2012 10th IEEE International Conference on Semiconductor Electronics

(ICSE 2012)

Kuala Lumpur, Malaysia
19-21 September 2012

IEEE Catalog Number: CFP12421-PRT
ISBN: 978-1-4673-2395-6

Copyright © 2012 by the Institute of Electrical and Electronic Engineers, Inc
All Rights Reserved

Copyright and Reprint Permissions: Abstracting is permitted with credit to the source. Libraries are permitted to photocopy beyond the limit of U.S. copyright law for private use of patrons those articles in this volume that carry a code at the bottom of the first page, provided the per-copy fee indicated in the code is paid through Copyright Clearance Center, 222 Rosewood Drive, Danvers, MA 01923.

For other copying, reprint or republication permission, write to IEEE Copyrights Manager, IEEE Service Center, 445 Hoes Lane, Piscataway, NJ 08854. All rights reserved.

***This publication is a representation of what appears in the IEEE Digital Libraries. Some format issues inherent in the e-media version may also appear in this print version.**

IEEE Catalog Number: CFP12421-PRT
ISBN 13:   978-1-4673-2395-6

Additional Copies of This Publication Are Available From:

Curran Associates, Inc
57 Morehouse Lane
Red Hook, NY 12571 USA
Phone: (845) 758-0400
Fax: (845) 758-2633
E-mail: curran@proceedings.com
Web: www.proceedings.com

Message from IEEE-ICSE2012 Conference Chair

Selamat Datang to Kuala Lumpur and IEEE-ICSE2012.

On behalf of the organizing committee, it is with great pleasure to welcome you to the 10th IEEE International Conference on Semiconductor Electronics 2012 (ICSE2012). Over the last twenty years, ICSE has become the preeminent international forum on semiconductor electronics embracing all aspects of the semiconductor technology from circuit device, modeling and simulation, photonics and sensor technology, MEMS technology, process and fabrication, packaging technology and manufacturing, failure analysis and reliability, material and devices and nanoelectronics.

This time, the conference offers six keynote lectures by distinguished persons in their own fields, 153 oral presentations and over 40 posters contributed by participants. We are proud to have Prof. Dr. Akhlesh Lakhtakia from PennState University, Prof. Dr. Edward Yi Chang from National Chiao Tung University, Taiwan R.O.C, the Director of EPFL's Institute of Microengineering (EPFL IMT), Switzerland Prof. Dr. Nico F. de Rooij, Prof. Dr. Yoon Soon Fatt from Nanyang Technology University Singapore and Prof. Dr. Ooi Boon Sew from KAUST Saudi Arabia as our keynote speakers. Besides that we are also delighted to have an Industrial Talk from Dr. Kamarulzaman Mohamed Zin, the CEO Silterra Malaysia Sdn Bhd who will talk about challenges and opportunities for IC design and fabrication in Malaysia. We hope participants would appreciate knowledge imparted by the lectures.

The conference proceedings provide views into the current advances in semiconductor electronics in the region. We are pleased to note the significant increase in number of submissions from the previous conference back in 2010 and we look forward to the presentations of our participants.

To all participants, I hope we can gain knowledge and benefits from the conference while making new contacts with other participants. To participants from overseas, I wish you a pleasant stay in this country and we will endeavor to make your stay here enjoyable.

Terima kasih.

Prof. Dato' Dr. Burhanuddin Yeop Majlis
Chair, 10th IEEE International Conference on Semiconductor Electronics

IEEE-ICSE 2012 Committee

Conference Chair:
Prof. Dato' Dr. Burhanuddin Yeop Majlis   *Universiti Kebangsaan Malaysia, Malaysia*
Co-Chair:
Prof. Dr. Ibrahim Ahmad   *Universiti Kebangsaan Malaysia, Malaysia*
Technical Chair:
Assoc. Prof. Dr. Roslina Mohd. Sidek   *Universiti Putra Malaysia, Malaysia*
Hon. Secretary:
Dr. P. Susthitha Menon   *Universiti Kebangsaan Malaysia, Malaysia*
Treasurer:
Mr. Azrif Manut   *Universiti Teknologi MARA, Malaysia*

International Advisory Committee

Prof. Dr. Edward Yi Chang, Taiwan
Prof. Dr. Hiroshi Iwai, Japan
Prof. Dr. Arokia Nathan, UK
Prof. Dr. Cary Yang, USA
Prof. Dr. Jong Duk Lee, Korea
Prof. Dr. John David, UK
Prof. Sima Dimitrijev, Australia
Prof. Dr. Yan Long, China
Prof. Dato' Dr. Muhammad Rasat Muhammad, Malaysia
Dr. Kamarulzaman Mohd Zain, Silterra, Malaysia
Dr. Gopi Kurup, TMR&D, Malaysia
Prof. Emeritus Dato' Dr. Muhammad Yahaya, Malaysia
Prof. Dr. Aklesh Lakhtakia, USA
Prof. Dr. Arshad Saleem Bhatti, Pakistan
Prof. Dr. Yoon Soon Fatt, Singapore
Prof. Dr. Kamran Eshraghiran, Australia
Prof. Dr. Nico Frans de Rooij, Switzerland
Prof. Dr. Vijay Arora, USA

Committee Member
Prof. Dr. Muhammad Mat Salleh   *Universiti Kebangsaan Malaysia*
Prof. Dr. Norani Muti Mohamed   *Universiti Teknologi Petronas, Malaysia*
Prof. Dr. Uda Hashim   *Universiti Malaysia Perlis, Malaysia*
Prof. Datin Dr. Saadah Abdul Rahman   *Universiti Malaya, Malaysia*
Prof. Dr. Razali Ismail   *Universiti Teknologi Malaysia, Malaysia*
Prof. Dr. A.H.M Zahirul Alam   *International Islamic University Malaysia, Malaysia*
Assoc. Prof. Dr. Norhayati Soin   *Universiti Malaya, Malaysia*
Assoc. Prof. Dr. Mohd Nizar Hamidon   *Universiti Putra Malaysia, Malaysia*
Assoc. Prof. Dr. Azlan Abdul Aziz   *Universiti Sains Malaysia, Malaysia*

Assoc. Prof. Zaliman Sauli	*Universiti Malaysia Perlis, Malaysia*
Assoc. Prof. Dr. Dee Chang Fu	*Universiti Kebangsaan Malaysia, Malaysia*
Dr. Mohd Haris MdKhir	*Universiti Teknologi Petronas, Malaysia*
Dr. Badariah Bais	*Universiti Kebangsaan Malaysia, Malaysia*
Dr. Mohd Sharizal Alias	*TMR&D Sdn. Bhd.*
Dr. Ismail Saad	*Universiti Malaysia Sabah, Malaysia*
Dr. Zainal Arif Burhanudin	*Universiti TeknologiPetronas, Malaysia*
Dr. Asrulnizam Abdul Manaf	*Universiti Sains Malaysia, Malaysia*
Dr. Norhana Arsad	*Universiti Kebangsaan Malaysia, Malaysia*
Mr. Rahman Wagiran	*Universiti Putra Malaysia, Malaysia*
Mrs. Maizatul Zolkapli	*Universiti Teknologi MARA, Malaysia*
Mr. Mohd Azizi Chik	*Silterra (M) Sdn. Bhd.*
Mrs. Anees Abdul Aziz	*Universiti Teknologi MARA, Malaysia*
Mrs. Norazreen Abdul Aziz	*Universiti Kebangsaan Malaysia, Malaysia*

Secretariat

Khairul Nisha Mohd. Kharuddin
Hayati Hussin
Aishah Fauthan
Shafii Abdul Wahab
Nor Nashrique Nasarudin
Farahiyah bt. Ali
Noraini Marsi
Hafzaliza Erny Zainal Abidin
Ummikalsom Abidin
Marianah Masrie

Secretariat Office

Secretariat of IEEE-ICSE2012
Electron Devices Malaysia Chapter
Institute of Microengineering and Nanoelectronics (IMEN)
Universiti Kebangsaan Malaysia, 43600 UKM Bangi, Selangor, MALAYSIA
Telephone +603 8921 7151/6987/7113
Fax +603 8925 9080
Email: edsmalaysia@gmail.com

Objective

The IEEE-ICSE is aimed at bringing together researchers from industry and academia to gather and explore various issues and trends in the field of micro and nano electronics. ICSE has become the prominent international forum on semiconductor electronics embracing all aspects of the semiconductor technology from circuits, device modeling and simulation, photonics and sensor technology.

Scopes of Conference

- MEMS/ NEMS
- Device Modeling and Simulation
- Nanoelectronics
- Device Physics and Characterization
- Opto-electronics and Photonics Technology
- Microwave Device and MMIC
- IC Packaging & Testing
- Reliability and Failure Analysis
- Semiconductor Manufacturing & Process
- VLSI Design
- Process Technology (CMOS, Bipolar, BiCMOS, GaAs, etc)
- Microelectronics Application in Product Development
- Electronics Materials and Device Fabrication

TABLE OF CONTENTS

Foreword by Conference Chairman	iii
Advisory and Organizing Committee	iv
Scope of Conference	vi
Table of Contents	vii

No.	Papers	Page

K1 **Nanotechnology and Metamaterials: Conceptualization and Intersection for New Opportunities**
Prof Dr Akhlesh Lakhtakia *(Charles Godfrey Binder (Endowed) Professor of Engineering Science and Mechanics, Pennstate University, USA)* — A1

K2 **InAs HEMT for Terahertz Applications**
Prof Dr Edward Yi Chang *(Distinguished Lecturer of the IEEE Electron Devices Society, National Chiao Tung University, Taiwan, R.O.C.)* — A2

K3 **Product Innovation Enabled by MEMS**
Prof Dr Nico F. de Rooij *(SAMLAB, Institute of Microengineering, EPFL STI, Switzerland)* — A3

K4 **Thermal Effects in InAs/GaAs Quantum Dot Vertical Cavity Surface Emitting Lasers**
Prof Dr Yoon Soon Fatt *(Nanyang Technological University, Singapore)* — A4

K5 **Broadband Semiconductor Lasers and Their Applications**
Prof Dr Ooi Boon Siew *(King Abdullah University of Science and Technology (KAUST))* — A5

I1 **Challenges & Opportunities for IC Design and Fabrication in Malaysia**
Dr Kamarulzaman Mohamed Zin *(CEO, Silterra Sdn. Bhd., Malaysia)* — A6

Cluster 1

1 **Statistical Modeling of Solar Cell using Taguchi Method and TCAD tool**
M. S. Bahrudin, S.F. Abdullah and I. Ahmad
Universiti Tenaga Nasional, MALAYSIA — 1

2 **Analysis of A MEMS Piezoelectric Microgenerator Using Network Placement Method**
Luay Yassin Taha, Burhanuddin Yeop Majlis, Ahmad Al Ali
Emirates Aviation College, UAE — 6

3 **Placement Effect of MEMS Thermoelectric Generator to Harvest Waste Heat using Shunt Configuration for Microprocessor**
Tai Zhi Ling and Ong Hang See
Universiti Tenaga Nasional, MALAYSIA — 10

4   **Effect of Co Catalyst on PECVD Growth of Carbon Nanotubes for NEMS Applications**   14
Mai Woon Lee, Nadia Md Razib, Aun Shih Teh, Daniel C. S. Bien, Soo Kien Chen and Shahrul Azam Abdullah
MIMOS Berhad, MALAYSIA

5   **A Study of Catalyst-based ZnO Nanowires Growth by PECVD**   18
Muhammad Aniq Shazni Mohammad Haniff, Khairul Anuar Abd Wahid, Wai Yee Lee, Hing Wah Lee, Saat Shukri Embong, Ishak Hj. Abd. Azid
MIMOS Berhad, MALAYSIA

6   **Recent Advancement in Microgap Electrode Fabrication by Conventional Photolithography Technique**   22
Q. Humayun and U. Hashim
Universiti Malaysia Perlis, MALAYSIA

7   **Numerical Simulation of Microfluidic Devices**   26
Uda Hashim, P.N A. Diyana and Tijjani Adam
Universiti Malaysia Perlis, MALAYSIA

8   **Effect of Dye Coating Duration of Eosin Y Synthesized on Nanorods Zinc Oxide Toward Hybrid Solar Cell**   30
Siti Aisyah Zawawi, Chi Chin Yap, Haslinda Abdul Hamid and Mohd Zaki Mohd Yusoff
Universiti Teknologi MARA, MALAYSIA

9   **Effect of Oxygen ratio to the Transparency and Conductivity of Nanostructured ZnO Thin Films deposited by RF Magnetron sputtering**   34
N.D. Md Sin, S. Ahmad, M.Z. Musa, M.H. Mamat, Abdul Aziz. A and M. Rusop
Universiti Teknologi MARA, MALAYSIA

10   **Enhanced Light Absorption in Bifacial Solar Cells**   38
Suhaila Sepeai, M.Y.Sulaiman, Saleem H.Zaidi, Kamaruzzaman Sopian
Universiti Kebangsaan Malaysia, MALAYSIA

11   **Infrared Emissivity of Co,Ni Co-Doped ZnO Powders by Solid-State Reaction**   42
Yinhua Yao and Quanxi Cao
Xidian University, P.R. CHINA

12   **Electric Field Induced Enhancement in Multisubband Electron Mobility in Strained GaAs/InGaAs Double Quantum Well Structures**   47
Trinath Sahu, Sangeeta Palo and Narayan Sahoo
National Institute of Science and Technology, INDIA

13   **Properties of Amorphous Carbon Thin Films with Nitrogen Incorporation by Aerosol-assisted CVD**   52
A.N. Fadzilah, K. Dayana, U. M. Noor and M. Rusop
University Teknologi MARA, MALAYSIA

14 **Aromatic and Non-Aromatic Solvent in Nanocomposited MEH-PPV:CNTs Thin Film for Organic Solar Cells** 57
M.S.P. Sarah, F.S.S. Zahid, Z. Zulkifli, U.M. Noor, M. Rusop
Universiti Technologi MARA, MALAYSIA

15 **Diameter Dependence of Spin Relaxation in SiGe nanowires** 61
Bhupesh Bishnoi, Vikas Nandal and Bahniman Ghosh
Indian Institute of Technology Kanpur, INDIA

16 **Effects of PMMA concentration on PMMA-based organic capacitor behavior** 65
L.N. Ismail, M. Khairizal, Z. Habibah, A.N. Arshad, M.H. Wahid, N.N. Hafizah, S.H. Herman and M.Rusop
Universiti Teknologi MARA, MALAYSIA

17 **Influence of Electron-Electron Scattering on Spin Relaxation Length in Single and Bilayer graphene** 69
Bhupesh Bishnoi, Dharmendra Hiranandani, Akshay Salimath, Vikas Nandal and Bahniman Ghosh
Indian Institute of Technology Kanpur, INDIA

18 **Properties of ZnO/MgO Multilayer Films as Insulating Layer Prepared by Sol-Gel Method** 73
Z. Habibah, A. N. Arshad, M. H. Wahid, L. N. Ismail, R. A. Bakar, M. H. Mamat, M. Rusop
Universiti Teknologi MARA, MALAYSIA

19 **AFM Images of Undoped Amorphous Carbon Thin Films Deposited by Bias-assisted Thermal-CVD** 78
A. Ishak, M. Amirul, and M. Rusop
Universiti Teknologi MARA, MALAYSIA

20 **Electrical Properties of ZnO/TiO$_2$ Nanocomposite Film Deposited by Simultaneous Radio-Frequency Magnetron Sputtering** 82
I. Saurdi, M.H. Mamat, M.H. Abdullah, M.Z. Musa, and M. Rusop
Universiti Teknologi MARA, MALAYSIA

21 **Scattering Effect in Silicon Nanowire Fin Field Effect Transistor** 86
Fatimah K. A. Hamid, Zaharah Johari, Wei Sun Leong, Norazlin Bahador, Munawar A Riyadi, Jatmiko E Suseno, S.Isaak, M. T. Ahmadi and Razali Ismail
Universiti Teknologi Malaysia, MALAYSIA

22 **Low Actuation Voltage through Three Electrodes Topology for RF MEMS Capacitive Switch** 90
Norhaslinawati Ramli, Othman Sidek and M.Amir
Universiti Kuala Lumpur - British Malaysian Institute, MALAYSIA

23 **Photoluminescence of Porous Silicon Nanostructures with Optimum Current Density of Photo-Electrochemical Anodisation** 94

M. Ain Zubaidah, N.A. Asli, M. Rusop, S. Abdullah
Universiti Teknologi MARA, MALAYSIA

24   **Simulation of Nanoscale Dual-channel Strained Si/Strained Si_{1-y} Ge_y/Relaxed Si_{1-x} Ge_x PMOSFET**   97
Yu Chan Thien, Eng Siew Kang and Razali Ismail
Universiti Teknologi MARA, MALAYSIA

25   **Electrical and Optical Properties of Iodine Doped Amorphous Carbon Thin Film by Thermal CVD**   102
K. Dayana, A. N. Fadzilah, U.M. Noor and M. Rusop
Universiti Teknologi MARA, MALAYSIA

26   **The Effect of Exposure Time and Development Time on Photoresist Thin Film in Micro/Nano Structure Formation**   107
Tijjani Adam and U. Hashim
Universiti Malaysia Perlis, MALAYSIA

27   **Substrate types and deposition pressure dependences of RF-magnetron sputtered Silicon thin films characteristics deposited at room temperature**   111
S. B. Hashim, N. H. Mahzan, S. H. Herman, R. Abu Bakar, U. Mohd Noor and M. Rusop
Universiti Teknologi MARA, MALAYSIA

28   **Simultaneous Study of Thermal and Optical Characteristics of Light-Emitting Diode**   115
Zhi-Yin Lee and Mutharasu Devarajan
Universiti Sains Malaysia, MALAYSIA

29   **The Octupole Microelectrode for dielectrophoretic trapping of single cells– Design and Simulation**   119
S.Noorjannah Ibrahim and Maan M. Alkaisi
University Of Canterbury, NEW ZEALAND

30   **Modeling and Simulation of Fluid Interactions with Bluff Body for Energy Harvesting Application**   123
M.S.Bhuyan, M.Othman, Sawal Hamid Md Ali, Burhanuddin Yeop Majlis and Md. Shabiul Islam
Universiti Kebangsaan Malaysia, MALAYSIA

31   **The Electrical Conductivity of Copper (I) Iodide (CuI) Thin Films Prepared by Mister Atomizer**   128
M.N. Amalina, A.R. Zainun, N.A. Rasheid and M.Rusop
Universiti Teknologi MARA, MALAYSIA

32   **Formation of copper oxide thin films from RF sputtered Cu thin film by ultra high pure boiled water**   132
Subramani Shanmugan and Devarajan Mutharasu
Universiti Sains Malaysia, MALAYSIA

33 **A Study on Lightly-Doped Cylindrical surrounding-gate 6H-SiC Nanowire FET** 137
Ru Han
Northwestern Polytechnical University, P.R. CHINA

34 **Synthesization of Carbon Nanotubes Using Single Stage Thermal CVD Method** 141
M. Maryam, A. B. Suriani, M.S. Shamsudin and M. Rusop
Universiti Teknologi MARA, MALAYSIA

35 **The Synthesis and Fabrication of Titanium Dioxide Nanowires-Based Biosensor** 145
Sharipah Nadzirah Syed Ahmad Ayob and U. Hashim
Universiti Malaysia Perlis, MALAYSIA

36 **The Effect of Surface Morphology to Photoluminescence Spectrum Porous Silicon** 149
M. H. Fadzilah Suhaimi, M. Ain Zubaidah, S. F. M. Yusop, M. Rusop and S. Abdullah
Universiti Teknologi MARA, MALAYSIA

37 **Influence of Heating Temperature on Electrical Photoconductivity of Nanocomposited Polymer-TiO$_2$ Thin Films for Organic Photovoltaic** 153
F.S.S.Zahid, M.S.P.Sarah, U.M.Noor and M.Rusop
Universiti Teknologi MARA, MALAYSIA

38 **Optical and Electrical Properties of ZnO and ZnO:TiO$_2$ Thin Films Prepared by Sol-Gel Spray-Spin Coating Technique** 158
C.M.Firdaus,M.Rusop,S.R.M.S.Baki and R.H.Salimin
Universiti Teknologi MARA, MALAYSIA

39 **Effect of Temperature Treatment on the Properties of ZnO Nanoparticle-Bi$_2$O$_3$-Mn$_2$O$_3$ Varistor Ceramics** 163
Rabab Khalid Sendi and Shahrom Mahmud
Universiti Sains Malaysia, MALAYSIA

40 **Optimization of Cantilever-Based MEMS Switch** 168
Mohammadmahdi Vakilian, Maryam Mousavi, Badariah Bais and Burhanuddin Yeop Majlis
Universiti Kebangsaan Malaysia, MALAYSIA

41 **Scaling Down the 32 nm Gate Length NMOS Transistor to 22 nm** 173
Afifah Maheran A.H., Menon, P.S., I. Ahmad, H.A. Elgomati, B.Y. Majlis and F. Salehuddin
Universiti Kebangsaan Malaysia, MALAYSIA

42 **Enhanced Performance Analysis of Vertical Strained- SiGeImpact Ionization MOSFET (VESIMOS)** 177
Ismail Saad, DivyaPogaku, Abu Bakar AR, Mohd Zuhir H., N. Bolong, Khairul A.M, Bablu Ghosh, Razali Ismail and U. Hashim

Universiti Malaysia Sabah, MALAYSIA

43 **Modulus and Thermal Properties of Free Standing PMMA/TiO2**   182
     Nanocomposite Films
     N.N. Hafizah, L.N. Ismail and M. Rusop
     Universiti Teknologi MARA, MALAYSIA

44 **Comparison of Mechanical Deflection and Maximum Stress of 3C SiC-**   186
     and Si-Based Pressure Sensor Diaphragms for Extreme Environment
     Noraini Marsi, Burhanuddin Yeop Majlis, Azrul Azlan Hamzah and Faisal
     Mohd-Yasin
     Universiti Kebangsaan Malaysia, MALAYSIA

45 **Study of ZnO Micro-gap on SiO$_2$/Si Substrate by Conventional**   191
     Lithography Method for pH Measurement
     K.L.Foo, U.Hashim, Haarindra Prasad s/o RajintraPrasat and M.Kashif
     Universiti Malaysia Perlis, MALAYSIA

46 **Design and Simulation of High Magnetic Gradient Device for Effective**   195
     Bioparticles Trapping
     Ummikalsom Abidin, Burhanuddin Yeop Majlis and Jumril Yunas
     Universiti Kebangsaan Malaysia, MALAYSIA

47 **Optical properties of zinc doped tin oxide synthesized by mechanochemical**   200
     processing
     Sharipah Nadzirah SAA, Azlan Zakaria, Mahesh Kumar Talari, Nurul
     Syahidah Sabri and U. Hashim
     Universiti Malaysia Perlis, MALAYSIA

48 **Design Study of Integrated Optical Transducer for Bioparticles Detection**   205
     Marianah Masrie, Burhanuddin Yeop Majlis, Jumril Yunas and P Susthitha
     Menon
     Universiti Kebangsaan Malaysia, MALAYSIA

49 **The Effect of Isopropyl Alcohol on Anisotropic Etched Silicon for the**   210
     Fabrication of Microheater Chamber
     Norihan Abdul Hamid, Burhanuddin Yeop Majlis, Jumril Yunas, and
     Mimiwaty Mohd Noor
     Universiti Kebangsaan Malaysia, MALAYSIA

50 **Controlled Growth of ZnO Nanostructures Prepared by Catalytic-**   214
     Immersion Method
     A.Azlinda, Z. Khusaimi and M. Rusop
     Universiti Teknologi MARA, MALAYSIA

51 **Optimization of Process Parameter Variation in 45nm p-channel**   219
     MOSFET using L18 Orthogonal Array
     F.Salehuddin, I.Ahmad, F.A.Hamid, A.Zaharim, Afifah Maheran A.Hamid,
     P.Susthitha Menon, H.A.Elgomati, B.Yeop Majlis and P.R.Apte
     Universiti Kebangsaan Malaysia, MALAYSIA

52   **A Simulation Study of The Effect Engineered Tunnel Barrier To The Floating Gate Flash Memory Devices**   224
Mohd Rosydi Zakaria, Uda Hashim, Ramzan Mat Ayub, Zarimawaty Zailan
Universiti Malaysia Perlis, MALAYSIA

53   **DC MEMS Switches with Self-x Features: Design, Simulation and Implementation Strategies**   229
Muhammad Akmal Johar, Pedro Torruella and Andreas König
University of Kaiserslautern, GERMANY

54   **Modeling of biomimetic flow sensor based fish dome shaped cupula using PDMS for underwater sensing**   234
Mohd Norzaidi Mat Nawi, Asrulnizam Abd Manaf, Mohd Rizal Arshad and Othman Sidek
Universiti Sains Malaysia, MALAYSIA

55   **Ultrasensitive Poly-Si Nanogap based on capacitive sensor for electrochemical detection**   238
Nazwa Taib, Uda Hashim, Thikra S.Dhahi, Ahmad Sudin, Nur Humaira Md Salleh, and Seng Teik Ten
Universiti Malaysia Perlis, MALAYSIA

56   **Optical properties of zinc oxide films growth on Si substrate via aqueous chemical growth**   242
Maria Abu Bakar, Muhammad Azmi Abdul Hamid, Azman Jalar and Roslinda Shamsudin
Universiti Kebangsaan Malaysia, MALAYSIA

57   **The Growth and Fabrication of High-Performance $In_{0.5}Ga_{0.5}As$ Metal-Oxide-Semiconductor Capacitor on GaAs Substrate by Metalorganic Chemical Vapor Deposition Method**   246
Hong Quan Nguyen, Hai Dang Trinh, Hung Wei Yu, Ching Hsiang Hsu, Chen Chen Chung, Binh Tinh Tran, Yuen Yee Wong, Thanh Hoa Phan Van, Quang Ho Luc, Diao Yuan Chiou, Chi Lang Nguyen, Chang Fu Dee and Edward Yi Chang
National Chiao Tung University, TAIWAN

58   **Investigation of Incorporating Dielectric Pocket (DP) on Vertical Strained-SiGe Impact Ionization MOSFET (VESIMOS-DP)**   249
Ismail Saad, Mohd. Zuhir H., Divya Pogaku, Abu Bakar AR, N. Bolong, Khairul A.M, Bablu Ghosh, Razali Ismail and U. Hashim
Universiti Malaysia Sabah, MALAYSIA

59   **Pierce Oscillator Circuit Topology for High Motional Resistance CMOS MEMS SAW Resonator**   254
Jamilah Karim, Anis Nurashikin Nordin, AHM Zahirul Alam and U.Hashim
Universiti Teknologi MARA, MALAYSIA

60   **Influence of LT-AlN Buffer Layers on Density of Threading Dislocation in AlGaN Layers**   259

xiii

H Meidia and S Mahajan
Universitas Multimedia Nusantara, INDONESIA

61 **Calibration Parameters in TCAD for Predictive MOSFET Device Simulations** 263
Muhamad Amri Ismail, Mohd Hezri Abu Bakar and Iskhandar Md Nasir
MIMOS Berhad, MALAYSIA

62 **Deposition of Titanium Dioxide (TiO$_2$) Thin Films Using In-house Nano-TiO$_2$ Powder** 267
M.Z. Sahdan, M.S. Alias, N. Nafarizal and U. Hashim
Universiti Tun Hussein Onn Malaysia, MALAYSIA

63 **T-Shaped Resonating Beam Pressure Sensor** 271
Y.Sujan, B.Vasuki and G.Uma
National Institute of Technology, Tiruchirappalli, INDIA

64 **Tuned Dual Beam Low Voltage RF MEMS Capacitive Switches for X – Band Applications** 276
E.S.Shajahan and Shankaranarayana M Bhat
National Institute of Technology Karnataka, INDIA

65 **Magnetic Force on a Magnetic Bead** 280
Alireza Bahadorimehr, Jafar Alvankarian and Burhanuddin Yeop Majlis
Universiti Kebangsaan Malaysia, MALAYSIA

66 **Polycrystalline p-β-FeSi$_2$(Al) on n-Si(100): Heterojunction Thin-Film Solar Cells** 285
A. Bag, S. Mallik, C. Mahata and C. K. Maiti
Indian Institute of Technology, INDIA

67 **An Investigation in the Impact of Structural Parameters on the Electrical Characteristics of Nanoscale Heterostructure p-MOSFETs** 288
Fatemeh Kohani Khoshkbijari, Reza Fouladi, Shiva Nejati, Reza Barkhordari, Reza Kohani Khoshkbijari and Shide Nejati
Islamic Azad University, Rasht Branch, IRAN

68 **Effect of Solution Concentration on the Morphology, Electrical, and Optical Properties of MEH-PPV Thin Films** 293
Shafinaz Sobihana Shariffudin, Nurhafizah Zainal Abidin, Nurul Zayana Yahya, Anees Abdul Aziz, Sukreen Hana Herman, and Mohamad Rusop
Universiti Teknologi MARA, MALAYSIA

69 **Temperature Effect on Quantum Capacitance Zig-Zag Graphene Nanoscrolls (ZGNS) (16,0)** 298
Afiq Hamzah, M.T.Ahmadi, Mohammad Javad Kiani, Fatimah. K. A. Hamid, Azlin Bahador and Razali Ismail
Universiti Teknologi Malaysia, MALAYSIA

70   **Effects of Annealing Temperature on Morphology and Crystallinity of**   302
Nitrogen Doped Zinc Oxide (ZnO:N) Nano Films
J. Karamdel, F. Razaghian, A. Hadi, C. F. Dee and B. Y. Majlis
Universiti Kebangsaan Malaysia, MALAYSIA

71   **Studies on the Growth of Alumina Nanoporous Film & Nanowires on**   306
Planar & Cylindrical Substrate
Tiong Teck Yaw, Abrar Ismardi, Dee Chang Fu and Burhannuddin Yeop
Majlis
Universiti Kebangsaan Malaysia, MALAYSIA

72   **The Role of Reactive Ion Etching(RIE) on Wirebond Formation: A Study**   311
on Successful Rate of Thermosonic Gold Wire on Aluminium
Bondpad
Sauli. Z., Retnasamy, V., Rahman, N.A.Z., Aziz, M.H.A., Razak, H.A. and
Palianysamy, M.
Universiti Malaysia Perlis, MALAYSIA

73   **Effect of Copper FAB Impact on Palladium Bond Pad**   316
Sauli. Z., Retnasamy, V., Taniselass, S., Norhaimi, W.M.W., Aziz, M.H.A. and
Hashim, M.N.
Universiti Malaysia Perlis, MALAYSIA

74   **Single Hole at Constrained Location for Stress Analysis in PCB Plate**   320
Bending
Sauli. Z., Retnasamy, V., Vengdasalam, K., Taniselass, S., Shapri, A.H.M. and
Vairavan, R.
Universiti Malaysia Perlis, MALAYSIA

75   **Design and Analysis of a Localised Environment Monitoring Sensor**   324
System
Asral Bahari Jambek, Lau Chyun Wenn and Uda Hashim
Universiti Malaysia Perlis, MALAYSIA

76   **Physical Characteristic of Room-Temperature Deposited Ti Thin Films by**   328
RF Magnetron Sputtering at Different RF Power
Z. Aznilinda, S.H. Herman, R.A.Bakar and M. Rusop
Universiti Teknologi MARA, MALAYSIA

77   **Issues and Challenges in Microfluidic Research Studies**   333
Jafar Alvankarian, Alireza Bahadorimehr, Benyamin Davaji and Burhanuddin
Yeop Majlis
Universiti Kebangsaan Malaysia, MALAYSIA

78   **Geometrical Characterization of Single Layer Silicon Based Piezoresistive**   336
Microcantilever using ANSYS
Mohd Hazrul Zakaria, Badariah Bais, Rosminazuin Ab. Rahim and
Burhanuddin Yeop Majlis
Universiti Kebangsaan Malaysia, MALAYSIA

79   **Induced Mass Change Technique for Glucose Detection in Microcantilever-based Sensors**   340
Mardhiah Mohd Nor, Badariah Bais, Norazreen Abd Aziz, Rosminazuin Ab. Rahim and Burhanuddin Yeop Majlis
Universiti Kebangsaan Malaysia, MALAYSIA

80   **Localized Surface Plasmon Resonance Sensor of Gold Nanoparticles for Detection Pesticides in Water**   344
Norhayati Abu Bakar, Akrajas Ali Umar, Muhamad Mat Salleh, Muhammad Yahaya and Burhanuddin Yeop Majlis
Universiti Kebangsaan Malaysia, MALAYSIA

81   **Electrical Characterization of Interdigital Electrode Based on Cyclic Voltammetry Performances**   348
Hafzaliza Erny Zainal Abidin, Azrul Azlan Hamzah and Burhanuddin Yeop Majlis
Universiti Kebangsaan Malaysia, MALAYSIA

82   **High Sensitivity Localized Surface Plasmon Resonance Sensor of Gold Nanoparticles : Surface Density Effect for Detection of Boric Acid**   352
Marlia Morsin, Akrajas Ali Umar, Muhamad Mat Salleh and Burhanuddin Yeop Majlis
Universiti Kebangsaan Malaysia, MALAYSIA

83   **Effect of Varying Thickness of Electroplated NiFe Film on Magnetic Properties**   357
Nadzril Sulaiman, Jumril Yunas and Burhanuddin Yeop Majlis
Universiti Kebangsaan Malaysia, MALAYSIA

84   **Micro-heater Filament on Polyimide Membrane for Gas Sensor Applications**   360
Mimiwaty Mohd Noor, Gandi Sugandi and Burhanuddin Yeop Majlis
Universiti Kebangsaan Malaysia, MALAYSIA

85   **Modeling of high-intensity expert system of the catalytic oxidation reactor of phosphorous gases**   363
Akzhigitova Meruyert, Eskendirov Sharipzhan and Umarova Zhanat
South-Kazakhstan State University, KAZAKHSTAN

Cluster 2

86   **Multiplication Gain and Excess Noise Factor in 4H-SiC APD**   366
C. C. Sun, A. H. You and E. K. Wong
Multimedia University, MALAYSIA

87   **All Optical Switch Using Ultra compact Multi Mode Interference Coupler**   370
Mehdi Tajaldini and Mohd Zubir Mat Jafri
Universiti Sains Malaysia, MALAYSIA

88   **Theoretical Triangular Quantum Well Model for AlGaN/GaN HEMT**   374
     Structure Used as Polar Liquid Sensor
     Sulaiman Rabbaa and Johan Stiens
     Vrije Universiteit Brussel, BELGIUM

89   **Demonstration of DC Current Sensing through Microfiber Knot**   378
     Resonator
     Azlan Sulaiman, Sulaiman Wadi Harun, Jalil. Md. Desa and Harith Ahmad
     Universiti Malaya, MALAYSIA

90   **Microfiber Coupler Devices**   381
     M. Z. Muhammad, A. A. Jasim, H. Ahmad and S.W. Harun
     University of Malaya, MALAYSIA

91   **Influence of Optical Power on Thermal Resistance Measurement for High**   384
     Power Infrared Emitter
     Chin-Peng Ching and Mutharasu Devarajan
     Universiti Sains Malaysia, MALAYSIA

92   **Analysis on Optical Properties for Various Types of Light Emitting Diode**   388
     Chin-Peng Ching, Zhi-Yin Lee, Sze-Yen Lee and Mutharasu Devarajan
     Universiti Sains Malaysia, MALAYSIA

93   **On the Waves in Circular Waveguides Containing Chiral Nihility**   392
     Metamaterial under PMC Boundary
     M.A. Baqir and P.K. Choudhury
     Universiti Kebangsaan Malaysia, MALAYSIA

94   **Improved Dead Time Response for Si Avalanche Photodiode**   396
     Norazlin Bahador, Fatimah K. A. Hamid, Afiq Hamzah, Suhaila Isaak and
     Razali Ismail
     Universiti Teknologi Malaysia, MALAYSIA

95   **Physical Effects from Etching Parameters of the Bragg Grating**   399
     Waveguide Fabricated on Porous Silicon Nanostructure
     Ahmad Afif Safwan Mohd Radzi, Shamsul Faez Mohd Yusop, Nurul Izrini
     Ikhsan, Mohamad Rusop and Saifollah Abdullah
     Universiti Teknologi MARA, MALAYSIA

96   **High Performance of a SOI-based Lateral PIN Photodiode Using SiGe/Si**   403
     Multilayer Quantum Well
     P.Susthitha Menon, S.Kalthom Tasirin, Ibrahim Ahmad, S.Fazlili Abdullah
     Universiti Kebangsaan Malaysia, MALAYSIA

97   **Compatibility Issues of Si Technology with Higher Band Gap Materials**   407
     for RF Applications
     Bablu K. Ghosh, Ismail Saad, Khairul Anuar Mohamad, Nurmin Bolong,
     Norfarariyanti Parimon, Afishah Alias and Mohd Zuhir Hamzah
     Universiti Malaysia Sabah, MALAYSIA

98 **RF Characteristics of AlGaN/GaN HEMTs under Different Temperatures** 411
Yu-Sheng Chiu, Jui-Chien Huang, Tai-Ming Lin, Yu-Ting Chou, Chung-Yu
Lu, Chia-Ta Chang and Edward Yi Chang
National Chiao Tung University, TAIWAN

99 **Investigation of Efficiency Droop in GaN-based UV LEDs with N-type** 414
AlGaN Underlayer
Shun-Kuei Yang, Po-Min Tu, Shih-Cheng Huang, Ya-wen Lin, Chih-Peng
Hsu, Jet-Rung Chang, and Chun-Yen Chang
Advanced Optoelectronic Technology Inc., TAIWAN

100 **Study of Efficiency Droop in InGaN-based Near-UV LEDs with** 418
Quaternary InAlGaN Barrier
Po-Min Tu, Shih-Cheng Huang, Ya-wen Lin, Shun-Kuei Yang, Chih-Peng
Hsu, Jet-Rung Chang, and Chun-Yen Chang
Advanced Optoelectronic Technology Inc., TAIWAN

101 **Modeling of SOI-based MRR by Coupled Mode Theory using Lateral** 422
Coupling Configuration
Hazura H., Menon, P.S, Burhanuddin Yeop Majlis, Hanim A.R, Mardiana B.,
Hasanah, L., Mulyanti, B., Mahmudin, D. and Wiranto, G.
Universiti Kebangsaan Malaysia, MALAYSIA

Cluster 3

102 **Fabrication and Characterization of Cu Pellet Using Powder Metallurgical** 426
Method
Chew Pei Yi, You Ah Heng and Vijayaram Thoguluva Raghavan
Multimedia University, MALAYSIA

103 **Gross Die Estimator's Caveats For ASIC Floorplanning** 431
Ang Boon Chong
PMC-Sierra, MALAYSIA

104 **Functional OBIRCH Strategy in Analyzing Complex Functional Failures** 436
Including Logic Failures
Gaojie Wen, Li Tian, Binghai Liu, Grace Song, Joe Yu and Winter Wang
Freescale Semiconductor (China) Limited, Tianjin, P.R. CHINA

105 **Leakage in CMOS Devices Induced by Pattern-Dependent Microloading** 440
Effect
Miao Wu, Winter Wang, Li Tian, Chunlei Wu and Diwei Fan
Freescale Semiconductor (China) Limited, Tianjin, P.R. CHINA

106 **Combined Emission with simulation technique to resolve unstable failure** 444
mode sample
DiWei Fan, Winter Wang, Li Tian, Miao Wu and ChunLei Wu
Freescale Semiconductor (China) Limited, Tianjin, P.R. CHINA

107 **PPTP: Pre-Post Terminal Propagation in Modern Fixed-Outline Soft** 448
Module VLSI Floorplanning Design

xviii

Chyi-Shiang Hoo, Kanesan Jeevan, Velappa Ganapathy and Harikrishnan Ramiah
Universiti Malaya, MALAYSIA

108 **A Compliant Lead-Free Solder Alloy**   453
Mohd Faizul Mohd Sabri, Dhafer Abdul-Ameer Shnawah, Irfan Anjum Badruddin and Suhana Binti Mohd Said
Universiti Malaya, MALAYSIA

109 **Temperature Cycling and Thermal Shock Correlation in DPAK & DSO Packages**   458
Lee Chai Ying and Cheong Choke Fei
Infineon Technologies (M) Sdn. Bhd., MALAYSIA

110 **Mechanism and Improvement of Breakdown Degradation Induced by Interface Charge in UHV Device**   462
Md. Imran Siddiqui, Abijith Prakash, Mohammed Sadique Anwar, Gene Sheu and P A Chen
Asia University, TAIWAN

111 **Design of a Low Voltage Charge Pump Circuit for RFID Tag**   466
Kang Cheng Wei, M. B. I. Reaz, Md. Syedul Amin, Jubayer Jalil and Labonnah F. Rahman
Universiti Kebangsaan Malaysia, MALAYSIA

112 **Novel Architecture of Pipeline Radix 22 SDF FFT Based on Digit-Slicing Technique**   470
Yazan Samir Algnabi, Furat A. Aldaamee, Rozita Teymourzadeh, Masuri Othman and Md Shabiul Islam
Universiti Kebangsaan Malaysia, MALAYSIA

113 **Methodology To Execute SPARC Binary of Silterra Memory Compiler 0.18µm Process Technology on x86 Architecture**   475
Raja Mohd Fuad Tengku Aziz, Rozaimah Baharim, Md Hanif Md Nasir, Rohaya Abdul Wahab, Nazaliza Othman, Nabihah Razali, Sharifah Saleh
MIMOS Berhad, MALAYSIA

114 **Simulated Annealing vs. Genetic Simulated Annealing for Automatic Transistor Sizing**   478
Nishant Singh and Bahniman Ghosh
Indian Institute of Technology Kanpur, INDIA

115 **Development of Automated Neighborhood Pattern Sensitive Fault Syndrome Generator for SRAM**   482
J.R. Rusli, R.M. Sidek and W.H. Wan Zuha
Universiti Putra Malaysia, MALAYSIA

116 **Frequency Reduction in Quantum Dot Cellular Automata**   486
Bhupesh Bishnoi, Diwakar Agrawal, Vikas Nandal, Akshay Salimath and Bahniman Ghosh

Indian Institute of Technology Kanpur, INDIA

117 **Design and Implementation of Reversible Logic Based Bidirectional Barrel Shifter**   490
O.Anjaneyulu, T.Pradeep and C.V.Krishna Reddy
Kakatiya University, INDIA

118 **Failure Analysis Case Studies on Open Defect**   495
Grace Song, Chunlei Wu, Joe Yu, Gaojie Wen and Winter Wang
Freescale Semiconductor (China) Limited, Tianjin, P.R. CHINA

119 **Thermal and Optical Analysis of Four-chip HPLED Package with Different Thermal Interface Material**   499
S.Y. Lee and M. Devarajan
Universiti Sains Malaysia, MALAYSIA

120 **A Study on the Effect of Test Vector Randomness on Test Length and its Fault Coverage**   503
Muhammad Sadiq Sahari, Abu Khari A'ain and Ian Grout
Universiti Teknologi Malaysia, MALAYSIA

121 **New Low Power Delay Element in Self Resetting Logic with Modified Gated Diffusion Input Technique**   507
Uma.Ramadass and P. Dhavachelvan
Pondicherry University, INDIA

122 **Multiply-Accumulate Instruction Set Extension in a Soft-core RISC Processor**   512
Ahmad Jamal Salim, Nur Raihana Samsudin, Sani Irwan Md Salim and Yewguan Soo
Universiti Teknikal Malaysia Melaka, MALAYSIA

123 **Adjustable Phase-Locked Loop with Independent Frequency Outputs**   517
Robert Freier, Hamam Maher Abd and Andreas König
Technische Universiẗat Kaiserslautern, GERMANY

124 **Capacitor-Grounded Electronically Tunable Voltage-Mode OTA-C Multifunction Filter with Three Inputs and Five Outputs**   522
Montree Kumngern and Kobchai Dejhan
King Mongkut's Institute of Technology Ladkrabang, THAILAND

125 **Design of Low Power, Low Jitter DLL Tested at all Five Corners to Avoid False Locking**   526
Himadri Singh Raghav, Sachin Maheshwari, Mola Srinivasarao and B. P. Singh
Mody Institute of Technology and Science, INDIA

126 **Effect of Damaged-Chip Infrared Emitter Package on Ge Substrate**   532
Wei Ching LIEW and Mutharasu DEVARAJAN
Universiti Sains Malaysia, MALAYSIA

127 **New Circuit Models of Complementary-Symmetry Class-AB and Class-B Push-Pull Amplifiers** 538
SachchidaNand Shukla, Beena Pandey and Susmrita Srivastava
Ram Manohar Lohia Avadh University, INDIA

128 **A Three-Stage Power Amplifier for WiMedia Ultra-Wideband Applications** 543
Zi-Yi, Lam, Yun-Fen, Yong, Sew-Kin, Wong, Chee-Pun, Ooi
Multimedia University, *MALAYSIA*

129 **High Speed Direct Digital Frequency Synthesizer with Pipelining Phase Accumulator Based on Brent-Kung Adder** 547
Salah Hasan Ibrahim, Sawal Hamid Md Ali and Md. Shabiul Islam
Universiti Kebangsaan Malaysia, MALAYSIA

130 **Design Optimization Platform for Synthesizable High Speed Digital Filters Using Retiming Technique** 551
Deepa Yagain, Vijaya Krishna A. and Sheetal Chennapnoor
Visveswaraya Technological University, INDIA

131 **The Design of DC Motor Driver for Solar Tracking Applications** 556
Zi-Yi, Lam, Sew-Kin, Wong, Wai-Leong, Pang and Chee-Pun, Ooi
Universiti Malaya, MALAYSIA

132 **Planar Dipole Antenna Design At 1800MHz Band Using Different Feeding Methods For GSM Application** 560
Waleed Ahmed AL Garidi, Norsuzlin Bt Mohad Sahar and Rozita Teymourzadeh
UCSI University, MALAYSIA

133 **Qualitative Study of a New Circuit Model of Small-signal Amplifier using Sziklai Pair in Compound Configuration** 565
SachchidaNand Shukla, Beena Pandey and Susmrita Srivastava
Ram Manohar Lohia Avadh University, INDIA

134 **Design of a 9-bit UART Module Based on Verilog HDL** 570
Nennie Farina Mahat
MIMOS Berhad, MALAYSIA

135 **A $\Delta\Sigma$ Modulator with 3-Bit, 37-Level Pre-Detective Dynamic Quantization** 574
Chien-Hung Kuo and Kuan-Hsun Wang
National Taiwan Normal University, TAIWAN

136 **Influences Study on MIM Capacitors' Reliability** 578
Chu Tsui Ping, Yang Peng, Tee Pei Ling
X-FAB Sarawak Sdn. Bhd., MALAYSIA

137 **Investigation of Crosstalk Impact on Channel Performance from IC package and Motherboard Breakout Routing** 583
Azri Husni Hasani, Aftanasar Md. Shahar, Ahmad Jalaluddin Yusof and Jackson Kong

Universiti Sains Malaysia, MALAYSIA

138 **Failure Mechanism and Improvement on Gate Oxide Failure at the Edge of LOCOS**
Lesley Wong Ying Ying, Deb Kumar Pal, Raymond Tan, Ng Hong Seng, Michaelina Ong, Tong Gee Hong and Wong Jian Sang
X-FAB Sarawak Sdn. Bhd., MALAYSIA
588

139 **Charge Collection Efficiency Measurement System Based On Field Programmable Gate Array Multipurpose Card**
Norizam Saad, Ishak Mansor, Muhammad Azmi Abdul Hamid, Azman Jalar and Roslinda Shamsudin
Universiti Kebangsaan Malaysia, MALAYSIA
592

140 **Development and Application of In-House High Voltage Power Supply For Atmospheric Pressure Plasma Treatment System**
Nafarizal Nayan, Mohammad Redzuan Zahariman, Mohd Fadzlie Bin Ahmad, Riyaz Ahmad Mohamed Ali, Mohd Zainizan Sahdan and Uda Hashim
Universiti Tun Hussein Onn Malaysia, MALAYSIA
596

141 **A New Design Methodology based on Particle Swarm Optimization (PSO) Algorithm for Multi-clad Single Mode Optical Fibers**
Shiva Nejati, Reza Barkhordari, Fatemeh Kohani Khoshkbijari, Reza Fouladi, Shide Nejati and Reza Kohani Khoshkbijari
University of Tabriz, IRAN
600

142 **Radiation Exposure induced Failure on Semiconductor Package Material**
Wan Yusmawati Wan Yusoff, Azman Jalar, Norinsan Kamil Othman, Irman Abdul Rahman, Roslinda Shamsudin and Muhammad Azmi Abdul Hamid
Universiti Kebangsaan Malaysia, MALAYSIA
604

143 **Automated Switching Mechanism for Multi-Standard RFID Transponder**
Teh Kim Ting, Khaw Mei Kum and Faisal Mohd-Yasin
Multimedia University, MALAYSIA
608

144 **Switched Inverter Comparator based 0.5 V Low Power 6 bit Flash ADC**
Rajeev Komar, M S Bhat and T Laxminidhi
National Institute of Technology Karnataka, INDIA
613

145 **A Ultra-Wideband, Downconversion Folded Mixer in 0.13-um CMOS Technology**
Xuelian Liu and John F. McDonald
Rensselaer Polytechnic Institute, USA
618

146 **Design of Single-Stage Folded-Cascode Gain Boost Amplifier for 14 bit 12.5Ms/S Pipelined Analog-to Digital Converter**
Xuelian Liu and John F. McDonald
Rensselaer Polytechnic Institute, USA
622

147 **Technique to Improve Visibility for Cycle Time Improvement in Semiconductor Manufacturing**
627

Syahril Ridzuan Ab Rahim, Ibrahim Ahmad and Mohd Azizi Chik
Universiti Tenaga Nasional, MALAYSIA

148 **Accessing AHB Bus using WISHBONE Master in SoC Design**   631
Muhamad Khairol Ab Rani and Mohd Zubir Khalid
MIMOS Berhad, MALAYSIA

149 **Tunable Loop Filter in Fractional-N Frequency Synthesizer for Wireless**   636
Applications
Gan Leong Kit, Fazrena Azlee Hamid and Syed Khaleel Ahmed
Universiti Tenaga Nasional, MALAYSIA

150 **Stress Analysis on Centric Through Hole PCB**   641
Sauli. Z., Retnasamy, V., Rahman, N.A.Z., Man, B., Nadzri, N.S. and
Vairavan, R.
Universiti Malaysia Perlis, MALAYSIA

151 **Very High Speed and Low Voltage Open-Loop Dual Edge Triggered**   645
Sample and Hold Circuit in 0.18μm CMOS Technology
Mohamad Hasan-Sagha and Mohsen Jalali
Shahed University, IRAN

152 **Development of Capacity Indices for Semiconductor Fabrication**   649
Mohd Azizi Chik, Kader Ibrahim, Mohd Hazmuni Saidin, Faizah Md Yusof, G.
Devandran and U. Hashim
Universiti Malaysia Perlis, MALAYSIA

153 **Oxidation on Copper Lead Frame Surface Which Leads to Package**   654
Delamination
Lai Chin Yung, Lee Chai Ying, Cheong Choke Fei, Aw Tiam Ann and Soellner
Norbert
Infineon Technologies (M) Sdn. Bhd., MALAYSIA

154 **Measurement and Characterization of Hot Carrier Safe Operating Area** 659
(HCI-SOA) in 24V n-type Lateral DMOS Transistors
N. Soin, S.S. Shahabuddin and K.K. Goh
Universiti Malaya, MALAYSIA

155 **Characteristic Analysis of 1024-Point Quantized Radix-2 FFT/IFFT**   664
Processor
Rozita Teymourzadeh, Memtode Jim Abigo and Mok Vee Hong
UCSI University, MALAYSIA

Cluster 4

156 **Effect of Platinum Catalyst Loading on Membrane Electrode Assembly**   669
(MEA) in Proton Exchange Membrane Fuel Cell (PEMFC)
Norfarhanim Mohd Zahari and Azlan Abdul Aziz
Universiti Sains Malaysia, MALAYSIA

157 **Optimized Flow Field Bipolar Plate Design in Proton Exchange Membrane Fuel Cell** 674
Mohd Ikhwan Mohd Isa and Azlan Abdul Aziz
Universiti Sains Malaysia, MALAYSIA

158 **Fabrication of Cu_2ZnSnS_4 Thin film Solar Cells by the Spin Coating technique** 678
M.A. Olopade, O.E. Awe, A.M. Awobode, A. Oberafo and M.G. Zebaze Kana
University of Lagos, NIGERIA

159 **Structural Study and Sensitivity Measurements of ZnO based Ammonia (NH_3) Sensor** 682
S. Ahmad, N. D. Md Sin, M. H. Mamat, M.Salina, M. N. Berhan and M. Rusop
Universiti Teknologi MARA, MALAYSIA

160 **Design and Characterization of Bandgap Voltage Reference** 686
Yuzman Yusoff, Hanif Che Lah, Nabihah Razali, Siti Noor Harun and Tan Kong Yew
MIMOS Berhad, MALAYSIA

161 **An Improved P+/N Diode Leakage Current in BiCMOS Technologies with Fluorine Co-implant** 690
Siti Zubaidah Md Saad, Tan Chan Lik, Marhanis Abu Othman, Poehle Holger and Sukreen Hana Herman
Infineon Technologies (Kulim) Sdn Bhd, MALAYSIA

162 **Oxygen Uptake During the MBE Growth of $Al_xGa_{1-x}As$ Epitaxial Layers** 694
A. A. RahmanOthman, B. F. Usher, A. Loykaew and D. Nelson
La Trobe University, AUSTRALIA

163 **Characterization of NBTI by Evaluation of Hydrogen Amount in the Si/SiO_2 Interface** 699
Surya Kris Amethystna, Karuna Nidhi, Shao-Ming Yang, Gene Sheu, Jung-Ruey Tsai and Md Imran Siddiqui
ASIA University, TAIWAN

164 **Effect of Annealing Duration on the Memristive Behavior of $Pt/TiO_2/ITO$ Memristive Device** 703
N.S Kamarozaman, Z. Aznilinda, S.H Herman, R. A Bakar and M. Rusop
Universiti Teknologi MARA, MALAYSIA

165 **Effective Heat Dissipation of High Power LEDs Mounted on MCPCBs with Different Thickness of Aluminium Substrates** 707
Soon Bee Law, Anithambigai Permal and Mutharasu Devarajan
Universiti Sains Malaysia, MALAYSIA

166 **Rhombohedral In$_2$O$_3$ Thin Films Preparation from In Metal Film using Oxygen Plasma** 711
Subramani Shanmugan, Devarajan Mutharasu and Ibrahim Kamarulazizi
Universiti Sains Malaysia, MALAYSIA

167 **Capacitive Micro-Sensor for the detection of dextrose** 716
Q.Humayun, U.Hashim and M.Kashif
Universiti Malaysia Perlis, MALAYSIA

168 **Thermoelectric Properties and Devices of p-type Bi$_{0.4}$Sb$_{1.6}$Se$_{2.4}$Te$_{0.6}$ and n-type Bi$_2$Se$_{0.6}$Te$_{2.4}$ Prepared By Solid State Microwave Synthesis** 720
Arej Kadhim Abbas, Arshad Hmood Abd Al Kadhim and Haslan Abu Hassan
Universiti Sains Malaysia, MALAYSIA

169 **Fabrication and Characterization of Pb$_{1-x}$Yb$_x$Te Based Alloy Thin Film Thermoelectric Generators Using Thermal Evaporation Method** 725
Arshad Hmood Abd Al Kadhim, Arej Kadhim Abbas and Haslan Abu Hassan
Universiti Sains Malaysia, MALAYSIA

170 **Growth and Fabrication of AlGaN/GaN HEMT on SiC Substrate** 729
Yuen-Yee Wong, Yu-Sheng Chiu, Tien-Tung Luong, Tai-Ming Lin, Yen-Teng Ho, Yue-Chin Lin, Edward Yi Chang
National Chiao Tung University, TAIWAN

171 **Structural, Morphological and Photoluminescence Studies of SnO$_2$ Microparticles** 733
Karkeng Lim, Muhammad Azmi Abdul Hamid, Roslinda Shamsudin, Azman Jalar and N. H. Al-Hardan
Universiti Kebangsaan Malaysia, MALAYSIA

172 **Fabrication and Characterization of IDE ZnO Thin Films Using Sol-Gel Method for PBS Solution Measurement** 736
K.L.Foo, U.Hashim, Haarindra Prasad s/o RajintraPrasat and M.Kashif
Universiti Malaysia Perlis, MALAYSIA

173 **Surface Defect on SiC Ohmic Contact During Thermal Annealing** 740
Izhan Abdullah, Azman Jalar, Mohammad Azmi Abdul Hamid, Ishak Mansor and Burhanuddin Yeop Majlis
Universiti Kebangsaan Malaysia, MALAYSIA

174 **Fabrication of AlGaN/GaN HEMTs with Slant Field Plates by Using Deep-UV Lithography** 744
Ting-En Hsieh, Lu-Che Huang, Yueh-Chin Lin, Chia-Hua Chang, Huan-Chung Wang and Edward Yi Chang
National Chiao-Tung University, TAIWAN

175 **Influence of Post Deposition Annealing Temperatures on Electrical Properties of Al₂O₃/InSb MOSCAPs** 747
Hai-Dang Trinh, Yue-Chin Lin, Edward Yi Chang, Hong-Quan Nguyen, Shin-Yuan Wang, Yuen-Yee Wong, Binh-Tinh Tran, Quang-Ho Luc, Chi-Lang Nguyen and Chang-Fu Dee
National Chiao Tung University, TAIWAN

176 **Organic Field-Effect Transistors for Nonvolatile Memory Devices using Charge-Acceptor Layers** 750
Khairul Anuar Mohamad, Afishah Alias, Ismail Saad, Bablu Kumar Gosh, Katsuhiro Uesugi and Hisashi Fukuda
Universiti Malaysia Sabah, MALAYSIA

177 **Nanoindentation Creep Analysis of Gold Ball Bond** 755
Muhammad Nubli Zulkifli, Azman Jalar, Shahrum Abdullah, Norinsan Kamil Othman and Muhammad Azmi Abdul Hamid
Universiti Kebangsaan Malaysia, MALAYSIA

178 **The Effects of Mixed Electroluminescent (EL) Polymer Layer Thickness on the Single Layer Organic Light Emitting Diode (OLED) Performance** 759
Mohd Shahrul Akram Mohd Mokhtar, Muhamad Mat Salleh, Akrajas Ali Umar and Muhammad Yahaya
Universiti Kebangsaan Malaysia, MALAYSIA

179 **Fabrication of CuGaO₂ Films by Sol-gel Method for UV Detector Application** 763
Afishah Alias, Khairul Anuar Mohamad, Bablu Kumar Gosh, Masato Sakamoto and Katsuhiro Uesugi
Universiti Malaysia Sabah, MALAYSIA

180 **The Parasitic Reaction During the MOCVD Growth of AlInN Material** 766
Wei-Ching Huang, Yuen-Yee Wong, Kusan-Shin Liu, Chi-Feng Hsieh and Edward Yi Chang
National Chiao Tung University, TAIWAN

181 **Analysis of Energy Harvesters for Powering a Wireless Sensor Node Device** 769
Asral Bahari Jambek, Choo Pey See and Uda Hashim
Universiti Malaysia Perlis, MALAYSIA

182 **Development of Microstructure on Polysilicon Substrate by Reactive Ion Etching (RIE) for Future Reproductivity of Nanogap** 774
Q.Humayun and U.Hashim
Universiti Malaysia Perlis, MALAYSIA

Nanotechnology and Metamaterials: Conceptualization and Intersection for New Opportunities.

Akhlesh Lakhtakia
Department of Engineering Science and Mechanics
Penn State University
University Park, PA 16802-6812, USA
Email: akhlesh@psu.edu

Although the worldwide nanotechnology market is currently estimated to be 2 trillion dollars, its continuing growth presents both opportunities and challenges. The opportunities are concentrated in the reliable and inexpensive fabrication of nanomaterials, integrated electronics and optoelectronics, bionanotechnology and nanomedicine, and nanometrology. The health impacts of proliferating nanotechnology continue to cause concerns; the convergence of nanotechnology, biotechnology, information technology, and cognition science could lead to reduced individual liberties; and the specter of a "nanodivide" between the rich and the poor still looms.

InAs HEMT for Terahertz Applications

Edward Yi Chang
Department of Materials Science and Engineering, & Department of Electronics Engineering,
National Chiao Tung University
1001 Ta-Hsueh Rd., Hsinchu,
Taiwan, R.O.C.
E-mail: edc@mail.nctu.edu.tw

InP based high indium concentration high-electron mobility transistors (HEMTs) are an attractive transistor technology for millimeter-wave and terahertz applications[1]. Ultra-short gate length (Lg) InAlAs/InGaAs HEMTs on InP substrate has been fabricated and demonstrated excellent RF-performance over the past decade[2,3]. Higher electron mobility and drift velocity can be realized by reducing of Lg and increasing the indium content in the $In_xGa_{1-x}As$ channel. Due to the high electron mobility, velocity and large conduction band offset in InAs, InAs-channel HEMT is promising for high speed and low power logic applications[4]. In this talk, $InAs/In_xGa_{1-x}As$ channel InP HEMTs are fabricated and evaluated for high-frequency applications.

The epitaxial structure of the InAs-channel HEMTs were grown by MBE on InP substrate. When Lg of the T-shaped gate was 80 nm. Superior drain current density of 1015 mA/mm was achieved with an extremely high transconductance (gm) of 1900 mS/mm when the drain voltage (VDS) was biased at 0.5 V, it indicated that the In-rich $In_xGa_{1-x}As$-channel HEMTs can be biased at a low VDS to reduce overall dc power consumption, while maintaining relatively high current density and gm. The S-parameter of the $In_xGa_{1-x}As$ HEMTs was measured using Cascade Microtech™ on-wafer probing system with vector network analyzer from 1 to 110 GHz. The current gain cutoff frequency (fT) and maximum oscillation frequency (fmax) of the device were extracted to be 393 GHz and 260 GHz at the low VDS = 0.5 V, respectively. Moreover, the dc power dissipation was 5.8 mW when the device was biased at peak fT and fmax.

Forty-nanometer InAs HEMT fabricated by two-step recess and Pt-buried gate were also fabricated and evaluated for low-noise and low-power sub millimeter-wave applications. Interestingly, a high fT of 663 GHz was obtained for the InAs HEMTs when the device was biased near the occurrence of impact ionization. To demonstrate the capability of the proposed device for low noise applications at high frequencies, noise measurement was performed on the device with DC bias set as the same for maximum fT. The overall minimum noise figure is below 2.5 dB with frequency ranging from 18 GHz to 64 GHz, and the corresponding associated gain is 7 dB at 64 GHz while the total DC power consumption remains as low as 4.3 mW. Overall, these superior results show great potential of 40-nm InAs HEMTs for ultra low-power, low-noise sub-millimeter wave and terahertz applications.

Product Innovation enabled by MEMS

Nico F. de Rooij
Institute of Microengineering
EPFL STI
Rue Jaquet-Droz 1
Case postale 526
CH-2002 Neuchâtel
Switzerland
Email :nico.derooij@epfl.ch

An overview will be given of current R&D activities at EPFL's SAMLAB and at CSEM focusing on product innovation exploiting MEMS as the key technology. In particular examples will be presented from a variety of applications dealing with life sciences, communication technology, advanced instrumentation, space, environmental monitoring and luxury goods. It will be demonstrated that MEMS has allowed successful product innovations.

Thermal Effects in InAs/GaAs Quantum Dot Vertical Cavity Surface Emitting Lasers

Yoon Soon Fatt
School of Electrical and Electronic Engineering
Nanyang Technological University
50 Nanyang Avenue
Singapore 639798
Rep. of Singapore
Email : ESFYOON@ntu.edu.sg

The performance of 1.3-µm InAs/GaAs quantum dot (QD) vertical cavity surface emitting lasers (VCSELs) is adversely affected by self-heating effect. In this talk, a self-consistent model comprising rate equations and thermal conduction equation will be presented to analyze the influence of self-heating on the carrier dynamics and output power of QD VCSELs. The simulation results indicate that the low output power is attributed to hole thermalization and escape due to the thin wetting layer. The hole confinement can be improved by increasing the number of QD layers and surface density, as well as adopting p-type modulation doping. The fabricated p-doped QD VCSELs exhibit high temperature stability in the threshold current. The highest output power of 0.435 mW and lowest threshold current of 1.2 mA under single-mode operation were achieved, with side mode suppression ratio (SMSR) of 34 dB at room temperature (RT). However, the output power is limited by the small-sized oxide apertures. To achieve both high output power and enhancement of the fundamental mode emission, a dielectric-free (DF) approach with surface-relief (SR) technique is applied in our device fabrication. Compared with the conventional dielectric-dependent (DD) method, the DF approach potentially reduces the fabrication cost and complexity. Moreover, with the same oxide aperture area, the differential resistance is reduced by 36.47% and output power is improved by 78.32% under continuous-wave (CW) operation. The output power increases up to 3.42 mW under pulsed operation with oxide aperture diameter of ~15 µm. The surface-relief technique effectively enhances the fundamental mode emission of the QD VCSEL.

Broadband Semiconductor Lasers and Their Applications

Ooi Boon Siew

Division of Physical Sciences and Engineering, King Abdullah University of Science & Technology (KAUST),
Thuwal 23955-6900, Saudi Arabia.
Email: boon.ooi@kaust.edu.sa

Self-assembled semiconductor quantum dots (QDs) constitute a class of nanoscale materials that provide fundamental advantages compared to the dominating 2-D quantum well (QW) structures in photonic device applications. QD devices based on InAs/GaAs material system operating at emission wavelengths between 1.0 and 1.3 μm have achieved a relative maturity and many outstanding performances such as low threshold current, high temperature stability, high gain and differential gain, have already been demonstrated. For long wavelength operations in the S-C-L communication bands, particularly for the 1.55 μm window, the InAs/InP QD and quantum-dash (Qdash) material systems have been seen as the most suitable material system. Recent attempts in extending the technology of self- assembled QD on InP substrate has led to the development of InAs-based Qdashes that give the wavelength spans of ground state (GS) transition over several bands of optical telecommunication windows between 1.4 and 2.0 μm. Apart from its superior characteristics as compared to conventional quantum well structures and apart from its predominant applications in optoelectronics industry, self-assembled QD/Qdash lasers have demonstrated a number of unique features like broad emission spectra which have been attributed to the carrier localization in non-interacting or spatially isolated dot/dash employing a highly inhomogeneous QD/Qdash structures. These novel semiconductor light emitters are particularly attractive for novel many practical imaging and sensor applications due to their compactness and relatively low energy requirement in comparison to other state-of-the-art broad-spectrum light sources.

In this talk, we will present the recent development of both GaAs- and InP-based QD and Qdash broadband lasers. The technology for growing these nanostructures, as well as the various technologies for engineering the bandgap of the QD/Qdash systems using epitaxy growth technique and postgrowth intermixing methods will be presented. Improvement of the QD/Qdash laser characteristics using various active region designs, and recent development in theoretical analysis of QD/Qdash laser, particularly, the effect of the quantum dot/dash inhomogeneity on laser characteristics will also be reviewed. At device level, we will focus our discussion on broad gain semiconductor optical amplifiers (SOA), mode-locked lasers, comb-laser and broadband lasers fabricated on these material systems. The applications of broadband lasers in telecommunications, biomedical imaging using optical coherence tomography (OCT), spectroscopy and sensing, etc, will be discussed.

978-1-4673-2395-6/12 $31.00 © 2012 IEEE

Challenges and Opportunities for IC Design and Fabrication in Malaysia

Kamarulzaman Mohamed Zin
Silterra Malaysia Sdn.Bhd.
Lot 8, Phase II, Kulim Hi-Tech Park
09000 Kulim, Kedah Darul Aman,Malaysia
Email: kamarulzaman@silterra.com

The Electrical and Electronics Industry in Malaysia has consistently been the highest contributor to the annual Malaysian exports despite the cyclical and volatile nature of the industry. For 2011, as in the previous years, it has even outperformed the Oil & Gas, Palm Oil and Chemicals & Chemical Products sectors of the economy, even if they are all combined.

It is however realised that the true measure of economic significance is the local value-adding of the Malaysian activity. For although the exports are high, the trade statistics also show that imports are also high. The gross numbers show that local value-adding is in the 15-25 % range; which present a significant scope (and challenge) for the local players to rise up to.

This paper describes the efforts that are currently underway to increase the local value-adding and challenges it faces to make it a sustainable industry transformation exercise.

Statistical Modeling of Solar Cell using Taguchi Method and TCAD Tool

M. S. Bahrudin, S.F. Abdullah and I. Ahmad
Center for Micro and Nano Engineering (CeMNE)
Department of Electronics and Communication Engineering, College of Engineering,
Universiti Tenaga Nasional,
Jalan IKRAM-UNITEN, 43000 Kajang, Selangor, Malaysia

Abstract- **This paper focuses on optimizing silicon based solar cell fabrication using Taguchi Optimization Method (TOM). Optimization focused on 3 parameters namely doping concentration of boron, creating phosphorus PN-junction and energy used for ion-implantation with 2 noise factors, Diffuse time and diffuse temperature. The aim is to have a shallow junction in order to decrease the recombination process but higher fill factor (FF) for better efficiency. Fabricating are done in computer simulation environment by Silvaco TCAD software that also conducting an electrical testing for measurement. Each factor (product from the parameters through TOM) has 2 levels of best values taken from the previous researches. In this research, L_8 orthogonal array consists of 8 set of different combination of experiment has been done. Optimized values are analyzed by finding Signal to Noise Ratio (SNR) of each experiment and applied it on Larger the Better (LTB) for highest FF and Smaller the Better (STB) for shallowest junction depth. Result reveal that boron at concentration of 5.0×10^{15} cm^{-3}, phosphorus at concentration of 2.0×10^{16} cm^{-3}, and energy at 10 keV gave a result of 0.3 um ~ 0.5 um for junction depth and stable FF value of 0.8 at any noise factor contributing efficiency of 15% to 16%. As a conclusion, TOM has achieved predicting the best solution for optimizing silicon solar cell fabrication.**

Index Terms—Solar cell, Taguchi Optimization Method, PN-junction, ion-implantation and Silvaco TCAD software.

I. INTRODUCTION

Silicon solar cells are basically a creation of n-type and p-type semiconductor material, combining the two create a PN-junction. The PN-junction layered with anti-reflecting coat (ARC) to trap more light into the cell and connected with the contact circuit that link the solar cell with the load.

Electron are made free to move around by energy from the absorbed light. Thus it generates an electron-hole pair [1]. This higher energy remained only for a short of time [2]. When this occurs, it settles back into an empty valance band state and effectively removes the hole. This process called recombination and it affected solar cell's efficiency.

Inside fabricating process, permitting high doping level increase open circuit voltage and lead to tolerance of impurities and defect [3]. But the doping profile has to be carefully controlled because heavy doping enhances a recombination and affecting the solar cell efficiencies [4].

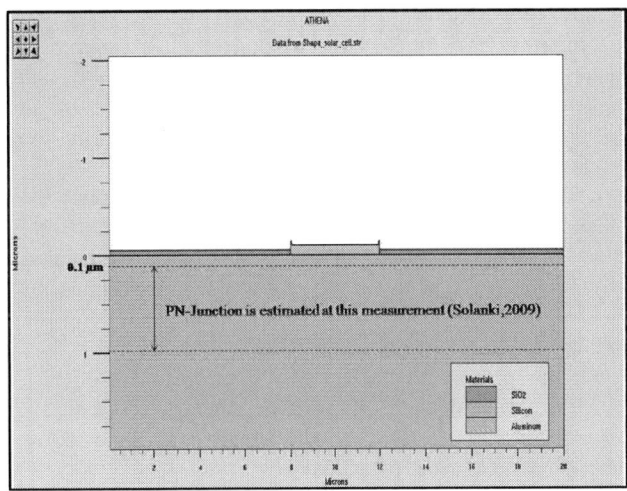

Fig. 1 Solar Cell structure and location of PN-Junction

The depth of the junction is measured from the first surface of solar cell as shown in Fig. 1. This is the place where the electrons diffuse into the p-type region and the hole diffuse into n-type region. They are then collected by the electrodes resulting difference voltage and when connected to load electric current flow [5]. Junction depth is range between $0.1\mu m$ to $1.0\mu m$ at 10 to 60 minutes diffusion time and 800 °C to 950°C temperature [1].

Most electron-hole pairs are generated at nearer to the surface from where the sunlight initially enters into the solar cell. The generation rate decay exponentially with the depth of the material [1].

This implies that the PN-junction must be as closed as possible to the surface. It is also implied that by reducing the junction depth will decrease the cell processing time ultimately leading to appreciable output [4]. Since the junction is not to be brought too close to the surface as it increase resistive loss [1] which will be effected the open circuit voltage (V_{oc}). Therefore optimum depth is introduced in this project for shallower junction depth.

FF, which is a ratio of a maximum electrical power available from the cell i.e. at the point of maximum current (J_m) and maximum voltage (V_m) to the product of V_{oc} and short circuit current (J_{sc}). Visually, it defined as the squareness of the I-V curve

The FF is mainly related to the resistive losses in a solar cell [1]. Ideally its value can be 100% correspond to I-V curve square but of course it is not feasible to have a square I-V curve as there always some losses, thus reduce the FF values. A good solar cell typically has FF values more or close to 0.80 (Julian, 2011 & Solanki, 2009).

II. TAGUCHI OPTIMIZATION METHOD

This study focused on optimizing the optimum value of fabricating and designs of the solar cell using Taguchi Optimization Method (TOM) . Created by Genichi Taguchi it is primarily used by industrial engineering for experiment planning to achieve in designing a product efficiently and high reliability. In this method, effects of various factors on design are calculated then by controlling dominant factor, the product is optimized in order to achieve better quality [6].

TOM is basically a systematically approach to study the process parameter of the experiment with small number of experiment. Inside TOM, the experiment is set into matrix form called orthogonal array. This matrix experiment is conducted using special matrices called orthogonal array [7]. In this project L_8 type of orthogonal was used.

Various process parameters setting are changed for each performed experiment. Following the conduct of the matrix experiment, all the data are set together and analyzed to determine the effect of the parameters.

III. CELL DESIGN AND L8 ORTHOGONAL ARRAY

The designs of solar cell were based on 2 models [8] and [9]. All designs was constructed using ATHENA, a silicon wafer with orientation of <1 0 0> and 50 μm of thickness.

As a group III material in periodic table, Boron with 3 covalent atoms made one electron out of four covalent bonds for silicon lattice missing, and the dopant atom became acceptors thus creating a p-type region.

For the n-type region, the phosphorous impurity is doped into the Silicon layer by ion implantation. Phosphorus is a group V material in periodic table with 5 valence electrons, hence made extra one electron in the Silicon lattice and the dopant atom become a donors.

The energy for ion implementation was set at 10 keV and 30 keV on both modeling process. The result on the example of fabricated process the solar cell is shown in Fig. 2.

It is noticed that the anode contact of the solar cell is not visible but it was located at the very bottom of the structure.

The top layer of the Silicon was oxidized to create a thin layer of Silicon Oxide (SiO_2). This was the ARC layer use to trap more light into the solar cell for higher efficiencies.

There were 3 parameters of interest that had been done in this project for optimization, hence produced various factors.

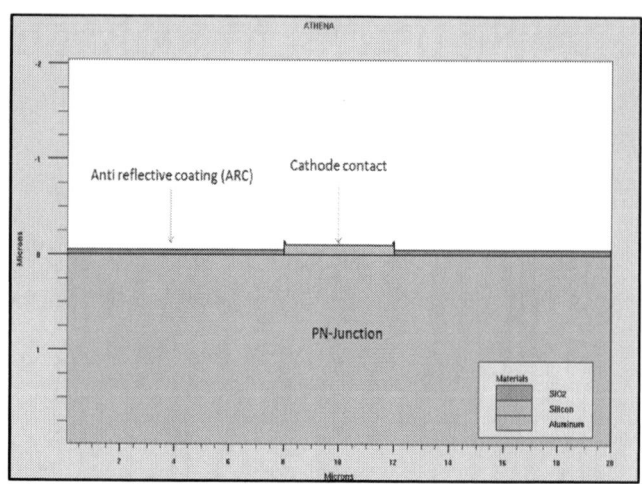

Fig. 2 Fabricated solar cell structures

In order to achieve better quality products, the dominant factors need to be optimized.

L_8 orthogonal array has been utilized in this project which consists of 3 control factors with 2 levels for each factor. TABLE 1 summarized the variance of values involved for the control factors and their respective level.

TABLE 1
CONTROL FACTORS AND THEIR RESPECTIVE LEVELS

Parameters (Control factors)	Level	
	1	2
Phosphorus concentration ($/cm^3$)	1.0×10^{14}	2.0×10^{16}
Boron concentration ($/cm^3$)	5.0×10^{15}	1.0×10^{19}
Energy (KeV)	10	30

Value for level 1 was taken from research done by [8] and level 2 was by [9].

The noise factor was defined by varying the diffusion temperature at 800 ºC to 950 ºC with 50 ºC increment and diffusion time at range of 10 to 30 minutes [3]. The repetitions are done 4 times and the combination setting are shown in TABLE 2.

Using L_8 orthogonal array, the control factors and respective level was made into 8 set of experiments. Each experiment has different combination of level showed in TABLE 3.

TABLE 2
COMBINATION OF NOISE FACTORS

Noise Factor	Repetitions			
	1	2	3	4
Diffuse time	10	15	20	25
Temperature	800	850	900	950

TABLE 3
VARIANT OF 3 FACTORS AND 2 LEVEL ON L₈ ORTHOGONAL ARRAY

Experiment Set No	Control Factors		
	Phosphorus concentration (/cm³)	Boron concentration (/cm³)	Energy (KeV)
1	1.0×10^{14}	5.0×10^{15}	10
2	1.0×10^{14}	5.0×10^{15}	30
3	1.0×10^{14}	1.0×10^{19}	10
4	1.0×10^{14}	1.0×10^{19}	30
5	2.0×10^{16}	5.0×10^{15}	10
6	2.0×10^{16}	5.0×10^{15}	30
7	2.0×10^{16}	1.0×10^{19}	10
8	2.0×10^{16}	1.0×10^{19}	30

IV. RESULT AND DISCUSSION

Upon completing all set of experiments. The result on Fill Factors and Junction Depths are showed in TABLE 4 and TABLE 5.

TABLE 4
RESULT ON FILL FACTOR

Experiment Set No.	Fill Factor (FF)			
	1	2	3	4
1	0.75052	0.75052	0.75053	0.75062
2	0.75122	0.75121	0.75114	0.75111
3	0.75090	0.75095	0.75103	0.75114
4	0.75098	0.75099	0.75088	0.75020
5	0.79838	0.79856	0.79856	0.79955
6	0.81286	0.81285	0.81282	0.81303
7	0.81314	0.81285	0.81307	0.81323
8	0.81303	0.81310	0.81277	0.81213

TABLE 5
RESULT ON JUNCTION DEPTHS

Experiment Set No.	Junction Depths			
	1	2	3	4
1	0.03418	0.04232	0.06129	0.11691
2	0.41714	0.41756	0.42029	0.43450
3	0.07874	0.11570	0.18775	0.30022
4	0.16137	0.18312	0.28100	0.49939
5	0.01871	0.02223	0.03046	0.05791
6	0.24675	0.24704	0.24890	0.32323
7	0.06175	0.09797	0.16873	0.26664
8	0.13850	0.16884	0.26458	0.46966

These result in both TABLE 4 and TABLE 5 are then attributed for calculating SNR. For FF, the SNR is determined by LTB equation [7] give by

$$\eta = -10 \log_{10}\left[\frac{1}{n}\sum\left(\frac{1}{Y_1^2} + \frac{1}{Y_2^2} + \cdots + \frac{1}{Y_n^2}\right)\right] \quad (1)$$

LTB is justified to be used for FF signal-to-noise ratio's calculation for its LTB type characteristic. Complied by [7] explanations, FF values are at its best when larger and performance (or efficiency) is progressively better when these values increase.

For Junction depth, the STB equations are used for calculating their SNR. STB equation [7] are given by

$$\eta = -10 \log_{10}\left[\frac{1}{n}\sum(Y_1^2 + Y_2^2 + \cdots + Y_n^2)\right] \quad (2)$$

STB are also justified to be use for Junction Depth signal-to-noise ration's calculation as explained by [7]. In this project, Junction Depth values are aimed to be smaller but can never take negative values. Also their critical performance a progressively increases as the value decreases. But as mentioned in introduction, having the junction depth value too small or too near to zero sacrifice the open circuit voltage so optimum value is tallied with the FF for the best result.

Applying Equation (1) and (2) for respective factor effects give SNR values. Both SNR result are represented side by side in TABLE 6.

TABLE 6
SNR VALUES ON FF AND JUNCTION DEPTH FOR EACH EXPERIMENTS

Experiment Set No.	Signal to Noise Ratio (dB)	
	Fill Factor (FF)	Junction Depth
1	-2.49197	17.82384
2	-2.48519	6.65842
3	-2.48635	10.73898
4	-2.49523	6.11412
5	-1.94291	23.86881
6	-1.79925	9.48113
7	-1.79557	11.99317
8	-1.80218	6.69561

From TABLE 5, the SNR values for each level of control factors on both factor effect are calculated. The results are respectively represented in TABLE 7 and TABLE 8.

TABLE 7
SNR VALUE FOR EACH LEVEL AND THEIR TOTAL MEAN FOR FILL FACTOR

Control Factors	Level		Total Mean
	1	2	
Phosphorus concentration (/cm³)	-2.48967	-1.83498	
Boron concentration (/cm³)	-2.17981	-2.1448	-2.16232
Energy (KeV)	-2.17920	-2.14545	

TABLE 8
SNR VALUES FOR EACH LEVEL AND THEIR TOTAL MEAN FOR JUNCTION DEPTH

Control Factors	Level		Total Mean of SNR
	1	2	
Phosphorus concentration (/cm³)	10.3338	13.00970	
Boron concentration (/cm³)	14.45810	8.88540	11.67176
Energy (KeV)	16.10620	7.23730	

The control factors are investigated through analysis of variance or ANOVA and the investigation are aimed to verify more accurately the optimum combination of the control factors that affect the result. The result of ANOVA is as in TABLE 9.

TABLE 9
RESULT ON ANOVA FOR BOTH FILL FACTOR AND JUNCTION DEPTH

	Control Factor	Degree Of Freedom	Sum of Square	Factor Effect on SNR (%)
Fill Factor	Phosphorus Concentration	1	0.4298	98.91
	Boron Concentration	1	0.0024	0.56
	Energy	1	0.0023	0.54
Junction Depth	Phosphorus Concentration	1	7.1601	6.31
	Boron Concentration	1	31.0537	26.57
	Energy	1	78.6570	67.30

The factor effect on SNR percentage indicate the priority of the factor where the factor with highest percentage have a great influence on the performance of the device. [10].

The result shows that Phosphorus concentration has the highest influence for higher FF at 98.91% almost dominating all other factor effects. The bigger concentration on the phosphorus applied the higher FF is achievable.

As for junction depth it is show that Energy plays the most principal factor as agreed by [4]. For shallower junction depth smaller energy should be applied during fabrication process. Other factor contributing to Junction depth is Boron concentration (26.57%) and Phosphorus concentration plays minor role.

The best value as predicted by TOM area is shown in TABLE 10.

TABLE 10
BEST SETTING AS PREDICTED BY TAGUCHI OPTIMIZATION METHOD.

Control Factors	Represented symbol	Best Value
Phosphorus concentration (/cm^3)	P2	2.0×10^{16}
Boron concentration (/cm^3)	B1	1.0×10^{19}
Energy (KeV)	E1	10

From the results on TABLE 10, final simulation on the Silicon solar cell is performed by using the best parameter values predicted by TOM and the results on the final simulation is shown in TABLE 11.

TABLE 11
RESULT FROM FINAL SIMULATION BASE ON BEST VALUES

Factor Effect	Repetitions			
	1	2	3	4
Fill Factor	0.80313	0.80304	0.802366	0.801262
Junction Depth	0.321135	0.321474	0.323665	0.417379

V. CONCLUSION

As a conclusion, TOM successfully achieved prediction of the optimum values of control factors in fabricating Silicon solar cell. It also accomplished determining the dominant factors that contributing both FF and Junction depth. Nevertheless, there always a room for fine tuning of the process parameter value in order to get the best result. For future works it is suggested that application of bigger orthogonal array like L_{18} and L_{24} with more factor level is used for best prediction and stronger result.

ACKNOWLEDGMENT

This work was financially supported by the Ministry of Higher Education (MoHE), Malaysia with the research grant no: 01101031 FRGS.

REFERENCES

[1] Chetan Singh Solanki, *Solar Photovoltaic: Fundamental Technologies And Application*. New Delhi: PHI Learning Limited, 2009.

[2] N.A. Rashied, M. Rusop M.H. Abdullah, "Layer Thickness Analysis of Silicon Solar Cell," *2010 Internation Conference on Electronic Devices, Systems and Application (ICEDSA2010)*, pp. 149-153, 2010.

[3] Tomas Markvart, Luis Castaner, *Solar cells : materials, manufacture and operation*. Oxford: Elsevier, 2005.

[4] Upadhyaya Ajay, Yelundur Vijay, and Rohatgi Ajeet, "High Efficiency Mono-Crystalline Solar Cells with Simple Manufacturable Technology," in *University Center of Excellence for Photovoltaics Conference Papers*, Atlanta, 2006.

[5] Safa O Kassap and Peter Capper, *Springer Handbook of Electronic and Photonic Material*. Saskatoon,Canada: Springer, 2006.

[6] Vinayak Hande, Maryam Shojaei Baghini, and Prakash Apte, "Design and Optimization of High Precision CMOS Voltage Reference Using Taguchi Orthogonal Array Technique," *International Sysmposium on Integrated Circuits*, pp. 576-578, 2011.

[7] Madhav Phadke, *Quality Engineering using Robust*

Design.: Pearson Education, 2008.

[8] K. Sopian, Y. Othman, H.R. Fallah F. Jahanshah, "pn junction depth impact on short circuit current of solar cell," *Science Direct*, pp. 1629-1633, 2009.

[9] H. John, G. Gary V. Frank, "PV Lesson Plan 1 - Solar Cells," Oregon, 2000.

[10] Ugur Esme, "Application of Taguchi Method for the Optimization of Resistance Spot Welding," *The Arabian Journal for Science and Engineering*, pp. 34:2B, 2009.

[11] Robert Castellano, *Solar Panel Processing*. Philadephia: Old City Publishing, 2010.

[12] Eddy Simoen Cor L. Claeys, *Radiation Effects In Advanced Semiconductor Materials And Devices*. Berlin: Springer, 2002.

Analysis of A MEMS Piezoelectric Microgenerator Using Network Placement Method

Luay Yassin Taha[1], *Member, IEEE,* Burhanuddin Yeop Majlis[2], *Member, IEEE*
and Ahmad Al Ali[3]
[1]Emirates Aviation College
Email:luaytaha59@hotmail.com
P. O. Box: 53044; Dubai, United Arab Emirates
[2]UKM University, Institute of Microengineering and Nanoelectronics (IMEN),
Universiti Kebangsaan Malaysia; 43600, Bangi, Selangor, Malaysia
Email: burhan@vlsi.eng.ukm.my
[3]Emirates Aviation College, P. O. Box: 53044; Dubai, United Arab Emirates
Email: dr.alali@emirates.com

Abstract— **This paper presents a new approach for analyzing a MEMS piezoelectric microgenerator based on the network placement method. First, new piezoelectric model is proposed by modifying the device model using its equivalent closed loop transfer function. The network placement method is then used to vary the microgenerator poles and zeros thus altering the generated voltage and power. The new model can be used to calculate the load that produces certain transient parameters. The model is simulated by applying a pulse force of 1 N amplitude and 1.2 ms width. A considerable improvement on the average voltage and apparent power curves is recorded by placing a Wein Bridge passive network having $R = 278.2$ MΩ and $C = 1$ nF. The optimum voltage and power are 3.8 V and 8.8 μVA, respectively.**

I. INTRODUCTION

In piezoelectric microgenerator terminology, the device can be modeled using the transformer equivalent circuit [1]–[4] shown in Fig.1. In our previous work, the transfer function $V(s)/F(s)$ is derived by referring to the primary side and applying electrical circuit analysis [5]:

$$\frac{V(s)}{F(s)} = \frac{C_m Z_L \phi}{b_3 s^3 + b_2 s^2 + b_1 s + b_o} \quad (1)$$

where:

$$b_3 = C'_p Z_L M_m C_m$$
$$b_2 = M_m C_m + C'_p Z_L B_m C_m$$
$$b_1 = B_m C_m + C'_p Z_L + \frac{d_{33} Z_L}{\phi}$$
$$b_0 = 1$$

$$(2)$$

It is clear from (1) and (2) that the piezoelectric microgenerator transfer function is altered by the electrical parameters (Z_L *and* $C'p$), the mechanical parameters (B_m, M_m and C_m), and (Φ). Varying these parameters result in changing the coefficients of the transfer function

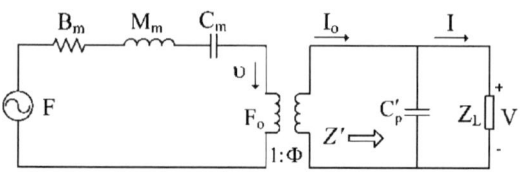

Fig.1 Piezoelectric microgenerator transformer model. (F = input force, B_m = damping of the piezoelectric element, M_m = mass of the piezoelectric element, C_m = Piezoelectric short circuit compliance, F_0 = net force transferred into electrical voltage, υ = velocity of the piezoelectric element in the z-axis direction, I_o = current passing through the piezoelectric element, C'_p = blocked capacitance, Φ = mechanical to electrical conversion ratio, Z_M = Piezoelectric mechanical impedance, I = output load current, Z_L = electrical Load impedance, V = generated output voltage).

and hence changing the positions of poles and zeros. These changes affect the transient response specifications such as the settling time, the overshoot and the steady state output voltage [6]. As the individual poles and zeros are calculated from the roots of the transfer function and since these roots depend upon the mixed electrical and mechanical parameters, the study of the effect of varying each parameter on the output voltage and power using the transformer modeling approach will be analytically difficult. However, different simulation and experimental approaches are recorded in previous work [1]-[5], [7]-[15]. In these approaches, some parameters such as the device dimensions and the piezoelectric material type are assumed to be fixed. The electrical parameter Z_L and the input vibrating frequency are varied to address the optimum voltage or power. The study of the effect of changing the device dimensions on the output voltage and power is studied in our previous work [4]-[5] and is also recorded in [9]-[10]. For a pulse applied force, the effect of varying device dimensions and material types on the output power is recorded in [16]-[17].

A general look at the above overview clearly shows the following. *First,* the transformer model is used as a two-port network to analyze the electromechanical circuit and address the optimum parameters. *Second*, the analysis is based on finding the device-equivalent model, which is an open-loop model. *Third*, due to the complexity of the piezoelectric generator transfer function the transformer model cannot be represented by a block diagram consisting of separate electrical and mechanical transfer functions; therefore, making it difficult to investigate a pole zero analysis.

The aim of this paper is to produce a new piezoelectric microgenerator closed-loop model called *the inner feedback loop (IFL)* model. The model's forward transfer function consists of clearly separated electrical and mechanical transfer functions. This feature drastically simplifies the analysis and design phases. The new model uses the pole placement method and passive network analysis and synthesis to study the effect of varying the electrical network poles and zeros on the positions of closed loop poles and zeros. This helps designers in selecting the device's electrical and mechanical components, resulting in an improvement in its output properties such as voltage and power.

II. THE INNER FEEDBACK LOOP MODEL

From Fig.1, the mechanical–electrical analogy, the electrical circuit analysis, and the Laplace transform method, the followings can be written:

$$\frac{F - F_o}{\upsilon} = Z_M = B_m + s M_m + \frac{1}{s C_m} \tag{3}$$

$$\frac{1}{\phi} = \frac{F_o}{V} = \frac{I_o}{\upsilon} \tag{4}$$

$$V = I_o Z' \tag{5}$$

$$Z' = Z_L // \frac{1}{s\, C_p'} = \frac{Z_L}{1 + s\, C_p' Z_L} \tag{6}$$

where Z' is the equivalent electrical impedance seen at the secondary side and s is the Laplace transform operator. From (3)–(6), and Fig.1, we constructed the new piezoelectric microgenerator model shown in Fig. 2.

The new model forms a negative feedback loop with a forward transfer function $G(s) = Z'/\phi Z_M$ and a feedback transfer function $H(s) = 1/\phi$. As this feedback is established without any external connection, we assigned the name *inner feedback loop (IFL)* model.

The *IFL* model has many merits. *First*, pole placement approach can be applied to the model to improve the device output parameters. *Second:* there is a complete separation between the mechanical and electrical transfer functions. *Third:* variations of Z_L and Z_M affect the open and closed loop zeros and poles.

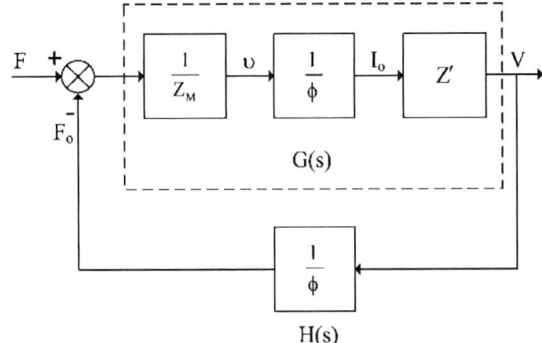

Fig. 2   The inner feedback loop (IFL) model

III. ANALYSIS USING NETWORK PLACEMENT METHOD

The network placement method is now applied to the new model to investigate the effect of varying the load impedance Z_L on the positions of open and closed loop poles and zeros. Fig. 3 illustrates typical root loci plotted for different passive networks selected in this paper. The device parameters are chosen according to Table I. The type of the piezoelectric element used is PZT-5H [18].

Higher order passive networks can be connected to place complex poles and zeros in the s-plane. This may result in a considerable shifting of the root loci to the right, therefore improving the time response and consequently the output voltage and power. Fig. 4 illustrates the result of placing a Wein bridge network.

The results shown in Fig. 3 and 4 cannot be obtained using the conventional piezoelectric microgenerator transformer model since the root locus method requires forward and feedback transfer functions. However, the new *IFL* model is a closed loop model which can used to obtain the root locus of the system. The *IFL* model has the advantage of identifying the dominant pole (DP) region thus simplifying the overall transfer function. The DP regions are illustrated in Fig. 3 and 4. Furthermore, all transient response parameters such as settling time, damping, and overshoot, can be easily extracted using the *IFL* model and Matlab tool such as *rltool*.

Next, a Matlab program is written to determine and plot the average voltage and power generated from the piezoelectric element type PZT-5H when a narrow pulse force of 1 N amplitude and 1.2 ms width is applied. The program uses both the open and *IFL* models. The type of connected load is the same as in section III. The loads are varied according to Table II.

TABLE I
PIEZOELECTRIC GENERATOR PARAMETERS

Parameter	Value	Unit
A	1.963×10^5	μm^2
t	500	μm
d_{33}	370×10^{-12}	C/m
s_{33}^E	15×10^{-12}	m/N
ε_r	1200	-
C_p'	0.58	pF

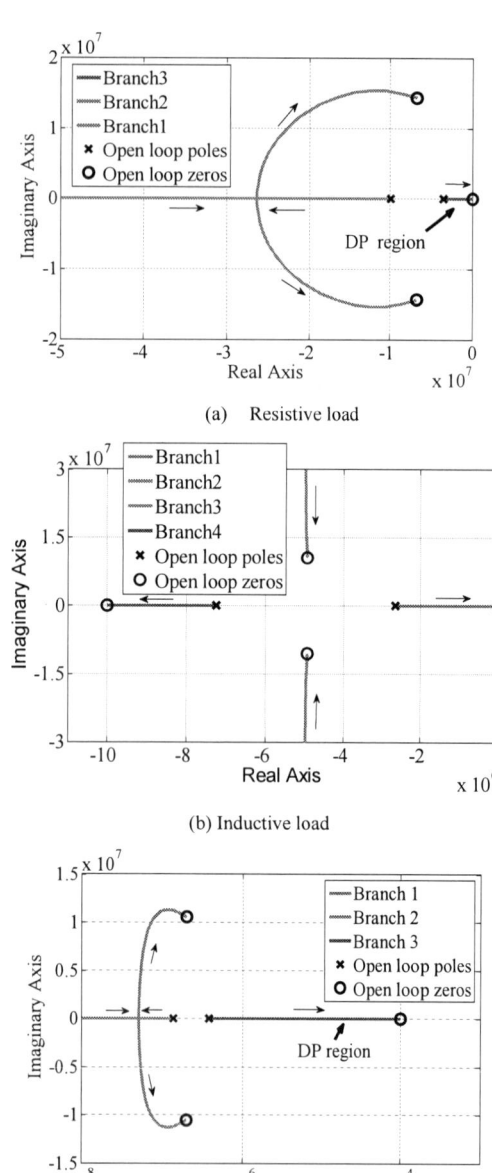

(a) Resistive load

(b) Inductive load

(c) Capacitive load

Fig. 3 Root locus of different connected networks and $B_m = 10$ Ns/m. (DP = dominant poles)

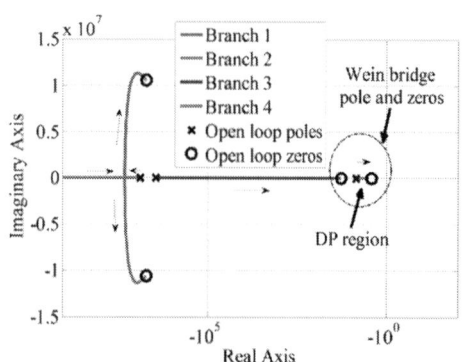

Fig. 4 Root locus of Wein bridge network, $B_m = 10$ Ns/m

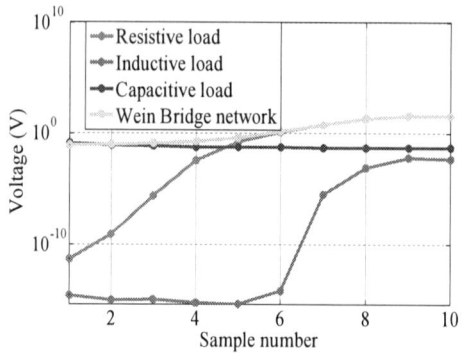

(a) Voltage design curves

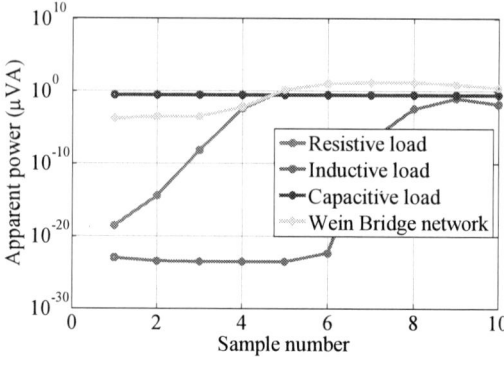

(b) Power design curves

Fig. 5 Design curves obtained from simulation using open and *IFL* models

IV. RESULTS AND DISCUSION

The results from simulation for both open and *IFL* models are identical and are plotted as shown in Fig. 5. These plots can be used for design purposes to identify the required load that produces certain voltage and power. A considerable improvement in the average voltage and power are recorded for Wein bridge network with $R = 278.2$ MΩ and $C = 1$ nF (sample 9). These values produce maximum voltage and power at 3.8 V and 8.8 μVA, respectively. Although both models produce the same results, the *IFL* model, ass explained in section III, can be used to simplify the analysis and design of the system using *rltool*. For example, it can be used extract the transient response parameters for certain output.

TABLE II
LOAD VARIATIONS USED IN SIMULATION

Sample Number	Resistive load	Inductive load	Capacitive load	Wein bridge network
	MΩ	mH	nF	MΩ
1	0.01	0.001	1	0.01
2	0.036	0.0046	2.15	0.036
3	0.13	0.0215	4.64	0.13
4	0.46	0.1	10	0.46
5	1.67	0.46	21.5	1.67
6	6	2.15	46.4	6
7	21.5	10	100	21.5
8	77.4	46.4	215	77.4
9	278.2	215	464	278.2
10	1000	1000	1000	1000

V. CONCLUSION

A new MEMS piezoelectric microgenerator model, named the *IFL* model, is presented in this paper. The model is powerful in analyzing the microgenerator based on the network placement approach. The connected load alters the device's poles and zeros resulting in considerable change in both output voltage and power. The *IFL* model can identify the dominant pole region thus simplifying the overall transfer function. The model has another advantage of calculating the load that produces certain output parameters. Optimum solution is found in this paper when a Wein bridge network is placed at the output terminals with R = 278.2 MΩ and C = 1 nF. These values produce maximum voltage and power at 3.8 V and 8.8 μVA, respectively.

REFERENCES

[1]   S. Roundy and P. K. Wright, "A piezoelectric vibration based generator for wireless electronics", *Smart Materials and Structures*, vol. 13, no. 5, pp. 1131–1142, 2004.

[2]   S. R. Platt, S. Farritor, and H. Haider, "On low-frequency electric power generation with PZT ceramics", *IEEE/ASME Trans. Mechatronics*, vol. 10, no. 2, pp. 240- 252, Apr. 2005.

[3]   E.M Sayed, F.P Dawson, and E.S. Rogers, "Analysis and modeling of a Rosen type piezoelectric transformer", in *Proc. IEEE Int. Conf. Power Electronics Specialists,* Vancouver, BC, Canada, June 2001, pp.1761-1766.

[4]   L. Y. Taha, B. Y. Majlis, and A. Al Ali, "Modeling and analysis of a transformer based MEMS Piezoelectric vibration type Microgenerator", in *Proc. IEEE Int. Conf. Semiconductor Electronics (ICSE)*, Kuala Lumpur, Malaysia, 2006, pp. 294-298.

[5]   L. Y. Taha, B. Y. Majlis, and A. Al Ali "Design of MEMS piezoelectric microgenerator using voltage and power optimization ", in *Proc. IEEE Int. Conf. Semiconductor Electronics (ICSE)*, Melaka, Malaysia, 2010, pp. 313-316.

[6]   K. Ogata, *Modern Control Engineering*, 4th Ed. Prentice- Hall, 2004.

[7]   M. Zhu, E. Worthington, and James Njuguna, "Analyses of power output of piezoelectric energy-harvesting devices directly connected to a load resistor using a coupled piezoelectric-circuit finite element method", *IEEE Trans. Ultrasonics, Ferroelectrics, and Freq. Control*, vol. 56, no. 7, Jul. 2009.

[8]   D. Koyama, and K. Nakamura, "Array configurations for higher power generation in piezoelectric energy harvesting", *Japanese Journal of Applied Physics*, vol. 49, no. 7, pp. 07HD04 1-5, 2010.

[9]   D. Koyama and K. Nakamura, "Electric power generation using a vibration of a polyurea piezoelectric thin film", in *Proc. IEEE Int. Sym. Ultrasonics*, Beijing, Nov. 2008, pp. 938-941.

[10]   J. C. Park, J. Y. Park, and Y. Lee, "Modeling and characterization of piezoelectric d33-mode MEMS energy harvester", *J. Microelectromechanical Syst.*, vol. 19, no. 5, pp. 1215-1222, Oct. 2010.

[11]   M. Renaud1, T. Sterken, A. Schmitz, P. Fiorini1, C. Van Hoof, and R. Puers, "Piezoelectric harvesters and MEMS technology: fabrication, modeling and measurements", in *Proc. 14th Int. Conf. Solid-State Sensors, Actuators and Microsystems*, Lyon, France, June 2007, pp. 891-894.

[12]   H. Chen, C. Jia, C. Zhang, Z.Wang, and C. Liu, "Power harvesting with PZT ceramics", in *Proc. IEEE Int. Sym. Circuits and Systems (ISCAS)*, New Orleans, LA, May 2007, pp. 557-560.

[13]   R. D'hulst and J. Driesen, "Power processing circuits for vibration-based energy harvesters", in *Proc. IEEE Int. Conf. Power Electronics Specialists*, Rhodes, Greece, Aug. 2008, pp. 2556-2562.

[14]   L. J Blystad, E. Halvorsenm, and S. Husa, "Simulation of a MEMS piezoelectric energy harvester including power conditioning and mechanical stoppers", Tech. Digest, PowerMEMS 2008, Sendai, Japan, Nov. 2008, pp. 237-240.

[15]   F. Lu, H.P Lee and S.P Lim, "Modeling and analysis of micro piezoelectric power generators for microelectromechanical-systems applications", *Smart Materials and Structures*, vol. 13, no.1, pp. 57-63, 2004.

[16]   C. Keawboonchuay and T.G Engel, "Maximum power generation in a piezoelectric pulse generator", *IEEE Trans. Plasma Science*, vol. 31, no. 1, pp. 123-128, 2003.

[17]   C. Keawboonchuay and T.G Engel, "Electrical power generation characteristics of piezoelectric generator under quasi-static and dynamic stress conditions", *IEEE Trans. Ultrasonic, Ferroelectrics, and Freq. Control*, vol. 50, no. 10, pp. 1377-1382, 2003.

[18]   Morgan Electro Ceramics, "*Introduction to piezoelectric ceramics*", Tech. inf., 2001.

Placement Effect of MEMS Thermoelectric Generator to Harvest Waste Heat using Shunt Configuration for Microprocessor

Tai Zhi Ling and Ong Hang See
Asia R&D Software and Firmware Development,
Western Digital Malaysia
Jalan SS8/6, Sungai Way, 47300 Petaling Jaya, Selangor
Department of Electrical and Electronic Engineering, University Tenaga Nasional
Km7,Jalan Kajang Puchong, Kajang, Selangor
Email: zhiling.tai@wdc.com, ong@uniten.edu.my

Abstract— Using the waste heat generated by the microprocessor is a prospect to improve the total energy efficiency of a microprocessor system. There is a need for an optimal design to harvest waste heat efficiently. This study investigated the effect of the placement of the TEG on the total energy harvested. The heat dissipation performance of the microprocessor relative to the placement of the TEG is also investigated. To evaluate the placement effect, the system is subjected to a non-uniform temperature analysis using numerical method. The thermoelectric energy conversion equations were derived from Seebeck effect. The simulation evaluates two types of heat spreader material, copper and pyrolytic graphite. The advantages and their shortfalls are discussed in the results.

Index Terms— Microprocessor, Energy harvesting, MEMS Thermoelectric

I. INTRODUCTION

According to Institute of Energy Economics, Japan through APEC, the primary world energy consumption is expected to grow up to 16,487 million tons in 2030 [1]. Malaysia, in the Tenth Malaysia Plan has committed to reduce the country energy consumption by 10 percent by 2020 [2]. An option is to recover the waste heat energy from the microprocessor. The energy that is recovered can be channeled to other application, hence improving the total energy efficiency. This paper focuses on modeling the placement of MEM's based Thermo Electric Generator (TEG) integrated into a microprocessor with a shunt configuration.

Microprocessor is used in various types of applications including datacenter environment. IDC Server Power and Cooling Expenses Report [3] forecast the total servers worldwide to grow to approximately 45 million by 2010. The increase number of servers deployed will increase the power consumption and heat dissipation as highlighted by researcher K. G. Brill [4]. Therefore, recovering the energy from the waste heat generated by the microprocessors in the datacenter environment is worth investigating.

The fundamental principle to convert thermal energy into electricity rests on the *Seebeck effect*. Seebeck effect describes an electromotive force (emf) that is produced when the

junction of two different materials are heated [5]. The concept of using TEG to harvest waste heat from the microprocessor based on Seebeck effect, was first proposed and patented by Suski [6]. Suski proposed to harvest the waste heat from the microprocessor by connecting the TEG thermally in serial with the microprocessor and the heat sink. However in [7], it was shown that Suski's configuration was inefficient and an alternative configuration was suggested by placing the TEG thermally in parallel with the microprocessor. The aforementioned alternative configuration requires an additional heat sink and suffers from convective heat losses during heat transfer process from the source. This is overcome in [8] by proposing the TEG to be integrated into the microprocessor as shown in Fig. 1.

In this paper, we have performed the modeling and analysis of the thermoelectric energy conversion using distance from the heat source as a parameter of study. The configuration will be used as shown in Fig. 1. An analytical model is also developed to estimate the amount of emf that is able to be generated by the TEG.

II. MODELING

Previous work presented in [9] has analyzed the thermoelectric energy conversion of the TEG using a nonuniform temperature distribution. However, the model did not cover system level analysis for TEG integrated in a shunt configuration. Furthermore, the placement effect of TEG is not discussed.

A. Geometric Configuration of the TEG and the System

We considered the configuration proposed by [8] where the TEG is integrated into the microprocessor as shown in Fig. 1. Fig. 1 shows the MEM's TEG is contacting the heat spreader (heat source) and the integrated heat spreader (heat sink). The heat spreader will transfer the heat from the microprocessor to the TEG. The integrated heat spreader (IHS) acts as a heat sink to dissipate the heat that flows through the TEG. Since the MEM's TEG is embedded in the microprocessor, convective heat loss will be minimal. We assume that the MEM's TEG have an ideal contact on both surfaces.

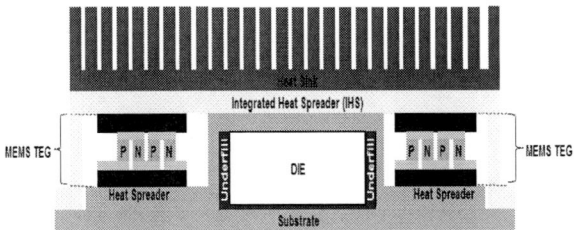

Fig. 1: MEMS TEG integrated in the processor.

B. Governing Equations

The derivation of the general equation for the simulation is described as follow. At steady state, the amount of heat entering the TEG will be equivalent to the heat dissipated from the TEG. At this state, a stable temperature gradient exists between the heat source and the heat sink. A general solution for an open circuit voltage is given by Equation (1):

$$V_{OC} = \sum_{k=1}^{N} \left(\alpha_p \Delta T_{k,p} - \alpha_n \Delta T_{k,n} \right) \qquad (1)$$

In Equation (1), V_{OC} is the open circuit voltage, N is the total number of p-n thermoelectric elements, α_p and α_n is the Seebeck coefficient of the p and n thermoelectric element respectively and $\Delta T_{k,p}$ and $\Delta T_{k,n}$ is the k-th element average temperature gradient for the p and n element respectively . The power received by the load from the TEG is given by Equation (2)

$$P_L = \left(\frac{\sum_{k=1}^{N} \left(\alpha_p \Delta T_{k,p} - \alpha_n \Delta T_{k,n} \right)}{R_L + R_{pn}} \right)^2 R_L \qquad (2)$$

In Equation (2), R_L and R_{pn} are the load resistance and the p-n thermoelectric element resistance respectively. In order to deliver the maximum load, the p-n thermoelectric element resistance and the load resistance need to be identical. The maximum power equation is defined by Equation (3)

$$P_{L,max} = \frac{\left(\sum_{k=1}^{N} \left(\alpha_p \Delta T_{k,p} - \alpha_n \Delta T_{k,n} \right) \right)^2}{4 R_L} \qquad (3)$$

The heat transfer from the microprocessor will use the thermal model proposed by [8]. The thermal model is based on a finite volume method define by Equation (4) with the heat diagram shown in Fig. 2:

$$T_{i,j,t+\Delta t} = T_{i,j,t} + \frac{2\Delta t}{(\rho_1 c_1 + \rho_2 c_2 + \rho_3 c_3 + \rho_4 c_4)\Delta x^2} [(k_1$$

$$+ k_8)T_{i-1,j,t} + (k_2 + k_3)T_{i,j-1,t} + (k_4 + k_5)T_{i+1,j,t}$$

$$+ (k_6 + k_7)T_{i,j+1,t} - \left(\sum_{n=1}^{8} k_n \right) T_{i,j,t}] \qquad (4)$$

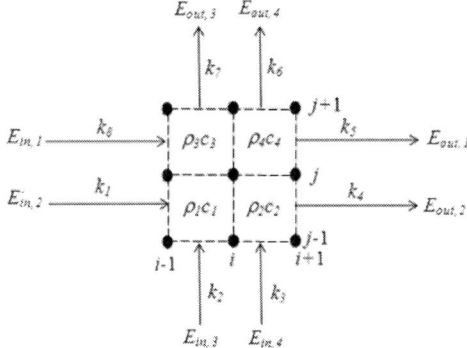

Fig. 2: Heat Diagram of a single segment

C. Simulation

The simulation model is based on Fig. 1 configuration which is composed of three major elements. The first element is the microprocessor which is based on Intel Pentium 4, 3.40GHz, LGA 775 microprocessor. The second element is the MEM's TEG, Micropelt MPG-D751. The third element is the heat spreader – two types of hears spreaders are used in this study, copper and pyrolytic graphite.

The boundary condition is assumed to be uniform with a temperature of 65°C at the microprocessor die. The heat sink surface is assumed to be uniform with an ambient temperature of 25°C. The remaining items in Fig. 1 such as the substrate and underfill are assumed to be an insulator with an adiabatic behavior in the model.

The mechanical properties that are critical in the simulation are described in the Table 1.

TABLE I
Mechanical Properties used in Simulation Model

Component	Parameter	Value
Micro-processor	Die Length*	13 mm
	Die Thickness*	1.1 mm
	Underfill gap(left and right)*	2 mm
	IHS nickel plating thickness*	0.1 mm
	IHS copper thickness*	0.8 mm
	Total Processor length*	29.2 mm
TEG	Substrate thickness [12]	0.525 mm
	TEG element thickness [12]	0.04 mm
	n-type element (Bi$_2$Te$_3$) [10]	-160 µV/K
	p-type element (Sb$_2$Te$_3$) [11]	195 µV/K
	TEG electrical resistance [12]	300 Ω
Heat Spreader	Spreader thickness	0.03 mm
	Copper ∥ thermal conductivity [13]	401 W/k-m
	Copper ⊥ thermal conductivity [13]	401 W/k-m
	PGS ∥ thermal conductivity [14]	1600 W/k-m
	PGS ⊥ thermal conductivity [13]	5.70 W/k-m

*Estimated from a Pentium 4 LGA 775 Microprocessor

The distance of the TEG from the underfill edge is varied from 0.1 mm to 2.4 mm. The model is simulated through MATLAB. Since the geometric configuration is symmetrical, only one of the symmetry is analyzed. The subsequent subsection will discuss on the results and the optimum distance of TEG from the heat source for different type of heat spreader.

III. RESULT AND DISCUSSION

A. Shunt Configuration TEG using Copper Heat Spreader

In this subsection, the TEG shunt configuration using copper heat spreader is analyzed. First, the effect of the placement in relation to the energy conversion is investigated as shown in Fig. 3. The energy conversion behavior towards the TEG distance can be divided to two parts. The first part indicated that a gradual energy conversion increase proportional to the distance. The second part meanwhile showed a gradual decrease proportional to the distance.

The first part is influenced by the TEG's cold surface (contacting the IHS). The TEG's cold surface temperature is contributed by the heat conducted through the TEG and the heat received from the IHS. The horizontal thermal resistance between the TEG's cold surface and the heat source will increase with the distance from the heat source. Therefore less heat will be received by the TEG's cold side from the IHS. This results in lower temperature on the TEG's cold surface. The lower temperature will increase the temperature gradient and energy generated.

The second part is influenced by the TEG's hot surface (contacting the heat spreader). The horizontal thermal resistance on the TEG's hot surface will also increase with the distance from the heat source. Therefore less heat will flow to the TEG's hot surface. As the distance of the TEG from the heat source continue to increase, the TEG's hot surface will have a lower temperature. As a result, the temperature gradient and energy generated will gradually decrease once the TEG's cold surface is unable to compensate for the temperature drop.

The highest amount of energy generated by the TEG integrated with copper spreader is 14.7uW at a distance of 0.5 mm from the source. The small amount of energy generated indicates that the average temperature gradient on the TEG is small.

Second, the microprocessor heat dissipation performance with relation to the TEG placement is explored as shown in Fig. 4. The temperature is observed to gradually increase as the TEG placement approach the heat source. This is because heat will be conducted from the heat spreader to the TEG's cold surface. As the TEG move closer to the source, the additional heat from the heat spreader will also increase. The presence of additional heat on the TEG's cold surface will reduce the IHS horizontal heat flow hence increasing the temperature of the microprocessor.

The heat spreader that is introduced has a maximum temperature increase of 0.4896°C and a temperature increase is 0.4892°C at the highest power generated.

Generally from Fig. 4 it is observed that the temperature change with respect to the TEG placement is negligible.

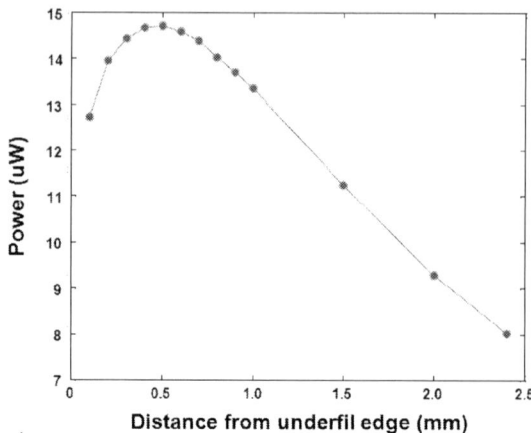

Fig. 3 Maximum power generated in TEG versus the distance from the underfill edge for copper spreader

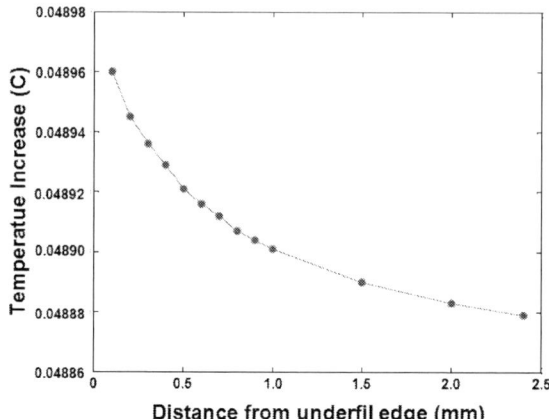

Fig. 4 Temperature increase in TEG versus the distance from the underfill edge for copper spreader

B. Shunt Configuration TEG using Pyrolytic Graphite Heat Spreader

In this setting, the TEG shunt configuration using pyrolytic graphite heat spreader is investigated. The effect of the TEG placement in relation to the energy conversion is shown in Fig. 5. Since the thermal conductivity of the pyrolytic graphite is approximately 4 times higher than copper the maximum amount of energy generated by the TEG is 220.4 uW at a distance of 0.6 mm. The optimum placement of the TEG is influence by the thermal conductivity. This is because there is a shift on the optimum distance of the TEG from the source between both copper and pyrolytic heat spreader.

The microprocessor heat dissipation is observed to increase non-linearly as it approaches the heat source. A temperature increase of 3.101°C and 3.098°C is observed for TEG placement corresponding to the maximum temperature and maximum power respectively.

Since the temperature change of the pyrolytic graphite spreader is relatively significant compared to copper spreader, care must be taken on the TEG placement while designing the heat sink. The results also show that the optimum placement of

TEG is dependent on the thermal conductivity of the heat spreader.

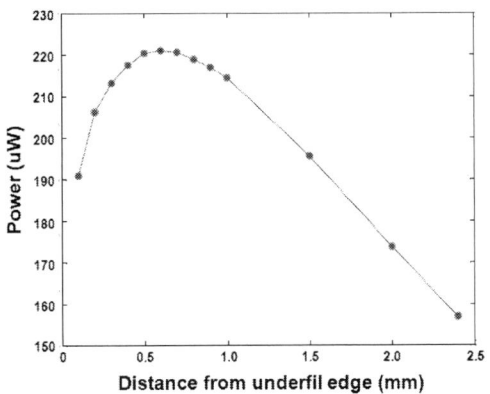

Fig. 5 Maximum power generated in TEG versus the distance from the underfill edge for pyrolytic graphite spreader

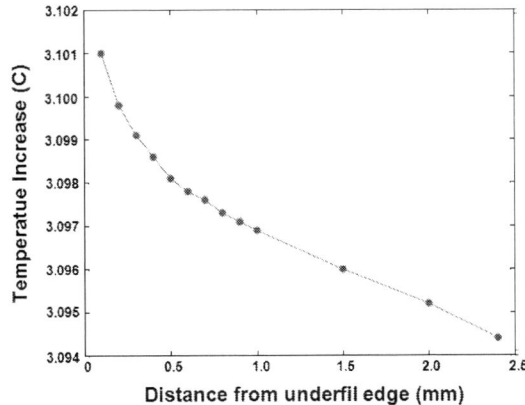

Fig. 6 Temperature increase in TEG versus the distance from the underfill edge for pyrolytic graphite spreader

IV. CONCLUSION

In this paper, we have presented a study on the effect of the TEG placement for microprocessor waste heat harvesting using shunt configuration proposed by [8]. We have analyzed two types of heat spreader which are copper and pyrolytic graphite integrated into the TEG. Result shows that pyrolytic graphite heat spreader has a better energy harvesting performance compare to copper. Care must be taken when integrating the TEG to the microprocessor as the placement will affect the heat dissipation performance

ACKNOWLEDGMENT

We would like to acknowledge the contribution of Teh Eng Lee from Altera Corporation Malaysia who has provided us with inputs and support on TEG simulation and analysis.

REFERENCES

[1] Agency for Natural Resources and Energy Ministry of Economy and Trade Industry, Japan, "Fiscal 2005 Annual Energy Report (Outline)" Jun 2006.
[2] APEC Secretariat, "APEC Energy Overview 2009". 2010.
[3] Jed Scaramella, "Worldwide Server Power and Cooling Expense 2006 - 2010 Forecast", IDC #203598, 2006
[4] K.G. Brill, "The invisible Crisis in the Data Center: The Economic Meltdown of Moore's Law", Uptime Institute, 2007
[5] H.J Goldsmid, Introduction to Thermoelectricity. Springer, 2009, ch. 1.
[6] Suski, Edward D., "Method and Apparatus for Recovering Power from Semiconductor Circuit Using Thermoelectric Device," US Patent #5,419780, 1995
[7] Solbrekken, G.L. et al., "Thermal management of portable electronic equipment using thermoelectric energy conversion," ITHERM., 2004.
[8] Tai Zhi Ling. et al., "Thermal Modeling for Harvesting Waste Heat From Microprocessor Using Shunt Configuration," ICPER., 2012.
[9] Yu Zhou et al., "Harvesting Wasted Heat in a Microprocessor Using Thermoelectric Generators: Modeling, Analysis and Measurements", Design, Automation and Test Europe, 2008
[10] H. Bottner et al., "New thermoelectric components using microsystem technologies," Journal of Microelectromechanical Systems, vol. 13, no. 3, pp. 414- 420, Jun. 2004.
[11] Wei-Chuan Feng et al., "Large-Scale Preparation of Ternary BiSbTe Films with enhanced Thermoelectric Properties using DC Magnetron Sputtering", Microsystem, Packaging, Assembly and Circuits Technology Conference, pp. 457-460, Oct. 2009
[12] MPG-D651,MPG-D751 Thin Film Thermogenerator and Sensing Device, Micropelt GmBH, Freiburg, Germany,
[13] Theodere L. Bergman et. al, (April,12 2011). Fundamentals of Heat and Mass Transfer (7th edition)
[14] PGS Graphite Sheet Type: EYG, Panasonic Industrial Company, Secaucus, New Jersey, 2012

Effect of Co Catalyst on PECVD Growth of Carbon Nanotubes for NEMS Applications

Mai Woon Lee[*,a], Nadia Md Razib[a], Aun Shih Teh[a], Daniel C. S. Bien[a], Soo Kien Chen[b],
Shahrul Azam Abdullah[c]

[a]MEMS, NEMS and Nanotechnology Division, MIMOS Berhad, Technology Park Malaysia, Kuala Lumpur
[b]Department of Physics, Faculty of Science, University Putra Malaysia, Serdang, Selangor
[c]Faculty of Mechanical Engineering, UniversitiTeknology Mara, Shah Alam, Selangor
*Corresponding Author, Email : mw.lee@mimos.my

Abstract- In this paper the effect of cobalt (Co) catalyst on the growth of carbon nanotubes (CNTs) was studied. CNTs were vertically grown by plasma enhanced chemical vapor deposition method (PECVD) at 700°C with various sputtered Co catalyst thicknesses. Experimental results shows that for carbon nanotube growth duration of 20 minutes, growth was only achieved with thinner catalyst layers but when the growth duration was doubled, high density of CNTS were also observed with thicker catalyst layers with taller nanotubes formed. The nucleation of the catalyst with various thicknesses was also studied as the absorption of the carbon feedstock is dependent on the initial size of the catalyst island.

Keywords: Carbon nanotubes; nanomaterials; chemical vapor deposition; PECVD, cobalt, nucleation.

I.   INTRODUCTION

Since its discovery by Ijima in 1991 [1], carbon nanotubes have undergone vigorous studies with intentions of utilizing its unique physical and chemical characteristics which stems from its nanoscale geometry and molecular structure. The material piques further interest due to its stable mechanical structure in the form of high strength, stiffness and high thermal stability. The nature of carbon nanotubes possessing either metallic or semiconducting behavior depending on its chirality extends further possible applications in the form of novel electronic devices.

In the past decade, the development of CNTs towards electronic applications has been growing significantly, in particular with the integration of nanotubes to micro-and nano-electromechanical devices (MEMS/NEMS). For example, carbon nanotube based sensors [2] have demonstrated improved device performances such as increased sensitivity, targeted selectivity, and shorter response time. Such improvements enable the development of low cost and portable gas and chemical sensor devices. Other examples of promising CNTs devices include torsion switches for nonvolatile memory devices [3]; supercapacitors [4]; pressure sensors [5]; and filtration element [6] for micro-/nano-fluidics.

The typical methods for synthesizing CNTs include laser ablation [7], arc discharge [8], and chemical vapor deposition (CVD) [9]. However, as CVD is more comparable with standard semiconductor fabrication, most efforts have focused on CVD as it allows a more compatible integration of CNTs

onto a silicon platform. In this paper, the research work focuses on the effect of cobalt (Co) catalyst towards the growth of carbon nanotubes via plasma enhanced chemical vapor deposition (PECVD) where the process is focused on the integration of CNTs onto NEMS devices fabricated on silicon. This is done at wafer level to establish a process in line with 8" wafer processing. This study analyses the effect of catalyst thickness and nucleated catalyst morphology on the CNT growth.

II.   EXPERIMENTAL DETAILS

The growth of CNTs was performed by Plasma Enhanced Chemical Vapor Deposition (PECVD) with the aid of Oxford Instruments Nanofab700 tailored for depositions up to 8" wafers. Acetylene (C_2H_2) was used as the carbon feedstock while cobalt (Co) was chosen as the catalyst material. Ammonia (NH_3) is added to C_2H_2 during the growth process to assist in the etching of by-products such as amorphous carbon [9]. The Co catalyst is prepared on thermally oxidized Si wafers. Prior to the deposition of Co, a thin TiN layer is deposited to act as a diffusion barrier. Both TiN and Co catalyst materials were deposited via RF-magnetron sputtering whereby the deposition pressure was maintained at 5×10^{-3} mbar for all layers. The sputtering was performed with a TF600 Edwards system using a rolling table to achieve depositions on 8" wafers. The thickness of the TiN layer was maintained at 10nm while the Co catalyst thicknesses were varied with thicknesses of 2, 4, 6, 8, 10 and 15nm. Prior to the growth of CNTs, the deposited catalysts samples were annealed for 10 minutes for nucleation into a seeding layer. Annealing is performed in 200W H_2 plasma ambient at 700°C with the H_2 flow of 100sccms and chamber pressure of 1000mTorr. Carbon nanotubes were grown on the annealed Co catalyst layers with a C2H2:NH3 flow of ratio of 20:80sccm at 700°C and chamber pressure of 1000mTorr for 20 minutes and 40 minutes respectively.

The nucleated Co catalysts were characterized to study its morphology and the correlation between the nucleated particle sizes with the initial Co thickness and its effects on the nanotube growth were determined. The morphology of the Co and CNT samples were characterized with a JEOL Field Emission Scanning Electron Microscope (JSM7500)

III. RESULTS AND DISCUSSIONS

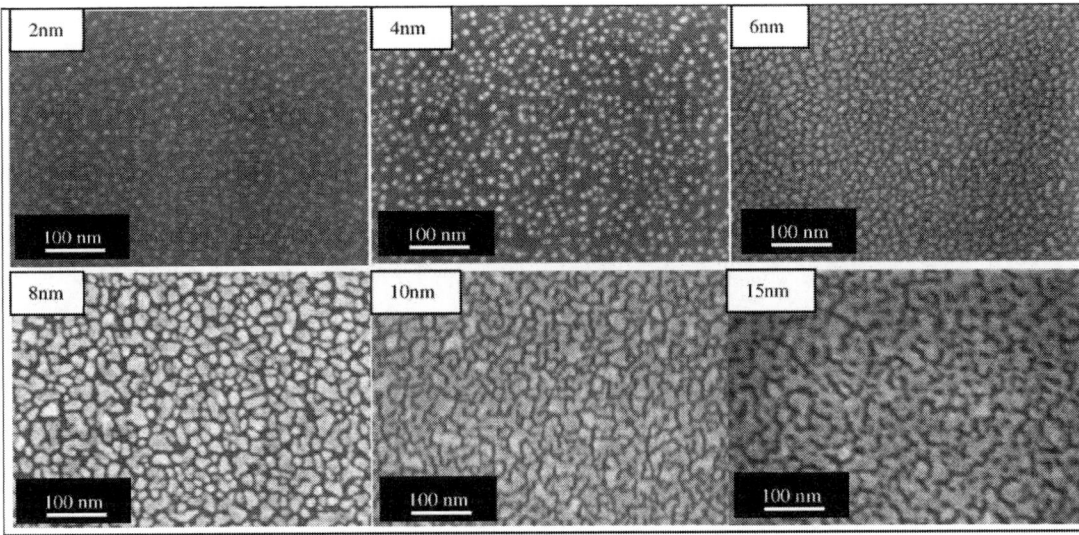

Figure 1 : FESEM images with varying Co catalyst thickness on TiN/Si after annealing at 700 °C in 1000 mTorr of H_2 for 10 minutes.

The formation of carbon nanotubes in this study are analysed based on electron micrograph images of the catalyst conditions and also of the nanotubes grown at different deposition times. Figure 1 illustrates the variation of the Co catalyst after having subjected to annealing conditions of 10 minutes at 700°C in H2 plasma ambient at a pressure of 1000mTorr. This annealing step is the nucleation step prior to the actual CNT deposition which results in the formation of a seeding layer in the form of Co nanoparticles. The seed layer resemble small islands which are formed due to the surface tension and compressive stress as a result of the mismatch in thermal expansion coefficients of Si and Co [10, 11] at elevated temperatures. Based on the images in Figure 1, the catalyst of 2, 4 and 6nm Co forms a well dispersed seed layer. For thicker initial catalyst layers the formation of the seed layer becomes incomplete resulting in irregular island sizes. The island sizes were estimated from the electron micrographs and are compiled in the graph of Figure 2. It was found that the average island diameter was larger when the initial catalyst layers were thicker.

Figure 3 illustrates the growth of nanotubes grown for corresponding Co catalyst layers and deposition times. The CNT deposition condition was fixed at C2H2:NH3 flow ratio of 20:80sccm at 700°C and chamber pressure of 1000mTorr with only the deposition duration varied for 20 minutes and 40 minutes. Based on a previous study [12] it was observed that the nanotubes deposited on TiN are of multiwalled carbon nanotubes (MWCNT) and forms via a base growth mechanism. Base growth is the dominant mechanism due to a strong interaction between the Co catalyst and TiN layer [13]. From the images, the 20 minute deposition revealed very little growth and almost no growth for the Co catalyst which was thicker than 4 nm. This variation in growth is attributed to the size of the seed/catalyst particles. Smaller seeds or particles from the thinner catalyst layer are expected to have greater energetics due to a higher surface to volume ratio. The absorption and saturation of the carbon feedstock on the smaller catalyst island occurs at a faster rate resulting in the early formation of nanotubes as seen with the 2, 4 and 6nm Co catalyst layer. However, as the deposition duration is extended further as seen with the 40 minutes deposition, a greater density of MWCNT formation is observed for the catalyst thickness of 6nm and above. Performing cross section images of the nanotubes (Figure 5) grown for 40 minutes revealed that the nanotubes were much longer and taller for the thicker Co catalyst. It is possible that with the thinner Co catalyst layers the supersaturation of the carbon feedstock occurs at such a greater rate which results in formation of a carbon shell around the Co catalyst [14] instead of the nanotube structure which in turn halts the nanotube formation. Based on this, a lower or controlled carbon feedstock supply would be required when utilising the thinner Co catalyst.

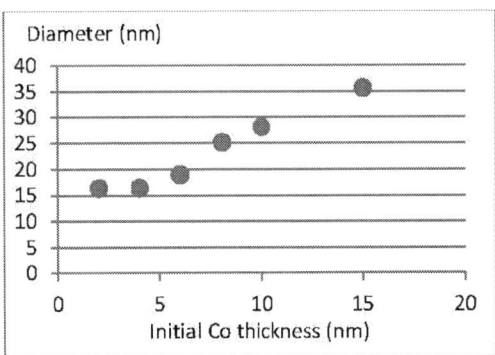

Figure 2 : The average Co island diameter after annealing versus initial Co layer thickness.

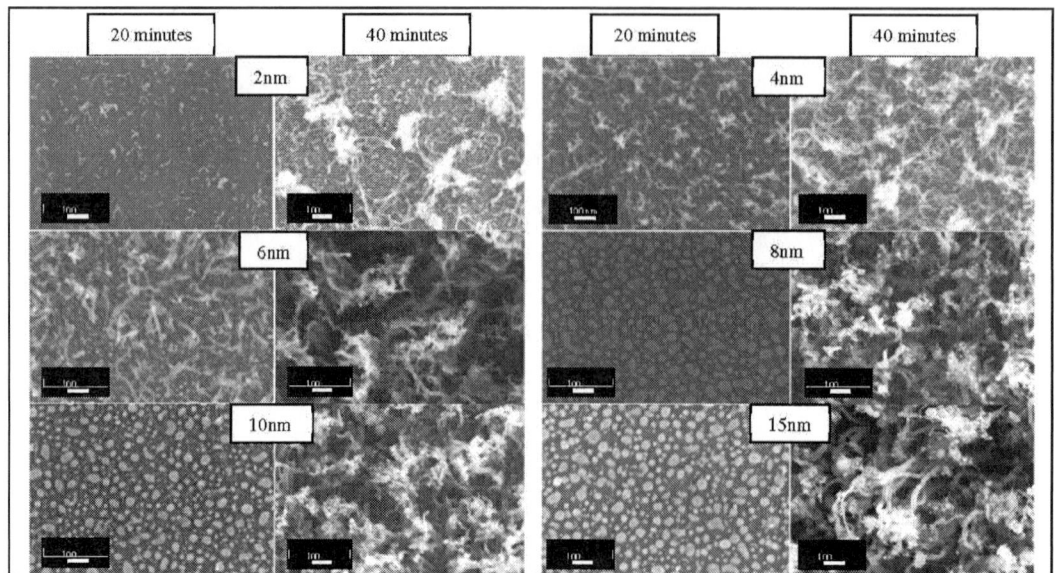

Figure 3 : Comparison of top view images with varying thickness of nanotubes grown for 20 minutes and 40 minutes. Standard growth conditions (C_2H_2:NH_3 = 20:80 sccm, temperature = 700 °C)

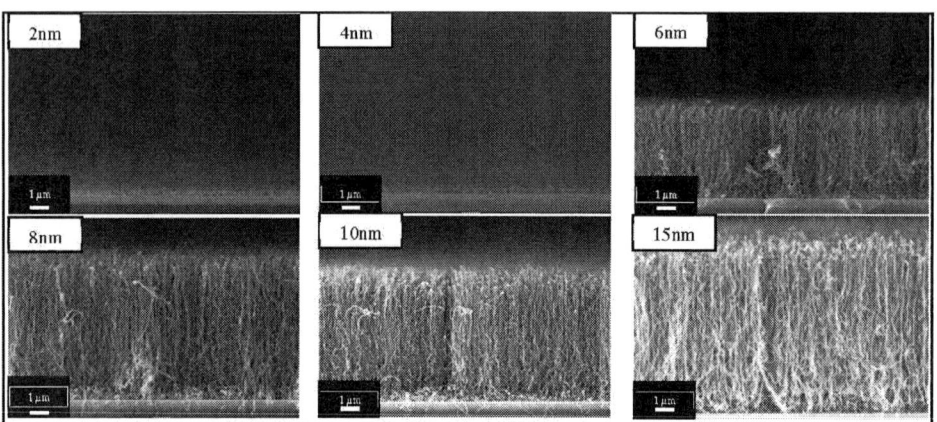

Figure 4 : FESEM images of nanotubes grown of various initial thicknesses shown in Figure 1. Standard growth conditions (C_2H_2:NH_3 = 20:80 sccm, temperature = 700 °C, time = 40 minutes)

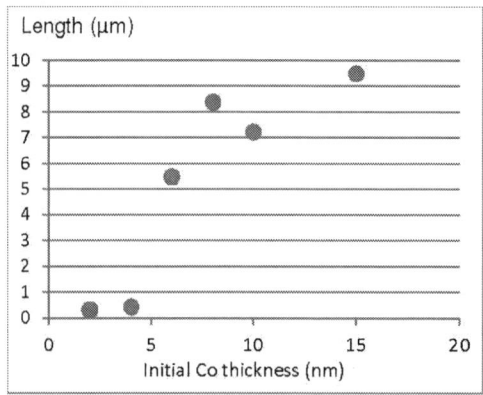

Figure 5 : The average nanotube length as a function of the Co thickness.

978-1-4673-2395-6/12 $31.00 © 2012 IEEE

It can be seen that the tubes are very long for the thicker catalyst layer with lengths up to 10μm. The results here indicates further that a thicker catalyst layer is sufficient to maintain the steady process of feedstock diffusion into the catalyst, supersaturation and nanotube formation resulting in taller MWCNTs as compared to the thinner catalyst layer. From Figure 4, only short nanotubes were found to have grown with the 2nm and 4nm of Co layer. Attempts to grow nanotubes on thinner layers of 0.5nm Co showed no growth. Figure 5 shows a clear relation of increased nanotube length with a thicker catalyst.

Another observation was that the lower portion of the MWCNTs appears to be not perfectly aligned and distorted. However, as the growth process is sustained the bunching of MWCNT results in vertically aligned tall MWCNTs.

IV. CONCLUSION

In summary, the process condition for the growth of vertically aligned carbon nanotubes by PECVD with variation of the Co catalyst layer thickness was studied. An annealing step is required to fragment the initial catalyst thin film into islands of nanoparticles required as the seeding layer. The results showed that for deposition duration of 20 minutes only nanotubes were found to grow from the thinner catalyst layers. However, with the deposition of 40 minutes high density MWCNTs were found to grow with the thicker Co catalyst layer. Based on the results, the absorption of the carbon feedstock is highly dependent on the initial size of the catalyst island. Taller MWCNT formation was observed to be more favorable with thicker catalyst layers.

ACKNOWLEDGMENTS

This research was supported by eScience funding 03-03004-SF0006 and 01-03-04-SF0027 together with National Nanotechnology Directorate funding NND/ND/(2)/TD11-012 under the Ministry of Science, Technology and Innovation (MOSTI), Malaysia.

REFERENCES

[1] S. Iijima, "Helical microtubules of graphitic carbon", *Nature*, **354**, p. 56-58, 1991.

[2] Y. Wang and J.T.W. Yeow, "A review of carbon nanotubes-based gas sensors", *Journal of Sensors*, 493904, 2009.

[3] W. Xiang and C. Lee, "Nanoelectromechanical torsion switch of low operation voltage for nonvolatile memory application", *Applied Physics Letters*, **96**, 193113, 2010.

[4] Y.Q. Jiang, Q. Zhou, and L. Lin, "Planar MEMS supercapacitor using carbon nanotube forests", *IEEE 2nd Int. Conf. on MEMS*, p. 587-590, 2009.

[5] K.S.Karimov, M.Saleem, Z.M.Karieva, A. Khan, T.A.Qasuriaand A Mateen, "A carbon nanotube-based pressure sensor", *PhysicaScripta*, **83**, 065703, 2011.

[6] O. Bakajin, N. Ben-barak, J. Peng and A. Noy, "Carbon nanotube based microfluidic elements for filtration and concentration", *7th Int. Conf. on Miniaturized Chemical and Biochemical Analysis Systems*, 2003.

[7] T. Guo, P. Nikolaev, A. Thess, D.T. Colbert and R.E. Smalley, "Catalytic growth of single-walled nanotubes by laser vaporization", *Chemical Physics Letters*, **243**, p. 49-54, 1995.

[8] T.W. Ebbesen and P.M. Ajayan, "Large-scale synthesis of carbon nanotubes", *Nature*, **358**, p. 220, 1992.

[9] M. Chhowalla, K.B.K. Teo, C. Ducati, N.L. Rupesinghe, G.A.J. Amaratunga, A.C. Ferrari, D. Roy, J. Robertson and W.I. Milne, "Growth process conditions of vertically aligned carbon nanotubes using plasma enhanced chemical vapor deposition", *Journal of Applied Physics*, **90**, 5308, 2001.

[10] C. Bower, O. Zhou, W. Zhu, D. J. Werder, and S. Jin, "Nucleation and growth of carbon nanotubes by microwave plasma chemical vapour deposition," *Appllied Physics Letters*, **77**, 2767, 2000.

[11] V. I. Merkulov, D. H. Lowndes, Y. Y. Wei, G. Eres, and E. Voelkl, "Patterned growth of individual and multiple vertically aligned carbon nanofibers," *Appllied Physics Letter*, **76**, 3555, 2000.

[12] A.S. Teh, D.C.S. Bien, R.M. Saman, S.K. Chen, K.S. Tan, H.W. Lee, " Multiwalled carbon nanotube growth mechanism on conductive and non-conductive barriers", *Advanced Materials Research*, **403-408**, p. 12011204, 2012.

[13] J-B. A. Kpetsu, P. Jedrzejowski, C. Côté, A. Sarkissian, P. Mérel, P. Laou, S. Paradis, S. Désilets, H. Liu, X. Sun, "Influence of Ni catalyst layer and TiN diffusion barrier on carbon nanotube growth rate," *Nanoscale Research Letters*, **5**. p. 539-544, 2010.

[14] C. Journet, M. Picher and V. Jourdain, "Carbon nanotube synthesis: from large-scale production to atom-by-atom growth", *Nanotechnology*, **23**, 142001, 2012.

A Study of Catalyst-based ZnO Nanowires Growth by PECVD

Muhammad Aniq Shazni Mohammad Haniff*,[a], Khairul Anuar Abd Wahid[a], Wai Yee Lee[a], Hing Wah Lee[a], Saat Shukri Embong[a], Ishak Hj. Abd. Azid[b]

[a]MEMS & NEMS Cluster,MIMOS Berhad,Technology Park Malaysia,Kuala Lumpur
[b]Department of Mechanical Engineering, Engineering Campus, Universiti Sains Malaysia, Pulau Pinang
*Corresponding Author, Email: aniq.haniff@mimos.my

Abstract – **This paper analyzed the effects of the temperature and plasma on the nucleation and growth of the zinc oxide (ZnO) nanowires. Uniformly distributed ZnO nanowires were successfully synthesized through plasma-enhanced chemical vapor deposition (PECVD) on silicon (100) substrate under different thickness of gold (Au) as the catalyst material. Experimental results were characterized using field emission scanning electron microscopy (FESEM) showed that the optimum growths of ZnO nanowires were achieved at a higher temperature of 650°C for Au catalyst thickness between range of 2 to 5 nm. It is demonstrated as well that the introduction of plasma affects both the size of the Au nanoparticles formed during the nucleation process and subsequently the density of the ZnO nanowires synthesized during the growth process.**

Keywords: **zinc oxide nanowires; nanomaterials; PECVD.**

I. INTRODUCTION

Nanostructured ZnO materials have attracted considerable interest in recent years due to its potential application to varieties of optical, electronics and sensor devices such as transparent conducting material [1], light emitting diodes [2], piezoelectric devices [3], sensors [4] and UV laser diodes [5]. One of the more common nanostructured ZnO materials utilised is in the form of ZnO nanowires which have been incorporated into devices to form NEMS sensors and actuators.

At the early phase of the research on ZnO nanowires, the solution and gas phases method has been commonly employed. The growth process for the solution phase method; typically referred to as the hydrothermal growth process [6] is carried out in a aqueous solutions while for the gas phase synthesis, one of the more established approaches utilised is the vapor-phase transport method as demonstrated by Z.W. Pan et. al [7]. Recently, researchers have shifted their focus on the synthesis of ZnO nanowires via the PECVD techniques as it has shown capability in forming nanowires with high uniformity at a lower temperature, pressure and power.

The synthesis of ZnO nanowires throught the PECVD method typically requires catalyst layer of various metal or metal oxide materials [8-9]. However, it has been demonstrated that highly uniform vertically-aligned ZnO nanorods can also be grown without the need of any catalyst material via a two-step growth mechanism by changing the oxygen concentration of the gas mixture during the nucleation and growth steps [10]. Meanwhile, Colin A. Wolden [11] varies the mixtures ratio of the diethylzinc (DEZn) and oxygen gas to examine the amount of oxygen required to fully combust the DEZn to promote the growth of the ZnO nanowires.

Based on the reported literatures, it is evident that various parameters; which includes the growth temperature and the presence of plasma; had influenced the growth of the ZnO nanowires synthesized through the PECVD techniques. In view of that, this paper tends to study the effects of the growth temperature and the presence of plasma on the growth characteristics of the ZnO nanowires synthesized via the PECVD technique for various catalyst thicknesses. The interaction between the different formation of the catalyst nanoparticles on the growth of the ZnO nanowires were also studied.

II. EXPERIMENTAL DETAIL

In this paper, ZnO nanowires growth is achieved by plasma-enhanced chemical vapor deposition (PECVD) on Si (100) substrate. Prior to the ZnO nanowires growth process, a thin layer of gold (Au); acting as the catalyst material; was deposited by RF-magnetron sputtering at a pressure of 5×10^{-3} mbar. The thicknesses of the deposited Au catalyst were varied at 2, 5, 10, 15 and 20nm. Once the catalyst material was prepared on the Si wafer, the growth of ZnO nanowires were performed through the PECVD with the aid of Oxford Instruments Nanofab700 for 8" wafer-scale depositions at different substrate temperature and plasma conditions. A short nucleation process involving annealing of the Au catalyst for 10 minutes is accomplished by pulsing 100 sccm of hydrogen gas at different temperatures of 550°C or 650°C, coupled with or without the presence of Ar plasma. The growth process for the ZnO nanowires will supersede once the nucleation process has completed. Mixtures of Diethylzinc (DEZn) and oxygen gas (O_2) were employed as the gas pre-cursors for the growth of ZnO nanowires via the PECVD technique. During the growth process, the temperature of the DEZn pre-cursor is maintained at -20°C with Ar flow rate of 25 sccm for DEZn dilution, which it will then be subsequently transported into the reaction chamber with 3sccm of Ar as carrier gas. As for the oxygen gas, it will be transported separately into the reaction chamber. For this work, ZnO nanowires were grown on the nucleated Au nanoparticles under continuous flow of DEZn vapor with O_2 flow rate of 100 sccm for 40 minutes in different

978-1-4673-2395-6/12 $31.00 © 2012 IEEE

temperatures (550°C or 650°C). Results for both the nucleation of the Au catalyst and growth of the ZnO nanowires at different substrate temperature and plasma conditions were then characterized with a JEOL JSM-7500F field emission scanning electron microscopy (FESEM).

III. RESULT AND DISCUSSIONS

A. Effect of temperature and plasma condition on nanoparticles formation

SEM images for nanoparticles formed during the nucleation process; with and without plasma; for temperatures of 550 °C and 650 °C, are illustrated in Figure 1 and Figure 2 respectively. Clearly, the sizes of nanoparticles formed are increasing with an increasing thickness of Au layers. However, a critical thickness is observed at Au thickness of 10 nm, whereby the catalyst was most dense with small homogeneous nanoparticles. Thinner layers (2 nm and 5 nm) showed more isolated, smaller size nanoparticles formation. Similarly, for thicker layers (15 nm and 20 nm), large size and cluster-like nanoparticles were observed. These observations can be explained as a function of Au consumption during the nucleation process. For thin layers (2 nm and 5 nm), Au layers were fully consumed or somewhat insufficient, thus nanoparticles formed were small and isolated. A 10 nm Au was observed to be the best condition for catalyst formation, in which adequate Au were formed into small and dense catalyst. At thicknesses above 10nm, the nucleation process were incomplete in excess of large amount of Au, therefore the formation of the nanoparticles were partially formed resulting in an irregular cluster, large-sized nanoparticles. This occurrence can be further described from the short annealing time in the nucleation process, which offer cause insufficient time for the nanoparticles to form in a complete manner.

Figure 3 further demonstrate the size of nanoparticles formation with the effects of the temperature and plasma treatment on the nucleation process. At a higher temperature of 650°C, the nanoparticles formed tend to be smaller than those formed at 550°C. This is most likely due to a faster melting activity at a higher temperature as compared to a lower temperature nucleation process. The effect of plasma treatment does not show a significant impact to the overall results.

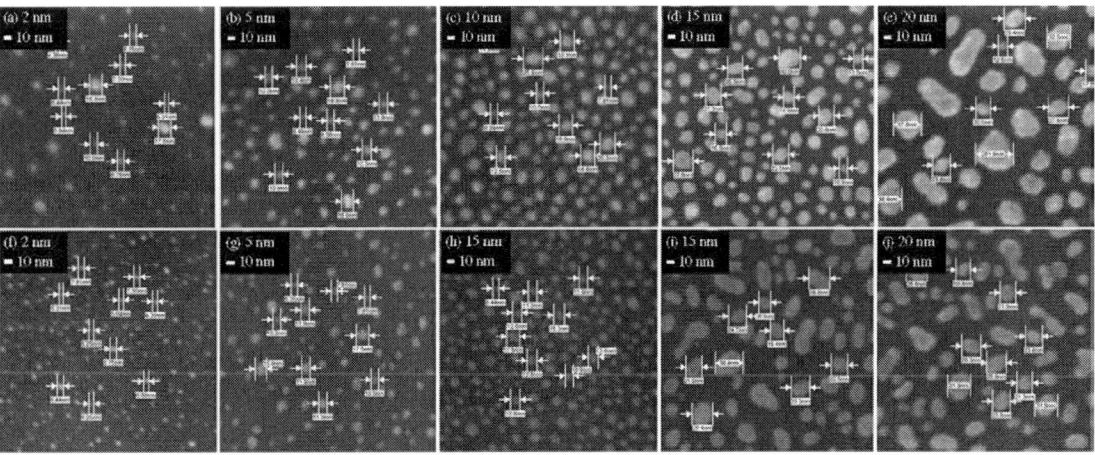

Figure 1. SEM images of Au nanoparticles after annealing at 550˚C for 10 minutes: (a–e) with plasma and (f–j) without plasma. The nucleation conditions (Pressure =1000 mTorr, H2 flow rate = 100 sccm, RF power = 100 W)

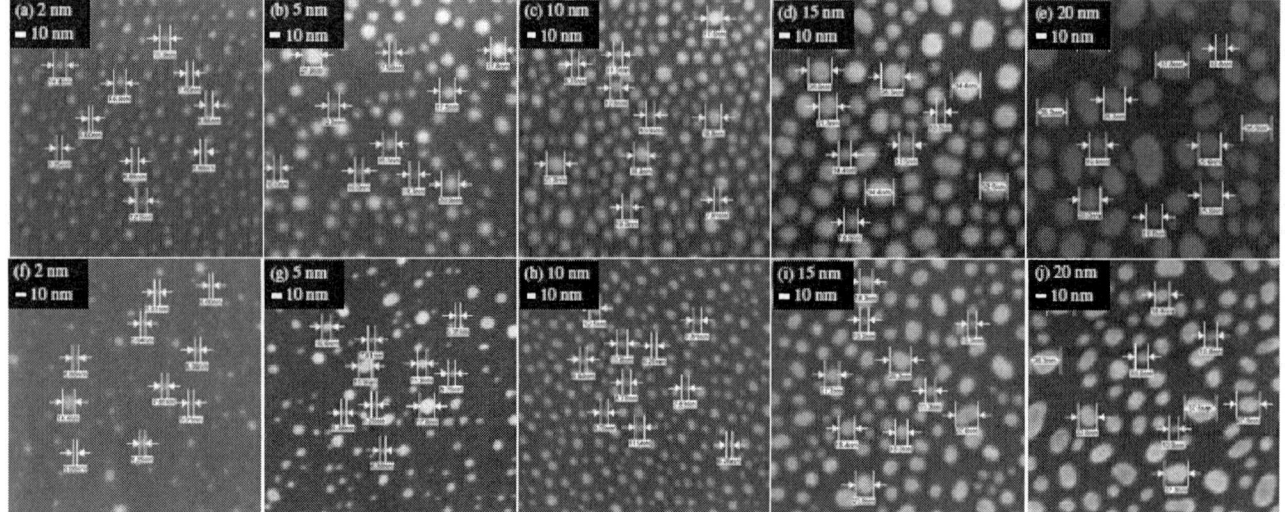

Figure 2. SEM images of Au nanoparticles after annealing at 650°C for 10 minutes: (a–e) with plasma and (f–j) without plasma. The nucleation conditions (Pressure=1000 mTorr, H2 flow rate = 100 sccm, RF power = 100 W)

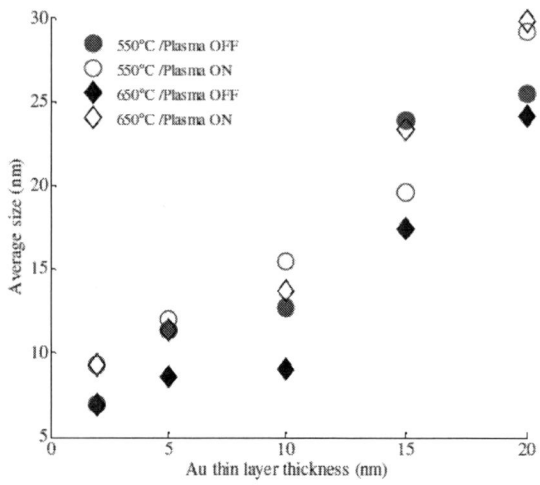

Figure 3. The average nanoparticles size as a function of Au thin layer thickness under different temperature and plasma conditions.

B. Effect Of Growth Temperature And Plasma Conditions On Zno Nanowires Growth

In this section, we will describe the result of ZnO nanowire grow based on the catalyst study as discussed previously. Figure 4 and 5 shows the SEM images of ZnO nanowires grown on Si(100) substrates at temperature of 650°C with and without the presence of plasma respectively. We would like to highlight that our results shows that there were no successful growth at lower temperature condition at 550°C. This is maybe cause by the fact that lowers growth temperature, the vapour has insignificant energy, and therefore the atoms are deposited exactly where they land thus creating a rough uneven layer [12]. The lack of vapour tends inhibit nanowire growth since the catalyst was unable to fully react with the mixtures of DEZn and O_2 gas pre-cursors. At 650°C, a highly-dense ZnO nanowire was observed only at thin layers of Au catalyst (2 nm and 5 nm). In contrast to our catalyst study previously, highly dense ZnO nanowires were only grown on catalyst with small size nanoparticles. Despite our best catalyst formation at 10 nm Au, we only manage to grow discrete nanowires as illustrated in Figure 4(c) and Figure 5(c). In addition, the ZnO nanowires grown had good uniformity in terms of the size, but were also randomly oriented on the substrate. This could be due to lack of lattice compatibility between the ZnO nanowires and Si (100) substrate [10]. For the thicker Au catalyst (15 nm and 20 nm), there were almost no growth exhibited. These phenomena can be related to the poor catalyst formation as described previously. With such a large nanoparticles, the vapour reaction could not form nanowires, but instead were forming layers of ZnO as observed in Figure 4(d)(e) and Figure 5(d)(e) respectively. Furthermore, we observed that for nanowire growth process, a better density of nanowire were achieved with the presence of plasma (refer Figure 4(a)(b) and Figure 5(a)(b)). Evidently, the presence of plasma would encourage better process conditions for nanowire growth.

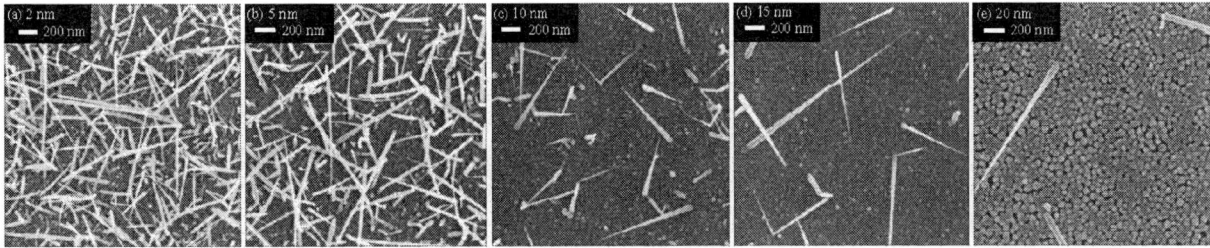

Figure 4. Top view of SEM images of ZnO nanowires grown on Si (100) substrates at growth temperature of 650°C for 40 minutes growth time from nucleated nanoparticles with plasma.

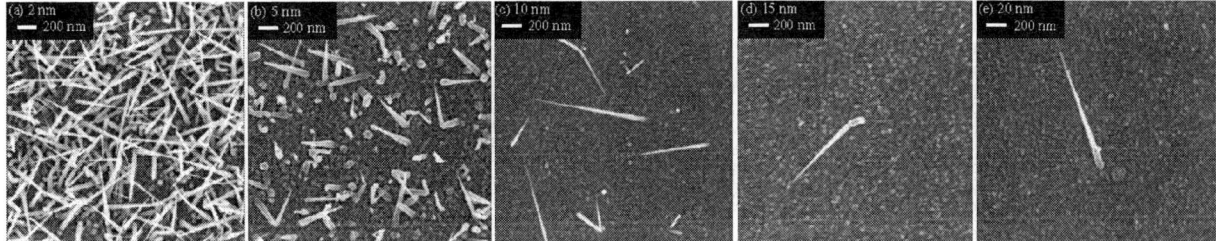

Figure 5. Top view of SEM images of ZnO nanowires grown on Si (100) substrates at growth temperature of 650°C for 40 minutes growth time from nucleated nanoparticles without plasma.

IV. CONCLUSION

The effects of the growth temperature and the effect of plasma in process conditions on the nucleation and growth of the ZnO nanowires were studied at various thicknesses of Au catalyst layer were presented. Results from the nucleation process showed that the size of the Au nanoparticles increases with the increasing Au thickness. Apparently, ZnO nanowires were only successfully grown on silicon (100) substrates at a temperature of 650°C. No nanowires growth was observed at 550°C. The growth of ZnO nanowires were attained only with thin Au catalyst. Thicker catalyst results in formation of ZnO layers. An enhanced density of ZnO nanowires were observed with the introduction of plasma in the growth process.

ACKNOWLEDGEMENT

This research was supported by National Nanotechnology Directorate funding NND/ND/ (2)/TD11-012 under the Ministry of Science, Technology and Innovation (MOSTI), Malaysia.

REFERENCES

[1] David S. Ginley and Clark Bright, "Transparent conducting oxides," *MRS bulletin 25*, **15** (2000) 15-18.

[2] S. J. Jiao, Z. Z. Zhang, Y. M. Lu, D. Z. Shen, B. Yao, J. Y. Zhang, B. H. Li, D. X. Zhao, X. W. Fan, and Z. K. Tang, "ZnO p-n junction light-emitting diodes fabricated on sapphire substrates," *Appl. Phys. Lett. 88*, **3** (2006) 031911-031911-3 .

[3] T. Itoh and T. Suga, "Force sensing microcantilever using sputtered zinc oxide thin film," *Appl. Phys. Lett. 64*, **37** (1994).

[4] G. Heiland, "Homogeneous semiconducting gas sensors," *Sensors and Actuators*, **2** (1982) 343–361.

[5] Toru Aoki, Yoshinori Hatanaka, and David C. Look, "ZnO diode fabricated by excimer-laser doping," *Appl. Phys. Lett. 76*, **22** (2000) 32573258.

[6] Bin Liu and and Hua Chun Zeng, "Hydrothermal Synthesis of ZnO Nanorods in the Diameter Regime of 50 nm," *Journal of the American Chemical Society 125*, **15** (2003) 4430-4431.

[7] Z. W. Pan, Z. R. Dai and Z. L. Wang, "Nanobelts of Semiconducting Oxides," *Science 291*, **5510** (2001) 1947-1949.

[8] Z. Zhu, T. Chen, Y. Gu, J. Warren and Richard M. Osgood, Jr., "Zinc Oxide Nanowires Grown by Vapor-Phase Transport Using Selected Metal Catalyst: A Comparative Study," *Chem. Mater 17*, **16** (2005) 4227–4234.

[9] T.Y. Kim, J.Y. Kim, M. Senthil Kumar, E.-K. Suh, and K.S. Nahm, "Influence of ambient gases on the morphology and photoluminescence of ZnO nanostructures synthesized with nickel oxide catalyst," *Journal of Crystal Growth 270*, **3-4** (2004) 491–497.

[10] X. Lui, X. Wu, H. Cao and R.P.H Chang, "Growth Mechanism and Properties of ZnO nanorods synthesized by Plasma-enhanced Chemical Vapor Deposition," *Journal of Applied Physics 95*, **6** (2004) 3141-3147.

[11] Colin A. Wolden, "The Role of Oxygen Dissociation in Plasma Enhanced Chemical Vapor Deposition of Zinc Oxide from Oxygen and Diethyl zinc," *Journal of Plasma Chemistry and Plasma Processing 25*, **2** (2005) 169.

[12] D.L. Smith, "Thin Film Deposition," Principles and Practice, McGraw-Hill, Inc, New York, (1995).

Recent Advancement in Microgap Electrode Fabrication by Conventional Photolithography Technique

Q. Humayun, U. Hashim

Nano Structure Lab-On-Chip Research Group, Institute Nano Electronic Engineering (INEE),
Universiti Malaysia Perlis (UniMap), 01000 Kangar, Perlis Malaysia
qhumayun2@gmail.com
uda@unimap.edu.my

Abstract-The biosensors which based on polysilicon material should be sensitive and selective for the detection of label free bio molecule. The objective of this research is to characterize, design, and fabricate polysilicon nanogap. The proposed device was designed initially in two masks by using AutoCAD software and then transferred to commercial chrome mask. The standard CMOS photolithography process coupled with ICP dry etching process is used for fabrication of the proposed nanogap. The fabrication process start by microgap formation and than by size expansion technique the microgap expand to nanogap and finally gold pad were deposited on polysilicon nanogap surface. Using TEM, FIB, FESEM and AFM the nanogap was inspected and characterized. At last the surface modification of polysilicon nanogap and DNA immobilization is done on polysilicon nanogap for the testing and validation by using real biological samples. The initial study demonstrates the fabrication of microgap of 1, 2, 3μm and 40.7, 43.5, 115.2, 335.9μm on polysilicon wafer samples.

Keywords: Polysilicon, microgap, chrome mask, Photolithography, size expansion, DNA immobilization

I. INTRODUCTION

Biosensor technologies have been widely studied. They have been based on electrochemical optical mass sensitive and acoustic wave transducers [1]. Recently, nanogap capacitors for direct electrical DNA detection without labeling like fluorescent, electrochemical intercalate magnetic, nanoparticles etc have been fabricated and used [2]. The dielectric detection mechanism (capacitance change) of a nanogap sensor promises fast and directing situ monitoring of DNA hybridization without a time consuming DNA labeling procedure [3]. The dielectric properties of molecules depend on electron transfer, atomic bonds, and the large-scale molecular structure [15]. Therefore, a sensing element silicon nanogap arrays fabricated in this research. The sensing element polysilicon nanogap integrated [16] with a reference probe electrodes (gold).The reference electrodes are formed on the silicon nanogap arrays. The proposed research describes fabrication of polysilicon nanogap for bio molecules detection. The sensor is proposed to design and fabricate using microgap expansion to nanogap coupled with metallic deposition to create ohmic contact for the conducting-probe, for this, gold [17] was chosen as two conduction probes, the reason why gold was preferred is, Gold

is resistant against almost all acids, even hydrochloric acid, The notable performance when used in analysis of many ionic species and the extraordinary affinity of thio [17] compounds for its surface make this electrode very suitable for this study taking other application in close proximity in consideration [20].

II. MATERIALS AND METHOD

A. Mask Design

In this research polysilicon wafer is used to fabricate the polysilicon nanogap structure. AutoCAD software was used for the electrode design with different gap sizes, 1μm, 2μm, 3μm having the length and width of 2.00mm and 2.50mm respectively (Fig.1A). The AutoCAD design mask (Fig.1A) is finally transferred to actual commercial soda lime glass photomask (Fig. 6A,B).

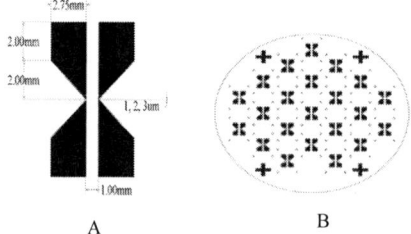

Figure.1. (A) Microgap electrode design specification (B) Microgap electrode AutoCAD designed masks.

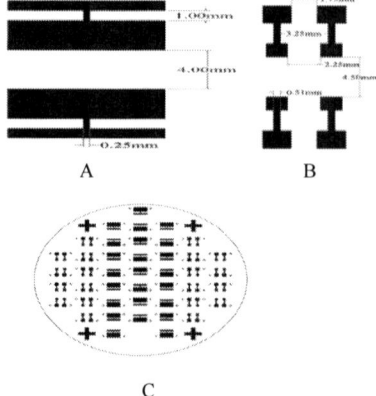

Figure2: (A) Design specification of common electrode padding (B) Design specification for Individual electrode padding. (C) AutoCAD design mask for common and individual electrode padding.

B. Micro to Nanogap Electrode Fabrication

The proposed process steps for the gold electrodes with polysilicon nanogap fabrication are as follow:clean the silicon wafer with IPA, ethanol, acetone and DI water in ultrasonic cleaner for 5 minutes followed by BOE in order to remove the native oxide layer. After cleaning 150 nm of silicon nitride (Si_3N_4) substrate on the silicon wafer and 200 nm Si layers, respectively, by using plasma-enhanced chemical vapor deposition (PECVD) equipment. Then, a layer of 135 nm polysilicon substrate as a hard mask using physical vapor deposition (PVD) equipment was performed to avoid damage to the silicon layer during the etching/RIE process. Next, in the photolithography process, a layer of positive photoresist1200 nm thick is first applied on the polysilicon substrate, and then exposed to ultraviolet light through mask 1 as shown in Fig. 3(E). After development only the unexposed resist would remain, and then the wet-etching process the polysilicon layer is performed before removing the resist Fig.3 (H). After that, the dry-etching process for a layer of silicon was done to fabricate the nanogap using the size expansion technique, then a layer of 100 nm gold substrate is deposited after a 30 nm Ti layer; the resist coating process before exposing mask2 and developing of the resist as seen in Fig. 3(M, N).Then the dry-etching process of the Ti/Au substrate is performed before removing the resist. Finally, the structure of common and individual gold electrode with a silicon nanogap is obtained as shown in Fig. 3(O,P,Q,R).

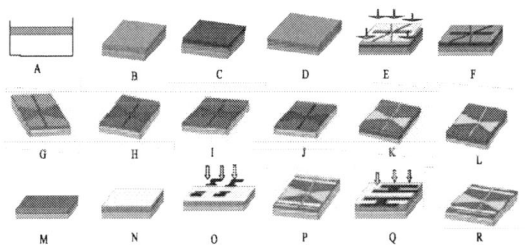

Figure3: Fabrication process flow of proposed Micro to nanogap electrode (A)Wafer Cleaning (B) Silicon Nitrite Si_3N_4 deposition (C) Poly Silicon deposition (D) Resist coating (E)Ultra Violet Exposure (F) Exposure Development (G) Ploy silicon dry etching (H) Resist Stripping (I) Poly Silicon microgap (J) Microgap to nanogap by size expansion thermal oxidation process (K) Wet etching process (L) Poly silicon nanogap (M) Titanium (Ti) deposition(N) Gold (Au) deposition (O) Alignment mark for individual gold padding (P) nanogap with individual gold padding (Q) Alignment mark for common gold padding (R) Nanogap with common gold padding.

III. SURFACE MODIFICATION

The surface of the fabricated nanogap is modified for the perfect attachment of bio molecule probe to appropriate locations within the nanostructure for DNA immobilization process.

A. Surface Modification Process for polysilicon Nanogap

The modification of silicon surface is, started with forming amino group on the surface of the silicon and followed by gold deposition on the amino group and finally followed by linking thiolated probe DNA. Below are the major steps in picture form.

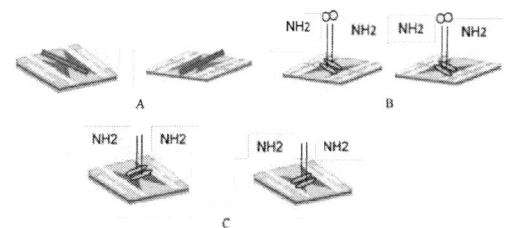

Figure4: Shows the polysilicon nanogap surface modification.

IV. DNA IMMOBILIZATION

In order to attach bimolecular, DNA and proteins in particular, onto a solid support, it is necessary to have a strong chemical binding chemistry. As we know, the most critical step while preparing a DNA biosensor is the immobilization of DNA probe on the surface of a sensing device such as an electrode. The amount of immobilized DNA probe will influence the accuracy, sensitivity, selectivity, and life of a DNA biosensor directly.[7] Because of the high surface-to-volume ratio and excellent biological compatibility, Nano-materials can enlarge the sensing surface area to increase the amount of immobilized DNA greatly, and the DNA mixed with Nano-materials can keep its biological activity well. Gold (Au) Nano-particles is used this research as previously and they play a key role in DNA bio-sensors. Thiol–Au linkages usually were used to bind Au Nano-particles covalently with solid electrodes or DNA because of the strong affinity of covalent bonds between sulphur atoms and gold atoms. Immobilized the DNA on the surface of Au Nano-particles which were co-immobilized at a gold electrodethrough4,4'-bis(methanethiol) biphenyl (MTP) molecules by an assembly process. Their results indicated that the Au nanoparticles modified gold electrode can enlarge the electrode surface area and enhance the amount of immobilized single stranded DNA (ssDNA) greatly in previous research, and the surface coverage value of DNA molecules decreased as the size of the gold Nano-particles increased [15].

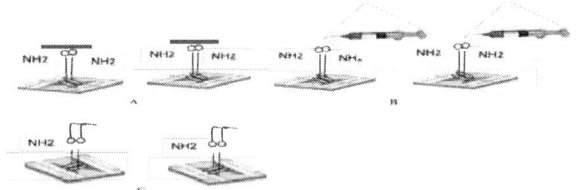

Figure5: Shows the DNA Immobilization at polysilicon nanogap surface.

V. RESULTS AND DISCUSSION

A. Soda lime quartz glass photo mask

Commercial soda lime glass photo masks are used in this research for a better photo masking process. It consists of 22 dies with 3 different microgap sizes. The designed mask was transferred on a 5x 5 inch soda lime quartz glass with a thickness of 3mm from Photonik (Singapore) Pte Ltd to conduct the research. The critical dimensions of the microgap and common, individual electrode padding are shown in Fig. 1(A) and 2(A,B).

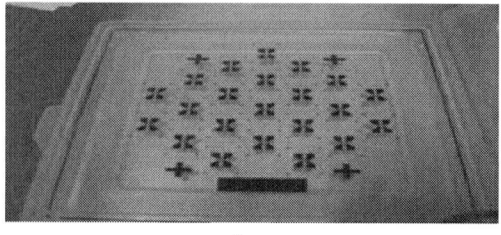

A

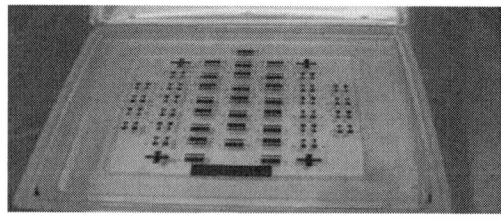

B

Figure6: Shows (A) the microgaps soda lime quartz glass mask (B) the electrode padding soda lime quartz glass mask.

B. Pattern transfer by exposure and development

For reliable transformation of microgap pattern to wafer the designed should be well specified to avoid from error to the critical dimension. The photoresist needs to have good resolution, high etch resistance and good adhesion. High resolution is the key to achieve successfully microgap pattern transfer. Without high etch resistance and good adhesion of the photoresist, the etching processes will most likely fail to meet and will cause intolerable error the thinner the photoresist film, the higher the resolution. However, the thinner the photoresist, the lower the etching resistance, this is one of the most critical trade-off between these two conflicting requirement which is considered during this research. Following are some SEM result obtained from developed microgap pattern transfer.

- Well developed pattern transfer to poly silicon deposited silicon wafer;It is clear from Fig.7A, B, C, poly silicon deposited silicon wafer samples that the pattern transferred to samples wafer same to the critical dimension. The micro gap sizes of the all positive photo resist coated samples wafer in Fig.7A,B,C are 1μm, 2μm, 3μm respectively which is exactly same to the critical dimension of the microgap electrode. Its shows that the developed wafer samples are suitable for the next dry etching process.

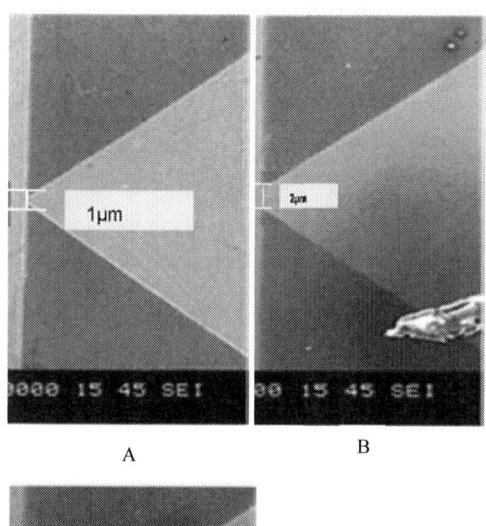

A   B

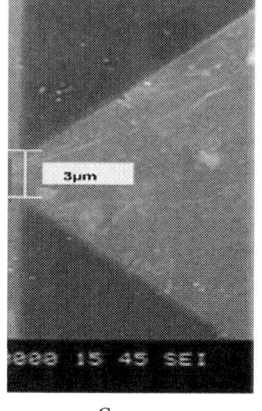

C

Figure7: (A,B,C) SEM images of well developed polysilicon deposited wafer samples.

- Partial developed pattern transfer poly silicon deposited silicon wafer; The sizes of microgaps patterned on polysilicon wafer samples in Fig.8A,B,C,D are not exactly same to the critical dimension due to partial developing process. The first difficulty faced during developing the silicon wafer sample as in Fig. 8A is due to irregular thickness of the resist because "thick" resist means a film thickness if the thickness is higher than the penetration depth of the exposure light this will cause non uniform and irregular exposure. When the exposure time is too low and the development time of positive resists is high than it increases the total dark erosion and if the exposure time is too low, and the development time increase too high exposure time values cause light scattering and diffraction in the resist film which deteriorates the resolution Fig.8B. The big lumps and coating failure which is shown in Fig.8C, D is caused due to uniform developing sample wafer.

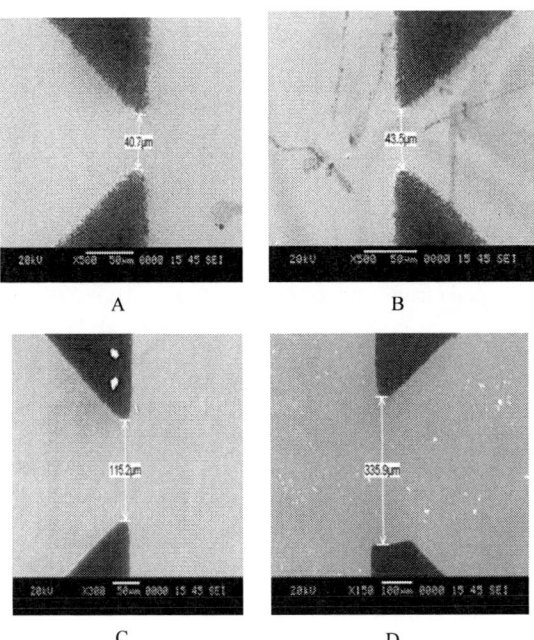

A   B

C   D

Figure8: SEM images of partial developed polysilicon wafer samples.

CONCLUSION

The proposed nanogap which is highly sensitive and selective with common and individual probe, designed, fabricated, characterized and optimized in this paper by standard CMOS conventional photolithography process coupled with ICP dry etching process respectively. The target application is the label-free detection for low concentration and single bio-molecule label-free detection that lead to any related diseases using in-vitro clinical samples. The proposed method has been experimentally demonstrated by fabricating microgaps of 1μm, 2μm, 3μm and 40.7 μm, 43.5 μm, 115.2 μm, 335.9μm dimensions. The next part of this research is to decrease the micro sizes gaps to nano sizes experimentally by expanding the micro sizes to nano by size expansion technique using thermal oxidation process. Future experiments will focus on building up nano-biosensors by coupling biomolecules within the nanogaps with dimensions close to their persistence length.

ACKNOWLEDGEMENT

The authors acknowledge the financial support from ministry of higher education (MOHE). The authors also would like to thank all of the team members and technical staff in the Institute of Nano Electronic Engineering.

REFERENCE

[1]   Likharev K 2003 Nano and Giga Challenges in Microelectronics ed J Greer, A Korkin and J Labanowski (Amsterdam: Elsevier).

[2]   Reed M A, Zhou C, Muller C J, Burgin T P and Tour J M 1996Conductance of a molecular junction Science 278 252–4.

[3]   Fischbein M D and Drndic M 2007 Sub-10 nm device fabrication in a transmission electron microscope Nano Lett. 7 1329–37.

[4]   Venkataraman L, Klare J E, Nuckolls C, Hybertsen M S and Steigerwald M L 2006 Dependence of single-molecule junction conductance on molecular conformation Nature 442 904–7.

[5]   Haag R, Rampi M A, Holmlin R E and White H S 1999Electrical breakdown of aliphatic and aromatic self-assembled monolayers used as nanometer-thick organic dielectrics J. Am. Chem. Soc. 121 7895–906.

[6]   Fischbein M D and Drndic M 2006 Nanogaps by direct lithography for high-resolution imaging and electronic characterization of nanostructures Appl. Phys. Lett.88 063116.

[7]   Gergel-Hakett N et al 2006 Effects of molecular environments on the electrical switching with memory of nitro-containing OPEs J. Vac. Sci. Technol. A 24 1243–8.

[8]   Long D P, Patterson C H, Moore M H, Seferos DS,Bazan G C and Kushmerick J G 2005 Magnetic directed assembly of molecular junctions Appl. Phys. Lett.86 153105.

[9]   Ramachandran G K et al 2003 Electron transport properties of a carotene molecule in a metal–(single molecule)–metal junction J. Phys. Chem. B 107 6162–9.

[10] Galperin M, Ratner M A and Nitzan A 2007 Molecular transport junctions: vibrational effects J. Phys.: Condens.Matter 19 103201.

[11] Datta S, Tian W, Hong S, Reifenberger R, Henderson J I and Kubiak C P 1997 Current–voltage characteristics of self-assembled monolayers by scanning tunneling microscoy Phys. Rev. Lett. 79 2530–3.

[12] Tian W, Datta S, Hong S, Reifenberger R, Henderson J I and Kubiak C P 1998 Conductance spectra of molecular wiresJ. Chem. Phys. 109 2874–82.

[13] Wiesendanger R and G¨untherodt H-J (ed) 1996 Scanning Tunneling Microscopy III: Theory of STM and RelatedScanning Probe Methods (New York: Springer).

[14] Bonnell D A (ed) 1993 Scanning Tunneling Microscopy and Spectroscopy: Theory, Techniques, and Applications (New York: VCH.

[15] Uda Hashim, Siti Fatimah Abd. Rahman, M. Nuzaihan Md. Nor, Shahrir Salleh," Design and Process Development of Silicon Nanowire Based DNA Biosensorusing Electron Beam Lithography", 2008 International Conference on Electronic Design, December 1-3, 2008, Penang, Malaysia.

[16] Th.S.Dhahi,U.Hashim,M.E.Ali,N.M.Ahmed, and T.Nazwa Fabrication of Lateral Polysilicon Gap of Less than 50nm Using Conventional Lithography Journal of Nano materials. Volume 2011, Article ID 250350.

[17] Dhahi, T. S.,Hashim,U.Ahmed, N.M. 2011. Improvement in Processing of Nano Structure Fabrication Using O2

[18] Dhahi.T.S, Hashim, U.Ahmed, N.M.Taib,A.M.2010.Areview on the Electrochemical Sensors and Biosensors Composed of Nanogaps as Sensing Material. J. Optoelectr. Adv. Materials 12(29):1857-1862.

[19] Dhahi. T.S, Hashim. U, Ahmed. N.M. 2011. Fabrication and Characterization of 50nm Silicon Nanogap Structure. J. Sci. Adv. Materials 3(2): 233–238.

[20] Dhahi.T.S, Hashim. U,Ali. M.E., Nazwa. T, Ahmed,N.M. 2011. Fabrication and characterization of lateral polysilicon gap less than 50nm using conventional lithography process. J. Nano Materials, Article in Press.

Numerical Simulation of Microfluidic Devices

Uda Hashim[1], P.N A. Diyana [2], Tijjani Adam[#1]

[1#]Institute Of Nano Electronic Engineering, Universiti Malaysia Perlis(UniMAP)
[2#]School of Microelectronic Engineering, Universiti Malaysia Perlis(UniMAP)
Perlis Indera Kayangan, Malaysia.
[1]pu3_aliaa@yahoo.com , [2]uda@unimap.edu.my,[1]tijjaniadam@yahoo.com

Abstract—**Microfluidic devices present a unique powerful platform for working with living cells. The length and volume scales of these devices in miniaturize system make it possible to perform detailed analyses with several advantages. The small volume facilitates response detection by effectively increasing the local concentration of the analyte. Microfluidic lab-on-chip technologies represent a revolution in laboratory experimentation, bringing the benefits of miniaturization to many researchers. This project aim on simulation of the fluid flow, the mixing concentration distribution to the interaction of protein and other biological reagents, fluid flow pressure, and fluid flow velocity field with designing Micromixer, Microchannel and Microchamber for Microfluidic devices using simulation software COMSOL Multiphysics 3.5. Fluid flow patterns, concentration distribution and velocity field were observed by using simulation software. Fluid flow patterns obtained at the junction and The component consists of a PDMS microchannel to give a continuous open circuit flow, a 0.7μm/s was obtained.**

Keywords— **microfluidics devices. Micro total analysis system (μTAS), laboratory-on-a-chip, miniaturized, fluid flow, mixing concentration distribution, fluid flow pressure, fluid flow velocity field, micromixer, microchannel, microchamber, simulation.**

I. INTRODUCTION

The potential of Microfluidic devices as a microchips has attracted many researchers as it offers several advantages and the applications either in a biotechnology, pharmaceuticals, agriculture health and many more. Manipulation of fluids, colloidal particles, or biomolecules on the microscale can be performed in integrated devices for sample pre-treatment and analyte detection. Such systems integrated in a single chip are known as micro-Total Analysis Systems (mTAS) or lab-on-a-chip devices. Present, more company and institution doing research and development based on miniaturization system. The advantages of microfluidics include, yet are not limited to, small sample volumes leading to greater efficiency of chemical reagents; low production costs per device thereby allowing for disposability; high throughput synthesis and screening of biological species and drug targets; parallel processing of samples[1-3]; fast sampling times; accurate and precise control of samples and reagents reducing the need for pipetting; low power consumptions; and versatile format for integration of various detection schemes thereby leading to greater density. Fluid flow is generally categorized into two flow regimes: laminar and turbulent. Laminar flow is characterized by smooth and constant fluid motion, whereas turbulent flow is characterized by vortices and flow fluctuations. Physically, the two regimes differ in terms of the relative importance of viscous and inertial forces. In an environment where the fluid flow is restrictedly laminar, mixing is largely dominated by passive molecular diffusion and advection. Diffusion is defined as the process of spreading molecules from a region of higher concentration to one of lower concentration by Brownian motion, which results in a gradual mixing of material.

Micromixer are generally designed with channel geometries that act as the mixing path and increase the mixing before flow through the channel. Channels perform the essential but simplest of tasks: transferring fluid from one device into another, usually as a continuous fluid flow. At present, microchannels are used to transport and mix biological materials such as proteins, DNA, cells, etc. or to send chemical samples from one place to the other [3,4]. To reduced the mixing path in micromixer by a simple narrowing of the mixing channel and therefore shortening of the diffusion length. Further reduction of the mixing time could be achieved by using a high flow rate, hence high Re,where a chaotic flow is expected [5]. In the development of micro total analysis systems, it is often necessary to achieve complete, uniform filling of relatively large microchambers, such as those needed for nucleic acid amplification or detection. A microfluidic device is made from varieties categories of components and in this paper will discuss three important components: micromixer, microchannel and microchamber. In this paper, the results observed for concentration distribution to determine the mixing point of the devices, The pressure for the fluid flow, and the velocity field to determine whether the fluid is flow in microchannel. Other than that, it will explain the numerical simulation process development to form Micromixer, Microchannel and Microchamber and characterization of the results.

II. Experimental Procedure

A. *Micromixer and Microchannel*

The constant below is for modeling 3 inlet and 1 outlet microchannel and micromixer. There are 3 inputs for inlet concentration, 2 inlets will be using high concentration and 1 inlet low concentration as Table 1, 2 and 3:

TABLE 1. Shows Constant Used IN Simulation 3 Inlet AND 1 Outlet Of Micromixer AND Microchannel

NAME	EXPRESSION	DESCRIPTION
rho	1e3[kg/m^3]	Density
eta	1e-3[Pa*s]	Viscosity
D	1e-10[m^2/s]	Diffusion constant
p0	2[Pa]	Pressure drop
c0	1[mol/m^3]	Inlet concentration
c1	2[mol/m^3]	Inlet concentration
alpha	0.5[(m^3/mol)^2]	Viscosity c^2-term prefactor

The geometry modelling in 2-D cross section of Micromixer is made with 3 inlet and 1 outlet as figure 2 (a). Figure 2(b) shows after extrude into 3-D cross section.

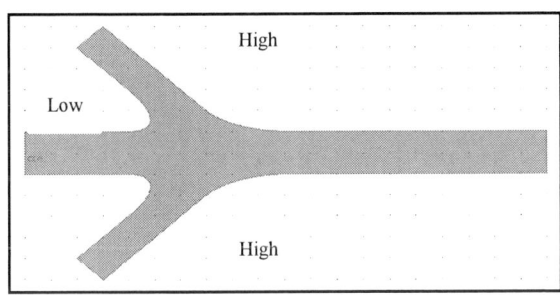

(a)

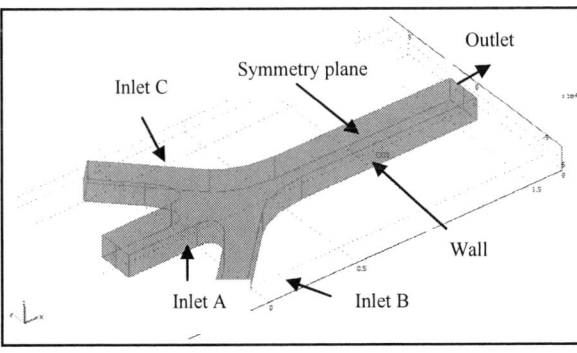

(b)

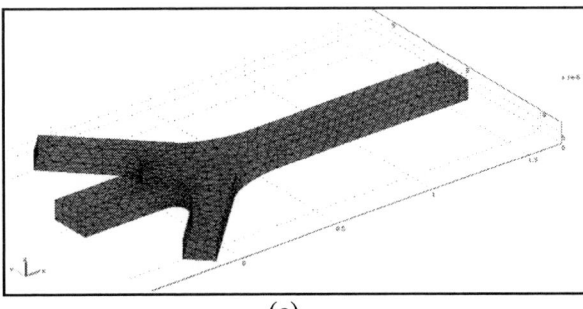

(c)

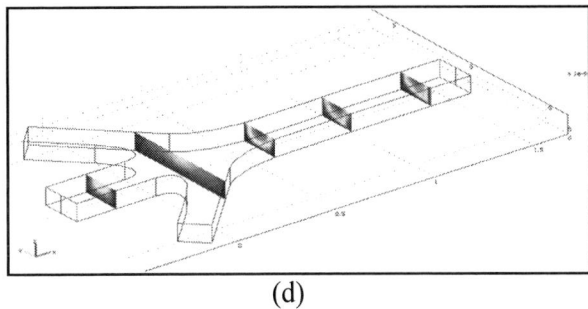

(d)

Fig1. (a) shows the geometry model in 2-D of 3 inlet and 1 outlet using high concentration and low concentration of Micromixer and Microchannel. (b) shows the Sub domain Settings and Boundary Settings used to specify each boundary condition of Micromixer and Microchannel in 3-D, (c) shows the Micromixer and Microchannel after Remesh into small units (d) Fig 2 (d) shows the figure after Solve the design of Micromixer and Microchannel

TABLE 2. Shows THE Boundary Conditions FOR Incompressible Navier-stokes (MMGLF) USED TO SPECIFY EACH Boundary Condition OF Micromixer AND Microchannel IN 3-D

Settings	Boundaries 1,5,10,12	Boundaries 19,20	Boundary 4	All Others
Boundary type	Inlet	Outlet	Symmetry boundary	Wall
Boundary condition	Pressure, no vicious stress	Pressure, no vicious stress		No slip
P_0	p0	0	-	-

TABLE 3. shows the Boundary Conditions for Convection and Diffusion (chcd) used to specify each boundary condition of Micromixer and Microchannel in 3-D.

Settings	Boundary 1,5	Boundary 10,12	Boundaries 19,20	All others
Type	Concentration	Concentration	Convective flux	Insulation / Symmetry
c_0	c0	c1	-	-

B. Microchamber

Simulation for Microchamber is done in two-dimensional cross section. The constant below is for modelling Microchamber:

TABLE 4. SHOWS CONSTANT USED IN SIMULATION 3 INLET AND 1 OUTLET OF MICROMIXER AND MICROCHANNEL.

NAME	EXPRESSION	DESCRIPTION
rho	1e3[kg/m^3]	Density
eta	1e-3[Pa*s]	Viscosity
D	1e-10[m^2/s]	Diffusion constant
p0	4[Pa]	Pressure drop
c0	1[mol/m^3]	Inlet concentration

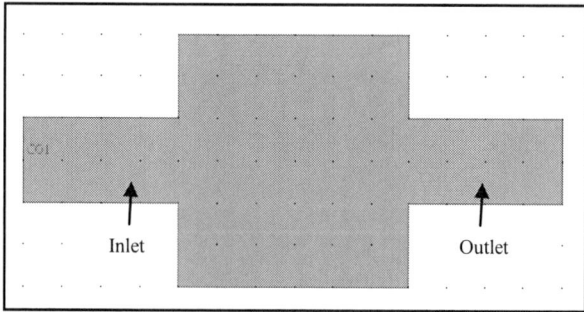

Fig2 shows the geometry model of Microchamber in 2-D of 1 inlet and 1 outlet using 1mol/m^3 concentration.

TABLE 5. SHOWS THE BOUNDARY CONDITIONS FOR INCOMPRESSIBLE NAVIER-STOKES (MMGLF) USED TO SPECIFY EACH BOUNDARY CONDITION OF MICROCHAMBER IN 2-D.

Settings	Boundary 1	Boundary 12	All Others
Boundary type	Inlet	Outlet	Wall
Boundary condition	Pressure, no vicious stress	Pressure, no vicious stress	No slip
P_0	p0	0	-

TABLE 6. SHOWS THE BOUNDARY CONDITIONS FOR CONVECTION AND DIFFUSION (CHCD) USED TO SPECIFY EACH BOUNDARY CONDITION OF MICROCHAMBER IN 2-D.

Settings	Boundary 1	Boundary 12	All others
Type	Concentration	Convection flux	Insulation / Symmetry
c_0	c0	-	-

III. RESULT AND DISCUSSION

A COMSOL Multiphysics software was used for Capillary flow experiments in a microchannel, the channel was formed by two parallel plates separated by distance H. The plate separation (H) is assumed to be precisely maintained through out the length of the channel covering a range of 8 mm to 10 mm as shown in fig2. Experiments have been conducted using Newtonian liquids (Navier-Stokes) for model flow. Fig11 Shows the variation of the velocity with length of the channel, the inertia point where the first time when the fluid is injected of typical runs with unstable fluid dynamics. At the inlet, are dual forces associated in driving the fluid, a gravitation force since the inlet is vertical (ρgH) and the forces due capillary, in the figure 1a, where the first point marked with pink arrow indicate the flow orientation is vertical where the velocity increases rapidly. The point shown with yellow arrow in figure1a showing distinct point where the fluid entered the channel and stabilizes for the flow, the fronts rises and stabilize and reaches the equilibrium rise height (H) at 1.28mm/s. The most interesting point is where a constant velocity is maintained through out the channel length with this velocity. A sudden upshot is experience due to the fluid leaving the channel indicate by convective flux and at the 1.3mm/s, Which is a very interesting result, to explain this, the difference in velocity between the stable state to convective influx is 0.02mm/s at least 4% higher. This indicates that all the fluids entered the channel leave the channel at the outlet without any disturbance hence, the channel will work perfectly without any clog of sample The results can be obtained in the form of graph from cross-section plot parameters after post processing and visualization steps.

A. Micromixer and Microchannel

The mixing process in this type of micromixer is obtained by guiding the three liquids to be mixed in contact through a flow-through channel. The mixing solely depends on diffusion of the species at the interface between the two liquids; hence the mixing is accomplished before the liquid flow through the Microchannel.

- Concentration Distribution

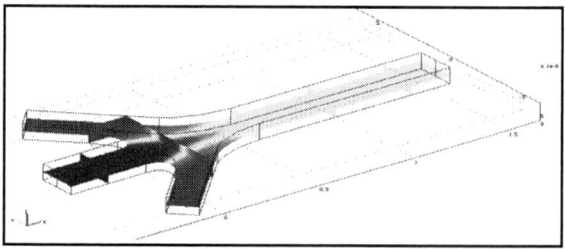

Fig 3. shows 3D view of the Micromixer and Microchannel after postprocessing and visualization process for Concentration distribution for a species with diffusivity 1·10−10m2/s. Because of the relatively large diffusion coefficient, the degree of mixing is almost perfect for the

lightest species. The highest concentration is set at 2.00 and the low concentration is set at 1.00. The mixing stablilize after reach 1.3 mol/m^3.

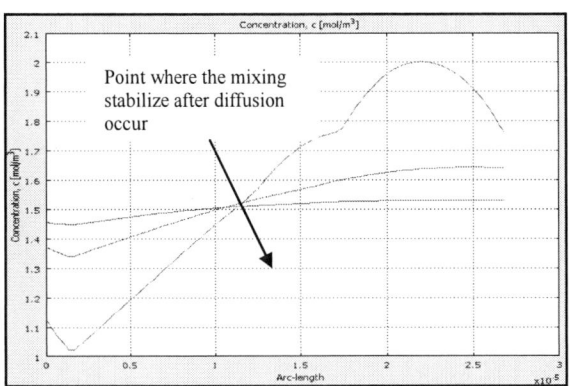

Fig4: Simulation software using COMSOL MULTIPHYSIC 3.5. Concentration Graph, C, for 3 input of inlet concentration, 2 inlet will be using high concentration; 2mol/m^3 and 1 inlet low concentration; 1 mol/m^3. The mixing occur at concentration of 1.3 mol/m^3 and at arc-length of 0.12µm.

At microscale level, fluid properties become increasingly controlled and provide the possibility of exerting control over the different processes carried out in a Microfluidic environment [9]. When designing a Microfluidic device, it is important to be able to mix chemicals quickly and easily before the chemicals reach the sensor that will be place at the Microchannel. The flow comes in the three inputs on the left and exits on the right. Diffusion occurs across the mid-plane in the device. The flow is basically straight down the device or known as laminar flow, except at the junction between three inlets region, with diffusion sideways, and there is no convection sideways. Thus, diffusion controls the mixing and determines how far the reagents can diffuse with each other [6]. The molecular diffusion, which is an extremely slow process; and various ingenious methods have been and are still actively developed to enhance mixing fig5.

- Velocity Field

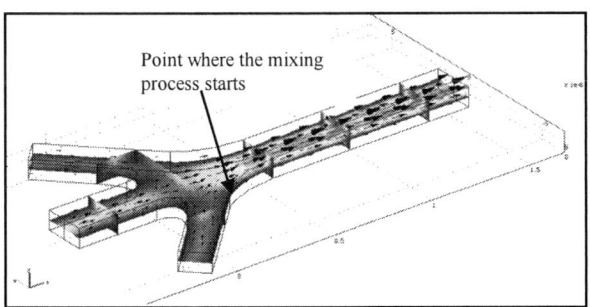

Fig5 shows the 3D view of Velocity Field for the Micromixer and Microchannel after postprocessing and visualization process. The mixing

occurs around 60%-70% at the junctions before the liquids flow through the Microchannel.

IV. CONCLUSION

In this paper, the design, simulation and characterize the Micromixer and Microchannel in three-dimensional; and Microchamber in two-dimensional are presented. For a better understanding of the fluid flow and the mixing characteristics in the mixer, a computational fluid, Comsol Multiphysics 3.5, using flow with species, to simulate three-dimensional of Micromixer and Microchannel; and two-dimensional of Microchamber to determine the flow fields, concentration and pressure as well as three-fluid mixing. All simulations were done in Comsol Multiphysics, 3.5 by first solving the In compressible Navier-Stokes equations and then Convection and Diffusion (chcd). The mixing performance is great for this experiment. And the chamber need to achieve a uniform filling such as those needed for nucleic acid amplification or detection.

ACKNOWLEDGMENT

The authors wish to thank Universiti Malaysia Perlis (UniMAP and Ministry of Higher Education Malaysia for giving FRGS grant to conduct this research in the Micro & Nano Fabrication Lab. Appreciation also goes to all the team members in the Institute of Nanoelectronic Engineering especially the Nano structure Lab On chip Research Group.

REFERENCES

[1] Lab-on-a-chip: *A micromixer to facilitate chemical and biochemical analyses*, Press Release, Paris, January 25,2002.

[2] H. M. Xia and S. Wan, (Jan-March 2007), *Development of a novel passive micromixer*, SIMTech technical reports, Mechanical Engineering Department, National University of Singapore, Singapore.

[3] Simon Jayaraj, Sangmo Kang, Yong Kweon Suh, *A Review on the Analysis and Experiment of Fluid Flow and Mixing in Micro-Channels*, Journal of Mechanical Science and Technology 21, (January 2007).

[4] Stefan Haeberle and Roland Zengerle, (2007), *Microfluidic platforms for lab-on-a-chip application*, RSC Publishing.

[5] Lorenzo Capretto, Wei Cheng, Martyn Hill, and Xunli Zhang, (2011), Micromixing Within Microfluidic Devices, School of Engineering Sciences, University of Southampton, Southampton SO17 1BJ, UK.

Effect of Dye Coating Duration of Eosin Y Synthesized on Nanorods Zinc Oxide Toward Hybrid Solar Cell

*Siti Aisyah Zawawi, **Chi Chin Yap, *Haslinda Abdul Hamid, *Mohd Zaki Mohd Yusoff
*Department of Applied Sciences
Universiti Teknologi MARA Pulau Pinang
13500 Permatang Pauh, Pulau Pinang, MALAYSIA
**School of Applied Physics, Faculty of Science and Technology,
Universiti Kebangsaan Malaysia 43600 UKM Bangi, Selangor, MALAYSIA
aisyahzawawi@gmail.com

Abstract— The efficiency of inverted organic solar cells become interested among researchers. The various parameters changed to improve the efficiency of cell.This study were conducted to investigate the effect of dye coating duration on the performance of inverted orgnic solar cell. The characterization of the sample was done by Scanning Electron Microscope test (SEM) for morphology characterization, UV-Vis-NIR test for optical properties and current voltage test (IV) to determine the efficiency. The ZnO nanorods-coated FTO substrates were immersed in the eosin Y dye solution with ethanol at $60^{\circ}C$ for 15, 60 and 120 minute. The power conversion efficiency of the solar cell increased with dye coating duration and reached an optimum value at dye coating duration of 60 minute. The device with dye coating duration of 60 minute exhibited the highest power conversion efficiency of 0.0001528 % with short I_{sc} 0.00765 mA/cm^2, V_{oc} of 0.192 V and fill factor of 0.104 %.

Keywords-hybrid solar cell; ZnO; eosin Y

I. INTRODUCTION

The demand on the alternative energy production is become more interest because of global demand for energy resources and environmental awareness of climate change. One of the alternative energy production is solar energy which is clean and safe process beside the materials used are cheap and widely available. For example is conventional solar cell. This solar cell is based on inorganic materials such as silicon. The efficiency of this conventional solar cells made from inorganic materials reached up to 24%, using expensive materials of high purity and energy intensive processing techniques [1]. To overcome this problem, organic solar cell and dye synthesized solar cell (DSSC) are created. Dye synthesized solar cell composed of a dyed oxide semiconductor proved as the most low-cost alternative for the effective conversion of light energy into electrical energy. Up to now, the conversion efficiency of the DSSC has reached to ~11% [2].

Basically, organic solar cell consist of three layers; indium tin oxide (ITO), polymer poly(3-hexylthiophene), P3HT with phenylC61 butyric acidmethylester, PCBM and aluminium, Al arranged ITO/P3HT:PCBM/Al. One of the limitting performance factors of this structure is the smaller internal electric field to attract charge toward electrode. After some research, this organic solar cell structured improved by adding poly(3,4-ethylenedioxythiophene), PEDOT and lithium fluoride, LiF, with improved structure ITO/PEDOT:PSS/P3HT:PCBM/LiF/Al. However, this structure not necessarily stable and get high efficiency. The power conversion efficiency of this solar cell is around ~2.11% [3]. This study is focus on the effect of eosin Y toward the performance of hybrid solar cell at difference dye-coating duration.

II. EXPERIMENTAL WORKS

Provision of research materials is done according to the steps and performed in the laboratory. This solar cell is synthesized using poly [2-methoxy-5-(2'-ethyl-heksilosi)-1.4-fenelin vinilin] (MEHPPV) act as donor and (6,6)-C6-phenyl-butyric acid methyl ester (PCBM) act as acceptor with structure of FTO/ZnO/MEHPPV:PCBM/Au. FTO substrate was prepared by cleaning the surface using zinc powder and mixture of 10 ml hydrocloric acid with 10 ml distilled water. It is important to clean all the contaminations which can cause short circuit during IV test. Then, the substrate were washed with aceton and 2-propanol in ultrasonic bath for 15 minutes respectively.

ZnO nanoparticles seeding on FTO glass substare via sol-gel and spin coating method. Sol-gel solution prepared using 0.4 g Zn asetate dihydrate and put into bottle contain magnet. Solution of 6 ml etanol and 0.1725 ml DEA prepared and dropped slowly into the bottle. Mix all the solution using digital stirrer machine for 30 minutes at $60^{\circ}C$ and 400 rpm. Then, wrap the solution with aluminium foil and keep for 24 hours. Growth of nanoparticle ZnO is prepared via spin

978-1-4673-2395-6/12 $31.00 © 2012 IEEE

coating method. 0.07 ml sol-gel solution was dropped into substare for three times. ZnO seed layer was prepared by anealling process to get uniform growth of nanarod and nanoparticle ZnO. The furnace was setted at 300°C for 1 hour. Substrate is then kept at room temperatrure. The hydrothermal growth of nanorod ZnO was prepared solution of 0.112 g hexamethylenetetramine (HMT) and 0.238 g zinc nitrate hexahydrate in 20 ml deionized water. Noted that the FTO-coated with ZnO nanoparticle is facing down.

After nanorod ZnO was growth, the substrate will dye coated with eosin Y solution. Eosin Y solution was prepared by 20 ml of ethanol and 0.00259 g eosin Y. Make sure the solution is totally dissolved. Then, the substrate is coated with eosin Y solution by hydrothermal process at 60°C for three difference dye coating duration (15 min, 60 min and 120 min). After hydrothermal process, the substrate was rinsed with ethanol before dried and stored in vacuum.

Polymer used in this research are MEHPPV and PCBM with ratio 1:2. Polymer solution was prepared by adding 2 ml of cloroform, 12 mg PCBM and 6 mg MEHPPV. The solution must totally dissolved and dark-brown color formed. Substrate are then coated with polymer layer using a spin coating method. 0.12 ml polymer solution is dropped onto the substrate using an adjustable pipette at 1000 rpm for 40 seconds. Lastly, the gold layer is coated. Gold is deposited on the substrate surface using electron evaporation with a thickness of 150 nm for about 2 hours. Quartz is used to obtain the absorption of light by using UV-Vis test.

III. RESULTS & DISCUSSION

Fig. 2-4 shows SEM images of dye-coated ZnO nanorods samples for b) 15 minute c) 60 minute d) 120 minute in eosin Y solution, respectively. An uncoated sample Fig. 1 is also included for comparison. Generally, we found that all the samples have better surface uniformity. It can be seen that the dye-coating duration changes the morphology of ZnO nanorods structure. Specifically, all the ZnO nanorods were invisible after the samples were dye-coating with eosin Y solution. The eosin Y agregates were started growth on ZnO nanorods at 15 minutes treatment. From the results, the size of pacth and amount of eosin Y have increased with the dye-coating duration, respectively. Absorption spectrum analysis on the samples has been studied by using ultraviolet (UV)-visible spectroscopy system. UV-visible spectroscopy involves the spectroscopy of the photons in the UV-visible region. The absorption in visible ranges directly affects the color of the chemicals involved. In this region of the electromagnetic spectrum, molecules undergo electronic transitions.

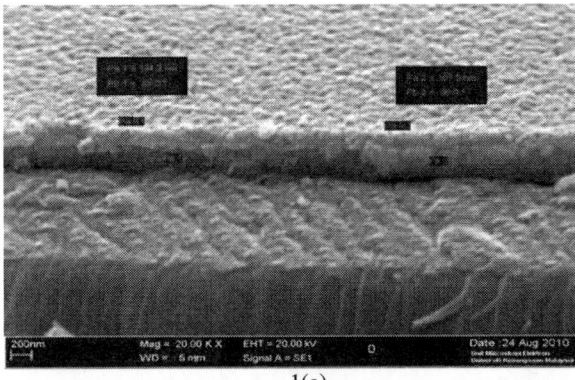

1(a)

1(b)

Figure 1: (a) Cross-section view (b) top-view of SEM images for ZnO nanorods without eosin Y

2(a)

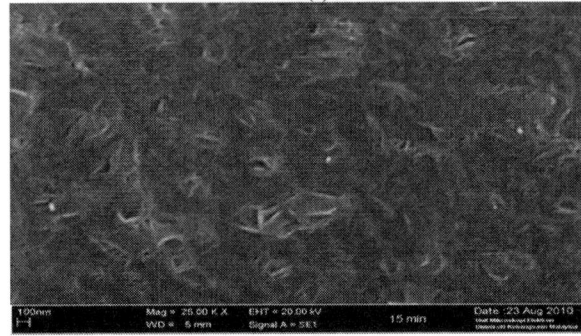

2(b)

Figure 2: (a) Cross-section and (b) top-view of SEM images for ZnO nanorods with dye-coating duration 15 minutes.

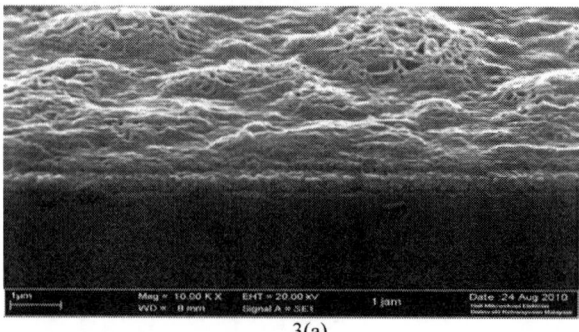

Figure 3: (a) Cross-section and (b) top-view of SEM images for ZnO nanorods with dye-coating duration 60 minutes.

Figure 4: (a) Cross-section and (b) top-view of SEM images for ZnO nanorods with dye-coating duration 120 minutes.

Figure 5 shows the absorption spectrum of ZnO nanorods with different dye-coating duration. It can be seen that the dye-coating process improves the light absorption of the samples. In agreement with the [4] and [5] dye-coated ZnO nanorods samples have maximum light absorption at range 400 nm – 600 nm compared to control sample. Moreover, we found that the absorption of eosin Y was increased with dye-coating duration. Result from Figure 5 therefore suggest that the differences in absorption characteristics could be partly explained by the quantity of coated-eosin Y on ZnO nanorods with various dye-coating duration. Sample with 120 minutes exhibited high absorption compared to other samples. This result was agreed with SEM results which showing that quantity of coated-eosin Y increased with dye-coating duration.

Figure 6 shows current-voltage (IV) characteristics of ZnO nanorods with various dye-coating duration under sun light 1.5 AM luminescence with power source $100 mW/cm^2$. short circuit current (I_{sc}), open circuit voltage (V_{oc}), maximum current (I_{max}), maximum voltage (V_{max}), fill factor (FF), and efficiency (η) can be determined from this graph. The 60 minutes and control samples exhibited maximum and minimum I_{sc} values of 0.00765 mA/cm^2 and 0.00196 mA/cm^2, respectively. However, for 120 minutes sample, the I_{sc} values was decreased. The dye-coated samples also have open circuit voltage (V_{oc}) values, with 60 and 15 minutes samples showed maximum and minimum values of 0.192 V and 0.086 V, respectively. The fill factor (FF) is essentially a measure of quality of the solar cell. It is calculated by comparing the maximum power to the theoretical power that would be output at both the open circuit voltage and short circuit current together.

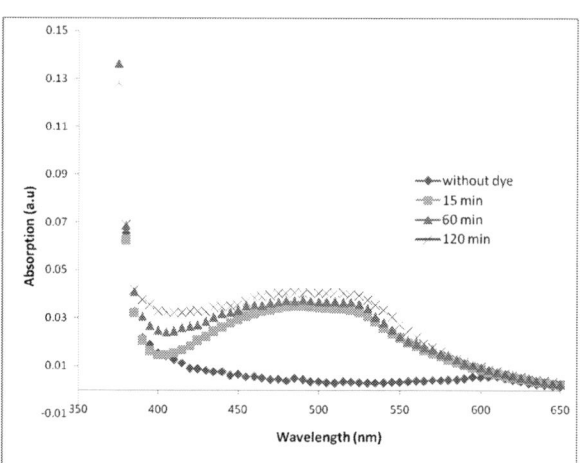

Figure 5: Absorption spectrum of ZnO nanorods with different dye-coating duration.

The maximum and minimum FF values were 0.509 and 0.104 for dye-coating 15 and 60 minutes, respectively. The efficiency From the FF results, the efficiency of solar cells can be calculated. The efficiency of control sample was 0.0000559%. Generally, dye-coated samples from 15 to 60 minutes started to increase except 120 minutes sample which exhibited degradation. 60 minutes sample has highest efficiency (0.000153%) compared to other samples. Table 1 shows the summary for all samples.

978-1-4673-2395-6/12 $31.00 © 2012 IEEE

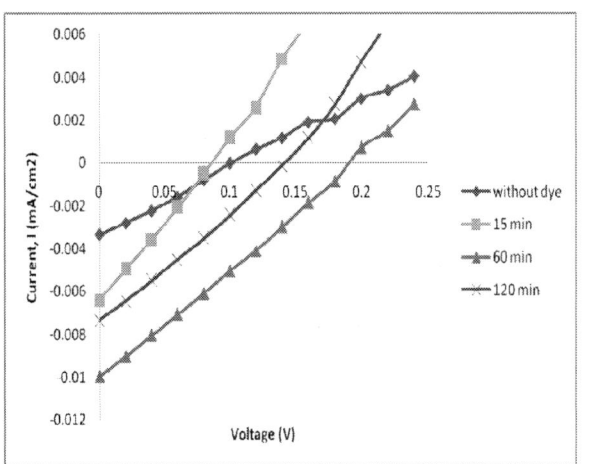

Figure 6: IV characteristic for ZnO nanorods with different dye-coating duration.

According to [6], short circuit current (I_{sc}) is directly proportional with number of dye-coated on ZnO nanorods surface. We suggest that this is mainly due to the existence of exciton and free charges. The excitons and free charges are increased when the optical absorption increased. However, at 120 minutes, the eosin Y patch is large as shown in Figure 4 (b). Results from the Figure 4 (b) will cause the disturbances of electron movement from heterojunction layer to ZnO nanorods and it will decrease the number of electron free [7]. The application of dye-coating on ZnO nanorods is also preventing the contact between ZnO nanorods surface and heterojunction. It results the degradation of electron-hole recombination between ZnO and polymer [8]. The degradation of electron-hole recombination will increase open circuit voltage of solar cell device.

TABLE 1
PARAMETER FOR EACH SAMPLE

Dye-coating duration Eosin Y (minute)	I_{sc} (mA/cm^2)	V_{oc} (V)	Fill Factor , FF (%)	Efficiency, η (%)
Control	0.00196	0.1	0.285	0.00005586
15	0.00288	0.086	0.509	0.000126
60	0.00765	0.192	0.104	0.0001528
120	0.00506	0.141	0.181	0.0001291

IV. CONCLUSION

Nanorod ZnO have been successfully grown on FTO substrate by sol-gel method. The structure of inverted hybrid solar cell is FTO/ZnO/MEHPPV:PCBM/Au. Eosin Y is one of an organic dye that applied together in photovoltaic power conversion and semiconductor, ZnO. ZnO film respond well when synthesized by eosin Y. Eosin Y will stick on ZnO film surface and the reduction of current when immersion period is extended. The morphology and optical properties of the sample have been analyzed by SEM and UV-Vis test respectively. The efficiency of the solar cell has been determined by IV graph. The optimum efficiency of the solar cell is 0.0001528 % at time duration 60 minutes.

ACKNOWLEDGMENT

The support from the Dana Kecemerlangan grant no: 600-RMI/ST/DANA 5/3/Dst (133/2011) and Universiti Kebangsaan Malaysia are gratefully acknowledged.

REFERENCES

[1] Gunes, S. & Sariciftci, N.S. 2007. Hybrid Solar Cells. *Inorganica Chimica Acta* 361: 581-588.
[2] Lee, Kun-Mu, Suryanarayanan, Vembu, Huang, Jen-Hsien., Thomas, K.R.J., Lin, J.T. & Ho, Kuo-Chuan. 2009. Enhancing the Performance of Dye-Sensitized Solar Cells Based on an Organic Dye by Incorporating TiO$_2$ Nanotube in a TiO$_2$ Nanoparticle Film. *Electrochimica Acta* 54: 4123-4130.
[3] Baek, Woon-Hyuk., Yang, Hyun, Yoon, Tae-Sik., Kang, C.J., Lee, Hyun Hoo & Kim, Yong-Sang. 2009. Effect of P3HT:PCBM Concentration in Solvent on Performances of Organic Solar Cells. *Solar Energy Materials & Solar Cells* 93: 1263-1267.
[4] Kim, Seok-Soon, Yum, Jun-Ho & Sung Yung-Eun. 2003. Improved Performance of a Dye-Sensitized Solar Cell Using a TiO$_2$/ZnO/Eosin Y Electrode. *Solar Energy Materials & Solar Cells* 79: 495–505.
[5] Hosono, Eiji, Fujihara, Shinobu & Kimura, Toshio. 2004. Synthesis, Structure and Photoelectrochemical Performance of Micro/Nano-Textured ZnO/Eosin Y *Electrodes. Electrochimica Acta* 49: 2287–22.
[6] Ghicov, A., Albu, S., Hahn, R., Kim, D., Stergiopoulos, T., Kunze, J., Schiller, C.A., Falaras, P. & Schmuki,P. 2009. TiO$_2$ Nanotubes in Dye-Sensitized Solar Cells: Critical Factors for the Conversion Efficiency. *Chem. Asian Journal* 4: 520-525.
[7] Postels, B., Kasprzak, A., Buergel, T., Bakin, A., Schlenker, E., Wehmann, H.H. & Waag, A. 2008. Dye-Sensitized Solar Cells on the Basis of ZnO Nanorods. *J. Korean Phys. Soc.* 53: 115-118.
[8] Thitima, Rattanavoravipa, Patcharee, Chareonsirithavorn, Takashi, Sagawa & Susumu, Yoshikawa. 2008. Efficient Electron Transfers in ZnO Nanorod Arrays with N719 Dye for Hybrid Solar Cells. *Solid-State Electronics* 53: 176-180.

Effect of Oxygen ratio to the transparency and conductivity of Nanostructured ZnO Thin Films deposited by RF Magnetron sputtering

*N.D. Md Sin, S. Ahmad, M.Z. Musa, M.H. Mamat, Abdul Aziz. A, M. Rusop

NANO-ElecTronic Centre, Faculty of Electrical Engineering,
Universiti Teknologi MARA, 40450 Shah Alam, Selangor, Malaysia
*e-mail: nordiyana86@yahoo.com, rusop@salam.uitm.edu.my

Abstract—**The effect of oxygen ratio to the transparency and conductivity of nanostructured ZnO thin films deposited by RF magnetron sputtering are investigated. The effect of oxygen ratio at (0~10 sccm) on the ZnO thin films have been investigated. The I-V measurement specify that at ratio 45: 5 sccm (Ar:O₂) show the highest conductivity. All films have show high highly transparent in the visible region which (>80 %). The average transmittance in the visible range is 350 nm-1200 nm. The image of nanocolumnar size increase as the insertion of oxygen ratio increase.**

Keywords-component; ZnO thin films; oxygen flow rate; electrical properties; optical properties; structural properties

I. INTRODUCTION

The credibility of ZnO is noted with high transparent and high conductive material. ZnO also has high potential for device fabrication because of their low resistance, high temperature stability, catalyst, transport layer, electrode transparent, and photoluminescence properties[1]. Various application of ZnO has been applied such as devices electronic [2], optoelectronic[3], and solar cell[4]. The advantage of R.F magnetron sputtering are highly uniform thin film, increase adhesion, highly crystalinity [5]. Oxygen vacancies can be prevented by introducing sufficient oxygen gas during deposition process. More over oxygen gas may worsen the properties of the thin film. T. Gosh et al reported the oxygen flow rate is important parameter that influence the quality of the thin film[6].

In this project, we prepared ZnO thin films at different ratio injection argon to oxygen deposited on glass substrate by applying heat during deposition. The effect of additional oxygen on the electrical, optical and morphology evolution of ZnO nanostructured thin films also have been investigated.

II. METHODOLOGY

The sputter chamber was pump down to 5×10^{-4} Pa using a turbo molecular pump. ZnO thin films were prepared on glass substrate at high temperature 500°C in pure Ar:O₂ (99.99%) gas ratio of 45 sccm of Ar to 0,3,5,7,10 sccm of O₂ was introduce into the chamber. The pressure of the system was maintained at 7mTorr during deposition with rf power 200 watt

for 1h. All sample were annealed for 1h (500 °C) in air ambient to increases the crystallinity of films. Then , the thin film were characterized using current-voltage (I-V) measurement (Keithley 2400) for electrical properties. All thin film were deposit with metal contact of gold (Au) for I-V measurement that deposited using thermal evaporator. The optical properties were characterized using UV-VIS spectrophotometer (JASCO 670). The structural properties has been characterized using field emmision scanning electron microscopy (FESEM) (JEOL JSM 6701F).

III. RESULTS AND DISCUSSIONS

A. Electrical properties

Fig. 1(a) show that ZnO thin films for all sample reveal ohmic characteristic. The current intensity at 45:0 sccm (Ar:O₂) RF is the highest compare to other thin films. The results. Fig. 2 show the resistivity of ZnO thin films at different injection ratio of oxygen gas. The conductivity of 45 sccm Ar without oxygen show the highest 13.20 Sm⁻¹. This is may be due to the ZnO thin film are approximately to metal Zn nature. The resistivity of the thin film was obtained using equation reporting by [1]. The conductivity of ZnO thin films at 0:45 sccm show the highest and lowest for resistivity. The carrier concentration decrease as the oxygen flow increase [6]. As the ratio of argon to oxygen increase thus increase the concentration of oxygen vacancies [7]. The electrical properties of the thin film were also investigated by calculating the resistivity of the thin film using equation (1) [1]:

$$\rho = \left(\frac{V}{I}\right)\left(\frac{wt}{x}\right) \qquad (1)$$

where ρ = resistivity,

V = voltage,

I = current,

w =width of Au,

t = thin film thickness and

x = distance between electrodes and

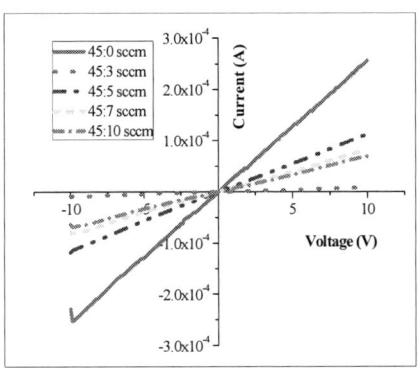

Figure 1. I-V curves of ZnO thin films deposited on glass substrates with different argon to oxygen ratio.

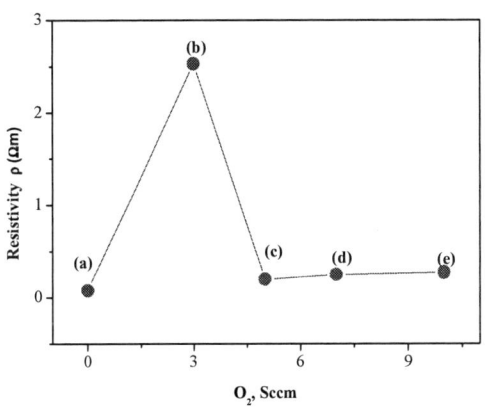

Figure 2. The film conductivity and resistivity of ZnO thin films at different argon to oxygen ratio (a) 45:0, (b) 45:3, (c) 45:5, (d) 45:7, and (e) 45:10 .

B. Optical properties

Fig. 3 demonstrate the transmittance spectra for ZnO thin films prepared at different injection ratio of gas . It was observed that the UV absorption edge occur at all the thin films were highly transparent in the visible region between 380 nm-1200 nm for all films are over 80 %. The transmittances of the sample is related to the energy band where the electron jump from valence band to conduction band to fill the conduction band states [8].

Figure 4 shows the graph of $(\alpha h v)^2$ vs. Photon Energy, hv plot for ZnO thin films optical band gap, Eg energy estimation. The Burstein-Moss effect explained the

broadening of band gap energy with the increasing of carrier concentration [9-10]. The value of optical band gap is taken

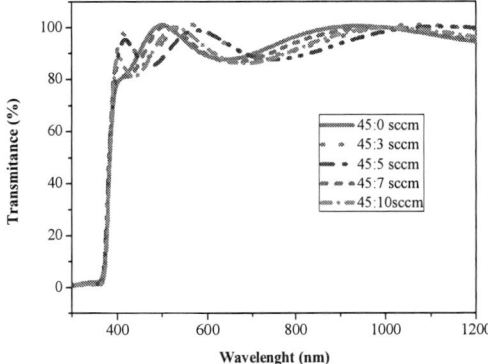

Figure 3. Transmittance spectra of ZnO thin films at various different argon to oxygen ratio.

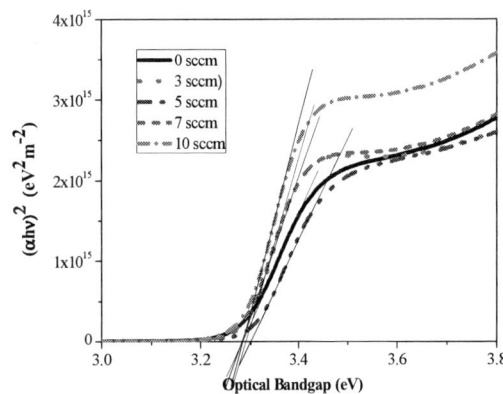

Figure 4. $(hv)^2$ vs hv graph for Optical Band gap, Eg   energy estimation of ZnO thin films for all sample.

TABLE I.   PROPERTIES OF ZNO THIN FILMS AT DIFFERENT SUBSTRATE TEMPERATURE.

Oxygen flow rate (sscm)	Optical Band gap Energy, Eg (eV)
0	3.280
3	3.284
5	3.310
7	3.281
10	3.289

from the linear portion of the curve were about 3.280 eV to 3.310 eV. At ratio gass 45:5 sample show the highest band gap which is 3.310 eV.

C. *Structural properties*

The FESEM image of ZnO thin films deposit at different injection gas ratio are shown in Fig. 3. The columnar size structure become denser as increased the injection oxygen increase. Fig.3 (a) show agglomerate structure of ZnO thin film which the pure ZnO without any oxygen injection. At Fig.3 (b) image show hole (circle in red) between nano columnar structure and fig.3 (c) the columnar structure became denser and fig.3 (d) and (e) the columnar structure increasing and also become denser.

The stoichiometry can be destroy as more oxygen was added to the discharge more low energy neutral oxygen might diffuse into the growing film. The good stoichiometry by construct with suitable oxygen content [11]. From this image of FESEM, the ratio of 45:5 is close to the stoichiometry. Beside by applying heat could also enhance the migration capability of the atom resulting better crystalline quality and better electrical properties [12].

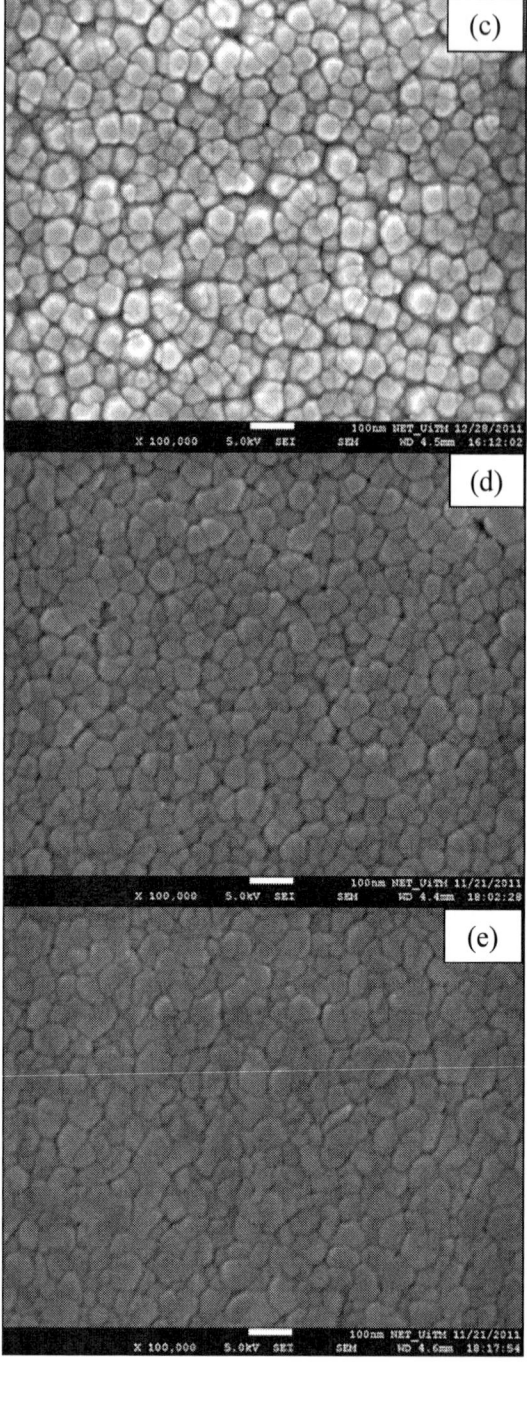

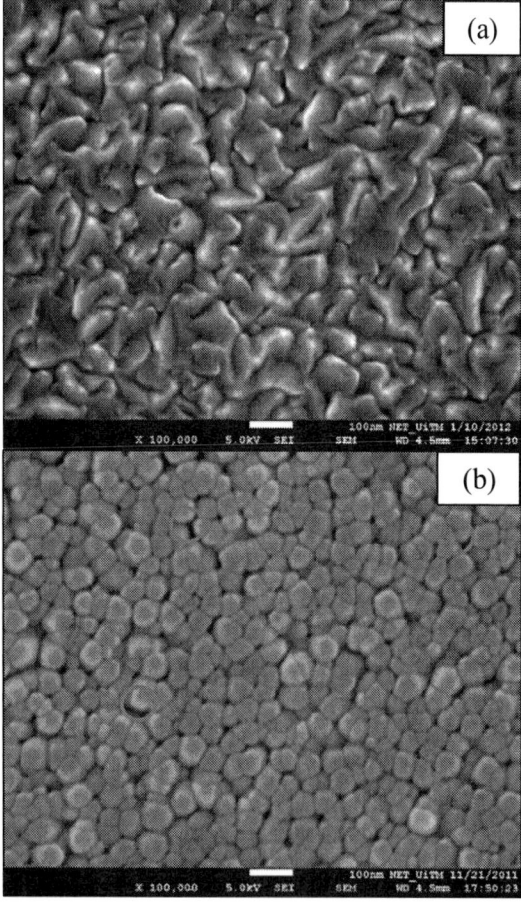

Figure 5. FESEM image of the ZnO thim films deposited on glass substrates deposited at different injection of gas ratio (a) 45:0 sccm (b) 45:3 sccm (c) 45:5 sccm (d) 45:7 sccm and (e) 45:10 sccm.

IV. CONCLUSION

The transparent ZnO thin fim have been successfully deposited at different injection have been successfully investigate. The IV measurement indicate ohmic nature for all thin films. At 45:0 sccm sample without oxygen show lowest resistivity and higher conductivity. All films have show highly transmittance at visible range (>80%) using UV-VIS spectrophotometer. The surface morphology from the FESEM image indicate that the columnar size structure become denser as increased the injection oxygen increase. From the image of FESEM, the ratio of 45:5 is close to the stoichiometry.

ACKNOWLEDGMENT

The author would like to express her thanks to Institute of Science and Faculty of Mechanical UiTM for providing the laboratory facilities. Thanks also to Research Management Institute (RMI) UiTM through the project 600-RMI/ST/DANA 5/3/Dst (392/2011) and Ministry of Higher Education (MOHE) for financial support.

REFERENCES

[1] Rajesh Das and Swati Ray," Zinc oxide—a transparent, conducting IR-reflector prepared by rf-magnetron sputtering" *J. Phys. D: Appl. Phys. 36* 152–155(2003).

[2] S.J. Lim, Soonju Kwon, H. Kim," ZnO thin films prepared by atomic layer deposition and rf sputtering as an active layer for thin film transistor" *Thin Solid Films 516* 1523–1528 (2008).

[3] P.T. Hsieh,Y.C. Chen, K.S. Kao, C.M. Wang,"Structural effect on UV emission properties of high-quality ZnO thin films deposited by RF magnetron sputtering" *Physica B 392* 332–336 (2007).

[4] Z.A. Wang, J.B. Chu, H.B. Zhu, Z. Sun, Y.W. Chen, S.M. Huang," Growth of ZnO:Al films by RF sputtering at room temperature for solar cell applications" *Solid-State Electronics 53* 1149–1153(2009).

[5] Liangxian Chen , Chengming Li, WeiLing Yin, Jinlong Liu, Lifu Hei, Fanxiu Lu," Effect of deposition temperature and quality of free-standing diamond substrates on the properties of RF sputtering ZnO films" *Diamond & Related Materials 20* 527–531 (2011).

[6] T. Ghosh, D. Basak," Effect of oxygen flow rate and radio-frequency power on the photoconductivity of highly ultraviolet sensitive ZnO thin films grown by magnetron sputtering" *Material Research Bulletin 46* 1975-1979(2011).

[7] Shu Jie, Li Yang and DongHua fan" Effects of Argon-Oxygen ratio and Sputtering Power on Opitcal Properties of ZnO:Al Thin Films fabricated by RF Magntron Sputtering" *Advanced Materials Research Vol. 216* pp 307-311(2011).

[8] Yihua Sun, Chenhui Li, Weihao Xiong, Caihua," Influnce of substrate temperature on the structural, electrical and optical properties of Al-doped ZnO films by RF magnetron sputtering" *Advanced Materials Research Vols.* 287-290 (2011).

[9] Ligang Ma, Xiaoqian Ai, Xinli Huang, Shuyi Ma," Effects of the substrate and oxygen partial pressure on the microstructures and optical properties of Ti-doped ZnO thin films" *Superlattices and microstructures 50* 703-712(2011).

[10] Keunbin Yim, Hyounwoo Kim, Chongmu Lee" Effects of the O_2/Ar gas flow ratio on the electrical and transmittance properties of ZnO:Al films deposited by RF magnetron sputtering" *J Electroceram 17:* 875–877(2006).

[11] Che-Wei Hsu, Tsung-Chieh Cheng, Chun-Hui Yang, Yi-Ling Shen, Jong-Shinn Wu, Sheng-Yao Wu "Effects of oxygen addition on physical properties of ZnO thin film grown by radio frequency reactive magnetron sputtering" *Journal of Alloys and Compounds 509* 1774–1776 (2011).

[12] Ma Junwei, Ran Feng, Xu Meihua, Ji Huijie, ' Influence of substrate temperature on the performance of zinc oxide thin film transistor' *Journal of semiconductor Vol. 32,* No. 4 (2011).

Enhanced Light Absorption in Bifacial Solar Cells

Suhaila Sepeai, M.Y.Sulaiman, Saleem H.Zaidi, Kamaruzzaman Sopian
Solar Energy Research Institute (SERI)
Universiti Kebangsaan Malaysia (UKM)
43600 UKM Bangi, Selangor, Malaysia
Email: suhailas@ukm.my/ suhaila_sepeai@yahoo.com

Abstract- **Solar cell is a semiconductor device that converts sunlight into electricity. Bifacial solar cell is a specially designed solar cell for the production of electricity from both sides of the solar cell. Bifacial solar cell became an active field of research making photovoltaic (PV) more competitive together with current efforts to increase the efficiency and lower material costs. In silicon (Si) solar cell, the inability to absorb all the incident sunlight fundamentally limits the Si solar cell performance. As the wafer is efficiently used by fabricate a bifacial solar cell to save on Si costs, efficiencies will become even lower due to incomplete optical absorption. Subwavelength surface texturing helps offset some of the absorption loss. To trapped more lights, anti reflecting coating (ARC) was studied. There are three type of texturing methods and two type of ARCs that has been studied in this research. According to optical properties of the textured Si wafer and ARC thin film, it was found that wet texturing and SiN are the best methods to improve the light absorption in bifacial solar cell. The efficiency obtained from the bifacial solar cell is 9.62% for front side and 4.5% for back surface.**

I. INTRODUCTION

Currently, close to 90% of the global photovoltaic (PV) production is based on crystalline silicon (Si) solar cell. In spite of the expensive manufacturing, the crystalline silicon solar cell still dominates the market due to the stable technology, abundant supply of silicon as a raw material, low ecological impact and no degradation in crystalline form [1]. Incomplete optical absorption is one of the key factors that fundamentally limit the Si solar cell performance [2]. In this research, we address to minimizing the reflection losses and improve the light absorption by deposition of anti reflective coating (ARC) and textured Si wafer.

Texturing of crystalline silicon (c-Si) has been agreed for the development of high efficiency c-Si solar cells. In PV device, texturing is used to enhance the amount of light absorbed into devices by reduce the light reflectance on the silicon surface. Beside that, texturing is used in order to increase the light trapping, and therefore increase the short circuit current and the efficiency in the solar cells. The geometry of the texture determines how the photons absorb in solar cell. There are a few geometry has been reported, for instance pyramid [3], inverted pyramids [4], microgroove and honeycomb [5].

For crystalline silicon, some sophisticated technique has been used for texturing, such as plasma etching [6,7], Reactive Ion Etching (RIE) [8]. The photolithographic patterning and isotropic etching technique is used to generate a square matrix of holes through a masking oxide in a square array [5]. Traditional method for texturing is anisotropic chemical etching. For pores with a size from 2 to 20 microns, the silicon was anodized in hydrofluoric acids [9]. This technique is widely used due to their dependence on silicon surface crystallographic orientation and can maintain wafer lifetime [10]. Electrochemical method is better than chemical method due to its current density, but the reflectance of silicon surface subjected to chemical texturing is higher than reflectance of inverted pyramid and microgroove [9]. Isotropic etching with acidic solution includes the formation of meso- and macro-porous structures on mc-Si that helps to minimize the grain-boundary delineation [3] and also remove metallic contamination [10].

Anti Reflective Coating (ARC) is the most essential layer in silicon solar cells. The main purpose of ARC is anti-reflection. This layer allows more photon absorption and reduces the light reflection. The second role of ARC is as surface passivation to solve the defect and impurities problem in solar cells. It is important to note that those defects and impurities act as recombination centers for the electrons and the holes created by external photons and degrade the efficiency of silicon solar cell [11]. The capability of ARC to reduce the recombination of charge carriers at the Si surface is well known for more than 20 years [12]. The third function is hydrogenation of the wafer if the layer undergoes a short thermal treatment after deposition. This hydrogenation induces passivation of defects and impurities in the bulk Si especially at the grain boundaries and can extensively increase the lifetime of minority charge carriers in the bulk of crystalline wafers [13].

In this paper, we discussed our approach to reduce the reflection losses in bifacial solar cell. For texturing, we had investigated three methods of texturing, namely chemically-etch, vapor etch and wet texturing. Meanwhile, we studied two types of ARC had been investigated in this research, namely Silicon Nitride (SiN) and Silicon Dioxide (SiO_2). SiN is deposited by plasma enhanced chemical vapor deposition

(PECVD) while for SiO_2, we used two types of deposition method, namely thermally grown oxide and PECVD.

II. EXPERIMENTAL

Bifacial solar cells with a configuration of n^+pp^+ with Aluminium Back Surface Field (Al-BSF) were designed. The cell has screen-printed front surface Ag and back surface Ag/Al contacts. Figure 1 is a schematic of the basic bifacial solar cell structure.

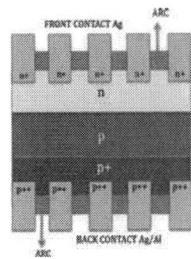

Fig. 1. Bifacial solar cell with n+pp+ structure

A p-type <100> Si wafer with a sheet resistivity ranging between 1 ohm/cm and 10 ohm/cm was used. The Si wafer was initially cleaned by dipping into solution of hydrofluoric acid (HF) and nitric acid (HNO_3) in a ratio of 1:100 for 10 minutes. After rinsing with deionized water, it was then dipped into HF and water (H_2O) in a ratio of 1:50 for 1 minute. The wafers were then subjected to the texturing process. In this research, we used three types of texturing process, namely chemical etching, vapor etching, and wet texturing. Chemical and vapor etching used the same solution, that is $HF:HNO_3$. For chemical etching, the Si wafer was immersed in that solution for 3 hours, while for vapor etching, the Si wafer was exposed to the vapor of the solution for 24 hours. For a wet texturing process, the solution of texturing process is $KOH:IPA:H_2O$, where IPA is iso-propil alcohol (IPA) in the ratio of 1:5:125. The texturing temperature was set at 70°C for 30 minutes. For the characterization of the the texturing, cross section and top view images by Scanning Electron Microscope (SEM), surface photovoltage measurement and reflectance measurement checked the outcome from that experiment.

After the texturing process, the wafers were subjected to the n-type diffusion procedure using gas-source phosphorous oxychloride ($POCl_3$) at a temperature of 908 C. The edges of the Si wafers were then mechanically diced to achieve edge isolation. For bifacial solar cells with Al- BSF, Al pastes were screen-printed onto the back side of the Si wafer. The paste was annealed at 150°C for 10 minutes prior to firing at a temperature of 830°C in a rapid thermal annealing (RTA) furnace to form Al-diffused p+ layer. Excess Al was removed by soaking in 100% hydrochloric acid (HCl) solution. Thus, n+pp+ structure was successfully fabricated.

Next, to determine an appropriate Anti Reflective Coating (ARC) to be used in bifacial solar cell, we had performed the characterization studies thermally grown Silicon Dioxide (SiO_2) and Silicon Nitride (SiN). SiN is deposited by plasma enhanced chemical vapor deposition (PECVD) at the temperature of 150 C, while for SiO_2, we used two types of deposition method, namely thermally grown oxide and PECVD- SiO_2. The ARC thin film has been characterized by surface photovoltage (SPV) measurement system. Finally, the metallization processes were carried out using screen printing of Ag and Ag/Al pastes by employing identical grid masks on the front and back surfaces, respectively. Screen-printed contacts were fired at 830°C to form ohmic front and back contacts. The finished solar cells were experimentally analyzed using light Current-Voltage (LIV) Measurement System.

III. RESULT AND DISCUSSION

We had performed characterization analyses after texturing process before proceed to device fabrication. Figure 2 shows the Scanning Electron Microscopy (SEM) images of (a) chemically-etched, (b) vapour etched and (c) wet texturing of Si wafer with a magnification of 10 000 and an operating voltage of 3 kV. From Figure 2 (a) and (b), we can see that the $HF:HNO_3$ that were used as chemical and vapour etch solution is not uniformly textured the Si wafer. This is due to the etched thickness that in a range of 1.7 m to 3.0 m. The texture pattern is not able to determine too. Compared to the wet texturing, it is clearly shown that this technique produces a uniform pyramid-like structure. The diameter of the pyramid is in nano size.

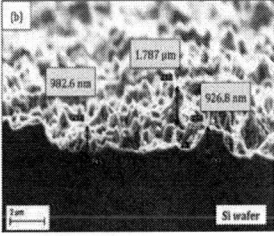

Fig. 2 SEM images of (a) chemically-etched, (b) vapor-etched and (c) wet texturing.

In order to justify the best texturing method, we study the reflectance measurement (Figure 3) of chemically-etched, vapor-etched and wet texturing. We compare the textured Si wafer with planar or un-textured Si wafer. From the figure, it is clearly shown that textured wafer reflect less light than un-textured or planar wafer in a visible light (range from 400-700 nm). The planar wafer shows the highest reflectance followed by vapor-etched, chemically-etched and wet texturing. We can conclude that wet texturing is shows the best results with least reflection of light as well as enhance the light absorption in solar cells.

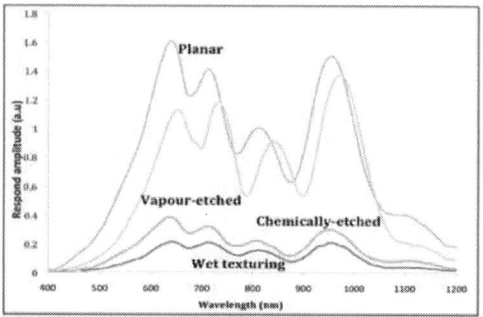

Fig. 3. I-V curve of solar cell with a variation on firing temperature, x = 650 – 880 °C

Figure 4 shows the surface photovoltage (SPV) meaurement of PECVD-SiN, thermally grown SiO_2 and PECVD-SiO_2. All measurements are done in an operating voltage of 50 V. From Figure 4, it can be seen that all ARC increases the SPV for the visible light range (400 nm – 700 nm). This means that blue light absorbs more compared to red light since small number of carriers are collected in the range of 400-500 nm. It is good to compare between thermally grown SiO_2 and PECVD-SiO_2. In visible light range, both of them show similar performance, but they differ in longer wavelength response. The thermally grown SiO_2 has a faster decay rate in 700 – 800 nm compared to PECVD- SiO_2. The thermally grown process consists of the transportation of oxygen to the surface, diffusion of the oxygen through the already grown oxide and finally the reaction of the oxygen with the silicon at the interface between silicon and silicon oxide. With growing oxide thickness, the growing rate slows down because the time of the diffusion through the oxide depends on its thickness. This process makes the degradation of thermally grown SiO_2 faster than PECVD-SiO_2. Meanwhile, SiN has shows the highest response amplitude compared to thermally grown SiO_2 and PECVD-SiO_2, means that a lot of minority carriers are collected on SiN surface. This indicates that SiN is a good passivation compared to SiO_2. Based on SPV performance, we decided to choose SiN as a surface passivation for bifacial solar cell.

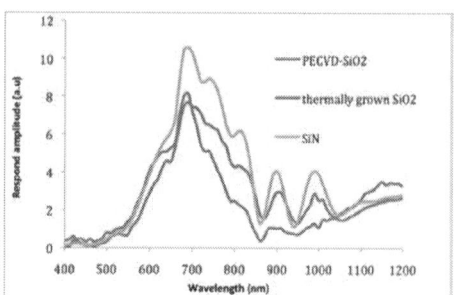

Fig. 4. Surface photovoltage (SPV) measurement of various anti reflecting coating (ARC).

After performing the texturing and ARC characterization on thin film, we applied the best result of texturing and ARC, namely wet texturing and SiN in our n+pp+ bifacial solar cells. Figure 5 shows the current-voltage (I-V) curve of bifacial solar cell for a front and back surface. Open circuit voltage (V_{oc}) obtained from the device was 580 mV and 560 mV for illumination from front and back sides respectively. The short circuit current (I_{sc}) are 0.47 A and 0.23 A for the front and rear sides; respectively. The efficiency obtained was 9.62% and 4.5% for front and back surface, respectively. It was found that the efficiency of back side is approximately half from the front side. The poor performance could be due to the insufficient effect of back surface field (BSF). Removing the fired Al from the Si wafer using hydrochloric acid (HCl) is a messy and hazardous process and therefore we think that this approach is not a good practice to obtain BSF.

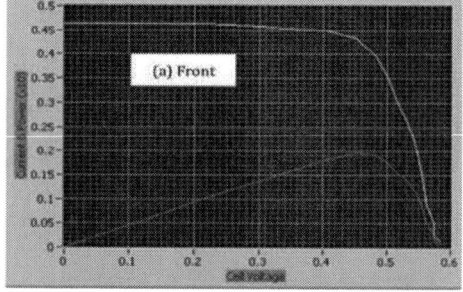

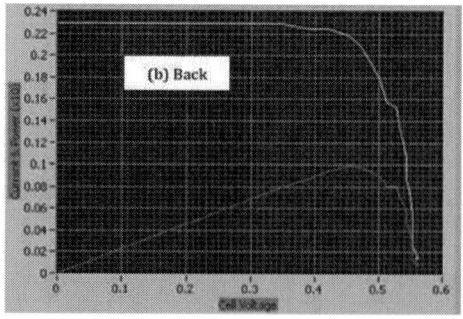

Fig. 5. I-V curve of bifacial solar cell for (a) front (a) and back (b) surfaces.

IV. CONCLUSION

Bifacial solar cells performance with a structure of n+pp+ had been investigated. The pyramid pattern from wet texturing and SiN has been choose for improve the light absorption in bifacial solar cell. The efficiency for front surface is 9.62%. Meanwhile 4.5% of efficiency is occurred from back surface of bifacial solar cell. As our intuition on the poor performance of back surface may be due to lack of BSF, for our future research, we plan to developed a simulation model of bifacial solar cells with a structure of n+pp+ using PC1D software where the simulated parameters are obtained from experiment. The study will increase understanding of the parameters that influence the bifacial solar cell performance. We also interested to develop a dry process for texturing and improve the device performance by double anti-reflective coating, namely SiN/SiO$_2$ stack.

ACKNOWLEDGMENT

This work has been carried out with the support of the Malaysia Ministry of Science, Technology and Innovation (MOSTI).

REFERENCES

[1] A.Goetzberger, C.Hebling, and H.Schock, "Photovoltaic material, history, status and outlook", *Materials Science and Engineering* , vol.40, pp. 1–46, 2003.

[2] J.Nelson, "The Physics of solar cells", Imperial College, 2003.

[3] U. Gangopadhyay , S.K. Dhungel, P.K. Basu, S.K. Dutta, H. Saha, and J.Yi, "Comparative study of different approaches of multicrystalline silicon texturing for solar cell fabrication", *Solar Energy Materials & Solar Cells*, vol. 91, pp. 285–289, 2007.

[4] M. Moreno n, D. Daineka, P. Roca, and I.Cabarrocas, "Plasma texturing for silicon solar cells: From pyramids to inverted pyramids-like structures", *Solar Energy Materials & Solar Cells*, vol. 94, pp.733–737, 2010.

[5] J. Zhao, A. Wang, P.Campbell, and M.A.Green, "A 19.8% Efficient Honeycomb Multicrystalline Silicon Solar Cell with Improved Light Trapping", *IEEE Transaction On Electron Devices*, Vol. 46 (10), Oct 1999.

[6] U.Kaiser, M.Kaise, and R.Schlinder, "Texture etching of multicrystalline silicon", *Proc. 10th Eenergy Photovoltaic Sol.Energy Conf.*, pp. 293-294, 1991.

[7] S.Narayanan, S.R.Wenham, and M.A. Green, "High Efficiency Polycrystalline Solar Cells", *Proc. 4th Photovoltaic Sol. Energy Conf.*, Sydney, Australia, pp. 111-116, 1989.

[8] S.H. Zaidi, D.S. Ruby, and J.M. Gee, "Characterization of Random Reactive Ion Etched-Textured Silicon Solar Cells", IEEE Transaction On Electron Devices, Vol. 48(6), June 2001.

[9] V.Y.Yerokhov, R.Hezel, M. Lipinski, R.Ciach, H.Nagel, A. Mylyanych, and P.Panek. "Cost effective methods of texturing for silicon solar cells", *Sol.Ener.Mat & Solar Cell*, vol. 72, pp. 291-298, 2002.

[10] M.Edwards, S.Bowden, U.Das, M.Burrows, "Effect of texturing and surface preparation on lifetime and cell performance in heterojunction silicon solar cells, *Solar Energy Materials & Solar Cells*, vol. 92, pp.1373– 1377, 2008.

[11] J.Kim and J.Hong, "Application of PECVD SiNx film to screen-printed multicrystalline silicon solar cell", *Journal of the Korean Physical Society*, vol. 44 (2), pp. 479–482, February 2004.

[12] T.Markvart, and L.Castaner, "Solar Cell: Material, Manufacture and Operation", Elsevier Science, 2005.

[13] W.Soppe, H.Rieffe and A.Weeber, "Bulk and Surface Passivation of Silicon Solar Cells Accomplished by Silicon Nitride Deposited on Industrial Scale by Microwave PECVD", Progress in PV: Research and Applications, Prog. Photovolt: Res. Appl. ,Vol.13, pp. 551–569,2005;

Infrared Emissivity of Co,Ni Co-Doped ZnO Powders by Solid-State Reaction

Yinhua Yao, Quanxi Cao
School of Technical Physics
Xidian University
Xi'an, Shaanxi, 710071, China
Email: yaoyinhua2009@126.com

Abstract- **This paper studies the infrared emissivity of $Zn_{0.99-x}Ni_{0.01}Co_xO$ (x=0.00, 0.01, 0.03, 0.05) powders prepared by solid-state reaction. The phase and morphology of samples were characterized by XRD and SEM. The UV absorption spectra and infrared emissivity at the wavelength of 8-14μm were measured by UV Spectrophotometer and IR-2 Dual-Band Emissivity Measuring Instrument. The peak of the impurity phase, NiO, is detected at temperatures below 1150℃. The optical band-gap decreases with increasing Co concentration and calcination temperature, which are much smaller than results in other studies. The infrared emissivity of $Zn_{0.99-x}Ni_{0.01}Co_xO$ falls with the increase of Co content. It descends to the minimum (0.772) at 1100℃ and then ascends when the temperature is raised.**

I. INTRODUCTION

ZnO-based diluted magnetic semiconductor (DMS) has evoked much theoretical and experimental work due to its enormous applications in spintronics, such as nonvolalatile storage, spin-valve transistors, and spin-based light-emitting diodes [1-4]. ZnO has a wide and direct band gap of 3.37eV at room temperature and a very large exciton binding energy of 60meV. It is a multifunctional material with wide applications in transparent conductive oxides [5], ultraviolet light-emitting devices [6], and gas sensor [7]. Up to now, a lot of experimental results have been obtained on transition metal(TM) doped ZnO, but little research results have been reported concerning the infrared emissivity of TM-doped ZnO. Due to the high transmittance and good conductive properties, doped ZnO can show excellent infrared properties. Because of the magnetism, TM-doped ZnO may display different infrared performances.

From our previous works [8], among TM-doped ZnO powders, ZnO:Co has the lowest emissivity in the same preparation condition. To investigate the impact of Co on the emissivity of ZnO:Ni, $Zn_{0.99-x}Ni_{0.01}Co_xO$ (x=0.00, 0.01, 0.03, 0.05) powders were synthesized by solid-state process in this paper. The effects of Co and Ni ions on the phase and morphology were studied by XRD and SEM. The infrared emissivities of samples with various Co contents and at different temperatures were discussed in the range of 8-14μm.

II. EXPERIMENT

The co-doped ZnO were synthesized by the conventional solid-state reaction process. According to certain proportions, ZnO(99.95%, Tianjin Fuchen Chemical Reagent Factory) were mixed with Co_2O_3 and Ni_2O_3(99.95%, Tianjin Fuchen Chemical Reagent Factory). With alcohol as the solvent, the mixed powders were ball milled in planetary mills (QM-1SP04). Then the mixtures were dried, grinded, and screened. In the end, the mixtures were calcined in furnaces (sx2-6-13 box-type resistance furnace, Shanghai Laboratory Electric Furnace Works). Besides pure ZnO at 1100℃, $Zn_{0.99-x}Ni_{0.01}Co_xO$ (x=0.00, 0.01, 0.03, 0.05) powders were prepared at 1000℃, 1050℃, 1100℃, 1150℃, and 1200℃.

The structure characterization of the samples was investigated by an X-ray diffractometer (DX-1000, Dandong Fangyuan Co., Ltd.). The particle size and morphology of the powders were observed by scanning electron microscope (SEM) (JSM 6360LV, JEOL). The UV spectra were analyzed by a spectrophotometer (V-570, JASCO). The emissivity was measured by an infrared emissivity measuring instrument (IR-2 Dual-Band Emissivity Measuring Instrument, Shanghai Chengbo Optoelectronic Technology Co., Ltd.).

978-1-4673-2395-6/12 $31.00 © 2012 IEEE

III. RESULTS AND DISCUSSION

A. XRD Analysis

By the software (MDI Jade 6), the XRD patterns in Fig. 1 confirm the change of Co_2O_3 and Ni_2O_3 into Co_3O_4 and NiO. From Fig. 2(a), only the diffraction peaks corresponding to the ZnO wurtzite structure, such as (100), (002), and (101), are observed in $Zn_{0.99-x}Ni_{0.01}Co_xO$. This implies that the impurities have occupied the sites of Zn ions, or exist as interstitial ions or atoms. But we can not exclude the existence of Co and Ni clusters or their oxide clusters whose amounts are too small to be detected. From Fig. 2(b), the characteristic peak of NiO is seen at $42.84°$ in $Zn_{0.94}Ni_{0.01}Co_{0.05}O$ below $1150℃$. Huang et al.[9] also observed the presence of NiO in $Zn_{0.95}Ni_{0.05}O$ nanoparticles synthesized by an ultrasonic assisted sol-gel process. Compared with the peak of NiO at $43.286°$ in the standard card PDF44-1159, the peak we obtained shifts to degrees. The intensities of impurity peaks are weakened and the crystalline quality of powders becomes better at elevated temperatures.

Increasing Co content from 0at% to 3at% causes the lattice parameter contraction. However, the parameter of Co-5at% ZnO climbs. The corresponding parameters are 0.52111nm, 0.52053nm, 0.51978nm, and 0.52044nm, which are smaller than that of pure ZnO (0.52197nm) at $1100℃$. This can be explained by the differences among their radiuses. Because Co^{3+}, Co^{2+}, and Ni^{2+} have smaller radiuses (0.063nm, 0.072nm, and 0.068nm) than Zn^{2+}(0.074nm)[10], the replacement of Zn^{2+} ions by these ions will reduce the lattice parameter. The increasing parameter of Co-5at% ZnO is ascribed to the formation of interstitial defects about Co or Zn

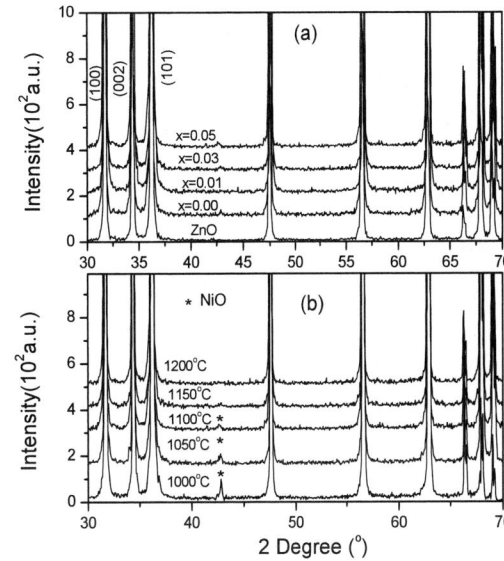

Fig. 2 XRD patterns of (a) $Zn_{0.99-x}Ni_{0.01}Co_xO$ (x=0.00-0.05) at $1100℃$ and (b) $Zn_{0.94}Ni_{0.01}Co_{0.05}O$ at different calcination temperatures

rather than the substitutional doping [11]. The lattice parameter dependent on the temperature has the same tendency as that with different Co contents. They are 0.52154nm, 0.52072nm, 0.52044nm, 0.52091nm, and 0.52100nm, with the minimum value at $1100℃$.

B. SEM Analysis

From the SEM images shown in Fig. 3(a) and 3(f), poor crystalline qualities are observed in ZnO and ZnO:Ni powders synthesized at $1100℃$. Aggregation of grains can also be seen in pure ZnO powders. Although the dispersibility of particles in ZnO:Ni becomes better, the particle size is small. In contrast, bigger particle sizes and better dispersibility of particles are obtained in ZnO co-doped by Co and Ni ions (see Fig. 3(b) to Fig. 3(d)).

C. UV Absorption Spectrum Analysis

The UV absorption spectra of pure and doped ZnO are depicted in Fig. 4. As the Co content and calcination temperature grow, the absorption, α, is enhanced, with three peaks located at 565nm, 610nm, and 660nm. And the incorporation of Co and Ni ions leads to a red shift of absorption edge. The optical band-gap, E_g, is determined from the absorption spectra using the equation: [12]

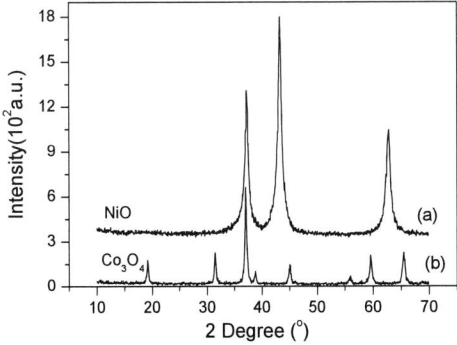

Fig. 1 XRD patterns of raw materials (a) Ni_2O_3 and (b) CO_2O_3

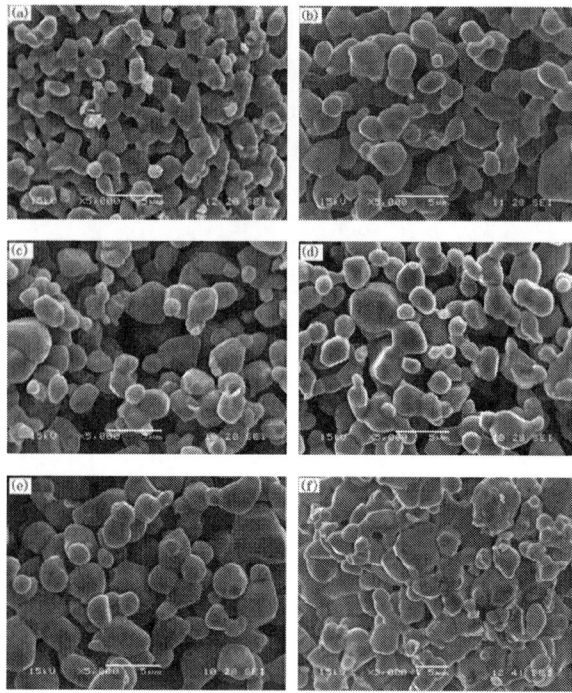

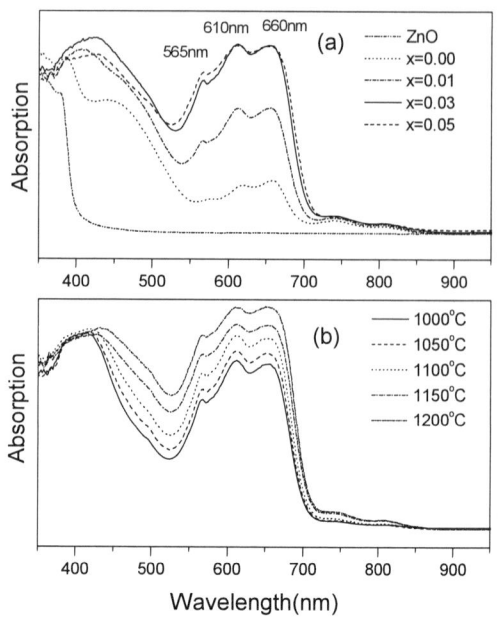

Fig. 4 UV absorption spectra of (a) $Zn_{0.99-x}Ni_{0.01}Co_xO$ (x=0.00-0.05) at 1100℃ and (b) $Zn_{0.94}Ni_{0.01}Co_{0.05}O$ at different calcination temperatures

Fig. 3 SEM images of (a)-(d) $Zn_{0.99-x}Ni_{0.01}Co_xO$ (x=0.00, 0.01, 0.03, 0.05) at 1100℃, (e) $Zn_{0.94}Ni_{0.01}Co_{0.05}O$ at 1150℃, and (f) pure ZnO

$$(\alpha h\nu)^2 = A(h\nu - E_g) \qquad (1)$$

where $A, h,$ and ν are constants, Planck constants, and the electromagnetic wave frequency. The optical band-gap is shown in Fig. 5. From Fig. 5(a), doped ZnO have smaller optical band-gaps than that of pure ZnO (3.1625eV) at 1100℃. Corresponding to the values of x varying from 0.00 to 0.05, E_g are 2.85eV, 2.275eV, 2.2375eV, and 2.4375eV, respectively. These E_g are not as big as those of ZnO:Ni films in the works of Snure et al.[1], as well as those of ZnO:Co obtained by Sarsari et al.[2]. They all observed that the band-gap decreased with rising Co content. On the contrary, the growing optical band-gap with the increase of Co concentration was obtained in $Zn_{1-x}Co_xO$ (x=0.00, 0.01, 0.05, 0.10) films prepared by Ozerov et al.[3].

When the temperature rises, E_g fall, which are 2.5750eV, 2.5250eV, 2.4375eV, 2.2375eV, and 2.1500eV, respectively. The same goes for the results of Maensiri et al.[4]. But they also obtained bigger optical band-gap (~3.2eV). Additionally, they thought that the peaks at 550nm, 610nm, and 660nm in absorption spectra were attributed to the substitution of Co^{2+}

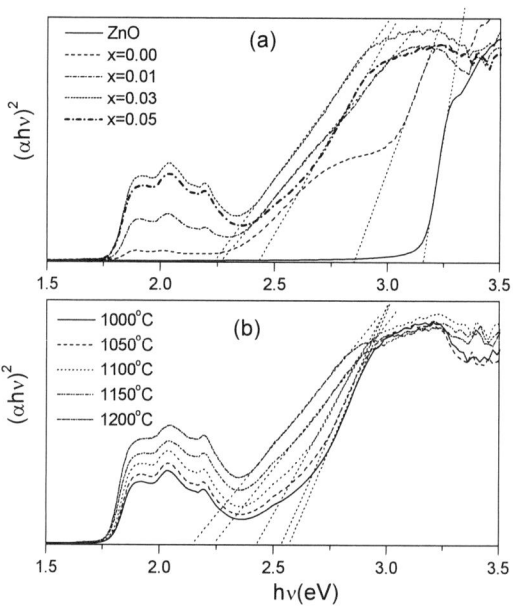

Fig. 5 Optical band-gaps of (a) $Zn_{0.99-x}Ni_{0.01}Co_xO$ (x=0.00-0.05) at 1100℃ and (b) $Zn_{0.94}Ni_{0.01}Co_{0.05}O$ at different calcination temperatures

ions for Zn^{2+} ions. X-ray photoelectron spectroscopic measurement confirms that Co^{2+} ions substitute for Zn^{2+} ions with 58.93 and 65.40 atomic mass, respectively [2]. But the chemical state of Co ions in ZnO needs to be further investigated. Nevertheless, this is enough to confirm the replacement of Zn ions by Co ions, which causes the changing band structure.

D. Infrared Emissivity Analysis

From Fig. 6(a), the emissivity of $Zn_{0.99}Ni_{0.01}O$ (x=0.00) powders is 0.91, much greater than that of pure ZnO(0.840) at 1100℃. This is in agreement with our previous results that the emissivity was improved by the introduction of Ni ions [8]. When the Co content rises, the emissivity of samples drops sharply and then slowly.

The effect of Co^{2+} ions in the sites of Zn^{2+} ions has been elaborated in previous works [8]. Because of the similar radiuses between Co^{2+} (0.072nm) and Zn^{2+} (0.074nm), Zn^{2+} ions can be replaced by the produced Co^{2+} ions during the calcination easily. In addition, CoO and ZnO have cobalt and oxygen vacancies, respectively. Consequently, occupying the sites Zn^{2+} ions, the Co^{2+} ions reduce the lattice defects and distortion so the crystalline quality gets better. This is consistent with the XRD and SEM analysis. As a result, carrier scattering is weakened and carrier mobility is enhanced. Besides, entering the ZnO lattice, Co ions improve the oxygen vacancy concentration and carrier concentration through the equation:

$$CoO \longrightarrow Co_{Zn} + V_o^{2+} + \frac{1}{2}O_2 + 2e \qquad (2)$$

Naeem et al.[13] also thought that carrier concentration could be increased by the incorporation of Co ions into ZnO lattice. From above, with higher Co content, more Co ions are incorporated as substitutional ions and the lattice parameter is decreased. Therefore, the emissivity is lessened by the increasing conductivity due to the introduction of Co ions.

From Fig. 6(b), the emissivity of $Zn_{0.94}Ni_{0.01}Co_{0.05}O$ decreases to 0.772 up to 1100℃ and then increases with the changing temperature. And the emissivity shows the same relationship as that between the lattice parameter and the temperature. At elevated temperatures, more Co ions are dissolved into ZnO lattice. Stated above, the implantation of Co ions lessens the emissivity. And the obtained better crystalline quality due to the reduction in the amount of NiO will reduce the defects and emissivity. On the other hand, the decreasing emissivity can be attributed to the magnetism as

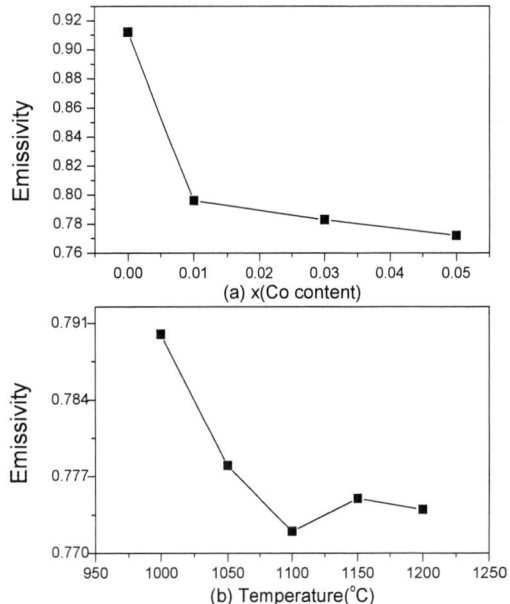

Fig. 6 Emissivities of (a) $Zn_{0.99-x}Ni_{0.01}Co_xO$ (x=0.00-0.05) at 1100℃ and (b) $Zn_{0.94}Ni_{0.01}Co_{0.05}O$ at different calcination temperatures

well. Huang et al.[9] reported that ferromagnetism was observed in ZnO:Ni with single phase while the samples turned to paramagnetic with the presence of NiO. Chen et al.[14] identified that the ferromagnetic thin films showed a high electron concentration, while the electron content of the thin films without ferromagnetism decreased rapidly. As well known, samples with high electron concentrations exhibit low emissivities. Thus, to some extent, it can be thought that ferromagnetism leads to a lower emissivity. However, from the lattice parameter analysis, higher temperatures can cause more interstitial defects which damage the symmetry of lattice and enhance the infrared absorption. Another point is that the formation of interstitial Co and Ni atoms or ions will contribute to lower carrier concentrations [15]. As a result, the emissivity is improved.

IV. CONCLUSION

$Zn_{0.99-x}Ni_{0.01}Co_xO$ (x=0.00, 0.01, 0.03, 0.05) powders are synthesized by solid-state reaction at various calcination temperatures. All the samples have the wurtzite structure. At temperatures below 1150℃, the peaks of the secondary phase, NiO, are detected. The optical band-gap decreases first and then increases with growing Co content. As the calcination temperature is elevated, the band-gap falls gradually. These values are much smaller than those in other studies. The emissivity of ZnO can be reduced by the replacement of Zn ions by Co ions. However, ZnO:Ni powders show a much higher emissivity(0.91) at 1100℃. Bigger x values reduce the emissivity of $Zn_{0.99-x}Ni_{0.01}Co_xO$ fabricated at 1100℃. The emissivity of $Zn_{0.94}Ni_{0.01}Co_{0.05}O$ falls to 0.772 at 1100℃ and jumps with rising calcination temperature. As expected, the emissivity is associated with the magnetism.

REFERENCES

[1] M. Snure, D. Kumar, and A. Tiwari, "Ferromagnetism in Ni-doped ZnO films: extrinsic or intrinsic?" Appl. Phys. Lett., vol. 94, issue. 1, pp.012510-2, Jan 2009.

[2] I.A. Sarsari, H. Salamati, P. Kameli, and F.S. Razavi, "Optical, structural, and magnetic properties of ZnO:Co nanoparticles prepared by a thermal treatment of ball milled precursors," J. Supercond. Nov. Magn., vol. 24, pp.2294-2295, Sept 2011.

[3] I. Ozerov, F. Chabre, and W. Marin, "Incorporation of cobalt into ZnO nanoclusters" Mater. Sci. Eng. C, vol. 25, issue. 5-8, pp.615-616, 2005.

[4] S. Maensiri, P. Laokul, J. Klinkaewnarong, and C. Thomas, "Structure and magnetic properties of Zn0.9Co0.1O nanorods synthesized by a simple sol-gel method using metal acetylacetonate and poly(vinyl alcohol)," Appl. Phys. A, vol. 94, pp.602-603, 2009.

[5] K.M. Lin, and P. Tsai, "Parametric study on preparation and characterization of ZnO:Al films by sol–gel method for solar cells," Mater. Sci. Eng. B, vol. 139, issue. 1, pp.81–82, Apr 2007.

[6] L. J. Mandalapu, Z. Yang, S. Chu, and J. L. Liu, "Ultraviolet emission from Sb-doped p-type ZnO based heterojunction light-emitting diodes," Appl. Phys. Lett., vol. 92, pp.122101-1, Mar 2008.

[7] S.M. Chou, L.G. Teoh, W.H. Lai, Y.H. Su, and M.H. Hon, "ZnO:Al thin film gas sensor for detection of ethanol vapor," Sens., vol. 6, pp.1420-1422, Oct 2006.

[8] Y.H. Yao, and Q.X. Cao "Infrared emissivity of transition elements doped ZnO," J.Cent.South Univ.Technol., in press.

[9] G.J. Huang, J.B. Wang, X.L. Zhong, G.C. Zhou, and H.L. Yan, "Synthesis, structure, and room-temperature ferromagnetism of Ni-doped ZnO nanoparticles," J. Mater. Sci. vol. 42, pp. 6464–6468, Apr 2007.

[10] Y. Zhou, Ceramic materials, 1st ed., Beijing: Science Press, 2004, pp.12,58-59.

[11] Y.S. Yu, G.Y. Kim, B.H. Min, and S.C. Kim, "Optical characteristics of Ge doped ZnO compound," J. Eur. Ceram. Soc., vol. 24, pp.1866, 2004.

[12] G.G. Valle, P. Hammer, S.H. Pulcinelli, and C.V. Santilli, "Transparent and conductive ZnO:Al thin films prepared by sol-gel dip-coating," J. Eur. Ceram. Soc., vol. 24, pp.1011, 2004.

[13] M. Naeem, S. K. Hasanain, and A. Mumtaz, "Electrical transport and optical studies of ferromagnetic cobalt doped ZnO nanoparticles exhibiting a metal-insulator transition," J. Phys.: Condens. Matter, vol. 20, issue. 2, pp.0252, 2008.

[14] J. Chen, G.J. Jin, and Y.Q. Ma, "Effect of oxygen vacancy defect on the magnetic properties of Co-doped ZnO diluted magnetic semiconductor," Acta Physica Sinica, vol. 58, issue. 4, pp.2708-2710, Apr 2009.

[15] H.X. Liu, H. Qiu, X.B. Chen, M.P. Yu, and M.W. Wang, "Structural and physical properties of ZnO:Al films grown on glass by direct current magnetron sputtering with the oblique target," Curr. Appl. Phys., vol. 9, issue. 6, pp.1217-1222, Feb 2009.

Electric Field Induced Enhancement in Multisubband Electron Mobility in Strained GaAs/InGaAs Double Quantum Well Structures

Trinath Sahu[1*], Sangeeta Palo[2] and Narayan Sahoo[3]

[1]National Institute of Science and Technology, Berhampur-761008, Odisha, India
[2]Department of ECE, Kalam Institute of Technology, Berhampur, India
[3]Department of Electronic Science, Berhampur University, Berhampur-760007, India
* Email : tsahu_bu@rediffmail.com

Abstract— **We study the effect of external electric field *F* in enhancing the multisubband electron mobility mediated by intersubband effects in a pseudomorphic *GaAs / InₓGa₁₋ₓAs* coupled double quantum well structure. An electric field *F* changes the potential profile of the structure which in turn amends the subband energy levels and wave functions. By varying, *F*, the occupation of different subbands can be changed. As a result the system can be transformed from double subband occupancy to single subband occupancy resulting enhancement in the mobility due to the suppression of intersubband interaction. We show that the effect of the doping concentration and hence the 2D-electron density on the electric field dependence of the subband mobility yields interesting results.**

I. INTRODUCTION

The effect of external electric field, perpendicular to the interface plane of a quantum well structure, changes the potential energy profile of the quantum well. Accordingly the subband energy levels and wave functions vary leading to changes in the occupation of different subband energy levels. Hence by varying the electric field *F* one can transform a system from double subband occupancy to single subband occupancy resulting in a large enhancement in mobility due to the suppression of intersubband effects. Attempts have been made to study the effect of electric field on the electron mobility in single and double quantum well structures [1,2]. However, very little work has been done on the study of the effect of the quantum well structure parameters on the field dependence of the multisubband mobility in coupled quantum well structures.

In the present work we study the effect doping concentration and hence the two-dimensional (2D-) electron density on the electric field dependence of the multisubband electron mobility in a coupled quantum well structure. We consider a *GaAs/InGaAs* double quantum well structure on which the field *F* is applied perpendicular to the interface plane. The low temperature mobility is assumed to be due to the ionized impurity (*imp-*) scattering, alloy disorder (*AL-*) scattering and interface roughness (*IR-*) scattering. The screening of the scattering potentials is

obtained by adopting the static dielectric response function formalism within the random phase approximation [3,4]. We analyse the changes in the intrasubband and intersubband scattering rate matrix elements as a function of the electric field *F*. We take different doping concentrations No to study their effect on the functional dependence of subband mobility upon the electric field *F*. We show that the interplay of different scattering mechanisms mediated by intersubband interaction leads to enhancement in mobility along with other interesting features. As long as the double subband is occupied, the *imp-*scattering is dominant. Whereas, in case of single subband occupancy, *IR-* and *AL-*scatterings are prominent. With increasing *Ns* the effect of *AL-*scattering is enhanced while that of *IR-*scattering is reduced. We show that large enhancement in mobility can be achieved through the application of an external electric field with suitable choice of the material parameters such as the doping concentration. Our results can be utilized for low temperature device applications.

II. THEORY

We consider a pseudomorphic *GaAs/InGaAs* double quantum well structure in which the side barriers (GaAs layers) are delta-doped with layers of Si. The width of the central barrier is *b* and *w* is the well width. The width of the doping layer is *d*. The doping concentration is N_0. The doping layers lie at a distance *s* from the corresponding interfaces. A schematic structure profile is presented in Fig.1. Because of diffusion, the electrons move to the adjacent *InGaAs* well layers. The impurity and electron concentration distributions $n_D(z)$ and $n(z)$ perpendicular to the interface plane, i.e., along the growth axis, (say z-axis), can be written respectively as [4] :

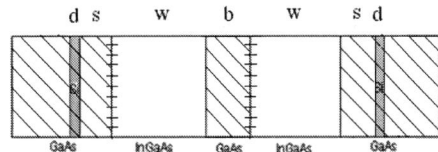

Fig.1 : Schematic diagram of the barrier delta-doped double quantum well structure

$$n_D(z) = \begin{cases} N_0 & (b/2+w+s) < |z| < (b/2+w+s+d) \\ 0 & otherwise \end{cases} \quad (1)$$

$$n(z) = \sum_n n_n |\psi_n(z)|^2 \qquad -\infty < z < +\infty \quad (2)$$

n_n is the number of electrons per unit area in the nth subband and $\psi_n(z)$ is the subband electron wave function [3]. At T = 0 ^oK , assuming all the donors are ionized, the surface electron density $N_s = 2N_0 d$, which can be related to the Fermi energy [3] :

$$N_s = m/(\pi\hbar^2) \sum_{n=1}^{N} (E_F - E_n) \; \theta \; (E_F - E_n) \quad (3)$$

N is the number of filled levels, θ is the Heaviside step function. E_n is the subband energy level and E_F is the Fermi energy. We obtain potential profile of the structure $V(z)$ using the variational technique adopted by us [2,4] in which the effect of the external electric field, applied perpendicular to the interface plane, is incorporated. We calculate the subband energy levels E_n and wave functions $\psi_n(z)$ by using multi-step potential approximation [5].

For a multisubband occupied system, the subband transport lifetime τ_n satisfies a coupled linear equation obtained from the Boltzmann equation [3]. The expression for transport life time τ_n can be written for a single occupied band (n=0) as [4]:

$$\frac{1}{\tau_0} = B_{00} \quad (4)$$

B_{00} is the intrasubband scattering rate matrix element. For double occupied subbands, τ_n can be written as [4]:

$$\frac{1}{\tau_0} = \frac{(B_{00}+C_{01})(B_{11}+C_{10})-D_{01}D_{10}}{(B_{11}+C_{10})+(E_{F1}/E_{F0})^{1/2} D_{01}}$$

$$\frac{1}{\tau_1} = \frac{(B_{00}+C_{01})(B_{11}+C_{10})-D_{01}D_{10}}{(B_{00}+C_{01})+(E_{F0}/E_{F1})^{1/2} D_{10}} \quad (5)$$

where C_{nm} and D_{nm} are the intersubband scattering rate matrix elements. The scattering rate matrix elements are expressed in terms of the screened scattering potentials . The screened alloy disorder scattering potential can be written as [4] :

$$|V_{nm}^{AL}(q)|^2 = \sum_j \left[a_j^3 (\delta V_j)^2 x_j (1-x_j)/4 \right] \times$$

$$\int_j dz \left| \sum_{n'm'} \varepsilon_{nm,n'm'}^{-1}(q) \Psi_{n'}(z) \Psi_{m'}(z) \right|^2 \quad (6)$$

The screened ionized impurity scattering potential [4] :

$$|V_{nm}^{imp}(q)|^2 = \frac{4\pi^2 e^4 N_0}{\varepsilon_0^2 q^2} \times$$

$$\left[\int_{-(b/2+w+s+d)}^{-(b/2+w+s)} dz_i \left| \sum_{n'm'} \varepsilon_{nm,n'm'}^{-1}(q) P_{n'm'}(q,z_i) \right|^2 \right. \quad (7)$$

$$\left. + \int_{(b/2+w+s)}^{b/2+w+s+d} dz_i \left| \sum_{n'm'} \varepsilon_{nm,n'm'}^{-1}(q) P_{n'm'}(q,z_i) \right|^2 \right]$$

where

$$P_{n'm'}(q,z_i) = \int_{-\infty}^{\infty} dz \; \psi_{n'}(z)\psi_{m'}(z) e^{-q|z-z_i|} \quad (8)$$

Similarly the interface roughness scattering potential [4]

$$|V_{nm}^{IR}(q)|^2 = V_b^2 \pi \Lambda^2 \Delta^2 e^{-q^2 \Lambda^2/4} \times$$

$$\sum_{bi} \left| \sum_{n'm'} \Psi_{n'}^*(z)\Psi_{m'}(z) \Big|_{z=z_{bi}} \varepsilon_{nm,n'm'}^{-1}(q) \right|^2 \quad (9)$$

j stands for number of wells, 'a_j', x_j and δV_j are the lattice constant, alloy fraction, and alloy scattering potential respectively of the strained well layer. Sum over bi is over the bottom sides of two wells of the structure leading to an asymmetric scattering potential [6]. The ionized impurity scattering occurs from the delta-doped layers lying in the side barriers. The alloy scattering occurs in the well regions. The interfaces lying towards the bottom sides (substrate sides) of the two wells of the structure are assumed to be rough, leading to an asymmetric IR-scattering potential [4, 6]. In Fig.1 the rough interfaces are depicted by marking (+) signs.

At T=0 ^{o} K, the subband mobility is related to the subband transport life time by the relation μ_n (E_F) = (e/m) $\tau_n(E_F)$. The mobility μ^R due to a certain scattering mechanism R (R= imp, IR or AL) can be written in terms of the subband mobility μ_n^R as : \

$$\mu^R = \frac{\sum_n n_n \mu_n^R}{\sum_n n_n} \quad (10)$$

where n_n is the number of electrons per unit are in the nth subband. The total mobility μ are calculated by using Matthiessen's rule [3, 4].

978-1-4673-2395-6/12 $31.00 © 2012 IEEE

III. RESULRS AND DISCUSSION

We analyse the effect of two-dimensional electron density Ns (in 10^{11} cm^{-2}) on the external electric field dependence of the subband mobility. The electric field F is applied perpendicular to the interface plane of a *GaAs / $In_xGa_{1-x}As$* pseudomorphic double quantum well structure. We take the well width w = 80 A°, spacer width s = 80 A°, delta-doping layer width d=20 A°, and the barrier width b= 40 A° . We take the alloy fraction x = 0.25 and calculate the conduction band offset of the *GaAs / $In_xGa_{1-x}As$* structure using the strain dependent band gap, V_b = 175 meV [4]. The strain dependent effective mass of the electron is m=0.061. We vary the doping concentration N_0 (in 10^{18} cm^{-3}) and two-dimensional electron density N_s =$2N_od$ (in 10^{11} cm^{-2}). The interface roughness parameters Δ = 2.83 A° and Λ = 100 A° and the alloy disorder scattering potential δV = 530 meV as adopted in [4].

In Fig.2 we present the results of subband mobility μ_n^{imp}, μ_n^{IR} and μ_n^{AL} (cm^2/Vsec) as a function of F (kV/cm) along the z-axis for Ns = 2.0 X 10^{11} cm^{-2}. The second subband is occupied up to F = 4.0 kV/cm. The subband mobilities are governed by the intrasubband and intersubband scattering rate matrix elements as described in [4]. For double subband occupancy, the *imp*-scattering is dominant. Whereas, for single subband occupancy, the *IR*-scattering is dominant.

We calculate the subband mobility as a function of F by taking different values of Ns. In Fig.3(a), 3(b) and 3(c), we present the results of μ^{imp}, μ^{IR} μ^{AL} and total mobility μ as a function of the external field F (in kV/cm) for doping concentrations No = 0.5, 1.0 and 1.5 (X 10^{18} cm^{-3}) , i.e., Ns = 2.0, 4.0 and 6.0 (X 10^{11} cm^{-2}) respectively. To start with we have two occupied subbands. As F increases transition from double subband to single subband occupancy occurs leading to sudden enhancement in mobility due to suppression of the intersubband effects. As Ns increases the value of F at which the transition occurs increases. For example, as Ns increases from 2.0, 4.0 to 6.0 the transition to single subband occupancy occurs at F = 4.5, 12.0, 18.0 kV/cm respectively.

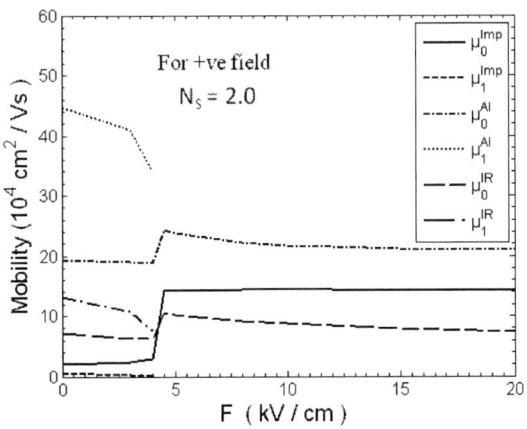

Fig.2 : Results of μ_n in units of 10^4cm^2 / Vs as a function of F(kV/cm) for Ns = 2.0 X 10^{11} cm^{-2}.

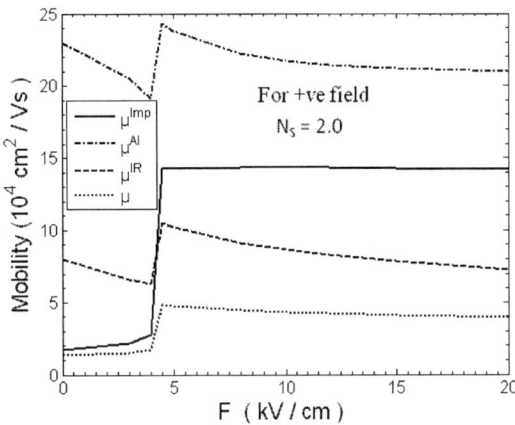

Fig.3(a) : Results of μ^{imp} μ^{IR} , μ^{AL} and μ in units of 10^4cm^2 / Vs as a function of F(kV/cm) for Ns = 2.0 X 10^{11} cm^{-2}

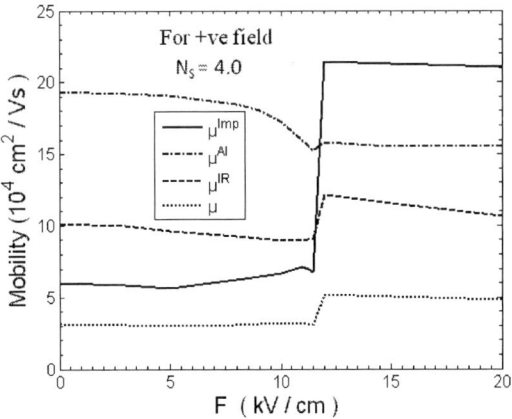

Fig.3(b) : Results of μ^{imp} μ^{IR} , μ^{AL} and μ in units of 10^4cm^2 / Vs as a function of F(kV/cm) for Ns = 4.0 X 10^{11} cm^{-2}.

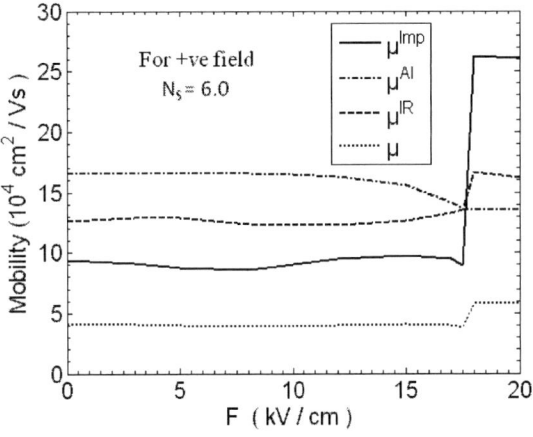

Fig.3© : Results of μ^{imp} μ^{IR} , μ^{AL} and μ in units of 10^4cm^2 / Vs as a function of F(kV/cm) for Ns = 6.0 X 10^{11} cm^{-2}.

We shall compare the interplay of different scattering mechanisms mediated by the intersubband effects in these structures. From Figs. 3(a), 3(b) and 3(c) we note that with increase in Ns, for double subband occupancy, the mobility is governed mostly by the *imp*- scattering and to an extent by *IR*-scattering. However, once the single subband is occupied, the mobility is influenced by the *IR-* and the *AL-* scatterings. As *Ns* increases, the effect of *AL -* scattering is enhanced while that of *IR-* scattering is decreased. Therefore for Ns=2 X10^{11} cm^{-2} the mobility is more influenced by *IR -* scattering. While for Ns=6 X10^{11} cm^{-2}, the mobility is mostly due to *AL*-scattering.

In Fig.4 we present the total mobility μ as a function of the applied electric field *F* (kV/cm) for different 2-D electron concentrations *Ns*. It is interesting to show that the rise in mobility at the transition point becomes less with increase in *Ns*. For Ns= 8.0 X 10^{11} cm^{-2} the system is double subband occupied up to *F* = 20 kV/cm and hence no rise in mobility is seen.

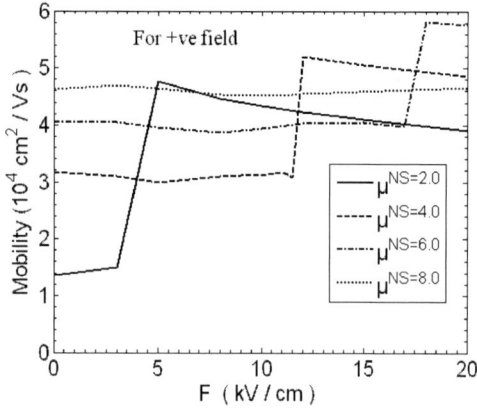

Fig.4 : Results of μ in units of 10^4 cm^2 / Vs as a function of F(kV/cm) for different Ns (in 10^{11} cm^{-2})

We change the direction of the field i.e., along the negative z-axis to study its effect on the mobility. In Figs. 5(a), 5(b) and 5(c), we plot the results of μ^{imp}, μ^{IR} μ^{AL} and μ as a function of *F* along the negative z- axis. We note that the imp- scattering and *AL*-scattering potentials are symmetric about the centre of the structure along the z-axis. Whereas the *IR*-scattering is asymmetric in nature since the interface planes lying towards the substrate sides of the wells are only assumed to be rough [6]. In the present work we consider the interfaces in the left side of the wells of the double quantum well structure as rough. As a result the change in the direction of the electric field alters the effect of the IR-scattering potential on mobility. In Figs. 3 and 5, the difference in mobility is therefore due to μ^{IR}. We show that more enhancement in mobility μ is obtained by applying the field *F* along the negative z-axis due to the asymmetric nature of the *IR*-scattering potential.

IV. CONCLUSION

The principal work involved in this paper is the study of the effect of 2-D electron density on the electric field induced enhancement of electron mobility. We consider a GaAs / In$_{0.25}$Ga$_{0.75}$As double quantum well structure where the outer barriers are delta- doped with Si. Effect of applied electric field, perpendicular to the interface

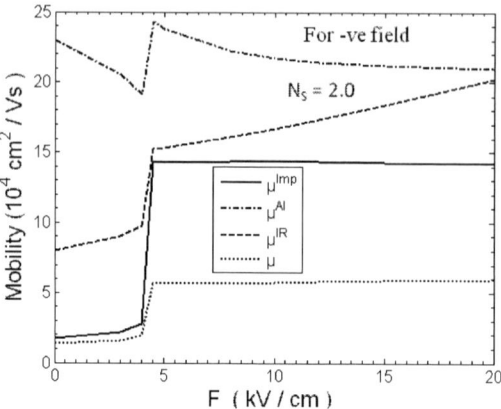

Fig.5(a) : Results of μ^{imp}, μ^{IR}, μ^{AL} and μ in units of 10^4 cm^2 / Vs as a function of F(kV/cm) along negative z-axis for Ns = 2.0 X 10^{11} cm^{-2}.

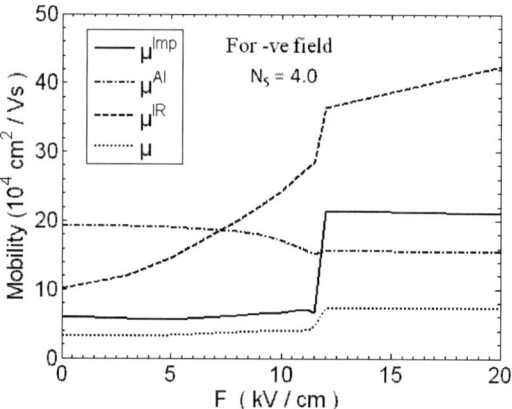

Fig.5(b) : Results of μ^{imp}, μ^{IR}, μ^{AL} and μ in units of 10^4 cm^2 / Vs as a function of F(kV/cm) along negative z-axis for Ns = 4.0 X 10^{11} cm^{-2}.

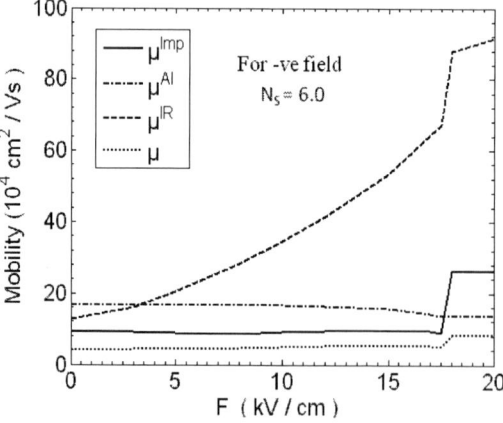

Fig.5© : Results of μ^{imp}, μ^{IR}, μ^{AL} and μ in units of 10^4 cm^2 / Vs as a function of F(kV/cm) along negative z-axis for Ns = 6.0 X 10^{11} cm^{-2}.

plane of the structure, amends the subband electron energy levels and wave functions. The occupation of a subband is also altered. In a multisubband occupied quantum well structure, the change in the occupation of subband energy levels greatly influences the mobility through the intersubband interactions. At the transition from double subband occupancy to single subband occupancy, enhancement of mobility occurs due to the suppression of intersubband effects. We show that the field induced enhancement in mobility is maximum for Ns. = $2.X10^{11}$cm^{-2}. As Ns increases the magnitude of enhancement in mobility decreases. Our results of μ can be utilised for low temperature device applications.

REFERENCES

[1] F. M. S. Lima, A. L. A. Fonseca, and O. A. C. Nunes, " Electric field effects on electron mobility in n-AlGaAs/GaAs/AlGaAs single asymmetric quantum wells",*J. Appl. Phys.* Vol. 92, 2002, pp-5296-5303.

[2] T. Sahu and K.A.Shore, "Effect of electric field on low temperature multisubband electron mobility in a coupled GaInP/GaAs quantum well structure", *J. Appl. Phys.* Vol.107, 2010, pp-113708(6).

[3] T. Ando, A.B. Fowler and F. Stern, "Electronic properties of two-dimensional systems", *Rev. Mod. Phys.* vol. 54, 1982, pp-437-672.

[4] P.K.Subudhi, S. Palo and T. Sahu, "Effect of strain on multisubband electron transport in GaAs/In$_x$Ga$_{1-x}$As coupled quantum well structures", *Superlattices and Microstructures*, vol. 51,2012, 430 - 442.

[5] Y.Ando and T.Itoh, " Calculation of transmission tunneling current across arbitrary potential barrier ", *J. Appl. Phys.* vol. 61, 1987, pp-1498-1502.

[6] S.K. Lyo, "Real space and energy representations for the interface roughness scattering in quantum well structures" *J. Phys.: Condens. Matter* vol. 13 , 2001, pp-1259 -1264.

Properties of Amorphous Carbon Thin Films with Nitrogen Incorporation by Aerosol-assisted CVD

[*1]A.N. Fadzilah, [1]K. Dayana, [1]U. M. Noor and [1]M. Rusop

[1]Nano-ElecTronic Centre (NET),

Faculty of Electrical Engineering, Universiti Teknologi Mara (UiTM),

40450 Shah Alam, Selangor, Malaysia

*Corresponding Author: nurfadzilahahmad@yahoo.com

Abstract. In this experiment, a new deposition technique was employed to deposit and to dope the amorphous carbon (a-C). Nitrogen doped amorphous carbon (a-C: N) thin films were prepared by Aerosol-assisted Chemical Vapor Deposition (AACVD) by varying the deposition time from 15 minutes to 60 minutes. The electrical and optical properties of deposited a-C: N thin films were characterized by current-voltage Solar Simulator system and UV-Vis-NIR spectroscope. Electrical characterization results in ohmic behavior with the optimum conductivity were indicated at sample deposited for 15 minutes. At visible range, the transmittance is high (above 80%) for sample deposited at 15 minutes and possess lower transmittance (60% to 85%) when deposition time increase up to 45 minutes. The absorption coefficient, α for a-C: N is reported to be ~$\times 10^5$ cm^{-1}. From the atomic force microscope characterization, surface morphology and roughness value was measured.

Keywords: Amorphous carbon; Nitrogen doped; Aerosol-assisted CVD; Electrical properties; Optical properties; Structural properties

1. INTRODUCTION

At present amorphous carbon (a-C) thin films was studied due to many superior properties such as high mechanical hardness, chemical inertness, and optical transparency [1]. It is able to accept dopants due to the semiconducting nature [2], has band gap from 0.0 eV to 5.5 eV [3] and the band gap can be tuned by adjusting the sp^2 and sp^3 of carbon bonding ratio [4] by doping techniques. The properties of a-C are dependent on the deposition condition for such the deposition time. The deposition parameter of a-C is also compatible with glass substrate that is cheap since the production can be as low as 350°C.

Researchers has fabricated the a-C by various techniques, such as pulsed laser deposition (PLD) method [3], pulsed filtered cathodic method [5] and filtered cathodic vacuum Arc method [6] that uses the expensive reactor and vacuum systems. On the other hand, the low cost chemical vapor deposition (CVD) technique which involves homogenous and heterogenous chemical reaction of gaseous reactants was also being utilized in the fabrication of a-C. However, some problems for such the lack of proper volatile precursor and the difficulty in controlling the stoichiometry of the deposition [7] is the limitation of using the CVD method. These problems were countered by modifying the conventional CVD process into aerosol-assisted CVD (AACVD), where this method uses the aerosol droplets to transport the precursor, with the aid of carrier gas.

The properties of semiconductor material can be modified such that in optical band gap and also the photoconductivity by effective doping technique. Several materials were reported as the suitable dopant material for a-C, such as Nitrogen gas and Phosphorous for n-type dopant, and Boron and Iodine for p-type dopant. Nitrogen doped amorphous carbon (n-C:N) is reported as a conductive material for electrical analysis. Several advantages include a low background current, a large potential window and inertness in chemically aggressive environments [8]. Other than that, the optical band gap and conductivity differ due to increment of electron concentration [9, 10]. To produce high electrical conductivity of a-C, doping was introduced Nitrogen (N) incorporated a-C shows better electrical conductivity and high photoconductivity [11].

This paper reports on the effect of N doped a-C as well as the effect of varying the deposition time of a-C: N by AACVD method. Electrical, optical and structural properties were discussed.

2. METHODS

Camphor ($C_{10}H_{16}O$) which consists both sp^2 and sp^3 carbon is an attracting new material for carbon-based preparation and nitrogen doped amorphous carbon (a-C: N) thin films were prepared using Aerosol-assisted Chemical Vapor Deposition (AACVD) system, where the AACVD consist of an aerosol-assisted system and CVD with two furnace systems. Camphor oil was heated at the aerosol-assisted system at 180°C to make sure the transformation of the camphor oil from liquid phase into vapours. A double furnace CVD was connected next to the aerosol-assisted setup, where the deposition took place. For nitrogen doped amorphous carbon (a-C: N), nitrogen (N$_2$) gas was utilized as the carrier gas and dopant gas.

The glass substrates were by acetone (C_3H_6O) and methanol (CH_3OH) for 10 min respectively using an ultrasonic cleaner (Power Sonic 405) to eliminate contaminations. Lastly the substrate was blown with nitrogen gas (N$_2$). Temperature of the furnace two was fixed at 500°C under N$_2$ atmosphere whereas furnace one is kept constant at 200^{0}C. The electrical, optical and structural properties of deposited a-C: N thin films were characterized by current-voltage Solar Simulator system,

UV-Vis-NIR spectroscope and atomic force microscope. In this experiment, effect of deposition time of a-C:N was investigated for 15 minutes (sample B), 30 minutes (sample C), 45 minutes (sample D) and 60 minutes (sample E) whereas as a reference, pure a-C thin film was deposited at 15 minutes (sample A).

3. RESULTS AND DISCUSSION

Nitrogen doped amorphous Carbon (a-C:N) films were deposited on glass substrate at 4 different deposition time, 15 minutes (sample B), 30 minutes (sample C), 45 minutes (sample D) and 60 minutes (sample E) whereas as a reference, pure a-C thin film was deposited at 15 minutes (sample A). The current-voltage (I-V) of the a-C thin films gives analysis on the electrical performance of the films when supply voltage was set from -10 to 10 V. The measurement was carried out using BUKOH KEIKI (CEP2000) Solar Simulator/Spectral Sensitivity Measurement by two probe method. Gold (Au) was used as the metal contact with 60nm thickness.

An ohmic graph was obtained as in figure 1 and the value of resistivity (ρ) and conductivity (σ) was calculated using equation (1) and (2) respectively, where R is the resistance, w is the width of the electrode, t is the thickness of the a-C thin film and L is the length of the electrodes. From figure 1, obviously noted that sample D has highest slope of ohmic graph followed by sample C, E, B and A. The thickness of the thin films was evaluated from the Surface Profiler and it was noted that as deposition time increased (sample B, C, D and E) the thickness increased from 30 nm to 90 nm.

$$\rho = (\frac{V}{I})(\frac{wt}{L}) \qquad \text{in unit } \Omega.\text{cm} \qquad (1)$$

$$\sigma = \frac{1}{\rho} \qquad \text{in unit S.cm}^{-1} \qquad (2)$$

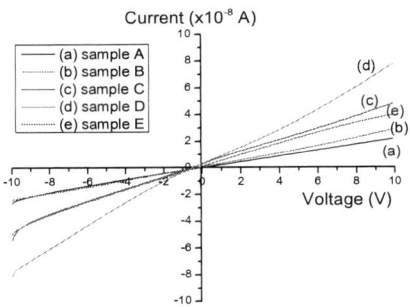

Figure 1. Current-voltage characteristics for a-C and a-C: N at different samples.

Electrical conductivity, σ was plotted, as in figure 2 and it was observed that when doping was introduced at sample B, it exhibit higher σ (7.19 x 10^{-4} Scm^{-1}) compared to sample A (5.06 x 10^{-4} Scm^{-1}). Conductivity from sample B to Sample C decrease gradually and has the second highest conductivity value, 6.76 x10^{-4} Scm^{-1} followed by sample D, 3.67 x 10^{-4} Scm^{-1} and sample E, 3.86 x 10^{-4} Scm^{-1}.

This conductivity value is higher in comparison with the reported value of O. S. Panwar et. al [12]. The difference in conductivity when doping introduced from sample A to

sample B is due to the successful doping which leads to decrease of spin density gap with N addition. Thus, the defect also decrease and higher the conductivity [13]. Another reason for the increment in conductivity is because when the N doping is introduced, the C-N$_x$ bonds were formed from the large amount of C and N atoms. As the impurity center the C-N$_x$ bonds has activated the films conductivity performance and increase the conductivity [14].

More, lower conductivity noted as deposition temperature increase is said to be effect from the improper growth of a-C as can be seen from the AFM images in figure 6. Huang et. al [15] discussed on the theoretical part of thickness relation towards the conductivity value for the semiconductor layer, where the conductivity is directly related to the carrier concentration and carrier mobility for sample at different thickness. The reduction of conductivity is said to be influenced by the reduction of the carrier mobility at increasing thickness.

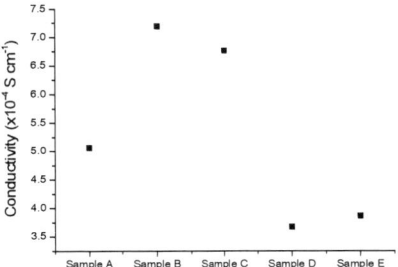

Figure 2. Conductivity for a-C and a-C:N at different samples.

For optical properties, a-C thin films deposited on the glass substrates was investigated by a Perkin Elmer LAMBDA 750 UV-vis-NIR spectroscope measurement. Figure 3 shows the optical transmittance versus wavelength of the a-C thin films deposited at different sample for pure a-C and a-C:N. It is observed that the optical transmittance at the visible wavelength for a-C:N decrease from sample B (80-90%), sample C (70-80%) and sample D (70-80%). However, for sample E with longest deposition time, the transmittance is high (80-90%) and as the comparison, pure a-C exhibit highest transmittance value that is more than 90%. According to Han et.al [16], the high considerable transmittance value for thin films is good for solar cell since the reflectance at the surface of the material is supressed.

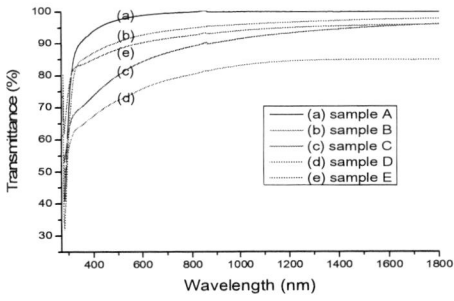

Figure 3. Transmittance spectrum for a-C and a-C:N at different samples.

The absorption coefficient (α) was calculated by equation (3) ;

$$\alpha = \frac{1}{t}\ln(\frac{1}{T}) \qquad (3)$$

where t is the thickness and T is the transmittance of the film. Based on figure 4, absorption coefficient at low energy is higher compared to the high energy. Further discussion on this is related to the π-π* transition. The localized π electrons inside a-C and the bonds between the π electron is inconsistent with any other allotrope of carbon. On the other hand, a-C contains high concentration of dangling bonds, which will affect the band gap value. Thus, since the localized π electrons play an important role in determining the optical properties of a-C, due to the reduced spatial overlapping of π and σ states wave overlapping, the contribution of σ-π* and π-σ* can be neglected [17, 18]. The similar range value of α had also been reported for a-C deposited by surface wave microwave plasma CVD (~10^2-10^5 cm^{-1}) [19, 20], ion beam sputtering (~10^4-10^5 cm^{-1}) [21].

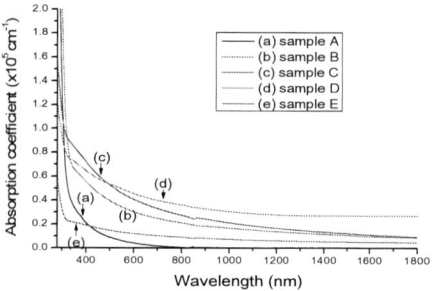

Figure 4. Absorption coefficient spectra for a-C and a-C:N at different samples.

The optical band gap of pure a-C and a-C:N thin film was deduced using the Tauc plot from Tauc relationship $(\alpha h v)^{\frac{1}{2}} = B(E_g - h v)$, where α is the absorption coefficient, B is the Tauc parameter and hv is the photon energy [22] as shown in Figure 5 in the absorption region (α~10^5 cm^{-1}). The optical band gaps (E_g) for amorphous semiconductors were obtained by The intercept of the Tauc's slope in the photon energy axis gives the optical gap, E_g as shown in figure 5.

For N_2 doping comparison at same parameter (sample A and B), sample A has higher band gap that is 1.4 eV whereas sample B, 0.5 eV. Theoretically, when ratio of sp^2 towards sp^3 increase, optical band gap decrease due to π-π* transition contribution. Here, there is evolution of π and π* peaks at the high energy edge of the valence band and also low energy edge at the conduction band. On the other hand, the optical band gap is also determined by the sp^2 aromatic cluster at the sample [17]. The small gap between the valence and conduction band that the thermal or other excitations can bridge the gap. The interaction between the lattice phonons and the free electrons and holes will also affect the band gap to a smaller extended. Thus it is said that that the N incorporation induced graphitization due to increase of sp^2 [3]. Besides, the presence of N atom acts as bridging atom between a-C clusters. This in turn will increase the size of sp^2 cluster, broadening the π-π* states and decrease the band gap [23].

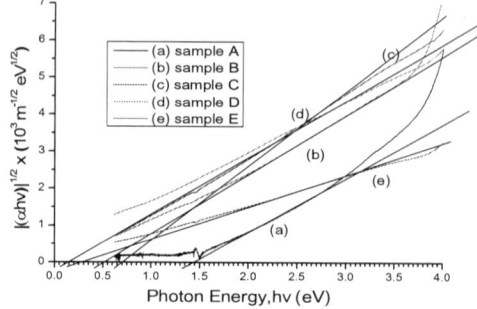

Figure 5. Tauc plot for a-C and a-C:N at different samples.

The measurement for structural properties was carried out by Atomic Force Microscopy (AFM). Figure 6 revealed the surface morphology of a-C films (a) sample A, (b) sample B, (c) sample C, (d) sample D and (e) sample E with the average roughness value 0.62nm, 22.5nm, 14.9nm, 8.93nm and 9.75nm respectively. The typical average roughness was estimated over 30μm x 30μm area. For a-C:N, less thicker thin film shows smaller and dispersed particles. However, the even thin films disappear as thickness increase up to 90nm. Moreover, there are some aggregation spot of sample E, which might be due to improper dispersion of carbon atoms.

The results as in table 1 show that the surface roughness and the growth rate decrease along with the increase in deposition temperature (up to sample E). For highest roughness at sample B, this implies that some particles has combines into large particle due to the enhancement of surface migration [24]. Park et.al suggested that the decrease in the surface roughness might be related to the increase of the energetic ion bombardment at the surface of thin film [25]. The different kinetic energy of carbon species generated during deposition cause the increase of roughness value. The high energy of carbon species would considerably enhance the surface mobility of the depositing atom, which then increase the roughness and decrease the compactness of the a-C films [19]. The roughness value which is smaller for sample deposited at longer deposition time is the consequence of the presence of high energetic carbon species with higher sp^3 ratio. Thus this explains the lower conductivity [26].

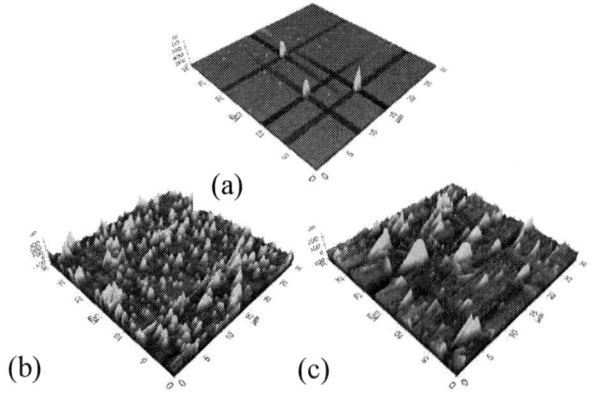

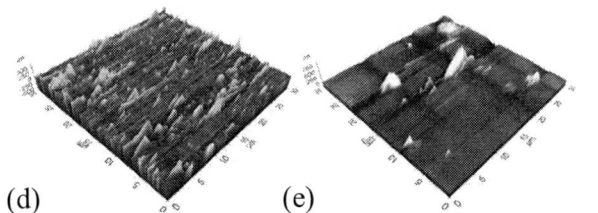

(d) (e)

Figure 6. AFM images for a-C and a-C:N at different samples.

Table 1. Details on roughness gap for a-C and a-C:N

Sample	Roughness (nm)	Growth rate (nm/min)
Sample A	0.62	2.0
Sample B	22.5	1.7
Sample C	14.9	1.4
Sample D	8.93	1.35
Sample E	9.75	1.5

4. CONCLUSIONS

Nitrogen doped amorphous carbon (a-C:N) thin films were prepared using Aerosol-assisted Chemical Vapor Deposition (AACVD) by varying the deposition time from 15 minutes to 60 minutes. The properties of a-C thin films have been investigated using standard measurement techniques. Electrical characterization results in ohmic behavior and the conductivity shows high value, $\sim 10^{-4}$ Scm^{-1}. The difference in conductivity at different sample is related to the surface morphology as measured by atomic force microscope. Optical measurement results in high transmittance (above 80%) for sample deposited at 15 minutes and possess lower transmittance (60% to 85%) when deposition time increase up to 45 minutes at the visible range. Absorption coefficient value for a-C:N is approximately $\sim 10^{5}$ cm^{-1} and surface roughness of the sample was estimated to be around 22.5 nm to 9.75 nm for a-C:N. The decrease in the surface roughness might be related to the increase of the energetic ion bombardment at the surface of thin film.

Acknowledgment

Greatest gratitude to Research Management Institute of Universiti Teknologi Mara (UiTM), Universiti Teknologi Mara (UiTM) and Ministry of Higher Education for the financial support.

References

[1] Y. Tang, Y. S. Li, Q. Yang, and A. Hirose, "Characterization of hydrogenated amorphous carbon thin films by end-Hall ion beam deposition," *Applied Surface Science,* vol. 257, pp. 4699-4705, 2011.

[2] J. Podder, M. Rusop, T. Soga, and T. Jimbo, "Boron doped amorphous carbon thin films grown by r.f. PECVD under different partial pressure," *Diamond and Related Materials,* vol. 14, pp. 1799-1804, 2005/12//.

[3] S. M. Mominuzzaman, M. Rusop, T. Soga, T. Jimbo, and M. Umeno, "Nitrogen doping in camphoric carbon films and its application to photovoltaic cell," *Solar Energy Materials and Solar Cells,* vol. 90, pp. 3238-3243, 2006.

[4] K. Abe and O. Eryu, "Optical properties of amorphous carbon films implanted with nitrogen," Nuclear Instruments and Methods in Physics Research Section B: Beam Interactions with Materials and Atoms, vol. 242, pp. 637-639, 2006.

[5] N. D. Baydogan, "Evaluation of optical properties of the amorphous carbon film on fused silica," *Materials Science and Engineering B,* vol. 107, pp. 70-77, 2004.

[6] O. S. Panwar, M. A. Khan, B. S. Satyanarayana, S. Kumar, and Ishpal, "Properties of boron and phosphorous incorporated tetrahedral amorphous carbon films grown using filtered cathodic vacuum arc process," Applied Surface Science, vol. 256, pp. 4383-4390, 2010.

[7] X. Hou and K. L. Choy, "Processing and Applications of Aerosol-Assisted Chemical Vapor Deposition," *Chemical Vapor Deposition,* vol. 12, pp. 583-596, 2006.

[8] A. Zeng, Y. Yin, M. Bilek, and D. McKenzie, "Ohmic contact to nitrogen doped amorphous carbon films," *Surface and Coatings Technology,* vol. 198, pp. 202-205, 2005.

[9] P. P, "Optical properties of amorphous carbons and their applications and perspectives in photonics," *Thin Solid Films,* vol. 519, pp. 3990-3996, 2011.

[10] B. Bhushan, "Nanotribology of Ultrathin and Hard Amorphous Carbon Films Springer Handbook of Nanotechnology," Springer Berlin Heidelberg, 2004, pp. 791-830.

[11] S. Liu, G. Wang, and Z. Wang, "Study of the conductivity of nitrogen doped tetrahedral amorphous carbon films," *Journal of Non-Crystalline Solids,* vol. 353, pp. 2796-2798, 2007.

[12] O. S. Panwar, M. A. Khan, B. S. Satyanarayana, S. Kumar, and Ishpal, "Properties of boron and phosphorous incorporated tetrahedral amorphous carbon films grown using filtered cathodic vacuum arc process," *Applied Surface Science,* vol. 256, pp. 4383-4390, 2010.

[13] M. Rusop, S. Adhikari, A. M. M. Omer, T. Soga, T. Jimbo, and M. Umeno, "Effects of methane gas flow rate on the optoelectrical properties of nitrogenated carbon thin films grown by surface wave microwave plasma chemical vapor deposition," *Diamond and Related Materials,* vol. 15, p. 6, 2006.

[14] J.-r. Xiao, X.-h. Li, and Z.-x. Wang, "Effects of nitrogen content on structure and electrical properties of nitrogen-doped fluorinated diamond-like carbon films," *Transactions of Nonferrous Metals Society of China,* vol. 19, pp. 1551-1555, 2009.

[15] L. Y. Huang and L. Meng, "Effects of film thickness on microstructure and electrical properties of the pyrite films," *Materials Science and Engineering: B,* vol. 137, pp. 310-314, 2007.

[16] K.-S. Han, J.-H. Shin, and H. Lee, "Enhanced transmittance of glass plates for solar cells using nano-imprint lithography," *Solar Energy Materials and Solar Cells,* vol. 94, pp. 583-587, 2010.

[17] R. Gharbi, M. Fathallah, N. Alzaied, E. Tresso, and A. Tagliaferro, "Hydrogen and nitrogen effects on optical and structural properties of amorphous carbon," *Materials Science and Engineering: C,* vol. 28, pp. 795-798, 2008.

[18] G. Fanchini, S. C. Ray, and A. Tagliaferro, "Density of electronic states in amorphous carbons," *Diamond and Related Materials,* vol. 12, pp. 891-899, 2003/7//.

[19] M. Rusop, S. Adhikari, A. M. M. Omer, T. Soga, T. Jimbo, and M. Umeno, "Effects of methane gas flow rate on the optoelectrical properties of nitrogenated carbon thin films grown by surface wave microwave plasma chemical vapor deposition," *Diamond and Related Materials,* vol. 15, pp. 371-377, 2006.

[20] S. Adhikari, S. Adhikary, A. M. M. Omer, M. Rusop, H. Uchida, T. Soga, and M. Umeno, "Optical and structural properties of amorphous carbon thin films deposited by microwave surface-wave plasma CVD," *Diamond and Related Materials,* vol. 15, pp. 188-192, 2006.

[21] K. M. Krishna, Y. Nukaya, T. Soga, T. Jimbo, and M. Umeno, "Solar cells based on carbon thin films," *Solar Energy Materials and Solar Cells,* vol. 65, pp. 163-170, 2001.

[22] T. J, "Optical properties and electronic structure of amorphous Ge and Si," *Materials Research Bulletin,* vol. 3, pp. 37-46, 1968.

[23] L. Valentini, J. M. Kenny, G. Carlotti, M. Guerrieri, G. Signorelli, L. Lozzi, and S. Santucci, "Effect of nitrogen addition on the elastic and structural properties of amorphous carbon thin films," *Thin Solid Films,* vol. 389, pp. 315-320, 2001.

[24] M. Rusop, S. M. Mominuzzaman, X. M. Tian, T. Soga, T. Jimbo, and M. Umeno, "Nitrogen doping and structural properties of amorphous carbon films deposited by pulsed laser ablation," *Applied Surface Science,* vol. 197–198, pp. 542-546, 2002.

[25] Y. S. Park and B. Hong, "Characteristics of sputtered amorphous carbon films prepared by a closed-field unbalanced magnetron sputtering method," *Journal of Non-Crystalline Solids,* vol. 354, pp. 5504-5508, 2008.

[26] M. Rusop, T. Kinugawa, T. Soga, and T. Jimbo, "Preparation and microstructure properties of tetrahedral amorphous carbon films by pulsed laser deposition using camphoric carbon target," *Diamond and Related Materials,* vol. 13, pp. 2174-2179, 2004.

Aromatic and Non-Aromatic Solvent in Nanocomposited MEH-PPV:CNTs Thin Film for Organic Solar Cells

M.S.P. Sarah, F.S.S. Zahid, Z. Zulkifli, U.M. Noor, M. Rusop
NANO-ElecTronic Centre, Faculty of Electrical Engineering,
Universiti Teknologi MARA, 40450 Shah Alam
e-mail: puteri_ajip@yahoo.com

Abstract- **This paper explained the effect of solvent used to dissolve MEH-PPV powder in preparing the nanocomposited MEH-PPV:CNTs. The ratio of MEH-PPV powder to the solvent used is 1:1 (20 mg/ml). The solvents involved were toluene and tetrahydrofuran (THF). The preparation of the MEH-PPV solution took 48 hours of stirring to ensure that the MEH-PPV powder was well dissolved. After 48 hours of stirring, the black powder of annealed carbon nanotubes (CNTs) was added to the polymer solution. Both polymer solutions were added with 1,2,3 and 4 wt% of CNTs respectively. Nanocomposited MEH-PPV:CNTs thin film which used THF as the solvent is labelled as S1 where else S2 used toluene. The electrical properties of the two sets of samples are characterized by means of current-voltage in dark and under illumination. Where else, optical properties are done by measuring the absorbance and photoluminescence of the samples. From the results, it indicates that the conductivity in S2 is higher compared to conductivity under illumination in S1. Thickness of S2 is higher compared to S1. This indicates that THF as the solvent can ensure the dispersity of the CNTs in the nanocomposites. The optical properties indicate that S1 and S2 showed no peak shifting. Meanwhile in photoluminescence, S1 showed better quenching than S2.**

Keywords: aromatic; non-aromatic; MEH-PPV;CNTs

I. INTRODUCTION

MEH-PPV is known as poly (2-methoxy-5-(2'-ethylhexyloxy)-1,4-phenylenevinylene). MEH-PPV is chosen as an object for the present study because of both its widespread usage throughout the polymer community and the particular sensitivity of its photo physics to the solid-state morphology. This conjugated polymer is known as a hole transport material [1] and it can be good excite generator in optoelectronic component [2]. It is also known as electron donor (p-type semiconductor polymer) with relatively low conductivity due to low hole and electron mobility when comparing to inorganic semiconductor material [3]. Normally, MEH-PPV was used in the organic light emitting diodes (OLED), sensors and an organic diode because of its good environmental stability, easier conductivity control and cheap in production in large quantities. MEH-PPV is a conjugated polymer which is known as a novel class of material that combine the optical and electronic properties of semiconducting with the processing advantages and mechanical properties of plastics [4] . MEH-PPV was affected

by the solvent choice [4]. It is functionalized with flexible side groups, become soluble in common organic solvents, such as tetrahydrofuran (THF), chloroform, xylene and chlorobenzene, and can be solution processed at room temperature into uniform and large area by spin-coating [5].

MEH-PPV alone device has low absorption which will directly influence the photocurrent of device [6, 7]. In order to improve the performance of MEH-PPV, it needs conductive filler like carbon nanotubes (CNTs). CNTs will help in providing lateral electrical conductivity for collecting current from the front surface of MEH-PPV thin film solar cells [1]. Since their discovery, carbon nanotubes have been extensively studied because of their unique structures and remarkable mechanical, electronic, magnetic, photonic, and transport properties, etc. The small diameter (scale of nanometers) and the long length (at order of microns) of CNTs lead to such large aspect ratios that it can act as ideal one-dimensional systems [8]. CNTs have been recognized as a potential candidate for the reinforcement of polymeric materials due to their excellent physical, mechanical properties, nanometer scale diameter and high aspect ratio [9]. However, the solubility and the aggregation of CNTs is a major problem [10] prior to vital solvent used.

Here we report, the fabrication of blend MEH-PPV:CNTs thin film towards the application of organic solar cells with various compositions of CNTs in different solvent to dissolve the MEH-PPV. The goal of this research is to highlight which solvent will produce a good performance of nanocomposited MEH-PPV in electrical and optical properties towards the application of organic solar cells.

II. METHODOLOGY

A. *Solution Preparation*

The materials that have been used in preparing the solutions are MEH-PPV (Mw = 40,000-70,000) with toluene and MEH-PPV (Mw = 40,000-70,000) with THF. The processes are divided into two, which is preparing the MEH-PPV solution with different solvent and preparing the nanocomposite solution. The first set of MEH-PPV used tetrahydrofuran (THF) are labelled as S1 where else the second set used toluene as S2. The polymer solutions are stirred for 48 hours to ensure the dispersity of the MEH-PPV in the solvents. Next, the nanocomposited MEH-PPV:CNTs solution by adding different compositions of annealed CNTs [11] to the polymer solution (0,1,2,3 and 4 wt%) are prepared. The CNTs with an average diameter <8 nm were chosen for this study to provide a large surface area-to-volume ratio.

B. Thin Film Deposition

Spin coating technique is used for the deposition process. The nanocomposited MEH-PPV:CNTs solution was deposited ontop of 2 cm x 2 cm glass. The samples were spin coated at 2000 rpm for 1 minute. Then it goes through drying process at 50 °C to evaporate the solvent.

C. Characterization of Samples

The deposited MEH-PPV:CNTs samples were divided into three characterizations. As a consequence, the thickness of the thin film varies with the escalating of the CNTs composition. The thicknesses of the thin film were measured by using Veeco Dektak 150 surface profiler. Optical characterization of the nanocomposite films was performed by UV-Visible spectrometer and photoluminescence. The UV-Visible absorbance was recorded with a Perkin–Elmer spectrometer while photoluminescence with Fluoro Max 3. It was taken at room temperature. The electrical measurements are done with a source measurement unit (Keithley 2400). The current-voltage measurement (in dark and under illumination) was done after depositing the metal contact, Au. The deposition of the metal contact is done by electron beam thermal evaporator. During the evaporation of metal a pressure of 6.12×10^{-4} Pa is typical.

III. RESULTS AND DISCUSSIONS

A. Physical Properties

Table I shows the thicknesses of the thin films with different solvent. Thicknesses were measured to verify the nano size of the thin film and to get the conductivity of the thin film. For both sets of samples, S1 and S2, the highest thickness value for the five samples recorded at 0 wt% of CNTs which is 143.53 nm for S1 and 62.48 for S2. Results obtained showed that there were decrements in thicknesses as the CNTs are increase. This is due to rearranging of atomic structure by the CNTs [12]. Therefore, the atomic rearrangement will compressed the atomic structure that will directly minimize the thicknesses of the thin film. We assumed that it is also due to undisperse of CNTs that cover the MEH-PPV surface.

Table I
Thickness of nanocomposited MEH-PPV:CNTs using THF and toluene

Concentration of CNTs (wt%)	0	1	2	3	4
Thickness nm (S1)	143.53	119.25	103.95	58.72	55.35
Thickness nm (S2)	62.48	60.68	59.52	58.54	57.34

The FESEM images show that S1 has the ability to hold the CNTs however, in S2 the CNTs agglomerates at the surface

prior to the MEH-PPV unable to cling to the CNTs. The high surface area of the CNTs also influence the surface energy to be high, hence the CNTs tend to agglomerate in both S1 and S2. It can be said that the degree of interchain interactions between MEH-PPV and CNTs is very much depends on the solvent used to dissolve the polymer and the polymer concentration [13].

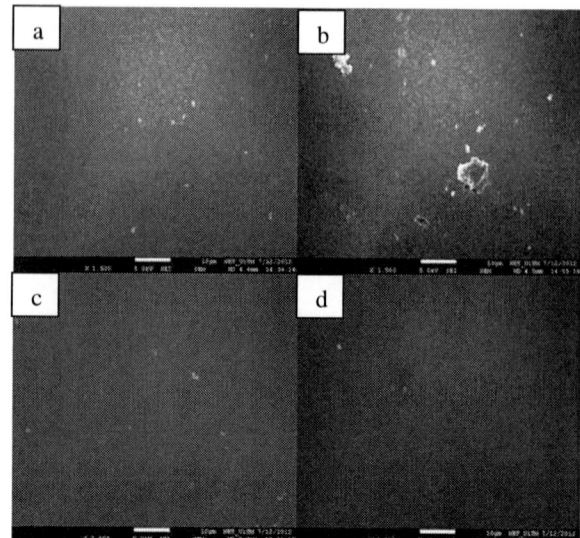

Fig. 1: FESEM images of nanocomposited MEH-PPV:CNTs for S1 a) 0 wt% and b) 4 wt%; S2 c) 0 wt% and d) 4 wt%

B. Electrical Properties

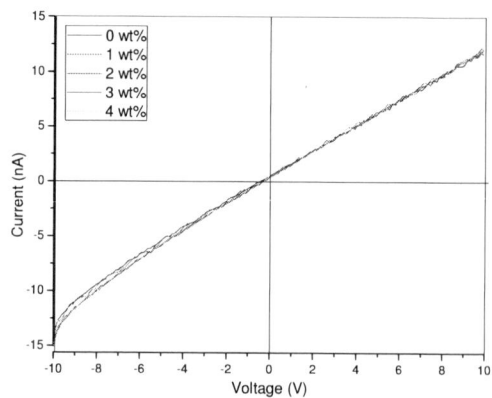

Fig. 2: Photo-current for various compositions in THF

Figs. 2 and 3 below shows the I-V characteristics under illumination for nanocomposited MEH-PPV:CNTs using THF and toluene. The I-V characteristics indicate that the metal contact Au and the organic semiconductor form ohmic contact for S1 and S2. Fig. 2 shows that as the concentration of CNTs increase, the current overlapped with each other. The current showed no significance of changes.

Where else, in Fig. 3, the current showed slightly improvement as the concentration increase. We assumed that using THF as the solvent has enable a low photo-current compared to toluene [14]. THF is known as a non-aromatic solvent which will directly hold back the contact between conjugated polymers with the CNTs [15]. Thin films deposited using THF in MEH-PPV/THF is not easy to vaporize since there are strong aggregation among the main chains of the MEH-PPV. The aggregation could not be disaggregated by heat. On the other hand, thin films depositing by vaporising the toluene solvent from MEH-PPV/toluene has weak interactions among the side chains. Therefore, it is easier for MEH-PPV disaggregated when heat is applied. Consequently, the difference in polymer solubility depends on the aggregation of certain amount of soluble molecular segments in each solvent [16, 17]. As the polymer is illuminated with photons of energy larger than the band gap, electron–hole pairs are generated.

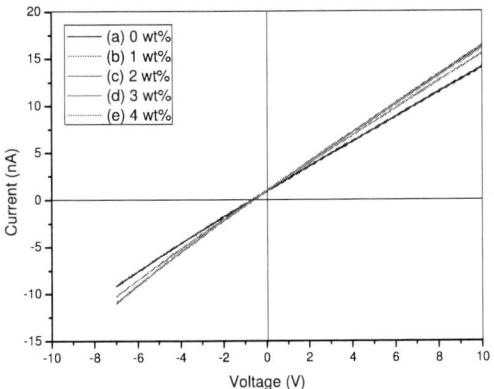

Fig. 3: Photo-current for various compositions in toluene

Fig. 4 shows the photoconductivity for S1 and S2. By comparing the graphs, it can be said that S2 gave the highest conductivity compared to S1. Conductivity of 0 wt% for S2 gave value of 0.022 S/m compared to conductivity in S1 with only around 0.0009 S/m. The nature of the charge transfer process depends on the optical properties of two materials as well as the surface properties of the nanoparticles [18] regardless that the nanoparticles or nantotubes are homogeneously distributed.

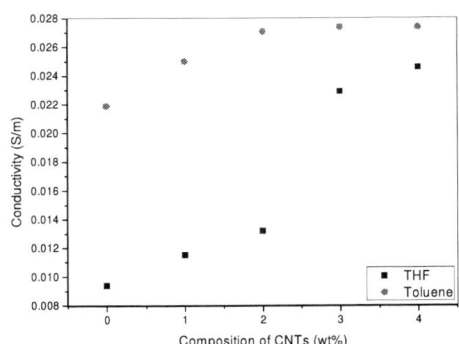

Fig. 4: Photoconductivity of nanocomposite MEH-PPV:CNTs in THF and toluene

C. Optical Properties

Under optical properties, this paper describes the absorbance and photoluminescence of the nanocomposites thin films. Fig. 5 and 6 shows the absorbance spectra for S1 and S2 respectively. In S2, 0 wt% showed the highest absorbance value where else in S1, 2 wt% showed the highest absorbance value. In optical viewpoint, S1 demonstrate better results compare to S2. We assumed that CNTs powder in S2 shows inhomogeneous distribution of the CNTs that have caused no improvement in absorbance prior to THF is good as solvent not only for MEH-PPV but also CNTs. The absorbance features are not significantly affected by the incorporation of the CNTs. UV–Vis electronic absorption spectra of MEH-PPV solutions show an intense absorption band at 500 nm whose peak position is practically independent of the solvent [17]. For both S1 and S2, there are no peak shifting. S1 showed sharp peaks while S2 showed broad peaks.

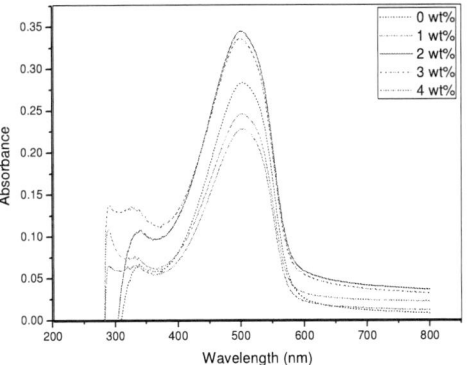

Fig. 5: Absorbance spectra for S1

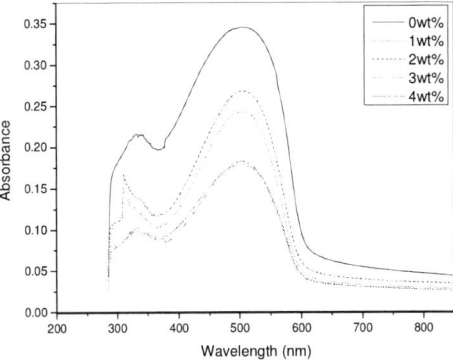

Fig. 6: Absorbance spectra for S2

Photoluminescence quenching in a nanocomposites is a useful indication of the degree of success of exciton dissociation either suitable or not to be used in organic solar cells. Figs. 7 and 8 show the PL spectra of nanocomposited MEH-PPV:CNTs for S1 and S2 respectively. Both figures showed PL emission of MEH-PPV under excitation at ~555 nm. A significant quenching of the emission intensity resulting from the infiltration of CNTs into the MEH-PPV can be observed in Fig. 7 compared to Fig. 8. This indicates the efficient charge separation at the MEH-PPV/CNTs interface for S1 is better compared to S2, which results in the production of electricity under illumination by light [19]. With existence of CNTs, it can be concluded that the decreased of PL intensity (in S1), the exciton dissociation is further improved. As in S2, the intensity of the nanocomposite MEH-PPV:CNTs is increasing compared to thin film without CNTs. This shows that the probability of recombination to occur is high prior to it will create electron-hole pair.

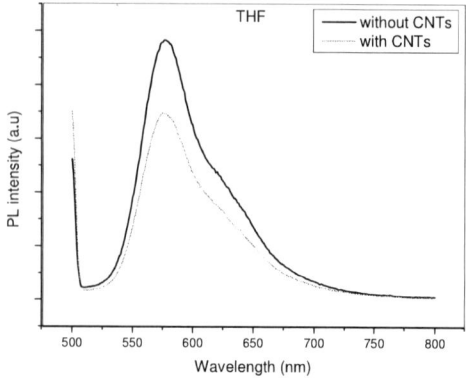

Fig. 7: Photoluminescence spectra for S1

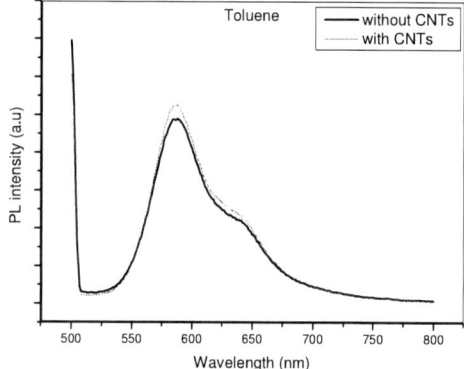

Fig. 8: Photoluminescence spectra for S2

CONCLUSION

As conclusion MEH-PPV properties is dependence of the solvent used. From our observation, using THF will not give any changes to current in MEH-PPV:CNTs nanocomposite. However, slight changes can be seen in samples using toluene. Electrical conductivity in S2 shows higher value than S1. Nevertheless, conductivity in S1 and S2 show increment in value as the concentration of CNTs is increased. In optical point of view, there are no peak shifting for both S1 and S2 but in S1 we can observed a better absorbance of light. S1 also showed efficient PL quenching in S1 compared to S2. We conclude that THF is suitable to be used in organic solar cells.

ACKNOWLEDGEMENT

This work was supported by Ministry of Higher Education (MOHE), Research Management Institute (RMI), UiTM (600-RMI/ST/DANA 5/3/Dst (399/2011) and MOSTI (e-Science grant: 06-01-01-SF0328) is grateful acknowledge. Also special thanks to all members of NANO-SciTech Centre.

REFERENCES

[1] A. R. Inigo, H. C. Chiu, W. Fann, Y. S. Huang, U. Jeng, C. Hsu, K. Y. Peng, and S. A. Chen, "Structure and charge transport properties in MEH-PPV," *Synthetic metals,* vol. 139, pp. 581-584, 2003.

[2] Y.-J. Choi, H.-H. Park, S. Golledge, and D. C. Johnson, "A study on the incorporation of ZnO nanoparticles into MEH-PPV based organic-inorganic hybrid solar cells," *Ceramics International,* vol. In Press, Accepted Manuscript, 2011.

[3] Satish Patil, Qianxi Lai, Filippo Marchioni, Miyong Jung, Zuhua Zhu, Y. Chen, and a. F. Wudl, "Dopant-configurable polymeric materials for electrically switchable devices," *Journal of Materials Chemistry,* vol. 16, pp. 4160–4164, 2006.

[4] H. Zhang, X. Lu, Y. Li, X. Ai, X. Zhang, and G. Yang, "Conformational transition of poly[2-methoxy-5-(2'-ethylhexoxy)-p-phenylene vinylene] in solutions: solvent-induced emitter change," *Journal of Photochemistry and Photobiology A: Chemistry,* vol. 147, pp. 15-23, 2002.

[5] D. Braun and A. J. Heeger, *Appl. Phys. Lett.,* 1991.

[6] I. Khatri and T. Soga, "Carbon Nanotubes Towards Polymer Solar Cell," *Carbon and Oxide Nanostructures,* pp. 101-123, 2011.

[7] S. Vedraine, P. Torchio, D. Duché, F. Flory, J.-J. Simon, J. Le Rouzo, and L. Escoubas, "Intrinsic absorption of plasmonic structures for organic solar cells," *Solar Energy Materials and Solar Cells,* vol. 95, pp. S57-S64, 2011.

[8] J. Zhao, X. Chen, and J. R. H. Xie, "Optical properties and photonic devices of doped carbon nanotubes," *Analytica Chimica Acta,* vol. 568, pp. 161-170, 2006.

[9] H. H. So, J. W. Cho, and N. G. Sahoo, "Effect of carbon nanotubes on mechanical and electrical properties of polyimide/carbon nanotubes nanocomposites," *European Polymer Journal,* vol. 43, pp. 3750-3756, 2007.

[10] L. Vaisman, H. D. Wagner, and G. Marom, "The role of surfactants in dispersion of carbon nanotubes," *Advances in Colloid and Interface Science,* vol. 128â€"130, pp. 37-46, 2006.

[11] PUTERI Sarah Mohamad Saad, MOHD Hazrin Zainal, Fazlinashatul Suhaidah Zahid, Zurita Zulkifli, Suriani Abu Bakar, and M. R. Mahmood, "Investigation on Annealed CNTs to the Electrical and Optical Properties of Nanocomposited MEH-PPV: CNTs Thin Film," *Advanced Materials Research* vol. 364, pp. 144-148, 2012.

[12] M. HARISH, "Processing and Study of Carbon Nanotube/ Polymer Nanocomposites and Polymer Electrolyte Materials," in *Department of Mechanical, Materials and Aerospace Engineering* Orlando, Florida University of Central Florida 2007.

[13] T.-Q. Nguyen, I. B. Martini, J. Liu, and B. J. Schwartz, "Controlling Interchain Interactions in Conjugated Polymers:â€‰ The Effects of Chain Morphology on Excitonâˆ'Exciton Annihilation and Aggregation in MEHâˆ'PPV Films," *The Journal of Physical Chemistry B,* vol. 104, pp. 237-255, 1999.

[14] J. Liu, Y. Shi, and Y. Yang, "Solvation-induced Morphology Effects on the Performance of Polymer-Based Photovoltaic Devices," *Advanced Functional Material,* vol. 11, 2001.

[15] E. Katz, D. Faiman, S. Tuladhar, J. Kroon, M. Wienk, T. Fromherz, F. Padinger, C. Brabec, and N. Sariciftci, "Temperature dependence for the photovoltaic device parameters of polymer-fullerene solar cells under operating conditions," *Journal of Applied Physics,* vol. 90, p. 5343, 2001.

[16] W.-C. Ou-Yang, T.-Y. Wu, and Y.-C. Lin, "Supramolecular Structure of Poly[2-methoxy-5- (2′-ethylhexyloxy) -1,4-phenylenevinylene] (MEH-PPV) Probed Using Wide-angle X-ray Diffraction and Photoluminescence," *Iranian Polymer Journal* vol. 18 pp. 453-464, 2009.

[17] S. Quan, F. Teng, Z. Xu, L. Qian, Y. Hou, Y. Wang, and X. Xu, "Solvent and concentration effects on fluorescence emission in MEH-PPV solution," *European Polymer Journal,* vol. 42, pp. 228-233, 2006.

[18] C. Ton-That, M. R. Phillips, and T.-P. Nguyen, "Blue shift in the luminescence spectra of MEH-PPV films containing ZnO nanoparticles," *Journal of Luminescence,* vol. 128, pp. 2031-2034, 2008.

[19] S. S. Kim, J. Jo, C. Chun, J. C. Hong, and D. Y. Kim, "Hybrid solar cells with ordered TiO2nanostructures and MEH-PPV," *Journal of Photochemistry and Photobiology A: Chemistry,* vol. 188, pp. 364-370, 2007.

Diameter dependence of spin relaxation in SiGe nanowires

Bhupesh Bishnoi, *Associate Member, IEEE,* Vikas Nandal and Bahniman Ghosh

Department of Electrical Engineering, Indian Institute of Technology Kanpur, Kanpur, India 208016
Email: bbishnoi@iitk.ac.in, bishnoi@ieee.org

Abstract- **In this paper, we use 1-D semi classical Monte Carlo Method to investigate spin polarized transport in SiGe nanowires (SiGeNWs) having Ge mole fraction 0.3. We use a multi-subbands semi classical Monte Carlo approach to model spin dephasing. Monte Carlo simulations have been widely adopted to study electron transport in devices and have recently been used in conjunction with spin density matrix calculations to model spin transport. Spin dephasing in SiGeNWs is caused due to D'yakonov-Perel (DP) relaxation and due to Elliott-Yafet (EY) relaxation. The spin polarization is studied along the length of the SiGeNWs. The ensemble averaged spin components variation has been studied for SiGeNWs along the nanowires length. The effect of variation of nanowires diameter on spin dephasing length has been studied. It is found that as the diameter of nanowires increases spin dephasing length also increases and become saturated beyond some value of diameter, this is due to the reduction in surface roughness scattering, which is a dominant contributor to the total scattering rate.**

I. INTRODUCTION

In recent past spintronics [1-2] has fascinated noteworthy concern from the researchers. This has resulted in a gradual increase in exploring the spin degree of freedom [3] as opposed to the charge degree of freedom. Semiconductor based spintronics can be used to integrate logic, communication and memory storage devices on a single chip thereby providing a multifunctional device [4]. Also spin, being a quantum operator, can be used for quantum computation [5]. The spintronics mechanism works on three basic principles, spin injection, spin transport and spin detection. SiGe offers the opportunity for more flexible band gap tuning than silicon technology by varying the Ge mole fraction in the alloy, the band gap can be very easily reduced as compared to Silicon and thus can provide many advantages over Si technology. Strained silicon is a layer of silicon in which the silicon atoms are stretched beyond their normal interatomic distance. This can be accomplished by putting the layer of silicon over a substrate of silicon germanium (SiGe).

II. MODEL

A full account of the Monte Carlo simulations and spin transport model is presented elsewhere [6-10]. In this paper we shall restrict ourselves to discussing only the necessary features of the model and the key modifications.

Fig.1 shows the geometry of the nanowires structure and the designation of the axes in accordance with the coordinate system chosen. The current flow is maintained by the application of a driving electric field Ex along the channel. Along with Ex, a transverse field is also present. The transverse field breaks the structural inversion symmetry which causes Rashba spin-orbit coupling [11-12]. The electron spin and momentum are coupled via spin-orbit interaction. The electron spin evolves under the influence of spin-orbit Hamiltonian which comprises of the Rashba interaction written as [13],

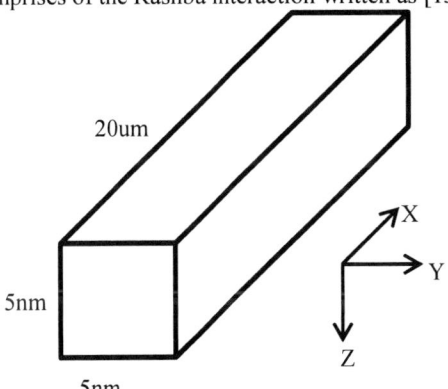

Fig.1 Geometry of the nanowires and the co-ordinate axes

$$H_R = -\eta k_x \sigma_y \tag{1}$$

Rashba coefficient η depends on the material and on the external transverse electric field and this dependence is explicit from the expression of η [14],

$$\eta = \frac{h}{2m^*} \frac{\Delta}{E_g} \frac{2E_g + \Delta}{(E_g + \Delta)(3E_g + 2\Delta)} e\vec{E} \tag{2}$$

Where Δ is the spin orbit splitting of the valence band, e is the electronic charge, m^* is the effective mass, E_g is the band gap and E is the transverse electric field.

The temporal evolution of the spin vector during the free flight time occurs in accordance with the following equation [14],

$$\frac{d\vec{S}}{dt} = \overline{\Omega} \times \vec{S} \tag{3}$$

The so-called "precession vector" $\overline{\Omega}$ has only the Rashba component due to the Rashba Hamiltonian, and can be written as,

$$\overline{\Omega}_R(k_x) = -\frac{2\eta k_x \hat{j}}{\hbar} \tag{4}$$

Using Eq. (3) and Eq. (4) and expressing spin vector

$$\vec{S} = S_x \hat{i} + S_y \hat{j} + S_z \hat{k} \tag{5}$$

We get the following relations for the individual components of spin,

$$\frac{dS_x}{dt} = -\frac{2\eta k_x S_z}{\hbar} \tag{6}$$

$$\frac{dS_y}{dt} = 0 \tag{7}$$

978-1-4673-2395-6/12 $31.00 © 2012 IEEE   61

$$\frac{dS_z}{dt} = \frac{2\eta k_x S_x}{\hbar} \qquad (8)$$

The scattering mechanisms taken into account in our simulations are acoustic phonon scattering, alloy scattering, ionized impurity scattering, optical phonon scattering and surface roughness scattering. The formulae for computation of scattering rates are taken from references [15, 16, 17, 18].

III. RESULTS

We have used the model described in the preceding section to simulate spin polarized electron transport in SiGeNWs. The nanowires structure considered has length 20um and varying cross-section from 4nm x 4nm to 12nm x 12nm. The transverse effective field is taken to be 100kV/cm. This effective field results in Rashba spin orbit coupling. Four subbands are considered in the simulations to account for the confinement along the two transverse directions. The higher subbands will be higher up in energy due to small transverse dimensions and hence for the purpose of the simulation they can be considered to be depopulated and thus are neglected. Moreover, moderate values of driving electric field (100kV/cm) are used in our simulation ensure that the majority of electrons are contained in the first four subbands. The energy levels of subbands are calculated using an infinite potential well approximation. The Rashba coefficient η is calculated using the Equation (2) for SiGe. The electrons are injected with a specific polarization from the source i.e. x=0. A time step of 0.02 fs was selected and the simulation run for 12 x 10^5 such time steps.

This allows the electrons to reach steady state. Data is recorded for the final 50,000 steps only. The ensemble average is calculated for each component of the spin vector for the last 50,000 steps at each point of the nanowires. The magnitude of the average spin vector is then computed using the expression,

$$|<S>(x,T)| = \sqrt{<S_x>^2 + <S_y>^2 + <S_z>^2} \qquad (9)$$

Spin dephasing length is defined as the distance from the source (x=0, from which the electrons are injected) where $|<S>|$ drops to 1/e times of its initial value of injection. In our simulations the electrons are injected with an initial polarization of 1 and hence the initial value of $|<S>|$ is 1.

TABLE I

E_z	100kV/cm
Dt	0.2×10^{-15} second
Nstep	2×10^5
Lx	20 μm
Nd	$2 \times 10^{25}/m^3$
T	300 k
Vs	0 V
Vd	2 V
Grid points	500
Sub-band	4
Valley	4

1. Decay of magnitude of ensemble averaged spin vector for SiGeNWs.

From the Figure 2, 3 and 4 we note that the decay of the magnitude of ensemble averaged spin vector along the length of SiGeNWs in X, Y, Z polarization direction with varying cross section of SiGe Nanowires.

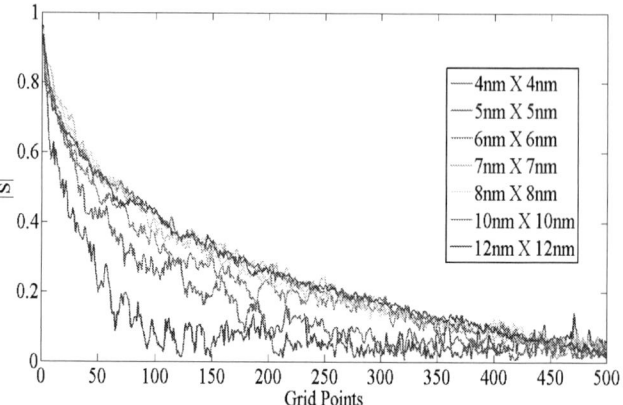

Fig.2. shows the decay of the magnitude of ensemble averaged spin along the length of SiGeNWs. Spin polarized electrons are injected along the X-direction

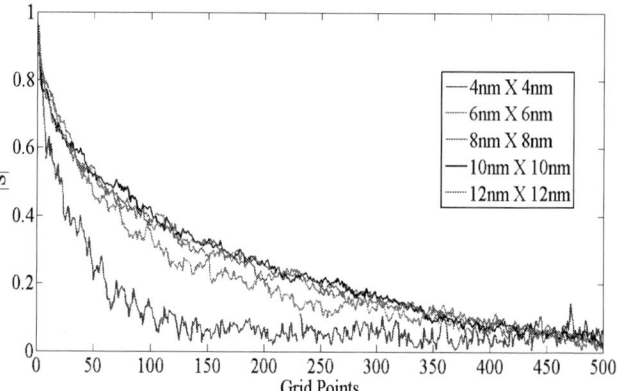

Fig.3. shows the decay of the magnitude of ensemble averaged spin along the length of SiGeNWs. Spin polarized electrons are injected along the Y-direction

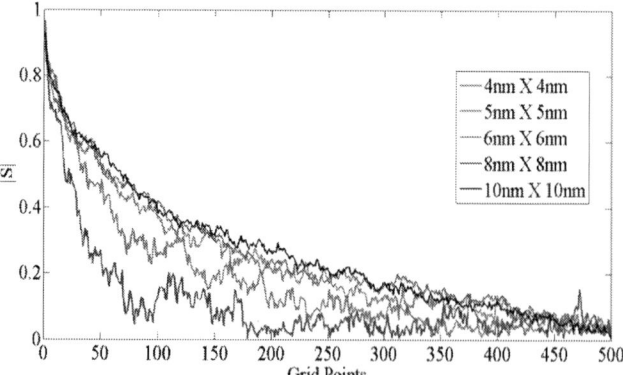

Fig.4. shows the decay of the magnitude of ensemble averaged spin along the length of SiGeNWs. Spin polarized electrons are injected along the Z-direction

2. Decay of components of spin in SiGeNWs

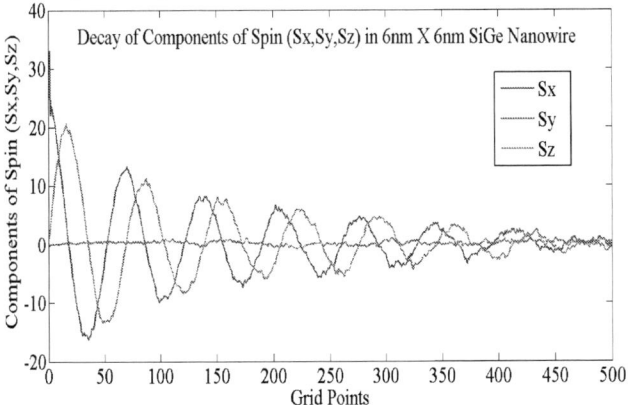

Fig.5. Decay of spin component along the length of SiGeNWs for injection polarization along the X-direction

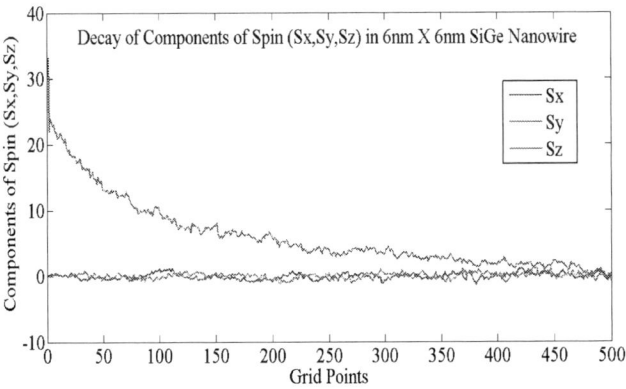

Fig.6. Decay of spin component along the length of SiGeNWs for injection polarization along the Y-direction

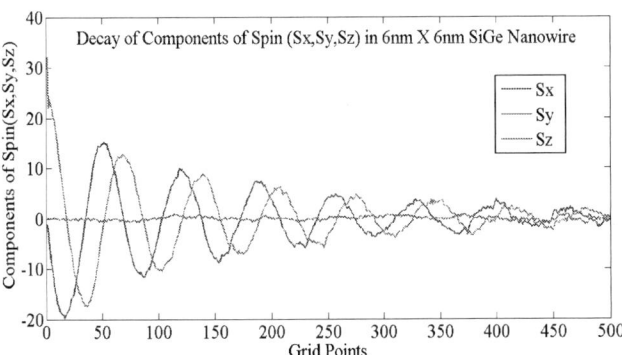

Fig.7. Decay of spin component along the length of SiGeNWs for injection polarization along the Z-direction

From the Figure 5, 6, 7 we note that the spatial decay of the spin components for SiGeNWs has an oscillatory component for the X, Y and Z polarized injection, respectively. Based upon the above reasoning, this leads us to the inference that the incoherent dynamics due to ensemble dephasing is weaker than the coherent dynamics in SiGeNWs. The coherent dynamics dominates because the spin precession vector is large due to large spin orbit coupling in SiGeNWs.

3. Variation of Spin dephasing length with diameter of SiGeNWs for different Spin injection

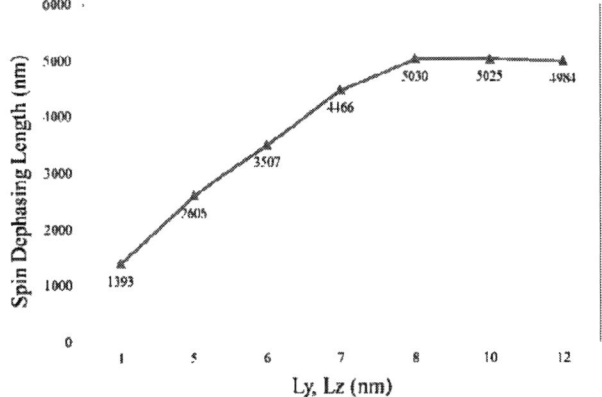

Fig.8. Variation of Spin Dephasing Length with Cross sectional area of SiGeNWs at 300K at a driving electric field of 100kV/cm with initial injection polarization along the X-direction

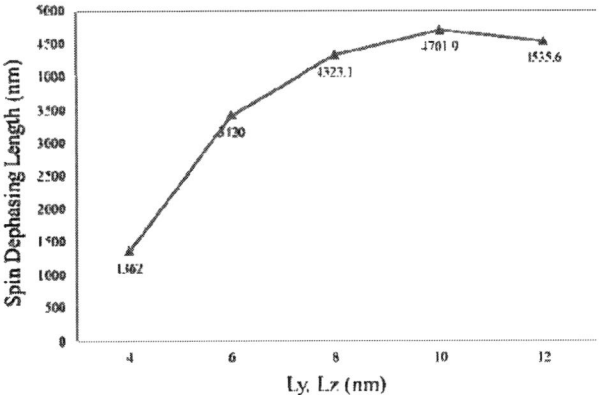

Fig.9. Variation of Spin Dephasing Length with Cross sectional area of SiGeNWs at 300K at a driving electric field of 100kV/cm with initial injection polarization along the Y-direction

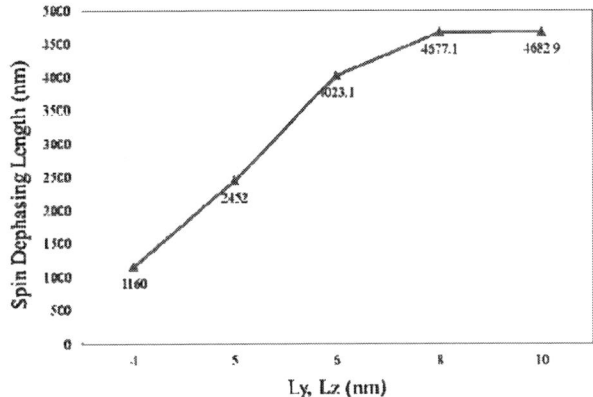

Fig.10. Variation of Spin Dephasing Length with Cross sectional area of SiGeNWs at 300K at a driving electric field of 100kV/cm with initial injection polarization along the Z-direction

IV. CONCLUSION

In this work, we studied spin dephasing in a nanowires. The study was conducted on SiGeNWs, in a bid to investigate their spin transport properties. The electrons were injected with initial polarization along the x, y, and z-direction separately. The decay of spin components is oscillatory for SiGeNWs. The effect of variation of nanowires diameter on spin relaxation length is also studied. From the Figure 8, 9 and 10 Spin Relaxation Length increases with cross sectional area from 4nm x 4nm to 8nm x 8nm because in this cross sectional area range SRS dominates as compared to Acoustic Phonon scattering and decrease monotonically hence Spin Relaxation Length increases. Spin Relaxation Length became constant with cross sectional area from 8nm X 8nm to 12nm X 12nm because in this cross sectional area range SRS became comparable to Acoustic Phonon scattering and hence Spin Relaxation Length became constant. Hence from spin transport view point SiGeNWs with cross sectional area of 10nm x 10nm is best suited as have maximum spin dephasing length and for longer distance electron can retain its spin information. We also found that ensemble averaged spin components and the steady state distribution of spin components vary with initial polarization direction.

REFERENCES

[1] Albert Fert, *"Thin Solid Films,"* 517(2008).

[2] S. Bandyopadhyay and M. Cahay, *Introduction to Spintronics* (CRC Press, March 2008).

[3] S. Das Sarma, *Am. Sci.* 89, 516(2001).

[4] X.F. Wang, P. Vasilopoulos, F.M. Peeters, "Spin-current modulation and Square-wave transmission through periodically stubbed electron Waveguides," *Phys. Rev. B* 65, 165217 (2002)

[5] R.G. Mani, W.B. Johnson, V. Narayanamurti, V. Privman, Y.H. Zhang, "Nuclear Spin Based Memory and Logic in Quantum Hall Semiconductor Nanostructures for Quantum Computing Applications," *Physica E 12*, 152 (2002).

[6] L.Kong, G.Du, Y.Wang, J.Kang,R. Han, X. Liu, Liu, "Simulation of Spin-polarized Transport in GaAs/GaAIAs Quantum Well Considering Intersubband Scattering by the Monte Carlo Method," *International Conference on Simulation of Semiconductor Processes and Devices*, 2005, pp.no.175-178, (2005).

[7] Pramanik, S., Bandyopadhyay, S., Cahay, M., "Spin transport in nanowires," *Third IEEE Conference on Nanotechnology*, pp.no.87-90 vol.2 (2003).

[8] S.Pramanik, S. Bandyopadhyay, M. Cahay, "Spin dephasing in quantum Wires," *Phys. Rev. B*, 68, 075313(2003).

[9] A. Bournel, P. Dollfus, P. Bruno, P. Hesto, "Spin polarized transport In1D and 2D semiconductor Heterostructures," *Materials Science Forum* (vol. 297 - 298), pp.no.205-212(1999).

[10] A.Kamra, B.Ghosh and T.K.Ghosh,"Spin relaxation due to electron-electron magnetic interaction in high Lande g-factor semiconductors," *J. Appl. Phys.* 108, 054505 (2010) .

[11] A.Kumar and B.Ghosh, (unpublished) arxiv: 1105.0173v1 [cond-mat.mes-hall] (2011) .

[12] Borzdov, A. V.; Pozdnyakov, D. V.; Borzdov, V. M.; Orlikovsky, A. A.; V'yurkov, V. V., "Effect of a transverse applied electric field on Electron drift velocity in a GaAs quantum wire: A Monte Carlo Simulation," *Russian Microelectronics* 39: 411-417 (2010).

[13] Y. A. Bychkov, E. I. Rashba, "Properties of a 2d electron gas with lifted Spectral degeneracy," *JETP Lett.* 39 (1984) 78–81.

[14] Charles Tahan and Robert Joynt, "Rashba spin-orbit coupling and spin Relaxation in silicon quantum wells," *Phys. Rev. B* 71, 075315 (2005).

[15] Ramayya, E. B., Vasileska, D., Goodnick, S. M. & Knezevic, "Electron Transport in silicon nanowires: The role of acoustic phonon confinement and surface roughness scattering," *J. Appl. Phys.* 104,063711 (2008).

[16] Lee,J., Spector,Harold N.,"Impurity-limited mobility of Semiconducting thin wire," *J. Appl. Phys.* 54, 3921.

[17] A.Kumar and B.Ghosh, "Spin Relaxation in Germanium Nanowires," (Unpublished) arxiv: 1106.4378 [cond-mat.mes-hall] (2011).

[18] Ben Yu-Kuang Hu and S. Das Sarma, "Self-consistent calculation of ionized impurity scattering in semiconductor quantum wires," *Phys. Rev. B* 48, 14388–14392 (1993).

Effects of PMMA concentration on PMMA-based organic capacitor behavior

L.N. Ismail[1]*, M. Khairizal[1], Z. Habibah[1], A.N. Arshad[2], M.H. Wahid[2], N.N. Hafizah[2], S.H. Herman[1] and M.Rusop[1]

[1]NANO-ElecTronic Centre (NET), Faculty of Electrical Engineering,
[2]Faculty of Applied Sciences,
Universiti Teknologi MARA (UiTM) Shah Alam
Selangor, MALAYSIA
E-mail*: lyly2909@gmail.com

Abstract— This paper reports the behavior of poly (methyl methacrylate) (PMMA) as insulator films. The PMMA concentrations were from 0.3, 0.6, 0.9 to 1.2 g. The electrical and dielectric properties results showed that PMMA concentration influenced the insulator behavior in the capacitor. The resistivity increased and the leakage current reduced from 10^{-2} A/cm^2 to 10^{-7} A/cm^2 when the PMMA concentrations were increased.

Keywords; PMMA, insulator, electrical properties, dielectric properties

I. INTRODUCTION

Poly (methyl methacrylate) (PMMA) is one of the most versatile polymeric materials that are being used as gate dielectric in many applications in micro electric and electro-optics areas. PMMA is chosen to be dielectric materials in organic thin film transistor (OTFT) and organic field effect transistor (OFET) because it has good insulation properties such as low dielectric constant and dielectric loss over a wide range of frequency [1, 2]. This polymer offers low costs, processability, possibility of functionalization. Its melting point is 160°C [3-5]. Besides having a low dielectric constant and dielectric loss, the surface morphology of dielectric also determine the device performance. The surface morphology has to be smooth and uniform because it affects the electronic transport properties which depends on the grain boundaries and defect located in the film [6-8].

Many researchers have characterized the properties of PMMA film, however the characterization only focusing on the optical [9], thermal [9, 10] and mechanical properties [9, 11]. J. H. Park et al. have characterized the dielectric properties of PMMA thin film by focusing on the effects of the polymer tacticity (isotactic, sydiotactic, and atatic) and they find that different type of tacticity give different dielectric properties. They have conclude that PMMA with isotactic have the best dielectric properties compared to other tacticity [12]. However, they have not discussed on the effect of the PMMA concentration to the electrical, dielectric and morphology properties of PMMA film. Therefore a systematic study of the electrical and dielectric behavior of PMMA with different concentration is needed.

In this paper, we report on the fabrication and characterization of metal-insulator-metal (MIM) capacitors using PMMA as the insulator. The effects of PMMA concentration on the electrical and dielectric properties of PMMA are investigated.

II. EXPERIMENTAL SETUP

PMMA powders (M_W of 120,000) and anhydrous solvent Toluene purchased from Sigma Aldrich were used to form the PMMA solution. Glass substrates were used and cleaned by standard process using acetone, methanol, de-ionized water (DI), in sequence, for 10 minutes, respectively and blown dry using nitrogen (N$_2$) gas.

The PMMA concentrations were varied from 0.3, 0.6, 0.9 to 1.2g were dissolved in toluene. The solutions were sonicated at temperature of 50 ^{O}C for 30 min before being stirred at room temperature for 24 hr. Aluminum (Al) with 60 nm thickness was deposited on the substrates by thermal evaporation to form a bottom contact. The PMMA solution was spin coated on Al/glass with the speed of 1500 rpm for 40sec. After the PMMA deposition, Al electrodes were evaporated as top contact (60nm thickness) through shadow mask to form the MIM capacitor having the structure of Al/PMMA/Al.

Device preparation and characterizations were done in normal ambient conditions. The electrical characterizations is done using metal-insulator-metal (MIM) configuration to evaluate the capacitance and leakage current densities of gate insulator as shown in Fig. 1. The electrical properties were measured using Keithley 2400 source meter. The film morphology was characterized using atomic force microscopy (AFM) (Park System XE-100). Film thickness measurements were performed by measuring the step height from the substrate using surface profiler (VEECO DEKTAK 150).

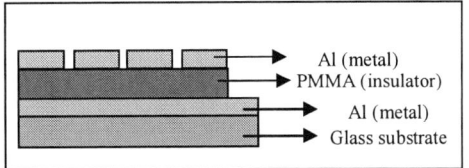

Fig. 1 Schematic diagram of MIM configuration

III.   RESULTS AND DISCUSSION

A. Electrical Characterization

Figure 2 show the thickness of PMMA thin film. The thickness of the samples is determined by averaging several measurements with 115 nm, 382 nm, 857 nm and 1162 nm for the sample of 0.3g, 0.6g, 0.9g and 1.2g respectively. The thickness is depending on the amount of PMMA concentration results in thicker films.

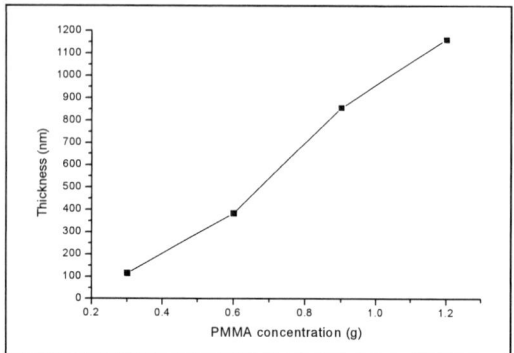

Fig. 2 Film thickness versus PMMA concentration

Figure 3 is shows the variation of the resistivity and conductivity for the PMMA thin film with respect to the different concentration of PMMA. Typically, for a material to be an insulator, it must have high resistivity ~ 10^{13} to 10^{15} Ωcm [4, 13] and low conductivity approximately 10^{-16} S/cm [14, 15]. The measured resistivity for all samples is approximately ~ 10^{6} Ωcm. It can be seen that the film with low concentration that is 0.3g of PMMA has the lowest resistivity and high electrical conductivity ~ 10^{-3} S/cm compare to other concentration. This indicates that the characteristics of PMMA thin film with a concentration of 0.3g are more to semiconductor which is ~ 10^{-4} S/cm [14] than an insulator. An abrupt drop can be seen in the conductivities from 10^{-3} to 10^{-8} S/cm when the concentration is increased. This value is nearly consistent with the magnitude of insulator even though it is much lower as mentioned earlier.

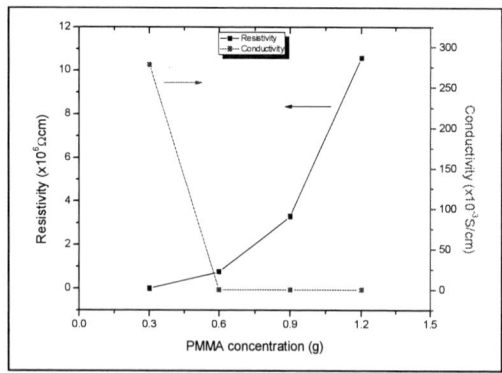

Fig.3 Resistivity and conductivity of PMMA film with different concentration

The current characteristics of the capacitors are shown in Fig. 4, 5, and 6. Figure 4 show the capacitor with PMMA concentration of 0.3g, Fig. 5 for 0.6g and Fig. 6 show the capacitor having 0.9 and 1.2 g of PMMA. The voltage applied to the Al electrode was swept from -10 to 10V while the other Al electrode was grounded. The measured current density for all samples is in the range of 10^{-2} to 10^{-7} A/cm^2 at the applied voltage 10V. As shown in Fig. 3, the film with 0.3g of PMMA with the thickness approximately 115 nm thick exhibited the highest leakage current about 10^{-2} A/cm^2 at the applied voltage of 5V. As the PMMA concentration increased, the leakage current was reduced by five orders of the magnitude at the same applied voltage. This is due to the increment of the film thickness. The leakage current density indicate that PMMA film with 0.3g concentration is not a suitable insulator to be used as dielectric in OFET or OTFT devices because it has high leakage current even though it has a very thin film. For thin film sample, the thickness become dependent and the electrical properties will progressively decrease. This is parallel with the finding by Paul et al where they have suggested that this must be due to the percolation phenomenon where at high concentration, the film experience a distinctive transition from high conductivity to a low conductivity. When the percolation threshold is reached, the transition from semiconductor to an insulator are occure wherein conductivity is reduce from 10^{-3} S/cm to 10^{-8} S/cm. They also conclude that when the thickness of the thin films is reduced, the electrical properties become thickness dependent [16].

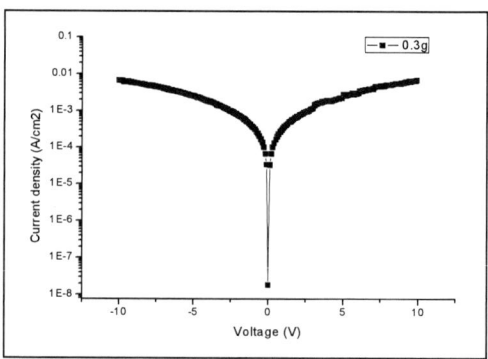

Fig.4 Leakage current density of PMMA thin film with 0.3g concentration

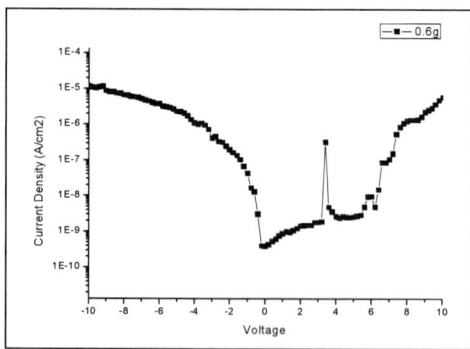

Fig. 5 Leakage current density of PMMA thin film with 0.6g concentration

978-1-4673-2395-6/12 $31.00 © 2012 IEEE

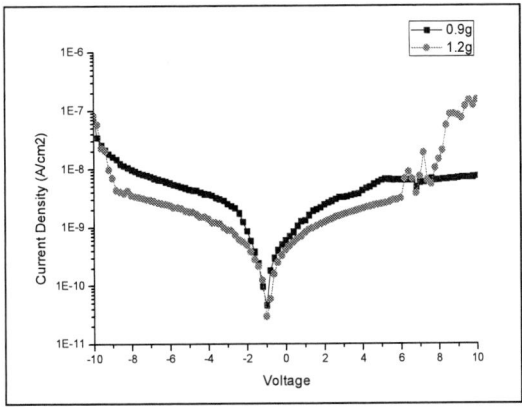

Fig. 6 Leakage current density of PMMA thin film with 0.9 and 1.2g concentration

Basically, dielectric constant of an organic material usually depends on the frequency and the temperature [13]. Figure 7 show the dielectric constant, k versus frequency for PMMA thin film. The inset (a) and (b) in Fig. 7 illustrate the dielectric constant versus PMMA concentration measured at frequency of 1 kHz and 1 MHz. It can be seen that the k values at frequency of 1 kHz for 0.3, 0.6, 0.9 and 1.2g are 3.9, 10.7, 4.5 and 25 respectively. The k values drastically reduced at higher frequency of 1 MHz. PMMA film with 0.3 g concentration give the lowest k value that is 0.6, meanwhile the k for 0.6, 0.9 and 1.2g are 2.9, 2.6 and 2.7, respectively. This value is similar that has been reported by as reported by the several researcher that is 2.6 ~ 3 [17-19]. Overall, the film with concentration of 0.6, 0.9 and 1.2g give the higher dielectric constant compare to 0.3g and it is suitable to be used as insulator. This is because high dielectric constant enhances the capacitor performance which can be seen from eq. 1.

$$C = \varepsilon_o \varepsilon_r \qquad (1)$$

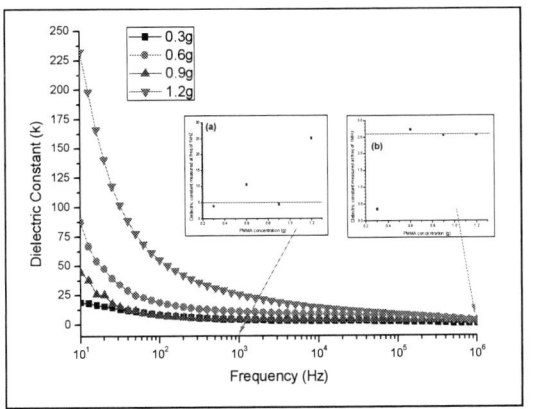

Fig. 7 Dielectric constant versus frequency with different amount of PMMA

The variation in capacitance density as a function of applied frequency is shown in Fig. 8. The measured maximum capacitance values at frequency of 10 Hz are 0.25, 0.98, 0.47 and 1.36µF/cm^2 for 0.3, 0.6, 0.9 and 1.2 g concentration, respectively. When the frequency increased to 1 kHz, the capacitance values are lower for all concentration that is 2.76 nF/cm^2, 1.38 nF/cm^2, 421 pF/cm^2 and 1.75 nF/cm^2 for 0.3, 0.6, 0.9 and 1.2 g respectively. At higher frequency (1 MHz), the capacitances drastically reduce and there are no significant changes between all samples. The capacitances measured at 1 MHz for 0.3, 0.6, 0.9 and 1.2g are 2.79, 3.91, 2.68 and 1.99 nF/cm^2 respectively.

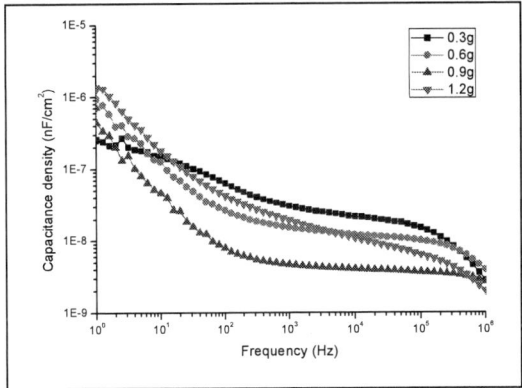

Fig. 8 Capacitance density of PMMA MIM with different concentration

The electrical behavior of the PMMA films with different concentration was analyzed by means of impedance plots in which the impedance Z is shown in complex plane with the reactance (Z") plotted against the resistance (Z'). Figure 9 shows the impedance curves for PMMA films with different PMMA concentration. It can be seen that the semicircles curve are form when the PMMA concentration are increased from 0.6 to 1.2g. As the concentration increased from 0.6g until 1.2g, the radii of semicircles became smaller indicating that the resistance decreased. However, the semicircle curve cannot be seen at 0.3g of PMMA due to higher impedances at lower frequency.

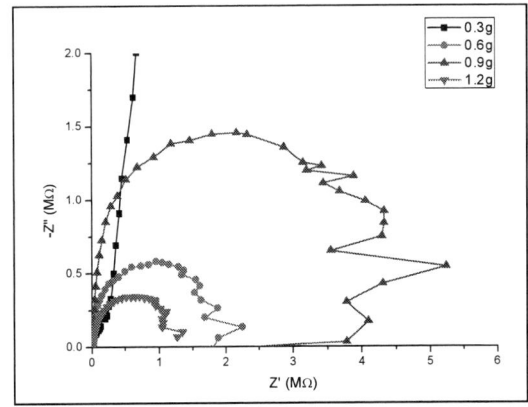

Fig.9 Impedance spectroscopy curves of PMMA film with different concentration

978-1-4673-2395-6/12 $31.00 © 2012 IEEE

IV. CONCLUSION

The behaviors of PMMA film with different concentration as insulator in a MIM capacitor have been characterized. PMMA film with 0.3g of PMMA shows poor insulator behavior because it has high leakage current and low resistivity compare to other concentration. PMMA film with 0.9g and 1.2g of PMMA show good insulator behavior. As summary, the electrical and dielectric properties results indicate that as the concentration is increased the PMMA thin film showed better insulator behavior.

ACKNOWLEDGMENT

The authors would like to acknowledge NANO-SciTech Centre (NST) and Microwave Technology Centre (MTC), UiTM Shah Alam for their support using their AFM and thermal evaporator equipment.

REFERENCES

[1] D. Dorranian, Z. Abedini, A. Hojabri, and M. Ghoranneviss, "Structural and Optical Characterization of PMMA Surface Treated in Low Power Nitrogen and Oxygen RF Plasmas," *Journal of Non-Oxide Glasses,* vol. 1, pp. 217-229, 2009.

[2] H. M. Ahmed and S. A. B. Aziz, "Dielectric Properties of Commercial non-Polar Polymers," *Journal of Zankoy Sulaimani,* vol. 11, pp. 1-8, 2008.

[3] X. Sun, X. Chen, X. Liu, and S. Qu, "Optical properties of poly(methyl methacrylate)–titania nanostructure thin films containing ellipsoid-shaped titania nanoparticles from ex-situ sol-gel method at low growth temperature," *Applied Physics B: Lasers and Optics,* pp. 1-8, 2010.

[4] R. R. Prakash, S. Pandiarajan, P. Venkatesh, and N. Kamaraj, "Performance analysis of PMMA - TiO2 nanocomposite dielectrics," in *Emerging Trends in Electrical and Computer Technology (ICETECT), 2011 International Conference on,* 2011, pp. 46-49.

[5] L. N. Ismail, F. A. Ahmad, H. Zulkefle, M. M. Zaihidi, M. H. Abdullah, S. H. Herman, U. M. Noor, B. Mahmood, and M. Rusop, "Optical Properties and Surface Morphology of PMMA: TiO2 Nanocomposite Thin Films," *Advanced Materials Research,* vol. 364, pp. 105-109, 2012.

[6] D. Knipp, P. Kumar, A. R. Völkel, and R. A. Street, "Influence of organic gate dielectrics on the performance of pentacene thin film transistors," *Synthetic Metals,* vol. 155, pp. 485-489, 2005.

[7] M. Mukherjee, B. Mukherjee, Y. Choi, K. Sim, J. Do, and S. Pyo, "Investigation of organic n-type field-effect transistor performance on the polymeric gate dielectrics," *Synthetic Metals,* vol. 160, pp. 504-509, 2010.

[8] C. Feng, T. Mei, and X. Hu, "Influence of trapping states at the dielectric-dielectric interface on the stability of organic field-effect transistors with bilayer gate dielectric," *Organic Electronics,* vol. 12, pp. 1304-1313, 2010.

[9] M. E. Nicho, C. H. García-Escobar, M. C. Arenas, P. Altuzar-Coello, R. Cruz-Silva, and M. Güizado-Rodríguez, "Influence of P3HT concentration on morphological, optical and electrical properties of P3HT/PS and P3HT/PMMA binary blends," *Materials Science and Engineering: B,* vol. In Press, Corrected Proof, 2011.

[10] X. Xu, H. Ming, P. Wang, Z. Liang, and Q. Zhang, "Multi-photon-absorption-induced birefringent grating in azobenzene-doped polymethyl methacrylate optical fibres," *Journal of Optics A: Pure and Applied Optics,* vol. 4, p. L5, 2002.

[11] F.-A. Zhang, D.-K. Lee, and T. J. Pinnavaia, "PMMA/mesoporous silica nanocomposites: effect of framework structure and pore size on thermomechanical properties," *Polymer Chemistry,* vol. 1, pp. 107-113, 2010.

[12] J. H. Park, D. K. Hwang, J. Lee, S. Im, and E. Kim, "Studies on poly (methyl methacrylate) dielectric layer for field effect transistor: Influence of polymer tacticity," *Thin Solid Films,* vol. 515, pp. 4041-4044, 2007.

[13] B. C. Shekar, J. Lee, and S.-W. Rhee, "Organic Thin Film Transistors: Materials, Processes and Devices," *Korean Journal Chemical Engineering,* vol. 21, pp. 267-285, 2004.

[14] K. Xu, D. Erricolo, M. Dutta, and M. A. Stroscio, "Electrical conductivity and dielectric properties of PMMA/graphite nanoplatelet ensembles," *Superlattices and Microstructures,* vol. 51, pp. 606-612, 2012.

[15] W. Zheng and S.-C. Wong, "Electrical conductivity and dielectric properties of PMMA/expanded graphite composites," *Composites Science and Technology,* vol. 63, pp. 225-235, 2003.

[16] P. J. King, T. M. Higgins, S. De, N. Nicoloso, and J. N. Coleman, "Percolation Effects in Supercapacitors with Thin, Transparent Carbon Nanotube Electrodes," *ACS Nano,* vol. 6, pp. 1732-1741, 2012.

[17] I. Mejia and M. Estrada, "Characterization of Polymethyl Methacrylate (PMMA) Layers for OTFTs Gate Dielectric," vol. 6th, pp. 375 - 377, 26 - 28 2006.

[18] T. S. Huang, Y. K. Su, and P. C. Wang, "Poly(methyl methacrylate) Dielectric Material Applied in Organic Thin Film Transistors," *Japanese Journal of Applied Physics,* vol. 47, pp. 3185 - 3188, 2008.

[19] M. Estrada, I. Mejia, A. Cerdeira, and B. I iguez, "MIS polymeric structures and OTFTs using PMMA on P3HT layers," *Solid-State Electronics,* vol. 52, pp. 53-59, 2008.

978-1-4673-2395-6/12 $31.00 © 2012 IEEE

Influence of Electron-Electron Scattering on Spin Relaxation Length in Single and Bilayer graphene

Bhupesh Bishnoi, *Associate Member, IEEE,* Dharmendra Hiranandani, Akshay Salimath,
Vikas Nandal and Bahniman Ghosh

Department of Electrical Engineering, Indian Institute of Technology Kanpur, Kanpur, India 208016
Email: bbishnoi@iitk.ac.in, bishnoi@ieee.org

Abstract- **Theoretical study of influence of electron-electron scattering on spin relaxation length in single layer graphene (SLG) and bilayer graphene (BLG) is done using ensemble semi classical Monte Carlo simulation. The comparison is made by including electron-electron interactions with electron-phonon interactions (acoustic and optical). The D'yakonov-Perel (DP) and Elliot-Yafet (EY) spin relaxation mechanisms are utilized in the Monte Carlo routines. The results are simulated with varying temperatures to show that e-e scattering holds significant importance as a scattering mechanism at low temperatures and gradually loses its importance in as we reach room temperature and above. We report highly contrasting effect of e-e scattering on SLG and BLG.**

Index Terms- **Single Layer and Bilayer Graphene, Monte Carlo Simulation, Scattering, Spin Relaxation**

I. INTRODUCTION

Graphene was experimentally discovered in 2004 by *Novoselov et al.* [1], since then there has been an enormous increase in interest in graphene related devices due to the many outstanding properties that the two dimensional sheet of carbon atoms displays. *Geim and MacDonald* [2] describe it as the "Ultimate Flatland". Its 2D honeycomb lattice is the most perfect of 2D electronic materials available in nature. It is exactly one atomic layer thick and the *sp2* hybridized carbon atoms are separated by a bond length of 1.42 Å and when stacked in multiple layers the interplanar distance is 3.35 Å [3, 4]. Graphene shows exceptionally high mobilities [5] and its bilayer has shown tremendous spin relaxation length [6, 7]. These properties make graphene a possible candidate for future use in applications like superconductors and spin-transistors [8].

The π orbitals between neighboring carbon atoms in the graphene sheet overlap and can be described by a tight-binding Hamiltonian [9]. The corners of the in equivalent dirac [10] points are located at the corners of the hexagonal Brillouin zone. The valance and conduction band at these points and the Fermi level passes through these and the band gap is zero [11]. The energy dispersion relationship for single layer graphene is that of mass less Dirac fermions [12, 13],

$$E = \hbar v_f |k| \qquad v_f = \frac{10^6 \, m}{s} \qquad (1)$$

Whereas on the other hand, bilayer graphene possess a band gap that is tunable using external electric field [14] and its

energy-dispersion follows a parabolic relation, with a very low effective mass [15,16].

$$E = \frac{\hbar^2 k^2}{2m_{eff}} \qquad (2)$$

The effective mass used in our model is $\sim 0.03 m_e$ [17]

One promising area of graphene's application is Spintronics. Long spin relaxation lengths which are in order of few micrometers in bilayer graphene [18] have been realized experimentally. Though spin relaxation length for SLG are much less than BLG, they are significantly large compared to the general magnetic materials used for spintronics devices presently [19]. Hence this has led to active pursuit of theoretical study of spin transport in graphene. Bahniman et al. have previously made use of semi classical Monte Carlo simulations to study graphene and other materials relevant for spintronics application [20, 21, 22]. However, it has been shown experimentally that spin current doesn't remain unaffected by carrier-carrier collision [23]. Hence it becomes crucial to consider the effects of e-e scattering, which are currently ignored and remain unaccounted for, on spin relaxation length in graphene and its bilayer.

In conventional charge carriers the ensemble velocity is conserved upon an event of e-e collision. It is a mere redistribution of momentum and is no direct consequence to current flow. This is because the charge quantity (-e) is quantized as same for both the electrons involved in collision. Unlike current, the particles undergoing collision can have two values (↑ and ↓) so that net spin of system need not be same after e-e collision [24]. Therefore, inter-carrier collisions deserve careful consideration in the determination of the spin transport properties.

II. ELECTRON ELECTRON SCATTERING

Goodnick and Lugli [25] discuss the inclusion of e-e scattering while considering the transport phenomenon in 3D systems. We make use of their model and modify it to use it for 2D systems such as single layer and bilayer graphene. Since calculation of e-e scattering rate for an electron requires summation over carrier distribution of all other electrons possessing anti-parallel spin the process is computationally almost impossible. As proposed by them, the maximum rate is instead utilized to determine the occurrence of electron-electron scattering and the actual value of calculated only when the process is selected. The maximum value of e-e scattering is calculated using,

$$\Gamma[7]_{max} = \frac{4\pi^2 e^4 m^* N_s}{\hbar^3 (\mathcal{E}_r \mathcal{E}_o)^2} \frac{1}{q_o^2} \qquad (3)$$

Where Ns is the total sheet density of electrons and εr is the dielectric constant for graphene and if EF denotes the Fermi energy of graphene, $q0$ is given by [26, 27]

$$q_o = \frac{e^2 m}{2\pi \mathcal{E}_r \hbar^2} \left\{ \exp\left(-\frac{E_f(T)}{k_b T} \right) - 1 \right\}^{-1} \qquad (4)$$

When e-e scattering is selected, a partner electron is chosen from ensemble by the approach which slightly differs from *Mosko-Moskova*. The methods they suggest are understandable in the light of wave nature considerations of electron wave function [28], which specify no information in coordinate space. But since we are working with a semiclassical model, there is a definite coordinate value associated with every electron. We aim to utilize this information and make the choice of the partner electron on the basis of least distance to the scattered electron. The scattering angle is chosen at random and the actual value of integrand is calculated. A random number is chosen between zero and maximum value, i.e. $\Gamma[7]_{max}$, the e-e scattering is rejected if the integrand is less than this number. In the case of rejection of e-e scattering, the electron and its partner are allowed to continue with their flights unaltered.

III. SIMULATION RESULTS

Jacobani et al [29, 30] give comprehensive review of application of Monte Carlo technique for modeling spin transport. Important assumption made here is that all electron interactions are confined to the lowest subband. The scattering mechanisms employed are acoustic and optical phonon scattering. The detailed description of model is provided in the previous works of *Bahniman et al.* [20, 21]. The model is incapable to perform at ballistic transport levels as we study diffusive transport only with Monte Carlo method. Hence, the highest mobility that we consider in our simulations is nearly $2m^2 V^{-1} S^{-1}$. Further works on ballistic hot transport are provided in [24].

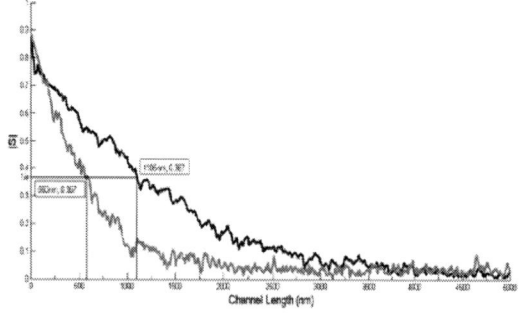

Fig.1. Spin relaxation length with (grey line) and without (blackline) e-e scattering for single layer grapheme at temperature 4K

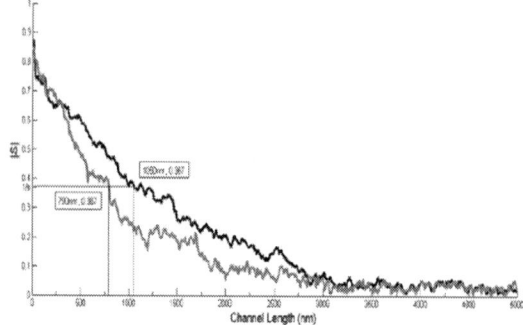

Fig.2. Spin relaxation length with (grey line) and without (black line) e-e scattering for single layer grapheme at temperature 77K

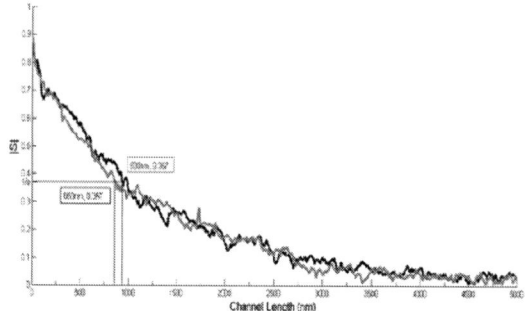

Fig.3. Spin relaxation length with (grey line) and without (black line) e-e scattering for single layer grapheme at temperature 200K

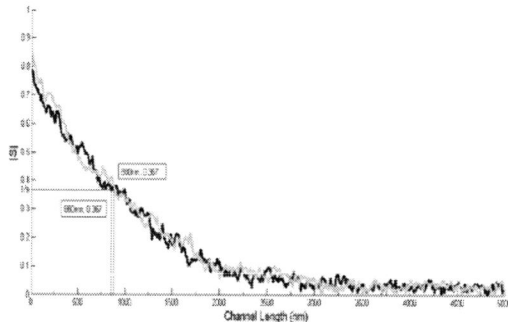

Fig.4. Spin relaxation length with (grey line) and without (black line) e-e scattering for single layer grapheme at temperature 300K

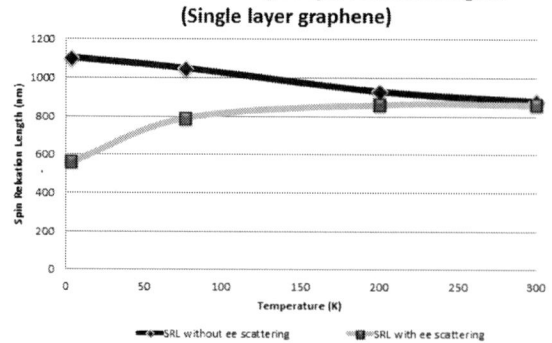

Fig.5. The decreasing effect of e-e scattering with temperature. Spin relaxation length with (grey line) and without (black line) e-e scattering for single layer graphene

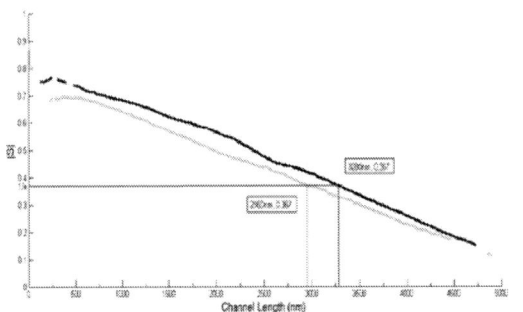

Fig.6. Spin relaxation length with (grey line) and without (black line) e-e scattering for bilayer graphene at temperature 4K

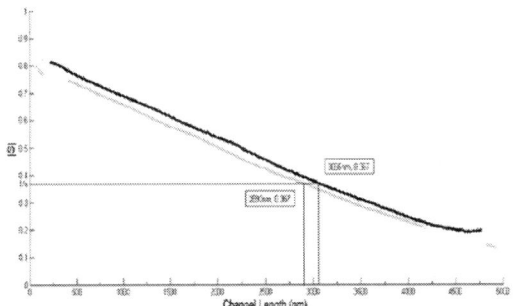

Fig.7. Spin relaxation length with (grey line) and without (black line) e-e scattering for bilayer grapheme at temperature 77K

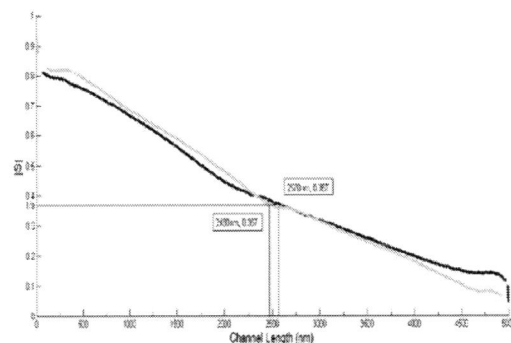

Fig.8. Spin relaxation length with (grey line) and without (black line) e-e scattering for bilayer grapheme at temperature 300K

Fig.9. The low effect of e-e scattering on spin relaxation length in bilayer graphene.Spin relaxation length with (grey line) and without (black line) e-e scattering for single layer grapheme

The simulation was run for single layer and bilayer graphene at varying temperatures. The spin relaxation lengths obtained are plotted. Figure 1-4 show the spin relaxation length with and without e-e scattering for temperatures 4K, 77K, 200K and 300K respectively for single layer graphene. Fig 5 shows the decreasing effect of e-e scattering with temperature. The effect of e-e scattering nearly dies out at300K. Thus showing it is operative at only low temperatures. Inclusion of e-e scattering thus becomes crucial for any accurate model spintronics in graphene. Fig 6-8 show spin relaxation length for temperatures 4K, 77K and 300K for bilayer graphene and Fig 9 shows the low effect of e-e scattering in bilayer graphene.

One intriguing result is indeed the comparatively lesser effect of e-e scattering on s pin relaxation in graphene and its bilayer. Whereas in the former it is responsible for a fall of around 45% in spin relaxation length at 4K, the latter saw a drop of just 10% at the same temperature. This can be due to two operations under process here. One that other scattering mechanisms operational in bilayer layer graphene are relatively stronger then e-e scattering or bilayer graphene is more robust towards carrier induced scattering. Understanding the exact reason of this phenomenon can pave new ways in discovering other novel properties and applications of bilayer graphene. Further theoretical and experimental research is needed to be done in this area.

IV. CONCLUSION

To summarize our simulation results, unlike mobility, the e- e scattering affects the spin relaxation length and a general drop is seen as e-e scattering is incorporated in the simulation model. But the drop in spin relaxation length is more prominent for single layer graphene than bilayer graphene, pointing towards its more affinity towards carrier scattering. Qualitatively, it should be noted that e-e scattering needs to be incorporated in any theoretical model of spintronics devices based on graphene so as to make it predict accurate results.

REFERENCES

[1]   K. S. Novoselov, A. K. Geim, S. V. Morozov, D. Jiang, Y. Zhang, S.V.Dubonos, I. V. Grigorieva, and A. A. Firsov. *Science*306, 666 (2004).
[2]   Geim, A. K., and A. H. MacDonald, *Phys. Today*60, 35 (2007)
[3]   RajiHeyrovska, arXiv:0804.4086 (2008)
[4]   C D Reddy, S Rajendran and K M Liew. *Nanotechnology* 17 864 (2006)
[5]   K.I. Bolotin, K.J. Sikes, Z. Jiang, M. Klima, G. Fudenberg, J. Hone, P.Kim,H.L. Stormer. *Solid State Commun.*146, 351 (2008)
[6]   W. Han, K. Pi, W. Bao, K. M. McCreary, Y. Li, W. H. Wang, C. N. Lau and R. K. Kawakami. *Appl. Phys. Lett.* 94, 222109 (2009)
[7]   N.Tombros, C.Jozsa, M. Popinciuc, H.T. Jonkman & B. J. van Wees,*Nature*448, 571-574 (2007)
[8]   S. A. Wolf, D. D. Awschalom, R. A. Buhrman, J. M. Daughton, S. von Molnar, M. L. Roukes, A. Y. Chtchelkanova, and D. M. Treger. *Science*294, 1488–1495 (2001)
[9]   E. H. Hwang, Ben Yu-Kuang Hu and S. Das Sarma. *Phys. Rev. B*76,115434 (2007)
[10] Y. Zhang, Y.W.Tan, H.L. Stormer &P. Kim. *Nature* 438, 201 (2005)
[11] A.H. Castro Neto, F. Guinea, N.M.R. Peres, K.S. Novoselov, and A. K.Geim, *Rev. Mod. Phys.*81, 109 (2009)
[12] C. Jacoboni and L. Reggiani, *Rev. Mod. Phys.*55, 645 (1983).
[13] P.R.Wallace.*Phys.Rev.*71,622-634(1947)

[14] Y. Zhang, T. Tang, C.Girit, Z.Hao, M.C. Martin, A. Zettl, M.F.Crommie, Y. R. Shen&F.Wang. *Nature* 459, 820-823 (2009)

[15] K. S. Novoselov, E. McCann,S. V. Morozov, V. I. Fal'ko, M. I.Katsnelson, U. Zeitler, D. Jiang, F. Schedin & A. K. Geim. *NaturePhysics* 2, 177-180 (2006)

[16] E. McCann and V. I. Fal'ko*Phys. Rev. Lett.* 96, 086805 (2006)

[17] J. Nilsson, A. H. Castro Neto, F. Guinea, and N. M. R. Peres. *Phys. Rev.Lett.* 97, 266801 (2006)

[18] M. Shiraishi, M. Ohishi, R. Nouchi, T. Nozaki, T. Shinjo, and Y.Suzuki,*Adv. Funct. Mater.* 19, 3711 (2009).

[19] N. Tombros, S. Tanabe, A. Veligura, C. Jozsa, M. Popinciuc, H. T.Jonkman, and B. J. Van Wees. *Phys. Rev. Lett.* 101, 046601 (2008)

[20] B. Ghosh and S. Misra.J. Appl. Phys. 110, 043711 (2011)

[21] B. Ghosh. *J. Appl. Phys.* 109, 013706 (2011)

[22] A. Kumar, M. W. Akram, and B. Ghosh. *AIP Advances* 2, 012165 (2012)

[23] C. P. Weber, N. Gedik, J. E. Moore, J. Orenstein, J. Stephens & D. D.Awschalom. *Nature* 437, 1330(2005)

[24] A. Kamra andB. Ghosh, *Journal of Applied Physics*109, 024501 (2011)

[25] S.M. Goodnick and P.Lugli. *Phys. Rev. B* 37, 2578–2588 (1988)

[26] T. Ando, A. B. Fowler and F. Stern, *Rev. Mod. Phys.* 54, 437 (1982)

[27] F. Stern. *Heterojunctions and Semiconductor Superlattices.* (Springer-Verlag, Berlin, 1986)

[28] M. Moko and A. Moková. *Phys. Rev. B* 44, 10794 (1991) [29] C. Jacoboni and P. Lugli, *The Monte Carlo Method for SemiconductorDevice Simulation* (Springer-Verlag, October 1989)

[30] C. Jacoboni and L. Reggiani, *Rev. Mod. Phys.*55, 645 (1983).

Properties of ZnO/MgO Multilayer Films as Insulating Layer Prepared by Sol-Gel Method

Z. Habibah[a,*], A. N. Arshad[b], M. H. Wahid[b], L. N. Ismail[a], R. A. Bakar[a], M. H. Mamat[a], M. Rusop [a,c,*]

[a]NANO-ElecTronic Centre (NET), Faculty of Electrical Engineering;
[b]Faculty of Applied Sciences;
[c]NANO-SciTech Centre (NST), Institute of Science;
Universiti Teknologi MARA, UiTM 40450 Shah Alam, Selangor, Malaysia

*Email: habibahzulkefle@yahoo.com, nanouitm@gmail.com

Abstract—**This work present the synthesis of ZnO/MgO multilayer films using spin coating technique with different MgO layer thickness which are 171nm, 238nm and 506nm. The influence of MgO thickness on the insulating layer properties was investigated. The best prepared insulating layer to be used as dielectrics was found to be ZnO/MgO film deposited using 238 nm MgO layer. This is due to it uniformity, low porosity, high resistivity (28.7 x 10^5 Ω.cm) and low leakage current (below 1E-7). The particle produced also in nanometer dimension which in the range of 42 to 84 nm that will lead to the improvement in the device characteristic.**

Keywords- ZnO/MgO; insulating layer; dielectric; leakage current; sol-gel

I. INTRODUCTION

Magnesium oxide, MgO has attracted attention in many applications such as in plasma display panel, optoelectronic devices and sensors [1, 2] . This is due to its excellent properties such as high temperature stability, chemical inertness, optical transparency, and high thermal conductivity [3]. Furthermore, MgO is the best candidate to be used as insulating layer material due to its low heat capacity and high melting point [4]. In addition, it also has high dielectric constant (~9.8), wide band gap (~7.8 eV), high breakdown field (12MV/cm) and low leakage current density [5-7].

Moreover, due to the excellent properties of Zinc oxide, ZnO such as large exciton binding energy of 60 meV at room temperature and wide bandgap [8, 9] making ZnO desirable to be used in many applications. As stated by N. Tjitra Salim et.al due to chemical stability and optical transparency of ZnO it could be considered to be used as dielectric [10]. Moreover, without introducing oxygen vacancies and doping materials ZnO has high resistivity which is up to $10^5\Omega$.cm [10, 11].

Recently deposition film in multilayer form has attracted much attention in device fabrication due to its ability to alter films properties [12, 13] and also enhance the device performance [14]. Multilayer film can be deposited using several techniques such as plasma enhance chemical vapor deposition, RF magnetron sputtering, hydrothermal method and sol-gel spin coating method. However, among these methods, spin coating has its own advantages which are ease of processing, cost effective, large area coating and stoichiometric control [15, 16].

Although many researches had been done on the deposition of ZnO/MgO multilayer films, the effect of the thickness on its insulating properties has not been explored. Therefore, this research is carried out to investigate the effect of different layer thickness of MgO on the electrical and structural properties of prepared ZnO/MgO insulating films via sol-gel method.

II. EXPERIMENTAL PROCEDURE

A. Material Used for Preparing Zinc Oxide, ZnO and Magnesium Oxide, MgO Solutions

Zinc acetate dehydrate, 2-methoxylethanol, mono-ethanolamine, magnesium acetate tetrahydrate, ethanol and nitric acid

B. Synthesis and Characterizations of ZnO/MgO Multilayer films

For zinc oxide solution preparation, zinc acetate tetrahydrate was dissolved in 2-methoxylethanol with addition of monoethanolamine while for magnesium oxide solution, magnesium acetate and nitric acid were dissolved in ethanol solution. Then both ZnO and MgO solutions was sonicated at 50°C for 30 and 20 minutes respectively. The solutions were then stirred and heat for 3 hours at 80°C. Aging process was then performed at room temperature for 24 hours.

Deposition of MgO films on ZnO layer film was conducted using spin coating method with rotating speed of 3200 rpm for 30 seconds. Then the deposited multilayer films were dried at 200°C for 10 minutes and these processes was repeated until the desired thickness was achieved. Finally the films were annealed at 500°C for 1 hour.

Electrical and structural characterizations were performed using two point probes I-V measurement (BUKOH KEIKI), surface profiler (VEECO), atomic force microscopy (AFM-Park System XE 100) and field emission scanning electron microscope (FESEM- JSM- J600F) respectively. The summary of the overall synthesis and characterizations process of ZnO/MgO multilayer films was shown in Fig.1.

978-1-4673-2395-6/12 $31.00 © 2012 IEEE

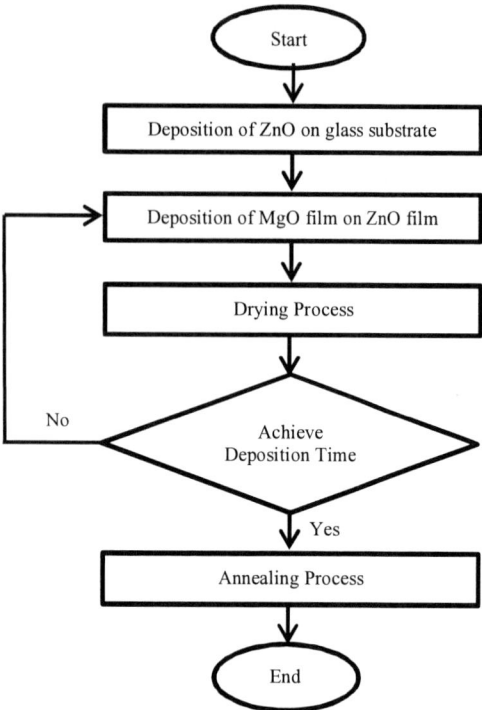

Fig. 1. Synthesis of ZnO/MgO multilayer film via sol-gel method.

III. RESULTS AND DISCUSSION

A. Effect on Electric Properties

Electrical behaviors of deposited ZnO/MgO multilayer films were investigated in terms of its current voltage (*I-V*) characteristic and leakage current density. *I-V* characteristic in Fig. 2 shows that the current is linearly increased with the applied voltage and from the slope it can be seen that the resistance of the multilayer film varied with the change of MgO film thickness.

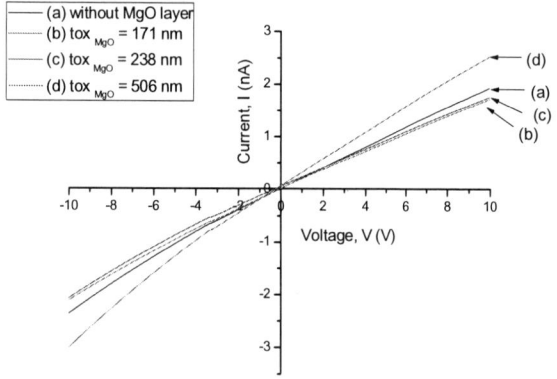

Fig. 2. Current voltage, *I-V* characteristics of deposited ZnO/MgO multilayer films.

The resistivity, ρ of multilayer films was then calculated using the following equation [17]:

$$\rho = \frac{wt}{l} R \qquad (1)$$

where R is the film resistance, w and l are the width and length of the metal contact respectively and t is the films thickness. The resistivity value of the deposited multilayer films is presented in Fig. 3. From the graph, it can be said that the MgO thickness has positive correlation with the multilayer films resistivity. This can be proven by Eqn. 1 where the resistivity is directly proportional to the thickness. Moreover, the plotted graph also shows increment in resistivity as the film is in multilayer form where the resistivity value increased from 8.72×10^4 Ω.cm up to $44.7 \times 10^4 \Omega$.cm for single layer ZnO and multilayer ZnO/MgO films respectively. This result is supported with the observation done by K. Ellmer et al. where their film resistivity increased up to three orders of magnitude when the film is in multilayer form [18].

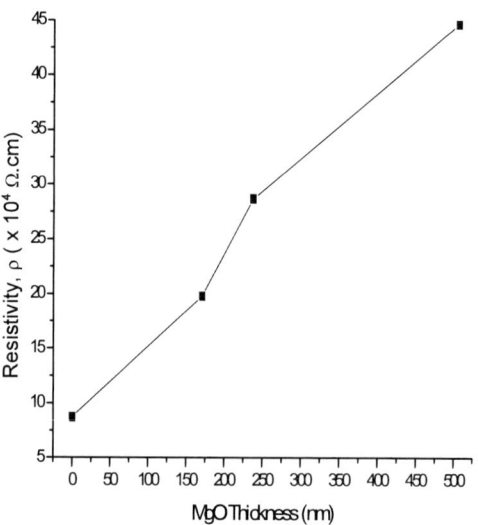

Fig. 3. Resistivity distribution for ZnO/MgO multilayer films at different MgO layer thickness.

In addition, the rise in resistivity might also due to the agglomerated particle and large grain boundary produced in the prepared multilayer films (Fig. 5) that will reduced the carrier mobility due to the carrier scattering [19].

Leakage current density, J of the deposited ZnO/MgO multilayer films was presented in Fig. 4. It can be seen that, the rise in MgO thickness resulting an increase in leakage current. The same observation has been reported by Somnath et al. where the thickness give negative effect to the leakage current in their MIM capacitor [20]. The leakage current density observed in this work was below 10^{-7} A.cm^{-2} at voltage 5V.

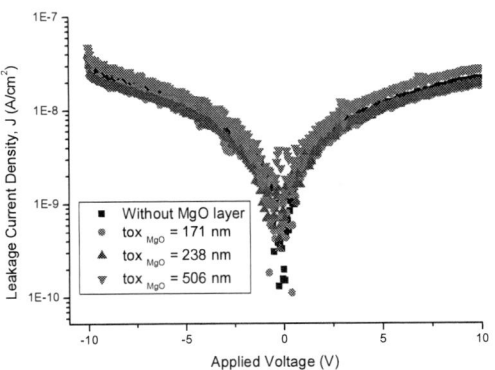

Fig. 4. *J-V* characteristics for different MgO thickness of multilayer films.

Rising in leakage current with the thickness might due to the non-uniform film produced at thicker thickness. This was supported by its surface topology and morphology shown in Fig. 5 where at thicker film, the large grain boundary produced will act as short path for the electron to flow from top to the bottom electrode. Additionally, as stated by Kun Ho Ahn et al. , the dependence of the leakage current with thickness is also due to the depletion region formed at the top and bottom electrode interfaces [21].

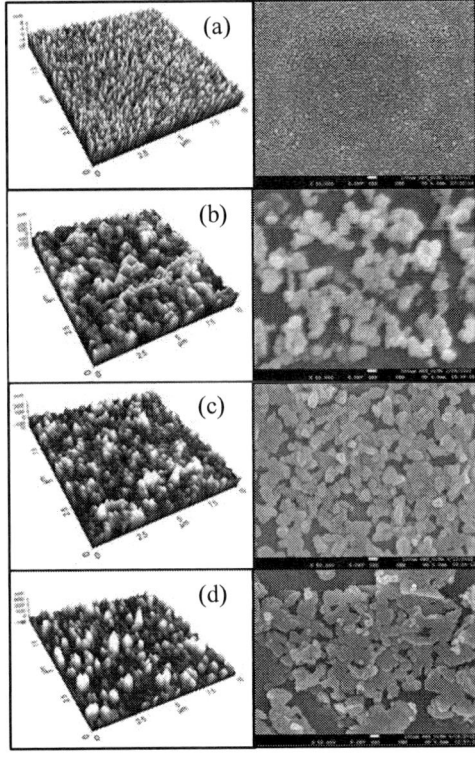

Fig. 5. Surface topography and morphology for (a) without MgO and multilayer film with various MgO thicknesses: (b) 171nm, (c) 238nm and (d) 506nm.

B. *Effects on Surface Topography and Morphology*

Surface topography of the deposited multilayer film observed by AFM (Park System-XE 100) respectively was shown in Fig. 5 and it revealed that film in ZnO/MgO multilayer form rougher surface compared to ZnO single layer. As reported by K. Kato et al. , the formation of rougher surfaces in their multilayer films was due to the high melting point of MgO layer [22]. Moreover as tabulated in Table 1, increased in MgO thickness was resulted in the formation of rougher films surface. Multilayer films prepared with MgO thickness of 238nm has denser, uniform and smaller roughness compared to MgO thickness of 171nm and 506nm.

Fig. 5 also shows surface morphology of ZnO/MgO multilayer films using FESEM (JSM-J600F) at 50k magnification. From the images, it can be observed that as the MgO film thickness increased, its causes changing in film growth mechanism and grain size. When the MgO thickness is 171nm, the grain boundary is clear and the grains are round in shape. Therefore it can be said that the growth mechanism at 171 nm thickness is in vertical growth [9].

However, at 238nm thickness of MgO layer, the film produced has smaller roughness compared to other prepared multilayers. This was due to the change in growth mechanism from vertical to lateral growth that contributes to the production of denser film [23]. This was proven by its AFM image (Fig. 5c) and the roughness value that was tabulated in Table 1. As further increased in MgO film thickness, lead to the formation of agglomerated particle with large grain boundary.

TABLE I. SURFACE ROUGHNESS OF DEPOSITED ZnO/MgO MULTILAYER FILMS

Thickness of MgO Layer Films (nm)	Average Surface Roughness, Ra (nm)	Root Mean Square Surface Roughness, Rq (nm)
0	1.484	1.878
171	44.029	56.322
238	32.779	41.710
506	38.020	50.011

One of the important criteria for an insulating material to be used as dielectrics film, the films must be good in it structural properties which including small surface roughness and less in porosity. Besides, it also should have less thickness and small particle size in order to improve the films properties that can lead to enhancement in the device performance.

From the surface morphology obtained, it illustrates that ZnO/MgO mutilayer film with 238 nm MgO thickness has smallest particle size which in range of 42 to 84 nm (Fig. 6). In addition, its grain boundary also the smallest compared to others.

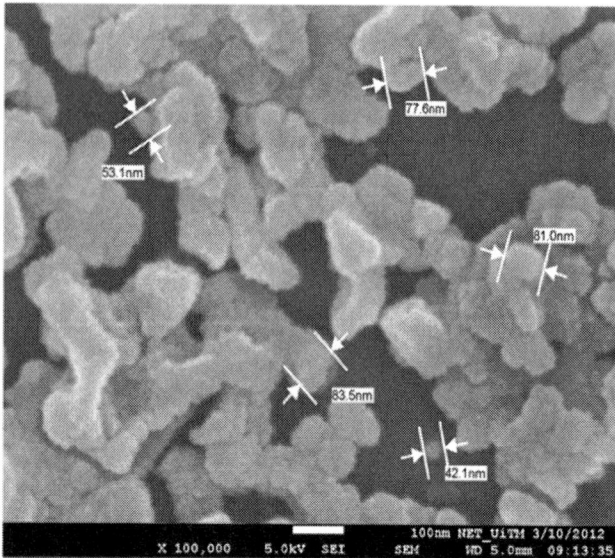

Fig. 6. FESEM image of deposited multilayer film with 238 nm MgO thickness.

IV. Conclusions

Based on the electrical and structural properties of deposited ZnO/MgO multilayer films obtained, it can be concluded that film with 238 nm MgO layer is the best prepared insulating film to be used as dielectric layer due to its high resistivity, low leakage current, uniformity, less porosity and the particles size is in nanometer dimension.

Acknowledgment

The authors would like to thanks Ministry of Higher Education, MOHE and Research Management Institute of Universiti Teknologi MARA for providing financial support under grant 600-RMI/ST/DANA 5/3/Dst (401/2011).

References

[1] N. Yasui, H. Nomura, and A. Ide-Ektessabi, "Characteristics of ion beam modified magnesium oxide films," *Thin Solid Films,* vol. 447-448, pp. 377-382, 2004.

[2] H. S. Jung, J.-K. Lee, J.-Y. Kim, and K. S. Hong, "Crystallization behaviors of nanosized MgO particles from magnesium alkoxides," *Journal of Colloid and Interface Science,* vol. 259, pp. 127-132, 2003.

[3] D.-Y. S. a. K.-N. Kim, "Electrical and optical properties of MgO films deposited on soda lime glass by a sol-gel process using magnesium acetate," *Journal of Ceramic Processing Research* vol. 10, pp. 536-540, 2009.

[4] A. M. E. Raj, M. Jayachandran, and C. Sanjeeviraja, "Fabrication techniques and material properties of dielectric MgO thin films--A status review," *CIRP Journal of Manufacturing Science and Technology,* vol. 2, pp. 92-113, 2010.

[5] H.-Y. Tsao, Y.-J. Lin, Y.-H. Chen, and H.-C. Chang, "Leakage currents through In/MgO/n-type Si/In structures," *Solid State Communications,* vol. 151, pp. 693-696, 2011.

[6] P. P. C. Bondoux, P. Belleville, F. Guillet, R. Jerisian, "Development of sol-gel MgO thin films for SiC insulation applications," *Material Science Forum,* vol. 457-460, pp. 1373-1376, 2004.

[7] G. H. P. Casey, E. O'Connor, R. D. Long, P. K. Hurley, "Growth and characterization of thin MgO layers on Si (100) surfaces," *Journal of Applied Physics: Conference Series,* 2008.

[8] J. Sengupta, R. K. Sahoo, K. K. Bardhan, and C. D. Mukherjee, "Influence of annealing temperature on the structural, topographical and optical properties of sol-gel derived ZnO thin films," *Materials Letters,* vol. 65, pp. 2572-2574, 2011.

[9] L. Xu, X. Li, Y. Chen, and F. Xu, "Structural and optical properties of ZnO thin films prepared by solâ€'gel method with different thickness," *Applied Surface Science,* vol. 257, pp. 4031-4037.

[10] K. C. A. N. Tjitra Salim, W. Gao, Z. W. Li, B. Wright, "ZnO as a dielectric for organic thin film transistor-based non-volatile memory," *Microelectronic Engineering,* vol. 86, pp. 2127-2131, 2009.

[11] M. Sahal, B. Hartiti, A. Ridah, M. Mollar, and B. MarÃ-, "Structural, electrical and optical properties of ZnO thin films deposited by solâ€'gel method," *Microelectronics Journal,* vol. 39, pp. 1425-1428, 2008.

[12] A. Kaushal and D. Kaur, "Pulsed laser deposition of transparent ZnO/MgO multilayers," *Journal of Alloys and Compounds,* vol. 509, pp. 200-205, 2011.

[13] Z.-y. Wang, L.-z. Hu, J. Zhao, H.-q. Zhang, and Z.-j. Wang, "The fabrication of ZnO/MgO multilayer films on Si (1 1 1) by PLD," *Vacuum,* vol. 80, pp. 977-980, 2006.

[14] J. Y. Cho, S. W. Shin, Y. B. Kwon, H.-K. Lee, K. U. Sim, H. S. Kim, J.-H. Moon, and J. H. Kim, "The effect of Mg0.1Zn0.9O layer thickness on optical band gap of ZnO/Mg0.1Zn0.9O nano-scale multilayer thin films prepared by pulsed laser deposition method," *Thin Solid Films,* vol. 519, pp. 4282-4285.

[15] Z. Jiwei, Z. Liangying, and Y. Xi, "The dielectric properties and optical propagation loss of c-axis oriented ZnO thin films deposited by sol–gel process," *Ceramics International,* vol. 26, pp. 883-885, 2000.

[16] Y. Li, L. Xu, X. Li, X. Shen, and A. Wang, "Effect of aging time of ZnO sol on the structural and optical properties of ZnO thin films prepared by sol–gel method," *Applied Surface Science,* vol. 256, pp. 4543-4547.

[17] D. K. Schroder, *Semiconductor material and device characterization* 3ed. New Jersey: John Wiley 2005.

[18] G. V. Klaus Ellmer, "Electrical transport parameters of heavily-doped zinc oxide and zinc magnesium oxide single and multilayer films heteroepitaxially grown on oxide single crystals " *Thin Solid Films,* vol. 496, pp. 104-111, 2006.

[19] J. Li, J.-H. Huang, W.-J. Song, Y.-L. Zhang, R.-Q. Tan, and Y. Yang, "Effects of post-annealing temperature on structural, optical, and electrical properties of $Mg_xZn_{1-x}O$ films by RF magnetron sputtering," *Journal of Crystal Growth,* vol. 314, pp. 136-140, 2011.

[20] T.-M. P. Somnath Mondal, "High-performance $Ni/Lu_2O_3/TaN$ metal-insulator-metal capacitors," *IEEE Device Letters,* vol. 32, pp. 1576-1578, 2011.

[21] S. S. K. Kun Ho Ahn, Sunggi Baik, "Thickness dependence of leakage current behavior in epitaxial $(Ba,Sr)TiO_3$ film capacitors," *Journal of Applied Physics,* vol. 93, pp. 1725-1729, 2003.

[22] K. Kato, H. Omoto, A. Takamatsu, and T. Tomioka, "Crystal growth of MgO thin films deposited on ZnO underlayers by magnetron sputtering," *Journal of Crystal Growth,* vol. 333, pp. 59-65.

[23] M. Sharma and R. M. Mehra, "Effect of thickness on structural, electrical, optical and magnetic properties of Co and Al doped ZnO films deposited by sol–gel route," *Applied Surface Science,* vol. 255, pp. 2527-2532, 2008.

AFM Images of Undoped Amorphous Carbon Thin Films Deposited by Bias-assisted Thermal-CVD

A. Ishak[1], M. Amirul, and M. Rusop[2]

NANO - ElecTronic Centre (NET), Faculty of Electrical Engineering,
Universiti Teknologi MARA (UiTM),
40450 Shah Alam, Selangor, Malaysia
Email: [1] ishakannuar@yahoo.com, [2] nanouitm@gmail.com

Abstract—**The undoped of amorphous carbon thin films were deposited by bias assisted thermal-CVD system at various deposition temperatures in the range 300°C to 500°C with fixed negative bias of -40V for 3 h deposition. The thin films were characterized by Atomic Force Microscopy (AFM), surface profiler, and I-V measurement. The results showed that the distributions of undoped thin films were more density when the temperature is increased at 400°C and 500°C for 3 h deposition times. The resistivity of undoped a-C thin film at 300°C, 350°C, 400°C and 500°C is 9.57x10^6 Ω/cm, 9.44x10^6 Ω/cm, 9.81x10^5 Ω/cm and 337738.124 Ω/cm respectively. The conductivity of thin films was increased by the increasing of temperature. The AFM images showed that, density and uniformity had correlated with the resistivity and conductivity of undoped amorphous carbon thin films for various temperatures.**

Keywords—*Amorphous carbon; Solar cell; Thermal-CVD; Negative bias; DC bias*

I. INTRODUCTION

For a decade, there is a trend to adapt solar cell from an alternative material that are abundantly in nature, low cost and non-toxic instead of using remarkably prominent silicon [1,2]. Accordingly, allotropes carbons as reported will be promised as a potential candidate for an alternative material in the future due to the abundantly in nature, suitability as a precursor, excellent photoconductivity, high optical absorption of visible light, and can be deposited on any inexpensive substrate. In addition, it is possible to form a very wide area of solar cell since it can be deposited directly from a kind of vapor phase growth onto noncrystalline substrate [3].

Carbon is found of having a wide band gap which is suitable for optical devices as well as for photovoltaic solar cells application. The band gap of some allotrope carbon such as amorphous carbon in the range of 0.1 eV to 5.5 eV could be tuned tailoring with energy band gap of photon by deposition process [4]. The unique ability to tune its energy band gap has offered a wide ability to absorb more photon energy from spectrum of light. However, there is a serious barrier to understanding of the physic of amorphous materials due to no equivalent reliable rule for theoretical analysis [5].

Meanwhile, the undoped amorphous carbon in contrast is reported as weakly p-type, complex structure and high density of intrinsic defect [6].Until now, various standard deposition techniques such as thermal-CVD, plasma enhanced chemical vapor deposition (PECVD), hot wire CVD, radio frequency CVD, ion beam sputtering (IBS), microwave surface-wave plasma (MW-SWP) CVD, filtered cathodic vacuum arc (FCVA) etc. are practically used [7,8]. It was reported that, dc bias would improve the deposition process by acceleration deposition rate, reduce contamination or maintain repeatable process as were done in many standards CVD [9,10].

In negative DC bias, electric field developed around the substrate was significantly influences the energy with which ions impinges upon growing film. The negatively substrate repels any negatively ions in chamber from being reach the surface [11]. By this technique, contamination might reject from being reach the substrate through repulsion ion [12]. Due to those advantages; a proposed to apply dc bias to support the process will be embarked.

II. EXPERIMENTAL DETAILS

we deposited the undoped amorphous carbon thin film on insulator glass substrates by using bias assisted thermal CVD at different temperatures. A fixed of negative bias - 40 V is set by 3 h of deposition times. Fig. 1 shows the schematic diagramm of bias assisted thermal CVD for deposition of amorphous carbon thin film.

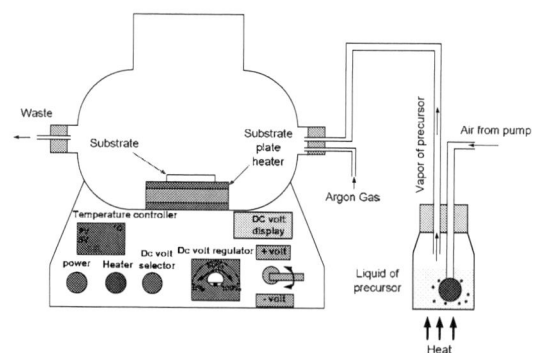

Figure 1: Schematic diagram of bias assisted thermal-CVD system

The glass substrates were cleaned with acetone (C_5H_6O) for remove the contaminated glass followed by methanol (CH_3OH) for remove acetone solution and deionized water for remove methanol solution for 15 min respectively in Ultrasonic Cleaner (power Sonic 405).

The cleaned of glass substrates are then placed in the chamber for 3 hours deposition. The chamber is heated with temperature starting from 300°C to 500°C and setup with a fixed dc bias of -40°C. A vaporized of ethanol precursor is pressure into the chamber by the air pump and heated at 50°C with the hot platter (Stuart CB162). The argon gas is used for carry the deposited charge and also contaminated gas.

The samples are then characterized by using atomic force microscope (AFM, XE-100 Park Systems) for structural properties, I-V measurement (Bukuh Keiki EP-200) for electrical properties and surface profiler (Vekko Dektak 150) for thickness of the thin film.

III. RESULTS AND DISCUSION

A. Surface morphology (2-dimensional imaging)

The surface morphology of amorphous carbon thin film deposited on glass substrates were characterized by atomic force microscopy (AFM, XE-100 Park Systems), as shown in Fig. 2.

Fig. 2 (a) shows that a very less of the deposited thin film on insulator glass is present. However, after the temperature increased from 350°C to 500°C, the particles are slowly increased by showing a biggest particle size at 400°C. However, the sizes of particle are decreased at temperature 500°C but more uniform and density compared with low temperatures as shown Fig. 2 (c) and (e).

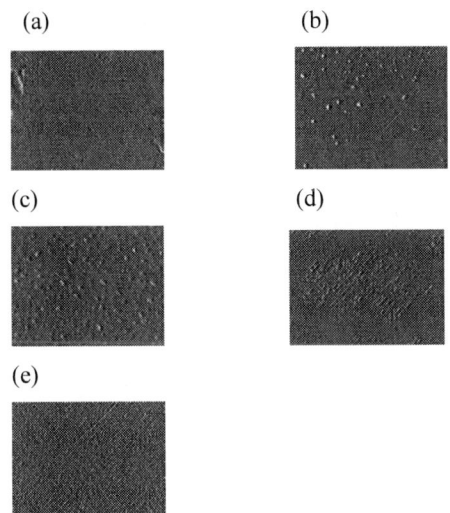

(a)   (b)

(c)   (d)

(e)

Figure 2: Two-dimensional view of the tapping mode image analysis from the top of the glass surface at different temperatures,-: (a) 300°C, (b) 350°C, (c) 400°C, (d) 450°C, (e) 500°C.

B. Line profile from section analysis

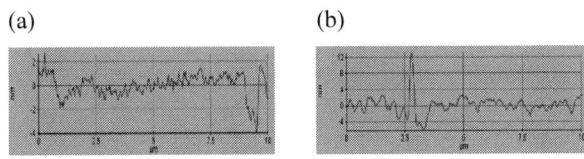

(a)   (b)

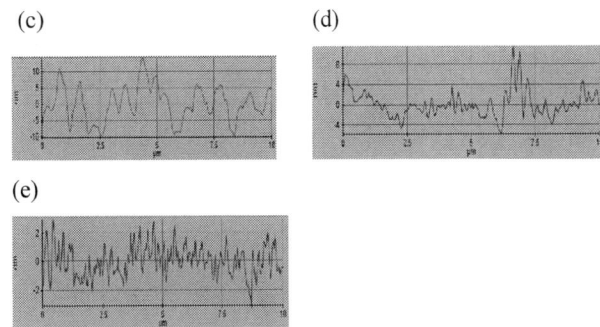

(c)   (d)

(e)

Figure 3: Line profile of amorphous carbon thin film deposited on glass substrates from sectional view analysis at different temperatures: (a) 300°C, (b) 350°C, (c) 400°C, (d) 450°C, (e) 500°C.

AFM cross-sectional analysis showed that the amorphous carbon thin films were uniformed in size and growth pattern, giving rise to a much rougher amorphous carbon as shown in Fig. (c) and (e). These results are correlated with the resistivity and conductivity of thin film measured by I-V measurement as shown in Fig. 7 and Fig. 8 and supported by two-dimensional view of surface morphology as shown in Fig. (c) and (e).

Fig. 3-(a), (b), and (d) show the nonuniform of thin film deposited on the insulator glass substrates. The low density and uniformity of the thin films compared with Fig. (c) and (d) contribute to the high resistivity and low conductivity of thin films as shown in Fig. 7 and Fig. 8.

C. Surface roughness via AFM

Table 1 shows that the surface roughness properties were analysed using AFM (XE-100, Park System). In Table 1, at temperature 400°C and 300°C, the surface has the highest and lowest roughness value of 10.328 nm 1.983 nm respectively. The surface roughness of thin film is increased from 300°C until 400°C and slowly dropped starting from 450°C to 500°C.

The surface roughness average is 2.783 nm at 500°C, 5.949 nm at 450°C, and 3.296 nm at 350°C. From this result, we can conclude that, the surface roughness increased until reached the optimum temperature of 400°C and drastically dropped when temperature at 450°C to 500°C. Nevertheless, the surface roughness did not effect on the resistivity and conductivity since the resistivity is decreased as the deposition temperature is increased as shown in Fig. 7 and Fig. 8 respectively.

Table 1: AFM Surface roughness properties of amorphous carbon thin film on glass substrates

Temperature (°C)	Root mean square Roughness, Rq (nm)	Surface Roughness Average, Ra (nm)
300 °C	2.821	1.983
350 °C	6.265	3.2963
400 °C	12.603	10.004
450 °C	8.503	5.949
500 °C	3.49	2.783

D. Surface topology from 3-dimensional AFM image analysis.

Figure 4-(a) to (e) show the topology images at which the amorphous carbon thin film deposited at different temperatures. It was shown that in Figure 3 (a), thin film do not properly deposited at lower temperature which might be due to less energy to form thin film. The thin film is showing an increasing of density of thin film at the temperature of 350°C to 400°C and dropped again from 450°C to 500°C. However, at 450°C to 500°C, the sharp of the thin films have smaller in size and but more density compared with 400°C. In conclusion, the results show that changing the temperature affects the uniformity and thickness of the amorphous carbon thin film.

(a)   (b)

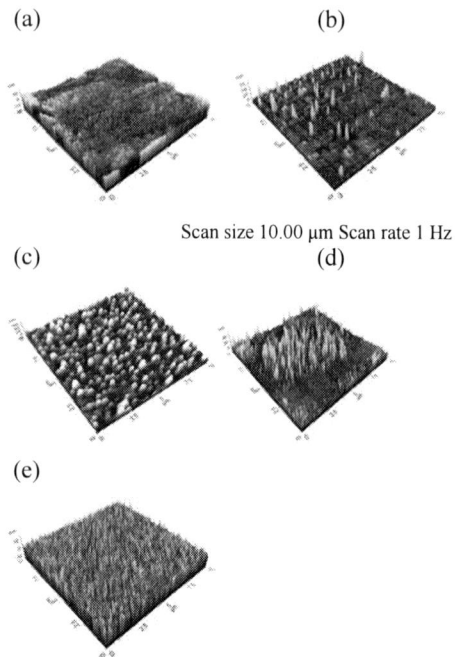

Scan size 10.00 μm Scan rate 1 Hz

(c)   (d)

(e)

Figure 4 Representative AFM 3-dimensional image surveys of amorphous carbon thin film grown on insulator glass substrates for 3 hours using negative bias of -40V at different temperatures: (a) 300°C, (b) 350°C, (c) 400°C, (d) 450°C (e) 500°C.

E. Surface profiler

Table 2 shows the relationship between the changing of temperature and thickness of amorphous carbon thin film characterized by surface profiler (Vekko Dektak 150). From Table 2, the highest thickness is found at 300°C by 230.78 nm and the lowest is 153.3 nm at 450°C. It can be concluded that by increasing the temperature, the thickness will be decreased until certain temperature at 450°C. However, at 500°C, the trend of decreasing thickness is suddenly changed.

Table 2: The thickness of undoped amorphous carbon thin film deposited at various temperature.

Temperature (°C)	Thickness (nm)
300 °C	230.78
350 °C	198.30
400 °C	172.75
450 °C	153.30
500 °C	220.54

F. Electrical Properties

The electrical properties of the undoped amorphous carbon thin films were characterized by using current-voltage (I-V) measurement whereas gold is used as a metal contact. Figure 6 shows the amorphous carbon thin film and gold metal are in ohmic contact. The value of resistivity and conductivity were calculated from using equation in (1) and (2) [13]. The resistance R is obtained from the average of resistance measure from -2 V to 2 V. the variable A is the area of the mask electrode, and L is the length between two measured mask electrodes as shown in figure 5.

$$\rho = \frac{RA}{L} \qquad (1)$$

$$\sigma = \frac{1}{\rho} \qquad (2)$$

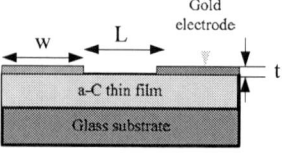

Figure 5: Cross sectional view of the a-C thin films on the glass substrate

Figure 7 and Figure 8 show the effect of various temperatures on the resistivity and conductivity between two measured mask electrodes. It is observed that from Figure 7, the resistivity is slightly decreased from 350°C until 350°C and drastically decreased starting from 350°C to 500°C. In contrast, the conductivity is flatted between 300°C and 350°C and drastically increased until 500°C as shown in Fig. 8.

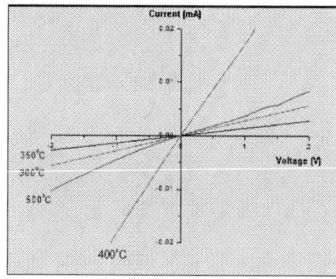

Figure 6 : I-V curves of amorphous thin films deposited at different temperatures

It was believed that, the structure of amorphous carbon has transformed from semiconductor to semi-metalic materials. Beside that, the density and uniformity has contributed the increasing of the conductivity as proved by the more density and uniformity in Fig. 4 (c) and Fig. 4 (e).The conductivity increases from 1.04454×10^{-7} cm/Ω to 2.67825×10^{-6} cm/Ω for undoped a-C thin film deposited from 300°C to 500°C. According to the Fig. 7, the resistivity is varied against various deposition temperatures. The resistivity for 350°C is 9.57×10^{6} Ω/cm and decreased gradually to 373378.124 Ω/cm at 500°C. This trend for resistivity and conductivity variation resembles the trend reported by Hussin et al [13] and A. N. Fadzilah et al [14].

978-1-4673-2395-6/12 $31.00 © 2012 IEEE

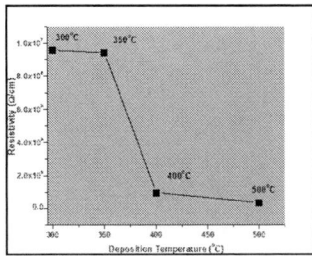

Figure 7: Resistivity of a-C thin films deposited on the glass substrate at different temperature

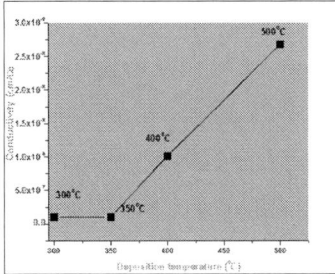

Figure 8: Conductivity of a-C thin films deposited on the glass substrate at different temperature

IV. CONCLUSION

The undoped of a-C thin films were successfully deposited on insulator glass substrates by using bias assisted thermal–CVD for 3 h deposition. The effect of the temperature on the structural and electrical properties of thin films was investigated by Atomic Force Microscopy (AFM), surface profiler and I-V Measurement. The results showed that at temperature of 400°C and 500°C, thin films became more uniform compared with others. The thickness is however, higher at 300°C and minimum at 400°C. The resistivity of undoped a-C thin film at 300°C, 350°C, 400°C and 500°C is 9.57x10^6 Ω/cm, 9.44x10^6 Ω/cm, 9.81x10^5 Ω/cm and 337738.124 Ω/cm respectively. The conductivity of thin films is increased by the increasing of temperature. The AFM images showed that, density and uniformity has correlated with the resistivity and conductivity of undoped a-C thin films at various temperatures. The less uniformity and density of undoped thin films provided less conductivity and high resistivity while the more uniform and high density contributed to the high conductivity and low resistivity.

ACKNOWLEDGMENT

This authors are grateful for Research Management Institute (RMI) Universiti Teknologi MARA (UiTM), and Ministry of Higher Education for the funding. The authors are also thankful to the staffs of NANO-ElecTronic Center (NET) and NANO-SciTech Centre of Universiti Teknologi MARA, UiTM for equipment and facilities.

REFERENCES

[1] Zheng Maxwell, Takei Kuniharu, Hsia Benjamin, Fang Hui, and Zhang Xiaobo, "Metal-catalyzed Crystallization of Amorphous Carbon to Graphene," *Applied Physics Letters*, vol. 96, pp. 63110-63110-3, 2010.

[2] S. M. Mominuzzaman, M. Rusop, T. Soga, T. Jimbo, and M. Umeno, "Photoresponse Characteristics of Nitrogen Doped Carbon/P-silicon Photovoltaic Cell," *IEEE 4th World on Conference on Photovoltaic Energy Conversion*, vol. 1, pp. 302-305, 2006.

[3] Hongwei Zhu, Jinquan Wei, Kunlin Wang, and Dehai Wu, "Applications of Carbon Materials In Photovoltaic Solar Cells," *Solar Energy Materials & Solar Cells*, vol. 93, pp. 1461-1470, 2009.

[4] Hiroki Akasaka, Takatoshi Yamada, and Naoto Ohtake, "Effect of Film Structure on Field Emission Properties of Nitrogen Doped Hydrogenated Amorphous Carbon Films," *Journal In Diamond & Related Materials*, vol. 18, pp. 423-425, 2009.

[5] Sudip Adhikari, Hare Ram Aryal, Dilip Chandra Ghimire, Golap Kalita, and Masayoshi Umeno, "Optical Band Gap Of Nitrogenated Amorphous Carbon Thin Films Synthesized By Microwave Surface Wave Plasma CVD," *Journal Of Diamond & Related Materials*, vol. 17, pp. 1666-1668, 2008.

[6] Dwivedi, Neeraj Kumar, Sushil Malik, and K. Hitendra, "Studied on Pure and Nitrogen-incorporated Hydrogenated Amorphous Carbon Thin Films and their Possible Application for Amorphous Silicon Solar Cells," *Journal Of Applied Physics*, vol. 111, pp. 14908-14916, Jan. 2012.

[7] O. Cubero, F. J. Haug,D. Fisher, and C. Ballif, "Reduction of the Phosphorous Cross-Contamination in n–i–p solar Cells Prepared in a Single-chamber PECVD Reactor," *Journal of Solar Energy Materials & Solar Cells*, vol. 95, pp. 606-610, 2011.

[8] C. Corbella, M. Rubio-Roy, E. Bertran, and J. L. Andujar, "Plasma Parameters of Pulsed-dc Discharge in Methane used to deposit Diamondlike Carbon Films," *Journal of Applied Physics*, vol.106, pp. 103302-10302-11, 2009.

[9] A. Mallikarjuna Reddy, A. Sivasankar Reddy, and P. Sreedhara Reddy, "Effect of Substrate Bias Voltage on the Physical Properties of dc Reactive Magnetron Sputtered NiO Thin Films," *Journal of Materials Chemistry and Physics*, vol. 125, pp. 434–439, 2011.

[10] J. H. Shim, N. H. Cho, and E. H. Lee, "The Effect of Negative Direct Current Bias on the Crystallization of nc-Si:H Films Prepared by Plasma Enhanced Chemical Vapor Deposition," *4th IEEE International Conference on Group IV Photonic*, 2007.

[11] A. Mallikarjuna Reddy, A. Sivasankar Reddy, and P. Sreedhara Reddy, "Effect of Substrate Bias Voltage on the Physical Properties of dc Reactive Magnetron Sputtered NiO Thin Films," *Journal of Materials Chemistry and Physics*, vol. 125, pp. 434–439, 2011.G. Eason, B. Noble, and I. N. Sneddon, "On certain integrals of Lipschitz-Hankel type involving products of Bessel functions," Phil. Trans. Roy. Soc. London, vol. A247, pp. 529–551, April 1955. (references)

[12] Jui-Yun Jao, Sheng Han, Chung-Chih Yen, Yu-Ching Liu, Li-Shin Chang, Chi-Lung Chang, and Han-C. Shih, "Bias Voltage Effect on the Structure and Property of the (Ti:Cu)-DLC," *Journal in Applied surface Science*, vol. 256, pp. 7490-7495, 2010.

[13] H. Hussin, F. Mohamad, S. M. A. Hanapiah, M. Muhammad, and M. Rusop, "Electrical properties of a-C Thin Film Deposited Using Methane Gas as Precursor," *International Conference on Electronic Devices, System and Aplication (ICEDSA2010)*, 2010.

[14] A. N. Fadzilah, and M. Rusop, " Effect of Deposition Temperature of Amorphous Carbon Thin Films," *International Conference on Electronic Devices, System and Aplication (ICEDSA)*, 2011.

Electrical Properties of ZnO/TiO₂ Nanocomposite Film Deposited by Simultaneous Radio-Frequency Magnetron Sputtering

I. Saurdi[1], M.H. Mamat, M.H. Abdullah, M.Z. Musa, and M. Rusop[2]
NANO - ElecTronic Centre (NET), Faculty of Electrical Engineering,
Universiti Teknologi MARA (UiTM),
40450 Shah Alam, Selangor, Malaysia
Email: [1] saurdy788@gmail.com, [2] nanouitm@gmail.com

Abstract — In this work, the ZnO/TiO₂ nanocomposite thin films were prepared by simultaneous Radio-Frequency Magnetron sputtering of ZnO and TiO₂ targets on glass substrates at different deposition times in the range of 30-75 minutes that increases the film thickness. The electrical and surface morphology were characterized by I-V measurement and atomic force microscopy (AFM) measurement respectively. The electrical characteristics indicate that the conductivity increases as the thickness increase due to the improvement in surface contact between particles and photocatalytic activity. High conductivity at 1.67×10^{-4} S/cm and lowest resistivity about 5.14×10^{4} Ω/cm have been obtained for 75 minutes deposition time. Atomic force microscopy (AFM) showed particle size of ZnO/TIO₂ thin films increases from 26nm to 50nm with an increasing in deposition time.

Keywords — *Zinc oxide; Titanium oxide, Deposition time; Sputtering; properties;*

I. INTRODUCTION

TiO₂ and ZnO are two most important wide-band gap semiconductors that have tunable physicochemical properties, high surface activities and cost effectiveness and non-toxicity [1]. Due to that, it can be used for photocatalysis, purification air and water [2] and also for different electronic devices such as gas sensors [3], field-effect transistors [4], dye sensitized solar cells [5], etc.

There are three main phases of TiO₂ in normal pressure which is rutile, anatase and brookite. Rutile has very low electron mobility as well as low photocatalytic activity as compared to anatase phase that has indirect band gap of 3.2 eV. Meawhile, brookite was rarely formed at room temperature. Zinc Oxide a well known II-IV compound semiconductor has a direct band gap 3.37eV and 60meV of free-exciton excitation energy at room temperature [6]. Moreover, the electron mobility in ZnO is higher than anatase. By mean of mixing these two semiconductors can enhance the activity of photocatalytic [7].There are several methods have been used to synthesize ZnO/TiO₂ nanocomposites: atomic layer deposition [8], electrospinning [9], direct current and radio-frequency (rf) magnetron sputtering [10], sol-gel [11].

In this paper, we report the effect of the different deposition times on the structural and electrical properties of ZnO/TiO₂ thin films grown on glass substrates by simultanous rf magnetron sputtering which rarely reported by other researchers. The properties of the ZnO/TiO₂ films deposited at different deposition times are systemically investigated by I-V Measurement (Bukuh Keiki CEP-2000) and atomic force microscopy (AFM, XE-100 Park Systems).

II. EXPERIMENTAL DETAILS

Microscope glass was used as substrates and was cut into a size of 2.5 x 2.5 cm. The surfaces were cleaned in the acetone, methanol and deionised water for 10 minutes in order to remove all the contamination. Then the cleaned glass substrates were dried using nitrogen gas blower.

ZnO/TiO₂ thin films were prepared using a SNTEK RSP 5004 RF magnetron sputtering unit with RF generator operating at 13.56 MHz by ZnO target (99.99%) and TiO₂ target (99.99%). The glass substrates were fixed onto the silicon wafer using polyimide tape. Before the deposition, the chamber was vacuumed to a base pressure less than 5×10^{-6} Torr. The ZnO/TiO₂ thin films were sputtered on glass substrates at different deposition times 30, 45, 60 and 75 minutes. Argon gas was used for sputtering at an argon flow rate of 50sccm. At the same time, the working pressure was maintained at 5mTorr. The ZnO/TiO₂ thin film was deposited at 150W rf power, while rotating the substrates at the speed of 5 rpm. After the deposition process, ZnO/TiO₂ thin film was annealed at 500^{0}C for 1hour.

The film thicknesses were measured using the surface profiler (Vekko Dektak 150). Then, the electrical properties were obtained by using I-V Measurement (Bukuh Keiki CEP-2000). The structural properties of ZnO/TiO₂ thin films were examined using Atomic force microscopy (AFM, XE-100 Park Systems).

III. RESULTS AND DISCUSION

A. Electrical Properties

The current voltage measurements (I-V measurements) of ZnO/TiO$_2$ thin films were obtained by using I-V Measurement (Bukuh Keiki CEP-2000). Figure 1 show the current increase when deposition time increase which is obeys ohm's law. The current will flow and increase accordingly to the magnitude of the voltage applied to the circuit as in equation (1).

$$V=IR \qquad (1)$$

V = voltage (V)

I = current (A)

R = resistance (Ω)

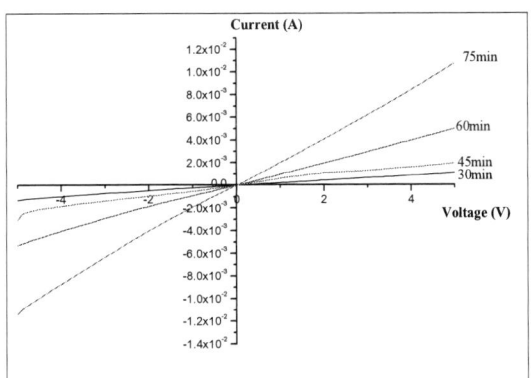

Fig.1 I-V characteristics of ZnO/TiO$_2$ thin film with different deposition times at (a) 30 (b) 45 (c) 60 (d) 75 minutes

However, there are some factors making the result in schottky response as shown in figure 1. These are may be due to the barrier of the ZnO/TiO$_2$ and metal contact [12], also due to the defect on the ZnO/TiO$_2$ interface. The thickness of ZnO/TiO$_2$ thin films deposited by simultaneous RF Magnetron sputtering with different deposition times is shown in Table I. The thickness increases as the deposition time increases. As the simultaneous RF-Magnetron sputtering deposited ZnO and TiO$_2$ at different times, the minimum and maximum thicknesses were 118.57nm and 382.4nm respectively. From the I-V graph and reported resistivity, conductivity, and thickness of ZnO/TiO$_2$ thin films have been calculated using the equation (2):

$$\rho = \left(\frac{V}{I}\right)\frac{wt}{l} = \frac{1}{\sigma} \qquad (2)$$

t = thickness of the nanostructured ZnO/TiO$_2$ thin film
w = width of electrode

(V/I) = electric resistance in ohm
x = gap between electrodes

TABLE I
ZnO/TiO$_2$ FILM THICKNESS AT DIFFERENT DEPOSITION TIMES

Deposition times	ZnO/TiO$_2$ thickness (nm)
30 minutes	118.57
45 minutes	227.59
60 minutes	290.27
75 minutes	382.39

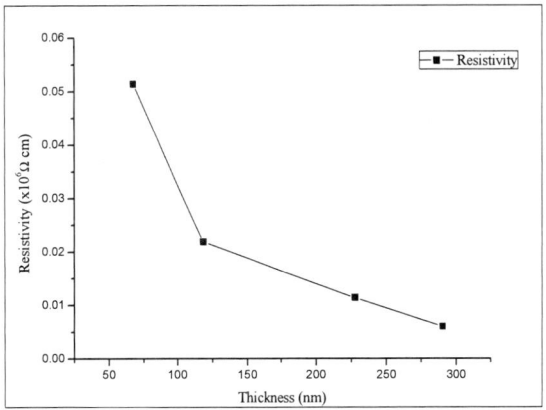

Fig.2 Resistivity of ZnO/TiO$_2$ thin film at different thickness

Figure 2 show the resistivity of ZnO/TiO$_2$ thin films. From the graph, it shown that the resistivity is decreasing when the thickness increase and then due to that the resistivity decreases when the deposition time increase. As a result, $5.14 \times 10^4\ \Omega/cm$ is the lowest resistivity at 30 minutes deposition time. The free carrier concentration increases and hence, decreases the resistivity. The resistivity is decreasing because of the better properties of ZnO/TiO$_2$ thin films as the thickness increases that show the increment of electron concentration at higher deposition time [13]. Moreover, increment of particle size can improve the surface contact between particles that produces better electron mobility in the thin films to reduce resistivity at higher deposition time [14-16]. Therefore, the nanocomposite of ZnO/TiO2 thin films deposited by simultaneous RF-Magnetron Sputtering that exhibiting high conductivity is possibly can be used in DSSC.

As shown in Figure 3, the conductivity of ZnO/TiO$_2$ thin films increases from 30 minutes to 75 minutes of simultaneous RF-Magnetron sputtering time that produce densely ZnO/TiO$_2$ thin films as represented in AFM images figure 4, while deposition time of 75 minutes shows a high electrical conductivity which is 1.67×10^{-4} S/cm. It means that the ability of ZnO/TiO$_2$ thin film to generate electric current is very high at 75 minutes.

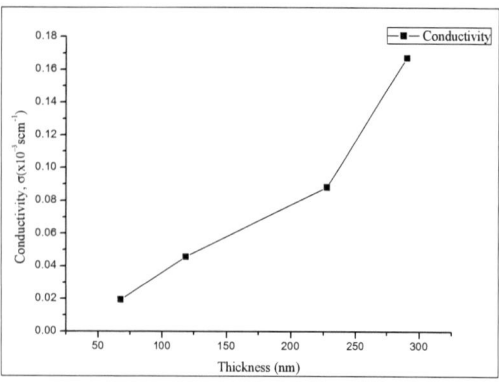

Fig.3 Conductivity of ZnO/TiO$_2$ thin film at different thickness

B. Structural Properties

Figure 4 represent AFM images of the structural of ZnO/TiO$_2$ thin films at different deposition times prepared by simultaneous RF-Megnetron sputtering. These images show that the changes of ZnO/TiO$_2$ particle sizes from 30 to 75 minutes of deposition whereby densely packed ZnO/TiO$_2$ thin films produced. Meanwhile, average roughness R$_a$ of deposited ZnO/TiO$_2$ thin film at different deposition times (30min, 45min, 60min, and 75min) were 0.795nm, 0.565nm, 0.233nm and 0.587nm respectively.

The particle sizes ZnO/TiO$_2$ thin films increases from 26.8nm to 50.6nm with increasing deposition time as represented in table I, which closely agree with results reported by G. Battalin et. al. [17]. These results support I-V characteristic in figure 1 and reported resistivity and conductivity in figure 2 and figure 3, that as increment of particle sizes improved the surface contact between particles and may caused better electron mobility and reduced resistivity at higher deposition time [14-16]. Due to that, the I-V characteristic in figure 1 is dependent on the deposition times that increased particle size in the thin film which might increase the surface contact between particles and improve packing density in the thin films [18].

TABLE II
ZnO/TiO$_2$ PARTICLES SIZE AT DIFFERENT DEPOSITION TIMES

Deposition times	Particle size (nm)
30 minutes	26.8
45 minutes	33.8
60 minutes	40.3
75 minutes	50.6

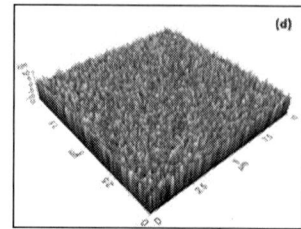

Fig.4 AFM images of ZnO/TiO$_2$ thin films at (a) 30min, (b) 45min
(c) 60min and (d) 75min

IV. CONCLUSIONS

The effect of the deposition time on the structural and electrical properties of ZnO/TiO$_2$ thin films was investigated by atomic force microscopy (AFM) and I-V measurement. The highest conductivity of ZnO/TiO$_2$ thin films were obtained by RF Magnetron Sputtering at deposition time at 75 minutes and the lowest resistivity 5.14×10^4 Ω/cm has been obtained at deposition time at 30 minutes. The study has shown that different thicknesses also affect the conductivity of thin films. As the deposition time increases, the thickness of the ZnO/TiO$_2$ films is thicker that improved the surface contact between particle sizes and then contribute to better electron mobility as well as conductivity.

ACKNOWLEDGEMENT

The author is grateful for the financial support from Research Management Institute (RMI) and the Ministry of Higher Education (MHOE).

REFERENCES

[1] M. Liao, C. Hsu and D. Chen., "Preparation and properties of amorphous titania-coated zinc oxide nanoparticles," Journal of Solid State Chemical, Vol. 179, pp. 2020-2026, 2006.

[2] H. Wang, Z. Wu, Y. Liu, Z. Sheng, " The characterization of ZnO-Anatase-Rutile three components semiconductor and enhanced photocatalytic activity of nitrogen oxides, "J.Mol Catal, A: Chem. Vol. 287, p. 176-181, 2008.

[3] R. Bhatt, H. Sankaranarayanan, C.S. Ferekides, D.H. Morel. 2. (Eds.), Proceedings of the 26th PVSC, vol. 171,Anaheim,California, pp.383,1997.

[4] M.S. Arnold, P. Avouris, Z.W. Pan, Z.L. Wang," Field-effect tansistor based on single semiconducting oxide nanobelts," Journal Physic Chemistry, Vol. 107, pp. 659-663, 2003

[5] D. wei, " Dye-sensitized solar cell review," Int. J. Mol. Sci. 2, Vol.11, pp1103-1113, 2010.

[6] G. Srinivasan, N. Gopalakrishnan, Y.S. Yu, R. Kesavamoorthy, J. Kumar, "Influence of post-deposition annealing on the structural and optical properties of ZnO thin films prepared by sol-gel and spin-coating method," *Superlattices and Microstructures,* vol. 43, pp. 112-119, 2008.

[7] C. Karunakaran, G. Abiramasundari, P. Gomathisankar, G. Manikandan and V. Anandi., "Preparation and characterization of ZnO–TiO2 nanocomposite for photocatalytic disinfection of bacteria and detoxification of cyanide under visible light," *Materials Research Bulletin,* Vol. 46, pp. 1586-1592.

[8] D.M. King, X. Liang, C.S. Carney, L.F. Hakim, P. Li, A. and W. Weimer., "Atomic Layer Deposition of UV-Absorbing ZnO Films on SiO2 Nanoparticles Using a Fluidized Bed Reactor," *Advance Functional Materials,* Vol. 18, pp. 607-615, 2008.

[9] S.H. Hwang, J. Song , Y. Jung , O. Y. Kweon, H. Song and J. Jang., "Electrospun ZnO/TiO2 composite nanofibers as a bactericidal agent," *Chemical Communication,* Vol. 47, pp. 9164-9166, 2011

[10] J.L Chung, J.C Chen, C.J Tseng., " Preparation of TiO2-doped ZnO films by radio frequency magnetron sputtering in ambient hydrogen–argon gas," *Applied Surface Science,* Vol. 255, pp. 2494-2499, 2008.

[11] J. Tian, L. Chen, J. Dai, X. Wang, Y. Yin and P. Wu, "Preparation and characterization of TiO2, ZnO, and TiO2/ZnO nanofilms via sol-gel process." *Ceramics International,* Vol. 35, pp2261-2270, 2009

[12] M. Ohyama, H. Kouzuka., " Sol-gel preparation of ZnO films with extremely preferred orientation along (002) plane from zinc acetate solution," *Thin solid Films,* Vol. 306, pp. 78-85, 1997.

[13] S. Benkara and S. Zerkout., "Preparation and characterization of ZnO nanorods grown into aligned TiO2 nanotube array," *Journal of Materials and Environmental Science,* Vol. 1, pp. 173-188, 2010.

[14] S.W. Xue, X.T. Zu, L.X. Shao, Z.L. Yuan, W.G. Zheng, X.D. Jiang and H. Deng, "Effects of annealing on optical properties of Znimplanted ZnO thin films," *Journal of Alloys and Compounds,* Vol. 458, pp. 572, 2008.

[15] J.F. Chang, M.H. Hon, "The effect of deposition temperature on the properties of Al-doped zinc oxide thin films," *Thin Solid Films,* Vol. 386, pp. 78-86, 2001

[16] S.H. Jeong, J.W. Lee, S.B. Lee and J.H. Boo., "Deposition of aluminum-doped zinc oxide films by RF Magnetron sputtering and study of their structural, electrical and optical properties," *Thin solid Films,* Vol. 435, pp. 78-82, 2003.

[17] G. Battaglin, V. Bello, E. Cattaruzza, F. Gonella, G. Mattei, P. Mazzoldi, R. Polloni, C. Sada, B.F. Scremin., RF Magnetron Co-Sputtering Deposition of Cu-based Nanocomposite Silica films for optical applications," *Journal of Non-Crystalline Solids,* Vol. 345, pp. 689-693, 2004.

[18] S. Youssef, P. Combette, J. Podlecki, R. A. Asmar and A. Foucaran., "Structural and Optical Characterization of ZnO Thin Films Deposited by Reactive rf Magnetron Sputtering," *Crystal Grown and Design,* Vol. 9, pp. 1088-1094, 2009

Scattering Effects in Silicon Nanowire Fin Field Effect Transistor

F. K. A. Hamid[1], J. F. Webb[2], Z. Johari[1], W. S. Leong[1], M. A Riyadi[1], M. T. Ahmadi[1] and R. Ismail[1]*

[1] Computational Nanoelectronics Research Group (CONE)
Universiti Teknologi Malaysia (UTM)
81310 UTM, Skudai, Johor, Malaysia
[2] Department of Mechanical, Materials and Manufacturing Engineering,
The University of Nottingham Malaysia Campus,
Jalan Broga, 43500 Semenyih, Selangor Darul Ehsan, Malaysia
Email: razali@fke.utm.my*, jeffwebb@physics.org

Abstract- **The velocity of an electron traveling from the source to drain of a field effect transistor can be much degraded by scattering effects. The scattering effects eventually become dominant as the devices are scaled down to the nanometer regime. In this paper, we propose a current-voltage (I-V) model for a two-dimensional Silicon Nanowire FinFET (SNWFinFET) which considers the scattering mechanism effects as well. Based on our simulated model, a notable scattering effect is observed and the I-V characteristics are in good agreement with experimental data. To evaluate the model, three parameters, temperature, channel length and drain voltage, were varied; the variation of each parameter has a significant effect on the I-V characteristics. These simulation results provide insights helpful for implementing SNWFinFETs as future devices, especially for high speed applications.**

Keywords: **FinFET, scaterring effect, nanoscale.**

I. INTRODUCTION

Over the past few decades, the Metal-Oxide-Semiconductor field-effect transistor (MOSFET) has become the basic building block for almost all computing devices. However as the MOSFET channel length is being scaled to the nanometer region, there are a lot of limitations which severely affect the device performance, as has been shown in the ITRS 2009 [1]. The MOSFET is expected to reach its channel length limits of 10 nm before 2020 [1] which means that new techniques will be required in the quest to sustain Moore's law. Thus, new materials such as carbon nanotubes and other graphene-based materials, as well as new device structures like Multi-Gate MOSFETs [2] are being considered as alternatives. A double gate is advantageous due to its robustness against short channel effects, high transconductance, almost ideal subthreshold swing and better mobility performance [3][4][5][6]. There have been a number of studies on the modeling and fabrication of double gate MOSFETs which shows promising characteristics [7]. However, FinFETs are favourable for fabricating double gate structures due to the much simpler fabrication process involoved. FinFETs can also be categorized as quasi planar. This is because they have fin shaped channels that are not fully planar [7]. Several means have been used to fabricate FinFETs [8][9]. They are easy to fabricate using conventional planar MOSFETs due the quasi planar structure, which is not much different from conventional MOSFETs [10][4]. The advantages of FinFETs are that they can suppress leakage current, have good electrostatic properties, and have high mobility [11]. Based on a FinFET channel shape and gate, devices with unique structures such as double FinFETs and omega FinFET have been developed [12][13]. In this study, a SINWFinET, as depicted in Fig. 1, is considered. Our main concern is to develop an I-V model which includes scattering effects for a one dimensional SNWFinFET. A model is proposed for one-dimensional electron motion resulting in scattering in the channel. Simulated results are confirmed by experimental data [14].

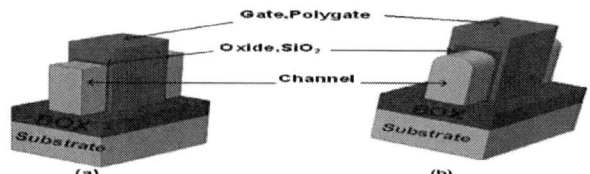

Fig 1:3-Dimensional depiction of (a) Standard FinFET (b) Proposed Silicon Nanowire FinFET in which the channel is a silicon nanowire.

II. ANALYTICAL MODEL

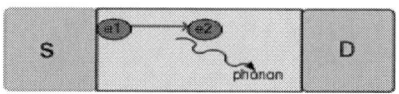

Fig 2: Velocity of electrons reduced after scattering

A drain current-voltage model based on scattering effects is considered, as depicted in Figure 2. First, let us consider the movement of a single electron, the total forced F_t exerted on a single electron can be expressed as.

$$F_t = q\varepsilon \qquad (1)$$

Where q is the electron charge and ε is the applied electric field. Scattering occurs when one or more of electrons moving from source to drain collide with other electrons. Thus, the force exerted on the electron will be reduced according to

---

978-1-4673-2395-6/12 $31.00 © 2012 IEEE   86

$$F_t = q\varepsilon - m\frac{\Delta v}{\Delta t} \qquad (2)$$

where m is the mass of an electron, v the average velocity, and Δt is the average time taken to move from source to drain. By using Newton's second law, the velocity for a single electron can be derived by integrating

$$m\frac{dv}{dt} = q\varepsilon - m\left(\frac{v - v_0}{t - t_0}\right) \qquad (3)$$

which results in

$$v = \frac{q\varepsilon t}{2m} + \frac{C_1}{t} \qquad (4)$$

The velocity for a single electron depends on an applied electric field, electron mass and time. This shows that the the less time taken for an electron to travel from source to drain, the higher speed of the device. If $t = \tau$ is the mean free time between between collisions and $v = v_{sat}$, we find that

$$v_{sat} = \frac{q\varepsilon\tau}{2m^*} + \frac{C_1}{\tau} \qquad (5)$$

and so C_1 is given by

$$C_1 = v_{sat}\frac{\tau}{t} - \frac{q\varepsilon\tau^2}{2m^*} \qquad (6)$$

By substituting the this expression for C_1 into equation (4), the final equation for the saturation velocity of a single electron is derived as

$$v = \frac{q\varepsilon t}{2m^*} + v_{sat}\frac{\tau}{t} - \frac{q\varepsilon\tau^2}{2m^*t} \qquad (7)$$

The electric field depends on the applied drain voltage and channel length such that

$$\varepsilon = \frac{V_{DS}}{L_{CH}} \qquad (8)$$

The average velocity is given by

$$v_{av} = v\int_0^\infty D(E)f(E)dE \qquad (9)$$

where $D(E)$ is the density of states (for one dimension) and $f(E)$ is the Fermi function. Putting in appropriate functions for D and f, this can be expressed by

$$v_{av} = v\left(\frac{1}{2\pi}\right)\left(\frac{2m^*}{\hbar}\right)^{1/2}\int_A^\infty \frac{E - E_C}{\left(1 + e^{(E-E_F)/k_BT}\right)}dE \qquad (10)$$

where n_A is the carrier concentration in the channel. Putting $\eta = (E_C - E_F)/k_bT$ and $x = (E - E_C)/k_BT$ in equation (11) gives

$$v_{av} = \left(\frac{v}{2\pi}\right)\left(\frac{2m^*k_bT}{\hbar}\right)^{1/2}\frac{\Gamma(1/2)}{\Gamma(1/2)}\int_0^\eta \frac{x^{-1/2}}{1 + e^{x-\eta}}dx \qquad (11)$$

The final drain equation can then be written as

$$I_D = n_A q v_{av} A \qquad (12)$$

Where A is the area given by

$$A = WL \qquad (13)$$

Where W and L are width and length of the device. The final drain equation can then be written as

$$I_D = (WL)\left(q\frac{v}{2\pi}\right)\left(\frac{2m^*k_bT}{\hbar}\right)^{1/2}\frac{\Gamma(1/2)}{\Gamma(1/2)}\int_0^\eta \frac{x^{-1/2}}{1 + e^{x-\eta}}dx$$

$$(14)$$

III. RESULTS AND DISCUSSION

To validate the proposed model, a comparison with experimental data is presented in this section. A one-dimensional drain current model was developed based on scattering effects for a SNWFinFET. The significant of these effects in the model could be seen when our simulated results agreed well with experimental data [14], which indicates the validity of this model. For the comparison, we assume that the parallel nano FinFETs is operating as a single channel, as in Fig. 3. The channel length used is $L_G = 65$ nm, with an oxide thickness of $t_{ox} = 3$nm. A small discrepancy was noticed. This may be attributed to the different channel shape between our model and that from which the experimental data was gathered. Besides this , the model is one dimensional, instead of 3 dimensional which is only an approximation to the actual situation [14].

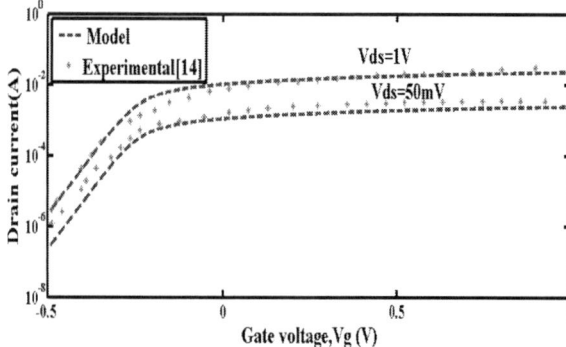

Fig 3. Current-voltage characteristics in a silicon nanowire finfet compared with experimental data (square line) taken from [14].

An investigation of the model was carried out by varying parameters such as channel length, temperature, and drain voltage. The simulation results show that there are some corresponding variations in the I-V characteristics. Temperature variations indicate that changes in the characteristics occur in the linear regions (Fig 4). In addition, it was found that the leakage current reduced when the temperature was decreased. This shows that this temperature response could result in an improvement in device performance by reducing the off-state current.

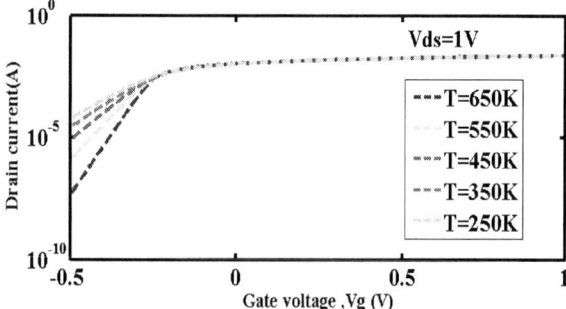

Fig 4. A variation of current-gate voltage in term of temperature

Such favorable effects on the I-V characteristics, both in the linear and saturation regions, are also apparent at reduced values for the channel length and drain voltage, as is evident in Figures 5 and 6. Thus the model shows that FinFet based devices can be scaled down further for high-performance device applications. Devices with smaller channel lengths could contribute to the making of higher speed devices due to the high on-current. Meanwhile, a higher drain value could reduce the electric field due to the gate being biased during the gate switching period. These positive effects on the electrical performance of the finFETS predicted by our model indicate that SNWFinFETs have a strong potential to be commercialized for use in future devices in the nanoelectronics industry.

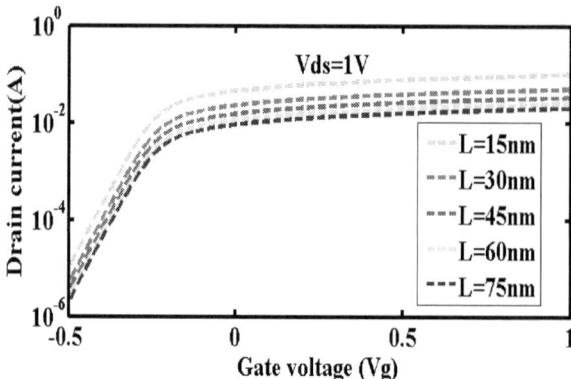

Fig 5. A variation of current-gate voltage in term of channel length

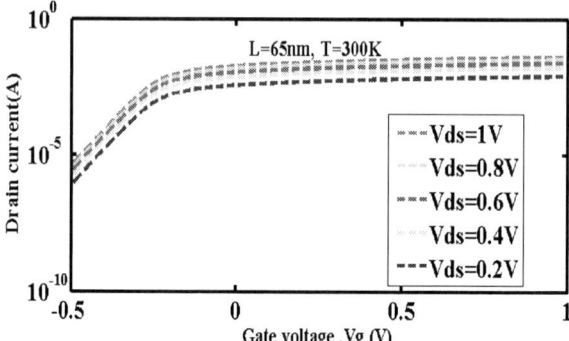

Fig 6. A variation of current-gate voltage in term of drain voltage

IV. CONCLUSION

Scattering effects play an important role in determining the current-voltage characteristics of SNWFinFETs, as is shown by our simulation results. The model developed has been used to investigate the influence temperature, drain voltage and channel length, on the I-V charactersitics; this has demonstrated the capability of FinFET devices, indicating that they are good candidates for commercialization.

ACKNOWLEDGEMENT

The authors would like to acknowledge financial support from MOSTI (Ministry of Science, Technology and Innovation), Malaysia; and Research University funding from MOHE (Ministry of Higher Education), Malaysia under Projects Q.J130000.7123.02H24 and Q.J130000.7123.02H04. Also we thank the Research Management Center of the University Teknologi Malaysia for providing excellent research environment in which to complete this work.

REFERENCES

[1] International Technology Roadmap for Semiconductors, 2009.

[2] P.-E. Gaillardon, J. Liu, I. O'Connor, and F. Clermidy, "Interconnection scheme and associated mapping method of reconfigurable cell matrices based on nanoscale devices," in Nanoscale Architectures, 2009. NANOARCH '09. IEEE/ACM International Symposium on, july 2009, pp. 69 –74.

[3] J. Kedzierski, D. Fried, E. Nowak, T. Kanarsky, J. Rankin, H. Hanafi, W. Natzle, D. Boyd, Y. Zhang, R. Roy, J. Newbury, C. Yu, Q. Yang, P. Saunders, C. Willets, A. Johnson, S. Cole, H. Young, N. Carpenter, D. Rakowski, B. Rainey, P. Cottrell, M. Ieong, and H.-S. Wong, "High performance symmetric-gate and cmos-compatible vt asymmetric-gate finfet devices," in Electron Devices Meeting, 2001. IEDM Technical Digest. International, 2001, pp. 19.5.1 –19.5.4.

[4] Y.-K. Choi, N. Lindert, P. Xuan, S. Tang, D. Ha, E. Anderson, T.-J. King, J. Bokor, and C. Hu, "Sub-20 nm cmos finfet technologies," in Electron Devices Meeting, 2001. IEDM Technical Digest. International, 2001, pp. 19.1.1 –19.1.4.

[5] X. Wu, P. Chan, and M. Chan, "Impacts of nonrectangular fin cross section on the electrical characteristics of finfet," Electron Devices, IEEE Transactions on, vol. 52, no. 1, pp. 63 – 68, jan. 2005.

[6] D. Frank, R. Dennard, E. Nowak, P. Solomon, Y. Taur, and H.- S. P. Wong, "Device scaling limits of si mosfets and their application dependencies," Proceedings of the IEEE, vol. 89, no. 3, pp. 259 –288, mar 2001.

[7] S. Tang, L. Chang, N. Lindert, Y.-K. Choi, W.-C. Lee, X. Huang, V. Subramanian, J. Bokor, T.-J. King, and C. Hu, "Finfet-a quasi-planar double-gate mosfet," in Solid-State Circuits Conference, 2001. Digest of Technical Papers. ISSCC. 2001 IEEE International, 2001, pp. 118 – 119,437.

[8] N. Lindert, Y.-K. Choi, L. Chang, E. Anderson, W. Lee, T.-J. King, J. Bokor, and C. Hu, "Quasi-planar nmos finfets with sub-100 nm gate lengths," in Device Research Conference ,2001.IEEE International, 2001, pp. 26 –27.

[9] N. Lindert, L. Chang, Y.-K. C. Y.-K. Choi, E. H. Anderson, W.-C. L. W.-C. Lee, T.-J. K. T.-J. King, J. Bokor, and C. H. C. Hu, "Sub-60-nm quasi-planar finfets fabricated using a simplified process," IEEE Electron Device Letters, vol. 22, no. 10, pp. 487–489, 2001.

[10] X. Huang, W.-C. Lee, C. Kuo, D. Hisamoto, L. Chang, J. Kedzierski,E. Anderson, H. Takeuchi, Y.-K. Choi, K. Asano, V. Subramanian, T.- J. King, J. Bokor, and C. Hu, "Sub 50-nm finfet: Pmos," in Electron Devices Meeting, 1999. IEDM Technical Digest. International, 1999,pp. 67 –70.

[11] N. Collaert, A. De Keersgieter, A. Dixit, I. Ferain, L.-S. Lai, D. Lenoble,A. Mercha, A. Nackaerts, B. Pawlak, R. Rooyackers, T. Schulz, K. Sar,N. Son, M. Van Dal, P. Verheyen, K. von Arnim, L. Witters, D. Meyer, S. Biesemans, and M. Jurczak, "Multi-gate devices for the 32nm technology node and beyond," in Solid State Device Research Conference, 2007. ESSDERC 2007. 37th European, sept. 2007, pp. 143 –146.

[12] Chien-Shao Tang; Shao-Ming Yu; Hong-Mu Chou; Jam-Wem Lee; Yiming Li; , "Simulation of electrical characteristics of surrounding-and omega-shaped-gate nanowire FinFETs," Nanotechnology, 2004. 4th IEEE Conference , aug. 2004, pp. 281- 283.

[13] X. Sun, V. Moroz, N. Damrongplasit, C. Shin, and T.-J. K. Liu,"Variation study of the planar ground-plane bulk mosfet, soi finfet, and trigate bulk mosfet designs," Electron Devices, IEEE Transactions, oct.2011, vol. 58, no. 10, pp. 3294 –3299,.

[14] Sato, S., Yeonghun, L., Kakushima, K., Ahmet, P., Ohmori, K., Natori, K., et al. (2010, 14-16 Sept. 2010). *Gate semi-around Si nanowire FET fabricated by conventional CMOS process with very high drivability.* Paper presented at the Solid-State Device Research Conference (ESSDERC), 2010 Proceedings of the European.

Low Actuation Voltage through Three Electrodes Topology for RF MEMS Capacitive Switch

Norhaslinawati Ramli[1], Othman Sidek[2], M.Amir[1]

[1]*Universiti Kuala Lumpur, British Malaysian Institute, Jln Sg. Pusu, Gombak, Selangor*
[1]norhaslinawati@bmi.unikl.edu.my, [1]dmamir@bmi.unikl.edu.my
[2]*Universiti Sains Malaysia, Kampus Kejuruteraan, Nibong Tebal, Pulau Pinang*
[2]othman@cedec.usm.my

Abstract— This paper presents the investigation results of low actuation voltage for RF MEMS capacitive switches using three electrodes topology. The main purpose of the investigation is to verify the reduction of actuation voltage as a result of applying three electrodes in RF MEMS. The new switch structure emphasizes three parallel electrodes instead of two electrodes as in previous structure. In the design stage all performance factors include beam width, beam length, area and beam thickness have been optimized to ensure the best physical dimension of the switch. The investigation of the performance was carried out using Architect Coventorware. Preliminary results shows that the actuation voltage of the three parallel electrodes switch gives very significant reduction of actuation voltage which is approximately half compared to other topologies or standard structure using two parallel electrodes.

Keywords— RF MEMS, Capacitive Switch, Actuation Voltage, Coplanar Waveguide, Electrostatic

I. INTRODUCTION

MEMS (micro-electro-mechanical system) is a micro dimension devices or system which consists of a combination of three components namely sensors, control unit and mechanical part. MEMS devices are fabricated on a substrate usually silicon by micro machining technique. There are two types of micromachining process called bulk and surface micromachining. In surface micromachining, layers are deposited on a substrate surface followed by the removal of the sacrificial layer in order to obtain the required structure [1].

There are many types of MEMS for different types of applications. One of them is RF (Radio Frequency) MEMS which is manned for RF integrated circuit. One of RF MEMS devices is RF MEMS switch. RF MEMS switches commonly use mechanical movement to exhibit a short circuit or open circuit in the RF transmission line [2]. Presently the RF MEMS switches become highly demand in replacing other RF switches which is p-i-n diode and field effect transistor. It shows good electrical performance characteristics such as high isolation, very low insertion loss, wide bandwidth operation and excellent linearity.

However, RF MEMS switch has high actuation voltage which is higher than the standard voltage of CMOS, usually 5V or less. This is absolutely not compatible with the CMOS circuit. Therefore the integration of MEMS with CMOS would be difficult and complicated in a single chip.

Hence the purpose of this research is to explore a new topology of switches design for RF MEMS which can facilitate to reduce the actuation voltage and make the integration of MEMS and CMOS in a single chip possible.

The next section of this paper highlights the previous works of MEMS and followed by standard design structure for MEMS switches. The third section is focusing on the simulation results and finally the conclusion of the findings.

II. LITERATURE REVIEW

Different design topologies for MEMS have been intensively introduced using different types of materials [3], [4], [5], new process flow [6], [7], and various architectures [8], [9], [10], [11]. The development activities are still actively in progress especially in the industries and research institution. The demand to produce compatible MEMS for CMOS circuitry is greatly focused and explored.

III. THE STRUCTURE AND OPERATION OF THE SWITCH

There are various types of RF MEMS switches and it can be classified either by contact mechanism (capacitive, metal-to-metal), anchor mechanism (cantilever, fixed-fixed beam) or actuation method (electrostatic, magnetic, piezoelectric, thermal). It also can be implemented using the most common configuration either in series or shunt structure. In a shunt switch, the transmission line is sandwiched between two ground line and the switch is turned ON by shorting the RF signal to the ground thus preventing the signal from crossing over the switch. Fig. 1 shows the switch structure in the up-state position.

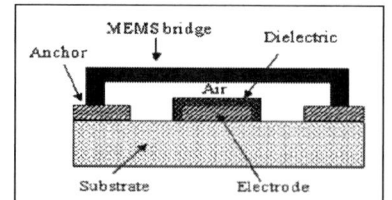

Fig.1. A shunt capacitive switch in the up-state position

978-1-4673-2395-6/12 $31.00 © 2012 IEEE

When a DC voltage is applied between the beam and actuation electrode as shown in Fig.2, an electrostatic given by equation (2) is created to pull the membrane down.

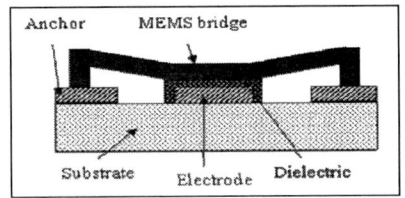

Fig. 2. A shunt capacitive switch in the down-state position

This force is exerted at the pull-down electrode and must be transferred to the contact point using the stiffness of the bridge.

$$F_e = -\frac{1}{2}\frac{\varepsilon_0 W w V^2}{g^2} \qquad (1)$$

The DC actuation voltage, at which the membrane collapsed and makes a contact with the dielectric layer, is called the pull-in voltage. For capacitive switch, where the gap is zero, an excellent contact must be achieved between the bridge and the dielectric layer. Now, the device is in the on-state and its capacitance, C_{on} is much larger.

IV. THE PROPOSED STRUCTURE

The design strategy involves two parts in RF MEMS switch. The first part is to design a switch with optimized spring constant and accurate physical dimension for reducing the actuation voltage as a result improving the RF performance. In second part, the design involves calibrating the impedance of coplanar waveguide transmission line at value of $Z_0 = 50\Omega$.

A. Proposed strategy in new switching structure

The investigation introduces three possibilities to achieve a reduction of the actuation voltage [12] which are described as follow:

i. First strategy is by enlarging the effective actuation area, $A = wW$ which could reduce the compactness of the switch integration. However the implication is the size becomes slightly bigger.

ii. Next is by decrementing of the initial gap, g_0. However the impact resulting poor RF performance (insertion loss and isolation).

iii. The third strategy is by reducing the effective spring constant k. The results show very significant outcomes which justify the best option for actual design implementation.

Therefore for this particular research, the third approach has been chosen. Together with this approach, a new switch structure has been proposed. The proposed structure is similar with the typical capacitive shunt switch. It consists of CPW (Coplanar waveguide) transmission lines and a fixed-fixed membrane that has been suspended from the anchor,

over a bottom electrode insulated by a dielectric. However, there is a change in a number of stacked electrodes. Instead of using two parallel electrodes, this proposed switch will have three stacked electrodes. The middle electrode will be movable and the top and bottom electrode will be maintained at a constant potential. Since there will be two forces acted on the beam, so it will be expected that the switch can be actuated with a lower voltage than normally required by a similar switch. The whole implementation for the new structure is illustrated as shown in Fig. 3.

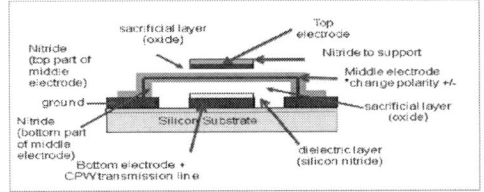

Fig. 3. Proposed switch with three parallel electrode structure

The mechanical design of most electrostatically based switches starts with considering the required dc actuation voltage. Equation (2) presents a formula given by Rebeiz [13] that can be used for calculating the pull-in voltage of fixed-fixed beam.

$$V_p = \sqrt{\frac{8kg_0^3}{27\varepsilon_0 wW}} \qquad (2)$$

where k is the effective spring constant of the moving structure in the direction of desired motion (typically z-direction), g_0 is the initial gap height between the switch and the electrode, ε_0 is the free-space permittivity, and w is the width of the beam and W is the width of the pull-down electrode. The effective spring constant, k depends on the geometrical dimensions of the metal beam and the Young's modulus of the material used. The spring constant for non-meander structure is given by (3) as below:

$$k_{nm} = 4Ew\left(\frac{t^3}{l^3}\right) \qquad (3)$$

The spring constant for meander structure is given by (4) as below:

$$k_m = \frac{Ew\left(\dfrac{t^3}{l^3}\right)}{1+\dfrac{l_s}{l_c}\left[\left(\dfrac{l_s}{l_c}\right)^2 + 12\left(\dfrac{1+v}{1+\left(\dfrac{w}{t}\right)^2}\right)\right]} \qquad (4)$$

Table 1 shows the formula used to calculate the spring constant for 1 Meander structure.

TABLE I
SPRING CONSTANT FORMULA FOR EACH STRUCTURE

Design Type	Spring constant formula
1 Meander Structure	$k_m = \dfrac{k_m k_{nm}}{k_m + 2(k_{nm})}$ (5)

V. COPLANAR WAVEGUIDE DESIGN

Coplanar waveguide (CPW) consists of a center conductor on a substrate with two grounds located parallel to it as shown in Fig. 4 below.

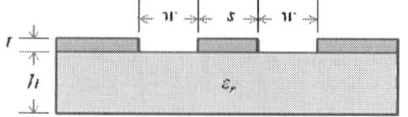

Fig. 4. Coplanar waveguide transmission lines

The formula for the characteristic impedance of coplanar waveguide is given by:

$$Z_o = \frac{30\pi^2}{\sqrt{\frac{(\varepsilon_r + 1)}{2}}} \left[\ln\left(2\frac{1+\sqrt{k}}{1-\sqrt{k}} \right) \right]^{-1} \quad (10)$$

where $k = \dfrac{s}{s+2w}$ with s = center strip width, w = slot width and ε_r = relative dielectric constant of the substrate.

VI. SWITCH DESIGN METHODOLOGY

A. Design Part

The switch structure that has been designed and simulated was shown in Fig. 5:

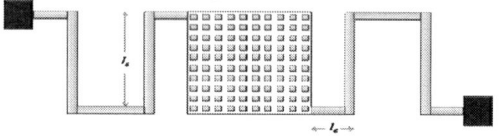

Design C

Fig. 5. 1-Meander support structure

All optimization factors include beam width, beam length, area and beam thickness was considered to get the best physical dimension of the switch. Table 2 shows the switch physical dimension for pulling-in the voltage for less or equal to 5V.

TABLE II
RF MEMS CAPACITIVE SWITCH PHYSICAL DIMENSION

Physical Dimension	1-Meander
Beam Length, l (μm)	-
Beam Length, ls (μm)	100
Beam Length, lc (μm)	50
Beam Thickness, t (μm)	1.7
Beam Width, w (μm)	5
Electrode width, W1=W2 (μm)	120
Air gap, g0 (μm)	3
Modulus Young, E (GPa)	70
Spring constant, k	0.1337
Pull-In Voltage, Vp (V)	2.9

B. Simulation Part

Advanced software tool known as Architect CoventorWare has been used in this research. The design flow starts by inserting the materials property parameters in material properties database. Next, the simulation process was declared in the process editor. The third step is to design by using the schematic capture. The designer has to select the suitable component form electromechanical library that representing for the plates, electrodes and beam. Finally the schematic capture that has been design was simulated and the results were plotted and analysed. The CAD is capable to display in 2D layout or 3D model.

VII. RESULT

TABLE III

RESULT OF PULL-IN VOLTAGE FOR 1-MEANDER SUPPORT STRUCTURE

Design	Calculation		Simulation	
	Standard	Proposed	Standard	Proposed
1 Meander	4.4V	2.5V	4.0V	2.0V

A slightly different exist between calculated and simulated value of pull-in voltage is due to dissimilar value of actuation area used. The actuation area is assume to be $(120\mu m)^2 - \left((5\mu m)^2 \times 100\right)$ in the calculation. However, it is not an accurate area utilized in the simulation.

A. Electromechanical Analysis

Fig. 6 and 7 show the electromechanical analysis of standard and the proposed struture. The pull-in voltage for three electrodes structure have been significantly reduced to half compared to switch with two electrodes structure with the value from 4V to 2V.

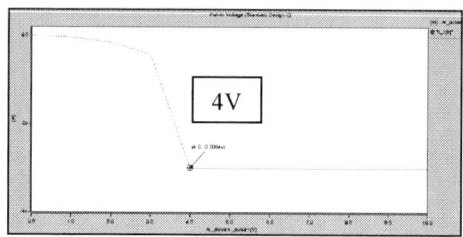

Fig.6. Pull-in voltage for 1-meander switch with 2 electrodes

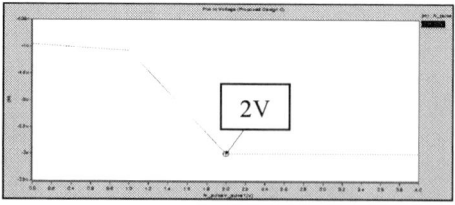

Fig. 7. Pull-in voltage for 1-meander switch with 3 electrodes

B. RF Electrical Analysis

Next simulation was focusing on RF performance of the MEMS switch in down-state and up-state position. The purpose is to investigate the insertion loss and isolation of the switch by obtaining the value of S11 and S21. Fig. 8 shows the results of the switch response for 1-meander switch with 3 electrodes. Fig. 8(a) indicates the response in up-state position. The value for the insertion loss (S21) at 40GHz is -0.3129dB and return loss (S11) at 40GHz is -11.685dB.

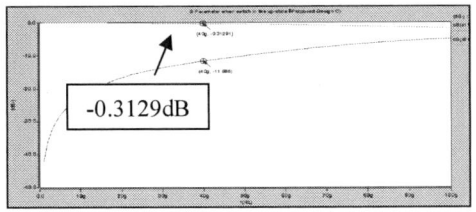

Fig. 8(a). RF Performance when switch in the up-state
(1-meander switch with 3 electrodes)

Fig. 8(b) shows the RF performance result at down-state position. The value for the insertion loss (S21) at 40GHz is -26.694dB and return loss (S11) at 40GHz is -0.1147dB.

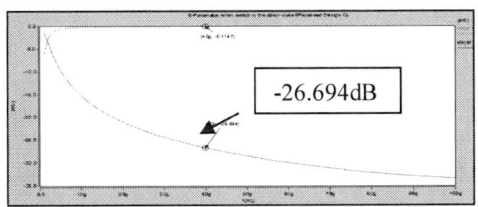

Fig. 8(b). RF Performance when switch in the down-state
(1-meander switch with 3 electrodes)

Table 4 summarize the insertion loss for 1-meander switch with 3 electrodes in up-state position. While Table 5 summarize the isolation results in the down-state position. The result shows that the RF performance for the three parallel electrode design has equal as the two parallel electrode structure. This indicates that the changes in number of parallel electrodes from two to three does not effect the RF performance of the switches.

TABLE IV
INSERTION LOSS (S21) WHEN SWITCH IN THE UP-STATE

Design	Standard	Proposed
1-meander	-0.14823dB	-0.3129dB

TABLE V
ISOLATION (S21) WHEN SWITCH IN THE DOWN-STATE

Design	Standard	Proposed
1-meander	-26.694dB	-26.694dB

VIII. CONCLUSIONS

The new topology of implementing three parallel electrodes in RF MEMS has been investigated and analyzed. In addition two different structures of switches were simulated to study the performance of actuation voltage .The results indicate that the actuation voltage of proposed RF MEMS switches has good reduction compared to the previous standard topology. It gives good reduction of actuation voltage which is half of the standard design. While for the overall performance the new topology has same results as two electrode design. This highlights that the new topology maintains the performance which is considered a minimum benchmark for this particular study. Finally the two findings provide one significant solution where the new design of parallel electrode is able to reduce the actuation voltage which is absolutely important for circuit integration with CMOS devices without jeopardizing the circuit performance.

REFERENCES

[1] H. S. Newman, "RF MEMS Switches and Application," in *40 th Annual International Reliability Physics Symposium* Dallas, Texas, 2002.

[2] G. M. Rebeiz, "RF MEMS Switches: Status of Technology," in *The 12 th International Conference On Solid State Sensors, Actuators and Microsystems, Boston*, 2003, pp. 1726-1729.

[3] B. B. Devarajan, S.K; and Ayazi, F., "Low Cost Low Actuation Voltage Copper RF MEMS Switches," pp. 1225-1228, 2002.

[4] G. P. Li, "On the Design and Fabrication of Electrostatic RF MEMS Switches," Dept. of Electrical and Computer Engineering, California University.

[5] F. M. Guo, Z. Q. Zhu, Y. F. Long, W. M. Wang, S. Z. Zhu, Z. S. Lai, N. Li, G. Q. Yang and W. Lu, "Study on low voltage actuated MEMS rf capacitive switches," *Sensors and Actuators A: Physical,* vol. 108, pp. 128-133, 2003.

[6] S. P. Pacheco, L. P. B. Katehi, and C. T. C. Nguyen, "Design of low actuation voltage RF MEMS switch," in *Microwave Symposium Digest, 2000 IEEE MTT-S International*, 2000, pp. 165-168 vol.1.

[7] J. P. H. L. D. Ching, C.L. Mao and C.W. Chyan, "Design and Fabrication of RF MEMS switch by the CMOS Process," *Tamkang Journal of Science and Engineering,* vol. 8, No.3, pp. 197-202, 2005.

[8] Y. L. H. H. X. Zhang, Z. Y. Xiao, D. M. Lou, N. Finch, J. Marchetti, D. Keating and V. Narashima, "Design of A Novel Bulk Micro-machined RF MEMS Switch," *International Conference on Micro and Nano System,* August 8-11, 2002.

[9] L. H. C. Y. L. LAI, "Design of Electrostatically Actuated MEMS Switches," *Colloids and Surfaces A : Physicochem eng. Aspects 313 - 314,* pp. 469-473, 2008.

[10] M. Song, J. Yin, X. He, and Y. Wang, "Design and analysis of a novel low actuation voltage capacitive RF MEMS switches," in *Nano/Micro Engineered and Molecular Systems, 2008. NEMS 2008. 3rd IEEE International Conference on*, 2008, pp. 235-238.

[11] S. Touati, N. Lorphelin, A. Kanciurzewski, R. Robin, A. S. Rollier, O. Millet, and K. Segueni, "Low actuation voltage totally free flexible RF MEMS switch with antistiction system," in *Design, Test, Integration and Packaging of MEMS/MOEMS, 2008. MEMS/MOEMS 2008. Symposium on*, 2008, pp. 66-70.

[12] J. Boussey, *Microsytems Technology Fabrication, Test aand Reliability.* London And Sterling, VA: Kogan Page Science, 2003.

[13] G. M. Rebeiz, *RF MEMS Theory, Design and Technology:* John Wiley and Sons, 2003.

Photoluminescence of Porous Silicon Nanostructures with Optimum Current Density of Photo-Electrochemical Anodisation

M. Ain Zubaidah[1,a], N.A. Asli[1], M. Rusop[2], S. Abdullah[1]

[1]Faculty of Applied Sciences,
[2]Faculty of Electrical Engineering,
Universiti Teknologi MARA,
40450 Shah Alam, Selangor, Malaysia.
[a]ainzubaidahmaslihan@yahoo.com

Abstract—P-type silicon wafer (<100> orientation; boron doping; $0.75 \sim 10$ Ωcm^{-1}) was used to prepare samples of porous silicon nanostructures. All the samples have been prepared by using photo-electrochemical anodization. A fixed etching time of 30 minutes and volume ratio of electrolyte, hydrofluoric acid 48% (HF48%) and absolute ethanol (C_2H_5OH), 1:1 were used for various current densities, J. There were sample A (J=10 mA/cm^2), sample B (J=20 mA/cm^2), sample C (J=30 mA/cm^2), sample D (J=40 mA/cm^2) and sample E (J=50 mA/cm^2). Photoluminescence (PL) spectra were investigated. Sample B gives the maximum peak position of PL spectrum at ~675 nm.

Index Terms—Photoluminescence, porous silicon nanostructures.

I. INTRODUCTION

In 1956, even though porous silicon (PSi) was first noticed by Uhlir [1], major interest in this material is more up to date. Electroluminescence has not been sufficiently investigated until present because it is rather complicated to obtain steady structures based on PSi with high emission efficiency [2]. PSi layers emiting visible light at room temperature has created much advantage due to its potential applications for fabrication of light emitting silicon devices integrated surrounded by main of the silicon technology [3]. As far as we know, attempts to correlate PL and Raman information taking into account the morphological parts of porous silicon nanostructures is still lacking [4].

In this study, porous silicon nanostructures (PSiNs) were prepared by photo-electrochemical anodisation and will be measured the PL emission as well as morphology through FESEM micrograph.

II. EXPERIMENTAL

P-type silicon wafer (<100> orientation; boron doping; $0.75\sim10$ Ωcm-1) was used to prepare sample of PSiNs. For photo-electrochemical anodization, a fixed volume ratio of electrolyte, 1:1, used was HF 48% and C_2H_5OH. Besides that, fixed etching time (30 minutes) was used for several of applied current density, J. There were sample A (J=10 mA/cm^2),

sample B (J=20 mA/cm^2), sample C (J=30 mA/cm^2), sample D (J=40 mA/cm^2) and sample E (J=50 mA/cm^2).

III. RESULT AND DISCUSSION

A. Structural Properties of Porous Silicon Nanostructures

Structural properties of porous silicon nanostructures (PSiNs) were studied by using Field Emission Scanning Electron Microscope (FFESEM). Based on the Fig. 1, it shows the various structural of PSiNs samples. Magnification of 200K X was used in order to capture all the sample pictures.

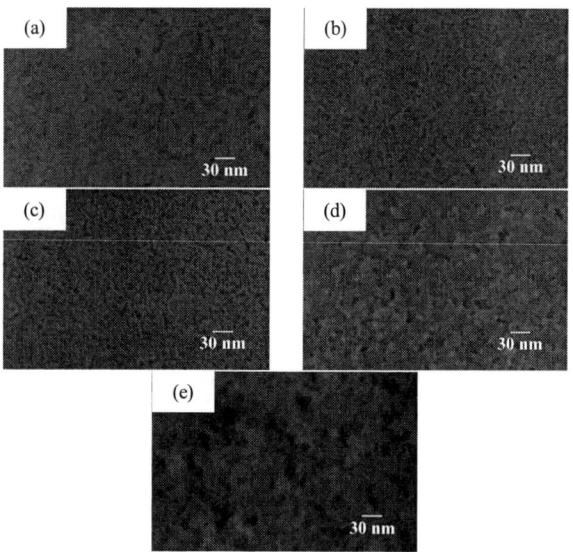

Fig. 1. Structural properties of porous silicon nanostructures sample at various current densities of (a) 10 mA/cm^2, b) 20 mA/cm^2, c) 30 mA/cm^2, d) 40 mA/cm^2 and e) 50 mA/cm^2.

As can be observed changes of growth pores formation of PSiNs samples from Fig. 1 is clearly significant. Referring to the Fig. 1a, sample A (J = 10 mA/cm^2) produces the lowest formation of pores on the silicon wafer surface. Fig. 1b shows the formation of pores increased for sample B (J = 20 mA/cm²). From Fig. 1c, for sample C with current density, J = 30 mA/cm², pores formed on the silicon wafer surface become clearly and the size of pores become larger than sample A and B. Very clearly pores have been seeing from the sample of D, J = 40 mA/cm² (Fig. 1d) and E, 50 mA/cm² (Fig. 1e). This is because, when the current density increased during the experiment, the formations of pores increased. Besides that, the pores size increased too. The bigger nanopores give meaning the smaller nanopillar in these samples.

B. Raman Spectroscopy

Based on Fig. 2, raman spectra also were measured for all PSiNs samples. The raman spectra were measured under room temperature at atmospheric pressure. Sample B with current density, J = 20 mA/cm², shows the smallest peak and shifted to low wavelength which is at 506.448 cm⁻¹. While PSiNs samples A, C, D and E with current densities, J = 10, 30, 40 and 50 mA/cm², show the Raman shift results at 519.661, 512.79, 515.961 and 518.076 cm⁻¹.

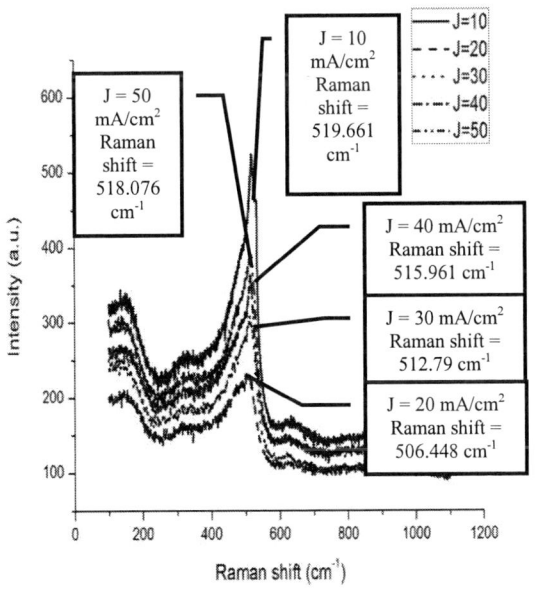

Fig. 2. Raman spectra of porous silicon nanostructures at various current densities (10, 20, 30, 40 and 50 mA/cm²).

From the Raman shift, crystallite sizes, L, of PSiNs samples were calculated. The crystallite sizes of PSiNs samples were calculated from equations given by,

$$\Delta\omega = 52.3 \, (0.543 \text{ nm} / L)1.586 \qquad (1)$$

$$\Delta\omega = \omega_{\text{(silicon wafer)}} - \omega_{\text{(porous silicon nanostructures)}} \qquad (2)$$

Where ω is the raman shift. Raman shift of silicon wafer, $\omega_{\text{silicon wafer}}$, is 520 cm⁻¹. After all the crystallite sizes were calculated, the results of crystallite sizes for each sample were listed in Table I.

TABLE I. CRYSTALLITE SIZE OF POROUS SILICON NANOSTRUCTURES AT VARIOUS CURRENT DENSITIES

Current density, J (mA/cm²)	Raman shift, ω (cm⁻¹)	Crystallite size, L (nm)
10	519.661	13.019
20	506.448	1.272
30	512.790	1.894
40	515.961	2.729
50	518.076	4.356

Table I shows the crystallite sizes of PSiNs samples. Sample B with current density, J = 20 mA/cm², shows the smallest crystallite sizes which is 1.272 nm compared to the rest of samples. Otherwise sample A with current density, J = 10 mA/cm², shows the largest crystallite size which is 13.019 nm. While crystallite sizes for samples C, D and E with current densities, J = 30, 40 and 50 mA/cm², are 1.894, 2.729 and 4.356 nm. When the crystallite size becomes very small, quantum confinement of phonons inside the particles relaxes [4]. However the Raman estimation value is not really reliable to quantify the nanosize level which can be observed the crystallite values in Table I.

C. Photoluminescence spectroscopy

Photoluminescence (PL) spectra for all PSiNs samples were characterized under room temperature at atmospheric pressure. Fig. 3 shows PL spectra of PSiNs samples. The PSiNs samples exhibited PL behavior in a wavelength at a range of 500-800 nm. The highest PL intensity was shows by sample B with current density, J = 20 mA/cm², at a wavelength of ~675 nm. While sample A with current density, J = 10 mA/cm², shows the lowest peak of PL spectrum.

Sample B with current density, J = 20 mA/cm² shows the highest peak positions of PL spectrum. Visible PL intensity at room temperature is attributed to the quantum confinement effects owing the presence of the aggregates in the etched layer of PSiNs [5, 6]. When PL intensity increases, it is caused by the total volume of the nanocrystallites on the surface of PSiNs [6-9]. The uniformity in the shape and size of the crystallites are always uncertain which is the major problem will contribute the results.

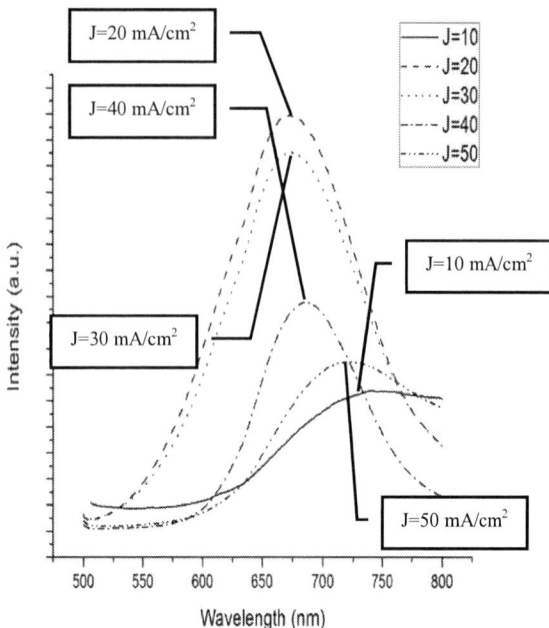

Fig. 3. Photoluminescence spectra of porous silicon nanostructures at various current densities (10, 20, 30, 40 and 50 mA/cm^2).

IV. CONCLUSION

As a conclusion, PSiNs samples were prepared by photo-electrochemical anodic etching of p-type silicon wafer (1 0 0) in electrolyte of HF 48% and ethanol. The PSiNs sample B with current density, J = 20 mA/cm^2, has the smallest crystallite size which is 1.272 nm. The PSiNs samples exhibited PL spectra in wavelength range of 500-800 nm with the maximum peak position at ~675 nm, which is attributed to quantum confinement efficiencies owing to the presence of the aggregates of the nanocrystalline silicon. Besides that, from this study, Raman and PL features are sensitive to the various applied current density during etching time process. The structural information determined from PL spectra generally agrees with Raman scattering results.

ACKNOWLEDGMENT

The authors would like to thank the staff of NANO-SciTech Centre for their technical support. This work was financially supported by Excellent Fund 600-RMI/ST/DANA 5/3/Dst (407/2011).

REFERENCES

[1] A. Uhlir, "Electrolytic shaping of germanium and silicon," The Bell System Technical Journal, vol. 35, pp. 333-347, 1956.

[2] S. A. Nepijko, D. N. Ievlev, D. B. Dan'ko, R. D. Fedorovich, W. Schulze, and G. Ertl, "Electroluminescence spectra of porous silicon as a function of the applied voltage," Physica B: Condensed Matter, vol. 299, pp. 32-35, 2001.

[3] G. Bomchil, A. Halimaoui, I. Sagnes, P. A. Badoz, I. Berbezier, P. Perret, B. Lambert, G. Vincent, L. Garchery, and J. L. Regolini, "Porous silicon: material properties, visible photo- and electroluminescence," Applied Surface Science, vol. 65–66, pp. 394-407, 1993.

[4] O. Ben Younes, M. Oueslati, and B. Bessaïs, "Anodisation-related structural variations of porous silicon nanostructures investigated by photoluminescence and Raman spectroscopy," Applied Surface Science, vol. 206, pp. 37-45, 2003.

[5] A. Bsiesy and J. C. Vial, "Voltage-tunable photo- and electroluminescence of porous silicon," Journal of Luminescence, vol. 70, pp. 310-319, 1996.

[6] D. A. Kim, J. H. Shim, and N. H. Cho, "PL and EL features of p-type porous silicon prepared by electrochemical anodic etching," Applied Surface Science, vol. 234, pp. 256-261, 2004.

[7] D. F. Timokhov and F. P. Timokhov, "Influence of injection level of charge carriers in nanostructured porous silicon on electroluminescence quantum efficiency," Microelectronic Engineering, vol. 81, pp. 288-292, 2005.

[8] F. Philippe M, "Photoluminescence and electroluminescence from porous silicon," Journal of Luminescence, vol. 70, pp. 294-309, 1996.

[9] E. Savir, J. Jedrzejewski, A. Many, Y. Goldstein, S. Z. Weisz, M. Gomez, L. F. Fonseca, and O. Resto, "Relation between electroluminescence and photoluminescence in porous silicon," Materials Science and Engineering: B, vol. 72, pp. 138-141, 2000.

Simulation of Nanoscale Dual-channel Strained Si/Strained $Si_{1-y}Ge_y$/Relaxed $Si_{1-x}Ge_x$ PMOSFET

Yu Chan Thien, Eng Siew Kang, Razali Ismail
Faculty of Electrical Engineering (FKE),
UniversitiTeknologi Malaysia (UTM),
81310 Skudai, Johor, Malaysia.
Email:razali@fke.utm.my

Abstract - **In this paper, the effects of several parameters on the threshold voltage of nanoscale dual-channel strained Si/Strained $Si_{1-y}Ge_y$/relaxed $Si_{1-x}Ge_x$ PMOSFET are investigated using SILVACO TCAD tools. The aspects discussed include strain induced at the channel, channel length, oxide thickness and substrate doping concentration. The electrical characteristics such as current-voltage relationship, subthreshold swing, drain induced barrier lowering and threshold voltage are investigated for 45nm channel length dual-channel strained Si/Strained $Si_{1-y}Ge_y$/relaxed $Si_{1-x}Ge_x$ PMOSFET. The quantum mechanical effects that arise in sub-nanometer regime are explained in detail. The simulated results show good agreement with the developed analytical model with the incorporation of quantum mechanical effects, showing the accuracy of the obtained results.**

I. INTRODUCTION

Aggressive geometry scaling on the MOSFET dimensions brought about advantages in three different aspects: functionality, performance and cost. As the MOSFETs technology advanced into the 22nm channel length regime, new materials are incorporated to prolong the lifetime of MOSFETs scaling. To sustain the Moore's Law to many generations, strained Si technology is adopted whereby a thin layer of silicon (Si) is pseudomorphically grown on top of relaxed $Si_{1-x}Ge_x$ layer, where x denoted as the germanium (Ge) fraction in the SiGe substrate. This technology is attracting more and more attention in semiconductor industry attributed to its potential to achieve enhanced performance in terms of enhanced carrier mobility, high field velocity, and carriers' velocity overshoot [1].

The latest attention in strained technology is given to dual-channel architecture. The strained Si/relaxed SiGe favors only the electron confinement and hence this structure is commonly used for n-type MOSFET.

Dual channel heterostructure consists of compressively strained $Si_{1-y}Ge_y$ on relaxed $Si_{1-x}Ge_x$ (y>x) capped with tensile strained Si. The cross section of dual channel heterostructure is illustrated as in Figure 1. The main advantages of this structure is the ability to increase both electrons and holes mobility simultaneously. Moreover, the same structure can be to fabricate both NMOSFETs and PMOSFETs which results in the reduction of manufacturing costs [2]. Hence, dual channel heterostructure MOSFETs have the potential to increase the performance for many more generations and leads to a new performance roadmap for the Si MOSFETs technology [3].

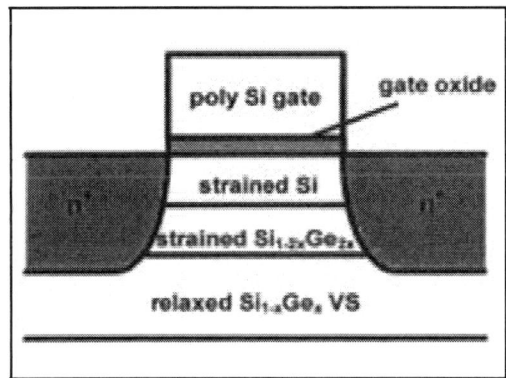

Figure 1: Cross Section of the dual-channel architecture of Si MOSFET.

II. I_d-V_{ds} CHARACTERISTICS

The substrate concentration for the device is 5×10^8 cm^{-3} with 4.62nm oxide thickness and 0.14µm junction depth. Figure 2 illustrates the comparison of the I_d-V_{ds} characteristics for 45nm and 200nm dual-channel heterostructure PMOSFET. It is

978-1-4673-2395-6/12 $31.00 © 2012 IEEE   97

illustrated that the curve acts as a resistor with linear relationship at a small value of V_{ds}. At this region, MOSFET is turned on as the channel formed to allow current to flow between source and drain. By increasing the value of V_{ds}, the drain current starts to saturate. At this point, pinch- off occurs whereby inversion layer density becomes very small. Hence we can observe that the drain current depends weakly on the drain voltage but is greatly affected by the gate voltage. The slope of the curve at this region almost dropped to zero. In an ideal saturation behavior, the I_d-V_{ds} curve should become constant at the saturation point. However, due to early effect of channel length modulation, a positive slope appears at saturation region whereby the drain current is still slightly dependent of the drain voltage.

Furthermore, it is observed for shorter channel length, in our case 45nm, the resistance decreases proportionally with channel length, thus resulting in stronger dependence of I_d to V_{ds}. The behavior of the I_d-V_{ds} curve for 45nm channel length deviates slightly compared to 200nm channel length due to the short channel effects. The device for 45nm channel length exhibits higher magnitude of saturation point due to velocity saturation whereby the carrier velocity had reached its maximum value. In the state of velocity saturation, the carrier velocity will no longer increase and thus the current in the channel is limited by this saturation value. Therefore the short channel device saturates at a lower drain current compare to the long channel device.

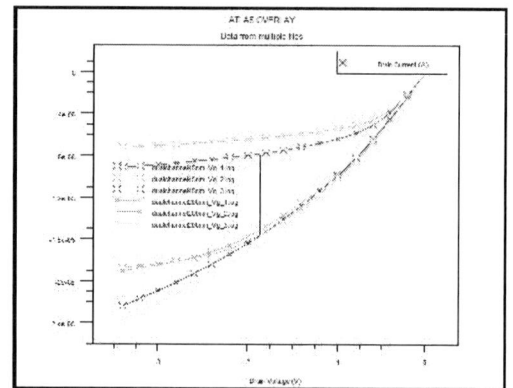

Figure 2: Comparison of I_d-V_{ds} for 45nm dual- 200nm dual-channel heterostructure PMOSFET.

The saturation slope and saturation point for each curve can be summarized as in Table 1.

TABLE 1
Extraction of saturation slope and saturation point at different gate voltage for 45nm and 200nm channel length dual channel heterostructure PMOSFET

		Channel Length	
		45nm	200nm
Vgs = - 1 V	Saturation Slope	6.93×10^{-7}	4.75×10^{-7}
	Saturation Point	-9.18×10^{-6}	-7.19×10^{-6}
Vgs = - 2 V	Saturation Slope	1.42×10^{-6}	1.02×10^{-6}
	Saturation Point	-3.87×10^{-5}	-3.58×10^{-5}
Vgs = - 3 V	Saturation Slope	3.71×10^{-6}	3.74×10^{-6}
	Saturation Point	-2.20×10^{-5}	-2.37×10^{-5}

III. I_d-V_{gs} CHARACTERISTICS

The relationship of drain current-gate voltage (I_d-V_{gs}) for 45nm and 200nm dual-channel heterostructure PMOSFET, with V_{ds} of -1V, -2V and -3V are illustrated respectively in Figure 3 and Figure 4. 35% of Ge fractionof strained $Si_{1-y}Ge_y$ concentration and oxide thickness of 4.0nm are used in the simulation.

Threshold voltage (V_{th}) is the gate voltage required to switch on the device by inducing a highly conductive channel which extends from source to drain. From the I_d-V_{gs} characteristic, the value of V_{th} is obtained as the gate voltage when the drain current started to transition from zero to nonzero value. It is demonstrated that lower gate voltage is required to induce the channel for 45nm device compared to the 200nm device. At a lower gate voltage, the 45nm dual-channel PMOSFET form an inversion layer at interface of oxide and substrate. In shorter channel length, the source and drain depletion region will extends into the channel depletion region. In addition, the channel depletion region charge will also connect to the charge of the source drain depletion region. This causes some of the channel to be depleted at zero gate bias. Hence, smaller gate voltage is required to create an inversion layer for the 45nm dual-channel heterostructure PMOSFET. For long channel device, the space charge under the gate is only influenced by the vertical electric field. This means that the channel depletion region is induced by the gate voltage only.

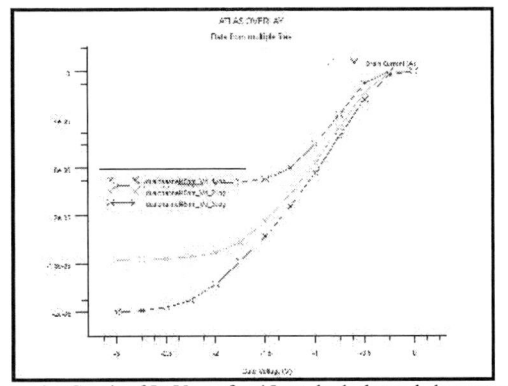

Figure 3: Graph of I_d-V_{gs} for 45nm dual-channel heterostructure PMOSFET.

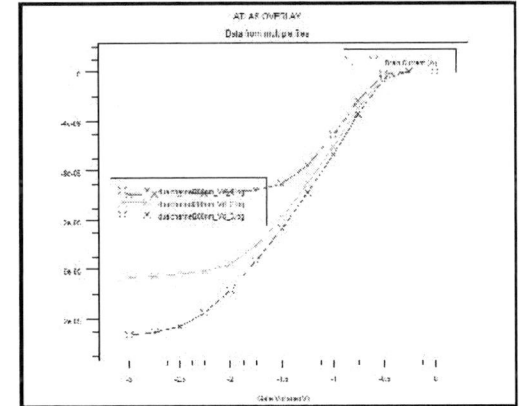

Figure 4: Graph of I_d-V_{gs} for 200nm dual-channel heterostructure PMOSFET.

IV. SUBTHRESHOLD SWING

Subthreshold swing is the key characteristics of the dual-channel PMOSFET performance. It is defined as the gate voltage which is required to change the subthreshold drain current by one decade. Smaller subthreshold swing is desired as this will provides an improvement in the ratio of the on and off current. Figure 5 shows the simulated log I_d-V_{gs} graph of the 45nm dual-channel heterostructure. The magnitude of the subthreshold swing is determined by the inverse of the subthreshold slope from the log I_d-V_{gs} graph. The relationship between the subthreshold swing and the strained Si thickness (T_{Si}) is depicted in Figure 6. We can observe that the subthreshold swing is directly proportional T_{Si}. When the gate voltage is low, the holes of dual-channel device travel in the buried strained $Si_{1-y}Ge_y$ layer. The holes confinement at the buried channel is increased as the T_{Si} is increased. Furthermore, if the applied gate voltage is high enough, the holes will eventually reside at the surface of the strained Si layer. In this case, increasing the T_{Si} will results in the increase of the valence band offset. Consequently, the concentration

of holes in the strained $Si_{1-y}Ge_y$ layer is increased and causes the depletion depth to reduce and the capacitance to increase. Thus the magnitude of the subthreshold swing is increased proportional to the T_{Si}.

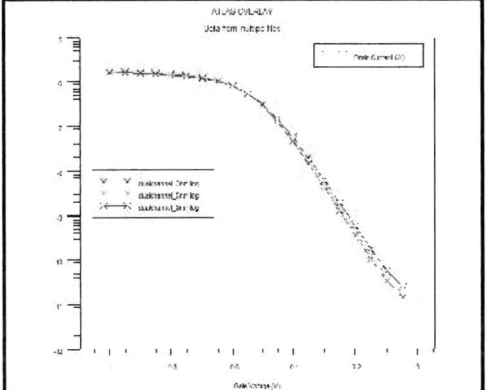

Figure 5: Graph of subthreshold slope, S for 45nm dual-channel strained Si PMOSFET for different strained Si thickness, T_{Si}.

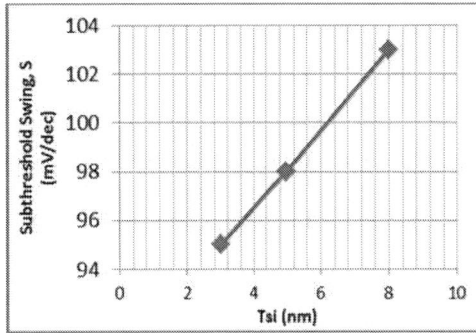

Figure 6: Variation of subthreshold swing with changes in strained Si thickness, T_{Si} 45nm channel length.

V. DRAIN INDUCED BARRIER LOWERING

Drain-Induced-Barrier-Lowering (DIBL) refers to the secondary effect of MOSFET which is defined by the reduction of V_{th} at higher drain voltages [3] and it plays an important role in determining the performance of MOSFET. Dual-channel device possess a better confinement of holes in the buried strained $Si_{1-y}Ge_y$ layer, which leads to smaller gate-controlled depletion charge and smaller electric field in its subsurface. This results in the suppression of DIBL and punchthrough of the device that enables the device to have enhanced performance and better scaling advantage.

The illustration of DIBL for 30nm to 300nm dual-channel heterostructure PMOSFET is shown in Figure 7. It is observed that the shorter channel length exhibits higher DIBL, which causes a lower output resistance. In the short channel device, the barrier of electron injection has become so near to each other that it is harder to turn off the device compares to the long channel device. Thus lower DIBL is observed for longer channel length compared to shorter channel length.

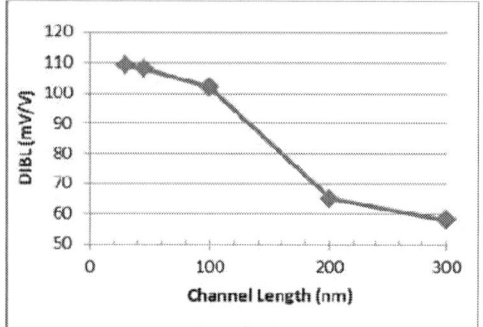

Figure 7: Graph of DIBL for different channel length of dual-channel heterostructure PMOSFETs.

VI. THRESHOLD VOLTAGE

A. Quantum Mechanical Effects

As the channel length is gradually shrink into nanometer regime, the quantum mechanical effects (QMEs) become dominant where the classical physics are no longer sufficient for the physics explanation. The combination of ultra thin oxide and high substrate doping concentration of the channel leads to a high electric field at the oxide/semiconductor interface [4] where the energy quantization of the carriers occur. In state-of-the-art nanoscale MOSFETs, at high electric field, the width of the well is small enough so that the carriers' motion is only in the direction normal to the Si/SiO_2. This quantization leads to the splitting of the continuous energy band and the energy levels of these carriers are grouped into several discrete subbands, with two dimensional density of state [4][5]. Furthermore, the carrier charge distribution and transport properties under the influences of QMEs are different from the classical case. According to the quantum physics, the sheet charge is effectively shifted away from the Si/SiO_2 interface towards the substrate by an amount of ΔZ, with ΔZ is the average inversion layer depth due to the law of quantum effects [6]. Hence, the incorporation of QMEs in the V_{th} is essential in order to attain higher accuracy in the device performance. A V_{th} model that takes into account the QMEs was carried out for dual channel heterostructure and these results are compared with the TCAD simulated results. The simulated V_{th} variation in this paper is valid only for $V_{th} < V_{ts}$, where V_{ts} is the threshold inversion induced in the strained Si channel. In the condition of $V_{th} < V_{ts}$, the strained Si layer acts as an

dielectric and decreases the total capacitance.

B. Effect of Ge Fraction in Strained $Si_{1-y}Ge_y$

The relationship between V_{th} and Ge fraction in strained $Si_{1-y}Ge_y$ for relaxed $Si_{0.85}Ge_{1.5}$ is illustrated in Figure 8. The simulation results track well with the analytical results for a set of similar parameters, indicating the accuracy of the obtained results. The shift of the V_{th} is attributed to the alteration in the energy band diagram of strained $Si_{1-y}Ge_y$. The splitting of the band degeneracy changes the relative masses of holes residing in the bands, resulting in the preferable occupation of holes in the heavy bands (HH) with higher energy compared unstrained Si. Hence, lesser gate voltage is required to induce the same level of holes in the channel. At the-state-of-art dual channel heterostructure, the Ge faction in the strained $Si_{1-y}Ge_y$ must be greater than the Ge fraction in the relaxed $Si_{1-x}Ge_x$ in order to induce the strain in the channel. Thus it is observed that there is no change in the V_{th} for less than 15% Ge fraction of the strained $Si_{1-y}Ge_y$. It is apparent that there is a significant discrepancy in the magnitude of V_{th} for 45nm analytical and simulated results, compared to the longer channel length. This phenomena is due to the domination of QMEs in short channel where most of the carriers that are responsible for the current transportation in the inversion layer reside in the lowest subband. Thus, higher gate voltage is required to induce the strong inversion.

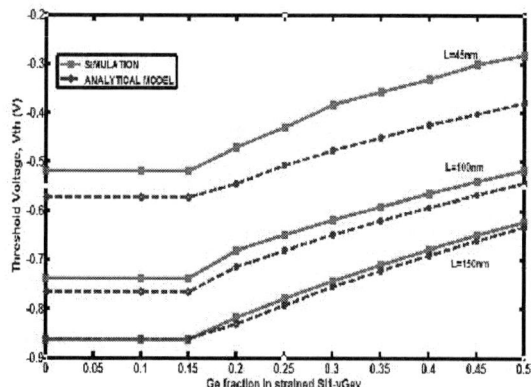

Figure 8: Variation of Vth with changes in the Ge mole fraction in strained $Si_{1-y}Ge_y$ for relaxed $Si_{8.5}Ge_{1.5}$.

C. Effect of Oxide Thickness

The relationship between the V_{th} and oxide thickness is depicted in Figure 9. Again, the simulation results show a good agreement with the analytical model by demonstrating higher V_{th} with increased oxide thickness because of the higher charge density in the depletion region formed under the channel which required larger gate voltage to overcome this charge in the strong inversion. Furthermore, it is apparent that 45nm channel length device exhibit huge V_{th} differences between the simulated and analytical results while small changes for 120nm devices. This can be understood by the fact that QMEs start to arise as the device dimension reach 45nm.

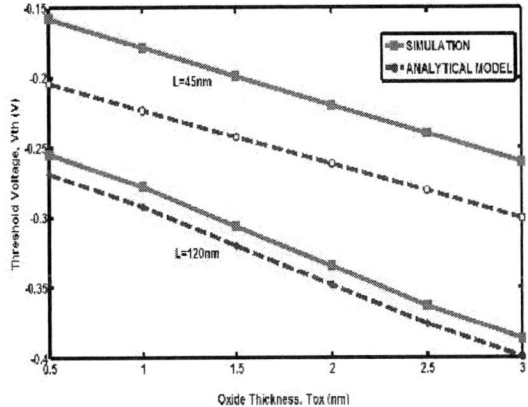

Figure 9: Variation of V_{th} with change in the oxide thickness.

D. Effect of Substrate Doping Concentration

The threshold voltage variation with changes in the substrate doping concentration for 0.5nm and 2nm oxide thickness is shown in Figure 10. The increment in the magnitude of threshold voltage with increasing doping concentration is attributed to the increased ionized charge that causes the carrier density to decrease. Furthermore, the combination of ultra thin oxide and high doping concentration causes carriers occupation in the lower subband and the splitting of the energy subbands. Additionally, higher doping concentration raises the flatband voltage, which in turn, required a larger gate voltage to reach the threshold inversion point.

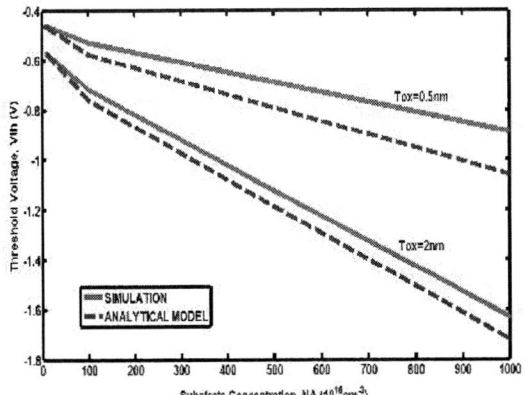

Figure 10: The effects of doping concentration on the variation of

threshold voltage

VII. CONCLUSION

The simulation of the dual-channel heterostructure PMOSFETs demonstrated enhanced performance of dual channel heterostructure. Besides holes mobility enhancement, dual-channel device also excel in suppressing short channel effects which is useful to sustain the scalability of MOSFET. In order to obtain the most optimized dual-channel electrical characterization, the incorporation of quantum mechanical effects that arise in the short channel devices is taken into account in the developed analytical model. The magnitude of threshold voltage obtained from the simulation track well with the analytical model. Significant discrepancies between the simulated and analytical results are observed due to the domination of quantum mechanical effects, where at this stage the classical mechanism is no longer sufficient to explain the underlying physics.

ACKNOWLEDGEMENT

The authors would like to thank the Research Management Centre (RMC) of Universiti Teknologi Malaysia (UTM) for providing excellent research environment in which to complete this work.

REFERENCES

[1] K. Rim, J. L. Hoyt and J. F. Gibbons. "Fabrication and analysis of deep submicron strained-Si NMOSFETs," *IEEE Transaction On Electron Devices*, Vol. 47, No.7, July 2000.

[2] B. Bindu, N. DasGupta, A. DasGupta. "Analytical model of drain current of strained-Si/strained $Si_{1-y}Ge_Y$/relaxed $Si_{1-x}Ge_x$ NMOSFETs and PMOSFETs for circuit simulation," *Solid-State Electronics*,Vol.50 ,2006.

[3] K. Chandrasekaran, X. Zhou and S. B. Chiah. "Physics-Based Scalable Threshold-Voltage Model for Strained-Silicon MOSFETs," *NSTI-Nanotech* , ISBN 0-9728422-8-4 Vol. 2, 2004.

[4] G.S. Jayadeva and A. DasGupta. "Compact Model of Short Channel MOSFETs Considering Quantum Mechanical Effects," *Solid-State-Electronics*, Vol. 53, pg 649-657, 2009

[5] J. He, M. Chan, X. Zhang and Y. Wang. " An Analytical Model to Account for Quantum Mechanical Effects on MOSFETs Using a Parabolic Potential Well Approximation", *IEEE Transaction on Electron Devices*, Vol. 53, No. 9, pg 2082-2090, 2006

[6] M. A. Karim and A. Haque. "*A Physically Based Accurate Model for Quantum Mechanical Correction to the Surface potential of Nanoscale MOSFETs,*" IEEE Transactions on Electron Devices, Vol. 57, No. 2, pg 496-502, 2010

[7] Y. L. Tsang, S. Charropadhyay, S. Uppal, E. Escobedo-Cousin, H. K. Ramakrishnan , S. H Olsen and A. G. O'Neill. "Modeling of the Threshold Voltage in Strained Si/Si1-xGex/Si1-yGey (x>y) CMOS Architectures," *IEEE Transactions on Electron Devices*, Vol. 54, No. 11, pg. 3040-3048, 2007

Electrical and Optical Properties of Iodine Doped Amorphous Carbon Thin Film by Thermal CVD

K. Dayana[1, a], A. N. Fadzilah[1], U.M. Noor[1] and M. Rusop[1,2,b]
[1]NANO-ElecTronic Centre (NET), Faculty of Electrical Engineering,
Universiti Teknologi MARA (UiTM), 40450 Shah Alam, Selangor, Malaysia
[2]NANO-SciTech Centre (NST), Institute of Science,
Universiti Teknologi MARA (UiTM), 40450 Shah Alam, Selangor, Malaysia
[a]dyna2171@gmail.com, [b]rusop@salam.uitm.edu.my

Abstract— **In this paper, iodine doped amorphous carbon thin films with varying the doping time were prepared onto glass and n-type silicon substrates by Thermal CVD technique. The optical and electrical properties of iodine doped amorphous carbon thin films were characterized by using UV-VIS-NIR spectroscopy and current-voltage (I-V) measurement respectively. The optical band gap of a-C thin films shows a reduction of the optical band gap upon iodine doping. The higher electrical conductivity was found to be at 10 min iodine doping. The structural properties of the films were studied by FESEM and FTIR. FESEM studies shows a uniform distribution of fine small particles on the iodine doped a-C thin films indicating deposition of iodine atoms on the substrates. FTIR measurement shows the effect of iodine on reduction on sp^3 bonded carbon in a-C thin film.**

Keywords-amorphous carbon; iodine doping; thermal CVD; thin film;

I. INTRODUCTION

Amorphous carbon is one of the most investigated materials and has attracted much attention due to its outstanding properties like inertness to any aggressive chemical, very hard hardness like sapphire, high thermal conductivity, low electrical resistivity and tunability of band gap from graphite to diamond[1, 2] . These advantages give a good opportunity for the potential technological applications of amorphous carbon (a-C) materials in a wide variety of electronic applications such as hard coating, electronics devices and photovoltaic therefore it has been studied extensively. a-C has been found to exhibits semiconducting electrical properties at ambient conditions [3]. The most common chemical bonds in a-C is sp^3 and sp^2 hybridizations and has been reported that a-C consists of sp^2 bonded cluster interconnected by a random network of sp^3 bonded atomic sites [4]. The undoped a-C thin films is p-type semiconductor and the usefulness of a-C for electronic applications has been greatly enhanced by recent demonstration of p-type doping by iodine and boron [5-8]. Many effective approaches have been carried out to improve the performance of carbon based solar cells. It is known that the performance of a-C based solar cell is still lower compared with silicon based solar cells, which results from the low charge carrier mobility and presence of high defect density by complex sp^3/sp^2 mixed structure[9, 10]. In this work, iodine doped a-C thin films were prepared by thermal chemical vapor deposition (CVD) technique onto the glass substrates. The aim of this work was to investigate the effect of iodine doping time on the electrical and optical properties by a simple thermal CVD technique.

II. EXPERIMENTAL DETAILS

The apparatus employed for deposition and doping process of amorphous carbon thin films consists of double furnaces, a quartz tube and water bubbler system as shown in Figure 1. The quartz tube which having an inner diameter of 2.75 cm acts as a container for deposition of the films and placed into the horizontal tubular furnaces. Argon gas which does not give any reaction with other chemical material is used as a carrier gas while camphor oil used as a precursor of carbon source to synthesize a-C thin film. For the deposition of a-C thin film, the combustion boat filled with 3ml camphor oil was placed inside the quartz tube at first furnace and heated at 200°C. The substrates were loaded into the second furnace at 550°C. Then, the quartz tube was purged with the argon gas for approximately 10 min. in order to flush out the air inside the quartz tube and to provide an inert environment during the deposition. The carrier gas was flowed towards the quartz tube from the first furnace to the second furnace with the gas flow rate maintained at 50 bubbles per minute and the thin films were deposited for 30min.

After that, the post-deposition iodine doping method was employed. The deposited a-C thin film was doped with iodine as a dopant by thermal CVD. For iodine doping process, the first furnace for heating iodine (in solid form) at 100°C while second furnace for controlling temperature of the deposited a-C thin films at 300°C. Iodine doped a-C (a-C:I) thin films were deposited on glass by thermal CVD with varied the doping time (10, 20 and 30 min.). Before deposition of a-C thin films, the glass substrates were effectively cleaned by the standard organic cleaning method. Acetone and methanol was used to clean oils and organics residues which appear on substrate surfaces using an ultrasonic cleaner for 10 min respectively. This is to avoid any possibility of contaminations during the deposition of thin films. Then, the substrate was blow with nitrogen gas. The resistivity of iodine doped a-C thin films was measured by the dark and illuminated current-voltage (I-V) characteristics and the cell configuration for I-V measurement was shown in Figure 2. The gold (Au) electrodes (thickness = 60 nm) which is a known conductor material were sputtered on top of the thin films deposited on glass substrate by sputter coater to achieve the ohmic contacts between the thin film and the top contact. The optical

978-1-4673-2395-6/12 $31.00 © 2012 IEEE

properties were conducted by UV-VIS-NIR spectroscopy while the structural and bonding properties were investigated by Field emission scanning electron microscopy (FESEM) and Fourier transform infrared spectroscopy (FTIR).

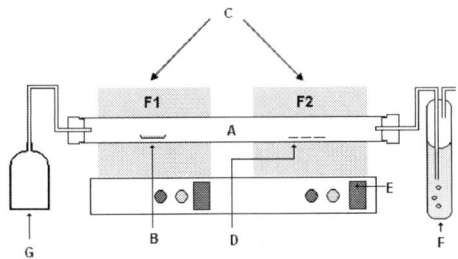

Figure 1. Schematic diagram of thermal CVD system (A) Quartz tube (B) Combustion boat (C) Furnaces (D) Substrate (E) Temperature controller (F) Water bubbler system (G) Argon gas cylinder [1]

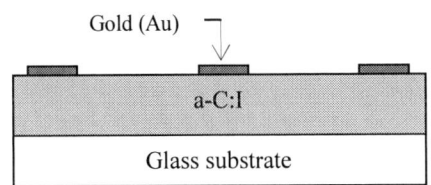

Figure 2. Schematic structure of iodine doped a-C (a-C:I) thin film

III. RESULT AND DISCUSSION

A. Determination of optical band gap

The UV-Vis-NIR spectroscopy analysis is one of the most used techniques in the literatures to measure the optical band gap of amorphous semiconductors. The optical band gaps of iodine doped a-C thin films deposited on glass substrates were estimated by measuring the optical transmittance as function of wavelength by a UV-Vis-NIR spectroscopy in the range 300-2000 nm at room temperature. The optical transmittance of iodine doped a-C thin films with different doping time as well as a-C thin film (undoped) is shown in Figure 3 for comparison. The optical band gaps of iodine doped a-C thin films were calculated by the Tauc formula as

$$(\alpha h v)^{1/2} = B(E_g - h v) \qquad (1)$$

Where B is the Tauc parameter, E the photon energy, α the absorption coefficient and E_g the band gap energy[11]. The optical band gap for the iodine doped a-C thin films with different doping time is shown in Figure 4. A transmittance of higher than 80% could be obtained for undoped a-C thin film (0 min). The result shows the optical transmittance decreased for a-C:I thin films which can be explained to be related to the presence of iodine atoms in the film [12]. The transmittance results indicates that there is an optimal condition of the iodine dopant to be maintained during the doping process of a-C thin films for the higher efficiency carbon based

solar cell fabrication. Figure 4 shows the plot of $(\alpha h v)^{1/2}$ as a function of photon energy (hv) and the summarized optical band gap obtained from before doping and after doping a-C thin films was shown in Table 1. Interestingly, the results show that the decrease of the optical band gap from 0.513 eV to 0.136 eV with doping time increase which can be explained caused by the sp^2 content in the a-C thin films increases and has less dangling bonds in the random carbon network due to the presence of the C-I bonding [2, 13]. This effect could be due to the induced graphitization in the a-C structure and thus leading to the narrow optical band gap. Our results also agree with the literature reported by M. M. Omer et.al. [14] , that the iodine in the a-C thin film deposited induced graphitization of the structure and helps to decrease the sp^3 content in the films.

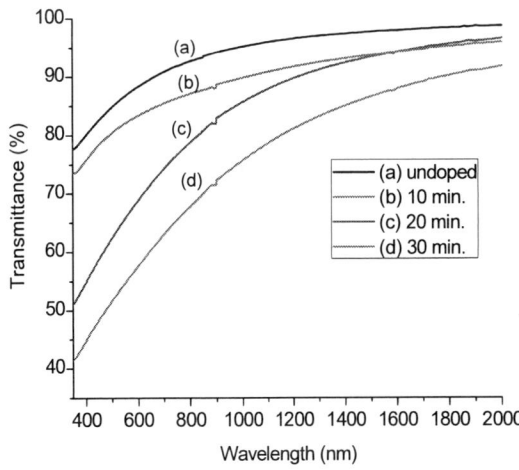

Figure 3. The optical transmittance of iodine doped a-C thin films

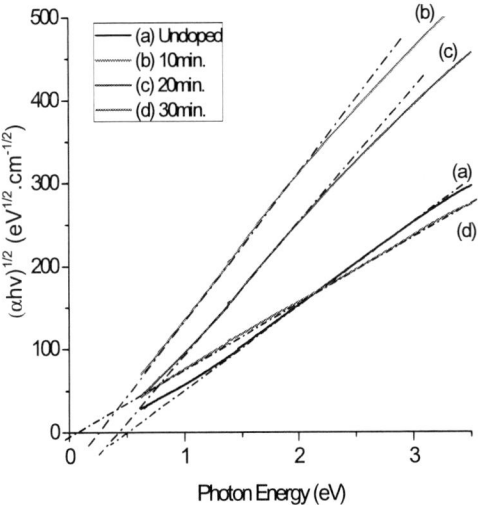

Figure 4. The Tauc plot of $(\alpha h v)^{1/2}$ as a function of photon energy for a-C thin films

Table 1: The optical band gap of a-C thin films doped with different doping time by thermal CVD

Sample	Optical band gap (eV)
Undoped	0.513
10 min	0.264
20 min	0.430
30 min	0.136

B. Electrical conductivity

Figure 5 illustrates the I-V characteristic of iodine doped a-C thin films deposited on glass substrates at room temperature. The current changed linearly with the voltage which indicates the ohmic behavior achieved between the thin films and electrode. The rectifying behavior has not been formed between the Au and a-C:I thin film because the work function of Au is larger than the a-C:I and the existence of sp^2–rich interfacial layer [15]. The electrical conductivity was calculated by I-V measurement data and the thin film thickness data by surface profiler. As shown in Figure 6, the electrical conductivity of the thin films was found to be higher up to 3.81×10^{-4} S.cm^{-1} after iodine doping compared to before iodine doping. The electrical conductivity and resistivity of undoped and iodine doped a-C thin films with different doping time were summarized in Table 2. It has been clearly observed that doping time at 10 min has the higher conductivity. However, as the doping time increased from 10min to 30min, the conductivity was decreased to 1.51×10^{-4} S.cm^{-1}. This is might due to the microstructural change of a-C thin film.

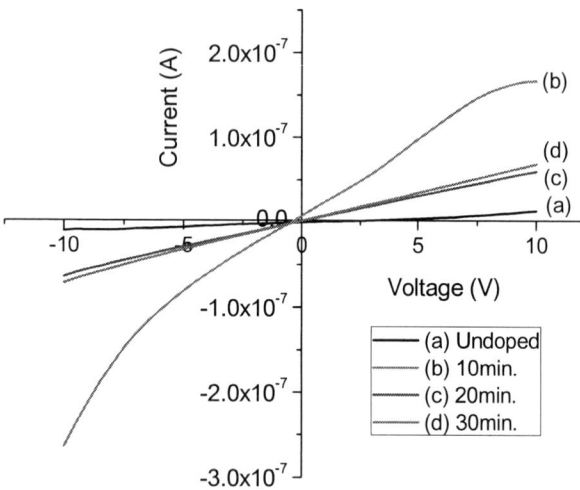

Figure 5. Current-Voltage (I-V) characteristic of iodine doped a-C (a-C:I) thin films

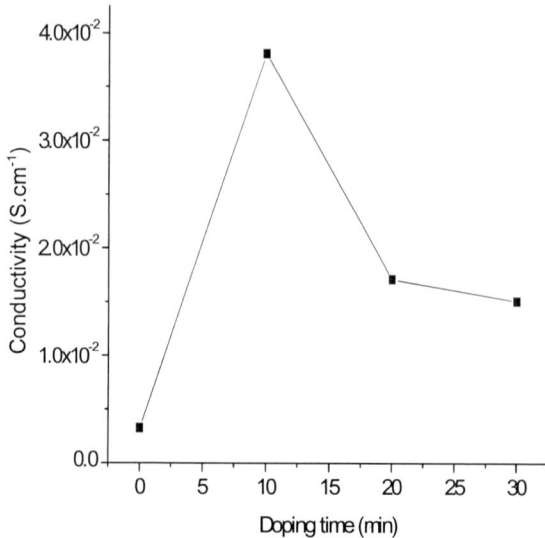

Figure 6. Conductivity of iodine doped a-C (a-C:I) thin films

Table 2: Electrical conductivity and resistivity of undoped and iodine doped a-C thin films with different doping time.

Doping time (min)	Conductivity (S.cm^{-1})	Resistivity (Ω.cm)
0	3.26×10^{-5}	3.06×10^4
10	3.81×10^{-4}	2.63×10^3
20	1.71×10^{-4}	5.84×10^3
30	1.51×10^{-4}	6.64×10^3

The electrical conductivity of the a-C thin films mainly occurs due to electrical transport in the valence band [16]. Due to the increasing of mobile holes, the electrical conductivity of the films was improved and the resistivity was decreased by iodine doping. The result indicated that iodine doping increased the electrical conductivity and showed that iodine dopant could help in neutralization of the dangling bonds. Therefore, it could be explain that the role of Iodine atoms in assisting carbon atoms to form sp^2 bonding and to facilitate graphitization which decreased the optical band gap in a-C thin film [5, 17].

C. Surface Morphology

The surface morphology of the undoped and iodine doped a-C thin films were studied by FESEM. All the micrographs are at the same magnification (50kX). Figure 7(a) shows the surface morphology of the undoped a-C thin film. It can observed that small discrete particles having irregular shape which indicated the amorphous nature of the samples. The iodine doped a-C thin film with doping time at 10min and 20min show a uniform distribution of fine small particle ranging from 25 to 40nm in diameter (Figure 7(b and c)). However, the surface morphology of iodine doped a-C thin film at 30min., it is obvious that the fine small particle almost disappear. From FESEM observation, this can be explained as the iodine atoms being deposited on the substrates after cooling. The thin film's morphology changed with the different doping time which could influence the charge carrier mobility and the electrical conductivity of thin films [18].

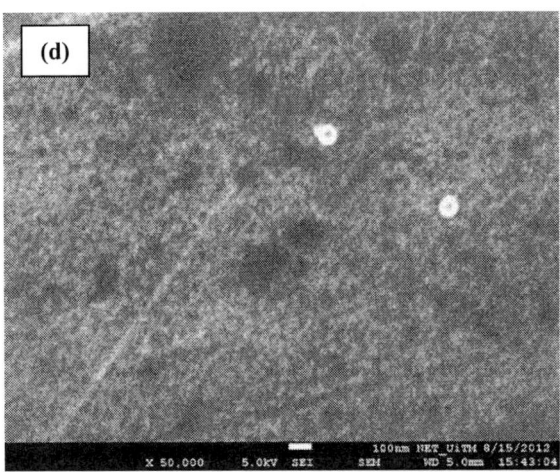

Figure 7. FESEM micrographs of a-C thin film and a-C:I thin film with different doping time. (a) undoped, (b) 10 min, (c) 20 min, (d) 30 min.

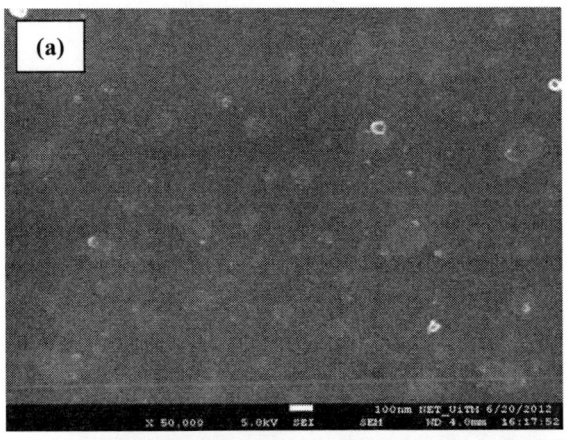

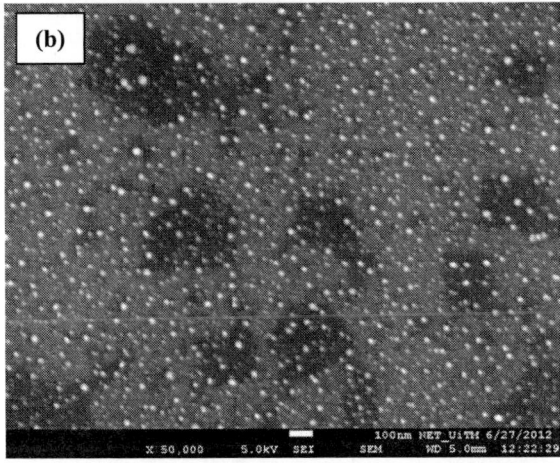

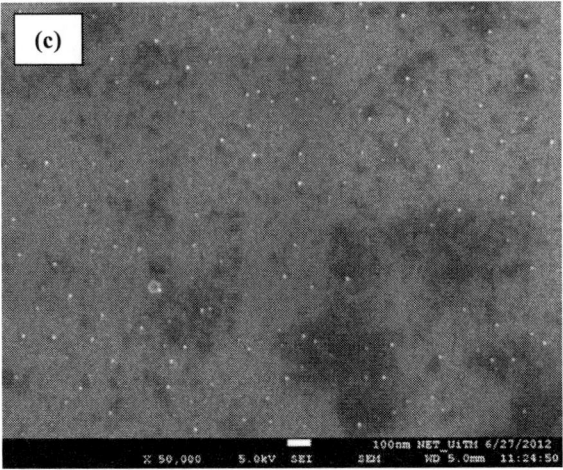

D. FTIR spectral analysis

In order to demonstrate the effect of iodine doping into a-C, the FTIR spectra were measured at room temperature. Figure 8 shows the FTIR spectra of the a-C thin films before and after iodine doping for 10 min in the region from 2000 to 3000 cm^{-1}. The observed peak FTIR spectra at 2926 and 2957 cm^{-1} are corresponding to sp^3-CH_2 asymmetric and sp^2-CH_2 olefinic bonds respectively which is the stretching vibrations of C-H groups. Our results also agree with the literature reported by M. Rusop et. al. [19], that the FTIR absorption for C-H stretching vibration on sp^2 bonded carbon is found in the 1950-3060 cm^{-1} range, while for sp^3 bonded carbon is found in the 2850-2945 cm^{-1} range. As shown in Figure 8, it is apparent that the FTIR spectra of the undoped a-C thin film shows higher C-H band in the range 2850-2945 cm^{-1} compared to the iodine doped a-C thin film. This result shows that iodine doping helps to decrease the sp^3 bonded carbon in a-C thin film, indicating the increase of conductivity of the a-C thin films.

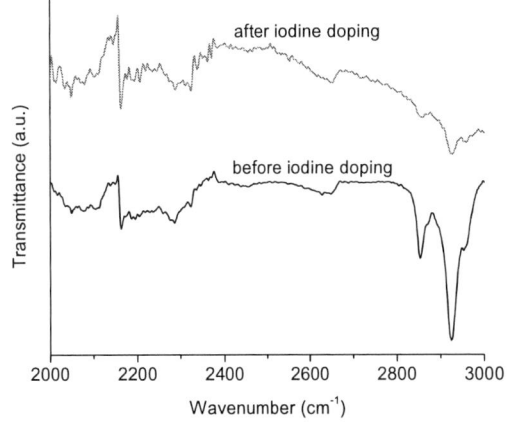

Figure 8. FTIR spectra of undoped and iodine doped a-C thin films for 10 min.

IV. CONCLUSION

The effect of doping time on the optical and electrical properties of iodine doped a-C thin films deposited by Thermal CVD technique has been investigated. The UV-Vis-NIR spectroscopy studies have been performed for the investigation of the optical band gap of a-C thin films and show a reduction of the optical band gap upon iodine doping. The optical band gap of a-C thin films decreased from 0.513 to 0.136 eV as the doping time increase which could be attributed to the formation of graphitization in the films. The electrical conductivity was increases by iodine doping. The higher electrical conductivity was found to be the lowest doping time which is 10 min. The structural properties were investigated to support the optical and electrical properties. FESEM of the iodine doped a-C thin film for 10min doping shows a uniform distribution of fine small particles indicating deposition of iodine atoms on the substrates. The results obtained from the above are well matched with the results evaluated from FTIR measurement that shows the effect of iodine on reduction on sp^3 bonded carbon in a-C thin film which responsible for a low optical band gap and a high value of the conductivity. These results show that there is the possibility of the improvement of the electrical properties by optimizing the iodine dopants for future prospects of high efficiency carbon solar cells.

ACKNOWLEDGMENT

The authors of this paper would like to express their deepest appreciation to UiTM, Research Excellence Fund (600-RMI/ST/DANA 5/3/Dst (396/2011)), Research Management Institute (RMI) and Ministry of Higher Education (MOHE) for the facilities and the financial support.

REFERENCES

[1] K. Dayana, A. N. Fadzilah, and M. Rusop, "Optical properties of amorphous carbon thin films deposited by thermal CVD using camphor oil," in *Research and Development (SCOReD), 2011 IEEE Student Conference on*, 2011, pp. 25-29.

[2] Ishpal, O. S. Panwar, M. Kumar, *et al.*, "Effect of ambient gaseous environment on the properties of amorphous carbon thin films," *Materials Chemistry and Physics*, vol. 125, pp. 558-567, 2011.

[3] C. W. Tan, S. Maziar, E. H. T. Teo, *et al.*, "Microstructure and through-film electrical characteristics of vertically aligned amorphous carbon films," *Diamond and Related Materials*, vol. 20, pp. 290-293, 2011.

[4] P. K. Chu and L. Li, "Characterization of amorphous and nanocrystalline carbon films," *Materials Chemistry and Physics*, vol. 96, pp. 253-277, 2006.

[5] L. Kumari, V. Prasad, and S. V. Subramanyam, "Effect of iodine incorporation on the electrical properties of amorphous conducting carbon films," *Carbon*, vol. 41, pp. 1841-1846, 2003.

[6] L. Kumari, S. V. Subramanyam, A. Gayen, *et al.*, "Characterization and thermal stability of iodinated amorphous conducting carbon films," *Thin Solid Films*, vol. 471, pp. 252-256, 2005.

[7] S. Adhikari, D. C. Ghimire, H. R. Aryal, *et al.*, "Boron-doped hydrogenated amorphous carbon films grown by surface-wave mode microwave plasma chemical vapor deposition," *Diamond and Related Materials*, vol. 15, pp. 1909-1912, 2006.

[8] C.-S. Park, S. G. Choi, J.-N. Jang, *et al.*, "Effect of boron and silicon doping on the surface and electrical properties of diamond like carbon films by magnetron sputtering technique," *Surface and Coatings Technology*, 2012.

[9] M. Rusop, S. Adhikari, A. M. M. Omer, *et al.*, "Effects of methane gas flow rate on the optoelectrical properties of nitrogenated carbon thin films grown by surface wave microwave plasma chemical vapor deposition," *Diamond and Related Materials*, vol. 15, pp. 371-377, 2006.

[10] A. M. M. Omer, S. Adhikari, S. Adhikary, *et al.*, "Iodine doping in amorphous carbon thin-films for optoelectronic devices," *Physica B: Condensed Matter*, vol. 376–377, pp. 316-319, 2006.

[11] M. Rusop, X. M. Tian, S. M. Mominuzzaman, *et al.*, "Photoelectrical properties of pulsed laser deposited boron doped p-carbon/n-silicon and phosphorus doped n-carbon/p-silicon heterojunction solar cells," *Solar Energy*, vol. 78, pp. 406-415, 2005.

[12] A. M. M. Omer, S. Adhikari, S. Adhikary, *et al.*, "Effects of iodine doping on optoelectronic properties of diamond-like carbon thin films deposited by microwave surface wave plasma CVD," *Diamond and Related Materials*, vol. 13, pp. 2136-2139, 2004.

[13] P. P, "Optical properties of amorphous carbons and their applications and perspectives in photonics," *Thin Solid Films*, vol. 519, pp. 3990-3996, 2011.

[14] A. M. M. Omer, S. Adhikari, S. Adhikary, *et al.*, "Electrical conductivity improvement by iodine doping for diamond-like carbon thin-films deposited by microwave surface wave plasma CVD," *Diamond and Related Materials*, vol. 15, pp. 645-648, 2006.

[15] X. M. Tian, M. Rusop, Y. Hayashi, *et al.*, "A photovoltaic cell from p-type boron-doped amorphous carbon film," *Solar Energy Materials and Solar Cells*, vol. 77, pp. 105-112, 2003.

[16] M. Umeno, S. Adhikary, H. Uchida, *et al.*, "Amorphous carbon thin film deposition by microwave surface-wave plasma CVD for photovoltaic solar cell," in *Photovoltaic Specialists Conference, 2005. Conference Record of the Thirty-first IEEE*, 2005, pp. 163-166.

[17] L. Klibanov, M. Allon-Alaluf, N. Croitoru, *et al.*, "Study of photoconductivity in thin amorphous diamond-like carbon (a:DLC) films prepared by r.f. glow discharge technique," *Diamond and Related Materials*, vol. 5, pp. 1414-1417, 1996.

[18] Z. Zhuo, F. Zhang, J. Wang, *et al.*, "Efficiency improvement of polymer solar cells by iodine doping," *Solid-State Electronics*, vol. 63, pp. 83-88, 2011.

[19] M. Rusop, T. Kinugawa, T. Soga, *et al.*, "Preparation and microstructure properties of tetrahedral amorphous carbon films by pulsed laser deposition using camphoric carbon target," *Diamond and Related Materials*, vol. 13, pp. 2174-2179, 2004.

978-1-4673-2395-6/12 $31.00 © 2012 IEEE

The Effect of Exposure Time and Development Time on Photoresist Thin Film in Micro/Nano Structure Formation

Tijjani Adam and U. Hashim
Nano Structure biosensor Research Group,
Institute of Nano Electronic Engineering (INEE),University Malaysia Perlis (UniMAP)
uda@unimap.edu.my, tijjaniadam@yahoo.com

Abstract—**precise transfer of pattern means guarantee in high repeatability and reliability, high throughput and low cost of ownership. By improving this resolution and alignment precision the minimum size can be further reduced to 1nm and beyond. The other important aspect of achieving minimum precised size is the photo resist must be very sensitive to the exposure light to achieve reasonable throughput. However, if the sensitivity is too high, other photoresist characteristics can be affected, including the resolution. Thus, the paper present a preliminary study on fundamentals of resist exposure and development mechanisms for fabrication of Micro- Nanowire formation, We demonstrated significance of considering process parameters such as quality of resist, soft bake, exposure time and intensity, and development time.**

Keywords- Photoresist; Minimum size; fabrication; Development; alignment; critical dimension

I   INTRODUCTION

With high aspect ratio and unique electronic and optical properties, Nanowire have a large potential for applications in Biomedical-electronic Nano devices, including using as sensor for life science examination, microwave amplifier driving circuits for active matrix liquid crystal displays [1], low energy consumption of Nanowire light emitting diode displays [2], and field emission transistors [3]. Besides, Nanowire also have an ideal shape to act as the probes of atomic force microscopes or scanning tunneling microscopes [4].Therefore, to fit the demand of practical applications, more techniques for controlling the shape, sizes and qualities of Nanowire efficiently thus become very important. Nanowire formed by various mechanisms is often randomly oriented [5-7]. Although, vertical alignment of as-grown Nanowire has been demonstrated [8-10], only limited material systems can be applied. Practical devices require having wires parallel to each other and possibly parallel or perpendicular to the substrate surface. The common way of fabricating aligned wires by photolithography and need to carefully design. The Nanowire fabrication needs more 15 patterning process steps; each one must be precisely align with previous requirements such as alignment which confirm with the critical dimensions to achieve successful pattern transfer for the whole device design. For technology of device with nano size, misalignment must be controlled to operate within tolerable error limit or total elimination of misalignment but in most cases, such precision needs automatic alignment systems. This is a very challenging, since not every researcher in this field can afford such device but with careful design procedure, researchers can eliminate error dimension and yet maintain low cost of fabrication [11- 14]. Hence, in this study, we demonstrated a good designed and fabrication method for realization of a minimum possible nano sizes wires, we have carried out the design of four wires of size 1nm, 2nm, 3nm, and 5nm which we were successfully established not only the design and fabrication parameters but we went further to establish some critical parameters such as resist coating, soft bake, exposure time and intensity, and development time used in realization of nano structures.

II.   MATERIALS AND METHOD

The study comprises of device design and fabrication the design comprises of two masks; The layout and specification of the design can be seen here in figure1, After the design, the device fabrication process follows which consists of Nanowire fabrication and gold pad formation, where the gold is used probing and the nano wire and sensor, the fabrication comprises of 15 photoligraphy processes, it begins with wafer preparation and this is done by cleaning the wafer with acetone first to remove all unwanted particle may sticked to the wafer surface, this process followed by depositing silicon nitride to isolate the silicon substrate from subsequent structures to be built thereafter. The nano structure formation was started by depositing polysilicon on the insulated substrate which is also followed by positive photoresist coating for pattern transfer process and after the photoresist coating, alignment exposure follows to realize the polysilicon microwire, after the microwire formation, the wire trimming process follows and this is done through plasma oxidation process. The gold pad formation was carried out on the trimmed Nanowire by first, depositing Titanium metal and thereafter followed by gold deposition the process flow is shown in figure2.

978-1-4673-2395-6/12 $31.00 © 2012 IEEE

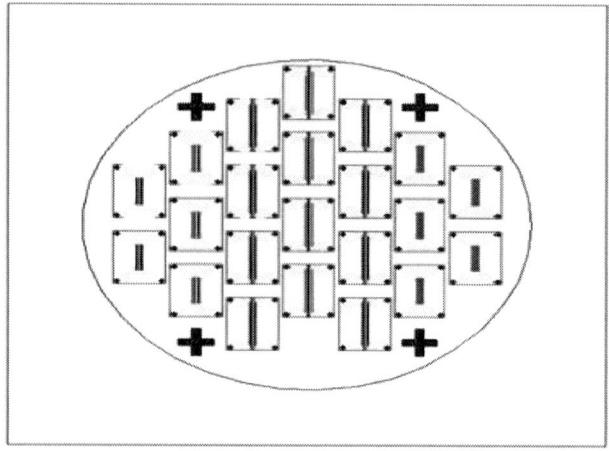

(a)

(b)

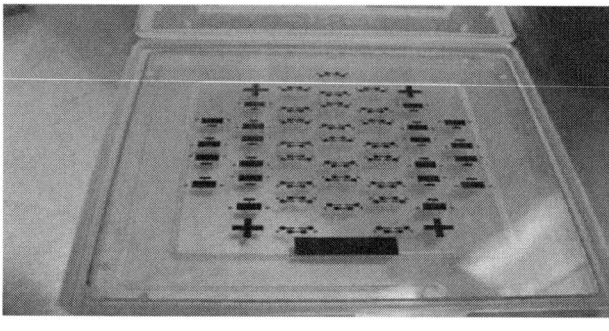

(c)

Fig.2. (a) Assembled mask design layout showing the combined device (a) wire mask (b) Electrode mask on chrome mask

A. Device Fabrication

The polysilicon Nanowire fabrication consist of five major photoligraphy processes: Photoresist coating, alignment, exposure, photoresist development, and etching process. At the beginning, The polysilicon wafer is cleaned and coated with a thin layer of positive photoresist (+PR), which we exposed to ultraviolet light through a mask (f) after the exposure, the

exposed part's cross-link broke down become softened due to the photochemical reaction called photosolubilization and later dissolved by developer while un exposed part remained on the wafer surface.

After the coating process, etching process was conducted to remove the unwanted part of polysilicon on the wafer surface to form polysilicon micro wire with the sizes of 1µm, 2µm and 3µm to confirm with our designed after the micro wire formation the trimming process follows by using plasma processes to required nanosizes this is done silicon dioxide grow by consuming polysilicon. The thickness of the polysilicon consumed depends on the total penetration of the oxide and this is controlled and limited by movement of the oxygen through the oxide-silicon interface.

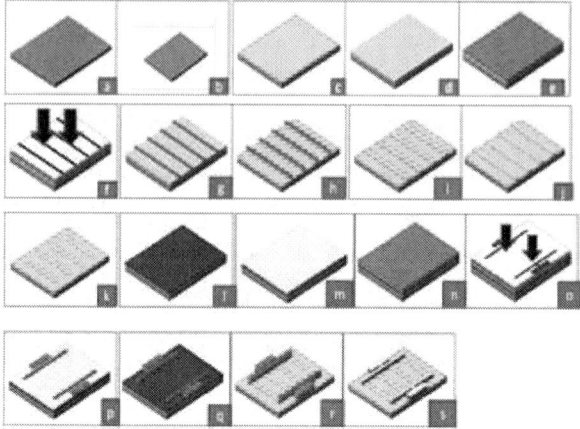

Fig.2. Fabrication process flow of Polysilicon Nanowire with common electrode (a) silicon wafer (b) wafer preparation (c) wafer insulation (Si_3N_4) (d) polysilicon deposition (e) Photo resist coating (f) Alignment and exposure (g) Resist development (h) Resist striping (i) polysilicon microwire (j) Trimming process by plasma oxidation (k) polysilicon Nanowire (i) Titanium (Ti) deposition for metallic connection (m) Gold deposition for contact formation (n) photoresist coating (o) Alignment and exposure (p) resist development (q) gold etching (r) resist stripping (s) common electrode polysilicon Nanowire

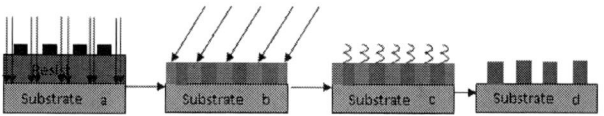

Fig.3: (a) Shows exposure using a positive mask for positive resist the unexposed areas finally remain (b) Shows the resists area which is exposed and soluble in developer (c) The hard bake crosslink's unexposed area, while the exposed area remains photoactive(d) After development, the areas unexposed in (a)now remain

B. Exposure and Development

We used samples with same viscosity through out the experiments with different set of the resist exposure and development. The resist was spin coated on polysilicon wafers with thicknesses of 0.3µm. The resist coated wafers were baked at 75 °C for 3 minutes on a hotplate. The exposure process is shown in figure8. Developing time was chosen at least twice the clearing time. Standard values were 20 s, 25 s and 30s, corresponding to the suitable exposure time. The exposure and

978-1-4673-2395-6/12 $31.00 © 2012 IEEE   108

development characteristics were measured using High power microscope. Before the sample subjected to hard bake, the exposed resist needs a certain time depending on the resist type and thickness and for purpose of this study we are using positive photoresist with thickness of 1μm and allowed it to stay for 3mins to outgas nitrogen (N2) formed during exposure. This will avoid bubbling (irregular developed structures and foaming of the resist by thermally activated N2. Nitrogen preferentially accumulates near locations with inferior resist adhesion to the substrate, which can also be optimized with substrate pre-treatment by a sufficient soft bake. Photo mask and resist: Reflection causes number of problems for photo resist pertaining, reflection from the under laying substrate can cause standing wave in the photoresist or scattering from the under laying substrate may lead to photoresist exposure in undesired area. The Gap between Photo mask and resist surface extends the diffraction pattern and therefore deteriorates the resolution therefore, the exposure process was done with the photo mask tightly closed with substrate to avoid possible(i)Particles in the resist caused by either insufficient clean room conditions, contaminated substrates, or expired photoresist(ii) bubbles in the resist film caused during dispensing, or an insufficient delay time after refilling/diluting/moving the resist(ii) mask contamination by particles, or resist from previous exposure steps

II   RESULT AND DISCUSSION

A preliminary study based on high power microscope measurements on fundamental resist exposure and development mechanisms are presented: When the exposure time is too low, the development time of positive resists is high which increases the total dark erosion and If the exposure time is too low, the development time increase too high exposure time values cause light scattering and diffraction in the resist film which deteriorates the resolution., while resists show an increased erosion of the exposed areas due too a weak cross-linking figure4a, the thickness of the resist is a very important factor be considered because "thick" resist means a film thickness much higher than the penetration depth of the exposure light this will cause Nonuniform exposure and irregular figure4b. Too high exposure time cause an undesired exposure by scattering, diffraction and reflection of the part of the resist which should not be exposed. As a consequence, too much resist is cleared during development figure4c. On different substrate, the resist film will show a different colour, especially in case of thin resist films, a change of the resist film thickness of only few 10 nm will cause an interference-based colour change of the resist film. Such a thickness change can be induced by solvent loss, resist ageing, changed temperature or air humidity, or modified coating equipment or parameters figure4h, A show big lumps and coating failures occurred figure4d may due to either expired resists or resists stored under the wrong (temperature) conditions, as well as resists diluted too strongly or diluted with unsuitable solvents, may form particles and we think from preliminary study this assumption cannot be excluded, in the next experiment we discovered that the ageing of the resist due to elevated temperature and high resist dilution ratio are caused of the

failure which discovered in figure4e, f and g . In early stage averaging, the coated resist reveals a silky resist surface figure 4g. With a serious consideration of steps which started from soft bake after coating produce a good result shown in figure 4i, since A soft bake too cool/too short keeps the remaining solvent concentration too high allowing the N_2 formed during exposure to form bubbles in the resist film. A soft bake temperature of 100°C for 1 minute/μm resist film thickness was used. The study also revealed that N_2 generated during exposure of positive tone and image reversal resists may form bubbles at locations of minor resist adhesion to the substrate. The substrate should be clean and The N_2 generated during exposure needs to dissipate from the resist film before its concentration becomes too high. If the exposure intensity is too high, the N_2 cannot timely outgas and therefore forms bubbles or resist cracks due to mechanical stress. Working with lowering intensity exposure will solve the problem figure4k into several steps with delays in-between and very thick resist films, it becomes more difficult for the N_2 formed during exposure to diffuse towards the resist film surface. Thus, bubble formation is much more pronounced during thick resist processing. For this reason, we used we recommend suited thick resists with a lower photo active compound concentration causing less N_2 during exposure.

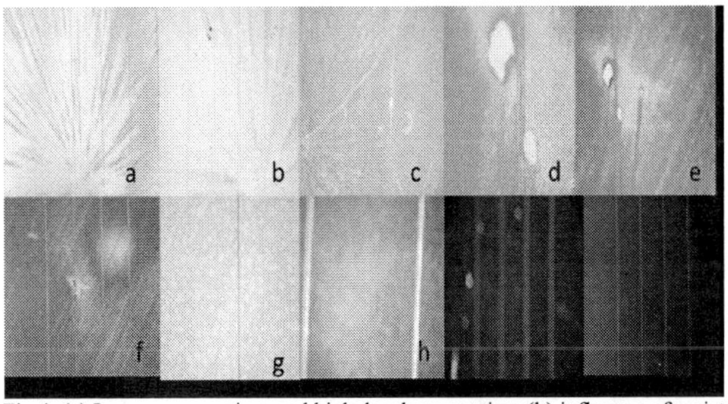

Fig.4. (a) Low exposure time and high development time (b) influence of resist thickness (c) undesired exposure caused by scattering, diffraction and reflection (d) show big lumps and coating failure (e, f and g) ageing of the resist due to elevated temperature and high resist dilution ratio (h) the resist film show a different color (i) High speed coated wafer to improve uniformity (j) lowering intensity exposure [14]

The other important aspect is the light spectrum, In the I-line stepper, the monochromatic light as interference effects that strongly influence the energy coupling in the resist film. This energy coupling causes the variation in dose-to-clear resist as resist thickness varies. However the variation is minimizing when the resist thickness is at a quarter lambdas divided by refractive index minimized when the resist thickness is between quarter and half of lambda divided by the refractive index. This maxima and minima values can be determines by investigating the swing curve. The swing curve is periodic and the curve descending as the resist thickness decreases. Thus, it is very important to find the swing curve to determine the change in line-width caused by resist thickness variation. In other words, small resist thickness variation can caused big variation in critical dimension (CD) depending on the resist thickness. Swing curve can also be described as the dose-to-clear versus thickness plot. In the experiment, the values for does-to-clear or clearing point were obtained by ranging both resist thickness. This mass sensitivity of fabricated device was obtained using simple electrical characterization set up. A current-voltage measurement were followed and the studies were performed at room temperatures, to evaluate the sensitivity of the polysilicon Nanowire, we took the measurement at four different time at 4hrs interval this is gain strong assessment and see the consistency of the ,for the different results reported here were measured at room temperature. A peak of the nanowire's difference signal as a function its dimension, to characterize the sensitivity, In this case, the dimension with the greatest sensitivity is when the Nanowire is smaller, the sensitivity of Nanowire is linearly dependent upon the on the thickness of the wire, this dependency suggests that that conductivity of the polysilicon nano wire (analogous to the conventional electrical conductor) is operating within these phenomena. In addition, these mass sensitivities could be significantly improved by further reducing the size of the nanowire's and It is expected that thinner nanowire's may further enhance the sensitivity of the devices due to an increased surface to volume ratio, which may lead to the realization of single biomolecules detection, for better understanding the sensitivity of the sensing element, more vigorous study is needed a parameter such resonant frequencies with concise resolution and Such resolution could conceivably allow the observation to single stranded DNA single to events of relatively more concentrated molecules and proteins sample. However, our experimental setup is currently not equipped with the advanced testing systems required for such experiments.

III CONCLUSION

The fabrication of Polysilicon Nanowire has some very important factors to be considered. The development of Nanoscience process technology is measured by the shrinking of the minimum feature size on the production wafer. The smaller the minimum feature size, the smaller the device can be produced but this is limited by the photolithography resolution. By the improving this resolution, the minimum size can be further reduced beyond the current attainable sizes and other important aspect of achieving minimum precised size is, the photo resist must be very sensitive to the exposure light to achieve reasonable throughput. However, if the sensitivity is too high, other photoresist other characteristics can be affected, including the resolution. To achieve complete pattern transfer, we came to realized that a perfect design is needed; the photoresist needs to have good resolution, high etch resistance and good adhesion. High resolution is the key to achieve successfully pattern transfer. Without high etch resistance and good adhesion of the PR, the next etch processes will most likely fail to meet process requirement and will cause intolerable error and we observed that the thinner the photoresist film, the higher the resolution. However, the thinner the photoresist, the lower the etching resistance, thus in

this study, it is clearly shown how important have good design on which all further processes depends. In summary, we have demonstrated that the fabrication of Nanowire from few nanometers to 1 nanometer is possible with simple conventional photoligraphy and sensitivity of NW biosensor largely depends on the surface to volume ratio of the sensing element.

ACKNOWLEDGEMENT

The authors wish to extend his sincere appreciation to Universiti Malaysia Perlis (UniMAP) for giving the opportunities to use the research facilities in the Bio-chip Fabrication and characterization Lab and Ministry of Science, Technology & Innovation (MOSTI) for providing us with grant to carry out this research. The appreciation also goes to all the team members in the Institute of Nanoelectronic Engineering especially in the Nano Biochip Research Group.

IV REFERENCE

[1] Cui JB, Gibson UJ. Electrodeposition and room temperature ferromagnetic anisotropy of Co and Ni-doped ZnO Nanowire arrays. Appl Phys Lett 2005; 87: 133108.

[2] Wang X, Song J, Li P, Ryou JH, Russell D, Christopher D,Summers J, Wang ZL. Growth of uniformly aligned ZnO nanowire heterojunction arrays on GaN, AlN, and Al0.5Ga0.5N Substrates. J Am Chem Soc 2005; 127: 7920-7923.

[3] Duan X, Huang Y, Cui Y, Wang J, Lieber CM. Indum phosphide nanowires as building blocks for nanoscale electronic and optoelectronic devices. Nature 2001; 409: 66-69

[4] Parthangal PM, Cavicchi RE, Zachariah MR. A universal approach to electrically connecting nanowire arrays using nanoparticles-application to a novel gas sensor architecture, Nanotechnology 2006; 17:3786-90.

[5] Xua CL, Qina DH, Lia H, Guoa Y, Xub T, Lia HL. Low temperature growth and optical properties of radial ZnO nanowires. Materials Lett 2004; 58: 3976-79.

[6] Yao BD, Chan YF, Wang N. Formation of ZnO nanostructures by a simple way of thermal evaporation. Appl Phys Lett 2002; 81: 757-759.

[7] Kim H, Wolfgang S. Zinc oxide nanowires on carbon nanotubes.Appl Phys Lett 2002; 81: 2085-2587.

[8] Levin I, Davydov A, Nikoobakht B, Sanford N. Growth habits and defects in ZnO nanowires grown on GaN/sapphire substrates. Appl Phys Lett 2005; 87: 103110.

[9] Wang X, Song J, Li P, Ryou JH, Russell D, Christopher D, Summers J, Wang ZL. Growth of uniformly aligned ZnO nanowire heterojunction arrays on GaN, AlN, and Al0.5Ga0.5N Substrates. J Am Chem Soc 2005; 127: 7920-7923.

[10] Zhang Y, Wang L, Liu X, Yan Y, Chen C, Zhu J. Synthesis of nano/micro zinc oxide rods and arrays by thermal evaporation approach on cylindrical shape substrate. J Phys Chem B 109: 13091-93.

[11] Uda Hashim, Siti Fatimah Abd. Rahman, M. Nuzaihan Md. Nor, Shahrir Salleh," Design and Process Development of Silicon Nanowire Based DNA Biosensorusing Electron Beam Lithography", 2008 International Conference on Electronic Design, December 1-3, 2008, Penang, Malaysia

[12] Bien DCS, Badaruddin SA, Saman RM: Method of fabricating nanowires, Malaysian Patent Office, MyIPO, PI 20097036

[13] Tijjani Adam, U.Hashim, Pei Ling Leow , Pei Song Chee and K. L. Foo "Fabrication of PDMS multi-layer microstructure: The electroosmosis mechanism in fluidics for life sciences "Enabling Science and Nanotechnology (Escinano), 2012 International Conference, 5-7 Jan. 2012, pp 1 – 4

[14] Tijjani Adam, U.Hashim, Pei Ling Leow , Pei Song Chee and K. L. Foo "Mask Design for the reproducible fabrication and reliable pattern transfer for polysilicon Nanowire "Enabling Science and Nanotechnology (Escinano), 2012 International Conference, 5-7 Jan. 2012, pp 1 – 4

Substrate types and deposition pressure dependences of RF-magnetron sputtered Silicon thin films characteristics deposited at room temperature

[1]S. B. Hashim, [1]N. H. Mahzan, [1]S. H. Herman, [1]R. Abu Bakar, [1]U. Mohd Noor, and [1,2]M. Rusop

[1]NANO-ElecTronic Centre (NET), Faculty of Electrical Engineering, Universiti Teknologi MARA (UiTM), 40450 Shah Alam, Selangor, Malaysia

[2]NANO-SciTEch Centre (NST), Institute of Science, Universiti Teknologi MARA (UiTM), 40450 Shah Alam, Selangor, Malaysia

shaiful_bakhtiar@yahoo.com

Abstract—**Silicon thin films was successfully deposited on glass and Teflon substrates at room temperature by using radiofrequency (RF) magnetron sputtering. The effect of deposition pressure on crystallinity and structural properties of the thin films on glass and Teflon substrates was studied. From Raman spectroscopy results it showed that the highest peak was around 512 cm^{-1} on Teflon substrate indicating the existing of poly-Si phase. The crystalline quality of the Si films improved with the increasing sputtering pressure, for films deposited on Teflon substrates. While, the crystalline quality of the films on glass substrates deteriorated with the increasing sputtering pressure. The different on surface roughness and thermal conductivity for both glass and Teflon substrates contribute on the different crystallinity for both substrates.**

Keywords: Silicon thin films; glass substrate; Teflon substrate; RF magnetron sputtering; Deposition pressure.

I. INTRODUCTION

Polycrystalline and nanocrystalline silicon (poly-Si, nc-Si) has received special attention for the use in large area devices, such as solar cells [1, 2] and thin film transistors [3, 4]. In general, nc-Si thin films are prepared by low chemical vapor deposition (LPCVD) [5] and hot-wire enhanced chemical vapor deposition (HWCVD) [6] methods. Disadvantages of CVD methods are the use of extremely toxic gases such as silane, diborane and phosphine. Meanwhile, for poly-Si thin films method such as solid phase crystallization (SPC) [7], excimer laser annealing (ELA) [8] and metal induced crystallization (MIC) [9] which convert amorphous silicon (a-Si) films into crystallized poly-Si films are been favorable. ELA method can be done at low temperature which is suitable for glass substrates but there are some problems such as non-uniformity of grain growth on large area glass substrates and expensive processing costs. Although the SPC produce uniform thin films, but longer processing time (4-64hours) and high process temperature (~650°C) are their drawbacks.

Magnetron sputtering is also promising methods for the preparation of silicon (Si) thin films, because the crystallinity of the poly-Si thin films can be controlled easily [10]. Thin films growth can be substantially modified by ion bombardment during sputtering [11]. In magnetron sputtering, there are few parameters can be controlled during this sputtering process to get good crystalline quality such as sputtering power, substrate bias power, temperature, pressure and mixing gas.

In modern day, flexible substrates have represented a new form of electronics large area circuit with rapidly rising and promising application such as flat panel displays (FPDs), solar cell and other disposable and wearable electronic devices. These are because of some advantages of the flexible substrates over the conventional Si and glass substrates, such as light weight, unbreakable, flexible and remarkable [12].

Higher temperature deposition is needed to growth both poly-Si and nc-Si thin films with good crystalline quality, so it difficult to growth both poly-Si and nc-Si films at lower temperature. However for future electronics application especially those using flexible substrates, lower depositions are needed because the melting point for flexible substrate is lower than conventional glass substrates.

In this study, we attempted to prepare poly-Si thin films at room temperature by a radiofrequency (RF) magnetron sputtering method from undoped silicon wafer target and investigated the crystallinity and structural properties of the films at different deposition pressure on different substrate. We present Teflon as flexible substrate and compared the effect on the crystallinity and structural properties with glass substrate. The crystallinity of thin films was characterized by Raman spectroscopy and structural properties were characterized by Field emission scanning electron microscopy (FESEM).

II. EXPERIMENTAL

Si thin films were deposited directly on Teflon and glass (slide glass) substrates at room temperature by using RF magnetron sputtering system using a pure (99.999%) n-type silicon target (4 inches diameter and 0.25 inches thickness). Prior to each deposition, the base sputtering pressure was evacuated down to (~ 10^{-7} mTorr), and for about 120 seconds,

978-1-4673-2395-6/12 $31.00 © 2012 IEEE

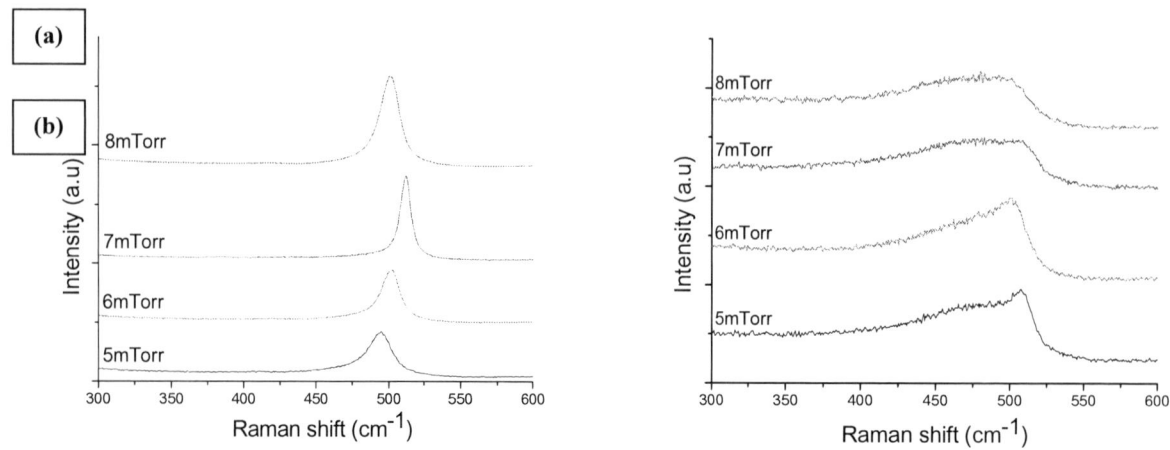

Fig. 1: Raman spectroscopy result for films deposited at different deposition pressure on glass (right) and Teflon (left) substrates

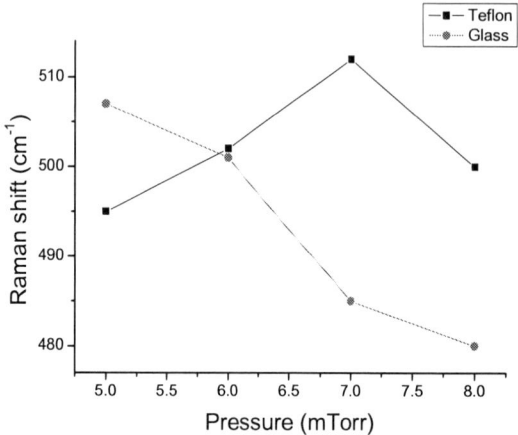

Fig. 2: Graph of Raman shift over deposition pressure for Teflon

the target was pre-sputtered to remove any impurity on the surface of the target. Prior to the sputtering deposition, the Teflon and glass substrates were cleaned by the conventional organic cleaning using acetone, ethanol and deionized water (DI water) in ultrasonic bath for 10 minutes respectively for each solution. During the deposition process, the RF sputtering power was maintained at 200W for 60 minutes. The pressure was varied from 5 mTorr to 8 mTorr in pure argon atmosphere.

Crystalline properties of thin films on Teflon and glass substrates were measured with a laser Raman scattering spectrometer using a semiconductor laser of 532 nm wavelength, over a range of Raman shifts from 300–600 cm^{-1}. Surface morphology of thin films was characterized with a FESEM.

III. RESULT AND DISCUSSION

The crystallization of the deposited Si films was measured by Raman spectroscopy. Fig. 1(a) and 1(b) shows Raman

scattering result of thin films deposited using deposition pressure of 5, 6, 7, and 8 mTorr on different substrates; (a) Teflon and (b) Glass. Fig. 1(a) clearly shows that for thin film deposited at 5 mTorr, the transverse optical (TO) peak located below 500 cm^{-1} which means that the thin film is not crystallized [13] under this condition. But when deposition pressure increase from 6 mTorr to 8 mTorr, the Si-Si TO peaks ranging from 502 to 512 cm^{-1} are observed, indicating that the thin film partially crystallized. Considering that the single crystalline Si peak is at 520 cm^{-1} [13], Fig. 1(a) indicates that the thin film deposited at 7 mTorr gives the highest crystallinitiy among the other with the Raman peak shift at around 512 cm^{-1}. It also indicating the existence of poly-Si phase for the sample deposited at 7 mTorr on Teflon substrate.

Fig. 1(b) shows films deposited on glass substrate, at 5 mTorr deposition it shows that the TO located at 508 cm^{-1} and when the deposition pressure at 6mTorr the TO decrease to 500 cm^{-1}. It can be seen clearly that, the sharp peak of TO at 5

978-1-4673-2395-6/12 $31.00 © 2012 IEEE

mTorr become broad when it reach 8 mTorr. So, the highest Raman spectra for thin films deposited on glass substrate is only 508 cm^{-1} which can be categorized as nc-Si thin films.

Fig. 2 shows the tendency graph of Raman shift over deposition pressure for Teflon and glass substrates. It shows that, for thin films deposited on Teflon substrates, the number of Raman shift increase with increasing deposition pressure indicating the improvement in crystalline quality. Meanwhile, for thin films deposited on glass substrates the number of Raman shift decrease with decreasing deposition pressure, suggesting an opposite phenomena. This shows that the type of the substrate has quite an effect on the crystal growth of the Si film. This is probably due to the different thermal conductivity between glass (0.8~1.4 W/cm.K) and Teflon (~0.25 W/cm.K) in which glass has higher thermal conductivity. Even though in this work the substrate is not heated intentionally, the heat from the plasma may be transferred to substrate thus providing energy to the arriving Si atoms to construct themselves to form crystalline structure. Further, the difference on surface roughness on both Teflon and glass substrates may contribute to the different crystallinity.

Considering crystallinity dependences on the sputtering pressure, theoretically, at lower deposition pressure, higher bombardment occurs and there are two effects on the deposited thin films, one is to damage the crystallize particles and the other is to enhance surface migration of the deposited particles and to promote the crystallization of the films. It is supposed that the damage become more serious when particles with higher kinetic energy bombard a film with higher quality. On the other hand, crystallization becomes more effective when the film quality is poorer [14]. This can be seen for the films on

Teflon substrate in which the crystallinity improves with the increasing pressure however degrades at higher pressure.

Fig. 3 shows SEM micrographs of Si films grown on glass and Teflon substrates at deposition pressure 5 mTorr and 8 mTorr. Based on the shape of thin film, there is not much can be differentiate either on glass or on Teflon substrates. Average cluster size for thin films grown on glass substrate at 5 mTorr is around 12 nm which similar for films grown on Teflon substrate. Meanwhile, at deposition pressure 8 mTorr, average cluster size on glass is around 14 nm and 16 nm for Teflon substrate. Theoretically, at lower deposition pressure, the sputtered materials have higher surface mobility and hence, a higher growth rate takes place. Thus, films deposited at lower deposition pressure should have surface with more compact structures due to higher growth rate [14]. It can be seen clearly from films grown on glass substrate, where the surface morphology transform from concise structured to not closely pack on the surface.

IV. CONCLUSION

In conclusion, Si films having a mixture of nanocrystalline and polycrystalline structures were deposited directly at room temperature by RF magnetron sputtering method. The effect of deposition pressure and the type of substrates on the crystallization and structural properties were investigated. We found that the crystalline quality of the Si films improved with the increasing sputtering pressure, for films deposited on Teflon substrates. Based on Raman spectroscopy result, Si film deposited at 7 mTorr gives the highest crystallinity. On the contrary, the crystalline quality of the films on glass substrates deteriorated with the increasing sputtering pressure.

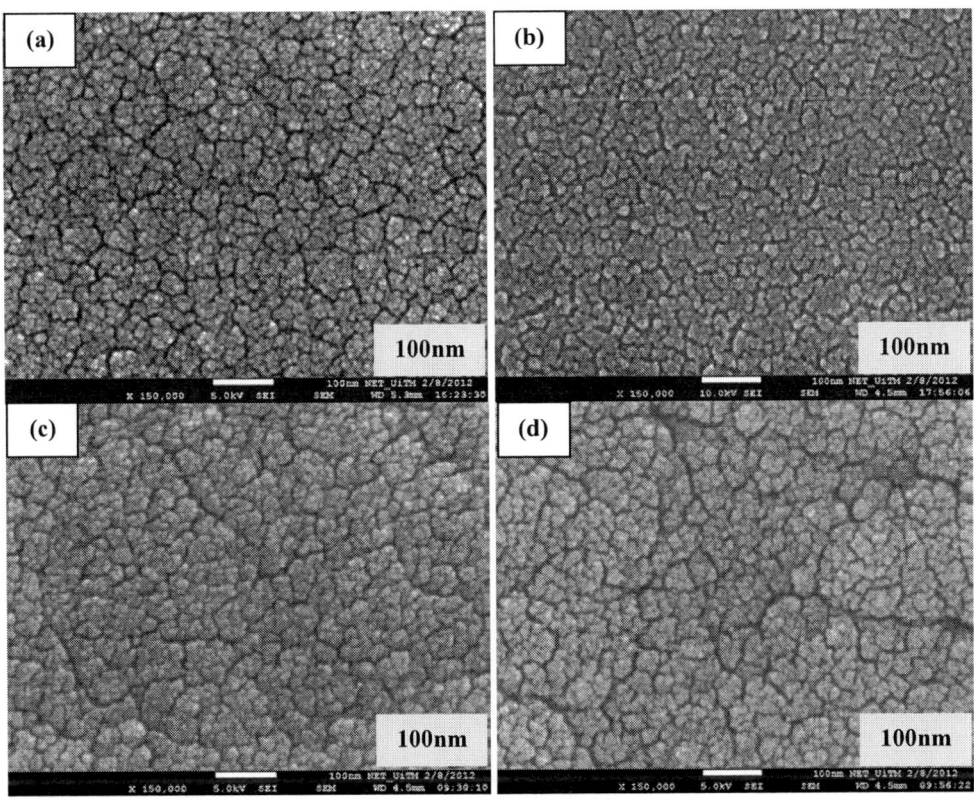

Fig. 3: SEM micrographs of Si films grown on glass substrate (a) 5mTorr; (b) 8mTorr; and Teflon substrate (c) 5mTorr and (d) 8mTorr.

The surface morphology of the Si films were not much different between both substrates and sputtering pressures, however we found a slight increase in the cluster size as the pressure increases. The surface roughness on both Teflon and glass substrates may contribute to the different crystallinity.

Acknowledgment

Authors would like to thank Fundamental Research Grant Scheme (FRGS) Research Management Institute (RMI) Universiti Teknologi Mara, Shah Alam (Project Code: FRGS/1/2012/TK02/UITM/03/8) for their financial support. The authors thank to the technicians and science officers at NET and NST for their kind support in this research.

References

[1] S. Gall, C.Becker, K.Y.Lee, T.Sontheimer and B.Rech, "Growth of polycrystalline silicon on glass for thin-film solar cells," Journal of Crystal Growth 312 (2010) 1277–1281.

[2] Ian Y.Y. Bu, Room temperature synthesis of nanocrystalline silicon by aluminium induced crystallization for solar cell applications, Vacuum 86 (2011) 106-110.

[3] Mutsumi Kimura, "Extraction of trap densities in entire bandgap of poly-Si thin-film transistors fabricated by solid-phase crystallization and dependence on process conditions of post annealing," Solid-State Electronics 63 (2011) 94–99

[4] Masaki Hara, High mobility bottom gate nanocrystalline-Si thin-film transistors, Thin Solid Films 519 (2011) 3922-3924.

[5] M. Matsumoto, Y. Inayoshi, M. Suemitsu, T. Yara, S. Nakajima, T. Uehara and Y. Toyoshima, " Low temperature growth of polycrystalline Si on polyethylene terephtalate (PET) films using pulsed-plasma CVD

under near atmospheric presuure," Thin Solid Films 516 (2008) 6673–6676.

[6] F. Villar, J. Escarré, A. Antony, M. Stella, F. Rojas, J.M. Asensi, J. Bertomeu and J. Andreu, "Nanocrystalline silicon thin films on PEN substrates," Thin Solid Films 516 (2008) 584–587.

[7] S. Gall, C.Becker, E.Conrad, P.Dogan, F.Fenske, B.Gorka, K.Y.Lee, B.Rau, F.Ruske and B.Rech, " Polycrystalline silicon thin-film solar cells on glass," Solar Energy Materials & Solar Cells 93 (2009) 1004–1008.

[8] Chil-Chyuan Kuo, " Dynamical resolidifaciton behavior of silicon thin films during frontside and backside excimer laser annealing," Optics and Lasers in Engineering 49 (2011) 804–810.

[9] Thanh Nga Nguyen, Van Duy Nguyen, Sungwook Jung and Junsin Yi, "The metal-induced crystallization of poly-Si and the mobility enhancement of thin film transistors fabricated on glass substrate," Microelectronic Engineering 87 (2010) 2163–2167.

[10] Y. Leconte, P. Marie, X. Portier, M. Lejeune and R. Rizk, " Pronounced crystallization of silicon layers deposited with high deposition rates at temperatures < 200°C," Thin Solid Films 427 (2003) 252.

[11] P.Reinig, V.Alex, F.Fenske, W.Fuhs and B.Selle, "Pulsed dc magnetron sputtering of microctystalline silicon," Thin Solid Films 86 (2002) 403-404.

[12] P.I. Hsu, M. Huang, S. Wagner, Z. Suo, J.C. Sturm, " Amorphous Si TFTs on plastically deformed spherical domes," Electron-Emissive Materials, Vacuum Microelectronics and Flat-Panel Displays, Materials Research Society Symposium Proceedings 621 (2000) Q8.6.1.

[13] Susumu Horita and Sukreen Hana, "Low-Temperature Crystallization of Silicon Films Directly Deposited on Glass Substrates Covered with Yttria-Stabilized Zirconia Layers, " Japanese Journal of Applied Physics 49 (2010) 105801.

[14] C.H. Tseng, W.H. Wang, H.C. Chang, C.P. Chou, C.Y. Hsu, Effects of sputtering pressure and Al buffer layer thickness on properties of AZO films grown by rf magnetron sputtering, Vacuum 85 (2010) 263-267.

Simultaneous Study of Thermal and Optical Characteristics of Light-Emitting Diode

*Zhi-Yin Lee, Mutharasu Devarajan
Nano-optoelectronic Lab, School of Physics,
Universiti Sains Malaysia,
11800 Penang, Malaysia
*E-mail: lzy10_phy032@student.usm.my

Abstract- **This paper deals with the simultaneous study of thermal and optical behaviors of light-emitting diode (LED) through various measuring conditions. The conditions such as different types of thermal interface materials (TIMs) and increases of driving current have been selected for detailed investigation. The results revealed that the measuring conditions led to a greater impact on thermal properties than optical properties of the LED. The determined junction-to-ambient thermal resistance and junction temperature were enhanced about 1.35% and 0.78%, respectively by replacing alumina to silver thermal compound. This indicates that the application of silver thermal compound provided a slightly weaker heat transfer from the packaged LED to ambient. On the other hand, it was only 0.5% deviation on the overall optical performance was found between both the TIMs. Despite, it was observed that the increase of driving current caused an augment in optical power, and adversely decreased the efficiency of the LED. Finally, the study of real thermal resistance of the LED was performed as to improve the accuracy of the measurement.**

Keywords: Light-emitting diode; real thermal resistance; thermal interface material; driving current

I. INTRODUCTION

Light-emitting diode (LED) is a sophisticated invention of solid-state lighting that including certain important criteria such as high brightness, efficiency, operating lifetime and low power consumption [1]. At high operating power condition, despite LED exhibits the excellent optical properties, the inherently poor thermal management may lead to LED degradation [2]. In addition, heat stored in the parasitic resistances of contact and cladding layers has a great impact on the heating of LED [3]. Various methods have been adopted to improve thermal transfer of an LED, such as selection of packaging material, application of an external heat sink with different design and orientation, micro-jet cooling and development of thermal interface materials (TIMs) [4, 5].

From the pass studies, concurrent characterizations on thermal and optical behaviors of LED have been introduced. The combined measurement of thermal and optical properties of LED provides real thermal metrics as well as optical parameters. This becomes an important indicator to evaluate LED performance [6, 7]. The distinctions between effective thermal resistance, R_{th} and real thermal resistance, R_{thr} have been explained [8]. Besides, there is a report on the improvement of the accuracy of thermal transient measurement. Authors [9] had presented the methodology to increase the accuracy of recorded function based on thermal material parameter measurement and parasitic heat flow path. Even though the heat flow path is assumed to be one-dimensional, parasitic paths always do exist.

In the present work, simultaneous study of the thermal and optical performance of an LED was carried out under the influence of different TIMs and increases of driving current. Next, detailed analyses on the measured R_{th} and R_{thr} were reported in order to accomplish this investigation.

II. THEORETICAL BACKGROUND

The heat flow is assumed to be one-dimensional if the heat lost to surrounding is insignificant and negligible [10]. Heat path can be characterized as the process of conduction and convection. In the case of an LED mounted on metal-core board (MCB), heat conduction is occurring from junction to MCB and heat convection is occurring from MCB to ambient. Fig. 1 shows the schematic diagram of an LED.

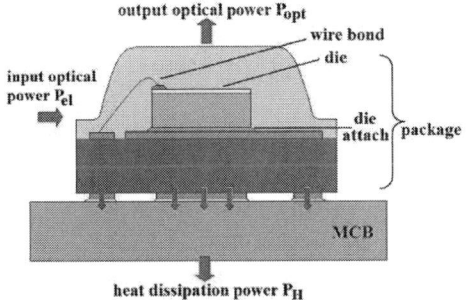

Fig. 1. The schematic diagram of LED.

A. Thermal transient measurement

Recording of the thermal transient by using Thermal Transient Tester (T3Ster, MicRed Ltd.) is a sophisticated approach to study the thermal characteristics of LED [12]. Response function either heating curve or cooling curve can be recorded in real-time. The supporting software is applied to evaluate the recorded functions into cumulative and differential structure functions. The former illustrates the sum of thermal capacitance, ΣC_{th} against the sum of thermal resistance, ΣR_{th}; and the latter determines by the derivative of the cumulative structure function. By referring to structure function, partial thermal resistance, die-attach failure and packaging failure can be described [13]. Thermal resistance of a semiconductor device is a measure of temperature difference by which it resists a heat flow and can be written as [11]

$$R_{thJA} = \frac{T_J - T_A}{P_H} \qquad (1)$$

where R_{thJA} represents thermal resistance from device junction to the specific environment; T_J represents junction temperature in the steady state test condition; T_A represents reference temperature for the specific environment; and P_H represents power dissipation in the device.

B. Optical characterization

In convectional semiconductor device, input electrical power, P_{el} is converted into output optical power, P_{opt} and heat dissipation power, P_H [14]. This is described as

$$P_{el} = P_{opt} + P_H \qquad (2)$$

TERALED provides the combined measurement of photometric/radiometric characterization of high-power LEDs [15]. Optical parameters such as efficiency, luminous intensity, chromaticity coordinates can be determined. Basically, it can be calculated by taking the ratio of P_{opt} to P_{el}, as described in

$$\eta = \frac{P_{opt}}{P_{el}} \qquad (3)$$

The efficiency, η of an LED [16] reduces as the operating temperature increases.

III. METHODOLOGY

In this work, a commercial warm-white LED mounted on MCB was tested. Simultaneous measurement was performed by using T3Ster and TERALED to investigate the thermal and optical properties of the LED. Fig. 2 illustrates the basic measurement setup.

Fig. 2. The basic measurement setup.

A specific fixture was used to connect both the T3Ster and TERALED. The LED was attached on a temperature-controlled holder by using TIM. Two types of TIMs were considered for comparison, which were alumina and silver thermal compounds. Measuring conditions such as driving current of 350mA and 700mA; and constant ambient temperature at (30°C ± 0.01°C) were set. Optical test of the LED was first conducted and at the same time response function (cooling curve) was captured. Supporting software of T3Ster-Master and Teraled-View were applied to evaluate the recorded information.

IV. RESULTS AND DISCUSSION

C. Thermal and optical characterizations of the LED

The differential structure function is presented in order to describe the heat flow manner of the LED. It is useful in terms of interpreting the partial thermal resistances which represented by the separation between each consecutive peak. Fig. 3 illustrates the evaluated differential structure functions for different TIMs and currents.

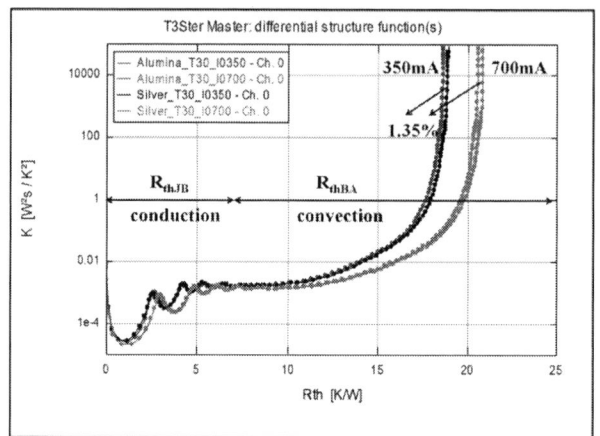

Fig. 3. Differential structure functions for different thermal interface materials and currents.

Heat flow of the LED is described as heat conduction and convection. The conduction process begins from the junction to MCB, known as junction-to-board thermal resistance R_{thJB}; while the convection process occurs from the MCB to ambient, known as board-to-ambient thermal resistance R_{thBA}. It was apparently shown that the evaluated functions were consistently right shifted as the driving current was increased from 350mA to 700mA.

TABLE I
JUNCTION-TO-AMBIENT THERMAL RESISTANCE AND JUNCTION TEMPERATURE FOR DIFFERENT THERMAL INTERFACE MATERIALS AND CURRENTS

Current, I (mA)	Junction-to-ambient thermal resistance, R_{thJA} (K/W)		Junction temperature, T_J (°C)	
	Alumina	Silver	Alumina	Silver
350	18.60	18.85	50.5	50.9
700	20.57	20.84	78.2	78.8

Table I shows the R_{thJA} and T_J for different measuring conditions. The results revealed that a monotonous difference on R_{thJA} and T_J between both the TIMs as the LED operated at the same driving current. In both the driving currents of 350mA and 700mA, the R_{thJA} and T_J were enhanced 1.35% and 0.78% respectively after replacing the alumina with silver thermal compound. The application of TIM is capable to eliminate the formation of micro-air gaps and to conform the imperfection between the contacting surfaces [16]. Thus, thermal contact resistance can be reduced and on the other hand, heat transfer can be improved [17]. However, the values showed that the presence of silver thermal compound has led to a slightly lower heat dissipation from the packaged LED into ambient than alumina thermal compound.

Generally, it was observed that the R_{thJA} for both the TIMs increases for about 10% as the driving current increases from 350mA to 700mA. Explanation can be made based on the amount of heat generated within the LED. Basically, heat is generated if the non-radiative recombination occurred at the p-n junction. In addition, more heat is produced as the P_{el} increases. Thus, this would lead to heat accumulation within the packaged LED if weak thermal conductivity material was used in the packaging. A further implication is illustrated by referring to the determined partial thermal resistances under different measuring conditions, as shown in Table II.

TABLE II

PARTIAL THERMAL RESISTANCES FOR DIFFERENT THERMAL INTERFACE MATERIALS AND CURRENTS

Current, I (mA)	Junction-to-board thermal resistance, R_{thJB} (K/W)		Board-to-ambient thermal resistance, R_{thBA} (K/W)	
	Alumina	Silver	Alumina	Silver
350	13.67	13.67	4.93	5.18
700	15.30	15.30	5.27	5.54

The values showed that in spite of the application of various TIMs has no effect on R_{thJB}, the increases of current would lead to a uniform shifting of R_{thJB}. This occurrence is believed due to the influence of P_{opt}. At the same operating current, similar enhancement of 5% on R_{thBA} was found from the alumina to silver thermal compound. The enhancement indicates that the rate of heat convection from the MCB region to ambient was reduced. It showed that silver tends to have lower thermal conductivity than alumina thermal compound. Therefore, the application of silver thermal compound caused a heat stored within the LED and has a difficulty to dissipate to ambient. This has described the reason for the LED with silver thermal compound showed higher T_J.

Next, the determined optical parameters for the LED were tabulated in Table III. The values showed that the application of different TIMs has brought to an insignificant effect on the LED optical characteristics. The calculated percentage of variation in terms of the optical parameters between both the TIMs was only 0.5% and thus, it is negligible. It was noticed that as the driving current increases, P_{opt} was slightly increased, where the LED tends to appear brighter. In spite of this, higher

driving current will cause an augment in T_J and adversely decrease the efficiency of the LED [18].

TABLE III

OPTICAL PARAMETERS FOR DIFFERENT THERMAL INTERFACE MATERIALS AND CURRENTS

Optical parameters	350mA		700mA	
	Alumina	Silver	Alumina	Silver
Electrical power, P_{el} (W)	1.10	1.10	2.34	2.33
Optical power, P_{opt} (mW)	311.6	312.9	490.0	492.5
Heat dissipation power, P_H (W)	0.787	0.785	1.84	1.84
Efficiency, η (lm/W)	89.8	89.8	66.9	67.1

D. Real thermal resistance, R_{thr}

As mentioned in the earlier part, the consistent shifting of the functions is believed due to the influence of P_{opt}. Thus, a detailed description of this consistency can be conducted by taking P_{opt} into the consideration to obtain the R_{thr}. By substituting (2) into (1), R_{thr} which represents the corrected thermal resistance of the LED is written as

$$R_{thr} = \frac{T_J - T_A}{P_{el} - P_{opt}} \qquad (4)$$

Fig. 4 illustrates the evaluated differential structure functions of R_{thr} for different measuring conditions. It can be seen that the shifting of the functions have been mostly eliminated.

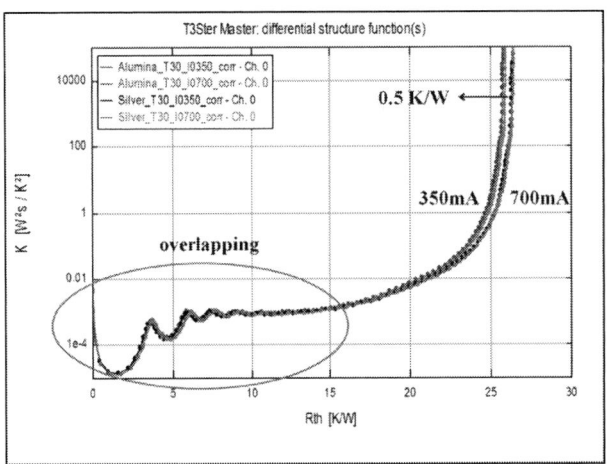

Fig. 4. Differential structure functions of R_{thr} for different thermal interface materials and currents.

In this case, the parasitic heat flow [9] and P_{opt} of the LED are necessary to be included to obtain the corrected thermal resistance or R_{thr}. This is an important step to differentiate the peak formation caused by noise, packaging failure or the existence of a material layer. Thus, the accuracy of the measurement has been improved and reliable.

V. CONCLUSION

Concurrent investigations of thermal and optical characteristics have been carried out on a commercial warm-white LED by using T3Ster and TERALED. The effects of different TIMs and increases of driving current on the LED performance have been studied. The experimental results demonstrated that the measuring conditions caused a greater influence on thermal properties than optical properties of the LED. In addition, a consistent shifting in the evaluated functions was apparently shown as the driving current increases. However, the occurrence was eliminated by including P_{opt} in the determination of R_{thr}. This allowed the accuracy of measurement to be improved. In order to prolong the operating lifetime and maintain the performance of the LED, the design of the internal thermal structure and the selection of packaging material need to be considered.

ACKNOWLEDGEMENT

This work was supported by Postgraduate Research Grant Scheme no. 1001/PFIZIK/ 834112 and Institute of Postgraduate Studies (IPS) of Universiti Sains Malaysia through Graduate Assistance Scheme.

REFERENCES

[1] J. Kovac, L. Peternai and O. Lengyel, "Advanced light emitting diodes structures for optoelectronic applications", *Thin Solid Films* 433, pp.22-26, 2003.

[2] L. Kim, G. W. Lee, W. J. Hwang, J. S. Yang, and M. W. Shin, "Thermal analysis and design of GaN-based LEDs for high power applications", *Phys. Stat. Sol.(c)*0, No.7, pp.2261-2264, 2003.

[3] E. Fred Schubert, *Light-Emitting Diodes*, 2nd ed., Cambrigde University Press, ISBN-13 978-0-521-86538-8, 2006, pp.101.

[4] C. J. Kobus and T. Oshio, "An experimental and theoretical investigation into the thermal performance characteristics of a staggered vertical pin fin array heat sink with assisting mixed convection in external and in-duct flow configurations", *Exp. Heat Transfer* 19:2, pp.129-148, 2006.

[5] S. Liu and X. Luo, *LED Packaging for Lighting Applications: Design, Manufacturing and Testing*, John Wiley and Sons, 2011, pp.2-24.

[6] L. Jayasinghe, T. Dong and N. Narendran, "Is the thermal resistance coefficient of high-power LEDs constant", *7th International Conference on Solid State Lighting*, Proceedings of SPIE 6669:666911.

[7] A. Poppe, G. Farkas and G. Horváth, "Electrical, thermal and optical characterization of power LED assemblies", *THERMINIC*, ISBN: 2-916187-04-9, 2006.

[8] G. Farkas, Q. V. V. Vader, A. Poppe, and G. Bognar, "Thermal investigation ogh power optical devices by transient testing", *IEEE Transaction on Components and Packaging Technologies*, vol. 28, No. 1, pp.45-50, 2005.

[9] M. Renz, *Member, IEEE*, A. Poppe, E. Kollár, S. Ress and V. Székely, "Increasing the accuracy of structure function based thermal material parameter measurements", *IEEE Transactions on Components and Packaging Technologies*, vol. 28, No. 1, pp.51-57, 2005.

[10] H. Chen, Y. Lo, Y. Gao, H. Zhang and Z. Chen, "The performance of compact thermal models for LED package", *Thermochimica Acta* 488, pp.33-38, 2009.

[11] EIA/JEDEC Standard No. 51-1.

[12] Thermal transient tester general overview, from http://www.mentor.com/products/mechanical/products/upload/t3ster.pdf

[13] M. Rencz, V. Szekely, A. Morelli and C. Villa, "Determining partial thermal resistances with transient measurements, and using the method to detect die attach discontinuities", *18th IEEE SEMI-THERM Symposium*, 0-7803-7327-8/02, pp.15-20, 2002.

[14] L. Kim and M. W. Shin, "Thermal resistance measurement of LED package with multichips", *IEEE Transaction on Components and Packaging Technologies*, vol. 30, No. 4, pp.632-636, 2007.

[15] Thermal and radiometric characterization of LEDs, from http://www.mentor.com/products/mechanical/products/upload/teraled.pdf

[16] D. Sahray, H. Shmueli, G. Ziskind and R. Letan, "Study and Optimization of Horizontal-Base Pin-Fin Heat Sinks in Natural Convection and Radiation", Department of Mechanical Engineering, Heat Transfer Laboratory, Ben-Gurion University of the Negev, *Journal of Heat Transfer*, vol. 132, pp.012503.1-012503.13, 2010.

[17] K. J. Puttlitz and P. Totta, *Area Array Interconnection Handbook*, Kluwer Academic Publishers, 2001, pp. 394-395.

[18] D. Wu and C. P. Wong, *Materials for Advanced Packaging*, Springer Science+Business Media, ISBN: 978-0-387-78218-8, 2009, pp.676.

The Octupole Microelectrode for dielectrophoretic trapping of single cells–Design and Simulation

S.Noorjannah Ibrahim[1,2] and Maan M. Alkaisi[1]

[1] The MacDiarmid Institute of Advanced Materials & Nanotechnology, Department of Electrical and Computer Engineering,
University Of Canterbury, Christchurch 8041, New Zealand
[2]Department of Electrical and Computer Engineering, Faculty of Engineering, International Islamic University Malaysia,
Kuala Lumpur, Malaysia
email: noorjannah@iium.edu.my,siti.ibrahim@pg.canterbury.ac.nz

Abstract- **Different microelectrode designs for a dielectrophoresis (DEP)-based lab-on-chip, have significant effect on the DEP force produced. Study on the microelectrode factor is essential, as the geometry and the numbers of microelectrode can influence particles and/or cells polarization. This paper presents one of the three new microelectrode designs, called the octupole microelectrode, in a study on the microelectrode factor of the DEP trapping force. The octupole pattern was constructed on a metal-insulator-metal layer structured on a Silicon Nitride(Si_3N_4) coated Silicon(Si) substrate. The first layer or back contact is made from a 20nm Nickel-Chromium(NiCr) and a 100nm gold (Au). Then, an insulator made of SU-8-2005 was spin-coated on the metal layer to create arrays of microcavities or cell traps. The third layer, where the octupole geometry was patterned, consists of 20nm NiCr and 100nm Au layers. The microcavities which were defined on the SU-8 layer, allows access to the back contact. Gradient of electric fields which represent the actual DEP trapping regions were profiled using COMSOL Multiphysics 3.5a software. Then, the microelectrode trapping ability was evaluated using polystyrene microbeads suspended in deionised (DI) water as the cell model. Results obtained from the experiment were in agreement with results from simulation studies where polystyrene microbeads concentrated at the trapping region and filled the microcavity.**

I. INTRODUCTION

The development of lab-on-chip (LOC) devices has evolved tremendously in recent years. A LOC device with trapping capability can benefit life science studies and is useful for real-time observation of single cell responses to chemical stimulus. It provides more control over labelling of individual cells for comparison purposes. One vital question however, is how to precisely localize single cells on the LOC devices. Dielectrophoresis (DEP) is one of the cell manipulation techniques used on the lab-on-chip devices. This method utilizes the polarization effect between the dielectric properties of cell and suspension medium, and the supplied AC signals to initiate cell movements. On a DEP-based device, the DEP forces are generated by microelectrodes fabricated on the LOC platform. The fabricated microelectrodes can be customized to accommodate the

electric fields non-uniformity, the particle sizes and the type of physical manipulations required. Some examples of the common microelectrode designs for DEP device are planar [1], quadruple [2], interdigitated (IDE) [3] and planar ring arrays [4-6].

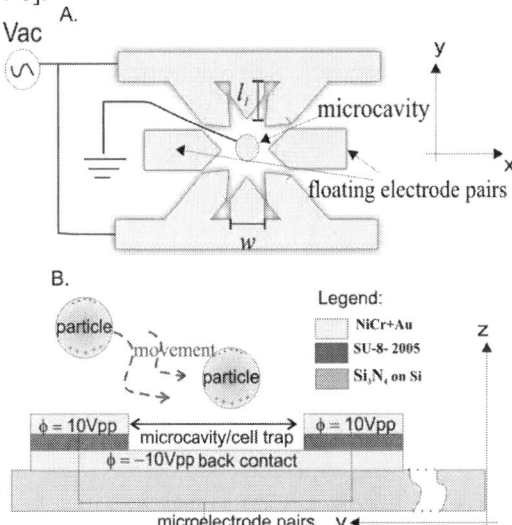

Fig. 1: a)The schematic of the octupole microelectrode pattern for trapping single cells. b) The cross section of the microelectrode structure.

This paper presents a new microelectrode design for trapping single cells, called the octupole microelectrode. The motivation behind this work is to design a LOC device that is capable of trapping single cells, with minimum chemical interventions on the cell's physical conditions. As illustrated in Fig. 1(a), the octupole design comprises of a pair of three-arm electrodes, two floating electrodes and a microcavity. The octupole microelectrode is structured on a multilayer LOC platform. As shown in Fig.1(b), the microcavity is made of an insulating material, SU-8-2005. The microcavity allows access to the back contact electrode at the bottom layer and acts as the cell trap. Meanwhile, the three-arm electrode and two floating electrode are patterned on the uppermost metal layer.

978-1-4673-2395-6/12 $31.00 © 2012 IEEE

Numerical simulations were conducted using COMSOL Multiphysics 3.5a software, to identify the DEP trapping regions on the device. In the simulations, electric field distributions that represent the actual DEP forces were solved in two-dimensional (2D) geometric domains with time-dependent electrostatics model mode. Results show that the V-shaped of octupole microelectrode tips produce high gradient of electric fields and generate strong DEP trapping forces on the LOC. Meanwhile the back contact creates a low DEP trapping force to anchor cell inside the trap. Hence, the trapping regions of the octupole design are from cumulative DEP forces generated from the electrode tips and the back contact. After fabrication, the microelectrode functionality was assessed using polystyrene microbeads suspended in DI water as the cell model. Experimental results obtained were in agreement with simulation results conducted earlier.

II. THE MICROELECTRODE DESIGN

The DEP force is described as polarization effects on cell due to non-uniform electric fields that manifests into translational movements. The octupole microelectrodes are designed to produce strong DEP force governed by [7] :

$$\langle F_{\text{DEP}} \rangle = \pi\, \varepsilon_m\, R^3 Re\, [K] \nabla |E|^2 \qquad (1)$$

where ε_m, R, $\nabla|E|^2$ and $Re[K(\omega)]$ denote the medium permittivity, the cell radius, gradient of electric fields and the real part of Claussius-Mosotti (CM) factor. The CM factor is frequency dependent and is defined by:

$$K(\omega) = \left(\frac{\varepsilon^*_p - \varepsilon^*_m}{\varepsilon^*_p + 2\varepsilon^*_m} \right) \qquad (2)$$

where $\varepsilon^*_{p/m}$ is the particle complex permittivity of cell and medium and defined as:

$$\varepsilon^*_{p/m} = \varepsilon_p - j\frac{\sigma_{p/m}}{\omega} \qquad (3)$$

where σ is the conductivity, ε is the permittivity of cell or medium respectively, ω is the angular frequency and $j = \sqrt{-1}$. Cell movements due to DEP force can be categorised into positive DEP (pDEP) i.e., movements towards high electric fields intensity or negative DEP (nDEP) i.e., movements towards low electric fields intensity. As described in (2), due to CM factor dependency on the frequency of AC signals, particle trapping is in nDEP polarization when $K(\omega) < 0$ or in pDEP polarization when $K(\omega) > 0$.

From (1), the gradient of electric fields or $\nabla|E|^2$ is proportional to the DEP force strength. Hence, in order to generate strong DEP forces, the designed microelectrode has to create high gradient of electric fields ∇E^2, on the LOC platform. This has points out the reason for designing eight electrode tips surrounding a microcavity, as the ∇E^2

concentrations from the octupole microelectrode manifest into strong DEP holding force. However, excessive exposure to electric fields can deteriorate cell viability on the LOC which leads to cell membrane rupture or cell death. Therefore, two floating electrodes are incorporated in the arrangement to reduce the electric field concentrations. The two floating electrodes generate induced electric fields [8] which are useful to reduce total amount of DEP forces exerted on cells and can minimize damages imposed on the cells.

III. METHODS AND MATERIALS

A. Simulation

The DEP forces generated by the octupole geometry were profiled using finite element analysis software, COMSOL Multiphysics 3.5a. Results were obtained by solving the Maxwell's equations in electrostatic approximations i.e., the dielectric properties were in ideal condition where materials were considered to have only permittivity and zero conductivity. The Laplace equation was simplified to $= -\nabla\varphi$, where φ is the supplied AC signals and therefore, the electric fields can be derived from:

$$D = \varepsilon.E \qquad (4)$$

where D, ε and E denote as the electric displacement fields, the permittivity of suspension medium and the electric fields respectively. Due to the multilayer structure, studies on the φ configuration between the octupole pattern and the microcavity with respect to its phase, were also conducted to estimate the performance of the microelectrode. Table 1 depicts parameters used in the numerical simulations.

TABLE1
PARAMETERS USED IN THE OCTUPOLE MICROELECTRODE SIMULATIONS [9].

Parameter	Polystyrene Microbeads
Relative Permittivity	2.55
Relative Conductivity	0.5 µS/m
Medium Permittivity	78.5
Medium Conductivity	1.7mS/m (DI water)
K calculated	-0.4760269 (nDEP)
Radius	5µm
Frequency	1MHz
Potential (φ)	10Vpp
Phase difference	0°,90°,270°,180°

B. Fabrication

Arrays of the octupole microelectrode were fabricated using photolithography technique on a metal-insulator-metal layer platform of Fig.1(b). The back contact or the bottom electrode is made of a 20nm NiCr and a 100nm Au deposited on top of a Si_3N_4 coated silicon substrate. Then, SU-8-2005 (a negative photoresist) was spin-coated on top of the back contact creating an approximately 5µm thick layer. Finally, the third layer consists of a 20nm NiCr and a 100nm Au, was deposited on top of the SU-8 layer. As illustrated in Fig.1(b), the sandwiched SU-8 layer separates the back contact from the microelectrode arrays on the third layer but allows access to the back contact only through arrays of microcavity. Using a positive resist AZ1518, the third layer was patterned with arrays of the design. Any unwanted area was removed by wet etching after the AZ1518 resist exposure and development. Then, the photoresist was stripped using photoresist remover leaving behind arrays of the microelectrode pattern. Fig.2 illustrates the octupole microelectrode fabricated on 15mmx15mm Si_3N_4 coated Si substrate.

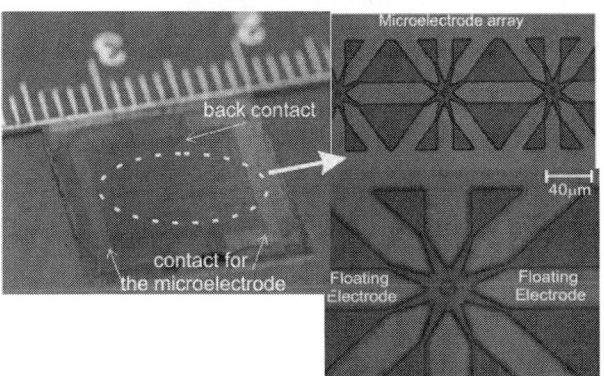

Fig.2: The fabricated 15mmx15mm LOC with arrays of octupole microelectrode.

C. Experiments

The octupole microelectrode trapping ability were tested using polystyrene microbeads (Polyscience Inc.) of 9-10 μm size, suspended in deionised (DI) water as the cell model. Before testing, the LOC of Fig.2 was placed on a glass holder for electrical connections. Then, a 10mmx10mmx1mm (Lenght x Width x Height) square spacer made of PDMS material was placed on top of the biochip to contain the suspension medium before dispersing the polystyrene microbeads using a micropitte. Then, a glass cover slip is placed on the spacer to level the medium.

The biochip was tested using 10 Vpp AC signals of various frequency (50 kHz to 5 MHz) supplied by a function generator (HP3312A). After the function generator was turned ON, movements of the suspended microbeads were observed using a microscope (Nikon eclipse 80i). Then, the movements were recorded using a camera (Nikon digital Sight DS-U1). Each video was recorded for 3 seconds every 3 minutes controlled by the ACT-2U software. Finally, the video segments were compiled using commercial video editing software before thoroughly studied.

IV. RESULTS AND DISCUSSION

D. Simulation Results

Fig.3 illustrates the electric fields and the DEP force profiles generated by the octupole microelectrode. The results were obtained by setting the φ between the microcavity and the microelectrode pattern with 180° phase different. The electric field intensities in Fig.3(a), indicate that the microcavity generate the low electric field region while the high electric field region occurs at the tips of microelectrodes. The DEP trapping region occurs at the central region of the octupole microelectrode i.e., surrounding the microcavity. Note that, DEP forces are also generated in between the floating electrodes pair and the three-arm electrodes that indicate the presence of induced electric fields generated by the floating electrodes.

Table 2 presents the DEP force values calculated between microelectrode tips and the microcavity edge i.e., the trapping region. The results show that the differences of φ between microelectrode pairs and microcavity can affect the trapping region. The maximum calculated DEP forces occurred when the back contact is 180° out of phase from the microelectrode pattern on the third layer. Meanwhile, the minimum DEP force was resulted from connecting φ of the same phase for both microcavity and the microelectrode pair. Interestingly, if the microcavity is not connected with φ, the DEP force is stronger than connecting the microcavity with the same φ of the same phase with the microelectrode pair.

TABLE 2

THE ESTIMATED DEP FORCE EXERTED ON MICROBEAD USING φ OF DIFFERENT PHASE CONNECTED TO THE BACK CONTACT.

Back contact (φ)	DEP Force (mean) (N)
No connection	2.75×10^{-14}
No phase shift	2.94×10^{-18}
90°	5.51×10^{-14}
180°	1.15×10^{-13}
270°	5.51×10^{-14}

TABLE 3

TRAPPING RESULTS USING THE OCTUPOLE MICROELECTRODE.

Frequency of φ	50 kHz	100 kHz	250 kHz	500 kHz	1 MHz	2 MHz	5 MHz
Trapping in the microcavity	No	No	No	No	No	Yes	No

E. Experimental Results

The functionality test of the octupole microelectrode was conducted using polystyrene microbeads. In the experiments, the microelectrode was supplied with φ that has 180° phase difference from the φ of the back contact. Table 3 depicts the results for trapping microbeads using the octupole design. Successful trapping of polystyrene microbead inside the microcavity was observed at 2MHz indicating nDEP force. Fig. 4 illustrates microbeads movement toward octupole trapping region at 2 MHz. During the first 60 seconds, microbeads moved toward the central region of the octupole microelectrode. The movements are due to the DEP forces generated at the tips of microelectrodes. At t= 300s of Fig.5(b) a microbead was successfully trapped inside the microcavity.

978-1-4673-2395-6/12 $31.00 © 2012 IEEE

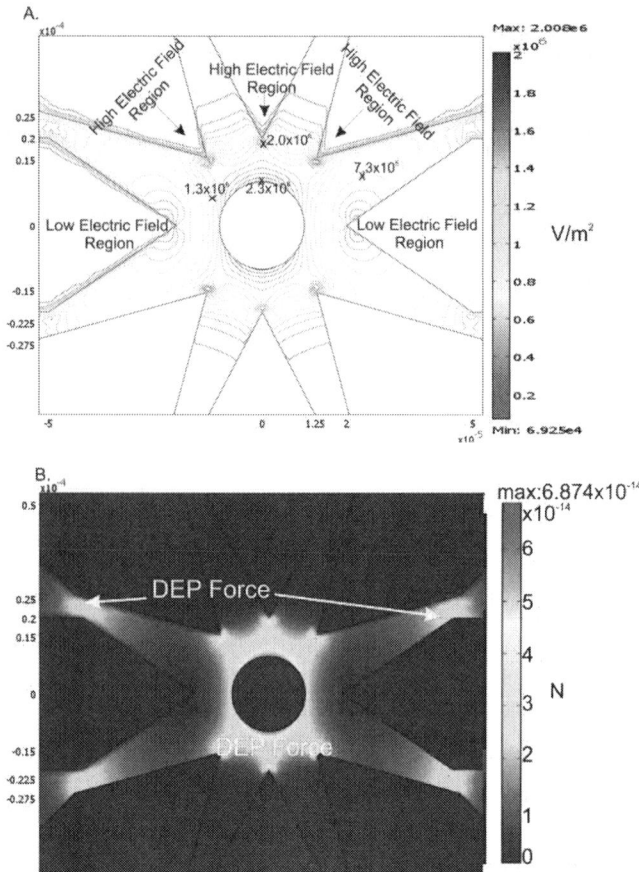

Fig. 3:a) Regions of high and low electric field intensity generated by the octupole microelectrode. b) The DEP trapping region.

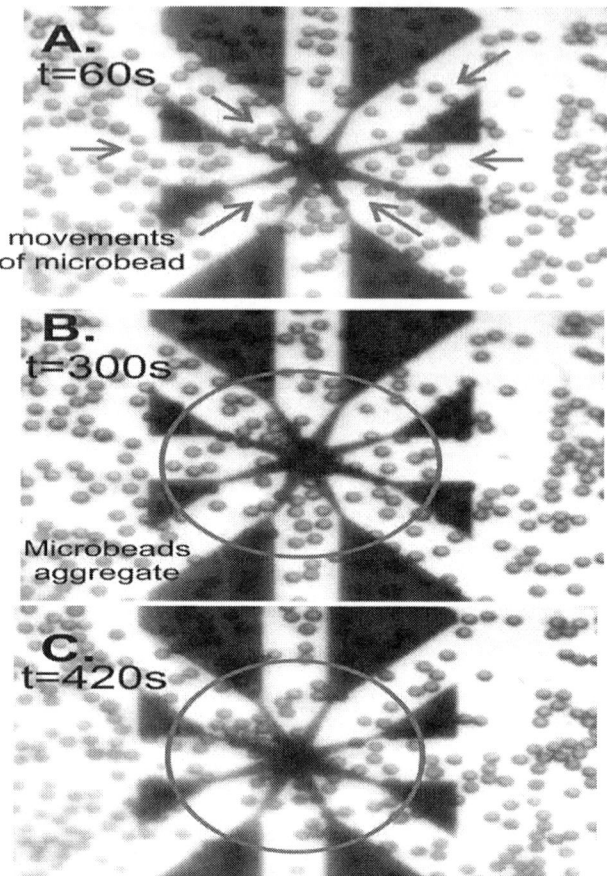

Fig. 4: Movement of microbeads toward the trapping region of the octupole microelectrode at 2MHz.

The microbeads however, tend to concentrate at the trapping region if the DEP forces were exerted for a longer period. Results from the experiment have demonstrated the octupole microelectrode's trapping ability and the presence of DEP trapping regions that surrounding the microcavity. Overall, microbeads movements to the trapping region are slow, due to the low conductivity (σ) of the DI water with measured σ = 1962 μS/m. From the results, it can be deduced that with the existence of high DEP force at the trapping region, single cells can be held properly and stay inside the microcavity.

V. CONCLUSION

Precise localization of cell on lab-on-chip devices can be realized using the DEP force. In this paper, the octupole microelectrode on the multilayer structured LOC has demonstrated its ability to trap single cells using polystyrene microbeads suspended in DI water. Connecting the SIBC biochip with AC signals that has 180° phase difference between microelectrode pair and microcavity, has resulted with successful cell trapping at 2 Mhz.

REFERENCE

[1] M. R. Tomkins, A. W. Jeffery, and A. Docoslis, "Observations and analysis of electrokinetically driven particle trapping in planar microelectrode arrays," *The Canadian Journal of Chemical Engineering,* vol. 86, p. 609, 2008.

[2] L. F. Hartley, K. V. I. S. Kaler, and R. Paul, "Quadrupole levitation of microscopic dielectric particles," *Journal of Electrostatics,* vol. 46, pp. 233-246, 1999.

[3] M. Hywel, *et al.,* "The dielectrophoretic and travelling wave forces generated by interdigitated electrode arrays: analytical solution using Fourier series," *Journal of Physics D: Applied Physics,* p. 2708, 2001.

[4] E. G. Cen, *et al.,* "A combined dielectrophoresis, traveling wave dielectrophoresis and electrorotation microchip for the manipulation and characterization of human malignant cells," *Journal of Microbiological Methods,* vol. 58, pp. 387-401, 2004.

[5] L. C. Hsiung, *et al.,* "A planar interdigitated ring electrode array via dielectrophoresis for uniform patterning of cells," *Biosensors and Bioelectronics,* vol. 24, pp. 869-875, 2008.

[6] L. Qian, *et al.,* "Integrated planar concentric ring dielectrophoretic (DEP) levitator," *Journal of Electrostatics,* vol. 55, pp. 65-79, 2002.

[7] H. A. Pohl, *Dielectrophoresis*: Cambridge University Press, 1978.

[8] S. Golan, *et al.,* "Floating electrode dielectrophoresis," *Electrophoresis,* vol. 27, pp. 4919-4926, 2006.

[9] M. P. Hughes, *Nanoelectromechanics in engineering and biology*: Boca Raton, Fl: CRC Press, 2003.

Modeling and Simulation of Fluid Interactions with Bluff Body for Energy Harvesting Application

M.S.Bhuyan[1], M.Othman[2], Sawal Hamid Md Ali[3], Burhanuddin Yeop Majlis[1] and Md. Shabiul Islam[1]

Institute of Microengineering and Nanoelectronics (IMEN)
Universiti Kebangsaan Malaysia (UKM)
43600 UKM Bangi, Selangor, Malaysia
[2]Ministry of Science, Technology and Innovation (MOSTI),
Federal Government Administrative Centre, 62662, Putrajaya, Malaysia.
[3]Department of Electrical, Electronic and Systems Engineering, Faculty of Engineering and Built Environment, Universiti Kebangsaan Malaysia, 43000 UKM, Bangi, Malaysia.
Email: shakir_dhaka@yahoo.com/shabiul@ukm.my

Abstract-This paper presents the study of modeling and simulation of fluid flow and bluff-body interactions in different fluid velocity in order to investigate the vortex induced vibration phenomena for energy harvesting application. The Strouhal number for two bluff-bodies is analyzed to identify the right bluff-body for optimized cantilever based energy harvester design. From vibration based energy harvesting standpoint, it is important to predict the frequency of vibrations at various fluid speeds and thereby identify the desirable resonances between the vibrations of bluff body structure and the vortex shedding. This study employs the use of COMSOL-multiphysics computational fluid dynamics software using the fluid structure interaction module. A Fast Fourier Transformation (FFT) in Matlab is performed on stationary bluff bodies lift force oscillation yielding the frequency of vortex shedding by taking the inverse of the difference between the time periods for each vortex pair. The main motive here is to seek a higher synchronized region of frequencies for the oscillation amplitudes for a range of fluid velocity and to calculate the lift and drag coefficients. The wake velocity profile is used to determine lift oscillation and calculate vortex shedding frequency for different Reynolds numbers. From two-dimensional, transient incompressible fluid flow simulation it is found that D-shaped bluff body has a comparatively higher lifting force than the cylinder-shape, suitable for optimized cantilever based energy harvester design in terms of a wide-range of lock-in.

I. INTRODUCTION

In recent years harvesting energy from fluid flow has gained increasing interest in the research community. Researchers have developed energy harvesting devices based on fluid induced vibration. Two types of device have been developed; one relies on the conversion of air motion to air steady oscillation, the other on the drag force of the airflow. In the first type of energy harvester's researchers uses Helmholtz resonators to convert airflow motion to steady air oscillation [1-2]. In the second type of energy harvester, the lift force of the fluid flow causes vibration on a piezoelectric cantilever or magneto-electric component which, in turn, converts kinetic energy of motion into electricity [3-4]. For vibration based

energy harvesting device, it is important to predict the frequency of vibrations at various fluid speeds and thereby identify the desirable resonances between the bluff-body structure and the vortex shedding. When a flowing fluid is unable to negotiate its way smoothly around a bluff object, the phenomenon of periodic vortex shedding occurs. Since fluid flow is viscous, significant boundary layer, separation appears on the bodies' surface, depending on the surface geometry. This separated layer that bounds the wake and free stream, tend to cause fluid rotation, since its outer side, in contact with the free stream, moves faster than its inner side, in contact with the wake. It is this rotation which then results in the formation of individual vortices. The larger vortex draws the smaller vortex across the wake causing the shear layers to interact, which effectively cuts-off the supply to the larger vortex, forming a vortex pair; this vortex is then shed from the rear of the bluff-body and carried downstream down the wake. The pattern of periodic, alternating vortex shedding that occurs in the flow behind the body is referred as von Karman's vortex street [5] as represented in Fig. 1. Depending on the characteristics of the flow, mainly the Reynolds number, different types of vortex streets may form.

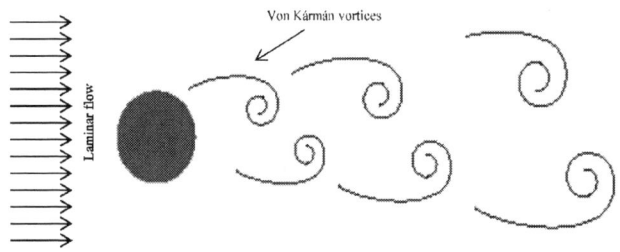

Fig.1. Von Kármán vortex streets forming in the wake of a Bluff-body.

When the pattern of shed vortices is not symmetrical about the body, which is the case in any vortex street, an irregular pressure distribution is formed on the upper and lower sides of the body, leads to periodic variations in the lift and drag forces.

The lift force has a period equal to the vortex shedding frequency perpendicular to the flow direction. Since the vortices are shed in a periodic manner, the resulting lift forces on the body also vary periodically with time. When a degree of freedom is allowed in the cross-flow direction, these variations in the lift force induce vibrations and therefore can induce oscillatory motion of the body. This occurrence alone would qualify as Vortex Induced Vibration (VIV) [6]. When the frequency of vortex shedding is close to the natural frequency of the body it is referred as "lock-in", the vortex shedding frequency actually shifts to match the bodies' natural frequency, and as a result, much larger amplitudes of vibration can occur that is relevant to the design of an optimized energy harvesting device. A cantilever beam attached to the bluff-body model is required to function at a wide range of fluid velocities in ambient conditions due to the unpredictable nature of fluid. This functionality can be examined by predicting the lock-in bandwidth, where the vortex shedding frequency locks into the natural frequency of the cantilever beam.

In this paper we investigated the vortex induced vibration for two different types of bluff-bodies namely cylindrical and D-shaped. The vortex formation is studied for a wide range of Reynolds number without a cantilever beam attached. Computational Fluid Dynamics (CFD) simulations were used to simulate a wide range of flow speeds and Fast Fourier Transformation in Matlab is performed on the lift force yielding the frequency of vortex shedding. The Strouhal number analysis for bluff-bodies will help in tackling the decision of which bluff-body shape to use in design of a fluid flow energy harvester.

II. FLUID-STRUCTURE INTERACTION BASICS AND PARAMETERS

A. Fluid motion

This section explains the theory behind the models including the interaction between fluids and structures, often named as Fluid-Structure Interaction (FSI). Assuming the fluid to be incompressible (material's density is constant or almost constant) the fluid motion is governed by the following Incompressible Navier-Stokes equations [7] and transient models can be solved.

$$\rho \frac{\partial u}{\partial t} - \nabla \cdot [\eta(\nabla u + (\nabla u)^T)] + \rho(u.\nabla)u + \nabla_p = F \quad (1)$$
$$\nabla . u = 0 \quad (2)$$

where, η is the dynamic viscosity, ρ is the density, u is the velocity field, p is the pressure, F is a volume force field such as gravity.

The viscous forces on a bluff-body are proportional to the gradient of the velocity field at the surface. A pair of reaction forces namely drags and lift force operators is used to compute the integrals of the viscous forces. The dimensionless drag and lift coefficients depend only on the Reynolds number and an object's shape, not its size. The coefficients are defined as:

$$C_D \quad \frac{D}{\rho U_{mean}^2 D} \quad (3)$$

$$C_L = \frac{2F_L}{\rho U_{mean}^2 D} \quad (4)$$

where, F_D and F_L are the drag and lift forces, ρ is the fluid's density, U_{mean} is the mean velocity, and D is the characteristic length of the bluff-body.

B. Reynolds Number

The dimensionless Reynolds number (Re) expresses the ratio of inertial forces to viscous forces. Re number scales the boundary layer thickness and transition from laminar to turbulent flow. The boundary layer is impelled about the bluff body by the inertia of the flow. Viscous friction at the model surface retards the boundary layer. By definition, the ratio of inertial force to viscous force in the boundary layer is

$$R_e = \frac{UD}{v} \quad (5)$$

Where, U, D and v are the free stream velocity, characteristic dimension of the bluff-body, and kinematic viscosity of the fluid, respectively. The main significance of Re number is that it is used to predict the nature of vortex shedding at various flow speeds.

C. Strouhal Number

The Strouhal number (St) is the dimensionless proportionality constant between the predominant frequency of vortex shedding and the free stream velocity divided by the bluff-body dimension. The Strouhal number is often approximated by a constant value. The frequency of vortex shedding becomes constant at lock-in and the free-stream velocity changes the value of the Strouhal number. For example, the Strouhal number for cylindrical bluff-bodies, for a wide range of Reynolds number, is approximated as $St = 0.21$ in most of the literature [6].

$$S_t = \frac{f_s D}{U} \quad (6)$$

Where, f_s is the frequency of vortex shedding. This nondimensional number is investigated for different bluff-body shapes in the simulation section of this work.

III. SIMULATIONS AND RESULTS

This section describes the simulations and results involving the integrated fluid structure interaction using COMSOL-multiphysics software for Circular and D-shape bluff-bodies with characteristic length of 0.1m and the Strouhal number calculation using Matlab. Each bluff-body shape studied for 9 different flow velocities. First the step-by-step procedure to create the model of this system is described then the analysis of the results obtained from the simulations is presented.

A. Model Definition

Our model examines unsteady, incompressible flow past a bluff-body placed in a rectangular channel at right angle to the oncoming fluid. The bluff-body is offset somewhat from the center of the flow to destabilize what otherwise would be steady-state symmetrical flow. Fluid enters from the left side with a parabolic velocity profile, passes over the bluff body and leaves through the right boundary as shown in Fig.2.

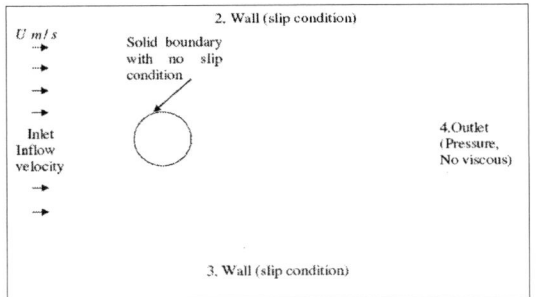

Fig.2. Diagrammatic representation of boundary conditions used in the CFD simulations.

The model computes the fluid's velocity components u = (u, v) in the x and y directions and its pressure p in the region defined by the geometry in the preceding Fig.2. The viscous forces on the cylinder are proportional to the gradient of the velocity field at the bluff-body surface. The velocity gradient on the boundary is evaluated by computing integrals of reaction forces operators to compute the time-varying forces on the bluff-body. The fluid properties and varying fluid parameters used in the simulations are tabulated in Table I.

TABLE I
FLUID PROPERTY RANGE AND VALUES

Velocity	0.3-2.7 m/s
Density	1.23[kg/m^3]
Dynamic Viscosity	1.79e-5[Pa*s]

B. Geometric Modeling

The COMSOL-multiphysics CAD tools were used to create the 2D composite solid objects using Boolean operations like union, intersection, and difference of the bluff-bodies. Fig. 3 shows the D-shape composite solid objects in COMSOL environment.

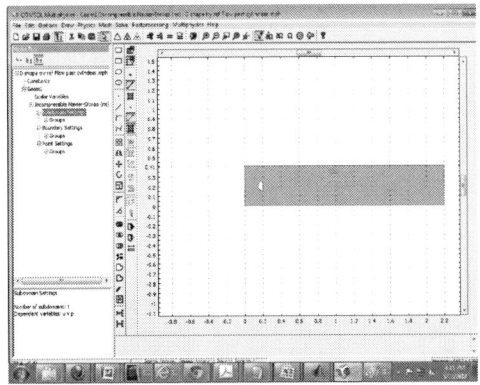

Fig.3. 2D composite solid objects in COMSOL.

C. Boundary Conditions

The inlet flow conditions taken into consideration in this study depend upon the Reynolds number value. The boundary condition for the inlet or the opening of the rectangular closed boundary system is the normal inflow velocity. The range of velocities used in this study is 0.3m/s to 2.7m/s. The wall feature represents wall boundaries in a fluid-flow simulation. On the longer sides of the rectangular boundary, the boundary condition at the wall has the with-slip boundary condition. This is to approximate the fact that the boundaries are assumed to be further apart in real world conditions and also to nullify any stress from the walls due to viscosity on the solid bluff-body setup. The slip conditions assume that there are no viscous effects at the slip wall and hence, no boundary layer develops. From a modeling point of view, this may be a reasonable approximation if the important effect of the wall is to prevent fluid from leaving the domain which means that there is no flow across the boundary and no viscous stress in the tangential direction. The boundary conditions applied on the solid bluff-bodies are with no-slip condition. The outlet boundary conditions used are pressure with no viscous stress.

D. Mesh Generation

A mesh is a partition of the geometry model into small units of simple shapes. This work follows the 2D meshing techniques. The mesh generator partitions the sub-domains into triangular mesh elements. The geometry being investigated uses mainly a free mesh consisting of triangular elements. Fig. 4 shows the meshing of the D-shaped bluff-body.

Fig.4. Mesh of the 2D D-shape bluff body.

E. Post Processing

Once the boundary conditions are set up and meshing is done the next step is to simulate the transient model. Fig. 5 shows as an example, the velocity contour plots for a D- shape bluff-body. The contour shows the incipient vortex at the top end and shed vortex at the bottom end in the wake of the bluff-body.

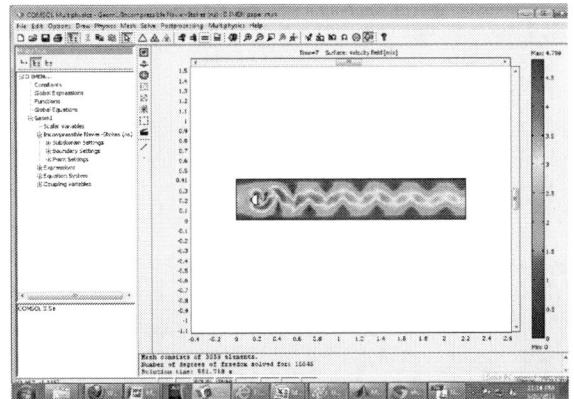

Fig.5. Velocity contour plots for a D-shape bluff-body in COMSOL.

Table II, shows the formation of wake vortices for the D-shape bluff-body for 9 different fluid velocities in the time-dependent simulation. At Re=<60, there is no flow separation. At higher Re=<90, downstream of the bluff- body, a fixed pair of vortices are seen in the wake of the bluff body and von Karman path is clearly visible. Until Re=270, a laminar vortex street regime is seen. However 300<Re, the vortex streets flows into a transition range of turbulence.

TABLE II
BLUFF-BODY WAKE (D-SHAPED)

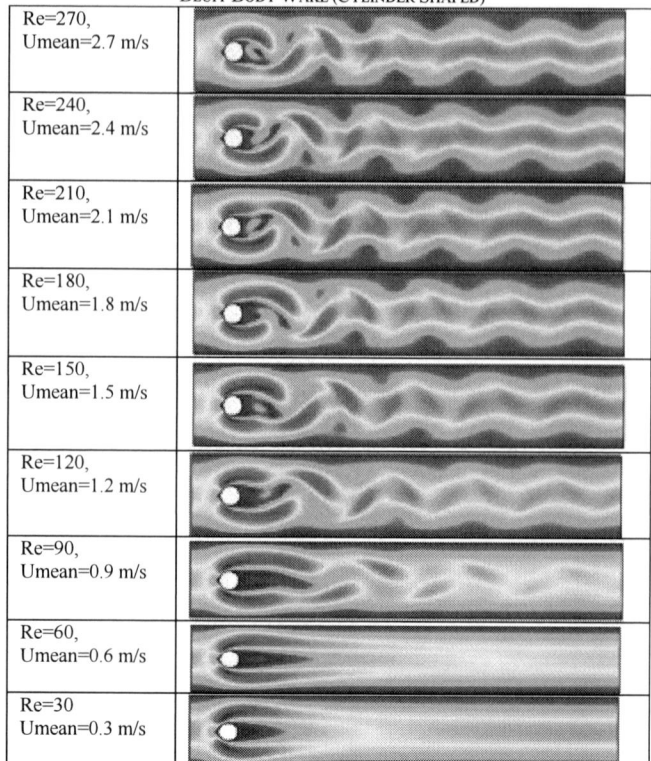

TABLE III
BLUFF BODY WAKE (CYLINDER SHAPED)

Table III, shows the formation of wake vortices for the Cylinder shape bluff-body for 9 different fluid velocities in the time-dependent simulation.

The quantities in the study are the drag and lift coefficients. These are calculated by integrating reaction force operators which the Lagrange multipliers is corresponding to the viscous forces and the pressure over the surface of the bluff body. Fig. 6 shows as an example, the oscillating lift force at Re=150 on the D-shape bluff body. The clear sinusoidal pattern in the plot is the proof of a sustained vortex shedding. Fig.7 shows FFT on the lift force oscillation that yields the frequency of vortex shedding. This plot is used to compute the correct value of Strouhal number using FFT in Matlab. A fairly pronounced spike at frequency equal to 6±0.5Hz shows the shedding frequency for Re=150.

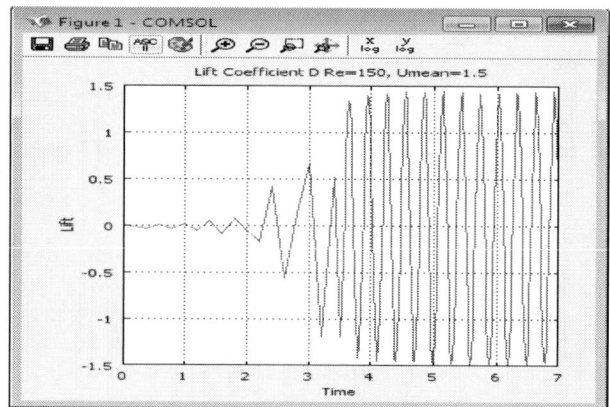

Fig.6. D-shape bluff-body lifts force oscillations.

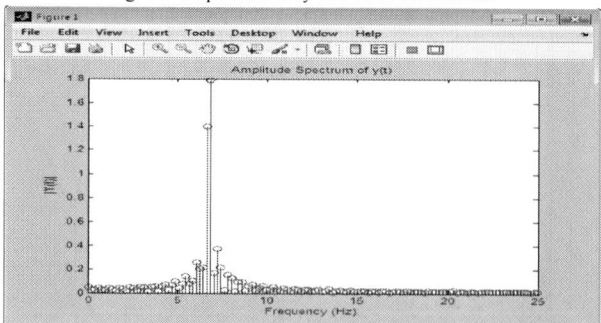

Fig.7. D shape bluff-body shedding frequency Re=270.

978-1-4673-2395-6/12 $31.00 © 2012 IEEE

TABLE IV
SIMULATION RESULTS

Umean	Re	D-Shape			Cylinder- Shape		
		Coff_drag	Coff_lift	St	Coff_drag	Coff_lift	St
0.3	30	5.649706	-0.003097	0	5.580709	0.011075	0
0.6	60	4.240458	0.021975	0	4.037306	-0.007536	0
0.9	90	4.069957	0.136415	0.1953125	3.480902	-0.08039	0.173611
1.2	120	4.493084	1.084495	0.2115885	3.296793	0.64853	0.195313
1.5	150	4.289937	-0.146683	0.2213542	3.139638	-0.526418	0.195313
1.8	180	4.474245	0.165702	0.2387153	3.137394	-1.04089	0.195313
2.1	210	5.072582	-1.357123	0.2418155	3.006547	0.28321	0.195313
2.4	240	5.753137	1.571753	0.2441406	3.081943	1.524873	0.203451
2.7	270	5.246778	1.334015	0.2531829	3.009086	0.45257	0.202546

Table IV summarize the total 18 simulations results yielding the Strouhal Number for the D-shape and cylinder shape bluff bodies for flow velocity range 0.3 to 2.7 m/s. The results matches fairly well with the experimental value as reported in the literature [8] where experimental studies have shown that most bluff-bodies have a Strouhal number around 0.2 for a wide range of Reynolds number. Our study also confirms these results through CFD simulations, showing Strouhal numbers for different bluff-body shapes in the Table IV.

Fig.8 shows how Strouhal number varies with respect to the Reynolds number. Both bluff-bodies show, for increasing Reynolds number, Strouhal number becomes a constant due to a gradual increase in the vortex shedding frequency. The statistics for the Strouhal number fluctuation range showing that the shedding frequency is not a constant value, but it fluctuates inside a frequency range as a function of the bluff-body shape and Reynolds Number. At low Reynolds number the constant linear increase in the Strouhal number is the region where a attached cantilever beam with the bluff-body will resonates with the vortex shedding frequency.

The lift force of the cylindrical bluff-body is comparatively lower than that of the D-shaped bluff-body due to a reduced shear surface force and flow separation. The physical significance is that the vortex shedding frequency of the D-shape bluff-body scales linearly with the velocity for the Strouhal number constant over a wide range of velocities. Hence, the lock-in will occur at this region.

IV. CONCLUSION

In this paper, vortex induced vibrations from two different stationary bluff bodies namely Cylindrical and D-shape for different flow regime velocity and dimensionless parameters: Strouhal number, Reynolds number were studied. The main conclusions obtained are: the vortex shedding frequency for a bluff- body in cross flow is not a constant value, but rather it fluctuates within a frequency range which is function of the body geometry and Reynolds number. This driving oscillating forces generated from the bluff-body induced by the von Kármán vortex streets will mechanically strain an attached piezoelectric cantilever at the trailing edge, and hence power could be generated. The D- shape wake phenomenon presented

the most fluctuant shedding frequency; therefore, the future optimized piezoelectric cantilever energy harvester design in terms of efficiency and over a wide-range of lock-in from this study is concluded to be the design with the D-shaped bluff body.

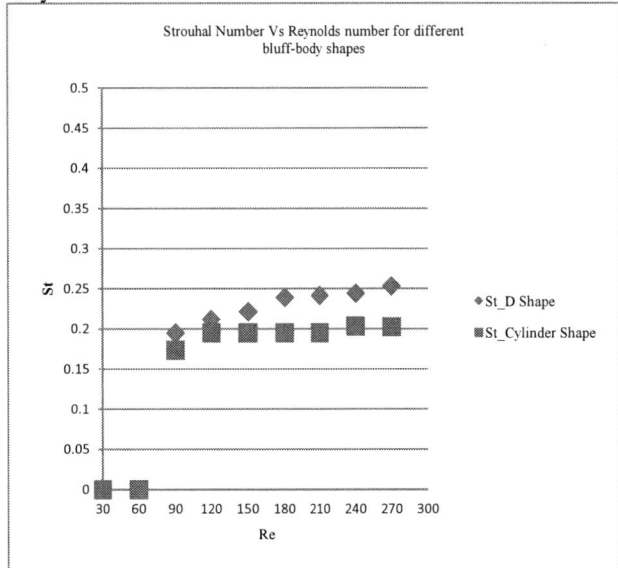

Fig. 8. Strouhal Number vs Reynolds number for D-shape and Cylinder shape bluff bodies.

REFERENCES

[1] Horowitz B. St. "Design and characterization of compliant backplate Helmholtz resonators" *MSc Thesis,* University of Florida. 2001.

[2] Seong-Hyok Kim et al., "An electromagnetic energy scavenger from direct airflow", *J. Micromech. Microeng.* vol. 19, 2009.

[3] Cuadras A. and Ovejas V. J., "Piezoelectricity from the wind", *Proc. Eurosensors XXII,* pp. 305–8, 2008.

[4] Akaydin H. D., Niell E,, and Yiannis A., "Energy harvesting from highly unsteady fluid flows using piezoelectric materials", *J. Intel. Mat. Sys. Struct.* vol. 21 1263–78., 2010.

[5] http://en.wikipedia.org/wiki/K%C3%A1rm%C3%A1n_vortex_street.

[6] Ming Zhaoa and Liang Chengb, "Numerical simulation of vortex-induced vibration of four circular cylinders in a square configuration,`` *Journal of Fluids and Structures,* vol. 31, 2012.

[7] http://en.wikipedia.org/wiki/Navier%E2%80%93Stokes_equations.

[8]. C.H.K. Williamson and G.L. Brown, "A Series in to represent the Strouhal-Reynolds number relationship of the cylinder wake," *J. Fluids Struct., v*ol. 12, 1998.

The Electrical Conductivity of Copper (I) Iodide (CuI) Thin Films Prepared by Mister Atomizer

M.N. Amalina[1,*], A.R. Zainun[1, 2], N.A. Rasheid[1], M.Rusop[1,*]

[1]NANO - ElecTronic Centre (NET), Faculty of Electrical Engineering, Universiti Teknologi MARA (UiTM), 40450 Shah Alam, Selangor, Malaysia

[2]Faculty of Electrical & Electronics Engineering, Universiti Malaysia Pahang (UMP), Lebuhraya Tun Razak, 26300 Kuantan, Pahang, Malaysia

[*]Email: amalina.muhamad@gmail.com, rusop@salam.uitm.edu.my

Abstract- In this paper, the copper (1) iodide (CuI) thin films were prepared by mister atomizer with different thickness. The effect of thickness of CuI thin films were done by varying the deposition flow rate and deposition time. The effects of thickness to its structural, electrical and optical properties were studied. The resistivity increases as the thickness of thin film increase with highest resistivity of 4.79 x 10^1 Ω cm. The transmittance for most of the samples was transparent of above 80% in the visible wavelength. The transmittance and absorption coefficient was measured and then the energy gap was determined which shows the direct transition of n=2. The maximum band gap observed here is 2.82 eV for the thickest thin films. The observation on effect of thickness in this study shows that the increasing of thin film thickness increased the resistivity while the absorption coefficient decrease with slight rise of band gap which due to the bulk grain properties for thick thin film.

Keywords – **Copper (I) Iodide; Mister Atomizer; Flow Rate; Electrical Properties; Optical Properties**

I. INTRODUCTION

Cuprous iodide (CuI) is a p-type wide band gap semiconductor with a great potential for optoelectronic applications. CuI is water insoluble solid with three crystalline phases which are α, β, γ. The low temperature of γ-CuI which is below 350°C with a cubic structure was chosen since it behaves as a p-type semiconductor with a large band gap of 3.1eV.The γ- CuI has potential application in light emitting diode, field emission display, organic catalyst, solid state dye sensitized solar cell as a hole transport material [1-3].

CuI thin film have been fabricated by a variety methods either physical or chemical type of deposition. Each method of deposition contributes to different properties of CuI thin films and is valuable for different application. The deposition includes using the pulse laser deposition (PLD), rf- dc coupled magnetron sputtering and etc [4, 5]. All the methods above are possible for preparing thin films with the size of nanometer. In our experiment, the mist atomization which a process similarly known as spray pyrolysis technique is used for the CuI thin film deposition. The process involves spraying precursor concentration solution onto the heated substrate and the constituents react to form the intended products. Spraying technique is chosen in this research work because it is simple, inexpensive since no vacuum condition needed and able to give high production rates attracts our attention to further studies on this deposition technology. There are several main prime requirements to obtain good quality thin film which are by optimizing the preparation condition such as the gas flow rate, solution concentration, substrate temperature and deposition time.

In the present work, we studied the electrical and optical properties of CuI thin films by using our own designed mister atomizer instrument. The observation made is on the effect of flow rate and deposition frequency. The effect of precursor concentration of CuI thin film using this method has been report elsewhere.

II. EXPERIMENTAL PROCEDURE

CuI particle were obtained by using mister atomizer which the CuI solution is atomized or break up into small droplets due to high forces created by the carrier gas. In this experiment, the pneumatic atomizer is considered. The three main components of the experimental setup are the steel pneumatic atomizer, hot plate and the chamber. The parameters investigated here is the flow rate and the frequency of the deposition. The precursor concentration, substrate temperature, distance between the atomizer and hot plate is kept constant throughout the experimental process. The CuI solution was prepared by mixing the CuI powder (ALDRICH 98%) and the acetonitrile as the solvent. All chemicals were used without any further purification. The CuI solution was released through a tiny outlet as mists with atomic magnitude before it reaches the substrate. The substrate temperature was fixed at 150 °C and was measured by using IR thermometer (Fluke 62 Mini IR). For each sample, the spraying was repeated for 10 and 5 times with 1 minute deposition and 5 minutes of heating process in the furnace at 150 °C. This is

978-1-4673-2395-6/12 $31.00 © 2012 IEEE

done to evaporate the solvent and at the same time to maintain the substrate temperature by having intermittence spraying technique. The spray rate was varied with three different flow rates which are 2 ml/min, 3.5 ml/min and 10 ml/min and the samples were labeled as FR1, FR2 and FR3 respectively in this research paper.

The characterization carried in this experiment is the surface morphology, electrical and optical properties. The thicknesses of the thin films have been characterized by using surface profiler (VEECO DEKTAK 150). The surface morphology of these CuI films was observed by field emission scanning electron microscope (JEOL JSM- J600F). The electrical study was carried out through two point probe I-V measurement by using solar simulator (CEP 2000). Gold was used as metal contacts for I-V measurement and deposited using sputter coater (EMITECH K550X). The optical properties were characterized by JASCO UV-VIS-NIR spectrophotometer at wavelengths between 400 nm- 800 nm. The uncoated glass slide was used as the reference sample. All of the measurements were done in room ambient.

III. RESULTS AND DISCUSSION

A. Electrical Properties

The electrical conductivity of CuI thin films was determined by investigating its behavior between metal and semiconductor junction by having the I-V measurement spectra and hence its resistivity. Fig. 1 shows the I-V spectra of CuI thin films at different flow rate with 5 times. The I–V measurements of the films were attained in the voltage range of -5V and 5V. The γ-phase CuI is known as semiconductor with p-type conductivity [6] . This is because the vacancies exist in CuI is in the cation sites. From each cation, there is one vacancy formed leading to p- type conductivity. Fig 1 and 2 shows linear relationship between the current and voltage. This linear characteristic designates that the gold which act as the metal contacts are ohmic between the semiconductor and the metal. An appropriate ohmic contact between the metal semiconductor junctions (M-S) will not significantly change the performance of the device. To get ohmic contact between metal and p- type semiconductor, the work function of metal must greater than the work function of the semiconductor. Ideally, the work function of the metal should be greater than the electron affinity and the band gap of the semiconductor. There are several reports on the value of work function of CuI. The work function of a material is greatly affected by the deposition technique, morphology of thin films and etc. The work function of CuI reported were between 4.86 eV to 5.4 eV [7, 8]. The work function of gold is said to be around (5.28 eV) [9]. Other metal contact which has the possibility to be used for ohmic behavior are platinum, Pt (5.63 eV) and paladium Pd, (5.40 eV) [9]. In solid state dye sensitized solar cells (SSDSSC), CuI is used as the hole transport layer replacing the redox couple electrolyte for the regeneration of dye. In SSDSSC, the CuI is in contact with counter electrode where the electron flows from the counter electrode to CuI. Therefore, the junction between the metal and semiconductor must not having a rectifying curve to avoid barrier between the two junction and thus easy for movement of electron from back electrode to CuI. The fabrication of SSDSSC that used CuI/ Au as the back electrode was done by several authors such as K. Tennakone et. al, Q.B. Meng et. al and many more

[10, 11]. From the I-V curves and reported thickness, the resistivity, ρ and conductivity, σ of CuI thin film were calculated using following Eqn (1) :

$$\rho = \left(\frac{V}{I}\right)\frac{wt}{l} = \frac{1}{\sigma} \qquad (1)$$

where V is supplied voltage, I is measured current, t is the film thickness, w is the electrode width and l is the length between electrodes.

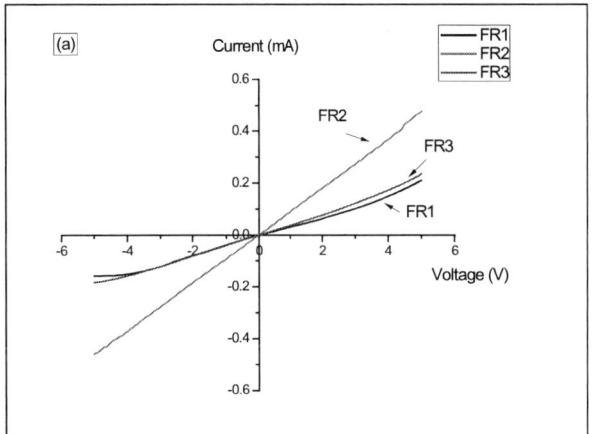

Fig. 1 I-V spectra of CuI thin films at different flow rate with 5 times deposition frequency (in dark condition)

The I-V measurements were done with two different measurements ambient which are in dark and illumination condition. The white bias light with intensity of 1000Wm^{-2} with illuminated area of 0.09cm^2 is used to observe the photoconductivity of CuI thin films and the data were recorded with solar simulator (CEP 2000). Fig. 3 and Fig. 4 show the plot of resistivity of CuI thin films at different flow rate with 5 times and 10 times deposition frequency. From the results, slight photo responses were observed for all the thin films since the resistivity value were less compared to the measurement carried in dark condition. The low electron carrier in the p-type CuI compared to holes carrier leads to small photo response properties. The photo response characteristic is very important for the n-type semiconductor especially for the application of solar cells.

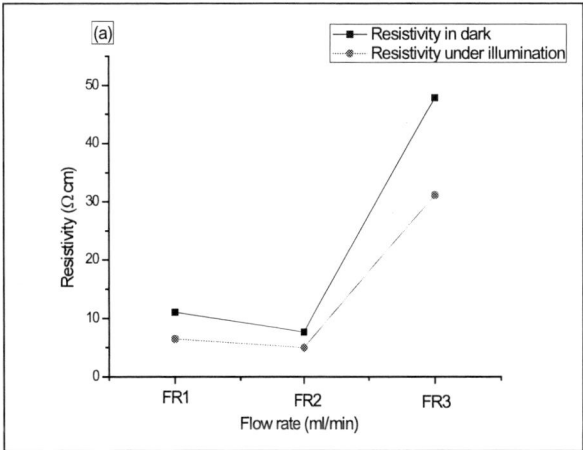

Fig. 3 Resistivity of CuI thin films at different flow rate with 5 times deposition frequency

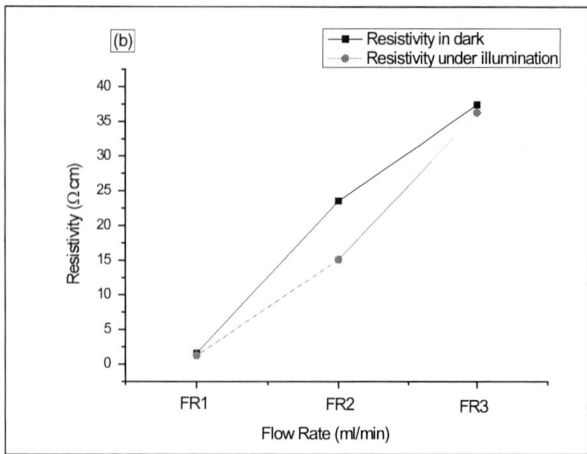

Fig. 4 Resistivity of CuI thin films at different flow rate with 10 times deposition frequency

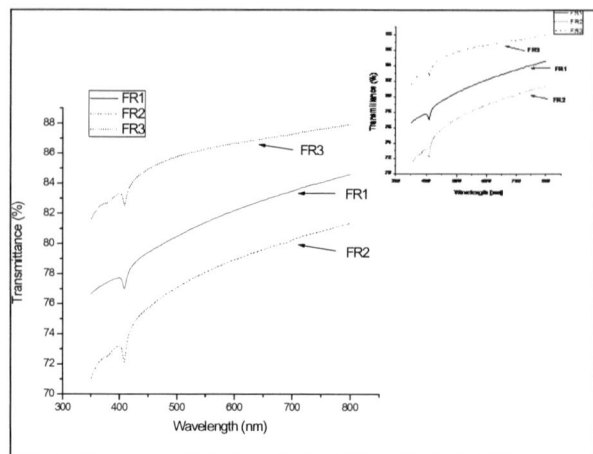

Fig. 5 Transmittance of CuI thin films at different flow rate with 5 times deposition frequency and 10 times deposition frequency (inset)

For both measurement conditions, the resistivity increases as the deposition flow rate (ml/min) increases. It can be clearly seen that, the resistivity increased for thicker thin films as shown in Fig. 4 for 10 times deposition frequency. This observation shown that despite the surface morphology, the thickness of a thin film greatly affect the electrical conductivity of a thin film. The movement of an electron on a thin film determined its electrical characteristics. Table I shows the relationship between the thickness and the resistivity value of CuI thin films. The thickness is increased from 98 nm up to 409 nm. The linear relationship between the thickness and the flow rate indicates that, as the flow rate increases the thickness increased. Generally the resistivity decrease as the film thickness increases, since the grain size is raised and lowers the grain boundaries as the thickness increases. The increase of resistivity at higher thickness was caused by the defect structure which is fully controlled by the bulk grain electrical properties [12]. Besides that, at higher spray rate (FR3, 10 ml/min) the films tend to be powdery in nature and therefore, we preferred to use lower flow rate.

B. *Optical Properties*

The purpose of our study is to use the CuI films in the opto-electronic devices such as solar cells. A good conductivity and optical transparency are highly favorable for the applications. Consequently, the choice of thin film in this research is dictated by the optimum electrical and optical properties. The optical transmission spectra of CuI thin films prepared on glass substrate at different flow rate and deposition frequency were shown in Fig. 5. All films exhibit high transparency in the visible range between 400–800 nm of the optical spectrum. The observed hump at 410 nm may be due to excitation of electrons from sub bands in the valence band to the conduction band. The transmittance spectra for most of the samples showed a good transparency in the visible range which is around (~ 80%). The lowest transmittance value was obtained at 63% for sample deposited at 10 times deposition frequency with 2ml/min.

The transmittance analysis in the locality of the fundamental absorption edge shows the variation of absorption coefficient, α with photon energy (hv) is in accordance to the following relation which implies direct band transition as shown in Eqn [2]:

$$A\,(\,hv - Eg)^{1/2} \qquad (2)$$

where α is the absorption coefficient, A is a constant, hv is the photon energy, Eg is the bandgap.

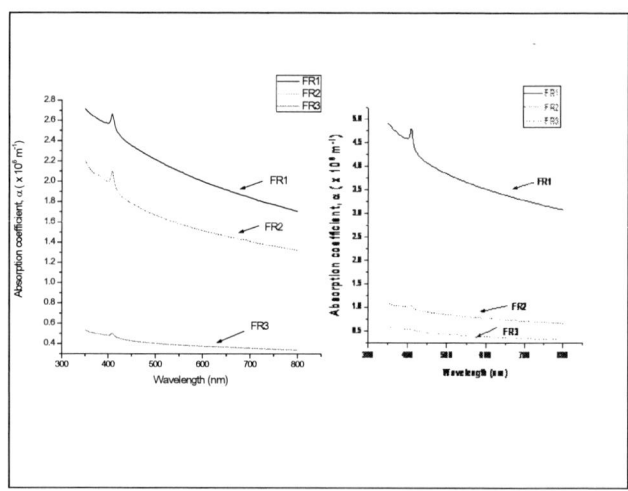

Fig. 6 Absorption coefficient of CuI thin films at different flow rate with 5 times and 10 times deposition frequency (inset)

The absorption coefficients of CuI films at different deposition condition are shown in Fig. 6. A close investigation of both figures reveals that the thick films which deposited at higher flow rate have lower absorption coefficient at high energy region than the thin films. This phenomenon might be due to bigger crystallites or grain sizes that can be obtained for thicker thin films.

TABLE I

THE RELATIONSHIP BETWEEN THICKNESS, RESISTIVITY, TRANSMITTANCE AND BAND GAP OF CuI THIN FILMS

Deposition Frequency	Sample Name	Thickness (nm)	Resistivity in dark ($\times 10^1 \Omega$ cm)	Resistivity under illumination ($\times 10^1 \Omega$ cm)	Transmittance (%)	Band gap (eV)
5 times	FR1	98	1.10	0.65	82	2.70
	FR2	156	0.76	0.50	79	2.80
	FR3	379	4.79	3.11	86	2.80
10 times	FR1	131	0.16	0.12	63	2.80
	FR2	242	2.35	1.51	82	2.80
	FR3	409	3.74	3.63	85	2.82

Therefore, the thicker thin films are closer to bulk crystalline CuI, but bigger grain sizes results in larger unfilled inter-granular volume so the absorption per unit thickness is reduced [13]. The band-gap energy E_g of CuI thin films for different flow rate was summarized in Table 1. It shows that the band gaps for all thin films are almost the same for all samples which are between 2.70 eV to 2.82 eV. The broadening effect can be understood based on the Burstein effect, which implies an increase in the Fermi level in the conduction band of semiconductors leading to widening of the optical band-gap. The increase Fermi level in the conduction band is due to increased carriers in the films. The value obtained in this research work is below the reported band gap of 3.1 eV obtained by other authors. The lower band gap compared to 3.1 eV helps in better utilization of the blue region of the solar spectrum which means at longer wavelength of visible spectrum.

IV. CONCLUSION

CuI thin films have been successfully deposited by mister atomizer at different deposition flow rate and deposition frequency. The thickness was found to be increased as the deposition flow rate and deposition frequency increased up to 409 nm. The electrical and optical properties of CuI thin films investigated here show its relationship between the thickness and its opto-electrical properties. The CuI thin films show an ohmic behavior between the gold metal contact and CuI thin films. Therefore only low voltage drop across the metal and semiconductor interface. The resistivity increased for thicker thin films because of the defect on bulk grain films. The transmittance spectra of most of CuI films show high transmittance of which around 80% in the visible range. The optical band gap obtained in this research is around 2.70 eV to 2.82 eV. The optimization of thickness for thin film properties is owing to the suitable resistivity, transmittance and optical band gap dependence on the required application.

ACKNOWLEDGEMENT

Many thanks to Universiti Tekonologi Mara (UiTM), Ministry of Higher Education (MOHE) and Research Management Institute (ERGS Fund (600-RMI/ST/ERGS 5/3/ (19/2011)), UiTM for the financial support.

REFERENCES

[1] J.-H. Lee, D.-S. Leem, and J.-J. Kim, "High performance top-emitting organic light-emitting diodes with copper iodide-doped hole injection layer," *Organic Electronics,* vol. 9, pp. 805-808, 2008.

[2] P. M. Sirimanne, T. Soga, and T. Jimbo, "Identification of various luminescence centers in CuI films by cathodoluminescence technique," *Journal of Luminescence,* vol. 105, pp. 105-109, 2003.

[3] V. P. S. Perera and K. Tennakone, "Recombination processes in dye-sensitized solid-state solar cells with CuI as the hole collector," *Solar Energy Materials and Solar Cells,* vol. 79, pp. 249-255, 2003.

[4] P. M. Sirimanne, M. Rusop, T. Shirata, T. Soga, and T. Jimbo, "Characterization of transparent conducting CuI thin films prepared by pulse laser deposition technique," *Chemical Physics Letters,* vol. 366, pp. 485-489, 2002.

[5] T. Tanaka, K. Kawabata, and M. Hirose, "Transparent, conductive CuI films prepared by rf-dc coupled magnetron sputtering," *Thin Solid Films,* vol. 281-282, pp. 179-181, 1996.

[6] A. Gruzintsev and W. Zagorodnev, "Temperature-dependent conductivity and photoconductivity of p-CuI crystals," *Semiconductors,* vol. 46, pp. 35-40.

[7] A. Zhukov, V. Gartman, D. Borisenko, M. Chernysheva, and A. Eliseev, "Measurements of work function of pristine and CuI doped carbon nanotubes," *Journal of Experimental and Theoretical Physics,* vol. 109, pp. 307-313, 2009.

[8] J.-H. Lee, D.-S. Leem, H.-J. Kim, and J.-J. Kim, "Effectiveness of p-dopants in an organic hole transporting material," *Applied Physics Letter,* vol. 94, p. 123306, 2009.

[9] M. Z. Musa, M. S. P. Sarah, S. S. Shariffudin, M. H. Mamat, and M. Rusop, "A study on ohmic contact of different metal contact materials on nanostructured titanium dioxide (TiO2) thin milm," in *Electronic Devices, Systems and Applications (ICEDSA), 2010 Intl Conf on,* pp. 412-414.

[10] G. R. R. A. K. K Tennakone, A R Kumarasinghe, K G U Wijayantha and P M Sirimanne, "A dye-sensitized nano-porous solid-state photovoltaic cell " *Semiconductor Science and Technology,* vol. 10, p. 1689, 1995.

[11] Q. B. Meng, K. Takahashi, X. T. Zhang, I. Sutanto, T. N. Rao, O. Sato, A. Fujishima, H. Watanabe, T. Nakamori, and M. Uragami, "Fabrication of an Efficient Solid-State Dye-Sensitized Solar Cell," *Langmuir,* vol. 19, pp. 3572-3574, 2012/05/30 2003.

[12] J. Bruneaux, H. Cachet, M. Froment, and A. Messad, "Structural, electrical and interfacial properties of sprayed SnO2 films," *Electrochimica Acta,* vol. 39, pp. 1251-1257, 1994.

[13] M. M. Abbas, A. A.-M. Shehab, A. K. Al-Samuraee, and N. A. Hassan, "Effect of Deposition Time on the Optical Characteristics of Chemically Deposited Nanostructure PbS Thin Films," *Energy Procedia,* vol. 6, pp. 241-250.

Formation of copper oxide thin films from RF sputtered Cu thin film by ultra high pure boiled water

Subramani Shanmugan* Devarajan Mutharasu

Nano Optoelectronics Research Laboratory, School of Physics, Universiti Sains Malaysia (USM), Minden, Pulau Penang, 11800, Malaysia

* Corresponding author E-mail: subashanmugan@gmail.com (S.Shanmugan)

Tel: +60-04-6533672; Fax: +60-04-6579150.

Abstract- **Copper oxide thin films were synthesized from rf sputtered Cu thin film using ultra high pure hot water. X ray diffraction studies confirmed the formation of mixed phases (Cubic Cu_2O and Monolithic CuO) after processing for 3 hrs duration. Transmittance studies also revealed the conversion of Cu into Copper oxide by showing high transmittance for processed film. The calculated band gap (2.6 – 2.69 eV) is higher than the bulk copper oxides (2.14 eV). PL spectra showed three dominant peaks at 390nm, 522nm and 680 nm related to nano structured Cu_2O. Raman spectra also confirmed the formation of nano crystals of Copper oxides.**

I. INTRODUCTION

Cu and Cu-based materials have a number of industrial applications at lower temperatures (300°C), which is partly due to the relatively good oxidation resistance in water-containing environments. However the expose of Cu surface in presence of moisture content air atmosphere cause the surface oxidation independent to the atmospheric temperature. Copper oxide (CO) is a proven p-type transparent conducting oxide (TCO) semiconductor for fabricating variety of devices. The Cu-O system contains two stable oxides at high temperatures, Cu_2O and CuO, where the change in Gibbs free energy for the reaction $2CuO(s) = Cu_2O + \frac{1}{2} O_2 (g)$ can be expressed as $\Delta G° = 146230 + 25.5T \log T - 85T$ (J/mol) [1]. Thus at the higher oxygen pressures, a two-phase scale is formed during oxidation of Cu, an outer CuO layer and an inner Cu_2O layer. Both the CuO (monoclinic) and Cu_2O (cubic) are p-type semiconductors with a band gap ranging 1.9 – 2.1 and 2.1 – 2.6 eV, respectively [2]. Cu_2O is one of the oldest semiconducting materials [3] that is used in solar cell applications [4], owing to the fact that the constituent materials are nontoxic and abundantly available on the earth.

The large excitonic binding energy (150 meV) [5] of Cu_2O allows the observation of a well-defined series of excitonic features in the absorption spectrum at low temperatures [6] Further, excitons in Cu_2O are long-lived (approximately 10 ms) and their motion through the solid can be coherent, thus it can be applied to photon coherence in a laser [7]. Its high optical absorption coefficient in the visible range (350 – 800 nm) and reasonably good electrical properties make it suitable for use in the fabrication of thin film solar cells, with a theoretically achievable efficiency of up to 13%. Besides, Cu_2O evolves potential in photochemical applications such as catalyst for water splitting [8-10].

The control of semiconductor size at the nanometer scale is an effective approach for enhancing optical and electrical properties, which is originated in quantum confinement and surface effects. A variety of methods such as electro deposition, thermal oxidation, chemical deposition, sputtering, plasma evaporation, sol–gel, molecular beam epitaxy, and chemical vapor deposition [11-19] have been employed to study the CO films. Chemical methods based on reactions in solution, such as the reduction method and the electrochemical method, etc., may be preferable from a production cost perspective. Copper metal does not react with water or steam since the Cu is less reactive. The aim of this paper is an attempt to covert metal Cu film into Copper oxide film by simple boiling in ultra high pure water. The structural and optical properties of the processed films are reported here.

II. EXPERIMENTAL TECHNIQUES

Copper films were deposited on soda lime glass substrates in Ar atmosphere at ambient temperature by RF Magnetron Sputtering (Edwards make, Model-Auto 500). Pure Cu (99.999%) and high pure argon gas were used as the sputtering target and the work gas, respectively. The base pressure of the chamber was ~2 x 10^{-7} torr. During the deposition, a gas flow rate of 16 sccm and a gas pressure P_{Ar} of 1.4 x 10^{-2} torr were employed. RF power of 110 W was used for all Cu coatings. To get the uniform thickness, rotary drive system was used and 25 RPM was fixed for all Cu film coatings. Sputtering duration was adjusted to yield Cu thicknesses of about 50 nm. All Cu films were coated at 0.023 nm/sec.

Ultra pure hot water treatment was performed at 95°C, using magnetic stirrer (resistivity more than 18.2 MΩ cm) for the specimen coated with metallic Cu thin film. The hot water treatment was performed at three different times (120 min, 150 min and 180min). The crystalline nature of the CuO thin films was investigated by using a high resolution X-ray diffraction (HRXRD, X'pert-PRO, Philips, Netherlands). A CuKα (k = 1.54056 Å) source was used, and the scanning range was between 2h = 20° and 80°. The transmittance in the visible range was measured using Shimadzu UV-1800 UV/Vis/NIR Scanning Spectrophotometer (Shimadzu). The extinction coefficient and band gap were obtained from the transmittance curve. PL and Raman spectra at room temperature of the samples were measured by Jobin Yvon

978-1-4673-2395-6/12 $31.00 © 2012 IEEE

HR 800 UV using 325 nm line of a He–Cd laser and Ar laser as the excitation source respectively. The sheet resistance of all the processed samples was measured using a four probe resistivity measurement system (Model: Changmin Tech CMT-SR2000N).

III. RESULTS AND DISCUSSIONS

3.1 Structural analysis

To determine the crystal structure of the processed films, the X ray diffraction (XRD) of the films was carried out and Fig. 1 represents the XRD patterns of hot water processed Cu thin films at various time durations. The XRD patterns indicate that the processed samples are structurally defected and hence amorphous like pattern were observed. According to the XRD patterns, the crystallographic phase of the films can be indexed to cubic structure of Cu_2O (JCPDS file no. 78-2076) and Cu (JCPDS file no. 03-1005), Monoclinic structure of CuO (JCPDS file no. 89-5898). Figure 1 also shows that metal Cu related peaks are observed at low process time (1hr). But these peaks slowly disappear and Cu_2O phase exists when the process time increases to 3 hrs. In addition to that, a CuO phase is also observed when the film processed at 3hr. The oxidation reaction of copper can be expressed as follows [20]:

$$4Cu + O_2 = 2Cu_2O \qquad (1)$$
$$2Cu_2O + O_2 = 4CuO \qquad (2)$$

When copper is oxidized as a result of O_2 addition, the major product is Cu_2O as shown in Reaction 1, and CuO forms slowly only through the second step of oxidation as in Reaction 2 in which Cu_2O serves as a precursor to CuO. In other words, CuO forms at a higher temperature or needs longer reaction time than Cu_2O.

Since the pH value of ultra high pure water is about 7, the copper (II) ion is the more common oxidation state [21] and will form complexes with hydroxide and carbonate ions [22]. But in our study, we didn't observe any hydroxide peaks in XRD analysis for all processed films. It seems to be the effect of process temperature (90°C) on control the formation of hydroxide groups while Cu reacts with water.

Based on these observations, it is concluded that the formation of Cu_2O and CuO phases is possible by the proposed hot water process method at long process time duration. Meanwhile, the intensity of (110) plane increases for the film processed at 2 hrs duration and decreases for 3 hrs duration. But the (110) peak is observed at higher 2θ than the standard value. The intensity of each peak varies with respective to process time. It seems that the high process time has the ability to increase the crystalline quality.

3.2 Optical Properties

Transmittance spectra of metal Cu and all processed films are recorded and given in Fig. 2. It clearly indicates that the conversion from metal Cu into Copper oxides occurs by observing high transmittance of processed samples than the as grown metal Cu thin film. The low transmittance (<35%) of the as-deposited films (due to absorption of Cu) is observed and the transmittance increases as process time increases. The transmittance of the films processed for 1 hr shows good transmittance than the others. The transmittance all processed films varies from 44 % to 88 % in the visible region. The transmittance range varies from 60 to 88, 54 to 86 and 44 to 84 in the visible region for 1 hr, 2 hrs and 3 hrs respectively. The absorption coefficient (α) was calculated from the transmittance spectra using the relation [23].

$$\alpha\,(h\upsilon) = 4\pi\,kf\,/\,\lambda \qquad (3)$$

The extinction coefficient (kf) coming out from the experimental measurements was then calculated from the relation.

$$kf = 2.303\,\lambda\,log\,(1/T_0)\,/\,4\pi t \qquad (4)$$

where T_0 is the transmittance, t is the film thickness and λ is the wavelength of the incident radiation. The calculated absorption coefficient was $> 10^5$. The extinction coefficient k is found to be in between 0.25 and 0.65 and shown in Fig.3. The film processed at 3 hrs duration shows high extinction coefficient than the other samples. The efficient optical band gap or width of the forbidden zone can be determined from the plot of $(\alpha h\upsilon)^2$ versus ($h\upsilon$) following the Mott and Davis expression for direct transitions [24]:

$$\alpha h\upsilon = B\,(h\upsilon - E_g)^n \qquad (5)$$

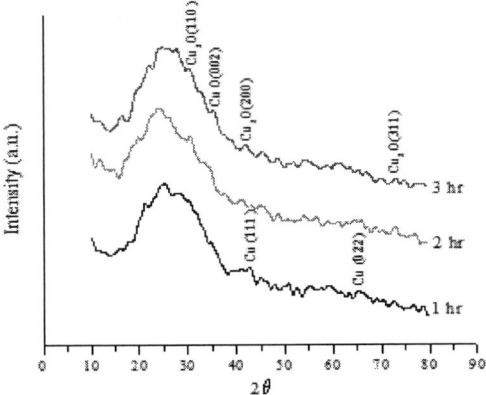

Fig. 1. XRD spectra of showing mixed Cu_2O and CuO phases.

Fig. 2. Transmittance spectra of as grown and hot water processed Cu thin films for various wavelength regions.

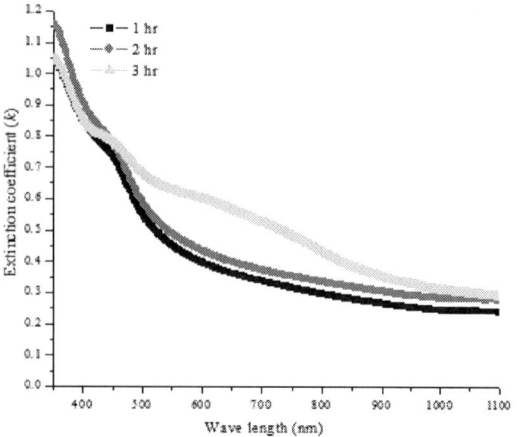

Fig. 3. Variation of extinction coefficient (k) of all processed thin films against wavelength.

where α is the absorption coefficient at the frequency v, n = 1/2 for allowed direct transitions, B is a constant and Eg is the band gap. The intersection of the slope of the linear part of the plot $(\alpha h v)^2$ vs. $h v$ gives the values of the Eg (in eV). The calculated band gap are plotted against photon energy and shown in Fig.4. The band gap of the processed nano structured Copper oxide thin films was measured by extrapolation which varies from 2.6 – 2.69 eV.The estimated values of Eg for samples treated at 1, 2 and 3 hrs durations are 2.60, 2.66 and 2.69 eV respectively. The absorption edge for all processed samples is blue shifted in comparison to bulk Cu_2O (Eg = 2.0 eV) [25]. The shift is due to the nanosize of the crystallites. The increase in the value of Eg for the processed samples for 2 hrs and 3 hrs, in comparison to sample treated for 1 hr is related to the decrease in crystallite size with process time. All the processed films show dual Eg values. In general, the presence of dual band gap occurs due to the mixed phase or poor crystallinity [26]. The same behavior is also reported by different authors for gallium doped zinc oxide films [27].

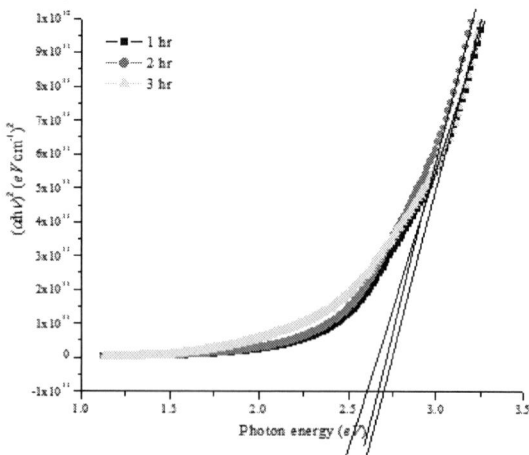

Fig. 4. Band gap variation of all Cu thin films processed at 95° C for various time.

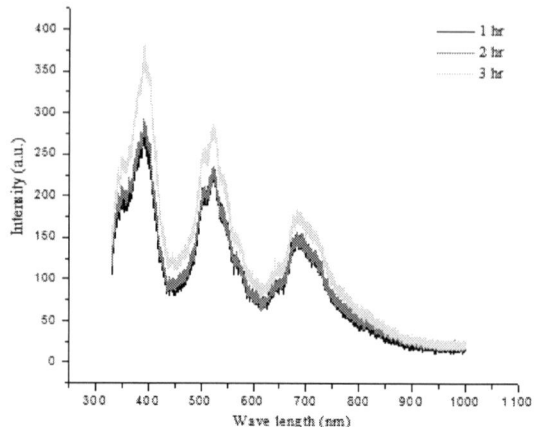

Fig. 5. Room temperature PL spectra of Cu thin films processed at 95° C for various time.

In this study, the observation is attributed to the relatively poor crystallinity of the films. It is also attributed to the formation of nano size particles on the processed thin films. The same behavior was also observed by R.K. Swarnkar et al. [28].

3. 3 Photoluminescence studies

PL spectra of all processed samples are recorded and shown in Fig.5. The strong UV emission band peaking at 390nm show good crystalline nature of the material and can be attributed to high NBE-to-DLE intensity ratio, resulting in detectable near-UV emission at room temperature [29]. This peak may be attributed to CuO phase (399 nm), but this could not be identified by XRD analysis for 1 hr and 2 hrs processed samples. The peak at 390 nm possibly originates from the existance of weak O – O bonding, which is likely a preexisting defect [30]. The intensity of 390 nm emission shows high value for the film processed at 3 hrs duration. Moreover the intensity of all peaks increases with the process time. A peak due to near band-edge emission is observed at ~680 nm (1.82 eV) for all processed films. This peak is most likely attributable to acceptor-related luminescence, taking account of the bandgap energy of Cu_2O (2.0 eV) [31]. This peak is probably due to transitions between $^3\Gamma_5^+$ valence band and $^2\Gamma_7^+$ conduction band in Cu_2O nano crystals [32-34].

The observed energy positions of band edge luminescence at 503 and 522 nm in all the samples are related to surface of nano structured with large concentration of defects and hence the hamper PL emission [35, 36]. The observed results are matched with the published results for the Eg ~2.46 and 2.38 eV of Cu_2O obtained by the different author. The observed peak at 503 is well matched with the results of Cu_2O published by R. S. Patil et al [37].

It is therefore suggested that the broad red emission peak at around 650nm is probably due to the overlap of the emission due to Cu nanocrystals and the transitions between $^3\Gamma_5^+$ valence band and $^2\Gamma_7^+$ conduction band in Cu_2O nanocrystals [37]. It is suggested that the process time has an effect of increasing a luminescence intensity of nanocrystalline Cu_2O [38].

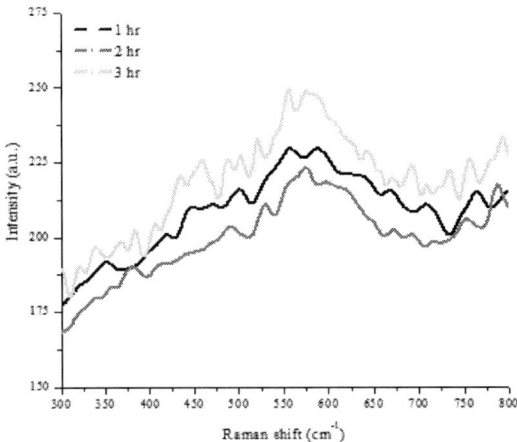

Fig. 6. Raman spectra of Copper oxide thin films processed at 95° C for various time.

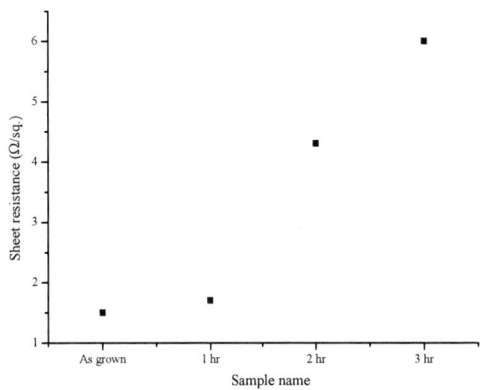

Fig. 7. Sheet resistance of as grown and hot water processed Cu thin film for various time.

3.4 Raman studies

The Raman spectra of all processed Cu thin film are displayed in Fig. 6. It is seen that a peak at around 570 cm^{-1} attributed to the Cu$_2$O phase [39,40]. A weak peak at around 458 cm^{-1} is also observed with the film processed at 3 hr duration. This peak is correspond to the CuO phase [41,42]. Raman spectrum of the processed film at 3 hr duration is in complete agreement with XRD results and reveals the formation of CuO when the process time increases.

It is well-known that the intensity of Raman scattering is directly proportional to the number of scattering centers present in the volume illuminated by the laser beam. Thus, it is concluded that both CuO and Cu$_2$O exist in samples processed for 3 hrs duration. We also note that the Cu$_2$O-related Raman peak does not disappear completely for sample processed at 1 hr duration whereas the Cu$_2$O diffraction peak in XRD cannot be observed. This difference may exist because Cu$_2$O does not crystallize or the amount of crystalline Cu$_2$O is low so that it cannot be detected by XRD. Another possible explanation is that Raman scattering is more sensitive to copper oxide compared with XRD, as reported by Gong et al. [40].

3.5 Resistivity

The sheet resistance of the prepared Cu thin film and hot water processed thin film were studied and the observed values are plotted in Fig. 7. It shows that the resistance of hot water processed Cu thin films shows high value compared to as grown thin film which confirms the formation of copper oxide on the surface.

IV. CONCLUSION

Copper oxide was successfully synthesized by boiling rf sputtered Cu thin films in ultra high pure water. The phases were indexed as mixed phases of Cu$_2$O and CuO, for the increase in process time. Improved crystalline quality is possible with long process time duration.

The observed band gaps of the processed thin films were higher than the reported value. PL and raman spectra showed the presence of structural defects in processed films. The increased sheet resistance of the film was an evidence of the formation of copper oxide by hot water process. Overall it is concluded that nano structured copper oxide could be synthesized by simple hot water process.

ACKNOWLEDGMENT

The authors would like to thank the Solid State Laboratory and Nano Optoelectronics Research Laboratory in School of physics who has provided the RF sputtering system to coat Zn metal films and the ICP-RIE, SEM, AFM and four probe apparatus to perform the plasma process and characterization for the prepared film respectively.

REFERENCES

[1] O. Kubaschewski, E.L.L. Evans and C.B. Alcock, "Metallurgical Thermochemistry," Pergamon Press, Elmsford, New York (1967).

[2] S.C. Ray, "Preparation of copper oxide thin film by the sol–gel-like dip technique and study of their structural and optical properties," Sol. Energy Mater. Sol. Cells, vol. 68, pp. 307-312, march 2001.

[3] L.O. Grondahl, "The Copper-Cuprous-Oxide Rectifier and Photoelectric Cell," Rev. Mod. Phys., vol. 5, pp. 141–168, 1933.

[4] B. P. Rai, "Cu$_2$O solar cells: A review," Solar Cells, vol. 25, pp. 265-272, 1998.

[5] A. O. Mousa, T. Akomolafe and M. J. Carter, "Production of cuprous oxide, a solar cell material, by thermal oxidation and a study of its physical and electrical properties," Sol. Energy Mater. Sol. Cells, vol. 51, pp. 305-316, 1988.

[6] K. Shindo, T. Goto and T. Anzai, "Exciton-LO Phonon Scattering in Cu$_2$O," J. Phys. Soc. Jpn., vol. 36, pp. 753-758, 1974.

[7] D. Snoke, "Coherent Exciton Waves," Science, vol. 273, pp. 1351–1352, 1996.

[8] A. E. Rakhshani, "Preparation, Characteristics and Photovoltaic Properties of Cuprous Oxide," Solid State Electron, vol. 29, pp. 7-17, 1986.

[9] M. Hara, T. Kondo, M. Komoda, S. Ikeda, K. Shinohara, et. al., 'Cu$_2$O as a Photocatalyst for Over All Water Splitting under Visible Light Irradiation,' Chem. Commun., vol. 1998, pp. 357-358, 1998.

[10] S. Ikeda, T. Takata, T. Kondo, G. Hitoki, M. Hara, et. al., 'Mechano-Catalytic Overall Water Splitting,' Chem. Commun., vol. 1998, pp. 2185-2186, 1998.

[11] P.E. de Jongh, D. Vanmaekelbergh and J.J. Kelly, "Cu_2O: A Catalyst for the Photochemical Decomposition of Water," Chem. Commun., vol. 1999, pp. 1069–70, 1999.

[12] E. M. Alkoy and P. J. Kelly, "The structure and properties of copper oxide and copper aluminium oxide coatings prepared by pulsed magnetron sputtering of powder targets," Vacuum, vol. 79, pp. 221-230, 2005.

[13] T. Mahalinga, J. S. P. Chitra, J. P. Chu, H. Moon, H. J. Kwon et. al., "Photoelectrochemical solar cell studies on electroplated cuprous oxide thin films," J. Mater. Sci., Mater. Electron., vol. 17, pp. 519-523, 2006.

[14] M. T. S. Nair, L. Guerrero, O. L. Arenas and P. K. Nair, "Chemically deposited copper oxide thin films: structural, optical and electrical characteristics," Appl. Surf. Sci., vol. 150, pp. 143-151, 1999.

[15] P. Samarasekara, M. A. Mallika Arachchi, A. S. Abeydeera, C. A. N. Fernando, A. S. Disanayake, et. al., "Photocurrent enhancement of d.c. sputtered copper oxide thin films," Bull. Mater. Sci., vol. 28, pp. 483-486, 2005.

[16] K. Akimoto, S. Ishizuka, M. Yanagita, Y. Nawa, G. K. Paul, et. al., "Thin film deposition of Cu_2O and application for solar cells," Sol. Energy, vol. 80, pp. 715-722, 2006.

[17] K. H. Yoon, W. J. Choi and D. H. Kang, "Photoelectrochemical properties of copper oxide thin films coated on an n-Si substrate," Thin Solid Films, vol. 372, pp. 250-256, 2003.

[18] K. Santra, C. K. Sarker, M. K. Mukherjee and B. Ghosh, "Copper oxide thin films grown by plasma evaporation method," Thin Solid Films, vol. 213, pp. 226-229, 1992.

[19] A. Y. Oral, E. Mensur, M. H. Aslan and E. Basaran, "The preparation of copper(II) oxide thin films and the study of their microstructures and optical properties," Mater. Chem. Phys., vol. 83, pp. 140-144, 2004.

[20] K. P. Muthe, J. C. Vyas, S. N. Narang, D. K. Aswal, S. K. Gupta, et. al., "A study of the CuO phase formation during thin film deposition by molecular beam epitaxy," Thin Solid Films, vol. 324, pp. 37-43, 1998.

[21] X. Jiang, T. Herricks, and Y. Xia, "CuO Nanowires can be Synthesized by Heating Copper Substrates in Air," Nano Lett., vol 2, pp. 1333-1338, 2002.

[22] US EPA (1995) "Effect of pH, DIC, orthophosphate and sulfate on drinking water cuprosolvency," Washington, DC, US Environmental Protection Agency, Office of Research and Development (EPA/600/R-95/085).

[23] W. Stumm and J.J. Morgan, "Aquatic chemistry," New York, Wiley Interscience (1996).

[24] H. Padmanabhasarma, V. Subramanian, N. Rangarajar and K.R. Muralli, "A Comparative study on CdSe synthesized at low temperature," Bull. Mater. Sci., vol. 18, pp. 875-881, 1995.

[25] N.F. Mott and E.A. Davis, "Electronic Processes in Non-Crystalline Materials," 2nd edition, Clarendon Press, Oxford, (1979).

[26] E. Rakshani, "Preparation, characteristics and photovoltaic properties of cuprous oxide—a review," Solid-State Electron., vol. 29, pp. 7-17, 1986.

[27] A. U. Mane and S. A. Shivashankar, "MOCVD of cobalt oxide thin films: dependence of growth, microstructure, and optical properties on the source of oxidation," J. Cryst. Growth, vol. 254, pp. 368-377, 2003.

[28] G. Goncalves, E. Elangovan, P. Barquinha, L. Pereira, R. Martins et. al., "Influence of post-annealing temperature on the properties exhibited by ITO, IZO and GZO thin films," Thin Solid Films, vol. 515, pp. 8562-8566, 2007.

[29] R.K. Swarnkar, S.C. Singh and R. Gopal, "Optical characterizations of copper oxide nanomaterial," ICOP 2009-International Conference on Optics and Photonics CSIO, Chandigarh, India, 30 Oct.-1 Nov. 2009.

[30] H. Cao, X. Qiu, Y. Liang, L. Zhang, M. Zhao et al., "Sol–Gel Template Synthesis and Photoluminescence of n- and p-Type Semiconductor Oxide Nanowires" Chem. Phys. Chem., vol. 7, pp. 497-501, 2006.

[31] H. Nishikawa, T. Shiroyama, R. Nakamura, Y. Ohki, K. Nagasawa et. al., "Photoluminescence from defect centers in high-purity silica glasses observed under 7.9-eV excitation' Phy. Rev. B: cond. matt., vol. 45, pp. 586-591, 1992.

[32] Y. Okamoto, S. Ishizuka, S. Kato, and T. Sakurai, "Passivation of defects in nitrogen-doped polycrystalline Cu_2O thin films by crown-ether cyanide treatment," Appl. Phy. Lett., vol. 82, pp. 7-17, 2003.

[33] M.I. Freedhoff and A.P. Marchetti, 'Quantum confinement in semiconductor nanocrystals, in: R.E. Hummel' P. Wissman (Eds.), Handbook of Optical Properties, 2, CRC Press, Boca Raton, (1997).

[34] A. Sengupta and J.Z. Zhang, Semiconductor nanoparticles, in: Z. Lin Wang, Y. Liu, Z. Zhang (Eds.), Handbook of Nanophase and Nanostructured Materials, Vol.3, Plenum Publishers, New York, (2003).

[35] Y. Liu, '"Excitons at high density in cuprous oxide and coupled quantum wells," Ph.D. thesis, University of Pittsburgh, 2004.

[36] K. Asai, T. Yamaki, K. Ishigure and H. Shibata, "Bombardment effect on electronic states in CdS fine particles," Thin Solid Films, vol. 277, pp. 169-174, 1996.

[37] K. Asai, T. Yamaki, S. Seki, K. Ishigure, and H. Shibata, "Surface treatment effect of ion irradiation on size-quantized semiconductor particles incorporated into LB films," Thin Solid Films, vol. 284, pp. 541-544, 1996.

[38] Y. Okamoto, S. Ishizuka, S. Kato, T. Sakurai, N. Fujiwara et. al., "Passivation of defects in nitrogen-doped polycrystalline Cu_2O thin films by crown-ether cyanide treatment," App.Phy. Lett., vol. 82, pp. 1060-1062, 2003.

[39] G. Niaura, "Surface-enhanced Raman spectroscopic observation of two kinds of adsorbed OH- ions at copper electrode," Electrochim. Acta, vol. 45, pp. 3507-3519, 2000.

[40] Y. S. Gong, C. Lee, and C. K. Yang, "Atomic force microscopy and Raman spectroscopy studies on the oxidation of Cu thin films," J. Appl. Phys., vol. 77, pp. 5422-5426, 1995.

[41] S. Guha, D. Peebles, and T. J. Wieting, "Zone-center (q=0) optical phonons in CuO studied by Raman and infrared spectroscopy," Phys. Rev. B, vol. 43, pp. 13092-13101, 1991.

[42] H. Hagemann, H. Bill, W. Sadowski, E. Walker, and M. Francois, "Raman spectra of single crystal CuO," Solid State Commun. vol. 73, pp. 447-451, 1990.

A Study on Lightly-Doped Cylindrical surrounding-gate 6H-SiC Nanowire FET

Ru Han

School of Computer Science and Engineering
Northwestern Polytechnical University
Xi'an, China
Email: hanru@nwpu.edu.cn

Abstract- The device characteristics of Cylindrical surrounding-gate (CSG) 6H-SiC NW FET is investigated in this paper. The results indicate that the surface potential, threshold voltage and the electric characteristics (transfer characteristics and output characteristics) is very sensitive to 6H-SiC nanowire radius, channel length, oxide thickness and temperature. The temperature dependence of CSG 6H-SiC NW FET is also discussed in this paper. When the nanowire radius is decreased, the minimum potential is lowered, the locations of minimum potential moves to the source side and the threshold voltage is increased. When the oxide thickness is increased, the locations of minimum potentials are not changed, but the minimum potentials themselves become larger and the threshold voltage become smaller. The minimum potential increases as the gate length decreases, but the threshold voltage decreased as the gate length decreases. With increasing temperature, the surface potential decreases and the location of minimum potential moves to the source side. The threshold voltage decreases monotonically with temperature. At strong inversion region, the drain current decrease as nanowire radius decreases, but increases as temperature decreases.

I. INTRODUCTION

Semiconducting one-dimensional (1D) objects such as nanowires are currently being intensively studied as a result of their importance in fundamental research and for their potential in the fabrication of nanoscale electronic, optoelectronic, and sensor devices [1]. The 1D structures can function both as active components of devices as well as interconnects and thus have potential to provide two of the most critical functions in any integrated nanoelectronic system [2].

At the same time, excellent physical properties (such as the wide bandgap, high breakdown electric field, high thermal conductivity, high electron drift velocity, high strength at elevated temperatures and physical stability) of silicon carbide (SiC) offer the opportunity to realize a breakthrough for its high power, high frequency and high temperature electronic applications. The native oxide is silicon dioxide, which makes SiC directly compatible with usual Si technology.

Therefore, SiC nanowires combine the properties of 1D materials with that of SiC and devices based on SiC NWs are expected to present concrete advantages [3-4]. Hence, research on 1D SiC nanowires is highlighted, both from the fundamental research standpoint and for potential application in nanodevices and nanocomposites.

On the other hand, with the reduction in channel length, short-channel effects (SCEs) and hot carrier effects (HCEs) impose a physical limit on the ultimate performance of traditional planar metal-oxide-semiconductor field effect transistors (MOSFETs) [5]. The cylindrical surrounding-gate (CSG) MOSFETs (CSG) offers the best control of SCEs and HCEs, and is considered one of the most promising devices for downscaling below 50 nm [5]. For SiC nanowires, the cylindrical surrounding-gate is more advantageous than any other gate structures.

To incorporate the advantages of SiC nanowire and CSG structures, in this paper we investigate the device characteristics and circuit behavior of CSG 6H-SiC NW FET. For circuit simulation, it is generally agreed that the use of surface potential can provide a good approximation to the physical behavior of transistors and yields better predictions of the performance of integrated circuits compared with other alternatives, particularly in the nanoscale and high-frequency operations [6]. So, the 3-D numerical device simulator ISE is employed to simulate the device characteristics, such as the surface potential, the fundamental device parameter threshold voltage and the electric characteristics (transfer characteristics and output characteristics) of CSG 6H-SiC NW FET.

II. STRUCTURE OF 6H-SIC NANOWIRE FET

Considering the cylindrical surrounding-gate 6H-SiC nanowire FET represented in Fig. 1, the two-dimensional potential described in the lightly doped 6H-SiC nanowire ($\approx 10^{16}$ cm^{-3}) is modeled in follow section. The source and drain regions are highly doped. In this paper, V_{gs} is the gate-to-source voltage, V_{ds} is the drain-to-source voltage, R is the radius of the 6H-SiC nanowire, t_{ox} is oxide thickness, L is the gate length, x is the channel direction and T is the temperature.

III. SIMULATION RESULTS AND DISCUSSION

A. Surface potential

It is well known that the surface potential based model had been chosen the next generation industry standard model of the bulk CMOS since December 2005 [6]. Therefore, we investigates the surface potential in this section.

Fig. 2 and Fig. 3 shows the surface potential profile by varying the 6H-SiC nanowire radius and gate oxide thickness. It is shown in Fig. 2 that when the nanowire radius is decreased, the minimum potential is lowered (implying that higher threshold voltage is obtained), and the locations of

978-1-4673-2395-6/12 $31.00 © 2012 IEEE

minimum potential moves to the source side. The oxide thickness of current state-of-the-art nanowire transistors is rather thick. The (equivalent) oxide thickness will have to be reduced to about 0.8 – 1nm if improved device performance has to be achieved. Therefore, we have considered insulator thickness ranging from 1 to 4 nm. It is indicated in Fig. 3 that when the oxide thickness is increased, the locations of minimum potentials are not changed, but the minimum potentials themselves become larger.

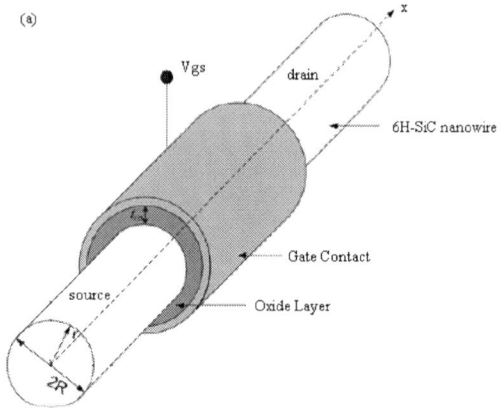

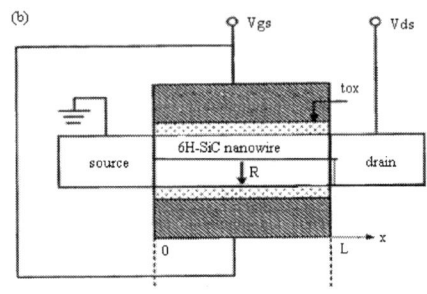

Fig. 1.   Schematic of the CSG 6H-SiC nanowire FET: (a) three dimension view, (b) Cross senction view

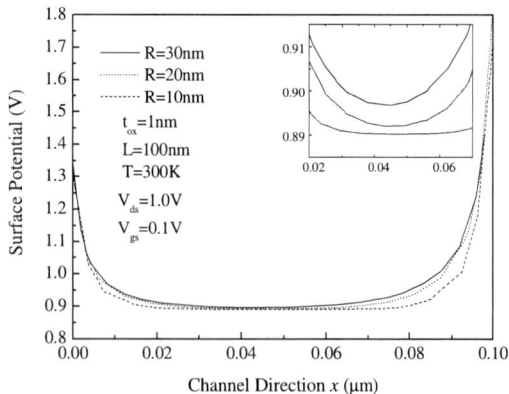

Fig. 2.   Surface potential along the channel versus channel direction x with different 6H-SiC nanowire radius

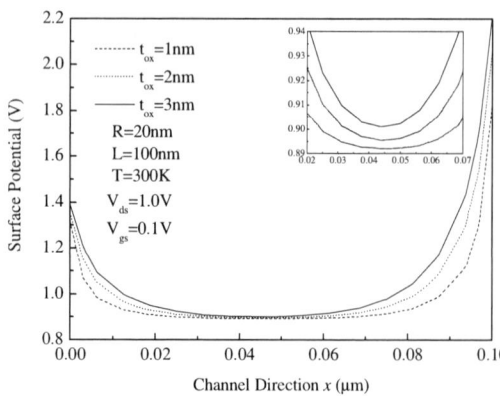

Fig. 3.   Surface potential along the channel versus channel direction x with different oxide thickness

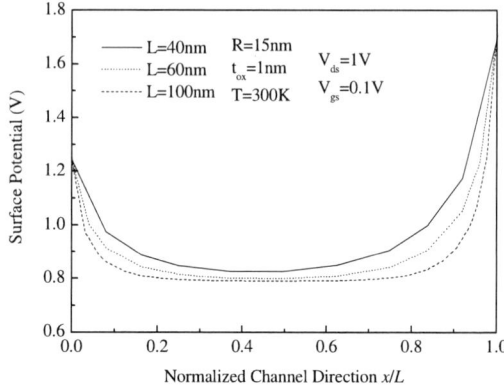

Fig. 4.   Surface potential along the channel versus normalized channel direction x/L with different gate lengths

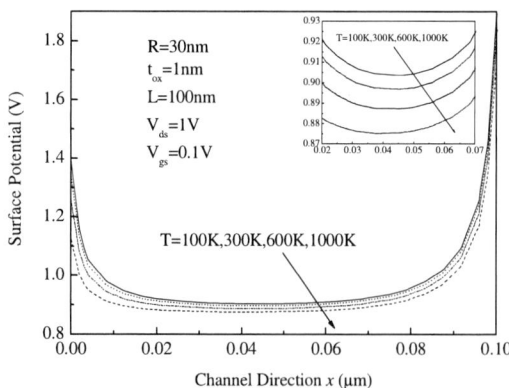

Fig. 5.   Surface potential along the channel versus channel direction x with different temperature

In order to model the impact of the gate length reduction on the characteristic of CSG 6H-SiC nanowire FET, the surface potential profile for different channel lengths is shown in Fig. 4. It is clearly evident from the figure that the minimum potential increases as gate length decreases.

The temperature dependence of the surface potential of CSG 6H-SiC nanowire FET is presented in Fig. 5. As it is evident from Fig. 5, with increasing temperature, the surface potential decreases and the location of minimum potential moves to the source side.

B. Threshold voltage

The threshold voltage is the most important electrical parameter for MOSFET modeling and characterization. The threshold voltage value is very sensitive to 6H-SiC nanowire radius, channel length, oxide thickness and temperature.

Fig. 6 and Fig. 7 shows the threshold voltage variation with different 6H-SiC nanowire radius and channel length. It is obviously that the threshold voltage decreases as 6H-SiC nanowire radius increases, but increases as channel length increases. Fig. 8 and Fig. 9 shows the threshold voltage variation with different oxide thickness and temperature. Fig. 8 indicates that the threshold voltage of the lightly doped CSG 6H-SiC NW MOSFET decreases with increasing gate oxide thickness, which is different from that in traditional bulk MOSFET. Fig. 9 shows that the threshold voltage decreases monotonically with temperature.

C. Drain current

Fig. 8 plots drain current versus V_{gs} curves for several different voltages, which are obtained from different nanowire radius and different oxide thickness. At the strong inversion region, the drain current decrease as nanowire radius decreases. The variation of the oxide thickness could influence the surface potential and the threshold voltage, therefore the oxide thickness dependence of drain current is the result of the surface potential and the threshold voltage.

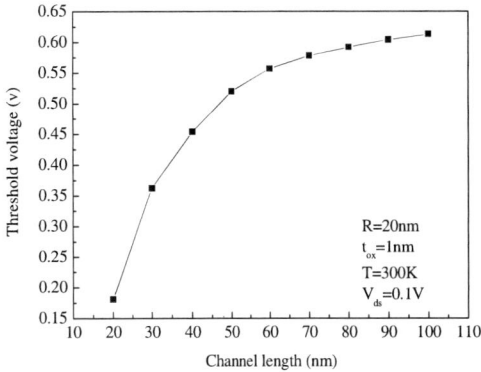

Fig. 7.   Threshold voltage versus channel length

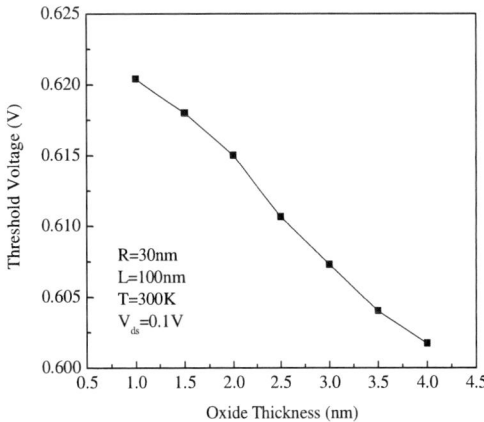

Fig. 8.   Threshold voltage versus oxide thickness

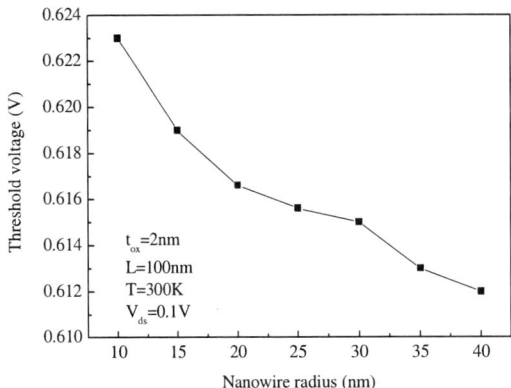

Fig. 6.   Threshold voltage versus 6H-SiC nanowire radius

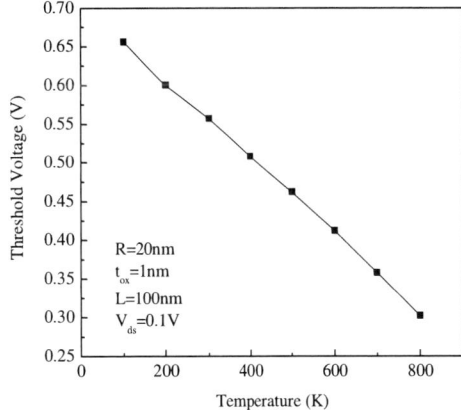

Fig. 9.   Threshold voltage versus temperature

978-1-4673-2395-6/12 $31.00 © 2012 IEEE   139

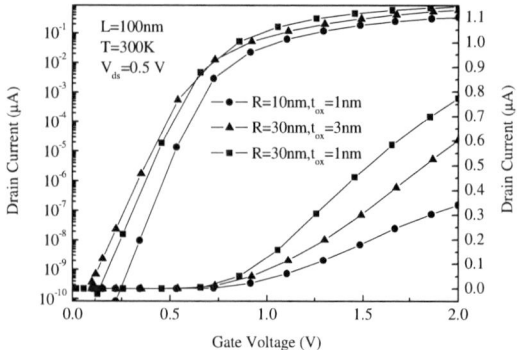

Fig. 10.   Drain current simulations versus gate voltage with different 6H-SiC nanowire radius and oxide thickness

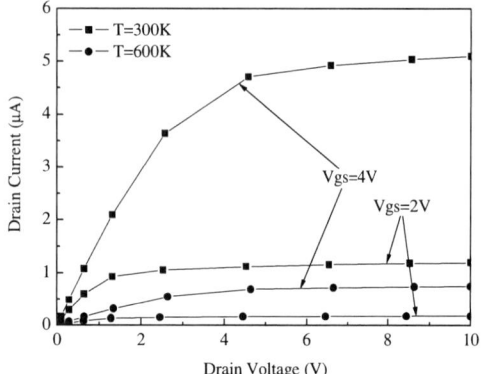

Fig. 11.   Drain current simulations versus drain voltage with different gate voltage and temperature

Fig. 9 plots drain current versus drain voltage for several different gate voltages. It is observed from the figure that the drain current decrease as temperature increases. From the figure, it is observed that when the $V_{ds} > V_{dsat}$ (saturation drain-source voltage) there is a steady increase of drain current for CSG 6H-SiC nanowire FET at 300K as compare to 600K. This increase is caused due to channel length modulation effect.

IV.   CONCLUSION

SiC nanowires are used for the reinforcement of various nanocomposite materials or as nanocontacts in harsh environments, mainly due to their superior mechanical properties and high electrical conductance. The device characteristics of CSG 6H-SiC NW FET is investigated in this paper.

The results   indicate that the surface potential, threshold voltage and the electric characteristics (transfer characteristics and output characteristics) is very sensitive to 6H-SiC nanowire radius, channel length, oxide thickness. The temperature dependence of CSG 6H-SiC NW FET is also discussed in this paper. When the nanowire radius is decreased, the minimum potential is lowered, the locations of minimum potential moves to the source side and the threshold voltage is increased. When the oxide thickness is increased, the locations of minimum potentials are not changed, but the minimum potentials themselves become larger and the threshold voltage become smaller. The minimum potential increases as the gate length decreases, but the threshold voltage decreased as the gate length decreases. With increasing temperature, the surface potential decreases and the location of minimum potential moves to the source side. The threshold voltage decreases monotonically with temperature. At strong inversion region, the drain current decrease as nanowire radius decreases, but increases as temperature decreases.

ACKNOWLEDGMENT

This work is supported by Northwestern Polytechnical University foundation for fundamental Research (Grant No. GCKY1001).

REFERENCES

[1]   K. Zekentes and K. Rogdakis, "SiC nanowires: material and devices" J. Phys. D: Appl. Phys., vol. 44, pp. 133001, 2011

[2]   M. Choueib, A. Ayari, P. Vincent, S. Perisanu, and S. T. Purcell, "Evidence for Poole–Frenkel conduction in individual SiC nanowires by field emission transport measurements" J. Appl. Phys., vol. 109, pp. 073709, 2011.

[3]   K. Rogdakis, S. Poli, E. Bano, K. Zekentes, and M. G. Pala, "Phonon- and surface-roughness-limited mobility of gate-all-around 3C-SiC and Si nanowire FETs" Nanotechnology, vol. 20, pp. 295202, 2009.

[4]   Shinobu Onoda, Takahiro Makino, Naoya Iwamoto, Gyorgy Vizkelethy, Kazutoshi Kojima, Shinji Nozaki, and Takeshi Ohshima, "Charge Enhancement Effects in 6H-SiC MOSFETs Induced by Heavy Ion Strike" IEEE Transactions on Nuclear Science, Vol. 57, pp. 3373, 2010

[5]   Cong Li, Yiqi Zhuang and Ru Han, "Cylindrical surrounding-gate MOSFETs with electrically induced source/drain extension" Microelectronics Journal, Vol. 42, pp. 341-346, 2011

[6]   Wei Bian, Jin He, Yadong Tao, Min Fang, and Jie Feng, "An Analytic Potential-Based Model for Undoped Nanoscale Surrounding-Gate MOSFETs" IEEE Trans. Electron Devices, vol. 54, no. 9, pp. 2293-2303, 2007.

Synthesization of Carbon Nanotubes Using Single Stage Thermal CVD Method

M. Maryam[1, 2, 3, a], A. B. Suriani[1, 4], M.S. Shamsudin[1, 2,c] and M. Rusop[1, 3,d]

[1] NANO-SciTech Centre, Institute Of Science;
[2] School of Physics and Material Studies, Faculty of Applied Sciences;
[3] NANO-ElecTronic Centre, Faculty of Electrical Engineering;
Universiti Teknologi MARA,
40450 Shah Alam, Selangor, Malaysia
[4] Department of Physics, Faculty of Science and Mathematics,
Universiti Pendidikan Sultan Idris,
35900 Tanjung Malim, Perak, Malaysia.
[a]mary_am_mohd@yahoo.com, [b]absuriani@yahoo.com, [c]nanopizza@rocketmail.com, [d]rusop@salam.uitm.edu.my

Abstract— **This paper will report on the synthesization of varieties of nanostructured carbon nanotubes such as bundles of vertically alligned CNTs (VACNTs), spaghetti-like CNTs and spiral-like structure of CNTs (SCNTs) which are potentially useful in various application. Multiwall carbon nanotubes (MWCNTs) were produced from palm oil precursor and ferrocene as catalyst source by this method at various deposition temperature ranging from 650-950oC with an increment of $50^{\circ}C$. Direct heat was used to vaporize both fixed parameter of precursor and catalyst placed in the middle of quartz tube. Field emission scanning electron microscopy was used to obtain the image of CNTs which showed different structures and diameters of CNTs relative to the deposition temperature of furnace. Raman Spectroscopy and Thermogravimetric analysis was also used to determine the purity of samples.**

Keywords-carbon nanotubes; palm oil; chemical vapor deposition

I. INTRODUCTION

Carbon Nanomaterials were found to be very useful in many applications such as energy conversion devices because unlike the conventional graphite phase, carbon nanostructures possess metallic or semiconductor properties that can induce catalysis by participating directly in the charge transfer process. Furthermore, the electrochemical properties of these materials facilitate modulation of their charge transfer properties and aid in the design of catalysts for hydrogenation, sensors, and fuel cells [1]. Various types of carbon nanostructures were succesfully synthesized in the form of nanotubes [2], nanocones [3], nanofibers[4], nanoballs[5], nanowires[6] and many more.

The most popular type of carbon nanostructures are carbon nanotubes which were discovered to have great properties [7]. Since then, more research had been done in this field due to its promising capability. Methods such as arc discharge [8], chemical vapor deposition or spray pyrolysis [9], laser ablation and many more were done to produce these

nanomaterials. However, in this experiment TCVD method was chosen due to it being the most inexpensive method and have higher probability of producing carbon nanotubes (CNTs) in large scale [10]. It was also found that variety of carbon nanomaterials were formed using metal catalysts and therefore Fe was chosen to be the metal cagvtalyst [8].

This paper will report on the production of different structures of multi wall carbon nanotubes such as bundles of vertically alligned CNTs (VACNTs), spaghetti-like structure of CNTs and spiral-like structure of CNTs (SCNTs) which are successfully synthesized by single zone Thermal CVD technique using palm oil (PO) as the precursor, ferrocene (Fe) as the catalyst and Nitrogen (Ni) as the carrier gas. Previous reports and studies done on CNTs from palm oil shows that by using this natural bio-hydrocarbon source, not only can we reduce the production cost; it can also be the green alternative for industrial scale production of CNTs. Fe as a catalyst also have a higher rate in producing higher quality and larger scale of CNTs [11-15].

II. EXPERIMENTAL PROCEDURE

CNTs were formed by single zone TCVD system (Figure 1). This method was based on the vaporization of both liquid PO precursor and Fe catalyst in a single stage furnace (PROTHERM PTF 14/38/250) equipped with a quartz tube of length 100cm and diameter 6cm. First of all, ferrocene and palm oil were weighed onto alumina boats. The boats were placed side by side in the middle of the furnace and nitrogen gas was flowed through the tube into a bubbler connected to the fume hood with a flow rate of 11 bubbles/10 sec. The reaction furnace was heated at deposition temperature of 650 to $950^{\circ}C$ for 1 hour and allowed to cooled down. Black substance was then collected from the wall of quartz tube and alumina boats. The powder like samples collected were then characterized by the field emission scanning electron microscope, FESEM (ZEISS Supra 40VP) operated at 5kV to evaluate the structure, diameter and identification of the

978-1-4673-2395-6/12 $31.00 © 2012 IEEE

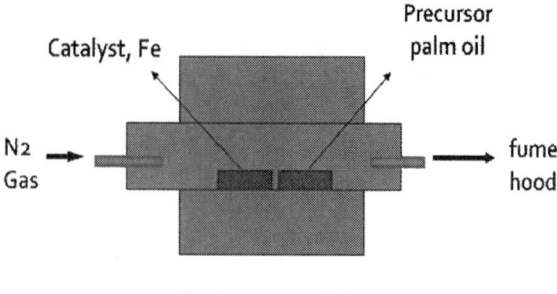

Single Furnace CVD

Figure 1.   Schematic diagram of TCVD system

sample. Raman spectra was obtained using micro-Raman spectroscopy (Horiba Jobin Yvon-DU420A-OE-325) with Ar⁺ ion of wavelength 514.5nm to determine the quality and impurities of sample and thermogravimetric analysis, TGA (Perkin Elmer Pyris 1 TGA) was done to determine the decomposition temperature and impurities of sample.

III.   RESULTS AND DISCUSSIONS

A.   FESEM images

Figure 2 showed the resulting images of samples collected from single zone TCVD method at different deposition temperature of (a) 650 °C, (b) & (c) 750 °C, (d) 850 °C, and (e) & (f) 950 °C. Carbon nanotubes and few amorphous carbon

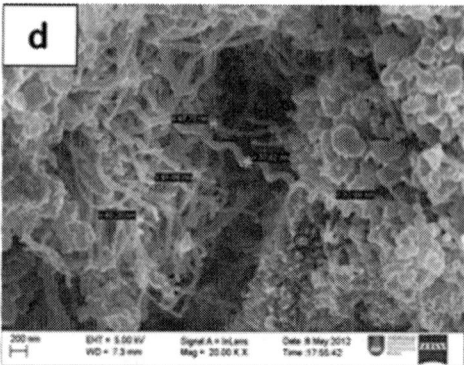

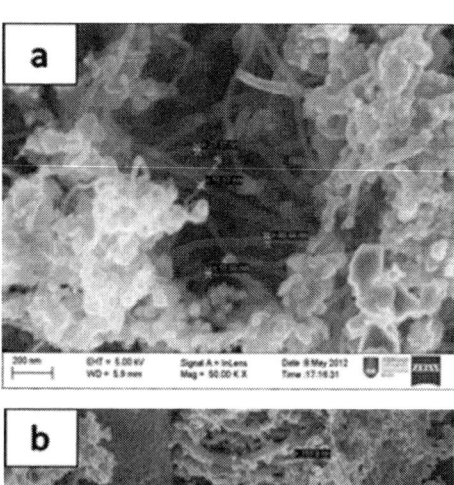

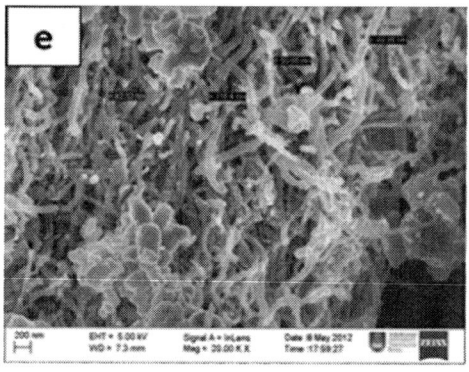

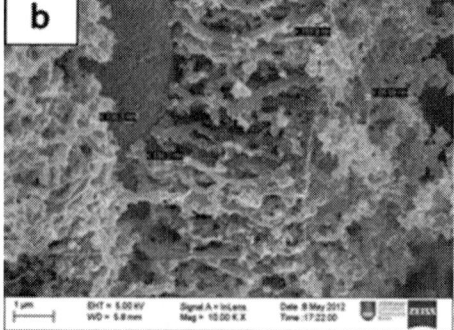

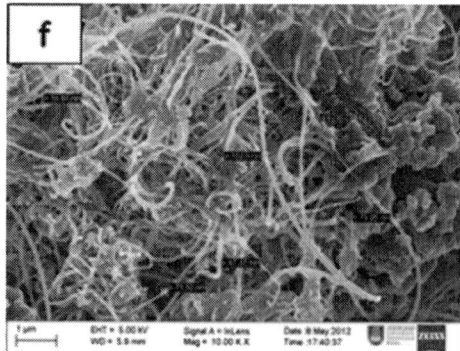

Figure 2.   FESEM images of CNTs from palm oil using floated catalyst TCVD method at deposition temperature of (a) 650 °C, (b) & (c) 750 °C (d) 850 °C, (e) & (f) 950 °C

were present at temperature (a) - (f). Apparently, bundles of spaghetti-like structures of CNTs were found at lower deposition temperature of 650°C with average diameter of ~24.2nm as seen in Fig. 2a. However, CNTs produced at temperature 750°C were found to be in bundles of vertically alligned CNTs (Fig. 2b) and few spiral-like structures of CNTs (Fig. 2c) with average diameters of ~121.35nm to ~158.45nm respectively. Sample at deposition temperature of 850 also produced a few spiral-like structure of CNTs at smaller average diameter of ~70.11nm. At highest deposition temperature of 950°C on the other hand, showed bundles of spaghetti-like structure of CNTs (Fig. 2e & 2f) with average diameter of ~60.64nm to ~77.83nm respectively. Based on the observation from the FESEM images of the samples, it can be discussed that lowest deposition temperature of 650°C produced the most narrow diameter of CNTs with the usual spaghetti-like structure. However, at temperature 750°C, different structures of CNTs were succesfully synthesized at larger diameters. It can be said that temperature below 650°C were insufficient enough to pyrolyse the precursor and catalyst resulting in few bundles of spaghetti-like structure of CNTs and some a-C were present with small diameters. As the temperature increases more than 750°C, the diameters of CNTs also decreases.

B. TGA analysis

Table 1 represented the data calculated from the TGA curve as seen in Figure 3. TGA curve shown in Figure 3 showed an initial weight loss at temperature around ~98.8°C to ~478.5°C which may be caused by the decomposition of residual hydrocarbon impurities. Significant weight loss were estimated around temperature range of ~480 to ~600°C due to the decomposition of CNTs. Remaining percentage of weight loss consists of the Fe catalyst and nonvolatile elements [14, 15]. It can be said that the highest purity of CNTs is at temperature of 900°C with total of 63.1 %.

TABLE I.   TGA DATA OF CNTs FROM PALM OIL AT TEMPERATURE OF 650 TO 950 °C

Samples (°C)	Initial weight loss (%)	Residual weight loss (%)	Purity of CNTs (%)
650	25.4	18.7	55.9
750	29.0	17.9	53.1
850	50.8	10.7	38.5
950	17.4	19.5	63.1

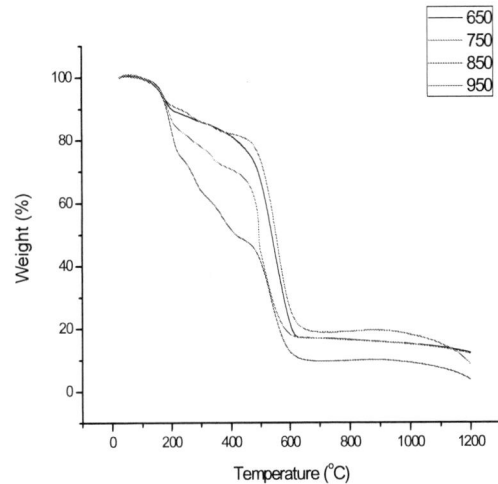

Figure 3.   TGA curves of CNTs from palm oil using floated catalyst TCVD method at deposition temperature of 650 °C, 750 °C, 850 °C and 950 °C

C. Raman Spectroscopy

By using microraman spectroscopy, the quality of samples can be calculated. The vibrational spectroscopic analysis of the samples were obtained and tabulated in the tables below:

TABLE II.   RAMAN PEAK POSITION AND INTENSITY RATIOS OF TEMPERATURE 650 TO 950 °C

Samples (°C)	Peak Positions (cm^{-1})		Integrated Intensity Ratio of D, G Bands (I_D/I_G) Ratio.
	D	G	
650	1347.8	1583.3	0.82
750	1349.7	1576.1	0.48
850	1347.5	1583.4	0.66
950	1351.3	1581.0	0.48

Table 2 represents the raman peak position and intensity ratios of samples at deposition temperature of 650 to 950°C. The peaks intensity ranging from ~1347.5-1351.3 cm^{-1} represented the disordered D line and peaks ranging from ~1576.1-1583.4 cm^{-1} represented the graphitic G line. The I_D/I_G ratio were calculated to estimate the variation of CNTs quality with precursor temperature. The average ratio for all samples are 0.61 which showed high quality of samples. Smaller ratio (<1.0) indicates that the CNT sample had a narrower distribution of defects while larger ratio shows broader distribution of defects. Figure 4 represents the microraman spectra of CNTs from palm oil using floated catalyst TCVD method at different deposition temperature of 650-950°C with increment rate of 50°C for multiple first order and second order peaks; D,G & G'.

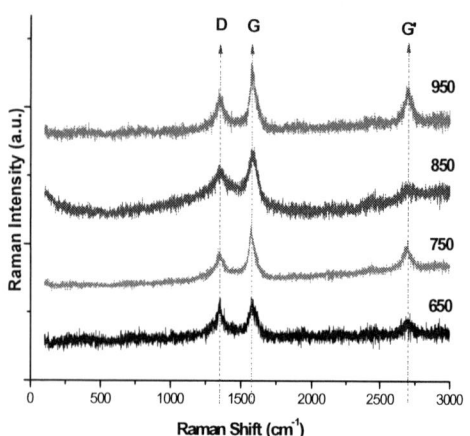

Figure 4. Microraman spectra of CNTs from palm oil using floated catalyst TCVD method at different deposition temperature of 650-950°C with increment rate of 50°C for multiple first order and second order peaks; D,G & G'

The results showed a clear trend of decreasing I_D/I_G ratio as the temperature increased with ratio of 0.82 at temperature 650°C and 0.48 at temperature 750°C. However, the ratio increased a bit at temperature 850°C and decreases again at highest temperature of 900°C with ratio 0.48.

IV. CONCLUSIONS

It can be concluded that by using the floated catalyst TCVD method, various structures of carbon nanotubes were successfully synthesized at deposition temperature ranging from 650-950°C. As seen from the FESEM image, the diameters of CNTs produced were smaller at lower deposition temperature and increased as the deposition temperature increased and finally decreased again at higher temperature resulting in more alligned and spiral-like structures. The diameters are as follow with 24.2, 121.3, 70.11, and 60.64 nm for temperature 650, 750, 850 and 950°C respectively. The a-C content also decreased as the temperature increased resulting in more CNTs content in samples. The microraman spectra study showed the presence multi wall carbon nanotubes with lower intensity ratio at higher deposition temperature with values of 0.82, 0.48, 0.66 and 0.48 for temperature 650, 750, 850 and 950 °C respectively. From TGA analysis, it was found that the CNTs yielded at temperature 650°C is 55.9%. Bundles of vertically alligned carbon nanotubes and spiral-like structure of CNTs were found to grow within temperature of 750-850°C yielding 53.1 and 38.5 % of CNT's purity. The best CNTs yield was at optimized deposition temperature of 950°C which showed

uniformed CNTs with diameter of ~60.64 nm within the bundles with highest purity calculated from the TGA curve of 63.1 % and lowest intensity ratio of 0.48.

ACKNOWLEDGMENT

The authors would like to thank Universiti Teknologi MARA for their funding and the Excellence Fund grant (project no 600-RMI/ST/DANA 5/3/Dst (494/2011)) and also Malaysian Government for their support and funding.

REFERENCES

[1] P. Kamat, "Carbon Nanomaterials: Building Blocks in Energy Conversion Devices," *The Electrochemical Society interface,* pp. 45-47, 2006.

[2] S. Iijima and T. Ichihashi, *Nature,* vol. 363, p. 603, 1991.

[3] Y. E. Lozovik and A. M. Popov, "Formation and growth of carbon nanostructures: fullerenes, nanoparticles, nanotubes and cones," *Physics-Uspekhi,* vol. 40, pp. 717-737, 1997.

[4] J. K. Chinthaginjala, S. Unnikrishnan, M. A. Smithers, G. A. M. Kip, and L. Lefferts, "Carbon nanofiber growth on thin rhodium layers," *Carbon,* vol. 50, pp. 1434-1437, 2012.

[5] X. J. Wu, X. Yuan, and L. G. Yu, "Preparation of Carbon Nanoball from Starch by Arc Discharge," *Advanced Materials Research,* vol. 476-478, pp. 1533-1536, 2012.

[6] K. C. Pingali, S. Deng, and D. A. Rockstraw, "Synthesis of Nanowires by Spray Pyrolysis," *Journal of Sensors,* vol. 2009, pp. 1-6, 2009.

[7] S. Iijima, "Nature," *London,* p. 354, 1991.

[8] C. H. Kiang, M. S. Dresselhaus, R. Beyers, and D. S. Bethune, "Vapor-phase self-assembly of carbon nanomaterials," *Chemical Physics Letters,* vol. 259, pp. 41-47, 1996.

[9] A. Aguilar-Elguézabal, W. Antúnez, G. Alonso, F. P. Delgado, F. Espinosa, and M. Miki-Yoshida, "Study of carbon nanotubes synthesis by spray pyrolysis and model of growth," *Diamond and Related Materials,* vol. 15, pp. 1329-1335, 2006.

[10] Z. E. Horváth, K. Kertész, L. Pethő, A. A. Koós, L. Tapasztó, Z. Vértesy, Z. Osváth, A. Darabont, P. Nemes-Incze, Z. Sárközi, and L. P. Biró, "Inexpensive, upscalable nanotube growth methods," *Current Applied Physics,* vol. 6, pp. 135-140, 2006.

[11] S. Abu Bakar, S. Muhamad, P. S. Mohamad Saad, S. Afif Mohd Zobir, R. Md Nor, Y. Mohd Siran, S. A. M. Rejab, A. Jaril Asis, S. Tahiruddin, S. Abdullah, and M. Rusop Mahmood, "The Effect of Precursor Vaporization Temperature on the Growth of Vertically Aligned Carbon Nanotubes Using Palm Oil," *Defect and Diffusion Forum,* vol. 312-315, pp. 906-911, 2011.

[12] A. B. Suriani, R. M. Nor, and M. Rusop, "Vertically aligned carbon nanotubes synthesized from waste cooking palm oil," *Jounal of ceramic Society of Japan,* vol. 118, pp. 963-966, 2010.

[13] S. Abu Bakar, S. Muhamad, P. S. Mohamad Saad, R. Md Nor, Y. Mohd Siran, S. A. M. Rejab, A. Jaril Asis, S. Tahiruddin, S. Abdullah, and M. Rusop Mahmood, "Effect of Temperature on the Growth of Vertically Aligned Carbon Nanotubes from Palm Oil," *Defect and Diffusion Forum,* vol. 312-315, pp. 900-905, 2011.

[14] M. S. Azmina, A. B. Suriani, A. N. Falina, M. Salina, J. Rosly, and M. Rusop, "Preparation of Palm Oil Based Carbon Nanotubes at Various Ferrocene Concentration," *Advanced Materials Research,* vol. 364, pp. 408-411, 2011.

[15] M. S. Azmina, A. B. Suriani, A. N. Falina, M. Salina, and M. Rusop, "Temperature Effects on the Production of Carbon Nanotubes from Palm Oil by Thermal Chemical Vapor Deposition Method," *Advanced Materials Research,* vol. 364, pp. 359-362, 2011.

The Synthesis and Fabrication of Titanium Dioxide Nanowires-Based Biosensor

Sharipah Nadzirah Syed Ahmad Ayob and U. Hashim
Nanostructure Lab-on-chip Research Group
Institute of Nano Electronic Engineering
Universiti Malaysia Perlis (UniMAP)
01000 Kangar, Perlis Malaysia.
E-mail: shnadzirahsaa@gmail.com, uda@unimap.edu.my

Abstract- **A better performance and low cost biosensor is believed can be formed when Titanium dioxide (TiO₂) nanowires working together with Aurum (Au) Inter Digitated Electrode (IDE). The use of TiO₂ nanowires for DNA application can be considered as a new exploration in research and development field. Bottom-up approach of the sol-gel method is used to grow TiO₂ nanowires while the topdown simple lithography method is used to fabricate Aurum (Au) interdigitated electrodes (IDEs). Thus, the combination of both bottom-up and top-down approaches are able to form a very high sensitivity of the sensor. Biomaterial DNA application on the modified surface of TiO₂ nanowires cause it to form a biosensor.**

Keywords: Titanium dioxide nanowires; Biosensors; Sol-gel method; Simple fabrication process.

I. INTRODUCTION

The problem of producing rapid and reliable detection of biomolecules is still considered a serious issue. Therefore, considerable attention is now being given to the development of biosensors that can provide continuous and express detection of biomolecules that suited for medical applications. Hence, biosensors must have the ability to support interchangeable biorecognition elements, miniaturization to allow automation and ease of operation at a competitive cost. Future development of sensor platforms will require significant improvements in sensitivity, specificity and parallelism in order to meet the future needs in a variety of fields. With the existence of titanium nanowires working together with interdigitated electrode (IDE) is expected can enhance the performance of biosensor. TiO₂ has attracted much attention in biosensing research and related applications due to their high sensitivity and specificity for rapid analyte detection. High sensitivity range is due to their wide energy band gap between 1.8 and 4.1 eV; where it depends on its crystal structure on that particular time since this material has a different crystalline structure at different temperature applied for annealing. Those structures are brookite, anatase and rutile [1]. Besides that, using TiO₂ as a main material with good biocompatibility allowed this material to be chosen as biomolecules detection in biological systems. TiO₂ provides high chemical stability, high resistance to deteriorate in acid or alkali, furthermore with the advantage of being non-poisonous, safe, easy to be made, not expensive and strong oxidation–reduction reactions.

Various methods within bottom-up approach can be used to synthesis nanowires such as chemical vapor deposition, electrochemical deposition and pressure injection [2]. However, sol-gel is one of the simplest and low cost methods to grow TiO₂ nanowires. Sol-gel is a process that transforms liquid like "solution" to "solid-like" gel with the existence of precursor. This method can be used to grow TiO₂ nanowires due to its excellent control on the purity and homogeneity of nanowires, which greatly diminishes the probability of defect formation. This is particularly important in dealing with TiO₂-based materials, where some nonstoichiometric defects such as oxygen vacancy and trivalent titanium are easily formed [3]. Besides that, it is very suitable for nanotechnology since it promises all gel products may contain nanoparticles or nanocomposites. Moreover, it is possible to produce materials of new compositions in high purity, high homogeneity, and to control particle size distributions in nano-scale level.

Nanowires working together with interdigitated electrode (IDE) can form a sensor with enhancement in sensitivity level. Fig. 1 shows the image of nanowires working with IDE. The IDE is composed of a set of interdigitated finger electrodes which are generally employed as an impedance or a capacitance biosensor for label-free and sensitive detection of biomolecules. Using IDE may give a lot of advantages such as high signal-to-noise ratio, fast detection due to the numerical number of electrodes and simple testing procedure [2]. This research is aimed to develop

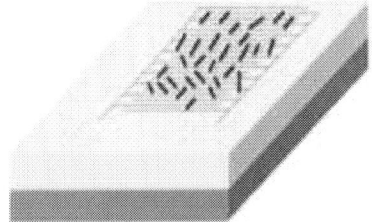

Fig. 1: TiO nanowires with IDEs

TiO₂ nanowires-based biosensor as a molecular sensing device which intimately couples a biological recognition element to an electrode transducer; also known as a biosensor with

978-1-4673-2395-6/12 $31.00 © 2012 IEEE

maximum sensitivity and very rapid detection of biomolecules especially cancer cells which used TiO_2 nanowires as a bridge between electrodes that expected to provide a better performance of the biosensor in terms of sensitivity, by low cost and low processing temperature of sol-gel method. Detail explanation of the biosensor formation will be clarified in methodology.

II. METHODOLOGY

To form a biosensor, three separate methods are required. Those are TiO_2 nanowires growth process, IDE fabrication process, a combination of TiO_2 nanowires with IDE, surface modification and DNA target hybridization. Fig. 2 shows roughly steps taken to form TiO_2 nanowires-based biosensor.

A. Interdigitated Electrode Fabrication

Aurum (Au) electrodes and pads are fabricated using simple lithography method on a silicon dioxide (SiO_2) substrate. The gaps of the electrode are varied from 1μm, 5μm and 10μm. After fabrication, it is ready to be electrically characterized and the result will be used to see changes of the electrical reading when DNA probe immobilized and target hybridized.

B. TiO₂ Nanowires Growth

Next, after Au electrodes are ready, TiO_2 nanowires are grown on a different SiO_2 substrate via simple sol-gel method. A seed layer or nanoparticles of TiO_2 film is required to grow nanowires. The particles size of film should be in nanoscale; so that, the wires grow will be in nano-size too. Seed layer of TiO_2 is spin coating first on SiO_2 substrate at certain speed. Then it will be dried on a hot plate and ready for the annealing process at 600°C. Characterization samples using Atomic Force Microscopy (AFM), X-Ray Diffraction (XRD) and Field Emission Scanning Electron Microscopy (FESEM) are required to identify its surface roughness, crystal structure also the physical appearances of particles respectively.

When the particle sizes are verified in nanoscale, nanowires are ready to be grown on it. The coated and annealed substrate is then placed into the Teflon-liner against the wall with an angle and the coated side will be kept face down. The growth process takes 150-220°C within 3-10 hours. The entire nanowires grow will undergo physical characterization using Scanning Electron Microscopy (SEM) to identify their active area size. To separate nanowires with its seed layer, ultrasonic bath with low frequency is used.

C. TiO₂ Nanowires-Based Biosensor

TiO_2 nanowires will be combined with IDE (as shown in Fig. 2c) by dropping nanowires on IDE. Then let the device dry in ambient temperature before electrical testing is done.

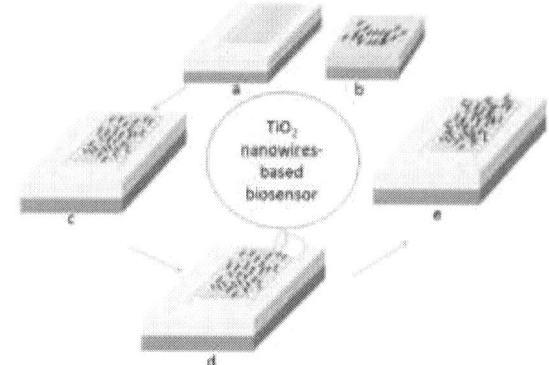

Fig. 2: Formation of TiO_2 nanowires-based biosensor. (a) IDE fabrication, (b) TiO_2 nanowires growth, (c) TiO_2 nanowires working together with IDE, (d) Surface modification of TiO_2 nanowires, (e) DNA target hybridization.

C. TiO₂ Nanowire Surface Modification and DNA Target Hybridization

Au will be doped on TiO_2 nanowires which act as an element to modify the nanowires surface by simply dropping Au nanoparticles on TiO_2 nanowires. Then single strand DNA forms when target DNA will be immobilized on the modified surface by dropping a marker known as thiol group. After that, DNA probe will be hybridize for DNA zipper-up forms double strands.

III. DISCUSSION

For this project, we are choosing an electrode with the shape of comb due to certain advantages. The IDE compose a set of interdigitated finger electrodes which are generally employed as an impedance or a capacitance biosensor for label-free and sensitive detection of biomolecules. With the numerical number of electrodes there, it provides high signal-to-noise ratio, fast detection and simple testing procedure [2]. The expected design of the IDE after fabrication can be seen in Fig. 2(a). Due to Jun Tamaki who studied the effect of electrode gap sizes of sensing properties, the sensitivity levels may improve once the gap sizes between electrodes are getting smaller [4]. For the electrical measurement, suppose the current values should be very small since the gap sizes are in micron. Besides that, a Schottky contact is expected when I-V testing will be done on Au pads since the IDEs is behaving like a capacitance. The sensing principle of this sensor is based on the changes in dielectric properties, charge distribution, and/or conductivity change that are brought on by antibody–antigen complex formed on the surface of the electrodes. Capacitive affinity biosensors can be constructed by measuring changes in the dielectric/surface properties when an analyte binds. For providing larger sensor surface, conductors can be made into a pattern of interdigitated fingers. Au itself has very good conductivity which makes it be a very popular material to be chosen especially in biomedical applications. It is believed that, when there is a change in the dielectric properties of the material between the electrodes, a change in the capacitance will occur and it is correlated to the bound antigen molecules and amount captured by antibodies on the surface, as well as between the electrodes.

An expected result of structure and sizes of TiO_2 nanowires can be relate with some result of Akshay Kumar who studied the growth of single-crystalline rutile TiO_2 nanowires. Roughly, the expected physical structure of TiO_2 seed layer and nanowires can be seen in Fig. 3.

The diameter and length size of wires may be in the range of ~90nm and ~4µm respectively. Since we are using different TiO_2 precursor, the expected time taken to grow nanowires with that size is quite longer than the author was done [5]. Suppose, the smaller the active area (diameter of TiO_2 nanowires), the higher the sensitivity levels towards biomolecules.

When both IDE and nanowires are combined together, it forms sensor with enhancing in sensitivity level. Referring to Guo, L., et al. [6] who is studied ZnO nanowires based UV detector, the I-V curve should be as in Fig. 4. Since ZnO and TiO_2 are known as metal oxide semiconductor materials, besides that those are from period 4 materials, thus we are expecting the performance of both materials must be slightly the same. The graphs also reflect the amount of currents are slightly decreased when the wider electrode gap have been used. For detecting of DNA, immobilization and hybridization of DNA on the surface of TiO_2 nanowires are required but before that, some modification will be made on the TiO_2 nanowires. Surface modification is the most critical part to be prepared and it aims to modify the charge on the surface of TiO_2 nanowires; so that, it can attract different ion of biomolecules toward it due to the electrostatic force occurs between modified surfaces of nanowires with any ionized biomolecules. The step of DNA immobilization is to create a hybridization probe. The labelled DNA probe is first denatured into single-stranded DNA (ssDNA). These ssDNA will immobilize on the TiO_2 nanowires. Immobilization of biological element on physical transducer is able to improve the sensitivity and long-lasting biosensor. During the immobilization process, the biological sensing element must be confined on the transducer and keep it from leaking out over the lifetime of biosensor, allow contact with the analyte solution, allow any product diffuse out of the immobilization layer and not denature the biologically active material. Hybridization is a process to apply analyte (DNA) on the immobilized surface. Then it causes the nanowires and DNA collides randomly. If both probe DNA and target DNA are correct matched, they will "zipper up" to form a duplex. Fig. 5 shows the image of DNA is zipped up. To see the effect of surface modification, DNA target immobilization also probe DNA hybridization towards TiO_2 nanowires-biosensor device, the changes in the I-V curve will be observed. Thus the expected I-V curves for those DNA applications should be slightly changed towards increment of the current amount.

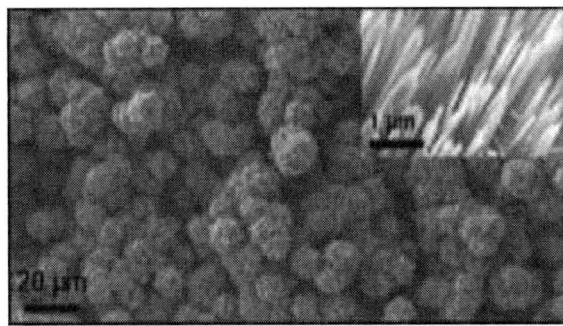

Fig. 3: SEM image of TiO_2 nanowires grown on SiO_2 substrate [5].

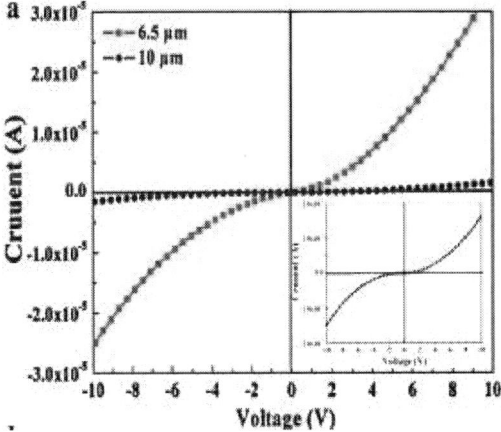

Fig. 4: I-V curves of ZnO nanowires-based device for 6.5µm and 10µm gaps electrodes. Inset is an enlarged image of 10µ gap I-V curve [6]

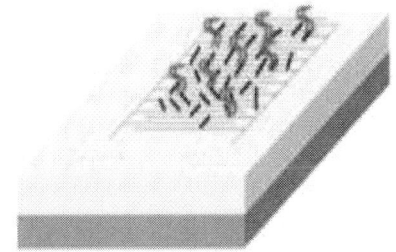

Fig. 5: DNA Target hybridization "zipper up" on TiO_2 nanowires

IV.   CONCLUSION

As a conclusion, TiO_2 nanowires-based biosensor where TiO_2 nanowires working as enhancer towards IDE sensor is expected to provide a better performance of the biosensor in terms of rapid and accurate identification of biomolecules.

REFERENCES

[1] Dr. A. Ahmad, Gul Hameed Awan, and Salman Aziz. *Synthesis and Applications of TiO2 nanoparticles*. in *Pakistan Engineering Congress, 70th Annual Session Proceedings*. 2007.

[2] Shankar, K.S. and A.K. Raychaudhuri, *Fabrication of nanowires of multicomponent oxides: Review of recent advances*. Materials Science and Engineering: C, 2005. 25(5â€"8): p. 738-751.

[3] Sheng, Y., et al., *Low-temperature deposition of the high-performance anatase-titania optical films via a modified solâ€"gel route*. Optical Materials, 2008. 30(8): p. 1310-1315.

[4] Tamaki, J., et al., *Effect of micro-gap electrode on sensing properties to dilute chlorine gas of indium oxide thin film microsensors*. Sensors and Actuators B: Chemical, 2006. 117(2): p. 353-358.

[5] Kumar, A., A.R. Madaria, and C. Zhou, *Growth of Aligned Single-Crystalline Rutile TiO2 Nanowires on Arbitrary Substrates and Their Application in Dye-Sensitized Solar Cells*. The Journal of Physical Chemistry C, 2010. 114(17): p. 7787-7792.

[6] Guo, L., et al., High responsivity ZnO nanowires based UV detector fabricated by the dielectrophoresis method. Sensors and Actuators B: Chemical, 2012. 166-167(0): p. 12-16.

The Effect of Surface Morphology to Photoluminescence Spectrum Porous Silicon

M. H. Fadzilah Suhaimi[1,2,*], M. Ain Zubaidah[1,2,*], S. F. M. Yusop[1,2], M. Rusop[1,3], S. Abdullah[1,2]

[1]NANO-SciTech Centre (NST), Institute of Science;
[2]Nano-Innovation Centre (Nano-IC), Faculty of Applied Sciences;
[3]NANO-ElecTronic Centre (NET), Faculty of Electrical Engineering;
Universiti Teknologi MARA (UiTM),
40450 Shah Alam, Selangor, Malaysia
*E-mail: mhusairifadzilah@yahoo.com

Abstract - **The porous silicon nanostructures was prepared by electrochemical etching of p-type silicon wafer. Porous silicon structure has good mechanical robustness, chemical stability, and compatibility with existing silicon technology. Therefore, it also has a wide area of potential applications such as waveguides, 1D photonic crystals, chemical sensors, biological sensor etc. Photoluminescences characteristics of porous silicon depend on their morphology because the size and distribution of pore its self will effect to their exciton energy level. The structure of porous silicon controlled by the parameters used during the experiment which know as experimental factor. These factors such as etching time, current density applied, temperature, doping concentration etc play an important role during the formation of porous. It will effect either to the thickness or porosity of sample. In this work, we select one of that factor to corellate which optical properties of porous silicon. We investigated the surface morphology by using Atomic Force Microscope (AFM) and photoluminescences using Photoluminescences (PL) spectrometer.**

I. INTRODUCTION

Porous silicon was fabricated by using the electrochemical anodization of silicon in a hydrofluoric acid (HF) based electrolyte. Porous silicon got a lot of attention as an interesting material due to its large ratio of surface area to volume [1], high chemical reactivity at room temperature [2] and potential compatibility with silicon integration technologies [3]. Since porous silicon discovered in 1956 by Uhlir, a type of application for porous silicon had explore like light emitting diode, photodetector, solar cell and sensor. Porous silicon is an interesting material in sensor field because it had a high sensitivity because of its very large specific surface area and possesses the capability in term of device integration [4, 5]. The surface structure of porous silicon plays an important role in sensor application such as Usually porous silicon used in sensing of gas [6], pH [7] and humidity [4]. It has been proved that the humid sensitivity of porous silicon depend the morphology structure, thickness and size distribution [4, 8].

The effect of HF concentration used as an electrolyte on physical and electronic properties (optical properties) being studied by Pushpendra Kumar et. Al [9] and porous silicon nanostuctures for photoluminescence device under various anodization conditions had been studied by M. Jayachandran et.al [10]. Based on previous study, anodization conditions not only effect the PSiNs physical properties, but also other properties like optical and electrical properties [9, 10]. In this paper, we test characterized samples using Photoluminescence (PL) spectrometer. The surface topography and roughness of porous silicon were investigated by using Atomic Force Microscopy (AFM XE-100 Park Systems).

II. EXPERIMENTAL

PSiN is prepared by using electrochemical etching method and [100] of p-type Si wafer with 0.4-2 Ω cm resistivity and 330 ± 40 μm thickness used as a sample. A square dimension approximately 2 cm on an edge is cleaved and placed in cell using a piece of aluminum foil as a back contact and a small O-ring to seal the wafer to the cell. The etching process is set up at various current densities (20 - 40 mA/cm^2) with a constant time etching, 20 minutes. Then the sample is prepared by anodizationethanoic hydrofluoric acid (HF) 48% electrolyte and absolute ethanol (C$_2$H$_5$OH) at ratio 1:1 with illumination by halogen lamp. The halogen lamp used to accelerate etching process during the preparation of sample. After the preparation of PSiN finished, the sample was clean with distilled water and it dried by using nitrogen gas.

Fig 1. Experimental set-ups with halogen lamp.

The surface topography of the porous silicon was investigated using Atomic Force Microscope (AFM) and photoluminescences using Photoluminescences (PL) spectrometer brand Modu – Laser PL which used He-Cd Laser 325 as a source.

III. RESULTS AND DISCUSSION

A. Surface Morphology

Pores in silicon form during the anodization etching depend on parameters used. Many theory suggested to explain the formation of pore on silicon surface [11-13]. Generally the porous silicon is a structure form when the silicon surface dissolution by HF with applied potential [14]. The equation during the formation porous silicon can be expressed as below:

$$Si + 2HF + 2h^+ \longrightarrow SiF_2 + 2H+,$$
$$SiF_2 + 4HF \longrightarrow H_2 + H_2SiF6. \qquad (2)$$

The hole will injected to silicon surface and accumulate at tip of pore like in figure 2. The etching rate determinde by the hole (h^+) accumulation in adjacent region of the electrolyte and Si atoms [15].

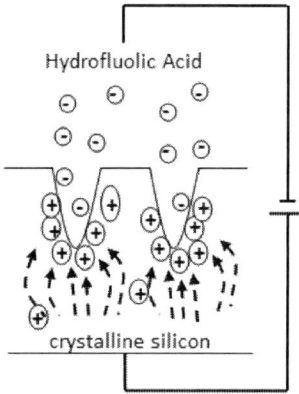

Fig 2. The injected of current (holes) into Si during the anodization etching process.

In this experiment, the current density was constant, so the etching rate also constant, so the changing or effect of morphology will cause by etching time. This factor was effect to thickness of porous layer and its surface morphology.

Atomic force microscope (AFM XE-100 Park Systems) was used to investigate the surface structure of PSiN. Figure 3 show the 3D AFM images (10 μm x 10 μm) of porous silicon at different etching time during the electrochemical etching process. The surface roughness can be calculated using formula (2) [16] below

$$Ra = \frac{A}{n} \qquad (2)$$

Where; Ra = surface roughness average, A= total peak height and n= total number of peaks

Figure 3 show the surface morphology of porous silicon etched at constant current density and different etching time. The differential of size of pore and pillar can be observed. The surface topography sample etched 20 (fig. 3(a)) and 40 minutes (fig. 3 (c)) have a lower porosity while sample 30 minutes etched had highly porous nature. The black and white region

on the surface represent the porous and pillar area form on silicon surface.

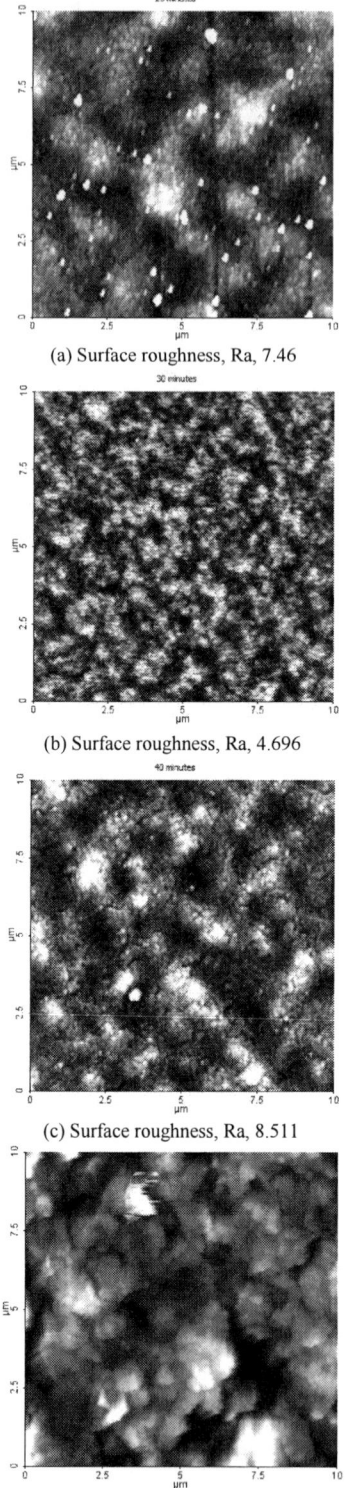

(a) Surface roughness, Ra, 7.46

(b) Surface roughness, Ra, 4.696

(c) Surface roughness, Ra, 8.511

(d) Surface roughness, Ra, 0.193

Fig 3. 3-D image of PSiNs with different etching time used: (a) 20 minutes (b) 30 minutes (c) 40 minutes and (d) 50 minutes

Fig (3b) has the lowest surface roughness compared to other samples because the high and thin of pillars are more. The number black area distribution are more but in small size. That proved that the porosity for this sample is high. Compare with other samples, the back area is bigger but at certain point only. When the etching time increases, the thickness of porous layer increase [17]. It was proved from the figure 3(a) to 3(c), the pillars of the PSiNs become sharper and in good arrangement expecially in figure 3(b).

The surface roughness decreases in figure 3(c) because the pillars of PSiNs continuously etched to form large diameter of pores. It will make the number of pillar decrease especially the higher pillar because its more expose to solution. The lower pillar leave on the surface and make the ratio of total high of pillar to total number is decrease.

B   Photoluminescences Spectrum

The PL emission spectra from porous silicon effected by etching time was correlated with porous silicon surface morphology determined by AFM. The intensity of the PL signal provides information on the quality of surfaces and interfaces. The PL spectrum shows a peak at around 660 to 650 nm know as a vesible region as show in figure 4 below. In the visible spectral range, changes at the low energy side of the broad PL band were observed. The peak position of porous silicon etched at 20 minutes and 40 minutes is almost same which at aroud 660 nm. This is related with the of sample which almost same in their surface roughness. For the sample 30 minutes, the peak shift to right which know as red region.

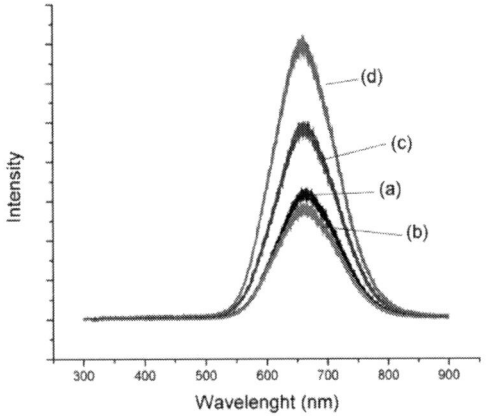

Fig.4. PL emission of porous silicon on different etching time; (a) 20 minutes (b) 30 minutes (c) 40 minutes and (d) 50 minutes

When the peak more to blue region, it is more better because we can produce high energy exciton. The intensity of luminence depend on exciton energy of porous silicon. An exciton form when a photon is absorbed by porous silicon. When the size of pillar become smaller, the surface area and surface energy increase and produce more sharp peak. The large broadening of the peak is caused by the size distribution of these pores. That why the samples etched at 20 minutes which had more distribution of pores produced broad of PL

spectrum. Sample etched 40 minutes had high intensity because high exciton energy. High intensity of PL spectrum for 40 minutes cause by the charge trapped at the wall of pillar after anodization process (Fig. 5). This region will provide high energy of exciton.

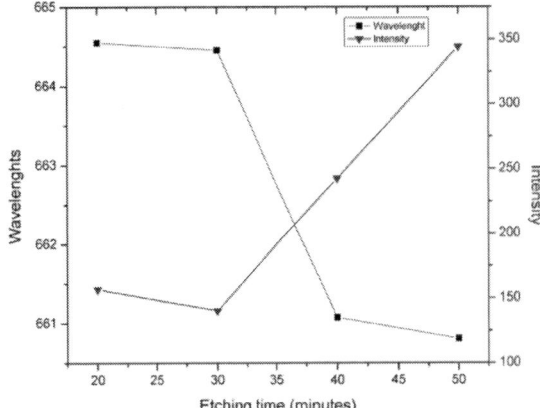

Fig. 5. The relationship between etching time and PL spectra produced.

Porous silicon prepared at the same etching time and certain range of current densities will produced similar morphology (using electron microscope), whereas obvious effect to morphology observed for sample prepared at different etching time [17]. When different time etching applied, different morphology produced, resulting in a shift of PL peak to blue region. That same with reported by Dian et al. [18].

IV.   CONCLUSION

In this paper, the relation between surface topography and photoluminencence of the PSiNs at different etching time were investigated. The morphology of anodized PS is dependent on anodization time applied. When the etching time increases, the porosity increases. Results show that the surface of the PSiNs affected the luminencences properties of PSiNs because the PL spectrum of samples show shifted happen and different intensity collected.

ACKNOWLEDGMENT

One of the authors (Mohd Husairi Fadzilah Suhaimi) is grateful to Universiti Teknologi MARA through Young Lecturer Skime for financial supports. Thanks to NANO-Scitech Centre and NANO-Electronic Centre colleagues for their helpful support and encouragements.

REFERENCES

[1]   G. Barillaro, A. Diligenti, G. Marola, and L. M. Strambini, "A silicon crystalline resistor with an adsorbing porous layer as gas sensor," *Sensors and Actuators B: Chemical,* vol. 105, pp. 278-282, 2005.

[2]   T. Islam and H. Saha, "Hysteresis compensation of a porous silicon relative humidity sensor using ANN technique," *Sensors and Actuators B: Chemical,* vol. 114, pp. 334-343, 2006.

[3] C. Baratto, G. Faglia, G. Sberveglieri, L. Boarino, A. M. Rossi, and G. Amato, "Front-side micromachined porous silicon nitrogen dioxide gas sensor," *Thin Solid Films,* vol. 391, pp. 261-264, 2001.

[4] S.-J. Kim, J.-Y. Park, S.-H. Lee, and S.-H. Yi, "Humidity sensors using porous silicon layer with mesa structure," *J. Phys. D: Appl. Phys. 33 1781,* vol. 33, pp. 1781–1784, 2000.

[5] S.-J. Kim, S.-H. Lee, and C.-J. Lee, "Organic vapour sensing by current response of porous silicon layer," *J. Phys. D: Appl. Phys. 33 1781* vol. 34, pp. 3505–3509, 2001.

[6] L. Boarino, C. Baratto, F. Geobaldo, G. Amato, E. Comini, A. M. Rossi, G. Faglia, G. Lérondel, and G. Sberveglieri, "NO2 monitoring at room temperature by a porous silicon gas sensor," *Materials Science and Engineering: B,* vol. 69–70, pp. 210-214, 2000.

[7] M. J. Schöning, A. Kurowski, M. Thust, P. Kordos, J. W. Schultze, and H. Lüth, "Capacitive microsensors for biochemical sensing based on porous silicon technology," *Sensors and Actuators B: Chemical,* vol. 64, pp. 59-64, 2000.

[8] G. D. Francia, M. D. Noce, V. L. Ferrara, L. Lancellotti, P. Morvillo, and L. Quercia, "Nanostructured porous silicon for gas sensor application," *Mater. Sci. Technol.,* vol. 18, pp. 767–771, 2002.

[9] P. Kumar, P. Lemmens, M. Ghosh, F. Ludwig, and M. Schilling, "Effect of HF Concentration on Physical and Electronic Properties of Electrochemically Formed Nanoporous Silicon," *Journal of Nanomaterials,* p. 7, 2009.

[10] M. Jayachandran, M. Paramasivam, K. R. Murali, D. C. Trivedi, and M. Raghavan, "SYNTHESIS OF POROUS SILICON NANOSTRUCTURES FOR PHOTOLUMINESCENT DEVICES," *Mater. Phys. Mech.,* vol. 4, pp. 143-147, 2001.

[11] M. I. J. Beale, J. D. Benjamin, M. J. Uren, N. G. Chew, and A. G. Cullis, "An experimental and theoretical study of the formation and microstructure of porous silicon," *Journal of Crystal Growth,* vol. 73, pp. 622-636, 1985.

[12] R. L. Smith, S.-F. Chuang, and S. D. Collins, "A theoretical model of the formation morphologies of porous silicon," *Journal of Electronic Materials,* vol. 17, pp. 533 - 541, 1988

[13] G. X. Zhang, "Porous Silicon: Morphology and Formation Mechanisms Modern Aspects of Electrochemistry." vol. 39, C. G. Vayenas, R. E. White, and M. E. Gamboa-Adelco, Eds., ed: Springer US, 2006, pp. 65-133.

[14] X. G. Zhang, "Morphology and Formation Mechanisms of Porous Silicon," *Journal of the Electrochemical Society,* vol. 151, p. C69, 2004.

[15] P. Kumar and P. Huber, "Effect of Etching Parameter on Pore Size and Porosity of Electrochemically Formed Nanoporous Silicon," *Journal of Nanomaterials,* vol. 2007, p. 4 pages, 2007.

[16] M. F. b. Achoi, M. N. b. Asiah, M. Rusop, and S. Abdullah, "The Effect of Growth Temperature on The Surface Properties of TiO_2 Nanostructures Grown on TiO_2 Templete," *Transactions of the Materials Reasearch Society of Japan,* vol. 36, pp. 273-279, 2011.

[17] S. D. MILANI, R. S. DARIANI, A. MORTEZAALI, V. DAADMEHR, and K. ROBBIE, "The correlation of morphology and surface resistance in porous silicon," *JOURNAL OF OPTOELECTRONICS AND ADVANCED MATERIALS,* vol. 8, pp. 1216 - 1220, 2006.

[18] J. Dian, A. Macek, D. Nižňanský, I. Němec, V. Vrkoslav, T. Chvojka, and I. Jelínek, "SEM and HRTEM study of porous silicon—relationship between fabrication, morphology and optical properties," *Applied Surface Science,* vol. 238, pp. 169-174, 2004.

Influence of Heating Temperature on Electrical Photoconductivity of Nanocomposited Polymer-TiO$_2$ Thin Films for Organic Photovoltaic

F.S.S.Zahid[1], M.S.P.Sarah[1], U.M.Noor[1] and M.Rusop[1]

[1] NANO-ElecTronic Centre, Faculty of Electrical Engineering,
Universiti Teknologi MARA (UiTM), 40450,
Shah Alam, Selangor, Malaysia
E-mail: fazlinazahid@gmail.com

Abstract- The influence of heating temperature on the properties of nanocomposited poly [2-methoxy 5-(2'-ethyl-hexyloxy)-1,4-phenylene vinylene (MEH-PPV) polymer and titanium dioxide (TiO$_2$) thin films has been investigated. The heating temperature was varied from 50°C, 75°C, 100°C, 125°C and as deposited thin films as reference sample. Compared with a pristine thin film (as deposited) the MEH-PPV: TiO$_2$ heated thin film shows an improved in current-voltage, conductivity and absorbance characteristics. The variation in heating temperature has affect to the reduction of series resistance of the thin film and therefore improves the photoconductivity from 2.134×10^{-6} to 6.656×10^{-6} S.cm^{-1}. The absorbance also shows an improvement as upon heating, the absorption spectrum of the MEH-PPV: TiO$_2$ nanocomposite undergoes a modification compared to as deposited thin film which remains unchanged. The absorption of the heated nanocmposited thin films shows a significant red-shifted in the wavelength region ascribed to MEH-PPV (500 nm) while the absorption due to TiO$_2$ (340 nm) does not change. Optimization of heating temperature plays a significant role for conductivity properties of nanocomposite MEH-PPV: TiO$_2$.

Keywords: MEH-PPV polymer; Titanium dioxide; Heating Temperature; Conductivity; Absorbance

I. INTRODUCTION

Photovoltaic (PV) cells are designed to convert available light into electrical energy. Conventional photovoltaic cells are usually made from silicon however the cost to fabricate photovoltaic cell by using silicon cost a lot of money. [1]. The introduction of organic photovoltaic (OPV) based on nanocomposited of electron donor and acceptor has receive much attention as a promising material to deliver future solution to the low cost-energy solution. However, the performance and efficiency of OPV is still low compared to other type of photovoltaic. In OPV, the efficiency is depend on the ability to absorb photons, generation of charge carrier and separation, and the ability to transport the separated charge to the respected electrode with minimal recombination rate.

Nonetheless, the nanocomposited polymer-nanoparticles based OPV can be employ to improve exciton dissociation [2, 3].

In this study, MEH-PPV has been chosen as potential conjugated polymer in organic photoactive matrices as it has higher absorption in visible range, easy charge generation under illumination and easy to be deposited on substrate at room temperature [4]. Whilst TiO$_2$ has been chosen to embed in MEH-PPV polymer matrices as it has higher absorption coefficient, good carrier mobility and excellent conductivity. Moreover, TiO$_2$ also shows the possibility to overcome polymer charge-transfer limitations by higher electron-injection of TiO$_2$ and this property make it as the suitable candidate for inorganic semiconductor nanoparticle to be introduced in MEH-PPV polymer [5].

There are some literature stated the polymer which as an active layer in solar cells, was gradually dependent and influenced under gradual increase in the heating temperature during a specific range [6]. In this paper we have investigated the effect of heating temperature on the nanocomposited MEH-PPV: TiO$_2$ thin films for OPV application. This is because heating temperature also plays a significant role in determining the performance of nanocomposited thin films properties.

II. EXPERIMENTAL PROCEDURE

A. Material and Solution

MEH-PPV or poly [2-methoxy 5-(2'-ethyl-hexyloxy)-1, 4- phenylene vinylene] (product of Aldrich) was used without further purification. MEH-PPV was dissolved in organic solvent 1, 2-dichlorobenzene and stirred for 48 hours. Then, TiO$_2$ nanopowder with particle size less than 25 nm (product of Aldrich) which has been purified at 450°C for 2 hours was mixed into MEH-PPV solution. The nanocomposited solution was sonicate in ultrasonic bath for 1 hour to improve the solubility between MEH-PPV and TiO$_2$.

978-1-4673-2395-6/12 $31.00 © 2012 IEEE

B.   Deposition of nanocomposited MEH-PPV:TiO₂ thin films

The glass substrates were sequentially cleaned in ultrasonic bath using acetone, methanol, rinsed with deionized water and finally dried in flowing nitrogen. The spin coating deposition method was performed at room temperature to deposit the nanocomposited MEH-PPV: TiO₂ thin films. Each time after deposition, the thin films were heated on a hot plate stirrer for 5 minutes to vaporize the solvent at 50, 75, 100 and 125°C. The as-deposited sample was using as the reference sample. The deposition process was repeated for 5 times to get the desired thickness. An approximate 60 nm thick Au metal contact was thermally deposited onto the active layer using a vacuum deposition system at pressure about ~5x10⁻⁴ Pa through a shadow mask.

C.   Characterization of thin films

The current-voltage (I-V) characteristics of the thin film under white light illumination were determined using a standard solar irradiation of 100 mW/cm² (AM 1.5) in range -5 to 5V. The optical absorption and transmittance of MEH-PPV: TiO₂ hybrid was measured by ultra violet/ visible/near-infrared (UV-Vis-NIR) spectrophotometer (JASCO V-670). The thickness was measured using surface profiler (VEECO DEKTAK 750). All measurements were performed soon after preparation of the thin films to avoid any changes in the thin films properties caused by aging.

III.   RESULTS AND DISCUSSIONS

A.   Electrical Properties

The influence of the heating temperature on the electrical properties of nanocomposited MEH-PPV: TiO₂ thin films can be observed in the current-voltage (I-V) curves in the dark and under illumination. As appeared on Figure 1, all nanocomposited thin films shows a linear relation indicating ohmic contact behavior between Au metal contact and nanocomposited thin films. It was found that the current values of the the thin films at respective voltage increased with heating temperature. The purpose of heating the thin films each time after deposition is to removed any residual solvents and also to enhance crystal structure in nanocomposited MEH-PPV: TiO₂ [7]. The highest heating temperature (125°C) gives the highest current value both in dark and under illumination to indicate the enhancement of electron conductivity as heating temperature increase.

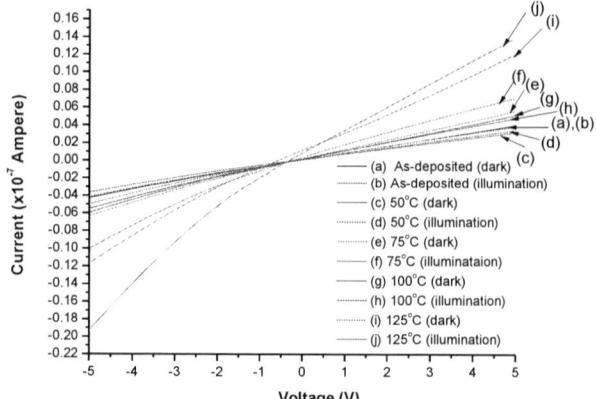

Figure 1. Current-Voltage of nanocomposited MEH-PPV:TiO₂ thin films in dark and under illumination.

From I-V curves, the resistivity, ρ and conductivity, σ of nanocomposited MEH-PPV: TiO₂ thin films were calculated using following equation (1) and (2) and the result is plotted in Figure 2.

$$\rho = (R.w.t)/(L) \qquad (1)$$

$$\sigma = \left(\frac{1}{\rho}\right) \qquad (2)$$

where:
R =Resistance
w= Metal contact width
t= Thin film's thickness
L= Distance between metal contact

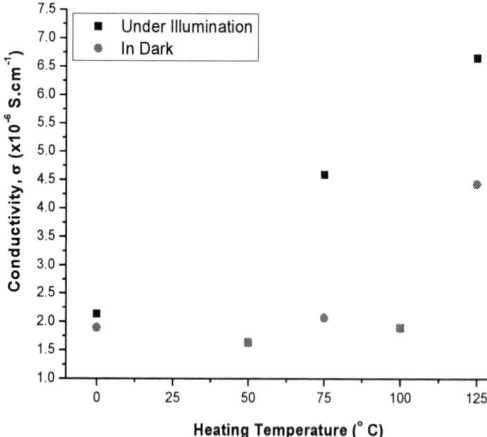

Figure 2. Conductivity of nanocomposited MEH-PPV: TiO₂ thin films in dark and under illumination.

It can be seen that there is a changes in the photoconductivity in comparison with the conductivity in dark for the nanocomposite thin films upon illumination. This indicates that an the nanocomposited thin films is response to illumination and efficient charge separation takes place at the MEH-PPV polymer and TiO_2 nanoparticles interfaces. The calculated conductivity under illumination (photoconductivity) in this work to be 2.134×10^{-6}, 1.637×10^{-6}, 4.592×10^{-6}, 1.893×10^{-6} and 6.656×10^{-6} S.cm^{-1} for nanocomposited thin films heated at 0°C, 50°C, 75°C, 100° C and 125°C respectively. The change in photoconductivity is due to a change in oxidation state of polymer where heating temperature increase specifically where heating temperature is 125°C [8]. From the literature, it is found that when polymer is heat to a temperature higher than its glass transition temperature ($T_g \approx 100$°C), the polymer chains become more mobile and degree of ordering of the polymer significantly increase thus the hole mobility subsequently increase [9]. Furthermore, at this temperature we can tell that the majority of organic solvent is removed from the thin film thus enhance the interfacial contacts between the blend of MEH-PPV and TiO_2. By heating the thin films above the glass transition temperature, the charge separation of the nanocomposited could be improved. It can be explained that such interfacial charge transfer and separation process can reduce the rate of excitons recombination because the process of charge transfer is much faster than recombination thus improve the conductivity [10, 11].

B.   Optical Properties

The optical properties play an important role in solar cells to know the absorption value and absorption wavelength as these are the factors that affect the total number of absorbed photons by a solar cells active layer. The number of absorbed photons in solar cells application increases when the active layer (MEH-PPV: TiO_2) thin films absorbs in the red or near infrared region of the solar spectrum where there is the largest number of the visible light photons. Figure 3 shows optical absorption spectra of nanocomposited MEH-PPV: TiO_2 thin films at different heating temperature in the wavelength range of 250-900 nm. From the figure, we can see that there are two peaks distinguished in the absorption spectrum of the nanocomposited thin films. The absorption spectra of MEH-PPV and TiO_2 exhibited a strong absorption at wavelengths of 510 nm in visible ranges and 350 nm in UV ranges respectively. The optimum absorption peak is found to be located at 500 nm due to electron transitions between non-localized bands (π-π*) of MEH-PPV polymer [12]. From the absorbance figure we can see that as the heating temperature increases, the peak of absorption decreases except for the temperature of 75°C. However, this result is contradicted with electrical properties as temperature at 125°C gives the highest photoconductivity. These phenomenons might be because of two reasons; 1) at 75°C, the organic solvent in the nanocomposited thin films are completely vaporized and capable to absorbed higher spectrum spectra and 2) at 125°C

the polymer MEH-PPV has degrade as polymer is very sensitive to higher temperature (> 100°C).

Moreover, a considerably red-shifted absorption peak can be noticed from the figure which may indicate interchain interactions as well as the extension of the conjugated polymer segments and widens the absorbance spectrum region and solar radiation [12]. This can explained as the heating temperature increase, the crystallites size of TiO_2 increase as well. As a result, the MEH-PPV polymer chains extend through the larger crystallites thus gives a longer conjugation length and lead to the red-shifted of the absorbance peak [10].

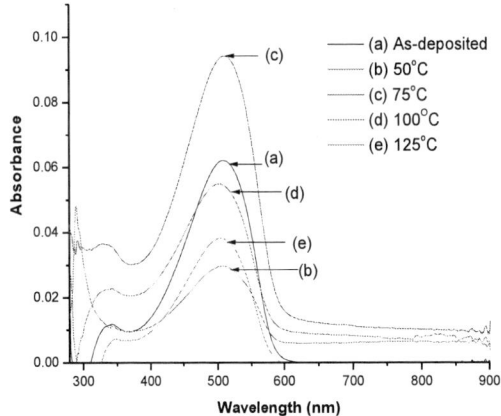

Figure 3. Absorbance spectra of nanohybrid MEH-PPV:TiO_2 thin films.

In order to find the optical band gap of MEH-PPV:TiO_2 thin films, the absorption coefficient of the nanocomposite thin films at different heating temperature have been calculated using Lambert's Law as shown in Equation (3):

$$\alpha = \left(\frac{1}{t}\right) \ln \left(\frac{100}{T}\right) \qquad (3)$$

where α is the absorption coefficient, t is the film thickness and T is the transmittance of the thin films. Since MEH-PPV is organic polymer and TiO_2 nanoparticle is a direct band gap, both of them has a direct transition of electron between the valence and conductance band for which the variation in the absorption coefficient with the photon energy. The relation of these direct band gap energy with the absorption coefficient and photon energy can conclude by plotting $(\alpha h\nu)^2$ against $h\nu$ where $h\nu$ is the photon energy. The optical band gap can be estimate by extrapolating the linear portion near the onset of absorption edge to the $h\nu$ axis. The extrapolation of the graph gave the optical band gap value as shown in Figure 4 below.

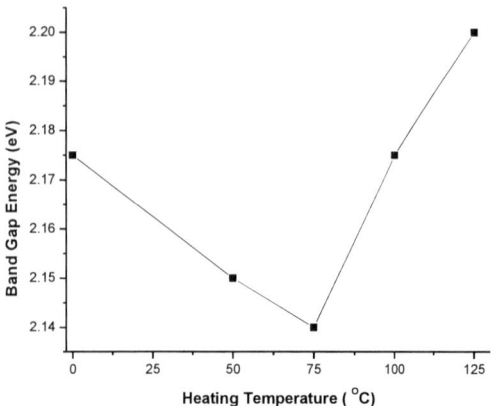

Figure 4. Band gap energy of nanocomposited MEH-PPV: TiO$_2$ thin films.

From the band energy, as heating temperature increases up to 75^{0}C the band energy decrease while at temperature 100°C and 125°C the band gap energy increases. It can be said that the decrement of band gap energy is related to polymer conjugation length. From the previous explanation, there is an improvement in conjugation length that leads to the alternation of single-double bonds of the carbon atoms in the MEH-PPV polymer and therefore, the π-π^* transition shift to lower energy [6]. Because of this reason, the energy band gap tends to decrease in nanocomposited MEH-PPV: TiO$_2$ thin films by heating temperature. The optimum and low optical band gap energy is important as lower the band gap energy; it is much easier for electrons to excite from valence band to conductance band in TiO$_2$ and from HOMO to LUMO in MEH-PPV thus gives a better performance to the nanocomposite thin films. The photoconductivity and optical band gap energy has been estimated and summarized in Table 1.

TABLE I
PHOTOCONDUCTIVITY AND OPTICAL PROPERTIES OF MEH-PPV: TiO$_2$ NANOOCMPOSITE THIN FILMS AT DIFFERENT HEATING TEMPERATURE.

Heating Temperature (^{o}C)	Photoconductivity (S.cm^{-1})	Band Gap Energy (eV)
As-deposited	2.134x10^{-6}	2.18
50	1.636x10^{-6}	2.15
75	4.592x10^{-6}	2.14
100	1.893x10^{-6}	2.18
125	6.656x10^{-6}	2.20

IV. CONCLUSION

In summary, we have presented the photoconductivity and optical properties of nanocomposited MEH-PPV: TiO$_2$ thin films at different heating temperature. nanoparticle at different weight ratio has been deposited and then characterized. In electrical properties, there is a slight change for I-V curves both in dark and under showing the influence of heating temperature towards the sense of light. For optical properties, the highest absorbance in visible range spectra indicates that these thin films are favourable for organic photovoltaic application. The optical band gap energy is found to decrease from 0 to 75 ^{o}C and increase when the heating temperature above 100 ^{0}C.

ACKNOWLEDGEMENT

The authors of this paper would like to express their deepest appreciation to MOSTI E-Science grant (06-01-01-SF0328), UiTM Excellent Fund (600-RMI/ST/DANA 5/3/Dst (399/2011)) and Research Management Institute (RMI), Universiti Teknologi MARA (UiTM), Shah Alam, Selangor, Malaysia for the financial support.

References

[1] J. Chandrasekaran, D. Nithyaprakash, K. B. Ajjan, S. Maruthamuthu, D. Manoharan, and S. Kumar, "Hybrid solar cell based on blending of organic and inorganic materials - An overview," *Renewable and Sustainable Energy Reviews,* vol. 15, pp. 1228-1238, 2010.

[2] A. J. Breeze, Z. Schlesinger, S. A. Carter, and P. J. Brock, "Charge transport in TiO2/MEH-PPV polymer photovoltaics," *Physical Review B,* vol. 64, p. 125205, 2001.

[3] Z. C. D. Chirvase, M. Knipper, J. Parisi, V. Dyakonov and J. C. Hummelen "Temperature dependent characteristics of poly□3 hexylthiophene□-fullerene based heterojunction organic solar cells," *Journal of Applied Physics,* vol. 90, pp. 3376-3383, 2003.

[4] P. M. Sirimanne, T. Shirata, L. Damodare, Y. Hayashi, T. Soga, and T. Jimbo, "An approach for utilization of organic polymer as a sensitizer in solid-state cells," *Solar Energy Materials and Solar Cells,* vol. 77, pp. 15-24, 2003.

[5] C. C. Oey, A. B. Djurisic, H. Wang, K. K. Y. Man, W. K. Chan, M. H. Xie, Y. H. Leung, A. Pandey, J. M. Nunzi, and P. C. Chui, "Polymer TiO 2 solar cells: TiO 2 interconnected network for improved cell performance," *Nanotechnology,* vol. 17, p. 706, 2006.

[6] Y. A. M. Ismail, T. Soga, and T. Jimbo, "Investigation of Annealing and Blend Concentration Effects of Organic Solar Cells Composed of Small Organic Dye and Fullerene Derivative," *Advances in OptoElectronics,* vol. 2011, 2011.

[7] Y.-M. Chang, W.-F. Su, and L. Wang, "Influence of photo-induced degradation on the optoelectronic properties of regioregular poly(3-hexylthiophene)," *Solar Energy Materials and Solar Cells,* vol. 92, pp. 761-765, 2008.

[8] Y. Kim, A. M. Ballantyne, J. Nelson, and D. D. C. Bradley, "Effects of thickness and thermal annealing of the PEDOT:PSS layer on the performance of polymer solar cells," *Organic Electronics,* vol. 10, pp. 205-209, 2009.

[9] J. J. Dittmer, E. A. Marseglia, and R. H. Friend, "Electron Trapping in Dye/Polymer Blend Photovoltaic Cells," *Advanced Materials,* vol. 12, pp. 1270-1274, 2000.

[10] E. Kymakis, E. Koudoumas, I. Franghiadakis, and G. A. J. Amaratunga, "Post-fabrication annealing effects in polymer-nanotube photovoltaic cells," *Journal of Physics D: Applied Physics,* vol. 39, p. 1058, 2006.

[11] A.-w. Tang, F. Teng, H. Jin, Y.-h. Gao, Y.-b. Hou, C.-j. Liang, and Y.-s. Wang, "Investigation on photoconductive properties of MEH-PPV/CdSe-nanocrystal nanocomposites," *Materials Letters,* vol. 61, pp. 2178-2181, 2007.

[12] S. L. M Nam, J Park, SW Kim and KK Lee, "Development of Hybrid Photovoltaic Cells by Incorporating $CuInS_2$ Quantum Dots into Organic Photoactive Layers," *Japanese Journal of Applied Physics,* vol. 50, 2011.

Optical and Electrical Properties of ZnO and ZnO: TiO₂ Thin Films Prepared by Sol-Gel Spray-Spin Coating Technique

C.M.Firdaus[a], M.Rusop[b], S.R.M.S.Baki[c], R.H.Salimin[d]

NANO-ElecTronic Centre (NET), Faculty of Electrical Engineering
Universiti Teknologi MARA (UiTM)
40450 Shah Alam, Selangor, Malaysia
Email:
[a]firdaus_chemat86@yahoo.com, [b]rusop@salam.uitm.edu.my, [c]sharizam@salam.uitm.edu.my, [d]rhidayah@salam.uitm.edu.my

Abstract— In this work, a ZnO and composite ZnO: TiO₂ thin films were conducted on glass prepared by sol-gel spray-spin coating technique. The performance and properties of nanostructured ZnO and ZnO: TiO₂ thin films at different thickness were investigated. The effect of thickness on the electrical properties and optical properties have been characterized by using 2-point probe KEITHLEY 2400 source meter and UV-Vis spectrophotometer respectively. From the current-voltage measurement it shows the conductivity of nanocomposite ZnO: TiO₂ thin films were higher compare to nanostructure ZnO thin films. The optical properties show the band gap for nanocomposite ZnO: TiO₂ decreases as the thickness increase.

Keywords-component; Nanostructured ZnO; Nanocomposite ZnO: TiO₂; Sol-Gel Spray-Spin coating; Thickness; Electrical Conductivity

I. INTRODUCTION

Many researchers have been study the performance of the Dye-Sensitized Solar Cells (DSSCs) by using other semiconducting metal oxide such as ZnO, TiO₂, CuI, and SnO₂ and prepared composite oxide films to explore the possibilities of enhancing the efficiency of DSSCs [1]. In DSSCs, the conversation of visible light to electricity is achieved by providing an effective path for electron transport and a higher surface area for dye adsorption to maximize light absorption [2-3]. ZnO has an exciton binding energy about 60meV at room temperature and has a wide band gap energy 3.37eV [4-5], making it frequently used in the production of dye-sensitized solar cells [6]. Higher electron mobility in ZnO thin films will reduce the electron recombination rate by making injection of photo excited electron into the conduction band became easier thus will enhance the performance of solar cell [7]. A lot of presented result based on ZnO thin films shows the performance of device is direct connected with the good crystalline structure quality ZnO thin film first [8-9]. Compare to ZnO, TiO₂ semiconductor also has similar band gap energy approximately 3.2 eV [10]. Both of these semiconductors have an excellent properties and extensive application and have attracted much attention of researchers. However, the ZnO and TiO₂ intrinsic

semiconductor characteristics are different which are ZnO is a direct band gap energy meanwhile TiO₂ is an indirect semiconductor [10-11]. In DSSCs, TiO₂ photoanode thin films have been demonstrated with a power conversion efficiency of 11% [2, 12]. TiO₂ is known to be a good candidate for the degradation of environmental contaminants due to its high photocatalytic activity, non-toxic nature, and stable in aqueous solution, and relatively low cost [10, 13]. However, TiO₂ thin film present two drawbacks during photocatalytics process which are the low use of solar spectrum and the relatively high electron-hole recombination rate [2-3, 10, 12, 14]. Since TiO₂ and ZnO have excellent properties and have similar band gap energy, this drawbacks can be possible overcame by composite these two nano-semiconductor to enhance the photocatalytic efficiency [10]. Nano-composited has been extensively study because they exhibit better structural, optical, and electrical properties. Generally, the activity of the photocatalytic was relying on the adapting condition of the composite materials. These composite materials will be manipulated the particles size, crystallographic phase and morphology of the nanocrystallite according to condition of method preparation [15]. There are several method preparations of ZnO and TiO₂ thin films that have been done by other researchers [16-17]. Among of this method, sol-gel has the advantages such as being not conglomerated, good uniformity of thin film, high purity, low temperature synthesis, easily reclaimed after reaction and it also easily controlled reaction condition [10, 16, 18]. In this work, the optical and electrical properties of ZnO and composite ZnO: TiO₂ thin film have been studied by preparing the thin films using spray-spin coating technique with various thickness. The effects of thickness on the optical and electrical properties of ZnO and composite ZnO: TiO₂ thin films have been reported.

II. EXPERIMENTAL

ZnO thin films with various thicknesses were deposited on glass substrates using sol-gel spray-spin coating technique. ZnO solution was prepared by dissolving zinc acetate dihydrate, monoethanolamine (MEA), diethanolamine (DEA) and sodium

978-1-4673-2395-6/12 $31.00 © 2012 IEEE

dodecyl sulfate (SDS) in absolute ethanol solution. TiO_2 was prepared by mixing together the Titanium (IV) isopropoxide (TiPP) as precursor, diethanolamine (DEA) as stabilizer, Triton X-100 as surfactant, ethanol absolute as solvent. The composite solution was prepared by mixing together the resultant solution of ZnO and TiO_2 with volume ratio 1:1. The deposition process by spray-spin coating technique was conducted at the speed of 500 rpm for 5 second and the substrate was kept at a distance of 25 cm below spray nozzle. The Argon gas was used as a carrier gas and the glass was spin while coating with spray of the prepared solution. Each time after deposition process, the films were pre-heated at 200 °C for 10 min to evaporate the solvent and remove organic residual. The spray-spin coating depositions were repeated from 1 to 5 times to increase film thickness. Then, the films were finally post heated at 500 °C for 1 hour in air using a furnace (Protherm).

III. RESULTS AND DISCUSSION

TABLE I
THE THICKNESS, BANDGAPS AND RESISTIVITIES VALUES FOR ZNO AND COMPOSITE ZNO:TIO₂ THIN FILMS.

	ZnO				ZnO:TiO₂		
Sample	Thickness (nm)	Bandgap (eV)	Resistivity (Ω.cm)	Sample	Thickness (nm)	Bandgap (eV)	Resistivity (Ω.cm)
a	896.73	2.81	0.85×10^6	a	890.49	2.71	0.87×10^6
b	2557.32	2.87	2.24×10^6	b	2424.77	2.97	2.13×10^6
c	3719.73	2.94	3.44×10^6	c	2457.40	2.52	2.05×10^6
d	4491.84	2.95	2.74×10^6	d	3350.13	2.51	2.13×10^6
e	4884.00	2.98	2.86×10^6	e	3919.62	2.88	2.25×10^6

A. Optical properties

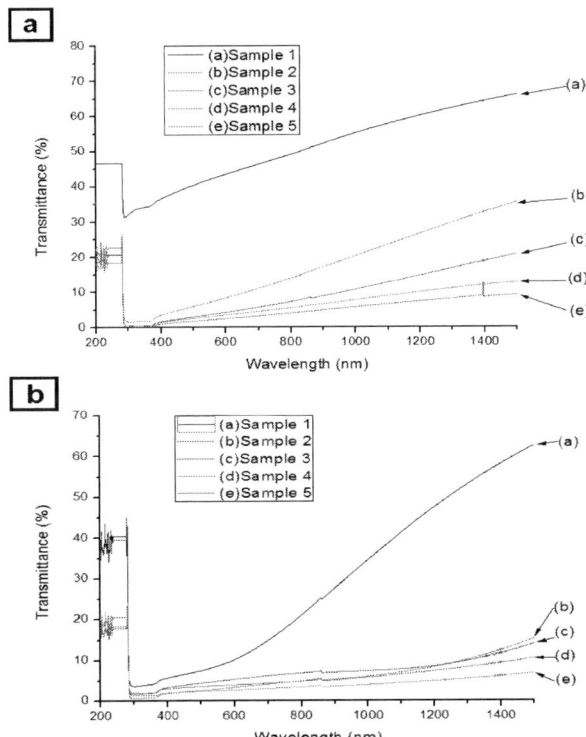

Fig. 1. Transmittance spectra of ZnO (a) and composite ZnO: TiO_2 (b) thin films with different thickness.

The optical transmittance, absorbance and the band gap values, (Eg) of ZnO thin films as a function of different thickness has been studied. Fig. 1 shows the optical transmittance through the different thickness of ZnO and composite ZnO: TiO_2 thin films are measured in the range of wavelength 200 to 1500 nm. The transmittance spectra revealed that all films in Fig. 1 (a) and (b) had low average transmittance between 1% – 48.9% and 1.8% – 5.8% within the visible region (400 - 800 nm). The transmission decreases sharply near the ultraviolet region about 380 nm for both of the samples due to the band gap absorption. From this figure, the transparency of films becomes decreases as the increase of thickness and surface roughness.

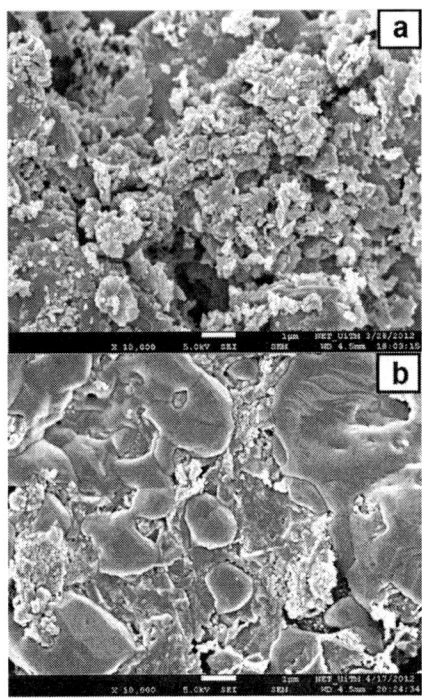

Fig. 2. FESEM images of ZnO (a) and composite ZnO: TiO_2 (b) thin films.

This result was supported by Fig.2. From the Fig. 2(a) and 2(b), it can been seen that the ZnO thin film has high surface roughness on surface morphology with porous and also posses agglomerate particles for the fifth thickness compare to surface morphology of composite ZnO: TiO_2 thin film improved for the fifth thickness that has batter surface, more compact, well distributed particles on surface morphology and also has densely pack of particles might be due to attraction between two different particles ZnO and TiO_2 materials. The low transmittance in visible range might be attributed to the possible light scattering on roughness surface morphology and high grain boundary that the thin films has, because the particles are less uniform dispersed on the glass substrate and forming large agglomerate particles. Other paper has report that the high transmittance in visible range could be achieved if the thin films have smoother surface morphology and less grain boundary [19-20].

978-1-4673-2395-6/12 $31.00 © 2012 IEEE

It is known that the band gap of ZnO is approximately 3.2 - 3.37 eV [5, 21]. It can be excited by photon with wavelengths below 387 nm. From Fig. 3(a) and (b), it shows the variance of absorption coefficient in the range 200 to 1500 nm for five different thicknesses. The exciton absorption is at about 370nm. It can be seen that samples from Fig. 3(b) have high absorption coefficient in visible range compare to Fig. 3(a).

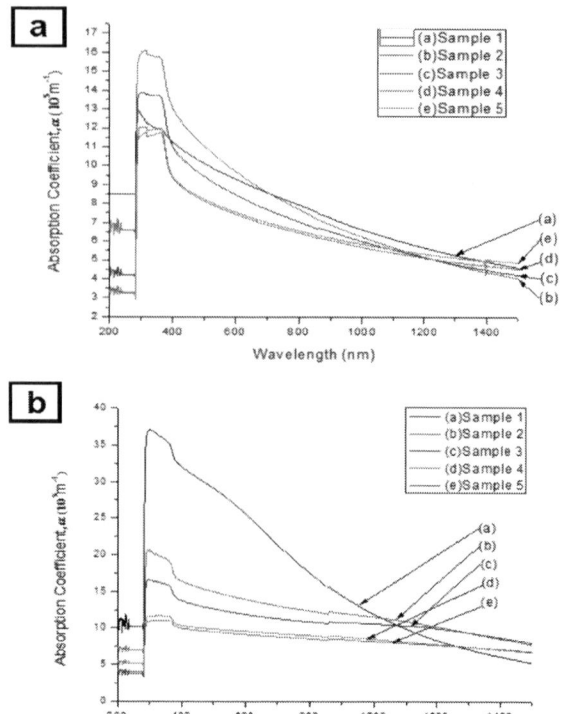

Fig. 3. Absorption coefficient of ZnO (a) and composite ZnO: TiO$_2$ (b) thin films with different thickness.

From the Fig. 3, a sharp decrease in absorption coefficient near band edge about 370 nm indicates the intrinsic optical band gap energy of ZnO (3.37 eV). The strong absorption in the UV region when the photon energy is used for electron excitation from valence band to conduction band which implies the light with photon energy need be higher or equal than optical band gap absorbed by the semiconductor, the electron will have adequate energy to jump from the valence band to conduction band [22].

The optical band gap values of the ZnO and composite ZnO: TiO$_2$ thin films were obtain from the transmission measurements by plotting $(\alpha h\nu)^2$ versus photon energy graphs where α is the absorption coefficient and hν is the photonic energy. From Fig. 4(a) and (b) and Table I, it shows the variation of ($\alpha h\nu$ vs. hν). The optical band gap energy for all samples in Fig. 4(a) for ZnO thin films are found increased from 2.81 to 2.98 eV. But for samples composite ZnO: TiO$_2$ thin films in Fig. 4(b) show increases in optical band gap from 2.71 to 2.97 eV but decreases after 2.97 to 2.88 eV when the thickness increases. This means by adding TiO$_2$ material into ZnO material, the band gap of ZnO become reduce and improve the minimum energy required for excitation electron. The

electron will become easily excite from the valence band to conduction band.

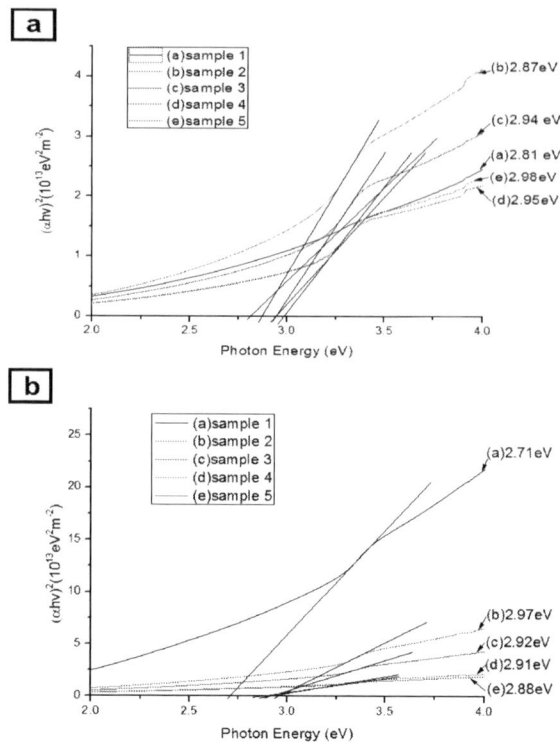

Fig. 4. Optical band gap energy of ZnO (a) and composite ZnO: TiO$_2$ (b) thin films with different thickness.

B. Electrical properties

Fig. 5(a) and (b) shows the result of Current-Voltage measurement for ZnO and composite ZnO: TiO$_2$ thin film with different thickness in ambient light with voltage supply ranging from -5 to 5V. The results show an almost linear I-V curve obtained from the measurement meaning that all the samples exhibit Ohmic behavior with Al metal contact. The electrical behavior of samples ZnO and composite ZnO: TiO$_2$ thin films with respect of their thickness were examined through measurement of electrical conductivity (σ).

From the Fig. 6, the conductivity of ZnO thin films was decreased when the thickness of thin film increase might be due to chemisorptions of oxygen at grain boundaries and also on the surface [23].The decrease of conductivity may attributed from the decreases of electron concentration resultant from chemisorptions of oxygen that causes the increasing of hole concentration [23]. The absorbed oxygen may produce potential barrier which hinder the electrical mobility. According to FESEM result, there are so many pores among the grains that it becomes more difficult for carriers to travel from one grain to another neighboring grain. From the FESEM result, as the thickness film increases, the pores created among grains increases. The pores will interfere the electron mobility to travel from grain to grain.

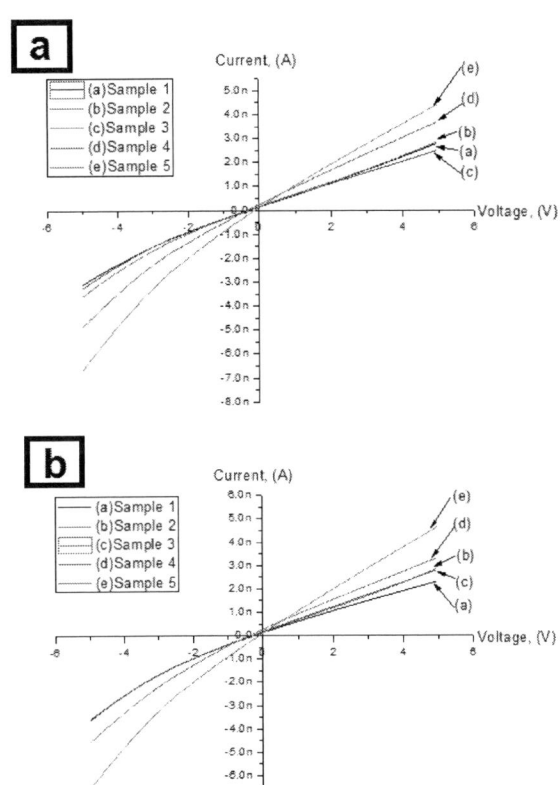

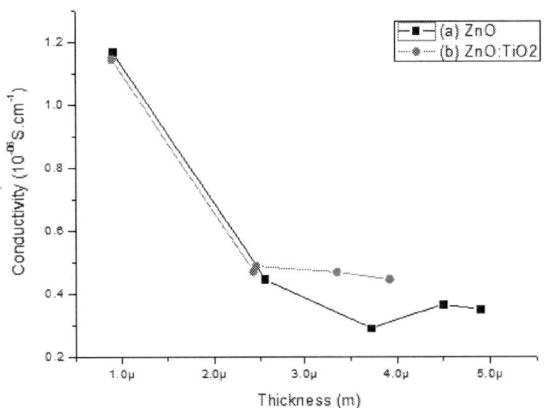

Fig. 6. Conductivity of ZnO (a) and composite ZnO: TiO$_2$ (b) thin films with different thickness

the band gap decreases when the thickness increases. The conductivity of ZnO found to be decreased as the thickness increase and conductivity for composite ZnO: TiO$_2$ thin films found to be higher than ZnO thin films.

ACKNOWLEDGMENT

I would like to express my deep sense of gratitude and appreciation to RMI, Universiti Teknologi MARA (UiTM) for financial support. This project was supported by Fundamental Research Fund Grant No. 600-RMI/FRGS 5/3 (56/2012).

Fig. 5. I-V curve of ZnO (a) and composite ZnO: TiO$_2$ (b) thin films with different thickness

This is agreed to other report [23-24] where the film thickness is affected the electrical properties. From the Fig. 6 also shows the conductivity of composite ZnO: TiO$_2$ decreased as the thickness increase but for comparison between these two results, the conductivity of composite ZnO: TiO$_2$ thin films were higher compare to ZnO thin film. This might be due to combination of two materials with different properties that will improve the performance of thin film. From FESEM result, the structural properties of composite ZnO: TiO$_2$ thin film shows the surface morphologies are homogeneous, less porous and the particles become more densely pack so it will not affect the electron mobility to travel among grain to grain.

IV. CONCLUSIONS

ZnO and ZnO: TiO$_2$ thin films with different thicknesses were prepared on glass substrate by using a sol-gel spray-spin coating method. The optical and electrical properties of ZnO thin films have been investigated and it is found to be influenced by the thickness of the film. The transmittance spectra revealed that all films had low average transmittance and transparency of film decreases with the increase of thickness and roughness. The optical band gap found to be increased with the increasing of the thin films thickness for ZnO thin films but for composite ZnO: TiO$_2$ thin films show

REFERENCES

[1] K. E. Kim, et al., "Enhancement in the performance of dye-sensitized solar cells containing ZnO-covered TiO2 electrodes prepared by thermal chemical vapor deposition," Solar Energy Materials and Solar Cells, vol. 91, pp. 366-370, 2007.

[2] J. B. Baxter and E. S. Aydil, "Dye-sensitized solar cells based on semiconductor morphologies with ZnO nanowires," Solar Energy Materials and Solar Cells, vol. 90, pp. 607-622, 2006.

[3] S. Yun, et al., "Improvement of ZnO nanorod-based dye-sensitized solar cell efficiency by Al-doping," Journal of Physics and Chemistry of Solids, vol. 71, pp. 1724-1731, 2010.

[4] Y. Zeng, et al., "Enhanced toluene sensing characteristics of TiO2-doped flowerlike ZnO nanostructures," Sensors and Actuators B: Chemical, vol. 140, pp. 73-78, 2009.

[5] P. Lv, et al., "I-V characteristics of ZnO/Cu2O thin film n-i-p heterojunction," Physica B: Condensed Matter, vol. 406, pp. 1253-1257, 2011.

[6] R. Velmurugan and M. Swaminathan, "An efficient nanostructured ZnO for dye sensitized degradation of Reactive Red 120 dye under solar light," Solar Energy Materials and Solar Cells, vol. 95, pp. 942-950, 2011.

[7] M. Giannouli and F. Spiliopoulou, "Effects of the morphology of nanostructured ZnO films on the efficiency of dye-sensitized solar cells," Renewable Energy, vol. 41, pp. 115-122, 2012.

[8] L. Xu, et al., "Effect of TiO2 buffer layer on the structural and optical properties of ZnO thin films deposited by E-beam evaporation and sol–gel method," Applied Surface Science, vol. 255, pp. 3230-3234, 2008.

[9] R. Romero, et al., "The effects of zinc acetate and zinc chloride precursors on the preferred crystalline orientation of ZnO and Al-doped ZnO thin films obtained by spray pyrolysis," Thin Solid Films, vol. 515, pp. 1942-1949, 2006.

[10] J. Tian, *et al.*, "Preparation and characterization of TiO2, ZnO, and TiO2/ZnO nanofilms via sol-gel process," *Ceramics International*, vol. 35, pp. 2261-2270, 2009.

[11] D. W. Kim, *et al.*, "Effects of heterojunction on photoelectrocatalytic properties of films," *International Journal of Hydrogen Energy*, vol. 32, pp. 3137-3140, 2007.

[12] Z. Liu, *et al.*, "Controlled synthesis of ZnO and TiO2 nanotubes by chemical method and their application in dye-sensitized solar cells," *Renewable Energy*, vol. 36, pp. 1177-1181, 2011.

[13] S. Chen, *et al.*, "Preparation, characterization and activity evaluation of p–n junction photocatalyst p-ZnO/n-TiO2," *Applied Surface Science*, vol. 255, pp. 2478-2484, 2008.

[14] Z. Zhang, *et al.*, "Preparation of photocatalytic nano-ZnO/TiO2 film and application for determination of chemical oxygen demand," *Talanta*, vol. 73, pp. 523-528, 2007.

[15] C.-h. Zhou, *et al.*, "Titanium dioxide sols synthesized by hydrothermal methods using tetrabutyl titanate as starting material and the application in dye sensitized solar cells," *Electrochimica Acta*, vol. 56, pp. 4308-4314, 2011.

[16] D. Raoufi and T. Raoufi, "The effect of heat treatment on the physical properties of sol-gel derived ZnO thin films," *Applied Surface Science*, vol. 255, pp. 5812-5817, 2009.

[17] M. Sasani Ghamsari and A. R. Bahramian, "High transparent sol-gel derived nanostructured TiO2 thin film," *Materials Letters*, vol. 62, pp. 361-364, 2008.

[18] M. H. Mamat, *et al.*, "Electrical characteristics of sol-gel derived aluminum doped zinc oxide thin films at different annealing temperatures," in *Electronic Devices, Systems and Applications (ICEDSA), 2010 Intl Conf on*, 2010, pp. 408-411.

[19] L. Xu, *et al.*, "Structural and optical properties of ZnO thin films prepared by sol–gel method with different thickness," *Applied Surface Science*, vol. 257, pp. 4031-4037, 2011.

[20] E. Bacaksiz, *et al.*, "Structural, optical and electrical properties of Al-doped ZnO microrods prepared by spray pyrolysis," *Thin Solid Films*, vol. 518, pp. 4076-4080, 2010.

[21] G.-Y. Zeng, *et al.*, "Characteristics of a dye-sensitized solar cell based on an anode combining ZnO nanostructures with vertically aligned carbon nanotubes," *Diamond and Related Materials*, vol. 19, pp. 1457-1460, 2010.

[22] M. H. Mamat, *et al.*, "Influence of doping concentrations on the aluminum doped zinc oxide thin films properties for ultraviolet photoconductive sensor applications," *Optical Materials*, vol. 32, pp. 696-699, 2010.

[23] A. Zhong, *et al.*, "Thickness effect on the evolution of morphology and optical properties of ZnO films," *Applied Surface Science*, vol. 257, pp. 4051-4055, 2011.

[24] T. Prasada Rao and M. C. Santhoshkumar, "Effect of thickness on structural, optical and electrical properties of nanostructured ZnO thin films by spray pyrolysis," *Applied Surface Science*, vol. 255, pp. 4579-4584, 2009.

Effect of Temperature Treatment on the Properties of ZnO Nanoparticle-Bi$_2$O$_3$-Mn$_2$O$_3$ Varistor Ceramics

Rabab Khalid Sendi and Shahrom Mahmud

Nano Optoelectronic Research (NOR) Lab, School of Physics, Universiti Sains Malaysia
11800 Minden, Pulau Pinang, Malaysia
E-mail: Last-name3@hotmail.com

Abstract- In the current study, 20 nm zinc oxide nanoparticles were used to make high-density ZnO discs doped with Bi$_2$O$_3$ and Mn$_2$O$_3$ via the conventional ceramic processing method. Different sintering temperatures were found to have significant impacts on the ZnO discs, especially on enhancing grain growth even at a low sintering temperature of only 980 °C. The strong solid-state reaction during sintering may be attributed to the high surface area of the 20 nm ZnO nanoparticles that promoted a strong surface reaction even at low sintering temperatures. Moreover, the sintering process also improved the grain crystallinity, as shown in the lowering of the intrinsic compressive stress based on the X-ray diffraction lattice constant and full-wave half-maximum data. The sintering temperatures also significantly influenced the electrical properties of the doped ZnO discs with a marked drop in the breakdown voltage from 335 V (sample at 980 °C) to 130 V (sample at 1220 °C). The resistivity also experienced a dramatic drop from 303.5 kΩ.cm (sample at 980 °C) to 180.7 kΩ.cm (sample at 1220 °C). The observed shift in the energy band-gap from a higher to a lower value may be attributed to the conversion of compressive stress to tensile stress with increasing sintering temperature. Therefore, the sintering process can be used as a new technique for controlling the breakdown voltage of doped ZnO discs made from ZnO nanoparticles with improved structural and optical properties.

I. INTRODUCTION

Considerable interest has focused on many oxide ceramic semiconductor materials based on ZnO [1], TiO$_2$ [2], and SnO$_2$ [3, 4] in recent years. These ceramics are used as varistors, which are semiconductor devices with linear (ohmic) symmetric current-voltage characteristics (CVC), a wide direct band-gap, and strong excitonic binding energy, making it a promising material for ultraviolet (UV) lasers with low thresholds [5], field-emission arrays [6, 7].

Certain ceramics consist of highly conductive grains with grain-boundary potential barriers that are formed during sintering [1, 8]. Although several metal oxides, such as Sb$_2$O$_3$, Bi$_2$O$_3$, Co$_2$O$_3$, NiO, Cr$_2$O$_3$, and Mn$_2$O$_3$, are used as additives, ceramic manufacturing, where Bi$_2$O$_3$ plays the essential role of imparting non-linear behavior to the material [9]. The quality of processing, such as the sintering temperature and heating and cooling rates, plays a major role in controlling and understanding the general properties of ZnO.

Sintering is a widely used and effective technique for improving the crystalline quality of ZnO. The sintering temperature plays an important role in controlling the intrinsic defects in ZnO and the properties of the samples. Recrystallization can occur at higher sintering temperatures,

and the concentration of the defects changes with the sintering temperature.

The electrical characteristics of a device are directly related to the electrical properties of the boundary and the grain size. In general, the current-voltage behavior is a grain boundary phenomenon because the number of the grain boundaries between electrodes has a significant effect on the non-linear characteristics of ZnO and on the breakdown voltage of ceramics. In the photoluminescence process, the luminescence from ZnO extends from the band-edge to the green or orange spectral range with a common broadband centered at 2.45 eV.

This study investigated the impact of the sintering temperature on doped ZnO discs made from 20 nm ZnO nanoparticles with respect to their structure, optical, and electrical characteristics.

II. EXPERIMENTAL DETAILS

A. Sample Preparation

ZnO discs were prepared via the conventional ceramic processing method involving ball milling, drying, pressing, and sintering. Oxide precursors of 99.9% purity were used. The composition consists of 99 mol% 20-nm ZnO + 0.5 mol% Bi$_2$O$_3$ + 0.5 mol% Mn$_2$O$_3$ powder.

The powder was blended with poly vinyl alcohol (PVA) by mixing with distilled water in a ball milling jar for 6 hours. The ZnO slurry was dried at 60 °C in air for 1 hour and then was granulated by sieving through a 20-mesh sieve. The resulting granules were used to make discs by pressing at 4 ton/cm^2 pressure. The green ZnO discs were 26 mm in diameter and 2 mm thick. Finally, the green discs were sintered at 980, 1060, 1140, 1220 °C in air for 3 h.

B. Characterization

The microstructure of the ZnO discs was studied using a scanning electron microscopy (SEM) system. The crystalline phases were studied using a high resolution X-ray diffractometer (XRD) equipment (PANalytical X' Pert PRO MED PW3040) with Cu K$_\alpha$ radiation =1.5406Å).

C. Electrical and Optical Ttesting

The I – V characteristics of the samples were measured using a high voltage source measure unit (KEITHLEY instruments 246 high voltage supply). The resistance R was evaluated from Voltage (V) - Current (I) characteristics in accordance with Ohm's Law. Room temperature photoluminescence (PL)

978-1-4673-2395-6/12 $31.00 © 2012 IEEE

spectra were obtained using a Jobin Yvon HR 800 UV spectrometer system.

III. RESULTS AND DISCUSSION

A. Microstructural Analysis

Figure 1 shows the SEM micrographs of doped ZnO nanoparticle-based discs at various sintering temperatures. These SEM images show that the surface morphology of the samples was strongly dependent on the sintering temperature, and the average grain size and density of the ZnO discs increased when the temperature reached the optimum density at 1220 °C. The ZnO grains grew bigger to approximately 11.5, 16.5, 23.6 and 35.4 μm as the sintering temperature increased to 980, 1060, 1140 and 1220 °C, respectively (Table 1).

The grains became more structured and resembled polygons or hexagons and their porosity became smaller as they grew bigger. A high sintering temperature is known to provide a larger driving force for internal atomic diffusion responsible for grain growth and pore elimination. The high densification of the ZnO discs at the high sintering temperatures can be attributed to the Bi_2O_3 and the Mn_2O_3 doping effects in the ZnO lattice, which caused the formation of oxygen vacancies during the sintering process. As the dopants ions substitute at the Zn ion site during the sintering process, oxygen vacancies are produced and contribute to an enhancement of the density of the ZnO discs.

An interesting observation is the clear presence of some spinel phases (1–4 μm), which may impede the growth of the ZnO grains during the sintering process. These spinel phases are usually located in groups but they may also be found individually between, sometimes even within, the ZnO grains. The liquid phase during sintering produces Bi-rich phases that are associated with the spinel particles in the intergranular region of the microstructure.

Another observation is the layer-by-layer growth of several grains at the extreme sintering temperatures at 1140 °C and 1220 °C. Judging from the SEM micrographs in Figure 1(c-d), the polycrystalline layer possessed a thickness of 80-120 nm. Since these multilayer growths were localized in only several grains, there must have been a high concentration of free zinc

in these localized grains in order for the secondary layer-growth to occur.

As can be seen in the EDX data in Figure 1, the different metallic oxides at different concentrations were detected in the samples at various sintering temperatures. The Bi_2O_3 and Mn_2O_3 had average concentration similar to the initial chemical nominal content.

Figure 2 shows the XRD spectra of the ZnO disc specimens characterized by strongest major peaks of (101), (100), (002), and (110) that arose from the ZnO layer and confirmed through the polycrystalline nature of the discs. Other peaks, namely, the ($Bi_{48}ZnO_{73}$) and (Mn_3O_4) phases, appeared as secondary phases. Doped ZnO discs are multiphase materials. These materials contain ZnO as a major phase. Spinal, pyrochlore, and some other phases were present in the specimen. The existence of these phases is dependent on the nature and amount of additives to ZnO and the processing parameters. Incorporation of these oxides forms atomic defects at the grain and grain boundary, with donor or donor-like defects dominating the depletion layer and acceptor and acceptor-like defects dominating the grain boundary states [10].

The intense (101) peak of the ZnO disc increased from $2\theta = 36.13°$ to $2\theta = 36.18°$ after sintering at 1060 °C, indicating the occurrence of residual stress in the ZnO disc. The (101) peak became narrower, less asymmetric, and gradually shifted to higher 2θ values with increasing sintering temperatures, resulting in the decrease in the lattice constant (Table 1). Moreover, the FWHM of the (101) diffraction peak decreased with increasing temperature (Table 1), suggesting that the sintering process improved the crystallization of the ZnO nanoparticles.

However, the FWHM of the ($Bi_{48}ZnO_{73}$) and (Mn_3O_4) increased with sintering temperature because of the structural defect and the impurity of the grain boundaries in the ZnO disc.

Table 1 summarizes the results from the SEM, XRD, and PL. The lattice constant "c" was calculated from equation (1) [11]:

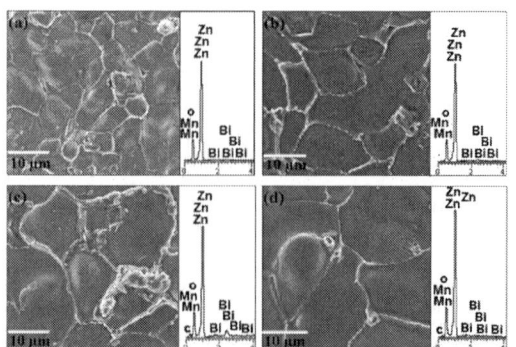

Fig. 1. Typical SEM images of varistor sintered at (a) 980 °C, (b) 1060 °C, (c) 1140 °C, and (d) 1220 °C temperatures.

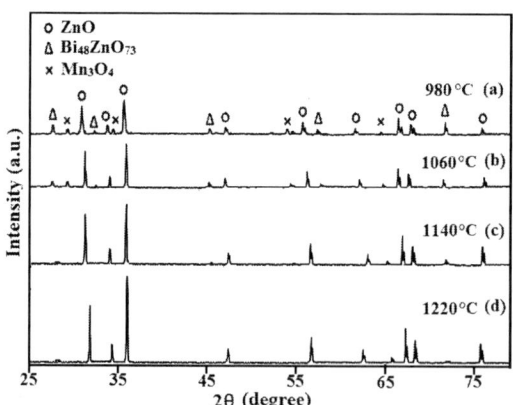

Fig. 2. XRD patterns of varistor at (a) 980 °C, (b) 1060 °C, (c) 1140 °C, and (d) 1220 °C temperatures.

$$\frac{1}{d^2} = \frac{3}{4}\left(\frac{h^2+hk+k^2}{a^2}\right) + \frac{l^2}{c^2} \qquad (1)$$

where (h, k, l) were evaluated from the XRD analysis. The lattice spacing d was calculated based on Bragg's equation.

The perpendicular parameters (c) of the ZnO sample were obtained from the peak positions of the ZnO (002) planes, and the results are summarized in Table 1. These parameters are comparable to the lattice constants c_0 of bulk ZnO at 5.206 Å [12]. Employing the data above, the stress (σ) in the ZnO disc was calculated using the strain model for hexagonal crystals as follows [13]:

$$\sigma = \left(\frac{2c_{13}^2 - c_{33}(c_{11}-c_{12})}{c_{13}}\right)\left(\frac{c_o-c}{c_o}\right) \qquad (2)$$

where σ is the mean stress in the ZnO sample, C_{11} = 209.7 GPa, C_{12} = 121.1 GPa, C_{13} = 105.1 GPa, and C_{33} = 210.9 GPa are the elastic stiffness constants of bulk ZnO (-453.6 GPa), c_0 is the lattice constant of strain-free bulk ZnO, and c is the lattice constant of the ZnO disc. The estimated values of stress "σ" in the ZnO disc for (002) plane grown at different sintering temperatures are listed in Table 1. The increasing order of the stress values is as follows: $\sigma(a) < \sigma(b) < \sigma(c) < \sigma(d)$. The negative value of the compressive stress indicates that the lattice constant is more elongated than the stress-free sample [14]. The positive value indicate a tensile stress generated by the stretching crystal size, indicating that the lattice constant decreased compared with that of the stress-free sample.

B. Optical Properties

Figure 3 shows the room-temperature PL spectra of varistor at different sintering temperatures. A characteristic UV peak at 359.53 nm to 380.46 nm and a broad emission in the red region with band peaks at 751 nm to 763 nm were observed. The UV and visible peak intensities varied with sintering temperature. The quality of the ZnO discs generally improved with increasing sintering temperature. The ZnO discs sintered at 1220 °C exhibited strong emission, good crystal quality with few structural defects and impurities, and excellent optical properties at approximately 380.46 nm in the ultraviolet region. The UV emission intensity also monotonically decreased when the temperature was decreased to 980 °C. This behavior can be understood by considering the formation of defects. At the sintering temperature of 980 °C, more defects responsible for nonradiative transition were introduced into the discs, causing a low UV emission peak. The higher sintering temperatures facilitated the migration of grain boundaries and promoted the coalescence of small crystals, thus favoring decreased concentration of nonradiative recombination centers. However, for the excitation spectrum of the red emission centered at 657 nm, a sharp absorption was observed near the band gap edge at 380 nm, suggesting that the main excitation mechanism for the red emission is an electron-hole pair via interband absorption.

Interestingly, the ultraviolet emission in the varistors in the current study was prevalent over the properties and crystal quality with several impurities and structural defects. This can be interpreted in terms of the increase in PL intensity with increased grain size and grain boundary in the sample. The increase in PL intensity can be attributed to the dependence of the PL spectra on the size distribution of the particles in the grains [15]. If the distribution is very wide, a large number of particles will be excited in the grains and the emission peak will occur very strongly. On the other hand, if the distribution is narrow, the emission peak will occur very broadly, as shown in the INSET in Figure 3.

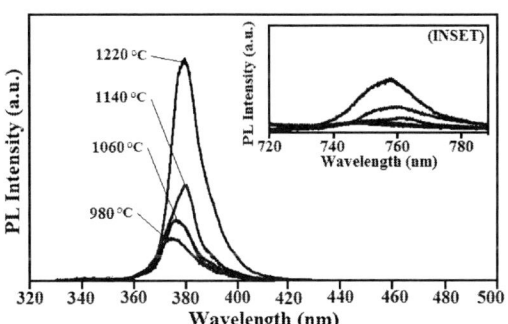

Fig. 3. PL spectra of varistor at different sintering temperatures in the UV region and in the visible region (INSET)

TABLE I
SUMMARIZED DATA FROM SEM IMAGES, XRD PATTERNS, AND PL SPECTRA

Sintering temperature	Density g/cm³	SEM Grain size (μm)	XRD 2θ (deg.)	a (Å)	c (Å)	Stress (σ) (G.Pa)	PL peak Wavelength (nm)	Energy band-gap (eV)
980 °C	5.48	11.5	36.13	3.255	5.220	-1.226	359.53	3.449
1060 °C	5.50	16.5	36.18	3.253	5.217	-0.959	371.53	3.338
1140 °C	5.57	23.6	36.21	3.252	5.205	0.086	377.93	3.281
1220 °C	5.63	35.4	36.24	3.250	5.203	0.263	380.46	3.259

As a result, the band-gaps of varistors decreased with increasing sintering temperature, and the tensile stress became more dominant with increasing grain size. The shift in the band-gap energy is related to its structural property. The inherent tensile strain in the ZnO disc can be relaxed by providing sufficient thermal energy, which lowers the band-gap energy. The band-gap was found to vary within the sintering temperature range of 980 °C to 1220 °C, which agrees with the reported value [15]. The variation in the energy band-gap as a function of sintering temperatures was measured (Table 1).

The observed shift in the energy band-gap from a higher to a lower value may be attributed to the conversion of the compressive stress to tensile stress with increasing sintering temperature. A decrease in the UV emission peak was observed at higher sintering temperatures as the compressive stress increased because of the increase in grain size. Hence, the variation in the band-gap with temperature significantly correlates with the stress in the discs, indicating that the increase in the sintering temperature induced an increase in the grain size and caused a reduction in the quantum size effect (or in the energy band-gap) in the doped ZnO nanoparticles [16].

C. Electrical Properties

Figure 4 shows the I-V characteristics of the varistor at different sintering temperatures. Current was applied and the residual voltage was observed.

The I-V curves in Figure 4 resemble a diode response with a high-resistivity region (left of breakdown voltage) and a low-resistivity region (right of breakdown voltage). The low resistivity region is referred to as the nonlinear region, whereas the nonlinearity coefficient (α) can be determined as follows, equation (3) [17]:

$$\alpha = \frac{\log (I_2 - I_1)}{\log (V_2 - V_1)} \qquad (3)$$

Table 2 summarizes the electrical data of the ZnO disc specimens. The reduction in breakdown voltage (V_b) from 335 V (sample at 980 °C) to 130 V (sample at 1220 °C) can be explained by the increase in the average grain size from

11.5 μm to 35.4 μm, resulting in lower numbers of grain boundaries for the 1220 °C sample and reduced "p-n junctions." Lower "p-n junctions" produced lower breakdown voltages, as shown in the I-V curves in Figure 4.

The addition of Bi_2O_3 to the ZnO led to the higher nonlinearity coefficient (α) of the discs. Thus, Bi_2O_3 has a significant effect on the grain boundary voltage barrier of the discs. Moreover, Mn_2O_3 is also a commonly used dopant in ZnO discs for building up the potential barrier in the grain boundary. Mn exhibited as a deep donor in ZnO and diminished the concentration of intrinsic donors at a high sintering temperature. When the specimens were immediately cooled to room temperature, the condition at the sintering temperature could be frozen and the concentration of the intrinsic donors would be low at room temperature [18,19]. Mn doping had a more significant effect on the grain boundary than on the grain, suggesting that the grain boundary was more resistive than the grain.

Another dramatic change was in the resistivity (ρ) at the high-resistivity region, in which the ρ of the sample sintered at 980 °C (303.5 kΩ.cm) dropped by more than thrice to 180.7 kΩ.cm for the 1220 °C sample (Figure 5). The surface sheet resistance also decreased from 19.71 MΩ/□ to 5.630 MΩ/□ as the sintering temperature increased , as shown in Figure 5. Thus, the extreme sintering temperatures at 1220 °C reduced the potential barrier of the "p-n junctions," resulting in the big drop in ρ. However, the nonlinear coefficient (α) was not significantly affected by the sintering temperature as only a small drop in α was observed (Table 2).

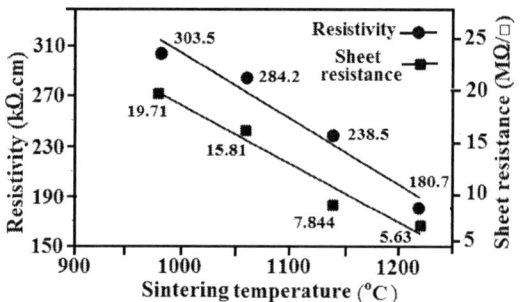

Fig. 5. Resistivity of varistor at different sintering temperatures and the sheet resistance

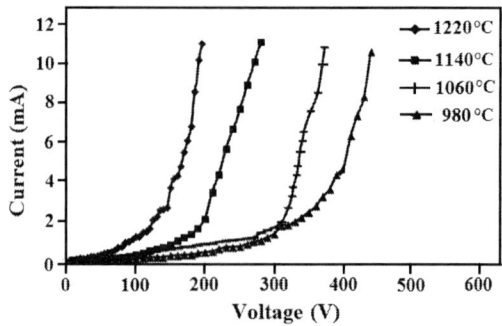

Fig. 4. Current-voltage characteristics of doped ZnO at different sintering temperatures

TABLE II
ELECTRICAL PROPERTIES OF SAMPLES SINTERED AT DIFFERENT TEMPERATURES

Sintering temperature	V_b (V)	α	ρ (kΩ.cm)	R_s MΩ/□
980 °C	335	57	303.5	19.71
1060 °C	292	50	284.2	15.81
1140 °C	219	44	238.5	7.844
1220 °C	130	40	180.7	5.630

IV. CONCLUSION

ZnO nanoparticles-Bi_2O_3-Mn_2O_3 varistors were prepared via a conventional ceramic processing method and were then sintered at different temperatures. SEM, EDX, and XRD methods were used for the characterization of the morphologies and crystal structures of the discs. The crystal quality of the ZnO discs was highly dependent on the sintering temperature. The increase in the sintering temperature from 980 °C to 1220 °C led to larger grain size and better crystallinity. The sintering process was also found to improve grain crystallinity, as shown in the lowering of intrinsic compressive stress based on the XRD lattice constant and FWHM data. The sintering temperature also significantly influence the electrical properties of the ZnO discs. The photoluminescence spectra of the doped ZnO nanoparticles showed a strong ultraviolet emission at approximately 381 nm with a broad red emission at approximately 750 nm. The observed shift in the energy band-gap from a higher to a lower value may be attributed to the conversion of the compressive stress to tensile stress with increasing sintering temperature. It can be concluded that the sintering process can be used as a new technique for controlling the grain size and different properties of doped ZnO discs made from ZnO nanoparticles with improved structural and optical properties.

ACKNOWLEDGMENT

This work was supported by Research University (Individual) Grant from Universiti Sains Malaysia. We express gratitude to the Cultural Mission of the Royal Embassy of Saudi Arabia. We further acknowledge the priceless assistance from the NOR Lab of USM.

REFERENCES

[1] M. Matsuoka, T. Masuyama and Y. Iida, "*Nonlinear electrical properties of zinc oxide ceramics*", suppl. J. Jap. Soc. Appl. Phys. 39 (1970) 94.

[2] J. F. Yan and W. W. Rhodes, "*Preparation and properties of TiO2 varistors*", Appl. Phys. Lett. 40 (1982) 536.

[3] A. B. Glot, A. M. Chakk, B. K. Chernyj and A. Ya. Yakunin, "*Dependence of the Electrical Conductivities of the Semiconductors ZnO-SnO2-Bi2O3 in the Temperatures and Additional Heats-Treatment procedure*", Inorganic Mater. 10 (1974) 1866-1868.

[4] A.B. Glot and A.P. Zlobin, "*Nonohmic conductivity of tin dioxide ceramics*". Inorganic Mater. 25 (1989) 274-276.

[5] L. Miao, S. Tanemura, H. Y. Yang, S. P. Lau, "*Synthesis and random laser application of ZnO nano-walls: a review*", 2009, J. Nanotech. 6, 723.

[6] C. X. Xu and X. W. Sun , "*Field emission from zinc oxide nanopins*", 2003, Appl. Phys. Lett. 83, 3806-3808.

[7] C. X. Xu, X. W. Sun, and B. J. Chen, 2004, "*Field Emission from gallium-doped zinc oxide nano fiber array*", Appl. Phys. Lett. 84, 1540–1542.

[8] R. Einzinger, "*Microcontact Measurement of ZnO Varistors*" ,Ber. Dt. Keram. Ges. 52 (1975) 244-245.

[9] D. R. Clarke, "*Varistor ceramics*" . J. Am. Ceram. Soc. 82 (1999) 485–502.

[10] F.A. Selim, T.K. Gupta, P.L. Hower, W.G. Carlson, "*Low voltage ZnO varistor: Device process and defect model*", J. Appl. Phys. 51 (1980) 765.

[11] S. Maniv, W. D. Westwood, E. Colombini, "*Pressure and angle of incidence effects in reactive planar magnetron sputtered ZnO layers*", J. Vac. Sci. Technol. 20 (1982), 162.

[12] R. R. Reeber, "*Lattice parameters of ZnO from 4.2$_0$ to 296$_0$ k*", J. Appl. Phys. 41 (1970) 5063.

[13] Z. Z. Zhi, Y. C. Liu, B. S. Li, X. T. Zhang, Y. M. Lu, D. Z. Shen, X. W. Fan, "*Effect of thermal annealing on ZnO films grown by plasma enhanced chemical vapour deposition from Zn(C2H5)2 and CO2 gas mixtures*", J. Phys. D: Appl. Phys. 36 (2003) 719–722.

[14] Shinji Okamotu, Yoshihiko Kanemitsu, Hiroji Hosukawa, Kei Mura Koshi and Shozo Yanagida, Solid State Commun. 105, 7 (1998).

[15] R. Kumar, N. Khare, V. Kumar, G.L. Bhalla, "*Effect of intrinsic stress on the optical properties of nanostructured ZnO thin films grown by rf magnetron sputtering*", 2008, Appl. Surf. Sci. 254, 6509–6513.

[16] A. Mahmood, N. Ahmed, Q. Raza , T. M. Khan, M. Mehmood , M. M. Hassan, N. Mahmood, "*Effect of thermal annealing on the structural and optical properties of ZnO thin films deposited by the reactive e-beam evaporation technique*", 2010, Phys. Scr. 82, 065.

[17] J. Han, P. Q. Mantas, and A. M. R. Senos, "*Effect of Al and Mn doping on the electrical conductivity of ZnO*", J. Eur. Ceram. Soc. 21 (2001) 1883–1886.

[18] J. Han, P. Q. Mantas, and A. M. R. Senos, "*Defect chemistry and electrical characteristics of undoped and Mn-doped ZnO*", J. Eur. Ceram. Soc. 22 (2002) 49– 59.

[19] R.L. Petritz, Phys. "*Theory of Photoconductivity in Semiconductor Films*", Rev. 104 (6) (1956). 1508.

Optimization of Cantilever-Based MEMS Switch

Mohammadmahdi Vakilian, Maryam Mousavi, Badariah Bais and Burhanuddin Yeop Majlis
Institute of Microengineering and Nanoelectronics (IMEN)
Universiti Kebangsaan Malaysia (UKM)
43600 UKM Bangi, Selangor, Malaysia
Email: burhan@vlsi.eng.ukm.my

Abstract- **In this paper, RF MEMS switches were studied. One of the applications of RF MEMS switches is in the reconfigurable antenna where it is used to replace the traditional switches such as FET and PIN switches. Among the RF MEMS switches, the cantilever-based MEMS switch is the most popular for its low cost and ease of the fabrication process. The 3D builder and Thermo-Electro-Mechanical (TEM) modules of the IntelliSuite software were used for the simulation of the cantilever-based MEMS switch. The effect of materials and beam geometry of the switch on the pull in voltage and the operating frequency were discussed. The simulation results showed that there is a trade-off between the pull in voltage and the operating frequency for optimizing the cantilever-based MEMS switch.**

I. INTRODUCTION

Micro-Electro-Mechanical Systems (MEMS) is a very diverse technology that can be classified as miniaturized devices with numerous functions. MEMS devices may contain elements such as sensors, actuators and embedded microcomputers for the development of new and enhanced systems. Being microstructures, these devices can be integrated from tiny functional elements into a whole new package for improved performance and effectiveness. Among some of the usual application areas for MEMS lies in the radio frequency and millimeter wave field, also known as the Radio Frequency Micro-Electro-Mechanical Systems (RF MEMS). New advances in the RF MEMS technology are currently paving the way for new generations of RF related devices such as the RF MEMS switches [1].

RF MEMS switches are utilized in reconfigurable antennas. This type of antenna can electronically modify the outgoing topology inside the same physical aperture [2]. However, an important part of reconfigurable antenna is actually the switch. As opposed to conventional switches i.e. transistors and diodes, MEMS switches in RF applications offer greater advantages such as small insertion loss, less energy consumption, excellent quality factor (Q), RF isolation and less expensive.

The designs for RF MEMS switches can be varied according to their usage; as either shunt or series and can move up and down or side to side. Whereas the MEMS switches can either be metal-to-metal connection or a capacitive connection switches. Their movements can be carried out via electrostatic, magnetostatic, piezoelectric or thermal actuation [3]. Prevalent types of MEMS switch include a metal cantilever, air bridge and other structures electrically arranged in series or parallel with the RF transmission line [4]. The most omnipresent structures in the MEMS area are cantilever beams. These

cantilevers are generally made out of silicon (Si), silicon nitride (SiN) or polymers [5]. The most important reasons to use of MEMS cantilevers are their low cost and easy to fabricate even in huge quantities.

In this paper, a cantilever-based RF MEMS switch for reconfigurable antenna was simulated. The aims of this study are to evaluate the effect of materials and the switch geometry on the pull-in voltage and the operating frequency.

II. LITERATURE REVIEW

A. The Reconfigurable Antenna

With the advances in technology, life is easier for the human race when information can be moved from one point to the other via wireless communications. This type of technology relies a lot on the transmitting capability of the antenna. Of late, the MEMS technology has been going through major advancement in designing and fabricating antennas for various purposes [6]. The ability of antenna to be reconfigured has some advantages for tuning its operating frequency to new setting and the reduction of its size [7]. This ability can be boosted by incorporating the MEMS switch to the antenna.

Reconfigurable antenna can electronically modify the outgoing topology inside the same physical aperture. This type of antenna is better suited to use with spatial multiplexing methods and space time block codes [8]. It is considerably lighter, less bulky and lower cost in comparison to conventional antennas. However, an important part of reconfigurable antenna is actually the switch used to connect all components of the antenna. The switches' insertion loss and isolation are the factors that determine the effectiveness of the whole array of reconfigurable antenna [9]. Lately, PIN diodes are utilized in reconfigurable antenna for RF switching which are important for steering the beam, band functions with dual-frequency and dual-polarized radiation. Nevertheless, using these diodes often resulted in elevated insertion loss due to their low Q at high frequencies. MEMS switches are the best replacement for the semiconductor devices since they boast higher Q as opposed to the semiconductor devices [10]. RF-MEMS switches are well suited to be used with reconfigurable antennas since these switches have very low switching loss. In huge antenna arrays, the broad bias network of the RF-MEMS switches will not obstruct and corrupt the antenna radiation pattern. Most importantly, the biasing network will not use up any power which is good for such antenna arrays [11]. In addition, the antenna itself is essentially a high Q structure and

978-1-4673-2395-6/12 $31.00 © 2012 IEEE

the frequencies of operation as well as radiation pattern play vital role in the reconfigurable antenna system [12].

Fig.1 shows the structure of a reconfigurable dual-band antenna containing two patches of basic microstrip resonator. When Patch 1 is operated at frequency f_1, a second frequency f_2 will resonate when Patch 1 and 2 are connected. The control over the length and size of the antenna is carried out by the MEMS switch in order to achieve a dual-band reconfigurable antenna.

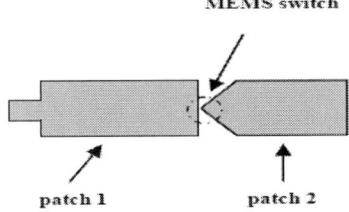

Fig. 1 A reconfigurable dual-band antenna [13]

B. RF MEMS Switch

The advances in MEMS technology have provided the almost perfect switches in terms of small insertion loss, good isolation, low power consumption and linearity. Thus, more researches are being carried out to develop improved components that have excellent RF properties with low actuation voltage [14].

Typically in recent microelectronic components, switching from 'on' to 'off' condition requires solid state devices to employ voltage potential for adjusting the carrier concentration in semiconductors. Modulating conductivity within the same channel of usually crystalline silicon substances could result in some performance degradations such as signal insertion loss, cut-off frequency and isolation. With these problems in mind, MEMS switches have been added with movable micro-mechanical elements to eliminate such problems. This is because in MEMS devices, switching from 'on' to 'off' can be done mechanically.

An RF switch is made up of two elements which are the mechanical element (actuator) as well as the electrical element. The energy behind the mechanical movement of switch is usually acquired from electrostatic, piezoelectric, magnetic and thermal mechanism. Electrostatic actuation is generally the ubiquitous choice for these reasons: i) in the electrostatic actuation, there is no DC current, thus there is no DC power consumption; ii) the electrodes of the actuator are miniaturized in size and similar in temperament with IC production since they are commonly lithographically defined using micro-fabrication processes; iii) electrostatic actuation can produce switching speed ranging from 1 to 200 μs which is faster than a thermal or magnetostatic actuation [15].

Apart from that, a mechanical switch is a very important component in a variety of microwave circuits. It can either move vertically or laterally and positioned in series or in shunt configurations in the electrical component.

Two standard contact RF MEMS categories are available; metal–metal contacts and metal–insulator–metal contacts. The metal–metal contacts can be found mostly in the cantilever

kind of switches and they are known as ohmic or direct contacts. A direct contact means that by touching metal to metal, a signal can be transmitted causing current to flow through the interface. As for metal–insulator–metal type of contacts (also known as capacitive or indirect contacts), they can be found in membrane type of MEMS switches [3]. All in all, 32 kinds of diverse RF switches can be assembled by using 4 actuation methods (electrostatic, magnetic, piezoelectric or thermal), 2 movement choices (vertically or laterally), 2 contact types (metal-metal or metal-insulator-metal) and 2 circuit configurations (series or shunt).

RF MEMS switches can be utilized in diverse applications such as switching and reconfigurable networks, transferable wireless systems as well as phased arrays in communication and radar systems [11].

Cantilever beams are typically made out of silicon (Si), silicon nitride (SiN) or polymers. Its fabrication steps normally require the cantilever component to be undercut in order to obtain it, either with a wet or dry anisotropic etching method [16]. Recently, researchers are working hard at developing cantilever arrays in the form of biosensors to be used in medical diagnostics [17]. MEMS cantilevers have also paved their way as radio frequency switches in reconfigurable antennas.

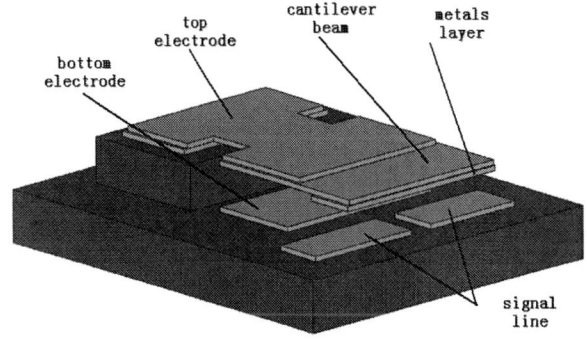

Fig. 2 Structure of RF MEMS [18]

Fig. 2 shows a typical broadside MEMS-series switch. It has one fixed end while another is layered with metal for connecting the microwave signal line. Apart from that, a metal-layer is also placed in the center of the cantilever beam, making it the capacitor's top pole. By applying a control voltage between the top and bottom poles, the electrostatic force will cause the cantilever beam to distort. With bigger control voltage that surpasses the threshold voltage, the metal layers on the switch and on the cantilever will connect the signal line. In the case when a broadside MEMS-series switch is closed, the signal will cross the switch only instead of the entire cantilever beam [18].

A range of different optimizing methods are available for MEMS switches. One of the important parameters for optimizing the RF MEMS switch is the pull in voltage. Wang et al. [19] studied the pull in voltage optimization of the cantilever MEMS switch via beam shape and dimension. Spasos et al. [20] proposed actuation pulse optimization of RF MEMS switch by using Taguchi's method. Another critical parameter in optimization of MEMS switch is the operating frequency of the switch and also materials selection is an important criterion for MEMS switch design. In this paper, the effect of materials and the switch geometry on the pull in voltage and the operating frequency of the switch by using IntelliSuite software will be discussed.

III. METHODOLOGY

The design of the switch used in this study is based on the switch designed by [19, 21] as shown in Fig. 3. The switch consists of bottom substrate made of Silicon. The substrate is then coated with a layer of silicon dioxide which acts as a dielectric. One end of the silicon dioxide layer is lightly coated with aluminium which acts as the bottom contact of the switch. A cantilever made of silicon dioxide and aluminium with a dimple of 20 μm x 20 μm in size is suspended above the substrate with a 1.7 μm air gap separating them.

A differential voltage is applied between the bottom section and the wide aluminium area on the left side of the cantilever. When the appropriate voltage is supplied, the cantilever will touch the ground plane and a closed circuit will be formed by the aluminium conductors on the right side.

In this study, the 3D builder and the Thermo-ElectroMechanical (TEM) analysis modules of the IntelliSuite software were used for the simulation of the cantilever-based MEMS switch. The 3D Builder module was utilized for designing of MEMS switch. The design was then transferred to the ThermoElectroMechanical (TEM) module to perform the simulation. The effect of materials on the displacement of the switch was investigated and then the effect of changing the beam dimensions on the pull in voltage and operating frequency of switch were obtained.

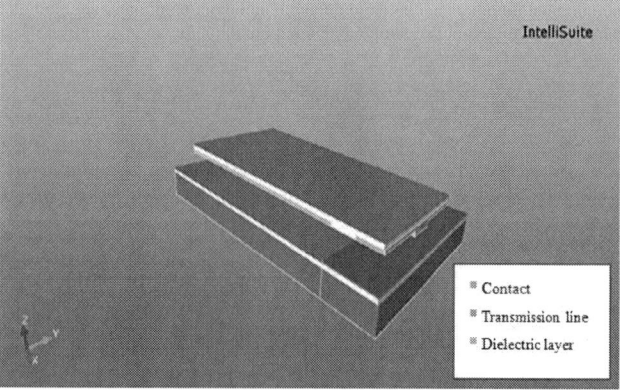

Fig. 3 Cantilever MEMS switch

IV. RESULTS AND DISCUSSION

A voltage of 23.4 V and 0 V were supplied to the conductor on the left side of the cantilever and the silicon section, respectively. The result of this analysis was shown as the z-displacement, as illustrated in Fig. 4.

It can be seen that maximum deflection has taken place on the right side of cantilever, with a z-displacement valued at -1.72555 microns which means the cantilever have touched the ground plane.

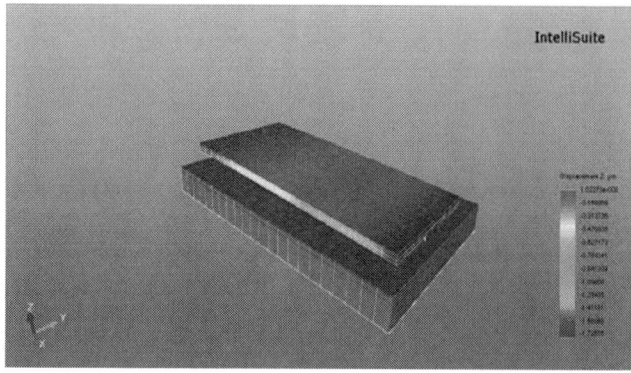

Fig. 4 Z-displacement result

In this section, we discuss on material selection for the cantilever-based switch and then the effect of beam dimensions on pull in voltage and operating frequency of the switch by using IntelliSuite software.

A. Material Selection

In this section, we have investigated some different materials for the cantilever MEMS switch.

TABLE I
Z-DISPLACEMENT BY USING VARIOUS MATERIALS AT 15 V APPLIED VOLTAGE

Contact	Transmission line	Dielectric layer	Z-displacement (μm)
Al	Al	Si_3N_4	-0.179218
Au	Al	Si_3N_4	-0.179244
Au	Au	Si_3N_4	-0.167387
Al	Au	Si_3N_4	-0.167388
Al	Al	SiO_2	-0.367521
Au	Au	SiO_2	-0.343638
Au	Al	SiO_2	-0.367514
Al	Au	SiO_2	-0.343644

As shown in Table I, different materials were used for contact, transmission line and dielectric layer of MEMS switch and z-displacement was obtained by applying 15 V for each state. Simulation results showed that Aluminium (Al) gave the highest z-displacement compared with gold (Au) when it was used for both contact and transmission line of the switch. Also the maximum z-displacement was occurred when SiO_2 was used as the dielectric layer compared with Si_3N_4.

B. Relationship between the Pull in Voltage and the Length of the Beam

Table II illustrates relationship between pull in voltage and length of the beam in various thicknesses with W=100 μm. When length of the beam is increased, pull in voltage will be decreased and by increasing thickness of the beam, pull in voltage will be increased.

TABLE II

PULL IN VOLTAGE AT VARIOUS BEAM LENGTHS AND THICKNESSES (W=100 μm)

Length of beam (μm)	Pull in voltage (V) (T=1.25μm)	Pull in voltage (V) (T=1.75μm)	Pull in voltage (V) (T=2.25μm)
200	15.9	26.3	38.3
210	14.1	23.4	34
220	12.7	20.9	30.4

C. Relationship between the Frequency of Switch and the Length of the Beam

Relationship between the frequency of switch and the length of the beam in various modes is shown in Table III. According to this table there is a great difference between mode 1 and mode 2, therefore it can be concluded that mode 1 is dominant mode.

TABLE III

FREQUENCY AT VARIOUS BEAM LENGTHS AND MODES (T=1.75, W=100 μm)

Length of the beam (μm)	Mode 1 (Hz)	Mode 2 (Hz)	Mode 3 (Hz)
200	29017.4	169602	240251
210	26136.2	160484	217954
220	23728.2	152312	198635

D. Relationship between the Mode 1 and the Length of the Beam

Table IV shows relationship between mode1 (operating frequency) and length of the beam in various thicknesses with W=100 μm. When length of the beam is increased, mode1 will be decreased and by increasing thickness of the beam, mode1 will be increased.

TABLE IV

MODE 1 AT VARIOUS BEAM LENGTHS AND THICKNESSES (W=100 μm)

Length of the beam (μm)	Mode 1 (Hz) (T=1.25μm)	Mode 1 (Hz) (T=1.75μm)	Mode 1 (Hz) (T=2.25μm)
200	20330.2	29017.4	37768.8
210	18380.9	26136.2	34110.3
220	16524.7	23728.2	30937.6

V. CONCLUSION AND FUTURE WORK

During this study, the focus has been on applying the RF MEMS switch in the reconfigurable antenna. RF MEMS switches definitely trumps over traditional switches due to their superior performance and low cost.

The IntelliSuite software, 3D builder module and ThermoElectroMechanical (TEM) analysis module were used for simulation of the cantilever-based MEMS switch.

As it was shown materials selection is an important criteria for MEMS switch design. Pull in voltage and operating frequency of switch are critical parameters in optimization of MEMS switch.

Simulation results showed that the pull in voltage was reduced for longer and thinner cantilever beams. However, the operating frequency was increased when the thickness of the cantilever beam was increased and also when the length of the

cantilever beam was decreased. Therefore, there is a trade-off between the pull in voltage and the operating frequency.

Following works are suggested to continue this research:

1)   During this research, a cantilever MEMS switch was effectively designed and optimized by studying different materials as well as the effect of changes in its beam dimensions. This research methodology is also relevant for the study of other types of switches.

2)   Two parameters were studied for optimizing the MEMS switch which were the pull-in voltage and operating frequency. It is suggested that other parameters e.g. switching speed and power handling should also be studied.

3)   At the end of this research, it has been proven that the optimized design of the RF MEMS switch is viable to use in real applications. Since the pull-in voltage has direct effect on life time of RF MEMS switch, our design is expected towards more practical utilization.

REFERENCES

[1]   F. M. Guo, et al., "The experimental model of micro-cantilever switch and its optimization model," in Nano/Micro Engineered and Molecular Systems (NEMS), 2010 5th IEEE International Conference on, 2010, pp. 1-4.

[2]   I. J. Hyeon, et al., "Package-Platformed Linear/Circular Polarization Reconfigurable Antenna Using an Integrated Silicon RF MEMS Switch," ETRI Journal, vol. 33, pp. 802-805, Oct 2011.

[3]   T. H. Lin, et al., "A study on the performance and reliability of magnetostatic actuated RF MEMS switches," Microelectronics Reliability, vol. 49, pp. 59-65, Jan 2009.

[4]   W. Han and R. J. Pryputniewicz, "Design, fabrication, and characterization of surface micromachined MEMS cantilever components," in sem-proceedings, 2007, pp. 1162-1170.

[5]   N. J. R. Muniraj, "MEMS based humidity sensor using Si cantilever beam for harsh environmental conditions," Microsystem Technologies, vol. 17, pp. 27-29, 2011.

[6]   Yulindon, et al., "Investigation of reconfigurability of dual-band microstrip patch antenna by utilizing MEMS switch," in Space Science and Communication, 2009. IconSpace 2009. International Conference on, 2009, pp. 111-114.

[7]   J. T. Bernhard, "Reconfigurable antennas," Synthesis Lectures on Antennas, vol. 2, pp. 1-66, 2007.

[8]   C. P. Sukumar, et al., "Link Performance Improvement Using Reconfigurable Multiantenna Systems," Antennas and Wireless Propagation Letters, IEEE, vol. 8, pp. 873-876, 2009.

[9]   J. Kiriazi, et al., "Reconfigurable dual-band dipole antenna on silicon using series MEMS switches," in Antennas and Propagation Society International Symposium, 2003. IEEE, 2003, pp. 403-406 vol.1.

[10]   J. Chang won, et al., "Reconfigurable scan-beam single-arm spiral antenna integrated with RF-MEMS switches," Antennas and Propagation, IEEE Transactions on, vol. 54, pp. 455-463, 2006.

[11]   G. M. Rebeiz, RF MEMS: Wiley Online Library, 2003.

[12]   D. Rodrigo, et al., "MEMS-reconfigurable antenna based on a multi-size pixelled geometry," in Antennas and Propagation (EuCAP), 2010 Proceedings of the Fourth European Conference on, 2010, pp. 1-4.

[13]   Yulindon, et al., "Reconfigurable Dual-Band Antenna Using MEMS for Wireless Application," presented at the Proc. 2008 Student Conference on Research and Development (SCOReD 2008), 2008.

[14]   J. Héctor, RF MEMS circuit design for wireless communications: Artech House Publishers, 2002.

[15]   Z. Yang, Contact material optimization and contact physics in metal-contact microelectromechanical systems (MEMS) switches: ProQuest, 2008.

[16]   C. W. Shong, et al., Science at the Nanoscale: An Introductory Textbook: Pan Stanford Pub, 2010.

[17]   R. A. Rahim, et al., "Design and analysis of MEMS piezoresistive SiO2 cantilever-based sensor with stress concentration region for biosensing

applications," in *Semiconductor Electronics, 2008. ICSE 2008. IEEE International Conference on*, 2008, pp. 211-215.

[18] G. Yanjue, *et al.*, "Design optimization of cantilever Beam MEMS switch," in *Industrial Mechatronics and Automation, 2009. ICIMA 2009. International Conference on*, 2009, pp. 28-31.

[19] Z. Wang, *et al.*, "Contact physics modeling and optimization design of RF-MEMS cantilever switches," in *Antennas and Propagation Society International Symposium, 2005 IEEE*, 2005, pp. 81-84 Vol. 1A.

[20] M. Spasos, *et al.*, "RF-MEMS switch actuation pulse optimization using Taguchi's method," *Microsystem Technologies-Micro-and Nanosystems-Information Storage and Processing Systems*, vol. 17, pp. 1351-1359, Aug 2011.

[21] *IntelliSuite® Training Manual User Documentation*: IntelliSense Inc, 2009.

Scaling Down of the 32 nm to 22 nm Gate Length NMOS Transistor

Afifah Maheran A.H.[1], Menon, P.S.[1], I. Ahmad[2], H.A. Elgomati[1], B.Y. Majlis[1], F. Salehuddin[2]

Institute of Microengineering and Nanoelectronics (IMEN)
Universiti Kebangsaan Malaysia (UKM)[1]
43600 Bangi, Selangor, Malaysia
Centre for Micro and Nano Engineering(CeMNE)[2]
College of Engineering
Universiti Tenaga Nasional (UNITEN)
Jalan IKRAM-UNITEN
43000 Kajang, Selangor, MALAYSIA
Email: susi@eng.ukm.my

Abstract- **In this paper, we provide the downscaling design and simulation of NMOS transistor with 22 nm gate length, based on the 32 nm design simulation from our previous research. A combination Titanium dioxide (TiO_2) was used as the high-k material and tungsten silicide (WSi_x) was used as the metal gate instead of SiO_2 dielectric from the 32 nm gate length device. The NMOS transistor was simulated using fabrication tool ATHENA and electrical characterization was simulated using ATLAS. The scale down ratio was used and the dimension of device was scaled down with minimal issues. Our simulation shows that the optimal value of threshold voltage (V_{th}) and leakage currents (I_{on} and I_{off}) was achieved according to specification in ITRS 2011. This provides a benchmark towards the fabrication of 22 nm NMOS in future work.**

Keywords: 22 nm NMOS, Scaling down ratio, high-k/metal gate, Silvaco.

I. INTRODUCTION

SiO_2 has been used as the gate dielectric material over decades, and the current device scaling trend requires the film thickness to be as thin as $t_{ox} = 1.5$ nm [1]. The most critical issue for the microelectronics industry is the need for higher permittivity (high-k) dielectrics to replace the silicon dioxide as a gate CMOS. There are number of high-k materials being proposed and analyzed as the replacement of silicon dioxide (SiO_2) in the next generation of metal–oxide–semiconductor field-effect transistor (MOSFETs). In order to design the new device with new materials in CMOS technology, a few crucial characteristics must be taken into account, such as short channel effects, leakage current, low sheet resistance and others. The International Technology Roadmap for Semiconductors (ITRS) gives a good reference to researchers as a guidance to scale down the size of the MOSFET transistor.

The continual downscaling of semiconductor devices into the nanoscale regime not only requires the usage of alternative material for SiO_2 as the gate dielectric but also requires new material for the polycrystalline silicon (poly-Si) gate electrodes.

Example of high-k materials that are being studied nowadays such as HfO_2, ZrO_2, TiO_2 and Al_2O_3 [2]. Most of the researchers are using HfO_2 as gate dielectric for future CMOS applications. In this design, we explore the usage of TiO_2 as the dielectric material since it has a very high dielectric constant, due to better thermo dynamical stability with silicon [3], while using WSi_x as the metal gate [4, 5].

In semiconductor industry, performing an actual fabrication is extremely expensive and time consuming [6]. Therefore the simulation process is one of the smart techniques and favourable to researchers to obtain the optimized design before proceeding to the actual fabrication. In this research, Silvaco Software Tools is used for the 22 nm gate length device.

The Ivy Bridge released by Intel in 2012 on its 22 nm technology node is the first volume production of microprocessors based on FinFETs [7]. Future device development is expected to focus on 3D, non-planar device architectures. However, the planar architectures are still on going to production because of the easy implementation in production and the new technology by using high-k/metal gate. While the FinFETs still have disadvantage such as self heating and dissipation problem in the device and highly set up cost [7].

As the downscaling of the transistor device reaches the submicron regime, the phenomena called short channel effect arises [8]. The short channel effect will cause changes in the electric field at the drain and source area and then influence the charge distribution [8].

The short-channel effects are attributed to two physical phenomena which are the limitation imposed on electron drift characteristics in the channel and the modification of the threshold voltage due to the shortening channel length. These

978-1-4673-2395-6/12 $31.00 © 2012 IEEE

will interrupt with the basic characteristics of transistor devices such as voltage threshold value and transistors' off-current due to the non-linearity of the physical and electrical characteristics of the materials used in a device. This will contribute to the difficulty and complexity of getting a good V_{th} value. This is because of the leakage current between the source and drain and also gate and the channel, causing the transistor to be 'ON' while no voltage is applied to the gate.

The aim of the research is to meet the ITRS 2011 specification for 22 nm gate length NMOS transistor with the V_{th} value of 0.306V and the lower of leakage currents (I_{on} = 100 µA/µm and I_{off} = 1582 nA/µm) [9].

II. SCALING DOWN

Threshold voltage is being the most critical part in determining the functionality of the device. As well as the leakage currents, it should be kept as low as possible in order to increase the speed of the device by shortening the time to accumulate charge in the channel for a transistor to be turn on. The NMOS transistor can be scaled down using the downscale rule [10], which is;

$$\beta = \frac{1}{\alpha} \qquad (1)$$

Since α = 32 nm / 22nm, where is the downscaling from 32 nm gate length to 22 nm, then β = 0.6875. By using the ratio value, the device dimension and mesh setting are down scaled with other parameter changes, especially on the gate length. Then to get the optimal value as that stated in ITRS 2011, the previous research [10,11] is considered a few parameters that can be adjusted in order to get the optimal V_{th}. Theses parameters are the Halo implantation, Source/Drain implant dose, Compensation Implantation dose, V_{th} adjustment implantation dose and others.

III. MATERIALS AND METHODS

The fabrication process steps are as follows. The substrate used for experiment was p- type silicon, <100> orientation. It is then followed by growing an oxide layer at the top of silicon bulk using dry oxygen. P-well implantation process was done using this oxide layer as a mask. This was done using Boron as dopant with a dose of 3.75×10^{12} ions/cm^2. The silicon wafer then has underwent the annealing process at 900°C in Nitrogen environment, and followed by dry oxygen in order to ensure that boron atoms being spread properly in the wafer. The following step was to produce Shallow Trench Isolator (STI) of 130 Å thickness [12].

In order to form the STI layer, the wafer was oxidized in dry oxygen for 25 minutes. Then, a 1000Å nitride layer was deposited on top of the oxide layer by applying low pressure chemical vapour deposition process (LPCVD), followed by a photo resist deposition with a thickness of 1.0µm. The trench depth of 3200 Å was achieved. Thereafter, a sacrificial oxide layer was grown and then etched followed by a sacrificial nitride layer whereby the trench is then completed.

Then the high-k material, TiO_2 was deposited for thickness of 2 nm [13] and it followed by etching to get the desired thickness and was adjusted to produce a 22 nm gate length. The next step was to implant the N well active area, in order to adjust the V_{th} value. The dosage for boron was 6.98×10^{12} atom/cm^2. Tungsten silicide (WSi_x) will then be deposited on the top of the bulk with thickness of 8 nm and etch accordingly to produce the gate contact point as desired [14].

Later on, Halo implantation then took place in order to get an optimum performance for NMOS device, Indium with the dose of 12.76×10^{12} ions/cm^2 was implant. The dosage was varied in order to get the optimum value [10, 11]. Then spacers were formed at each of the polysilicon sides, namely source and drain regions respectively. Side wall spacers was used as a mask for source and drain implantation [15]. Then, there are source-drain implantations, Arsenic was firstly implanted with dose of 5.1×10^{13} ions/cm^2, followed by phosphorous with dose of 1.5×10^{12} ions/cm^2 to ensure the smooth current flow in the device.

The next process was the development of 0.5 µm Borophosphosilicate Glass (BPSG) layer [16]. This layer acts as a pre metal dielectric (PMD). After BPSG deposition, the wafer has undergoes annealing process at temperature of 950°C [16]. The next process was compensation implantations using phosphorous, with the dose of 3.71×10^{13} ions/cm^2 [17]. Then, aluminium layer was deposited on top of the structure, next it was etched accordingly to form the metal contact for source and drain. At this stage the transistor design is completed.

Figure 1(a) and Figure 1(b) shows the device structure and the zoom up figure of 22 nm gate length of high k/metal gate NMOS transistor respectively while Figure 2 show the doping profile that successful designed. Then, the transistor undergoes electrical characteristic measurement in order to find the threshold voltage (V_{th}) by reference to ITRS 2011 [9].

978-1-4673-2395-6/12 $31.00 © 2012 IEEE

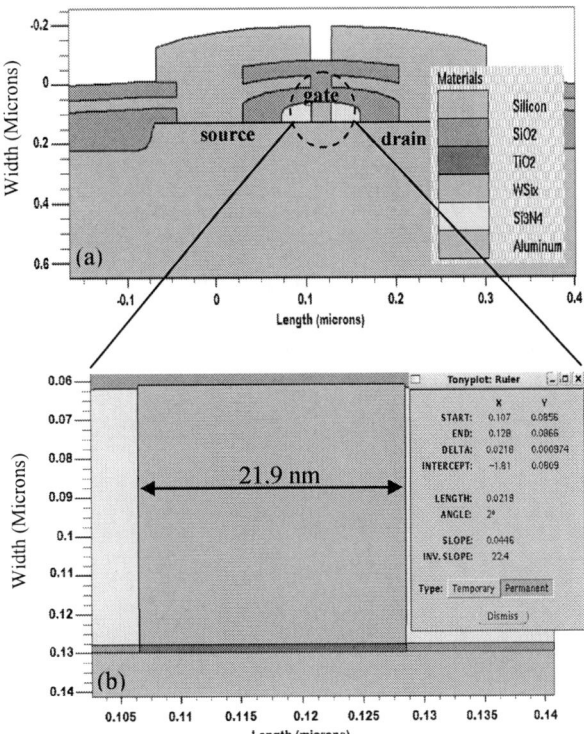

Fig. 1. (a) Completed NMOS transistor with 22 nm gate length;
(b) Zoom up of gate length of 22 nm NMOS technology

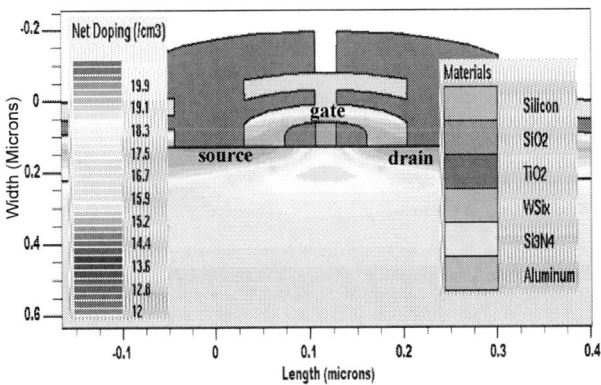

Fig. 2. The doping profile of the 22 nm gate length NMOS transistor

IV. RESULTS AND DISCUSSION

Finally, we succeed in obtaining the values which result in good V_{th} values upon various adjustments, as well as lower leakage currents value. Table I shows the design parameters that have been optimized in reference to previous work and the new parameter to design the 22 nm NMOS transistor [18].

TABLE I
DESIGN PARAMETER OF 22 NM GATE LENGTH NMOS TRANSISTOR

Type	Unit	32 nm NMOS	22 nm NMOS
Gate Material		SiO$_2$/poly	TiO$_2$/WSi$_x$
Halo Implantation	atom/cm^3	12.75 x 10^{12}	12.76 x 10^{13}
V_{th} Adjust Implant	atom/cm^3	1.75 x 10^{11}	6.98 x 10^{12}
S/D Implant	atom/cm^3	5.0 x 10^{13}	5.1 x 10^{13}
Compensation Implant	atom/cm^3	3.7 x 10^{13}	3.71 x 10^{13}

Table II lists the simulated values versus ITRS prediction for V_{th}, I_{on} and I_{off} in 22 nm gate length CMOS technology. It is clearly shown that our design simulation is close to the ITRS prediction and our NMOS values of I_{on} and I_{off} are respectively closed and much lower than the prediction. The lower the value of I_{on} and I_{off}, resulted the better device performance.

TABLE II
SIMULATION RESULT VERSUS ITRS 2011 PREDICTION

Parameter	Unit	Simulation	ITRS 2011 Prediction
V_{th}	V	0.3058	0.3060
I_{on}	μA/μm	7.90215 x 10^{-5}	1.0 x 10^{-6}
I_{off}	nA/μm	6.63916 x 10^{-10}	1.582 x 10^{-6}

While in result of characteristic simulation of the NMOS device, Figure 3 and Figure 4 shows the graph of Drain current (I_d) versus Gate voltage (V_g) and Drain current (I_d) versus Drain voltage (V_d) respectively.

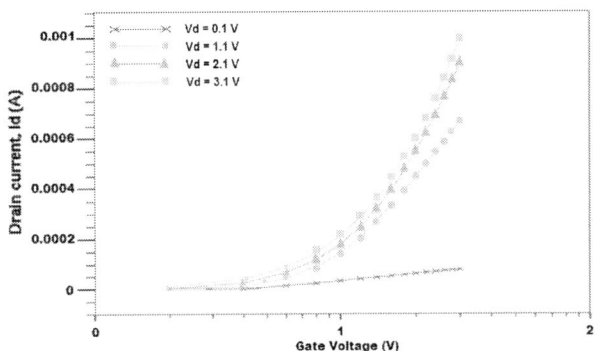

Fig. 3. Drain current (I_d) versus Gate voltage (V_g)

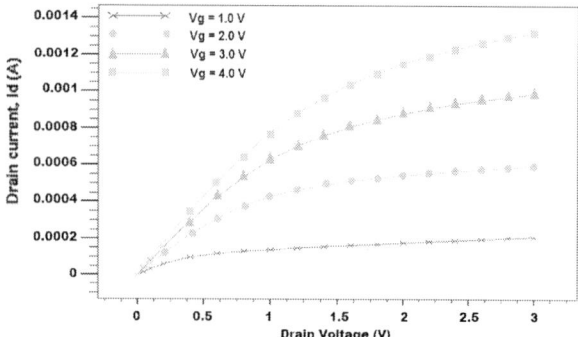

Fig. 4. Drain current (I_d) versus Drain voltage (V_d)

By using the information gathered from this simulation, all the parameters can be used for the next research in order to optimize best parameter to design the 22 nm transistor. In future work, Taguchi Method can be used as the analysis tools to find the optimum parameters of the device.

CONCLUSION

As a conclusion, downscaling to 22 nm gate length of the NMOS transistor was successful. The Halo implantation, Source/Drain implant dose, Compensation Implantation dose and V_t adjustment implantation dose are being used as the adjustment parameter. Our simulation shows that the optimal value of threshold voltage (V_{th}), leakage currents (I_{on} and I_{off}) was achieved according to specification in ITRS 2011.

ACKNOWLEDGMENT

The authors would like to thank to the Institute of Microengineering and Nanoelectronics (IMEN), Universiti Kebangsaan Malaysia (UKM), Centre for Micro and Nano Engineering (CeMNE), College of Engineering Universiti Tenaga Nasional (UNITEN), University Teknikal Malaysia Melaka (UTeM) and Ministry of Higher Education for the support, facilities and financial support throughout the project.

REFERENCES

[1] M.K. Bera and C.K. Maiti, "Reliability of UltraThin ZrO₂ Film on Strained-Si," Microelectronics Reliability 48, pp.682-692, 2008.

[2] Hei Wong and Hiroshi Iwai, "On the Scaling Issues and High-k Replacement of Ultrathin Gate Dielectrics for Nanoscale MOS Transistors," Microelectronic Engineering 83, pp.1867–1904, 2006.

[3] Shashank N, S Basak and R K Nahar, "Design and Simulation of Nano Scale High-k Based MOSFETs with Ploy Silicon and Metal Gate Electrodes," International Journal of Advancements in Technology, pp.252-261, 2010.

[4] J. Widiez, M. Vinet, B. Guillaumot, T. Poiroux, D. Lafond, P. Holliger, 0. Weber, V. Barral, B. Previtali, F. Martin, M. Mouis and S. Deleonibus, "Fully Depleted SOI MOSFETs with WSix Metal Gate on HfO2 gate Dielectric," IEEE International SOI Conference Proceedings, pp.161-162, 2006.

[5] R. Gassilloud, F. Martin, C. Leroux, M. Hopstaken, X. Garros, M. Cassé, Gilles Reimbold, Thierry Billon, Daniel Bensahel, "MOCVD Fluorine Free WSix Metal Gate Electrode on High-k Dielectric for NMOS Technology," Microelectronic Engineering 86, pp.263–267, 2009.

[6] H.A.Elgomati, B.Y.Majlis, I.Ahmad, F.Salehuddin, F.A.Hamid, T.Z. Mohamad, P.R.Apte, "Statistical Optimization for Process Parameters to Reduce Variability of 32nm PMOS Transistor Threshold Voltage", International Journal of the Physical Sciences Vol. 6(10), pp. 2372-2379, 2011.

[7] Andrew Marshall, Circuit design with FinFET processes, July 2006 http://www.cmoset.com/uploads/Marshall.pdf

[8] S.M Sze, Semiconductor Devices Physics and Technology, Second Edition , John Wiley & Sons, USA, 2002, pp.199-204.

[9] ITRS, 2011, *www.ITRS2011.net*.

[10] F.Salehuddin, I. Ahmad, F.A. Hamid and A. Zaharim, "Characterization and Optimizations of Silicide Thickness in 45nm pMOS Device," ICEDSA. 2010.

[11] H.A. Elgomati, B.Y.Majlis, I. Ahmad, F. Salehuddin, F.A. Hamid, A. Zaharim, T. Ziad Mohamad and P.R.Apte, "Investigation Of The Effect for 32nm PMOS Transistor and Optimizing Using Taguchi Method," Asian Journal of Applied Sciences, 2011.

[12] J.W. Sleight, I Lauer, O Dokumaci, D M Fried, D Guo, B Haran, S Narasimha, C Sheraw, D Singh, M Steigerwalt, X Wang, P Oldiges, D Sadana, C Y Sung, W Haensch, M Khare, "Challenges and Opportunities for High Performance 32nm CMOS Technology," IEDM Tech Digital, 2006.

[13] M.K. Bera and C.K. Maiti, "Electrical Properties of Sio₂/Tio₂ High-K Gate Dielectric Stack," Materials Science in Semiconductor Processing, pp.909–917, 2006.

[14] L. Pereira, P. Barquinha, E. Fortunato and R. Martins, "Low Temperature High K Dielectric on Poly-Si Tfts," Journal of Non-Crystalline Solids 354 pp. 2534–2537. 2008.

[15] J. Laeng, Zahid A. Khan and S.Y. Khu, "Optimizing Flexible Behaviour of Bow Prototype Using Taguchi Approach", Journal of Applied Sciences 6 (3):pp.622-630, 2006.

[16] Sarcona G.T., M. Stewart, and M.K Hatalis, "Polysilicon Thin Film Transistor Using Self_Aligned Cobalt and Nickel Silicide Source and Drain Contacts", IEEE Electron Device lett, vol.20, 1999

[17] Hashim, U., "Statistical Design of Ultra-Thin SiO2 for Nanodevices", Sains Malaysia 38(4), pp.553-557, 2009

[18] H.A.Elgomati, B.Y.Majlis, I.Ahmad, F.Salehuddin, F.A.Hamid, A.Zaharim and P.R.Apte, "Application of Taguchi Method in the Optimization of Process Variation for 32nm CMOS Technology," Australian Journal of Basic and Applied Sciences, 5(7), pp.346-355, 2011.

978-1-4673-2395-6/12 $31.00 © 2012 IEEE

Enhanced Performance Analysis of Vertical Strained-SiGeImpact Ionization MOSFET (VESIMOS)

Ismail Saad[1*], DivyaPogaku[1], Abu Bakar AR[1], Mohd Zuhir H.[1], N. Bolong[1], Khairul A.M[1], Bablu Ghosh[1], Razali Ismail[2], U. Hashim[3]

[1] Nano Engineering & Material (NEMs) Research Group, School of Engineering & IT, Universiti Malaysia Sabah, 88999, Kota Kinabalu, Sabah
[2] Computational Nanoelectronics (CONE) Research Group, Faculty of Electrical Engineering, UniversitiTeknologi Malaysia, Skudai 81310, Malaysia
[3] Institute of Nano Electronic Engineering (INEE), Universiti Malaysia Perlis (UniMAP), 01000 Kangar, Perlis, Malaysia
[*]email: ismail_s@ums.edu.my / ismailsaad07@gmail.com

Abstract—The Vertical Strained Silicon Germanium (SiGe) Impact Ionization MOSFET (VESIMOS) has been successfully developed and analyzed in this paper. VESIMOS device integrates vertical structure concept of Impact Ionization MOSFET (IMOS) and strained technology. The transfer characteristics of VESIMOS revealed an inverse proportionality of supply voltage, V_D and sub-threshold, S due to lower breakdown strength of Ge content. However, the Sis in direct proportion to the leakage current. The S=10mV/dec was successfully obtained at threshold voltage, V_T=0.9V, with V_D=1.75V. This V_T is 40% lower than $V_T$for Si-vertical IMOS.The output characteristics goes into saturation for $V_D$more than 2.5V, attributed to the presence of Ge that has high and symmetric impact ionization rates. Electron mobility wasimproved by 40% compared to Si-vertical IMOS and an increase in strain will also increase mobility and reduce further the V_T. However, the increase in strain layer thickness, T_{SiGe}, resulted in an increase of $V_T$and lowered the mobility. This is due to the strain relaxation in the SiGe layer. Finally, at high source-drain doping concentration, S/D=2×10^{18}/cm^3, the $V_T$dropped to 0.88V, with $V_D$of 1.75V. This is due to high electric field effect in the channel at high doping concentration, which is contrary to the doping effects of conventional MOSFET.

I. INTRODUCTION

Continuous scaling of MOSFETs has leads inexorably toward fundamental physical limits. A severe difficulty is the limitation of the sub-threshold slope *S* [which is defined as*dVG/d*(log *ID*)] of a MOSFET. It is governed by kT/q and hasa theoretical limit of 60 mV/dec at room temperature [1]. Accordingly, the supply voltage V_D andthreshold voltage V_T had to be reduced significantly to reduce thedynamic power maintain the magnitude of the saturation currentand assure a reliable operation for the devices [2-4].This limit for the sub-threshold slope can be overcome by lateralimpactionization MOSFETs (IMOS) [5-10], which utilized drift mechanisms of carriers rather than diffusion. However, the IMOS needs a highsupply voltage and suffers from reliability problems due to hot carriers effect [11-13]. The vertical IMOS, which isa planar-doped barrier MOSFET with a floating body, has beenintroduced and investigated [14-18]. The deviceis capable

toreduce supply voltage, mitigating thehot electrons damages almost completely [14-16] and showscapability of working properly under high temperatures [17].This vertical IMOS is not based on avalanche breakdown like thelateral IMOS. Instead, the holes generated by impact ionizationcharge the floating p-body and cause a dynamic reduction of thethreshold voltage, which leads to an extremely fast rising draincurrent in the sub-threshold region. The vertical IMOS, however, has also limitations including aremarkable hysteresis and still high V_D.In [19-22] a vertical IMOS with a strained SiGe layerwas introduced and investigated. The simulation results showedthat a reduction of both threshold voltage and supply voltagewas obtained by integrating such a strained SiGe layer.

This paper present an enhanced analysis performance of Vertical Strained-SiGe Impact Ionization MOSFET (VESIMOS). The correlation between supply voltage, sub-threshold, leakage current and saturation currentwas revealed based on device transfer characteristics and output characteristics. Enhancement of electron mobility and strain composition effects towards threshold voltage was successfully obtained. Finally, the effect of strain layer thickness,T_{SiGe} and doping concentration had merge to give enhance performance of VESIMOS as a potential candidate for future nanoelectronics device.

II. DEVICE SIMULATION

Figure 1 shows the detailed cross-sections of the device structure which simulated for performance analysis of the Vertical strained-SiGe Impact Ionization MOSFET (VESIMOS) using Silvaco's package [23]. Thisstructure comprises a source and a drain region with n+ doping,an intrinsic channel containing a highly doped dp+ layer and twosided gates. The high doping of the delta layer, which creates alarge potential barrier, makes it possible to achieve high electricfields in the intrinsic zone near the drain without applying a veryhigh drain–source voltage [20].Careful selection of delta layer thicknessis recommended as to have good sub-threshold slopes. Hence, an optimum value of delta layer thickness and doping was chosen to get desired sub-

978-1-4673-2395-6/12 $31.00 © 2012 IEEE

threshold slopes. The intrinsic-Si, i-Si regions reduce the lateral electric field near source and the drain [24]. Therefore, an optimum thickness for i-Si region has been chosen, so that they could effectively reduce the lateral electric fields. Due to the presence of i-Si regions between highly doped S/D regions, the impurity scattering is also reduced.The strained layer thickness was 20nm. However, the strained layer thickness was varied to examine the device performance. The increase in strained layer thickness is related to the strain relaxation. Hence, varying this parameter is essential to understand the behavior of the device under strain.

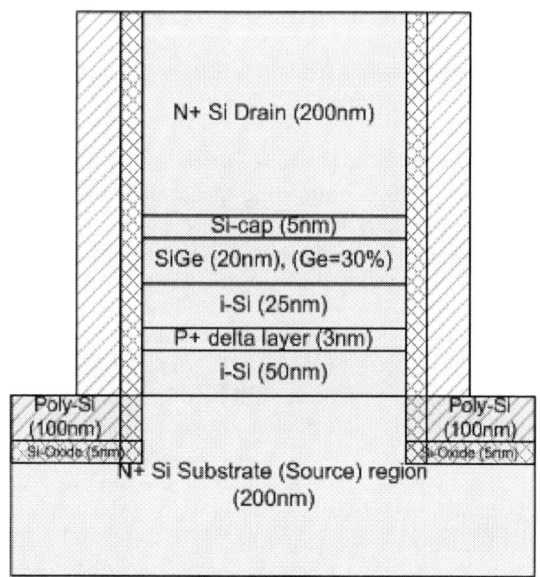

Fig. 1 VESIMOSdevice structure with dimensions, d=0.8μm, L=0.5μm, respective layer thickness of source, drain, δp⁺, i- Si, SiGe and Ge=30%

The Ge mole fraction is related to the amount of strain in the SiGe layer. The Ge mole fraction used was 30%, initially. Later, the mole fraction was varied from 10% to a maximum mole fraction of 50%. The lower the mole fraction, the lower is the strain. Hence, by varying the amount of strain, the device performance was analyzed.The Source was n-doped with Antimony with a doping concentration of 2.08×10^{18} /cm³. The Drain was also n-doped with Phosphorus with a doping concentration of 2.08×10^{18} /cm³. The high doped S/D doping was chosen, as the device concept was based on impact ionization. VESIMOS is an impact ionization device with drift current mechanism, which requires high electric fields. Both the drift current and the electric fields depend on the doping concentrations. Hence, high doping concentrations are imperative for obtaining better device characteristics.

The electrical characteristics of devices were done by solving Poisson's equation and continuity equation numerically and self-consistently within explicitly defined meshes of the devices [23]. The electrical potential energy and electronic band structures can be computed by using Poisson's equation. Continuity equations for electrons and holes are then used to

calculate the current densities of electrons and holes. Boltzmann transport framework is used in solving these two equations. The relationship between the current density of electrons/holes and carrier concentration as well as quasi-Fermi potential is exhibited during this self-consistent process. Thedrift–diffusion(DD)transportmodelwiththe Boltzmann carrier transportframeworkwasused,as it is ableto predict I–V characteristicsofDG-MOSFET [25]. Evenfornanoscalesize >10 nm, Granzneretal. [26] have shownthatfor DG-MOSFET's currentcharacteristics,theDDandMonte-Carlo simulation resultsproducedexcellentagreement,whileRen and Lundstorm [27] and RhewandLundstorm [28] have revealedthat the DDmodelcanpredict I–V characteristics ofshort-channel MOS devicesmorerealisticallythantheenergy-balance(EB) model.The Selberherr's[29] impact ionization modelwasemployed, which is a local impact ionization model. Selberherr's model is recommended for most cases of device simulation.

III. DEVICE PERFORMANCE ANALYSIS

VESIMOS device works in three different modes: conventional MOSFET, Impact Ionization (II) and Bipolar (BJT) mode. The transfer characteristic is examined by biasing the drain voltage, V_D and ramping the gate voltage, V_G at defined bias steps.Figure 2 shows the I_D-V_G characteristics of the VESIMOS.

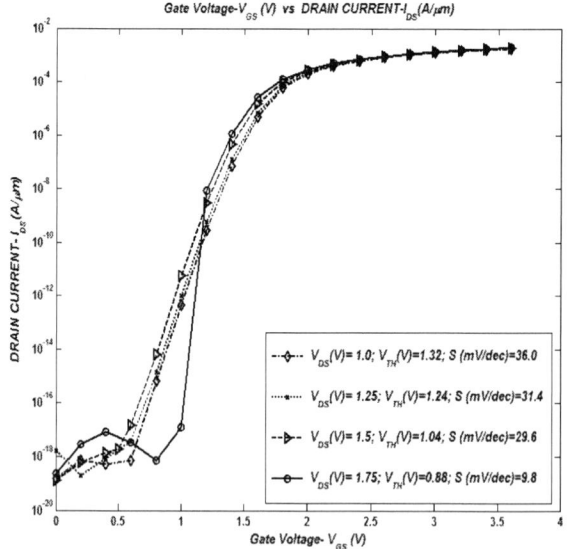

Fig.2 Transfer Characteristics, I_D-V_G of VESIMOS for $Si_{0.7}Ge_{0.3}$, S/D doping $=2.0 \times 10^{18}$/cm³.

Figure 2 shows an inverse proportionality between the supply voltage, V_D and the sub-threshold voltage, S.This is due to the electrons having enough energy to cross the potential barrier only at high supply voltage, which is V_D=1.75V. However, beyond this value, the device undergoes breakdown, due to lower breakdown strength of Ge. Hence, it can be concluded that VESIMOS works well for lower supply voltage, which has

978-1-4673-2395-6/12 $31.00 © 2012 IEEE

overcome the problem faced by conventional IMOS devices [14-16].As we know, sub-threshold voltage is in direct proportion to the leakage currents. Hence, lower sub-threshold voltage ensures low leakage currents. For VESIMOS, a good sub-threshold voltage, S=10mV/decade was obtained at threshold voltage of 0.9V, with a supply voltage of 1.75V. This $V_T$is found to be 40% lower than V_T for Si-vertical IMOS [14]. In addition, theS value is much lower than the conventional MOSFET limit due to the impact ionization mechanism of VESIMOS. The I_{ON}/I_{OFF} ratio observed approximately 10^{12}.

Table 1 VESIMOSV_{TH} and S at different modes of operation

Mode	V_{DS} (V)	V_{TH}(V)	S (mV/decade)
Conventional	≤1.25	1.24	31
Impact Ionization	>1.25	0.88	9.8
Bipolar	>1.80	-	-

Table 1 shows the threshold voltage and sub-threshold voltage in various modes of operation. It shows that for V_{DS}>1.80V, the device, instead of going to the bipolar mode, as in Si vertical I-MOS device, itundergoes breakdown. This was one of the limitations observed with VESIMOS device. The low breakdown voltages of this device can be attributed to the low bandgap of the Ge compared to other semiconductor materials [30]. It's also the inherent properties of Germanium (Ge), that it has a breakdown strength that is two times lower than the Si. Nevertheless, in planar doped barrier FET (PDBFET) with impact ionization devices, it was observed that for V_{DS}>1.5V, the device undergoes avalanche breakdown [31]. Based on this comparison, it can be seen that strained SiGe vertical I-MOS has a higher breakdown voltage.

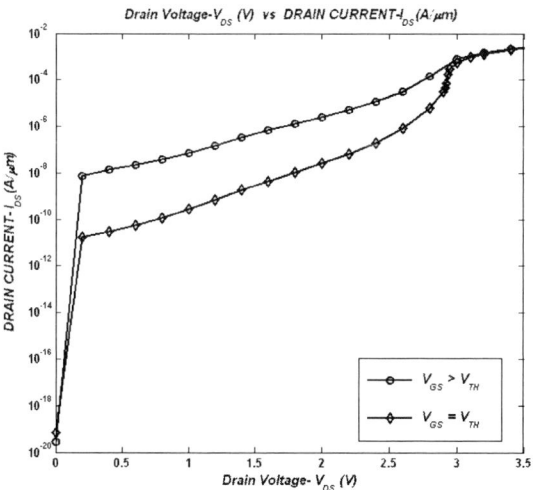

Fig.3 Output characteristics of VESIMOS device

Figure 3 shows the output characteristics of VESIMOS. It can be seen that the drain current rises sharply initially, and then increases gradually, before going into saturation for V_{DS}>

2.5V. The sharp rise in drain current can be attributed to the presence of Ge. Germanium has high and symmetric impact ionization rates ($\alpha_N \approx \alpha_P$), which ensures that the transition from OFF state to the ON state is abrupt [6]. Carrier mobilities in strained SiGe layer are different from that of Si [32] as a result of local distortion of band extreme due to strain effects, as well as due to the alloy scattering on the carriers. The mobilities in strained layer also depend on the transport direction, either parallel to the original SiGe growth interface or in the perpendicular direction [32]. In VESIMOS, on applying the bias, the electrons are transported towards the drain end and the holes are transported in the opposite direction towards the source end [21].

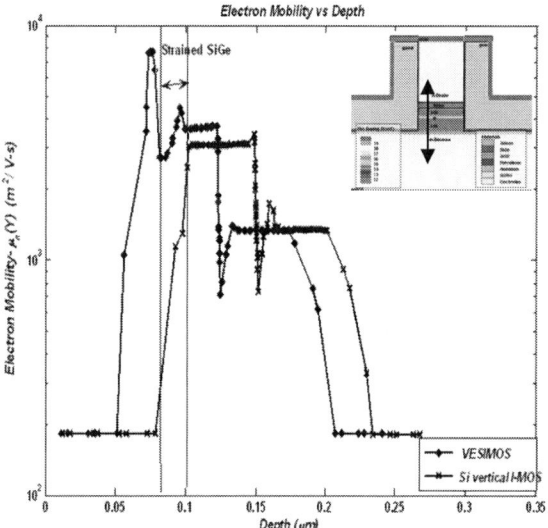

Fig. 4 Comparison of electron mobility profiles of VESIMOS and relaxed Si vertical I-MOS

A comparison between the electron mobility profiles, on application of bias, of both Si vertical I-MOS and VESIMOS were done as shown in figure 4. It can be observed that the electron mobility is higher in the strained SiGe layer. The electron mobility in the strained $Si_{0.7}Ge_{0.3}$layer ($\sim$4200m^2/V-s), was found to be increased by 40% in comparison to relaxed Si vertical IMOS (3000 m^2/V-s).Due to the splitting of the valleys in conduction band into lower four-fold and higher two fold states, the electron mobility becomes dependent on the in-plane and out-of plane directions. In the in-plane direction (in the plane of growth), the heavy longitudinal electron mass leads to lower electron mobility, while, in the out-of plane direction (out-of plane growth), the effective electron mass is reduced and hence, the mobility increases.

Figure 5 shows the reduction in threshold voltages with increasing strain for various drain biases. At high strain, the bandgap reduction is higher. Hence, low threshold voltage value is obtained, such that for Ge=50%, the threshold voltage is about 0.72V. In addition, the effect of strain on electron mobility was also analyzed, which revealed that, with increasing Ge content would increases the mobility of the electrons, linearly.Although with increasing strain would increases the mobility and reduces threshold voltage, it is not preferred to increase the strain much further, as it results in bandgap degradation [33].

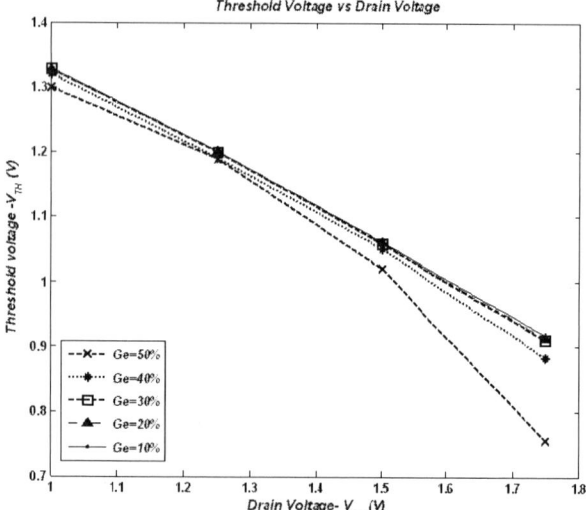

Fig. 5Variation of threshold voltage with Ge content

Figure 6 shows the variation of SiGe layer thickness on the threshold voltage. A threshold voltage increment of approximately 5-7% was observed, with increasing T_{SiGe} layer.This can be explained by the degree of strain relaxation (α)as a result of increasing thickness defined by equation 1.

$$\alpha = \frac{a_{SiGe(x)}(strained) - a_{Si}}{a_{SiGe(x)}(fullyrelaxed) - a_{Si}} \times 100 \quad (1)$$

whereas$_{SiGe(x)}$ is an average in-plane lattice constant of the strained epi SiGe layer, a_{Si} is the lattice constant of the substrate and $a_{SiGe(x)}$ is the average lattice constant of a fully relaxed SiGe layer. Thus, it shows that smaller $\alpha$is obtained by increasing the compressive strain in the SiGe layer.

Figure 7 show the threshold voltage decreases with increasing doping concentration. This is in contrary to the doping effects of MOSFET, where the threshold voltage increases with increasing doping concentration. Hence, the I-MOS devices have this unique characteristic of showing reduced threshold voltages with increase in doping concentration. This effect was observed as a consequence of high electric field in the channel region [10], at high doping concentrations and hence, aids in reducing the threshold voltages. All simulated data was validated with experimental data in order to verify the accuracy of the simulation work. As no experimental data available for Vertical SiGe IMOS, the vertical Si IMOS data was used [15]. It can be observed that the simulated data is synchronous with the experimental data and is in close approximation of 0 to 5% range as shown in figure 8.

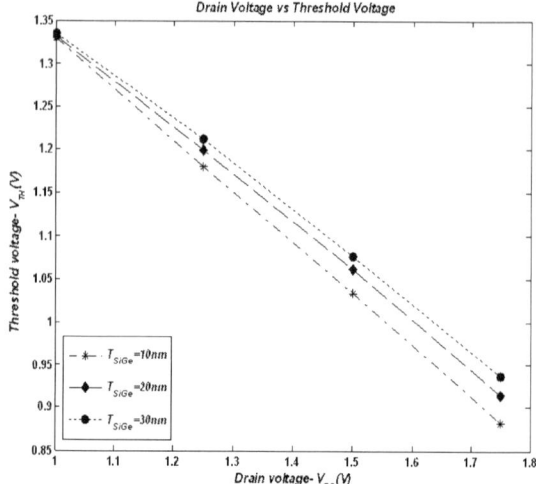

Fig. 6Variation of threshold voltage with Si$_{0.7}$Ge$_{0.3}$layer thickness

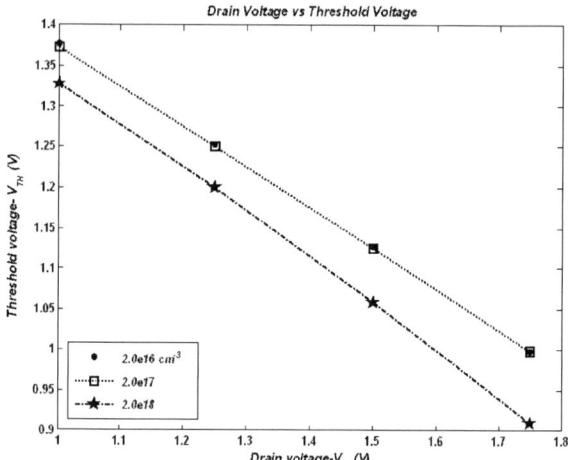

Fig. 7Variation of threshold voltage with Ge content

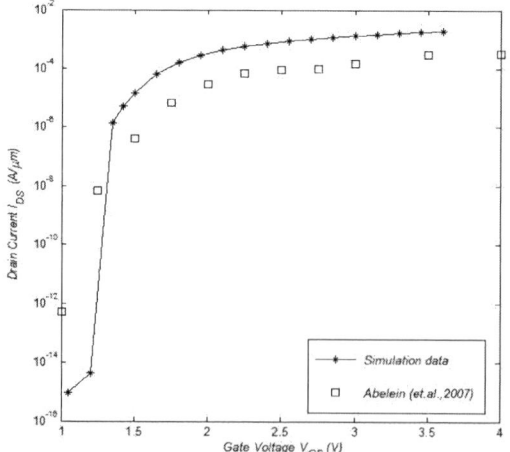

Fig. 8Validation of simulated data with experimental data (Abelein*et.al.*,2007) of Si vertical I-MOS device.

IV. Conclusion

The superior performances of Vertical Strained-SiGe Impact Ionization MOSFET (VESIMOS) were successfully analyzed. VESIMOS device integrate vertical concept of IMOS device with strained SiGe technology. The electrical characteristics of the VESIMOS by varying the amount of strain, the strained layer thickness, and the S/D doping concentrations were obtained using Silvaco TCAD tools. Moreover, 40% improvementof electron mobility was revealed, besides an improvement in obtaining lower threshold voltage and supply voltage. The threshold voltage is better for VESIMOS (V_{TH}=0.88V) than the vertical Si IMOS and strained SiGe IMOS. The sub-threshold voltage (S=9.8 mV/dec) in VESIMOS is also lower than the Si-vertical IMOS (S=20mV/dec). The VESIMOS also shows good I_{on}/I_{off} ratio of 10^{12}.Based on these results, VESIMOS was projectedas the most prominent candidate for nanoelectronics device.

Acknowledgement

The authors would like to acknowledge the financial support from ERGS (ERGS0002-TK-1/2011) fund of MOHE andE-Science fund (03-01-10-SF0175) of MOSTI Malaysia. The author is thankful to the UniversitiMalaysia Sabah (UMS) for providing excellent research environment in which to complete this work.

References

[1] International Roadmap Committee.The international technology roadmap for semiconductors.<public.itrs.net>.

[2] Morifuji E, Yoshida T, Kanda M, Matsuda S, Yamada S, Matsuoka F. Supply andthreshold-voltage trends for scaled logic and SRAM MOSFETs. IEEE TransElectron Dev 2006;53:1427–32.

[3] Khakifirooz A, Antoniadis DA. MOSFET performance scaling – part I: historicaltrends. IEEE Trans Electron Dev 2008;55:1391–400.

[4] Khakifirooz A, Antoniadis DA. MOSFET performance scaling – part II: futuredirections. IEEE Trans Electron Dev 2008;55:1401–8.

[5] Gopalakrishnan K, Griffin PB, Plummer JD.I-MOS: a novel semiconductor device with a substhreshold slope lower than kt/q.In: IEEE tech dig IEDM; 2002. p. 289–92.

[6] Gopalakrishnan K, Griffin PB, Plummer JD. Impact ionization MOS (I-MOS) –part I: device and circuit simulations. IEEE Trans Electron Dev2005;52(1):69–76.

[7] Gopalakrishnan K, Woo R, Jungemann C, Griffin PB, Plummer JD. Impact ionization MOS (I-MOS) – part II: experimental results. IEEE Trans Electron Dev 2005;52(1):77–84.

[8] Choi WY, Song JY, Lee JD, Park YJ, Park B-G. 100-nm n-/p-channel I-MOS using a novel self-aligned structure. IEEE Electron DevLett 2005;26(4):261–3.

[9] Choi WY, Song JY, Lee JD, Park YJ, Park B-G. 70-nm impact-ionization metal–oxide–semiconductor (I-MOS) devices integrated with tunneling field-effect transistors (TFETs). In: IEEE tech dig IEDM; 2005. p. 955–8.

[10] Choi WY. Effect of device parameters on the breakdown voltage of impact ionization metal–oxide–semiconductor devices.Jpn J ApplPhys, 2009;48:040203-1–3-3.

[11] Charbuillet C, Monfray S, Dubois E, Bouillon P, Judong F, Skotnicki T. Highcurrent drive in ultra-short impact-ionization MOS (I-MOS) devices. In: IEDM;2006. p. 1–4.

[12] Onal C, Woo R, Koh H-Y, Griffin P, Plummer J. A novel depletion-IMOS (DIMOS)device with improved reliability and reduced operating voltage. IEEE ElectronDevLett 2009;30(1):64–7.

[13] Savio A, Monfray S, Charbuillet C, Skotnicki T. On the limitations of silicon for IMOSintegration. IEEE Trans Electron Dev 2009;56(5):1110–7.

[14] Abelein U, Born M, Bhuwalka K, Schindler M, Schmidt M, Sulima T, et al. Anovel vertical impact ionization MOSFET (I-MOS) concept.In: Proc 25thinternational conference on microelectronics (MIEL '2006), Niš, Serbia; 2006.p. 127–9.

[15] Abelein U, Born M, Bhuwalka K, Schindler M, Schlosser M, Sulima T, et al.Improved reliability by reduction of hot-electron damage in the verticalimpact-ionization MOSFET (I-MOS). IEEE Electron DevLett 2007;28(1):65–7.

[16] Abelein U, Assmuth A, Iskra P, Schindler M, Sulima T, Eisele I. Doping profiledependence of the vertical impact ionization MOSFET's (I-MOS) performance.Solid-State Electron 2007;51:1405–11.

[17] Abelein U, Assmuth A, Iskra P, Reinl M, Schlosser M, Sulima T et al. Vertical40 nm impact ionization MOSFET (I-MOS) for high temperature applications.In: Proc 26th international conference on microelectronics (MIEL '2008), Niš,Serbia; 2008. p. 287–90.

[18] Kraus R, Jungemann C. Investigation of the vertical IMOS-transistor by devicesimulation. In: Proc ULIS, international conference on ultimate integration ofsilicon, vol. 10; 2009. p. 281–4

[19] Dinh TV, Kraus R, Jungemann C. Investigation of the performance of strained-SiGe vertical IMOS-transistors. In: Proc ESSDERC; 2009. p. 165–8.

[20] Dinh TV, Kraus R, Jungemann C. Investigation of the performance of strained-SiGe vertical IMOS-transistors.Solid-State Electronics 54 (2010) p. 942–949.

[21] P.Divya andIsmail Saad, Feasibility study of Integrated Vertical and Lateral IMOS and TFET devices for Nano-scale Transistors. IEEE International Conference On Semiconductor Electronics (ICSE 2010), ICSE2010 Proc. 2010.

[22] DivyaRavindra, Ismail Saad. Effects of S/D doping concentrations on strained SiGe vertical I-MOS device characteristics, ICECT2011, Kanyakumari, India, DOI: 978-1-4244-8679-3/11/$26.00 ©2011.

[23] Atlas and Athena user Manual Device and Process Simulation Software, Silvaco International, 2005.

[24] Scheinert, S., Paasch, G., Kittler, M., Nuernbergk, D., Mau, H., Schwiers, F. (1998). Requirements and Restrictions in Optimizing Homogenous and Planar Doped Barrier Vertical MOSFETs,International Caracas Conference on Devices, Circuits and Systems, ICCDCS-1998, Isla de Margarita.

[25] N.D. Jankovic, G.A. Armstrong, Comparative analysis of the DC performance of DG MOSFETs on highly-doped and near-intrinsic silicon layers, Microelectronics. J. 35 (2004) 647–653.

[26] R. Granzner, et al., On the suitability of DD and HD models for the simulation of nanometer double-gate MOSFETs, Physica E 19 (2003) 33–38.

[27] Z. Ren, M. Lundstrom, Simulation of nanoscale MOSFETs: a scattering theory interpretation, SuperlatticesMicrostruct. 27 (2000) 177–189.

[28] J. Rhew, M. Lundstrom, Drift–diffusion equation for ballistic transport in nanoscale metal-oxide-semiconductor field effect transistors, J. Appl. Phys. 92 (2002) 5196.

[29] Selberherr S., "Analysis and Simulation of Semiconductor Devices", Springer-Verlag, Wien-New York. 1984.

[30] Zoolfakar, A.S., and Ahmad, A., 2009. Hole mobility enhancement using Strained Si, SiGe technology. *5th International Colloquium on Signal Processing & Its Applications (CSPA)*, 4-6 March.

[31] Born, M., Abelein, U., Bhuwalka, K.K., Schindler, M., Schmidt, M., Ludsteck, A., Schulze, J., Eisele, I. 2005. Sub-50 nm high performance PDBFET with impact ionization, *International Conference on Silicon Epitaxy and Heterostructures- ICSI-4*, Japan, 23-26 May.

[32] Cressler John D., GuofuNiu, 2002. Silicon- Germanium Heterojunction Bipolar Transistors.Artech House Publishers. Boston, London. pp: 42-43.

[33] Kuhn, K.J., Murthy, A., Kotlyar, R., Kuhn M., 2010.Past, Present and Future: SiGe and CMOS transistor scaling, Electrochemical Society, Intel Corporation.

Modulus and Thermal Properties of Free Standing PMMA/TiO₂ Nanocomposite Films

N.N. Hafizah[1, 2,*], L.N. Ismail[3] and M. Rusop[1, 3]

[1]NANO-SciTech Centre (NST), Institute of Science (IOS);
[2]Faculty of Applied Sciences
[3]NANO-ElecTronic Centre (NET), Faculty of Electrical Engineering;
Universiti Teknologi MARA (UiTM), 40450 Shah Alam, Selangor, Malaysia
Email: *n.noor_hafizah@yahoo.com

Abstract—**TiO₂ nanopowder (0-20 w%) has been introduced in the PMMA matrix using the single step of sonication method to produce the PMMA/TiO₂ nanocomposite films in the free standing form. The PMMA and TiO₂ solution were mixed in the sonication bath. The effects on Tg, thermal degradation and also modulus of the PMMA/TiO₂ nanocomposites films were investigated. The properties of the nanocomposite films were depending on the dispersion of TiO₂ nanopowder in the PMMA matrix. The results of differential scanning calorimetry (DSC), thermo gravimetric analysis (TGA), and dynamic mechanical analysis (DMTA) were discussed in this paper.**

Keywords;PMMA/TiO₂ nanocomposites; free standing films; thermal; modulus.

I. INTRODUCTION

Composite materials have been widely used in our daily life applications for many years. Polymer composite systems have largely scale for industry and research field due to many advantages such as light weight, design flexibility and also processability [1-2]. The composite also has been developed for chemical, physical and also mechanical functions so their purpose becomes much larger. Because of the improvements on these composite materials, the industries are growing on automotive, aerospace, electronics and biotechnology [3-4].

In composites, two materials are mixed to make the improvement in the mechanical properties. Many new composite materials have potential to combine to obtain the desired properties of the composite system. The properties is depends not only upon the properties of the individual component phases, but also upon their interaction, polymer phase morphology and interfacial properties with the other component [5-6].

Poly (methyl methacrylate) (PMMA) has been widely used in the preparation of the nanocomposite to form the polymer nanocomposite. It has been broadly used in optical devices due to its excellent optical properties and processability [7-8]. However, its thermal and mechanical properties have limited applications. The achievable solution is by addition of nanoscale silica or titania particles to the polymer [5,7,9]. Polymer nanocomposite has widespread application especially in special engineering materials such as aerospace industry, automotive and civil engineering structures due to their outstanding mechanical properties. Polymer composite

theories were well-known on properties of composite constituents, volume fraction, shape, matrix-inclusion interface [10].

In this study, polymer PMMA is used as a matrix and titanium dioxide (TiO₂) nanopowder act as a filler was mixed into the PMMA by using sonication method. Sonication method is commonly used in the area of nanotechnology for evenly to disperse the nanoparticles in the liquids. In this method, an extra force comes from the mechanical shock waves directly exerted on the protrusions of the nanostructure. Cavitations are normally used to explain the creation of high pressure and heating temperature that produced during the sonication process. [11-12] The PMMA/TiO₂ nanocomposite films were characterized using differential scanning calorimetry (DSC), thermo gravimetric analysis (TGA), and dynamic mechanical analysis (DMTA) for their thermal and mechanical properties.

II. EXPERIMENTAL

In the preparation of free standing PMMA/TiO₂ nanocomposites film, the sonication method was used. PMMA/TiO₂ nanocomposite solution was prepared in two different beakers which are beaker A and B. Beaker A contains 10 ml toluene (ChemAR) solution was then added with 0.6 g PMMA (ALDRICH Chemicals) in powder form. Meanwhile in the beaker B that also contain 10 ml toluene solutions was added with 1 wt% of TiO₂ nanopowder (<100 nm, ALDRICH Chemicals). Both of the solutions were then sonicated in the sonication bath for 30 min. Solution A and B were then mixed together in the 250 ml beaker, and was sonicated for 1 hr to obtained the homogenous mixture. The mixture of the solution was then moved to solution casting process where the solution was pour into the Petri dish in a fume hood for 24 hr to evaporate the solvent. The nancomposite film in free standing form were named as T-0, T-1, T-5, T-10, T-15 and T-20 for 0, 1, 5, 10, 15 and 20 wt% TiO₂ nanopowder respectively. The samples were characterized using differential scanning calorimetric (DSC), thermo gravimetric analysis (TGA), and dynamic mechanical analysis (DMTA) and the results were discussed.

III. RESULTS AND DISCUSSIONS

We make use of DSC to measure the Tg of our samples. Tg or glass transition temperature is the temperature at which the polymer molecules undergo relaxation from a glassy state to rubbery state. The value of glass transition temperature was taken as the midpoint of the glass transition level. The thermo scan of DSC results for free standing PMMA/TiO$_2$ nanocomposite films were sketched. The Fig. 1 and the tabulated data were listed in the TABLE 1.

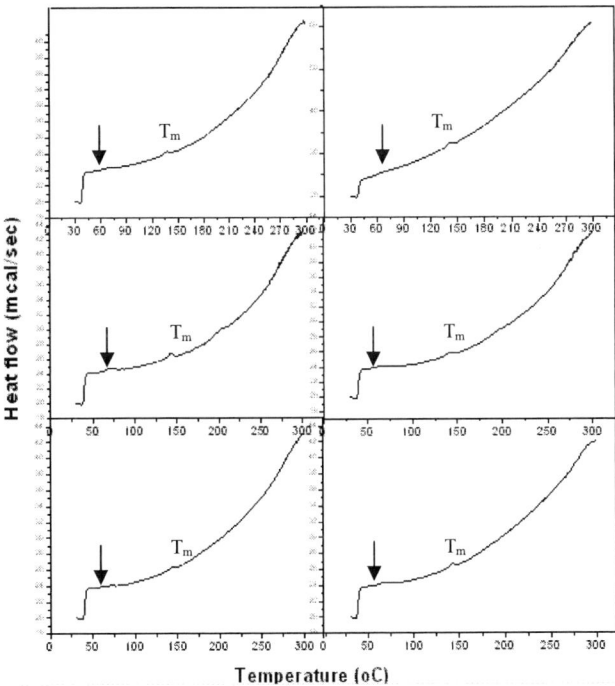

Figure 1: DSC thermograms of the sample (a) T-0, (b) T-1, (c) T-5, (d) T-10, (e) T-15 and (f) T-20

From the Figure 1, the results obtained are unexpected. The Tg obtained is low for all the samples. It can be found that the sample of T-0 that contains 0 w% TiO$_2$ nanopowder has a single Tg at 59.43 °C. However, when the TiO$_2$ nanopowder was added in the PMMA, there is little increase in the Tg reading, but the value is still low compared with the previous study obtained. The storage and loss modulus, Tg, pendulum hardness and activation energy of the thermal degradation of the nanocomposites supposed to be increased with the increasing TiO$_2$ content in the PMMA-TiO$_2$ nanocomposites, as reported in the previous study [13]. From the Fig. 1, the Tg for the sample T-5 exhibit the single Tg but with little different of glass transition of 66.73 °C. Meanwhile, the Tg for the sample T-10 has little increase with the reading of 67.5 °C and Tg for the sample T-20 and T-25 is 66.2 °C and 61.4 °C respectively. The results indicate that the phase separation is still existed in the nanocomposite samples. Also could be seen from the DSC thermograms results that the Tg for the free standing PMMA/TiO$_2$ nanocomposite materials was higher than that of the neat PMMA, which indicated that the incorporation of the filler due to existence of the strong interaction s between the TiO$_2$ filler with the polymer matrix. These strong interactions limit the movement of the polymer chains segments due to increase rigidity structure of the polymer [14-15].

The DSC was measured from room temperature of 27 to 300 °C. The sample was start heating at the room temperature of 27 °C. After a certain temperature, the graph was shifted upwards suddenly. At this time, the sample getting more heat flow which means that the heat capacity of the polymer increases. This happen because the sample has just gone through the glass transition. The changes in the heat capacity occur at the glass transition. Therefore, the DSC reading shows the Tg of the sample of 59.4 °C for the PMMA sample which is labeled with an arrow as shown in the Fig. 1. When more heat is applied to the sample, the sample has reached the melting temperature of Tm, and the polymer crystal in the films begin to fall apart, that is they melt. At this time, the polymer crystals melt and absorb heat.

TABLE 1: Data of Tg for free standing PMMA/TiO$_2$ nanocomposite films

Sample Name	TiO$_2$ Weight Percentage (w %)	Tg (°C)
T-0	0	59.4
T-1	1	66.7
T-5	5	67.5
T-10	10	60.4
T-15	15	66.2
T-20	20	61.4

From the Tg obtained, there is not much different in the values of Tg for each sample. Enis et al in their study also observed that Tg of PMMA/TiO$_2$ nanocomposites decrease approximately 20 °C. In the study, they synthesized the PMMA/TiO$_2$ nanocomposites by using 6-PAA. They reported that the reducing of Tg is due to the palmitoyl chains in TiO$_2$/6-PAA nanoparticles that act as a plasticizer and hence, reducing the Tg of the nanocomposite [16]. Meanwhile, some researcher suggested that the decrease of Tg was associated with the significant formation of the agglomerates and poor interaction. The decrease of the Tg also was attributed to the increased in the length of the polymer chains that grown from the nanoparticles which is due to increased polymerization time [17]. Meanwhile, some researchers reported that the addition of inorganic nanoparticles clearly leads to increase in Tg [18-22].

TGA was used to determine the thermal behavior of the PMMA/TiO$_2$ nanocomposite samples. The influence of nanocomposites on thermal behavior was studied by comparing the degradation stability with the synthesized reference of PMMA. The corresponding thermograms are illustrated in Fig. 4. The thermograms show that PMMA thermally decomposes by depolymerization [23]. In the previous study, the TGA curve for pure PMMA is lower than the curve for PMMA nanocomposite sample [24-27]. In the sample of PMMA/TiO$_2$ nnanocomposite films, the thermal degradation occurs in three stages which are at about 120, 220 and 300 °C for the sample T-5, T-10, T-15 and T-20 and the additional small stage at around 380-430 °C was observed in the T-1 sample . The

PMMA nanocomposites show slightly less thermal stability at 120 °C which is probably due to water evaporating. Kashiwagi et al synthesized the PMMA nanocomposites by in situ radical polymerization of methylmethacrylate (MMA), with colloidal silica (ca. 12 nm) to study the effects of nanoscale silica particles on the physical properties and flammability properties of PMMA nanocomposite. They found that the less thermal stability at 200 °C was detected [28]. The change is also considered to be the disappearance of the low molecular weight of polymer.

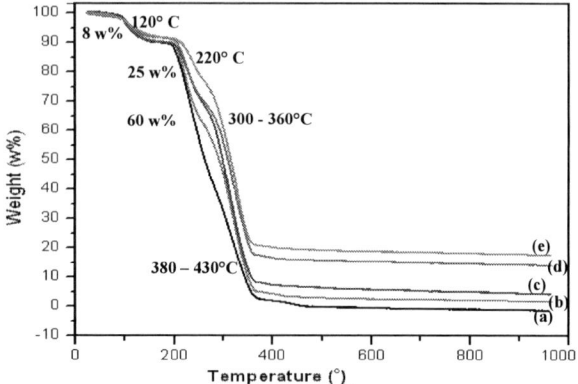

Figure 2: TGA thermograms of the sample (a) T-0, (b) T-1, (c) T-5, (d) T-10, (e) T-15 and (f) T-20

From the TGA thermograms, the most intense peak was detected for all samples is approximately at 300 °C. Sharma et al reported, the most intense peak is observed at around this stage is due to depolymerization initiated by the random chain scission [29]. Meanwhile, others have reported that the stage at approximately between 300–450 °C is corresponding to the decomposition of the main chain of PMMA [28]. Above 500°C, there is no residue remains.

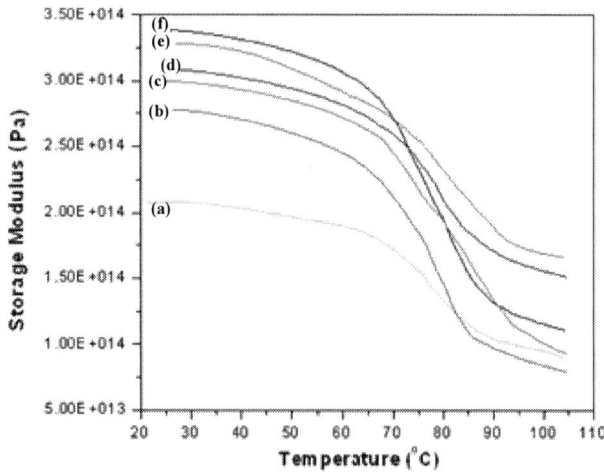

Figure 3: Storage modulus curve of the sample (a) T-0, (b) T-1, (c) T-5, (d) T-10, (e) T-15 and (f) T-20

Further, DMA was used to measure the viscoelastic and mechanical properties of the polymer nanocomposites. Fig. 3

shows the temperature versus storage modulus of the samples. The E' versus T curve of the sample T-0 that contain 0 w% TiO_2 shows the value of the sample has the lowest initial value at around 2.00×10^{14}. Meanwhile, sample of T-1, T-5, T-10, T-15 and T-20 show the higher initial value than T-0 which is at around 2.55×10^{14} to 3.50×10^{14}. Note that the storage modulus of PMMA was increased significantly by the addition of TiO_2 nanoparticles in the polymer.

IV. CONCLUSION

The $PMMA/TiO_2$ nanocomposite films were successfully synthesized using sonication method. The 5 w% TiO_2 nanopowder in the PMMA composite shows the highest value of Tg with 67.5 °C among other samples. The sample without TiO_2 shows low Tg with only 59.4 °C than the samples that contain TiO_2. The thermal and modulus value also show the increase in its value with the increase of TiO_2 content. The possible applications of using TiO_2 as filler in the nanocomposite are as flame retardant, semiconductor materials, electrical devices and also in photovoltaic application.

ACKNOWLEDGMENT

The authors would like to thank to NANO-ElecTronic centre (NET) for the laboratory facilities.

REFERENCES

[1] J.H. Park, and S.C. Jana, "The relationship between nano- and micro-structures and mechanical properties in PMMA-epoxy-nanoclay composites," *Polymer*, vol. 44, pp.2091-2100, March 2003.

[2] C.L. Wu, M.Q. Zhang, M.Z. Rong and K. Friedrich, "Tensile performance improvement of low nanoparticles filled-polypropylene composites," *Composites Science and Technology*, vol. 62, pp.1327-1340, August 2002.

[3] V.M.F. Evora and A. Shukla, "Fabrication, characterization, and dynamic behavior of polyester/TiO_2 nanocomposites," *Materials Science and Engineering: A*, vol. 361, pp.358–366, November 2003.

[4] Y. Iwahori, S. Ishiwata, T. Sumizawa and T. Ishikawa, "Mechanical properties improvements in two-phase and three-phase composites using carbon nano-fiber dispersed resin," *Composites Part A: Applied Science and Manufacturing*, vol. 36, pp.1430-1439, October 2005.

[5] T. Kashiwagi, A.B. Morgan, J.M. Antonucci, M.R. Van Landingham, R.H. Harris, W.H Awad and J.R. Shields, "Thermal and flammability properties of a silicapoly(methyl methacrylate) nanocomposite," *Journal of Applied Polymer Science*, vol. 89, pp.2072-2078, August 2003.

[6] B. Morgan, J.M. Antonucci, M.R. van Landingham, R.H Harris and T. Kashiwagi, "Thermal and flammability properties of a silica-PMMA nanocomposite," *Polymer Materials Science and Engineering*, vol. 83, pp.57-58, 2003.

[7] Y.Y Yu, and C W.C. hen, "Transparent organic–inorganic hybrid thin films prepared from acrylic polymer and aqueous monodispersed colloidal silica," *Materials Chemistry and Physics*, in press, 2003.

[8] L.J. Bian, X.F. Qian, J. Yin, Z.K Zhu, and Q.H. Lu, "Preparation and luminescence properties of the $PMMA/SiO_2/EuL_3.2H_2O$ hybrids by a sol-gel method," *Materials Science and Engineering*, vol. 100, pp.53-58, 2003.

[9] Y. Y. Yu, C. Y. Chen, and W. C. Chen, "Synthesis and characterization of organic-inorganic hybrid thin films from poly(acrylic) and monodispersed colloidal silica," *Polymer*, vol. 44, vol. 3, pp. 593–601, December 2002.

[10] J Jordan, K.I Jacob, R Tannenbaum and M.A Sharaf, Jasiuk I, "Experimental trends in polymer nanocomposites: a review," *Materials Science and Engineering: A.*; vol. 393, pp.1-11, February 2005.

[11] Dojalisa Sahua, B.S. Acharyaa, B.P. Baga, Th. Basanta Singhb and R.K. Gartiac, "Probing the surface states in nano ZnO powder synthesized by sonication method: Photo and thermo-luminescence studies," *Journal of Luminescence,* vol. 130, pp.1371–1378, August 2010.

[12] Tao Gao, Qiuhong Li, and Taihong Wang, "Sonochemical Synthesis, Optical Properties, and Electrical Properties of Core/Shell-Type ZnO Nanorod/CdS Nanoparticle Composites," *Chemistry of Materials,* 2005, vol. 17, pp.887–892, January 2005.

[13] Z. Ling, L. Zhongshi, F.A. Wenjun, P. Tianyou, "A novel polymethyl methacrylate (PMMA)-TiO₂ nanocomposite and its thermal and photic stability," *Journal of Natural Sciences,* vol. 11, pp.415-418, 2006.

[14] C. Landry, B. Coltrain and B. Brady, "In situ Polymerization of Tetraethoxysilane in Poly(methyl methacrylate", *Morphology and Dynamic Mechanical Properties,* Polymer, vol. 33, pp.1486-1495, 1992.

[15] M. Mulder., "Basic Principles of Membrane Technology," *Kluwer Academic Publisher,* Netherland, 1991.

[16] D. Gersappe, "Molecular Mechanisms of Failure in Polymer Nanocomposites," *Physical Review Letters,* vol. 89, pp.1-4, 058301-1-4, 2002.

[17] L.M. Hamming, R. Qiao, P.B. Messersmith, L.C. Brinson, "Effects of dispersion and interfacial modification on the macroscale properties of TiO2 polymer–matrix nanocomposites," *Composites Science and Technology,* vol. 69, pp.1880-1886, 2009.

[18] Enis Dz unuzovic, Katarina Jeremic and Jovan M. Nedeljkovic, "In situ radical polymerization of methyl methacrylate in a solution of surface modified TiO₂ and nanoparticles," *European Polymer Journal,* vol. 43, pp.3719–3726, September 2007.

[19] M. Yang and Y. Dan, "Preparation and characterization of poly(methyl methacrylate)/titanium oxide composite particles," *Colloid Polym Sci,* vol. 284, pp.243–250, 2005.

[20] C.C. Weng and K.H. Wei, "Selective distribution of surface modified TiO₂ nanoparticles in polystyrene-b-poly(methyl methacrylate) diblock copolymer," *Chemistry of Materials,* vol. 15, pp.2936–2941, 2003.

[21] A. Laachachi, M. Cochez, M. Ferriol, J.M. Lopez-Cuesta and E. Leroy, "Influence of TiO₂ and Fe₂O₃ fillers on the thermal properties of poly(methyl methacrylate) (PMMA)," *Mater Lett,* vol. 59 , pp.36–39, August 2005.

[22] M. Marinovic-Cincovic, Z. V. Saponjic, V. Djokovic, S. K. Milonjic and J.M. Nedeljkovic, "The influence of hematite nanocrystals on the thermal stability of polystyrene," *Polymer Degradation and Stability,* vol. 91, pp.313–316, February 2006.

[23] Technology," *Kluwer Academic Publisher,* Netherland, 1991.

[24] McNeill IC, "A study of the thermal degradation of methyl methacrylate polymers and copolymers by thermal volatilization analysis," *European Polymer Journal,* vol. 4, pp.21–30, 1968.

[25] Amit Chatterjee, "Properties improvement of PMMA using nano TiO₂," *Journal of Applied Polymer Science,* vol. 118, pp.2890-2897, December 2010.

[26] Enis Dz unuzovic, Katarina Jeremic and Jovan M. Nedeljkovic, "In situ radical polymerization of methyl methacrylate in a solution of surface modified TiO₂ and nanoparticles," *Macromolecular Nanotechnology, European Polymer Journal,* vol. 43, pp.3719–3726, 2007.

[27] Jun Zhang, Shengcheng Luo and Linlin Gui, "Poly(methyl methacrylate)-titania hybrid materials by sol gel processing," *Journal of Material Science,* vol. 32, pp.1469-1472, Mac 1997.

[28] A. Laachachi, E. Leroy, M. Cochez, M. ferrol and J.M. Lopez cuesta, "Use of oxide nanoparticles and organoclay to improve thermal stability and fire retardancy of poly(methyl methacrylate)," *Polymer Degradation and Stability,* vol. 89, pp.344-352, August 2005.

[29] T. Kashiwagi, A.B. Morgan, J.M. Antonucci,; M.R. VanLandingham, R.H. Jr. Harris, W.H. Awad, J.R. Shields, "Thermal and flammability properties of a silica-poly(methylmethacrylate) nanocomposite," *Journal of Applied Polymer Science,* vol. 89, pp.2072-2078, 2003.

[30] R.K. Sharma, M.C. Bhatnagar and G.L. Sharma, "Mechanism in Nb doped titania oxygen gas sensor," *Sensors and Actuators B: Chemical,* vol. 46, pp.194–201, May 1998.

Comparison of Mechanical Deflection and Maximum Stress of 3C SiC- and Si-Based Pressure Sensor Diaphragms for Extreme Environment

[a]Noraini Marsi, [a]Burhanuddin Yeop Majlis, *SMIEEE*, [a]Azrul Azlan Hamzah and [b]Faisal Mohd-Yasin, *SMIEEE*

[a]Institute of Microengineering and Nanoelectronics (IMEN)
Universiti Kebangsaan Malaysia (UKM)
43600 UKM Bangi, Selangor, Malaysia
[b]Queensland Micro- and Nanotechnology Centre (QMNC)
Griffith University, 4111, Brisbane, QLD, Australia
Email: burhan@vlsi.eng.ukm.my

Abstract-The design of a capacitive-sensing pressure sensor for extreme environment is proposed in this project. The movable diaphragm (top plate) is made of either cubic silicon carbide (3C-SiC) or Silicon (Si), while the fix diaphragm (bottom plate) is made of Si. This paper specifically compares the mechanical performance of the movable diaphragm utilizing both materials. Two important parameters associated with the behavior of the diaphragm are examined, namely the maximum deflection and maximum stress, and they are simulated at a pressure of 0-100 MPa, and at temperature of 27-1000 °C. The graphs of maximum deflection and stress *vs* pressures at different temperatures and thicknesses are plotted to summarize the data. SiC diaphragm has lower deflection and stress compares to Si diaphragm at different thicknesses, pressures and temperatures. Then, a linear regression analysis is performed to determine the R-square value. It is shown from these analyses that SiC diaphragm exhibits better linear behavior compares to Si diaphragm. Generally, this work proves that SiC is a better material over Si for the development of a pressure sensor at extreme environment.

Keywords: silicon carbide (SiC), pressure sensor, harsh environment, diaphragm

I. INTRODUCTION

The operating temperature inside the gas turbine engine is typically greater than 300˚C [1]. However, the silicon-based MEMS pressure sensor could only operate below this point due to the limitation of its material properties. In such extreme environment, SiC is the best material to replace Si because it can operates up to 1000˚C [2]. In addition, SiC has more advantages at this environment due to its chemical inertness and corrosion resistance, and an extremely low coefficient of thermal expansion and high Young's Modulus. The latter is a good mechanical advantage because that makes SiC more immune to stress effects from deflection and thermal shock. Electrically speaking, SiC has an excellent properties compares to Si as well, namely a larger bandgap (2.3-3.4eV), a higher breakdown field (30×10^5 V/cm), a higher thermal conductivity (3.2 - 4.9 W/cm K), and a higher saturation velocity (2×10^7 cm/s) [3]. This paper will demonstrate the superiority of SiC material over Si as the pressure sensor diaphragm through a series of simulations.

II. THEORY AND DESIGN

A. Design of a capacitive pressure sensor

A structural model for the pressure sensor is shown in Figure 1. It consists of two components to make the parallel-plate structure; the first is the top diaphragm as the movable plate, and the second is a bottom substrate as a fixed plate. The pressure signal comes from the top and generates the stress-deformation of the movable diaphragm, which modulates the sealed air gap length. The corresponding differential capacitance value between the top and bottom plates is measured as the electrical output. In this design, the square shape diaphragm with the area of 400 μm² is employed. The thickness of the top movable diaphragm is simulated at the following points: 0.05μm, 0.1μm, 0.2μm, 0.3 μm, 0.5μm and 1.0μm to study the stress deformation of both SiC and Si membranes at different thicknesses.

TABLE I MATERIAL PROPERTIES OF TESTED MATERIALS [4]

Material properties	Silicon	SiC
Young's modulus (GPa)	1.69	4.70
Poisson ratio	0.30	0.22
Density (kg/μm³)	2.5×10^{-15}	3.2×10^{-15}
TCE Integral Form (1/K)	2.50×10^{-6}	2.30×10^{-6}
Thermal Conductivity (pW/umK)	$1.48 \times 10^{+8}$	$5.0 \times 10^{+8}$
Specific Heat (PJ/kgK)	7.12×10^{14}	1.34×10^{15}

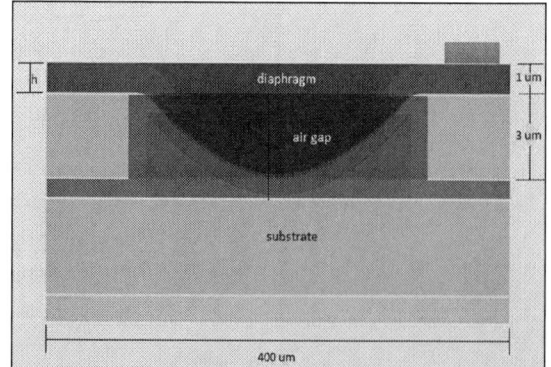

Fig. 1. Structural model for capacitive pressure sensor

B. Theory of operation of the square deflection and maximum stress

For a two dimensional diaphragm with a pressure load, p, the differential equation for the displacement of the diaphragm can be derived by analyzing the balance conditions for forces and bending moments in an elemental area of diaphragm, $dxdy$. The general equation for displacement $w(x, y)$ is found to be the differential equation of diaphragm for displacement is given as equation (1), where we measure the point at the center of the diaphragm due to pressure applied on its surface:

$$D\left[\frac{\partial^4 w}{\partial x^4} + 2\frac{\partial^4 w}{\partial x^2 \partial y^2} + \frac{\partial^4 w}{\partial y^4}\right] + h\rho \frac{\partial^2 w}{\partial t^2} = p(x, y) \quad (1)$$

This equation is based on time dependence, so it can be used for frequency analysis. D is the flexural rigidity, h is the diaphragm's thickness and ρ is density of the diaphragm material. If the pressure p is uniform then the steady displacement can be expressed as equation (2):

$$D\left[\frac{\partial^4 w}{\partial x^4} + 2\frac{\partial^4 w}{\partial x^2 \partial y^2} + \frac{\partial^4 w}{\partial y^4}\right] = p \quad (2)$$

For a square diaphragm with a side length of $2a$ as shown in Figure 2, the simplest expression of displacement for a pressure p can be expressed by equation (3) [5]:

$$w(x, y) = \frac{1}{47} p \frac{a^4}{D}\left(1 - \frac{x^2}{a^2}\right)^2 \left(1 - \frac{y^2}{a^2}\right)^2 \quad (3)$$

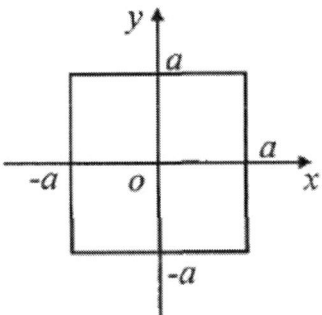

Fig. 2. Structure geometries of a square diaphragm

Since the deflection of the diaphragm is higher than its thickness, the strain in the middle plane of the diaphragm could be neglected. Thus the diaphragm stresses as well as the maximum stresses components for normal stress on surfaces perpendicular to x axis (σ_x) and normal stress on surfaces perpendicular to y axis (σ_y) were considered. The equations for diaphragm stresses are as follows equation (4) and (5):

$$\sigma_x = -\frac{E}{1-v^2}\left[\frac{\partial^2 w}{\partial x^2} + v\frac{\partial^2 w}{\partial y^2}\right] - \frac{E\alpha\Delta T}{1-v} \quad (4)$$

$$\sigma_y = -\frac{E}{1-v^2}\left[\frac{\partial^2 w}{\partial y^2} + v\frac{\partial^2 w}{\partial x^2}\right] - \frac{E\alpha\Delta T}{1-v} \quad (5)$$

Note that E is Young's modulus, v is Poisson's ratio of the diaphragm material, w is vertical deflection of the diaphragm, α equal to $\varepsilon/\Delta T$, where ε is strain and ΔT is temperature change with respect to reference temperature.

III. SIMULATION AND RESULTS

A. Simulation Settings

CoventorWare ver.2008 simulation software is used in this research to design, simulate and modify the performance of MEMS capacitive pressure sensor. It has three main components: Architect, Designer and Analyzer [6]. Architect is used to design the schematic of the capacitive pressure sensor. Designer is used to build the 3D design of the diaphragm. Analyzer is used to analyze the diaphragm deflection with given pressure and temperature.

Si and 3C-SiC materials with their respected electrical and mechanical properties are used to simulate the movable diaphragm (top plate). The anode (top plate) and cathode (bottom plate) separation is set to be 3 μm, and the range of applied pressures is between 0 to 100MPa. The design steps include selecting the substrate layer and wet etching of the backside of the substrate to create the membrane. The next process is mesh creation for the device that is essential to allow the Analyzer to do the pressure and temperature analysis. Both are performed using CoventorWare mechanical solver MemMech. Figure 3 and Figure 4 shows the diaphragm deformation and maximum stress under a series of different pressures, respectively.

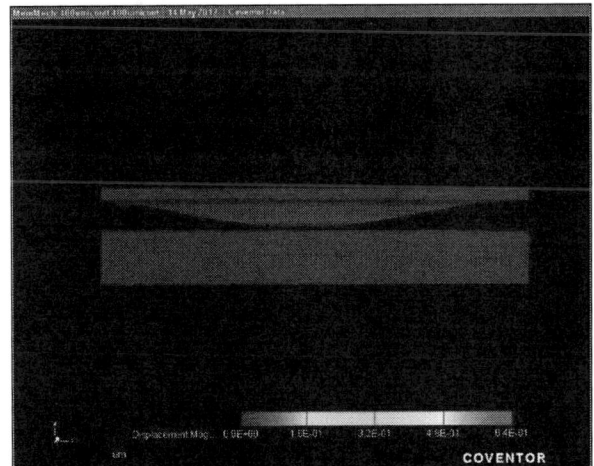

Fig. 3. Deflection of diaphragm by applied pressure

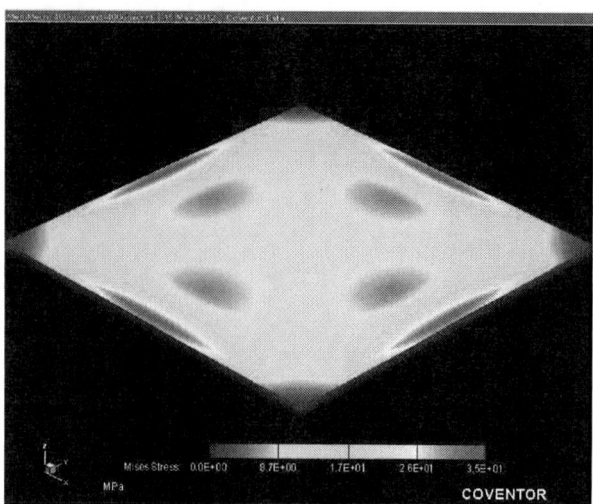

Fig. 4. Maximum stress of diaphragm by applied pressure

Figure 5 shows the 3D model of the diaphragm using mapped bricks mesh generated by Designer [7]. There are three steps involved. Firstly is the substrate step, secondly is the silicon planar fill with an anistropic front side wet etch, and thirdly Si/SiC top diaphragm by using stack material modeling action. The substrate layer material is silicon. The wet etch properties are 3 μm depth and 54.7° degree angle of the silicon substrate. The highest stress-deflection of the diaphragm for an applied pressure of 100MPa is at the center of the diaphragm. The resulting diaphragm stress-deflection at 3 μm considered as the touch-mode deflection. But in this paper, we are focuses on the deflection of the diaphragm only.

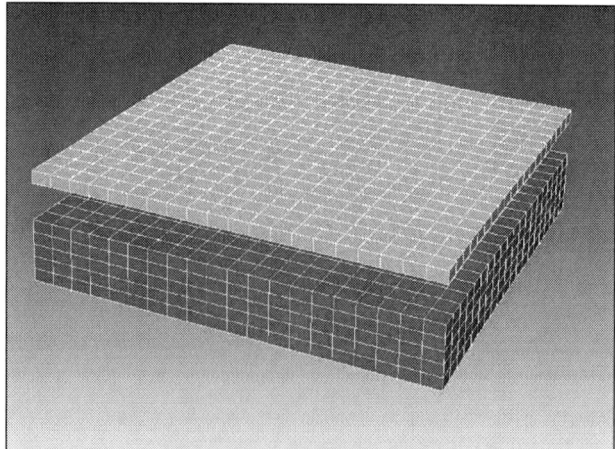

Fig. 5. 3D diaphragm model using mapped brick mesh with relevant boundary conditions.

B. Results and discussion

The capacitive pressure sensor is subjected to a range of temperatures namely 27°C, 300°C, 700°C and 1000°C. The reference temperature is at 27°C, where we assume that the diaphragm is stress free. It is also assumed that by heating the

diaphragm rapidly, the deflection is achieved without relaxation. Symmetric boundary conditions have been chosen for the planes at x = 200μm and y = 200μm, where the pressure is applied at the center of the diaphragm.

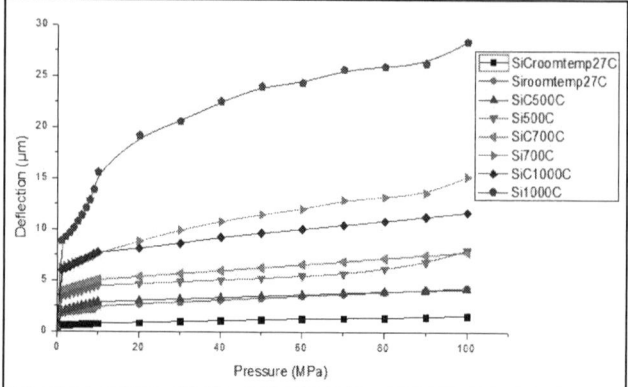

Fig. 6.Deflection at differential pressures and temperatures with the diaphragm thickness of 1μm

TABLE II:
LINEAR REGRESSION ANALYSIS FOR DEFLECTION RESULTS AT DIFFERENTIAL
TEMPERATURES WITH THICKNESS OF 1μM

Temperature (°C)	SiC		Si	
	Equation	R^2	Equation	R^2
27	y=0.008x + 0.74	1.00	y=0.020x + 2.30	1.00
500	y=0.014x + 2.77	1.00	y=0.033x + 3.81	0.87
700	y=0.03x + 4.78	1.00	y=0.075x + 7.44	0.98
1000	y=0.043x + 7.38	1.00	y=0.122x + 16.54	0.92

Figure 6 and Table II showed the deflection of the 1 μm diaphragm made of SiC and Si under a differential temperatures. At room temperature, the deflection at the center of the diaphragm for both SiC and Si increases linearity (R-squared value = 1.00) from 0-100MPa. It is observed that the elastic limit of both materials can sustain 100 MPa at room temperature.

However, this linear trend does not hold for Si at high temperatures. The deflection of Si diaphragm increases non-linearly for 500°C, 700°C and 1000°Cwith R-squared value at 0.87, 0.98 and 0.92, respectively. On the other hand, the deflection of the SiC diaphragm increases at high temperatures up to 1000°C, all with R-square value of 1.

The maximum deflection of the center of the diaphragm for SiC and silicon at 100MPa yields predicted result, where Si diaphragm deflects more compares to SiC diaphragm. At room temperature, the deflection for both Si and SiC diaphragm are 1.6 μm and 4.3 μm, respectively. At 1000°C, both Si/SiC diaphragms deflect 11.7μm/28.4μm, respectively. The Young modulus of SiC is approximately 3 times higher compares to Si, and hence, the SiC deflection is three time less [8]. It must also be pointed out that the silicon diaphragm shows non-linear deflection at high temperatures because of the stress due the stretching of the diaphragm [9]. In other words, Si diaphragm cannot sustain the loading pressure at such extreme environment.

978-1-4673-2395-6/12 $31.00 © 2012 IEEE

For the capacitive pressure sensor to be used continuously over a long period of time, the diaphragm must be capable of responding to applied pressure and extremely temperature, and must retain its elastic property [10]. This analysis indicates that SiC diaphragm is indeed is much better suited for extreme environments than Si.

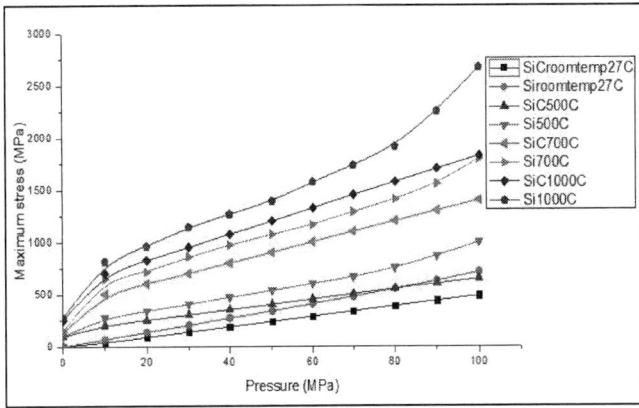

Fig. 8. Results of maximum stress atdifferential temperature with the thickness diaphragm for 0.2 μm

TABLE IV
LINEAR REGRESSION ANALYSIS FOR MAXIMUM STRESS AT DIFFERENT THICKNESS
AT 500°C

Temperature (°C)	SiC		Si	
	Equation	R^2	Equation	R^2
27	y=5x - 15	1.000	y=7.098x - 13.2	0.998
500	y=5x + 150	1.000	y=7.643x + 168.7	0.981
700	y=10x + 400	1.000	y=12.13x + 480.2	0.985
1000	y=12.5x + 575	0.996	y=18.92x + 531.2	0.959

The mechanical characterization to investigate the maximum stress on the diaphragm is performed using "Mises Stress" simulation. Figure 8 and Table IV show the maximum stress of a 200 nm thick diaphragm at the temperatures of 27°C, 500°C, 700°C and 1000°C.In theory as shown in equation (4) and (5), the stress diaphragm is proportional to the pressure and temperature. The stress distribution increases with the increment of pressure and temperature, however the stresses are less by using SiC material compare to Si. The temperature shows a significant influence of the diaphragm on the stress state.

The main point from Figure 8 is the fact that the SiC diaphragm has more linear stresses versus pressure as compares to Si diaphragm at all temperatures, as shown in Table IV. Amazingly, even at 1000°C the SiC diaphragm shows linear graph (R-squared = 0.996), which Si diaphragm shows a non-linear maximum stress (R-squared value = 0.959).

These results show that the maximum stress of the SiC diaphragm at 1825 MPa is much below the yield strength of the SiC material at 3440 MPa. Similarly, the maximum stress for silicon diaphragm is 2000 MPa, whereas the yield strength for silicon is 2673 MPa [13]. Further analysis will be carried out to investigate this finding.

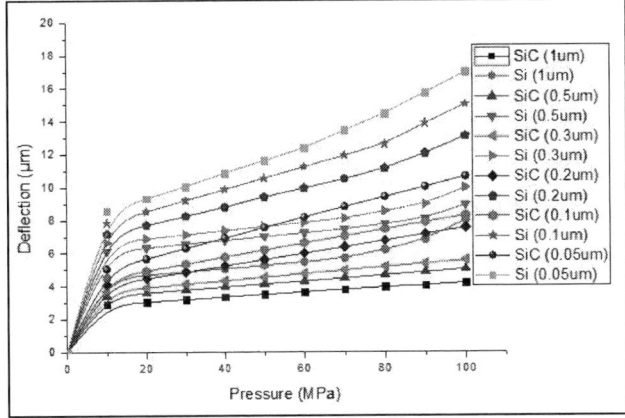

Fig. 7. Results of deflection at differential thickness of diaphragm at 500°C with thickness diaphragm of 0.05μm, 0.1μm, 0.2μm and 1.0μm

TABLE III
LINEAR REGRESSION ANALYSIS FOR DEFLECTION DIFFERENT THICKNESS AT
500°C

Thickness (μm)	SiC		Si	
	Equation	R^2	Equation	R^2
0.05	y=0.062x + 4.43	1.000	y=0.091x + 7.32	0.988
0.10	y=0.042x + 4.09	1.000	y=0.076x + 6.87	0.987
0.20	y=0.038x + 3.70	1.000	y=0.063x + 6.33	0.987
0.30	y=0.020x + 3.50	0.999	y=0.032x + 6.11	0.933
0.50	y=0.017x + 3.23	0.999	y=0.027x + 5.70	0.949
1.00	y=0.014x + 2.77	0.998	y=0.033x + 3.81	0.873

Figure 7 and Table III show the diaphragm deflection at 500°C at the different thicknesses. As shown in equation (3), the deflection is inversely proportional to the thickness. Moreover, Si diaphragms have larger deflection (16.96 μm) compare to SiC diaphragm (10.66μm) at the thickness of 50nm. More importantly, the SiC diaphragm has more linear deflection compares to Si diaphragm at all thicknesses, as shown in Table III. We also observe the mechanical failure of Si diaphragm at 1000°C, which is due to the degradation of the surface of SiC diaphragm [11]. On the other hand, show consistent linearity throughout different thicknesses at such high pressure and temperature as recorded in Table III. It implies that very thin diaphragm below 1 μm of the capacitive pressure sensor should be use SiC material indicating a very high mechanical strength of the diaphragm that can able to survive a very high pressure and temperature without fracture [12]. These deflection results are useful in real fabrication practical applications by specified the parameter of the diaphragm.

IV. CONCLUSION

This paper simulates and compares the performances of a parallel-plate model of a pressure sensor, where the top movable plate is made of either 3C-SiC or Si. The movable square diaphragm has the area of 400 μm² with six different thicknesses (0.05, 0.1, 0.2, 0.3, 0.5 μm and 1.0μm). The mechanical stress-deflection is performed by loading high pressure and temperature up to 100 MPa and 1000°C, respectively. The 1 μm thick SiC diaphragm exhibits close linearity graph with R-squared value of 1, and the maximum deflection of 11.7 μm, whereas the Si diaphragm of the same thickness exhibits non-linear behavior with R-square value of 0.92, and maximum deflection of 28.4 μm at the applied pressure and temperature of 100MPa and 1000°C, respectively. SiC diaphragm consistently shows better linearity (in term of R-square values) over Si diaphragm with the stress simulation as well. The 200 nm thick diaphragm shows the R-square value of 0.996 at the extreme temperatures and pressures. Finally, it is shown that the maximum stresses of both SiC and Si are lower than the yield strength of both materials.

ACKNOWLEDGMENT

The authors would like to thank the Institute Microengineering and Nanoelectronics (IMEN) of Universiti Kebangsaan Malaysia (UKM), Queensland Micro- and Nanotechnology Centre (QMNC) of Griffith University for providing the resources and facilities to perform this project. This work was performed in part at the Queensland node of the Australian National Fabrication Facility, a company established under the National Collaborative Research Infrastructure Strategy to provide nano and microfabrication facilities for Australia's researchers.

REFERENCES

[1] R. S. Okojie, G. M. Beheim, G. J. Saad and E. Savrun, *Characteristic of a hermetic 6H-SiC pressure sensor at 600°C*, Conference and Exposition, 2001.

[2] Y. Hezarjaribi, M. N. Hamidon, S. H. Keshmiri and A. R. Bahadorimerhr, *Capacitive pressure sensors based on MEMS, operating in harsh environments*, ICSH proceedings, 2008.

[3] M. Mehregany and C. A. Zorman, *SiC MEMS: opportunities and challenges for applications in harsh environments*, Journal of Elsevier, Thin solid film, 1999, pp. 355-356.

[4] R. Pratap and A. Arunkumar, *Material selection for MEMS devices*, Indian Journal of Pure and Applied Physics, 2007, pp. 358-367

[5] S. Timshenko and S. Woinosky-Krieger, *Theory of Plates and Shells*, McGraw Hill Classic Textbook Ressue, 1987.

[6] I. A. Ali, *Micromachining techniques for fabrication micro and nano structures*, 2009, pp. 227-298.

[7] M. Young, *The Technical Writer's Handbook.*Mill Valley, CA: University Science, 1989.

[8] Shanmugavalli, M. Uma, G. Vasuki and M. Umapathy, *Design and simulation of MEMS using interval analysis.* Journal of physics: Conference series 34, 2006, pp. 601-605.

[9] D. J. Yiung, J. Du, C. A. Zorman and W. H. Ko, *High-temperature single-crystal 3C-SiC capacitive pressure sensor*, IEEE sensors journal, vol.4, 2004, pp. 464-470.

[10] Ganji, B. A. and Majlis, B. Y. (2009). *Design and fabrication of a new MEMS Capacitive Microphone Using a Perforated Aluminum Diaphragm.* Sensors and Actuators A: Physical , 149: pp.29-37.

[11] E. Tobin, M. Magida, S. Kishner and M. Krim, *Design, fabrication and test of a meter-class reaction bonded SiC mirror blank*, SPIE vol.2543.

[12] Kudimi, J. M. R., MohdYasin, F. and Dimitrijev, S., SiC-Based Piezoelectric Energy Harvester for Extreme Environment

[13] The Engineering Toolbox, *Elastic Properties and Young Modulus for Some Materials* retrieved April 20, 2012 from http://www.engineeringtoolbox.com/young-modulus-d_417.html

Study of ZnO Micro-gap on SiO₂/Si Substrate by Conventional Lithography Method for pH Measurement

K.L.Foo[1], U.Hashim[1], Haarindra Prasad s/o RajintraPrasat[2] and M.Kashif[1]

[1] Nano Biochip Research Group, Institute of Nano Electronic Engineering (INEE), UniversitiMalaysia Perlis (UniMAP), 01000 Kangar, Perlis, Malaysia

[2] Microelectronic Engineering, University Malaysia Perlis (UNIMAP), 01000 Kangar, Perlis , Malaysia

Email: elitefoo@yahoo.com

Abstract- **ZnO films, type of the metal-oxide semiconductor promised a wide range of application. ZnO prepared from zinc acetate dehydrate acted as a precursor and IPA acted as a solvent exhibit high crytallinity with the hexagonal wurzite structure. The ZnO films with the grains uniformly distributed on the substrate was deposited using low-cost sol-gel technique. In this paper, the zinc oxide thin films are further used for the formation of micro gap device using conventional fabrication process. The influence of surface morphologies and uniformity distribution of ZnO nanoparticles on the substrate had been investigated using FESEM, whereby the crystallization and structure types of ZnO was determined using XRD. FTIR study was used to determine the chemical compound existed on the ZnO films with the SiO2/Si acted as a substrate. The electrical characteristic of the ZnO microp gap with different pH had been tested using source meter.**

I. INTRODUCTION

Metal-oxide-semiconductor films, which have been widely studied by the research all around the world, have currently received considerable attention in various applications. This is due to the advantages of their electrical and optical properties. ZnO films, one of the metal-oxide-semiconductor films, have a wide band-gap (3.37eV) and high exciton binding energy (60meV). Due to those characteristic, ZnO films are widely used in various type of sensor, such as bio-molecule sensor[1-2], ultraviolet detector[3] and chemical and gas sensors[4-5]. Besides that, its transparent conductive oxide film is very suitable applied in optoelectronic device, like light emitting diodes (LED)[6] and solar cell[7-8] applications.

ZnO films can be prepared with various types of method, including sol-gel spin coating[9-10], physical vapor deposition[11], chemical vapor deposition (CVD)[12], spray pyrolysis[13], sputtering[14], as well as ink-jet printing[5]. ZnO films is also the key factor for the formation of the ZnO nanostructure, like nanowires (NWs)[15], nanopores[9], nanorods[16], nanobelts[17], nanorings[18], nanocables[19] and nanotubes[10, 20], nanocolumns[21], nanocombs[22] and nanoneedles[23]. Due to the low-cost, chemical composition control, low-temperature annealing and homogeneity of the sol solution, sol-gel spin coating technique had been chosen in this project.

Micro-gap, which is defined as formation of gap in between 1- 100 μm ranges, has been widely studied on these days. Various types of material, ranging from insulator, semiconductor until conductor are used for the formation of the micro-gap. Due to the potential application of the micro-gap in electrochemical sensor[24], bimolecular sensor[25] and power saving device [26], it received enormous research attentions nowadays.

In this paper, a ZnO semiconductor thin films micro-gap had been successfully fabricated using conventional lithography process. Prior electrical characterization of ZnO micro-gap with source meter, the surface morphologies and microstructures of the ZnO films was studied using field emission scanning electron microscope (FESEM), while X-ray diffraction (XRD) and Fourier transform infrared spectroscopy (FTIR) would be used to confirm the crystallinity and structural properties of the ZnO films. Besides that, the electrical properties to the ZnO micro-gap device under different condition had been studied using source meter.

II. METHODS AND MATERIALS

The films were deposited by sol-gel low-cost method on SiO₂/Si substrate. The ZnO solution was synthesized by taking Zn ions from the zinc acetate material which is free from chlorin ion. Therefore, the ZnO sol was prepared by using zinc acetate dehydrate [Zn (CH₃COO)₂•2H₂O] as a precursor, isopropanol (IPA) as solvent and monoethanolamine (MEA) as a stabilizer. The zinc acetate dehydrate was first dissolved in 100ml of IPA to form a ZnO solution. The concentration of the solution was maintained at 0.2M. The mixed solution was stirred on a hot plate with the temperature fixed at 60°C for 20 minutes to make sure the ZnO powder was completely dissolved in the IPA solution. Then, the resultant solution was continuing stirred at 60°C for another 2 hours. During this period, MEA was dropped drop by drop into the solution as to yield a clear and homogenous ZnO aqueous solution. The solution was then left for 24 hours at room temperature for aging process before it could be used for thin film deposition process.

In this project, p-type Silison (100) was used as the substrate. The substrate was first cleaned with 70% concentrated of nitric acid (HNO_3), buffered oxide etch (BOE), acetone and rinsed with deionized water (DIW). Then wet oxidation process was used to grow an oxide layer on the cleaned silicon surface. For wet oxidation process, the substrate was put in the furnace at 1000°C with the water vapor continuously flowed into the tube furnace for 1 hours. The reason of this process was to get an isolator layer between silicon wafer and ZnO films. The final thickness of the oxide layer was ~180nm.

ZnO solution was deposited onto the substrate right after the oxidation process for avoiding the contamination of the substrate. Sol-gel technique was used in this deposition process, whereby the cleaned substrate placed on the spin coater and a few drops of ZnO solution were dropped on the substrate by using pipet. The ZnO solution was spin on the substrate with the speed 3000rpm for 30 s. The deposited substrates were then dried on the hot-plate at 150°C for 10 minutes. Finally, The ZnO films were annealed in the furnace at 500°C for 2 hours as to get the crystallization of ZnO

Fabricated ZnO thin films were further used for the ZnO micro-gap formation using conventional lithography process. In this project, two masks were be used. Both masks were designed by AutoCAD software and printed onto chrome glass surfaces. First mask was used for the formation of the micro-gap while the second mask was used for the aluminium (Al) electrode formation. Both ZnO thin films and aluminium electrode were etched using wet etching with alum etcher and 2% of diluted hydrochloric acid (HCl), respectively. The fabrication process is shown in Fig. 1.

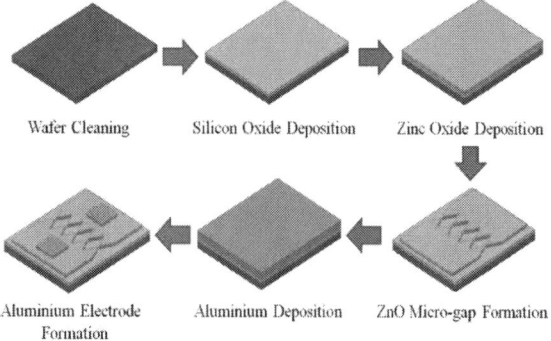

Fig. 1. ZnO micro-gap fabrication process

The morphologies of the ZnO films were characterized using field emission scanning electron microscope (FESEM, Hitachi SU-70). The crystal structure of sol-gel derived ZnO film was examined with X-ray diffraction (XRD, Bruker D8). The Fourier transform infrared spectroscopy (FTIR, Perkin Elermen 400) was used to confirm the chemical compound and hexagonal wurtzite structure of the ZnO films. After the formation of the ZnO micro gap, the size of the gap was measured using high power microscope (HPM, Olympus BX51) Finally, the electrical properties of the fabricated ZnO micro-gap was characterized with source meter (keithley 2400) and dielectric analyzer (Alpha high dielectric analyzer from Novo Control).

III. RESULTS AND DISCUSSION

The FESEM photograph of the deposited and annealed ZnO films had been analyzed. The top view of FESEM of the films, prepared using sol-gel method with the IPA as a solvent, is shown in Fig. 2. From the result, it shows that the size of the grains is in nanometer size (approximately 50nm). It can be seen that the distribution of the grains are very uniform on the SiO_2/Si substrate.

The phase evaluation, crystallinity and crystallographic had been performed and studied with XRD measurement as shown in Fig. 3. All the diffraction peaks in the result is according to the standard card (JCPDS 36-1451). From the XRD analysis, it clearly shows that the ZnO films are polycrystalline with the hexagonal wurzite structure, which had been investigated by the previous researcher[27]. From Fig. 2, it indicates that the characteristic peaks with high intensities appears has preferential growth alone (100), (002) and (101). This sharp and narrow of those diffraction peaks prove that the sol-gel fabricated ZnO films exhibit high crytallinity. The XRD measurement also shows some others low peaks appeared at (102), (110), (103) and (112).

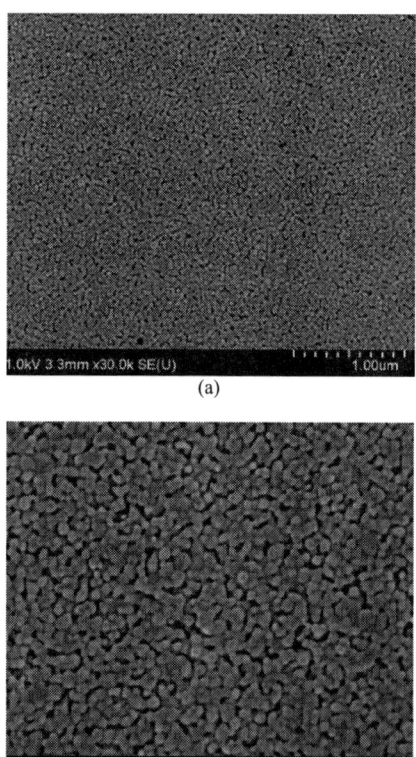

Fig. 2. Top view images of FESEM (a) 30k magnification and (b) 70k magnification

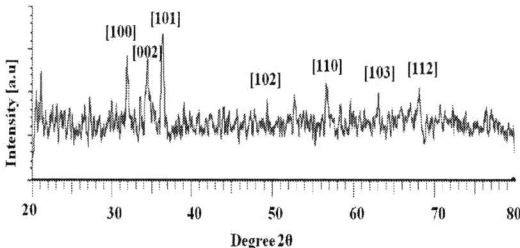

Fig. 3. X-ray diffraction pattern of ZnO films

The result of the compound chemical bonds had been indicated in FTIR spectra. The sharp and intense band at ~460cm^{-1}, which is shown in Fig. 4 is due to the absorption spectra of Zn-O groups[28]. This absorption spectra shows that these thin films are hexagonal wurtzite structure, which also proved from above XRD graph. The absorption spectra at ~611cm^{-1} is assigned to the existence of the local vibration of the substitution carbon in the Si crystal lattice[29]. Whereas the sharp peaks at ~1902cm^{-1} is assigned to the Si-O stretching vibration[30].

Fig. 5 shows the HPM image of ZnO micro-gap fabricated using conventional lithography process. The image shows that ZnO micro-gap formed using chrome mask and exposed with 365nm ultra-violet (UV) light giving a shape profile. The smallest gate that obtained in this experiment is 19.34µm.

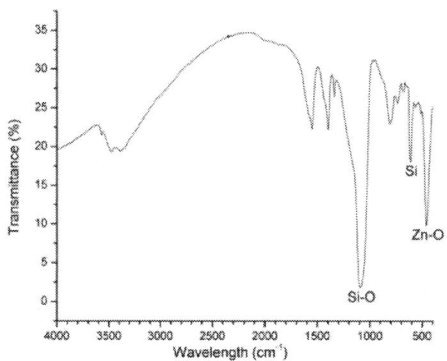

Fig. 4. FTIR spectra of the ZnO samples

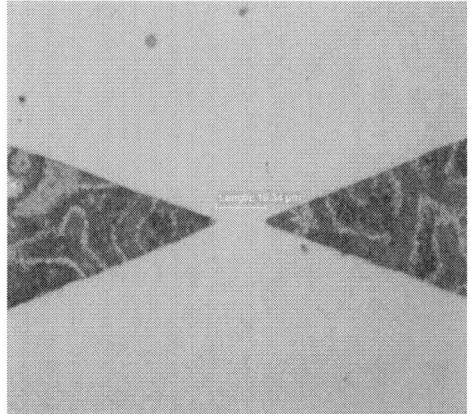

Fig. 5. HPM image of ZnO micro gap with magnification of 50X

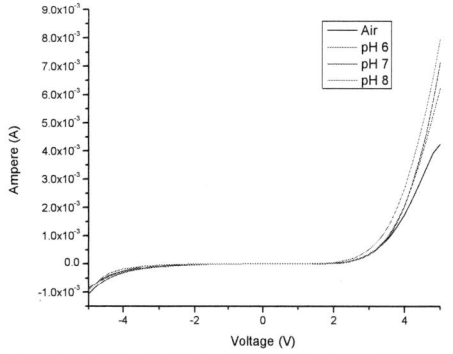

Fig. 6. I-V characteristic for the ZnO micro-gap in air condition and different pH value.

The current–voltage (I-V) characteristics of ZnO heterojunction thin film prepared under IPA solvent was studied using source meter. The heterojunction demonstrates rectifying behavior with a typical forward-to-reverse current in the voltage of -5 to 5 V. The result (Fig. 6) shows that ZnO micro-gap acted like a back-to-back diode. In this experiment, the ZnO micro-gap had been tested under air condition and different pH value, ranging from pH6 to pH8. The result indicated that the change of the forward bias current under different condition is more obvious if compare to the reverse bias current. Therefore, forward bias current is studied in this experiment. The result shows that the ZnO micro-gap device gives the lowest current (4mA, 5V) in air condition. This is believed due to the no ion exhibited in the air and therefore electron is hardly flowed between the gaps. When pH solution dropped between the gaps, the current is increased due to the [H]$^{+}$ and [OH]$^{-}$ ions appeared in the solution. By the way, pH 6 gives lower value of current (6mA, 5V) if compare to pH 7 (7mA, 5V) and pH 8 (8mA, 5V) It can be concluded that the increasing of the pH value will increase the current value of the device. In the other words, the current value of the device would increase while the solution changing from more acidic to more alkali or vice versa.

IV. CONCLUSIONS

By sol-gel spin coating technique, a high crytallinity with the hexagonal wurzite structure of ZnO films had been prepared. According to the FESEM result, the sol-gel technique also shows that the grains of ZnO films were in nano-size and were uniformly deposited on the substrate. The shape and narrow peaks shown by XRD indicted that the fabricated ZnO is highly crystalline, while the FTIR study also shows that the ZnO thin films are hexagonal wurtzite structure. For device electrical testing, ZnO micro-gap giving higher current value in alkali solution if compared to the acidic solution.

ACKNOWLEDGEMENTS

The authors are very grateful to ministry of high education (MOHE) for providing the FRGS grand to conduct this research. Besides that, the authors also would like to thank all of the team members in the Institute of Nano Electronic

Engineering especially the member in Nano Biochip Research Group.

REFERENCES

[1] A. Fulati, *et al.*, "An intracellular glucose biosensor based on nanoflake ZnO," *Sensors and Actuators B: Chemical*, vol. 150, pp. 673-680, 2010.

[2] S. M. Usman Ali, *et al.*, "A fast and sensitive potentiometric glucose microsensor based on glucose oxidase coated ZnO nanowires grown on a thin silver wire," *Sensors and Actuators B: Chemical*, vol. 145, pp. 869-874, 2010.

[3] G. Chai, *et al.*, "Crossed zinc oxide nanorods for ultraviolet radiation detection," *Sensors and Actuators A: Physical*, vol. 150, pp. 184-187, 2009.

[4] C. Baratto, *et al.*, "ZnO nanocrystals by chemical route for optical gas sensing," in *Sensors, 2008 IEEE*, 2008, pp. 1293-1296.

[5] W. Shen, *et al.*, "The preparation of ZnO based gas-sensing thin films by ink-jet printing method," *Thin Solid Films*, vol. 483, pp. 382-387, 2005.

[6] K. Kim, *et al.*, "Light-emitting diodes composed of n-ZnO and p-Si nanowires constructed on plastic substrates by dielectrophoresis," *Solid State Sciences*, vol. 13, pp. 1735-1739, 2011.

[7] K. Matsubara, *et al.*, "ZnO transparent conducting films deposited by pulsed laser deposition for solar cell applications," *Thin Solid Films*, vol. 431-432, pp. 369-372, 2003.

[8] E. Guillen, *et al.*, "ZnO solar cells with an indoline sensitizer: a comparison between nanoparticulate films and electrodeposited nanowire arrays," *Energy & Environmental Science*, vol. 4, pp. 3400-3407, 2011.

[9] M. Kashif, *et al.*, "ZnO nanoporous structure growth, optical and structural characterization by aqueous solution route," in *Enabling Science and Nanotechnology (ESciNano), 2010 International Conference on*, pp. 1-1.

[10] S. M. U. Ali, *et al.*, "Functionalised zinc oxide nanotube arrays as electrochemical sensors for the selective determination of glucose," *Micro & Nano Letters, IET*, vol. 6, pp. 609-613, 2011.

[11] L. Wang, *et al.*, "Synthesis of well-aligned ZnO nanowires by simple physical vapor deposition on c-oriented ZnO thin films without catalysts or additives," *Applied Physics Letters*, vol. 86, pp. 024108-024108-3, 2005.

[12] B. S. Li, *et al.*, "Effects of RF power on properties of ZnO thin films grown on Si (0??) substrate by plasma enhanced chemical vapor deposition," *Journal of Crystal Growth*, vol. 249, pp. 179-185, 2003.

[13] A. Ashour, *et al.*, "Physical properties of ZnO thin films deposited by spray pyrolysis technique," *Applied Surface Science*, vol. 252, pp. 7844-7848, 2006.

[14] B. Deng, *et al.*, "AFM characterization of nonwoven material functionalized by ZnO sputter coating," *Materials Characterization*, vol. 58, pp. 854-858, 2007.

[15] M. H. Huang, *et al.*, "Catalytic Growth of Zinc Oxide Nanowires by Vapor Transport," *Advanced Materials*, vol. 13, pp. 113-116, 2001.

[16] G.-n. He, *et al.*, "Positive temperature coefficient of resistance of single ZnO nanorods," *Nanotechnology*, vol. 22, p. 065304, 2011.

[17] X. Y. Kong and Z. L. Wang, "Polar-surface dominated ZnO nanobelts and the electrostatic energy induced nanohelixes, nanosprings, and nanospirals," *Applied Physics Letters*, vol. 84, pp. 975-977, 2004.

[18] X. Y. Kong, *et al.*, "Single-Crystal Nanorings Formed by Epitaxial Self-Coiling of Polar Nanobelts," vol. 303, ed, 2004, pp. 1348-1351.

[19] S. Kim, *et al.*, "Fabrication of Zn/ZnO nanocables through thermal oxidation of Zn nanowires grown by RF magnetron sputtering," *Journal of Crystal Growth*, vol. 290, pp. 485-489, 2006.

[20] X. Ren, *et al.*, "Fabrication of ZnO nanotubes with ultrathin wall by electrodeposition method," *Materials Letters*, vol. 62, pp. 3114-3116, 2008.

[21] J. Y. Park, *et al.*, "Synthesis and electrical properties of aligned ZnO nanocolumns," *Composites Part B: Engineering*, vol. 37, pp. 408-412, 2006.

[22] Y. Yang, *et al.*, "Effective photoluminescence modification of ZnO nanocombs by plasma immersion ion implantation," in *Nanoelectronics Conference, 2008. INEC 2008. 2nd IEEE International*, 2008, pp. 20-24.

[23] J. Zhang, *et al.*, "Fabrication, structural characterization and the photoluminescence properties of ZnO nanoneedle arrays," *Physica E: Low-dimensional Systems and Nanostructures*, vol. 27, pp. 302-307, 2005.

[24] T. S. Dhahi, *et al.*, "Electrical characterization of in-house fabricated polysilicon micro-gap for yeast concentration measurement," *Journal of Engineering and Technology Research*, vol. 3(8), pp. 246-254, 2011.

[25] X. Chen, *et al.*, "Electrical nanogap devices for biosensing," *Materials Today*, vol. 13, pp. 28-41, 2010.

[26] T. h. S. Dhahi, *et al.*, "Fabrication and Characterization of 50 nm Silicon Nano-Gap Structures," *Science of Advanced Materials*, vol. 3, pp. 233-238, 2011.

[27] S. Flickyngerova, *et al.*, "Structural and optical properties of sputtered ZnO thin films," *Applied Surface Science*, vol. 254, pp. 3643-3647, 2008.

[28] R. Singh, *et al.*, "Growth and characterization of high resistivity c-axis oriented ZnO films on different substrates by RF magnetron sputtering for MEMS applications," *Journal of Materials Science*, vol. 42, pp. 4675-4683, 2007.

[29] Y. Sun, *et al.*, "Characterization of excess carbon in cubic SiC films by infrared absorption," *Journal of Applied Physics*, vol. 85, pp. 3377-3379, 1999.

[30] N. Laidani, *et al.*, "Spectroscopic characterization of thermally treated carbon-rich $Si1-xCx$ films," *Thin Solid Films*, vol. 223, pp. 114-121, 1993.

Design and Simulation of High Magnetic Gradient Device for Effective Bioparticles Trapping

Ummikalsom Abidin, Burhanuddin Yeop Majlis, *SMIEEE* and Jumril Yunas, *MIEEE*
Institute of Microengineering and Nanoelectronics (IMEN)
Universiti Kebangsaan Malaysia (UKM)
43600 UKM Bangi, Selangor, Malaysia
Email: burhan@vlsi.eng.ukm.my

Abstract- **In this work, a design and simulation of high magnetic gradient device for effective bioparticles trapping is reported. The planar square-shaped microcoil and a V-shaped nickel iron (NiFe) alloy core is designed to guide and confine the magnetic flux lines through its small tip area and thus enhance the magnetic flux density and its gradient. The effects of core structure and coil parameters are analyzed using Finite element analysis (FEA) of two dimensional axial symmetry modeling. The simulation results revealed that the V-shaped magnetic core has significantly increased the magnetic flux density, its gradient and the magnetic force affecting on the beads sample. The highest magnetic flux density value, B_{norm} is 66 mT is achieved for microcoil turns of N = 20, thickness of h = 5 μm, width and spacing of w = s = 50 μm and on tip surface area of 1 μm². Furthermore, a maximum magnetic force value of F_m = 1700 pN which is much higher than the drag force experienced by the magnetic beads in the microchannel has also been observed. Therefore, a promising effective trapping of the magnetic beads in the microfluidic channel is enable with this high magnetic gradient device design.**

I. Introduction

Separation of bioparticles or biological cells i.e. viruses, bacteria, blood cells, cancer cells are crucial in clinical diagnostics, cell therapeutics and biological studies. In magnetic separation technique, magnetic field with high gradient is used to capture the bioparticles that have been labeled with magnetic beads.

In current Lab on a Chip (LoC) systems, a surface micromachined planar wire coil carrying electrical current has been the most interesting part that has been utilized in obtaining controllable and well defined magnetic field in isolating bioparticles [1, 2, 3]. The planar microcoil is chosen due to its compactness, simple fabrication process and stacking or sandwich structure possibility [4].

High magnetic field and its gradient are generated by spiral- and square-shaped planar microcoil due to the summation of total magnetic flux density at its centre [5, 6, 7]. Therefore, a high magnetic gradient is possible to be generated from a wire wound around a tapered core structure or a needle tip [8, 9]. It was reported previously that a combination of microcoil and magnetic pillar has been used for trapping 1 to 5 μm diameter magnetic beads [10]. Integrated micro-separator with soft magnetic micro-pillars has also been design and tested for lymphocytes cells trapping and separation [11]. In addition, a

three-dimensional (3D) microfabricated permalloy core tips has been design, fabricated and characterized for effectively separate bioparticles tagged with magnetic beads [12]. Complexity of the design and its fabrication processes are some of the previous design disadvantages. Furthermore, Joule heating effect and maintaining a close distance between the magnetic source and the microchannel contributed to the drawbacks of the preceding works.

This paper presents a new design concept in generating high strength magnetic flux and its gradient using planar micro coil with V-shaped NiFe permalloy magnetic core structure. The proposed design is expected to produce high magnetic flux gradient and its force and hence reducing the magnetic losses in the microfluidics channel. A comparison of the magnetic force from the microcoil design with and without V-shaped magnetic core using Finite Element Analyses (FEA) will justify this new design concept which is designed to be used as bioparticles separator in LoC system.

II. Magnetic Separation Concept

In trapping magnetic particles within a fluid volume, an inhomogeneous magnetic field with high field gradient is required [9]. The principle of the magnetic separation of magnetic particles in the continuous microfluidics flow involves the interaction of magnetic and hydrodynamics forces. The magnetic force developed on a magnetic particle of volume, $V = \frac{3}{4} \pi r^3$, with different magnetic susceptibility $\Delta\chi$ (χ_p for particle and χ_m for the buffer medium) and the strength and gradient of the magnetic flux density B can be theoretically calculated as follow,

$$\vec{F}_m = \frac{V\Delta\chi}{\mu_0}(\vec{B}\cdot\vec{\nabla})\vec{B} = \frac{V\Delta\chi}{2\mu_0}\vec{\nabla}(\vec{B}\cdot\vec{B}) \tag{1}$$

In the direction of flow, the x-component of the force on the magnetic particles can be written as

$$F_{m,x} = \frac{V\Delta\chi}{\mu_0}\left(B_x\frac{\partial}{\partial x} + B_y\frac{\partial}{\partial y} + B_z\frac{\partial}{\partial z}\right)B_x \tag{2}$$

In two dimensional (2D) axial symmetrical system, the magnetic force along the radial axis becomes

$$F_{m,r} = \frac{V\Delta\chi}{\mu_0}\left(B_r\frac{\partial B_r}{\partial r} + B_z\frac{\partial B_r}{\partial z}\right) \tag{3}$$

A magnetic particle of radius, R experiences hydrodynamics drag force in the microfluidics channel flow of

$$\vec{F}_d = 6\pi\eta R(v_p - v_{medium}) \qquad (4)$$

where v_p is the particles velocity, v_{medium} is the fluid velocity and η is the fluid viscosity. For a 1.4 μm radius magnetic particle flowing in a fluid of viscosity 10^{-3} N.s/m^2 and velocity of 10^{-4} m/s, the drag force is calculated to be F_d = 2.64 pN. For that reason and in order to successfully trap the magnetic particles in the continuous microfluidics channel flow, the magnetic force must be greater from the drag force experience by the magnetic particle

$$\vec{F}_m > \vec{F}_d \qquad (5)$$

III. DESIGN CONCEPT

The cross-sectional, front- and back-side view of the device is shown in Fig. 1. It consists of three main parts, namely on silicon planar coil as magnetic field generator, silicon based through hole soft magnetic core as magnetic flux guidance and PDMS base microfluidic channel. Silicon is chosen as the substrate due to the microfabrication processes compatibility and its excellent heat sink capability. The novelty of the proposed design is the V-shaped soft magnetic core structure while the small surface area magnetic core tip guides, concentrates and thus intensifies the magnetic flux lines. The V-shaped core structure of <100> silicon through hole is formed using the anisotropic etching process.

In this work, a nitride mask of 0.92 x 0.92 mm^2 area is designed to enable the forming of smallest core tip. Soft magnetic material of nickel iron (NiFe) alloy is selected to be electroplated in the V-shaped core structure. The selection of NiFe as the magnetic core is due to its high permeability and low coercivity value. The high permeability will provide an easy magnetic flux linkage and thus intensify the magnetic field. The low coercivity value is important to make the device easily to be magnetized and demagnetized with the effect of current pass through the microcoil. Low coercivity value is also required for LoC magnetic separator for better trapping and release of the bioparticles tagged with the magnetic beads.

An electromagnetic field will be generated through a metal microcoil carrying electrical current. A square-shaped planar microcoil design has been chosen due to the high magnetic field generation from the summation of each individual microcoil. Copper is chosen as the microcoil metal conductor due to its low in cost, ease of sputtering process and high conductivity value, ρ = 5.998 x 10^7 S/m.

In realizing the micromagnetic system as a LoC magnetic separator, a polydimetylsiloxane (PDMS) microfluidic channel will be bonded to the silicon chip back surface. The microfluidic channel is to provide the mean of fluid and bioparticles transport. PDMS is chosen as the microchannel materials due to its relatively cheap and easy fabrication process. In realizing the micromagnetic system as a LoC magnetic separator, a polydimetylsiloxane (PDMS)

microfluidic channel will be bonded to the silicon chip back surface. The microfluidic channel is to provide the mean of fluid and bioparticles transport. PDMS is chosen as the microchannel materials due to its relatively cheap and easy fabrication process. In addition, biocompatibility, disposability and optically transparent properties are also the advantages of using PDMS materials for realization of the microfluidics channel [13].

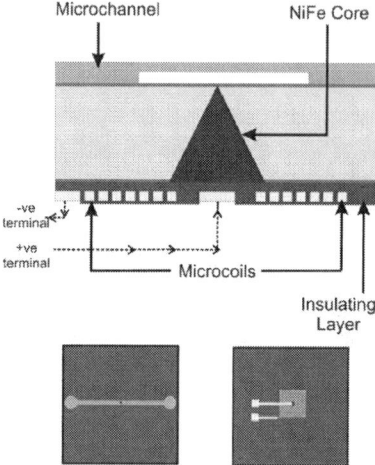

Fig. 1. (a) Cross-sectional view of the integrated device (b) Front-view of the chip showing the microchannel and the tip of the magnetic core (c) back-view of the chip showing the microcoil

The basic design structures of the microfluidics channel are the inlet port, microchannel and the outlet port. The continuous flow in the microfluidic will be control using the micro-syringe pump that provides the range of microliter flow per minute. Flow rate in the microchannel need to be control to ensure laminar low Reynold number flow and sufficient drag force during the magnetic trapping of the bioparticles.

IV. MODELING AND SIMULATION

Accurate analytical solution in determining the magnetic flux density and its gradient from the microcoil is a difficult task due to the structure shape and equations complexity. In this work, Finite Element Analysis (FEA) has been used to model and simulate the proposed design in obtaining the magnetic flux density and its gradient. Furthermore, the magnetic force on the micro magnetic beads is also possible to be calculated. In order to reduce the computational limitation in meshing and solving the problem, a simplification of the three-dimensional (3D) model to two-dimensional (2D) axial symmetry geometry model has been made [14]. An approximation of square-shaped to the spiral-shaped planar microcoil has also been made in this study. The schematic of the model is as shown in Fig. 4. The modeled microelectromagnet consists of copper microcoil, V-shaped magnetic core structure made of NiFe and an air space box domain surrounding the structures. In this model, the axial symmetry boundary is along r = 0 and the air box boundaries condition are set to be magnetically and electrically insulated. The NiFe relative permeability used is $\mu_r = 3000$. All the materials properties are chosen to be linear and isotropic. In addition, a static magnetic field analysis using Ampere's Law is used in solving the magnetic field with current injection of $I = 100$ mA. The geometrical parameters used in this analysis are shown as in Table I.

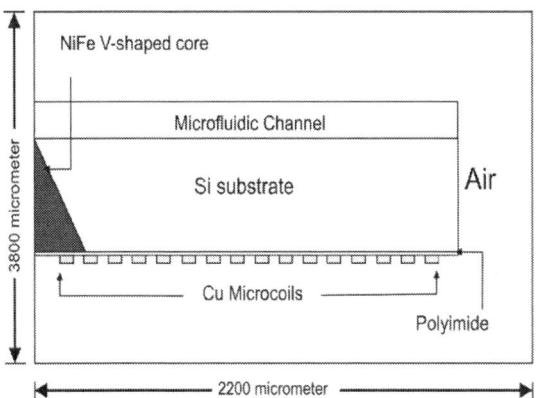

Fig. 4. 2D axial symmetry modeling with Finite Element Analysis (FEA)

TABLE I
GEOMETRICAL PARAMETERS USED IN THE DESIGN AND MODELING

Geometrical Parameters		Value	Unit
Substrate	Si		
Thickness	t	650	μm
Microcoil	Cu		
Width	w	50	μm
Spacing	s	50	μm
Thickness	t	5	μm
Winding	N	20	
Outer radius	r_o	2300	μm
Magnetic Core	NiFe		
Tip surface area	A	1	μm²
Height	h	650	μm

The simulation is conducted with physics controlled meshing of extremely fine elements. The 2D axial symmetry analysis uses a low processor memory and faster solution convergence.

V. RESULTS AND DISCUSSION

This section will discuss the magnetic flux density generated from the microcoil. The r-coordinate is the radial distance and the z-coordinate is the axis that is perpendicular to the axis of the channel. Z-coordinate of $z = 0$ is corresponding to the top of the microcoil and tip of the V-shaped core respectively.

Planar microcoil generates a periodic profile of magnetic flux density, B_{norm} along the radial axis as shown in Fig. 5. This trend is expected because of the electrical current flowing through the microcoil metal structure. In addition, the highest B_{norm} of 2.65 mT is observed at the top surface of the first

microcoil structure. This effect is due to the accumulation of the magnetic flux from each of the spiral-shaped microcoil design. A decaying trend of B_{norm} magnitude is also observed along the z-coordinate above the microcoil.

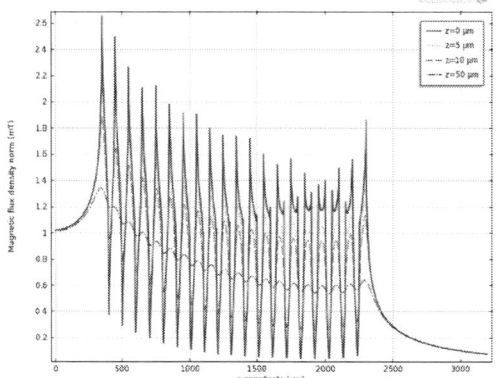

Fig. 5. Normal component of the magnetic flux density, B_{norm} along the r-coordinate axis at different height above the microcoil structure

At $z = 5$μm above the microcoil, a 21 percent dropped of B_{norm} value is calculated. At the farthest distance above the microcoil which is at $z = 50$ μm, 51 percent reduction of B_{norm} value has been determined. The gradient of the B_{norm} along the z-coordinate distance above the microcoil can be clearly seen from Fig. 6. For the microcoil design without core integration, a steepest fall in the gradient value is at 5 μm above the microcoil. Further z-coordinate distance away from the microcoil, the gradient descended deliberately. This considerable declining trend of the magnetic flux with respect to the height above the microcoil has also been observed by the work of other researchers [3, 15].

Even though the spiral-shaped design microcoil is possible to generate maximum value of the magnetic flux density and its gradient, however the values are consider very low and not sufficient in overcoming the drag force on the magnetic beads in the continuous microfluidics channel flow. Increasing the number of microcoil in the design will only create a Joule heating problem in the microchannel flow and thus affecting the viability of the bioparticles tagged with the magnetic beads.

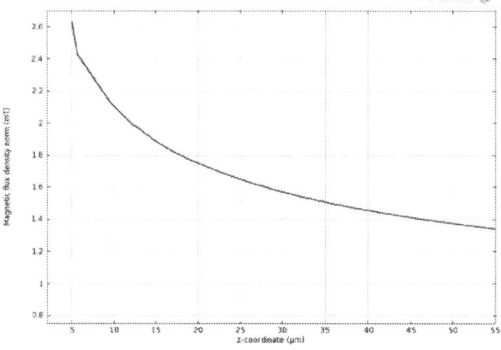

Fig. 6. Normal component of the magnetic flux density, B_{norm} along the z-coordinate axis above the microcoil structure

As mentioned previously, high permeability V-shaped magnetic core is use to concentrate and increase the magnetic potential generated by planar microcoil. In the FEA analysis, it is confirmed that the V-shaped magnetic core has drastically increased the magnetic flux density generation. Fig. 7 shows the characteristics of normal magnetic flux density of V-shaped core tip. The maximal B_{norm} generated from this new proposed design is 66 mT. The maximal value B_{norm} is observed at the core tip surface of 1 μm^2. Therefore, this result has proven that the V-shaped magnetic core tip has guides, concentrates and greatly intensifies the magnetic flux density through it small tip surface area. This effect is expected from the magnetic field theory where the magnetic flux density is defined as total magnetic flux divided by the cross-sectional area of which it flows. Furthermore, the high magnetic flux density generation is contributed by the high permeability of NiFe core which provides ease flux linkage for the magnetic flux lines to pass through. The high peak profile is corresponding to the tip corner effect which is at a distance of 0.56 μm from the centre.

A significant dropped of the magnetic flux density value has been observed with greater z-coordinate distance from the V-shaped core surface area. A great loss of B_{norm} of up to 91 percent is observed at $z = 5$ μm ($B_{norm} = 6$ mT) comparing at the tip surface, $z = 0$ μm B_{norm} value of 66 mT. Therefore, it is crucial to bond the microfluidics channel at the closest proximity onto the magnetic core surface tip as proposed in the design concept.

A decreasing profile of the B_{norm} with respect to the height above the V-shaped is clearly seen in Fig. 8. The abrupt gradient descendent is observed at the first $z = 5$ μm height above the magnetic core tip. On the other hand, a gradual dropped of the B_{norm} gradient is observed from 5 μm to 50 μm distance above the tip surface. The inset picture of Fig. 8 shows the magnetic flux density contour plot above the magnetic core tip.

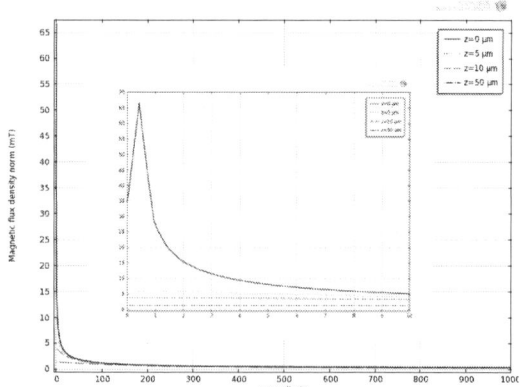

Fig. 7. Normal component of the magnetic flux density, B_{norm} along the r-coordinate axis at different height above the microcoil structure

The contour plot confirmed the significant declined in the magnetic flux density gradient above the tip. In addition, a high magnetic flux density gradient is observed at the sharp tip of the NiFe magnetic core.

Fig. 9 shows the comparison of the B_{norm} values from the microcoil with and without the V-shaped core effect. The design of microcoil with V-shaped core effect has enhanced the magnetic flux density by 25 times comparing the microcoil without core design.

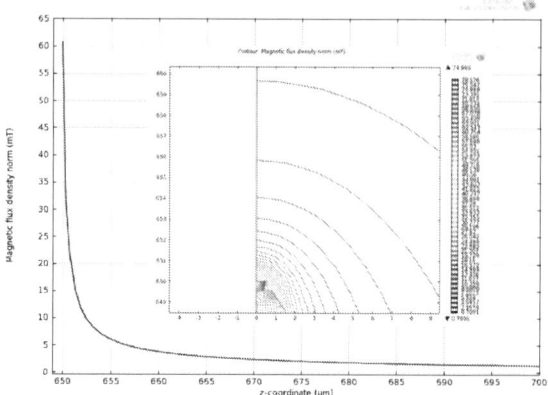

Fig. 8. Normal component of the magnetic flux density, B_{norm} along the z-coordinate axis above the microcoil structure

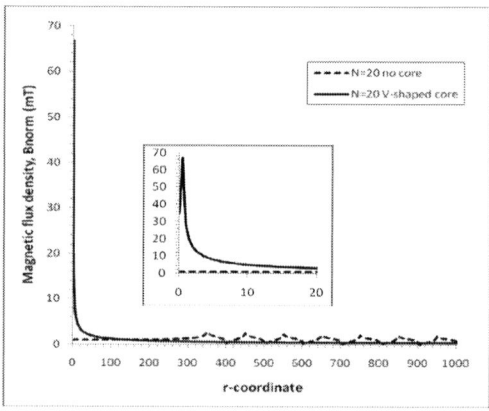

Fig. 9. Comparison of the magnetic flux density normal component, B_{norm} of the microcoil with and without V-shaped magnetic core

The most important parameter in magnetic beads trapping is the magnetic force value. As previously discussed, the magnetic force is the proportional to the summation of the magnetic flux density and its gradient. Fig. 10 shows the graph of magnetic force on the 2.8 μm diameter bead with respect to the r-coordinate for microcoil with and without V-shaped core. As clearly seen, the effect of V-shaped core has intensified the F_m on the beads. The maximum magnetic force $F_{m,max}$ value for microcoil with V-shaped core is 1700 pN. On the other hand, a small $F_{m,max}$ value of 0.034 pN is obtained for the microcoil without core. As the drag on the magnetic bead has been calculated as $F_d = 2.64$ pN, the magnetic force obtained by the V-shaped core LoC magnetic separator is considered far more effective in trapping the magnetic beads.

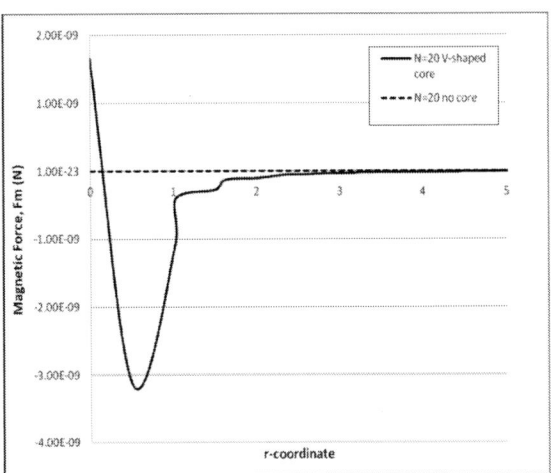

Fig. 10. Comparison of the magnetic flux density normal component, B_{norm} of the microcoil with and without V-shaped magnetic core

VI. CONCLUSION

In this work, a high magnetic flux density device has been designed and simulated. The proposed design is considered as a new innovation in generating a high magnetic gradient due to the exploitation of the V-shaped NiFe magnetic core as the flux guidance and concentrator in the microfluidic channel. The FEA simulation result has revealed that high permeability V-shaped core structure significantly enhanced the magnetic flux density at its very small tip area. The highest value of the magnetic flux density is found to be 66 mT for $w = l = s = 50$ μm, $N = 20$ and $I = 100$ mA at tip surface area of 1 μm^2. The simulation results also show that there is significant decreased of the magnetic flux density gradient value with respect to the z-distance from the tip surface. For that reason, the microfluidic channel must be brought as close as possible to the tip of the magnetic core in order to minimize the magnetic flux density gradient losses and enhance the bioparticles trapping.

The magnetic force value obtained for the 2.8 μm diameter magnetic bead is $F_m = 1700$ pN which is much greater that the drag force experience by the bead in the microfluidics channel flow. Therefore, efficient trapping of the magnetic beads is expected to be realized. In conclusion, a new and novel concept of high magnetic flux density generation device has been proposed. This new design is believed to be functioned as a Lab-on-chip magnetic separator in trapping and separating bioparticles.

ACKNOWLEDGMENT

I would like to thank Ministry of Higher Education of Malaysia and Universiti Kebangsaan Malaysia for the research grant under the project NND/ND/(1)/TD11-002 (Development of lab-on chip for peripheral blood stem cell isolation and rapid detection of tropical diseases from blood).

REFERENCES

[1] K.-H. Han and A. B. Frazier, "Paramagnetic capture mode magnetophoretic microseparator for high efficiency blood cell separations," *Lab on a Chip*, vol. 6, pp. 265-273, 2006.

[2] J. Kim, H. H. Lee, U. Steinfield, and H.Seidel, "Fast Capturing on Micromagnetic Cell Sorter," *Sensors Journal, IEEE*, vol. 9, pp. 908-913, 2009.

[3] Q. Ramadan, V. Samper, D. P. Poenar, and C. Yu,"An integrated microfluidic platform for magnetic microbeads separation and confinement," *Biosensors and Bioelectronics*, vol. 21, pp. 1693-1702, 2006.

[4] J. Yunas, A. A. Hamzah, and B. Y. Majlis, " Fabrication and characterization of surface micromachined stacked transformer on glass substrate," *Microelectronics Engineering*, vol. 86 (10) pp. 2020-2025, 2009.

[5] J.-W. Choi, T. M. Liakopoulos, and C. H. Ahn, "An on-chip magnetic bead separator using spiral electromagnets with semi-encapsulated permalloy," *Biosensors and Bioelectronics*, vol. 16, pp. 409-416, 2001.

[6] Q. Ramadan, "On-chip micro-electromagnets for magnetic-based bio-molecules separation," *Journal of Magnetism and Magnetic Materials*, vol. 281, pp. 150-172, 2004.

[7] K. Smistrup, O. Hansen, H. Bruus, and M. Hansen, "Magnetic separation in microfluidic systems using microfabricated electromagnets—experiments and simulations," *Journal of Magnetism and Magnetic Materials*, vol. 293, pp. 597-604, 2005.

[8] M. Barbic, J. J. Mock, A. P. Gray, and S. Schultz,"Scanning probe electromagnetic tweezers," *Applied Physics Letters*, vol. 79, p. 1897, 2001.

[9] N. Pamme, J. Eijkel, and A. Manz, "On-chip free-flow magnetophoresis: Separation and detection of mixtures of magnetic particles in continuous flow," *Journal of Magnetism and Magnetic Materials*, vol. 307, pp. 237-244, 2006.

[10] Q. Ramadan, D. P. Poenar, and C. Yu, "Customized trapping of magnetic particles," *Microfluidics and Nanofluidics*, vol. 6, pp. 53-62, 2008.

[11] T. Dong, Q. Su, Z. Yang, F. Karlsen, H. Jakobsen, E. B. Egeland, and S. Hjelseth, "Fully integrated micro-separator with soft-magnetic micro-pillar arrays for filtrating lymphocytes," 2010, pp. 6522-6526.

[12] R. Rong, J.-W. Choi, and C. H. Ahn, "An on-chip magnetic bead separator for biocell sorting, " *Journal of Micromechanics and Microengineering*, vol. 16, pp. 2783-2790, 2006.

[13] G. M. Whitesides, "The origins and the future of microfluidics," *Nature*, vol. 442, pp. 368-373, 2006.

[14] M. Shafique and M. F. Hansen, "Modeling the Magnetic Field From a Microelectromagnet in FEMLAB," in *Proceedings of the Nordic Matlab Conference*, 2003, pp. 209-214.

[15] R. Fulcrand, D. Jugieu, C. Escriba, A. Bancaud, D. Bourrier, A. Boukabache, and A. M. Gué, "Development of a flexible microfluidic system integrating magnetic micro-actuators for trapping biological species," *Journal of Micromechanics and Microengineering*, vol. 19, p. 105019, 2009.

Optical properties of zinc doped tin oxide synthesized by mechanochemical processing

Sharipah Nadzirah SAA[*1], Azlan Zakaria[**2], Mahesh Kumar Talari[**2], Nurul Syahidah Sabri[**2] and U. Hashim[*1]

[*1]Nanostructure Lab-on-chip Research Group, Institute of Nano Electronic Engineering,Universiti Malaysia Perlis (UniMAP), 01000 Kangar, PerlisMalaysia.

[**2]Faculty of Applied Sciences, Universiti Teknologi Mara (UiTM), Shah Alam, Selangor, Malaysia

E-mail: [*1]shnadzirahsaa@gmail.com, [**2]azlan515@salam.uitm.edu.my

Abstract- **Simple doping process of Zinc (Zn) with Tin oxide (SnO_2) was successfully prepared by the mechanochemical process, followed by a heat treatment and leaching. The raw materials are Tin Chloride ($SnCl_2$) as a base element, Sodium Carbonate (Na_2CO_3) as an oxidizer and Sodium chloride (NaCl) as a diluent, while Zinc Chloride ($ZnCl_2$) as a doping element. In this paper, we with fix wavelength investigated the effect of different concentration of Zn dopant on the optical properties of SnO_2. The chemical formula of doped SnO_2 will be $Sn_{1-x}Zn_xO_2$. The peaks of the X-ray Diffraction (XRD) prove that all Zn had been successfully doped into SnO_2 host. The average sizes of crystalline were around 25 to 44 mm and it was calculated by using Scherrer's equation. While Ultraviolet-visible spectroscopy (UV-VIS) and photoluminescence were used to analyze the energy gap differences of pure SnO_2 and doped SnO_2 with varying Zn concentration. Results show that increasing in the concentration can decrease the volume of samples. Red shift in energy gap (E_g) with an increasing Zn concentration ($x \leq 0.06$) could be attributed to the Burstein-Moss effect that relates the E_g with the crystallite size of SnO_2 formed before and after the doping process. The blue shift in E_g at $x>0.06$ was possibly due to excess oxygen and may also be affected by sudden increases in crystallite size. The emission intensity was changed inversely with the E_g.**

Keywords: Zinc doped tin oxide; Mechanochemical; and Optical properties.

I. INTRODUCTION

Tin oxide (SnO_2) is an n-type semiconductor with stable structure [1] with wide energy gap, $E_g>3.0$ eV [2]. This material is suit well as transparent conducting oxide (TCO), oxidation catalyst, solid state gas sensing material, and negative electrodes for lithium batteries [1]. These applications are due to it novelties such as high transparency, high conduction, high chemical stability even at high temperatures and high sensitivity for gas-sensing applications [1]. Zn was

used as a dopant in order to form stable bonds [3]. Nanosized SnO_2 has been produced via several methods; including sol-gel [4], spray pyrolysis [2], pulsed laser ablation [5], chemical vapour deposition [6], sputtering and many others [7]. However these techniques often cause a very high degree of agglomeration. Thus mechanochemical processing is an alternative method for the production of nanosize materials, where separated nanoparticles can be produced and it is the only process that is simple, cheap and convenient method suitable for large scale production of nanoparticles [1]. Recently, a diluents of the reaction (salt), has been added to the starting materials during the mechanochemical processing. The main purpose is to separate the nanoparticles, prevent their subsequent growth or the synthesized powder from being significantly agglomerated [8]. In this work, the effect of varying Zn concentration in doped SnO_2 nanoparticles on the optical properties prepared by mechanochemical synthesis was reported and the results were discussed in detail.

II. METHODOLOGY

Four main materials need to be measured by using an electronic balance before start doping process wereTin Chloride (SnCl2), Sodium Carbonate (Na_2CO_3), Zinc Chloride ($ZnCl_2$) and Sodium chloride (NaCl). $SnCl_2$ was used as a base element, Na_2CO_3 as an oxidizer, NaCl as a diluent and $ZnCl_2$ as a doping element. The mixing of these powders was known as a simple doping process. Different concentrations of Zn were used to dope with the SnO_2. Zn doped SnO_2 nanoparticles with a balanced chemical equation of $Sn_{1-x}Zn_xO_2$, where $x = 0, 0.02, 0.04, 0.06, 0.08, 0.10$. The powders were heated in a vacuum oven at $150^{\circ}C$ within 45minutes to drive out the majority of solvent inside. The starting material powder was sealed in a 250 ml zirconium oxide vial along with 10 zirconium oxide balls of 20 mm diameter using a ball-to-powder weight ratio of 10:1. Then the mixed powder was milled in a planetary mill for 5 hours at 500 rpm.The milled powder was then heat-treated at $600^{\circ}C$ inside a furnace within 2 hours to get pure SnO_2 by removing carbon dioxide. Samples then were leached out by distilled water to extract samples from NaCl and produce nanocrystalline SnO_2.

The phases and crystallite size of the samples were characterized and calculated by X-ray diffraction (XRD) analysis (Cu-Kα radiation). The average of crystallite size (D) for the SnO_2 was calculated from the data of XRD peaks based on the Scherrer's equation. Ultraviolet Visible spectroscopy (UV-Vis) technique was used for the optical properties characterization of the samples. Energy gap (E_g) value of the samples was determined from Tauc plots. Photoluminescence (PL) was used to characterize the emission spectra intensity of SnO_2 when doped with Zn.

III. DISCUSSION

SnO_2 has been formed when $SnCl_2$ was mixed with Na_2CO_3 and NaCl. The chemical equation involves for the formation of SnO_2 is shown below:

$SnCl_2 + Na_2CO_3 + 6NaCl \longrightarrow SnCO_3 + 8NaCl$ (after milling)

(After heat treatment) $\longrightarrow SnO + 8NaCl$ (remove carbon dioxide, CO_2)

(After leaching) $\longrightarrow SnO$ (remove NaCl)

Finally, the SnO formed were shifted to SnO_2 once it reacts with oxygen in the atmosphere. The oxidation begins at the surface of Sn particles. As a result, SnO also SnO_2 are nucleated on the surface of Sn particles and continuously growing in the nanoparticles to allow oxygen diffuse into the SnO particles. The chemical equation of $Sn_{1-x}Zn_xO_2$ was formed once SnO is doped with Zn. "x" represents the concentration of Zn dopant introduced into SnO_2 in mole unit.

There was a change in colour from white powder (before milling process) to brown nanopowder (after milling process). However, after annealing under $600^{\circ}C$, the powder samples were turned to white again. The changes were affected by the interaction of SnO with oxygen from the environment. In fact, the quantity of SnO_2powder samples experience

TABLE I
ORIENTATION AND 2 THETA READINGS FOR for $Sn_{0.10}Zn_{0.00}O_2$

ORIENTATION	2 THETA (DEGREES)
(110)	26.6
(101)	33.9
(200)	38.0
(111)	38.8
(211)	51.7
(220)	54.7
(002)	57.8
(310)	61.9
(112)	64.7
(301)	66.0

extremely reduced in almost three quarter after annealing. This is because CO_2 gas had been successfully released.

An XRD pattern of the doping SnO_2 with Zn nanopowder is shown in Figure 1. The XRD peaks of the samples were indexed and they are corresponding to the wurtzite structure (JCPD files, No. 88-0287). Table I shows its orientation with peaks. The intensity of the (110) peak increased relatively rapidly compared to that of the (211) peak indicates that the SnO_2 nanopowders were crystallized into a perfect polycrystalline structure with orientations along the (110) plane [9] with no impurity existence. The 2θ (diffraction angle) were shifted towards increment as the concentration of Zn were increased. This phenomenon reflects all samples were successfully doped [10] and it can be proved by the shrinking of lattice constant with increasing Zn content as shown in Figure 2. Lattice constant was calculated from the XRD peak positions using least square method. The strongly shrinkage of the lattice constant could be due to smaller ionic radius of Sn^{2+} (1.10 Å) [11] substituted in place of Zn^{2+} (0.74 Å) site [3]. Since the ionic radius of Zn^{2+} is less than that of Sn^{2+}, Zn atoms are most likely to be located in Sn positions in the crystallite bulk. Figure 2 reflects the decreasing values of lattice parameters of a and c and indirectly it causes the volume of the SnO_2 shrinking.

Figure 3 shows the crystallite size of $Sn_{1-x}Zn_xO_2$ nanoparticles calculated using the Scherrer's formula in Equation (1) from the full width at half maximum (FWHM) of the XRD peaks as a function of the Zn content.

$$d = \frac{k\lambda}{B\cos\theta} \qquad (1)\ [12]$$

where d is the mean crystallite size, k is a constant usually equal to 0.9, λ the wavelength of Cu Kα ($\lambda = 1.5405$ Å), B the full width at half maximum intensity of the peak (FWHM) in radian and θ is Bragg's diffraction angle [3]. Due to the figure, the crystalline sizes were increased from. $Zn_{0.00}$ until $Zn_{0.06}$. This is suggested to be caused by increasing in diffusion rate with the addition of dopant.Then, suddenly the size was decreased at $Zn_{0.08}$ and $Zn_{0.10}$. It was occurring because when SnO_2 is doped with Zn, some quantity of Zn atom tends to locate in or near crystal boundary regions which courage the growth of the crystals during the reaction in the ball mill, resulting in increasing crystallite size [13]. Table II gives the values of crystalline size for each dopant concentration of Zn. The nanosized of the crystallites form may give advantages for high gas sensing applications.

Further confirmation of the incorporation of Zn into SnO_2 lattice was provided by monotonic reduction in band gap determined from optical absorption spectra shown in Figure 4 by using Tauc plots of $(\alpha h\nu)^2$ versus $h\nu$. Figure 4 shows the result of energy gap with varying Zn concentrations. The result was observed from the Lambda 35Uv-Vis system. The E_g was evaluated from the intercept of the linear portion of the curve of $(\alpha h\nu)^2$ versus $h\nu$ plots using the relation $(\alpha h\nu)^2 = A(h\nu - E_g)$, where $h\nu$ is the photon energy and α is the absorption coefficient [13].The band gap decreased from 3.760 eV, 3.62 eV, 3.33 eV and 3s.281 eV for $Sn_{0.10}Zn_{0.00}O_2$,

978-1-4673-2395-6/12 $31.00 © 2012 IEEE   201

$Sn_{0.08}Zn_{0.02}O_2$, $Sn_{0.06}Zn_{0.04}O_2$, and $Sn_{0.04}Zn_{0.06}O_2$ samples respectively. Then it was increased at $Sn_{0.02}Zn_{0.08}O_2$ and $Sn_{0.00}Zn_{0.10}O_2$ from 3.675eV to 3.761 eV respectively.

The red shift absorption edge or decrement of the energy gap (from 3.76 to 3.281 eV) at the first four concentrations was explained by the Brus formula which explains about the confinement effect. In the quantum confinement range, the band gap is inversely proportional to the crystallite size of particles [14]. This can be proved by the Brus formula in Eq. (2) given below:

$$Eg = E_g^{bulck} + \frac{\hbar^2\pi^2}{2r^2}\left(\frac{1}{m_e}+\frac{1}{m_h}\right) - \frac{1.8}{4\pi\varepsilon\varepsilon_o r}e^2$$
$$- \frac{\hbar^2\pi^2}{\hbar^2(4\pi\varepsilon\varepsilon_o r)^2}\left(\frac{1}{m_e}+\frac{1}{m_h}\right)^{-1} \qquad (2)$$

where E_g^{bulck} is the bulk semiconductor band gap energy, E_g is the absorption band gap of nano-semiconductor particles, r is the particle radius, m_e is the effective mass of the electrons = $0.24m_o$, m_h is the effective mass of the holes = $0.45m_o$, ε is the relative permittivity = 3.7, ε_o is the permittivity of free space, $\hbar$ is Planck's constant, and e is the charge of the electron. Thus, this formula reflects the relation between the crystalline sizes (r) with the energy gap. Once the crystallite size of SnO_2 was increased, the band gap of the particle will be decreased. This phenomenon can be clearly compared in Figure 3 and Figure 4.

However, the increment of the optical bandgap energy of Zn doped SnO_2 nanoparticles from 3.675 to 3.761eV at $Zn_{0.08}$ and $Zn_{0.10}$ respectively can be attributed to the Moss–Burstein effect. The Burstein-Moss shift Eq. (3) can be seen as below:

$$\Delta E_g = \left(\frac{\hbar^2}{2m^*}\right)(3\pi^2 n_e)^{\frac{2}{3}} \qquad (3)$$

where ΔE_g is the shift of the doped semiconductor with respect to the host semiconductor, m^* is the reduced effective mass, $\hbar$ is Plank constant and n_e is the free carrier concentration. The blue shift of the absorption edge or bandgap widening (BGW) is due to the increment of donor atom or Zn concentration. As the Zn concentration increased, the donor states push the Fermi level higher in energy. Therefore, we can observe an increase in the measured band gap as the Zn concentration increase [15] and the total measured band gap is the summation of actual band gap with Moss-Burstein shift. (Measured band gap = Actual band gap + Moss-Burstein shift [16].

The parabolic shape of the energy gap of SnO_2 result is almost similar with J. S. Bhat et.al [17], people who studied the influence of Zn doping on electrical and optical properties of multilayered SnO_2 thin films via sol-gel method. Figure 5 shows the PL results synthesized by mechanochemical process. Fluorescent LS55 with the wavelength ranges were 190-1100 nm and the bandwidths applied were 0.5-4 nm had been used to identify the energy gap result. From these emission spectra, it is clearly observed that the fluorescence intensities ratio transitions increase gradually along with the addition of Zn^{2+} doping from $Zn_{0.00}$, $Zn_{0.02}$, $Zn_{0.04}$ and $Zn_{0.06}$ with the values of the emission intensity were 113.462a.u, 114.983 a.u, 128.449a.u and 148.043a.u respectively. After that, the

TABLE II
CRYSTALLINE SIZE OF $Sn_{1-x}Zn_xO_2$ PARTICLES.

Zn concentration, x	Crystalline size (nm)
0.00	25.636
0.02	31.916
0.04	40.156
0.06	42.581
0.08	28.431
0.10	24.806

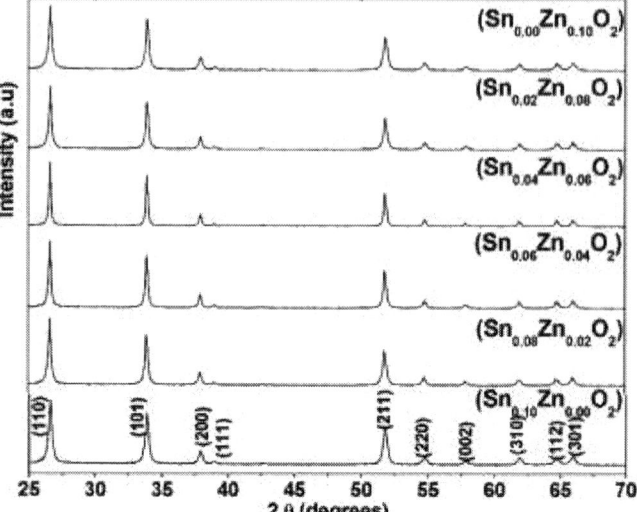

Figure 1: XRD patterns of $Sn_{1-x}Zn_xO_2$

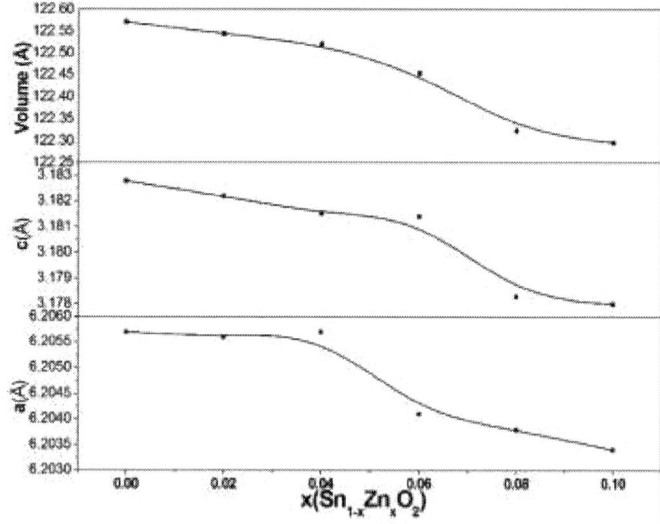

Figure 2: Lattice structure of $Sn_{1-x}Zn_xO_2$

intensity was decreased at $Zn_{0.08}$ with the intensity was 137.576a.u and increased again at $Zn0.10$ with 145.905a.u. The reason to enhance the luminescent intensity spectra may be attributed to many aspects.

The doping process of Zn with SnO_2 causes higher

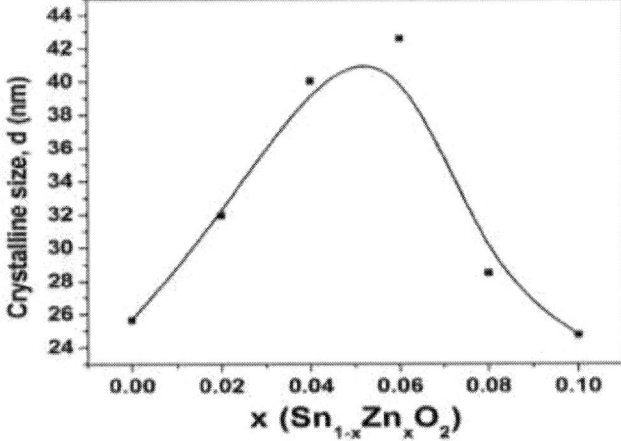

Figure 3: Crystalline size of $Sn_{1-x}Zn_xO_2$ particles.

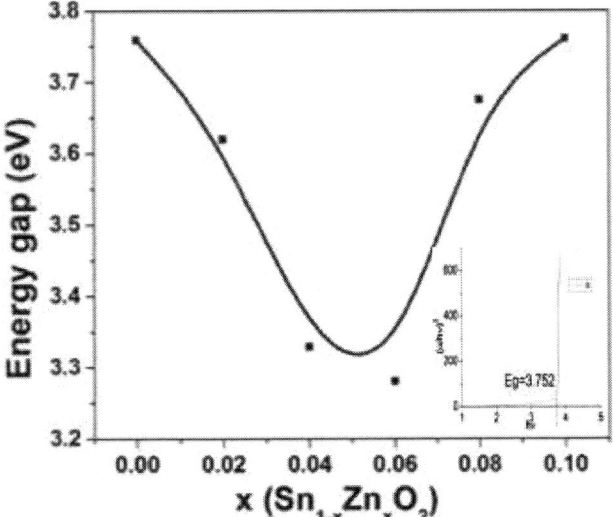

Figure 4: The energy gap of $Sn_{1-x}Zn_xO_2$ nanoparticles and Inset shows a plot of $(\alpha h\nu)^2$ versus $h\nu$.

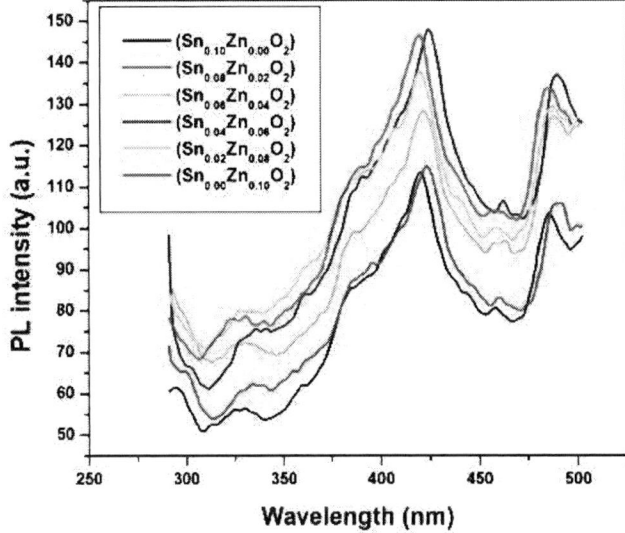

Figure 5 The emission intensity of different $Sn_{1-x}Zn_xO_2$ with fixed wavelength.

is relatively high possible due to the presence of unreacted Zn [20].

IV. CONCLUSION

As a conclusion, Zn doped SnO_2 have been successfully prepared by mechanochemical process with different concentration ranging from $Zn_{0.00}$ to $Zn_{0.10}$. The XRD gives the fact that Zn substitution into the SnO_2 nanoparticles can be confirmed by the shifting of the peaks and shrinkage of the lattice constant with increasing Zn content. Uv-Vis result can be explained by Brus and Burstein-Moss shift formula that is inversed with the crystalline size graph. The result of PL can be attributed to a combined effect of crystallite size reduction and compressive stress as a result of Zn doping in SnO_2 nanoparticles.

ACKNOWLEDGMENT

This work was supported by Faculty of Applied Sciences in Universiti Teknologi Mara and Institute of Nano Electronic Engineering in Universiti Malaysia Perlis.

strain was generated due to incorporation of Zn^{2+} ions inSnO_2 lattices. Such a strain increases crystalline defects due to Zn doping and consequently the intensity of the corresponding PL peak

increases. It is well known that the band gap of semi-conducting materials increase with decreasing in crystallite size. Hence the relative energy difference between the defect levels causing PL is also expected to increase as observed here [18]. It is also reported to be caused by compressive strain within the material. In the present case, Zn^{2+} incorporated into the SnO_2 lattice results in compressive strain due to smaller ionic radius of Zn^{2+} ions compared to Sn^{2+} ions. Hence, the PL 'blue-shift' observed can be attributed to a combined effect of crystallite size reduction and compressive stress as a result of Zn doping in SnO_2 nanoparticles [19]. The phenomenon of increasing relatively high PL intensity also can be referred to the interaction of phonon with exciton at these peaks almost disappeared after Zn doped SnO_2 even though it has similar size as that of nanoparticles. The intensity of the $Zn_{0.10}$emission

REFERENCES

[1] Yang, H., et al., Synthesis of tin oxide nanoparticles by mechanochemical reaction. Journal of Alloys and Compounds, 2004. 363(1–2): p. 276-279.

[2] Patil, G., et al., Synthesis, characterization and gas sensing performance of SnO<sub>2</sub> thin films prepared by spray pyrolysis. Bulletin of Materials Science. 34(1): p. 1-9.

[3] Torabi, M. and S.K. Sadrnezhaad, Electrochemical evaluation of nanocrystalline Zn-doped tin oxides as anodes for lithium ion microbatteries. Journal of Power Sources. 196(1): p. 399-404.

[4] Wu, Y.C., et al., Synthesis of tin oxide nanosized crystals embedded in silica matrix through sol–gel process using alkoxide precursors. Journal of Non-Crystalline Solids, 2009. 355(16–17): p. 951-959.

[5] Ponce, L., et al., Preparation and characterization of tin oxide thin films by pulsed laser deposition. AIP Conference Proceedings, 1996. 378(1): p. 165-168.

[6] Liu, X., et al., Preparation of Tin Oxide Self-assembly Nanostructures by Chemical Vapor Deposition. Acta Physico-Chimica Sinica, 2007. 23(3): p. 361-366.

[7] Cukrov, L.M., T. Tsuzuki, and P.G. McCormick, SnO2 nanoparticles prepared by mechanochemical processing. Scripta Materialia, 2001. 44(8â€"9): p. 1787-1790.

[8] Ao, W., et al., Mechanochemical synthesis of zinc oxide nanocrystalline. Powder Technology, 2006. 168(3): p. 148-151.

[9] Jeong, J., et al., Photoluminescence properties of SnO2 thin films grown by thermal CVD. Solid State Communications, 2003. 127(9): p. 595-597.

[10] Bilgin, V., et al., The effect of Zn concentration on some physical properties of tin oxide films obtained by ultrasonic spray pyrolysis. Materials Letters, 2004. 58(29): p. 3686-3693.

[11] Shi, Z.M., et al., The phase transformation behaviors of Sn2+-doped Titania gels. Journal of Non-Crystalline Solids, 2007. 353(22â€"23): p. 2171-2178.

[12] Sutti, A., et al., Inverse opal gas sensors: Zn(II)-doped tin dioxide systems for low temperature detection of pollutant gases. Sensors and Actuators B: Chemical, 2008. 130(1): p. 567-573.

[13] Sabri, N.S., A.K. Yahya, and M.K. Talari. Effect of Aluminium Doping on Structural and Optical Properties of ZnO Nanoparticles Prepared by Mechanochemical Synthesis. AIP Conference Proceedings [cited 1328 1]; 268-270].

[14] Gondal, M.A., Q.A. Drmosh, and T.A. Saleh, Preparation and characterization of SnO2 nanoparticles using high power pulsed laser. Applied Surface Science. 256(23): p. 7067-7070.

[15] Hudait, M.K., P. Modak, and S.B. Krupanidhi, Si incorporation and Bursteinâ€"Moss shift in n-type GaAs. Materials Science and Engineering: B, 1999. 60(1): p. 1-11.

[16] Grundmann, M., The Physics of Semiconductors. Second Edition ed. An Introduction Including Nanophysics and Applications, ed. P.W.T. Rhodes, P.H.E. Stanley, and P.R. Needs. 2010: Springer.

[17] Bhat, J., K. Maddani, and A. Karguppikar, Influence of Zn doping on electrical and optical properties of multilayered tin oxide thin films. Bulletin of Materials Science, 2006. 29(3): p. 331-337.

[18] Zhang, H., et al., Luminescence properties of Li+ doped nanosized SnO2:Eu. Journal of Luminescence, 2005. 115(1â€"2): p. 7-12.

[19] Rani, S., et al., Structure, microstructure and photoluminescence properties of Fe doped SnO2 thin films. Solid State Communications, 2007. 141(4): p. 214-218.

[20] Kundu, S. and P.K. Biswas, Synthesis and photoluminescence study of nanostructured solâ€"gel Mn(II) doped indium tin oxide films on silica glass. Chemical Physics Letters, 2006. 432(4â€"6): p. 508-512.

Design Study of Integrated Optical Transducer for Bioparticles Detection

Marianah Masrie, Burhanuddin Yeop Majlis, *SMIEEE*, Jumril Yunas, *MIEEE* and P Susthitha Menon, *MIEEE*
Institute of Microengineering and Nanoelectronics (IMEN)
Universiti Kebangsaan Malaysia (UKM)
43600 UKM Bangi, Selangor, Malaysia
Email: Burhan@vlsi.eng.ukm.my

Abstract- **This paper reports the design study of optical transducer for bioparticles detector integrated in Lab-on-Chip (LoC) system. Optical detector has been identified as potential device for real time detection of bioparticles because of its simple and compact structure, high sensitivity and easy of integration with microfluidic system. In this study, the absorbance detection using optical transducer is selected due to its label free nature and relatively simple implementation. The analysis work is focused on the characteristics of PIN photodetector simulated using Atlas and Athena software. It is observed that the lateral SOI based PIN photodiode exhibits the behavior of silicon PIN photodiode characteristics. The result also shows that Peak intensity at wavelength of 400nm was achieved and therefore, it is suitable to be used as bioparticles detector in UV/Vis spectrum.**

I. INTRODUCTION

In recent years, there is an increasing demand for highly sensitive and portable devices that is compact and able for integration with other part of Lab-on-Chip components. This kind of devices can provide rapid biological sample analysis in biomedical diagnosis. The importance of rapid and precise bioparticles detection is necessary for the medical diagnosis, treatment and prevention of diseases.

The evolution of research in bioparticles detection was started in late 1960's when some of the conventional bioparticles detector devices were fabricated [1]. Currently, many different approaches have been used in bioparticles detection such as a conventional method using culturing, using antibody and biosensing techniques [2]. Now, there is a growing need for developing microfluidic and bioparticles detection based technology in lab-on-a-chip device that able to fast detect, low cost and portable sensing of biological and chemical samples. These samples are no longer to be sent to the Centres for Disease Control and laboratory for further study since this technology may provide on-site laboratory analysis solution. Therefore, the application for these lab-on-a-chip devices may be applied in medical diagnostics. One of the detection principles is employing an optical transduction method which is identified as higher potential for real time detection detection of bioparticles because it is simple, highly sensitive and easy to be integrated with microfluidic system [3].

This paper discusses the idea of concepts for the MEMS devices design which employ microfluidics integrated with optical transducer. Apart from that, it also includes the previous works on bioparticles detection using optical method for both label and label free detection.

II. PAST RESEARCH WORKS ON BIOPARTICLES DETECTION USING OPTICAL METHOD

In general, there are two types of methods for bioparticles detection using optical transduction; with label detection and label-free detection [4]. The material that is used to label the particle is normally fluorescence tags where the intensity of the fluorescence indicates the presence of the target particles. On the other hand, in label-free detection target particles are not labeled or unmodified and are detected in their natural forms.

A. With Label detection method

One of the labeling methods in particles sensing that apply optical method as the transducer is flow cytometer. In this method, particles that are labeled with fluorescent dye emit light when it is excited with a light source. It has three main systems: one or more laser light sources and a sensing system that comprises the sample/flow chamber and optical assembly, a hydraulic system that controls the passage of cells through the sensing system, and a computer system that collects data and performs analytical routines on the electrical signals relayed from the sensing system.

Due to a number of recent technological advances, a hand-held flow cytometer can be achieved by using MEMs technology. Barat et al. has reported a microfluidic cytometer that performs simultaneous optical and electrical characterization of single particle [5]. The device consists of solid state laser as the light source and the fluorescence was detected by integrated optical fibers coupled to photo multipliers tube (PMT). The cytometer managed to identifies four different sizes of beads in the range of size from 10 to 25 micrometer by identifying the intensities in the fluorescent beads.

The usage of laser source as the light source requires additional optical component alignment to focus the light is now replaced with light emitting diode (LED). Grafton et al

978-1-4673-2395-6/12 $31.00 © 2012 IEEE

[6] demonstrated a valve-less, on-chip magnetic sorting of immunomagnetically labeled white blood cells, bright Qdot labeling of lymphocytes, and counting of labeled white blood cells. The microfluidic cytometer was composed with UltraViolet (UV) LED source which illuminate the white blood cells and fluorescence was detected by silicon photomultipliers. For the prototyping of the microfluidic devices, PDMS were designed and fabricated using standard photolithography and soft lithography techniques.

A low cost and low power consumption of microfluidic LOC system is developed for portable cytometers for a wide range of applications [7]. It comes with planar design with low layer thickness and implemented LED as light source and a PIN photodiode which reduce component cost and power consumption. PMT is expensive and sensitive component since it requires high quality optical filter to separate the fluorescence from the excitation light. PIN photodiode do not have this limitation and much cheaper. The device managed to identify up to 600 particles per second.

There is a drawback behind the advancement of these fluorescence-label devices which is time consuming, complex and expensive since the bioparticles need to be labeled using fluorescence markers [8]. In addition, this method is not suitable for the biophysical characterization of living cells due to the modification of their natural forms.

B. Label Free Detection method

In contrast to the previous method, label free detection method has the potential to deliver high quality, high information content detection to bioparticles [9]. Light absorption principle in the range from blue, uv/Vis to infrared spectrum is a nature label free for direct detection of bioparticles that can overcome the problem in fluorescence-labeling method. It provides the information about the structure of molecules where it measures the vibrational levels of the material. When the light source illuminates the molecules, some of the absorbed energy changes the vibrational levels of the bond within a molecule and the energy with this change is different depending on the structure of the molecule [10].

Chandrasekaran et al. developed a hybrid integrated optical microfluidic system in order to study a single molecule and enzymatic reactions in both chemical and biological reaction through the optical absorption technique [11]. A visible wavelength was chosen for absorption experiment in a glass microfluidic channel. This type lab-on-a-chip device can be applied in the area of nanotechnology since it can provide in situ biomedical diagnosis of micro and nano level species, point of care testing and rapid pathogenic detections.

A microfluidic device with integrated waveguides and a long path length detection cell for UV/Vis absorbance detection is reported by Petersen et al. [12]. Visible and ultraviolet wavelength region are chosen since a large number of molecules absorb light in these wavelength. The microfabricated device was connected to the laser light source and detector by optical fibers waveguide [13]. A band pass

filter was included to exclude any fluorescent light from the sample.

A lab on chip integrated with optical transducer was developed by Balslev et al. [14]. The dye laser was fabricated on-chip as well as embedded photodiodes in the silicon substrate. The device employed dye laser which emit light to waveguides that brought the light into the cuvette for absorbance measurement. The core planar waveguide was formed by SU-8 layer instead of coupling of optical fibers to the microchip since it is expensive and coupling losses are a significant problem. The light was detected by the photodiode at the other side of the cuvette. The structure of the device is shown in Fig. 3. It managed to perform an absorbance measurement on two different concentrations of xylenol orange dye.

Bulteel et al. [15] demonstrated a DNA concentration measurement in tubes using UV SOI lateral PIN photodiodes. The microsystem consists of a UV LED as the light source in which the light at precise wavelength transmitted through the quartz container of DNA. The photodiode managed to measure responses for DNA concentrations in the range from 400 ng/μL to 4 pg/μL and correlate bacteria concentrations from 6.10^{11} spores/ml to 6.10^7 spores/ml.

III. NEW PROPOSE DESIGN FOR BIOPARTICLES DETECTION USING OPTICAL METHOD

A. Theory of concept

Optical detection at low wavelength has many applications in medical diagnostics such as bacteria and protein detection and DNA concentration measurement. These biological molecules can absorb light when excited at low wavelength in the range from ultraviolet (UV) to visible (Vis) spectrum. DNA, RNA and proteins has strong absorption of UV light at the wavelength around 260nm and 280nm [16].

The illustration of absorption light by biological molecules is shown in Figure 1. Light (photon) from light source is positioned directly to microfluidic channel. When the particles pass through the channel, the incident light is scattered and absorbed by the particles reducing the intensity of the transmitted light [17]. This change light level is measured by a photodetector and recorded as series of pulses.

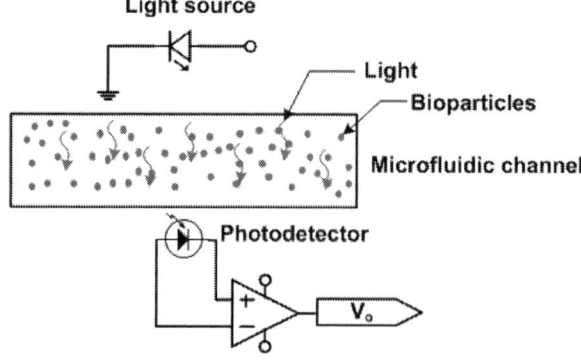

Fig. 1. Optical detection using absorbance method

B. Concept design

The MEMS device for bioparticles detection consists of a microfluidic chip with integrated optical transducer comprises of a UltraViolet light emitting diode (UV LED) as the light source and a lateral thin film SOI PIN photodiode as the photodetector as illustrated in Fig 2(a). Polydimethylsiloxane (*PDMS*) is chosen as the material for the fabrication of the microfluidic since it is optically transparent and suitable for UV/VIS absorbance detection [18].

The lateral SOI PIN photodiode design as published in [19, 20] is used as the reference where in this application is bonded with silicon microfluidic and is placed just below the microfluidic channel. An anti reflection coating (ARC) of Silicon Oxide (SiO2) is deposited with an appropriate thickness below the PDMS microfluidic to ensure minimal reflectivity at the surface of the diode. The UV LED is attached on top of the microfluidic channel and basically covers the microfluidic channel as shown in Figure 2(b).

absorbed and produces electron-hole pairs which are transported to the fingers and collected.

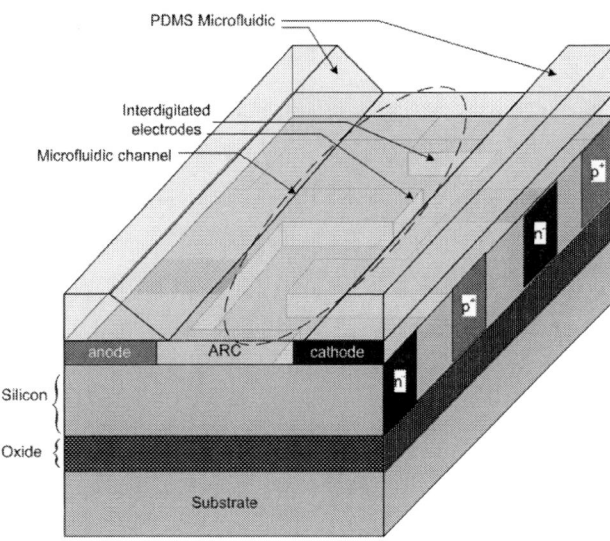

Fig. 3. Schematic 3D view of the microfluidic chip bonded with lateral SOI PIN photodiode.

C. Photodiode characteristics

The simulation on photodiode characteristics is done based on the previous report in [21, 22]. The cross section of lateral PIN photodiode is shown in Figure 4. The thickness of 80 nm-thick top silicon layer is chosen due to the optimum absorption of light emission for bioparticles detection in the blue and UV range. The isolation layer by oxide provides high speed application since it has very low dark current, high responsivity and quantum efficiency [19] . The other dimensions as shown in Figure 4 provide the thickness for the buried oxide at 390nm and the anti reflection coating (ARC) at 50 nm and also the doping concentrations for lightly and heavily doped at the top silicon layer. The length 8μm of the intrinsic region is used in the simulation since it gives an optimum results for the photogenerated hole-electron pairs, dark current and electron-hole recombination [23].

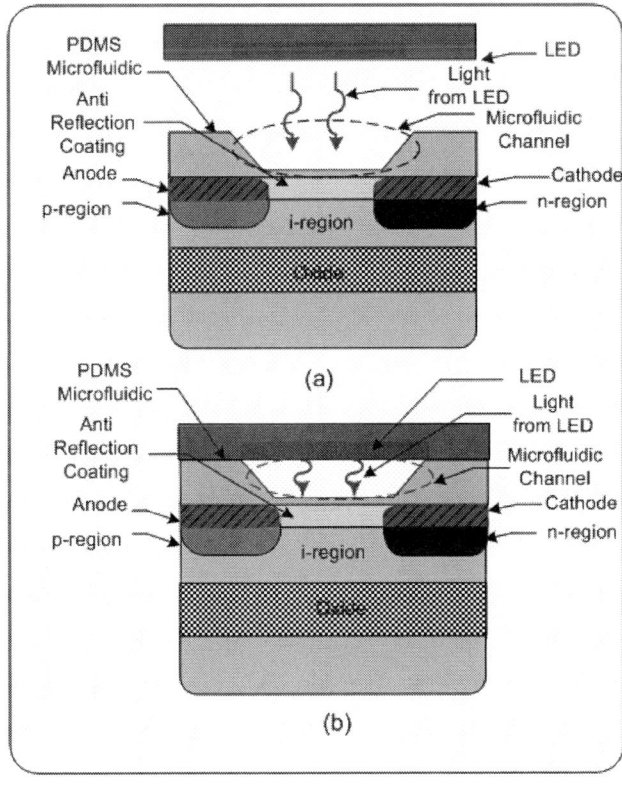

Fig. 2. Cross – sectional outline; excitation light directly into the photodiode through microfluidic

Figure 3 shows the 3D schematic view of the device with the lateral SOI PIN photodiode consists of interdigitated patterns with associated electrodes from the cathode and the anode. Light (photon) reaching the area between electrodes is

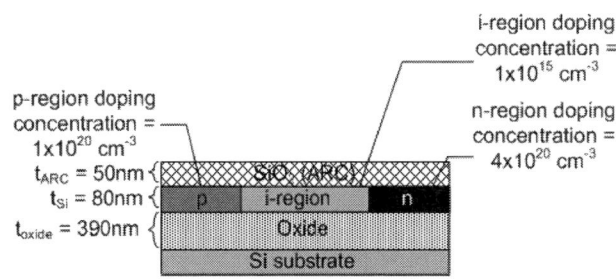

Fig. 4. Cross-section indicating thickness and other dimensions

IV. RESULTS AND DISCUSSION

This paper provides a preliminary study of an analysis of the 2D model of lateral PIN photodiode structure with and without light illumination, electrical and optical characteristics for bioparticles detection. It was simulated using Athena and Atlas from SILVACO software.

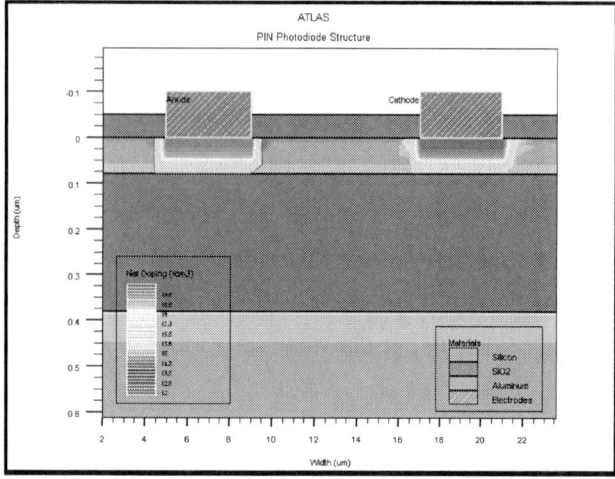

Fig. 5. Lateral PIN photodiode structure

The structure of the PIN photodiode detector is shown in Figure 5. The intrinsic region was uniformly doped with n-type at 1×10^{15} cm^{-3} with heavily doped p$^+$ and n$^+$ regions at the top silicon layer of the structure. An oxide layer which acted as a mask was deposited on top of the silicon layer. The junction depth was measured using extract statement and the result of the depth was equivalent to 0.0666019 μm. Shockley Read Hall (SRH) and Auger recombination mechanisms have been specified in order to see the effects on the quantum efficiency of photodiode.

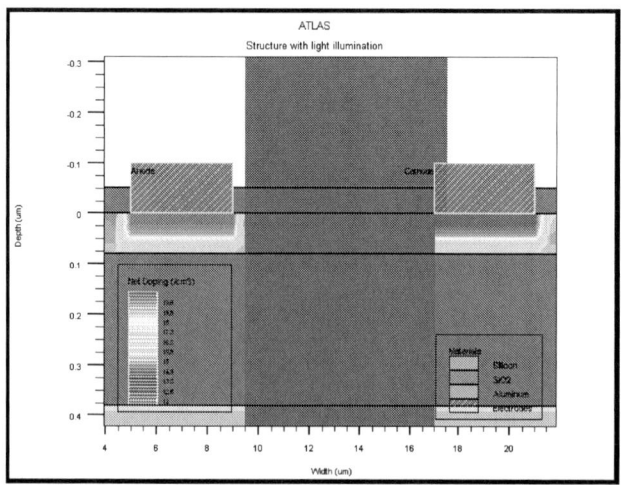

Fig. 6. PIN photodiode structure with light illumination

The DC simulation in order to obtain the forward bias (current-voltage) I-V curves was done without light illumination and with light rays directed at 90° within one micron above the photodiode to the front surface with coordinate point at (9.0,-1.0) as shown in Figure 6. There are 3 curves plotted to compare the I-V without illumination and with light intensity at1000, 100, 10 and 1 W/cm^2. The turn on voltage is found approximately at 1V with forward bias voltage at 2V as depicted at the curve respectively.

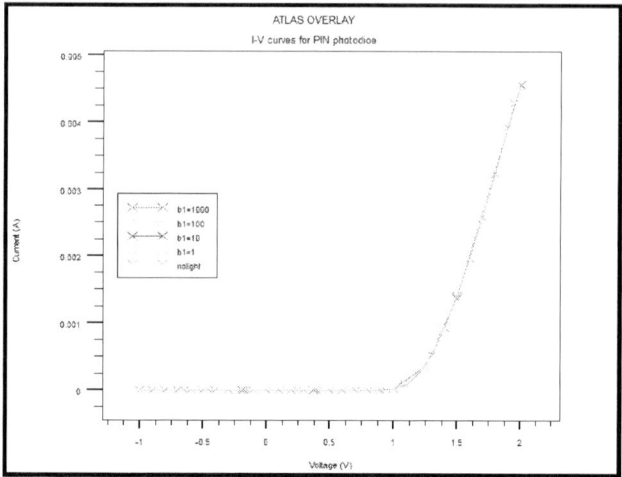

Fig. 7. I-V curves (forward bias) of the PIN photodiode

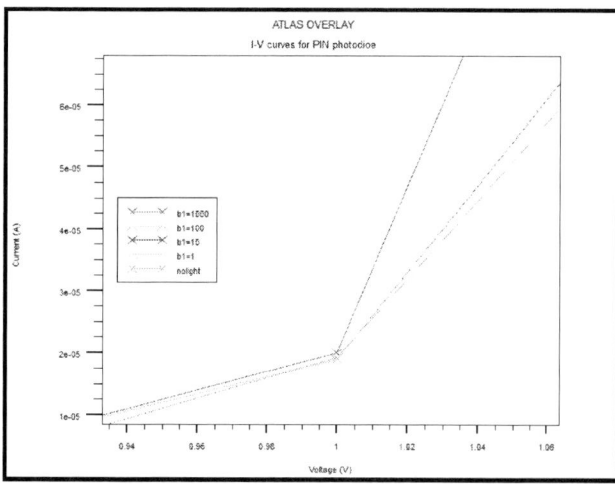

Fig. 8. I-V curves with different light intensity

As the light intensity increased from lowest to highest value, it can be seen that the photocurrent is also increased right after the turn on voltage. This can be observed clearly in Fig 8. As shown in Fig 9, the peak responsivity of 0.117 A/W is found at 400 nm and the structure is blind when it reach longer wavelength. This is due to low level of photon energy in the photodiode structure. Based on the result, it shows that the photodetector exhibits good quantum efficiency below wavelength about 400nm.

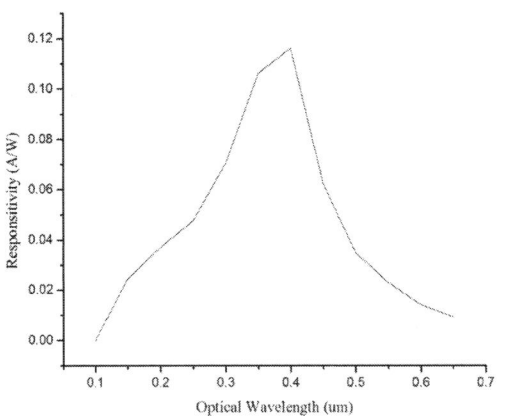

Fig. 9. Responsivity for various spectrums

V. CONCLUSIONS

The development of medical diagnostic devices as part of lab on a chip component has realized the demand of point of care diagnosis needs. The label free optical detection of bioparticles has been used to overcome the complexity of detection method . Therefore, the new propose design for bioparticles detection using optical transduction has been discussed in this paper. An SOI based lateral PIN photodiode was designedfor the detection of bioparticles in UV/VIS spectrums. The preliminary study of integrated optical transducer for bioparticles detection has been successfully conducted by simulating the photodetector - lateral SOI photodiode. Based on the simulation results, the PIN photodiode exhibits the behavior of silicon PIN photodiode characteristics and therefore, it is suitable to be used in bioparticles detection in UV/Vis spectrum.

ACKNOWLEDGMENT

This work is supported by Universiti Kebangsaan Malaysia (UKM) under the project NND/ND/(1)/TD11-002 (Development of lab-on chip for peripheral blood stem cell isolation and rapid detection of tropical diseases from blood).

REFERENCES

[1] H. M. Shapiro and R. C. Leif, *Practical flow cytometry* vol. 4: Wiley Online Library, 2003.

[2] C. Jae-Woo, A. Pu, and D. Psaltis, "Bacteria detection in a microfluidic channel utilizing electromagnetic cellular polarization and optical scattering," in *LEOS Summer Topical Meetings, 2006 Digest of the*, 2006, pp. 17-18.

[3] D.-Q. Huo, Z. Liu, C.-J. Hou, J. Yang, X.-G. Luo, H.-B. Fa, J.-L. Dong, Y.-C. Zhang, G.-P. Zhang, and J.-J. Li, "Recent Advances on Optical Detection Methods and Techniques for Cell-based Microfluidic Systems," *Chinese Journal of Analytical Chemistry,* vol. 38, pp. 1357-1365, 2010.

[4] X. Fan, I. M. White, S. I. Shopova, H. Zhu, J. D. Suter, and Y. Sun, "Sensitive optical biosensors for unlabeled targets: A review," *Analytica Chimica Acta,* vol. 620, pp. 8-26, 2008.

[5] D. Barat, D. Spencer, G. Benazzi, M. C. Mowlem, and H. Morgan, "Simultaneous high speed optical and impedance analysis of single particles with a microfluidic cytometer," *Lab on a Chip,* 2011.

[6] M. M. G. Grafton, T. Maleki, M. D. Zordan, L. M. Reece, R. Byrnes, A. Jones, P. Todd, and J. F. Leary, "Microfluidic MEMS hand-held flow cytometer," 2011, p. 79290C.

[7] S. W. Kettlitz, S. Valouch, W. Sittel, and U. Lemmer, "Flexible planar microfluidic chip employing a light emitting diode and a PIN-photodiode for portable flow cytometers," *Lab on a Chip,* 2012.

[8] A. Liu, H. Huang, L. Chin, Y. Yu, and X. Li, "Label-free detection with micro optical fluidic systems (MOFS): a review," *Analytical and Bioanalytical Chemistry,* vol. 391, pp. 2443-2452, 2008.

[9] M. A. Cooper, "Label-free screening of bio-molecular interactions," *Analytical and Bioanalytical Chemistry,* vol. 377, pp. 834-842, 2003.

[10] R. A. Yotter, L. A. Lee, and D. M. Wilson, "Sensor technologies for monitoring metabolic activity in single cells-part I: optical methods," *Sensors Journal, IEEE,* vol. 4, pp. 395-411, 2004.

[11] A. Chandrasekaran and M. Packirisamy, "Absorption detection of enzymatic reaction using optical microfluidics based intermittent flow microreactor system," 2006, pp. 137-143.

[12] N. J. Petersen, K. B. Mogensen, and J. P. Kutter, "Performance of an in-plane detection cell with integrated waveguides for UV/Vis absorbance measurements on microfluidic separation devices," *Electrophoresis,* vol. 23, pp. 3528-3536, 2002.

[13] K. B. Mogensen, N. J. Petersen, J. Hübner, and J. P. Kutter, "Monolithic integration of optical waveguides for absorbance detection in microfabricated electrophoresis devices," *Electrophoresis,* vol. 22, pp. 3930-3938, 2001.

[14] S. Balslev, A. M. Jorgensen, B. Bilenberg, K. B. Mogensen, D. Snakenborg, O. Geschke, J. P. Kutter, and A. Kristensen, "Lab-on-a-chip with integrated optical transducers," *Lab on a Chip,* vol. 6, pp. 213-217, 2006.

[15] O. Bulteel, V. Overstraeten-Schlogel, P. Dupuis, and D. Flandre, "Complete microsystem using SOI photodiode for DNA concentration measurement," 2010, pp. 142-145.

[16] A. Karczemska and A. Sokolowska, "Materials for DNA sequencing chip," *Journal of Wide Bandgap Materials,* vol. 9, pp. 243-259, 2002.

[17] J. Wu and M. Gu, "Microfluidic sensing: state of the art fabrication and detection techniques," *Journal of Biomedical Optics,* vol. 16, p. 080901, 2011.

[18] J. C. McDonald, D. C. Duffy, J. R. Anderson, D. T. Chiu, H. Wu, O. J. A. Schueller, and G. M. Whitesides, "Fabrication of microfluidic systems in poly (dimethylsiloxane)," *Electrophoresis,* vol. 21, pp. 27-40, 2000.

[19] M. de Souza, O. Bulteel, D. Flandre, and M. A. Pavanello, "Temperature and Silicon Film Thickness Influence on the Operation of Lateral SOI PIN Photodiodes for Detection of Short Wavelengths," *Journal of Integrated Circuits and Systems,* vol. 6, pp. 107-113, 2011.

[20] P. Menon, K. Kandiah, A. Ehsan, and S. Shaari, "Concentration-dependent minority carrier lifetime in an In0. 53Ga0. 47As interdigitated lateral PIN photodiode model based on spin-on chemical fabrication methodology," *International Journal of Numerical Modelling: Electronic Networks, Devices and Fields,* vol. 24, pp. 465-477, 2011.

[21] O. Bulteel, P. Dupuis, S. Jeumont, L. Irenge, J. Ambroise, B. Macq, J. L. Gala, and D. Flandre, "Low-cost miniaturized UV photosensor for direct measurement of DNA concentration within a closed tube container," 2009, pp. 1057-1061.

[22] P. S. Menon, K. Kandiaha, A. A. Ehsana, and S. Shaaria, "An interdigitated diffusion-based In0. 53Ga0. 47As lateral PIN photodiode," 2008, pp. 68380C-1.

[23] O. Bulteel and D. Flandre, "Optimization of blue/UV sensors using pin photodiodes in thin-film SOI technology," 2009.

The Effect of Isopropyl Alcohol on Anisotropic Etched Silicon for the Fabrication of Microheater Chamber

Norihan Abdul Hamid, *MIEEE*, Burhanuddin Yeop Majlis, *SMIEEE*, Jumril Yunas, *MIEEE*, and
Mimiwaty Mohd Noor, *MIEEE*
Institute of Microengineering and Nanoelectronics (IMEN)
Universiti Kebangsaan Malaysia (UKM)
43600 UKM Bangi, Selangor, Malaysia
Email: burhan@vlsi.ukm.my

Abstract- In bulk micromachining technology, anisotropic etching process has been one of the most popular processes in creating 3-dimensional MEMS structure, due to its simple and low cost process techniques. This paper presents the investigation of isopropyl alcohol (IPA) effect on anisotropic etched silicon surface of microheater chamber. The aim of the study is to find the optimal etch solution composition which will produce a smooth etched surface, low lateral etch (undercutting) effect and controllable etching rate. The etching process was carried out at with various potassium hydroxide (KOH) concentrations by adding various IPA compositions in the etchant solution. The effects of the solution were observed for several temperature conditions ranging from 50°C to 80°C. From the experimental results, it was observed that surface roughness and etch rate are highly dependent on the temperature, etchant composition and IPA concentrations in the solution. It can also be concluded that the addition of an appropriate IPA concentration provide a simple method in achieving a smooth and controlled etching of silicon substrate that plays an important factor in the fabrication of micro-heater chamber.

I. INTRODUCTION

Anisotropic wet etching has been employed for many years as a key technology in creating 3-dimensional silicon micro structures. A lot of studies have been conducted to explain the anisotropic etching mechanism in terms of fastest process, temperature control, solution concentration, orientation dependencies, surface roughness and many more [1-9].

All anisotropic etchants are aqueous alkaline solution that can be either organic or inorganic [7]. An inorganic aqueous solution of KOH is the most popular and commonly used in anisotropic wet etching technique because of its simple implementation, excellent repeatability and low production cost [4, 8]. However, KOH solution may cause undesirable roughness on the etched silicon surface [10], and difficulties in controlling the etching rates, especially in lateral etch direction [11, 12].

Addition of IPA in KOH solution is one alternative to overcome the problems, Sundaram [2] indicated that an addition of IPA can improve the smoothness of etchant surfaces. By using the Raman spectrum method, Palik[13] found that IPA is not participating in KOH etching. It only moderate the reaction of KOH and decrease the etching rate which results in smoother etchant surface for (100) and (110) of silicon [10]. An addition of more than one hydroxyl group

still cannot influence the etching anisotropy however the best surface morphologies can be achieved with the saturated KOH/IPA solution [10, 14].

In this paper, various composition of KOH and IPA additive were analyzed in order to find the optimum composition of KOH/IPA to get smooth and uniform etchant surfaces, high etch rate and controlled etching process. Smooth and uniform etchant surfaces are important for the final KOH to sustain the thin film membrane fixed in the position and avoid the membrane peel off or dilute into KOH solution.

II. ETCHING MECHANISM

The etching rate by KOH strongly depends on the crystallographic orientations of the Si material to produce the 3D micro-structures. The multi step reduction-oxidation reaction mechanism for silicon etching in potassium hydroxide (KOH) was proposed by Seidel [7]. The oxidation of silicon by hydroxide ion is shown in equation (1)[15],

$$Si + 2OH^- \rightarrow Si(OH)_2^{++} + 4e^- \quad (1)$$

The four electron produced by equation (1) reduced water molecules near the Si surfaces to produce hydroxide ion and hydrogen gas as equation,

$$4H_2O + 4e^- \rightarrow 4OH^- + 2H_2 \quad (2)$$

The negative charged silicon surface tends to repel aqueous anions. The Si (OH)₄ molecules float away from the Si surfaces. This molecule however, is unstable in a solution with pH greater than 12. A high pH stable complex, depicted in equation (2) is formed by

$$Si(OH)_2^{++} + 4OH^- \rightarrow SiO_2(OH)_2^{2-} + 2H_2O \quad (3)$$

which leads to the overall reaction,

$$Si + 2OH^- + 2H_2O \rightarrow SiO_2(OH)_2^{2-} + 2H_2 \quad (4)$$

Silicon reacts with water and OH- ion and produces hydroxide ion and hydrogen gas bubbles. The dependency of the etching rate on the crystallographic orientation is due the differences in the number of dangling bond at the surfaces and of the atomic step structures.

In general, an addition of IPA leads to a decrease in the etching rate. Maximum etch rate occurred at a KOH concentration of 10-15 weight percent when no alcohol added and 30% with alcohol [13].

III. EXPERIMENTAL SET-UP

A uniform and smooth surfaces of silicon is important at the beginning of an etching process to prevent the peel-off of the nitride membrane and dilute into solution during the final KOH, due to the remaining thin nitride layer. Although KOH act very slow on the nitride layer the long etching time and repeatedly process of KOH may also attack the membrane layer gradually, resulting to an uncontrolled device dimension.

The schematic fabrication process of silicon based micro chamber and thin film membrane is depicted in Fig. 1. A standard <100> p-type Si wafer, with thickness of 680 μm and a 200 nm thin double coated nitrate was used as the starting material. An AZ 4620 photo resist layer was coated on the top side of the wafer and repeated on the bottom surface. Various sizes of nitride window array were patterned on the back side of the silicon nitride layer by photolithography. Then, buffered hydrofluoric acid (BOE) etching was used to remove the exposed nitride layer. The remaining nitride layer will then act as a mask for the silicon etching in KOH etchant.

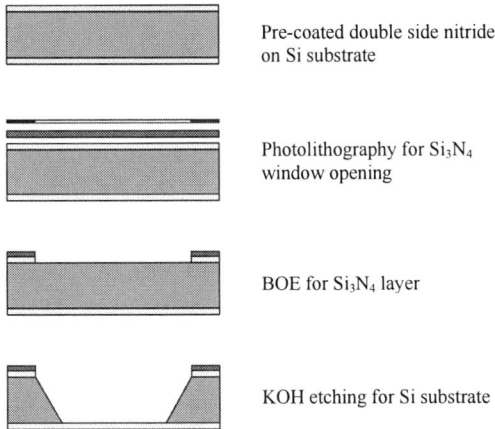

Pre-coated double side nitride on Si substrate

Photolithography for Si_3N_4 window opening

BOE for Si_3N_4 layer

KOH etching for Si substrate

Fig. 1. Process flow for the fabrication of micro chamber and thin film membrane structure

TABLE I
COMPOSITION RATIO BETWEEN KOH, DI WATER AND IPA

Composition ratio	Minimum (wt.%)	Maximum (wt.%)
KOH:DI water	10:90	90:10
KOH:IPA:DI water	10:10:80	50:50:50

The sample were cut into small pieces and etched in KOH solution. Etching solution was prepared using different KOH concentration diluted with DI water with and without IPA. The etching process produces a required depth chamber

capped by the thin film membrane. Table 1 illustrates the range of ratio composition between KOH with or without IPA diluted with DI water.

All experiments were varied within the range of 10 to 50 wt.% KOH concentration according to the pellet weight. Taking into account the natural evaporation of IPA during the etching process [10, 11] and boiling point of IPA is 80.37°C , therefore the temperature selected in this experiment was within 50-80°C. The sample prepared was immersed in respective concentration and temperature for maximum 3 hour. However, the sample will be taken out from the etchant solution in every 1 hour for property measurements and surface morphology observations.

IV. RESULTS AND DISCUSSION

Previous studies by [2, 7, 14] indicated that silicon etch rate decreases with the increase of IPA concentration in etch solution. The same situation prove the statement as depicted in Fig. 2, where it is shown that at 10% concentration of IPA, the etch rate is higher compared to the one at 50% concentration. From the same figure, it also remarks that 30% is the maximum etching rate due decreasing of etch rate at higher IPA concentration. Since function of IPA is to moderate the etching reaction [10], therefore by increasing the concentration of IPA, etch rate will also decrease respectively.

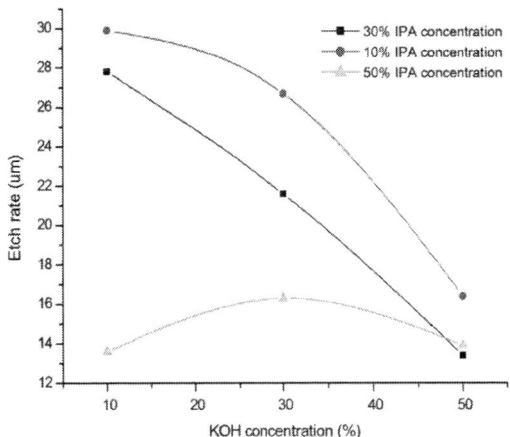

Fig. 2. KOH concentration versus etch rate at temperature of 60°C

Fig. 3 shows various concentrations of KOH with and without 10 wt% of IPA in a varying temperature. From Fig. 3, it can be observed that the etch rate increases as the temperature increases. In addition, as mentioned in [7, 13] when the concentration of KOH solution increases, the etch rate will decrease significantly. Fig. 3 also indicates that when the etch rate is higher, the morphology of silicon surface become rougher and vice versa. From Fig. 3, we can conclude that miscibility of 10 wt.% IPA in the 50 wt% KOH solution will increase the etch rate and give the rougher surface at the higher temperature. Therefore, lower temperature should be considered to improve the smoothness. The same results can

978-1-4673-2395-6/12 $31.00 © 2012 IEEE   211

also be observed when different concentration of KOH mixes in 30 wt.% and 50 wt% of IPA.

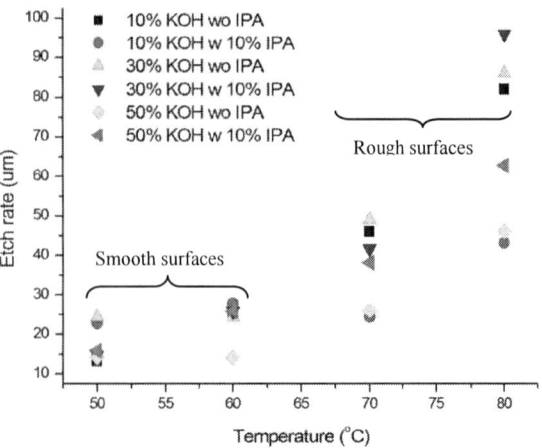

Fig.3. Etch rate versus temperature in KOH with and without 10 wt% IPA

Fig. 3 also indicates that etch rate of KOH solution without IPA is higher compared to the solution with IPA. This is due to the presence of IPA in KOH solution that can absorb the molecule on the solution surfaces, which in turn, reduces the etching rate in KOH solution [12]. In the other way [16] molecules of water that surrounding the alcohol molecule are released and disappear, which ultimately slow down the etching process.

Images of etched surface at differential weight percentage of IPA and temperature are illustrated in Fig. 4. The comparison shows that morphology of lower temperature is smoother compare than higher temperature whereby physical texture of (a) to (f) is much smaller compare with (g) to (l). As mention at the earlier of study, a uniform and smooth surface is a crucial part at the final stage of etching. Hence, if the texture and surface produced in (g) to (l) tendency of membrane to release to KOH solution is higher compare from (a) to (f)

V. CONLUSION

Etching rate with various KOH concentrations, temperatures and IPA compositions were studied and discussed. From this study, it is concluded that uniform and smooth surfaces of silicon can–be achieved with lower etching temperature and higher KOH concentration.

The presence of IPA in KOH solution helps in reducing the etch rate and produce a smoother silicon surfaces. Although there is no reaction when IPA mixed with KOH solution miscibility of IPA reduces with the increases of KOH solution concentration and decrease the etch rate respectively.

ACKNOWLEDGMENT

This work was supported by UKM via the grant NND/ND/(1)/TD11-002 (Development of lab-on chip for peripheral blood stem cell isolation and rapid detection of tropical diseases from blood).

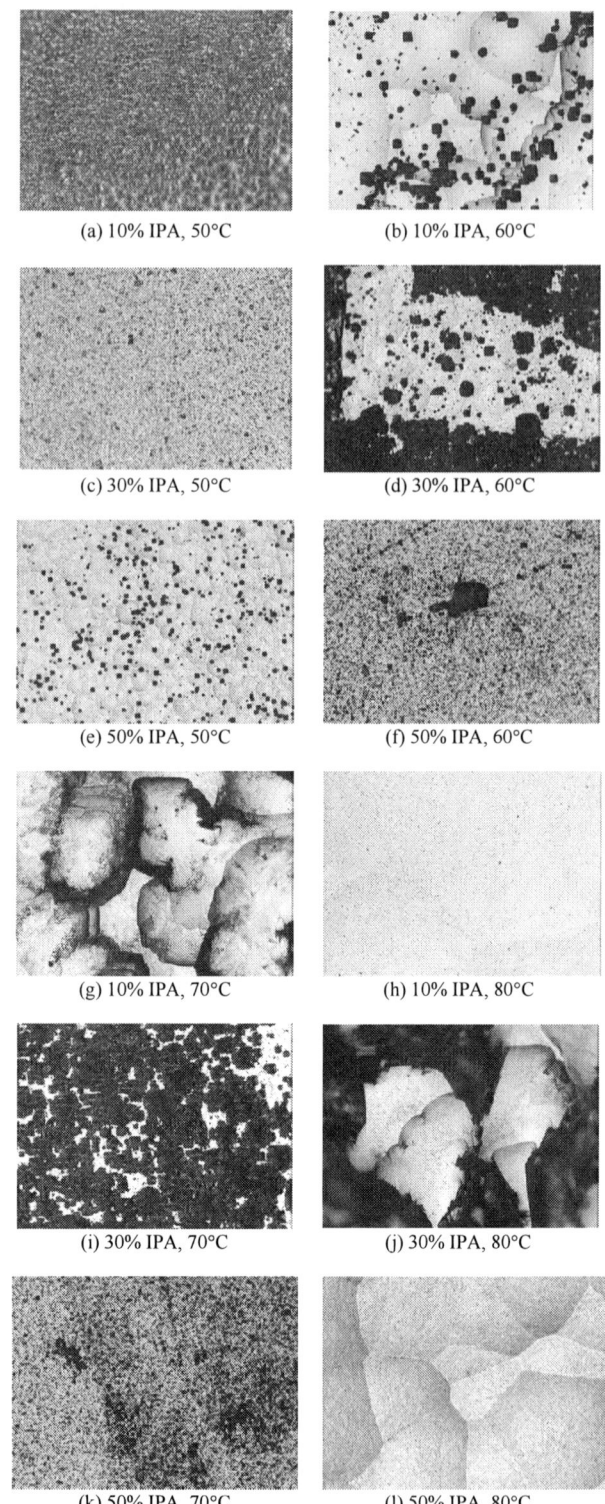

Fig. 4. The morphologies of surfaces etched in 30% wt.KOH at various wt% of IPA and temperature within 3 hour of etching period

REFERENCES

[1] H. Tanaka, S. Yamashita, Y. Abe, M. Shikida and K. Sato, "Fast etching of silicon with a smooth surface in high temperature range near the boiling point of KOH solution", *J. Sen. & Act. A*, 114, 2004, pp. 516-520

[2] K. B Sundaram, A. Vijayakumar, G. Subramanian, "Smooth etching of silicon using TMAH and Isopropyl alcohol for MEMS applications", *J. Microelectronic and Engineering*, 77, 2005, pp. 230-241

[3] M. M. Noor, B. Bais and B. Y. Majlis, "The Effects of Temperature and KOH Concentration on Silicon Etching Rate and Membrane Surface Roughness", IEEE Proc. ICSE 2004.

[4] C. R. Yang, P Y Chen, C. H yang, Y. C Chiou, R. T. Lee, "Effect of various ion typed surfactants on silicon anisotropic etching properties in KOH and TMAH solutions", *J. Sen. & Act, A*. 119, 2005, pp.271-281

[5] K. Sato, M. Shikida, Y. Matsushima, T. Yamashiro, K Asaumi , Y. Iriye and M. Yamamoto, "Characterization of orientation-dependent etching properties of single crystal silicon: effects of KOH concentration", *J. Sen. & Act. A*, 64, 1998, pp 87-93.

[6] N. Soin,B. Y. Majlis, "Realization of perfect silicon corrugated diaphragm using KOH etching", IEEE Proc, ICSE, 2004

[7] H. Seidel, L. Csepregi, A. Heuberger and H. Baumgartel, "Anisotropic etching of crystalline silicon in alkaline solutions", *J. Electrochem. Soc.,* Vol 137, 1990, pp. 3612-3625

[8] M. Shikida, K. Sato, K. Tokoro and D. Uchiawa "Differences in anisotropic etching properties of KOH and TMAH solutions", *J. Sen. & Act. A,* 80, 2000 pp. 179-188

[9] Soin. N, B. Y. Majlis, "Development of perfect silicon corrugated diaphragm using anisotropic etching", J. Microelect. Eng. 83, 2006, pp. 1438-1441

[10] W. J. Cho, W. Kuo, Chin, C. T. Kuo, "Effects of alcoholic moderators on isotropic etching of silicon on aqueous potassium hydroxide solutions", *J. Sen & Act. A*, 116, 2004, pp, 357-368

[11] I. Zubel, "Silicon anisotropic in alkaline solution III, on the possibility of spatial structure formingin the course of Si (100) anisotropic etching in KOH and KOH + IPA solutions", *J. Sen & Act. A*, 84, 2000, pp. 163-171.

[12] I. Zubel, I. Barycka, K. Kotowska "Silicon anisotropic in alkaline solution IV: the effect of organic and inorganic agent on silicon anisotropic etching process", *J. Sen, & Act ,* 84, 2000, pp.163-171.

[13] E. D. Palik, H.F. Gray, P. B. Klein, "A raman study of etching silicon in aqueous KOH", *J. Electrochem. Soc. 130*, 1983, pp. 956-959.

[14] I. Zubel, K. Rola and M. Kramkowska, "The effect of isopropyl alcohol concentration on etching process of Si-substrates in KOH solutions", *J. Sen. & Act. A*, 171, 2011, pp 436-445

[15] J. S. Starzynski,"Etch rate and etch selectivity of p++ doped silion and undoped Silicon and dielectric film in KOH ethylene-glycol-water solution", Electrochem. Soc. Proceeding, 2004.

[16] Y. F. Yano,"Correlation between surface and bulk structures of alcohol-water mixtures", *J. Colloid Interface Sci*, 284, 2005, pp.255-259

Controlled Growth of ZnO Nanostructures Prepared by Catalytic-Immersion Method

A.Azlinda,[1,2,*] Z. Khusaimi,[1,2] M. Rusop,[1,3]

[1]NANO-SciTech Centre (NST), Institute of Science;
[2]School of Physics and Material Studies, Faculty of Applied Sciences;
[3]NANO-ElecTronic Centre (NET), Faculty of Electrical Engineering;
Universiti Teknologi MARA (UiTM), 40450 Shah Alam, Selangor, Malaysia.
*E-mail: azlindazz@gmail.com, rusop@salam.uitm.edu.my

Abstract— This paper presents an efficient method to prepare ZnO nanostructure via immersion of Si substrate with catalyst assistance (gold) (Au/Si) in the mixture of zinc nitrate hexahydrate ($Zn(NO_3)_2.6H_2O$) and urea (CH_4N_2O). The effect of precursor concentration ranging from 0.01 to 0.6 (molar ratio 1:1) was evaluated in this study. Solution-immersion method was adopted with the intention to develop a large area deposition at low-temperature. As concentration increase, the morphologies changed from rod (diameter ~208 nm and length ~ 920) to accumulated nano-sheets that consist of many pores. The structural, morphology and photoluminescence effect of changing the precursor concentration on the synthesization of ZnO films were investigated by X-ray diffractometer (XRD), field emission scanning electron microscope (FESEM) and room temperature photoluminescence (PL) measurement, respectively. A unique development of size and growth orientation is seemingly affected by the change of the precursor concentration.

Keywords- Immersion; catalyst; precursor concentration; zinc nitrate hexahydrate; urea; photoluminescence

I. INTRODUCTION

Nowadays, ZnO nanostructures with small dimensions have been comprehensively receiving growing interest for nanodevices applications. This is due to their peculiar and fascinating properties as a wide band gap semiconductor of 3.37 eV with large exciton binding energy approximately 60 meV at room temperature [1]. These unique properties of ZnO nanostructures are attributed by its nanoscale shapes and size. ZnO nanostructures is a versatile material that are rich with various configurations such as nanorods, nanowires, nanotubes, nanoflower, nanosheet, etc [2-6] which have been successfully fabricated by a variety of growth conditions. Due to these properties, ZnO is one of the most promising materials for electronic applications such as light-emitting diodes (LEDs), UV photoconductive sensor, chemical sensors and solar cell [7-10].

It has been reported that, ZnO nanostructures can be successfully achieved by various deposition techniques such as spray pyrolysis, chemical vapor deposition (CVD), thermal treatment, sol-gel spin coating, immersion [11-16], that have been used to generate ZnO nanostructures. Among the aforementioned techniques, immersion technique is preferred due to its simplicity, faster operation, less power consumption, and lower cost than the other techniques.

It is well known that the properties of ZnO produced by immersion method are dependent on its preparation parameter such as a material concentration, deposition temperature and time. Concentration plays an important role in controlling the morphologies and size of a ZnO nanostructure. In this work, we study the influence of precursor concentration ranging from 0.01 to 0.6 M together with urea as a stabilizer using solution-immersion method. Stabiliser was used to shrink hasty precipitation and agglomeration of ZnO particle and aid to control the formation of ZnO at nanometer range. However, controlled synthesis of ZnO nanostructures using urea's stabiliser still needs further investigation.

II. EXPERIMENTAL DETAILS

A. Growth process

A mixture of zinc nitrate hexahydrate ($Zn(NO_3)_2.6H_2O$) (purity 99.9 %), and urea (CH_4N_2O) was used as a precursor and stabilizer respectively to form Zn^{2+} aqueous solution. Substrate used in this work prepared by using p-type Si wafer (100), was sputter coated for 60 s using gold target in argon plasma. Then it was annealed at 500 °C for 30 min to ensure good adherence and improved crystallographic quality. The preparation is as described in our previous paper [15]. The concentrations of the solutions varied from 0.01 M to 0.60 M. The clear solutions were used after 24 h of ageing. The prepared substrates (Au/Si) were immersed in prepared Zn^{2+} solution for 4 h. The deposited films on the substrates (ZnO/Au/Si) were annealed at 500 °C for 1 h in ambient. Subsequently, the samples were taken out after cooling it down to room temperature.

B. Characterization

The ZnO/Au/Si was characterized with a field-emission scanning electron microscope (FESEM) using ZEISS Supra 40VP, to determine the surface morphology of the samples. The photoluminescence and structural properties were analysing using photoluminescence spectroscopy (PL) Horiba Jobin Yvon HR800 using helium-cadmium (He-Cd) laser source and X-ray diffractometer, Rigaku RINT 2200/Ultima IV.

III. RESULTS AND DISCUSSIONS

A. Photoluminescence spectroscopy

Figure 1 shows room temperature PL spectra of ZnO nanostructures with various precursor concentrations using excitation wavelength of 325 nm. PL spectrum of a seven samples shows the existence of two prominent peaks. A violet emission peak and broad peaks at visible emission were observed. The obtained spectra exhibits similar pattern for all the samples but giving the differences in the intensity and position of peaks depending on parameters set during samples synthesization. The as-deposited samples analyzed by PL display a strong violet emission (402-416 nm) with low visible emission determining the quality of the crystal with low defect [17]. It is observed that the peak position shifted to the blue region. Generally, blue-shift of PL spectra is attributed due to the reduction of size in ZnO particle. This phenomenon corresponds to the quantum confinement effect [18]. The shrinkage in ZnO size takes place when size of nanocrystals grown is smaller than its corresponding de Broglie wavelength or the Bohr radius [19].

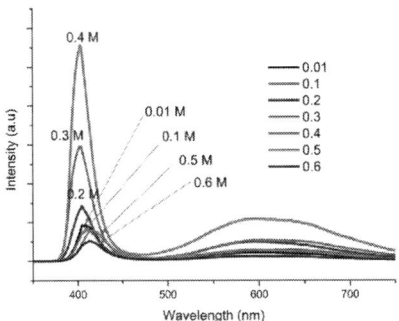

Figure 1: Room temperature PL spectra of ZnO/Au/Si at various precursor concentrations.

Generally, broad and strong visible emission (Fig.2) refers to the level of defect concentration with different types of defect [20]. Zhao et al. [21]reported that the broad emission in the visible region is linked to the oxygen defect (major-carrier traps), a result of the recombination of a photogenerated hole with an electron occupying the vacant oxygen site formed during synthesis of ZnO. However, the defect states in visible region are also attributed by other types of defect, e.g. zinc vacancy, Zn interstitial, antisite oxygen, etc [21-22].

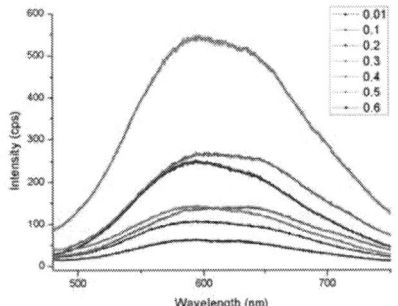

Figure 2: visible emission PL spectra of ZnO/Au/Si at various precursor concentrations

The energy band gap (eV) for each precursor concentration of ZnO samples at the highest intensity were calculated as shown in Table 1. It was found that the samples with 0.30 M and 0.40 M of precursor concentration gave the highest band gap energy of 3.09 eV while 0.50 gave the lowest energy band gap of 2.99 eV. The transition of shapes from nanorods to nanoflowers (as found from FESEM studies), may have affected to give highest Eg, these phenomena probably due to the higher presence of sharp tips of ZnO that improve the violet emission of PL. Furthermore, from XRD data, 0.40 M concentration was the smallest crystallite size which is probably why its UV peak is the most intense.

TABLE 1
THE PEAKS POSITION AND ITS CORRESPONDING ENERGY BAND GAP (Eg) VALUE OF ZNO NANOSTRUCTURES AS A FUNCTION OF VARIOUS PRECURSOR CONCENTRATIONS

Precursor concentration (M)	λ (nm)	Eg (eV)
0.01	408	3.05
0.10	409	3.04
0.20	404	3.08
0.30	402	3.09
0.40	402	3.09
0.50	416	2.99
0.60	413	3.01

B. Morphologies and structures

Generally, it was observed that ZnO nanostructures were successfully formed on Si substrate (100) covered by 6 nm layer of gold (Au) as a seeded catalyst throughout the synthesis parameters. The morphology and surface density of ZnO produced were observed to be highly dependent on precursor concentration.

Figure 3(a) – (g) show FESEM micrographs of ZnO rods, flowers, spheres and flakes like structures grown on gold-seeded Si substrate by immersion process as a function of different concentration of Zn^{2+} precursor solution. From the micrograph, it was found that the micro-rods appeared at 0.01 M of Zn^{2+} precursor solution.

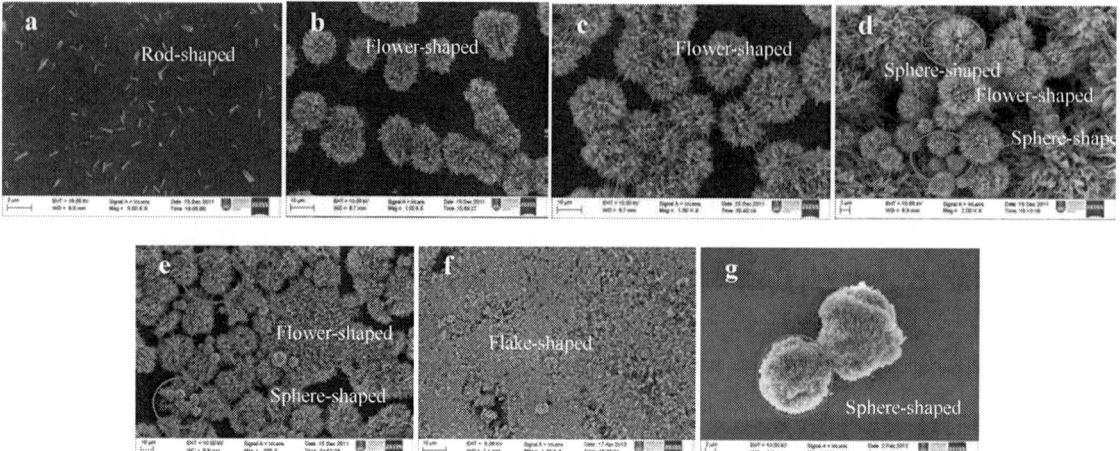

Figure 3: FESEM micrograph of ZnO/Au/Si at (a) 0.01 M, (b) 0.10 M, (c) 0.20 M, (d) 0.30 M, (e) 0.40 M, (f) 0.50 M and (g) 0.60 M concentration.

The rods with diameter of ~208 nm and length of ~920 nm grew parallel to the substrate. The formation of rod-shaped at low precursor concentration is similar with the previous study reported by Kamaruddin et.at. [23], Khusaimi et.al [24], and Peiro et.al. [25]. It was suggested that further investigation needed to be done to optimize the growth of rods with smaller diameter and better alignment. With further increase in concentration, the micro-rod disappeared and ZnO micro-flowers were formed instead. A cluster of ZnO micro-flower with diameter in range of 22 - 25 µm was formed at 0.10 M until 0.40 M concentration as shown in **Figure 3 (b) – (e)**. In contrast, the population of ZnO flower-shaped seems to be dispersed and denser as concentration increase. The clusters of ZnO micro-flower with serrated broad nano-sheets that consist of many pores can be clearly seen at 50,000 x magnification as shown in the inset. The loss of volatile gases such as H_2O and CO_2 could be attributed to the formation of pores during the heat treatment [26]. By increasing the concentration to 0.30 M and 0.40 M, growth of ZnO micro-spheres was observed. However, ZnO micro-flower still dominates the substrate surface. When the concentration of the ZnO precursor was increased to 0.50 M, nano-flakes consist of pores structures were formed instead and covered almost the entire surface of ZnO sample. The structure with uniform growth and high surface area are found to have applications as UV photoconductive sensor [27].

When 0.60 M of precursor concentration was employed, the spheres structure appears. These spheres consist of agglomerates particle that are not well dispersed and grew like an island on the substrate surface.

Figure 4 are X-ray diffraction of ZnO/Au/Si grown at different precursor concentration. The spectra showed strong peak at (100), (002), (101) and weak peaks at (102) and (110). The results is in good arrangement with standard ZnO (JCPDS 36-1451).

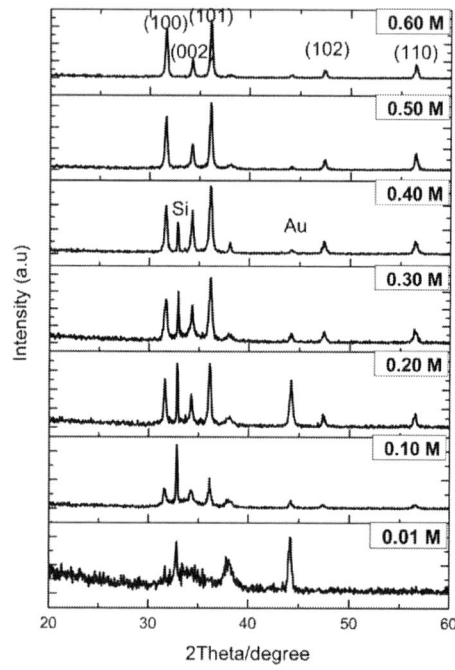

Figure 4 XRD spectra of ZnO/Au/Si samples

The intensity of ZnO (100), (002) and (101) lattice orientation for each precursor concentrations are revealed in Figure 5. By increasing the precursor concentrations, the intensity of [100], [002] and [101] planes increases significantly. However [002] plane show an optimum intensity at 0.40 M before it start to decrease as concentration increase.

XRD data were further analysed by comparing the full width at half maximum (FWHM) of [100], [002] and [101] lattice planes and the calculated crystallite size of ZnO as revealed in Table 2. The average crystallite sizes of ZnO on 0.10 M, 0.20 M, 0.40 M, 0.50 M, and 0.60 M Zn^{2+} concentration are 33.7, 34.3, 31.3, 30.3, 31.3 and 33.3 nm, respectively. Due to the absence of ZnO XRD diffraction peak at 0.01 M, the calculation for crystallite size was neglected.

Crystallite size on 0.40 M of Zn^{2+} solution concentration was smallest while largest crystallite size provided by 0.20 M. This result suggested that 0.40 M Zn^{2+} solution concentration grown on Au/Si were the best precursor concentration as in agreement with PL result shown in Figure 1.

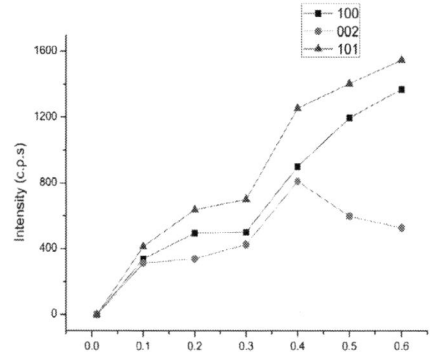

Figure 4 Intensity of XRD lattice phase of (100), (002) and (101) at various precursor concentrations

TABLE 2

CALCULATED FWHM AND CRYSTALLITE SIZE OF ZNO/AU/SI AT (100) PHASE FOR VARIOUS PRECURSOR CONCENTRATIONS

Precursor concentration (M)	FWHM (deg)	Crystallite Size (nm)	Average Crystallite Size (nm)
0.01	-	-	-
0.10	100=0.2681	32	
	002=0.3723	23	33.7
	101=0.1902	46	
0.20	100=0.2541	34	
	002=0.2225	39	34.3
	101=0.2904	30	
0.30	100=0.3431	25	
	002=0.2010	43	31.3
	101=0.3393	26	
0.40	100=0.2923	29	
	002=0.2393	36	30.3
	101=0.3356	26	
0.50	100=0.2804	31	
	002=0.2778	31	31.3
	101=0.2693	32	
0.60	100=0.2833	30	
	002=0.2338	37	33.3
	101=0.2679	33	

IV. CONCLUSIONS

In summary, ZnO nanostructures were successfully fabricated on Au/Si by solution-immersion method using the mixture of zinc nitrate hexahydrate and urea. The ZnO/Au/Si samples analyzed by PL displayed a strong violet emission (402-416 nm) with low visible emission which determines the quality of the crystal with low defect, while the blue-shift of PL spectra is corresponds to quantum confinement effect. Energy band gap of PL spectra shows that ZnO/Au/Si at 0.30 M and

0.40 M of is the highest band gap energy of 3.09 eV while 0.50 is the lowest energy band gap of 2.99 eV.

FESEM micrographs of ZnO rods, flowers, spheres and flakes like structures grown on gold-seeded Si substrate were observed as a function of different concentration of Zn^{2+} precursor solution. The rod-shaped were formed at low precursor concentration with diameter of ~208 nm and length of ~920 nm. The morphology of ZnO/Au/Si significantly was affected by changing the precursor concentrations. From XRD data, the calculated crystallite size on 0.40 M of Zn^{2+} solution concentration was smallest while largest crystallite size provided by 0.20 M.

Based on PL properties, suggesting that 0.40 M concentration was the best precursor concentration to grow ZnO/Au/Si by solution-immersion method.

ACKNOWLEDGEMENT

The author would like to thank Universiti Teknologi MARA (UiTM) (Excellent Fund 600-RMI/ST/DANA 5/3/Dst (406 /2011) for their financial support. The authors would also like to thank the faculty of Applied Sciences (Mr. Hayub) for the use of their FESEM.

REFERENCES

[1] C. Liu, Y. Masuda, Y. Wu, and O. Takai, "A simple route for growing thin films of uniform ZnO nanorod arrays on functionalized Si surfaces," Thin Solid Films, vol. 503 pp.110– 114, 2006.

[2] Z. Khusaimi, S. Amizam, H. A. Rafaie, M. H. Mamat, M. Z. Sahdan, N. Abdullah, and M. Rusop, "Growth pattern of zinc oxide nanorods on gold coated silicon surfaces," in AIP Conference Proceedings, 2009, pp. 867-871.

[3] G. S. Wua, Y. L. Zhuang, Z. Q. Lin, X. Y. Yuan, T. Xie, and L. D. Zhang, "Synthesis and photoluminescence of Dydoped ZnO nanowires," Physica E, vol. 31, pp. 5-8, 2006.

[4] G. S. Wu, T. Xie, X. Y. Yuan, Y. Li, L. Yang, Y. H. Xiao, and L. D. Zhang, "Controlled synthesis of ZnO nanowires or nanotubes via sol–gel template process," Solid State Communications, vol. 134, pp. 485–489, 2005.

[5] Y. Cao, B. Liu, R. Huang, Z. Xia, and S. Ge, "Flash synthesis of flower-like ZnO nanostructures by microwaveinduced combustion process," Materials Letters, vol. 65, pp. 160-163, 2011.

[6] K. Kakiuchi, E. Hosono, T. Kimura, H. Imai, and S. Fujihara, "Fabrication of mesoporous ZnO nanosheets from precursor templates grown in aqueous solutions," Journal of Sol-Gel Science and Technology, vol. 39, pp. 63-72, 2006.

[7] S. Kishwar, K. ul Hasan, N. H. Alvi, P. Klason, O. Nur, and M. Willander, "A comparative study of the electrodeposition and the aqueous chemical growth techniques for the utilization of ZnO nanorods on p-GaN for white light emitting diodes," Superlattices and Microstructures, vol. 49, pp. 32-42, 2011.

[8] M. H. Mamat, M. Z. Sahdan, Z. Khusaimi, A. Z. Ahmed, S. Abdullah, and M. Rusop, "Influence of doping concentrations on the aluminum doped zinc oxide thin films properties for ultraviolet photoconductive sensor applications," Optical Materials, vol. 32, pp. 696-699, 2010.

[9] D. Barreca, D. Bekermann, E. Comini, A. Devi, R. A. Fischer, A. Gasparotto, C. Maccato, G. Sberveglieri, and E. Tondello, "1D ZnO nano-assemblies by Plasma-CVD as chemical sensors for flammable and toxic gases," Sensors and Actuators B: Chemical, vol. 149, pp. 1-7, 2010.

[10] L. Lu, R. Li, K. Fan, and T. Peng, "Effects of annealing conditions on the photoelectrochemical properties of dyesensitized solar cells made with ZnO nanoparticles," Solar Energy, vol. 84, pp. 844-853, 2010.

[11] T. Tani, L. Madler, and S. E. Pratsinis, "Homogeneous ZnO nanoparticles by flame spray pyrolysis," Journal of Nanoparticle Research, vol. 4, pp. 337–343, 2002.

[12] S. Music, D. Dragcevic, M. Maljkovic, and S. Popovic, "Influence of chemical synthesis on the crystallization and properties of zinc oxide," Materials Chemistry and Physics, vol. 77, pp. 521-530, 2003.

[13] P. Singh, A. Kumar, A. Kaushal, D. Kaur, A. Pandey, and R. N. Goyal, "In situ high temperature XRD studies of ZnO nanopowder prepared via cost effective ultrasonic mist chemical vapour deposition," Bull. Mater. Sci, vol. 31, pp. 573–577, 2008.

[14] M. H. Mamat, M. Z. Sahdan, S. Amizam, H. A. Rafaie, Z. Khusaimi, and M. Rusop, "Al doped zno thin film based ultraviolet photo-conductive sensor prepared by sol-gel spincoating method," in AIP Conference Proceedings, 2009, pp. 591-595.

[15] A. Azlinda, Z. Khusaimi, S. Abdullah, and M. Rusop, "Characterization of Urea versus HMTA in the Preparation of Zinc Oxide Nanostructures by Solution-Immersion Method Grown on Gold-Seeded Silicon Substrate," Advanced Materials Research, vol. 364, pp. 45-49, 2012.

[16] Z. Khusaimi, S. Amizam, H. A. Rafaie, M. H. Mamat, N. Abdullah, and M. Rusop, "A Surface Morphology Study On The Effect Of Annealing Temperature To Nanostructured ZnO And Its Reaction Mechanism In Solution Method. ,"2009.

[17] A. George, S. K. Sharma, S. Chawla, M. M. Malik, and M. S. Qureshi, "Detailed of X-ray diffraction and photoluminescence studies of Ce doped ZnO nanocrystals," Journal of Alloys and Compounds, vol. In Press, Corrected Proof, 2011.

[18] N. A. K. Aznan and M. R. Johan, "Quantum Size Effect in ZnO Nanoparticles via Mechanical Milling," Journal of Nanomaterials, vol. 2012, pp. 1-4, 2012.

[19] Z. Khusaimi, "Synthesis and Characterisation of Low- Dimensional Zinc Oxide Nanostructures by Solution-Immersion and Mist-Atomisation ",

Faculty of Applied Sciences Universiti Teknologi MARA, Selangor, Malaysia, 2012.

[20] S. Wei, J. Lian, and H. Wu, "Annealing effect on the photoluminescence properties of ZnO nanorod array prepared by a PLD-assistant wet chemical method," Materials Characterization, vol. 61, pp. 1239-1244, 2010.

[21] X. Zhao, J. Y. Lee, C.-R. Kim, J. Heo, C. M. Shin, J.-Y. Leem, H. Ryu, J.-H. Chang, H. C. Lee, and W.-G. Jung, "Dependence of the properties of hydrothermally grown ZnO on precursor concentration," Physica E: Lowdimensional Systems and Nanostructures, vol. 41, pp. 1423-1426, 2009.

[22] S. H. Mousavi, H. Haratizadeh, and H. Minaee, "The effect of morphology and doping on photoluminescence of ZnO nanostructures," Optics Communications, vol. 284, pp. 3558-3561, 2011.

[23] S. A. Kamaruddin, M. Z. Sahdan, K. Y. Chan, M. Rusop, and H. Saim, "Effect of precursor concentration on the structural and optical properties of ZnO nanostructures," Physica Status Solidi (A) Applications and Materials, vol.207, pp. 1596-1599, 2010.

[24] Z. Khusaimi, S. Amizam, M. H.Mamat, M. Z. Sahdan, M. K. Ahmad, N. Abdullah, and M. Rusop, "Controlled growth of zinc oxide nanorods by aqueous-solution method,"Synthesis and Reactivity in Inorganic, Metal-Organic and Nano-Metal Chemistry, vol. 40, pp. 190-194, 2010.

[25] A. M. Peiro, C. Domingo, J. Peral, X. Domènech, E. Vigil, M. A. Hernández-Fenollosa, M. Mollar, B. Marí, and J. A. Ayllón, "Nanostructured zinc oxide films grown from microwave activated aqueous solutions," Thin Solid Films, vol. 483, pp. 79-83, 2005.

[26] A. Lei, B. Qu, W. Zhou, Y. Wang, Q. Zhang, and B. Zou, "Facile synthesis and enhanced photocatalytic activity of hierarchical porous ZnO microspheres," Materials Letters, vol. 66, pp. 72-75, 2012.

[27] M. H. Mamat, Z. Khusaimi, M. Z. Musa, M. F. Malek, and M. Rusop, "Fabrication of ultraviolet photoconductive sensor using a novel aluminium-doped zinc oxide nanorodnanoflake network thin film prepared via ultrasonic-assisted sol-gel and immersion methods," Sensors and Actuators A: Physical, vol. 171, pp. 241-247, 2011.

Optimization of Process Parameter Variation in 45nm p-channel MOSFET using L_{18} Orthogonal Array

F.Salehuddin[1,2], I.Ahmad[1],*Member,IEEE*, F.A.Hamid[1],*Member,IEEE*, A.Zaharim[3],*Member,IEEE*,
Afifah Maheran A.Hamid[2,3], P.Susthitha Menon[3],*Member,IEEE*, H.A.Elgomati[3],
B.Yeop Majlis[3],*Member,IEEE* and P.R.Apte[4],*Member,IEEE*.

Universiti Tenaga Nasional (UNITEN), Kajang, Selangor, MALAYSIA[1]
Universiti Teknikal Malaysia Melaka (UTeM), Melaka, MALAYSIA[2]
Universiti Kebangsaan Malaysia (UKM), MALAYSIA[3]
Indian Institute of Technology (IIT), INDIA[4]
Email: fauziyah@utem.edu.my

Abstract— In this study, orthogonal array of L_{18} in Taguchi method was used to optimize the process parameters variance on threshold voltage (V_{TH}) in 45nm p-channel Metal Oxide Semiconductor Field Effect Transistor (MOSFET) device. The signal-to-noise (S/N) ratio and analysis of variance (ANOVA) are employed to study the performance characteristics of the PMOS device. There are eight process parameters (control factors) were varied for 2 and 3 levels to performed 18 experiments. Whereas, the two noise factors were varied for 2 levels to get four readings of V_{TH} for every row of experiment. V_{TH} results were used as the evaluation variable. This work was done using TCAD simulator, consisting of a process simulator, ATHENA and device simulator, ATLAS. These two simulators were combined with L_{18} Orthogonal Array to aid in design and optimize the process parameters. The predicted values of the process parameters were verified successfully with ATHENA and ATLAS's simulator. In PMOS device, V_{TH} implant dose (26%) and compensate implant dose (26%) were the major factors affecting the threshold voltage. While S/D Implant was identified as an adjustment factor in PMOS device. These adjustment factors have been used to get the nominal values of threshold voltage for PMOS device closer to -0.289V.

Keywords— NMOS Device, Threshold Voltage, Leakage Current, Taguchi Method.

I. INTRODUCTION

The scaling of transistors to smaller dimensions has therefore profound effect on the manufacturing yield and reliability of integrated circuit [1,2]. During chip manufacturing, random process variations affect all transistor dimensions: length, width, junction depths, oxide thickness etc., and become a greater percentage of overall transistor size as the transistor shrinks [3,4]. In 1974, an analysis of systematic variation was performed by Schemmert and Zimmer with their paper on threshold-voltage sensitivity [5]. Their research looked into the effect that the oxide thickness and implantation energy had on the threshold voltage of MOS devices.

In order to limit the impact of variations, statistical treatment of random variation of devices (statistical design) is becoming increasingly important. The increasing statistical

variation in the process parameters has emerged as a serious problem in the circuit design and can cause significant effect in the MOS transistor [6]. The technique to identify semiconductor process parameters whose variability would impact most on the device characteristics is realized through a process using Taguchi Method. The Taguchi method involves an analysis that reveals which of the factors are most effective in reaching the goals and the directions in which these factors should be adjusted to improving the results [7]. This method uses a special design of orthogonal arrays to study the entire process parameter space with only a small number of experiments. Using an orthogonal array to design the experiment could help the designers to study the influence of multiple controllable factors on the average of quality characteristics and the variations in a fast and economic way [8].

II. METHODOLOGY

Sample used in these experiments were <100> orienting and p-type (boron doped) silicon wafers [9] with density of 7.0×10^{14} atom/cm^2. Most of the process steps or 65nm to 32nm MOSFET devices were similar [9,10]. Retrograde P-well was created starting with developing a 200Å oxide layer on the top of substrate. The oxide layer was grown to prevent contamination of the substrate. The silicon wafer then had undergone the annealing process to ensure that amorphous phosphorus atoms are being spread properly in the wafer. This process was also to improve the doping intensity across the substrate and to strengthen the structure. Next, Shallow Trench Isolation (STI) was developed to isolate neighbouring transistor. A 130Å stress buffer was grown on the wafers with 25 minutes at 900°C in dry oxygen. Then, a 1350Å nitride layer was deposited using the Low Pressure Chemical Vapour Deposition (LPCVD) process. This thin nitride layer was acted as the mask when silicon was etched to expose the STI area. Photo resistor layer was then deposited on the wafers, and unnecessary part will be etched using the Reactive Ion Etching (RIE) process. An oxide layer was grown on the

978-1-4673-2395-6/12 $31.00 © 2012 IEEE

trench sides to eliminate any impurity from entering the silicon substrate [11]. Chemical Mechanical polishing (CMP) was then applied to eliminate extra oxide on the wafers. Lastly, Phosoro-Silicate-Glass (PSG) layer was developed in 15 minutes with $900^{\circ}C$ at the top of substrate. A sacrificial oxide layer was then grown and etched to eliminate defects on the surface. Next process was to growth the gate oxide layer. The silicon wafer was oxidized in dry oxygen for short time at 100ms. The short time is needed to ensure 1.1nm of gate oxide thickness (T_{ox}) was grown [12]. The V_{TH} adjust implantation was performed to implant Boron Difluoride (BF_2) at P-well active area in order to adjust the V_{TH} of the device. The polysilicon gate was then deposited on the substrate and defined followed by the halo implantation.

In order to get an optimum performance, arsenic implantation with 18° tilt angle was performed in PMOS device. Halo implantation was followed by depositing sidewall spacers. Sidewall spacers were then used as a mask for source/drain implantation. Boron atom was implanted at a desired concentration to ensure the smooth current flow in PMOS device. Silicide layer was formed and then annealed on the top of polysilicon. The next step in this process was deposited of Boron Phosphor Silicate Glass (BPSG) layer. This layer will be acted as Premetal Dielectric (PMD), which is the first layer deposited on the wafer surface when a transistor was produced. This transistor was then connected with aluminum metal. After this process, the second aluminum layer was deposited on the top of the Intel-Metal Dielectric (IMD) and unwanted aluminum was etched to develop the contacts [13]. The procedure was completed after the metallization and etching were performed for the electrode formation, and the bonding pads were opened. Once the devices were built with ATHENA, the complete devices can be simulated in ATLAS to provide specific characteristics such as the I_D versus V_{GS} curve. The threshold voltage (V_{TH}) can be extracted from that curve.

A. Taguchi Orthogonal L_9 Array Method

For the study and modelling of variability of this device performance parameters, eight relevant process parameters were identified as the source of uncontrollable variation for 45nm PMOS technology, as shown in Table I.

TABLE I
PROCESS PARAMETERS AND THEIR LEVELS

Sym	Process Parameter	Unit	Level 1	Level 2	Level 3
A	Oxide Growth Temp	$^{\circ}C$	815	820	-
B	Vth Implant Dose	atom cm^{-2}	1.75E11	1.80E11	1.85E11
C	Vth Implant Energy	keV	4.5	5	5.5
D	Pocket-Halo Implant Dose	atom cm^{-2}	2.60E13	2.65E13	2.70E13
E	Pocket-Halo Implant Tilt	Degree	18	20	22
F	S/D Implant Dose	atom cm^{-2}	6.85E13	6.90E13	6.95E13
G	S/D Implant Energy	keV	11	11.5	12
H	Compensate Implant Dose	atom cm^{-2}	2.50E13	2.55E13	2.60E13

Meanwhile, the two noise factors are sacrificial oxide temperature and annealing process temperature. These noise factors were varied for 2 levels to get four readings of threshold voltage (V_{TH}) for every row of experiment. The values of the noise factor at the different levels are listed in Table II. In this research, an L_{18} ($2^1 \times 3^7$) orthogonal array which has 18 experiments was used.

TABLE II
NOISE FACTORS AND THEIR LEVELS

Symbol	Noise Factor	Unit	Level 1	Level 2
N	Sacrificial Oxide Temp.	$^{\circ}C$	900 (N_1)	905 (N_2)
M	Annealing Process Temp.	$^{\circ}C$	910 (M_1)	915 (M_2)

III. RESULT AND DISCUSSION

The results of V_{TH} were analyzed and processed with Taguchi Method to get the optimal design. The optimized results from Taguchi Method were simulated in order to verify the predicted optimal design.

A. Analysis of process parameters variation on Threshold Voltage

Eighteen different experiments in PMOS device was performed using the design parameter combinations in the specified orthogonal array table. Four specimens were simulated for each of the parameter combinations. The completed response for V_{TH} data is shown in Table III.

After eighteen experiments of L_{18} array have been done, the next step is to determine, which control factors can gave more effect to a device characteristics. Signal-to-noise (S/N) ratio was used to easily find out the optimal process parameters and analyze the experimental data. The S/N ratio for each level of process parameters is computed based on the S/N analysis. Regardless of the category of the performance characteristic, the larger S/N ratio corresponds to the better performance characteristic [14]. Therefore, the optimal level of the process parameters is the level with the highest S/N ratio [15, 16].

TABLE III
V_{TH} VALUES FOR PMOS DEVICE

Exp. No	Threshold Voltage (Volt)			
	Vth1	Vth2	Vth3	Vth4
1	-0.2179	-0.2168	-0.2174	-0.2177
2	-0.3550	-0.3382	-0.3540	-0.3548
3	-0.4119	-0.3943	-0.4113	-0.4117
4	-0.2921	-0.2743	-0.2740	-0.2919
5	-0.2626	-0.2625	-0.2714	-0.2625
6	-0.4004	-0.3839	-0.3996	-0.4003
7	-0.2896	-0.2794	-0.2890	-0.2894
8	-0.3718	-0.3560	-0.3552	-0.3716
9	-0.2927	-0.2827	-0.2929	-0.2927
10	-0.3839	-0.3835	-0.3837	-0.3837
11	-0.2613	-0.2606	-0.2409	-0.2611
12	-0.3117	-0.3192	-0.3289	-0.3115
13	-0.3548	-0.3587	-0.3545	-0.3546
14	-0.2742	-0.2742	-0.2816	-0.2741
15	-0.3053	-0.2965	-0.3046	-0.3051
16	-0.2462	-0.2467	-0.2464	-0.2461
17	-0.3804	-0.3884	-0.3716	-0.3803
18	-0.3177	-0.3173	-0.3173	-0.3175

In this research, threshold voltage of the 45nm device belongs to the nominal-the-best quality characteristics. This S/N Ratio (SNR) is selected to get threshold voltage value closer or equal to a given target value (-0.289V), which is also known as nominal value [16]. The S/N Ratio (Nominal-the-best), η can be expressed as [7]:

$$\eta = 10 Log_{10} \left[\frac{\mu^2}{\sigma^2} \right] \qquad (1)$$

Where:

$$\qquad (2)$$

$$\sigma^2 = \frac{\sum_{i=1}^{n} (Y_i - \mu)^2}{n-1} \qquad (3)$$

While n is number of tests and Y_i the experimental value of the threshold voltage, μ is mean and σ is variance. In the nominal-the-best, there are two types of factor to find which are dominant and adjustment factors. By applying (1)-(3), SNR (η) for PMOS device was calculated and given in Table IV [7].

The SNR for each level of the process parameters is summarized in Table V. In addition, the overall mean SNR for the nine experiments is also calculated. Basically, the larger the S/N ratio, the quality characteristic for the threshold voltage is better [8,17].

TABLE IV
MEAN, VARIANCE AND S/N RATIOS FOR PMOS DEVICE

Exp. No.	Mean	Variance	S/N Ratio (Mean)	S/N Ratio (Nominal-the-Best)
1	-0.217	2.30E-07	-13.25	53.13
2	-0.351	6.74E-05	-09.11	32.61
3	-0.407	7.52E-05	-07.80	33.44
4	-0.283	1.06E-04	-10.96	28.78
5	-0.265	1.97E-05	-11.54	35.52
6	-0.396	6.57E-05	-08.04	33.78
7	-0.287	2.47E-05	-10.85	35.22
8	-0.364	8.65E-05	-08.79	31.84
9	-0.290	2.53E-05	-10.74	35.22
10	-0.384	2.67E-08	-08.32	67.42
11	-0.256	1.01E-04	-11.84	28.12
12	-0.318	6.74E-05	-09.96	31.76
13	-0.356	4.15E-06	-08.98	44.84
14	-0.276	1.38E-05	-11.18	37.42
15	-0.303	1.81E-05	-10.37	37.04
16	-0.246	7.00E-08	-12.17	59.38
17	-0.380	4.71E-05	-08.40	34.87
18	-0.317	3.67E-08	-09.97	64.39

B. Analysis of Variance (ANOVA)

The analysis of variance (ANOVA) is a common statistical technique to determine the percent contribution of each factor for results of the experiment. It is also can be used to investigate which of the process parameters significantly affect the performance characteristics. It calculates parameters known as sum of squares (SSQ), degree of freedom (DF), variance, F-value and percentage of each factor. The result of ANOVA for the PMOS device is presented in Table VI.

According to these analyses, the most dominant factors for S/N Ratio are factor C (V_{TH} implant energy – 26%) and factor H (Compensate implant dose – 26%). Therefore, these factors should be set as a dominant factor and can not be used as an adjustment factors. Whereas factor G (S/D implant energy) was described as an adjustment factor because it has the large effect on mean (41%) and small effect on variance (1%) if compare with other factors. The analysis of average performance showed that the optimum condition is $A_2, B_3, C_1, D_3, F_2, H_1$. Because factor E (Halo implant tilt angle) and factor G (S/D implant energy) and were found not significant (pooled) in threshold voltage, there could be set at any level [7]. The full recommendation for optimization is $A_2, B_3, C_1, D_3, E_1, F_2, G_1, H_1$ i.e. oxide growth temperature at level 2, V_{TH} implant dose at level 3, V_{TH} implant energy at level 1, halo implant dose at level 3, halo implant tilt at level 1, S/D implant dose at level 2, S/D implant energy at level 1 and compensate implant dose at level 1.

TABLE V
S/N RESPONSES FOR VTH IN PMOS DEVICE

Sym	Process Parameter	S/N Ratio (dB)			Max - Min
		Level 1	Level 2	Level 3	
A	Oxide Growth Temp	35.50	45.03	-	9.53
B	Vth Implant Dose	41.08	36.23	43.49	7.26
C	Vth Implant Energy	48.13	33.40	39.27	14.73
D	Pocket-Halo Implant Dose	36.19	40.72	43.88	7.69
E	Pocket-Halo Implant Tilt	42.01	36.90	41.88	5.11
F	S/D Implant Dose	38.37	44.30	38.12	6.18
G	S/D Implant Energy	42.07	40.16	38.57	3.5
H	Compensate Implant Dose	48.22	38.99	33.58	14.64

Overall mean of SNR = 40.26 dB

The Pareto analysis of effect with parameters V_{TH} (PMOS) using Taguchi Method and Response Surface Methodology (RSM) [17] are shown in Fig. 1 and Fig. 2 respectively. According to Fig.1, it can be seen that for the response, V_{TH}, G (S/D implant energy-16%), F (S/D implant dose-14%) and A (Oxide growth temperature-10%) and H (Compensate implant dose-10%) are the top four significant parameters.

From the Pareto plots of V_{TH} for PMOS device using these two methods, it can be seen that the V_{TH} implant energy has an influential role on the threshold voltage of the devices. Table VII summarizes the most significant process steps that influence V_{TH} for MOSFET devices using RSM and Taguchi Method.

978-1-4673-2395-6/12 $31.00 © 2012 IEEE

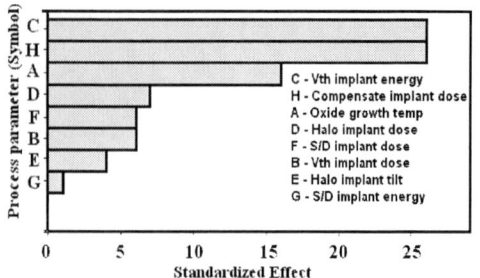

Fig. 1: Pareto plot for PMOS Device using Taguchi Method

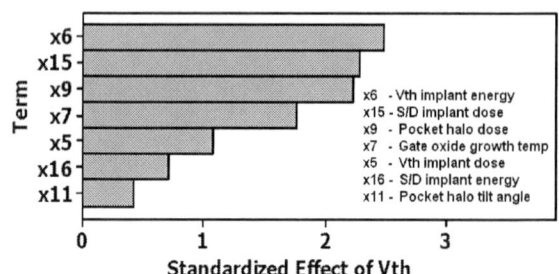

Fig. 2: Pareto plot for PMOS Device using RSM

TABLE VI
RESULTS OF ANOVA FOR PMOS DEVICE

Symbol	Process Parameter	DF	Sum of Square	Mean square	F-Value	Factor Effect on SNR (%)	Factor Effect on Mean (%)
A	Oxide Growth Temperature	1	408	408	16	16	0
B	Vth Implant Dose	2	164	82	3	6	0
C	Vth Implant Energy	2	660	330	13	26	12
D	Pocket-Halo Implant Dose	2	179	90	3	7.0	17
E	Pocket-Halo Implant Tilt	2	102	51	2	4	18
F	S/D Implant Dose	2	147	74	3	6	10
G	S/D Implant Energy	2	37	18	1	1	41
H	Compensate Implant Dose	2	658	329	13	26	2

[a]At least 95% confidence

TABLE VII
THE MOST SIGNIFICANT PROCESS PARAMETER THAT IMPACT VTH
FOR PMOS DEVICE

Rank	Method	
	Taguchi Method [my work]	RSM [18]
1	Vth Implant Energy	Vth Implant Energy
2	Oxide Growth Temp	S/D Implant Dose
3	Halo Implant Dose	Halo Implant Dose
4	S/D Implant Dose	Oxide Growth Temp
5	Vth Implant Dose	Vth Implant Dose
6	Halo Implant Tilt	S/D Implant Energy
7	S/D Implant Energy	Halo Implant Tilt

TABLE VIII
BEST SETTING OF THE PROCESS PARAMETERS

Sym	Process Parameter	Unit	Best Value
A	Oxide Growth Temperature	^{o}C	820
B	Vth Implant Dose	atom cm^{-2}	1.85E11
C	Vth Implant Energy	keV	4.5
D	Pocket-Halo Implant Dose	atom cm^{-2}	2.70E13
E	Pocket-Halo Implant Tilt	Degree	22
F	S/D Implant Dose	atom cm^{-2}	6.90E13
G	S/D Implant Energy	keV	11
H	Compensate Implant Dose	atom cm^{-2}	2.50E13

TABLE IX
RESULTS OF THE CONFIRMATION EXPERIMENT FOR VTH

Threshold Voltage (Volts)				SNR (Mean)	SNR (Nominal-the-Best)
Vth1	Vth2	Vth3	Vth4		
-0.2935	-0.2940	-0.2935	-0.2933	-10.65	59.85

C. Confirmation Test

The confirmation test is used to verify the estimated result with the experimental results. Best setting of the process parameters for PMOS device that had effects on V_{TH} which had been suggested by Taguchi Method is shown in Table VIII. In here, the confirmation test was required in PMOS device because the optimum combination of parameters and their levels i.e. $A_2, B_3, C_1, D_3, E_1, F_2, G_1, H_1$ respectively did not correspond to any experiment of the orthogonal array. The result of the final simulation for this device is shown in Table IX.

Before the optimization approaches, the best S/N ratio (Nominal-the-best) is 67.42dB at row of experiment no. 10 (Please refer Table IV). Whereas the variance is 0.163mV and mean for threshold voltage is -0.384V. The percent different of this threshold voltage value from the nominal value, -0.289V [12] is higher (32.6%).

After the optimization approaches, the S/N Ratio (Nominal-the-best) and S/N Ratio (Mean) of threshold voltage for PMOS device are 59.85 dB and -10.65 dB respectively. These values are within the predicted range. For S/N Ratio (Nominal-the-best), 59.85 dB is within predicted range S/N ratio of 85.50 to 57.95 dB (71.73 ± 13.77dB). While for S/N Ratio (Mean), -10.65 dB is within predicted range S/N ratio of -10.73 to -9.26 dB (-10.00 ± 0.73 dB).

These show that Taguchi Method can predict the optimum solution in finding the 45 nm PMOS fabrication recipe with appropriate threshold voltage value. The variance and threshold voltage for the device after optimization approaches are 0.298mV and 0.294V respectively. Although the variance is slightly increase from the previous value (0.163mV) but threshold voltage value is closer to the nominal value (target). The V_{TH} value after the optimization approach is just 1.73% different from the target. Table X shows the simulated values versus ITRS 2009 prediction for V_{TH} in 45nm PMOS device. It is clearly shown that our design simulation is closer to International Technology Roadmap for Semiconductor (ITRS) 2009 prediction [12].

TABLE X
SIMULATION VERSUS ITRS 2009 PREDICTION

Device	Simulation	ITRS 2009 Prediction [12]
PMOS	-0.294V	-0.289V

The closer the quality characteristic value to the target, the better the product quality will be [16]. At row of experiment no. 9 (Table V), V_{TH} value (-0.290V) is exactly same with the target but the variance is high (5.03mV) and S/N ratio is small (35.22dB). The S/N ratio value is also not within predicted range S/N ratio of 85.50 to 57.95 dB.

IV. CONCLUSIONS

Through this paper, the main factors that affect the response characteristics of 45nm PMOS device was found, together with the optimal factor levels. There were shown that V_{TH} implant energy and compensate implant dose were identified as the most dominant or significant factors for S/N Ratio in PMOS device. While S/D Implant energy was identified as an adjustment factor in this device. The adjustment factor has been used to get the nominal (target) value of threshold voltage for PMOS device closer to -0.289V. The percent different of threshold voltage value from the target after the optimization approach is just 1.73%. This value is closer with International Technology Roadmap for Semiconductor (ITRS) 2009 prediction. It can be shown that the optimum solution in achieving the desired transistor was successfully predicted by using Taguchi Method.

ACKNOWLEDGMENT

The authors would like to thanks to the Ministry of Higher Education (MOHE) for their financial support and the Universiti Teknikal Malaysia Melaka (UTeM) for the moral support throughout the project.

REFERENCES

[1] K.Kuhn, C.Kenyon, A.Kornfeld, M.Liu, A.Maheshwari, W.Kai Shih, S.Sivakumar, G.Taylor, P.VanDerVoorn, and K.Zawadzki, (2008). Managing process variation in intel's 45nm CMOS technology, Intel Technology Journal, Vol. 12, No.2, pp. 93–109.

[2] A. Chin and S. McAlister, (2005). The power of fundamental scaling: Beyond the power consumption challenge and the scaling roadmap, IEEE Circuit and Device Magnetic, pp. 27–35.

[3] S.Borkar, (2005). Designing reliable system from unreliable components: the challenges of transistor variability and degradation, IEEE Microelectronics, Vol. 26, No. 6, pp. 10–16.

[4] S.Springer et al., (2006). Modelling of variation in submicron CMOS ULSI technologies, IEEE Transactions on Electron Devices, Vol. 53, No. 9, pp. 2168–2178.

[5] W.Schemmert and G.Zimmer, (1974). Threshold-voltage sensitivity of ion- implanted MOS transistors due to process variations, Electronics Letter, Vol. 10, No. 9, pp. 151–152.

[6] Nassif, S.R., Strojwas, A.J., Director, S.W., (1984). FABRICS II: A statistically based IC fabrication process simulator, IEEE Transactions on Computer-Aided Design of Integrated Circuits and Systems 3, pp. 40-46.

[7] Phadke, Madhav S., (1998). Quality Engineering Using Robust Design, Pearson Education, Inc. and Dorling Kindersley Publishing, Inc.

[8] Esme U., (2009). Application of Taguchi Method for the Optimization of Resistance Spot Welding Process, The Arabian Journal for Science and Engineering, Vol. 34, No. 2B.

[9] F.Salehuddin, I.Ahamd, F.A.Hamid, and A.Zaharim, (2011). Influence of halo and source/drain implantation on threshold voltage in 45nm pmos device, Australian Journal Basic Applied Sciences, Vol. 5, No.1, pp. 55–61.

[10] H.Elgomati, B.Y.Majlis, I.Ahmad, F.Salehuddin, F.A.Hamid, A.Zaharim, and P.R.Apte, (2011). Application of Taguchi Method in the optimization of process variation for 32nm cmos technology, Australian Journal Basic Applied Science, Vol. 5, No.7, pp. 346–355.

[11] F.Salehuddin, I.Ahmad, F.A.Hamid, and A.Zaharim, (2011). Optimization of process parameter variability in 45 nm pmos device using taguchi method, Journal of Applied Sciences, Vol. 11, No. 7, pp. 1261–1266.

[12] ITRS 2010 Report; http://www.itrs.net

[13] Fauziyah Salehuddin, Ibrahim Ahmad, Fazrena Azlee Hamid, Azami Zaharim, (2009). Application of Taguchi Method in Optimization of Gate Oxide and Silicide Thickness for 45nm nMOS Device, International Journal of Engineering & Technology (IJET), Vol. 9, No. 10, pp. 94-98.

[14] G.P.Syros, (2003). Die casting process optimization using Taguchi methods, Journal of Materials Processing Technology 135, pp. 68-74.

[15] Abdullah H., Jurait J., Lennie A., Nopiah Z.M., Ahmad I., (2009). Simulation of Fabrication Process VDMOSFET Transistor Using Silvaco Software, European Journal of Scientific Research, ISSN 1450-216X, Vol. 29, No. 4, pp. 461-470.

[16] Naidu, N.V.R., (2008). Mathematical model for quality cost optimization, Robotics and Computer-integrated Manufacturing, The 17th International Conference on Flexible Automation and Intelligent Manufacturing, Vol. 24, No. 6, pp. 811-815.

[17] Nalbant M., Gokkaya H., Sur G., (2007). Application of Taguchi method in the optimization of cutting parameters for surface roughness turning, International Journal of Materials and Design, Vol. 28, pp. 1379-138.

[18] Ziyad Al Tarawneh, (2010. An analysis of SEU robustness of C-element structures implemented in bulk CMOS and SOI technologies, International Conf. on Microelectronics (ICM 2010), pp. 280-283.

A Simulation Study of The Effect Engineered Tunnel Barrier To The Floating Gate Flash Memory Devices

Mohd Rosydi Zakaria, Uda Hashim, Ramzan Mat Ayub, Zarimawaty Zailan
Institute of Nano Electronic Engineering (INEE),
Universiti Malaysia Perlis, Malaysia

Abstract- **Flash memory is a device that is used as a tool to store data electrically. The main advantage of this device is in the non-volatility which can store data without power supply, thus make the device very popular in broad application. Conventional Flash memory generally uses single tunnel oxide with a thickness of 7 nm to 10 nm as a tunnel barrier. In order to obtain good device performance, the thickness of the tunnel barrier must be reduced. If the thickness of the oxide is reduced below than 5 nm, device performance will be better but suffer from problems such as current leakage and data retention. To overcome this problem, a technique identified as Engineered Tunnel Barrier is used to replace the single oxide used in conventional flash memory. The programming characteristic of memories with different tunnel barrier stacks single layer oxide, symmetric layer and asymmetric layer dielectric are investigated using TCAD simulator. The T-suprem-4 was used for device process fabrication and MEDICI simulator used for electrical characteristics. From theoretical, confirmed that the memory with the multilayer tunnel barrier exhibits better programming characteristics in term of, programming tunneling current, programming speed and programming voltage.**

***Keyword*s:** Engineered tunnel barrier, Floating gate Flash Memory, Non-volatile memory device.

I. INTRODUCTION

Dynamic RAM (DRAM), static RAM (SRAM) and flash memory are three types of memory devices are most commonly used in semiconductor product today [1]. Flash memory is a one of a non-volatile device that can be electrically erased and reprogrammed. Because of that, this device is becoming increasingly popular as mass storage media for devices such as digital cameras, camcorders and USB [2].

Conventional Flash memory used Silicon Oxide (SiO_2) as a tunnel barrier due to its excellent interface properties with Silicon [3]. Conventional flash memory facing several serious limitation examples is non-scalability of the tunnel oxide and the thickness of the tunnel oxide reached their limit <5nm [4]. If the tunnel oxide is decreased below than <5nm, the operation

voltage could be reduced and the programming speed has also improved for several second. However, reducing the tunnel oxide will affect to the leakage current and this case will destroy the retention time. From the above problem, conventional tunnel barrier flash device cannot meet the requirement of NAND flash.

Engineered Tunnel Barrier (ETB) is one of an alternative to overcome above problems and are expected to reduce the programming voltage and allow greatly improve the programming speed. ETB also possible to turn on and turn off independently compare the single layer dielectric [5]. The concept of ETB is based on multiple barrier dielectric stack is used as a tunnel barrier in flash device to replace a single tunnel oxide barrier. The ETB stack can potentially achieve smaller programming speed and programming voltage compares the single oxide barrier [6]. This because, the ETB is consists of multilayer dielectric stack with low-κ/high-κ dielectric combination as shown in figure 1. In this paper the Engineered Tunnel Barrier Flash memory (ETB-FM) using VARIOT-type [7] was simulated by stacking thin layer of oxide and nitride. The writing/Erase tunneling current and programming speed was evaluated and analyzed.

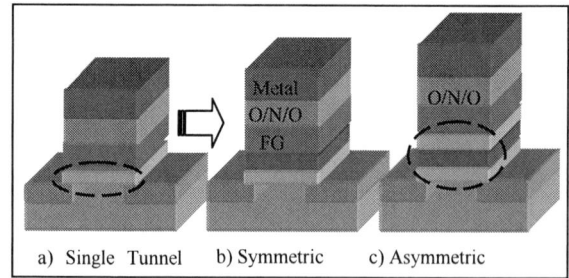

Fig.1. The conventional single tunnel oxide replaces by an engineered tunnel barrier multilayer dielectric

978-1-4673-2395-6/12 $31.00 © 2012 IEEE

II. METHODOLOGY

In this work, we use technology CAD tools to study the potential of Flash memory. Taurus Supreme-4 (TS-4) is being used to simulate the device process of flash memory and MEDICI is being used to simulate the electrical characteristic device.

In the first half of the research, we perform simulations of conventional flash to optimize the single tunnel oxide and the effect to device performance was studied. In the second part, we implement engineered tunnel barriers based on the simulation results. We simulate the asymmetric (3-Layer) and symmetric layer (2-Layer) with same EOT of conventional which aspect better performance. From the design, the gate 1 (Floating Gate) and Gate 2 (Control gate) gate length sizes are kept the same. The channel gate length is fixed to be 2.5µm, while the thickness of polysilicon is at 250µm and 150µm for control gate and floating gate respectively.

The EOT of Inter-Poly Dielectric (IPD) is 15nm with stacks of Oxide/Nitride/Oxide, thickness (3.5/7/2) nm. For this design, LOCOS is used as the device isolation. Therefore, for this part, the thickness of the tunnel barrier with Equivalent Oxide Thickness (EOT) is 8nm made with a combination of stacks (SiO_2/Si_3N_4) and $(SiO_2/Si_3N_4/SiO_2)$ creating a 2 layer and 3 layer, to replace a single oxide. For materials of high-κ dielectric, Si_3N_4 is used because it has high dielectric properties of 7eV [8]. All parameters used to simulate the device structure are summarized in table I.

TABLE I
PARAMETER USED TO SIMULATE CELL FLASH MEMORY STRUCTURE

Parameter	Value
Tunnel Barrier (nm) EOT=8nm	i. 8, ii. (5/5.5), iii.(3.5/4.5/2)
Channel Lengh (µm)	2.5
Tunnel Window (µm)	0.3x0.3
Source /Drain Doping cm^{-3}	$1x10^{15}$

III. RESULT AND DISCUSSION

Results and discussion will cover the behavior of tunneling efficiency and programming speed for three characterize sample. Critical issues such as, programming tunneling current, programming speed during writing and erasing are investigated through TCAD simulation by using Fowler-Nordheim model.

A. Device Structure

Figure 2 shows the device structure after simulating used a Tsuprem-4.

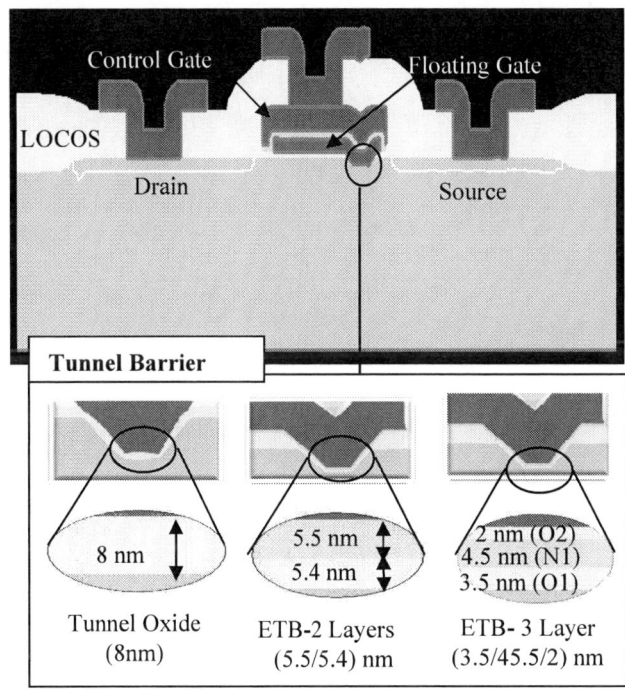

Fig. 2. Device structure of Flash memory and Tunnel Barrier Stacks

B. Programming Tunneling Current

Fowler-Nordheim tunneling mechanism is widely used for writing and erasing a flash memory device [9]. The writing tunneling current for ETB can extract from figure J_{FN} vs V_{CG} shown in figure 3. In order to see the improvement of ETB to the flash memory device, the FN current density for the conventional tunnel oxide with same physical thickness EOT 8nm also plot in the same graph. The writing was carried out with V_{CG}=20V.

It is observed that the three layers of ETB ONO layer with symmetric stack 3.5/4.5/2 nm has a higher tunneling efficiency than asymmetric 2-layer (5/5.5) nm stacks and single oxide 8nm. The tunneling maximum current for ONO stacks is ($1x10^{-2}$ A/cm) compare $1x10^{-5}$ for single layer oxide.

978-1-4673-2395-6/12 $31.00 © 2012 IEEE

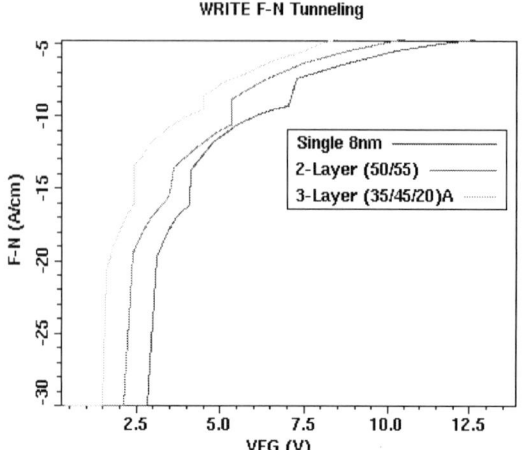

Fig. 3. F-N Tunneling current through three different tunnel barrier

The erase simulation results after changing the tunnel barrier between single oxide and Engineered Tunnel Barrier are plotted in figure 4. For erase, we also plot the F-N tunneling erase through the single oxide for comparison. It clearly shows that the erase through the ETB is better than single oxide even though same EOT are given. From the figure the erase tunneling current increases when the tunnel barriers are improved to 3-Layer tunnel barrier. The tunneling current for three tunnel barriers are in the range of 10^{-9} A/cm for engineered tunnel barrier while 10^{-10} A/cm for conventional flash. From the simulation result, tunneling current during erase has not change much compared during writing. At same Eox, the tunneling efficiency has changed about one order of magnitude.

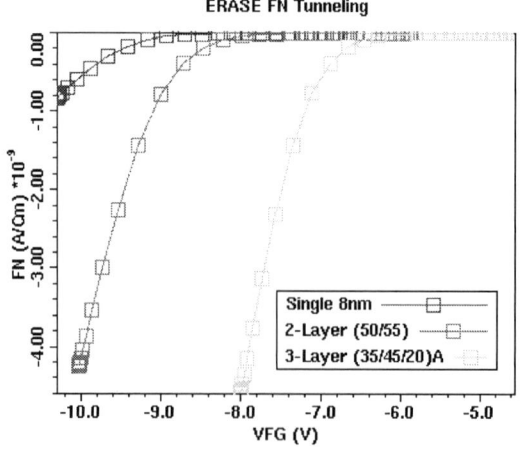

Fig. 4. Simulated result for erase with different tunnel barrier.

From both results, shows that the writing and erasing tunneling is increasing when improve the tunnel single oxide to engineered tunnel barrier. It is shown that, the ETB can overcome the low programming tunneling current in conventional flash.

C. Programming Speed

Figure 5 shows the simulated Vth versus programming time for different tunnel barrier structure. The charging was carried out with V_{CG}=20V. From figure, when increases the tunnel barrier from single oxide to ETB it improves programming time. At the same programming time (e.g., 1ms) a V_{th} for ETB is increasing about 2V compare other layers. However, when setting the shift threshold voltage at a constant $V_{th} = 5V$, the programming time increases from about 100µs for single SiO_2 to about 10µm for the (5/5.5) nm SiO_2/Si_3N_4 stacked and 1µs for (3.5/4.5/2)nm $SiO_2/Si_3N_4/SiO_2$ stacks. As a result, for 3 layered, tunnel barrier show apparent faster programming time as compared to 2-layer and single layer. shown in figure 5. From the result, the programming speed produced for the three layers is 5µs, for the second layer it is about 50µs, while for a single layer of about 400µs. The result is shown within writing the programming speed is about 1µs for 3-Layer tunnel barrier. It is clear that, increasing of various layer barriers is much faster than a single layer in terms of programming speed. This is evidently as in the ITRS 2009 standard that by using multilayer tunnel barrier can increase the programming speed around 1µs.

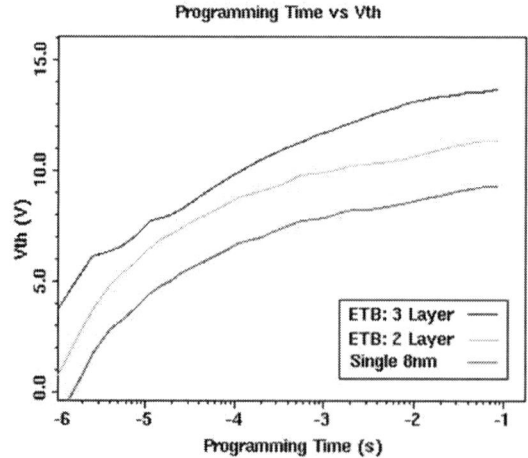

Fig. 5. Simulate of writing transient of Floating Gate Flash Memory with different tunnel barrier

The erase operation however is influenced by the tunnel barrier variation as shown in figure 6. From the figure, the erase time increases after converting the tunnel oxide to engineer tunnel barrier. At the same programming time (e.g., 100μm) Vth for ETB increases to about 1V. From both results, it can be seen that Vth decrease logarithmically with time.

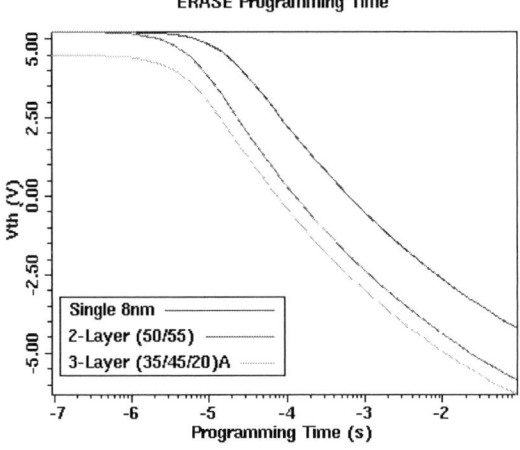

Fig. 6. Simulate of Erase transient of Floating Gate Flash Memory with different tunnel barrier

The programming speed for tunnel barrier characteristics can be calculated from the slope dv/dt change along the higher value to a lower value from the modeling study and result shown in figure 5. From the result, the programming speed produced for the three layers is 5μs, for the second layer it is about 50μs, while for a single layer of about 400μs. The result is shown within writing the programming speed is about 1μs for 3-Layer tunnel barrier. It is clear that, increasing of various layer barriers is much faster than a single layer in terms of programming speed. This is evidently as in the ITRS 2009

standard that by using multilayer tunnel barrier can increase the programming speed around 1μs.

The comparison for flash memory device performance between ETB and single tunnel oxide is summarized in Table 3. From table, Programming current, programming speed and programming voltage result has improve after changing the single oxide to engineered tunnel barrier.

IV. CONCLUSION

In summary, Engineered Tunnel Barrier was simulated in order to overcome the weaknesses of conventional Flash Memory. For engineered Tunnel Barrier technique, two engineered tunnel barrier approach namely the asymmetric and symmetric are with as high-κ (Si_3N_4) already successfully simulated using TS-4. The performance of the device is observed to be improved as well as when the engineered tunnel barrier used to replace single oxide. The result shows the ETB technique has about one order magnitude higher in tunneling efficiency at the same EOT than single tunnel oxide. The programming speed can achieved to 1μm for engineered tunnel barrier and the programming voltage can reduce less than 15V.

ACKNOWLEDGMENT

The authors wish to thank Universiti Malaysia Perlis (UniMAP), ministry of science, technology & innovation (mosti) and ministry of higher education (mohe) for giving the opportunities to do this research in the Micro & Nano Fabrication Cleanroom. The appreciation also goes to all the team members in the Institute of Nanoelectronics Engineering (INEE) especially in the NVM Research Group.

TABLE 3:
SUMMARY RESULT OF FLASH MEMORY DEVICE

	ETB (Sio2/SiN4/SiO2)	ETB (Sio2/SiN4)	Conventional Sio2 barrier
Physical Thickness	3.5nm/4.5nm/2nm	5.5nm/4.4nm	8nm
Programming Speed (W/E)	1μs/100μs	50μs/100μs	400μs /1ms
Programming voltage (W/E)	+12V/-12V	+15V/-15V	+17V/-17V
FN Gate current (A/cm)	1×10^{1}	1×10^{-1}	1×10^{-2}

REFERENCES

[1] Andy Chung and Jamal Deen and Jeong-Soo Lee and, M.M., "Nanoscale memory devices". IOP Publishing, 2010. 21(41) p. 412001.

[2] Aritome, S. *"Advanced flash memory technology and trends for file storage application".* in *Electron Devices Meeting, 2000. IEDM Technical Digest. International.* 2000.

[3] Gehring, A. and S. Selberherr, *"Modeling of tunneling current and gate dielectric reliability for nonvolatile memory devices".* Device and Materials Reliability, IEEE Transactions on, 2004. **4**(3): p. 306-319.

[4] Bez, R., *"Innovative technologies for high density non-volatile semiconductor memories".* Microelectronic Engineering, 2005. **80**(0): p. 249-255.

[5] Likharev, K.K., *"Layered tunnel barriers for nonvolatile memory devices".* Applied Physics Letters, 1998. **73**(15): p. 2137-2139.

[6] Cho, J.J.a.W.-J., *"Tunnel Barrier engineering for non-volatile memory".* journal of semiconductor technology and science, 2008. **8**.

[7] Govoreanu, B., et al., *"VARIOT: a novel multilayer tunnel barrier concept for low-voltage nonvolatile memory devices".* Electron Device Letters, IEEE, 2003. **24**(2): p. 99-101.

[8] Govoreanu, B., D.P. Brunco, and J. Van Houdt, *"Scaling down the interpoly dielectric for next generation Flash memory: Challenges and opportunities".* Solid-State Electronics, 2005. **49**(11): p. 1841-1848.

[9] Pavan, P., et al., *"Flash memory cells-an overview".* Proceedings of the IEEE, 1997. **85**(8): p. 1248-1271.

DC MEMS Switches with Self-x Features: Design, Simulation and Implementation Strategies

Muhammad Akmal Johar, Pedro Torruella and Andreas König
Institute of Integrated Sensor System
Department of Electrical and Computer Engineering
Technische Universitaet Kaiserslautern
67655 Kaiserslautern, Germany
Email: johar@eit.uni-kl.de

Abstract—**Successful MEMS products depend on device reliability to perform their function. While many MEMS products have been introduced into the mass market, a lot of work is now concentrated on adding functionality and reliability. This paper will layout the idea of the implementation of self-x features in MEMS DC switches to improve these characteristics. It introduces the self-monitoring and self-repairing functions at component level. Special additional structures have been added into the MEMS DC Switch device to give these additional functionality. Our MEMS switches are designed using the MetalMUMPs technology offered by MEMSCAP. Our switch proposal uses lateral switching movement with metal to metal contact, with a gap of $10\mu m$. It is equipped with two moveable structures with electrostatic actuation as primary source of force. Heat actuators are used for self-repairing procedures. Our initial simulation results show an actuation voltage of around 92V for the electrostatic actuation and a resonance frequency of 6.724kHz.**

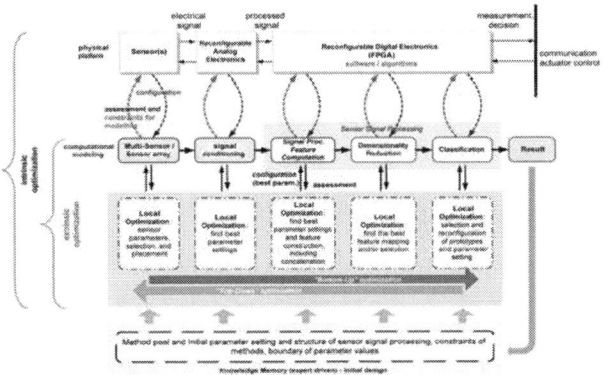

Fig. 1. Recofigurability in sensory system [6]

I. INTRODUCTION

MEMS products have become an important component in nowadays consumer electronics. The ability to convert mechanical information to electrical signals in such smaller package opens the possibility to create more efficient products which are smaller in size and cheaper in price. MEMS accelerometers and gyroscopes are examples of MEMS Sensors that have been successful in the market. They allow the implementation of many consumer devices with special sensing features like game consoles, digital cameras and mobile phones.

DC MEMS switches or relays are another MEMS components that are entering into the mass market. Several possible application areas can benefit from MEMS switches. Recent research work reported [1] the possibility to reintroduce the MEMS switches to replace the CMOS switches for the low power digital logic circuit. This is because the relay can perform as an ideal switch with on/off switching behaviour and zero off-state leakage current (I_{off}). Since logic circuit require high switching frequency and low voltage operation, a Nano-electro-mechanical (NEM) switch has been proposed to meet these requirements [2].

Another potential area is in the area of industrial automation. The switches are used to reconfigure the analogue circuits for example in automated test equipment(ATE). The switching frequency requirements are moderate but a higher power handling is needed. Generally three options of switches

are available: electromechanical, reed relays and MEMS. A recent survey reported that the demands for MEMS switches are increasing to replace the older technology starting from the year 2011 [3]. They are able to match the power handling and switching speed with small size and low cost [4].

Reconfigurable analog circuits using MEMS switches can also be used in cyber-physical systems. They can help to increase reliability of a complex CPS where the interconnections between embedded and physical systems are enormous. This is in line with our research group goals in pursuing a dynamical sensor system with self-x capability as shown in Fig. 1. The group has been working on the reconfiguration capability at various stages of sensor system [6], [7]. Our previous work also suggested self-x features in embedded systems for AMR sensor systems using MEMS switches [8]. In that paper, MEMS switches have been used as part of calibration actuator to enable the self-monitoring and self-repairing for AMR sensors to guarantee optimum performance.

As a continuation from previous work, the implementation of self-x features in MEMS level as shown in Fig. 2 is reported. This paper proposed switch designs that are equipped with self-monitoring and self-repairing features. Self-monitoring gives the capability to monitor the function of the switches using indirect measurement. As stiction is a common problem in MEMS devices, the ability to monitor the switches functionality is important for these MEMS switches. As for

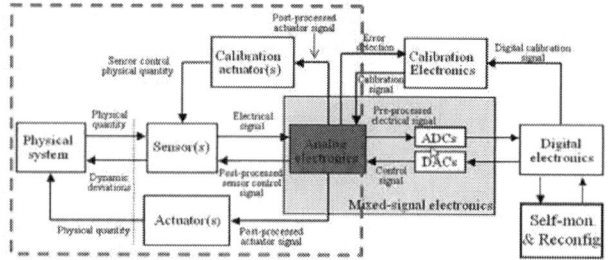

Fig. 2. Self-x implementation in a)the sensor level [8] b)calibration actuator level

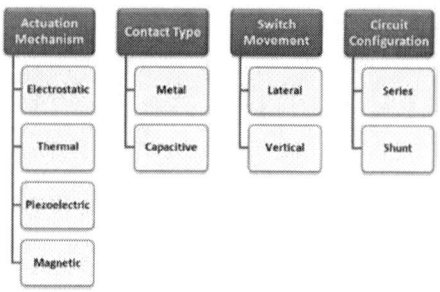

Fig. 3. Possible design configuration of MEMS switches

self-repairing, it is the capability to repair the switch function in the case of failure. Several strategies will be discussed in this paper concerning this matter.

II. FABRICATION TECHNOLOGY

Getting access to the latest MEMS fabrication technology is difficult and involves high costs. New technology usually will cost more and these high costs will scare away the industry player from using MEMS technology. So our approach is to use a well proven technology that has a lower cost which is offered by the MEMSCAP company. Among several technologies offered by them, we choose MetalMUMPs technology in our design.

MetalMUMPs is a good candidate to built MEMS switches due to several advantages. The main structure of the devices are built using a metal layer from nickel. As nickel is a conductive material, nitride layers have been used to act as an isolator and at the same time to provide a mechanical link between internal structures. The standard nickel layer with a thickness of $20\mu m$ has low resistance and good mechanical properties. This layer can be thickened up to $30\mu m$ in order to have larger effective area for electrostatic actuation. This opens the possibility to design lateral MEMS switches. Using the standard MetalMUMPS process two actuation methods are possible, namely electrostatic actuation and thermal actuation. Table I explains in details the layer description of the Metal-MUMPs process.

III. MEMS SWITCH DESIGN CONFIGURATION AND PARAMETERS

When designing MEMS switches, these can be configured based on four design parameters as shown in Fig. 3. The proposed switches in this work are built from two types of actuators: electrostatic and thermal. The lateral switch movement has been chosen to allow a metal-to-metal contact. Also, the switches can be connected in series.

One of the design challenges when using lateral movement is to overcome the minimum gap between nickel structures which is $10\mu m$ [9]. Typical configurations of switches contain one moveable conductor to make contact with two static signal lines. Such a huge gap requires high actuation voltage. In order to eliminate this problem, the proposed switches have two moveable MEMS structures. Each moveable structure will only require to travel half of the gap and thus will reduce the actuation voltage.

In this paper, two concept designs are presented. The first concept design is 'Single Pole-Single Throw' (SPST) MEMS switch. It has one terminal for source and drain. It is built by two identical moveable structures. Each structure consists of a one direction of electrostatic actuator, two pairs of heat actuators, signal line: source and drain ports, self-monitoring structure and switch contact area. The second concept design is 'Double Poles-Double Throws' DPDT MEMS switch. This switch has two separate sources and drains. However, it can become Single Pole-Double Throws (SPDT) switch if both sources are short circuit. It has three moveable structures, namely center, left and right mass. The left and right mass are identical to the component of SPST switches. The center mass

TABLE I
METALMUMPS LAYER DESCRIPTION

Layer	Thickness (μm)	Material
Isolation Oxide	2	Thermal Oxide
Oxide 1	0.5	PSG
Nitride 1	0.35	Silicon Nitride
Poly	0.7	Polysilicon
Nitride 2	0.35	Silicon Nitride
Oxide 2	0.35	PSG
Anchor Metal	0.035	10nm Cr + 25nm Pt
Plating Base	0.55	500nm Cu + 50nm Ti
Metal	20.5	$20\mu m$ Ni + $0.5\mu m$ Au
Sidewall Metal	1-3	Gold

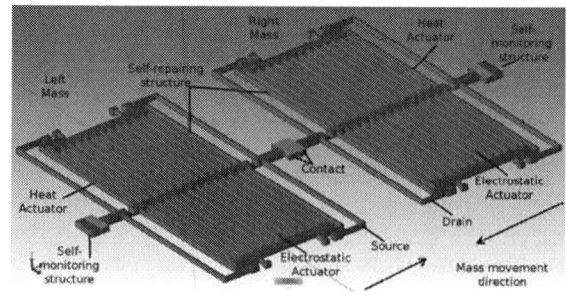

Fig. 4. General construction of SPST MEMS Switch

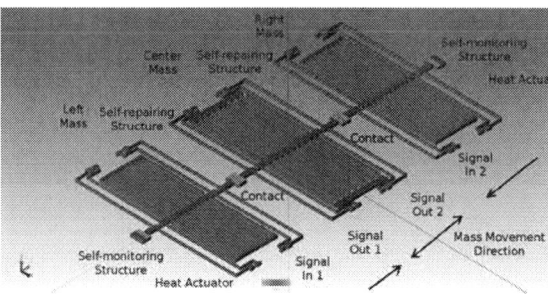

Fig. 5. General topology of DPDT MEMS switches

has bidirectional electrostatic actuators. Both of the source ports are attached to the center mass, while the drain ports are connected to left and right mass each. Only one signal line can be activated at one time. The general construction of both switches are in Fig.4 and Fig.5.

IV. MEMS SWITCHES MODE OF OPERATION

The MEMS structure is equipped with two types of actuators namely an electrostatic actuator and thermal actuation. Both actuators can be activated in several modes either in the independent mode or hybrid mode. Electrostatic actuation has the advantage of low power consumption during operation, high speed switching and easy design implementation. However it lacks of force that is essential to reduce the contact resistance. Thermal actuation however provides higher force but with slower response, a little more power consumption and small displacement.

The key of our proposal is that the designer can choose between different modes of operation, by coherently turning on or off the different actuators. A summary of operation is shown in Table II. Thus the application engineers have the freedom to operate in those modes that are suitable for their application. For example, mobile devices that have a limited power source will surely avoid using thermal actuators since it will consume more power; on the other hand, test circuits could ensure good contact by using them.

In mode A, only the electrostatic actuation is activated while the heat actuator only serves as support to the switch structure. This mode can give high switching frequencies and very small current consumption during operation. The setbacks are only that our switch structure requires actuation voltage of around 90V.

In mode B, only the thermal actuation is activated. In this mode, the electrostatic actuator can be used for self-monitoring

TABLE II
MODE OF OPERATION FOR SELF-X MEMS SWITCHES

| Operation Mode | Actuation | | Remarks |
	Electrostatic	Thermal	
A	Active	Passive	Driven by one actuator
B	Passive	Active	Produced slow response
C	Active	Active	In hybrid mode thermal actuator active first

purposes. The movement of the moveable structure causes the gap between the comb structure to become smaller. Thus we can sense the difference in capacitance between on or off state of the switch.

Meanwhile mode C represents a hybrid mode where both actuators are active in a smart way. First the thermal actuator is used to close the gap between switch contact about half from the initial gap. Then the electrostatic actuator will activate and make a full switch contact. Since the contact gap is smaller then the actuation voltage will be lower. The advantage of this mode is that it can reduce power consumption of the thermal actuator and also allow to have a smaller size for the driving structure of the electrostatic actuator. Using the right strategy, the combination of these actuators will compensate their respective weaknesses.

V. SELF-X CONCEPTS AND IMPLEMENTATION

Stiction is a common problem of MEMS structures. Stiction is caused when two movable parts are stick together, making the device to malfunction. It can happen either during the dice release process, which will affect the manufacturing yield, or during operation. The later can be fatal to the end user, specially when the device is used in applications with a high level of safety. Even more, since physical contact does take place during switching operation, tear and wear at the contact area are another concern. Therefore it is desirable to implement techniques that allow the system to improve its reliability and fault tolerance.

A. Self-monitoring Concept and Design

Self-monitoring comprises the monitoring of the movable structure's condition. In our proposal, capacitive changes between the movable and static parts of the device are measured. Two reference states, representing 'OFF' and 'ON', are sensed and recorded. When in 'ON' state, the gap of the equivalent capacitor will be smaller, thus producing a higher than reference value. Both measurements are recorded during normal operation. In case of stiction, the sensed capacitance will be that as in the 'ON' state, but, without any actuation voltage being applied. Therefore a faulty condition could be identified.

B. Self-repairing Concept and Design

Self-repairing is another feature that can significantly improve the device reliability. Implementing the self-monitoring alone is not adequate in most cases. When the monitoring system detects a faulty event, sending a notification message alone is not enough without taking certain action. Without self-repairing capability, the action normally requires external intervention which normally takes time and increases system complexity. A built-in subsystem that is able to repair will further increase the device reliability.

In this MEMS design, the implementation of self-repairing is done by using thermal actuation. The thermal actuation has been chosen because it can produce largest forces amongst other micro-actuators and also due to its simplicity in design. Each moveable structure has two pairs of thermal actuators

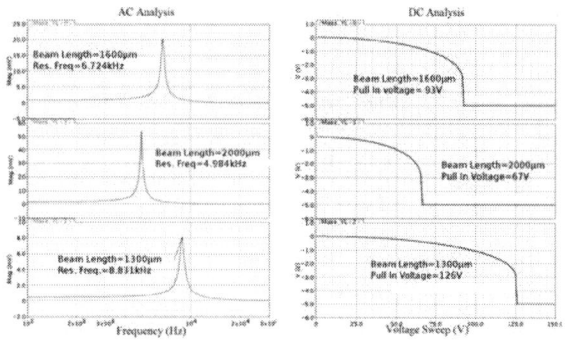

Fig. 6. Voltage actuation and resonance frequency for SPST MEMS switch with various spring length

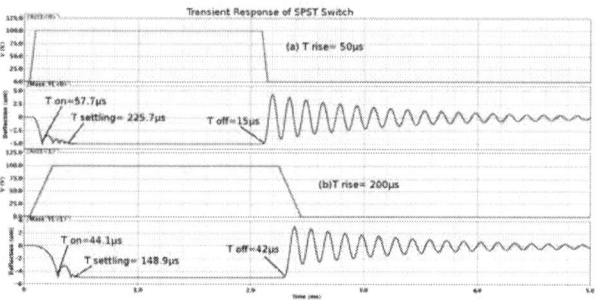

Fig. 7. Transient response for SPST MEMS switch with a) $50\mu s$ t_{rise} and b) $200\mu s$ t_{rise}

TABLE III
SIMULATION RESULT OF ACTUATION VOLTAGE AND RESONANCE
FREQUENCY WITH VARIOUS SPRING LENGTH

Spring length [μm]	Actuation Voltage [V]	Resonance Frequency [kHz]
2000	67	4.984
1600	93	6.724
1300	126	8.831

namely pair A and pair B. Pair A actuator acts in the direction of closing the switch, while pair B will move in the opposite direction providing forces separating the switch contact in case of stiction.

VI. RESULT AND DISCUSSION

In evaluating MEMS switches performances, several design parameters have been considered. Two of the most important design requirements are high resonance frequency and low voltage actuation. Unlike CMOS transistors that have no moveable structure, the maximum switching speed of a mechanical device depends only on the resonance frequency of the structure. A small structure with high stiffness leads to a higher resonance frequency. On the other hand, the need of having low pull-in actuation voltage leads to a large area for the driving circuit and also a small spring constant. As both requirements are contradictory, an optimum value is needed to satisfy both.

Switching time values are another design parameters that have been considered. The time taken to change the state of the switch has been measured. Ideally fast switching characteristics are desired. Three parameters of switching time have been evaluated. Firstly, T_{On} is the time for the switch to make contact after the voltage actuation reaching the required value. Secondly, T_{Off} is the time taken by the switch to change state after the actuation voltage are no longer active. Finally, $T_{Settling}$ represent the time taken by the MEMS switch to reach settling condition after bouncing effects.

In this work we report the structural analysis of the suggested MEMS switch structure. The DC Simulation analysis will give the information regarding the actuation voltage of the electrostatic actuator. The optimum structure size can be obtained with the desired actuation voltage. Then, AC Simulation analysis give the resonance frequency of the MEMS switch structure. Identifying the resonance frequency will define the working region of switches. Finally, the transient analysis is performed to study the detail structure behaviour during switching movement especially the contact bounce phenomena.

The SPST MEMS switch has voltage actuation of 92V. From frequency response analysis, the resonance frequency of

the structure is at 6.72kHz. Based on the iteration simulation result, in order to get lower actuation voltage, the spring constant, k needs to be low. However, MEMS structures with a low spring constant lead to lower resonance frequency. As a higher resonance frequency and low voltage actuation are desired, a moderate value is taken for this design. Table III shows the variant result of voltage actuation and resonance frequency.

The initial results of the transient response show the behaviour of the switch during state transition. Based on the result, T_{On} took little longer compared with T_{Off}. This is because during T_{On}, the incremental actuation force needs to overcome the mechanical spring force of the structure in order to pull in the switch. During T_{Off}, the spring force pulls off the structure rather quick after the actuation force is diminished.

The bouncing effects of the switch also can be seen in the transient analysis. For simulations with a rise time of the voltage source set to $50\mu s$, the switch takes about $225.7\mu s$ after first contact to settle. This is due to the presence of an inelastic collision between two movable parts. This bouncing effect is normal to any mechanical switch even at micro size level. One way to ease this problem is to apply a slower rise of the actuation voltage. By implementing this, the full contact action can be achieved faster. Simulation results show that for a rise time of the voltage supply set to $200\mu s$ the time required to reach the settling point is reduced to about $148.9\mu s$. Graph 7 shows the improvement of the switch behaviour at different values of rise time.

For the DPDT Switch, our analysis concentrates on the center mass as the left and the right parts are identical to the SPST switches. Since the center mass is larger, it has a slightly different characteristic. The actuation voltage for the

TABLE IV
SPST AND DPDT DESIGN SPECIFICATION

Design Name Properties[Unit]	SPST Switch	DPDT Switch
Overall Length [μm]	2315	3315
Overall Width [μm]	2200	2200
Electrostatic Actuator	1-Dir	2-Dir
Length [μm]	700	700
Width [μm]	510	1070
Heat Actuator		
Length [μm]	1600	1600
Width [μm]	10	10
Voltage Actuation [V]	Left=92 Right=92	Left-100 Right=100 Center=96
AC Analysis Resonance Freq [kHz]	Left=6.72 Right=6.93	Left=6.92 Right=6.94 Center=6.00
Switching Performances at t_{rise}=200μs		
t_{On} [μs]	44.1	51.5
t_{Off} [μs]	42	32
$t_{Settling}$ [μs]	148.9	204.5

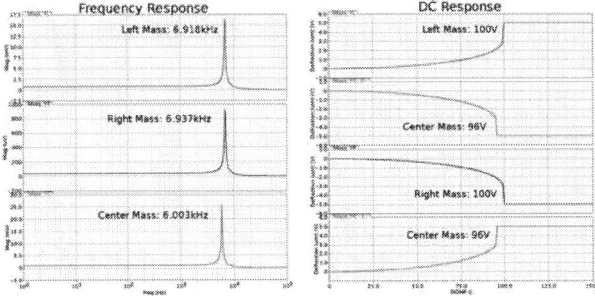

Fig. 8. Voltage actuation and resonance Frequency for DPDT MEMS switch

center mass is 96 V with a resonance frequency of 6 kHz. As for the left and right mass, the actuation voltage is 100V and the resonance frequency is 6.937kHz. Similar to SPST design, the bouncing effect is also visible especially in the center mass. By using the same approach we compared the effect between different rise times of the voltage supply. At t_{rise}= 50μs, the center mass takes 523.9μs to settle. The best result can be obtained when the t_{rise} is set to 200 μs and

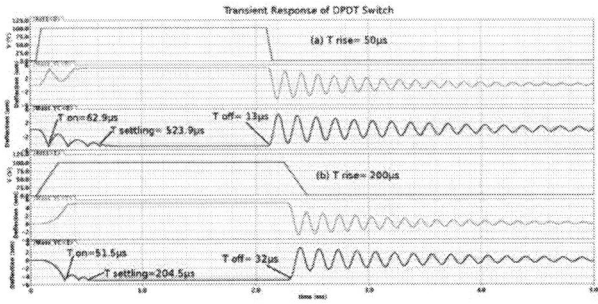

Fig. 9. Transient response for DPDT MEMS switch with a) 50μs t_{rise} and b) 200μs t_{rise}

producing a settling time for center mass of 204.5 μs. The detailed result can be seen on graph 8 and graph 9.

VII. CONCLUSION AND FUTURE WORK

In this work initial analysis results of MEMS switch structures are presented. The proposed design requirements were achieved. Further analysis in terms of parasitic capacitance and total resistance of the switch will be pursued. A lot of improvement in terms of the performance can be done. For our future work, we will investigate the suggestion to improve our MEMS design. A possible approach is to combine electrostatic and thermal actuation. However, this must be done in an appropriate way so that the power consumption stays minimal. We believe that with the help of the thermal actuator, the switch will settle a in much shorter time.

Our vision is to put as many MEMS switches as possible in a single die. Based on our current size, we are able to put 12 SPST switches and 6 DPDT switches in a single die with size of 1cm x 1cm. These array of switches are better than current MEMS switches on the market which is a single switch on the single die.

Our future work will also include the real implementation of self-x features in these MEMS switches. Since this paper only explained the concept and the idea for self-x implementation, we will continue to work for realization of this concept and idea. We also planning to explore other MEMS technology for example Imec's SiGe-MEMS technology that has possibility on combining MEMS structure and CMOS circuit on a single die.

REFERENCES

[1] Hei Kam; Tsu-Jae King Liu; Stojanovic, V.; Markovic, D.; Alon, E.; , "Design, Optimization, and Scaling of MEM Relays for Ultra-Low-Power Digital Logic," Electron Devices, IEEE Transactions on , vol.58, no.1, pp.236-250, Jan. 2011 doi: 10.1109/TED.2010.2082545.

[2] Nathanael, R.; Pott, V.; Hei Kam; Jaeseok Jeon; Tsu-Jae King Liu; , "4-terminal relay technology for complementary logic," Electron Devices Meeting (IEDM), 2009 IEEE International , vol., no., pp.1-4, 7-9 Dec. 2009 doi: 10.1109/IEDM.2009.5424383

[3] John, Williamson, "Automated Test Equipment: An Overview of MEMS and Non-MEMS Switching Options", published in MEMS Investor Journal April 2011.

[4] J.Maciel, S. Majumder, R. Morrison and J. Lampen,"Lifetime Characteristic of Ohmic MEMS Switches", Radant MEMS Publications, www.radantmems.com

[5] Edward A. Lee, "Cyber Physical System: Design Challenges", Technical Report N0. UCB/EECS-2008-8

[6] Iswandy, K.; Koenig, A. Methodology, Algorithms, and Emerging Tool for Automated Design of Intelligent Integrated Multi-Sensor Systems. Algorithms 2009, 2, 1368-1409.

[7] Peter Messiha Mehanny Tawdross, Bio-Inspired Circuit Sizing and Trimming Methods for Dynamically Reconfigurable Sensor Electronics in Industrial Embedded Systems. Kaiserslautern 2011, ISBN 978-3-941438-62-0

[8] Johar, Muhammad Akmal; Freier, Robert; Koenig, Andreas; , "Adding self-x capabilities to AMR sensors as a first step towards dependable embedded systems," Intelligent Solutions in Embedded Systems (WISES), 2011 Proceedings of the Ninth Workshop on , vol., no., pp.41-46, 7-8 July 2011

[9] A. Cowen et al., "MetalMUMPs Design Handbook- a MUMPs process", Revision 2.0, MEMSCAP, 2002-2006.

[10] A. Cao, P. Yuen and L. Liwei, "Microrelays With Bidirectional Electrothermal Electromagnetic Actuators and Liquid Metal Wetted Contacts" Journal of Micromechanical Systems, Vol. 16, pg:700-708 June 2007

978-1-4673-2395-6/12 $31.00 © 2012 IEEE

Modeling of biomimetic flow sensor based fish dome shaped cupula using PDMS for underwater sensing

Mohd Norzaidi Mat Nawi[1], Asrulnizam Abd Manaf[1,2], Member, IEEE, Mohd Rizal Arshad[1], Member, IEEE, Othman Sidek[2], Member,IEEE

[1]Underwater Robotics Research Group (URRG), School of Electrical and Electronic engineering
[2]Collaborative microelectronic Design Excellence Center (CEDEC),
Engineering Campus, Universiti Sains Malaysia,
14300 Nibong Tebal, Pulau Pinang, Malaysia
Email: norzaidiurrg@gmail.com

Abstract- **This paper presents the initial modeling on the biomimetic flow sensor based fish dome-shaped cupula for underwater sensing. Fish depend on this cupula to monitor the flow fields especially for maneuvering and survival underwater. We proposed the design structure and the principle of sensing using microchannel which is consist of liquid and electrolyte. PDMS material was chosen because it is easy to deform and soft. By using a computational fluid dynamic and finite element method, the optimal performance was obtained by optimizing the geometrical dimension of the radius and thickness of the dome. The sensor performance is measured on the basis of displacement and strain. The sensitivity of the sensor has been investigated by using different radius and thickness of the dome. Dome with a radius of 0.2mm until 1.2mm was chosen for this study. The resulting in a maximum displacement is 0.27µm and the strain is 3.98E-4 for a flow rate of 1m/s. Simulation results show that the sensitivity of the dome is a maximum when the radius of the dome at the maximum and the thickness of the dome at the minimum.**

Keywords: PDMS, dome shape cupula, biomimetic flow sensor

I. INTRODUCTION

Fish were distributed with mechanosensors along the length of the body called lateral line. Lateral line system consists of 100 to 1000 sense organs called neuromasts that are usually appeared as faint lines running lengthwise down each side. Superficial neuromasts located in the skin in direct contact with the stream while canal neuromasts exist in the sub-epidermal canals connecting pore openings on the skin surface as shown in Fig. 1[1]. In the recent years, many types of the flow sensor based artificial lateral line system including hair cell sensor have been developed using different material and types of sensing. The general structure of the previous flow sensor consists of single hair cell that perpendicular to the substrate and using strain gage or piezoresistive to measure the strain due to the deflection of hair cell [2,3,4]. The limitation of the previous flow sensor is the hair cell cannot be performed in roughness flow condition because it only survives 55° of deflections [4]. Based on this problem, we proposed the dome-shaped flow sensor and will be implemented with the microfluidic technology as a sensing element in order to achieve the miniaturize sensor and high sensitivity of the sensor.

Microfluidic technology has attracted attention because of their potential in chemical and biochemical engineering. Microfabrication techniques allow researchers to control the physical and biochemical environment [5]. Microfluidic systems also have potential for wide application in miniaturization, and it will lead to many benefits, including lower costs in the manufacturing, use and disposal. In addition, smaller channels improve resolution and reduce the overall size of the device, but also make the detection of small vessels to be more challenging, and more sensitive to the adsorption [6]. In designing the flow sensor, material is the most important thing that needs to be considered. In this modeling, Polydimethylsiloxane (PDMS) is proposed. PDMS is a silicon-based polymer usually used for the fabrication of microdevices due to its interesting feature related to our desired application, including biocompatible, gas permeable, deform, chemically inert, and exhibit low auto fluorescence. In addition, it is a cheap material that can be easily attached to the glass, making it user-friendly materials that allow the user to create any type of geometry using mold-replication technology [7].

In this paper, we concentrate on the modeling of PDMS dome shaped structure using computational fluid dynamic (CFD) and finite element method (FEM) approach for underwater sensing applications.

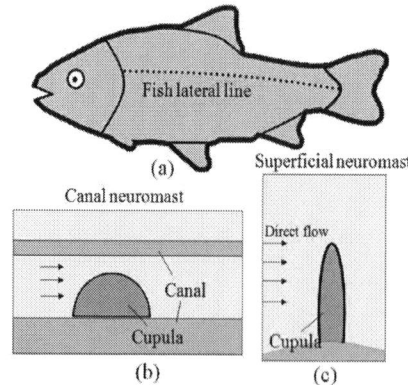

Fig. 1. Schematic of Fish lateral line (a) location of the lateral line system; (b) canal neuromast; (c) superficial neuromast.

978-1-4673-2395-6/12 $31.00 © 2012 IEEE

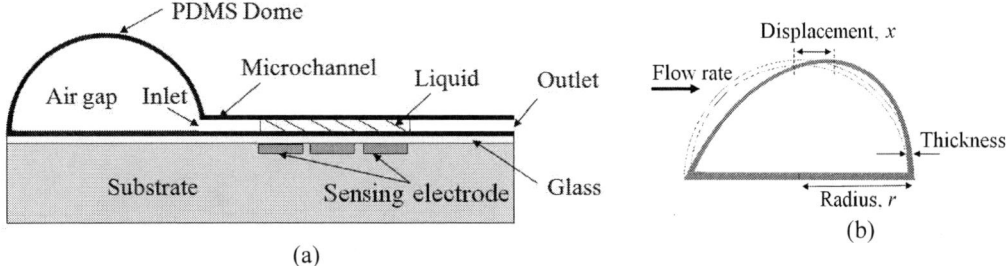

Fig. 2. Scheme structure of the sensor; (a) The biomimetic flow sensor consists of PDMS dome and microchannel
(b) PDMS dome moved at certain displacement

I. SENSOR STRUCTURE

The structure of the biomimetic flow sensor is based on the dome-shaped cupula in fish body and the microchannel which is contained of liquid and electrode as a sensing element, as shown in Fig 2(a). When the flow rate was applied, the fluid will produce the drag force on the dome surface. Based on the properties of PDMS in Table I, the structure is easily deformed where the structure will give some displacement, Fig 2(b). At initial condition, the pressure at inlet and outlet of microchannel was the same. Because of the displacement of the dome, the inlet pressure is changing and gives the effect to the liquid inside the channels. The moving liquid is measured by using sensing electrode by implementation of electrical double layer capacitance. The selection of correct liquid and the dimension of channel also contribute to increased sensitivity of the sensor. The properties of liquid are low surface tensioning and high-polarity solution is important to ensure the smoothly movement in PDMS channel. The potential liquids are methanol, propylene carbonate and polyurethane [9]. For the complete system, the change of capacitance on the electrode will be converted to the voltage by using charge-balanced capacitance-voltage (C-V) conversion readout circuitry [10].

TABLE I
PROPERTIES OF PDMS [8]

Material	Density (kg/m³)	Elastic Modulus (kPa)	Poisson's ratio
PDMS (1:10 mixing ratio)	920	750	0.5

III. METHODOLOGY

The structure design was modeled by using different radius and thickness in order to study the performance of sensor based on dome-shaped geometry. The detail of the parameter is listed in the Table II. Denote that the flow rate as u_o and the radius of the dome as r. At certain flow rate, the Reynold number is given by

$$Re = \frac{u_o x}{v} \tag{1}$$

where x is the leading edge and v is the kinematic viscosity. The drag force acting on the dome surface is obtained

$$F_D = C_D (\frac{1}{2} \rho u^2 dA) \tag{2}$$

where the ρ, C_D and A are the density of water, drag coefficient and surface area that facing the flow. The C_D is depended on the area of the object where the surface area can be calculated using equation below

$$A = \pi r^2 + \pi r t \tag{3}$$

The π and t is constant equal to 3.14159 and 150μm respectively.

TABLE II
LIST OF PARAMETERS FOR DOME-SHAPED FLOW SENSOR

Parameter	Dimension
Microchannel dimension	150μm x 150μm
Dome radius, r	r
Dome height, h	$r + 150$μm

For the simulation process, the package in ANSYS 12.1 was used to model the structure including CFD software FLUENT and FEM software Mechanical APDL. FLUENT is commercial software based on finite volume method that it is enabled to calculate the laminar/turbulent, 2D/3D, steady/unsteady flows, and to run on a personal computer. The FLUENT solves the equation of the mass and momentum conservation equations in terms of primitive variables velocity and pressure for a constant property fluid [11].

$$\nabla \cdot U = 0 \tag{4}$$

$$\frac{\partial U}{\partial t} + U \cdot \nabla U = -\frac{1}{\rho} \nabla P + v \nabla^2 U \tag{5}$$

where ρ and v are the density and kinematic viscosity of water respectively. U is the vector velocity and P is the pressure. The CFD FLUENT will analyze the drag force acting on the surface of dome structure. The geometrical model of the dome is created using Fluent's Gambit pre-processing tool. The boundary condition such as inlet velocity, wall and outlet pressure was set. The several of flow rate were applied perpendicular to the substrate in range 0.1 to 1 m/s. The result from the Fluent is substituted into the model in Mechanical APDL Ansys for finite element method (FEM) solution. The purpose is to do the structural analysis such as to analyze the maximum strain and displacement due to the flow rate.

IV. RESULT AND DISCUSSION

The flow parameter such as velocity and pressure was determined using the CFD software by study the effect of dome dimension to the sensor performance. The iteration for simulation was stopped at 300 because has already met the criterion, and it is limited to 1E-5 for convergence criterion. Fig. 3 shows the comparison between four different radius of dome 0.6mm, 0.8mm, 1mm and 1.2mm. It is clearly shown that maximum radius 1.2mm provide higher drag force compare with others. For the structural analysis, the PDMS property in Table I was used in FEM software ANSYS. The FEM helps to study the sensor structure effect such as strain distribution and displacement based on the flow rate. The comparison cupula structure between the dome shaped, and hair cell is the surface area for the dome-shaped is bigger than the hair cell. According to the equation (2), the increase of area will increase the drag force acting on the sensor. Therefore, the consideration parameter is the radius and the thickness of the dome.

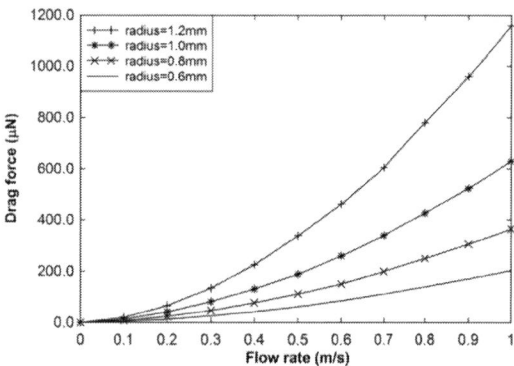

Fig. 3. Drag force versus flow rates with different radius of the dome

The Fig. 4 shows the dome-shaped model in ANSYS for strain distribution and displacement vector for with radius 1mm. The bar bottoms of the figure shows the distribution for strain and displacement respectively. For the strain distribution, the red color was the highest strain which is equal to 2.98E-4 (E-6). Also same for the displacement distribution where the maximum displacement is 0.172μm. From the result of ANSYS simulation, it can be seen that both dome displacement and the maximum strain showed the same trend of improvement in which dome radius increase and thickness of dome decrease as shown in Fig 5 and Fig 6. From the graph, we can see the maximum performance for the highest radius and smallest thickness of the dome. The PDMS dome cannot be broken or failure because the PDMS is soft and deformable.

The purpose of microchannel is to convert the displacement based on the flow rate to the fluid movement in channel. Then, the capacitance was used to measure the fluid changes and will be connected to the readout circuitry. Therefore, the selection of electrolyte in microchannel is also important in producing the effect of capacitive sensing purposes.

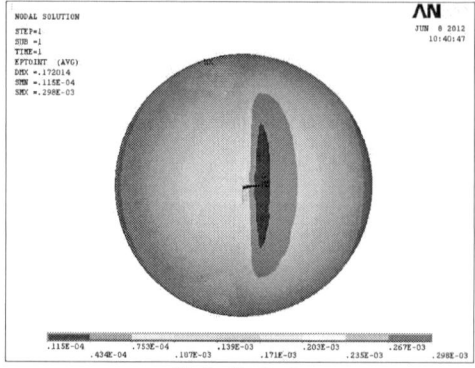

Strain distribution

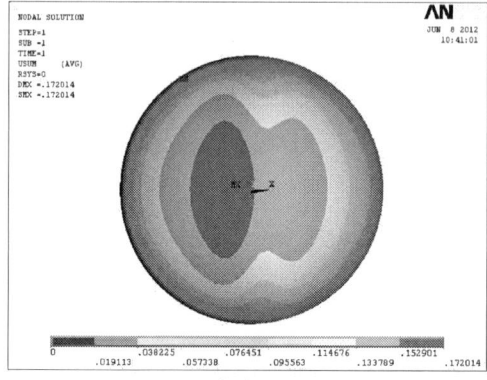

Displacement

Fig. 4. Top view of flow sensor for flow rate 0.1m/s. The max strain 2.98E-4 (E-6) and displacement 0.172μm.

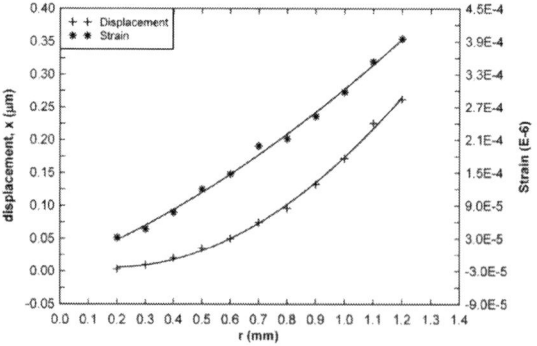

Fig. 5. Deflection and strain as functions of dome shape radius. The thick of the dome was fixed to 40μm.

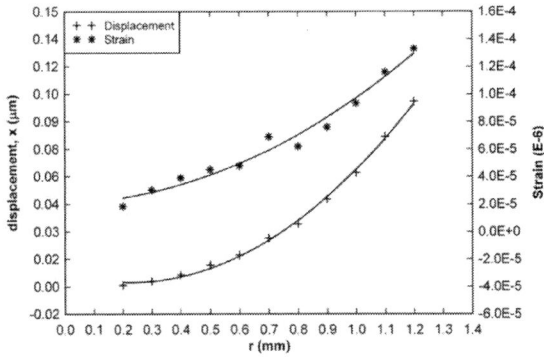

Fig. 6. Deflection and strain as functions of dome shape radius. The thick of the dome was fixed to 100μm

V. CONCLUSION

The simulation was done by using CFD and FEM approach. Based on the simulation response with different radius and thickness to the flow rate, the PDMS dome with 1.2mm radius and 40μm thickness was suggested for the future fabrication. The effect of acting drag force, strain and displacement of the dome were analyzed and discussed. The highest sensitivity can be achieved by the increase the radius of dome and decrease the thickness of the dome.

ACKNOWLEDGMENT

The author is a USM fellowship holder. This study was supported by short term grant number 304/PELECT/60310023 and grant number 1001/PELECT/814168.

REFERENCES

[1] M.J. Montgomery, S. Coombs, M. Halstead, "Biology of the mechanosensory lateral line in fishes". Reviews in Fish Biology and Fisheries, vol.5, pp. 399-416, 1995.

[2] Z. Fan, J. Chen, J. Zou, D. Bullen, C. Liu, F. Delcomyn, "Design and fabrication of artificial lateral line flow sensors ". Journal of Micromechanics and Microengineering vol. 12, pp. 655-661, 2002.

[3] J. Chen, J. Engel, and C. Liu, "Development of Polymer-Based Artificial Haircell Using Surface Micromachining and 3D Assembly." 12th International Conference on Solid-State Sensors, Actuators and Microsystems, Boston, MA, 2003.

[4] Y. Yang, N. Chen, C. Tucker, J. Engel, S. Pandya, and C. Liu, "From artificial hair cell sensor to artificial lateral line system: development and application," in Proceedings of the 20th IEEE International Conference on Micro Electro Mechanical Systems (MEMS '07), pp. 577–580, January 2007..

[5] E. Verpoorte, N.F. De Rooij, "Microfluidics meets MEMS," Proc. IEEE, vol. 91, pp. 930–953, 2003.

[6] J.C. McDonald, D.C. Duffy, J.R. Anderson, D.T. Chiu, H. Wu, O.J. Schueller, G.M. Whitesides, "Fabrication of Microfluidic Systems in Poly(dimethylsiloxane)". Electrophoresis, vol. 21, pp. 27–40, 2000.

[7] V. Senez, E. Lennon, S. Ostrovidov, T. Yamamoto, H. Fujita, Y. Sakai, T. Fujii, "Integrated 3-D Silicon Electrodes for Electrochemical Sensing in Microfluidic Environments: Application to Single-Cell Characterization," Sensors Journal, IEEE , vol.8, no.5, pp.548-557, May 2008.

[8] J. Tong, C.A. Simmons, Y. Sun, "Precision patterning of PDMS membranes and applications". J. Micromech. Microeng. Vol. 18, pp. 1-5, 2008.

[9] A.A. Manaf, K. Nakamura and Y. Matsumoto, "One-Side-Electrode-Type Fluid-Based Inclinometer Combined with CMOS circuitry", Sensors and Materials, Vol. 19, No.7(2007), pp. 417-434.

[10] A.A. Manaf and Y. Matsumoto, "Low voltage charge-balanced capacitance-voltage conversion circuit for one-side-electrode-type flui-based inclination sensor", Solid-State Electronics, Vol. 53 (2009), pp. 63-69

[11] C. Barbier and J.A.C. Humphrey. "Drag force acting on a neuromast in the fish lateral line trunk canal. I. Numerical modelling of external–internal flow coupling", Journal of The Royal Society. Interface, vol. 6, no.36, pp. 627-640, 2009.

Ultrasensitive Poly-Si Nanogap based on capacitive sensor for electrochemical detection

Nazwa Taib, Uda Hashim, Thikra S.Dhahi, Ahmad Sudin, Nur Humaira Md Salleh, and Seng Teik Ten

Institute of Nanoelectronic Engineering (INEE)
Universiti Malaysia Perlis (UniMAP)
Malaysian Agricultural Research and Development Institute (MARDI)
Lot 106,108&110, Tingkat 1 Block A, Taman Pertiwi Indah,
Jalan Kangar-Alor Setar, Seriab, 01000 Kangar Perlis Malaysia.
Email: nazawa_nice@yahoo.com

Abstract- **Nanogap has become a new emerging subject to researcher around the globe for its complementary study with biomolecule and ionic level detection. Higher sensitivity and selectivity in sensing device attained by nanogap has resolved many limitations faced by previous bulk device. For this project, we present a novelty device provide ultrasensitive label-free detection using real time measurement. The fabrication of device involves only conventional lithographic process and simple dry oxidation process. By using 42 nm nanogap devices, the sensitivity of nanogap were tested by dropping 10μL of different level pH and NaCl concentrations onto the target. Experiments were performed by sweeping frequencies from 1 Hz to 1 MHz at room temperature with 30 mV input signal(0 V,DC, Offset). The effects of excitation frequency on capacitance sampling were analyzed.**

Keywords: Nanogap, lithography, ionic capacitance, thermal oxidation

I. INTRODUCTION

Nanogap, namely, a pair of electrodes with nanometer gap sizes is fundamental blocks for the fabrication of nanometer-sizes device and circuit. It is an important tool to examine material properties at nanoscale scale, even at the molecular scale[1]. Nanogap is kind of simple integrated technology system that capable to tune electrical properties from sample solution and biomolecule sample. The biomimetic characteristic found in nanogap enhances the sensitivity towards biomolecule and ionic sample. The electrode structure activated using a lower voltage and frequency to minimize the electrode polarization effect due to parasitic capacitance from solution conductance. Despite all these characteristics, nanogap emerges as the promising device to characterize DNA and other biomolecule activities that provide new insight in biomedical field[2]. Current research provide new insight of nanogap development by functionalize its surface electrodes and used for glucose sensing [3]. Generally, transducer can confer sensitivity to sample liquid conductance. The most common used transducers are electrochemical. Electrochemical detection offer many magnificent features such as higher sensitivity,

least expensive, easy handling and low detection limit. . The electrical signal is directly converted into the electronic domain, therefore, allowing simple instrumentation and design of compact systems[4]. Electrochemical recognition may target current, potential, conductance and/or capacitance measurement. Capacitance may be readily measured in nanogap design using techniques such as dielectric spectra. However, when dealing with solution sampling on metal electrode, there is one barrier for accurate reading. When potential voltage applied to the electrode/ solution sample interface, a double layer charges will takes place. At equilibrium state, nanogap electrode having good conductivity cannot support electric filed causing all charge in the electrode accumulates on the surface. Consequently, two opposite polarity charge get arrange themselves on the solution electrode interface forming the double layer of charges. OHP to the bulk solution [5-6]. To overcome this problem, gap must be downsized to less than 100 nm to overlap the Electrical Double Layer. This solution will produce accurate result by eliminating the noise from parasitic capacitance and ionic conductivity.

In recent years, there has been proposed a nanogap less than 100nm to reduce double layer capacitance contributions so to increase the sensitivity of the device. The device is proven having novelty by immunosensing the glycoprotein laminin, which is clinically relevant to kidney disease at concentration of 0.5μg/ml [7]. By looking at how nanogap size effecting the case sensitivity of device, it was clearly proven that gap less that 100nm has promised the excellent platform to detect wider aspect of biomolecules. Nanogap also has been used to study how water behaves within nanoscale using nanogap biosensor device [8]. The research has shown the capability of nanogap sensor to investigate the mobility of ions through the various water structures.

In this article, we present an array of Poly-Si nanogap device less than 50nm that were used to detect the first performance of device in changes of pH sampling and different concentrations of NaCl in terms capacitance value with excitation frequency.

978-1-4673-2395-6/12 $31.00 © 2012 IEEE

II. EXPERIMENTAL

A. Device Fabrication

Prior to fabrication process, silicon substrates (100) 4 inch size need to be all set. Before applying further process onto wafer, some properties like thickness (Si thickness) and sheet resistance have been checked. The silicon substrates were cleaned and rinsed with deionized water. The deposition of Si_3N_4 were employed by using Plasma-enhanced chemical vapor deposition (PECVD).Plasma nitrides always contain a large amount of hydrogen which provides the enhancement in electrical conductivity, stability and mechanical stress of thin layer of wafer. By using low pressure chemical vapor deposition (LPCVD) machine, about 250 nm layer of poly-si has been deposited on top of nitrate film at 400'C with 80 sccm silane gases supplied. The thickness of poly-Si is designed to be thicker to bear the loading of compression stress from Aluminium deposition plus to increase the value of capacitor. Theoretically, for every 1µm of SiO2 grown, about 0.46 µm of silicon is consumed. The thickness required to support the expansion of nanogap pattern were initially estimated based on kinetics of thermal oxidation principles. It is estimated that for every 1µm of SiO2 grown, about 0.46µm of silicon is consumed. Each layer thickness of deposition was checked by Spectro-photometer and Hawk 3D Nanoprofiler. Aluminium is required to be deposited about 150nm by using thermal evaporator machine. For further reason, aluminium was acting as a hard mask to bear the inflexible ion bombardment during poly-Si RIE process. Then, for the conventional photolithography process, a positive photoresist were coated onto the flat surface of the substrates. The thickness could be estimated by adding 3 drops of photoresist (PR1-2000A) can provide about 1000nm. During photolithography process, the substrates were exposed to 10s through mask 1 shown in fig. 1(b). After development, aluminium layer was removed in the alum etch medium. The substrates were soft baked for 20s to remove the residue solvent used in the development. A permanent pattern is thus were created in the substrate after the removal of photoresist step by using acetone. Poly-Si nanogap pattern on the substrates were dry etched by using RIE recipe of 50sccm of SF6 , 10sccm of O2 ,1.0Pa of pressure, 250 bias and etch for 10s.Then, completely etched the aluminium by using alum etches to get final pattern of nanogap structure. The dry oxidation were carried out at 1000 °C for different time ranging starting from 40mins until the oxidation growth were stop. The poly-Si nanogap patterns were expanded to the closest gap size. Same procedures were applied to fabrication of pad using mask 2 in fig.2. The Ti/Au will be deposited about 60nm and 100nm respectively. By using another pad chrome mask, photolithography process was employed to produce the Ti/Au pad for electrical characterization of nanogap electrodes.

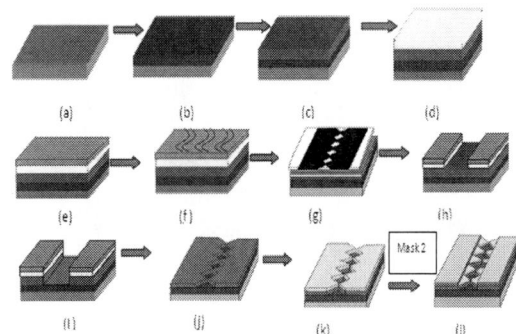

Fig. 1. Process flow; (a) Starting Material, (b)Deposit Si3N4, (c)Deposit poly-silicon, (d) Deposit Al, (e) Resist coating, (f) Soft bake, (g) Exposure mask, (h) Develop resist (i) Polysilicon RIE, (j) Alum etch and stripe resist, ion(k) Dry oxidation, (l) Poly-silicon nanogap pattern with pad Pt/Au fabrication(Electrical checking of the device can be performed on the fabricated pad)(Repeat step (a) to (j) for mask 2). Fig. 3 shows the circuit after serial impedance is measured, a simple resistor model is developed representing the substrate and polysilicon layer. The capacitor also found in series to describe the device with no liquid test.

b. Sample preparation

To provide clear condition of electrical characterization, the frequency used was in the range of 100 Hz to 1 MHz with the constant 30 mV of AC voltage .Because the nanogap was designed as a capacitive sensor, dielectric analyzer were programmed to measure capacitance and loss tangent (ratio of capacitive reactance to resistance).The capacitance reading were studied at different excitation of frequency. Then, the volume effect towards capacitance and time were examined to get clear view of effectiveness of dropping sample onto target electrode. In this study, 10 different levels of pH solution were tested. The results were combined to see the different capacitance for each level solution. Next, various concentrations of Sodium Chloride (NaCl) in deionized water have been tested. Different mass of NaCl were used to get various concentration of solution tested. The molar mass and volume of NaCl is constant in this calculation while the molarity is the case to figure out. Figure 3.11 shows the example of probing system and the target sample.

The solution sample was dropped carefully using pipette onto target between the two nanogap electrodes. Consequently, the result can be tuned out automatically through the capacitance graph from dielectric analyzer interface. To check the next sample, device was cleaned using deionized water and filter paper. Blower was used to ensure no dirt and dust on the sample. Device was kept static with the probe needle each time the sample was located. At this stage, the environment must be silent and avoid any vibration around the characterization system. The test was done for, deionized water and various concentration of sample solution. Figure 2 shows the nanogap device with 42 nm gap size used in detection.

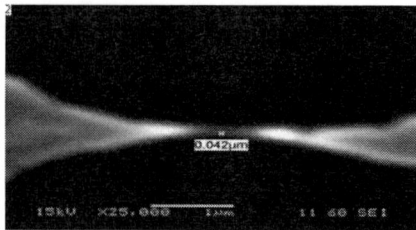

Fig. 2 : SEM by JEOL : Device shows nanogap cavity at 42 nm using in electrochemical testing sample.

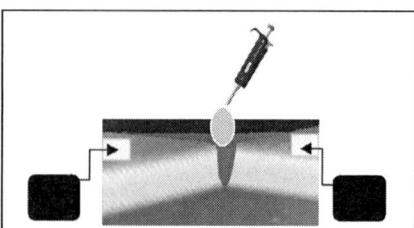

Fig. 3: Schematic image of characterization system. The solution sample was dropped onto the target and the output reading was tuned via dielectric analyzer interface.

The solution sample was dropped carefully using pipette onto target between the two nanogap electrodes as shown in figure 3. Consequently, the result can be tuned out automatically through the capacitance graph from dielectric analyzer interface. To check the next sample, device was cleaned using deionized water and filter paper. Blower was used to ensure no dirt and dust on the sample. Device was kept static with the probe needle each time the sample was located. At this stage, the environment must be silent and avoid any vibration around the characterization system. The test was done for dry capacitance, deionized water and various concentration of sample solution.

III. RESULTS AND DISCUSSIONS

The first set of experiments investigated the effect of excitation frequency on capacitance sampling, where the capacitance was measured at regular intervals between 0.05 Hz. In the experiment, figure 3 showed the capacitance value range from 19nF-36 nF respectively. It can be seen that the acidic sample showed the decrease trend for capacity value while alkalic show the highest reading among them. Capacitance–Frequency (C–F) characterization of nanogap by connecting a two-point probe method were done by using a Keithley 4200 semiconductor characterization system .Experiments were performed for dry measured capacitance for typically by sweeping frequencies from 1 Hz to 1 MHz at room temperature with 30 mV input signal(0 V,DC, Offset). The sensor converts a change in measured into a change of capacitance.

Since a capacitor consist of two electrodes separate by a dielectric, the capacitance change can be caused by the changes in the dielectric between two fixed electrodes [9]. The ionic relaxation and dispersion can be studied during dielectric changes between the nanogap electrodes in terms of capacitance and frequency reading. Thus, capacitance reading variation according to relaxation and resonance process and is usually observed in the frequency range 10^2-10^{10} Hz.[10]

pH	Capacitance value (nF)	Eps" (Tan Delta)
1	46	8.6105e-01 - 6.682e +02
2	42	1.2208e+0 - 6.3043e+02
3	39	1.3985e +0 - 7.1235e+02
4	36	11.158e+0 - 5.3258e+02
5	34	5.0789e+0 - 5.5789e+02
6	31	1.3797e+0 - 5.4453e+02
7	29	1.1503e+0 - 7.3601e+02
8	28	1.1904e+0 - 6.622e+02
9	27	1.098e+0 - 9.7070e+02
10	22	4.8546e+0 - 5.475e+02

Table 1: The capacitance and loss tangent values for different level of pH

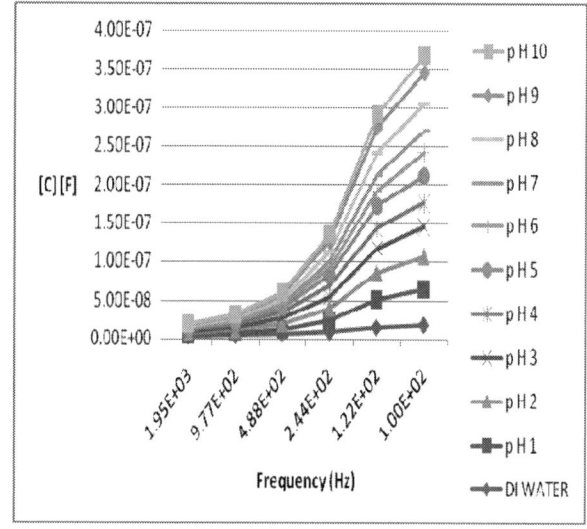

Fig. 4: Capacitance-Frequency relationship of 10 different level of pH using device A

Table 1 and figure 4 shows the relationship between capacitance and loss tangent for 10 different pH solution samples. In a consequent, the capacitance value is stable during 10 measurements taken. The measurement is repeated for 3 times for each pH solution being characterized.

Fortunately, the result able to maintain and almost uniform result will be obtained for each sample respectively.

Capacitance value in the range of 22- 46 nF were obtained and los tangent in the range of 1 to 01 indicated that small losses of dielectric occurred in nanogap capacitive sensor.

Molarity (M)	Capacitance value (nF)	Eps" (Tan Delta)
0.1	26	1.2595e+0- 6.014 e+02
0.3	28	1.239e+0 - 6.832e+02
0.5	29	4.485e-01 - 7.138e+02
0.7	35	1.1848e-01 - 6.04e+02
0.9	38	9.33e-01 - 8.8074e+02
1.1	47	1.2659e+ - 9.197e+02

Table 2 :The capacitance and loss tangent values for different concentrations of NaCl

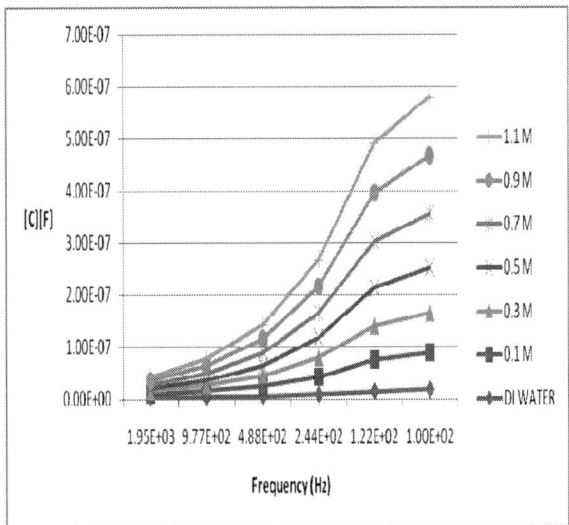

Fig. 5: Capacitance-Frequency relationship of 6 different concentrations of Sodium Chloride (NaCl) using nanogap device

From the table 2 and figure 5 , noted that the greatest changes are in the frequency range between 1000 Hz to 100 Hz, But in this project nanogap capacitance value indicated at 100 Hz. Higher than 100 Hz , the response is dominated by charge accumulation at the electrodes. The loss peak is also shifted higher in frequency and the small peak at 100 Hz. The different reading for each concentration in the range of 2 nF-9 nF which is desirable in this project to show the sensitivity of device. Whenever the test has conducted twice for the same sample, the results configure almost similar.

IV. CONCLUSIONS

The capability of nanogap device in tuning different capacitance values from different concentration of NaCl and pH in uniform way has proved that nanogap was very receptive to any ionic and molecular activity in sample solution. Capacitance measured in the range of 20 nF to 60 nf were acquired during sensitivity test and stable for each measurement. Results were uniforms even repeated many times. Matter concerning on repeatability of device can be automatically resolved. This simple testing system required less time sample handling, scalable and reliable for studying sample solution.

.

ACKNOWLEDGMENT

We are grateful for fruitful discussions with our collaborators at the Institute of Nano Electronic Engineering (INEE) at University Malaysia Perlis (UniMAP). This work was supported by INEE at UniMAP and MARDI through the Nano Technology project. The views expressed in this publication are those of the authors and do not necessarily reflect the official view of the funding agencies on the subject.

REFERENCES

[1] W. H. Tao Li, Daoben Zhu, "Nanogap Electrodes. Advance," *Advance Material,* vol. 22, pp. 286-300, 2010.

[2] J. S. Ho Kwan Kang, Dino Di Carlo,Yang-Kyu Choi, Luke P.Lee, "Planar nanogap capacitorarrays on quartz for optical and dieelctric bioassays," in *7th International conference on miniaturized Chemical and biochemical analysis system,* Squay Valley ,California USA, 2003, pp. 697-700.

[3] F. C. Y. U.S.Dinish, Ajay Agarwal, Malini Olivo, "Development of highly reproducible nanogap SERS substrates; comparative performance analysis and its application for glucose sensing," *Biosensors and Bioelectronics,* vol. 26, pp. 1987-1992, 2011.

[4] D. S. o. J. Hong, M.-I. Park, J. Choi, T.S. Kim, G. Im, S. Kim, Y.E. Pak and K. No, "dielectric biosensor using the capacitance change with AC frequency integrated on glass substrates.," *Japanese Journal of Applied Physics,* vol. 43, pp. 5639-5645, 2004.

[5] D. M. Kolb, "Reconstruction phenomena at metal-electrolyte interfaces," *Progress in surface science* vol. 51, pp. 109-174, 1996.

[6] W. Schmickler, "Electronic effects in the electric double layer," *Chemical Reviews,* vol. 96, pp. 3177-3200., 1996.

[7] H. K. Dino Di Carlo, Xianghui Zeng, Ki-Hun Jeong, Luke P.Lee, "Nanogap-based dielectric immunosensing " in *TRANSDUCERS'03 the 12th IEEE international conference on solid state sensors ,actuators and microsystem,* 2003, pp. 1180-1183.

[8] Z. G. Xing Chen, Gui-Mei Yang, Jie Li, Min-Qiang Li, Jin-Huai Liu, Xing-Jiu Huang. (2010, Electrical Nanogap devices for biosensing. *Materialstoday 13(11),* 28-41.

[9] S.M.Sze, Ed., *Semiconductor sensors.* John Wiley &Sons. Inc, 1994, p.^pp. Pages.

978-1-4673-2395-6/12 $31.00 © 2012 IEEE

Optical properties of zinc oxide films growth on Si substrate via aqueous chemical growth

Maria Abu Bakar[1], Muhammad Azmi Abdul Hamid[1,*], Azman Jalar, *Member, IEEE*[2] & Roslinda Shamsudin[1]

[1]School of Applied Physics, Faculty of Sciences and Technology
[2]Institute of Microengineering and Nanoelectronics (IMEN)
Universiti Kebangsaan Malaysia, 43600 Bangi, Selangor
Email: azmi@ukm.my

Abstract- **This paper presents a methodology to synthesized zinc oxide (ZnO) films by an aqueous chemical growth based technique. Structural, morphology and optical properties of ZnO films were characterized using X-ray diffractometry (XRD), field emission scanning electron microscopy (FESEM) and UV-Visible spectroscopy. This paper aims to study the effect of growth time on the structural, morphology and optical properties of ZnO films. XRD patterns showed a well-defined (101) peak that indicates the crystalline hexagonal ZnO structure. FESEM results revealed hexagonal rods structures growth on Si substrate from 1.5 h to 3 h growth time. The optical band gap for ZnO film has increased from 3.80 eV to 3.90 eV with the increasing of growth time. This result shows that the growth time have influences on the structural, morphology and optical properties of ZnO films.**

I. INTRODUCTION

Zinc oxide (ZnO) has received much attention due to its various applications such as dye sensitized solar cells [1], light-emitting diodes [2], humidity sensor [3], and surface acoustic wave (SAW) devices [4]. ZnO has a wide bandgap of 3.37 eV and a large exciton binding energy of 60 meV which is higher than thermal energy at room temperature, 27 meV [5]. There are many deposition techniques to prepared ZnO films such as thermal evaporation [6], molecular beam epitaxy (MBE) [7], chemical vapor deposition [8] and dc magnetron sputtering [9]. However, these growth techniques require higher temperature, pressure and expensive apparatus.

Recently, chemical solution route has become a promising approach for the large scale production of nano/microscale materials because of it is less expensive, simple, fast, and requires a low growth temperature [10]. Up to now, aqueous chemical growth (ACG) approaches have gained in importance because of low temperature growth technique and cost efficient.

In the present work, we report on the growth of ZnO film on Si substrate via aqueous chemical growth method. The aim of this paper is to study the effect of different growth time process from 1.5 hour to 3 hour on ZnO structure, morphology and its optical properties. The structural and morphological of the ZnO films were carried out using X-ray diffraction (XRD) and field emission scanning electron microscope (FESEM).

The optical energy band gaps were calculated from absorption spectra using UV-Visible spectroscopy.

II. MATERIALS AND METHOD

ZnO films were grown by the aqueous chemical growth method. The experimental setup for the growth of ZnO films is shown in Fig. 1.

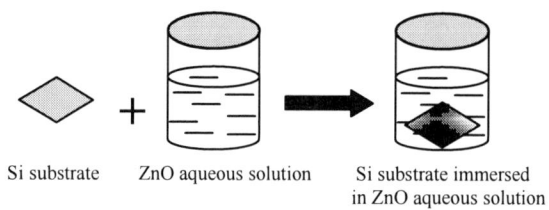

Si substrate   ZnO aqueous solution   Si substrate immersed in ZnO aqueous solution

Fig. 1. Schematic diagram of the experimental setup for the growth of ZnO films on Si substrate

The Si substrates used in the experiment were cleaned with acetone for 15 min. ZnO aqueous solution was prepared using an equimolar (0.05 M) of zinc nitrate hexahydrate $(Zn(NO_3)_2.6H_2O)$ and hexamethylenetetramine $(C_6H_{12}N_4)$. The mixed solution was magnetically stirred until complete dissolution for one hour. The Si substrates were then immersed in the prepared solution and heated at 90 °C for different growth time; 1.5 hour, 2 hour, 2.5 hour and 3 hour in an oven without any stirring. Subsequently, the samples were removed and washed thoroughly with deionized water to eliminate any residual salts and were then dried in air.

The samples were analyzed using field emission scanning electron microscopy (FESEM-SUPRA 55VP) to determine the morphology of the ZnO nanostructures. The phase composition and crystal structure were determined by X-ray diffractometer (XRD-Bruker D8 Advance) with CuKα radiation source (40 kV, 40 mA, λ = 0.15406 nm). The optical absorbance of the ZnO film was recorded using UV/VIS/NIR spectrometer (LAMBDA 900). The optical absorbance measurement of the samples was carried out at room temperature with the wavelength range of 300-800 nm.

978-1-4673-2395-6/12 $31.00 © 2012 IEEE

III. RESULT AND DISCUSSION

Fig. 2 shows the XRD pattern of the ZnO films grown on the Si substrates in aqueous solution at 90°C for 1.5 h to 3 h. As can be seen from the figure, all samples gave similar XRD pattern indicating the ZnO films are in high crystallinity. In particular, it was observed that all the high intensity peaks, (100), (002), (102), (110) and including the strong (101) peak are assigned to typical wurzite hexagonal structure of ZnO according to the reference data (JPCDS card No. 36-1451, a = 0.325 nm, c = 0.521 nm). No other diffraction peaks are found in any of our samples. It is clear from Fig. 1 that the (101) peak increases with the increasing of growth time. The grain size, t of ZnO films was calculated using the Debye-Scherrer equation as follows [11]:

$$t = k\lambda / \beta \cos \theta \qquad (1)$$

where t is the crystallite size, k =1, λ is the X-rays wavelength (0.15406 nm), β is full width half maximum (FWHM) in radians, and θ is the Bragg diffraction angle. The grain size of the as deposited films was estimated about 34 nm for ZnO films growth for 1.5 h, 36 nm for 2 h and 2.5 h and increased to 37 nm for 3 h growth time. The increasing of the growth time clearly has contributed to increase the ZnO grain size, indicating better crystallinity of the ZnO films grown at longer periods.

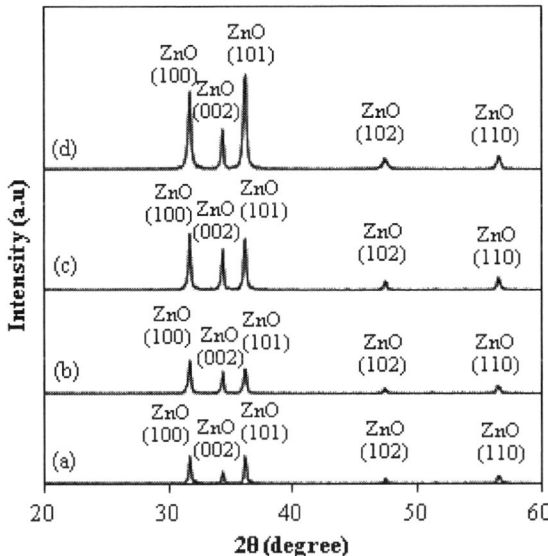

Fig. 2. X-ray diffraction patterns of ZnO grown on the Si substrate at different growth time: (a) 1.5 h, (b) 2 h, (c) 2.5 h and (d) 3 h.

The morphology of prepared ZnO was determined by using FESEM. Fig. 3 shows the typical SEM images of the ZnO nanostructures at different growth time.

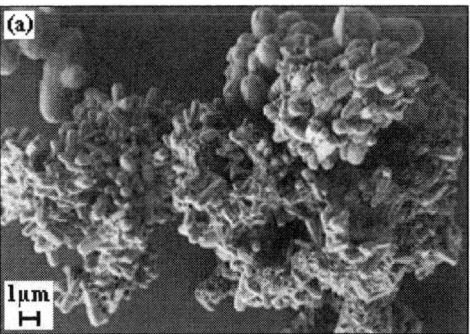

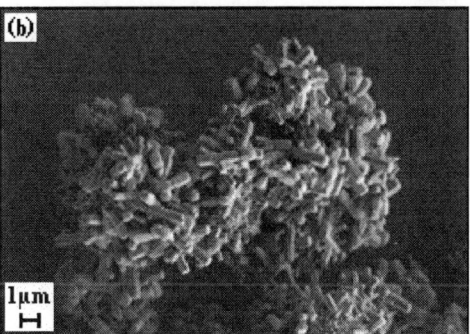

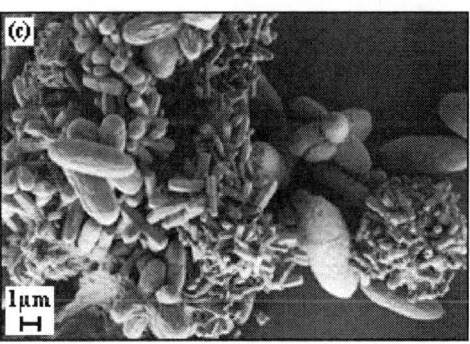

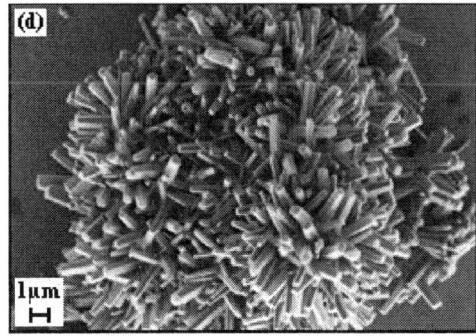

Fig. 3. FESEM micrographs of the ZnO on Si at different growth time: (a) 1.5 h, (b) 2 h, (c) 2.5 h and (d) 3 h.

These ZnO nanostructures were observed to have a random distribution of branched hexagonal rods. Understandably, with the increase of the time, it appears that the growth is self-assemble and the hexagonal rods started to grow and branched from the center on the top of the oval-like ZnO nanostructures.

These branched hexagonal rods were found to start grown from 1.5 h growth time process (Figure 1a). Clearly, it was observed that the ZnO hexagonal rods size increase with the increasing of growth time. This is because of "Ostwald ripening" as reported previously by Krichevsky and Stavans [12].

Fig. 4 shows the absorption spectrum in the wavelength range from 300 nm to 800 nm of ZnO samples at 90°C for different growth time. Clearly seen from the figure, all samples gave similar absorption spectrum.

The optical band gap energy in Fig. 5 shows of ZnO samples deposited for 1.5 h to 3 h. The optical band gap energy of ZnO film was calculated using Tauc's method from absorbance spectra [13]. The values of the optical band gap energy (E_g) are determined by extrapolating the linear portion of the curves to $(ahv)^2 = 0$.

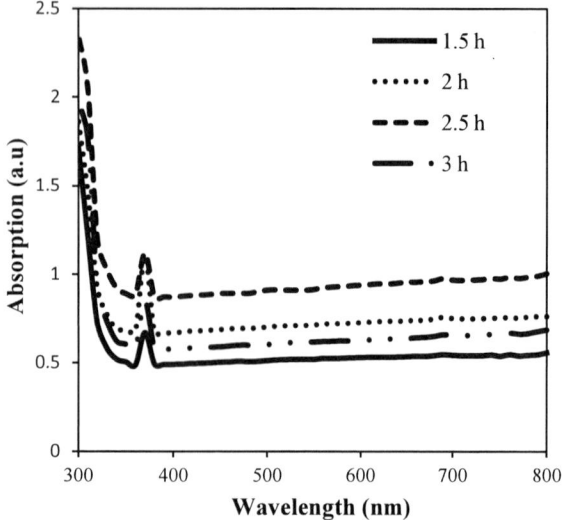

Fig. 4. Absorption graphs of ZnO grown on the Si at different growth time.

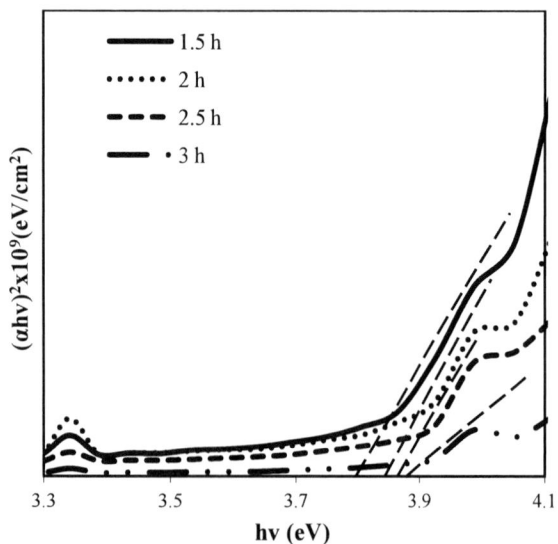

Fig. 5. Plot of $(\alpha hv)^2$ vs. hv of ZnO grown on the Si at different growth time.

The optical band gap for ZnO sample deposited for 1.5 h growth time was found to be 3.80 eV, which increased to 3.86 eV, 3.88 eV and 3.90 eV for 2 h, 2.5 h and 3 h growth time, respectively. The increased of optical band gap in our system may attributed to the enlarged of grain size which in agreement with Ting et al. [14].

IV. CONCLUSIONS

In summary, we has demonstrated a simple method for synthesis ZnO films deposited on Si substrate for 1.5-3 h growth time process. The grain size range of the as deposited films was estimated about 34 nm to 37 nm, which gradually increased with the increasing of growth time. This indicates that for longer growth time period, the ZnO appeared to have better crystalline. The optical band gap for ZnO calculated from absorption spectra ranged between 3.86 to 3.90 eV, which higher the values obtained for bulk ZnO crystals, 3.37 eV. The results show that the growth time has some influence on the ZnO morphology, structure and optical properties.

ACKNOWLEDGEMENT

The authors acknowledge financial support from Universiti Kebangsaan Malaysia (research funding: UKM-RRR1-07-FRGS0257-2010 and OUP-2012-012).

REFERENCES

[1] R.C. Pawar, J.S. Shaikh, P.S. Shinde and P.S. Patil, "Dye sensitized solar cells based on zinc oxide bottle brush," Materials Letters, vol. 65, pp.2235–2237, 2011.

[2] Q. Qiao, B.H. Li, C.X. Shan, J.S. Liu, J. Yu, X.H. Xie, Z.Z. Zhang, T.B. Ji, Y. Jia and D.Z. Shen," Light-emitting diodes fabricated from small-size ZnO quantum dots," Materials Letters, vol. 74 , pp. 104–106, 2012.

[3] P.K. Kannan, R.Saraswathi and J.B.B. Rayappan, "A highly sensitive humidity sensor based on DC reactive magnetron sputtered zinc oxide thin film," Sensors and Actuators A, vol. 164, pp. 8–14, 2010.

[4] D.T. Phan and G.S. Chung, "The effect of post-annealing on surface acoustic wave devices based on ZnO thin films prepared by magnetron sputtering," Applied Surface Science, vol. 257, pp 4339–4343, 2011.

[5] Y.Chen, X.L.Xu, G.H.Zhang, H.Xue and S.Y.Ma, "A comparative study of the microstructures and optical properties of Cu- and Ag-doped ZnO thin films," Physica B, vol. 404, pp.3645–3649, 2009.

[6] S.N.F. Hasim, M.A.A. Hamid, R. Shamsudin and A. Jalar, "Synthesis and characterization of ZnO thin films by thermal evaporation," Journal of Physics and Chemistry of Solids, vol. 70, pp 1501–1504, 2009.

[7] C. Wang, Z. Chen, H. Hu and D. Zhang, "Effect of the oxygen pressure on the microstructure and optical properties of ZnO films prepared by laser molecular beam epitaxy," Physica B, vol. 404, pp 4075–4082, 2009.

[8] F.S.S. Chien, C.R. Wang, Y.L. Chan, H.L. Lin, M.H. Chen and R.J.Wu, "Fast-response ozone sensor with ZnO nanorods grown by chemical vapor deposition," Sensors and Actuators B, vol. 144, pp 120–125, 2010.

[9] A.Tanusevski and V.Georgieva, "Optical and electrical properties of nanocrystal zinc oxide films prepared by dc magnetron sputtering at different sputtering pressures," Applied Surface Science, vol. 256, pp 5056–5060, 2010.

[10] L. Vayssieres, "Growth of arrayed nanorods and nanowires of ZnO from aqueous solutions," Adv. Mater. **15**, pp.464-466, 2003.

[11] R. Elilarassi and G.Chandrasekaran, "Microstructural and photoluminescence properties of Co-doped ZnO films fabricated using

a simple solution growth method," Materials Science in Semiconductor Processing, vol.14, pp.179–183, 2011.

[12] O. Krichevsky and J. Stavans, "Correlated Ostwald Ripening in Two Dimensions," Physical Review Letters, vol. 70, pp. 1473-1476, 1993.

[13] S.E. Rodil, S. Muhl, S. Maca and A.C. Ferrari, "Optical gap in carbon nitride films," Thin Solid Films, vol. 433, pp 119–125, 2003.

[14] C.C. Ting, C.H. Li, C.Y. Kuo, C.C. Hsu, H.C. Wang and M.H. Yang, "Compact and vertically-aligned ZnO nanorod thin films by the low-temperature solution method," Thin Solid Films vol. 518, pp 4156–4162, 2010.

The Growth and Fabrication of High-Performance $In_{0.5}Ga_{0.5}As$ Metal-Oxide-Semiconductor Capacitor on GaAs Substrate by Metalorganic Chemical Vapor Deposition Method.

Hong Quan Nguyen[1], Hai Dang Trinh[1], Hung Wei Yu[1], Ching Hsiang Hsu[1], Chen Chen Chung[1], Binh Tinh Tran[1], Yuen Yee Wong[1], Thanh Hoa Phan Van[1], Quang Ho Luc[1], Diao Yuan Chiou[1], Chi Lang Nguyen[1], Chang Fu Dee[1], and Edward Yi Chang,[1,2,*] *Senior member, IEEE*

[1]Department of Materials Science and Engineering, National Chiao Tung University, 1001 University Road, Hsinchu 300, Taiwan
[2]Department of Electronics Engineering, National Chiao Tung University, 1001 University Road, Hsinchu 300, Taiwan
Email: edc@mail.nctu.edu.tw

Abstract- **Growth conditions have investigated for growing high quality $In_{0.3}Ga_{0.7}As$ and $In_{0.5}Ga_{0.5}As$ on GaAs substrate by metalorganic chemical vapor deposition method. Annihilation reactions between threading dislocations observed by transmission electron microscopy are experimental evidences to confirm threading dislocations had been blocked in $In_xGa_{1-x}As$ buffer layers. A high quality smooth surface $In_{0.5}Ga_{0.5}As$ epi-film with threading dislocation density of $2x10^6$ cm^{-2} was achieved at growth temperature of 490 ^oC. Metal-oxide-semiconductor capacitor devices fabricated on $In_{0.5}Ga_{0.5}As/GaAs$ perform nice capacitance-voltage response, with small frequency dispersion. The conductance contours indicate that the Fermi level moves freely to the lower part of the InGaAs bandgap without pinning.**

I. INTRODUCTION

Future complementary metal-oxide-semiconductor (CMOS) device technologies will require the integration of higher carrier mobility materials to increase drive current capability [1]. The ternary compounds of $In_xGa_{1-x}As$ ($x\approx0.5$) are potentially suitable for the device applications due to its high electron mobility and low band gap properties. Currently, most of the devices are designed based on $In_{0.53}Ga_{0.47}As$ grown on InP substrates by molecular beam epitaxial (MBE) method [2, 3]. A major advantage of using InP as a substrate is that it is ideal for growing lattice-matched InGaAs/InP. However, InP substrates are more expensive, fragile, available in smaller sizes, and have less mature processing technology as compared to GaAs substrates [4]. Integration of III-V compounds such as InGaAs on Si substrates has received strongly interest in recent years [5, 6], because the successful integration of the III-V compounds on Si substrate is promising for less expensive high-speed low power logic devices. However, the large lattice and thermal expansion coefficient mismatches between $In_xGa_{1-x}As$ compounds and Si substrate obstruct the growth of a high-quality epi-layer, especially by using metal organic chemical vapor deposition (MOCVD) method. These difficulties have been being addressed by using III-V materials [7, 8] or Si_xGe_{1-x} alloy [9] as buffers in order to reduce dislocation propagating to active layer. Most efforts aim at high-quality GaAs layers on

Si, then, the GaAs layers will play as alternative wafer platform for growing other III-V compounds. However, high device performance III-V based heterostructures on Si has not been commercial successful. Therefore, enabling $In_xGa_{1-x}As$ ($x\approx0.5$)-based devices on GaAs substrate is desired, making them one step closer to future goal of large-scale integration of devices on Si substrate. In this paper, we investigate the grow conditions to achieve high-quality $In_{0.5}Ga_{0.5}As$ epi-film on GaAs substrate by MOCVD method. The high-quality $In_{0.5}Ga_{0.5}As$ film allows fabricating high performance $In_{0.5}Ga_{0.5}As$-based MOSCAP devices.

II. EXPERIMENTS

InGaAs compound was growth by metalorganic chemical vapor deposition (MOCVD-EMCORE D180). Samples were grown on epi-ready GaAs(001) substrates with 6° off-cut toward the [110] direction. Group-III precursors of trimethylindium (TMIn) and trimethylgallium (TMGa) and the group-V precursor of pure arsine (AsH3) were used. Monosilane (SiH4) was used as n- and p-type doping sources. The total pressure in the reactor was kept at 70 Torr. Growth temperature was vaired from 490 to 620 ^oC. $In_xGa_{1-x}As$ step-graded layers with a parabola-like composition profile were used as buffer layers to block threading dislocation (TD) [10]. The indium composition and degree of relaxation were determined with a high-resolution X-ray diffractometer (HRXRD). The surface texture and roughness were examined by atomic force microscopy (AFM). The dislocation densities were characterized by cross-sectional transmission electron microscopy (TEM). For MOSCAP device fabrication, initial InGaAs/GaAs wafer was degreased in acetone and isopropanol for 2 min each. The sample was then dipped into HCl 4% solution for 2 min, followed by rinsing in deionized (DI) water and N_2 blowing before loading into the atomic layer deposition (ALD) chamber (Cambridge NanoTech Fiji 202 DSC) within 5 min. In ALD chamber, 10 trimethyl aluminum (TMA)/Ar pulses were used for pre-cleaning, followed by the deposition of 9 nm Al_2O_3 at 250^oC using TMA and water as precursors.

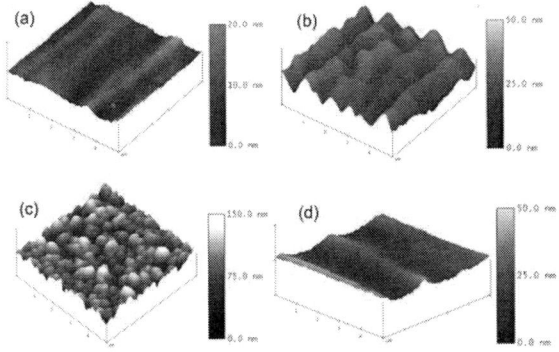

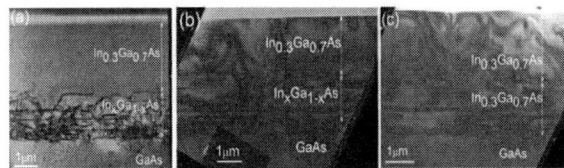

Fig. 2. Cross-sectional TEM images of $In_{0.3}Ga_{0.7}As$ epi-films grown on 6^0 off-cut GaAs substrate grown at (a) 490 °C, (b) 550 °C, and (c) 620 °C.

Fig. 1. AFM images of $In_{0.3}Ga_{0.7}As$ films grown on GaAs substrate at (a) 490 °C, (b) 550 °C, and 620 °C; and (d) is AFM of $In_{0.5}Ga_{0.5}As$ on GaAs substrate grown at 490 °C.

After that, the sample was post deposition annealed (PDA) at 500°C in N_2 for 5 min. Ni/Au gate metal was formed via lithography/e-beam deposition/lift-off steps. Finally, backside Au/Ge/Ni/Au ohmic contact was e-beam deposited, followed by post metal annealing (PMA) at 300°C in N_2 for 5 min. Capacitance-voltage (C-V), conductance-voltage (G-V) characteristic of samples were measured by using an HP4284A LCR meter, while gate leakage current (I-V) characteristic was acquired using an HP4200 meter.

III. RESULTS AND DISCUSSION

Fig. 1(a)-1(c) represents the AFM images of $In_{0.3}Ga_{0.7}As$ epi-films grown on GaAs substrate using $In_xGa_{1-x}As$ step-graded buffer layers at 490, 550 and 620 °C, respectively. It is observed growth temperature strongly affects to surface roughness and surface morphology of the epi-films. A smooth surface with root mean square (RMS) roughness of 2 nm and crosshatch pattern (Fig. 1(a)) was obtained at growth temperature of 490 °C. The surface roughness of epi-film was increased to 6.5 and 23 nm when growth temperature increases to 550 (Fig. 1(b)) and 620 °C (Fig. 1(c)), respectively. The poorer surface morphology of samples grown at higher growth temperatures was due to the increase in the growth rate and TDs propagation from buffer layers to epi-film.

Crystal quality of $In_{0.3}Ga_{0.7}As$ epi-films were examined by TEM measurement. Fig. 2(a)-2(c) shows cross-sectional TEM images of the epi-films grown at 490, 550 and 620 °C, respectively. All samples were grown using the same step-graded buffer layer design. In these structures, 1.5 μm of $In_xGa_{1-x}As$ buffer layers was grown first on GaAs substrate followed by 2 μm $In_{0.3}Ga_{0.7}As$ epi-layer. It is clearly shown in Fig. 2(a) that TDs was confined within the buffer layer when growth temperature of 490 °C was used, resulting in almost no TD extension into the epi-layer on top. Detail calculation shows that TD density in the epi-layer was about 10^6 cm^{-2}. In contrast, as shown in Fig. 2(b) and 2(c), the growth at higher temperatures results in worse crystal quality. TDs could not be confined well within the buffer layers. They elongated or propagated to the top free surface, result is TD densities were 5×10^7 cm^{-2} and 10^8 cm^{-2} for epi-films grown at 550 and 620 °C,

respectively. The high TD density extended from the bottom layers has worsened the surface morphology of the epi-layer as shown in Fig. 1(b)-1(c).

The growth of high-quality, smooth surface $In_{0.3}Ga_{0.7}As$ epi-film allows us to achieve a high-quality $In_{0.5}Ga_{0.5}As$ on GaAs substrate. Fig. 3(a) shows a cross-sectional TEM mage of 1.5 μm n-type-doped $In_{0.5}Ga_{0.5}As$ grown continuously on $In_{0.3}Ga_{0.7}As$ using $In_xGa_{1-x}As$ (x=0.34÷0.49) buffer layers. It is shown that no TDs were observed on the $In_{0.5}Ga_{0.5}As$ epi-layer. The plan-view TEM image (not shown) indicates that a TD density of 2×10^6 cm^{-2} in $In_{0.5}Ga_{0.5}As$ was achieved. The growth of high indium composition $In_{0.5}Ga_{0.5}As$ also caused increased RSM roughness of the film to 3.5 nm [Fig. 1(d)].

Fig. 4 shows the asymmetric (115) ω-2θ scans performed by double-axis HRXRD of $In_{0.3}Ga_{0.7}As$ and $In_{0.5}Ga_{0.5}As$ epi-layers grown on GaAs substrates at 490 °C. The indium composition and degree of relaxation of epi-films were estimated through the peak separation between films and substrate [10]. The calculated values of out-of-plane ($a_\perp$) and in-plane ($a_\parallel$) lattice constants of $In_{0.3}Ga_{0.7}As$ film were around 5.772 and 5.767 Å, and the out-of-plane ($a_\perp$) and in-plane ($a_\parallel$) lattice constants of $In_{0.5}Ga_{0.5}As$ film were 5.8562 and 5.8542 Å. These parameters indicate that the indium composition of the epi-layers was about 29.5% and 50%, respectively, and the epi-layers were nearly fully relaxed on GaAs substrate.

Fig. 5 (a) shows the multi-frequency C-V responses of 9nm Al_2O_3/n-$In_{0.5}Ga_{0.5}As$/GaAs. The C-V of MOSCAP shows a nice performance with distinct accumulation, depletion, and inversion regions. The frequency dispersion in accumulation regime of $In_{0.5}Ga_{0.5}As$ MOSCAP is dramatically reduces as compared with the report on low In content MOVCD InGaAs/GaAs MOSCAPs devices [11].

To study in detail the performance of Al_2O_3/$In_{0.5}Ga_{0.5}As$ MOSCAP, conductance methods were performed. The two dimensional contour plot of parallel conductance as a function of bias voltage and measurement frequency of the sample is

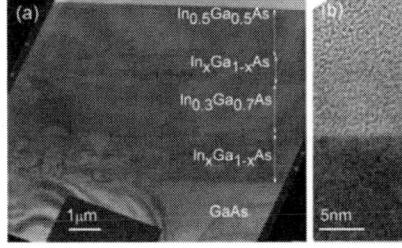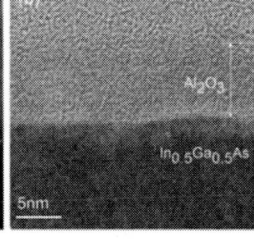

Fig. 3. (a) Cross-sectional TEM images of $In_{0.5}Ga_{0.5}As$ epi-films grown on GaAs substrate, (b) HR-TEM image of Al_2O_3/$In_{0.5}Ga_{0.5}As$ MOSCAP

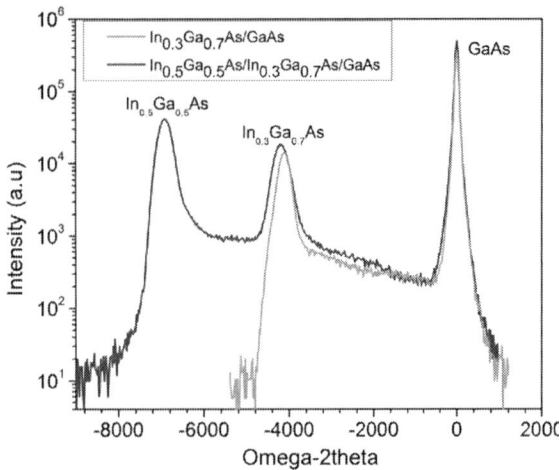

Fig. 4. XRD data from asymmetry (115) ω-2θ scan of $In_{0.3}Ga_{0.7}As$ and $In_{0.5}Ga_{0.5}As$ epi-films on GaAs substrate.

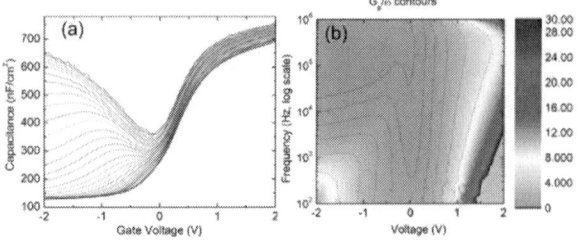

Fig. 5. (a) Multi-frequency C-V curves of 9nm Al_2O_3/MOVCD n-$In_{0.5}Ga_{0.5}As$/GaAs, and (b) Conductance contours shows clearly the free movement of the Fermi level.

shown in Fig. 5(b). From the figure, a frequency-dependent shift in the conductance peak maximum with gate voltage is observed in whole range of measured frequency (100Hz ÷ 1 MHz), indicating that the Fermi level moves freely to the lower part of the InGaAs bandgap without pinning.

IV. CONCLUSION

The effects of growth temperature on the surface morphology and the crystal quality of the $In_{0.3}Ga_{0.7}As$ epi-layer have been investigated. A smooth surface with a rms roughness of 2 nm, has been obtained at growth temperature of 490 °C. TEM measurement indicated that TD density in $In_{0.3}Ga_{0.7}As$ film was about $1x10^6$ cm^{-2}. The optimum growth condition and buffer design allow us to grow 50% indium of InGaAs on GaAs substrate with low TD density. The MOSCAP devices fabricated on $In_{0.5}Ga_{0.5}As$/GaAs film show a good performance with nice C-V response and low frequency dispersion. The conductance map indicates that Fermi level moves freely to the lower part of the InGaAs bandgap without pinning.

ACKNOWLEGEMENT

The authors are thankful for the assistance and support of the National Science Council, Taiwan, R.O.C., under Contract Nos. NSC 98-2120-M-009-010 and NSC. 99-2221-E-009-170-MY3.

REFERENCES

[1] Peng Chen, et al., "High crystalline-quality III-V layer transfer onto Si substrate", *Appl. Phys. Lett.* 92, pp 092107-1-3, 2008.

[2] L. K. Chu, et. al, "Low interfacial trap density and sub-nm equivalent oxide thickness in $In_{0.53}Ga_{0.47}As$ (001) metal-oxide-semiconductor devices using molecular beam deposited HfO_2/Al_2O_3 as gate dielectrics," *Appl. Phys. Lett.* 99, 042908, 2011.

[3] Roman Engel-Herbert, Yoontae Hwang, Joël Cagnon, and Susanne Stemmer, "Metal-oxide-semiconductor capacitors with ZrO_2dielectrics grown on $In_{0.53}Ga_{0.47}As$ by chemical beam deposition" *Appl. Phys. Lett.* 95, 062908, 2009.

[4] Y. M. Kim, M. Dahlstrom, S. Lee, M. J. W. Rodwell, and A. C. Gossard, "InP/$In_{0.53}Ga_{0.47}As$/InP double heterojunction bipolar transistors on GaAs substrates using InP metamorphic buffer layer", *Solid-State Electronics* 46, pp. 1541-1544, 2002.

[5] N. Mukherjee, et al., "MOVPE III-V material growth on silicon substrates and its comparision to MBE for future high performance low power logic applications" *IEDM* 11, 6.3.1, 2011.

[6] N. Waldron, D-H. Kim, and J. A. del Alamo, "A Self-Aligned InGaAs HEMT Architecture for Logic Applications," *IEEE Trans. Electron Devices* 57, pp. 297-304,2010.

[7] R. J. W. Hill, et al., "Self-aligned III-V MOSFETs heterointegrated on a 200 mm Si substrate using an industry standard process flow," IEDM 10, pp. 6.2.1-6.2.4, 2010.

[8] Suman Datta, et al., "Ultrahigh-Speed 0.5 V Supply Voltage $In_{0.7}Ga_{0.3}As$ Quantum-Well Transistors on Silicon Substrate," *IEEE Electron. Device Lett.* 28, pp. 685-687, 2007.

[9] M. E. Groenert, et al., " Monolithic integration of room-temperature cw GaAs/AlGaAs lasers on Si substrates via relaxed graded GeSi buffer layers," *J. Appl. Phys.* 93, pp. 362-367, 2003.

[10] Hong Quan Nguyen, et al,. "Threading dislocation blocking in metamorphic InGaAs/GaAs for growing high-quality In0:5Ga0:5As and In0:3Ga0:7As on GaAs Substrate by using metal organic chemical vapor deposition," *Applied Physics Express* 5, pp. 055503, 2012.

[11] É. O'Connor, et al. "Temperature and frequency dependent electrical characterization of HfO2 / $In_xGa_{1-x}As$ interfaces using capacitance-voltage and conductance methods," *Appl. Phys. Lett.* 94, pp. 102902, 2009.

Investigation of Incorporating Dielectric Pocket (DP) on Vertical Strained-SiGe Impact Ionization MOSFET (VESIMOS-DP)

Ismail Saad[1*], Mohd. Zuhir H.[1], Divya Pogaku[1], Abu Bakar AR[1], N. Bolong[1], Khairul A.M[1], Bablu Ghosh[1], Razali Ismail[2], U. Hashim[3]

[1] Nano Engineering & Material (NEMs) Research Group, School of Engineering & IT, Universiti Malaysia Sabah, 88999, Kota Kinabalu, Sabah

[2] Computational Nanoelectronics (CONE) Research Group, Faculty of Electrical Engineering, Universiti Teknologi Malaysia, Skudai 81310, Malaysia

[3] Institute of Nano Electronic Engineering (INEE), Universiti Malaysia Perlis (UniMAP), 01000 Kangar, Perlis, Malaysia
*email: ismail_s@ums.edu.my / ismailsaad07@gmail.com

Abstract—The Vertical Strained Silicon Germanium (SiGe) Impact Ionization MOSFET with Dielectric Pocket (VESIMOS-DP) has been successfully developed and analyzed in this paper. Due to the DP layer, improve stability of threshold voltage, VT was found for VESIMOS-DP device of various DP size ranging from 20nm to 80nm. The stability is due to the reducing charge sharing effects between source and drain region. However, the presence of DP layer has introduced another potential barrier in addition to δp+ triangular potential barrier. Thus, increased amount of gate source voltage for lowering both barriers and allows the electron to move from source to drain. Accordingly, slight different and consistency of VESIMOS-DP sub-threshold value as compared to VESIMOS has revealed to give advantages for incorporating DP layer near the drain end. Moreover, the DP layer has suppressed the parasitic bipolar transistor effect with higher breakdown voltage as compared to without DP layer.

I. INTRODUCTION

In order to face theoretical limit of 60 mV/dec of sub-threshold slope, S [1-4], a device that work on the principle of impact ionization MOSFETs (IMOS) has been developed successfully [5-10]. However, this lateral IMOS needs a high supply voltage and suffers from reliability problems due to hot carrier effect [11-13]. The vertical IMOS, which is a planar-doped barrier MOSFET with a floating body has been introduced and investigated [14-18]. The device is capable to reduce supply voltage, mitigating the hot electrons damages almost completely [14-16] and shows capability of working properly under high temperatures [17]. This vertical IMOS is not based on avalanche breakdown like the lateral IMOS. Instead, the holes generated by impact ionization charge the floating p-body and cause a dynamic reduction of the threshold voltage, which leads to an extremely fast rising drain current in the sub-threshold region. The vertical IMOS, however, has also limitations including a remarkable hysteresis and still high V_D. In [19-22] a vertical IMOS with a strained SiGe layer was introduced and investigated. The simulation results showed that a reduction of both threshold voltage and supply voltage was

obtained by integrating such a strained SiGe layer. However, this device suffers from low breakdown voltage that limits its applicability in digital and analog circuit. This is due to the parasitic bipolar transistors (PBT) effect, since the npn structure acts as a bipolar transistor with floating base as shown in fig.1 [23]. At high drain voltages (> 2.5V), the gate loses its control over the drain current due to sufficiently higher impact ionization rates.

This paper presents an investigation of incorporating dielectric pocket (DP) into Vertical Strained-SiGe Impact Ionization MOSFET (VESIMOS-DP) for suppressing the effect of PBT in getting higher breakdown voltage. The introduction of a DP concept was first proposed as an alternative to pocket ion implantation in order to overcome short channel effects (SCE) in conventional lateral MOSFET [24]. The DP serves a number of functions in vertical MOSFET structure [25-27]. First, it greatly reduces the influence of the large area PBT effects and prevents encroachment of the doping from the extrinsic source, thus reducing bulk punch-through effects. Furthermore, it reduces charge sharing effects associated with the source and improves threshold voltage control.

II. DEVICE SIMULATION

Figure 1 shows the detailed cross-sections of the device structure which simulated for performance analysis of the Vertical Strained-SiGe Impact Ionization MOSFET with dielectric pocket (VESIMOS-DP) using Silvaco's package [23]. This structure comprises a source and a drain region with n+ doping, an intrinsic channel containing a highly doped p+ layer (Boron = $4 \times 10^{19}/cm^3$) and two sided gates. The high doping of the delta layer which creates a large potential barrier, makes it possible to achieve high electric fields in the intrinsic zone near the drain without applying a very high drain–source voltage [20]. Careful selection of the delta layer thickness is recommended as to have good sub-threshold slopes. Hence, an optimum value of delta layer thickness and doping was chosen

---

978-1-4673-2395-6/12 $31.00 © 2012 IEEE

to get desired sub-threshold slopes. The intrinsic-Si, i-Si regions reduce the lateral electric field near source and the drain [24]. Therefore, an optimum thickness for i-Si region has been chosen, so that they could effectively reduce the lateral electric fields. Due to the presence of i-Si regions between highly doped S/D regions, the impurity scattering is also reduced. The strained layer thickness was 20nm with Ge=30%. The DP layer thickness was also 20nm. However, the DP layer thickness was varied to examine its effects towards device performance. The DP layer was also sandwiched with intrinsic Silicon caps with 5nm thickness. This Si-cap acts the same function as i-Si to improve the stability of the overall device.

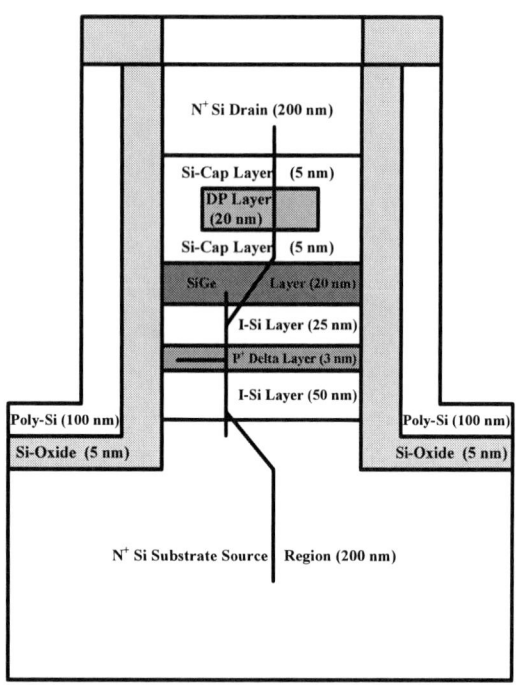

Fig. 1 VESIMOS-DP device structure with respective layer thickness of source, drain, δp+, i- Si, SiGe (Ge=30%), Si-cap and DP

The Source was n-doped with Antimony with a doping concentration of 2.08×10^{18} /cm^3. The Drain was also n-doped with Phosphorus with a doping concentration of 2.08×10^{18} /cm^3. The high doped S/D doping was chosen as the device concept was based on impact ionization. VESIMOS is an impact ionization device with drift current mechanism, which requires high electric fields. Both the drift current and the electric fields depend on the doping concentrations. Hence, high doping concentrations are imperative for obtaining better device characteristics.

The electrical characteristics of devices were done by solving Poisson's equation and continuity equation numerically and self-consistently within explicitly defined meshes of the devices [23]. The electrical potential energy and electronic band structures can be computed using Poisson's equation.

Continuity equations for electrons and holes are then used to calculate the current densities of electrons and holes. Boltzmann transport framework is used in solving these two equations. The relationship between the current density of electrons/holes and carrier concentration as well as quasi-Fermi potential is exhibited during this self-consistent process. The drift–diffusion (DD) transport model with the Boltzmann carrier transport framework was used, as it is able to predict I–V characteristics of DG-MOSFET [25]. Even for nanoscale size > 10 nm, Granzneretal [26] have shown that for DG-MOSFET's current characteristics, the DD and Monte-Carlo simulation results produced excellent agreement. While Ren and Lundstorm [27] and Rhew and Lundstorm [28] have revealed that the DD model can predict I–V characteristics of short-channel MOS devices more realistically than the energy-balance (EB) model. The Selberherr's [29] impact ionization model was employed, which is a local impact ionization model. Selberherr's model is recommended for most cases of device simulation.

III. RESULT AND DISCUSSIONS

Figure 2 shows the SIMS profile, showing the doping concentrations in the drain, substrate and the intrinsic regions, respectively. It also shows the SiGe layer thickness of 20nm and the Ge content is 30%. The δp+ layer thickness is 3nm and the intrinsic-Si layers are 50nm thick. The DP layer thickness is 20nm. This DP layer length size will be varying accordingly as to evaluate its effects on the device performance. The antimony doping at the source region is shown as 2.0×10^{18}/cm^3, the Phosphorus doping at the drain side is 2.0×10^{18} /cm^3, the boron doping in the δp+ is 4.0×10^{20} /cm^3 and the intrinsic-Si layer doping is approximately 1.0×10^{14} /cm^3.

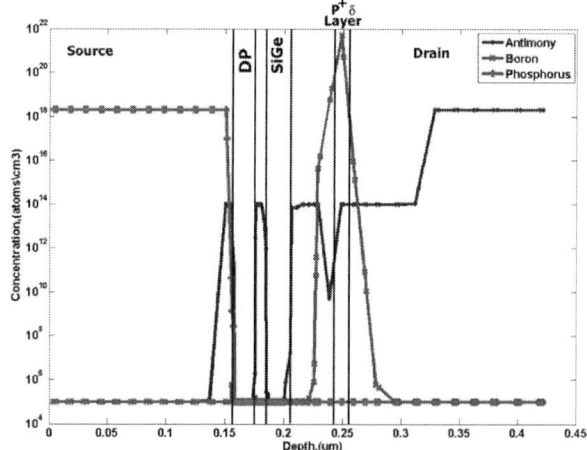

Fig. 2 SIMS profile showing the doping concentration in different regions of the VESIMOS-DP structure

VESIMOS-DP device works in three different modes: conventional MOSFET, Impact Ionization (II) and Bipolar (BJT) mode. The transfer characteristic is examined by biasing the drain voltage, V_D and ramping the gate voltage, V_G at defined bias steps. Figure 3 shows the comparison of I_D-V_G characteristics between VESIMOS-DP and VESIMOS without DP layer.

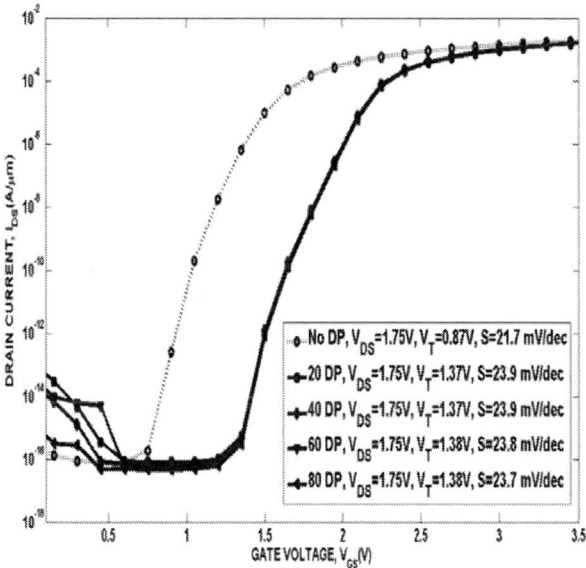

Fig. 3 Transfer Characteristics, I_D-V_G of VESIMOS-DP and VESIMOS without DP for $Si_{0.7}Ge_{0.3}$, S/D doping =$2.0 \times 10^{18}/cm^3$, V_{DS}=1.75V.

Figure 3 revealed that both devices works well for lower power supply voltage (V_{DS}=1.75V), which has overcome the problem faced by conventional IMOS devices [14-16]. Due to the DP layer, improve stability of threshold voltage, V_T was found for VESIMOS-DP device of various size ranging from 20nm to 80nm. This stable V_T=1.37V obtained since the vicinity of DP layer near the drain end has reduce charge sharing between source and drain. However, as a trade-off very low VT=0.87V was expected for VESIMOS since the presence of DP layer has introduced another potential barrier in addition to $\delta p+$ triangular potential barrier. Thus, increase amount of gate-source voltage needed for lowering both barriers and allows the electron to move from source to drain. Accordingly, a lower threshold voltage and lower leakage currents was observed for VESIMOS as an advantage of triangular potential barrier [16]. However, for VESIMOS device without DP sub-threshold voltage, S=21.7 mV/decade was obtained at threshold voltage of 0.87V with a supply voltage of 1.75V. The slight different and consistency of VESIMOS-DP sub-threshold value (S=23.7 mV/dec) as compared to VESIMOS has given advantages for incorporating DP layer near the drain end.

This analysis can be summarized as in Table 1, which shows the comparison of threshold, sub-threshold and breakdown voltage, BV obtained for both VESIMOS and VESIMOS-DP

Table 1: VESIMOS V_{TH} and S at different modes of operation

Parameter	No DP	DP
V_{TH}(V)	0.87	1.37
S (mV/decade)	21.7	23.7
B_V(V)	2.8	3.6

Table 1 also depicted that higher breakdown voltage, Bv was obtained for VESIMOS-DP. The Bv = 2.5V and 3.5V for VESIMOS and VESIMOS-DP respectively can be seen from figure 4. Figure 4 shows that for VDS > 1.80V, the device, instead of going to the bipolar mode as in Si vertical I-MOS device, it undergoes breakdown. In bipolar mode, the n+ source region acts as emitter, the n+ drain acts as collector and $\delta p+$ layer as base as the symbol draw in figure 1. The holes generated by impact ionization acts as base currents and drain source current increases as the current of the parasitic bipolar transistor (PBT) is switched on. This additional current amplification mechanism contributes to lower sub-threshold slope, but with a certain hysteresis [20] due to PBT effect. In addition, the low breakdown voltages of VESIMOS can be attributed to the low bandgap of the Ge compared to other semiconductor materials [35]. It is also the inherent properties of Germanium (Ge) which has breakdown strength two times lower than the Si.

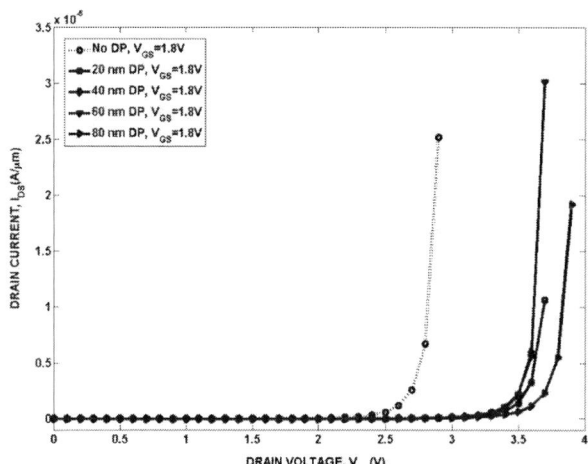

Fig. 4 Comparison of output characteristics between VESIMOS and VESIMOS-DP device at VGS=1.8V (Impact Ionization mode)

978-1-4673-2395-6/12 $31.00 © 2012 IEEE

However, with the vicinity of DP layer at the drain region for VESIMOS-DP device the effect of PBT has been suppressed. The advantages of using DP layer for PBT suppression was known for Si vertical MOSFET [25-27] and applied here for the same purpose. Figure 5 shows the output characteristics of VESIMOS and VESIMOS-DP at V_{GS} = 3.5V and 2.5V respectively. It can be seen that the drain current rises sharply initially and then increases gradually before going into breakdown for V_{DS} > 2.5V for VESIMOS and V_{DS} > 3.5V for VESIMOS-DP. It can be observed that for all size of DP (20 to 80nm), almost identical breakdown voltage was obtained that merits further study on it effect towards device performance.

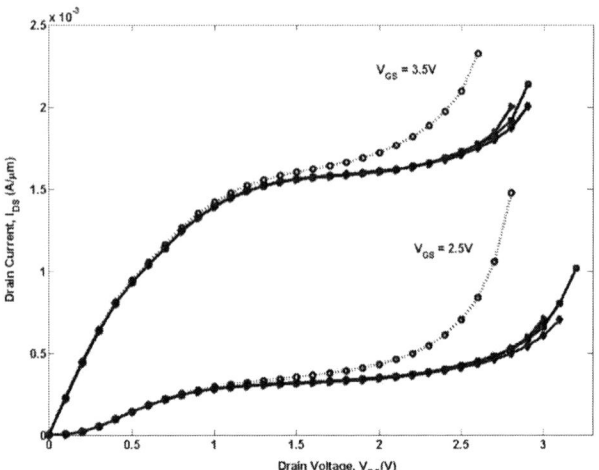

Fig. 5 Comparison of output characteristics between VESIMOS and VESIMOS-DP device at V_{GS}=3.5V and 2.5V (Bipolar mode)

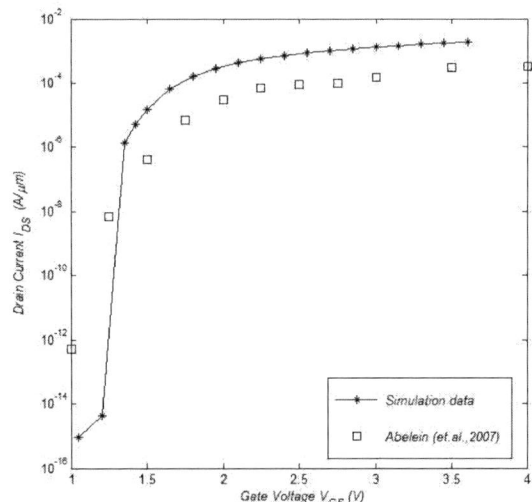

Fig. 6 Validation of simulated data with experimental data (Abelein *et.al.*, 2007) of Si vertical I-MOS device.

All simulated data was validated with experimental data in order to verify the accuracy of the simulation work. As no experimental data available for Vertical SiGe IMOS, the vertical Si IMOS data was used [15]. It can be observed that the simulated data is synchronous with the experimental data and is in close approximation of 0% to 5% range as shown in figure 6.

IV. CONCLUSION

The investigation on the effects of incorporating dielectric pocket (DP) into Vertical Strained-SiGe Impact Ionization MOSFET (VESIMOS-DP) was successfully analyzed. The comparison of electrical characteristics between VESIMOS and VESIMOS-DP by varying the DP size was obtained using Silvaco TCAD tools. Due to DP layer, improve stability of threshold voltage of various size ranging from 20nm to 80nm was found. The stability is due to reduce charge sharing between source and drain region. However, as a trade-off very low threshold voltage was expected for VESIMOS since the presence of DP layer has introduced another potential barrier in addition to δp+ triangular potential barrier. Thus, increase amount of gate-source voltage needed for lowering both barriers and allows the electron to move from source to drain. The slight different and consistency of VESIMOS-DP sub-threshold value (S = 23.7 mV/dec) as compared to VESIMOS has given advantages for incorporating DP layer near the drain end. Moreover, the DP layer has suppressed the parasitic bipolar transistor effect with higher breakdown voltage as compared to without DP layer. Based on these results, VESIMOS-DP merits more studies on its DP size effects towards device performance.

ACKNOWLEDGEMENT

The authors would like to acknowledge the financial support from ERGS (ERGS0002-TK-1/2011) fund of MOHE and E-Science fund (03-01-10-SF0175) of MOSTI Malaysia. The author is thankful to the University Malaysia Sabah (UMS) for providing excellent research environment in which to complete this work.

REFERENCES

[1] International Roadmap Committee. The international technology roadmap for semiconductors. <public.itrs.net>.

[2] Morifuji E, Yoshida T, Kanda M, Matsuda S, Yamada S, Matsuoka F. Supply and threshold-voltage trends for scaled logic and SRAM MOSFETs. IEEE Trans Electron Dev 2006;53:1427–32.

[3] Khakifirooz A, Antoniadis DA. MOSFET performance scaling – part I: historical trends. IEEE Trans Electron Dev 2008;55:1391–400.

[4] Khakifirooz A, Antoniadis DA. MOSFET performance scaling – part II: future directions. IEEE Trans Electron Dev 2008;55:1401–8.

[5] Gopalakrishnan K, Griffin PB, Plummer JD. I-MOS: a novel semiconductor device with a substhreshold slope lower than kt/q. In: IEEE tech dig IEDM; 2002. p. 289–92.

[6] Gopalakrishnan K, Griffin PB, Plummer JD. Impact ionization MOS (I-MOS) –part I: device and circuit simulations. IEEE Trans Electron Dev2005;52(1):69–76.

[7] Gopalakrishnan K, Woo R, Jungemann C, Griffin PB, Plummer JD. Impact ionization MOS (I-MOS) – part II: experimental results. IEEE Trans Electron Dev 2005;52(1):77–84.

[8] Choi WY, Song JY, Lee JD, Park YJ, Park B-G. 100-nm n-/p-channel I-MOS using a novel self-aligned structure. IEEE Electron Dev Lett 2005;26(4):261–3.

[9] Choi WY, Song JY, Lee JD, Park YJ, Park B-G. 70-nm impact-Ionization metal–oxide–semiconductor (I-MOS) devices integrated with tunneling field-effect transistors (TFETs). In: IEEE tech dig EDM; 2005. p. 955–8.

[10] Choi WY. Effect of device parameters on the breakdown voltage of impact ionization metal–oxide–semiconductor devices. Jpn J Appl Phys, 2009;48:040203-1–3-3.

[11] Charbuillet C, Monfray S, Dubois E, Bouillon P, Judong F, Skotnicki T. High current drive in ultra-short impact-ionization MOS (I-MOS) devices. In: IEDM; 2006. p. 1–4.

[12] Onal C, Woo R, Koh H-Y, Griffin P, Plummer J. A novel depletion-IMOS (DIMOS) device with improved reliability and reduced operating voltage. IEEE Electron Dev Lett 2009;30(1):64–7.

[13] Savio A, Monfray S, Charbuillet C, Skotnicki T. On the limitations of silicon for IMOS integration.IEEE Trans Electron Dev 2009;56(5):1110–7.

[14] Abelein U, Born M, Bhuwalka K, Schindler M, Schmidt M, Sulima T, et al. A novel vertical impact ionization MOSFET (I-MOS) concept. In: Proc 25ᵗʰ international conference on microelectronics (MIEL '2006), Niš, Serbia; 2006. p. 127–9.

[15] Abelein U, Born M, Bhuwalka K, Schindler M, Schlosser M, Sulima T, et al. Improved reliability by reduction of hot-electron damage in the vertical impact-ionization MOSFET (I-MOS). IEEE Electron Dev Lett 2007;28(1):65–7.

[16] Abelein U, Assmuth A, Iskra P, Schindler M, Sulima T, Eisele I. Doping profile dependence of the vertical impact ionization MOSFET's (I-MOS) performance. Solid-State Electron 2007;51:1405–11.

[17] Abelein U, Assmuth A, Iskra P, Reinl M, Schlosser M, Sulima T et al. Vertical 40 nm impact ionization MOSFET (I-MOS) for high temperature applications. In: Proc 26th international conference on microelectronics (MIEL '2008), Niš, Serbia; 2008. p. 287–90.

[18] Kraus R, Jungemann C. Investigation of the vertical IMOS-transistor by device simulation. In: Proc ULIS, international conference on ultimate integration of silicon, vol. 10; 2009. p. 281–4

[19] Dinh TV, Kraus R, Jungemann C. Investigation of the performance of strained-SiGe vertical IMOS-transistors. In: Proc ESSDERC; 2009. p. 165–8.

[20] Dinh TV, Kraus R, Jungemann C. Investigation of the performance of strained-SiGe vertical IMOS-transistors. Solid-State Electronics 54 (2010) p. 942–949.

[21] P.Divya and Ismail Saad, Feasibility study of Integrated Vertical and Lateral IMOS and TFET devices for Nano-scale Transistors. IEEE International Conference On Semiconductor Electronics (ICSE 2010), ICSE2010 Proc. 2010.

[22] Divya Ravindra, Ismail Saad. Effects of S/D doping concentrations on strained SiGe vertical I-MOS device characteristics, ICECT2011, Kanyakumari, India, DOI: 978-1-4244-8679-3/11/$26.00 ©2011.

[23] Kraus, R., Jungemann, C., 2009. Investigation of the vertical IMOS- transistor by device simulation, International Conference on Ultimate Integration of Silicon, ULIS-2009, Aachen, 18-20 March

[24] M. Jurczak, T. Skotnicki, R. Gwoziecki, M. Paoli, B. Tormen, P. Ribot, D. Dutartre, S. Monfray, and J. Galvier. Dielectric Pockets – A New Concept of the Junctions for Deca-Nanometric CMOS Devices. *IEEE Trans. Electron Devices*, 48(8):1770–1775, August 2001.

[25] S.K. Jayanarayanan et.al. A Novel 50nm vertical MOSFET with a dielectric pocket. Solid-State Electronics, 50 (2006) 897-900.

[26] D. Donaghy, S. Hall, C. H. de Groot, V. D. Kunz, and P. Ashburn. Design of 50-nm vertical MOSFET incorporating a dielectric pocket. *IEEE Trans. Electron Devices*, 51(1):158–161, January 2004.

[27] D.C Donaghy, S. Hall, V.D. Kunz, C.H. de Groot, and P. Ashburn. Investigating 50nm channel lenght vertical MOSFETs containing a dielectric pocket, in a circuit environment. *ESSDERC Conf. Proc.*,pages 499–502, 2002.

[28] Atlas and Athena user Manual Device and Process Simulation Software, Silvaco International, 2005.

[29] Scheinert, S., Paasch, G., Kittler, M., Nuernbergk, D., Mau, H., Schwiers, F. (1998). Requirements and Restrictions in Optimizing Homogenous and Planar Doped Barrier Vertical MOSFETs, International Caracas Conference on Devices, Circuits and Systems, ICCDCS-1998, Isla de Margarita.

[30] N.D. Jankovic, G.A. Armstrong, Comparative analysis of the DC performance of DG MOSFETs on highly-doped and near-intrinsic silicon layers, Microelectronics. J. 35 (2004) 647–653.

[31] R. Granzner, et al., On the suitability of DD and HD models for the simulation of nanometer double-gate MOSFETs, Physica E 19 (2003) 33–38.

[32] Z. Ren, M. Lundstrom, Simulation of nanoscale MOSFETs: a scattering theory interpretation, Superlattices Microstruct. 27 (2000) 177–189.

[33] J. Rhew, M. Lundstrom, Drift–diffusion equation for ballistic transport in nanoscale metal-oxide-semiconductor field effect transistors, J. Appl. Phys. 92 (2002) 5196.

[34] Selberherr S., "Analysis and Simulation of Semiconductor Devices", Springer-Verlag, Wien-New York. 1984.

[35] Zoolfakar, A.S., and Ahmad, A., 2009. Hole mobility enhancement using Strained Si, SiGe technology. *5ᵗʰ International Colloquium on Signal Processing & Its Applications (CSPA)*, 4-6 March.

Pierce Oscillator Circuit Topology for High Motional Resistance CMOS MEMS SAW Resonator

Jamilah Karim[1,2], Anis Nurashikin Nordin[1], AHM Zahirul Alam[1], U.Hashim[3]

[1]Electrical and Computer Engineering Department
Faculty of Engineering
IIUM, Gombak.
[2]Electrical Engineering Faculty
University Technology MARA
Shah Alam, Selangor
[3]Institute of Nano Electronic Engineering (INEE)
University Malaysia Perlis

[1]*Abstract*— **This paper presents the design and simulation for a pierce oscillator using CMOS MEMS SAW resonator. The MEMS resonator utilizes surface acoustic waves to generate resonant frequencies of 600MHz and 900MHz. The MEMS resonator is fully compatible with CMOS technology, allowing the possibility of full integration with circuits. The pierce circuit topology was chosen as a sustaining circuit, connected to the resonator to form an oscillator. For simulation purposes, the CMOS MEMS SAW resonator was modeled using its RLC equivalent circuit. The oscillator produces transient oscillation of 300mV peak to peak voltage. The phase noise performance for 600MHz oscillator is -70dBc/Hz at 100 kHz and consume 1.17mW power. The 900MHz oscillator has achieved -63dBc/Hz phase noise at 100 kHz offset frequency and consume about 1.62mW power.**

Keywords-component; CMOS MEMS Oscillator, Pierce Oscillator, MEMS SAW resonator

I. INTRODUCTION

Modern transceivers need a local oscillator to generate a sinusoidal signal as its carrier frequency. Traditionally the implementation of communication systems used multiple discrete devices such as the receiver, transmitter, and the oscillator [1]. Thus each device was optimized locally. Advancement in the complementary metal oxide semiconductor (CMOS) process sometimes in the late 1990's, has lead to the integration of external components and multiple devices onto a single chip [1]. Unfortunately, the oscillator is one of the hardest blocks in the systems to integrate on silicon since it has to be fabricated on piezoelectric substrates. This quartz based oscillators have excellent performance.

MEMS based oscillator has provide great solutions for fully integration systems. Full integrated systems consume minimum power, small in size and reduce production cost. Although MEMS based oscillator is still new, it shows promise

with measured Q factors > 1000 and operating frequencies in GHz range [2-5].

This paper will present the design and simulation of CMOS MEMS oscillator based on MEMS SAW resonator. Two resonators with different resonant frequencies are used. Section II begins with general discussion of CMOS MEMS SAW resonator. Section III explains about the implementation of MEMS oscillator and the discussion of the output from the simulation and Section IV concludes the paper.

II. CMOS MEMS SAW RESONATOR

A. Resonator Implementation

The two port CMOS MEMS SAW resonators used in this oscillator design was obtained from [6]. These resonators consist of reflectors and delay line. Reflectors are the key component of SAW resonator, as these electrodes restrain the outward propagating acoustic waves within the cavity, thus reducing its loss. The delay lines are formed by using interdigital arrays of electrodes. These transducers produce spatially non uniform time varying electric fields when excited with a sinusoidal electrical signal. The electric fields generate local stresses at the surface of piezoelectric material, initiating elastic waves. Symmetrical interdigital transducers, IDT_1 and IDT_2 have greater efficiency than single port IDTs and can act both as transmitter and receiver. Figure 1 shows the two port SAW resonator.

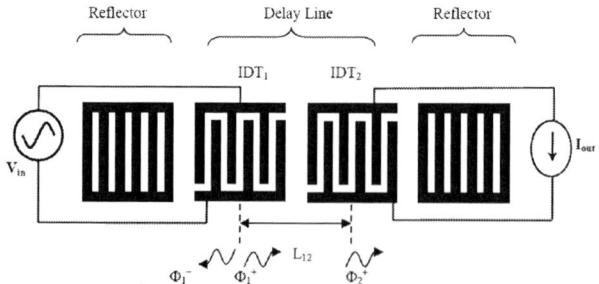

Figure 1: Schematic of a two-port SAW resonator[7].

This work was supported by the Exploratory Research Grant: ERGS 11-009-009 provided by the Ministry of Higher Education of Malaysia.

978-1-4673-2395-6/12 $31.00 © 2012 IEEE

The operating principle of SAW device is based on the piezoelectric effect. When a microwave voltage input is applied at the transmitting (input) IDT, it generates a propagating acoustic wave on the surface of the substrate [7]. This propagating acoustic wave in turn produces an electric field localized at the surface which can be detected and translated back into an electrical signal at the output IDT port. Thus with existence of reflector, the resonance frequency are created, which contain the acoustic waves within the cavity.

Figure 2 shows the implementation of SAW resonator structure in standard CMOS. The IDTs and the reflectors are placed underneath the piezoelectric layer (ZnO) and were implemented using CMOS Metal 2. Metal 1 was used as a ground shield. This implementation result in the acoustic wave to propagate above the IDTs. The cross-section of the structural layers for CMOS SAW resonator is shown in Figure 3. For ease of integration, the Si substrate was used in this CMOS SAW resonator.

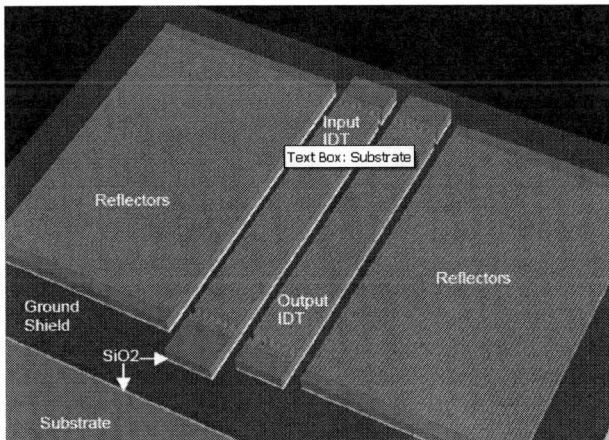

Figure 2: Implementation of a two-port SAW resonator structure in CMOS. The interdigital transducers (IDT) and reflectors are implemented using CMOS Metal 2. The ground shield is implemented using Metal 1[6].

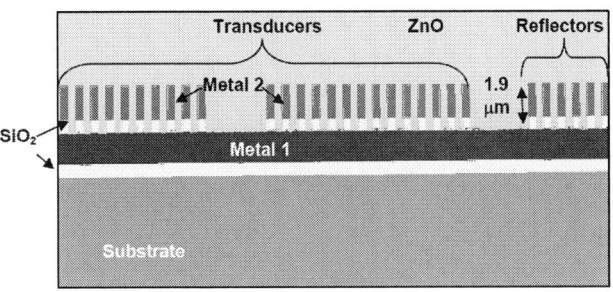

Figure 3: The Cross-section of CMOS SAW resonator utilizing two CMOS metal layers[6].

The resonator to be used in wireless applications must have precise resonant frequency, low insertion losses and high quality factor.

B. Electrical Model of Resonator

Figure 4 shows the equivalent circuit model of a SAW resonator. This circuit can be divided into two parts: acoustic and parasitic components. The acoustic component can be described electrically using circuit elements R_x, C_x, L_x, and C_f. The R_x is also known as the 'motional resistance", is a key factor in determining quality factor and insertion losses of the device. The existence of parasitics which is represented by C_1 and C_2 in the circuit model can degrade the performance of the resonator. This parasitic component exists due to the structure of the SAW resonator's transducer, formed using a composite of Al and SiO_2 layers [6].

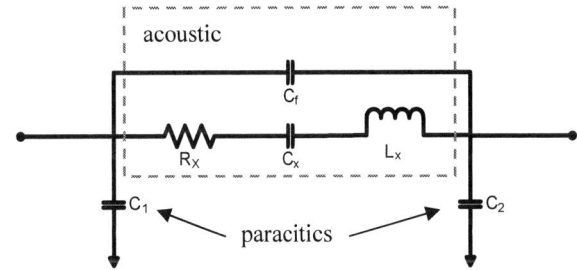

Figure 4: SAW resonator equivalent circuit model[8]

TABLE1
Measurement Result of CMOS MEMS SAW resonator[7]

Element	Resonator 1	Resonator 2
R_x (kΩ)	8.95	3.347
L_x (mH)	0.350	4.715
C_x (aF)	199	6.05
C_f (fF)	24.9	85
f_s (GHz)	0.603	0.942
Q	285	94
IL	-36dB	-23.7dB

Table 1 shows the parameter of the resonator used to test the pierce oscillator circuit topology. All these parameter were extracted from the S21 measurement. Using the parameter from table 1, the following simulated admittance plots were obtained for resonator 1 (Reson 1) and resonator 2 (Reson 2). The insertion losses for each of the resonators are -36dB and -23.7dB respectively.

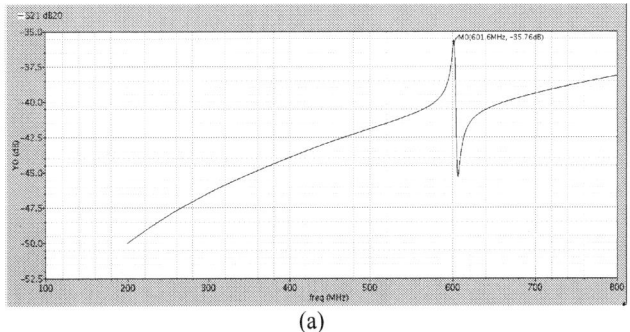

(a)

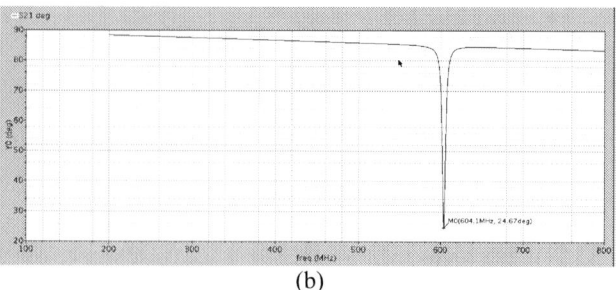

(b)

Figure 5: Simulated admittance plot of CMOS MEMS SAW resonator at 603MHz (a) Magnitude and (b) Phase

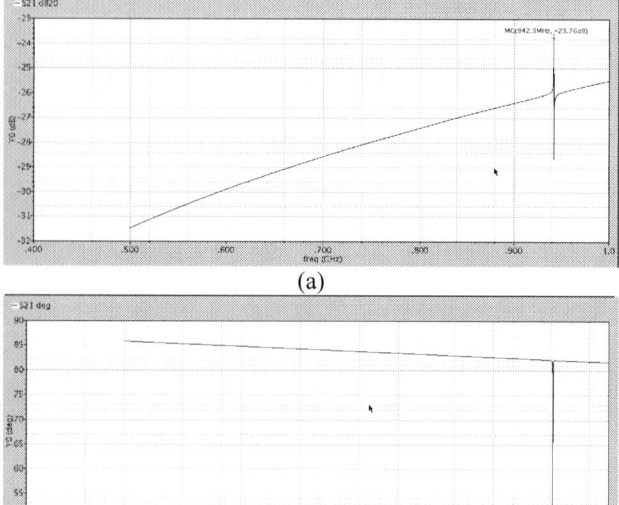

(a)

(b)

Figure 6: Simulated admittance plot of CMOS MEMS SAW resonator at 942MHz. (a) Magnitude, (b) Phase

Resonator 1 has the highest insertion loss due to the highest motional resistance it have.

III. PIERCE OSCILLATOR CIRCUIT DESIGN AND SIMULATION OUTPUT

A. Sustaining Circuit

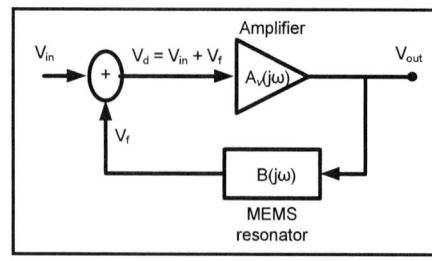

Figure 7: Basic Feedback Oscillator Circuit

The performance of the MEMS based oscillator is not only dependent on resonator but also the supporting circuit. To

provide the oscillation normally the MEMS resonator will be embedded in the feedback loop of an amplifier as in Figure 7. The chosen amplifier circuit must be compatible with the relatively high motional resistance MEMS SAW resonator. Thus the pierce circuit topology is a logical choice since it has simple and straightforward biasing, sustained high frequency stability and can achieve very low phase noise. To guarantee start up and sustainable oscillation; the pierce circuit topology must have sufficient gain and also 0° or 360° phase shift. Equation (1) and (2) show the Barkhausen criteria for the oscillation.

$$|\beta A_v| \geq 1 \tag{1}$$

$$\angle \beta A_v = 0°, \pm 360° \tag{2}$$

Figure 8 presents the transistor level design of pierce oscillator circuit which was adopted from [2]. The MEMS SAW resonator is represented by its equivalent electrical circuit, RLC. A three stages circuit is required in order to achieve the needed gain and phase shift. Transistors T_1-T_3, represent a single stage pierce circuit with T_1 as the main transistor which provides the critical transconductance for oscillation. T_1 is biased by transistor T_2. Besides providing the bias resistance to the gate of T_1, T_3 also provides enough resistance to ensure the oscillation is sustained. Transistors T_4 through T_9 are added and become the second and third stages in order to ensure that the total phase shift of the oscillator circuit is 0° or 360°.

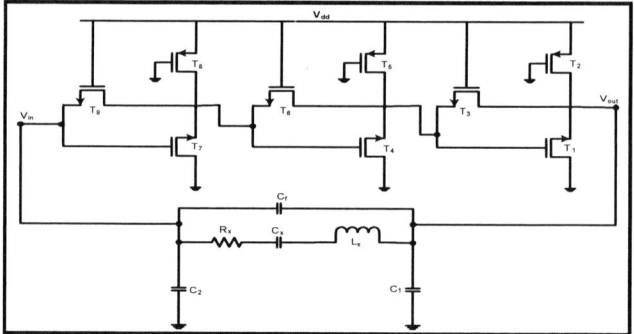

Figure 8: Detailed schematic circuit of pierce oscillator circuit topology

And gain for the amplifier is:

$$A_v = g_{m1} R_L \frac{C_2}{C_1}$$

$$R_L = 1 \Bigg/ \left(R_x \omega_o^2 C_T^2 \left(\frac{(C_1 + C_2)^2}{C_1^2} \right) \right) + \left[\left(\frac{(C_1 + C_2)^2}{C_1^2} \right) \frac{\omega_o C_T}{Q_{cap}} \right]$$
$$+ \frac{\frac{(C_1 + C_2)^2}{C_1^2}}{R_{fb}} + \frac{1}{r_{o1}} + \frac{1}{r_{o2}}$$

978-1-4673-2395-6/12 $31.00 © 2012 IEEE   256

B. Simulation Result

TABLE2
DC analysis parameter of main Transistor for different resonator

	OSC 1	OSC 2
	T_1, T_4, T_7	T_1, T_4, T_7
I_d (uA)	657.19	789.06
V_{ds} (mV)	592.35	683.91
g_m (mS)	7.65	6.17
Power consume	1.17mW	1.62mW

Both resonators 1 and 2 work well with pierce oscillator circuit topology. Table 2 show the DC analysis result obtained for both oscillators. OSC 1, requires higher transconductance g_m compare to OSC 2, which has higher motional resistance R_x.

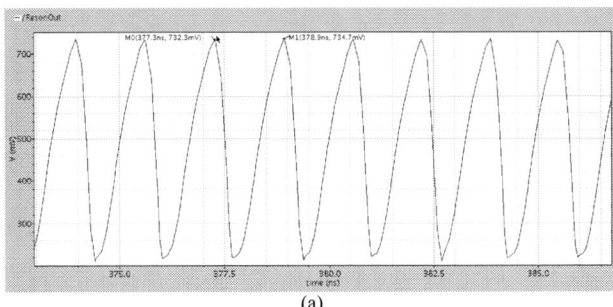

(a)

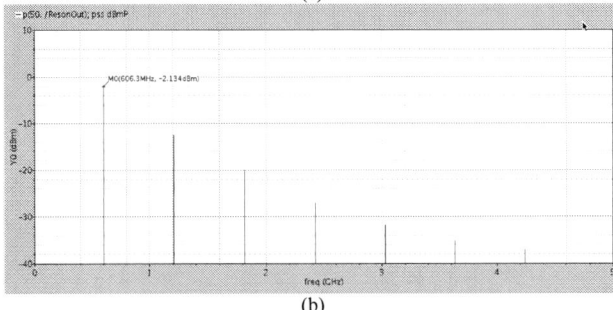

(b)

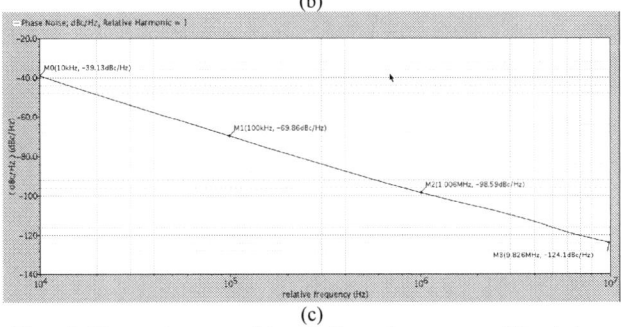

(c)

Figure 9: The transient output (a), periodic steady state output (b) and phase noise output for OSC 1 (c)

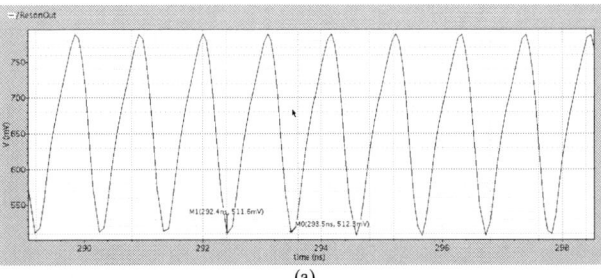

(a)

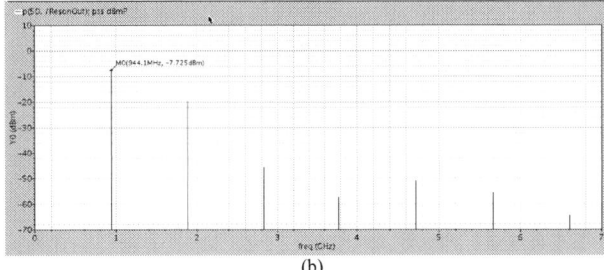

(b)

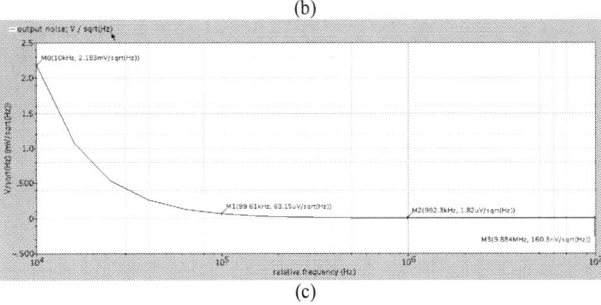

(c)

Figure 10: (a) The transient output, (b) periodic steady state output and (c) phase noise output for OSC 2

Figure 9 and Figure 10 shows the transient output, periodic steady state output and phase noise output for OSC1 and OSC 2 respectively. The periodic steady state output shows the highest spectrum at 606.3MHz for OSC 1, with signal power equals to -2.135dBm. OSC 2 shows the highest signal power at 949.5MHz. OSC 1 has better phase noise performance at 100 kHz offset frequency compare to OSC 2. Better phase noise is necessary for OSC 1 and OSC 2 to be used as a local oscillator in communication systems. As an example the phase noise of oscillator for Bluetooth technology is -110dBc/Hz @ 500 kHz offset frequency. GSM850 and GSM900 require -118dBc/Hz at 600 kHz offset frequency[1] . OSC 1 has better noise performance compare to work done in [9].

IV CONCLUSION

The pierce circuit topology has been chosen as a sustaining circuit for the oscillator based on MEMS SAW resonator. The circuit has been design and simulated using silterra CMOS 0.18um technology. The simulation result obtained shows that this topology worked well with both resonators. The phase noise performance obtained for OSC 1 and OSC 2 are -70dBc/Hz and -63dBc/Hz at 100 kHz offset frequency.

REFERENCES

[1] M. Hammes, C. Kranz, and D. Seippel, "Deep submicron CMOS technology enables system-on-chip for wireless communications ICs," *Communications Magazine, IEEE,* vol. 46, pp. 154-161, 2008.

[2] Z. Chengjie, J. Van Der Spiegel, and G. Piazza, "1.05-GHz CMOS oscillator based on lateral- field-excited piezoelectric AlN contour-mode MEMS resonators," *Ultrasonics, Ferroelectrics and Frequency Control, IEEE Transactions on,* vol. 57, pp. 82-87, 2010.

[3] Z. Chengjie, N. Sinha, J. Van der Spiegel, and G. Piazza, "Multifrequency Pierce Oscillators Based on Piezoelectric AlN Contour-Mode MEMS Technology," *Microelectromechanical Systems, Journal of,* vol. 19, pp. 570-580, 2010.

[4] S. S. Rai and B. P. Otis, "A 600uW BAW-Tuned Quadrature VCO Using Source Degenerated Coupling," *Solid-State Circuits, IEEE Journal of,* vol. 43, pp. 300-305, 2008.

[5] H. M. Lavasani, P. Wanling, B. Harrington, R. Abdolvand, and F. Ayazi, "A 76 dB, 1.7 GHz 0.18um CMOS Tunable TIA Using Broadband Current Pre-Amplifier for High Frequency Lateral MEMS Oscillators," *Solid-State Circuits, IEEE Journal of,* vol. 46, pp. 224-235, 2011.

[6] A. N. Nordin and M. E. Zaghloul, "Modeling and Fabrication of CMOS Surface Acoustic Wave Resonators," *Microwave Theory and Techniques, IEEE Transactions on,* vol. 55, pp. 992-1001, 2007.

[7] A. N. Nordin, "Design, Implementation and Characterization of Temperature Compensated SAW Resonators in CMOS Technology for RF Oscillators " in *School of Engineering and Applied Science*: The george Washington University, 2008, p. 221.

[8] "IEEE Standard Definitions and Methods of Measurement for Piezoelectric Vibrators," *IEEE Std No.177,* p. 1, 1966.

[9] A. N. N. Jamilah Karim, AHM Zahirul Alam, "Design of a Pierce Oscillator For CMOS SAW Resonator," in *4th International Conference on Computer and Communication Engineering (ICCCE'2012),* Kuala Lumpur, Malaysia, 2012, p. 4.

Influence of LT-AlN Buffer Layers on Density of Threading Dislocation in AlGaN Layers

H Meidia[a] and S Mahajan[b]

[a]Department of Computer Engineering, Universitas Multimedia Nusantara, Scientia Garden, Jl, Boulevard Gading Serpong, Tangerang, Indonesia.
Email:hira.meidia@um.ac.id
[b]Department of Chemical Engineering and Materials Science, University of California, 3001 Ghausi Hall, Davis, California, 95616-5294, USA.
Email:smahajan@ucdavis.edu

Abstract- **To assess the influence of low temperature (LT) AlN buffer layers on the density of threading dislocations, we deposited multilayer structures consisting of AlN and AlGaN using metal organic chemical vapor deposition. We used two types of composite substrate in this study: (1) ELOG GaN/sapphire, and (2) sapphire. The resulting multilayer structures were examined in cross section by transmission electron microscopy. We show that the deposition of low temperature AlN buffer on highly perfect ELOG GaN leads to defective AlN/GaN interfaces. Some of the dislocations associated with these interfaces evolve into threading dislocations in AlGaN layers. We also demonstrate that multilayer of AlN buffers are only moderately effective in reducing the density of TDs into overgrown AlGaN layers.**

I. INTRODUCTION

Group III-nitride semiconductor materials give the spectral range from UV to visible and to infrared, with unique properties for optoelectronic applications such as ultraviolet light emitting diodes (LED). AlGaN-based material has been used to generate ultraviolet (UV) light sources that corresponding to wavelength of 200-360nm. The UV optical devices are usually used in medical applications, high-density optical storage and chemical process [1, 2, 3]. The production of thicker AlGaN film with high aluminium composition is very desirable for these applications. It also important to achieve a high quality crystal AlGaN that is crack-free with a low dislocation density. However the growth of AlGaN directly on sapphire or GaN can cause the high-density crack network due to the mismatch of thermal expansion coefficients and lattice parameter of AlGaN and sapphire or GaN.

Khan et al reported that the 3μm thick of crack free $Al_xGa_{1-x}N$ (x>0.2) layer grown on sapphire is achieved by depositing AlN/AlGaN superlattices (SLs) to control strain [4]. They found that $Al_xGa_{1-x}N$ (x>0.2) layers are under biaxial tensile strain and AlN/AlGaN SLs are under biaxial compressive strain. Therefore by growing SLs to the top of $Al_xGa_{1-x}N$ (x>0.2), the cracking can be prevented. Koide et al also reported that a thin low temperature (LT) AlN buffer layer was used to grow the high-crystalline quality of AlGaN film on

sapphire with surface free from crack by using metal organic vapor phase epitaxy (MOVPE) [5]. The LT-AlN buffer layer is not only used to reduce the residual strain, but also to perform the nucleation center and to promote the lateral growth to facilitate the lattice mismatch between the epitaxy layer and sapphire. Although AlGaN layer is deposited on sapphire using LT-AlN, it still contains high degree of threading dislocations (TDs) density. It is well recognized that the presence of a high density of TDs has deleterious effects on the performance and reliability of optoelectronic devices. To decrease these TDs density, the LT-buffer interlayer has been inserted in between AlGaN layers [6]. The LT-interlayer method adjusts the film stress and reduces TDs density. LT-AlN interlayers were found to be more effective than LT-GaN interlayer in suppressing cracking by reducing tensile stress [7, 8]. The nucleation layers (NLs) of LT AlN are relatively smooth compare to NLs LT GaN although the strain may exist at the AlN/sapphire interface. However, the mechanism for dislocation reduction has not yet understood.

In this paper, we report our transmission electron microscopy (TEM) observations of a TDs in the AlGaN layer with the growth of LT-AlN. We show that the deposition of LT-AlN in growth of AlGaN layer on high quality ELOG GaN results in a high density of TDs in the overgrowth. We also found the insertion of four periods LT-AlN buffer layers in between AlGaN layers grown on sapphire is moderately effective in reducing the TDs density.

II. EXPERIMENTAL DETAILS

For this study, we grew two types of samples by using conventional MOVPE process; the growth pressure was kept constant at 76Torr. First sample is 1.2μm $Al_xGa_{1-x}N$ (x=0.22) grew on 20nm LT-AlN/ELOG-GaN/sapphire. Second sample consisted of four periods of 20nm LT-AlN/450nm $Al_xGa_{1-x}N$ (x=0.35) layers grown on sapphire. AlN and AlGaN layers were deposited at 650°C and 1050°C respectively. To study the influence of LT-AlN on the density of TDs in the AlGaN layer, we used transmission electron microscopy (TEM) technique.

The TEM samples were prepared in cross sectional configuration. The samples were prepared by standard mechanical grinding and ion milling method. TEM was performed at 400kV on a JEM-4000EX microscope.

III. RESULTS AND DISCUSSIONS

To investigate the structural properties of the samples, ie $1.2\mu m$ $Al_{0.22}Ga_{0.78}N$/ 20nm LT-AlN/ ELOG-GaN/ sapphire and four periods of 20nm LT-AlN/ 450nm $Al_{0.35}Ga_{0.65}N$/ sapphire, cross-sectional TEM observations were performed. Figure 1a and 1b show WBDF images of ELOG GaN/LT-AlN/AlGaN under (01-10) and (0002) reflections respectively. The LT-AlN is deposited as a nucleation layers (NLs) on ELOG-GaN, some of dislocation evolved to TDs in AlGaN layer. These TDs seem to originate from the LT AlN/AlGaN interfaces. A dense array of dislocation is visible at LT AlN/AlGaN interface. By applying the general rule $\mathbf{g.b}=0$ and $\mathbf{g.b}x\mathbf{u}=0$ for contrast extinction of a dislocation of Burgers vector $\mathbf{b}$ and line direction $\mathbf{u}$ imaged at reflection $\mathbf{g}$. It shown that the primary TDs have $\mathbf{c}$ and $\mathbf{c+a}$ types and some of dislocations are in the form of half-loops. These half loops have been observed to appear without reaching the surface. The pairs of dislocation with opposite b vectors tend to annihilate each other when more material was deposited during the growth. Therefore the TDs density seems reduced with increasing distance from the LT-AlN/AlGaN interface. It is shown that the reduction of the TDs density at the top surface region could be due to "dislocation bending".

The dislocation density in AlGaN overgrowth is increased to $10^9/cm^2$ after depositing LT-AlN layer. It shows that the deposition of 20nm LT-AlN buffer layer on highly perfect ELOG GaN leads to defective AlN/AlGaN interface and produces TDs in the AlGaN layer. Lorenz et al [8] and Gonsalves et al [9] have shown the role of LT-AlN as a nucleation layer in the optimization the epitaxial overgrowth. The AlN is roughening by developing surface undulation through surface relief. The growth parameter, such as the thickness, deposition temperature of AlN buffer layer, the growth temperature of epitaxial layer and etc, are crucial to obtain the high quality films. LT-AlN with thickness 20 nm appears to enhance the structural properties of the columnar growth structures in AlGaN [10].

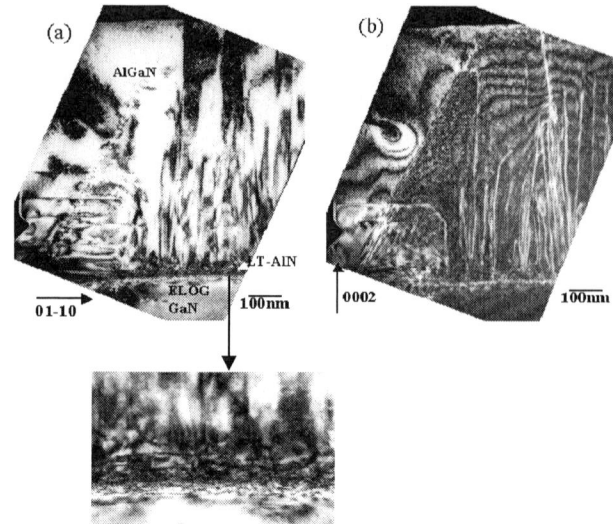

Fig 1. WBDF images of ELOG GaN/LT-AlN/ $Al_{0.22}Ga_{0.78}N$ at two different reflections (a) <01-10> and (b) <0002>

Figure 2 shows a cross-sectional of weak beam dark field (WBDF) image at low magnification of four periods of 20nm LT-AlN/450nm $Al_{0.35}Ga_{0.65}N$ layers grown on sapphire taken under <0002> reflection. Due to the large lattice mismatch between sapphire substrate and AlN, the high density of TDs exists in LT-AlN layer. These dislocations are extended into overgrown AlGaN layer. In order to reduce the density of TDs into overgrown AlGaN layer, LT-AlN was inserted. The lattice parameter of AlN is relatively small in-plane, therefore the AlGaN layer is compressed by the LT-AlN layer and the misfit between the LT-AlN and AlGaN becomes more rigorous. Then the second LT-AlN layer was grown on the top of this AlGaN layer. Figure 2 indicates the slightly reduction of TDs density in AlGaN layer by increasing the number of LT-AlN layers. The TDs density in the first and second AlGaN layer is about $3.0x10^{10}/cm^2$ and $9.0x10^9/cm^2$ respectively. The density of TDs in the third and fourth AlGaN layers is $5.0x10^9/cm^2$ and $2.0x10^9/cm^2$ respectively. Furthermore this figure shows a high density of loops as far as 200nm from LT AlN buffer layer in each AlGaN layer. However most of TDs propagate through the LT AlN layers; it indicates that the three interlayer of LT-AlN are not really effective in reducing the dislocation densities. The TD density in the fourth AlGaN layer is only ten times less than the TD density in the first AlGaN layer.

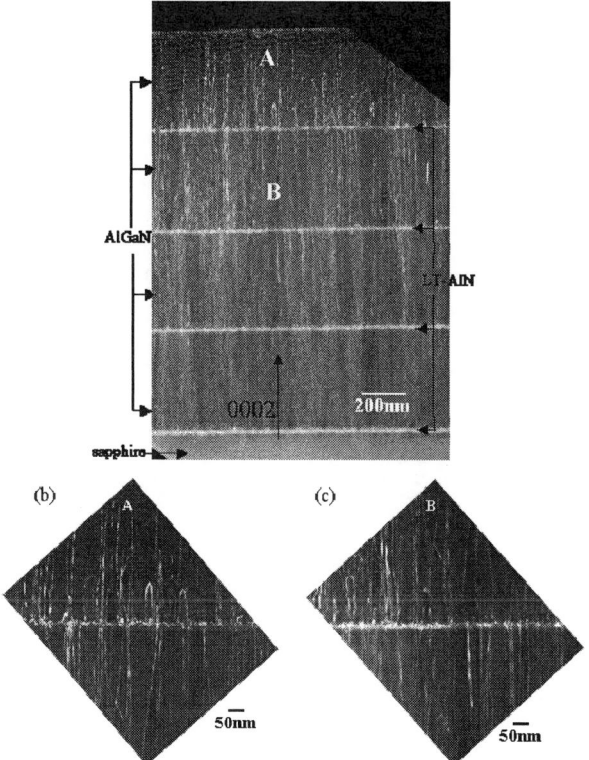

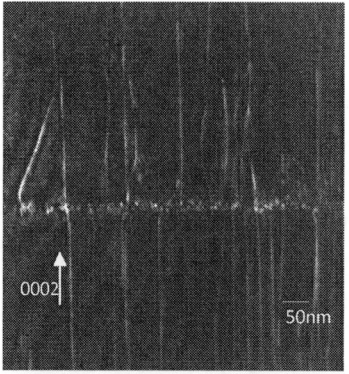

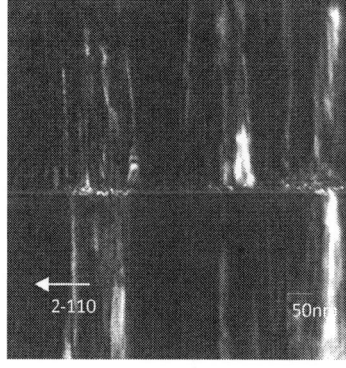

Fig. 2 (a) WBDF images of the four periods of LT-AlN/ Al$_{0.35}$Ga$_{0.65}$N with the magnification region A and B show on (b) and (c) respectively.

Fig. 3: WBDF images of Al$_{0.35}$Ga$_{0.65}$N /LT-AlN/ Al$_{0.35}$Ga$_{0.65}$N at two different reflections (a) 0002 and (b) 2-110

Figure 3a and 3b present a cross-sectional WBDF images of Al$_{0.35}$Ga$_{0.65}$N/ LT-AlN/ Al$_{0.35}$Ga$_{0.65}$N at two different reflections, which are <0002> and <2-110> respectively. It is shown the presence of a, c and c+a types of TDs in the AlGaN layers. Figure 3a also shows some of TDs are carried on from AlGaN to LT-AlN and continued to AlGaN layer. Although some of these TDs are appeared to generate from LT-AlN layer. The LT-AlN layer is moderately effective in reducing the density of TDs. There are some TDs that may be introduced at the interface due to the lattice mismatch between AlN and AlGaN. The origins of these TDs might be from the growth faults in NLs [11] as the partial dislocations associated with the faults in the NLs compose the major source of basal plane dislocations, which may develop into TDs by self-glide and climb. The TD density is very high near the interface and decreases with increasing film thickness.

IV. SUMMARY

The introduction of LT-AlN on highly perfect ELOG-GaN leads to imperfect LT-AlN/ ELOG-GaN interface. It also introduces TDs in AlGaN that deposited on the top of LT-AlN layer.

The TDs evolve at the GaN/LT-AlN and LT-AlN/ Al$_{0.22}$Ga$_{0.78}$N interfaces. The insertions of LT-AlN buffer interlayer in Al$_{0.35}$Ga$_{0.65}$N layer are moderately effective in reducing the density of TDs. It shows that the TDs density in the Al$_{0.35}$Ga$_{0.65}$N layer decreases by one order of magnitude after the incorporation of three layers of LT-AlN in Al$_{0.35}$Ga$_{0.65}$N layer. The work on AlN as a nucleation layer or a buffer layer in AlGaN has to pursue in order to get a high quality crystal with a low TDs density.

REFERENCES

[1] A. Kimoshita, H. Hirayama, M. Ainoya, Y. Auyagi, A. Hirata, "Room Temperature Operation at 333 nm of Al$_{0.03}$Ga$_{0.97}$N/Al$_{0.25}$Ga$_{0.75}$N Quantum Well Light Emitting Diodes with Mg-doped Superlattice Layers", App. Phys. Lett., 77, 175, 2000.

[2] J.Han, M.H. Crawford, R.J. Shul, J.J.Figrel, M.Banas, L.Zhang, Y.K.Song, H.Zhou and A.V. Nurmikko, "AlGaN/GaN Quantum Well Ultraviolet Light Emitting Diodes", App. Phys. Lett., 73, 1688, 1998.

[3] T.Nishida, H.Saito and N.Kobayashi, "Submilliwatt Operation of AlGaN-based Ultraviolet Light Emitting Diode using Short-period Alloy Superlattice", App. Phys. Lett., 78, 399, 2001.

[4] J.P.Zhang, H.M.Wang, M.E.Gaevski, C.Q.Chen, Q.Fareed, J.W.Yang, G.Simin and M.A.Khan, "Crack-free Thick AlGaN Grown on Sapphire

Using AlN/AlGaN Superlattices for Strain Management", App. Phys. Lett., 80, 3542, 2002.

[5] Y.Koide, N.Itoh, K.Itoh, N. Sawaki and I. Akasaki, "Effect of AlN Buffer Layer on AlGaN/α-Al₂O₃ Heteroepitaxial Growth by Metalorganic Vapor Phase Epitaxy", Jpn. J. Appl.Phys., 27, 1156, 1988.

[6] H.Amano, M. Iwaya, N.Hayashi, T.Kashima, S.Nitta, C.Wetzel and I.Akasaki, "Control of Dislocations and Stress in AlGaN on Sapphire Using a Low Temperature Interlayer", Phys. Stat. Sol (b), 216, 683, 1999.

[7] H.Amano, M. Iwaya, T.Kashima, M.Katsuragawa, I.Akasaki, J.Han, S. Hearne, J.A.Floro, E. Chason and J.Figiel, "Stress and Defect Control in GaN Using Low Temperature Interlayers", Jpn. J. Appl. Phys., 37, L1540 1998.

[8] K.Lorenz, M.Gonsalves, Wook Kim, V.Narayanan and S. Mahajan, "Comparative study of GaN and AlN nucleation layers and their role in growth of GaN on sapphire by metalorganic chemical vapor deposition", App. Phys. Lett., 77, 3391, 2000.

[9] M. Gonsalves, W. Kim, V. Narayanan, S. Mahajan, Influence of AlN nucleation layer growth conditions on quality of GaN layers deposited on (0 0 0 1) sapphire", J. Crystal Growth, 240, 347, 2002.

[10] X.L. Wang, D.G. Zhao, X.Y. Li, H.M. Gong, H. Yang, J.W. Liang, "The effects of LT AlN buffer thickness on the properties of high Al composition AlGaN epilayers", Materials Letters 60, 3693–3696, 2006.

[11] V. Narayanan, K. Lorenz, W. Kim, S. Mahajan, "Gallium nitride epitaxy on (0001) sapphire", Philosophical Magazine A, 82, 885, 2002.

Calibration Parameters in TCAD for Predictive MOSFET Device Simulations

Muhamad Amri Ismail, Mohd Hezri Abu Bakar and Iskhandar Md Nasir

MIMOS Wafer Fab, MIMOS Berhad,
Technology Park Malaysia, 57000 Kuala Lumpur, MALAYSIA
Email: amris@mimos.my, mhezri.abakar@mimos.my, iskhand@mimos.my

Abstract— **Predictive TCAD tool is crucial for several reasons such as to provide pre-silicon data, shorten the technology development cycle and reduce the fabrication cost. This paper presents a methodology for TCAD advanced calibration of MOSFET particularly on critical electrical parameters during device simulations. A few physical device model parameters have been experimented to solve the inaccuracy issues due to the default values. The comparisons between measured and simulated data of electrical parameters are presented for the verification purpose. It is proven that modifying the surface mobility, high-field saturation and band-to-band models had been successful in significantly improved the TCAD accuracy.**

I. INTRODUCTION

Process and device simulations with Technology Computer Aided Design (TCAD) have gained a lot of interests from semiconductor communities with respect to its several powerful features. One of them is its capability to predict the device behaviour from the provided process flow where the simulation results agree with the actual one. With the ever increases demand of technology or product time-to-market, this predictive capability provides a huge advantages to the silicon foundries in order to meet the scheduled timeline [1].

Parallel with the gaining interest of predictive TCAD tool, there are many approaches of calibrating the tool have been studied and presented in the literatures [2], [3]. From the calibration of Secondary Ion Mass Spectrometry (SIMS) profile for accurate prediction of process simulation by Park *et al*, the study supported with literature by Aoyama *et al* in calibrating the TCAD considering the effect of statistical process variations. Consideration of statistical analysis in TCAD provides good benefit for both predictability and preparation of pre-silicon data for smaller technology nodes. The literatures are not only focused on the bulk-CMOS technologies but they also involve recent and advanced device materials and structures such as Germanium, Silicon Carbide and FinFET [4]-[6].

Generally those reports are referring to the smooth link between process and device simulations resulting to a good correlation between measured and simulated electrical parameters. The motivation of this paper lies in the authors experience that even the well calibrated process simulation did not guarantee that the same case will go to the device simulation. This paper presents a method for TCAD advanced calibration of MOSFET device simulation particularly on important electrical parameters namely threshold voltage (V_{th}),

saturation current (I_{dsat}) and off-state leakage current (I_{off}). The remainder of the paper is structured as follows. Section II explains the selection of physical parameters needed in the calibration process. Section III presents the calibration procedures follows with the discussion of experimental outcomes in Section IV. The paper is concluded in Section V.

II. IDENTIFICATION OF CALIBRATION PARAMETERS

Considering the fact that the focus of this paper is mostly on the calibration of TCAD device simulations, it is assumes that the TCAD process simulations are fully calibrated in the first place. The calibration of actual process flow to the TCAD process file needs to consider both 1-D and 2-D simulations. Various SIMS profiles from different process steps such as well, channel, lightly doped drain (LDD) and source/drain implantations are the golden reference in 1-D calibration process [2]. Besides that, sheet resistance data at every process steps are other important parameters for the calibration of 1-D process simulation. Accuracy of 1-D process simulation is vital because it gives a good starting point for later 2-D calibration especially on the extension and deep source/drain regions. For 2-D process calibration, Transmission Electron Microscope (TEM) pictures are required where measured physical process parameters such as transistor's gate length and oxide thickness are inspected against simulation results. Another important assumption in the early stage is the generated meshes are in a good pattern to ensure the correct simulated outcome with respect to device scalability. Adaptive meshing strategy is proposed for that purpose because it has the capability to generate more mesh profiles during simulation stage which contributes to more accurate refinements.

After the completion of TCAD process calibration, then the next step is the calibration of device characteristics. Simulation file which contains various device setups is the prerequisite for any TCAD device simulation. These device setups are actually the standard device models used to describe the physical behaviour of MOSFET such as carrier transport, band gap narrowing, quantization effect, carrier mobility and recombination-generation models [7]. Our main objective is to identify what are the relevant calibration parameters from those mentioned standard device models that need to be considered so that the simulated electrical parameters match with the measured data. There are a few carrier transport models available in standard TCAD tool such as drift-diffusion, thermodynamic and hydrodynamic.

978-1-4673-2395-6/12 $31.00 © 2012 IEEE

Selection of transport model is straight forward for MOSFET where drift-diffusion equation is typically used while thermodynamic and hydrodynamic equations are suitable for more advanced simulations such high power and high voltage devices [8]. The same thing goes to band gap narrowing and quantization models where Slotboom model and density gradient models, respectively, are used for MOSFET device. In fact, these drift-diffusion, Slotboom and density gradient models have been complementing each other as a basis of the MOSFET device and related to general constants such as Boltzmann and Poisson, so there is no need for further calibrating those parameters.

In contrast, the suitability of carrier mobility and recombination-generation models are more foundry-specific and the parameters need to be corrected based on the provided model parameters during TCAD device simulation [9]. For mobility model, the inversion layer mobility equation with Lombardi model is given by

$$\mu_{ac} = \frac{B}{F_\perp} + \frac{\left(C \times \left(\frac{N_D + N_A}{N_0}\right)\right)^\lambda}{F_\perp (T/300K)^K}$$

(1)

where $F_\perp$ is a local field component, T is the temperature node, K is the temperature dependence parameter while N_D and N_A are for donor and acceptor concentrations, respectively. There are four fitting parameters available in (1) such as B, C, N_0 and λ. In order to reduce the time for calibration process, only one fitting parameter is necessary and parameter C is proposed as being suggested by Darwish *et al* due to its sensitivity to the impurity concentration [9].

Another important parameter under carrier mobility model is high-field saturation which is the phenomenon when the carrier drift velocity is no longer relative to electric field. The recommended velocity model (V_{sat}) for silicon device at height electric fields is given by the following relationship [7]:

$$V_{sat} = V_{sat0} \left(\frac{300K}{T}\right)^{V_{satexp}}$$

(2)

where V_{sat0} is saturation velocity at nominal temperature and V_{satexp} is the temperature dependence parameter for saturation velocity. Since this work is basically concerns at room temperature analysis, then V_{sat0} is the only candidate for the calibration parameter. Therefore there are two parameters have been identified for mobility model namely parameter C in (1) and V_{sat0} in (2).

For recombination-generation mechanism, there are a few physical device models involve such as band-to-band, Shockley-Read-Hall (SRH) and Auger and these models are provided in most of the available TCAD device simulator. With regards to our aim in determining the most significant model for the simplification, band-to-band model is proposed for further calibration because this model dominates the recombination mechanism in direct bandgap semiconductor

[10]. Whereas SRH model is typically represents recombination through defects while Auger is mainly for heavily doped devices. Band-to-band model itself contains several theories but we are concentrated with Schenk fitting parameters available in [7].

III. EXPERIMENT

The electrical parameters of NMOS device are measured with auto parametric tester namely Agilent 4073UX system. The device contains a few variations of channel length (L) with fixed channel width (W) from MIMOS's 0.35 um CMOS process. Single site measurement from one golden wafer is performed in order to ensure that the measurement is representing the typical case data. Device characteristics such as V_{th}, I_{dsat}, and I_{off} are required where these data will be the golden reference during the calibration exercise.

On the other hand, the process and device simulations are performed using Sentaurus TCAD tool from Synopsys. Fig. 1 shows the details flow diagram of TCAD device calibration with related approaches proposed in this work. The first approach attempted in simulating the TCAD device simulation file using defaults physical model parameters provided in the standard TCAD package. From the first approach, if the simulation of targeted parameters is not matching the measurement result, the default device models need to be calibrated.

The second approach was aimed at calibrating the default models of carrier mobility, specifically on normal Lombardi and high-field saturation models. As being described during the previous section, there are two parameters involve in this step namely parameter C and V_{sat0} given by

$$C = \text{Calibration factor 1} \times \text{Default model}$$ (3)

$$V_{sat0} = \text{Calibration factor 2} \times \text{Default model}$$ (4)

where calibration factor 1 and 2 is respectively for for parameter C in (1) and V_{sat0} in (2). The simulation with calibration factor was then examined versus the measurement to ascertain the accuracy of new calibrated model file. If the result still unsatisfactorily fit the target, then the calibration flow continues with the third approach.

The third approach was concerned at calibrating both carrier mobility and recombination-generation models to the default model parameters. Therefore (3) and (4) are required and combined with calibration of band-to-band model parameters, specifically parameter A in [7] of Schenk model. This is the last step for optimization of the related calibration factors. Any persistence of discrepancies between measured and simulated after the last step will needs further revision of the provided calibrated TCAD process file.

IV. RESULT AND DISCUSSION

In order to verify the proposed methodology, we observed the TCAD simulation result against the actual measurement data. Table I sums up the detail comparison of measured and simulated device characteristics, obtained from each step of

978-1-4673-2395-6/12 $31.00 © 2012 IEEE

calibration approaches for the shortest and the largest transistors. The dimension for shortest device is $L=0.35$um and $W=20$um while for largest device is $L=20$um and $W=20$um.

Although there is some improvement for I_{off} parameter with the calibration of mobility models but it still about 60% off the measurement data. Finally the data shows that only after the third approach, all of the simulated electrical parameter could accurately match with the actual measured data.

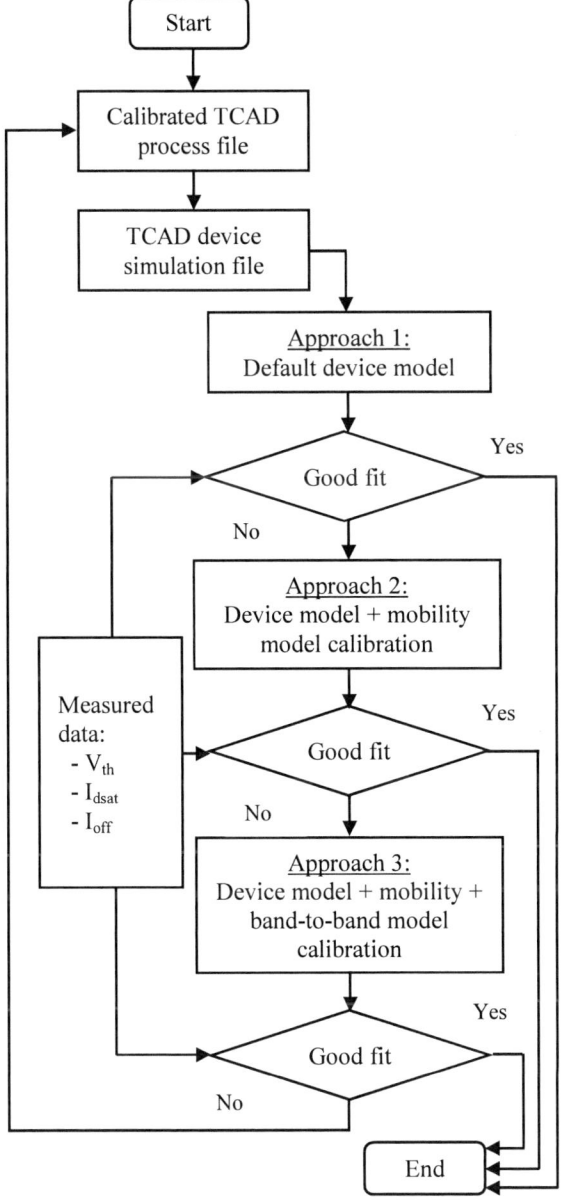

Fig. 1. Flowchart for TCAD device calibration

TABLE I
COMPARISON OF MEASURED VERSUS SIMULATED DATA

Electrical parameters	Results	Short	Large
V_{th} (V)	Measured	0.560	0.410
	Approach 1	0.552	0.435
	Approach 2	0.558	0.408
	Approach 3	0.558	0.408
I_{dsat} (A)	Measured	9.89e-03	4.82e-04
	Approach 1	5.41e-03	1.48e-04
	Approach 2	1.06e-02	4.86e-04
	Approach 3	1.01e-02	4.84e-04
I_{off} (A)	Measured	1.24e-11	1.81e-11
	Approach 1	5.29e-12	1.97e-12
	Approach 2	7.44e-12	2.22e-12
	Approach 3	1.33e-11	1.65e-11

The results of measurement versus simulation data obtained from various device geometries are shown respectively in Fig. 2, Fig. 3 and Fig. 4 for V_{th}, I_{dsat} and I_{off} parameters. The default model is the simulation result from approach 1 while the calibrated model is the simulation result from approach 3, which is representing this report.

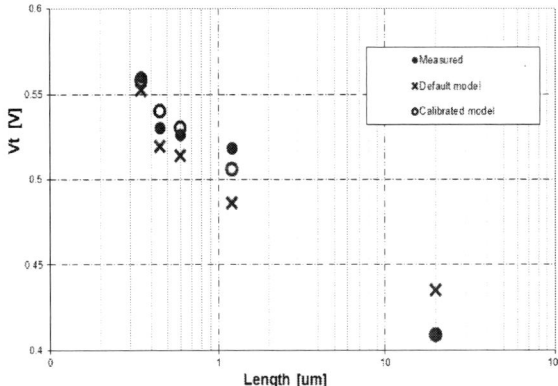

Fig. 2. V_{th} vs. gate length

The data shows that the default physical device model could only managed to accurately match the V_{th} value while there is a significant error of 80% and 130% respectively for I_{dsat} and I_{off}. The simulation results from the second approach provide better accuracy compared to the first approach, where V_{th} and I_{dsat} parameters are now in an acceptable tolerance.

Fig. 2 generally shows that the default model could considerably match V_{th} measured data in an acceptable range for calibrated TCAD process without device model calibration. But Fig. 3 and Fig. 4 shows the major concern in using default model since it might seriously underestimates the actual values of I_{dsat} and I_{off}. The calibrated model from this work produced a favourable fitting between measured and simulated results for all the critical electrical parameters.

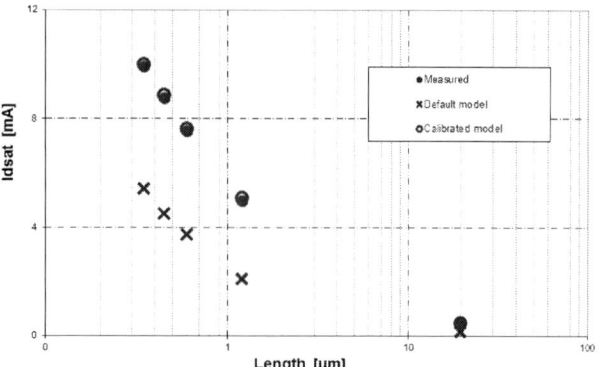

Fig. 3. I_{dsat} vs. gate length

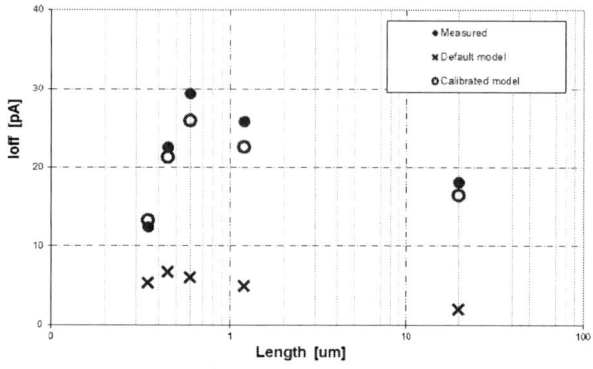

Fig. 4. I_{off} vs. gate length

These results show the important of having the calibration parameters for the standard physical device models available in the TCAD tool. In short, mobility model calibration is critical for on-state current characteristics while the calibration of both mobility and band-to-band models is important for both on-state and off-state current characteristics. The accuracy of I_{off} simulation from TCAD is crucial considering the difficulties of controlling that parameter in recent advanced nanotechnologies especially to the low power applications.

V. CONCLUSIONS

This paper has presented a methodology to calibrate the TCAD tool especially on the device simulation part with the consideration of various calibration parameters namely surface mobility, high-field saturation and band-to-band models. Those calibration parameters are important in order to guarantee that the simulated electrical parameters could accurately match with the actual measured data. Surface mobility and high-field saturation models are important for the predictive on-state current characteristics while band-to-band model is critical for leakage current variations. We hope this work would facilitate other TCAD users in selecting appropriate parameters during calibrating the critical electrical parameters with TCAD device simulations.

REFERENCES

[1] R. W. Dutton, and A. J. Strojwas, "Perspectives of technology and technology-driven CAD," *IEEE Trans. on Computer-Aided Design of Integrated Circuits and Systems*, vol. 19, no. 12, pp. 1544-1560, Dec. 2000.

[2] H. Park, M. Bafleur, L. Borucki, C. Sughama, T. Zirkle, and A. Wild, "Systematic calibration of process simulations for predictive TCAD," *Proc. of IEEE International Conference on Simulation of Semiconductor Processes and Devices*, pp. 273-275, Sept. 1997.

[3] K. Aoyama, H. Kunitomo, K. Tsuneno, K. Sato, K. Mori, and H. Masuda, "Rigorous statistical process variation analysis for quarter-um CMOS with advanced TCAD metrology," *Proc. of IEEE International Workshop on Statistical Methodology*, pp. 8-10, Jan 1997.

[4] G. Hellings, G. Eneman, R. Krom, B. De Jaeger, J. Mitard, A. De Keersgieter, T. Hoffman, M. Meuris, and K. De Meyer, "Electrical TCAD simulations of a Germanium pMOSFET technology," *IEEE Trans. on Electron Device*, vol. 57, no. 10, pp. 2539-2546, Oct. 2010.

[5] M. Philip, and A. O'Neil, "Calibration of 4H-SiC TCAD models and material parameters," *Proc. of IEEE Conference on Optoelectronic and Microelectronic Materials and Devices*, pp. 137-140, Dec. 2006.

[6] B. Raj, A. K. Saxena, and S. Dasgupta, "Quantum inversion charge and drain current analysis for double gate FinFET device: Analytical Modeling and TCAD simulation approach," *Proc. of IEEE European Symposium on Computer Modeling and Simulation*, pp. 526-530, Nov. 2010.

[7] *Sentaurus Device User Guide, Synopsys* Inc., Dec. 2010.

[8] T. Grasser, T. Ting-Wei, H. Kosina, and S. Selberherr, "A review of hydrodynamic and energy-transport models for semiconductor device simulation," *Proceeding of the IEEE*, vol. 91, no. 2, pp. 251-274, Feb. 2003.

[9] M. N. Darwish, J. L. Lentz, M. R. Pinto, P. M. Zeitzoff, T. J. Krutsick, and H. V. Hong, "An improved electron and hole mobility model for general purpose device simulation," *IEEE Trans. on Electron Device*, vol. 44, no. 9, pp. 1529-1538, Sept. 1997.

[10] A. Heigl, A. Schenk, and G. Wachutka, "Correction to the Schenk model of band-to-band tunneling in silicon applied to the simulation of Nanowire tunneling transistors," *Proc. of IEEE International Workshop on Computational Electronics*, pp. 1-4, May 2009.

978-1-4673-2395-6/12 $31.00 © 2012 IEEE

Deposition of Titanium Dioxide (TiO$_2$) Thin Films Using In-house Nano-TiO$_2$ Powder

[1]M.Z. Sahdan, [1]M.S. Alias, [1]N. Nafarizal and [2]U. Hashim

[1]Microelectronic and Nanotechnology-Shamsuddin Research Centre, Faculty of Electrical and Electronic Engineering, Universiti Tun Hussein Onn Malaysia, 86400 Batu Pahat, Johor, Malaysia
[2]Institut Nano-Engineering, Universiti Malaysia Perlis, 01000 Kangar, Perlis, Malaysia
Email: zainizno@gmail.com

Abstract – The purpose of this study is to fabricate uniform TiO2 thin films using in-house TiO2 nanopowder. The nanopowder was obtained from a tin mining waste (Ilmenite) and its concentration was optimized during fabrication. The TiO2 thin films were characterized by an atomic force microscope (AFM), a surface profiler, an X-ray diffractometer, an Ultra violet- Visible (UV-Vis) and a current-voltage (I-V) measurement system. The relation of the uniformity and the properties of the TiO2 thin films will be discussed in detail in this paper.

Keywords: *Titanium dioxide (TiO2), ilmenite, thin films, sol-gel, atomic force microscope (AFM).*

I. INTRODUCTION

Titanium dioxide or known as TiO$_2$ has become one of the researched semiconductor material because of the promising result in the photocatalytic oxidation. TiO$_2$ has wide band gap energy around 3.2 eV or have photon energy (hv) <390 nm that is transparent to visible light and has excellent optical transmittance [1]. It has high refractive index and good insulating properties, and as a result it is widely used as protective layer for very large scale integrated circuits and for manufactures of optical elements. Besides that, Titanium dioxide is made of TiO$_2$ octahedral. The structural of octahedral gives three different polymorphs of Titanium dioxide that is anatase, rutile and brooktile [2].

Anatase and rutile are the most researched polymorphs but anatase is proven recently as the most photocatalytically active of the three polymorphs. Furthermore, basically anatase transform into rutile under the heat between 600 $^{\circ}$C and 700°C. TiO$_2$ has many structures but it depends on its polymorphs [3]. If they are anatase and rutile, then the structure is tetragonal but if it is brooktile, the structure is orthorhombic [4-5]. Nowadays TiO$_2$ has wide application from paint to sunscreen to food coloring.

The price of TiO$_2$ powder in the market for fabrication of TiO$_2$ thin films is a way too high, but we can bring it into a solution. In this research, TiO$_2$ powder obtained from tin mining waste (Ilmenite) is introduced which will reduce the

cost up to 100 times. However, it is an issue whether TiO$_2$ obtained from Ilmenite can be a functionalized material or not. Moreover, it is questionable whether the TiO$_2$ can be fabricated into a uniform thin film. Using sol-gel, the challenge is to control the process parameters which influence the material properties (electrical and optical). Therefore, process optimization is required to enhance the electrical and optical properties.

II. METHODOLOGY

An aqueous solution was prepared from in-house nano-TiO$_2$ powder by varying its mass from 1g, 0.4g, 0.1g and 0.01g in 30 ml of ethanol and mixed with 6 ml acetic acid as the catalyst. Then, the solution was stirred on a magnetic stirrer at 700 r.p.m. at room temperature for 20 hours. ITO was cleaned using acetone in ultrasonic cleaner for 5 minutes before rinsed with de-ionized water. The samples were dried under atmospheric ambient at room temperature.

The thin films were fabricated using a spin coater using 2 steps procedure (1000 r.p.m. for 30s and 3000 r.p.m. for 60s). Each layer of the film requires 10 drops of TiO$_2$ solution and we prepared 10 layers for each sample. Then, the samples were annealed at 500°C for anatase phase. The annealing process was carried out at 1 hour and he samples underwent slow cooling at room temperature at about 5 minutes.

The samples were characterized using an atomic force microscopy (Park System XE-100) for topological observation, a thickness Profiler (KL-Tenko) for thickness measurement, a UV-Vis spectrometer (Varian) for the optical characterization and an *I-V* measurement system (Keithley 2400) for the electrical measurement.

III. RESULT AND ANALYSIS

The thickness of the sample using mass concentration of 1g, 0.4g, 0.1g and 0.05 g are 427 nm, 57 nm, 54 nm and 17 nm, respectively. Fig.1 (a)-(d) indicates the TiO$_2$ thin films when the nano-TiO$_2$ powder concentration was reduced from 1g to 0.05g.

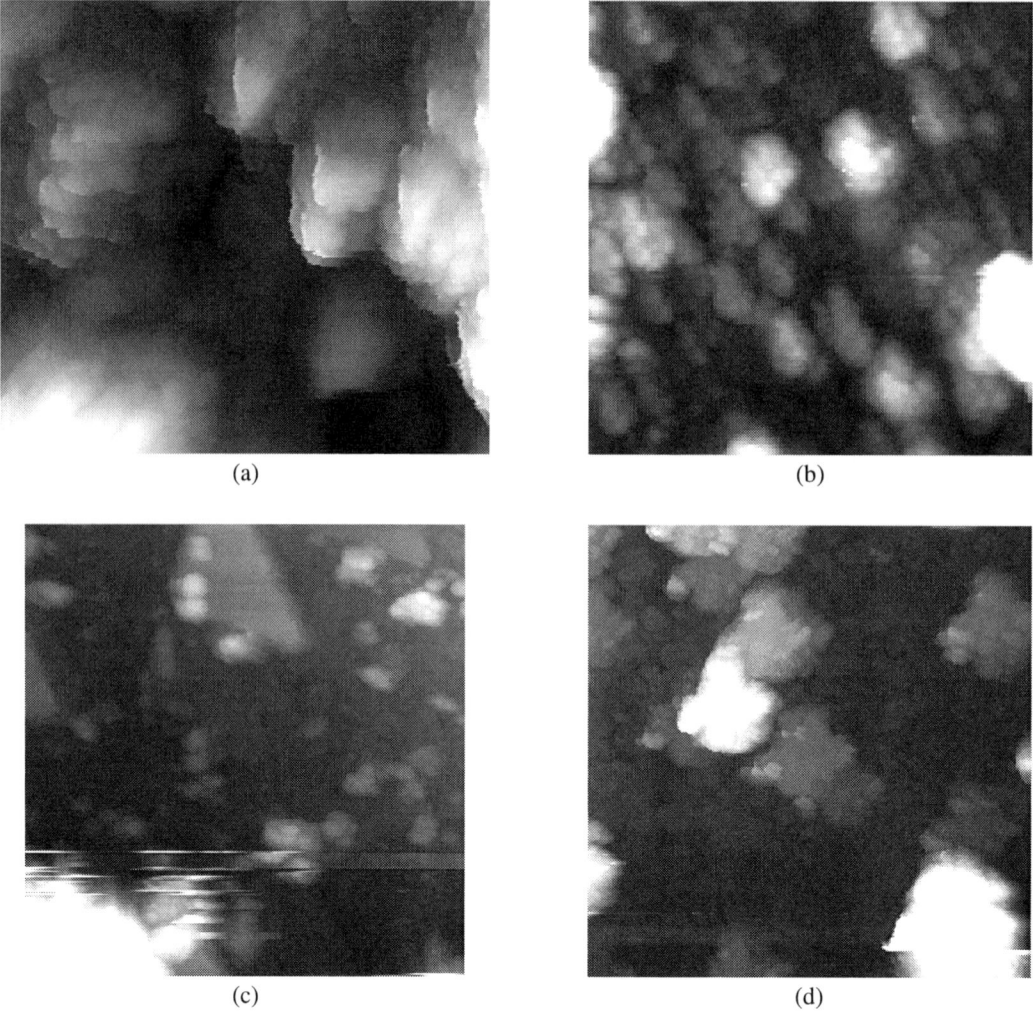

(a)   (b)

(c)   (d)

Fig. 1   The AFM topography of TiO$_2$ thin films by varying the nano-TiO2 powder mass concentrations (a) 1g; (b) 0.4g; (c) 0.1g; (d) 0.05g

As the concentration of the nano-TiO$_2$ powder reduced, the particles size which constructs the thin film also reduced. It is observed that TiO$_2$ thin film deposited using 0.4g has the optimum uniformity.

The transmittance spectra were measured using a UV-Vis spectrometer. Fig. 2 shows the transmittance spectra of the TiO$_2$ thin films deposited using different nano-TiO$_2$ powder mass concentration. Generally, TiO$_2$ has high absorption at wavelength of 360nm due to its optical band gap. As the mass concentration of nano-TiO2 powder decreased, the optical transmittance increased. There are two reasons of explaining this changes; (i) since the thickness of the film reduced, the transmittance increases, and (ii) as the particles size become smaller, the transmittance became higher. However, the transmittance for 0.4g at wavelength below 400 nm depicted a

different behavior compared to the other samples. The reason is still unknown.

Fig. 4 and 5 show the *I-V* characteristics of the TiO$_2$ samples which indicate Schottky and Ohmic response, respectively. The Schottky contact may resulted from the large different of conduction band between Platinum (Pt) and TiO2 which has the value of 5.6 eV and 4.4 eV, respectively. Therefore, for an electron to be excited from the conduction band of Pt to the conduction band of TiO2, it requires energy greater than 1.2 eV. However, this phenomenon is not happening for 0.1g and 0.05g. Instead of Schottky, an Ohmic response was produced. This may be due to reduce in the thickness of the film which causes electron tunneling between the Pt and TiO$_2$.

978-1-4673-2395-6/12 $31.00 © 2012 IEEE

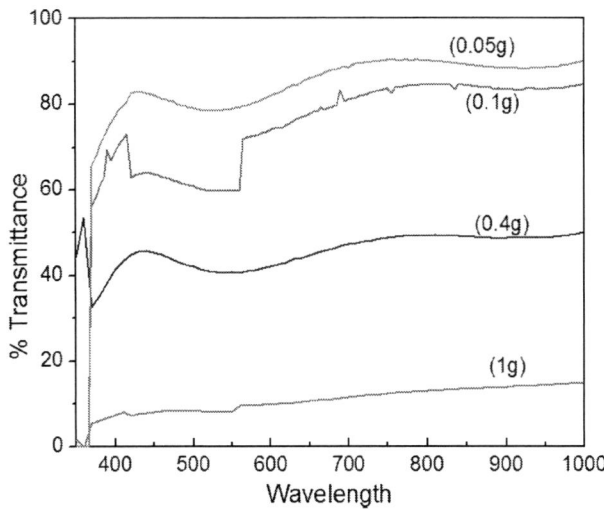

Fig. 2   The transmittance spectra of TiO_2 thin films

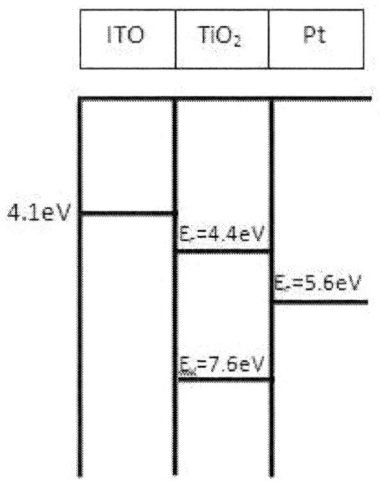

Fig. 3   The band off-set of ITO/TiO_2/Pt contact

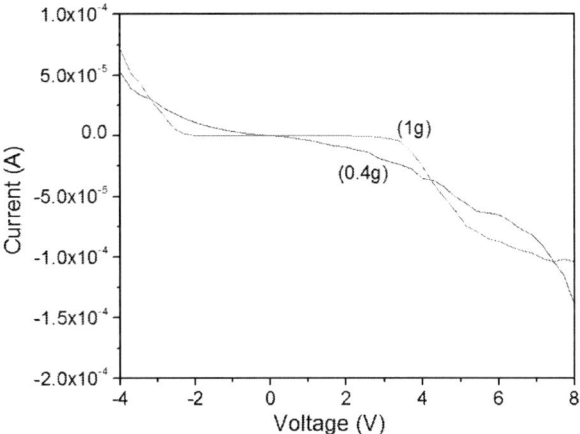

Fig. 4   The Schottky behavior of TiO_2 thin films

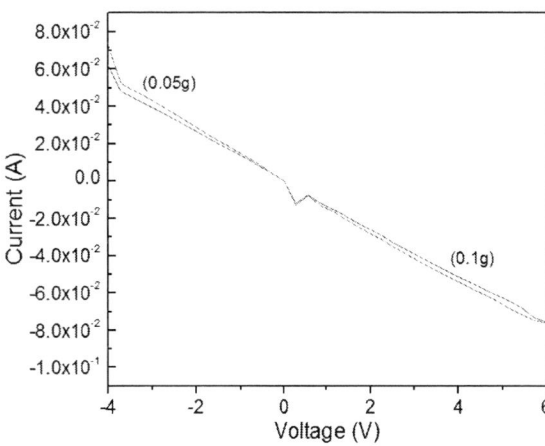

Fig. 5   The Ohmic behavior of TiO_2 thin films

IV. CONCLUSION

As a conclusion, Ilmenite is a good material to fabricate uniform TiO_2 thin films although its purity is only 95.07 %. It can be used as a functionalized material whether for electronic or optoelectronic applications. Furthermore, spin coating is the suitable fabrication technique in sol-gel method. From the characterization analysis, we can conclude that sample deposited using 0.4 g of nano-TiO_2 powder have the best characteristic for both electrical and optical properties. There a few recommendations for a better result. Ethanol is not a good solvent. So, using other different solvent which has higher solubility may improve the results. The acetic acid which act as a catalyst may also need a process optimization which would produce more uniform thin film.

V. ACKNOWLEDGEMENT

A special thanks to Mint-SRC, Mechanical Engineering Laboratory and Science Chemistry Laboratory of Universiti Tun Hussein Onn Malaysia for providing the equipment needed.

VI. REFERENCES

[1] L A. K. Sharma, R. K. Tareja, U. Wilker, and W. Schade, Appl. Surf. Sci. 206, 137 (2003).

[2] Jamieson J, Olinger B (1969). "Pressure Temperature Studies of Anatase, Brookite, Rutile and TiO2 II: A discussion". *Mineralogical Notes* 54: 1477.

[3] D. Luca, L. S. Hsu, Department of Physics, National Chang-Hua University of Education, "Structural Evolution and Optical Properties of TiO_2 Thin Films Prepared by Thermaloxidation of PLD Ti Films"

[4] Hanaor, D.; Sorrell, C. (2011). "Review of the anatase to rutile phase transformation". *Journal of Materials science* 46 (4): 855–874.

[5] Di Paola, A; Addamo, M. Bellardita, M. Cazzanelli, E. Palmisano, L. (2007). "Preparation of photocatalytic brookite thin films". *Thin solid films* 515 (7–8): 3527–3529.

T-Shaped Resonating Beam Pressure Sensor

Y.Sujan, B.Vasuki, G.Uma

Department of Instrumentation and Control Engineering,
National Institute of Technology, Tiruchirappalli 620 015, India.
sujan.1053@gmail.com, bvas@nitt.edu, guma@nitt.edu.

Abstract- **A resonant pressure sensor with a provision for incorporating appropriate actuation and detection mechanism using fully surface micromachined process is presented in this paper. The pressure sensing element is a T-shaped resonating beam encapsulated beneath a polysilicon rectangular diaphragm and fully enclosed in a sealed vacuum cavity. This design enables high pressure sensitivity, miniature chip size and as well as good environmental isolation. The proposed structure is evaluated for pressure sensitivity theoretically and numerically using MEMS CAD tool CoventorWare. The pressure sensitivity of the sensor with a T-shaped 1.2 μm thick beam is found to be 10.2%/bar with the beam resonance frequency of about 792 kHz.**

Keywords- **T-shaped beam; Pressure sensor; MEMS; Surface micromachining.**

I. INTRODUCTION

Micro pressure sensors are the second most important, fabricated and sold today next to temperature sensors. A wide variety of applications of micro pressure sensors are found in biomedical, automotive, aerospace, process and industrial control applications. Capacitive micro pressure sensors as in [1-3] show the advantages of higher sensitivity, insensitive to temperature, more robust structure, and low power consumption suitable for bio-medical applications, but they require precise and complex interfacing circuit for their small capacitance due to miniature size. Piezoresistive micro pressure sensors can be easily fabricated since the sensing element, pizeoresistance can be produced by doping boron ions on silicon. The first surface micromachined piezoresistive pressure sensor for blood pressure measurements has been commercialized as in [4]. These sensors have high linearity but very sensitive to temperature change and temperature compensation circuits have to be employed. Optical micro pressure sensors have the advantages of high resolution, high sensitivity and immune to electromagnetic interference when compared to electrical sensors which usually employ capacitive or piezoresistive detection. However temperature compensation is necessary for high accuracy measurement. Resonant micro sensors for pressure measurements have an increasing importance in

all fields of engineering applications, due to its advantages like high stability, high resolution, insensitive to temperature and quasi digital output. These sensors consist of a resonating element attached to a non vibrating diaphragm or the resonating element is diaphragm itself. In these sensors the stiffness variation of the resonating element attached to the diaphragm as in [5] or the stiffness variation of the vibrating diaphragm is converted into resonance frequency shift which can be sensed using piezoelectric, electrostatic, optical and piezoresistive based secondary sensors giving a direct measure of pressure. There are several resonant pressure sensors using MEMS technology have been proposed as in [6, 7].

An earlier reported, all surface micromachined resonant pressure sensor as in [8] has one end of the resonating beam suspended beneath a pressure sensitive rectangular diaphragm and the other end to the edge of the vacuum sealed side wall cavity with electrostatic excitation and piezoresistive detection. The structure proposed as in [8] is modified to achieve higher sensitivity as in [9] and verified theoretically. The main drawback of the proposed sensor design is that, in the sensing mechanism the electrical connections should be made on the diaphragm that could cause stiffening effect which may counteract the increased sensitivity. In this paper, a modified design for the sensor structure is proposed with appropriate actuation and sensing mechanism to avoid the drawback of the previous design as in [9]. Surface micromachining process is used in the sensor design to meet the great demand for small chip requirements for applications where miniaturization is required.

II. PRINCIPLE OF OPERATION

In this sensor design, the resonating beam is suspended at both of its ends by rigid supports to a polysilicon diaphragm as in [9] with a T-shape extension of width 2μm at the centre of the beam, in any one side of the wall cavity and in the same plane. The extended beam, with the fixed-fixed beam act as a resonant structure with a provision of piezoelectric deposition on the T- shape extension for resonant detection; thereby eliminates the

978-1-4673-2395-6/12 $31.00 © 2012 IEEE

electrical connections on the diaphragm. The electrical connections can be done through the T shaped extension on the top surface of the substrate. The beam can be set to resonance by providing an electrostatic ac excitation by placing a bottom electrode on the substrate. Fig.1 shows the 3D model of the sensor. The entire T-beam is enclosed in a sealed walled cavity and this vacuum encapsulation inside the cavity functions as a vacuum pressure reference.

During the design of the sensor, it should be ensured that the beam resonance frequency is lower than the resonance frequency of the diaphragm as in [8]. This constraint arises due to the fact that diaphragm need to be isolated from the beam vibration via the beam attachment to the diaphragm. With the applied pressure, the diaphragm deflects and causes a tensile axial force to act on opposite ends of the T-beam. Thus the T-beam gets strained which shifts its resonance frequency, which is a measure of pressure and can be detected by the piezoelectric layer on the extended part of the beam and with the closed loop electronics for resonant sensors proposed as in [10, 11].

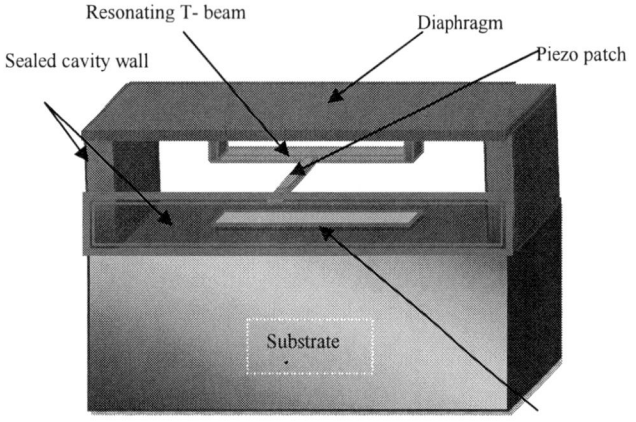

Fig. 1. 3D layout of the sensor structure

III. RESONANCE FREQUENCY OF THE BEAM

The resonant frequency of the resonating beam attached to the pressure sensitive rectangular shaped diaphragm at one end and the other end attached to the edge of the cavity is derived as in [8] is given as

$$ f = \frac{H}{2\pi L^2} \sqrt{42 \frac{E}{\rho}\left(1 + \frac{2}{7}\varepsilon \frac{L^2}{H^2}\right)} \qquad (1) $$

where E is the Young's Modulus , ρ is the mass density, H is the thickness of the beam, L is the length of the beam and ε is the total strain equal to the sum of built in strain and the strain developed in the beam due to applied pressure. The resonant frequency (f_T) of resonating beam with T shape extension given in (3) is derived from modified Euler Bernoulli equation given in (2) using Rayleigh-Ritz method as in [12, 13].

$$ EI \frac{\partial^4 w(x)}{\partial x^4} + \rho A \frac{\partial^2 w(x)}{\partial t^2} + Kw(x) = 0 \qquad (2) $$

$$ f_T = \frac{1}{2\pi}\sqrt{504 \frac{EI}{mL^4}\left(1 + 0.00488K\frac{L^3}{EI}\right)} \qquad (3) $$

where I is the moment of inertia along the cross section and m is the mass per unit length, A is the cross sectional area, K is the stiffness constant of the extended part of the beam and $w(x)$ represents the transverse deflection of the beam. The deflection assumption $w(x)$ is given by (4) which satisfy all the boundary conditions as stated in (5) and (6).

$$ w(x) = c\,x^2(x-L)^2 \quad \text{for } 0 \le x \le L \qquad (4) $$

$$ w(x) = 0 \qquad \text{for x=0 and x=L} \qquad (5) $$

$$ \frac{\partial w(x)}{\partial x} = 0 \qquad \text{for x=0 and x=L} \qquad (6) $$

The resonant frequency $(f_T)_{load}$ of resonating beam with T shape extension with applied axial load in terms of strain given in (8) is derived from modified Euler Bernoulli given in (7) using Rayleigh-Ritz method as in [12, 13].

$$ EI \frac{\partial^4 w(x)}{\partial x^4} + \rho A \frac{\partial^2 w(x)}{\partial t^2} - p\frac{\partial^2 w(x)}{\partial x^2} + Kw(x) = 0 \qquad (7) $$

$$ (f_T)_{load} = \frac{1}{2\pi}\sqrt{504 \frac{EI}{mL^4}\left(1 + 0.00488K\frac{L^3}{EI}\right) + 12\frac{E}{\rho L^2}\varepsilon} \qquad (8) $$

The pressure induced strain in the resonating beam with T- shape extension is calculated theoretically by using the large deflection theory of rectangular plates with the assumptions (i) When no pressure is applied the plate remains flat, (ii) All the edges of the plate are clamped and

(iii) The residual stress is neglected since it is determined by the fabrication process. The out of plane deflection of the diaphragm with the applied pressure load is given as

$$
\begin{aligned}
w(x,y) = w_0 & [(1-\frac{x^2}{a^2})(1-\frac{y^2}{b^2}) \\
& (1+A\frac{x^2}{a^2}+B\frac{y^2}{b^2}+C\frac{x^4}{a^4}+D\frac{y^4}{b^4}+E\frac{x^2}{a^2}\frac{y^2}{b^2})]+ \\
& 12\frac{w_0^3}{h^2}[(1-\frac{x^2}{a^2})(1-\frac{y^2}{b^2}) \\
& (T\frac{x^2}{a^2}+U\frac{y^2}{b^2}+V\frac{x^4}{a^4}+W\frac{y^4}{b^4}+Z\frac{x^2}{a^2}\frac{y^2}{b^2})]
\end{aligned}
\tag{9}
$$

where w_0 is the central deflection of the diaphragm, $2a$ is the width, $2b$ is the length of the diaphragm and A, B, C, D, E, T, U, V, W, Z are numerical constants as in [14].

The pressure induced elongation in the beam can be related to the angle (α) between the horizontal plane of the diaphragm (under no pressure load) and the diaphragm with pressure load. Equation (9) is used in calculating (α) as in [15] and the elongation (Δx) in the resonating T-shaped beam is given as

$$
\Delta x = d \tan(\alpha)
\tag{10}
$$

where d is the height of the beam to diaphragm attachment point. From the elongation, strain values are calculated and substituted in (8) to get the corresponding theoretical resonance frequencies of the T-shaped beam for applied pressure in the range of 0.1 bar to 1 bar.

IV. EVALUATION THROUGH NUMERICAL SIMULATION

The resonating beam attached to the diaphragm as in [9] is modelled using Designer in MEMS CAD tool CoventorWare. The length and width of the resonating beam is optimised to have maximum pressure sensitivity by keeping thickness constant. In the simulations, the length of the beam is varied from 90μm to 140μm in steps of 10μm by keeping the width at 40μm; the pressure sensitivity (%/bar) and strain sensitivity (micro strain/bar) are estimated for an applied pressure range of 0.1 bar to 1 bar. To improve the sensitivity of the beam further, numerical simulations are performed by varying the width of the beam from 40μm to 10μm by keeping the length at 130μm. Pressure sensitivity and strain sensitivity are estimated for the same input pressure range. The maximum pressure sensitivity is obtained for the length of 130μm and width of 10μm as shown in Fig. 2. For this optimum

length and width of the resonating beam, an extension of the same thickness as of the resonating beam is attached, by keeping its other end fixed to any one of the side wall cavity to get the required T-shaped resonating beam.

Modal analysis is performed on the T-shaped beam using Memmech analyzer in Designer module of CoventorWare and the first two mode frequencies under no load conditions are obtained as shown in Fig. 3. It is observed that in the first mode frequency, the beam alone oscillates laterally as required and hence the first mode frequency is preferred for sensor design.

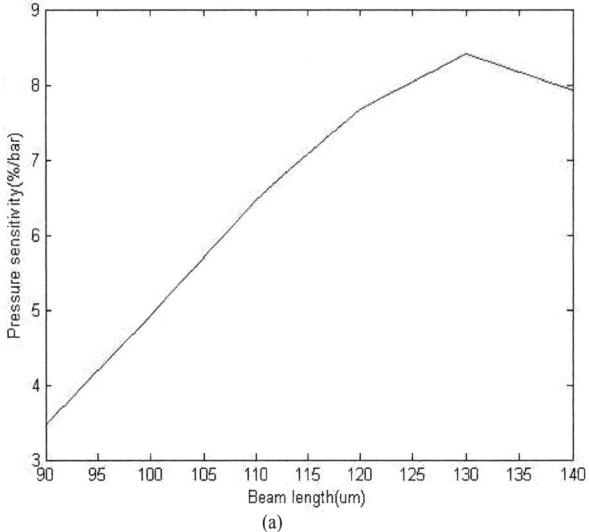

(a)

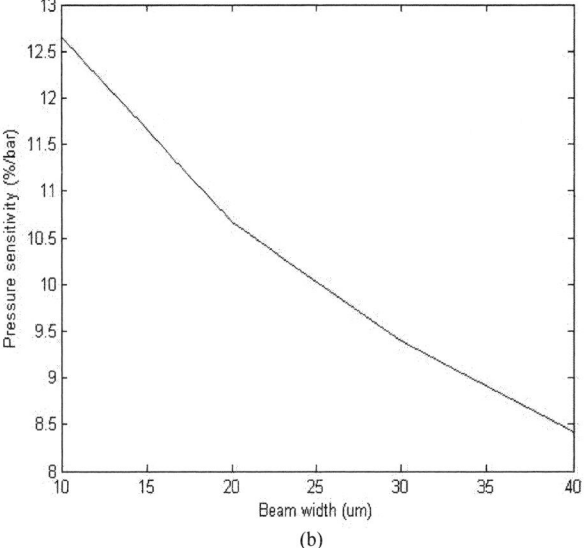

(b)

Fig. 2. Results for pressure sensitivity (Simulation)
(a) For variation in beam length (b) For variation in beam width.

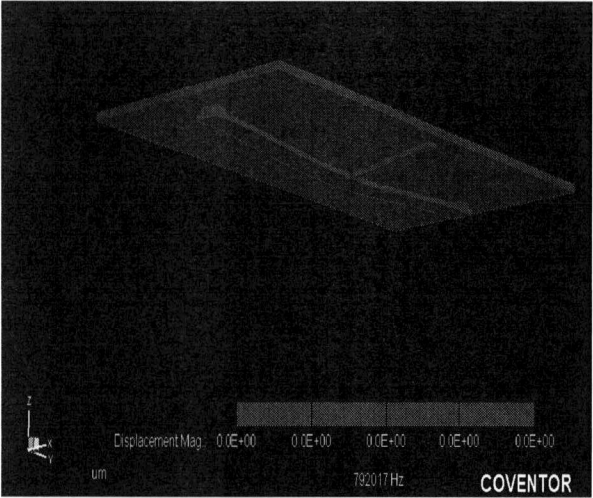

(a) First mode

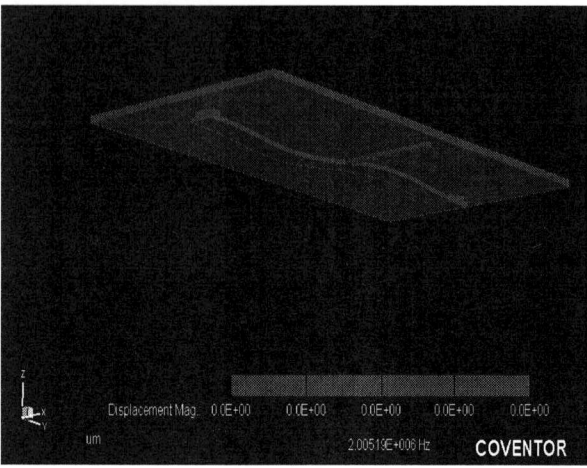

(b) Second mode

Fig. 3. Modal analysis of the T-shaped beam showing the first two modes

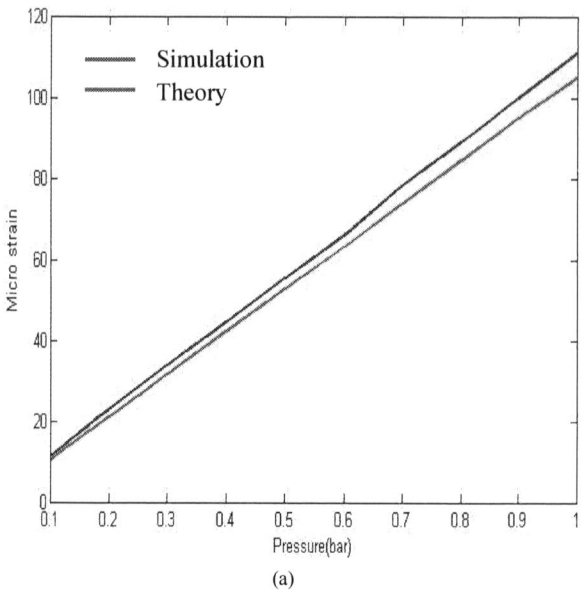

(a)

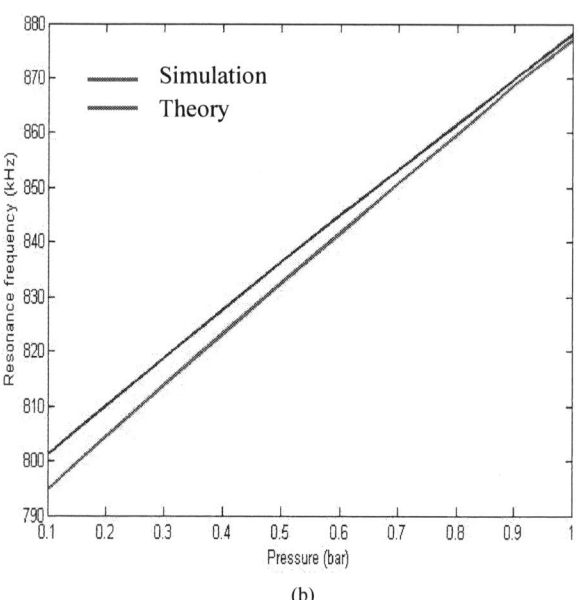

(b)

Fig. 4. Comparison of analytical and simulation results (a) showing variation in strain (b) showing variation in resonance frequency

The variation in resonance frequency and strain in the T-beam is carried out by varying the applied pressure in the range of 0.1 bar to 1 bar in steps of 0.1 bar using modal and stress analysis. The variation in the resonance frequency and strain in the T-shaped beam are estimated and are matching with the resonance frequency and strain values computed as in (8) and (10) as shown in Fig. 4. The variation in resonance frequency obtained in numerical simulation is in agreement with resonant frequency computed analytically. The sensor is having a linearity of 0.0758 % within the specified pressure range and a resolution of 0.0005 bar.

The T-shaped resonating beam, with piezoelectric patch placed on its top surface for sensing, and an electrode for electrostatic actuation is modelled using Saber Architect. The layout of Saber Architect model and the 3D schematic are shown in Fig. 5. Small signal ac analysis is performed under no load conditions and its first mode frequency is found to be 798 kHz which is close to the frequency obtained from the simulation in Designer module and the analytical expression given in (8).

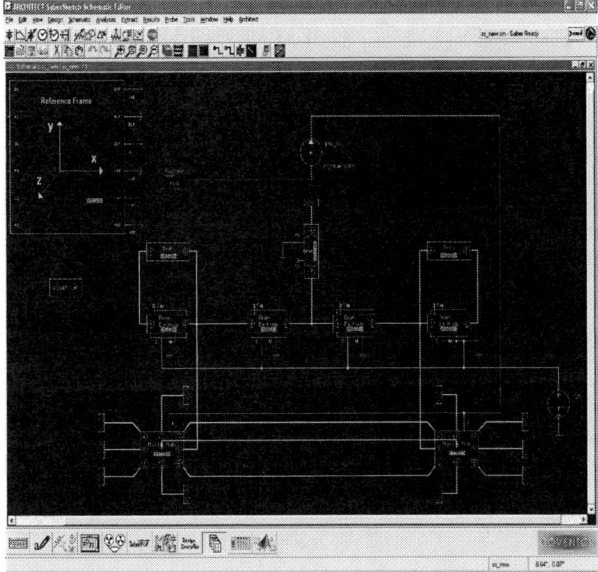

(a)

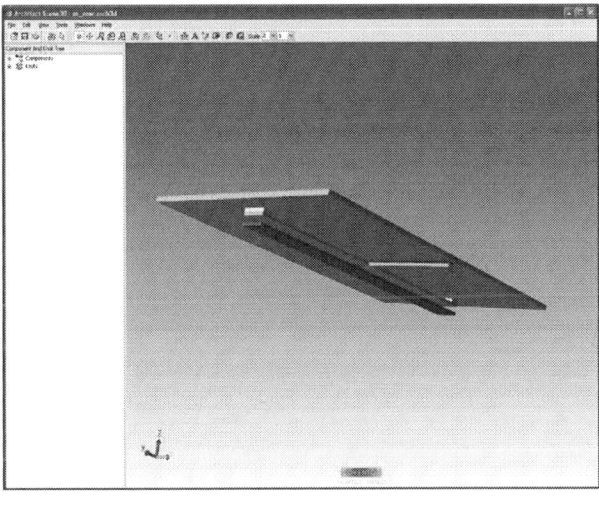

(b)

Fig. 5. Saber architect (a) 2D layout of the sensor (b) 3D schematic of the sensor

V. CONCLUSION

A resonant pressure sensor structure with a rectangular diaphragm and a T- shaped resonating beam is designed, modelled and simulated using MEMS CAD tool CoventorWare and verified theoretically. The sensor design is simple and can be fabricated by available surface micromachining techniques. The resonance frequency of the modified resonating beam is found to be 792 kHz and the sensor pressure sensitivity is found to be 10.2%/bar which is higher than the one reported in [8].

REFERENCES

[1]. Robert B. McIntosh, Philip E. Mauger, and Steven R. Patterson," Capacitive transducers with curved electrodes," *IEEE Sensors Journal*, vol. 6(1),pp.125-138, February 2006.

[2]. Ezzat G. Bakhoum and Marvin H. M. Cheng,"Capacitive Pressure Sensor with very large dynamic range," *IEEE Transactions on components and packaging technology*, vol. 33(1), pp.79-83, March 2010.

[3]. Ezzat G. Bakhoum and Marvin H. M. Cheng ,"A Novel Capacitive Pressure Sensor," *J. Microelectromechanical systems*, vol. 19(3), pp.443-450, June 2010.

[4]. Kalvesten.E, Smith.L, Tenerz.L, and Stemme.G, "The first surface micromachined pressure sensor for cardiovascular pressure measurements, "in *Proc. MEMS 98*, pp.574 – 579, 1998.

[5]. Jacques Mandle,Olivier Lefort and Andre Migeon ,"A new micromachined silicon high accuracy pressure sensor," *Sensors and Actuators* A 46-47, pp.129-132,1995.

[6]. Lee Dong-Weon and Choi Young-Soo ," A novel pressure sensor with a PDMS diaphragm," *J.Microelectronic Eng.* Vol. 85,pp. 1054–1058, January 2008

[7]. Olfatnia.M, Xu.T, Miao.J.M, Ong.L.S, Jing.X.M, Norford.L ," Piezoelectric circular microdiaphragm based pressure sensors," *Sensors and Actuators* A 163, pp.32–36, June 2010.

[8]. Melvas P, Ka¨lvesten. E, Stemme. G , "A Surface micromachined resonant beam pressure sensing structure," *J. Microelectromechanical systems*, vol. 10(4),pp. 498-502, December 2001.

[9]. Sujan.Y, Vasuki.B,Uma.G,Umapathy.M, Harihara Krishnan.S ," Sensitivity Enhancement of Surface Micromachined Resonant Pressure Sensor," *Proceedings of ISSS 2012,* IISc Bangalore, January 2012.

[10]. Suresh.K, Uma.G, Umapathy.M , "A new resonance based method for the measurement of non magnetic conducting sheet thickness," *IEEE Transactions on Instrumentation and Measurement*, vol. 60(12), pp.3892 – 3897, December 2011.

[11]. Santhosh Kumar B.V.M.P, Suresh.K, Varun Kumar.U, Uma.G and Umapathy.M , " Design and simulation of resonance based DC current sensor," *International Journal of Interaction and Multiscale Mechanics*,vol. 3(3),pp. 257-266, 2010.

[12]. Beards.C.F, *Structural Vibration: Analysis and Damping*. Butterworth-Heinemann publishers, Linacre House, Jordan Hill, Oxford OX2 8DP, 200 Wheeler Road, Burlington,1996.

[13]. Meirovitch Leonard, *Fundamentals of vibration*. McGraw Hill publications, Inc. 1221, Avenue of the Americas, New York, NY, 2001.

[14]. Hooke.R, "Approximate analysis of the large deflection elastic behaviour of clamped, uniformly loaded, rectangular plates," *J. Mech. Eng. Sci.*, vol. 11(3),pp. 256–268, November 1969.

[15]. Melvas Patrik, kalvesten Edvard. Enoksson Peter, Stemme Goran, "A free hanging strain- gauge for ultra miniaturized pressure sensors," *Sensors and Actuators* A 97-98, pp.75-88, October 2002.

Tuned Dual Beam Low Voltage RF MEMS Capacitive Switches for X – Band Applications

E.S.Shajahan[1], Shankaranarayana M Bhat[2]
Department of Electronics and Communication Engineering
National Institute of Technology Karnataka
Mangalore, India
[1]shajes2007@gmail.com, [2]msbhat@ieee.org

Abstract—**This paper presents a low voltage, low loss tuned RF MEMS (Radio Frequency Micro Electro Mechanical Systems) capacitive shunt switches for use in X – band. The tunable switch is designed using two shunt beams with meander springs. The switch achieved low actuation voltage along with small up state capacitance. Simulation using CoventorWare shows the actuation voltage as 7.5 Volts and up state capacitance of 47fF. HFSS simulation reveals the insertion loss in the range of (0.1 – 0.2) dB and up state return loss better than -25 dB in the X-band (8-12 GHz). The switch offers down state isolation of 60 dB at 12 GHz and is better than 40 dB in the frequency range 8-25 GHz.**

I. INTRODUCTION

Recent research has demonstrated beyond doubt the role of low loss RF MEM (Radio Frequency Micro Electro Mechanical) switches in millimeter and microwave applications [1]. This is because of the numerous advantages RF MEMS switches have over their semiconductor counterparts (PIN and FET switches) in terms of their low power dissipation, transmission loss, high quality factor, better isolation and very low inter modulation products [2]. Disadvantages include low switching speeds (10-20μs), high actuation voltages (15-60 V) and hot switching in high power applications and failure due to stiction phenomenon. Despite better performance over other competing technologies such as PIN or FET switches, reliability issues hamper the commercialization of RF MEMS capacitive switches.

This study discusses the design of low voltage, low loss tuned Micro Electro Mechanical (MEM) switches for use in X – band. To reduce the actuation voltage, low spring constant meander springs have been employed to connect the capacitor plates to the anchors. Vast majority of the switches in the literature typically require pull-in voltage of the order of 40-100 V. These ranges are quite challenging for handheld devices and wireless devices that rely on low voltage power supplies. Goldsmith *et al.* [4] have experimentally observed significant lifetime improvement for every volt drop in pull-in voltage. Thus, reducing the actuation voltage not only broaden the range of applications but also improves their performance.

II. DESCRIPTION OF THE X – BAND RF MEM SWITCHES

The dual beam tuned switch is constructed over CPW transmission line with a characteristic impedance of 50 Ω. The tuned switch is designed to be a low voltage, low loss and high isolation RF switch for operation in the X – band (8-12 GHz).

The switch is found to have satisfactory RF characteristics in K_u band as well. This tunable switch design employs two thin metal membranes with meander springs for switching action. The MEMS switches are separated by a high impedance transmission line of characteristic impedance 60 Ω. The transmission line separating the switches can be made shorter if its impedance is higher than at the ports. The length as well as the impedance of the transmission line can be adjusted such that the reflections from the two switches are out of phase and thus gets cancelled at the input port thereby improving the up state return loss.

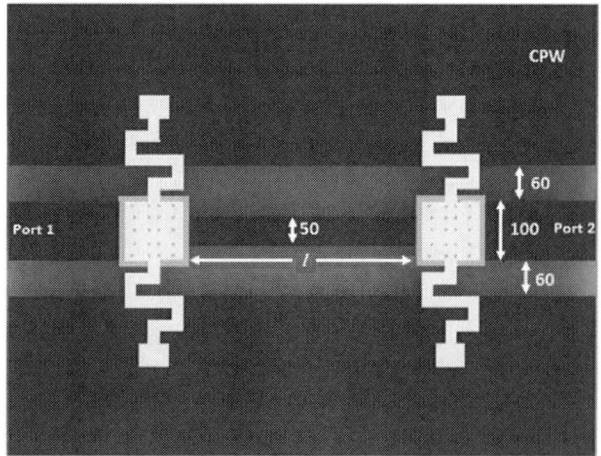

Fig. 1. Dual beam tuned switch – top view

Fig. 1 shows the top view of the dual beam switch. Meandered beams have been used not just to lower the actuation voltage alone but to lower the LC resonant frequency to the X-band. In MEM switches, thin metal membranes are made to shunt the CPW transmission line. A thin dielectric layer over the transmission line and below the beam provides a capacitive path for the RF signal to ground. Unlike in solid state switches, here in MEMS, switching action is due to the mechanical deformation of the metal membrane.

III. SPRING CONSTANT AND ACTUATION VOLTAGE

A. Switch Design

The mechanical design of the electrostatically actuated switches is dictated by the required actuation voltage. When

978-1-4673-2395-6/12 $31.00 © 2012 IEEE

designing switches with low actuation voltage, the choice of membrane material and of the support design is critical. Equation (1) showed below presents a widely cited formula for calculating the pull-in voltage (V_p) for fixed-fixed beams [1]

$$V_P = \sqrt{\frac{8 K_{eff} g^3}{27 \varepsilon_0 A}} \qquad (1)$$

K_{eff} is the effective spring constant of the movable structure, g is the air gap, ε_0 is the free-space permittivity and A is the switch area. A meandered spring supported beam shown in Fig. 3 is used to reduce the effective spring constant. DC simulation shows the pull-in voltage as 7.5 V

B. Spring Design

As shown in Fig., the switch is connected to the anchors through four meander springs, two on each side. The effective spring constant, K_{eff}, of the entire MEMS switch can be determined by combining the simple spring equations. Since the mathematical details have been analyzed in [1] and [2], just the equations are presented here. Fig. 2 shows the dimensions of single meander spring of spring constant, K_m

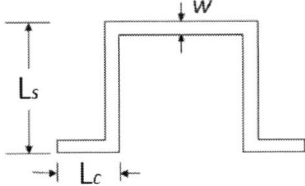

Fig. 2. Meander spring dimensions

For the meandered spring, spring constant, K_m [1] is given by

$$K_m = \frac{E w \left(\dfrac{t}{L_c}\right)^3}{1 + \dfrac{L_s}{L_c}\left[\left(\dfrac{L_s}{L_c}\right)^2 + 12 \dfrac{1+v}{1+\left(\dfrac{w}{t}\right)^2}\right]} \qquad (2)$$

where E is Young's modulus, v is poisons ratio, w is width of meander, L_s is overall width of meander and L_c is the distance from end of spring to start of meander. Non meandered spring constant K_{n-m} [2] is given by

$$K_{n-m} = 32 E W \left(\frac{t}{L}\right)^3 \qquad (3)$$

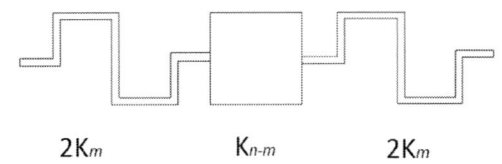

$2K_m \qquad K_{n-m} \qquad 2K_m$

Fig. 3. Switch with meanders

The effective spring constant, K_{eff} of the membrane shown in Fig. 3 can be derived as in [2]

$$K_{eff} = \frac{K_m K_{n-m}}{K_m + 4 K_{n-m}} \qquad (4)$$

All the necessary dimensions and material constants of this design are given in Table I and Table II

TABLE I

PARAMETERS AND DIMENSIONS OF THE SWITCH

Parameters	Values
CPW dimensions (µm)	60/100/60
Width of the membrane W (µm)	100
Distance between membranes X (µm)	400
Dimension of Dielectric (µm)	120x120
Dielectric constant of Si_3N_4 (ε_r)	7.4
Thickness of membrane, t (µm)	1.0
Thickness of Dielectric, T (µm)	0.1
Air gap, g (µm)	3.0
Relative dielectric constant of Si substrate	11.7
Young's modulus of gold (E)	80 GPa
Poisson's ratio of gold (v)	0.44

TABLE II

DIMENSIONS OF THE MEANDER SPRING

Dimension	Values
Overall width L_S (µm)	60
Length L_C (µm)	20
Width of the spring w (µm)	20
Thickness of the meander spring t (µm)	1
Length of the non-meander spring L (µm)	120

IV. CIRCUIT MODEL OF THE TUNED SWITCH

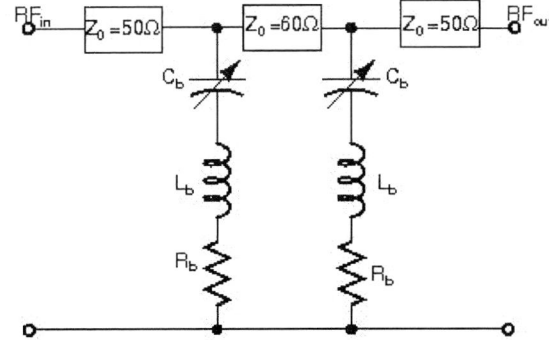

Fig. 4. CLR circuit model of the dual beam switch

978-1-4673-2395-6/12 $31.00 © 2012 IEEE

Fig. 4 shows the circuit model of the two beam switch. The model is constructed using two 50Ω transmission lines at the ports, two shunt CLR branches for modeling the beams and a short section of high impedance transmission line of 60Ω for separating the shunt branches. In this design, meanders supporting the beam introduce additional inductance thus increasing the beam inductance. To operate the switch in X–band, the LC resonant frequency is decreased by increasing the beam inductance. The LC resonant frequency is given by [5]

$$f = \frac{1}{2\pi\sqrt{L_b C_b}} \qquad (1)$$

Resonance frequency in the vicinity of 35GHz for single beam switches is brought down to X - band frequency by employing meandered spring beam geometry. This is inductive tuning [7]. The switch is designed for a down capacitance C_d of 6.7 pF and the extracted beam inductance L_b from the isolation characteristics is 24pH. Thus it is evident that LC resonant frequency can be reduced by increasing bridge inductance even if the capacitance is kept the same due to process limitations. Increased inductance increases Q resulting in reduced bandwidth and increased isolation around the resonant frequency

V.   S- PARAMETERS OF THE SWITCH

The simulated S-parameters of the designed dual beam switch are presented in Fig. 5- Fig.7. Simulation has been carried out for various physical lengths of the high impedance

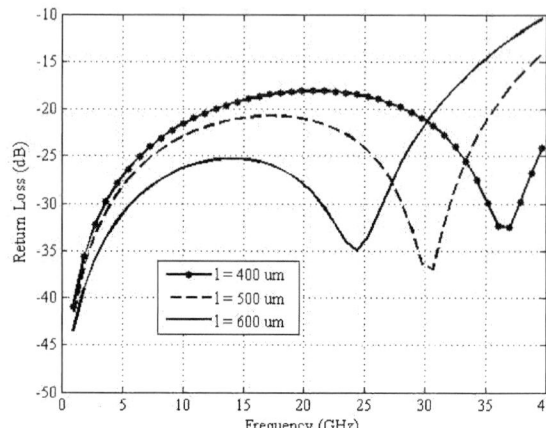

Fig. 6.   Return Loss in Up- state

transmission line between the membranes. Table III shows the results of the RF and DC simulation. The insertion loss (S_{12}) of the switch in the up state is shown in Fig. 5. S_{12} is 0.1-0.15 dB in X-band for a high impedance transmission line length of 600 μm between the membranes. This dual beam design gives a reflection coefficient of (S11 in up-state)   less than -25 dB in X- band. The frequency of cancellation of reflections from the switches for transmission line lengths of 400, 500 and 600 μm are found to be 24.4, 30.7 and 33.4 GHz respectively. This is illustrated in Fig. 6.

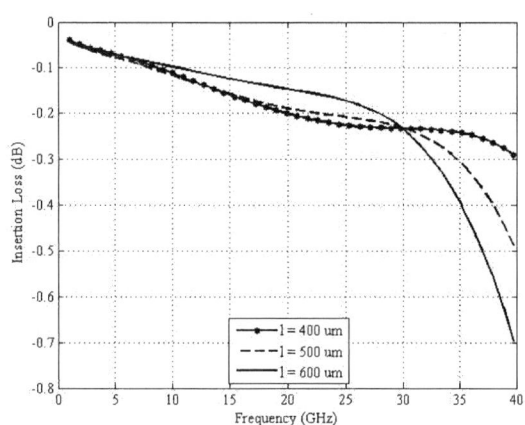

Fig. 5. Insertion Loss in Up- state

TABLE III
S - PARAMETERS OF THE DESIGNED SWITCH

S11 Up-State	S12 Up-State	S12 Down-State	Frequency GHz	Pull-in voltage
> 25 dB	< 0.2 dB	> 45 dB	8-20	7.5 V

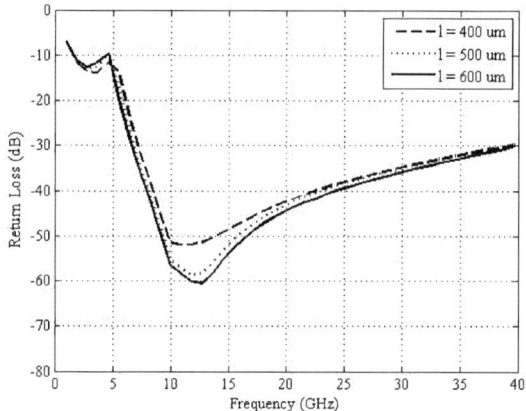

Fig. 7. Isolation

A capacitance ratio (C_d/C_u) of over 100 is achieved with 3 μm air gap between the membrane and transmission line which results in isolation better than -50 dB in X-band. The isolation of the switch is limited by the downstate capacitance (C_d) for frequencies less than LC resonance and by the beam inductance (L_b) for frequencies higher than the resonant frequency. The bandwidth of the switch is specified by isolation performance dictated by these values.

VI. Loss Characteristics with Substrate Resistivity

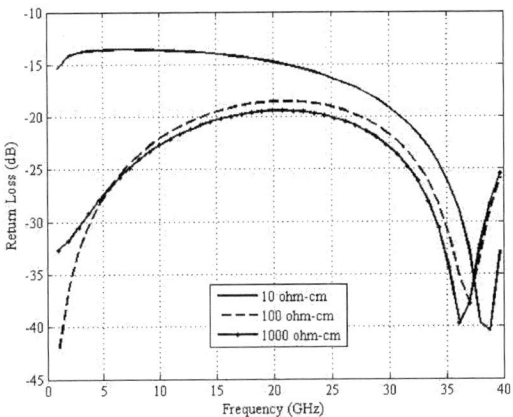

Fig. 8. Return Loss with substrate resistivity

Insertion loss in RF MEMS switches includes substrate loss, conductor loss due to skin effect and reflections due to mismatch. Radiation losses are significant only when the device dimensions become comparable to one tenth of the RF signal wavelength. Conductor loss due to skin effect can be minimized by making the CPW conductor thickness greater than twice the skin depth. Since skin depth of gold at 10 GHz is 0.7 μm, CPW conductor thickness is chosen to be 2μm in this work. It has been observed from simulation that substrate resistivity influences the S–parameters. It is highly likely that the signal penetrates deep into the substrate and gets absorbed if it is highly doped. This is substantiated by the simulation results. The total loss tangent [3], $\tan \delta_t$, of a lossy dielectric medium is given by

$$\tan \delta_t = \tan \delta_l + \tan \delta_d$$

$\tan \delta_l$ is the intrinsic loss from the polarization loss of the intrinsic silicon substrate and $\tan \delta_d$ is the extrinsic loss due to the finite conductivity of the silicon substrate.

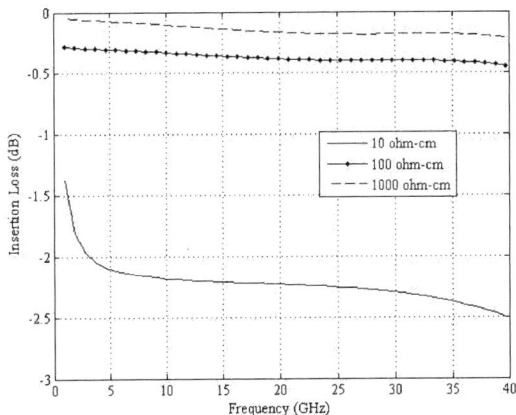

Fig. 9. Insertion loss with resistivity

Therefore, the total loss increases with conductivity of the substrate. The insertion loss is 2-2.5 dB and up-state return loss is better than -15 dB with a substrate resistivity of 10 Ω-cm and the insertion loss improves to 0.1-0.2 dB with a return loss better than -20 dB when the substrate resistivity is increased to 1000 ohm-cm as shown in Fig.6 - Fig.7. From these results, it is inferred that for millimeter wave (mm-wave) switching applications, low loss, and low dielectric constant substrates like glass or quartz would yield better loss characteristics. Another possible alternative is to use micro-machined high resistive silicon (1000 Ω-cm) [6].

VII. Conclusion

A low voltage low loss dual beam tuned RF MEM capacitive shunt switch suited for applications in the X- band (8-12 GHz) is presented. Meandered spring supported beam is used to lower the actuation voltage (7.5 V) and the increased beam inductance reduced the LC resonant frequency to X-band range. Thus actuation voltage of 7.5 V is achieved with an air gap of 3 μm between the CPW conductor and the membrane. The length of the high impedance transmission line separating the beams is fixed to be 600 μm for better RF characteristics in the X-band.

RF simulation shows that the designed tuned switch could achieve high isolation in down state and better return loss in upstate. Results reveal an insertion loss of 0.1-0.2 dB and return loss in excess of -25 dB in up-state and isolation better than -50 dB in X- band. Possible reasons for observed losses are discussed in detail. Detailed RF simulations are being carried out to arrive at better results. Work is also underway with regard to fabricating the device with the final goal of demonstrating the switch for real applications. The results of these studies will be reported in literature in near future.

References

[1] Gabriel M. Rebeiz, "RF MEMS: Theory, Design and Technology," 2003 John Wiley & Sons, Inc.

[2] Gere J., "Mechanics of materials," Fifth Edition. Thompson – Engineering, 2003.

[3] M.S.Giridhar, et al, "Design Fabrication and Testing of a Bulk Silicon Micromachined RF MEMS switch," International conference on Smart material Structures and Systems January 04-07 2012,Bangalore,India

[4] C. Goldsmith, J. Ehmke, A. Malczewski, B. Pillans, S. Eshelman, Z. Yao, J. Brank, and M. Eberly, " Lifetime characterization of capacitive RF MEMS switches," IEEE MTT-S Int.Microwave Symp.Dig.,vol1,June 2001,pp.227-230.

[5] Jeremy B. Muldavin, Gabriel M. Rebeiz, "High-Isolation CPW MEMS Shunt Switches---Part 1: Modeling," IEEE Transaction on Microwave Theory and Techniques," Vol., 48, No.6, June 2000

[6] Takehiko Makita, Isao Tamai, Shohei Seki, "Coplanar Waveguides on High-Resistivity Silicon Substrates with Attenuation lower than 1 dB/mm for Microwave and Millimeter-wave bands," IEEE Transactions on Electronic Devices, Vol.,58, No.3, March 2011

[7] K. Topalli, et al. "Empirical formulation of Bridge Inductance in Inductively Tuned RF MEMS Shunt Switches," Progress In Electromagnetic Research. PIER 97, 343-356, 2009 5-8, 2009.

Magnetic Force on a Magnetic Bead

Alireza Bahadorimehr, Jafar Alvankarian, Burhanuddin Yeop Majlis

Institute of Microengineering and Nanoelectronics (IMEN)
Universiti Kebangsaan Malaysia (UKM)
43600 UKM Bangi, Selangor, Malaysia
Email: burhan@vlsi.ukm.my

Abstract- **Micron-size magnetic particles are utilized in different bio-applications. Many in-vivo and in-vitro analysis use these particles because of their interesting abilities to attach to biomolecules. The magnetically actuation of magnetic beads is important to confine the area of the specified cell analysis. Electromagnets and permanent magnets are used for different applications to capture these particles. Therefore exerting optimum force on magnetic particles is crucial in micron size analysis. In this paper we present the magnetic force calculation on a single magnetic bead using magnetic related equations from both permanent and electromagnet sources. It is shown that the magnetic force exerted on larger particles is more than smaller ones.**

I. Introduction

Magnetism and microfluidics concepts are not new fields; however, they have been combined in recent years. In microfluidics, electric fields have long been utilized in different literatures for separation, trapping, pumping and many other purposes [1]. With advent of microfluidic systems, nanomaterials and nanoparticles have become an amazing research trend. Among all these particles, the functional magnetic micro and nanoparticles have been widely utilized for diagnostic and therapeutic applications. Magnetic forces are now being used in several microfluidic applications. In biomedical applications, magnetic microparticles mostly are used in the form of magnetic beads which have been polymerized. These beads consist of nanoparticles attached together in a suitable polymer matrix. The ability of the magnetically manipulation of these particles are used to investigate various bioreactions by keeping them stable using permanent magnets or electromagnets. The selection of the type of magnetic beads, their size, and their polymer matrix depends to the biomolecule of interest (e.g., specific cells, proteins or DNA sequence) and proteins that are investigated. In the fabrication and design of these beads, biocompatibility and biodegradability are considered as two of the main factors. Meanwhile, the size and shape of the magnetic beads are controlled for specific applications.

Magnetic nanoparticles offer attractive advantages for their applications in biomedicine [2]. First, their sizes could be controlled in the range of a few nanometers to tens of nanometers, so they could be smaller than or comparable to a cell (10–100 μm), a virus (20–450 nm), a protein (5–50 nm) or a gene (2 nm wide and 10–100 nm long). Therefore they can get close to, or enter a biological entity of interest.

Permanent magnets or electromagnets are utilized for trapping of the functionalized particles in desired points in in-vivo and in-vitro

applications. Various fabrication processes are applied using different substrate materials to produce accurate magnetic field generation systems. The magnetic field magnitude and its gradient are important factors in trapping of the magnetic particles. The increasing distance from the magnet source, permanent magnet or electromagnet, reduces the magnetic field rapidly. Therefore, an accurate calculation is required for obtaining desired magnetic force on the particles in the fluidic vessels, capillaries, tissues, or microchannels.

In this paper the magnetic force on micro/nano particles using combined Finite Element Method (FEM) and analytical methods are discussed in details.

II. Theory

To realize how a magnetic field is used to transport and manipulate magnetic particles, it is essential to understand that a magnetic field gradient is required to exert a translation force on particles (a uniform field gives rise to a torque, but no translational action) [2].

The magnetic force on a particle depends on the magnetic moment m and gradient of the magnetic flux density B and is calculated according to $\vec{F}_{mag} = \nabla(\vec{m}.\vec{B})$. For small magnetic particles with single domains the magnetic force is determined by the absolute value of the field gradient [3].

$$\vec{F}_{mag} = \left(\left(m.\frac{\vec{B}}{B} \right).\nabla \right) \tag{1}$$

$$\vec{F}_{mag} = m.\nabla B \tag{2}$$

$$m = M_s V \tag{3}$$

where m and B are the absolute values of $\vec{m}$ and $\vec{B}$, V is the volume of the particle. The magnetization M_s for magnetite, Fe_3O_4, and maghemite, γ-Fe_2O_3, are 480 kA/m and 380 kA/m respectively. For instance, magnetite particles with 10nm diameter and magnetic flux density gradient of 110 T/m experience the force of 2.8×10^{-17} N.

$$\vec{F}_{mag} = M_s V \nabla B = 480000[A/m]\frac{4}{3}\pi(5\times10^{-9}[m])^3 \times 110[T/m]$$

$$= 2.8\times10^{-17}[A.mT] = 2.8\times10^{-17}[A.m.\frac{N}{A.m} = N]$$

However, for larger particles the magnetic force on magnetic particles is described using electromagnetic field theory. In this case, the magnetic force on a particle in a magnetic field depends on the volume of particle (V), the magnetic susceptibility difference between the particle and its surrounding medium ($\Delta\chi$), amplitude and the gradient of magnetic flux density which is written by [4]:

$$\vec{F}_{mag} = \frac{V \Delta \chi}{\mu_0} (\vec{B} \cdot \vec{\nabla}) \vec{B} = \frac{V \Delta \chi}{2\mu_0} \vec{\nabla} (\vec{B} \cdot \vec{B}) \qquad (4)$$

For instance, when the magnetic particles move in a fluid channel in the x direction, the x component of the force becomes:

$$F_{mag,x} = \frac{V \Delta \chi}{\mu_0} \left(B_x \frac{\partial}{\partial x} + B_y \frac{\partial}{\partial y} + B_z \frac{\partial}{\partial z} \right) B_x \qquad (5)$$

Which in circular case with r-z plane can be written as [5]:

$$F_{mag,r} = \frac{V \Delta \chi}{\mu_0} \left(B_r \frac{\partial B_r}{\partial r} + B_z \frac{\partial B_r}{\partial z} \right) \qquad (6)$$

From these equations it is obtained that in a homogeneous magnetic field where the gradient of magnetic field is zero, there is no force exerted on particles (Magnetic field of a Helmholtz coil in the uniform regions cannot be used for magnetic bead capturing purposes). The forces on magnetic beads are typically in the range of few tens of fN to few tens of pN.

In these equations the term $\Delta\chi = \chi_p - \chi_m$ is the difference between magnetic susceptibility of the particle, χ_p, and its surrounding buffer, χ_m. The magnetic susceptibility of diamagnetic objects is negative. Therefore, for a diamagnetic particle ($\chi_p < 0$) in a diamagnetic medium ($\chi_m < 0$) the term $\Delta\chi$ can be positive or negative which cause the particle to repel from or attract to the magnetic field respectively. In many cases, water is used as the buffer for magnetic particles ($\chi_{water} = -9 \times 10^{-6}$). The magnetic susceptibility of water is negligible compare to paramagnetic materials with $\chi_p > 0$. Ferromagnetic materials have the largest magnetic susceptibility ($\chi_p >> 0$) and are attracted easily to magnetic fields.

It is obvious that by calculation of the magnetic flux density of a magnet source, the force on particles is obtained using these equations. Here, for permanent and electromagnets, FEM is used for calculation of magnetic field.

II. Magnetic field of a coil

Due to special structure of current carrying planar microcoils, precise calculation of the magnetic field is complicated and time consuming. The finite element method (FEM) is employed to calculate the magnetic field. The forces on magnetic particles from this analysis are analytically calculated using different equations. This magnetic field problem is a nonlinear Maxwell's equation for magnetic flux density calculations. For instance, in spiral planar coils, the coil is replaced by a concentric coil and the effect of each of these separate coils is considered for the entire device. Therefore, as a consequence of the nonlinearity of this problem, the superposition principle which has been used in [6] cannot be applied.

A. Single turn coil

A single turn circular coil with a current of 1A and square cross section of 25 μm by 25 μm with radius of 300 μm is considered in the following 3D simulation (Fig. 1 and 2).

To estimate the magnetic force on beads, when one turn of coil is utilized, the magnetic field at the middle of four finite square conductors is calculated by the Biot–Savart law (Fig. 3 (a)). A single turn circular current loop is shown in Fig. 3 (b). In microsystems, the

scale of the section of conductors becomes comparable with that of the device, so that the conductor is no more considered as a line. Thus, the section is divided to segments, and the magnetic field is calculated by integrating each contribution. Biot-Savart law is used for calculation of magnetic flux density produced by a short segment of wire which is formulated as:

$$d\vec{B} = \frac{\mu_0 I}{4\pi} \left(\frac{d\vec{l}' \times \vec{R}}{R^3} \right) \qquad (7)$$

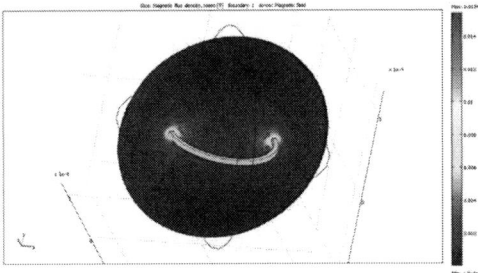

Fig. 1. Single turn coil magnetic flux density at 100 μm above coil

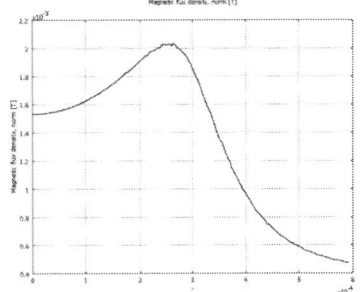

Fig. 2. Magnetic flux density at 100 μm distance above coil

where $\mu_0 = 4\pi \times 10^{-7}$ (H/m) is the permeability of vacuum, I is the current passing the wire segment, $d\vec{l}'$ is the wire segment in direction of current, $\vec{R}$ points from the short segment of wire to the observation point.

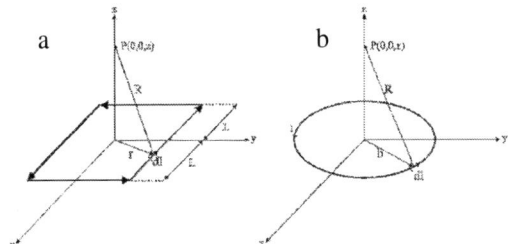

Fig. 3. Square and circular current loop for magnetic field calculation

The magnetic flux density of single turn square and circular loops can be calculated using:

$$\vec{B}_{square\_coil} = \vec{a}_z \frac{2\mu_0 I L^2}{\pi (L^2 + z^2) \sqrt{(2L^2 + z^2)}} \qquad (8)$$

$$\vec{B}_{circular\_coil} = \vec{a}_z \frac{\mu_0 I b^2}{2(z^2 + b^2)^{3/2}} \qquad (9)$$

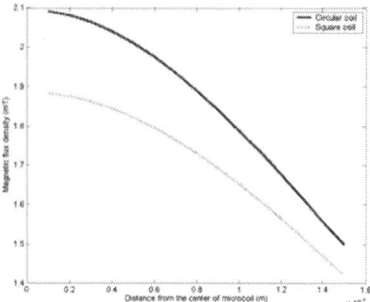

Fig. 4. Calculated magnetic flux density of square and circular coils at their center line

In (8) and (9), for a single turn microcoil with I=1 A current, the magnetic flux density at the distance of 100 μm above the coil surface (z=100 μm) and at the center of it (x=0, y=0) is 1.65 mT and 1.79 mT for square and circular coils respectively (L=300 μm, b=300 μm). Magnetic flux density decreases quickly with increasing distance from the microcoil. Results of analytic calculation with MATLAB software is illustrated in Fig. 4. The slopes of the graphs with data fitting option in MATLAB and first derivative of the fitting equation are -5.5 T/m and -4.5 T/m for circular and square loops respectively after the distance of 100 μm from the coil where the graphs are approximately linear. It means that by increasing of each 100 μm distance from the center of the coil, the magnetic flux density reduces 550 μT and 450 μT for circular and square coils respectively. It is obvious that magnetic flux density at the center of circular coils is more than square loops. However, the reduction rate of the magnetic flux density in square types is less than circular types.

B. Multiple turn coil

For a circular multiple turn coil we used a 2D axisymmetric simulation in COMSOL software. In this simulation each coil turn has its specific loop voltage for obtaining 1 A current for all the turns by using the well-known $R = \rho \ell / A$ equation and the revolving option in the software.

These points show the increase in one of the terms in (6). Magnetic flux density from the microelectromagnet coil decreases quickly with increasing distance from the coil surface. The magnetic field reduction rate from center to the outside areas of the coil is shown in Fig. 6 (b). This figure displays the magnetic flux density variations along the channel at 50 μm above the coil (Fig. 6 (a) and (b)). It shows that the local maxima and minima points are based on the value of magnetic flux density along the channel in these areas. Moreover, the reason of the maximum force at the center regions is the maximum magnetic flux density in z direction in these areas since the B_r part is zero. In the center area of the microcoil the term $B_z \frac{\partial B_r}{\partial z}$ from (6) is at its maximum, however, in each local point above the coil turns, $B_r \frac{\partial B_r}{\partial r}$ is predominant. Therefore, above each turn and also at the center of the mocrocoil there are maximum magnetic flux points. The number of points that magnetic particles can be captured is reduced by increasing the distance from the coil surface which is one of the significant factors in thickness optimization of layers in between.

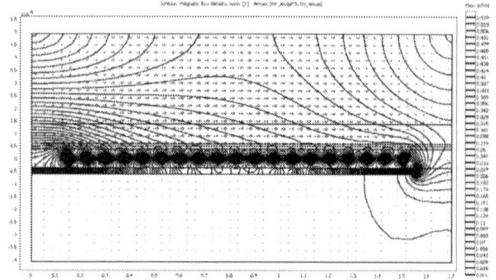

Fig. 5. Simulation results for magnetic flux density

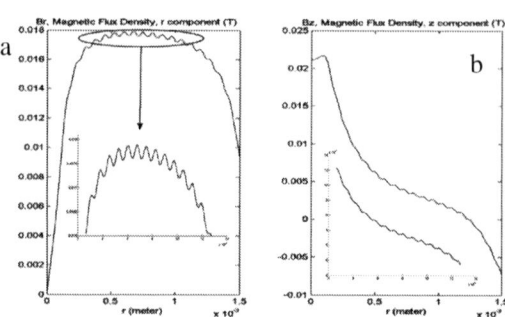

Fig. 6. Magnetic flux density along the channel at 50 μm distance from the coil(a) r-component (b) z-component

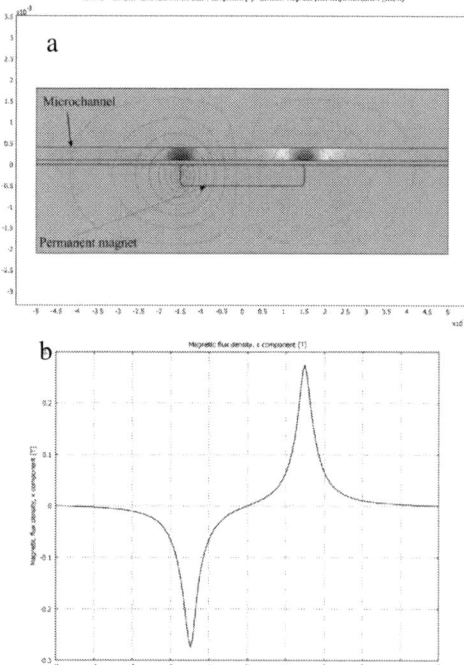

Fig. 7. (a) Simulation results for a permanent magnet beneath the microchannel, (b) Magnetic flux density

III. Magnetic field of a permanent magnet

In many researches permanent magnet is used because of its high magnetic field. The Neodymium magnets are usually used in capturing different micro and nano particles. Here, the simulation of rare earth neodymium-iron-boron (NdFeB) permanent magnet is performed to compare the difference between microcoil magnetic field strength and strong permanent magnets. The fields due to the permanent magnet are calculated using FEM. Then, forces on particles based on the fields are calculated using (5). In Fig. 7(a) the permanent magnet is placed 50 μm beneath the microfluidic channel. As shown in this figure, the magnetic field is maximum at the edges of the magnet and reduced at the center.

Fig. 7(b) illustrates the magnetic flux density and its gradient at 100 μm above the magnet surface. The magnetic flux density can reach up to 280 mT at the edges, where in microcoil this is 18 mT above the coils and 22 mT at the center. The magnetic flux density has its minimum value ($B_x=0$) at the center of the magnet. Moreover, the gradient of magnetic flux density can reach up to 900 T/m at the edges and 20 T/m at the center.

IV. Magnetic force on a magnetic bead

Two different types of magnetic beads with 2.8 μm (Fig. 8 (a)) and 750 nm (Fig. 8 (b)) diameters from Dynabeads and chemicell were selected for this simulation. The magnetic force on two types of magnetic beads with diameters of 2.8 μm and 750 nm are shown in Fig. 9 (a) and (b) respectively. In this figure a multiple turn coil with 19 turns and 1 A current with circular cross section wire with diameter of 25 μm is utilized. For magnetic force calculation, the magnetic field was simulated using COMSOL software and subsequently using (6) the force was calculated. It is obvious that the force on larger particles is much higher than smaller ones. One of the reasons of this result is due to the volume of large particles which are much more than smaller ones which has direct effect on force according to (6). The volumetric magnetic susceptibility is another factor that leads force increase on larger particles. The same results are shown in Fig. 10 for NdFeB permanent magnet.

In Fig. 9 the force on 750 nm and 2.8 μm particles are illustrated. The magnetic force on 2.8 μm particles at 200 μm distance from the magnet surface is shown in Fig. 9 (a). The maximum force is at the edges of the magnet which is 55 pN and reduces to 0 at the center of the magnet. It shows that this force is too much larger (40 times) than the force exerted on particles in the maximum points in microcoil. However, at the centers of the channel electromagnetic coil force surpasses from the permanent magnet. The same results for 750 nm particles are shown in Figure 9 (b). The capturing efficiency is much higher than the microcoil due to the use of strong rare magnets. Again, it is obvious that in both types of magnets the larger particles experience maximum force.

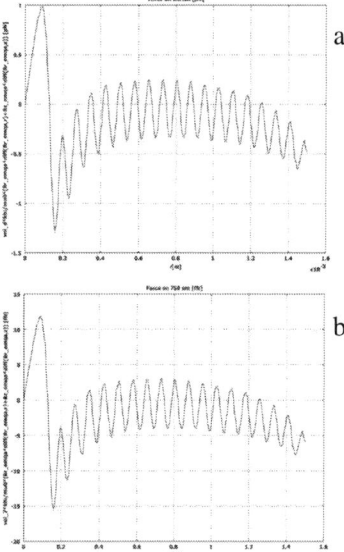

Fig. 9. Force on particles with diameters of (a) 2.8 μm, (b) 750 nm at 50 μm above the microcoil surface

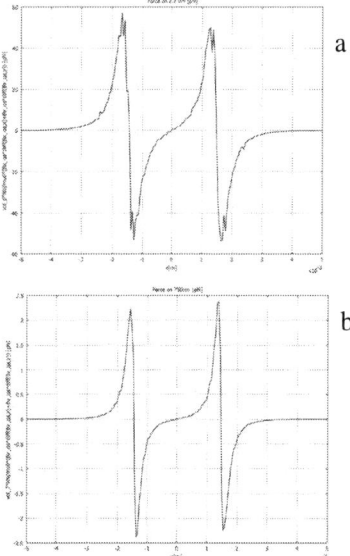

Fig. 10. Force on particles with diameters of (a) 2.8 μm (b) 750 nm, in 200 μm above the surface of the permanent magnet

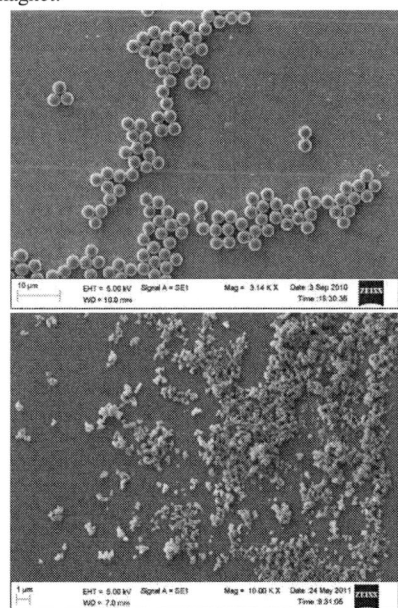

Fig. 8. SEM image of magnetic beads with diameters of (a) 2.8 μm Dynabeads and (b) 750 nm Chemicell

V. CONCLUSION

In this paper, the magnetic force on micro/nano particles using combined FEM and analytical methods was discussed in details. A single and multiple turn coil were presented and the forces on particles was calculated. Furthermore, permanent magnet was simulated and the magnetic flux density and its gradient for magnetic force calculation were described. The larger force on larger particles was observed using the Maxwell equations because of their larger volume and higher magnetic susceptibility.

REFERENCES

[1] P. Gravesen, *et al.*, "Microfluidics-a review," *Journal of Micromechanics and Microengineering*, vol. 3, p. 168, 1993.

[2] Q. A. Pankhurst, *et al.*, "Applications of magnetic nanoparticles in biomedicine," *Journal of Physics D: Applied Physics*, vol. 36, p. R167, 2003.

[3] T. H. Boyer, "The force on a magnetic dipole," *American Journal of Physics*, vol. 56, pp. 688–692, 1988.

[4] M. A. M. Gijs, "Magnetic bead handling on-chip: new opportunities for analytical applications," *Microfluidics and Nanofluidics*, vol. 1, pp. 22-40, 2004.

[5] J. W. Choi, *et al.*, "A new magnetic bead-based, filterless bio-separator with planar electromagnet surfaces for integrated bio-detection systems," *Sensors and Actuators B: Chemical*, vol. 68, pp. 34-39, 2000.

[6] A. Beyzavi and N. T. Nguyen, "Modeling and optimization of planar microcoils," *Journal of Micromechanics and Microengineering*, vol. 18, p. 095018, 2008.

Polycrystalline p-β-FeSi$_2$(Al) on n-Si(100): Heterojunction Thin-Film Solar Cells

A. Bag[*], S. Mallik, C. Mahata and C. K. Maiti

Dept. of Electronics and ECE, Indian Institute of Technology,
Kharagpur 721302, India
Phone: +91-3222-281475 Fax: +91-32222-55303 [*]Email: abag@iitkgp.ac.in

Abstract- **Photovoltaic properties of solar cells based on aluminum (Al) alloyed polycrystalline p-type β-phase iron disilicide [p-β-FeSi$_2$(Al)]/n-type Si(100) bi-layer heterojunctions are reported. Current density-voltage and photo response characteristics were measured at room temperature. Short circuit current density of ~26 mA/cm^2 and open-circuit voltage of ~335 mV were obtained for p-β-FeSi$_2$(Al)/p$^+$- Si/n-Si(100) cells with indium-tin-oxide (ITO) as top electrode. Under air mass (AM) 1.5 illumination, the cell showed a conversion efficiency of 3.0%. The device showed a series resistance of 103.95 Ω, a high shunt resistance of 761 Ω, and an ideality factor of 1.135, resulting in a fill factor of 34.4%.**

I. INTRODUCTION

Semiconducting β-FeSi$_2$, made from ubiquitous and nontoxic materials, is expected to be used in optical devices including thin film solar cells as it has a high absorption coefficient (α >10^5 cm^{-1}, at photon energies higher than 1 eV), a direct energy band-gap of 0.85 eV, chemical stability at temperatures as high as 937 °C, high thermoelectric power (Seebeck coefficient of k ~ 10^{-4} /K) and high resistance against the humidity, chemical attacks and oxidization [1-7]. It is a novel photovoltaic material with reported theoretical energy conversion efficiency about 16-23% [8]. However, up to now, there are only a few reports on β-FeSi$_2$ solar cells. The highest conversion efficiency was reported only 3.7%, obtained from a β-FeSi$_2$/Si heterojunction formed by a facing-target sputtering (FTS) technique on p-Si(111) [9].The poor features of β-FeSi$_2$ solar cells might be due to the low quality of β-FeSi$_2$ films and rough interface between the β-FeSi$_2$ films and the Si substrate, which is mainly induced by the significant interdiffusion of Fe and Si atoms during β-FeSi$_2$ formation. It is well known that the diffused Fe impurities in Si form deep energy levels, depending on the quenching rate, are located at 0.10, 0.33, 0.40, 0.43 and 0.52 eV above E$_v$ and act as efficient trap centers for photo-generated carriers. This results in high series resistivity and large leakage current of β-FeSi$_2$/Si heterojunction devices [9-11].

Recently, heterojunction solar cells with Al-alloyed polycrystalline p-type β-FeSi$_2$ on n-Si(100) were investigated by Dalapati *et al.* with a ~80 nm thick β-FeSi$_2$(Al) layer subjected to RTA in nitrogen (N$_2$) ambient at temperatures of 650–700 °C for 60 s [12]. In this study, we have fabricated ITO/p-β-FeSi$_2$(Al)/p$^+$-Si/n-Si(100) heterojunction solar cells with ~40 nm thick β-FeSi$_2$(Al) layer with a thin Al interlayer (~10nm) on n-Si(100) substrates. For the first time, an energy conversion efficiency as high as 3.0% was obtained from a p-β-FeSi$_2$(Al)/p$^+$-Si/n-Si heterojunction solar cell under illumination of AM 1.5, 100 mW/cm^2.

II. EXPERIMENTAL

The photovoltaic devices were fabricated on n- type Si(100) substrates, with resistivity of ~5 Ω-cm. Prior to deposition of FeSi$_2$(Al)/Al films, the substrates were cleaned using standard RCA process followed by etching in 1% dilute hydrofluoric acid (HF) solution to remove the native oxide layer. After cleaning, the substrates were immediately loaded into a magnetron sputtering chamber. The chamber was then evacuated to a base pressure of 4.0×10^{-6} mbar. A layer of ~40 nm thick Al-containing amorphous FeSi$_2$ (Al) was deposited at room temperature by co-sputtering of stoichiometric FeSi$_2$ target and pure Al target in Ar ambient at a working pressure of 3.33×10^{-3} mbar. A thin Al interlayer (~10 nm) was deposited prior to deposition of amorphous-FeSi$_2$(Al) layer. The samples were then subjected to RTA in N$_2$ ambient (with 1×10^3 cm^3/min flow rate) at temperature of 650 °C with ramp rate 10.8 °C/s and dwell time 2 min (ULVAC-RIKO, MILA-3000). Heterojunction solar cells were fabricated by sputter deposition of ITO (~100 nm thick) at room temperature on p-β-FeSi$_2$(Al) as top electrode (area ~0.086 cm^2) and by evaporating Al layer as Ohmic contact at the back-side of n-Si substrate. Current density-voltage (J-V) characteristics of the devices in dark and under air mass (AM) 1.5 illumination (XES-151S) with a power density of 100 mW/cm^2 were measured using Agilent B1500A semiconductor device analyzer. The polycrystalline β phase formation after RTA at ≥650 °C was confirmed by x-ray diffraction (results not shown). Sharp film-substrate interfaces and film thickness were confirmed by field emission scanning electron microscopy (FESEM).

III. RESULTS AND DISCUSSION

Fig. 1(a) shows the basic structure of ITO/p-β-FeSi$_2$(Al)/ p$^+$-Si/ n-Si(100) heterojunction solar cells with Al back electrode. During RTA, the Al interlayer was mostly dissolved into FeSi$_2$ to replace Si, which resulted in the formation of the heavily Al-doped epitaxial-Si (p$^+$-Si) interfacial layer by the expelled Si. It was reported by Dalapati *et al.* [12] that open circuit voltage increased considerably once thin Al layer was deposited prior to deposition of amorphous-FeSi$_2$(Al) layer. The improvement is due to the formation of epitaxial Al-containing p$^+$- Si at p-β-FeSi$_2$(Al)/ n-Si(100) interface and suppressed

978-1-4673-2395-6/12 $31.00 © 2012 IEEE   285

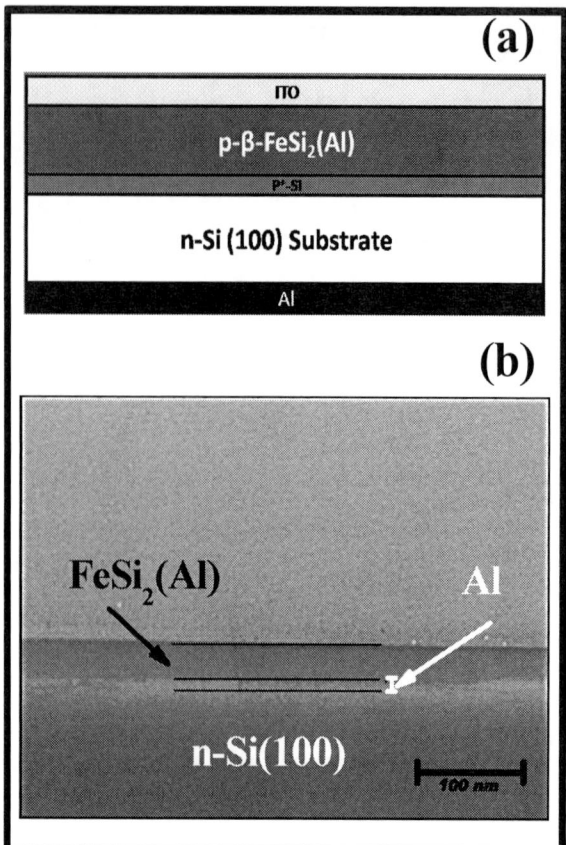

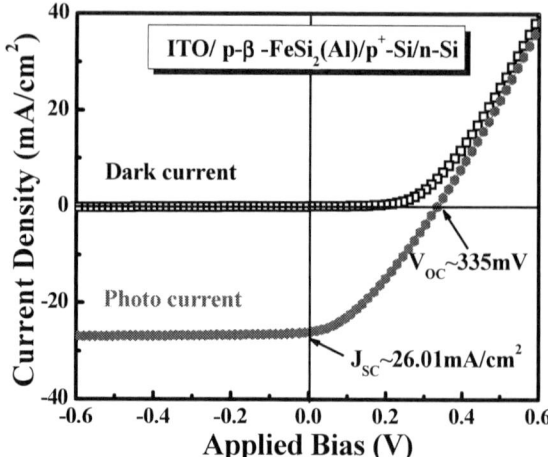

Fig. 2 Current density-voltage (J-V) characteristics of ITO/ p-β-FeSi$_2$(Al)/ p$^+$-Si/ n-Si(100) heterojunction solar cell in dark (open symbol) and under AM 1.5 illumination (100 mW/cm^2) simulated solar irradiation condition (solid symbol).

$$FF = \left(\frac{I_m \times V_m}{I_{sc} \times V_{oc}} \right) \qquad (1)$$

and

$$\eta = \left(\frac{V_{oc} \times I_{sc} \times FF}{P_{in}} \right) \qquad (2)$$

where V_{OC} is the open-circuit voltage, I_{SC} is the short circuit current, FF is the fill factor, η is the power conversion efficiency and P_{in} is the incident light power density, I_m and V_m are the current and voltage at the maximum power point. The small FF indicates high series resistance (R_{sh}) of the device [13]. For further investigation, the precise measurement of all the parameters of solar cell have been done by analyzing the J–V curve, shown in Fig. 2 under air mass 1.5 illumination using the model proposed by Ishibashi et al. [14]. From the slope of the I–V curve at $I = I_{SC}$ ($V = 0$), the shunt resistance (R_{sh}) has been independently derived. Fig. 3 shows the experimental data

Fig. 1. (a) Schematic diagram of ITO/p-β-FeSi$_2$(Al)/p$^+$-Si/n-Si(100) heterojunction solar cell with Al back electrode. (b) Cross-sectional FESEM image of FeSi$_2$(Al)/Al film deposited on n-Si(100) substrate.

back-diffusion of photo generated electrons into ITO.

The cross-sectional FESEM image of FeSi$_2$(Al) film of thickness of ~40 nm with ~10 nm thick Al inter layer is shown in Fig. 1(b). The film is continuous and almost uniform in thickness. A sharp interface between the FeSi$_2$(Al) film, Al layer and the Si substrate was confirmed from the FESEM image. The as-deposited films exhibited smooth surfaces.

PV properties of the ITO /p-β-FeSi$_2$(Al)/p$^+$-Si/n-Si(100) heterojunctions were measured using a solar simulator (XES-151S) with a power density of 100 mW/cm^2. The J–V characteristics of the device both under dark and AM 1.5 (100 mW/cm^2) simulated solar irradiation condition is displayed in Fig. 2. The device exhibits an open-circuit voltage (V_{OC}) of 0.335 volt, a short-circuit current density (J_{SC}) of 26.01 mA/cm^2 under illuminated condition. The fill factor (FF) and the power conversion efficiency (η) were estimated to be 34.44% and 3.0%, respectively. The fill factor (FF) and the power conversion efficiency (η) of the device was measured using the relations:

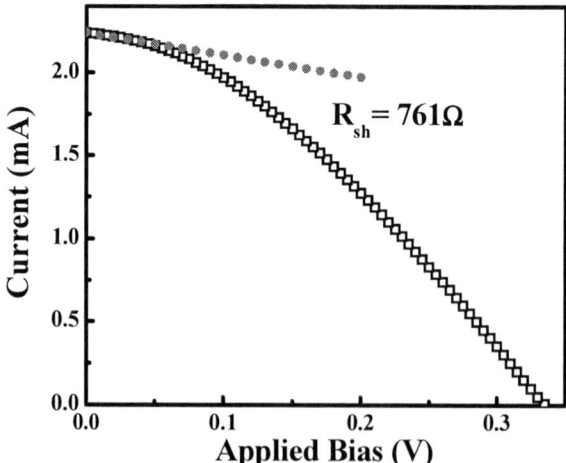

Fig. 3 Analysis of the I–V curve under illumination for determination of device shunt resistance (R$_{sh}$).

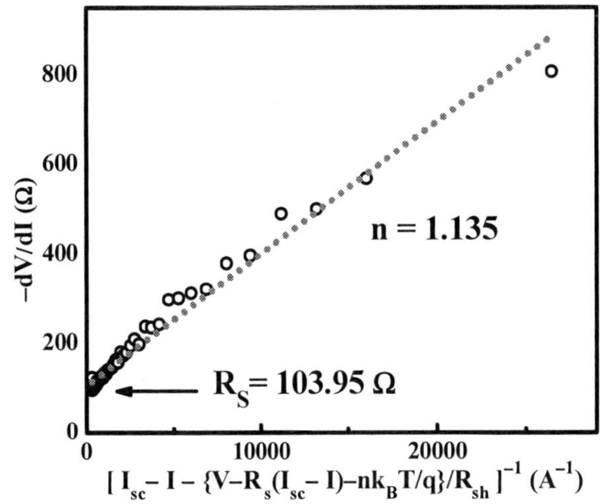

Fig. 4 Plot of $(-dV/dI)$ as a function of $[I_{sc} - I - \{V - R_s (I_{sc} - I) - nk_BT/q\}/R_{sh}]^{-1}$ for estimation of device series resistance (R_s) and ideality factor (n).

together with the calculated R_{sh} of the device. The device shows a shunt resistance (R_{sh}) of 761 Ω.

Fig. 4 shows a plot of $(-dV/dI)$ as a function of $[I_{SC} - I - \{V - R_S (I_{SC} - I) - nk_BT/q\}/R_{sh}]^{-1}$, where I_{SC}, R_S, R_{sh}, q, n, k_B and T are the short circuit current, series resistance, shunt resistance, electron charge, the ideality factor, the Boltzmann constant and the temperature, respectively. The plot of $(-dV/dI)$ exhibits a good linearity in agreement with the model. Other parameters have been estimated from the y-intercept and the slope of the plot. The device shows a series resistance (R_S) of 103.95 Ω with an ideality factor (n) of 1.135. Low ideality factor (<2.0) confirms reduced recombination losses within the device [15]. The saturation current (I_0) and photocurrent (I_{Ph}) of the cell was measured using the following relations [14]:

$$I_0 \sim \left(I_{sc} - \frac{V_{oc} - R_s I_{sc}}{R_{sh}} \right) \exp\left(-\frac{qV_{oc}}{nk_BT} \right) \qquad (3)$$

and $$I_{ph} = I_0 \left\{ \exp\left(\frac{qV_{oc}}{nk_BT}\right) - 1 \right\} + \frac{V_{oc}}{R_{sh}} \qquad (4)$$

The I_0 and I_{Ph} of the cell have been estimated to be 29.85 nA and 2.98 mA, respectively.

IV. CONCLUSIONS

In conclusion, we have fabricated polycrystalline p-β-FeSi₂(Al) heterojunction thin film solar cell on n-Si(100). This study has shown that p-β-FeSi₂/n-Si heterojunctions are promising new candidate for future photovoltaic applications.

Further improvement of the device performance may be achieved by optimizing the β-FeSi₂(Al) thickness and by introducing an electron-blocking layer between ITO and p-β-FeSi₂(Al).

ACKNOWLEDGEMENT

The authors acknowledge DST, New Delhi (project sanction number: DST/TM/SERI/2k11/100/(G)) for supporting the solar cell work reported here.

REFERENCES

[1] M. C. Bost, and J. E. Mahan, "Optical properties of semiconducting iron disilicide thin films," *J. Appl. Phys.*, vol. 58, no. 7, pp. 2696-2703, October 1985.

[2] M. C. Bost, and J. E. Mahan, "A clarification of the index of refraction of betairon disilicide," *J. Appl. Phys.*, vol. 64, no. 4, pp. 2034-2037, August 1988.

[3] C. A. Dimitriadis, J. H. Werner, S. Logothetidis, M. Stutzmann, J. Weber, and R. Nesper, "Electronic properties of semiconducting FeSi₂ films," *J. Appl. Phys.*, vol. 68, no. 4, pp. 1726-1734, August 1990.

[4] K. Lefki, and P. Muret, "Photoelectric study of FeSi₂ on silicon: Optical threshold as a function of temperature," *J. Appl. Phys.*, vol. 74, no. 2, pp. 1138-1142, July 1993.

[5] Y. Makita, T. Ootsuka, Y. Fukuzawa, N. Otogawa, H. Abe, L. Zhengxin, and Y. Nakayama, "β-FeSi₂ as a Kankyo (environmentally friendly) semiconductor for solar cells in the space application," *Proc. SPIE*, vol. 6197, pp. 619700, May 2006.

[6] R.H. Bube, "Photovoltaic Materials," *Imperial College Press*, Amsterdam, 1998, ch. 1.

[7] Z. Yang, K. P. Homewood, M. S. Finney, M. A. Harry, and K. J. Reeson, "Optical absorption study of ion beam synthesized polycrystalline semiconducting FeSi₂," *J. Appl. Phys.*, vol. 78, no. 3, pp. 1958-1963, August 1995.

[8] M. Powalla, and K. Herz, "Co-evaporated thin films of semiconducting β-FeSi₂," *Appl. Surf. Sci.*, vol. 65/66, pp. 482-488, March 1993.

[9] Z. Liu, S. Wang, N. Otogawa, Y. Suzuki, M. Osamura, Y. Fukuzawa, T. Ootsuka, Y. Nakayama, H. Tanoue, and Y. Makita, "A thin-film solar cell of high-quality β-FeSi₂/Si heterojunction prepared by sputtering," *Sol. Energy Mater. Sol. Cells*, vol. 90, no. 3, pp. 276-282, February 2006.

[10] K. Wünstel, and P. Wagner, "Interstitial Iron and Iron-Acceptor Pairs in Silicon," *Appl. Phys. A*, vol. 27, no. 4, pp. 207-212, April 1982.

[11] T. Suemasu, T. Fujii, K. Takakura, and F. Hasegawa, "Dependence of photoluminescence from β-FeSi₂ and induced deep levels in Si on the size of β-FeSi₂ balls embedded in Si crystals," *Thin Solid Films*, vol. 381, no. 2, pp. 209-213, January 2001.

[12] G. K. Dalapati, S. L. Liew, A. S. W. Wong, Y. Chai, S. Y. Chiam, and D. Z. Chi, "Photovoltaic characteristics of p-β-FeSi₂(Al)/n-Si(100) heterojunction solar cells and the effects of interfacial engineering," *Appl. Phys. Lett.*, vol. 98, no. 1, pp. 013507-1-013507-3, January 2011.

[13] M. Shaban , K. Nakashima , W. Yokoyama, and T. Yoshitake, "Photovoltaic Properties of n-type β-FeSi₂/p-type Si Heterojunctions," *Jpn. J. Appl. Phys.*, vol. 46, no. 27, pp. L667-L669, July 2007.

[14] K. Ishibashi, Y. Kimura, and M. Niwano, "An extensively valid and stable method for derivation of all parameters of a solar cell from a single current-voltage characteristic," *J. Appl. Phys.*, vol. 103, no. 9, pp. 094507-1-094507-7, May 2008.

[15] A. Moliton, and J. M. Nunzi, "Review: How to model the behaviour of organic photovoltaic cells," *Polym. Int.*, vol. 55, no. 6, pp. 583-600, March 2006.

An Investigation in the Impact of Structural Parameters on the Electrical Characteristics of Nanoscale Heterostructure p-MOSFETs

Fatemeh Kohani Khoshkbijari, Reza Fouladi, Shiva Nejati, Reza Barkhordari, Reza Kohani Khoshkbijari, Shide Nejati

Device Simulation and Modeling Lab

Sama Technical and Vocational Training College, Islamic Azad University, Rasht Branch

Rasht, IRAN

f.kohani@ieee.org

Abstract— Over the past few decades, CMOS has proved to be the choice device in the fabrication of the high density integrated circuits. However, in this technology the device performance is degraded primarily due to mobility limitation in PMOSFET. One way to elevate this problem is to alter electronic properties of the channel region using strained layers. In this paper, we propose a novel Heterosructure PMOSFETs with optimum Ge content in SiGe layer. This investigation proves that an increase in Ge mole fraction reduces threshold voltage on Si/SiGe interface, while threshold voltage on Si/SiO2 is increased. As the Ge mole fraction is increased the gate capacitance also will increase. The results provide useful guide lines for optimizing Nanoscale Heterostructure for low power applications.

Keywords; Nanoscale, MOSFET, Tensile-Strained Si, Compressively Strained SiGe

I. INTRODUCTION

The past several years have witnessed rapid growth in the study of strained silicon due to its potential ability to improve the performance of very large scale integrated (VLSI) circuits independent of geometric scaling [1]-[2]. However, the practical benefit of scaling is declining as physical and economic limits are approached, and novel solutions are increasingly being sought. The incorporation of new materials, from the interconnect level (Cu, Low-K), to the gate stack (high-K dielectrics, metal gate electrodes), and even the substrate (strained silicon, silicon-on-insulator (SOI) wafers) is emerging as an important way to continue to improve circuit performance. Strain improves MOSFET drive currents by fundamentally altering the band structure of the channel and can therefore enhanced performance even at aggressively scaled channel length. Such band-engineered heterostructures can be optimized to allow mobility enhancement factors over bulk Si of ~ 2 for electrons and as high as ~ 10 for holes [3]. In this paper different types of the Heterostructure transistors has been investigated and a novel Si/SiGe Heterostructure pMOSFET has been proposed.

II. HETEROSTRUCTURE PMOSFET WITH SI/SIGE CHANNEL

PMOS transistors with Si/SiGe heterostructure are classified into 3 types:

1- Compressively strained $Si_{1-x}Ge_x$ buried-channel PMOSFET with Si cap layer

2- Tensile-strained Si/relaxed $Si_{1-x}Ge_x$ surface-channel PMOSFET

3- Dual channel tensile-strained-Si/compressively strained $Si_{1-y}Ge_y$/ relaxed- $Si_{1-x}Ge_x$ PMOSFET

A. Compressively Strained $Si_{1-x}Ge_x$ Buried-Channel PMOSFET with Si Cap Layer

A schematic diagram of a Si/SiGe/Si p-channel MOSFET is shown in Fig. 1, where V_G, V_D and V_B represent the gate, drain, and bulk voltage, respectively. It may be noted that for the p-channel MOSFET, V_G and V_D are negative voltages while V_B is usually positive. The device consists of an SiO_2 gate dielectric layer of thickness t_{ox}, an Si cap layer of thickness t_{cap}, an SiGe layer of thickness t_{SiGe} and Si buffer layer of thickness t_{buff}, all of which are undoped. These layers are grown on an n-type substrate, which is uniformly doped with a concentration of N_D. Another important parameter for this device is the Ge mole fraction (x) in $Si_{1-x}Ge_x$, which not only determines the valance-band discontinuity (ΔE_V) at the $Si/Si_{1-x}Ge_x$ heterointerface but also the mobility of holes in the 2-DHG. The depletion-layer with (W) in the doped silicon substrate widens as V_G is made increasingly negative and reaches maximum value ($W=W_{MAX}$) at $V_G = V_{TH}$, when strong inversion occurs at the top heterointerface (x = t_{cap}). With a further increase in | V_G |, the sheet hole concentration in the 2-DHG in SiGe at the Si/SiGe heterointerface (ρ_s^H) increases, resulting in an increased potential drop across the strained silicon and oxide layers. Fig. 2 shows the energy band diagram of the metal / SiO_2 / Si / Strained-SiGe / Si (undoped) /n-Si structure at this condition.

978-1-4673-2395-6/12 $31.00 © 2012 IEEE

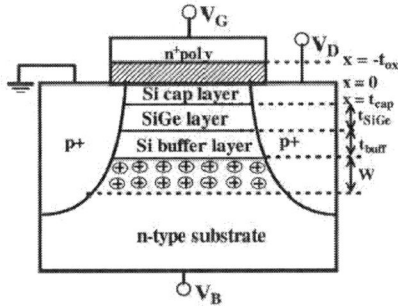

Fig. 1.   Cross-sectional view of an Si/SiGe/Si p-channel MOSFET[4]

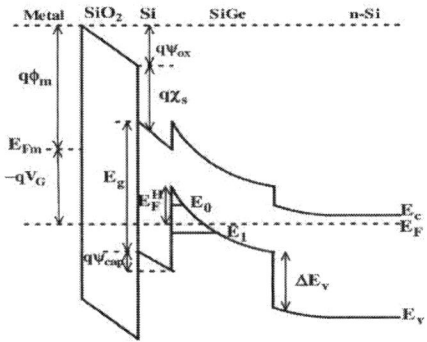

Fig. 2.   Energy band diagram of metal / SiO2 / Si / Strained-SiGe / Si(undoped) /n-Si structure[4]

B. Tensile-Strained Si/ Relaxed $Si_{1-x}Ge_x$ Surface-Channel PMOSFET

Single-Channel heterostructures, consisting of strained-Si layer grown on a relaxed $Si_{1-x}Ge_x$ buffer, are typically grown in two steps. First, the relaxed buffer is deposited at high temperature in order to insure maximum relaxation. Finally, the strained-Si layer itself is grown (Fig. 3). Since the lattice consist of the relaxed $Si_{1-x}Ge_x$ is larger than that of Si, the thin Si film grows in a state of biaxial tension. Tensile strained silicon layers are useful for electron channels of high mobility in nMOSFETs. Hole transport is improved in both tensile strained-Si and compressively strained-SiGe compared with bulk-Si. Modifications to the electronic band structure and a reduction of the hole effective mass have been found to increase the mobility in strained-SiGe by five times [5].

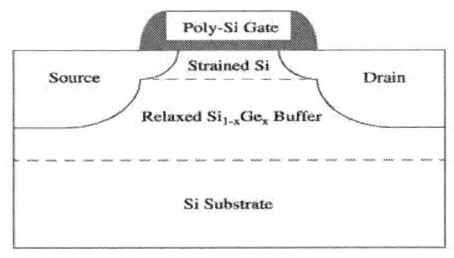

Fig. 3.   Tensile strained-Si/ relaxed $Si_{1-x}Ge_x$/ bulk-Si MOSFET[6]

Fig.4 illustrates the energy band diagram of strained Si pMOSFET.

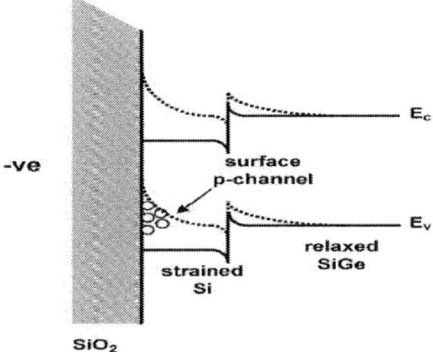

Fig. 4.   Energy Band Diagram of tensile strained-Si/ relaxed $Si_{1-x}Ge_x$/ bulk-Si MOSFET[7]

There are some challenges in using strain to enhance the performance of CMOS devices and many of these are related to the critical thickness of a strained-layer [8]. If a strained layer is grown above the critical thickness, strained relaxes with the introduction of misfit defects at the strained-Si-SiGe heterointerface and the enhanced transport properties arising from the strain are lost. Therefore a dual channel structure with a compressive strained-SiGe layer followed by a tensile strained-Si layer on a single-relaxed SiGe (virtual substrate) has been proposed.

C. Dual Channel Tensile-Strained-Sii / Compressively Strained $Si_{1-y}Ge_y$/ Relaxed- $Si_{1-x}Ge_x$ PMOSFET

The schematic diagram of a strained-Si/compressively strained $Si_{1-y}Ge_y$/ relaxed- $Si_{1-x}Ge_x$ (X<Y) MOSFET is shown in Fig. 5. The source, drain and gate polysilicon are n^+ doped for n-channel MOSFET and p^+ doped for p-channel MOSFET.

The device consists of a SiO_2 gate dielectric layer of thickness t_{OX}, a strained-Si cap layer of thickness t_{cap}, a strained-$Si_{1-y}Ge_y$ layer of thickness t_{SiGe} and relaxed- $Si_{1-x}Ge_x$ buffer layer of thickness t_{buff}, all of which are undoped. These layers are grown on an n-type substrate for a p-channel MOSFET. Another important parameter for this device is the Ge mole fraction (Y) in strained $Si_{1-y}Ge_y$, which not only determines the valance band discontinuity (ΔE_V) at the Si/SiGe heterointerface but also the mobility of holes in the p-channel. In p-channel MOSFET, the depletion layer width (W) in the doped silicon substrate widens as $|V_G|$ is made increasingly negative and reaches a maximum value (W = W_{max}) at V_G = V_{TH}, when strong inversion occurs at the top heterointerface (x = t_{cap}). With further increase in $|V_G|$, the sheet hole concentration in the strained- $Si_{1-y}Ge_y$ at the Si/SiGe heterointerface (ρ_s^H) increases, resulting in increased potential drop across strained-Si and oxide layers. At V_G = V_{TS}, the bands bend sufficiently to create inversion at the Si/SiO_2 interface. When $|V_G|>|V_{TS}|$, any increase in the negative gate voltage results in an increase in the sheet hole concentration at the Si/SiO_2 interface (ρ_s^S), while the sheet hole concentration at the Si/SiGe interface remains essentially fixed at its peak value of $\rho_{s\,max}^{H}$.

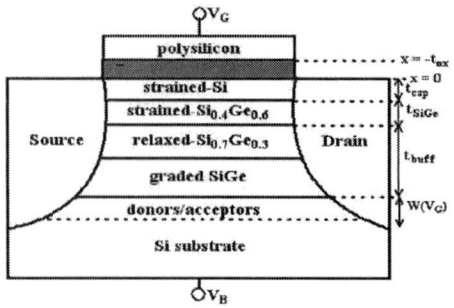

Fig. 5.   Schematic diagram of a strained-Si/strained-SiGe/relaxed SiGe MOSFET[9]

Fig. 6 shows the energy band diagram of the structure under this condition. In this region of operation, the drain current is due to the flow of holes in the Si/SiGe heterointerface as well as the Si/SiO$_2$ interface.

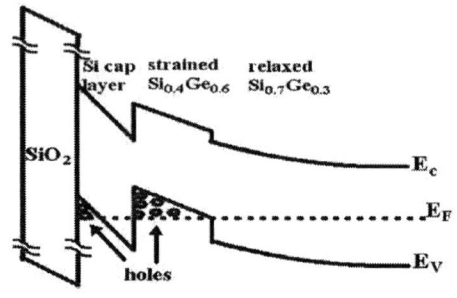

Fig. 6.   Energy band diagram of metal/SiO$_2$/strained-Si/strained-SiGe/relaxed SiGe p-MOSFET[10]

In this structure, tensile strain is created in Si layer. Thus, it is appropriate for n-channel MOSFETs. Depletion layer width in doped Si substrate is increased and hit its zenith (W=W$_{max}$) when gate voltage enhances in n-channel MOSFET. In this condition, inversion layer occurs at SiO$_2$/Si interface. Any increase in gate voltage results in an increase in the sheet electron concentration at strained Si. Thus, drain current is improved.

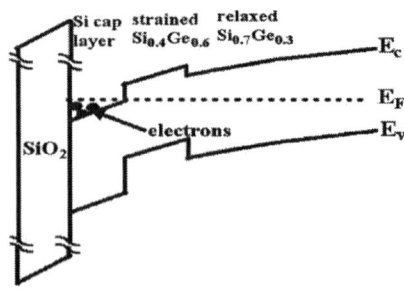

Fig. 7.   Energy band diagram of strained-Si/strained-SiGe n-MOSFET[10]

In Fig.7 the energy band diagram of Strained-Si/ Strained-SiGe nMOSFET is shown. It is obvious that electron accumulation in lower band energy (in Strained Si) is more noticeable because of energy bands bend.

III.   PROPOSED HETEROSTRUCTURE PMOSFET SIMULATION

Ge mole fraction in SiGe channel has an important role in performance of heterostructure transistors. In this paper we investigate the impact of Ge mole fraction on the electrical characteristics of a novel heterostructre pMOSFET.

A.   Device Structure

Fig.8 shows the schematic diagram of a Si/SiGe/Si p-MOSFET. The gate length and gate oxide thickness are considered as 150 nm and 4 nm respectively. The silicon cap thickness and SiGe layer thickness are 1 nm and 4 nm. The bulk doping in this work is 3×10^{18} cm^{-3}.

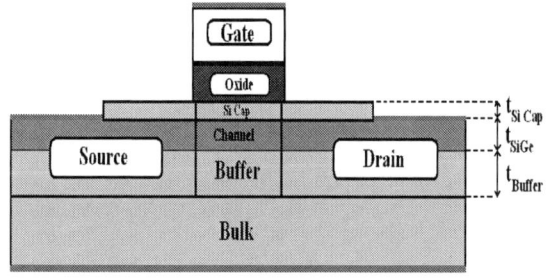

Fig. 8.   Schematic of simulated device

We have considered following model in our simulation: tunneling and thermionic emission at hetero interface, Shockley-Read-Hall (SRH) for recombination. Moreover, the mobility models include doping dependence, high field saturation and transverse field dependence [11].

IV.   SIMULATION RESULTS

A.   The Energy Bands Diagram of Simulated Device

The conduction and valance bands of structure when inversion layer is formed in Si/SiGe interface is depicted in Fig.9. A significant enhancement in carriers speed is due to created quantum well in Si/SiGe interface.

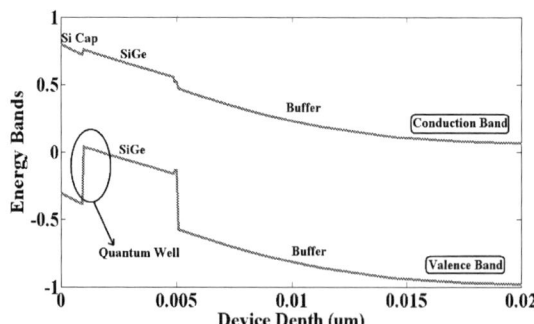

Fig. 9.   Energy Band diagram of proposed device when inversion layer is formed in Si/SiGe interface

Fig.10 shows the bands structure diagram when the inversion layer is created in SiO₂/Si interface. In this condition gate voltage is greater than V_{TS} so the hole concentration increases in SiO₂/Si interface.

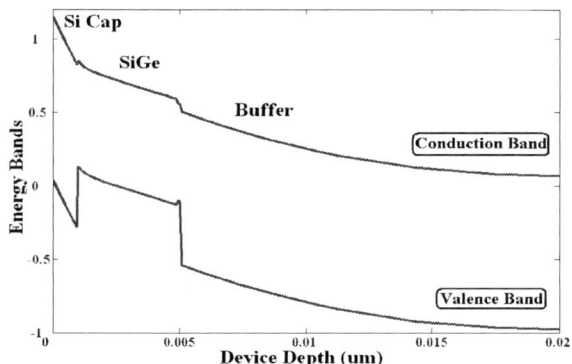

Fig. 10. Energy Band diagram of simulated device when the inversion layer is created in SiO₂/Si interface

B. The Impact of Ge Mole Fraction on Threshold Voltage

The dependency of the threshold voltages on the germanium content (x) in the Si/Si$_{1-x}$Ge$_x$ layer is shown in Fig. 11. It is observed that the absolute value of the threshold voltage V_{TH} decreases as Ge mole fraction is increased. This is due to the increase in the valence band discontinuity at the Si/SiGe heterointerface. Consequently the channel forms at a lower gate voltage. Increasing Ge mole fraction also yields larger |V_{TS}|. It is preferable to prevent the parallel conduction at SiO₂/Si interface. There are other parameters affecting threshold voltage which are not considered in this paper. For example layer thicknesses and substrate doping concentration affect threshold voltage. Controlling the magnitude of these two threshold voltages is very important in low power applications.

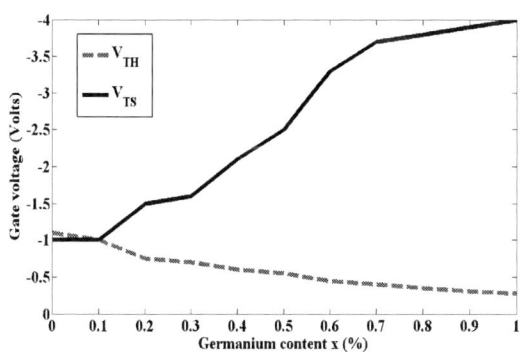

Fig. 11. The impact of Ge mole fraction on threshold voltages

Fig. 12 shows the effect of mole fraction on the drain current. As Ge mole fraction is increased, the hole effective mass is decreased. Therefore carrier mobility is increased. This results in an increase in drain current. It is interesting to note that by increasing mole fraction above 0.6 the rate of increase in drain current is reduced. Because of the complicated and warped valence structure of strained Si, it is difficult to explain

the reason for hole mobility enhancement [12]. However from Fig. 12 it may be confirmed that mobility enhancement continues to increase with x > 0.6.

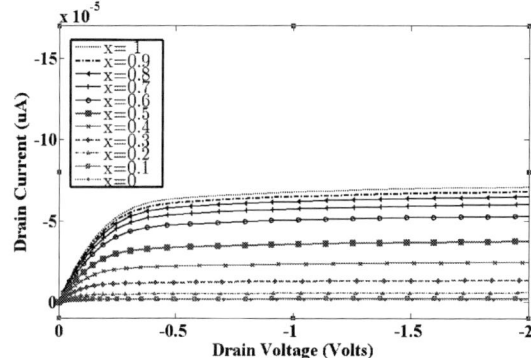

Fig. 12.   The impact of Ge content (x) on drain current

C. The Impact of Ge Content on Gate Capacitance

In order to find the gate capacitance, AC analysis was carried out. Fig. 13 shows gate capacitance (C_{gate}) as a function of Ge mole fraction. We note that as Ge mole fraction increased, C_{gate} increases. This is mainly attributed to an increase in hole concentration and leads to the increase in drain current. For more fraction larger than 0.6 the C_{gate} becomes almost flat which confirms the data in Fig. 12.

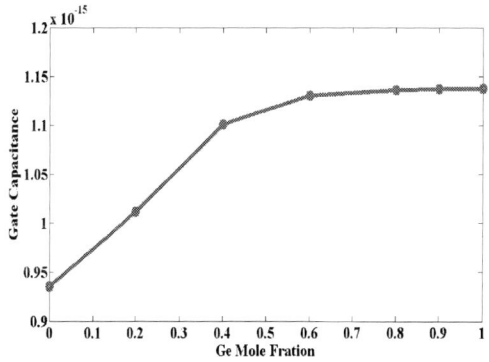

Fig. 13.   The impact of Ge content (x) on C_{gate}

V.   CONCLUSION

In this paper, different types of heterostructure p-MOSFET and their performance was investigated. Furthermore, the effect of Ge mole fraction on electrical characteristics of a proposed heterostructure pMOSFETs was surveyed. We have shown that as the Ge mole fraction is increased, V_{TH} is decreased while V_{TS} is increased. Increasing Ge mole fraction also causes increase in drain current and C_{gate}. The results of these studies are useful for design optimization of heterostructure devices for low power applications.

REFERENCES

[1] D. A. Antoniadis, "MOSFET scalability and new frontier devices", Digest of Symposium on VLSI Technology, Honolulu, HI, pp.2-5, 2002.

[2] J. L. Hoyt, H. M. Nayef, S. Eguchi, I. Aberg, G. Xia, T. Drake, E. A. Fitzgerald, and D. A. Antoniadis, " Strained Si MOSFET technology", Technical Digest-International Electron Devices Meeting, San Francisco, CA, pp.23-26, 2002.

[3] S. Thompson et al., "A 90-nm logic technology featuring strained-silicon", IEEE Trans. on Electron Devices, vol. 51, No. 11, pp. 1790-1797, Nov.2004.

[4] B. Bindu, N. DasGupta and A. DasGupta, "Analytical model of drain current of Si/SiGe heterostructure p-channel MOSFETs for circuit simulation" , IEEE Trans. on Electron Devices, vol.53, No.6, pp.1441-1419, Mar 2002.

[5] M. T. Currie, C. W. Leitz, T. A. Langdo, G. Taraschi, D. A. Antoniadis, and E. A. Fitzgerald, " Carrier motilities and process stability of strained Si n- and p-MOSFETs on SiGe virtual substrates", Journal of Vacuum. Sci. Technology B vol 19, No.6, pp 2268-2279, Nov-Dec2001.

[6] W. Zhang and J. G. Fossum, "On the threshold voltage of strained-Si/Si$_{1-x}$Ge$_x$ MOSFETs", IEEE Trans. on Electron Devices, vol. 52, No. 2, pp 263-268, Feb 2005.

[7] S.H.Olsen et al. , "Study of single- and dual-channel designs for high-performance strained-Si-SiGe n-MOSFETs", IEEE Trans. on Electron Devices, vol. 51, No. 7, pp. 1245-1253, July.2004.

[8] J. W. Matthews and A. E. Blakeslee, "Defects in epitaxial multilayers", J. Cryst. Growth, vol.27, pp 118-125, 1974.

[9] J. Jung, M. L. Lee, S. Yu, E. A. Fitzgerald and D. A. Antoniadis, "Implementation of both high-hole and electron mobility in strained Si/strained Si$_{1-y}$Ge$_y$ on relaxed Si$_{1-x}$Ge$_x$ (x < y) virtual substrate", IEEE Electron Devices Letters, vol. 24, pp 460-462, July 2003

[10] B. Bindu, N. DasGupta and A. DasGupta, "Analytical model of drain current of strained-Si/strained-Si$_{1-Y}$Ge$_Y$ /relaxed-Si$_{1-X}$Ge$_X$ NMOSFETs and PMOSFETs for circuit simulation" , Solid State Electronics, vol. 50, pp.448-455, Mar 2006.

[11] ISE-TCAD Dessis 7.0 Manual

[12] S. Takagi, T. Mizuno, T. Tezuka, N. Sugiyama, S. Nakaharai, T. Numata, J. Koga and K. Uchida, "Sub-band structure engineering for advanced CMOS channels", Solid-State Electronics, vol. 49, pp. 315-320, 2005.

Effect of Solution Concentration on the Morphology, Electrical, and Optical Properties of MEH-PPV Thin Films

Shafinaz Sobihana Shariffudin, *Member, IEEE*, Nurhafizah Zainal Abidin, Nurul Zayana Yahya, Anees Abdul Aziz, *Member, IEEE,* Sukreen Hana Herman, *Member, IEEE*, and Mohamad Rusop, *Member, IEEE*
NANO-Electronic Centre (NET), Faculty of Electrical Engineering, Universiti Teknologi MARA,
40450 Shah Alam, Selangor, Malaysia.
Email: sobihana@gmail.com

Abstract - Polymer and organic light-emitting diodes (OLEDs), photovoltaic cells and field effect transistor are being focused towards commercialization. Poly[2-methoxy-5-(2′-ethyl-hexyloxy)-1,4-phenylene vinylene]or known as MEH-PPV, have been extensively studied for OLED applications. In this study, MEH-PPV powders were dissolved in toluene and 1, 2-dichlorobenzene (1, 2-DCB) at different solutions concentration and deposited using spin coating technique. Surface morphologies of the films were obtained using Atom Force Microscope (AFM) and thickness using surface profiler, while the electrical properties were investigated using 2-point probe. The optical properties were characterized using UV-Visible-NIR (UV-VIS-NIR), and photoluminescence (PL) spectrometer. The surface roughness, Ra of the films increased with the concentration for both types of solvent, however for films in toluene, as the concentration increased above 5mlmg^{-1}, the roughness decreased which is caused by the aggregation of the MEH-PPV. Films' thicknesses increased as well as the resistivity for both types of solvents. However, for MEH-PPV thin films in 1, 2-DCB show better conductivity compared to toluene. PL spectra of films using toluene are red-shifted due to longer conjugation length, and also show higher intensity compared to films using 1,2 DCB.

Keywords—Polymer (MEH-PPV); solvents concentration; morphology; electrical properties; optical properties

I. INTRODUCTION

Conjugated polymers have been remarkably studied to be used in electronics and optoelectronics devices such as light emitting diodes[1] and solar cells [2]. Burroughes et. al. were the first group to report on electroluminescence in conjugated polymers, using (1, 4--phenylene vinylene), PPV as the active layer [3]. However, one disadvantage of using PPV is that it is insoluble in common organic solvents such as tetrahydrofuran (THF), chloroform, xylene, chlorobenzene (CB) and toluene, which makes it difficult to process to become thin films [4].

One of derivatives with alkoxy on phenyl group known as poly[2-methoxy-5-(2′-ethyl-hexyloxy)-1,4-phenylene vinylene] or more known as MEH-PPV exhibits greater solubility in common organic solvents. MEH-PPV can be easily dissolved in non-aromatic and aromatic solvents to exhibit different electrical and optical properties [5]. According to B. Kang et. al, the organic solvents can control the semiconducting properties of the MEH-PPV thin films, either it shows electron

transporting (n-type) using aromatic solvents or hole transporting (p-type) behavior using non-aromatic solvents [6]. Many researchers have compared the properties of the MEH-PPV thin films between these two solvents in their studies, however none had compared between the same groups of solvents. So here we chose to compare the aromatic solvents which will be used as the active layer in our device later on.

Spin coating technique is mostly applied in synthesizing polymer film as it can produce uniform films on large area. MEH-PPV is one of the conjugated polymers that have Newtonian rheological properties where the film thickness obtained by spin coating can be affected by polymer solution concentration, solution viscosity and spin speed [7].

The objective of this study is to investigate the characteristic of MEH-PPV in terms of its morphology, electrical and optical properties. Two types of solvents were used to dissolve the MEH-PPV which are toluene and 1, 2-dichlorobenzene (1, 2-DCB) to produce p-type polymer. Using both solvents, the concentration of the MEH-PPV were varied from 2 to 7mgml^{-1}.

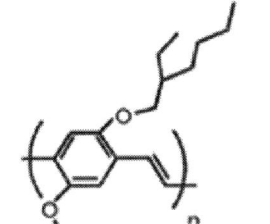

Figure 1. Molecular structure of MEH-PPV

I. EXPERIMENTAL

A. Solution Preparation

MEH-PPV with an average molecular weight of 40,000 to 70,000 was purchased from Sigma-Aldrich and was used without further purification. The MEH-PPV were dissolved in toluene and 1,2 DCB with varies concentration of 2mgml^{-1} to 7mgml^{-1} ml. The solutions were stirred for 48 hours at room temperature to ensure the reaction of the chemicals dissolved completely. The reaction mixture became red/orange during the addition.

B. Thin Film Formation by Spin Coating

The thin film was formed by spin coating on pre-cleaned glass substrates at room temperature. The spin speed was fixed at 2000 rpm with a spin time of 60 seconds. After each coating, the film was then dried at 50 °C for 10 minutes. The spin coating process was repeated for 3 times in order to achieve desired thickness of thin film.

C. Characterization

Surface morphology of the MEH-PPV thin films were obtained using Atomic Force Spectroscopy (AFM, model: Park System XE-100), while the thickness were measured using surface profiler (model: Veeco Dektak 750). For the electrical properties, current-voltage (I-V) measurements were done using two point probe solar simulator (model: Bukoh Keiki CEP 2000) within voltage range of -10 to 10 V. Room temperature photoluminescence (PL) spectra were measured using PL spectrometer (model: Horiba Jobin Yvon Fluoromax 3), and transmittance spectra were measured using UV-Vis spectrophotometer (model: JASCO UV-Vis) for the optical properties.

II. RESULTS AND DISCUSSIONS

A. Atomic Force Spectroscopy (AFM) Surface Morphology and Film Thickness

Surface morphologies of the MEH-PPV thin films at different concentration were studied using AFM and are shown in Figure 3 for films using toluene, and Figure 4 for films using 1,2 DCB solvents. For the case of thin films using solution dissolved in toluene, it can be seen that the surface of the thin film becomes more uniform as the concentration of the MEH-PPV increased. When the concentration of the MEH-PPV is more than 5mgml-1, the MEH-PPV becomes more aggregately spread. While for MEH-PPV in 1,2 DCB, it is observed that the surfaces of the thin films are smoother and more uniform than the thin films in toluene. This is also proven by the surface roughness, Ra of the thin films which is shown in Table 1.

Table 2 shows the thin film average thickness at different concentration of MEH-PPV solutions. As reported by others, spin coated conjugated polymers for optoelectronic devices typically have a thickness in the sub-micron range [5]. In this study, the range of thin film thickness obtained were between 50-300nm.

TABLE 1
SURFACE ROUGHNESS OF THIN FILM

Thin Film Concentration (mgml^{-1})		2	3	4	5	6	7
Surface Roughness, Ra (nm)	Toluene	2.68	3.68	3.92	4.48	3.51	3.25
	1,2 DCB	1.91	1.99	1.33	1.85	2.24	2.74

TABLE 2
MEH-PPV THIN FILMS AVERAGE THICKNESS AT DIFFERENT CONCENTRATIONS

Thin Film Concentration (mgml^{-1})		2	3	4	5	6	7
Average thickness (nm)	Toluene	49.5	67.8	100.7	131.0	195.8	281.9
	1,2 DCB	53.9	75.2	99.81	110.5	170.0	236.2

The thickness of thin film both for toluene and 1, 2-DCB increased with the solutions concentration. This phenomenon can be explained by equation 1:

$$h_f = kx_{1,0}\omega^{-\beta} \tag{1}$$

The final thin film thickness, h_f after spin coating of a polymer solution was found to be correlated to the initial polymer weight fraction, $x_{1,0}$ and the spin speed, ω [7, 8]. k is the constant that depends on the concentration solution viscosity and other properties of the polymer and solvent. In this study, the spin speed was fixed at 2000 rpm while the concentrations of solutions were varied. Therefore, in this case, as the concentration increased, the value of k is increased, hence thickness of thin film also increased.

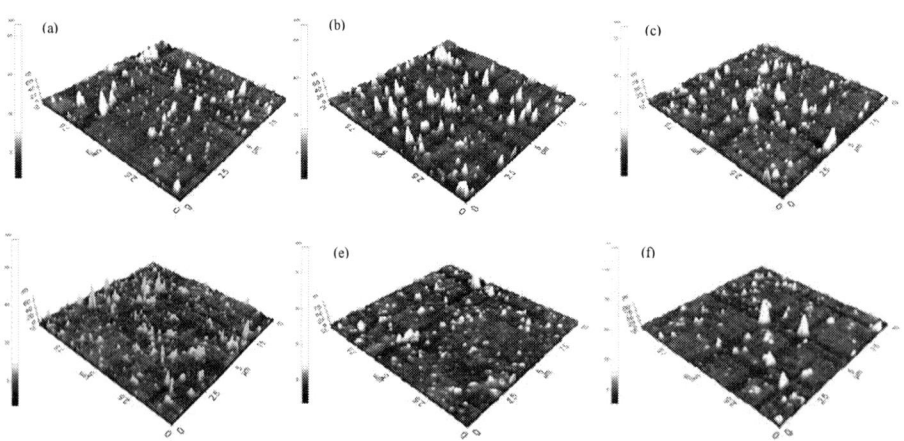

Figure 2. AFM surface morphology of MEH-PPV thin films in toluene at (a) 2 mgml^{-1} (b) 3 mgml^{-1} (c) 4 mgml^{-1} (d) 5 mgml^{-1} (e) 6 mgml^{-1} (f) 7 mgml^{-1}

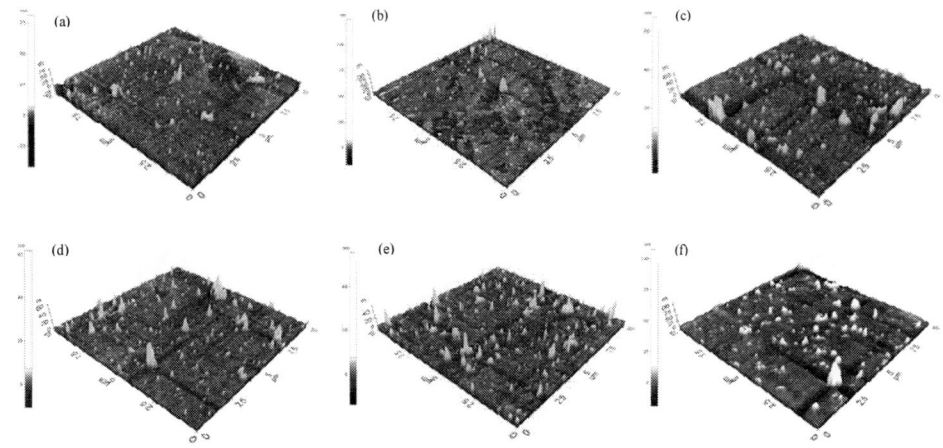

Figure 3. AFM surface morphology of MEH-PPV thin films in 1,2 DCB at (a) 2 mgml^{-1} (b) 3 mgml^{-1} (c) 4 mgml^{-1} (d) 5 mgml^{-1} (e) 6 mgml^{-1}(f) 7 mgml^{-1}

B. Electrical Properties of MEH-PPV Thin Films

Gold were firstly deposited by thermal evaporation on the surface of the thin films with a thickness of about 10 nm; which act as metal contact. I-V measurements were done using a two-point probe at room temperature.

Resistivity and conductivity of the thin films were calculated and presented in Figure 4 for films in toluene and Figure 5 for films in 1,2 DCB. It is observed that for both types of solvents, the resistivity increased with the concentration of the MEH-PPV. However comparing films between toluene and 1,2 DCB, resistivity value of films using 1,2 DCB are much lower than toluene.

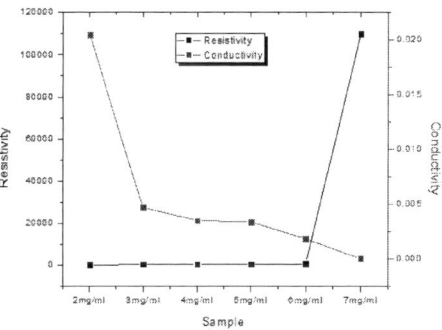

Figure 5. Resistivity and conductivity of MEH PPV in 1, 2-DCB.

Therefore, as the thin film thickness increased due to increase of concentration, the resistivity also increased for both toluene and 1, 2-DCB.

C. Optical Properties of MEH-PPV Thin Films

1) Photoluminescence (PL) Spectra

Room temperature PL spectra of toluene and 1, 2-DCB are presented in Figure 6(a) and 6(b). The range of wavelength in the measurement is between 450 nm to 700 nm. PL spectra of toluene gives higher peak at range between 550nm to 600nm compare to 1, 2-DCB. The PL peak for 1, 2-DCB increased as the concentration of thin film is increased. Same goes for toluene however, at 6 mgml^{-1} and 7 mgml^{-1} concentration of toluene, the peak quenched when compared to the films with concentration of 5 mgml^{-1} toluene. It is found that the PL intensities correlated to the surface roughness, which shows the same pattern.

The peak of PL from toluene is accompanied with small shoulder located at 650 nm. The shoulder exhibited when aggregation is occurred and grows further as concentration increased caused by the increasing aggregation [9]. According to Nguyen et. al, the degree of aggregation is dependent on both concentration and types of solvents used [10].

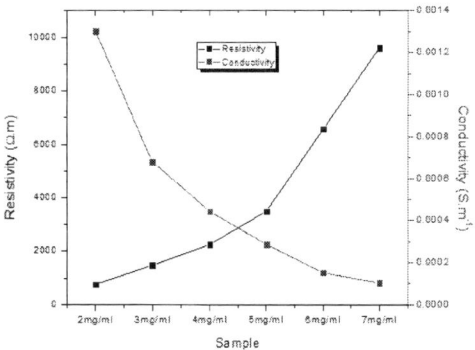

Figure 4. Resistivity and conductivity of MEH PPV in toluene.

The most possible explanation for this phenomenon is because of the increment of the thickness of the thin films with increasing of the concentration. As we know, resistivity is calculated using equation 2 as below:

$$\rho = R \frac{w \times t}{l} \qquad (2)$$

where R is the resistance of the films, w is the width, t is the thickness, and l is the length between the metal contacts.

978-1-4673-2395-6/12 $31.00 © 2012 IEEE

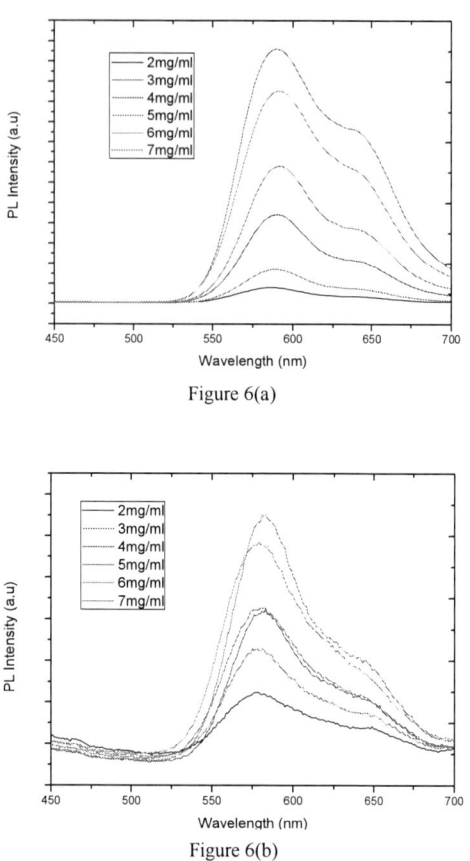

Figure 6(a)

Figure 6(b)

Figure 6. PL spectra of MEH-PPV in (a) toluene and (b) 1, 2-DCB

A broader PL spectra exhibited by toluene compared to 1, 2-DCB due to its greater conformational disorder. Besides, MEH-PPV film prepared from toluene is more red-shifted compare to 1, 2-DCB. The shifting can be explained by considering that the conjugation length of polymer changes in different solvents caused by conformational changes of the polymer chain [10, 11]. Shorter average length of conjugation length will tend to exhibit blue shifted spectra. Thus, in this case, can be said that the polymer in toluene has longer conjugation length compared to 1, 2-DCB. In addition, the enhancement of red portion of film's PL can also be attributed by the unwanted aggregation of the MEH-PPV in toluene which also leads to decrease of PL efficiency at higher solution's concentration [12].

2) UV-Vis Absorption

As shown in Figure 7(a) and 7(b), the absorbance of the MEH-PPV films are significantly affected by the solvents and concentration of the solutions. The highest absorption peak of toluene in the wavelength range of 450 nm to 550 nm is given by the film with concentration of 5 mgml⁻¹. The peaks observed to be decreased as the concentration increased to 6 mgml⁻¹ and 7 mgml⁻¹. However for films using 1, 2-DCB, the absorption of the films increased with the solution concentration.

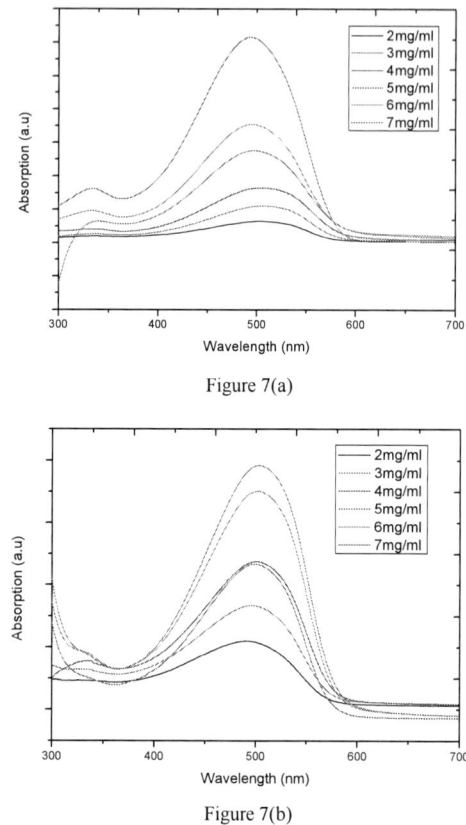

Figure 7(a)

Figure 7(b)

Figure 7. Absorption for (a) toluene and (b) 1, 2-DCB

The absorption coefficient of MEH-PPV thin film in different solvents can be obtained from the transmittance spectra as indicated in Figure 7 and calculated by using Lambert's Law as in equation 3:

$$\alpha = \frac{1}{t} ln \left(\frac{100}{T} \right)$$

(3)

where α is the absorption coefficient, t is the thin film thickness (nm) and T is the transmittance of the thin film. The determination of optical band gap energy of thin film was obtained by applying Tauc's Plot formula by extrapolation of the linear portion to the photon energy axis.

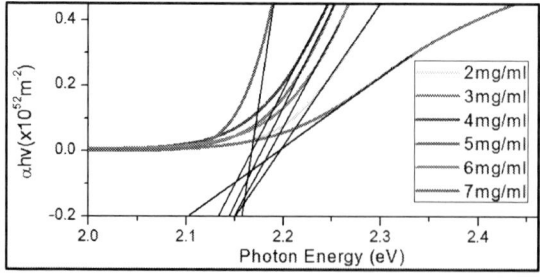

Figure 8(a)

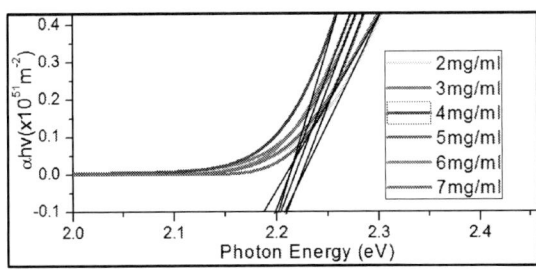

Figure 8(b)

Figure 8. Tauc's plot determined by transmittance data samples from (a) toluene and (b) 1, 2-DCB

Figure 8 shows the Tauc's plot determined by transmittance data samples and Table 2 shows the value of the band gap for each thin film from toluene and 1,2-DCB. From the results, both thin films from toluene and 1, 2-DCB has the highest band gap energy at 5 mgml^{-1} of solution concentration. These data also shows that by average, 1, 2-DCB requires higher energy to excite the electron form valence band to conduction band. Lower optical band gaps of films in toluene also proved that MEH-PPV in toluene has longer conjugation segments [13].

TABLE 3
OPTICAL BAND GAP OF MEH-PPV AT DIFFERENT SOLUTION CONCENTRATIONS

Thin Film Concentration (mgml^{-1})		2	3	4	5	6	7
Optical band gap (eV)	Toluene	2.15	2.14	2.13	2.16	2.16	2.10
	1,2 DCB	2.21	2.20	2.20	2.21	2.20	2.19

IV. CONCLUSIONS

The morphology, electrical and optical properties of MEH-PPV thin films with different solvents and concentration investigated. The surface roughness observed by AFM indicate that thin films from toluene was rougher than 1, 2-DCB. For film in toluene, roughness of film above 5mgml^{-1} was attributed from the aggregation of the films. From the electrical properties of MEH-PPV thin film, it can be concluded that 1, 2-DCB offers better conductivity compared to toluene. In terms of optical properties, PL spectra show that for both solvents, the intensity increased with the concentration. However the spectra become quenched as the concentration of MEH-PPV in toluene increased above 5 mgml^{-1}. The overall PL intensity of films in toluene is higher and more red shifted compared to films in 1, 2-DCB. The same phenomena was also observed from the UV-Vis absorption spectra. A broader PL spectra exhibit by toluene compared to 1, 2-DCB. Moreover, MEH-PPV films prepared from toluene is more red-shifted compared to 1, 2-DCB. Lower optical band gap for films in toluene were attributed to the longer conjugation length of the MEH-PPV.

ACKNOWLEDGEMENT

The authors would like to thank Universiti Teknologi MARA and Ministry of Higher Education, Malaysia for supporting this research.

REFERENCES

[1] M. S. AlSalhi, J. Alam, L. A. Dass, and M. Raja, "Recent Advances in Conjugated Polymers for Light Emitting Devices," *International Journal of Molecular Sciences*, vol. 12, pp. 2036-2054, 2011.

[2] J. Liu, W. Wang, H. Yu, Z. Wu, J. Peng, and Y. Cao, "Surface ligand effects in MEH-PPV/TiO2 hybrid solar cells," *Solar Energy Materials and Solar Cells*, vol. 92, pp. 1403-1409, 2008.

[3] J. H. Burroughes, D. D. C. Bradley, A. R. Brown, R. N. Marks, K. Mackay, R. H. Friend, P. L. Burns, and A. B. Holmes, "Light-emitting diodes based on conjugated polymers," *Nature*, vol. 347, pp. 539-541, 1990.

[4] A. Kraft, A. C. Grimsdale, and A. B. Holmes, "Electroluminescent Conjugated Polymers - Seeing Polymers in a New Light," *Angewandte Chemie, International Edition*, vol. 37, p. 402, 1998.

[5] N. Juhari, W. H. A. Majid, and Z. A. Ibrahim, "Structural and Optical Studies of MEH-PPV using Two Different Solvents Prepared by Spin-coating Technique," *Journal of Solid State Science and Technology*, vol. 15, pp. 141-146, 2007.

[6] B. Kang, Y. Yang, L. Wang, and Y. Qiu, "Solvent induced semiconductor type conversion of MEH-PPV investigated by surface photovoltage spectra," *Displays*, vol. 25, pp. 57-60, 2004.

[7] C. C. Chang, C. L. Pai, W. C. Chen, and S. A. Jenekhe, "Spin Coating of Conjugated Polymers for Electronic and Optoelectronic Applications," *Thin Solid Films*, vol. 479, pp. 254-260, 2005.

[8] P. C. Sukanek, "Dependence of film thickness on speed in spin coating," *Journal of the Electrochemical Society*, vol. 138, p. 1712, 1991.

[9] W. C. Ou-Yang, T. Y. Wu, and Y. C. Lin, "Supramolecular Structure of Poly [2-methoxy-5-(2'-ethylhexyloxy)-1, 4-phenylenevinylene](MEH-PPV) Probed Using Wide-angle X-ray Diffraction and Photoluminescence," *Iranian Polymer Journal*, vol. 18, pp. 453-464, 2009.

[10] T. Q. Nguyen, V. Doan, and B. J. Schwartz, "Conjugated polymer aggregates in solution: Control of interchain interactions," *The Journal of chemical physics*, vol. 110, p. 4068, 1999.

[11] S. Quan, F. Teng, Z. Xu, L. Qian, Y. Hou, Y. Wang, and X. Xu, "Solvent and concentration effects on fluorescence emission in MEH-PPV solution," *European Polymer Journal*, vol. 42, pp. 228-233, 2006.

[12] S. H. Chen, A. C. Su, Y. F. Huang, C. H. Su, G. Y. Peng, and S. A. Chen, "Supramolecular Aggregation in Bulk Poly(2-methoxy-5-(2'-ethylhexyloxy)-1,4- phenylenevinylene)," *Macromolecules*, vol. 35, pp. 4229-4232, 2012/06/17 2002.

[13] F. Kong, S. Zhang, C. Yang, and R. Yuan, "Interchain excited states in annealed poly [2-methoxy-5-(2'-ethyl-hexyloxy)-phenylene vinylene] films," *Materials Letters*, vol. 60, pp. 3887-3890, 2006.

978-1-4673-2395-6/12 $31.00 © 2012 IEEE

Temperature Effect on Quantum Capacitance Zig-Zag Graphene Nanoscrolls (ZGNS) (16,0)

Afiq Hamzah[1], M.T.Ahmadi*[1], Mohammad Javad Kiani[1,2], Fatimah. K. A. Hamid[1], Azlin Bahador[1], Razali Ismail[1]

[1]Department of Electronic Engineering, Faculty of Electrical Engineering, Universiti Teknologi Malaysia, Johor Bahru, 81310 Johor, Malaysia
[2]Department of Electrical Engineering, Islamic Azad University, Yasooj branch, Yasooj, Iran.
Tel No.: +6075535246, Fax No.: +607-5566272,
*Email address: taghi@fke.utm.my

Abstract - **Device scaling of the electronic devices has brings the dominancy of quantum effect in nano-size device characterization. This paper presented the first band analytical model of the quantum capacitance for (16,0) zig-zag graphene nanoscroll (ZGNS). The derivation of quantum capacitance is based on the differentiation of carrier density towards the Fermi energy. The Taylor's series expansion is employed on parabolic energy band structure so that it can be modified in the form of Fermi Intergal. Owing to its unique geometry structure that provides high intercalation area, it is expected that ZGNS exhibit high quantum capacitance.**

I. INTRODUCTION

A Graphene nanoscroll (GNS) is a novel carbon-based structure which is made by rolling a layer of graphene sheet in the form of Archimedean type spiral [1-3]. It is also known as the "Swiss roll" graphene due to its spiral shape structure [4]. In distinct to the carbon nanotube (CNT), GNS has an open edge along the translational axis. Since GNS is a scroll graphene layer, the chirality can be describe as $\vec{C} = n\vec{a_1} + m\vec{a_2}$ where the nomenclature can be describe as zigzag for $\theta = 0$, armchair for $\theta = 90°$, and chiral for $0 < \theta < 90°$ [1] as shown in Figure 1(B). GNS exhibit fascinating properties due to its novel structure [4-7] and it cannot be determined uniquely by its chirality (n,m) [4] since the tuneable core size that alters its properties that signify the GNS properties dependence on its geometry structure [6, 7]. Moreover, GNS interlayer galleries can be used to intercalate with dopants and the diameter can be expand to accommodate the volume of the dopant-layer interactions [8-11]. Because of its novel structure, researchers speculate the variety of promising applications to be implemented using the GNS especially as energy storage devices [11-14]. Despite rapid advances research in carbon-based material [15] transport properties, the carrier statistics and electrostatic temperature dependence properties such as quantum capacitance remain unexplored for GNS.

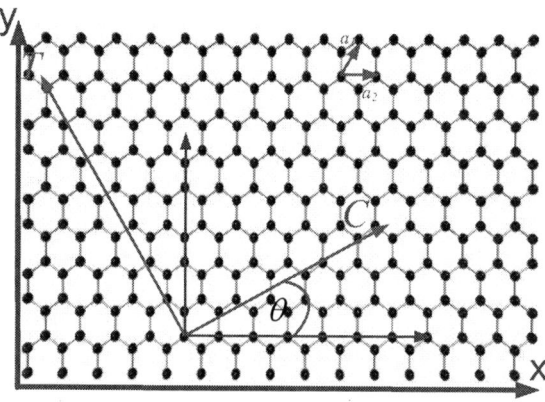

Fig. 1. Shows a graphene layer as the precursor to the formation of GNS where θ is the scroll angle with respect to the xy axes, C and T is the scroll vector and translational vector respectively.

Owing to the GNS geometry and temperature dependence properties, the quantum capacitance of a semiconducting zigzag GNS (ZGNS) at (16,0) chirality in a semi-classical regime is reported. The quantum capacitance is derived based on the parabolic energy dispersion approximation. In this paper, the ZGNS of chirality (16,0) is used and the energy dispersion is equated as [4]

$$E_{ZGNS}(k_x) = \pm t\sqrt{1 + 4cos\left(\frac{3k_x a_{cc}}{2}\right)cos\left(\frac{2\pi v - \theta}{2n}\right) + 4cos^2\left(\frac{2\pi v - \theta}{2n}\right)} \quad (1)$$

where the circumferential direction is around the y-axis, thus $\sqrt{3}k_y a_{cc}/2 = (2\pi v - \theta)/2n$ is replaced into the graphene general equation to form Eq. (1) and θ is the only parameter that control the geometry of the GNS which also modulate the energy gap [6] and t is the nearest neighbour C-C overlap energy which is between 2.5eV to 3.2eV. As a consequence to Chen's et al., $\theta = 1.9242$ for ZGNS (16,0). k_x is the wave vector along the x-axis and a_{cc} is the length of carbon-carbon atom and n is the chirality of ZGNS (n,m). The derivation of the quantum capacitance is based on the approximation of Maxwell-Boltzmann within 5% agreement with Fermi-Dirac probability distribution law in parabolic function is employed for estimating Fermi level for classical regime of 1D ZGNS and high carrier density effect with probability of occupation equal to one in band energy is approximated for degenerate regime.

II. QUANTUM CAPACITANCE

Capacitance is an essential parameter that helps to gain physical insights in determining the MOSFET characteristics and the electrical properties such as carrier statistic, conductance and mobility [16, 17]. There have been numbers of research on GNS characteristic through molecular dynamic simulation and fabrication process [5, 18-20]. Numerical simulation does give accurate result but very time consuming for fast circuit simulation [17]. Thus by analytically model the quantum capacitance of GNS, could provide the accessibility on its device physics and the device characteristic performance can be measured [16, 21, 22]. Because of its interlayer galleries which able to intercalated with dopant atoms, signify the concept of generating electricity from solar energy [8, 23]. As the interconnect capacitance increased due to the quantum confinement and CMOS scaling, it can also store the harvested photon energy without the battery materials to be integrated on-chip. C_q is the quantum capacitance which was significantly affected by the quantum confinement into a nanoscale size device. Thus, the application of the quantum capacitance has been considered for nanoscale devices modelling especially in the carbon-based material [16, 21, 24]. The quantum capacitance limit is dominant in one dimensional (1D) device compare to the conventional two-dimensional (2D) and three-dimensional (3D) structures. As for the case of GNS, research shows that it exhibit 1D properties [1, 4]. A general expression for quantum capacitance is

$$ C_q = \frac{\partial Q}{\partial V} = \frac{e^2 \partial n}{\partial E} \ , \tag{2} $$

where $\partial Q = e \cdot \partial n$ is the change in charge per unit length. $\partial V = \partial E / e$ is the voltage differential applied to the device and e is the magnitude of an electron charge. For 1D devices, the numbers of electrons/cm with energies between E and E+dE is established as $D(E)f(E, E_F)\,dE$. Thus the total of carrier concentration within a band can be obtained by integrating the Fermi-Dirac distribution function against energy that equated as [17]

$$ n = \int_{E_C}^{E_F} D(E) f(E, E_F)\,dE \tag{3} $$

whereby $D(E)$ is the density of states (DOS) for the ZGNS. Since ZGNS is a confined 1D structure, hence the carrier density is defined as [25]

$$ n_{1D}(E_F) = N_{1D} F_{-1/2}(\eta_F) \ . \tag{4} $$

N_{1D} is the 1D effective DOS and $F_{-1/2}(\eta_F)$ is Fermi integral of order minus half and $\eta_F = (E_F - E_C)/K_B T$ is the normalized Fermi energy, The effective DOS for ZGNS (16,0) is $N_{1D} = 2\sqrt{K_B T}/(3a_{cc}\sqrt{8\pi t})$. In order to fulfil the constraints for the calculations and physical interpretation, the first unfilled energy level (conduction band) at $T = 0\,K$ is being considered where the probability for electrons to occupy is high. Then, by employing the Taylor's series expansion, Eq. (1) is approximated to be

$$ E_{zz(16,0)}(k_x) = \pm t\left(0.5 - \frac{9k_x^2 a_{cc}^2}{8}\right). \tag{5} $$

The quantum capacitance only considered within a parabolic band. Eq. (5) indicate that the parabolic energy dispersion for (16,0) ZGNS as illustrated in Figure 2.

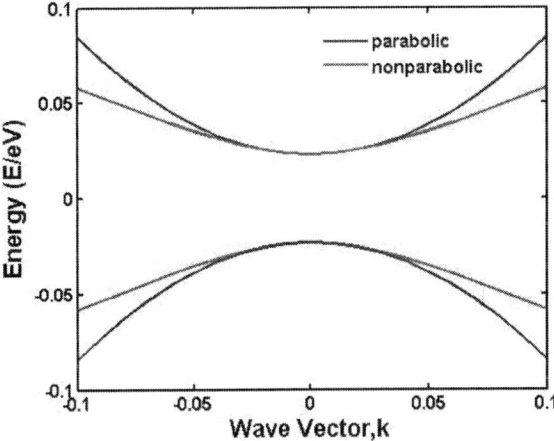

Fig. 2. The energy dispersion of ZGNS near the minimum energy is parabolic which will be useful for order determination of known Fermi integral form in calculating the quantum capacitance.

By differentiating energy with respect to the wave vector, the DOS (normalized per unit length) for infinite length of ZGNS is obtained as in Eq. (6)

$$ D(E) = \frac{\Delta n}{\Delta E L_x} = \frac{2}{3\pi a_{cc}\sqrt{8t}} (E - 0.5t)^{-\frac{1}{2}}. \tag{6} $$

The total summation of the product of DOS and Fermi-Dirac distribution function is obtained as in Eq. (7) by resolving in terms of η where $\eta = (E_F - 0.5t)/K_B T$ and $x = (E - 0.5t)/K_B T$. The carrier concentration to be the total numbers of electron resides within the conduction band. K_B is the

Boltzmann constant and T is a temperature in Kelvin. The expression for the total DOS is the same for the conduction band and valence band and exhibits Van Hove singularities at energies E is depicted as the bottom of the sub-band (assumed one) [25].

$$n = \frac{2\sqrt{K_B T}}{3\pi a_{cc}\sqrt{8t}} \int_{E_C}^{E_{top}} \frac{(x)^{-\frac{1}{2}}}{\exp(x-\eta)+1} dx \ . \tag{7}$$

Hence, the quantum capacitance at a Dirac point or charge neutrality is obtained by the differentiation of carrier concentration against the states of energy level which is

$$C_Q = \frac{2e^2}{3\pi a_{cc}\sqrt{8t}} \frac{(x)^{-\frac{1}{2}}}{\exp(x-\eta)+1} \tag{8}$$

In the non-degenerate regime, "1" from the denominator is neglected. At room temperature (T=300K), the states within the conduction band (allowed band) are partially filled, thus the exponential part of Eq. (8) is big enough for "1" to be neglect. Thus in non-degenerate regime, the quantum capacitance is obtained as

$$C_{QND} = \frac{2e^2 (x)^{-\frac{1}{2}}}{3\pi a_{cc}\sqrt{8t}} \exp(x-\eta). \tag{9}$$

While for degenerate regime, the exponential part is very small because of the probability for electron to fill all the available states up to Fermi level one is "1", signifying that there are no available states within the conduction band for $E - E_F < 3K_B T$. Therefore the quantum capacitance within degenerate regime is

$$C_{QD} = \frac{2e^2 (x)^{-\frac{1}{2}}}{3\pi a_{cc}\sqrt{8t}}. \tag{10}$$

From the result obtained in Figure 3, at low concentration the non-degenerate regime can be approximate using Maxwell-Boltzmann approximation within 5 per cent of each other, which is valid for approximately $\eta < -1$. It is also shows that quantum capacitance operating in the degenerate regime where it reach degenerate limit at $\eta > 9$ for ZGNS (16,0) at $\theta = 1.9242$. At this particular condition, it can be seen that for ZGNS (16,0) at ground state, the quantum capacitance has reach its limit at $\cong 1.7 \times 10^{-10}$F/m @ 170 pF/m. The dominancy of the quantum capacitance can be describe at different concentration level. In degenerate regime, quantum capacitance has more influence for the fact that carrier concentration within the conduction band exceeds the density of states. Therefore the Fermi level resides in the conduction band.

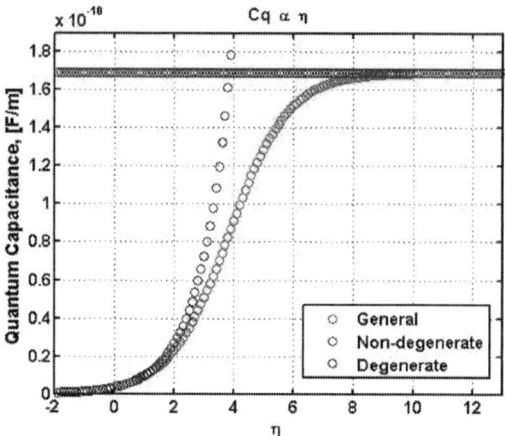

Fig. 3. The validity of the approximation on both classical and degenerate regime based on the general quantum capacitance curve.

The temperature dependence on the quantum capacitance for the ZGNS (16,0) is shown in Figure 4 where the non-degenerate regime has cause the normalized Fermi level, η to decreased (shifted to the left) by increasing the temperature. The distance between Fermi energy to the conductance band is decreased. By intuition, the electron can be easily elevated to the conduction band by the energy provided by the temperature so that the electron moves freely within the 1D GNS structure to conduct current. Degenerate regime is theoretically not affected by the temperature due to the assumption of the high carrier density effects that cause the probability of the occupation for conduction band is one. Thereby, the quantum capacitance in degenerate regime is temperature independent and according to Kliros et al. that the temperature independent is due to its Fermi energies level higher than the broadening parameter [26]. Furthermore, as the temperature increased the quantum capacitance reach its limit at low η. Since the quantum capacitance is a function of the total charge, the electron concentration is vastly increased by the temperature increment causing the quantum capacitance reach its limit faster.

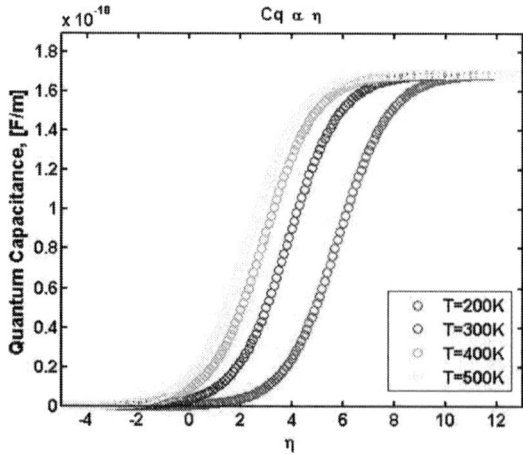

Fig. 6. The quantum capacitance temperature dependence for ZGNS (16,0).

III. CONCLUSION

As device is being shrunk into nanoscale topology, quantum mechanics play important role in determining the carrier transport. In 1D structure, it is very important to consider the quantum transport in analytical model. Quantum capacitance is one of the parameter that needs to be considered. In this paper we model the quantum capacitance by applying the first derivative of Taylor's expansion approximation on energy dispersion relation. The Maxwell-Boltzmann's distribution law is employed at low concentration where Fermi level within a range of $3K_BT$ for a validity of $\eta < -1$ where the Maxwell-Boltzmann probability function assumption is in accordance with the Fermi-Dirac distribution function. A simple expression is used as the quantum capacitance reach degeneracy at $170\,pF/m$ for $\eta < 3K_BT$. By having the quantum capacitance model, one can predict the efficiency of its storage capability at different temperature condition.

ACKNOWLEDGEMENT

Authors would like to acknowledge the financial support from Research University grant of the Ministry of Higher Education (MOHE), Malaysia under Projects Q.J130000.7123.02H24 and Q.J130000.7123.02H04. Also thanks to the Research Management Center (RMC) of Universiti Teknologi Malaysia (UTM) for providing excellent research environment in which to complete this work.

REFERENCES

[1] S. F. Braga, V. R. Coluci, S. B. Legoas, R. Giro, D. S. Galvao, and R. H. Baughman, "Structure and dynamics of carbon nanoscrolls," *Nano Letters,* vol. 4, pp. 881-884, 2004.

[2] L. M. Viculis, J. J. Mack, and R. B. Kaner, "A chemical route to carbon nanoscrolls," *Science,* vol. 299, pp. 1361-1361, 2003.

[3] M. Grundmann, "Nanoscroll formation from strained layer heterostructures," *Applied Physics Letters,* vol. 83, pp. 2444-2446, 2003.

[4] Y. Chen, J. Lu, and Z. X. Gao, "Structural and electronic study of nanoscrolls rolled up by a single graphene sheet," *Journal of Physical Chemistry C,* vol. 111, pp. 1625-1630, 2007.

[5] A. K. Schaper, H. Q. Hou, M. S. Wang, Y. Bando, and D. Golberg, "Observations of the electrical behaviour of catalytically grown scrolled graphene," *Carbon,* vol. 49, pp. 1821-1828, 2011.

[6] T. S. Li, M. F. Lin, and J. Y. Wu, "The effect of a transverse electric field on the electronic properties of an armchair carbon nanoscroll," *Philosophical Magazine,* vol. 91, pp. 1557-1567, 2011.

[7] T. S. Li, M. F. Lin, S. C. Chang, and H. C. Chung, "Optical excitations in carbon nanoscrolls," *Physical Chemistry Chemical Physics,* vol. 13, pp. 6138-6144, 2011.

[8] X. H. Shi, N. M. Pugno, and H. J. Gao, "Tunable Core Size of Carbon Nanoscrolls," *Journal of Computational and Theoretical Nanoscience,* vol. 7, pp. 517-521, 2010.

[9] X. H. Shi, Y. Cheng, N. M. Pugno, and H. J. Gao, "Tunable Water Channels with Carbon Nanoscrolls," *Small,* vol. 6, pp. 739-744, 2010.

[10] X. A. Peng, J. Zhou, W. C. Wang, and D. P. Cao, "Computer simulation for storage of methane and capture of carbon dioxide in carbon nanoscrolls by expansion of interlayer spacing," *Carbon,* vol. 48, pp. 3760-3768, 2010.

[11] G. Mpourmpakis, E. Tylianakis, and G. E. Froudakis, "Carbon nanoscrolls: A promising material for hydrogen storage," *Nano Letters,* vol. 7, pp. 1893-1897, 2007.

[12] V. R. Coluci, S. F. Braga, R. H. Baughman, and D. S. Galvao, "Prediction of the hydrogen storage capacity of carbon nanoscrolls," *Physical Review B,* vol. 75, 2007.

[13] S. F. Braga, V. R. Coluci, R. H. Baughman, and D. S. Galvao, "Hydrogen storage in carbon nanoscrolls: An atomistic molecular dynamics study," *Chemical Physics Letters,* vol. 441, pp. 78-82, 2007.

[14] V. R. Coluci, S. F. Braga, R. H. Baughman, and D. S. Galvao, "Hydrogen storage in carbon nanoscrolls: A molecular dynamics study," *Hydrogen Cycle-Generation, Storage and Fuel Cells,* vol. 885, pp. 153-158, 2006.

[15] T. Fang, A. Konar, H. Xing, and D. Jena, "Carrier Statistics and Quantum Capacitance of Graphene Sheets and Ribbons," *Applied Physics Letters,* vol. 91, p. 3, 2007.

[16] M. T. Ahmadi, J. F. Webb, N. A. Amin, S. M. Mousavi, H. Sadeghi, M. R. Neilchiyan, and R. Ismail, "CARBON NANOTUBE CAPACITANCE MODEL IN DEGENERATE AND NONDEGENERATE REGIMES," in *Proceedings of the Fourth Global Conference on Power Control and Optimization.* vol. 1337, N. W. J. F. V. P. Barsoum, Ed., ed, 2011, pp. 173-176.

[17] J. Liang, D. Akinwande, and H. S. P. Wong, "Carrier density and quantum capacitance for semiconducting carbon nanotubes," *Journal of Applied Physics,* vol. 104, Sep 15 2008.

[18] H. Q. Zhou, C. Y. Qiu, H. C. Yang, F. Yu, M. J. Chen, L. J. Hu, Y. J. Guo, and L. F. Sun, "Raman spectra and temperature-dependent Raman scattering of carbon nanoscrolls," *Chemical Physics Letters,* vol. 501, pp. 475-479, 2011.

[19] J. Zheng, H. T. Liu, B. Wu, Y. L. Guo, T. Wu, G. Yu, Y. Q. Liu, and D. B. Zhu, "Production of High-Quality Carbon Nanoscrolls with Microwave Spark Assistance in Liquid Nitrogen," *Advanced Materials,* vol. 23, pp. 2460-+, Jun 3 2011.

[20] D. Xia, Q. Z. Xue, J. Xie, H. J. Chen, C. Lv, F. Besenbacher, and M. D. Dong, "Fabrication of Carbon Nanoscrolls from Monolayer Graphene," *Small,* vol. 6, pp. 2010-2019, 2010.

[21] V. Parkash and A. K. Goel, "Quantum Capacitance Extraction for Carbon Nanotube Interconnects," *Nanoscale Research Letters,* vol. 5, pp. 1424-1430, Sep 2010.

[22] J. Xia, F. Chen, J. Li, and N. Tao, "Measurement of the quantum capacitance of graphene," *Nature Nanotechnology,* vol. 4, pp. 505-509, Aug 2009.

[23] H. Zhu, J. Wei, K. Wang, and D. Wu, "Applications of carbon materials in photovoltaic solar cells," *Solar Energy Materials and Solar Cells,* vol. 93, pp. 1461-1470, Sep 2009.

[24] L. Wei, D. J. Frank, L. Chang, and H. S. P. Wong, "Noniterative Compact Modeling for Intrinsic Carbon-Nanotube FETs: Quantum Capacitance and Ballistic Transport," *Ieee Transactions on Electron Devices,* vol. 58, pp. 2456-2465, Aug 2011.

[25] M. Lundstrom and J. Guo, "Basic Concept," in *Nanoscale transistors: device physics, modeling and simulation,* ed 233 Spring Street, New York, NY 10013, USA: Springer Science+Business Media, Inc, 2006 pp. 1-10.

[26] G. S. Kliros, "Quantum capacitance of bilayer graphene," in *Semiconductor Conference (CAS), 2010 International,* 2010, pp. 69-72.

[27] X. Xie, L. Ju, X. F. Feng, Y. H. Sun, R. F. Zhou, K. Liu, S. S. Fan, Q. L. Li, and K. L. Jiang, "Controlled Fabrication of High-Quality Carbon Nanoscrolls from Monolayer Graphene," *Nano Letters,* vol. 9, pp. 2565-2570, 2009.

Effects of Annealing Temperature on Morphology and Crystallinity of Nitrogen Doped Zinc Oxide (ZnO:N) Nano Films

J. Karamdel [1], F. Razaghian [1], A. Hadi [2], C. F. Dee [3] and B. Y. Majlis [3]

[1] Electrical Department, Faculty of Engineering, Islamic Azad University-South Tehran Branch
No. 209 North Iranshahr ave. Tehran, Iran
[2] Department of Electrical Engineering, Islamic Azad University, Science And Research Branch, Tehran, Iran
[3] Institute of Microengineering and Nanoelectronics (IMEN), Universiti Kebangsaan Malaysia,
43600 Bangi, Selangor, Malaysia
j-karamdel@azad.ac.ir and burhan@vlsi.eng.ukm.my

Abstract: **Semiconductor of ZnO has been extensively researched in recent years for its extraordinary properties. ZnO is naturally an n-type semiconductor and due to asymmetric doping limitations, it is difficult to obtain p-type ZnO. In this work the deposited nitrogen doped zinc oxide nano films by reactive magnetron sputtering technique, were treated using conventional thermal annealing, while, the annealing temperature were varied from 300 °C to 800 °C in a mixture of nitrogen and oxygen ambient. The surface morphology, Crystallinity and electrical characteristics of prepared films have been investigated with respect to the temperature of annealing process. The XRD spectra of samples before and after annealing processes confirmed the deposition of wurtzite crystalline structures of ZnO. However, the annealed samples exhibited smaller FWHM compared to un-annealed ones, which confirms better crystalline structure of annealed films. Moreover, un-annealed specimens showed n-type conductivity with an electron concentration of 2.5×10^{16} cm^{-3}, while the annealed samples exhibited p-type behavior with a hole concentration of 8.2×10^{15} cm^{-3}.**

Keywords: **Thin film, doped ZnO, annealing, conductivity**

I. INTRODUCTION

Zinc oxide (ZnO) thin films have been widely used in various electronic and optoelectronic devices such as thin film transistors, light emitting diode, UV lasers, solar cells, flat panel displays and buffer layers, because of the combination of extraordinary electrical properties and high transparency[1-6]. For realization of such applications, both n-type and p-type semiconductors are needed [7]. Due to intrinsic defects such as oxygen vacancy (V_O) and zinc interstitial (Zn_i), pure ZnO is an n-type semiconductor and different electron conductivity can be achieved by adding appropriate dopants. Although, many research works have been done on fabrication of p-type ZnO nanostructures, still, reliable generation of p-type ZnO is a big challenge in this field [7-10].

Nitrogen is known as the potential material as acceptor dopants in ZnO nanofilms [10-13]. Beside of some reports on successful fabrication and characterization of p-type ZnO:N thin films [11-13], some researchers have reported that, the solubility rate of N is low and also it is unstable with variation of temperature. Even it is reported that N doped ZnO behave as n-type semiconductor [10, 15]. Therefore, in order to utilize ZnO:N in device applications, more investigations and study are needed to clarify the low solubility and thermal instability of ZnO:N nano films.

In this work, thin films of ZnO:N were deposited on Si substrate by reactive magnetron sputtering technique and then were annealed at different temperatures using conventional thermal annealing process. Subsequently, the morphology and electrical characteristics of deposited films were investigated with respect to the temperature of annealing process.

II. EXPERIMENTAL

The ZnO:N specimens were made by depositing a 80 nm thin layer of ZnO:N on a 500 μm thick n-type Si wafer (100) using reactive RF magnetron sputtering method. The source of ZnO was a pure ZnO target (99.9%) and the sputtering gases were a mixture of high purity argon and nitrogen. The cleaned Si wafers, by normal cleaning process, were loaded into the sputtering chamber. Subsequently, the chamber was evacuated to base pressure of 3×10^{-6} Torr using a rotary and turbo molecular pump. Since the dopant source is nitrogen gas, various concentration of N doped ZnO film is obtainable by control of nitrogen gas. The typical values of sputtering parameters are shown in table I.

TABLE I
SPUTTERING PARAMETERS

Substrate temperature	200°C
Targets	Zinc oxide (99.9%)
Gas flow rate	Ar (4 sccm) N2 (12-14 sccm)
Base pressure	2×10^{-5} torr
RF Power	150 W
Sputtering pressure	17 mtorr
Deposition time	20 minutes

In order to stabilize and activate the doped N atoms inside the crystalline ZnO lattice, the sputtered samples were annealed at different temperature and conditions in a horizontal quartz furnace. The temperature of annealing was varied from 300°C to 800°C in a mixture of nitrogen and oxygen ambient. Surface topology and morphology of the thin films were investigated by atomic force microscopy (AFM) (Veeco di Dimension V), and field emission scanning electron microscopy (FESEM) (Zeiss- SUPRA 55VP), respectively. To examine elemental composition of layers, energy dispersive x-ray (EDX) instrument (Oxford Inca Penta FETx3) that has been attached to FESEM was utilized. X-ray diffractometer (XRD) (Panalytical Diffractometer system- XPERT-PRO) was used to investigate the quality of crystalline nano films. Moreover, Hall Effect measurement was performed for understanding the conductivity type and carrier concentration of sputtered and annealed samples.

III. RESULTS AND DISCUSSIONS

The surfaces of as sputtered and annealed samples with different temperatures were investigated by FESEM. Fig. 1(a) to Fig. 1(d) show the FESEM micrographs of sputtered ZnO:N, annealed ZnO:N at 400 °C , annealed ZnO:N at 600 °C and annealed ZnO:N at 800°C, respectively. These figures demonstrate that the films have a smooth surface and are formed from small grains. In addition, it can be obviously seen that by increasing the annealing temperature up to 800°C, the average grain sizes become bigger. That is, the average grain size for un-annealed sample was measured around 15 nm, while, for annealed sample at 800°C was measured around 30 nm.

Shown in Fig. 2 is the surface topography of samples, which are imaged by AFM. The results exhibit smooth and uniform surface for all specimens, verifying the achieved results from the FESEM. The average roughness of samples was measured 0.97 nm, 1.47 nm and 2.21 nm for un-annealed, annealed at 400°C and annealed at 600°C, respectively.

The EDX spectra of samples revealed the incorporation of N atoms inside the crystalline structure of ZnO. The amount of nitrogen in the samples were decreased dramatically after annealing at higher temperatures, in which, the measured percentage of N (atomic) in the sputtered samples were 13%, while it decreased to less than 1% after annealing at 800 °C. This intensive reduction in the N concentration of the ZnO:N thin films via annealing process has been reported in our previous paper [10].

In the XRD spectra of both annealed and un-annealed specimens, which are shown in Fig. 3, only the (002) peak were observed. It confirms that the deposited ZnO:N films are single crystalline with the wurtzite hexagonal phase and preferentially grow along the c-axis direction.

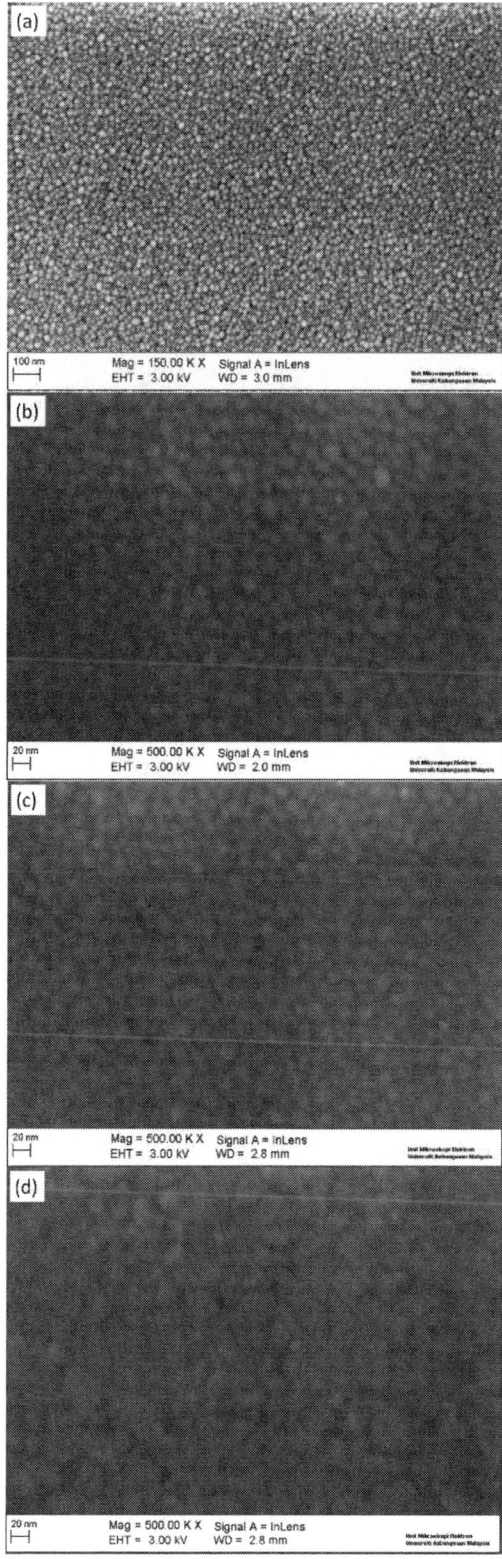

Fig. 1. Top-view FESEM images of the ZnO:N thin film on Si substrates (a) un-annealed sample (b) annealed at 400 °C (b) annealed at 600 °C and (c) annealed at 800 °C.

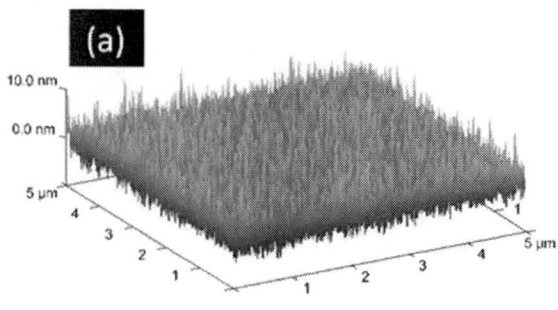

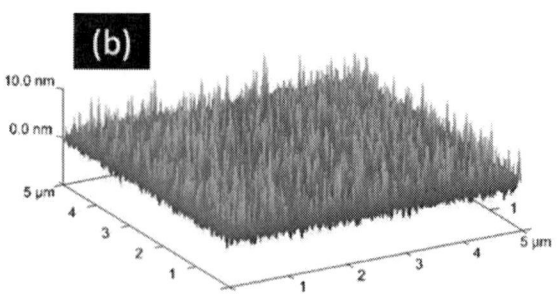

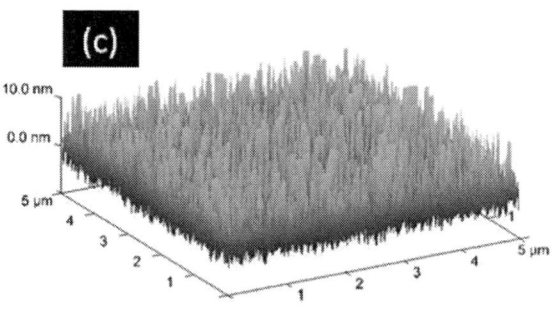

Fig. 2. AFM results of (a) sputtered sample, (b) annealed sample at 400 °C and (b) annealed sample at 600°C, show that the roughness of sample increase with temperature

More investigation on the spectra of samples revealed that the annealed samples exhibited smaller full width at half maximum (FWHM) (around 0.59 °2Theta) compared to the un-annealed one (around 0.68 °2Theta). Moreover, the (002) peak position of annealed sample at 800 °C was slightly shifted towards higher angles. Higher peak intensity and shift in position of (002) peak can be assigned to better crystallinity with more relaxation that is caused by thermal annealing.

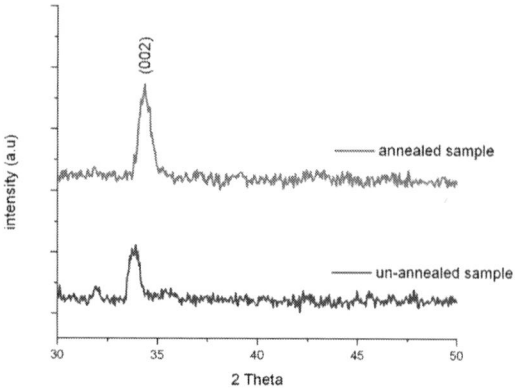

Fig. 3. XRD spectra of un-annealed and annealed ZnO:N thin films

Hall Effect measurements were performed to investigate the conductivity type and electrical characteristics of ZnO:N thin films for un-annealed and annealed films under different conditions. The results revealed that un-annealed N doped ZnO films and also annealed samples with temperature less than 600°C behaved as n-type semiconductor with an electron concentration of 2.5×10^{16} cm^{-3}, while the annealed samples showed p-type conductivity with a hole concentration of 8.2×10^{15} cm^{-3}.

IV. CONCLUSION

The N doped ZnO thin films were deposited on Si substrate by reactive magnetron sputtering and then were annealed by conventional thermal annealing methods. AFM, FESEM, XRD and Hall Effect measurements were performed to characterize the samples. All specimens showed a smooth surface with roughness of 0.97 nm and 2.21 nm for un-annealed sample and annealed sample at 600°C, respectively. The average grain size for un-annealed sample was measured around 15 nm, while, for annealed sample at 800°C was measured around 30 nm. The XRD spectra of samples confirmed that increasing the annealing temperature improve the crystallinity of samples, in which, the corresponding peaks have higher intensity with smaller FWHM. Hall Effect measurements showed n-type conductivity for un-annealed samples and annealed samples with annealing temperature of less than 600°C, while the annealed films showed p-type conductivity with a hole concentration of 8.2×10^{15} cm^{-3}.

ACKNOWLEDGEMENT

The authors wish to express their hearty thanks to Islamic Azad University-South Tehran Branch of Iran and the Ministry of higher education of Malaysia for their financial supports.

978-1-4673-2395-6/12 $31.00 © 2012 IEEE

REFERENCES

[1] J. Poortmans and V. Arkhipov, "Thin film solar cells fabrication, characterization and applications", 2006 Wiley

[2] M. Sasani Ghamsari, M. Vafaee, "Sol–gel derived zinc oxide buffer layer for use in random laser media", Mater. Lett. 62 (2008) 1754–1756.

[3] Javad Karamdel, A. Hadi, C. F. Dee and B. Y. Majlis, "Effects of substrate on surface morphology, crystallinity and photoluminescence properties of sputtered ZnO nano films", Ionics (2012) 18:203 – 207

[4] Z. L. Wang, "ZnO nanowire and nanobelt platform for nanotechnology", Mater. Sci. Eng. R 64 (2009) 33–71.

[4] Y.Y. Kim, C.H. Ahn, S.W. Kang, B.H. Kong, S.K. Mohanta, H.K. Cho, J.Y. Lee, H.S. Kim, "Temperature dependence of ZnO thin films grown on Si substrate", J. Mater. Sci.: Mater. Electron 19 (2008) 749–754.

[5] Y. Zhang , H. Zhang , X. Li , L. Dong and X. Zhong, Nanotechnology 21 095606 (2010)

[6] P.G. Li, X. Wang, W.H. Tang, Journal of Alloys and Compounds 479 (2009), P. 634-637

[7] S. Chakrabarti, B. Doggett, R. O'Haire, E. McGlynn, M.O. Henry, A. Meaney, J.-P. Mosnier, p-type conduction above room temperature in nitrogen-doped ZnO thin film grown by plasma-assisted pulsed laser deposition, Electronics Letters 42 (2006) 20.

[8] L. L. Yang, Z. Z. Ye, L. P. Zhu, Y. J. Zeng, Y. F. Lu, and B. H. ZHAO, "Fabrication of p-Type ZnO Thin Films via DC Reactive Magnetron Sputtering by Using Na as the Dopant Source" Journal of Elec. Materials, Vol. 36 4 498-501 (2007)

[9] Y.G. Wang, S.P. Lau, X.H. Zhang, H.W. Lee, H.H. Hng, B.K. Tay, "Observations of nitrogen-related photoluminescence bands from nitrogen-doped ZnO films", J. Crystal Growth 252 (2003) 265–269.

[10] J. Karamdel, C. F. Dee, and B. Y. Majlis, 'Characterization and aging effect study of nitrogen-doped ZnO nanofilm', Appl. Surface Science 256 (2010) 6164–6167

[11] A. Sandip Gangil, A. Nakamura, Y. Ichikawa, K. Yamamoto, J. Ishihara, T. Aoki, J. Temmyo, "P-type nitrogen-doped ZnO thin films on sapphire 1120 substrates by remote-plasma-enhanced metalorganic chemical vapor deposition", J. Crystal Growth 298 (2007) 486–490.

[12] A. Nakagawa, F. Masuoka, S. Chiba, H. Endo, K. Megro, Y. Kashiwaba, T. Ojima, K. Aota, I. Niikura, Y. Kashiwaba, "Photoluminescence properties of nitrogendoped ZnO films deposited on ZnO single crystal substrates by the plasmaassisted reactive evaporation method", Appl. Surf. Sci. 254 (October (1)) (2007) 164–166.

[13] M. Zheng, J. Wu, "One-step synthesis of nitrogen-doped ZnO nanocrystallites and their properties", Appl. Surf. Sci. 255 (2009) 5656–5661.

[14] H.W. Liang, Y.M. Lu, D.Z. Shen, Y.C. Liu, J.F. Yan, C.X. Shan, B.H. Li, Z.Z. Zhang, J.Y. Zhang, X.W. Fan, "P-type ZnO thin films prepared by plasma molecular beam epitaxy using radical NO", Phys. Status Solidi, A Appl. Res. 202 (2005) 1060– 1065.

[15] L. L. Kerr, X. Li, M. Canepa, A. J. Sommer, "Raman analysis of nitrogen doped ZnO", Thin Solid Films 515 (2007) 5282–5286.

Studies on the Growth of Alumina Nanoporous Film & Nanowires on Planar & Cylindrical Substrate

Tiong Teck Yaw, Abrar Ismardi, Dee Chang Fu, Burhannuddin Yeop Majlis, Member, IEEE

Institute of Microengineering and Nanoelectronics (IMEN)
Universiti Kebangsaan Malaysia (UKM)
43600 UKM Bangi, Selangor, Malaysia
Email: t_yawt05@yahoo.com, cfdee@ukm.my

Abstract- **2D-Planar and 3D-cylindrical nanoporous anodic aluminium oxide was fabricated by anodization of high purity aluminium. High yield of alumina nanowires were prepared from planar and cylindrical nanoporous aluminium oxide film by chemical etching in H_3PO_4 aqueous. The tensile stress at the aluminium oxide/aluminium interface corresponds to the formation of parallel cracks on the cylindrical aluminium oxide film. A possible mechanism for the formation of alumina nanopores and nanowires on planar and cylindrical substrate was proposed.**

I. INTRODUCTION

The synthesis of nanostructures [1-5] has received much interest due to their unique properties and potential applications in electronics [6], optoelectronics [7], magnetic memory [8] and sensors [9]. The precisely aligned and ordered nanostructures are significant in the fabrication of nanoelectronic devices. Up to date, high technology such as e-beam and ion-beam lithography is among the significant facilities used to fabricate a precise and well-controlled nanoelectronic device. There is a major drawback with the use of these sophisticated facilities to fabricate a nanodevice, where as it is very costly, time consuming, limited to laboratory scale and limited to two dimensional (2D) nanostructures. Therefore new techniques such as nano-imprinting [10], self-assembly [11] and template assisted growth [12] were developed to offer a cost effective, simple and easy controllable method for synthesis of multi-dimensional nanostructures and well aligned nanostructures. Among the template assisted growth methods, nanoporous anodic aluminium oxide (AAO) film has been employed as one of the frequently used template for synthesizing different dimensional of nanostructures over a macroscopic surface area [13-15]. The previous studies have proven that the pore diameter, distribution and density of the AAO template can be precisely controlled from few nanometers to several hundred nanometers by using the suitable electrolyte, voltage or current and well controlled reaction temperature [16-18].

On the other hand, alumina nanowires (NWs) have drawn considerable attention due to their large surface area, high strength, dielectric constant and chemical stability. It is interested in some possible applications in future nanotechnology such as water purification system [19], catalysis [20] and orthopaedic [21]. Alumina NWs have been extensively synthesized through chemical etching on AAO film since it is the most facile and promising method [22, 23].

The alumina NWs grown from this method shows high uniformity and subjected to high yield production.

To date, most of the studies have been done on synthesizing alumina nanopores and NWs were on planar platform. From our point of view, the synthesis of alumina nanopores and NWs on three dimensional (3D) substrates may enhance its performance in photonics, filtration and catalysis application. In this paper, we demonstrate the fabrication of AAO film and alumina NWs on both 2D-planar and 3D-cylindrical aluminium substrates. The possible formation mechanism of AAO film and alumina NWs on planar and cylindrical substrate will be further discussed.

II. EXPERIMENTAL METHOD

In this work, high purity (99.9 %) aluminium wires with the diameter of 0.15 cm were used as a starting material. Some of the aluminium rods were flattened in order to serve a flat surface for the anodization process to take place. Before anodize, the samples were ultrasonically cleaned in acetone and methanol, and then were annealed in the tube furnace at 400 °C for 3 hours with flowing nitrogen. The annealing process was subjected to remove the mechanical stress of the samples. Both cylindrical and flatten rods were polished by using typical metal polisher (Brasso) to a mirror finishing. The samples were cleaned with acetone and methanol followed by rinsing in de-ionized water for several times. The samples were then dipped in sodium hydroxide (1.0 M) for 1 min to eliminate surface oxide. Nail polish was applied on both curved sides of the flatted sample to ensure the growth of AAO only occur on the flat surface. Finally, anodization was performed in slowly stirred sulfuric acid (1.8 M) at a constant temperature of 25 °C for 30 min. The constant potential of 20 V was applied with platinum stripe serve as a cathode. The samples were further anodized for 90 min under the same condition without removing the first anodized layer. The as-prepared AAO on both samples were etched with 5 % phosphoric acid (H_3PO_4) for 30 min at 25 °C. The morphologies of the obtained samples were observed by field effect scanning electron microscopy (FESEM).

III. RESULT & DISCUSSION

A. Nanoporous Alumina Film

An AAO film was obtained by anodizing both the planar and cylindrical aluminium in sulfuric acid. Fig. 1 shows the

FESEM images of the as-grown AAO film surface for both 2D and 3D samples.

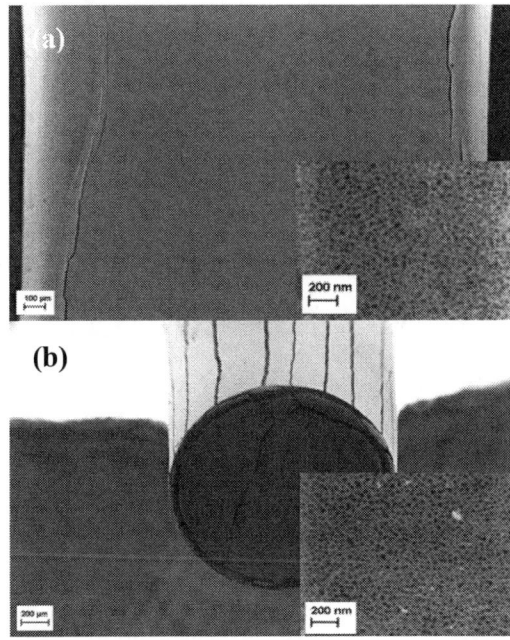

Fig. 1. FESEM images of the AAO film obtained on (a) planar surface and (b) cylindrical surface.

As shown in Fig.1 (a), there was no crack observed on the as-grown planar AAO film except at both edge of this sample. Since both the curved sides of this aluminium substrate were covered by the nail polish, the aluminium under the covered areas was not being anodized. During the anodization, the aluminium which was exposed to the selected electrolyte was oxidized to form AAO film. The oxidation process happened on the aluminium surface was mainly performed by the migration of oxygen containing ions from the electrolyte. Since the formed alumina is Al_2O_3 so the atomic density of the aluminium in the AAO film is lower than in metallic aluminium substrate. This manifests that the volume of the AAO film expands to about twice of the original volume of the aluminium. This volume expansion has lead to the compressive stress at the AAO/aluminium interface at the bottom, thus the continuous growing of AAO film has pushed the film that was formed earlier vertically upwards. Subsequently, it caused the cracks to form at both the curved sides along AAO/aluminium interface. On the other hand, there were many cracks observed along the AAO film grown on cylindrical sample shown in Fig. 1 (b). This image reveal that there was no any crack formed perpendicularly to the surface of cylinder sample but all of the cracks have been found to be longitudinal to the surface. The insets of Fig. 1 (a) and (b) shows the identical well defined alumina porous layer grown on both samples. The average pore diameters obtained were 30 nm. Even though there were many cracks being found on the cylindrical AAO film but they did not affect the pore formation and distribution.

The AAO porous film obtained on cylindrical substrate after the first and second anodization was shown in Fig. 2. As we were unable to see all of the cracks that were formed on the whole cylindrical AAO porous film from the figure, we were then decided to analyse it by observing the changes occurred on the upper part of the cylindrical sample.

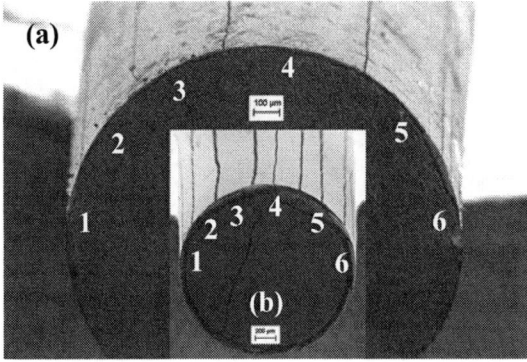

Fig. 2. FESEM images of the AAO porous film obtained on cylindrical substrate after anodizing at (a) 30 min and (b) 30 min + 90min

From Fig. 2 (a), we can clearly see that the cylindrical AAO porous film was divided into 6 parts which was caused by the cracks formed after 30 min of anodization. Compared to Fig. 2 (a), it was obviously seen that more cracks occur on the upper part of the cylindrical AAO film as shown in Fig. 2 (b). After the extended anodization, there were new cracks observed at part 1, 4, 5 and 6, and this has divided the upper part of the cylindrical AAO film into 10 parts. This manifest the anodizing duration is one of the factors that affect the crack formation on the cylindrical AAO film. From the analysis, we observed that the part with bigger surface area formed by the cracks after 30 min of anodization conserved higher tensile stress than part 2 and 3. Prolong of anodization process has led to thicker AAO film formation since more aluminium from the substrate was oxidized. This means the volume of AAO film was further increased. Since the continuous growing of AAO film pushes the existing film vertically upwards, thicker AAO film induces higher tensile stress that acting on it. The film cracks when this stress reaches its critical point. As we see in Fig 2. (b), new cracks were formed on part 1, 4, 5 and 6 since the critical stress has reached its maximum on these parts. The critical stress here refers to the large tensile stress that acting on the AAO film, which is able to separate the AAO film from the strong molecular bonding between each other. We can relate that the thicker the cylindrical AAO film, more cracks will be induced on it.

The micro-crack formed on the cylindrical AAO film is shown in Fig. 3. As observed, the bottom of the AAO film was partially separated about 6 μm apart. We can see that both sides of the separated AAO film have matching edges as labelled by the white dash ellipses in the image, Thus, it is confirmed that the gap seen between the both sides of the AAO porous film was formed due to the cracking effect but not by the dissolution of any defect parts of the AAO porous

978-1-4673-2395-6/12 $31.00 © 2012 IEEE   307

film in the electrolyte during the anodization process nor by the peeling off of void AAO film.

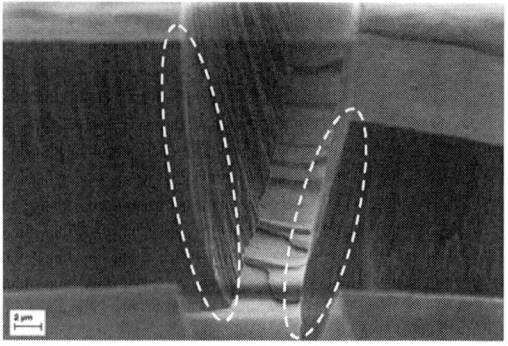

Fig. 3. FESEM image of a micro-crack formed at cylindrical AAO film

Fig. 4 shows the nanopores distribution of the pores formed on the top and bottom of the cylindrical AAO film which have been obtained before and after the formation of cracks. The randomly distributed nanopores were shown in Fig. 4 (a). It consisted of more pores compared to Fig. 4 (b).

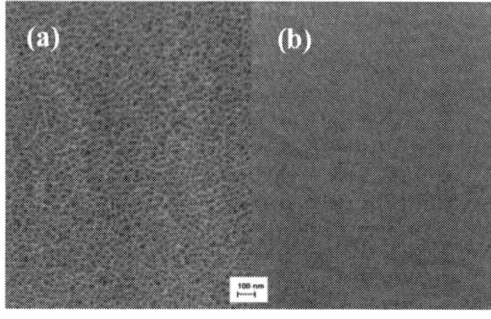

Fig. 4. FESEM images of cylindrical AAO film (a) before and (b) after crack

This phenomenon could be explained as followed. In the beginning, the initial pore arrangement was very disorder because the formations of the pores were not stable initially due to the unfixed current distribution. It was focused locally on fluctuations of the surface. After a period, self-organization occurred upon steady current distribution and repulsive forces formed between neighboring pores during the long-anodization. As a result, better aligned and close-packed pores arrays were obtained at the interface between the porous alumina layer and the aluminum substrate. Following by crack formation, it left the better distributed and aligned pores pattern behind. Subsequently, the new pores will be anodized following these self-ordered pre-patterns left on it. Since all the parameters were kept constant along the anodization, the pores quantity at the bottom would not much differ from the top ones. During the long anodization, the distribution and alignment of the pores on planar substrate will be shifted but the quantity of pores may not decrease much. In this experiment, we found that the pores quantities at the bottom decayed a lot. This implies that some discontinuous growth of the initial pores happened before it

reached the bottom. In term of nanodimension, the surface area on the top of the cylindrical sample is larger than the bottom. The changes of the surface area caused reduction on the pores quantity since the pores diameter and the wall thickness were fixed with constant parameters. The repulsive force between the pores is important to maintain the aligned pores distribution. As the surface area become smaller at the bottom, the repulsive force from the pores caused some of the pores to stop growing thus effectuate to a lot of pores unable to reach or to stay connected with the bottom pores.

Schematic in Fig. 5 shows the possible formation mechanism of the cracks on cylindrical AAO film during the anodization process. The outwardly expanding AAO film layer creates high tensile stress on the film itself and at AAO/aluminium interface due to the volume expansion [24]. As a consequence, cracks are form to minimize this stress.

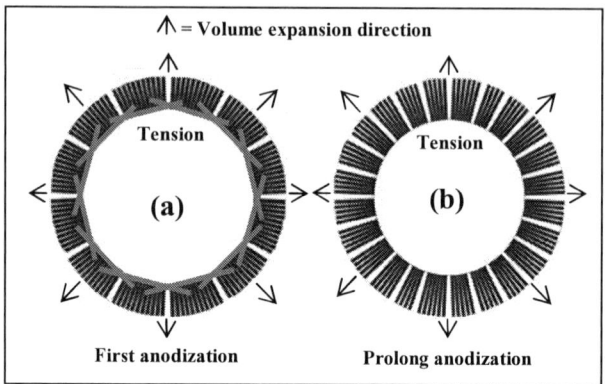

Fig. 5. The formation of micro-crack on cylindrical AAO film due to the tensile stress at the interface of AAO/Aluminium

From Fig. 2, we found that the crack do not occurs randomly but it follows some sequences which were shown in Fig. 5. This schematic diagram can be used to illustrate the possible cracks formation on a cylindrical AAO film. Initially the magnitude of tensile stress acting on the whole AAO film is equally divided because the rate of AAO film growth is the same around the cylinder sample. However, when the AAO film keeps on growing and expanding from all direction, there will be some areas which are concentrated with higher stress. In the illustration shown in Fig. 5, we assume the cylindrical samples are perfect circles. As displayed in Fig. 5 (a), there are 12 tangents (grey line) which are differ 30° side by side to each other on a perfect circle, they represent the tensile stress endured by the AAO film, which were concentrated at certain areas equally throughout the circle. This was expected to happen as the stress was uniformly divided on AAO film to keep it in the original shape. The overlapping points of the tangents indicate the areas which withstanding the highest tensile stress when the AAO film keeps expanding. This area can be called as critical point. Once the tensile stress occurred exceeded that the AAO film can withstand, this extreme stress will pull the AAO film to separate apart from its original circle position into 12 different parts as to minimize the stress acting on all parts. Therefore several cracks will form on the cylindrical AAO film simultaneously with the same distance to each other. As we observed from this work, the cracks were formed simultaneously on the cylindrical AAO film but not with the same distance to each other. This might be the effect from the used of non perfect circle in our cylindrical sample, resulted to the stress acting on the AAO film is not totally equal. From the results gained after the prolong anodization, we realise that another set of crack will be formed again at the critical point on each separated part if the extreme tensile stress acting on them is high enough to break them from their strong molecular ionic bonding. We hypothesized that the twelve existing parts that divide the cylindrical AAO film will further divide following with the prolong anodization, and that will happen in between the cracks, but not from the remaining cracks.

B. Alumina Nanowires

Fig. 6 reveals that planar AAO film has a smoother surface and higher uniformity compared to the surface of cylindrical AAO film after it was chemical etching.

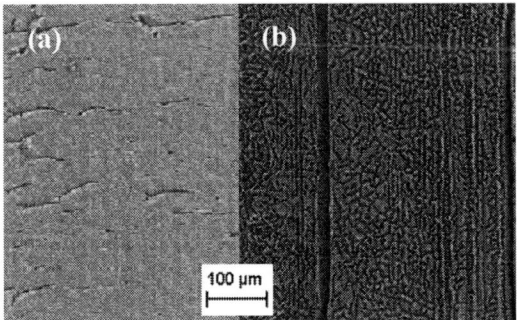

Fig. 6. FESEM images of (a) planar AAO film and (b) cylindrical AAO film after etching in 5 % H_3PO_4 for 30 min

The initially smooth cylindrical AAO porous film surface as seen in the inset of Fig. 1 (b) has turned into undulating hills structure after immersed in acid etchant.

Images in Fig. 7 (a) and (b) indicate high yield of alumina NWs formed after the etching process has been performed on both 2D and 3D AAO film. There were some slits and crumple surface observed on the planar sample shown in Fig. 7 (a). The alumina NWs obtained were from the part of the AAO film left after some part of its pores wall has been etched away thus resulted in the lost of pores wall which was holding the pores to each other. In fact, some spaces formed in between the alumina NWs from one another instead of closely pack. These NWs were unable to withstand with their own weight and hence collapsed upon each other. As a result, the slits and crumple surface was formed. From Fig.7 (b), the surface of bulk NWs formed is different compared to Fig. 7 (a). The surface looked rough and uneven. Previous studies have shown the growth of nanopores on planar sample is almost straight from the top to the bottom with a little change of pores distribution and alignment, but there will not be many changes in the pores quantity. However, this result shows the growth of nanopores quantity on cylindrical sample

was reduced from the top to the bottom because of the narrowing in surface area as discussed above earlier. This resulted to some of the pores from top were unable to reach the bottom and their growth was terminated at somewhere in the middle of the AAO film. After chemical etching, some part of the alumina NWs formed on the cylindrical AAO film left the substrate because they are not connected to the AAO pores wall at the bottom anymore. This had bring to the formation of a lots of spaces in between the alumina NWs, when those NWs collapse and lean on each other, they left gaps on the bundle of NWs formed and made it to look not as compact and as smooth as the alumina NWs obtained on the planar sample.

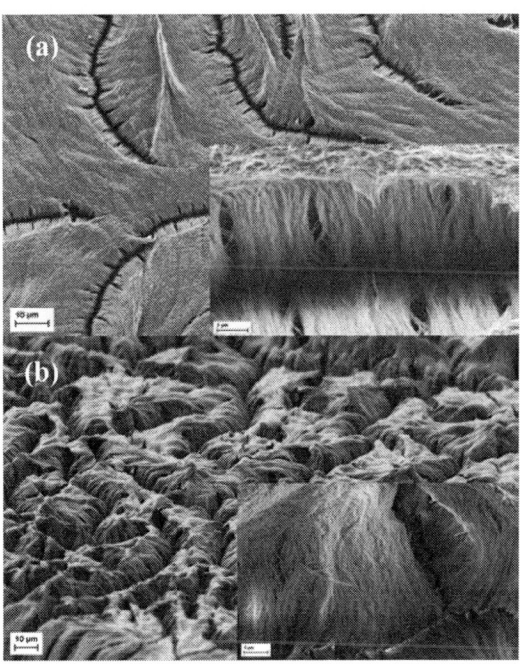

Fig. 7. FESEM images of alumina NWs formed by (a) planar AAO film and (b) cylindrical AAO film after etching

Magnified view of both images in Fig. 7 shown very fine alumina NWs formed either on planar or cylindrical nanoporous AAO film. They have the same structure and those NWs were found to lean against one another after they lose their support from the neighbouring walls. An average diameter of the NWs obtained from both AAO film was 20 nm.

The NWs formation mechanism during the etching process has been displayed schematically in Fig. 8. The ideal nanopores of AAO film consists of a densely packed array of hexagonal shape cells with the columnar pores in the cell centre as shown in Fig. 8 (a). There was a study found that the alumina cell wall surrounded the pores were contaminated with acid anions which was used in the anodization. While pure alumina was found to be formed near to the cell boundary [25]. Compared to pure alumina which is harder and with high chemical stability, the contaminated alumina could be etched or dissolved relatively easier by dilute H_3PO_4

acid. This corresponds to the rapid widening of the original pores of AAO film as shown in Fig. 8 (b). The pores enlargement process lead to wall thinning and the wall was totally dissolved by the etchant after 30 min etching in this experiment. The pure alumina found at the cell boundary (triple point) was left after the process [23]. Eventually an array of alumina NWs was formed by the pure alumina which remained at the centre of the three surrounding nanopores, they were found to lean on each other since they had lose support from the neighbouring walls shown in Fig.8 (c, d).

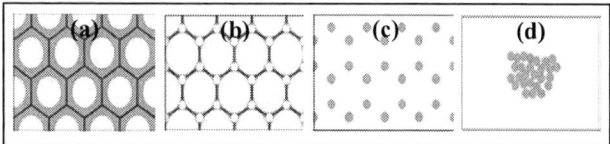

Fig. 8. Schematic diagram of the formation of alumina NWs

IV. CONCLUSION

2D-planar and 3D-cylindrical AAO films were fabricated by using a simple and cost effective anodization method. The presented results shown the existence of tensile stress due to the volume expansion of AAO film during the anodization had induced micro cracks to form on the cylindrical AAO film. Prolong anodizing duration will results in more cracks formation on the cylindrical AAO film. The cracks formed were due to the alleviation of overload tensile stress that was acting on the cylindrical AAO film. The model of the cracks formation was proposed based on our observation from this experiment. On the other hand, high yield of alumina NWs were successfully prepared by simply carry out the chemical etching on the as-prepared planar and cylindrical AAO film with 5 % H_3PO_4 for 30 min. Compact alumina NWs were obtained on planar AAO film while alumina NWs with lower density were formed on cylindrical AAO film. It was caused by more discontinuous pore formation on this AAO film. Slowing down the rate of AAO formation by reducing the electrolyte concentration, anodization temperature and potential or using other electrolyte could lower the tensile stress on cylindrical AAO film thus micro crack problem on the cylindrical AAO film might be eliminated.

ACKNOWLEDGMENT

This work has been carried out with the support of University Research Grant from Universiti Kebangsaan Malaysia with code: UKM-GUP-2011-378.

REFERENCES

[1] T. Qiu, X.L. Wu, Y.F. Mei, G.J. Wan, Paul K. Chu and G.G. Siu, "From Si nanotubes to nanowires: Synthesis, characterization, and self-assembly," *J. Crystal Growth*, vol. 277(1–4), pp.143-148, 2005.

[2] A. Ismardi, C.F. Dee, H. Abdullah, B.Y. Majlis and M.M. salleh, "Synthesis and characterisation of Zn-Sn-In-O quaternary nanostructure system," *Mater. Res. Innov.*, vol. 15(2): pp. 173-175, 2011.

[3] T.Y. Tiong, M. Yahaya, C.F. Dee, K.P. Lim, B.Y. Majlis and C.H. Sow, "Influence of growth temperature on SnO₂ nanowires," *Mater. Res. Innov.*, vol. 13(3), pp.203-206, 2009.

[4] J. Karamdel, C.F. Dee, K.G. Saw, B. Varghese, C.H. Sow, I. Ahmad and B.Y. Majlis, "Synthesis and characterization of well-aligned catalyst-free phosphorus-doped ZnO nanowires," *J. Alloy Compd.*, vol. 512(1), pp. 68-72, 2012.

[5] J. Karamdel, C.F. Dee and B.Y. Majlis, "Characterization and aging effect study of nitrogen-doped ZnO nanofilm," *Appl. Surf. Sci.*, vol. 256(21), pp.6164–6167, 2010.

[6] B. Yu, and M. Meyyappan, "Nanotechnology: Role in emerging nanoelectronics," *Solid-State Electron.*, vol. 50(4), pp.536-544, 2006.

[7] V. Sgobba and D.M. Guldi, "Carbon nanotubes-electronic /electrochemical properties and application for nanoelectronics and photonics," *Chem. Soc. Rev.*, vol. 38(1), pp.165-184, 2009.

[8] S.S.P. Parkin, C. Kaiser, A. Panchula, P.M. Rice, B. Hughes, M. Samant and S.H. Yang, "Giant tunnelling magnetoresistance at room temperature with MgO (100) tunnel barriers," *Nature Mater.*, vol. 3, pp.862-867, 2004.

[9] T.Y. Tiong, C.F. Dee, M.M. Salleh, B.Y. Majlis and M. Yahaya, "Butane Sensing Property of Si-ZnO Nanowires p-n junction," *Adv. Mat. Res.*, vol. 364, pp.260-265, 2012.

[10] M. Konijn, M.M. Alkaisi and R.J. Blaikie, "Nanoimprint lithography of sub-100 nm 3D structures," *Microelectron. Eng.*, vol. 78-79, pp. 653-658, 2005.

[11] B.X. Wang, Z. Rozynek, J.O. Fossum, K.D. Knudsen and Y.D. Yu, "Guided self-assembly of nanostructured titanium oxide," *Nanotechnology*, vol.23, pp.279502, 2012.

[12] A. Huczko, "Template-based synthesis of nanomaterials," *Appl. Phys. A: Mater. Sci. Process.*, vol. 70(4), pp.365-376, 2000.

[13] L.J. Pan, H. Qiu, C.M. Dou, Y. Li, L. Pu, J.B. Xu and Y. Shi, "Conducting Polymer Nanostructures: Template Synthesis and Applications in Energy Storage," *Int. J. Mol. Sci.*, vol. 11(7), pp.2636-2657, 2010.

[14] P.J. Zhang, J.T. Chen, R.F. Zhuo, L. Xu, Q.H. Lu, X. Ji, P.X. Yan and Z.G. Wu, "Carbon nanodot arrays grown as replicas of specially widened anodic aluminum oxide pore arrays," *Appl. Surf. Sci.*, vol. 255(8), pp.4456-4460, 2009.

[15] J. Sarkar, G.G. Khan and A. Basumallick, "Nanowires: properties, applications and synthesis via porous anodic aluminium oxide template," *Bull. Mater. Sci.*, vol. 30(3), pp.271-290, 2007.

[16] L.A. Meier, A.E. Alvarez, D.R. Salinas and M.C. del Barrio, "A clean method to obtain a porous alumina template," *Mater. Lett.*, vol. 70(0), pp.119-121, 2012.

[17] J.P. Zhang, J.E. Kielbasa and D.L. Carroll, "Controllable fabrication of porous alumina templates for nanostructures synthesis," *Mater. Chem. Phys.*, vol. 122(1), pp.295-300, 2010.

[18] C.Y. Han, G.A. Willing, Z.L. Xiao and H.H. Wang, "Control of the Anodic Aluminum Oxide Barrier Layer Opening Process by Wet Chemical Etching," *Langmuir*, vol. 23, pp.1564-1568, 2007.

[19] D.J. Yang, B.Paul, W.J. Xu, Y. Yuan, E.M. Liu, X.B. Ke, R.M. Wellard, C. Guo, Y. Xu, Y.H. Sun and H.Y. Zhu, "Alumina nanofibers grafted with functional groups: A new design in efficient sorbents for removal of toxic contaminants from water," *Water Res.*, vol. 44(3), pp.741-750, 2010.

[20] W.L. Suchanek, "Hydrothermal Synthesis of Alpha Alumina (α-Al₂O₃) Powders: Study of the Processing Variables and Growth Mechanisms," *J. Am. Ceram. Soc.*, vol. 93(2), pp.399-412, 2010.

[21] M. Sato and T.J. Webstera, "Nanobiotechnology: implication for the future of nanotechnology in orthopedic applications," *Expert. Rev. Med. Devices*, vol. 1(1), pp.105-114, 2004.

[22] X.J. Xu, G.T. Fei, L.Q. Zhu andX.W. Wang, "A facile approach to the formation of the alumina nanostructures from anodic alumina membranes," *Mater. Lett.*, vol. 60(19), pp.2331-2334, 2006.

[23] Z.L. Xiao, C.Y. Han, U. Welp, H.H. Wang, W.K. Kwok, G.A. Willing, J.M. Hiller, R.E. Cook, D.J. Miller and G.W. Crabtree, "Fabrication of Alumina Nanotubes and Nanowires by Etching Porous Alumina Membrane," *Nano Lett.*, vol. 2(11), pp.1293-1297, 2002.

[24] B.Y. Yoo, R.K. Hendricks, M. Ozkan and N.V. Myung, "Three-dimensional alumina nanotemplate," *Electrochim. Acta*, vol. 51(17), pp.3543-3550, 2006.

[25] G.E. Thompson and G.C. Wood, *Treatise on Materials Science and Technology*, vol. 23, New York: Academic Press, 1983, pp.205-329.

The Role of Reactive Ion Etching(RIE) on Wirebond Formation: A Study on Successful Rate of Thermosonic Gold Wire on Aluminium Bondpad

[1]Sauli. Z. *, [1]Retnasamy, V., [1]Rahman, N.A.Z., [1]Aziz, M.H.A., [1]Razak, H.A. & [1]Palianysamy, M.

[1]School of Microelectronic Engineering, Universiti Malaysia Perlis (UniMAP),
Kampus Alam Pauh Putra, Perlis, Malaysia.
*Corresponding email: zaliman@unimap.edu.my

Abstract—**Wire bond has been an important tool in the world of microelectronic interconnections. The effect of Reactive Ion Etching (RIE) on the successful rate of thermosonic bonding using gold wire on aluminium pad is studied. Surface morphology images from Atomic Force Microscopy (AFM) are used as correlation comparison. In this work wire bonding adhesion is studied on two different surface conditions, which are treated with RIE and the other without RIE treatment. In this experiment only the bonding time was varied for each set of experiments. Results of the wire bond from both samples, with and without RIE were compared. The RIE treated surfaces yield better adhesion results in this work.**

Keywords-**Thermosonic; bonding time; reactive ion etching.**

I. INTRODUCTION

Wire bonding is a technique that will link microelectronic interconnections between an integrated circuit and lead frame by using wire as connector. Currently many semiconductor packages use thermosonic gold wire bonding, such as in optoelectronic devices and micro-electromechanical systems. It is because gold wire bonding have benefits such as self-cleaning, high yield rate, flexibility and reliability[1]. To improve further in terms of bondability and yield rate of wire bond, plasma as pre-treatment has been introduced [2].

Plasma is an ionized gas with equal number of positive and negative charges. It involves three types of collision which is ionization(generates and sustains the plasma), excitation(causes plasma glow) and disassociation(creates free radicals). These collisions will generate chemically reactive species and react with the materials being etched to form volatile by-products.

This pre-treatment method uses only efficient resources and apparently improves the manufacturability, reliability and yield for advanced semiconductor packages. The removal of the acquired amount of material from substrates will contribute to the improvement of the surface uniformity and directly to the wire bonding yield rate [2].

Plasma etching is extremely sensitive to many variables, making etch result inconsistent and irreproducible. Therefore, important plasma parameters and their influences will be treated [3]. The overall etching process is controlled by the absorption of the etchant on the surface, the reaction rate at the surface and desorption of the products [4]. Controlling the RIE process with a balance set of parameters are important to produce a desired surface morphology of aluminum bond pad.

As a key element, aluminium plays an important role in the formation of intermetallic bonding with gold, which is used as the bond wire to form the interconnections[5]. Therefore, surface morphology of aluminium will affect the contact area when the gold wire is bonded. The actual contact takes place only over a small area [6]. In this case, bonding time is varied to improve the contact area which will directly give a better wire bond result.

Another important factor that contributes to the bondability and contact area is the ultrasonic vibration produced by the capillary. The vibration disrupts the contaminations and oxide layer on the surface of the ball and the surface of the bond pad during the wire bonding. Based on microslip theory towards bonding, the vibration on the capillary also supplies a dynamic force causing the capillary tip to move in a reciprocating motion. The results of the interaction are a contact produced from both normal and tangential forces [7].

Ultrasonic vibration affect begins from peripheral penetrating into the sample where the depth is dependent on the bonding time. The Mindlin's microslip theory considered the contact between two spheres pressed into contact with relative elastic displacements between them caused by the action of an oscillating tangential load. In the proposed model, the central region was referred to as stationary (non-slip) region, while the slip occurred in peripheral areas[8]. The theory concludes that the bonding time does affect the quality of wire bond based on area of contact.

II. EXPERIMENTAL METHOD

Samples used in this experimental procedures are 2 same type wafers. Wafers are prepared and proceeded to metalization process, where aluminium layer is deposited on the wafer using the Physical Vapour Deposition (PVD) machine. Then, the photolithography process is done to form the bond pads. One of the wafer is directly put into RIE

978-1-4673-2395-6/12 $31.00 © 2012 IEEE

treatment to prevent oxidation caused by the exposure to the environment. Whereas, the other wafer is proceeded to wire bonding without RIE tratment. For the wafer which is treated using RIE, the parameters of RIE is given as in Table 1.

Wire bonding parameters are set up as it is shown in Table 2. Overall experimental procedures are shown in a flow chart as in Fig. 1.

TABLE 1
REACTIVE ION ETCHING PARAMETER

Time	20s
Temperature	50°C
RF Power	40w
CF4 Gas	10ccm
O2 Gas	5ccm

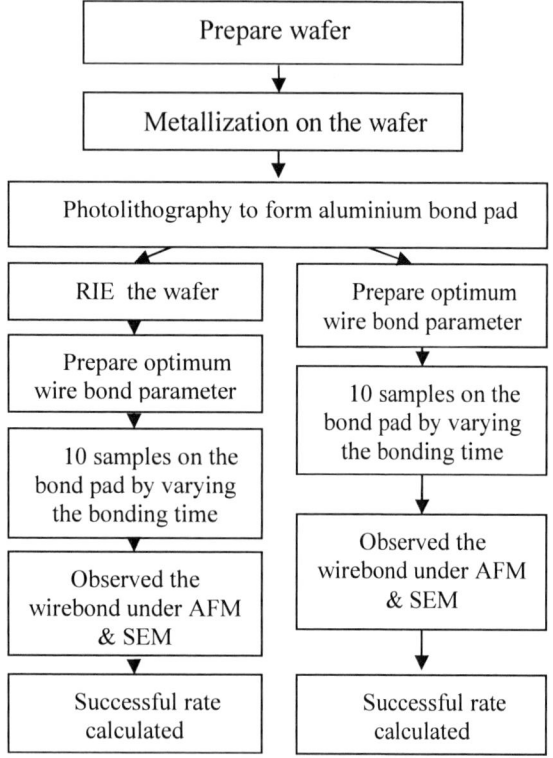

Fig. 1.Experimental process

TABLE 2
WIREBONDING PARAMETER

Ball		Wedge	
Temperature		150°C	
Power	1.99	Power	1.52
Force	2	Time	3ms
		Force	1.8

The samples are carried out with the wire bonding process. Below are the steps of wire bonding process:

1. A gold wire is fed through a hole in the capillary, and an electronic flame off spark melts the wire, forming a ball at the end of the wire.

2. The ball is pressed against a bond pad when the capillary tool is lowered. The interface temperature rises caused by the heated work holder; the ultrasonic energy is applied and the ball bond is formed.

3. The capillary tool is raised, leaving the ball bonded to the surface, and forming the wire loop as it moves towards the second bond position.

4. The capillary tool is lowered to the second bond pad, to make a second bond. This bond is also called a stitch bond, crescent or tail bond.

5. After the second bond is formed, the capillary tool is raised and a wire clamp above the capillary tool pulls and breaks the wire free.

The process was repeatedly done with total of 100 samples which differs in term of the bonding time. Each wafer has 10 different wire bonds which is varied by the bonding time. The results were recorded and the images was taken using High Power Optical Microscopy. The results is also correlated using the images from Atomic Force Microscopy (AFM). Successful rate of the wire bonding process was calculated using the recorded result.

III. RESULT & DISCUSSION

The results from the samples that were not treated by RIE before the wire bonding process is tabulated in Table 3. Sucessful rate of the wirebonding process is also plotted in a graph as shown in Fig. 2. From the results, the sucessful rate of the wire bonding shows inconsistency. This is probably because of the surface morphology of the aluminum layer which does not contribute to a sucessful wire bond. The rate was only high when the bonding time is high which is at 8 and 9 ms. The longer bonding time gives a window of more oxide layer removed at the bonding area. This is because the ultrasonic vibration at the cappilarry will penetrate through as much as possible to get the closest to the proper bonding area where a stronger intermetallic bond can be formed. This phenomena can be related to Mindlin's microslip theory [7].

TABLE 3
RESULT OF WIREBOND SUCCESFUL RATE WITHOUT RIE PROCESS

Time (ms)	Without Reactive Ion Etching										Successful rate (%)
	Sample										
	1	2	3	4	5	6	7	8	9	10	
0	F	√	X	X	X	√	X	√	X	√	40
1	√	√	√	√	√	X	X	H	X	H	50
2	√	X	√	√	√	H	X	√	√	√	70
3	√	√	√	√	√	B	√	X	√	√	80
4	X	X	X	√	X	√	√	X	√	F	40
5	√	√	X	√	X	√	√	√	√	√	80
6	X	F	√	√	√	√	X	√	√	X	60
7	√	√	√	√	√	√	√	√	X	X	80
8	√	√	B	√	X	√	√	H	√	√	70
9	X	√	B	√	√	√	√	X	√	√	70
Total Successful Rate (%)											64

√: Complete Wirebond (ball & wedge)
X: Failure (non effect on aluminum pad)
H: Failure (heel break)
B: Failure (neck break/ball stick)
F: Failure (ball lift up/footprint)

TABLE 4
RESULT OF WIREBOND SUCCESFUL RATE WITH RIE PROCESS

Time (ms)	With Reactive Ion Etching										Successful rate (%)
	Sample										
	1	2	3	4	5	6	7	8	9	10	
0	√	√	√	√	√	√	√	√	√	√	100
1	√	H	√	√	X	√	√	√	√	√	80
2	√	√	√	√	√	H	√	√	√	√	90
3	√	H	√	√	√	B	√	√	√	√	80
4	√	√	√	√	√	√	√	√	√	√	100
5	√	√	√	√	√	H	√	√	√	√	90
6	√	√	H	√	√	√	√	√	√	√	90
7	√	√	√	√	√	√	√	√	H	√	90
8	√	√	√	√	√	√	√	√	√	H	90
9	√	H	√	√	√	√	√	√	√	√	90
Total Successful Rate (%)											90

√: Complete Wirebond (ball & wedge)
X: Failure (non effect on aluminum pad)
H: Failure (heel break)
B: Failure (neck break/ball stick)
F: Failure (ball lift up/footprint)

using RIE gives a high chances of sucessful wire bond on the sample.

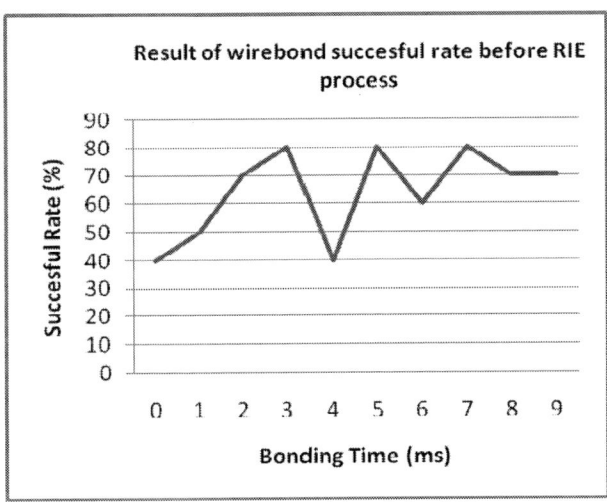

Fig. 2.Result of wirebond succesful rate before RIE process

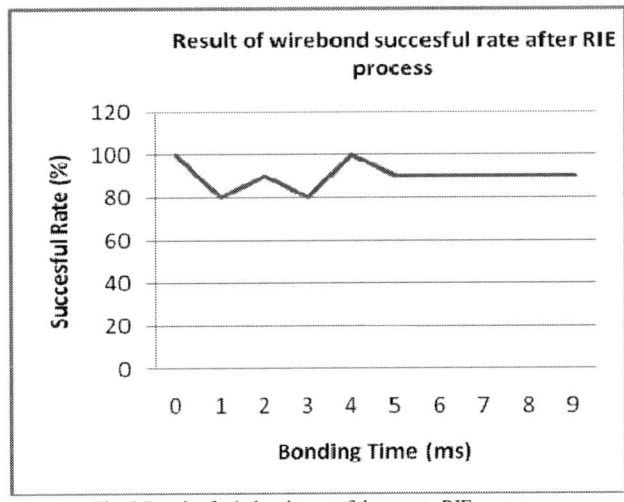

Fig. 3.Result of wirebond succesful rate rate RIE process

The results of the samples that were treated by RIE before the wire bonding process is shown in Table 4. The successful rate is plotted in a graph as in Fig. 3. Comparisons shows clearly that the succesful rates are consistent with the samples that had been treated using RIE. The rate is just between 80% and 100% for all the 10 samples and it is 90% of success from the bonding time of 5 ms to 9 ms. From this consistent sucessful rate, it can be deduced that the samples that is treated with RIE has a better chances of successful wire bonding. Altough the preventation method had been taken, there are high chances of oxide formation on the samples because of the exposure to the environment and the experimental procedures. Removal of this oxide layer by

The images from the High Power Microscope show a better justification on the calculated values. Below, Fig. 4 and Fig. 5 show the images of wire bond on certain samples. The wire bond images from the samples which were not treated using RIE shows the successful bonding, failure in terms of heel break, neck break and footprint. In this case, it is focused on the footprint failure. This is because the other failures do not involve the surface morphology. Whereas the results from the samples that had undergone RIE treatment has none of footprints. Footprint failure is because the bond ball fails to form a proper intermetallic bond with the bonding area or the bonding area is too small which prevents a successful bonding and leaves a footprint.

The other failures such as neck break are because of the damage on the gold wire itself. Whereas, heel break is probably because of incorrect bond placement. Both this failures does not involve the surface morphology of the pad.

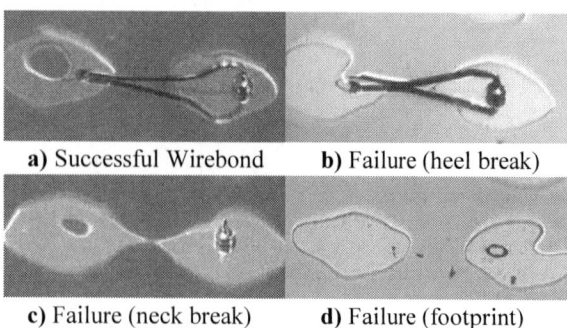

a) Successful Wirebond   **b)** Failure (heel break)

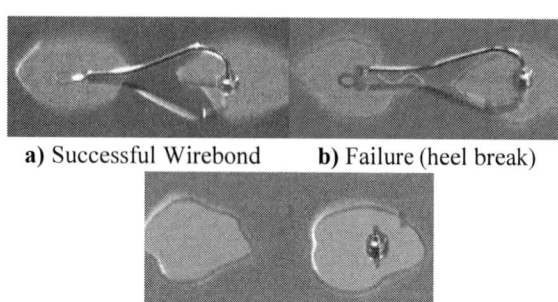

c) Failure (neck break)   **d)** Failure (footprint)

Fig. 4. High power microscope for wirebond image before RIE

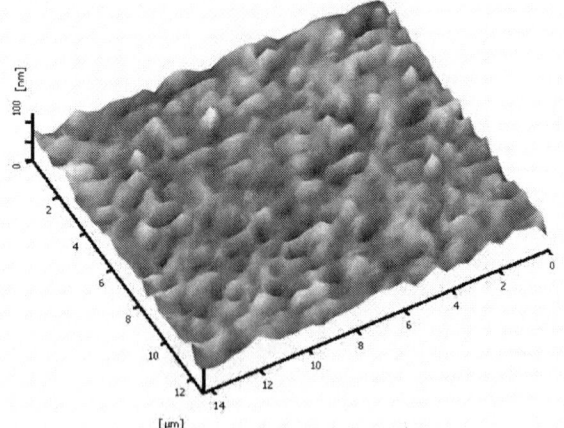

a) Successful Wirebond   **b)** Failure (heel break)

c) Failure (neck break)

Fig. 5. High power microscope for wirebond image after RIE

The AFM 3D images are shown in Fig. 6 and Fig. 7. From the surface images, it can be seen that the sample that had been treated using RIE shows a slightly smoother surface compared to the other sample which is not treated with RIE. The smoother surface is another reason for a better bonding area. This justifies the successful rate of the samples that had gone through the etching process using RIE.

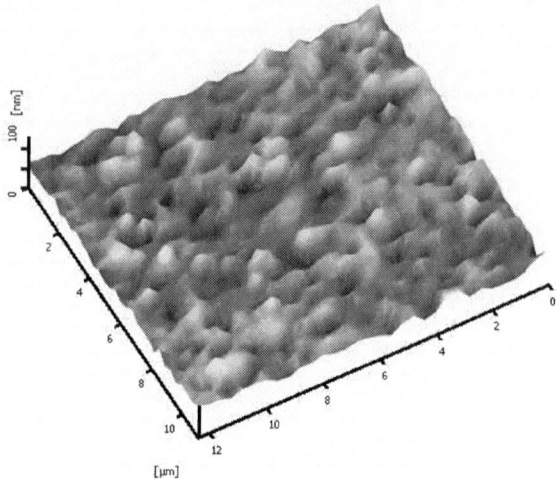

Fig. 7. AFM 3D image after RIE process

For a better understanding of the surface morphology, the grain size was studied for both sample types. The images are shown in Fig. 8 and Fig. 9 below. The resulting grain size of the sample that is not treated using RIE records 7.794×10^5 nm^2 whereas the sample that had been treated gives a grain size of 6.334×10^5 nm^2. This proves that the surface of the layer is smoother when it is etched using RIE. The possibility for a stronger bond is higher because there will be more bonding area when the surface is smoother.

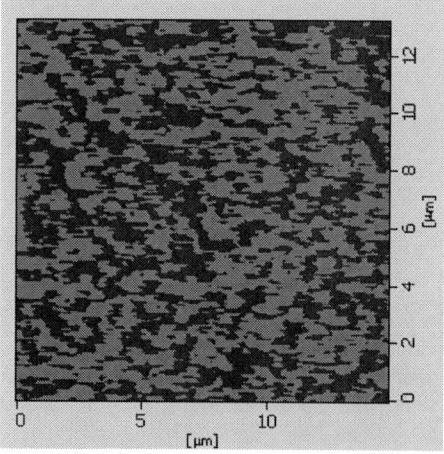

Fig. 8. AFM grain size before RIE process

978-1-4673-2395-6/12 $31.00 © 2012 IEEE

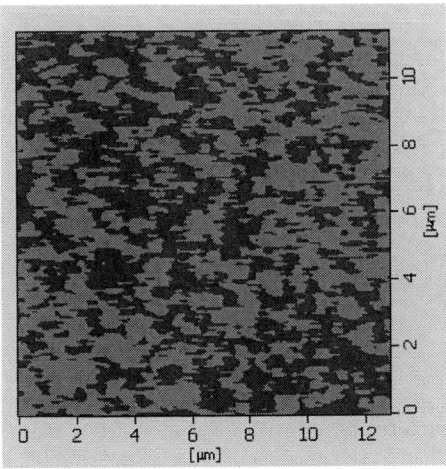

Fig. 9.AFM grain size after RIE process

IV. CONCLUSION

The successful rate of thermosonic gold wire bond on the aluminum bond pad with samples treated with RIE and withot treatment investigated by bonding 10 samples of wire bond. The bonding time was varied from 0 ms to 9 ms. Bonding parameters was constant except the bonding time which was varied. This is to observe the effect of the bonding time on the different surface morphology of aluminium bond pad. The informations gathered from High Power Microscope and AFM were used to corellate with the calculated successful rate. The major findings are summarized as follows:

- Lower bonding time produces lower successful rate in thermosonic gold wire bonding. In exception, there are still 100% sucess with the samples that were treated using RIE process which means larger bonding area is still possible at lower bonding time.

- The samples that were treated using RIE has a 90% successful rate in thermosonic gold wire bonding which is higher compared to 64% from the samples which is not treated using RIE.

- Successful rate of samples treated by RIE have earlier consistency starting at 5 ms compared with samples that is not treated with RIE which has consistency only after 8 ms.

- The grain size of the treated samples are lower(6.334×10^5 nm^2)than the samples that is not treated with RIE which are higher(7.794×10^5 nm^2). This gives a larger bonding area with a shorter bonding time.

- The bonding time can be optimized for a higher sucessful rate of the wire bonding by using RIE treatment before the wire bonding process. This will increase the yield and reduce the process time.

REFERENCES

[1] Harman G.G, *Wire Bonding in Microelectronics: Material, Processes Reliability, and Yield*, 2 ed. New York: McGraw-Hill, 1998.

[2] J. D. Getty, "Considerations for the Use of Plasma in Pre-Wire Bond Application."

[3] Henri Jansent, Han Gardeniers, Meint de Boer, Miko Elwenspoek and Jan Fluitman, (1995), "A survey on the reactive ion etching of silicon in microt echnology".

[4] Hwaiyu Geng, "Semiconductor Manufacturing Handbook", McGraw-Hill, 2005,pp 3.2-3.5, 21.33-21.34

[5] E. A. Amerasekera and F. N. Najm, "Failure Mechanisms in Semiconductor Devices," 1997.

[6] D. H. Bu ey, "Surface Effects in Adhesion, Friction, Wear, and Lubrication," 1981.

[7] Hui Xu, Changqing Liu, Vadim V. Silberschmidt, Zhong Chen, Jun Wei, Initial bond formation in thermosonic gold ball bonding on aluminium metallization pads, Journal of Materials Processing Technology, Volume 210, Issue 8, 1 June 2010, Pages 1035-1042

[8] Mindlin, R.D., 1949. Compliance of elastic bodies in contact. J.Appl. Mech. 71,259–268.

Effect of Copper FAB Impact on Palladium Bond Pad

[1]Sauli. Z.*, [1]Retnasamy, V., [1]Taniselass, S., [1]Norhaimi, W.M.W., [1]Aziz, M.H.A. & [1]Hashim, M.N.

[1]School of Microelectronic Engineering, Universiti Malaysia Perlis (UniMAP),
Kampus Alam Pauh Putra, Perlis, Malaysia.
*Email: zaliman@unimap.edu.my

Abstract- **In this study, the initial impact stage of the copper free air ball (FAB) towards the palladium bond pad had been analyzed by using simulation. The effect of copper FAB impact on palladium bond pad was evaluated by simulation using the commercial computational software Ansys. The impact forces of copper FAB were varied from 1N to 5N. The stress response on copper FAB, lead frame and bond pad were obtained at various bond force were. The simulation results showed that when the applied force increases, the stress level on all these three parts increased. Highest stress response was recorded at impact force of 5N.**

*Keywords-***Copper wire; copper FAB ;palladium bond pad**

I INTRODUCTION

The adoption of copper wire in high volume wire bonding production has proven successful, as seen in tremendous increase in copper wire consumption in the past years. This increase largely came from the conversion from gold to copper wire in fine pitch devices which previously had been more conservative due to concerns related to copper wire reliability [1]. Fine pitch applications with bond pad openings of not more than 45um using a wire diameter lesser than 0.8mil require more stringent control on bonding qualities. Specially designed copper hardware that maintains an oxygen free environment, as well as more elaborate wire bonding processes were required to meet the challenges in copper wire bonding, some had succeeded but often with compromises to production throughput and mean-time-between-assist (MTBA)[2]. Palladium-coated copper (Pd-Cu) wire has emerged as an alternative solution to bare Cu wire especially in 0.8mil and even finer wire diameter applications. Its ability to form uniformly shaped free-air-balls (FABs) with nitrogen instead of forming gas, has better bondability on lead surfaces, as well as its resistance to oxidation and corrosion tend to offset the higher cost of Pd-Cu wire compared to bare Cu wire. The advantages of Pd-Cu wire are clearly seen in terms of production throughput and MTBA. Both bare Cu and Pd-Cu wires have been widely studied in terms of their hardness and FAB formation. Second bond bondability has also been compared with Pd-Cu wire demonstrating higher stitch strength and a wider process window without the use of elaborate bonding processes. For first bond, both bare Cu and Pd-Cu wires are seeing challenges in pad damage control in the form of peeling and cracking, with Pd-Cu wire showing smaller working process windows in most cases. Much attention has been put in studying the higher hardness of Pd-Cu wire compared to pure Cu wire, though the difference is not too significant. FABs generated from Pd-Cu wire have been studied with respect to various factors such as electric-flame off (EFO) parameters, types of inert gas, and the gas flow rate used to maintain the oxygen free environment. Correlation between these parameters and the distribution of Pd on the surface of a FAB was used to explain the different degrees of hardness seen. It has been claimed that both EFO parameters and the type of inert gas used affect the mixing of Pd into the FAB and result in different FAB hardness. The control of FAB formation by means of EFO parameters and inert gas used showed a certain influence on the first bond process window.Simulation are ulitzed as a guidance to understand and characterize the wire bonding process[4-6]. In this report, simulation were carried out to understand the impact of copper in FAB on palladium bond pad. This understanding helps in setting up a robust process with a special attention to process parameters that are critical in controlling key bonding performance like ball size, shear, stitch pull strength, padsplash, and pad damage. The objective of this report to evaluate the stress response during copper FAB impact on palladium bond pad.

II METHODOLOGY

Simulation is used as a guideline to understand the impact stress of copper FAB on palladium bond pad during wire bonding process. The simulation is conducted with the following assumptions.

1) The temperature of the FAB is assumed to be same as the substrate .

2) The FAB is assumed to be a rate-dependent elastic plastic
material during the bonding process. The bond pad and the copper lead frame are elastic plastic material.

3) The ultrasonic energy effect is not included in this simulation. Only impact stage is simulated.

The commercially available finite element code Ansys was used as a simulation tool. A 3D model was developed to simulate the Fab impact on the bond pad. The wire diameter used is 25µm. The diameter of the copper FAB is 70µm. The thickness of the palladium bond pad used 1.5µm and the thickness of copper lead frame is 3µm. Fig. 1 illustrates the 3D model.

978-1-4673-2395-6/12 $31.00 © 2012 IEEE

A 10-node Quadratic Tetrahedron solid element(SOLID 187) were utilized to design the 3-D model. Surface-to-surface target element (TRAGE170) and contact element (CONTA174) pair were used simulate the contact between the copper fab and the palladium bond pad . The focus of this simulation is placed at the initial impact stage only. To allow a sufficient accuracy, a fine meshing density was used. The 3D model consisted of 199582 elements and 297393 nodes.Therefore, the better and more automated the meshing tools, the better the solution. The material properties of the 3D model is presented in Table 1.

In this study, the impact force of the Cu Fab is varied from 1 N to 5N to investigate the stress response due to fab impact on the bond pad.

Fig. 1. The 3D CopperFab Impact model

TABLE 1
MATERIAL PROPERTIES

Material	Density, ρ (g/cm^3)	Young's modulus,E (Pa)	Poisson ratio, v
Copper	8.94	130 x 10^9	0.34
Palladium	12.023	121 x 10^9	0.39

III RESULTS & DISCUSSION

In this experiment, the impact of copper free air ball (FAB) on the palladium bond pad based on the different force had been studied and the analysis was done by using the software Ansys. In general, there are three types of wire bonding processes which are thermosonic bonding, ultrasonic bonding and thermocompression bonding. All these techniques utilize applied force and temperature in forming the bonding. Since this study only investigates the amount of stress produced based to the applied force, the temperature parameter is neglected and all the other factors that could contribute to the amount of stress such as size of the ball and thickness of both bond pad and leadframe were set to be constant.In this study,

the force is appliedfrom 1N until 5N to simulate the free air ball impact towards bond pad. . The bond pad and lead frame were constrained to resist any movement during impact stage.

Based on this set up, analysis was conducted and the results were gained. According to the figures in the Table 2, it can be conclude that the depth of damage on the ball, bond pad and leadframe increases as the applied force increases. When small force which is 1N is applied, the damages were more like cratering on the surface[3]. However, when the force reached 5N, it can be clearly seen that the damage on ball, bond pad and leadframe was deep and the cratering was very obvious as shown in Fig. 2.

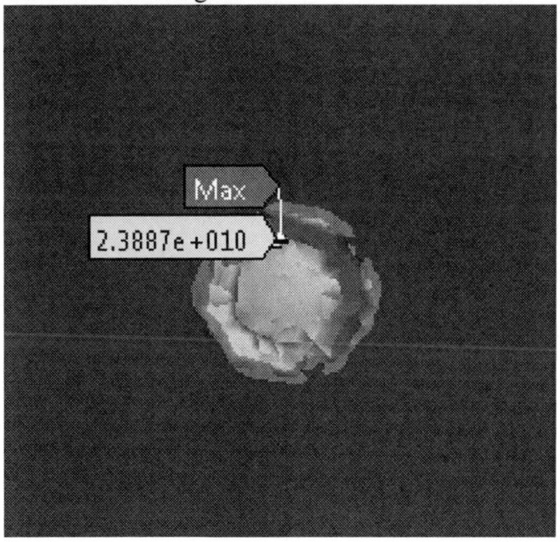

Max
2.3887e+010

Fig. 2. The cratering at palladium bond pad at impact force of 5N

The measurement of stress which indicates the impact due to applied force was tabulated in Table 2 and the finding shows that the stress on the ball, bond pad and leadframe increases as the force increases.

TABLE 2
SIMULATION RESULTS

Force (N)	Ball Stress (Pa)	Bond Pad Stress(Pa)	Lead Frame Stress(Pa)
1	3.8198e+009	4.4494e+009	3.5553e+009
2	7.2282+009	9.3185e+009	1.4472e+010
3	2.0773e+010	2.3887e+010	2.5569e+010
4	1.6066e+010	1.8184e+010	1.4472e+010
5	2.0773e+010	2.3887e+010	2.5569e+010

The stress reaches the peek when 3N force is applied and then fluctuated at high stress level as the force increases. The trend is similar for ball, bond pad and leadframe which can be observed based on Fig. 3, Fig. 4, Fig. 5.

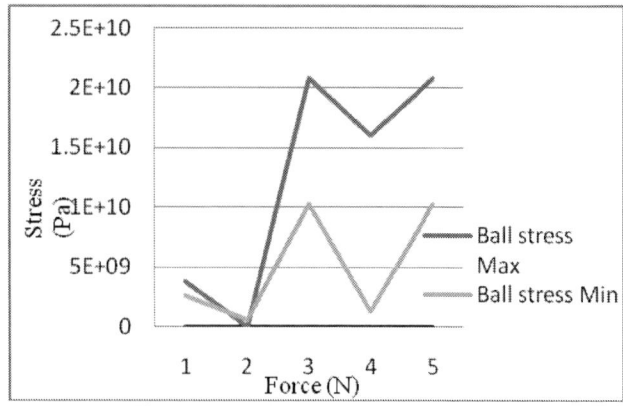

Fig. 3. Stress on Copper FAB versus applied force

In comparison, the stress on the ball is smaller compared to the stress on the bond pad and the stress on the leadframe is smaller compared to the stress on both ball and bond pad for a constant applied force. This indicates that the bond pad receives the most impact in the form of stress in the bonding process. This is very much related to the types of materials that had been used. In this experiment, copper wire had been used replacing the commonly used gold wire. Though copper has the advantages as the gold wire, however it has its drawbacks where it will be easily corrode and it is a hard material. Pure copper is twice as hard as pure gold. Therefore, additional 20% of the ultrasonic energy will be required in the bonding process which could lead to the damage of the bond pad[2].

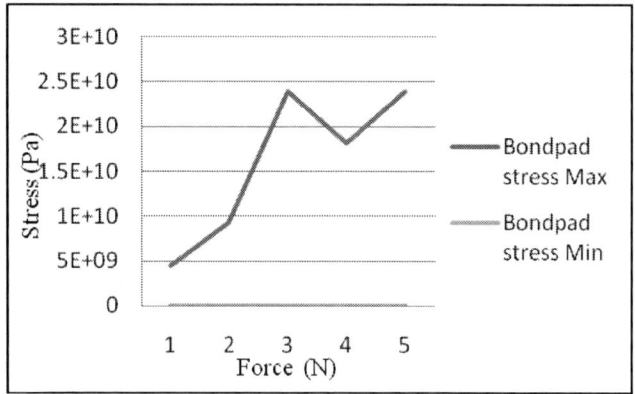

Fig. 4. Stress on the bond pad versus applied force

That is part of the reason why aluminum bond pad is not suitable when copper is used as the bonding material. Though the bond pad of this experiment had been replaced with palladium, this material is still considered as softer than the copper as shown in the figure below. As can be seen in Table 2 , the difference between the stress on the ball and the stress on the bond pad is not very large. This indicates the palladium is a strong material which is suitable be used as bond pad for copper bonding because it can withstand the force applied from the copper ball. Therefore less damage can be produced on the bond pad though large amount of force is applied.

Stress on the copper leadframe will be the smallest because most of the stress in the bonding process is absorbed towards the bond pad. The small stress on the leadframe is due to the acting of force from the bond pad which is in the opposite direction of the fixed force that holds the leadframe.

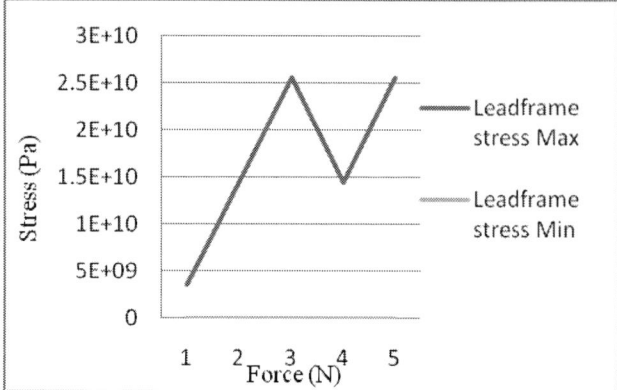

Fig.5.Stress on the leadframe versus applied force

IV CONCLUSION

The Wire bonding is one the most crucial process in the semiconductor industry. Since gold pricing is increasing, studies had been carried out to replace the gold wire with the copper wire. Since copper is twice as harder as the gold, therefore the bond pad also had to be replaced from aluminum to other materials. In this study, the impact of the copper free air ball towards the palladium hadbeen analyzed. The impact is studied based on the stress level on the copper ball, palladium bond pad and copper leadframe. As the applied force increases, the stress level on all these three parts increases as well until reaches the peak. For the given diameter of FAB and thickness of bond pad and leadframe, the peak stress level is reached when 3N of force is applied and the stress fluctuate at high level as the applied forces increase. The leadframe obtained the smallest stress compared to copper ball and palladium bond pad. The most stress is seen on the bond pad but however the difference between stress on the copper ball and palladium bond pad is not large . Palladium is slightly softer than copper and therefore the stress were more felt on the bond pad.

ACKNOWLEDMENT

The author would like to thank and acknowledge School of Microelectronic Engineering, Universiti Malaysia Perlis for their support and facilities

REFERENCES

[1] Zhong ZW. Wire bonding using copper wire. MicroelectronInt 2009:10–6.

[2] Zhaohui Chen, Yong Liu and Sheng Liu, "Modeling of copper wire bonding process on high power LEDs", Microelectronics Reliability 51, 2011, p. 171–178.

[3] Hsiang-Chen Hsu, Wei-Yao Chang, Chang-Lin Yeh and Yi-Shao Lai, "Characteristic of copper wire and transient analysis on wirebonding process", Microelectronics Reliability 51, 2011, p. 179–186.

[4] Yuan, C.; Weltevreden, E.; van dan Akker, P.; Kregting, R.; de Vreugd, J.; Zhang, G.Q.; , "FE modeling of Cu wire bond process and reliability," Thermal, Mechanical and Multi-Physics Simulation and Experiments in Microelectronics and Microsystems (EuroSimE), 2011

[5] Qiuxiao Qian; Yong liu; Timwah Luk; Irving, S.; , "Wire bonding capillary profile and bonding process parameter optimization simulation," Thermal, Mechanical and Multi-Physics Simulation and Experiments in Microelectronics and Micro-Systems, 2008. EuroSimE

[6] Yong Liu; Irving, S.; Timwah Luk; , "Thermosonic Wire Bonding Process Simulation and Bond Pad Over Active Stress Analysis," Electronics Packaging Manufacturing, IEEE Transactions on , vol.31, no.1, pp.61-71, Jan. 2008

Single Hole at Constrained Location for Stress Analysis in PCB Plate Bending

[1]Sauli. Z.*, [1]Retnasamy, V.,[1]Vengdasalam, K., [1]Taniselass, S., [1]Shapri, A.H.M.&[1]Vairavan, R.

[1]School of Microelectronic Engineering, Universiti Malaysia Perlis (UniMAP),

Kampus Alam,Pauh Putra Perlis, Malaysia.

*Corresponding email: zaliman@unimap.edu.my

Abstract–In this paper, simulation on the bending process of PCB during depaneling was done. The stress response during the bending process was evaluated using a computational program Ansys version 11. Two PCB plate model were developed:one of the model with single hole and other model without hole. The stress responce of the two models were then compared. From the simulation, it has been observed that the value of stress response of the PCB increases with increasing displacement height. The PCB model with single hole exhibited higher stress compared to the PCB model without hole. Highest stress response was obtained at the displacement height of 5cm for both models.

Keywords-Printed circuit board(PCB); depaneling process; single hole

I INTRODUCTION

Currently, the swift advancement of the electronics industry necessitates the electronic manufactured goods to be light weight, diminutive sized and compact. Hence, the miniaturized electronic components such as surface mount components are place on printed circuit board which functions as a back bone support during packaging assembly [1]. The printed circuit board (PCB) is utilized to sustain and provide connections to the passive and active components by means of conductive conduit. The intention of PCB is to electrically interconnect all the components, to provide power and dissipate the heat generated by the mounted components [2]. Mass production of electronic devices is done and these devices are produced in a batch form. During mass production, the PCBs undergo board level assembly. Then, this process is followed by a depaneling process which individualizes the PCBs.

The PCB plates are subjected to mechanical bending process during the depaneling process. The mechanical bending process may induce stress on the PCBs which may damage and reduce the reliability of components mounted on the PCBs [3,4]. The depaneling process causes movement between the mounted components and the board which will damage the solder joint of the components [5,6]

. Therefore, the stress induced by the PCB plates during the bending process and ways to improvise the stress distributions have to be investigated.

As a result, the bending process of the PCB plate is simulated in this study. The commercial computational program, Ansys is utilized for the simulation process. The stress response during bending process is examined. To evaluate the stress response, two types of PCB design were used, one model with hole and one model without hole. There stress responses of both models were compared. The bending process was replicated by using a displacement scheme, where one end of the PCB plate is constrained and the displacement is applied at the other end. The displacements height of the PCB plates was ranged from 1cm to 5cm.

II METHODOLOGY

The bending process was simulated using Ansys version 11 software. Two 3 dimensional models, one model resembles a PCB plate without hole and the other model resembles a PCB plate with a single hole located at the near the left end of the PCB plate. These models were developed using 20 Node Quadratic Hexahedron (Solid 186). The 3D models were then meshed with 69867 number of elements for the PCB plate without hole and 69308 number of elements for the PCB plate with single hole. The dimension of the PCB plate model is 175mm x 250mm and the thickness is 1mm. As for the single hole model, the diameter of the hole is 30 mm and it is placed at the front part of the left side, 15mm from the constrained area of the model. The models are shown in Fig.1.

The material properties used in the analysis is shown in Table 1.

TABLE 1
MATERIAL PROPERTIES OF FR4

Material	FR4
Young Modulus ,E (Pa)	22×10^9
Poison Ratio	0.136
Density (kg/m³)	1850

For the both 3D model, the loading and boundary conditions are applied as follows:

i. Displacement

As the design is in three dimension, there will be x, y and z axis. The displacement height is loaded at the right end side of the pcb plate, towards the positive y axis direction. The value of displacement height are in range of 1cm to 5cm.

ii. Fixed Support

The left end side of the PCB plate is contrained by applying fixed support at the negative x-axis. This results in a fixed end of the PCB plate with zero movements.

The Von Mises Stress is used as a criterion in this study to evaluate the stress response of the PCB plates during the bending process.

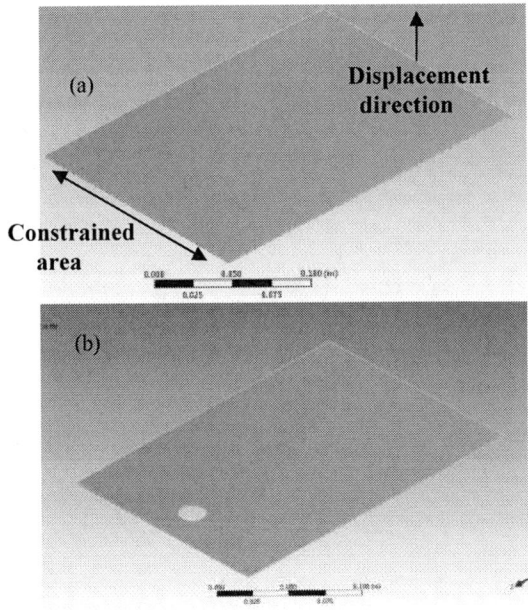

Fig.1. The 3D model of PCB plates, (a)Without hole and (b) with single hole

III RESULTS AND DISCUSSION

The stress response of the PCB plates during bending process was simulated using Ansys software. In this study, two types of PCB plates were modeled and simulated which are without hole and with single hole. The effects of the PCB geometry on the stress response were investigated. The displacement height of 1cm, 2cm, 3cm, 4cm, and 5cm were applied to both PCB plates. Based on the Fig. 2., the maximum stress on both PCB plates are different based on the PCB plate geometry. For PCB plate without hole, the maximum stress onto the board is lower compared the PCB plate with single hole. For displacement height of 1cm, the stress response of PCB plate without hole is 534500 Pa and the stress response of PCB plate with single hole is 803420 Pa.

Based on the Fig. 3., the maximum stress on both PCB plates are different based on the PCB plates geometry. For PCB plate without hole, the maximum stress response the plate is lower compared the PCB plate with single hole. For 2cm displacement height, the stress response of PCB plate without hole is 1069000Pa and the stress response of PCB plate with single hole is 1606800Pa.

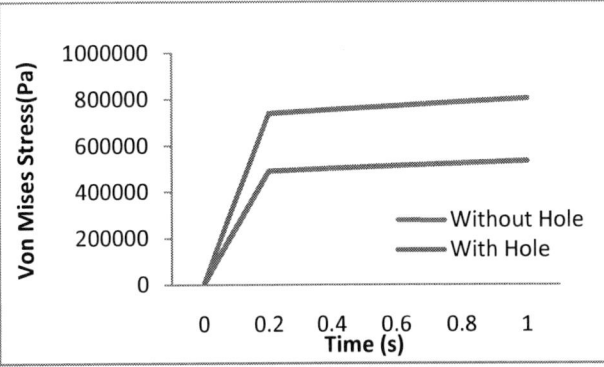

Fig. 2. Stress vs Time for displacement height of 1cm

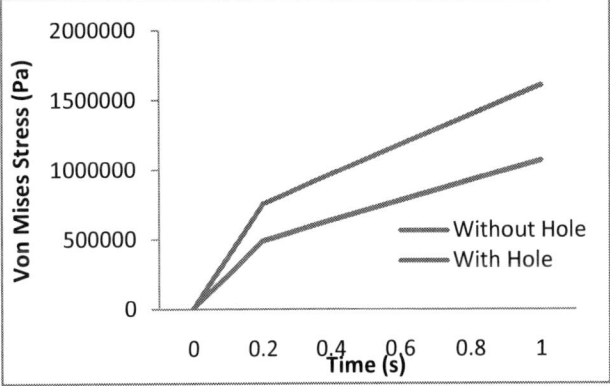

Fig. 3. Stress vs Time for displacement height of 2cm

Based on the Fig. 4., it is observed that the stress response is increasing with an increase in displacement height. For 3cm displacement, the stress response of PCB Board without hole is 1603500 Pa and the stress response of PCB Board with single hole is 2410300Pa.

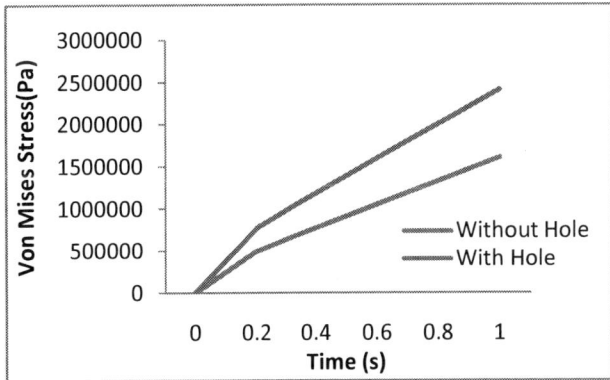

Fig. 4. Stress vs Time for displacement height of 3cm

Based on the Fig. 5., for 4cm displacement height the stress response of PCB plate without hole is 2138000Pa and the stress onto of PCB plate with single hole is 3213700Pa.

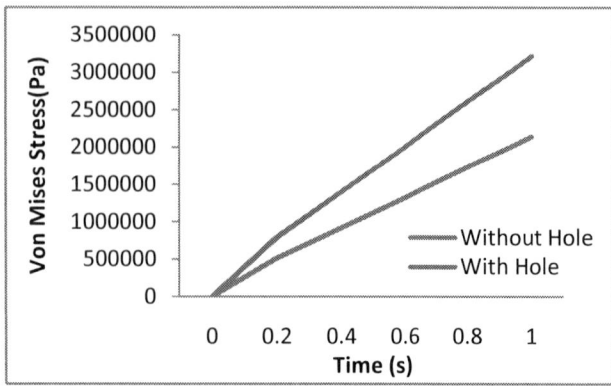

Fig. 5. Stress vs Time for displacement height of 4cm

Based on the Fig. 6., for 5cm displacement height, the stress response of PCB plate without hole is 2672500 Pa and the stress response of PCB plate with hole is 4017100 Pa.

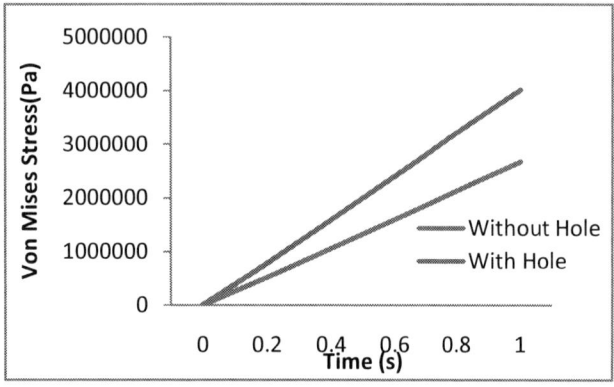

Fig. 6. Stress vs Time for displacement height of 5cm

Hence, by comparing the over all results, the PCB plate with hole exhibited higher stress response compared to the PCB plate without hole. During the depaneling process, a curvature bend exist on the PCB plates due to the mechanical bending[3]. When PCB plate experiences the bending motion, stress is induced on the opposite side of the PCB plates which distributes along the width and length direction of the PCB plates[7]. This can be seen in the stress contor of PCB model without hole in Fig7(a). The distribution of the stress along the width and length direction of the PCB may induce damage to the solder joints of the mounted components. The high stress induced in this region may induce crack on the solder joints if the bending effect is prolonged[3]. Hence, a single hole is placed near the constrained area of the PCB to focus the stress distribution area to the hole structure as shown in Fig 7(b) where high stress is induced at the periphery of the hole structure. The hole structure acts as a weak link since it could not repulse the opposing stress and hence the stress distribution will be directed to the hole structure. This will reduce the stress distribution along the width and length direction of the PCB plate which can be vizualized when both the Fig.7(a) and (b) is compared.Therefore the damage

induced to the solder joints of the mounted components can be reduced as exhibited by the simulated results. Therefore, this explains why the PCB plate with single hole exhibits higher stress response compared to the PCB plate without hole at the constrained area.

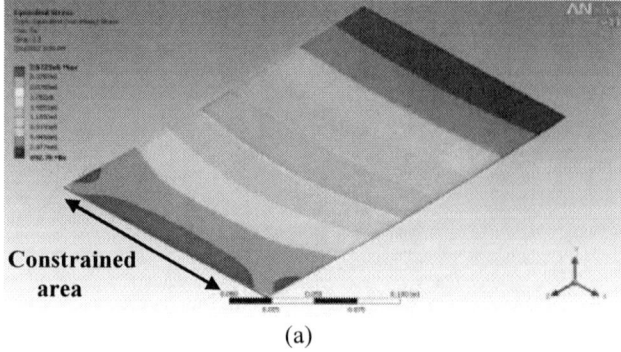

(a)

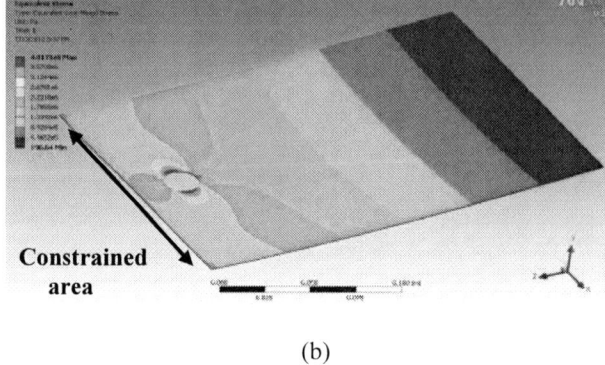

(b)

Fig. 7.Stress contour of the PCB plates. (a) without hole and (b) with single hole

IV CONCLUSION

In this study, the stress analysis during bending process was simulated using Ansys. The analysis was performed using 3D Models resembiling PCB plates with single hole and without hole. From the simulation, it has been observed that the value of stress response of the PCB plates was increasing with increased displacement height. The with single hole PCB model exhibited higher stress compared to the without hole PCB model. The single hole structure was utilized to focus the stress distribution on one area. This will reduce the stress distribution along the width and length direction of the PCB plate.Therefore the damage induced to the solder joints of the mounted components can be reduced. Highest stress response was obtained at the displacement height of 5cm for both models.

ACKNOWLEDMENT

The author would like to thank and acknowledge School of Microelectronic Engineering, Universiti Malaysia Perlis for their support and facilities.

REFERENCES

[1] W. Hai-feng, et al., "Life Predict and Simulation of the Copper Wire in Flexible Printed Circuit Board," in Measuring Technology and Mechatronics Automation (ICMTMA), 2011 Third International Conference on, 2011, pp. 470-472.

[2] Tummala Rao. R., Fundamentals of Microsystems Packaging, Mc Graw Hill International Edition, 2004.

[3] D. Lau, et al., "Experimental testing and computational stress analysis of printed circuit board for the failure prediction of passive components under the depaneling load condition," in Electronic Components and Technology Conference, 2005.Proceedings.55th, 2005, pp. 1783-1791 Vol. 2.

[4] S. W. Ricky Lee and Dennis Lau, "Correlation between the Strain on the Printed Circuit Board and the Stress in Chips for the Failure Prediction of Passive Components," Proceedings of IMECE2005, ASME,Orlando, Florida USA, November 5-11, 2005.

[5] Luan, J.E., Tee, T.Y., Pek, E., Lim, C.T., and Zhong, Z.W., "Modal Analysis and Dynamic Responses of Board Level Drop Test," 5th EPTC Conference Proc., Singapore, 2003, pp. 233-243.

[6] S. W. Ricky Lee, Dennis Lau, Mabel Tsang, Jeffery Lo, Fu Lifong, Jin Jiwen, Liu Sang, "Experimental Testing and Computational Stress Analysis of Printed Circuit Board for the Failure Prediction of Passive Components under the Depaneling Load Condition," Fifty Fifth IEEE ECTC Conference, Wyndham Palace Resort & Spa. Orlando, FL USA, 2005, Vol 2, pp. 1783-1792.

[7] T. Tong Yan, et al., "Novel numerical and experimental analysis of dynamic responses under board level drop test," in Thermal and Mechanical Simulation and Experiments in Microelectronics and Microsystems, 2004. EuroSimE 2004. Proceedings of the 5th International Conference on, 2004, pp. 133-140.

Design and Analysis of a Localised Environment Monitoring Sensor System

Asral Bahari Jambek[1], Lau Chyun Wenn[2], Uda Hashim[3]

[1, 2]School of Microelectronic Engineering, Universiti Malaysia Perlis
[3]Institute of Nanoelectronic Engineering, Universiti Malaysia Perlis
[1]asral@unimap.edu.my, [2]lauchyunwenn@gmail.com, [3]uda@unimap.edu.my

Abstract- **This paper discusses the design of a localised environment monitoring system. The system consists of a temperature sensor, a humidity sensor, a global positioning system (GPS), a microcontroller and an LCD display. The temperature and humidity sensors measure the surrounding environment, while the GPS collects the current longitude, latitude and altitude position of the system. This paper highlights the verification method to ensure the accuracy of the system when performing the sensing operations. Each sensor is compared against the existing commercial device under various environmental conditions. The results show that the system can measure the temperature, humidity and location information, within an accuracy range of 96% to 99%, when it is benchmarked against the commercial devices.**

Keywords: environment sensor device, temperature, humidity, GPS, microcontroller.

I. INTRODUCTION

Environment monitoring devices are important in current and future monitoring applications, such as in precision agriculture [1], automobiles [2], machines, home automation [3] and buildings [4]. In animal tracking, for instance, the animal might travel from one place to another. In order to study the behaviour of the animal, it is important to collect information about the ambient condition and the animal's location. To collect these data, a reliable sensing device must be developed and attached to the animal. This device must be able to measure and store the data accurately so that the data can be used to study the subject's behaviour.

When designing the environment sensor devices, one of the main challenges is to obtain accurate measured data. In this work, we develop an environment sensor system that can collect temperature, humidity and location information. The objective of this work is to ensure that the device can accurately and reliably perform the required measurements.

The rest of the paper is arranged as follows. First, Section II will discuss the literature related to existing environment monitoring systems. Section II will discuss the method used to design our localised environment monitoring system. Section IV will highlight the results obtained from the experiments. Finally, Section V will conclude the paper.

II. LITERATURE REVIEW

In [5], information, such as temperature collection, tri-axial and GPS data, is collected to measure the comfort levels found in public transportation. The system consists of a microcontroller, an accelerometer, a temperature sensor and a GPS system. By collecting travelling pattern information, the driver's behaviour can be studied. Furthermore, the road conditions can also be assessed to ensure better comfort for the driver and to provide information that could be used to improve road maintenance.

Paper [6] discusses designing data acquisition for three-dimensional electromagnetic explorations for oil and gas detection. The design consists of data acquisition, a digital signal processing module (DSP), an embedded controller and a GPS system. While the embedded controller performs the main control operation of the system, the DSP module performs the FFT, decimation, filtering, correlation and custom data processing required by the electromagnetic exploration [6].

To improve the ability of drivers to deliver frozen products to the consumer market, paper [7] discusses developing a cold-chain temperature monitoring system. The system is equipped with RFID tags, a temperature sensor and a GPS system. The RFID tags provide a continuous record of temperature data, time and data storage. By implementing the system, various aspects of the services can be monitored and improved, such as product quality, management and logistics.

In [8], an animal tracking system is built to monitor the behaviour and migration patterns of swamp deer. The system collects important information, such as temperature, humidity, GPS, head orientation and ambient light. Once the data is collected, it is transmitted to the base station through wireless communication. The system is built using a microcontroller, temperature and humidity sensors, an accelerometer, a GPS receiver, a Li-Ion battery and a solar panel. The solar panel harvests the energy from ambient light and stores the energy in the rechargeable battery.

III. METHODOLOGY

This section discusses the design of the localised environment monitoring system in detail. This system consists of five main components: a temperature sensor, a humidity sensor, a GPS receiver, an LCD display and a controller, as shown in Figure 1. In this design, the microcontroller is the main component. Its main function is to receive input data from various sensors, process the measured data and output it to the LCD display. In this work, a PIC16F887 microcontroller from a microchip was selected. It is a low-power CMOS 8-bit MCU based on high performance RISC architecture [9]. The device has a built-in analog-to-digital converter (ADC) that converts analog

978-1-4673-2395-6/12 $31.00 © 2012 IEEE

signals from the sensor to a 10-bit binary. This eliminates the need for an external ADC in the system.

An SN-HMD sensor is used to measure the temperature and humidity [10]. That sensor is selected since it has good accuracy and a wide measuring range that is suitable for the targeted application. The device produces a linear output voltage as the temperature and humidity change. For the temperature, each change of 1°C will result in a change of 0.1V at the output, whereas for the humidity sensor, a 10% change in the air humidity results in a 0.35 V change in the output voltage. All output pins from the sensor are connected to the PIC microcontroller's I/O through pin RA0 and RA1.

In this system, the location data is collected using the GPS receiver model EB-85A [11]. The GPS receiver will receive signals from satellites, calculate the current location information and output the longitude, latitude and altitude information. These data are outputted in NMEA format and are transfer to the microcontroller's universal asynchronous receiver transmitter (UART) port at a rate of 38400 bits per second [11]. The overall schematic diagram of the system is shown in Figure 2.

In order to ensure that the device can operate properly, the microcontroller has to be programmed to read data from multiple sources of sensors. In this work, the microcontroller will first read the data from the temperature and humidity sensors before displaying the value. In the next phase, the microcontroller will read the data from the GPS receiver, extract the location information and then display the latitude, longitude and altitude of the current position. This process will then be repeated every 10 seconds until the microcontroller receives an interrupt signal from the user. Figures 3 and 4 show the flow chart of how the microcontroller reads the output from the temperature, humidity and GPS sensors.

IV. RESULTS AND DISCUSSION

This section discusses the results obtained in this work. First, we will discuss the results obtained from calibrating each of the components used in this work. Next, the results obtained from integrating all the modules will be explained. Lastly, the verification of the sensor node will be discussed.

Tables 1 and 2 show the results from measuring the temperature and humidity, respectively. The microcontroller measures the voltage output from these sensors every 5 seconds. Since the analog voltage output from the sensor varies all the time, due to variations in the environment, 10 values will be taken and the average value will be calculated. This is done to ensure that a stable value can be obtained and displayed on the LCD. The average value of the measured voltage will be converted to Celsius (°C) using the formula given in the datasheet [10]. To ensure high accuracy, the measured temperature is calibrated against a mercury thermometer.

Table 1 shows the results obtained from measuring the temperature using our system. In this experiment, the temperature is measured between 31.5 °C and 38 °C. This measurement is taken under different conditions, namely, indoor, shaded and direct sunlight. As shown in the Table, the measured temperature obtained using our

system give 97.52% accuracy when compared to the results obtained using a mercury thermometer. For the humidity sensor, the measurement we obtained using our system is compared against the results obtained from using a commercial humidity sensor with a humidity range of 40% to 62%. As seen in the Table, the data collected by the humidity sensor in our system shows 99.35% accuracy compared to the commercial humidity sensor.

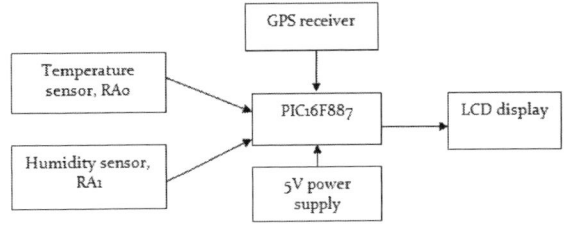

Figure 1: Block diagram of the localised environment monitoring system.

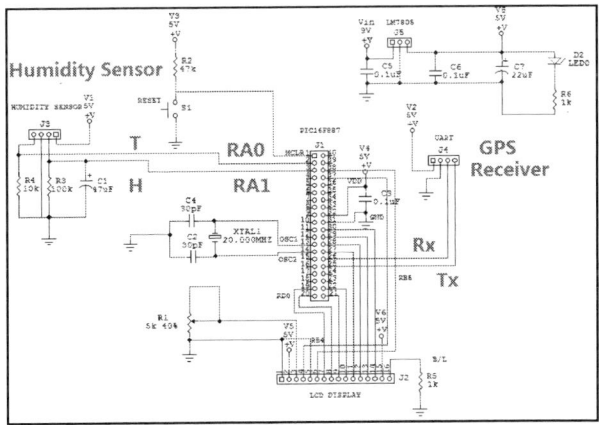

Figure 2: Schematic diagram of the complete localised environment monitoring system.

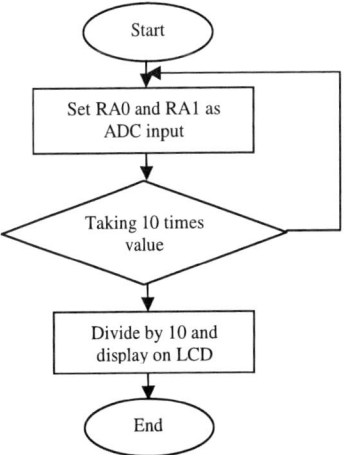

Figure 3: Flow chart for how the microcontroller measures humidity and temperature.

978-1-4673-2395-6/12 $31.00 © 2012 IEEE

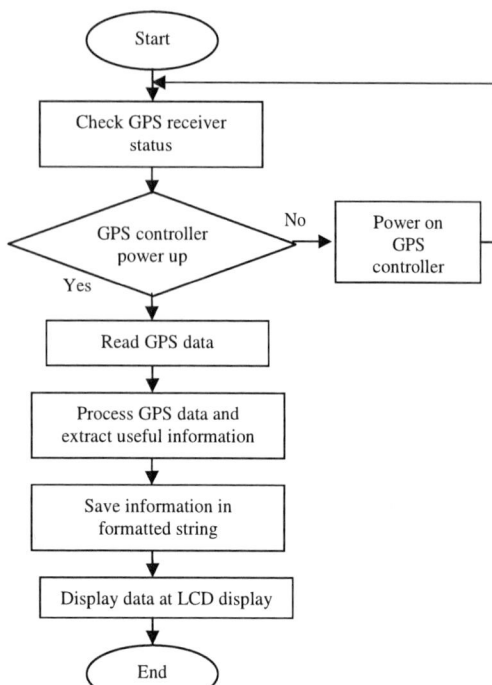

Figure 4: Flowchart for how the microcontroller reads the GPS output data.

To verify the operation of the GPS receiver, NMEA data obtained during the experiment were inputted into the microcontroller through UART before being displayed on the LCD. The location data output from our system is compared against the data measured by a commercial GPS device embedded in a smart phone (model Samsung GT-S5570). The smart phone is an Android version 2.2.1 and it uses the GPS Locator Utility application to display the GPS information. In this experiment, latitude, longitude and altitude data were measured while travelling from Ulu Pauh, Perlis to Kampung Wai, Kuala Perlis. This route covers a distance of 18.6 km. Figure 7 shows the traveling map taken from Google Map. On the map, the point A–J shows the position where the location information is recorded. The data output from our system and the smart phone were recorded at the same time.

Tables 3 and 4 show the comparison between the data collected from our system against the data collected using the smart phone. As shown in the Table, the latitude and longitude information recorded by our device matches the information recorded by the smart phone with 99% accuracy. For altitude information, our system gives 96% accuracy compared to the commercial device. Figure 5 and Figure 6 show the displayed value of the LCD during the temperature, humidity and location measurements.

Figure 5: Temperature and humidity readings displayed on the LCD.

Figure 6: GPS receiver output displayed on the LCD.

Table 1: Data comparisons between the measured temperature sensor and the mercury thermometer

Mercury Thermometer	Temperature Sensor		Accuracy (%)
Temperature (°C)	Measured Voltage (V)	Displayed Temperature (°C)	
31.5	2.65	31.0	98.41
32.0	2.65	31.0	96.88
32.5	2.75	32.0	98.46
35.0	2.90	34.0	97.14
36.0	3.00	35.0	97.22
37.0	3.10	36.0	97.30
38.0	3.20	37.0	97.37
38.0	3.20	37.0	97.37
Average			97.52

Table 2: Measured humidity values

Commercial Device	Humidity Sensor		Accuracy in %
Humidity (%RH)	Measured Voltage (V)	Displayed Humidity (%RH)	
40.0	1.66	39.9	99.77
41.0	1.70	40.7	99.37
42.0	1.71	41.8	99.42
43.0	1.75	43.0	99.96
44.0	1.78	44.0	99.93
50.0	1.95	49.6	99.24
51.0	2.00	50.9	99.78
61.0	2.30	60.2	98.73
62.0	2.35	60.7	97.91
Average			99.35

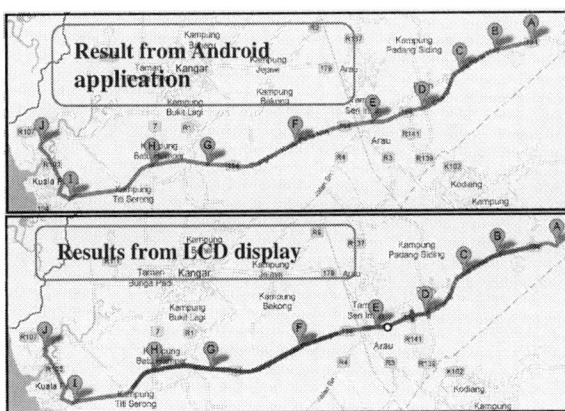

Figure 7: Traveling point for the GPS measurement.

Table 3: Latitude and longitude data comparison

Point	Result output using commercial GPS		Results displayed on the LCD display		Average Error (%)
	Latitude	Longitude	Latitude	Longitude	
A	6.45637	100.35602	6.45673	100.35626	0.0029
B	6.45238	100.33083	6.45266	100.33125	0.0024
C	6.44437	100.31595	6.44445	100.31663	0.0010
D	6.44474	100.2996	6.42861	100.29998	-0.1250
E	6.42257	100.27842	6.42304	100.27863	0.0038
F	6.41401	100.24478	6.41446	100.24543	0.0038
G	6.40552	100.2068	6.40565	100.20697	0.0011
H	6.40497	100.18223	6.40554	100.18251	0.0046
I	6.39134	100.14726	6.3914	100.14803	0.0009
J	6.41379	100.13495	6.41414	100.13523	0.0029
Average					-0.0102

Table 4: Altitude data comparison

Point	Altitude (feet)		Accuracy (%)
	Result from commercial device	Displayed Altitude on LCD	
A	164	169	96.95
B	100	95	95.00
C	95	100	94.74
D	87	90	96.55
E	72	76	94.44
F	59	60	98.31
G	56	55	98.21
H	78	82	94.87
I	55	58	94.55
J	42	41	97.62
		Average:	96.12

V. CONCLUSION

This paper discusses the design and analysis of a localised environment monitoring system. The system consists of a temperature sensor, a humidity sensor, a GPS system, a microcontroller and an LCD display. In order to achieve good accuracy, rigorous experiments were conducted to verify the operation of each component of the system. From our experiments, our system is able to collect data with 96% to 99% accuracy when compared to the existing commercial devices. The next step of this work will include an evaluation of our system's wireless capability, as well as determining how to reduce the system's size and minimise its power consumption so that it can be used effectively in a wireless sensor node application.

REFERENCES

[1] T. Hamrita, J. Durrence, G. Vellidis, "Precision Farming Practice", Industry Applications Magazine, IEEE, Volume 15, Issue 2, 2009 , pp 34 – 42.

[2] Tavares, J., Velez, F.J., Ferro, J.M. "Application of wireless sensor networks to automobiles", Measurement Science Review, Vol. 8, Secion 3, pp: 65-70, 2008.

[3] "ZigBee Home Automation Public Application Profile", ZigBee Alliance, Rev 25, Ver 1.0, 2007.

[4] Yu, Y., Prasanna, V.K., Krishnamachari, B., "Information Processing and Routing in Wireless Sensor Networks", World Scientific Publishing Co., 2006.

[5] J.C. Castellanos, A.A.Susin, F. Fruett, "Embedded sensor system and techniques to evaluate the comfort in public transportation", The 14th International IEEE Conference on Intelligent Transportation Systems (ITSC), 2011 , pp 1858 - 1863

[6] Rujun Chen; He Zhangxiang; Qiu Jieting; He Lanfang; Cai Zixing, "Distributed data acquisition unit based on GPS and ZigBee for electromagnetic exploration", IEEE Instrumentation and Measurement Technology Conference (I2MTC), 2010 , pp: 981 – 985.

[7] Bo Yan; Danyu Lee, "Application of RFID in cold chain temperature monitoring system", ISECS International Colloquium on Computing, Communication, Control, and Management, 2009. CCCM 2009.
Volume: 2,2009 , pp: 258 – 261.

[8] Jain, V.R.; Bagree, R.; Kumar, A.; Ranjan, P., "wildCENSE: GPS based animal tracking system", ISSNIP 2008. International Conference on Intelligent Sensors, Sensor Networks and Information Processing, 2008 , pp: 617 – 622.

[9] "PICmicro Family Tree", PIC16F Seminar Presentation http://www.microchip.com.tw/PDF/2004_spring/PIC16F%20semin ar%20presentation.pdf

[10] Cytron Technologyies, "SN-HMD Humidity Sensor User's Manual V1.2", April 2009.

[11] Datasheet "Smart GPS receiver Model EB-85A", Etek Navigation Inc, 28 Nov 2006.

Physical Characteristic of Room-Temperature Deposited Ti Thin Films by RF Magnetron Sputtering at Different RF Power

Z. Aznilinda *[a], S.H. Herman**[a], R.A.Bakar[a], and M. Rusop[ab]

[a] NANO-Electronic Centre (NET), Faculty of Electrical Engineering, Universiti Teknologi MARA (UiTM),
40450 Shah Alam, Malaysia
[b] NANO-SciTech Centre (NST), Institute of Science (IOS), Universiti Teknologi MARA (UiTM),
40450 Shah Alam, Malaysia
*aznilinda.zainuddin@yahoo.com, **hana1617@salam.uitm.edu.my

Abstract— **Ti thin films of various thicknesses were grown on glass substrates by using RF magnetron sputtering technique with sputtering power varied from 50W to 300W. The thickness of the thin films are measured using surface profiler KLA Tencor P-6 and it is observed that the thickness increased as the sputter power increased. Sputtering rate increases form 1.59nm/min to 8.77nm/min as the sputter power increases from 50W to 300W. Atomic force microscopy (AFM) was used to study the surface roughness and surface topography of the Ti thin films. The surface roughness is also proportional to the sputter RF power. FESEM analysis revealed that the particle size transform from dense agglomeration particle to bigger particle size with voids in between as the increase of RF power. The growth of the Ti on glass is in columnar structure and the RF power place a big role in order to modify a structure of a Ti thin film.**

Keywords- RF Magnetron Sputtering; room temperature; Ti; deposition rate, surface topography.

I. INTRODUCTION

Titanium (Ti) is a silver-gray non-toxic material with an atomic number of 22, 47.87 of atomic weight , 1,660 °C melting point and 3,287 ° C boiling point. It is the ninth most abundant element in the earth's crust and lithosphere, and the fourth most abundant metallic element [1]. In nature, titanium is found in the forms of rutile (titanium dioxide, TiO_2) and ilmenite (titanium iron oxide, FeTiO3) [2].

This work was supported by the Excellence Fund of Research Management Institute (RMI –UiTM) Project Code: (Project Code: FRGS/1/ 2012/TK02/UITM/ 03/8) and the Ministry of Higher Education (MHOE) Malaysia.

Ti is well known due to its excellent physical properties which are light and strong. Titanium's corrosion rate is so low even in seawater [3], and due to this, is has been used to build ships, propeller, rigging and parts that are unshielded to the sea water. Ti metal is used as an alloying agent with metals such as aluminum and iron, and used in aircraft, military, automotive and aerospace where strong, high-temperature stability and lightweight are needed.

Ti which is also a biocompatible material [4] has the ability to bind with human bone and due to that is used for a number of medical and dental purposes such as bone plates, hip or knee joints, dental implants and crowns. Ti is also known to be a good material for electronics device fabrication such as microsensor and microelectromechanical systems [5],

With all these benefits and applications of Ti, there have been a lot of studies done on the deposition technique for Ti which includes the surface morphology and nanostructure properties of the deposited Ti.

Radio Frequency (RF) Magnetron sputtering method, which is a well-known physical vapor deposition (PVD) technique has been broadly used in depositing metal contact or thin film [6]. RF magnetron sputtering method offers deposition of thin film at room temperature with high deposition rates. It also produce high adherence to substrate quality.

The thickness and crystallite size of a thin film can be controlled by the power and pressure during sputtering. The thin film thickness can be minimized by reducing the power and/or increasing the pressure [7-11]

The thin film physical properties and deposition rate are highly reliant on the sputtering conditions, such as substrate temperature, substrate bias, argon gas flow, sputtering power and also sputtering time [12].

In the present work, we will study on the physical properties and the deposition rate of Ti deposited by this method with different RF power.

II. EXPERIMENTAL PROCEDURE

Titanium was deposited on slide glass for 1 hour by using RF magnetron sputtering as a function of RF power (50 W, 100 W, 150 W, 200 W, 250 W and 300 W) at room temperature. The Rf Sputtering system was vacuumed to a background pressure of 5.0 x10^{-7} Torr in order to remove and minimize any remaining gas particles in the system chamber. A high-purity Ti (99.99% Ti, 4-inch diameter) target was used at working pressure of 5mTorr and in argon and oxygen gas flow rate of 50:00. The Ti target was pre-sputtered for 15 min in order to clean the target surface from any impurities. The target-substrate distance was at 13cm during the whole deposition process. Detailed sputtering conditions are illustrated in Table I.

The substrates were subjected to ultrasonic cleaning in acetone for 10 min. It is then rinsed with deionized water and next ultrasonically cleaned in methanol for 10 min. the last step is to rinse the substrates with deionized (DI) water and being dried with nitrogen gas.

The thickness of the thin films is measured using surface profiler (SP, KLA Tencor P-6) and from the thickness, the deposition rate is calculated. The surface topology is analyzed by using an atomic force microscopy (AFM, XE-100 Park System). The surface topology images were taken at 1um x 1um scan area and at 3 different spot, the surface roughness is then being measured. Field Emission Scanning Electron Microscope (FESEM, JEOL JSM 7600 F) analysis is used to observe the particle structure on surface view and also cross-sectional view.

III. RESULT AND DISCUSSION

The average thickness of the thin films measured is plotted in Fig. 1. It shows that the film thickness and deposition rate is proportional to the sputtering power. During the deposition process, as RF power increases, it provides substantial energy to the Argon (Ar) ions and results in extensive bombarded ion onto the target. Concurrently, the ions which received a very high kinetic energy increases the amount of ejected target atoms [13]. These two phenomena are linearly proportional to the sputtering power consequently affected the changes in deposition rate.

In addition, thickness and deposition rate is also contributed by the power density [14]. When the RF power increases, its power density increases from 2.47 W/cm^2 at 50 W to 7.40 W/cm^2 at 150 W and 14.80 W/cm^2 at which leads to the higher plasma density in the chamber hence increases the number of atoms collision and atoms deposited on the substrate.

Fig. 2 shows the FESEM images of the Ti thin film's surface morphology deposited at 50 W, 100 W, 150 W, 200 W, 250 W and 300 W respectively. The thin films deposited at 50W revealed fine grains morphology with smaller grain structure as compared to the thin film deposited at 300 W which have bigger grain structure with voids.

TABLE 1
DEPOSITION CONDITIONS FOR THE RF MAGNETRON
SPUTTERING OF Ti THIN FILMS

Substrate	Slide glass 2x2cm
Target	Ti 99.99% purity ; 50.8mm diameter
Gas	Ar(99.995%);O$_2$ (99.995%)
Base Pressure	5.0 x10^{-7} Torr
Deposition hour	1hour
Substrate rotate vertical axis	30rpm
Substrate temperature	Room temperature
Oxygen-argon ratio	50:00
RF power	50W - 300W

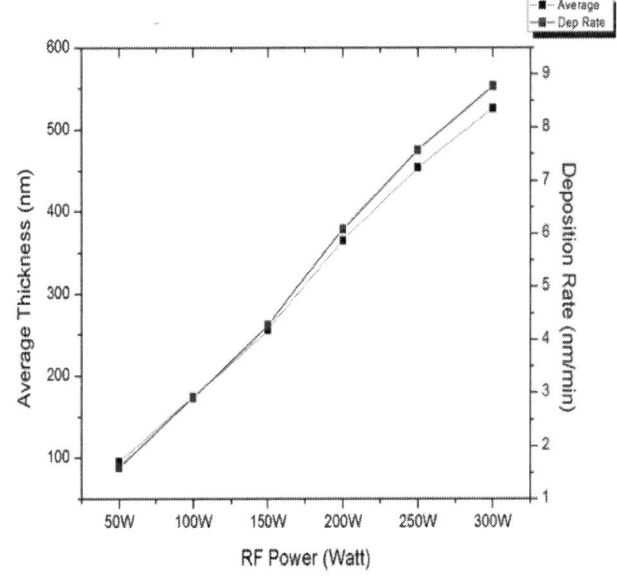

Fig. 1 Dependent of the thickness and deposition rate on the RF sputtering power.

This has proven that the increase in sputtering power changed the size of the grains and also the morphology of the grains to be less dense with voided boundaries. It is also observed that the particle structure transform from agglomeration of particles to a porous structure which can also been seen from the cross section images in Fig. 3. As the sputter power increase, the high energy carry by the ions collides on the target's surface and giving the ability to eject bigger particle size from the target [15].

Fig. 2 FESEM surface images of Ti deposited at RF sputtering power of 50 W ,100 W, 150 W,200 W,250 W and 300 W.

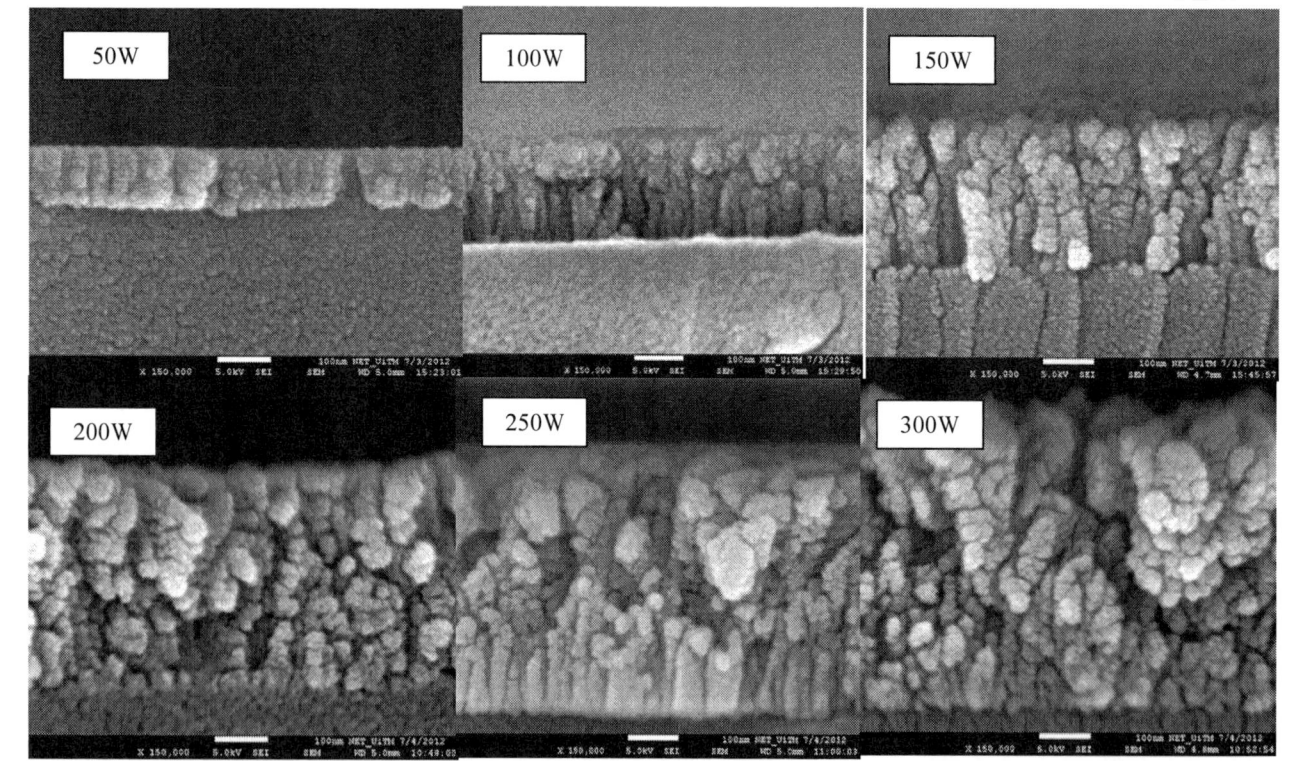

Fig.3 FESEM cross-section images of Ti deposited at RF sputtering power of 50 W ,100 W, 150 W,200 W,250 W and 300 W.

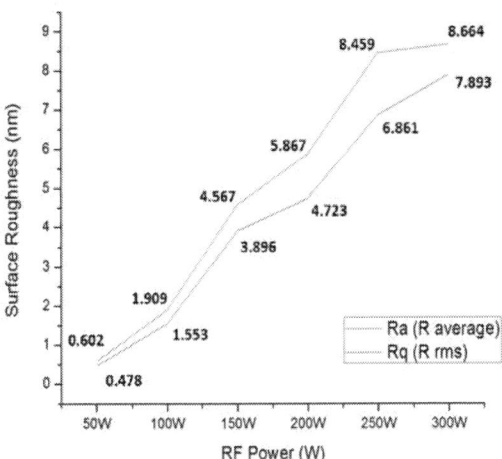

Fig.4 The Roughness (Ra) and Rrms (Rq) of the Ti thin films surface sputtered at different RF Power.

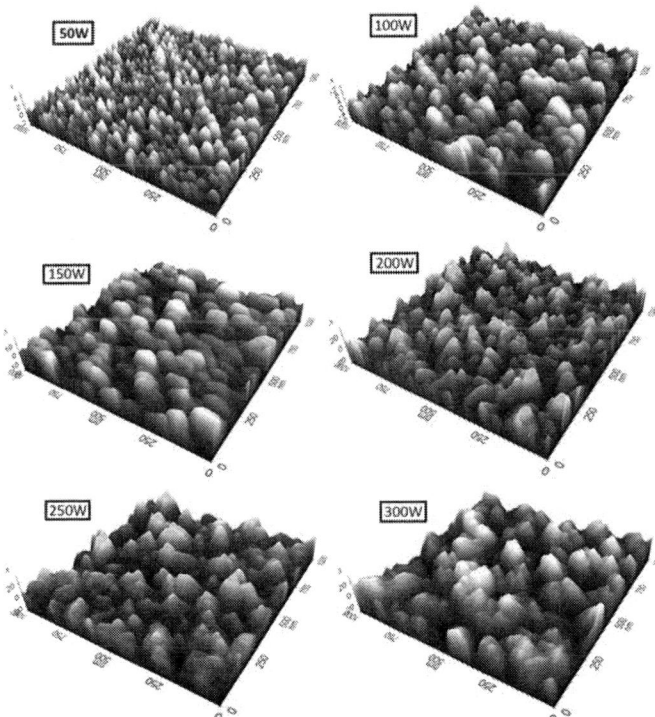

Fig.5 AFM images at scan area of 1um x 1um of Ti thin films deposited at sputtered power 50 W ,100 W, 150 W,200 W,250 W and 300 W.

As this happen with a higher sputtering rate, it provides a deposition of bigger grain size and the existence of voids between the particles.

At lower RF power, the sputtered or ejected atoms from the target will be smaller due to the low energy carried by the argon ions. And with the smaller size of particles ejected, it produce a more dense thin film. The cross section image in Fig. 3, we also observe that the growth of the Ti thin film is in columnar structure.

The surface topography of the sputtered Ti was studied by AFM. Topography scans were carried out in the noncontact mode under ambient temperature. Fig. 5 shows the AFM images that were collected at scan area of 1um x 1um on the Ti sputtered on glass substrates at RF power 50 W to 300 W. The figure manifests the changes in the surface grain size which it is bigger as the sputter power increases. The trend of the thin film's surface average roughness (Ra) and root mean square roughness (Rq) in Fig. 4 shows the proportional relationship between the surface roughness and the sputtering RF power.

The roughness increases as the RF power increases is due to the higher bombardment of the highly energetic sputtered particles and ions to the deposited film surface. This is also due to increases of adatom mobility as the increases of RF power giving bigger particle of atom being ejected and deposited on the substrate surface and hence increases the surface roughness [16].

IV. CONCLUSION

In this study, we have successfully deposited Ti thin films on glass substrates at a room temperature by using RF magnetron sputtering technique. We observe how the sputtering power affected the thin film thickness, deposition rate, surface topography and particle structure of Ti films. The change in sputtering power is proportional to the particle size, film thickness, deposition rate and also surface roughness. The changes in sputtering power also exhibit different structure of the particles and grains. Generally, higher sputtering power (up to 300W) can lead to increase of deposition rate, grain size and surface roughness. We conclude that the Ti thin film thickness and its physical structure can be optimize and modify by properly controlling the sputtering RF power during the deposition.

ACKNOWLEDGMENT

The authors would like to thank the technicians and science officers at NET and NST for their kind support in this research.

REFERENCES

[1] D.Knittel, *Titanium and titanium alloys*, 3 ed.: John Wiley & Son., 1983.

[2] W. Zhang, *et al.*, "A literature review of titanium metallurgical processes," *Hydrometallurgy*, vol. 108, pp. 177-188, 2011.

[3] B. D. Craig, *Handbook of Corrosion Data*, 2 ed., 1995.

[4] D. Williams, "Titanium Epitome of Biocompatibility or Cause of Concern," *The Journal of Bone and Joint Surgery*, vol. 76-B, p. 766, May 1994 1994.

[5] D. Resnik, *et al.*, "Experimental study of heat-treated thin film Ti/Pt heater and temperature sensor properties on a Si microfluidic platform," *Journal of Micromechanics and Microengineering*, vol. 21, p. 025025, 2011.

[6] X. Yu, *et al.*, "Fabrication and structural characterization of metal films coated on cenosphere particles by magnetron sputtering deposition," *Applied Surface Science*, vol. 253, pp. 7082-7088, 2007.

[7] J. L. Andújar, *et al.*, "Effects of gas pressure and r.f. power on the growth and properties of magnetron sputter deposited amorphous carbon thin films," *Diamond and Related Materials*, vol. 11, pp. 1005-1009, 2002.

[8] J.-A. Jeong and H.-K. Kim, "Thickness effect of RF sputtered TiO2 passivating layer on the performance of dye-sensitized solar cells," *Solar Energy Materials and Solar Cells*, vol. 95, pp. 344-348, 2011.

[9] Y. Jin, *et al.*, "Effect of sputtering power on surface topography of dc magnetron sputtered Ti thin films observed by AFM," *Applied Surface Science*, vol. 255, pp. 4673-4679, 2009.

[10] J. Keun Seo, *et al.*, "The effect of deposition RF power on the SiC passivation layer synthesized by an RF magnetron sputtering method," *Journal of Crystal Growth*, vol. 326, pp. 183-185, 2011.

[11] D. K. Kim and H. B. Kim, "Room temperature deposition of Al-doped ZnO thin films on glass by RF magnetron sputtering under different Ar gas pressure," *Journal of Alloys and Compounds*, vol. 509, pp. 421-425, 2011.

[12] P. Singh and D. Kaur, "Room temperature growth of nanocrystalline anatase TiO2 thin films by dc magnetron sputtering," *Physica B: Condensed Matter*, vol. 405, pp. 1258-1266, 2010.

[13] K.-Y. Chan and B.-S. Teo, "Sputtering power and deposition pressure effects on the electrical and structural properties of copper thin films," *Journal of Materials Science*, vol. 40, pp. 5971-5981, 2005.

[14] A. Matsuda, *et al.*, "Control of plasma chemistry for preparing highly stabilized amorphous silicon at high growth rate," *Solar Energy Materials and Solar Cells*, vol. 78, pp. 3-26, 2003.

[15] J. Zhou, *et al.*, "Influence and determinative factors of ion-to-atom arrival ratio in unbalanced magnetron sputtering systems," *Journal of University of Science and Technology Beijing, Mineral, Metallurgy, Material*, vol. 15, pp. 775-781, 2008.

[16] V. Chawla, *et al.*, "Microstructural characterizations of magnetron sputtered Ti films on glass substrate," *Journal of Materials Processing Technology*, vol. 209, pp. 3444-3451, 2009.

Issues and Challenges in Microfluidic Research Studies

Jafar Alvankarian, Alireza Bahadorimehr, Benyamin Davaji, and Burhanuddin Yeop Majlis, *Senior Member, IEEE*

Institute of Microengineering and Nanoelectronics (IMEN)
Universiti Kebangsaan Malaysia (UKM)
43600 UKM Bangi, Selangor, Malaysia
Email: burhan@vlsi.eng.ukm.my

Abstract- **This article investigates the common problems of fabrication and testing of microfluidic devices. Compatibility of the main substrate properties with the materials and experiments involved is very important for performing a test using a microfluidic device. Methods of solving the problems associated with the fabrication techniques involved are critical in creating a new microfluidic device. Specific preparatory steps need to be considered before microfluidic testing and during experimental works to obtain acceptable conclusion.**

I. INTRODUCTION

The goal in this paper is to highlight some of the problems we have observed during prototyping and testing of the new microfluidic devices for biological analysis applications. So many materials and fabrication techniques have been introduced by different research groups on developing new microfluidic devices for biological applications[1, 2]. But there are less published articles on the problems and the associated solution techniques as most of the presented results are focused on the successful tests.

Some of the issues are related to the selected material. Chemical compatibility, biocompatibility, mechanical properties, thermal properties and optical properties need to satisfy the requirements of the tests. Over-swelling under severe chemical exposure, structure deformation under extreme thermal condition, toxic effects on the living cells are some of the problems related to substrates specially polymers. Glass and silicon are the materials most suitable and robust in this case. Moreover, there are other problems appearing during fabrication process. Complexity of the processes, failing of microstructures and local defects in the microstructures are issues we have seen on polymeric and non-polymeric substrate.

Finally, testing process can be problematic if not enough and precise preparatory consideration applied. Device bonding or fluidic interconnect failure under high fluidic pressure is a common trouble involved in the microfluidic devices. In this paper, we refer to the problems we have faced during different fabrication processes such as defects appear in the polymer structures due to the trapped bubbles, or defects in the fabrication process of the silicon structures as well as the fluidic leakage during testing a microfluidic device.

II. MATERIAL SELECTION

Fig. 1 show a single layer of the microstructure fabricated using lithograph technique. The main material for the fabrication has been polyurethane methacrylate which is demonstrating elastomeric behaviors [1]. Although this material presents a good bonding strength as other elastomeric substrates such as PDMS but it also suffers from swelling with chemicals.

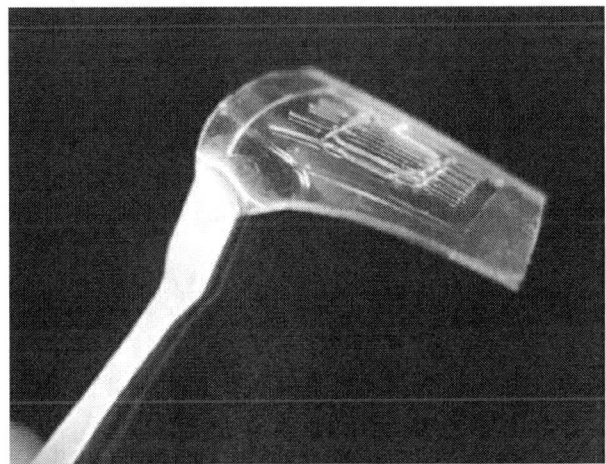

Fig. 1 Flexibility of the elastomeric substrates in rapid prototyping of microfluidic devices[3]

One of the main candidates for realization of microfluidic devices is silicon. The most important reason is the developments in microelectronic industry. Due to the existing standard machines, materials and protocols, many of the fabricated microfluidic devices are fabricated based on silicon substrate. Undoubtedly, microfluidic devices may still be somewhat expensive using silicon substrates, but the use of cheaper materials are increasing. The biocompatibility of silicon substrate is another important factor in biomedical applications especially in vivo experiments. Biocompatibility of silicon-based microfluidic devices is relatively not well understood. Indeed, the surface biocompatibility of silicon is poor and interaction between Si-based microfluidic devices and human body is not desirable [4].

Glass has several advantages as a fluidic substrate: i) it is optically transparent over a wide range of wavelengths, ii) its

978-1-4673-2395-6/12 $31.00 © 2012 IEEE

surface is hydrophilic, iii) its surface chemistry is well understood, iv) there are numerous etching methods and v) modification of surface is performed with various techniques [5]. Glass in most cases is the material of choice for the chemists and biologists, but the isotropic nature of its wet etching makes it difficult to create deep channels in glass. Also the fragility of this material is another issue[6].

III. FABRICATION PROCESS

Fig. 2 shows the scanning electron microscopy of a microstructure made of the elastomeric material in a photolithograph process. In this photo, the microstructure has been partly delaminated from the base substrate which has caused failure of the process. The step of the fabrication has to be iterated several times to solve the issue.

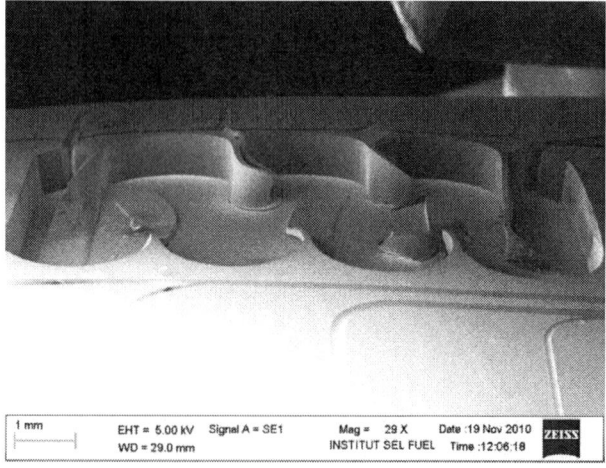

Fig. 2 Delamination of the microstructure from the base substrate during lithography process.

Fabrication of microfluidics using Polydimethylsiloxane (PDMS) is the most growing method in this field. There are lots of advantages of using this polymer which often encourage researchers to use this material as their first choice. There are a lot of research in chemistry and biomedical applications which use this material without reporting any problem. However, there are also many researchers that point to the problems of this polymer for bio applications [7-9]. These groups prefer to work with silicon and glass substrates instead of PDMS. Here, both cons and pros of PDMS as the main microfluidic biochip fabrication material are discussed, since in this thesis also PDMS is the main material for magnetic bead capturing. PDMS in microfluidics is the most well-known material due to many advantages as follows:

- Easy to pattern by molding process and soft lithography: It is easy to fabricate and also reproduce many devices from one master.
- Optically transparent from 240 nm to 100 nm wavelengths

- Flexibility and elasticity: Compare to glass there is no fragility concerns for breaking problems. Moreover, the elasticity of this material makes it an excellent choice for thin diaphragms and membranes with low young's modulus (360-870 kPa depend to the mixing ratio) and high deflection.
- Nontoxic and Gas permeable: Permeability is the product of solubility (partition) of a gas in polymer and its diffusivity. PDMS has high permeability compare to other siloxane groups. Therefore, for cell based analysis is one of the best choices.
- Inexpensive material and low cost procedure: Its disposability is not concerned.
- Biocompatible
- Fast prototyping material
- Excellent bonding properties using different techniques for surface treatments. It can easily bond to itself or glass.

Replication of PDMS structures for mass manufacturing is not reliable because of the following issues:
- Deformation of PDMS under pressure
- Swelling effect of PDMS in the presence of solvents is inevitable.

Some of the researchers due to these drawbacks of the most popular material in microfluidic fields recommend using of each PDMS chip only once [10].

IV. TESTING

Fig. 3 shows the optical photography of a microfluidic device under test. In this device, five inlet and outlet ports have been included in the design. One of the important issues is the device functioning failure due to the leakage from the fluidic interconnect as demonstrated in this figure or delamination of two layers of the microfluidic device. To solve this problem, either the fabrication process need to be improved or device need to be manufactured again for performing the desired functions.

The issues of PDMS in all microfluidic devices also should be considered. There are some problems using PDMS which are as follows:
- It operates only in moderate temperatures
- Hydrophobic surface and adsorption problems: The sample adsorption to the surface of the channels due to migration of siloxane groups to the surface of PDMS is inevitable. Unfortunately, some fluorescent dyes, hydrocarbon solvents, and many of proteins adhere to the surface of the PDMS. Some magnetic and non-magnetic beads also have the adsorption properties. Modification of the surface is applied to reduce this effect. The hydrophilicity revert back to hydrophobicity only few hours after surface activation and in some literatures for 3-6 days [11].
- Surface roughness: It depends to the fabrication of master and molding process. Larger area for adsorption and air bubble production occurs when the roughness

increase. Moreover, local eddy currents and pressure losses are inevitable.

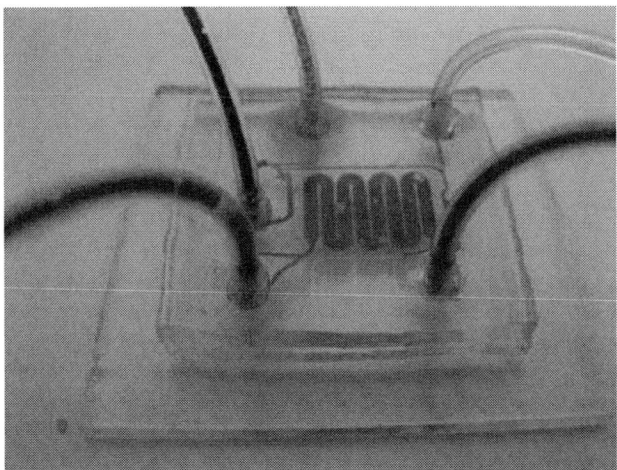

Fig. 3 Leakage of the fluidic from the fluidic interconnection during device testing.

V. CONCLUSION

In this paper, we highlighted some of the issues related to the microfluidic research works as our experience and also mentioned in the literature in dealing with fabrication of new microfluidic devices. Being familiar with this type of problems is fundamental for the beginner and also for further improvement of the microfluidic research studies. Solving these problems effectively is important for the success of the ongoing researches.

ACKNOWLEDGMENT

This work was supported by the Universiti Kebangsaan Malaysia through the grant funded under "PROJEK ARUS PERDANA", UKM-AP-NBT-10-5-2009.

REFERENCES

[1] J. Alvankarian and B. Y. Majlis, "A new UV-curing elastomeric substrate for rapid prototyping of microfluidic devices," *Journal of Micromechanics and Microengineering,* vol. 22, p. 035006, 2012.

[2] J. Alvankarian, M. Damghanian, and B. Y. Majlis, "Thick-Film Deposition of High-Viscous Liquid Photopolymer," presented at the NEMS/MEMS Technology and Devices - ICMAT2011Singapore, 2011.

[3] J. Alvankarian and B. Y. Majlis, "Low cost prototyping of microfluidic structure," presented at the International Conference on Semiconductor Electronics (ICSE), 2010 IEEE Melaka, Malaysia, 2010.

[4] X. Liu, R. K. Y. Fu, P. K. Chu, and C. Ding, "Mechanism of apatite formation on hydrogen plasma-implanted single-crystal silicon," *Applied Physics Letters,* vol. 85, p. 3623, 2004.

[5] G. T. Roman, T. Hlaus, K. J. Bass, T. G. Seelhammer, and C. T. Culbertson, "Sol-gel modified poly (dimethylsiloxane) microfluidic devices with high electroosmotic mobilities and hydrophilic channel wall characteristics," *Analytical Chemistry,* vol. 77, pp. 1414-1422, 2005.

[6] A. Bahadorimehr and B. Y. Majlis, "Fabrication of Glass-based Microfluidic Devices with Photoresist as Mask," *Electronics and Electrical Engineering,* vol. 116, pp. 45-48, 2011.

[7] G. K. Toworfe, R. J. Composto, C. S. Adams, I. M. Shapiro, and P. Ducheyne, "Fibronectin adsorption on surface-activated poly (dimethylsiloxane) and its effect on cellular function," *Journal of Biomedical Materials Research Part A,* vol. 71, pp. 449-461, 2004.

[8] J. M. Bruder, N. C. Monu, M. W. Harrison, and D. Hoffman-Kim, "Fabrication of polymeric replicas of cell surfaces with nanoscale resolution," *Langmuir,* vol. 22, pp. 8266-8270, 2006.

[9] E. K. F. Yim, R. M. Reano, S. W. Pang, A. F. Yee, C. S. Chen, and K. W. Leong, "Nanopattern-induced changes in morphology and motility of smooth muscle cells," *Biomaterials,* vol. 26, pp. 5405-5413, 2005.

[10] J. Monahan, A. A. Gewirth, and R. G. Nuzzo, "Indirect fluorescence detection of simple sugars via high-pH electrophoresis in poly (dimethylsiloxane) microfluidic chips," *Electrophoresis,* vol. 23, pp. 2347-2354, 2002.

[11] R. Mukhopadhyay, "When PDMS isn't the best," *Analytical Chemistry,* vol. 79, pp. 3248-3253, 2007.

Geometrical Characterization of Single Layer Silicon Based Piezoresistive Microcantilever using ANSYS

Mohd Hazrul Zakaria[1], Badariah Bais[1,2], *Member, IEEE*, Rosminazuin Ab. Rahim[3] and Burhanuddin Yeop Majlis[1,2],
Senior Member IEEE
[1]Dept. of Electrical Electronic & Systems Engineering, Faculty of Engineering & Built Environment,
[2]Institute of Microengineering and Nanoelectronics (IMEN)
Universiti Kebangsaan Malaysia (UKM)
43600 UKM Bangi, Selangor, Malaysia
[3]Faculty of Engineering, International Islamic University Malaysia,
50728 Kuala Lumpur, Malaysia
Email: badariah@vlsi.eng.ukm.my

Abstract- **In this paper, characterization on the geometrical aspects of a single layer, p-doped silicon based piezoresistive microcantilever using finite element method is presented. The displacement and the von Mises stress obtained from the simulation were observed by varying the geometries of the microcantilever namely the thickness, length, width and the distance between the piezoresistor legs. The sensitivity of the microcantilever was then calculated and tabulated. From the simulation results, it can be shown that the displacement and sensitivity of the single layer piezoresistive microcantilever is comparable to the dual layer counterpart with the thinner microcantilever resulted in a maximum displacement and sensitivity, compared to other geometrical factors.**

I. INTRODUCTION

Microcantilever has been widely researched due to various effective applications as sensor elements, simple design of microcantilever structure, and advantage of existing silicon batch fabrication through semiconductor technology. Piezoresistive microcantilever sensor is one of the promising mechanical microsensors that have been around for many years. This cantilever-based device offers high sensitivity, high selectivity, easy to fabricate and can be easily integrated with on-chip electronics circuitry [1].

Currently, most of the piezoresistive microcantilever sensors were designed in a dual-layer comprising of piezoresistor incorporated on top of a microcantilever. The piezoresistor can be made from a polysilicon or a doped, single-crystalline silicon while the microcantilever from silicon nitride or silicon dioxide [2-5]. This dual layer design has been effective in optimizing the piezoresistance effect and sensitivity of the device [6-7]. However, the incorporation of the piezoresistor on the microcantilever requires additional sequence in the fabrication process which will incur additional cost and manufacturing time.

In this paper, a new piezoresistive microcantilever design that incorporate both piezoresistor and the microcantilever in a single layer, is simulated and characterized. This new design utilizes a p-doped silicon material that will act as both piezoresistor and microcantilever. Characterization was done by varying the geometrical aspects of the piezoresistive microcantilever namely, the thickness, length, width and the distance between the piezoresistors.

II. DESIGN AND SIMULATION

The design of a single layer piezoresistive microcantilever is based on a dual-leg microcantilever as shown in Fig. 1. The microcantilever is designed with the starting dimension of 100 μm in width (w), 250 μm in length (l), 1 μm in thickness (t) and 40 μm in the distance between the piezoresistor legs (d). The geometrical dimensions were varied and the effect of each of the parameters on the displacement, von Mises stress and sensitivity was observed. The simulations were performed using a multi-purpose finite element simulator, ANSYS.

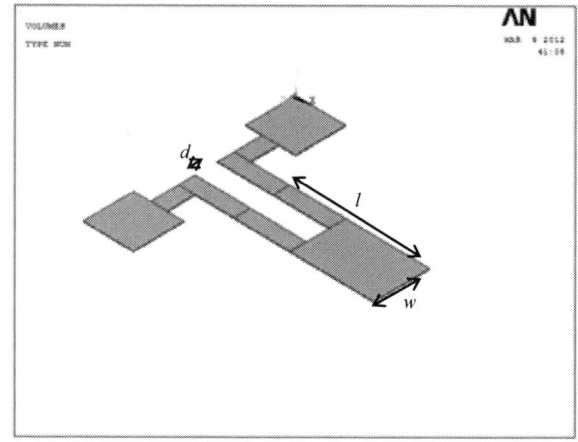

Fig. 1 3D Model of the piezoresistive microcantilever design

For the simulation, element SOLID227, a 3-D 10-node coupled-field solid tetrahedral structure is used. Appropriate meshing and boundary condition is done on the model, before the uniform load ranging from 1 to 10 μN is applied.
Table 1 summarized the material properties used in the simulation done with ANSYS:

TABLE 1
MATERIAL PROPERTIES OF SI

Parameter	Value
Young's Modulus (MPa)	1.30191×10^5
Poisson's Ratio	0.278
Piezoresitive Coefficient (MPa^{-1})	$\Pi_{11} : 6.60 \times 10^{-5}$ $\Pi_{12} : -1.10 \times 10^{-5}$ $\Pi_{44} : 1.381 \times 10^{-3}$

In this work, the sensitivity is calculated and expressed in terms of the resultant fractional resistance change given by:

$$\frac{\Delta R}{R} = \pi_l \sigma_{max} \qquad (1)$$

where π_l is the longitudinal piezoresistive coefficient of silicon at the operating temperature and σ_{max} is the maximum stress.

III. RESULTS AND DISCUSSION

Fig. 2 shows the simulation output of ANSYS software in terms of displacement and von Mises stress values. Colour gradient indicates the maximum or minimum values given by the simulation.

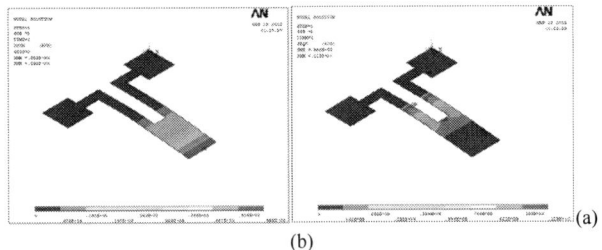

(b)   (a)

Fig. 2 ANSYS simulation showing (a) displacement and (b) von Mises stress of the microcantilever

When uniform load are applied to the microcantilever, it replicates target analytes binding onto the microcantilever surface. This will cause different stress levels generated between the top and bottom layer of the microcantilever, thus, resulting in the displacement of the microcantilever The highest displacement can be seen at the free end of the microcantilever as shown in Fig. 2(a) while the maximum stress point is located at the fixed end of the microcantilever as shown in Fig. 2(b).

The displacement and the von Mises stress of the single layer microcantilever with the original dimensions are shown in Fig. 3. From Fig. 3, one can see the linear relationship between displacement and von Mises stress values with the applied load on the microcantilever. These will also serve as the reference values for the geometrical variations.

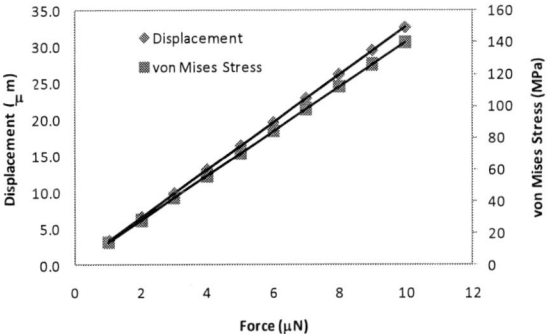

Fig. 3 Displacement and von Mises stress of the reference microcantilever with different load levels.

The effect of the thickness of the microcantilever on the displacement, von Mises stress and the sensitivity is shown in Fig. 4 and Fig. 5, respectively. The simulations were done with the fixed applied uniform load of 2 µN. From Fig. 4, it can be shown that when the thickness of the microcantilever decreases, the displacement and von Mises stress values increase. The sensitivity of the microcantilever is determined using Eq. (1) and plotted where the maximum stress value is taken from the simulation. As the proposed structure consists of a single layer, having thinner structure will result in higher displacement and sensitivity. This is consistent with the findings in [2]. However, the minimum thickness allowable will be determined by the fabrication process and equipment limitations.

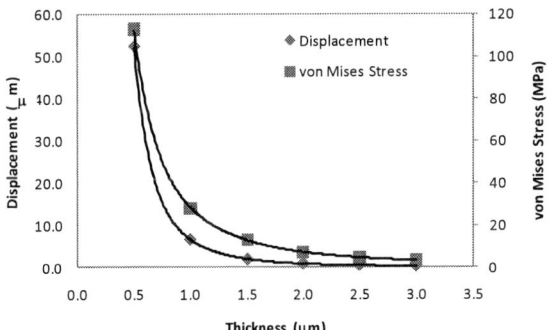

Fig. 4 Displacement and von Mises stress vs. microcantilever thickness

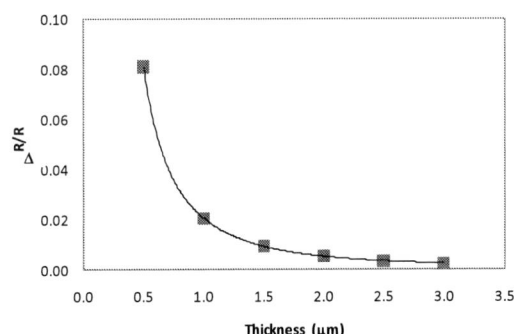

Fig. 5 Sensitivity vs. microcantilever thickness

Finite element analysis results of the second geometrical factor is shown in Fig. 6 and Fig. 7. When uniform load is applied to the microcantilever structure, its displacement increases from the fixed end to the free end of the microcantilever. As the length of the microcantilever is increased, the displacement and von Mises stress also increases. Hence, the sensitivity also increases.

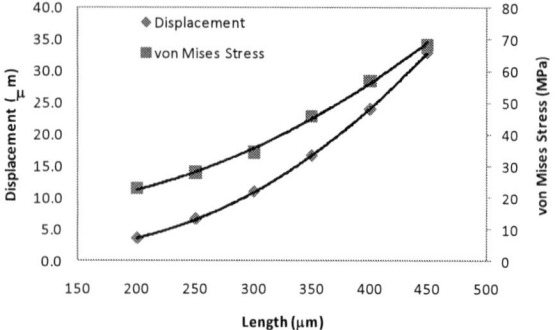

Fig. 6 Displacement and von Mises stress vs. microcantilever length

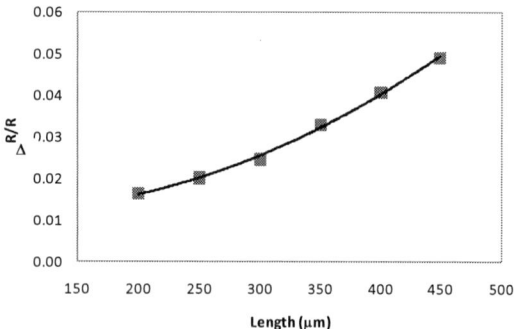

Fig. 7 Sensitivity vs. microcantilever length

The next geometrical factor is the microcantilever width. Finite element analysis results of this parameter showed that as the width increases, both displacement and sensitivity decrease as shown in Fig. 8 and Fig. 9, respectively. This is due to the fact that when the microcantilever width increases, the spring constant of the microcantilever also increases which results in the decrease of displacement and sensitivity values.

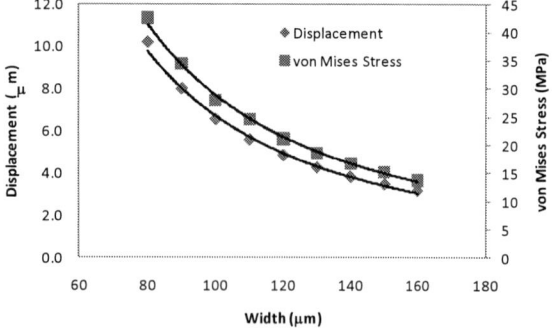

Fig. 8 Displacement and von Mises stress vs. microcantilever width

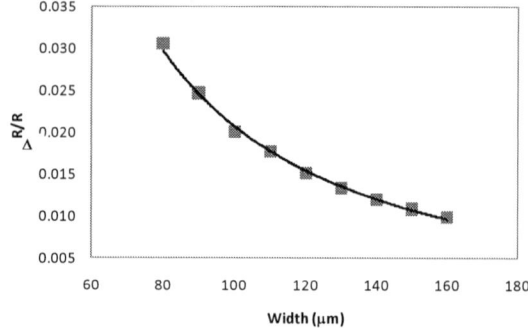

Fig. 9 Sensitivity vs. microcantilever width

Finite element analysis results obtained from varying the distance between piezoresistor legs also resulted in desirable response as shown in Fig. 10 and Fig. 11. As the distance between the piezoresistive legs increases, both displacement and von Mises stress values also increases. This is because thinner piezoresistor legs will reduce the spring constant of the microcantilever. Hence, the sensitivity also increases.

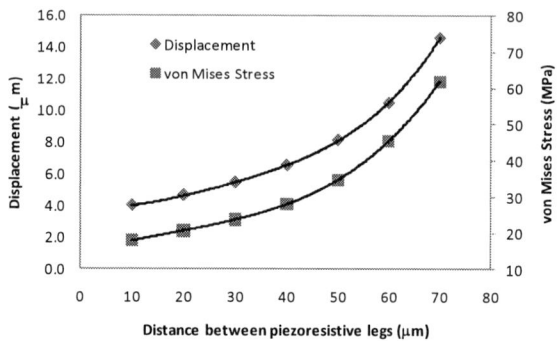

Fig. 10 Displacement and von Mises stress vs. distance between piezoresistor legs

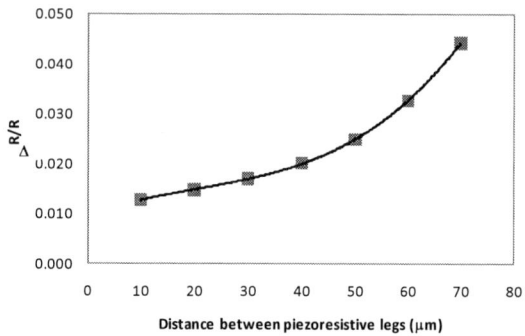

Fig. 11 Sensitivity vs. distance between piezoresistor legs

From this study, one can conclude that that the thickness of the microcantilever has the most significant impact on the displacement and the sensitivity of single layer piezoresistive microcantilever. One can also conclude that displacement and sensitivity of the single layer piezoresistive microcantilever is comparable to the dual layer counterpart as done in [7].

IV. CONCLUSION

This paper highlight the characterization of a new single layer, p-doped silicon based piezoresistive microcantilever using finite element method. The geometries of the microcantilever were varied and the displacement and the von Mises stress were simulated and the sensitivity is calculated and tabulated. Simulation results show that the displacement and sensitivity of the piezoresistive microcantilever are highly affected by the geometry of the microcantilever with the thickness of the microcantilever being the major contributing factor. Silicon is known to have large value of Young's Modulus i.e. approximately 130 GPa. Thus, using piezoresistive materials with lower modulus can further improve the microcantilever performance and sensitivity.

ACKNOWLEDGEMENT

This work is supported by Research Grant: UKM-RRR1-07-FRGS0258-2010 (Synthesization and Characterization of Glucose-sensitive Hydrogel).

V. REFERENCES

[1] S. K. Vashist, "A Review of Microcantilevers for Sensing Applications", *J. Nanotechnol* vol. 3, pp. 1-15, 2007.

[2] V. Chivukula, M. Wang, H. F. Ji, A. Khaliq, J. Fang and K. Varahramyan, "Simulation of SiO_2-based piezoresistive microcantilevers", *Sensors and Actuators A: Physical*, vol.125, issue 2, pp. 526–533, January 2006.

[3] Y. Tang, J. Fang, X. Yan, and H. F. Ji, "Fabrication and characterization of SiO_2 microcantilever for microsensor application," *Sensors and Actuators B*, vol. 97, pp. 109–113, 2004.

[4] D. Saya, P. Belaubre, F. Mathieu, D. Lagrange, J. B. Pourciel and C. Bergaud, "Si-piezoresistivemicrocantilevers for highly integrated parallel force detection applications", *Sensors and Actuators A123-124*, pp.23–29, 2005.

[5] J.A. Harley, T.W. Kenny, "High-sensitivity piezoresistive cantilevers under1000 °A thick", *Appl. Phys. Lett.*, 75, pp.289–291, 1999.

[6] M. Yang, X. Zhang, K. Vafai, C.S. Ozkan, "High sensitivity piezoresistive cantilever design and optimization for analyte–receptor binding", *J. Micromech. Microeng.*, 13, pp.864–872, 2003.

[7] R. Ab. Rahim, B. Bais and B. Yeop Majlis, "Design and analysis of MEMS piezoresistive SiO_2 cantilever-based sensor with stress concentration region for biosensing applications", *Proc. IEEE International Conference on Semiconductor Electronics (ICSE2008)*, pp.211-215, Nov. 2008.

978-1-4673-2395-6/12 $31.00 © 2012 IEEE

Induced Mass Change Technique for Glucose Detection in Microcantilever-based Sensors

Mardhiah Mohd Nor [1], Badariah Bais[1,2], *Member, IEEE*, Norazreen Abd Aziz[1,2], Rosminazuin Ab. Rahim[3], and Burhanuddin Yeop Majlis[1,2], *Senior Member IEEE*

[1]Institute of Microengineering and Nanoelectronics (IMEN)
[2]Dept. of Electrical Electronic & Systems Engineering, Faculty of Engineering & Built Environment,
Universiti Kebangsaan Malaysia (UKM)
43600 UKM Bangi, Selangor, Malaysia
[3]Dept. of Electrical Electrical and Computer Engineering, Kulliyyah of Engineering,
International Islamic University Malaysia (IIUM)
Email: badariah@vlsi.eng.ukm.my

Abstract- Microcantilever-based glucose sensor is a state-of-art sensor that has been widely investigated for biosensing applications. This work focuses on the preliminary study of correlation of the glucose reaction and the induced force on the microcantilever sensor by measuring the induced mass changes due to the glucose reaction. The induced force is then used to test the fabricated microcantilever sensor for glucose detection. Gold-coated glass slides were used as the immobilization sites for glucose oxidase enzyme. The reaction of the glucose oxidase enzyme and ß-D glucose was observed by injecting 5 mM glucose solution on 0.3 mL immobilized enzyme on three samples. The resulted mass changes, ranging from 151.87 mg to 180. 14 mg for the three samples, indicate a good correlation between the enzyme-glucose reaction and the induced mass change. The induced forces of 0.28 µN to 0.29 µN, converted from the mass changes of the glucose-enzyme reaction, is in a range of interactive force values commonly found in biochemical detection applications.

I. INTRODUCTION

Nowadays, microcantilever has been widely used as transducers in chemical sensing systems. This is primarily because microcantilever is one of the easiest micromechanical systems devices that can be easily micromachined and mass-produced. Microcantilever normally found in the atomic force, microscopy and have excellent potential as sensors due to several notable advantages which include ultrasensitivity, ease of mass production and low cost [1]. Microcantilevers that are sensitive, reliable, fast label-free detection of protein/ protein conformations make them capable to detect any marker proteins related to diseases even at low concentration. Moreover, microcantilevers are able to recognize the specific protein conformations and/or reversible conformation changes of polymers/proteins enzyme.

Glucose sensor is one of the most popular microcantilever sensors that has been used in biosensing applications such as in clinical diagnosis and biochemical study [2]. The sensing was organized normally on the biologically elements such as enzyme, antibody, protein and glucose because they are low cost, simple, easy to use and rapid detection [3]. The glucose detection was derived from the reaction of glucose oxidase enzyme (GOx) with glucose by immobilizing a layer of GOx on the surface of the microcantilever [4,5]. The enzyme reaction in the presence of glucose resulted in the stress change in the microcantilever which, in turn, caused the microcantilever to deflect. Therefore, the amount of mechanical deflection in the microcantilever is equivalent to the induced force in the glucose oxidase enzyme and glucose reaction.

One of the methods to determine the equivalent force in the glucose oxidase enzyme and glucose reaction is by measuring the induced mass change. A mass under the influence of the earth's gravitational field is a straightforward and simple way to produce a force, which is referred to by the term deadweight [6]. The measured mass change was then converted into load or force (N) change using the equation $F = m \times g$ where m is the mass from the GOx – glucose reaction and g is the gravitational force which is equivalent to 9.8 m/s^2. Therefore, this work investigates experimentally the induced mass change caused by the glucose oxidase enzyme and glucose reaction and the converted force is utilized in testing the microcantilever sensor for glucose detection.

II. EXPERIMENTAL PROCEDURE

A. Materials

The materials and reagents that were used in this experiment, include glucose oxidase (GOx) (type VII-S) which was purified from *Aspergillus niger,* bovine serum albumin (BSA), glutaraldehyde (GA) in 50% aqueous solution, ß-*D*-Glucose, phosphate buffer solution (PB, 100mM, pH 7) and sodium acetate buffer, pH 5.1. For weighing a small amount of reagents, a micro balance with 0.0001 g readability, Mettler Toledo XS 105 Dual Range was used.

B. Immobilization of Glucose Oxidase

Immobilization is a process to improve the condition of the enzyme. The main advantages of using this process is that it is easy and economical as immobilized enzyme is easy to remove. There are three types of immobilization, which are adsorption on glass, alginate or matrix, entrapment and cross linking. Cross linking is the most effective method in immobilization while adsorption is the slowest. In this work, GOx was immobilized by cross-linking it with GA in the presence of BSA. GOx is derived from Aspergillus Niger which consists of 2 equal subunits with a molecular weight of 80 kDa for each. Prior to the immobilization process, 20 mg glucose oxidase (GOx) was dissolved in a 1 mL of 50mM sodium acetate buffer before being added into a sequence of 1 mL of PB solution with 5 mg of BSA and 40μL of 50% GA. A 0.5 mL capacity pipette was used to withdraw 0.3 mL of the enzyme solution onto the glass slide before the sample was stored in a refrigerator and dried at 4^0C overnight. For this study, Au deposited glass slide with a dimension of 18 mm x 18 mm was used as the substrate and the immobilization site of the enzyme-glucose reaction. Prior to Au sputtering, the glass slide was cleaned in an ultrasonic bath of acetone and methanol, subsequently. The deposition of 100 nm thick Au layer was realized through Au sputtering process.

C. Loading/release of Glucose

Prior to glucose deposition step, the glucose solution was initially prepared by adding 5 mM of ß-D glucose in 1 mL PB solution. A volume of 0.3 mL glucose solution was then deposited using a pipette onto the sample's surface. It should be noted that the sample was weighed at the end of each discrete process to identify the mass difference due to the enzyme-glucose reaction. It is also worth noted that the amount or volume of the reagents used were kept constant throughout the procedure to ensure accurate mass change measurement.

D. Glucose Detection on Microcantilever Sensor

The glucose detection was derived from the reaction of glucose oxidase enzyme (GOx) with glucose by immobilizing a layer of GOx on the surface of cantilever and the mechanical bending will be detected by the enzyme reaction in the presence of glucose. From the previous simulation work [7], a range of 1 to 10 μN was suggested as the suitable applied load for biosensing applications. The induced force converted from the induced mass change caused by the glucose oxidase enzyme and glucose reaction is used as the applied load onto the microcantilever. At this initial measurement stage, by applying load on the suspended microcantilever sensor, the correlation between the stress change due to the presence of load and the resulting resistance was observed. The observation on the direct conversion of the mechanical response into the electrical response in the PRM sensor will paves the way forward for the utilization of microcantilever-based sensor in any stress induced applications.

III. RESULTS AND DISCUSSION

Fig. 1 shows the observations of mass changes due to enzyme-glucose reaction. Overall, the induced mass increases linearly from glucose oxidase enzyme immobilization to glucose solution deposition. This mass change is due to the enzyme-glucose reaction on the immobilization site of the glass slides. The observed mass change is in the range of 190 mg for all the samples as shown in Fig. 1. It should be mentioned that the glucose oxidase enzyme is highly selective towards glucose due to its high specificity. As reported by [2], GOx possesses a very high specificity toward glucose (5×10^4 greater activity) compared with its specificity for any other sugars (e.g., D-fructose or D-mannose). The authors also carried out an investigation on the specificity of GOx-functionalized microcantilever towards D-fructose and D-mannose. They concluded that the common interferences for glucose detection have shown no effect on the measurement of blood glucose level. The same observation was found out by [8] in which they observed that analytes other than glucose such as bovine serum albumin, sucrose, galactose and deoxyglucose showed no specific response to the enzymatic microcantilever. However, they noticed a slight response of the GOx-functionalized microcantilever-based sensor towards mannose but at a very minimal value. Therefore, based on these observations on the GOx specificity towards glucose, it can be concluded that our observation on the enzyme-glucose reaction is valid and represents the actual enzyme-glucose reaction.

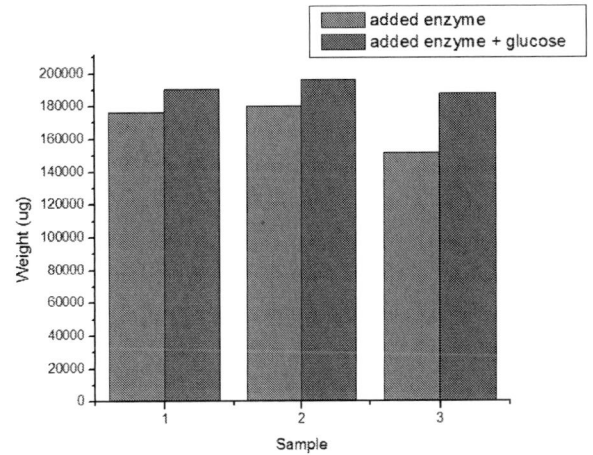

Fig. 1 Mass changes due to enzyme-glucose reaction

Table 1 shows the results of the equivalent force translated from the induced mass change of the glucose-enzyme reaction. As shown in Table 1, the measured induced force due to glucose-enzyme reaction is about 0.3 μN. This value matches up well with the range of interactive force values commonly found in biochemical detection applications, with the lower end is at 0.25 μN [8]. The same force range was also reported by [9] on the development of vibration sensor and [10] in their work on microcantilever-based force sensor.

TABLE I
MASS CHANGE OBSERVATION

	Mass measurement	Sample 1	Sample 2	Sample 3
i.	Au sputtered sample [mg]	176.48	180.14	151.87
ii.	Immobilized GOx + (i) [mg]	521.47	451.15	451.02
iii.	Glucose deposited + (ii) [mg]	540.50	470.74	469.81
iv.	Mass change (iii) − (ii) [μg]	190.30	195.95	187.82
v.	$F = m.g$ [μN]	1864.97	1920.31	1840.60
vi.	Pressure, $P = F / A$ [μN/m^2]	5756088	5926882	5680874
vii.	F on microcantilever [μN]	0.281	0.289	0.277

(b)

Fig. 2 The resultant resistance change due to the presence of applied force in the range of one-tenth μN (a) without the load applied (b) with load applied

From the glucose oxidase – glucose reaction investigation, it was found out that induced load for the glucose sensing reaction is in the range of one-tenth μN. For the observation of mechanical-electrical signals conversion purpose, the micron-Newton range load is approximately represented by applying a 0.03 mg mass on the surface of the microcantilever sensor. As reported by [6], 0.03 mg mass is approximately equivalent to 0.3 μN force. During the measurement, the resultant resistance change, which is due to the applied load, was observed by connecting the micromanipulator's probe to the device. The induced resistance change due to the applied load was observed by comparing the resistance values before and after load applied. As shown in Fig. 2, the initial measured resistance value at no load condition of 54.1 Ω is changed to 54.4 Ω which gives a sensitivity value of 184.8 x 10^{-6}. From this observation, it can be concluded that the mechanical response of the microcantilever's bending due to the applied load can be directly converted into an electrical response in terms of the resistance change in the Wheatstone bridge configuration.

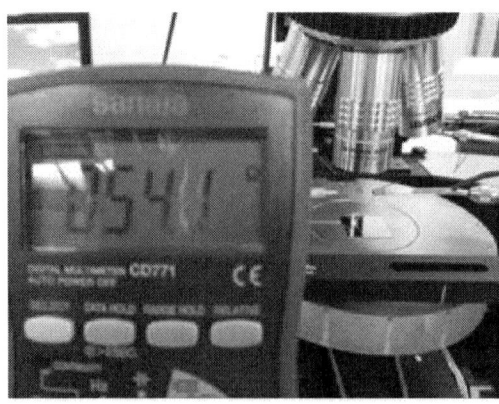

(a)

IV. CONCLUSION

In conclusion, the equivalent force in the glucose oxidase enzyme and glucose reaction is determined by measuring the induced mass change. The force is then used to test the fabricated microcantilever sensor. The resulted mass changes of the three samples indicate a good correlation between the enzyme-glucose reaction and the induced mass change. The converted interactive force values conform to the force commonly found in biochemical detection applications, with the lower end is at 0.25 μN. Based on the observations on the GOx specificity towards glucose, it can be concluded that the measurement of GOx-glucose reaction is valid and represents the actual enzyme-glucose reaction. The functionality of the fabricated microcantilever sensor is tested for glucose sensing reaction in which an equivalent reaction force in the range of one-tenth μN is represented on the fabricated sensor. The observation on the reaction shows a good response in which the resistance of the microcantilever sensor changes with the presence of applied load or stress.

ACKNOWLEDGEMENT

This work is supported by Research Grant: UKM-RRR1-07-FRGS0258-2010 (Synthesization and Characterization of Glucose-sensitive Hydrogel).

REFERENCES

[1] Youzheng Zhou, Qi Zhang, Zheyao Wang, LitianLiu, A Silicon Microcantilever Biosensor Fabricated on SOI Wafers Using a Two-Step Releasing Approach, Proc. IEEE Sensors (2008) 227-230.

[2] K.R. Buchapudi, X. Huang, X. Yang, H.F. Ji, T. Thundat, Microcantilever Biosensors for Chemical and Bioorganisms, Analyst 136 (2011) 1539-1556.

[3] J. Pei, F. Tian, T. Thundad, Glucose Biosensor Based on Microcantilever, Analytical Chemistry 76 (2004) 292-297

[4] S.M. Yang, T.I. Yin, Design and Analysis of Piezoresistive Microcantilever for Surface Stress Measurement in Biochemical Sensor, Sens. Actuator B: Chemical 120, 2007, 736-744.

[5] S.M. Yang, T.I. Yin, C. Chang, Development of A Double-Microcantilever for Surface Stress Measurement in Microsensors, Sens. Actuator B: Chemical 121, 2007, 545-551.

[6] M.S. Kim, J.R Pratt, Si Traceability: Current Status and Future Trends for Forces Below 10 Micronewtons. Measurement 43, 2010, 169-182.

[7] Bais, B., Rahim, R. A. & Majlis, B. Y. 2011. Finite element and system level analyses of piezoresistive microcantilever for biosensing applications. *Australian Journal of Basic and Applied Sciences* 5(12): 1038-1046.

[8] A. Subramanian, P. Oden, S. Kennel, K. Jacobson, R. Warmack, T. Thundat, M. Doktycz, Glucose Biosensing Using an Enzyme-Coated Microcantilever, Applied Physics Letters 81 (2002) 385.

[9] D.R. Baselt, G.U. Lee, K.M. Hansen, L.A. Chrisey, R. Colton, A High-Sensitivity Micromachined Biosensor, Proc. IEEE 85 (1997) 672-680.

[10] I. Behrens, L. Doering, E. Peiner, Piezoresistive Cantilever as Portable Micro Force Calibration Standard. J. Micromech. Microeng. 13 (2003) 171-177.

978-1-4673-2395-6/12 $31.00 © 2012 IEEE

Localized Surface Plasmon Resonance Sensor of Gold Nanoparticles for Detection Pesticides in Water

Norhayati Abu Bakar[1], Akrajas Ali Umar[1], *Member, IEEE*, Muhamad Mat Salleh[1*], *Member, IEEE*, Muhammad Yahaya[2], *Member, IEEE*, and Burhanuddin Yeop Majlis[1], *Senior Member, IEEE*

[1]Institute of Microengineering and Nanoelectronics (IMEN),
Universiti Kebangsaan Malaysia (UKM)
43600 UKM Bangi, Selangor, Malaysia
[2]School of Applied Physics, Faculty of Science and Technology,
Universiti Kebangsaan Malaysia (UKM)
43600 UKM Bangi, Selangor, Malaysia
*Email: mms@ukm.my

Abstract- **This paper reports the utilization of plasmonic properties of Gold Nanoparticles (GNPs) as an optical sensor to detect pesticides in water. The pesticides with trade name of Ridomil G MZ 68 WP fungicide and Water-Dispersible Granules WG insecticide were used in this study. Spherical GNPs of the average size of ca. 31±7 nm were grown on quartz substrate using seed mediated growth. An optical sensor system was setup, comprises a tungsten lamp light source, a duplex fiber optic probe, a spectrometer and a sensor chamber. Detection of pesticides was done by comparing the Localized Surface Plasmon Resonance (LSPR) spectra of the GNPs film immersed in the deionised water and in pesticides solutions by varying the concentration of pesticides solutions from 2.5 µg/L to 2500 µg/L. The LSPR spectra of GNPs sample were very sensitive to the presence of pesticides where the spectra intensities are increases with the concentrations of the pesticides solutions.**

I. INTRODUCTION

Pesticides are widely used in agriculture for improving the productivity and quality of crops. Pesticides have chemical substances that belong to different chemical groups, such as organophosphates, carbamates, organochlorine, nitro compounds, pyrethroids and amides. These chemical used to kill unwanted organisms for protecting crops from damages by insects, weeds and disease. Unfortunately, the pesticides which act as carcinogen agents may also harmful to humans and animals [1]. Hence, widely used pesticides may cause environmental problems such as contamination of water, soil and vegetation [2]. It is important to detect pesticides residues in food and water.

Detection of pesticides normally used chemical techniques such as chromatography and mass spectroscopy which require tedious sample pretreatments, highly qualified technicians and sophisticated instruments [3]. Hence, it is urgently need to develop a portable, faster and users-friendly technique to detect pesticides in water. Recently, sensor-based optical sensing technique has received growing interest since current development of optical technology is able to produce low cost and potable equipments [4]. One of the most attractive of the optical method is based on Localized Surface Plasmon Resonance (LSPR) sensing technique.

Localized Surface Plasmon Resonance (LSPR) is an interesting phenomenon of metal nanoparticles arises when their conduction electrons interact with electromagnetic radiations such as light waves [5]. Since the interaction is occurred at the surface of nanoparticles, the LSPR properties are very sensitive with size, shape of metal nanoparticles and the surrounding medium [6]. Hence, the LSPR properties can be utilized as optical sensors to detect a particular material. The rapid progress gold and silver nanoparticles synthesis technologies has attracted intense research works to develop the applications of LSPR as biosensors [7,8] and other chemical sensors.

This paper reports the synthesis Gold Nanoparticles (GNPs) using a wet chemical approach, namely the seed mediated growth technique, and study their plasmonic properties in the pesticides solutions in water. The sensitivity of the LSPR of the GNPs samples towards pesticides may be utilized as sensors to detect these toxic chemicals.

II. EXPERIMENTS

GNPs were grown on quartz substrate by a seed mediated growth technique. The seed mediated growth technique consists of two processes, namely seeding and growth processes. For seeding process, seed solution was firstly prepared by mixing 0.01M $HAuCl_4.3H_2O$, 0.01M trisodium citrate and 0.1M $NaBH_4$. The gold nanoseeds with 3 to 5 nm will be obtained using this recipe. The nanoseeds were attached onto the surface by simply immersing the substrate into the seed solution. Finally, the nanoseeds film was taken out and rinsed with pure water. Then the sample was dried by purging with nitrogen gas and annealed at 200°C. Subsequently, the nanoseeds were grown with variation of growth time in the growth solution. The solution consists of 0.1M CTAB, 0.01M $HAuCl_4.3H_2O$ and 0.1M L(+)-ascorbic acid. For this study, the substrate was taken out from growth solution after two hours, rinsed with pure water, purged with nitrogen and then annealed at 200°C in ambient air.

Pesticides were received from Malaysian Agricultural Research and Development Institute (MARDI) and directly used without any further purification process. The pesticides with trade name of Ridomil G MZ 68 WP fungicide and

978-1-4673-2395-6/12 $31.00 © 2012 IEEE

Water-Dispersible Granules WG insecticide were used in this study. The active ingredients for Ridomil G MZ 68 WP are phenylamide and dithiocarbamate, and Water-Dispersible Granules WG is acibenzolar. The pesticides solutions in deionized water were prepared with twelve concentrations of 2.5, 5, 10, 15, 20, 25, 50, 75, 150, 250, 1250 and 2500 μg/L. The solutions were made homogenous using an ultrasonic cleaner machine.

Detection of pesticides was studied using an optical sensor system as shown in Fig. 1. The system comprises a light source that was provided by tungsten lamp, a duplex fiber optic, a spectrometer HR2000 and a sensor chamber that contains GNPs sample immersed in the test solution. The sample was illuminated with light that was transmitted using one of the fiber arms. The reflected light from the GNPs sample was collected by the other fiber arm to the spectrometer and recorded as the absorption spectrum of the GNPs sample. The sensing property of GNPs was observed through comparing the absorption spectra of the sample when first dipped in deionized water and then in the pesticides solutions.

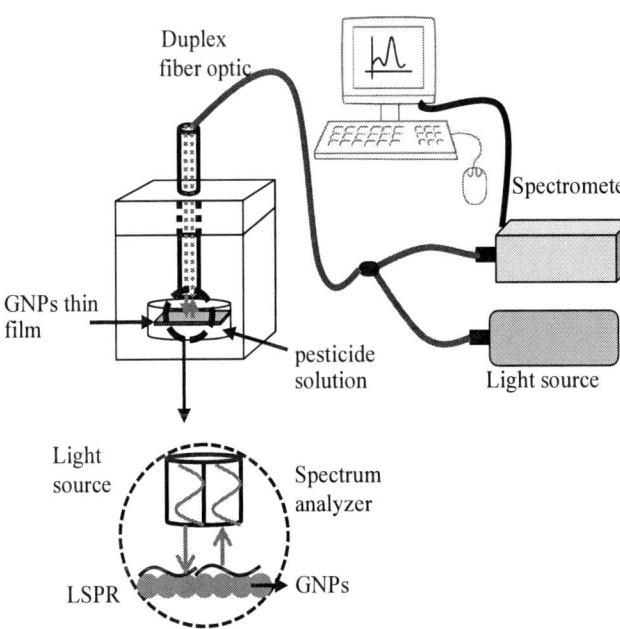

Fig. 1: An optical sensor system for pesticides detection

III. RESULTS AND DISCUSSION

The successful of GNPs grown on quartz substrate surface was examined using FESEM image as shown in Fig. 2. The size and particles distribution density were calculated based on the FESEM image using Image J software. This software calculates area nanoparticles at 1024 x 710 pixels with total area of FESEM image is 3.4 μm². As can be seen from the image, the nanoparticles were found to be spherical with size particles are ca. 31±7 nm. The calculation also revealed the average particles density is 189 count/μm².

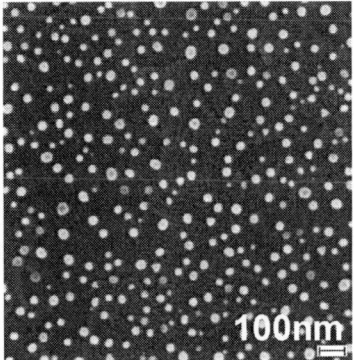

Fig. 2: FESEM image of gold nanoparticles

The plasmonic properties of the GNPs samples were studied using an optical fiber spectrometer by simply illuminating the samples with light via a fiber optic and then recorded and analyzing the reflected light from the samples. Fig. 3 showed the optical absorption spectra of GNPs under different environment conditions namely air, water and Ridomil G MZ 68 WP solution. Each spectrum consists of a single absorption peak which is identified as the Localized Surface Plasmon Resonance (LSPR) of spherical GNPs. The LSPR peak position of GNPs in air is 524 nm. The peak was shifted to 539 nm when the medium replaced with water and the pesticides solution. The shift of the LSPR spectra is caused by changes of the medium refractive index as described by the following relationship [9]:

$$\Delta\lambda = m(n_{adsorbate} - n_{medium})[1 - \exp(-2d/l_d)] \quad (1)$$

Where m is the sensitivity factor (in nm per refractive index unit (RIU)), $n_{adsorbate}$ and n_{medium} are the refractive indices (in RIU) of the adsorbate and medium surrounding the nanoparticle, respectively, d is the effective thickness of the adsorbate layer (in nm), and l_d is the electromagnetic field decay length (in nm).

It is also shown in Fig. 3 that the intensity of the LSPR spectra is increased with the change of medium from air to pesticides solution. These responses are related to the change of dielectric properties of the medium as explained from the theoretical model of Mie theory of the optical properties of metallic nanospheres as expressed by this relationship [10]:

$$E(\lambda) = \frac{24\pi N_A a^3 \varepsilon_m^{2/3}}{\lambda \ln(10)}\left(\frac{\varepsilon_i}{\left(\varepsilon_r + 2\varepsilon_m\right)^2 + \varepsilon_i^2}\right) \quad (2)$$

Where $E(\lambda)$ is the extinction which is, in turn, equal to the sum of absorption and Rayleigh scattering, N_A is the areal density of nanoparticles, a is the radius of the metallic nanosphere, ε_m is the dielectric constant of the medium surrounding the metallic nanosphere, λ is the wavelength of the absorbing radiation, ε_r is the real portion of the metallic nanosphere's dielectric function, and ε_i is the imaginary portion of the metallic nanosphere's dielectric function.

original concentration of the solution. Hence, the sensing response towards Water-Dispersible Granules WG solutions at high solution concentration is not linear and also not repeatable.

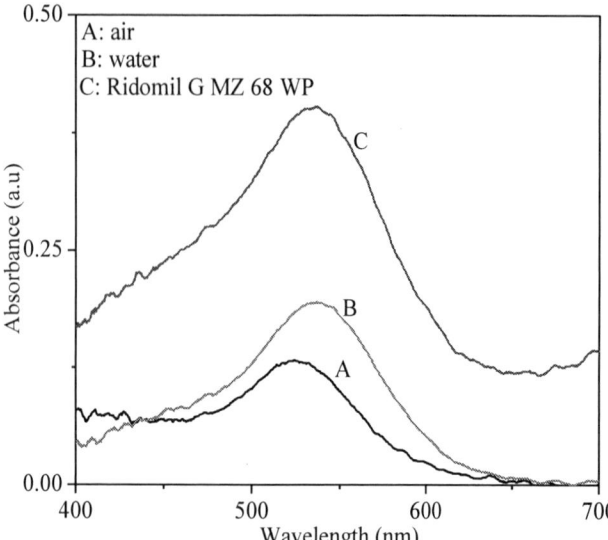

Fig. 3: LSPR response of gold nanoparticles film in air, water and 20 μg/L of Ridomil G MZ 68 WP fungicide.

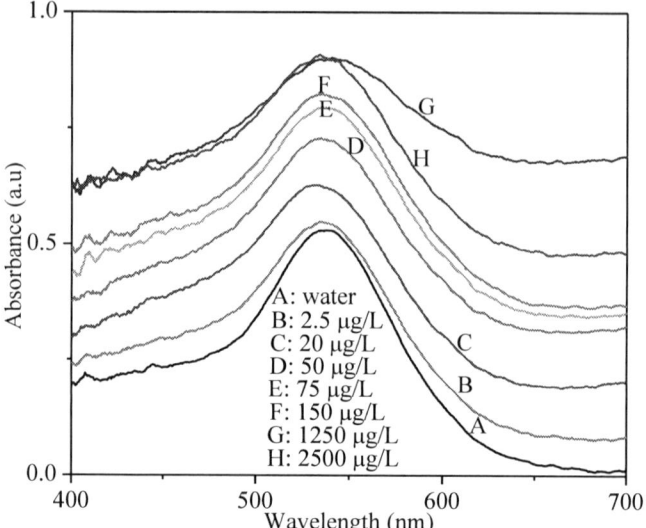

Fig. 4: LSPR response of gold nanoparticles in seven concentrations of Ridomil G MZ 68 WP fungicide.

Sensing response of a GNPs plasmonic sensor is normally based on the shift of the LSPR spectra when the reference medium, such as air or water, is replaced with the analyte as medium. However, in this work, we found that the spectral shift from water to pesticides medium is very small. As alternative, we used the change of resonance peak intensities in water and pesticides as sensing sensitivity parameter of the sensor.

The sensing responses of GNPs samples toward variation concentrations of pesticides; Ridomil G MZ 68 WP and Water-Dispersible Granules WG in water, were studied by varying the concentration from 2.5 to 2500 μg/L. The results are shown in Fig. 4 and 5 for Ridomil G MZ 68 WP and Water-Dispersible Granules WG respectively. It can be observed that intensity of LSPR peak is increased with increasing of pesticides concentrations. This sensing response is satisfied with theoretical analysis where the spectral intensity increases with the refractive index of medium, hence the concentrations of the pesticides solutions.

Fig. 6 displays the calibration curves of the GNPs sensor towards two pesticides solutions, i.e. Ridomil G MZ 68 WP and Water-Dispersible Granules WG. The sensing response towards Ridomil G MZ 68 WP has shown excellent linearity with linear correlation (r) of 0.95. However, the linearity sensing response towards Water-Dispersible Granules WG solutions is only satisfied for the solutions concentrations from 5 to 25 μg/L. At higher solution concentrations, some of the pesticide particles are accumulated together to form grains or granules which are not dissolve in water. This will reduce the

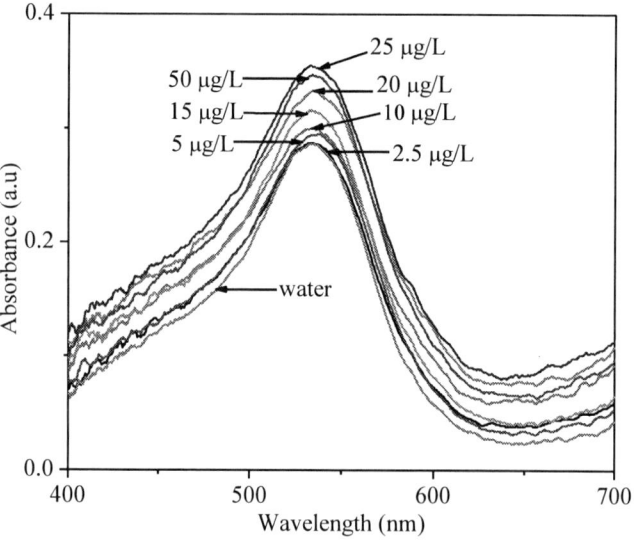

Fig. 5: LSPR spectra of gold nanoparticles in seven concentrations of Water-Dispersible Granules WG insecticide.

The calibration curves in Fig. 6 may be used to estimate the amount of pesticides residues in water. The sensitivity of the GNPs sample towards pesticides is depended on the type and physical properties of pesticides. We may extend further our analysis on the LSPR spectral curves in Fig. 4 and Fig. 5 to study the selectivity property of the sensors.

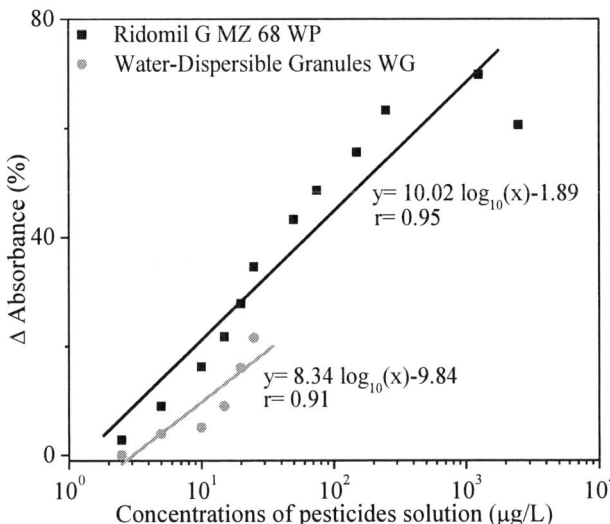

Fig. 6: The percentage of LSPR peaks increased toward
pesticides concentrations

IV. Conclusions

In this study we have successful grown GNPs on quartz
substrate using seed mediated growth. The plasmonic property
of GNPs is sensitive towards the presence of pesticides in
water. Hence, potentially be used as LSPR sensor for
monitoring pesticides contamination in water. There are linear
relationships between pesticides concentrations and LSPR
parameters. However, these relationships are limited for
concentrations of Ridomil G MZ 68 WP fungicide from 2.5 to
1250 µg/L and Water-Dispersible Granules WG insecticide
from 5 to 25 µg/L.

Acknowledgments

This work has been supported by the Malaysian Ministry of
Higher Education and Universiti Kebangsaan Malaysia.

References

[1] S. Mavrikou et al., "Assesment of organophosphate and carbamate
pesticides residue in cigarette tobacco with a novel cell biosensor,"
Sensors, vol. 8, pp.2818-2832, 2008.

[2] I. Akca, C. Tuncer, A. Gumer and I. Saruhan, "Residual toxicity of 8
different insecticides on honey bee (Apis mellifera Hymenoptera:
Apidae)," *J. Anim. Vet. Adv.*, vol. 8, pp.436-440, 2009.

[3] S. Liu, L. Yuan, X. Yue, Z. Zheng and Z. Tang, "Recent advances in
nanosensors for organophosphate pesticide detection," *Adv. Powder
Tech.*, vol. 19, pp.419–441, 2008.

[4] C. Massie, G. Stewart, G. McGregor and J. R. Gilchrist, "Design of a
portable optical sensor for methane gas detection," *Sensor. Actuat. B-
Chem.*, vol. 113, pp. 830-836, 2006.

[5] J. Zhao et al., "Interaction of plasmon and molecular resonances for
Rhodamine 6G adsorbed on silver nanoparticles" *J. Am. Chem. Soc.*, vol.
129, pp.7647-7656, 2007.

[6] H. Chen, X. Kou, Z. Yang, W. Ni, and J. Wang, "Shape- and size-
dependent refractive index sensitivity of gold nanoparticles" *Langmuir*,
vol. 24, pp. 5233-5237, 2008.

[7] M. S. Mehand, and B. Srinivasan, "Increasing throughput of surface
plasmon resonance-based biosensors by multiple analyte injections" *J.
Mol. Recognit.*, vol. 25, pp. 208-215, 2012.

[8] M. Piliarik et al., "High-resolution biosensor based on localized surface
plasmons" *Opt. Express*, vol. 20, pp. 672-680, 2012.

[9] J. N. Anker et al., "Biosensing with plasmonic nanosensors" *Nat. Mater.*,
vol. 7, pp.442-453, 2008.

[10] U. Kreibig and M. Vollmer, *Optical properties of metal clusters*, 1st ed.,
vol. 25, Springer Series in Materials Science: Springer, 1995.

Electrical Characterization of Interdigital Electrode Based on Cyclic Voltammetry Performances

Hafzaliza Erny Zainal Abidin, Azrul Azlan Hamzah and Burhanuddin Yeop Majlis, *SMIEEE*

Institute of Microengineering and Nanoelectronics (IMEN)
Universiti Kebangsaan Malaysia (UKM)
43600 UKM Bangi, Selangor, Malaysia
Email: azlanhamzah@ukm.my

Abstract- **Cyclic voltammetry is a common characterization technique for electrochemical reactions of electroactive species. Cyclic voltammetry is an important characteristic to determine the performances of energy storage devices such as battery and supercapacitor. In this paper we focus on the interdigital electrode with the various physical parameters such as gap, width and length of electrodes to increase the current response based on cyclic voltammetry performances. 2-D cyclic voltammetry time dependent model was developed. The interdigital electrode array consists of working electrode and counter electrode and the reaction at the working electrode is described by concentration dependent Butler-Volmer kinetics. The counter electrode is modeled so that it acts like an ideal reference electrode. In the design, the counter electrode with respect to ground is set to zero. The applied cell potential range from -0.5V to 0.5V. To explore the performances on cyclic voltammetry with different physical parameters of interdigital electrode, Comsol Multiphysics ver.4.2a is used to design the interdigital electrode, and the cyclic voltammetry was plotted. The maximum current response achieved is 1.6 A/m corresponding to gap of electrodes of 250 um. For the width and length of the electrodes, the maximum current response is 0.85 A/m and 1.6 A/m respectively corresponding the width of electrodes of 50 um and the length of electrodes of 700 um.**

Keywords: Interdigital structure; Cyclic Voltammetry; Working Electrode; Counter Electrode

I. INTRODUCTION

Cyclic Voltammetry techniques commonly involved the application of a potential to an electrode and the monitoring of the resulting current flowing through the electrochemical cell and commonly used to characterize the microelectrode to give a better performances of the current response [1]. Cyclic voltammetry one of the electrochemical properties in energy storage devices such as supercapacitor, batteries and fuel cells to give some information of current response when the potential is applied [2]. Supercapacitor has the potential to be an excellent power storage material for renewable energy as it has high power density, unlimited number of recharge cycles, and rapid discharge [3]. In cyclic voltammetry, the power density and capacitance can be obtained to determine the performances of the energy storage devices. A cyclic voltammetry measures the current response of an electrochemical cell that consists of two electrodes in contact

with electrolyte interface in response to a linear sweep potentials applied to the working electrode. A structure used frequently in combination with the redox–cycling process effect is the interdigital array micro electrode. The interdigital electrode has recently received a lot of attention [4]. Besides, the interdigital electrode is widely used because of the major effect on the electrical current performances [5]. At interdigital electrode array, the current at working electrode flowing through a counter electrode is monitored as a triangular excitation when the potential is applied to the working electrode [6]. To improve the current response, the physical parameters of interdigital electrode such as gap, width and length are varied. The cyclic voltammetry performances of the interdigital electrode at various gap, width and length were investigated.

II. THEORY

In an electrochemical reaction, the bidirectional reaction can be formulated as:

$$A + e \underset{\leftarrow}{\overset{\rightarrow}{}} B \qquad (1)$$

The forward and backward reactions are characterized by the rate constant k_f and k_b respectively. The forward and backward reaction rates on a single electrode are described as:

$$N_f = k_f . C_O \qquad (2)$$

$$N_b = k_b . C_R \qquad (3)$$

Where N_f and N_b is the mass flow density expressed as the number of moles crossing the unit surface in the time unit.

These reaction rate constants at the electrodes are described by the Butler-Volmer reaction kinetics.

$$k_f = k_s e - \alpha \left(E - E_0 \right) \frac{F}{RT} \qquad (4)$$

$$k_b = k_s e (1 - \alpha)(E - E_0)\frac{F}{RT} \qquad (5)$$

where k_s is the standard rate constant and α the transfer coefficient. In symmetrical reactions, α has a value 0.5. E is the applied voltage on the electrode interface and E_0 has a value 0 V. For F, R and T, it is a Faraday constant (96485 C/mol), molar gas constant (8.3144 J mol^{-1} K^{-1}) and temperature (298 K) respectively.

Additionally, the relationship for current, potential and concentration can be summarized as follows:

$$\frac{i}{nFA} = k0\{c_0 e[-\alpha\theta] - c_{Re}[(1-\alpha)\theta]\} \qquad (6)$$

Where $\theta = nF(E - E_0)/RT$

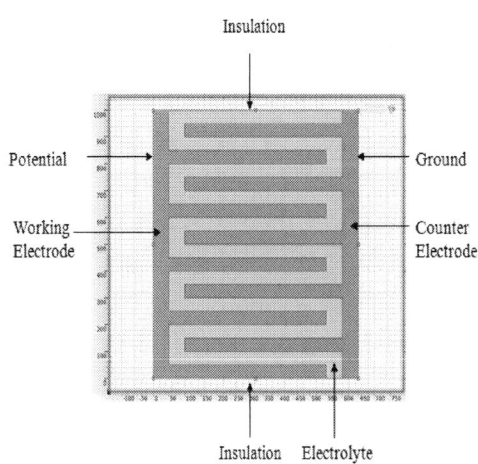

Fig. 1. 2-D structure interdigital electrode and boundary settings

IV. METHOD OF SIMULATION

In Comsol Multiphysics ver. 4.2a, the Secondary Current Distribution and Transport of Diluted Species are chosen as the Application module. Cyclic Voltammetry 2-D time dependent model was developed. The transport of a reduced and an oxidized species is described by time-dependent mass transfer principles for diffusion under dilute conditions.

In this model, it consists of two electrode cell. The reaction at the working electrode is described by concentration dependent Butler-Volmer kinetics. The counter electrode is model so that it acts like an ideal reference electrode. The applied cell voltage sets in range -0.5V to 0.5V.

The boundary conditions define the interface between the electrode and electrolyte. In our case, we set one of the boundaries as ground and the applied electrode potential was implemented using the built-in function 1[V]*Scan (t/1[s])-0.5[V]. Fig.1 shown, gives the schematic of the model boundaries. Different geometries were drawn for different dimensions such as width, gap and length of the electrode. For width and gap, the values are selected between 50 um to 250 um and for the length, the values are between 500 um to 700 um. The current responses were simulated plotted using the post processing feature in Comsol Multiphysics.

V. RESULT AND DISCUSSION

The 2-D model was drawn with the settings as shown in Fig.1. A structure of interdigital electrode has been made with different geometries such as width, gap and length. The redox reaction on an electrode depends on multiple factors such as surface concentration of the reactant and reaction constant.

The surface concentration of redox reaction is shown in Fig. 2. The maximum is surface concentration is illustrated. From the Fig.2, the maximum concentration is 0.4815 mol/m^3 at t = 1s. The current flow depends directly proportional on the flux of material to the electrode surface. When oxidation and reduction process occur at the surface, the increased concentration provides the force for its diffusion toward the bulk of the solution. Diffusion flux can be determined the rate movement of molecules across a unit area.

The flux at the electrode surface controls the rate of reaction, and thus the faradaic current flowing in the cell. In the bulk solution, concentration gradients are generally small and ionic migration carries most of the current. The ionic migration will be occurring when the potential applied in the electrolyte. The current can be defined a quantitative measure of how fast a species is being redox process at the electrode surface. The value of this current is affected by many additional factors such as redox species, material of the electrode and the number of electrons transferred. When concentration gradient decreases, flux also decreases due to the decrease in current responses at cyclic voltammetry performances.

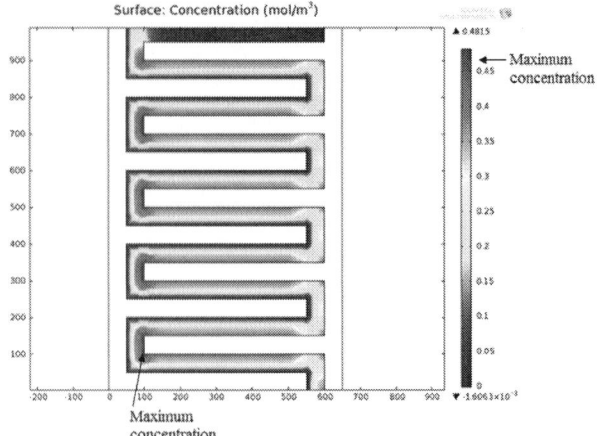

Fig. 2. Surface concentration of redox process (mol/m³) for interdigital electrode L=500 um, W=50 um and G= 50 um

For all interdigital electrode, the cyclic voltammetry performances were simulated with the applied potential (E_{cell}) at working electrode between -0.5 V to 0.5 V. From the 2-D analyses, cyclic voltammetry for the interdigital electrode at various gap values between 50 um to 250 um are plotted in Fig.3. From Fig.3, we can see, the higher value of gap, give a higher effect the response of current. The value for maximum response current with gap = 250 um is 1.6 A/m. Note that we presented current at working electrode per depth of electrode data because the modelling is done in 2-D. The increasing value of gap causes space between electrodes to increase. This allows the higher amount of the active species produced at the working electrode and electrolyte. At working electrode, the reduction and oxidation occured at the approriate applied potential and will be generate the current.

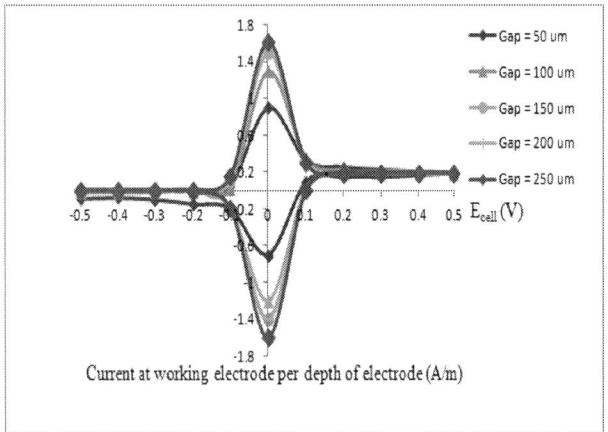

Fig. 3. Cyclic Voltammetry for various values of the gap interdigital electrode

From the 2-D analyses, cyclic voltammetry for the interdigital electrode at various width values between 50 um to 250 um as shown in Fig. 4. From Fig.4, we can see, the smaller width, the higher response of current. The value of current response is 0.9 A/m. For the various value of width,

we can see only slight changes for the current responses and smaller effect at the current responses.

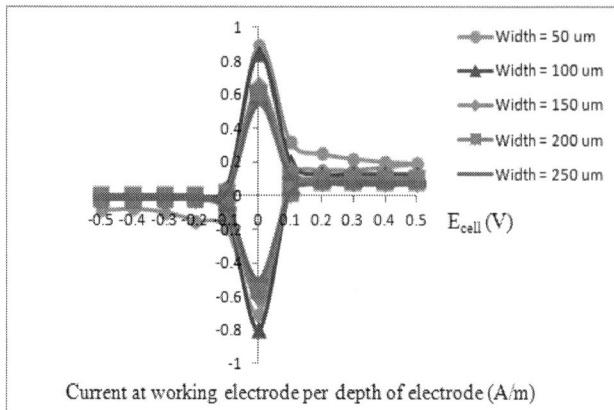

Fig. 4. Cyclic Voltammetry for various values of the width interdigital electrode

From the 2-D analyses, cyclic voltammetry for the interdigital electrode at various length values between 500 um to 700 um as shown in Fig. 5. From Fig.5, it can be seen that, the higher value of length, the higher effect the response of the current. The maximum of the current response is 1.6 A/m corresponding to the length 700 um of the electrodes. The increasing length of the electrodes increases the surface area which enable a higher redox activity between the electrodes. The increasing the length of the electrodes, increasing the mass transport activity to generate the maximum current responses on cyclic voltammetry performances.

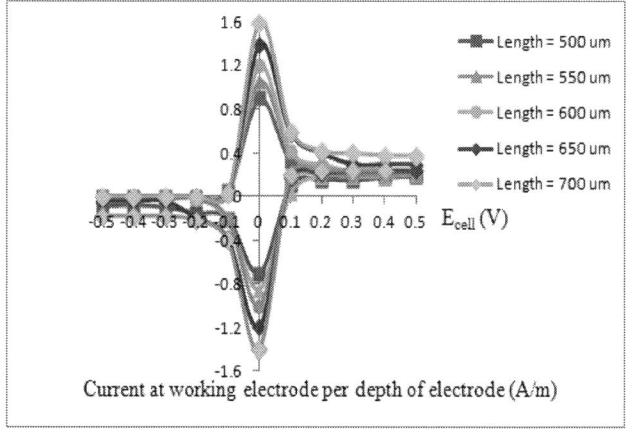

Fig. 5. Cyclic Voltammetry for various values of the length interdigital electrode

V. CONCLUSION

From the Figures 3,4 and 5, we can conclude that gap and length of the electrodes give a higher effect for the current response while the width of the electrodes to have a smaller current response. It was observed that increase the gap and length of the electrodes provided sufficiently the higher surface areas for active species take place at the electrode surface. The maximum current response achieved is 1.6 A/m corresponding to gap of electrodes of 250 um. For the width and length of the electrodes, the maximum current response is 0.85 A/m and 1.6 A/m respectively corresponding the width of electrodes of 50 um and the length of electrodes of 700 um. The effect of the materials for the interdigital electrodes will be explored in our future work to give a high impact at the current response to improve the cyclic voltammetry performances.

ACKNOWLEDGMENT

The authors would like to thank UKM-GUP-2011-380 for supporting this project.

REFERENCES

[1] J. Guo and E. Lindner, "Cyclic Voltammetry at Shallow Recessed Microdisc Electrode: Theoretical and Experimental Study," *J.Electroanal Chem,* 2009, pp. 180-184.

[2] H. Wang, L. Pilon, "Physical interpretation of Cyclic Voltammetry for Measuring electric double layer capacitances", *Electrochemica Acta,* 64(2012), pp. 130-139, 2012.

[3] H. E. Zainal Abidin, A. A. Hamzah and B. Y. Majlis, "Design of Interdigital Structured Supercapacitor," Proceeding *of the Regional Symposium on Micro and Nanoelectronics*, vol. A247, September 2011.

[4] X. Yang and G. Zhang, "Diffusion – Controlled Redox Cycling at Nanoscale Interdigitated Electrodes," Proceedings of the COMSOL Multiphysics User's Conference, 2005.

[5] K. V. Singh, A. M. Whited, Y. Ragineni and etc, "3D nanogap interdigitated electrode array biosensors, 2010, pp.1493-1502.

[6] H. J. Kwon and E. Akyiano, "Simulation of cyclic voltammetry of ferrocyanide/ferricyanide redox reaction in the EQCM Sensor," Proceedings of the COMSOL Multiphysics User's Conference, 2011.

High Sensitivity Localized Surface Plasmon Resonance Sensor of Gold Nanoparticles : Surface Density Effect for Detection of Boric Acid

Marlia Morsin, Akrajas Ali Umar*, *Member*, IEEE, Muhamad Mat Salleh*, *Member*, IEEE and Burhanuddin Yeop Majlis, *Senior Member*, IEEE

Institute of Microengineering and Nanoelectronics (IMEN),
Universiti Kebangsaan Malaysia,
43600 UKM Bangi, Selangor, Malaysia.
Email : akrajas@ukm.my, mms@ukm.my

Abstract: **This paper presents the study on localized surface plasmon resonance (LSPR) of gold nanoparticles (AuNPs) sensor and its detection towards the presence of boric acid (H_3BO_3). Boric acid is a substance commonly used as consumer products and also been used in food preparations to increase elasticity, better texture and crispiness. However, this chemical was claimed as non-permitted preservative and additive in food preparation. In this study, spherical AuNPs were grown on quartz substrate using Seed Mediated Growth Technique (SMGT) with variations seeding process. The plasmonic responses of the AuNPs in deionized water and in various concentration of boric acid solution based on optical absorption spectra of the AUNPs samples were recorded. The changes of the resonance peak intensity and position of the spectra were used as sensing parameters. It was found that the surface density of nanoparticles grown on the AuNPs samples affected the plasmonic responses and the sensing sensitivity of the AuNPs samples.**

Key words: Gold nanoparticles (AuNPs), boric acid, localized surface plasmon resonance (LSPR), Optical Sensor

I. INTRODUCTION

A noble metal plasma nanoparticle such as gold (Au) and silver (Ag) have attracted the attention of scientists in the field of electronic, sensing and non - linear optic. It displays strong and unique localized surface plasmon resonance (LSPR) in the visible and near infra – red region [1] and also have sensitive plasmon resonance characteristics to the change in the dielectric of surrounding medium [2]. LSPR is a phenomenon resulting from the interaction between incident light and surface electrons in a conduction band. LSPR is strongly dependent on surface density [3], interparticle spacing [1], size and shape [1-2] of metal nanoparticles structure which are the key parameters in determining and enhancing their function and potential applications. Recently, LSPR of metal plasma nanoparticles has been widely used to detect various types of chemicals by direct method [4].

A water - soluble boron compound namely boric acid (H_3BO_3) have been used as pesticide products to kills fungi, weeds, algae, insects, mites and molds since 1948 [5]. Also, boric acid is one of the frequent preservatives and additives used in various foods processing especially by small scale producers. Foods such as noodles, seafood product, dairy products and meat products were added with boric acid during preparation process to control freshness and enhance colour, texture and flavour [6-7]. However, this salt - like white powder is poisonous and declared unsafe to use as food additives by FAO/WHO Expert Committee [8]. Potential lethal boric acid doses are 3 - 6 g for infants and 15 – 20 g for adults [6] while exposure to large amount of boric acids over long period can produce toxic symptoms which include diarrhoea, vomiting and abdominal pain [6, 9]. Boric acid is a weak acid and the detection of this chemical is normally done indirectly by laborious chemical techniques [10]. These detection techniques such as mannitol titration, colorimetric and spectrophotometric [6] are less sensitive, requires a huge amount of sample and complex test procedures.

This paper reports a study on localized surface plasmon resonance (LSPR) of gold nanoparticles (AuNPs) response toward boric acid. The spherical AuNPs was syntheses using seed mediated growth method with variation seeding process to observe its surface density effect and then used as sensing material in an optical sensor system. Au is chosen as sensing materials due to its inert nature and biocompatibility [11, 12].

II. METHODOLOGY

A. Synthesis of AuNPs

The gold nanoparticles (AuNPs) were grown on quartz substrate using seed mediated growth method as previously reported [13] with minor modifications. The chemicals used for the synthesis were hydrogen tetrachloroaurate ($HAuCl_4.3H_2O$), poly-l-lysine, trisodium citrate ($C_6H_5Na_3O_7$), sodium tetraborohydride ($NaBH_4$) ascorbic acid and cethyltrimethy ammonium boromide (CTAB). These chemicals were purchased from Sigma-Aldrich except trisodium citrate, sodium tetraborohydride and ascorbic acid which were obtained from Wako Pure Chemical Ltd. All these chemicals were used as received. The solutions of these chemicals were prepared using deionized (DI) water with resistivity around 18.2 MΩcm.

In order to grow gold nanoparticles on quartz substrate, two types of solutions were prepared namely seed solution and growth solution. The seed solution was prepared by mixing 0.5 ml of 0.01 M $HAuCl_4$ with 0.5 ml of 0.01 M trisodium citrate and 20 ml DI water. After that, 0.5 ml of 0.1 M cold aqua

NaBH$_4$ was added into the solution. Before the seeding process, the quartz substrate was immersed in a solution of 5% poly-l-lysine for 30 minutes to impose a positive charge on the substrate surface. Then, the substrate was rinsed with DI water before immersing into gold nanoseeds solution. The nanoseeds can easily attach onto the positively charged substrate due to the electrostatic interaction with negative charges from gold nanoparticles. All processes were repeated two, three and four times to obtain difference density nanoseed particles on the substrate surface. Finally, the substrate was rinsed and dried before proceeding with growth process. The growth solution was prepared by dissolving 20 ml of 0.1 M CTAB – capped seed using ultrasonic. Then, 0.5 ml of 0.01 M HAuCl$_4$ and 0.1 ml of 0.1 M ascorbic acid were added into the solution. Next, the seeds were immersed into the growth solution for 2 hours. After the growth process, the quartz substrate colour changes to violet red indicating the formation of gold nanoparticles. Prior to characterization process, the substrate was rinsed and dried.

The gold nanoparticles were then used as sensing material in an optical sensor system.

B. Setup of Boric Acid Sensor

An optical sensor system was setup to study the optical response of gold nanoparticles (AuNPs) toward boric acid. As shown in Fig. 1 the setup consists of a sensor chamber, a light source (LS-1 tungsten halogen lamp), duplex fiber optical probe system, a USB-2000 Ocean Optics spectrometer and a computer with OOIBase32 software as spectrum analyzer. The light source beam was transmitted by one of the fiber arm toward AuNPs sample, and the reflected light was transmitted by the other fiber arm to the spectrometer. The optical responses of the AuNPs sample were recorded as the absorption spectra of light at first when the sample immersed in deionized (DI) water and then immersed in boric acid solution. Boric acid solution was prepared by dissolving the white boric acid powder purchased from R&M Chemicals with DI water under ultrasonic treatment.

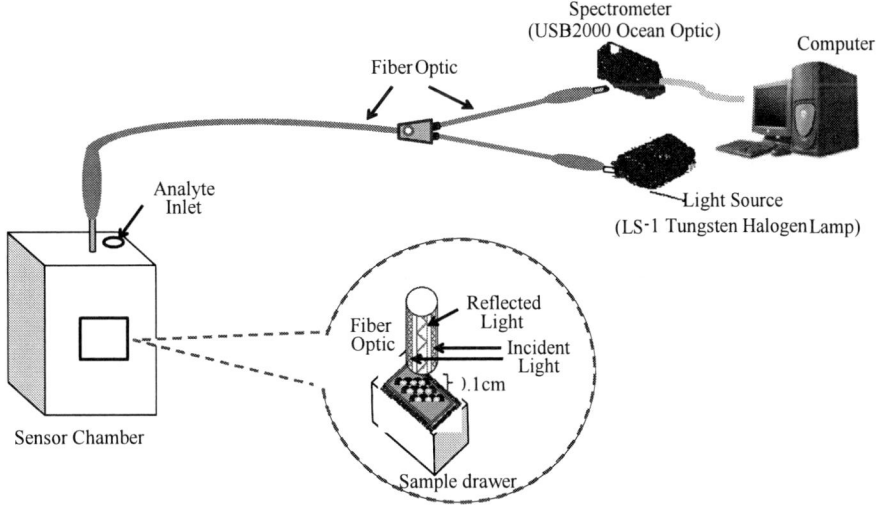

Fig. 1. Setup for sensor system to detect boric acid

III. Results and Discussion

The gold nanoparticles (AuNPs) have been grown using seed mediated growth method. The visible optical absorption spectrum of the sample was used to confirm the formation of AuNPs on the quartz substrate. Fig. 2 shows the UV - Vis absorption spectra of AuNPs for various seeding process. It was observed that the spectrum contains a single absorption peaks from 530 nm - 550 nm . The spectrum tends to shift right when the seeding process was increased. Instead of that, the intensity of the spectrum increases with the increment of the seeding process, This situation can be understood by the generation of more nanoparticles on the quarts substrate. This absorption band agrees with previous observations of the localized surface plasmon resonance (LSPR) spectrum of spherical shape gold nanoparticles [10, 14 - 15].

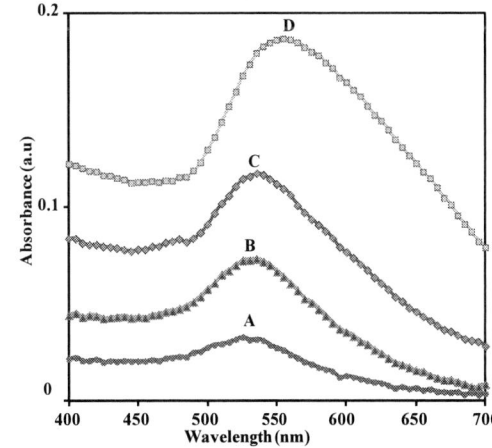

Fig. 2. Optical absorption spectrum using UV-VIS for (A) 1, (B) 2, (C) 3 and (D) 4 times seeding process in the air medium.

The morphology of the gold nanoparticles (AuNPs) were characterized using FESEM with accelerating voltage of 3.0kV under 50k magnification. Fig. 3 shows the FESEM image of AuNPs that have grown from solution by the seed mediated growth method in 2 hours. Figure 3 (A) and (B) shows the spherical grown gold nanoparticles on the substrate. For Figure 3 (C), the spherical gold nanoparticles tend to aggregate with the increment of seeds. Therefore, for 4 times repeated seeding process as shown in Figure 3 (D), the spherical nanoparticles start to aggregate and network after gold formation. These gold nanoparticles grown on the substrate were deviated from single spherical geometry.

The trend of size and surface density are shown in Fig.4. Form the graph, we can see that the surface density showed a linear relationship with the increment of gold nanoseeds while the size of AuNPs after 2 hours growth was slightly decreased. Hence, we can conclude that the increase of seeding process was effective to promote higher amount attachment of GNPs on the surface. It also found that the size of AuNPs for 1, 2, 3 and 4 repeated seeding process are 25±9 nm, 18±7 nm, 16±8 nm and 15±10 nm. The same source of gold seeds was used until all the seeding processes complete.

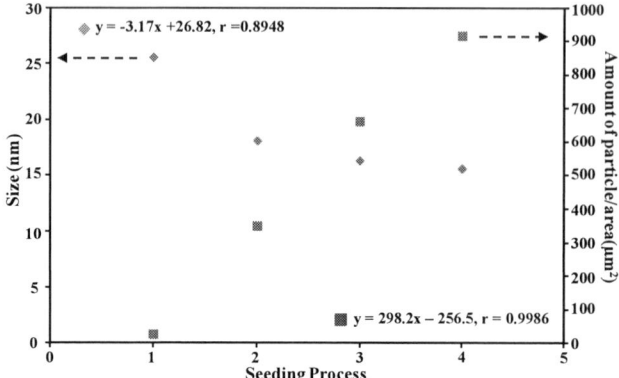

Fig. 4.   The size and surface density of AuNPs grown in quartz substrate for various repeated seeding process after growth.

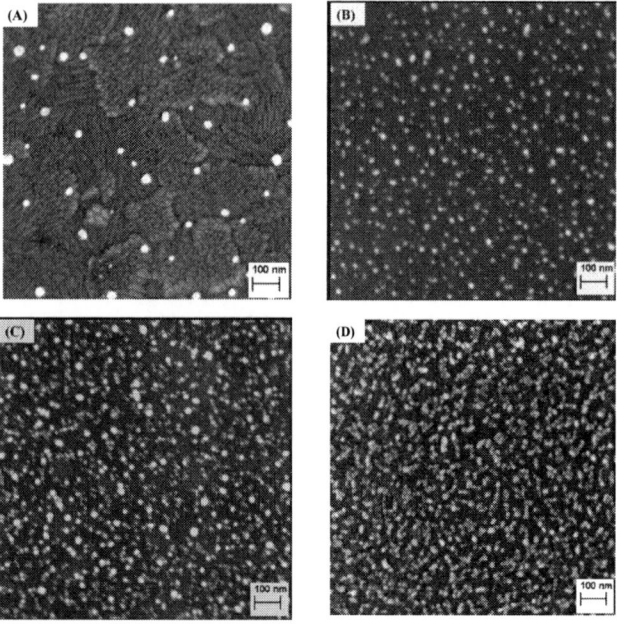

Fig. 3 : FESEM image of AuNPs grown on quartz substrate for (A) 1, (B) 2, (C) 3 and (D) 4 repeated seeding process for 30 minutes each.

For the first seeding process, the gold seeds were attached on the substrate and was repeated for the next seeding process. When more seeds were generated on the substrate, the seeds generated by 2 ,3 ,4 repeated seeding process tends to grow on the original seeds. The lack of space caused the generated seeds to grow on the original seeds. The smaller the seeds caused the smaller size of AuNPs after the growth process.

Using the optical sensor system as shown in Fig. 1, the optical response of gold nanoparticles (AuNPs) towards boric acid was measured. The LSPR spectra of AuNPs when the sample was immersed in the deionized (DI) water and then, in 10 mM boric acid solution for all 4 seeding process shows that the resonance peak position and its intensity of the LSPR spectrum of the sample were changed as the water medium was replaced by the boric acid solution. Fig. 5 (A) and (B) depicts the example for sample 2 and 3 seeding repeated process. It was observed that the peak position of the spectrum in water are around 545nm and red-shifted by 1 nm in the presence boric acid. Meanwhile, the LSPR spectrum increased with the change of medium from water to boric acid. These spectral changes are related to the change of refractive index of the medium [10, 16]. Besides, for the intensity change, the delta absorbance for 2 seeding repeated process was 0.11 and 0.14 for 3 seeding repeated process.

Hence, the change of resonance peak position and its intensity when the water medium is replaced by the boric acid solution may be used as sensing sensitivity parameters of gold nanoparticles sensor based on LSPR. Also, from the graph, we can see a pattern which is with the increment of seeding process, the intensity of the LSPR spectrum were increased while the peak position of the spectrum remain. This situation can be explained by Mie theory [17] which is the density of the particles will give effect to absorbance. In this case, the AuNPs density was increased because of the mediums used for all seeding process were same (water and 10mM boric acid).

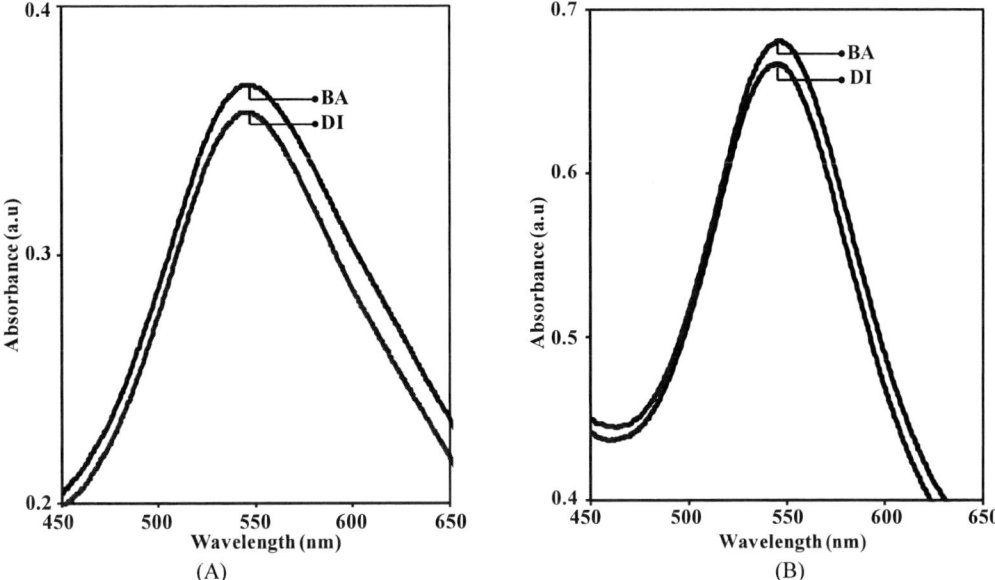

Fig. 5. LSPR spectrum of spherical gold nanoparticles in (A) deionized (DI) water and (B) 1 mM boric acid

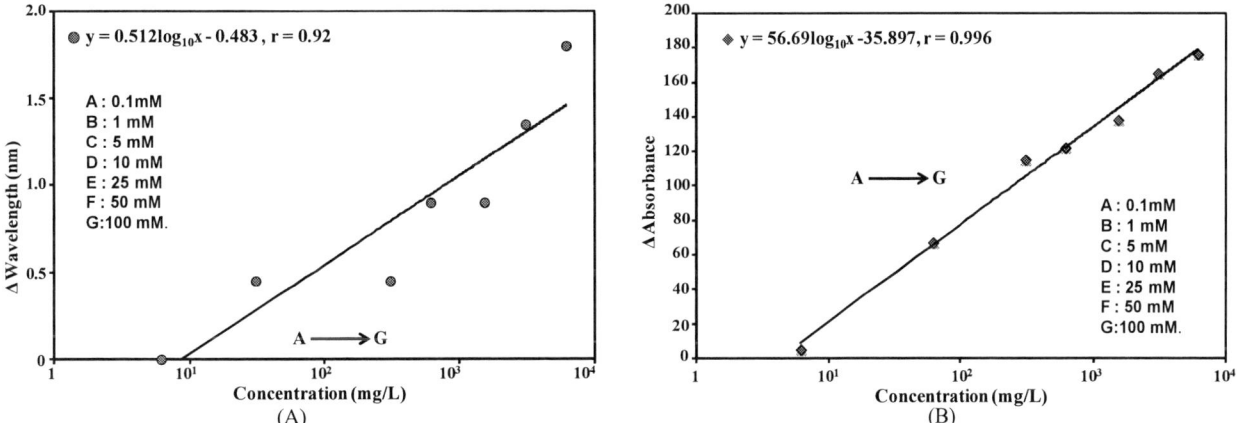

Fig. 6. Plasmonic response of spherical gold nanoparticles to the variation concentration of boric acid; based on the change of the LSPR spectra; i.e. (A) the shift resonance peak positions and(B) their intensities

In order to study the sensitivity of AuNPs towards various concentrations of boric acid solutions, samples from 2 repeated seeding process was taken. The boric acid solutions were prepared with seven concentrations of 0.1 mM, 1 mM, 5 mM, 10 mM, 25 mM, 50 mM 100 mM. Fig. 6 shows the sensing response toward the variation of boric acid concentrations based on the change of the LSPR spectra ; i.e. (A) the shift resonance peak positions and (B) their intensities. It appears that the sensing responses are linear logarithm with the boric acid solution concentrations, where the linear correlation coefficients (r) is greater than 0.90.

IV. CONCLUSION

This investigation shows that the LSPR of gold nanoparticles grown from solution by seed mediated method is sensitive toward the presence of boric acid. The sensing parameters are based on changing the resonance peak position and its intensity showed a linear relationship with the boric acid concentrations. Also, it was found that the surface density of nanoparticles grown on the AuNPs samples affected the plasmonic responses and then sensing sensitivity of the AuNPs samples.

ACKNOWLEDGEMENT

The first author would like to acknowledge Universiti Tun Hussien Onn for SLAB - MOHE fellowship. This work was supported by Universiti Kebangsaan Malaysia under research university grant.

REFERENCES

[1] S. K. Ghosh and T. Pal, "Interparticle Coupling Effect on the Surface Plasmon Resonance of Gold Nanoparticles: From Theory to Applications", *Chemical Reviews* vol. 107 (11), pp. 4797- 4862, 2007.

[2] W.A. Murray and W. L. Barnes, "Plasmonic Materials"; *Advanced Materials* vol .19 (22), pp. 3771-3782 ,2007

[3] M.H. Tu, T. Sun and K.T.V. Grattan, "Optimization of the Gold – Nanoparticle - based Optical Fibre Surface Plasmon Resonance (SPR) – based Sensors", *Sensors and Actuators B: Chemical* vol. 164 (1), pp. 43 – 53, 2012

[4] K.M. Mayer and J.H. Hafner, "Localized Surface Plasmon Resonance Sensors", *Chemical Reviews* vol. 111 (6), pp. 3828-3857, 2011

[5] National Pesticide Information Center , " *Boric Acid (Technical Fact Sheet)*", Oregon State University, 2001

[6] A.S. See, A.B. Salleh, F.A. Bakar, N. Yusof, A.S. Abdulamir and L.Y Heng, "Risk and Health Effect of Boric Acid", *American Journal of Applied Sciences* vol. 7(5), pp 620-627, 2010.

[7] P.H Yiu, J. See, A. Rajan and C. J. Bong, "Boric Acid Levels in Fresh Noodles and Fish Balls", *American Journal of Agricultural and Biological Sciences* vol. 3 (2), pp. 476-481, 2008

[8] FAO/WHO Expert Committee on Food Additives, "*Report of the First Session of the Joint FAO/WHO Codex Alimentarius Commission*", Appendix H.1, 1963

[9] T.L. Litovitz, W. Klein – Schwartz, G.M. Oderda and B.F. Schmitz, "Clinical Manifestations of Toxicity in a Series of 784 Boric Acid Ingestions", *The American Journal of Emergency Medicine* vol. 6 (3), pp. 209-213, 1988

[10] B. Sepúlveda, P.C. Angelomé, L.M. Lechuga and L. M. Liz-Marzán, "LSPR - based Nanobiosensors", *NanoToday* vol 4 (3), pp. 244 – 251, 2009

[11] E. Petryayeva and U.J Krull, "Localized Surface Plasmon Resonance: Nanostructures, Bioassay and Biosensing – A Review", *Analytica Chimica Acta* vol.706 (1), pp. 8-24, 2011

[12] W.H. Low, "Boric Acid: Its Detection and Determination in Large or Small Amounts*", Journal of the American Chemical Society* vol. 28 (7),pp. 807-823, 1906

[13] Marlia Morsin, M.M. Salleh, A.A Umar, " Detection of Boric Acid using Localized Surface Plasmon Resonanace Sensor of Gold Nanoparticles, *The 14th International Meeting on Chemical Sensor*, pp .1418 - 1412, 2012

[14] S. Nengsih, A.A. Umar, M.M. Salleh and M. Yahaya, " Detection of Formaldehyde in Water: A Shape-Effect on the Plasmonic Sensing Properties of the Gold Nanoparticles", *Sensors* vol. 12(8), pp. 10309-10325, 2012

[15] S. Nengsih, A.A. Umar, M.M. Salleh and M. Oyama, "Detection of Volatile Organic Compound Gas using Localized Surface Plasmon Resonance of Gold Nanoparticles", *Sains Malaysiana* vol. 40(3), pp. 231-235, 2011

[16] J. N. Anker, W. P. Hall, O. Lyandres, N. C. Shah, J. Zhao and R.P.V. Duyne, "Biosensing with Plasmonic Nanosensors", *Nature Materials* vol.7, pp. 442-453, 2008

[17] U. Kreiberg and M. Vollmerr, " Optical Properties of Metal Clusters" 1st ed. vol. 25, German, Springer,1995

Effect of Varying Thickness of Electroplated NiFe Film on Magnetic Properties

Nadzril Sulaiman, Jumril Yunas and Burhanuddin Yeop Majlis, *Member, IEEE*
Institute of Microengineering and Nanoelectronics (IMEN)
Universiti Kebangsaan Malaysia (UKM)
43600 UKM Bangi, Selangor, Malaysia
Email: burhan@vlsi.eng.ukm.my

Abstract- The thickness variations of NiFe based magnetic core and its effects on magnetic properties have been investigated. The thin film core material is fabricated by using electroplating that may have direct effect on the performance on magnetic device. In this work, the main interest is on varying thickness of NiFe film that is being used as magnetic core material for a micro-scaled magnetometer. The magnetic properties of interest are coercivity and saturation magnetization. From the experiments, the relationship between the thickness of electroplated NiFe film and coercivity and saturation magnetization are determined. The low coercivity and low saturation magnetization can be achieved by increasing the core thickness. This relationship can be used to improve the functional of a micro magnetometer.

I. INTRODUCTION

Magnetic core is used in many devices that require concentration of magnetic flux or magnetic field lines in its operations. Such devices include transformer, inductor and magnetometer. In recent years, numerous magnetic-based electronic devices have been made into smaller scale. This has led into the development of many miniaturized devices such as transformers [1], inductors [2, 3], and also magnetometers [4, 5]. There are many types of magnetometer based on different principle of operation. Among many, fluxgate is considered as one of the most popular type due to its versatility and toughness.

Besides a set of driving and sensing coil, fluxgate magnetometer requires a core which is made of ferromagnetic material in order to function. This ferromagnetic material must possessed certain magnetic properties such as low coercivity, low magnetization saturation, and high permeability.

Many types of material have been reported to be used as magnetic core. These includes the NiFe [6-8], ferrites and amorphous based materials [9, 10]. NiFe is the most common material simply because of its excellent magnetic properties and also can be easily fabricated.

Various methods in the fabrication of magnetic core have been reported such as MBE [11], MOCVD [12] and electrodeposition [5-8]. In comparing between the stated methods, electrodeposition is considered to be practical. This is mainly due to its simplicity, low cost and ability to produce thick feature.

Geometrical characteristic of magnetic core such as thickness may have effects on certain magnetic properties. Based on this

assumption, analyses are conducted to observe the effect of varying the NiFe film thickness on magnetic properties such as coercivity and saturation magnetization.

In this paper, the fabrication of NiFe magnetic core with various thicknesses is presented. These magnetic cores are then analyzed using variable sample magnetometer (VSM) to observe the hysteresis characteristic which reveals its coercivity and saturation magnetization. The results from these analyses should display the relationship between the varying thickness of electroplated NiFe film and magnetic properties particularly coercivity and saturation magnetization.

II. MAGNETIC CORE FABRICATION

The magnetic core is fabricated using dc electrodeposition method. This method utilizes the effect of material ion transfers between anode and cathode terminals through an electrolyte solution. The common setup in dc electrodeposition consists of dc power supply with two terminals which are anode and cathode. The target material is located at anode terminal while the sample to be electroplated is placed at cathode terminal. Fig. 1 illustrates the electrodeposition setup.

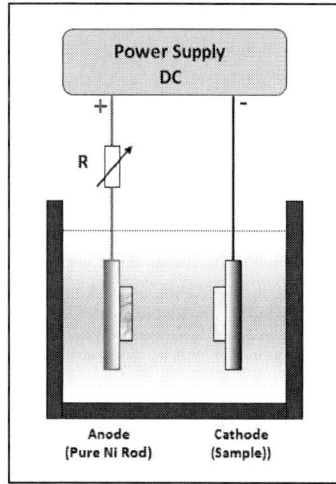

Fig. 1. Electrodeposition setup consisting of dc power supply, anode and cathode terminals

In the experiment, the magnetic core test samples were prepared on ceramic substrate using standard UV photolithography technique which creates pattern for electrodeposition process. The pattern is square-shaped with dimension of 3 mm². The film thickness depends on the thickness of photoresists. Hence, with multiple coatings of AZ 4620 photoresist could produce samples with thickness ranging from 30 µm to 120 µm. These samples were then sputter coated with 100 nm thick Au as seed layer and lifted-off to realize the square pattern.

In order to electroplate NiFe, anode terminal is connected to pure nickel rod while the square-shaped sample is connected to cathode terminal. The electrolyte solution is of sulfamate based which is made of a mixture of chemicals such as Nickel Sulfate (20%), Ferrous Sulfate (1.4%), Sodium Chloride (2.5%), Orthoboric Acid (4%), Sodium Salicylate (2%) and Saccharine (0.5%). These percentages represent the proportion with respect to 1000 ml of deionized water.

The electrodeposition is conducted at room temperature which is approximately 28 °C. The dc power supply provides voltage at 11.2V while applied current is at 2 mA. Current density is maintained at 22.2 mA/cm². The average rate of electrodeposition is measured to be 320 nm/min. Fig. 2 plots the electroplated NiFe film thickness against time.

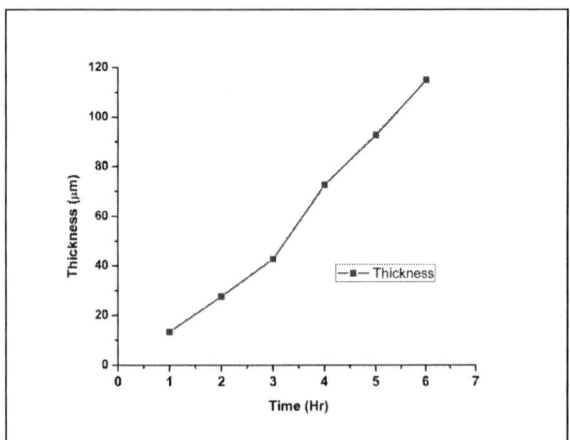

Fig. 2. Electrodeposition rate of NiFe magnetic core

III. MAGNETIC CORE CHARACTERIZATION

The characterization of the electroplated NiFe magnetic core are conducted using vibration sample magnetometer (VSM) to determined relevant magnetic properties i.e. coercivity and saturation magnetization. Through the data generated by VSM, hysteresis curve of the electroplated NiFe magnetic core can be plotted which is shown in Fig. 3. It can be seen that the Fig. 3, the magnetization (M) to reach saturation is 36 emu/g.

Preferable characteristic of magnetic core for a fluxgate magnetometer is having high permeability, low coercivity and low saturation magnetization. Due to the properties of nickel and iron, NiFe alloy is recognized to possess high permeability while coercivity and saturation magnetization depends on the geometry of the magnetic core [13].

Characterization processes are conducted on each sample at different thickness to observe the effect of film thickness on coercivity and saturation magnetization.

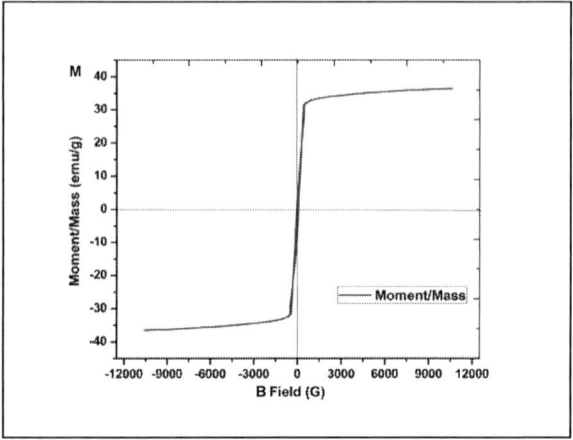

Fig. 3. Hysteresis curve of NiFe magnetic core

IV. RESULTS AND DISCUSSION

A number of samples with different thicknesses have been prepared and analyzed using VSM. From the VSM, hysteresis of the electroplated NiFe magnetic core is plotted as shown in Fig. 3. The hysteresis curve have characteristic such as very narrow loop. This imply that the material is capable of very fast magnetize and demagnetize operations besides reducing the possibility of wasted energy during that period. This is expected since the material electroplated is NiFe which is known to have such hysteresis curve [14].

Each sample undergoes VSM analysis to observe its coercivity and saturation magnetization. Fig. 4 shows the plot of coercivity and saturation magnetization with respect to NiFe film thickness.

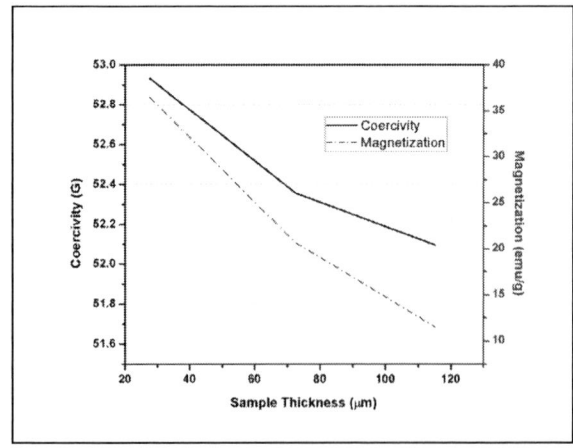

Fig.4. Coercivity and saturation magnetization against NiFe film thickness

978-1-4673-2395-6/12 $31.00 © 2012 IEEE

From Fig. 4, it can be seen that the relationships of both coercivity and saturation magnetization with film thickness are of inverse proportional nature. In other words, both the coercivity and saturation magnetization tend to decrease as the film thickness increase.

This finding may suggest that preferable conditions such as lower coercivity and lower saturation magnetization can be achieved by increasing the thickness of electroplated NiFe film. Thicker film may also increase the sensitivity and operational range of magnetometer [15]. Therefore, the statement "the thicker the magnetic core, the better the performance of magnetometer" can be established.

V. CONCLUSION

The analyses of finding the relationship between the thickness of the electroplated NiFe magnetic core with coercivity and saturation magnetization have been conducted. From the results, it can be concluded that lower coercivity and lower saturation magnetization can be made possible by increasing the magnetic core thickness. Thick magnetic core also has the advantage of increasing the sensitivity and operational range hence improving the functionality of the magnetometer.

ACKNOWLEDGMENT

The authors would like to thank Institute of Microengineering and Nanoelectronics and Universiti Kebangsaan Malaysia for providing assistance and facilities.

REFERENCES

[1] Jumril Yunas, A. A. Hamzah Burhanuddin Yeop Majlis, "Fabrication and characterization of surface micromachined stacked transformer on glass substrate," *Microelectronic Engineering* vol. 10, pp. 2020-2025, 2009.

[2] M. G. A. Chong H. Ahn, "Micromachined planar inductors on silicon wafers for MEMS applications," *IEEE Transactions on Industrial Electronics,* vol. 45, pp. 866-876, 1998.

[3] B. Y. M. J. Yunas, A. A. Hamzah, B. Bais, "Design, fabrication and testing of sandwich coupled inductors," *Microsyst. Technol.,* pp. 739-744, 2012.

[4] P. K. Pavel Ripka, "Portable fluxgate magnetometer," *Sensors and Actuators A,* vol. 68, pp. 286-289, 1998.

[5] J.-S. H. Hae-Seok Park, Won-Youl Choi, Dong-Sik Shim, Kyoung-Won Na, Sang-On Choi, "Development of micro-fluxgate sensors with electroplated magnetic cores for electronic compass," *Sensors and Actuators A,* vol. 114, pp. 224-229, 2004.

[6] G. L. Chun-Lei Kang, Jian-Zhong Yang, Long-Hua Liu, Ying Xiong, Yang-Chao Tian, "Electroplating a magnetic core for micro fluxgate sensor," *Microsyst. Technol,* vol. 15, pp. 413-419, 2009.

[7] S. N. Jean-Marie Quemper, J.P. Gilles, J.P. Grandchamp, A. Bosseboeuf, T. Bourouina, E. Dufour-Gergam, "Permalloy electroplating through photoresist molds," *Sensor and Actuators* vol. 74, pp. 1-4, 1999.

[8] J. T. R. F.E. Rasmussen, P.T. Tang, O. Hansen, S. Bouwstra, "Electroplating and characterization of cobalt-nickel-iron and nickel-iron for magnetic microsystems applications," *Sensor and Actuators A,* vol. 92, pp. 242-248, 2001.

[9] G. C. S. Luiz Carlos de Carvalho Benyosef, Mauricio Bochner, "Optimization of the magnetic properties of materials for fluxgate sensors," *Materials Research,* vol. 11, pp. 145-149, 2008.

[10] F. V. Predrag M. Drljača, Pavel Kejik, Radivoje S. Popović, "Advanced process of the magnetic core integration for the microfluxgate magnetometer," *Sensors and Actuators A,* vol. 129, pp. 58-61, 2006.

[11] A. M. Z. K. Rook, J.O. Artman, D.E. Laughlin, R.M. Chrenko, "Multilayer Permalloy Films Grown by Molecular Beam Epitaxy," *Journal of Applied Physics,* vol. 69, pp. 698-703, 1991.

[12] C. L. R. Peter E. Oliver, Anthony D. Pitt, John M. Keen, Michael C.L. ward, Matthew E. G. Tilsley, Nigel A. Smith, Brian Cockayne and I. Rex Harris, "Growth of Iron, Nickel, and Permalloy Thin Films by MOCVD for use in Magnetoresistive Sensors," *Chemical Vapor Deposition* vol. 3, pp. 97 - 101, 1997.

[13] P. Ripka, Ed., *Magnetic Sensors and Magnetometers.* Norwood, MA: Artech House, Inc., 2001, p.^pp. Pages.

[14] R. C. O. Handley, *Modern Magnetic Materials Principles and Applications.* New York: John Wiley & Sons, Inc., 2000.

[15] C. H. A. Trifon M. Liakopoulos, "A micro-fluxgate magnetic sensor using micromachined planar solenoid," *Sensors and Actuators,* vol. 77, pp. 66-72, 1999.

Micro-heater Filament on Polyimide Membrane for Gas Sensor Applications

Mimiwaty Mohd Noor, Gandi Sugandi and Burhanuddin Yeop Majlis, *Senior Member, IEEE*
Institute of Microengineering and Nanoelectronics (IMEN)
Universiti Kebangsaan Malaysia (UKM)
43600 UKM Bangi, Selangor, Malaysia
Email: burhan@eng.ukm.my

Abstract - **Micro-heater filaments on polyimide membrane for micro gas sensors are fabricated and measured. Resistance as well as power consumption are calculated and temperature of the filament is then extracted. The polyimide membrane is supported by silicon frame that has been etched from the back side by bulk micromachining in a potassium hydroxide (KOH) solution. The filament is made of platinum with a thickness of 300 nm. The membrane thickness is 20 μm. It has an area of 3.5 mm x 3.5 mm and the area covered by the heater filament is 750 μm x 750 μm. The filament shows power consumption of 243 mW at temperature of 115 °C.**

I. INTRODUCTION

In general, a gas sensor comprises of a substrate provided with micro-heater and sensor electrodes and a gas sensitive layer. A heater filament functions to heat the sensitive layer and electrodes to measure the resistivity of the sensing film. To reduce power consumption, the sensing layer can be deposited on a thin membrane realized on a silicon substrate using the bulk micromachining technique [1-3].

Platinum is the most widely used filament material due to its good linear thermal coefficient and high stability at elevated temperature [4].

A thin membrane is applied as mechanical support for the micro-heater and the sensing film. It is also responsible for thermal isolation and uniformity as well as low-power consumption [2]. In most cases, the membrane is made of either silicon nitride or silicon oxide thin films [5]. Several investigations have been conducted on micro heaters that utilize polyimide as the membrane materials.

The advantage of a polyimide membrane is that heat loss to the substrate is less than for silicon dioxide or silicon nitride membranes [6]. Furthermore, robustness and flexibility is increased when fabricated on polyimide as compared to micro-hotplates on silicon with membranes made of dielectric layers [7].

In this work, we present the fabrication of micro-heater filament on polyimide membrane. The membrane has an area of 3.5 mm x 3.5 mm, and the area covered by the heater is 750 μm x 750 μm. Photolithography, sputtering and wet etching are the processes used in realization of the micro-heater.

II. FABRICATION PROCESS

A schematic structure and dimensions of the fabricated micro-heater are shown in Fig. 1(a) and 1(b) respectively.

As a starting material, 650 μm silicon wafer coated with 200 nm silicon nitride was used. First, the silicon membrane was fabricated using anisotropic wet etching on the back side until 50-100 mm thick membrane was left. Polyimide was spun onto the substrate and then cured at 350 °C for 30 minutes [8]. Then, 300 nm thick platinum filament and temperature sensor were patterned. In the final step, the silicon wafer was etched by anisotropic etching in KOH at 80 °C [9] from the backside to release the polyimide membranes.

Photos of the fabricated micro-heater are shown in Fig. 2.

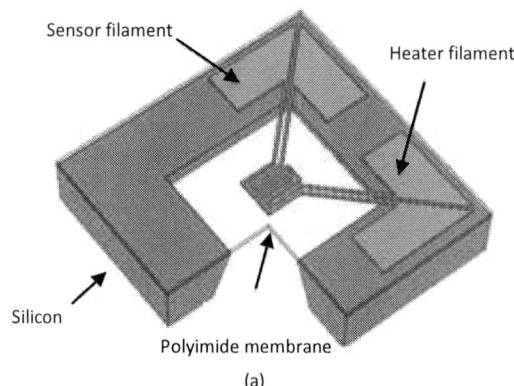

(a)

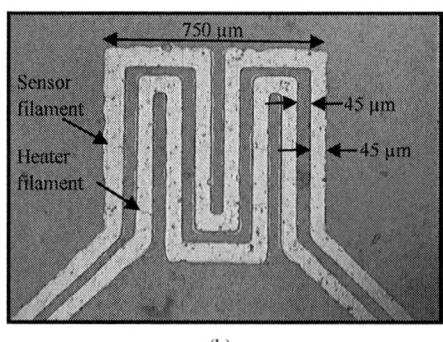

(b)

Fig. 1. (a) Schematic structure of the micro-heater. (b) Image of fabricated heater filament and sensor filament.

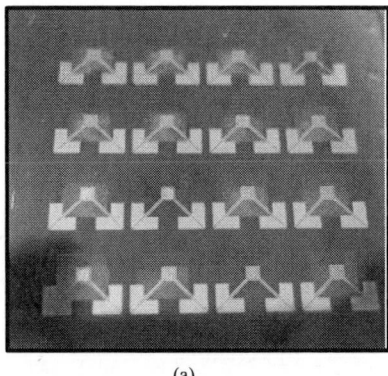

(a)

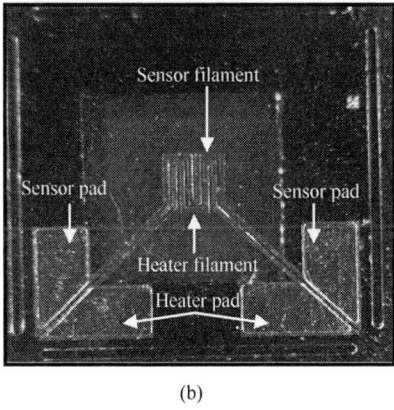

(b)

Fig. 2. (a) A photograph of an array fabricated micro-heater (b) Close up of the micro-heater

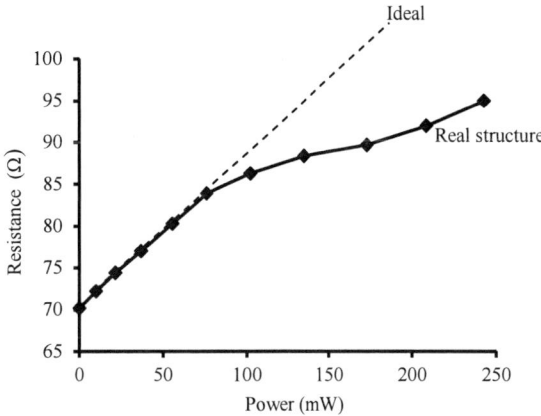

Fig. 3. Resistance as a function of applied power.

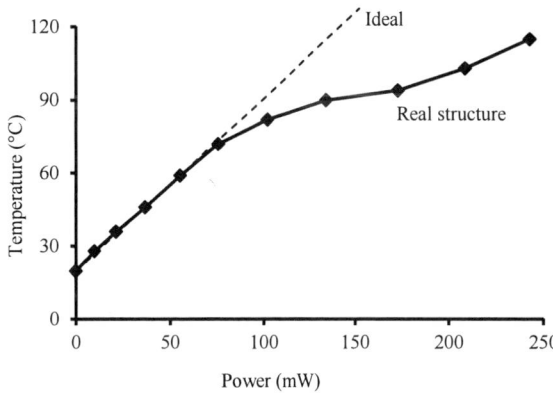

Fig. 4. Temperature as a function of the applied power

III. RESULTS

It is known that the dependence of the specific resistance on temperature can be expressed by the equation (1),

$$R_T = R_a [1 + \alpha_a(T - T_a)] \tag{1}$$

where R and R_a are the resistance values at temperatures T and T_a, respectively, T_a the ambient temperature and α_a the temperature coefficients of resistance at T_a [10,11].

During the measurement, the heater filament was connected to a dc power supply (ISO-TECH, IPS 3202). The applied voltage of the device was then increased from 0 to 6.0 V with steps of 0.5 V. The resistance of temperature sensor was measured using a digital multimeter (SANWA CD771) and recorded as a function of heating power as shown in Fig. 3. It shows that the resistance increases with increament the input power linearly up to 90 mW.

The temperature coefficient of resistance (TCR), α_a, of platinum at 20 °C is 0.003729 °C^{-1} [12].

With this value and resistance measured, the extracted temperature by using Eq. (1) as a function of supplied power is shown in Fig. 4. It also shows temperature increases with increasing the input power linearly up to 90 mW.

Both Figures illustrate a non linear association after a 90mW power input. This may be due to the amount of vertical heat radiation becoming significant after 90 mW as heat losses due radiation increases with heated membrane area [13].

However, temperature increased linearly with increasing the resistance following the linear relationship of equation (1). A linear association between temperature and voltage was also demonstrated upon voltage change.

The variation of temperature with resistance of the temperature sensor and temperature with input voltage of the heater filament are shown in Fig. 5 and Fig. 6 respectively.

From Fig. 5, the sensitivity, S = ΔT/ΔR, was calculated. S = 3.8 °C/Ω. This means that every 1 Ω increase in resistance, the temperature rises by 3.8 °C.

From Fig. 6, the variation of temperature with input voltage is 18.3 °C/V. This shows that every 1 volt increase in voltage, the temperature rises by 18.3 °C.

The micro-heater filament was found to have the power consumption of 243 mW at temperature of 115 °C before the filament malfunctions at the input voltage of 6.0 V in room ambient.

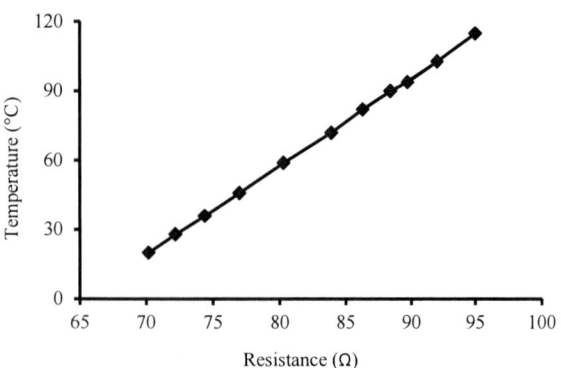

Fig. 5. The variation of temperature with filament resistance

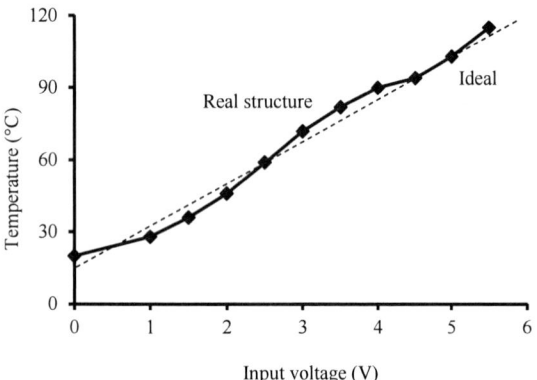

Fig. 6. The dependence of temperature on input voltage.

IV. CONCLUSION

In conclusion micro-heater filaments on polyimide membrane has been fabricated and characterized. Resistance as well as power consumption were calculated and temperature of the filament was extracted. The filament was made of platinum with a thickness of 300 nm. The membrane thickness was 20 μm. It has an area of 3.5 mm x 3.5 mm, and the area covered by the heater was 750 μm x 750 μm. The maximum filament temperature obtained was 115 °C with the power consumption was 243 mW. The sensitivity, $S = \Delta T/\Delta R$ of the micro-heater filament was about 3.8 °C/Ω and the variation of temperature with input voltage was 18.3 °C/V.

ACKNOWLEDGEMENT

This work has been supported by Institute of Microengineering and Nanoelectronics (IMEN), Universiti Kebangsaan Malaysia (UKM).

REFERENCES

[1] M. Heule and L.J. Gauckler, Sens. Actuators B, 2002, in press.
[2] Jean Laconte, Cedric Dupont, Denis Flandre and Jean-Pierre Raskin, "SOI CMOS compatible low-power microheater optimization for the fabrication of smart gas sensor", IEEE Sensor Journal, Vol. 4, No. 5, (2004) pp.669-680.
[3] G.C. Cardinali, L.Dori, M. Fiorini, P. Maccagnani and V. Liberali, "An integrated microstructure with temperature control for gas sensors", Proc. 21st International Conference on Microelectronics, Vol. 2 (1997) pp.515-518.
[4] Jian Wei Gong, Quan Fang Chen, Ming Ren Lian, Nen Chin Liu and Claude Daoust, " Temperature feedback control for improving the stability of a semiconductor metal oxide (SMO) gas sensor", IEEE Sensors Journal, Vol. 6, No. 1, (2006) pp. 139-145.
[5] Danick Briand, "Thermally isolated microelectronic devices for gas sensing applications" unpublished.
[6] M. Aslam and J.V. Hatfield, "Fabrication of thin film microheater for gas sensors on polyimide membrane", 2003, pp. 389-392
[7] D. Briand, S. Colin, J. Courbat, S. Raible, J. Kappler and N. F.de Rooij, "Integration of MOX gas sensors on polyimide hotplates", Sensors and Actuators B 130 (2008), pp. 430-435.
[8] G. Sugandi and B. Y. Majlis, "Fabrication of MEMS based microspeaker using bulk micromachining technique", Advances Material Research, 254, (2011) pp. 11-174
[9] Mimiwaty Mohd Noor, Badariah Bais and Burhanuddin Yeop Majlis, "The effects of temperature and KOH concentration on silicon etching rate and membrane surface roughness", Proc. 2002 IEEE International Conference on Semiconductor Electronics, pp. 524-528.
[10] Yong Shin Kim, "Microheater-integrated single gas sensor array chip fabricated on flexible polyimide substrate", Sensors and Actuators B 114 (2006), pp.410-417.
[11] A. Scorzoni, M. Baroncini and P. Placidi, "On the relationship between the temperature coefficient of resistance and the thermal conductance of integrated metal resistors", Sensors and Actuators A 116 (2004), pp. 137-144.
[12] http://allaboutcircuits.com/vol_1/chpt_12/6.html.
[13] M. Baroncini, P. Placidi, A. Scorzoni, G.C. Cardinali, L. Dori and S. Nicoletti, "Characterization of an embedded micro-heater for gas sensors applications"

Modeling of high-intensity expert system of the catalytic oxidation reactor of phosphorous gases

Akzhigitova Meruyert, Eskendirov Sharipzhan, Umarova Zhanat
South-Kazakhstan State University (SKSU)
Tauke-khan Avenue 5, Shymkent, Kazakhstan
e-mail: akzhigitova80@mail.ru

Abstract- The article considers mathematical statement of the problem of creating chemical- technological device and chemical-technological system. The invention relates to the utilization of aspiration gases in the workshops for the production of phosphorus and phosphorus-containing substances.

Catalytic reactor of phosphorus-containing gases oxidation to utilize aspiration gases in the workshops for the production of phosphorus and phosphorus-containing substances has been developed.

The present invention - a catalytic reactor of oxidation of phosphorus containing gases is intended mainly for the utilization of phosphine in waste gas emissions in the workshops, tanks of milk, department of clay milk production of thermal treatment workshop of raw. Along the way, phosphoric anhydride and hydrogen fluoride are utilized too.

I.   INTRODUCTION

Intensification of production is before everything else provided by creating corresponding highly efficient technological equipment, beginning with individual   built-in process units and ending with chemical technological aggregates, complexes and systems

The model describes the utilization of aspiration gases in the workshops for the production of phosphorus and phosphorus-containing   substances.
Catalytic reactor, containing iron body of rectangular in shape of columnar type, has branch pipes for inlet and outlet of gas in four units with cartridges filled with catalyst balls and feeder of irrigating fluid passing through the unit, separated from one another by water seal in the form of bubble cap plate. The use of in parallel operating four units provides even wear of catalyst and removable cartridges with a catalyst provide their quick replacement   that   does   not   require   disassembly of the whole reactor   to   replace   the   catalyst.
Application of the proposed design of the reactor allows high treatment of waste gases from phosphorus, harmful gas components to human health, as well as deriving weak phosphoric acid. The reactor is simple to manufacture and isn't difficult to maintain.

Reactor is known for decontamination of organic impurities in the gas emissions of chemical, coke-chemical, metallurgic and other enterprises. The reactor is a tower-type structure, which includes mixing air heat, heat exchanger, built-in mixing air heater and gas burner, at the top catalyst case with platinum catalyst for deep thermo catalytic oxidation is placed. (A.S. 1060214 USSR. Reactor for catalytic processes / A.M . Sychev, V.S.Genkin, S.I.Melnikov and etc. / / Discoveries. Inventions.1983. № 46.p.18,19).

The disadvantage of this reactor is high energy consumption, complexity in manufacture, large hydraulic resistance and the difficulty   in   replacing   the   catalyst.
The objective of the present invention is to develop the design of the catalytic reactor for the oxidation of waste gases of phosphorous production with a high degree of ecological purification.

The technical result is effective treatment of aspiration gases of phosphorus production.

The set problem is solved by the fact that the catalytic reactor is built up as a column-type device, rectangular in shape, with four units, separated by a water seal and containing removable cartridge with the catalyst.

II.   STATEMENT OF THE PROBLEM

Mathematical statement of the problem of creating both individual chemical technological device (CTD) and chemical technological system (CTS) is in whole common to them and consists in stating a problem of multicriteria optimization with the given set of target functions $\overrightarrow{W}$, determining requirements for a designer to being created object, and a constraint vector of two types: restriction of equality type $\overrightarrow{F}(\overrightarrow{Z})=0$, corresponding to the full mathematical model of being constructed object, and restrictions of inequality type $\overrightarrow{Z}_{min} \leq \overrightarrow{Z} \leq \overrightarrow{Z}_{max}$, corresponding to the conditions of physical realizability of the object and engineering assignment for its construction:

$$\overrightarrow{W}(\overrightarrow{Z}) = \overrightarrow{W}(\overrightarrow{X}, \overrightarrow{Y}, \overrightarrow{T}, \overrightarrow{K}, \overrightarrow{H}, \overrightarrow{M}) \rightarrow extr, \quad (1)$$

$$\overrightarrow{F}(\overrightarrow{Z}) = \overrightarrow{F}(\overrightarrow{X}, \overrightarrow{Y}, \overrightarrow{T}, \overrightarrow{K}, \overrightarrow{H}, \overrightarrow{M}) = 0, \quad (2)$$

$$\overrightarrow{Z}_{min} \leq \overrightarrow{Z} \leq \overrightarrow{Z}_{max} \quad (3)$$

where $\overrightarrow{F} = \overrightarrow{F}(f_1, f_2, ...f_n)$ - vector-function of functional operator of the object, that is,   system of equations of its mathematical model; $\overrightarrow{Z}$ – vector of varying variable, restricted from above and below on the basis of conditions of physical, technological and constructible realizability of the technological system and engineering assignment; $\overrightarrow{T}$ – a vector of the operator of technological effect on the processed medium; $\overrightarrow{K}$ – a vector of constructible parameters of the object; $\overrightarrow{H}$ – a vector of restrictions for the construction; $\overrightarrow{M}$ – a vector of engineering assignment requirements; $\overrightarrow{X}$ – a vector of parameters of input flows, coming into technological device or system; $\overrightarrow{X}$ – a vector of parameters of output flows – derived products.

Mathematical model of the designed object $\overrightarrow{F}(\overrightarrow{Z}) = 0$ is the most important constituent of mathematical setting of the problem. It is essential that its structure for individual CTD will be one, and for all CTS – another one [2].

As a mathematical model of built-in-process module, heterogeneous-catalytic reactor of phosphine oxidation developed by us, we will mention generalized mathematical model, describing both stochastic and determinate properties of polydisperse physico-chemical systems [3]:

$$\frac{\partial p(x,y,t)}{\partial t} + \sum_{i=1}^{3} \frac{\partial}{\partial x_i}[V_i(x,t)p(x,y,t) + \frac{\partial}{\partial \tau}[\frac{\partial \tau}{\partial t}p(x,y,t)] + \frac{\partial}{\partial l}[\frac{\partial l}{\partial \tau}p(x,y,t)] +$$

$$+ \sum_{k=1}^{n} \frac{\partial}{\partial c_k}[I_k p(x,y,t)] + \frac{\partial}{\partial T}[p(x,y,t)\sum_{j=1}^{n}\frac{\Delta H_j}{C_p}I_j] + \frac{\partial}{\partial \rho}[\frac{\partial \rho}{\partial t}p(x,y,t)] +$$

$$+ \frac{\partial}{\partial \mu}[\frac{\partial \mu}{\partial t}p(x,y,t)] = q[p(x,y,t),t].$$

(4)

where: $\quad y = (\tau, l, \tau_1, \tau_1, \tau_2, ..., \tau_n, T, \rho, \mu) \quad ;$

$I_k = \dfrac{dC_k}{dt}$ – rate of chemical reaction in dispersed phase along k-key component; n – a number of key components, reacting in dispersed phase; ΔH_j – thermal effect of j-reaction; N – a number of reactions in dispersed phase; C_p – volumetric heat capacity of dispersed phase.

The expression $q[p(x,y,t),t]$ is defined by the mechanism of interactions of particles among themselves, as well as by the presence of external sources and runs-off, that is characterizes the rate, appearance and disappearance of new particles at the moment of the time t of particles with coordinates x, y. As internal coordinates such physico-chemical characteristics as particle residence time in the device τ, typical linear size of the particle l, concentration of the key component in the particle C_k, temperature T, density ρ, viscosity μ .have been accepted.

The waste gases of phosphorous production contain high-toxic compounds of sulfur, phosphorus and fluorine, such as hydrogen sulfide, phosphoric anhydride, phosphorous and hydrogen fluoride. Currently, for treatment of these gases wet cleaning methods using water, lime and soda solutions are employed. However, according to the stated method of treatment only the gases containing phosphoric anhydride and hydrogen fluoride are exposed, and the most toxic gas -hydrogen phosphide (phosphine) is emitted into the atmosphere. The present invention -a catalytic reactor of oxidation of phosphorus containing gases is intended mainly for the utilization of phosphine in waste gas emissions in the workshops, tanks of milk, department of clay milk production of thermal treatment workshop of raw. Along the way, phosphoric anhydride and hydrogen fluoride are utilized too.

In real aspiration gases of phosphorus production on the average: phosphine - 80mg/m3, hydrogen fluoride - 10mg/m3 and phosphoric anhydride - 120mg/m3 are contained. The proposed reactor purifies from phosphoric anhydride and hydrogen fluoride for 68-70%, from phosphine for 85%. The most effective ones are the nickel-carbon catalysts containing 10% NiO and palladium-carbon containing 0,5% PdO. The temperature in the catalyst zone at the temperature of reflux water of $15-25^0$C does not exceed $30-35^0$C. The gas velocity in the apparatus of 2-3 m / s provides the above-mentioned oxidation of phosphorus-containing components in the gas.

III. RESULTS

The structure of mathematical model for individual CTD: material-energetic flows come into CTD entry, which are presented by input vector $\vec{X} = (x_1, x_2, ... x_n)$ with the given

physico-chemical properties $(a_i^1, a_i^2, ... a_i^c), i = \overline{1,n}$.

CTD output is characterized by output flows vector

$\vec{Y} = (y_1, y_2, ... y_m)$ with required physico-chemical properties $(b_j^1, b_j^2, ... b_j^c), j = \overline{1,m}$. The vector $\vec{X}$ of input flows is subject to purposeful physico-chemical changes in the vector $\vec{Y}$, which are formalized as an operator of engineering effect $\vec{T}$. The latter one is convergency (superimposition) of the simplest operators of various physico-chemical natures (mechanical $\vec{T}_M$, heat $\vec{T}_T$, hydrodynamic $\vec{T}_\Gamma$, diffusion $\vec{T}_Д$, phase change $\vec{T}_\Phi$, chemical change $\vec{T}_X$ and etc.). Each of the constituents of operators $\vec{T}_i$ requires specific conditions of their realization. So, mechanical effect can be realized as vibration, crosspartition, mixing and etc.; phase change as evaporation, condensation, dissolution, crystallization and etc., heat one – as heating and cooling; chemical one – as a chemical reaction, flowing in homogeneous (liquid, gas) medium, heterogeneous medium (liquid–gas– solid), on the surface of the catalyzer and etc., diffusion one – as molecular or turbulent diffusion and etc.

The proposed design of catalytic oxidation reactor is shown in figure 1.

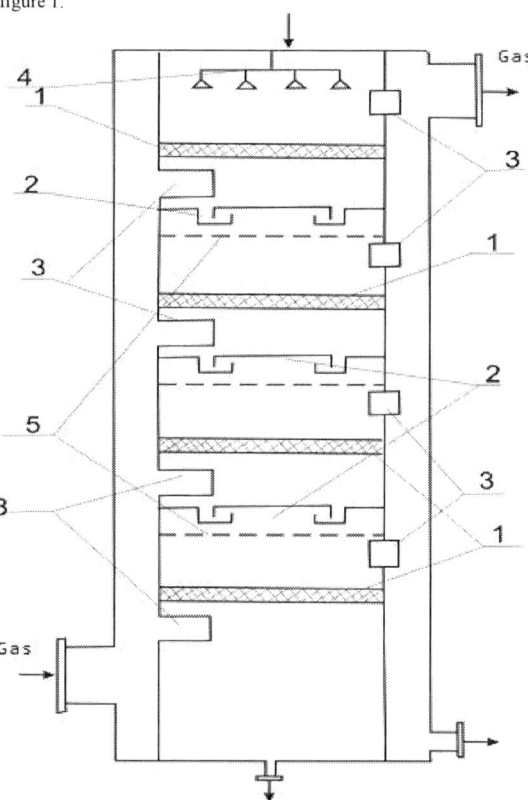

Fig.1 The proposed design of catalytic oxidation reactor

The catalytic reactor consists of a rectangular iron body of 1 column-type with four units with cartridges of palladium-carbon catalyst 2, 3, 4, 5 in the form of balls with a diameter of 5-10 mm., depending on the volume of gas supply; gas supply is conducted by gas ducts 6, 7, 8, 9, in each unit, the output of the reacted gas occurs through the ducts 10, 11, 12, 13; and each unit is separated from one another by water seals 14, 15, 16 in the form of bubble cap plates to prevent reverse gas flow; irrigating fluid supply is realized by the sprinklers 17 to the upper unit of the reactor; at the lower unit of the reactor fluid passing through the water seal is put into perforated distribution plates 18, 19, 20 for even irrigation of the lower units; in the lower part of the columns there are branch pipes for fluid withdrawal 21, 22; to the rear side of the column along the surface the heat exchanger 23 is adjacent to the refrigerating fluid.

The reactor of the proposed design works as follows: the incoming gas in a wide range of speeds from 0,5 to 5m/sec is distributed through four ducts 6, 7, 8, 9 and goes to the bottom of the contact device, of each unit with the catalyst cartridges 2, 3, 4 5, in which the catalyst balls are sprinkled with thickness of bulk bedfrom 5 to 10 cm, depending on the speed of the gas supply and the percentage of harmful phosphorus-containing components in the gas. After passing of the catalyst bed, phosphorus components are oxidized and absorbed by irrigated liquid, becoming weak phosphoric acid. Replacement of depleted catalyst is carried out by fluting of cartridges from the slots, the pouring of a new catalyst and sending depleted one to the regeneration. Irrigated liquid may be water or weak phosphoric acid [4].

The proposed design in the invention of the catalytic oxidation reactor of phosphorus containing gases has the following advantages of the existing analogs:
- removable catalyst cartridges, allowing easy replacement of the catalyst,
  and the possibility to adjust the thickness of the catalyst bed;
-low hydraulic resistance;
-low energy consumption;
-easiness in manufacture;
- high efficiency of utilization of phosphine gas.

The use of the very catalytic reactor allows effective treatment of emissions of aspiration gases from the plants and workshops producing phosphorus and phosphorus-containing substances with the aim of environmental cleanup of gas wastes into the environment. The most toxic and poorly utilizable gas is phosphine forming in all industries associated with the use of phosphorus.

IY. CONCLUSION

The catalytic reactor of oxidation of phosphorous containing gases containing a vertically placed iron body, with input and output branch pipes of gas, irrigation devices, different from that it has a case rectangular in shape, divided into four units with removable cartridges with a catalyst having water seal between has four removable cartridges with a catalyst in four units, separated by water seal cap, devices for even irrigation of catalyst units, feeders and output of gas and emptying the used fluid, which has a heat exchanger.

The waste gases of phosphorous production contain high-toxic compounds of sulfur, phosphorus and fluorine, such as hydrogen sulfide, phosphoric anhydride, phosphorous and hydrogen fluoride. Catalytic reactor containing iron body of rectangular in shape of columnar type has branch pipes for inlet and outlet of gas in four units with cartridges filled with balls of catalyst and feeder of Irrigating fluid passing through the unit, separated from one another by water seal in the form of bubble cap plate. The use of in parallel operating four units provides even wear of catalyst and removable cartridges with a catalyst provide their quick replacement that does not require disassembly of the whole reactor to replace the catalyst.

This reactor allows utilizing 85% of phosphine and 68-70% of phosphorus anhydride and hydrogen fluoride.

REFERENCES

[1] Zadorski V.M. Intensification of chemico-technological process on the basis of system approach. – Kiev: Techniques, 1989.-210 p.

[2] Kardashev G.A. Physical methods of intensification of chemical technology processes. – M.: Chemistry, 1990.-206 p.

[3] Yunusova D.U. Modeling gas-liquid catalytic processes.-Shymkent.-Pub.M.Auezov SKSU, 2007.-p.122, ISBN 9965-898-70-9

[4] Materialy N.K. Перспективы развития химической переработки горючих ископаемых. – Moscow: 2010.- p.56.

Multiplication Gain and Excess Noise Factor in 4H-SiC APD

C. C. Sun[a,*], A. H. You[b] and E. K. Wong[b]

[a]Centre for Diploma Program, Multimedia University, Jalan Ayer Keroh Lama, 75450 Melaka, Malaysia.
[b]Faculty of Engineering and Technology, Multimedia University, Jalan Ayer Keroh Lama, 75450 Melaka, Malaysia.
*Email: ccsun@mmu.edu.my

Abstract—4H-SiC is an attractive material for ultraviolet detection owing to its wide band gap and a matured material technology. This paper reports the mean multiplication gain and excess noise factor with electron- and hole-initiated multiplication of a thin 4H-SiC APDs. The impact ionization coefficients for electron (α) and hole (β) using Monte Carlo method over an electric field ranging from 2000 kV/cm up to 5000 kV/cm are simulated in our work. The results show that $\beta > \alpha$, and the ratio remains large even at very high electric field region. The electric field dependence of the impact ionization coefficients equations have been deduced from our model. Based on these equations, the avalanche breakdown voltage, multiplication gain and excess noise factor at different avalanche width have been investigated by considering the effect of dead space. As the width is increasing, the breakdown voltage and multiplication gain also increases proportionally. We observe a significantly higher multiplication gain for hole- than that of electron-initiated multiplication. The excess noise factor of electron-initiated multiplication is greater than that of hole-initiated due to higher number of feedback carriers. Thus, pure hole injection is necessary in order to ensure low excess noise in 4H-SiC APD owing to the large $\beta > \alpha$.

Index Terms— Avalanche photodiodes (APDs), impact ionization coefficients, multiplication gain, excess noise factor.

I. INTRODUCTION

Over the years, wide band gap semiconductor material such as silicon carbide (SiC) and III-nitrides has become the promising candidate for high power and high frequency applications. However, due to the lack of native III-nitrides substrates, SiC is technologically more mature and stems as the promising alternative materials, especially under the influence of high electric field operation. Among different polytype structures of SiC, 4H-polytype poses the largest energy bandgap, thus, has received extensive interest in high voltage device applications. With the ability of operating under elevated temperature with high thermal conductivity and high saturation velocity, 4H-SiC has been chosen for the fabrication of optical sensing devices in ultraviolet (UV) regime [1-3]. The widely differing ionization coefficients have proven to be an important parameters in realising high performance avalanche

photodiodes (APDs) [4,5]. A large ratio of hole to electron impact ionization coefficient ($k = \beta/\alpha$) is essential in producing high multiplication gain with low excess noise factor which arise during the avalanche process.

Significant efforts had been carried out by many researchers to investigate the impact ionization coefficients, multiplication gain and excess noise factor of 4H-SiC [1-5]. The first impact ionization coefficients in 4H-SiC were published by Konstantinov *et al.* [4] from DC mode photomultiplication measurement on three different APD structures. Guo *et al.* [6] had characterized the 4H-SiC UV APD by using separate absorption and multiplication (SAM) method. Their device structure is such that doping and thickness of multiplication layer is designed to limit the breakdown and reach through voltage. They reported a gain higher than 1000 without edge breakdown. Ng *et al.* [1] and Loh *et al.* [5] had investigated the multiplication gain and excess noise factor of 4H-SiC APD through photomultiplication measurement in thin and thick *i*-region respectively. For the earlier author, the effect of nonlocal impact ionization is taken into consideration since dead space effect is significant in thin devices. They concluded that apart from pure hole-initiated multiplication, the inclusion of dead space does play an important role in reducing the excess noise factor. On the other hand, the later reported the low field impact ionization coefficients and multiplication characteristics by interpolating a local model. Recently, Nguyen *et al.* [7] had determined the impact ionization coefficients of electrons and holes through optical beam induced current (OBIC) measurements. They also simulated the breakdown voltage by considering the multiplication coefficient in space charge region in their model. The breakdown voltage is simulated at 59.5 V for 0.104 µm avalanche width.

However, it can be noted that most of the experimental results [4-7] are mix carrier injection process, which in turn can caused a significant uncertainty towards the reported data. Furthermore, not much simulation work is performed thus far in 4H-SiC. In this work, a thorough study of impact ionization coefficients, multiplication gain and excess noise factor for

978-1-4673-2395-6/12 $31.00 © 2012 IEEE

both electron- and hole- initiated multiplication is simulated by using Monte Carlo (MC) method. The model is able to perform for either pure electron- or pure hole- initiated multiplication. This is an extension of the previous work which had been carried in the earlier stage [8,9].

II. IMPACT IONIZATION COEFFICIENTS

Impact ionization is a process that occurs between conduction and valence bands in generating new electron-hole pairs. During this event, an electron (hole) in the valence (conduction) which had gain sufficient energy from the electric field will be promoted to conduction (valence) band. This electron (hole) will collide with the lattice structure of conduction (valence) band and an electron-hole pair is created. Thus, a total of two electrons (holes) and one hole (electron) are generated. The process repeatedly occurs until the electron (hole) had successfully left the avalanche region.

The impact ionization coefficients for electron and hole in 4H-SiC had been investigated in our previous work [9]. These parameters are extracted from the number of times the electrons and holes scatter due to impact ionization, n and the distance (l_{ei} and l_{hi} for electrons and holes respectively) they had travelled. The impact ionization coefficients are computed by taking the reciprocal of the averaging distance given as:

$$\alpha = \left[\sum_{i=1}^{n} l_{ei} \middle/ n \right]^{-1} \text{cm}^{-1} \quad \text{and} \quad \beta = \left[\sum_{i=1}^{n} l_{hi} \middle/ n \right]^{-1} \text{cm}^{-1} \quad (1)$$

Fig. 1 shows the electron and hole impact ionization coefficients as a function of inverse electric field obtained from (1). The impact ionization coefficient for hole is

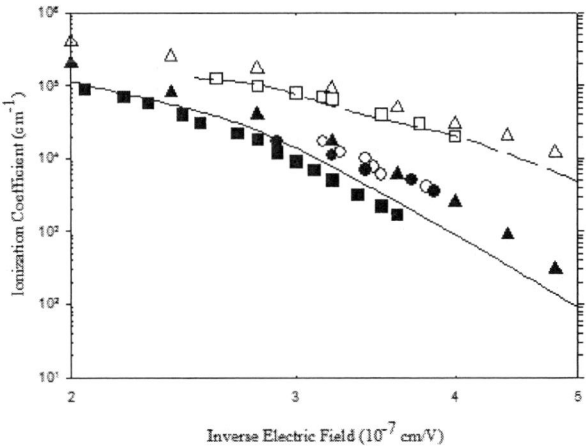

Fig.1. α (solid line) and β (dash line) of this work are compared with those by Raghunathan *et al.* [10] (α : ●, β : ○), Konstantinov *et al.* [4] (α : ■, β : □), and Loh *et al.* [5] (α : ▲, β : △).

significantly higher than that of electron. This shows that holes dominate the multiplication process.

Typical practices usually express impact ionization coefficients as a function of electric field for easier estimation of breakdown voltage and avalanche width at breakdown. With reference to the impact ionization curve plotted in Fig. 1, the electric field dependent impact ionization coefficient expressions are deduced. This is done based on the comparison of the general equations and parameterized value of previous researchers. The best fitted expression obtained in this work is given as (2) and (3), where E is the electric field.

$$\alpha = 1.803 \times 10^6 \exp\left[-\left(\frac{1.352 \times 10^7}{E} \right)^{1.20} \right] \text{cm}^{-1} \quad (2)$$

$$\beta = 1.861 \times 10^6 \exp\left[-\left(\frac{9.986 \times 10^6}{E} \right)^{1.11} \right] \text{cm}^{-1} \quad (3)$$

III. MULTIPLICATION GAIN AND EXCESS NOISE FACTOR

The multiplication in our model is initiated either with single electron or hole injection into the avalanche region from $x = 0$ to w. For electron-initiated multiplication, an electron which is injected at $x = 0$ will travel in the positive x-direction throughout the avalanche region. Conversely, hole-initiated multiplication begins with a hole being injected at $x = w$ which will travel in the negative x-direction in the avalanche region.

The multiplication gain and excess noise factor of 4H-SiC are simulated based on (2) and (3). The model considers the random ionization path length theory (l_e and l_h for electron and hole respectively) [11] with uniform electric field, written as

$$l_e = d_e - \ln(r)/\alpha \quad (4)$$

$$l_h = d_h - \ln(r)/\beta \quad (5)$$

where d_e and d_h are the dead space for electron and hole respectively. Multiplication gain is calculated by taking the averaging of number of trials (n) generated by a specific random number, r. The equations used for the calculation of multiplication gain $\langle M \rangle$ and excess noise factor, F are given as follow

$$\langle M \rangle = \sum_{i=1}^{n} M_i \middle/ n \quad (6)$$

$$F = \sum_{i=1}^{n} M_i^2 \middle/ n \langle M \rangle^2 \quad (7)$$

In this work, the breakdown voltage, multiplication gain and excess noise factor with different avalanche width are investigated. To validate our model, the multiplication gain and excess noise factor are simulated at the avalanche width reported by other researchers [1, 6, 12]. The electron- and hole-initiated multiplication gain versus breakdown voltage obtained

978-1-4673-2395-6/12 $31.00 © 2012 IEEE

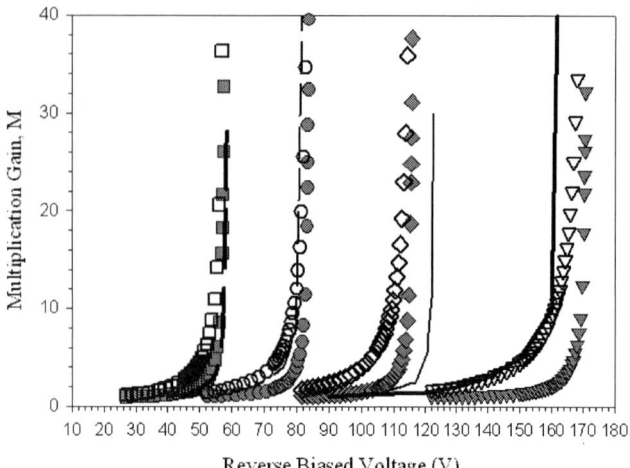

Fig. 2. Electron- (filled symbol) and hole- (open symbol) initiated multiplication gain as a function of reverse-biased voltage with w = 0.1 µm (square), 0.18 µm (circle), 0.285 µm (diamond), and 0.485 µm (down triangle), of this work are compared with Ng *et al.* [1] (w = 0.1 µm: short dash line and w = 0.285 µm: thin solid line), Guo *et al.* [6] (w = 0.18 µm: long dash line), and Zhou *et al.* [12] (w = 0.48 µm: thick solid line).

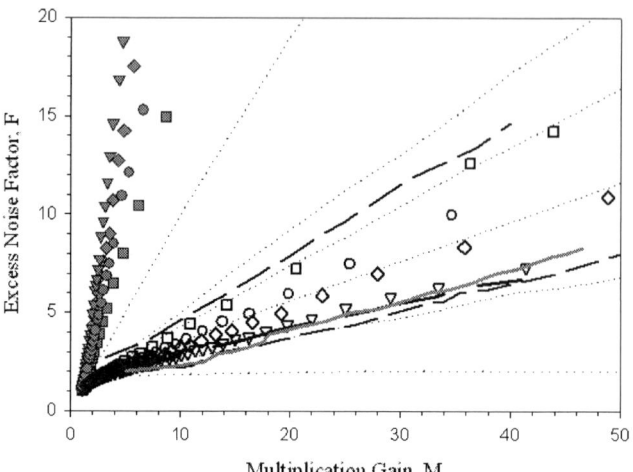

Fig. 3. Electron- (filled symbol) and hole- (open symbol) initiated multiplication gain as a function of reverse-biased voltage with w = 0.1 µm (square), 0.18 µm (circle), 0.285 µm (diamond), and 0.485 µm (down triangle), of this work are compared with Ng *et al.* [1] (w = 0.1 µm: short dash line and w = 0.285 µm: solid line), Guo *et al.* [6] (w = 0.18 µm: long dash line), and Zhou *et al.* [13] (w = 0.25 µm: dash-dot-dot line). Dotted line indicates the k value calculated from McIntyre equation.

from the model are as shown in Fig. 2. Our simulation results agree with the experimental results reported by other researchers [1,6,12]. It shows unambiguously that hole-initiated process gives a much higher multiplication gain than that of electron-initiated process at the same applied voltage. This is due to the higher hole impact ionization coefficient of 4H-SiC, as shown in Fig. 1. It can be noted that hole-initiated process has softer threshold than the electron-initiated process in earlier work [9]. When a higher voltage is applied, carriers gain more energy from the electric field, allowing more impact ionization to occur. With larger avalanche width, these carriers are able to undergo more impact ionization and it leads to higher multiplication gain and breakdown voltage. Thus, the device length is proportional to the applied voltage.

Fig. 3 shows the resultant excess noise factor for electron- and hole-initiated process obtained in this work. It is apparent that electrons produce much higher noise than holes during impact ionization. This occurs due to the energy band structure of 4H-SiC where conduction bands have the higher anisotropy than the valence bands [4]. Hence, electrons compensate higher energy loss due to tunnelling in order to reach the threshold energy, which in turn lead to the increasing excess noise. This phenomenon is more significant in thicker avalanche region. The thicker device results in higher noise being generated due to higher number of feedback holes in the avalanche region.

The results reported by Ng *et al.* [1] and Zhou *et al.* are included for reference. Ng *et al.* use HeCd laser light illuminated at different wavelength onto the device structure at

w = 0.1 µm and w = 0.285 µm respectively. They reported a very low excess noise with the k value close to 0.1. In this work, we obtained that hole-initiated process produce lower excess noise in thicker avalanche region, as opposed to electron-initiated multiplication described earlier. At w = 0.485 µm, electron initiated multiplication produce very high excess noise whereas the hole-initiated multiplication produce very low noise at k < 0.15. We justify our results as it is related to the energy band structure of 4H-SiC. The continuous energy spectrum of valence bands make holes to be impact ionised more easily, especially when a higher voltage is applied. This make lesser feedback of minority carriers (electrons), thus resulting in a reduction of excess noise factor [3].

IV. CONCLUSION

A MC model is applied to study the avalanche characteristics of 4H-SiC APD. The multiplication gain and excess noise factor under different avalanche width for electron- and hole-initiated multiplication are performed. It is observed that hole-initiated multiplication gives higher multiplication gain with low excess noise as compare to electron-initiated multiplication. This is the benefit of the large β/α ratio which is crucial in realizing a high performance APD. A high multiplication gain, large breakdown voltage and low excess noise in large avalanche width SiC-4H APDs are reported for hole-initiated multiplication.

ACKNOWLEDGEMENT

This work is supported by FRGS fund 2/2010, MoHE.

REFERENCES

[1] B. K.Ng, P. R. David, R. C. Tozer, G. J. Rees, F. Yan, J. H. Zhao, and M. Weiner, "Nonlocal effects in thin 4H-SiC UV avalanche photodiodes," IEEE Trans. Electron Devices, vol. 50, no. 18, pp. 1724-1732, Aug. 2003.

[2] X. Bai, D. Mcintosh, H. Liu, and J. C. Campbell, "Ultraviolet single photon detection with Geiger-mode 4H-SiC Avalanche photodiodes," IEEE Photon. Technol. Lett., vol. 19, no. 22, pp.1822-1824, Nov. 2007.

[3] J. P. R. David, and C. H. Tan, "Material considerations for avalanche photodiodes," IEEE J. of Selected Topics in Quantum Electron., vol. 14, no. 4, pp.998-1009, July/Aug. 2008.

[4] A. O. Konstantinov, Q. Wahab, N. Nordell, and U. Lindefelt, "Study of avalanche breakdown and impact ionization in 4H silicon carbide," J. Electron. Mater., vol. 27, no. 4, pp. 335-341, Apr. 1998.

[5] W. S. Loh, B. K. ng, J. S. ng, S. I. Soloviev, H. Y. Cha, P. M. Sandvik, C. M. Johnson, and J. P. R. David, "Impact ionization coefficients in 4H-SiC," IEEE Trans. Electron Devices, vol. 55, no. 8, pp. 1984-1990, Aug. 2008.

[6] X. Guo, L. B. Rowland, G. T. Dunne, J. A. Fronheiser, P. M. Sandvik, A. L. Beck, and J. C. Campbell, "Demonstration of ultraviolet separation absorption and multiplication 4H-SiC avalanche photodiodes," IEEE Photon. Technol. Let., vol. 18, no. 1, pp. 136-138, Jan. 2006.

[7] D. M. Nguyen, C. Raynaud, N. Dheilly, M. Lazar, D. Tournier, P. Brosselard, and D. Planson, "Experimental determination of impact ionization coefficients in 4H-SiC," Diamond & Rel. Mater., vol. 20, pp. 395-397, Jan. 2011.

[8] C.C. Sun, A. H. You, and E. K. Wong, "Monte Carlo Simulation of Electron Transport in 4H- and 6H-SiC," AIP Conf. Proc. 1250, pp. 281-284, 2010.

[9] C.C. Sun, A. H. You, and E. K. Wong, "Impact ionization coefficients of electrons and holes in 4H-SiC," AIP Conf. Proc., 1328, pp.277-280, 2011.

[10] R. Raghunathan, and B. J. baliga, "Temperature dependence of hole impact ionization coefficients in 4H and 6H-SiC," Solid State Electron., vol. 43, no. 2, pp.199-211, Feb. 1999.

[11] A. H. You and D. S. Ong, 'Avalanche multiplication and noise characteristics of thin InP p^+-i-n^+ diodes', Japan. J. Appl. Phys., vol. 43, no.11A, pp. 7399-7404, 2004.

[12] Q. Zhou, H. D. Liu, D. C. McIntosh, C. Hu, X. Zheng, and J. C. Campbell, "Proton-implantation-isolated 4H-SiC avalanche pgotodiodes," IEEE Photon. Technol. Lett., vol. 21, no. 23, pp.1734-1736, Dec. 2009.

[13] Q. Zhou, D. C. McIntosh, H. D. Liu, and J. C. Campbell, "Proton-implantation-isolated separate absorption charge and multiplication 4H-SiC avalanche photodiodes," IEEE Photon. Technol. Lett., vol. 23, no. 5, pp.299-301, March 2011.

978-1-4673-2395-6/12 $31.00 © 2012 IEEE

All Optical Switch Using Ultra compact Multi Mode Interference Coupler

Mehdi Tajaldini, Mohd Zubir Mat Jafri
School of Physics
University Sains Malaysia (USM)
11800 USM, Penang, Malaysia

Abstract- **This paper proposes an approach for accessing optical switching performance designed for signal processing purposes through one continuous short MMI coupler in a nonlinear regime. The Switching operations are studied in lengths of less than$100\mu m$. In solving the set of coupled mode nonlinear propagation equations for multimode waveguide, as well as the possibility of measured electric field at output waveguides on different input electric fields and MMI lengths, the central waveguide length is optimized to obtain a smaller length of possible switching. Therefore, two input electric fields are presented, where the MMI coupler operates each state of switching when launched with either one. Finally two states of switching performance are simulated, and switching states appear on each output. The results show very low cross-talk and insertion loss.**

Keyword: MMI, switch, multimode Interference, insertion loss, cross-talk

I. INTRODUCTION

Multimode interference couplers, as the most important types of MMI structures, are the key elements of photonic integrated circuits [1]. MMI couplers have broad applications in photonic complex circuits especially when operated in the nonlinear regime [2], such as, wavelength (de)multiplexers [3,4], optical power splitters/combiners [5,6], optical switches [7-11] and optical gates[12]. It can also be used as a part of complex devices which contribute to the processing in the integrated photonic circuits [13], such as micro ring resonator [14] and Mach Zender switch [15]. As mentioned above, one of the operations of MMI is optical switching, which is achieved through a continuous MMI and it is performed using several methods, such as thermal [10] and electro optic [8]. Recently scientist, have focused on the design of devices with all-optical performance because the offer mentioned approaches fail to meet the important aspects such as, short dimension, high speed and bandwidth for medium, high speed for switching operation and user ability to choose the number of input and output channels. Therefore, several all optical switches have been proposed to show high speed and bandwidth [7,9,10]. However, issues in size, which are necessary in all optical integrated circuits and telecommunications, remain non feasible. Most of the efforts are concentrated on the design of the all optical switch with the smallest dimension for micro integrated circuits. In addition the switch was designed to operate as a component of complex devices with a micro dimension. However, an MMI approximately a few millimeters in length is not suitable in photonic integrated design due to a very essential goal in photonic development: the production of shorter devices for compact circuits. With the development an all-optical ultra compact switch using an individual MMI, a large number of switches in a circuit can be used, even if this circuit consists of several devices, each with several MMI switches.

In this paper we address this problem by proposing the shortest MMI switch in the nonlinear regime. This regime would be helpful in obtaining the desirable result, as well as low insertion loss and cross-talk. We propose a method based on Mode Propagation Analysis method (MPA) that is able to introduce a procedure for achieving the M×N switching and switching for 2×2 MMI with various input intensities are demonstrated. In this work; switching is accomplished using phase modulation as well as self focusing in effect of Kerr nonlinear effect in a multimode interference region, which leads to a different self imaging process and guide modes interference acting in small lengths. We also present the Kerr nonlinear effect, which reduces the distance of the first self imaging in the multimode interference waveguide to reduce and which allows it to switch with a smaller length. In the second section, we theoretically investigate electromagnetic wave propagation through the Kerr nonlinear effect in a multimode interference waveguide, and measuring the electric field at the output facet. In the third section, we demonstrate the design and simulation result for wave propagation, lengths optimization, and switching operation. The results are discussed in the fourth section.

II. WAVE PROPAGATION IN MULTIMODE INTERFERENCE WAVEGUIDE IN NONLINEAR REGIME

The MMI coupler is introduced to the photonic device as the simplest structure. Although this device has very broad applications in the integrated photonic circuits and telecommunications, these applications increase with the appearance of nonlinear effects due to the change in the modes of electric field in terms of amplitudes or phases. This application exchanges energy between modes. This advantage leads to an ability to control the wave propagation in the medium, contributing to signal processing in all-optical functions.

The central region of MMI coupler is the multimode waveguide. The access waveguides which are usually single mode are fixed at the input and output facets of the multimode waveguide.

Figure1 shows the schematic structure of the (2×2) MMI coupler, where L_{MMI} and W_{MMI} are the length and the effective width of MMI region respectively. n_{MMI} and n_c represent the effective refractive index of the MMI and cladding regions, respectively.

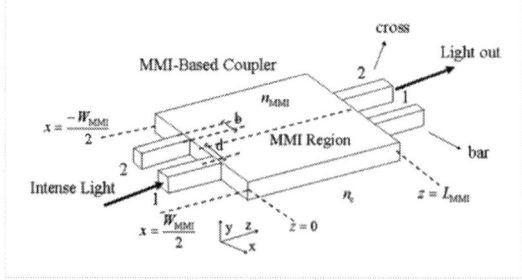

Fig. 1. Schematic structure of (2×2) MMI coupler.

The performance of these devices depends on the interference of guided modes, where the complete constructive interference contributes to the formation of the single or multiple self-images at precise distances in the input facet. The interference property of the MMI waveguide depends on the refractive indices of the core and cladding regions of the Multimode waveguide. In other words, by varying the refractive index in the core region, changing the interference type is possible. In fact, by imposing the intense light into the multimode region, the optical features of this region can be varied. In this case, the response of the region is changed from the linear to the nonlinear regime. The Kerr nonlinear effect plays an important role in the propagating of light in MMI devices. By studying this effect in the multimode interference couplers and applying the obtained results, we can design all-optical switch to have very small MMIs. In this section, we theoretically study the Kerr nonlinear effect in a small MMI coupler by discussing the nonlinear wave equation in the central region. Then, we discuss the electric field in the output single mode waveguides.

When an intense input light launches into the MMI region from the input waveguide, the refractive index of the region changes by an amount that is proportional to the intensity of the input light. In fact, varying the intensity of the input light produces a nonlinear change in the refractive index of the MMI region. The change of the refractive index leads to a change in interference of the modes and in the self imaging. In other words, light propagation in MMI region is changed. Notably, the MMI region is isotropic, and the field in this region is a superposition of all the modal fields of the MMI. In linear regime, the amplitude of the modal fields is constant; however in the nonlinear regime, the amplitude of the modes changes with the direction of propagation and exchange energy with each other (in the z direction). The field distribution of the light in MMI region is expressed by

$$E(x,z,t) = \sum_{\nu=0}^{n} A_\nu(z) e^{j\gamma_\nu x} e^{j\beta_\nu z} e^{(j\omega t + \phi_0)} . \quad (1)$$

Where ν is the mode number, $A_\nu(z)$ is the amplitude of the νth mode, γ_ν and β_ν are lateral and longitudinal propagation constants of the νth mode, respectively.

With enhanced Kerr nonlinear effect in the central region of MMI coupler, the set of mode propagation equations in nonlinear regime is expressed by [16]

$$2j\gamma_\nu \frac{dA_\nu(z)}{dz} = -\frac{3\chi_e^{(3)}\omega^2}{c^2} \sum_p \sum_q C_{\nu(p,q)} A_p A_q A_{p+q-\nu}^* . \quad (2)$$

By solving the set of ν coupled equations of the field amplitudes ($\nu = 0, \pm1, \pm2, \pm3, ...$), the field amplitude $A_\nu(z)$ of the modes is obtained. Consequently, by using the (1) the field in MMI region will be obtained.

The field profile of the guided modes, in general, consists of a superposition of sine and cosine light waves, with zero fields at the boundaries of the guiding region. In addition we assume that there is negligible penetration of the fields in the cladding layer. Under these assumptions, Ref.[1]the lateral propagation constant $\gamma_{x,\nu}$ can be expressed as

$$\gamma_{x,\nu} = \frac{(\nu+1)\pi}{W_{MMI}} . \quad (3)$$

Setting Eq. 11 into dispersion equation yields the longitudinal propagation constant β_ν as [1]

$$\beta_\nu = \left[k_0^2 n_{MMI}^2 - \left(\frac{(\nu+1)\pi}{W_{MMI}} \right)^2 \right]^{1/2} \quad (4)$$

The number of guided modes is determined by using the total internal reflection condition [17].

$$\nu \le \frac{2n_{MMI} W_{MMI}}{\lambda_0} \left(1 - \frac{n_c^2}{n_{MMI}^2} \right)^{1/2} - 1 \quad (5)$$

When the light profile $E(x,z)$ enters the MMI waveguide through the input waveguides (at z=0), it is decomposed into the modal fields of the MMI waveguide. Therefore, various modal combinations are excited, as follows:

$$E(x,0) = \sum_\nu A_\nu \varphi_\nu(x,0) . \quad (6)$$

The input light field has the profile of the input waveguide mode. The field profile of the input and output waveguides (which are single-mode) are determined by using the EIM method [18] as

$$E_{in}(x,z,t) = E_{0,in} \cos[k_x(x - (-1)^{i+1}d) + \phi_0] e^{j(\beta_{in}z)} \quad (6)$$

$$E_{out}(x,z,t) = E_{0r} \cos[k_x(x - (-1)^{m+1}d) + \phi_0] e^{j(\beta_{in}z)} \quad (7)$$

Where $i = 1,2$ and $m = 1,2$ are the number of input and output waveguides, respectively. The excitation coefficients of the modes are determined by using the overlap integral between the input field profile and the profile of the νth mode at z=0, As

$$A_\nu = \frac{\int e^{j\gamma_\nu x} E^*_i(x,0) dx}{\left[\int |e^{j\gamma_\nu x}|^2 dx \int |E_i(x,0)|^2 dx \right]^{1/2}} \quad (8)$$

The field amplitude in output port is obtained by evaluating the overlap integral between the profile of an output waveguide and the profile of the excited modes of the MMI region at z=L [16].

$$E_{0r,\mathbf{m},\nu} = \frac{\int A_\nu(L) e^{j\gamma_\nu x} \cos[k_x(x - (-1)^{m+1}d) + \phi_0]}{\left[\int |e^{j\gamma_\nu x}|^2 dx \int |\cos[k_x(x - (-1)^{m+1}d) + \phi_0]|^2 dx \right]^{1/2}} \quad (9)$$

III. SIMULATION RESULT

The most important purpose is to propose the shortest switch, thus the MMI coupler is studied to find the shortest length. As such, in a situation where the variation of the switching state is required in the input intensity variation, at least, two input intensities for switching performance are considered. In this section, the two states of switching are simulated, and the procedure for accessing the switching length and input electric field amplitudes are discussed (all results and solving equation have been done with Mathematica software). Our considerations are limited to the MMI coupler with the following structure, $n_c = 1$, $n_{MMI} = 1.5$, $W_{MMI} = 10\mu m$, $d = 4.5\mu m$, $b = 1\mu m$, $\chi_e^{(3)} = 2.8 \times 10^{-18} m^2/w$ at $\lambda_0 = 1.55\mu m$. Fig. 2 shows the calculated intensity profiles along the length of the MMI for the nonlinear regime while input intensity is $56 w/\mu m^2$. For every distance, the light and dark positions are related to the constructive and destructive interference of the guided modes of the region, respectively. The figure clearly shows that the light propagation in the nonlinear case (with high input optical intensity) is very different from that in the linear case (with low input intensity). The first self imaging at linear region with above characteristics is formed as mirror image at $387.097\mu m$ and the direct image at $774.199\mu m$ [1]. Fig. 2 shows that the first direct self image is formed at $180\mu m$. Two smaller images, similar to the self image can be seen at $90\mu m$ and $150\mu m$. These differences with linear region are the results of self focusing and the phase modulation effect. We investigate small-length MMI operation using Kerr nonlinear effect because it decreases self-imaging length, whereas the bases of the MMI coupler are its self-imaging properties. We found that, by using the Kerr nonlinear effect in the MMI couplers, we can control the self-imaging along the length of the coupler. In the nonlinear regime, the position of the constructive interference changes longitudinally and transversely through the variation in the input intensity. This characteristic enables us to model an all-optical switch for the MMI coupler.

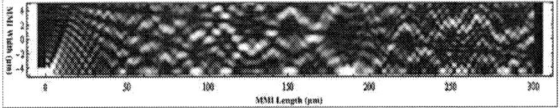

Fig. 2. Intensity distribution of the light along the length of the MMI in nonlinear regime, $E_i = 1.87 \times 10^8 N/C$, $L_{MMI} = 300\mu m$.

Clearly, switching at all lengths would be impossible because switching can only be performed at specific lengths, as it is discussed below.

The electric field of the bar and cross waveguides was calculated in a long interval of length, with approximately six input field amplitudes. Then, the bar and cross field amplitudes were compared. Figures 3 and 4 show the results.

In Fig. 3 and 4, the Normalized electric field at the bar and cross waveguide is shown in terms of length variation in the individual magnitude of the electric input field amplitudes, which are equal to $5.101 \times 10^8 \, N/C$ and $6.006 \times 10^8 \, N/C$, respectively. The blue line shows the electric field at primordium of the bar waveguide, whereas the red line shows this at primordium of the cross waveguide. In comparison to these curves we observe that, in particular lengths; one curve falls whereas another arises. For instance, in fig. 3, for lengths 60, 74 and 98μm, the red curve, which shows the value of the electric field amplitudes in the cross waveguide, moves nearer the X axis, whereas the blue curve related to the bar waveguide shows larger values at above lengths. In addition, the opposite situation is observed at the lengths of 38, 58, 80 and 86μm, whereas, the similar situation is observed in Fig. 4 but in different length.

In other word, one state of the switching has been occurring at several lengths, as shown in Fig. 3 and 4.

In fact, in order to access to desirable switching length, the appropriate input fields must be found so that they inversely convert the switching state at any of the abovementioned length. For instance, at length of 30μm with $E_i = 5.101 \times 10^8 \, N/C$ field amplitude, the entire output intensity switches to the bar waveguide; if another input field is found which can switch the intensity to the cross waveguide, thus switch can be introduced in terms of two insertion input field which make each state of switching.

In fact, we tested the switching operation at any length which are the switching candidates; the output electric field at the primordium of the bar and cross waveguides were calculated in terms of input electric field variation in the interval which consist of 10^5 measured point, which begins with a threshold field of nonlinear effect in this medium. We could easily confirm that not all lengths induce switching except length of 80μm which meets two state of switching on two diffrent input electric field and the best input fields are $5.101 \times 10^8, 6.006 \times 10^8 \, N/C$ for switching the wave to the cross and bar waveguides, respectively, as it is demonstrated in Fig. 3 and 4.

The first state switches the light to the cross output. The second performs in the opposite in terms of output choice. The first state is applicable when the field amplitude of insertion light is $5.101 \times 10^8 \, N/C$; for the second state, it is $6.006 \times 10^8 \, N/C$.

Two states of switching are simulated, in which the length of central region of switch is 80μm, as shown in Figs. 7 and 8. In Fig. 7, the electric field throughout the MMI coupler is demonstrated in $E_i = 5.101 \times 10^8 \, N/C$. As observed, the electric field in the bar waveguide has a rather high peak, whereas no peak exists at the cross waveguide. This situation shows a state of MMI switch, while the insertion loss and cross-talk are 0.12dB, -28dB respectively.

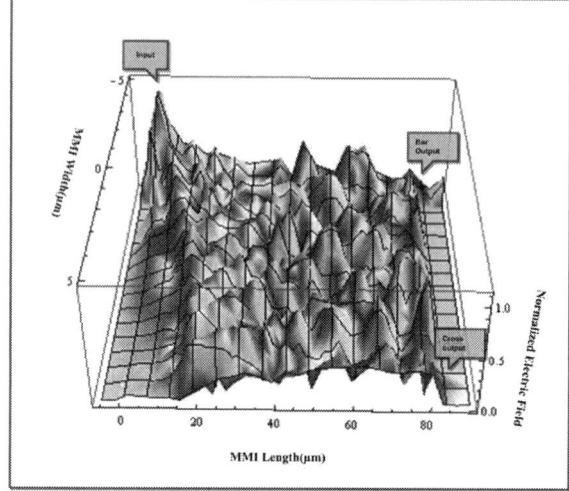

Fig. 7. 3D-Plot of electric field is an MMI coupler in situation of switching the wave to the bar waveguide, $E_i = 5.101 \times 10^8 \, N/C$.

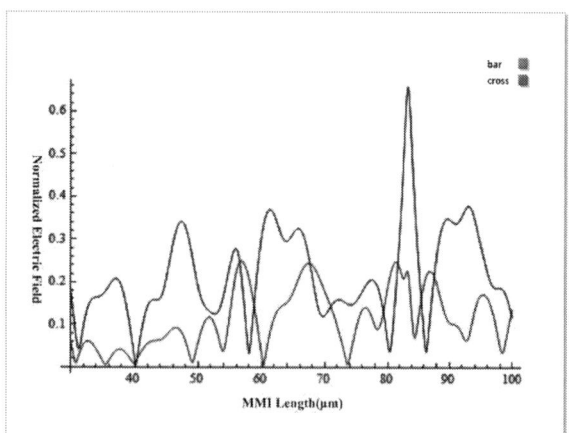

Fig. 3. Normalized electric fields at outputs (bar and cross) in terms of variation of length: $E_i = 5.101 \times 10^8 \, N/C$.

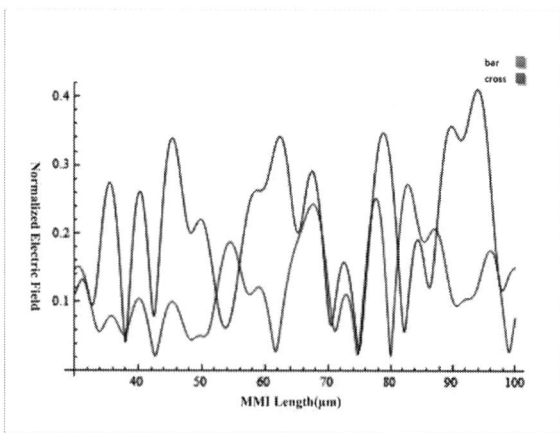

Fig. 4. Fig. 6. Normalized electric fields at outputs (bar and cross) in terms of variation of length, $E_i = 6.006 \times 10^8 \, N/C$.

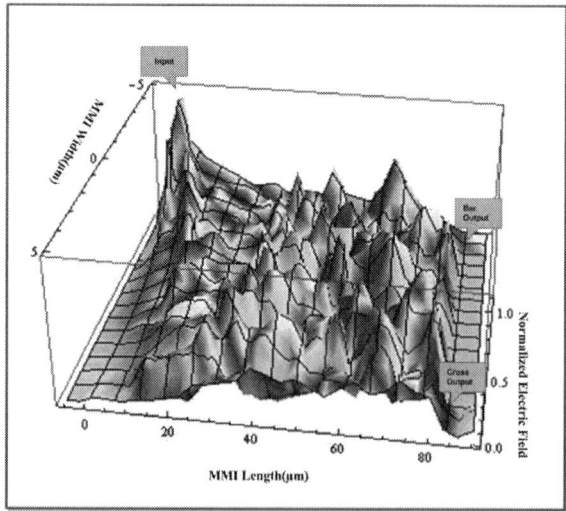

Fig. 8. 3D-Plot of electric field is an MMI coupler in situation of switching the wave to the cross waveguide, $E_i = 6.006 \times 10^8 \, N/C$.

The other state is demonstrated in Fig. 8. This figure demonstrates electric field in the MMI switch, whereas the input electric field amplitude is, $E_i = 6.006 \times 10^8 \, N/C$. As shown in the figure, there is a high peak at the cross waveguide against the field amplitude in bar waveguide. This shows the switching to the bar waveguide, whereas the insertion loss and cross-talk are 0.13dB, -31dB, respectively

In these two simulated figures, the wave inserts are from left input. However, if the wave is inserted from right input, no change occurs in results due to the symmetry between inputs and also output waveguides.

Our procedure for designing the all-optical switch is also very useful for designing an M×N switch. Thus, the above explorations are conducted for $M/2$ times, Which corresponds to our work for 2× 2 switch. In addition, N curves are required, instead of the two curves in any figure for each output. Thus a unique length and $M/2 \times N$ desirable input amplitude field are found and each amplitude creates two states of switching due to the symmetry.

IV. CONCLUSIONS

In this paper, a model of all optical switch with a small dimension MMI coupler was presented. A simple approach to access to the switching at desirable intervals of length was introduced, as well as the response for a $M \times N$ switch for $M \times N$ states. This switching occurs due to the Kerr nonlinear effect because this effect is able to exchange the constructive or destructive interference with each other. Thus, our approach, does not focus on the self-imaging properties of the MMI; rather, switching performance is based on the length, which should be smaller than the first self-image's length. In fact, variations on the input electric field cause the multimode waveguide to change the configuration of the propagation due to the Kerr nonlinear effect. In finding the desirable input field for any stage of switching, we were able to simulate the switch for the operation for bar or cross waveguides, which were unified from the last section. The insertion loss and cross-talk were very low due to the small dimensions of the coupler. In addition, the recorded bandwidth and speed were very high.

For future works, we intend to investigate the enhancement of the two photon absorption effect in order to conduct the wave in a specific way in semiconductor medium.

REFERENCES

[1] L. Soldano and E. Pennings, "Optical multi-mode interference devices based on self-imaging: Principles and applications," J. Light. Technol., Vol. 13, pp.615–627, April. 1995.

[2] M. Magana, D. Modato, R. Morandoti, S. Linden, J. Aitchison, "Kerr Nonlinear Effect in ALGaAs Multimode Waveguides", Applied Phisics Letters, Vol. 85, No. 16, 2004, pp.3390-3392.

[3] Z. Zhu,W. Ye, J. Ji, X. Yua and C. Zen, "Wavelength Demultiplexer Based on Multimode Interference Effect in Photonic crystals", Physics LettersA , VOl. 372, 2008, pp.2534-2538.

[4] X. Yueyu and H. Sailing, "A new Design Approach to MMI-based (de)Multiplexers" ,Optic Communications, Vol. 239, 2004, pp.85-90.

[5] A. Ferreras, F. Roriguez, L. Miguel and F. Hernandez, "Usful Formulas for Multimode Interference Power Splitter/Combiner Design", IEEE, Photonics Technology Letters, Vol. 5, No. 1, 1993.

[6]M. Swillam, M. Bakr and X. Li, "Efficient Design of Integrated Wideband Polarization splitter/Combiner" ,Journal of Light Wave Technology, Vol. 28, No. 8, 2010, pp.1176-1183.

[7] A. Bahrami and A. Rostami, "A proposal for1×8 All Optical Switch Using Multimode Interference", Optica aplicata, Vol. XLI, NO. 1, 2011.

[8] G. Singh, R. Yadav and V. Janyani, "Multimode Interference(MMI) Coupler Based All Optical Switch: Design, Applications and Performance Analyis", International journal of Recent Trends in Engineering, Vol. 1, NO. 3, 2009.

[9] D. Arrioja, N. Bickel, R. Selvas and P. Likamawa, "MMI-based 2×2 photonic switch", Proc. of SPIE, Vol. 6013, 2005.

[10] F. Wang, J. Yang, L. Chen, X. Jiang and M. Wang, "Optical Switch Based on Multimode Interference Coupler", IEEE Photonics Technology Letters, Vol. 18, No. 2, 2006.

[11] A. Bahrami, S. Mohammadnejad and A. Rostami, "All-Optical Multimode Interference Switch Using Nonlinear Directional Coupler as a Passive Phase Shifter" , Fiber and Integrated Optic, pp.139-150, 2011.

[12] L. Thanh, "ALL-Optical NAND and AND Gates Based on 3×3 General Interference Multimode Interference Couplers", VNU Journal of Science-physics , Vol. 26, 2010, pp.107-113.

[13] L. Xu, X. Leijtens, B. Docter, T. Vries, E. Smalbrugge, F. Karouta and M. Smit, "MMi-RerlectorAnavel on-Chip Reflector for Photonic Integrated Circuits", ECOC, 2009, pp.20-24.

[14] D. Xu, A. Densmore, P. Waldron, J. Lapointe, E. Post, A. Delags, J. Seigfried, P. Cheben, J. Schmid and B. Lamontagne, "High Bandwidth SOI Photonic Wire Ring Resonator Using MMI Coupler", Optic Express, Vol. 15, No. 6, 2007.

[15] J. Kim, H. Kim, E. Sim, K. Kim, O. Kwen and K. Oh, "Filter-free Wavelength Conversion Using Mach-Zender Interferometer with Integrated Multimode Interference Semiconductor Optical Amplifirs", ETRI Journal, Vol.26, No. 4, 2004

[16] M. Tajaddini and A. Mahmoodi, "Comparison of Linear and Nonlinear Regimes in MMI Couplers", SASTECH, Iran, 1312-419, May. 2011.

[17] G. Euliss, "Temporal Characteristics and Scaling Conditions of Multimode Interference Coupler", Journal of Light wave Technology, Vol. 1, No. 7, 1999, pp. 1206-1210.

[18] K. Izuka, "Elements of Photonics, Volume II: For Fiber and Integrated optics". New York, Wiley, 2002.

Theoretical Triangular Quantum Well Model for AlGaN/GaN HEMT Structure Used as Polar Liquid Sensor

Sulaiman Rabbaa and Johan Stiens

Laboratory of Micro- and Photon Electronics (LAMI), Dept. of Electronics and Informatics, Vrije Universiteit Brussel (VUB),
Pleinlaan 2, 1050-Brussels, Belgium
e-mail: srabbaa @ etro.vub.ac.be

Abstract— **A triangular quantum well model is introduced to investigate a doped AlGaN/GaN high electron mobility transistor (HEMT) structure as a sensor for polar liquids. We calculate the drain current of the transistor as a function of the dipole moment of the polar liquid. The results show good agreement with experimental measurements for different polar liquids. It is also found that the device has large linear sensitivity by detecting the change of drain current with dipole moment and therefore it can distinguish molecules with slightly different dipole moments. The device can be extended to sense biomolecules (such as proteins) with very large dipole moments.**

Keywords- HEMTs;triangular quantum well; AlGaN/GaN heterostructure; chemical sensors; polar liquids.

I. INTRODUCTION

A considerable progress had been made in the technology of gallium nitride (GaN) based HEMTs during the first decade of the 21th century. GaN has exceptional chemical and physical stability and so it is ideal for immobilization of bio-molecules. It features spontaneous (P_{sp}) and piezoelectric (P_{pz}) polarization with a non-toxic surface content compared to arsenide (As) [1, 2]. Materials like GaN with a large band gap and with previously distinguished features are alternative options for biosensors, particularly at high temperatures and in harsh environmental conditions which are not suitable for silicon-based sensors [3, 4]. For example, GaN-based sensors can be operated at elevated temperature of 600 ºC, whereas Si-based devices are limited to temperatures below 350 ºC. This is due to the large band gap and the low intrinsic carrier concentration in GaN, especially at high temperatures [5]. The response time of sensing using GaN is about several hundred milliseconds compared with several tens of minutes in GaAs-based sensors. This response delay in GaAs HEMTs can be ascribed to the coated layer necessary for the device fabrication, whereas GaN-based sensors do not require surface passivation and so it has an immediate and effective response to molecules with different dipole moments [6].

The global sensor market grows to cover many needs and purposes. Chemical sensors include detection of gases, pH and polar liquids. Examples of biosensors applications are the detection of kidney injury molecules, glucose detection in exhaled breath condensate and the detection of biomolecules such as DNA [7].

In the present contribution we apply a theoretical model to calculate the two-dimensional electron gas (2DEG) induced in AlGaN/GaN structure before and after exposing the device surface to polar liquids. We, then, determine the drain current for each liquid at different values of drain and gate voltages. Results show that the current can be quantitatively modulated by the magnitude of liquid dipole moment. We discuss how a HEMT will be effective for sensing molecules with very large dipole moments. We compare our results with experimental measurements and we find good agreement. In the present work we don't consider the effect of changing the pH of the liquid on the drain current. This will be studied later taking into account the change of dipole moment with pH.

II. DEVICE STRUCTURE

A cross section of a HEMT is shown in Fig. 1. The

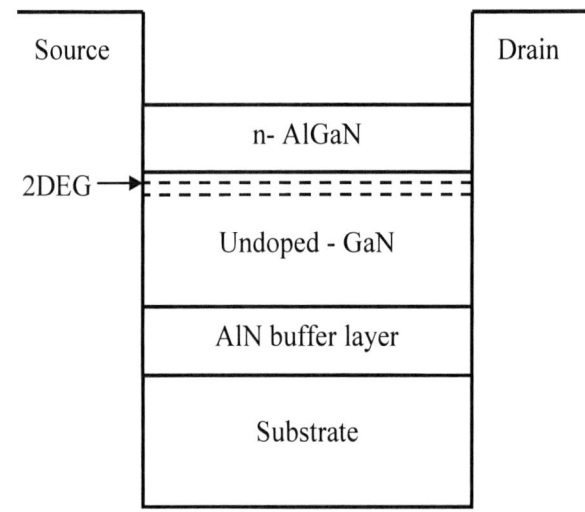

Fig. 1. Schematic diagram of AlGaN/GaN HEMT structure

epitaxial layers are usually grown by various techniques such as molecular beam epitaxy (MBE), Metalorganic vapor phase epitaxy (MOVPE) and metalorganic chemical vapor deposition (MOCVD) on a substrate, typically sapphire or SiC. The thermal expansion of sapphire is close to aluminum nitride ceramics used in high temperature sensors fabrication. The growth sequence is designed as follows: aluminum nitride

(AlN) buffer layer followed by undoped GaN of thickness 0.6-3 μm (typically 1 μm) and doped aluminum gallium nitride (AlGaN) barrier layer. The buffer layer improves the structural quality of the undoped-GaN layer and it produces uniform Ga-face polarity of the III-nitride epilayer across the entire substrate [2, 4].

The metallic gate of the normal transistor is removed. The transistor surface is subjected to a surface treatment to become suitable for sensing purposes, For example, wet chemical and plasma etching [8]. The surface is encapsulated by a passivation material such as silicon nitride or thermal oxide with only the gate region opened to allow the polar liquids to reach across the surface [9]. The surface should be modified to achieve hydrophilic and hydrophobic behavior, which becomes important for the measurements in droplets [10].

III. THEORETICAL MODEL

The model we will use to determine the characteristics of AlGaN/GaN HEMT sensors for polar liquids is the triangular quantum well model for the transistor channel as explained in [2]. This triangular potential well is one of the most common geometries that are considered to take into account quantum effects in the HEMT. It is properly the most investigated in all heterostructures. The model was used in many researches and it exhibited good results compared with experiments such as in [2], and to the authors knowledge the first time in combination with liquid sensing.

As shown in Fig. 1, a 2DEG channel with high electron density is formed at the AlGaN/GaN heterointerface due to the induced polarization in the heterostructure. A spontaneous polarization results from the cation and anion positions in the lattice, and a piezoelectric polarization results from the lattice mismatch between GaN and AlGaN layers. The large piezoelectric effect makes nitride HEMTs excellent candidates for piezoelectric-related applications. The sheet charge density N_s of the 2DEG as a function of the position x in the channel can be obtained by solving the one dimensional Poisson's equation and is given as [2, 11]:

$$N_s(x) = \frac{\varepsilon_0 \varepsilon(m)}{q(d + \Delta d)}\left(V_g - V_{th} - V_c(x) - \frac{E_F}{q}\right), \quad (1)$$

where $\varepsilon(m)= 9.5 - 0.5\,m$ is the dielectric constant of $Al_mGa_{1-m}N$, ε_0 is the permittivity of free space, m is the Al-mole fraction, q is the electron charge, d is the AlGaN barrier thickness, Δd is the 2DEG average position from the hetero-interface, V_g is the gate voltage, $V_c(x)$ is the channel potential at a certain distance x from the source due to the drain-source potential, E_F is the Fermi energy, and V_{th} is the threshold voltage given by [2]:

$$V_{th}(m) = \phi_b - \frac{\Delta E_c}{q} - \frac{qN_d d^2}{2\varepsilon(m)} - \frac{|\sigma(m)|}{\varepsilon(m)}(d + \Delta d), \quad (2)$$

where σ(m) is the total polarization (= $P_{sp}+P_{pz}$), $\phi_b(m)= 1.3+ 0.84\,m$ is the Schottky barrier, $\Delta E_c= 0.7[E_g(m) - E_g(0)]$ is the conduction band discontinuity between AlGaN and GaN, with $E_g(m)=6.13m+3.42(1-m) -m(1-m)$ is the band gap of AlGaN, and N_d is the doping concentration of AlGaN layer. This equation is valid also for undoped HEMT: we just substitute zero for the term N_d.

When the HEMT surface is exposed to a polar liquid, the adsorption of polar molecules causes a potential drop ΔV at the AlGaN surface / liquid interface which can be determined by the following Helmholtz model [3, 4]:

$$\Delta V = \frac{n_s p(\cos\theta)}{\varepsilon\varepsilon_0}, \quad (3)$$

where n_s is the dipole density per unit area, p is the liquid dipole moment in Debye (D), θ is the angle between the dipole and the surface normal and ε is the dielectric constant of the liquid.

To determine the channel current I_{ds} between drain and source, we use the analytical model given in [11]. It can be expressed as:

$$I_{ds} = Zqv_d(x)N_s(x), \quad (4)$$

where Z is the gate width, $v_d(x)$ is the position dependence electron drift velocity.

I. RESULTS AND DISCUSSIONS

Exposing the transistor surface to a polar liquid leads to a change in the surface charge concentration, which produces a reduction in the surface potential at the semiconductor/liquid interface as given in (3). As a result, the sheet charge density in the channel decreases, and the drain-source current is modulated as shown in (4). The sensing mechanism for molecules dipole moment is represented by the partial compensation of the 2DEG with the charge on the device surface which is produced as a result of interaction between the surface and the liquid dipoles [12].

To compare our results with experiments given in [6], we consider the same device characteristics: aluminum composition of the doped AlGaN layer m=0.23, AlGaN thickness d=22 nm, the gate length L=10 μm and the gate width is Z=500 μm.

Fig. 2 shows the drain-source current as a function of p/ε for liquids at different values of drain-source voltages.

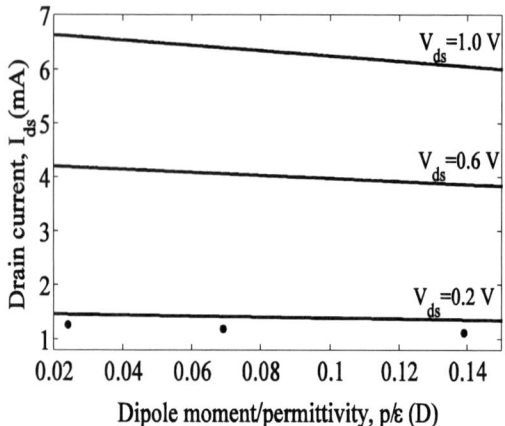

Fig. 2 Drain current as a function of p/ε at different drain-source voltages and $V_g = -0.5$ V (filled circle points represent experimental results)

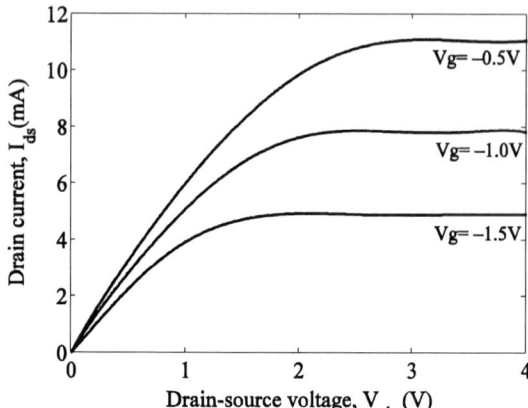

Fig. 4 Current-voltage characteristics at different gate voltages

Liquids with a high value of p/ε produce a large surface potential drop according to (3). This drop causes reduction in the sheet charge density and then the current decreases as expected by (4). The sensitivity of the device can be represented by the magnitude of change in current with respect to change in the ratio p/ε (slopes of lines). We get a sensitivity of 0.90, 2.83 and 4.90 mA/Debye at drain voltages of 0.2, 0.6 and 1.0 V, respectively. The points in Fig. 2 near the lower line represent the experimental measurements of current for acetone (p/ε=0.139 Debye), ethanol (p/ε=0.069 Debye) and water (p/ε=0.024 Debye) at V_{ds}=0.2 V and V_{gs}= − 0.5 V as done in [6]. We notice a good agreement between our calculations and those measurements.

In Fig. 3 we describe the current-voltage characteristics of the HEMT with and without exposing to polar liquids. The curves consist of two regions: linear (Ohmic mode) and saturation (active mode) region. We notice that liquids with smaller p/ε values have a wider linear region because they

have less effect on the surface potential of the device according to (3).

Fig. 4 represents the current-voltage characteristics of the device at different gate potentials when its surface is exposed to acetone. We notice that the curves in Fig. 3 and Fig. 4 are similar to those of a common transistor.

In Fig. 5 we plot the drain current versus p/ε at different gate potentials with fixed drain-source voltage. If we extrapolate the line with $V_g = -1.5$ V, as an example, it will intersect with the x-axis at a certain value of p/ε, which means that the current is zero and the device stops detecting the effect of liquids with larger p/ε values. To solve this problem, we change the gate potential (for example to − 1.0 V) and we notice that the intersection of the line with the x-axis will be at larger p/ε value. This technique improves the ability of the device to sense molecules of very large dipole moments such as biomolecules.

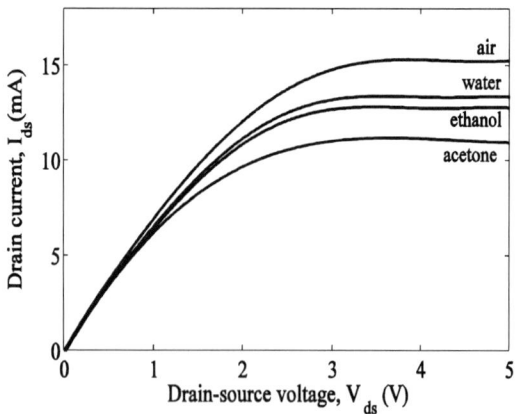

Fig. 3 Current-voltage characteristics of HEMT exposed to air and polar liquids at $V_g = -0.5$ V

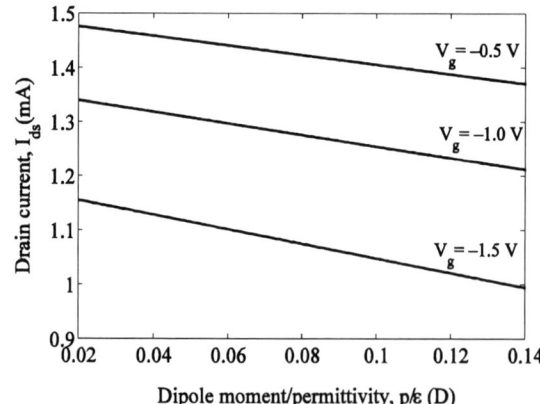

Fig. 5 Drain current as a function of p/ε at different gate voltages and $V_{ds} = 0.2$ V

II. CONCLUSIONS

We have shown that the HEMT is sensitive to the changes in electrostatic conditions of the gate when it is exposed to polar liquids. By drawing the current versus the dipole moment, we notice a linear response for polar liquids. This means that we can extrapolate the results to measure unknown dipole moments. The calculations exhibit a good sensitivity due to the high mobility and the large sheet density in the channel. We got results in good agreement with the experimental measurements using the same device characteristics.

The drain source current in the channel can be controlled by the change in drain-source voltage and/or the gate voltage which can be changed by applying a reference electrode to the liquid.

The calculations suggest the possibility of functionalizing the HEMT surface as a sensor for biomolecules with very large dipole moment by increasing the gate potential. The excellent stability of the AlGaN surface will minimize degradation of adsorbed molecules.

ACKNOWLEDGMENT

This work has been funded by the FWO-Vlaanderen project "FWOAL600" of the Flemish region in Belgium.

REFERENCES

[1] S. Rabbaa, W. Vandermeiren, and J. Stiens, "Longitudinal Optical Phonon-Plasmon Interaction in Ga-Group V Compounds for IR and THz Applications", Proc. of the 15th Annual Symp. of the IEEE Photonics Benelux Chapter, TU Delft, Netherlands, pp 257-60, November 2010.

[2] S. Rabbaa, and J. Stiens, "Charge Density and Plasmon Modes in a Triangular Quantum Well Model for Doped and Undoped Gated AlGaN/GaN HEMTs", J. Phys. D: Appl. Phys. , vol. 44, pp 325103, July 2011.

[3] Wang Soo Jeat, et al, "Fabrication and Characterization of GaN-Based Two Terminal Devices for Liquid Sensing", IOP Conf. Series: Materials Science and Engineering , vol. 17, pp 012024, 2011.

[4] Mastura Shafinaz Zainal Abidin, et al, "Gateless-FET Undoped AlGaN/GaN HEMT Structure for Liquid-Phase Sensor" IEEE Inter. Conf. ICSE2010 Proc. (Melaka, Malaysia) , pp R961, 2004.

[5] S. J. Pearton, et al, " GaN-based diodes and transistors for chemical, gas, biological and pressure sensing", J. Phys.: Condens. Matter , vol. 16, pp 031101, 2008.

[6] T. Kokawa, T. Sato, H. Hasegawa and T. Hashizume, "Liquid-phase sensors using open-gate AlGaN/GaN high electron mobility transistor structure", J. Vac. Sci. Technol. B , vol. 24, pp 1972-76, 2006.

[7] B. S. Kang, H. T. Wang, F. Ren, and S. J. Pearton, "Electrical Detection of Biomaterials Using AlGaN/GaN High Electron Mobility Transitors", J. Appl. Phys. , vol. 104, pp 031101, 2008.

[8] F. Ren, and S. Pearton, Semiconductor Device-Based sensors for Gas, Chemical, and Biomedical Applications, CRC Press Taylor and Francis Group, New York, pp. 11, 2011

[9] R. Mehandru, et al, "AlGaN/GaN HEMT Based Liquid Sensors", Solid State Electronics, vol. 48, pp.351-352, 2004.

[10] C. Wood, and D. Jena, Polarization Effects in Semiconductors from Ab Initio Theory to Device Applications, Springer, New York, pp. 77, 2008.

[11] S. Ehsan Abtahi Hosseini, and S. Ebrahim Hosseini, "An Accurate Analytical Model for Current-Voltage Characteristics and TransConductance of Al_mGa_{1-m}/GaN MODFETs", IEEE Inter. Conf. EDSSC (Xia, Chaina), pp 412-415, 2009.

[12] R. Neuberger, G. Muller, O. Ambacher, and M. Stutzmann, "High-Electron-Mobility AlGaN/GaN Transitors (HEMTs) for Fluid Monitoring Applications", Phys. Stat. Sol. (a) , vol. 185, pp 85-89, 2001.

Demonstration of DC Current Sensing through Microfiber Knot Resonator

Azlan Sulaiman[1,2], Sulaiman Wadi Harun[2,3], Jalil. Md. Desa[1] and Harith Ahmad[3]

[1]Telekom Research & Development TM Innovation Center,
Lingkaran Teknokrat Timur
63000 Cyberjaya Selangor Malaysia

[2]Department of Electrical Engineering University Malaya
50603 Kuala Lumpur Malaysia.

[3]Photonic Research Center Department of Physics
University of Malaya
50603 Kuala Lumpur Malaysia
Email: azlan27@gmail.com

Abstract – **A compact current sensor is demonstrated using a Microfiber Knot Resonator (MKR) which is obtained by knotting silica microfiber fabricated from flame brush technique. With the assistance of a copper wire touching the circumference of the ring, resonant wavelength inside the MKR can be tuned by injecting electric current into the copper wire. The wavelength shift is due to the thermally induced optical phase shifts attributable to the heat produced by the flow of the current. It is shown experimentally that the wavelength shift is linearly proportional to square of the amount of current with a tuning slope of 90 pm/A^2. Compared to the previous work that place the copper in the loop of the MKR, this technique is simple, more sensitive and robust.**

Index Terms – microfiber, current sensor, knot resonator.

I. INTRODUCTION

Microfibers have attracted growing interest recently due to their interesting optical properties, which can be used to develop low-cost, miniaturized and all-fiber based optical devices for various applications [1-4]. For instance, many research efforts have focused on the development of microfiber based optical resonators that can serve as optical filters, which have many potential applications in compact tuneable fiber laser [5] and sensors [6-7]. These devices are very sensitive to a change in the surrounding refractive index due to the large evanescent field that propagates inside the fiber and thus they can be used in various optical sensors. To date, a variety of fiber optic sensors have been demonstrated [8]. For instance, fiber-optic based current sensors based on either Faraday or thermal effect [9]. The former is capable of measuring electrical currents remotely, but the device requires a long fiber due to the extremely small Verdet constant of silica. The latter needs a short length of fiber but requires complex manufacturing techniques to coat fibers with the

metals. In this paper, a new approach is proposed to enhance the sensitivity of the MKR-based current sensor. In this approach, a copper wire is placed in contact with the outer surface of the MKR loop to induce the wavelength shift as a result of the heat emitted by the current flowing in the wire. Analytical and experimental analysis specifically on the characteristic of the resonant wavelength of MKR is presented.

II. EXPERIMENTAL AND THEORETICAL ANALYSIS

The MKR was fabricated as follows: first, a microfiber was fabricated using the so-called flame brush technique. The microfiber was made by flame brush technique; heating and stretching uncoated fiber until ~2µm waist diameter is acquired [10]. The microfiber was then cut into 2 unequal parts, where the longer part of the microfiber was twisted to form a large loop and the end of the microfiber is inserted inside the loop to form a knot. The loop diameter decreased gradually when the fiber ends were pulled. Another part of the microfiber was coupled to the knot by van der Walls force to collect the light transmitted out from the knot by means of evanescent coupling. At least ~3 mm of coupling length between two microfibers is required to achieve sufficient van der Waals attraction force to keep them attached together. The knot configuration is sustained by elastic-bend-induced tensile force and friction at the inter-twisted area. Therefore, it offers higher structural stability as compared to a micrometer loop resonator, thus ensuring a strong contact at the coupling region and maintaining a consistent resonance condition of the resonator. Fig. 1 shows the experimental setup of the MKR-based current sensor where both SMF ends are connected to an amplified spontaneous emission (ASE) source and optical spectrum analyser (OSA). A copper wire with a radius of 0.16 mm is connected to the power supply and placed horizontally on the translation stage. Then it is gently lowered so that it touches firmly on the outer radius of the MKR as illustrated in

978-1-4673-2395-6/12 $31.00 © 2012 IEEE

the figure. The launched ASE light oscillates in the resonator to generate an optical resonance in multiple 2π of phase shift. The current flowing inside the copper wire generates heat which eventually transfers into the microfiber in contact with the copper wire.

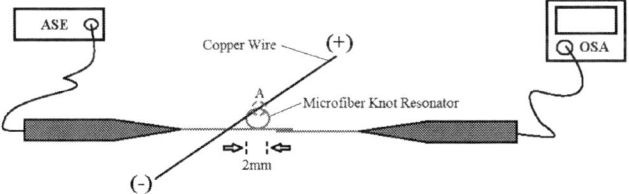

Fig. 1: The optical microscope image of MKR and the schematic illustration of MKR touched on a copper rod.

The electrical power dissipated through the copper wire is given by

$$P_I = I^2 \rho L / A \qquad (1)$$

where I is the current, ρ is resistivity of the copper wire, L is the length and A is the cross sectional area of the copper wire. It is intuitive that the flow of current through a wire will cause the temperature of the wire to increase. The temperature rise in the copper wire is directly proportional to the power, which lead to:

$$\Delta T \approx I^2 \rho L / AK \qquad (2)$$

where K is the dissipation constant. The temperature variation affects both the effective refractive index and the loop length L, of the MKR, which shifts the phase of the oscillating light leading to spectral shift. The relation between the spectral shift and the temperature change is expressed as:

$$\Delta\lambda / \lambda = (\alpha + \beta) \Delta T \qquad (3)$$

where α and β are the thermal expansion coefficient (TEC) and thermal-optic coefficient (TOC) of the microfiber respectively [12]. Based on equations (2) and (3), the relationship between the wavelength shift and the conducting current can be expressed by

$$\Delta\lambda / \lambda \approx (\alpha + \beta) I^2 \rho L / AK \qquad (4)$$

III. RESULT AND DISCUSSION

The optical characteristic of the proposed MKR-based current sensor is investigated by launching a broadband source from ASE into the device and measuring the transmitted light by an OSA while increasing the amount of DC current flowing through the copper wire. The optical resonance is generated when light travelling in the MKR acquires a phase shift multiple of 2π when the current flows through the copper wire, heat is produced in the wire causing temperature to change. Since the MKR is in contact with the copper wire, any temperature changes will influence the refractive index and the optical path length of the MKR. Fig. 3 shows the change of resonant spectrum of the MKR, which is in contact with the copper wire at different current loadings. In the experiment, the DC current is loaded with different values in the range of 0 to 2.0 A in step of approximately 0.5 A. For each change in the current value, 10s settling time is given in order to stabilize the wavelength shift before the spectrum is recorded.

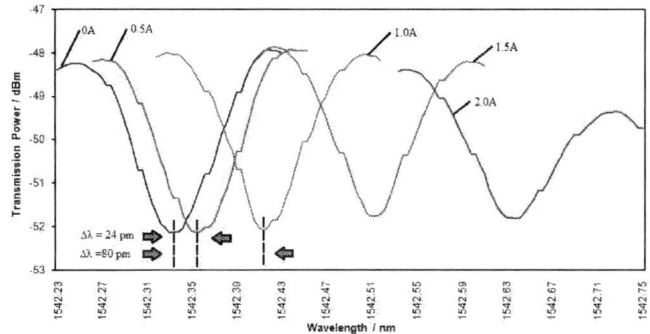

Fig. 2: Resonant wavelength of the MKR at various injected DC currents.

As shown in Fig. 2, the resonant wavelength shifts to a longer wavelength with the increase in conducting current of the copper wire. For instance, at loading currents of 0.5 and 1.0 A, the resonant wavelengths shifts to a longer wavelength by 28 pm and 80 pm respectively, in comparison with the initial spectrum. The spectrum returns to the original state as the current is turned off. It is also observed that the FSR and extinction ratio remain unchanged at 0.17 nm and 4.5 dB respectively when the value of the flowing current is below 1.0 A. However, when the current is increased up to 2.0 A, the extinction ratio is obtained at 3.4 dB, which is a reduction of 24 % from the original wave and the FSR slightly increases to 0.18 nm. The reduction of the extinction ratio and the small increment of the FSR are observed may be due to the deterioration of van der Waals bonding, which introduces slip at the coupling region as the temperature increases above the threshold level of 1.0 A.

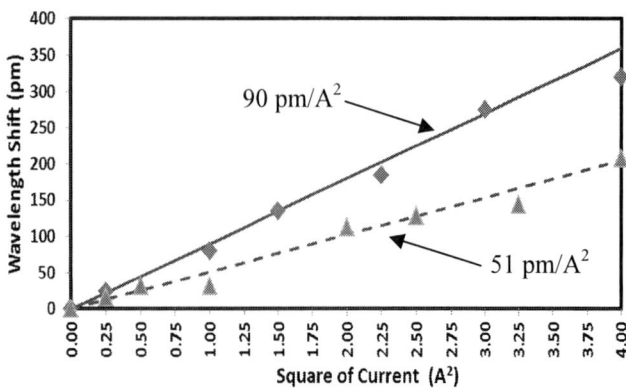

Fig. 3: wavelength shift against the square of the injected current.

In an earlier work, Guo [11] demonstrated that the shift of the resonant wavelength of microfiber loop resonator (MLR) linearly increases with the increase of the injected current with a slope of 26.5 pm/A. Later, Lim et. al. [12] demonstrated that the wavelength shift is actually proportional to the square of electric current which is in good agreement with Equation (3). In addition, a slope efficiency of 51.3 pm/A^2 is obtained. In this work, the wavelength shift is plotted against square of current for the proposed sensor and the result is shown in Fig. 3. The plot shows the linear relation with a slope efficiency of 90 pm/A^2, which is so much higher than those of both earlier works. In the proposed work, the copper wire is touching the outer ring area of MKR as compared to the previous work, which the copper wire touches the inner ring. Since the peak of evanescent field is near to the outer ring area due to the bending effect, the phase shift is more pronounced if the effective refractive index is increased at this region compared to that of the case of inner region. This makes the proposed sensor is more sensitive compared to the conventional approaches. This technique is also less complicated compared to earlier techniques. In both earlier techniques, the copper wire is wrapped by the microfiber ring or knot and thus the microfiber is easily damaged and broken during the process.

IV. CONCLUSION

A compact and robust current sensor using a microfiber knot resonator (MKR) is demonstrated with the assistance of a copper wire. The copper wire is left touching the outer diameter area of the MKR to increase the sensitivity of this sensor. The resonant wavelength change is based on the thermally induced optical phase shift in the MKR due to the heat produced by the flow of electric current over a short transit length. It is shown that the wavelength shift is linearly proportional to the square of the amount of current with the tuning slope of 90 pm/A^2.

V. REFERENCES

[1]. L. Tong, R. R. Gattass, J. B. Ashcom, S. He, J. Lou, M. Shen, I. Maxwell, and E. Mazur, "Subwavelength-diameter silica wires for low-loss optical wave guiding," Nature, Vol 426, pp 816–819, 2003.

[2]. G. Brambilla, V. Finazzi, and D. J. Richardson, "Ultra-low-loss optical fiber nanotapers," Optics Express, Vol. 12, pp. 2258-2263 (2004).

[3]. M. Sumetsky, Y. Dulashko, J. M. Fini, A. Hale, and D. J. DiGiovanni, "The Microfiber Loop Resonator: Theory, Experiment, and Application," J. Lightwave Technol. Vol. 24, No. 1, pp. 242-250 (2006).

[4]. Sulaiman, A.; Harun, S. W.; Arof, H.; Ahmad, H.; , "Compact and Tunable Erbium-Doped Fiber Laser With Microfiber Mach–Zehnder Interferometer," Quantum Electronics, IEEE Journal of , vol.48, no.9, pp.1165-1168, Sept. 2012.

[5]. Sulaiman, A.; Harun, S. W.; Ahmad, F.; Norizan, S. F.; Ahmad, H.; , "Electrically Tunable Microfiber Knot Resonator Based Erbium-Doped Fiber Laser," Quantum Electronics, IEEE Journal of , vol.48, no.4, pp.443-446, 2012

[6]. M. Sumetsky, "Basic Elements for Microfiber Photonics: Micro/Nanofibers and Microfiber Coil Resonators," J. Lightwave Technol., vol. 26, no. 1, pp. 21-27, 2008.

[7]. Y. Li and L. Tong, "Mach-Zehnder interferometers assembled with optical microfibers or nanofibers," Opt. Lett., Vol. 33, pp. 303-305 (2008)

[8]. M. Yasin, S. W. harun, H. A. Abdul-Rashid, Kusminarto, Karyono and H. Ahmad, "The performance of a fiber optic displacement sensor for different types of probes and targets," Laser Phys. Lett., vol. 5, pp. 55-58, 2008.

[9]. G. A. Sanders, J. N. Blake, A. H. Rose, F. Rahmatian, and C. Herdman, "Commercialization of Fiber-Optic Current and Voltage Sensors at NxtPhase," 15th Optical Fiber Sensors Conference, Portland, OR, May 2002, pp. 31-34.

[10]. Sulaiman, A.; Harun, S.W.; Desa, J.M.; Lim, K.S.; Zamzuri, A.K.; Ahmad, H.; , "Fabrication and characterization of optical microfiber structures," Communications (APCC), 2011 17th Asia-Pacific Conference on , vol., no., pp.330-333, 2-5, 2011

[11]. X. Guo, Y. Li, X. Jiang, L. Tong, "Demonstration of critical coupling in microfiber loops wrapped around a copper rod," Applied Physics Letters, vol. 91, no. 7, pp. 073512 - 073512-3, 2007.

[12]. K. S. Lim, S. W. Harun, S. S. A. Damanhuri, A. A. Jasim, C. K. Tio, and H. Ahmad, "Current sensor based on microfiber knot resonator," Sensors and Actuators A: Physical, Vol. 67, No. 1, pp. 60-62, (2011).

Microfiber Coupler Devices

M. Z. Muhammad[1], A. A. Jasim[1], H. Ahmad[2] and S.W. Harun[1,2]

[1]Department of Electrical Engineering, University of Malaya 50603 Kuala Lumpur, Malaysia
[2]Photonics Research Center, Department of Physics, University of Malaya, 50603 Kuala Lumpur, Malaysia
Email: swharun@gmail.com

Abstract-A 2x2 microfiber coupler is demonstrated by laterally fusing and tapering two optical fibers using a flame brushing technique. The coupler has an overlapping length of 40mm with a uniform waist of around 5 μm. The coupler is then used to demonstrate a microfiber knot resonator (MKR) coupler. It is obtained by forming a knot within the coupling region of a microfiber coupler with a 50:50 splitting ratio. With an MKR structure, the coupler produces a resonant response at both output ports. The free spectral range (FSR) of the output spectrum from both ports is obtained at 0.2 nm at a knot diameter of 260 μm. The resonance extinction ratio (RER) of the device varies from 2 to 6 dB while the calculated Q factor and finesse are ~25646 and 3.3 respectively at both output ports.

I. INTRODUCTION

Microfibers have attracted considerable interests in recent years, as they exhibit a number of exciting properties such as large evanescent field, strong confinement, easy configurability and high robustness [1, 2]. These properties are advantageous for a wide range of applications including high-sensitivity optical sensors, nonlinear optics, atom trapping, micro/nano-scale photonic devices and for evanescent coupling to planar waveguides or microcavities. Recently, Jung et. al. demonstrated that sub-wavelength optical microwires can be used as an efficient element for higher-order mode filtering in multimode waveguides, creating effectively an endless single mode operation in conventional optical fibers [3]. The stable and low-loss single-mode operation obtained both at short wavelengths and over a wide spectral range is suitable for applications in high performance fiber lasers, sensors, photolithography, and optical coherence tomography (OCT) systems. However, single-mode output from a single fiber strand is not sufficient to fulfill all technical demands within these fields making the development of multi-port devices an important requirement [4].

In this paper, fabrication and characterization of fused-typed microfiber coupler devices are presented. A bi-conical 2×2 MKR coupler structure is also demonstrated by forming a knot from a microfiber coupler. The microfiber coupler is also used to form a microsphere at the coupling region.

II. FABRICATION OF MICROFIBER COUPLER

A microfiber coupler is made by laterally fusing and tapering two optical fibers as shown in Fig. 1. In this experiment, a standard telecom optical fiber (Corning SMF-28) was used to make a low noise microfiber coupler with the aid of the well-established single stage "flame-brushing" technique [2]. In the fabrication process, two fibers are brought into close proximity after the protective plastic

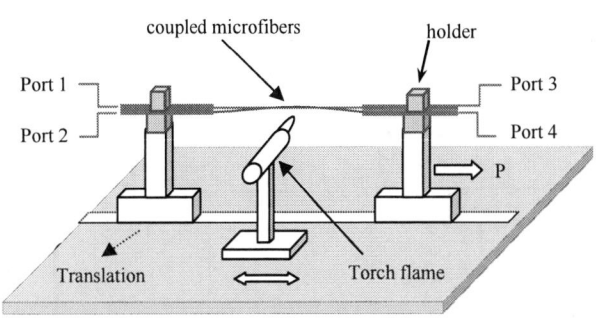

Fig. 1. Fabrication setup for microfiber coupler

jacket is removed. Then, both fibers are twisted at two different locations to make overlapping contact. Then, while heated by an oxy-butane torch, the fibers are fused and stretched. The longitudinal profile of the conical transition tapers was achieved by reliable control of the hot zone and precise movement of the translation stages. During the tapering process, the coupling ratio is being monitored in real time by using a 1550 nm light source and power meters. The heating and pulling processes are stopped at the moment of achieving the desired coupling ratio.

III. CHARACTERISTIC OF THE FABRICATED MICROFIBER COUPLER

An Erbium amplified spontaneous emission (ASE) source is used to couple light into one of the ports of the microfiber-based coupler, and the output spectrum is measured by an optical spectrum analyzer (OSA) from the output ports. Figs. 2(a) and 2(b) show the spectral response of the two output ports of the coupler when the ASE is injected from port 1 and port 2 respectively. As shown in these figures, an equal splitting of the output power into the two output ports is obtained, resulting in a 50:50 coupler. However, a slight spectral oscillation and amplitude modulation is also observed. The throughput loss of the coupler is quite acceptable which is around 8 dB between port 1 and port 3 due to the polarization effect. The excess loss is also observed at around -4.5 dB. This coupler has a broad range of single-mode optical operation in two output ports and is directly applicable to the high performance fiber lasers, fiber sensors, optical coherence tomography (OCT) and fiber test & measurement systems.

978-1-4673-2395-6/12 $31.00 © 2012 IEEE

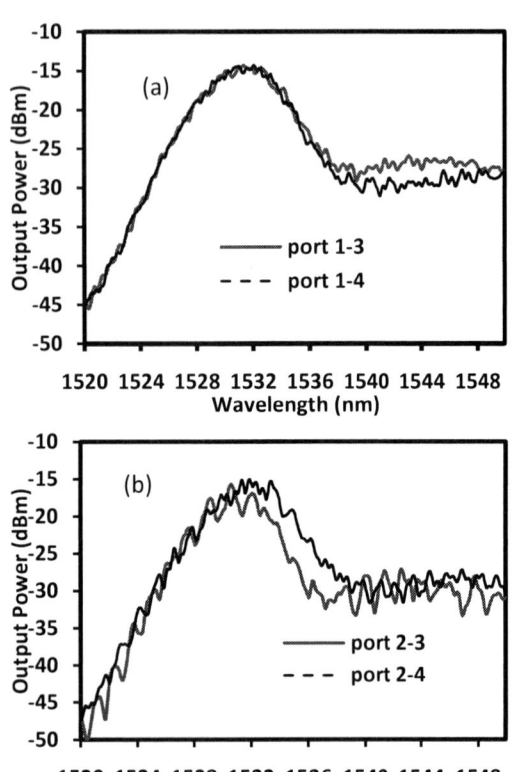

Fig. 2. Output spectra from the two output ports of the fabricated coupler when the ASE is injected from (a) port 1 and (b) port 2.

IV. MKR COUPLER

The 50/50 microfiber coupler was then cut into 2 unequal parts, where the longer part of the coupling region was twisted to form a large loop and the end of the microfiber is inserted inside the loop to form a knot. The required loop diameter is obtained by gradually pulling one of the fiber ends. The shorter part of the coupler was coupled to the knot by van der Walls force to collect the light transmitted out from the knot by means of evanescent coupling. At least ~3 mm of overlapping length between the two coupled microfibers is required to achieve sufficient van der Waals attraction force to keep them together. The knot configuration is sustained by elastic-bend-induced tensile force and friction at the inter twisted area and thus ensuring the structural stability of the device. The output spectrum of the MKR is characterized by using an ASE source in conjunction with an OSA. Fig. 3 shows the microscope image of the MKR coupler with a loop diameter of 260 μm, which was fabricated using two coupled microfibers with a diameter of around 5 μm.

As an ASE is injected through one of the two input ports, it is coupled into the neighboring microfiber at the coupling region. The light is then split into two after entering the knot resonator. A portion of the light oscillates and interferes in the loop to generate the resonant response at the output of the resonator before it is separated or split into the two output ports at the end of the hybrid structure.

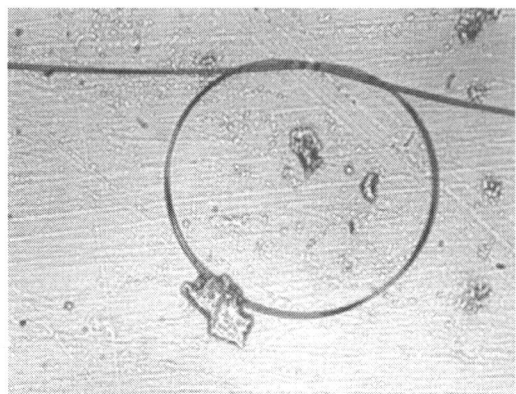

Fig. 3. Microscope image of the fabricated MKR coupler.

The overlapping microfibers guide light with the evanescent field extending outside the microfiber.This evanescent field depends on the wavelength of operation, diameter of the overlapping region of the coupler and the surrounding medium. Fig. 4 shows the output transmission power of the resonant response at both output ports. As shown in the figure, wavelength dependent filtering responses are obtained at both output ports, irrespective whether the broadband source is injected from the first port or the second port. The additional power loss of approximately 5 dB which is considered normal during the fabrication of the MKR is incurred as compared to a coupler device without the knot structure. The free spectral range (FSR) or the spacing between two peaks of the resonant wavelength is measured to be the same at around 0.2nm for all spectra irrespective of which input or output signal ports. The spacing is strongly dependent on the knot diameter which is around 260 μm. The resonance extinction ratio (RER) of the device varies from 2 to 6 dB. The calculated of Q factor and finesse of the MKR are ~25646 and 3.3 respectively at both ports.

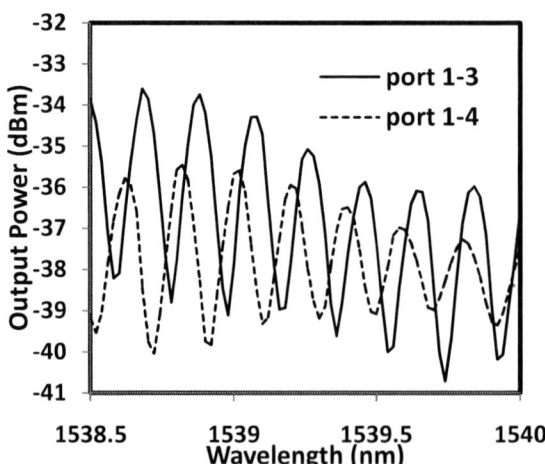

Fig. 4. Output transmission spectrum for the MKR coupler device

V. Conclusion

A microfiber coupler is fabricated to generate two output resonance spectra, which is then used to construct MKR. The knot is formed within the coupling region of a microfiber coupler, which is obtained by laterally fusing and tapering two optical fibers using a flame-brushing technique. The coupler has a 50:50 splitting ratio and an overlapping length of 40mm with a uniform waist of around 5 μm. The hybrid device produces an output comb with the same FSR of 0.2 nm at each port with knot diameter of 260 μm. The RER of the device varies from 2 to 6 dB while the calculated Q factor and finesse are ~25646 and 3.3 respectively at both output ports.

Acknowledgment

This project was funded by the Ministry of Science, Technology and Innovation (MOSTI) under Brain-Gain program (Grant No. 5302031067) and the Ministry of Higher Education (MOHE) under HIRG (Grant No: HIR-MOHE D000009-16001).

References

[1] M. Sumetsky, Y. Dulashko, J. M. Fini, A. Hale, and D. J. DiGiovanni, "The microfiber loop resonator: theory, experiment, and application," J. Lightwave Technol., vol. 24, no. 1, pp. 242–250 (2006).

[2] S. W. Harun, K. S. Lim, S. S. A. Damanhuri, H. Ahmad, "Microfiber loop resonator based temperature sensor", J. Europ. Opt. Soc. Rap. Public., vol. 6, pp. 11026, (2011).

[3] Y. Jung, G. Brambilla, D. J. Richardson, "Broadband single-mode operation of standard optical fibers by using a sub-wavelength optical wire filter," Optics Express, vol. 16, pp.14661-14667, (2008).

[4] Y. Jung, G. Brambilla, D. J. Richardson," Optical microfiber coupler for broadband single mode Operation," Optics express, vol. 17, pp.5273-5278, (2009).

[5] K. S. Lim, A. A. Jasim, S. S. A. Damanhuri, S. W. Harun, B. M. A. Rahman, and H. Ahmad, "Resonance condition of a microfiber knot resonator immersed in liquids," Appl. Opt., vol. 50, pp. 5912-5916 (2011).

[6] F. Xu, P. Horak, and G. Brambilla, "Conical and biconical ultra-high-Q optical-fiber nanowire microcoil resonator," Appl. Opt., vol. 46, pp. 570–573 (2007).

[7] X. S. Jiang, Y. Chen, G. Vienne, and L. M. Tong, "All-fiber add-drop filters based on microfiber knot resonators," Opt. Lett., vol. 32, pp. 1710–1712 (2007).

[8] X. S. Jiang, Q. H. Song, L. Xu, J. Fu, and L. M. Tong, "Microfiber knot dye laser based on the evanescent-wave-coupled gain," Appl. Phys. Lett., vol. 90, pp. 233501 (2007).

[\9] G. Vienne, Y. H. Li, L. M. Tong, and P. Grelu, "Observation of a nonlinear microfiber resonator," Opt. Lett., vol. 33, pp. 1500–1502 (2008).

[10] K. S. Lim, S.W.Harun, S.S.A.Damanhuri, A.A.Jasim, C.K.Tio, H.Ahmad," Current sensor based on microfiber knot resonator", Sens. Actuators A: Phys, \ vol. 167, pp. 60-62, (2011).

Influence of Optical Power on Thermal Resistance Measurement for High Power Infrared Emitter

Chin-Peng Ching[*] and Mutharasu Devarajan
Nano Optoelectronics Research (NOR) Lab, School of Physics,
Universiti Sains Malaysia, 11800 USM, Pulau Pinang, Malaysia.
[*]E-mail: ccp10_phy033@student.usm.my

Abstract- **This work signifies the importance of optical power in determining the real thermal resistance value for high power infrared (IR) emitter. Thermal transient measurement and optical test have been employed to study the influence of optical power on thermal resistance determination. For measurement, the IR emitter is driven at constant current of 1.0A with ambient temperature of 24.6°C under still air condition. From the findings, real junction-to-board thermal resistance Rth_{JBR} obtained from the structure function with optical power consideration is 7.52±0.01 K/W. However, the electrical junction-to-board thermal resistance Rth_{JBE} value (4.87±0.01 K/W) is much lower when optical power is not taken into consideration. It is found that structure function considering optical power offers higher values of real thermal resistance compared to that without optical power consideration.**

Keywords- Thermal transient measurement, optical power, real junction-to-board thermal resistance, structure function, high power infrared emitter.

I. Introduction

Light-emitting diodes (LEDs) are becoming an alternative for traditional light source as technology of LEDs have caused significant impacts on illumination industry and increasing market [1,2]. LEDs are developing fast due to their improved performance in optical power, efficiency and reliability [3-7]. Recently, high power infrared emitter has been used extensively in infrared illumination for cameras, driver assistance systems and also as remote controls for consumer products.

Since the optical and electrical characteristics of LEDs are strongly dependent on diode junction temperature, thermal management of solid state devices is of great interest especially for high power LEDs that involve large amount of power dissipation [8]. Subsequently, poor thermal management will cause shorter lifetime, lower efficiency, junction breakdown and hence catastrophic device failure [9,10].

This paper presents the influence of optical power value on thermal resistance determination for a high power IR emitter. Comparison of structure functions with and without optical power consideration has been made to study the importance of optical power in determining the real junction-to-board thermal resistance value. Structure function with optical power consideration will yield a higher real thermal resistance compared to that without optical power consideration that provides a lower electrical thermal resistance value.

II. Theoretical Background

Thermal transient method is a recording of the thermal step response functions such as heating and cooling curves [11]. Evaluation of these curves can be either in the form of a few discrete values or continuous time-constant spectrum [12]. Thermal transient response functions are usually plotted on logarithmic time axis as it is highly characteristic that temperature changes in an extremely wide time range [13].

Structure function is a thermal resistance and capacitance map of a heat flow path. Cumulative structure function is defined as the sum of the thermal capacitance C_Σ with respect to the sum of the thermal resistance R_Σ measured from excitation point toward the ambience [14]. Whereas differential structure function is defined as the derivative of the cumulative thermal capacitance with respect to the cumulative thermal resistance [15].

According to JEDEC standard [16], thermal resistance of a semiconductor device is commonly defined as,

$$R_{thJX} = \frac{T_J - T_X}{P_h} \qquad (1)$$

where R_{thJX} is the thermal resistance from device junction to a specific point, T_J is junction temperature, T_X is reference temperature at specific point and P_h is power dissipated.

Electrical thermal resistance is defined without taking into consideration of energy being dissipated; however, real thermal resistance considers the optical power being emitted.

Electrical thermal resistance,
$$R_{thE} = \frac{\Delta T}{P_{el}} \qquad (2)$$

Real thermal resistance,
$$R_{thR} = \frac{\Delta T}{P_{el} - P_{op}} \qquad (3)$$

where P_{el} is input power, P_{op} is optical power and ΔT is temperature difference between the junction and a reference point [17].

Optical power which is a radiometric parameter is generally know as radiant flux or radiant power of a LED. It is defined as the energy (Q) radiated by a source per unit of time (t).

$$P_{op} = \frac{dQ}{dt} \qquad (4)$$

where the unit is watt, W [18].

978-1-4673-2395-6/12 $31.00 © 2012 IEEE

III. EXPERIMENTAL WORK

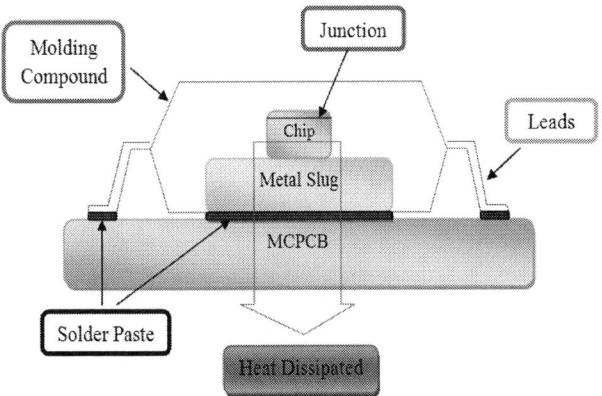

Fig. 1. Schematic diagram of a high power infrared emitter soldered on MCPCB.

For thermal characterization, a commercial high power IR emitter has been soldered on MCPCB as shown in Fig. 1. At first, the IR emitter went through calibration to obtain temperature sensitive parameter which was the K-factor. In calibration, the ambient temperature was varied from 40°C to 80°C with a spacing of 10°C and a small value of sensor current, 1.0mA was then applied to the IR emitter at each temperature. K-factor of 1.83mV/°C was obtained through the gradient of the graph as shown in Fig. 2.

By using thermal transient tester (T3Ster), the IR emitter has been driven by forward current of 1.0A at an ambient temperature of 25.0 ± 1.0°C under still air condition. Then, the IR emitter was heated for 5-10 min to reach thermal equilibrium with ambient temperature environment and allowed to cool down. After that, the IR emitter went through optical measurement to obtain optical power. In order to obtain real thermal resistance values, optical power obtained was used to correct the power dissipation for the computation of structure functions by using T3Ster evaluation software.

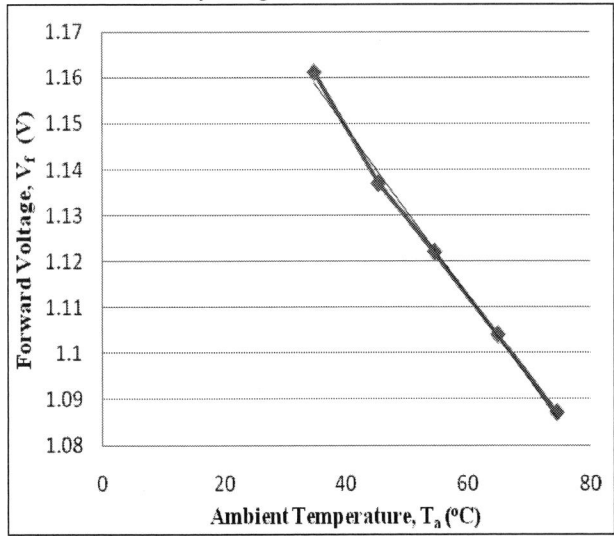

Fig. 2. Forward voltage versus ambient temperature for calibration measurement.

IV. RESULTS AND DISCUSSION

Differential and cumulative structure functions have been extensively used for identification of different regions in a heat flow path of a sample. Fig. 3 shows an example of differential structure function of a complete heat flow path for the IR emitter. By assuming heat is flowing in one-dimensional direction, heat generated at the junction will spread towards the end of the sample then dissipate to ambience. In the figure, the left-hand side of the structure function refers to the junction in a chip whereas the right-hand side denotes the ambience. Region I (from origin to first peak) represents partial thermal resistance from junction to metal slug, Rth_{JS} of the sample. Region II refers to the partial thermal resistance from junction to interface material, Rth_{JI}. Region III denotes partial thermal resistance from junction to board, Rth_{JB} of the sample and Region IV describes partial thermal resistance from junction to ambient, Rth_{JA}.

Fig. 4 illustrates the comparison of differential structure functions with and without optical power consideration recorded at constant current of 1.0A and ambient temperature of around 25.0°C under still air condition. As shown in the figure, real junction-to-board thermal resistance Rth_{JBR} obtained from the structure function with optical power consideration is 7.52 ± 0.01 K/W. However, the electrical junction-to-board thermal resistance Rth_{JBE}, which is 4.87 ± 0.01 K/W, is much lower when optical power is not taken into consideration. It is obvious that the structure function which considering optical power offers higher thermal resistance values compared to that without optical power consideration. The comparison of thermal resistance values with and without optical power consideration are tabulated in Table 1.

From the table, when optical power is not taken into consideration, the electrical junction-to-board thermal resistance Rth_{JBE} obtained is 35.2% lower than its real value. Correspondingly, percentage of difference between the electrical and real junction-to-slug thermal resistance Rth_{JS}, junction-to-interface thermal resistance Rth_{JI} and junction-to-ambient thermal resistance Rth_{JA} are 38.5%, 33.8% and 34.5% respectively.

TABLE I

COMPARISON OF REAL AND ELECTRICAL THERMAL RESISTANCE VALUES

Thermal resistance values	Real thermal resistance with optical power consideration, K/W	Electrical thermal resistance without optical power consideration, K/W
Junction-to-slug thermal resistance, Rth_{JS}	1.56	0.96
Junction-to-interface thermal resistance, Rth_{JI}	3.73	2.47
Junction-to-board thermal resistance, Rth_{JB}	7.52	4.87
Junction-to-ambient thermal resistance, Rth_{JA}	46.7	30.6

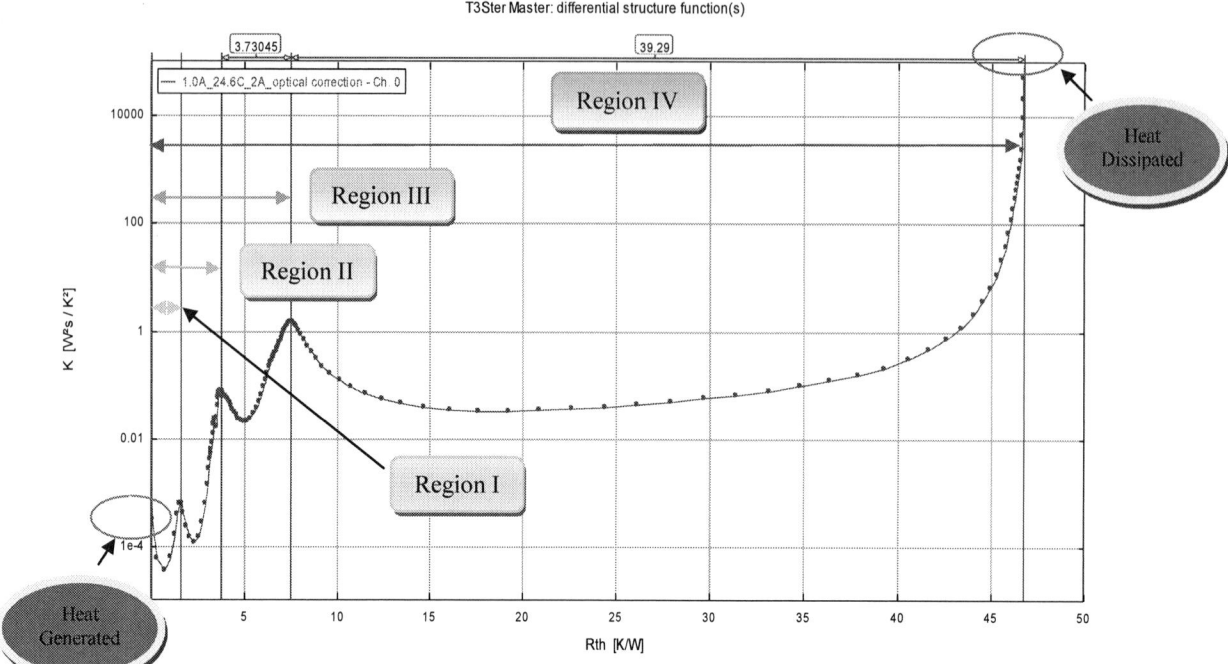

Fig. 3. An example of differential structure function of a heat flow path for the IR emitter.

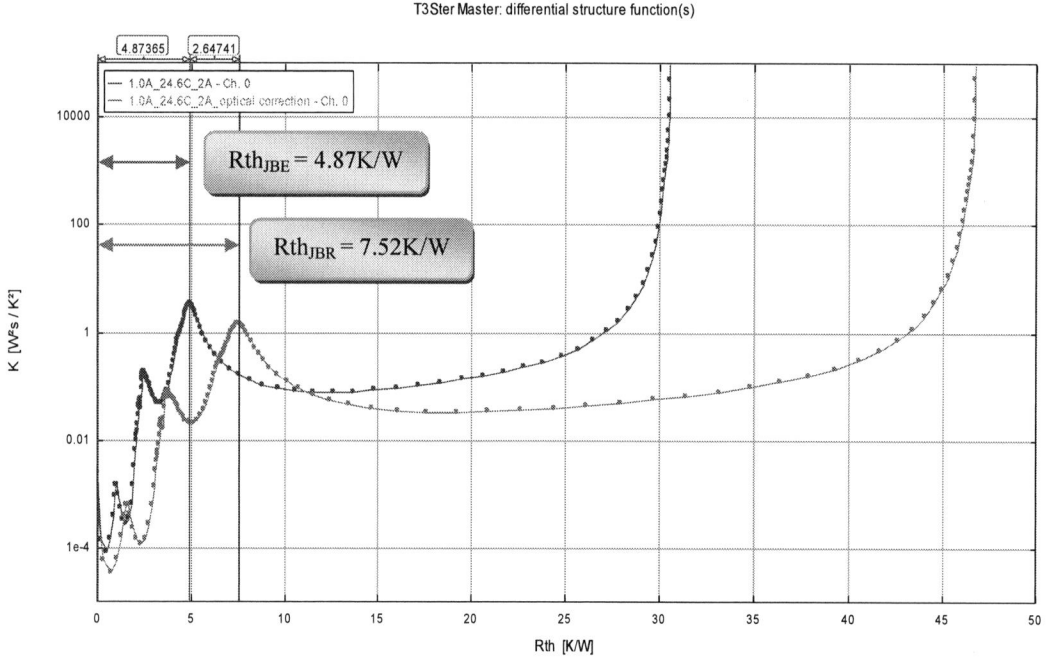

Fig. 4. Comparison of structure functions with and without optical power consideration for the IR emitter.

In reality, once electrical power is applied, optical power will be emitted as light source and heat will be dissipated [19]. In order to obtain real thermal resistance values, real thermal resistance equation that takes into consideration of optical power and heat dissipated should be involved since electrical thermal resistance considers only power supplied. Therefore, influence of optical power in the determination of real thermal resistance could be well explained by using (2) and (3).

Electrical thermal resistance, $\qquad R_{thE} = \dfrac{\Delta T}{P_{el}}$

Real thermal resistance,

$$R_{thR} = \frac{\Delta T}{P_{el} - P_{op}}$$

According to (2) and (3), when optical power is taken into consideration, real thermal resistance value obtained would be higher than the electrical thermal resistance. Correspondingly, neglecting optical power in the measurement will yield a much lower electrical thermal resistance value. This shows that it is necessary to include optical power for the computation of structure functions to obtain accurate thermal resistance values [20]. Since about 30-40% of the input power will leave the system as light for IR emitter, neglecting optical power in thermal resistance determination will yield a value which is much lower than the reality.

V. SUMMARY AND CONCLUSION

In this paper, thermal transient measurement and optical test have been employed to signify the importance of optical power in the determination of real thermal resistance values for the high power IR emitter. It is found that the real and electrical junction-to-board thermal resistance Rth_{JB} values are 7.52±0.01 K/W and 4.87±0.01 K/W respectively. When optical power is not taken into consideration, the electrical junction-to-board thermal resistance Rth_{JBE} obtained is 35.2% lower than its real value. It is proved that optical power has significant influence on the determination of real thermal resistance values. Hence, it is necessary to consider optical power in the evaluation of structure functions that provide accurate real thermal resistances.

ACKNOWLEDGMENT

Ching would like to thank Institute of Postgraduate Studies (IPS) of Universiti Sains Malaysia for offering USM Fellowship and PRGS grant no. 1001/PFIZIK/834073.

REFERENCES

[1] P. Hanselaer, A. Keppens, S. Forment, W. R. Ryckaert and G. Deconinck, "A New Integrating Sphere Design for Spectral Radiant Flux Determination of Light-Emitting Diodes", Measurement Science and Technology, 20, 2009, 095111.

[2] Lianqiao Yang, Jianzheng Hu and Moo Whan Shin, "Degradation of High Power LEDs at Dynamic Working Conditions", Solid-State Electronics, 53, 2009, 567-570.

[3] Gilbert Held, Introduction to Light Emitting Diode Technology and Applications, Taylor & Francis Group, New York, 2009.

[4] Bahaa E. A. Saleh and Malvin Carl Teich, Fundamentals of Photonics, second ed., John Wiley & Sons, Inc., New York, 2007.

[5] Ting Cheng, Xiaobing Luo, Suyi Huang and Sheng Liu, "Thermal Analysis and Optimization of Multiple LED Packaging Based on a General Analytical Solution", International Journal of Thermal Science, 49, 2010, 196-201.

[6] Jianzheng Hu, Lianqiao Yang and Moo Whan Shin, "Electrical, Optical and Thermal Degradation of High Power GaN/InGaN Light-Emitting Diodes", Journal of Physics D: Applied Physics, 41, 2008.

[7] Peter Baureis, "Compact Modeling of Electrical, Thermal and Optical LED Behavior", Proceedings of ESSDERC, 2005.

[8] C. P. Ching, M. Devarajan and W. C. Liew, "Thermal Characterization of a High Power Infrared Emitter as a Function of Input Current", 2nd International Conference on Photonics, 2011, pp. 134-138.

[9] Chun-Jen Weng, "Advanced Thermal Enhancement and Management of LED Packages", International Communications in Heat and Mass Transfer, 36, 2009, 245-248.

[10] Adam Christensen and Samuel Graham, "Thermal Effects in Packaging High Power Light Emitting Diode Arrays", Applied Thermal Engineering, 29, 2009, 364-371.

[11] V. Székely, M. Rencz and B. Courtois, "A Step Forward in the Transient Thermal Characterization of Chips and Packages", Microelectronics Reliability, 39, 1999, 89-96.

[12] V. Székely, "Enhancing Reliability with Thermal Transient Testing", Microelectronics Reliability, 42, 2002, 629-640.

[13] V. Székely and M. Rencz, "Increasing the Accuracy of Thermal Transient Measurements", IEEE Transactions on Components and Packaging Technologies, vol. 25, 2002, pp. 539-546.

[14] M. Rencz and V. Székely, "Measuring Partial Thermal Resistances in a Heat-Flow Path", IEEE Transactions on Components and Packaging Technologies, vol. 25, 2002, pp. 547-553.

[15] M. Rencz and V. Székely, "Structure Function Evaluation of Stacked Dies", 20th IEEE Semi-Therm Symposium, 2004, pp. 50-54.

[16] EIA/JEDEC Standard No. 51-1.

[17] Lianqiao Yang, Jianzheng Hu, Lan Kim and Moo Whan Shin, "Variation of Thermal Resistance with Input Power in LEDs", Phys. Status Solidi C, 3, 2006, 2187-2190.

[18] F. Graham Smith and Terry A. King, Optics and Photonics: An Introduction, John Wiley & Sons, Ltd., 2000.

[19] Lan Kim, Moo Whan Shin, "Thermal Resistance Measurement of LED Package with Multichips", IEEE Transactions on Components and Packaging Technologies, vol. 30, 2007, pp. 632-636.

[20] Lalith Jayasinghe, Tianming Dong and Nadarajah Narendran, "Is the Thermal Resistance Coefficient of High-power LEDs Constant?", Seventh International Conference on Solid State Lighting, Proceedings of SPIE, vol. 6669, 2007, pp. 666911.

Analysis on Optical Properties for Various Types of Light Emitting Diode

Chin-Peng Ching[*], Zhi-Yin Lee, Sze-Yen Lee and Mutharasu Devarajan
Nano Optoelectronics Research (NOR) Lab, School of Physics,
Universiti Sains Malaysia, 11800 USM, Pulau Pinang, Malaysia.
[*]E-mail: ccp10_phy033@student.usm.my

Abstract- **In this paper, comparison on wall-plug efficiency for the infrared (IR) emitter, warm-white LED and multi-chips LED has been studied. The main objective of this work is to provide a specific reference on wall-plug efficiency between various types of LED. Optical measurement is carried out at constant current of 1.0A under ambient temperature of 26.0 ± 1.0°C. It is found that IR emitter offers the highest wall-plug efficiency (34.5%) compared to warm-white LED (18.4%) and multi-chips LED (29.3%). In addition, influence of forward current on optical power for all the LEDs has also been discussed. As a result, optical power increases with increasing of forward current.**

Keywords- Wall-plug efficiency, optical power, input power, forward current, IR emitter, warm-white LED, multi-chips LED.

I. INTRODUCTION

Light-emitting diodes (LEDs) are considered as an alternative for traditional light source as technology of LEDs have caused significant impacts on illumination industry and increasing market [1,2]. LEDs are developing fast because of their improved performance in optical power, efficiency and reliability [3-5].

The first practical LED introduced in 1962 had a luminous efficacy of only 0.15lm/W [6]. Since then, rapid development of LEDs resulted in improvement in luminous efficacy which is mainly due to the improved materials and designs from worldwide effort to create the most efficient light source [7]. From the past review, the researchers used Taguchi method to monitor the increase in external quantum efficiencies of LED while attempting to restrain the limited forward voltage value [8]. However, optical and electrical characteristics are strongly dependent on diode junction temperature which in turn is determined by forward current, heat sink and ambient temperature. Since spectral radiant flux is the primary optical characteristics for LED that determining luminous flux and color, efforts have been made to investigate the LED spectral flux and its variation with junction temperature [9].

In this paper, input power and optical power for the infrared (IR) emitter, warm-white LED and multi-chips LED have been obtained at constant input current of 1.0A under room temperature of 26.0 ± 1.0°C. Then, we make comparison on wall-plug efficiency for all the LEDs. Likewise, variation of optical power as a function of forward current has been studied to investigate the dependence of optical power on forward current. Besides, influence of forward current on wall-plug efficiency has also been reported.

II. THEORETICAL BACKGROUND

According to [10,11], radiometry is a measurement of optical radiation, which is electromagnetic radiation in the frequency range between 3×10^{11} Hz and 3×10^{16} Hz. This range corresponds to wavelengths between 10 nm and 1000 μm, and includes the regions commonly called the ultraviolet, visible, and infrared. Typical radiometric units include watt (radiant flux), watt per steradian (radiant intensity), watt per square meter (irradiance), and watt per square meter per steradian (radiance).

Photometry is a measurement of light, which is defined as electromagnetic radiation detectable by the human eye. It is thus restricted to the visible region (wavelength range from 360 nm to 830 nm), and all the quantities are weighted by the spectral response of the eye. Typical photometric units include lumen (luminous flux), candela (luminous intensity), lux (illuminance), and candela per square meter (luminance).

The difference between radiometry and photometry is that radiometry includes the entire optical radiation spectrum while photometry deals with the visible spectrum weighted by the response of the eye.

A. Optical Power and Wall-Plug Efficiency

For LED as a highly efficient electronic-to-photonic device, once electrical power is applied, optical power will be emitted as light source and heat will be dissipated to the ambience.

$$P_{el} = P_{op} + P_h \qquad (1)$$

where P_{el} is input electrical power, P_{op} is output optical power and P_h is heat power dissipation [12].

Optical power or simply known as radiant flux or radiant power is the energy (Q) radiated by a source per unit of time (t), expressed as

$$P_{op} = \frac{dQ}{dt} \qquad (2)$$

where unit of optical power is watt, W [13].

Wall-plug efficiency which is also called power-conversion efficiency is a measure of performance for solid-state lighting devices. Wall-plug efficiency is defined at the

978-1-4673-2395-6/12 $31.00 © 2012 IEEE

ratio of emitted optical power to the applied electrical power [14]. It is usually expressed in percentage.

$$WPE = \frac{P_{op}}{P_{el}} \tag{3}$$

III. EXPERIMENTAL WORK

Three commercial units of high power IR emitter, warm-white LED and multi-chips LED have been soldered on MCPCB as shown in Fig. 1. Each LED was driven by constant forward current of 1.0A under ambient temperature of $26.0 \pm 1.0^\circ$C to obtain input power and optical power values. For warm-white and multi-chips LEDs, measurement was carried out by using Thermal and Radiometric Characterization of LED (TERALED) measurement set. Since TERALED could only determine the optical properties for visible LEDs, optical characterization of IR emitter was carried out by using the different set of specific equipment. Subsequently, wall-plug efficiency for each LED was determined by using the input power and optical power obtained through the testing. In addition, optical power values for each LED driven with varying forward current (0.2A, 0.4A, 0.6A, 0.8A and 1.0A) at constant ambient temperature were determined to study the variation of optical power as a function of forward current. Similarly, wall-plug efficiency calculated from (3) has been used to plot against forward current to investigate the dependence of wall-plug efficiency on forward current.

(a)

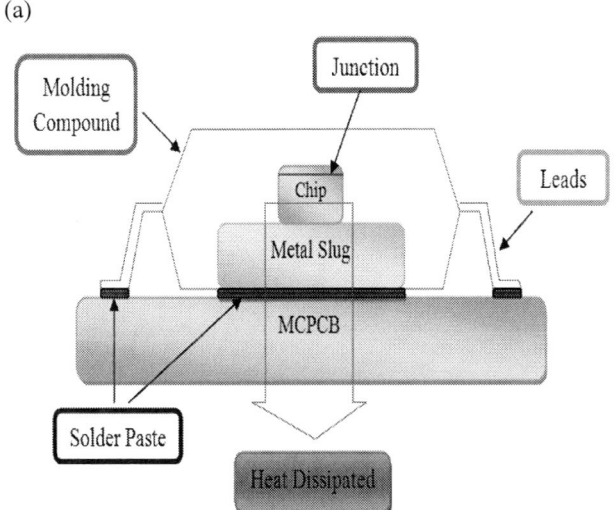

(b)

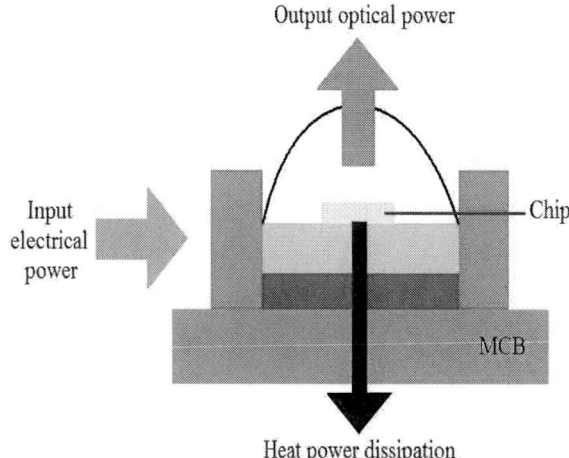

(c)

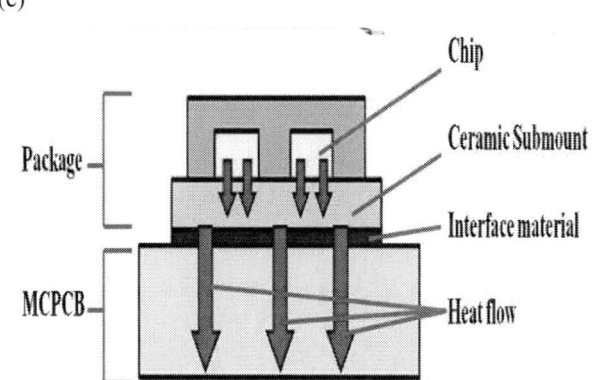

Fig. 1. Schematic diagram of high power (a) IR emitter, (b) warm-white LED, (c) multi-chips LED soldered on MCPCB.

IV. RESULTS AND DISCUSSION

In this section, the recorded data for the comparison of wall-plug efficiency for the IR emitter, warm-white LED and multi-chips LED has been presented. Besides the variation of optical power as a function of forward current, influence of forward current on wall-plug efficiency would also be discussed in sequence.

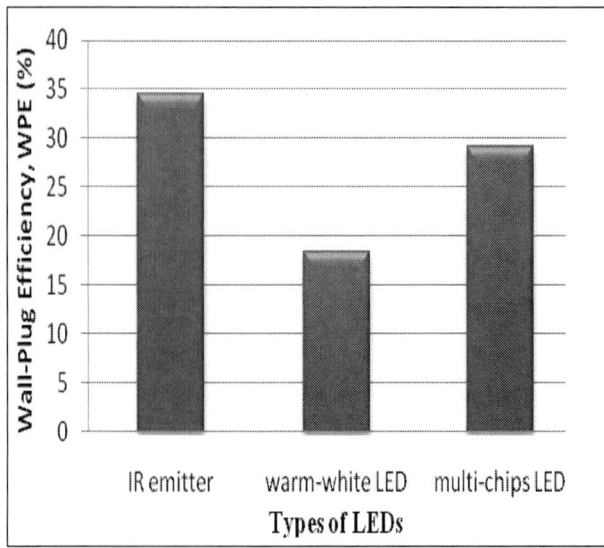

Fig. 2. Comparison of wall-plug efficiency for IR emitter, warm-white LED and multi-chips LED.

Fig. 2 presents the comparison of wall-plug efficiency for the various types of LED recorded at constant current of 1.0A under ambient temperature of $26.0 \pm 1.0°C$. In the figure, it is obvious that IR emitter offers the highest wall-plug efficiency, follows by multi-chips LED and warm-white LED. Warm-white LED shows the lowest value among the others. It means that higher input power is required in order to produce similar brightness as IR emitter and multi-chips LED. The power loss is probably happened during the phosphor convertion process as to generate warm-white light source. The generated light was trapped under the lens and lead to the occurrence of lower optical power [15].The calculated wall-plug efficiency for all the LEDs is tabulated in Table 1.

TABLE I

WALL-PLUG EFFICIENCY FOR THE LEDS

Types of LEDs	Input Power, P_{el} (W)	Optical Power, P_{op} (W)	Wall-Plug Efficiency, WPE (%)
IR emitter	1.62	0.559	34.5
Warm-white LED	3.52	0.647	18.4
Multi-chips LED	3.11	0.910	29.3

From the table above, the obtained wall-plug efficiencies for the IR emitter, warm-white LED and multi-chips LED are 34.5%, 18.4% and 29.3% respectively. Thus, IR emitter could effectively convert the 34.5% of the input power into optical power compared to warm-white and multi-chips LEDs which have lower wall-plug efficiencies. It is also found that the variation in wall-plug efficiency is due to the different chip technology used for the LEDs. According to [16], thin film technology allows the almost similar type of IR emitter

offering stellar performance and achieving wall-plug efficiency of approximately 35%.

Fig. 3 illustrates the variation of optical power as a function of forward current for all the LEDs recorded at varying forward current under constant ambient temperature. It is observed that optical power increases monotonously with increasing of forward current. This might be due to the increase of electrons into a specially designed recombination region, where they will recombine with excess holes [17] and hence, causes augment in the emission of radiant energy from recombination process [18]. Therefore, increase in the forward current will greatly enhance the recombination process and further lead to high emission of optical power.

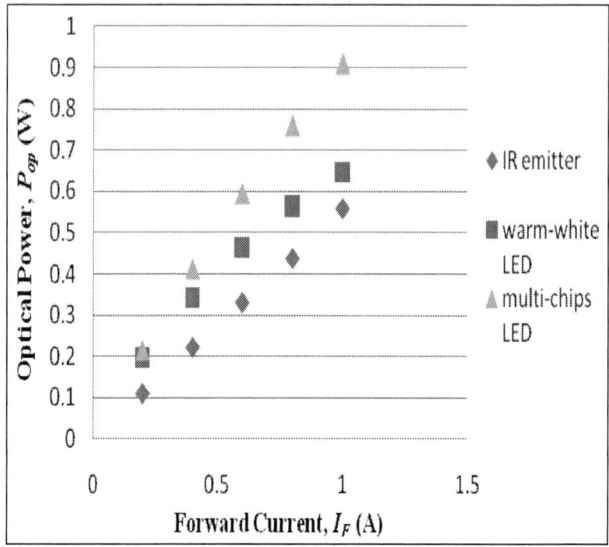

Fig. 3. Optical power as a function of forward current for the IR emitter, warm-white LED and multi-chips LED.

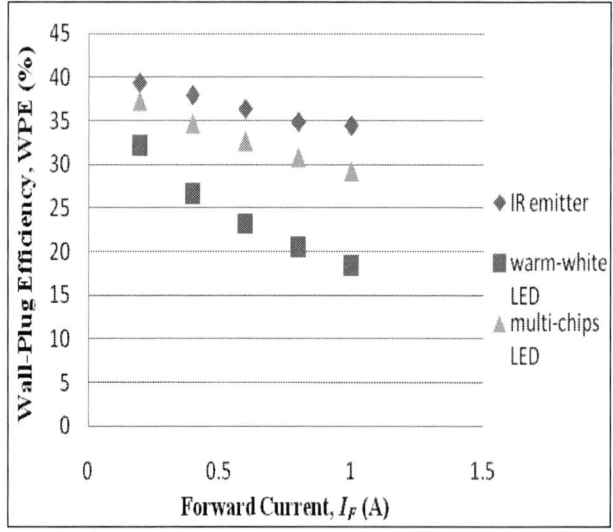

Fig. 4. Influence of forward current on wall-plug efficiency for the different types of LED.

Fig. 4 displays the influence of forward current on the wall-plug efficiency for the various types of LED recorded at varying forward current and constant ambient temperature. From the figure, wall-plug efficiency decreases gradually as forward current increases. Using (3), by assuming optical power is constant in this case, wall-plug efficiency is inversely proportional to the input power. As a result, increase in forward current causes input power to increase and hence, wall-plug efficiency will proportionally decrease [19]. Some researchers reported that this phenomenon might be due to the increment of forward voltage [20].

In Fig. 4, the maximum value of wall-plug efficiency achieved for the IR emitter might up to 40%. In case of warm-white and multi-chips LEDs, the highest wall-plug efficiencies are around 32% and 37% respectively. Even though LEDs play an important role in lighting systems nowadays, an average of 60-70% of the input power will be dissipated as heat to the ambience [21]. Heat generated will be conducted from the p-n junction of the LED package to the ambience in the direction opposing to the light output [22]. Since, the heat generated will cause excessive rise in junction temperature, and subsequently cause degradation of the LED package. Therefore, junction temperature is the key of reliability of an LED package.

V. SUMMARY AND CONCLUSION

The optical properties for the IR emitter, warm-white LED and multi-chips LED have been analyzed. At constant current of 1.0A under ambient temperature of $26.0 \pm 1.0^\circ$C, it is found that IR emitter offers the highest wall-plug efficiency, 34.5% compared to warm-white LED and multi-chips LED with much lower values, 18.4% and 29.3% respectively. By increasing the forward current, optical power increases whereas wall-plug efficiency would decrease. In conclusion, the maximum value of wall-plug efficiency achieved for the IR emitter, warm-white LED and multi-chips LED might up to 40%, 32% and 37% respectively.

ACKNOWLEDGMENT

The authors would like to thank Institute of Postgraduate Studies (IPS) of Universiti Sains Malaysia for the financial support through USM Fellowship and Graduate Assistant Scheme. In addition, this work was also supported by PRGS Grant no. 1001/PFIZIK/834073 and 1001/PFIZIK/834112.

REFERENCES

[1] P. Hanselaer, A. Keppens, S. Forment, W. R. Ryckaert and G. Deconinck, "A New Integrating Sphere Design for Spectral Radiant Flux Determination of Light-Emitting Diodes", Measurement Science and Technology, 20, 2009, 095111.

[2] Lianqiao Yang, Jianzheng Hu and Moo Whan Shin, "Degradation of High Power LEDs at Dynamic Working Conditions", Solid-State Electronics, 53, 2009, 567-570.

[3] Jianzheng Hu, Lianqiao Yang and Moo Whan Shin, "Electrical, Optical and Thermal Degradation of High Power GaN/InGaN Light-Emitting Diodes", Journal of Physics D: Applied Physics, 41, 2008.

[4] Ting Cheng, Xiaobing Luo, Suyi Huang and Sheng Liu, "Thermal Analysis and Optimization of Multiple LED Packaging Based on a General Analytical Solution", International Journal of Thermal Science, 49, 2010, 196-201.

[5] Peter Baureis, "Compact Modeling of Electrical, Thermal and Optical LED Behavior", Proceedings of ESSDERC, 2005.

[6] Holonyak N. J. and Bevacqua S. F., "Coherent (visible) Light Emission from Ga(As$_{1-x}$P$_x$) Junctions", Applied Physics Letter, 1, 1962, 82-83.

[7] Miran Burmen, Franjo Pernus and Bostjan Likar, "LED Light Sources: A Survey of Quality-Affecting Factors and Methods for Their Assessment", Measurement Science and Technology, 19, 2008, 122002.

[8] R. M. Lin, J. C. Li and T. E. Nee, "Brightness Improvement and Limited Forward Voltage of the AlGaInP MQW LED with Wet-Oxidation by Taguchi Method", IEEE Conference on Emerging Technologies-Nanoelectronics, 2006, pp. 245-248.

[9] A. Keppens, W. R. Ryckaert, G. Deconinck and P. Hanselaer, "Modeling High Power Light-Emitting Diode Spectra and Their Variation with Junction Temperature", Journal of Applied Physics, 108, 2010, 043104.

[10] Yoshi Ohno, "OSA Handbook of Optics, Volume III Visual Optics and Vision Chapter for Photometry and Radiometry", 1999, pp. 10-13.

[11] Ian Ashdown, "Photometry and Radiometry: A Tour Guide for Computer Graphics Enthusiasts", 2002.

[12] Lan Kim and Moo Whan Shin, "Thermal Resistance Measurement of LED Package with Multichips", IEEE Transactions on Components and Packaging Technologies, 30, 2007, pp. 632-636.

[13] F. Graham Smith and Terry A. King, Optics and Photonics: An Introduction, John Wiley & Sons, Ltd., 2000.

[14] B. E. A. Saleh and M. C. Teich, Fundamentals of Photonics, second ed., John Wiley & Sons, Inc., 2007.

[15] N.T. Tran and F.G. Shi, "Studies of Phosphor Concentration and Thickness for Phosphor-Based White Light-Emitting-Diodes", Journal of Lightwave Technology, 26, 2008, 3556-3559.

[16] http://my.mouser.com/osramirdragon/

[17] Robert L. Boylestad and Louis Nashelsky, Electronic Devices and Circuit Theory, ninth ed., Prentice Hall, 2006.

[18] Z. Y. Lee and M. Devarajan, "Optical Characterization of Different LED Packages at Various Biasing Current", IEEE Symposium on Industrial Electronics and Applications, 2011, pp. 424-427.

[19] C. P. Ching, M. Devarajan and W. C. Liew, "Thermal Characterization of a High Power Infrared Emitter as a Function of Input Current", 2nd International Conference on Photonics, 2011, pp. 134-138.

[20] Yukio Narukawa, Masahiko Sano, Masatsugu Ichikawa, Shunsuke Minato, Takahiko Sakamoto, Takao Yamada and Takashi Mukai, "Improvement of luminous efficiency in white light emitting diodes by reducing a forward-bias voltage", Japanese Journal of Applied Physics, 46, 2007, L963-L965.

[21] A. Christensen and S. Graham, "Thermal Effects in Packaging High Power Light Emitting Diode Arrays", Applied Thermal Engineering, 29, 2008, 364-371.

[22] S. Y. Lee and M. Devarajan, "Thermal Analysis of Multi-Chip LED Package with Different Position and Ambient Temperatures", 2nd International Conference on Photonics, 2011, pp. 269-272.

On the Waves in Circular Waveguides Containing Chiral Nihility Metamaterial under PMC Boundary

M.A. Baqir and P.K. Choudhury[*], *Senior Member, IEEE*
Institute of Microengineering and Nanoelectronics (IMEN)
Universiti Kebangsaan Malaysia
43600 UKM Bangi, Selangor, Malaysia
[*]E-mail: pankaj@ukm.my

Abstract—**Analytical investigation has been presented of the propagation of electromagnetic waves in a circular optical waveguide containing chiral nihility metamaterial. The waveguide under consideration has the inner region as free-space and the outer region as the chiral nihility metamaterial. The outer region of guide is coated with perfect magnetic conductor (PMC). It has been found that no net magnetic field exists in the chiral nihility region of the guide whereas the same exists in non-nihility (i.e. the free-space) region. The expressions corresponding to fields and power have been derived with the emphasis on the power pattern corresponding to the excitation of transverse electric modes in the guide.**

Keywords—*EM wave propagation, metamaterials, chiral fibers.*

I. INTRODUCTION

Investigators have reported recently about nano-engineered metamaterials, which find versatile technological applications [1,2]. These materials are designed in a way to interact with electromagnetic (EM) waves with a suitable control – the fundamental need for the fabrication of several EM devices [3,4]. Metamaterials are composites – the artificially designed structures by assembling tiny scale objects together [2], which serve to replace the atom and molecules of the conventional materials. This way, a composite structure can exhibit particular EM properties which cannot be observed in naturally occurring or chemically synthesized materials, making thereby the properties of *metamaterials* extended beyond that usually found in naturally occurring materials.

One of the interesting properties of metamaterials is to possess the characteristics of negative refraction and reflection of EM waves. Within the context, chiral metamaterials have been of great technological interest owing the ability to achieve negative refraction through the control over the chirality parameter [3–6]. It has been reported that the EM waves, when reflected by isotropic chiral media, unusual negative reflection occurs at the interface between the chiral medium and a perfectly conducting plane under the condition of strong chirality parameter. That is, the incident wave and one of the reflected eigenwaves lie in the same side of the interface normal [6].

The phenomenon of negative refraction and reflection can also be achieved by using the property of nihility in chiral medium. A medium with nihility is essentially a dielectric one wherein both the permittivity and the permeability are zero simultaneously [7], and wave propagation cannot be realized. Tretyakove *et al.* [8] implemented the concept of nihility to isotropic chiral mediums, and it was found that an unbounded chiral nihility medium can support two circularly polarized modes with one of them as backward mode. Later on chiral nihility medium attracted the attention of many researchers [9–12].

Within the context, an interesting contribution in the area of chiral nihility mediums has been the phenomenon of negative reflection of uniform plane wave by a planner perfect electric conductor (PEC) interface placed in chiral nihility medium [6]. In this arrangement, the electric fields of the incident and the reflected plane waves cancel each other yielding thereby net zero electric field. That is, the propagation of power remains non-existing in chiral nihility medium if the PEC interface affects the fields. This motivated to introduce waveguides with chiral nihility metamaterial for the purpose of confinement of power in the non-nihility region of the guide. Relevant studies have been covered for parallel, rectangular and circular waveguides [11,12].

In this communication, we present the study of the propagation of fields in (apparently) a three-layer circular optical waveguide containing free-space in the core region and chiral nihility metamaterial as the clad, which is coated outside with a perfect magnetic conductor (PMC). We investigate the behavior of fields in the chiral nihility region and the ordinary dielectric region of the guide, and study the behavior of the flow of power under the situation of transverse electric (TE) mode excitation.

II. ANALYTICAL TREATMENT

We consider a waveguide of circular cross-section having the core radius as a and the outer radius as b, as shown in fig. 1. The length of the waveguide is assumed to be infinitely extended. Within the guide, the region with $r \leq a$ is a free-space (having the permittivity and the permeability values as μ_0 and ε_0, respectively), and that with $b \geq r > a$ is made of chiral nihility metamaterial characterized by the parameters as $\varepsilon = 0$, $\mu = 0$ and $\kappa = 0$, κ being the chirality admittance parameter. Furthermore, the z-axis of the cylindrical polar coordinate system (r, ϕ, z) is assumed to coincide with the optical axis of the guide. We consider the field as harmonic in time t and the coordinate z, i.e. of the form $\exp(j\omega t)$, the mention of which is suppressed throughout the text. The constitutive relations of an isotropic, lossless and reciprocal chiral medium are generally given as [13]

$$D = \varepsilon E - j\kappa\sqrt{\varepsilon_0\mu_0}H \tag{1a}$$

and $\quad B = \mu H + j\kappa\sqrt{\varepsilon_0\mu_0}E, \tag{1b}$

where the symbols have their usual meanings as stated before.

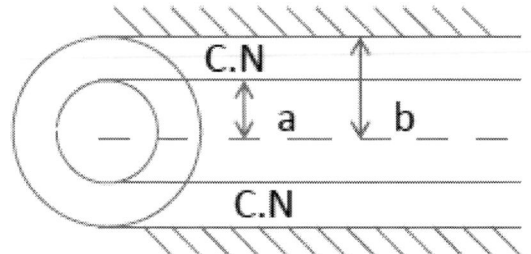

Fig. 1. Illustration of the circular waveguide made of chiral nihility metamaterial with a PMC boundary.

As described earlier, chiral nihility is considered as the special case corresponding to a chiral medium under the conditions when $\varepsilon \to 0$, $\mu \to 0$ and $\kappa = 0$. Under the situation, the constitutive relations for a chiral nihility metamaterial would be stated as

$$D = -j\kappa\sqrt{\varepsilon_0\mu_0}H \tag{2a}$$

and $\quad B = j\kappa\sqrt{\varepsilon_0\mu_0}E. \tag{2b}$

In chiral nihility metamaterials, of the propagating two modes, one is right circularly polarized (RCP,+) and the other one is left circularly polarized (LCP,−), and their wave numbers can be represented as $k_\pm = \pm\kappa k_0$. As such, the field in chiral nihility metamaterial may be decomposed as [13]

$$E = E_+ + E_- \tag{3a}$$

and $\quad H = \dfrac{j}{\eta}(E_+ + E_-), \tag{3b}$

where the parameter η is the impedance of the chirality nihility metamaterial, and can be represented as

$$\eta = \lim_{\substack{\mu \to 0 \\ \varepsilon \to 0}} \sqrt{\frac{\mu}{\varepsilon}}. \tag{4}$$

We now consider that the TE modes are excited into the free-space section of the guide. In that case, the solutions to the wave equation in this region can be written as

$$E_{z0} = 0, \tag{5a}$$

$$H_{z0} = B_{1n}J_n(k_{0r}r)e^{jn\phi}, \tag{5b}$$

$$E_{r0} = \frac{j\omega\mu}{k_{0r}}B_{1n}J_n(k_{0r}r)e^{jn\phi}, \tag{6a}$$

$$H_{r0} = -\frac{\beta n}{k_{0r}}B_{1n}J_n'(k_{0r}r)e^{jn\phi}, \tag{6b}$$

$$E_{\phi0} = \frac{j\omega\mu}{k_{0r}}B_{1n}J_n'(k_{0r}r)e^{jn\phi}, \tag{7a}$$

and $\quad H_{\phi0} = \dfrac{\beta n}{k_{0r}}B_{1n}J_n(k_{0r}r)e^{jn\phi}. \tag{7b}$

In eqs. (5), (6) and (7), n is the azimuthal index which can take discrete values and B_{1n} is a coefficient. Also, we assumed

$$k_{0r} = \sqrt{k_0^2 - \beta^2}. \tag{8}$$

Considering the fields in the chiral nihility region of the guide, as stated earlier, the incident wave splits into two components as (RCP,+) and (LCP,−) in nihility metamaterials. In this region, the longitudinal components of the electromagnetic field $E_\pm$ can be written as

$$E_{+z1} = C_{1n}J_n(k_{r+}r)e^{jn\phi} + C_{2n}Y_n(k_{r+}r)e^{jn\phi} \tag{9a}$$

and $\quad E_{-z1} = D_{1n}J_n(k_{r-}r)e^{jn\phi} + D_{2n}Y_n(k_{r-}r)e^{jn\phi}, \tag{9b}$

taking into account the oscillatory nature of fields in this region; $J_n(\bullet)$ and $Y_n(\bullet)$ being Bessel functions of the first and the second kinds, respectively. Further, C_{1n}, C_{2n}, D_{1n} and D_{2n} are the unknown coefficients to be determined by the boundary conditions. Also, the parameter $k_{r\pm}$ is given as

$$k_{r\pm} = \sqrt{k^2 - \beta^2} = k_0\sqrt{\kappa^2 - (\beta/k_0)^2}. \tag{10}$$

Now, following the analysis in ref. [10], the transverse components in the chiral nihility region of the guide can be expressed as

$$E_{r1} = (C_{1n} + D_{1n})\left\{\frac{jn\kappa k_0}{k_r^2 r}J_n(k_r r) - \frac{j\beta}{k_r}J_n'(k_r r)\right\}e^{jn\phi}$$
$$+ (C_{2n} + D_{2n})\left\{\frac{jn\kappa k_0}{k_r^2 r}Y_n(k_r r) - \frac{j\beta}{k_r}Y_n'(k_r r)\right\}e^{jn\phi}, \tag{11a}$$

$$H_{r1} = \frac{j}{\eta_1}(C_{1n} - D_{1n})\left\{\frac{jn\kappa k_0}{k_r^2 r}J_n(k_r r) - \frac{j\beta}{k_r}J_n'(k_r r)\right\}e^{jn\phi}$$
$$+ \frac{j}{\eta_1}(C_{2n} - D_{2n})\left\{\frac{jn\kappa k_0}{k_r^2 r}Y_n(k_r r) - \frac{j\beta}{k_r}Y_n'(k_r r)\right\}e^{jn\phi}, \tag{11b}$$

$$E_{\phi 1} = \left(C_{1n} + D_{1n}\right)\left\{\frac{\beta n}{k_r^2 r}J_n\left(k_r r\right) - \frac{\kappa k_0}{k_r}J_n'\left(k_r r\right)\right\}e^{jn\phi}$$

$$+ \left(C_{2n} + D_{2n}\right)\left\{\frac{\beta n}{k_r^2 r}Y_n\left(k_r r\right) - \frac{\kappa k_0}{k_r}Y_n'\left(k_r r\right)\right\}e^{jn\phi}, \quad (12a)$$

and
$$H_{\phi 1} = \frac{j}{\eta_1}\left(C_{1n} - D_{1n}\right)\left\{\frac{\beta n}{k_r^2 r}J_n\left(k_r r\right) - \frac{\kappa k_0}{k_r}J_n'\left(k_r r\right)\right\}e^{jn\phi}$$

$$+ \frac{j}{\eta_1}\left(C_{2n} - D_{2n}\right)\left\{\frac{\beta n}{k_r^2 r}Y_n\left(k_r r\right) - \frac{\kappa k_0}{k_r}Y_n'\left(k_r r\right)\right\}e^{jn\phi}. \quad (12b)$$

Thus, by the use of eqs. (9), (11) and (12), the axial components of the E- and the H-fields in the chiral nihility region can be written as

$$E_{z1} = \left(C_{1n} + D_{1n}\right)J_n\left(k_r r\right)e^{jn\phi} + \left(C_{1n} + D_{1n}\right)Y_n\left(k_r r\right)e^{jn\phi} \quad (13a)$$

and
$$H_{z1} = \frac{j}{\eta_1}\left(C_{1n} - D_{1n}\right)J_n\left(k_r r\right)e^{jn\phi} + \frac{j}{\eta_1}\left(C_{1n} - D_{1n}\right)Y_n\left(k_r r\right)e^{jn\phi}. \quad (13b)$$

As discussed earlier, in the case of a PMC waveguide, we assume that the wall of the guide at the interface $r = b$ is a perfect magnetic conductor, which leads the situation of non-existing magnetic fields in the region with PMC.

Now, in the attempt to evaluate the unknown coefficients, we apply the suitable boundary conditions at the layer interfaces, which ultimately yields their forms as

$$C_{1n} = D_{1n} = \Theta B_{1n} \quad (14)$$
and $$C_{2n} = D_{2n} = \Phi B_{1n}. \quad (15)$$

In eqs. (14) and (15), the used symbols have their meanings as

$$\Theta = \frac{k_r Y_n(k_r a)\left\{J_n(k_r a)Y_n(k_r b) - J_n(k_r b)Y_n(k_r a)\right\}}{\kappa k_0^2 J_n(k_{0r}a)Y_n(k_r b)\left\{J_n'(k_r a)Y_n(k_r a) - J_n(k_r a)Y_n'(k_r a)\right\}} \quad (16)$$

and $$\Phi = \Theta\frac{J_n(k_{0r}a)}{Y_n(k_{0r}a)}C_{1n}. \quad (17)$$

By substituting the coefficients as, determined by eqs. (14) and (15), back into field eqs. (11), (12) and (13), it can be demonstrated the net magnetic field inside the chiral nihility becomes zero. This essentially provides that there will be vanishing power inside the chiral nihility region, i.e.

$$P_{cn} = 0. \quad (18)$$

Now, following some tedious mathematical steps [14], it can be shown that the power inside the core of non-nihility region can be given as

$$P_c = 2\pi\left[\frac{na^2}{2}\left\{J_n^2\left(k_{0r}a\right) - J_{n-1}\left(k_{0r}a\right)J_{n-1}\left(k_{0r}a\right)\right\}\right.$$

$$+ \frac{n}{k_{0r}\Gamma^2(n+1)}2^{-2n+1}(k_{0r}a)^{2n}\,{}_2F_3\left(n, n+0.5;\right.$$

$$n+1, n+1, 2n+1; -k_{0r}^2 a^2\right) + \frac{1}{2}\left\{J_{n+1}^2\left(k_{0r}a\right)\right.$$

$$\left. - J_n\left(k_{0r}a\right)J_{n+2}\left(k_{0r}a\right)\right\} - \frac{2^{-(2n+1)}na(k_{0r}a)^{2n+1}}{k_{0r}}$$

$$\left. \frac{{}_2F_3\left(n, n+0.5; n+1, n+1, 2n+1; -k_{0r}^2 a^2\right)}{(n+1)\Gamma(n+1)\Gamma(n+2)}\right] \quad (19)$$

III. RESULTS AND DISCUSSION

So far we have deduced the expressions corresponding to the field and power inside the chiral nihility as well as the non-nihility regions of the guide. It has been found that the net magnetic field remains zero inside the chiral nihility region, which is owing to the negative refraction attained inside the chiral nihility medium when a perfect conductor is placed outside of it. This is due to the phenomenon of backward wave produced at the PMC interface canceling out the incident wave. Similarly happens in the case of guides with a coating of PEC, wherein the net electric field becomes zero inside the chiral nihility medium. However, the detailed description of that is beyond the scope of the paper.

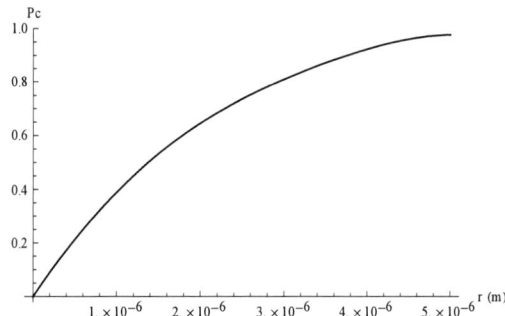

Fig. 2. Illustration of the power inside non-nihility region of guide corresponding to the excitation of TE_{01} mode.

As stated before, the core section of the guide structure is free-space. In our computations, we made an attempt to investigate the flow of power in the core of the guide corresponding to the excitation of TE_{01} and TE_{11} modes. We consider the core region as having a radius of 5 μm and the value of the operating wavelength as 1.55 μm. Figures 2 and 3, respectively, illustrate the plots of the normalized power against the core radius r. We observe that, corresponding to the case of

TE$_{01}$ mode, the power initially increases, and reaches a maximum around 4 μm radius, and then it becomes almost saturated with increasing radial distance. This may be attributed to the reason that the core section is bounded by the PMC coated nihility medium which possesses the property of suppressing the flow of power, and the effect is observable in this case as the power becomes saturated in the vicinity of the core boundary without exhibiting further increase.

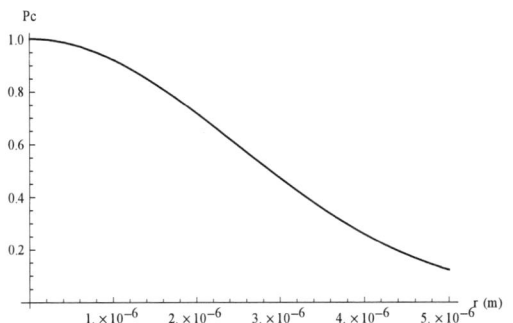

Fig. 3. Illustration of the power inside non-nihility region of guide corresponding to the TE$_{11}$ mode.

Corresponding to the case of the first higher order TE$_{11}$ mode, the power is found to be highly confined within the central region of the core of the guide structure, and it starts decreasing with the increase in radial distance. We find that the effect of the existence of the PMC coated nihility medium is more pronounced in this case as the power drops gradually with the increase in r. As such, the ultimate effect of the existence of the nihility medium is to suppress the flow of power.

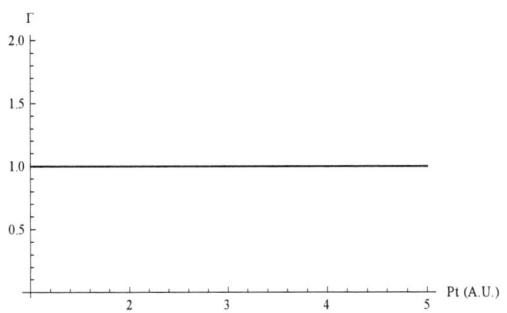

Fig. 4. Plot of Γ against the total power P_t in the guide.

Considering the total power P_t in the guide, which is the algebraic sum of the power P_c in the non-nihility core section and the power P_{cn} in the chiral nihility section, we give a look at the plot of the ratio Γ of P_c and P_t, as shown in Fig. 4. We observe that the plot remains simply as a straight line, which is owing to the reason that the power does not exist in the nihility region of the waveguide as the forward and the backward waves

cancel each other. Thus, the total power essentially remains confined in the non-nihility region of the guide.

IV. CONCLUSION

Analytical treatment has been made of a circular optical waveguide containing chiral nihility metamaterial having PMC coating in respect of its propagation characteristics. Expressions corresponding to the electric/magnetic fields as well as power inside the guide are derived with the conclusion that, for a PMC waveguide, the net magnetic field remains zero inside the chiral nihility region whereas both (the electric and the magnetic fields) exist in achiral region of the guide. Power patterns of the guide reveal that the power inside the core is suppressed near the chiral-achiral boundary, and the effect of nihility remains more prominent in the case of higher order transverse electric modes.

ACKNOWLEDGEMENT

The authors are thankful to Prof. Burhanuddin Yeop Majlis, the Director of IMEN (Universiti Kebangsaan Malaysia), for constant encouragement and help.

REFERENCES

[1] J.B. Pendry, "Negative refraction makes a perfect lens," Phys. Rev. Lett., vol. 85, 2000, pp. 3966–3969.

[2] E.S. Poutrina Larouche and D.R. Smith, "A metamaterial based parametric oscillator," Opt. Comm., vol. 283, 2010, pp. 1640–1646.

[3] T.G. Mackay and A. Lakhtakia, "Plane waves with negative phase velocity in Faraday chiral mediums," Phys. Rev. E, vol. 69, 2004, pp. 026602-1–9.

[4] T.G. Mackay, "Plane waves with negative phase velocity in isotropic chiral mediums," Microw. and Opt. Technol. Lett., vol. 45, 2005, pp. 120–121.

[5] T.G. Mackay and A. Lakhtakia, "Simultaneous negative- and positive-phase-velocity propagation in an isotropic chiral medium," Microw. and Opt. Technol. Lett., vol. 49, 2007, pp. 1245–1246.

[6] C. Zhang and T.J. Cui, "Spatial dispersion and energy in strong chiral medium," Opt. Exp., vol. 15, 2007, pp. 5114–5119.

[7] A. Lakhtakia, "An electromagnetic trinity from 'negative permittivity' and 'negative permeability'," Int. J. Infrared and Millimeter. Waves, vol. 23, 2002, pp. 813–818.

[8] S. Tretyakov, I. Nefedov, A. Sihvola, S. Masloviski, and C. Simovski, "Wave and energy in chiral nihility," J. Electromagn. Waves and Appl., vol. 17, 2003, pp. 695–706.

[9] Q.A. Naqvi, "Fractional dual solutions to the Maxwell equations in chiral nihility medium," Opt. Comm., vol. 282, 2009, pp. 2016–2018.

[10] J. Dong, "Exotic characteristics of power propagation in the chiral nihility fiber," Prog. Electromagn. Res., vol. 99, 2009, pp. 163–178.

[11] M.A. Baqir, A.A. Syed, and Q.A. Naqvi, "Electromagnetic fields in a circular waveguide containing chiral nihility metamaterial," Prog. Electromagn. Res. M, vol. 16, 2011, pp. 85–93.

[12] Q.T. Cheng, J. Cui and C. Zhang, "Waves in planar waveguide containing chiral nihility metamaterial," Opt. Comm., vol. 276, 2007, pp. 317–321.

[13] I.V. Lindell, A.H. Sihvola, S.A. Tretyakov, and A.J. Viitanen, Electromagnetic waves in chiral and bi-isotropic media, Artech House, Boston, 1994.

[14] K.Y. Lim, P. K. Choudhury, and Z. Yusoff, "Chirofibers with helical windings – An analytical investigation," Optik, vol. 121, 2011, pp. 980–987.

Improved Dead Time Response for Si Avalanche Photodiode

Norazlin Bahador, Fatimah K. A. Hamid, Afiq Hamzah, Suhaila Isaak, Razali Ismail

Computational Nanoelectronics Research Group (CoNE)

Universiti Teknologi Malaysia (UTM)

81310 UTM, Skudai, Johor, Malaysia

Email: razali@fke.utm.my

Abstract- **An improved model in defining the dead time (t_D) response for Silicon Avalanche Photodiode (Si APD) is presented in this paper. The generation of a mathematical model for Si APD is used to characterize the performance of passively quenched Si APD. For this purpose, quenching time (tq) and recharging time (tr) models are implemented to obtain t_D less than 50 ns. Therefore, the value of quenching resistor (R_L) should be minimized to reduce the recharge duration of time constant R_LC. By concerning various values of R_L and the influence of parasitic capacitance (C), tr and tq are analyzed in defining t_D for Si APD, which is working in Geiger's mode.**

Keywords: Avalanche photodiode, quenching resistor, recharging time, Geiger mode, dead time.

I. INTRODUCTION

In response to a single photon arriving, the detector must provide a high-output signal due to an internal high multiplication mechanism to allow easy recognition of the event by subsequent electronic circuits. A Si APD is a p-n junction device that is operated at high reverse bias voltages where the avalanche multiplication takes place [1]. Avalanche multiplication is whereby each free electron or holes causes a large number of free electrons and holes by impact ionization. Si APD can detect the very low intensity light signals and absorption of a single photon can initiate a strong avalanche current [5]. In recent years, sensors capable of detecting single photons are required for imaging systems, for example, astronomy, laser ranging, optical time-domain reflectometry (OTDR), single molecule detection, fluorescence decay and biomedical imaging [1].

These progress interests have influenced the demand of design optimization of a passive quenching circuit for Si APD to prior fabrication by numerical model simulation. The simulation analysis aims at relating both the VB and t_D to be relevant quenching circuit parameters, such as RL and C.

II. Si APD SIMULATION MODEL

The operation of Si APD with avalanche quenching circuits generated from the equivalent circuit model of the p-n junction, and the numerical analysis for Si APD avalanche quenching reports in this paper are based on a passive quenching circuit. Passive quenching is the simplest way to quench and reset the Si APD by connecting a single high load quenching resistor,

R_L between the bias voltage and the cathode of Si APD [1, 2, 6]. To ensure reliable quenching of breakdown, the value of R_L should be very large. The disadvantage of passive quenching is that it is very slow due to the large recovery tr and influences the size (duration) of the t_D, which limits the maximum count rate of the SPAD. The capacitance of Si APD capacitance (CD), R_L and parasitic capacitance (CP) all contribute to a large t_D. In practice, the maximum count rates achievable with off-chip passive quenching circuits do not exceeded 100,000 Cps [6, 10, 11]. In addition, passively quenched Si APDs do not have a well-defined t_D because of the slow increase of the detection probability during the recharge process [6, 4]. However, most of the developments of passively quenched Si APD is experimentally characterized. In this paper, the generation of a numerical model analysis based on passively quenched Si APD is discussed to obtain a small value of t_D, hence high count rates.

Si APD is biased above breakdown, V_B and gets triggered then current keeps flowing until the avalanche process is quenched by lowering the bias voltage down to V_B or below. The operating voltage must be restored after the t_D in order to make the Si APD able to detect another photon. Figure 1 show configuration of a passive quenching circuit for photon detection[5].

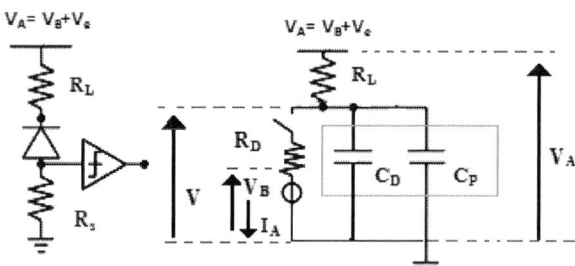

Fig.1: Configuration of passive quenching circuit

In addition, total capacitance is given by C=C_D+C_P and Ve are the excess voltage. When V=V_A, no current flows in the diode. From the passive quenching circuit above, before the arrival of a photon the avalanche current, I_A and the voltage, V change with time according to:

$$I_A(t) = \frac{V_e}{R_D} \exp\left(-\frac{t}{R_D C}\right) \tag{1}$$

$$V(t) = V_B + V_e \exp\left(-\frac{t}{R_D C}\right) \tag{2}$$

978-1-4673-2395-6/12 $31.00 © 2012 IEEE   396

During this operation, the avalanche current and the voltage across the diode follow respectively equation (1) and (2) until the quenching state. At quenching state, V is assumed equal to V_B, and C start to be recharged across R_L. The experimental values for lowering the bias voltages are implemented by using various value of R_L with fixed total capacitance value of 100 fF.

The recharge process of time-constant is given by:

$$V(t) = V_B + V_e \left[1 - \exp \left(-\frac{t}{R_L C} \right) \right] \qquad (3)$$

In order to model the quenching parameter, the definition of the initial conditions before the first carrier penetrates the multiplication region (t<0), all the currents are equal to zero as shown in Figure 1 and SPAD in waiting state. The voltage V is equal to V_A. At t=0, the first carrier entered the multiplication region can be expressed as in (4) and (5). The avalanche current, I_A:

$$I_A(0) = \frac{q.v_{sat}}{w} \qquad (4)$$

$$V(0) = V_A \qquad (5)$$

Velocity saturation in SI, v_{sat}: 1×10^7 cm/s
Electronic charge, q : 1.6×10^{-9} C

The value from initial condition will be used to solve the differential equation (6). Since, the passive quenching photoelectrons' detection probability increases with the bias voltage during the slow recharge, dead time is determined by the values of the series resistor and the depletion layer capacitance of the diode [8].

III. NUMERICAL MODEL OF SPAD

The circuit operation of SPAD during quenching and recharging time can be expressed by using a numerical model for biasing a voltage below breakdown voltage, V_B. Quenching process requires analysis of voltage near V_B. As shown in Figure 1, the differential equation for equivalent schematic of SPAD is equal to:

$$\frac{dV(t)}{dt} = \frac{V_A}{R_L C} - \frac{V(t)}{R_L C} - \frac{I_A(t)}{C} \qquad (6)$$

By using a definition of initial condition equation (5), and substitute equation (1) and (2) into (6), the quenching moment of voltage below V_B is established as follows:

$$V(t) = \left(\frac{V_A - V_B}{R_L C} \right) t + V_E \exp \left(\frac{-t}{R_D C} \right) + \frac{V_e R_D}{R_L} \exp \left(\frac{-t}{R_D C} \right) + A * (constant)$$

$$A = V_A - \left(V_e + \frac{V_e}{R_L} \right) \qquad (7)$$

The final implementations of a numerical model during quenching are expressed in (7), the graph in Figure 2 shows the simulation of voltage during quenching with varies RL. The implementations of a numerical model are proven when comparing within the empirical data [11] as shown in Figure 3. For comparison's purpose, quenching occurs between numerical model and experimental data in the range of (0.7±0.05) ns. There no significant differences between quenching for develop model and experiment data. Analysis are done considering a dead time response of the device, which

are the dead times of the SPAD, Δtdt is then defined as the sum of the quenching duration Δtq and the recharge duration Δtr, where for efficient quenching, Δtq<< Δtr.

IV. RESULTS AND DISCUSSION

The time occurrences for quenching are fast enough in a nanosecond. Usually, to measure dead time the quenching time can be ignored since its quite small compared to recharge process. From analysis, it had shown that the main contributions of dead time are during recharging process. Hence, the higher value of RL would increase a time for detection of the photon to occur in a passive quenching circuit. Therefore, the detection of an event becomes slower and less sensitive. Figure 4 shown the recharge process based on varies RL from equation (3).

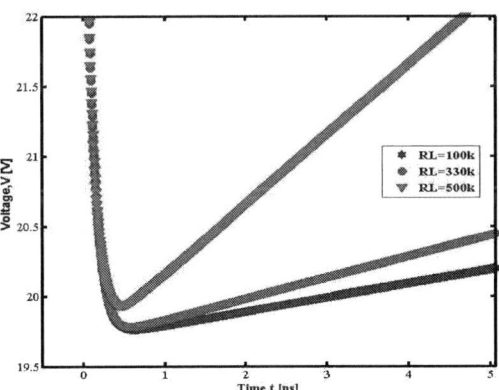

Fig. 2: Simulation of voltage for various quenching resistor value.

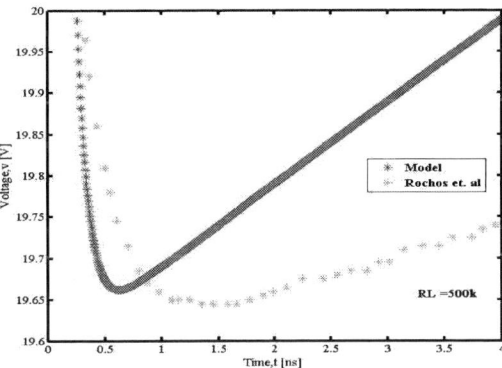

Fig. 3: Comparisons quenching moment between numerical model and experimental data.

The device is recharged via the R_L, it becomes increasingly biased beyond its breakdown voltage, so it can detect the next photon arrival. The quenching time constant, Δtq is set by the total capacitance, C=100fF and by R_D and R_L in parallel can be simplify by R_D= 1kΩ,

$$\Delta tq = \left(\left(\frac{R_D * R_L}{R_D + R_L} \right) \right) \cong R_D C \qquad (8)$$

Therefore, as can be seen, the higher value of R_L will provide the increase dead time response for the device performance. Table 1 show dead time, t_D simulation with varies R_L. The result indicated the dead time of less than 50ns can be achieved by concerning a value of R_L and C in the device.

TABLE I
DEAD TIME, ΔT_{DT} SIMULATION

Quenching Resistor, R_L (kΩ)	Recharge duration, Δtr (ns)	Dead Time, Δt_{dt} (ns)
100	7.035	7.135
330	23.28	23.38
500	35.20	35.30

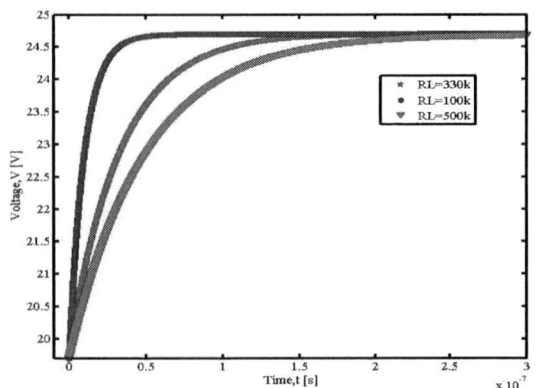

Fig. 4: Simulation of recharge process for varies quenching resistor value.

V. CONCLUSION

A numerical model for passively quenched Si APD was designed and the simulation characterization was performed in MATLAB . By optimizing the value of t_q and t_r diameter for various values of R_L, has allowed the simulation of the behaviour of the passive quenching circuit during the design phase.

ACKNOWLEDGEMENT

The authors acknowledge the Ministry of Higher Education Malaysia and the administration of Universiti Teknologi Malaysia (UTM) and Research Management Centre (RMC) for the project financial support through Institutional Grant vote number 77331 and 4P026.

REFERENCES

[1] S. Cova, A. Longoni, and A. Andreoni, "Towards picoseconds resolution with singlephoton avalanche diodes", Rev. Sci. Instruments, 52(3): pp. 408_412, 1981.

[2] W.J. Kindt, N.H Shahrjerdy, H.W van Zeijil, "A silicon avalanche photodiode for single optical photon counting in the Geiger Mode," Sensors and Actuators A: 60, pp.98-102,1997.

[3] Wang Yifeng,Ma Yu, "Development of Quenching Circuits of Single Photon Avalanche Diodes," DianziKeji - Electronic Science and Technology. Vol. 24, no. 4, pp. 113-118. Apr 2011.

[4] L. Neri, S. Tudisco, F. Musumeci, A.Scordino, G. Fallica, M. Mazzillo, M.Zimbone, "Dead Time of Single Photon Avalanche Diodes," Nuclear Physics B (Proc. Suppl.),215(2011),pp.291–293, 2011.

[5] S. Tisa, F. Zappa, A. Tosi, S. Cova, " Electronics for single photon avalanche diode arrays," Sensors and Actuators A 140,pp 113–122, 2007.

[6] S. Cova, M. Ghioni, A. Lacaita, C. Samori, and F. Zappa, "Avalanche photodiodes and quenching circuits for single-photon detection," Applied Optics, Vol. 35, Issue 12, pp. 1956-1976, April 1996.

[7] Li-Clang Li and Lloyd M. Davis, "Single photon avalanche diode for single molecule detection," Rev. Scl. lnatrum., Vol. 64, No. 6, pp. 1524-1529, June 1993.

[8] Alberto Dalla Mora, Alberto Tosi, Member, IEEE, Simone Tisa, and Franco Zappa, Senior Member, IEEE, "Single-Photon Avalanche Diode Model for Circuit Simulations," IEEE Photonics Technology Letters, Vol. 19, No. 23, December 1, 2007.

[9] S. A. Castelletto, I .P. Degiovanni, V. Schettiniy and A. L. Migdall, " Reduced Deadtime And Higher Rate Photon-Counting Detection Using A Multiplexed Detector Array," Journal of Modern Optics, Vol. 54, No. 2–3, pp 337–352,2007

[10] A.Rochas ,P.A.Besse., R.S. Popovic, "Actively recharged single photon counting avalanche photodiode integrated in an industrial CMOS process, "Sensors and Actuators A:Physical, Vol.110, p. 124-129, 2004.

[11] A.Rochos, "A. Rochas, "Single Photon Avalanche Diodes in CMOS Technology,"École Polytechnique Fédérale de Lausanne, Ph.D. Thesis, 2003.

Physical Effects from Etching Parameters of the Bragg Grating Waveguide Fabricated on Porous Silicon Nanostructure

Ahmad Afif Safwan Mohd Radzi*[1,2], Shamsul Faez Mohd Yusop [1,2], Nurul Izrini Ikhsan[1], Mohamad Rusop[2,3] and Saifollah Abdullah[1,2]

[1]Faculty of Applied Sciences
[2]NANO-SciTech Centre, Institute of Science
[3]NANO-ElecTronic Centre, Faculty of Electrical Engineering
Universiti Teknologi MARA, 40450 Shah Alam, Selangor D.E, Malaysia.
*Corresponding author email: afifsafwan7788@gmail.com

Abstract— **Multilayer structure of Bragg Grating Waveguide (BGW), porous silicon (PSi)-based was fabricated and characterized. The BGW was directly etched on a PSi-based planar waveguide. Adjustment of parameters for the electrochemical process will let the realization of multilayer properties of PSi. Fabricated BGW structure depends on thickness and layers of porous structure, and also average pore size. It is well known from previous study that the modulation of multilayer PSi much affected by the HF concentration of electrolyte, etching time, and current density applied during the electrochemical etching process. Surface homogeneity and layer uniformity are also the scope of study and both are much relying on those factors. The average refractive index, n and pore sizes for the multilayer structure were determined and the comparison of the results based from the study was shown. Fabricated BGW on PSi is now intensely investigated for application as an optical sensor for chemical substances.**

Keywords-Porous silicon; multilayer; refractive index; Bragg grating waveguide; optical sensor.

I. INTRODUCTION

Since the discovery of photoluminescence from porous silicon by Canham et al., intense researches have been done to investigate other electrical and optical properties of porous silicon. Multilayer porous silicon has been investigated due to its optical properties which can behave such as Bragg mirror, optical micro cavity and Fabry-Perot interferometer [1]. The structure of BGW is very much relying on certain parameters during the electrochemical etching process. Since etching parameters such as the HF electrolyte concentration, current density and time of etching playing such a big role during the fabrication of porous silicon, those factors would determine the structure's surface morphology uniformity and layer homogeneity of the fabricated BGW [2]. The specifications, drying technique and doping type of the silicon wafer also affects overall structure of porous silicon during etching. In order to produce a good gas sensor base from porous silicon, pore size distribution of porous silicon will be emphasized. This is because the pore is situated at the surface of structure and is the place where gas will be trapped and condensed inside. To control the pore size, several parameters during etching process must be varied. The pore size would increase if the HF acid concentration decreased, current density increased and etching

time increased [3]. Capillary effect also occurred due to the skeleton structure of the porous silicon itself. This effect is very interesting because it realize the properties of a gas sensing device [4]. Different doping of silicon will yield different orientation to the surface and reaction to the etching parameters. For example n-type silicon will produce higher porosity porous silicon while a p-type would react inversely when both electrochemically etched with same method [5]. The pore structure will never overlap its neighboring pore and also has the constant hole etching rate which will never overlying the upper etched layer. This is due to the existence of depletion region between pore tips and the directional pore growth [3]. Thus a uniform and complex multilayer porous structure can be fabricated without any doubt if the structure may collapse [6]. By using the electrochemical etching technique, it allows a simpler and cheaper method to produce a BGW structure with the advantage of producing multilayer thin films consisting different refractive indexes and thicknesses in a single substrate without the requirement of deposition process.

II. EXPERIMENTAL DETAILS

Porous silicon (PSi) is conventionally produced by using the electrochemical etching method on a crystalline silicon wafer. Electrolyte is the medium for chemical etching reaction to occur. The electrolyte is a mixture of hydrofluoric (HF 48%) rich acid-based solution and absolute ethanol (C_2H_5OH 99%). The mixture were composed with ratio HF:Ethanol of 1:1. Crystalline silicon substrate used was a highly doped p-type silicon wafer, <1 0 0> oriented, and 525μm thickness. Electrochemical cell used was Teflon cell with aluminium plate as the base. A tungsten rod served as cathode of the electrolytic cell while Silicon wafer was mounted at bottom of the cell. Usually bottom of the sample was coated first with aluminium, but we just use aluminium contact for anode. In order to produce multilayer porous silicon, the etching current density and etching time were varied during etching process so that the realization of thin films with different porosity can be achieved. There are 3 types of sample fabricated for this study. All samples were first etched on the polished surface with current density of 10mA for a period of time and continued with second current density of 20mA with the same period of time. For type 1, porous silicon was etched with 30 seconds etching time for each layer and 20 minutes for the whole process. While for type

978-1-4673-2395-6/12 $31.00 © 2012 IEEE

2 the etching time was 20 seconds for each layer and 14 minutes for the whole process. For type 3 each layer was etched for 10 seconds. For structural view, samples were characterized by FESEM (JEOL JSM 7600F and ZEISS 77 SUPRA 40VP) and atomic force microscopy (AFM, XE-100 Park Systems). For the ellipsometric study, Ellipsometer L 116S (Gaertner Scientific Corp.), PL intensity (Horiba Jobin Yvon 79 DU420A-OE-325) and reflectance spectra from UV-Vis (PerkinElmer Lambda 750).

III. RESULTS AND DISCUSSION

Porous silicon structure formed once the chemical reaction between surface of silicon wafer and electrolyte started. The chemical reaction can be expressed as below:

$$Si + 2HF + 2h^+ \longrightarrow SiF_2 + 2H^+,$$

$$SiF_2 + 4HF \longrightarrow H_2 + H_2SiF_6. \quad [2]$$

Porosity determination from reaction is depending on the etching rate. h^+ ion is the hole charge which shows the magnitude of etching. From literature, pore size distribution can be classified in standard measurement prepared by IUPAC. For structure with pore width less than 2nm, the type of pore can be classified as micro. For pore width about 2-50nm, it is in the range of mesopore. For pore width of 50nm and above, the pore is in macro size [3].

A. Ellipsometric Study of Multilayer Porous Siliocn

Table 1 displays the thickness and refractive index values of fabricated multilayer structure. The incident angle of laser source (He-Ne) used was set at variable angles and average thickness for each corresponding thicknesses were measured.

TABLE I
THICKNESS AND REFRACTIVE INDEX VALUES FROM MEASUREMENT.

Sample	Thickness, t (nm)	Refractive index, n	
		n_H	n_L
Type 1	t_1=341, t_2=375	1.719	2.012
Type 2	t_1=298, t_2=275	1.859	2.109
Type 3	t_1=163, t_2=164	1.412	1.538

t_1 = thickness for the first alternating layer, t_2= thickness for the second alternating layer
n_H = refractive index of high porosity layer, n_L= refractive index of low porosity layer

The overall corresponding thickness of porous silicon layer on crystalline silicon measured was 14.33 μm for type 1, 11.48 μm for type 2 and 6.5 μm for type 3. These results were calculated by effective medium approximation (EMA) from ellipsometer. Refractive index is directly related with the porosity of silicon nanostructure. Alternating higher and lower porosity layer within the periodic layers will provide alternating values of refractive index. The refractive index for high porosity layer is referred to n_H while for lower porosity is n_L. It is known from reference [3] that high porosity structure will have lower refractive index value since the structure consisted of larger void area. To fabricate a BGW by using thin films, alternating thin films with low and high refractive index is strongly recommended. Thus by modulating the etching time and current density, we can see the alternating pattern existed with a strong capability to be applied as a waveguide.

B. Structural Study of Multilayer Porous Silicon

Surface morphology investigation is a major part of sensor device fabrication. The interaction between substance and sensor first happened on the surface of a sensor itself. Gas sensor with high surface area would provide a high sensitivity and very crucial for sensing method. Table 2 displays the average pore size for all samples. From Fig. 1, we can see the pore size distributions for type 1, type 2 and type 3 are utterly different. Fig. 1(a) shows pore sizes which are slightly bigger compared to Fig. 1(c) and 1(e). In Fig. 1(c), the pore size is smaller and well distributed on the surface and smaller pore size distribution from Fig. 1(e). Thus sample type 2 and 3 should provide better sensing efficiency due to smaller pore and grain size.

TABLE II
AVERAGE PORE SIZE OF SAMPLES

Sample	Average Pore Width	Standard IUPAC Pore Size
Type 1	20-50 nm	Mesopore
Type 2	15-30 nm	Mesopore
Type 3	5-15 nm	Mesopore

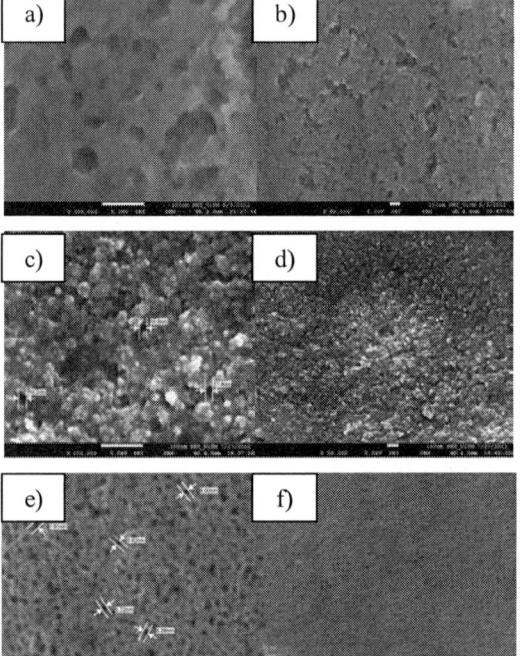

Fig. 1. Figure shows the top view of Type 1 ((a) and (b)), Type 2 ((c) and (d)) and Type 3 ((e) and (f)).

Fig. 2 shows the 3D view from AFM where (a) is for type 1, (b) for type 2 and (c) for type 3. We can see the peaks in 2(b) and 2(c) are well distributed compare to 2(a). This is because the pores on type 2 and 3 are much narrower and well distributed resulting better peak formation. While in 2(a) the peaks are less uniform because the pores are bigger, thus the nanocrystalline structures formation on the surface are less.

978-1-4673-2395-6/12 $31.00 © 2012 IEEE

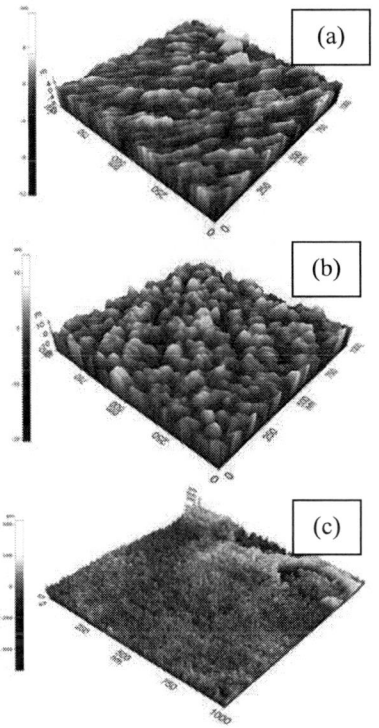

Fig. 2. Three-Dimensional surface view from AFM.

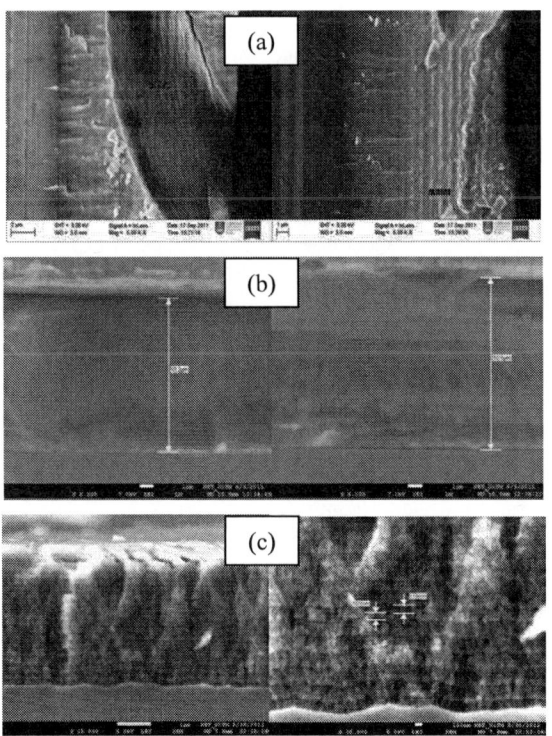

Fig. 3. Figures show the cross sectional view for (a) Type 1, (b) Type 2 and (c) Type 3.

In terms of roughness, higher peaks formation would produce rougher surface morphology because the number of skeletal structures on the surface is significantly more. The roughness for 2(a) is 1.847 nm, 2(b) is 2.640 nm and 2(c) 10.517 nm. Here we can conclude that with less peak formation the surface roughness of a sample can be reduced. Another crucial part in fabricating a BGW is the multilayer thin films formation. PSi multilayer was produced by etching technique. Compare to deposition technique, etching is cheaper and faster fabrication method. The main difference between etching and deposition is in etching, the thin films formed inside the substrate while in deposition, the films formed on the substrate. Multilayer structure using etching process has the same view and properties as we can in Fig. 3. Here we can see the cross sectional view of multilayer PSi consisted of layers with alternating high and low refractive index. From Fig. 3, the whole thicknesses determined for both samples are 15.5 μm for type 1, 10.2 μm for type 2 and 4.85 μm for type 3. With all samples has the same number of layers, thickness of each layer can be calculated. The approximate average thickness of each layer in type 1 is 387.5 nm and type 2 is 255 nm while 120 nm for type 3. Here we can see the apparent difference between the thicknesses from both samples. It is obvious that type 1, which has longer etching time, is slightly thicker than type 2 and 3. The realization of multilayer stacking of PSi structure can be seen and verified. From ellipsometric study, we can see the comparison of thickness values for both characterization methods.

C. Photoluminscence (PL) Spectroscopy

Optical properties are a major scope for this study. To get a good optical sensor, PL can describe the optical response for it. From room temperature PL graph in Fig. 4, the value of full width half maximum FWHM for type 1 is 101.31 and type 2 is 100.66. The difference between those FWHMs is 0.65 which is very low. Grain size estimation can be made where broader FWHM will have bigger grain size. The lattice structure of nanocrystalline silicon determines the PL response from both samples.

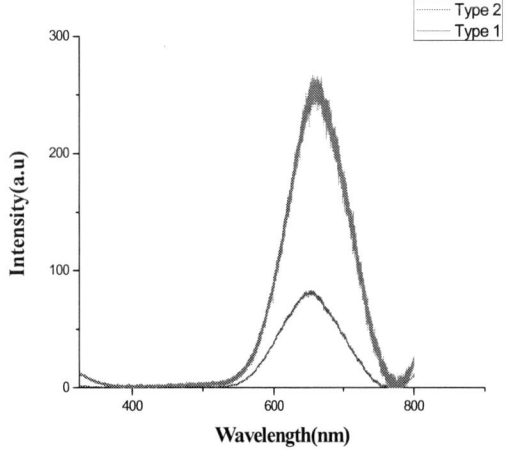

Fig. 4. PL Intensity for Type 1 and Type 2.

Type 1 has a slightly longer wavelength peak due to the fact that the optical phonon response within the crystals structure was degraded because of high porosity [7]. The carrier confinement within the lattice was limited based from quantum confinement. Because smaller grain size, such that the nanoscale silicon pillars are narrower, better PL emission detected as we can see from type 2. Wavelength shift indicates that band gap energy for sample 1 is lower than sample 2. This is because as crystalline structure decreases, the quantized carrier phonon moves toward lower energy emission [7]. The quantized carrier was affected by alternating thickness and porosity hence affecting the PL. This shift giving information that the quantum confinement of carrier in the crystalline is dependent on the structure of the sample itself. In terms of intensity, obviously type 1 is higher compare to type 2. Intensity value for type 1 is 247.79 and type 2 is 79.07. From PL intensity, we can describe the optical efficiency of a sample in terms of light sensitivity. With high intensity, such as type 1, the optical response from the surface is much better. BGW is a light reflector, thus low absorbance and high reflectance from the optical structure is a must.

D. UV-VIS Spectroscopy

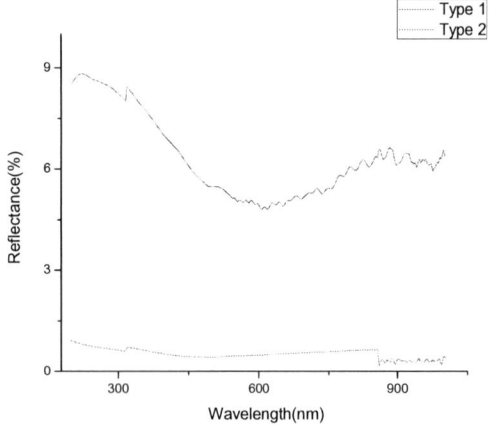

Fig. 5. Reflectance spectra from sample 1 and sample 2.

From Fig. 5, the graphs show very low reflectance from BGW due to thin multilayer film with alternating refractive index. The alternating refractive index then produces a total internal reflection effect which confine and guiding light passing through it and may have increase the light's optical path length. This effect also explains about the fringes pattern produced from both graphs [7]. The low reflectance also affected by scattering based on Rayleigh scattering [8]. The high refractive index contrast between alternating layers and the substrate would produce very high light scattering and change its propagation path. Light scattering also can be described by the roughness measurement from AFM. type 1 has a low roughness value compare to type 2. This concludes

that rough surface scatter incident light from the surface and disturbing reflected light from the multilayer structure.

IV. CONCLUSION

BGW fabricated on PSi was successfully prepared by electrochemical etching method. The physical properties of BGW have been investigated and verified. The multilayer PSi nanostructure with alternating refractive index and thickness determined. From FESEM images, the multilayer PSi can be seen and BGW structure was verified. From Ellipsometer, the average thickness per layer for type 1 is 341.54~375.16 nm and FESEM is 387.5 nm. For type 2, thickness measurement from Ellipsometer is average 275.29~298.71 nm and FESEM is 255 nm. Layer Thickness for Type 3 from Ellisometer is 163~164 nm and 120 nm from FESEM. AFM analysis shows that surface roughness is contributed by the pillar/skeletal structure on the surface. The PL spectroscopy determined that intensity for type 1 is much higher than sample 2 thus possessed better efficiency as a BGW and optical reflector. From reflectance spectra, clearly the reflectance of both samples is low due to the light scattering on the surface and rayleigh scattering within the multilayer structure. Multilayer PSi has the properties of an optical gas sensor with high optical efficiency and the next step for this study should be discussing about the gas sensing method of this device.

ACKNOWLEDGMENT

The author would like to thank Universiti Teknologi MARA (UiTM) Malaysia and Ministry of Higher Education Malaysia for the financial support. Also special thanks to Nano-Optoelectronics Research and Technology Laboratory USM for the ellipsometer measurement.

REFERENCES

[1] E.K Squire, P.A. Snow, P.St.J. Russell, L.T. Canham, A.J. Simons, C.L. Reeves, D.J. Wallis, "Light Emission from Highly Reflective Porous Silicon Multilayer Structures", Journal of Porous Materials 7 (2000) pp. 209–213.

[2] Pushpendra Kumar and Patrick Huber, "Effect of Etching Parameter on Pore Size and Porosity of Electrochemically Formed Nanoporous Silicon", Journal of Nanomaterials, Volume 2007, (2007).

[3] O. Bisi, Stefano Ossicini, L. Pavesi, "Porous silicon: A Quantum Sponge Structure for Silicon Based Optoelectronics" Surface Science Reports 38 (2000) pp. 1-126.

[4] Lei Zhen-Kun, Kang Yi-Lan, Qiu Yu, Hu Ming, Cen Hao, "Experimental Study of Capillary Effect in Porous Silicon Using Micro-Raman Spectroscopy and X-Ray Diffraction" Chinese Physical Society, 21, No. 7 (2004) pp. 1377-1381.

[5] A. G. Nassiopoulou, "Porous Silicon For Sensor Applications" Nanostructured and Advanced Materials, (2005) pp.189–204.

[6] Shu-Zee A. Lo and Thomas E. Murphy, "Porous Silicon Based Terahertz Bragg Grating Filter" Optical Society of America (2009).

[7] R. S. Dubey and D. K. Gautam, "Synthesis and Characterization of Nanocrystalline Porous Silicon Layer for Solar Cells Applications" Journal of Optoelectronic and Biomedical Materials 1 (2009), p. 8-14

[8] Tomas Svensson, Zhijian Shen, "Laser spectroscopy of Gas Confined in Nanoporous Materials" Applied Physics Letters 96, (2010).

High Performance of a SOI-based Lateral PIN Photodiode Using SiGe/Si Multilayer Quantum Well

[1]P.Susthitha Menon, *Member, IEEE*, [2]S.Kalthom Tasirin, [2]Ibrahim Ahmad, *Senior Member IEEE*, [2]S.Fazlili Abdullah

[1]Institute of Microengineering and Nanoelectronics (IMEN), Universiti Kebangsaan Malaysia
43600 Bangi, Selangor, Malaysia
[2]Department of Electronics and Communication Engineering, College of Engineering, Universiti Tenaga Nasional, Jalan
IKRAM-UNITEN, 43000 Kajang, Selangor, Malaysia
E-mail: susi@eng.ukm.my, sitikalthom@gmail.com

ABSTRACT - Silicon–on-insulator (SOI) based SiGe quantum well infrared pin photodiode has the potential of being a serious candidate for applications in sensing applications as well as in optical fiber communications. The present work investigates the performance of a virtual lateral PIN photodiode with a SiGe/Si multi-quantum well structure. In this paper, 5 periods of stacked SiGe quantum wells were grown on Si(100). A lateral PIN photodiode consisting of the SiGe/Si multi-quantum well layers as the active absorption layer with intensity response in the 700-1600 nm wavelength range was demonstrated. The results obtained for responsivity, total quantum efficiency and frequency response were 0.89 A/W, 71% and 21 GHz respectively for design parameters of intrinsic region length of 6 μm, photoabsorption layer thickness of 50 μm, incident optical power of 1 mW/cm^2 and bias voltage of 3 V. As a conclusion, the SiGe/Si multi-quantum well solution in achieving the desired high performance photodiode was achieved.

Keywords; SiGe/Si multi quantum well (MQW), PIN photodiode, Silvaco, SOI

I. INTRODUCTION

The advantage of compatibility with the VLSI technology and low cost makes Silicon–on-Insulator (SOI) based photodiodes a very important research project for the application in sensing as well as optical fiber communications. With thin layers, the electrical current leakage that degrades the performance of the device can be reduced.

Due to the strong confinement effects of electron and holes, quantum wells exhibit novel physical properties leading to important applications in microelectronic and optoelectronics. Quantum wells are essentially suitable for infrared photo-detection because of their high sensitivity to normal incidence lights and low dark current [1]. The advantage of using SiGe for optoelectronic devices, include the low defect density of the material, that it can enhance operations even at room temperatures [2]. Also a SiGe multi-quantum well devices device's operating wavelength can be tuned over a range of 1.3 μm to 1.55 μm making them ideal choices for sensing and optical fiber communications [2]. Besides that, the SiGe quantum well has lower dark current and high frequency performance compared with bulk Silicon [3]. The SiGe offers the opportunity for more flexible band gap tuning than silicon-only technology for being a heterojunction

technology with an adjustable band gap [3]. The lateral PIN photodiodes can be fabricated in this technology [4], giving incident light direct access to the active intrinsic region of the device.

In the literature, SiGe/Si quantum well photodiode were designed with 20 multi quantum wells (MQW) of Ge/Si bilayers and 25 nm of Si spacer layers. The device exhibited responsivity of 0.158 mA/W at 1.31 μm with an applied voltage of 1-V [1]. In previous research [5], the results of frequency response was up to 40GHz at 1300 nm to 1550 nm. Sood et. al [6] reported at -1V and wavelength of 1550 nm, the 5 x 10 μm device with Ge quantum wells were alternated within 10 nm to 30 nm Si barrier layers and had a responsivity of 0.18 A/W. The photodiode with a SiGe/Si MQW thicknesses of 5 and 25 nm respectively,exhibited responsivity of 0.3 A/W at a small bias of 2V [7].

In this paper, for the first time the lateral PIN photodiode based on $Si_{0.5}Ge_{0.5}$/Si multi-quantum well on a SOI substrate as a high performance photodetector is discussed. This paper investigates the SiGe/Si multi-quantum well structure giving the effects on the electrical and optical characteristics of a lateral PIN photodiode based on the device which was developed previously on a pure silicon substrate [8], [9], [10], [13]]. A lateral photodiode design is capable of reducing parasitic capacitances and hence maximize the bandwidth ..

In this paper, numerical modeling was used to design and characteristize the device. In terms of simulation for the fabrication and device electrical characterization, ATHENA and ATLAS software from Silvaco Int. were used respectively. In this way a more physical insight into the device operation can be obtained. The solution of Poisson-Fermi_Schroedinger equations is needed for the calculation of the hole density in the quantum well of SiGe/Si PIN photodiode for quantum well thicknesses of less than 10 nm [11], [12]. The models in bulk material which were used are the Shockley-Read-Hall (SRH) recombination model, the concentration dependent mobility model, the Auger model and the Fermi-Dirac statistic model [8], [9], [10], [13]

II. METHODOLOGY

The Si lateral pin-photodiode was simulated on a silicon substrate (n$^+$-type) with thicknesses of 50 μm using ATHENA

978-1-4673-2395-6/12 $31.00 © 2012 IEEE

software from Silvaco Int. Then, the SiO_2 buffer layer was fabricated on the Silicon substrate with thicknesses of 500 nm. In order to fabricate the device, 5 periods of $Si_{0.5}Ge_{0.5}/Si$ multi-quantum well were fabricated onto a (100) n^+-type Si substrate. The thicknesses of the SiGe with a Ge content of 50% and Si layers are 5 and 20 nm, respectively Then, the n-well was developed using phosphorus diffusion with dopant concentration of 2.02×10^{19} cm^{-3} on the left side of the photodiode with diffusion temperature of 1000 ^{0}C for 50 seconds. Whilst the p-well was diffused with boron concentration of 8.09×10^{19} cm^{-3} on the other side of the photodiode with temperature of 1000 ^{0}C for 120 seconds. SiO_2 layer with thickness of 280 nm was deposited on the silicon substrate to act as a passivation layer. The electrode contacts with thickness and length of 500 nm and 6000 nm respectively, were processed by depositing aluminum on the n-well and p-well areas of the silicon photodiode.

The responsivity of a pin-photodiode is given by [10]:

$$\mathcal{R} = \frac{I_T}{I_S}\left(\frac{\lambda}{1.24}\right) \qquad (1)$$

where I_s is the source photocurrent, I_T is the cathode current and λ is the optical wavelength. Calculation for the total quantum efficiency (%) is given by equation (2) [8] :

$$\eta_{total}\,(\%) = \frac{I_{cathode}}{I_{source\ photo\ current}} \cdot 100\% \qquad (2)$$

The frequency response is defined as [10] :

$$f_{-3dB} = 20^* \log\left(\frac{I_R}{I_{RO}}\right) \qquad (3)$$

where I_R is the real cathode current and I_{Ro} is the real component current.

III. RESULTS AND DISCUSSION

The developed device structure is shown in Fig. 1. The p+ region resides beneath the anode electrode whereas the n+ region resides below the cathode electrode. The area between the p+ and n+ region is the intrinsic (or *i*) region. The device's electrical and optical characteristics were executed using the ATLAS module from Silvaco Int.

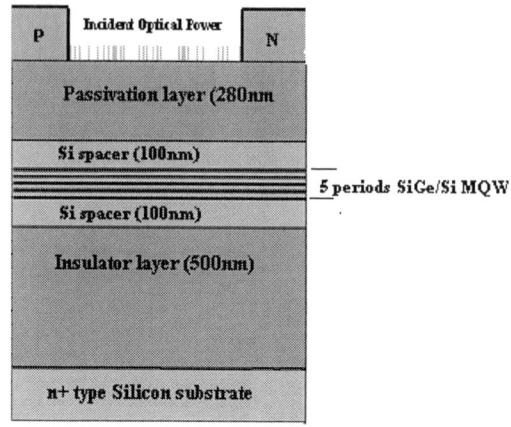

Fig. 1: The device structure of SiGe/Si MQW lateral pin-photodiode

A nanostructured array of a material with smaller bandgap than the host material would allow for additional absorption of light because of an effective bandgap tuning as is necessary to increase the efficiency of the multi-junction photodiode.

Fig. 2 shows compares the dark IV curve of SiGe/Si MQW photodiode versus the bulk Si photodiode.

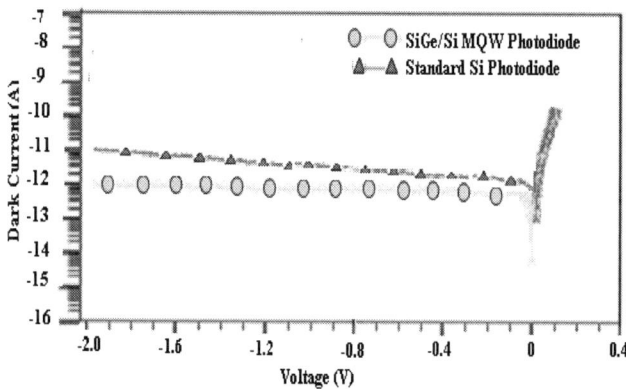

Fig. 2: Dark current of SiGe/Si MQW PIN photodiode and bulk silicon PIN photodiode without intensity optical power.

The dark current I-V curve was measured up to a bias voltage of 2V as shown in Fig. 2. The low dark current are obtained at SiGe/Si MQW PIN photodiode.

The optical current-voltage (I-V) characteristics were extracted when an optical power of 1 mW/cm^2 and another power of 10 mW/ cm^2 were used. The result at a bias voltage of 2 V is shown in Fig. 3. The photocurrent increases more rapidly with reverse bias in the SiGe MQW photodiode due to the enhanced electric field in both the SiGe MQW and the intrinsic region. In this voltage range, the electric field across the MQW and the intrinsic region is large enough to initiate higher recombination rates which photocurrent to increase sharply [2].

978-1-4673-2395-6/12 $31.00 © 2012 IEEE   404

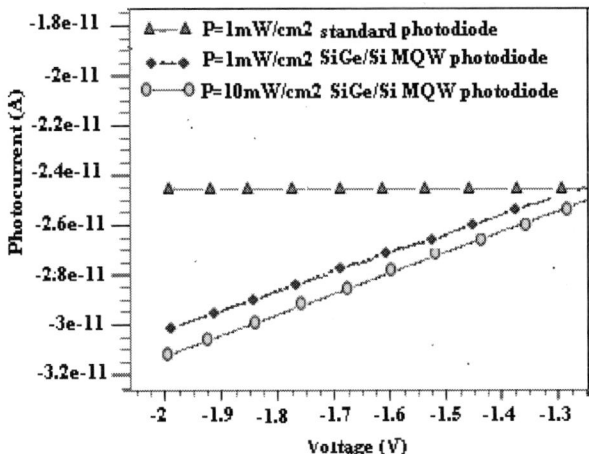

Fig. 3: Photocurrent of SiGe/Si MQW PIN photodiode and a standard Silicon photodiode at 1 mW/cm² intensity optical power.

Fig. 4 shows the responsivity as a function of reverse bias with optical wavelength from 400 nm until 1600 nm. At a small bias of 3 V, the SiGe/Si MQW and bulk Silicon PIN photodiode for the device with 6 μm active area diameter produced responsivities of 0.85 A/W at 800 nm and 0.62 A/W at 800 nm, respectively [13]. The SiGe/Si MQW photodiode also achieved high responsivity of 0.89 A/W at the infrared wavelength of 1550 nm.The total quantum efficiency for the SIGe/Si MQW PIN photodiode is 71%. The result of responsivity for SiGe/Si multi-quantum layer photodiode increases because of the recombination rate in the SiGe layer. The responsivity could be further increased by increasing the MQW layer thickness. The spectral responsivity of the fabricated optical devices can be tuned by the applied reverse bias [14].

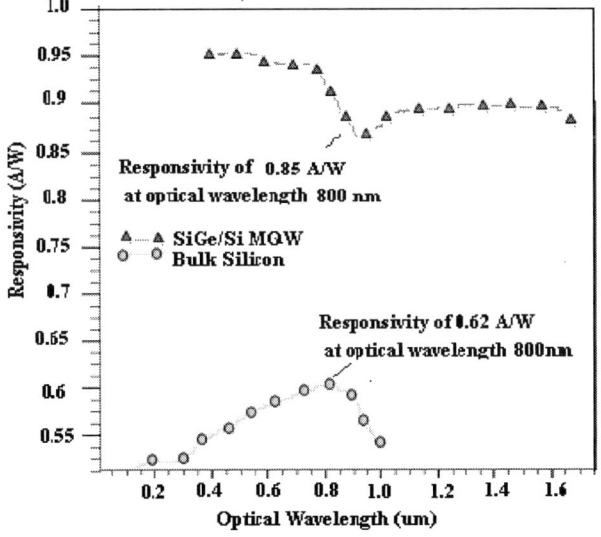

Fig. 4: Responsivity of SiGe/Si MQW and bulk Silicon lateral pin photodiode after optimization.

The simulation results for the frequency response of the SiGe/Si MQW PIN photodiode and bulk Silicon lateral PIN photodiode was 15 GHz at optical wavelength of 800 nm and 13.1 GHz at optical wavelength of 800 nm, respectively as shown in Fig 5 [13]. The SiGe/Si MQW PIN PD also achieved -3dB frequency of 21 GHz at infrared wavelength of 1550 nm. The SiGe quantum well can potentially operate up to a device speed of 50 GHz with sufficiently high electric field intensity [15].

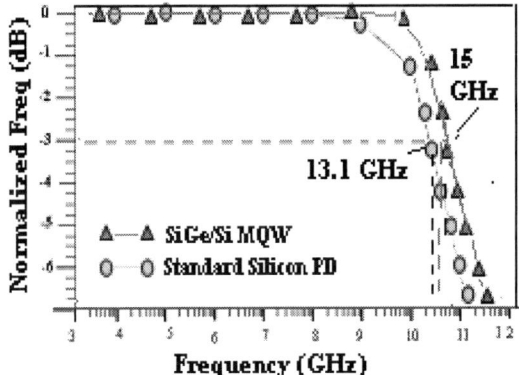

Fig. 5: The -3dB frequency response at optical wavelength 800 nm.

IV. CONCLUSION

As a conclusion, the SiGe/Si multi quantum well lateral PIN photodiode with a high performance in the visible and near infrared spectrum (400 – 1600 nm) is reported. The origin of detection is due to optical absorption in the multi quantum well region in the intrinsic layer. The PIN photodiode device structure is similar to PIN photodiode developed in previous work with the addition of SOI substrates and MQW layers in the intrinsic region. The fabrication of SiGe/Si multi quantum well photodiode is compatible with complementary metal-oxide semiconductor (CMOS) technology processes, which makes the device reliable and suitable for CMOS integration. In order to make a faster and smaller device in the future, we need to optimize the design of the device including the layer thicknesses, SiGe composition and the number of MQW.

ACKNOWLEDGEMENT

The authors would like to thank the Public Service Department of Malaysia (JPA) for their financial support, Photonic Lab, Institute of Microengineering and Nanoelectronics (IMEN), Universiti Kebangsaan Malaysia (UKM) and Pusat Latihan Teknologi Tinggi, Taiping (ADTEC) for the technical and moral support throughout the project. Universiti Kebangsaan Malaysia is also acknowledged for supporting this work under grant Industri-2012-017.

REFERENCES

[1] Rongshan Wei, Ning Deng, Minsheng Wang, Shuang Zhang, Peiyi Chen, Litian Liu and Jing Zhang," Study self assembled Ge quantum dot infrared photodetectors," Proc. Of the IEEE (2006).

[2] Silvaco int," Enhanced silicon light emission intensity with multiple SiGe quantum well structure," a journal for process and device engineers, Vol. 16, No. 5, May 2006.

[3] Wikipedia website : http://en.wikipedia.org/wiki/Silicon-germanium

[4] A. Apsel, E. Culurciello, A. Andreou, and K. Aliberti, "Thin film pin photodiodes for optoelectronic silicon on sapphire CMOS," in *ISCAS'03. Proceedings of the 2003 International Symposium on Circuits and Systems*, vol. 4, May 2003, pp. 908 – 911.

[5] Russell M. Kurts, Ranjit D. Pradhan, Alexander V. Parfenov, Jason Holmstedt and Vladimir Esterkin," High speed nanotechnology based photodetector," Proc. Of SPIE, Vol. 5925 (2005).

[6] Ashok K. Sood, Robert A. Richwine and Yash R.Puri," Development of low dark current SiGe-detector arrays for visible-NIR imaging sensor," Proc. Of the SPIE, Vol. 7298 (2009).

[7] Po-Hsing Sun, Shu-Tong Chang, Yu-Chun Chen and Hongchin Lin,"A SiGe/Si multiple quantum well avalanche photodetector," Solid-State Electronics 54 (2010) 1216-1220.

[8] Menon, P. S. 2005. Pembangunan diodfoto planar p-i-n silikon (Development of silicon-based p-i-n photodiode), MSc Thesis. Universiti Kebangsaan Malaysia.

[9] Menon, P. S. & Shaari, S. 2005. Surface versus lateral illumination effects on an interdigitated Si planar PIN photodiode (Poster). *Proceedings of the SPIE Symposium on Optics and Photonics: Infrared and Photoelectronic Imagers and Detector Devices, 2005, San Diego, USA*, Volume 5881: art. no. 58810S, pp. 1-8.

[10] Menon, P. S. K. Kandiah. A. A. Ehsan & S. Shaari. 2011. Concentration-dependent minority carrier lifetime in an In(0.53)Ga(0.47)As interdigitated lateral PIN photodiode based on spin-on chemical fabrication methodology. *International Journal of Numerical Modelling: Electronic Networks, Devices and Fields.* 24(5):465-477.

[11] Tania Tasmin, Nicolas Rouger, Guangrui Xia, Lukas Chrostowski, and Nicolas A. F. Jaeger," Design of a 1550 nm SiGe/Si quantum well optical modulator," Proc. Of SPIE Vol. 7750, May 2011.

[12] Marris, D., Cassan, E., and Vivien, L..," Response time analysis of SiGe/Si modulation doped multiple quantum well structures for optical modulation," Journal of Applied Physics 96, 6109-6112 (Dec 2004).

[13] S.K. Tasirin, P.S. Menon, I. Ahamd, S.F. Abdullah. 2012. Optimization of Process Parameters For Si Lateral PIN Photodiode. *International Conference of Mathematical Aplications in Engineering.*

[14] Efe Onaran, M. Cengiz Onbasli, Alper Yesilyurt, Hyun Yong Yu, Ammar M. Nayfeh and Ali K. Okyay," Silicon-Gemarnium multi quantu well photodetectors in the near infrared," Vol. 20, No. 7 / OPTICS EXPRESS 7608, March 2012.

[15] Yiwen Rong, Yangsi Ge, Yijie Huo, Marco Fiorentino, Michael R. T. Tan, Theodore I, Tomasz J. Ochalski, Guillaume Huyet and James S. Harris, "Quantum-confined stark effect in Ge/SiGe quantum wells on Si," IEEE journal of selected topics in quantum electronics, Vol. 16, No. 1, Jan/Feb 2010.

Compatibility Issues of Si Technology with Higher Band Gap Materials for RF Applications

Bablu K. Ghosh[1], Ismail Saad[1], Khairul Anuar Mohamad[1], Nurmin Bolong[1], Norfarariyanti Parimon[1], Afishah Alias[2] and Mohd Zuhir Hamzah[1]

[1]School of Engineering and IT, Universiti Malaysia, Sabah
[2]School of Science and Technology Universiti Malaysia, Sabah
Jalan UMS 88400 Kota-kinabalu, Sabah, Malaysia
e-mail: ghoshbab@ums.edu.my

Abstract- **Now-a-days microwave (MW) electronics is an extremely rapid developing field in semiconductor electronics. For the present and near future demand, research is progressing for the development of high speed and high power density RF devices fabrication beside main stream Si based research. For mixed and RF signal performance, Si still has limitations for it further scaling due to excess leakage of current and low trans-conductance or fT and f $_{Max}$. So, suitable alternative materials device fabrication is potential. In this paper doping profile of GaAs channel and compositional (% of Al) variation in AlGaAs layer for schottky contact for AlGaAs/GaAs compound semiconductor (CS) based HEMT (high electron mobility transistor) is evaluated. Besides that, gate oxide thickness effects on covalent bonded Si based nMOS gate turn on time and ON/OFF current have also been evaluated. It appears that increasing Al composition in AlGaAs and more doping in GaAs layer enhances channel trans-conductance while low % of Al in AlGaAs layer and low doping in GaAs layer increases ON/OFF current ratio (reduces leakage current). In case of Si MOS, decreasing oxide layer thickness, transconductance is increased but ON/OFF current ratio is decreased (increase leakage current).**

I. INTRODUCTION

Si based matured microelectronics in the VLSI technology for memory and processor appears as mainstream technology (MT). In respect of MT, power electronics, high power density RF devices and optoelectronics as well as sensor technology is also emerging very fast for the present and near future demand [1-4]. Latest advancement in hetro-junction of compound semiconductor devices based on Si wafer brings the technology steps ahead [2-4]. There is also a scope to possible integration of such CS based RF device with Si CMOS technology for compactness and to minimize different parasitic effects and phase mismatch issue [4]. Epitaxial grown III-Vs CS on CMOS-compatible Si substrates is the focus of the technology. Due to higher band gap compound semiconductor, it can be operated at high temperatures with higher breakdown voltage and high electron mobility even at higher dislocation densities that appears due to lattice mismatch with wafer [1-4]. It can also be doped more heavily than Si, enabling fabrication of devices with low dissipation loss and higher speed. Due to persist iconicity, it polarization impact positively to make 2DEG (two dimensional electron

gas AlGaN/GaN or AlGaAs/GaAs) is prospective for sensor applications [5].

In case of Radar & Satellite communication links operating at frequencies ranging from 100 MHz to 90 GHz, have large power requirements [2-4]. Currently amplifiers are using Si technology that is roughly 10% efficient; 90% of the power that goes into a transistor is wasted as heat as leakage effect [2-4]. So, for high power and RF device fabrication, it is very important to assure the device with higher power density/ breakdown voltage and less prone to dissipation or heat generation. GaAs and GaN as wider band gap materials are capable to handle higher frequency with less junction temperature [2-3]. Though SiGe based device can operate even at higher frequency than GaAs based device but it is unable to handle high power densities. For integration of RF circuit with logic CMOS; low leakage current and high speed is desired for logic but for RF circuit, high carrier mobility, high drive current and breakdown voltage is required. In that purpose, as a high-mobility or peak saturation velocity semiconductor, gallium arsenide (GaAs) based RF device is a best and gallium nitride (GaN) based RF power device has the highest power density is suitable for base station applications. Even longer channel length CS device response very fast compared to tiny channel length Si based device [2-3].

II. EXPERIMENTAL

TCAD software based AlGaAs/GaAs HEMT (high electron mobility transistor) on GaAS substrate and Si based n-MOS have been designed in similar channel length (0.55μm) for RF device application. The GaAs 250 Å channel thickness and doping profile are shown in the model below. The construction of the device and the grid is made by using ATLAS syntax. The schottky contact region is depleted n-AlGaAs with highest carrier concentrations reduces the barrier height.

978-1-4673-2395-6/12 $31.00 © 2012 IEEE

III. RESULT AND DISCUSSION

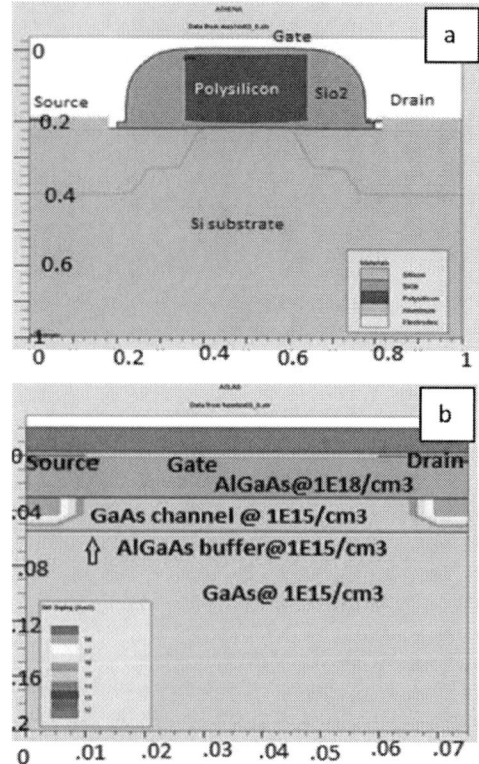

Fig. 1 1x1 μm Si based n-MOS (a) and 0.75x 0.20 μm GaAs based AlGaAs/GaAs HEMT (b) structure developed by using TCAD software

Fig.1 shows the simulated Si based nMOS and GaAs based AlGaAs/GaAs HEMT. Trans-conductance of the MOSFET decides its gain and is proportional to electron or hole mobility (depending on device type), depends on drain voltages, Vd. As MOSFET size (I) is reduced, the fields in the channel increase. As channel lengths are reduced without proportional reduction in drain voltage, raising the electric field in the channel, the result is velocity saturation of the carriers, limiting the current and the trans-conductance.

$$v_d = \frac{\mu}{E} \quad and \quad E = \frac{Vd}{l} \qquad (1)$$

$$c_{ox} = \frac{\varepsilon_s A}{t_{ox}} \quad and \quad Q = c_{ox} Vgs \quad (2)$$

Switching time is roughly proportional to the gate capacitance. In case of Si technology, it is usually expected that smaller transistors switch faster and linear. The device dimensions are the transistor length, width, and the oxide thickness, each (used to) scale with a factor of 0.7(say) per node. This way, the transistor channel resistance does not change with scaling, while gate capacitance is cut by a factor of 0.7 due to reduction of area/thickness factor by 0.7. Hence, the RC delay of the transistor scales with a factor of 0.7. For high frequency application, the input impedance must be required higher for possible high fan-out effect. Fixed gate area, reduction of

oxide layer thickness enhances the capacitance, input impedance and trans-conductance but it also increases the leakage current consequently wastage of lot of power [3]. This limitation of application of high drain voltage, reduction of channel length, and t_{ox} are the main limitations for the application of Si technology as high power density RF device.

The device merits include the transit frequency at unity current gain fT and the gain at higher frequencies i.e, maximum frequency of oscillation fMAX at unity power gain. For fMAX assessment, noise margin, breakdown voltage and capacitor density of the transistors are also important parameters those also assess figure of merit. Scaling performance of RF devices has less importance while for logical devices; it is the most important issue. RF devices and circuits must meet many other performance specifications and parameters that do not scale in the same manner as mainstream Si based CMOS digital devices [6].

$$f_T = \frac{Gm}{2\pi(Cgs+Cgd)} \qquad (3)$$

$$f_{max} = \frac{f_T}{2\sqrt{\{Gds\,(Rs+Rg)+2\pi f_T Cgd Rg\}}} \qquad (4)$$

To make faster device, channel length must to be reduced but reduction of channel length much impact on leakage current. If oxide layer thickness is increased then the gate capacitance is decreased, gate response must be faster but due to reduction of trans-conductance, f_T is also decreased. Increment of f_T is only possible by decreasing channel length and oxide layer thickness. But decreasing channel length and gate oxide layer thickness enhances leakage current as a result wastage of power is increased. Due to high specific resistance of gate polySi materials, the gate resistance, Rg becomes high leading to decrease f_{max}. This trade-off for Si based nMOS device is still not so attractive for very high frequency RF devices application. Avoidance of such junction oxide materials and poly Si at the gate, high carrier concentrated n- AlGaAs contact layer with metal for AlGaAs/GaAs HEMT could the able to perform efficiently over longer range of frequency. Hetro-junction can avoid substantial leakage problem thus it may be effective to increase ON/OFF current ratio.

TCAD software based simulation is done for the comparison between AlGaAs/GaAs HEMT (high electron mobility transistor) and covalent bonded Si based nMOS. The GaAs 250 Å channel thickness and doping profile are shown in the above model in Fig-1. From the result it appears that whatever the composition of Al, the impact of increment of doping enhances channel conductivity as well as trans-conductance. Higher percentage of Al ensures more depletion and more carriers accumulate at junction interface result in increased drain current with similar Vgs and Vds as shown in Fig. 2(a). So, doping in channel layer ensures higher sheet carrier density at high % of Al while different Band gap materials junction avoids regenerative effect and leakage problem thus increases ON/OFF current ratio (~10^{12}) as compared to Si based nMOS device ON/OFF current ratio (2x10^7) as shown in Fig-2 (b) and (c) respectively.

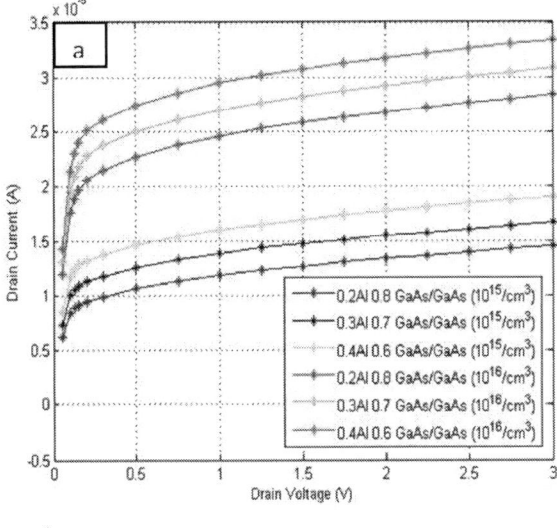

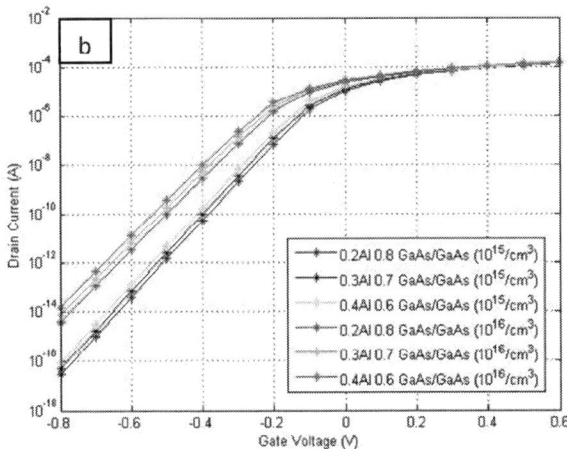

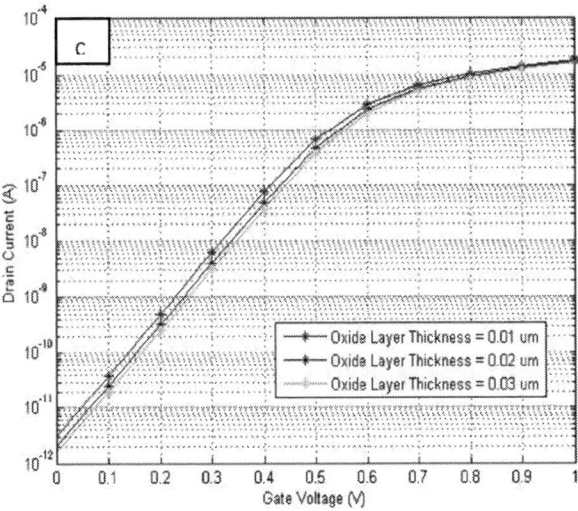

Fig.2 VD-ID characteristic curve for AlGaAs/GaAs HEMT (a) ; Vgs –log ID characteristic curve for AlGaAs/GaAs HEMT (b) and Si based nMOS (c) respectively

Small gate voltage as well as VT can swing the ID much more due to variation of sheet carrier density that is so sensitive for % of Al composition in AlGaAs layer and channel doping concentration is certainly suitable for sensor applications [5]. The shift of two branches of data as shown in Fig-2(a) is due to different doping profile in GaAs channel area while the trend of increment of drain current or calculated (graph is not shown here) transconductance (400 mS/mm for channel doping @ 1016/cm3 to 300 mS/mm for channel doping @ 1015/cm3) is very similar for increment of Al compositional factor in the AlGaAs layer.

For Si nMOS, decreasing oxide layer thickness, the generation of carrier in the channel is also increased as a result the drain current or calculated (graph is not shown here) channel trans-conductance (40 mS/mm to 50 mS/mm) is increased and the gate turn on (time delay) is found to be increased 45 pSec to 60 pSec (graph is not shown here). It appears that due to thinner oxide layer, increasing capacitance and hence turns on transient/response time, Γ = RC is increased since gate poly silicon has higher specific resistance. Though the increment of trans-conductance for increasing capacitance has very less effect on transition frequency, f_T (as refer to equation-3) since gate capacitance is also increased as the area is remained unchanged. Such complex property is completely unavailable for the AlGaAs/GaAs based CSHEMT. Due to high specific resistance of gate poly Si material, Si based device can be attributed to less efficient even at similar gate/channel length and for the superior electron transport properties of AlGaAs/GaAs CS- HEMT must enhances f_T and f_{max} (as refer to equation 3 and 4). The replacement of poly Si gates in conventional Si CMOS by low specific resistance materials and introduction of high dielectric constant materials may be effective approach for the reduction of channel length in order to make faster device and at the same time it may avoids the excess leakage current. The effect of such possible variation might be possible solution.

IV. CONCLUSION

In AlGaAs/GaAs HEMT structure, increasing % of Al composition in depleted n-AlGaAs layer and the GaAs channel doping, the trans- conductance is increased while ON/OFF current ratio is decreased or vice-versa. In Si nMOS increasing trend of ON/OFF current ratio by increasing oxide layer thickness favor to reduce gate turn on (delay time) but trans- conductance is ultimately decreased. Comparing to Si technology, much higher level of trans- conductance and ON/OFF current ratio is observed for HEMT technology. The poly Si gates in conventional Si MOS and relatively low permittivity dielectric materials are the main reasons to make such wide variation.

ACKNOWLEDGEMENTS

I would like to thank our colleagues specially NEM group in SKTM, UMS. I would like to thank the UMS authority for supporting me financially to attend the conference.

978-1-4673-2395-6/12 $31.00 © 2012 IEEE

REFERENCES

[1]. Yuki Niiyama a, Zhongda Li b, T. Paul Chow b, Jiang Li a, Takehiko Nomura a, Sadahiro Kato, Solid-State Electronics 56, 73 (2011)

[2]. Michael S. Shur, GaN based electronic devices, chapter-5, P 61-86 (1999)

[3] Frank Schwierz, index-978-1-4244-2186-2/08/$25.00 ©2008 IEEE

[4] Jin Wook Chung, Bin Lu and Tomás Palacios, index 978-1-4244-5191-3/09/$26.00 ©2009 IEEE

[5] H.T. Wang, *et al.*, *Appl Phys Lett* **90,** . 252109 (2007)

[6] C.E. Weitzel and K.E. Moore, Journal of electronic materials, Vol27, No 4, 1998

RF Characteristics of AlGaN/GaN HEMTs under Different Temperatures

Yu-Sheng Chiu, Jui-Chien Huang, Tai-Ming Lin, Yu-Ting Chou, Chung-Yu Lu, Chia-Ta Chang, Edward Yi

Chang, *Senior Member, IEEE*

Department of Materials Science and Engineering
National Chiao Tung University
1001 University Rd., Hsinchu, 30010 Taiwan
Email: laurance0319@yahoo.com.tw

Abstract- We present the Rf characteristics of 0.7-µm gate length n-GaN/AlGaN/GaN high-electron mobility transistors (HEMTs) with different source-drain spacing tested under different temperatures.. The 7-µm source-drain spacing device demonstrated 800 mA/mm drain current density and 257 mS/mm tranceconductance, and the 5-µm source-drain spacing device demonstrated 700 mA/mm drain current density and 260 mS/mm tranconductance. The 7-µm source-drain spacing device was measured at room temperatures of 25 ℃ and -40 ℃, the current gain (f_T) were 18 GHz and 21GHz and the maximum oscillation (f_{max}(U)) frequency were 63 GHz-and 87 GHz, respectively The f_T was nearly linearly dependent on the temperature. As operating temperature increased from -40 ℃ to 50 ℃, the f_T dropped more dramatically for the 5-µm SD spacing device than for the 7-µm device. The f_{max} characteristic of 5-µm SD spacing device decreases more dramatically above 125 ℃ than the 7-µm SD spacing device. This phenomenon might be due to stronger phonon scattering for shorter channel device at high temperatures.

I. INTRODUCTION

GaN material has attracted many interests in recent years due to its unique physical properties, such as wide band-gap (3.4 eV), high electron mobility (1500 $cm^2V^{-1}s^{-1}$), high breakdown electrical field (5×10^6 Vcm^{-1}) and high peak saturation electron velocity (3×10^7 cms^{-1}) [1-4]. As a result, GaN devices are expected to posses high voltage and high current characteristics and is suitable for high frequency and high power electronic device applications.

Many progresses have been made in GaN HEMT power devices in recent years [1]. However, when the devices were operated under high voltage and high current conditions, , they would generate lots of heat and degrade device performances [2-4], such as the degradations of current density, current gain (f_T) and maximum oscillation (f_{max} (U)) cut-off frequency, etc. The mechanism of degradation related to the operation temperature is rarely studied.

In this paper, we study the DC and RF performances of the GaN HENT devices on high thermally conductive SiC substrate tested under various temperatures. The RF characteristics of the devices with 5-µm and 7-µm source-drain spacing are discussed.

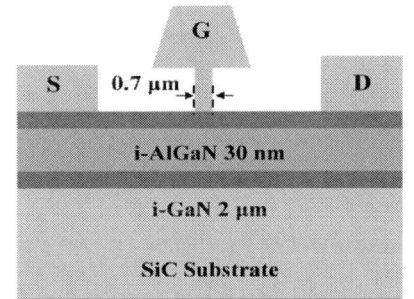

Fig. 1. Cross section of the 0.7-µm-gate GaN/AlGaN/GaN HEMT

II. EXPERIMENTAL

The wafer used in this paper was grown by metal-organic chemical vapor deposition (MOCVD) system on 3-inch semi-insulating SiC substrate. The epitaxial structure consisted of a 2-µm-thick undoped GaN buffer layer, 1-nm AlN spacer layer, a 30-nm-thick undoped $Al_{0.22}GaN_{0.75}N$ barrier layer and a 5-nm-thick Si-doped (3×10^{18} cm^{-3}) GaN cap layer, as shown in Fig. 1.

First, the ohmic contact to the nAlGaN/GaN structure was achieved by depositing Ti (20-nm)/Al (120-nm)/ Ni (25-nm)/ Au (100-nm) metal stacks using electron-beam evaporation and followed by high temperature annealing at 800°C in ambient N_2. After ohmic formation, a Cl_2-based gas plasma dry-etch was used to define the active region on the wafer by using the ICP-RIE. The low ohmic contact resistance and sheet resistance of 0.75 Ω · mm and 286 Ω/□ were obtained using the TLM method, respectively. A tri-layer resist system of PMMA/P(MMA-MAA)/PMMA were exposed for the photolithography by deep-UV light. The Ni (20-nm)/Au (300-nm) were used as gate metal. The gate length and gate width of the T-shaped gate were 0.7-µm -and 100-µm, respectively. In order to avoid the surface state effect, a SiNx passivation layer was deposited with 100-nm thickness by using the plasma enhanced chemical vapor deposition (PECVD). There are two different sizes of source-drain spacing, 5-and 7-µm, on this wafer.

978-1-4673-2395-6/12 $31.00 © 2012 IEEE

III. RESULTS AND DISCUSSION

Fig. 2 shows the current-voltage characteristics of $0.7 \times 100 \ \mu m^2$ devices with 7-μm source-drain (SD) spacing. As observed from the plot, the fabricated GaN HEMT shows transconductance of 257 mS/mm and drain current density of 800 mA/mm at the drain bias of 5 V. The threshold voltage was 3.2 V. Fig. 3 shows the current-voltage characteristics of the same wafer with SD spacing of 5 μm. As observed from the plot, the fabricated GaN HEMT shows transconductance of 260 mS/mm and drain current density of 700 mA/mm at the drain bias of 5 V. The threshold voltage of the device was 2.8 V.

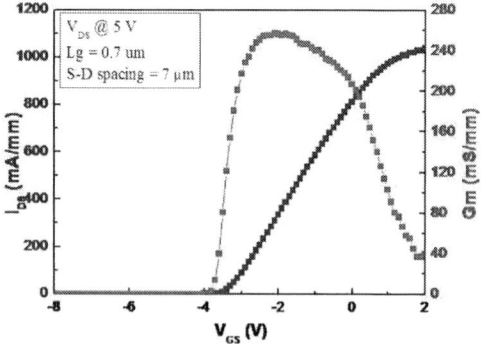

Fig. 2. Current-voltage characteristics at V_{DS} 5 V of $0.7 \times 100 \ \mu m^2$ device with 7-μm source-drain spacing.

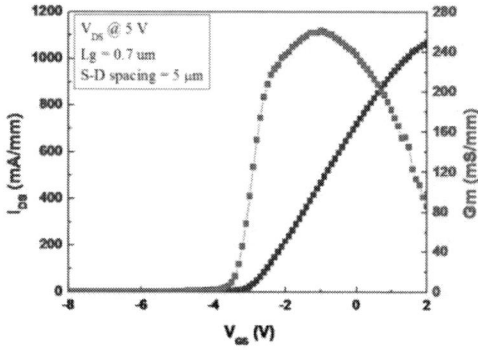

Fig. 3. Current-voltage characteristics at V_{DS} 5 V of $0.7 \times 100 \ \mu m^2$ device with 5-μm source-drain spacing.

Fig. 4 shows the s-parameters measured under various temperatures (-40 ℃~200 ℃) for the 7-μm source-drain spacing device. The Gain vs Frequency plots of GaN HEMT with temperatures at -40 ℃, 25 ℃ and 200 ℃ are shown. Moreover, the linear drop of the Gain with temperature of GaN HEMT is clearly observed.

Fig.5 summarizes the s-parameters results. As the device measured at room temperature (25 ℃), the current gain (f_T) and maximum oscillation ($f_{max}(U)$) frequency are 18 GHz and 63 GHz, respectively. Moreover, the f_T and f_{max} are 21 GHz and 87 GHz at -40 ℃. As the device was measured at

200 ℃, the f_T and f_{max} dramatically deceased to 11 GHz and 30 GHz. The f_T linearly degraded as the temperature rises. C. H. Oxley *et al* [2] and DARWISH *et al* [3] have demonstrated that carrier mobility decreases as the test temperature increases. Therefore the phenomenon shown in Fig.4 can be explained by the mobility decreased when the temperature was increased, resulting in the degradation of the device transconductance. Regarding to the f_T reduction ratio from -40 ℃ to 50 ℃, for the 5-μm SD devices, the f_T slope of the 7-μm SD spacing device was -0.0376; the f_T slope of the 5-μm SD spacing device was -0.0567. The 5-μm SD spacing device degraded faster than the 7-μm SD spacing device. It is expected that f_{max} of the 5-μm SD spacing device will also degrade faster. As judged from Fig. 4, the f_{max} characteristic of 5-μm SD spacing device decreases more dramatically above the 125 ℃ than the 7-μm SD spacing device. It might be due to stronger phonon scattering in the shorter channel at high temperature condition.

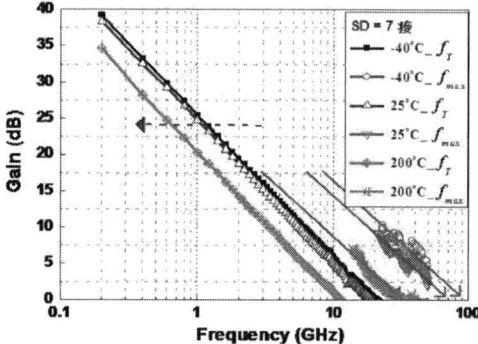

Fig. 4. The Gain vs frequency plot of GaN HEMT at -40, 25 and 200 ℃.

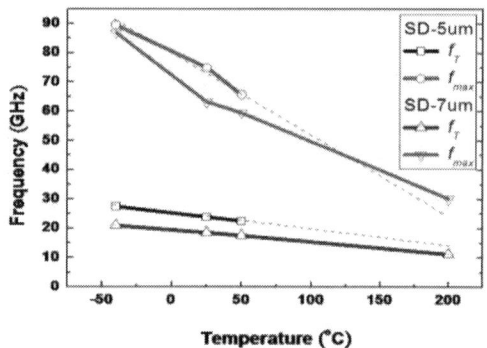

Fig. 5. The f_T and f_{max} of 5-and 7-μm source-drain spacing GaN devices measured at different temperature.

IV. CONCLUSIONS

As the GaN device was operated at high voltage and high current, it would generate lots of heat from the device itself and degrades the device performances. The 7-μm SD spacing device demonstrated 800 mA/mm drain current density and 257 mS/mm transconductance, and the 5-μm SD device demonstrated 700 mA/mm drain current density and 260 mS/mm transconductance. The f_T degradation was nearly linearly dependent on the temperature. As operating temperature increased from -40 ℃ to 50 ℃, the f_T dropped more dramatically for the 5-μm SD spacing device than for the 7-μm device. The f_{max} characteristic of 5-μm SD spacing device decreases more dramatically above the 125 ℃ than the 7-μm SD spacing device. It might be due to stronger phonon scattering in the channel at higher temperatures.

ACKNOWLEDGMENT

The authors acknowledge the high-frequency technology division of the NDL (NCTU, TAIWAN) for the help of S-parametes measurement.

REFERENCES

[1] L. Shen *et al,."*AlGaN/AlN/GaN high-power microwave HEMT," *IEEE Electron Devices Letter,* vol. 22 ,pp 457-459 , 2001.

[2] C. H. Oxley *et al.,* "On the temperature and carrier density dependence of electron saturation velocity in an AlGaN/GaN HEMT" *IEEE Trans. Electron Devices*, vol.53, pp 565-567, 2006.

[3] DARWISH *et al.,* " Dependence of GaN HEMT Millimeter-Wave Performance on Temperature," *IEEE Trans. MTT*, vol. 57, pp. 3205-3211, 2009.

[4] S. Arulkumaran *et al.,* "Temperature dependent microwave performance of AlGaN/GaN high-electron-mobility transistors on high-resistivity silicon substrate," *Thin Solid Films*, vol. 515, Issue 10, pp. 4517- 4521, 2007.

978-1-4673-2395-6/12 $31.00 © 2012 IEEE

Investigation of Efficiency Droop in GaN-based UV LEDs with N-type AlGaN Underlayer

Shun-Kuei Yang[*a], Po-Min Tu[a,b], Shih-Cheng Huang[a], Ya-wen Lin[a], Chih-Peng Hsu[a],
Jet-Rung Chang[b], and Chun-Yen Chang[b], *Life Fellow, IEEE*

[a]Advanced Optoelectronic Technology Inc., Hsinchu 30352, Taiwan.
[b]Department of Electronic Engineering, National Chiao Tung University,
1001 University Rd., Hsinchu, 30010 Taiwan
Email: stan.yang@aot.com.tw

Abstract- The efficiency droop in InGaN-based 380nm UV light emitting device (LED) with n-GaN and n-AlGaN underlayer grown on sapphire substrate by metal-organic chemical vapor deposition (MOCVD) was investigated. From simulation result of high resolution x-ray diffraction (HRXRD) ω-2θ curve by using dynamical diffraction theory, the Al composition in the n-AlGaN layer was determined to be about 3%. The experimental results of temperature dependent photoluminescence (PL) demonstrated that the internal quantum efficiency (IQE) of n-GaN and n-AlGaN UV-LEDs are 43% and 39%, respectively, which are corresponding to an injected carrier density of 8.5×10^{17} #/cm^3. It could be explained that the crystal quality of n-GaN is better than of n-AlGaN. In addition, the observation of pit density from atomic force microscopy (AFM) surface morphology is consistent with the interpretation. It was well-known that the pits appearing on the surface in the virtue of the threading dislocations. Thus, it means that defects induce the non-radiative centers and deteriorate the IQE of the UV-LED with n-AlGaN underlayer

I.

II. INTRODUCTION

Recently, the GaN-based ultraviolet light emitting diodes (UV-LEDs) have been focused on as one of the most important targets for the pumping source to develop white-light LEDs. However, it is difficult to fabricate near-UV LEDs with high efficiency, because the external quantum efficiency (EQE) decreases drastically below the wavelength of 400 nm [1]. This is due to the smaller InN mole fluctuation with reduced indium composition in the near-UV quantum wells (QWs), and thus less localized energy states lead to lower efficiency of the near-UV LEDs [2, 3]. Moreover, crystalline quality and light absorption of GaN are significant for short wavelength near-UV LEDs [4, 5].

Besides, most of the light will be trapped inside the GaN-based LED, resulting in the low light extraction efficiency [6]. Once the light trapped inside, the LED will be reabsorbed eventually. On the other hand, the junction temperature, the temperature of the active region, is a critical parameter and affects quantum efficiency of device, maximum output power, reliability, and other parameters. In general, heat can be generated in the ohmic contact, cladding layers, and the non-radiative recombination in the active region [7].

In this study, we proposed to remove the conventional n-type GaN below InGaN multi-quantum-wells (MQWs) and replaced it with an n-type $Al_{0.03}Ga_{0.97}N$ to improve the optical performance in the GaN-based UV LEDs.

III. EXPERIMENTAL

The samples in this study were grown on c-plane 2" sapphire substrates by using an atmospheric-pressure metal organic chemical vapor deposition (AP-MOCVD SR4000) system. The metalorganic compounds of trimethylgallium, trimethylaluminum, trimethylindium and ammonia (NH3) were employed as the reactant source ma-terials for Ga, Al, In, and N, respectively. Silane and bis-cyclopentadienyl magnesium (Cp$_2$Mg) were used as the sources for n-type and p-type dopants, respectively. Prior to the growth, the sapphire substrates were thermal cleaned in hydrogen ambient at 1100°C.

As shown in Fig.1, the UV LED structure with InGaN/AlGaN multi-quantum-well (MQW) consisted of a 30-nm-thick low-temperature (500°C) GaN nucleation layer (GaN NL), a 2-μm-thick undoped GaN epilayer (u-GaN), a 2-μm-thick Si-doped n-layer, an InGaN/AlGaN multiple quantum wells (MQWs) active layer, a 15-nm-thick Mg-doped AlGaN electron blocking layer (p-AlGaN) and a 0.2-μm-thick Mg-doped GaN contact layer (p-GaN). In this study, Si-doped GaN and Si-doped $Al_{0.03}Ga_{0.97}N$ n-type layer were grown and denoted as UV-LED with n-GaN and UV-LED with n-AlGaN, respectively.

Temperature dependent PL and Power-dependent PL were used to determine internal quantum efficiency (IQE) and internal electric field (IEF), respectively. In addition, these samples were also characterized by atomic force microscopy (AFM) to reveal the surface morphology. Electrical and optical characteristics were measured by electroluminescence (EL). Their junction temperatures were extracted from the current-voltage (I–V) curves measured under DC current condition for a broad temperature range, from 30 °C to 140 °C.

Finally, the UV LED wafers were processed into mesa-type chips (size: 1mm×1mm) and mounted on epoxy-free metal can (TO-39). The output power of the UV LED chips was measured using an integrated sphere detector.

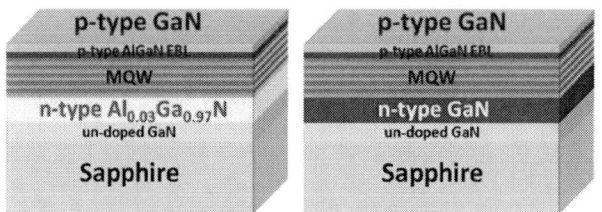

Fig. 1. Schematic illustrations of GaN UV-LEDs with n-AlGaN and n-GaN.

IV. RESULTS AND DISCUSSION

Firstly, the surface morphologies of UV-LEDs with n-AlGaN and n-GaN were investigated by AFM, as shown in Fig. 2 (a) and (b). The root mean square (RMS) value of the UV-LEDs with n-AlGaN and n-GaN were 1.36 nm and 0.60 nm, respectively. While, it can be clearly seen that, the pit density of the UV-LEDs with n-AlGaN and n-GaN were 6.5×10^9 cm^{-2} and 4×10^9 cm^{-2}, respectively. It have been reported that the pits appearing on the surface correspond to the threading dislocations. Fig. 2 (c) and (d) shows the dependence of the IQE of UV-LED with n-GaN and UV-LED with n-AlGaN as a function of injected carrier density at 15K and 300K. As the excitation energy increases, the IQE of UV-LED with n-GaN is usually higher than UV-LED with n-AlGaN at 300K. The experimental results demonstrated that the IQE are 43% and 39%, respectively, which are corresponding to an injected carrier density of 8.5×10^{17} #/cm^3 [8]. It could be explained by the crystal quality of n-GaN is better than that of n-AlGaN.

In addition, the observation of pit density of AFM morphology is consistent with the interpretation. It was well-known that the pits appearing on the surface in the virtue of the threading dislocations. Thus, it means that defect-induced non-radiative centers deteriorated IQE of LED with n-AlGaN layer. Therefore, it can be believed that the crystalline quality of the UV-LED with n-GaN is slightly better than the UV-LED with n-AlGaN.

Fig. 3(a) shows the injection current versus voltage (I-V) characteristics of the both LEDs. Under a forward current of 350 mA, the forward voltage was 3.92 and 4.07 V for UV-LED with n-GaN and UV-LED with n-AlGaN. A little high forward voltage of UV-LED with n-AlGaN can be attributed to the higher Al content compare to the UV-LED with n-GaN, thus increase the series resistance in the device. The (I-V) characteristics of both are almost the same. The light output power versus injection current (L-I) characteristics of both UV-LEDs are also shown in Fig. 3(b). The light output powers are 60mW and 63mW with the injection current at 350 mA for UV-LEDs with n-GaN and n-AlGaN, respectively. The light output powers are quite similar for both LEDs at 350 mA. However, when the injection current is increased to 600 mA, the output power of UV-LED with n-AlGaN has is much better than UV-LED with n-GaN. The light output powers for UV-LEDs with n-GaN and n-AlGaN are 70 mW and 86 mW, individually. There exhibits 22% enhancement in UV-LED with n-AlGaN compared to UV-LED with n-GaN at 600mA. This indicates that UV-LED with n-AlGaN had higher efficiency at high injection current. Fig. 3(c) shows the normalized efficiency curves as a function of forward current for the two samples. The UV LED with n-GaN attain to the normalized efficiency maximize with the forward current at 200mA and begin to decrease. On the other hand, the UV LED with n-AlGaN attain to the normalized efficiency maximize with the forward current at 250mA and keep the normalized efficiency maximize to the forward current at 350 mA then start to decrease.

From AFM and IQE measurements, it revealed that the crystal quality of the UV-LED with n-GaN is slightly better than UV-LED with n-AlGaN. Therefore, the light output power of UV-LED with n-GaN was slightly higher below 250 mA. For the UV LED with n-GaN, when the injection current exceeds 600mA, the efficiency is reduced to 33% of its maximum value. In contrast, the UV LED with n-AlGaN exhibit only 20% efficiency droop when increasing the injection current to 600 mA.

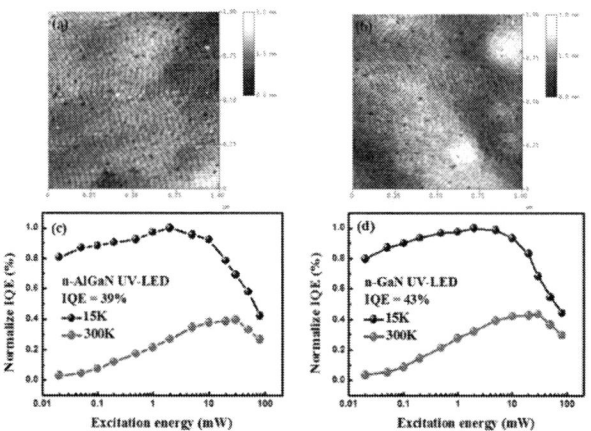

Fig. 2. AFM analysis over 1x1 μm² for (a) UV-LED with n-AlGaN (b) UV-LED with n-GaN. The IQE of UV-LEDs with (c) n-AlGaN and (d) n-GaN as a function of excitation power at 15K and 300 K.

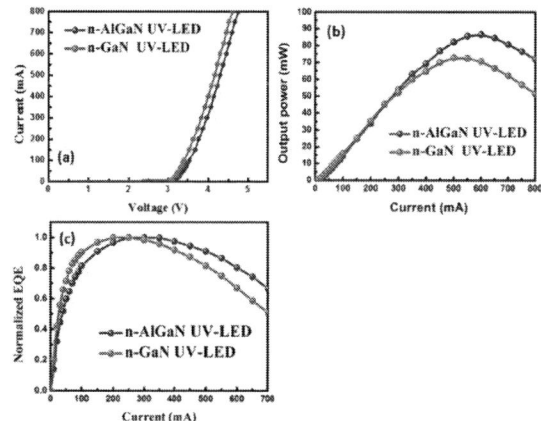

Fig. 3. (a) Injection current versus voltage and (b) the light output power versus injection current characteristics for UV-LED with n-AlGaN and UV-LED with n-GaN (c) Normalized EQE curve for UV-LED with n-AlGaN and UV-LED with n-GaN.

These results indicate that the UV LED with n-AlGaN not only enhance the total light output power but successfully improve the efficiency at high injection current. In general, the ideal semiconductor has a zero band-to-band absorption coefficient at the bandgap energy (E=Eg).

However, the absorption strength in the real semiconductor, for below-bandgap light, can be expressed in terms of exponentially decaying absorption strength. The absorption coefficient, α of the $Al_{0.03}Ga_{0.97}N$ and GaN films are 8×10^2 and 1.2×10^3 cm^{-1} at 380nm, respectively [9]. The transmission percentage of AlGaN and GaN are 85% and 77%, respectively. The light absorbing magnitude of GaN is 8% more than AlGaN at 380nm, which is significantly related to the peak shift result. In other ward, the droop effect issue has been improved in ultraviolet LED.

Fig. 4 shows the emission peak energy under different excitation power. It was found that the peak of the emission spectra of the LEDs shift to higher energy when the excitation power increased. The blue shift of both UV-LEDs with increasing excitation power may be explained by the carrier screening of the QCSE resulting from piezoelectric fields. In addition, the IEF was fitted by the peak energy shift under various injection levels at RT [10]. Our result shows the IEF of the both UV-LEDs was 0.198 MV/cm. Therefore, it was clarified that, in this case, QCSE is not the main reason to influence the characteristic of the light output power and efficiency droop.

Fig. 5(a) depicts the variation of the emission peak versus injection current characteristics of these UV-LEDs. It was shown that the initial peak positions of both UV-LEDs were located at about 380 nm, and the appearance of monotonic red-shift was revealed with increasing current. It was recognized that the junction temperature increased as the injection current increased. The red-shift magnitudes of UV LEDs with n-GaN and n-AlGaN were 6.6 and 4.3nm, respectively.

This implied that the phenomenon of red-shift is more serious in UV LED with n-GaN. Moreover, the full width at half maximum (FWHM) of both UV-LEDs increased with increasing current as shown in Fig. 5(b), and the FWHM variations of UV LEDs with n-GaN and n-AlGaN were 3.5 nm and 2.6 nm, respectively.

In our experiment, the difference between each sample is mainly in the Al mole fraction in the n-type layer. In previous results, no matter in crystalline quality, piezoelectric field and IQE, there are no immensely distinction between UV-LEDs with n-AlGaN and n-GaN.

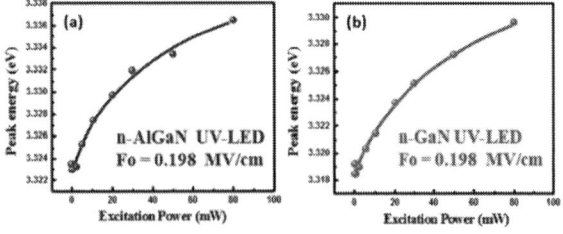

Fig. 4. Emission energy of UV-LED with (a) n-AlGaN and (b) n-GaN at different power density.

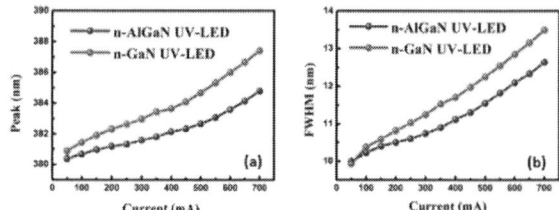

Fig. 5. EL characteristics of (a) peak shift and (b) FWHM versus injection current for UV-LEDs with n-GaN and n-AlGaN.

However, from EL spectrum peak shift result, the different self-heating phenomena of UV-LEDs have been observed [11]. Therefore, we suggested that the heat is generated from the absorption of light in the n-type layer, leading to the broadening of FWHM and red-shift. As a matter of fact, the self-absorption of emitting light in LEDs will generate heat, which means the performance of LEDs will deteriorate. Fortunately, the larger band gap of $Al_{0.03}Ga_{0.97}N$ can significantly suppress the self-absorption effect in LEDs. Thus, these results indicated that the self-heating phenomenon of UV LED with n-GaN is much serious than UV LED with n-AlGaN.

Fig. 6 shows the junction temperature versus injection current characteristics of both UV-LEDs. The junction temperature of the UV-LED with n-GaN is always higher than UV-LED with n-AlGaN. Also, it can be clearly seen that the ΔT increases as increasing the injection current. The observation was consistent with EL analyses, which showed that the existence of temperature difference between n-type $Al_{0.03}Ga_{0.97}N$ and GaN. Consequently, self-heating effect will occur and lead to a quantum efficiency droop.

As well-known, a semiconductor material absorbs photon energy when the photon energy is larger than the band-gap energy, whereas the semiconductor material is transparent when photon energy is smaller than the band-gap energy [9].

By Urbach tail law, a semiconductor material also can absorb below band-gap light even though the absorption coefficient is small. From energy point of view, the absorbed light transfers to thermal energy will influence on the performance of LED device.

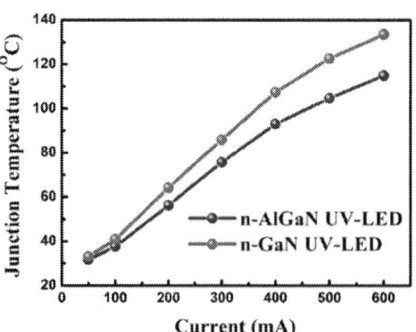

Fig. 6. The junction temperature as a function of injection current of UV-LEDs with n-AlGaN and n-GaN.

V. CONCLUSIONS

The quantum efficiency droop in InGaN-based 380nm UV LED with n-GaN and n-$Al_{0.03}Ga_{0.97}N$ underlayer grown on sapphire substrate by metal-organic chemical vapor deposition (MOCVD) was investigated. The measurements of temperature dependent PL and AFM revealed that the crystal quality of UV-LED with n-GaN is slightly better than that with n-AlGaN. Nevertheless, the output power of n-AlGaN UV-LED was enhanced from 70 mW to 86 mW (about 22%) as the injection current was increased to 600 mA. The n-AlGaN UV-LED exhibits only 20% quantum efficiency droop and 22% power enhancement at high injection current, we attributed this improvement can be less self-absoption by replacing n-GaN nuderlayer with n-AlGaN.

REFERENCES

[1] H. Hirayama, "Quaternary InAlGaN-based high efficiency ultraviolet light-emitting diodes," *J. Appl. Phys.*, vol. 97, pp. 091101−091119, 2005.

[2] I. H. Ho and G. B. Stringfellow, "Solid phase immiscibility in GaInN," *Appl. Phys. Lett.*, vol. 69, pp. 2701−2703, 1996.

[3] T. Mukai and S. Nakamura, "Ultraviolet InGaN and GaN Single-Quantum-Well-Structure Light-Emitting Diodes Grown on Epitaxially Laterally Overgrown GaN Substrates." *Jpn. J. Appl. Phys.*, vol. 38, pp. 5735−5739, 1999.

[4] R. H. Horng, W. K. Wang, S. C. Huang, S. Y. Huang, S. H. Lin, C. F. Lin, and D. S. Wuu, "Growth and characterization of 380-nm InGaN/AlGaN LEDs grown on patterned sapphire substrates," *J. Cryst. Growth*, vol. 298, pp. 219−222, 2007.

[5] D. Morita, M. Yamamoto, K. Akaishi, K. Matoba, K. Yasutomo, Y. Kasai, M. Sano, S. i. Nagahama and T. Mukai, "Watt-Class High-Output-Power 365 nm Ultraviolet Light-Emitting Diodes," *Jpn. J. Appl. Phys.*, vol. 43, pp. 5945−5950, 2004.

[6] T. Fujii, Y. Gao, R. Sharma, E. L. Hu, S. P. DenBaars, and S. Nakamura, "Increase in the extraction efficiency of GaN-based light-emitting diodes via surface roughening," *Appl. Phys. Lett.*, vol. 84, pp. 855−857, 2004.

[7] K. C. Yung, H. Liem, H. S. Choy, and W. K. Lun, "Degradation mechanism beyond device self-heating in high power light-emitting diodes," *J. Appl. Phys.*, vol. 109, pp. 094509−094514, 2011.

[8] Y. J. Lee, C. H. Chiu, C. C. Ke, P. C. Lin, T. C. Lu, H. C. Kuo, and S. C. Wang, "Study of the Excitation Power Dependent Internal Quantum Efficiency in InGaN/GaN LEDs Grown on Patterned Sapphire Substrate," *IEEE Journal of Selected Topics in Quantum Electronics*, vol. 15, pp. 1137−1143, 2009.

[9] D. Brunner, H. Angerer, E. Bustarret, F. Freudenberg, R. Hopler, R. Dimitrov, O. Ambacher, and M. Stutzmann, "Optical constants of epitaxial AlGaN films and their temperature dependence," *J. Appl. Phys.*, vol. 82, pp. 5090−5096, 1997.

[10] A. Chtanov, T. Baars, and M. Gal, "Excitation-intensity-dependent photoluminescence in semiconductor quantum wells due to internal electric fields," *Phys. Rev. B*, vol. 53, pp. 4704−4707, 1996.

[11] Y. Yang, X. A. Cao, and C. Yan, "Investigation of the Nonthermal Mechanism of Efficiency Rolloff in InGaN Light-Emitting Diodes," *IEEE Trans Electron Devices*, vol. 55, pp. 1771−1775, 2008.

Study of Efficiency Droop in InGaN-based Near-UV LEDs with Quaternary InAlGaN Barrier

Po-Min Tu[*a,b], Shih-Cheng Huang[a], Ya-wen Lin[a], Shun-Kuei Yang[a], Chih-Peng Hsu[a],
Jet-Rung Chang[b], and Chun-Yen Chang[b], *Life Fellow, IEEE*

[a]Advanced Optoelectronic Technology Inc., Hsinchu 30352, Taiwan.
[b]Department of Electronic Engineering, National Chiao Tung University,
1001 University Rd., Hsinchu, 30010 Taiwan
Email: bomin.tu@gmail.com

Abstract- **In this study, we demonstrate high efficient near-UV LEDs by replacing low-temperature AlGaN by InAlGaN barrier in active region. The efficiency droop in InGaN-based near-UV LED with AlGaN and InAlGaN barrier is investigated. High-resolution x-ray diffraction (HRXRD) and transmission electron microscopy (TEM) measurements show the two barriers are consistent with the lattice, and smooth morphology of quaternary InAlGaN layer can be observed in atomic force microscopy (AFM). Electroluminescence results indicate that the light performance of quaternary LEDs can be enhanced by 25 % and 55 % at 350 mA and 1000mA, respectively. Furthermore, simulations show that quaternary LEDs exhibit 62 % higher radiative recombination rate and low efficiency degradation of 13 % at a high injection current. We attribute this improvement to increasing of carrier concentration and more uniform redistribution of carriers.**

I. INTRODUCTION

GaN-based near-ultraviolet light emitting devices (LEDs) have attracted great attention in last few years due to its potential applications in photo-catalytic deodorizing such as air conditioner [1], biological, medical and environmental instrumentation, resin curing, UV light source, and there have been interests in solid-state lighting by using near-UV LEDs light for the phosphor-converting source [2, 3].

However, it is difficult to fabricate near-UV LEDs with high efficiency, because the internal quantum efficiency (IQE) decreases drastically under the low indium composition [4-6]. Moreover, crystalline quality and light absorption of GaN are significant for short wavelength near-UV LEDs [7, 8]. It's well known that in low indium content InGaN based-quantum wells, AlGaN barrier is necessary for carrier confinement. But the two materials of AlGaN and InGaN are very different in growth temperature which affects strongly on the quality of material and device performances.

To improve the quantum efficiency of the InGaN-based LEDs, previous reports used InAlGaN in the quantum barrier instead of AlGaN or GaN for polarization, strain, material quality, and interfacial abruptness issues [9-12]. However, by introducing of indium in AlGaN without increase aluminum content will cause the enhancement of the quantum confined Stark effect and other band gap issues. In this letter, the InAlGaN barrier was not for lattice or band gap matched in

InGaN QW but matched in optimized AlGaN barrier, for a fair investigation on the light output and efficiency current droop characteristics.

In this study, we demonstrate high efficient near-UV LEDs by replacing AlGaN by InAlGaN barrier in active region. Furthermore, the efficiency droop characteristics and optical properties of high efficient near-UV LEDs have been measured and investigated by APSYS.

II. EXPERIMENTAL

All samples used in this study were grown on 2-inch c-plane sapphire substrates using an atmospheric-pressure metal organic chemical vapor deposition (AP-MOCVD) in a Taiyo Nippon Sanso SR4000 reactor system.

The conventional structure is as follows. A 500 °C low temperature (LT) 30-nm-thick GaN nucleation layer was deposited, followed by a 1-μm-thick un-doped GaN layer and a 2.5-μm-thick n-type $Al_{0.02}Ga_{0.98}N$ layer grown at 1150 °C. A ten-period InGaN/AlGaN multi-quantum-well (MQW) active region was grown at 830 °C. Subsequently, a 15-nm-thick Mg-doped $Al_{0.3}Ga_{0.7}N$ and a 10-nm-thick Mg-doped $Al_{0.1}Ga_{0.9}N$ electron-blocking layers (EBL) were grown at 1050 °C, followed by a 60-nm-thick Mg-doped GaN contact layer grown at 1030 °C.

The quaternary structure of InGaN/InAlGaN MQW was almost identical to that of the InGaN/AlGaN MQW LED, the only difference was that we used InAlGaN instead of AlGaN as the barrier layers in the active region. Here, the MQW active region consisted of ten periods of 2.6-nm-thick un-doped $In_{0.025}Ga_{0.975}N$ well layers and 11.7-nm-thick Si-doped $In_{0.0085}Al_{0.1112}Ga_{0.8803}N$ or $Al_{0.08}Ga_{0.92}N$ quantum barrier layers growth on n- $Al_{0.02}Ga_{0.98}N$/ud-GaN/Sapphire. The schematic of InGaN/AlGaN and InGaN/InAlGaN MQWs structure are shown in Fig. 1 (a).

To probe the detailed properties of epitaxial layers, a 50-nm AlGaN and InAlGaN single heteroepitaxial layers were also deposited on n-AlGaN/ud-GaN/Sapphire substrate. The schematic of AlGaN and InAlGaN single heteroepitaxial layers are shown in Fig. 1 (b).

The mole fractions of Al and In in MQWs were identified by High-resolution double crystal x-ray diffraction (DCXRD) using Cu Kα as source. These samples were also

978-1-4673-2395-6/12 $31.00 © 2012 IEEE

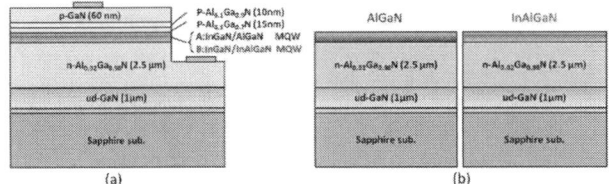

Fig. 1. Schematic of (a) InGaN/AlGaN and InGaN/InAlGaN MQWs structures, and (b) AlGaN and InAlGaN single heteroepitaxial layers.

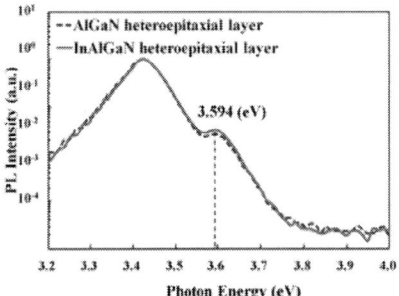

Fig. 2. Room-temperature PL spectra of AlGaN and InAlGaN single heteroepitaxial layers.

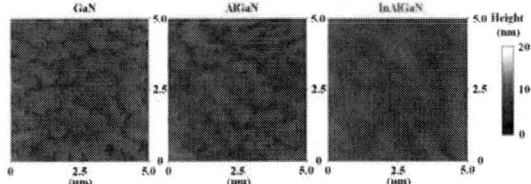

Fig. 3. Surface morphology of top-view AFM images over 5×5 μm² of LT GaN, AlGaN, and InAlGaN layer.

characterized by photoluminescence (PL), atomic force microscopy (AFM) and transmission electron microscopy (TEM) to reveal the optical property, surface morphology and MQWs structure, respectively.

Finally, the UV LED wafers were processed into mesa-type chips and packaged on epoxy-free metal cans. The output power of the UV LED was measured using an integrated sphere detector and tested at room temperature with currents up to 1 A. Testing is done in pulsed mode with 100 μs pulses and a 1 % duty cycle to prevent self-heating, because the thermal time constant of the LEDs is in the millisecond range [13]. In this paper, the optical and electrical properties of InGaN/InAlGaN and conventional InGaN/AlGaN MQW LEDs are numerically calculated using the APSYS simulation software [14].

III. RESULTS AND DISCUSSION

A. Investigation of Optical Property and Surface Morphology

Fig. 2 shows that the PL emission energy of these two samples are very close (~ 3.594 eV) and the peak intensity of InAlGaN is slightly higher than AlGaN. The strong PL emission is attributed to the better crystal quality [15]. Fig. 3 shows the surface morphology of LT GaN, AlGaN and InAlGaN single heteroepitaxial layers with the same thickness about 50-nm. The root-mean-square (RMS) roughness measured by AFM is about 0.851 nm, 0.813 nm and 0.595 nm, respectively. The relatively high roughness of AlGaN single heteroepitaxial layer can mainly be attributed to the low deposition temperature of 830°C necessary for the adjacent InGaN well. As shown in Fig. 3, the dimension of each pit in GaN and AlGaN is slightly larger than in InAlGaN layer. The relatively small pits of LT InAlGaN layer can mainly be attributed to the smaller tensile strain in LT AlGaN or LT GaN by inserting the isoelectronic In atoms.

B. Compositions and Thicknesses Analysis

Fig. 4 shows the HRXRD (ω-2θ) curves in the (004) and (105) reflections of GaN, AlGaN, and InAlGaN single heteroepitaxial layers. The HRXRD results show that the locations of right side peaks of AlGaN and InAlGaN layers are very close. It proves that the lattices in those two samples of ternary and quaternary material are matched. It is worth mentioning for the asymmetric spectrum in the center.

This is because the single heteroepitaxial layers are grown on an n-AlGaN/ud-GaN/Sapphire substrate, and the peak in center and the merged peak around -300 arcsec indicate the 2.5-μm $Al_{0.02}Ga_{0.98}N$ and 1-μm GaN, respectively. Experiments have shown that the presence of In leads to a smooth morphology with better crystal quality and optical properties, and this result is due to the interaction between In atoms and screw dislocations [16].

Fig. 5 (a) shows the HRXRD (ω-2θ) curves in the (002) reflections of InGaN/AlGaN and InGaN/InAlGaN MQWs. The results show that the locations of multiple satellite peaks of InGaN/AlGaN and InGaN/InAlGaN MQWs are very close. This indicates that the thickness of barrier layer in these two samples is matched, and it is quite consistent with the measured values of 11.7 nm from HRTEM images as shown in Fig. 5 (b) and (c). In addition to experimentally estimate the In and Al composition in the MQWs, we simulate the HRXRD (ω-2θ) curve by using dynamical diffraction theory. The In composition in the QWs was determined to be about 2.5 %, where the thickness of the well was about 2.6 nm. The compositions of ternary and quaternary barriers were $Al_{0.08}Ga_{0.92}N$ and $In_{0.0085}Al_{0.1112}Ga_{0.8803}N$, respectively. Besides, the growth rates of well and barrier were estimated about 0.329 and 0.308 Å/s, respectively.

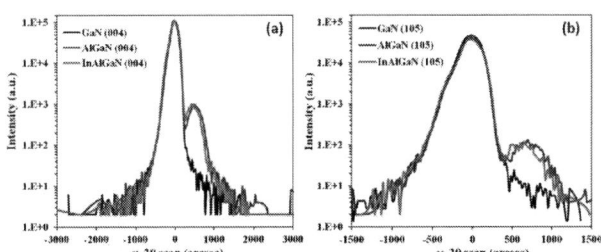

Fig. 4. HRXRD (ω-2θ) curves in the (a) (004) and (b) (105) reflections of GaN, AlGaN, and InAlGaN single heteroepitaxial layers.

978-1-4673-2395-6/12 $31.00 © 2012 IEEE

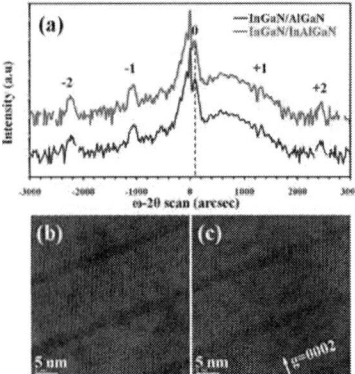

Fig. 5. (a) HRXRD (ω-2θ) curves in the (002) reflections of InGaN/AlGaN and InGaN/InAlGaN MQWs, and Cross-sectional TEM images of (b) InGaN/AlGaN and (c) InGaN/InAlGaN MQWs.

C. Current-dependent Intensity and Efficiency

Fig. 6 (a) shows the optical properties of light output power–current–voltage (L-I-V) characteristics for the AlGaN and InAlGaN barrier near-UV LEDs. The forward voltage was 3.89 and 3.98 V for InGaN/AlGaN and InGaN/InAlGaN MQWs near-UV LED at a forward current of 350 mA, respectively. A little high forward voltage of InAlGaN barrier LED can be attributed to the higher Al content compare to the AlGaN barrier, thus increase the series resistance in the device. The light output power of InGaN-based near-UV LED with the InAlGaN barrier is higher by 25 % and 55 % than the AlGaN barrier at 350 mA and 1000 mA, respectively. Fig. 6 (b)-(d) show the images of epi-wafer and mesa-type chip. Besides, the wavelength is nearly constant about 380 nm over the entire current range.

D. Theoretical Analysis by APSYS

In order to investigate the physical origin of efficiency droop in these near-UV LEDs, we performed a simulation of the above structures by using the APSYS simulation software. Commonly accepted Shockley-Read-Hall recombination lifetime (about ~6 ns) and Auger recombination coefficient (about ~10^{-30} cm^{6}s^{-1}) are used in the simulations. Other material parameters of the semiconductors used in the simulation can be found in Ref. [17].

In addition, we can reduce the effect of spontaneous and piezoelectric polarizations, because of lattice match condition in barrier between AlGaN and InAlGaN. The total polarization fields in different combination of materials can be obtained through the calculation, and the results of $In_{0.025}Ga_{0.975}N$, $Al_{0.08}Ga_{0.92}N$ and $In_{0.0085}Al_{0.1112}Ga_{0.8803}N$ are -0.0305, -0.0391 and -0.0398 (C·m^{-2}), respectively [18]. Besides, a different band-offset ratio from 6:4 to 7:3 is used in this simulation for introducing of indium in AlGaN. We can know that under the same energy band of barrier, the band-offset ratio from 6:4 to 7:3 will lead higher conduction-band and lower valence-band between well and barrier. This is useful for electron confinement and hole distribution in low indium content InGaN-based near-UV LEDs.

Fig. 7 shows the normalized efficiency curves of experimental (open circles) and simulated (solid lines) as a function of forward current for the two samples. For the InGaN/AlGaN near-UV LEDs, when the injection current exceeds 1000 mA, the efficiency is reduced to 66 % of its maximum value. In contrast, InGaN/InAlGaN near-UV LEDs exhibit only 13 % efficiency droop when we increase the injection current to 1000 mA. The reduction of efficiency droop is quite clear and the current at maximum efficiency shifts from 150 to 400 mA. It can clearly be seen that the droop behavior is dominated by hole mobility, and we find the efficiency curve will nearest to the experimental result when electron and hole mobility of InAlGaN is about 1.8 and 2.5 times the value of AlGaN. Finally, the results of the EQE droop simulation of both different structures are in good agreement with the experimental data as shown in Fig. 7.

We performed the numerical simulation with different parameters in band-offset ratio and carrier mobility, listed in Table I.

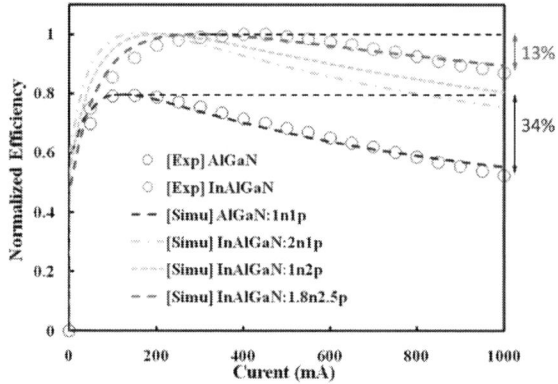

Fig. 7. Simulation results of normalized IQE under different carrier mobility.

TABLE I

SIMULATION PARAMETERS IN BAND-OFFSET RATIO AND CARRIER MOBILITY.

Items	Band offset ratio	Electron mobility (cm²/V⁻¹s⁻¹)	Hole mobility (cm²/V⁻¹s⁻¹)
InGaN/AlGaN MQWs	6 : 4	354	2
InGaN/InAlGaN MQWs	7 : 3	642	5

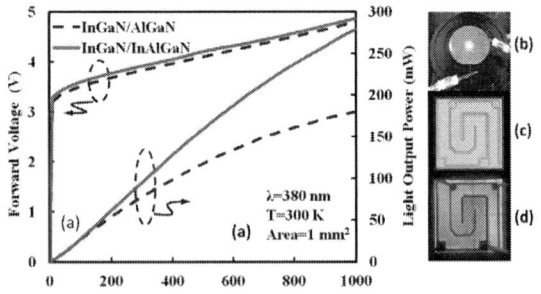

Fig. 6. (a) L-I-V curves of the LEDs with AlGaN (dash) and InAlGaN (solid) barrier. (b) 2-inch near-UV LED epi-wafer under 100 mA, (c) the mesa-type near-UV chip and (d) chip image under 350 mA driving current.

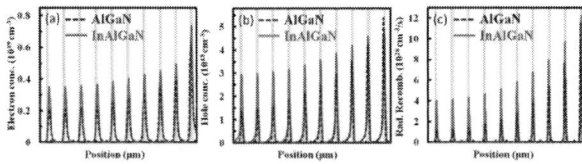

Fig. 8. Distribution of (a) Electron (b) Hole concentrations, and (c) Radiative recombination rates concentrations of the LEDs with AlGaN and InAlGaN barrier under a high current density of 100 A/cm².

Fig. 8 shows the calculated carrier distribution in these near-UV LEDs structure under a high forward current density of 100 A/cm² by APSYS. When we apply the corresponding band-offset ratio and the carrier mobility in InGaN/InAlGaN MQWs, the electron and hole concentration increases in the QW by about 26 % and 35 %, respectively, and the distribution of carrier becomes more uniform than InGaN/AlGaN case. Under high current density, the carrier distribution of both electrons and holes determines how efficient the photon-emission process will be. As shown in Fig. 8, the peak-to-peak carrier ratio of InAlGaN barrier sample is reduced due to better carrier transportation, and this is also more obvious in the hole distribution. The direct consequence is the increasing radiative recombination rate and thus the light output is expected to rise. On the other hand, in the traditional AlGaN barrier samples, the holes are locally concentrated in the first quantum well which causes the unbalanced distribution between different types of carriers, and thus leads to reduction of radiative recombination rate. Comparing electrons and holes, holes suffer more as a result of this nonuniformity due to their large effective mass and low mobility. Thus, our InAlGaN design can reduce the carrier leakage and increase electron-hole pair radiative recombination simultaneously, especially for the distribution of holes.

IV. CONCLUSIONS

In summary, we fabricated and compared the performance of LEDs of InGaN-based near-UV MQWs active region with ternary AlGaN and quaternary InAlGaN barrier layers. Measurements show the two barriers are consistent with the lattice, and smooth morphology of quaternary InAlGaN layer can be observed in AFM. EL results indicate that the light performance can be enhanced effectively when the conventional LT AlGaN barrier layers are replaced by the InAlGaN barrier layers. Furthermore, simulation results show that near-UV LEDs with InAlGaN barrier exhibit about
62 % higher radiative recombination rate and low efficiency droop of 13 % at a high injection current. We attribute this improvement to increasing of carrier concentration and more uniform redistribution of carriers.

REFERENCES

[1] A. Sandhu, "The future of ultraviolet LEDs," *Nature Photonics*, vol. 1, p. 38, 2007.

[2] Y. S. Tang, S. F. Hu, C. C. Lin, N. C. Bagkar, and R. S. Liu, "Thermally stable luminescence of KSrPO₄:Eu²⁺ phosphor for white light UV light-emitting diodes," *Appl. Phys. Lett.*, vol. 90, pp. 151108-1–151108-3, 2007.

[3] Y. C. Chiu, W. R. Liu, C. K. Chang, C. C. Liao, Y. T. Yeh, S. M. Jang, and T. M. Chen, "Ca₂PO₄Cl : Eu²⁺: an intense near-ultraviolet converting blue phosphor for white light-emitting diodes," *J. Mater. Chem.*, vol. 20, pp. 1755–1758, 2010.

[4] H. Hirayama, "Quaternary InAlGaN-based high-efficiency ultraviolet light-emitting diodes," *J. Appl. Phys.*, vol. 97, pp. 091101-1–091101-19, 2005.

[5] I. H. Ho and G. B. Stringfellow, "Solid phase immiscibility in GaInN," *Appl. Phys. Lett.*, vol. 69, pp. 2701–2703, 1996.

[6] T. Mukai and S. Nakamura, "Ultraviolet InGaN and GaN single quantum well structure light-emitting diodes grown on epitaxially laterally overgrown GaN substrates," *Jpn. J. Appl. Phys.*, vol. 38, pp. 5735–5739, 1999.

[7] R. H. Horng, W. K. Wang, S. C. Huang, S. Y. Huang, S. H. Lin, C. F. Lin, and D. S. Wuu, "Growth and characterization of 380-nm InGaN/AlGaN LEDs grown on patterned sapphire substrates," *J. Cryst. Growth*, vol. 298, pp. 219–222, 2007.

[8] D. Morita, M. Yamamoto, K. Akaishi, K. Matoba, K. Yasutomo, Y. Kasai, M. Sano, S. i. Nagahama, and T. Mukai, "Watt-class high-output-power 365 nm ultraviolet light-emitting diodes," *Jpn. J. Appl. Phys.*, vol. 43, pp. 5945–5950, 2004.

[9] A. Knauer, H. Wenzel, T. Kolbe, S. Einfeldt, M. Weyers, M. Kneissl, and G. Tränkle, "Effect of the barrier composition on the polarization fields in near UV InGaN light emitting diodes," *Appl. Phys. Lett.*, vol. 92, pp. 191912-1–191912-3, 2008.

[10] M. F. Schubert, J. Xu, J. K. Kim, E. F. Schubert, M. H. Kim, S. Yoon, S.M. Lee, C. Sone, T. Sakong, and Y. Park, "Polarization-matched GaInN/AlGaInN multi-quantum-well light-emitting diodes with reduced efficiency droop," *Appl. Phys. Lett.*, vol. 93, pp. 041102-1–041102-3, 2008.

[11] J. J. Wu, G. Y. Zhang, X. L. Liu, Q. S. Zhu, Z. G. Wang, Q. J. Jia, and L.P. Guo, "Effect of an indium-doped barrier on enhanced near-ultraviolet emission from InGaN/AlGaN:In multiple quantum wells grown on Si(111)," *Nanotechnology*, vol. 18, pp. 015402-1–015402-5, 2007.

[12] S. H. Baek, J. O. Kim, M. K. Kwon, I. K. Park, S. I. Na, J. Y. Kim, B. J.Kim, and S. J. Park, "Enhanced Carrier Confinement in AlInGaN–InGaN Quantum Wells in Near Ultraviolet Light-Emitting Diodes," *IEEE Photon. Technol. Lett.*, vol. 18, pp. 1276–1278, 2006.

[13] Q. Shan, Q. Dai, S. Chhajed, J. Cho, and E. F. Schubert, "Analysis of thermal properties of GaInN light-emitting diodes and laser diodes," *J. Appl. Phys.*, vol. 108, pp. 084504-1–084504-8, 2010.

[14] APSYS by Crosslight Software Inc., Burnaby, Canada: http://www.crosslight.com.

[15] J. J. Wu, X. X. Han, J. M. Li, H. Y. Wei, G. W. Cong, X. L. Liu, Q. S. Zhu, Z. G. Wang, Q. J. Jia, L. P. Guo, T. D. Hu, and H. H. Wang, "Crack control in GaN grown on silicon (111) using In doped low-temperature AlGaN interlayer by metalorganic chemical vapor deposition," *Optical Materials*, vol. 28, pp. 1227–1231, 2006.

[16] S. Yamaguchi, M. Kariya, T. Kashima, S. Nitta, M. Kosaki, Y. Yukawa, H. Amano, and I. Akasaki, "Control of strain in GaN using an In doping-induced hardening effect," *Phys. Rev. B*, vol. 64, pp. 035318-1–035318-5, 2001.

[17] H. Amano, N. Sawaki, I. Akasaki, and Y. Toyoda, "P-Type Conduction in Mg-Doped GaN Treated with Low-Energy Electron Beam Irradiation (LEEBI)," *Jpn. J. Appl. Phys.*, vol. 28, pp. L2112–L2114, 1989.

[18] M. H. Kim, M. F. Schubert, Q. Dai, J. K. Kim, E. F. Schubert, J. Piprek, and Y. Park, "Origin of efficiency droop in GaN-based light-emitting diodes," *Appl. Phys. Lett.*, vol. 91, pp. 183507-1–183507-3, 2007.

978-1-4673-2395-6/12 $31.00 © 2012 IEEE

Modeling of SOI-based MRR by Coupled Mode Theory using Lateral Coupling Configuration

[1]Hazura H., [1]Menon, P.S, *MIEEE*, [1]Burhanuddin Yeop Majlis, *SMIEEE*, [1]Hanim A.R, [1]Mardiana B., [2]Hasanah, L.,
[3]Mulyanti, B., [4]Mahmudin, D., [4]Wiranto, G., *MIEEE*,

[1]Institute of Microengineering and Nanoelectronics (IMEN),
Universiti Kebangsaan Malaysia (UKM), 43600 UKM Bangi, Selangor, Malaysia.
[2]Department of Physics Education, Faculty of Mathematics and Natural Sciences Education,
[3]Department of Electrical Engineering,
Indonesia University of Education (UPI), Jalan Dr. Setiabudhi 207, Bandung 40154, Indonesia.
[4]Research Centre for Electronics and Telecommunications,
Indonesian Institute of Sciences (LIPI), Jl. Sangkuriang, Bandung 40135, Indonesia.
susi@eng.ukm.my

Abstract-We present the modeling of a first order waveguide-coupled microring resonator (MRR) by coupled mode theory (CMT) using transfer matrix model. The design topology is based on the lateral coupling configuration and single mode propagation which is integrated on a Silicon-on-Insulator (SOI) platform. Performance parameters including Free Spectral Range (FSR) and Quality Factor (Q-factor) are investigated. For verification, we compare these results with the results obtained from the Finite Difference Time Domain (FDTD) commercially available software. We found that both results agree well with each other.

I. INTRODUCTION

Optical waveguide-based devices using evanescent wave are currently applied for a variety of applications. One of the most promising devices is the microring resonator (MRR), which can be employed as a wavelength filter[1], multiplexer[2], sensor[3] and modulator[4].

The device can be modeled by several methods such as Finite Difference Time Domain (FDTD) method [5] and conformal transformation method [6]. Withal, both methods are time consuming. The research on modeling of the optical waveguides structures by Coupled Mode Theory (CMT) has been started in the early 1970's by researchers such as Yariv [7] and Snyder [8]. CMT provides the interaction description between the circulating waves in the microring and the bus waveguides. The fraction of power exchanged between waveguides is highly dependent on the coupling coefficient, which will determine the microring resonator performance such as Free Spectral Range (FSR) and Quality Factor (Q-Factor).

To date, many studies have been performed to develop the modeling of microring resonators incorporated in the transfer matrix model, which involves complicated calculations and numerous parameter assumptions. We refine each parameter to provide a more understandable model. Based on the model, we studied the effect of major physical characteristic variation on the device performance.

FDTD simulation from RSoft is adopted to validate the accuracy of the analytical model based on the transfer matrix method and CMT.

II. THEORY

A. Device Design

Fig. 1 depicts schematically the MRR-based optical filter under study which comprises of a ring waveguide closely coupled to double straight bus waveguides (Fig. 1(a)) and the cross section of the waveguide structure is shown in Fig. 1(b). The bus waveguides serve as evanescent light input and output couplers, while the ring waveguide acts as the wavelength selective element. *R* depicts the ring radius, *gap* is the separation gap between straight and ring waveguide, *W* is the waveguide width and *H* is the total waveguide height. The fully etched waveguide structure is considered throughout this study.

An example of the spectral response of the microring resonator is shown in Fig.2. The microring will be at the ON-resonance state, when the optical path length of a roundtrip is a multiple of the effective wavelength. The light wave will then be transferred to the drop port. On the contrary, if the resonance condition is not complied with, the light wave will be delivered to the through port.

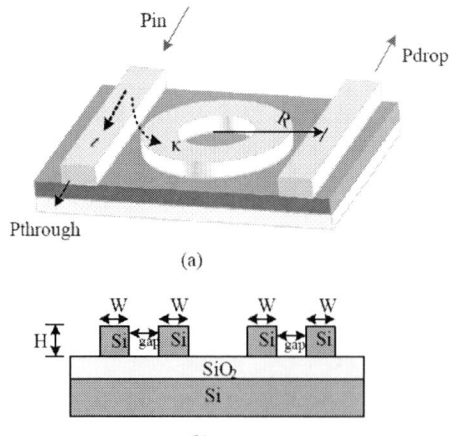

Figure 1. (a) Layout and (b) cross section of the proposed MRR.

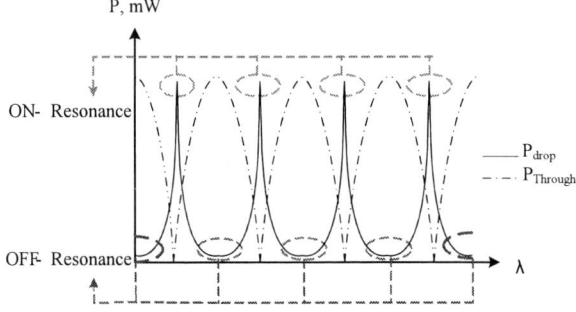

Figure 2. An example of MRR's response spectrum.

$$\tilde{\beta} = \frac{2\pi}{\lambda_o} n_{eff} - j\frac{\alpha}{2} \tag{3}$$

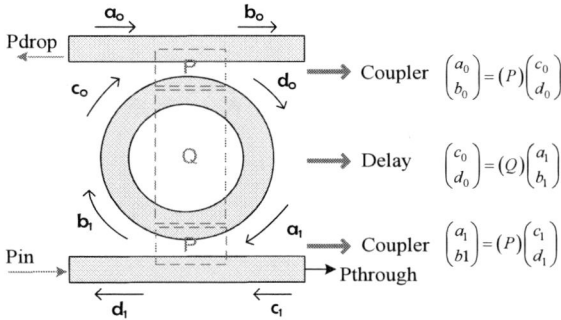

$$\text{Coupler} \quad \begin{pmatrix} a_0 \\ b_0 \end{pmatrix} = (P)\begin{pmatrix} c_0 \\ d_0 \end{pmatrix}$$

$$\text{Delay} \quad \begin{pmatrix} c_0 \\ d_0 \end{pmatrix} = (Q)\begin{pmatrix} a_1 \\ b_1 \end{pmatrix}$$

$$\text{Coupler} \quad \begin{pmatrix} a_1 \\ b1 \end{pmatrix} = (P)\begin{pmatrix} c_1 \\ d_1 \end{pmatrix}$$

Figure 3. Visualization of the couplers and delay unit from microring resonator .

The substrate cladding is composed of air (n=1.0 at 1.55μm), while the silicon-on-insulator (SOI) platform is chosen for the bus and ring waveguide formation. Lateral coupling is chosen between the bus waveguides and ring waveguides due to fabrication limitation.

Here, α is the loss per unit length in the microring, λ_o is the free space wavelength and n_{eff} is the effective refractive index. The transfer matrix between two bus waveguides is:

$$\begin{pmatrix} a_0 \\ b_0 \end{pmatrix} = (PQP)\begin{pmatrix} c_1 \\ d_1 \end{pmatrix} \equiv (M)\begin{pmatrix} c_1 \\ d_1 \end{pmatrix} \equiv \begin{pmatrix} m_{11} & m_{12} \\ m_{21} & m_{22} \end{pmatrix}\begin{pmatrix} c_1 \\ d_1 \end{pmatrix} \tag{4}$$

B. Coupled Mode Theory and Transfer Matrix Method

One of the most fundamental approaches in the photonic device modeling and design is the coupled mode theory (CMT). This theory describes the expressions for the output transmission in terms of coupling coefficient, effective refractive index, waveguide losses and ring radius. We consider the MRR-based device configuration, comprising of a single ring coupled to a double bus waveguides (please refer to Fig.1). A transfer matrix method (TMM) is integrated with CMT to develop the transfer function of multiple optical waveguides, in our case from input bus waveguide to microring waveguide and then from the microring waveguide to the output or drop port waveguide. The matrix chain form of the TMM eases the signal transmission formulations and is practical to synthesize higher cascaded MRRs.

The basic elements in coupled microring optical waveguides are very similar to directional couplers and delay units. For instance, a single microring coupled to two bus waveguides can be decomposed into an input coupler, a delay unit, and an output coupler, as visualized in Fig. 3.

The transmission of the input and output coupler, P can be calculated by [9]:

$$P = \frac{1}{\kappa}\begin{pmatrix} 1 & t \\ -1 & t^* \end{pmatrix} \tag{1}$$

In Eq. (1), κ is the normalized coupling coefficient of the coupler and t is the transmission coefficient, respectively. The optical phase delay and the waveguide loss, Q is given by:

$$Q = \frac{1}{\kappa}\begin{pmatrix} 0 & e^{-i\tilde{\beta}\pi R} \\ e^{i\tilde{\beta}\pi R} & 0 \end{pmatrix} \tag{2}$$

where R is the ring radius and $\tilde{\beta}$ is the propagation constant which is equal to:

Considering only 1 input, c_1 will be zero, hence the final transfer functions for the through port |T| and drop port |D| signals are [10]:

$$|T| = \frac{b_o}{a_o} = \frac{m_{22}}{m_{12}} = \frac{\sqrt{1-\kappa_{in}} - \sqrt{1-\kappa_{out}}e^{-i\tilde{\beta}2\pi R}}{1 - \sqrt{(1-\kappa_{in})(1-\kappa_{out})}e^{-i\tilde{\beta}2\pi R}} \tag{5}$$

$$|D| = \frac{d_1}{a_o} = \frac{1}{m_{12}} = \frac{-\sqrt{\kappa_{in}\kappa_{out}}e^{-i\tilde{\beta}\pi R}}{1 - \sqrt{(1-\kappa_{in})(1-\kappa_{out})}e^{-i\tilde{\beta}2\pi R}} \tag{6}$$

Assuming that $\kappa=\kappa_{in}=\kappa_{out}$, eq.(5) and eq.(6) can be simplified as eq.(7) and eq. (8), respectively.

$$|T| = \frac{b_o}{a_o} = \frac{m_{22}}{m_{12}} = \frac{\sqrt{1-\kappa} - \sqrt{1-\kappa}e^{-i\tilde{\beta}2\pi R}}{1 - \sqrt{(1-\kappa)(1-\kappa)}e^{-i\tilde{\beta}2\pi R}} \tag{7}$$

$$|D| = \frac{d_1}{a_o} = \frac{1}{m_{12}} = \frac{-\sqrt{\kappa^2}e^{-i\tilde{\beta}\pi R}}{1 - \sqrt{(1-\kappa)(1-\kappa)}e^{-i\tilde{\beta}2\pi R}} \tag{8}$$

For symmetrical lateral coupling waveguide structures, κ is analytically computed according to the theory developed in [11]:

$$\kappa = \frac{2\eta^2\gamma \exp.(-\gamma.gap)}{\tilde{\beta}(W + 2/\gamma)(\eta^2 + \gamma^2)} \tag{9}$$

where gap is the distance between the bus waveguide and the microring waveguide, while η and γ are as follows:

978-1-4673-2395-6/12 $31.00 © 2012 IEEE

$$\eta = \sqrt{n_{core}^2 k_o^2 - \widetilde{\beta}^2} \qquad (10)$$

$$\gamma = \sqrt{\widetilde{\beta}^2 - n_{clad}^2 k_o^2} \qquad (11)$$

It is noted that $k_o = (2\pi/\lambda_o)$, n_{core} is the refractive index of the core waveguide and n_{clad} is the refractive index of cladding.

The performance of the coupled microring can be evaluated for the FSR and Q-factor. FSR is the frequency separation between two successive resonances and is given by[12]:

$$FSR \approx \frac{\lambda_0^2}{n_g(\lambda)L_{eff}} \qquad (12)$$

where n_g is the group refractive index.

Meanwhile, the Q- factor of the mth resonance is estimated by computing the ratio of the center wavelength to the -3dB bandwidth.

III. RESULTS AND DISCUSSION

The waveguide width, W, the waveguide height, H and the free space wavelength considered in the following cases are 300nm, 550nm and 1550nm, respectively. Fig. 4 depicts the computed effective refractive index, n_{eff} of the bus waveguide and the coupling coefficient for TE mode of the microring resonator. The ring radius of 6μm and the gap separation of 100nm is considered in the theoretical calculations. The results show that for a gap of 100nm, the effective refractive index is 2.34 and the coupling coefficient, κ is 0.861. The effective refractive index obtained from the simulation is 2.36. The deviation between the theoretical calculation for CMT and simulation is therefore less than 0.9%.

Additionally, the coupling coefficient from the simulation is 0.848, which shows only a small discrepancy between both results. We also note that the separation distance between the bus waveguide and the microring waveguide is inversely proportional to the coupling coefficient. It can be concluded that the gap separation plays an important role in optimizing the performance of the microring resonator.

From the CMT analysis, the FSR values were calculated and compared with the rigorous FDTD simulations as shown in Fig. 5. The ring radius values were varied and the FSR was investigated. Results indicate that the FSR is highly dependent on the ring radius. Moreover, it can be noted that the theoretical results agree with the simulation observations. As an example, the highest difference is for R=4μm with FSR of 25nm (simulated) and FSR of 21 nm (theoretical) with 16% deviation and the smallest is 3% difference with R=6μm where FSR$_{simulated}$=16nm and FSR$_{theoretical}$=15.5nm. The different is possibly due to the discretized nature of the numerically simulated structure. Furthermore, Fig.6 shows that the separation gap has no significant effect on the FSR in which the values remain as we increase the gap size, while the radius is fixed with 6μm.

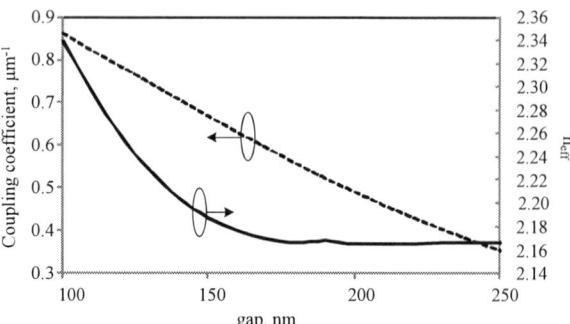

Figure 4. Theoretical coupling coefficient and effective refractive index values using CMT.

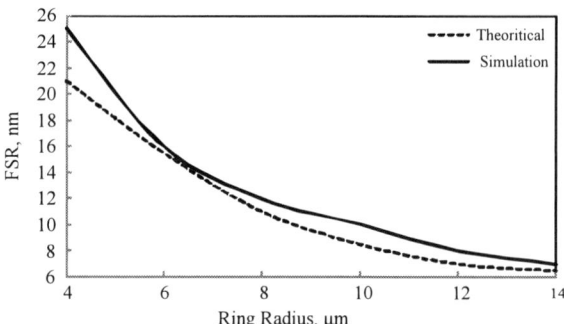

Figure 5. Theoretical and simulated values of FSR vs ring radius.

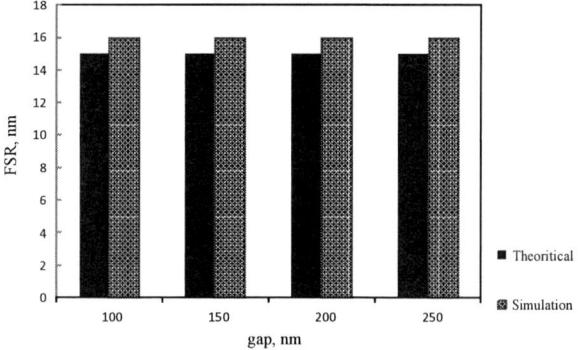

Figure 6. Theoretical and simulated values of FSR vs separation gap.

Fig. 7 plots the influence of gap separation on the Q-factor. It is shown that larger Q-factor is produced at larger separation gaps. The simulations predict a more linear response as compared to the theoretical analysis. This is may be due to human error while observing the data for calculating the theoretical Q-factor value while in simulation, the Q-factor is automatically calculated. The observable Q-factor difference at the separation gap of 0.13μm is 28 which contributes to 2.4% in percentage, while the largest deviation is 5.4%, which can be considered relatively low.

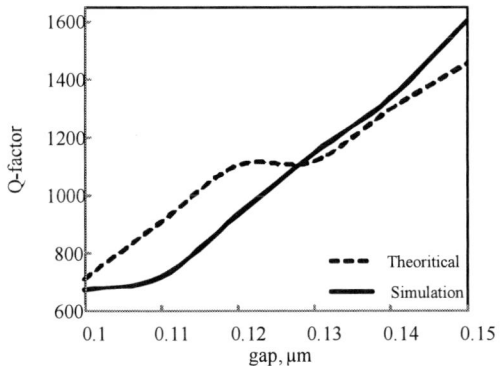

Figure 7. Q-factor results from theoretical calculations and simulations.

IV. CONCLUSIONS

We have presented the modeling of a lateral configuration microring resonator by Coupled Mode Theory. The influence of the microring parameters such as ring radius and separation gap on the coupling coefficient, FSR and Q-factor have been discussed. In the best case, the Q-factor reaches 1500 and the largest FSR is 16nm. We achieve a good match between the theoretical modeling by CMT and by FDTD simulations, which shows that this method can be reasonably accepted.

ACKNOWLEDGMENT

The authors would like to acknowledge Prof Dr Sahbudin Shaari from IMEN, UKM for his input on Device Modeling. The authors would also like to thank Universiti Teknikal Malaysia Melaka (UTeM) and Malaysian Ministry of Higher Education (MOHE) for the support. This research is supported by funding from Universiti Kebangsaan Malaysia, Industri-2011-015 and GUP-2012-012. UPI and LIPI are also acknowledged for their contribution via grant no 529/H40.8/TU/2012.

REFERENCES

[1] Hazura Haroon, Mardiana Bidin, Hanim Abdul Razak, Menon P.S, N. Arsad, "Impact of Coupled Resonator Geometry on Silicon-on-Insulator Wavelength Filter Characteristics", *IEEE Proc. of Regional Symposium on Micro and Nanoelectronics (RSM 2011)*, Kota Kinabalu, pp. 380-382, , 2011.

[2] Hazura Haroon, Sahbudin Shaari, P. S. Menon, B. Mardiana, A. R. Hanim, N. Arsad, B. Y. Majlis,W. M. Mukhtar and Huda Abdullah, "Design and characterization of multiple coupled microring based wavelength demultiplexer in silicon–on–insulator (SOI)", *Journal of Nonlinear Optical Physics & Materials*, 2012; 21(1): 1250004-1-8.

[3] C.Y Chao, W. Fung, and L.J Guo, "Polymer microring resonators for biochemical sensing applications", *Optics Express*, 2010, vol 18(2), pp.393-400.

[4] S. Shaari, A. R. Hanim, B. Mardiana, H. Hazura and P. S. Menon, "Modeling and analysis of lateral doping region translation variation on optical modulator performance", in *the 4th Asian Physics Symposium AIP Conference Proceedings*, 1325, September 2010, pp. 297–300.

[5] J. V. Hryniewicz, P. P. Absil, B. E. Little, R. A. Wilson, and P. T. Ho, "Higher order filter response in coupled microring resonators, " *IEEE Photonics Technol. Lett.* Vol. 12(3), 2000, pp.320–322.

[6] A. Belarouci, K. B. Hill, Y. Liu, Y. Xiong, T. Chang, and A. E. Craig, "Design and modeling of waveguide-coupled microring resonator," *J. Lumin*, vol (94),2001, pp. 35–38.

[7] Syner, A.W, "Coupled mode theory for optical fiber", *J.Opt. Soc.Am*, vol.62, 1972, pp.1267-1277.

[8] Yariv, A., "Coupled-mode theory for guided- wave optics," *IEEE J.Quantum Electron*, vol.9, 1973, pp.919-933.

[9] A. Yariv, "Universal relations for coupling of optical power between microresonators and dielectric waveguides," *Electronic Letters*, vol. 36, no. 4, 2000, pp. 321-322.

[10] Zen Pheng, "Coupled Multiple Micro-Resonators Design And Active Semiconductor Micro-Resonator Fabrication", Thesis Dissertation, University Of Southern California, 2007.

[11] B.E. Little, S.T. Chu, H.A Haus, J.Foresi, and J-P. Laine, "Microring Resonator Channel Dropping Filter", *J.of Lightwave Technol.*, vol. 15, No. 6, 1997, pp. 998-105.

[12] P. Rabiei, W.H. Steier, Z.Cheng, and L.R Dalton, "Polymer micro-ring filters and modulators," *J. Lightwave Technol.*, vol, 20, 2002, pp. 1968-1975.

Fabrication and Characterization of Cu Pellet Using Powder Metallurgical Method

Chew Pei Yi [1,2], You Ah Heng [1], Vijayaram Thoguluva Raghavan [1]

[1]Faculty of Engineering and Technology, Multimedia University, Jalan Ayer Keroh Lama, 75450 Melaka, Malaysia.
[2]Infineon Technologies (M) Sdn Bhd, Batu Berendam, 75350 Melaka, Malaysia
*Email: PeiYi.Chew@infineon.com

Abstract - **Pure Cu is prepared using by Powder Metallurgical (PM) method in order to study the impact of different sintering environment (with vacuum, N_2 and $N_2 + H_2$) and at different temperatures (1010 °C and 1080 °C) towards its surface and inner morphology, crystal structure and electrical properties. These samples are compacted at a constant pressure of 280 MPa in order to form a pellet with a dimension of 17.7 mm in diameter and about 10–15 mm in height. Cu pellet in vacuum sintering environment is detected with pores in the inner and outer surface of the pellet formed with PM method under SEM. PM method is able to form Cu pellet with electrical conductivity of up to ~ $1.50 \times 10^7 \, \Omega^{-1} \, m^{-1}$ which is within the electrical conductive range of Cu for lead frame and other electronic applications. This can be further explained through XRD analysis to understand the crystal structure and the SEM results on the grain structure in the Cu pellets.**

I. INTRODUCTION

Copper (Cu) is widely used in a large quantities in electrical applications such as electrical wires and transformers. Cu is also widely used in semiconductor industry as lead frame. The main function of the Cu lead frame is to connect the chip electrically with the outside world. Cu is the ideal leadframe base material due to its good electrical and thermal conductivity. Pure Cu is usually doped with other elements such as Zn, Sn, Ni & P etc to improve the disadvantages of pure Cu such as poor tensile strength, machinability and easily oxidized in order to form lead frames. The price of Cu has been rising steadily in recent years. Cu alloy lead frame had initially used when the Cu price was low in 2000 till the price now in year 2011 which is 4.5 times higher in comparison [1]. This is because the Cu is obtained from Cu ores which contains less than 1.5 % of Cu by weight. Therefore, with the increasing Cu price in the foreseeable future and coupled with the above disadvantages from Cu, it has become very critical and crucial to find an alternative material and preparation method for producing Cu lead frame.

Powder Metallurgy [2] (PM) is a technique which involves homogenous blending and mixing of the powder materials. This mixed powder is then compacted at room temperature and sintered in selected environments for densification of the pellets. PM method is a simple metal preparation process compared to conventional melting

involving milling, rolling and etc techniques in achieving the mechanical properties and electrical conductivity of a product. Upadhyaya et al [3] had reported a study of sintering of $Cu-Al_2O_3$ composites through blending and mechanical alloying PM routes. The study showed that there is an influence of compaction pressure on the mechanical properties of the samples. As the compaction pressure increases, both sintered density and hardness increase as well too. With the increase of Al_2O_3 powder content, hardness will generally increase as well but with an associated loss in electrical conductivity. It shows that PM method is able to give good mechanical properties but low electrical conductivity on the sample produced. However, F. Yi et al [4] had reported the microstructure and electrical conductivity of aluminum alloy foams. Aluminum alloy foams with different densities and cell diameters have been fabricated by using powder metallurgy technique. The result shows that electrical conductivity is in the range of 0.981 to 1.003×10^6 $(\Omega m)^{-1}$ of foams increases with increase of relative density while cell diameter has a minor influence on the electrical conductivity of foams. As compared to the Cu lead frame which is produced through melting, casting, and thermo-mechanical treatment (TMT) method, the electrical conductivity can be achieved in the range of 30 to 80 %IACS [5].

The conventional melting method is a complex process as compared to PM method. As melting method involved of melting ingots at about 1084 °C and the molten of Cu alloy is then poured and cast into block shape and soaked to solidified. It then proceeds with the TMT method which involves hot work, milling, cold work, annealing and slitting. After formation into a Cu alloy sheet, it then proceeds for stamping, etching and plating in order to form into a desirable commercial Cu lead frame. Therefore there is wastage of Cu alloy from stamping and etching process. While for PM method, the amount of Cu alloy powder is weighed accordingly, compressed into a desired lead frame design with a designed cavity and subsequently a final product is produced after sintering process [6]. It shows no wastage of Cu alloy with PM method.

This paper mainly describes the powder metallurgy preparation method and the evaluation on different sintering environments and temperatures in order to find the suitable and best condition to produce Cu pellets which have the best compactness, purity of phases and comparable electrical conductivity with a commercial Cu lead frame.

978-1-4673-2395-6/12 $31.00 © 2012 IEEE

The Cu pellets will be subjected for further characterizations such as SEM, EDX, XRD, electrical conductivity, density and porosity measurements.

II. EXPERIMENTAL METHOD

The study of pure Cu at different sintering environments and temperatures using powder metallurgy approach is hereby presented. The desired amount of Cu powder is weighed accordingly to the weight of 5 gm. The 5 gm of Cu powder is poured into the die hole of the pressing jig. Then the powder mixtures are compacted at pressure of 280 MPa using a manual hydraulic press as shown in Fig. 1. A Cu pellet with the dimension of 17.7 mm in diameter and 15 mm in height is produced. Finally, the pellets are sintered in a furnace at 3 different sintering environments and temperatures as shown in Table I.

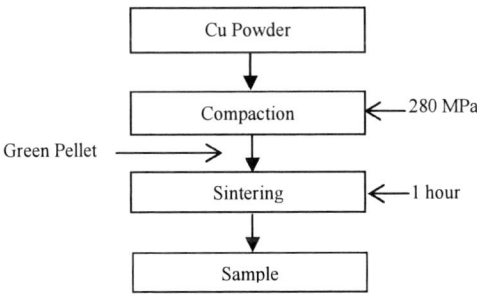

Fig. 1: Flow chart of Cu pellet with PM method

The pellet is sintered with the rate of 10 °C/min for all samples in an enclosed environment with the sintering duration of 1 hour with either 1010 °C or 1080 °C with different sintering environments.

TABLE I
THE SINTERING TEMPERATURES AND ENVIRONMENTS FOR FOUR SAMPLES

Sample	Sintered Temperature (°C)	Sintering Environment
S1	1010	No Gas
S2	1010	Nitrogen Gas (N_2)
S3	1010	Forming Gas ($N_2 + H_2$)
S4	1080	Forming Gas ($N_2 + H_2$)

Pure Cu and commercial lead frame both have the same value of density of 8.94 g/cm³. The densities and porosities are evaluated by using the water displacement method in our work. Water displacement is a method using Archimedean principle [7]. The sintered sample is weighed in air and then evacuated in desiccators before water was added to the infiltrate open pores. The sample is balanced with a wire hanger and is set to zero. Subsequently the suspended weight of sample is obtained. The sample is carefully blotted after removing from water and re-weighed

in air. Sintered density of sample is calculated with the formula below [8]:

$$\text{Sintered Density} = \frac{W_d}{(W_d - W_{ss})} \times \rho_o$$

, where W_d = weight of sample in air
W_{ss} = weight of saturated sample submerged in water

Porosities of sample can be calculated with the formula given by [9];

$$\text{Open Porosity} = \frac{W_s - W_d}{W_s - W_{ss}} \times 100\%$$

, where W_s = weight of sample suspended in water

Besides that, the surface morphology observations of the inner and outer pores have been carried out using scanning electron microscope (SEM). The electrical conductivity is measured using Ohm's law circuitry setup in order to obtain the conductivity of the pellets. Furthermore, the samples are then subjected to X-ray Diffraction (XRD) to study the crystal structure and phases that are present in the samples. The sample is then placed on a glass slide and then transferred into the XRD chamber. The diffraction patterns of the sample tested with Bragg's Law will be then matched with the International Centre for Diffraction Data (ICDD) database software in computer in order to determine the unknown phases that are observed in the sample.

III. RESULTS AND DISCUSSION

SEM micrographs show different grain structures under different sintering environments and temperatures. Sample S1 is observed and found to have incomplete grain growth as in Fig. 2(a). This is because in the early stage of the sintering process only inter-particle bridging through surface diffusion occurred. Moreover, the sintering mechanism had not yet reached boundary or lattice diffusion stage whereby no neck growth or pore elimination and densification occur. This is mainly because grain boundaries are not successfully formed within each neck, leaving every interstice between particles becoming a pore in the pellet. The average grain size is 5.69×10^{-6} m for sample S1. In Figs. 2(b), (c) and (d), the micrographs show a complete grain growth effect after sintering with either N_2 or $N_2 + H_2$ sintering environment. The samples S2, S3 and S4 have an average grain size of 9.23×10^{-6}, 1.12×10^{-5} and 2.35×10^{-4} m respectively.

Samples S3 and S4 with $N_2 + H_2$ gases show a total elimination of the pores from the surface of the Cu pellets, while some pores are still present on the surface of Sample S2. The creation of pores on the Cu pellet surface could be attributed to the formation of cuprous oxide at low temperatures (~ 700 °C) and then the decomposition of pores at higher temperature, ~ 1000 °C [10].

The decomposition of oxygen from the surface produces the Cu and oxygen, and subsequently creating the pores. Sample S2 with N_2 gas contained the highest oxygen level of 7.81 wt% compared to Samples S3 and S4 as shown in Table 1. The high level of oxygen is due to Cu and O bonding to form Cuprite oxide phase in the pellet which can be shown in XRD results. However, the Samples S3 and S4 with $N_2 + H_2$ gases contained the lowest oxygen levels which are 0 and 2.35 wt%, respectively. These samples show comparable result with the commercial Cu lead frame in term of oxygen content. But when at vacuum, Sample S1 contains two times of the oxygen level of 14.04 wt%. Cu pellet with high oxygen level would tend to form copper oxide which is brittle and porous. This formation of copper oxide should be avoided as it would strongly impact the mechanical, physical and electrical properties of the samples. Based on the SEM micrograph and EDX analysis, the degree of oxygen level and the porosity of sintered sample are ranked in decreasing order from vacuum, N_2, $N_2 + H_2$ regardless of any sintering temperature.

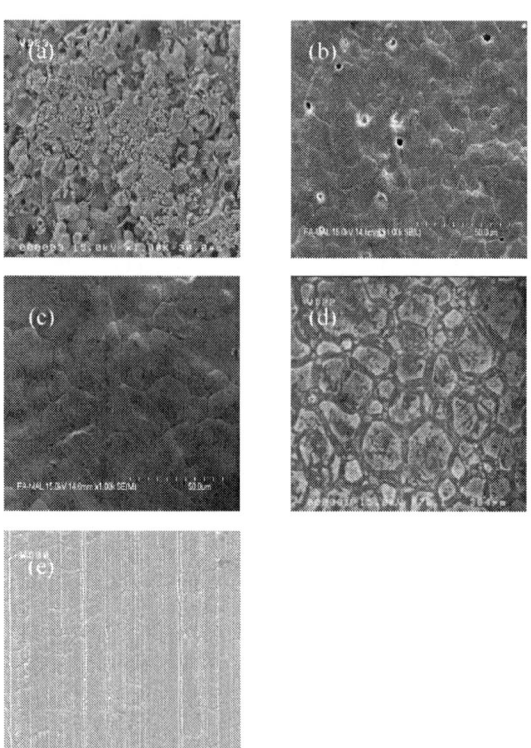

Fig. 2: SEM micrographs on the surface of the sintered Cu pellet in (a) vacuum at 1010 °C, Sample S1 (b) N_2 gas at 1010 °C, Sample S2 (c) $N_2 + H_2$ gases at 1010 °C, Sample S3 (d) $N_2 + H_2$ gases at 1080 °C, Sample S4 and (e) Commercial Cu lead frame.

TABLE II
EDX RESULTS OF SAMPLES S1, S2, S3 AND S4 COMPARE WITH COMMERCIAL CU LEAD FRAME.

	Composition (wt%)					
	C	O	Cu	W	Si	Cr
Commercial Cu Lead frame	9.79	-	89.8	-	-	0.42
S1	6.34	14.0	77.7	1.88	-	-
S2	4.51	7.81	87.7	-	-	-
S3	6.40	-	93.5	-	-	-
S4	5.53	2.35	93.7	-	1.17	-

XRD analysis is to identify the phase in the Cu pellets that sintered under vacuum, N_2 and $N_2 + H_2$ conditions in Table II. Scherrer developed a method for performing this crystallite size analysis based on the full width at half maximum (FWHM) of peaks [11]. The crystallite size can be determined through XRD by Debye Scherrer's formulation [12]. In order to achieve accurate crystallite size the peak should be selected below 45° (i.e 2θ).

$$D = \frac{k\lambda}{\beta cos\theta}$$

, where D = crystallite size, Å
k = crystalline shape factor, 0.9
λ = X-ray wavelength, 1.5418 Å for CuK_{α}
θ = angle of the peak in degree
β = X-ray diffraction broadening, rad

This XRD measurement uses the Cu to create K-alpha X-rays with the wavelength of 1.5418Å so that Cu pellets phases and crystallite size that are present can be determined.

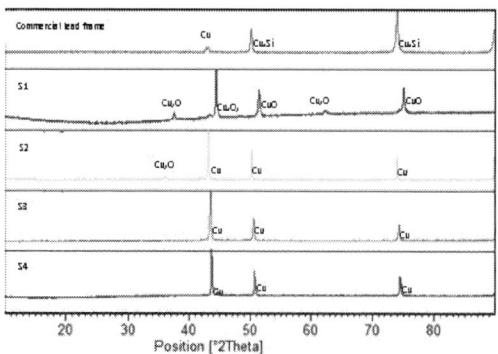

Fig. 3: The phases that presence in the Cu pellets, S1 to S4 and commercial Cu lead frame.

Cu pellets produced in different sintering environments show the presence of different phases as shown in Fig. 3. Sample S1 shows the presence of 3 different phases in Cu pellet, namely CuO, Cu_2O and Cu_2O_3. This means that there is no Cu phase present in this pellet. A high oxygen level is detected under EDX analysis. Paramelaconite structure (Cu_4O_3) shows the highest peak compared to all other composites. Cu_4O_3 has a tetragonal crystal structure which is brittle and black in color.

While for sample S2, the graph shows the presence of Cu_2O and Cu phases. These peaks do not match the peaks from commercial Cu lead frame as there is an extra Cu_2O phase present at the $2\theta = 35.9311°$ position. Samples S3 and S4 show comparable peaks as reference where it only shows Cu phase as the peaks matches the peaks of the reference. The peak of both reference and Sample S3 matches that in the database ICDD 01-070-3038 [13]. The database shows that this a face-centered-cubic structure because the space group of this Fm-3m is based on Hermann-Mauguin notation. This means that Sample S3 sintered at 1010 °C and $N_2 + H_2$ sintering environments has comparable phases as commercial Cu lead frame where only Cu phase was present in the sample.

Fig. 4 shows that the densities of Cu pellets have steadily increased from 6.06 g/cm^3 to 8.19 g/cm^3 from Samples S1 to S4. These samples are sintered in vacuum (Sample S1), N_2 (Sample S2) and $N_2 + H_2$ (Samples S3 and S4). As a comparison of the Cu pellets with the commercial Cu lead frame, it is observed that Samples S3 and S4 are quite desirable as the densities are approaching the commercial Cu lead frame density. This shows that sintering of Cu pellet with $N_2 + H_2$ at the optimum temperature of 1080 °C contributes to the densification of the pellet by PM method. Hence, good densification of Cu pellets come in the sequence of vacuum at 1010 °C, N_2 at 1010 °C, $N_2 + H_2$ at 1010 °C and $N_2 + H_2$ at 1080 °C. Although the density of Sample S4 is near to the commercial Cu lead frame density (produced using the conventional melting method), but the different methodology of forming Sample S4 causes the presence of minor pores.

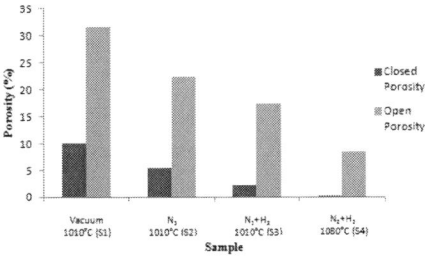

Fig. 4: The densities for Cu pellets with different sintering gas mixtures and temperatures.

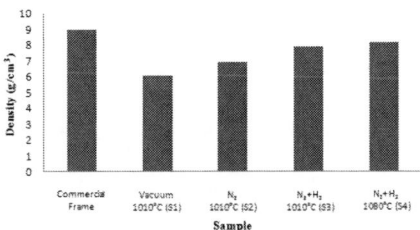

Fig. 5: The open and closed porosities of Cu pellets.

The open and closed porosities are measured by using Archimedes's principle to analyze the closed packed particle stacking and the compactness of the pellets formed through compression and also the optimum sintering temperature. Fig. 5 shows a trend of decreasing open porosities from 31.6 % to 8.4 % from Samples S1 to S4. The average open porosity in these four Cu pellets is considerably high. This is due to the friction of the tool and the pellet generated during the release of the Cu pellet from the die set. Sample S4 is the only sample with the lowest value of the open and closed porosities. This is mainly because the sintering process had reached the final stage of sintering where it involves the isolation of pores. The elimination of porosity and the shrinkage of pores to a limited size or total disappearance have occurred as in Sample 4 [14]. The graph shows the decreasing trend of the closed porosity from 10 % to 0.19 % from Samples S1 to S4. Samples S3 and S4 have low closed porosity in the range of 0.19 – 2.21 %. It shows the grains are quite closed packed and dense because the pores have been eliminated. The samples shrink with significant grain growth occurring as shown under SEM micrograph in Fig. 2. This shows that sintering in an appropriate environment will prevent oxidation in the Cu pellet that in turn will resist the reduction of porosity.

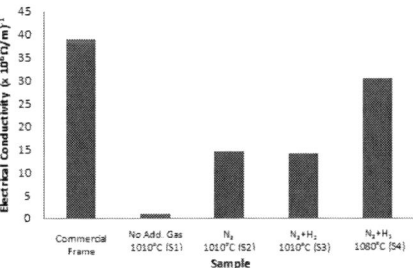

Fig. 6: The electrical conductivities of Cu pellets.

Fig. 6 shows that there is an increase of electrical conductivity in between 8.47×10^5 and 1.44×10^7 Ω^{-1} m^{-1} for Samples S1 and S2. The poor electrical conductivity of Sample S1 is due to the poor densification and high porosity of this Cu pellet resulting in the increase of resistivity. Furthermore, the presence of Cu oxide in sample S1 subsequently reduces the real contact area between the Cu.

The electrical conductivity is found almost equivalent for Samples S2 and S3 with 1.44×10^7 and $1.40 \times 10^7 \Omega^{-1} m^{-1}$ respectively. This is mainly because both the samples are sintered at the same temperature of 1010 °C and have comparable values of density, porosity and less oxidation layer in the Cu pellets. This also means that electrical conductivity is not influenced by the sintering environment regardless of N_2 or $N_2 + H_2$ gas is being used. However, the electrical conductivity of Sample 4 is $3.04 \times 10^7 \Omega^{-1} m^{-1}$ (i.e. 60.8 % IACS) even though Samples S3 and S4 are prepared under the same sintering environment. The difference between these two samples is that Sample S3 is sintered at 1010 °C while Sample S4 is at 1080 °C which is approaching the Cu melting point at 1084°C. When the Cu pellet reaches the melting point of Cu, it will have good bonding as it has already melted to form a solidified Cu structure. This is also related on the low porosity, high density and good electrical conductivity of sample S4 as compared to other samples. The contact particles give good conducting path with less pores. It is worth to note that Sample S4 has a comparable electrical conductivity with the commercial Cu lead frame of $3.94 \times 10^7 \Omega^{-1} m^{-1}$ (i.e. 70.8 % IACS).

IV. CONCLUSION

Powder Metallurgy (PM) method is a simple metal preparation process as compared to conventional melting involving milling, rolling and etc techniques in achieving the mechanical and electrical conductivity of a product. Sample 4 which is prepared with PM method is able to achieve electrical conductivity of $3.04 \times 10^7 \Omega^{-1} m^{-1}$ (i.e. 60.8 % IACS) with an optimum sintering temperature 1080°C which is reaching the Cu melting point at 1084°C. In addition, Sample 4 is desirable as the density, 8.19g/cm³, is approaching to the commercial Cu lead frame density with low porosity percentage. During the sintering of Cu pellet, a control environment is crucial in order to minimize the oxidation level and avoidance the formation of Cu oxide phases.

ACKNOWLEDGMENT

I would like to thank Infineon Technologies (M) Sdn Bhd for sponsoring my research study.

REFERENCES

[1] London Metal Exchange, *The World Centre For Non-Ferrous Metal Trading, 2003-2010*, UK.
[2] P.Y. Chew, S. Zahi, A.H. You, P.S. Lim, and M.C. Ng, "Preparation of Cu and fly ash composite by powder metallurgy Technique", *AIP Conference Proceedings,* 1328, 2010, 208-210.
[3] A.Upadhyaya and G.S.Upadhyaya, "Sintering of Cu-Al2O3 composites through blending and mechanical alloying powder metallurgy routes", *J. Materials & Design*, 16(1), 1995, 41-45.
[4] Y. Feng, H. Zheng, Z. Zhu, F. Zua, "The microstructure and electrical conductivity of aluminum alloy foams", 78, 2002, pp 196-201.
[5] Y. Tomioka, and J. Miyake, "A Copper Alloy Development for Leadframe", *Nippon Mining and Metals Co.,Ltd.,* pp 433-436.
[6] P. C. Angelo, R. Subramaniam, "Powder Metallurgy: Science, Technology and Applications", *PHI Learning Private Limited*, 2008, pp 8.
[7] J. S. Reed, "Principles of Ceramics Processing", *New York: Wiley*, 1995, pp 118-120.
[8] Manual of Weighing Applications- Part I Density, 1999, 21-22.
[9] J. Matějíček, B. Kolman, J. Dubský, K. Neufuss, N. Hopkins, and J. Zwick, "Alternative methods for determination of composition and porosity in abradable materials", *Materials Characterization,* Vol 57, 2006, pp 17–29.
[10] H. Ghasemi, M.A. Faghihi Sani, A.H. Kokabi and Z. Riazi, "Alumina-Copper Eutectic Bond Strength: Contribution of Preoxidation, Cuprous Oxides Particles and Pores", *Transaction B: Mechanical Engineering*, vol 16, No. 3, pp 263-268.
[11] R. Harrington, R. B. Neder, and J. B. Parise, "The nature of x-ray scattering from geo-nanoparticles: Practical considerations of the use of the Debye equation and the pair distribution function for structure analysis", *Chemical Geology*, 2011, pp 1-7.
[12] J. P. Enrıquez, and X. Mathew, "XRD study of the grain growth in CdTe films annealed at different temperatures", *Solar Energy Materials & Solar Cells*, 81, 2004, pp 363–369.
[13] E.A. Owen, E.L. Yates, *Philos. Mag.*, vol 15, 1933, pp 472
[14] G.Y. Onada, Jr. L. L. Hench, "Ceramic Processing Before Firing", *John Wiley & Sons, Inc. N.Y.*

Gross Die Estimator's Caveats For ASIC Floorplanning

Ang Boon Chong
PMC-Sierra,
1-18-12, SUNTECH@Penang Cybercity,
Lintang Mayang Pasir 3,
11950 Bayan Baru, Penang
Email:boonchong.ang@pmc-sierra.com

Abstract—**Floorplanning is essential elements in ASIC chip design. As design complexity increase with various IP packaging on single die, a good floorplanning will help to drive the product cost lower as well as reduce the design hiccup. For optimum product cost estimation, die count estimation is essential element based on floorplanning feedback. Gross Die Estimator is the common equation used to derive the product cost estimation in new product planning phase. The intend of this paper is to analyze the prior art to ASIC floorplanning from die analysis perspective, Gross Die Estimator and the caveats of gross die estimator usage.**

Index Terms—**Die Estimator, ASIC Floorplanning**

I. INTRODUCTION

For fullchip ASIC floorplanning methodology, it can be done either top down approach or bottom up approach. For most of the fullchip ASIC floorplanning methodology deployed, it is bottom up approach where individual IP owner provide the respective IP area, width, height and necessary constraint to fullchip ASIC planner to assemble the IPs and derive the product cost.

The comparison between top-down approach and bottom-up approach full chip floorplanning is as shown in table 1

Table 1. Floorplanning Methodology

Floorplanning Methodology	top-down	bottom-up
IP key parameter	area	width and height
area efficiency	√	
mask layer	√	
Planning duration		√
Product cost	√	

For bottom-up fullchip implementation, it has the advantage of faster turn around time(TAT) on planning duration. Such implementation is best to deploy on low volume none flagship product where its primary objective is to defend market share leadership. It is typically leveraging on the existing physical IPs in house with minimum product mask layer changes on mature fabrication technology. For new flagship product implementation with bottom up implementation approach, it may not yields better area efficiency and mask layer primary due lack of inter IP routing congestion visibility. Typically solutions deployed to resolve inter IP routing congestion for bottom-up floorplanning are adding extra routing channel to resolve the inter IP routing congestion or adding extra mask layer in order to minimize the product schedule push out.

For top-down fullchip ASIC floorplanning implementation, it is critical on flagship product as well as new product definition on advance technology where product cost and key performance parametric are equally important. For top down full chip implementation, individual IP aspect ratio is derived from top level where floor-plan is tuned properly with less dummy area overhead. The mask layer utilization of top down approach can yield lower mask cost as the inter IP routing congestion can be better predicted as well as macro pin placement is assigned with top level routing visibility. Hence it reduces the possibility of bad routing congestion at full chip assembly due to poor macro pin placement. However top down full chip implementation requires much longer product planning duration as there is no commercial design tools available in the market that can auto adjust hardmacro IP's width and height , hardmacro IP pin assignment for better die area as well as product routability. Top-down fullchip implementation serves better area efficiency, mask layers as well as product cost to protect market share leadership at trade off of longer planning phase but smoother implementation phase.

This paper is intended to share the prior art to fullchip floorplanning from die count estimator perspective. It will study the impact of gross die estimator[1] versus the physical die estimator solution. For the physical die estimator, inter die spacing,die width, die height, wafer diameter, notch height and wafer edge exclusion width is taken into consideration during gross die count per wafer estimation [3],[7]-[8]. The gross die count extracted does not include manufacturing yield impact cover multi product wafer(MPW) shuttle consideration or 3D-ICs [4], [5] with die count estimator taken into consideration. Due to design complexity, wafer size, fab process, technology node and signoff PVT, it may not impacts all ASIC floorplanners but do hope the readers may benefit from this sharing.

I. TRADITIONAL DIE ANALYSIS

For traditional gross die per wafer estimation, it can be approximate with the following equation[1]

GDPW =(wafer area/die area) – (wafer circumference/die diagonal length) -----Eq 1

$$GDPW = d\pi(d/4S - 1/(\sqrt{2}S))$$

where d denote wafer diameter in mm and S denote IC size in mm².

For the equation 1 above, the impact of die aspect ratio to die count is not taken into account.

Traditionally fullchip floorplanning always target the floorplan aspect ratio of 1. As a result, there is always a possibility of area redundancy introduced as shown in figure 2. For those redundant area, it is typically filled with decap, spare gates and etc.

Gross die estimator as shown in equation 1 has been widely adopted in engineer die count calculation routine. The objective of die count impact 1 analysis is to understand reliability of gross die estimator equation versus physical die count estimator. For this study, the range of aspect ratio choosen ranging from 1 to 0.01 and the area of IC sizes choosen is ranging from 30mm² to 500mm². Table 2 shows the width and height of IC sizes with selective aspect ratio applied.

[6]. It is not intended to cover physical die count estimator construction formula [1], [3],[7]-[8]. It is also not intended to

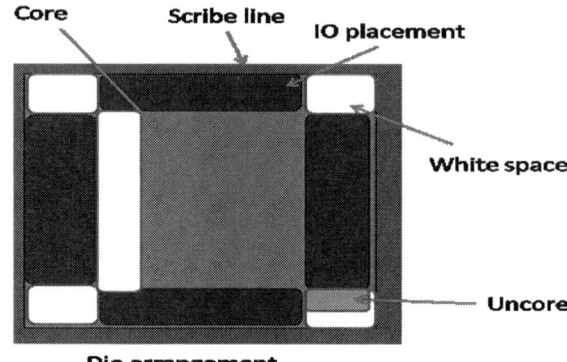

Figure 2. Die Arrangement

The objective of this paper is to explore the caveats of gross die estimator[1] with fixed floorplan aspect ratio of 1 from the actual die count estimation perspective. Hopefully, the value of physical die count estimator with various floorplan aspect ratio presented in this paper will benefit the readers as well.

II. DIE COUNT IMPACT

A. Die Count Impact 1: Die Estimator

From table 2, it is observed that for larger IC size range with bad aspect ratio of 0.01, the height of die is still within foundry manufacturability constraint.

The die count analysis for various IC sizes and chip's aspect ratio on 300mm wafer shown in table 3.

Table 3. Die Count Analysis With Respect To Floorplan Aspect Ratio

Wafer Diamater (mm)	IC size (mm2)	GDPW	aspect ratio 0.01	aspect ratio 0.05	aspect ratio 0.1	aspect ratio 0.2	aspect ratio 0.3	aspect ratio 0.4	aspect ratio 0.5	aspect ratio 0.6	aspect ratio 0.7	aspect ratio 0.8	aspect ratio 0.9	aspect ratio 1
300	30	2235	1887	2111	2177	2207	2239	2225	2245	2245	2241	2253	2247	2241
300	50	1319	1003	1239	1277	1289	1319	1315	1331	1319	1323	1339	1337	1313
300	100	640	281	547	613	633	635	635	653	643	651	645	653	657
300	200	306	187	237	269	307	303	309	301	305	317	307	311	317
300	300	197	141	171	185	195	199	195	197	201	203	205	207	205
300	400	143	111	121	115	133	145	147	145	149	143	147	147	145
300	500	112	89	55	99	109	113	115	113	113	117	115	119	121

From table 3, it is observed that gross die estimator overestimate the die count on bad floorplan aspect ratio of 0.05 and aspect ratio of 0.01 in general. The value of the gross die estimator is as shown in column GDPW (gross die per wafer) in table 3.

The normalized physical die count versus gross die estimator equation is as shown in figure 3.

Table 2. Die Width And Height With Various Floorplan Aspect Ratio

	Aspect Ratio 1		Aspect Ratio 0.8		Aspect Ratio 0.5		Aspect Ratio 0.3		Aspect Ratio 0.2		Aspect Ratio 0.1		Aspect Ratio 0.05		Aspect Ratio 0.01	
IC Size (mm2)	width (mm)	height (mm)	width (mm)	height (mm)	width (mm)	height (mm)	width (mm)	height (mm)	width (mm)	height (mm)	width (mm)	height (mm)	width (mm)	height (mm)	width (mm)	height (mm)
30	5.48	5.48	6.12	4.90	7.75	3.87	10.00	3.00	12.25	2.45	17.32	1.73	24.49	1.22	54.77	0.55
40	6.32	6.32	7.07	5.66	8.94	4.47	11.55	3.46	14.14	2.83	20.00	2.00	28.28	1.41	63.25	0.63
100	10.00	10.00	11.18	8.94	14.14	7.07	18.26	5.48	22.36	4.47	31.62	3.16	44.72	2.24	100.00	1.00
200	14.14	14.14	15.81	12.65	20.00	10.00	25.82	7.75	31.62	6.32	44.72	4.47	63.25	3.16	141.42	1.41
300	17.32	17.32	19.36	15.49	24.49	12.25	31.62	9.49	38.73	7.75	54.77	5.48	77.46	3.87	173.21	1.73
400	20.00	20.00	22.36	17.89	28.28	14.14	36.51	10.95	44.72	8.94	63.25	6.32	89.44	4.47	200.00	2.00
500	22.36	22.36	25.00	20.00	31.62	15.81	40.82	12.25	50.00	10.00	70.71	7.07	100.00	5.00	223.61	2.24

978-1-4673-2395-6/12 $31.00 © 2012 IEEE

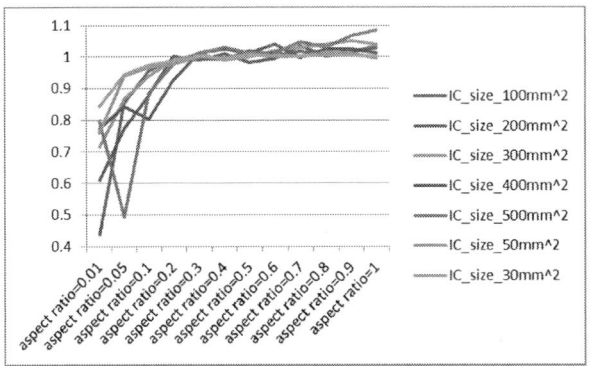

Figure 3. Normalized Physical Die Count Versus Gross Die Estimator Count

From figure 3, it is observed aspect ratio of fullchip floorplan closer to value of 1, the better the correlation of the physical die count estimator with gross die estimator equation for various IC sizes on 300mm diameter wafer.

For smaller IC size, gross die estimator equation show tight correlation with actual physical count with wider range of chip full chip floorplan aspect ratio as shown in figure 4.

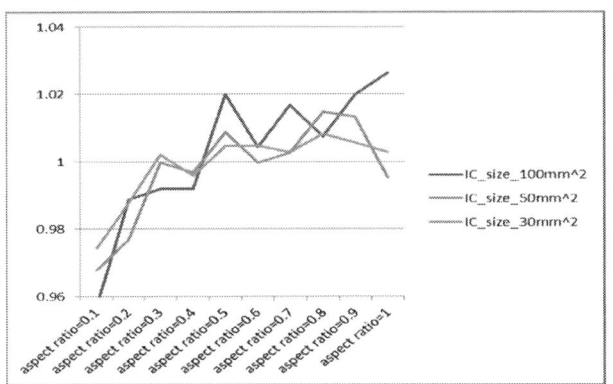

Figure 4. Normalized Physical Die Count Versus Gross Die Estimator Count On Smaller IC Sizes

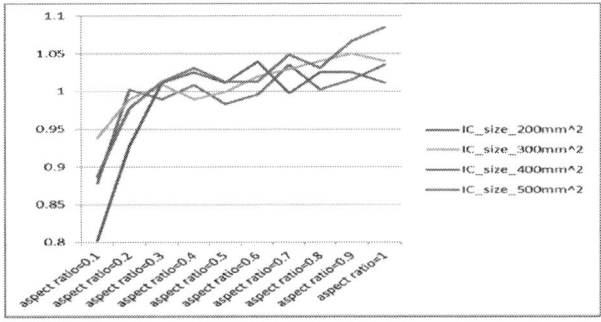

Figure 5. Normalized Physical Die Count Versus Gross Die Estimator Count On Larger IC Sizes

Keynote from die count impact 1 is gross die count estimator can reasonably estimate the die count per wafer on smaller IC sizes where impact of full chip floorplan aspect ratio is within reasonable range from 0.1 to 1 (ideal fullchip aspect ratio value)

A. Die Count Impact 2: Area Grow

In typical design, there is always a risk of fullchip die size grow after floorplan lock down period for various reasons such as extra IP features request for product competitiveness or design rule changes from foundry. In such occurrence, the floorplan

Table 5. Area Increment Percentage

Die X-Pitch x+0mm		x+0.1mm		x+0.2mm		x+0.3mm		x+0.4mm		x+0.5mm		x+1mm	
Base IC size (mm^2)	Effective Die Area (mm^2)	Area Increment (%)	Effective Die Area (mm^2)	Area Increment (%)	Effective Die Area (mm^2)	Area Increment (%)	Effective Die Area (mm^2)	Area Increment (%)	Effective Die Area (mm^2)	Area Increment (%)	Effective Die Area (mm^2)	Area Increment t (%)	Effective Die Area (mm^2)	Area Increment (%)
50	50.00	0.00	50.71	1.41	51.41	2.83	52.12	4.24	52.83	5.66	53.54	7.07	57.07	14.14
100	100.00	0.00	101.00	1.00	102.00	2.00	103.00	3.00	104.00	4.00	105.00	5.00	110.00	10.00
200	200.00	0.00	201.41	0.71	202.83	1.41	204.24	2.12	205.66	2.83	207.07	3.54	214.14	7.07
300	300.00	0.00	301.73	0.58	303.46	1.15	305.20	1.73	306.93	2.31	308.66	2.89	317.32	5.77
400	400.00	0.00	402.00	0.50	404.00	1.00	406.00	1.50	408.00	2.00	410.00	2.50	420.00	5.00

From table 5, the area increment percentage with fix x-pitch percentage is shown. For small x-pitch grown, the effective area grow observed is less than 10% in general.

The normalized gross die estimator count reduction rate, physical die count reduction rate as well as effective die size increment percentage due to fullchip X-pitch increment with

For larger IC size, gross die estimator show strong miscorrelation with actual physical die count on wafer due to aspect ratio impact as shown in figure 5.

changes is always limit to uni-direction changes such as width increment to minimize the project push out impact.

Table 4 shows the die count estimation from gross die count estimator as well as physical die estimator on 300mm diameter wafer for various IC sizes with fixed fullchip X pitch increment. For the physical die count calculation used in table 4, it is generated with chip floorplan aspect ratio of 1.

Table 4. Area Grow Impact On Die Count

Die X-Pitch x+0mm		x+0.1mm		x+0.2mm		x+0.3mm		x+0.4mm		x+0.5mm		x+1mm	
Base IC size (mm^2)	GDPW	Physical Die Count	GDPW	Physical Die Count	GDPW	Physical Die Count	GDPW	Physical Die Count	GDPW	Physical Die Count	GDPW	Physical Die Count	GDPW	Physical Die Count
50	1319	1313	1300	1309	1282	1289	1264	1275	1246	1255	1229	1243	1150	1165
100	640	657	634	641	627	637	621	625	614	615	608	615	579	591
200	306	317	304	309	302	299	299	299	297	299	285	287	285	287
300	197	205	196	201	195	201	193	199	192	199	191	199	185	187
400	143	145	143	145	142	145	141	145	140	145	139	141	136	141

respect to baseline on various IC sizes is as shown in figure 6 histogram.

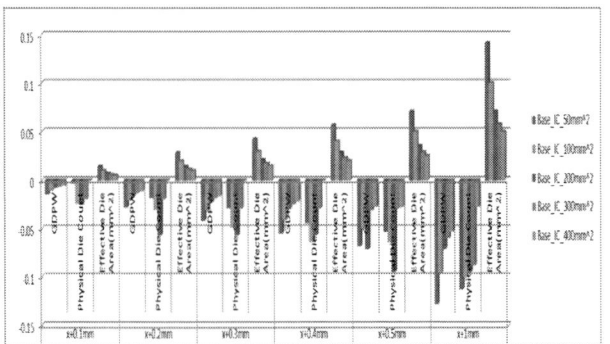

Figure 6. Normalized Gross Die Estimator Count, Physical Die Count Reduction Rate, Area Increment Percentage With Baseline Due to X-Pitch Increment

From figure 6, it is concludes that normalized gross die count reduction rate show tight correlation with normalized physical die count reduction rate. For small fixed pitch increment with aspect ratio changes taken into account, the die count reduction is linear with respect to the full chip die area increment.

Keynote from die count impact 2 is gross die count estimator can reasonably estimate die size reduction rate for small fixed fullchip X-pitch increment with die area as well as aspect ratio changes impact taken into consideration.

A. *Die Count Impact 3: Die Area Overhead*

In the event where die area overhead need to be budgeted in full chip area to cater for tentative IP size growth due to extra features required or extra resource count for wider product total available market (TAM) value, the estimation of die count reduction impact from die count estimator is as shown in table 6.

Table 6. Die Area Overhead Impact On Die Count

Die Overhead Percentage(%)		0		10		20		30	
Wafer Diameter (mm)	Base IC size (mm²)	GDPW	Physical Die Count	GDPW	Physical Die Count	GDPW	Physical Die Count	GDPW	Physical Die Count
300	50	1319	1313	1195	1201	1092	1101	1005	1017
300	100	640	657	579	585	528	537	485	489
300	200	306	317	276	277	252	261	231	241
300	300	197	205	178	177	161	177	147	149
300	400	143	145	129	137	117	121	107	113

All the physical die count used in table 6 is generated with fixed floorplan aspect ratio of 1.

The normalized histogram comparing gross die count reduction rate from gross die count estimator and physical die count reduction rate due to area overhead impact for various IC sizes is as shown in figure 7.

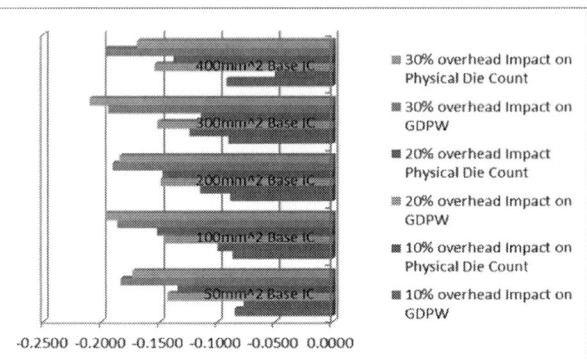

Figure 7. Normalized Gross Die Estimator Count, Physical Die Count Reduction Rate With Baseline Due to Area Overhead Impact

From figure 7, for 10% area overhead, gross die estimator equation has high miscorrelation on high area base IC sizes such as 400mm² base IC size. For 20% area overhead scenario, gross die estimator shows good die count reduction rate prediction for most IC base size as compare to physical die count reduction rate extracted, with exception on 300mm² base IC size. For 30% area overhead scenario, gross die estimator equation is able to predict the die count reduction rate across major base IC sizes.

Keynote from die count impact 3 is gross die count estimator equation may not reasonably predict die size reduction rate due to large area overhead.

B. *Die Count Impact 4: Alternative Die Count Estimator*

For further simplification on gross die size estimator, engineers tend to further simplify the gross die size estimation equation to

GDPW =scaling coefficient*(wafer area/die area) -----Eq 2

$$GDPW = Kd\pi(\frac{d}{4S})$$

where d denote wafer diameter in mm,K denote scaling coefficient and S denote IC size in mm².
The typical value of K applied on equation 2 is 0.9

Table 7 shows the die count extract for various IC sizes on 300mm diameter wafer with aspect ratio of 1 as well as theoretical maximum die count expectation which derive through equation 3 below

Max Die Count= wafer area/die area -----Eq 3

978-1-4673-2395-6/12 $31.00 © 2012 IEEE   434

Table 7. Die Count Estimation

IC size(mm^2)	GDPW	aspect ratio=1	max Die Count
30	2235	2241	2356
50	1319	1313	1414
100	640	657	707
200	306	317	353
300	197	205	236
400	143	145	177
500	112	121	141

The scaling ratio of physical die count versus maximum die count extracted with equation 3 as well as normalized gross die count estimator from equation 1 with maximum die count from equation 3 is as shown in figure 8.

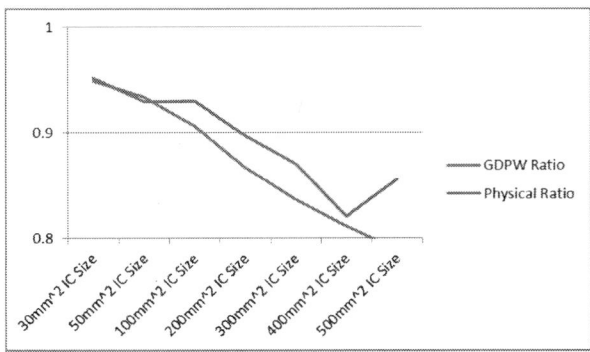

Figure 8. Die Count Scaling Ratio Analysis

From figure 8, it is observed that typical scaling ratio of 0.9 on equation 2 yield conservative estimation for smaller IC sizes range but optimistic result on higher IC base area size range above 100mm^2.

Keynote from die count impact 4 is gross die count estimation of equation 2 with scaling ratio of 0.9 may result in optimistic die count estimation on higher die area range.

III. SUMMARY

Gross Die Estimator can reasonably estimate die count per wafer on smaller IC sizes where aspect ratio impact is insignificant.

ACKNOWLEDGEMENTS

Thanks to Chong Thow Pang for the support given.

REFERENCES

[1] Gross die count estimator equation , http://en.wikipedia.org/wiki/Wafer_(electronics)
[2] Robert Doering, Yosio Nishi, Handbook of Semiconductor Manufacturing Technology, 2nd Edition
[3] D. K, de Vries, "Investigation of Gross Die per Wafer Formula," IEEE Trans. On Semiconductor Manufacturing, vol. 18(1), pp. 136-139, 2005

[4] K. Banerjee, S. J. Souri, P. Kapur, K.C. Saraswat, "3-D ICS: A Novel Chip Design for Deep-Submicrometer Interconnect Performance and Systems-on-Chip Integration, " Proceeding of the IEEE, vol. 89(5), pp. 602-633, 2001
[5] J. A. Burns, B. F. Aull, C.K. Chen, C. L. Chen, C.L. Keast, J.M. Knecht, V. Suntharalingam, K. Warner, P.W. Wyatt, D.RW. Yost, "A Wafer-Scale 3-D Circuit Integration Technology," IEEE Transactions on Electron Devices, vol. 53(10), pp. 2507-2516, 2006
[6] C. Hess and L.H. Weiland, "Extraction of Wafer-Level Defect Density Distribution to Improve Yield Prediction," IEEE Transactions on Semiconductor Manufacturing, vol. 12, pp. 175-183, 1999.
[7] G.D Croft, R. L. Lomenick, D.L. Youngblood, J.M. Johnston, "Die counting algorithm for yield modeling and die per wafer optimization," in Proc. SPIE, vol 3216, 1997, pp. 186-196
[8] A.V, Ferris-Prabhu, "An algebraic expression to count the number of chips on wafer," IEEE Circuits and Devices Mag, pp. 37-39, Jan 1989

Functional OBIRCH Strategy in Analyzing Complex Functional Failures Including Logic Failures

Gaojie Wen, Li Tian, Binghai Liu, Grace Song, Joe Yu, Winter Wang,
Freescale Semiconductor (China) Limited, Tianjin
Phone: (+86) No.22 85686020/13682170769   Email: b16245@freescale.com

Abstract- OBIRCH (Optical Beam Induced Resistance Change) analysis was usually used to analyze direct leakage or short failure. But when met complex functional failures, we only know the leakage existed and higher current was consumed, but we couldn't confirm the leakage path or which device was the main failed one. An efficient method was presented by performing OBIRCH analysis in function mode with different kinds of setup condition to trace the abnormal current. It could detect the current path and indicate the failed device directly without much microprobe work. It was also helpful in analyzing the logic failures without test pattern.

I. Introduction

OBIRCH is an efficient tool in failure analysis. It could analyze the leakage and short failures and located the failed device directly. But as the typical size of IC is becoming smaller, failure analysis is facing new challenge, especially when meet complex functional failure cases which microprobe must be performed. So how to use the OBIRCH efficiently is important in future failure analysis.

Usually, OBIRCH would be used only if leakage or short was found on pins or on a device. Voltage was biased on two pins or two nodes of leakage path, abnormal resistance change spot would be found on the failed device. But when met complex function failures, failed signal was found which was pulled down to a lower level, but we didn't find where the leakage was due to the many possible paths. Detailed microprobe would take long time and wouldn't assure to succeed. What's worse, if the failure was in logic, it would be more complicated due to the sea of Gates in it.

Functional OBIRCH could be helpful in tracing the failed signal and finally found the root cause of the failure. The most important point was that when there was a current consumption failure or abnormal lower level signal failure (control signal was normal), there should be leakage existed, no matter on metal lines or on device. The leakage path would be active in function mode even we didn't know where it is. After set the failed sample in failed mode on EVB (Electronic Verification Board), OBIRCH could be used to bias proper voltage on failed signal to reveal defect location or leakage current path which could provide analysis direction for further analysis.

Three kinds of cases were presented in this paper to show how the functional OBIRCH method could be helpful in analyzing the complex function failure cases.

II. Experiment And Results

A. Case 1

One case of SPI communication failure was analyzed. The failed sample was a mixed signal IC which contained analog module and logic module. As the SPI communication was mainly carried out in logic module, logic failure was highly suspected.

Post chemical decapsulation, EMMI analysis was performed on failed sample based on the SPI communication failure, but no abnormal emitted spot was found when compared with reference, so no useful analysis direction was provided by EMMI analysis. After studied schematic and layout, microprobe analysis was performed. Finally, the main failed signal was isolated on sclki signal which was an input signal of logic module. The sclki signal of failed sample was lower (2.1V) than reference (5.5V) and its control signal was normal as reference (Fig.1).

As no leakage was found between sclki signal and related signals including GND in schematic, we didn't know where the leakage path was. There was no direction for further microprobe analysis. After checked the signal on layout, it had a long metal line in the die wiring (Fig.2). The failure could be deeply isolated by cutting the failed signal using FIB (Focused Ion Beam), but this method had high risk to damage the defect and other metal lines due to the small metal line size.

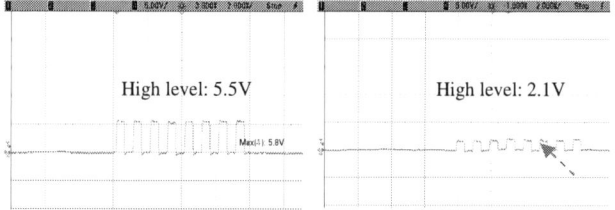

| Reference | Failed sample |

Fig. 1. Sclki signal on reference and failed sample.

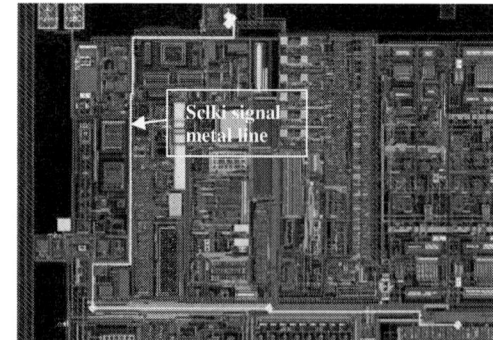

Fig. 2. Layout image of sclki signal wiring in die layout

Functional OBIRCH could be used when met this condition. If we forced a normal signal on the failed signal metal line and the leakage path should also pull down this signal. Much more current would be consumed which could be traced by OBIRCH. First of all, set up the EVB to make the sample in function mode. Then, OBIRCH was connected with the failed sclki signal through a 100k resistor and biased a normal 5V DC voltage. The proper resistor was used to adjust the current which should be within the control setting of OBIRCH machine. After powered on the EVB, SPI command was sent to the failed sample and traced the leakage current path using OBIRCH lock in mode (Fig.3).

Abnormal resistance change spot was found between sclki metal 1 line and another metal 1 line by function OBIRCH analysis. After checked related layout, the sclki signal metal line should be connected with miso_low signal metal line by a particle or something else. Two suspected metal lines were measured using curve tracer and abnormal high leakage was found. Since the sclki signal and miso_low signal had no relationship with each other in functional mode, the defect couldn't be found only by routine microprobe analysis. Although high magnification optical inspection focused on the suspected resistance change spot didn't reveal anomaly, the defect location was determined quickly in spite of the long wiring in die layout by functional OBIRCH analysis. Cross section was performed focused on the abnormal resistance change spot using FIB and metal bridge defect was found (Fig.4).

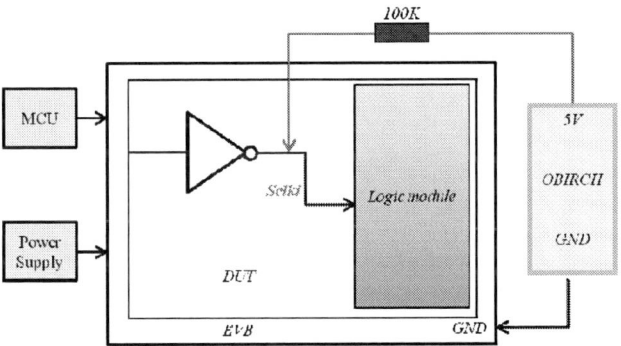

Fig. 3. Set up condition of functional OBIRCH

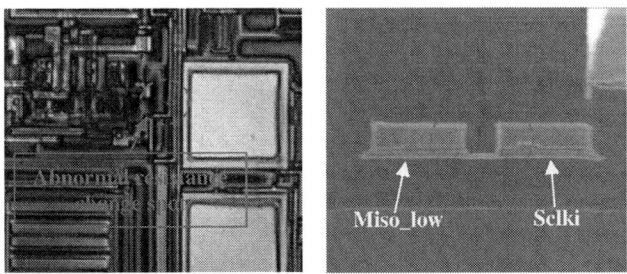

Fig. 4. Resistance change spot got by function OBIRCH and cross section image of metal bridge defect.

Based on this kind of special functional failure case, function OBIRCH was efficient in finding the root cause and could reduce much microprobe time.

B.   Case 2

Many ICs had the sleep/stop mode to save energy consumption. For many cases, we only got sleep/stop function mode current high failure and no other failure was found by ATE (Auto Test Equipment). It meant that one internal device failed and consumed much current, but it didn't affect the whole function. When met functional sleep/stop mode current consumption high failure, OBIRCH couldn't be performed directly as it was not a DC leakage but a functional leakage which contained thousands of MOSFETs and transistors in it. If EMMI analysis couldn't be helpful on pointing out the failed device, microprobe analysis would took long cycle time and couldn't surely to find the root cause. If the failed device was in logic block, it was really a difficult task to find the single failed device in SOG (Sea Of Gates).

A customer returned sample only failed on high current consumption in sleep mode and no other failure was found (Table I). Since many blocks would be disabled in sleep mode, the suspected failed block could be determined by studying the spec. EMMI analysis didn't reveal abnormal emitted spot post decap. As there was no other functional failure, no failed signal could be traced to perform microprobe. Deep failure analysis was difficult to perform.

If a device was failed (except open failure), it would consume current and OBIRCH could find it by proper setup condition. To analyze the current failure, FIB was used to cut the supply voltage signal metal line of internal active block. Source Meter was used to force the normal voltage to the block to check the current consumption in sleep mode. The failed high current consumption was found in logic module. That meant a device of logic was failed and consumed current in sleep mode. Microprobe in sea of gates was nearly impossible. To find the single failed device, functional OBIRCH was efficient in analyzing the logic failure from the current aspect.

TABLE I
COMPARE OF LOGIC MODULE CURRENT CONSUMPTION ON DIFFERENT MODE

Sample	Normal Mode Current Consumption (uA)	Sleep Mode Current Consumption (uA)
Failed Part:	46.2	35
Reference Part:	11.8	0.02

The failed sample was setup well on EVB; OBIRCH was used to bias a normal supply voltage to logic module through probe needle. The OBIRCH GND and signal GND should be tied together with the power supply GND to protect the electro-static discharge damage on die surface (Fig.5).

Functional OBIRCH revealed abnormal resistance change spot on a NMOS in logic module (Fig.6). It should be the NMOS failed and consumed current in sleep mode. Deep microprobe focused on the suspected NMOS revealed leakage between Gate and Source. Physical analysis focused on the abnormal resistance change spot revealed gate oxide rupture wafer defect (Fig.7).

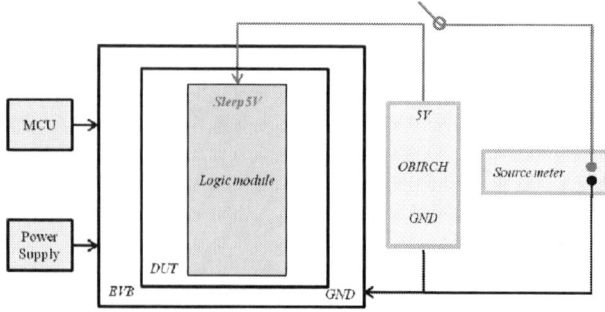

Fig. 5. Functional OBIRCH setup condition when analyzing the logic module.

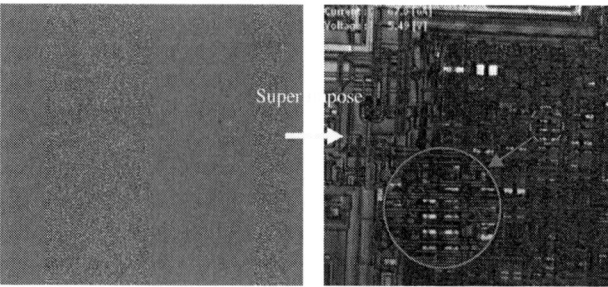

Fig. 6. Functional OBIRCH revealed abnormal resistance change spot in logic module

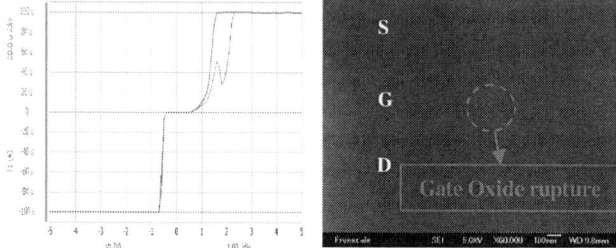

Fig. 7. Leakage IV curve between Gate and Source of failed device and Gox rupture was found on hot spot location.

Functional OBIRCH strategy was proved to be very efficient in analyzing the current failure; it also presented a useful method to analyze the logic failure.

C. Case 3

Sometimes we could get the abnormal emitted spot in logic, but as the device was too small that we couldn't justify if the emitted spot was a defect or saturated MOS caused by failed signal, we need to double confirm it by probe the IV character on Gate-Source or Gate-Drain on the single device. Usually we need to make test pad using FIB. But if the Gate, Drain or Source of suspected device was poly or active area which we couldn't probe, we couldn't double confirm the failed device deeply. The root cause may be lost and success rate couldn't be assured. Functional OBIRCH could also be helpful in double confirm the failed sample without microprobe on the suspected failed device.

One more sleep current high failure was analyzed and abnormal emitted spot was found in logic by EMMI analysis. But after studied the related layout, the suspected failed PMOS

was in an OR2 module and the source and drain were on active area which we couldn't probe (Fig.8).

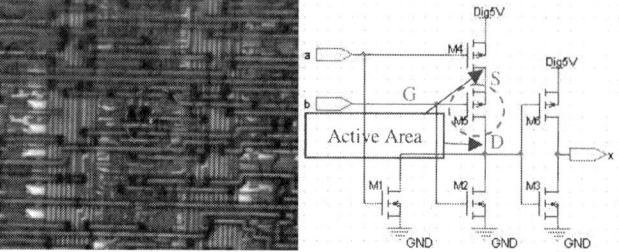

Fig. 8. EMMI image revealed abnormal emitted spot in logic module and the suspected failed MOS in OR2 of logic.

As the device in logic block was very small that we couldn't make sure if the device was really failed or only emitted in saturated state caused by failed signal. To assure our success rate, deprocess couldn't be performed directly only based on EMMI result. OBIRCH should be used to confirm the real failed device.

Functional OBIRCH was set up after we isolated the logic module completely by FIB cut. First of all, Source Meter was used to trace the current consumption in logic to see if the current was within the limit of OBIRCH. The result revealed that logic consumed higher current than reference sample (Failed: 150uA; Reference: 0.03uA) and function OBIRCH was feasible. Then, we connected the OBIRCH bias voltage to dig5V signal which was the power supply signal of logic module. After connected other related wires on EVB and set the failed sample into sleep mode, OBIRCH was turned on to get the abnormal resistance change spot focused on logic module.

OBIRCH analysis revealed the same abnormal resistance change spot on the suspected failed MOS in logic (Fig.9). OBIRCH image indicated that the abnormal emitted spot was not a saturated MOS but a real defect location. So the failed device was double confirmed both by EMMI analysis and OBIRCH analysis. Deprocess focused on hot spot revealed gate oxide rupture wafer defect (Fig.10).

Fig. 9. Functional OBIRCH revealed the same abnormal resistance change spot in logic module.

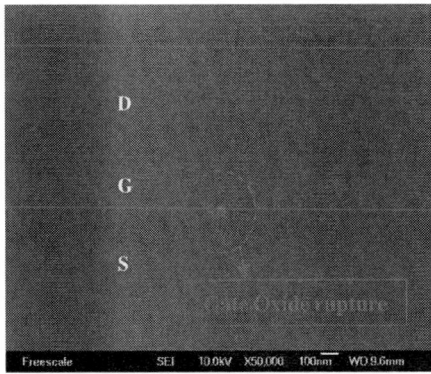

Fig. 10. SEM image revealed gate oxide rupture on the resistance change spot in logic module.

Due to the small size of device in logic module and layout arrangement, we may not be able to measure IV curve focused on single device. Functional OBIRCH strategy could help to confirm the real failed device and improve our success rate on logic failures.

III. SUMMARY

As typical size of integrated circuit is becoming smaller, it is difficult for failure analysis due to limitation on microprobe analysis. If only main failed signal was found and no leakage path was observed, the success rate couldn't be assured if we couldn't get abnormal resistance change spot on failed device.

Functional OBIRCH strategy could provide an efficient way from the current consumption aspect and failed signal aspect to find the root cause directly. It was also helpful in analyzing the logic failure without much microprobe work. This method can reduce much schematic analysis work and probe work and so much analysis cycle time was reduced significantly.

ACKNOWLEDGMENT

Thanks physical failure analysis team of Tian Jin failure analysis lab for their great efforts in cases analysis. Also thanks to other colleagues of electronic failure analysis team of Tian Jin FA lab who gave much assistance in case analyzing.

REFERENCE

[1] Ling, D.T.M. , Kuan, M. , Kwong Weng Yee , Cheong, D. , "Application of IR-OBIRCH to the failure analysis of CMOS integrated circuits", IPFA 2003, pp.86-91.
[2] Farisal Abdullah, Nafarizal Nayan, Muhammad Mahadi Abdul Jamil, "Failure analysis using IDD current leakage and photo localization for gate oxide defect of CMOS VLSI", SCOReD 2010, pp.329-333.
[3] Wu Chunlei, Linda Zhai, Winter Wang, "Defect Localization Using Photon Emission Microscopy Analysis with the Combination of OBIRCH Analysis",IPFA2010, pp.1-6.
[4] Gaojie Wen, Binghai Liu, Winter Wang, "Failure Isolation Using FIB Assist Photon Emission Microscopy Analysis and Microprobe Analysis", IPFA2011, pp.140-143.
[5] Dongwoo Lee, David Blaauw, Dennis Sylvester, "Gate Oxide Leakage Current Analysis and Reduction for VLSI Circuits", VLSI systems, VOL. 12, NO. 2, FEB 2004. pp.155-166.
[6] S.J. Cho, T.E. Kim, J.K. Hong, "Logic Failure Analysis 65/45nm Device Using RCI & Nano Scale Probe", IPFA2009, pp.50-53.

978-1-4673-2395-6/12 $31.00 © 2012 IEEE

Leakage in CMOS Devices Induced by Pattern-Dependent Microloading Effect

Miao Wu, Winter Wang, Li Tian, Chunlei Wu, Diwei Fan

Freescale Semiconductor (China) Limited, No.15, Xinghua Avenue, Xiqing Economic Development area, Tianjin, China
Email: b28231@freescale.com

Abstract- **The difference of pattern density in photomask may cause different photo resist profile, and then etch rate or implant profile can be impacted. This is called the microloading effect. Some special topological structures in chip layout design may result in serious leakage in CMOS devices due to the microloading effect. In this case, the solution is often a mixture of design change and process optimization. A product with such potential risk structure is selected to study the leakage mechanism by microloading effect during lithography and implant process. The result indicates that photo resist profile is impacted dramatically due to the microloading effect, which results in doping agent injecting in wrong area during implant process. Lithography process window is affected as well. Finally solution is presented for the leakage induced by microloading effect.**

Keywords- Microloading effect; CMOS; Lithography; Photo resist

I. INTRODUCTION

From CMOS technology invented in the 1960s [1], IC technology is developing rapidly more than half century. Higher density and performance and lower power consumption have been achieved by reduction of threshold voltage, channel length and gate oxide thickness. Together, semiconductor manufacturing process is kept innovating to scale down CMOS device to meet higher and higher requirement. Consequently, leakage mechanism in deep submicrometer regimes is becoming a significant research topic [2].

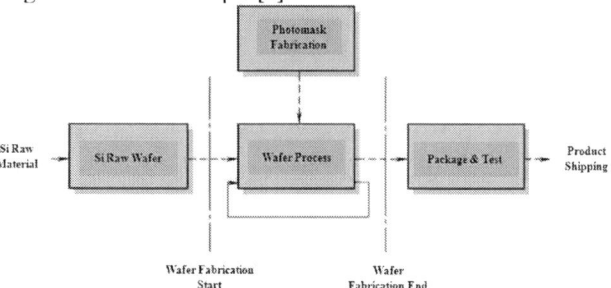

Fig. 1. Schematic diagram of procedure for an IC product manufacture

Along with technology progress, currently the most common business mode of semiconductor company is developed to an IFM company (Integrated Fabless Manufacturer) for product design, a foundry for chip fabrication and SAT suppliers (Subcontract Assembly and Test) for package and test. Consequently, yield of an IC product is determined by all the parts, especially IFM and foundry. The former provides photomask for chip fabrication, and the latter makes sure chip fabricated successfully after hundreds of complex process steps. Figure 1 shows a whole procedure to manufacture an IC

product. Usually people are used to divide IC manufacturing into frontend and backend. The front end means wafer process, which includes steps from raw wafer to chip probe. The backend stands for steps including package and final test. Figure 2 shows key process in wafer process.

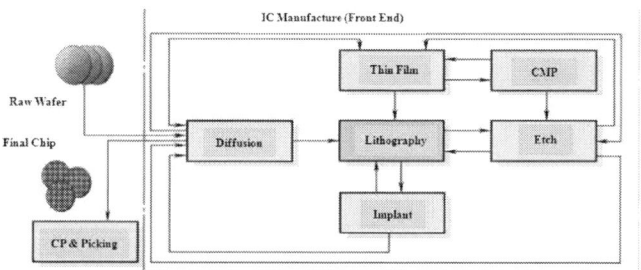

Fig. 2. Schematic diagram of key process in IC manufacture frontend

The mentioned business mode requires much tighter integration between the IFM, the foundries and the SAT suppliers [3]. Oppositely, an IC designer is not possible to know wafer process technique as well as a process engineer in a foundry. The process tool limitation, some side effect related with photomask pattern and some pattern dependent design bug which will kill wafer process window is probably not counted into physical design or layout design by the designer. This mismatch between design house and foundry may result in a disaster during wafer process for New Product Introduction (NPI). Usually, high yield dropout comes from leakage test, such as leakage at off state (Ioff) and leakage at sleep state (Isleep). It is very difficult for a foundry failure analysis engineer to identify failing device at die level, because complex bench test setup is impossible to perform at die level. Lot of time has to be waste to wait for package failing sample and then perform further failure analysis at package level. Such situation is a deadly issue for a NPI product.

In this paper, we present one of the famous side effects, called the microloading effect. Its mechanism is described as that the difference of pattern density in photomask may cause different PR (photo resist) profile, and then etch rate or implant profile can be impacted [4]. The leakage mechanism by the microloading effect during lithography and implant process is studied. Several solutions are put forward and experimental data reveals they can work well.

II. LEAKAGE MECHANISM

Currently, most of ICs can process mix signals, which are fabricated by CMOS process. Figure 3 shows a typical cross section schematic diagram of devices fabricated by CMOS process. In these ICs, analog and logic circuits often have

978-1-4673-2395-6/12 $31.00 © 2012 IEEE

similar proportion in the topology. To achieve higher performance and complex function, logic area is beyond hundred micrometers in geometry, and includes many inverters, flip-flops and other digital circuits. All of them consist of PMOS and NMOS, which are arranged alternately in the topology, as shown in Figure 4.

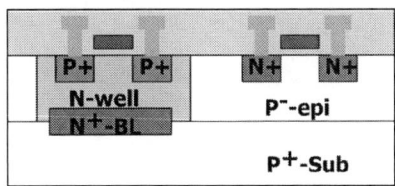

Fig. 3. Cross section schematic diagram of devices fabricated by CMOS process

From process point of view, these PMOS and NMOS are formed by a series of steps including lithography, implant, oxide growth, film deposition and etch process. In the case as shown in Figure 4, large size PR is coated to cover all patterns in the logic area before implant process.

Fig. 4. Topology of PMOS and NMOS in the logic area

Taking N-Well implant as an example, Figure 5 shows N-Well PR layout from topview. The area covered by box is to be performed N-Well implant, and other area is PR to block N-Well implant. We can easily detect that the PR in the middle of logic area, called as dense PR, has much smaller size than the PR at the edge of the logic area, called as isolated PR. The difference of PR size is determined by layout pattern, which may cause unexpected result during lithography process. Unfortunately, the problem is easily ignored by layout designer.

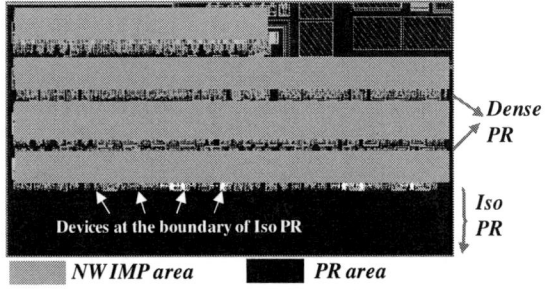

Fig. 5. Topology of N-Well PR in the logic area

During lithography process, the PR need be hardened by a series of hot and cool processes after PR coating. The isolated PR has huge size, which results in its higher fluidity than dense PR. The different fluidity can induce obvious PR profile difference if PR thickness is large enough, which can be treated as side effect of the pattern dependent microloading effect. Figure 5 shows a typical topology of layout design, which has

high potential risk to lose chip function due to pattern dependent microloading effect. The crux is not only the profile difference between dense and isolated PR, but also the devices existing at the edge of the isolated PR. These devices potentially cannot be protected intact by isolated PR during implant process.

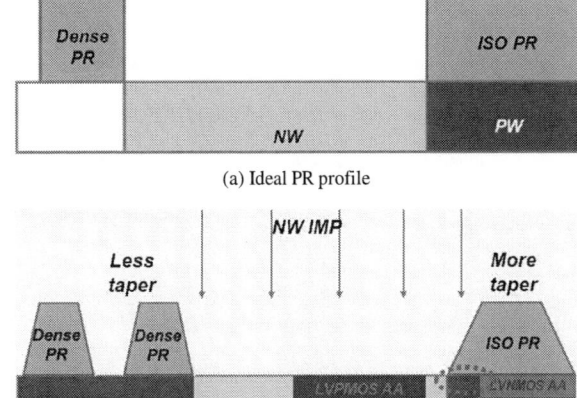

Fig. 6. Simplified failure mechanism of microloading effect during lithography and implant process

The simplified failure mechanism is revealed in Figure 6. If there is no any side effect caused by tool or process limitation, ideal PR profile should be coated as that in Figure 6 (a). However, lots of factors can affect the final result, and some of them are not expected by us, such as the microloading effect. After taking the effect into account, we can get the PR profile as that in Figure 6 (b). Due to more taper PR side, some doping agent for N-Well implant area will be injected into the area for P-Well implant. In this case, the boundary of P-N Well junction will not be as perfect as what it is expected to be, and finally serious leakage can be induced.

Such leakage failure is absolutely a tough task for failure analysis engineers, not only for the reason of setup difficulty at wafer level, but also for the hotspot capturing machines, such as PEM (Photon Emission Microscopy) and OBIRCH (Optical Beam Induced Resistance Change). These machines either cannot provide any hotspots, or reveal too many hotspots. The reason is leakage coming from a very big area, involving so many devices that no unique device can be highlighted. Furthermore, DFA (Data Failure Analysis) cannot provide helpful hints, neither. Tool commonality and correlation analysis [5] cannot capture any meaningful results due to following reasons. Firstly, its root cause is a design bug. Secondly, the design bug can be enhanced by wafer process shift, such as lithography overlay and PR thickness. DFA probably makes the problem more complex.

III. EXPERIMENT

A. PEM Technique for Hotspots Capturing

As important step in the failure analysis, PEM is necessary to localize which circuits are involved in the failure. As shown in Figure 7, many emission spots are captured, just like what we mentioned above. Several circuits are involved, which are listed

978-1-4673-2395-6/12 $31.00 © 2012 IEEE

as 1~4. All of the circuits have a commonality that is large size of NW PR and devices existing at the edge of the isolated PR.

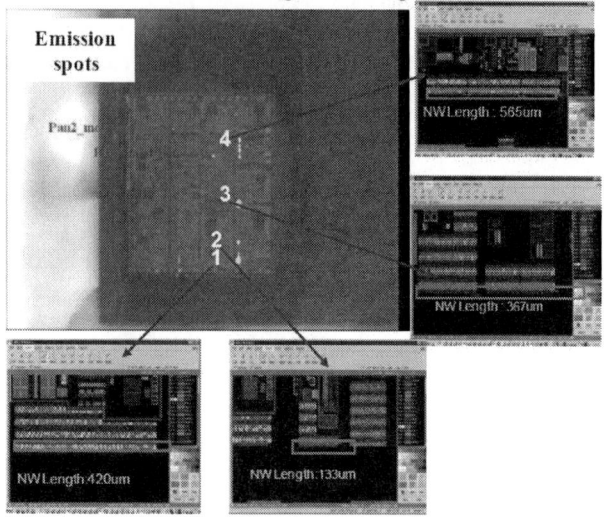

Fig. 7. PEM images for failing part

B. Physical Failure Analysis

Thanks to cross section SEM (Scanning Electron Microscopy), we inspect the rough profile of boundary between N-Well and P-Well after chip processed by wet chemical method. Figure 8 shows obvious angle difference of the boundary profile, which is suspected induced by different PR slope. We believe that such difference can result in unexpected effect on PN junction formed by Well implant, and consequently leakage is induced.

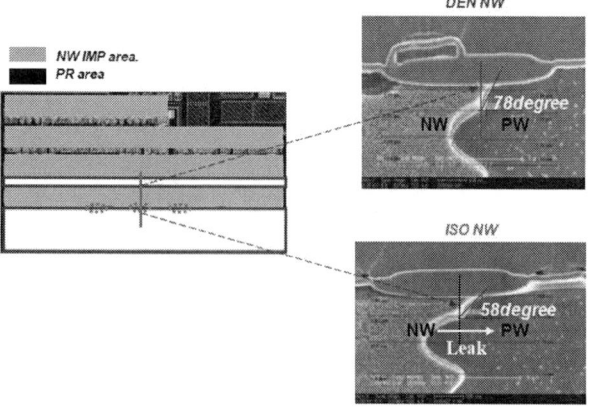

Fig. 8. Cross section SEM images for failing part

Of course, as direct evidence, inline PR profile is checked just after N-Well lithography process finished. Obviously different slope between dense PR and isolated PR is found, just as shown in Figure 9. The isolated PR cannot protect the area close to its boundary as perfect as dense PR does during implant process. N type agent will inject into the area which should have been injected by P type agent. Therefore, counter doping is formed at the weakly protected area. The chip will be screened out by Ioff or Isleep test for sure.

White band (WB) is defined to characterize the PR slope, as expressed by formula (1). Larger WB means more taper PR profile. We need decrease WB value to weaken microloading effect as possible as we can.

$$WB = (m - l)/2 \qquad (1)$$

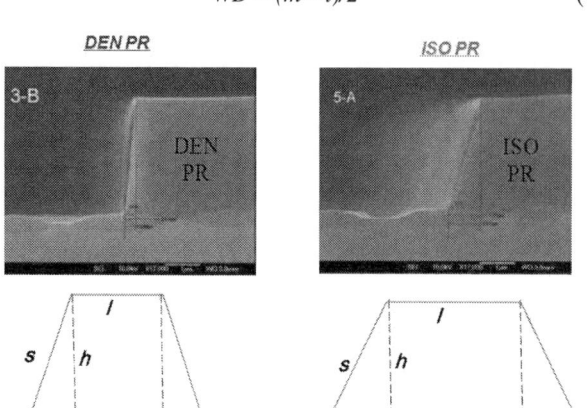

Fig. 9. Cross section SEM images of photo resist for failing part

C. Side Effect on Lithograph Process Window

During the analysis on the lithography tool comparison, we find a phenomenon that wafers processed by N-Well photo tool B are seriously suffered by the leakage issue, as shown in Figure 10. The composite yield map for the wafers reveals by reticle pattern, which is an obvious signal to tell us lithography process contributing to the failure.

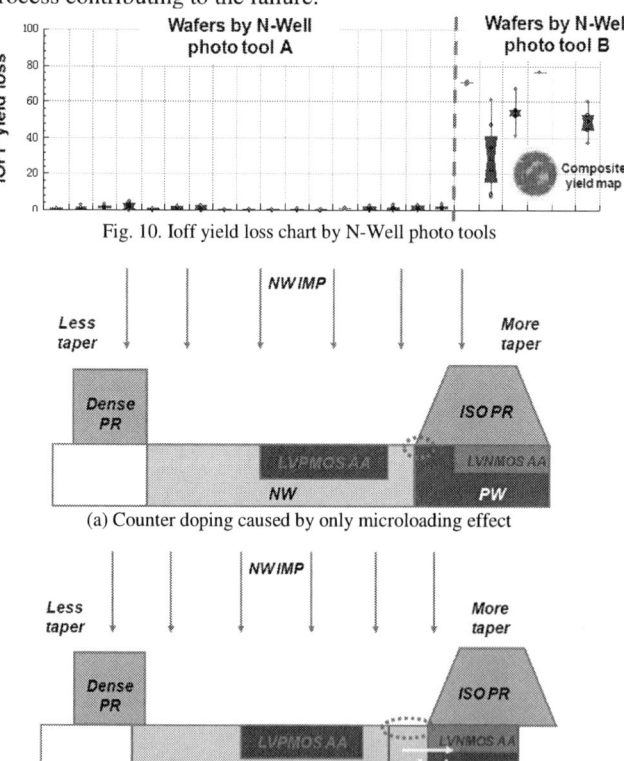

Fig. 10. Ioff yield loss chart by N-Well photo tools

(a) Counter doping caused by only microloading effect

(b) Counter doping area extended by microloading effect and overlay shift

Fig. 11. Simplified failure mechanism of microloading effect and overlay shift during lithography and implant process

DFA find the difference between the two tools. B has a little worse overlay performance than A. That is not to say B is a bad process tool, because no yield difference between A and B is observed on another product processed by same flow. The root cause is that the structure mentioned above consuming much lithography process window. For a product without the risk structure, overlay from -0.12um to 0.12um will not cause any yield issue. However, it may be a big problem for a product with the risk structure.

Figure 11 shows the mechanism of overlay enhancing microloading effect. The counter doping area is extended when microloading effect and overlay shift are working together. The existing microloading effect requests more critical overlay control during lithography process. In other words, microloading effect decreases overlay window of lithography process.

IV. SOLUTION

Since microloading effect is a famous problem, many researchers present solutions. Mainly we can simply divide them into two groups, one is process change, and the other is design change. Reduction of photo resist is proved useful to improve microloading effect [6]. However, it belongs to process change, which is not expected by many final customers. Another option seems more acceptable, that is adding dummy pattern in N-Well reticle, as shown in Figure 12.

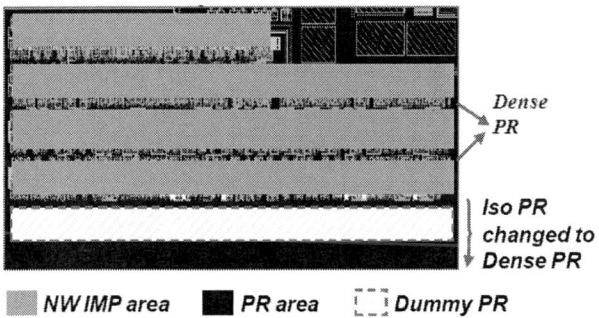

Fig. 12. Layout diagram of solution by adding dummy PR

By adding dummy PR pattern in the open area, isolated PR is changed to dense PR. The PR profile will become better for the boundary devices. This solution need update N-Well photomask, and its advantage is that it does not request any process change. Some measurements are performed to check the efficiency of the solution, as shown in Table 1. WB, overlay window and yield indicate that adding dummy PR in the N-Well photomask can work. PR profile becomes better around 50%, and overlay window increase around 70%. The most important is yield ramping up to 90%.

TABLE I
COMPARISON BETWEEN NEW AND OLD N-WELL PHOTOMASK

	WB	Overlay Window	Yield
New N-Well Mask	0.4um	-0.18um~0.18um	90%
Old N-Well Mask	0.9um	-0.11um~0.11um	30%

In other experiments, we try to skip hard bake during photo resist coating process to reduce fluidity of photo resist, and adjust focus and energy during exposure process to change PR profile. However, no obvious effect can be observed, and all of them belong to process change, which means high risk for final customers.

V. SUMMARY

The difference of pattern density in photomask may cause different photo resist profile, and then etch rate or implant profile can be impacted. This is called as microloading effect. We find that a structure is easily impacted during lithography and implant process due to the microloading effect.

After a series of analyses, we find that photo resist profile is affected by the pattern dependent microloading effect. PR with huge size is more oblique than small PR. In consequent implant process, counter doping happens due to implant agents penetrating the oblique PR. The effect will be enhanced when overlay performance is bad.

Some solutions are presented. Adding dummy PR pattern in blank area is proved to be the better option due to no process change involved, and effective data supported.

ACKNOWLEDGMENT

The author would like to thank all members from PA LAB of Freescale Semiconductor Limited in Tianjin, for their insight and helpful suggestions.

REFERENCES

[1] Wanlass, F. M, "Low Stand-By Power Complementary Field Effect Circuity" U. S. Patent 3,356,858 (Filed June 18, 1963. Issued December 5, 1967).

[2] KAUSHIK ROY, SAIBAL MUKHOPADHYAY, "Leakage Current Mechanisms and Leakage Reduction Techniques in Deep-Submicrometer CMOS Circuits", PROCEEDINGS OF THE IEEE, VOL. 91, NO. 2, FEBRUARY 2003, pp. 305-327.

[3] A.G.Street, "Failure Analysis in the Integrated Fabless Manufacturer (IFM) Environment," IEEE Proceedings of 16th IPFA- 2009.

[4] Hedlund, C., Blom, H. O., "Microloading effect in reactive ion etching," Journal of Vacuum Science & Technology A: Vacuum, Surfaces, and Films, Jul 1994, Volume: 12, Issue: 4, pp. 1962-1965.

[5] Wu Miao, "A Novel Data Analysis Methodology in Failure Analysis," ISDEA 2012, pp. 988-991.

[6] Azuma, T., Ohiwa, T., Okumura, K., "Impact of reduced resist thickness on deep ultraviolet lithography," Journal of Vacuum Science & Technology B: Microelectronics and Nanometer Structures, Nov 1996, Volume: 14, Issue: 6, pp. 4246-4251.

Combined Emission with simulation technique to resolve unstable failure mode sample

DiWei Fan, Winter Wang, Li Tian, Miao Wu, ChunLei Wu

Freescale Semiconductor (China) Limited

No.15, Xinghua Avenue, Xiqing Economic Development area, Tianjin, China

Email: b25961@freescale.com

Abstract – Failure analysis (FA) of semiconductor should base on a specific failure mode. But the failure mode has potential risk that it may change due to it is unstable (caused by weak defect or voltage stress etc). The change of unstable failure mode can occur in every stage of FA flow. If the change happens, typical FA flow cannot be continued base on the failure mode any more. The change of unstable failure mode in different stage will impact the final FA result deeply. In this situation, combined failed device simulation with emission result is a good choice to resolve unstable failure mode sample. This paper introduces the solution that combined Emission technique with simulation to resolve the unstable failure mode sample.

Keywords-Failure analysis; Failure mode; failed device simulation; Emission;

I. INTRODUCTION

The Failure Analysis (FA) process consists of five main steps, namely: failure validation, fault localization, sample preparation and defect tracing, defect characterization and root cause determination [1]. Failure validation is the first step of FA and basic of FA as well. Every FA technique should be performed base on a specific failure mode. Different FA techniques are used to resolve different failure modes. Generally, there are two categories of main FA techniques. One is passive techniques (no active stress on sample) and the other is active techniques (voltage stress or physical damage are added). Emission [2] is a typical technique of passive techniques. The active techniques include IR-OBIRCH, XIVA [3], Microprobe, and so on. The unstable failure mode can change in either passive techniques or active techniques. Most of the times, due to the impact of sample by passive techniques is less possible than active techniques, the unstable failure mode always change in active techniques analysis stage.

In this paper, the change of unstable failure mode occurs after emission analysis. When the failure mode change occurs the FA cannot be continued base on previous analysis result. And the variational failure mode cannot be used as analysis clew any more. In this situation, failed device simulation is a good choice for FA. One special failure mode can be induced by several possibilities. The function of simulation is to find these possibilities. But only find these possibilities is not enough. The Emission analysis result is important either. It can provide much information of the failed sample before failure mode change. Combined emission result with simulation result, even if the failure mode change occurs the FA still can be continued.

II. EMISSION METHODOLGY AND SIMULATION APPLICATION

During FA process, Emission analysis technique is a very important step. Generally, Emission is performed both on good sample and failed sample. Different emission spots are indicated on the background image as Emission result. The Emission result can provide analysis clew for FA analyzer. There are two categories of Emission source mechanism. One is generated by scattering of field accelerated carriers (F-PE). Mobile charge carriers are accelerated and gain kinetic energy in an electrical field. They relax scattering, accompanied by light emission with a broad spectral distribution from near IR into visible region. This is an intra-band process because it is all happening in the same energy band. This effect is most commonly used in Emission techniques for CMOS circuits. Most Emission spots in microelectronic failure analysis are electrical field induced, as described in below table.

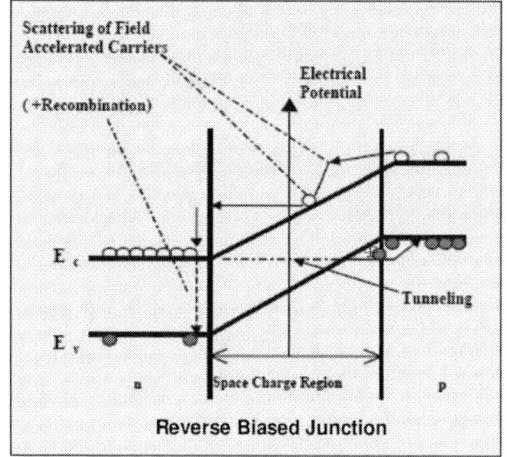

Fig.1. Scattering of Carriers Accelerated by Electric Field Voltage

TABLE I.
CLASSIFICATION OF F-PE

F-PE	Space Charge Region	Reverse Biased Junction (7V)
		Silicon Leakage Currents
		MOS Transistors / saturated mode
		ESD Protection Breakdown
		Bipolar Transistors / active mode (B-C reverse based)
	Locally High Current	Gate Oxide Defect / Leakage
	Fowler-Nordheim Current	Gate oxide Leakage Current (FLASH, EEPROM)

The other is generated by radiant electron-hole recombination (R-PE). Recombination involves carriers from conductance and valence band. It is called an inter-band process. These different Emission classifications can map to relative status of device. Emission result can be analyzed base on this information.

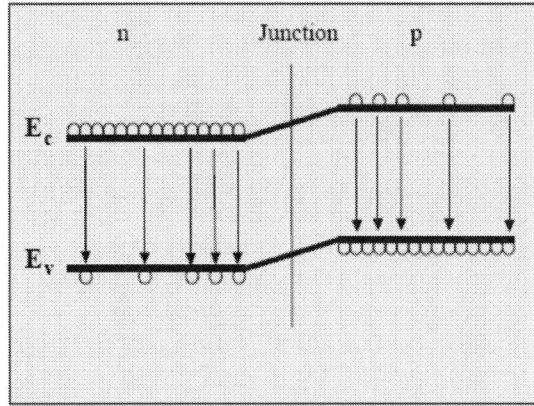

Fig.2. Radiant electron-hole recombination

TABLE II.
CLASSIFICATION OF R-PE

R-PE	E-H-Recombination	Forward biased Junction
		Bipolar Transistors / saturated mode (B-C forward based)
		Latch up (B-C forward based)

Failed device simulation is a special method of FA. For a specific failure mode, some suspected failed devices can be isolated after schematic and layout study. In this situation, simulation can be performed to confirm the suspicion. Analyzer simulate the failed devices by add similar condition (short, open, leakage etc.) at the suspected devices of good sample. Then observe that if the failure mode on good sample is same as failed sample. If the generated failure mode is same as failed sample the suspected failed device can be confirmed. Note that the simulation result may be many possibilities, but the actual failed device is only one.

Emission technique can confirm some devices' status, and simulation can find possibilities of failure mode. Combined Emission with simulation is a good solution in FA for unstable failure mode sample.

III. FAILURE MODE CHANGE SAMPLE ANALYSIS

The failed device under test (DUT) [4] is from eswitch family of freescale semiconductor, fabricated by 0.25μm "SmartMos8MV+HDTMOS5" wafer technology.

There are two status of DUT's five outputs, one is turn on status and other is turn off status. When DUT is working on turn on status, the relative output is pulled up by Vpwr input. In the opposite, when DUT is working on turn off status, the relative output cannot be pulled up by Vpwr input. The Vpwr input voltage range has a limitation. In this case, the Vpwr

input voltage is equal to 13v. The failed DUT's outputs are different with good sample. All outputs of failed DUT are abnormal whether in turn on status or turn off status. The voltage of outputs is shown as table 3. "HS" is the name of output.

Base on the failure mode, Emission technique is performed on both good sample and failed sample. Due to the failure mode is unstable. After Emission technique analysis, it is changed. The failed sample's outputs voltage are same as good sample.

TABLE III.
OUTPUT VOLTAGE OF GOOD SAMPLE AND FAILED SAMPLE

Vpwr=13v. turn off:

	HS1	HS2	HS3	HS4	HS5
Ref	Roughly 6.3v	Roughly 6.3v	Roughly 6.3v	Roughly 6.3v	Roughly 6.3v
CQI	Roughly 0v	Roughly 0v	Roughly 0v	Roughly 0v	Roughly 0v

Vpwr=13v. turn on:

	HS1	HS2	HS3	HS4	HS5
Ref	Roughly 12.9v	Roughly 12.9v	Roughly 12.9v	Roughly 12.9v	Roughly 12.9v
CQI	Roughly 0v	Roughly 0v	Roughly 0v	Roughly 0v	Roughly 0v

TABLE IV.
FAILURE MODE CHANGED

Vpwr=13v. turn off:

	HS1	HS2	HS3	HS4	HS5
Ref	Roughly 6.3v	Roughly 6.3v	Roughly 6.3v	Roughly 6.3v	Roughly 6.3v
CQI	Roughly 6.3v	Roughly 6.3v	Roughly 6.3v	Roughly 6.3v	Roughly 6.3v

Vpwr=13v. turn on:

	HS1	HS2	HS3	HS4	HS5
Ref	Roughly 12.9v	Roughly 12.9v	Roughly 12.9v	Roughly 12.9v	Roughly 12.9v
CQI	Roughly 12.9v	Roughly 12.9v	Roughly 12.9v	Roughly 12.9v	Roughly 12.9v

The next step of typical FA technique cannot be continued due to the failure mode disappears. Only Emission result is the analysis clew now. The low power microscope Emission result is shown as figure3. From the emission result, we can find that there are many different Emission spots between good sample and failed sample. Base on the schematic and layout study, we can conclude that the block "c_reg" failed maybe can induce these different emission spots between good sample and failed sample. So the block "c_reg" is highly suspected.

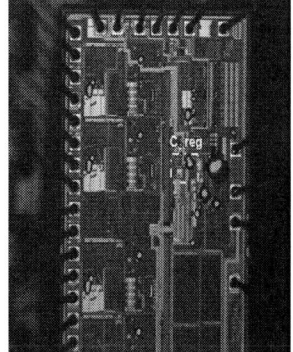 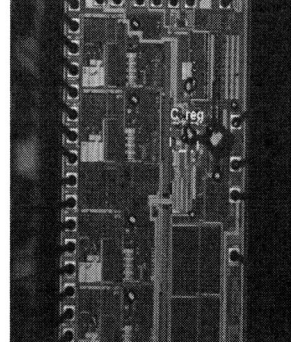

Good sample Mag:50x   Failed sample Mag:50x

Fig.3. Emission result on good sample and failed sample

978-1-4673-2395-6/12 $31.00 © 2012 IEEE

Simulation is performed base on the schematic and layout overlay study. The schematic diagram of "c_reg" is shown as figure 4. The output of "c_reg" is alim3p3. If alim3p3 works in normal mode, it will provide a 3.3v power supply to other blocks. But if alim3p3 is abnormal, it cannot provide enough power sources to the HS driver blocks. Thus, the voltage of outputs (HS) is abnormal.

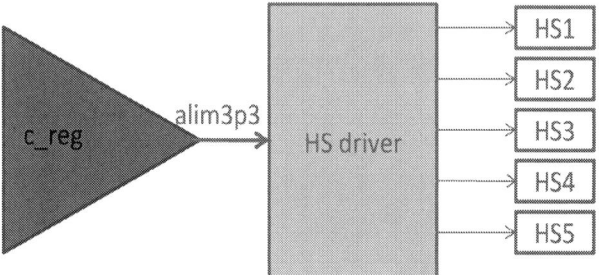

Fig.4. Schematic diagram of "c_reg"

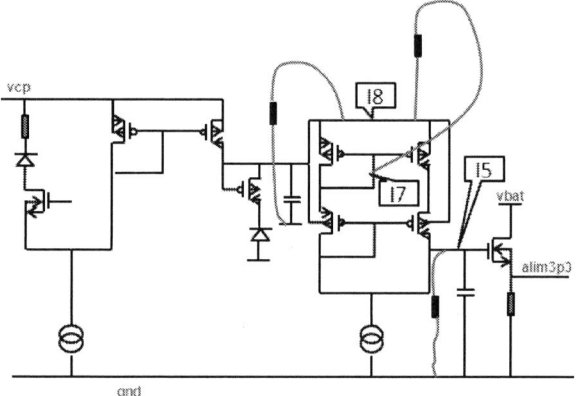

Fig.5. Diagram of suspected failure path

Base on the description of suspected failure path above (figure 5), making voltage of alim3p3 lower is the purpose of simulation. After measure the voltage of internal test pad (I5, I7, and I8) by microprobe[5] (see figure 6) on good sample, we can find that I5 equal to 5.3v, I7 equal to 3.3v and I8 equal to 6.4v. The simulation is performed on these three test pad. The voltage of alim3p3 is relative to a NMOS and the I5 is the gate of the NMOS. If voltage of I5 is equal to 5.3v as good sample, the NMOS turn on and alim3p3 is pulled up by vbat. The voltage of I5 decides the NMOS status and finally decides the voltage of alim3p3. To simulate the failed status on I5, we tie I5 to gnd by microprobe technique. And add different resistors in the simulation circuit. The simulation result is that the voltage of alim3p3 is changed when add different resistors, but the variational voltage is always lower than good sample. After simulation performed on I5, the simulation is performed on I8. I8 is the power source of I5, if the PMOS between I8 and I5 turn on, the I5 is pulled up by I8. If voltage of I8 is lower than good sample, even if the PMOS turn on, the I5 still cannot be pulled up to the same voltage as good sample. To simulate the failed status of this style, we tie I8 to gnd and add resistors

between I8 and gnd. The simulation result is that the voltage of alim3p3 is lower. I7 is simulated as same as I5 and I8. Tie I7 to I8 and finally find that the voltage of alim3p3 is lower than good sample either. From the simulation result, we can find that if I5 or I8 lower than good sample, or I7 higher than good sample, the alim3p3 will be lower than good sample and finally impact the outputs voltage. It can match the suspicion before.

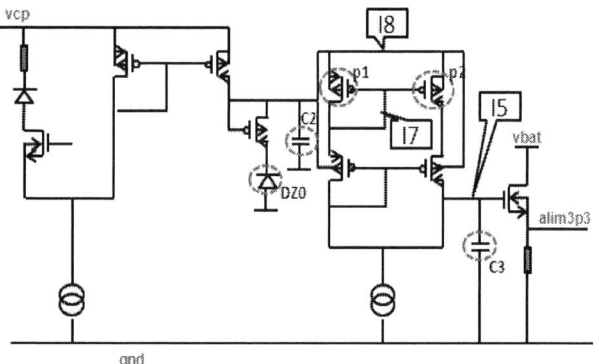

Fig.6. Simulation is performed on "c_reg"

Base on the simulation result, we can conclude that there are some possible defect can induce the failure. If DZ0 and C2 are failed, I8 can be pulled down by gnd so that the voltage of I8 is lower than good sample. Similarly, if p1 and p2 were failed, I8 and I7 are impacted. To make I5 lower than good sample, we can assume that C3 is failed.

Fig.7. Suspected device base on simulation results

In normal FA flow, we can measure the voltage of these suspected nodes by microprobe technique. But due to the unstable failure mode changes after Emission technique analysis, the voltages of these suspected nodes also change. Even if we get the voltages by microprobe technique, they are useless. So now, we must get more information about the failure mode from emission result. From the Emission result in low power microscope, we can only isolate the defect in block "c_reg". From the Emission result in high power microscope, we can get more analysis clew of block "c_reg". See figure 8.

978-1-4673-2395-6/12 $31.00 © 2012 IEEE   446

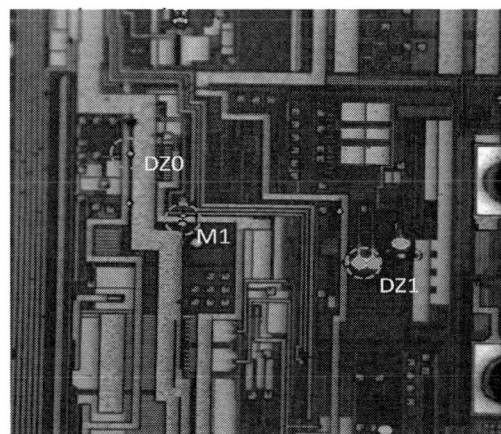

Fig.8. Emission result in high power microscope

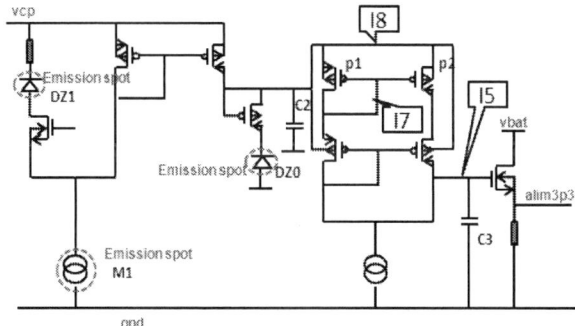

Fig.9. Emission spot in block "c_reg"

The Emission spots in block "c_reg" are shown as figure 9. The emission spots on DZ1, M1 and DZ0 is same as good sample. From the Emission spots on DZ1 and M1, we can find that DZ1 works in reverse biased breakdown status. It means that voltage of vcp should be normal. And the Emission spot on DZ0 can prove that voltage of I8 is normal. From the analysis information above, we can conclude that DZ0, C2 are normal. Simultaneously, p1 and p2 are normal. Because if p1 and p2 abnormal, it will impact voltages of I7 and I8. So now, combined Emission technique result and simulation information, the capacitor C3 in block "c_reg" is the most possible failed device. Base on this conclusion, physical de-process is performed focus on C3. Finally, dielectric rupture is observed as we expect. See figure 10.

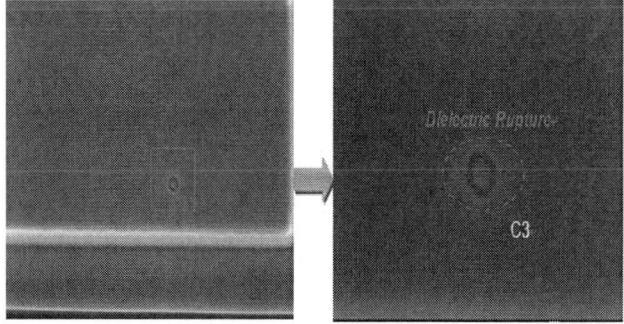

Fig.10. Dielectric rupture on C3

IV. SUMMARY

A specific failure mode is the premise of typical FA process, but some of the times, Due to the failure mode is unstable, it may change during FA. There are many reasons can induce the change. In this paper, we introduce a method to resolve the unstable failure mode case. This method should use Emission technique and simulation. Simlation technique can provide serveral possible failed devices, and then combined with emission result, choose the most possible failed device by a process of elimination. And finally, physical analysis on the unit can confirm the defect on the device.

ACKNOWLEDGEMENT

For this work and study, I would like to thank for cooperation with Miao Wu and help from ChunLei Wu and Winter Wang. Meanwhile I sincerely appreciate my colleagues in Tianjin product analysis laboratory of Freescale semiconductor.

REFERENCE

[1] Kudva SM, Clark R, Vallett D, "The SEMATECH Failure Analysis Roadmap," ISTFA 1995, pp. 1-5.

[2] JCH Pharng, DSH Chan, "A Review of Near Infrared Photon Emission Microscopy and Spectroscopy" Proceedings of 11th IPFA 2004, Taiwan, pp.255-261

[3] JCH Pharng, DSH Chan, "A Review of Laser Induced Techniques for Microelectronic Failure Analysis" Proceedings of 11th IPFA 2004, Taiwan, pp.255-261

[4] Jinglong Li, Zhai L, "A Novel Method to Analyze Analog and Mixed Mode" IPFA 2011, pp.1-4.

[5] DiWei Fan, Miao Wu, "Novel Failure Isolation Techniques for Circuits Sensitive to Microprobe" ISDEA 2012, pp.401-404.

PPTP: Pre-Post Terminal Propagation in Modern Fixed-Outline Soft Module VLSI Floorplanning Design

Chyi-Shiang Hoo*, Kanesan Jeevan, Velappa Ganapathy, Harikrishnan Ramiah
Department of Electrical Engineering, Faculty of Engineering,
University of Malaya (UM)
Lembah Pantai, 50603 Kuala Lumpur, Malaysia
Email: francioshoo@siswa.um.edu.my

Abstract—From the point of view of the industry, floorplanning is very important in VLSI chip physical design because it will deeply affect the time-to-market and the quality of the product. A new floorplanning algorithm namely Pre-Post Terminal Propagation (PPTP) has been proposed to handle soft module floorplanning by employing multilevel framework. Pre-Terminal Propagation is employed at the root node only in order to increase more possible partitionings, as introduction of TP at every level will restrict the partitioning results. However, non-inclusion of TP at the subsequent partitioning will lead to the lack of information about the external pins at every level in the tree. Hence, Post-Terminal Propagation is adapted to compensate this deficiency. PPTP gives improved optimal HPWL solutions and faster runtimes for soft module floorplanning based on Gigascale Systems Research Center (GSRC) benchmarks. The results obtained establish that PPTP is a high performance floorplanner as compared to other state-of-the-art floorplanning algorithms. This indicates this makes PPTP more suitable for industrial VLSI physical design implementation.

Keywords-Computer-Aided Design; Floorplanning; Terminal Propagation; VLSI; Circuit Layout.

I. INTRODUCTION

The physical design of VLSI chip has become more and more complicated and complex as the scale of the chip is decreasing due to the advancement in nanoscale IC technologies. This is because of the Moore's law [1][2] which has been steadily maintained in the commercial IC design. The density for a particular IC chip is highly depending on the electronic design automation (EDA) tool as well as the assumptions made, constraints considered and framework selection by the software developers. However, there are some crucial challenges existing in the classical outline-free floorplanning, which are limitations in packing-driven floorplanning [3], difficulty in choosing the coefficients in Pareto front/ non-dominated frontier (NDF) [4], low scalability, time consuming, and inconsideration of essential constraints such as fixed-outline [5].

Hence, as discussed by Chen et al. [6], the hierarchical [7] and multilevel floorplanning frameworks are preferred in modern floorplanning, rather than flat frameworks. The works in [5] and [8] have reviewed and revealed the new trend of modern EDA floorplanning concerning the elimination of classical floorplanning bottlenecks as discussed above. Adya et al. [9] have pointed out that the fixed-outline constraint

[10][11] and optimization of half-perimeter wirelength (HPWL) are the necessary parts in practical floorplanning. In [9], evolutionary turn points of the modern floorplanning are introduced where the minimal whitespace requirement is considered as a constraint instead of an objective. This simplifies the modern floorplanning by prioritizing the wirelength minimization. This is because the whitespace assigned can be used for other design purposes such as buffer insertion [12], level shifter positioning [13], thermal legalization, etc. Fixed-outline constraint causes modern floorplanning more difficult than conventional outline-free type, even in area-driven floorplanning [11].

A. Recent Works

Recent outperformed works on EDA floorplanning are employing hierarchical/multilevel frameworks. Defer [14] generated the generalized slicing tree by using the state-of-the-art hMetis [19] hypergraph partitioning method and the orientation of the modules are deferred to a certain level before they are combined via dynamic programming enumerative packing (DPEP) and refined. IMF [6] is the first floorplanner which proposed the "Λ-shaped" multilevel framework which imply connectivity-driven hMetis [19] and Accelerative Fixed-outline (AFF) packing techniques, with no interference in between. In [15], the whitespace fundamentals in top-down hierarchy are proposed so that whitespace can be controlled and predicted. However, there is no empirical comparison with other state-of-the-art floorplanning algorithms. PATOMA [16] employed purely cutsize-driven top-down approach followed by legalization. UFO [17] introduced a 2-ways multilevel framework. During the global distribution, circle and push-pull (PP) model is used to distribute the circles by considering the wirelength only. Meanwhile during the local legalization, the modules geometrical information is formulated. SAFFOA [18] introduced a SA-based top-down Ordered Quadtree hierarchical framework to handle purely soft modules floorplanning problems. Fixed-outline constraint is considered in all these recent modern floorplanning algorithms.

The rest of this paper is organized as follows. Section II shows the algorithm flow of PPTP and discusses the techniques used in PPTP. Section III compares the proposed PPTP with other recent fixed-outline floorplanning algorithms. Finally, this paper ends with the conclusion drawn in section IV.

978-1-4673-2395-6/12 $31.00 © 2012 IEEE

```
Algorithm flow of PPTP
Begin
    Step 1) Top-down partitioning algorithm
    Step 2) Bottom-up geometric formulation
    Step 3) Top-down recursive rotation/flipping
End
```

Fig. 1.   Pseudocode on algorithm flow of PPTP.

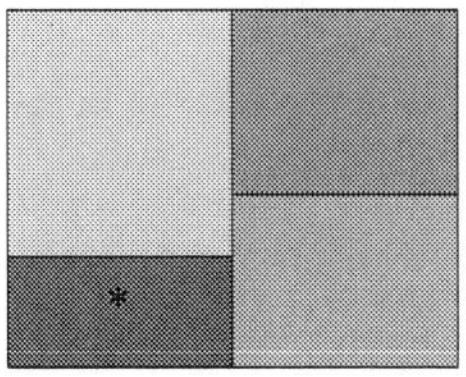

Fig. 2.   Outline Violation/aspect ratio outmatch in partition *.

II.   ALGORITHM FLOW OF PPTP

This paper presents a fast, high-quality, and scalable fixed-outline floorplanner, namely PPTP which can handle soft module modern floorplanning efficiently and effectively. Essentially, PPTP consists of three steps as shown in Fig. 1.

A. Partitioning Step (Pre-Terminal Propagation)

PPTP employs a slicing tree to represent the floorplan layout where the orientation of subcircuits (left-right/top-bottom) and cutline of the subcircuits are determined during the top-down recursive min-cut algorithm. There are three crucial factors in the partitioning stage:

1)   hMetis Hypergraph Partitioning

As discussed in section I, hierarchical frameworks are employed to handle large scale floorplanning problems as hierarchical framework can slit the floorplan problem into k-subproblems recursively and speed up the solution search space. Though flat framework can explore the solutions, it causes an extremely high cost. Thus this stage divides the floorplan circuit into two subcircuits recursively, by minimizing the min-cut/interconnections between the subcircuits until the number of modules in circuits has become four or less. In PPTP, the state-of-the-art hypergraph partitioner, namely hMetis [19] is adapted to construct the generalized slicing tree, until the number of modules is four or less. In order to make sure all the soft module dimensions fit to the constraints, two near square constraints are generated by cutting the outline appositively at every level (horizontal followed with vertical or vice versa).

2)   Area-driven Linear Ordering Partitioning

When the number of modules in a particular level is equal to four or less, an area-driven linear ordering partitioning is applied in order to reduce the fixed-outline constraint violation. The modules will be sorted into a linear descending order by referring to the areas of modules. If the number of module is four, the two largest modules will be put into the same partition, so do the other two. Meanwhile, if the number of modules is three, the largest module will be put into a single partition, so do the other two.   This is because when hypergraph partitioner generates two subfloorplans with similar weights, there is a high possibility that the largest module is grouped with the smallest module. This will cause the aspect ratio violation if the smallest module has to be fitted

into the flat outline which is caused by cutline, as shown in Fig. 2. Thus, the smallest module will be assigned the minimum sustainable aspect ratio as well as its dimensions, which infract the required outline constraint. This will degrade the overall floorplan in terms of outline constraint, area tolerance as well as wirelength optimization.

3)   Pre-Terminal Propagation

Terminal Propagation (TP) [20] is introduced at the first level of this partitioning stage although TP can be applied at every levels of a tree theoretically. However, in this work, TP is not applied at every level with the purpose of increasing the possible partition outcomes as TP restricts the possible partitioning results. On top of that, when the size of partition decreases, the effect of TP on the floorplan reduces. The TP involved during hypergraph partitioning is called as Pre-TP.

B. Geometrical Formulation Step

In this stage, the dimensions of the soft modules are determined based on the outline constraint and the soft modules with fixed dimensions will be orientated based on the cutline. The cutline direction is always parallel to the shorter side of a region in order to generate a pair of near-to-square subregions as this will maintain the aspect ratios of the subregions from exceeding the maximum aspect ratio of the soft modules at the leaf nodes.

C. Rotation/Flipping Step (Post-Terminal Propagation)

Terminal propagation [20] only considers the one third of the IO pads at the root node and at every level all the modules in a partition are considered to be concentrated on the center of the partition. The preferable location of a module and the subfloorplan order are unpredictable due to lack of TP at every level. Hence, Post-Terminal Propagation (Post-TP) is introduced to compensate the lack of global knowledge during hypergraph partitioning. After the final floorplan layout is obtained, the first three level partitions will be rotated through 180 degree or flipped one-by-one in order to try out all the possibility of best orientation. It is possible to try rotation/flipping for all partitions at every level but this will expensive in terms of computational complexity and runtime.

TABLE I.   RESULTS OF PPTP ON GSRC SOFT MODULE FLOORPLANNING PROBLEMS.

Problem	AR = 1:1		AR = 2:1		AR = 3:1		T_{run}
	HPWL	WS	HPWL	WS	HPWL	WS	(sec)
n10a	35140	0	36022.2	0	38046.8	0	0.05
n10b	41732.5	0	42117.9	0	45836.1	0	0.07
n10c	38865.5	0.1	40859.7	0	41818.3	0	0.08
n30a	104443	0	105145	0.72	112605	0.08	0.14
n30b	106800	0	114994	0	124645	0	0.15
n30c	131727	0	138248	0	137048	0	0.13
n50a	133989	0.06	135770	0.47	146708	0	0.17
n50b	150827	0.45	160243	0.37	170797	0.35	0.20
n50c	147462	0	155851	0.08	165129	0.09	0.18
n100a	212914	0.32	213235	0.53	223668	0.48	0.21
n100b	205948	0.13	213306	0.60	227920	0.49	0.22
n100c	207569	0.52	212597	0	222315	0.65	0.20
n200a	373712	0.48	383305	0.66	405305	0.34	0.53
n200b	376726	0.76	413417	0.84	413417	0.84	0.49
n200c	353991	0.96	381917	0.79	381917	0.79	0.50
n300	490601	0.43	507789	0.61	532582	0.44	0.75

978-1-4673-2395-6/12 $31.00 © 2012 IEEE

III. Empirical Validations and Discussions

PPTP is designed as a fixed-outline soft module floorplanner with 1% utilization rate as whitespace is no more an objective but a constraint. PPTP was compiled using *gcc* on a Linux Ubuntu PC platform with Intel Pentium IV 2.4-GHz CPU and 256-MB memory. The HPWL and runtimes of PPTP are obtained after 25 simulation runs. The GSRC [21] benchmarks are used as the standard for comparisons. The maximum dimension ratio of a soft module is found by calculating the ratio of the maximum dimension to the minimum dimension of the particular module and the minimum dimension ratio is the reciprocal of the maximum dimension ratio. In this work, for the ease of understanding, some abbreviations are introduced below:

AR	Aspect ratio
WS	Relative whitespace (%)
HPWL	Half-perimeter wirelength
nWL	Normalized HPWL
T_{run}	Runtime
nT_{run}	Normalized T_{run}

AR is the ratio of width to the height of the floorplan layout while WS is calculated by finding the ratio of the difference between the minimum area of the rectangle which covers all the modules and the summation of all the modules' areas, to the minimum area of the rectangle which covers all the modules. To provide a relative comparison, normalization of results is used to calibrate and compare PPTP results with other benchmark algorithms and the normalization is defined as the ratio of the results of other floorplanning algorithms to the PPTP results.

In this work, the near optimal HPWL, relative whitespace and runtime of all the GSRC soft module problems in five different aspect ratio constraints are shown in Table I. Amongst all the 80 results, the relative whitespaces of all cases are less than 1% which is the whitespace tolerance being assigned to the floorplanner. Xilinx® had explained that FPGA-based platform design demands shorter period of time in physical design so that the designers can modify, apprehend, analyze and implement their design in FPGAs rapidly [22]. This indicates that modern EDA floorplanning tool requires not only robustness but short timeframe also. The low increment of HPWL with respect to the aspect ratio of the floorplan layout indicates the efficiency and low dependency of PPTP to the aspect ratio of the chip. The near optimal HPWLs as well as the fast runtimes show that PPTP is a robust floorplanner which can search for near optimal solution within a restricted period of time.

The HPWL comparisons are made based on the following state-of-the-art floorplanning/placement algorithms: Defer [14], PATOMA [16], UFO [17], and SAFFOA [18], as shown in Table II. The normalized values of HPWLs given in the table bring out the comparisons more vividly. From Table II, it is very obvious that PPTP outperforms in terms of HPWL as compared to other state-of-the-art algorithms. PPTP shows improved results in 36 cases in HPWL as compared to 48 cases of other floorplanning algorithms. It should be noted that in Table II, although UFO shows a slightly lower average normalized HPWL value of 0.99 as compared to PPTP, the normalized runtime of UFO indicates that PPTP performs much better in terms of runtimes ranging from 60 to 6290

times faster than UFO and this is illustrated in Table III. Fig. 3 runtime results of PPTP are compared with Defer and PATOMA as other floorplanning algorithms show excessive normalized runtimes as compared to PPTP. In terms of HPWL, PPTP shows comparable results as that of Defer, but it is obvious that Defer runtimes are relatively higher as compared to PPTP when the number of modules in the floorplanning problems increases and is shown in Fig. 3. It is because in PPTP, there is an ineluctable constant execution period for the floorplanner to do the Post-TP algorithms.

TABLE II. HPWL COMPARISONS ON GSRC SOFT MODULE BENCHMARKS.

Floorplanners		Defer		PATOMA		UFO		SAFFOA		PPTP
Circuits	AR	HPWL	nWL	HPWL	nWL	HPWL	nWL	HPWL	nWL	HPWL
n10a	1	-	-	52258	1.48	36 398	1.04	-	-	35140
	2	-	-	-	-	-	-	-	-	36022.2
	3	-	-	-	-	-	-	-	-	38046.8
n30a	1	-	-	156921	1.50	102100	0.98	-	-	104443
	2	-	-	-	-	-	-	-	-	105145
	3	-	-	-	-	-	-	-	-	112605
n50a	1	-	-	180115	1.34	124300	0.93	-	-	133989
	2	-	-	-	-	-	-	-	-	135770
	3	-	-	-	-	-	-	-	-	112605
n100a	1	196457	0.92	283452	1.33	195200	0.92	263200	1.24	212914
	2	217686	1.02	-	-	214430	1.01	281509	1.32	213235
	3	235702	1.05	-	-	235210	1.05	296228	1.32	223668
n200a	1	354885	0.95	505716	1.35	346660	0.93	480014	1.28	373712
	2	380470	0.99	-	-	381320	0.99	520802	1.36	383305
	3	410464	1.01	-	-	406610	1.01	536040	1.32	405305
n300	1	476508	0.97	566242	1.15	476560	0.97	554240	1.13	490601
	2	514764	1.01	-	-	510020	1.00	615713	1.21	507789
	3	551610	1.04	-	-	541240	1.02	617554	1.16	532582
Average Normalized HPWL		**1.00**		**1.36**		**0.99**		**1.26**		**1.00**

TABLE III. RUNTIME COMPARISONS ON GSRC SOFT MODULE BENCHMARKS.

Floorplanners		Defer		PATOMA		UFO		SAFFOA		PPTP
Circuits	AR	T_{run}	nT_{run}	T_{run}	nT_{run}	T_{run}	nT_{run}	T_{run}	nT_{run}	T_{run}
n10a	1	-	-	1	20	3	60	-	-	0.05
	2	-	-	-	-	-	-	-	-	0.07
	3	-	-	-	-	-	-	-	-	0.08
n30a	1	-	-	1	7.14	87	621.4	-	-	0.14
	2	-	-	-	-	-	-	-	-	0.15
	3	-	-	-	-	-	-	-	-	0.13
n50a	1	-	-	1	5.88	204	1200	-	-	0.17
	2	-	-	-	-	-	-	-	-	0.20
	3	-	-	-	-	-	-	-	-	0.18
n100a	1	0.09	0.42	2	9.52	677	3223.8	158	752.4	0.21
	2	0.09	0.41	-	-	-	-	-	-	0.22
	3	0.09	0.45	-	-	-	-	-	-	0.20
n200a	1	0.18	0.33	3	5.66	3306	6237.7	690	1301.9	0.53
	2	0.19	0.36	-	-	-	-	-	-	0.49
	3	0.19	0.39	-	-	-	-	-	-	0.50
n300	1	0.78	1.04	4	5.33	4718	6290.7	1563	2084	0.75
	2	0.96	1.30	-	-	-	-	-	-	0.74
	3	0.97	1.34	-	-	-	-	-	-	0.72
Average Normalized T_{run}		**0.67**		**8.92**		**2938.9**		**1379.4**		**1.00**

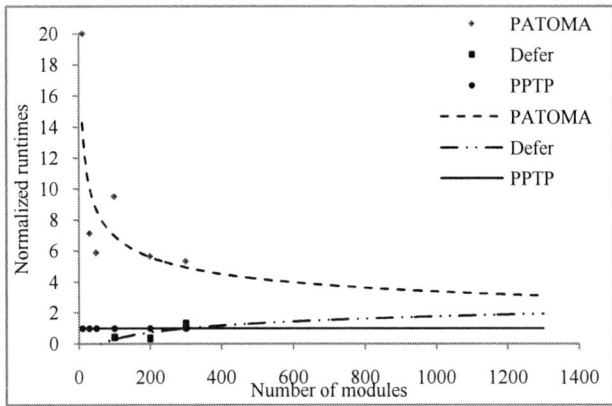

Fig. 3.   Normalized runtime comparisons with Defer and PATOMA.

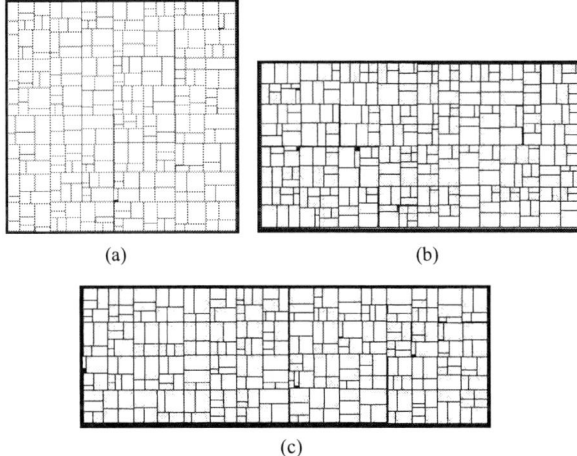

(a)   (b)

(c)

Fig. 4.   *n*300 soft module floorplan layout by PPTP: (a) AR = 1:1, (b) AR = 1:1 and (c) AR = 1:1.

Table III shows the runtime comparisons of the state-of-the art floorplanning/placement algorithms: Defer [14], PATOMA [16], UFO [17], and SAFFOA [18] and the runtimes are normalized for ease of comparisons. It is to be noted that PPTP is implemented on a PC with a slowest speed processor amongst all these state-of-the-art algorithms. Although Defer shows a lower average normalized runtime as compared to PPTP, the runtimes of Defer is relatively higher than PPTP when the number of modules is increasing. To conclude, it is proven that PPTP is capable of handling large-scale floorplanning problems with shorter runtimes. Its potential to be implemented in industry as the optimization tool for the reduction of HPWL and runtimes is always the priority in the industrial EDA tools. Fig. 4 shows the best results of *n*300 benchmark placement layouts using PPTP for different aspect ratios.

IV.   CONCLUSION

VLSI floorplanning design will affect the final chip layout directly and thus it has high impact on the performance of VLSI chip. In this paper, a fast and robust floorplanner, namely PPTP is proposed to handle fixed-outline soft module floorplanning. Pre-Terminal Propagation (Pre-TP) is

employed at the root node only as it will increase the computational complexity and runtime of the floorplan if it is applied at all the levels. However, non-inclusion of TP at the subsequent partitioning will lead to the lack of information about the external pins at every level in the tree. Hence, Post-Terminal Propagation (Post-TP) is adapted to compensate the deficiency. Experimental results have shown the efficiency, effectiveness, scalability, high-speed and robustness of the proposed floorplanner with various predefined aspect ratios. For future work, it is proposed to extend PPTP to handle some large scaled hard module standard cells and hybrid module (soft and hard) macro cell placements by adapting other tools.

ACKNOWLEDGEMENT

The authors would like to acknowledge ICT Research Cluster, University of Malaya, Kuala Lumpur, Malaysia for the research fund support.

REFERENCES

[1]   G. E. Moore, "Cramming more components onto integrated circuits", *Proceedings of the IEEE*, vol. 86, no. 1, pp. 4, Jan. 1998.

[2]   G. E. Moore, "Progress in digital integrated electronics", *IEEE Press on IEDM Tech. Digest*, 1975, pp. 11-13.

[3]   H. Murata, K. Fujiyoshi, S. Nakatake, and Y. Kajitani, "VLSI module placement based on rectangle-packing by the sequence-pair", *IEEE Transactions on Computer-Aided Design of Integrated Circuits and Systems*, vol. 15, no. 12, pp. 1518-1524, Dec.1996.

[4]   A. Jagannathan, S. W. Hur and J. Lillis, "A fast algorithm for context-aware buffer insertion", *ACM Transactions on Design Automation Electronic Systems (TODAES)*, vol. 7, no. 1, pp. 368-373, Jan.2002

[5]   A. B. Kahng, "Classical floorplanning harmful?", in *Proceedings of International Symposium on Physical Design*, 2000, pp. 207–213.

[6]   T. C. Chen, Y. W. Chang, and S. C. Lin, "A new multilevel framework for large-scale interconnect-driven floorplanning", *IEEE Transactions on Computer-Aided Design of Integrated Circuits and Systems*, vol. 27, no. 2, pp. 286-294, Feb. 2007

[7]   C.-S. Hoo, H.-C. Yeo, K. Jeevan, V. Ganapathy, H. Ramiah and I. A. Badruddin, Hierarchical Congregated Ant System for Bottom-up VLSI Placements, *Engineering Applications of Artificial Intelligence*, 2012, DOI: 10.1016/j.engappai.2012.04.007

[8]   A. E. Caldwell, A. B. Kahng and I. Markov, "Can recursive bisection alone produce routable placements?", in *Proceedings on Design Automation Conference*, 2000, pp. 477-482.

[9]   S. N. Adya and I. L. Markov, "Fixed-outline floorplanning: enabling hierarchical design", *IEEE Transactions on Very Large Scale Integration (VLSI) Systems*, vol. 11, no. 6, pp. 1120-1135, Dec. 2003.

[10] S. Chen, and T. Yoshimura, "Fixed-outline floorplanning: Block-position enumeration and a new method for calculating area costs", *IEEE Transactions on Computer-Aided Design of Integrated Circuits and Systems*, vol. 27, no. 5, pp. 858-871, May 2008.

[11] D. S. Chen, C. T. Lin, Y. W. Wang, and C. H. Cheng, "Fixed-outline floorplanning using robust evolutionary search", *Engineering Applications of Artificial Intelligence*, vol. 20, no. 6, pp. 821-830, 2007.

[12] J. Cong, T. Kong and D. Z. Pan, "Buffer block planning for interconnect-driven floorplanning", in *Proceedings of IEEE/ACM International Conference on Computer-Aided Design*, 1999, pp. 358-262.

[13] Q. Ma, Z. Qian, E. F. Y. Young, H. Zhou, "MSV-driven floorplanning", *IEEE Transactions on Computer-Aided Design of Integrated Circuits and Systems*, vol. 30, no. 8, pp. 1152-1162, 2011.

[14] J. Z. Yan and C. Chu, "Defer: Deferred decision making enabled fixed-outline floorplanning algorithm", *IEEE Transactions on Computer-Aided Design of Integrated Circuits and Systems*, vol. 29, no. 3, pp. 367-381, March 2010.

[15] A. E. Caldwell, A. B. Hahng and I. L. Markov, "Hierarchical whitespace allocation in top-down placement", *IEEE Transactions on Computer-*

Aided Design of Integrated Circuits and Systems, vol. 22, no. 11, pp. 1550-1556, 2003.

[16] J. Cong, M. Romesis, J. R. Shinnerl, "Fast floorplanning by look-ahead enabled recursive bipartitioning", *IEEE Transactions on Computer-Aided Design of Integrated Circuits and Systems*, vol. 25, no. 9, pp. 1719-1732, 2006.

[17] J. M. Lin and Z. X. Hung, "UFO: Unified convex optimization algorithm for fixed-outline floorplanning considering pre-placed modules", *IEEE Transactions on Computer-Aided Design of Integrated Circuits and Systems*, vol. 30, no. 7, pp. 1034-1044.

[18] O. He, S. Dong, J. Bian, S. Goto, and C. K. Cheng, "A novel fixed-outline floorplanner with zero deadspace for hierarchical design", in *Proceedings of IEEE/ACM International Conference on Computer-Aided Design*, 2008, pp. 16-23.

[19] G. Karypis and V. Kumar, "Multilevel k-way hypergraph partitioning", in *Proceedings of Design Automation Conference*, 1999, pp. 343-348.

[20] A. E. Dunlop, and B. W. Kerninghan, "A procedure for placement of standard-cell VLSI circuits," *IEEE Transactions on Computer-Aided Design of Integrated Circuits and Systems*, vol. 4, no. 1, pp. 92-98, 1985.

[21] J. Rabaey. (2005) Gigabyte Systems Research Center [online]. Available at: http://www.cse.ucsc.edu/research/surf/GSRC/progress.html

[22] L. Stanley. (2004) Xilinx acquires hier design, brings industry's fastest, most robust design flow to FPGA designers, Xilinx Press Releases [online]. Available at:

http://www.xilinx.com/prs_rls/xil_corp/0468_hierdesign.htm.

A Compliant Lead-Free Solder Alloy

Mohd Faizul Mohd Sabri, Dhafer Abdul-Ameer Shnawah, Irfan Anjum Badruddin, Suhana Binti Mohd Said

Department of Mechanical Engineering
University of Malaya (UM)
50603 Kuala Lumpur, Malaysia
Email: msfaizul@gmail.com

Abstract- **The effects of 0.5 wt.% Fe addition on the microstructural and mechanical properties of the Sn-1Ag-0.5Cu (SAC105) solder alloy were investigated. The addition of Fe leads to the formation of large FeSn$_2$ intermetallic compound (IMC) particles located in the eutectic regions besides the small Ag$_3$Sn and Cu$_6$Sn$_5$ IMC particles. The addition of Fe also leads to enlarge the primary β-Sn grains and diminish the eutectic regions. The formation of large FeSn$_2$ IMC particles together with the presence of large primary β-Sn grains leads to significantly lower elastic modulus and yield strength values for Fe-bearing SAC105 solder alloy. Moreover, the presence of large primary β-Sn grains causes the Fe-bearing solder alloy to maintain the total elongation at the level of SAC105 solder alloy. This effect can increase the bulk compliance and plastic energy dissipation ability of the solder joint, which play an important role in drop impact performance enhancement.**

I. INTRODUCTION

Increasing environmental and health concern over the toxicity of lead has provided a driving force to ban the use of Pb-Sn solders, and has stimulated to develop lead-free solder alloy [1-4]. The high-Ag-content Sn-Ag-Cu alloys such as Sn-4wt.%Ag-0.5wt.%Cu (SAC405) or Sn-3wt.%Ag-0.5wt.%Cu (SAC305) have been considered promising replacements for the Sn-Pb solder alloy for microelectronics applications based on low available melting temperature, near eutectic composition, and good cyclic fatigue properties [5]. However, due to the rigidity of the high Ag-content Sn-Ag-Cu alloys compared with the Sn-Pb solder alloy, more drop and high impact failures have been observed for these replacement alloys in mobile products such as personal data assistants, cellular phones, and notebook computers, which all require good drop impact reliability [6-11]. The root cause of the poor drop impact reliability of the high-Ag-content SAC alloys lies in the bulk alloy properties. These high-Ag-content alloys have a relatively high elastic modulus and yield strength and small elongation, which in turn result in a stiff bulk solders. Consequently, these stiff bulk solders more readily transfer dynamic stresses to the solder/substrate interface during drop impact loading conditions. The IMC layers formed at solder/substrate interface are of low ductility and exhibit brittle failure [12-14]. Moreover, the cost competitiveness of the high-Ag-content Sn-Ag-Cu alloys is a weak point due to the high cost of Ag [15]. Hence, there is a demand for soft, highly compliant, and more inexpensive alloy as a replacement for the high-Ag-content Sn-Ag-Cu alloys in drop and high impact applications.

Low-Ag-content Sn-Ag-Cu alloys such as Sn-1wt.%Ag-0.5wt.%Cu (SAC105) have been considered as a solution for resolving both issues. [16]. Reducing the Ag content of the Sn-Ag-Cu alloy has been shown to reduce the elastic modulus and yield strength and increase the elongation. This, in turn increases the elastic compliance and plastic energy dissipation ability of the bulk solder, which are key factors to enhancing the drop resistance [17]. To further enhance the drop impact resistance, a family of low-Ag-content Sn-Ag-Cu alloys doped with a fourth alloying element such as Ni, Mn, Ce,Ti, In, Cr, and Al have been investigated [13,18,19]. Generally, the formation and growth of the interface IMC layer, the compliance of the bulk solder, and the wetting properties are the most influential factors in drop impact tests. Fe is a low-cost element, and the use of Fe is considered to be environmentally friendly because Fe is a nonhazardous material. Moreover, recent studies have revealed that Fe-bearing solder alloys significantly reduce the interface IMC layer growth and increase the shear strength of the solder joint [20-25]. The effect of Fe is expected to be significant because the Fe has little solubility in the β-Sn matrix (and vice versa) below 200°C [26-29]. Moreover, the FeSn$_2$ phase may precipitate in the solder [20-25,29]. Finally, the addition of Fe could modify the wetting behavior [30].Therefore, this work investigates the effects of 0.5 wt.% of Fe on the microstructural and mechanical properties of the Sn-1Ag-0.5Cu (SAC105) solder alloy.

II. EXPERIMENTAL PROCEDURES

Bulk solder specimens of Sn-1Ag-0.5Cu (SAC105) and Sn-1Ag-0.5Cu-0.5Fe (SAC105-0.5Fe) with flat dog-bone shapes were used in this study. The dimensions of the gauge sections of the tensile test specimens were 5.0 mm thick × 5.0 mm wide × 21mm long. The alloys were prepared by melting pure ingots of Sn, Ag, Cu, and Fe in an induction furnace at more than 1000°C for 40 min. Then, the molten alloys were mixed with liquid pure Sn in a melting furnace at 290-300°C for 60 min. Subsequently, the molten alloys were cast to disk shaped ingots and sent to a third party lab (SGS) to verify the Fe element concentration. Chemical composition analyses were carried out to determine the exact composition of the casting ingots. Then, the molten alloys were poured into stainless steel molds that were pre-heated at 120-130°C, and the molds were air cooled naturally to room temperature (25°C).

978-1-4673-2395-6/12 $31.00 © 2012 IEEE

The solder bar was set onto a testing grip at two ends of the specimen using a universal tester. An extensometer was secured onto the specimen surface to measure the strain of the solder. In this study, a length of 10 mm was used as a gauge length. The tensile force applied to the specimen was measured by a load cell for stress calculation. A composite channel was used to measure the strain directly from an extensometer during the first part of the test (10%), and then, the strain was calculated from the extension of the crosshead for the remainder of the test. Tensile tests were conducted at room temperature (25°C) and at a constant strain rate of 10^{-3} s^{-1} to investigate the effects of the added Fe on the mechanical properties of the solder, such as the elastic modulus, yield stress, ultimate tensile strength (UTS), and elongation. A scanning electron microscope (SEM) with a backscattered electron detector was used to examine the microstructures. Additionally, energy dispersive X-ray spectroscopy was adopted to determine the phase compositions. Electron backscatter diffraction analysis was also carried out to determine the IMC phases. To obtain the microstructure, the solder samples were prepared by dicing, resin molding, grinding and polishing. The samples were ground with four grades of SiC paper (# 800, #1200, #2400 and #4000) and then mechanically polished with a diamond suspension (3 μm). Finally, the specimens were polished with a colloidal silica suspension (0.04 μm).

III. RESULTS and DISCUSSION

A. Tensile Test

The stress-strain curves of the tested solder alloys are shown in Fig. 1. As illustrated, adding Fe to the SAC105 solder alloy has a significant effect on the mechanical properties. Fig. 2 shows the elastic modulus, the 0.2% proof stresses, the ultimate tensile strengths (UTS), and the total elongation of the tested solder alloys. The addition of Fe decreases the elastic modulus, yield strength, and UTS, whilst the total elongation is still maintained at the level of SAC105 solder alloy. These changes can increase the bulk compliance and the plastic energy dissipation ability of the solder joints. Fig. 3 shows the elastic part of the stress-strain curve of the SAC105 and Fe-bearing SAC105 solder alloys. As illustrated, the Fe-bearing SAC105 solder alloy possesses a higher elastic compliance than the SAC105 solder alloy without Fe; as a result, the stress on the Fe-bearing SAC105 bulk solder is lower than that on the SAC105 bulk solder at the same strain. Thus, the Fe-bearing SAC105 bulk solder with a higher elastic compliance is expected to exhibit longer strain to failure than the SAC105 bulk solder under high strain loading conditions. Consequently, the Fe-bearing SAC105 bulk solder are expected to dissipate more drop impact energy compared with the SAC105 bulk solder.

B. Bulk Alloy Microstructure

Fig. 4 shows the as-cast microstructures of the SAC105 and SAC105-0.5Fe bulk solders. The as-cast microstructure of the SAC105 solder alloy is composed of primary β-Sn grains and

eutectic regions that consist of two IMC particles dispersed within the Sn-rich matrix, as shown in Fig. 4a. Using EBSD analysis, the bright IMC particles (approximately 0.11–0.72 μm) are identified as Ag$_3$Sn, and the gray IMC particles (approximately 0.46–2.90 μm) are identified as Cu$_6$Sn$_5$. These results are consistent with those of other studies [15,17]. The as-cast microstructure of the Fe-bearing SAC105 solder alloy consists of large primary β-Sn grains and eutectic regions of three distinct types of IMC particles dispersed in the Sn-rich matrix, as shown in Fig. 4b.

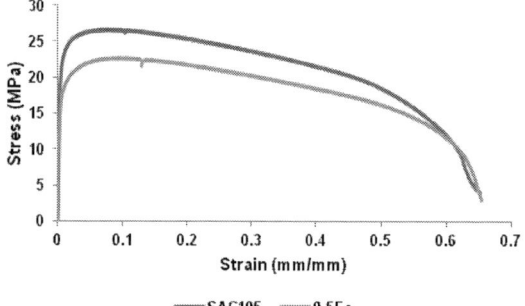

Fig. 1. Stress-strain curves of the SAC105 and Fe-bearing SAC105 solder alloys

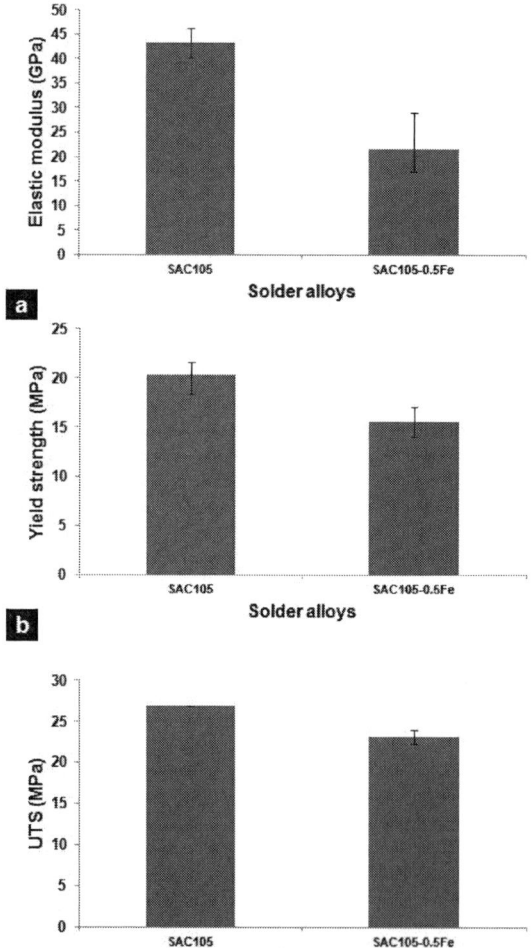

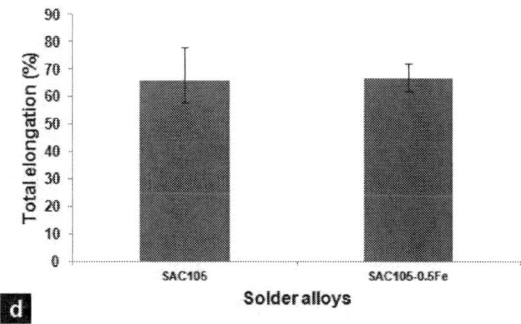

Fig. 2. The elastic part of stress-strain curves of as-cast SAC105 and Fe-bearing SAC105 solder alloys

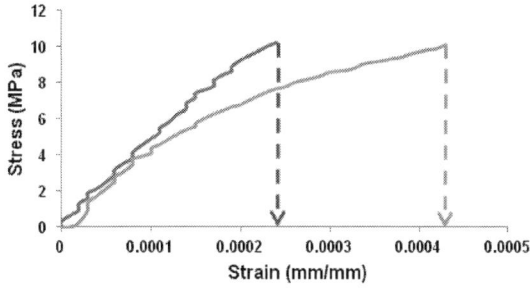

Fig. 3. The elastic part of stress–strain curves of as-cast SAC105 and Fe-bearing SAC105 solder alloys

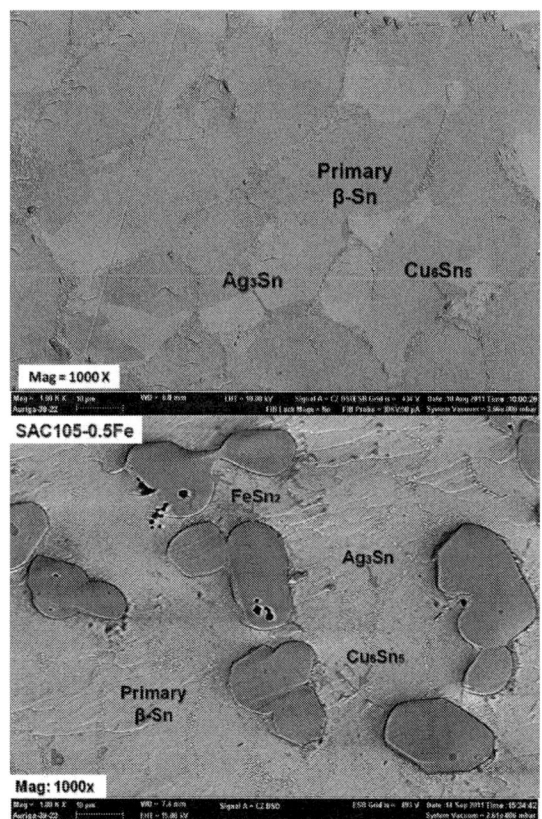

Fig. 4. SEM micrographs of (a) SAC105 solder and (b) Fe-bearing SAC105 solder alloys

The EDS analysis results indicate that the large gray particles (approximately 7.50–21.29 μm) are Fe-Snn75.26 (Fig. 5a), the fine bright particles (approximately 0.095–0.68 μm) are Ag47.49-Sn50.90-Fe01.61 (Fig. 5b), and the small gray particles (approximately 0.61–3.81 μm) are Cu39.42-Sn56.70-Fe03.88 (Fig. 5c). Using EBSD analysis, the large gray particles are identified as $FeSn_2$ IMC, the fine bright particles are identified as Ag_3Sn IMC with a small amount of Fe, and the small gray particles are identified as Cu_6Sn_5 IMC with a small amount of Fe. It is well known that, to a large extent, the microstructural characteristics of an alloy determine its mechanical performance. The Ag_3Sn and Cu_6Sn_5 phases are well known to possess a much higher strength than the bulk material in the SAC alloys whereas the primary β-Sn phase has the lowest elastic modulus and lowest yield strength among the bulk constituent phases of the SAC alloys [12, 31].

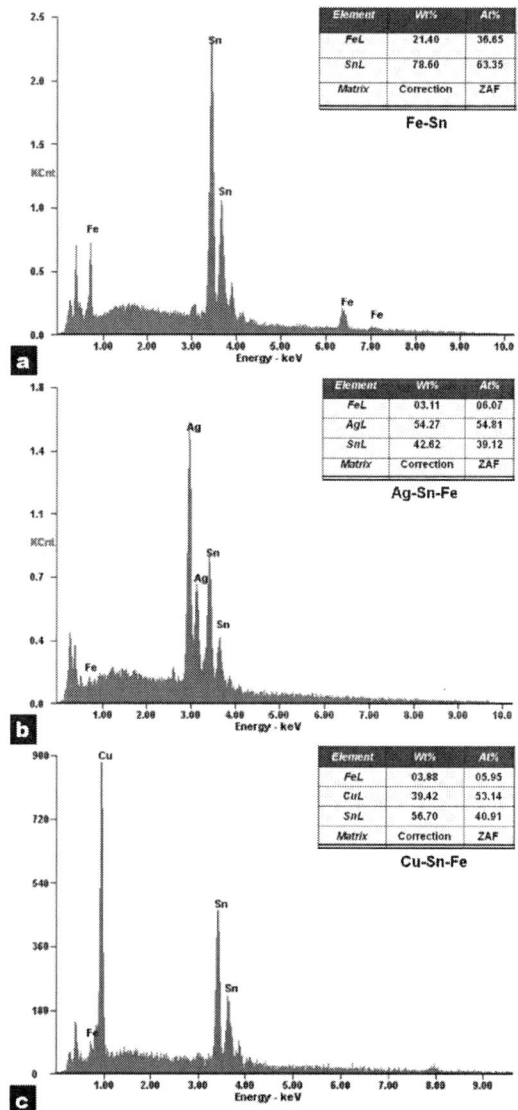

Fig. 5. EDS analysis results of the IMC particles in the Fe-bearing SAC105 bulk solders

Hence, the large amounts of the Ag_3Sn and Cu_6Sn_5 phases increase the elastic modulus and yield strength, which produce a stiff bulk solder. On the other hand, the high fraction of the primary β-Sn phase reduces the elastic modulus and yield strength, which produce a soft and highly compliant bulk solder. However, the volume fraction of Cu_6Sn_5 in the SAC105 solder alloy is much smaller than that of Ag_3Sn. Hence, the effect of Cu_6Sn_5 on the mechanical properties is smaller than that of the Ag_3Sn. Therefore, it is sufficient to consider the effect of Ag_3Sn only. The microstructure of the SAC105 solder alloy, which is shown in Fig. 4a, consists of large primary β-Sn grains and fine Ag_3Sn IMC particles sparsely distributed within the eutectic regions. The large size of primary β-Sn grains together with the sparsely distributed Ag_3Sn IMC particles causes the SAC105 solder alloy to exhibit a low elastic modulus and yield strength and a large elongation, which produce a soft and highly compliant bulk solder. In turn, these properties cause the SAC105 solder alloy to exhibit high bulk compliance solder joint which reduces the dynamic stress transferred to the interface IMC layers during drop impact loading conditions.

The 0.5 wt.% Fe addition to the SAC105 bulk solder leads to the formation of large $FeSn_2$ IMC particles. These large $FeSn_2$ IMC particles are sparsely distributed within the microstructure, as shown in Fig. 6, located in the eutectic regions besides the small Ag_3Sn and Cu_6Sn_5 IMC particles. Moreover, the 0.5 wt.% Fe addition leads to enlarge the primary β-Sn grains and diminish the eutectic regions. Fine particles in alloys are well known to effectively impede dislocation movement and produce an alloy with greater yield strength. When these particles grow in size, the yield strength decreases. In addition, when the coherence of the particles within the matrix is gradually lost with particle growth, the yield strength is further decreased [32]. Thus, the formation of large $FeSn_2$ IMC particles together with the presence of large primary β-Sn grains causes the Fe-bearing SAC105 solder to significantly reduce the elastics modulus and yield strength. Moreover, the presence of large primary β-Sn grains causes the Fe-bearing solder to maintain the total elongation at the level of SAC105. It is believed that the formation of large circular $FeSn_2$ IMC particles located in the eutectic regions does not affect the plastic deformation of the primary β-Sn.

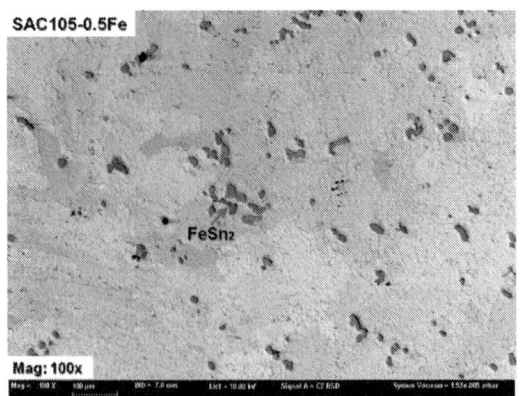

Fig. 6. SEM micrographs of Fe-bearing SAC105 solder alloy

IV. CONCLUSION

1. The addition of 0.5 wt.% Fe leads to the formation of large $FeSn_2$ IMC particles located in the eutectic regions besides the Ag_3Sn and Cu_6Sn_5 IMC particles.
2. The addition of 0.5 wt.% Fe leads to enlarge the primary β-Sn grains and diminish the eutectic regions.
3. The formation of large $FeSn_2$ IMC particles together with the presence of large primary β-Sn grains has a significant reduction on the elastic modulus and yield strength.
4. The presence of large primary β-Sn grains causes the Fe-bearing solder to maintain the total elongation at the level of SAC105 solder alloy.
5. The Fe addition to the SAC105 bulk solder alloy can further increases the bulk compliance of the solder joints, which play an important role in drop impact performance enhancement.

REFERENCES

[1] M. Abtew, G. Selvaduray, "Lead-free solders in microelectronics," *Mater Sci Eng R,* vol. 27, pp. 95-141, 2000.

[2] M.P. Renavikar, N. Patel, A. Dani, V. Wakharkar, G. Arrigotti, V. Vasudevan, O. Bchir, A. P. Alur, C. K. Gurumurthy, R. W. Stage, " Materials technology for environmentally green micro-electronic packaging," Intel Technol J, vol. 12, pp. 1-16, 2008.

[3] M. Amagai, M. Watanabe, M. Omiya, K. Kishimoto, T. Shibuya, "Mechanical characterization of Sn-Ag based lead-free solders," Microelectron Reliab, vol. 42, pp. 951-966, 2002.

[4] R. Plieninger, M. Dittes, K. Pressel, "Modern IC packaging trends and their reliability implications," Microelectron Reliab, vol. 46, pp. 868-1873, 2006.

[5] I.E. Anderson, J.C. Foley, B.A. Cook, J. Harringa, R. L. Terpstra, O. Unal, "Alloying effects in near-eutectic Sn-Ag-Cu solder alloys for improved microstructural stability," J Electron Mater, vol. 30, pp. 050-1059, 2001.

[6] D.A.-A. Shnawah, M.F.M. Sabri I.A. Badruddin, "A review on thermal cycling and drop impact reliability of SAC solder joint in portable electronic products," Microelectron Reliab, vol. 52, pp. 90-99, 2012.

[7] D.A.-A. Shnawah, M.F.M. Sabri I.A. Badruddin, S.B.M. Said and F.X. Che, "The bulk alloy microstructure and mechanical properties of Sn–1Ag–0.5Cu–xAl solders (x = 0, 0.1 and 0.2 wt. %)," J Mater Sci: Mater Electron, in press.

[8] D.A.-A. Shnawah, M.F.M. Sabri I.A. Badruddin, and F.X. Che (2012)," The bulk alloy microstructure and tensile properties of Sn-1Ag-0.5Cu-xAl lead-free solder alloys (x=0, 1, 1.5 and 2 wt.%)," Microelectron Int, in press.

[9] Wong EH, Seah SKW, Shim VPW. A review of board level solder joints for mobile applications. Microelectron Reliab 2008; 48:1747-1758

[10] E.H. Wong, S.K.W. Seah, W.D.V. Driel, J.F.J.M. Caers, N. Owens, Y.-S. Lai, "Advances in the drop-impact reliability of solder joints for mobile applications," Microelectron Reliab, vol. 49, pp. 139-149, 2009.

[11] Zhang B, Ding H, Sheng X. Reliability study of board-level lead-free interconnections under sequential thermal cycling and drop impact. Microelectron Reliab 2009; 49: 530-536.

[12] D. Kim, D. Suh, T. Millard, H. Kim, C. Kumar, M. Zhu, "Evaluation of high compliant low Ag solder alloys on OSP as a drop solution for the 2 level Pb-free interconnection." IEEE.

[13] D.A.-A. Shnawah, S.B.M. Said, M.F.M. Sabri I.A. Badruddin, and F.X. Che, "Microstructure and Tensile Properties of Sn-1Ag-0.5Cu Solder Alloy bearing Al for Electronics Applications." J Electron Mater, vol. 41, pp. 2073- 2082, 2012.

[14] F.X. Che, W.H. Zhu, E.S.W. Poh, X.W Zhang, X.R. Zhang, "The study of mechanical properties of Sn-Ag-Cu lead-free solders with different Ag contents and Ni doping under different strain rates and temperatures." J Alloy Compd, vol. 507, pp. 215–224, 2010.

[15] A.M. Yu, J.W. Jang, J.K. Kim, J.H. Lee, M.K. Kim, "Improved reliability of Sn-Ag-Cu-In solder alloy by the addition of minor elements." IEEE.

[16] W. Kittidacha, A. Kanjanavikat, K. Vattananiyom, "Effect of SAC alloy composition on drop and temp cycle reliability of BGA with NiAu pad finish." IEEE.

[17] Y. Kariya, T. Hossi, S. Terashima, M. Tanaka, M. Otsuka, "Effect of silver content on the shear fatigue properties of Sn-Ag-Cu flip-chip interconnect," J Electron Mater, vol. 33, pp. 321-328, 2004.

[18] D.A.-A. Shnawah, M.F.M. Sabri, I.A. Badruddin, S. Said, "A review on effect of minor alloying elements on thermal cycling and drop impact reliability of low-Ag Sn-Ag-Cu solder joints," Microelectron Int, vol. 29, pp. 47-57, 2012.

[19] D.A.-A. Shnawah, S.B.M. Said, M.F.M. Sabri I.A. Badruddin, and F.X. Che, "High-Reliability Low-Ag-Content Sn-Ag-Cu Solder Joints for Electronics Applications," J Electron Mater, in press.

[20] I.E. Anderson, J.H. Harringa, "Elevated temperature aging of solder joints based on Sn-Ag-Cu: Effects on joint microstructure and shear strength," J Electron Mater, vol.33, pp. 1485- 1496, 2004.

[21] I.E. Anderson, B.A. Cook, J. Harringa, R.L. Terpstra, "Microstructural modifications and properties of Sn-Ag-Cu solder joints induced by alloying," J Electron Mater, vol. 31, pp. 1166-1174, 2002.

[22] D.A.-A. Shnawah, S.B.M. Said, M.F.M. Sabri I.A. Badruddin, and F.X. Che, "Microstructure, mechanical, and thermal properties of the Sn-1Ag-0.5Cu solder alloy bearing Fe for electronics applications," Mater. Sci. Eng. A, vol. 551, pp. 160-168, 2012.

[23] S. Choi, J.P. Lucas, K.N. Subramanian, T.R. Bieler, "Formation and growth of interfacial intermetallic layers in eutectic Sn-Ag solder and its composite solder joints," J Mater Sci Mater Electron, vol. 11, pp.497-502, 2000.

[24] D.A.-A. Shnawah, S.B.M. Said, M.F.M. Sabri I.A. Badruddin, and F.X. Che, "Novel Fe-containing Sn–1Ag–0.5Cu lead-free solder alloy with further enhanced elastic compliance and plastic energy dissipation ability for mobile products," Microelectron Reliab, in press.

[25] I.E. Anderson, "Development of Sn–Ag–Cu and Sn-Ag-Cu-X alloys for Pb-free electronic solders applications," J Mater Sci Mater Electron, vol.18, pp. 55-76, 2007.

[26] T.B. Massalski, *Binary alloy phase diagrams*, Materials Park, OH, USA: American Society for Metals, 1986.

[27] K.S. Kim, S.H. Huh, K. Suganuma, "Effects of fourth alloying additive on microstructures and tensile properties of Sn-Ag-Cu alloy and joints with Cu," Microelectron Reliab, vol. 3, pp.259-267, 2003.

[28] L. Gao, S. Xue, L. Zhang, Z. Sheng, F. Ji, W. Dai, S.-I. Yu, G. Zeng, "Effect of alloying elements on properties and microstructures of SnAgCu solders," Microelectron. Eng., vol. 87, pp. 2025-2034, 2010.

[29] M. Hutter, R. Schmidt, P. Zerrer, S. Rauschenbach, K. Wittke, W. Scheel , H. Reichl, "Effects of additional elements (Fe, Co, Al) on SnAgCu solder joints." IEEE.

[30] P. Zerrer, A. Fix, M. Hutter, H. Reichl, "Solidification and wetting behaviour of SnAgCu solder alloyed by reactive metal organic flux," Solder Surf Mt Technol, vol. 22, pp. 19-25, 2010.

[31] D. Suh, D.W. Kim, P. Liu, H. Kim, J.A. Weninger, C.M. Kumar, A. Prasad, B.W. Grimsley, H.B. Tejada, "Effects of Ag content on fracture resistance of Sn-Ag-Cu lead-free solders under high strain rate conditions," Mater Sci Eng A, vol. 46-461, pp. 595–603, 2007.

[32] G.E. Dieter, Mechanical metallurgy, 2nd ed. Tokyo: McGraw-Hill, 1976.

Temperature Cycling and Thermal Shock Correlation in DPAK & DSO Packages

Lee Chai Ying & Cheong Choke Fei
Infineon Technologies (M) Sdn. Bhd.
P.O. 52, Free Trade Zone,
Batu Berendam, 75350, Melaka, Malaysia.
Email: chaiying.lee@infineon.com

Abstract- Short time to market is important in ensuring the competitiveness of a new semiconductor product. With the need to pass the listed qualification reliability stress test, temperature cycling (TC) is one of the stress tests that required the most time. Therefore, in order to speed up the stress test duration of TC, an initiative to assess the possibility of replacing TC with thermal shock (TS) in development stage was proposed. TS apply the similar failure mechanism model of Coffin Manson as TC. The main differences of both tests used in this study were the medium of the test environment (air-to-air in TC & liquid-to-liquid in TS), dwell time, time to reach specified temperature and load transfer time. With the conscious of the differences, a correlation study of TC to TS was carried out using DPAK and DSO packages as test vehicles. Positive results were reported in this correlation for the defined failure mode of the de-lamination percentage (%) at the interfaces of die pad to EMC, die-attach to lead frame/die pad. As a conclusion, TC H condition is correlated to 2x TS and TC C condition for die pad de-lamination in DPAK and DSO packages. These have been implemented in development stage of a project.

I. INTRODUCTION

With the intention to speed up the development time of new product introduction to market, an initiative to assess the possibility of replacing Temperature cycling (TC) to Thermal Shock (TS) was proposed. In the standard requirements for product qualification, a minimum of TC 1000 cycles with no failures detection is one of the criterions for product release. However, this replacement study is only applicable in development stage, and not for the replacement of TC in standard product qualification due to the obligation to international standards (e.g JEDEC).

TC is a highly-accepted test method to prove the component's reliability and the failure rate is monitored against the number of cycles [1]. The failure mode of interface de-lamination that is able to detect in TC is basically caused by the Coefficients of Thermal Expansion (CTE) mismatch of materials in the tested system. The mismatch causes the different expansion or contraction of the materials at different rates leading to significant stresses

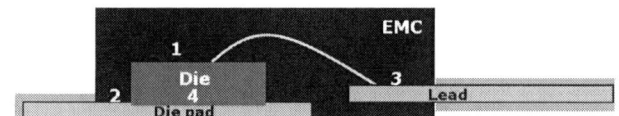

Fig. I. Sectional illustration of TO 252 DPAK with the interfaces identification: 1. Die top to Epoxy Moulding Compound (EMC); 2. Die pad to EMC; 3. Lead to EMC; 4. Die attach to die pad.

at the interface [2]. In this study, the focused failure mode for TC and TS correlation was package de-lamination. Figure I gives the defined 4 de-lamination interfaces. Other failure mechanisms that can be accelerated by TC are die cracking, package cracking, neck/heel/wire breaks and bond lifting will not be covered in this study. Besides, the physical de-lamination of the package, electrical test performance especially in DVSD test parameter was also collected for the correlation investigation to die attach performance in TO 252 DPAK package.

II. EXPERIMENTAL SET UP

A. Stress Test Conditions

3 stress test conditions were employed in this study mainly due to the machine availability in-house for convenient sample handling. Two TC conditions and one TS condition were applied and the details are given in Table 1. Referring to R. Pufall's findings, the acceleration of TC to TS is approximately 2 times or more [1]. Thus, in this correlation study is designed in the ratio of 2TC: TS in number of cycles as shown in Figure II.

B. Test Vehicles Selection Criteria

The test vehicles selection in this study is crucial to reveal the targeted failure modes across the stress test duration to see the correlation. Among the packages available in the development team, DPAK and DSO packages were selected based on the criteria below:
  a) Critical chip to die pad ratio with narrow clearance.
  b) Thick die thickness for higher CTE mismatches possibility.
  c) Origin of the samples must be from the same assembly lot.

978-1-4673-2395-6/12 $31.00 © 2012 IEEE

TABLE I
Test Condition For Stress Test Used

Stress Test Condition	Temperature Gradient	Dwell Time (min)	Load Transfer Time (min)	Time to reach specified temperature (min)
TC C	-65 to 150	10	<1 min	15
TC H	-55 to 150	10	<1 min	15
TS	-55 to 150	2	<10 sec	5

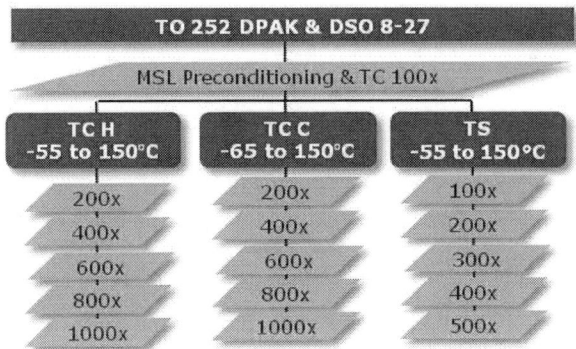

Fig. II. Stress test intervals designed.

C. Analytical Methods for De-lamination Quantification

a. SAM for De-lamination Assessment

Gen 5 Sonoscan Scanning Acoustic Microscopy (SAM) was used for the de-lamination assessment before and after each interval with minimum 10pcs of samples. Fig. III shows the examples of scanning result at the interested 4 interfaces.

b. Image J Setting Condition & De-lamination % Calculation

Image J software was used to calculate the de-lamination percentage regardless of severity (red shows more severe de-lamination and yellow gives less severe de-lamination). Three representative units were taken from each group for de-lamination percentage calculation. The definition of representativeness is given in the guidelines below that were used as the setting condition in Image J in order to get reasonable data:

a) Representativeness: The highest and lowest de-lamination percentages were not included for calculation. This was because some of the de-lamination severity reached saturation at certain interval and no de-lamination propagation observed.

b) Constant Selection Boundary: The exact same selected dimension was applied for all the units

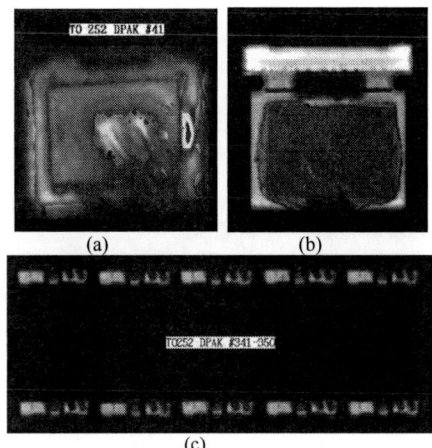

Fig. III. (a): The interfaces of die top/EMC & die pad/EMC were captured together in the same SAM image. The red arrow shows the minor de-lamination at the die pad/EMC interface; (b) SAM image taken from the back of the package for die attach de-lamination performance. No de-lamination was detected in this particular unit; (c) The SAM image focused at the leads of 10 units at once with no de-lamination detected.

calculated as shown in Figure IV (a). As the results of de-lamination percentage depend on the area selected, the consistency of dimension selected is crucial in getting valid data.

c) Similar Brightness & Contrast setting: ~90% brightness and contrast were used for the selected images as to compensate the darker (red) area of de-lamination (as shown in Figure IV (b) & (c)).

c. DVSD Electrical Testing

TO 252 DPAK samples were also taken for Delta Voltage Source Drain (DVSD) electrical testing with the attempt to correlate with die attach performance. The electrical test was conducted using thermal tester.

III RESULTS AND DISCUSSION

A. Die Pad De-lamination Correlation For TO 252 DPAK

Average value was calculated from the de-lamination assessment using Image J software. The data was then plotted into line graph as shown in Fig. V. From the graph, increased de-lamination percentage trend is observed for all the stress test conditions. Positive correlations are also demonstrated in the TC C condition and TS to TC H condition. The de-lamination percentage of 200x TC C is similar to 200x TS, also it is equivalent to the de-lamination percentage of TC H in the range of 71-74%. The accelerated factor is approximately 2 times of TC C condition and TS to 1x TC H condition in terms of number of cycles.

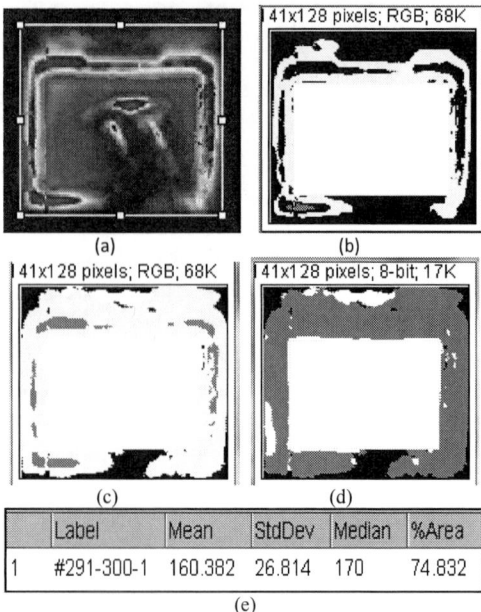

(a)

(b)

(c)

(d)

	Label	Mean	StdDev	Median	%Area
1	#291-300-1	160.382	26.814	170	74.832

(e)

Fig. IV (a): Same selection boundary for all units calculation; (b) Normal brightness and contrast image before adjustment; (c) ~90% brightness and contrast image after adjustment to compensate the dark (red) area of de-lamination; (d) The image ready for de-lamination percentage calculation after threshold adjustment in the software; (e) The calculated result of 74.832% de-lamination percentage.

B. Die Pad De-lamination Correlation For DSO 8-27

Similar observation is given in DSO 8-27 package where good correlation of de-lamination percentage was demonstrated. As shown in Fig. VI, the de-lamination percentage is increased from 40% to 78% for all the stress test conditions. 200x of TC C and TS are showing quite the same die pad de-lamination percentage of 45-47% as detected in TC H condition. The correlation shown in TC C and TS are in 50% reduction of number of cycles to TC H.

C. Die Attach De-lamination Correlation For TO 252 DPAK

Minor die attach de-lamination was detected in TO 252 DPAK package. However, the increased trend of the shown de-lamination for die attach is observed and the possible correlation is conducted in the similar way as done in die pad de-lamination. Fig. VII shown the results of die attach de-lamination percentage only at 2 points in the designed intervals. No die-attach de-lamination is observed at number of cycles below 300x for TS and 600x for TC H. TC C only shows die attach de-lamination at 400x unlike TS. This observation might indicate TS condition is somehow more stringent to the die attach material than TC C due to its shorter dwell time.

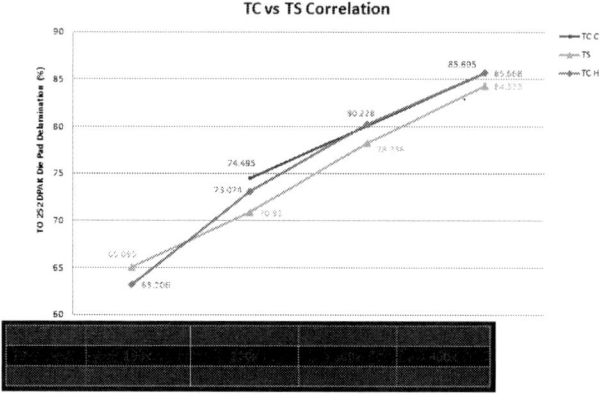

Fig. V Correlation results of die pad de-lamination percentage for TC C, TS and TC H condition in TO 252 DPAK with cycles given in x-axis.

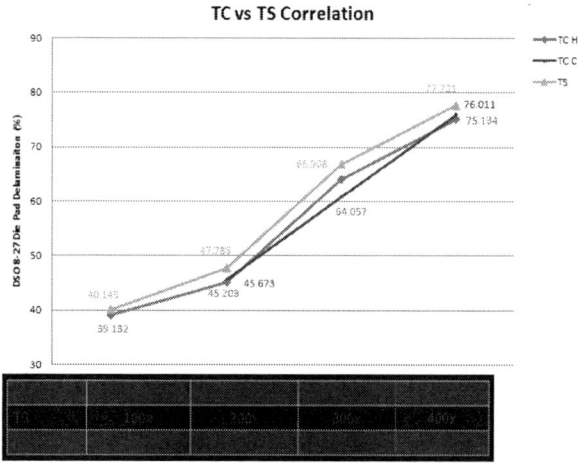

Fig. VI Die pad de-lamination percentage result for TC C, TS and TC H conditions with the cycles is given in x-axis.

The die-attach de-lamination percentage is correlated the same increasing trend as in die pad de-lamination against the increased number of cycles. The de-lamination percentage of TC C and TS shown in the range of 14-16% as detected in TC H condition. Again, 50% reduction in number cycles for TC C and TS to TC H is demonstrated.

Additional observations from the results are reported in this correlation study. As shown in Fig. V and Fig. VII, TC C seems to have similar failure behaviour to TC H for TO 252 DPAK package based on the observation of smaller deviation of de-lamination percentage. In the test vehicles of DSO 8-27, TS tends to have similar failure behaviour to TC H according to the die pad de-lamination percentage values. The different observations as mentioned is believe due to the different die attach materials used in the package.

978-1-4673-2395-6/12 $31.00 © 2012 IEEE

D. Die Attach De-lamination Correlation For DSO 8-27

Unlike the above mentioned correlation, no significant trend is observed in the die attach de-lamination percentage for DSO package. Fig. VIII shows no obvious die attach de-lamination is detected for all stress test conditions regardless of number of cycles. Thus, no possible de-lamination correlation is obtainable in this case.

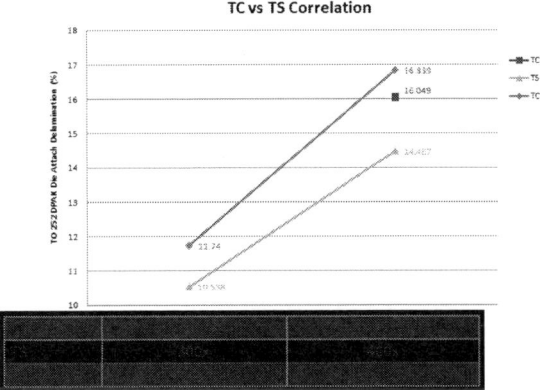

Fig. VII Die attach de-lamination percentage results for TC C, TS and TC H conditions.

E. Results of De-lamination at Other Interfaces

Minor de-lamination is detected for lead de-lamination at TO 252 DPAK, but no correlation is visible. Besides, no de-lamination is detected in the SAM assessment at the interface of die top to EMC for both packages. This observation might indicate good quality of adhesion at this interface.

F. Results of DVSD Electrical Testing

Sample size of 10 for all the intervals of TC C, TS and TC H stress test were tested to see the DVSD performance correlation to die-attach de-lamination percentage. Weibull distribution of DVSD failure probability is plotted as shown in Fig. IX. From the plot, DVSD of TC C fails first, followed by TS and lastly failure shown in TC H. The β values of the conditions show similar failure behaviour ($\beta>1$) for all the stress test condition, where TC C gives 3.8159, TS shows 1.3685 and TC H has 2.9353. The values of $\beta>1$ observation suggests that the failure is occurred at wear-out phase of die-attach material that might be shown in die attach de-lamination in this case.

As refer to the die-attach de-lamination percentage in Fig. VII, TC C shows the higher die attach de-lamination percentage than TS at 400x cycles. This phenomenon is consistent in the DVSD test performance, where TC C first gives the failure than TS. Therefore, it is acceptable that the DVSD test performance depends on the die attach de-lamination percentage. Higher the die attach de-lamination percentage might leads to higher DVSD failure rate.

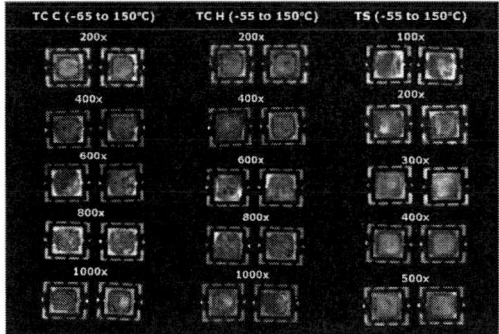

Fig. VIII No die attach de-lamination is detected in TC C, TS and TC H conditions for DSO 8-27.

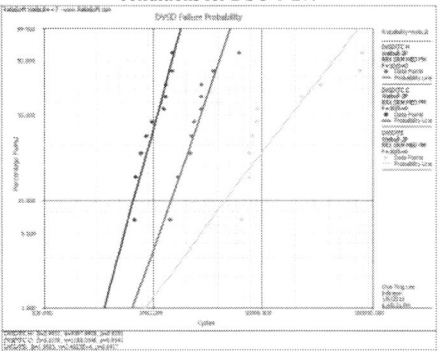

Fig. IX DVSD test performance of TC C, TS and TC H conditions for TO 252 DPAK.

IV. CONCLUSION

Positive correlation results in this study suggest the possibility of replacing TC with TS for DPAK & DSO packages with the focused failure mode of de-lamination at the interface of die pad to EMC and die attach to die pad/lead frame. This possible replacement is recommended to be held in development phase and not in product qualification due to obligation to international JEDEC standards.

FUTURE WORK & ACKNOWLEDGEMENT

Further work or continuation for this study is still needed for correlation validity and data verification. Besides, the inclusion of material analysis or simulation correlation to failure behaviours that are shown in TC C, TS and TC H is also recommended with more intervals of number of cycles. Besides, the authors would like to thank Mr. Adrian Tong SH for management support & reliability lab colleagues.

REFERENCES

[1] R. Pufall, W. Kanert, S. Aresu & M. Goroll, "Reduction of Test Effort. Looking for More Acceleration for reliable components for automotive application," *Microelectronics Reliability,* vol. 48, pp. 1490-1493, 2008.

[2] Alex MacDiarmid, Quanterion Solutions Inc.,"Temperature Cycling Failures-Part Two of Two," *The Journal of The Reliability Information Analysis Center*, July 2011.

Mechanism and Improvement of Breakdown Degradation Induced by Interface Charge in UHV Device

Md. Imran Siddiqui[a], Abijith Prakash[a], Mohammed Sadique Anwar[a], Gene Sheu[a], P A Chen[b]

[a]Department of Computer Science and Information Engineering, Asia University, Taichung, Taiwan, ROC
[b]Nuvoton Technology Corporation
500, Lioufeng Rd., Wufeng, Taichung 41354, Taiwan, R. O. C
Tel: +886-4-2332-3456 #1784, Fax: +886-4-23316699
Email: mdimrans1@gmail.com

Abstract- This paper presents an innovative p-top engineering to simulate and optimize the breakdown degradations in different regions of the interdigitated layout such as source center(SC), drain center(DC), and flat region of an Ultra high voltage(UHV) device. In manufacturing of UHV device, breakdown voltage degradation takes place due to interface charges, current crowding and breakdown degradation was also observed at wafer-stage with temperature stress resulted from package level reliability tests. Optimizations are done to sustain high breakdown voltage by varying the p-top mask design to investigate the interface charge effect on breakdown. ESD test is also conducted to show the difference in interface charges after stress. A better stability has been obtained for maximum p-top length structure with respect to breakdown and ESD testing.

I. INTRODUCTION

Double-RESURF technology is very well known for obtaining high breakdown voltages [1-2]. Optimizations and manufacturing of this technology is important due to various factors affecting its performance. One of the important factor is the breakdown voltage degradation due to charge balance issues, ESD strike and layout effects reported earlier [3]. The effect of drift length variations on breakdown characteristics is significant for an UHV device [4-5]. P-top engineering also being an important factor for D-RESURF technology must be optimized to avoid charge balance issues [6]. We investigate the effect of interface charge on device degradation. The p-top length variations and its effect on breakdown and interface charges have been demonstrated. Different regions of the layout like flat, source center and drain center are studied for their reliability on interface charges. It's been observed that for lower p-top lengths the degradation phenomenon due to interface charges is more severe. Varying p-top length to get better stability with respect to breakdown and interface charges has been the ploy for our analysis.

Previous studies have shown high-field stresses induce oxide-trapped charges in the oxide film that can result in threshold voltage shifts, excess leakage currents, and degradation of oxide breakdown [8-12]. Assessing ESD threats at various stages of manufacturing is important [13]. We report the impact of a 2kV HBM strike on interface charges for the proposed UHV device Flat region.

Technology Computer Aided Design (TCAD) tool such as Sdevice are used to simulate the ESD characteristics of the present device structure. Transient and circuit simulation are used in order to create the HBM pulse. Deliberate insertion of interface charges by Sdevice simulator is done in order to see effect of changes in the same after ESD strike. The output interface profiles are plotted by the Sdevice simulator and distribution of interface charges at source side and drain side are evident.

II. RESULTS AND DISCUSSION

A. Device Structure

Fig. 1 depicts our proposed LDMOS structure based on high resistivity (100 ohm-cm) p-type <100> wafer. The figure shows different regions of the device such as N-Epi, N-drift, P-well (p-type well), N-drift (n-type drift), P-top and source/drain/bulk regions. N-Epi region of 5μm is grown over low doped P-Substrate. Subsequently the drift region implant is performed followed by high temperature drive in. P-top mask design and optimized p-top dose are chosen for maintaining charge balance condition in the LDMOS structure. P-top implantation after pad oxidation is done to complete the RESURF structure. A thick field oxide acts as isolation between the active regions of the device.

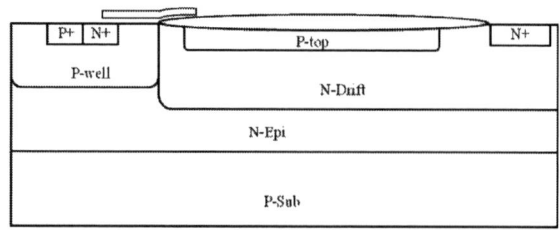

Fig. 1. A schematic cross section of LDMOS with P-top structure.

The gate region is formed by 800Å oxide and n-type doped polysilicon. The P+ and N+ implants for source/drain regions follow with aluminum contacts for connection.

The p-top of the above device is varied and several optimizations are done to obtain the desired breakdown characteristics in the flat region, source center and drain center of the device [14]. Source center and drain center curvatures are simulated using Medici simulator.

B. Interface charge analysis: Flat region, source center and drain center

The breakdown voltage degradation due to interface charges has been demonstrated in this paper. Taking 55um p-top as an initial case, we observe the breakdown characteristics. To meet the desired specification of breakdown when interface charges are inserted, different lengths of p-top are considered.

Working with 55um p-top structure to meet the desired specifications was the initial step. In order to have desired breakdown specification the device must satisfy and sustain breakdown in all the three cases of the device structure. Observing the interface characteristics at 55um we see degradation phenomenon to be very abrupt when interface charges are increased.

Fig. 2 shows the breakdown degradation in case of shorter lengths (55um, 65um) p-top structures (Flat region). As it shows the breakdown degradation has abrupt changes when the interface charges are increased to higher values. As we vary the p-top lengths to higher values (70um, 75um, 80um) it shows lesser degradation. Breakdown voltage drops to about 720V at 1.5e11/cm² interface charges in case of flat region in 55um case and improves up to 950V in 80um case.

The source center and drain center curvature regions have been simulated. Fig.3 shows source center region degradation with respect to different p-top lengths. Its breakdown at 1.5e11/cm² is minimum 660 V for 55um case and maximum 880V for 80um.

Similarly drain center region is simulated at different interface charges and 55um, 65um, and 70um p-top lengths which showed abrupt degradation down to 590V~ 600V respectively. Fig. 4 shows its breakdown sloping down at different interface charges.

Considering 1.5e11/cm² to be the maximum criteria for which breakdown decreases we plot the breakdown vs. p-top lengths of these regions of the devices is shown together in the same scale in Fig. 5. As seen the breakdown of these regions improves as the interface charges are increased. As shown

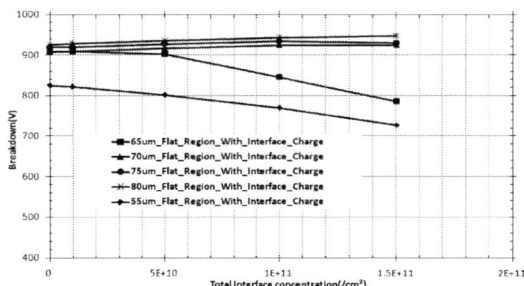

Fig. 2.Breakdown versus total Interface traps concentration with different p-top lengths for Flat region.

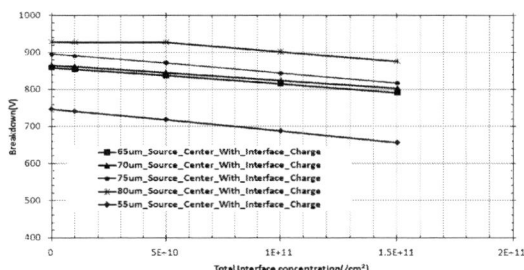

Fig. 3. Breakdown versus total Interface trap concentration with different p-top lengths for SC region.

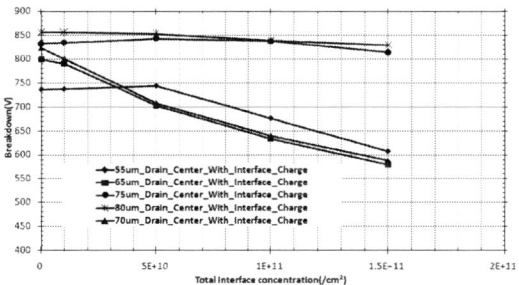

Fig. 4. Breakdown versus total Interface trap concentration with different p-top lengths for DC region.

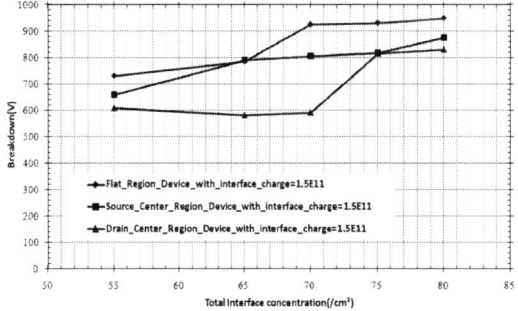

Fig. 5. Breakdown versus p-top lengths for interface trap concentration=1.5e11/cm².

978-1-4673-2395-6/12 $31.00 © 2012 IEEE   463

flat region, source center and drain center having 950V, 875V and 830V respectively at $1.5e11/cm^2$ interface charge concentration.

C. ESD

A study on the cause of interface charges has been done. A variation of interface charges after an ESD strike of 2kV is shown. Deliberate insertion of input interface charges is done by Sdevice simulator with a charge concentration of $1e8/cm^2$. The distribution of charges at source side and drain side regions have been investigated after an ESD strike. Device simulations are carried out to gain physical insight and to determine the main region where interface charges are spread out after the ESD strike. The measurements are taken with the device gate, source and bulk connected to ground and drain with ESD strike of 2kV.

Note: Source center and source side are different entities. Source side is w.r.t to ESD study.

The interface distribution at source side and drain side of the device after ESD strike is shown in Fig. 6 & Fig. 7.The peak interface charge distribution takes place for the least p-top length (only at source side). As shown in Fig. 6 more variation is seen near the source side region in comparison to the drain side region distribution which is shown in Fig. 7. The peak interface concentration takes place at the bird's beak area near the drain which can be correlated to peak points which is also shown in Fig. 7. The ratio of variation in interface charge concentration when compared to 55um & 80 um p-top is more near the source side compared to the drain side. This shows source side of the device being more prone to degradation due to interface charges. Interface charges generated after ESD can also affect the breakdown characteristics of the UHV device. Further the impact ionization points after breakdown is discussed for different interface charges.

As the input interface charges increase the impact ionization shifts toward the source side. Hence the interface charge distribution is more at the source side. The impact ionization shift for a device structure showing flat region with 55um p-top length is shown in Fig. 8 for different interface charges. The device depicted is after breakdown and its impact ionization points are shown.

Also Fig. 9 & Fig. 10 show the interface charge distribution at the drain side and source side before and after ESD strike taking 70um p-top as an example (The interface charge distribution before ESD remains same for different lengths of p-top due to same approach used for interface charge insertion during simulation).

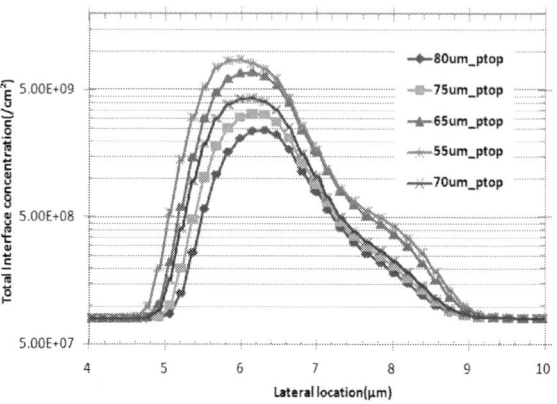

Fig. 6.Total output Interface charge distribution at the source side after ESD strike of 2kV and input interface charge concentration of $1e8/cm^2$

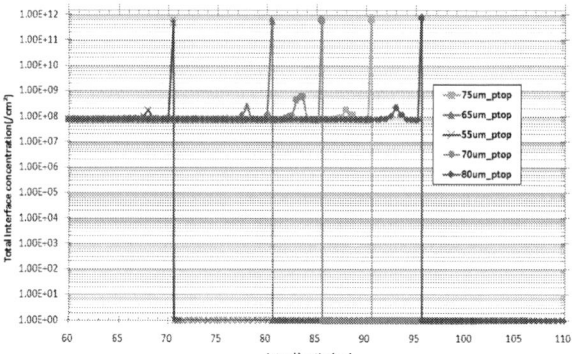

Fig. 7. Total output Interface charge distribution at the drain side after ESD strike of 2kV and input interface charge concentration of $1e8/cm^2$.

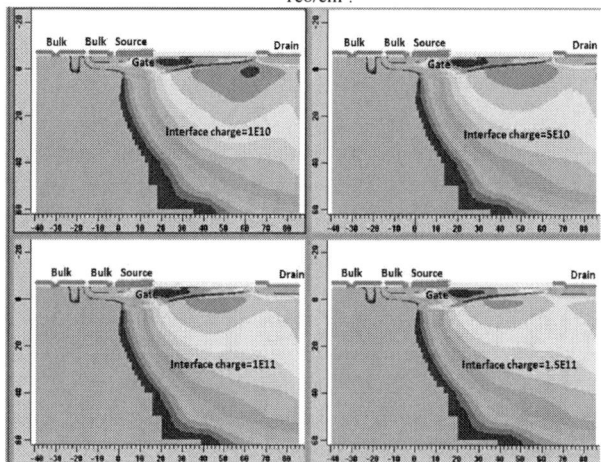

Fig. 8.Cross-sectional view of 55um P-top length Flat region device showing Impact ionization shift with increasing interface charge concentration.

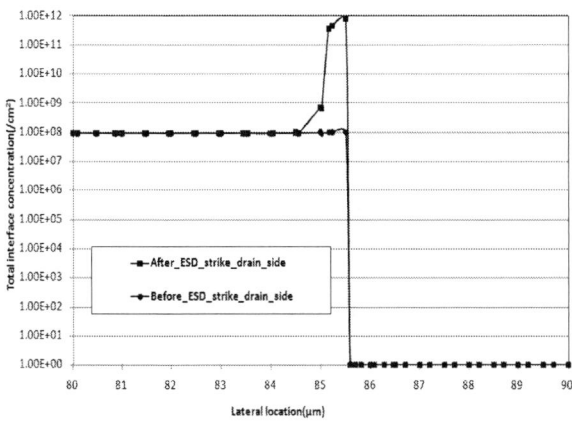

Fig. 9. Interface charge distribution of a 70um p-top structure at drain side before and after ESD strike.

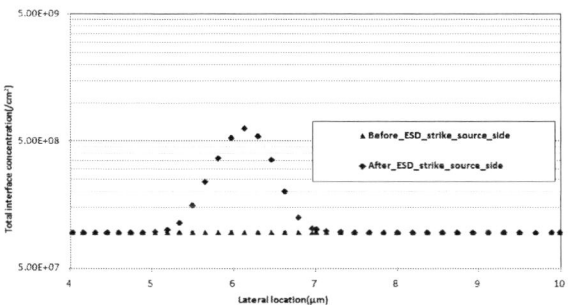

Fig. 10. Interface charge distribution of a 70um p-top structure at source side before and after ESD strike.

III. CONCLUSION

In this paper, we have demonstrated the interface charge effects on the breakdown of the proposed LDMOS structure. Varying the p-top lengths to higher values shows better breakdown stability. The flat region, source center and drain center regions have been simulated to check for their stability on insertion of interface charges. Maximum p-top length 80um proves to have better stability compared to others. The ESD strike causes variation in interface charges near the source side and drain side. Peak output interface charge concentration is seen at the source side and drain side of the device. Variation of interface charge concentration at source side is different for different p-top lengths while it's a constant peak for drain side region. This variation at the source side might cause the degradation in breakdown after

ESD stress. The impact ionization also shifts towards the source side as the interface charge concentration increases.

ACKNOWLEDGMENT

The authors would like to thank Nuvoton Technology Corporation engineers for their support and helpful discussions.

REFERENCES

[1] J. appels, M.c Collet, P. Hart, H. Vaes, J.Verhoeven, "Thinlayer HV-devices," Philips Journal Research. 35, 1980, pp. 1-13.

[2] AdriaanW.Ludikhuize,"A review of RESURF technology,"ISPSD'2000, pp. 11-18.

[3] Tsung-Yi, P. Y. Chiang, C. C. Huang, P. C. Huang, K. H. Huo, R. Y. Su, J. R. Shih, Fu-Hsin Chen, Clair Chen, Ken Chen, C. C. Chen, S. L. Hsu, Mingo Liu, J. Gong, Chun-Lin Tsai , "Mobile Charge Induced Breakdown Instability in 700V LDMOSFET," International Symposium on VLSI technology, system and applications, 21-23 April 2008, P.105-108.

[4] S.Merchant, E. Arnold, H. Baumgart, R. Egloff, T. Letavic, S. Mukherjee, and H. Pein, "Dependence of breakdown voltage on drift length and buried oxide thickness in SOI RESURF LDMOS transistors," in Proc. 5th Int. Symp. Power Semiconductor Devices and IC's, 1993, pp. 124–128.

[5] L. Vestling, J. Olsson, and K.-H. Eklund, "Drift Region Optimization of Lateral RESURF Devices," Solid-State Electronics, Vol. 46, No. 8, pp. 1177-1184 (2002).

[6] Bo Zhang, Wenlian Wang, Wanjun Chen, Zehong Li, and Zhaoji Li, "High-Voltage LDMOS with Charge-Balanced Surface Low On-Resistance Path Layer," Electron Device Letters, IEEE, Aug. 2009, Vol.30 Issue: 8, pp.849-851.

[7] M. S. Liang, C. Chang, Y. T. Yeow, C. Hu, and R. W. Brodersen, "MOSFET degradation due to stressing of thin oxide," IEEE Trans. Electron Devices, vol. ED-31, no. 9, pp. 1238–1244, Sep. 1984.

[8] M. S. Liang, S. Haddad, W. Cox, and S. Cagnina, "Degradation of very thin gate oxide MOS devices under dynamic high field/current stress," in IEDM Tech. Dig., 1986, pp. 394–398.

[9] M. Kimura and H. Koyama,"Stress-induced low-level leakage mechanism in ultrathin silicon dioxide films caused by neutral oxide trap generation," in Proc. IEEE Int. Reliab. Phys. Symp., Apr. 1994, pp. 167–172.

[10] R. S. Scott, N. A. Dumin, T. W. Hughes, D. J. Dumin, and B. T. Moore, "Properties of high-voltage stress generated traps in thin silicon oxide," IEEE Trans. Electron Devices, vol. 43, no. 7, pp. 1133–1143, Jul. 1996.

[11] J.C.Tseng and J.G.Hwu*, "Effects of Electrostatic Discharge (ESD) High-Field Current Impulse on Oxide Breakdown," Journal of Applied Physics, Vol.101, No.1, PP. 014103-1~014103-6, Jan. 2007.

[12] J.C.Tseng and J.G.Hwu*, "Oxide Trapped Charges Induced by Electrostatic Discharge (ESD) Impulse Stress," IEEE Transactions on Electron Devices, Vol.54, No.7, PP.1666~1671, Jul. 2007.

[13] Hugh Hyatt, ESD: Standards, Threats and System Hardness Fallacies, in EOS/ESD Symposium Proceedings, 2002, pp. 175-181.

[14] Hutomo Suryo Wasisto, Gene Sheu, Shao-Ming Yang, Rudy Octavius Sihombing, "A Novel 800V Multiple RESURF LDMOS Utilizing Linear P-top Rings," TENCON 2010-2010, 21-24 Nov. 2010,pp.75-79.

978-1-4673-2395-6/12 $31.00 © 2012 IEEE

Design of a Low Voltage Charge Pump Circuit for RFID Tag

Kang Cheng Wei, M. B. I. Reaz, Md. Syedul Amin, Jubayer Jalil, Labonnah F. Rahman

Department of Electrical, Electronic and Systems Engineering

UniversitiKebangsaan Malaysia

43600 UKM Bangi, Selangor, Malaysia

Email:mamun.reaz@gmail.com

Abstract – **Charge pump circuit is widely used in many systems due to its low power consumption, high performance, small area and low current drivability. This paper presents a low-voltage, high performance charge pump circuit suitable for low-voltage applications such as EEPROM of Radio Frequency Identification (RFID) tag. Designed in 0.18-μm CMOS process, the proposed charge pump circuit is able to pump an input voltage of 1.8V to a measured output of 5.95V through 20MHz clock signal with each pumping capacitor of 0.1pF and smoothing capacitor of 0.1pF at the output. Simulation result shows that the proposed charged pump circuit offers higher pumping gain compared with the existing charge pump circuit. Besides the RFID tag, the charge pump circuit can also be used in other memory circuits.**

Keywords – charge pump, CMOS, EEPROM, NVM, RFID

I. INTRODUCTION

RFID is a very familiar and emerging technology which is being used everywhere nowadays. Storing and reading the data without getting in touch with or involving contact between the tag and reader makes RFID technology a great application. An RFID tag, also known as transponder is attached to an object for identification and tracking by storing the identification data. Transponder can be applied to or attached into a product, animal, or even a person [1, 2]. The tag contains an electronic microchip which is fabricated as a low power integrated circuit (IC). The tag memory may consist of ROM, RAM, non-volatile memory (EEPROM or Flash memory) and data buffers depending on the device functionality [3].

Embedded Non-Volatile Memory (NVM) is mostly used as tag memory. The NVM has received much attention as it can be broadly applied into RFID tag, System on Chip (SoC), microcontroller unit, FPGA systems etc. On the contrary, fabrication of NVM such as EEPROM and Flash memory requires special multipolysilicon processes and multioxidation for thin SiO2 layers. Many masks are needed which result in lower yield, higher cost, lower reliability and longer process turnaround time compared to standard CMOS technology [4]. Many researchers [4-7] took these challenges and developed NVM in a standard CMOS logic process since they have the advantages of lower cost and lower power dissipation. On the other hand, the maintenance and endurance characteristics due to the nMOS tunneling junction [4] or the single-ended memory cell architecture with a too thin oxide [6] are unsatisfactory. It has large bit/area and consumes much power as each bit cell includes its own high voltage switch [5, 7].

Hence, an internal high-voltage generator circuits such as voltage doubler or charge pumps are required to supply these high voltages [8].

Charge pump circuit is applied in NVM, which functioned to generate a higher dc voltage from the supply voltage. As such, there will not be any necessity to use an external voltage regulator, which avoided the need of an external rail [9]. Charge pump circuit is widely used in many systems due to its low power consumption, high energy efficiency, small area and low current drivability. These features are suitable for the NVM in Radio Frequency Identification (RFID) tags [10]. They are useful to EEPROM or Flash memories, DC-DC converters and power management chips. Basically, charge pump circuit is a capacitor based circuit, where voltage is pumped-up stage by stage depending on each stage voltage gain. Most of the charge pumps are based on the Dickson charge pump circuit [10,11]. Diode connected nMOS and pumping capacitors are the main components of this charge pump circuit. Since there are threshold voltage and the body effect of the nMOS, the circuit needs to sustain large voltage loss, which leads to low voltage gain per stage. To develop higher pumping efficiency, charge pump circuit is modified by Jongshinet et al. [12]. However, threshold voltage and parasitic capacitance are still limiting the pumping efficiency [13].

This paper presents an improved charge pump circuit which is suitable for the NVM in RFID tags. It has four pumping stage without the parasitic capacitor to minimize the power dissipation. This paper is organized with the architecture and operation principles of different charge pump circuits, following by design and simulation of modified circuit. Finally, the design is compared with other charge pump circuits and conclusion is drawn.

II. ARCHITECTURE

In this section, the existing charge pump circuits, such as Luca Mensiet et al. [14] with new integrated charge pump architecture using dynamic biasing of pass transistors, Jingjing Che et al. [15] with ultra-low-voltage low-power charge pump are discussed.

A. Dynamic Biasing of Pass Transistors (DBPT)

This charge pump architecture is based on pMOS pass transistor with dynamic biasing of gates and bodies. The voltage loss due to the device threshold is removed by

978-1-4673-2395-6/12 $31.00 © 2012 IEEE

controlling the gate and body voltages of each pass transistor, leading to negligible voltage drop through each pumping stage.

In Fig. 1, the charge transfer block is behaving as a diode of the traditional Dickson charge pump. Similarly, the whole charge pump architecture is regulated by two complementary clock signals. The active control of Charge Transfer MOSFET (CTM) is based on two auxiliary pMOS M_{b1} and M_{b2}. This active control can reduce the body effect onto the threshold voltage of each charge transfer block [14].

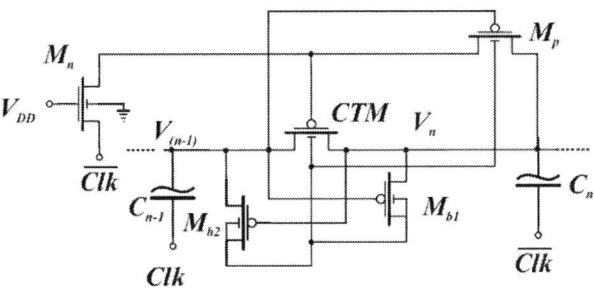

Fig. 1. DBPT charge transfer block architecture per stage

B. Ultra-Low-Voltage Low-Power (ULVLP)

Jingjing Cheet et al. proposed a new charge pump by using two symmetrical branches and pMOS-only transfer transistors per stage. Fig. 2 shows the schematic of the new proposed four-stage charge pump, which contains charge transfer branch A and branch B. The difference between these two branches is the clock signals, where their phases are in opposite to each other. Thereby symmetric nodes in branch A and branch B are charged alternately and both have the same amplitude but with opposite phase. There are two steps in this design. The first part is by using the voltage difference of symmetrical nodes to control the transistors. The second part is using only pMOS to overcome the body effect [15].

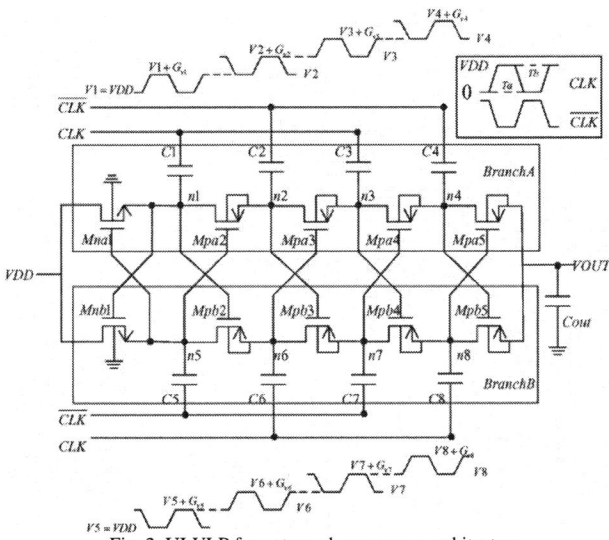

Fig. 2. ULVLP four-stage charge pump architecture

III. PROPOSED CHARGE PUMP CIRCUIT

In this work, pass transistors MNs and MPs have been used to dynamically control the inputs for the CTSs. Hence, the CTSs in this modified circuit can be turned on easily by the backward control mechanism. At the same time, they also can be turned off completely when required. To boost the charge, both clock signals CLKA and CLKB are out-of-phase but with the amplitudes of VDD.

Fig. 3 shows the proposed charge pump architecture. The operation of the improved CTS charge pump circuit is explained as below. For condition where CLKA is high and CLKB is low, voltages at node 1 and node 2 are the same, V_2 while voltage at node 3 is above $2\Delta V$ of voltage at node 1. If

$$2\Delta V > V_{tp} \text{ and } 2\Delta V > V_{tn}(V_2) \tag{1}$$

where V_{tp} is the threshold voltage of pMOS (MPs).

Both MP2 and MS2 are turned on by the voltage at node 3. At this phase, MN2 is always off since the gate-to-source voltage is zero.

For condition where CLKA goes low and CLKB goes high, voltage at node 1 is V_1. Both the voltages at node 2 and node 3 are above $2\Delta V$. If

$$2\Delta V > V_{tn}(V_1) \tag{2}$$

where V_{tn} is the threshold voltage of nMOS (MNs).

In this period, MN2 is turned on and MP2 is turned off as MS2 is free from the control of node 3. For successful operation of this modified charge pump circuit, both the equations 1 and 2 are necessary. For each pMOS, drain node of each individual well is connected. When CLKA or CLKB goes from high to low, the charges at the CTSs gate node during the short period of transition can be injected into the well.

IV. SIMULATION RESULT

The design of the proposed charge pump is verified using ELDONET simulator of Mentor Graphics. Generally, the voltage amplitudes of CLKA and CLKB are same as the power supply voltage (VDD). The simulation parameters used is at pumping clock frequency 20MHz; pumping capacitor with 0.1pF; smoothing capacitor with 0.1pF; input voltage (VDD) ranged from 1.8V to 3.3V; number of pumping stages ranged from 2 stages to 8 stages.

Fig. 4 shows the measured output voltage for the proposed four-stage charge pump schematics. It is observed that the proposed charge pump circuit can pump a low power supply voltage 1.8V up to 5.95V output voltage.

Comparison between proposed charge pump with existing charge pump is shown in Fig. 5. We can observe that the proposed charge pump giving the highest output voltage, which leads to higher pumping gain for each stage compared to the DBPT and ULVLP.

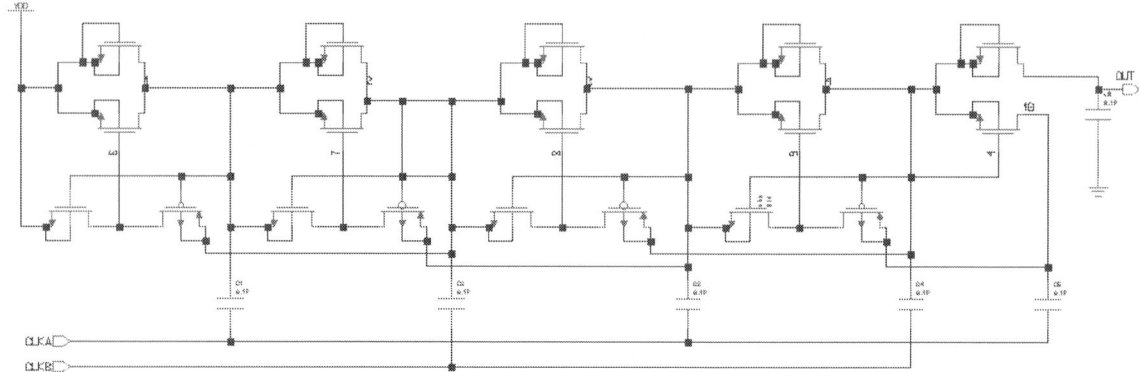

Fig. 3. Proposed four-stage charge pump circuit using dynamic CTS

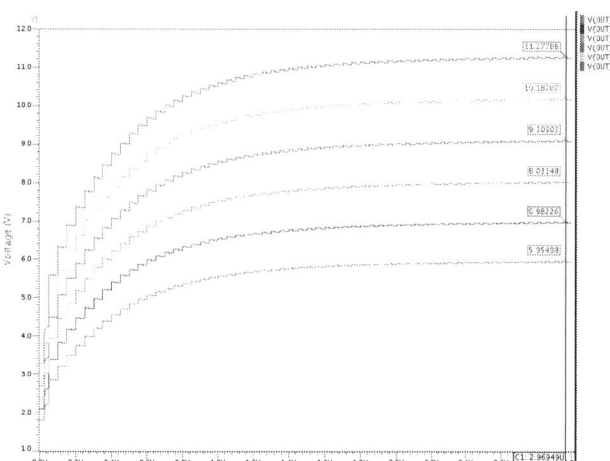

Fig. 4. Output voltages for proposed four-stage charge pump schematics with $1.8V \leq V_{DD} \leq 3.3V$

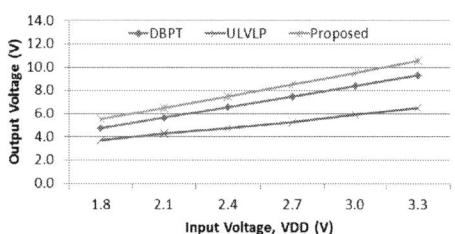

Fig. 5. Measured output voltages for four-stage charge pumps with $1.8V \leq V_{DD} \leq 3.3V$

Besides that, this paper also present the simulated output voltages for comparing the performance of DBPT, ULVLP and proposed charge pump circuit corresponding to the number of stages. From Fig. 6, Fig. 7 and Fig. 8, we can find that the proposed charge pump still giving the highest output voltage compared to the other two.

Fig. 9 shows the layout design for proposed four-stage charge pump using CEDEC 0.18-μm CMOS technology. The size of this layout design is 40μm X 60μm. This layout is designed for the ease to cascade more pumping stage to the desired output voltage for further development. Fig. 10 shows

the measured output voltage for the proposed four-stage charge pump layout design. It is found that the layout simulation can pump a low power supply voltage of 1.8V up to 5.07V output voltage. The layout simulation output voltage has decreased for 14% compared to schematic simulation as shown in Fig. 11. Schematic simulation is considering ideal case while layout design simulation is considering the real case which including parasitic capacitance, reverse charge sharing or the body effect, which lead to the decrease of output voltage. Yet, the proposed four-stage charge pump layout design able to boost a low input voltage 1.8V to 5.07V.

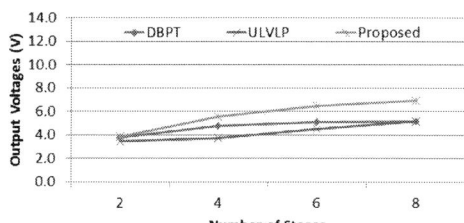

Fig. 6. Measured output voltages for $V_{DD} = 1.8V$

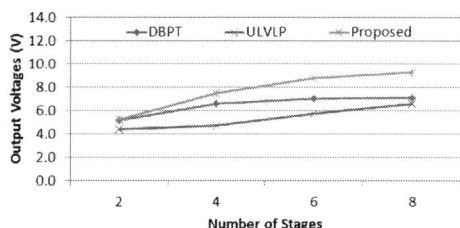

Fig. 7. Measured output voltages for $V_{DD} = 2.4V$

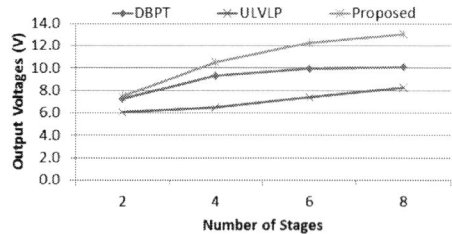

Fig. 8. Measured output voltages for $V_{DD} = 3.3V$

978-1-4673-2395-6/12 $31.00 © 2012 IEEE

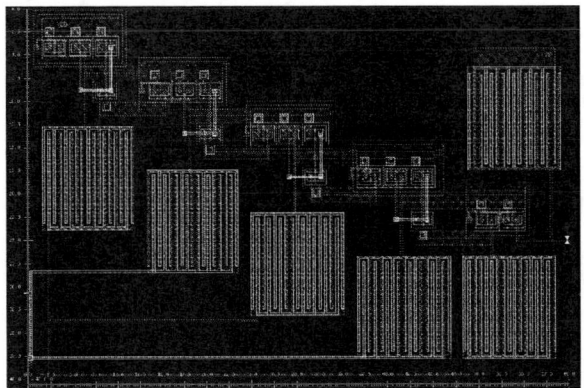

Fig. 9. Layout design for proposed four-stage charge pump using CEDEC 0.18-µm CMOS technology

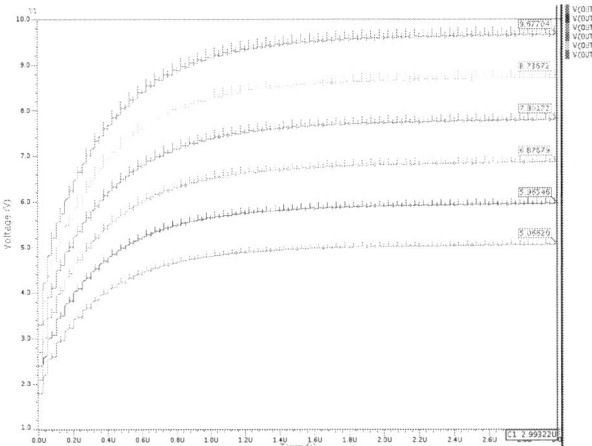

Fig. 10. Output voltages for proposed four-stage charge pump layout design with 1.8V≤ V$_{DD}$ ≤3.3V

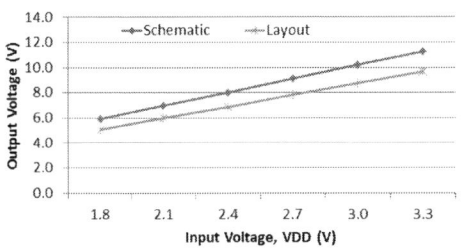

Fig.11. Comparison between schematic simulation and layout simulation for proposed four-stage charge pump with 1.8V≤ V$_{DD}$ ≤3.3V

V. CONCLUSION

An improved charge pump using CTS to direct charge flow with better voltage pumping gain has been presented in this paper. The limitation imposed by the diode-configured output stage is mitigated by pumping the output stage with clock of enhanced charge pump circuit. The modified charge pump is capable of working in low-voltage devices with minimum parasitic capacitance. From the simulation result, it is found

that the proposed four-stage charge pump circuit with each pumping capacitor of 0.1pF is able to produce 5.95V output for 1.8V power supply voltage (VDD). This proposed low voltage charge pump circuit is suitable for low voltage applications such as RFID tag EEPROM to generate the desired high output voltage.

REFERENCES

[1] M. J. Uddin, M.I. Ibrahimy, M.B.I. Reaz and A.N. Nordin, "Design and Application of Radio Frequency Identification Systems," European Journal of Scientific Research, vol. 33, no. 3, pp. 438-453, 2009.

[2] L.F Rahman, M.B.I. Reaz, M.A.M, Ali, M. Marufuzzaman and M.R. Alam, "Beyond the WiFi: Introducing RFID system using IPv6," in Proceedings of the Kaleidoscope: Beyond the Internet? – Innovations for Future Networks and Services, ITU-T, pp:1-4, 13-15 Dec. 2010, Pune, India.

[3] F. Mohd-Yasin, M.K Khaw and M. B. I. Reaz, "Radio Freqiency Identification: Evolution of Transponder Circuit Design," Microwave Journal, vol. 49, no. 6, pp. 56, 2006.

[4] K. Ohsaki, N. Asamoto and S. Takagaki, "A single poly EEPROM cell structure for use in standard CMOS processes," IEEE Journal of Solid-State Circuits, vol. 29, no. 3, pp. 311-316, Mar. 1994.

[5] D.X. Zhao, N.Y.W. Xu, L.W. Yang and J.Y. Wang, "Low-power, Single-poly, Non-volatile Memory for Passive RFID Tags," Chinese Journal of Semiconductors, Vol. 29, pp. 99-104, 2008.

[6] B. Wang, H. Nguyen, Y. Ma and R. Paulsen, "Highly Reliable 90-nm Logic Multitime Programmable NVM Cells Using Novel Work-Function-Engineered Tunneling Devices," IEEE Transactions on Electron Devices, vol.54, no. 9, pp. 2526-2530, Sept. 2007.

[7] J. Raszka, M. Advani, V. Tiwari, L. Varisco, N.D. Hacobian, A. Mittal, M. Han, A. Shirdel and A. Shubat, "Embedded flash memory for security applications in a 0.13um CMOS logic process," in Proceedings of the International Solid-State Circuits Conference (ISSCC), IEEE, 15-19 Feb. 2004.

[8] M.G. Mohammad, M.J. Ahmad and M.B. Al-Bakheet, "Switched positive/negative charge pump design using standard CMOS transistors," IET Circuits Devices Syst., vol.4, no. 1, pp.57-66, Jan. 2010.

[9] D. Somasekhar, B. Srinivasan,. G. Pandya, F. Hamzaoglu, M. Khellah, T. Karnik and K. Zhang, "Multi-Phase 1 Ghz Voltage Doubler Charge Pump in 32nm Logic Process," IEEE Journal of Solid-State Circuits, vol. 45, no.4, pp.751, Apr. 2010.

[10] J.F. Dickson, "On-chip high-voltage generation in MNOS integrated circuits using an improved voltage multiplier technique," IEEE Journal of Solid-State Circuits, vol. 11, no.3, pp.374-378, Jun. 1976.

[11] W. Henru and C. Yuhua, "A charge pump circuit design based on a 0.35um BCD technology for high voltage driver applications," in Proceedings of the 9th International Conference on Solid-State and Integrated-Circuit Technology (ICSICT), pp.2035, 30-23 Oct. 2008.

[12] J. Shin; I.Y. Chung; Y. J. Park; Hong and S. Min, "A new charge pump without degradation in threshold voltage due to body effect [memory applications]," IEEE Journal of Solid-State Circuits, vol. 35, no. 8, pp:1227-1230, Aug. 2000.

[13] L.D. Sheng, Z. Cheng, Z. Fan, Deng Min, Embedded EEPROM Memory Achieving Lower Power – New design of EEPROM memory for RFID tag IC," Circuits and Devices Magazine, IEEE, vol.22, no.6, pp.53-59, Nov.-Dec. 2006.

[14] L. Mensi, L. Colalongo, A. Richelli and Z.K. Vajna,"A New Integrated Charge Pump Architecture using Dynamic Biasing of Pass Transistors," in Proceedings of ESSCIRC, pp.85-88, Sept. 2005.

[15] J. Che, C. Zhang, Z. Liu, Z. Wang and Z. Wang, "Ultra-Low-Voltage Low-Power Charge Pump for Solar Energy Harvesting System," Communications, Circuits and System (ICCCAS), pp. 674-677, July 2009.

Novel Architecture of Pipeline Radix 2^2 SDF FFT Based on Digit-Slicing Technique

[1]Yazan Samir Algnabi, [2]Furat A. Aldaamee, [3]Rozita Teymourzadeh, [1]Masuri Othman and [1]Md Shabiul Islam
[1]Institute of Microengineering & Nanoelectronics (IMEN), Universiti Kebangsaan Malaysia (UKM),
43600 UKM Bangi, Selangor, Malaysia
[2]SCHOOL OF COMPUTER SCIENCE & IT, Linton University College, Bandar Universiti Teknologi Legenda
71700, Mantin, N. Sembilan, Malaysia
[3]Faculty of Engineering, Technology and Built Environment, Electrical & Electronic Engineering department,
UCSI University, 56000 Kuala Lumpur, Malaysia.
yazansamir@yahoo.com , furatali@legendagroup.edu.my, rozita@ucsi.edu.my, masuri@mosti.gov.my, shabiul@ukm.my

Abstract --**The prevalent need for very high speed digital signals processing in wireless communications has driven communications system to higher levels of performance. The objective of this paper is to propose a novel structure for efficient implementation of the Fast Fourier Transform (FFT) processor to meet the requirements of high speed wireless communication system standards. Based on the algorithm, architecture analysis, a design of pipeline Radix 2^2 SDF FFT processor based on the digit-slicing Multiplier-Less technique is proposed. Furthermore, this paper proposes an optimal constant multiplication arithmetic design to multiply a fixed point input by means of one of the several present twiddle factor constants. The proposed architecture was simulated using MATLAB software and the Field Programmable Gate Array (FPGA) Virtex 4 was targeted to synthesise the proposed architecture. The design was tested in real hardware of TLA5201 logic analyzer and the ISE synthesis report resulted in high speeds of 669.277 MHz with a total equivalent gate count of 14,854. This is a significant improvement over the Radix 2^2 DIF SDF FFT processor from which it can be concluded that the proposed pipeline Radix 2^2 DIF SDF FFT processor based on digit-slicing multiplier-less is capable of solving the problems that affect the capabilities of most high speed wireless communication systems in FFT and possesses huge potential for future research.**

Keywords -- Digit-Slicing Multiplier-Less, Constant Multiplication Arithmetic, Radix 2^2 DIF SDF, Fast Fourier Transform

I. INTRODUCTION

FFT plays an important role in many digital signals processing (DSP) applications such as communication systems and image processing. It is an efficient algorithm to compute the discrete Fourier transform (DFT) and it's inverse. The DFT is the perhaps the most important procedure in data analysis, system design and implementation [1]. The challenge in FFT hardware implementation is the speed functionality of the multiplier unit. In order to reduce the complexity of the FFT calculations, many modules were developed [2-11]. However, in order to implement the FFT processor as a system on a chip (SOC), ASIC implementation and FPGA prototyping were considered. Recently, FPGA has become an applicable option to direct hardware solution performance in real time applications. However, this paper will concentrate on FPGA implementation

of high multiplier-less FFT processors using the shift and add technique. Since multiplication causes high delay propagation in FFT calculation, the new digital-slicing technique is applied to build a novel architecture for a multiplier-less FFT processor. The motivation of this research was inspired by [12-14, 18]. Meanwhile, the study of the digit-slicing FFT in DSP applications has been introduced by [15]. This research uses a similar digit-slicing technique with those put forth by [15] but differs in regards to the use of a different algorithm, architecture and different platform, which helps to improve the performance and achieve higher speeds.

II. DIGIT SLICING ARCHITECTURE

The concept of digit-slicing is that any complex number, F, can be sliced into smaller blocks, each having a shorter word length, p, as shown in the following equations [14].

$$F = \sum_{k=0}^{b-1} (2^{p-1})^k FR_k + j \sum_{k=0}^{b-1} (2^{p-1})^k FI_k \quad (1)$$

$$FR_k = -(2^{p-1})FR_{(p-1),i} + \sum_{i=0}^{p-2} (2^i) FR_{k,i} \quad (2)$$

$$FI_k = -(2^{p-1})FI_{(p-1),i} + \sum_{i=0}^{p-2} (2^i) FI_{k,i} \quad (3)$$

In this equation $FI_{k,i}$ and $FR_{k,I}$ have values which are either zero or one. Any value whose absolute value is less than one can be represented in two's complement as:

$$x = \left[\sum_{k=0}^{b-1} 2^{pk} X_k \right] 2^{-(pb-1)} \quad (4)$$

Here x is any number with an absolute value less than one and x is sliced into b blocks, each block being p bits wide.

$$X_k = \sum_{j=0}^{p-1} 2^j X_{k,j} \quad (5)$$

III. RADIX 2^2 SDF FFT ALGORITHM

The Radix 2^2 FFT algorithm has the same multiplicative complexity as Radix 4 but retains the butterfly structure of the

Radix 2 algorithm [16]. In this algorithm, the first two steps of the decomposition of Radix 2 DIT-FFT are analysed and a common factor algorithm is used to illustrate.

$$X[k] = \sum_{n=0}^{N-1} x[n]W_N^{nk}, \quad k = 0,1,\dots\dots, \quad N-1 \tag{6}$$

In Eq. 6 the index n and k decomposed as:

$$n = < \frac{N}{2}n_1 + \frac{N}{4}n_2 + n_3 >_N \tag{7}$$

$$k = < k_1 + 2k_2 + 4k_3 >_N \tag{8}$$

The total value of n and k is N. When the above substations are applied to (6) the DFT definition can be written as the:

$$X[k_1 + 2k_2 + 4k_3] =$$
$$\sum_{n_3=0}^{(N/4)-1} \sum_{n_2=0}^{1} \sum_{n_1=0}^{1} x\left[\frac{N}{2}n_1 + \frac{N}{4}n_2 + n_3\right] \times$$
$$W_N^{\left(\frac{N}{2}n_1 + \frac{N}{4}n_2 + n_3\right)(k_1 + 2k_2 + 4k_3)} \tag{9}$$

$$X[k_1 + 2k_2 + 4k_3] =$$
$$\sum_{n_3=0}^{(N/4)-1} \sum_{n_2=0}^{1} \left[B_{N/2}^{k_1}\left[\frac{N}{4}n_2 + n_3\right] W_n^{\left(\frac{N}{4}n_2 + n_3\right)} \right] \times$$
$$W_N^{\left(\frac{N}{4}n_2 + n_3\right)(2k_2 + 4k_3)} \tag{10}$$

Where,

$$B_{N/2}^{k_1} = x\left[\frac{N}{4}n_2 + n_3\right] + (-1)^{k_1} x\left[\frac{N}{4}n_2 + n_3 + \frac{N}{2}\right] \tag{11}$$

For the normal Radix-2 DIF FFT algorithm, the expression in the brackets is computed first as part of the first stage in (10). However, in the Radix 2^2 FFT algorithm, the idea is to reconstruct the first stage and the second stage twiddle factors [16].

$$W_N^{\left(\frac{N}{4}n_2 + n_3\right)k_1} \quad W_N^{\left(\frac{N}{4}n_2 + n_3\right)(2k_2 + 4k_3)}$$
$$= W_N^{Nn_2k_3} W_N^{Nn_2(k_1 + 2k_2)} W_N^{n_3(k_1 + 2k_2)} W_N^{4n_3k_3} \tag{12}$$
$$= (-j)^{n_2(k_1 + 2k_2)} W_N^{n_3(k_1 + 2k_2)} W_N^{4n_3k_3}$$

Here the last twiddle factor in (12) can be rewritten as:

$$W_N^{4n_3k_3} = e^{\frac{-j2\pi}{N}(4n_3k_3)} = e^{\frac{-j2\pi}{4N}(n_3k_3)} = W_{N/4}^{n_3k_3} \tag{13}$$

By applying (12), (13) and (10) and expanding the summation over n_2, the result is a DFT definition with an FFT length that is four times shorter.

$$X[k_1 + 2k_2 + 4k_3] =$$
$$\sum_{n_3=0}^{(N/4)-1} \left[H[k_1 + 2k_2 + 4k_3] W_n^{n_3(k_1 + 2k_2)} \right] W_N^{n_3k_3} \tag{14}$$

Where,

$$H[k_1 + 2k_2 + 4k_3] =$$
$$\left[x(n_3) + (-1)^{k1} x\left(n_3 + \frac{N}{2}\right) \right] +$$
$$(-j)^{(k_1 + 2k_2)} \left[x\left(n_3 + \frac{N}{4}\right) + (-1)^{k1} x\left(n_3 + \frac{3N}{4}\right) \right] \tag{15}$$

Equation 14 is known as the Radix 2^2 FFT algorithm. Fig. 2 shows the butterfly signal flow graph for radix 2^2 FFT algorithms.

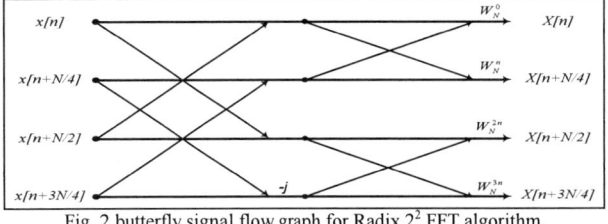

Fig. 2 butterfly signal flow graph for Radix 2^2 FFT algorithm.

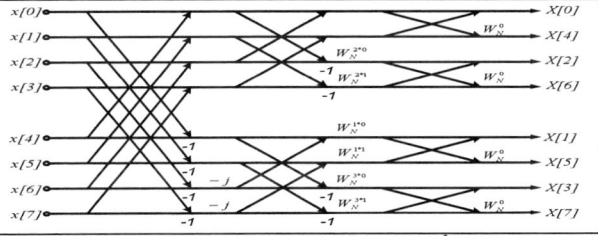

Fig. 3 Signal flow graph for an 8-Point Radix 2^2 DIF FFT

Fig. 3 shows the Radix 2^2 algorithm for an 8 point FFT. The complex multiplication in the first stage will be multiplication with $(-j)$, which means swapping the real with the imaginary and sign inversion. One complex multiplier can be reduced for 8-point FFT implementation.

From (15), each stage in Radix 2^2 SDF FFT consists of Butterfly I, Butterfly II, Complex multipliers with twiddle factors. Butterfly I calculates the input data flow, butterfly II calculates the output data flow from Butterfly I, then multiplies the twiddle factors with the output data from Butterfly II to get the result of the current stage. Fig. 4 shows the structure of 8 point Radix 2^2 SDF FFT.

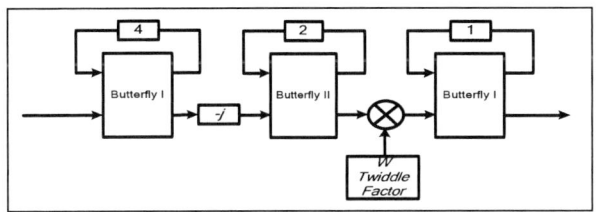

Fig. 4 8-Point Radix 2^2 SDF DIF FFT Structure

A. Butterfly I Structure

Fig. 5 shows the Butterfly I structure. The input A_r, A_i for this butterfly comes from the previous component which is the twiddle factor multiplier except that the first stage comes from the FFT input data. The output data B_r, B_i goes to the next stage which is normally the Butterfly II. The control signal $C1$

has two options $C1=0$ in which the multiplexers direct the input data to the feedback registers until they are filled. The other option is $C1=1$ in which the multiplexers select the output of the adders and subtracters.

B. Butterfly II Structure

Fig. 6 shows the Butterfly II structure. The input data B_r, B_i comes from the previous component, Butterfly I. The output data from the Butterfly II are E_r, E_i, F_r and F_i. E_r, E_i and are fed to the next component, normally twiddle factor multiplier. The F_r and F_i go to the feedback registers.

The multiplication by $-j$ involves swapping between real and imaginary parts and sign inversion. The swapping is handled by the multiplexers Swap-MUX efficiently and the sign inversion is handled by switching between the adding and the subtracting operations by means of the Swap-MUX. The control signals $C1$ and $C2$ will be one when there is a need for multiplication by $-j$, therefore the real and imaginary data will swap and the adding and subtracting operations will switch.

In order to not lose any precision, the divide by 2 is used where the word lengths imply successive growth as the data goes through the adder, subtracter and multiplier operations. Rounding off has been also applied to reduce the scaling errors.

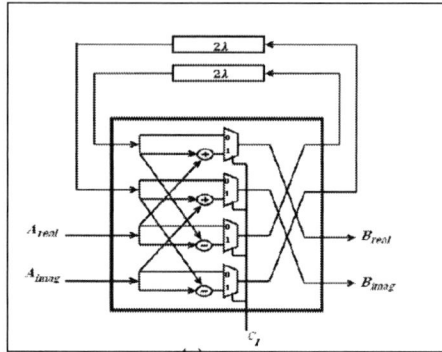

Fig. 5 Butterfly I

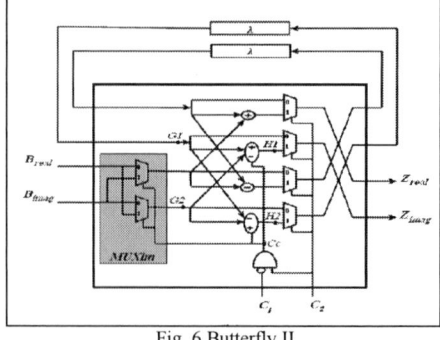

Fig. 6 Butterfly II

C. Digit Slicing Complex Multiplier Less

Complex multiplier can be realized by digit-slicing multiplier-less and real adder [17] based on (16) as shown in Fig. 7.

$$(a_r+ja_i)(b_r+jb_i)=\{b_r[a_r-a_i]+a_i[b_r-b_i]\}+j\{b_i(a_r+a_i)+a_i(b_r-b_i)\} \quad (16)$$

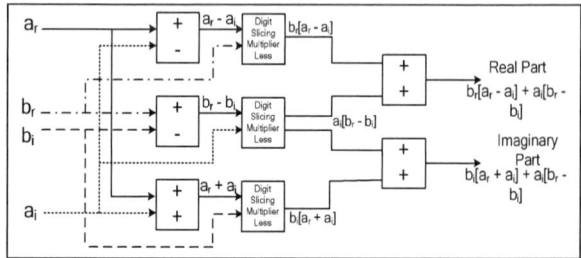

Fig. 7 Complex multiplier with three real multiplier structures

IV. SHIFT AND ADD DIGIT SLICING MULTIPLIER LESS

The proposed design slices the input data into four blocks with each block carrying four bits. By considering the input data for the multiplier are A and B with the word-length of 16 bits two's complement fixed point signed number with 15 bits fraction. The digit slicing architecture is applied for the input A as shown in Fig 8. There are four different cases for the multiplication between the four bits and the twiddle factors. Fig. 8 shows the block diagram of the digit-slicing multiplier less using the shift and addition technique.

Because of the shifts operation according to the digit slicing algorithm, the twiddle factors will store with right shifts by 6 which means that the ROM for storing the twiddle factors will be 10 bits width only not 16 bits. As mentioned in (4) and (5) the digit-slicing algorithm for this case will be:

$$A_3 . B = \sum_{j=0}^{3} 2^j A_{3,j} . B \qquad A_2 . B = \sum_{j=0}^{3} 2^j A_{2,j} . B$$

$$A_1 . B = \sum_{j=0}^{3} 2^j A_{1,j} . B \qquad A_0 . B = \sum_{j=0}^{3} 2^j A_{0,j} . B$$

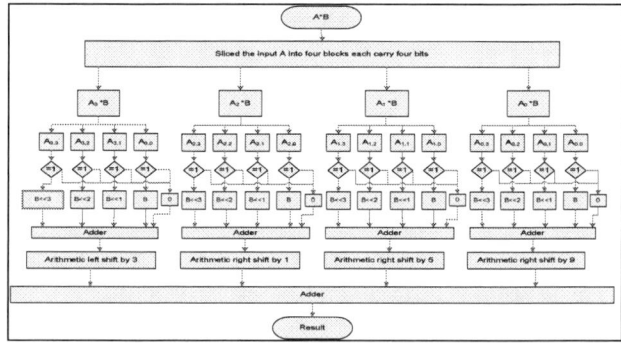

Fig. 8 The structure of Digit Slicing Multiplier Less with shift and add

V. IMPLEMENTATION RESULT

The proposed design of the pipeline Radix-2^2 DIF SDF FFT processor based on digit-slicing multiplier-less has been implemented using Matlab to prove and check the result for all stages as shown in Fig. 9. The design has been coded in Verilog HDL and tested in real hardware using Xilinx Virtex-4 FPGA as shown in Fig. 10 and Fig. 12.

In addition, the Modelsim XE-III was used to get the simulation result of the proposed design as shown in Fig. 11

and the technology schematic of top level proposed module shown in Fig.13. Table 1 shows the synthesis results with a comparison with conventional 8-point FFT processor for the prior art.

TABLE I
FFT COMPARISON

Xilinx Virtex- 4 FPGA	Total gate count	Max.Freq. MHz
Conventional 8 point FFT [9]	77,418	200
Design Proposed by [9]	16,580	400
Proposed Design	**14,854**	**669.277**

VI. CONCLUSION

This study presented the FPGA Implementation of the pipeline Radix 2^2 DIF SDF-FFT processor based on digit-slicing multiplier-less. The implementation has been coded in Verilog HDL and was tested on Xilinx Virtex-4 FPGA prototyping board. A maximum clock frequency of 669.277 MHz with a total equivalent gate count of 14,854 have been obtained from the synthesis report for the 8 point pipeline Radix 2^2 DIF SDF FFT which is 3.35 times faster than the conventional butterfly and requires 20% of the conventional butterfly area. It can be concluded that the proposed design is capable of solving problems that affect the capabilities of communications capability in FFT and possesses huge potential for future research.

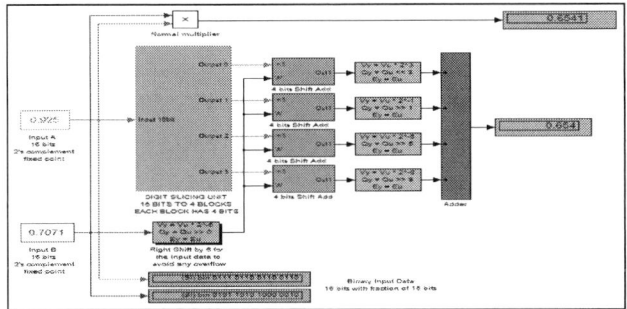

Fig. 9 Simulation of the 8-point pipeline digit slicing Radix 2^2 SDF-FFT

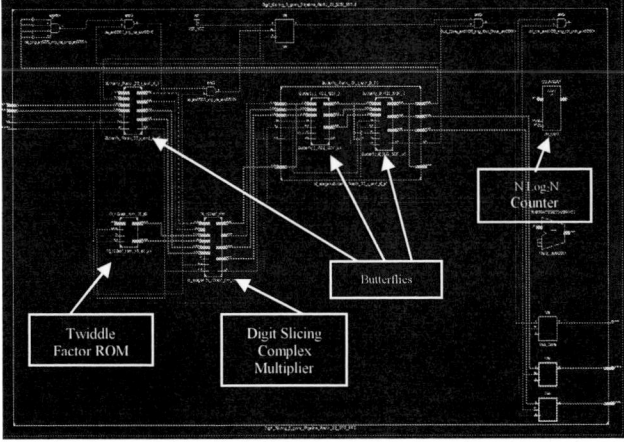

Fig. 10 Internal behavioral layout of top level Module of the 8-point pipeline digit slicing radix 22 SDF FFT

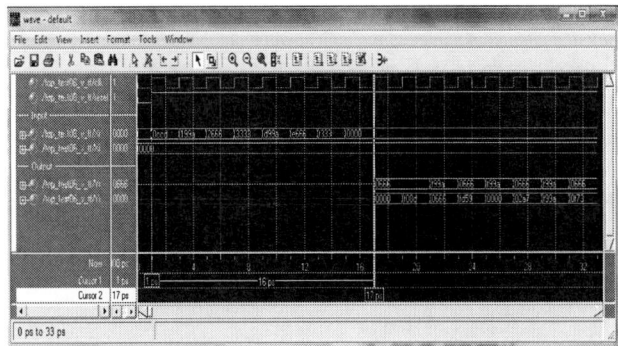

Fig. 11 Behavioral Simulation of the proposed design

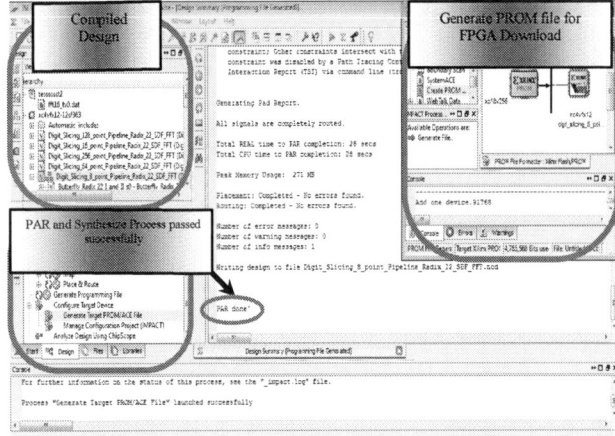

Fig. 12 FPGA synthesize report of the proposed design

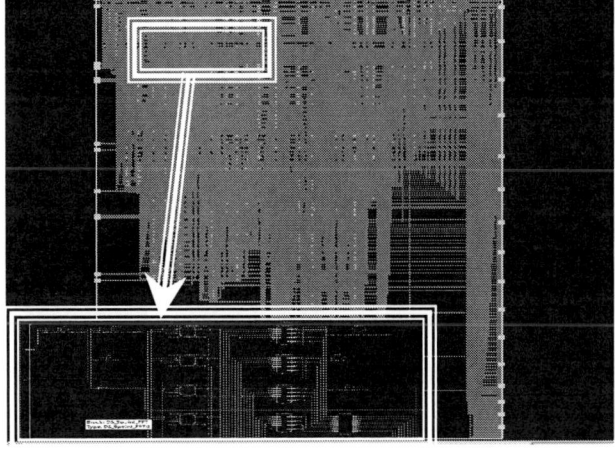

Fig. 13 Technology Schematic of top level Module of 8-point pipeline Digit Slicing radix-2^2 SDF FFT.

REFERENCES

[1] A. V. Oppenheim, R. W. Schafer, and J. R. Buck, Discrete-time signal processing, 2 ed., N.J.: Prentice Hall, 1999.

[2] G. D. Bergland, "A radix-eight fast-Fourier transform subroutine for real-valued series.," IEEE Trans. Audio Electroacoust, vol. 17, pp. 138-144, 1969.

[3] R. C. Singleton, "An algorithm for computing the mixed radix fast Fourier transform," Audio and Electroacoustics, IEEE Transactions on vol. 17, pp. 93-103, 1969.

[4] D. P. Kolba and T. W. Parks, "A prime factor FFT algorithm using high-speed convolution," IEEE Trans Acoust. Speech, Signal Process, vol. 25, pp. 281-294, 1977.

[5] A. R. Varkonyi-Koczy, "A recursive Fast Fourier Transform algorithm," IEEE Trans.Circuits, vol. 42, pp. 614-616, 1995.

[6] Y.Wang,Y,Y.J.Tang,J.G.Chung,S.Song, "Novel memory reference reduction methods for FFT implementation on DSP processors," IEEE Trans. Signal Process, vol. 55, pp. 2338-2349, 2007.

[7] Y. Zhou,J.M. Noras, S. J. Shephend, "Novel design of multiplier-less FFT processors," Signal Proc., vol. 87, pp. 1402-1407, 2007.

[8] T. Sansaloni, A. P´erez-Pascual, V. Torres, and J. Valls, "Efficient pipeline FFT processors for WLAN MIMO-OFDM systems," Electronics Letters, vol. 41, pp. 1043–1044, 2005.

[9] B. Mahmud and M. Othman, "FPGA implementation of a canonical signed digit multiplier-less based FFT Processor for wireless communication applications," in ICSE2006 Proc Kuala Lumpur, Malaysia, 2006, pp. 641-645.

[10] B. M. Baas, "A Low-Power, High-Performance,1024-Point FFT Processor," IEEE JOURNAL OF SOLID-STATE CIRCUITS,, vol. 34, pp. 380-387, 1999.

[11] Y. P. Hsu and S. Y. Lin, "Parallel-computing approach for FFT implementation on Digital Signal Processor (DSP)," World Acad. Sci., Eng. Technol., vol. 42, pp. 587-591, 2008.

[12] M. A. B. Nun and M. E. Woodward, "A modular approach to the hardware implementation of digital filters " Radio and Electronic Engineer vol. 46, pp. 393 - 400 1976.

[13] A. Peled and B. Liu, Digital signal processing : theory, design, and implementation. New York: Wiley, 1976.

[14] Z. A. M. Sharrif, "Digit slicing architecture for real time digital filters." vol. Ph.D UK: Loughborough University, 1980.

[15] S. A. Samad, A. Ragoub, M. Othman, and Z. A. M. Shariff, "Implementation of a high speed Fast Fourier Transform VLSI chip " Microelectronics Journal, vol. 29, pp. 881-887 1998.

[16] S. He and M. Torkelson, "A new approach to pipeline FFT processor," in Parallel Processing Symposium, Proceedings of IPPS '96, The 10th International Honolulu, HI, 1996, pp. 766 - 770

[17] Rozita Teymourzadeh, Yazan Samir Algenabi, Nooshin Mahdavi, Masuri Bin Othman. On-Chip Implementation of High Resolution High Speed Floating Point Adder/Subtractor with Reducing Mean Latency for OFDM. American Journal of Engineering and Applied Sciences. 3(1): 25-30. ISSN: 1941-7020. 2010.

[18] Yazan Samir Algenabi,Rozita Teymourzadeh, Masuri Othman & Md Shabiul Islam, 2011. FPGA Implementation of pipeline Digit-Slicing Multiplier-Less Radix 2^2 DIF SDF Butterfly for Fourier Transform Structure. *IEEE European Conference on Antennas and propagation.* pp 4168- 4172

Methodology To Execute SPARC Binary of Silterra Memory Compiler 0.18um Process Technology on x86 Architecture

Raja Mohd Fuad Tengku Aziz, Rozaimah Baharim, Md Hanif Md Nasir, Rohaya Abdul Wahab, Nazaliza Othman,
Nabihah Razali, Sharifah Saleh
Integrated Circuit Development (ICD)
MIMOS Berhad
57000 Technology Park Malaysia (TPM), Kuala Lumpur, Malaysia
Email: rozaimah@mimos.my

Abstract— This paper addresses the issue of Silterra memory compiler 0.18 micrometer (um) technology that can only be run with Solaris operating system with Scalable Processor Architecture (SPARC). However, a lot of technologies in software applications had moved to x86 platforms. SUN, which produced SPARC machines, also had merged into ORACLE, a database expert company. Today it is hard to find support for SPARC machines, as most developers now are moving to x86 operating system with Intel architecture. This paper proposes a solution by using QEMU approach. QEMU stands for "Quick Emulator" which emulates the role of SPARC machines that installs Solaris operating system with SPARC architecture on Linux machines. Now, the Silterra memory compiler 0.18um process can be executed on the virtual machine, as it recognizes the platform as Solaris SPARC architecture. This methodology not only for memory compiler of 0.18um Silterra but also can be use for others SPARC binaries.

I. INTRODUCTION

Evolution of technology decreases the dependency on hardware requirements. The revolution of open source software and competition with other operating system such as Red Hat Linux has forced Sun, the manufacturer of SPARC machines, to start supporting other platform, especially Intel [1]. However, some applications such as Silterra Memory Compiler 0.18um process technology do not have binaries for Intel architecture. This limits the hardware requirement to only machines with SPARC processor.

CMOS memory compiler has become more and more the standard offer that fulfils customers' requirements for embedded applications. The Memory Compiler is intended to be used when a certain functionality and size of memory are required. Silterra is one of the foundries that offers memory compiler with varies technologies. However for memory compiler 0.18um process technology, the foundry only provides SPARC binary format, which means it can only be run with Solaris operating system with SPARC architecture. Considering this problem regards to hardware limitation it can be solved by using QEMU emulator.

QEMU is a generic and open source machine emulator and virtualizer that enable a single computer as a host to fulfil several roles by running multiple guest operating systems (OS). It stands for "Quick Emulator "and was written by Fabrice Bellard and is free software [2]. Various parts are released under different GNU General Public License version 2-compatible licenses. These include the GNU Lesser General Public License (GNU LGPL) or permissive licenses such as the BSD license [3]. QEMU also relies on dynamic binary translation to achieve a reasonable speed and easy to port to new host CPU architectures.

Based on the capability of QEMU to run a virtual machine with SPARC architecture on an Intel machine, users do not require a SPARC machine to run Memory Compiler or other applications that can only be run on SPARC architecture. Due to lack of support by administrators for Sun operating system and to reduce cost, QEMU will be used to resolve the problem.

This paper will discuss the methodology to execute SPARC binary of Silterra Memory Compiler 0.18um on x86 architecture using QEMU.

II. SYSTEM EMULATION WITH QEMU APPROACH

Fig. 1 shows the process flow to run Silterra Memory Compiler 0.18um process technology by using QEMU approach. First the end user connects to the x86 machine (host) using secure shell (SSH) client in order to execute a QEMU commands. Then QEMU emulates SPARC architecture and now the machine can be viewed as a SPARC virtual machine (guest). Since the machine is running on SPARC architecture which supports the SPARC binaries, the end user can now generate the Memory Compiler 0.18um process technology. The generated files from the memory compiler will be stored in the SPARC virtual machine before those files are transferred to the host machine using file transfer protocol (FTP).

978-1-4673-2395-6/12 $31.00 © 2012 IEEE

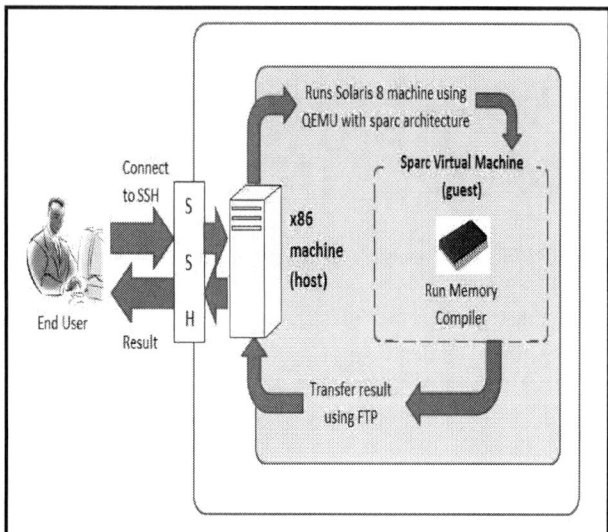

Fig.1. Running Memory Compiler using QEMU Approach

III. SYSTEM EMULATION OPERATING MODE

In this section, we will explain how we use QEMU system emulation operating mode in our methodology. QEMU supports two operating modes: user-mode emulation and system-mode emulation. User-mode emulation allows a process built for one CPU to be executed on another (performing dynamic translation of the instructions for the host CPU and converting Linux system calls appropriately). System-mode emulation allows emulation of a full system, including processor and assorted peripherals [4]. Software running in this system sees a computer that can be quite different from the host system. Fig.2 shows the structure of QEMU system operating mode in our x86 machine where the Silterra Memory Compiler 0.18um process can be generated in SPARC virtual machine.

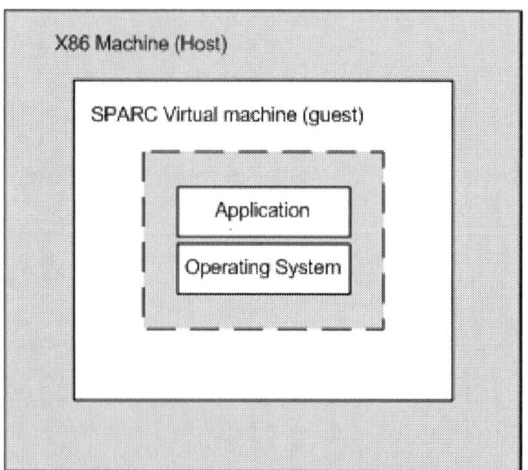

Fig.2. Structure of QEMU System Operating Mode

IV. TRANSFERRING FILES BETWEEN QEMU GUEST AND HOST USING TUN/TAP

QEMU integrates the guest and host system with several options, thus allowing the file to transfer among each other. Currently, there are five options that can be executed to transfer the files between QEMU guest and host. These options are:

- User mode networking.
- Using option *–tftp* for file transfer.
- Using option *–redir* for connection from Host OS to Guest OS.
- Using TUN/TAP interfaces.
- By mounting the hard disk image *–raw* image.

Each of the option has its own setup requirement. However, after trying all the options, only virtual network kennel devices (TUN/TAP) interfaces manage to communicate between QEMU guest and host in this scenario.

Using TUN/TAP interface method, the QEMU Virtual Machine opens a pre-allocated TUN or TAP device on the host and uses that interface to transfer data to the guest OS.

V. METHODOLOGY TO EXECUTE MEMORY COMPILER

This implementation is using Ubuntu as the host operating system and Solaris SPARC 8 as the guest operating system. The host machine is using x86 machine. Fig. 3 shows process flow to run Memory Compiler using QEMU approach.

First QEMU is installed on the host machine in order to emulate the SPARC architecture, allowing applications compiled for one architecture to be run on another.

Once QEMU has been installed, it should be ready to run a guest operating system from a disc image. This image is a file that represents the data on a hard disc. From the perspective of the guest operating system, it actually is a hard disc, and it can create its own file system on the virtual disc.

The next step is to use proprietary Basic Input Output System (BIOS) before installing Solaris SPARC 8. Here we used SPARCstation 5 (SS5) as BIOS to match with the operating system [5]. Upon completion, we need to telnet to local host where the Solaris operating system will boot up. Right after the boot up process, the disk is formatted so that it is compatible with SS5.

Then we install uml-utilities and bridge-utilities to bring up the TUN/TAP interface where TUN is used with routing whereas TAP is used to create a network bridge. Now host and guest are able to communicate via the interface.

Next we connect to Solaris8 by using host with new user account. The virtual machine is available now and Silterra Memory Compiler 0.18um process technology SPARC binary can be downloaded. All the binary files are extracted and then we run the commands to generate the memory compiler. All the files generated should be copied from host via ftp. Detail implementations are available in [6].

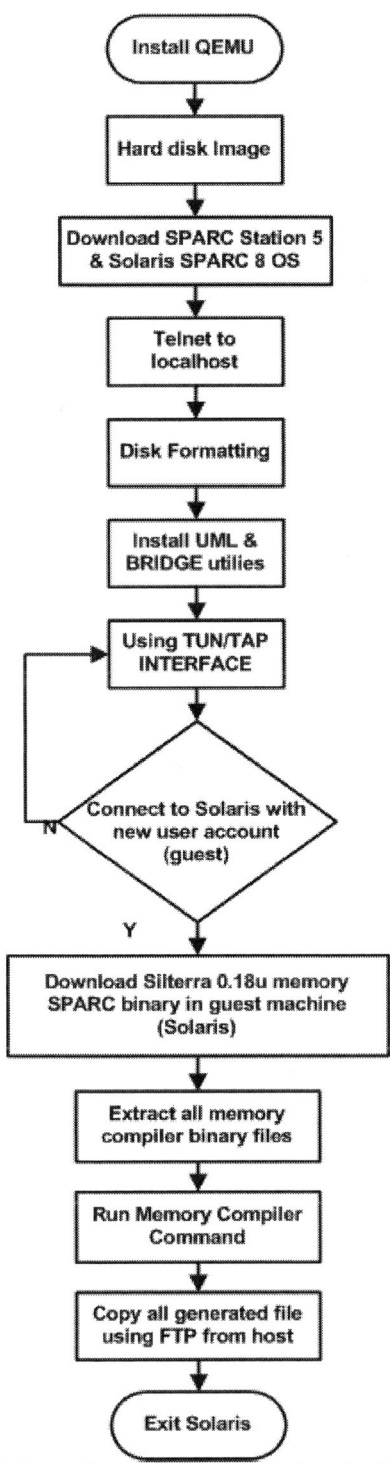

Fig.3. Process flow to run Memory Compiler using QEMU

and data similarity between two machines and it supposed to be equivalent. For layout file SRAM1_8192x64.gds2, the design evaluation has been done by executing design rules check (DRC) and layout versus schematic (LVS) process flow.

TABLE 1
OUTPUT DATA COMPARISON BETWEEN x86 AND SOLARIS MACHINE

Output Files	x86	Solaris
SRAM1_8192x64.gds2	1. File size: 6612992 byte. 2. DRC: Passed 3. LVS: Passed	1. File size: 6612992 byte. 2. DRC: Passed 3. LVS: Passed
SRAM1_8192x64_ant.l ef	1. File size: 4878 byte 2.Data similarity: Passed	1. File size: 4878 byte 2. Data similarity: Passed
SRAM1_8192x64_fast _syn.lib	1.File size: 26673 byte. 2. Data similarity: Different on date and time.	1. File size: 4878 byte 2. Data similarity: Different on date and time.
SRAM1_8192x64_fast. tlf	1.File size: 26673 byte. 2. Data similarity: Different on date and time.	1. File size: 26673 byte. 2. Data similarity: Different on date and time.
SRAM1_8192x64_fast. tlf	1.File size: 26673 byte. 2. Data similarity: Different on date and time.	1. File size: 26673 byte. 2. Data similarity: Different on date and time.
SRAM1_8192x64.v	1. File size: 41749 byte. 2. Data similarity: Different on date and time.	1. File size: 41749 byte. 2. Data similarity: Different on date and time.

VI. CONCLUSIONS

Using the above mentioned method, it is thus possible to generate Memory Compiler 0.18um process technology using QEMU approach. QEMU emulates SPARC architecture to create a SPARC virtual platform on a Solaris 8 operating system hence gives great advantages to end user because he/she can run any software on x86 machines without having to rely on any hardware requirements (e.g. SPARC machine). In addition, apart from creating a virtual platform to be used to perform this scenario, this methodology can also be used for others SPARC binaries application.

REFERENCES

[1] Hardware and Software. Engineered to Work Together. [Online]. Available: http://www.oracle.com/us/sun/index.htm
[2] Fabrice Bellard, "QEMU, a Fast and Portable Dynamic Translator," FREENIX Track: 2005 USENIX Annual Technical Conference.
[3] QEMU, open source processor emulator. [Online]. Available: http://www.qemu.org/
[4] System Emulation With QEMU. [Online]. Available: http://www.ibm.com/developerworks/linux/library/l-qemu/
[5] The SPARC Architecture Manual Version 8. [Online]. Available: http://www.sparc.org/standards/V8.pdf
[6] Brezular's Technical Blog. [Online]. Available: http://brezular.wordpress.com

V. DATA COMPARISON

To verify and confirm that the output files from x86 are similar with Solaris, the list of data comparisons has been done at Table 1. Here, SRAM1_8192x64 has been taken as an example. Each of the output files will be compared on file size

978-1-4673-2395-6/12 $31.00 © 2012 IEEE

Simulated Annealing vs. Genetic Simulated Annealing for Automatic Transistor Sizing

Nishant Singh and Bahniman Ghosh*
Department of Electrical Engineering,
Indian Institute of Technology Kanpur,
Kanpur 208016, India.
Email: *bahniman@iitk.ac.in

Abstract - Transistor size optimization is an important aspect of circuit design. Small and non-complex circuits can be designed easily using manual calculations and circuit simulations. But, as the complexity of circuits increases, manual design becomes too difficult and time consuming. Therefore, tools and techniques for automatic transistor sizing are of great importance in the area of circuit design. The goal of this paper is to implement Genetic Simulated Annealing algorithm as a tool for transistor sizing, and compare its performance with Simulated Annealing, one of the most popular optimization algorithm in use today. The algorithms have been tested on four different digital circuits and the results collated and compared in this paper.

I. INTRODUCTION

Circuit design involves balancing the speed of the circuit with the power consumed. The challenge is to minimize both power and delay. In MOS circuits, this is mostly done by comprehensive transistor sizing. Doing so manually is iterative, slow, tedious and error-prone.

For a simple design, an experienced designer can produce a decent design in a few trials on paper, but it is not that easy for others. The updating of transistor widths from one iteration to the next relies on human intuition. This is complicated because, in a complex circuit, sizing a transistor to speed up one signal path may slow down another due to the capacitive loading effects of path interactions. To add to that, deep sub-micron transistors have very complex mathematical models which are difficult to handle manually. The only option to get an accurate design is to use a simulator, and it is easier and less time consuming to interface an optimization code with the simulator.

However, the complexity of modern circuits means that every simulation of the circuit takes a long time. If too many simulations are involved, the automated process will take a very long time. Such algorithms, when given less time, will be unable to produce good results. Also, in complicated circuits, the number of parameters involved is very large or the range of sizes is wide, so the search domain is very large. This slows the algorithm down, and affects the final result. The need, therefore, is to develop algorithms for circuit design that can produce better results in lower number of runs while searching in a large domain.

In this paper, we started with one of the most popular optimization algorithms, Simulated Annealing. We analyzed its performance in automatic transistor sizing and looked into ways to improve the performance. Eventually, we looked towards hybrid algorithms which could overcome the shortfalls of Simulated Annealing. We decided to hybridize Simulated Annealing with Genetic Algorithms. The result was Genetic Simulated Annealing, which we developed and tested extensively.

II. ALGORITHMS

A. Simulated Annealing

Simulated annealing (SA) is a generic probabilistic metaheuristic for the global optimization problem of locating a good approximation to the global optimum of a given function in a large search space. It is normally used in problems where the search space is discrete. The strength of simulated annealing lies in the ability to get out of local minima and locate the global minimum in an uneven result domain.

The algorithm takes its inspiration from the metallurgical process of annealing. Metal is heated to a high temperature, and then allowed to cool in a slow and controlled manner. The aim there is to increase crystal size and reduce defects. The heat allows the atoms to escape local minima and wander randomly through states of higher energy, the slow cooling increases their chances of finding configurations with lower energy than the initial one.

Simulated annealing, by analogy, attempts to replace the current solution with a random solution from its neighborhood. If the new solution is a downhill move, i.e., it is a better solution, it is accepted. If, however, it is an uphill move, it may be allowed with a probability that depends on the magnitude of the uphill move and a global parameter, T. As the temperature is reduced over time, the probability of uphill moves decreases and at T=0, it becomes zero.

In SA, the existing solution is given a small perturbation to generate a new solution. The difference in the *cost functions*, $\triangle E$, is calculated as:

$$\triangle E = E_f - E_i$$

where,

E_f = Cost function of the new solution

E_i = Cost function of the old solution

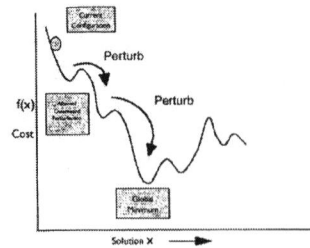

Fig.1 Simulated Annealing [17]

If the value of $\triangle E$ is negative, the solution is accepted. If it is positive, the solution is accepted with the probability:

$$ P = \frac{1}{(1 + e^{\frac{\triangle E}{K_B T}})} $$

where,

$$ K_B = \text{Boltzmann Constant} $$

As the temperature decreases, the value of P decreases, thereby reducing the probability of uphill movements. Finally, the solution settles to the lowest state if the final value of T is sufficiently low.

The method used in this paper is called the homogeneous algorithm. Here, the temperature is decreased after a certain number of iterations have been completed. So, the parameters to be controlled are initial and final temperatures, the cooling schedule and the number of iterations at any given temperature.

SA has the ability to get out of local minima, but the random search pattern makes it inefficient when the number of variables involved is large, or the range of transistor sizes is wide. To improve the search, we decided to use Genetic Algorithm (GA) in place of random neighborhood search. GA gave us the ability to tune the searches and make the whole algorithm more efficient. The result is a hybrid algorithm called Genetic Simulated Annealing.

B. Genetic Simulated Annealing

Genetic Simulated Annealing (GSA) is a hybrid algorithm that combines Simulated Annealing and Genetic Algorithm. It uses GA to select a new solution, and then uses SA to decide whether to select the new solution or not.

The technique implemented in this project is called **Binary GSA**. Each solution is represented as a *chromosome*, made up of streams of bits, each of which is called a *gene*. A set of chromosomes is called a *population*.

GSA starts by selecting two *parents* out of the population. Each parent is selected from the population using a technique called **Tournament Selection**. We choose four chromosomes at random from the population. We then compare their *fitness functions*. The chromosome with the best fitness function is selected as the first parent. The procedure is repeated with the remaining population to generate the second parent.

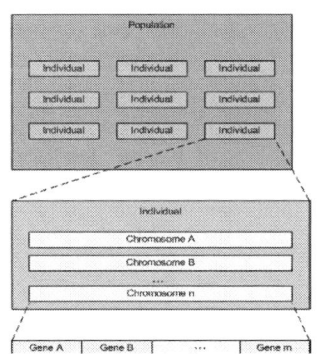

Fig.2 Concept of population [18]

Fig.3 Individual chromosomes [18]

Once the parents are selected, we perform *crossover*. Sometimes called *recombination*, it combines the attributes of both the parents to produce offspring that inherits attributes from one or both the parents. There are various crossover techniques used in GA. We are using a technique called **Multi-point Crossover**. The parent chromosomes are aligned side-by-side, and each boundary of genes can be used as a crossover point. The genes are then exchanged between parents along the crossover points with a finite cross over probability, and the resultant chromosomes are referred to as the *children*.

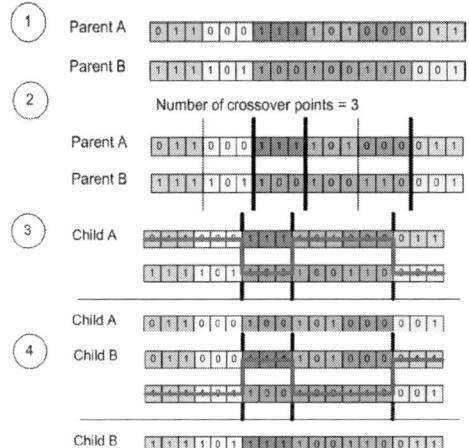

Fig.4 Multi-point crossover [18]

Once we have the two children, we use SA to compare the parents and the children. The comparison is done between the parent and the child who share the same first gene. In the illustration, for example, we will compare Parent A with Child A and Parent B with Child B. The new solution, selected as per SA rules is added back to the population, and the procedure is repeated.

The benefit of GSA over SA lies in the way the new solution is selected for comparison. In SA, the process was completely random. In GSA, the evolutionary method helps concentrate the best genes in the pool into the population. SA helps overcome GA's tendency to get trapped in local minima. Thus, the two parent algorithms help cover each other's weaknesses, and at the same time retain their strengths, thereby making a good optimization algorithm.

III. TEST CIRCUITS

The testing for both the algorithms was done on four digital circuits using BSIM3 models in Synopsys® HSPICE 2006. The optimization codes were written in Perl, which invoked HSPICE with the desired values of transistor sizes and extracted the parameters required from the output file. Both algorithms were given roughly the same number of iterations on each circuit and the same cooling schedule, and the final value of the *objective function*, a function of some circuit parameters, was used for comparison of performance.

A. NAND Gate

Our first test circuit was a three input CMOS Nand gate using TSMC 180nm model. There are six transistors used in the circuit and hence there are six variables to optimize. The circuit has been optimized in terms of the delay and power consumption. The detailed result for the two algorithms is shown in Table 1. Following is the objective function used for the optimization:

$$O.F = 0.9 \times \frac{T_d}{T_{d-ref}} + 0.1 \times \frac{P_{avg}}{P_{avg-ref}}$$

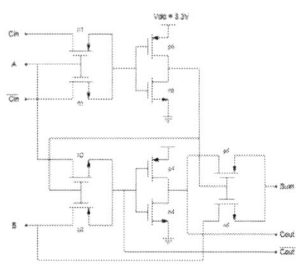

Fig.5 Three input CMOS Nand gate

B. Adder

The second circuit was a one bit complementary level restoring carry adder using TSMC 350nm model. The detailed results are shown in Table 2. The objective function used is:

$$O.F. = Power \times Delay$$

Fig. 6 One bit Adder

C. CMOS Buffer Chain

The third circuit optimized was a buffer chain of 4 CMOS inverters. IBM 130nm technology has been used to make the circuit. Variables which were optimized were: Ratio of (W/L) PMOS to (W/L) NMOS and (W/L) NMOS of last three inverters. The detailed results are shown in Table 3. The objective function used is shown below:

$$O.F = 0.45 \times \left(\frac{T_{d-rise}}{T_{d-rise-ref}} + \frac{T_{d-fall}}{T_{d-fall-ref}} \right) + 0.1$$
$$\times (\frac{P_{avg}}{P_{avg-ref}})$$

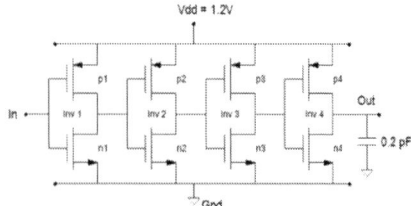

Fig.7 Four transistor CMOS Buffer Chain

D. XOR Gate

The final digital circuit optimized is a simple two input XOR gate using TSMC 180nm model. Variables optimized were the W/L ratios of all the transistors. The detailed results are shown in Table 4. The objective function used is:

$$O.F = 0.9 \times \frac{T_d}{T_{d-ref}} + 0.1 \times \frac{P_{avg}}{P_{avg-ref}}$$

Fig.8 Two input XOR Gate

IV. RESULTS

There is a notable improvement in the value of the objective function on using GSA. For the NAND gate, the best result for SA was 1.2992 while GSA attained a best result of 0.7058. For the Adder, the best result for SA was 0.43809 while for GSA, it was 0.01753. For the CMOS buffer chain, the best result reached by SA was 0.73364 while GSA reached 0.34598. For the XOR Gate, SA reached 0.9721 while GSA reached 0.6199. The detailed results with the number of runs, average values for the best three runs and standard deviations are given in the appendix.

V. DISCUSSION

The performance gain of GSA compared to SA is very promising. The gain is more pronounced in circuits like CMOS Buffer, where the transistor size range is very wide, or the Adder, where the number of transistors to be sized is large. The evolutionary selection process helps narrow down the search for minima/maxima. In simulated annealing, the search process is rather random while in GSA, the search process incorporates the "genes" from the best results in previous iterations. This helps make the search more focused. The simulated annealing part of the circuit helps push the search out of local minima.

The idea for GSA has been in theory for a long time ([6], [7], [8], [9], [11], [12] and [13]). But the implementation to circuit design has not been done before. Also, the other versions of the hybrid algorithm have been somewhat different to our implementation.

There is always the risk of annealing algorithms missing the global minima by jumping out of it at high temperatures if the valley is not steep enough, but we took care of that by tracking the absolute minima hit by the algorithm. We avoided that by using a variable to track the absolute minima, thereby avoiding any interference with the algorithm. Only at the very end, we would compare the two minima reported, one by the algorithm, and one by the tracking variable and use the lower value. In our tests with standard optimization functions and the circuit optimization problems, this small tweak made an observable difference almost every single time in case of SA, and several times in GSA.

VI. CONCLUSION

GSA can produce a better result in the same number of iterations as SA, so, for the same result, it will take a smaller number of iterations, which was our aim. GSA is a more complex algorithm compared to SA, but since our bottleneck is the speed with which simulations take place, there is no effect on the overall speed as long as the number of simulations is kept in check. What we lose in terms of complexity, we more than recover in terms of final results. The main issue with GSA is that it is more memory intensive than SA. But with memory becoming cheaper, and the fact that we can control the population size and encoding of each variable to reduce memory consumption, the problem can be alleviated.

ACKNOWLEDGMENT

The authors acknowledge the Ministry of Human Resource Development (MHRD) of India for funding this work.

REFERENCES

[1] A. R. Conn et al.: "Optimization of Custom CMOS circuits by Transistor Sizing," in *Proc. IEEE Int. Conf. Computer-Aided Design*, 1996, pp.174–180

[2] Lance A. Glasser, Lennox P. J. Hoyte: "Delay and Power Optimization in VLSI Circuits", in *Proceedings of the IEEE Design Automation Conference*, 1984

[3] Jyuo-Min Shyu, Alberto Sangiovanni-Vincentelli, John P. Fishburn, Alfred E. Dunlop: "Optimization-Based Transistor Sizing", in *IEEE Journal of Solid-State Circuits*, Vol. 23. No. 2, April 1988

[4] S. Kirkpatrick, C. D. Gelatt, Jr., M. P. Vecchi: "Optimization by Simulated Annealing", in *Science 13 May 1983*, Volume 220, Number 4598

[5] Dimitris Bertsimas, John Tsitsiklis: "Simulated Annealing", in *Statistical Science*, Vol.8, No. 1, 10-15, 1993

[6] Xin Yao: "Optimization by Genetic Annealing", in *Proc. of Second Australian Conf. on Neural Networks*, pp. 94-97, 1991

[7] Samir W. Mahfoud, David E. Goldberg: "Parallel recombinative simulated annealing: A genetic algorithm", in *Parallel Computing 21*, 1-28, 1995

[8] Dan Adler: "Genetic Algorithms and Simulated Annealing: A Marriage Proposal", in *IEEE Conference on Neural Networks 1993*, pp. 1104-1109, 1993

[9] Feng-Tse Lin, Cheng-Yan Kao, and Ching-Chi Hsu: "Applying the Genetic Approach to Simulated Annealing in Solving Some NP-Hard Problems", in *IEEE Transactions on Systems, Man, and Cybernetics*, Vol. 23. NO. 6, November/December 1993

[10] El-Hosseini, M.A.; Hassanien, A.E.; Abraham, A.; Al-Qaheri, H.; , "Genetic Annealing Optimization: Design and Real World Applications," in *Intelligent Systems Design and Applications, 2008. ISDA '08. Eighth International Conference on* , vol.1, no., pp.183-188, 26-28 Nov. 2008

[11] Garcia-Martinez, C., Lozano, M.: "Simulated annealing based on local genetic search," in *Evolutionary Computation, 2009. CEC '09. IEEE Congress on* , vol., no., pp.2569-2576, 18-21 May 2009

[12] Hui Huang, Siyu Bai, Xiaoqian Du, Nansheng Pang: "Application of Genetic Annealing Algorithm in Multi-resource Balanced Optimization considering maneuver time", in *Electronic and Mechanical Engineering and Information Technology (EMEIT), 2011 International Conference on* , vol.6, no., pp.2876-2879, 12-14 Aug. 2011

[13] Jing Jiang, Boxue Tan, Lidong Meng, Lin Jiang: "Genetic and simulated annealing algorithm based on chaos variables," in *Automation and Logistics, 2009. ICAL '09. IEEE International Conference on* , vol., no., pp.424-427, 5-7 Aug. 2009

[14] Robert Rogenmoser, Hubert Kaeslin, Tobias Blickle: "Stochastic Methods for Transistor Size Optimization of CMOS VLSI Circuits", in *Proc. 4th Int. Conf. Parallel Problem Solving from Nature (PPSN IV) (Lecture Notes in Computer Science, vol. 1141), W. Ebeling, I. Rechenberg, H.P. Schwefel, and H.-M. Voigt*, Eds. Berlin, Germany: Springer-Verlag,1996

[15] Carlos A. Coello Coello, David A. Van Veldhuizen, Gary B. Lamont: "Evolutionary Algorithms for Solving Multi-Objective Problems", *ISBN:0-306-46762-3*

[16] S. Rajasekaran, G.A. Vijayalakshmi Pai: "Neural Networks, Fuzzy Logic, and Genetic Algorithms", *ISBN:978-81203-2186-1*

Image sources:

[17] Performance groups-based fast simulated annealing for improving speed and quality of VLSI circuit placement: http://www.freepatentsonline.com/6725437.html

[18] Genetic Algorithm: http://www.xatlantis.ch/education/genetic_algorithm.html

APPENDIX

TABLE 1
NAND GATE

Algorithm	Total Number of Iterations	Best Value of Objective Function	Mean of the Three Best Values	Standard Deviation
SA	10310	1.2992	1.3289	0.0374
GSA	10311	0.7058	0.7183	0.024

TABLE 2
ADDER

Algorithm	Total Number of Iterations	Best Value of Objective Function	Mean of the Three Best Values	Standard Deviation
SA	1201	0.43809	0.47065	0.04326
GSA	1193	0.01753	0.03251	0.0331

TABLE 3
CMOS BUFFER CHAIN

Algorithm	Total Number of Iterations	Best Value of Objective Function	Mean of the Three Best Values	Standard Deviation
SA	2001	0.73364	0.78784	0.06958
GSA	1991	0.34598	0.34798	0.00264

TABLE 4
XOR GATE

Algorithm	Total Number of Iterations	Best Value of Objective Function	Mean of the Three Best Values	Standard Deviation
SA	1501	0.9721	0.9893	0.0194
GSA	1495	0.6199	0.6326	0.0113

Development of Automated Neighborhood Pattern Sensitive Fault Syndrome Generator for SRAM

J.R. Rusli[1], R.M. Sidek[2] and W.H. Wan Zuha[2]

[1]Department of Electronics
[1]Universiti Kuala Lumpur British Malaysia Institute
[2]Department of Electrical and Electronic Engineering
Universiti Putra Malaysia
43400 UPM Serdang, Selangor, Malaysia
Email: [1]julie@bmi.unikl.edu.my, [2]roslina@eng.upm.edu.my

Abstract– **With the increasing complexity of memory devices, fault diagnosis is becoming as important as fault detection. Fault diagnosis is to locate and identify type of fault. One of the memory faults is Neighborhood Pattern Sensitive Faults (NPSF) which is one of the faults that are hard to test due to higher number of cells to be tested at one time. To improve the process of analyzing NPSF detection and to generate the fault syndrome for NPSF diagnosis, an Automated NPSF Syndrome Generator (ANPSFSG) is developed. A proven March algorithms are used in this generator to verify the efficiency of this generator by producing the fault coverage and diagnostic resolution. A user-friendly Graphical User Interface (GUI) of ANPSFSG is also developed by using Microsoft Visual Basic software to load the algorithm under test and display the results.**

*Keyword; **SRAM, NPSF, ANPSFSG, GUI, March Algorithm, memory diagnosis, memory testing, memory fault syndrome.***

I. INTRODUCTION

The new trend of system-on-chip (SoC) technology indicates memory as the main component occupying major area of the system [1]. SoC designs contain increasing number and variety of embedded memories such as random access memory (RAM), read only memory (ROM) and registers file memories. As a result, more faults are likely to occur hence intensify the importance of efficient memory testing. Furthermore, the increasing size of the memory chip also results in the testing of more memory bits. This will indirectly increase the manufacturing testing time and manufacturing cost as well as dropping yield because of the more complicated manufacturing process [2,3,4]. Therefore, the need for optimum testing during memory development phase is extremely critical.

Memory testing is divided into two main categories which are electrical testing and functional testing. The parametric testing under the electrical testing categories is on voltage, current and frequency. To ensure a good memory device produced during manufacturing time this testing method need to be applied. In functional testing, a memory is treated as a gray box and the test pattern will be used for fault detection [2]. The generation of test pattern is done by writing and reading of each memory cell thus with the large size of memory device, the writing and reading require more time. An effective test is a test that detects high fault coverage with less test pattern [5]. The test patterns which represent the fault types will be generated for the use of Automatic Test Equipment (ATE) or Built-In-Self-Test (BIST).

Effective test algorithms contribute to the reduction and minimization of testing time. March algorithms are the most used algorithms during memory testing [3]. This is due to their linearity, less complexity and high fault coverage compared to the traditional testing algorithms such as Checkerboard, GALPAT, Walking 1/0, Sliding diagonal and Butterfly[6]. Unlike March test, the traditional testing approaches are not based on functional fault model. In March based algorithms, improvement can be done easily with functional fault model.

Functional Fault Model (FFM) is defined as the functional behavior of a faulty memory. Among the memory FFMs are Neighborhood Pattern Sensitive Fault (NPSF), Coupling Fault (CF), Stuck-At Fault (SAF) and Transition Fault (TF). SAF and TF are the single cell types of faults whereas the CF and NPSF are for two cells and multiple cell types of faults, respectively. This paper focuses on NPSF that have three types of faults which are the Active Neighborhood Pattern Sensitive Fault (ANPSF), Passive Neighborhood Pattern Sensitive Fault (PNPSF) and Static Neighborhood Pattern Sensitive Fault (SNPSF). The detection of these faults has been proposed such as in [7] using MT-R4CF algorithm.

March algorithms that have been incorporated with diagnostic capability can ease the process of debugging failures by identifying the fault type and fault location in memory cell array. Diagnostic capability is becoming a major concern with the rapid development of semiconductor memories since it can improve the manufacturing yield by reducing the number of faulty memory device and testing time. Fault diagnosis procedure is done by distinguishing detected fault using fault syndromes concept. A good diagnostic algorithm should be able to distinguish all the detected faults. In this paper the term 'fault syndrome' will be used to represent the set of read position. Good diagnostic resolution relies on the fault syndromes that represent each detected fault.

Although work on March diagnosis algorithm has been proposed on faults such as SAF and CF [8], work on NPSF has only been reported in [6,9]. Many existing memory test

978-1-4673-2395-6/12 $31.00 © 2012 IEEE

algorithms have not been investigated for their capability to detect and diagnose NPSF. New test algorithms may also require NPSF detection and diagnosis. Since NPSF involves multiple cells, the analysis is complex and thus, manual fault analysis is no longer practical. To improve the process of analyzing NPSF detection and to generate the fault syndrome for NPSF diagnosis, an Automated NPSF Syndrome Generator (ANPSFSG) is developed as discussed in the next section.

II. Methodology

The process of developing this tool is shown in Fig. 1. Suitable March algorithms were selected based on their capability to detect and diagnose NPSF. The algorithms are March 17N, March 12N, MarchPS 23N and March 160N. A set of procedures were applied to analyse the faults and to generate the NPSF database. An ANPSFSG tool is then developed by using the Visual Basic (VB) software with entire Graphical User Interface (GUI) element. The database will be used by ANPSFSG tool for producing the fault coverage and diagnostic resolution. The efficiency results produced by the tool will be compared with the known results to validate the tool. The details of each step in Figure 1 are discussed in the following subsections.

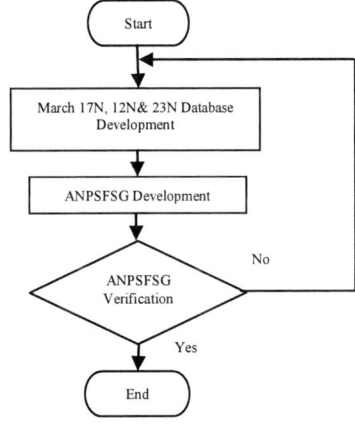

Fig. 1. Development flow of ANPSFSG

A. Database Development

The development of database is derived from the manual analysis detection and diagnosis procedure of March operation sequence as explained in [10]. In the database, the detail fault types to be detected by each element will be recorded. Since the ANPSF requires multiple data background to get 100% fault coverage, the data background will also be included in the database as discussed in [11,12]. The final compilation of database will be converted into Microsoft Access. The database is the main element in this ANPSFSG tool, where the results of the fault analysis were applied. This database mainly describes the fault detected by all possible sub march and identifies

which read operation detected the fault. The development of the database is described in the procedure in Fig. 2. The generated database is stored in Microsoft Access and is as shown in Fig. 3.

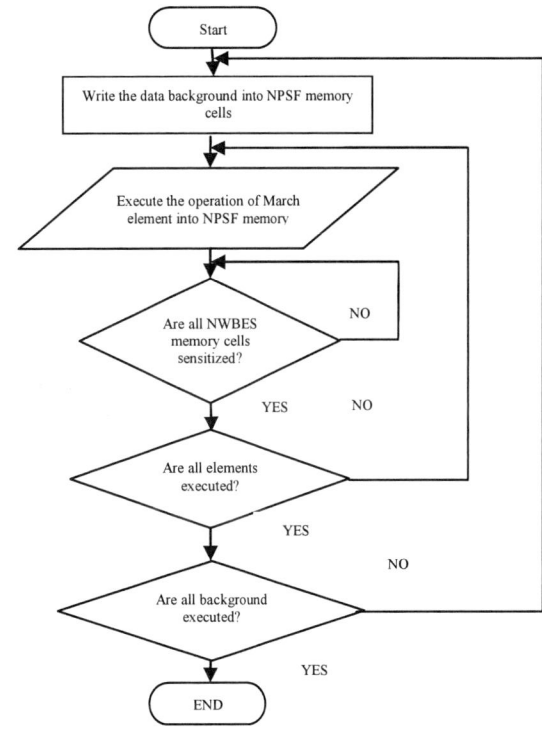

Fig. 2. Procedure of Database development

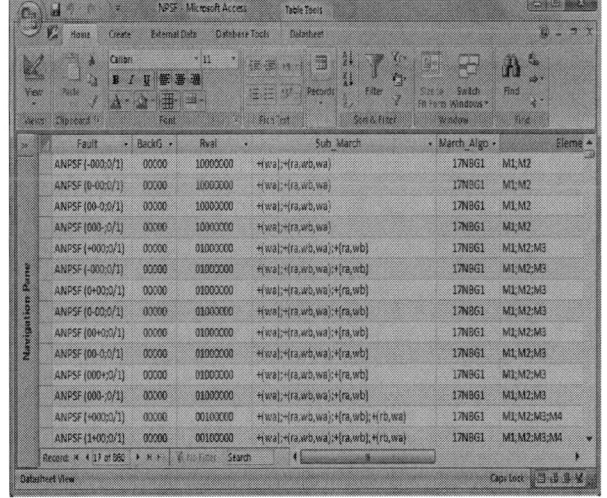

Fig. 3. Database of March 17N

B. ANPSFSG Development

Figure 4 presents the overall architecture of ANPSFSG. This Object Oriented generator will read the two main inputs from the user. The user needs to load the March element that needs to be tested and then choose the data background that is required for the testing. The tool will search through the database according to the loaded March element and selected data background. The process will generate a list of entire possible march elements together with the fault coverage in the form of 'RTable'. A Fault Syndrome table will list the total fault coverage together with fault syndrome for each of the fault. The core algorithm for the ANPSFSG is shown in Fig. 5. User-friendly clickable icons were designed for users to display the Fault Syndrome table and to calculate the diagnostic resolution.

C. Validation of ANPSFSG

In order to validate the effectiveness of the developed ANPSFSG, March 17N is used as a benchmark to ensure that the fault coverage and the fault diagnostic generated by the tools agree with the results of the same algorithms in [6]. After the tool has been verified for detection and diagnostic of NPSF using this March 17N, the database is expanded to cover the detection and diagnostic of March 12N and MarchPS 23N. Fig. 7 illustrates the Graphical User Interface (GUI) of the tool where desired March algorithms and data backgrounds can be inserted as inputs. After executed the program, the fault coverage and the diagnostic resolution can be displayed as outputs as shown in Fig. 8.

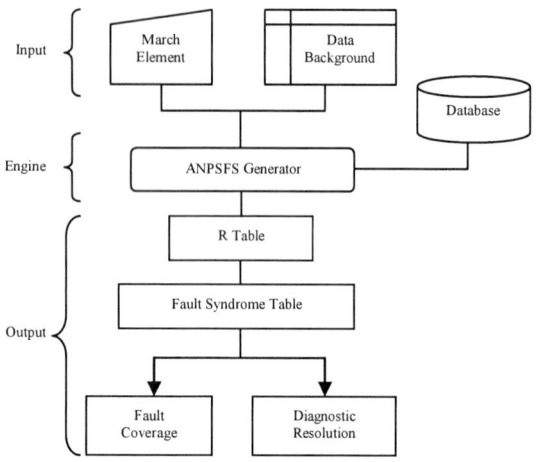

Fig. 4. ANPSFSG Architecture

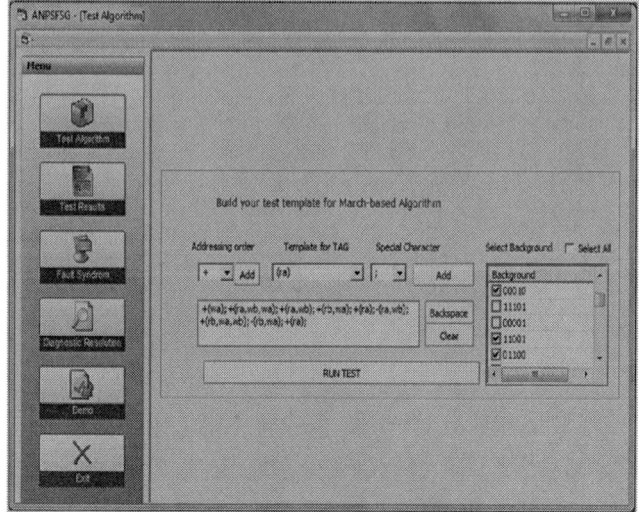

Fig. 7. ANPSFSG Test Algorithm Windows

```
for each element begin
    if (sub_March > 0) and ( R<> 1) then
        begin
            if (sub_March) detect NPSFs and (R<>1) ;
                Run compare (element,Rval,BackG);
        R_value set to 1;
        until all element compared;
            fault syndrome table generated;
            calculate Fault coverage;
            if Diagnostic resolution button press;
                Diagnostic resolution percentage display;
            end if;
        elseif
            begin
                if (sub_March) not detect NPSFs and (R<>0) then;
                    R_value all set to 0;
                    Fault syndrome table not generated;
                end if;
        end if
endfor
```

Fig. 5. Core algorithm for ANPSFSG

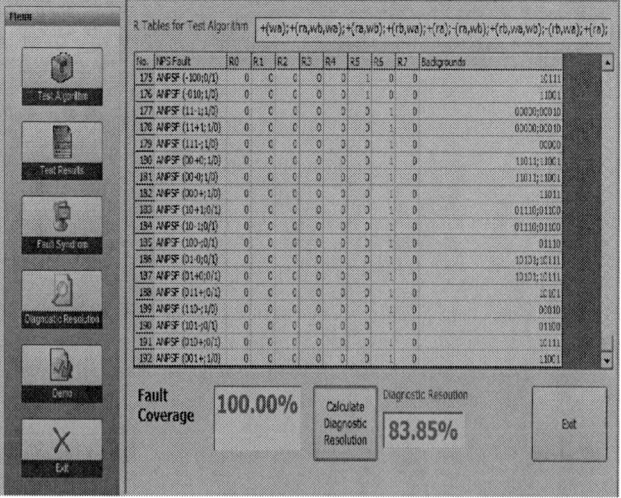

Fig. 8. Fault coverage and Diagnostic Resolution for March 17N

IV. RESULTS AND DISCUSSION

The effectiveness of the ANPSFSG tool to detect and diagnose NPSF by using selected March algorithm are summarized and compared with RAMSES tool as shown in Table I [13,14]. Using manual analysis, one fault type can be detected by different elements of the algorithms hence a single fault can have redundant fault syndrome. Using the ANPSFSG, the redundancy is eliminated and the detection of elements for a single fault is compiled such that a single fault syndrome representing the entire elements that can detect the single fault is produced. In terms of fault coverage, March 12N, March 17N and March 160N has 100% fault coverage for NPSF detection. However, from these three algorithms, March 12N has shorter test length hence makes it more ideal and suitable for NPSF detection. For the diagnose of NPSF, March 17N has higher diagnostic resolution compared to March 12N which is at 83.9% compared to March 12N at only 60.4% as shown in Table I. For MarchPS 23N the diagnostic resolution and the fault coverage is only cover for 33% of the NPSF. Although it has the shortest test length, the results only covers the diagnostic of PNPSF and SNPSF. The tool has shown that March 160N which is the combination of March 17N and additional March 6N gives the best result where 100% diagnostic of NPSF can be achieved. However compared to March 12N, March 17N and MarchPS 23N the test length of March 160N is much longer which can be a disadvantage in term of testing time as longer test length requires longer test time.

TABLE I
VALIDATION OF ANPSFSG TOOL

March Algorithm	March 17N	March 12N	MarchPS 23N	March 160N
NPSF Coverage	ANPSF PNPSF SNPSF	ANPSF PNPSF SNPSF	PNPSF SNPSF	ANPSF PNPSF SNPSF
Total NPSF Detected	100%	100%	33%	100%
NPSF Diagnostic Resolution	84%	60%	33%	100%
Test Length	136N	96N	92N	160N
Number of Background	8	8	4	12
Simulator Tools	RAMSES ANPSFSG	ANPSFSG	ANPSFSG	RAMSES ANPSFSG

V. CONCLUSION

The fault detection and diagnosis of NPSF have been successfully analyzed by using four different March algorithms which are March 17N, March 12N, MarchPS 23N and March 160N. Automated approach to analyze the detection and diagnosis of NPSF by using selected March algorithms has been successfully demonstrated. In order to improve the effectiveness of this generator, the coverage of type 2 neighborhood need to be considering as well as the other type of RAM circuit faults. By having these features, the developed generator can provide a platform to ease the process of diagnosing and selecting March algorithm for an optimum testing result. In addition, the tool can help to analyze andproduce a good March-based algorithm with a better faults

coverage and diagnostic resolution especially for the detection and diagnosis of NPSF in shorter time compared to using traditional manual analysis. For further improvement on this generator, a new embedded memory Build in Selft Test(BIST) can be developed [14,15].

ACKNOWLEDGMENTS

Appreciation goes to the Faculty of Engineering UPM for providing the facilities needed to undertake this project and UniKL-BMI for funding my research work.

REFERENCES

[1] ITRS,2007 *International Technology Roadmap for Semiconductors 2007*, "System Driver. Retrieve on 2007" from www.itrs.net/links/ 2007itrs/2007_chapters/ 2007 System Drivers"

[2] Li, J.F. and Huang, C.D., "An Efficient Diagnosis Scheme for Random Access Memories" *Proc. of the 13th Asian Test Symposium* 0-7695-2235-1/04 pp. 277-282, 2004

[3] Van de Goor, A.J., " Testing semiconductor memories: theory and practice" Netherlands: ComTex, 1999.

[4] Wang, L.T., Wu C.W., and Wen, X. "VLSI Test Principles and Architectures: Design for Testability", San Fancisco: Morgan Kaufmann, 2006.

[5] Mourad, S. and Zorian, Y., "Principles of Testing Electronic Systems. Canada", John Wiley & Sons, 2000.

[6] Cheng, K.L., Tsai, M.F. and Wu, C.W., "Neighborhood pattern-sensitive fault testing and diagnostics for random-access memories", *IEEE Trans. On Computer-Aided Design of Integrated Circuits and Systems*, vol. 21, pp. 1328-1336, 2002.

[7] Cascaval, P., Bennett, S. and Hutanu, C., "Efficient March Tests for a Reduced 3-Coupling and 4-Coupling Faults in Random-Access Memories," *Journal of Electronic Testing: Theory and Applications*, 20, 227–243, 2004.

[8] Li, J.F. and Liang, C.K., Huang, C.T and Wu, C,W., "March-based RAM diagnosis algorithms for stuck-at and coupling faults" *Test Conference, Proc. International* pp.758 – 767, 2001.

[9] Cheng, K.L., Tsai, M.F. and Wu, C.W.,"Efficient neighborhood pattern-sensitive fault test algorithms for semiconductor memories", *Proc. VLSI Test Symposium*, pp. 225-230, 2001.

[10] R.R. Julie, W.H. Wan Zuha and R.M. Sidek, "12N test procedure for NPSF testing and diagnosis for SRAMs" *Proc. IEEE International Conference on Semiconductor Electronics*, pp. 430-435, 2008.

[11] Yarmolik, V., Klimets, Y. and Demidenko, S., "March PS (23N) test for DRAM pattern-sensitive faults" *Proc. Seventh Asian Test Symposium* pp. 354-357, 1998.

[12] Huzum, C., Cascaval, P., "A Multibackground March Test For All Static Simple Neighborhood Pattern-Sensitive Faults in RAMs" *System Theory, Control and Computing(ICSTCC),15th International Con.* pp.1-6, 2011.

[13] Wu, C.F., Huang, C.T., Wang, C.W., Cheng, K.L. and Wu, C.W. "Error Catch and Analysis for Semiconductor Memories Using March Tests" *IEEE/ACM International Con. 10.1109/ ICCAD*, pp. 468 – 471, 2000.

[14] Wu, F.W., Huang, C.T., Cheng, K.L and Wu, C.W., "Fault Simulation and Test Algorithm Generation", *IEEE Transactions On Computer-Aided Design of Integrated Circuits and Systems* vol.21, no.4, pp.480-490, 2002.

[15] De Carvalho, M., Bernardi, P.,Sonza Reorda,, " Optimized Embedded Memory Diagnosis", *Design and Diagnostics of Electronic Circuits & Systems (DDECS),IEEE 14th International Symposium* p.p: 347-352, 2011.

[16] Bernardi, P., Grosso, M., Sonza Reorda, M., Zhang, Y., "A Programmable BIST for DRAM testing and diagnosis", *Test Conference , 2010 IEEE International*, p.p: 1-10, 2010.

978-1-4673-2395-6/12 $31.00 © 2012 IEEE

Frequency Reduction in Quantum Dot Cellular Automata

Bhupesh Bishnoi, *Associate Member, IEEE,* Diwakar Agrawal, Vikas Nandal,
Akshay Salimath and Bahniman Ghosh
Department of Electrical Engineering, Indian Institute of Technology Kanpur, Kanpur, India 208016
Email: bbishnoi@iitk.ac.in, bishnoi@ieee.org

Abstract- **In this paper, we have discussed the frequency aspects of the Quantum dot cellular automata (QCA) circuits. Techniques to operate a QCA circuit at a frequency which is a fraction of the universal clock frequency are presented. For the purpose many intermediate circuits like periodic signal generator and master-slave flip-flop circuits are designed. Prospective applications are suggested and their implementation details are discussed.**

I. INTRODUCTION

Quantum Cellular Automata [1][2][3][4][5] is being seen as a technology with the calibre to replace the CMOS technology which is reaching the physical limit of the device size [6]. Various advancements have been made in this field which enable QCA devices to do almost all the functions that can be done by the CMOS devices. Conventional QCA devices use the quantum dots and electron-electron repulsion to transfer the data and operate on it. A QCA cell consists of four quantum dots having two electrons. Due to repulsion between the electrons, electrons take diagonal positions [7]. The two diagonal positions represent state '0' and '1'. Various geometric configurations of QCA cells can perform different computational functions [8].

In this paper we have identified a new problem and tried to provide a solution for the same. The frequency of operation is a very important factor in operation of any circuit. In many digital and analog circuits there are parts which have different frequency of operation than the rest of the circuit. Use of counters is an example of the same. In QCA circuits, transfer of data from one part of the circuits to another depends on the frequency of the QCA clock and hence one solution for having a different frequency is to change the clock frequency. The problem with that is that in this case designer will have to divide the whole circuits into parts in which the QCA clock will have different frequencies and use of a universal clock will not be possible.

A lot of work has been done for frequency divider circuits for CMOS circuits, but by far such circuits have not been discussed in QCA. We, in this chapter, have presented circuits for generating the signals of frequency which are a fraction of QCA clock frequency and are similar to CMOS clock in nature.

Clocking signals of lower frequency are generally used for sampling purposes. We have considered a sampling problem (discussed in section V) and tried to create a solution for the

same within QCA technology. In the process we designed and used latches and master-slave edge triggered flip-flops.

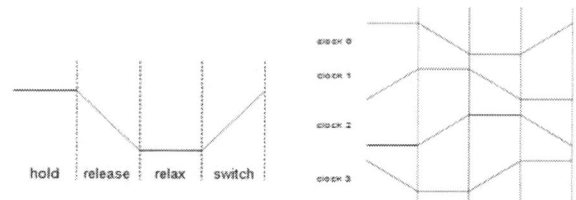

Fig.1. Four phase clocking mechanism for QCA devices [7]

QCA devices use a unique clocking mechanism [9]. QCA clocks work by changing the tunnelling barrier among the quantum dots. Figure 1 shows the different phases of a QCA clock and phase difference between various QCA clocks.

In this paper we have tried to create circuits which can generate the signals which are fraction of the QCA clock frequency. These signals are used to implement the functions like sampling, with the help of periodic signal generator and edge triggered circuits. Each circuit presented is simulated and verified using QCADesigner [10][11] version 2.0.3.

II. PERIODIC SIGNAL GENERATOR

As discussed in the Section I above, the QCA circuits have clock which synchronize and control the data/information flow. Here we present a method to operate circuits at frequency which is fraction of the QCA clock frequency. By QCA clock frequency we mean the frequency of all four phases of the clock taken together.

To achieve the above goal we first have to create a periodic signal which has a frequency f given as follows

$$f = \frac{f_0}{2n}$$

Where f_0= QCA clock frequency and n=1, 2, 3…..

We can achieve this goal by using an astable multivibrator [12] having odd number of inverters as shown in figure 2.

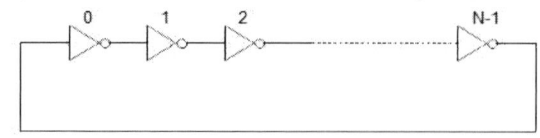

Fig.2. Astable multivibrator [12]

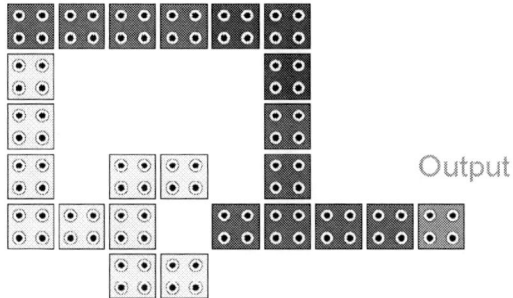

Fig.3. QCA layout of a single inverter Astable multivibrator

Figure 3 shows the QCA implementation of astable multivibrator circuit with a single inverter. Thus for a frequency output of $f0/2n$ the number of clock zones in the feedback loop must be $4*n$. Figure 4 shows the output of the circuit at two different runs. We shall discuss the difference in the outputs in the next section.

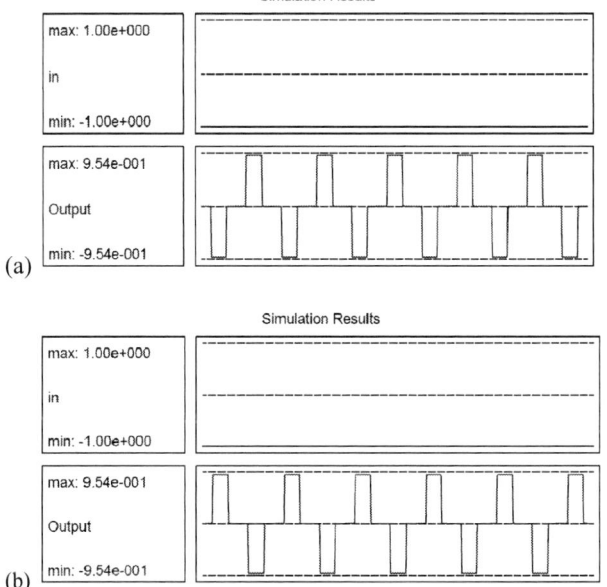

Fig.4. Output of circuit in figure 3 starting with (a) bit '0' (b) bit '1'

III. DETERMINISTIC DESIGN

The circuit described in Figure 3 is not deterministic. It means that since the initial state can be '0' as well as '1', the output can be any one of the figures shown in figure 4. Figure 4 shows the output of two different runs of circuit in figure 3 showing different starting bits. Hence to make the circuit deterministic we use the multiplexer circuit [13]

Whose output is fed back to one of its inputs through an inverter and other input is fixed to the required initial value. The delay of the feedback path will determine the frequency of

the output signal. Figure 5 shows the QCA layout of a circuit which generates an output with frequency 1/4th of the QCA clock frequency.

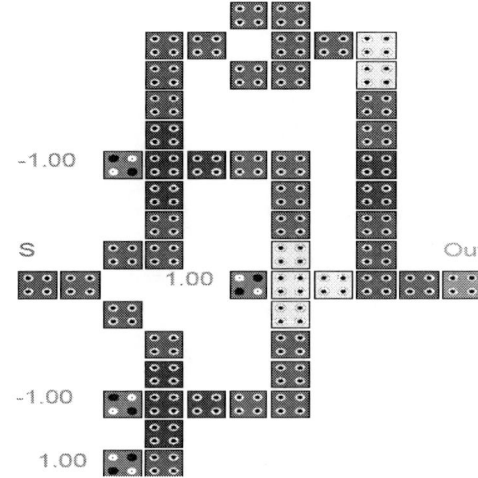

Fig.5. Periodic deterministic signal generator for frequency $f_0/4$

Similarly a signal for different values of n can be generated just by changing the number of clock phases in the feedback loop. To ensure the initial value of the output signal to be of the required value, the select bit of the multiplexer is initialized with the '1' or '0' for n number of times in the starting to make the periodic signal deterministic. For example for an $f_0/4$ signal, the select bit is initialized with '1' for first two clock periods and then it is '0' afterwards. Note that 4 clock phases are by default present in the part of multiplexer which is used in the loop. Figure 6 shows the output of the circuit which is of frequency $f_0/4$.

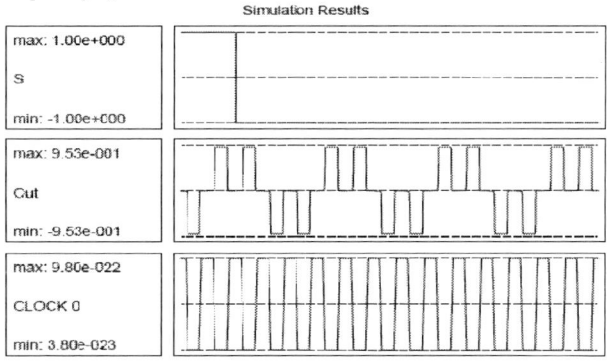

Fig.6. A periodic signal of frequency $f_0/4$

IV. LATCHES AND REGISTERS

In this section we have designed QCA latches and edge triggered flip-flops. We will see in the next section that they will be very useful for the purpose of the frequency reduction applications like sampling and low frequency counters. Figure 7 shows the circuit layouts for positive and negative latches. Positive latch transfers the input to the output whenever the value of 'S' is '1'. Rest of the time it latches the same value at

978-1-4673-2395-6/12 $31.00 © 2012 IEEE

output. Negative latch does the same thing except that the input is transferred to the output when value of 'S' is '0'.

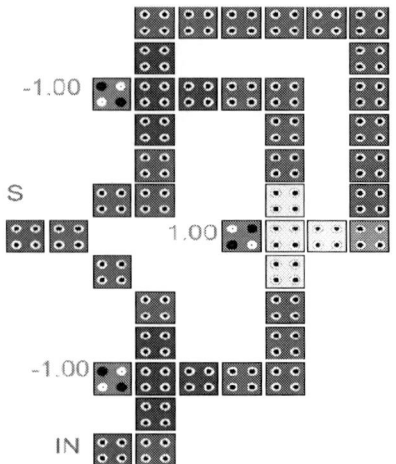

Fig.7. (a) a Positive QCA latch

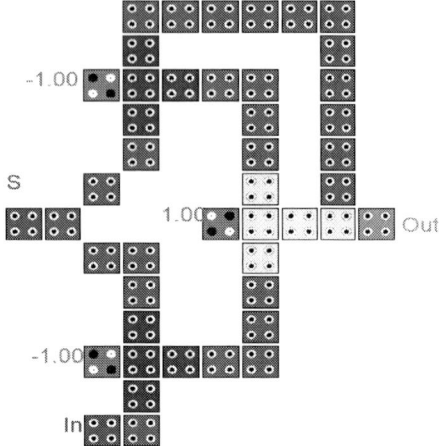

Fig.7. (b) Negative QCA latch

The series combination of positive and negative latches will form a master slave edge triggered flip flop. The type of the latter latch decides the type of edge triggered circuit. If the combination is positive latch followed by negative latch, then the circuit is negative edge triggered flip flop. For the reverse connection the circuit is positive edge triggered [14][15]. We have used a negative edge triggered flip flop in section V for sampling an incoming data stream.

V. APPLICATION AND DISCUSSION

In this section we will try to explain above discussion with the help of an application. In communication systems many a times we require to sample bits of an incoming data stream. Figure 8 shows a data separator sampling every other bit and creating two different bit streams of half frequency.

To achieve this we will require a signal that is of half the frequency of the inherent QCA clock. The circuit shown in figure 5 will be useful here. We shall also use an edge triggered register circuit made from series combination of positive and negative latches.

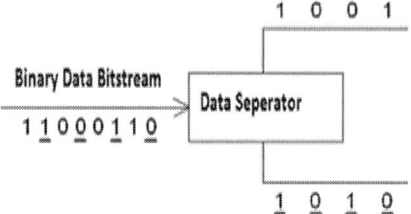

Fig.8. A data separator

Now, we will discuss an example to sample every forth bit of an incoming data stream. To sample every forth bit of a data stream we will require a signal which is of 1/4th of the frequency of the input stream (same as the QCA clock frequency). We want to sample the data stream once every four universal clock cycle. We shall use a negative edge-triggered master slave flip flop for the purpose. Figure 9 shows the layout of a negative edge triggered circuit. This also has a periodic signal generator of frequency f0/4 at the S input.

Circuit in figure 9 consists of three parts, which are signal generator, positive latch and negative latch. Positive and negative latches are connected in series making it a negative edge triggered flip flop. The signal generator generates the signal of desired frequency i.e. $f_0/4$. It is to be noted that the output of the periodic signal generator is fed in both latches with same clock phase difference i.e. the latency of the 'S' signal of both the latches is same.

For a negative edge triggered master slave flip flop the positive latch takes in the data when the signal (S) value is positive, But it is transferred to the negative latch (and hence to output) only when the next value of the signal is negative. Hence the input goes to output only once in one period of signal 'S'. Table 1 shows the simulation results of the negative edge triggered circuit for a particular set of inputs. Figure 10 shows the output of the circuit for an input sequence given in table 1.

Table1. Inputs and outputs for circuit in fig.9

A/B	-	1	1	0	0	1	1	0	0	1	1	0	0	1	1
Input	-	1	0	1	1	1	0	1	1	1	0				
Out1	-	-	1	0	0	0	1	0	0	0	1	0	0	0	
Out final	-	-	-	-	0	0	0	0	0	0	0	0	0	0	0

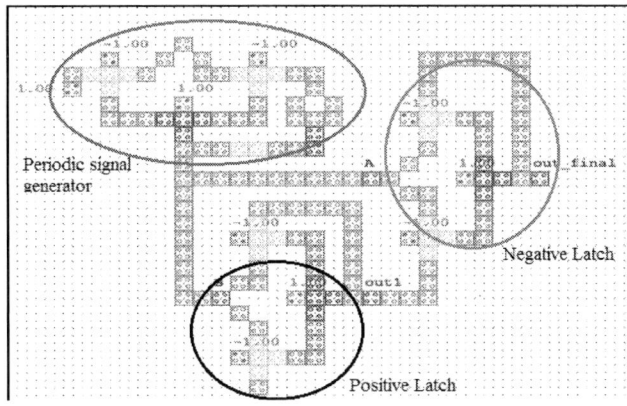

Fig.9. A negative edge triggered master slave flip flop with f0/4 periodic signal generator

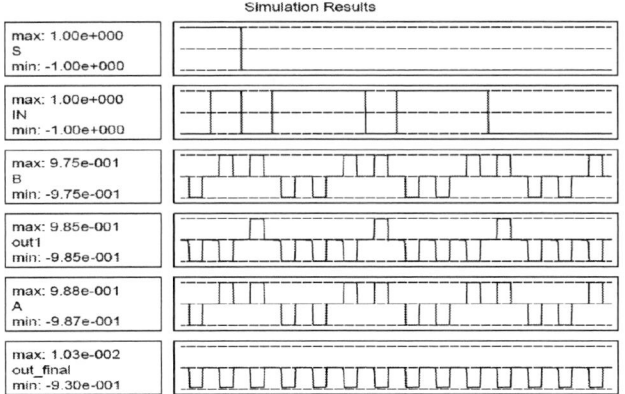

Fig.10. Output of figure 9 for inputs shown in table 1

It should be noted that first bit in the S, first two bits in out1 and first four bits in out_final should not be considered in the judgment since the effect of the input will appear in the output only after the latency period of the circuits; which is equal to 1 for signal generation circuit, 2 for out1 (1 for signal generator + 1 for positive latch). For out_final first 4 cycles output is meaningless because a negative latch's first meaningful output will be when the negative cycle arrives, hence the meaningful values will be after 4 clock periods.

Table 1 show that the input data stream had a '0' at every fourth place. The negative edge triggered circuit samples the input stream at every negative edge which is 0 at every negative edge (A/B value going from 1 to 0). Hence the output is all '0'. If we have to sample every forth bit of the data stream with the starting bit number 1 then we will just replicate the same circuit but the input to select bits of the latches will be delayed by 1, 2 and 3 clock periods respectively.

VI. CONCLUSION

We have tried to address a new problem of frequency reduction within QCA technology. Such circuits which can generate a signal of reduced frequency can be used in various situations like low frequency counters, sampling problems etc. We have showed different ways to generate periodic signals of different frequencies which are fraction of QCA clock frequency and imitate a similar behavior as CMOS clocks. First, a periodic signal generator using the astable multivibrator was used which had the shortcoming of being non deterministic. To solve this problem we used a multiplexer and created a deterministic periodic signal by having various delays in the feedback loop created in the multiplexer circuit. The different delays caused output to be of different frequencies. Circuit for master slave edge triggered flip-flop was presented and simulation results were shown. These circuits were combined together to solve the problem of sampling an incoming binary data stream.

REFERENCES

[1] C.S. Lent , P. D. Tougaw and W. Parod, "Quantum cellular automata," Nanotechnology, vol. 4, no.1, pp.49-57, Jan. 1993
[2] C. S. Lent and P. D. Tougaw. A device architecture for computing with quantum dots. *Proceedings of the IEEE*, 85:541, 1997.
[3] P.D. Tougaw, C.S. Lent, "Logical devices implemented using quantum cellular automata", J. Appl. Phys., 75 (3) 1818, 1994
[4] M. Macucci, Quantum Cellular Automata, Imperial college Press, 2006
[5] F. Lombardi, J. Huang, X. Ma, M. Momenzadeh, M. Ottavi, L. Schiano,and V. Vankamamidi, "Design and Test of Digital Circuits by Quantum-Dot Cellular Automata", F. Lombardi and J. Huang, Eds. Artech House, 2008.
[6] International Technology Roadmap for Semiconductors, Executive summary, http://www.itrs.net/Links/2005ITRS/ExecSum2005.pdf, 2005 Edition
[7] V.Vankamamidi, M. Ottavi and F. Lombardi, "Clocking and Cell Placement for QCA," *Nanotechnology, 2006. IEEE-NANO 2006. Sixth IEEE Conference on* , vol.1, no., pp. 343- 346
[8] A. Gin, P. D. Tougaw and S. Williams, "An alternative geometry for quantum cellular automata," *J. Appl. Phys.*, vol. 85, pp. 8281, 1999.
[9] K. Hennessy, C. S. Lent, "Clocking of Molecular Quantum-Dot Cellular Automata," *Journal of Vacuum Science and Technology*, vol 19(5), 2001, pp. 1752-1755.
[10] http://www.mina.ubc.ca/qcadesigner consulted 20 Nov. 2011.
[11] K. Walus, T. J. Dysart, G. A. Jullien, and R. A. Budiman. "QCADesigner: a rapid design and simulation tool for quantum-dot cellular automata". *Nanotechnology, IEEE Transactions on*, 3(1):26–31, 2004.
[12] A. C. Jan, M. Rabaey and B. Nikolic, "Digital Integrated Circuits". Upper Saddle River, NJ, Prentice-Hall, 2003.
[13] M. Askari and M. Taghizadeh "Logic Circuit Design in Nano-Scale using Quantum-Dot Cellular Automata", *European Journal of Scientific Research* Vol.48 No.3, pp.516-526, 2011.
[14]N. Weste and D. Harris, "CMOS VLSI Design". Boston, MA: AddisonWesley, 2005.
[15] V. Stojanovic and V.G. Oklobdzija, "Comparative analysis of master-slave latches and flip-flops for high-performance and low-power systems," *Solid-State Circuits, IEEE Journal of* , vol.34, no.4, pp.536-548, Apr 1999

Design and Implementation of Reversible Logic Based Bidirectional Barrel Shifter

O.Anjaneyulu[1],*Member, IEEE*, T.Pradeep[2], C.V.Krishna Reddy[3],*Member, IEEE*
KITS,Warangal[12], NNRESGI-Hyd[3]
Andhra Pradesh, India
Email: anjaneyulu_o@yahoo.com[1], Pradeep.thumma@yahoo.com[2], cvkreddy2@gmail.com[3]

Abstract—Embedded digital signal processors and general purpose processors will use barrel shifters to manipulate data. This paper will present the design and implementation of the barrel shifter that performs logical shift right, arithmetic shift right, rotate right, logical shift left, arithmetic shift left, and rotate left operations. The main objective of the upcoming designs is to increase the performance without proportional increase in power consumption. In this regard reversible logic has become most popular technology in the field of low power computing, optical computing, quantum computing and other computing technologies. Device scaling is limited by the power dissipation; and demands better power optimizations methods. Techniques like Energy recovery, Reversible Logic are becoming more and more prominent special optimization techniques in Low Power VLSI designs. Rotating and data shifting are required in many operations such as logical and arithmetic operations, indexing and address decoding etc. The feynman gate will remove the fanout. By comparing the quantum cost, number of ancilla bits and number of garbage outputs the design is evaluated. The performance characteristics of the proposed design are evaluated, and the transistor cost, Garbage outputs and Quantum Cost are also calculated. The performance characteristics analysis is carried out in Xilinx environment.

Keywords- barrel shifters, quantum cost, ancilla bits, verilog, garbage output, nanotechnology, quantum computing, fredkin gate, feynman gate.

I. INTRODUCTION

Rotating and shifting data is required in several applications including variable-length coding, arithmetic operations, and bit-indexing. Consequently, barrel shifters are capable of shifting or rotating data in a single cycle and are commonly found in both digital signal processors and general purpose processors. In reversible system information is not erased. Thus in reversible gates number of inputs and outputs are equal which means that the input stage can always be retained from the output stage. If a bit is erased in an irreversible circuit then it will dissipate $kTln2$ joules of heat energy where k is the Boltzmann's constant and T is the absolute temperature of environment [4]. There won't be dissipation of $kTln2$ joules of heat energy if the operations are performed in reversible manner based on reversible logic circuits [3]. Based on this observation, Bennett [3] showed, for a reversible computer the heat dissipation is exactly *kTln1* which is logically zero. Reversible logic also has the applications in emerging nanotechnologies such as quantum

dot cellular automata, quantum computing, optical computing and low power computing, etc.

Low power design also plays a significant role in high-performance integrated circuits such as microprocessors and other high-speed digital computational circuits. The power consumption in microprocessors is projected to grow linearly in proportion to their die size and clock frequency. Various cooling systems have been introduced to reduce the heat from power dissipation and keep the chip temperature at an admissible level. This in turn has increased the packaging cost, which results in large revenue.

II. BASIC REVERSIBLE GATES

A *Reversible Gate* is an n-input, n-output circuit. To maintain the reversibility property of reversible logic gates several dummy output signals are needed to be produced in order to equal the number of input to that of output. These signals are commonly known as *Garbage Outputs*. The quantum cost of reversible gate is equal to the number of 1x1 and 2x2 reversible gates needed to design a 3x3 reversible gate. The quantum cost of all 1x1 and 2x2 reversible gates are considered as unity [8], [7], [2]. The 3x3 reversible gates are designed from 1x1 NOT gate, and 2x2 reversible gates such as Controlled-V and Controlled-V+, the Feynman gate which is also known as Controlled NOT gate.

A NOT gate is 1x1 gate represented as shown in Fig. 1. Its quantum cost is unity since it is a 1x1 gate.

$$A \quad \oplus \quad P = \bar{A}$$

Fig. 1. NOT GATE

The input vector, *Iv* and output vector, *Ov* for 2*2 *Feynman Gate (FE)* is defined as follows: *Iv = (A, B)* and *Ov = (P = A* and *Q=A ^B)*. Feynman gates are typically used as copying gates. If *Iv = (A, B=0)* then *Ov = (P =A* and *Q=A)*. Fanout is not allowed in reversible logic. Feynman gate is helpful in this regard as it can be used for copying the signal by which it avoids the fanout problem as shown in Fig.2(c).

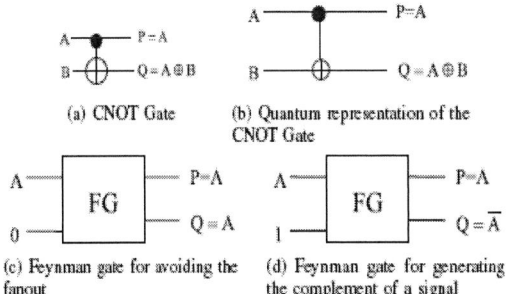

(a) CNOT Gate   (b) Quantum representation of the CNOT Gate

(c) Feynman gate for avoiding the fanout   (d) Feynman gate for generating the complement of a signal

Fig. 2. CNOT gate, its quantum implementation and its useful properties

The input and output vector for 3*3 ***Fredkin gate (FR)*** [1] are defined as follows: $Iv = (A, B, C)$ and $Ov = (P=A, Q= A'B \wedge AC$ and $R = A'C \wedge AB)$. Figure 3(a) shows the block diagram of a Fredkin gate. A Fredkin gate can work as 2:1 MUX, as it is able to swap its other two inputs depending on the value of its first input. Thus when A=0 the outputs P and Q will be directly connected to inputs A and B and if A=1 the inputs B and C will be swapped resulting in the value of the outputs as Q=C and R=B.

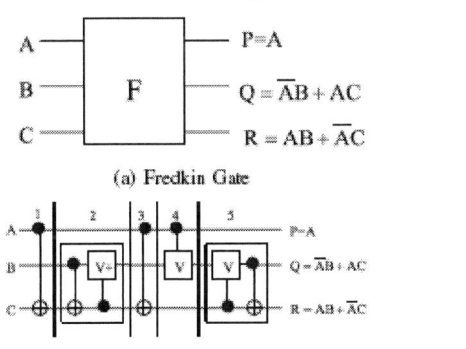

(a) Fredkin Gate

(b) Quantum representation of the Fredkin Gate

Fig. 3. Fredkin Gate and its quantum implementation

The quantum implementation of a Fredkin gate with a quantum cost of 5 is shown in Figure 3(b). In Fig. 3(b) each dotted rectangle is equivalent to a 2x2 Feynman gate and the quantum cost of each dotted rectangle is considered as 1 [8]. The same assumption is used for calculating the quantum cost of the Fredkin gate. Thus, the quantum cost of the Fredkin gate is 5 as it consists of 2 dotted rectangles, 1 Controlled-V gate and 2 CNOT gate.

III. BARREL SHIFTER

A Barrel shifter is an '*n*' input and '*n*' output combinational logic circuit in which k select lines controls the bit shift operation. Barrel shifter can be unidirectional allowing data to be shifted only to left (or right), or bi-directional which provides data to be rotated or shifted in both the directions. Among the different designs of barrel shifter, the logarithmic barrel shifter is most widely used because of its simple design, less area and the elimination of the decoder circuitry. The proposed work presents the design and implementation of reversible bidirectional arithmetic and logical barrel shifter that can perform six operations: logical right shift, arithmetic right shift, right rotate, logical left shift, arithmetic left shift and left rotate. The existing shifter is complex in design and requires large number of gates. As a result the total number of garbage outputs is high. Thus there is great room for improving the circuit complexity, total number of gates and garbage outputs, delay and quantum cost.

IV. PROPOSED REVERSIBLE BIDIRECTIONAL BARREL SHIFTER

The proposed design of reversible bidirectional barrel shifter can perform logical right & left shifting, arithmetic right & left shifting, rotating right & left operations. Table I shows that for different values of control signals sra, sla, rot and left the operations that can be performed by a (8,3) reversible bidirectional arithmetic and logical shifter.

TABLE I
OPERATION PERFORMED BY A (N, K) REVERSIBLE BIDIRECTIONAL BARREL SHIFTER

Operation performed	Control signal values			
Logical right shift	Left=0	Rot=0	Sra=0	Sla=0
Arithmetic right shift	Left=0	Rot =0	Sra=1	Sla=0
Rotate right	Left=0	Rot=1	Sra=0	Sla=0
Logical left shift	Left=1	Rot=0	Sra=0	Sla=0
Arithmetic left shift	Left=1	Rot=0	Sra=0	Sla=1
Rotate left	Left=1	Rot=1	Sra=0	Sla=0

In this design, the input data is represented as i7, i6, i5, i4, i3, i2, i1, i0 while the shift value is controlled by select signals represented as S2S1S0 and the output data is obtained as shown in Table II .

TABLE II
SHIFT AND ROTATE OPERATION OUTPUT FOR K = 3

Operation	Y
3-bit shift right logical	0 0 0 a7a6a5a4a3
3-bit shift right arithmetic	a7a7a7a7a6a5a4a3
3-bit rotate right	a2a1a0a7a6a5a4a3
3-bit shift left logical	a4a3a2a1a0 0 0 0
3-bit shift left arithmetic	a7a3a2a1a0 0 0 0
3-bit rotate left	a4a3a2a1a0a7a6a5

The proposed reversible bidirectional arithmetic and logical barrel shifter design approach is illustrated as shown in Fig. 4 with an example of a (8,3) barrel shifter. The barrel shifter performs the various operations such as logical right shift, logical left shift, rotate left etc. depending on the values of sra, sla rot and left control signals.

The design of a reversible barrel shifter can be divided into six modules: (i) Data reversal control unit-I, (ii) Arithmetic right shift control unit, (iii) Shifter or rotation unit which consists of three sub-modules that performs Stage I, Stage II and Stage III operations, (iv) Rotation unit, (v) Arithmetic left shift control unit, (vi) Data reversal control unit-II.

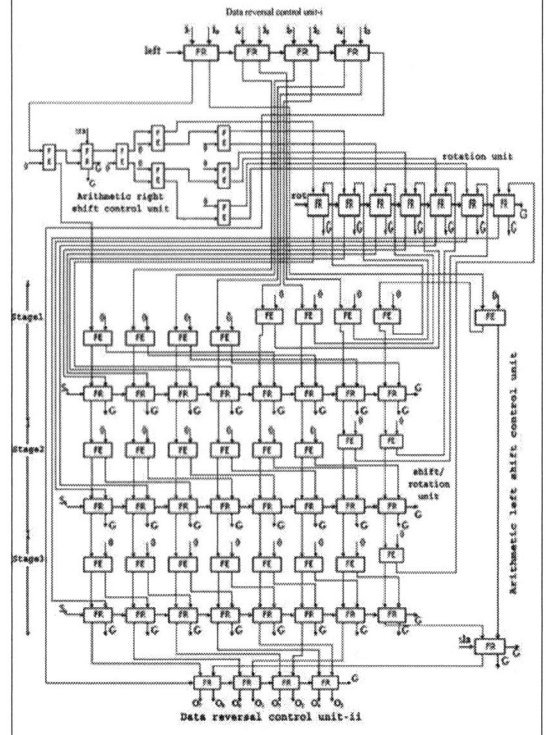

Fig. 4. Proposed (8,3) reversible bidirectional barrel shifter
*FE represents Feynman Gates, FR represents Fredkin gates and G represents the garbage outputs

V. PERFORMANCE ANALIZATION

To avoid the fanout problem, in the proposed design Feynman gate is used. Chains of n/2 Fredkin gates are used in data reversal unit-I and data reversal unit-II. The arithmetic right shift control unit uses one Fredkin and   -1 Feynman gates. Chain of n Fredkin gates and n Feynman gates are used in shifter or rotation unit at each stage. Rotation unit fredkin gates for m=0 to (k-1) for each stage. One Fredkin gate and one Feynman gate is used in arithmetic left shift control unit.

A. Ancilla input Bits

The table III shows the number of ancilla bits required to design a reversible bidirectional barrel shifter for different values of n and k.   +(n*k) Feynman gates are required to design a (n,k) reversible bidirectional barrel shifter. Each Feynman gate requires one ancilla input bit to copy the input

data. Additionally, the Fredkin gate used in arithmetic right shift control unit requires one ancilla bit.

TABLE III
ANCILLA INPUTS IN (N, K) REVERSIBLE BIDIRECTIONAL BARREL SHIFTER

n/k	n=4	n=8	n=16	n=32	n=64
K=2	13	21	37	69	133
K=3		33	57	105	201
K=4			81	145	273
K=5				193	353
K=6					449

B. Quantum Cost

Table IV shows the quantum cost for a reversible bidirectional barrel shifter for different *n* and *k* values. The number of Feynman and Fredkin gates used will decide the quantum cost of (*n, k*) reversible bidirectional barrel shifter. The quantum cost of the Feynman gate is considered as one, while the quantum cost of the Fredkin gate is considered as five.

TABLE IV
QUANTUM COST OF (N, K) REVERSIBLE BIDIRECTIONAL BARREL SHIFTER

n/k	n=4	n=8	n=16	n=32	n=64
K=2	137	165	301	573	1117
K=3		237	421	789	1525
K=4			565	1029	1957
K=5				1317	2437
K=6					3013

C. Garbage Outputs

Each Fredkin gate in the chain of n Fredkin gates produces atleast one garbage output except the last Fredkin gate which produces two garbage outputs. Two garbage outputs are produced by Fredkin gate which is used in the design of arithmetic left shift control unit and arithmetic right shift control unit. One garbage output is produced by last Fredkin gate of the data reversal control unit-II as the control signal left cannot be utilized further.

TABLE V
GARBAGE OUTPUTS IN (N, K) REVERSIBLE BIDIRECTIONAL BARREL SHIFTER

n/k	n=4	n=8	n=16	n=32	n=64
K=2	19	27	43	75	139
K=3		40	64	112	208
K=4			89	153	281
K=5				202	362
K=6					459

VII. IMPLEMENTATION AND RESULTS

The proposed design is functionally verified and the results are verified. The timing report was obtained after obtaining the netlist for the structural model of the digital implementation. The transistor cost was found in analog flow. Functionality was verified in Xilinx.

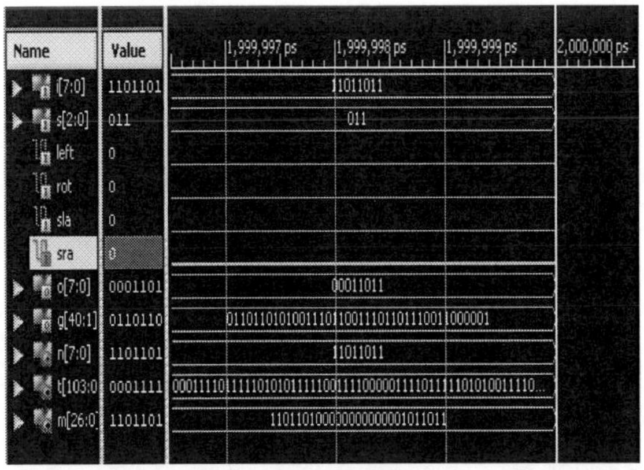

(a)

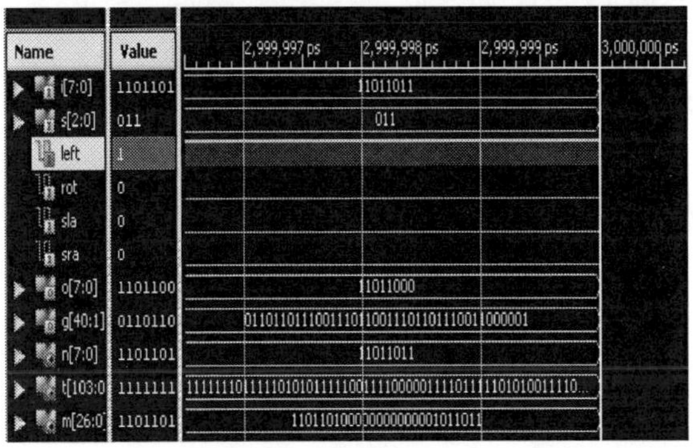

(b)

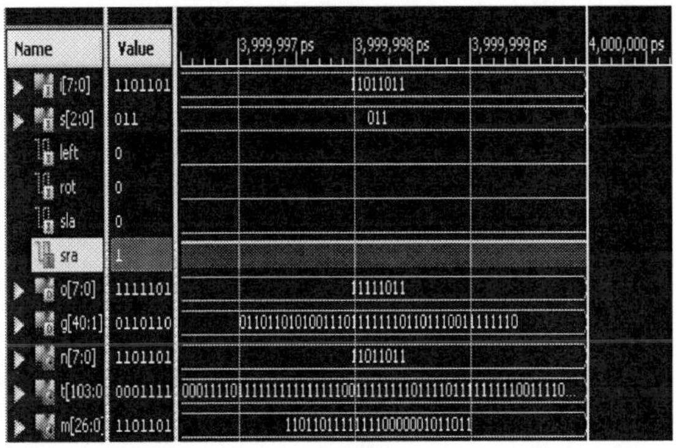

(c)

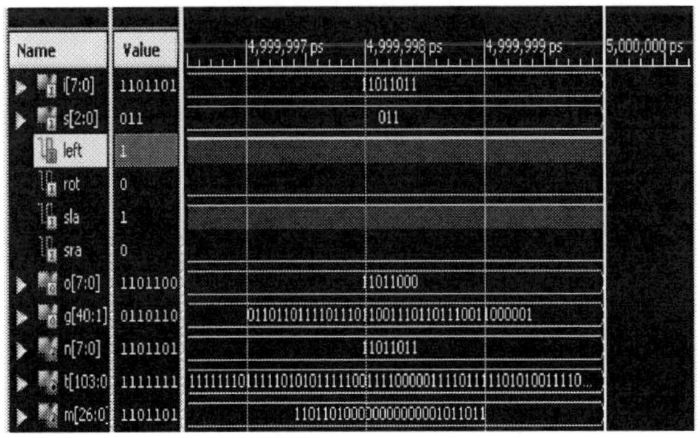

(d)

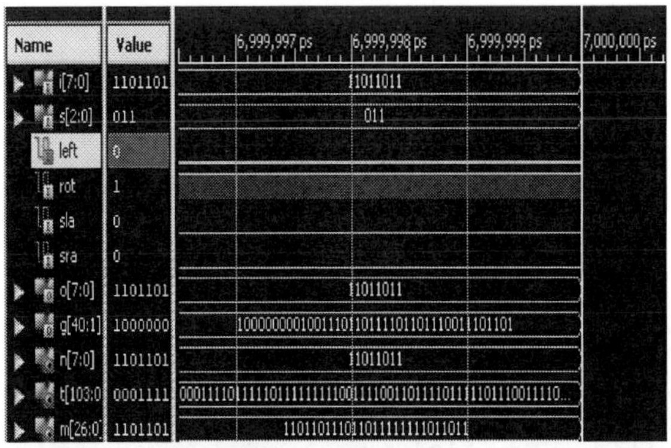

(e)

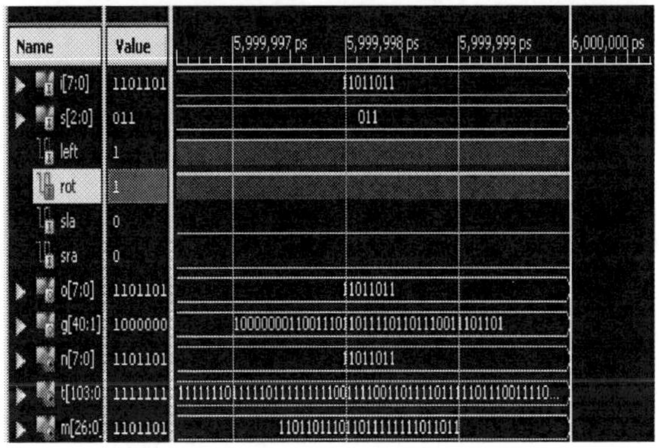

(f)

Figure.5: (a)-(f): A 8-bit bi-directional barrel shifter simulation result for all combinations outputs.

The universal barrel shifter is built by adding the rotation unit and the verification results are as follows. The schematic is a test circuit for input combination S=011, a square wave is applied to control signals to test for all the operations.

For the HDL structural design, the test vectors for excitation has been provided, and the response is as shown in Figure 5. Here the input reference vector is i=11011011.

Synthesis report

Final Results

RTL Top Level Output File Name	: modbar.ngr
Top Level Output File Name	: modbar
Output Format	: NGC
Optimization Goal	: Speed
Keep Hierarchy	: No

Design Statistics

# IOs	: 63

Cell Usage :

#	BELS	: 95
#	LUT2	: 8
#	LUT3	: 31
#	LUT4	: 51

#	MUXF5	: 5
#	IO Buffers	: 63
#	IBUF	: 15
#	OBUF	: 48

Timing constraints

Delay: 12.638ns (Levels of Logic = 9)
Source: left (PAD)
Destination: o<5> (PAD)

Data Path: left to o<5>

Cell: in->out	Gate fanout	Net Delay	Delay	Logical Name (Net Name)
IBUF:I->O	23	1.106	1.174	left_IBUF (left_IBUF)
LUT3:I0->O	20	0.612	0.940	f1/Mxor_e_Result1 (n<7>)
LUT4:I3->O	3	0.612	0.454	f37/Mxor_e_Result 1 (m<15>)
LUT4:I3->O	3	0.612	0.454	f15/Mxor_e_Result _and00011 (f15/Mxor_e_Resul t_and0001)
LUT4:I3->O	2	0.612	0.532	f23/f_and00011 (f23/f_and0001)
LUT4:I0->O	4	0.612	0.502	f33/e_and000011 (N3)
LUT4:I3->O	1	0.612	0.000	f33/Mxor_f_Result 1 (f33/Mxor_f_Result)
MUXF5:I1->O	1	0.278	0.357	f33/Mxor_f_Result _f5 (o_2_OBUF)
OBUF:I->O	3.169			o_2_OBUF (o<2>)

VII. CONCLUSION

In this paper An Efficient Design of Reversible Logic Based Bidirectional Barrel Shifter has been proposed and implemented. The design of the proposed bidirectional shifter is done using Fredkin gates and Feynman gates. The number of garbage outputs, the number of ancilla inputs and the quantum cost of the (n,k) reversible bidirectional barrel shifter increase more rapidly by varying n and keeping k as a constant compared to the designs in which n is kept as a constant while k is varied. The functional verification of the proposed design of the reversible barrel shifters are performed through simulations using the Verilog HDL flow in Xilinx for reversible circuits. The design of bidirectional barrel shifter is been evaluated in terms of garbage outputs, ancilla inputs and the quantum cost. The proposed design of reversible bidirectional barrel shifter can perform logical right shifting, arithmetic right shifting, rotating right, logical left shifting, arithmetic left shifting and rotating left operations.

REFERENCES

[1] S. Kotiyal, H. Thapliyal, and N. Ranganathan "Design of a reversible bidirectional barrel shifter," in *Proc. of the 11th IEEE International Conference on Nanotechnology,* Portland, Oregon, USA, Aug 2011, pp. 463-468.

[2] I. Hashmi and H. Babu, "An efficient design of a reversible barrel shifter," in *VLSI Design, 2010. VLSID '10. 23rd International Conference on,* Jan 2010, pp. 93–98.

[3] S.Gorgin and A. Kaivani, "Reversible barrel shifters," in Proc. 2007 Intl. Conf. on Computer Systems and Applications, Amman, May 2007, pp. 479–483.

[4] C.H. Bennett, "Logical reversibility of computation," *IBM J. Research and Development,* vol. 17, pp. 525–532, Nov. 1973.

[5] A. Peres, "Reversible logic and quantum computers," *Phys. Rev. A, Gen. Phys.,* vol. 32, no. 6, pp. 3266–3276, Dec. 1985.

[6] S. Kotiyal, H. Thapliyal, and N. Ranganathan, "Design of a ternary barrel shifter using multiple-valued reversible logic," in *Proc. of the 10th IEEE International Conference on Nanotechnology,* Seoul, Korea, Aug. 2010, pp. 1104–1108.

[7] D. Maslov and D. M. Miller, "Comparison of the cost metrics for reversible and quantum logic synthesis," http://arxiv.org/abs/quantph/0511008, 2006.

[8] H. Thapliyal and N. Ranganathan, "Design of reversible sequential circuits optimizing quantum cost, delay and garbage outputs," *ACM Journal of Emerging Technologies in Computing Systems,* vol. 6, no. 4, pp. 14:1–14:35, Dec. 2010.

Failure Analysis Case Studies on Open Defect

Grace Song, Chunlei Wu, Joe Yu, Gaojie Wen, Winter Wang

Freescale Semiconductor (China) Limited, Tianjin

Email: r64994@freescale.com

Abstract- **Failure analysis (FA) plays an important role in the integrated circuit industry. In FA cases, leakage and open are two main types of electrical faults. Optical beam induced resistance change (OBIRCH) technique is a very effective method in locating the leakage related defects or at least pointing out the abnormal leakage current path. But for open related defects, it seems that there is no common and effective method which can identify and locate the defect. In this paper, several open defect related failure analysis cases are presented. Some typical electrical and physical signatures of open fault and some analysis methods are discussed.**

I. INTRODUCTION

Failure analysis (FA) plays an important role in the integrated circuit industry. FA process consists of failure mode verification and electrical analysis, fault localization and physical failure analysis. Fault localization is the most critical step in which techniques such as emission microscopy (EMMI) and optical beam induced resistance change (OBIRCH) are employed to determine the physical location of the defect [1].

For most defects which can result in a leakage fault in the circuitry, OBIRCH is a very effective technique in locating the physical defect or at least pointing out the abnormal leakage current path. But for an open fault, it seems that there is no common and effective method which can identify and locate the defect.

In this paper, four open defect related real cases were presented. During the failure analysis procedure, circuit study, signal transient response analysis, I (V) measurement, passive voltage contrast (PVC) and OBIRCH technique as well, were all employed to identify and locate the physical defect.

II. CASE STUDIES

A. Case 1

For some cases, the voltage transient response and I (V) characteristic may give a direct indication of an open defect.

In this real case, an analog-digital mixed signal IC was returned from customer due to SO pin output abnormal. Failure verification in FA Lab confirmed the reject unit failed at serial peripheral interface (SPI) related failure. The SO pin always wrongly reported open load fault of two outputs. Post chemical decapsulation, EMMI analysis was performed on the unit while it was in the failing status. One abnormal distributing emitting site was detected on the drain side of one transistor in SPI related block, which indicated that the transistor was working in saturated mode. Layout & schematic study showed the gate of this transistor was signal "hof01_en" and it was the output of a 2NAND gate (Fig.1). Circuit study showed logic one on this signal can enable ON state open load detection of the two failed outputs. That is to say, abnormal higher voltage level on signal

"hof01_en" may result in the final failure. Microprobing analysis confirmed this signal was 0V on a reference unit. But for the reject unit, when probe needle was put on this "hof01_en" node, the unit would functionally recover and it measured 0V on "hof01_en". So the real voltage level of "hof01_en" signal for the reject unit could not be obtained under this normal static test condition. To further study the state of "hof01_en", the voltage transient response of it was measured. A falling edge and a rising edge were applied to an input signal which would affect "hof01_en", respectively. It could be observed that, as a response to the input signal, "hof01_en" could fall down or rise up. But the falling down time of it was longer than normal, while the rising time of it was normal (Fig.2). This result indicated that there may be some defect preventing "hof01_en" from quickly falling down. This could also explain why during steady state voltage measurement the unit would recover when probe needle put on "hof01_en" node, since the probe needle worked as a load between "hof01_en" and Vss and helped pull down this signal. I (V) measurement was then performed between "hof01_en" and Vss. Abnormal characteristic was confirmed. This anomaly showed that the parasitical body diode characteristic from Vss to "hof01_en" was absent (Fig.3), which indicated an open circuit in this current path (Fig.1). Detailed optical visual inspection revealed abnormal contact/missing contact defect at transistor N1 drain side and transistor N0 source side (Fig.4).

Although it was difficult to know the steady state voltage of the fault signal at first due to probe recoverable issue, it was easier to see the difference when the voltage transient response time was considered. The time delay is always an indication of an open fault. The I (V) curve could also confirm the existence of an open defect, as long as the defect is located on the current path indicated by the I (V) characteristic.

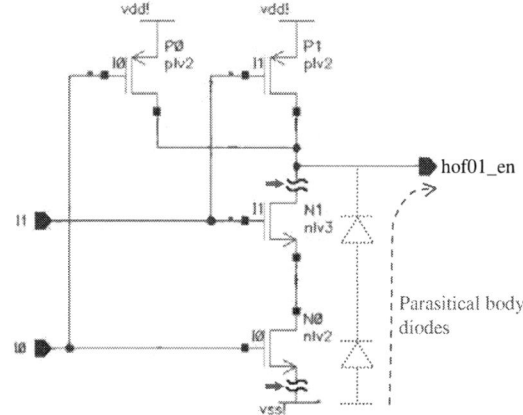

Fig. 1. Schematic of 2NAND circuitry. Red arrows point at the open defect locations.

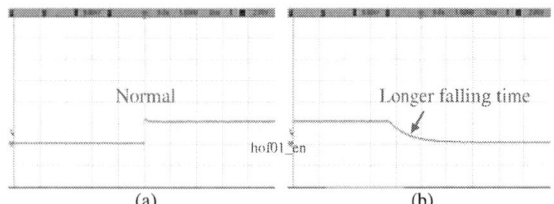

Fig. 2. Signal "hof01_en" rising edge (a) and falling edge (b).

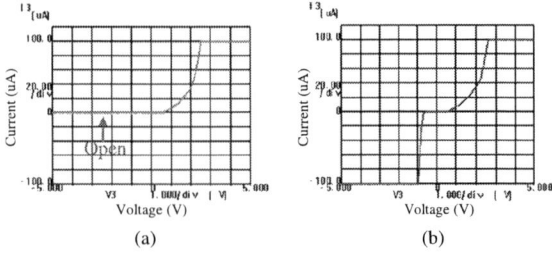

Fig. 3. I (V) curve between "hof01_en" and Vss for the failed unit (a) and a reference unit (b).

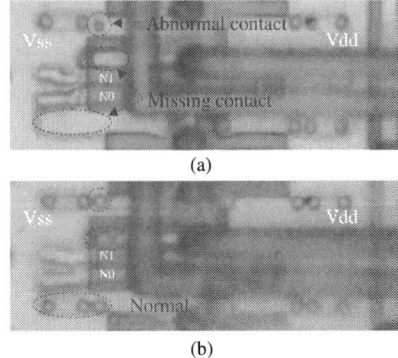

Fig. 4. Optical images of the failed unit (a) and a reference unit (b).

B. Case 2

Passive voltage contrast (PVC) is a very effective method in locating leakages and opens in circuitry [2]. Contact-level ion-beam induced PVC means the experiment was performed in a focused ion beam (FIB) system just after the metal1 layer was lapped off and the interlayer dielectric was still kept. There are three types of contacts which land on different kinds of structures, namely poly gate, NMOS and PMOS active regions, respectively. Poly contacts appear the darkest. This is due to the accumulation of positive charge when the electron beam scans the sample surface. Poly, being isolated from the substrate by gate oxide, does not drain off the beam induced charge and as a result recaptures low energy secondary electrons. Consequently, there is a decrease in the yield of secondary electrons which gives the poly contacts a dark contrast. In the case of the NMOS active regions contacts, the diode formed by the n+ active region and the p-well is always reverse biased. The accumulated charge from the primary beam scanning on the surface cannot drain off to the substrate which inhibits more secondary electrons from being emitted; hence, the n-diffusion contacts have a dark contrast. However, these contacts are slightly brighter than the poly contacts because unlike poly these are not completely isolated from the substrate by a dielectric. In the case of PMOS

active contacts, if the accumulated positive charge on the p+ region is greater than Vt, the diode will turn on and the positive charge will drain off through the n-well to the substrate. This will result in a higher yield of secondary electrons, giving the p-diffusion contacts the brightest contrast [3, 4].

In this case, contact-level ion-beam induced PVC analysis was employed successfully in identifying an open contact in a D flip-flop structure.

The reject unit also failed at SPI related failure. But for this unit, EMMI analysis did not detect any abnormal emitting site. Further fault localization was performed based on circuit study and microprobing. Finally, output Q of one D flip-flop was observed to be abnormally stuck high, which was confirmed as the cause of the final failure. The inputs of the D flip-flop were all confirmed normal. To further isolate the failure, Q signal was cut off from its downstream circuits with FIB tool. Post this cut, Q signal still stuck high at the D flip-flop side. So the fault can be isolated inside the D flip-flop. Fig.5 shows the schematic of the D flip-flop. I (V) measurement on Q node did not reveal any abnormal characteristic on it. To further locate the fault inside the D flip-flop, contact-level ion-beam induced PVC analysis was performed. The unit was deprocessed by face lapping method to contact level and was inspected in FIB system. As compared with an adjacent reference D flip-flop block, an abnormal dark contact was observed in the failed D flip-flop (Fig.6). Layout study showed this contact is connected to PMOS active, which should be bright during FIB inspection. The dark-looking in the PVC analysis means this contact was not well connected to PMOS active. FIB cross section on this contact confirmed the open defect (Fig.7).

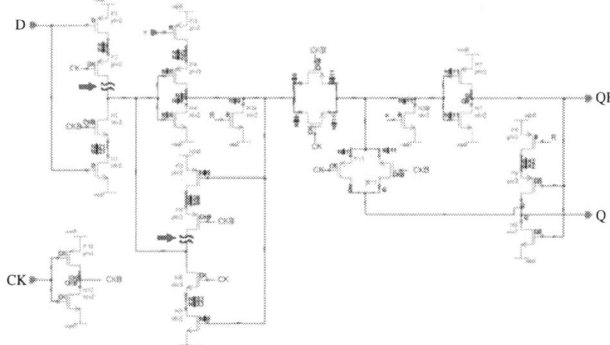

Fig. 5. Schematic of the D flip-flop. Red arrows point at the open defect locations. (Note: one contact open in layout results in two locations open in the schematic.)

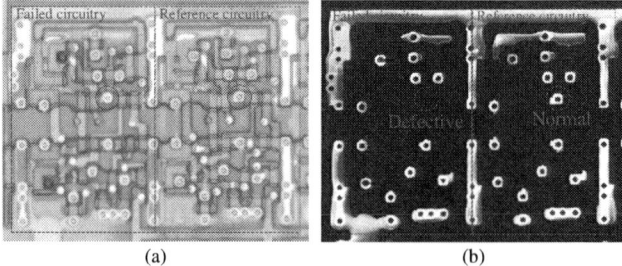

Fig. 6. Contact-level optical image (a) and the PVC image (b).

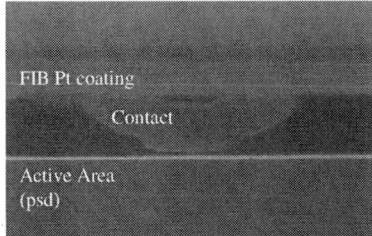

FIB image, Mag: 80kx

Fig. 7. FIB cross section image of the defective contact.

C. Case 3

Sometimes, even at the sample preparation step, there may be a hint pointing to an open defect. Again, this is an application of PVC theory.

In this case, the reject unit failed at interrupt related failure. No interrupt could be generated for a certain input of the unit. EMMI analysis on this unit did not detect any abnormal emitting site. Further fault localization was based on schematic study and microprobing. To facilitate microprobing, a necessary sample preparation step is to expose a window as test pad on the metal line of related signals by locally milling off the passivation with FIB tool. For milling status real-time monitoring, FIB system provides an end point detection (EPD) function. During the milling, when the primary gallium (Ga$^+$) ion beam reaches the metal layer, the EPD current would rise up suddenly, which means large amount of second electrons are detected and the metal is well exposed. This characteristic is used to determine when to stop the milling at the right depth. During this step, it was observed that when milling off the passivation on metal2 line of "intfb" signal at test pad 1 (TP1) location, the EPD current was abnormally lower than normal when milling depth was the same (Fig.8a). Another test pad TP2 was then milled on the same signal but on another metal 2 line (Fig.8b). For this test pad, the EPD current was normal. To further understand this phenomenon, circuit layout was carefully studied. It can be confirmed that the right terminals of the two metal2 lines where TP1 and TP2 lie on are connected to two duplicate blocks named as input_sp6 and input_sp5, respectively. Each of them only connects to one NMOS drain active region in each input_sp block. The left terminals of the metal2 lines connect to a same metal1 line and then to several other blocks (Fig.9 & Fig.10). According to the ion-beam induced PVC theory, for NMOS active connections, the accumulated charge from the primary beam scanning on the surface cannot drain off to the substrate which inhibits more secondary electrons from being emitted [4]. During the milling, when a metal line is exposed to the primary ion beam scanning, the one which only connects to NMOS active region would emit fewer secondary electrons than the one connects to many blocks where PMOS connections must be contained. So the lower EPD current at TP1 location probably meant that its connection to PMOS active was loss and only NMOS connection was remained. That is to say, there was high possibility that the via as marked in Fig.9 was open. Further microprobing analysis confirmed this hypothesis. The open defect at this via on "intfb" signal was the cause of the interrupt failure on the reject unit.

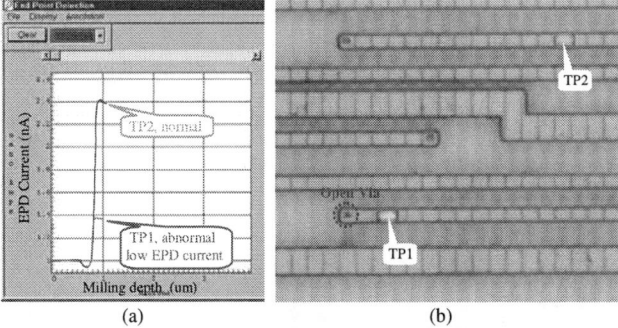

(a)   (b)

Fig. 8. FIB EPD curves (a) and optical image showing TP1 & TP2 location (b).

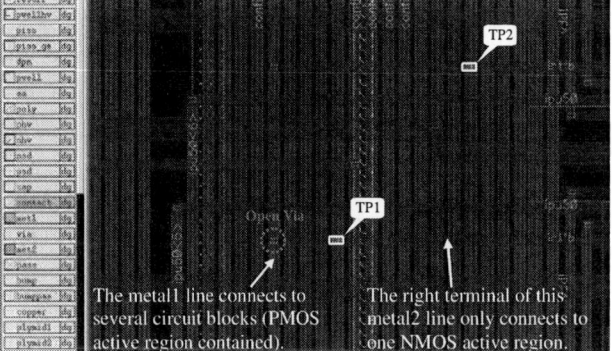

Fig. 9. Layout view.

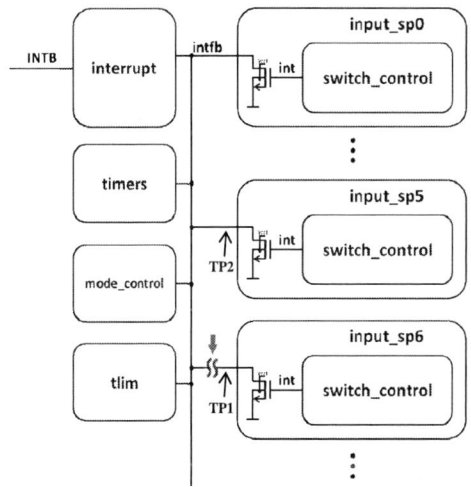

Fig. 10. Simplified schematic view. Red arrow points at the open defect location.

D. Case 4

For a NMOS transistor with its gate floating, the capacitive coupling between drain-gate and gate-source forms a capacitive voltage divider. When the drain voltage is sufficiently high, then the divider allows the gate voltage to exceed threshold and the transistor conducts [5].

This is a typical case that an open (i.e. floating) gate resulted in a NMOS transistor drain to source conductive (leakage).

978-1-4673-2395-6/12 $31.00 © 2012 IEEE   497

In this case, the reject unit failed functionally. EMMI analysis on this unit did not detect any abnormal emitting site. Microprobing analysis was performed based on schematic study. It was revealed that signal "A" voltage was abnormally lower than normal. This was confirmed as the cause of the final failure. I (V) measurement revealed signal "A" to Vss leakage (Fig.11). OBIRCH analysis was then performed on the unit. Abnormal resistance change location was detected, which indicated an abnormal current path through a NMOS transistor M125 (Fig.12a). M125 drain is connected to signal "A" and source to Vss. This kind of drain to source conductive for a NMOS transistor might be caused by a gate floating defect. Detailed visual inspection revealed the gate contacts of M125 were abnormal (Fig.12b&c). Besides, some of the drain and source contacts were also seems abnormal. FIB cross section was performed on M125 gate contacts. They were confirmed open because there was no over etch between the contacts and the poly below it (Fig.13). It was this open defect on the gate that resulted in drain (signal "A") to source (Vss) conductive and finally caused the final functional failure of the unit.

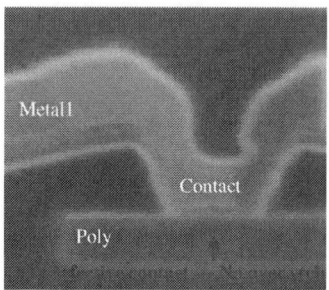

FIB image, Mag: 65kx

Fig. 13. FIB cross section image showing one of the defective gate contacts.

III. SUMMARY

Four open defect related failure analysis cases were presented. The purpose is to share some experiences on the analysis procedure and method on this kind of cases. For case 1, voltage transient response and I (V) curve pointed to an open defect in the circuitry. For case 2 & case 3, PVC was employed to identify and locate the open defect. Case 4 was an example of open gate induced NMOS transistor drain to source conductive.

Although there is no common methodology for identifying and locating the open defect, circuit study, signal transient response analysis, I (V) curve, PVC and OBIRCH technique as well, all can give a clue to the physical defect location.

ACKNOWLEDGMENT

Authors would like to thank Steven Che, Xuezhu Wang, Xiaocui Li and Changyan Qi of Product Analysis Lab in Freescale Semiconductor (China) Limited. Great support was given by them on physical failure analysis.

REFERENCES

[1] J. H. Lee, Y.S. Huang, D.H. Su, "Wafer Level Failure Analysis Process Flow", Microelectronic Failure Analysis Desk Reference 5th ed, pp.39-41.

[2] Yuan Chen, Xiaowen Zhang, "Focused Ion Beam Technology and Application in Failure Analysis", 2010 11th International Conference on Electronic Packaging Technology & High Density Packaging, pp. 957-960.

[3] Z. G. Song, G. Qian, J. Y. Dai, Z. R. Guo, S. K. Loh, C. S. Teh and S. Redkar, "Application of Contact-level Ion-beam Induced Passive Voltage Contrast in Failure Analysis of Static Random Access Memory", Proceedings of 8'" IPFA 2001, Singapore, pp. 103-106.

[4] H. Seiler, "Secondary Electron Emission in the Scanning Electron Microscope", Journal of Applied Physics, 54 (11), November, 1983, pp.R1-R18.

[5] Jerry Soden, Jaume Segura, and Charles F. Hawkins, "Electronics and Failure Analysis", Microelectronic Failure Analysis Desk Reference 5th ed, pp.89-109.

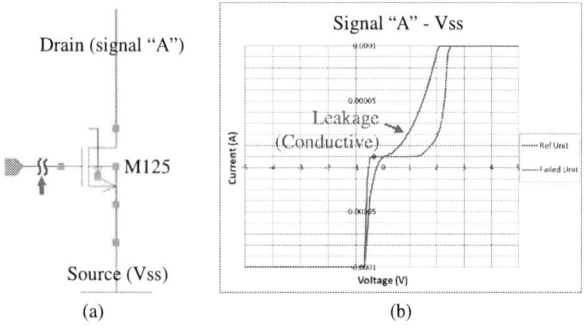

Fig. 11. Schematic (a) and the I (V) curve (b). Red arrow in schematic points at the open defect location.

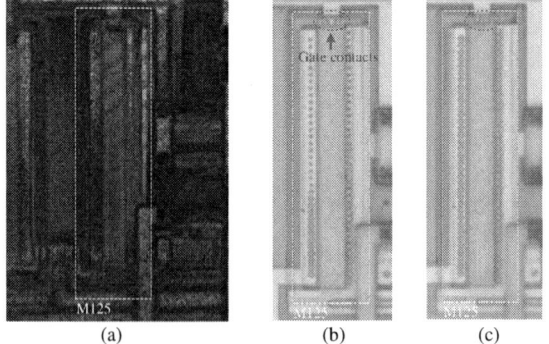

Fig. 12. OBIRCH image (a) and optical images of the failed unit (b) and a reference unit (c).

Thermal and Optical Analysis of Four-chip HPLED Package with Different Thermal Interface Material

S.Y. Lee[1], M. Devarajan[2]

Nano Optoelectronics Research (NOR) Lab, School of Physics,
Universiti Sains Malaysia (USM),
11800 USM, Pulau Pinang, Malaysia.
[1]Email: lsy10_phy038@student.usm.my

Abstract- **This paper is focused on thermal and optical properties of four-chip HPLED package with different thermal interface materials. The package is attached to MCPCB via solder paste and thermal compound. Through cumulative structure function, solder point of HPLED package is determined as 2.955K/W. The photometric properties of these two HBLED packages are evaluated by using TERALED. The total thermal resistance of the HPLED-S and HPLED-C is 10.68K/W and 13.00K/W respectively. Results show that the junction temperatures of HPLED with solder paste is lower than HPLED with thermal compound. Optical power, wall plug efficiency and luminous flux of both packages under various ambient temperature and forward current are discussed in this study. The HPLED-C produces more luminance than HPLED-S, but high thermal resistance of the HPLED-C induces the rise in junction temperature therefore increases the degradation rate of wall plug efficiency of the package.**

Keywords- HPLED, thermal resistance, junction temperature.

I. INTRODUCTION

Nowadays, solid state lighting or the LEDs become rife in many electronics applications and tend to replace conventional lighting [1-2]. For single chip LED package, luminance produced is not sufficient for general illumination compare with incandescent and fluorescent lamp. In order to fulfil the market requirement for higher luminance, HPLED and multi-chip LED (MCLED) are manufactured. Although HPLED and MCLED emit high luminance but load of heat also generated at the p-n junction. As the junction temperature of LED increases, lifetime of LED drops and LED package undergoes a dominant wavelength shift because of the yellowing of the epoxy lens [3]. Catastrophic failure of an LED package can be eliminated by controlling the rise of junction temperature. Therefore heat management in an LED package is crucial to maintain its reliability and durability.

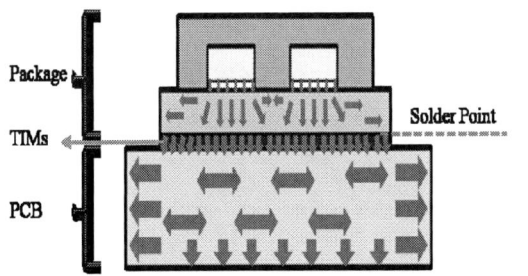

Fig 1. Schematic diagram of a heat transfer model of a board-level HPLED.

As a way to reduce the heat stored in the junction, LED package is attached to a printed circuit board (PCB) which functions as a heat sink. Since the thermal conductivity of air is low, interface material with higher thermal conductivity is used between LED package and PCB. Heat flow path in a typical LED package can be described by 1-dimensional heat flow model as shown in Fig 1. The red arrows in the Fig. 1 show the direction of heat flow. The low thermal conductivity of epoxy lens cause minor thermal energy is dissipated through radiation hence can be disregarded. From Fig. 1, thermal resistance in each region will affect the overall spreading resistance of the board-level LED package.

Thermal interface materials play an important role in heat dissipation. It is estimated around 50% of the total thermal resistance of a typical LED package is measured in this region [4]. G. De May studied that thermal interface materials (TIMs) have influence thermal resistance, thermal impedance, and thermal behaviour of an electronic package [5]. Some researchers work on the thermal effects of TIMs by using thermal transient testing and thermal simulation [6-7].

This paper proposes to investigate optical characterization of 4-chip HPLED package with different interface materials; solder paste and thermal compound. Thermal transient measurements were carried out for both packages under various ambient temperatures and input current by using TERALED system.

II. THEORETICAL BACKGROUND

Materials used to construct LED have their own thermal resistivity therefore total thermal resistance in a package is the sum of the individual thermal resistances [8]. Over few ten years, thermal transient tester [T3ster, MicRed] with software [T3ster Master] have been proved that can be used for analyzed heat-flow path of LED as well as deriving the resistance of any thermal interfaces [9]. These thermal behaviour can be described by using resistor-capacitor (RC) network. The transient thermal response in an LED package is modelled through the R-C network and transformed into structure function [10-11].

Fig. 2 shows a cumulative structure function of the four-chip HPLED with the different interface material. The x-axis represents thermal resistance while y-axis is thermal capacitance. Both curves diverge at the solder point which as 2.960K/W. This represents the total thermal resistance of the HPLED is 2.960K/W.

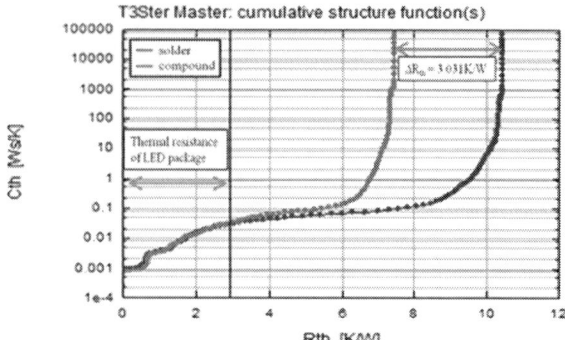

Fig. 2. Cumulative structure function of HPLED with different TIMs.

Refer to Fig. 2, the total resistance of board-level HPLED package with solder paste is observed as 3.031K/W lower than thermal compound. This is because the thermal conductivity of solder paste is 50.9W/mK and higher than the thermal compound which is only 2.4W/mK.. Heat conduction in the interface layer can be explained by the Fourier equation [12]:

$$q = -kA\frac{dT}{dx} \tag{1}$$

where q is the heat flow, k is thermal conductivity, A is the cross sectional area, and $\frac{dT}{dx}$ is the temperature gradient in the direction of heat flow.

From (1), thermal conductivity is proportional to heat flow, hence material with high thermal conductivity enable efficient heat dissipation.

According to JEDEC standard, thermal resistance of a single chip is the ratio between the rise in junction temperature and the heat dissipation. After modifying, average thermal resistance of a multi-chip LED package with identical geometry can be defined as:

$$R_{th} = \frac{T_j - T_a}{P_{el} - P_{opt}} \tag{2}$$

where T_j is junction temperature, T_a is ambient temperature, P_{el} is electrical power, P_{opt} is optical power and $P_{el}-P_{opt}$ is the heat dissipation.

III. EXPERIMENTAL METHOD

Integrating sphere, photometric detector and T3ster (TERALED) with power booster and Agilent power source are used in this experiment. With these apparatuses, thermal and optical properties of HPLED can be measured simultaneously [13].

Two four-chip high power warm white LED packages with different interface materials; HPLED-S and HPLED-C are prepared. HPLED-S is the package which uses solder paste while HPLED-C uses thermal compound. Both packages are calibrated to determine the relationship between junction temperature change and temperature sensitive parameter which is known as K-factor. The calibration process is important in order to ensure accuracy of the measurement [14]. In this process, 10.0mA sensor current is applied to the LED and calibration curve is recorded at each 5^0C. The K factor can be calculated by [15]:

$$K = \left|\frac{T_1 - T_2}{V_1 - V_2}\right| \tag{3}$$

For thermal transient measurement, each board-level HPLED package is powered by 0.8A, 1.2A and 1.6A input current. The temperature is increased in order of 35^0C from 25^0C to 85^0C. The package is heated to reach a steady state then cooling transient curve is recorded for 500s. The four chips in LED package are connected in parallel which means each chip will pass through a quarter of the total current.

IV. RESULT AND DISCUSSION

The measured sensitivity or inverse of K factor for HPLED-S and HPLED-C is 1.76mV/^{0}C and 1.86mV/^{0}C respectively. The sensitivity of LED represents the drops in forward voltage as junction temperature increases 1^0C in the particular LED. The electrical, thermal and optical characteristics for both packages are listed in Table 1. The recorded average total thermal resistance from junction to ambient for HPLED-S is 10.68K/W while for the HPLED-C is 13.00K/W.

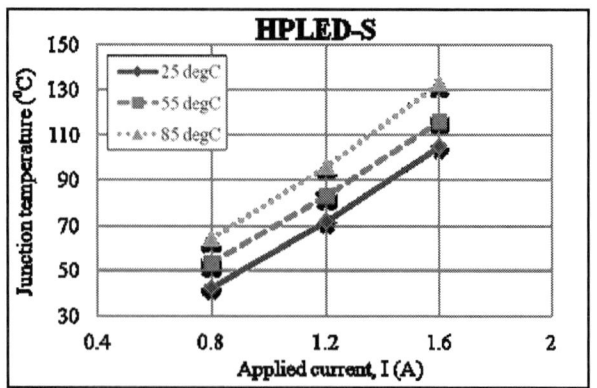 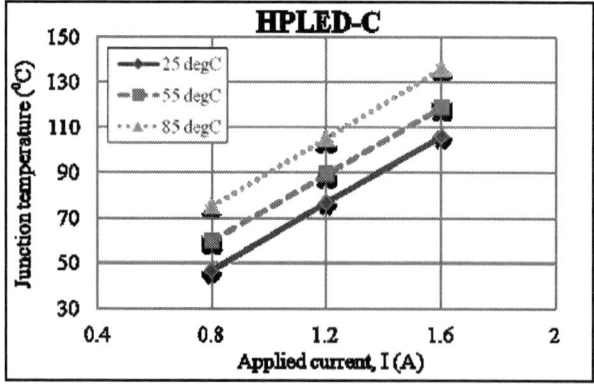

Fig. 3. Measured junction temperature as a function of forward current for (a) HPLED-S and (b) HPLED-C.

TABLE I

ELECTRICAL, THERMAL AND OPTICAL PROPERTIES OF HPLED-S AND HPLED-C WITH VARIOUS AMBIENT TEMPERATURE AND FORWARD CURRENT

T_a (^{0}C)	I_f (A)	HPLED-S					HPLED-C				
		P_{opt} (W)	P_{heat} (W)	T_j (^{0}C)	Φ (lm)	CCK(K)	P_{opt} (W)	P_{heat} (W)	T_j (^{0}C)	Φ (lm)	CCK(K)
25	0.8	0.77	1.68	42.3	242.2	3020.0	0.83	1.64	46.6	255.7	3042.0
	1.2	1.07	2.7	53.2	335.8	3079.7	1.13	2.65	59.8	352.2	3113.7
	1.6	1.32	3.79	63.9	416.0	3161.0	1.37	3.75	75.0	431.6	3198.0
55	0.8	0.73	1.69	71.8	226.1	3052.6	0.79	1.64	76.7	244.4	3061.1
	1.2	0.99	2.72	83.1	310.1	3123.0	1.06	2.66	89.4	331.9	3185.2
	1.6	1.2	3.83	95.8	377.8	3201.8	1.26	3.79	105.0	395.7	3282.0
85	0.8	0.64	1.74	105.0	201.5	3065.1	0.73	1.66	106.0	225.4	3131.4
	1.2	0.87	2.78	116.0	273.2	3171.9	0.95	2.71	119.0	297.8	3275.1
	1.6	1.03	3.93	133.0	328.0	3289.5	1.10	3.87	136.0	344.3	3425.5

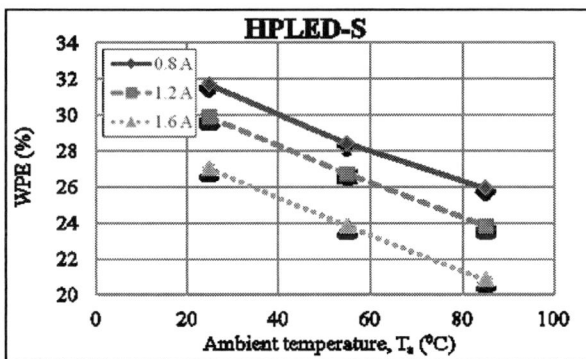 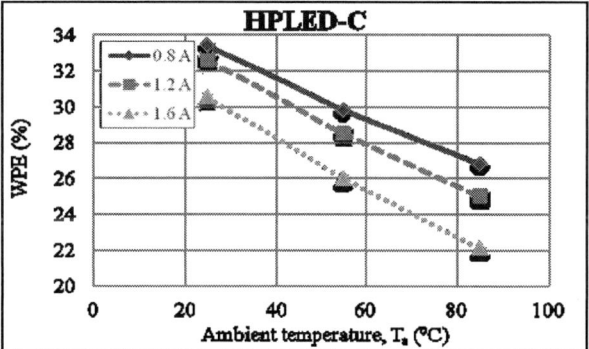

Fig.4. Relationship between Wallplug efficiency (WPE) and ambient temperature for (a) HPLED-S and (b) HPLED-C

Junction temperatures of the tested package can be calculated by referring to (2). The junction temperatures of each package are shown in Fig. 3. The junction temperature of both packages increases with the applied current and ambient temperature. The junction temperatures of HPLED-S are lesser than HPLED-C due to lower thermal resistance in the HPLED-S package.

However, it can be noted that the optical power of the HPLED-C is better than HPLED-S from Table 1. Optical powers for both package decrease with ambient temperatures. As both packages are powered by 1.2A constant current, the optical power for HPLED-S and HPLED-C decreases 0.2W and 0.18W respectively with the increase 60^0C in the ambient temperature.

Fig. 4 shows the wall plug efficiency (WPE) for HPLED-S and HPLED-C packages under various conditions . The wall plug efficiency or power efficiency can be defined as [16]:

$$WPE = \frac{P_{opt}}{P_{el}} \qquad (4)$$

The WPE of HPLED-C is higher than HPLED-S but the gradient of the curves is more steeper with the increasing of ambient temperature. This means the degradation effect of optical power with the ambient temperature of HPLED-C is more than HPLED-S. This is because of higher thermal resistance and junction temperature. The degradation effect would become more obvious when the device is operated for a long time.

Increase in junction temperature HPLED package cause luminance of the package decreases and shifts in the colour temperature. These effects are shown in Table 1. Luminous flux is obtained from[16]:

$$\Phi_{lum} = 683 \frac{lm}{W} \int_\lambda V(\lambda)P(\lambda)d\lambda \qquad (5)$$

where $P(\lambda)$ is the optical power per unit wavelength and $V(\lambda)$ is the eye sensitivity function.

Since luminous flux is proportional to optical power, decrease in optical power also cause reduction in luminous flux.

The function of TIMs in LED packaging is to increase the rate of heat flow to ambient by eliminating the air gap between LED package and MCPCB. Even the thickness of TIMs only few micrometers or millimeters, but has great influence on thermal and optical characteristics of the LED.

V. CONCLUSION

The effects of thermal interface material on photometric properties of warm white HPLED with four chips were studied.

The junction temperature of HPLED increases with ambient temperature but optical power behaves in contrast with junction temperature. Reduction in optical power of HPLED packages causes decrease in wall plug efficiency and luminous flux. HBLED-C able to produce more optical power but higher thermal resistance increase the rate of degradation. Therefore, choosing a suitable TIMs is crucial to maintain a typical LED's quality.

ACKNOWLEDGEMENT

The author, S.Y. Lee would like to thank the Institute of Post Graduate Studies (IPS), USM for providing Graduate Assistant Scheme.

REFERENCES

[1] Seong-Jin Kim, "Vertical chip of GaN-based blue light-emitting diode", Elsevier Solid-State Electronics, vol. 49, issue 7 (2005) 1153-1157.

[2] Jeung-Mo Kang, Jeong-Hyeon Choi, Du-Hyun Kim, Jae-Wook Kim, "Fabrication and Thermal Analysis of Wafer-LevelLight-Emitting Diode Packages", IEEE Electron Device Letter, vol.28, no. 7 (2008) 1118-1120.

[3] Nadarajah Narendran, Yimin Gu, "Life of LED-Based White Light Sources", Journal of Display Technology, vol. 1, no 1 (2005) 167-171.

[4] B. Smith, T. Brunschwiler, B. Michel, "Utility of Transient Testing to CharacterizeThermal Interface Materials", THERMINIC (2007).

[5] G. De Mey, J. Pilarski, M. Wójcik, etl., "Influence of interface materials on the thermal impedance of electronic packages", Elsevier International Communication in Heat and Mass Transfer, vol. 36, issue 3 (2009) 210-212.

[6] Ralph Schacht, Daniel May, Bernhard Wunderle, etl., "Characterization of Thermal Interface Materials to Support Thermal Simulation", THERMINIC (2006).

[7] Anithambigai Permal, Teeba Nadarajah, Dinash Kandasamy, etl. "Influence of Thermal Interface Material on the Thermal Resistance of High Power LED", IEEE 2nd International Conference on Photonics (2011) 1-5.

[8] Xingcun Colin Tong, Advanced Materials for Thermal management of Electronic packaging, Springer; 1st Edition, (2011) 305-370.

[9] V. Szekely, "Enhancing Reliability with Thermal Transient Testing", Microelectronics Reliability, vol. 42 (2002) 629-640.

[10] Oliver Steffens, Péter Szabó, Michael Lenz, Gábor Farkas, "Thermal Transient Characterization Methodology for Single-Die and Stacked Structures", 21st IEEE SEMI-THERM Symposium (2005).

[11] Andras Vass-Varnai, Shan Gao, Zoltan Sarkany, Jongman Kim, etl., "Issues in Junction-to-Case Thermal Characterization of Power Packages with Large Surface Area", 26th IEEE SEMI-THERM Symposium (2010) 158-164.

[12] Rao R. Tummala, Fundamental of Microsystems Packaging, International Edition, (2001).

[13] Andras Poppe, Gabor Molnar, Tamas Temesvolgyi, "Temperature Dependent Thermal Resistance on Power LED Assemblies and A Way to Cope with It", SEMI-THERM (2010) 283-288.

[14] Dr. John W Sofia, "Electrical Temperature Measurement Using Semiconductor", (1997) http://www.electronicscooling.com/1997/01/electrical-temperature-measurement-using-semiconductors/ (online available 1 Jan 2012).

[15] Lianqiao Yang, Jianzheng Hu, Lan Kim, etl. "Thermal Analysis of GaN-Based Light Emitting Diodes with Different Chip Sizes", IEEE Transactions on device and materials reliability, vol. 8, no.3 (2008) 571-575.

[16] E. Fred Schubert, "Light Emitting Diodes", 2nd Edition, (2007).

A Study on the Effect of Test Vector Randomness on Test Length and its Fault Coverage

[1]Muhammad Sadiq Sahari[*], [1]Abu Khari A'ain[#] and [2]Ian Grout[@],
[1]Department of Microelectronics and Computer Engineering,
Faculty of Electrical Engineering,
Universiti Teknologi Malaysia (UTM).
[2]Department of Electronics and Computer Engineering,
University of Limerick.
Email : [*]sdqnt8_que@yahoo.com.my, [#]abu@fke.utm.my and [@]ian.grout@ul.ie

Abstract- **This paper presents a study on the impact of test sequence randomness and the fault coverage (FC) it could produce through the use of a modified structure of the conventional linear feedback shift register (LFSR). By using double input signals, the modified LFSR can control the number of test patterns generated and also prevents the sequences from being stuck in all zeroes state. Fault simulations on ISCAS'85 benchmark circuits show that a high FC for combinational logic circuits has been obtained. Another observation is that the modified structure could achieve high FC with a smaller test sequence compared to other reported test pattern generation (TPG) techniques.**

I. INTRODUCTION

The main issue in test pattern generation (TPG) is to provide the most efficient test pattern generation method which resulted in high coverage. Other than high FC, testing time is also a significant factor that must be considered and proven to be a challenging task for the test technique. Too many test vectors required is amongst the bottlenecks of a test procedure. There have been many researchs carried out to understand why certain test patterns could produce high FC while others fail to do so.

The pseudo-random pattern generator (PRPG) is the most common technique employed to generate test patterns for pseudorandom testing. Random testing concept does not consider the previous (history of) test vectors which have been used in the production of the next test vectors, which results in these vectors capturing the same faults as the previous vectors. Thus the accumulated FC is low. The antirandom [1, 6] method was introduced to overcome this problem. The antirandom method does consider the history of the test patterns which have been generated when producing subsequent test vectors. It also employs maximum distance between every test vectors which results in a high FC and better randomness. This paper undertakes further research from [3] to study the effect of randomness when the test sequence is modified.

II. TPG UNDER STUDY

The n-bit structure of the TPG under study is presented in Fig. 1. It is a modified structure from the normal PRPG by adding an exclusive-NOR gate, a 2-to-1 multiplexer and an additional input signal (a selector) to the multiplexer. Thus, the TPG works with two input signals SEL and CLK. The CLK frequency is set to be the same as the conventional LFSR clock frequency. The input signal, SEL, is connected to a 2-to-1 multiplexer as an input (the selector) and is set to operate at a lower frequency than the register clock, CLK. By incorporating exclusive-NOR gates, an all zeroes state in the test vector sequence can be now be generated.

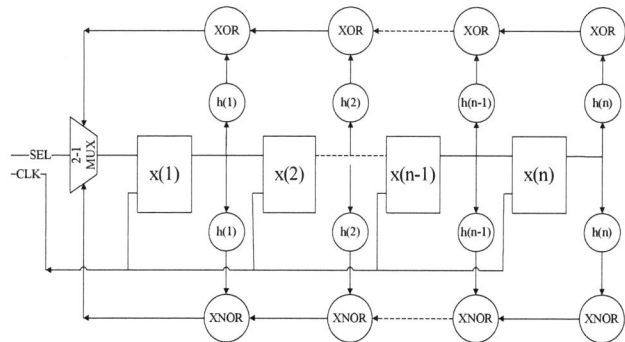

Fig. 1. *n-bit structure of the modified TPG method*

Like the conventional LFSR, the characteristic polynomial of the modified TPG is given as: $h_n x^n + h_{n-1} x^{n-1} + \ldots + h_0$ where all the terms still carry the same meaning as in normal LFSR. The modification results in the following equation:

978-1-4673-2395-6/12 $31.00 © 2012 IEEE

$$x_0^{i+1} = \overline{SEL} \bullet \left(h_1 x_0^i \oplus h_2 x_1^i \ldots \oplus h_{n-1} x_{n-2}^i \oplus h_n x_{n-1}^i \right)$$
$$+ SEL \bullet \overline{\left(h_1 x_0^i \oplus h_2 x_1^i \ldots \oplus h_{n-1} x_{n-2}^i \oplus h_n x_{n-1}^i \right)}$$
$$x_1^{i+1} = x_0^i$$
$$x_2^{i+1} = x_1^i$$
$$\vdots$$
$$x_{n-1}^{i+1} = x_{n-2}^i \qquad\qquad (1)$$

In equation 1, $x_0^{i+1}, x_1^{i+1}, x_2^{i+1}, \ldots, x_{n-1}^{i+1}$ represent the state of the modified TPG at time i+1 and $x_0^i, x_1^i, \ldots, x_{n-2}^i, x_{n-1}^i$ represent the modified TPG state at the time i. The TPG can now generate longer sequences than $2^n - 1$ due to the inclusion of the second input signal, SEL. The maximum sequence length depends on the ratio of the two input signal frequencies. Thus, the following equation can be used to obtain the maximum sequence length of the test sequence:

$$Max = (FreqRatio - 1)(2^{n+1} - 2) \qquad\qquad (2)$$

Where,
FreqRatio is the ratio between CLK and SEL and it must be greater than 1 (CLK: SEL >1)

III. DEGREE OF RANDOMNESS

A test sequence can be considered random if the ones and zeroes are evenly distributed in space and time, and if there is very little correlation between one sequence and the next [1]. Fig. 2 shows the comparison of randomness of the first 20 test sequences for 16-input sequences generated using PRPG and the modified TPG methods. Both TPG methods start with seed 0000000000000010. The state-time diagram for the first 20 time units are represented in black and white, where a black spot indicates '1' and white indicates '0'. In [1], it was proven that maximum distance between test vectors can generate more random sequences and produce high FC. The test sequence of the modified TPG method shows slightly more random compared to the test sequence from PRPG method. This is due to the repeating test vectors from the modified TPG disturb the correlation between one pattern and the next.

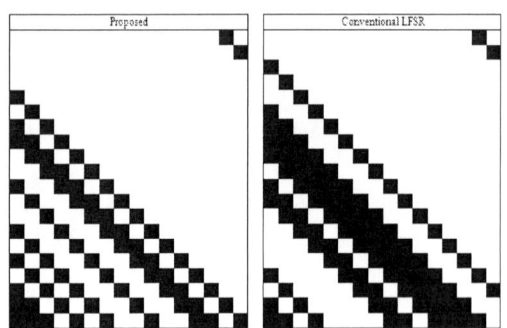

Fig. 2. *Comparison of randomness between the modified method and conventional LFSR-based method (PRPG)*

IV. SIMULATION RESULT

ISCAS'85 benchmark circuits have been employed as CUTs in order to evaluate the effectiveness of the modified TPG method. Table I presents the characteristic of ISCAS'85 combinational benchmark circuits such as circuit function, number of gates, number of primary input (PI), number of primary output (PO) and the total number of faults.

TABLE I
CHARACTERISTIC OF BENCHMARK CIRCUIT FOR ISCAS'85

Circuit	Gates	PI's	PO's	Faults
C17	6	5	2	22
C432	160	36	7	524
C499	202	41	32	758
C880	383	60	26	942
C1355	546	41	32	1574
C1908	880	33	25	1879
C2670	1193	233	140	2747
C3540	1669	50	22	3428
C5315	2307	178	123	5350
C6288	2406	32	32	7744
C7552	3512	207	108	7550

A. Comparison with PRPG

The comparison of FC between the PRPG and the modified TPG methods for the C5315 circuit is shown in Fig. 3.
From the graph, the FC of the modified TPG method sharply increases after 100th test vectors but PRPG needs to wait till 1000th test vectors to show a major increment of the FC. Thus, the modified TPG method produces the highest FC with less test time taken; that is 95.78% after 1000 test vectors have been applied while PRPG gives 92.09% FC after 10000 test vectors applied.

Fig. 4 shows the comparison of FC between the modified TPG and PRPG using C7552 circuit. The modified TPG method produces the highest FC with less test time taken; that is 90.80% after 400 test vectors have been applied while PRPG gives 90.53% FC after 4000 test vectors applied. The figure also indicates that the modified TPG provides higher FC even when the test is sampled at early test vectors.

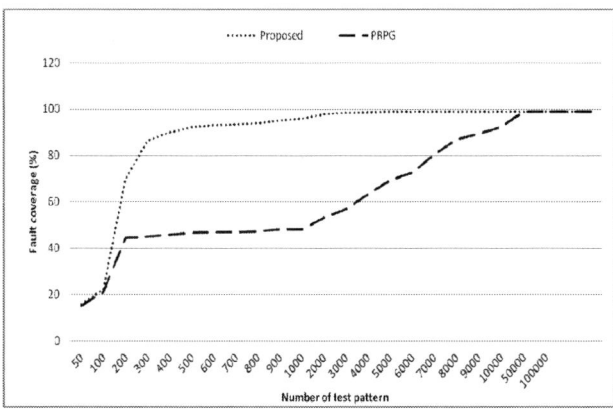

Fig. 3. *Comparison of FC between modified TPG and PRPG using C5315 (ISCAS'85) combinational benchmark circuit.*

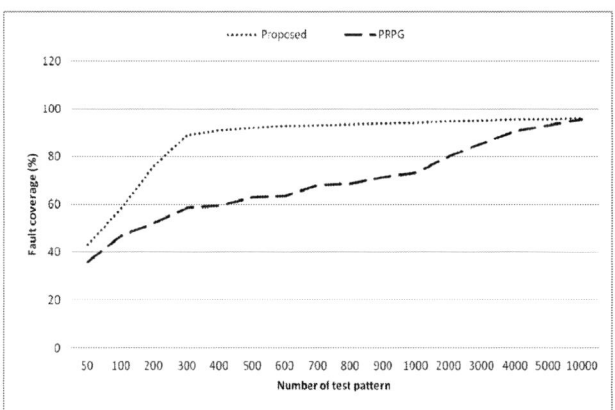

Fig. 4. *Comparison of FC between modified TPG and PRPG using C7552 (ISCAS'85) combinational benchmark circuit.*

Table II presents a summary of FC obtained using the modified TPG method and PRPG method on combinational CUTs. Certain circuits, such as C432, C880 and C5315 circuits show a major improvement in FC, no. of test vectors (TV) and test time (CPU time). Other circuits show only a minor improvement in FC, test length and test time. CPU time presents the total computational time in seconds needed to

obtain the result on a 2.93 GHz Intel Core 2 Duo CPU. Both methods show an acceptable time to generate a number of test vectors for all tested circuits. From Table II, the modified TPG can obtain the same FC using less test vectors and less test time (for C17, C6288, C432, C499, C1355, C880, C5315) or use the same number of test vectors and test time to obtain higher FC for (C3540, C7552, C2670).

TABLE II
COMPARISON WITH PRPG METHOD

Circuit	PRPG			Modified TPG		
	CPU time (s)	TV	FC (%)	CPU time (s)	TV	FC (%)
C17	0.25	10	100	0.25	7	100
C6288	2.0	175	99.56	1.8	103	99.56
C432	1.7	785	99.24	1.3	222	99.24
C499	1.7	757	98.94	1.5	640	98.94
C1355	2.5	2220	99.49	2.0	1548	99.49
C1908	3.7	3420	98.94	3.5	2906	99.15
C3540	4.4	5000	95.83	4.1	5000	95.86
C880	5.3	9000	100	3.8	6000	100
C5315	125	60000	98.90	18	8000	98.90
C7552	33	10000	95.60	33	10000	96.80
C2670	2000	1M	92.47	2000	1M	95.23

B.   Comparison with previous TPG methods.

Table III shows the comparison of number of test vector (TV) and FC between the modified TPG and other TPG methods [2, 3, 4 and 5]. From the table, the modified TPG method obtains higher FC using the same number of test vectors for C7552 and C2670 circuits. For other circuits, the modified TPG uses less number of test vectors to produce even higher FC than other TPG methods.

V. CONCLUSION

This paper reveals that a simple modification on basic PRPG produces more efficient test sequence. The produced test sequence is more random which result it being able to produce high FC. The test result could be sampled early which reduces test time, whilst retaining high FC. Another attractive feature of the modified TPG method is the ease of circuit modification and its' capability to extend the test length sequences depending on the frequency ratio between the two entry clock signals.

TABLE I
COMPARISON WITH PREVIOUS TPG METHODS [2, 3, 4 AND 5].

Circuit	Modified TPG		LPTPG [2]		TPG [3]		GA-LPTPG [4]		GAITPG [5]	
	TV	FC (%)	TV	FC (%)	TV	FC (%)	TV	FC (%)	TV	FC (%)
C432	222	99.24	2000	93.75	576	92.87	1882	98.89	2000	98.89
C499	640	98.94	1600	95.30	64	51.63	-	-	1600	98.81
C880	6000	100	10000	98.59	2112	98.84	9270	98.83	10000	98.83
C1908	997	96.65	2000	94.16	2688	81.24	1783	94.20	2000	94.20
C3540	4000	95.80	4000	92.54	416	91.11	4125	94.34	4000	94.34
C1355	1548	99.49	4000	93.39	-	-	4118	99.49	4000	99.49
C6288	103	99.56	2000	99.34	-	-	1783	99.56	2000	99.56
C7552	10000	96.80	10000	91.86	-	-	-	-	-	-
C5315	8000	98.90	15000	98.75	-	-	-	-	-	-
C2670	10000	88.79	10000	84.72	-	-	-	-	-	-

ACKNOWLEDGMENT

Authors are grateful for financial support from Research Management Centre (RMC) of Universiti Teknologi Malaysia through FRGS Grant vot. no: 4F040.

REFERENCES

[1] Shen Hui Wu, Yashwant K. Malaiya, A.P Jayakumana.: 'Antirandom vs. pseudorandom testing', Proc. of International Conference on Computer Design: VLSI in Computers and Processors, October 1998, pp. 221–223.

[2] He Ronghui, Li Xiaowei and Gong Yunzhan.: 'A low power BIST TPG design', Proc. of 5th International Conference on ASIC, October 2003, pp. 1136–1139.

[3] Jinyi Zhang, Qingfeng Zhang, Jiao Li.: 'A Novel TPG Method for Reducing BIST Test-Vector Size', IEEE International Symposium on High Density packaging and Microsystem Integration, 2007, pp. 1-4.

[4] Tan Enmin, Song Shengdong and Shi Wenkang.: 'Power Reduction in BIST Design Based on Genetic Algorithm and Vector-Inserted TPG', Proc. of 8th International Conference on Electronic Measurement and Instruments, July 2007, pp. 533-537.

[5] Enmin Tan and Li Wang.: 'A built-in self-test design with low power consumption based on genetic algorithm', Proc. of International Conference on Electronic Measurement & Instruments, August 2009, pp. 526-529.

[6] Shen Hui Wu, Sridhar Jandhyala, Yashwant K. Malaiya, and Anura P. Jayasumana, 'Antirandom Testing: A Distance-Based Approach', VLSI Design, vol. 2008, Article ID 165709, pp. 1-9.

New Low Power Delay Element in Self Resetting Logic with Modified Gated Diffusion Input Technique

Uma.Ramadass and P. Dhavachelvan
Department of Computer Science, School of Engineering,
Pondicherry University, Puducherry,
India.

Abstract- **Low power design has become one of the primary focuses in digital VLSI circuits, especially in clocked devices like microprocessor and portable devices. Optimization of several devices for speed and power is a significant issue in low-voltage and low-power applications. These issues can be overcome by incorporating Gated Diffusion Input (GDI) technique. Now-a-days dynamic circuits are becoming increasingly popular because of the speed advantage over static CMOS logic circuits. A fundamental difficulty with dynamic circuits is the monotonicity requirement and difficulties like charge sharing feed through, charge leakage, single-event upsets, etc. These issues can be eliminated using Self-reset logic (SRL). This logic provides a design solution where the clocking overhead is minimized. So the tradeoff between speed and power can be achieved through SRL and GDI technique. A new family of Modified self-reset logic (SRL) cells implemented with modified GDI technique is presented in this paper. The implementations proposed in this work are clocked storage element like D-FF in SRL with Modified Gate Diffusion Input Technique. This technique allows reducing power consumption and delay of digital circuits, while maintaining low complexity of logic design. Delay and power has been evaluated by Tanner simulator using TSMC 0.250μm technology. The simulation results reveal better delay and power performance of proposed delay elements as compared to existing dynamic, GDI cell and CMOS at 0.250μm technology.**

Keywords: ASIC, Self-resetting Logic, Modified Gate Diffusion Input technique, GDI cells, charge leakage, single-event upsets.

I. INTRODUCTION

LOW-POWER design has become a critical issue in VLSI design, especially for portable devices and high-density systems. For such submicron CMOS technology area, topology selection, power dissipation and speed are very important aspect especially for designing Clocked Storage Element (CSE), adder circuits and MAC unit for high-speed and low-energy design like portable batteries and microprocessors. To achieve the reduction of power consumption, optimizations are required at various levels of the design steps, such as the algorithm, architecture, logic, circuit and process technology. This paper considers the transistor logic level approach for low power digital design. Generally, the use of Pass Transistor Logic (PTL) [1] has many advantages over the CMOS design due the

reduced transistor count and smaller node capacitances thus decreasing the required area, rise/fall times and power dissipation. However, this scheme suffers from leakage problems because the input inverter is not the full swing signal and it becomes worse when the supply voltage drops. The Complementary Pass-Transistor Logic (CPL) and Swing Restored Pass-Transistor Logic (SRPL) [2-4], resolve the leakage and voltage swing. However, this logic operates in slow speed due to the regenerative property of the latch-type restoring circuit. The Gate Diffusion Input (GDI) [5, 6] is a lowest power design technique which offers improved logic swing and less static power dissipation. Using this technique several logic functions can be implemented using less number of transistor counts. This method is suitable for design of fast, low-power circuits, using a reduced number of transistors (as compared to TG and CMOS). Though GDI technique offers low power, less transistor count and high speed, the major challenges occurs in the fabrication process. The GDI technique requires twin-well CMOS or Silicon on Insulator (SOI) process to realize a chip which increases the complexity as well as the cost of fabrication.

Conventional CMOS has been a technique of choice in most digital design. The major advantage of CMOS circuit over single polarity MOS circuits (pass transistor), is that the static power dissipation is very small and produces minimal leakage [7]. However, the power dissipation of a CMOS device depends on the operating frequency. If the frequency of input signal increases, the CMOS devices dissipate more power. The main limitations of CMOS is slow speed when compare to PT and dynamic circuits. As the input capacitances of a CMOS gate are greater than the input capacitances of PTL the resistance and propagation delay is higher when compare to other logic family.

As input frequencies of a device increase, Dynamic circuits (domino logic) is becoming the circuit style of choice to implement critical paths because it offers significantly faster switching speeds than circuit styles [8-10]. A fundamental difficulty with dynamic circuits is the monotonicity requirement. In the design of dynamic logic circuits numerous difficulties may arise like charge sharing, feedthrough, charge leakage, single-event upsets, etc. A special dynamic logic circuit which resolves these issues is called Self- Resetting Logic (SRL) [11]. SLR is a commonly used piece of circuitry that automatically precharge themselves (i.e., reset themselves). In this logic the signal being propagated is buffered and used as the precharge or reset signal. By using a buffered form of the input, the input loading is kept almost as low as in normal dynamic logic while local generation of the reset assures that it is properly timed and only occurs when needed after a prescribed delay. One of the advantages of self-resetting logic is that when data present at evaluation does not require dynamic node to discharge, the precharge device is not active hence reduces power [12].The purpose of this paper is to propose new primitive cells using Self-resetting logic and Modified Gate Diffusion Input technique for high speed and low power circuits. The proposed structure will eliminates the issues of dynamic and static logic.

The proposed primitive structure is designed to operate in the 3-5V range with 250 nm process parameter. The organization of the paper is as follows: The section II, describes the preliminaries of SRL, GDI and MGDI. Section III, presents the proposed primitive cells in Self-resetting Logic with Modified Gate Diffusion Input (SRLMGDI). Section IV presents the implementation of DFF in SRLMGDI. Section V presents the discussion. Finally the conclusion is presented in section VI.

II. PRELIMINARY

A. Self-Resetting Logic (SRL)

A basic structure of a self-reset logic is shown in Fig.1. In the domino case, the clock is used to operate the circuit. In the self-resetting case, the output is fed back to the precharge control input and, after a specified time delay, the pull-up is reactivated. There is an NMOS sub block where the logic function performed by the gate is implemented which is represented as NMOS_LF through which the input data's are loaded.

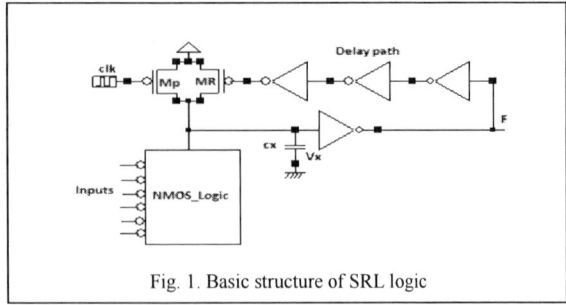

Fig. 1. Basic structure of SRL logic

The output of the gate F provides a pulse if the logic function becomes true. This output is buffered and it is connected to PMOS structure to precharge. The delay line is implemented as a series of inverters. The signals that propagate through these circuits are pulses. The width of the pulses must be controlled carefully or else there may

be contention between NMOS and PMOS devices, or even worst, oscillations may occur. In the circuit MP,MR and VSGR represents the precharge pull-up, reset pull-up and gate-source voltage of resetting transistor. During precharge phase clk=0, the transistor MP turns ON and the pull down network is OFF. Therefore the capacitor is charged to VDD. During evaluation phase clk=1, the transistor MP turns OFF and the pull down network is ON and evaluates the logic function. Therefore the capacitor is discharged making MR active which allows I_{DR} to flow and recharge C_X back up to a voltage of V_x=VDD.

B. Gate Diffusion Input (GDI)

The basic primitive of GDI cell consists of nMOS and pMOS as shown in Fig 2. A basic GDI cell contains four terminals – G (common gate input of nMOS and pMOS transistors), P (the outer diffusion node of pMOS transistor), N (the outer diffusion node of nMOS transistor), and D (common diffusion node of both transistors) [5]. Table 1 show how a simple change of the input configuration of the simple GDI cell corresponds to different Boolean functions. Referring to Table1 most of the functions are realized using the function F1 and F2 since they are possible to realize using CMOS p-well process.

TABLE I
LOGIC FUNCTION IMPLEMENTED WITH GDI TECHNIQUE

N	P	G	OUT	Function
'0'	B	A	$\overline{A}B$	F1
B	'1'	A	$\overline{A}+B$	F2
'1'	B	A	A+B	OR
B	'0'	A	AB	AND
C	B	A	$\overline{A}B+AC$	MUX
'0'	'1'	A	$\overline{A}$	NOT

Table II
LOGIC FUNCTION IMPLEMENTED WITH MGDI TECHNIQUE

N	P	G	OUT	Function
A	B	B	A+B	OR
B	A	A	AB	AND
B	A	C	$\overline{C}A+CB$	MUX
'0'	'1'	A	$\overline{A}$	NOT

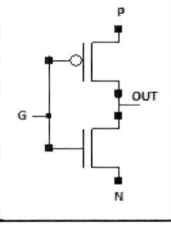

Fig. 2. Basic GDI cell

B. Modified Gate Diffusion Input (GDI)

In the basic structure of GDI cell the N diffusion node and P diffusion node act as a source and sink. Thereby there in no direct impedance path between VDD and GND as in the case of CMOS logic. Therefore this structure will considerably reduce the effect of dynamic short circuit power dissipation. The other advantage of this scheme, it requires lesser area to realize the logic functionality. While considering the basic structure of AND and OR gate of GDI the P diffusion (AND gate) is connected to GND and the N diffusion (OR gate) is connected to VDD (In Table

978-1-4673-2395-6/12 $31.00 © 2012 IEEE

1). This type of logic structure produces a slight degradation at the output voltage. To obtain the full swing voltage the proposed primitive cells has a modification in existing GDI technique. In the proposed cells of AND gate the P diffusion and Gate terminal are connected to 'A' input. Similarly for OR gate the N diffusion and gate terminal are connected to 'B' input. This modification produces less power consumption and high V_{OH}, while apparently maintaining the same reduced transistor count. Fig 3 shows the construction of modified basic gates of AND, OR, NAND, NOR, XOR, XNOR and MUX. The modified GDI primitive logic function (MGDI) is shown in Table 2.

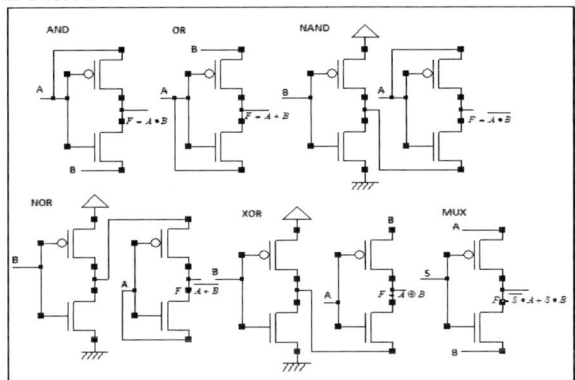

Fig. 3. Primitive cells in MGDI

The MGDI at the lower stack is used to perform the logic function, instead of NMOS block realizing the function in the conventional SRL. The output is buffered and it is connected to PMOS structure to precharge. The delay line is implemented as a series of inverters. The signals that propagate through these circuits are pulses. The width of the pulses can be increased or decreased by the insertion of inverters in the delay path. In this topology only one inverter is included in the precharge and evaluation path. The proposed SRL-MGDI primitives are simulated using Tanner EDA with BSIM3v3 250nm technology with supply voltage ranging from 0V to 5V in steps of 0.2V. All the primitive gates are simulated with multiple design corners to verify that operation across variations in device characteristics and environment. The test bed is supplied with a nominal voltage of 5V in steps of 0.2V and it is invoked with the technology library file Generic 025 and it is specified with TT, FF, FS and SS conditions. The W/L ratios of both nMOS and pMOS transistors are taken as 2.5/0.25μm. These widths were selected to ensure that the transistor threshold drop should be minimal and can only slightly impact the actual high output. To establish an unbiased testing environment, the simulations have been carried out using a comprehensive input signal pattern, which covers every possible transition for a logic gate. All the primitive structures are implemented in CMOS, PT (Pass Transistor), GDI, MGDI, SRL and SRL-MGDI with the same set up, providing the same temperature, biasing, aspect ratio and testing condition. All transitions from an input combination to another have been tested, and the delay at each transition has been measured. The average has been reported as the cell delay. The power consumption is also measured for these input patterns and its average power has been reported in Table 3.

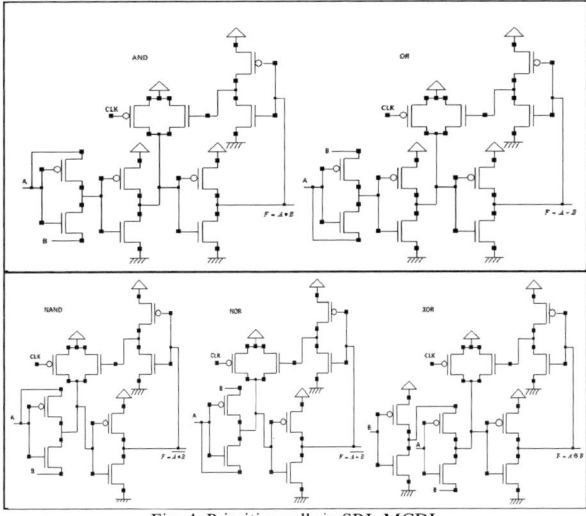

Fig. 4. Primitive cells in SRL-MGDI

IV. PROPOSED D-FF IN SRL-MGDI AND SIMULATION RESULTS

A novel implementation of a SRL-MGDI Flip-flop is shown in Fig. 5. It is based on the Master-Slave connection using NAND primitives. The enable of master and slave latch are triggered through the inverter. The first latch is called the master latch. The master latch is enabled when Clk = 0 and follows the primary input D. When Clk is a 1, the master latch is disabled but the second latch, called the slave latch, is enabled so that the output from the master latch is transferred to the slave latch. The slave latch is enabled all the while that Clk = 1, but its content changes only at the beginning of the cycle, that is, only at the rising edge of the signal because once Clk is 1, the master latch is disabled and so the input to the slave latch will not change. The reference circuits that are compared with the proposed model include (a) modified C^2MOS [14], (b) PowerPC [15], (c) HLFF [16], (d) DSTC [17] and GDI-FF [18]. The proposed SRL-MGDI FF is simulated using Tanner EDA with BSIM3v3 250nm technology with supply voltage ranging from 0V to 5V in steps of 0.2V. The simulated output of SLR-MGDI FF with the references [14-18] is presented in Table 4.

TABLE IV
PERFORMANCE OF D-FF

FF topology	Transist or Count	CLK-Q [ps]	Power [μW]	Delay [ps]	PD [fJ]
C^2MOS [14]	24	200.5	201.12	555.34	111.6
PowerPC [15]	20	250.12	220.23	523.41	115.3
HLFF [16]	20	187.62	320.4	420.75	134.3
DSTC [17]	12	320.34	240.51	501.34	120.28
GDI FF [18]	18	199.23	209.16	456.12	95.059
Proposed SLR-MGDI	46	165.25	110.24	325.3	35.861

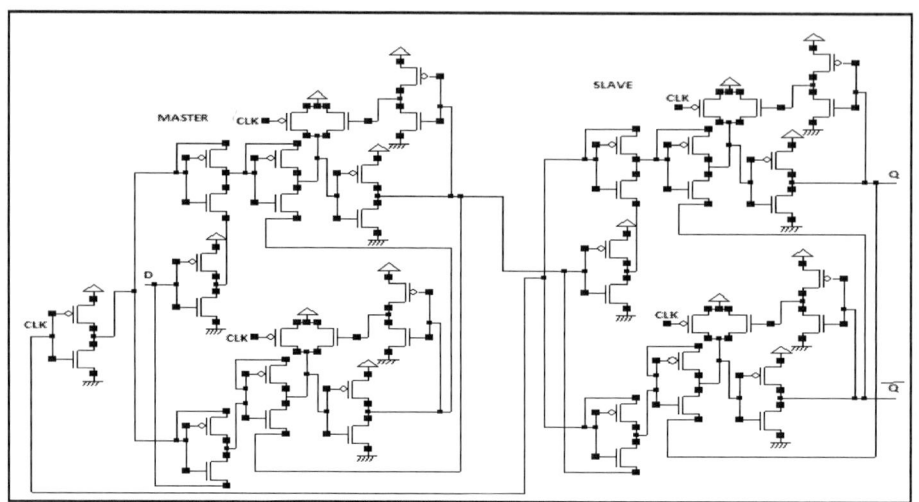

Fig 5 Proposed DFF using SRLMGDI

TABLE III
SIMULATION RESULT OF DIFFERENT LOGIC PRIMITIVE CELLS

AND	Parameters	CMOS	PT	GDI	MGDI	SRL	SRLGDI	SRLMGDI
	Delay (ns)	0.81913	.43316	0.87851	0.87851	0.71675	1.4628	0.78116
	Power (µW)	98.34	110.1	39.62	30.67	76.29	40.74	20.05
	Transistor count	6	4	2	2	8	10	10
OR	Delay (ns)	2.225	2.033	2.2912	2.2896	2.6003	2.7081	2.6119
	Power (µW)	56.33	57.66	48.91	45.09	28.16	21.21	20.13
	Transistor count	6	4	2	2	8	10	10
NAND	Delay (ns)	.83506	.23465	1.0489	.96212	1.5793	1.2346	1.2146
	Power (µW)	99.83	115.2	40.98	39.07	34.20	26.40	25.43
	Transistor count	4	6	4	4	10	8	8
NOR	Delay (ns)	2.2510	2.2346	2.2952	2.1988	2.3995	1.2346	1.2146
	Power (µW)	55.07	73.88	44.61	43.34	33.26	26.15	25.98
	Transistor count	4	6	4	4	10	8	8
XOR/ XNOR	Delay (ns)	6.7488	5.0836	5.5357	5.0357	6.2346	6.1795	5.1429
	Power (µW)	112.12	137.26	94.21	69.57	79.82	87.88	72.51
	Transistor count	12	6	4	4	14	10	10

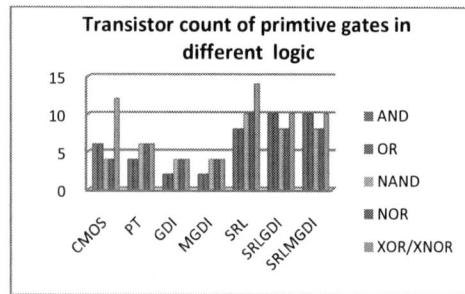

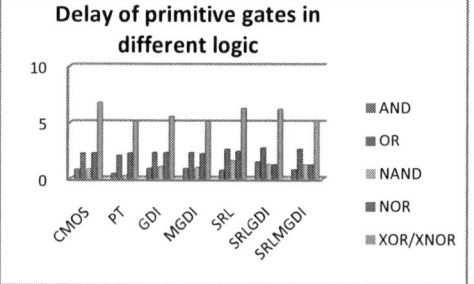

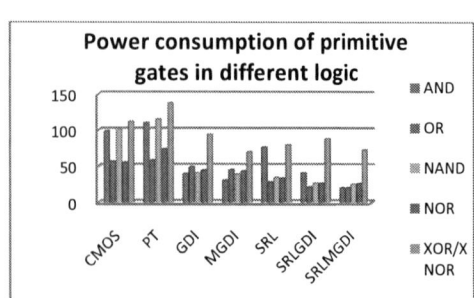

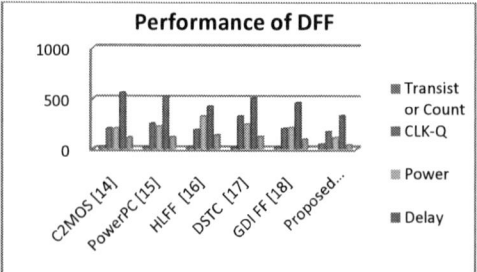

Fig. 6. Performance of primitive cells and DFF using SRL-MGDI

V. DISCUSSION

A new family of self-reset logic (SRL) cells implemented with modified GDI technique is presented in section III. A new low-power, high-speed Delay element is proposed using SRL-MGDI NAND gates. Its performances have been analyzed and reported in section IV. Using Self resetting logic the signal being propagated is buffered and used as the precharge or reset signal thereby keeping the input load as low, to minimize the loading effect. The bottom NMOS logic has been modified with Gate diffusion Input. This structure eliminates the number of transistors required to implement logic functionality, thereby reducing the transistor count as well as dynamic power dissipation. In conventional CMOS the upper PMOS stack and lower NMOS stack are connected to VDD and GND, which provides a low impedance path between VDD and GND. This situation is eliminated in MGDI so the dynamic power dissipation is less when comparing CMOS and NMOS. So by combining the features of SLR and MGDI the circuit produces high speed and low power output with slight increase in transistor count. Graph in Fig 6 shows the primitive gates implemented with CMOS, PT, GDI, MGDI, SRLGDI and SRLMGDI in terms of transistor count, delay and power, and performance of DFF. When comparing the transistor count SRLMGDI occupies significantly higher when compare to CMOS, GDI and PT. While comparing the delay aspect CMOS and SRL produces the maximum delay when compare to SRLMGDI. For power dissipation aspect the least occurs for the propose model about 62% while comparing the other logic families. The maximum power dissipation occurs for pass transistor logic. On comparing the DFF implemented with modified C^2MOS, PowerPC, HLFF, DSTC, GDI FF and Propose SRLMGDI FF the power reduction produces about 68% when comparing to the existing flip-flops.

VI. CONCLUSION

A new family of self-reset logic (SRL) cells implemented with modified GDI technique has been presented. The performance of this proposed primitive cells presents 62% of power reduction when compare to other logic families with slight increase in transistor count. With this proposed primitive cell a new low-power, high-speed Delay element has been implemented. The proposed circuit has a simple structure, based on Master-Slave principle, and contains 46 transistors. The performance issues like delay, transistor count, power and power-delay product has been analyzed with the existing structures like modified C^2MOS, PowerPC, HLFF, DSTC, GDI FF. The proposed DFF produces about 68% of power consumption when comparing to the existing flip-flops. Delay and power has been evaluated by Tanner simulator using TSMC 0.250 technologies. The simulation results reveal better delay and power performance of proposed delay element as compared to existing dynamic, GDI cell and CMOS at 0.250µm technologies.

REFERENCES

[1] Chatzigeorgiou & S. Nikolaidis, "Modelling the operation of pass transistor and CPL gates", INT . J . ELECTRONICS, 2001, VO L. 88, NO. 9, 97 7 - 100 0

[2] Yano, K., Yamanaka, T., Nishida, T., Saito, M., Shimohigash, K., and Shimizu, A., "A 3.8-ns CMOS 16x16 multiplier using complementary pass-transistor logic", IEEE Journal of Solid-state Circuits, 25, 388-394, 1990.

[3] Parameswar , A ., Hara , H., Sakurai, T., "A high speed , low power, swing restore d pass-transistor logic based multiply and accumulate circuit for multimedia applications", Custom Integrated Circuits Conference, pp . 278 – 281, 1994.

[4] Yano, K., Sasaki, Y., Rikino, K. , and Sek i, K., Top-down pass-transistor logic design . IEEE Journal of Solid-State Circuit s, 3 1 , 792- 803, 1996.

[5] Arkadiy Morgenshtein, Alexander Fish, and Israel A. Wagner, "Gate-Diffusion Input (GDI): A Power-Efficient Method for Digital Combinatorial Circuits", IEEE Transactions On Very Large Scale Integration (VLSI) Systems, VOL. 10, NO. 5, October 2002.

[6] Adarsh Kumar Agrawal, S. Wairya, R.K. Nagaria and S. Tiwari, "A New Mixed Gate Diffusion Input Full Adder Topology for High Speed Low Power Digital Circuits", World Applied Sciences Journal 7 (Special Issue of Computer & IT): 138-144, 2009

[7] L. BISDOUNIS, D. GOUVETAS & O. KOUFOPAVLOU, "A comparative study of CMOS circuit design styles for low-power high-speed VLSI circuits", INT. J. ELECTRONICS, 1998, VOL. 84, NO. 6, 599- 613.

[8] G. Yee and C. Sechen, " Clock-delayed domino for adder and combinational logic design" in proc.IEEE/ACM Int. Conf. Computer Design, Oct., 1996, pp. 332-337.

[9] P. Ng, P. T. Balsara, and D. Steiss, ―Performance of CMOS Differential Circuits,□ IEEE J. of Solid-State Circuits, vol. 31, no. 6, pp. 841-846, June 1996.

[10] P. Srivastava, A. Pua, and L. Welch, .Issues in the Design of Domino Logic Circuits, Proceedings of the IEEE Great Lakes Symposium on VLSI, pp. 108-112, February 1998.

[11] Woo Jin Kim, Yong-Bin Kim, ―A Localized Self-Resetting Gate Design Methodology for Low Power□ IEEE 2001.

[12] M. E. Litvin and S. Mourad, ―Self-reset logic for fast arithmetic applications, IEEE Transactions on Very Large Scale Integration Systems, vol. 13, no. 4, pp. 462–475, 2005.

[13] R.Uma, "4-Bit Fast Adder Design: Topology and Layout with Self-Resetting Logic for Low Power VLSI Circuits", International Journal of Advanced Engineering Sciences and Technology, Vol No. 7, Issue No. 2, 197 – 205, 2011.

[14] V. Stojanovic and V.G. Oklobdzija, "Comparative Analysis of Master–Slave Latches and Flip-Flops for High-Performance and Low-Power Systems", IEEE J. Solid-State Circuits, vol. 34, no. 4, April 1999.

[15] G. Gerosa, S. Gary, C. Dietz, P. Dac, K. Hoover, J. Alvarez, H. Sanchez, P. Ippolito, N. Tai, S. Litch, J. Eno, J. Golab, N. Vanderschaaf, and J. Kahle, "A 2.2 W, 80 MHz superscalar RISC microprocessor," IEEE J. Solid-State Circuits, vol. 29, pp. 1440–1452, December 1994.

[16] H. Partovi, R. Burd, U. Salim, F. Weber, L. DiGregorio, and D. Draper, "Flow-through latch and edge-triggered flip-flop hybrid elements," ISSCC Dig. Tech. Papers, pp. 138–139, February 1996.

[17] J. Yuan and C. Svensson, "New single-clock CMOS latches and flipflops with improved speed and power savings," IEEE J. Solid-State Circuits, vol. 32, January 1997.

[18] Arkadiy Morgenshtein, Alexander Fish and Israel A. Wagner, "An Efficient Implementation of D-Flip-Flop using the GDI Technique" IEEE, 2004.

Multiply-Accumulate Instruction Set Extension in a Soft-core RISC Processor

Ahmad Jamal Salim, Nur Raihana Samsudin, Sani Irwan Md Salim, Yewguan Soo
Faculty of Electronics and Computer Engineering
Universiti Teknikal Malaysia Melaka
Melaka, Malaysia
shaj@utem.edu.my, m021100037@student.utem.edu.my

Abstract— **Application Specific Instruction Set Processor (ASIP) design is known to offer optimum performance and flexibility in a processor performance although with limited application. Implementing the processor on Field Programmable Gate Array (FPGA) further extend the opportunity to reconfigure the architecture instantaneously. In this paper, the instruction set extension approach is implemented on a simple 8-bit soft-core RISC processor to enhance the processor capability by adding new instruction set that can allow it to perform basic digital signal processing (DSP) algorithm. Creation of new instruction set is achieved by modifying the processor's architecture using Hardware Description Language (HDL). For verification purposes, a multiply-accumulate (MAC) instruction is created in addition to existing RISC instructions. The MAC instruction set, which is the fundamental operation of DSP algorithms, involved 8x8 bit multiplication and the accumulation result is stored in two 8-bit register-pair. The new instruction set must adhere to the current instruction set architecture (ISA) in order to ensure the new instruction is fully compatible to the existing architecture. The instruction is successfully tested through execution of RISC processor on FPGA chip and correct output has been observed from the MAC instruction. The results show that through instruction set extension approach, a low-end RISC processor is capable to execute more complex instructions just by reconfiguring the instruction set to match the specific system requirement. The approach also offers flexibility in instruction extension and the resource is only limited to the constraint of the FPGA chip where the processor resides.**

Keywords-ASIP, RISC, multiply-accumulate

I. INTRODUCTION

It is a known fact that Application Specific Instruction Set Processor (ASIP) provides the middle ground in speed and flexibility when dealing with processor design. Generally, ASIP offers considerably speed performance although ASIC is still the best option when speed is the main design requirement. On the other hand, ASIP also has the flexibility of general-purpose processor [1], [2] by accommodating additional instruction set and on-field programming through it design. Flexibility can be achieved by introducing customized instruction set through instruction set extensions that is tailor-made to perform specific application optimally.

Instruction set extensions can be categorized into two approaches, which is complete customization and partial customization [2]. Complete customization requires a complete build from ground up on instruction set architecture of a processor to suit certain applications requirement. Meanwhile, partial customization involves extension of existing instruction set architecture where a limited number of new instruction set is added which is tuned to the specific functions [3]. Fundamentally, both approaches targeted a trimmed down processor design that only accommodate the most useful instruction set for the specific applications and discarded the others in order to maximize execution time.

RISC processor features a simple store/load architecture [4] that essentially offers highly optimized set of instruction. With each instruction is simplified and mostly executed in one clock cycle, RISC processor provides the best platform in implementing instruction set extension approaches. Combination of several basic instructions to perform a specific function e.g. multiply-accumulate operation can be accomplished in optimized performance, even for low-end RISC processor. Other than that, RISC architecture also incorporates a large number of registers to minimize the interaction between the processor and memory.

The emergence of reconfigurable architecture over the last two decades prompted the idea of combining conventional processor on top of a reconfigurable platform such as FPGA [5]. Normally, a general-purpose processor acts as the main controller that synchronizes and control several other modules or co-processors in an integrated embedded system [6]. Field programmable feature that is utilized in FPGA has made reconfiguring the processor architecture [7] easily done on the fly which ultimately reduces the cost while optimizing performance.

On a wider scope, implementation of a simple RISC processor on a FPGA platform would enable more opportunities to enhance the processor's capability by adoption of ASIP design methodologies [8], [9]. The term soft-core processor refers to a processor architecture that is synthesizable and resides in FPGA. Soft-core processor support generation-time configuration options to allow designers to trade off performance and cost. In this platform, the internal architecture of RISC processor is described in Hardware Description

Language (HDL) [10] that make it possible to make modification on memory allocation, register array and instruction decoder. Therefore, the designers are in control to determine the performance trade-off, resource utilization and detail configuration on the required instruction set.

This paper demonstrates the capability of a soft-core RISC processor in configuring new instruction set by modifying the processor's architecture. For testing purposes, multiply-accumulate (MAC) instruction is chosen to be evaluated because MAC performance is considered important to fundamental DSP algorithm such as FIR filters or Fast Fourier Transform (FFT) [11].

II. METHODOLOGY

A. MAC Operation

In the 8-bit RISC processor architecture, multiply-accumulate operation is implemented as an instruction set whereby the multiplier and adder functions are configured inside the processor's ALU. A signed two's complement multiplier and full adder supporting two's complement arithmetic calculation is utilized in the MAC operation. Two clock cycle are required in implementing both multiplication and accumulation process. The multiplier provides a double precision result of two single precision operands in the first clock cycle then the accumulator perform arithmetic function with double precision input and output during the next clock cycle.

Fig. 1 shows the basic MAC architecture. In this operation, the multiplication involves 8-bit multiplicand and 8-bit multiplier that reside in a register. The output would be in 16-bit wide and stored in a register-pair. As the register-pair is updated when MAC is executed, the overflow bit will be used to check the result of the operation.

B. Instruction Set Architecture (ISA)

For RISC processor, the instruction set width is set to 12-bit wide encompassing opcode and literal bit or file register or memory address. The formatting also supported direct and indirect addressing mode in its implementation. Instructions' opcode are varied between 3-6 bits depending on its mode of operation. The next corresponding bits in the instruction code are file register, literal value or bit address. Fig. 2 shows the overall instruction format of the RISC processor. The new MAC instruction set is configured according to the ISA. As MAC instruction involves registers or memory address, the instruction is operated in byte-oriented file register.

III. IMPLEMENTATION FLOW

The RISC processor in this paper essentially is an 8-bit RISC processor using Harvard architecture. Therefore, all program instructions will reside in ROM while other registers and variables are located in RAM. With the instruction set architecture is set, the Instruction Decoder module is configured to match the ISA and ensure the decoded bits are

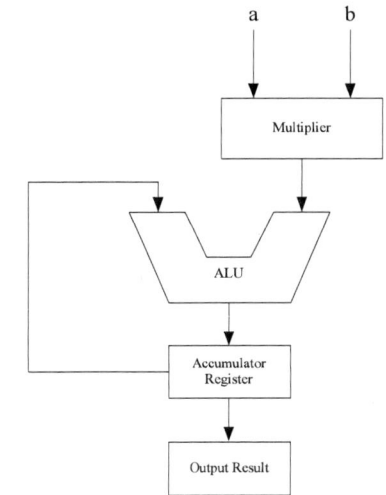

Fig. 1. Basic MAC Hardware Architecture

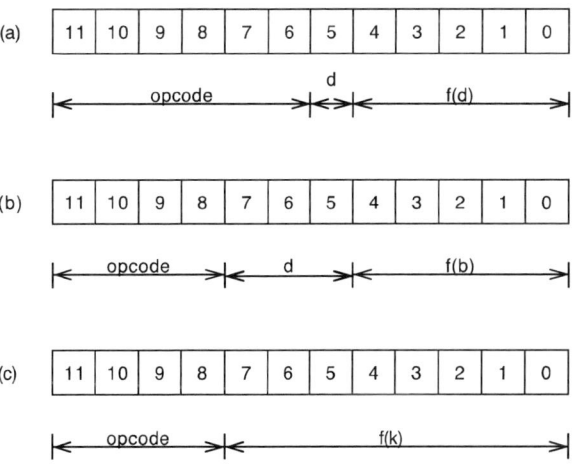

Fig. 2. General Instruction Format (a) Byte-oriented File Register Operation, (b) Bit-oriented File Register Operation and (c) Literal Value Operation

correctly translated. The MAC algorithm is programmed using HDL in ALU modules in order to perform multiplication and accumulation consecutively. A testbench is setup to simulate the newly added instruction set in FPGA environment.

The new MAC instruction set is inserted in assembly language programming file. The ASM file is then compiled to generate the hexadecimal file (.HEX), which is then converted to the coefficient file (.COE). The coefficient file is used as the initial data to be loaded to the ROM modules, as shown in Fig. 3. During the operation, ALU fetched the MAC instruction from Instruction Decoder modules and then execute the corresponding MAC sequences as per programmed. All modules are synthesized and simulated to validate the MAC operation's results.

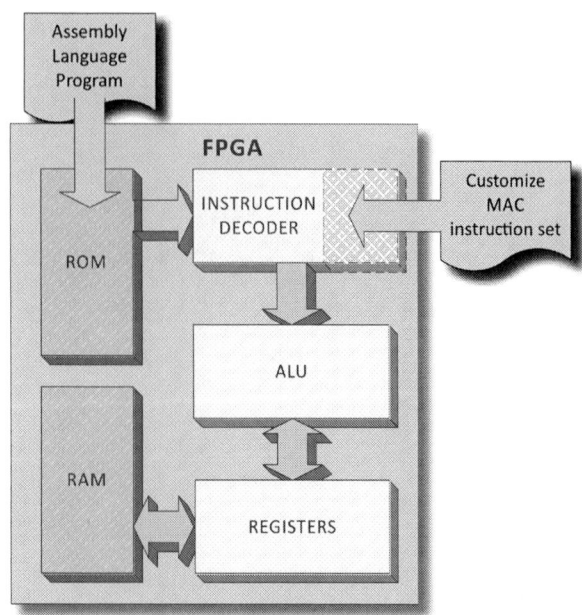

Fig. 3. RISC Processor Block Modules

IV. RESULTS AND DISCUSSION

Fig. 4 shows the assembly language program that is developed to call the MAC operation. The MAC instruction is executed in a continuous loop to verify its operation upon several execution cycles. The assembly language is compiled to generate the hexadecimal file. The coefficient file (.COE) is then transferred in ROM. Therefore, all instruction code are resided and fetched in ROM upon being decoded by the instruction decoder module.

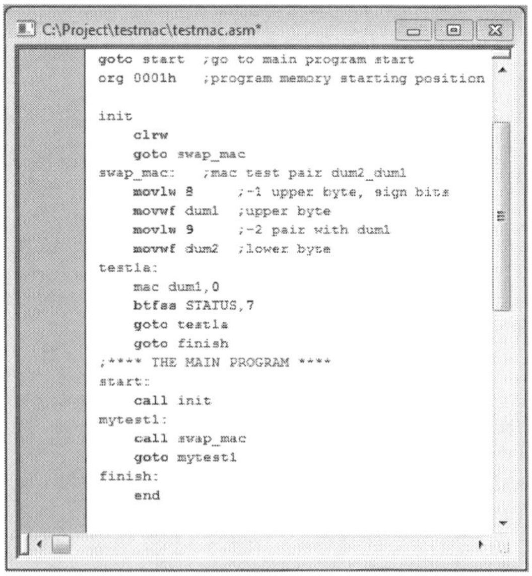

```
C:\Project\testmac\testmac.asm*

    goto start   ;go to main program start
    org 0001h    ;program memory starting position

init
    clrw
    goto swap_mac
swap_mac:    ;mac test pair dum2_dum1
    movlw 8      ;-1 upper byte, sign bits
    movwf dum1   ;upper byte
    movlw 9      ;-2 pair with dum1
    movwf dum2   ;lower byte
testla:
    mac dum1,0
    btfss STATUS,7
    goto testla
    goto finish
;**** THE MAIN PROGRAM ****
start:
    call init
mytest1:
    call swap_mac
    goto mytest1
finish:
    end
```

Fig. 4. Assembly Language Program for MAC instruction execution

The assembly language program is initiated by assigning starting value for FSR and w register; designated as dum1 and dum2. Therefore, for the first iteration of MAC instruction involving signed number -1 multiply by -2. During MAC instruction execution, the results are stored in FSR (dum1) and working register. To verify the output for subsequence iterations, the MAC instruction is set on a continuous loop. Overflow flag in STATUS register is checked regularly to avoid final sum error on the MAC results.

The overall result of MAC architecture is shown in Fig. 5. Input a and b obtained its values from working register (w) and FSR respectively. Two's complement is performed for negative number and stored as a1 and b1. Multiplication operation is executed and the result is stored as m1. Two's complement is performed again to reflect the negative sign and the result is stored as m2. The accumulate operation is done by adding m2 with previous value of m2, named as accu register. The 16-bit accumulation result is also stored in register pair y and y2 where register y contained the higher 8 bit data while y2 contained the lower 8 bit data. Indirectly, the data in register y and y2 is relayed back to w and FSR. Therefore, the cycle is restarted again upon the execution of the MAC instruction.

In the FPGA implementation, all RISC processor modules are instantiated and synthesized using the Xilinx Spartan-3AN platform with Xilinx ISE integrated environment. ISim is utilized as the simulation platform. Registers outputs are observed during the MAC instruction execution cycle. The overall MAC operation sequences are shown in Fig. 6.

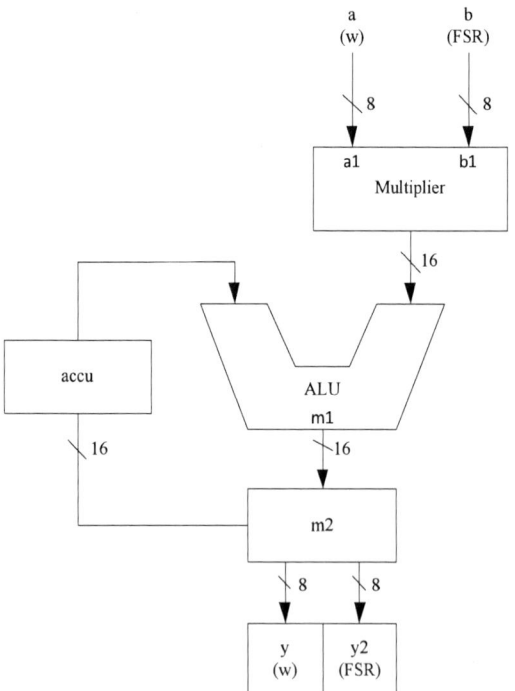

Fig. 5. MAC hardware implementation

978-1-4673-2395-6/12 $31.00 © 2012 IEEE

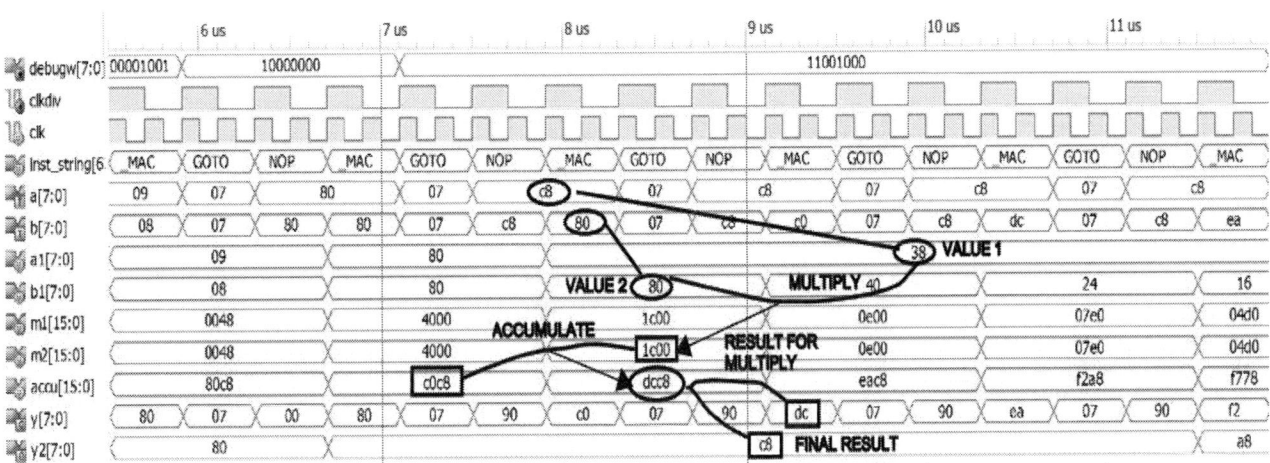

Fig. 6. Results of MAC Instruction Implementation

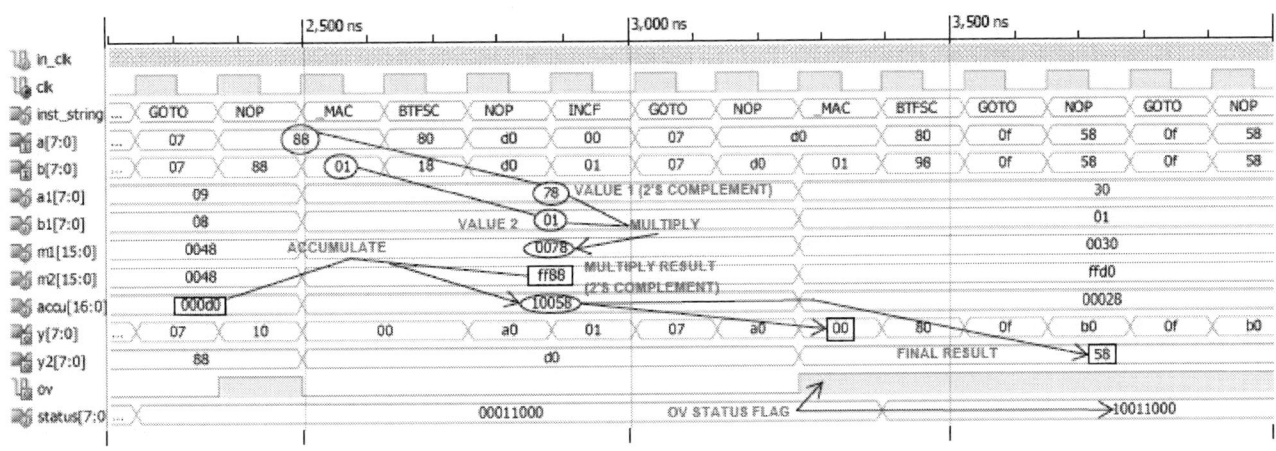

Fig. 7. Results of MAC Instruction Implementation with Overflow Status Bit Indicator

In cycle starting at 7.9 µs, when the MAC instruction is executed, value a (C8) is obtained from previous iteration of y2 while value b (80) from the previous iteration of y. Then, as both are negative numbers, 2's complement of those values is done and stored at value a1 (38) and value b1 (80). The multiplication operation is done by a1 and b1 whereby the result (1C00) is stored in m1. Value of m2 is maintained at 1C00 because a positive result is obtained from the two negative numbers multiplication; hence 2's complement is not required. Next, accumulate operation is performed by adding value of m2 (1C00) with value of accu (C0C8), which is the value of (y,y2) from previous MAC iteration. The accumulation result (DCC8) is updated to accu register and separated to two 8-bit data and stored as y (DC) and y2 (C8). At the same time, the value of y and y2 are also linked to w and FSR registers to be used in the next MAC instruction iteration. After running through several cycles, the overall observation has shown that the output of MAC operations turned out correctly. All respective registers in FSR and working register are updated and both multiplication and accumulation are executed accurately.

Further inspection have been done on the MAC instruction operation's accuracy particularly involving overflow value during the accumulate calculation. Fig. 7 shows the MAC instruction implementation whereby an overflow bit is used as an indicator to detect if any overflow has occurred during the accumulation process, on a different timescale. In this case, value a (88) is multiplied with value b (01). The accu register contained (00d0), which is acquired from the previous iteration. When the MAC operation is executed, the value of m2 (ff88) indicates that the multiplication result is a negative number. Next, the accumulate operation is performed by adding value of m2 (ff88) with the value of accu (00d0) resulting an overflow (10058) at the accu register. Although the value (0058) is updated at registers (y,y2) during the next iteration, concurrently the overflow bit in the STATUS register is set thus indicating that overflow has occurred and the result is inaccurate. As stipulated in assembly program, MAC operation loop is ended once the overflow bit is set. This routine is important to ensure that the MAC operation results are remained accurate within the 16-bit number range throughout all iterations.

978-1-4673-2395-6/12 $31.00 © 2012 IEEE   515

V. CONCLUSION

This paper described the capability of a simple RISC processor architecture to execute a new instruction set by instruction set extension approach. The extension involves modification of the instruction decoder and the ALU modules in accordance to the ISA of the processor. The newly added instruction is verified through assembly language program that is executed by the RISC processor. Having a soft-core RISC processor that is synthesized on a FPGA platform enables the architecture to be more flexible and reconfigurable. In addition, the successful operation of MAC instruction has provided a good basis for more complex application development in FPGA-based system especially in digital signal processing through ASIP methodologies.

ACKNOWLEDGMENT

The author would like to thank Universiti Teknikal Malaysia Melaka and Ministry of Higher Education Malaysia for the financial support given through the research grant number FRGS/2012/FKEKK/TK02/02/1/F00126.

REFERENCES

[1] J. Ball, "Designing Soft-Core Processors for FPGAs Processor Design," in Processor Design: System-on-Chip Computing for ASICs and FPGAs, J. Nurmi, Ed., 1st ed: Springer Netherlands, 2007, pp. 229-256.

[2] C. Galuzzi and K. Bertels, "The Instruction-Set Extension Problem: A Survey," ACM Trans. Reconfigurable Technol. Syst., vol. 4, pp. 1-28, 2011.

[3] L. Barthe, L. V. Cargnini, P. Benoit, and L. Torres, "The SecretBlaze: A Configurable and Cost-Effective Open-Source Soft-Core Processor," in IEEE International Symposium on Parallel and Distributed Processing Workshops and Phd Forum (IPDPSW), 2011, pp. 310-313.

[4] P. S. Mane, I. Gupta, and M. K. Vasantha, "Implementation of RISC Processor on FPGA," in IEEE International Conference on Industrial Technology 2006, pp. 2096-2100.

[5] J. G. Tong, I. D. L. Anderson, and M. A. S. Khalid, "Soft-Core Processors for Embedded Systems," in International Conference on Microelectronics, 2006, pp. 170-173.

[6] A. J. Salim, M. Othman, and M. A. M. Ali, "The Development of a Programmable DSC Chip: UKM8051DSC," European Journal of Scientific Research, vol. 19, pp. 350-361, 2008.

[7] P. Yiannacouras, J. G. Steffan, and J. Rose, "Exploration and Customization of FPGA-Based Soft Processors," IEEE Transactions on Computer-Aided Design of Integrated Circuits and Systems,, vol. 26, pp. 266-277, 2007.

[8] J. S. Lee and M. H. Sunwoo, "Design of New DSP Instructions and Their Hardware Architecture for High-Speed FFT," The Journal of VLSI Signal Processing, vol. 33, pp. 247-254, 2003.

[9] W. Wenxiang, L. Ling, Z. Guangfei, L. Dong, and Q. Ji, "An Application Specific Instruction Set Processor optimized for FFT," in IEEE 54th International Midwest Symposium on Circuits and Systems (MWSCAS), 2011, pp. 1-4.

[10] T. Coonan. (20 December 2011). Verilog Synthetic PIC. Available: http://www.mindspring.com/~tcoonan/newpic.html

[11] J. McAllister, "FPGA-based DSP," in Handbook of Signal Processing Systems, S. S. Bhattacharyya, E. F. Deprettere, R. Leupers, and J. Takala, Eds., 1st ed: Springer US, 2010, pp. 363-392.

Adjustable Phase-Locked Loop with Independent Frequency Outputs

Robert Freier, Hamam Maher Abd, Andreas König

Institute of Integrated Sensor Systems

Technische Universität Kaiserslautern

D-67663 Kaiserslautern, Germany

Email: freier@eit.uni-kl.de

Abstract—Phase-locked loops (PLLs) are used to synchronize the frequency of an oscillator with a reference. This work explains the architecture of a new multi-purpose PLL that can deliver different types of output signals which can be set independently to different frequencies ranging from 61 Hz up to 4 MHz. Possible outputs are square wave signals, sinusoidal signals and single voltage pulses with adjustable width. If no external oscillator is available, this PLL can also operate autonomously without reference. To reduce power consumption, the complete circuit can be sent to standby mode. Additionally, the whole structure is fully integrable in ICs and does not require external parts.

I. INTRODUCTION

PLLs are employed in a wide field of applications, ranging from telecommunications to control systems and measurement. Examples are signal modulation [1], synchronization [2] and clock recovery [3]. Generally, the task of a PLL is to deliver a frequency which is multiple, fractional or equal to a reference oscillator.

The featured circuit is developed in the framework of self-x sensor systems for a generic mixed-signal sensor conditioning IC with reconfiguration features. Self-x is substitutional for, e.g., self-monitoring, -calibrating, -trimming, and -repairing [4] and is one possible way to dependable systems in mixed-signal electronics (Figure 1). In this environment, flexible systems are needed with exceptional requirements that cannot be met by common designs. Existing PLLs usually generate one square wave output signal, where sometimes the frequency can be changed by programming or by applying external parts like a resistor [5] [6]. For the designated IC this won't be sufficient, as it should operate at different clock speeds and perform capacitive and impedance measurements at the same time. For this purpose two additional signals are needed that can be set independently from the clock, preferably a sinusoidal signal and a single voltage pulse. The setting of the frequencies for these purposes should be in the range from sub-kilohertz to more than one megahertz. As a reference a 1MHz crystal oscillator is intended to be used, but the structure should also be capable of working without any external reference. Further desired features are applicability in low-power environments as well as employment in different chips that have specific demands on PLLs, like diverse frequencies and variable reference oscillators. The PLL should be usable as a multi-purpose standard cell in 0.35µm CMOS technology for future ICs, especially in the domain of self-x systems.

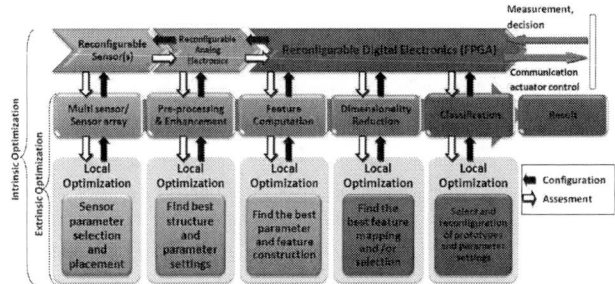

Fig. 1. Enhanced design methodology for intelligent sensor systems with possible implementation of self-x features [4]

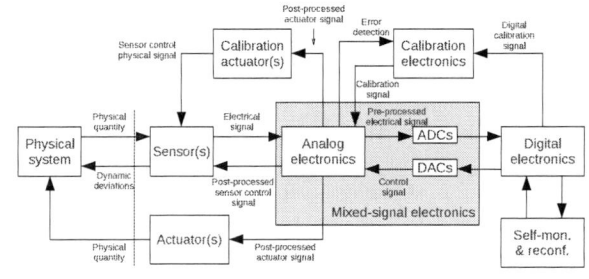

Fig. 2. Implementation of self-x features in sensor systems [7]

The design of these systems mostly relies on programmable or reconfigurable hardware, with the intention to adjust or modify characteristics of the circuits or the system itself during operation (Figure 2). These attributes, namely changing features while operating, are provided in the presented PLL. The basic functionality is shown in Figure 3.

II. CONCEPT OF A RECONFIGURABLE PLL

The basic operating principle of this novel PLL is similar to that of others. A voltage controlled oscillator (VCO) is put in relation to a reference oscillator through a closed loop control. A phase frequency detector (PFD) compares the feedback signal to the input and accordingly drives a charge pump (CP) to source or sink pulse-width modulated current pulses. Those

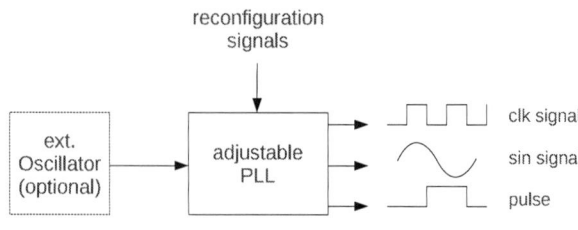

Fig. 3. Function of the presented PLL

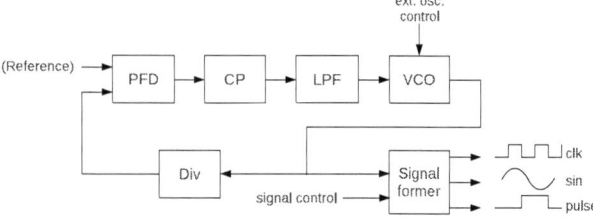

Fig. 4. Block diagram of the adjustable PLL with optional external reference

pulses are fed into a low-pass filter (LPF) which actually fulfills two tasks at the same time. On the one hand it acts as the controller to stabilize the feedback system, on the other hand it converts the current pulses to a smooth voltage signal that regulates the frequency of the VCO by either speeding it up or slowing it down. The feedback loop contains a divider which makes the VCO run at a higher frequency than the actual reference.

What sets this structure apart from existing PLLs is the connection of the VCO to programmable signal forming circuits that produce the desired outputs, where the output forms can be changed by control signals. These circuits will be explained in the next section. Another control signal determines whether the external oscillator shall be used or not. The whole structure is illustrated in Figure 4.

III. DESIGN AND IMPLEMENTATION

This section will give detailed explanations about the design of the used components. Figure 5 shows a block diagram of the PLL's architecture.

The core piece of the PLL is the voltage controlled oscillator. It will provide the oscillation needed for the output signals and its accuracy in frequency is mainly responsible for the quality of the outputs. The VCO is a ring oscillator type designed for differential operation (Figure 6), because this method is less sensitive towards deviations in supply and substrate voltage compared to single-ended approaches. Inside the ring oscillator, to set the cycle time, the circulating signal is delayed by five identical elements that are designed as differential amplifiers with replica circuit [8]. The principle is shown in Figure 7. To alter the speed of the oscillator, a voltage

signal adjusts the lateral current of the operational amplifiers inside the ring elements to change their delay time, resulting in a different frequency of the VCO (Figure 8). The VCO's output is used to generate the output signals of the PLL; it is also connected to the feedback of the control loop, where the frequency is divided by four in order to make the VCO run four times the speed of the external crystal oscillator.

The phase frequency detector is the device that compares the system's present frequency to that of an external reference, if any is used. Its design is a standard D flip-flop topology taken from [9], see Figure 9. As the output of the PFD is a voltage signal, but for the loop filter actually a current is needed, the signal is converted by a charge pump (Figure 10) to a pulse-width modulated current output.

With a PWM signal presenting the deviation between the VCO's actual and desired frequency, the speed of the ring oscillator has to be adjusted. Because the VCO requires a voltage signal to control the frequency, the current pulsed signal has to be converted to a smoothened voltage signal by a low-pass filter. The low-pass filter (LPF) is a crucial element inside a PLL, as it not only generates the control voltage for the VCO but also acts as the controller of the feedback system. Wrong LPF design will lead to instability of the control system, making the PLL useless. With proper dimensioning, speed and stability can be optimized. In this case, a passive second order LPF turned out to be sufficient and has been integrated (Figure 11). For other filters, the possibility is given to connect them externally.

As mentioned before, this PLL should be able to generate signals even if no external oscillator is available. To achieve this, the VCO must run at constant speed without a feedback control that regulates the oscillation. The frequency of the VCO depends primarily on the control voltage that is normally taken from the LPF. If this voltage is not available because there is no external oscillator, it will be replaced with a signal from a reference, namely a beta multiplier voltage reference

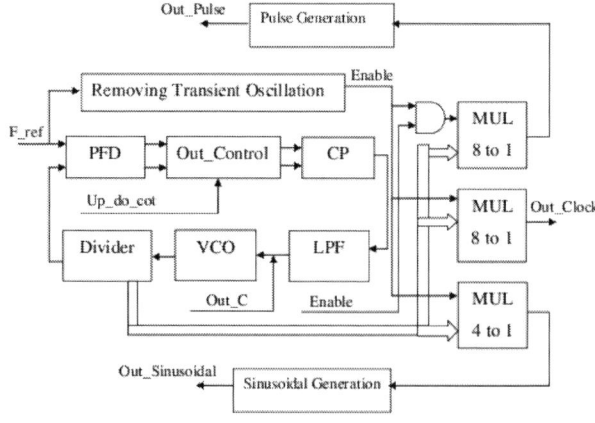

Fig. 5. Architecture of the phase-locked loop

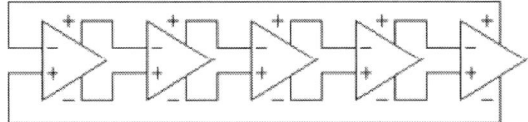

Fig. 6.   Principle of the used differential mode ring oscillator

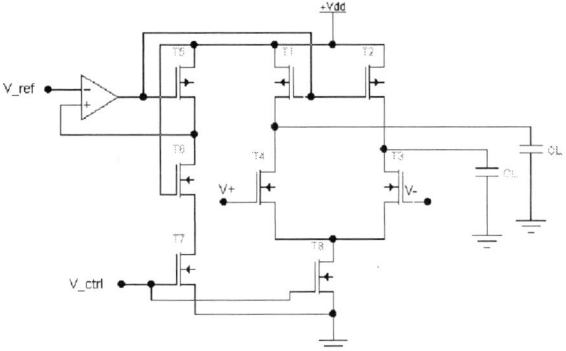

Fig. 7.   Operational amplifier with replica circuit

[10]. This action ensures the frequency of the VCO to be fixed at 4MHz, even without a reference oscillator.

The VCO generates a square wave signal at a constant frequency of 4MHz. But the required outputs should have variable, independent frequencies at the same time and different forms as well. In order to receive different frequencies from one oscillator, the output is fed into several divider circuits. Each divider, 16 in total that work like an asynchronous counter, cuts the input frequency in half, so the input of each divider is twice the frequency of the output. With this measure, frequencies from 4MHz down to around 60Hz can be obtained by selecting the desired output of the corresponding divider.

Still there is only one type of square wave signal. As there should be three output signals, three multiplexers are used to

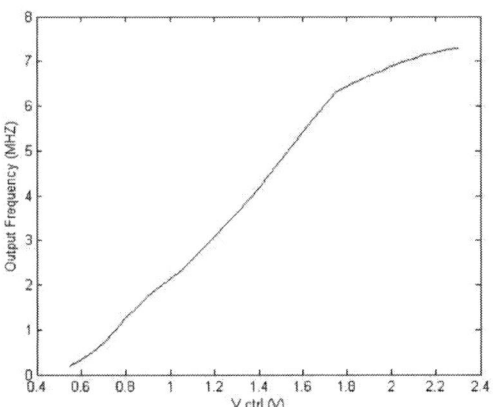

Fig. 8.   Dependence of input voltage and VCO frequency

choose independently from the different frequencies. One of the multiplexers delivers the square wave output, the remaining two are used for the sinusoidal signal and the voltage pulse output. To get a sine wave signal from a square wave, the following circumstance is considered: in frequency domain, an ideal square wave signal contains its fundamental frequency and an infinite number of harmonics. A sine wave is only at one frequency, in this case it should be the fundamental of the square wave. Removing the harmonics, i.e., filtering, will result in a sinusoidal signal. That means, by applying a low-pass filter, a sine wave can be taken from the signal output of the second multiplexer. The higher the order of the filter, the better the quality of the signal. For different signal frequencies also different filters are needed. Because this PLL is desired not to rely on any external parts, active low-pass filters for four different frequencies are integrated within the circuit. To keep the size of the layout within acceptable extent, second order Gm-C filters are used as a trade-off between physical dimension and output quality (Figure 12). If an improved THD is desired, also external filters of higher order can be attached. Last of the three signal types is a voltage pulse that can be adjusted in width and outputted on demand. For this purpose, the circuit shown in Figure 13 generates a pulse when an enable signal is given, where the pulse width equals the period of the input signal. By that the pulse width can be set between 0.24μm and 16384μm. The circuit is also connected to a multiplexer, so the duration of the pulse can be set by selecting the input frequency. The function of the RC filter is to remove the disturbance that occurs when powering up the PLL, to prevent unwanted pulse generation.

Additional features of the designed PLL include a power-down mode, which sets the complete circuit to standby when it is not needed. Power consumption can thereby be significantly reduced, which is crucial for applications that rely on battery supply. Furthermore, the PLL does not give flawed outputs while being turned on or resuming from standby. During start-up, a specific time is needed until the PLL reaches its set point and the VCO runs at the targeted frequency. This time span is called the lock time of the PLL and causes transient oscillation at the output. The lock time is a constant value that depends on the system's transfer function, in this case it is around 20μs. To prevent that behavior, the output is disabled during that time for 32μs until the PLL is ready and has definitely reached a steady state.

The layout of the PLL has been implemented in 0.35μm CMOS technology and can be seen in Figures 14 and 15. The area consumption of the whole design, including filters, is less than 0.9mm^2.

IV. SIMULATION RESULTS

Functionality has been verified by simulation for nominal and randomly distributed parameter values and temperatures between -40°C and 80°C. Table I shows all possible output signal forms and frequencies without external parts or with just a 1MHz reference oscillator. In Figure 16 the signals for one sample configuration are displayed. Total power consumption,

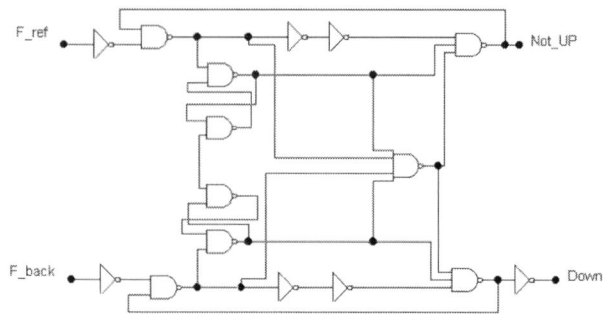

Fig. 9. Topology of the phase frequency detector

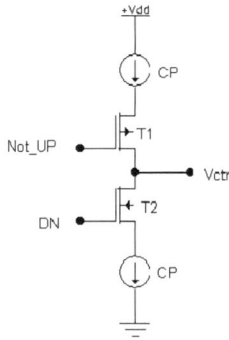

Fig. 10. Charge pump that is connected to the PFD

including the low-pass filters for sinusoidal signal generation, is less than 13mW during operation and lower than 3.5μW in standby mode.

V. CONCLUSION AND OUTLOOK

The reported phase-locked loop has been developed for the demands of mixed-signal circuits and SoCs that have varying requirements and should be fully integrable. In contrast to existing approaches, external components are not required and different output signal types can be set independently at the same time. Performance can even be increased or adjusted when it is negligible to use external parts. Currently the PLL's function has only been proven on post-layout simulation level. The next step will be to manufacture and test samples in a

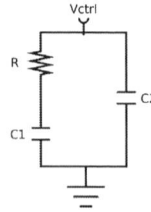

Fig. 11. Integrated second order low pass filter; it can optionally be exchanged with external filter

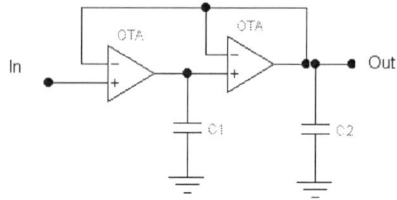

Fig. 12. Second order Gm-C filter

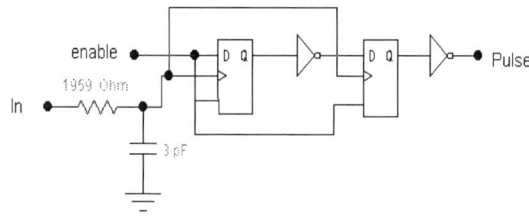

Fig. 13. Circuit for aperiodic pulse generation

0.35μm CMOS technology, where the PLL will be employed within the framework of a designated sensor interface IC.

REFERENCES

[1] M. Jung, A. Ferizi, T. Ussmueller, G. Fischer, R. Weigel, *A 9.7 mW 0.13 μm CMOS PLL for use in wireless sensor nodes*, IEEE 54th International Midwest Symposium on Circuits and Systems, Aug. 7-10, 2011.

[2] Fang Xiong, Wang Yue, Li Ming, Wang Ke, Lei Wanjun, *A novel PLL for grid synchronization of power electronic converters in unbalanced and variable-frequency environment*, 2nd IEEE International Symposium on Power Electronics for Distributed Generation Systems, pp. 466-471, June 16-18, 2010.

[3] N. Tall, N. Dehaese, S. Bourdel, O. Fourquin, R. Vauche, J. Gaubert, *Low-power clock and data recovery circuit for IR-UWB receiver power management*, 9th International Multi-Conference on Systems, Signals and Devices, March 20-23, 2012.

[4] K. Iswandy, A. König, *Methodology, Algorithms, and Emerging Tool for Automated Design of Intelligent Integrated Multi-Sensor Systems*, Algorithms 2009, 2, 1368-1409.

Fig. 14. Complete layout of PLL with filter circuits, dimensions are 1.01mm × 0.86mm

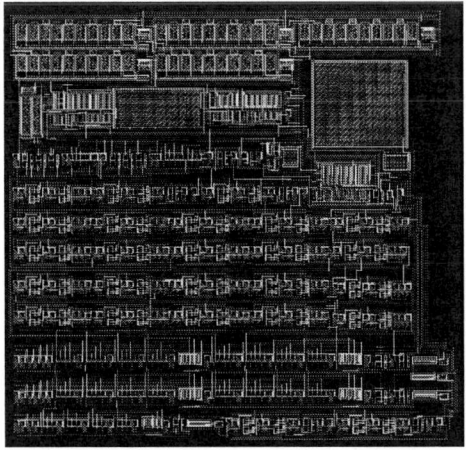

Fig. 15.   Layout of PLL without filter circuits

Square Wave [kHz]	Sine Wave [kHz]	Pulse Width [µm]
4000	-	0.24
2000	-	0.49
1000	-	0.99
500	-	2
125	125	8
15.625	15.625	64
0.997	0.997	1024
0.061	0.061	16384

TABLE I
POSSIBLE OUTPUT SIGNAL FORMS

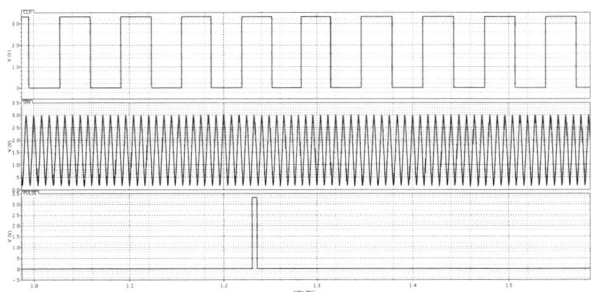

Fig. 16.   Simulation results for 15.625kHz square wave signal, 125kHz sine wave signal and 8µs voltage pulse

[5] Philips, *Data Sheet 74HC/HCT4046A Phase-locked-loop with VCO*, http://www.nxp.com/acrobat_download/datasheets/74HC_HCT4046A_CNV_2.pdf

[6] Linear Technology, *LTC6900 Low Power, 1kHz to 20MHz Resistor Set SOT-23 Oscillator*, cds.linear.com/docs/Datasheet/6900fa.pdf

[7] Peter M. Tawdross, *Bio-Inspired Circuit Sizing and Trimming Methods for Dynamically Reconfigurable Sensor Electronics in Industrial Embedded Systems*, Technische Universität Kaiserslautern, 2007.

[8] Behzad Razavi, *Design of Analog CMOS Integrated Circuits*, McGraw-Hill, 2001.

[9] R. Jacob Baker, *CMOS Circuit Design, Layout, and Simulation*, Wiley-IEEE Press, 2010.

[10] S.S. Prasad, Pradip Mandal, *A CMOS Beta Multiplier Voltage Reference with Improved Temperature Performance and Silicon Tunability*, Proceedings of the 17th International Conference on VLSI Design, 2004.

Capacitor-Grounded Electronically Tunable Voltage-Mode OTA-C Multifunction Filter with Three Inputs and Five Outputs

Montree Kumngern and Kobchai Dejhan
Department of Telecommunications Engineering, Faculty of Engineering,
King Mongkut's Institute of Technology Ladkrabang, Bangkok 10520, Thailand
E-mail: kkmontre@kmitl.ac.th

Abstract–**This paper presents an electronically tunable versatile multifunction voltage-mode biquadratic filter with three inputs and five outputs employing five operational transconductance amplifiers, two grounded capacitors and one grounded resistor. The use of grounded capacitors makes the circuit highly suitable for integrated circuit implementation. The proposed filter can simultaneously realize low-pass, band-pass, high-pass, band-stop and all-pass voltage responses by appropriately connecting the input and output terminals. The natural frequency and the quality factor can be orthogonally controlled. No inverting-type input signals are required for realizing all the filter responses, and all the incremental sensitivities are low. PSPICE simulation results are given to confirm the presented theory.**

Keywords–**Biquadratic filter, voltage-mode circuit, operational transconductance amplifier, analog signal processing**

I. INTRODUCTION

A universal filters is used because the circuit provides second-order filters, i.e., low-pass (LP), band-pass (BP), high-pass (HP), band-stop (BS) and all-pass (AP), as for example in phase-locked loop FM stereo demodulation, touch-tone telephone tone decode, cross-over network used in a three-way high-fidelity loudspeaker and higher-order filters [1]. As a result, a number of universal filters based on different design techniques have been developed in the literature; see, for example, [2]-[22]. In [2]-[4], universal filters using current conveyors were proposed. However, these filters suffer from the lack of the electronic tunability.

Operational transconductance amplifiers (OTAs) have exhibited some advantages in the circuit design. The OTA provides an electronic tunability, a wide tunable range and powerful ability to generate various circuits. Moreover, OTA based circuits require no resistors and, therefore, are suitable for integrated circuit (IC) implementation [5]. Many voltage-mode universal filters using OTAs have already been reported [6]-[22]. Considering the number of input and output ports, these filters can be classified into three categories: (i) a single-input, multiple-output (SIMO) type [6]-[10], (ii) a multiple-input, single-output (MISO) type [11]-[17] and (iii) a multiple-input, multiple-output (MIMO) type [18]-[21]. Generally, the SIMO filters can simultaneously realize three basic filter functions, i.e., low-pass (LP), band-pass (BP), and high-pass

(HP), at a time without altering the connection way of the circuits and without input signal matching. However, for the realizations of all-pass (AP) and band-stop (BS) functions, additional addition and subtraction circuits are usually required. The MISO filter can realize multifunction outputs by altering the way in which the input signals are connected. On the other hand, in comparison with the SIMO filter, the MISO and MIMO configurations provide a variety of circuit characteristics with different input voltage and usually do not require any parameter matching conditions. In addition, MISO and MIMO filters may lead to a reduction in the number of active elements used; hence the power consumption and area of chip can be reduced. Moreover, to realize a larger variety of filter functions such as inverting or non-inverting-type functions, the MISO and MIMO configurations seem to be more suitable than the SIMO configuration. As the reported OTA-based MISO and MIMO filtering circuits, circuit in [11] contains minimum elements and enjoys low sensitivities, but only BP and LP responses have the advantage of high input impedance. The circuits in [11], [13], [19], [20] enjoy the advantage of high input impedance, but they use two kinds of active components (i.e., OTA and op-amp, OTA and CCII). The circuits contain only OTAs that enjoy a variety of circuit characteristics with different input voltages were proposed in [17], [18], [20], [21] but these reported filters suffer from the use of floating capacitors and/or component-matching condition for realizing five standard filter functions.

In this paper, a new electronically tunable versatile MIMO voltage-mode universal filter using five OTAs, two grounded capacitors and one grounded resistors is presented. The circuit employs only a kind of active component which is ideal for IC implementation. The filter can realize LP, BP, HP, BS and AP filter responses without inverting-type input signals and component-matching condition requirements. Also the natural frequency and the quality factor can be orthogonally controlled. The active and passive sensitivities are low. PSPICE simulation results are performed to confirm the theoretical analysis.

II. PROPOSED CIRCUIT

The circuit symbol of the OTA is shown in Fig. 1. Its characteristic can be described by

978-1-4673-2395-6/12 $31.00 © 2012 IEEE

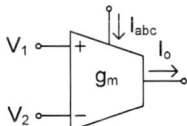

Figure 1. Circuit symbol of OTA.

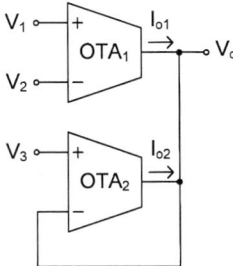

Figure 2. The addition/subtraction voltage using OTAs.

$$I_o = g_m(V_1 - V_2) \qquad (1)$$

where I_o is the output current, g_m is the transconductance gain, V_1 and V_2 are the non-inverting voltage and the inverting input voltage, respectively. The addition/subtraction voltage using OTAs is shown in Fig. 2. Assuming, two OTAs in Fig. 2 are identical, the voltage output V_o can be expressed as [23]

$$V_o = V_1 - V_2 + V_3 . \qquad (2)$$

The realization of the voltage-mode universal filter using OTAs is shown in Fig. 3. The circuit consists of five OTAs, two grounded capacitors and one grounded resistors. Since all the capacitors are grounded, thus the circuit is beneficial to an IC implementation [24]. If V_{in1}, V_{in2} and V_{in3} are the input signal voltages, the five output voltages of the proposed filter can be expressed as:

$$V_{o1} = \{ sC_2 g_{m1} g_{m3} V_{in1} + sC_2 g_{m1} g_{m3} V_{in2} \\ + (sC_2 g_{m1} g_{m2} + g_{m1} g_{m2} g_{m3}) \}/D(s) \qquad (3)$$

$$V_{o2} = \{ g_{m1} g_{m2} g_{m3} V_{in1} + g_{m1} g_{m2} g_{m3} V_{in2} \\ - sC_1 g_{m2} g_{m3} V_{in1} \}/D(s) \qquad (4)$$

$$V_{o3} = \{ (s^2 C_1 C_2 g_{m3} + g_{m1} g_{m2} g_{m3}) V_{in1} - sC_2 g_{m1} g_{m2} V_{in2} \\ + s^2 C_1 C_2 g_{m2} V_{in3} \}/D(s) \qquad (5)$$

$$V_{o4} = \{ -s^2 C_1 C_2 g_{m3} V_{in1} + (sC_2 g_{m1} g_{m2} + g_{m1} g_{m2} g_{m3}) V_{in2} \\ - (s^2 C_1 C_2 g_{m2} + sC_1 g_{m2} g_{m3}) V_{in3} \}/D(s) \qquad (6)$$

$$V_{o5} = (R_1 g_{m1}) \{ -s^2 C_1 C_2 g_{m3} V_{in1} - s^2 C_1 C_2 g_{m3} V_{in2} \\ - (s^2 C_1 C_2 g_{m2} + sC_1 g_{m2} g_{m3}) V_{in3} \}/D(s) \qquad (7)$$

when $D(s) = s^2 C_1 C_2 g_{m3} + sC_2 g_{m1} g_{m2} + g_{m1} g_{m2} g_{m3}$. From (3)-(7), it can be summarized as

Case I. From (3)-(7), if $V_{in1}=V_{in}$ and $V_{in2}=V_{in3}=0$ (grounded), it can be expressed that: BP when $V_{o1}=V_{out}$, LP when $V_{o2}=V_{out}$, BS when $V_{o3}=V_{out}$, HP when $V_{o4}=V_{out}$ or $V_{o5}=V_{out}$. Also from (5) if $V_{in3}=0$ and $V_{in1}=V_{in2}=V_{in}$, the AP transfer function can be obtained as V_{o3}.

Case II. From (4)-(5), it can be expressed that:

- LP: $V_{in1}=V_{in}$ or $V_{in2}=V_{in}$, $V_{in3}=0$ and $V_{o2}=V_{out}$.
- BP: $V_{in2}=V_{in}$, $V_{in1}=V_{in3}=0$ and $V_{o3}=V_{out}$.
- HP: $V_{in3}=V_{in}$, $V_{in1}=V_{in2}=0$ and $V_{o3}=V_{out}$.
- BS: $V_{in1}=V_{in}$, $V_{in2}=V_{in3}=0$ and $V_{o3}=V_{out}$.
- AP: $V_{in1}=V_{in2}=V_{in}$, $V_{in3}=0$ and $V_{o3}=V_{out}$.

Thus, the proposed filter can realize LP, BP, HP, BS and AP biquadratic filters by appropriately connecting the input and output terminals. It should be noted that no inverting-type input signals are required for realization five types of standard biquadratic function. Also the inputs V_{in2} and V_{in3} are connected to the high-input impedance, leading to easy cascadability without the need of any supplementary. The natural frequency (ω_o) and the quality factor (Q) of the proposed filter can be expressed as

$$\omega_o = \sqrt{\frac{g_{m1} g_{m2}}{C_1 C_2}} \qquad (8)$$

$$Q = g_{m3} \sqrt{\frac{C_1}{g_{m1} g_{m2} C_2}} . \qquad (9)$$

From (8)-(9), letting $g_{m1}=g_{m2}=g_m$, the parameter ω_o can be tuned electronically by changing the value of g_m while the parameter Q can be given by setting g_{m3}. This means that the parameters ω_o and Q can be provided the orthogonal control. The active and passive sensitivities of the parameters ω_o and Q are shown as Table I. From this table, the active and passive sensitivities are within unity in magnitude. Thus, the active and passive sensitivities are low. The CMOS implementation of OTAs is shown in Fig. 4. Assuming MOS transistors M_1 and M_2 are matched and operated in saturation, the transconductance gain (g_m) can be expressed by

$$g_m = \sqrt{\mu C_{ox}(W/L) I_{abc}} \qquad (10)$$

where I_{abc} is the bias current, μ is the carrier mobility, C_{ox} is the gate oxide capacitance per unit area, W and L are the channel width and length, respectively. From (10), the g_m can be tuned by varying the biasing current I_{abc}.

TABLE I. SENSITIVITIES OF CIRCUIT COMPONENTS

X	$S_x^{\omega_0}$	S_x^Q
g_{m1}	0.5	-0.5
g_{m2}	0.5	-0.5
g_{m3}	0.0	-1.0
C_1	-0.5	0.5
C_2	-0.5	-0.5

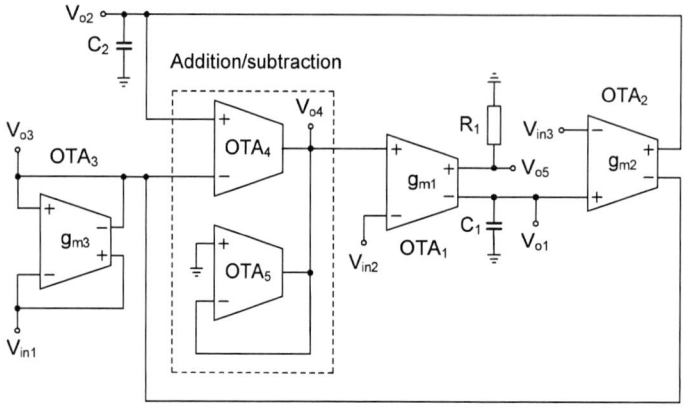

Figure 3. Proposed voltage-mode multifunction filter with three inputs and five outputs.

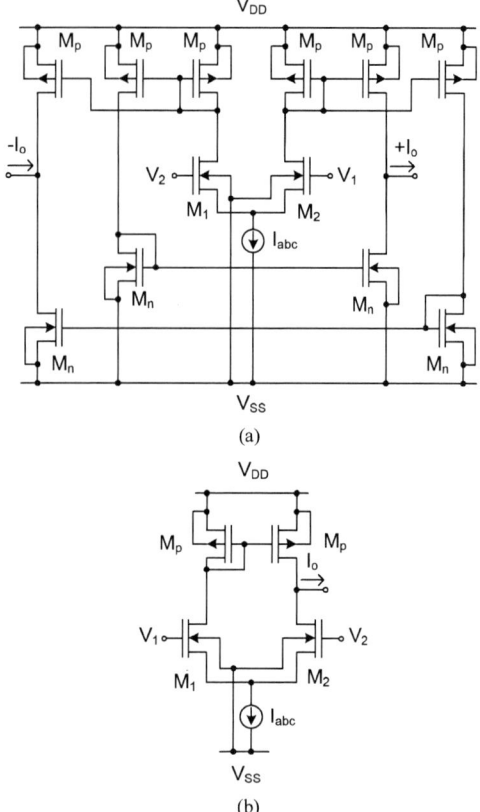

(a)

(b)

Figure 4. CMOS implementation of the OTAs: (a) for OTA_1-OTA_3; (b) for OTA_4-OTA_5.

III. SIMULATION RESULTS

The proposed universal filter was verified through PSPICE simulators using the circuit in Fig. 3 with 0.35 μm CMOS process from TSMC. The OTA in Fig. 4 was used. The aspect ratio of transistors are W/L = 10 μm/ 1 μm for pMOS devices and W/L = 5 μm/1 μm for nMOS devices [24]. The biasing currents for OTA_4 to OTA_5 are fixed as 20 μA. The power supplies are given as $V_{DD} = -V_{SS} = 1.65$ V.

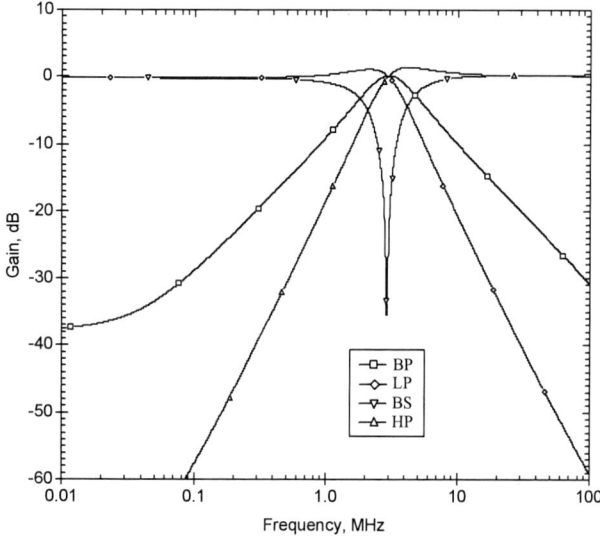

Figure 5. Simulated BP, LP, BS and HP responses.

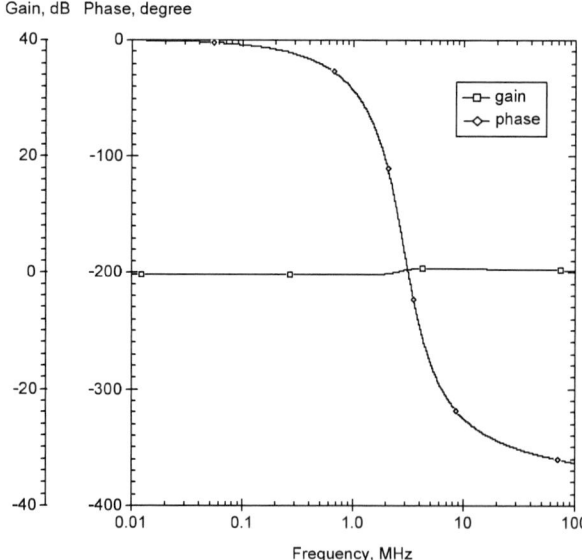

Figure 6. Simulated gain and phase responses of AP filter.

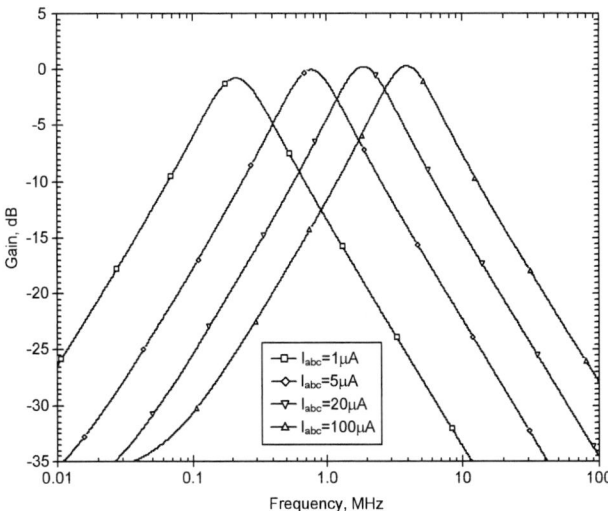

Figure 7. Simulated frequency responses of BP filter when I_{abc} is varied.

As an example design, the capacitors $C_1 = C_2 = 10$ pF, $R_1 = 5.5$ kΩ and the biasing currents $I_{abc1} = I_{abc2} = I_{abc3} = 50$ μA ($g_m = 181.97$ μS) are given. This setting has been designed to obtain the BP, LP, BS, HP and AP filter responses with $f_o \cong 2.89$ MHz and $Q \cong 1$. The simulated responses of the BP, LP, BS and HP of the proposed filter are shown in Fig. 5. In this figure, the pole frequency of 2.88 MHz and the power consumption of 1.39 mW are obtained. Fig. 6 shows the simulated frequency responses of the gain and phase characteristics of the AP filter. Fig. 7 shows the simulated a BP filter response when the biasing currents I_{abc} (i.e., $I_{abc} = I_{abc1} = I_{abc2} = I_{abc3}$) were simultaneously adjusted for the values of 1, 5, 20 and 100 μA, while keeping $C_1 = C_2 = 10$ pF. This result is confirmed by equation (8).

IV. CONCLUSIONS

This paper presents a new versatile voltage-mode three inputs and five outputs universal filter using five OTAs, two grounded capacitors and one grounded resistor which is very suitable for IC implementation. The proposed filter can realize the LP, BP, HP, BS and AP voltage responses without inverting-type input signal requirements. The natural frequency and the quality factor can be orthogonally controlled. Also, the parameters ω_o can be tuned electronically by adjusting the biasing currents of OTAs. The workability of the proposed filter is confirmed by PSPICE simulations.

REFERENCES

[1] C. K. Alexander and M. N. O. Sadiku, Fundamental of Electric Circuits, New York, McGraw-Hill, 2004.

[2] H.-P. Chen and S.-S. Shen, "A versatile universal capacitor-grounded voltage-mode filter using DVCCs," ETRI Journal, vol. 29, pp. 470-476, 2007.

[3] C.-N. Lee and C.-M. Chang, "Single FDCCII-based mixed-mode biquad filter with eight outputs," International Journal of Electronics and Communications, vol. 63, pp. 736-742, 2009.

[4] J. Koton, N. Herencsar, and K. Vrba, "KHN-equivalent voltage-mode filters using universal voltage conveyors," International Journal of Electronics and Communications, vol. 65, pp. 154-160, 2011.

[5] E. Sanchez-Sinencio, R. L. Geiger, and H. Nevarez-Lozano, "Generation of continuous-time two integrator loop OTA filter structure," IEEE Transactions on Circuits and Systems, vol. CAS-35, pp. 936-949, 1988.

[6] C. M. Chang, "New multifunction OTA-C biquads," IEEE Transactions on Circuits and Systems-II, vol. 46, pp. 820-824, 1999.

[7] T. Tsukutani, M. Higashimura, N. Takahashi, Y. Sumi, and Y. Fukui, "Novel voltage-mode biquad without external passive element," International Journal of Electronics, vol. 88, pp. 13-22, 2001.

[8] J.-W. Horng, "Voltage-mode universal biquadratic filter with one input and five outputs using OTAs," International Journal of Electronics, vol. 89, pp. 729-737, 2002.

[9] M. Kumngern, "New electronically tunable voltage-mode lowpass, highpass, bandpass filter using simple OTAs," International Journal of Computer and Electrical Engineering vol. 3, pp. 649-653, 2011.

[10] M. Kumngern and K. Dejhan, "Voltage-mode low-pass, high-pass, band-pass biquad filter using simple CMOS OTAs," in Proceedings of IEEE International Instrumentation and Measurement Technology Conference 2009 (I2MTC 2009), Singapore, 2009, pp. 924-927.

[11] I. A. Khan, M. T. Ahmed, and N. Minhaj, "A simple realization scheme for OTA-C universal biquadratic filter," International Journal of Electronics, vol. 72, pp. 419-429, 1992.

[12] T. Tsukutani, Y. Sumi, Y. Kinugasa, M. Higashimura, and Y. Fukui, "Versatile voltage-mode active-only biquad circuits with loss-less and lossy integrators," International Journal of Electronics, vol. 91, pp. 525-536, 2004.

[13] J.-W. Horng, "High input impedance voltage-mode universal biquadratic filter using two OTAs and one CCII," International Journal of Electronics, vol. 90, pp. 183-191, 2003.

[14] M. T. Abuelma'atti and A. Bentrcia, "A novel mixed-mode OTA-C universal filter," International Journal of Electronics, vol. 92, pp. 375-383, 2005.

[15] M. Kumngern, B. Knobnob, and K. Dejhan, "Electronically tunable high-input impedance voltage-mode universal biquadratic filter based on simple CMOS OTAs," International Journal of Electronics and Communications, vol. 64, pp. 934-939, 2010.

[16] C.-N. Lee, "Multiple-mode OTA-C universal biquad filters," Circuits, Systems and Signal Processing, vol. 29, pp. 263-274, 2010.

[17] M. Kumngern and K. Dejhan, "Electronically tunable voltage-mode universal filter with three-input single-output," in Proceedings of International Conference on Electronics Devices, Systems & Applications 2010 (ICEDSA 2010), Kuala Lumpur, Malaysia, 2010, pp. 317-322.

[18] J. Wu and C.-Y. Xie, "New multifunction active filter using OTAs," International Journal of Electronics, vol. 74, pp. 235-239, 1993.

[19] T. Tsukutani, M. Higashimura, Y. Sumi, and Y. Fukui, "Voltage-mode active-only biquad," International Journal of Electronics, vol. 87, pp. 1435-1442, 2000.

[20] T. Tsukutani, M. Higashimura, N. Takahashi, Y. Sumi, and Y. Fukui, "Novel voltage-mode biquad using only active devices," International Journal of Electronics, vol. 88, pp. 339-346, 2001.

[21] J.-W. Horng, "Voltage-mode universal biquadratic filter using two OTAs," Active and Passive Electronic Components, vol. 27, pp. 85-89, 2004.

[22] S.-H. Tu, C.-M. Chang, J. N. Ross, and M. N. S. Swamy, "Analytical synthesis of current-mode high-order single-ended-input OTA and equal-capacitor elliptic filter structures with the minimum number of components" IEEE Transactions on Circuits and Systems-I, vol. 54, pp. 2195-2210, 2007.

[23] R. R. Torrance, T. R. Viswanathan, and J. V. Hanson, "CMOS voltage to current transducers," IEEE Transactions on Circuits and Systems, vol. CAS-32, pp. 1097-1104, 1985.

[24] M. Bhusan and R. W. Newcomb, "Grounding of capacitors in integrated circuits," Electronics Letters, vol. 3, pp. 148-149, 1967.

Design of Low Power, Low Jitter DLL Tested at all Five Corners to Avoid False Locking

Himadri Singh Raghav[#1],Member, IEEE Sachin Maheshwari[*1],Student Member, IEEE, Mola Srinivasarao[*1] and Prof. B. P. Singh[*2], Member, IEEE

[#1,*2]Faculty of Engineering and Technology (FET),
Mody Institute of Technology and Science (MITS),
Lakshmangarh-332311, District Sikar, Rajasthan, India,
Email: himadri.singh.raghav@gmail.com, bpsingh@ieee.org
[*1]Birla Institute of Technology and Science, Pilani
Electrical and Electronics Engineering Department,
Pilani-333031, District Jhunjhunu, Rajasthan, India,
Email: sachin.mahe@gmail.com, ms02467@gmail.com

Abstract- **A modified Phase Selection Circuit, a modified Phase Frequency Detector and a modified Voltage Controlled Delay Line is proposed to improve the Delay Locked Loops (DLL) locking time, lock range and the jitter performance. Also the DLL presented in this paper has a wide-range frequency operation. A modified Phase Selection circuit is designed in order to operate DLL over wide frequency range and completely solve the false locking problem. Also a Modified Phase Frequency detector circuit has been designed to reduce the phase error as well as dead-zone situation. The proposed DLL design is simulated in Cadence Spectre using TSMC 180nm CMOS Technology and 1.8V power supply voltage operate correctly when the input clock frequency is changed from 84 to 800MHz and generate ten-phase clocks within just one clock cycle. The simulation is performed for all five process corners. The DLL consumes maximum power of 6.85mW at 800MHz working at FF corner, whereas, the maximum peak-to-peak jitter is 4ps at 84MHz working at FS corner. Both maximum power and jitter is measured at temperature and voltage of -40⁰C and 1.98V.**

Keywords-delay locked loop; false locking; phase selection circuit; phase frequency detector; phase error; dead-zone; peak-to-peak jitter.

I. INTRODUCTION

In nowadays, more and more applications, such as local oscillator and clock generator are employed using DLLs. A DLL based clock generation has proven itself a viable alternative in today's clock design segment over traditional PLL based design [1] [2] by offering lower design complexity [3] [4] [5] with satisfying low power budget. Another important application of a DLL is for the purpose of clock de-skewing in synchronous data transfer among communication chips. In high speed communication reducing clock skew [6] [7] [8] has become increasingly important. A DLL for this application requires fast lock time and excellent phase alignment between the reference signal and the corrected output signal [9]. Thus, to acquired high speed, the DLL requires low jitter, high bandwidth, low noise sensitivity, and low power consumption [10].

In this paper, we proposed a modified phase selection circuit, a modified phase frequency detector circuit and a modified voltage controlled delay line circuit to improve a DLL locking time, lock range and the jitter performance.

Moreover, the modified architecture of DLL completely eliminates the locking problem. Also the modified DLL architecture was simulated for all five process corner to check the effects of process, voltage and temperature variations. The rest of the paper is organized as follows. Section II summarizes the background and motivation of the work. The circuit description of DLL is described in detail in section III. Section IV describes the simulation results and observations, and finally section VI concludes the paper.

II. BACKGROUND AND MOTIVATION OF THE WORK

For a DLL, locking time, lock range and jitter performance are the most important metrics. Generally, locking time is related to the speed of the phase frequency detector (PFD), the magnitude of the charging or discharging current in the charge pump (CP), and the overall delay loop bandwidth. Lock range refers to the maximum and minimum delays of the voltage controlled delay line (VCDL), which set the range in which the delay of the VCDL can be varied. Phase noise, or time jitter, is the random variations of the period or phase of a clock signal. Fig. 1 shows the architecture of DLL [11] which suffers from slight false locking problem.

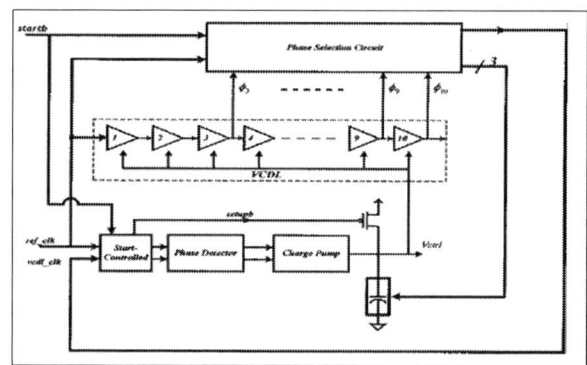

Fig. 1. Architecture of the DLL.

A. False locking problem

Whenever the reference clock is closer to any one of the phases generated by the VCDL, false locking problem occurs in the phase selection circuit of fig. 1.

978-1-4673-2395-6/12 $31.00 © 2012 IEEE   526

In the timing diagram of fig. 2 first waveform is the reference clock and second waveform is the phaseN clock, which is coming from VCDL. The phase selection circuit in fig. 1 selects this as the vcdl_clk. When phaseN clock is selected, it is leading with respect to reference clock and it goes to start controlled circuit via a multiplexer. When it passes through multiplexer it is delayed by T_{mux} delay as shown in fig.2. So, the output of the multiplexer gives a clock which is lagging with respect to reference clock. Thus, it gets locked to the next reference clock edge as shown in fig.2.

Thus, this motivated for a modified phase selection circuit which resolves the above problem. Moreover, the PFD in fig. 1 has also been designed to reduce the phase error, power dissipation, jitter and as well dead-zone situation.

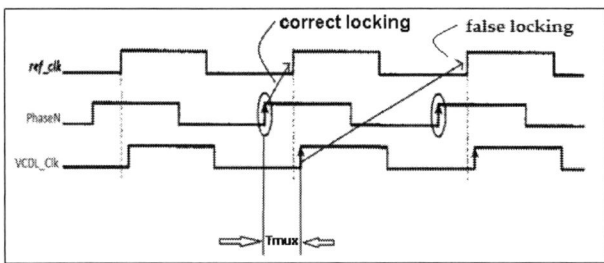

Fig. 2 . False Locking problem timing diagram

III. THE CIRCUITS DESCRIPTION OF DLL

The block diagram of the proposed DLL architecture is shown in fig. 3 consists of the following modified units.

- Phase selection circuit
- Phase frequency detector (PFD)
- Voltage controlled delay line (VCDL)
- Charge Pump (CP)

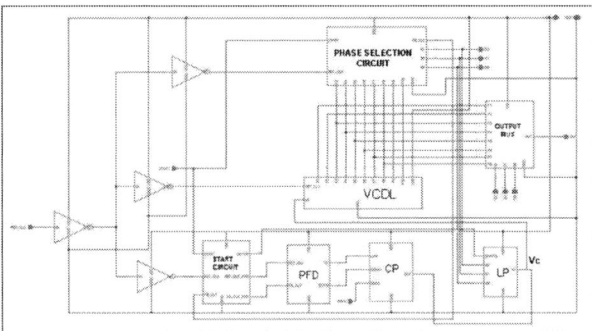

Fig. 3 Block diagram of DLL

A. Phase Selection Circuit

The Phase selection circuit consists of two blocks: an edge detector and a multiplexer with a decoder as shown in fig. 4. Initially, the start signal in fig. 4 is set to low to reset the edge detector outputs (i.e., d3~ d10) and the delay of the VCDL is set to its minimum value. When the start signal goes high, the edge detector will detect the rising edge of input signals in sequence during the next two rising edges of ref_clk. The vcdl_clk will be low until the selected phase is chosen. After the vcdl_clk is decided, the DLL will start the locking

process. By the decoder, signals (d3 ~ d10) are decoded to generate 3-bit control signals, which switch the number of capacitors used in the loop filter for tuning the loop bandwidth.

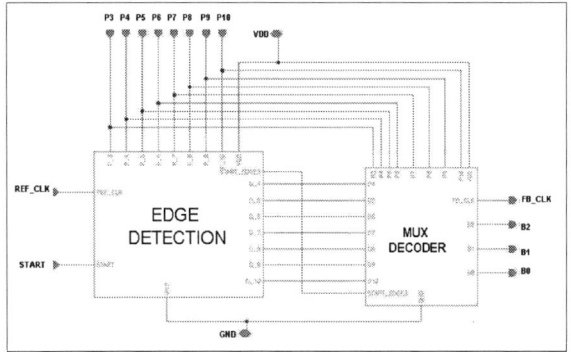

Fig.4. Block diagram of the phase selection circuit.

To overcome the problem of false locking of fig.1 the edge detection circuit in phase selection is modified and it is shown in fig.5. The basic idea is whenever the phaseN clock is leading the reference clock less than by T_{mux} delay then phase (N-1) clock should be selected as vcdl_clk instead of phaseN clock. To ensure this condition buffers is added in an edge detection circuit. The delay due to this added buffer is denoted by T_{buffer} as shown in fig.5. Delay element circuit is shown in Fig. 6 of modified schematic of latch N. Fig.7 shows a block level of 8:1 multiplexer with a decoder. The multiplexer is designed by using transmission gates so that it gives minimum delay from input to output. It also reduces feedback loop delay in the DLL circuit.

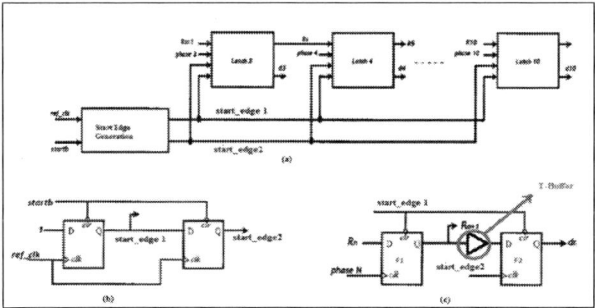

Fig.5 Schematic of edge detection circuit. (a) Edge detection circuits. (b) Start edge generation. (c) Modified Latch N.

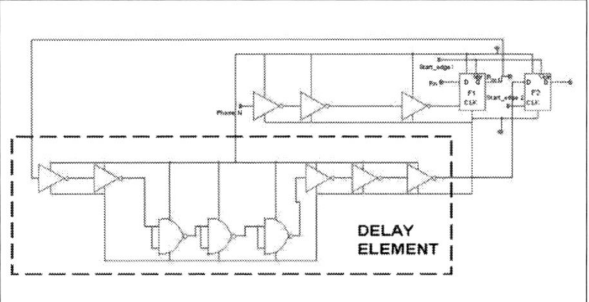

Fig.6 Modified Schematic of Latch N

The clock to second flip flop (F2) in fig.5 is start_edge2, is coming with respect to reference clock. So the delay between reference clock and start_edge2 is T_{phase}. The minimum delay between phaseN and reference clock is T_{mux}. The hold time of second flip flop (F2) in fig.5 is $T_{hold, F2}$ and CLK-Q delay of first flip flop (F1) in fig.5 is $T_{clk, F1}$. Thus, to ensure no false locking condition the two flip flops in fig.5 that is, F1 and F2 should satisfy the hold time condition which is given below.

$$T_{CLK, F1} + T_{buffer} > T_{mux} + T_{phase} + T_{hold, F2} \qquad (1)$$

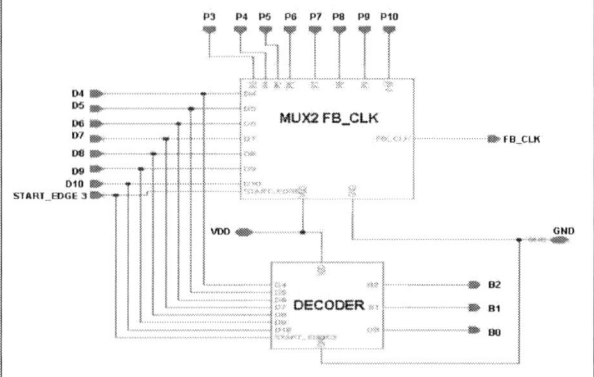

Fig. 7 Schematic of Multiplexer with a decoder.

B. Phase frequency detector (PFD)

The schematic of the Phase and Frequency Detector circuit is shown in fig. 8. In this DLL, the dynamic logic style PFD is adopted to avoid the dead-zone problem and improve the operating speed.

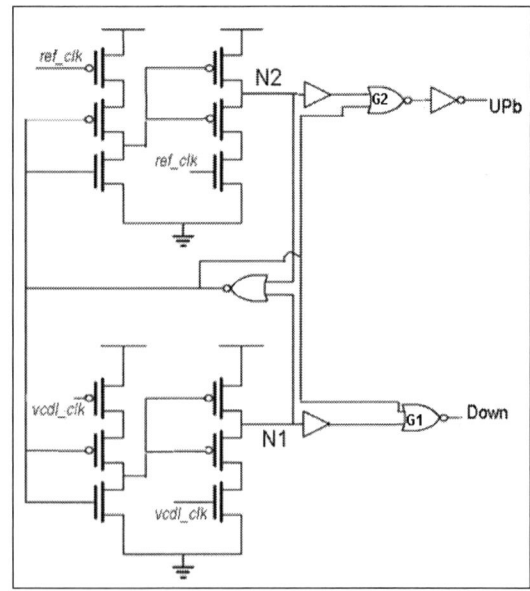

Fig. 8 Schematic of the modified PFD circuit

PFD has two outputs: UPb and DOWN, Since the NOR gate connected to the reset of the latches, both outputs cannot be high at the same time. When input reference clock ($\Phi 1$) leads delayed clock ($\Phi 2$)(vcdl_clk), the average value of UPb will be proportional to the phase difference and DOWN will

be zero. When $\Phi 1$ lags $\Phi 2$, the average value of DOWN reports the phase difference and UPb will be one. When the inputs are in phase, UPb is high and DOWN is low.

The PFD operation is explained by using the timing diagram as shown in fig.9. In this case vcdl_clk is leading with respect to ref_clk. When vcdl_clk rising edge comes the node N1 (in fig.8) goes to zero level and it goes back to V_{DD} when there is a rising edge on the ref_clk. The node N1 waveform is shown in fig.9. When there is a rising edge on reference clock node N2 (in fig.8) goes to zero level and it goes back to V_{DD} when there is a reset signal. This reset signal is generated by NOR gate and having NOR gate delay, so N1 and N2 node signals also should delayed by same amount. These delayed N1 and reset signals passed through G1 NOR gate. Similarly delayed N2 and reset are passed through G2 NOR gate. These nor gates eliminate the unwanted switching present on the UPb signal.

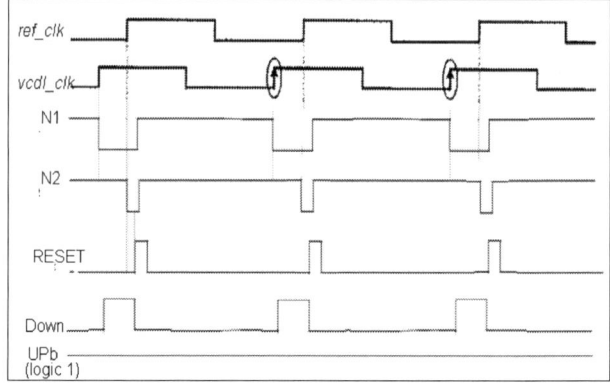

Fig. 9. Phase Frequency Detector timing diagram.

The operating frequency of this modified PFD is limited because the forward path delay from PFD input to output is more. The dead zone occurs when the loop is in a lock mode and the charge pump output does not adjust for small changes in input signals at the PFD. Any width of the dead zone directly translates to jitter in the DLL and must be avoided. However, the output signals of the PFD are active one at a time, not both. This gives better jitter and phase error performance. There is no need to take care about routing matches between UPb and DOWN signals, because only one signal is active at a time.

C. Voltage Controlled Delay Line (VCDL)

The VCDL takes two inputs: a control voltage and a clock. The output is a clock of the same frequency as the input, but phase shifted by some amount proportional to the control voltage. The schematic of the VCDL is shown in fig. 10. Fig. 11and fig. 12 shows the schematic and transfer function of a VCDL using two cascaded current starved inverter (CSI) as a delay element.

The delay elements are similar to Schmitt trigger circuits, with the current bias. It is split into upper (PMOS) and lower (NMOS) segments that are functional complements of each other. Transistors in the middle are the main switching device, while top and bottom transistors acts as a current

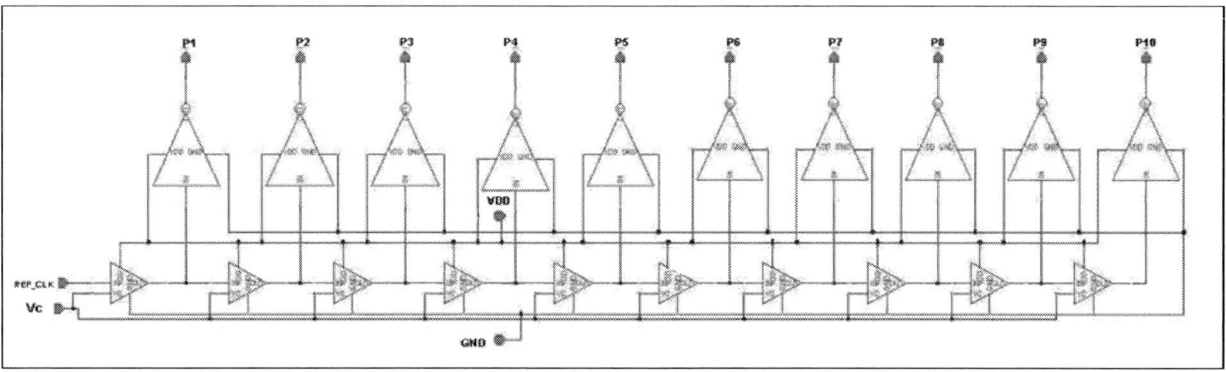

Fig. 10 Schematic of voltage controlled delay line (VCDL)

steering device. The maximum current supplied to the inverters is controlled by the bias voltage Vc. On the contrary, when the control voltage is high enough to turn on the current source/sink, the output clock becomes a delayed version of the reference signal. After insertion, the delay range is nearly unchanged and the control voltage can be full swing tuning. Also buffer has been added at the output of the delay element to prevent load on the delay cell and output clock waveform distortion due to the delay element. The delay is inversely related to the control voltage. These non-linear gain curves are one of the design challenges of a DLL. The VCDL contains 10 delay cells and thus the phase difference between two adjacent delay cells is about 2π / (no. of delay cells selected) when the lock is achieved.

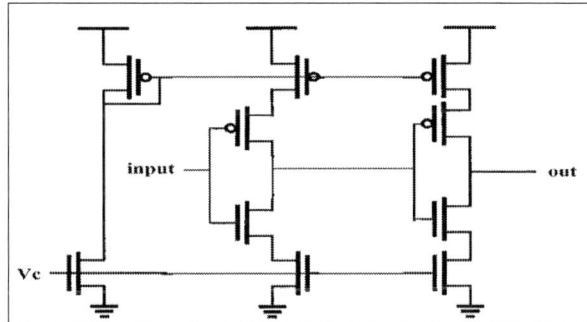

Fig. 11 Schematic of the delay cell

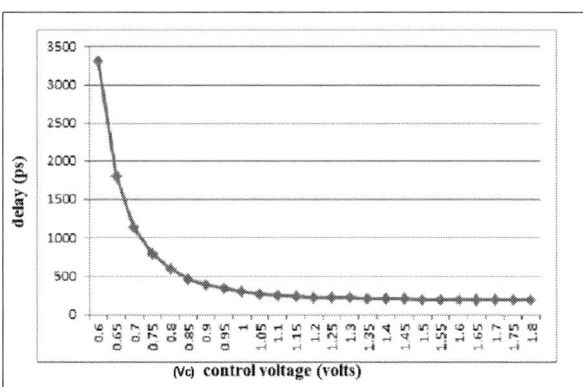

Fig. 12. Simulated transfer curve of the delay cell

D. Charge Pump

Fig. 13 shows the schematic of charge pimp. The value of current source, Ip used is 10µA. Use minimum size transistors for UPb and DOWN signals to reduce feed through effect caused by gate to drain capacitance. The loop bandwidth, ω_N can be expressed as;

$$\omega_N = \frac{I_{CH} \cdot K_{VCDL}}{T_{REF} \cdot C} \qquad (2)$$

A wider loop bandwidth can be used to achieve fast acquisition time, but the jitter performance will be degraded. Hence, the following tradeoff design guideline was suggested;

$$\frac{\omega_N}{\omega_{REF}} = \frac{I_{CH} \cdot K_{VCDL}}{2\pi \cdot C} \leq \frac{1}{10} \qquad (3)$$

where,

$\omega_{REF} = 2\pi / T_{REF}$, T_{REF} - time period of the reference clock
K_{VCDL} - Gain of the Delay Line
I_{CH} – Charge Pump Current,
C – Loop Filter capacitor,
ω_N – DLL loop Band width

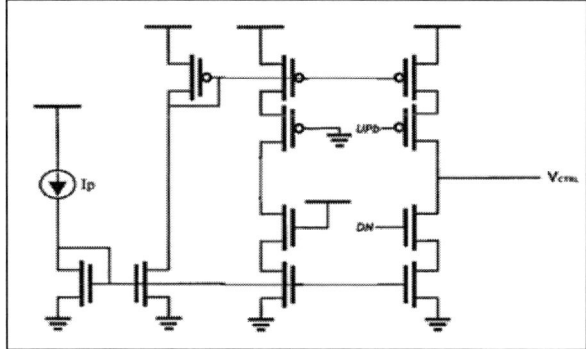

Fig. 13. Schematic of the charge-pump circuit

When the input frequency is higher, the phase selection circuit will select the smaller number of delay cells and K_{VCDL} will become smaller. In order to have an adequate loop bandwidth for the DLL, the capacitances used in the loop filter must become smaller. The 3-bit control signals generated from the phase selection circuit will switch the number of capacitors in the loop filter, as shown in fig. 14, depending on the selected phase. When decoder control bits are "000" loop filter selects 400fF capacitor. Every step the capacitor value increases by 200fF. Finally when decoder control bits are "111" loop filter selects 2pF capacitor.

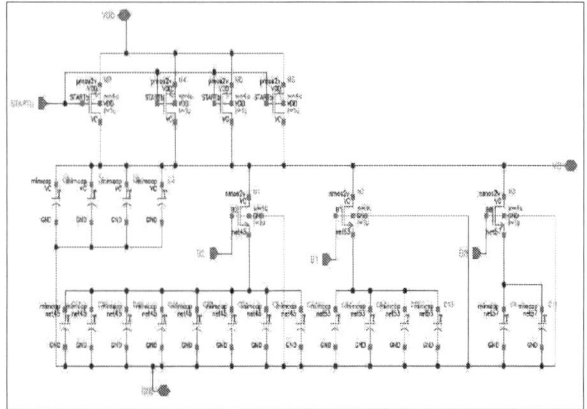

Fig. 14 Schematic of the loop filter circuit

E   Other Circuits

The Start control circuit is included before the PFD has two advantages: the circuit is simple, and the duty cycle of ref_clk and vcdl_clk is not required to be exactly 50%, only falling edge is used in start controlled circuit. The Output Multiplexor selects one phase out of eight phases depending on three control signals. Three control bits are select lines for the multiplexer. The valid selection phases are decided by the phase selection circuit decoder bits.

IV   SIMULATION RESULTS

The proposed DLL was simulated for all five corners. The simulated result shows that the DLL can operate in the frequency range of 84 MHz to 800 MHz.

The proposed DLL has low cycle-to-cycle and rms jitter for both frequencies when compared to other architecture. Also the maximum peak-to-peak jitter is 4ps in FS corner at 84MHz. The maximum power consumed by the DLL is 6.85mW at 800MHz. The lock time is 224 clock cycles for 800MHz and approximately 38 clock cycles for 84MHz. The DLL has a wide-operational range and a fixed latency of one clock cycle. Table I and table II summarizes the observation taken for maximum and minimum operating frequency. The proposed DLL is also tested at different corners and temperature. It has a maximum rms jitter of 5.15ps at FF corners at high frequency and 3.51ps at FS corner at low frequency which is very low when compared with other architecture. Similarly, the power consumed by the proposed DLL is 6.85mW at FF corner at high frequency which is very less when compared with the two designs.

TABLE I.   PERFORMANCE SUMMARY AT HIGH FREQUENCY

| | Process Corners | | | | |
	SS	TT	FF	FS	SF
Temperature	120⁰C	27⁰C	-40⁰C	-40⁰C	120⁰C
Voltage (V)	1.62	1.8	1.98	1.98	1.62
Frequency (MHz)	800	800	800	800	800
Peak-to-peak jitter	1.4ps	0.238ps	0.233ps	0.29ps	0.253fs
RMS jitter	15.7fs	3.7fs	5.15ps	4.46ps	3fs
Cycle-to-cycle jitter(fs)	37	8	20	20	10
Rise time (ns)	100	60	30	40	90
Fall time (ns)	110	60	40	50	80
Duty cycle (%)	49.6	45.6	45.6	44.8	46.4
Lock time (ns)	200	265	230	235	280
Power (mW)	2.75	5	6.85	6.06	3.6
Phase offset (ps)	2.60	10.21	11.62	59.06	45.35
Dead zone (ps)	3.57	10.25	11.10	59.43	45.35

TABLE II.   PERFORMANCE SUMMARY AT LOW FREQUENCY

| | Process Corners | | | | |
	SS	TT	FF	FS	SF
Temperature	120⁰C	27⁰C	-40⁰C	-40⁰C	120⁰C
Voltage (V)	1.62	1.8	1.98	1.98	1.62
Frequency (MHz)	84	84	84	84	84
Peak-to-peak jitter(ps)	2	2	3	4	1.14
RMS jitter	0.282ps	11.76fs	5.36fs	3.51ps	0.39fs
Cycle-to-cycle jitter(fs)	20	10	15	30	20
Rise time (ns)	100	100	100	90	100
Fall time (ns)	100	100	100	100	100
Duty cycle (%)	50	50	50.83	50.83	50
Lock time (ns)	450	400	370	400	425
Power (mW)	0.62	1.04	1.745	1.27	0.707
Phase offset (ps)	127.1	1.11	102.25	72.13	53.46
Dead zone (ps)	127.3	1.10	102.25	72.00	53.46

TABLE III.   PERFORMANCE COMPARISON

	This Work	[8]	[11]
Technology	180nm CMOS	180nm CMOS	350nm CMOS
Supply Voltage	1.8V @ 27⁰C	1.8V ±10%	3.3V
Frequency	84~800MHz	50~150MHz	6~130MHz
Peak-to-peak jitter	2ps @84 MHz 0.238ps@800MHz	58ps@100MHz	210ps@6MHz 24.3ps@130MHz
RMS jitter	11.76fs@84MHz 3.7fs@800MHz	8.69ps@100MHz	24.77ps@6MHz 3.3ps@130MHz
Cycle-to-cycle jitter	10fs @ 84MHz 8fs @ 800MHz	N/A	N/A
Lock time	~224 clock cycles	~200 clock cycles	~1130 clock cycles
Power	5mW@800MHz	15mW@100MHz	132mW@130MHz
Phase offset	10.21ps@800MHz	N/A	N/A
Dead zone	10.25ps@800MHz	N/A	N/A
Active Area	N/A	327.46μm x 116.16μm	880 μm x515 μm (without pads)
Chip Area	N/A	1 mm²	N/A

978-1-4673-2395-6/12 $31.00 © 2012 IEEE

The performance comparison in shown in table III. The Analog DLL [8] is not suitable for wide range operation though it has comparatively fast locking time. The DLL [11] has the problem of false locking which has been rectified in this paper and shows a much better results when it is compared. Thus, the proposed DLL has higher operating frequency range, low jitter and consumes less power. The proposed DLL has also been tested at all five corners with the temperature variation from -40^0 to 120^0 Celsius.

The Digital DLL [13] uses a different architecture to design ultra wide-range DLL in 65nm CMOS technology but fails to show jitter performance. The DLL [12] with edge combiner is used to increase the output frequency range. But if the wide input operation range is required the DLL will fail to operate.

V   CONCLUSION

A DLL tested at all five corners having low power and low jitter without false locking problem is proposed. First, the multiphase outputs of the VCDL are all sent to the modified phase selection circuit. Then the modified phase selection circuit will automatically select one of the delayed outputs to feedback. As a result, this DLL can operate over a wide range without suffering from false locking problems. The start-controlled circuit used here completely eliminated the requirement of exactly 50% duty cycle for ref_clk and vcdl_clk. The cycle-to-cycle and rms jitter performances are all in acceptable range and the latency is just one clock cycle. The maximum power consumption of the complete DLL is 6.85mW with a lock time of 224 clock cycles at 800MHz.

ACKNOWLEDGMENT

The authors are grateful to the lab facility at Electronics & Communication Engineering Department of Mody Institute of Technology & Science, Lakshmangarh where most of the work is carried out and is also thankful to Department of Electrical & Electronics Engineering of Birla Institute of Technology & Science, Pilani for the support to finish this work successfully.

REFERENCES

[1] B. Razavi, "Monolithic Phase-Locked Loops and Clock Recovery Circuits: Theory and Design. Piscataway", NJ: IEEE Press, 1996.

[2] R. E. Best, "Phase-Locked Loops: Theory, Design and Applications" NY: McGraw-Hill, 1998.

[3] Arto Rantala, David Gomes Martins and Markku berg, "A DLL clock generator for a high speed A/D-converter with 1 ps jitter and skew calibrator with 1 ps precision in 0.35 m CMOS", Journal of Analog Integrated Circuits and Signal Processing, Vol. 50, Issue 1, 69-79, Jan 2007.

[4] Mesgarzadeh, B. and Alvandpour, A., "A Low-Power Digital DLL-Based Clock Generator in Open-Loop Mode", IEEE Journal of Solid State Circuits, Vol. 44, Issue 7, 1907-1913, Jul 2009.

[5] Chulwoo Kim, In-Chul Hwang and Sung-Mo Kang, "A low-power small-area 7.28-ps-jitter 1-GHz DLL-based clock generator", IEEE Journal of Solid State Circuits, Vol. 37, Issue 11, 1414-1420, Nov 2002.

[6] C. T. Lu, H. H. Hsieh and L. H. Lu, "A 0.6V Low-Power Wide-Range Delay-Locked Loop in 0.18μm CMOS", IEEE Microwave and Wireless Components Letters, Vol. 19, No. 10, October 2009.

[7] A. Ghaffari, and A. Abrishamifar, "A Wide-Range Delay-Locked Loop with a new Lock-Detect Circuit," 13th IEEE International Conference on Electronics, Circuits and Systems, pp. 1168-1171, December 2006.

[8] S. Lip-Kai, M. S. Sulaiman, and Z. Yusoff, "Fast-Lock Dual Charge Pump Analog DLL using Improved Phase Frequency Detector," International Symposium on VLSI Design, Automation and Test, pp. 1-5, Malaysia, 2007.

[9] S.-I. Liu, J.-H. Lee and H.-W. Tsao, "Low-Power Clock De-skew Buffer for High-Speed Digital Circuits", IEEE J. Solid-State Circuits, vol. 34, pp. 554-558, Apr. 1999.

[10] Lee, M.-J.E., Dally, W.J., Greer, T., Hiok-Tiaq Ng, Farjad-Rad, R., Poulton, J. and Senthinathan, R., "Jitter transfer characteristics of delay locked loops - theories and design techniques", IEEE Journal of Solid-State Circuits, Issue 4, pp. 614 – 621, Apr. 2003.

[11] H. H. Chang, J. W. Lin, C. Y. Yang and S. I. Liu, "A wide-range delay locked loop with a fixed latency of one clock cycle," IEEE Journal of Solid-State Circuits, vol. 37, pp. 1021–1027, Aug. 2002.

[12] Po-Chun Huang, Chi-Jih Shih, Yu-Chang Tsai and Kuo-Hsing Cheng, "A Phase Error Calibration DLL with Edge Combiner for Wide-Range Operation", IEEE 9th international Conference on New Circuits and Systems(NEWCAS), pp. 1 - 4,June 2011.

[13] Ching-Che Chung and Chia-Lin Chang, "A Wide-Range All-Digital Delay-Locked Loop in 65nm CMOS Technology", International Symp International Symposium on VLSI Design, Automation and Test, pp. 66 - 69, April 2010.

978-1-4673-2395-6/12 $31.00 © 2012 IEEE

Effect of Damaged-Chip Infrared Emitter Package on Ge Substrate

Wei Ching LIEW[*], Mutharasu DEVARAJAN

Nano-Optoelectronics Research and Technology Laboratory (NOR-LAB)
School of Physics, University of Science Malaysia (USM),
11800 Penang, Malaysia.
[*]E-mail: liew.wching@gmail.com

Abstract- **Die cracking is the occurrence of fracture(s) in or on any part of the die of a semiconductor device. Failure caused by die cracking is one of the major concerns in packaging design and reliability. In this paper, fracture is inflicted onto the chip of a few units of p-Ge infrared emitting diode (IRED) package. Failure analysis such as curve tracing, decapsulation and visual inspection as well as SEM and EDX was performed on the units to characterize the fracture pattern and size. The results show dependence of the current-voltage (I-V) characteristic on the severity of the fracture inflicted onto the chip.**

I. INTRODUCTION

Light emitting diodes (LEDs) and infrared emitting diodes (IRED) have long been adopted in various electronic systems. Today, besides general illumination, LEDs open the way to new applications and markets in various different fields with a broad spectrum of requirements which includes the agriculture and medical applications [1-3]. In general, LEDs provide high reliability, durability and ability to operate even in harsh condition, energy-efficient, and lifetime of more than 50,000 hours can be reached. These factors have attracted considerable interests in LEDs especially in replacing incandescent and fluorescent lighting.

The IRED/LED-chip is the central element of the IRED/LED that generates light in a p-n junction by electron-hole recombination. The active zone is a complex structure of epitaxial layers. For different colours, different material combinations are used: InAlGaP – red, InGaN – blue, GaAlAs – IR, AlGaN – UV [4-6]. During normal operation, optical performance of IREDs gradually decreases with lifetime. Performance decrease is usually caused by growing defects in the epitaxy layers or on their boundaries, resulting in an increase of non-radiating recombination and a decrease of optical efficiency. In high power IREDs, heat generated within the device package itself must be extracted to improve power conversion efficiencies [7]. With the increase of injection current densities and decrease of die packaging size, IREDs are exposed to vulnerabilities to the generation of large heat fluxes during applied power cycling. Large heat fluxes and insufficient heat dissipation will ultimately lead to detrimental self-heating effects, reduction in light output, and inefficient

short and long term operation [8 – 10]. This will directly lead to accelerated ageing of the device. This is also due to low quality of epitaxy layers as well as extremely high junction temperature. A catastrophic defect like a sudden failure can be caused by electrostatic discharge (ESD) or electrical over stress (EOS) [11] due to electrical overload resulting in a serious damage of the epitaxy layer [12].

Besides heat and growing defects on the die being the main factors of device degradation/failure, poor workmanship in manufacturing and unfavorable operational conditions may reduce the reliability significantly. These, include component failures due to die/chip cracking, is one of the concerns in device packaging design and reliability [13]. Such failures typically are not detected until electrical testing is performed at the end of the process, and thus, making it challenging to identify where and how such damage occur [14]. To avoid failures or to achieve fast resolutions of existing problems, a good knowledge of the failure mechanisms and suitable analytical methods are required [15-16].

Various papers on degradation and failure mechanism of organic LEDs (OLEDs) and high power LEDs which includes visible-light LEDs (e.g. white, blue and red-light LEDs) [17-19] have been reported. However, only a number of papers have reported on degradation/failure analysis of IRED [20] which includes spectroscopic emission microscopy, cathodoluminescence and electrical stresses [21]. In this study, the defect/damage study of infrared emitters on Germanium substrate is reported, which consists of an evaluation on the severity of damage on the chip towards the electrical behaviour of the infrared emitter.

II. METHODOLOGY AND CHARACTERISATION FLOW

The device used in this study is a p-Ge IRED. This device has a maximum power consumption of 1.8W and able to withstand a maximum forward current of 1A. All the units are tested and data-logged beforehand to ensure normal behavior.

Due to unavailability of failed/damaged-ready units, the damages have to be created manually. Hence, fracture was inflicted onto the chip of a few random good units from the same product lots. This is created by tapping a clean and sharp

978-1-4673-2395-6/12 $31.00 © 2012 IEEE

needle onto the chip as shown in Fig. 1 with different amount of force. The fractures are done under a high magnification microscope to avoid damaging the wire-bonding of the device.

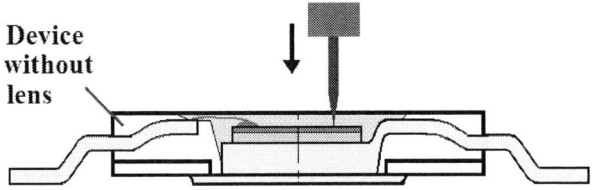

Fig. 1. Schematic of how the fracture is inflicted onto the chip.

A. Curve Tracing

Curve tracing was then performed on all the 'fractured' units and a good unit at ambient temperature ($25\pm1^{\circ}$C) and low current (100mA) to avoid further damage to the chip. Curve tracing is done to compare and to observe the change in their current-voltage (I-V) characteristic. The 4 units exhibiting high deviation and low deviation (2 each) from their normal I-V behaviour and a good unit are selected for further damages analysis and characterization.

From Fig. 2, Unit 2 & 3 shows a high deviation, Unit 4 shows a minimal deviation and Unit 5 shows a medium deviation from their normal I-V behavior which is shown by the good unit (line in blue).

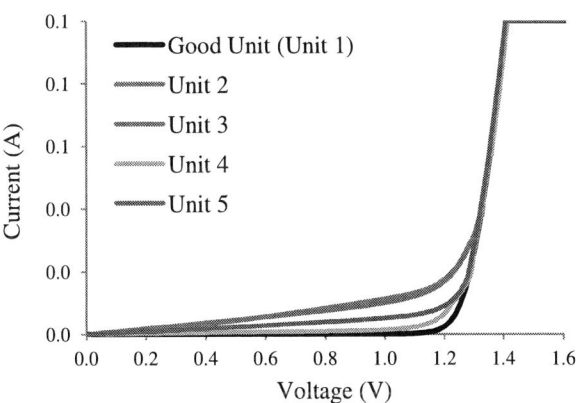

Fig. 2. Current-voltage (I-V) curves for the selected fractured units and good unit.

B. Decapsulation and Internal Visual Inspection

The optical microscope provides a quick overview survey of the die backside with virtually no sample preparations required [15]. However in this case, the fractured units must be decapsulated first as the damages during the fracturing process causes uneven cuts in the silicone resin, and hence, causing reflection of light from the microscope towards different angle.

This will result in a dark spot of image on the fractured location.

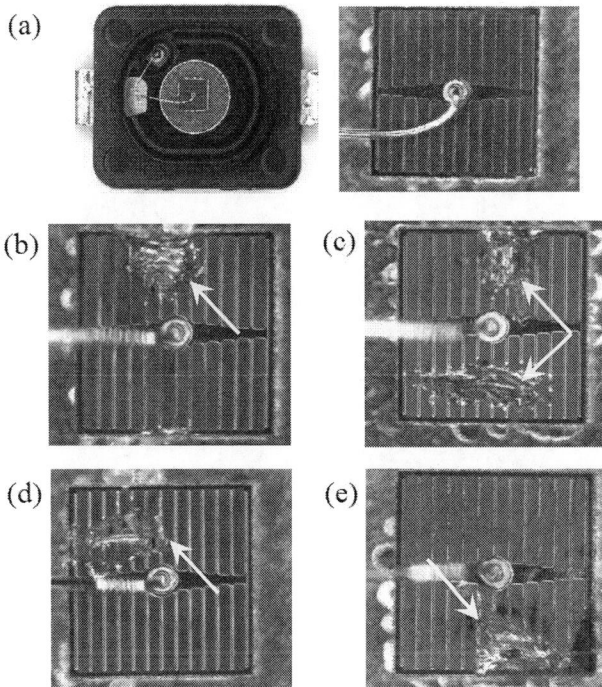

Fig. 3. Optical images of the chip topside with the fracture/damage (pointed with arrow) under high power microscope for (a) Unit 1 which is the good unit, (b) Unit 2 and (c) Unit 3 with high I-V deviation, and (c) Unit 4 with low I-V deviation and (d) Unit 5 with medium I-V deviation.

The 4 units exhibiting large I-V deviation and minimal deviation from their normal characteristic are chosen for chemical decapsulation before undergoing internal visual inspection and SEM. The units are soaked in Dynasolve solvent for approximately 3 hours to chemically remove the silicone resin layer that is protecting the chip. These procedures are also repeated on the good unit (without fracture) for comparison. Fig. 3 shows optical images of fracture severity on the chip of the decapsulated units captured using a high power microscope.

C. Scanning Electron Microscope (SEM) and Energy Dispersive X-Ray (EDX) Analysis

The SEM is used to capture high magnification micrographs of the die scratches, die cracks and to observe for underfill (UF) cracks. Energy Dispersive X-ray (EDX) commonly is used coherently to detect the presence of foreign material at the area of interest [15]. The EDX analysis system works as an integrated feature of SEM.

All of the 5 units are tested in SEM and EDX for fracture characteristics at 10kV for top view. Fig. 4 shows the electron micrographs of these 5 units.

978-1-4673-2395-6/12 $31.00 © 2012 IEEE

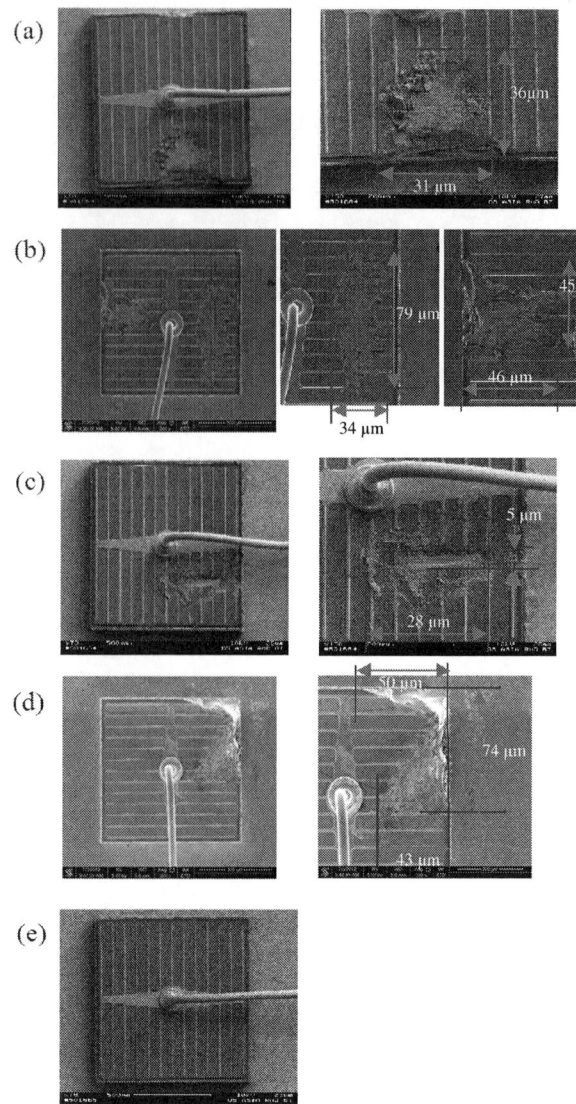

Fig. 4. Electron micrograph of all 5 units from top view and fracture size inflicted.

III. DISCUSSION

The effects of fracture towards the device I-V characteristic in this study, as shown in Fig. 2, are divided into a high deviation and low deviation in the I-V curve. In this paper, the focus is on these both types of deviation in comparison with a good unit without any fracture inflicted onto it. Unit 2 and 3 are the units exhibiting a high deviation in the I-V curve. Deviation in the horizontal section (below 1.2V) causes the curve to increase in gradient for that first section, and hence, showing a soft knee in the I-V curve. This soft knee signifies the effect of parallel resistance (shunt), as according to the Ohm's Law in (1), the proportional increase of current as the

potential difference increases proving the existence of resistance in unit 1.

$$V = I \cdot R \qquad (1)$$

where V is the potential difference measure across 2 points, I is current flowing in the circuit/conductor and R is resistance of the circuit/conductor.

Parallel resistance is said to be caused by any channel that bypasses the p-n junction due to damaged regions of the p-n junction or by surface imperfections [22, 23]. In other words, the short circuit is when a device exhibits an extremely low resistance [24], which clearly signifies the existence of shunt in the p-n junction as according to (2). Therefore, as the voltage increases, the low resistance in the 'damaged-chip' unit causes a higher gradient in the I-V curve (soft-knee effect), as compared to before fracturing is done.

$$R_{total} = \frac{R_1 \, R_2 \, ... \, R_n}{R_1 + R_2 + ... + R_n} \qquad (2)$$

where R_{total} is total resistance in a parallel circuit, R_1 and R_2 are individual resistance in the parallel circuit, and R_n is n-th individual resistance in the parallel circuit.

As for the case of unit 1 and 2, the damage inflicted onto the chip is large. Despite the depth of the damage however, a large region of passivation is damaged due to fracture which caused an increase in surface related leakage current and surface dangling bonds which will further induce more non-radiative surface recombination. Besides that, damage in large portion of passivation which protects the surface of the active layer from chemical and/or electronic degradation due to oxidation will also cause an increase density of electronically active surface states due to the increase in surface dangling bonds [25]. Besides that, damage in the passivation layer also expedite failure processes either by decreasing the threshold stress for nucleation of damage such as hillocks, or/and by increasing the diffusivity in the metallizations [26].

According to the I-V curve (under forward bias), the low current-voltage region (horizontal) is dominated by leakage currents whereas the high current-voltage region (vertical) is strongly influence by the internal series resistance within the diode [27]. As shown in Fig. 2, the increase of gradient in the low current-voltage region of unit 2 and 3 as indirectly indicates the increase of leakage current, which is consistent to the effect of large and shallow damages on the passivation region.

At the same time, semiconductors such as this IRED, also uses thin traces of metal to conduct current, and hence subject to high current densities. Under this condition, the material forming the interconnection actually 'moves'. This phenomenon, known as electromigration, would cause several different kinds of failure in the narrow ohmic contact (interconnect). Electromigration is generally considered to be the result of momentum transfer of electrons, which move in the applied electric field, to the ions which make up the lattice of the interconnect material. When electrons are conducted through a metal, their interactions with imperfections in the lattice will cause scattering, especially when an atom is out of place. The scattering produces by thermal energy will cause the atoms to vibrate, which is source of resistance in metals.

In these test samples, high current densities are common in the ohmic contacts and current spreader. A variation of microstructural parameters (e.g., grain size distribution, inclination of grain boudaries with respect to electron flow) will cause a non-uniform distribution of atomic flow rate, where the number of atoms flowing into the damaged region is not equal to the number of atoms flowing out per unit time, and hence, there will be mass accumulation, leading to formation of hillocks or mass depletion which leads to formation of voids.

At the damage location, the large damage which causes the damage to several spreader fins causes the current density to increase at the edges of the damaged vicinity itself due to the reduction in the cross sectional area of the conductor (contact) region. These will cause an increase of localize current density, known as current crowding, around the edges of the damage which lead to the deposition of migrated ions in the lumps of particles and debris from the damaged layers that lined the edges of the fracture as shown in Fig. 5 [28-30].

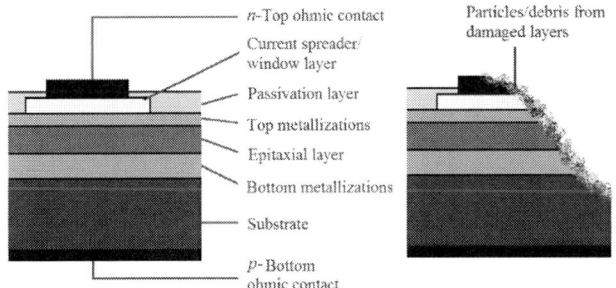

Fig. 5. A typical cross section of the damage area with particles and debris lining the edges.

Since junction temperature is proportional to the square of current density, the current crowding effect leads to a local temperature rise around the damaged region that will in turn, further accelerates the formation of hillocks. Besides that, increase in junction temperature due to the non-radiative surface recombination on the exposed surface of the epi-layer also exacerbated the current crowding effect. The whole process (as described by Fig. 6) continues till the hillock is large enough to create an alternative path for electron flow (shunt) through the *pn* junction. This in turn causes a short

circuit and a high deviation of I-V curve as shown by unit 2 and 3.

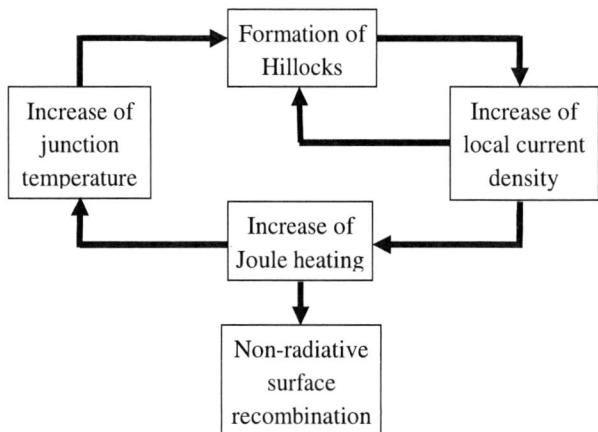

Fig. 6. Thermal acceleration loop during electromigration.

As for unit 4, it exhibits minimal/insignificant deviation in the I-V curve (Fig. 2); there is a soft knee formed in the I-V curve. This is most porbably due to the damage most probably due to depletion region generation and recombination. Carrier generation and recombination in the depletion region was not taken into account by Shockley due to the presence of trap levels in the depletion region in practical diodes. These carrier generation and recombination will cause an excess current for forward bias due to the recombination of minority carriers in the depletion region. However, this recombination current dominates only at low voltages, and hence, only a small effect (soft knee) in the I-V curve can be seen. Diffusion current in turn dominates at high voltages [23].

As for unit 5, it exhibits less deviation in the I-V curve (Fig. 2) as compared to unit 2 and 3 most probably due to the location of the damage region on the chip. From the optical images in Fig. 3, it can be seen that the damages on unit 4 was close to the center of the chip whereas the damage on unit 5 were mostly on the edges of the chip which does not carry many contacts. Hence, this causes less electromigration effect on unit 4, and therefore, depicting less deviation in the I-V curve.

IV. CONCLUSION

The effect towards the I-V curve of two different severity of damage on Ge-substrate IRED is first reported. Although unit 2 is more severely fractured, both unit 2 and 3 exhibits a significant deviation in the I-V curve, portraying a soft knee in the curve which is caused by the creation of shunt in the p-n junction. Unit 4 however, which has a small damage, in terms

978-1-4673-2395-6/12 $31.00 © 2012 IEEE

of total area size, only shows a very minimal significant deviation in the I-V curve as compared to the curve obtained before the fracture is inflicted where as unit 5 shows a medium deviation. With the findings from this work, we are able to clarify the effect of the damages are related to the passivation layer on the top of the IRED die. We would continue studying the effects of the damage on thermal behaviour which includes device thermal resistance and junction temperature in our future work. It has been proven in many other studies that the passivation layer is important in protecting the IRED die surface from chemical and/or electrical degradation due to oxidation and surface recombination as well as improving IRED light extraction efficiency.

ACKNOWLEDGMENT

Liew would like to thank University Science Malaysia (USM) for a USM Fellowship.

REFERENCES

[1] N. S. Takahashi, S. Fujiwara, K. Kohno, E. Shibano & S. Kurita, "AlGaInP/AlGaAs Double Heterostructure Light Emitting Diode Grown by Liquid Phase Epitaxy", Journal of Crystal Growth, Vol. 137, 1994, pp. 240 – 244.

[2] L. W. Ji, Y. K. Su, S. J. Chang, C. S. Chang, L. W. Wu, W. C. Lai, X. L. Du & H. Chen, "InGaN/GaN Multi-Quantum Dot Light Emitting Diodes", Journal of Crystal Growth, Vol. 263, 2004, pp. 114 – 118.

[3] T. Wang, Y. H. Liu, Y. B. Lee, Y. Izumi, J. P. Ao, J. Bai, H. D. Li & S. Sakai, "Fabrication of High Performance of AlGaN/GaN-based UV Light-Emitting Diodes", Journal of Crystal Growth, Vol. 235, 2002, pp. 177 – 182.

[4] Y. Guo & J. H. Zhao, "A Practical Die Stress Model and Its Applications in Flip-Chip Packages", Inter. Society Conf. on Thermomechanical Phenomena In Electronic System, Las Vegas, Nevada, 2000, pp. 393 – 399.

[5] P. Su, B. Khan & M. Ding, "An Evaluation of Die Crack Risk of Over-moulded Packages Due To External Impact", Electronic Components and Tech. Conf., Las Vegas, Nevada, 2010, pp. 1631 – 1636.

[6] F. J. Foo, M. C. Ong & M. Y. Tay, "Failure Analysis of Die Crack In Lidless Packages", 17th International Symposium on the Physical and Failure Analysis of Intergrated Circuits, 2010.

[7] N. Rada & G. Triplett, "Thermal and Spectral Analysis of Self-Heating Effects in High Power LEDs", Solid-State Electronics, Vol. 54(4), 2010, pp. 378 – 381.

[8] L. Jayasinghe, Y. Gu & N. Narendran, "Characterization of Thermal Resistance Coefficient of High-Power LEDs", 6th International Conference on Solid-State Lighting Proceedings of SPIE 6337, San Diego, CA, USA, 2006, pp. 63370-V.

[9] N. Narendran, Y. Gu, J. P. Freyssinier, H. Yu & L. Deng, "Solid-State Lighting: Failure Analysis of White LEDs", J. Crystal Growth, Vol 268, 2004, pp. 449 – 456.

[10] X. Zhou, J. He, L. S. Liao, M. Lu, X. M. Ding, X. Y. Hou, X. M. Zhang, X. Q. He & S. T Lee, "Real-time Observation of Temperature Rise And Thermal Breakdown Processes In Organic LEDs Using An IR Imaging And Analysis System", Advance Material, Vol 12(4), 2000, pp. 265 – 269.

[11] S. Gardonio, L. Gregoratti, P. Melpignano, L. Aballe, V. Biondo, R. Zamboni, M. Murgio, S. Caria & M. Kiskinova, "Degradation of

Organic Light-Emitting Diodes Under Different Environment at High Drive Conditions", Organic Electronics, Vol. 8(1), 2007, pp. 37 – 43.

[12] Y. Ryu, D. D Baldocchi, J. Verfaillie, S. Ma, M. Falk, I. Ruiz-Mercado, T. Hehn and O. Sonnentag, "Testing the Performance of A Novel Reflectance Sensor, Built With Light Emitting Diodes (LEDs), to Monitor Ecosystem Metabolism, Structure And Function", Agricultural And Forest Meteorology, Vol 150 (12), 2010, pp. 1597 – 1606.

[13] D. Pogany, N. Seliger, M. Litzenberger, H. Gossner, M. Stecher, T. Müller-Lynch, W. Werner & E. Gornik, "Damage Analysis In Smart-Power Technology Electrostatic Discharge (ESD) Protection Devices", Microelectronics Reliability, Vol 39, 1999, pp. 1143 – 1148.

[14] LED Failure Modes and Methods for Analysis. Retrieved on 14th June 2011 from http://www.led-professional.com/technology/measurement-simulation/led-failure-modes-and-methods-for-analysis.

[15] Y. Oshima, R. D. Coutts, N. M. Badlani, R. M Healey, T. Kubo and D. Amiel, "Effect of Light Emitting Diode Therapy (LED) on the development of osteoarthritis (OA) In A Rabbit Model", Biomedicine & Pharmacotherapy, Vol 65(3), 2011, pp. 224 – 229.

[16] Y. J. Jiang, X. Z. Song, T. F. Lam & S. Samatitchon, "FBGA Die Crack Issue Analysis", International Symposium on High Density Packaging and Microsystem Integration, Shanghai, 2007.

[17] A. Uddin, A. C. Wei & T. G. Andersson, "Study of Degradation Mechanism of Blue Light Emitting Diodes", Thin Solid Films, Vol. 483, 2005, pp. 378 – 381.

[18] S. C. Yang, P. Lin, C. P. Wang, S. B. Huang, C. L. Chen, P. F. Chiang, A. T. Lee & M. T. Chu, "Failure And Degradation Mechanism of High Power White Light Emitting Diodes", Microelectronics Reliability, Vol. 50(7), 2010, pp. 959 – 964.

[19] T. V. Torchinskaya, "Kinetics of Red AlGaAs Light-Emitting Diodes Degradation", Microelectronics Reliability, Vol. 33(6), 1993, pp. 845 – 857.

[20] K. L. Pey, W. K. Chim, L. S. Koh, Y. Y. Liu & S. Y. C. Chew, "An Application of Spectroscopic Emission Microscopy And Cathodoluminescence To The Failure Analysis of Nearinfrared Light Emitting Diodes", Microelectronics Reliability, Vol. 35(6), 1995, pp. 935 – 946.

[21] J. R. Van Der Ziel & R. D. Dupuis, "Degradation of GaAs Lasers and Light-Emitting Diodes on Silicon Substrates", Materials Science and Engineering, Vol. 1(1), 1988, pp. 37 – 45.

[22] A. Pieslinger, K. Plaetzer, C. Oberdanner, J. Berlanda, H. Mair, B. Krammer & T. Kiesslich, "Characterization of A Simple And Homogeneous Irradiation Device Based On Light Emitting Diodes: A Possible Low-Cost Supplement To Conventional Light Sources For Photodynamic Treatment", Medical Laser Application, Vol. 21(4), 2006, pp. 277 – 283.

[23] E. F. Schubert, "Light Emitting Diodes", 2nd ed. United Kingdom: Cambridge University Press, 2006, pp 1 – 26.

[24] P. L. Martin, "Electronic Failure Analysis Handbook", New York: McGraw-Hill, 1999, pp. 1.1 – 1.14.

[25] T. A. Railkar, A. P. Malshe, R. Bhide, S. Hullavarad, & S. V. Bhoraskar, "Different Modes of Surface Passivation of GaAs", International Microelectronics And Packaging Society (IMAPS') Southern California (SoCal) Chapter Symposium and Exhibition, 2001. Retrieved on 23 June 2012 from http://nepp.nasa.gov/docuploads/9CFA5E09-1BB7-45FA-9DA0E5961CE8D055/Railkar_DifferentModesSurface.pdf.

[26] C. A. Ross, J. S. Drewery, R. E. Somekh & J. E. Evetts, "Electromigration and Mechanical Stress In Aluminium Conductor Tracks Passivated By Anodisation", Journal of Electronic Materials, Vol 19(9), 1990, pp. 911 – 918.

[27] A. Keppens, D. De Smeyter, W. R. Ryckaert, G. Deconinck & P. Hansalaer, "Thermal Characterisation of Single-Die and Multi-Die High Power Light Emitting Diodes", 8th International Conference on Solid State Lighting (Proc. SPIE), San Diego, California, Vol. 7058, 2008, pp. 70580H-70580H-12.

[28] P. S. Ho & T. Kwok, "Electromigration in Metals", Reports on Progress in Physics, Vol. 50, 1989, pp. 301.

[29] H. C. Liu & S. P. Murarka, "Modeling of Temperature Increase Due to Joule Heating During Electromigration Measurements", Center for Integrated Electronics and Electronics Manufacturing, Materials Research Society (MRS) Sym. Proc., San Francisco, California, 1996, pp. 113 – 119.

[30] I. A. Blech & E. Kinsbron, "Electromigration In Thin Gold Films On Molybdenum Surfaces", Thin Solid Films, Vol. 25(2), 1975, pp. 327 – 334.

New Circuit Models of Complementary-Symmetry Class-AB and Class-B Push-Pull Amplifiers

SachchidaNand Shukla, Beena Pandey and Susmrita Srivastava
Department of Physics and Electronics
Dr. Ram Manohar Lohia Avadh University
Faizabad - 224001, U.P., India
Email: sachida_shukla@yahoo.co.in

Abstract- **New circuit models of 'Complementary-Symmetry Class-AB and Class-B Push-Pull Amplifiers' are proposed. The proposed Class-AB amplifier circuit, configured by matched Darlington pairs, possesses 62.83% efficiency with low harmonic distortion (0.94%). However the proposed Class-B amplifier that is configured by using Triple Darlington configuration of matched BJTs, produces improved efficiency (73.28%) on the cost of enhanced harmonic distortion (1.23%). The proposed circuit model of Class-B Amplifier is perhaps a fresh approach. Variation of efficiency and total harmonic distortion as a function of biasing resistances and temperature are also observed. Proposed power amplifier may be quite useful for transmitters and other analog communication circuits.**
Index Terms- **Power Amplifiers, Class-B and Class-AB Amplifiers, Triple Darlington amplifiers**

I. INTRODUCTION

The overall system efficiency of a communication transmitter is predominantly determined by power amplifier [1]. It is a circuit that converts DC input power into RF/microwave output power [2] with a goal to obtain high efficiency and low harmonic distortion. Classically, design of high efficiency power amplifier centers around the use of push-pull configuration of Class-B and Class-AB amplifiers [2], [3]. In a push-pull Class-B amplifier, each half of the amplifier conducts for one-half cycle of the RF signal (operating cycle 180°) and the power efficiency reaches as high as 78.5% on the cost of linearity [1], [4]. However, in push-pull Class-AB power amplifiers, the transistor becomes ON for more than half a cycle (like class B), but less than a full cycle (like class A) of the input signal [5]-[7]. Efficiency of Class-AB amplifiers lie between 50% to 78.5% while the output signal swing occurs between 180° to 360° [4], [7], [8]. The real devices operated in Class-AB or Class-B push-pull amplifiers does not change abruptly from cut-off mode to a linear mode [5]-[8]. Instead, the transition is gradual and nonlinear which generates the effect of crossover distortion [5], [9], [10]-[11] . This can be minimized by biasing the active devices to produce a small quiescent collector current [6] or by using 'Complementary-Symmetry' configuration but with a disadvantage that the circuit utilizes two independent complementary power supplies to feed the symmetrical sections of the circuit [4].

In fact, abundant research papers and books are available to explore Class-AB and Class-B push-pull amplifiers (with or without Complimentary-Symmetry configurations) having 1 KHz input frequency, low order load resistance (in the range of few Ohms) high efficiency and minimum THD as a favourite circuit for radios, stereos and communication transmitters [1]-[2], [4], [6], [11]-[12]. Sequence of such amplifiers with high load resistances (in the range of KΩ) have also been developed using monolithic or valve based circuits [5], [13]-[15].

Here, in the present manuscript authors developed such class of circuits, initially with high load resistance, at fundamental level using active devices. Simultaneously, the use of Darlington pairs in power amplifier circuits are less studied [4], [11]-[12] and the use of Triple Darlington configuration [16]-[17] of BJTs in push-pull circuits is a new approach. The present manuscript deals with the qualitative performance of the new circuits of 'Complementary-Symmetry Class-AB and Class-B Push-pull Amplifiers'. Variation of efficiency and total harmonic distortion with various biasing resistances, maximum voltage gain, maximum power gain and temperature are elaborately discussed.

II. EXPERIMANTAL CIRCUITS

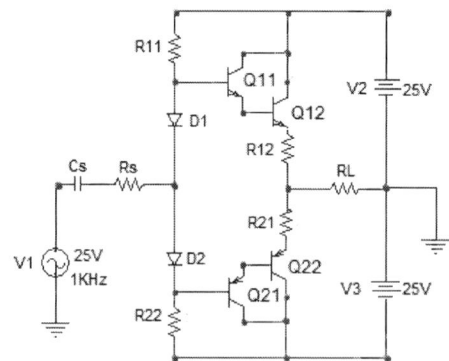

Fig.1. Proposed Class-AB amplifier

The proposed Class-AB amplifier (Fig.1) that is configured by matched Darlington pairs possesses symmetrical biasing resistances R12-R21 at its output section. However the proposed Class-B amplifier (Fig.2) that is configured by Triple Darlington units [16]-[17] of BJTs, use R12-R21 at its input section.

Both the proposed circuits utilize symmetrical compensating or biasing diodes [4] in their input halves which

978-1-4673-2395-6/12 $31.00 © 2012 IEEE   538

provide additional biasing to the matched transistor networks. Table-I summarizes the detail of circuit components used in both the circuits.

TABLE I
DETAIL OF CIRCUIT COMPONENTS

Components	Proposed Class-AB Amplifier	Proposed Class-B Amplifier
Q11, Q12	Q2N2222 (NPN, β=255.9)	Q2N2222 (NPN, β=255.9)
Q13	Not Available	Q2N2222 (NPN, β=255.9)
Q21, Q22	Q2N2907A (PNP, β=231.7)	Q2N2907A (PNP, β=231.7)
Q23	Not Available	Q2N2907A (PNP, β=231.7)
C_S	10μF	10μF
R_S	100Ω	100Ω
R_{11}, R_{22}	300KΩ	300KΩ
R_{12}, R_{21}	20KΩ (at Output Section)	20KΩ (at Input Section)
R_L	100KΩ	10KΩ
D_1, D_2	D1N4002	D1N4002
Biasing Source V_2	+25V DC	+25V DC
Biasing Source V_3	-25V DC	-25V DC
Input AC Signal V_1 for fair output	25V, 1KHz	25V, 1KHz

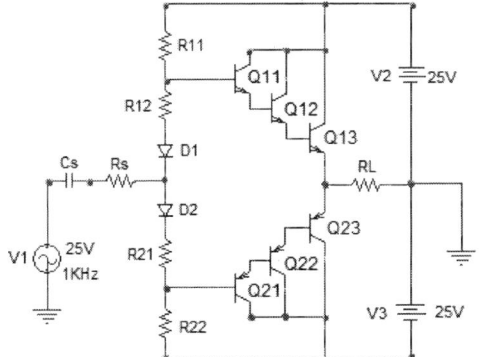

Fig.2. Proposed Class-B amplifier

Respective amplifiers are biased with ±25V DC power supply and observations are obtained for 25V AC signal source at 1KHz frequency through PSpice simulation software [12]-[13], [16], [18](Student version 9.2).

III. RESULTS AND DISCUSSIONS

Frequency response and efficiency for both the proposed amplifiers of Fig.1 and Fig.2 is estimated with the help of established theories [4], [12]-[13], [16].

With 62.83% efficiency and a frequency response (maximum voltage gain 0.826 and band-width 181.98MHz) like Class-A amplifier, the proposed circuit of Fig.1 is categorized as Class-AB amplifier [4]. Similarly, with 73.28% efficiency and a frequency response (maximum voltage gain 0.935 and band-width 1.928MHz) identical to Class-B amplifier, the proposed circuit of Fig.2 is identified as Class-B amplifier [4], [12]-[13].

Proposed Class-AB and Class-B amplifier circuits show significant variation with load resistance R_L. Increasing load resistance of the Class-AB amplifier to 500KΩ, efficiency reaches to 71.39% while reducing R_L to 50KΩ it declines to 53.80% and at about 10Ω the efficiency dips down to a non-

significant value. However for proposed Class-B amplifier, efficiency is found maximum at 10KΩ value of R_L which reduced to 72.98% at R_L=50KΩ and 67.31% at R_L=10Ω. Thus due to versatile behaviour with R_L, the proposed Class-B amplifier ensures its wide range of application as audio power amplifier [12]-[13].

Both the proposed circuits are having simple structure, wide bandwidth, capability to operate at high voltages [19] in wide range of load resistances and their efficiencies remain almost unaltered with respect to any change in R_S, C_S and input frequency [12]-[13].

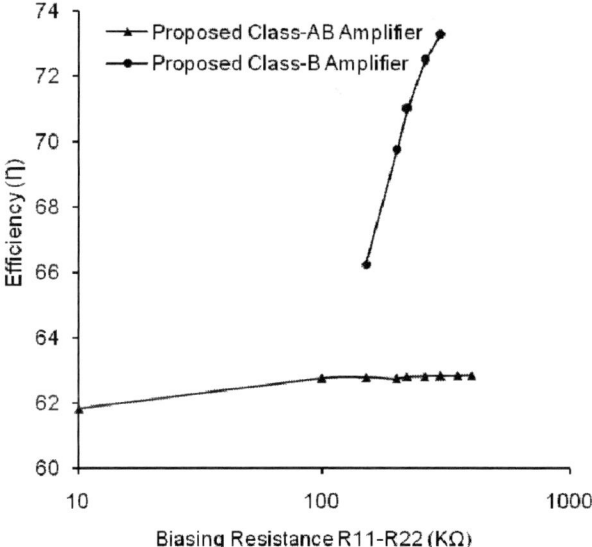

Fig.3. Variation of efficiency with biasing resistance R11-R22

Fig.3 shows the variation of efficiency with symmetrical biasing resistances R11-R22 for proposed amplifiers. It is noticed for Class-AB amplifier that the efficiency slightly increases at primary values of R11-R22 but at higher values it almost saturates. However for proposed Class-B amplifier, the efficiency graph rapidly increases with symmetrical biasing resistances R11-R22 within the limits of 150-300KΩ.

The nature of efficiency graphs with respect to R11-R22 in Fig.3 may be associated with the driving voltages of transistor units [12]. At increasing values of R11-R22, the driving voltage for Class-B amplifier rapidly falls (compared to that of Class-AB amplifier) which causes rapid increment in efficiency.

It is observed that in different resistance range of R12-R21 proposed amplifiers show almost similar response on efficiency scale (i.e. efficiency decreases non-linearly for increasing values of R12-R21). The respective observations are plotted in Fig.4. Figure suggests that the efficiency of Class-B amplifier decreases with a higher slope than that of Class-AB amplifier at increasing R12-R21. The efficiency of Class-B amplifier is also found higher than Class-AB amplifier at every value of R12-R21.

For proposed Class-B amplifier, when R12-R21 increases, the driving voltage to the matched active units also increases.

This brings matched units of transistors into saturation which causes reduction in output voltage and hence efficiency. On the other hand, due to parallel resistance network of R12, R21 and R_L at output section in the proposed Class-AB amplifier, effective load of the circuit decreases and hence the output voltage and efficiency.

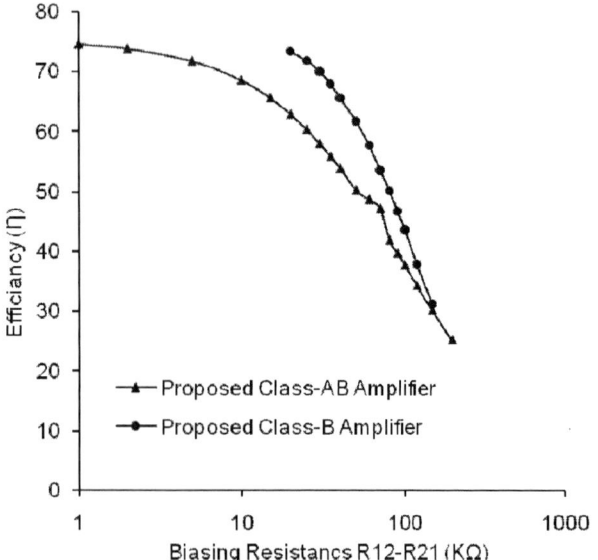

Fig.4. Variation of efficiency with biasing resistances R12-R21

Authors also attempted to study the variation of maximum power gain with symmetrical biasing resistances R11-R22 and R12-R21. The respective observations are listed in Table-II and Table-III.

Observations in Table-II suggest that for Class-AB amplifier, power gain slightly improves with increasing R11-R22. However for Class-B amplifier, it increases at rising values of R11-R22 with an achievement that the circuit shows a power gain greater than unity for 250-300 KΩ but produces distortion beyond 300KΩ. This suggests that the range of R11-R22 for the practical use of proposed Class-B amplifier should be 250-300 KΩ [4].

TABLE II
VARIATION OF MAXIMUM POWER GAIN WITH BIASING RESISTANCES R11-R22 (KΩ)

R11-R22	Class-AB Amplifier	Class-B Amplifier
150	0.285	0.633
200	0.331	0.891
250	0.366	1.152
300	0.394	1.399
350	0.417	Distortion
400	0.436	Distortion

TABLE III
VARIATION OF MAXIMUM POWER GAIN WITH BIASING RESISTANCES R12-R21 (KΩ)

R12-R21	Class-AB Amplifier	Class-B Amplifier
15	0.407	Distortion
20	0.394	1.398
40	0.349	1.214
80	0.280	1.014
100	0.253	0.919
120	0.230	0.840
150	0.201	0.744

Similarly, observations in Table-III suggests that for both the amplifiers, power gain decreases at increasing R12-R21. But the significant achievement is that the Class-B amplifier possesses a power gain greater than unity for 15-80KΩ which is also the advisable range of R12-R21 for this amplifier [4].

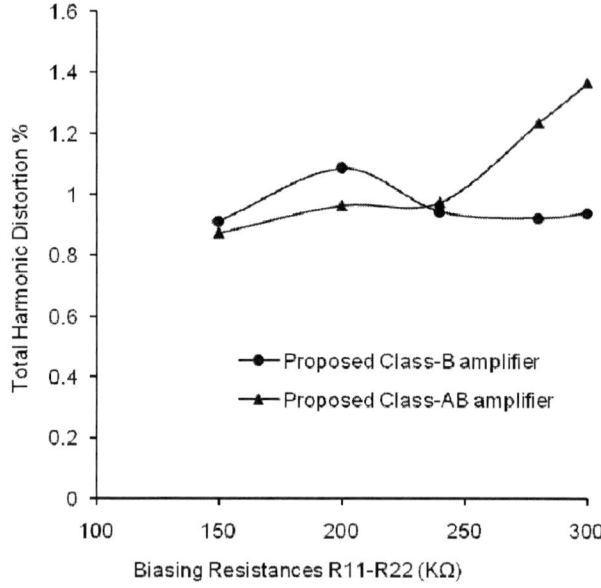

Fig.5. Variation of THD with biasing resistances R11-R22

Total Harmonic Distortion (THD) percentage for Class-AB and Class-B amplifiers are also calculated using standard formulae [4], [11]-[13]. Respective results show that THD of proposed Class-AB amplifier (0.94%) is less than that of proposed Class-B amplifier (1.23%). It is perhaps each half of the Class-AB amplifier circuit conducts for more than one half cycles while Class-B conducts for half cycle of the input signal. The higher operating cycle of Class-AB amplifier causes in reduction of cross-over distortion [4], [11]-[12].

Effect of biasing resistances R11-R22 on THD for proposed amplifiers is also observed. Respective plots are depicted in Fig.5. Figure shows that the THD for proposed Class-B amplifier has a rising tendency at increasing values of R11-R22. This tendency becomes intense at higher values of R11-R22 while at lower resistance values (up to 240KΩ), THD rises with slow pace. However in case of Class-AB amplifier, THD rises with R11-R22 up to 200KΩ, then falls up to 240KΩ and finally acquires a saturation tendency.

Effect of biasing resistances R12-R21 on THD for proposed amplifiers is depicted in Fig.6. It is observed here that the THD for proposed Class-B amplifier significantly rises with biasing resistances R11-R22. However in case of Class-AB amplifier, THD gradually falls with R12-R21 up to 120KΩ but beyond this critical limit, it starts rising.

It is obeserved for proposed Class-B amplifier that the efficiency and THD both gradually increases with R11-R22 and are maximum at 300KΩ (Fig.3 and Fig.5). This means, the efficiency of the proposed Class-B amplifier increases on the cost of THD. However, for proposed Class-AB amplifier the efficiency is found maximum with subsequent reduction in THD at 300KΩ value of R11-R22 (Fig.3 and Fig.5).

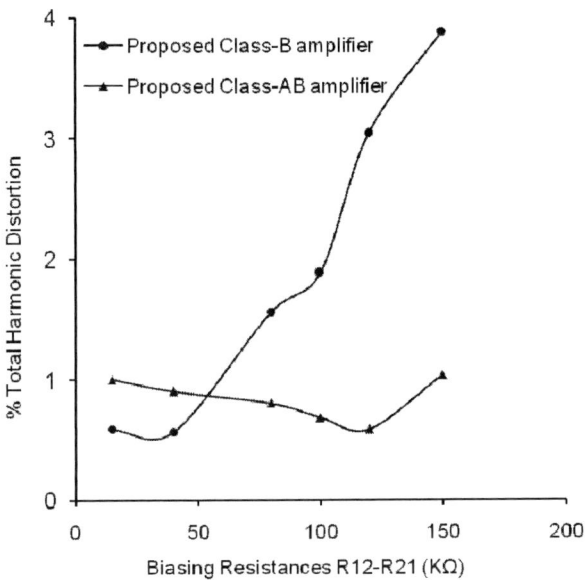

Fig.6. Variation of THD with biasing resistances R12-R21

Effect of load resistance on THD is also observed for both the proposed amplifiers. It is found that increasing load resistance of Class-AB amplifier to 500KΩ, THD rises to 1.01% while reducing R_L to 50KΩ it goes up to 1.08% and at 10Ω, it climbs up to 1.46%. However for proposed Class-B amplifier, THD reaches to a maximum 1.547% at 10Ω value of R_L which reduced to 1.02% at R_L=500KΩ.

Temperature dependency of both the amplifiers are also analyzed on the qualitative scale and respective observations are listed in Table-IV. Various parameters mentioned in Table-IV are showing that the maximum power gain, efficiency and harmonic distortion for both the proposed amplifiers bear almost non-significant variation with temperature (in certain range) which resembles with the prime feature of power amplifiers [4], [11]-[13].

TABLE IV

VARIATION OF MAXIMUM POWER GAIN, EFFICIENCY AND THD WITH TEMPERATURE

Temp °C	Class-AB Amplifier			Class-B Amplifier		
	$A_{P(max)}$	PAE (η%)	THD (%)	$A_{P(max)}$	PAE (η%)	THD (%)
0	0.393	62.511	1.094	1.423	73.071	1.106
10	0.393	62.593	1.088	1.415	73.676	1.055
25	0.394	62.731	0.993	1.399	73.275	1.044
50	0.395	62.973	0.863	1.413	73.209	1.154
80	0.394	63.004	0.679	1.421	73.058	1.411
100	NA	NA	0.657	1.423	73.373	1.257
125	NA	NA	0.697	1.420	73.379	0.943

IV. CONCLUSIONS

The proposed amplifiers with observed efficiencies and THDs are contemporary in their respective classes. Use of Triple Darlington transistor topology in proposed Class-B power amplifier is a new approach. The versatile behaviour of

this amplifier with R_L provides an extra dimension to the proposed circuit which suggests its effective use as an audio power amplifier at lower (10Ω) to higher (500KΩ) order load resistances.

Proposed Class-AB amplifier can be converted into high efficiency proposed Class-B amplifier by applying some minor changes. This conversion enhances the efficiency of Class-B amplifier at the cost of nominal increase in THD but at certain resistance values of symmetrical biasing resistances, the power gain of the proposed Class-B amplifier is found to be greater than unity. Efficiency, THD and maximum power gain of both the proposed amplifiers increase with R11-R22 and decrease with R12-R21 but the efficiency and THD do not vary significantly with temperature.

The qualitative properties explore proposed circuits suitable for low/high power consuming audio systems as well as transmitter stage of communication systems.

REFERENCES

[1] C. Trask , High Efficiency Broad band linear push-pull power amplifiers using linearity augmentation, IEEE International Symposium on Circuits and Systems, Vol-2, 2002, p. 432

[2] F. H. Raab , P. Asbeck , S. Cripps, P. B. Kenington et. al., Power Amplifiers and Transmitters for RF and Microwave, IEEE Transactions on Microwave Theory and Techniques, Vol.50, No.3, March 2002, p-814

[3] J. S. Lim, H. S. Kim, J. S. Park, D. Ahn and S. Nam, A Power Amplifier with efficiency Improved Using Defected Ground Structure, IEEE Microwave and Wireless Components Letters, Vol. 11, No. 4, April 2001, p-170

[4] R. L. Boylestad and L. Nashelsky, Electronic Devices and Circuit Theory, Pearson Education Asia, 2008, p-761

[5] A. Beohar and R. K. Baghel, A Class-AB Amplifier with a Reduced Crossover Distortion for Real Time Video Applications, International Journal of Engineering Science and Technology, Vol. 2(7), 2010, p-2783

[6] S. Kophon, P. Uthansakul, M. Uthansakul and S. Cheedket, Efficiency Improvement of Power Amplifier Class AB Push-Pull using Invert Doherty Combined with Negative Feedback, International Conference on Communication Engineering and Networks IPCSIT Vol.19, 2011, p-62 (IACSIT Press, Singapore)

[7] K. K. Mohammed, R. B. Mohammed, Linearization of Power Amplifier Class AB Using Cartesian Feedback, International Multi-Conference on Systems 7th, Signals and Devices, 2010.

[8] Z. Wang, Wideband Class-AB (Push-Pull) Current Amplifier in CMOS Technology, Electronics Letters, April 12, 1990, Vol.26, No.8, p-543-545

[9] A. T. Mendez, H.J. Aguilar, R. F. Leal et.al., Low harmonic distortion in single ended and push-pull class-E power amplifiers by using slotted micro-strip lines, AEU-International Journal of Electronics and Communications, Vol. 64 (1), 2010, p-66

[10] V. Paidi, R. Shouxuan Xie Coffie, B. Moran et. al., High linearity and high efficiency of class-b power amplifiers in GaN HEMT technology, IEEE Transaction on Microwave Theory and Techniques, 2003 Vol-51, Issue-2 p- 643

[11] R. K. Tiwari, and J. Mishra, A new model for distortion-less push-pull amplifier, Bulletin of Pure and Applied Sciences, Vol.29D (No.1), 2010, p- 53

[12] B. Pandey, S. Srivastava, S. N. Tiwari, J. Singh and S. N. Shukla, Complementary-Symmetry Class-B Push-Pull Amplifier with Improved Efficiency and Reduced Harmonic Distortion, Journal of Ultra Scientist of Physical Sciences Vol.23(2)B, 2011, p 353

[13] S. Srivastava, B. Pandey, S. N. Tiwari, J. Singh and S. N. Shukla, *Qualitative Analysis of MOS Based Complementary-Symmetry Class-B Push-Pull Amplifier with Improved Efficiency*, Journal of Ultra Scientist of Physical Sciences, Vol. 23(3)B, 2011, p.703-708

[14] L. Butler, *RF Power amplifiers-tank circuits & output coupling*, Amateur Radio, May 1988 (updated on Sept. 8, 2007)

[15] N. L. Frank, F. M. Patrick, L. Verdeyen and M. C. S. Willy, A CMOS Large-swing Low-distortion Three-stage Class-AB power amplifier, IEEE Journal of Solid State Circuits, Vol.25, No.1, Feb. 1990, p-265

[16] S. N. Tiwari, B. Pandey, S. N. Shukla, *Qualitative Analysis of Two Distinct Wide Band Triple Darlington Amplifiers*, Journal of Ultra Scientist of Physical Sciences, Vol.21(1), 2009, p 117

[17] D. A. Hodges, *Darlington's contributions to transistor circuit design*, IEEE Transactions on circuits and systems-1, Vol. 46, No. 1, 1999, p 102-104

[18] M. H. Rashid, *Introduction to PSpice Using OrCAD for Circuits and Electronics*, Pearson Education, 3rd Ed., 2004, p. 255-300

[19] P. S. Manhas, S. Sharma, L. K. Mangotra, and K. K. S. Jamwal, *New low- voltage class-AB current conveyor II for analog applications*, Indian Journal of Pure & Applied Physics, Vol.47 April 2009, pp 306-309

A Three-Stage Power Amplifier for WiMedia Ultra-Wideband Applications

Zi-Yi, Lam, *Member, IEEE*, Yun-Fen, Yong, Sew-Kin, Wong, *Member, IEEE*, Chee-Pun, Ooi
Faculty of Engineering (FOE)
Multimedia University, Persiaran Multimedia
63100 Cyberjaya, Selangor Darul Ehsan, Malaysia
Email: zylam@mmu.edu.my (corresponding author), yfyong@mmu.edu.my, skwong@mmu.edu.my, cpooi@mmu.edu.my

Abstract- **A three-stage 0.18μm CMOS power amplifier (PA) targeted for 3.1 to 4.8 GHz WiMedia Ultra-Wideband (UWB) system is discussed in this paper. The first and the last stage of the proposed PA are the conventional common source (CS) inductive degeneration while the inter-stage is a common gate (CG) structure with LC matching components. With careful optimization, the inter-stage CG and LC matching components enhances the gain of the amplifier over a wideband range while maintaining a reasonable output power of the proposed PA. Off-chip LC components are used to improve the overall input and output return losses of the proposed PA. The proposed PA has a simulated maximum power gain of 19.3 dB, minimum input return loss of 19.7 dB, maximum output return loss of 8.5 dB, reverse isolation of at least 36 dB, maximum output 1 dB gain compression of 10 dBm and power efficiency of more than 30 %, while dissipating a DC power of 20 mW from a 1.5 V supply. Simulation results gathered in this work can serve as a guideline for wideband PAs design, especially in radio frequency (RF) transmitter for mobile devices and consumer products.**

I. INTRODUCTION

At present, the Ultra-wideband (UWB) system is one of the promising technologies for Wireless Personal Area Network (WPAN). Compared to Bluetooth and WiMax, UWB has the capability of providing a high data-rate transfer with an effective range up to 20 meters and wide frequency range with extreme low power consumption [1]. WiMedia Alliance [2] led by Intel, Panasonic and Motorola has proposed the first industry UWB standard. For the first generation UWB system (WiMedia version 1.1) deployment, WiMedia UWB will use the low frequency band of 3.1 to 4.8 GHz (Band Group 1) as a mandatory mode to transmit short range multimedia files at a data-rate up to 480 Mbps. This group has three frequency bands, 3.168 to 3.696 GHz with a center frequency of 3.432 GHz (Band 1), 3.696 to 4.224 GHz with a center frequency of 3.96 GHz (Band 2) and 4.224 to 4.752 GHz with a center frequency of 4.488 GHz (Band 3). As suggested by WiMedia Alliance [2], UWB could be used as general USB cable replacement (also known as Wireless USB) and short-range high data rate communications between personal computer (PC), mobile devices, tablet PCs and digital consumer electronic products.

In radio frequency (RF) transmitter, the power amplifier (PA) is the most critical building blocks as it must fulfill stringent requirements such as broadband input and output matching, high power gain, maximum output power and optimum efficiency while maintaining low power dissipation. At present, various techniques have been used in the implementation of ultra-wideband power amplifiers. Among that have been reported are the current reused technique, resistive shunt feedback topology and the passive LC matching networks [3-7]. In practice, the techniques used in designing wideband PA depend on the features of the amplifier such as output power level, linearity and power efficiency. In this work, the proposed PA relies on common source (CS) and common gate (CG) topologies with external off-chip LC components in order to achieve wideband matching (both the input and output) while maintaining an optimized gain. In this work, Agilent Technologies's Advanced Design System (ADS) software is used to simulate, analyze and optimize the proposed PA before CMOS IC fabrications.

This paper illustrates the design, analysis and simulation of a three-stage CS inductive degenerated with inter-stage CG LC matching power amplifiers for WiMedia UWB transmitter using Silterra CMOS technology. 'Circuit Design and Description' describes the overview for the proposed PA which covers the circuit schematic descriptions. The simulation results and chip layout will be shown in 'Simulation Results'. Finally, the conclusion of this work is presented in 'Conclusion'.

II. CIRCUIT DESIGN AND DESCRIPTION

The proposed three-stage CS inductive degeneration and inter-stage CG LC matching PA circuit schematic is shown in Fig. 1. The supply voltage into this circuit is 1.5 V. With careful optimization, the width and length of NMOS field-effect transistor in all the three stages (M_1, M_2 and M_3) are set to 160μm and 0.18μm respectively. Common-source (CS) amplifier is put in the first stage due to its high power gain and efficiency compared to common-drain (CD) and common-gate (CG) circuitries [8]. In the CS stages, the source degeneration inductors L_{s_1} and L_{s_3} need to be chosen carefully since they decrease stability and gain of the PA [8]. In this work, the values of L_{s_1} and L_{s_3} are designed to be small enough (approximately 0.5 nH) so that it could be

978-1-4673-2395-6/12 $31.00 © 2012 IEEE   543

easily replaced by a bondwire inductance in order to reduce the chip area when necessary. The gate inductor in the first-

stage, L_{g_1} is needed to reduce capacitive effect of the input impedance of transistor M_1 [8].

Next, the inter-stage inductor L_{i_1} and capacitor C_{i_1} are added to provide a conjugate impedance matching between the CS stage (first stage) and CG stage (in which the impedances of these two stages are always capacitive) [8]. Due to this matching, the overall gain of the three-stage PA has increased significantly compared to the conventional amplifier without inductor L_{i_1} and capacitor C_{i_1} [9].

A shunt peaking inductor L_{o_2} with a value of 4.5 nH is used to resonate with the sum of parasitic capacitances at the drain of transistor M_2 and input of the third stage (gate of transistor M_3) around 4 GHz. At the output of the last stage, a large value of peaking inductance L_{o_3} is required in order to compensate the power consumption. Both the L_{o_2} and L_{o_3} inductors are also used as RF choke to the DC voltage supply. On top of that, a large value of R_{bias} (approximately 2 kΩ) is used as RF choke to provide RF signal isolation from the input. In order to provide sufficient RF shunting, four large on-chip capacitors (C_{b1} to C_{b4}) are included in the circuit. Finally, dc blockings are provided by capacitors C_{in_1} and C_{o_3}. External input and output matching circuitries are used to further improve the input and output return losses of the proposed PA. The proposed PA is simulated and optimized with Agilent Technologies's Advanced Design System (ADS) software before CMOS IC layout and fabrication. BSIM (Berkeley Short-channel IGFET Model) signal model version 3.3 is used for the CMOS transistor modelling. The on-chip spiral inductors, metal-fingered capacitors, pads, interconnects, bondwire inductance $L_{bondwire}$ and off-chip LC matching, L_{im1}, C_{im1}, C_{im2}, L_{om1}, C_{om1} and C_{om2} components are modeled by RLC equivalent networks in the circuit schematics.

III. SIMULATION RESULTS

The simulated S-parameter results are shown in Fig. 2 and 3. The proposed PA achieved a maximum gain of 19.3 dB at 3.75 GHz, compared to a three-stage PA without inter-stage LC matching components (without L_{i_1} and C_{i_1}) which only produces a gain of 14.8 dB. The proposed PA has a minimum input return loss ($|S_{11}|$) of approximately −19.7 dB at 4.75 GHz, while maintaining a maximum output return loss ($|S_{22}|$) of −8.5 dB across 3 to 5 GHz. Over the frequency of 1 to 8 GHz, the proposed PA obtained a high reverse isolation ($|S_{12}|$), with a minimum value of −36 dB. The simulated input and output 1 dB gain compression (P_{1dB}) and power added efficiency (PAE) for the proposed PA are illustrated in Fig. 4

and 5 respectively. As shown in Fig. 4, the simulated output P_{1dB} at 3.432, 3.96 and 4.488 GHz are 9.3 dBm, 8.4 dBm and 10 dBm respectively. When operating at output P_{1dB}, the proposed PA has an overall PAE of 27 %, 20.3 % and 22 % at 3.432, 3.96 and 4.488 GHz respectively, as depicted in Fig. 5. It also has the potential of reaching up to more than 30 % of PAE at higher output power in non-linear region.

Another important parameter used to test the functionality of a PA is by generating an output spectrum of the PA when co-simulated with WiMedia's MB-OFDM UWB modulated signal. In order to perform this co-simulation, pre-configured UWB Design Guide available in Agilent ADS, is carried out together with the designed PA circuit schematics. Simulations that are supported by this design guide include spectrum, power, constellation and EVM (Error Vector Magnitude). The simulated output spectrums before and after the PA's amplification are shown in Fig. 6 and 7 respectively. Here, the input data rate and channel power (P_{in}) are set to 320 Mbps and −12 dBm, assuming that the PA is operating near the linear region. The UWB spectrum at Band 2 (3.696 to 4.224 GHz) together with UWB SEM (Spectrum Emission Mask) is shown in Fig. 6, while the UWB spectrum for the whole mandatory mode (3.168 to 4.752 GHz) is depicted in Fig. 7. Overall, the proposed PA achieved satisfactory results of SEM and output channel power at Band 1, 2 and 3 of 7.27, 7.81 and 7.73 dBm respectively. Table 1 shows the simulation summary and comparison with other literatures. The die micrograph for the proposed PA (occupying an area of 1.4 mm × 1.6 mm) designed using Mentor Graphics IC Station software and fabricated in Silterra 0.18 μm CMOS technology is shown in Fig. 8.

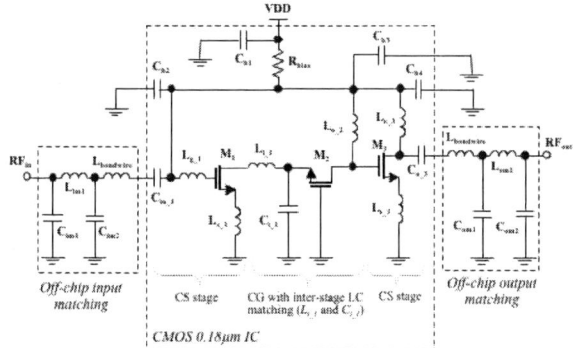

Fig. 1. Schematic of the proposed three-stage UWB PA

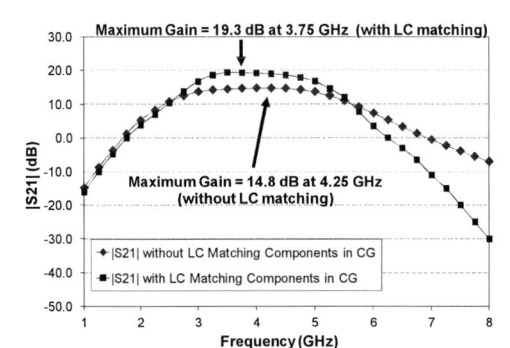

Fig. 2. Simulated gainc|S21| for the proposed PA (with and without LC matching components in the CG stage)

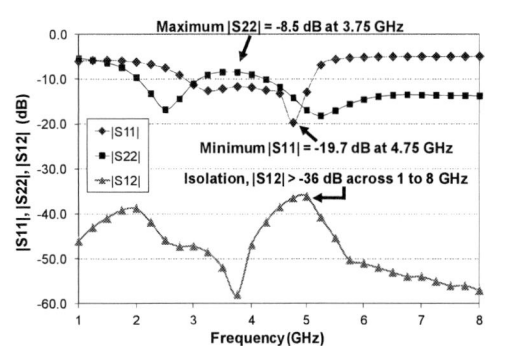

Fig. 3. Simulated input return loss |S11|, output return loss |S22| and reverse isolation |S12| for the proposed PA

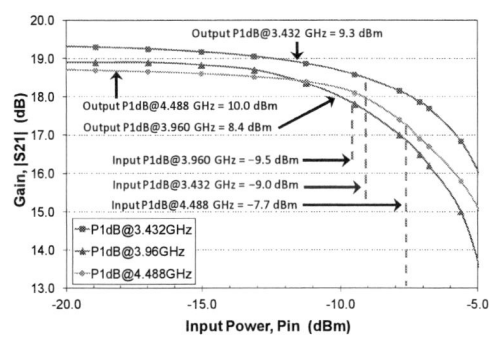

Fig. 4. Simulated 1 dB gain compression (P1dB) for the proposed PA at 3.432, 3.96 and 4.488 GHz

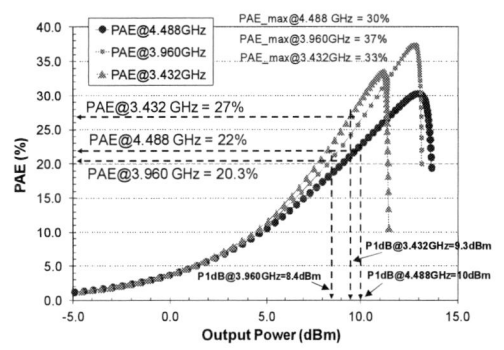

Fig. 5. Simulated power added efficiency (PAE) for the proposed PA at 3.432, 3.96 and 4.488 GHz

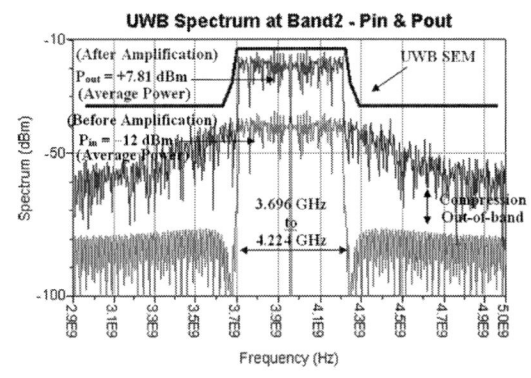

Fig. 6. Simulated output spectrum at MB-OFDM UWB Band 2 (3.696 to 4.224 GHz). Input data rate and channel power are 320 Mbps and −12 dBm

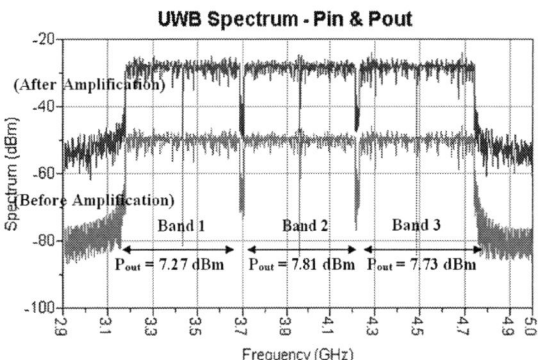

Fig. 7. Simulated output spectrum over the three bands (3.168 to 4.752 GHz)

TABLE I
COMPARISON OF WIMEDIA'S UWB CMOS PA PERFORMANCES

References	[3]	[4]	[6]	[7]	This work (Simulation)
Frequency (GHz)	3 to 5	3.1 to 10.6	3 to 4.8	3 to 7	3.1 to 4.8
S_{11} (dB)	< -10	< -9	< -10	< -6	< -8.5
S_{22} (dB)	< -8	< -8	< -10	< -7	< -12
Maximum S_{21} (dB)	19	15	17.5	14.5	19.3
Maximum Output P_{1dB} (dBm)	-4.2	0	1.25	7	10
Maximum PAE (%)	N/A	N/A	11	N/A	> 30
Power Dissipation (mW)	25	25.2	35	24	20

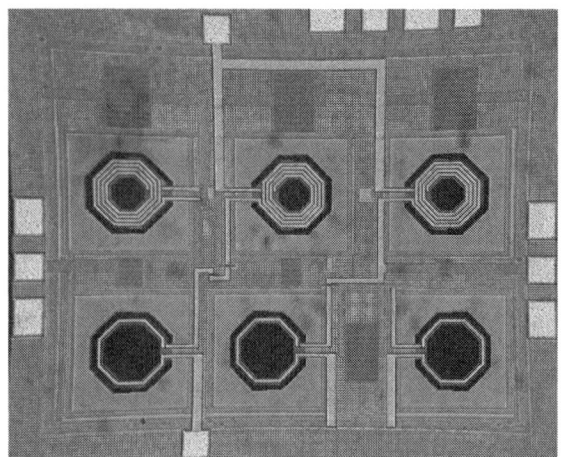

Fig. 8. Die micrograph of the proposed three-stage PA (1.4 mm × 1.6 mm)

IV. CONCLUTION

A three-stage 0.18μm CMOS WiMedia UWB PA for lower band UWB system (3.1−4.8 GHz) is designed, simulated and discussed in this paper. Using the CS-CG-CS topology, the proposed PA achieve a simulated maximum gain of 19.3 dB, maximum output 1 dB gain compression of 10 dBm and high power efficiency at 1dB gain compression of approximately 27%, while consuming 20 mW at 1.5 V DC supply. Having such feasibilities, this PA can be used as reference design for immediate multi-stage WiMedia UWB PA implementation. The proposed circuit occupies an area of 1.4 mm × 1.6 mm.

REFERENCES

[1] FCC, "Final rule of the Federal Communication Commission, 47 CFR Part 15, Sec. 503", Federal Register, vol. 67, no. 95, May 2002.

[2] "Multiband OFDM physical layer specification," WiMedia Alliance, 2005.

[3] S. Jose, H-J. Lee, D. Ha and S.S. Choi, "A low-power CMOS power amplifier for Ultra Wideband (UWB) applications," in Proc. IEEE International Symposium on Circuit and Systems, 2005, pp.5111-5114.

[4] C.H. Han, W.W. Zhi and K.M. Gin, "A low power CMOS Full-Band UWB power amplifier using wideband RLC matching method," in Proc. IEEE Electron Devices and Solid-state Circuit Conf., 2005, pp.223-236.

[5] C. Lu, A-V. Pham, and M. Shaw, "A CMOS power amplifier for full-band UWB transmitter," in Proc. IEEE RFIC Symp., 2006, pp.397-400.

[6] R-L. Wang, Y-K. Su and C-H. Liu, "3~5 GHz Cascoded UWB power amplifier," in Proc. IEEE Asia Pacific Conference on Circuit and Systems, 2006, pp.367-369.

[7] S.A.Z. Murad, R.K. Pokharel, A.I.A. Galal, R. Sapawi, H. Kanaya and K. Yoshida, "An excellent gain flatness 3.0-7.0 GHz CMOS PA for UWB applications," IEEE Microwave and Wireless Components Letters, vol. 20, no. 9, pp.510-512, Sep. 2010.

[8] T.H. Lee, The Design of CMOS Radio-Frequency Integrated Circuit, 2nd ed., Cambridge Univ. Press, New York, 2004.

[9] C. Zhang, D. Huang and D. Lou, "Optimization of cascode CMOS low noise amplifier using interstage matching network," in Proc. IEEE Conference on Electron Devices and Solid-state Circuit, 2003, pp.465-468.

High Speed Direct Digital Frequency Synthesizer with Pipelining Phase Accumulator Based on Brent-Kung Adder

Salah Hasan Ibrahim[1], Sawal Hamid Md Ali[1], Md. Shabiul Islam[2]

[1]Department of Electrical, Electronic & System Engineering, [2]Insitute of Microengineering and Nanoelectronics (IMEN),
Faculty of Engineering and Built Environment
University Kebangsaan Malaysia (UKM),
43600 UKM, Bangi, Selangor, Malaysia.
E-mail: mr.salah65@yahoo.com, sawal@eng.ukm.my, shabiul@ukm.my

Abstract – **This paper presents a high speed direct digital frequency synthesizer (DDFS) using pipelining phase accumulator (PA) with a modified parallel prefix adder based on Brent-Kung (BK) adder. The proposed 32-bit phase accumulator design consists of four pipeline stages, with 8-bit Registers and modifying 8-bit Brent-Kung adder in each stage with carries ripple between the stages. The proposed architecture with modifying 8-bit Brent-Kung adder has been implemented on Cyclone III FPGA kit. A comparison with conventional phase accumulator that using ripple carry adder (RCA) has been made and the results shown that the proposed architecture performs 24.9% faster than the conventional phase accumulator.**

Keywords: **direct digital frequency synthesizer (DDFS), phase accumulator (PA), Brent-Kung adder (BK), ripple carry adder (RCA).**

I.   INTRODUCTION

A high speed, high frequency resolution and fast frequency channel has became part of the requirements to design a new DDFS systems especially in communication and radar systems. Pipeline technique has been widely used to achieve higher speed phase accumulator (PA)[1]. Pipeline PA with a high bit-input are designed to achieve high speed and high frequency resolution [2-4]. 11- bit PA with one bit adder in each pipelining stages are used in[5], a large numbers of DFFs used in this design leads to high power consumption.

The output frequency of the DDFS is a analog sine wave and it can be calculated based on equation (1)

$$f_{out} = \frac{FCW}{2^{N-1}} \times f_{clk} \text{ [6]} \qquad (1)$$

Where FCW is a frequency control word, fclk – the clock frequency and N- is a bit-input number.

The standard DDFS consist of Phase accumulator (PA) with registers and adder, lookup table (LUT) or phase to amplitude convertor, digital to analog convertor (DAC) which the amplitude sine wave signals converted to analog signals waves and the low pass filter (LPF) to remove the unwonted harmonic, as shown in Fig. 1.

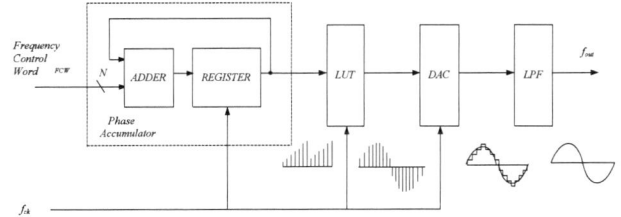

Fig. 1. Block diagram of Direct Digital frequency Synthesizer

PA has been designed with a high bit input number [7-8], to achieve high frequency resolution, where the relationship is between the clock frequency and the number of input bits is given by equation (2)

$$f_{res} = \frac{f_{clk}}{2^N} \qquad (2)$$

Pipelined carry lookahead adder design in quantum-dot cellular automata technology[9]. A large circuits design that used the QCA technology has disadvantages in delay and size due to wiring. Pipelined carry-dependent sum adder with the self-checking structure was proposed with (4-bit, 8-bit, 16-bit

978-1-4673-2395-6/12 $31.00 © 2012 IEEE

and 32-bit) adder's design [4]. Comparison result of the carry-dependent sum adder with the RCA revealed that the area and the delay can be reduced.

High frequency resolution in DDFS requires PA with a high bit inputs number. This will increase the complexity of the circuits design and the power consumption. Dividing the number of bits into several stages in pipelining technique is the best approach to overcome the complexity, power issues and increasing the speed of the PA by the number of the pipelining stages.

This paper presents high speed DDFS using pipeline PA design based on modified BK adder, exploiting the advantages of the pipelining technique , high speed and the simplify design of the modified BK adder in each of the pipeline stages.

This paper is organized as follows: Section II consists of phase accumulator architecture. Section III describes the circuit implementation of the 32-bit phase accumulator design. and section IV discusses the experimental results. Conclusion is presented in section V.

II. PHASE ACCUMULATOR ARCHITECTURE

The proposed PA design consists of four pipelining stages, each stages includes 8-bit registers (consists of DFFs) and a modified BK adder. The carries output of the modified BK adders ripple between the pipelining stages after one DFF delay, to maintain the timing synchronization in PA circuit. The block diagram of the 32-bit pipelining based PA using 8-bit modifying BK adder (in each pipelining stage) is shown in Fig.2.

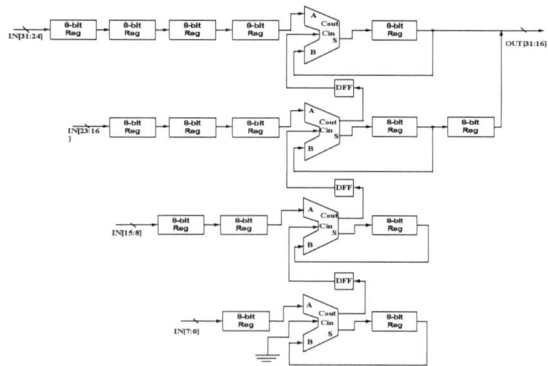

Fig. 2. Block diagram of 32-bit pipeline Phase Accumulator design

A. Brent-Kung Adder

In order to achieve a high speed circuit with less complex design, a parallel adder is a good candidate[10]. Brent-kung adder is a parallel prefix adder and suitable for PA designs with high bit input numbers. Propagate (Exclusive OR) and generate (AND gate) functions (pi, gi) of the BK adder consists of x and y –bits input, and written as equation (3) and (4)

$$p_i = x_i \oplus y_i \qquad (3)$$
$$g_i = x_i \times y_i \qquad (4)$$

Where p_i is a propagate function and g_i is a generate function.

In this PA design, the carries ripple between the pipelining stages, we modified the BK adder by removed the generate function (g_0) of the first bit of BK adder, the propagate function (p_0) with the x_0 input of the first bit and the carry input (cin) are connected to (2×1 Multiplexer) in the first bit of BK adder. The carry out of the first bit can achieve by the equation in (5).

$$C_1 = (x \times \bar{p}) + (c_{in} \times p) \qquad (5)$$

C_1 is a carry out of the first bit BK adder. Basic component of Brent-Kung (BK) adder is a Brent-Kung cells (BK), and it's a sum of higher- lower generate and propagate function [11]. Brent-Kung cells (BK) equation is sum of G and P as shown in Fig. 3.

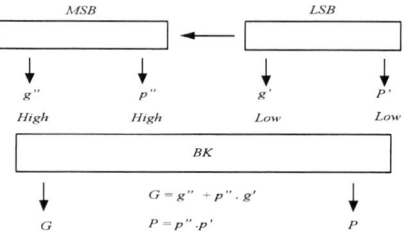

Fig. 3. Blok diagram of the Brent-kung cell

The carry out equations for the 8-bit modifying BK adder is:

$$C_{out} = g_7 + p_7.G_5 \qquad (6)$$

The zoom view of 8-bit modifying BK adder shows in Fig. 4.

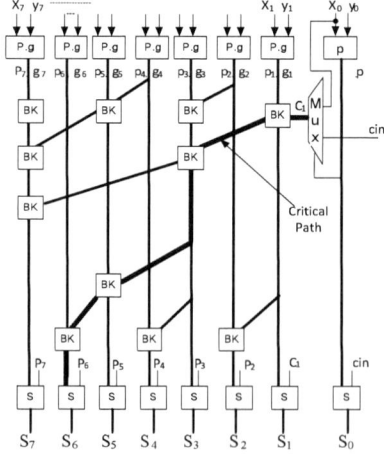

Fig.4. Zoom view of 8-bit modifying Brent-Kung adder

III. CIRCUIT IMPLEMENTATION OF 32-BIT PA DESIGN

The 32-bit phase accumulator has been completely designed with Verilog Hardware Description Language to be implemented on Cyclone III FPGA kit. The RTL view of the design is shown in Fig. 5. The total gate delay of the proposed PA design involves 17 gate delays (with a 12-gate delay for the modified BK adder and four 8-bit Reg. delays for the four pipelining stages and one SUM gate delay).

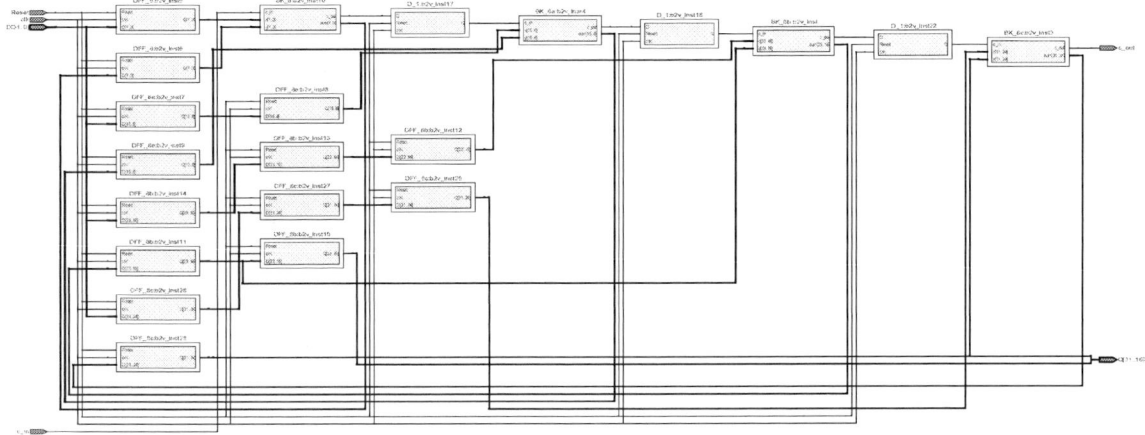

Fig. 5. RTL view of 32-bit pipeline PA design with 8-bit modifying BK

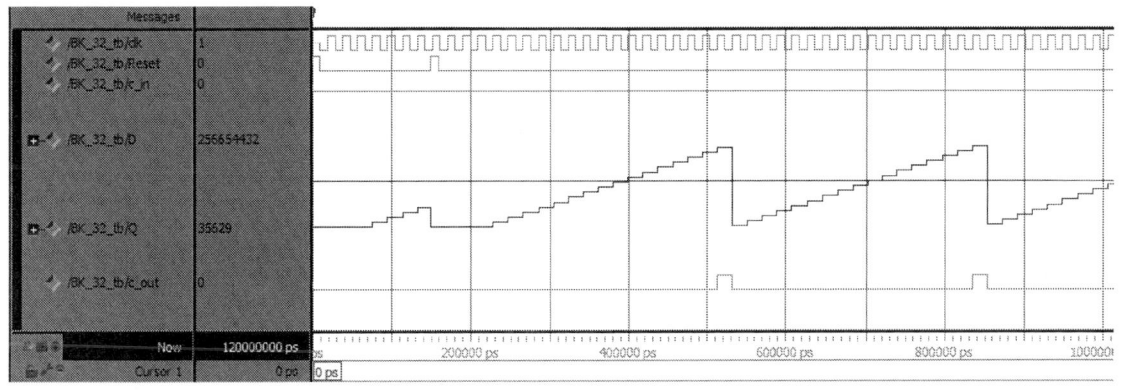

Fig.6. RTL simulation of 32-bit PA design modified 8-bit BK adder

A. RTL Simulation

The RTL or functional simulation of 32-bit PA with modified 8-bit BK adder using the ModelSim ALTERA as shown in Fig. 6, shows that the Q output data appears after five clock delays. This is due to the delays from four pipelining stages and one SUM operation in the final stage. Changing the Retest input to (1) makes the data output value return to zero, retuned the Reset input value to zero again, the data output start counting after five clock delay . Figure 5 shows the functional simulation of the 32-bit PA.

IV. EXPERIMENTAL RESULTS

The proposed PA has been compared with a conventional 32-bit pipelining PA based on 8-bit RCA adder. The maximum operation frequency of the proposed PA design with CLA adder is 254.91MHz, In the RCA based PA, the signal is propagated sequentially; the propagation delay for the 1-bit full adder includes two gate delays and one carry-out.For 8-bit RCA, the total delay is $2 \times 8 + 1 = 17$, and the 32-bit pipelining PA with 8-bit RCA will hage $17 + 5 = 22$ gate delays. 5 gate delays come from 4 pipelining stages and one SUM operation, for this the operation frequency of the PA design with RCA adder is 191.31MHz,   TABLE I. shows the performance comparisons between the proposed architecture, the RCA based architecture and the PA of the previous work.

TABLE I.

COMPARISON RESULTS OF MAXIMUM OPERATING FREQUENCY (F_{MAX}),
PROPAGATION DELAY (T_P), NUMBER OF REGISTERS AND LOGIC CELLS

PA Name & No. of Bits	No. of bit-stages	Max. Operation Freq.(MHz)	Propagation delay ($\square$ sec)	Registers	Logic Cells
BK-32	(8-4)	254.91	3.92	115	135
RCA-32	(8-4)	191.31	5.22	117	129
PA -32 RCA WMED2005[2]	(4-8)	91.78	10.89	200	217

V. CONCLUSION

The proposed 32-bit pipelined phase accumulator PA with modified BK adder has been successfully designed and simulated. The modified BK adder used in the architecture improves the operation speed of the PA and the same time maintains relatively small area consumption. A comparison has been made with conventional design that based on RCA adder, and PA designed of previous work. The results showed that the proposed architecture perform (2.77) times faster than the PA of the previous work and 24.9% faster than the conventional PA design.

REFERENCES

[1] B. S. Jensen, *et al.*, "Twelve-bit 20-GHz reduced size pipeline accumulator in 0.25 µm SiGe:C technology for direct digital synthesiser applications," *Circuits, Devices & Systems, IET*, vol. 6, pp. 19-27, 2012.

[2] I. Horowitz and G. S. La Rue, "Parallel phase accumulator architecture for DDFS," in *Microelectronics and Electron Devices, 2005. WMED '05. 2005 IEEE Workshop on*, 2005, pp. 63-66.

[3] M. Chappell and A. McEwan, "A low power high speed accumulator for DDFS applications," in *Circuits and Systems, 2004. ISCAS '04. Proceedings of the 2004 International Symposium on*, 2004, pp. II-797-800 Vol.2.

[4] L. Ming, *et al.*, "A Design of Pipelined Carry-dependent Sum Adder With its Self-checking Structure," in *Test Symposium, 2006. ATS '06. 15th Asian*, 2006, pp. 189-194.

[5] G. Xueyang, *et al.*, "An 11-Bit 8.6 GHz Direct Digital Synthesizer MMIC With 10-Bit Segmented Sine-Weighted DAC," *Solid-State Circuits, IEEE Journal of*, vol. 45, pp. 300-313, 2010.

[6] E. McCune, "Direct digital frequency synthesizer with designable stepsize," in *Radio and Wireless Symposium (RWS), 2010 IEEE*, 2010, pp. 356-359.

[7] O. I. Polikarovskykh and V. E. Havronskyy, "The New Type of Phase Accumulator for DDS," in *Microwave & Telecommunication Technology, 2007. CriMiCo 2007. 17th International Crimean Conference*, 2007, pp. 267-268.

[8] G. Xueyang, *et al.*, "24-Bit 5.0 GHz Direct Digital Synthesizer RFIC With Direct Digital Modulations in 0.13 µm SiGe BiCMOS Technology," *Solid-State Circuits, IEEE Journal of*, vol. 45, pp. 944-954, 2010.

[9] C. Heumpil and E. E. Swartzlander, "Pipelined Carry Lookahead Adder Design in Quantum-dot Cellular Automata," in *Signals, Systems and Computers, 2005. Conference Record of the Thirty-Ninth Asilomar Conference on*, 2005, pp. 1191-1195.

[10] R. P. Brent and H. T. Kung, "A Regular Layout for Parallel Adders," *Computers, IEEE Transactions on*, vol. C-31, pp. 260-264, 1982.

[11] K. Bazargan. (2006, *EE 5324 – VLSI Design II (University of Minnesota ed.)*.

978-1-4673-2395-6/12 $31.00 © 2012 IEEE

Design Optimization Platform for Synthesizable High Speed Digital Filters Using Retiming Technique

Deepa Yagain[1], Dr. Vijaya Krishna A.[2] , Sheetal Chennapnoor[3]

Department of E&C (VLSI Design & Embedded Systems)
People's Education Society Institute of Technology
Bangalore-560 085, Karnataka, INDIA
Email: [1]deepa.yagain@gmail.com, [2]vijayakrishna@gmail.com, [3]sheetal.chennapnoor@gmail.com

Abstract—**In signal processing applications the time critical sections are iterative and recursive and requires various optimization techniques for performance enhancement. Most of these applications require each iteration to be executed under a specific time constraint associated with the data input rate. Using optimization techniques like retiming, we achieve the desired performance. Digital filters are the most common blocks in signal processing applications and they can be represented by synchronous data-flow graphs (DFGs). Applying retiming techniques on the synchronous data flow graphs results in obtaining high speed digital circuits. Retiming is the process of rearranging the storage elements in the circuit to reduce the cycle time without changing its functionality. In this paper, a single optimization environment is developed for retiming the DSP filter blocks using cutset and clock period minimization techniques. Cutset retiming is specially used for filters designed for single processor systems. An optimized digital filter circuit is obtained after retiming from design optimization environment. Also, the HDL(Hardware Description Language) code of the optimized filter circuit is automatically generated and micro-architectural optimizations like usage of parallel prefix tree adders, supply voltage scaling are done at structural level of the circuit which still enhances the filter design performance.**

Keywords-Retiming,High-Level Synthesis[HLS], Cutset retiming technique, Retiming for minimum clock period technique, Floyd-Warshall algorithm, Data Flow Graphs, Parallel prefix Ling adder

I. INTRODUCTION

In recent times, the advancement in Digital signal processing arena has revolutionized the domain of applications like speech compression, telecommunication and data processing. At any given time, the real-time signal processing requirements impose several challenges while implementing the High-Level Synthesis [1] of DSP systems. These circuits should satisfy the enforced sampling rate constraints of the real time DSP applications and must require less area and power consumption. These DSP blocks like filter structures need to be transformed for high speed, low power and low area implementations [2]. Retiming [3][4] has many advantages in synchronous circuit design like reducing the clock period of the circuit, reducing the number of registers in the circuit, reducing the power consumption of the circuit and in overall it is an integral part of logic synthesis and optimization flow. If the filter design is for single processor

systems then cutset retiming method is useful as it allows folding of the circuit after retiming. Clock period minimization method will reduce the clock period and the digital filter block designed from this method is meant for multiprocessor systems. Reducing the computation time of the critical path and generating foldable designs are the main aim of the undertaken work. Here cutset technique and clock period minimization technique are used as retiming algorithms [5].

If any design fails to meet the target specifications at the circuit-level, then we may have to make the changes at the system level. This then necessitates re-writing the RTL (Register Transfer Level) which becomes very tedious if design cycle re-iterates due to various reasons. By automating the generation of optimized digital filter blocks using retiming and in-turn its RTL code, effort involved is reduced and also circuit-level measurements like power, area and throughput for these digital filters from synthesis results can be directly obtained. With the developed optimization environment, designers can conveniently explore the solution space of possible architectures and also analyse the tradeoffs in the energy-area-performance space. This work proposes such an optimization environment designed using MATLAB/Simulink with easy-to-use graphical platform for better user interface.

II. DESIGN AND ANALYSIS

The DSP Filter Blocks which have to be optimized and synthesized has to be described in some way. Use of Data Flow Graphs (DFG) for representing the filter blocks is an easier and efficient approach. A DFG [6] is a directed graph G (V, E) with set of nodes / vertices V and set of edges E. The set of nodes V are subdivided into computational nodes, input and output nodes, and conditional nodes:

1) Computational nodes: where actual computations are performed.
2) Input/output nodes: for the communication with the outer world.
3) Conditional nodes: where data dependent decision are taken.

An edge e having (vi, vj), transports stream of tokens from node vi to vj. The edge indicates a data dependency between node vi and vj, which means that vj need the result of

978-1-4673-2395-6/12 $31.00 © 2012 IEEE

the computation performed on vi before it can start its own computation. In the present work, design entry is done by using DFGs[7] and matrices are in-turn generated from the DFGs. These matrices are further used as inputs in optimization environment. A program is written for computation of critical path and shortest path[6] for the matrices. For example, Consider a Bi-Quad filter[8] block as shown in Figure 1(a).The DFG for this Bi-Quad filter is as shown in Figure 1(b).

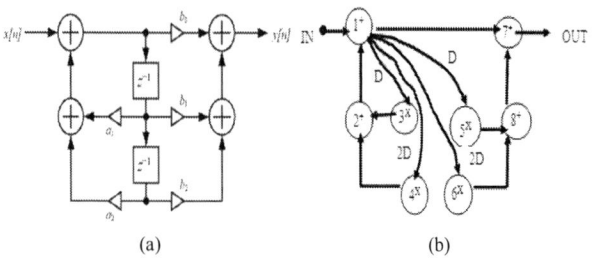

(a)   (b)

Figure 1.   Circuit if Bi-Quad filter

Digital filter must be converted to matrices depending on the connectivity information and node set. Here multiplication operation is assumed to take 2 time units and addition is assumed to take 1 time unit. The obtained matrix information and DFG for considered Bi-Quad filter is as shown in Figure 2 and 3.

```
a =[1 1; 2 1; 3 2; 4 2; 5 2; 6 2; 7 1; 8 1];
f =[inf inf   1   2   1   2   0  inf;
      0 inf inf inf inf inf inf  inf;
    inf   0 inf inf inf inf inf  inf;
    inf   0 inf inf inf inf inf  inf;
    inf inf inf inf inf inf inf    0;
    inf inf inf inf inf inf inf    0;
    inf inf inf inf inf inf inf  inf;
    inf inf inf inf inf inf   0  inf];
```

Figure 2.   Matrices of Bi-Quad filter

A program is written in MATLAB to obtain DFG, critical path and shortest path. The longest path between any two nodes with zero delay in between is called a critical path. Any changes in this path can affect the performance of a sequential digital design. The critical path is an important design parameter for digital filter implementations. The approach used to do this is to calculate the longest delay path between every possible pair of nodes, and keep the absolute longest delay path as the critical path. Critical path is found by taking every node as a starting point and traversing till the end, with weights of every node included. The computation complexity in this path dictates the clock period of the circuit. The generated graph and critical path for the considered Bi-quad filter is given in Figure 3.

The shortest path computation is based on Floyd-Warshall algorithm. This algorithm has a complexity of $O(n^3)$ in time and $O(n^2)$ in space, where n is the number of nodes in the DFG. Consider the vertices of graph G as V = {1, 2,..., n} and a subset {1, 2,..., k} of vertices V for some k.

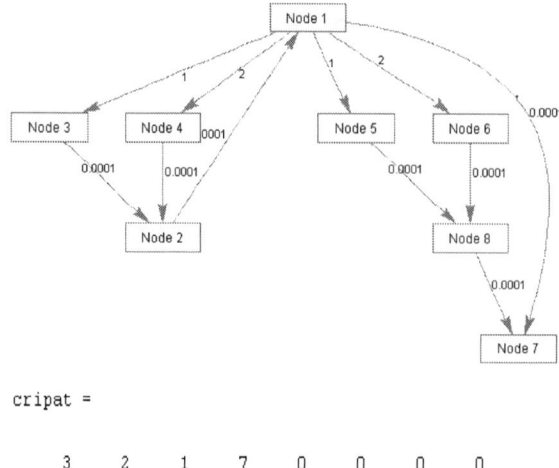

cripat =

   3   2   1   7   0   0   0   0

Figure 3. DFG and critical path generated by design optimization platform for Bi-Quad filter

For any pair of vertices i, j $\in$ V, consider all paths from i to j whose intermediate vertices can be drawn from {1, 2,..., k}. Let p be a minimum-weight path among them. The Floyd-Warshall algorithm exploits a relationship between different paths from i to j with all intermediate vertices in the set {1, 2,..., k}. The relationship depends on whether or not k is an intermediate vertex of path p. Figure 4 shows the matrix for the generated shortest path.

Biograph object with 8 nodes and 11 edges.

q =

1	1	1	2	1	2	0	1
0	1	1	2	1	2	0	1
0	0	1	2	1	2	0	1
0	0	1	2	1	2	0	1
Inf	Inf	Inf	Inf	Inf	Inf	0	0
Inf	Inf	Inf	Inf	Inf	Inf	0	0
Inf	Inf	Inf	Inf	Inf	Inf	Inf	Inf
Inf	Inf	Inf	Inf	Inf	Inf	0	Inf

Figure 4.   Matrix for the generated shortest path

These critical path and shortest path computations are further used for retiming the digital design blocks. This is applicable for any kind of DSP filter block set. Retiming can be performed using cutset technique or clock period minimization technique. User can choose the technique depending on the computational complexity and the design requirement.

II a): Retiming using cutset technique: Set of edges can be removed from the graph to create two disconnected sub graphs. If the two disconnected sub graphs are labeled as G1 and G2 then cutset retiming consists of adding k delays to each edge from G1 to G2 and removing k delays for each edge from G2 to G1.

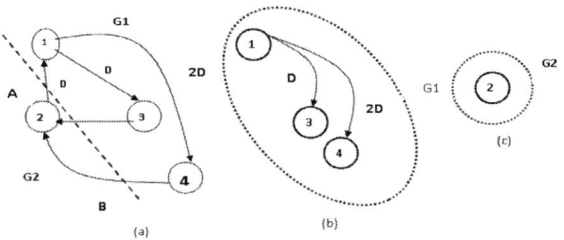

Figure 5.   (a) DFG with cutset, (b) cutset Graph G1, (c) cutset Graph G2

Cutset retiming is a special case of retiming where each node in the sub graph G1 has the retiming value j and each node in the sub graph G2 has the retiming value j+k. The value of 'j' is unimportant due to retiming property. For feasibility of the retimed graph, $w(e) \geq 0$ must hold for all edges e in Gr where w is the weight and r indicated the number of sub graph.

Let e12 → represent an edge from G1 to G2.

   e21→ represent an edge from G2 to G1.

Since cutset retiming adds k delays to each edge from G1 to G2:

$$w(e12) \geq 0 \quad => \quad w(e12)+k \geq 0. \qquad (1)$$

Similarly k delays are subtracted from each edge e21 from G2 to G1:

$$w(e21) \geq 0 \quad => \quad w(e21)-k \geq 0. \qquad (2)$$

Equation (1) and (2) is the condition on k for cutset retiming to give a feasible solution. The result of the cutset retiming in this case is found by adding one delay to each edge incident into the node 2 & subtracting '1' delay from each edge outgoing from node 2. Since we are adding and subtracting one delay element, this is called one node cutset technique. This is a special case of cutset retiming technique. The design optimization environment considers all the nodes and picks the best solution in the available solutions and gives the DFG of retimed circuit. Consider second order IIR filter shown in Figure 6 as an example.

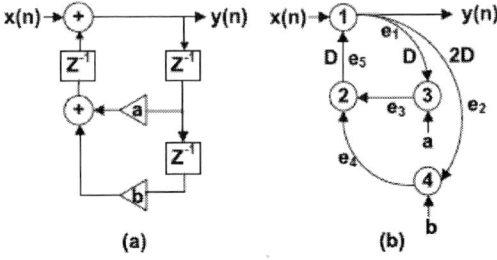

Figure 6.   (a) Circuit of 2nd order IIR filter (b) DFG

One node cutset Retiming is performed for the given graph. Results such as DFG and comparative chart showing the clock period before and after retiming are obtained.

After retiming it is observed that time period of the circuit is reduced from 3 to 2. Since clock period gets reduced, the frequency of the clock is increased there by increasing the operation speed of the circuit. The observed outputs are shown in Figure 7.

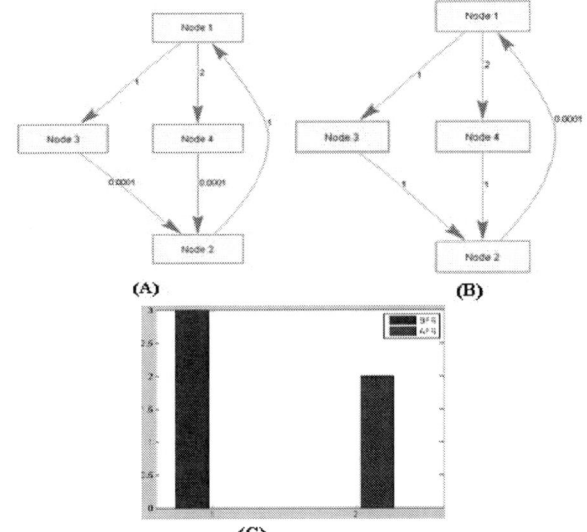

Figure 7.   (A),(B)DFG of IIR 2nd order filter before and after retiming(C)Graph showing reduction in the clock period after retiming

Cutset retiming technique may not always give the optimal solution. But the computational complexity is less. This method is applied whenever we need to fold the circuit for single processor systems. Retiming operation will not change the functionality of the circuit since the iteration bound of the circuit remains unaffected.

II b): Retiming using Clock period minimization technique: The retiming algorithm for clock period minimization is much more efficient than cutset retiming algorithm in terms of clock frequency improvement. But it has higher computational complexity of $O(n^3 \log n)$. The algorithm starts by building a new graph from the original DFG. The new graph can give us a set of inequalities called the critical path constraints. The original DFG also presents a set of equalities called the feasibility constraints. A constraint graph can be built from the critical path constraints and the feasibility constraints. The retiming values for each node can be derived by applying a Floyd-Warshall shortest path algorithm to the constraint graph. The weight for each edge in the retimed DFG can be calculated using the original weight and the retiming values of the two nodes connected by this edge. The description of the algorithm is presented below.

1)Calculate $M = tmax_n$, where n represents the number of nodes in the original DFG G and tmax is the maximum computation time of all the nodes in the DFG.

2) A new DFG G* can be created from G. G* has the same nodes and edges as G. For each edge in G*, the edge weight is w*(e)=Mw(e)-t(u), where w(e) is the edge weight of the same edge in G, t(u) is the computation time of the node initiating this edge.

3) We then apply the Floyd-Warshall algorithm to compute S^*_{UV}, which represents the shortest path from node U to node V.

4) From S^*_{UV}, W_{UV} and D_{UV} are calculated. If $U \neq V$, then

$$W_{UV} = \left\lceil \frac{S^*_{UV}}{M} \right\rceil \quad \text{and} \quad D_{UV} = MW_{UV} - S^*_{UV} + t(V) .$$

If U=V, then $W_{UV}=0$ and $D_{UV} = t(u)$. Here, t(u) and t(V) represent the computation times of node U and node V respectively.

5) We then find the maximum value of D_{UV} and the minimum valued of D_{UV}. We check all the possible clock periods starting from maximum value of D_{UV} to minimum valued of D_{UV} one by one. If we find a clock period, that can give us a feasible solution, we stop and we find the minimal clock period. The solution contains the retiming values for all nodes. We can derive the retimed DFG from the original DFG and the retiming values. This algorithm can efficiently compute the minimal clock period for a DFG. It dramatically reduces the designers' work on finding the minimal clock period. Consider 3rd order IIR filter as an example. Retiming is performed for the inputs which are given in Figure 8 and results such as DFG after the retiming and comparative chart showing clock time units before and after retiming are obtained. After retiming, it is observed that the clock period of the circuit is reduced from 7 to 4. Since clock period gets reduced, the frequency of the clock gets increased there by increasing the operation speed of the circuit. The results show that without any change in the functionality, the optimization platform will explore the hidden concurrency and increase the circuit performance. The two retiming techniques which are used can increase the number of registers after retiming.

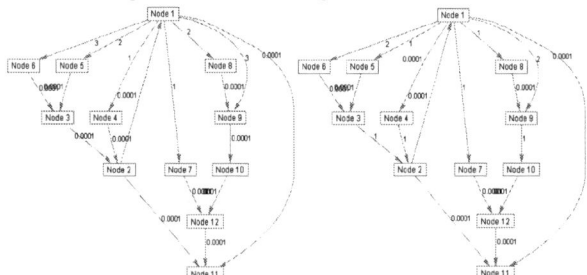

Figure 8. (a)DFG plot before retiming (b)DFG plot after retiming of 3rd order IIR Filter

III b) Parallel prefix Adders: The basic components of any digital filters are adders, multipliers and delay elements. The main constraints of all adders are their speed, performance, power consumption and die area. Parallel-prefix adders [9] offer a highly efficient solution to the binary addition problem. Here, the design of high speed, parallel-prefix adders such as Brent-Kung, Sklansky, Kogge- Stone and Ling adders by

Kogge-Stone implementation are done using CMOS logic and transmission gate logic. It is found Ling adder using Transmission gate for Kogge-Stone parallel prefix adders performs very much better[12]. Hence instead of normal ripple adders, Ling adders by Kogge-Stone implementation are used in retimed digital filter circuits which increase the speed.

III. SIMULATIONS AND RESULTS

IIR cascaded lattice filter is used as an application example for generating simulation results. Consider the schematic of Cascaded Lattice Filter shown in Figure 9. Here Digital filter model is converted in to matrices and retiming is done using one node cutset technique or clock period minimization technique.

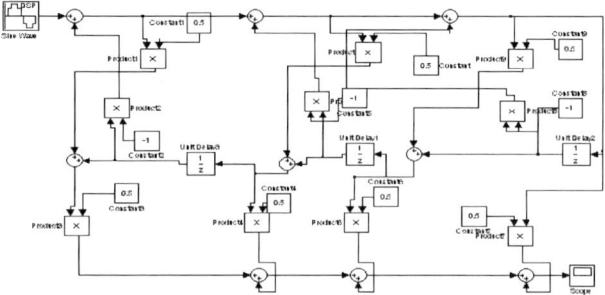

Figure 9. Circuit of IIR Cascaded Lattice filter before retiming

These retiming methods increase the register count slightly after retiming. The clock period gets minimized from 12 units to 11 units after cutset retiming technique and to 7 units after clock period minimization technique. Here clock period minimization technique is more efficient but it is more computationally complex. The graphs for the same are shown in Figure 10.

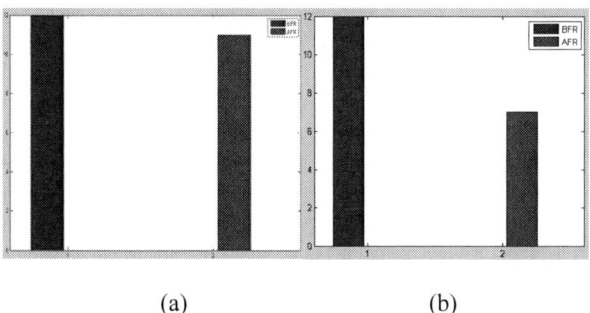

(a)   (b)

Figure 10. Bar Graph showing clock period minimization Before and after one node cutset and Clock period minimization technique retiming.

A script is written as a part of the optimization environment which generates synthesisable HDL code of the retiming circuit automatically. Consider an IIR lattice filter structure after retiming. The RTL generated using this design optimization environment is as shown in Figure 11. It is seen that the post synthesis results of lattice filter matches with the matlab results with 2% error. In the synthesized filter, we make use of Kogge-Stone parallel prefix Ling adder for adders

978-1-4673-2395-6/12 $31.00 © 2012 IEEE

since it is found to be very efficient in terms of power consumption, area and speed[12].

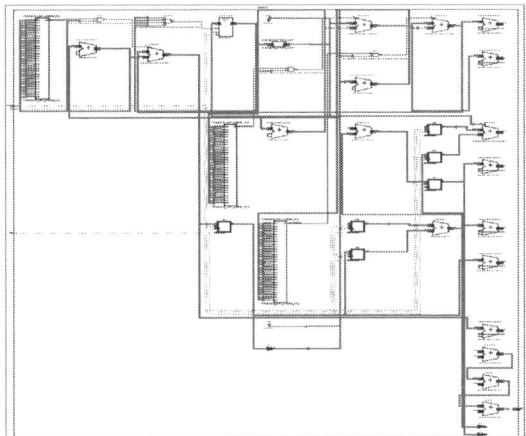

Figure 11. Schematic of the retimed IIR Cascaded Lattice filter after synthesis of generated HDL.

The synthesis of lattice filter is performed on Sparten3E XC3S100E device. The number of slices used by filter after cutset retiming technique is 190 and the number of slices after clock period minimisation technique is 220.

Different filter structures are designed and synthesised using the design optimization platform. They are compared w.r.t. the clock period (before and after retiming) and register count (before and after retiming). The results are shown in Figure 12 and 13. It is observed that after retiming the clock period gets reduced. The register count gets altered depending on the filter's iteration bound.

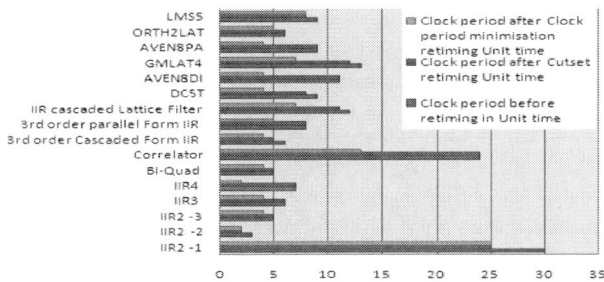

Figure 12. clock period before and after retiming for various digital filter blocks.

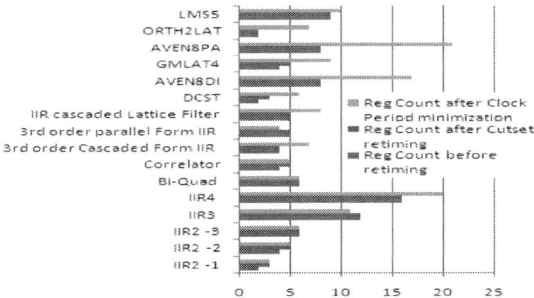

Figure 13. Register count before and after retiming for various digital filter blocks.

Along with all these, if 45 nm design library is used instead of 90nm and 180nm for the filter structures, the voltage gets minimised due to which saving in power can be obtained.

IV. CONCLUSIONS

In this paper, an optimization environment is developed using retiming technique for digital filters. Foldable circuits can be generated using the cutset method and circuit speed can be enhanced using clock period minimization technique. Along with optimization using retiming, it can also generate the synthesizable HDL of the retimed circuit efficiently and automatically which can be further analysed W.R.T. power, area and delay. Due to this automation, more of designer's time and effort will be saved. Even if reiteration of the design cycle due to specification change happens, time taken to reiterate is very less due to this optimization platform. Further, due to parallel prefix adders in the filter circuits, it is seen that performance improves. Future implementation of this work can be designing of retiming algorithm based on evolutionary computation where number of registers also gets minimised along with the clock period. Also, using register minimization retiming technique, the number of registers can be minimised.

REFERENCES

[1]. Daniel D. Gajski, Lognath Ramachandran "IEEE Design & Test," volume 11, Issue 4 (Oct 1994), Publishers: IEEE computer society press, Los Alamitos, CA,USA ,ISSN: 0740-7475,pp-44-54.

[2]. A. Chandrakasan, S. Sheng, and R. Brodersen, "Low-power CMOS digital design," IEEE J. Solid-State Circuits, vol. 27, pp. 473–484, Apr. 1992.

[3]. Monteiro, S. Devadas, and A. Ghosh, "Retiming sequential circuits for low power," in Proc. IEEE Int. Conf. Computer Aided Design, 1993,pp. 398–402.

[4]. T. C. Denk and K. K. Parhi, "Exhaustive scheduling and retiming of digital signal processing systems," IEEE Trans. Circuits Syst. II, vol. 45, pp. 821–838, July 1998.

[5]. Simon .S & Hofner .J "Retiming algorithms for Multiplexer circuits ," Technical Report TUM- LNS-TR-94-8, Institute for Network Theory & circuit Design , Technical University Munich,1994.

[6]. L. E. Lucke and K. K. Parhi, "Data-flow transformations for critical path time reduction in DSP synthesis," IEEE Trans. Computer-Aided Design, vol. 12, pp. 1063–1068, July 1993.

[7]. Lucke, L.E.; Brown, A.P.; Parhi, K.K".Circuits and Systems," IEEE International Sympoisum on Vol.4 , Issue , 11-14 Jun 1991 ,pp-2351 - 2354 .

[8]. J.Tow and Y.L.Kuo, "Coupled Biquad Active Filters" in proc. IEEE Int.Symp. Circuit Theory, IEEE Catalog No.72CH0594-2CT, pp.164-8

[9]. High-Speed Parallel-Prefix VLSI Ling Adders Giorgos Dimitrakopoulos and Dimitris Nikolos, IEEE transactions on computers, Vol. 54, No. 2, February 2005

[10]. N. L. Passos, E. H.-M. Sha, and S. C. Bass, "Optimizing DSP flow graphs via schedule-based multidimensional retiming," IEEE Trans. Signal Processing, vol. 44, pp. 150–155, Jan. 1996.

[11]. N. L. Passos and E. H.-M. Sha, "Synchronous circuit optimization via multidimensional retiming," IEEE Trans. Circuits Syst. II, vol. 43, pp.507–519, July 1996.

[12]. Akansha Baliga, Deepa Yagain, "Design of High Speed Adders Using CMOS and Transmission Gates in Submicron Technology: A Comparative Study," icetet, pp.284-289, 2011 Fourth International Conference on Emerging Trends in Engineering & Technology, 2011

The Design of DC Motor Driver for Solar Tracking Applications

Zi-Yi, Lam, *Member, IEEE*, Sew-Kin, Wong, *Member, IEEE*, Wai-Leong, Pang, Chee-Pun, Ooi

Faculty of Engineering (FOE)
Multimedia University, Persiaran Multimedia
63100 Cyberjaya, Selangor Darul Ehsan, Malaysia
Email: zylam@mmu.edu.my (corresponding author), skwong@mmu.edu.my, wlpang@mmu.edu.my, cpooi@mmu.edu.my

Abstract - **Solar trackers rely on a direct-current (DC) motor driver circuit to control the movement of the solar panel. However, conventional DC motor drivers used in solar tracking system do not provide any options for speed and torque control. Hence, the fixed speed of the DC motor leads to either too fast or too slow tracking movement. Usually, the output torque is set to the maximum. If the load (solar panels' weight) is small, the maximum torque is not fully utilized and therefore energy is wasted. So, a fully adjustable DC motor driver for solar tracking system shows great potential in commercialization because highly energy efficient DC motor driver circuit can improves the total efficiency of the solar tracker. Fewer solar panels are needed when they are attached to high efficiency solar tracking system, which will be reflected to lower system cost. Therefore, this enhanced motor driver will bring significant impact to the solar energy industry. The proposed DC motor driver is fully adjustable in term of speed and torque. The speed and torque of the motor is directly proportional to the output voltage and output current of the proposed DC motor driver, respectively. Adaptive controlling of the output voltage and current are possible by installing algorithm in the microcontroller of the DC motor driver and it can be reprogrammed according to the requirement. The speed control makes the solar tracking system to track the Sun more accurately and the torque control saves energy.**

I. INTRODUCTION

Solar tracking system is used to track the position of sun in order to get the maximum energy by aligned the solar panel perpendicular to the incidence sunlight [1, 2, 3, 12, 13]. Solar panels need to move along with the direction of sunlight during the tracking process. A DC motor driver is needed to control the directions as well as the turning speed of the solar panel. The maximum torque of the DC motor decides the total amount of solar panels can be mounted on the tracking system. The common limitation of the DC motor driver is the speed is fixed and the default torque is set to maximum. Logically, it is not preferable to set the motor in high turning speed since Sun tracking is a slow process. While setting the motor to operate at maximum torque causes the system to consume the highest amount of power constantly, regardless of the amount of load attached. It will bring significant impact on the efficiency of the system, since the energy consumed by the system is generated by the solar panel itself. Research has been done that DC-DC buck converter is used to step down the voltage in order to reduce the turning speed of solar panel [4].

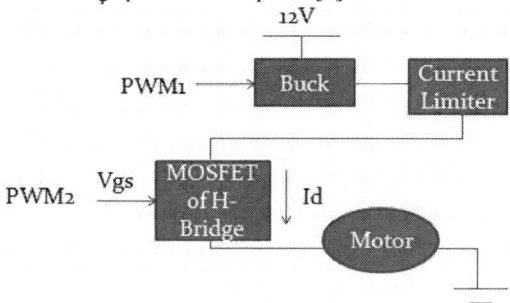

Fig.1. Block diagram of motor driver

However, the system reported has single speed only without torque control. System proposed by V. Gupta managed to control the torque by changing the duty cycle of Pulse Width Modulation (PWM) signal that feed into the low-side switching elements of the H-bridge circuit [5]. The proposed DC motor driver shows that it is possible to make both speed and torque adjustable. It brings great advantage to the solar tracking system as the tracking speed can be optimized according to the tracking condition. Moreover, power consumption of the tracking system can be reduced greatly since the DC motor driver supplies the amount of torque just sufficient to move solar panels attached. An addition current limiting circuit has been added in the motor driver in order to limit the maximum amount of current that can flow through the DC motor to avoid motor stall. Fig. 1 shows the block diagram of the proposed DC motor driver.

II. CIRCUIT DESIGN AND DESCRIPTION

A. Characteristic of DC motor

The characteristic of a DC motor is govern by the equation (1):

$$E_A = V_A - I_A R_A \qquad (1)$$

Where V_A is the applied voltage across the DC motor, I_A is the armature current, R_A is the armature resistance and E_A is the internal generated voltage. In normal operating condition, the armature current will increase with the applied voltage and the internal generated voltage remains constant. Turning speed of motor is directly proportional with internal generated voltage, showing in the equation (2) [6, 14, 15]:

$$E_A = K\emptyset w \qquad (2)$$

Where $\emptyset$ is magnetic flux, w is the turning speed of DC motor and K is a constant.

The proposed DC motor driver limits the amount of current flowing into the motor. In other words, it can maintains the armature current at a constant value although applied voltage increased. When the armature current remains constant, the internal generated voltage and therefore the turning speed will increase propotional to the voltage applied. Torque is directly proportional with the armature current, so the ability in controlling the armature current made torque control possible, as shown in the equation (3) [7]:

$$\tau_{induced} = K\emptyset I_A \qquad (3)$$

Where $\tau_{induced}$ is the induced torque, $\emptyset$ is magnetic flux, I_A is armature current and K is a constant. Whereas changing the flow direction of current across the DC motor will change the turning direction of the DC motor [8].

B. Speed Control of DC Motor

The voltage across the DC motor must be step down by using the DC-DC buck converter in order to reduce the turning speed of the solar panel to get accurate sun tracking. Fig. 2 shows the circuit of DC-DC buck converter. M1 is P12PF06 P-channel MOSFET as the switching element and M2 is IRF630 N-channel MOSFET as a driver to drive the P-channel MOSFET. Since PIC microcontroller cannot produce a negative voltage to turn on the P-channel MOSFET directly. Therefore, the N-channel MOSFET is use to turn on the P-channel MOSFET. The relationship between the input voltage and the output voltage of the DC-DC buck converter are given by [9]:

$$V_o = DV_{in} \qquad (4)$$

Where V_o is the output voltage, V_{in} is input voltage and D is the duty cycle of PWM signal. The output voltage can be step down by vary the duty cycle of the PWM signal that feed into M2.

C. Torque and Direction Control of DC Motor

Fig. 3 shows the H-bridge circuit. M1 and M2 are P12PF06 P-channel MOSFET as the high-side switching elements. M3 and M4 are IRF630 N-channel MOSFET as the low-side switching elements. Another two IRF630 N-channel MOSFET, M5 and M6 are needed as the driver to turn on the M1 and M2. The torque of the motor can be controlled by control the current flow through the motor as the torque is directly proportional to the current across it. This can be adjust by vary the duty cycle of PWM signal which feed into M5 or M6. Base on [10], increase the drain to source voltage (V_{ds}), will increase the drain current (I_d) of the MOSFET as well. Changing the duty cycle of PWM signal will change the average output voltage which is known as V_{gs}, when there are changes in V_{gs}, I_d which is the current that flow through the motor will change as well. This is how the torque of the motor being controlled.

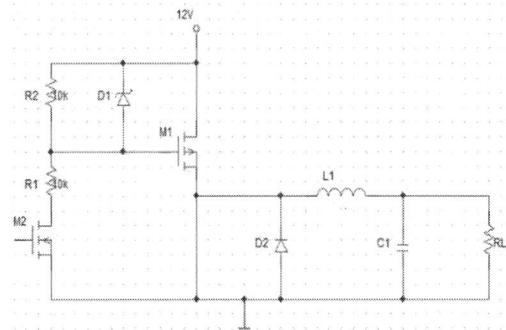

Fig. 2. DC-DC buck converter circuit

D. Current Limiting Circuit

Current limiter circuit is used to limit the current flowing through. This circuit is added right after the power supply to the H-bridge circuit to prevent the motor being stalled and protect the MOSFET. As shown in the data sheet of the DC motor, the stalled current is 3A. So the design of this current limiting circuit is to limit the current that above 3A as shown in Fig. 4. Basically, this circuit consists of two op-amps. U1 as differential amplifier which is use to amplified the voltage drop across the sense resistor (Rs) since the Rs is very small and the voltage across it is also very small. U2 used as a comparator to compare the input voltage with the reference voltage. If the current flow across is too high, which means that the output voltage of U1 is higher than the reference voltage, there will be output coming out from U2 and turn on the transistor Q1. As Q1 being turn on, the PWM signal will be grounded instead of feed into the MOSFET of H-bridge. So there will be no current flowing through the MOSFET as well as the motor. It prevents the motor being stalled, protect the MOSFET and saves the current consumption.

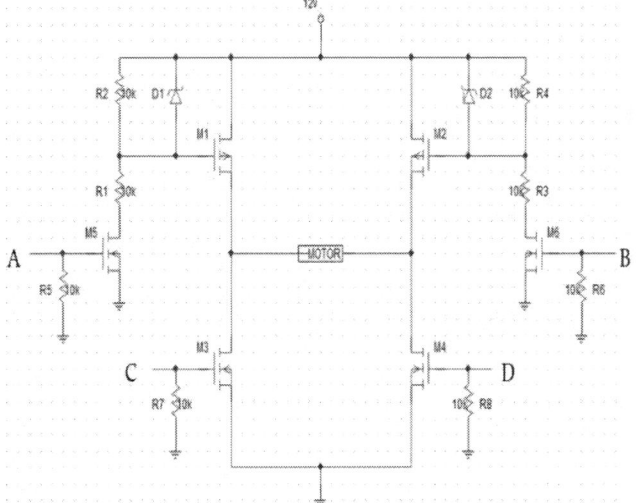

Fig. 3. H-bridge circuit using power MOSFET

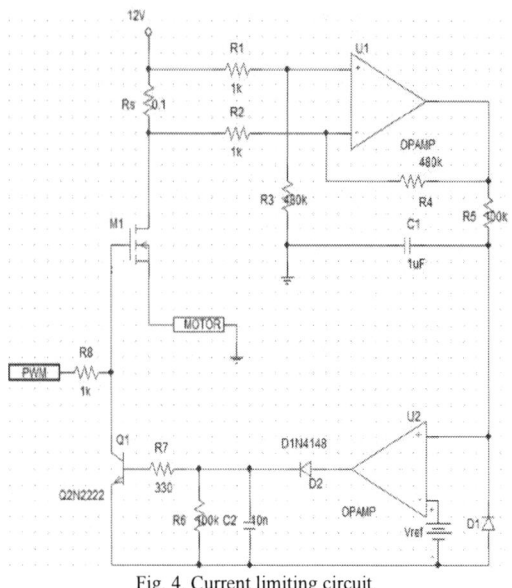

Fig. 4. Current limiting circuit.

III. RESULTS AND DISCUSSION

The revolution per minutes (rpm) of the DC motor is measured by using the digital stroboscope tachometer every time there are changes in output voltage. Fig. 5 shows the speed of DC motor changed with the output voltage of the driver. The results show that the driver manged to control the motor speed in a linear manner. Fig 6 shows the changes of the driver output power corresponding to various speed of the motor. As the speed of motor increase, the output power will remain closed to 5W. Which means that the driver output power has the ability to remain constant throughout the speed range. This characteristic enables the motor to turn slowly without lossing power, which is very important in promoting higher efficiency of solar tracking.

Current flow through the motor is measured under two conditions, i.e. with load and without load. The shaft of the DC motor is attached with weight of 130 grams as a load. Fig. 7 shows the relationship between the driver output current and duty cycle when there is a load attached. The results showed that the driver managed to control the output current linearly, and therefore the torque by using PWM control signal. This enables the solar tracking system to work properly later on when there is a need to mount more solar panels on it. When there is no load mounted on the motor, the driver only consume current maximum at 0.05A throughout the duty cycle range as shown in Fig. 8. Most of the time, solar tracking system is installed in a remote rural area where power grid is nowhere to be found. This will be a very

important power saving feature that helps in conserving the energy stored in the battery bank and allows the system to work in longer hours per full-charge cycle. To summuries, the performance differences between conventional and the proposed motor driver for solar tracking applications is listed in Table 1.

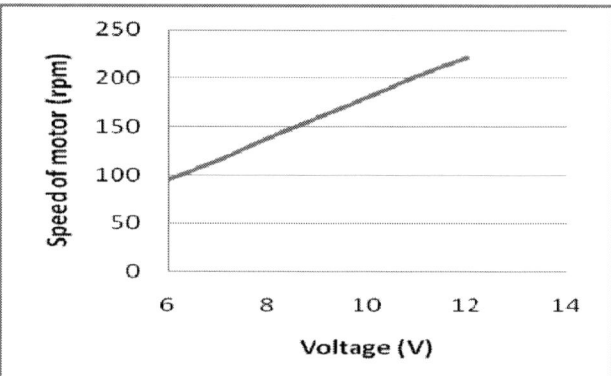

Fig. 5. Speed of motor when output voltage varies

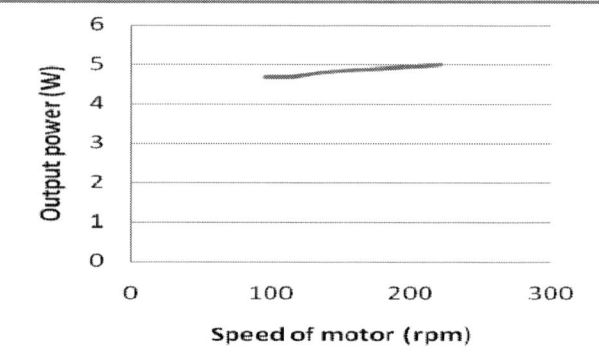

Fig. 6. Output power when speed of motor varies

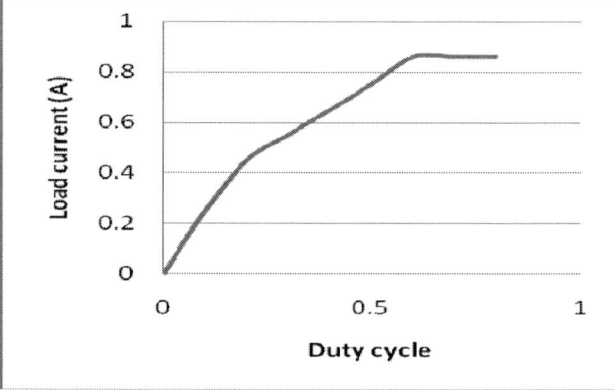

Fig. 7. Ouput current changes with the duty cycle when load is attached

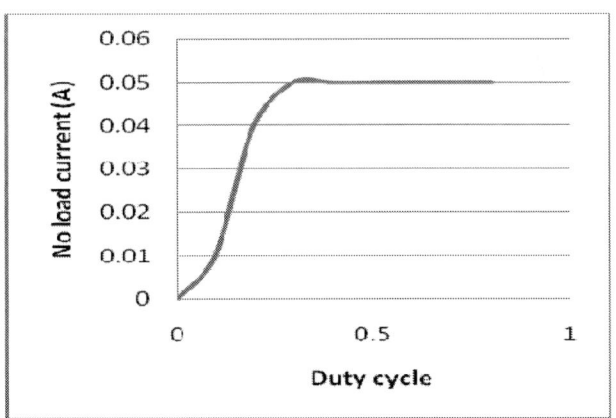

Fig. 8. Ouput current changes with the duty cycle when no load is attached

Table 1. Performance comparison between conventional and the proposed motor driver for solar tracking applications [16].

Conventional motor driver	Adaptive motor driver
Single speed control	Variable speed control
Either too fast or too slow (tracking speed)	Tracking speed can be programed to the needs (PID control is possible)
Full-torque (no matter how much load)	Adjustable torque according to the load attached
Operating at full torque, consumes maximum amount of energy	Giving only adequate amount of torque depending on the load attached, saves energy

IV. CONCLUSION

With the combination of the carefully designed torque, speed and direction control circuit, it can form a better motor driver for solar tracking system. This motor driver circuit is builded based on the idea to control the cost as low as possible. Therefore, the list of components used is simple and easy to be found in the market. Although the cost is low, the resultant driver delivers excellent capability of controling the speed and torque in a linear manner. The driver managed to control the current and the voltage of the output at the same time and therefore it can controls the resultant power transfer to the motor. The motor is only consumes maximum of 0.05A when it is idle. This stable and easy control definitely will shorten the development period for engineers to design the solar tracking system.

REFERENCES

[1] Nader Barsoum and Pandian Vasant, "Simplified Solar Tracking Prototype," Transaction in Controllers and Energy, (2010).

[2] Kh.S. Karimov, et al., "A simple photo-voltaic tracking system," Solar Energy Materials & Solar Cells, pp. 49-59 (2004).

[3] Shubhajit RoyChowdhury and HiranmaySaha, "Maximum power point tracking of partially shaded solar photovoltaic arrays," Solar Energy Materials & Solar Cells, pp. 1441-1447 (2010).

[4] Yie-Tone Chen and Cing-Hong Chen, "A DC-DC Buck Converter Chip with Integrated PWM/PFM Hybrid-Mode Control Circuit," Department of Electrical Engineering, pp. 181 – 186 (2009).

[5] Vibhor Gupta, "Working and Analysis of the H – Bridge Motor Driver Circuit Designed for," University Institute of Engineering and Technology, pp. 441-444 (2010).

[6] Aung Zaw Latt and Ni Ni Win, "Variable Speed Drive of Single Phase Induction Motor Using Frequency Control Method," Education Technology and Computer, pp. 30-34 (2009).

[7] Ayetül Gelen and Saffet Ayasun, "Effects of PWM Chopper Drive on the Torque-Speed Characteristic of DC Motor," Universities Power Engineering Conference, pp. 1-4 (2008).

[8] Ibrahim Sefa, et al., "Application of one-axis sun tracking system," Energy Conversion and Management, pp. 2709-2718 (2009).

[9] Shulin Liu, Jian Liu,Yinling Yang and Jiuming Zhong, "Design of intrinsically safe buck DC/DC converters," Electrical Machines and Systems, pp. 1327 - 1331 (2005).

[10] STMicroelectronics datasheet.

[11] Cytron technologies datasheet.

[12] Ahmed, M.M. and Sulaiman, M. "Design and proper sizing of solar energy schemes for electricity production in Malaysia" Power Engineering Conference, pp. 268-271, (2005).

[13] J. Rizk and Y. Chaiko. "Solar Tracking System: More Efficient Use of Solar Panels," Proceedings of World Academy of Science, Engineering and Technology, pp. 2-3, (2008).

[14] Wai Phyo Aung, "Analysis on Modeling and Simulink of DC Motor and its Driving System Used for Wheeled Mobile Robot", World Academy of Science, Engineering and Technology, pp. 32 (2007).

[15] R. Sahu and G. A. Rincon-Mora, "A High-Efficiency, Dual-Mode, Dynamic, Buck-Boost Power Supply IC for Portable Applications" , 2005 18th International Conference on VLSI Design, pp.858-861, 3-7 January, (2005).

[16] R. Valentine, "Motor Control Electronics Handbook", McGraw-Hill Handbooks, United States, 1998, pp. 59-83.

Planar Dipole Antenna Design At 1800MHz Band Using Different Feeding Methods For GSM Application

Waleed Ahmed AL Garidi, Norsuzlin Bt Mohad Sahar, Rozita Teymourzadeh, CEng. *Member IEEE/IET*
Faculty of Engineering, Technology & Built Environment
UCSI University, 56000, Malaysia
Email: rozita@ucsi.edu.my

Abstract- **This research work focuses on the design and simulation of planar dipole antenna for 1800MHZ Band for Global System Mobile GSM application using Computer Software Technology CST studio software. The antenna is structured on fire resistance FR4 substrate with relative constant of 4.3 S/m. Two types of feeding configuration are designed to feed the antenna in order to match 50 Ω transmission lines which are via-hole integrated balun and quarter wavelength open stub. The via-hole is capable to provide maximum return loss of -25dB, bandwidth of 18.4% and the voltage standing wave ratio (VSWR) of 1.116 V at optimum dimension of length 59mm and width 4mm; the bandwidth is improved 25% to 30% by extending the width of the antenna 8 mm to 10 mm followed by deterioration of return loss value to -15dB. While the open stub at length of 67 mm, width of 6 mm and height 1.6mm will provide max return loss of -47.88dB and bandwidth of 17% with VSWR 1.008 << 2. Way that the antenna substrate has influenced the performance of the antenna. The lower relative constant will result the higher return lows, narrower bandwidth and better radiation pattern in trade-off the resonant length Via-hole and then the quarter wave open stub are most convenient for practical implementation.**

Keywords-- Antenna, GSM, Planner, Dipole, VSWR, MMIC, Microstrip

I. INTRODUCTION

Antenna is playing significant role in the wireless communication systems. Radio frequency and Microwave has direct effect on antenna for trance-receiving the signals and antenna will transmit it through the electro-magnetite wave in free space [1]. In order to design antenna in the bandwidth of 900MHz – 1800 MHz in the industry of cellular network such as GSM network, the IEEE Standard [1] (Std 145-1983) is approached accordingly.

The trends of the mobile phone technology has been dramatically decreased the weight and size of the communication equipment. Microstrip dipole antennas consist of a thin sheet of low loss insulating called the dielectric substrate and it is completely covered with a metal on one side called the ground plane.

From the other side is partly metalized where the circuit of antenna is printed [2]. In particular, planar dipole-type exhibits many attractive features, such as a simple structure, inexpensive easy integration with monolithic microwave integrated circuits (MMIC), low-profile, comfortable to planar and non-planar surfaces. Therefore, it works best on air and portable application. Micro-strip dipole moment is attractive because they basically own large feature like simple analysis and manufacturing and its attractive radiation pattern particularly low between-polarized rays.

II. ANTENNA GEOMETRY DESIGN

The desired frequency that the dipole antenna is designed to operate on is 1800MHz GSM band. The substrate material has been used is FR4 with dielectric constant of 4.3 thickness of 1.6mm.

The length of the antenna is the parameter that controls the resonant frequency. As result, the length is treated as half wave dipole, which is given by the following formula:

$$L = \frac{\lambda}{2} \qquad (1)$$

$$\lambda = \frac{C}{F} \qquad (2)$$

Where L is dipole length, λ is the wavelength, C is the speed of light in free space and F is the frequency of operation [3].

Table 1 shows the range of the dimensions parameters and table 2 presents the calculated values.

TABLE I
DESIGN RECOMMENDATION AND RESTRICTION [2]

Parameters Symbol	Recommendation	Restriction
Width (W)	$0.05\lambda \le W \le 0.1\lambda$	$0.05 \le \frac{w}{h} \le 20$
Length (L)	$L \ge 0.48\lambda$	$T/W \le 0.5$
Height (H)	$h \le 0.02\lambda$	$T/h \le 0.5$
Thickness (T)	$T \triangleleft \lambda$	$\varepsilon r \ge 1$

TABLE II
CALCAULATED VALUES OF ANTENNA PARAMETERS

Material Type	ε_r	ε_e	$\lambda_{d\,mm}$	L-mm	w-mm	h-mm
FR4	4.3	2.65	104	52	6.25	2.08
Arlon AD300	3	2	118	59	7.08	2.36
Rogers RT5880	2.2	1.6	132	65.88	7.92	2.64

However, the speed of the signal when it propagates in free space varies as it propagates in medium and that is because of the effective dielectric constant of the microstrip substrate associated with fringing fields [4]. Therefore, the above equations have been modified to the following:

$$\lambda = \frac{c}{f\sqrt{\varepsilon}} \tag{3}$$

$$L = \frac{3\lambda}{4} \tag{4}$$

$$L = \frac{2\lambda}{3} \tag{5}$$

In most practical design, the wavelength of the printed dipole is treated as approximately in medium with equivalent to the average of that in free space, which is considered as alternative method for better determination of the length. Besides, the antenna parameters differ widely for different types of materials [5]. The feeding method is designed using integrated microstrip balun.

Planar antennas are integrated with other microwave circuitry. Therefore, the feeding techniques need to be designed and analyzed properly to reach between a good antenna performance and efficient circuit design. The word balun is a narrowing for "balanced to unbalanced" [5]. Table 3 has the practical parameters of the antenna being designed. At 1800 MHz band the antenna wavelength is 89.9 mm given by (3), the corresponding resonant length is 67mm for dipole with open stub and 60mm for dipole with via-hole by (4) and (5). Fig. 1 shows the open stub feeding while Fig. 2 illustrates the dipole antenna with via-hole feed.

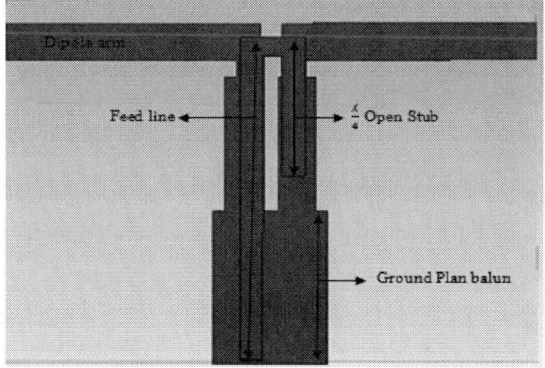

Fig. 1. Dipole with quarter wavelength open stub feed

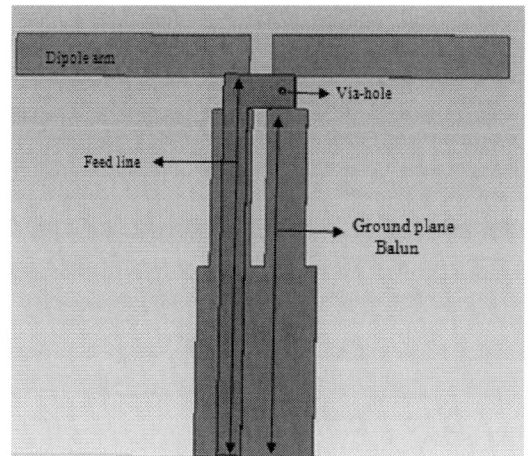

Fig. 2. Dipole with Via-hole feed

TABLE III
DIMENSIONS OF QUARTER WAVLENGTH AND VIA-HOLE

Parameter	Specifications
PCB Substrate	FR-4, h=1.6mm, ε_r =4.3, tanδ= 0.002.
Dipole arm	L= 67mm, W=6mm, gap g = 3mm
Microstrip Balun	L= 22mm, L=3mm, w=5mm, w= 3mm
Ground Plane	L = 25mm, w= 15mm.
Feed Line	L=25mm, w=3mm, Open stub=25mm
Dipole arm	Length L= 60mm, Width W=6mm, gap g = 3mm.
Via-hole	R= 0.375mm

III. PRINCIPLES OF BALUN OPERATION AND DESIGN

The main objective of balun design is to provide the system with balance transition between the antenna and its feeding circuit. However, as the rest of the system was then overturned as one part of antenna attached to external plate while the other connects to the internal connector. On the edge attached to the shield, the current can pass through over the outside of the coaxial cable. Baluns changes the tent onto the antenna ports the same magnitude on each but across stage, these tensions caused the same amount of current flow to the outside of the coaxial cable. This type of arrangement will let the feed point to have the same phase as the point of the top shield [6].

Fig. 1 shows quarter wavelength open stub feeds the dipole strips by creating virtual short circuit across the center of the dipole. This arrangement provides further possibilities for reactance compensation of the balance load [7]. The special feature of this method is that it does not require physical direct connection top of substrate. Therefore, it is commonly used to feed microstrip dipole antenna because it requires no soldered or plated through connection [8].

IV. SIMULATION RESULT

A. Quartered wavelength open stub method

The result shows the s-parameter as function of frequency. The plot of S11 shown in Fig. 3 is used to determine whether the antenna is a single band and operating at the desired resonant frequency also the bandwidth can be calculated for the corresponding frequency band. It provides bandwidth of 17% of the resonance frequency with maximum return loss of -47.88 dB as shown in Fig. 3. Fig. 4 illustrates how the resonant frequency is shifted down as the length reduces while Table 4 is a summary of the resulting measurements of different lengths. The balance feeding is illustrated by the value of VSWR in Fig. 5 while Fig. 6 shows the radiation pattern of the antenna.

TABLE IV
COMPARISON FOR DIFFERENT LENGTHS AT W=6 mm

Length	Z11	VSWR	RL	BW	Directivity
63mm	45-i 4	1.1013	-26dB	16%	1.7dB
65mm	48-i 2.2	1.05	-32dB	16.5%	1.8dB
67mm	51-i0.216	1.0081	-47.dB	17%	2dB

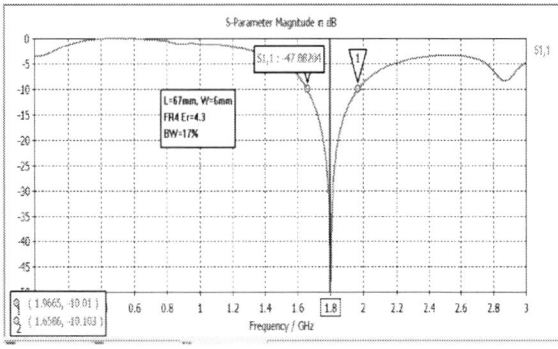

Fig. 3. Return Loss Plot at 1800MHz of antenna

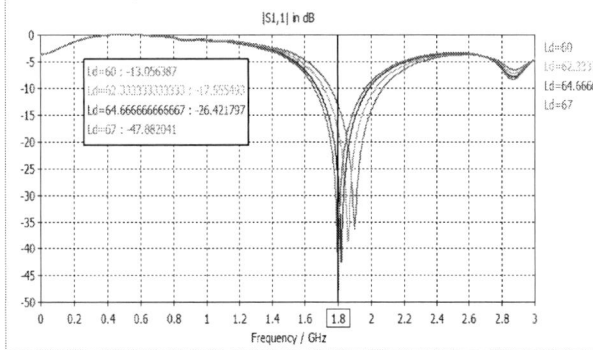

Fig. 4 Return Loss for different lengths of the dipole

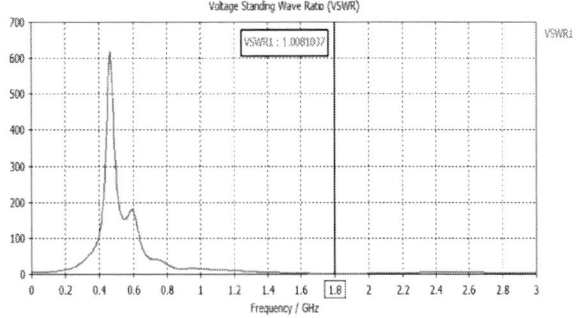

Fig. 5. VSWR of feeding circuit

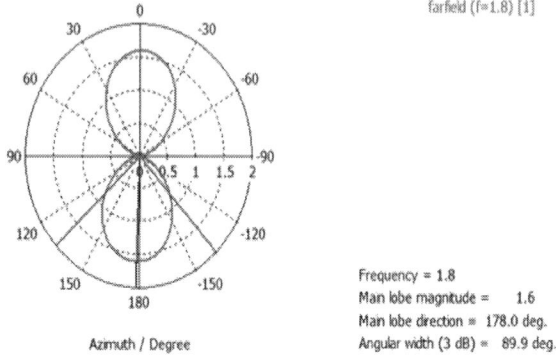

Fig. 6. Directivity in the Azimuth plane

B. Via-hole feeding method

The dipole radiator is modeled on the top of substrate and adopted with the integrated balun to form a ground plane as a return bath for the current distributions. Since this feeding is through via from the top to the bottom physically connecting the feed line to the center of the dipole to make the formation of center fed the simulations results differs for different widths of the antenna and Table 5 summarize the differences as well as Fig. 8. Besides, the antenna is resonating at 1.8GHz as shown in Fig. 7 with sufficient balance between antenna and its feeding circuit which is determined by the value of VSWR in Fig. 9. From Fig. 10, it can be seen that the antenna has omnidirectional radiation pattern.

TABLE V
COMPARED RESULTS FOR DIFFERENT WIDTHS

Width	Impedance	VSWR	Return Loss	Bandwidth %
8mm	43.614	1.255	-18.95dB	24.5%
7mm	44.43	1.216	-20.22 dB	22.5%
6mm	45.3	1.18	-21.6 dB	20.6%
5mm	46.17	1.17	-22.74	19.7%

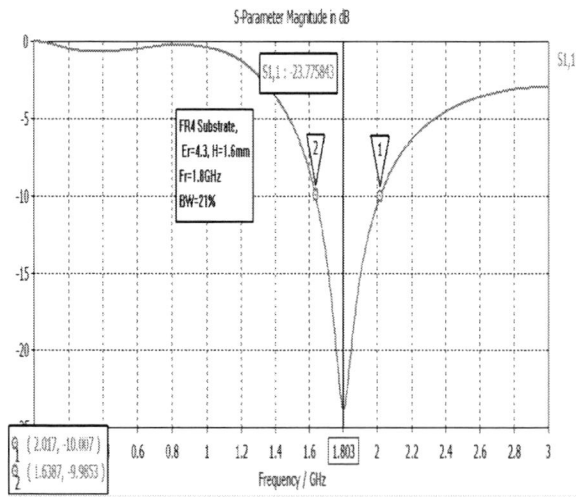

Fig. 7. Return Loss 1800MHz of antenna

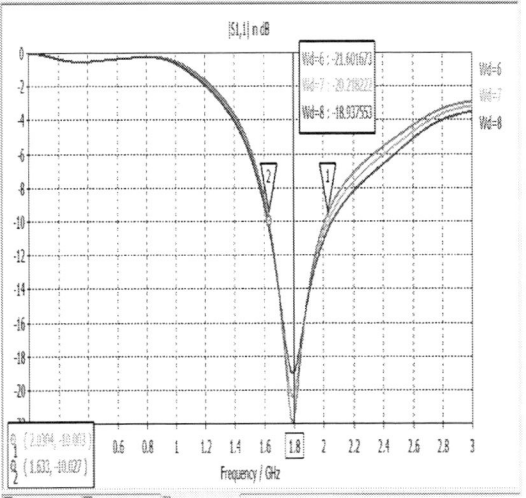

Fig. 8 . Return Loss for different widths of the antenna

Since the polarization of the antenna is horizontal, the E-plane coincides with the azimuth plane and the H-field coincides with elevation plane as shown in Fig. 11 and Fig. 12. The new features in this design in comparison to previous research works is that it has planar structure which can easily integrated to array form to provide greater directivity and high efficiency without causing substantial power losses in the feeding circuit since each element is match to its feeding circuit impedance. In term of bandwidth, this antenna has bandwidth range from 18.4% to 25% while previous search has bandwidth from 11.53% to 13.22%[3]. This advantage is a result of choosing the right configuration of the feeding circuit.

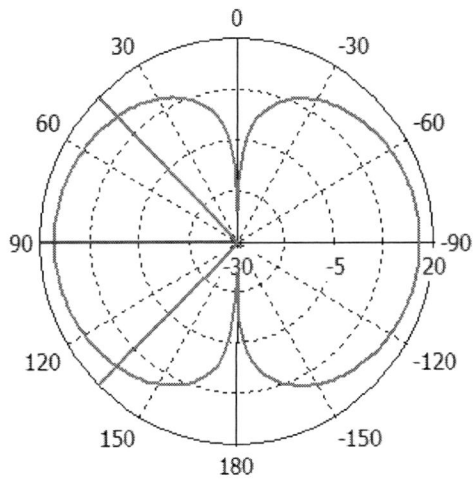

Fig. 11. E-field Radiation at 1.8GHz

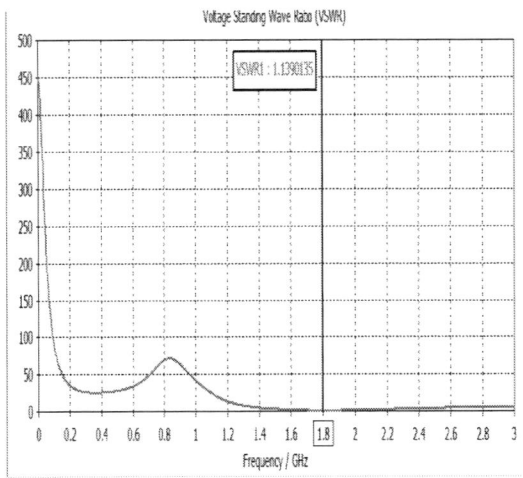

Fig. 9. VSWR of the feed circuit

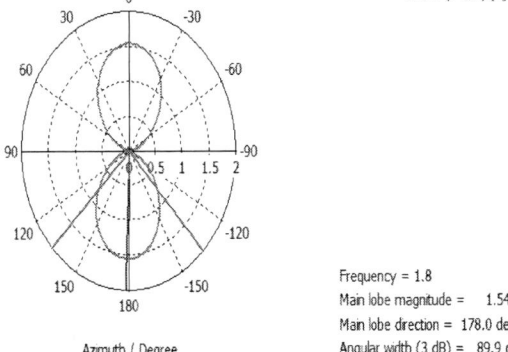

Fig. 10. Directivity of the dipole antenna

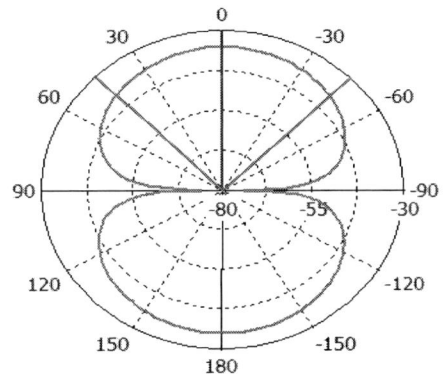

Fig. 12 . H-field Radiation at 1.8GHz

978-1-4673-2395-6/12 $31.00 © 2012 IEEE

V. DISCUSSION

Throughout the entire simulation results it has been seen how the antenna performance behaves as function of its geometry parameters. Besides, the feeding technique is important factor that makes substantial effects on the simulation results.

First, the dipole with integrated via-hole balun feed is being used as balance to unbalanced transformer and impedance transformer from coaxial line to the two printed dipoles strip via physical connection. This method is capable to provide us with a bandwidth of 18.4% With maximum return loss of -25dB as shown in Fig. 8 at length =59 mm, width=4 mm and height=1.6 mm. the bandwidth can be increased to 24.5% by changing the width of the antenna from 4mm to 8mm. However, the size will increase and other impacts such as reducing the return loss and increasing the value of VSWR.

In term of practical implementation, it is considered one of the most common suitable due to its compact size and acceptable performance but it requires soldering which may cause junction radiation.

The second part is about open stub integrated balun. This method does not require direct connection between the feed line and the dipole. Therefore, a virtual short circuit is created between the open stub and the dipole strip making a center fed current distribution [7].

The optimized dimension for optimum performance are at length=65 mm, width=6 mm and height=1.6mm. The optimum results are bandwidth of 17 percent with minimum VSWR1.008 v also maximum return loss of -47.88dB. The radiation pattern of this method is unidirectional and has main lobe directivity of 2 dB in the elevation plane with angular width of 89 deg and has the same magnitude on the azimuth plane. For implementation, it is very convenient and suitable since it does not require soldering though it has narrow bandwidth but again it depends on the application requirements.

V. CONCLUSION

This proposed work focuses on the design and simulation of planar dipole antenna for 1800MHz Band for GSM application using CST studio software. The antenna was structured on FR4 substrate with relative constant of 4.3. Two types of feeding configuration have been designed to feed the antenna in order to match 50 ohm transmission line.

These types are via-hole integrated balun, and quarter wavelength open stub. As results, the dimension of the antenna and simulation results such as return loss, bandwidth and voltage standing wave ratio (VSWR) has been analyzed. For instance, the via-hole is capable to provide maximum return loss of -25dB, bandwidth of 18.4% and VSWR of 1.116 v at length of 59 mm and width 4 mm; it is bandwidth can be improved to 25% and 30%. While the open stub at length of 67mm, width of 6mm and height 1.6mm can provide max return loss of -47.88dB and bandwidth of 17% with VSWR 1.008 << 2 and also has better radiation pattern.

REFERENCES

[1] Balanis C. A.. 2005.Antenna theory Analysis and design, 3rd edition. Wileterscience publication 3rd. ISBN: 0-471-66782-x.

[2] Garg, R., Bhartia, P., Bahl, I. & Ittipiboon, A. 2001. Microstrip antenna Design handbook. Artech House Boston London. ISBN 0-89006-513-6

[3] Jamaluddin M. H., Rahim, M. K. A., Abd. Aziz, M. Z. A. & Asrokin, A. 2005. Microstrip dipole antenna analysis with different width and length at 2.4 GHz. Asia Pacific conferences on applied electromagnetic, pp. 41-44. DOI: 0-7803-9431-3

[4] Pozar D.M. 2004. Microwave Engineering 3rd edition. , Wiley publication 3rd. ISBN10: 0471448788-

[5] X. Li, Yang, L., Gong, S.X. & Yang, Y.J. 2009. Dual-Band And Wideband Design of A Printed Dipole Antenna Integrated With Dual-Band Balun. *Progress In Electromagnetics Research Letters*, Vol. 6, pp:165-174.

[6] Kuo L.C. , Chuang H.R., Kan Y.-C. & Huang T.-C.2007. A Study of Planar Printed Dipole Antennas For Wireless Communication Application. *J. of Electromagnet. Waves and Appl.*, Vol. , NO 5 .pp: 637-652

[7] Huey-Ru Chuang & Liang-Chen Kuo. 2003. Design Analysis of a 2.4-GHz Polarization-Diversity Printed Dipole Antenna With Integrated Balun and Polarization Switch Circuit For WLAN and Wireless Communication Application. *IEEE Transaction On Microwave Theory and Techniques*,Vol.51(2):374-381, DOI:10.1101/TMT

[8] Shuchita Saxena & Kanaujia, B.K. 2011. Design And Simulation of A Dual Band GapCoupled ANnular Ring Microstrip Antenna, *International Journal of Advances in Engineering & Technology* Vol. 1(2):151-158. ISSN: 2231-1963

[9] Tang, T., , Li, C.M. & Lin C.Y. 2011, Printed Dipole Antenna With back To back Asymmetric Dual-c-Shape Uniform Strip For DTV Application. Progress In Electromagnetics Research C, Vol. 22, pp.73-83.

[10] Oltman G. & Huebner D. A. Member. 1986. Electromagnetically Coupled Microstrip Dipoles, *IEEE Transaction Antennas And Propagation*, Vol. 29(1):151-157. DOI: 0018-926X/81/0100-0151

[11] Shingo Tanaka, Yongho Kim & Hisashi Morishita. 2008. Wideband Planar Folded Dipole Antenna With Self-balance Impedance Property, *IEEE Transaction on Antennas and Propagation*, Vol. 56(5):1223-1228. DOI: 10.1109/TAP

Qualitative Study of a New Circuit Model of Small-signal Amplifier using Sziklai Pair in Compound Configuration

SachchidaNand Shukla, Beena Pandey and Susmrita Srivastava
Department of Physics and Electronics
Dr. Ram Manohar Lohia Avadh University
Faizabad - 224001, U.P., India
Email: sachida_shukla@yahoo.co.in

Abstract- **A new circuit model of small-signal wide-band voltage amplifier using Sziklai pair is proposed for the first time. The proposed amplifier circuit uses two PNP and one NPN transistors and two additional biasing resistances in its configuration. The upper half of the triple transistor system with transistors Q1 and Q2 constitute a Sziklai pair while the lower half with Q2 and Q3 forms complementary Sziklai pair. The over-all performance of the proposed amplifier circuit is found to be superior than conventional or modified small-signal Darlington and Sziklai pair amplifiers and as a rare feature the proposed circuit simultaneously produces high voltage gain and wide bandwidth. Poor response of small-signal Darlington pair amplifiers at higher frequencies and narrow bandwidth problem of small-signal Sziklai pair amplifier is found to be absent in the proposed amplifier. Variations in voltage gain as a function of frequency and different biasing resistances, bandwidth and total harmonic distortion of the amplifier is also pursued. Proposed amplifier may be useful for various analog communication applications.**

Index terms- **Small Signal amplifiers, Sziklai Amplifiers, Complimentary Darlington Pair Amplifiers**

I. INTRODUCTION

RC coupled Darlington pair amplifiers are popularly used to amplify small voltage signals in the range of milli-volts [1]-[7]. These amplifiers are usually known for high β value but suffer with the problem of poor response at higher frequencies [3]-[6]. Numerous books and research papers explored the usefulness of Darlington pair amplifiers [1]-[7] but least efforts are made to configure small-signal Sziklai pair amplifiers [8]-[9].

Sziklai pair [8], named after its inventor George Sziklai, is a composite unit of two bipolar transistors of opposite polarities (one NPN and other PNP transistor) and sometimes known as 'Compound feedback pair' or 'Complimentary Darlington pair'. Polarity of Sziklai pair unit is always determined by the *driver* transistor. For example, a PNP driver with NPN output transistor behaves like a PNP transistor and vice versa [8]-[9]. The current gain factor (β) of Sziklai pair is slightly less than Darlington pair topology, because it has a small amount of in-built negative feedback which reduces current gain [9]. However Sziklai pairs hold a better linearity than Darlington pairs when used in linear circuits. Another major advantage of Sziklai pair over the Darlington pair is that the base turn-on voltage is only half of the Darlington's turn-on voltage [8]-[9].

In electronics industry, Sziklai pairs are normally used in the push–pull output stage of power amplifiers or to configure quasi-complementary-symmetry power amplifiers [1], [8]. Contrary to this, authors had already developed a small-signal Sziklai pair amplifier [10] which was found to produce high voltage gain and removed the problem of poor frequency response of Darlington pair amplifier at higher frequency but having a narrow bandwidth. In the present manuscript, authors explore a new circuit model of small signal voltage amplifier using Sziklai pair in triple-transistor based compound configuration that simultaneously produces high voltage gain and wide bandwidth with fair response at higher frequencies.

II. EXPERIMENTAL CIRCUITS

The present study starts with a commonly used CC/CE Darlington pair amplifier having two NPN (Q2N2222 with β=255.9) transistors Q1 and Q2 in its paired unit. This amplifier circuit is depicted in Fig.1.

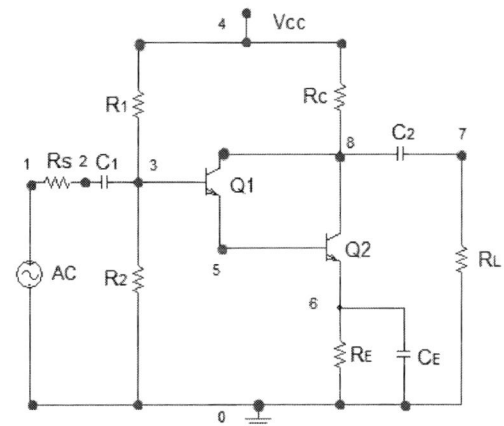

Fig.1. Darlington pair amplifier

On the other hand, in Sziklai pair amplifier [10], as shown in Fig.2, Q1 transistor of the Darlington's amplifier is replaced by a PNP (Q2N2907A with β=231.7) transistor. In addition, an extra biasing resistance R_D between collector of transistor Q1 and ground is introduced and emitter of upper PNP transistor Q1 is directly connected with DC supply voltage.

978-1-4673-2395-6/12 $31.00 © 2012 IEEE

TABLE I
BIASING PARAMETERS AND CONFIGURATIONAL DETAILS

Components	Darlington's Amplifier	Sziklai pair Amplifier	Proposed amplifier
Q1	Q2N2222	Q2N2907A	Q2N2907A
Q2	Q2N2222	Q2N2222	Q2N2222
Q3	Unavailable	Unavailable	Q2N2907A
R_S	500Ω	500Ω	500Ω
R_1	47KΩ	33KΩ	33KΩ
R_2	5KΩ	100KΩ	100KΩ
R_C	10KΩ	10KΩ	10KΩ
R_E	2KΩ	2KΩ	2KΩ
R_{D1}	Unavailable	500Ω	500Ω
R_{D2}	Unavailable	Unavailable	500Ω
R_L	10KΩ	10KΩ	10KΩ
C_1 , C_2	1μf	1μf	1μf
C_E	10μf	0.1μf	0.1μf
DC supply	+15V	+18V	+18V
Input AC signal	1-10mV (1KHz)	10-30mV (1KHz)	10-30mV (1KHz)

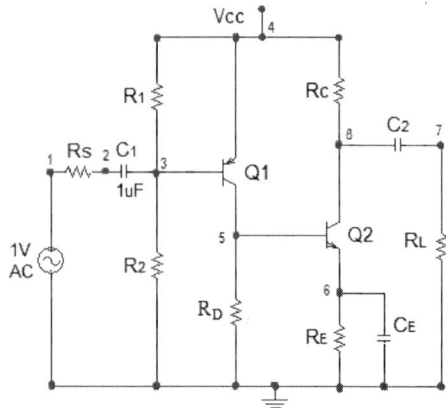

Fig.2. Sziklai pair amplifier

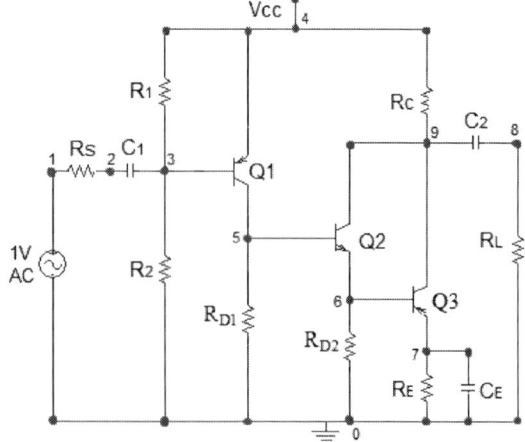

Fig.3. Proposed amplifier

However, the proposed amplifier, depicted in Fig.3, is obtained by adding an extra PNP transistor Q3 and biasing resistance R_{D2} in the circuit of Sziklai pair amplifier. The unit of transistors Q1 and Q2 in the proposed amplifier circuit (Fig.3) constitutes Sziklai pair while that of Q2 and Q3 jointly forms a complementary Sziklai pair. All the three amplifier circuits are properly biased using potential divider network with biasing parameters as described in Table-I.

Respective observations are made by feeding the amplifier circuits with 1V AC input signal source from which, a small and distortion less AC signal of 1mV for Darlington pair amplifier (Fig.1) while 10mV for Sziklai pair amplifier (Fig.2) and proposed amplifier (Fig.3) at 1KHz frequency is drawn as input for amplification purpose. All the observations mentioned in the present manuscript are furnished through PSpice simulation software [4], [6], [10]-[11] (Student version 9.2).

III. RESULTS AND DISCUSSIONS

The amplifier of Fig.1 is found to provide undistorted output up to 10mV AC input signal at 1KHz frequency while rest two amplifiers of Fig.2 and Fig.3 produce distortion-less results from 10 to 30mV AC input at similar frequency.

Variation of maximum voltage gain as a function of frequency for all the three amplifiers is depicted in Fig.4. It is found that the Darlington pair amplifier produces 16.98 maximum voltage gain and 102.058 KHz bandwidth (lower cut-off frequency f_L= 79.913 Hz and upper cut-off frequency f_H = 102.138 KHz) with a poor response at higher frequencies [4], [6], [10]. However, Sziklai pair amplifier produces 102.309 maximum voltage gain with narrow bandwidth of 4.80 KHz (with f_L=224.453Hz and f_H=5.0556 KHz) [10] whereas the proposed amplifier produces significantly enhanced maximum voltage gain of 177.006 with wide bandwidth of 1.2091MHz (with lower cut-off frequency f_L=547.54Hz and upper cut-off frequency f_H=1.2097MHz).

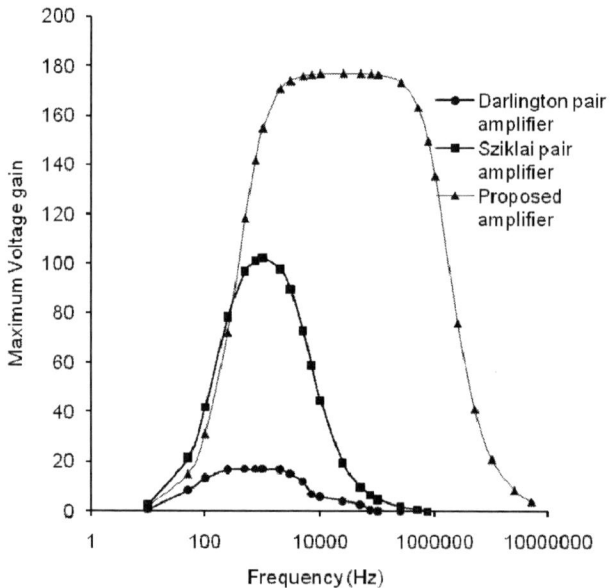

Fig.4. Variation of Maximum voltage gain with frequency

It is also found that Darlington pair amplifier delivers 1.7708 μA peak output current and 17.824mV peak output

978-1-4673-2395-6/12 $31.00 © 2012 IEEE

voltage [4], [6], [10], Sziklai pair amplifier crops 110.6µA peak output current and 1.106 volts peak output voltage [10] while the proposed amplifier delivers 155.435µA peak output current and 1.554 volt peak output voltage at the biasing parameters mentioned in Table-I. The maximum current gain of Darlington, Sziklai pair and proposed amplifiers are also measured and found to be 8.51, 7.345 and 5.059 respectively [4], [6], [10]. The output voltage waveform of Darlington pair and Sziklai pair amplifiers show phase-reversal whereas for proposed amplifier it shows a 228.6° phase difference with applied AC input.

Transistors' small-signal amplification parameters α and β corresponding to the Darlington pair, Sziklai pair and Complementary Sziklai pair topologies in Fig.1, Fig.2 and Fig.3 respectively are also estimated using standard formulae [1], [2], [10] and depicted in Table-II while Table-III mentions the transistors' driver voltages V_D.

The estimated α and β values in Table-II corresponding to the composite transistors unit of Darlington pair, Sziklai pair and proposed amplifier are found adequately in accordance with the prescribed range for small-signal amplifiers [1], [2], [10], [12]. It is further found that the values of α and β are continuously decreasing from Darlington Pair to proposed amplifier circuit.

It is to be mentioned that in absence of added resistance R_{D2} or on the simultaneous removal of R_{D1} and R_{D2} from the proposed circuit, the output voltage/current waveforms of the amplifier distorted badly while on removing only R_{D1} from the proposed circuit, the positive half cycle of the output waveform clipped off. Similarly, another interesting feature of proposed amplifier is obtained when the biasing resistance R_1 is removed from the circuit. This yields enhancement in maximum voltage gain to 188.97 value, peak output current to 168.30 µA, peak output voltage to 1.683 volts but reduction in maximum current gain to a 4.9398 value and bandwidth to 1.073MHz. On the other hand, if R_2 is made absent from the proposed circuit, the output voltage/current waveforms again distorted badly.

TABLE II

α AND β PARAMETERS BASED ON SIMULATION RESULTS

Configuration	Composite Transistor Unit	
	B	A
Darlington pair	65996.61	0.999
Sziklai pair	254.49	0.996
Proposed amplifier	214.99	0.995

TABLE III:

TRANSISTOR DRIVING VOLTAGES (V_D) BASED ON SIMULATION RESULTS

Configuration	V_D for Q1	V_D for Q2	V_D for Q3
Darlington pair	1.44 V	0.97 V	NA
Sziklai pair	17.2 V	13.8 V	NA
Proposed amplifier	17.18 V	9.41 V	8.48 V

This suggests that perhaps the simultaneous inclusion of added resistances R_{D1} and R_{D2} in the proposed amplifier circuits and the high driving voltages corresponding to the transistors Q1, Q2 and Q3 (Table-III) is responsible for the simultaneous increase in the voltage gain [3]-[4] and

bandwidth. The proposed amplifier is also found to effectively remove the problem of poor response of a conventional Darlington pair amplifier at higher frequencies as well as the narrow bandwidth problem of the small-signal Sziklai Pair amplifier [4], [6]-[7], [10].

TABLE IV

VARIATION OF MAXIMUM VOLTAGE (A_{VG}) AND CURRENT GAINS (A_{IG}) WITH TEMPERATURE

Temperature °C	Darlington's amplifier		Sziklai pair amplifier		Proposed amplifier	
	A_{VG}	A_{IG}	A_{VG}	A_{IG}	A_{VG}	A_{IG}
-30	8.95	4.48	83.43	5.74	152.22	3.88
-20	10.55	5.29	86.93	6.03	156.96	4.09
-10	12.07	6.05	90.36	6.31	161.52	4.29
0	13.51	6.77	93.70	6.59	165.93	4.50
10	14.86	7.45	96.96	6.87	170.16	4.71
27	16.98	8.51	102.31	7.34	177.00	5.06
50	19.53	9.79	109.12	7.96	185.54	5.53
80	22.38	11.22	117.22	8.73	195.50	6.12

TABLE V

VARIATION OF BANDWIDTH (IN KHZ) WITH TEMPERATURE

Temperature °C	Darlington pair amplifier	Sziklai pair amplifier	Proposed amplifier
-30	99	4.80	1555.38
-20	99	4.80	1462.09
-10	99	4.80	1401.91
0	99	4.80	1347.92
10	99	4.80	1292.93
27	99	4.80	1208.55
50	99	4.80	1106.27
80	99	4.80	981.12

Total Harmonic Distortion (THD) percentage is also calculated for the Darlington pair, Sziklai pair and the proposed amplifier circuit using standard formula [1]-[2], [10]. The THD for Darlington pair amplifier is estimated for 10 significant harmonic terms and found to be 0.734% while that for Sziklai pair amplifier is estimated for 8 significant harmonic terms and found to be 1.72%. Similarly for proposed amplifier it is estimated for 6 significant harmonic terms and found to be 4.154%. Hence conclusively, the voltage gain and bandwidth of the proposed amplifier significantaly enhances on the cost of THD.

It may be suggested here that the inclusion of additional biasing resistances (R_D in Sziklai pair amplifier while R_{D1} and R_{D2} in proposed amplifier) in the circuits of Fig.2 and Fig.3 improves the voltage gain performance of the respective circuits but simultaneously increases the harmonic distortion.

Variation of maximum voltage and current gains and bandwidth with temperature is also measured and listed in Tables-IV and Table-V respectively. For Darlington pair amplifier, the bandwidth remains unchanged but both varieties of gains increases with rising temperature. The similar situation persists for Sziklai pair amplifier [10]. On the other hand for proposed amplifier both varieties of gains increases with rising temperature but bandwidth gradually decrease. This observation verifies the usual behaviour of transistor parameter h_{FE} with temperature [10], [13].

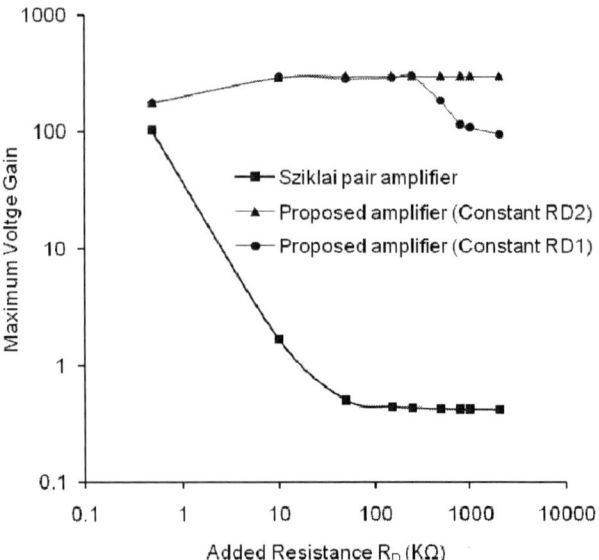

Fig.5. Variation of Maximum voltage gain with R_D

The variation of maximum voltage gain as a function of added resistances is shown in Fig.5. The maxim of the voltage gain corresponding to added resistance R_D for Sziklai Pair amplifier is observed at $R_D=0.5K\Omega$. The overall property is that the maximum voltage gain linearly decreases up to $R_D=50K\Omega$ and thereafter it tend towards saturation. Thus the Sziklai Pair amplifier is found to produce considerable response at $R_D=0.5K\Omega$.

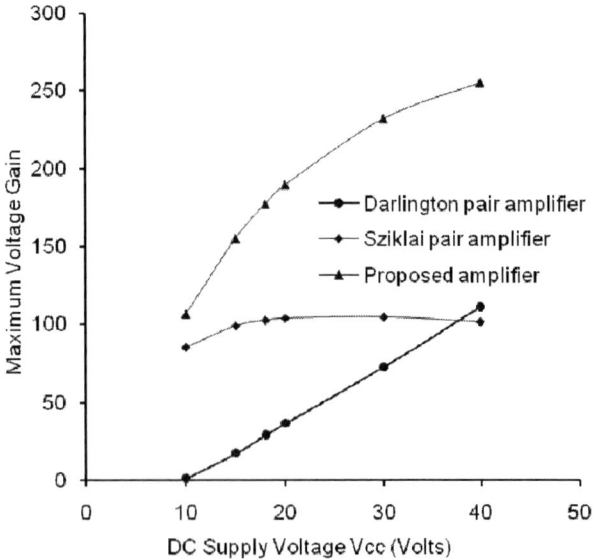

Fig.6. Variation of Maximum voltage gain with V_{CC}

Whereas in proposed amplifier, maximum voltage gain increases with increasing values of R_{D1} (keeping R_{D2} constant) from $0.5K\Omega$ to $50K\Omega$ thereafter it almost tends towards

saturation. On the other hand maximum voltage gain increases with increasing values of R_{D2} (keeping R_{D1} constant) up to $150K\Omega$ then starts decreasing.

Variation of maximum voltage gain with DC supply voltage is depicted in Fig.6. It is observed for proposed amplifier that the maximum voltage gain increases almost exponentially with increasing values of biasing voltage up to 40V. On the other hand the maximum voltage gain for Sziklai pair amplifier has a nonlinear rising tendency for increasing values of DC supply voltage up to 20 volts and beyond this critical limit it decreases with a slow pace [10]. However the maximum voltage gain for Darlington Pair amplifier possesses linear rising tendency with DC supply voltage. All the three amplifiers under discussion are found to respond fairly up to 40V of V_{CC}.

Variation of maximum voltage gain as a function of R_E for all the three amplifiers is traced in Fig.-7. The maximum voltage gain for Darlington pair amplifier has decreasing tendency (almost exponentially) at increasing values of R_E. However, the voltage gain for Sziklai pair amplifier increases with R_E and respective response curve is found to be inverted replica of the curve corresponding to Darlington pair amplifier. On the other hand, the maximum voltage gain for the proposed amplifier is found independent of any change in emitter resistance R_E. Sziklai pair amplifier fairly responds up to $25K\Omega$ value of R_E while Darlington pair and proposed amplifiers respond up to $50K\Omega$ value of R_E.

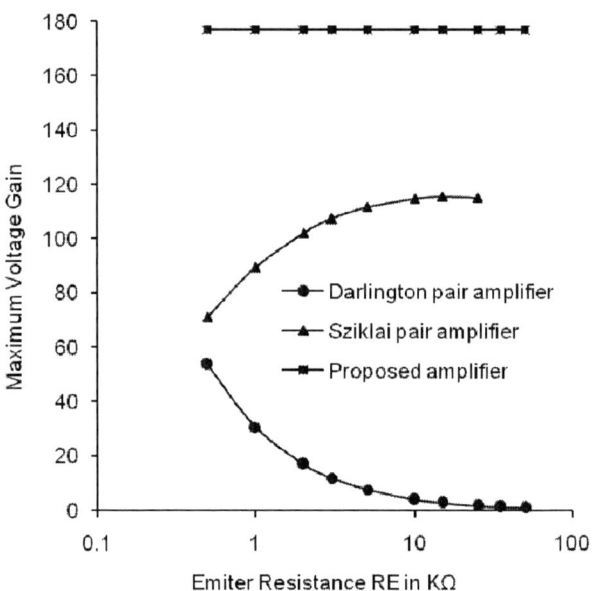

Fig.7. Variation of Maximum voltage gain with R_E

Variations of maximum voltage gain with collector resistance R_C and load resistance R_L are also estimated (not shown in form of figures). It is found that maximum voltage gain has a nonlinear rising tendency for increasing values of collector resistance R_C for all the three amplifiers up to $10K\Omega$ and beyond this critical limit, the voltage gain gradually acquires a saturation tendency. Darlington pair, Sziklai pair and the proposed amplifier circuits performs fairly below $40K\Omega$ of R_C.

Similarly for R_L, it is observed that voltage gain value rises up linearly in low resistance range up to $50K\Omega$ value of R_L but for higher R_L values it gradually acquires a sustained level. This rising and saturation of the voltage gain with R_L is well in accordance of the usual behaviour of small signal amplifiers [2]-[4], [6]-[7], [10], [12].

It is further observed that the basic nature of the variation of maximum voltage gain with R_L or R_C for all the three amplifier circuits are similar but the overall voltage gain of the proposed amplifier is always found higher than that of Darlington pair and Sziklai pair amplifier at every value of R_L or R_C.

IV. Conclusions

Sziklai pair topology is popularly used to design quasi-complimentary-symmetry push-pull Class-B power amplifiers but here it is explored to design a small-signal amplifier.

The proposed amplifier effectively removes the problem of poor response of conventional Darlington pair amplifiers at higher frequencies and narrow bandwidth problem of small-signal Sziklai pair amplifier.

The proposed amplifier shows a considerable response for additional biasing resistances R_{D1} and R_{D2} up to $1K\Omega$ value and simultaneously produces high voltage gain and wide bandwidth with a current gain greater than unity. The optimum performance of the proposed amplifier is received for 10-40 volts of DC supply voltage and the maximum voltage gain remains unaffected for any change in emitter resistance R_E.

References

[1] R. L. Boylestad and L. Nashelsky, *Electronic Devices and Circuit Theory*, Pearson Education Asia, 9th ed., 2008, p-299, 304, 681,

[2] A. Bell David, *Electronic devices and circuit*, Prentice Hall of India, 3rd ed., 2002, p-687

[3] A. M. H. Sayed ElAhl, M. M. E. Fahmi, S. N. Mohammad, *Qualitative analysis of high frequency performance of modified Darlington pair*, Solid State Electronics, 46, 2002, p-593

[4] S. N. Tiwari and S. N. Shukla, *Qualitative Analysis of Small Signal Modified Darlington Pair and Triple Darlington Amplifiers*, Bulletin of Pure and Applied Science, 28D, No.1, 2009, p-01

[5] T. A. Chris and G. M. Robert, *A New Wide-Band Darlington Amplifier*, IEEE Journal of Solid State Circuits, 24, No. 4, 1989, p-1105

[6] S. N. Tiwari, A. K. Dwivedi and S. N. Shukla, *Qualitative Analysis of Modified Darlington Amplifier*, Journal of Ultra Scientist of Physical Sciences, 20 No.3, 2008, p-625

[7] S. N. Tiwari, B. Pandey, A. K. Dwivedi, and S. N. Shukla, *Development of Small-Signal Amplifiers by Placing BJT and JFET in Darlington Pair Configuration*, Journal of Ultra Scientist of Physical Sciences, 21, No.3, 2009, p-509

[8] G. C. Sziklai, *Push-pull complementary type transistor amplifier.*, U.S. Patent 2,762,870, September 11, 1956

[9] P. Horowitz, H. Winfield, *The Art of Electronics*, Cambridge University Press. ISBN 0-521-37095-7, 1989

[10] B. Pandey, S. Srivastava, S. N. Tiwari, J. Singh and S. N. Shukla, *Qualitative Analysis of Small Signal Modified Sziklai Pair Amplifier*, Indian Journal of Pure and Applied Physics, 50, 2012, p-272

[11] M. H. Rashid, *Introduction to P.Spice Using OrCAD for Circuits and Electronics*, Pearson Education, 3rd Ed., 2004, p-255

[12] A. Motayed, T. E. Browne, A. I. Onuorah and S. N. Mohammad, *Experimental studies of frequency response and related properties of small signal bipolar junction transistor amplifier*, Solid State Electronics, 45, 2001, p-325

[13] A. G. Barua and B. Tiru, *Variation of width of the hysteresis loop with temperature in an emitter-coupled Schmitt trigger*, Indian Journal of Pure & Applied Physics, 44, 2006, p-482

Design of a 9-bit UART Module Based on Verilog HDL

Nennie Farina Mahat, *Member, IEEE*

Integrated Circuit Development (ICD)

MIMOS Berhad

Technology Park Malaysia

Bukit Jalil, 57000 Kuala Lumpur

Email: farina.mahat@mimos.my

Abstract— **Universal Asynchronous Receiver Transmitter (UART) is widely used in data communication process especially for its advantages of high reliability, long distance and low cost. In this paper, we present the design of 9-bit UART modules based on Verilog HDL. This design features automatic address identification in the character itself. We have implemented the VLSI design of the module and pass data between the proposed 9-bit UART module with a host CPU. The design consists of receiver module, transmitter module, prescaler module and asynchronous FIFOs. We have explained the functions of each individual sub-modules and how the design works in simulation.**

I. INTRODUCTION

UART is a Universal Asynchronous Receiver Transmitter that performs parallel-to-serial conversion on data character received from the host processor into serial data stream, and serial-to-parallel conversion on serial data bits received from serial device to the host processor. The RS-232 is applied to serial data communication as a standard to be complied with. Besides RS-232, RS-422 and RS-485 standards have been commonly applied into UART chip nowadays. These standards offer more reliable communication over much longer distances compared to RS-232.

A complete data frame format for UART consists of a start bit '0', 5-8 bits data, optional parity bit and stop bit '1'. The stop bit can be in length of 1, 1.5 or 2 bits [1]. Fig. 1 below shows the data frame format of a UART. While in idle state, serial data line will be in logic '1' state. A start bit '0' at the beginning of the data frame will cause a falling edge on the serial data line. This marks the detection of a data character. The idea of start bit and stop bit in UART is to achieve data synchronization [2-3]. An optional parity bit can be in odd parity or even parity. Odd parity means that sum of all bits gives an odd number, while even parity means sum of all bits gives an even number. The serial data frame is shifted out with the least significant bit (LSB) first.

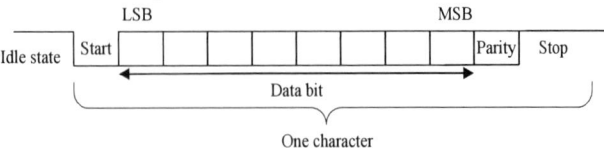

Fig. 1 UART data frame format

II. 9-BIT NETWORK

In a transmission, every Slave devices will search every character transmitted for address byte and try to match with its unique address. This results in a lot of wasted processing time for Slave devices. In a 9-bit network, UART uses the ninth bit of a character to differentiate between an address or a data character [4]. By proposing the ninth bit for address byte indication, Slave devices are able to distinguish an address byte, compare the address and decide whether to accept or discard the following data bytes. This reduces the processing time of the Slave's CPU [5].

The parity bit is set to logic '1' to indicate an address character and set to logic '0' to indicate a data character. Processor will poll for the parity bit and if it is set, the processor will try to match the address with its own. If the address matches, the following data bytes are received. If the address matching failed, the processor will ignore the following data bytes.

The 9-bit UART design proposed in this paper configures the data frame format to have a start bit '0', 8-bit address or data byte, sticky parity '1' or '0' and 1 bit of stop bit '1'. Fig. 2(a) shows an address character with parity '1' and Gig. 2(b) shows a data character with parity '0'. Data communication is in the order that address character to be received or transferred first, followed by one or more data characters.

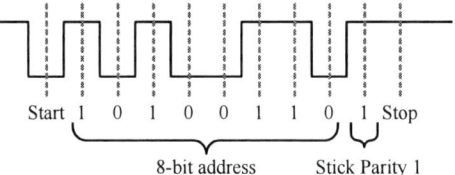

Fig. 2(a) Address character

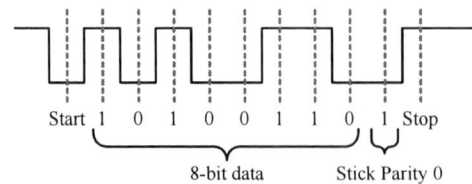

Fig. 2(b) Data character

III. Detailed Description of the Sub-modules

The 9-bit UART design proposed consists of the basic sub-modules of a UART which are the receiver, transmitter and baud rate generator that we called prescaler [6-7]. In addition to that, this design has internal buffers in both receiver and transmitter. Since this design operates in serial clock domain and interfaces with parallel clock domain of the processor, we implement the buffers by using asynchronous FIFOs. The asynchronous FIFO design provides smooth data transfer between two different clock domains. Fig. 3 below illustrates the overall block diagram of the 9-bit UART.

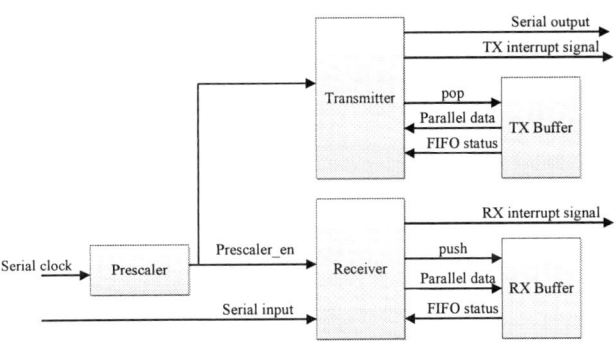

Fig. 3 Overall block diagram of 9-bit UART

A. Receiver

During reception, UART will be in listening mode and always sense for a start bit. A start bit is detected when there is a transition from logic '1' to logic '0' on the serial data line. UART receiver module will then capture the address byte and check the stop bit. If the stop bit is correct, the address captured will be compared with broadcast address and UART own address. UART will determine whether to receive or ignore the incoming data bytes. Receiver module also checks for the integrity of the start bit and stop bit. If a false start bit or stop bit is detected, reception will be terminated and an interrupt will be sent to the processor to read the remaining data, if available, in the RX buffer.

In this paper, we use finite state machine to implement these steps. The receiver finite state machine has five states; RX_ILDE, RX_START, RX_DATA, RX_PARITY and RX_STOP as shown in Fig. 4 below. The state machine changes state at every 16 baud clock cycles, which is equal to 16 prescaler_en pulses.

RX_IDLE State: When UART receiver module is reset, the state machine will be in this state. The state machine will wait for start bit detection on the serial data line. A start bit detection is identified when there is a fall transition on the serial data line from idle state '1' into logic '0'. In order not to detect a false bit, the state machine will only consider startbit_detect signal that has been synchronised to the serial clock domain. A correct start bit detection will cause the state machine to go into RX_START.

RX_START State: In this state, the state machine will wait for 16 baud clock cycles before going into the next state, RX_DATA.

RX_DATA State: The state machine will sample the character received at the most ideal time, which is at the middle of the bit. Each bit is then being stored into an internal register rx_data to form a complete 8-bit data. When a complete character of 8 bit has been received, the state machine will go into RX_PARITY.

RX_PARITY State: In this state, the state machine samples the parity bit at the middle of the bit and determines whether the character received is an address byte or a data byte,

RX_STOP State: Similar as in the previous states, the state machine will sample the stop bit at the midpoint. In this state, state machine will do stop bit error checking and address comparison. If the is no stop bit detected, that is the sampled bit is logic '0', receiver module interrupts the processor for error detection, skip address comparison and state machine will go into RX_IDLE state. If the state machine detects a correct stop bit, it proceeds with address matching by comparing the received address byte with its own address and broadcast address. A matched address will assert match_flag throughout the whole reception process so that when state machine goes into this state again while receiving data byte with parity bit '0', receiver will save the data byte into internal RX buffer, provided that the sampled stop bit is also correct.

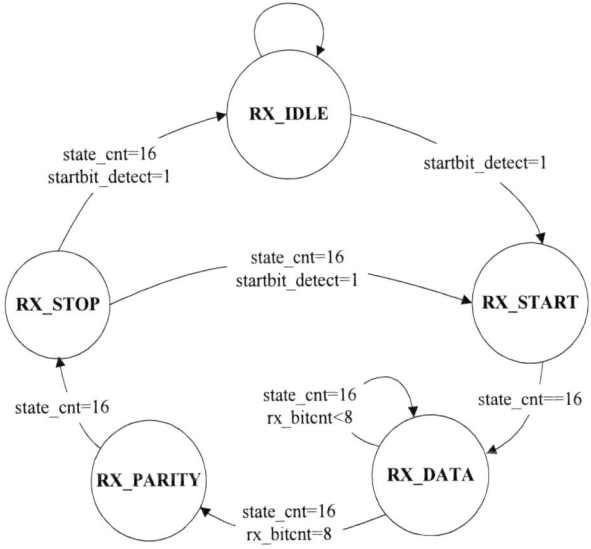

Fig. 4 Receiver module state machine

B. Transmitter

UART will be in transmission mode when TXMODE is enabled in the configuration register of the UART. The processor sets the destination address, data byte(s) to be sent and other transmission settings. Transmission data bytes are saved into an internal TX buffer before being processed for transmission.

Transmitter module converts the address and data received from the processor into serial bits, and adds in start bit of '0', parity bit of '1' for address byte or '0' for data byte, and stop bit of '1'. The address can be a broadcast address or a unique address that belongs to a specific Slave device. Therefore the following data bytes can be a broadcast message meant for all Slave devices or a specific command for a specific Slave device. When an address byte has been transferred successfully, UART transmit module will continue to send all data byte(s) available in the TX buffer until the buffer is empty.

An internal finite state machine is used to transmit the complete 11-bits character, bit-by-bit. The transition of each state will take place at every 16 baud clock cycles, which is equal to 16 cycles of prescaler_en. There are five states in the transmitter finite state machine; TX_ILDE, TX_START, TX_DATA, TX_PARITY and TX_STOP as shown in Fig. 5 below.

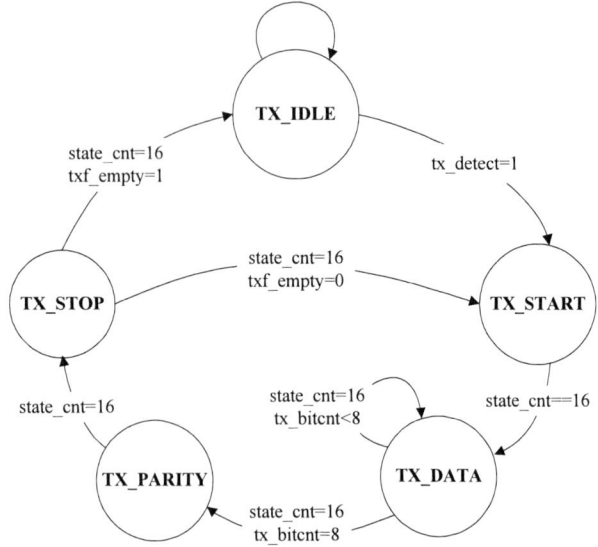

Fig. 5 Transmitter module state machine

TX_IDLE State: When UART is being reset, state machine will be in this state. Transmitter module will wait for tx_detect pulse that indicates data is ready to be transmitted. This enable pulse is coming from the processor and being synchronised to the serial clock.

TX_START State: In this state, transmitter module will send out a start bit '0' on the serial data line. State machine will be in this state for 16 baud clock cycles before going into TX_DATA state.

TX_DATA State: In this state, the state machine loads the internal tx_data register with data to be transmitted. For the first transmission, tx_data register will be loaded with a destination address. The state machine transmits the destination address bit by bit, starting from the least significant bit in tx_data[0] until the most significant bit in tx_data[7].

TX_PARITY State: For an address byte, the state machine transmits out parity bit '1' whereas for a data byte, it will transmit out parity bit '0'.

TX_STOP State: When a complete frame of data has been successfully transmitted out together with the parity bit, the state machine will go into this state to complete the transmission by sending out a stop bit of '1'. The state

machine then checks whether there is another data byte in the TX buffer to be sent out. If there is, state machine will go back to TX_START and repeats all the states until it goes back into this state again. If the TX buffer is empty, it means there is no more data byte to be transferred; therefore the state machine will go into TX_IDLE state and wait for another tx_detect pulse.

C. Prescaler Module

This 9-bit UART design can operate at any defined serial clock frequency and at the same time follows the desired baud rate. Here is where the prescaler module comes in hand where it acts as a baud rate generator module that calculates the divider factor. The divider factor will then generate prescaler_en pulse that acts as a baud clock. The prescaler module is used in this design for the transmission and reception to be at the desired baud rate. Fig. 6 below shows the timing diagram for transmission with prescaler ratio of 12. The serial clock is set to 32MHz and the desired baud clock rate is 115200bps. The calculation to get prescaler divide ratio is shown below:

$$Divide\ ratio = \frac{Serial\ clock\ frequency}{16 * Baud\ rate}$$

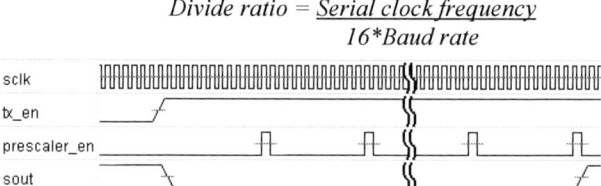

Fig. 6 Timing diagram for transmission with prescaler ratio of 12

D. Transmitter and Receiver Buffers

In serial data transmission, a huge stream of data can come in too fast and UART may not be quick enough to process what it is receiving. This causes data overrun which then results in data loss or other serious errors. This is why we need flow control in the UART where it will halt the flow of data bytes until UART is ready to receive more data.

There are many ways to implement flow control mechanism. One of the older methods used back then is to add bunch of zeros while receiving line is busy. This method is called padding. UART will ignore the zeros while it processes current data. To make it efficient, we need to provide just enough number of zeros and figuring this out is not easy.

Another method is to use RTS (Request to Send) and CTS (Clear to Send) flow control mechanism which is part of RS-232 standard. These two lines allow both sender and receiver to communicate and alert each other on their status. When sender has data to be sent, it will assert RTS to ask receiver whether it is ready to receive data or not. If receiver is free, it will reply by asserting CTS and sender will start transferring data. But if receiver is busy, CTS line will maintain low and sender needs to halt the data and wait until CTS goes high. This method however requires additional wires to the UART design.

Most UART nowadays has internal buffer with the size big enough to hold the data and that is less likely to overrun. The buffer can be set to interrupt the processor when the buffer reaches certain threshold level. In this paper, we are using this method to do flow control. Asynchronous FIFOs are being sued as an internal TX buffer and RX buffer in transmitter and receiver module respectively. The asynchronous FIFO allows smooth data transmission between two different clock domains and provides reliable empty and full status [8-10].

During reception, receiver module writes the received data bytes into RX buffer using serial clock. When RX buffer is almost full, receiver will interrupt the processor to start reading the data (using parallel clock) before RX buffer starts to overflow. While in transmission, processor writes data into TX buffer using parallel clock. When there is at least one data byte inside TX buffer, transmitter will start reading the data in serial clock, process the data byte and transmit out serial bits. If TX buffer is almost full, an interrupt is sent to processor to halt writing into the buffer.

IV. CONCLUSIONS

We have simulated the design using Synopsys' VCS Simulation tool. Fig. 7 shows the simulation waveform of the receiver module. In the simulation, receiver module is in idle state after being reset. When a start bit is detected, state machine is being activated. Parity bit received is logic '1'; therefore this indicates an address byte. The receiver captures the data address of 8'h01 and save it in rx_data register to be compared with UART's own address and broadcast address.

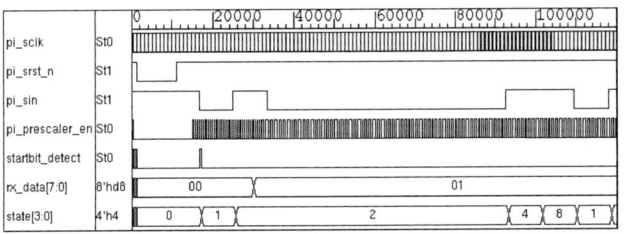

Fig. 7 Simulation waveform of the receiver

Fig. 8 below shows the simulation waveform of the transmitter module. In this simulation, UART is going to transmit out a data byte of 8'hCB. The transmitter module adds in start bit of '0', shifted data byte as '11010011', sets parity bit to '0' and 1 stop bit of '1'. From the waveform, we can see that the output on the serial data line, po_sout is correct.

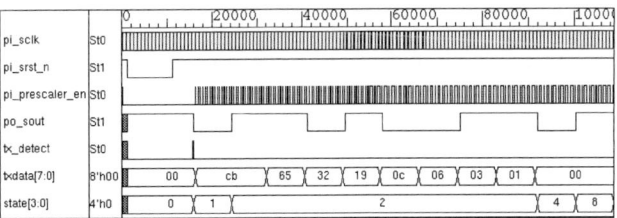

Fig. 8 Simulation waveform of the transmitter

V. CONCLUSIONS

In this paper, a modified UART design is proposed with automatic address indication, which is called 9-bit UART. The 9-bit UART design is implemented using Verilog HDL and simulated to see the functionality of each sub-modules and the result. This design shows that using the ninth bit method gives advantage of saving the UART processing time by comparing the address and decides whether to receive or ignore the incoming data packets. The stop bit error checking mechanism in this design also offers data integrity checking. With all the features mentioned, it adds to the flexibility, stability and reliability to the normal UART design that is widely being used.

REFERENCES

[1] J. Norhuzaimin, and H.H. Maimun, "The design of high speed UART," *Asia Pac. Conf. on Appl. Electromagnetics (APACE 2005),* Johor, Malaysia, Dec. 2005.

[2] C. He, Y. Xia, and L. Wang, "A universal asynchronous receiver transmitter design," *Int'l Conf. on Elect. Comm. and Control (ICECC 2011),* Ningbo, China, Sept. 2011.

[3] Y. Wang, and K. Song, "A new approach to realize UART," *Int'l Conf. on Elect. and Mech. Eng. and IT (EMEIT 2011),* Harbin, Heilongjiang, China, Aug. 2011.

[4] *Using a 9-bit Software UART with Stellaris® Microcontrollers Application Note (AN01280),* Texas Instruments, P.O. Box 655303, Dallas, Texas 75265, USA, Aug. 2010.

[5] *Zilog Z8 Encore! XP® 9-bit UART Implementation Application Note (AN014602-1207),* Zilog Inc., Dec. 2007.

[6] Z. Zhang, and W. Wu, "UART integration in OR1200 based SoC Design," *2nd Int'l Conf. on Comp. Eng. and Tech. (ICCET 2010),* Chengdu, China, Apr. 2010.

[7] Y. Fang, and X. Chen, "Design and simulation of UART serial communication module based on VHDL," *3rd Int'l Workshop on Intel. Sys. and App. (ISA 2011),* Wuhan, China, May 2011.

[8] C.E. Cummings, "Simulation and synthesis techniques for asynchronous FIFO design," *Synopsys User Group (SNUG),* San Jose, USA, 2002.

[9] X. Wang, and J. Nurmi, "A RTL asynchronous FIFO design using modified micropipeline," *The 10th Biennial Baltic Elect. Conf. (BEC 2006),* Tallinn, Estonia, Oct. 2006.

[10] X. Wang, T. Ahonen, and J. Nurmi, "A synthesizable RTL design of asynchronous FIFO," in *Proc. Int'l Sympo. on SoC 2004,* Tampere, Finland, Nov. 2004.

A ΔΣ Modulator with 3-Bit, 37-Level Pre-Detective Dynamic Quantization

Chien-Hung Kuo, *Member, IEEE* and Kuan-Hsun Wang
Dept. of Applied Electronics Technology
National Taiwan Normal University
Taipei, Taiwan, R.O.C.
E-mail: chk@ntnu.edu.tw

Abstract— In this paper, a high-resolution delta-sigma modulator with a pre-detective dynamic quantizer is proposed. A 37-level quantization can be achieved by using only a 3-bit quantizer in the proposed dynamic quantizer. In the proposed structure, a signal detector is added at the input of the presented modulator to pre-detect the magnitude of the sampled input and switch the dynamic quantizer to the corresponding quantization range. With the proposed technique, the quantization level can be greatly increased, and the number of comparators will hence be substantially reduced for a high-level quantization. The resulting resolution of delta-sigma modulators can thus be significantly promoted without consuming much power and area. The proposed delta-sigma modulator is implemented in a TSMC 0.18-μm 1P6M CMOS process. The signal-to-noise plus distortion ratio is 101.2 dB in a signal band of 25 kHz. The power consumption is 1.68 mW at a 1.8 V supply voltage.

I. INTRODUCTION

In general, the resolution of delta-sigma (ΔΣ) modulators can be promoted by increasing its order to enhance the degree of noise shaping [1]. However, for the single-loop structure, too many integrators will cause the ΔΣ modulator to be prone to unstable. For the multi-stage noise-shaping (MASH) structure [2], the components mismatch between analog and digital parts in error cancellation circuit will bring a great challenge to designers.

Another way to promote the performance of ΔΣ modulators is to reduce the quantization noise by increasing the bit number of quantizer [3]. Unfortunately, the number of comparators required in the multi-bit quantizer would hence be greatly increased with the bit number by a power of two. Consequently, for high-resolution data conversion, much more power and area cost would be consumed in the multi-bit ΔΣ modulator.

In this paper, we propose a new dynamic quantizer with tunable reference voltages to increase and extend the quantization levels without increasing the modulator order and comparator number. A 37-level quantization is achieved by using a 3-bit quantizer and a simple amplitude detector. The resulting performance of the ΔΣ modulator will thus be effectively promoted. In feedback paths of the modulator, a modified clocked averaging algorithm (mCLA) [4] is devised to simplify the dynamic element matching circuitry for reducing the mismatch noise caused by DAC components.

This work was supported by the National Science Council of Taiwan, R.O.C., under contract NSC 100-2221-E-003-017-MY3.

In the section II, a ΔΣ modulator structure with a dynamic quantizer will be introduced [5]. The simple amplitude detector and tunable reference voltages for dynamic quantizer will also be explained [6]. In section III, the detail circuit of the proposed subcircuits including simple amplitude detector, mode selector, dynamic quantizer with tunable references will be illustrated. In section IV, the simulation results of the proposed modulator with dynamic quantizer will be presented. Conclusion is in section V.

II. DELTA-SIGMA MODULATOR ARCHITECTURE

A. Principle of Proposed Dynamic Quantizer

In the general multibit quantizer, the step size, Δ_N, of quantization is inversely proportional to the number of comparators and proportional to the difference of two reference voltages. It can be expressed as

$$\Delta_N = \frac{V_{refp} - V_{refn}}{2^N - 1} \quad \text{for } N > 1 \quad (1)$$

where, N is the bit number, and the denominator denotes the number of comparators. In theory, the quantization noise can be reduced by increasing the bit number of quantizer. However, the resulting power consumption will also be substantially raised due to the large number of comparators.

In this paper, a new dynamic quantizer for high performance multibit ΔΣ modulators is proposed, as shown in Fig. 1(a). A small voltage swing range between positive and negative references, $V_{px}-V_{nx}$, where, x=1, 2, …, 8, for the multibit quantizer is chosen to decrease the power of quantization noise. Furthermore, to allow the large signal quantization can be achieved, this small voltage range can be tuned with the variation of the input magnitude, as shown in Fig. 1(b). The bold line sinusoidal wave represents the input signal, and the quantization range of the multibit quantizer is noted by Q_x. In this paper, a conventional 3-bit quantizer is adopted to be the basic quantization cell of the proposed dynamic quantizer. That is, the multibit quantizer can achieve one of quantizations, Q_1, Q_2, …, and Q_8. In other words, if we can know the magnitude of input signal in advance [7], the positive and negative references will be narrowed to the vicinity of the input voltage. A fine-level multibit quantization will then be performed to the input signal. The quantization noise power would thus be reduced and a high performance ΔΣ modulator could be resulted.

978-1-4673-2395-6/12 $31.00 © 2012 IEEE

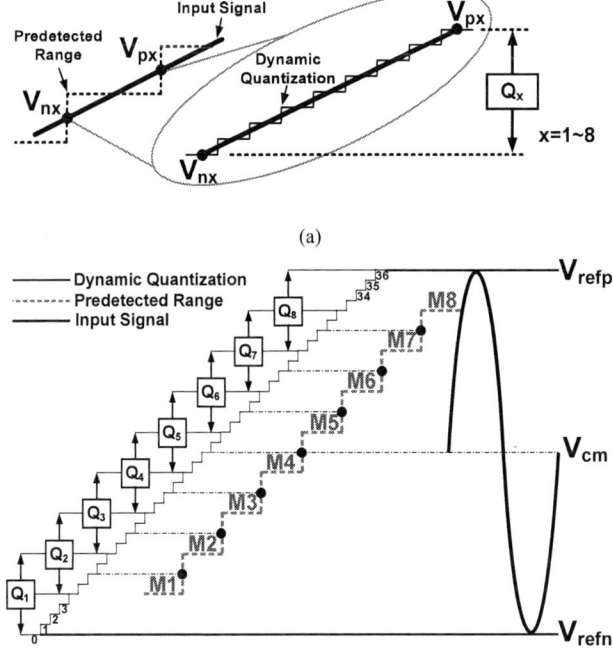

(a)

(b)

Fig. 1. (a) Dynamic quantizer having adjustable reference voltages, V_{px} and V_{nx}, where x=1, 2, ..., 8. (b) Principle of dynamic quantization.

In general, a 5-bit, 33-level quantization can be performed by the combination of the small-range quantizations Q_1, Q_3, Q_5, and Q_7. However, since the output of the second integrator includes signal and quantization noise, rapid switchings among these small-range quantizations would be required that brings a great challenge in the circuit design. Therefore, the designed quantization ranges will be overlapped to cover the inevitable voltage fluctuation to avoid the unnecessary switchings among small-range quantizations. In the presented dynamic quantizer, the quantizations Q_2, Q_4, Q_6 and Q_8 will be added to prevent incorrect quantizations from degrading the performance. Due to the addition of four overlapped quantization ranges, a 3-bit, 37-level quantization can be achieved in the proposed dynamic quantizer.

B. Architecture

The architecture of the proposed CIFB $\Delta\Sigma$ modulator with pre-detective dynamic quantizer is shown in Fig. 2. Because of the inherent feedbacks existed in the $\Delta\Sigma$ modulator, delaying integrators are often adopted to accept longer propagation time of feedbacks and avoid too much loading from causing the need of large slew rate of operational amplifiers. Thus, there will be two clock cycle delays from the signal input to the quantization. In this time period, the signal magnitude can be roughly detected. The detector signals will then activate the mode selector to choose one set of reference voltages for the multibit quantizer according to outputs M1, M2, ..., and M8, as the grey steps shown in Fig. 1(b). The fine quantization will be accomplished when the signal arrives at the dynamic quantizer. Finally, the output of pre-detector and dynamic quantizer signal

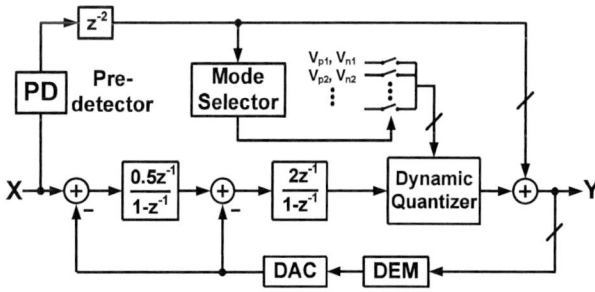

Fig. 2. Linear model of the $\Delta\Sigma$ modulator with pre-detective dynamic quantizer.

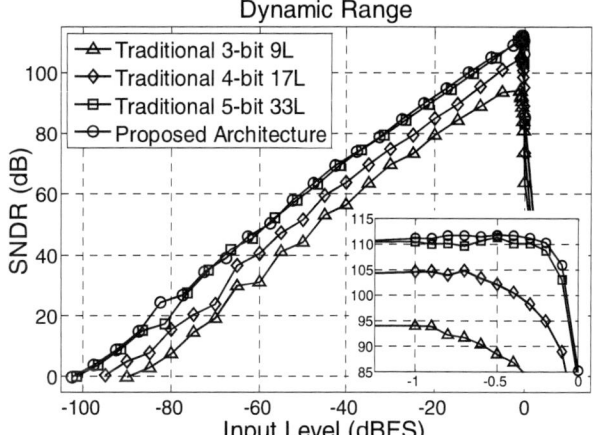

Fig. 3. Comparison of dynamic range of the poposed and traditional multibit $\Delta\Sigma$ modulators.

will be combined to produce a 3-bit, 37-level pre-detective dynamic quantization.

C. Dynamic Range

When the proposed multibit $\Delta\Sigma$ modulator with dynamic quantization is simulated by MATLAB, the resulting input dynamic range (DR) is shown in Fig. 3. The simulation results of other traditional multibit $\Delta\Sigma$ modulators are also presented for comparison. It can be seen that the proposed $\Delta\Sigma$ modulator presents a better DR than the 3-bit, 9-level and 5-bit, 33-level ones by about 10 dB and 1 dB, respectively. That is, the proposed $\Delta\Sigma$ modulator has the similar number of components in the traditional 3-bit, 9-level one, but exhibiting a good performance compatible to the traditional 5-bit, 33-level modulators [8].

III. CIRCUIT IMPLEMENTATION

A. Detector and Mode Selector

The conventional 3-bit quantizer is utilized to be the signal detector in the presented $\Delta\Sigma$ modulator, as shown in Fig. 4(a), where, V_{cm} is the common-mode voltage, V_{refp} and V_{refn} denote the maximum amplitude of the input signal range. Symbols qn, n=1, 2, ..., 7, are the output signal of the detector. After the input voltage is detected, the output signal of the detector will be transmitted to the mode selector to generate controlled

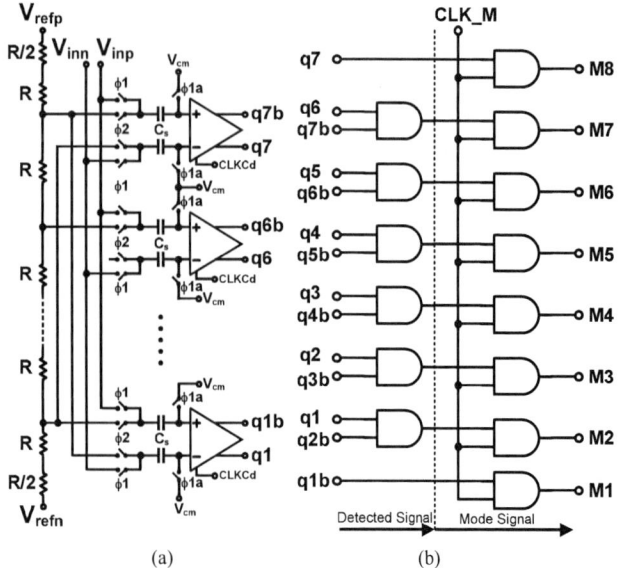

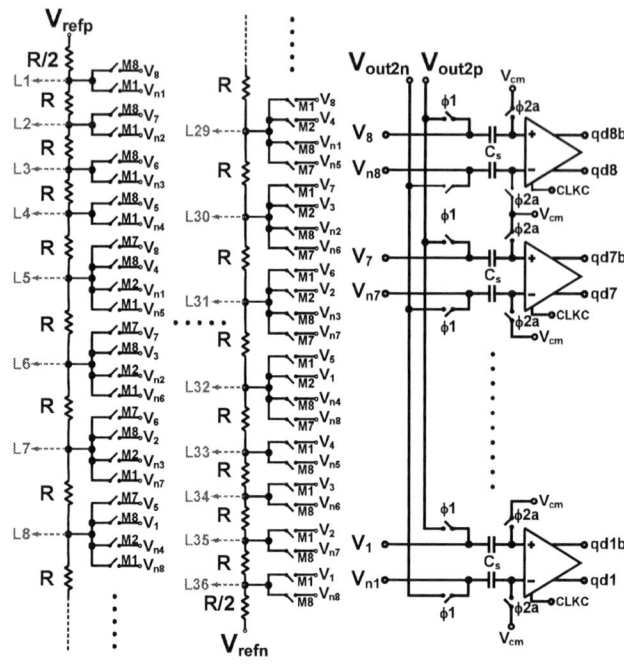

Fig. 4. Schematics of (a) detector and (b) mode selector.

signals, M1, M2, ..., M8, as shown in Fig. 4(b). When the CLK_M is on, the mode selector can record rough input levels and produce one mode at a time to switch reference voltages from the resister string of dynamic quantizer. One set of positive and negative reference voltages will thus be chosen in the proposed dynamic quantizer.

B. Dynamic Quantizer

The proposed dynamic quantizer is shown in Fig. 5. The main part of the architecture is a conventional 3-bit quantizer. The resister string of the conventional multibit quantizer is terminated by two constant reference voltages. However, the quantization range of the dynamic quantizer will be changed according to the magnitude of input signal, and different reference voltages will thus be required for the resister string. Once the output of the mode selector is determined, eight references, which are closer to the signal voltage, will be switched to the 3-bit quantizer before signal arrived at the

Fig. 5. Schematic of dynamic quantizer.

dynamic quantizer.

For example, when the detector output q7=1, the output signal M8 of mode selector will be high. Moreover, the sixteen differential references, $L_1 \sim L_8$ and $L_{36} \sim L_{29}$, on the resister string will be distributed to different comparison levels, $V_8 \sim V_1$ and $V_{n8} \sim V_{n1}$, of the 3-bit quantizer. When the detector output q6=1, output M7 of the mode selector will be high. The references starting from the fifth highest level of voltages, $L_5 \sim L_{12}$ and $L_{32} \sim L_{25}$, will be connected to the 3-bit quantizer to perform the following signal quantization. Finally, the fine-level output of dynamic quantizer will be produced.

C. Delta-Sigma Modulator with Pre-detective Dynamic Quantizer

The schematic of the proposed $\Delta\Sigma$ modulator with pre-

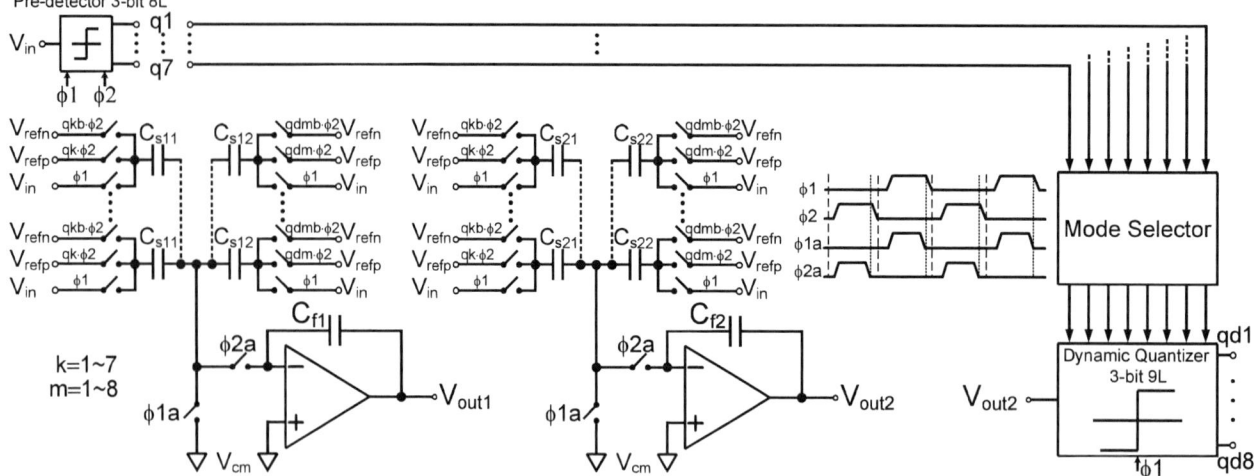

Fig. 6. Schematic of the proposed modualtor with pre-detective dynamic quantizer.

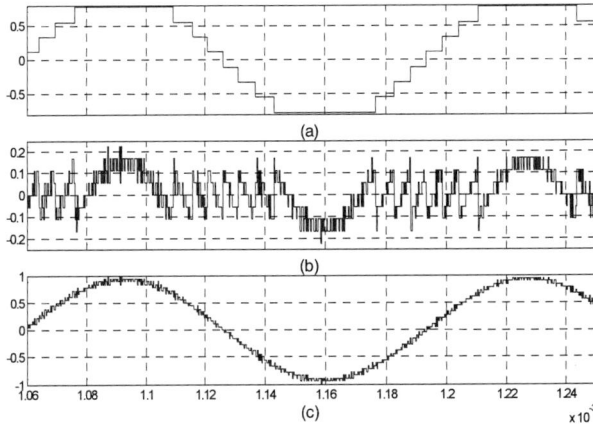

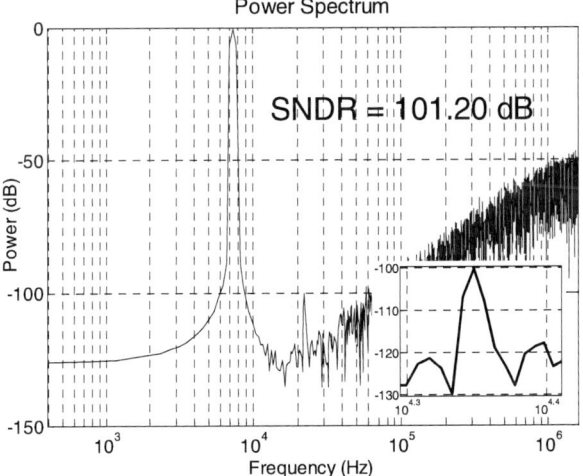

Fig. 7. (a) The output signal of pre-detector (b) The output signal of dynamic quantizer (c) The combination of above signal.

Fig. 8. Output spectrum of the proposed modualtor with pre-detective dynamic quantizer.

The transient outputs of the pre-detector and dynamic quantizer are shown in Fig. 7(a) and 7(b). The composite signal of these two components is plotted in Fig. 7(c). A 37-level quantization-step output has been successfully verified by HSPICE. The 8192-FFT output spectrum is obtained at a 3.2 MHz frequency, as shown in Fig. 8. The SNDR is 101.2 dB in a 25 kHz signal bandwidth. The power consumption is 1.68 mW at a 1.8 V supply voltage. The performance summary is listed in Table I.

TABLE I.
PERFORMANCE SUMMARY OF THE PROPOSED PRE-DETECTED $\Delta\Sigma$ MODULATOR WITH DYNAMIC QUANTIZER

Technology	0.18 μm 1P6M CMOS Technology
Supply Voltage	1.8 V
Signal Bandwidth	25 kHz
Clock Rate	3.2 MHz
Oversampling Ratio	64
Peak SNDR	101.2 dB
Dynamic Range	102 dB
Power Dissipation	1.68 mW
ENOB	16.52 bits

V. CONCLUSION

In this paper, a $\Delta\Sigma$ modulator with pre-detective dynamic quantizer is proposed. In the proposed dynamic quantizer scheme, only a 3-bit quantizer is used to perform a 37-level quantization. A significant reduction in the number of comparators is accomplished compared to traditional quantizers. The resulting power consumption would also be reduced.

ACKNOWLEDGMENT

This research was supported in part by a grant of finance support by the National Science Council (NSC) of Taiwan, R. O. C.

REFERENCES

[1] J. Ron, S. Byun, Y. Choi, H. Roh, Y. G. Kim, and J. K. Kwon, "A 0.9-V 60-μW 1-Bit Fourth-Order Delta–Sigma Modulator with 83-dB Dynamic Range," IEEE J. Solid-State Circuits, vol. 43, no. 2, pp. 361–370, Feb. 2008.

[2] A. Gharbiya, and D. A. Johns, "A 12-bit 3.125 MHz Bandwidth 0-3 MASH Delta-Sigma Modulator," IEEE J. Solid-State Circuits, vol. 44, no. 7, pp. 2010–2018, July. 2009.

[3] R. Schreier, and G.C. Temes, Understanding delta-sigma data converters, NJ: IEEE Press, 2005.

[4] L. R. Carley. "A Noise-Shaping Coder Topology for 15+ Bit Converters," IEEE J. Solid-State Circuits. vol. 24, no. 2, April. 1989.

[5] F. Colodro and A. Torralba, "Continuous-time sigma-delta modulator with a fast tracking quantizer and reduced number of comparators," IEEE Trans. Circuits Syst. I, Reg. Papers, vol. 57, no. 9, pp. 2413–2425, Sep. 2010.

[6] S. Pesenti, P. Clement, and M. Kayal, "Reducing the number of comparators in Multibit Delta Sigma Modulators," IEEE Transactions on Circuits and Systems I, vol. 55, no. 4, pp. 1011–1022, May 2008.

[7] A. Gharbiya, and D. A. Johns, "On the implementation of input feedforward delta-sigma modulators," IEEE Trans. Circuits Syst. II, Express Briefs, vol. 53, no. 6, pp. 453-457, June 2006.

[8] J. M. de la Rosa, "Sigma-Delta modulators: Tutorial overview, design guide, and state-of-the-art survey," IEEE Trans. Circuits Syst. I, Reg. Papers, vol. 58, no. 1, pp. 1–21, Jan. 2011.

detective dynamic quantizer is shown in Fig. 6. Single-ended structure is depicted here for simplicity. The corresponding clocks are also shown in the figure. At phase $\phi1$, signal will be sampled. At phase $\phi2$, input and DAC feedback signals will be integrated to the integrator. The switches near the input terminals of opamp will be turned off earlier to reduce the charge injection effect.

In feedforward path, because of propagation delay introduced by pre-detector, only one clock cycle delay after pre-detector is added to fit the requirement of signal timing in real implementation. The capacitances used in the first and second integrators are C_{s11}=0.6 pF, C_{s12}=0.15 pF, C_{s21}=0.2 pF, and C_{s22}=0.05 pF. The integrating capacitances are C_{f1}=10.8 pF and C_{f2}=0.9 pF.

IV. SIMULATION RESULTS

The proposed prototype is fabricated in TSMC 0.18-μm 1P6M standard CMOS process. A -0.6 dBFS signal is applied to the proposed $\Delta\Sigma$ modulator to perform the following test.

Influences Study on MIM Capacitors' Reliability

Chu Tsui Ping, Yang Peng, Tee Pei Ling
Technology development, X-FAB Sarawak Sdn. Bhd.
Sama Jaya Free Industrial Zone, 93350 Kuching, Sarawak, Malaysia.
Email: tsuiping.chu@xfab.com

Abstract – **Reliability assessment tests are used to evaluate the quality of different process schemes of MIM capacitors. Typically, VRAMP tests can be used to check for extrinsics; which are common and popular method used for evaluating yield issues and early life failures (in which the product failures in ppm level); while TDDB tests are used to determine the intrinsic quality of the capacitor dielectrics; thus the lifetime will be extrapolated accordingly from its dependency from accelerated tests at different higher stress conditions down to the corresponding use condition.**

In this paper, we will summarize the different approaches of making high capacitance density MIM capacitors in two dimensions – thickness or area: that is to say to achieve with thinner dielectrics or by using stack layers (i.e. single, double, or even triple stacked up in parallel). Comparisons are made in terms of breakdown voltages, leakages, linearity, RF Q-factors and mainly the reliability impact: extrinsics level and intrinsic lifetime even with the same capacitance density. Further Study on various process variants with reliability assessment has also been done, and will be discussed here as well.

Key words: high capacitance MIM; TDDB

I. INTRODUCTION

IN the past several CMOS generations, dielectric integrity check has already been one of the critical reliability tests assessments. Enormous studies have been carried out in the area of gate oxide integrity; while just a few and limited literature focus on some other dielectrics reliability performance, in particular, in this paper, we focus more on the commonly usage of MIM (metal-insulator-metal) capacitors as one of the popular components in RF/mixed-signal applications.

For the characterization of the quality of dielectric films, it is essential to have measurement methods available which can give a measure of dielectric reliability in a relatively short time. Stress biases are usually highly accelerated and cause destructive dielectric breakdown. A good understanding of the stress methods and the various measured parameters is essential to draw correct conclusions for the lifetime of the dielectric at operating conditions. MIM capacitor is a vertical structure which includes two electrodes, the dielectric and the routing. The weakest link in that construction will determine the measured lifetime. MIM dielectric might be affected by inhomogeneities which are not typical for gate dielectrics. Thickness variation due to electrode surface roughness or mechanical stress divergences also play a role. Achieving high capacitance MIM without jeopardizing all the other related quality factors of the capacitor in particular its reliability will be a challenge.

This paper will report and summarize the different approaches of achieving high capacitance MIM. Comparison in terms of major electrical parametric performance, and reliability assessments. Furthermore, various MIM capacitors' process influence study will also be discussed, especially the dependency on TDDB reliability performance.

Session II, introduction of different approaches in achieving high capacitance MIM, and its corresponding electrical characteristics. Session III, reliability performance comparison on different process approaches. Session IV other related characterization. Session V, stated some process evaluation impacts on TDDB results, and finally Session VI conclusion and outlook.

II. ELECTRICAL CHARACTERIZATION

TWO approaches in achieving high capacitance: say for MIM capacitors in this context by varying the two different dimensions- thickness and area. By a simple stacking up the process layers to increase the capacitor areas connected in parallel, the total capacitance can be doubled or tripled; on the other hand, to simplify the process, it is also easier to reduce the corresponding dielectric thickness by half or one-third to achieve the respective same target goals on capacitance.

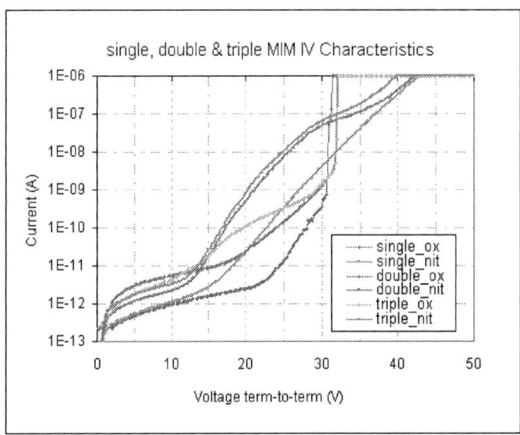

Fig.1a I-V characteristics of single, double, triple stack layers of MIM capacitors on two different dielectrics (oxide and nitride).

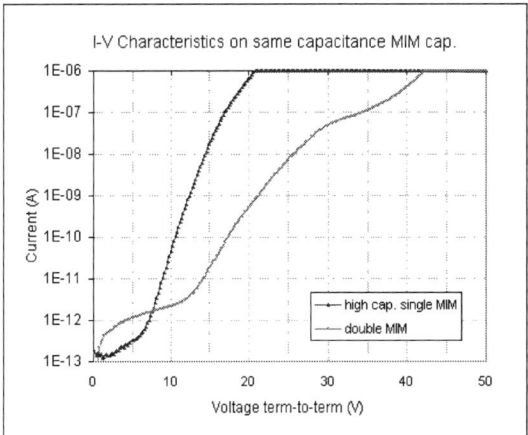

Fig.1b I-V characteristics of same capacitance density of single MIM (varying dielectric thickness) and double MIM capacitors.

Fig.1a showed two different commonly used dielectric materials oxides and nitrides in MIM capacitors; breakdown voltages are higher for nitride than for oxide, while the leakages within 15-30V is higher for nitride. In addition, with stacking layers of MIM (same area of single layer stacking up) breakdown are almost or slightly lower than that of single MIM layer, while leakages are in general higher as well with an order of magnitude difference. Fig.1b showed the I-V characteristics of the same capacitance density from a single MIM layer with reduced half thickness of dielectrics and a double stack MIM layers (same dielectric material, standard thickness); breakdown voltages are almost one-third for the single MIM than the stack, as it's strongly dependent on the dielectric thickness; while leakages are also high from 10V onwards.

Positive results on breakdown voltage as well as leakage from stack layers when complexity in process is not an issue. Tradeoff of good performance and cost of production maybe the concern.

III. RELIABILITY ASSESSMENTS

Tight distributions in VRAMP without extrinsics is key for early life reliability performance. Fig.2 showed all the variants have good performance. With stacking layers of MIM say double MIM, with similar capacitance density as the single high Cap. MIM (reduced dielectrics thickness), breakdown voltages are higher [thickness influence] while variation are comparatively tight. With the same dielectric thickness used, single MIM possess a higher breakdown than respective double and triple stack layers. [area influences]; larger area has a comparatively higher chance of earlier breakdown (conduction path established).

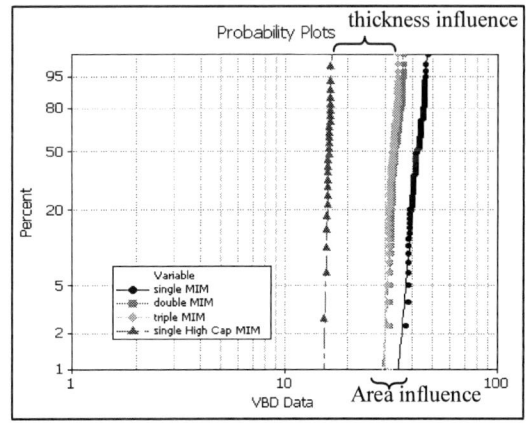

Fig.2 VBD distribution of double stack layers of MIM capacitor compared to single layer of reduced dielectric thickness (maintain same capacitance density). Intrinsic failure distributions obtained among the variables.

In general, using the dielectrics breakdown failure models; with ramp voltages applied, hot energetic carriers will be generated with tunneling leakages increase, more and more traps created until a percolation conduction path exist between the two MIM electrode terminals and thus breakdown happened. So the corresponding time-dependent-dielectric-breakdown (TDDB) depends on the level of breakdown voltages. The expectations, will be with the thinner the dielectrics, the lower the lifetime.

Experiment done on dielectric thickness splits to verify the theory, while keep all other process factors in the capacitor module remain the same.

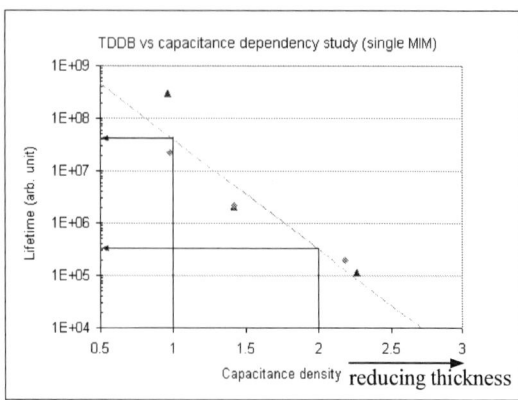

Fig.3 Lifetime dependency on capacitance density with respect to its dielectric thickness of single layer of MIM capacitor. (TDDB tests done at 175°C and lifetime extracted at standard reference area at 100ppm level). [Blue and pink points refer to different process variants in capacitor process module].

Figure 3 shows the relationship of the TDDB lifetime extraction at the use operating condition on the single layer MIM device with the capacitance density by varying the dielectric thickness. The thinner the thickness, the higher the capacitance with the lower the lifetime as expected from the breakdown failure mechanism model. The level of lifetime degradation with a double of capacitance density, reveals about a 2-orders of magnitude's difference. This is induced from the dielectric thickness difference in a single MIM layer.

On the other hand, by stacking up with two or three MIM layers (i.e. using same area for every layer repeat the process up) in order to increase its capacitance density by double or triple; the lifetimes do not seem to degrade much as compared to that of reducing its dielectric thickness. The influences are made from the area scaling base on the same random defect density from the Poisson distribution. The difference in lifetimes is small as the time to build up a conduction path lead to a dielectric breakdown depends more on dielectric thickness rather than the capacitor area because the electric field is higher with the thickness reduced by half while it remains unchanged even area is increased double.

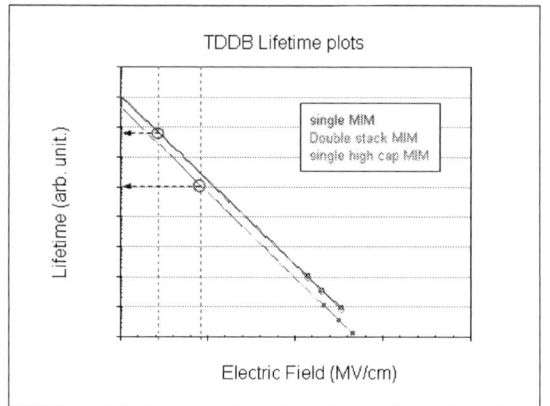

Fig.4 TDDB lifetime projections from various MIM (single MIM, double stack MIM and single high cap MIM). The latter two schemes possess the same capacitance density. [Tests are characterized at 175°C.]

Figure 4 reveals the TDDB lifetime projections from accelerated stress tests to the same use operating conditions with the different dielectric materials' thickness for the standard and high capacitance target, thus the corresponding electric fields are different (thin dielectric have a higher electric field criteria). Area is double for the double stack MIM compared to the single MIM; however lifetimes do not differ much. Single high capacitance MIM possess the same capacitance density as the double stack one, however the lifetime degrade more as the dielectric is with a higher electric field environment. This is a trade-off of cost of operation/process complexity than the reliability performance in general.

IV. OTHER CHARACTERISATION

Other related electrical properties of MIM also have been characterized accordingly for comparison on the difference approaches of

Table.I Summary of key electrical parameters performance among various MIM scheme.

MIM SCHEME	LINEARITY		RF	MATCHING
/KEY PARAMETERS	VC1 (ppm/V)	VC2 (ppm/V²)	Q	AC (%µm)
SINGLE	15	3.5	200	0.4
DOUBLE STACK	~ 0-3	3.5	100	0.28
SINGLE HIGH CAP	120	35	95	0.34

increasing the capacitance density. Table.I showed some advantages in terms of linearity in comparison for making double stack layers than using a reduced dielectric thickness of single high capacitance MIM. Linearity is found to be inversely proportional to its thickness; so using a reduced (say by half) dielectric thickness, [1]. RF Q-factor is roughly proportional to ~1/C; so it's quite comparable between the two approaches. Matching is considered to be comparable good; double stack with twice the capacitor area will be having better matching in a scaling factor of $1/\sqrt{2}$. [2]

V. PROCESS INFLUENCES STUDY

Besides the different construction approaches of integrating a high capacitance MIM discussed in the previous sessions, evaluation experiments have also been carried out on various process factors – such as dielectrics materials, dielectrics thickness, materials properties e.g. RI (reflective index), deposition RF power, capacitor metal scheme, bottom electrode metal process, capacitor pattern etching as well as post etch polymer removal cleaning processes, etc … to determine any reliability impact. In general, single MIM process with the different process variants are studied, and all reliability tests are characterized and assessed at the maximum operating temperature: 175°C.

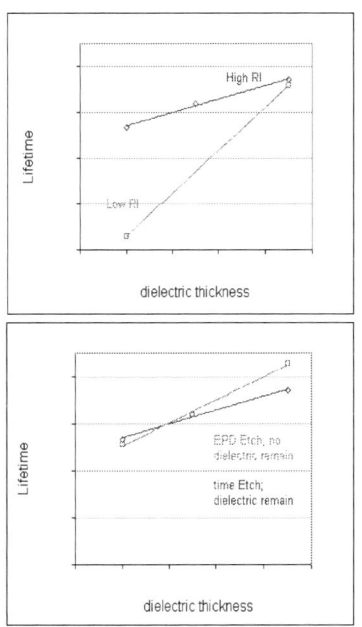

Fig.5. TDDB lifetime projections interaction plots between dielectric thickness dielectric materials properties as well as capacitor metal etch process.

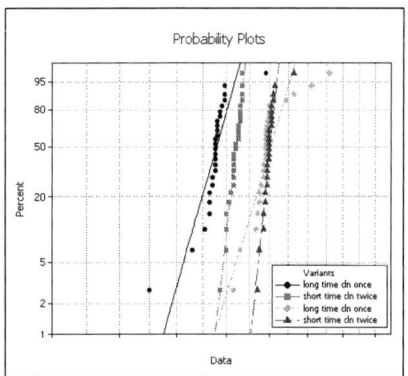

Fig.6. Effects of post etch polymer removal cleaning efficiency on VRAMP Weibull distribution.

Key factors impact on reliability lifetimes (extracted from accelerated constant voltages stress) are found to be mainly the dielectric thickness whereas the secondary with the dielectric material properties, for example, as illustrated in Fig.5, strong interactions seen for the material properties itself with thickness while the capacitor metal etch process does not have big differences in lifetimes.

Tighter distributions in VRAMP tests [3] seen in Fig.6 for the shorter time cleaning twice condition at the post etch polymer removal clean process. In terms of TDDB lifetimes, however, there is not much significant differences obtained from the different approaches of cleaning process.

Other factors reported elsewhere[4], such as the capacitor top electrode and the bottom electrode metal processes, have also been studied but with less influences seen from our evaluation experiments. Both the breakdown voltages and TDDB lifetimes were characterized for different dielectric materials as well. Nitride was found to be possessed a bit better in terms of both the breakdown voltages (as seen from the IV characteristics) and lifetimes than oxide as the capacitor dielectrics (higher k dielectric constant, thus thicker layer can achieve the same capacitances).

VI. CONCLUSION & OUTLOOK

In summary, comparison in terms of electrical characteristics and reliability assessments are done between varying different dimensions and the process construction approaches in achieving high capacitance MIM capacitors. There are pros and cons among the different approaches of the process scheme. Respective process schemes can be chosen dependent on the product application needs, reliability requirements and the cost of operations accordingly. In addition, process influences are studied comprehensively while the impact was found to be significant from the interactions of the dielectric material properties itself in the case with thin dielectric layer (possess the high capacitance density but a lower TDDB lifetime in general). Therefore, a reasonable "thick" dielectrics (high-k will be a must) is required to meet the reliability goal in achieving high capacitance together.

ACKNOWLEDGEMENT

The authors would like to thank their management, characterization team, Quality Reliability group and operation modules department for their support, technical advice in making contribution to this paper.

REFERENCES

[1] C.C. Ho, B-S. Chiou, "Reliability Analysis Using Weibull Distribution on the Breakdown of MIM Capacitors" Microsystems, Packaging, Assembly and Circuits Technology, 2007. IMPACT 2007. International, pp. 270-273.

[2] T. P. Chu, P. Yang, Evie Kho, Y.K. Ang and S.H. Tia, "Linearity Improvement on MIM Capacitors" IEEE ECS Transactions, 34(1) CSTIC 2011, pp. 119-124.

[3] T. P. Chu, P. L. Tee and Günter Lau, "Analog Matching properties Process Dependency on MIM Capacitors" Proceeding of IEEE-Regional Symposium on Micro and Nano-electronics RSM 2011 -conference.

[4] Inoue, et al. "Surface Control of Bottom Electrode in Ultra-Thin SiN Metal-Insulator-Metal Decoupling Capacitors for High Speed Processors" Japanese Journal of Applied Physics, Volume 46, Issue 4B, pp1968 (2007)

Investigation of Crosstalk Impact on Channel Performance from IC package and Motherboard Breakout Routing

Azri Husni Hasani[1], Aftanasar Md. Shahar[1], Ahmad Jalaluddin Yusof[2], and Jackson Kong[2],
[1] School of EE, Universiti Sains Malaysia,
14300 Nibong Tebal, SPS, Penang, Malaysia.
Email: azri.husnix.hasani@intel.com
[2] Intel Microelectronics (M) Sdn. Bhd.,
FIZ, 11900 Bayan Lepas, Penang, Malaysia
Email: ahmad.jalaluddin.bin.yusof@intel.com

Abstract- **Crosstalk is one of the signal integrity (SI) issues which is critical especially in systems with high operating speed and high circuit density. In this paper, the impact of crosstalk on channel performance is investigated based on the IC package and motherboard breakout routing. Differential pair spacing and coupling length were manipulated for the IC package breakout routing and motherboard breakout routing respectively. First, the crosstalk behavior on the breakout routing models are investigated in frequency and time domain. Next, the breakout routing models are used in the channel analysis, along with other components that form a channel. The crosstalk impact towards the channel is presented in terms of eye diagram parameters. Some conclusions are presented and can be used as a guideline for future circuit designers.**

I. INTRODUCTION

Due to the development of technology, circuits nowadays are designed in high densities which correspond to Moore's Law which stated that the number of transistors on a chip doubles roughly every two years [1]. Apart of that, higher clock operating frequency is used to make the system faster and more capable [2]. These two factors caused a signal integrity issue known as crosstalk to become more severe.

Crosstalk refers to the unintended coupling between adjacent transmission lines. It is an electromagnetic phenomenon caused by the capacitive and inductive coupling between the adjacent lines [3]. This issue may degrade system performance. Hence, it is essential to suppress the crosstalk to an acceptable margin. To achieve this objective, several methods have been implemented by high-speed PCB designers such as increasing the separating distance between the lines, reducing the coupling length of the lines, include

guard rings between adjacent lines, reducing the distance between the lines and the reference plane, and etc. [4, 5].

In this paper, the impact of crosstalk from IC package and motherboard breakout routing towards the channel is investigated. A channel is defined as the bidirectional data path of a signal from input to output. For the IC package, the differential pair spacing is varied whereby for the motherboard breakout routing, the coupling length is varied. The crosstalk on the breakout routing models were investigated prior to the crosstalk on the channel for correlation purpose.

In the next section, the types of crosstalk are presented to provide brief theoretical aspects of crosstalk. Then, the structure of the simulation models used in this work is presented. In section four, the simulation results in frequency and time domain is presented with respect to the types of crosstalk. The channel analysis comes next and finally, the conclusion of the crosstalk investigation.

II. NEXT AND FEXT

Crosstalk is associated to two mechanisms which is the capacitive coupling and the inductive coupling between adjacent transmission lines [3]. These mechanisms cause coupling of signal on the victim line which is known as crosstalk. The crosstalk that appears on the victim line can be differentiated by the location; near end and far end. Therefore, there are two widely known types of crosstalk such as the near end crosstalk (NEXT) and the far end crosstalk (FEXT). The NEXT is the total crosstalk seen at the near end of the victim line and FEXT is the total crosstalk seen at the far end of the victim line. To observe the NEXT and FEXT, a typical single-ended crosstalk configuration and waveform is depicted in Fig. 3.

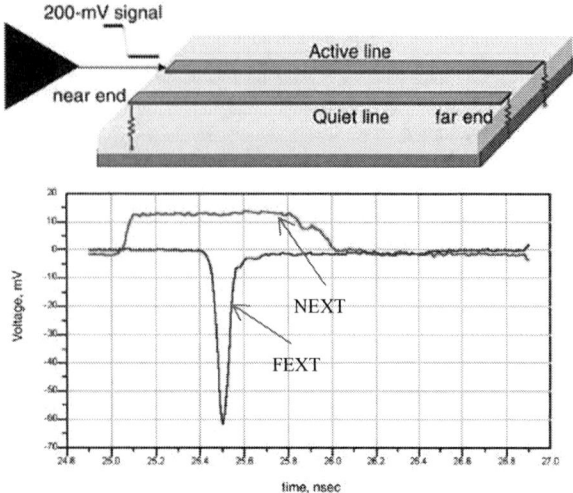

Fig. 1. Measurement configuration (top) and crosstalk on the quite (victim) line (bottom) [6]

In Fig. 3, the NEXT rises to a constant value instantly and stays for a period equals to twice the time [6]. This constant value is dependent on the separation distance of the lines. Before the NEXT reaches this value, it increases proportionally with the coupling length of the lines as long as the coupling length is less than the critical length [7].

The FEXT appears after the time delay, with quite a large magnitude. It appears only for a brief time but is large enough to degrade the SI of the system. This type of crosstalk is directly proportional to the coupling length of the lines. Mathematically, the NEXT and FEXT is expressed in [7] as

$$\text{NEXT} = \frac{1}{4}\left(\frac{C_{mL}}{C_L} + \frac{L_{mL}}{L_L}\right) \tag{1}$$

$$\text{FEXT} = \frac{len}{T_r} \times \frac{1}{2v}\left(\frac{C_{mL}}{C_L} + \frac{L_{mL}}{L_L}\right) \tag{2}$$

where

C_{mL} = mutual capacitance per unit length
C_L = capacitance per unit length
L_{mL} = mutual inductance per unit length
L_L = inductance per unit length
len = coupling length of the lines
T_r = signal rise time
v = speed of signal on the line

From (1) and (2), the NEXT and FEXT are directly proportional to the mutual capacitance and mutual inductance between the adjacent lines. Since the mutual capacitance and mutual inductance between adjacent lines increases when the lines are closer [4], it can be stated that the NEXT and FEXT are directly proportional to the distance between the lines. From (2), the FEXT also increases when the coupling length increases and when the rise time of the signal shrinks [7]. From these relationships, NEXT can be reduced by increasing the separating distance between the lines whereby FEXT can be reduced by increasing the separating distance, reducing the coupling length, and using signals with slower rise time [7].

III. SIMULATION MODELS

The breakout routing of IC package and motherboard are modeled using Ansys's HFSS. The modeling approach was different between the breakout routings where the IC package breakout routing is extracted from a 2D package layout whereby the motherboard breakout routing is modeled based on Fig. 4. Both the breakout routing models were designed to have the characteristic impedance of 50 Ω. This is to reduce reflections from unmatched transmission lines. The substrate material used for the breakout routing models has the dielectric constant equals to 3.41 and the loss tangent equals to 0.018 at 5 GHz.

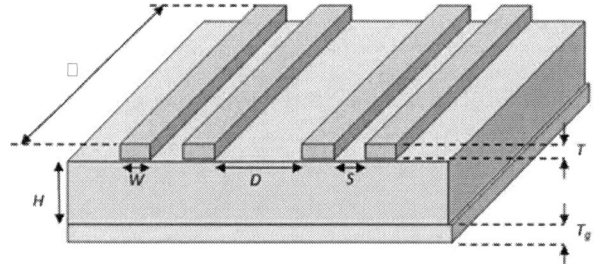

Fig. 4. Cross section and geometry of microstrip transmission line

For the channel analysis, the simulation is conducted using HSPICE. Therefore, the breakout routing models were converted into spice models to be included in the channel. The channel consists of other components such as balls, vias, AC capacitors, connectors, and etc.

To conduct the channel analysis for IC package, the breakout routing section in the channel at package level is replaced with the IC package breakout routing model. The same method is also used for the motherboard channel analysis where the motherboard breakout routing model is placed at the breakout routing section at board level. This enables us to investigate the crosstalk impact caused by the breakout routing models.

IV. SIMULATION RESULTS

Frequency Domain Analysis

In frequency domain, Ansys's HFSS is used to obtain the crosstalk behavior on the breakout routing models for the frequency range 50 MHz to 20 GHz. Fig. 5 shows the NEXT reading on the IC package breakout routing model. The crosstalk reading for differential pairs spacing 18 μm is denoted by continuous red color line, 27 μm denoted by dotted blue color line, and 36 μm denoted by dashed black color line.

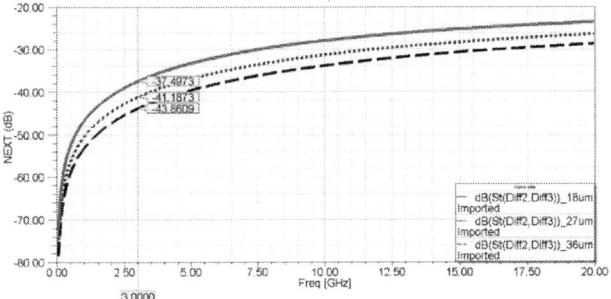

Fig. 5. NEXT versus frequency for IC package breakout routing model

As shown in Fig. 5, the NEXT is larger for smaller differential pair spacing. 18 μm spacing exhibits the largest NEXT followed by 27 μm, and 36 μm. Besides that, the NEXT for all spacing increases as the frequency becomes higher. This suggests that for a high operating frequency, it is necessary to increase the spacing distance between the differential pairs. The FEXT reading on the IC package breakout routing model is shown in Fig. 6. It behaves similarly to the case of NEXT, where it is larger for smaller spacing and increases with frequency.

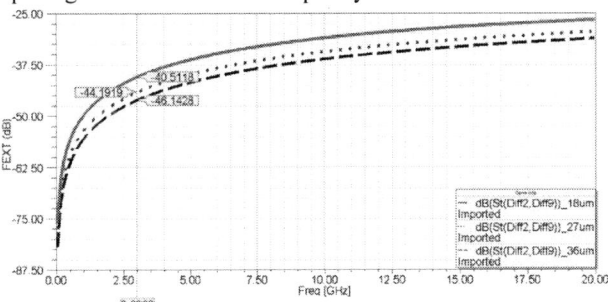

Fig. 6. FEXT versus frequency for IC package breakout routing model

For the motherboard breakout routing model, the NEXT is shown in Fig. 7. The colors represent different coupling length where for the NEXT, continuous red line represents 500 mils coupling length, dotted orange lines represents 1000

mils, black dotted dashed line represents 1500 mils, and blue dashed line represents 2000 mils.

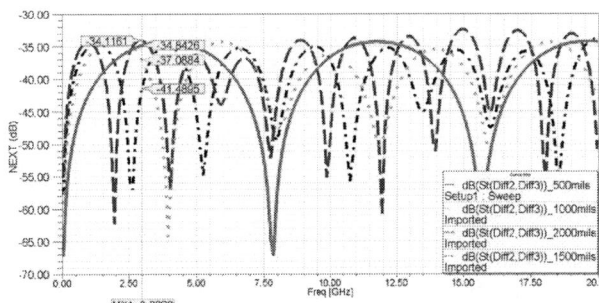

Fig. 7. NEXT versus frequency for motherboard breakout routing model

Compared to the NEXT on the IC package breakout routing previously, the NEXT on the motherboard breakout routing does not increase with frequency. The maximum magnitude is almost equal between different coupling lengths. However, different coupling length produces different NEXT behavior where longer coupling lengths have dips that occur more frequently compared to shorter coupling lengths. Fig. 8 shows the FEXT on the motherboard breakout routing model, with colors that represents similar coupling length as the NEXT except blue dashed line represents 1500 mils and black dotted dashed line represents 2000 mils. The FEXT has a consistent behavior, where longer coupling lengths exhibits larger FEXT. It also increases when the frequency becomes higher.

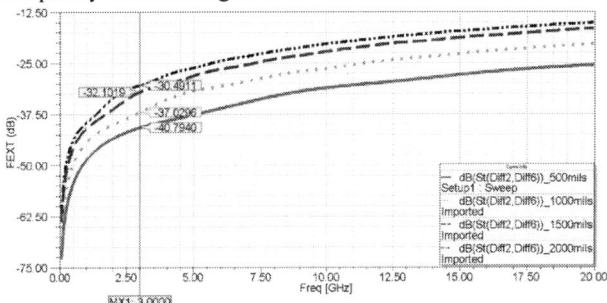

Fig. 8. FEXT versus frequency for motherboard breakout routing model

Time Domain Analysis

For the time domain analysis, the breakout routing models were converted into spice models. Then, they were used in Ansys's Designer to observe the step response of the breakout routing models. The aggressor pair was fed with a step input (1 V peak-to-peak) with signal rise time equals to 35 ps. The NEXT on the IC package breakout routing model is shown in Fig. 9. The continuous red line denotes the crosstalk reading for differential spacing 18 μm, blue dotted line for 27 μm, and black dashed line for 36 μm.

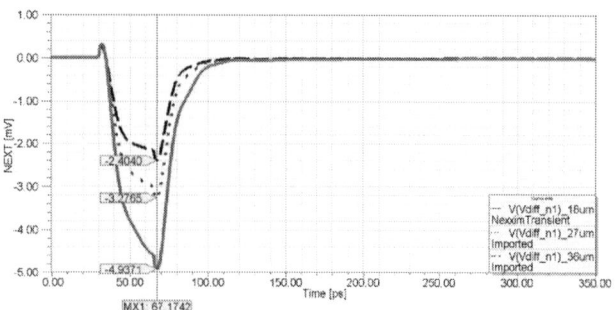

Fig.9. NEXT versus time for IC package breakout routing model

The NEXT behavior is similar as in the frequency domain previously where it increases when the spacing is smaller. Differential pair spacing 18 μm exhibits the largest NEXT magnitude compared to the other two spacing. The FEXT as shown in Fig. 10 also behaves similarly to NEXT where it is higher when the differential pair spacing distance is closer.

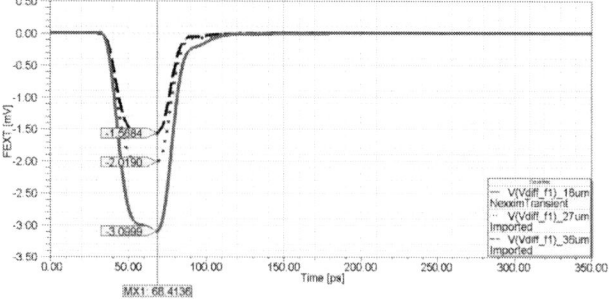

Fig. 10. FEXT versus time for IC package breakout routing model

For the motherboard breakout routing model, the NEXT in time domain is shown in Fig. 11. Continuous red line denotes 500 mils coupling length, dotted orange line for 1000 mils, dashed blue line for 1500 mils, and dotted dashed black line for 2000 mils.

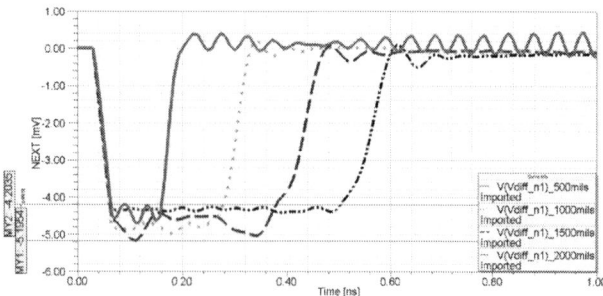

Fig.11. NEXT versus time for motherboard breakout routing model

In terms of magnitude, the NEXT is almost equal for all coupling lengths. A slight variation was observed due to the ripples which occur because of imperfect terminations. However, longer coupling lengths produce wider NEXT waveform. In [7], Bogatin stated that the NEXT magnitude will only increase with the coupling length if the coupling length is shorter than the critical length. The FEXT as shown

in Fig. 12 is significantly dependent on the coupling length of the pairs. 500 mils coupling length exhibits the smallest FEXT magnitude compared to the others. It can also be observed that the FEXT appears at different time for each coupling length. This is because the FEXT only appears when the coupled signal reaches the far-end of the victim pair [7].

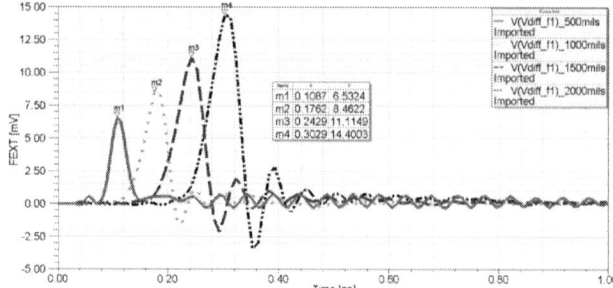

Fig. 12.FEXT versus time for motherboard breakout routing model

From the frequency and time domain analysis, it was found that the NEXT and FEXT coincides with (1) and (2). Besides that, the behavior of FEXT and NEXT on the breakout routing models were observed. These observations provided the first impression on the crosstalk impact towards the channel performance.

Channel Analysis

From the breakout routing models, both differential pair spacing and coupling length affects the crosstalk that appears on the victim pair. However, their impact towards the full channel might differ since extra components are taken into account. The results from the channel analysis were obtained in terms of eye diagram parameters such as the voltage margin and the timing margin.

A. IC Package

For the IC package channel analysis, the voltage margin and timing margin from differential pair spacing 18 μm is made as the reference. The margins from 27 μm and 36 μm were compared to the reference to observe the impact of differential pair increments towards the channel. Table 1 shows the margins improvement for the IC package channel analysis.

TABLE 1
EYE DIAGRAM MARGIN IMPROVEMENTS FOR IC PACKAGE CHANNEL ANALYSIS

Differential pair spacing (μm)	Voltage margin improvement (%)	Timing margin improvement (%)
18	ref	ref
27	0.8	0.6
36	1.0	0.6

When differential pair spacing 27 μm is implemented, the voltage margin and timing margin is found to improve by 0.8

percent and 0.6 percent respectively compared to the reference. However, the impact of differential spacing 36 μm is almost not visible. Compared to the reference, 36 μm differential spacing only improves the voltage margin by one percent, whereby the timing margin improves by 0.6 percent from the reference. From these observations, 27 μm provides better margin improvement compared to 36 μm.

B. Motherboard

For the motherboard channel analysis, the margins from 500 mils coupling length were made as the reference. Then, the results from other coupling lengths were compared to the reference to observe the impact of coupling length increments toward the channel. Table 2 shows the margins improvement for the motherboard channel analysis.

TABLE 2
EYE DIAGRAM MARGIN IMPROVEMENTS FOR MOTHERBOARD CHANNEL
ANALYSIS

Differential pair coupling length (mils)	Voltage margin improvement (%)	Timing margin improvement (%)
500	ref	ref
1000	-5.7	18.6
1500	-2.9	18.6
2000	-8.6	40.0

From Table 2, the voltage margins were found to be negative. This means the voltage margins become worse compared to the reference. For 1000 mils coupling length, the voltage margin is found to reduce by 5.7 percent but the timing margin improves by 18.6 percent. The same occurs for 1500 mils coupling length where the voltage margin reduces by 2.9 percent. The timing margin is found to improve by the same percentage as 1000 mils which is 18.6 percent. For 2000 mils coupling length, the voltage margin is worst where it reduces by 8.6 percent, but the timing margin improves the most at 40 percent.

From the results, it was found that there is no distinctive pattern on the margins. First, the voltage margin is found to reduce when the coupling length becomes longer. However, the reduction is not proportional to the coupling length. Second, the timing margin is found to improve despite the coupling length increments. It improves by 18.6 percent for 1000 mils and 1500 mils and 40 percent at 2000 mils coupling length.

The reason why there is no distinctive pattern with respect to the coupling length increments is because the channel analysis utilizes multilayer transmission line configuration. As explained in [7, 8], the FEXT which is dependent on the coupling length almost equals to zero when the trace is surrounded by uniform dielectric. Therefore, in the motherboard channel analysis, the margins reduction is mainly because of NEXT. From the frequency and time domain analysis previously, the NEXT magnitude did not increase with coupling lengths. Furthermore, ripples are observed in time domain due to imperfect terminations. This coincides with the irregular patterns from Table 2.

V. CONCLUSION

The investigation of crosstalk impact on channel performance was conducted with respect to IC package and motherboard breakout routing. For the IC package breakout routing, the three breakout routing models with differential pair spacing 18 μm, 27 μm, and 36 μm were used. From these models, the crosstalk reading from the frequency and time domain are affected by the differential pair spacing. However, in the channel analysis, it was found that the spacing only caused a small impact towards the channel performance. No significant improvements were observed on the channel results from two times the spacing increment.

In the motherboard case, the differential coupling length was increased from 500 mils to 2000 mils. In frequency and time domain, these coupling length increments does not really affect the NEXT. However, the FEXT is directly proportional to the coupling length. In channel analysis, the increments did not affect the channel performance significantly as the crosstalk impact from 500 mils to 2000 mils is almost similar. This is also due to the fact that the channel is surrounded by uniform dielectric. Hence, causing the FEXT to reduce to almost zero.

REFERENCES

[1] Moore G. E. (1965) Cramming more Components onto Integrated Circuits. Electronics, 38(8).
[2] Schauer B. (2008). Multicore Processors – A Necessity. Proquest, pp. 1-14.
[3] Xiaosong J. and Runjing Z. (2007) Crosstalk Analysis and Simulation in High-Speed PCB Design. 8th International Conference on Electronic Measurement and Instruments, Xi'an. Pp. 2-437 – 2-440.
[4] Johnson H. and Graham M. (1993) High-Speed Digital Design: A Handbook of Black Magic. New Jersey: Prentice Hall
[5] Montrose M. I. (2000) Printed Circuit Board Design Techniques for EMC Compliance. New York: IEEE.
[6] Hall S. H., Hall G. W., and McCall J. A. (2000) High-Speed Digital System Design-A Handbook of Interconnect Theory and Design Practices. New York: Wiley and Sons
[7] Bogatin E. (2003) Signal Integrity-Simplified. New Jersey: Prentice Hall
[8] Mbairi F. D. (2008) High-Frequency Transmission Lines Crosstalk Reduction using Spacing Rules. IEEE Transactions on Components and Packaging Technologies. 31(3), pp. 601-610.

Failure Mechanism and Improvement On Gate Oxide Failure At The Edge of LOCOS

Lesley Wong Ying Ying, Deb Kumar Pal, Raymond Tan, Ng Hong Seng, Michaelina Ong, Tong Gee Hong, Wong Jian Sang

X-FAB Sarawak Sdn. Bhd.
1 Silicon Drive, Sama Jaya Free Industrial Zone
93350 Kuching, Sarawak, Malaysia
Email: lesley.wong@xfab.com

Abstract **Gate oxide early breakdown was investigated. It was verified that the gate oxide quality is good and failure was due to extrinsic causes. The failure, which was localized at the edge of LOCOS was similar to Kooi effect. However, investigations showed that it was due to nitridation occured during high temperature nitrogen anneal. Investigation methods to find the root cause of failure were explained. Alternative methods to solve the failure were explored; including thickening the sacrificial oxide layer and changing the nitrogen anneal process sequence. Final solution was chosen based on PCM stress test, QBD and TDDB result with minimal process change.**

I. INTRODUCTION

The gate oxide breakdown failure occurred intermittently when gate oxide was processed with deep well isolation scheme and with breakdown voltages as low as 0.1V. Refer Fig. 1. The problematic process was using LOCOS isolation, with deep well implant driven by high temperature nitrogen anneal.

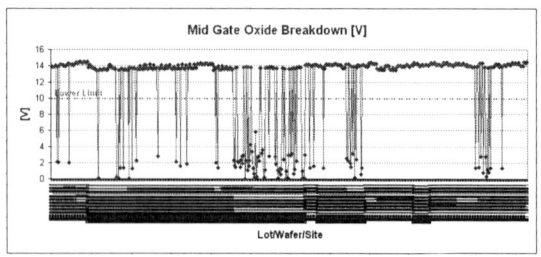

Fig. 1 Gate Oxide Breakdown trend showing many failure points as low as 0.1V

Repeated Qbd (charge-to-breakdown) tests increased the number of failures on the affected wafers, indicating that there was degradation in gate oxide performance with bias. The failure map in Fig. 2 was random and showed no wafer dependency.

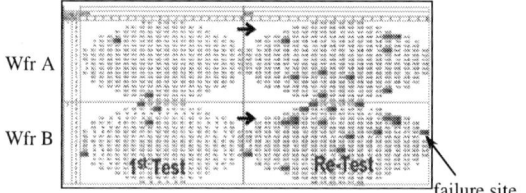

Fig. 2 Retest map shows more random failure than 1st test map

Oxide quality was verified problem-free by checking short loop wafers. So, it was apparent that failure was due to extrinsic causes. Reliability testing further confirmed the validity of failure.

II. PHYSICAL ANALYSIS

Several physical analyses had been conducted to understand the failure mechanism. Delayering the gate oxide breakdown structure for both good (~14V) and bad samples (~2.7V) showed random pinholes of ~0.5um in diameter at the edge of LOCOS. The physical observation was in agreement with the electrical result, where good sample had weak points which degraded in repeating test. Refer Fig. 3 for the localized failure.

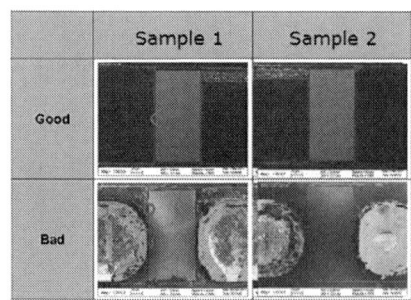

Fig. 3 Delayer result showed pinholes localized at many of the edges of LOCOS for both good and bad samples as shown by the red circles

Chemical treatment and dry etch tests showed signs of degradation along the edge of LOCOS (also known as bird's beak). It was also found that the pinholes extend to the substrate with chemical treatment. Other areas in the gate oxide capacitor were not affected. It was apparent that the whole length of the LOCOS edge was the weak region, vulnerable to chemical attack as well as other form of stress Refer Fig. 4.

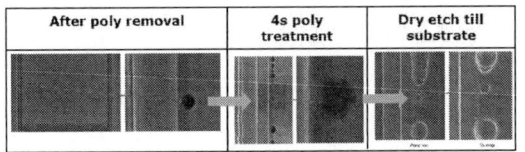

Fig. 4 Chemical treatment and dry etching on the same sample confirmed that the edge of LOCOS was the weak region

OBRICH analysis further confirmed that the fault was localized at the edge of LOCOS. TEM images exhibited clearly that failure mode was due to gate oxide thinning. Refer Fig. 5. The image on the oxide thinning described clearly the reason for oxide degradation during repeated testing.

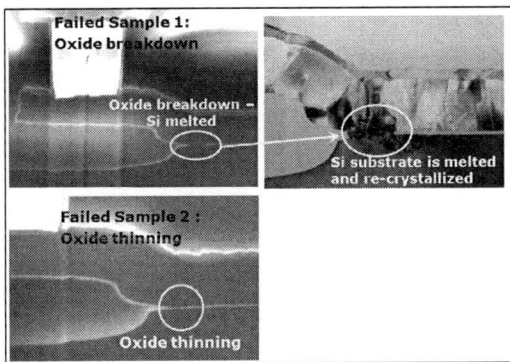

Fig. 5 TEM images showed that the failure was due to the localized oxide thinning at the edge of LOCOS

The TEM images indicated two possibilities which can cause the localized oxide thinning: (i) hardening of Kooi defect or (ii) nitridation during high temperature drive-in with nitrogen.

III. HYPOTHESIS

A. *Hypothesis on the hardening of Kooi defect*

A simplified diagram of the *Kooi effect* (which is also known as *white ribbon effect*) is shown in Fig. 6.

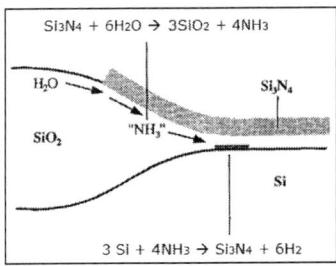

Fig. 6 Two reactions which take place during wet oxidation, causing Kooi effect to occur

A typical LOCOS thickness ranges from 400nm to 1500nm for 0.35um to 1.0um technologies. In order to grow the thick field oxide film, wet oxidation process is used for its fast oxide growth rate. Unfortunately, 2 side-reactions will take place during wet oxidation. At temperatures of $900^{\circ}C$ and above, the H_2O vapor will react with the masking nitride to form ammonia [1]. This can be illustrated with the chemical reaction below:

$$Si_3N_4 + 6H_2O \rightarrow 3SiO_2 + 4NH_3 \qquad (1)$$

The ammonia will permeate through the thin oxide at the edge of the LOCOS and react with the silicon dioxide and silicon to form an oxynitride compound:

$$Si + NH_3 + SiO_2 \rightarrow Si_xO_yN_z + H_2 \qquad (2)$$

This oxynitride compound then impedes the growth of the subsequent gate oxide, which will lead to early gate oxide breakdown failure. Since it is an oxynitride compound, the bonding with silicon is very strong and hot phosphoric cleaning during nitride removal is unable to remove this. To solve this problem, a sacrificial oxide layer is grown to lift up the oxynitride compound, followed by a HF strip. Refer Fig. 7. However, for this problematic process, the sacrificial oxidation was followed by a high temperature anneal before HF strip, which may harden the Kooi defect. The subsequent HF strip was then unable to totally remove the Kooi defect.

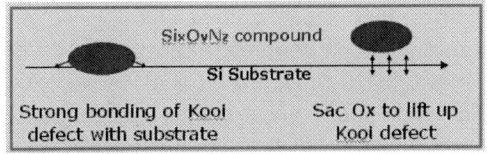

Fig. 7 To remove Kooi defect, a sacrificial oxide is grown to lift up the oxynitride compound, followed by HF strip

978-1-4673-2395-6/12 $31.00 © 2012 IEEE   589

To verify this hypothesis, a series of experiments were performed:

a. A thicker oxide layer was grown to lift up the Kooi defect, followed by HF strip for total oxide removal. After that, the sacrificial oxidation and anneal will take place.

b. An initial dry oxidation step was introduced during field oxidation to study the impact of denser oxide on retarding the diffusion of ammonia.

c. The last evaluation was adding HCL (hydrogen chloride) during wet oxidation [2]. This study is meant to verify whether HCL can limit the formation of ammonia by forming ammonium chloride (NH_4CL) as in the chemical reaction below.

$$HCL + NH_3 \rightarrow NH_4CL \qquad (3)$$

B. Hypothesis on nitridation during deep well anneal drive with nitrogen

The field oxide thinning at the edge of the LOCOS region is a natural behavior of this process due to limited supply of oxygen near the nitride edge as well as the compressive stress in the field oxide. This stress which is due to volumetric expansion reduces the surface reaction rate [3].

After the sacrificial oxidation, the field oxide thinning effect at the bird's beak region is still obvious. Refer Fig. 8.

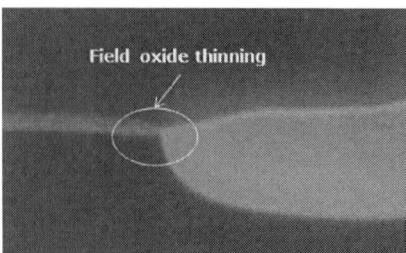

Fig. 8 Field oxide thinning effect is still obvious even after the growth of sacrificial oxide

During high temperature (>900^0C) N_2 anneal for more than an hour, the thin oxide at the edge of LOCOS will undergo thermal decomposition easier than other areas of the field oxide [4]. Silicon dioxide (SiO_2) at the bird's beak area will decompose into silicon monoxide (SiO), which can evaporate easily, as shown in equation (4).

$$Si + SiO_2 \rightarrow 2SiO \qquad (4)$$

In other words, the oxide at the bird's beak region can be partially or totally depleted, allowing nitridation to occur easily during the nitrogen anneal. The nitrided defect could not be removed easily and will retard the growth of the subsequent gate oxide.

To verify this hypothesis, two evaluations were performed:

a. To increase sacrificial oxide thickness.
b. Moving the deep well nitrogen anneal from after sacrificial oxidation to after field oxidation as shown in Fig. 9.

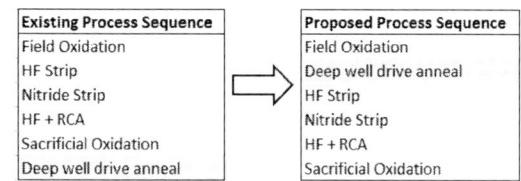

Fig. 9 Evaluation on shifting deep well drive anneal right after field oxidation

IV. RESULTS AND DISCUSSION

The assessment of gate oxide performance was based on characterized PCM stress test, QBD and TDDB (time-dependent dielectric breakdown) tests.

For hypothesis 1, the result for all the splits did not show any improvement. The result indicated that Kooi defect was not the root cause for failure. Refer Fig. 10.

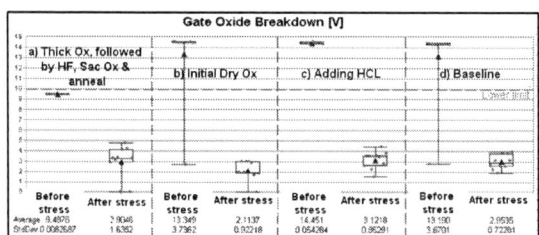

Fig. 10 PCM stress tests show no improvement from any of the evaluations, indicating that Kooi defect was not the root cause

For hypothesis 2, both splits showed improved breakdown voltages after PCM stressed test. This indicated that the gate oxide failure was due to the nitridation occurring during nitrogen anneal. Refer Fig. 11.

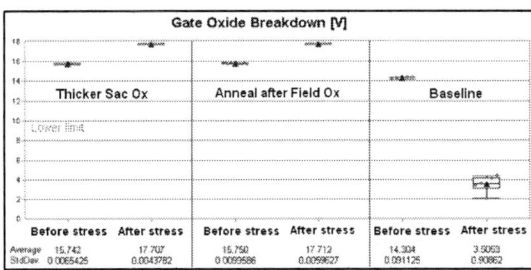

Fig. 11 Both splits showed good result, indicating that gate oxide failure was due to nitridation occur during anneal

Further analysis on the PCM parameters showed that device threshold voltages and saturation currents were adversely impacted for the thicker sacrificial oxide split. This was due to implant profile change with thicker screen oxide. For anneal right after field oxidation, the drift on PCM parameters was negligible.

Therefore, it was decided to shift the nitrogen anneal right after field oxidation. Refer Fig. 9. This condition was chosen as the optimal solution since process change was minimal and no adverse effects were observed.

The proposed solution showed more than 3 orders of improvement in QBD tests when compared to the baseline process. Refer Fig. 12.

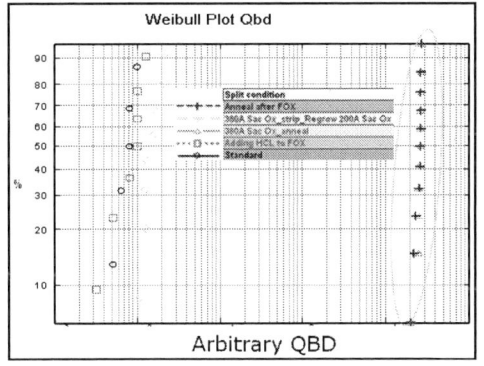

Fig. 12 Proposed solution showed improvement in QBD result as shown on the green circled.

The proposed solution also showed significant improvement in TDDB result. It improved the lifetime by 2 orders compared to the baseline process. Refer Fig. 13.

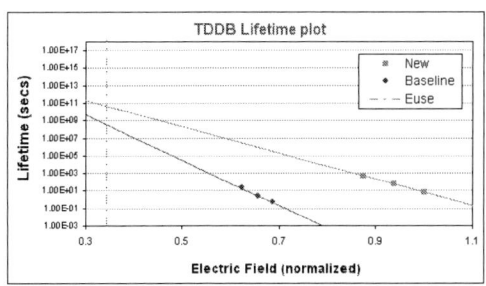

Fig. 13 Proposed solution showed improvement in TDDB

IV. CONCLUSION

Gate oxide failure was caused by the nitridation at the edge of LOCOS due to high temperature nitrogen anneal, leading to gate oxide thinning. It was learnt that anneal should be done with thicker oxide to reduce the nitridation effect. An optimal solution was found for the oxide thinning by shifting the nitrogen anneal step from post-sacrificial oxidation to post-field oxidation.

ACKNOWLEDGEMENT

The author would like to thank Brendan Bold, Paul Stribley and Raik Litzenberg of X-FAB for the support and guidance on this study.

REFERENCES

[1]   H. R. Huff, H. Tsuya, U Gosele, *The History of LOCOS,* Semiconductor Silicon 1998, vol. 1, pp 200-214, 1998.

[2]   William C. Young, *Oxidation of Silicon Wafers To Eliminate White Ribbon,* United States Patent, Aug 30, 1983.

[3]   Hwasik Park, Peter Smeys, Zakir H. Sahul, Krishna C. Saraswat, *Quasi-3-Dimensional Modelling of Submicron LOCOS Structures,* IEEE Transactions on Semiconductor Manufacturing, Vol. 8, No.4 , Nov 1995.

[4]   R Tromp, G.W Rubloff, F.K LeGoues, *High temperature SiO_2 Decomposition at the SiO_2/Si interface,* Physical Review Letters, 1985.

[5]   J. Cunningham, *Prevention of nitridation in longer drive recipe,* XFAB internal technical report, 2003.

[6]   T.A Shankoff, T.T Sheng, S.E Haszko, R.B Marcus, T.E Smith, *Bird's Beak configuration and Elimination of Gate Oxide Thinning Produced During Selective Oxidation ,* J. Electrochem Soc. Solid State Science and Technology , 1980.

[7]   T.T Sheng, *From White Ribbon to black belt: A direct observation of the Kooi effect masking film,* Journal of Applied Physics, 1994.

Charge collection efficiency measurement system based on field programmable gate array multipurpose card

Norizam Saad, Ishak Mansor, Muhammad Azmi Abdul Hamid, Azman Jalar and Roslinda Shamsudin

Faculty of Science & Technology (FST)
Universiti Kebangsaan Malaysia (UKM)
43600 UKM Bangi, Selangor, Malaysia
Email: nrm_mint@yahoo.com

Abstract—A charge collection efficiency (CCE) measurement system has been developed for CCE measurement of semiconductor nuclear detector using a nuclear counting system interface to a PC for data acquisition. A Field Programmable Gate Array (FPGA) based multipurpose card used for system development due to analog digital converter (ADC) and multichannel analyzer (MCA) functions needed for data acquisition available on board . This multipurpose card also portable and can easily interfaced to a computer. A data acquisition of CCE measurement system namely CCE Measurement V.1.0 also developed using LabVIEW 8.6. The CCE measurement system developed is a user friendly measurement system as it contains detector information, experimental, analysis and documentation. So, measurement results can automatically prepared during system execution. A CCE measurement of CdZnTe semiconductor detector AMPTEK XR-100T-CZT exposed to Cs-137 with 10.45 µCi activity with an increasing operating voltage proved that this developed measurement system is functional and ready to use.

Index Terms—charge collection efficiency (CCE), field programmable gate array (FPGA), nuclear counting system, semiconductor radiation detector

I. INTRODUCTION

Charge collection efficiency (CCE) of a semiconductor nuclear detector is a ratio of collected charge to the charge produced by the particle in the detector [1]. It is an important parameter used in semiconductor nuclear detector characterization as it explains the ability of detectors to convert all the incoming nuclear radiation into measureable signal.

CCE values are measured using a CCE measurement system that is basically consists of a nuclear detection system connected to a data acquisition system. The CCE measurement systems in other research institutes [2, 3, 4, 5, 6] use different approaches that are either partially similar or totally different from each other. None of these systems are a field programmable gate array (FPGA) based measurement system.

This paper discussed a CCE measurement system developed using a plug and play FPGA multipurpose card to interface a

nuclear counting system to computer. The developed CCE measurement system is a low cost measurement system as the system is an ordinary nuclear counting system interfaced to a computer. This system differ from other nuclear counting system by the usage of low cost FPGA multipurpose card and customize data acquisitision programme. The FPGA multipurpose card is used due to an analog digital converter (ADC) and multichannel analyzer (MCA) function available on this FPGA multipurpose card that are useful for data acquisition [7]. The usage of this plug and play FPGA multipurpose card makes a portable CCE measurement system that any counting system can be transformed into CCE measurement system by the usage of this FPGA multipurpose card and customize data acquisitision programme.

A customize data acquisition programme namely CCE Measurement V.1.0 is a developed data acquisition program using LabVIEW 8.6. It is developed to retrieve an ADC and MCA information to investigate nuclear detection ability in response with high operating voltage.

Results of from the developed CCE measurement system should be kept as record to monitor the detection ability of semiconductor nuclear detector. Therefore, this CCE measurement is a useful tool in quality assurance (QA) and quality control (QC) of semiconductor nuclear detector.

II. EXPERIMENTAL METHOD

A CCE measurement system based on FPGA multipurpose card basically a nuclear counting system interfaced to a computer by FPGA multipurpose card. A nuclear counting system as shown in Fig. 1 consists of a sample, a detector connected to a high voltage power supply (HVPS) to provide operating voltage, a pre-amplifier Ortec 142A and an amplifier Ortec 570.

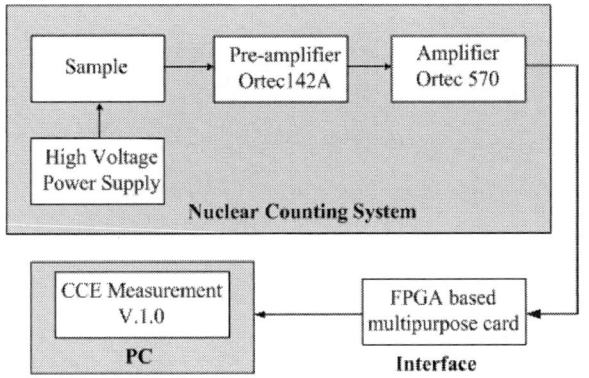

Fig. 1. Schematic diagram of the developed CCE measurement system.

FPGA multipurpose card in the system has an ADC and MCA functions to enable the developed CCE Measurement V.1.0 retrieved detector's operating voltage information peak channel information of nuclear radiation detected for further analysis.

CCE measurement system development been divided into two main tasks, interfacing a nuclear counting system to a PC and developing programming for data acquisition. Interfacing a nuclear system to a computer as described in Fig.2 involved C programming to enable access of ADC and MCA functions in FPGA multipurpose card [8]. A CCE Measurement V.1.0, a developed data acquisition program then acquired data from these functions for further analysis. This program is a structured user-friendly program developed using LabVIEW 8.6 for the purpose of CCE calculation and analysis. This program has been divided into two parts, data entry and the main program consisting of setting and configuration, counting and analysis. The completed CCE measurement system then tested with pulse generator and regulated power supply to substitute both detector and high voltage power supply (HVPS) to check the functionality of the developed system.

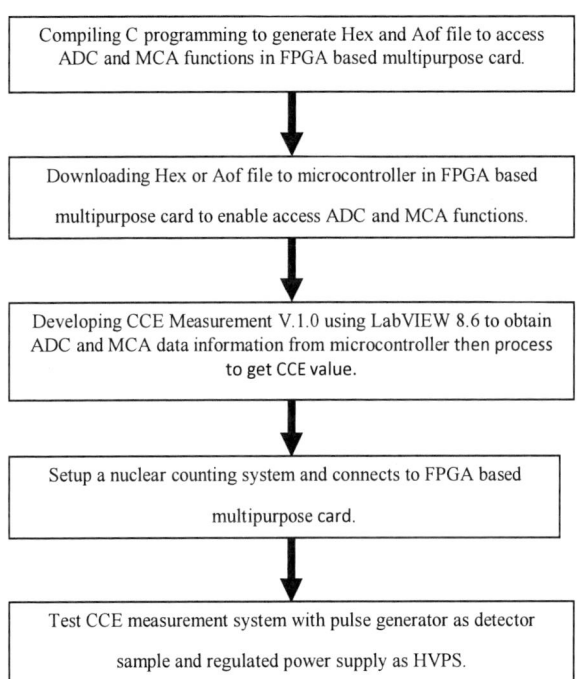

Fig. 2. Process in CCE measurement system development.

The CCE measurement system developed to measure CCE of a semiconductor nuclear detector. Therefore the functional CCE measurement system is tested to measure the CCE of a CdZnTe nuclear detector AMPTEK XR-100T-CZT for the operating voltage started from 50 V then gradually increased for 50 V until 500 V. This detector is used for testing due to its room operating behavior and portable. As this CCE measurement of this system applicable for any semiconductor nuclear detector, any nuclear source can be used for measuring purpose. But the nuclear source selection depends on the detection capability of the semiconductor nuclear detector. Therefore, operator of this measurement system has to refer the semiconductor nuclear detector manuals or datasheet.

A gamma emitter Cs-137 nuclear source with 10.45 μCi activity is used in CdZnTe nuclear detector CCE measurement. Cs-137 nuclear source is selected due to its single energy spectrum at 661.646 keV. This CCE measurement system calculated the CCE value from the peak location of the energy spectrum, therefore a single energy nuclear source such as Cs-137 used for CCE measurement.

III. RESULTS AND DISCUSSION

The developed CCE measurement system in Fig. 3 equipped with a data acquisition program in Fig. 4 consists of 3 main pages, which are introduction, detector information and main program. The introduction page as shown in Fig. 4 while the detector information page is for user's to fill detector's information needed for datasheet and generated report. Main program page consist of setting and configuration page, counting page as shown in Fig. 5 and analysis page. Setting and configuration page is to set parameters needed for CCE calculation, counting page is for counting nuclear source activity and measuring detector's CCE and analysis page in Fig. 6 is for analyzing CCE measured and documentation. These pages can be access from buttons on the left hand side in Fig. 5.

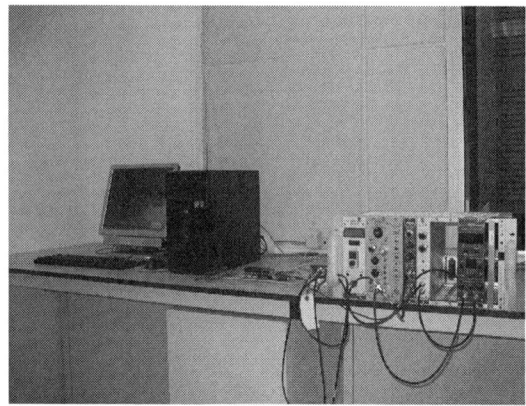

Fig. 3 The developed CCE measurement system.

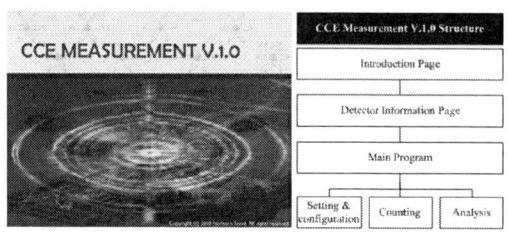

Fig. 4. Structures of CCE Measurement V.1.0.

CCE measurement of CdZnTe nuclear detector and Cs-137 activity shows in Fig. 5. This measurement system able to display Cs-137 activity both in counts and counts persecond (cps). The 661.646 keV spectrum of Cs-137 is clearly displayed in Fig. 5 and distinguishable. Detector's operating voltage, CCE information and counting status also displayed in this counting page.

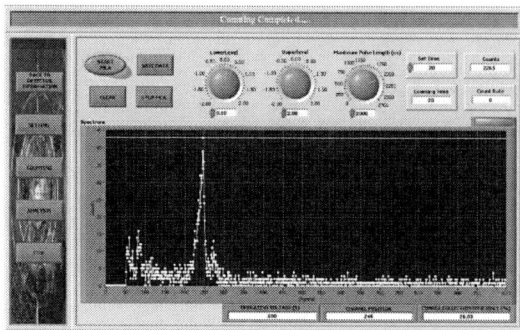

Fig. 5. Spectrum of Cs-137 displayed on main program page.

The CdZnTe nuclear detector's CCE measurement analysis in Fig. 6 able plotted graph of the measured CCE versus operating voltage. Operating voltage and CCE value information of the active cursor on this graph also available. This page also able to display the best operating voltage referred to highest CCE values.

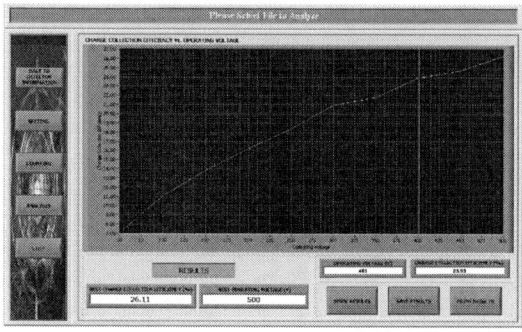

Fig. 6. CCE measurement of CdZnTe nuclear detector analysis.

The CCE measurement results of CdZnTe nuclear detector AMPTEK XR-100T-CZT is programmed to be available in datasheet and report format shown in Fig. 7 and Fig. 8. These results prepared by save results and print results buttons in Fig. 6.

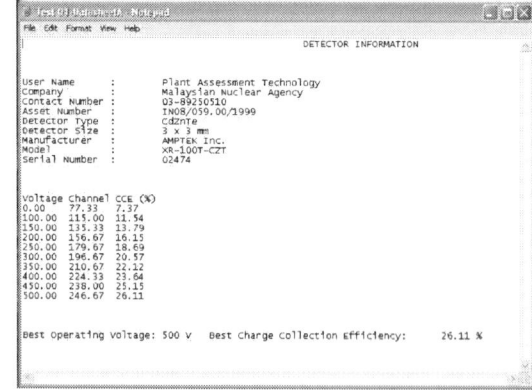

Fig. 7. CCE measurement of CdZnTe nuclear detector in datasheet format.

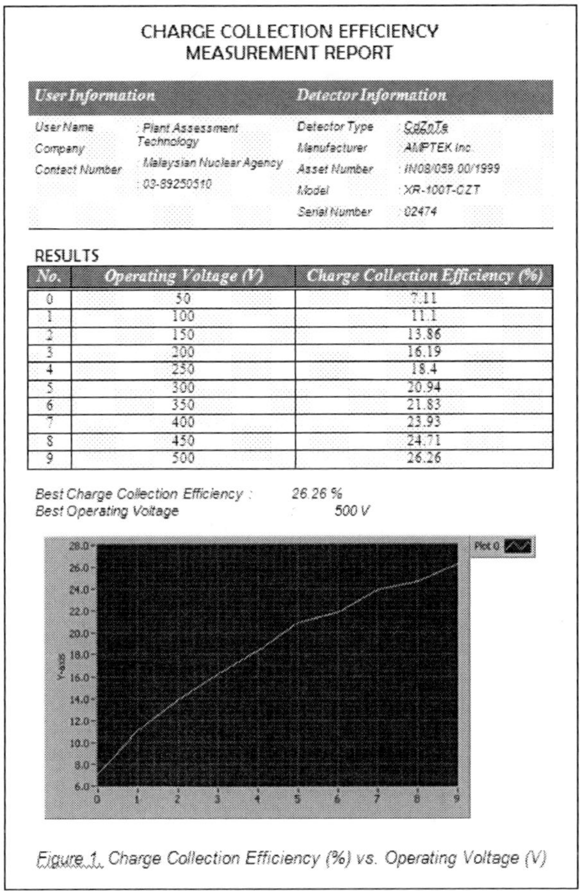

CHARGE COLLECTION EFFICIENCY MEASUREMENT REPORT

User Information		Detector Information	
User Name	: Plant Assessment Technology	Detector Type	: CdZnTe
Company	: Malaysian Nuclear Agency	Manufacturer	: AMPTEK Inc.
Contact Number	: 03-89250510	Asset Number	: IN08/059.00/1999
		Model	: XR-100T-CZT
		Serial Number	: 02474

RESULTS

No.	Operating Voltage (V)	Charge Collection Efficiency (%)
0	50	7.11
1	100	11.1
2	150	13.86
3	200	16.19
4	250	18.4
5	300	20.94
6	350	21.83
7	400	23.93
8	450	24.71
9	500	26.26

Best Charge Collection Efficiency : 26.26 %
Best Operating Voltage : 500 V

Figure 1. Charge Collection Efficiency (%) vs. Operating Voltage (V)

Fig. 8. CCE measurement of CdZnTe nuclear detector in report format.

CCE measurement datasheet and report, both in Fig. 7 and Fig. 8 are achievable after CCE analysis. Both results contain detector's information and measurement results. Only a report format that is for user shows measured CCE versus operating voltage.

IV. CONCLUSION

The developed CCE measurement system based on FPGA multipurpose card meet the research objective. Generated datasheet and reports functions of this CCE measurement are very helpful to user as it makes the documentation task easier and faster. Record process is faster and efficient with the generated datasheet and this data applicable to used in commercial software for example Microsoft Excel and Origin.

This system is also applicable for semiconductor nuclear detector characterization and QA/QC of semiconductor nuclear detector. A front end enhancement of this system to accommodate low current output from semiconductor radiation detector such as SiC and GaAs creates a a room temperature semiconductor nuclear detector research facilities.

ACKNOWLEDGMENT

Appreciation is expressed to the Ministry of Science, Technology and Innovation for funding and instrumentation, and to the Automation Center and Plant Assessment Testing Group of the Malaysian Nuclear Agency and the Universiti Kebangsaan Malaysia for their support.

REFERENCES

[1] L. Berluti, C. Canali, A. Castaldini, A. Cavallini, A. Centronio, S. D'Auria, C. Del Papa, C. Lanzieri, G. Mattei, F. Nava, M. Proia, P. Rinaldi, and A. Zichichi, "Gallium arsenide particle detector: a study of the active region and charge collection efficiency," *Nucl. Instr. Meth A*, vol. 354, no. 2-3, pp. 364-367, Jan 1995

[2] M.J. Harrison, A. Kargar, and D.S. McGregor, "Charge collection characteristics of frish collar CdZnTe gamma ray spectrometers," *Nucl. Instr. Meth. A*, vol. 579, no. 1, pp. 134-137, Aug. 2001.

[3] F. G. Hartjes, "Characterisation system to measure the charge collection properties of prototype solid state detectors," National Institute for Nuclear Physics and High Energy Physics, Amsterdam, Tech. Rep., Jul. 2005.

[4] M.D. Napoli, F. Giacoppo, G. Raciti, and E. Rapisarda, "Study of charge collection efficiency in 4H-SiC schottky diodes with ^{12}C ions," *Nucl. Instr. Meth. A*, vol. 608, no. 1, pp. 80-85, Sept. 2009.

[5] P.J. Sellin, A.W. Davies, F. Boroumand, A. Lohstroh, M.E. Özsan, J. Parkin, and M. Veale, "IBIC charcterisation of charge transport in CdTe:Cl," *Fizika I Tekhnika Poluprovodnikov*, vol. 41, no. 4, pp. 411-416, Oct. 2006.

[6] M.K. Peterson, R.F. Hurley, K. Arya, C. Betancourt, M. Bruzzi, B. Colby, M. Gerling, C. Meyer, J. Pixley, T. Rice, H.F. –W. Sadrozinski, M. Scaringella, J. Bernardini, L. Borello, F. Fiori, and A. Messineo, "Determination of the charge collection efficiency in neutron irradiated silicon detectors," *IEEE Trans. Nucl. Sci.*, vol. 56, no. 6, pp. 3828-3833, Dec. 2009.

[7] I. Mansor, N. Saad, M. Mohd. Ibrahim, H. Rongen, H. Kaufmann, and K. Sulaiman, "Refurbishment of the whole body counter bed type ND7500 using a rectangular NaI scintillation detector and development of integrated system software," in *Proc. of a Technical Meeting*, Vienna, 2006, pp. 121-128.

[8] H. Rongen, "IAEA/RCA Regional training course on a data acquisition and control module applicable in refurbishment of nuclear instrument," in IAEA/RCA Regional training course, Malaysian Institute for Nuclear Technology Research, Bangi, Jul. 7 – 18, 2003.

Development and application of in-house high voltage power supply for atmospheric pressure plasma treatment system

Nafarizal Nayan[1,2], *Member, IEEE,* Mohammad Redzuan Zahariman[2], Mohd Fadzlie Bin Ahmad[2],
Riyaz Ahmad Mohamed Ali[1,2], Mohd Zainizan Sahdan[1,2], Uda Hashim[3] *Member, IEEE*
[1]Microelectronic and Nanoelectronic – Shamsuddin Research Centre (MiNT-SRC),
[2]Department of Electronic Engineering, Faculty of Electrical and Electronics Engineering
Universiti Tun Hussein Onn Malaysia,
86400 Parit Raja,Batu Pahat, Johor
[3]Institute of Nano Electronic Engineering (INEE)
Universiti Malaysia Perlis,
01000 Seriab, Kangar Perlis
Email: nafarizal@gmail.com, nafa@uthm.edu.my

Abstract- **Atmospheric pressure plasma is now being widely developed for simple surface treatment process and for fast medical tools sterilization. Plasma surface modification involves the interaction of the plasma generated excited species with a solid interface or coatings. The previous vacuum plasma system is not applicable and very costly. In the present project, we have developed the high voltage power supply and atmospheric pressure plasma using dielectric barrier discharge concept. The high voltage power supply was developed using a simple 555 timer and car's ignition coil. Then, we investigate the plasma surface modification effect from the contact angle measurement and evaluate the roughness using surface profiler. We found that the contact angle decreased with the exposure time and surface roughness changed when exposed with atmospheric pressure plasma. It has been understood that a film coating will be create on glass and silicon surface when expose with atmospheric pressure plasma system in water vapor environment.**

I. INTRODUCTION

Atmospheric pressure plasma has high energy electrons that lead to the generation of chemically active species such as ozone. These highly reactive species are applicable for cleaning of organic dust on material surface, semiconductor substrate surface modification, and medical tools sterilization. In addition, a relatively low temperature in the plasma is suitable for the irradiation on thermal sensitive materials such as plastic tools without damage and stress. Basically, atmospheric pressure plasma is non-equilibrium plasmas produced by the dielectric barrier discharge are very attractive for various industrial applications because of their low-cost, high speed and the ability to operate without vacuum.

Plasma surface modification involves the interaction of the plasma generated excited species with a solid interface. Plasma surface activation means usually plasma treatment when non-polymerizing working gases are used. The chemical surface modification is initiated by the radical reaction of plasma species and the polymer surface. In atmospheric pressure plasma, the surface activation typically takes places with oxygen containing gas mixtures such as ambient air. The advantage of the plasma treatment is the ability to change the surface properties of the most external layers of the material without modifying its bulk characteristics. It means there is no change in mechanical properties too.

In addition, the previous system to clean medical tools need a high vacuum system where is very costly. The high vacuum is normally obtained using the combination of turbo molecular pump backed by rotary pump. The disadvantage by using high vacuum system is not an environment friendly due to the noise produced during pumping. So, development of plasma using atmospheric pressure plasma is very effective. The system is generally consisting of two parallel plate electrode and dielectric plate. A very high voltage is supply to electrode to produce plasma.

In the present work, we have designed and developed our own high voltage pulse generator for atmospheric pressure plasma system. Then, we design and develop the atmospheric pressure plasma system by using dielectric barrier discharge concept. Finally, we investigate the usage of atmospheric pressure plasma system for surface modification application.

II. HIGH VOLTAGE POWER SUPPLY

Figure 1 shows the block diagram of our pulse high voltage power supply. We use a simple 555 timer circuit and ignition coil to produce a high voltage output. The circuit was firstly designed using Multisim software and their output was simulated, as shown in figure 2 and 3.

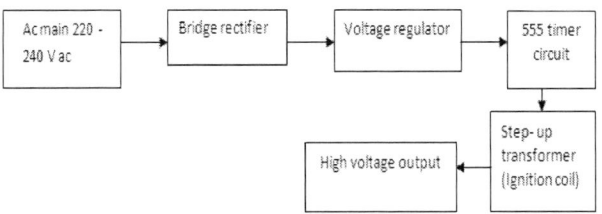

Fig. 1. Block diagram of developed high voltage power supply

The simulated output from the high voltage circuit diagram is shown in figure 3. It shows that the output sparks from ignition coil appear in the oscilloscope and the range of output voltage in 18-22 kV.

III. ATMOSPHERIC PRESSURE PLASMA

Figure 4 show the atmospheric pressure plasma using the dielectric barrier discharge concept. The high voltage power supply was connected to the upper metal electrode. The the ground electrode below was covered by quartz glass with 10 mm thickness. The gap between the upper electrode and the quartz glass was created using a microscope glass with thickness of 5 mm.

Prior to the surface treatment process by our in-house dielectric barrier discharge system, we clean the substrate using acetone and ultrasonic cleaner for glass substrate and using acetone, buffer oxide etching and ultrasonic cleaner for silicon wafer. This is to ensure that there is no oxide layer on the top of substrate. One can make sure the hydrophobic of silicon substrate after the cleaning process by dropping water on silicon substrate. Silicon is hydrophobic and silicon dioxide is hydrophilic.

The glass and silicon substrate was then inserted into the gap between the upper electrode and the lower electrode. The plasma was generated using high voltage power supply and exposed to the substrate. The treatment time was varied from 2-17 minutes. Although the plasma color is very dim, one can see small spark coming from the upper electrode to the substrate surface in dark room. Their color is slightly purple.

For the analysis, we used contact angle analysis and surface rougher analysis. The contact angle was evaluated from the image taken from a simple digital camera at specific angle. Then the surface roughness was evaluated using the KL Tencor surface profile system. Even this technique is very simple and rough, we hope that we could give some information of the effect of our plasma exposure to the substrate surface. These results may give us some knowledge on the surface tension activities before and after the plasma treatment.

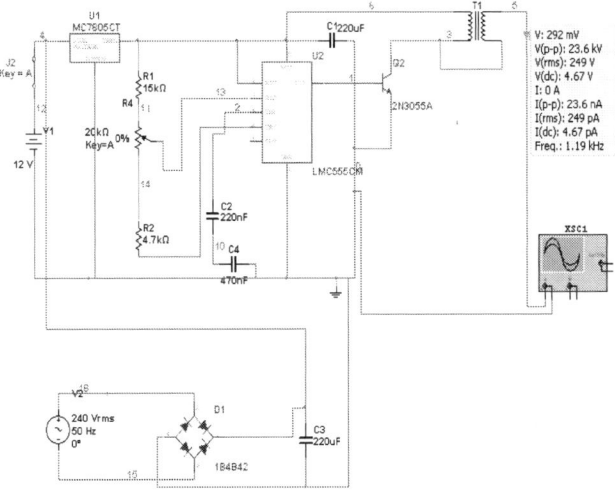

Fig. 2. High voltage circuit in Multisim software

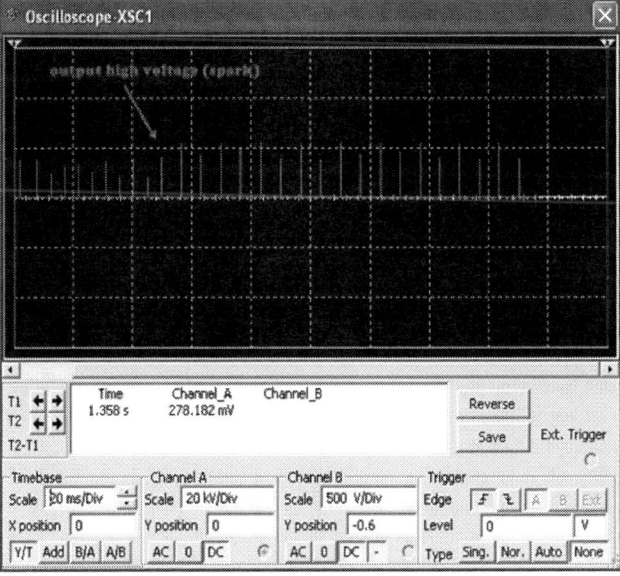

Fig. 3. Output from high voltage circuit simulation

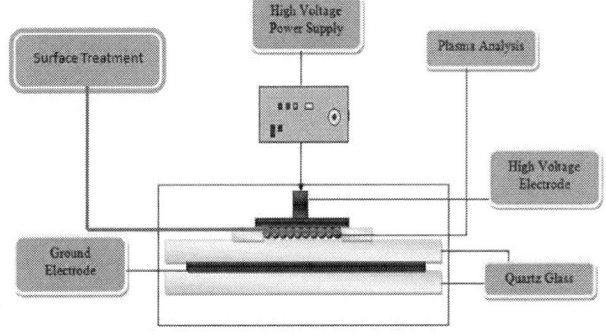

Fig. 4. Dielectric barrier discharge system for surface treatment application

IV. EXPERIMENTAL RESULTS AND DISCUSSIONS

For surface treatment process, the exposure time or treatment time was varied from 2 -17 minutes. We expose the substrate at a duty ratio of 66.7% where 2 minute "ON" and 1 minute "OFF". This is to cool down the circuit and maintain the discharge condition. A water drop from a pipette was observed before and after the treatment on glass and silicon substrate. A water drops behavior on the substrate surface is very instructive. If the water spreads out over the surface this shows the water molecules are more attracted to the surface than to itself. We call this condition as hydrophilic. The attractive quality of a surface is determined by its surface tension. If the water forms a spherical drop, this shows the substrate surface tension is less than that of the water. We call this condition as hydrophobic. As an example the water molecules would rather stick together than attach to the polymer surface molecules.

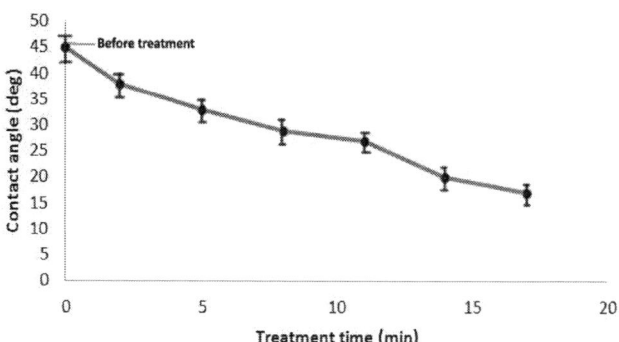

Fig. 6. Effect of DBD plasma exposure time on water contact angle on glass substrate

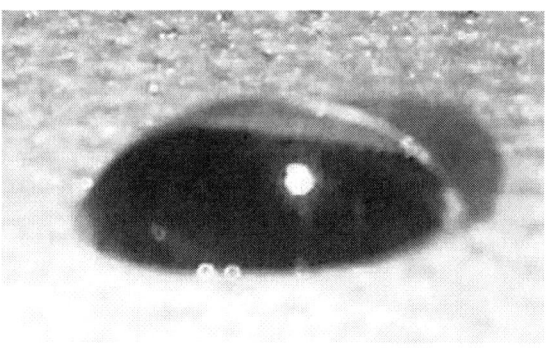

a) Water drop on an untreated, low energy surface

b)   Water drop on the plasma treated surface

Fig. 5. Surface tension effect of plasma before (a) and after (b) plasma treatment on glass substrate

Figure 6 shows the result of contact angle measurements. Freshly cleaned glass surfaces have a high surface energy and are wettable as shown at 0 min. However, they have a tendency to adsorb organic contamination from the ambient environment. After the treatment procedure, the result show that the contact angle decreased with the treatment time. It has been reported that the adsorbed organic contaminant molecules, generally less than full monolayer coverage of order nm thickness, will generate a heterogeneous wetability. After using atmospheric pressure plasma system for treatment, non uniform glass coatings will be formed. Therefore it will reduce the contact angle by increasing the glass surface wetability.

The use of atmospheric pressure plasma treatment instead of the widely used low pressure vacuum plasma treatment can increase the handling capacity of large area glass substrates, can increase the throughput by in line processing under the atmospheric environment, and can decrease the equipment cost. The surface of glass become hydrophilic after treatment using atmospheric pressure plasma. The higher the tension, the better the thin film will stick provided the surface is not degraded. Then the surface treatment process was repeated on silicon substrate using the similar procedure and parameter.

Figure 7 shows the image of water drop on silicon substrate after the plasma treatment. Before expose plasma on silicon surface, silicon have low energy surface with a contact angle of 40 degree. After 2 min. of exposure time, the surface tension has drastically changed to a very hydrophilic in their surface. Then after 2min., the results were almost the same. That is because we can assume there are some thin layer or coatings has form on silicon surface and this is prove that plasma have make a surface modification surface.

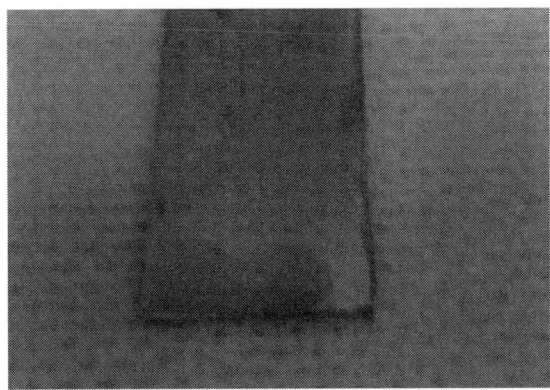

Fig. 7. Surface tension effect of after plasma treatment on silicon

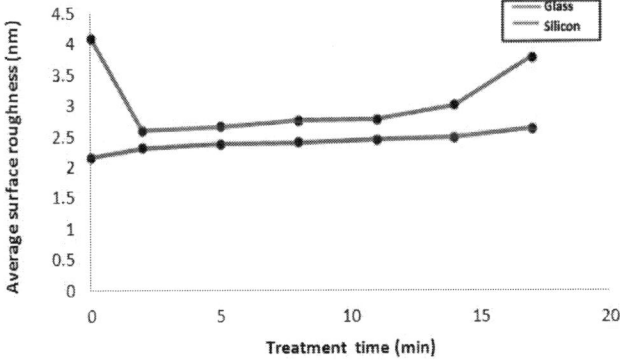

Fig. 8. Average surface roughness on glass and silicon surface evaluated by surface profiler meter

The wettability of silicon substrate surface is may be due to the increase in the hydroxyl –OH groups density, which are responsible for the hydrophilic properties. This is significant due to the presence of water vapor in the working environment.

Finally, the surface roughness was evaluated using surface profiler. The results are shown in figure 8. As shown in the figure 8, the surface roughness for silicon substrate was almost the same. On the other hand, the surface roughness for glass substrate decreased with the treatment time. This is consistent with the contact angle result.

IV. CONCLUSIONS

We have successfully developed the high voltage power supply and applied it to produce atmospheric pressure plasma system. The atmospheric pressure plasma system was used for the surface treatment on glass and silicon substrate. We have shown that the surface modification occurred when we expose the substrate surface with our in-house dielectric barrier discharge system. The potential reason is due to the existence of monolayer surface on the substrate surface. The results show the potential development of larger scale plasma treatment system for surface modification application. The result of treatment video will be shown during the conference.

ACKNOWLEDGMENT

The authors would like to express their thanks all the members of Microelectronic and Nanotechnology – Shamsuddin Research Centre (MiNT-SRC) of Universiti Tun Hussein Onn Malaysia (UTHM) for their support and kind suggestions during the project. This project has been sponsored partially under the Projek Sarjana Muda (PSM) of Faculty of Electrical and Electronic Engineering of UTHM, Short Term Research Grant of UTHM and FRGS of Ministry of Higher Education Malaysia.

REFERENCES

[1] Ochs D., Schroeder J., Cord B., Scherer J.: Surf. Coat. Technol. *142-144*, 767, 2001.

[2] Ratna Ika Putri, Ika Noer Syamsiana and La Choviya Hawa, "Design of High Voltage Pulse Generator for Pasteurization by Pulse Electric Field (PEF), Malang State Politechnic, East Java, Indonesia, 2010.

[3] F.D. Egitto and L.J.Matienzo,"Plasma Modification of Polymer Surfaces for Adhesion Improvement," *IBM Journal of Research and Development*, p. 423, July 1994.

[4] L.Wood, C. Fairfield et al., "Plasma Cleaning of Chip Scale Packages for Improvement of Wire Bond Strength," *Chip Scale Package Seminar*, December 2000.

[5] E. M. Liston, L. Martinu and M. R. Wertheimer "Plasma Surface Modification of Polymers for improved Adhesion: A Critical Review;" *Journal of Adhesion Science and Technology*; Vol. 7, No. 10, pp. 1091-1127; 1993.

[6] F.D.Egitto and L.J. Matienzo "Plasma Modification of Polymer Surfaces;" *36th Annual Technical Conference Society of Vacuum Coaters*, pp. 10-22, 1993.

[7] R.W. Burger and L.J. Gerenser, "Understanding the Formation and Properties of Metal/Polymer Interfaces via Spectroscopic Studies of Chemical Bonding" *34th Annual Technical Conference Society of Vacuum Coaters*, pp. 162-168, 1991.

A New Design Methodology based on Particle Swarm Optimization (PSO) algorithm for Multi-clad single mode optical fibers

Shiva Nejati, Reza Barkhordari, Fatemeh Kohani Khoshkbijari, Reza Fouladi, Shide Nejati, Reza Kohani Khoshkbijari

Photonic and Nanocrystal Research Lab. (PNRL)
Sama Technical and Vocational Training College, Islamic Azad University, Rasht Branch
Rasht, IRAN
E-mail: s.s.nejati@gmail.com

Abstract- **A new design methodology based on particle swarm optimization (PSO) algorithm for single mode multi-clad optical fiber is presented. Using numerical simulations, we evaluate fitness function with nonzero Gaussian parameter and propose optimal fitness function for multi-clad single mode fiber design. By use of a weighted fitness function with non-zero Gaussian parameter for the first time, we have decreased dispersion, dispersion slope and bending loss at least 10 times at λ=1.55 μm simultaneously. (-2.3104e-6 ps/km/nm, 0.0512 ps/km/nm and 2.3742e-5 dB/ are achieved for dispersion, dispersion slope and bending loss respectively). Also in comparison with last works, the simulation results have shown that PSO is so better than genetic algorithm (GA) in fiber design.**

I. INTRODUCTION

Fiber-based networks form a key part of international communication system. Recently multi-channel optical communication systems such as optical time division multiplexing (OTDM) and dense wavelength division multiplexing (DWDM) are required in industry thus, with considering these applications providing a large bandwidth and high speed communication possibility, using optical fibers is highly interesting [1].

However these fibers suffer from some disadvantages such as dispersion, dispersion slope and bending loss. To obtain a suitable fiber for mentioned purposed all these factors should be improved. Therefore many structures had presented [2-5]. Recently some works based on genetic algorithm (GA) were done [6, 7]. In these works by use of GA, the Optical and geometrical optimal parameters for a considered fiber were achieved.

In this work, a new fiber design approach is proposed. For the first time we use Particle Swarm Optimization (PSO) algorithm and consider three principle parameters including dispersion, dispersion slope and bending loss with different weights in predefined fitness function. It is expected to obtain better outcomes relative to GA because in PSO, each particle remembers its previous best value and the neighborhood best one; so it has a storage capability more efficiently than GA.

Organization of the paper is as follow:

In section II mathematical formulations is presented. In section III a brief review of PSO is expressed and optimization algorithm is considered. Simulation results are presented and discussed in section IV. Finally the paper ends with a short conclusion.

II. MATHEMATICAL FORMULATIONS

In this section, mathematical formulations that commonly used to calculate fiber characteristics including dispersion, dispersion slope and bending loss are presented. In fig.1 refractive index of considered fiber is shown.

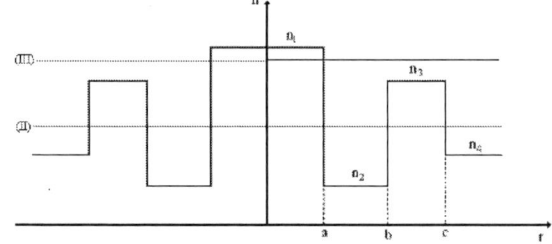

Fig. 1. Refractive Index of Considered Fiber

According to the traditional LP approximation [1] in wave equation propagating and boundary conditions, the characteristic equations for the proposed structure are achieved. These equations contain Bessel functions which defined as follows:

$$U_1 = a\sqrt{k_0^2 n_1^2 - \beta^2}, \quad U_2 = b\sqrt{\beta^2 - k_0^2 n_3^2},$$

$$U_3 = b\sqrt{k_0^2 n_3^2 - \beta^2}, \quad W_2 = a\sqrt{k_0^2 n_2^2 - \beta^2},$$

$$W_3 = a\sqrt{\beta^2 - k_0^2 n_2^2}, \quad W_4 = c\sqrt{\beta^2 - k_0^2 n_4^2}.$$

$$\overline{U_2} = \frac{1}{P}U_2, \quad \overline{U_3} = \frac{1}{P}U_3,$$

$$\overline{W_2} = \frac{P}{Q}W_2, \quad \overline{W_3} = \frac{P}{Q}W_3, \tag{1}$$

Where β is propagation wave vector of guided modes, k_0 is the wave number in vacuum, P and Q are the geometrical parameters defined as follows:

$$P = \frac{b}{c}, Q = \frac{a}{c}. \tag{2}$$

The optical parameters are defined as

$$R_1 = \frac{n_1 - n_3}{n_3 - n_2}, \quad R_2 = \frac{n_2 - n_4}{n_3 - n_2}, \quad \Delta = \frac{n_1^2 - n_4^2}{2n_4^2} \approx \frac{n_1 - n_4}{n_4}. \tag{3}$$

And the normalized frequency is defined as follow:

$$V = k_0 a\sqrt{n_1^2 - n_4^2} . \tag{4}$$

Dispersion and dispersion slope and bending loss are given by the following relations[1].

$$D = -\frac{\lambda}{c}\frac{d^2 n_4}{d\lambda^2}[1 + \Delta\frac{d(VB)}{dV}] - \frac{N_4}{c}\frac{\Delta}{\lambda}V\frac{d^2(VB)}{dV^2}, \tag{5}$$

$$S = -\frac{\lambda}{c}\frac{d^3 n_4}{d\lambda^3}\left[1 + \Delta\frac{d(VB)}{dV}\right] - \frac{1}{c}\frac{d^2 n_4}{d\lambda^2}$$
$$\left[1 + \Delta\frac{d(VB)}{dV}\right] + \frac{N_4}{c}\left(\frac{\Delta}{\lambda^2}\right)V^2\frac{d^3(VB)}{dV^3} \tag{6}$$
$$+ 2\frac{N_4}{c}\frac{\Delta}{\lambda^2}V\frac{d^2(VB)}{dV^2} + 2\frac{\Delta}{c}\frac{d^2 n_4}{d\lambda^2}V\frac{d^2(VB)}{dV^2},$$

$$2\alpha = \frac{\sqrt{\pi}F^2}{2P} \cdot \frac{ce^{\left(\frac{-4\Delta w^3}{3cv^2}R\right)}}{w\left(\frac{wR}{c} + \frac{v^2}{2\Delta w}\right)^{\frac{1}{2}}} \tag{7}$$

Where, $N_4 = n_4 - \lambda\frac{dn_4}{d\lambda}$ is the group index of the outer cladding layer. Also R is bending radius (in our simulations R=1).

III. OPTIMIZATION TECHNIQUE

In this section, optimization method is defined, which is PSO with a weighted fitness function. PSO optimizes an objective function by undertaking a population-based search. The population consists of potential solutions, named particles, which are a metaphor of birds in flocks. These particles are randomly initialized and freely fly across the multidimensional search space. During flight, each particle updates its own velocity and position based on the best experience of its own and the entire population. The updating policy drives the particle swarm to move toward the region with the higher objective function value, and eventually all particles will gather around the point with the highest objective value [8].

$$FF = \sum_i e^{-\frac{(\lambda-\lambda_0)^2}{2\sigma^2}}[C_1 D(\lambda) + C_2 S(\lambda) + C_3 BL(\lambda)] \tag{8}$$

We considered three parameters for fiber design: Dispersion, Dispersion slope and Bending loss .The fitness function was as follows where C_is are existence weight of each parameter in fitness function.

D, S and BL are dispersion, dispersion slope and bending loss coefficients respectively. For analysis the effect of C_is, simulation for three values of σ with three different combinations of C_is is done.

IV. SIMULATION RESULTS AND DISCUSSION

The simulation is done for three different value of σ. The results are discussed as follow:

- Simulation for $\sigma = 0$:

After analysis, results are illustrated in table I. It is shown that the best values for dispersion and dispersion slope are derived with [0.1, 1, 0.1] weight constants. However in case of bending loss, the derived value for [0.1, 0.1,1] is the best.

TABLE I
DISPERION, DISPERSION SLOPE AND BENDING LOSS VALUES FOR $\sigma = 0$ AT λ=1.55μm

[C1 C2 C3]	D (ps/km/nm)	S (ps/km/nm²)	BL(db/m)
[1,0.1,0.1]	0.1625	0.1543	0.1671
[0.1,1,0.1]	-0.0037	0.0881	0.0025
[0.1,0.1,1]	0.0617	0.1513	0.0006

The system parameters which obtained optimally are given in Table II.

TABLE II
OPTIMIZED SYSTEM PARAMETERS FOR $\sigma = 0$

[C1 C2 C3]	a(μm)	P	Q	R₁	R₂	Δ
[1 0.1 0.1]	2.4705	0.7121	0.3449	1.5769	0.7305	0.0083
[0.1 1 0.1]	2.2874	0.6364	0.3699	6.4260	0.2540	0.0086
[0.1 0.1 1]	2.4159	0.6673	0.3416	1.5971	0.6918	0.0094

- Simulation for $\sigma = 1.2256 \times 10^{-8}$

Considering the nonzero Gaussian parameter, numerical simulations are done and according to table III, similar to the previous sate, the best values for dispersion and dispersion slope obtained in [0.1, 1, 0.1] and for bending loss, the derived value for [0.1, 0.1,1] is the best. The system parameters which obtained optimally are given in Table IV.

TABLE III
DISPERION, DISPERSION SLOPE AND BENDING LOSS VALUES FOR $\sigma = 1.2256 \times 10^{-8}$
AT $\lambda = 1.55 \mu m$

[C1 C2 C3]	D (ps/km/nm)	S(ps/km/nm²)	BL(dB/m)
[1,0.1,0.1]	$3.8832e^{-4}$	0.1611	0.0150
[0.1,1,0.1]	$-2.3104e^{-6}$	0.0512	0.0078
[0.1,0.1,1]	6.5065e-5	0.0813	$2.3742e^{-5}$

TABLE IV
OPTIMIZED SYSTEM PARAMETERS FOR $\sigma = 1.2256 \times 10^{-8}$

[C1 C2 C3]	a(μm)	P	Q	R₁	R₂	Δ
[1 0.1 0.1]	2.3073	0.7175	0.3106	6.1407	0.2587	0.0084
[0.1 1 0.1]	2.2856	0.6481	0.4906	7.5265	0.9852	0.0086
[0.1 0.1 1]	2.4566	0.6598	0.4102	1.8138	0.2941	0.0089

- Simulation for $\sigma = 3.6935 \times 10^{-8}$

With this value for Gaussian parameter according to table V, best results for dispersion and dispersion slope derived with [0.1, 1, 0.1] and for bending loss with [0.1, 0.1, 1] as before. The system parameters which obtained optimally are given in Table VI.

TABLE V
DISPERION, DISPERSION SLOPE AND BENDING LOSS VALUES FOR $\sigma = 3.6935 \times 10^{-8}$
AT $\lambda = 1.55 \mu m$

[C1 C2 C3]	D (ps/km/nm)	S (ps/km/nm²)	BL(db/m)
[1,0.1,0.1]	-0.0027	0.0916	0.1005
[0.1,1,0.1]	$7.2334e^{-5}$	0.0748	0.0123
[0.1,0.1,1]	$6.5065e^{-4}$	0.0813	$2.9216e^{-5}$

TABLE VI
OPTIMIZED SYSTEM PARAMETERS FOR $\sigma = 3.6935 \times 10^{-8}$

[C1 C2 C3]	a(μm)	P	Q	R₁	R₂	Δ
[1 0.1 0.1]	2.4116	0.7080	0.3850	3.3682	0.3495	0.0077
[0.1 1 0.1]	2.3806	0.5953	0.3860	3.4047	0.4146	0.0081
[0.1 0.1 1]	2.4566	0.6598	0.4102	1.8138	0.2941	0.0089

Comparing these three parts, shows that Gaussian parameter has a critical impact on dispersion, dispersion slope and bending loss. According to simulation results the best values for these factors are obtained for $\sigma = 1.2256 \times 10^{-8}$. The effects of weight constants on dispersion, dispersion slope and bend loss are shown in fig.2, fig.3 and fig.4.

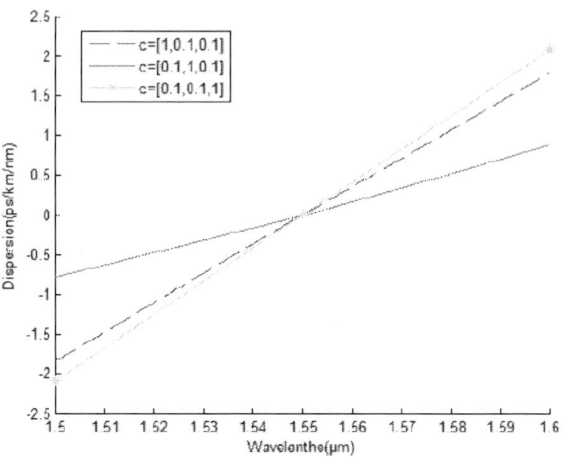

Fig. 2. Dispersion vs. Wavelength for $\sigma = 1.2256 \times 10^{-8}$

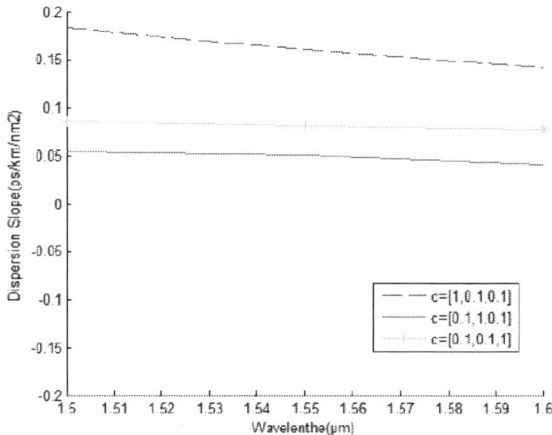

Fig. 3. Dispersion Slope vs. Wavelength for $\sigma = 1.2256 \times 10^{-8}$

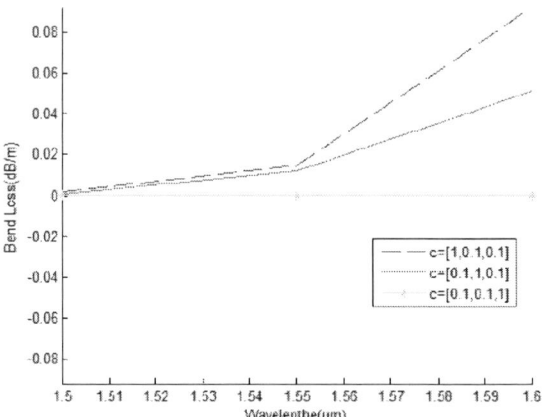

Fig. 4. Bending Loss vs. Wavelength for $\sigma = 1.2256 \times 10^{-8}$

CONCLUSION

In this paper we proposed a new designing method for single mode multi-clad optical fiber by use of PSO. As it was expected, in compare with GA, better results were derived. In term of fitness function, non-zero Gaussian parameter and weighted design factors led to a significant improvement in dispersion, dispersion slope and bending loss.

REFERENCES

[1] A. Ghatak, and K. Thyagarajan, "Introduction to Fiber Optics," Cambridge University Press, 1998.

[2] C.Zhao, R.Peng, Z.Tang, Y.Ye, Le.Shen, and D.Fan, "Modal fields and bending loss analyses of three-layer large flattened mode fibers," Optics Communications, vol.6, Issue 1, 2006.

[3] A.Rostami, and M.Savadi Oskouei, "Investigation of Chromatic Dispersion and Pulse Broadening Factor of Two New Multi-clad Optical Fibers," IJCSNS, vol.6, No.8, August 2006.

[4] Ali Rostami and Morteza Savadi Oskouei, "Investigation of dispersion characteristic in MI- and MII-type single mode optical fibers", Optics Communications J., available online, 2006.

[5] S. Makouei, A. Rostami and M. Savadi, "The Study of Bending Loss and Mode Field Diameter in Two Multi-clad Single Mode Optical Fibers" IJCSNS International Journal of Computer Science and Network Security, VOL.7 No.2, February 2007.

[6] M.Savadi Oskouei, S.Makouei, A.Rostami, and Z.D.Koozeh kanani, "Proposal for optical fiber designs with ultrahigh effective area and small bending loss applicable to long haul communications," Applied Optics, vol.46, No.25, pp. 6330-6339, 2007.

[7] M.Savadi-Oskouei, A.Rostami, and S.Makouei, " A novel fiber design strategy for simultaneously introducing ultra small dispersion and dispersion slope using Genetic algorithm," Wiley InterScience Euro. Trans. Telecomms, vol. 20, pp. 37-47, 2009.

[8] R.Shivakumar, and R.Lakshimapathi, "Implementation of innovative Bio inspired GA and PSO algorithm for controller design considering steam GT dynamics," IJCSI, vol.7, Issue 1, No.3, January 2010.

Radiation Exposure induced Failure on Semiconductor Package Material

Wan Yusmawati Wan Yusoff[1], Azman Jalar, *Member, IEEE*[1, *], Norinsan Kamil Othman[1, 2], Irman Abdul Rahman[2],
Roslinda Shamsudin[2], Muhammad Azmi Abdul Hamid[2]

[1]Institute of Microengineering and Nanoelectronics (IMEN)
[2]School of Applied Physics Faculty of Science and Technology
Universiti Kebangsaan Malaysia, 43600 UKM Bangi
Selangor Darul Ehsan, Malaysia
Email: azmn@ukm.my

Abstract- **Semiconductor industry has progressed towards the creation of packages with sub-micron technology. Quad-Flat No-Lead (QFN) package is among latest form semiconductor package in submicron size scale. Recent trends in digitization era lead to electronic package application in many fields including radioactive environment. In the development of semiconductor technology, the ability to predict and eventually to prevent failures of microelectronics is becoming increasingly important. This paper presents an effect of 1.33 MeV gamma irradiation induced failure in semiconductor packages. The packages were exposed to gamma radiation from a Cobalt-60 source with varying doses from 5 Gy to 50 000 Gy with an operating dose rate of 2.54 kGy/h. In this investigation, as-received packages were used as control samples. Following exposure to gamma ray, the in-house fabricated QFN package then subjected to Scanning Acoustic Microscope (CSAM) and X-ray Imaging System (3D CT scan X-ray) to check the influence of radiation exposure on the package. Detail analysis exhibited that the increment of exposure dose influenced the occurrence of the wire sweep, delamination and cracks. The delamination occurs at the silicon die and leadframe region and the cracks were observed at the die surface. The gamma irradiation is believed to induce the failure in QFN package.**

I. INTRODUCTION

Gamma radiation is the most energetic form of electromagnetic radiation with a very short wavelength. This radiation is a source of choice to observe the simulated effect of ionizing radiation on a component or material [1]. During the interaction with matter, gamma ray has sufficient energy to remove tightly bound orbital electrons of atom and alter the chemical bonding of the materials [2] and permanently modified the semiconductor structure [3].

Quad-Flat No-Lead (QFN) is the most popular semiconductor package. This package meet the demand of many new portable application by applying the concept of miniaturization to produce lightweight, multifunction and thin profile with superior electrical and thermal performance. In a stacked die package, their wire bonds usually posse's longer and higher interconnection as compare to conventional package. In the electronic package, there are various types of failure such as delamination, wire sweep, scratch on the body

and wire cracking, package cracks and cracking the silicon die [4,5]. The delamination is a core mode of failure [6] that may lead to various problems. The deflection of wire bonds can seriously cause wire crossover and circuit shorting [7]. Hence, every component and connection involved in this package must be maintained for effectiveness, durability and reliability of the package.

Consumer package encompasses every field of microelectronics applications, including radiation related industries such as radiography radiation used in medical and health sciences and non-destructive testing. According to Summer et al. [8], solid state electronic device will occur the displacement damage when exposed to radiation. Numerous studies have been done on the effect of radiation on microelectronic package and device. However, the radiation effects on semiconductor package have not been verified. Therefore, this present study concentrates on the effect of gamma irradiation towards the failure of QFN package.

II. MATERIALS AND METHODS

A QFN stacked die package was developed by researchers from Advanced Semiconductor Packaging (ASPAC) Research Group of Universiti Kebangsaan Malaysia, Malaysia. The dimension of package used is 7.00 mm x 7.00 mm x 0.85 mm with the die size of 5.00 mm x 5.00 mm x 0.15 mm and the top die size was 3.3 mm x 3.3 mm. This package was designed with the thickness of epoxy is 0.0254 mm. This package types has no lead exposed from mold and gold wire was used for interconnection. The package then exposed to gamma irradiation was performed at room temperature by using Gamma Cell (Excel 220), which uses a Co-60 source at dose rate of 2.54 kGy/h and varying dose of 5 Gy, 50 Gy, 500 Gy, 5000 Gy and 50000 Gy. In this investigation, a Scanning Acoustic Microscope (CSAM) and 3D CT Scan X-ray (HMX CT-160Xi) imaging system were used to detect the influenced of radiation exposure on the package. The 3D CT scan works on the principle that X-rays will penetrate the mold compound, while CSAM used the principle that detected reflected ultrasonic waves emitted to the sample to enable the internal condition of semiconductor package.

978-1-4673-2395-6/12 $31.00 © 2012 IEEE

III. RESULTS AND DISCUSSION

After exposure, all samples were analyzed using CSAM equipment in order to examine delamination effect. Fig. 1 shows the CSAM images obtained for the as-received and irradiated package with the dose of 50000 Gy. It found that there was no delamination effect for as-received package. However, the delamination effect was found for the package with 50000 Gy irradiation dose. It also found that the delamination occurred between the leadframe and die. This was confirmed with the signal in Fig. 1 (a) showed 80% signal for as-received package, while for package with exposure dose of 50000 Gy the signal intensity was only 10%. Based on the signals obtained, the low percentage of signal indicates the occurrence of delamination in that region of package. From this finding, the higher exposure dose gave lower durability in the QFN package as compared to as-received package. High radiation dose exposure causes the weakening of bond between the materials in packages that promote the occurrence of delamination.

the three-dimensional (3D) image was used to check the crack failure of the QFN package. For each package, a 2D image was captured before and after gamma irradiation. Then, the obtained image was analyzed to determine the initial of wire deflection and subsequent wire deflection change. Fig. 2 shows a 2D image (top view) before and after gamma irradiation. From these images, the wire deflection (indicates arrow in Fig. 2) happened for the wire bond which is placed at the edge of the silicon die. It is because of the wire length in this part is longest than the others. Fig. 3 shows the wire deflection measurements using 3D CT scan X-ray machine. The average change in the deflection of the wire with the exposure dose is plotted in Fig. 4. As shown in the graph, the overall view shows that the wire deflection of the irradiated package has change in comparison to the as-received package. From this result, it is clearly seen that the wire sweep was altered by exposure to gamma radiation. The change is believed due to wire sweep microstructure changes on the package.

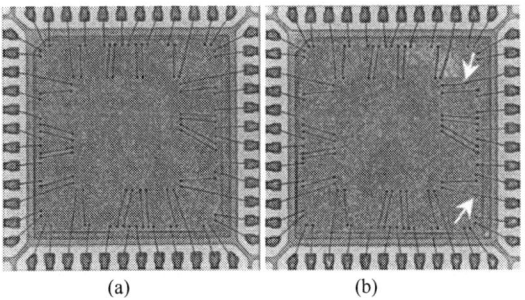

(a)   (b)

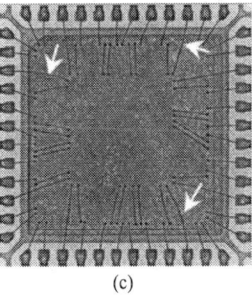

(c)

Fig. 2: 2D image of wire deflection (a) as-received (b) 500 Gy and (c) 50000 Gy

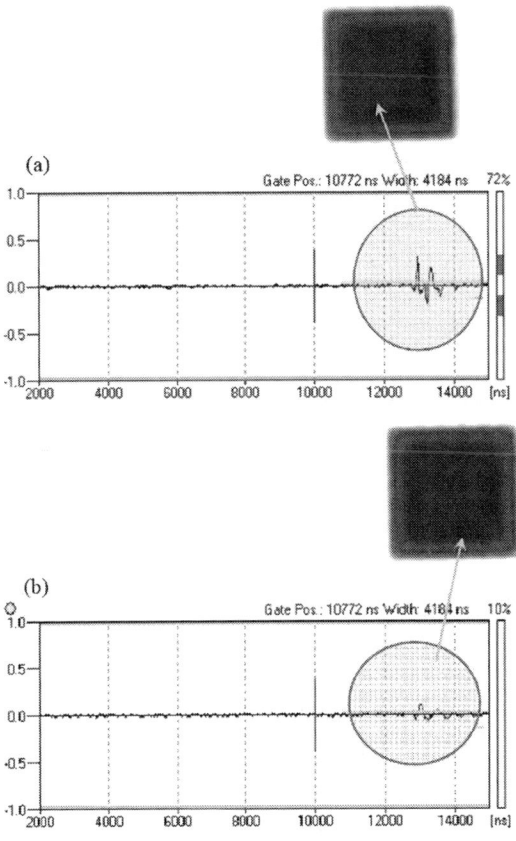

Fig. 1: The comparison signal between (a) as-received and (b) 50000 Gy

After complete the CSAM analysis, all entire packages then subjected to 3D X-ray to inspect internal discontinuity of package due to irradiation. In this study, the two-dimensional (2D) image was used to determine the deflection of wire while

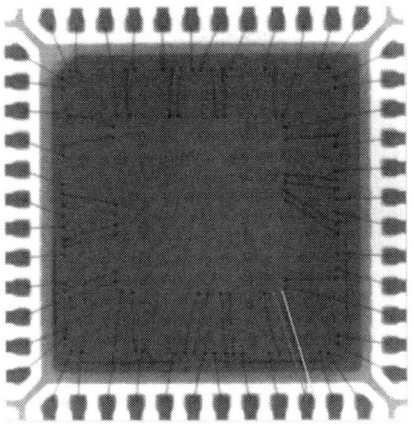

Fig. 3: The wire deflection measurements using 3D X-ray machine

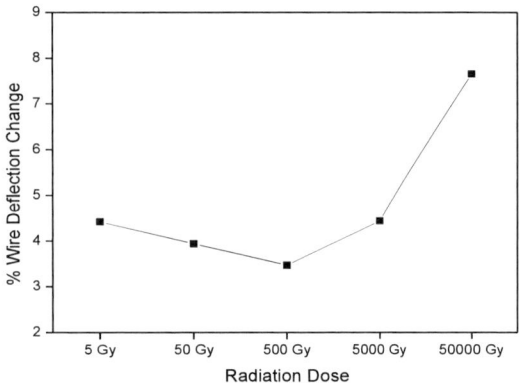

Fig. 4: The wire deflection change versus exposure dose

Fig. 5 shows 3D image for the QFN packages. During the observation, there are no sign of failure for the as-received package. However, the image shows the presence of cracks for the package exposed to a dose of 50000 Gy. This suggests that the formation of these cracks were due to gamma irradiation. These cracks occurred at the surface of silicon die as indicates by circle in Fig. 5 (a).

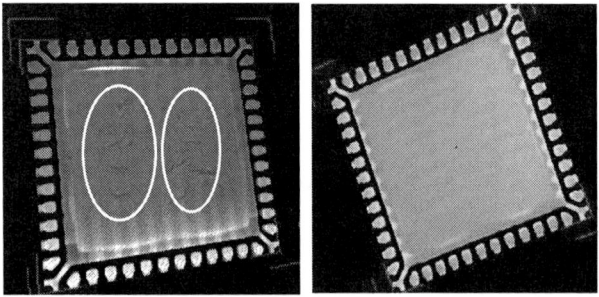

Fig. 5: The 3D image of (a) irradiated (50.0 kGy) and (b) as-received QFN

Exposure to gamma radiation induces radiation damage by virtue of collisions between the gamma ray and the atom constituting the material, called displacement damage [8,9]. The energetic gamma radiation interacts with both electrons and atoms in materials by elastic collision. These atomic collisions may finally compromise the durability of materials [10]. The gamma ray has sufficient energy to displace atoms from their original site. Each of photon has an energy related to the frequency, f given by Plank's relation:

$$E_{\gamma} = hf = h\frac{c}{\lambda} \qquad (1)$$

Where $h = 6.626 \times 10^{-34}$ Js is the Plank's constant, c is the speed of light and λ is the wavelength. It has a momentum, P given by de Broglie relation:

$$P = \frac{h}{\lambda} @ P = \frac{E}{c} \qquad (2)$$

From the above results, it can be summarized that the structure of the QFN stacked die package was altered after exposure to gamma radiation. It is believed that the differences in thermal expansion coefficients between the materials produce an interfacial stress, providing the driving force for the occurrence of delamination [11]. Due to the large differences in thermal mechanical properties, failure can easily occur at higher pressures, especially at locations adjacent to the interface. This higher pressure initiates microcracks near the interface. The crack phenomenon is believed to be due to the presence of the flaws at the die surface which is nature for brittle materials [12]. The existence of small (intrinsic) delamination leads to the occurrence of large delamination [13]. Initial delamination can be regarded as crack tip which occurs in the plastic zone. This fracture requires a large force to break the plastic zone and the plastic zone of crack propagation can be broken up and eventually cause cracking occurs at the die QFN package. Exposure of silicon to gamma ray is resulted in displacement damage. This occurs when silicon is exposed to gamma rays, Si atoms will shift from the substitution site to form Frenkel pairs which cause displacement damage. There are three types of interaction between gamma ray with matter namely pair production, photo electric effect and Compton effect. Compton Effect with energy of 1 MeV will produce isolated single defects when vacancies/interstitials act each other in silicon [14]. The use of ceramics in the production of mold compound can cause changes in the volume of crystal point defects and defect evolution in microstructure when exposed to gamma rays [15]. Gamma ray exposure resulted of revolution and movement of defects during the displacement cascade. In addition, gold wire (metal) can result in the displacement cascade and defect cluster after exposure to gamma radiation. This produces the change in the density of defect cluster. The changed in wire deflection is might be closely related to the change of microstructure of QFN sample.

IV. CONCLUSION

The effect of gamma irradiation on the failure of QFN stacked die package developed by ASPAC, UKM, Malaysia was investigated. The higher the dose applied to the package, the easier the package failure. This is supported by the observation of wire deflection and cracks, determined by using CT scan and delamination by using CSAM. Failure such as cracks, delamination and deflection of the wire is related to microstructure change. Gamma exposure plays a significant role in the change of the package material.

ACKNOWLEDGEMENT

The authors gratefully acknowledge the support provided by UKM (research fund: UKM-GGPM-089-2010, UKM-RRR1-07-FRGS0257-2010, ERGS/1/2011/STG/UKM/02/10 and OUP-2012-120) and Ministry of Science, Technology & Innovation, Malaysia (MOSTI).

REFERENCES

[1] C. Cevdet, "Radiation Effects in Wide-bandgap Semiconductors," in *Focus on Semiconductor Research*, Edited B.E. Thomas, Nova Science Publishers Inc., New York, 2005, pp.135-182.

[2] Y.H.A. Fawzy, A.E.H. Ali, G.F. El-Maghraby, and R.M. Radwan, "Gamma irradiation effect on the thermal stability, optical and electrical properties of acrylic acid/metyl methacrylate copolymer films". *World J. Condensed Matter Phys*, pp. 12-18, February 2011.

[3] P. Krasinski, D. Makowski, and B. Mukherjee, "Portable gamma and neutron radiation dosimeter reader.*" IEEE Nuclear Science Sym. Conf. Record*, October 19-25; Dresden, Germany, 2008.

[4] S. Abdullah, M.F. Abdullah, A.K. Ariffin, and A. Jalar, "Effect of leadframe oxidation on the reliability of a quad flat no-lead package", J. Electron. Packag. Transactions of the ASME 2009, 131 (3), pp. 0310021-0310027, September 2009.

[5] S. Abdullah, A.K. Ariffin, C.K.E. Nizwan, M.F. Abdullah, A. Jalar, and M.F.M. Yunoh, "Failure analysis of a semiconductor packaging using the signal processing approach", Int. J. Mod. Phys. B, 24(1-2), pp. 175-182, January 2010.

[6] M. Amagai, "Mechanical reliability in electronic packaging" Microelectron. Reliab. 42, pp. 607-627, March 2002.

[7] Y.F. Yao, T.Y. Hin, and K.H. Chua, "Improving the deflection of wire bonds in stacked chip scale package (CSP)", Microelectron. Reliab. 43, pp. 2039-2045, Disember 2003.

[8] G.P. Summers, E.A. Burke, and M.A. Xapsos, "Displacement damage analogs to ionizing radiation effects", Radiat. Meas. 24(1), pp. 1-8, January 1995.

[9] J. Kwon, G. Lee, and C. Shin, "Multiscale modelling of radiation effects on materials: pressure vessel embrittlement", Nucl Eng Tech, 41(1), pp. 11-20, February 2009.

[10] Y. Zhang, I-T. Bae, and W.J. Weber, "Atomic collision and ionization effects in oxides", Nucl. Instrum. Meth B, 266, 2828-2833, June 2008.

[11] V.K. Khanna, "Adhesion-delamination phenomeno at the surface and interfaces in microelectronics and MEMS structure and packages devices". J. Phys. D: Appl. Phys., 44(3), 034004, January 2011.

[12] S. Abdullah, M.F.M. Yunoh, A. Jalar, and M.F. Abdullah, "Hardness test on an epoxy mold compounds of a Quad Flat No Lead package using the depth sensing nanoindentation". Adv Mat Res, 146-147, pp. 1000-1003, 2011.

[13] J.A. Newman, W.T. Riddell, and R.S. Piascik, "A threshold fatigue crack closure model: Part 1-model development", Fatigue & Fracture of Engineering Materials & Structure 26(7), pp. 603-614, June 2003.

[14] Z. Li, "Radiation damage effects in Si materials and detectors and rad-hard Si detectors for SLHC". IOP Publishing Ltd and SISSA, March 2009, doi:10.1088/1748-0221/4/03/P03011.

[15] C. Kinoshita, "Microstructural Evolution of Irradiated Ceramics," in *Radiation Effects in Solids*, K.E. Sickafus, E.A. Kotomin, and B.P. Uberuaga, Eds, Springer, Netherlands, 2007, pp. 193-232.

Automated Switching Mechanism for Multi-Standard RFID Transponder

Teh Kim Ting and Khaw Mei Kum
Faculty of Engineering
Multimedia University
Cyberjaya, Malaysia
mkkhaw@mmu.edu.my

Faisal Mohd-Yasin
Queensland Micro- and Nanotechnology Centre
Griffith University
Brisbane, Australia
f.mohd-yasin@griffith.edu.au

Abstract— **This paper presents an automated switching mechanism for a multi-standard modulator operating at high frequency (13.56 MHz) passive Radio Frequency Identification (RFID) transponder. The design adheres to the automatic switching of ISO 14443 (with a data rate of 106 kbps), and ISO 15693 Single Subcarrier (with data rate of 6.62 kbps and 26.48 kbps). The concept of the automated selection of the two different protocols is based on the incoming RF amplitudes; taking advantage of the different reading distances of the two protocols. The switch design is made up of envelope detector, CMOS operational amplifier, comparator with hysteresis, CMOS voltage reference and a 2-to-1 multiplexer. Simulation results verified that the proposed switch design is capable of switching to the desired ISO standard accordingly in a multi-standard transponder. This simulation work is designed and verified by using TSMC 0.18μm CMOS technology with an operating voltage of 1.8V.**

I. INTRODUCTION

In recent years, contactless Radio Frequency Identification (RFID) has proliferated in many service industries, purchasing and distribution logistics industries, manufacturing companies and material flow systems. This RFID technology uses electromagnetic waves for automatic identification (auto ID) of people or objects. The aim of most auto-ID systems is to increase flexibility and efficiency, reduce data entry errors, labor cost as well as prevent internal theft. In an RFID system, the transponder or tag is a data-carrying device integrated on the object to be tracked, whereas the reader or interrogator is a data capture device. Naturally, a multi-standard transponder will be an advantage for wider range of applications due to their flexible operational distances and transmission methods, while maintaining similar costs. From a technical paper discussing on the multi-standard compatible RFID transponder [1], the selectors chosen to decide on the type of standard used by the data transfer from the transponder to the reader were previously done manually.

The goal of this work is to provide an automated switching mechanism that selects the desired protocol between the transponder and the interrogator, in a multi-standard modulator working in 13.56 MHz of a passive RFID transponder. Table 1 shows the summary and comparisons between the proximity standard of ISO 14443 and the vicinity standard of ISO 15693 Single Subcarrier. The major difference between both standards is that proximity cards need to be within 10cm range while vicinity cards can be read within 50cm range, up to 1m.

Therefore, the relationship between the recognition range of transponder to the reader and the voltage induced at the transponder are utilized for the automated switching concept. Particularly, wider distance between transponder and reader results in smaller amplitude of voltage received by the reader [3]-[4].

Based on a multi-standard 13.56MHz RFID system research paper, the measured induced voltage using ISO 14443 Type B commercial tag at a recognition distance of 10cm is 2.644V [2]. Hence, this voltage is used as the reference voltage to switch between the ISO 14443 and ISO 15693 standards signals accordingly. The desired ISO standard protocol is selected automatically, in accordance with the comparison of the received RF input signal against the reference voltage using a comparator circuit.

TABLE I
SUMMARY AND COMPARISONS BETWEEN SPECIFICATIONS OF ISO 14443 AND ISO 15693 [1]-[2]

ISO Standard	Proximity cards		Vicinity card
	14443A	14443B	15693 (Single Subcarriers)
Operating frequency, f_c	13.56MHz ± 7kHz		
Data Coding (Reader to tag)	ASK 100% (OOK) Modified Miller	ASK 10% NRZ	ASK 10% or 100% PPM
Data Coding (Tag to reader)	106kbps data rate: ASK 100% (OOK) Manchester	106kbps data rate: BPSK NRZ-L	ASK Manchester
	212 kbps, 424 kbps, 848 kbps data rate: BPSK NRZ-L		
Subcarrier frequency, f_s	847.5kHz		423.75kHz
Read or write range	Up to 10cm		Up to 1m
	~2cm	~10cm	~20cm
Required Bandwidth	1.7MHz		1MHz
Q-factor of TX antenna	Less than 8		13.6
I_{mag}	1mA	25mA	~220mA
V_{peak}	0.57V	2.16V	7.41V
$V_{induced}$ at tag	~2.644V (Based on 14443B at recognition distance of 10cm)		1.6833V (reader distance of~ 50cm)

The architecture of the automated switch design for passive RFID transponder is described in Section II, while Section III demonstrates the functions of each building block. Simulation results to validate the performance of the automated switch are shown in Section IV. Finally, the conclusion is stated in section V.

II. CIRCUIT ARCHITECTURE

Figure 1 shows the proposed block diagram of the automated switch design for the multi-standard HF passive RFID transponder. The automated switching concept is made up of envelope detector, CMOS operational amplifier, comparator with hysteresis, CMOS voltage reference and a 2-to-1 multiplexer.

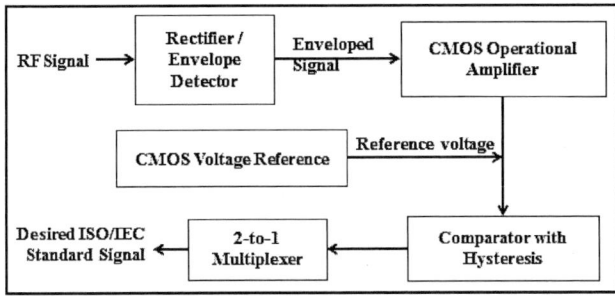

Fig. 1. Block diagram of an automated switch design for a multi-standard passive RFID transponder

According to the block diagram, the working principle of the switching mechanism commences with the detection and conversion of high frequency RF signal into the equivalent amplitude of enveloped signal. After the envelope detection, the amplitude of the enveloped RF signal is decreased. Therefore, a CMOS operational amplifier is required to provide the desired amount of amplification to the enveloped signal. In addition, some filtering effect which is capable of eliminating the high frequency ripple voltage contained in the enveloped signal is performed in this stage as well.

Subsequently, the amplified signal is fed into a comparator circuit. In particular, the comparator converts its input signal, V_{in} to a digital pulse of either a high (V_{DD}) or a low (ground) output, in accordance with the reference voltage V_{REF}. In this switch design, a comparator with hysteresis is designed so that the circuit is immune to the high frequency noise and jitter. Moreover, in order to reduce the number of off-chip component, a comparator with internal hysteresis is designed.

To supply a constant reference voltage to the comparator, a CMOS voltage reference circuit is included. Since the desired standard is selected based on whether the input signal is greater or lesser than reference voltage, it is essential for the V_{REF} to be stable and independent of the temperature and supply voltage variations. Particularly, as V_{in} is greater than V_{REF}, low output signal will be produced and hence it will refer to ISO14443 standard. In contrast, if V_{in} is smaller than V_{REF}, then the high output signal will correspond to the standard of ISO15693.

The last part of the switch design is a 2-to-1 multiplexer. Figure 2 illustrates the proposed integration of switch design in the multi-standard RFID transponder protocol. By referring to Figure 2, the subcarrier signal of 847.5 kHz which represents ISO14443 as well as subcarrier signal of 423.75 kHz denotes

ISO15693 from the frequency divider will be utilized as the input signals, IN_0 and IN_1 of the 2-to-1 multiplexer respectively. Furthermore, the output of the comparator circuit acts as the selector signal of the multiplexer. Therefore, when selector signal is low, IN_0 is output, so ISO14443 signal will be selected automatically. On the contrary, IN_1 signal that is corresponds to ISO15693 standard will be produced when the selector signal is high.

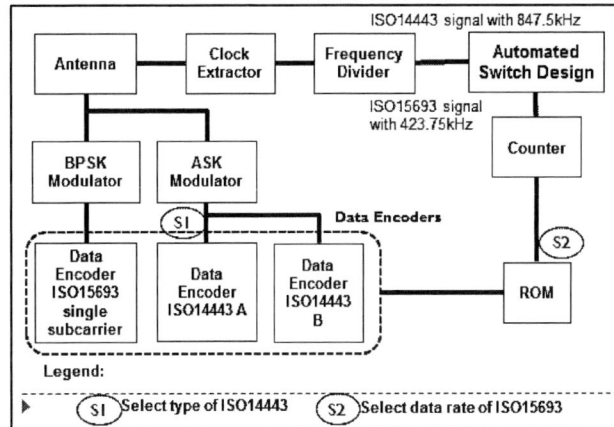

Fig. 2. Proposed integration of automated switch mechanism in multi-standard RFID transponder protocol

III. FUNCTIONAL BLOCKS

A. Complete Envelope Detector with Cascaded Doubler Cell

The peak of the radio frequency input signal is track by an envelope detector circuit as output signal. Since the conventional envelope detector with diode-connected PMOS, capacitor and resistor is inappropriate to be used in a high frequency circuit design [5], the complete envelope detector with cascaded doubler cell as shown in Figure 3 is designed. As it is desirable for the comparison level to be at zero voltage, the cascaded doubler cell acts as the negative DC voltage which replaces the ground terminal [6].

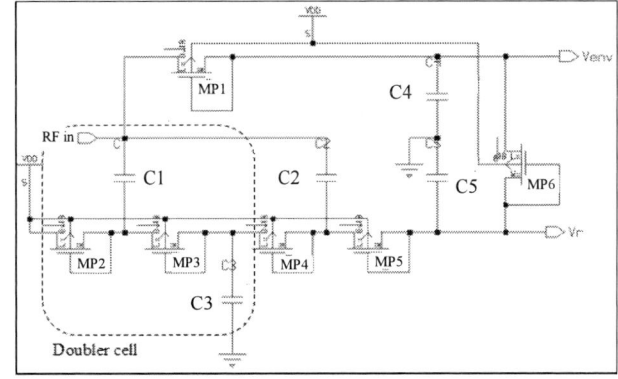

Fig. 3. Complete Envelope Detector Circuit

978-1-4673-2395-6/12 $31.00 © 2012 IEEE

B. CMOS Operational Amplifier

Figure 4 shows the actual configuration of a two-stage, internally compensated non-inverting CMOS operational amplifier (op-amp). Theoretically, this op-amp is capable of providing good voltage gain, good common-mode range and good output swing [7]. The two-stage op-amp comprises a differential stage and a gain stage. In the differential stage, differential amplification and differential to single-ended conversion are achieved, whereas additional gain is provided by the gain stage. A compensation capacitor, C2 is included between differential stage and gain stage to ensure the steadiness of op-amp for the feedback. Moreover, in case of a non-inverting op-amp, the gain can be defined as,

$$\text{Voltage gain}, A = 1 + \frac{R_2}{R_4} \qquad (1)$$

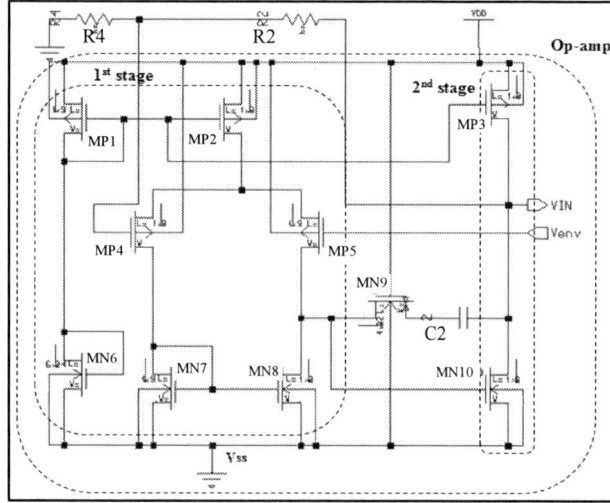

Fig. 4. CMOS Operational Amplifier

C. Comparator with Internal Hysteresis

The configuration of a comparator with internal hysteresis is shown in Figure 5. The hysteresis effect is provided by a latch structure comprising of MP2, MP3, MP4 and MP5. Besides, a bias circuit which composed of MN4 and a current source is included to supply a stable voltage to the gate of MN6. Particularly, larger value of current source will result in higher power dissipation but better output performance. Therefore, appropriate value of current source is required to compromise power consumption and circuit performance.

At the output stage, an inverter formed by MP6 and MN5 is utilized to output a binary signal, based on the comparison between V_{in} and V_{REF}. As V_{in} is larger than V_{REF}, then MP6 is turned off while MN5 is turned on, hence output is pulled to logic low of 0V. Alternatively, when V_{in} is smaller than V_{REF}, MP6 is turned on and MN5 is turned off, thus, the output voltage is pulled up and fixed as V_{DD} or 1.8V.

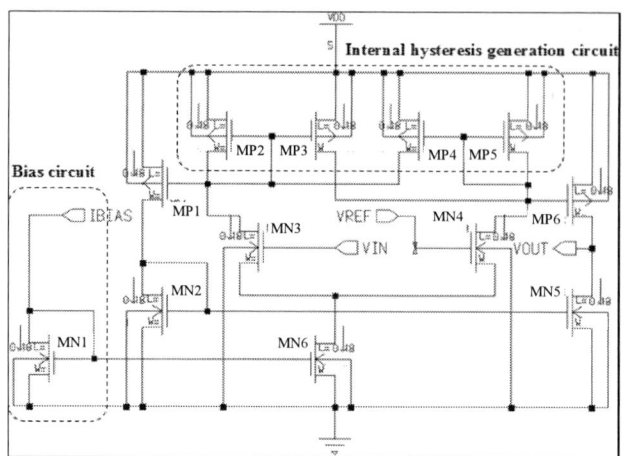

Fig. 5. Comparator with Internal Hysteresis

D. CMOS Voltage Reference

Figure 6 illustrates the configuration of a voltage reference circuit. Since the desired ISO standards are selected based on the comparison of the input signal with a reference voltage, it is essential for V_{REF} to be independent of temperature variation, power supply variation, load variation and other operating conditions. The output reference voltage can be defined as [8],

$$V_{REF} = \frac{V_{GS7}}{\frac{R_1}{R_2}+1} + \frac{V_{GS8}}{\frac{R_2}{R_1}+1} \qquad (2)$$

According to the equation 2, V_{REF} can be controlled by adjusting the ratio of R_1 and R_2 and thus the sensitivity of temperature coefficient of the output voltage against the change of R_1 and R_2 ratio is minimized. Besides, specific amplification through the implementation of an op-amp as well as voltage-controlled voltage source (VCVS) will be applied to the resulted reference voltage, in order to obtain final reference voltage of 2.644V that is required by the comparator circuit.

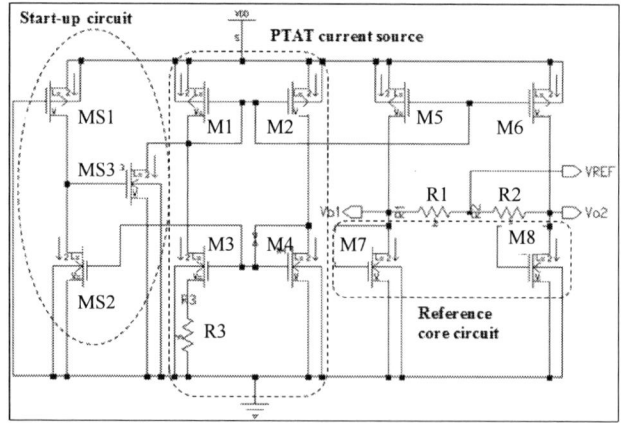

Fig. 6. CMOS Voltage Reference Circuit

E. 2-to-1 Multiplexer

A 2-to-1 multiplexer acts as a selector that selects between two available choices. The circuit of a pass-transistor based 2-to-1 multiplexer is shown in Figure 7. The pass transistors are capable of preventing threshold voltage loss, which will sequentially minimize the possibility of static power consumption [9]. As mentioned in Section II, the subcarrier signals for ISO 14443 and ISO 15693, which generated by the frequency divider of multi-standard RFID transponder protocol, are utilized as the input signals, IN_0 and IN_1 of multiplexer respectively.

In addition, the output signal from the comparator circuit is employed as the selector signal, SELECTOR and the output signal is denoted as MUX_OUT. When SELECTOR is low, IN_0 is output as MUX_OUT, thus ISO14443 signal can be selected automatically. On the other hand, IN_1 signal that is corresponded to ISO15693 standard is produced when selector signal is high.

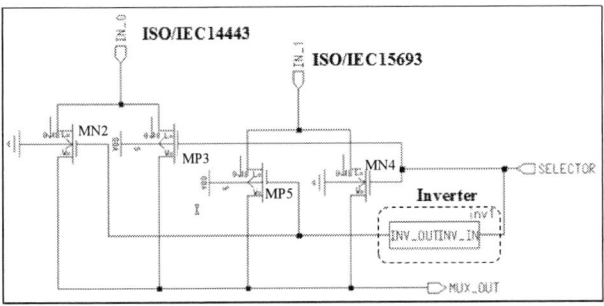

Fig. 7.   2-to-1 Multiplexer Circuit

IV.   RESULTS AND DISCUSSIONS

In order to test for the functionality of the proposed switch design, RF input signal with magnitude of 1.8V and 3.3V are used as the inputs to the design.

Figure 8 shows the simulated reference voltage over a temperature range of -50°C to 150°C. The variation in reference voltages for temperature below -48°C are approximately 8.49mV and 4.67mV for RF input signal of 1.8V and 3.3V respectively. Since V_{REF} is fixed over temperature range of -48°C to 150°C, a constant and temperature-stable reference voltage is obtained.

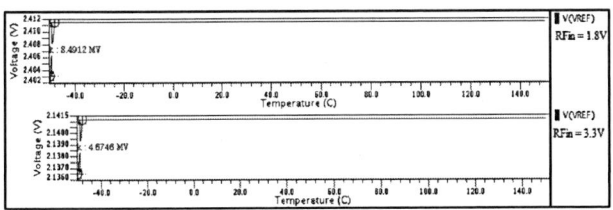

Fig. 8.   Simulated temperature characteristics of output reference voltage

Figure 9 and Figure 10 show the simulated waveforms for the automated switch with RF input signal of 1.8V and 3.3V respectively. According to Figure 9, the RF_{IN} signal of 1.8V is fed into the envelope detector circuit and the peak of the RF_{IN} signal is tracked as V_{ENV} signal. Next, V_{ENV} is applied into a two-stage op-amp circuit and the amplified signal is denoted as V_{IN}. The reference voltage V_{REF} that is required in the comparator is generated by a CMOS voltage reference circuit. Since V_{IN} (~1.8V) is smaller than the V_{REF} (~2.644V), the output voltage represented by SELECTOR signal is at logic high state and fixed at V_{DD} of 1.8V. At final stage, as SELECTOR signal is at logic high state, the IN_1 signal corresponded to ISO15693 standard is selected as MUX_OUT signal automatically.

On the contrary, referring to Figure 10, since V_{IN} signal (~3.2957V) is greater than V_{REF} signal (~2.644V), the output voltage represented by SELECTOR signal is at logic low state and stay at 0V. Therefore, at the final stage of switch design, as the SELECTOR signal is at logic low state, IN_0 signal that corresponding to ISO14443 standard is selected as MUX_OUT signal automatically.

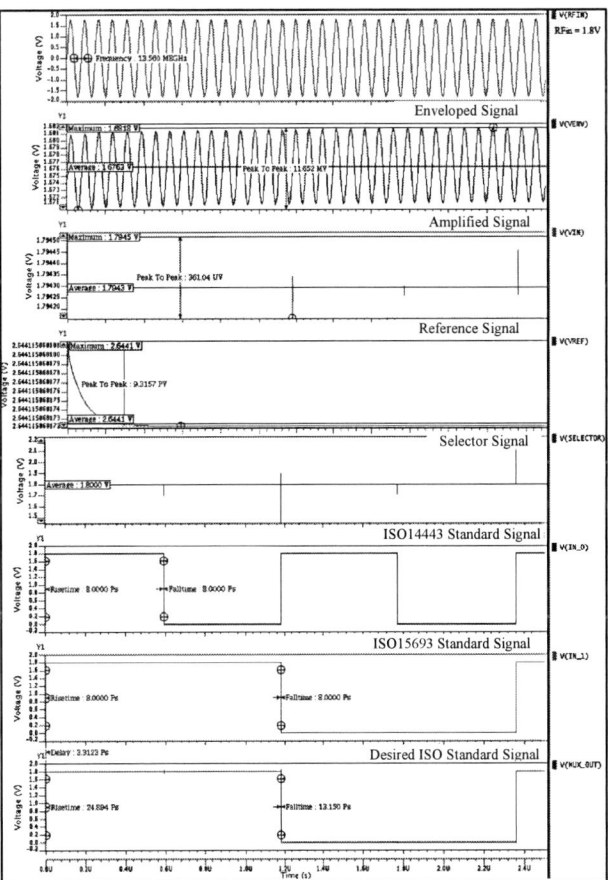

Fig. 9.   Simulated input and output waveforms for automated switch design with RF input signal of 1.8V

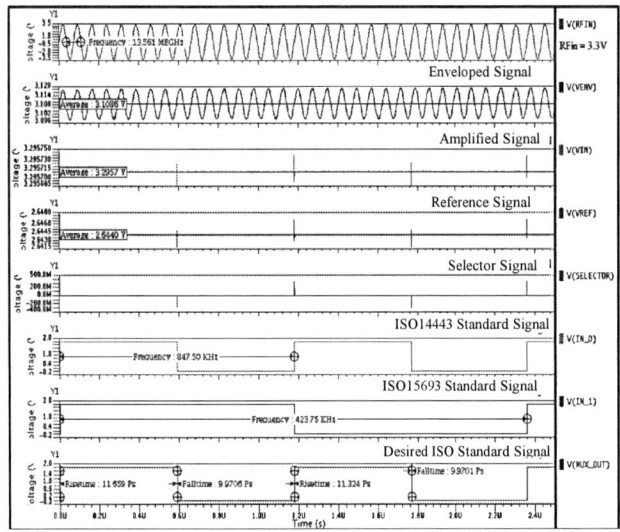

Fig. 10. Simulated input and output waveforms for automated switch design with RF input signal of 3.3V

proposed switch design is capable of switching to the desired ISO standard accordingly in a multi-standard transponder protocol. Since this automated selection of the two different protocols are based on the incoming RF amplitudes, therefore it will work in most situations except when we are receiving weak signals from the reader to the vicinity card/transponder.

V. CONCLUSION

The design of an automated switching mechanism for a multi-standard 13.56 MHz transponder has been presented. The proposed automated switch design conforms to the standards of ISO 14443 and ISO 15693 Single Subcarrier, working in HF of passive RFID transponder. The automated switch design was designed and validated using Mentor Graphics employing TSMC 0.18μm technology. Simulation results verified that the

REFERENCES

[1] Y. F. Lim, M. K. Khaw and F. Mohd-Yasin, "A Multi-Standard Modulator for High Frequency Passive RFID Transponder", Circuits and Systems (MWSCAS), 2011 IEEE 54th International Midwest Symposium on, 23 September 2011.

[2] Y. H. Kim, Y. C. Choi, M. W. Seo, S. S. Yoo, and H. J. Yoo, "A CMOS Transceiver for a Multistandard13.56-MHz RFID Reader SoC", IEEE Transactions on Industrial Electronics, vol. 57, No. 5, 5 May 2010.

[3] A. Shameli, A. Safarian, A. Rofougaran, M. Rofougaran and F. D. Flaviis, "An RFID System with Fully Integrated Transponder", University of California and Broadcom Corporation.

[4] C. Metzger, A. Ilic, P. Bourquin, F. Michahelles, E. Fleisch, "Distance-sensitive High Frequency RFID Systems".

[5] N. M. Karim, "Design of an Integrated Demodulator for a 13.56MHz RFID reader", IEEE Symposium on Industrial Electronics and Applications, 4 October 2009.

[6] P N. A. Fahsyar and N. Soin, "CMOS Implementation of Envelope Detector Circuit in 0.18μm Process", ICSE Proc. 2010.

[7] P. K. Alli, "Testing a CMOS Operational Amplifier Circuit Using a combinational of Oscillation and I_{DDQ} Test Methods.", Master thesis, Agriculture and Mechanical College, Louisiana State University, 2004.

[8] L. Najafizadeh, "Voltage References Using Mutual Compensation of Mobility and Threshold Voltage Temperature Effects", University of Alberta, Spring 2004.

[9] Jan M. Rabaey, A. Chandrakasan, and B. Nikolic, "Digital Integrated Circuits: A Design Perspective", 2nd Edition, Prentice Hall, 2003.

Switched Inverter Comparator based 0.5 V Low Power 6 bit Flash ADC

Rajeev Komar[1], M S Bhat[2], T Laxminidhi[3]
Department of Electronics and Communication Engineering
National Institute of Technology Karnataka
Mangalore, India – 575025
Email: [1]rtkomar@gmail.com, [2]msbhat@ieee.org, [3]laxminidhi_t@yahoo.com

Abstract- **This paper presents an ultra low power 6 bit Flash ADC designed in 180 nm CMOS technology for ultra low power applications. The design uses inverter based comparators to reduce the silicon area and power requirement. A novel clock delaying technique is used to power on the three stages of the comparator which work in series. This reduces the power consumption and increases speed of operation. Fat tree architecture is used to design the digital encoder. The power supply used for the design is 0.5 V and the sampling rate is 50 MS/s. The design consumes ultra low power of 600 µW and spans a very small area of 0.164 mm². In literature this is found to be the lowest for 6 bit ADCs in 180 nm with sampling frequency of 5 MS/s or above. The SNDR remains above 31.5 dB in the whole input frequency range of 0 to 25 MHz. The ADC has maximum DNL of 0.85 LSB and maximum INL of 1 LSB. The FOM of the ADC is found to be 0.39 pJ/conv.**

Keywords- **Flash ADC, Inverter comparator, Low power, Low voltage, Fat tree encoder**

I. INTRODUCTION

The exponential growth in the usage of handheld, battery operated devices like cell phones, tablets, laptops and portable medical instruments has made power consumption one of the most prominent design criterion. Also as the fabrication cost of the chip is directly proportional to its area, there is a dire need for power efficient designs consuming a small area. Although process scaling and integration of all the necessary electronic circuits of diverse functions onto a single chip, called system on a chip (SoC), have reduced the total power consumption and required area of the devices and brought in great performance improvements, clever design choices can bring in further improvements in terms of power, speed and area.

Analog to digital converters are an essential part of any device which interacts with real world. The flash architecture of the ADC is the fastest one which is used as a building block for many architectures like folding and interpolating, pipelined, subranging ADCs etc. But the exponential increase in area and power consumption with resolution has limited the use of flash ADC in ultra low power applications.

The proposed ADC intends to fill the gap by proposing a 6 bit flash ADC for ultra low power applications which uses a small area. The present work builds upon our work presented in [1] and improves it in a number of ways with layout results. As

power consumption reduces quadratically with the reduction of voltage [2], 0.5 V was chosen as the supply voltage. A novel technique of delayed clocks activating different stages of a modified inverter based comparator is used to reduce the power consumption and area when compared to OTA and Op Amp based comparators. Unlike threshold inverter quantizer (TIQ) which has a small fraction of the supply voltage as the input range, the proposed ADC has rail to rail input range. We have used low threshold MOSFETs in the design for the ultra low voltage operation of 0.5 V. The reference input for the ADC is generated from a resistor ladder network. The thermometer coded outputs of the comparators are converted into one-hot-code and the final digital output code is generated using a fat tree encoder.

In section 2 detailed explanation of the proposed comparator design is given. In section 3 the design of the 6 bit flash ADC using inverter comparator has been explained. The details of layout design and extracted layout simulation results are given in section 4 and section 5 concludes the paper.

II. DESIGN OF THE PROPOSED COMPARATOR

The use of inverter as a comparator was proposed in [3]. But because of its low gain and high parasitics inverter based comparator has not been exploited fully.

A scheme for voltage comparison using Inverter based comparator is shown in Fig. 1. In the figure, V_{IN} is the input signal, kV_{LSB} is the reference voltage of the kth comparator and ΔV is the voltage difference between the two points on the characteristics (see Fig. 2) where gain of the inverter is equal to -1. The use of inverter as a comparator, shown in Fig.1, can be explained with respect to the two phases of the clock, φ and $\overline{\varphi}$. S_1, S_2 and S_3 are three switches and C_{INV} is the sampling capacitor.

Figure 2 shows the transfer characteristics of the inverter. In this figure, V_{mid} is the voltage when the output of the inverter is equal to the input.

During φ,
switches S_1 and S_3 are closed and S_2 is opened. The voltage across the capacitor V_C can be expressed in terms of V_{IN} and V_{mid} of the inverter as,

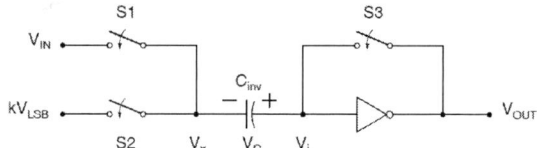

Fig. 1. Inverter comparator

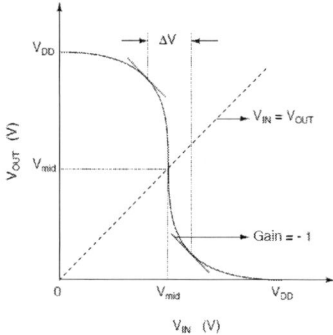

Fig. 2. CMOS inverter characteristics

$$V_I = V_{mid}$$
$$V_C = V_I - V_X$$
$$= V_{mid} - V_{IN} \quad (1)$$

During $\overline{\varphi}$,

switches S_1 and S_3 are opened, S_2 is closed connecting kV_{LSB} to V_X. Now, the input of the inverter V_I is calculated with the help of (1) as,

$$V_I = V_C + kV_{LSB}$$
$$= (V_{mid} - V_{IN}) + kV_{LSB}$$
$$= V_{mid} - (V_{IN} - kV_{LSB}) \quad (2)$$

Now, depending on the value of $(V_{IN} - kV_{LSB})$, the output of the inverter will be decided as follows,

$$(V_{IN} - kV_{LSB}) > \frac{\Delta V}{2} \Rightarrow V_I < V_{mid} - \frac{\Delta V}{2}$$

$$V_I < V_{mid} - \frac{\Delta V}{2} \Rightarrow V_{OUT} = '1'$$

$$(V_{IN} - kV_{LSB}) < -\frac{\Delta V}{2} \Rightarrow V_I > V_{mid} + \frac{\Delta V}{2}$$

$$V_I > V_{mid} + \frac{\Delta V}{2} \Rightarrow V_{OUT} = '0' \quad (3)$$

The low gain of the inverter and the parasitics associated with the sampling capacitor, switches and the switching kickback noise hinder the performance of the inverter comparator thereby making it unsuitable for low voltage and high speed operation. These limitations are overcome by adding a second inverter stage which is designed to increase the gain and clocked cleverly to nullify the effect of kickback noise and parasitics. A third stage of back to back inverters is added to further increase the gain and to hold the comparator output for digitization. The proposed comparator design is shown in Fig. 3. Figure 4 shows the clock signals used in Fig. 3.

Stage 1 of the comparator works as explained above. When $\overline{\varphi}$ goes high, the reference voltages are connected to the comparators. Hence there is a sudden current flow from the resistor ladder network to all the comparators of the ADC to charge the respective capacitors C_{inv}. This disturbs the reference voltages. The altered reference voltages would change the actual $(V_{IN} - kV_{LSB})$ values temporarily which may start driving the inverter Inv1 towards wrong output. Actual $(V_{IN} - kV_{LSB})$ values will be slowly restored once the C_{inv} capacitors are charged. But the output of the Inv1 will take much longer time to reflect the correct output, making the entire operation very slow. To overcome this, Stage 2 is clocked after a while, allowing reference voltages and kickback noises to settle. When φ_{d1} is applied, it connects the output of Inv2 to its input driving V_{O1} and V_{O2} to V_{mid}. This brings back the disturbed output of Inv1 to V_{mid}. When φ_{d1} goes low, Inv1 gives the correct output according to the settled values of $(V_{IN} - kV_{LSB})$. This output is further enhanced by Inv2. Now φ_{d2} is applied to the back to back connected inverters shorting their output to input making them equal to V_{mid}. When φ_{d3} is applied, the back to back inverters drive the Inv2 output to either supply voltage or ground resulting in high speed operation. They also act as a latch and hold the value till φ_{d2} goes high in the next cycle. This eliminates the need for a separate latch to hold the comparator outputs for the digital encoding.

III. ADC DESIGN

The block diagram of the ADC is shown in Fig. 5. The individual blocks are explained in the following subsections.

A. Reference Generation

The reference voltage for the comparators is generated by a resistor ladder network. The resistor ladder has been split into two parts: one generating the reference voltages for the odd numbered comparators (k = 1, 3, 5 etc.) and the other for even numbered comparators (k = 2, 4, 6 etc.). This reduces the RC delay of the resistor ladder and makes laying out the ADC easier. The individual resistor values are selected for optimum speed and power consumption.

B. Clock Generation

As explained in section II, the different stages of the

978-1-4673-2395-6/12 $31.00 © 2012 IEEE   614

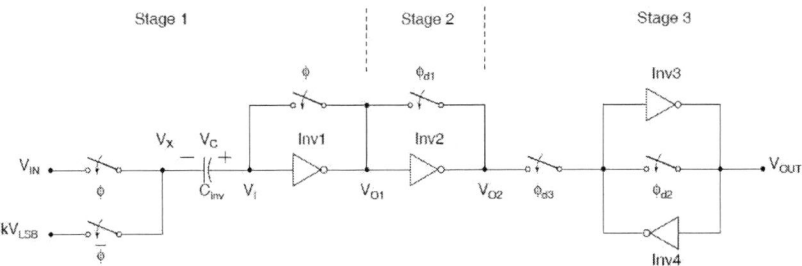

Fig. 3. Proposed Inverter Comparator

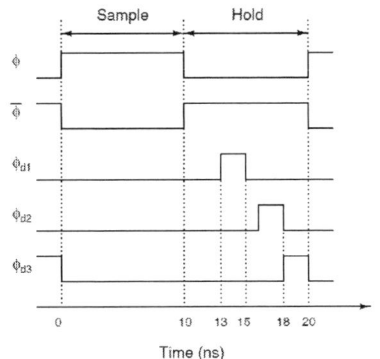

Fig. 4. Clock signals for the proposed design

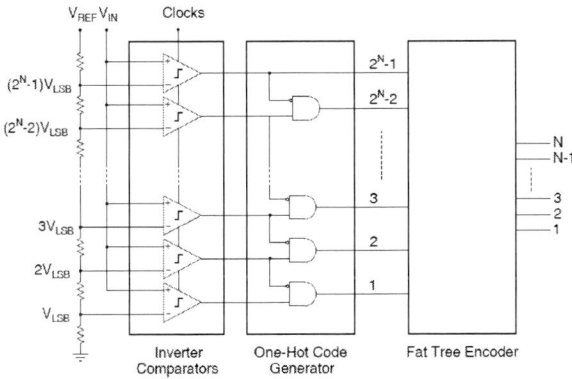

Fig. 5. Block diagram of the ADC

tree architecture is used. Fat tree architecture is one of the fastest architectures and it uses relatively small area when compared to commonly used ROM encoder [4]. The drawback is that, laying out a fat tree encoder is challenging as it has a tree structure and not as regular as a ROM encoder. For feeding the inputs of the fat tree encoder, thermometer code will have to be converted into one-hot-code. This can be easily achieved by using a NOT and AND gate combination as shown in Fig. 5. The one-hot-code is given as the input to the fat tree encoder. The fat tree encoder converts the one-hot-code into binary code using multiple tree structures.

IV. LAYOUT DETAILS AND SIMULATION RESULTS

The whole design has been laid out in UMC 180 nm CMOS technology. For the ultra low voltage operation of 0.5 V, low V_{th} transistors have been used in the design. Careful sizing of the inverters in the comparators, using low V_{th} PMOS as sampling capacitor C_{inv} and the optimum sizing of the switches along with the split resistor network have improved the overall ADC performance over [1]. Figure 6 shows a single inverter comparator layout along with its dimensions. Figure 7 shows the complete layout of the ADC.

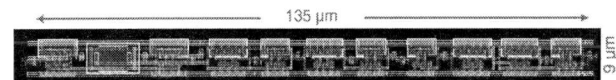

Fig. 6. Layout of a single inverter comparator

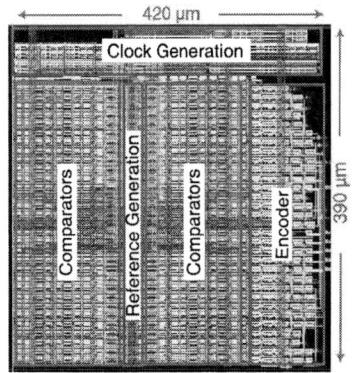

Fig. 7. Layout of the ADC

comparators are clocked cleverly to nullify the effect of reference variation and kickback noises achieving high speed operation. φ is generated from the system clock using buffers. $\overline{\varphi}$ is generated by inverting φ. φ_{d1} is generated by delaying φ and ANDing the delayed version with $\overline{\varphi}$. φ_{d2} and φ_{d3} are the delayed versions of φ_{d1}. Inverter buffers are used as delay elements in the design as they occupy small area and consume very low power.

C. Encoder

The output of the comparators will be thermometer coded. For converting the thermometer code to 6 bit binary code fat

The total area consumed by the ADC is 0.164 mm². The measured differential non linearity (DNL) and integral non linearity (INL) of the ADC are shown in Fig. 9. The maximum DNL and INL are found to be 0.85 LSB and 1 LSB respectively. There are no missing codes in the output. The dependency of the spurious free dynamic range (SFDR) and signal to noise and distortion ratio (SNDR) on input frequency is shown in Fig. 9 for the sampling frequency of 50 MS/s. Figure 10 shows the power spectrum of the reconstructed output of the ADC for the input frequency of 23.83 MHz. At zero frequency, the DC component superimposed on the input can be seen. Power spectrum has been calculated by performing 128 point FFT of the output. A 96% full scale input (480 mV peak to peak) has been used for all calculations. At the input frequency of 23.83 MHz, SNDR is found to be 31.5 dB giving effective number of bits (ENOB) as 4.97. The SNDR remains above this value for all the lower input frequencies. The performance of the ADC against temperature is shown in Fig. 11 and Fig. 12. The power consumption increases with temperature due to increase in leakage currents but the SNDR is not affected. The SNDR variation against process variation is shown in Fig. 13. All these simulations have been carried out at Nyquist frequency as that depicts the worst case scenario. The ADC consumes an ultra low power of 600 μW at room temperature.

The popular figure of merit (FOM) used to compare the ADCs is calculated as [5],

$$FoM = \frac{Power}{2 \cdot f_{in} \cdot 2^{ENOB}} \quad \text{(fJ/conv)} \quad (4)$$

The FOM of the ADC is 0.39 pJ/conv. Table I summarizes the overall performance of the ADC.

Table II compares the performance of the present work with recently published state of the art designs. The work presented in this paper is better than almost all the 180 nm designs in terms of power consumption, speed and area requirement. It is also better than many of the 130 nm and some of the 90 nm and 65 nm designs which have the advantage of lower power consumption, lower parasitics and smaller area.

TABLE I
PERFORMANCE SUMMARY OF THE ADC

Parameter	Value
Technology	UMC 180 nm CMOS, Low V_{th} transistors
Supply Voltage	0.5 V
Resolution	6 bit
Sampling Frequency	50 MS/s
Input Range	490 mV peak to peak
SFDR at 23.83 MHz	36.9 dB
SNDR at 23.83 MHz	31.5 dB
ENOB at 23.83 MHz	4.97 bits
Maximum DNL, INL	0.85 LSB, 1 LSB
Total Power	600 μW
FOM	0.39 pJ/conv
Total Area	0.164 mm²

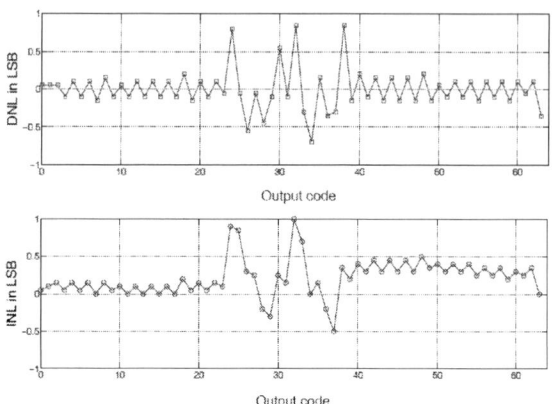

Fig. 8. DNL and INL plot of the ADC

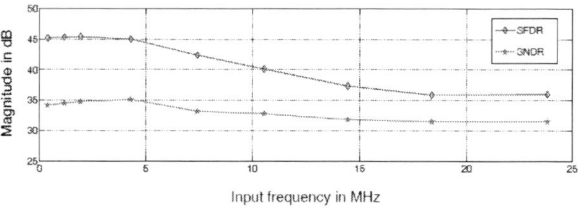

Fig. 9. SFDR and SNDR variation of the ADC with input frequency

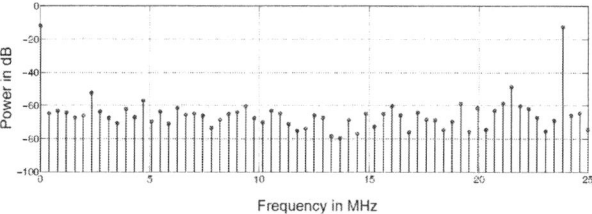

Fig. 10. ADC power spectrum with input frequency of 23.83 MHz

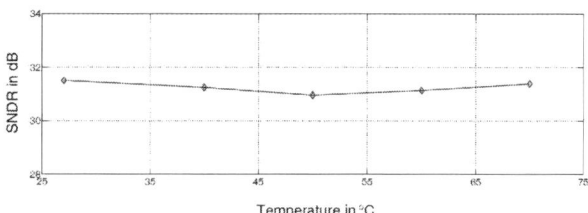

Fig. 11. SNDR variation of the ADC with temperature

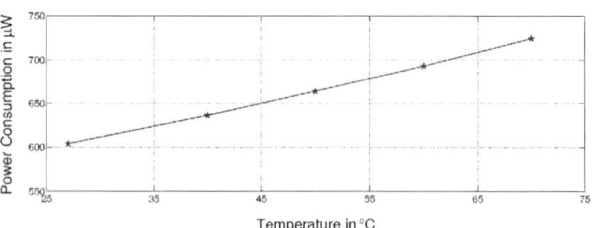

Fig. 12. Power consumption of the ADC with temperature

TABLE II
COMPARISON WITH LOW VOLTAGE STATE OF THE ART DESIGNS

Reference	Type	V_{DD} (V)	Resolution (bits)	Speed (MS/s)	Power (mW)	ENOB	Area (mm^2)	Technology (nm)	FOM (pJ/conv)
[6] - Apr/2009	Flash	1.2	6	100	4	5.63	0.13	65	0.8
[7] - 2010	Flash	0.35	4	200	0.54	3.03	0.0075	90	0.33
[8] - Nov/2008	Inverter Flash	0.6	5	60	1.3	4.36	0.11	90	1.06
[9] - Feb/2011	Pipeline	1.2	10	40	15.6	9.01	1.76	130	0.22
[10] - Aug/2011	Pipeline	1.2	10	60	23	9.16	1.84	130	0.67
[11] - Jan/2011	Flash	1.8	4	700	5.56	3.77		180	0.46
[12] - Nov/2009	Flash	0.4	6	0.4	0.0017	5.05	1.96	180	0.125
[13] - Feb/2009	Folding	1.2	8	80	30	6.94	1	180	3.1
[14] - July/2008	Pipeline	1.8	8	200	22	7.24	0.32	180	0.74
[15] - Mar/2010	Pipeline	1.8	10	60	18	8.6	0.84	180	1.15
[16] - May/2010	Pipeline	1.8	10	100	31	8.5	1.28	180	0.85
[17] - Jan/2010	Pipeline	1.8	11	40	21	10.5	2.1	180	0.18
Proposed Work	Inverter Flash	0.5	6	50	0.6	4.97	0.164	180	0.39

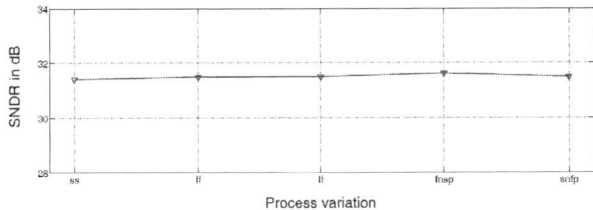

Fig. 13. SNDR variation of the ADC with process variation

V. CONCLUSION

As it can be seen from the literature, there have not been many low voltage ultra low power ADC designs in 180 nm technology. The present work has a relatively better FOM consuming ultra low power and very small area while achieving respectable speed of 50 MS/s. The ADC can be used in battery operated devices where power efficiency is very important criterion. It can also be used as a building block in the design of high resolution pipelined, folding-interpolating and subranging ADCs. The porting of the design to a lower technology node can bring in further improvement in the performance of the ADC.

REFERENCES

[1] R. Komar, M. Bhat, and T. Laxminidhi, "A 0.5 V 300μW 50MS/s 180nm 6bit Flash ADC using inverter based comparators," in *2012 International Conference on Advances in Engineering, Science and Management (ICAESM)*. IEEE, 2012, pp. 331–335.

[2] J. Rabaey, A. Chandrakasan, and B. Nikoli´c, *Digital Integrated Circuits, 2/e.* Pearson Education, 2003.

[3] A. Dingwall, "Monolithic Expandable 6 Bit 20MHz CMOS/SOS A/D Converter," *IEEE Journal of Solid-State Circuits*, vol. 14, no. 6, pp. 926–932, 1979.

[4] D. Lee, J. Yoo, K. Choi, and J. Ghaznavi, "Fat tree encoder design for ultra-high speed flash A/D converters," in *The 2002 45th Midwest Symposium on Circuits and Systems, 2002 (MWSCAS-2002)*, vol. 2. IEEE, 2002, pp. II–87.

[5] R. Plassche, *CMOS integrated analog-to-digital and digital-to-analog converters*, ser. Kluwer international series in engineering and computer science. Kluwer Academic Publishers, 2003.

[6] C. Chen, M. Le, and K. Kim, "A low power 6-bit flash ADC with reference voltage and common-mode calibration," *IEEE Journal of Solid-State Circuits*, vol. 44, no. 4, pp. 1041–1046, 2009.

[7] Y. Lin, Y. Lien, and S. Chang, "A 0.35-1 V 0.2-3 GS/s 4-bit low-power flash ADC for a solar-powered wireless module," in *2010 International Symposium on VLSI Design Automation and Test (VLSI-DAT)*. IEEE, 2010, pp. 299–302.

[8] J. Proesel and L. Pileggi, "A 0.6-to-1V inverter-based 5-bit flash ADC in 90nm digital CMOS," in *IEEE Custom Integrated Circuits Conference, 2008. CICC 2008.* IEEE, 2008, pp. 153–156.

[9] G. Shu, Y. Guo, J. Ren, M. Fan, and F. Ye, "A power-efficient 10-bit 40-MS/s subsampling pipelined CMOS analog-to-digital converter," *Analog Integrated Circuits and Signal Processing*, vol. 67, no. 1, pp. 95–102, 2011.

[10] J. Ruiz-Amaya, M. Delgado-Restituto, and ´A. Rodr´ıguez-V´azquez, "A 1.2 V 10-bit 60-MS/s 23 mW CMOS pipeline ADC with 0.67 pJ/conversion-step and on-chip reference voltages generator," *Analog Integrated Circuits and Signal Processing*, pp. 1–11, 2011.

[11] G. Torfs, Z. Li, J. Bauwelinck, X. Yin, G. Van der Plas, and J. Vandewege, "Low-power 4-bit flash analogue to digital converter for ranging applications," *Electronics letters*, vol. 47, no. 1, p. 20, 2011.

[12] D. Daly and A. Chandrakasan, "A 6-bit, 0.2 V to 0.9 V highly digital flash ADC with comparator redundancy," *IEEE Journal of Solid-State Circuits*, vol. 44, no. 11, pp. 3030–3038, 2009.

[13] H. Movahedian Attar and M. Bakhtiar, "Low voltage low power 8-bit folding/interpolating ADC with rail-to-rail input range," *Analog Integrated Circuits and Signal Processing*, vol. 61, no. 2, pp. 181–189, 2009.

[14] S. Jiang, M. Do, K. Yeo, and W. Lim, "An 8-bit 200-Msample/s pipelined ADC with mixed-mode front-end s/h circuit," *IEEE Transactions on Circuits and Systems I: Regular Papers*, vol. 55, no. 6, pp. 1430–1440, 2008.

[15] J. Lin, S. Chang, C. Liu, and C. Huang, "A 10-bit 60-MS/s low-power pipelined ADC with split-capacitor CDS technique," *IEEE Transactions on Circuits and Systems II: Express Briefs*, vol. 57, no. 3, pp. 163–167, 2010.

[16] M. Kim, J. Kim, T. Lee, and C. Kim, "10-bit 100-MS/s Pipelined ADC Using Input-Swapped Opamp Sharing and Self-Calibrated V/I Converter," *IEEE Transactions on Very Large Scale Integration (VLSI) Systems*, vol. 19, no. 8, pp. 1438–1447, 2011.

[17] M. Fan, J. Ren, N. Li, F. Ye, and J. Xu, "A 1.8-V 11-bit 40-MS/s 21-mW pipelined ADC," *Analog Integrated Circuits and Signal Processing*, vol. 63, no. 3, pp. 495–501, 2010.

A Ultra-Wideband, Downconversion Folded Mixer in 0.13-um CMOS Technology

Xuelian Liu, Student Member, IEEE, John F. McDonald, Life Senior Member, IEEE

Department of Electrical, Computer, &Systems Engineering, Rensselaer Polytechnic Institute

110 8th Street Troy, NY, 12180

Email: liux11@rpi.edu

Abstract- **In this paper, a double-balanced common gate folded Gilbert mixer with low voltage supply and low power consumption is presented in 0.13um CMOS technology. This mixer features the common gate gm stage following by a folded LO switch stage. These two stages can be biased independently to achieve gain, linearity and noise figure specification in a wideband range. The simulation shows this mixer can achieve 6.3~11.8dB conversion gain, 9.8~15.2dB noise figure, 39dBm~48dBm second-order intermodulation intercept point(OIP2), and around -7.5dBm Input Referred 1dB compression point between 2.5GHz~7.5GHz with 7mW power consumption under 1.5V voltage supply**

I. INTRODUCTION

Mixers are key components in both receivers and transmitters. Mixers translate signals from one frequency band to another. The output of the mixer consists of multiple images of the mixers input signal where each image is shifted up or down by multiples of the local oscillator (LO) frequency. The most important mixer output signals are usually the signals translated up and down by one LO frequency.

The CMOS direct-conversion receiver IC for multi-standard wireless systems has been intensively studied [1][2]. One of the key building blocks in the front direct-conversion receiver is a down-conversion mixer where low noise property and high linearity are required under low voltage operation. Linearity is quite an important issue because nonlinearity causes many problems such as gain compression, cross modulation and intermodulation, etc. As the scaling down of CMOS technology, the noise performance of circuit degrades since a shorter gate length introduces more flicker noise [3][4] (1/f noise). Another important issue of mixer is DC offset [5], which is mainly attributed to even-order nonlinearities and self-mixing of LO leakage to RF port.

There are mainly two kinds of mixer: active mixer like Gilbert cell mixer and passive mixer. The main difference between mixer types is that the active mixer consumes power, whereas the passive one, not including the gm-stage, does not [6]. From the RF metrics point of view, the main difference is in the linearity performance and the required LO drive level. Although the passive mixer has many benefits like higher linearity, lower power consumption and better way to suppress 1/f noise due to no DC current in LO switch than active mixer, active mixer can achieve higher gain than passive mixer as well as wider bandwidth, better

port to port isolation. So, it is preferable to use an active mixer as a direct down-conversion mixer (the 1st mixer in the Rx chain), or otherwise overall system noise figure would be degraded. The active mixer would be more suitable for use where the signal is healthy.

This paper presents a wideband down-conversion folded [7][8][9] active mixer design and implementation using a 0.13-um IBM RFCMOS process. This mixer achieves good linearity, high conversion gain and low NF over a wideband range with low power consumption and voltage supply.

II. DOUBLE BALANCED FOLDED SWITCH MIXER DESIGN

The schematic of the common gate double balanced folded switch mixer is shown in Fig.1. The NMOS differential pair, M1 and M2, is used as the input transconductance stage (g_m stage) . The PMOS LO switches (M3~M6) are folded with respect to the g_m stage. The PMOS is used as the switch transistors because it will contribute less flicker noise than NMOS[2].

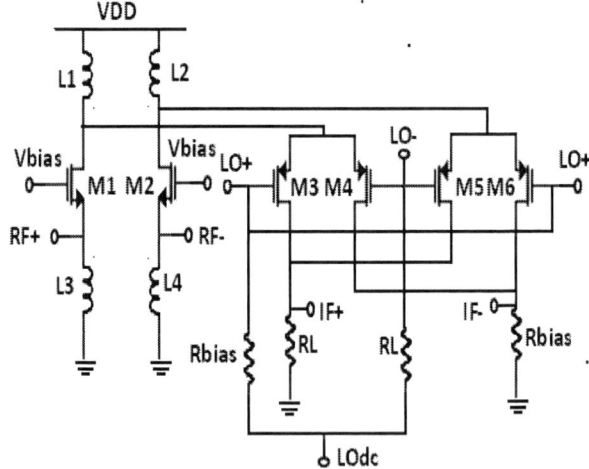

Fig.1 Double balanced folded switch mixer

The folded cascade architecture gives better performance under low voltage supply. It allows the independent bias current for the gm state and the LO switches. The bias current for the gm state should be large enough to achieve the decent gain, NF, and linearity, while the current in the LO switches should be low enough to suppress DC offset, thermal and 1/f noise.

A Gain and Noise

For the common gate input stage, the gate-source and gate-drain parasitic capacitances of the MOSFET M1 and M2 can be absorbed into the LC tank by L3 and L4 and resonated out. The Rin can be expressed as equation (1)

$$R_{in} = \frac{1}{g_m + g_{mb}}\left(\frac{r_{ds} + R_{L\prime}}{r_{ds}}\right) \quad (1)$$

where $R_{L\prime}$ is the equivalent shunt resistance introduced by the finite quality factor, Q, of the resonant load L1 and L2, which present a high impedance from 2.5GHz and 7.5GHz such that the current of gm state will flow into the LO switches[10].

Neglecting the FET gate-induced noise for FETs, assuming an input-matching condition Rs = R_{in}, neglecting the noise generated by inductor loss and taking the body effect into account, the NF of the proposed transconductance stage can be expressed as

$$F = 1 + \left(\frac{\gamma}{\alpha}\right)\left(\frac{1}{1+X}\right)\left(\frac{r_{ds} + R_L}{r_{ds}}\right) \quad (2)$$

where γ/α is usually 2/3 for long-channel device [2] and is used to account 2 for the short-channel device effect [2]. $\alpha = \frac{g_m}{g_{d0}}$ where g_{d0} is drain–source conductance at Vds=0 . $X = \frac{g_{mb}}{g_m}$, which is the ratio of the backgate transconductance to that of the MOS transistor. If $r_{ds} \to \infty$, $F = 1 + \left(\frac{\gamma}{\alpha}\right)\left(\frac{1}{1+X}\right)$.

As state in the [4], there are also other source of noise in the mixer mainly coming from LO switch stage, including noise from the LO Port and thermal noise generated in the switching pair, flicker-noise. Large LO amplitude increases the conversion gain and reduces the noise contribution of the switching pair and the LO port. After a certain value, the conversion gain of the switching pair reaches its maximum value 2/π and the noise contribution of the switching pair becomes negligible. Further increase of LO amplitude does not reduce the noise figure considerably and it even increases the noise coming from the LO port. So, the bias of LO switch should be suitable to achieve good current communication as well as suppress noise.

The total gain of the mixer combined g_m stage and LO switch stage is

$$G \approx \frac{2}{\pi} * g_m R_L \quad (3)$$

B Linearity

The non-linearity of RF gm stage determines the intermodulation characters of the folded mixer.

For the long channel device model, we can calculate ΔI_{RF} as a function of V_{RF}, $\Delta I_{RF} = \alpha 1 * V_{RF} + \alpha 2 * V_{RF}^2 + \alpha 3 * V_{RF}^3 + \cdots$, using the following equation (4~8) for the g_m stage.

$$I_{RF+} = k/2(V_{gs1} - V_t)^2, I_{RF-} = k/2(V_{gs2} - V_t)^2 \quad (4)$$
$$I_{RF+} = I_{ss} + \Delta I_{RF}, I_{RF-} = I_{ss} - \Delta I_{RF} \quad (5)$$
$$V_{RF} = V_{gs1} - V_{gs2} \quad (6)$$
$$V_{RF} = \sqrt{\frac{2*I_{ss}}{K}}\left(\sqrt{1 + \frac{\Delta I_{RF}}{I_{ss}}} - \sqrt{1 - \frac{\Delta I_{RF}}{I_{ss}}}\right) \quad (7)$$
$$\Delta I_{RF} = I_{ss} * \sqrt{\frac{k*V_{RF}^2}{2*I_{ss}}}\left(1 - \frac{kV_{RF}^2}{8I_{ss}}\right) \quad (8)$$

So, $\alpha 1 = \sqrt{\frac{K*I_{ss}}{2}}$, $\alpha 2 = 0$, $\alpha 3 = -\frac{k}{16}\sqrt{\frac{k}{2I_{ss}}}$, $A_{ip3} = \sqrt{\frac{4}{3}}\sqrt{\frac{\alpha 1}{\alpha 3}} = 8*\sqrt{\frac{1}{3}}\sqrt{\frac{I_{ss}}{k}} = 8*\sqrt{\frac{1}{3}}(V_{gs1} - V_t)$, where A_{ip3} is the IIP3 amplitude . As we can see from the expression of A_{ip3}, we need to increase the overdriving voltage of transistor M1 and M2 in order to improve the linearity.

Based on the discussion of noise figure, linearity and gain above, we know that high bias current improves the driver-stage transconductance and therefore the conversion gain and noise figure. We also set enough overdrive voltage for the transistor M1 and M2 in order to achieve the high linearity. For the LO stage, the Vgs of the LO switches(M3~M6) is set near Vt to achieve a low bias current while the size of the transistors of the switching pair must be large enough to complete current commutation. Because of the small bias current in the LO switch, relative large load resistance (RL) can be used to increase the gain without eating too much voltage headroom.

III. SIMULATION RESULTS

The mixer is designed in 130nm RFCMOS process with 1.5V power supply. Fig.2 shows the conversion gain vs the power of the LO signal (plo) of the proposed mixer when LO=4.5G and RF=4.51G. When Plo=13dBm, the maximal value of the conversion gain is equal to 11.82dB. Fig.3 shows frequency dependence of conversion gain when LO

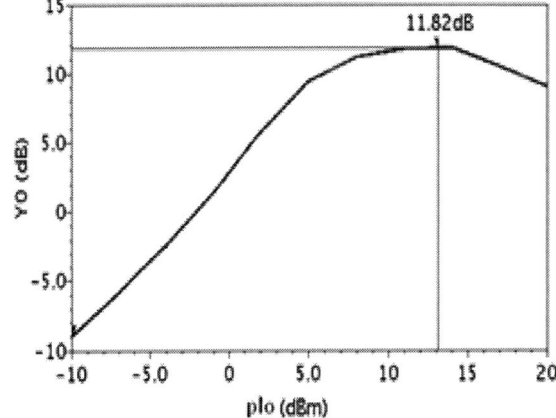

Fig.2 Effect of LO amplitude on CG

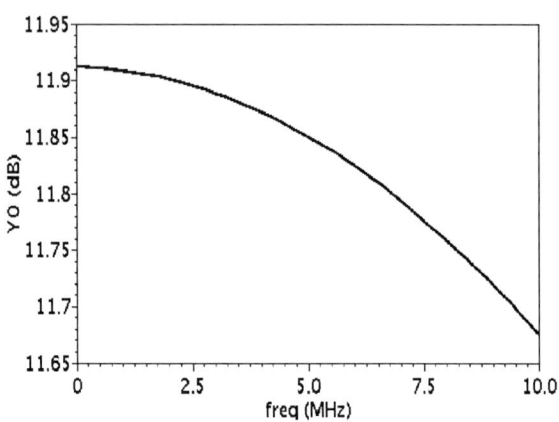

Fig.3 Conversion Gain when IF is in [0 10] MHz and LO frequency is 4.5GHz

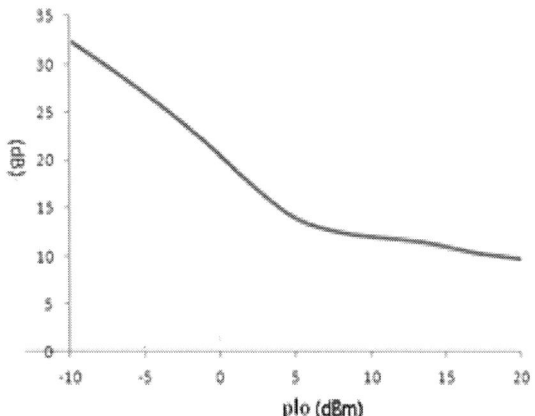

Fig.4 Effect of LO amplitude on noise figure

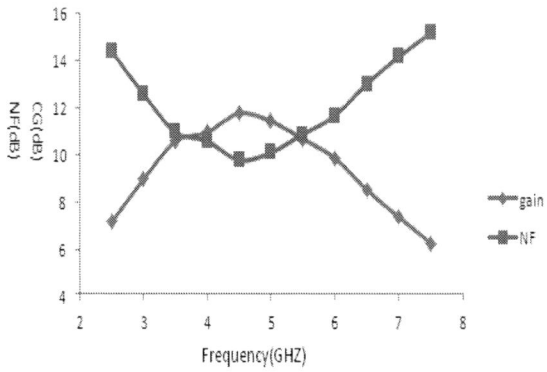

Fig.5 Frequency dependence of conversion Gain. IF frequency is 10MHz,Sweep LO Frequency

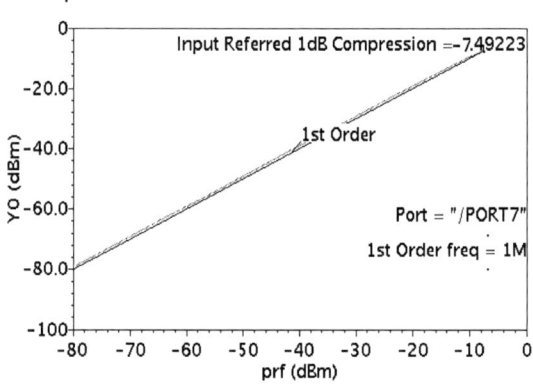

Fig.6 Input Referred 1dB compression point of the mixer. RF=4.501GHz and LO= 4.5GHz

is set at -20dBm to model low-power interferes. The OIP2 of the mixer is shown in the Fig. 7.

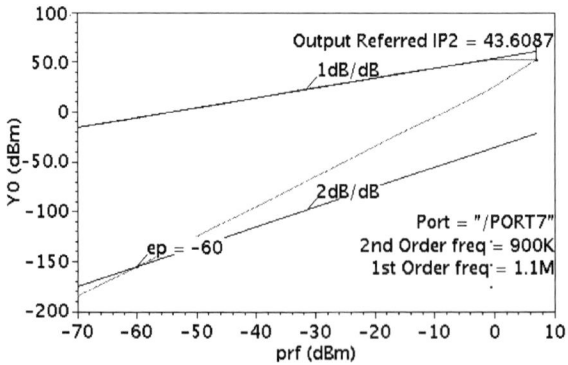

Fig.7 OIP2 extrapolation plot

frequency is 4.5GHz, RF frequency sweeps from 4.5GHz to 4.51GHz. After RF and LO signals pass through this down-conversion mixer, the IF frequency is from 0MHz to 10MHz.

Fig.4 shows DSB NF of the proposed mixer under different power of the LO signal (plo). It is noted that NF of the proposed mixer is 9.8dB@10MHz when LO=4.5G and RF=4.51G.

The simulated frequency responses of the mixer CG, NF are shown in Fig.5. Both CG and NF exhibits the best performance, 11.8 dB and 9.8dB, respectively, near 4.5GHz where the effective choke impedance reaches its peak value.

Fig.6 shows p-1dB character of this mixer at RF=4.501G and LO=4.5G. The simulation shows that the mixer achieves -7.5dBm input 1 dB compression point.

Two tones test is performed to characterize linearity performance of mixer. Two tone signals at 4.501GHz and 4.503GHz are applied to the RF input while the LO frequency is set to 4.5GHz. The two-two RF signal strength

978-1-4673-2395-6/12 $31.00 © 2012 IEEE   620

The same tones test method also execute at other frequency point from 2.5GHz to 7.5GHz. The OIP2 lies in the range [39dBm, 48dBm] during this frequency range [2.5GHz, 7.5GHz]

The summary of the design is shown in the following Table I.

The comparison of this design and the design in [9] is shown in Table II.

Table I. Post-layout Mixer performance summary

Parameter	Value
Technology	0.13um CMOS
Power	7mW at 1.5V
Frequency	2.5GHz~7.5GHz
CG	6.3~11.8dB
NF	9.8~15.2dB
OIP2	39dBm~48dBm
P1dB Input	-7~ -8 dBm

Table II. Performance contrast of mixers

Parameter	[9]	Value
Technology	0.13um CMOS	0.13um CMOS
Power	5.8mW at 1.2V	7mW at 1.5V
Frequency	3.0GHz~7.0GHz	2.5GHz~7.5GHz
CG	5.3-8.2dB	6.3~11.8dB
NF	9.6-13.5dB	9.8~15.2dB
OIP2	--	39dBm~48dBm
P1dB Input	--	-7~ -8 dBm

IV. CONCLUSION

Using the folded Gilbert mixer architecture, this mixer achieves high linearity and conversion gain, low noise figure in a wideband range using 0.13um CMOs technology. This mixer is suitable to be used as the direct down-conversion mixer in the direct-conversion receiver IC for multi-standard wireless systems.

REFERENCES

[1] B. Razavi, "Design Considerations for Direct-Conversion Receivers," IEEE Trans. Circuits and Systems II, vol.44, pp.428-435, June 1997.

[2] T. H. Lee, The Design of CMOS Radio-Frequency Integrated Circuits. Cambridge, U.K.: Cambridge Univ. Press, 1998.

[3] S. Chehrazi , R. Bagheri and A. Abidi "Noise in passive FET Mixers: a simple physical model", Proc. IEEE Custom Integrated Circuits Conf. (CICC'04), pp.375 2004

[4] Terrovitis, M.T.; Meyer, R.G.; , "Noise in current-commutating CMOS mixers," Solid-State Circuits, IEEE Journal of , vol.34, no.6, pp.772-783, Jun 1999

[5] R.Svitec et al., "DC offsets in direct-conversion receivers: characterization and implications," IEEE Microwave Magazin, pp.76-86,sep.2005

[6] Voltti, M.; Koivi, T.; Tiiliharju, E.; , "Comparison of active and passive mixers," Circuit Theory and Design, 2007. ECCTD 2007. 18th European Conference on , vol., no., pp.890-893, 27-30 Aug. 2007

[7] V.Vidojkovic et al, "A low-voltage folded-switching mixer in 0.18-um CMOS," in Proc.ISCAS, pp.300-303,2003.

[8] E.Abou-Allam et al., " Low-voltage 1.9GHz Front-End Receiver in 0.5-um CMOS Technology," IEEE J. Solid-State Circuits, vol.36,no.10,pp.1434-1443,Oct.2001

[9] Kihwa Choi; Dong Hun Shin; Yue, C.P.; , "A 1.2-V, 5.8-mW, Ultra-Wideband Folded Mixer in 0.13-μm CMOS," Radio Frequency Integrated Circuits (RFIC) Symposium, 2007 IEEE , vol., no., pp.489-492, 3-5 June 2007

[10] X. Guan and A. Hajimiri, "A 24-GHz CMOS front-end," IEEE J. SolidState Circuits, vol. 39, pp. 368–373, Feb. 2003

Design of Single-Stage Folded-Cascode Gain Boost Amplifier for 14bit 12.5Ms/S Pipelined Analog-to Digital Converter

Xuelian Liu, Student Member, IEEE， John F. McDonald, Life Senior Member, IEEE
Department of Electrical, Computer, &Systems Engineering, Rensselaer Polytechnic Institute
110 8th Street Troy, NY, 12180
Email: liux11@rpi.edu

Abstract - **This paper describes the design and simulation of a fully-differential, high gain, high speed CMOS Operational Transconductance Amplifier (OTA). The op-amp is designed for unity gain sampler stage of 14bit 12.5Ms/s pipeline analog-to digital converter. The design is implemented using a folding cascode topology with the addition of gain boosting amplifiers for increased gain. Common-mode feedback (CMFB) is used to stable the designed OTA against temperature and other process variations. This design has been implemented in 0.13μm IBM RF mixed signal CMOS Technology. The Spectre simulation shows the DC gain of 91.5 dB and a unity-gain frequency of 714.5MHz with phase margin of 62° (double 7.5-pF load) while consuming 9 mW power. For the normal corner, the settling time to 1/2 LSB of 14bit A/D converter accuracy is 40 ns.**

I INTRODUCTION

Speed and accuracy are two of the most important properties of amplifiers used in high performance pipeline A/D converters [1]. The settling speed mainly depends on the unity-gain frequency and a single pole settling time while high settling accuracy is due to high DC gain of the op-amp circuits. However, optimizing amplifiers for speed and gain usually leads to contradictory demand. The high-gain requirement leads to multistage designs with long-channel devices biased at low current levels, whereas the high-speed requirement calls for a single stage design with short channel devices biased at high current levels [2].

Two stage design or multistage design is capable of achieving a high gain and large output swing in [3], however, other stages introduce poles at a low frequency that affects the frequency response by lowering the unity gain frequency and phase margin. Another approach is the cascoding technology, which is the widely used to achieve high gain compared to 2-stage designs because of its superior frequency response. However, there are headroom problems while trying to cascode more transistors in a stack with limit power supply [4]. As a result, The DC gain of the cascode version is limited around 40-50dB, which is not good enough, in modern processes with short channel devices and an effective gate-driving voltage of several hundreds of millivolts. Other approaches such as dynamic biasing of transconductance amplifier [5], positive-feedback transconductance amplifier [6] were proposed, but the gain and unity gain frequency of those gain boosting techniques are not enough for the recent

submicron CMOS circuit applications.

One technique, which helps to increase the op-amp DC gain without sacrificing the unity-gain frequency, is the gain-boosting technique introduced in [7] and firstly applied to the folded cascode op-amp in [2] in 1990 by K. Bult and G. Geelen. The folded-cascode topology offers large output swing and has good performance on common mode input range even though it consumes more power. The purpose of this paper is to implement fully differential gain boost cascode op-amp topology in the unity gain sample stage of 14bit 12.5Ms/s Pipeline A/D converter.

This paper is organized as following. The gain boosting technique is explained in section II. The design process to meet the specification is discussed in section III. The implementation of main stage amplifier, boost stage amplifier, SC CMFB using IBM RF technology is discussed in the section IV. The simulation results are shown in section V. Finally, the conclusion is given in section VI.

II GAIN BOOSTING TECHNIQUE

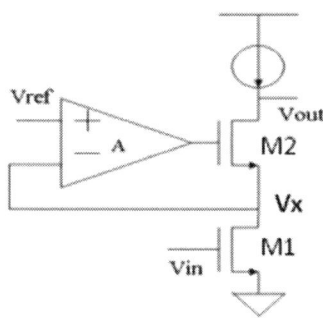

Fig 1. Cascode gain stage with gain boost enhancement

As shown in Fig. 1[8], the idea of gain boosting is based on negative feedback loop to set the drain voltage of M2[3]. M1 and M2 forms the main cascode amplifier. A is a gain-boosted amplifier. Negative feedback drives the gate of M2 until Vx has the same value as Vref. Therefore, the variation of Vout has much less effect on Vx, because A "regulates" this voltage. This topology is usually called "regulated cascode" or "active cascode". With the smaller

978-1-4673-2395-6/12 $31.00 © 2012 IEEE

variation of Vx due to the change of Vout, the output current becomes less sensitive to the voltage variation at Vout compared with conventional cascode structure. Therefore the output impedance increases as shown in equation (1):

$$Rout = (Gm2Ro2(A + 1) + 1)Ro1 + Ro2$$
$$\approx Gm2 * Ro2 * A * Ro1 \qquad (1)$$

Therefore, the DC gain can be increased several orders of magnitude to (2):

$$Aout = Gm1Ro1(Gm2Ro2(A + 1) + 1) \approx$$
$$Gm2 * Ro2 * A * Gm1 * Ro1 \qquad (2)$$

The DC gain without gain boost is shown in equation (3):

$$Aout = Gm2 * Ro2 * Gm1 * Ro1 \qquad (3)$$

which is A times less than the gain of boosted amplifier.

Fully differential folded-cascode main amplifier with fully differential gain boost amplifier is chosen to design OTA as shown in Fig 2. It offers relatively larger output swing and better single-pole roll-off frequency response than non-folded-cascode architecture with gain-boosting. The fully differential architecture is designed as boost amplifier rather than single-ended architecture because the single-ended architecture has some disadvantages such as extra pole from an internal current mirror and noise generated by the biasing circuit. However, fully differential amplifier requires common-mode feedback (CMFB) circuit.

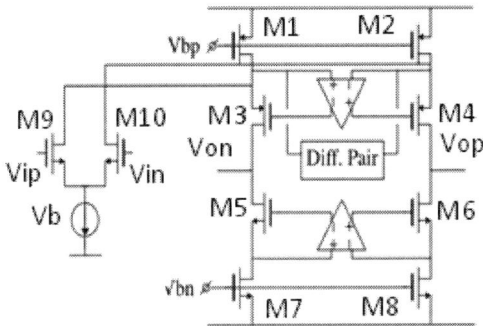

Fig 2. Fully differential folded cascode Op Amp with fully differential gain boosting amplifiers

III OTA DESOGM SPECIFICATIONS

The close loop configuration of this amplifier used as unity gain sampler stage of pipeline A/D Converter is shown in the Fig.3[9].

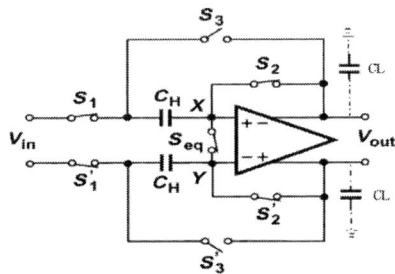

Fig.3 Differential realization of unity-gain sampler

A DC Gain

The closed loop gain of this amplifier can be determined by equation (4)

$$A_{CL} = \frac{A_{OL}}{1 + \beta A_{OL}} \qquad (4)$$

Where A_{OL} is the open loop gain, β is the feedback factor and is equal to 1 in this unity-gain sampler. In order to achieve the 14bit accuracy, we need to be able to amplifier 1/2 LSB of A/D converter. The open loop gain can estimate to be $A\_OL \geq 1/\beta * 2^{(N+1)}$, which is ≥ 90dB.

B. Noise

For a folded-cascode amplifier which is used as main amplifier in gain boosting topology, the total noise factor is Nf $=2*(N_p + N_n + 1)$, where $N_p = \frac{g_{mp}}{g_{mi}} = \frac{4I_b}{V_{dsat,p}} / \frac{2I_b}{V_{dsat,i}} = \frac{2V_{dsat,i}}{V_{dsat,p}}$ for these PMOS current sources and $N_n = \frac{g_{mn}}{g_{mi}} = \frac{2I_b}{V_{dsat,n}} / \frac{2I_b}{V_{dsat,i}} = \frac{V_{dsat,i}}{V_{dsat,n}}$ for the NMOS current sources. In order to minimize Nf, we need to reduce the gate overdrive of the diff. pair $V_{dsat,i}$ in the input and increase the gate overdrive of the current sources pair $V_{dsat,n}$ and $V_{dsat,p}$. However, there is tradeoff between the overdrive voltage, transistor size and output swings. The relative small gate overdrive of the diff. pair means the input transistors are relatively large; the relative large gate overdrive of the current source pair reduces the output swing. In our design, we set that the gate overdrive voltage is 240-mV of the current source and 80mv of the input diff., which is 1/3 of that of the current source. The gate overdrive voltage for the cascode pair is 80mV, which is also 1/3 of that of the current source in order to achieve higher output voltage swing. Based on the overdrive voltage we pick up in our design, the Nf $=2*(N_p + N_n + 1)=4$.

C Dynamic Range

In the A/D converter, the quantization step is Δ. The peak "signal-to-quantization noise ratio", SQNR, for sinusoidal inputs: is $6.02N + 1.76$dB, where N is the quantization bits [10]. For 14 bit A/D converter, SQNR=86dB. When we consider the circuit generated noise, the Dynamic range for the 14bit A/D Converter is 88 dB to meet the specification requirement.

The input dynamic range is defined as

$$DR_{input} = \frac{P_{signal}}{P_{noise}} = \frac{1}{8} \times \frac{\overline{V}_{i,pp}^2}{\overline{V}_{i,n}^2} = 10^{8.8}, \qquad (5)$$

Where $\overline{V}_{i,n}^2 = KT/C_H$ and $V_{i,pp}$ is the input swing.

The output dynamic range is defined as

$$DR_{output} = \frac{P_{signal}}{P_{noise}} = \frac{1}{8} \times \frac{\overline{V}_{o,pp}^2}{\overline{V}_{o,n}^2} = 10^{8.8}, \qquad (6)$$

Where $\overline{V}_{o,n}^2 = Nf*1/\beta*KT/C_{Leff}$ and $V_{o,pp}$ is the output swing. Nf is the noise figure, which is determined by the overdriven voltage of transistors in the circuit as discussed before. Based on the equation (5) and (6), the CH and CL capacitor value in the Fig.3 can be determined. In this design, the CL is chosen as 7.5pH, CH is chosen as 5pH to meet the dynamic range requirement.

D. Setting speed and small signal bandwidth

Because the A/D converter will work up to 12.5Ms/s, so each period is 80ns. It is switch-capacitor circuit with non-overlap clock, the setting time should be no more than 40ns. A single pole frequency response results in a single slope of -20 log(e)=-8.7 dB/τ for settling. Here assume slewing is not present. The settling accuracy is determined by the transient loop-gain achieved by the amplifier. The circuit needs to settle to < 1/2 LSB in 40 ns. With a feedback factor β = 1, the following equation (7) can be deduced and τ can be calculated.

$$\frac{-8.7dB}{\tau} * 10ns = 20 \log(\beta * \frac{1}{2} LSB) \qquad (7)$$

Because $\tau = \frac{1}{\beta * \omega_u}$, $\omega_u = g_m / C_{Leff}$, we can calculate g_m. Finally, use the square-law relationship $g_m = 2I_d / V_{dsat}$, we get to the biasing current needed, Id.

IV CIRCUIT IMPLEMENTATION

Based on the discussion in the last section, the design is implemented in IBM 8rf CMOS process as below.

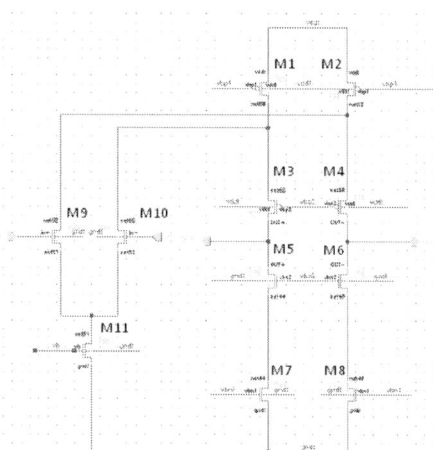

Fig 4. Folded cascode architecture

A Main Amplifier

The folded cascode topology provides high output swings and is ideally suited to operate in low voltage supply circuits [11]. Fig.4 shows a standard folded cascode single stage amplifier, which is used as the main amplifier in the gain boost topology. The gain and phase plot of the main amplifier is shown in Fig.5

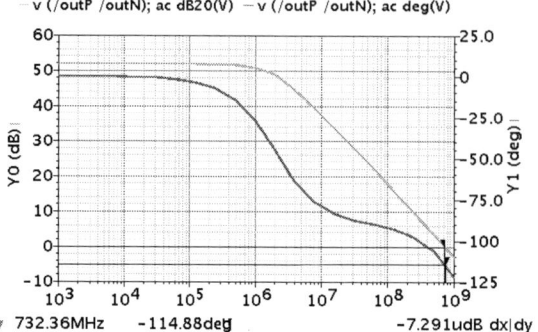

Fig.5 Gain and Phase plot of main amplifier

B Boosting Amplifier

The gain boosting stage increases the effective output resistance by the gain of the boosting amplifier, therefore increasing the overall gain of the folded cascode design. Since the gain of the main folding cascode architecture is approximately 52 dB then the gain of the gain boosting amplifier is required to be greater than 38 dB in order to achieve an overall gain of 90dB to meet the signal to noise ration specification requirement for 14bit A/D converter. The NMOS inputs folded cascode, which is the same architecture as the circuit in Fig.4, is used for the upper transistors and PMOS inputs folded cascode as shown in Fig.6 for the lower transistors to get a maximum voltage swing. Because the boost amplifier doesn't need to drive high capacitance, it can be designed in much lower current and consumer much less power than main amplifier. PMOS input boosting amplifier achieved a gain of 42 dB for each amplifier as shown in Fig.7. The current values in both the main amplifier and boost amplifier must be chosen carefully to ensure good setting performance.

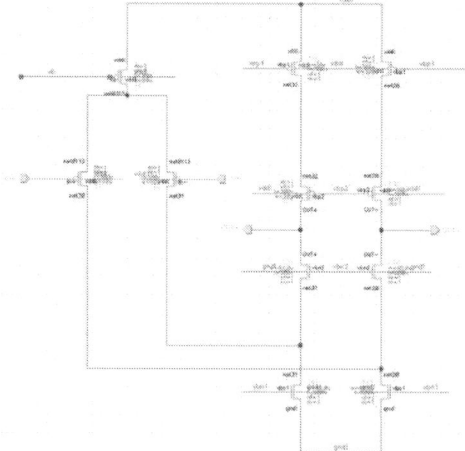

Fig 6. PMOS Input Boosting Amplifier

978-1-4673-2395-6/12 $31.00 © 2012 IEEE

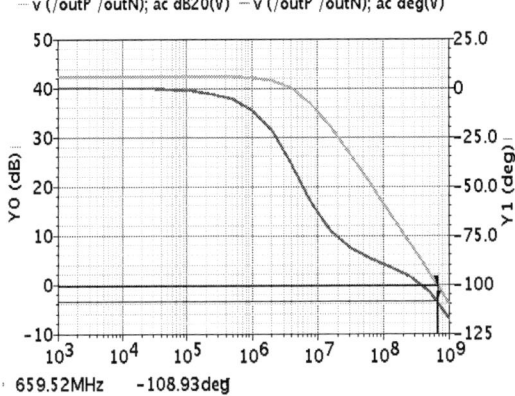

Fig.7 Gain and Phase plot of the Boost Amplifier

C CMFB Circuit Design

Fully differential op-amp must have a common-mode feedback(CMFB) circuit to get a common mode ouput voltage that is immune to variations in the process as well as temperature. Because the opamp is used in A/D converter, nonoverlapping phase clocks are available. The SC-CMFB is used which is shown in Fig.8 [12] because SC-CMFB consumes less power than continuous time CMFB[13]. It consists of four capacitors and six switches. During clock phase $\emptyset_2$, C_1 is connected to C_2. The dc voltage across C_2 is determined by C_1. During clock phase $\emptyset_1$, C_1 is charged to $V_{cmref} - V_{bias}$ and capacitor C_2 generate the control voltage V_b, level shifting the average output voltage by $V_{cmref} - V_{bias}$. The expression for V_b is as equation (8)

$$V_b = \frac{V_{op} + V_{on}}{2} - V_{dc}, \qquad (8)$$

where $V_{dc} = V_{cmref} - V_{bias}$. Because $V_{cm} = (V_{op} + V_{on})/2$, when the circuit reaches steady-state, if the input gate-source voltage(V_b) is precisely defined according to equation (9), $V_{cm} = V_b + V_{cmref} - V_{bias}$ where $V_b = V_{bias}$ typically. The whole circuit acts like a simple low-pass filter having a dc input voltage $V_{dc} = V_{cmref} - V_{bias}$.

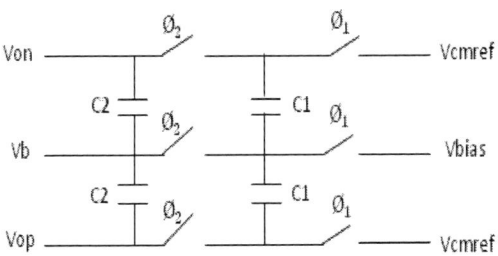

Fig.8 switched capacitance common mode feedback circuit

The output of each boosting amplifier does not need high swing, thus, a continuous time CMFB circuit is used. The advantage of the continuous time CMFB is that its speed is fast.

IV. SIMULATION RESULT

The gain-boosted folded-cascode op amp is designed by integrating the circuit discussed above according to Fig.2 topology, implemented in 0.13μm process with 1.8V power supply and simulated with Cadence Spectre. The load capacitance is 7.5pF. In Fig.9 AC simulation shows that the main amplifier has a dc gain of more than 91.55 dB. The phase margin is 62.38. The unity GBW is 714.5MHz.

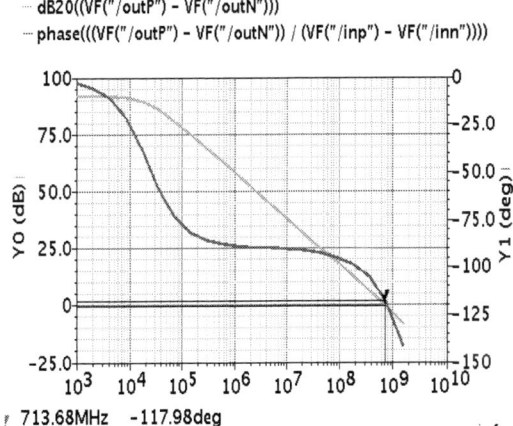

Fig.9 Gain and Phase plot of the Boost Amplifier

With a unity gain configuration, the settling time is measured when driving 7.5pF capacitor. The transient response of the test circuit is shown in Fig.10

The summary of the design is shown in the following tables.

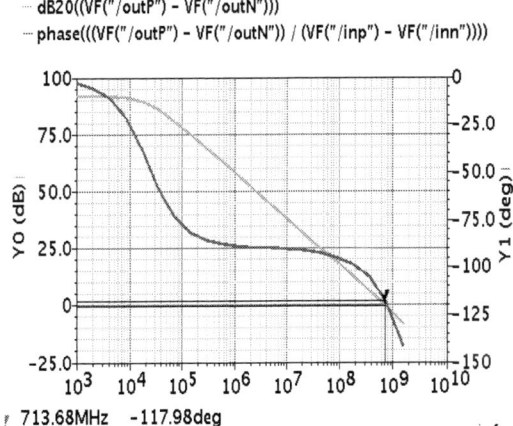

Fig.10 Simulated transient response of Op Amp

Table I. Op-Amp performance summary

Parameter	Value
Technology	0.13um CMOS
Power	9mW at 1.8V
CL	7.5PF
DC Gain	91.5dB
GBW	714.5MHz
Phase Margin	62 degree
Setting time to 1/2 LSB	40ns

V.   CONCLUSION

A single-stage folded cascode gain-boosted CMOS OTA has been designed and simulated using 0.13um RF CMOS technology. In this design, a differential amplifier is applied as gain-boost devices. The designed op-amp can meet specification of unity gain sampler stage for 14bit A/D converter.

REFERENCES

[1] K.Uyttenhove, M.S.J. Steyaert. Speed-power- accuracy tradeoff in high-speed CMOS ADCs. Circuits and Systems II, 2002, vol.49 , pp.280-283.

[2] K. Bult and G. J. G. M. Geelen. A fast-settling CMOS op amp for SC circuits with 90-dB DC gain. IEEE Journal of Solid State Circuits,25(6):1379–1384, Dec. 1990.

[3] Razavi, B. Design of Analog CMOS Integrated Circuits. McGraw-Hill, Boston, 2000.

[4] H. Hoara et al,. "A CMOS programmable self-calibratind 13-bit eight-channel data acquisition peripheral," IEEE J. of Solid-State Circuits, vol. SC-22, pp.930-938, Dec. 1987.

[5] M. A. Copeland and J. M. Rabaey, "Dynamic amplifier for MOS technology," Electron Lett., coll. 15, pp. 301-302, May 1979.

[6] C. A. Laber and P. R. Gray. "A positive-feedback transconductance amplifier with application to high-frequency, high-Q CMOS switched-capacitor filters," IEEE J. of Solid-State Circuits, vol. 23, no. 6, pp. 1370-1378, Dec. 1988.

[7] B. J. Hosticka, "Dynamic CMOS amplifiers." IEEE J. of Solid-State Circuits, vol. SC-14, no. 6, pp.1111-1114, Dec. 1979.

[8] X. Jiang, S. Seo, and Y. Lu, Final Project Report: A CMOS Single Stage Fully Differential Op-Amp With 120 dB DC Gain EECS Dept., Univ. of Michigan, Ann Arbor, MI, Fall, 2003, EECS 413 Fall 2003 Final Project Report.

[9] Yun Chiu, Ken Wojciechowski, "A gain-boosted 90-dB dynamic range fast settling OTA with 7.8 mW power consumption," UC Berkeley EE240 Final Project Report, 2000

[10] B. Razavi, Principles of Data Conversion System Design. New York: IEEE Press, 1995

[11] Mallya, S.; Nevin, J.H.; , "Design procedures for a fully differential folded-cascode CMOS operational amplifier," Solid-State Circuits, IEEE Journal of , vol.24, no.6, pp.1737-1740, Dec 1989

[12] Choksi, O.; Carley, L.R.; , "Analysis of switched-capacitor common-mode feedback circuit," Circuits and Systems II: Analog and Digital Signal Processing, IEEE Transactions on , vol.50, no.12, pp. 906-917, Dec. 2003

[13] D. Johns and K. Martin, Analog integrated circuit design, 287, 1997.A/D converter, IEEE Journal of Solid-state circuits,

978-1-4673-2395-6/12 $31.00 © 2012 IEEE

Technique to Improve Visibility for Cycle Time Improvement in Semiconductor Manufacturing

Syahril Ridzuan Ab Rahim, Ibrahim Ahmad & *Mohd Azizi Chik
Department of Electronics & Communication,
Universiti Tenaga Nasional (UNITEN),
43000 Kajang, Selangor, Malaysia.
*Silterra Malaysia Sdn. Bhd
Lot 8, Phase 2, Kulim Hi-Tech Park,
0900 Kulim, Kedah, Malaysia.
Email: syahril_ridzuan@uniten.edu.my, mohd_azizi@silterra.com.

Abstract – **Cycle time for a product is one of the key performance indicators in semiconductor manufacturing. Reduction of cycle time will shorten product time to market, increase throughput, reduce operational cost and develop customer trust. Semiconductor manufacturing that process 40,000 to 50,000 work-in-progresses (WIP), usually takes 50 to 70 days, 300 to 400 equipments and 300 to 900 steps to complete. Thus, any task related to manual data collection to make indices reports or analysis usually needs high resources requirements to spend for manual work and risk for mistake. In the modern facility of semiconductor fabrication, a system like Manufacturing Execution Systems (MES) was implemented to ease the process and operation traceability. The information is well kept in the appropriate databases. Many applications then are integrated with MES database to perform indices reports. In this paper, the improve method for data collection related to cycle time improvement is introduced. In this approach, the automated systems was developed using existing Advance Productivity Family (APF) programming platform to collecting the data. The system is integrated between MES and APF to have the real time data collection and analysis. In the systems, manual data collection is replaced with respective automated data transfer from real situation in the manufacturing environment. This program then able to shows real root caused with proper relational charting to display real problem for engineering to prioritize and resolve respectively. As a result, 39% reduction of cycle time gained by implementing this technique. The system has successfully implemented and supports the cycle time reduction.**

Index Terms—Advance Productivity Family (APF), Work In progress (WIP), Manufacturing Execution Systems (MES).

I. INTRODUCTION

Today demand to produce product with competitive cycle time is one for the key factors for business decision [1]. Many approaches for cycle time improvement have been published. Most of the papers are discussed cycle time improvement through improving throughput of the equipments [1]. In order to improve cycle time in semiconductor manufacturing, more detail data to plot trend and relational between issues are needed [2]. Key factors that influence cycle time are WIP quantity and its mixed, bottlenecks equipment availability and throughput, rework rates, others key equipments availability, setups and dispatching rules[1]-[3]. Normally high cycle time (CT) is contributed due to unnecessary waiting time, unscheduled downtime, not effective dispatching rule, rework and hold due to related engineering requirements [7]. In this analysis, more focus will be addressed to improve waiting time and hold time. Waiting time is when the WIP is idle. Hold time is when the product is not allow to process next step, mainly reserve for engineering activities to data validation, wait for recipe setup, wait for merge with other wafer and customer input. Most cases when product release from hold, it will require go to inspection step. Further in this paper will discuss technique and works that have done to reveal issue, causes and potential solution to improve cycle time visible to the operational team.

II. SYSTEMS ARCHITECTURE

SilTerra have existing MES systems that currently tracked the information for products. Its MES systems have been designed for easy integration with respective application for detail data collection. This research takes the advantages and utilizing its existing systems architecture to establish new computer program for cycle time improvement exploration. In this approach, few databases have been setup, and coding to collect related cycle time caused is established.

Fig.1 shows system architecture and explanation of how technique to improve visibility for cycle time improvement is implemented.

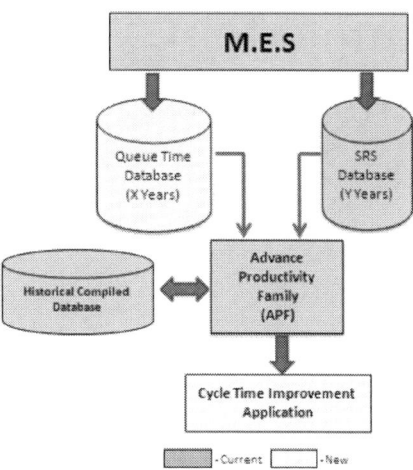

Fig. 1. System Architecture

A. Queue Time Database

This database stores all information regarding product information starting from first process steps until completed process step. All of the information was extracted from the MES database. Below is the list of information grabbed from this database;

TABLE I.
EXAMPLE QUEUE TIME DATABASE INFORMATION

Information Description	Information Description
LotID	LotID Identical name for every lot's (Code group of products)
StepName	Name of the processing step's
TrackIn Time	Date and time of product going to the process step's.
TrackOut Time	Date and time of product completed process step's.
Last TrackOut Time	Previous completed step's TrackOut Time
Equipment ID	Equipment Identification

B. SRS Database

SRS database contains many information regards WIP. However, in this approach only selected data regards caused of high cycle time is needed to extract from SilTerra Reporting Systems (SRS) database. This techniques help to provide fast data transfer for automated data compilation.

TABLE II.
EXAMPLE SRS DATABASE INFORMATION

Information Description	Information Description
Memo	Reason of high Cycle Time
Resources	Name a group of equipment
Date	Date of Activity
Uptime	Percent of Equipment avaibility
Utilization	Percent of utilizing the equipment

C. Advance Productivity Family (APF)

APF is needed as platform to code a program to calculate cycle time and consolidate various database sources to understand cause of high cycle time [5]. This is where automated data compilation replaced manual data collection in MES installed Semiconductor Manufacturing. Please refer to Fig. 1 for details. The code can be schedule automatically using one of the APF function to collect and to perform data snapshot at the respective time. The code is written to perform automated data integration and filtration for proper caused of high cycle time and search for responsible person that holding the responsibility of current issue. The information generated in this coding then integrated with Historical Compiled database.

D. Historical Compiled Data Database

Previous results from APF coding are stored in this database. An additional coding in APF then used this database to make an automated follow up with respective person that responsible for the action required. The automated follow up will end automatically when the issue resolved.

E. Alert and Notification

The alert and notification application is coded from APF platform. This is where an automated notification and escalation is made to respective responsible person and to its supervisor. The notification is done through automated reports update and email.

III. DATA ANALYSIS

In this paper, the scope for data analysis is started with capturing WIP waiting time at the respective limit. The information is gathered automatically from MES systems. Furthermore, coding that developed in APF will filter and capture respecting waiting time WIP. In the process, more data association and indexing with information like equipment, person that responsible for high waiting time, processing steps, module or area, and time stamp for the activities. Further explain in the fig 2. Finally the results from this analysis then publish to the respective personal for further action taken.

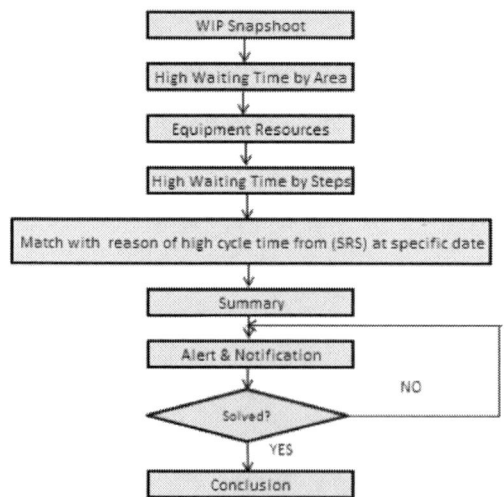

Fig. 2. Analyzing Method Process Flow Diagram

The definition of criticality of the equipment or steps related to cycle time is define to formula of waiting time x quantity that actually impacted. The formula is call impact factors and it's summarized in the equation 1 [4].

$$\begin{array}{ccc} \text{Impact} & = & \dfrac{90^{th}\ \text{Percentile}}{\text{Waiting Time}} \quad X\ \text{TrackIn Quantity} \\ \text{Factor} & & \end{array} \quad (1)$$

The highest impact is important and will sorted first. As an example the highest waiting time from the chart below is at Etch (ETH) area and followed by cleaning (CLN) area.

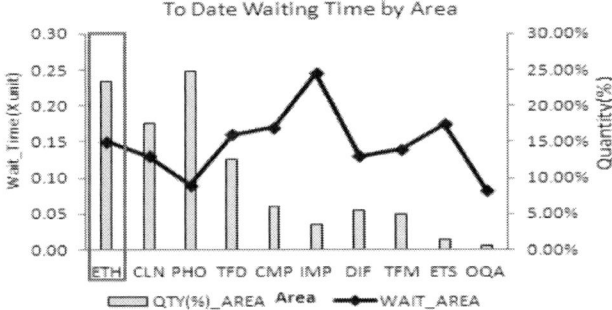

Resources are defined as a group of equipments that capable to process several types of steps. Front End (FE) resources are defined as resources that process to make transistors, capacitor and insulator. The processes begin with step "Oxidation Cleaning" until "Inter Layer Dielectric Planarization Annealing". While back end (BE) resources are define as resources that process to make metal connector. The processes continue with step "Lithography Contact Inspection" until "Outgoing Quality Assurance Packaging Inspection". Analysis is only choosed either front end or back end resources of the respective area based on the highest impact factor. As an example, fig. 4 below shows that the highest waiting time by resources for backend etching area is at Damascene Trench Etching and followed by Damascene Via-Etching.

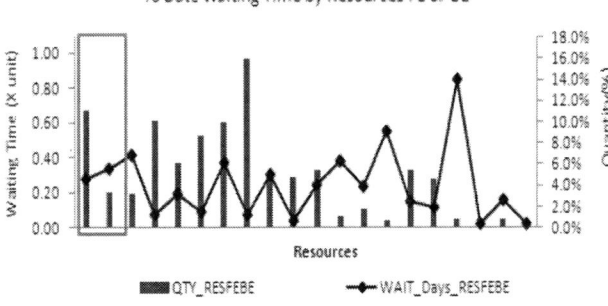

Fig. 4. To Date Waiting Time by Resources FE or BE

Every resource is process identical steps. At the same time, every product has a recorded actual processing time at particular steps. From the top two highest waiting times by resources from the previous analysis, drilldown analysis is performed to know the problem steps and grabbed the date of

bad situation. As an example, highest waiting time by steps for resources Damascene Trench Etching is at step Re-distribution Etching Layer and Damascene with Bi-layer Resist as in fig. 5 below .

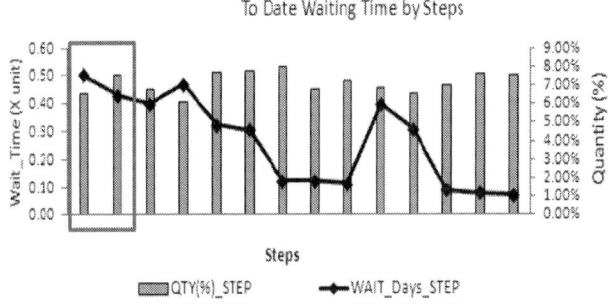

Fig. 5. Monthly Waiting time By Steps

Every day manufacturing engineers recorded their line problems in the SRS database through "SRS application". The problems date can be match with SRS data to find out the root cause of the problems based on the date and resources. For an example, the highest impact factor waiting time by resources is at Damascene Via-Etching is occur on day "2X" due to lack of avaibility at that resources.

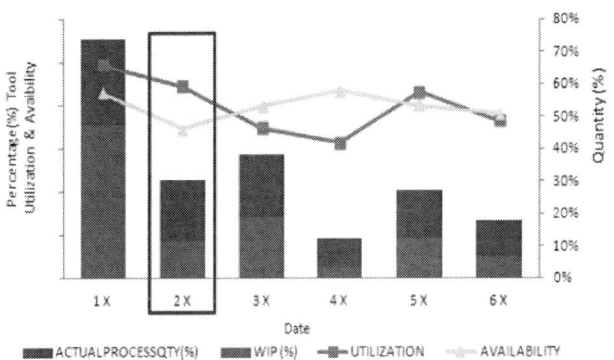

Fig. 6. Damascene Trench Etching Daily WIP, ActualProcessQty, Utilization and Avaibility

Finally, the result from the analysis is summarized and compile in the one PDF format as documentation. The sequence of the result arrangement is as in fig. 7 below. Result from the analysis arranged in sequence for solid explanation of the root cause that contribute to high waiting time at certain resource and date.

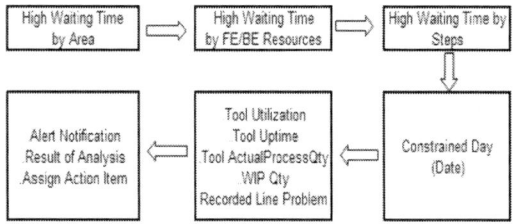

Fig. 7. Sequence of Result Documentation

Analysis result sent out to the respective personal for further action taken immediately. As an example, auto mail is sending the result of the analysis directly to manufacturing, engineering and management team for action taken to solve the problem that occur at Copper-Dual Damascene Trench Etching resources.

IV.   RESULT AND EMPIRICAL IMPLEMENTATION

This system was integrating with the production line and shows some improvement in the cycle time as presented in the fig. 7 below. There is an improvement about 39 % since year 2011.

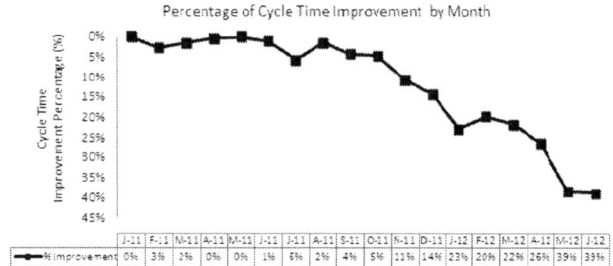

Fig. 8. Percentage of Cycle Time Improvement

V.   CONCLUSION

As a conclusion, manual analysis and data collection difficulties was eliminated by using this automated technique. This technique will help to improve the visibility for cycle time improvement in semiconductor manufacturing by identifying the causes of this problem. Manufacturing, Engineering and Management personal easily gets the information of the production problem by reading the real time report from this system.

REFERENCES

[1] Kader Ibrahim, Mohd Azizi Chik and Uda Hashim, "Variability Due to Tool Configurations That Impacts Overall Capacity in Wafer Fabrication Facility," 11th Asia Pacific Industrial Engineering & Management Systems Conference, 2010.

[2] Yair Meidan, Boaz Lerner, Gad Rabinowitz, and Michael Hassoun, "Cycle-Time Key Factor Identification and Prediction in Semiconductor Manufacturing Using Machine Learning and Data Mining" , IEEE Trans. Semicond. Manuf., vol. 24, no. 2, May 2011.

[3] Ingy A. El-Khouly, Khaled S. El-Kilany, and Aziz E.- Sayed, "Effective Scheduling of Semiconductor Manufacturing", World Academy of Science, Engineering and Technology, vol. 79, 2011.

[4] Chung-Jen Kuo, Chen-Fu Chien, Jan-Daw Chen, and Member, "Manufacturing Intelligence to Exploit the Value of Production and Tool Data to Reduce Cycle Time", IEEE Trans. Actions On Automations Science And Eng., vol. 8, no. 1, January 2011.

[5] Mohd Azizi Chik, Yeo Eng Teck, Mahalil Amin Abd Malek, and Mohd Hafidz Saidi, "Comprehensive Sequencing Dispatching Method for Identified Bottleneck Tool – Photolithography Process", NSM, Perlis, Malaysia, pp.19-21. 2003.

[6] Kader Ibrahim, Mohd Azizi Chik, Wan Shamsir Nizam, Nyioh Li Fern, and Nor Farahidah Za'bah, "Efficient lot Batching Systems for Furnace Operation", ASMC 14th Conference, Munich, Germany, pp. 322-324, 2003.

[7] Mohd Azizi Chik, Mohd Hazmuni bin Saidin, and Uda bin Hashim, "Industrial Engineering Roles in Semiconductor Fabrication", APIEM 11th Conference, December 2010.

Accessing AHB Bus using WISHBONE Master in SoC Design

Muhamad Khairol Ab Rani[#1], Mohd Zubir Khalid[#2]

Integrated Circuit Development

MIMOS Berhad

Kuala Lumpur, Malaysia

[1]muhamad.khairol@mimos.my

[2]mohd.zubir@mimos.my

Abstract— **An IP (Intellectual Property) based SoC (System-on-Chips) is getting popular among designers as it allows for a faster development cycle for SoC production. However, each IP may use different bus interface causing compatibility issues during design integration. A WISHBONE bus and an AHB (Advanced High Performance Bus) are among commonly used bus interfaces for many IP cores. This paper describes the conversion operation from WISHBONE Bus protocol into an AHB bus protocol. This is to allow an Open RISC Micro Controller Unit (ORMCU), a master device which uses WISHBONE bus protocols, to communicate and control all other devices (slaves) that use AHB bus protocols. The design is a WISHBONE-to-AHB Bridge, which consist of a WISHBONE slave and an AHB master inside one module. The simulation results confirm that the bridge is able to handle communication from a WISHBONE master in an AHB system.**

Keywords— **SoC, IP, AHB Bus, WISHBONE Bus, WISHBONE-to-AHB Bridge, OpenCores, Open RISC**

I. INTRODUCTION

The WISHBONE SoC interconnection architecture for portable IP cores is an open source and a flexible design methodology for use with semiconductor IP cores. Its purpose is to foster design reuse by reducing SoC integration problems. The aim is to allow the connection of differing cores to each other inside a chip. This is accomplished by creating a common interface between IP cores. It improves the portability and reliability of the system, and results in faster time-to-market for the end user [1] [4].

The AHB bus is a system bus used in an AMBA bus system along with Advance Peripheral Bus (APB) - a peripheral interconnect bus that is a simpler version of AHB. The AHB implements the features required for high-performance, high clock frequency systems including burst transfers, split transactions, single cycle bus master handover, single clock edge operation, non-tristate implementation and wider data bus configurations (64/128 bits) [2].

This paper focuses on porting an Open RISC Micro Controller Unit (ORMCU), which possesses a WISHBONE master and interface it into an AHB system, as shown in Fig. 1. We analyze the differences and similarities between the two buses and provide a way to port the WISHBONE master into

the AHB system using WISHBONE-to-AHB Bridge. The bridge behaves like a WISHBONE slave, and at the same time as an AHB master.

In order to evaluate the bridge, we design a WISHBONE master model, and use the AHB slave verification IP, in a VCS verification environment. The ORMCU is used as the master in our SoC, which is based on AMBA bus system.

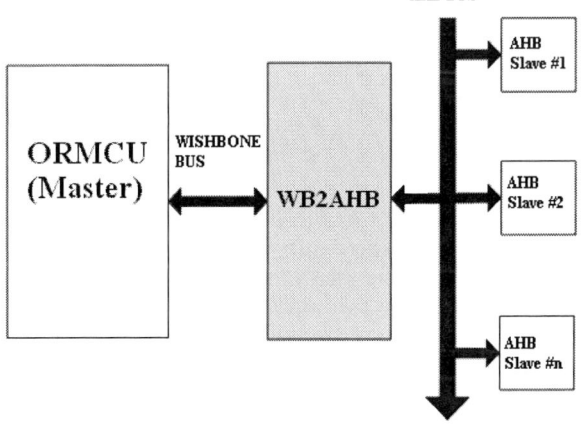

Fig.1 SoC architecture with WISHBONE and AHB busses

II. WISHBONE AND AHB

A. WISHBONE Bus Feature

The WISHBONE interconnection makes SoC and design reuse easy by creating a standard data exchange protocol [1]. Features of this technology include:

- Simple, compact, logical IP core hardware interfaces that require very few logic gates.
- Full set of popular data transfer bus protocols including: READ/WRITE cycle, burst transfer cycle and RMW (read-modify-write) cycle.
- The handshaking protocol allows each IP core to throttle its data transfer speed.
- Supports single clock data transfers.
- Supports normal cycle termination, retry termination and termination due to error.

978-1-4673-2395-6/12 $31.00 © 2012 IEEE

- Supports various IP core interconnection means, including: point-to-point, shared bus, crossbar switch, data flow interconnection, and off chip.
- Synchronous design assures portability, simplicity and ease of use.
- Independent of hardware technology (FPGA, ASIC, etc.).

B. AHB Bus Feature

The AHB is an AMBA bus which is intended to address the requirements of high-performance synthesizable designs [2]. Features of this technology include:
- High performance
- Pipelined operation
- Burst transfers
- Multiple bus masters
- Split transactions

C. Comparison of Both Buses

Comparison between the two buses is made in respect of interconnection and signals.

IP Cores Interconnection: The WISHBONE variable interconnection allows the system integrator to change the way that IP cores connect to each other. There are four defined types of WISHBONE interconnects, which are Point-to-point, data flow, shared bus and crossbar switch [1]. As for the AMBA AHB bus protocol, it is designed to be used with a central multiplexor interconnection scheme. Using this scheme all bus masters drive out the address and control signals indicating the transfer they wish to perform and the arbiter determines which master has its address and control signals routed to all of the slaves. A central decoder is also required to control the read data and response signal multiplexer, which selects the appropriate signals from the slave that is involved in the transfer [2].

Signals of Slave Device IP cores: For both Wishbone bus and AHB bus, the master signals are different from the slave signals. These slave signals are divided into reset, clock, response, data, address and control as shown in Table I.

III. PORTING FROM WISHBONE BUS TO AHB BUS

The WB2AHB Bridge is behaving as a WISHBONE slave, as well as AHB master. So, the architecture has a WISHBONE master that connects to the slave part of the bridge, and a number of AHB slaves that connect to the master part of the bridge as shown in Fig. 1.

Both WISHBONE and AHB support a single transfer. However, for burst transfer, not all AHB bursts transfer types are implemented in WISHBONE protocol.

Since ORMCU is a master device that uses WISHBONE protocol, it will use "incrementing burst of unspecified length" type of transfer to achieve the same type of transaction for the unsupported AHB burst transfers.

The master device is using a WISHBONE protocol while the slave devices are using AHB protocol. Therefore, the designed RTL module will convert WISHBONE protocol, into AHB protocol and not the other way.

TABLE I

SIGNAL OF WISHBONE AND AHB BUS

	WISHBONE [1]	AHB [2]
Reset	RST: Active HIGH, Synchronous	HRESETn: Active LOW
Clock	CLK: Coordinates activities	HCLK: Bus clock
Response	ACK: indicates okay	HRESP: Indicate status of the transfer – okay, error, retry and split.
	ERR: indicates an error	
	RTY: indicates retry is required	HREADY: transfer done and okay
Data	DAT_I: Data bus, input	HRDATA: Read data bus, input
	DAT_O: Data bus, output	HWDATA: Write data bus, output
Address	ADR: Address bus	HADDR: Address bus
Control	SEL: Byte(s) select input	HSELx: Slave select
	STB: Strobe input	HTRANS: Transfer type
	CYC: Bus valid	HSIZE: Transfer size
	WE: The write enable input	HWRITE: Transfer direction
	BTE: Burst Type Extension	HBURST: Burst type
	CTI: Cycle Type Identifier	HTRANS: Transfer type

We create four Finite State Machine (FSM) stages in the design; START, WRITE, READ and EOB (End of Burst). The START state is entered when WB_CYC and WB_STB is set to HIGH. The WRITE and READ states may cover single and burst cycle transfer. The EOB state is used to indicate the end of burst cycle, or just a single cycle.

There are four types of transfer from WISHBONE interface. They are: (1) Basic transfer/ single cycle transfer; (2) End of burst transfer; (3) Constant address burst transfer; and (4) Incrementing burst transfer. Table II shows the transfer type of AHB that are supported by WISHBONE [1]. The constant address burst transfer is normally used for accessing a FIFO. For incrementing burst transfer, the length is not specified in the WISHBONE protocol, and the transfer is stopped by end-of-burst indicator through the CTI signal.

For a single write transfer, the difference between WISHBONE protocol and AHB protocol is that in WISHBONE protocol, data, address and control signals are asserted at the same time. Meanwhile, for AHB protocol, address and control signals are asserted at first clock cycle and data are asserted in the next cycle. The difference is illustrated in waveform in Fig. 2.

978-1-4673-2395-6/12 $31.00 © 2012 IEEE

TABLE II

AHB Transfer Type For WISHBONE

AHB Transfer type	WISHBONE
Single Transfer	Supported
Incrementing burst of unspecified length	Supported
4-beat wrapping burst	Supported
4-beat incrementing burst	Not supported
8-beat wrapping burst	Supported
8-beat incrementing burst	Not supported
16-beat wrapping burst	Supported
16-beat incrementing burst	Not supported

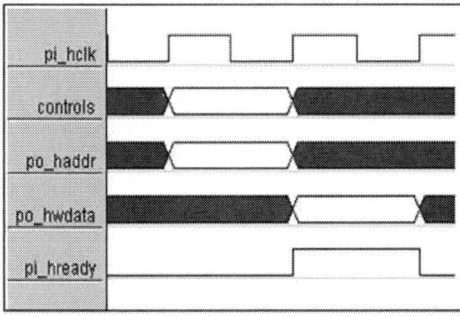

AHB Protocol

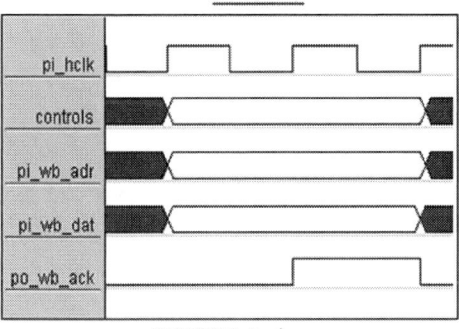

WISHBONE Protocol

Fig.2 WISHBONE versus AHB for single transfer protocol

The value of HTRANS is set accordingly to indicate non-sequential (NONSEQ) and sequential (SEQ) transfer of the burst cycle transfer. For a single cycle or the first transfer of a burst, HTRANS is set to NONSEQ. The HTRANS is set at idle when there is no transfer. The HTRANS is set as busy when the system is in WRITE or READ state, while either WB_CYC or WB_STB is LOW.

The AHB address (HADDR) is generated based on WB_CTI and WB_BTE values. The WB_SEL is decoded using binary to decimal decoder to produce the least significant bit of the AHB address. During START state, the AHB uses address given by the WISHBONE master. During WRITE and READ states, the FSM generates address and control signals for the AHB's address phase. For burst mode, data phase of the current burst will overlap with the address

phase of the next burst. AHB burst (HBURST) is similar to WISHBONE burst type extension (WB_BTE). They are 1-to-1 matching. For other HBURST value, it is implemented as a linear burst in WISHBONE protocol, as shown in Table III.

The Table IV shows how the WB_SEL signal is mapped onto HSIZE signal. Signal HSIZE indicates the size of transfers, which is 8 bits, 16 bits or 32 bits. Its value is determined by summing up the number of valid bytes indicated by WB_SEL.

For Single /Linear cycle, the AHB address output signal is taken from WISHBONE address input signal. In burst cycle mode (or incremental mode), the module reads only the first WISHBONE address. For the following data transfer, the module increments the address based on its own calculation, depending on whether it is linear increment, 4-beat wrap increment, 8-beat wrap increment or 16 bit wrap increment. The address is wrapped as shown in the Table V.

TABLE III

Relation Between WB_CTI, WB_BTE and HBURST

WB_CTI	WB_BTE	HBURST	Descriptions
3'b000	2'b00	3'b000	Single burst
3'b001	2'b00	3'b000	Constant address burst
3'b010	2'b00	3'b001	Incrementing burst of unspecified length / linear burst
	2'b01	3'b010	4-beat wrap burst
	2'b10	3'b100	8-beat wrap burst
	2'b11	3'b110	16-beat wrap burst

TABLE IV

Relation between WB_SEL and HSIZE

WB_SEL	HSIZE	Descriptions
4'b0000	Invalid	Use default size (32 bits)
4'b0001, 4'b0010, 4'b0100, 4'b1000	0	8 bits/ 1 byte
4'b0011, 4'b1100	1	16 bits / half word
4'b1111	2	32 bits / 1 word (default)
4'b0111	Invalid	Use default size (32 bits)
Others	-	Determined by summing up each WB_SEL valid bit.

TABLE V

Wrap Size Address Increments [1]

Starting address' LSBs	Linear	Wrap-4
3'b000	0-1-2-3-4-5-6-7	0-1-2-3-4-5-6-7
3'b001	1-2-3-4-5-6-7-8	1-2-3-0-5-6-7-4
3'b010	2-3-4-5-6-7-8-9	2-3-0-1-6-7-4-5
3'b011	3-4-5-6-7-8-9-A	3-0-1-2-7-4-5-6
3'b100	4-5-6-7-8-9-A-B	4-5-6-7-8-9-A-B
3'b101	5-6-7-8-9-A-B-C	5-6-7-4-9-A-B-8
3'b110	6-7-8-9-A-B-C-D	6-7-4-5-A-B-8-9
3'b111	7-8-9-A-B-C-D-E	7-4-5-6-B-8-9-A

IV. SIMULATION

We build AMBA bus environment using the Synopsys Designware Library and design a WISHBONE Master model with System Verilog, then we get the simulation results from Synopsys VCS (Verilog Compiled Simulator). We use the same AHB clock for WISHBONE clock.

The tests include classic read/write cycle; burst read/write cycle with constant, incremental and wrapped addresses; access to WAIT transaction state from WISHBONE master side; and access to WAIT transaction state from AHB slave side.

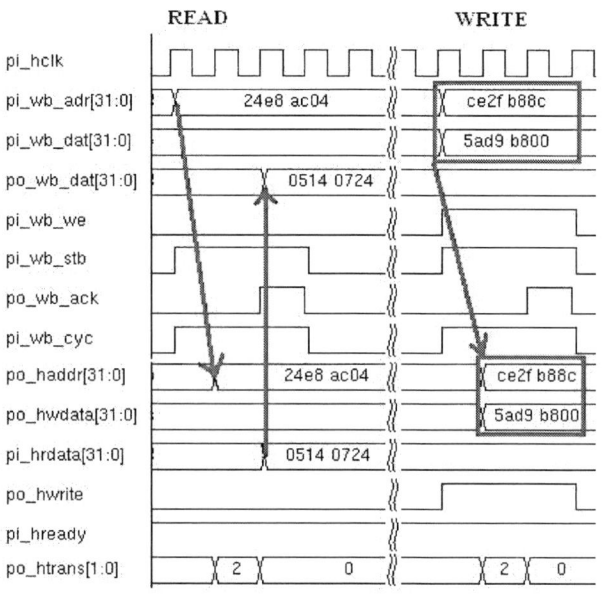

Fig. 3 Classic WRITE and READ

Fig. 3 shows a classic WRITE and READ by a WISHBONE master through the WISHBONE-to-AHB Bridge. The WRITE command is indicated by signal pi_wb_we in HIGH state. It is active when pi_wb_stb and pi_wb_cyc are also in HIGH state. It requires two clock cycles to complete the writing or reading, based on asserted po_wb_ack signal after the command is given. If compare to the waveform in Fig. 2, the po_wb_ack signal takes extra clock cycle because it needs to wait the AHB protocol to complete.

Fig. 4 shows a waveform of burst READ using linear incremental command without any WAIT transaction state. Initially, the po_haddr copies the first address value "d5bb3260" from pi_wb_adr. The AHB transfer type is set to indicate non-sequential transfer (po_htrans=2). After that, the next po_haddr "d5bb3264" is calculated based on the pi_wb_bte and pi_wb_cti values (refer Table III). The calculated po_haddr signal is asserted one cycle earlier than the actual WISHBONE address signal as shown in the Fig. 4. The AHB transfer type is set to indicate sequential transfer (po_htrans=3). The burst is terminated when ORMCU sends end-of-burst transfer, indicated by pi_wb_cti=7. Signal

po_wb_ack indicates the data at po_wb_dat is valid for the address given by the WISHBONE. The WISHBONE data and addresses are aligned in the same clock cycle as required by the WISHBONE protocol.

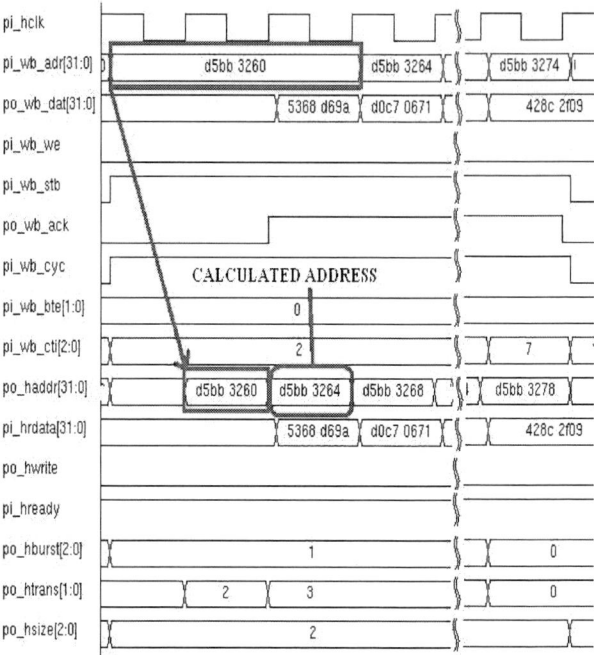

Fig. 4 Burst READ with linear incremental address

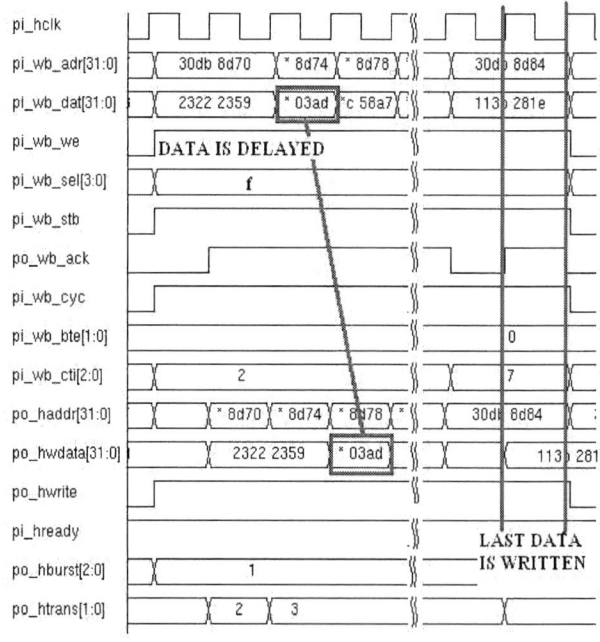

Fig. 5 Burst WRITE with linear incremental address

Fig. 5 shows a waveform of burst WRITE using linear incremental command without any WAIT transaction state. The address manipulation is similar to burst READ function as in Fig. 4. For writing function, the output data, po_hwdata are delayed one clock cycle to mimic an AHB transfer protocol. When the burst is to be terminated, the ORMCU sends end-of-burst indicator. The acknowledge signal, po_wb_ack is set to LOW until the writing is complete. Signal po_wb_ack indicates the data at pi_wb_dat is successfully written into the given address location. The AHB data and addresses are aligned in a manner of AHB protocol for burst transfer.

Fig. 6 shows a waveform of burst WRITE with WRAP8 incremental address and waiting state by the WISHBONE master. The WRAP8 incremental address is set by the WISHBONE master through pi_wb_cti and pi_wb_bte signal (refer Table III). The bridge passes the information to slave through po_hburst signal. When the master deactivates pi_wb_stb while pi_wb_cyc is still active, the bridge sets po_htrans as busy to inform the slave to wait. The red square boxes show the time of data "c9ffe563" is successfully transmitted for address "ceeb5e2c".

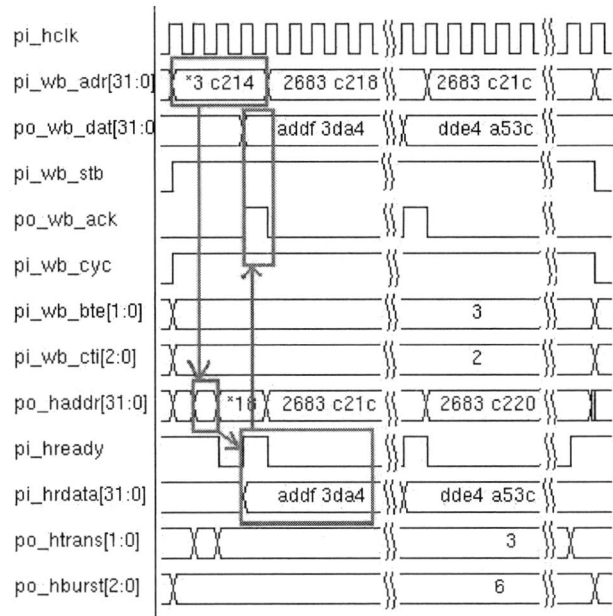

Fig. 7 Burst READ with WRAP16 incremental address and waiting state

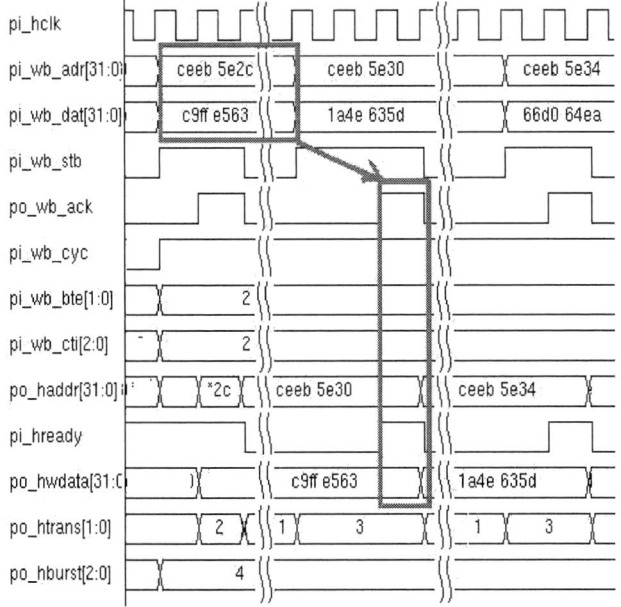

Fig. 6 Burst WRITE with WRAP8 incremental address and waiting state

Fig. 7 shows a waveform of burst READ with WRAP16 incremental address and waiting state by the AHB slave. The WRAP16 incremental address is set by the WISHBONE master through pi_wb_cti and pi_wb_bte signal (refer Table III). The bridge passes the information to slave through po_hburst signal. In the waveform, pi_hready is set to LOW by AHB slave to force the WISHBONE-to-AHB Bridge to enter WAIT state while both pi_wb_stb and pi_wb_cyc is still active. The po_wb_ack signal is held LOW until the slave gives the ready signal. From the waveform, the read data "addf3da4" is the data for address "2683c214", and not for the address "2683c218".

V. CONCLUSIONS

All simulation results above have proven the bridge is functioning correctly for both WISHBONE and AHB protocols. Our design has achieved initial success. We will test it with real physical IP in FPGA and improve its functions in the future.

WISHBONE and AMBA have their own architecture, thus their buses work differently. However, by bridging the busses, both architectures can work together. In addition to WISHBONE-to-AHB convertor, we can also create an AHB-to-WISHBONE convertor. This new bridge can be used if AHB master is used in WISHBONE architecture [3].

ACKNOWLEDGMENT

The author wishes to thank all colleagues who helped in building up the verification environment and provided technical support.

REFERENCES

[1] OPENCORES.ORG. WISHBONE System-On-Chip (SoC) Interconnection Architecture for Portable IP Cores, Revision B.3, 2002.9
[2] AMBA Specification, Rev 2.0, ARM Limited 1999
[3] Cao Fan, Chen Lan, and Yi Bo, "Designing WISHBONE to AMBA Wrapper", IEEE, 2009
[4] Xu Xing, Chen Zezong, Jiang Jing, and Ke Hengyu, "Porting from Wishbone Bus to Avalon Bus in SoC Design", ICEMI 2007, IEEE, 2007
[5] OpenCores, http://www.opencores.org

Tunable Loop Filter in Fractional-N Frequency Synthesizer for Wireless Applications

Gan Leong Kit[1], Fazrena Azlee Hamid[2], Syed Khaleel Ahmed[3]
[1]Department of Electronics and Communication Engineering, Universiti Tenaga Nasional,
Putrajaya Campus, Malaysia
[1]leong_kit0721@yahoo.com, [2]fazrena@uniten.edu.my, [3]syedkhaleel@uniten.edu.my

Abstract **A tunable loop filter in fractional-N frequency synthesizer is proposed and analyzed. The proposed concept allows designer to fine tune the loop bandwidth based on current injection into the loop filter. Fine tuning feature can be achieved with the utilization of a voltage-controlled transconductance component. The analysis and simulation results are presented for this technique. It shows that the frequency synthesizer employing the proposed tunable loop filter allows designer to have a fine control on the loop bandwidth and lock time of the system.**

Keywords: Phase locked loop, fractional-N frequency synthesizers, loop filter

I. INTRODUCTION

Phase Locked Loop (PLL) based synthesizers generate the desired periodic signal with high spectral purity and accurately defined frequencies. In radio-frequency (RF) transceiver, the considerations when one designs frequency synthesizers are frequency accuracy, phase noise, sidebands, switching time and sensitivity to noise [1]. The quality of the frequency synthesizer is highly dependent on the type of PLL used in the system.

Communication systems often pose severe requirements on the spectral purity of the tuning system local oscillator (LO) signal which exhibits finite phase noise. When the phase noise from LO signal is superimposed onto an adjacent channel signal during the down-conversion process, reciprocal mixing occurs. Reciprocal mixing decreases the receiver's selectivity and disturbs the reception of weak signals by the tail of interferer. In recent trend of communication products, the frequency synthesizers are commonly turned off when the system enters sleep mode, in order to prolong battery life and to reduce overall power consumption. Hence, it is vital for the output of frequency synthesizer to acquire lock state in the shortest duration of time when emerging from sleeping mode.

Fast settling time and good phase noise frequency synthesizers are essential building blocks of modern communication system. The limited bandwidth available to each user leads to smaller channel spacing in wireless systems. This has further challenged designers to consider the tradeoff between settling time and phase noise when designing loop filter.

In view of the above discussion, fractional-N frequency synthesizer is a key enabling technology and has become very popular in a range of RF applications due to the fact that they allow large reference frequency to achieve a small frequency resolution. In this paper, we enhance the flexibility of designing synthesizer by incorporating tunable loop filter into the system. In particular, we will introduce voltage-controlled tunable loop filter topology which provides loop bandwidth tunable capability.

Section II provides a short background discussion on loop parameters involved in loop filter design for wireless applications. We will then introduce and describe the proposed voltage-controlled tunable loop filter approach in Section III and describe its key attributes. Section IV presents the behavioral simulation results using Advanced Design System (ADS) and its analysis. Finally, Section V concludes the paper.

II. BACKGROUND

The key component used to achieve PLL transfer function is the loop filter. The choices of loop filter architecture and circuit component values are usually a very balanced compromise between both system static noise and dynamic performance. In particular, the loop filter implements desired poles and zeros for realizing a given open loop transfer function and, in combination with the other components in the PLL such as phase frequency detector (PFD), charge pump (CP), voltage controlled oscillator (VCO) and dividers, they also set the overall open loop gain of the system. A loop filter is needed within a PLL to attenuate the high frequency components and filter noise included in the error voltage which is mainly from the PFD. Obviously, the loop filter also acts as a current to voltage converter to obtain the controlling voltage which ultimately adjusts the frequency of VCO. The attenuation strength of the loop filter determines the voltage signal purity feeding to the VCO.

The main consideration when one designs a loop filter is the tradeoff between lock time and phase noise of the PLL system. Lock time defines the ability of the loop to change frequencies quickly. If the loop bandwidth is very low, the changes in the output control voltage will take place slowly, and the VCO will not be able to change its frequency fast. On the contrary, a high loop bandwidth allows the changes to happen faster. However, high loop bandwidth system allows more undesired high frequency signals that pass through the system and results in higher in-band phase noise. The phase noise in a PLL strongly impacts the quality of a transceiver system.

The charge pump PLL [2] structure is prevalent as the PLL topology of choice extends across a wide range of applications.

978-1-4673-2395-6/12 $31.00 © 2012 IEEE

It offers a seemingly simple implementation which can achieve low power operation, and can be applied to both integer-N and fractional-N frequency synthesizers. Over the years, various loop filter topologies have been explored to mitigate the tradeoffs between settling time and phase noise, power and area efficient in the charge-pump PLL architecture. In essence, prior works are active loop filter design topology, switch loop bandwidth, switch capacitor and switch resistor base technique [3] − [10]. In this paper, we proposed a new architecture of controlling the PLL loop bandwidth, i.e. using voltage controlled transconductance [11]. In the conventional variable-bandwidth PLL, fast lock is achieved through the switching of dumping resistance and increasing the charge pump current at the same time [3] [4]. The fast lock approach has brought several disadvantages. First, the in-band phase noise will increase since the fast lock requires that a higher current is switched in during frequency acquisition, this requires that the PLL is run in less than the highest current mode. Secondly, the fast lock approach is not working well for third or higher order filter [3]. In our tunable loop filter, the transconductance in the loop filter can be varied electronically and serve as an additional approach to alter the loop bandwidth. This allows designers to explore a larger design space in terms of varying the bandwidth.

III. PROPOSED TUNABLE LOOP FILTER

In this paper, a Type-II fractional-N frequency synthesizer with a second order tunable loop filter is proposed. The basic theory and transfer functions of the circuit are presented in the following. Further extending the tunable loop filter to a higher order is possible. However, it is beyond the scope of this paper to analyze higher order tunable filter. Practical circuits for the proposed tunable loop filter with the simulation results are discussed later in Section IV.

Fig. 1 shows the basic circuit which illustrates the operating principle of the proposed tunable loop filter. It consists of a voltage-controlled current source which injects its current $g_m V_\pi$ into the loop filter. The loop filter transfer function will be defined as the change in the voltage at the tuning port of the VCO divided by the current at the charge pump that caused it.

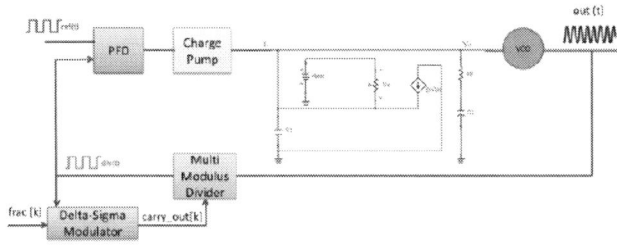

Fig. 1: Proposed Tunable Loop Filter

In the case of a second order loop filter, it is simply the impedance. The proposed tunable loop filter has a transfer function of:

$$\frac{V_0}{I_i + g_m V_\pi} = \frac{sR_2C_2 + 1}{R_2C_1C_2s^2 + (C_1 + C_2)s} \tag{1}$$

where g_m is the transconductance of the current source.

The output voltage of the loop filter is to be expressed as

$$V_0 = \frac{sR_2C_2 + 1}{R_2C_1C_2s^2 + (C_1 + C_2)s}(I_i + g_m V_\pi) \tag{2}$$

It is obvious that from (2), the output voltage of the tunable loop filter is a function of both charge pump current, I_i and bias voltage of the voltage-controlled current source, V_{bias}. The effective output voltage of the loop filter can be controlled electronically (voltage-controlled) by using a voltage-controlled transconductance, g_m. Such a voltage-controlled transconductance can be implemented using BJT or FET amplifiers whose transconductance is controlled linearly by an external bias voltage [11].

One very interesting usage of this new fine bandwidth-tuning protocol is to change the bandwidth with a fixed charge-pump current. This provides an alternative option and flexibility for designers to fine tune the loop bandwidth in a smaller step size and optimizes the performance needed for specific application. Note that the transconductance can be fine tuned with the voltage supply from a digital to analog converter (DAC), which can be controlled by the microcontroller for instance. Depending on specific design goals, if a wider bandwidth variation is required, the combination methods of varying charge pump current, switching loop components and voltage controlled capacitance can be used at the same time with care.

In order to cover wideband frequencies in modern communication systems, the flexibility of varying the loop bandwidth becomes an important and distinctive feature. Instead of optimizing the discrete components to alter the bandwidth to support a particular band of frequencies, a designer can tune the loop filter by fine tuning the biasing voltage of the transconductance.

IV. PRACTICAL CIRCUIT AND SIMULATION RESULTS

Fig. 2 shows a practical circuit for the implementation of the tunable loop filter. In this paper, the common collector BJT configuration is used to implement the voltage-controlled current source. As the focus of this paper is on loop filter design, the other PLL blocks are modeled using behavioral modeling technique. CMOS technology is commonly used to implement other blocks in the PLL due to its low phase noise. The transconductance is controlled by the bias voltage which is applied to the base of the transistor through the resistance R_B.

978-1-4673-2395-6/12 $31.00 © 2012 IEEE

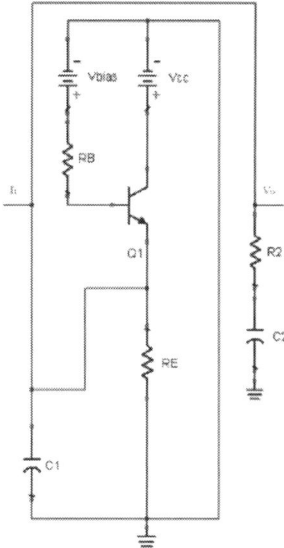

Fig. 2: A Practical Tunable Loop Filter Circuit

The simplified AC equivalent for the proposed circuit is shown in Fig. 3 where R_E and r_o are not shown because of their relative high resistance. The transfer function of the practical circuit is the same as that derived from (1) except the V_π is voltage divided of R_B and r_π, (3).

$$V_\pi = \left(\frac{r_\pi}{r_\pi + R_B}\right)V_{bias} \qquad (3)$$

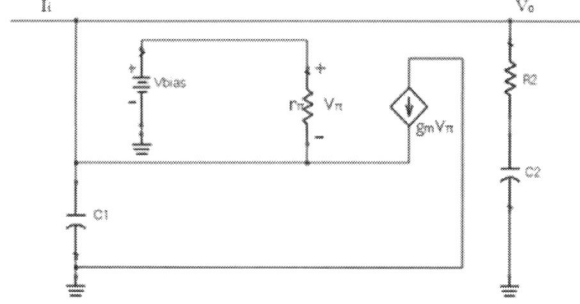

Fig. 3: Simplified AC Equivalent Circuit

Equation (4) shows the relationship between the transconductance of the transistor and other parameters, which can be obtained from the DC analysis

$$g_m = \frac{I_C}{V_T} = \frac{(V_{bias} - 0.7)}{(R_E + \frac{R_B}{\beta})V_T} \qquad (4)$$

where V_T is the thermal voltage ($V_T = \frac{kT}{e} = 0.026V$ at room temperature), β is the current gain of the transistor, and I_C is the DC collector current. The range over which the transconductance can be varied is determined by the range of the bias voltage, V_{bias} and the values of R_E, R_B and β provided that the transistor will not enter saturation region [11].

The circuit in Fig. 2 is incorporated into the fractional-N frequency synthesizer behavioral simulation. The synthesizer specifications in this simulation are tabulated in Table I below.

TABLE I
DESIGN PARAMETERS OF FREQUENCY SYNTHESIZER

Parameter	Value	Unit
Center Frequency, F_c	3020.1	MHz
Reference Frequency, F_{ref}	10	MHz
VCO Gain, K_v	28.504	MHz/V
Divide Ratio, N_o	302.01	-
Charge Pump Current, I_d	2.6	mA
Loop Filter Capacitor, C_1	0.075	nF
Loop Filter Capacitor, C_2	1.8	nF
Loop Filter Resistor, R_2	2.7	kOhm
Base Resistor, R_B	0.5	kOhm
Emitter Resistor, R_E	100	kOhm
V_{cc}	5	V

The bias voltage is swept across 1V to 3V with a step size of 0.1V. The range was also chosen with the reason of forbidding the transistor to operate in cutoff and saturation region. Fig.4 and Fig.5 presents the open loop frequency response and phase response respectively.

The bias voltage is swept across 1V to 3V with a step size of 0.1V. The range was also chosen with the reason of forbidding the transistor to operate in cutoff and saturation region. Fig.4 and Fig.5 presents the open loop frequency response and phase response respectively.

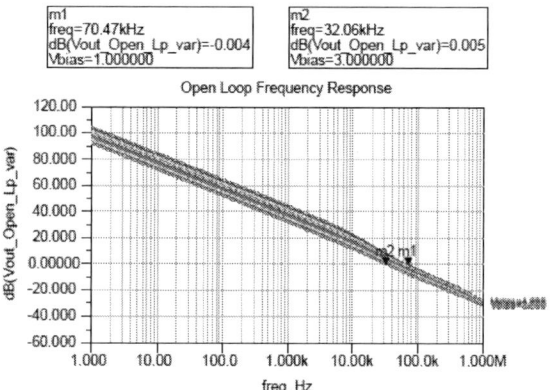

Fig. 4: Open Loop Frequency Response

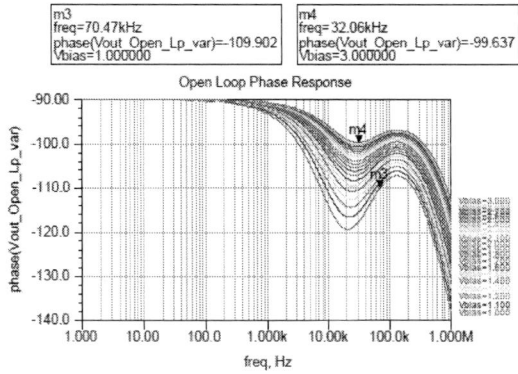

Fig. 5: Phase Frequency Response

From Fig.4, we observed that the open loop bandwidths vary as expected. As discussed earlier, the resolution of the loop bandwidth tuning is a function of the DAC resolution. As shown in marker1 (m1) and marker2 (m2), the open loop bandwidths vary from 70.47 kHz to 32.06 kHz for bias voltage of 1V and 3V respectively. From the phase response in Fig.5, the phase margin can be computed by taking the difference of 180° and the phase at the unity gain frequency, which for both of the extreme ends is 70° and 80.4° respectively. In this design, the phase margin variation with V_{bias} within the range of 1V and 3V is approximately 10°.

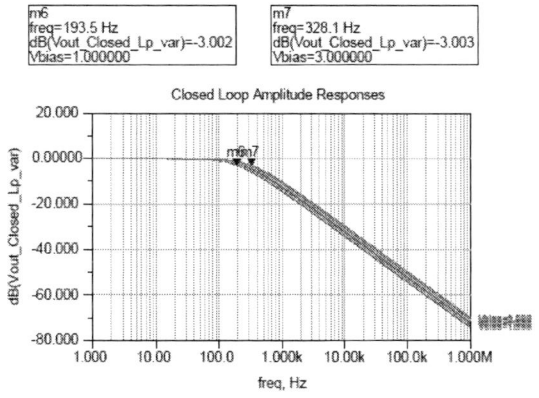

Fig. 6: Closed Loop Frequency Response

Let us now analyze the PLL system in a closed loop manner. Fig.6 shows the closed loop frequency response with the V_{bias} sweeping identical as the open loop response simulation. As portrayed in the plot, the closed loop bandwidth is proportional with the bias voltage of the transistor. The -3 dB frequency for $V_{bias} = 1V$ and $V_{bias} = 3V$ is shown in marker 6 (m6) and marker 7 (m7) respectively. Moreover, from the closed loop curves, we can know that the system has a low damping factor as there are no peaks in the response.

Finally, the transient response was simulated across $V_{bias} = 1V$ to $V_{bias} = 3V$ with a step size of 0.2V as depicted in Fig.7. In this simulation setup, it begins with the PLL in a locked state. At the time of pre-determined delay, the divide by N ratio is changed to force a step in the output frequency. In this case, the step input is 10 which caused the output frequency jump from 3020.1 MHz to 3120.1 MHz. Note that the system settles within the shortest time frame when V_{bias} is at the highest value (3V in this case) which correlates with our previous analysis. With the proper choice of current injection into the loop filter, the system is able to settle in an optimum time. In this simulation, the PLL managed to lock within tens of microseconds with quite a high frequency jump.

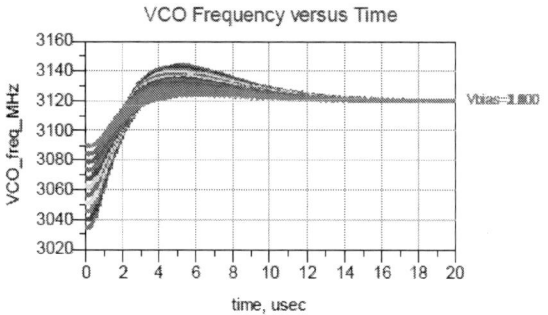

Fig. 7: Transient Response

Through the analysis above, we can see that the proposed system can be implemented in a very stable manner and yet can achieve a fast settling time. In order to further characterize the tunable loop filter, a DC analysis has been done for the transistor operation in this system. The current I_B, I_C and I_E are tabulated in the Table II below. The current consumption of this additional circuitry is very low (in the scale of μA) and hence the power consumption is low as well. With the additional power consumption, a designer has an option to optimize the loop bandwidth without changing components value. In view of this benefit, the design cycle time would definitely be reduced.

TABLE II:
DC ANALYSIS OF TRANSISTOR OPERATION

V_{bias}	I_B (nA)	I_C (μA)	I_E (μA)
1.000	31.440	5.000	5.031
1.200	42.780	6.909	6.952
1.400	54.140	8.383	8.892
1.600	65.500	10.780	10.840
1.800	76.870	12.730	12.800
2.000	88.250	14.680	14.770
2.200	99.640	16.640	16.740
2.400	111.000	18.600	18.710
2.600	122.400	20.560	20.680
2.800	133.900	22.530	22.660
3.000	145.300	24.500	24.640

V.   CONCLUSION

This paper described a fine tuning loop filter in fractional-N frequency synthesizer. Stability analysis and transient response were presented. The proposed design allows PLL designer to adjust the loop bandwidth of the frequency synthesizer in order have a good control and in the noise and dynamic performance within the constraints of all other parameters within the system.

REFERENCES

[1] B. Razavi, "Challenges in the design of frequency synthesizers for wireless applications," in *Proc. 1997 IEEE Custom Integrated Circuits Conf.*, May 1997, pp. 395-402.

[2] F. Gardner, "Charge-Pump Phase-Lock Loops," *IEEE Trans. Commun.*, vol. COM-28, pp.1849-1858, Nov. 1980.

[3] D. Banerjee, *PLL Performance, Simulation, and Design*, 4th ed. Santa Clara, CA: National Semiconductor, 2005.

[4] Valenta, V., Villegas, M., Baudoin, G., "Analysis of a PLL Based Frequency Synthesizer using Switched Loop Bandwidth for Mobile WiMAX," *Radioelektronika, 2008 18th International Conference*, September 2008.

[5] Vaucher, C.S., "An Adaptive PLL Tuning System Architecture Combining High Spectral Purity and Fast Settling Time," *IEEE J. Solid-State Circuits*, vol. 35, no. 4, pp.490 - 502, Feb. 2000.

[6] Thoka, S., Geiger, R.L., "Fast-Swithching Adaptive Bandwidth Frequency Synthesizer using a Loop Filter with Switched Zero-Resistor Array," *2005 IEEE International Symposium on Circuits and Systems (ISCAS)*, July 2005.

[7] Kyoungho Woo, Yong Liu, Eunsoo Nam, Donhee Ham, "Fast-Lock Hybrid PLL Combining Fractional-N and Integer-N Modes of Differing Bandwidths," *IEEE J. Solid-State Circuits*, vol. 43, no. 2, pp.379 - 389, Feb. 2008.

[8] Woo-Yeol Shin, Manho Kim, Gi-Moon Hong, Suhwan Kim, "A Fast-Acquisition PLL using Split Half-Duty Sampled Feedforward Loop Filter" *IEEE Trans. Consumer Electronics*, vol. 56, no.3, pp. 1856-1859, Aug. 2010.

[9] KELIU, S., SANCHEZ-SINENCIO, E. *CMOS PLL Synthesizers: Analysis and Design*. Springer, 2005.

[10] Tsukutani, T., Sumi, Y., Higashimura, M., Fukui, Y., "Electronically Tunable Low-Pass Filter in PLL Frequency Synthesizer," *Proc. 2001 IVEC*, pp. 55-59, 2001.

[11] Mansour I. Abbadi, Abdel-rahman M. Jaradat, "Artificial Voltage-Controlled Capacitance and Inductance using Voltage-Controlled Transconductance," *International Journal of Electrical and Computer Engineering*, vol. 3, no.11, pp. 756-759, 2008.

Stress Analysis on Centric Through Hole PCB

[1]Sauli. Z., [1]Retnasamy, V.[*], [1]Rahman, N.A.Z.,[1]Man, B., [1]Nadzri, N.S. &[1]Vairavan, R.

[1]School of Microelectronic Engineering, Universiti Malaysia Perlis (UniMAP),

Kampus Alam, Pauh Putra Perlis, Malaysia.

*Corresponding email: vc.sundres@gmail.com

Abstract–In this study, the bending process of PCB during depaneling is simulated. The stress response during the bending process is evaluated using a computational program Ansys. Two PCB plate model were developed:One of the model with centric through hole and other model without hole. The stress response of the two models are compared. From the simulation, it has been observed that the value of stress response of the PCB increases with increasing displacement height. The without hole PCB model exhibited a slightly higher stress response compared to the with centric through hole PCB model. Highest stress response for both models were recorded at the displacement height of 5cm.

Keywords-Printed circuit board(PCB); depaneling process; centric through holes

I INTRODUCTION

In the electronic industry, the printed circuit board (PCB) is the moral fiber for surface mounted components during packaging assembly. The numbers of PCBs production has increased due to the rapid growth in the electronic technologies [1].The printed circuit board operates as an instinctive support and it connects passive and active components by means of conductive conduit. The purposes of PCB board are as follows, 1.To electrically interconnect the components, 2.To provide mechanical support to the electronic components and 3. To provide power and dissipate the heat generated by mounted components [2].

Currently, all the electronic devices are produced in a batch form due to mass production. After completing the board level assembly, the printed circuit boards are subjected to depaneling process which individualizes the PCBs. During the depaneling process, the PCBs plates are subjected to mechanical bending process. This bending process may induce stress on the PCBs which may damage and reduce the reliability of the PCBs [3,4]. There are some works done based on computational system to detect the defects of mounted components on PCB during PCB handling process [5,6]. Hence, there is a necessity to evaluate the stress response of a PCB plate during the bending process.

Therefore, in this study, the bending process of the PCB plate is simulated using a commercial computational program, Ansys. The stress response during bending process is examined. To evaluate the stress response, two types of PCB designed were used, one model with centric through hole and one model without hole. The centric through hole model was utilized to observed if it has any impact on the stress response of the PCB plates. The stress responses of

both models were compared. The bending process was replicated by using a displacement scheme, where one end of the PCB plate is constrained and the displacement is applied at the other end. The displacements heights of the PCB plates were ranged from 1cm to 5cm.

II METHODOLOGY

A commercial computational program Ansys was utilized to perform the simulation. Two 3D models were developed using 20 Node Quadratic Hexahedron (Solid 186). One model will feature a PCB plate without hole and the other model will feature a PCB plate with a centric through hole. The 3D models were then meshed with 69867 number of elements for the PCB plate without hole model and 68872 number of elements for the PCB plate with centric through hole model. The dimension of the PCB plate models are 175mm vs 250mm and the thickness is 1mm. As for the centric through hole model, the diameter of the hole is 30 mm and it is placed at the center of the model. The 3D model is shown in Fig. 1. The material properties used in the analysis is shown in Table 1.

TABLE 1

MATERIAL PROPERTIES OF FR4

Material	FR4
Young Modulus ,E (Pa)	22×10^{9}
Poison Ratio	0.136
Density (kg/m³)	1850

In the both 3D model, the loading and the boundary conditions were applied as follows:

   i.   Displacement

       As the design is in three dimension, there will be x,y and z axis. the displacement is given at the right end of the pcb plate in the positive y direction . The value of displacement height are in range of 1cm to 5cm.

   ii.   Fixed Support

       The left end of the pcb plate is contrained by applying fixed support. This will result in no movement at this end of the pcb plate.

In this study, the stress response of the PCB plates during the bending process was evaluated by using the Von Mises Stress as a criterion.

978-1-4673-2395-6/12 $31.00 © 2012 IEEE

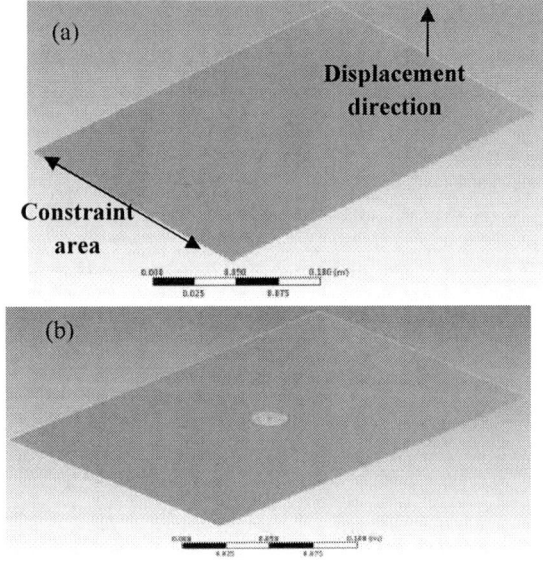

Fig. 1. The 3D model of PCB plates, (a)Without hole and (b) with centric through hole

III RESULTS AND DISCUSSION

Ansys software was used to analyze the stress response of the PCB plates. In this study, two types of PCB plates were modeled and simulated which are without hole and with centric through hole. The effects of the PCB geometry on the stress response were investigated. The displacement height of 1cm, 2cm, 3cm, 4cm, and 5cm were applied to both PCB plates.

As shown in Fig. 2. ,it can be observed that the maximum stress on both PCB plates are different based on the PCB plate characteristics. For PCB plate without hole, the maximum stress onto the board is slighty higher compared the PCB plate with centric through hole which produces less maximum stress. For displacement height of 1cm the stress onto of PCB plate without hole is 534500 Pa and the stress onto of PCB plate with centric through hole is 529690 Pa

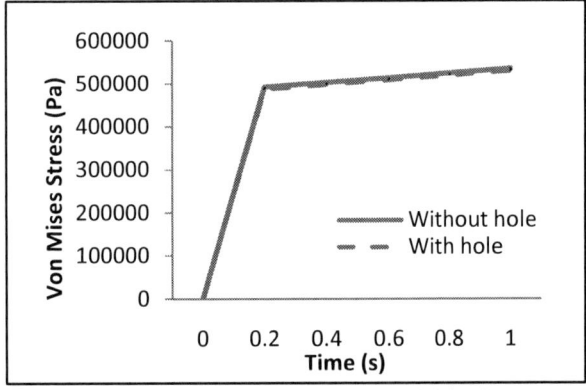

Fig. 2. Stress vs Time for displacement height of 1cm

From the Fig. 3., the maximum stress obtained on both PCB plate are different based on the PCB plate characteristics. For PCB plate without hole, the maximum stress onto the plate is slighty higher compared the PCB plate with centric through hole. For 2cm of displacement height, the stress response of PCB plate without hole is 1069000Pa and the stress response of PCB plate with centric through hole is 1059400Pa.

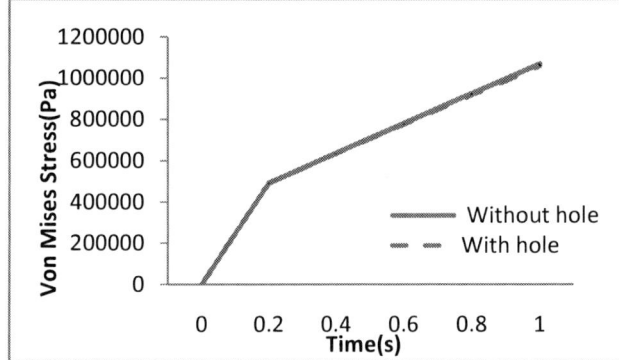

Fig. 3. Stress vs Time for displacement height of 2cm

Based on the Fig. 4, it is observed that the stress response is increasing with the increment of displacement height. For 3cm displacement height, the stress response of PCB plate without hole is 1603500 Pa and the stress response of PCB plate with centric through hole is 1589100Pa.

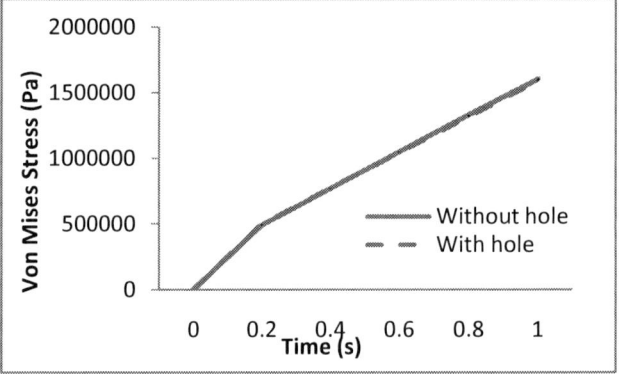

Fig 4. Stress vs Time for displacement height of 3cm

For the displacement height of 4cm , the stress response of PCB plate without hole is 2138000Pa and the stress onto of PCB Board with centric through hole is 2118800Pa as illustarted in Fig. 5.

978-1-4673-2395-6/12 $31.00 © 2012 IEEE

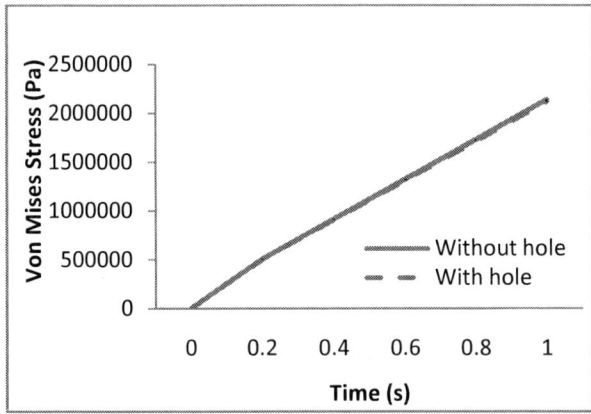

Fig 5. Stress vs Time for displacement height of 4cm

Based on the Fig. 6, for 5cm of displacement height, the stress response of PCB plate without hole is 2672500 Pa and the stress response of PCB plate with centric through hole is 2648500 Pa.

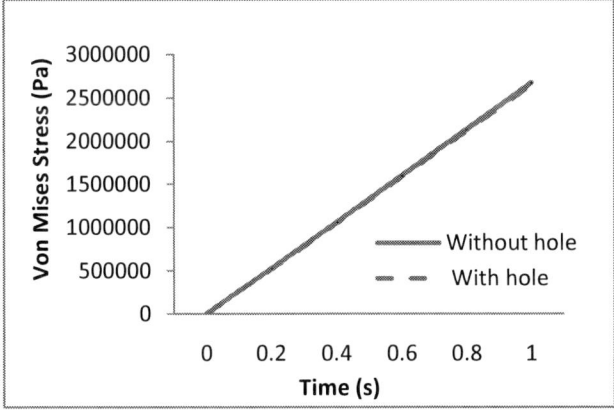

Fig 6. Stress vs Time for displacement height of 5cm

Hence, by comparing the over all results, the PCB plate with without hole exhibited a slightly higher stress response compared to the PCB plate with centric through hole. The mechanical bending exerts a curvature bend on the PCB plate structure [3]. During bending process, stress is exerted at the opposite of the PCB plate which is dispersed along the width and length direction of the PCB plates[7] which can be observed in Fig7(a). The high stress region along the width and length direction of the PCB may invoke damage to the solder joints of the mounted components. This may lead to solder joint crack formation[3]. Hence, a single centric through hole was placed at the center of the PCB. This was done to evalute if the hole structure assited in reducing the high stress region near the constarined are of the PCB. However, the hole structure had little effect in reducing the high stress region near the constarined area. This is due to the distance and position of the hole structure which is futher away from the constrained area which can be visualized in. Fig 7(b) . Therefore , it can be concluded that the position of the hole may have a significant effect if it was placed near the

contstained area of the PCB plate as the hole structure can function as a weak link in the PCB plate and the stress distribution can be solely focus on the hole structure. This may reduce the along the width and length direction of the PCB.

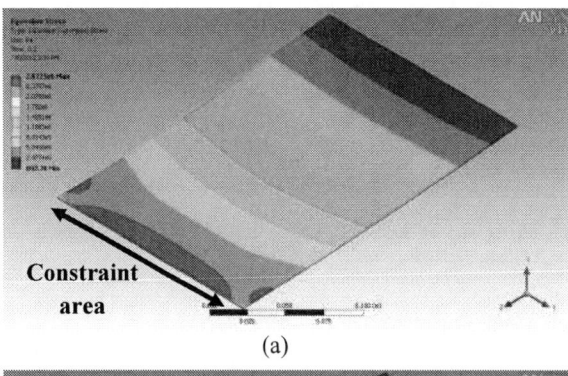

(a)

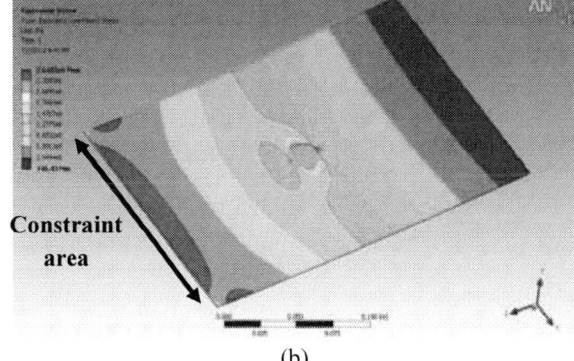

(b)

Fig 7. Stress contour of the PCB plates. (a) without hole and (b) with centric through hole

IV. CONCLUSION

The stress response between plate PCB with centric through hole and without hole were simulated in this study. The simulation were executed using Ansys version 11. From the simulation, it has been observed that the value of stress response of the PCB increases with increasing displacement height for both PCB models. However, the without hole PCB model exhibited a slightly higher stress response when compared to the centric through hole PCB model. The hole structure had a less significant effect on the stress response of the PCB plate. This may be due to the position of the hole structure which is futher away from the constrained area of the PCB plate. Highest stress response was obtained at the displacement height of 5cm for both models.

ACKNOWLEDMENT

The author would like to thank and acknowledge School of Microelectronic Engineering, Universiti Malaysia Perlis for their support and facilities.

REFERENCES

[1] R. C. Mata, et al., "Reverse engineering for obsolete single layer printed circuit board (PCB)," in Computing & Informatics, 2006. ICOCI '06. International Conference on, 2006, pp. 1-7.

[2] Tummala Rao. R., Fundamentals of Microsystems Packaging, Mc Graw Hill International Edition, 2004.

[3] D. Lau, et al., "Experimental testing and computational stress analysis of printed circuit board for the failure prediction of passive components under the depaneling load condition," in Electronic Components and Technology Conference, 2005.Proceedings.55th, 2005, pp. 1783-1791 Vol. 2.

[4] S. W. Ricky Lee and Dennis Lau, "Correlation between the Strain on the Printed Circuit Board and the Stress in Chips for the Failure Prediction of Passive Components," Proceedings of IMECE2005, ASME, Orlando, Florida USA, November 5-11, 2005.

[5] F. R Leta,. et al., "Computer Vision System for Printed Circuit Board Inspection", 19th International Congress of Mechanical Engineering, Brasília, Brazil, 2007.

[6] F. R. Leta and F. F. Feliciano, "Computational system to detect defects in mounted and bare PCB Based on connectivity and image correlation," in Systems, Signals and Image Processing, 2008. IWSSIP 2008. 15th International Conference on, 2008, pp. 331-334.

[7] T. Tong Yan, et al., "Novel numerical and experimental analysis of dynamic responses under board level drop test," in Thermal and Mechanical Simulation and Experiments in Microelectronics and Microsystems, 2004. EuroSimE 2004. Proceedings of the 5th International Conference on, 2004, pp. 133-140.

Very High Speed and Low Voltage Open-Loop Dual Edge Triggered Sample and Hold Circuit in 0.18μm CMOS Technology

Mohamad Hasan-Sagha and Mohsen Jalali
Electrical and Computer Engineering Department
Shahed University
Tehran, Iran
Email: m.hasansagha@shahed.ac.ir

Abstract—**A very high speed and low voltage open-loop dual edge triggered sample and hold (S/H) circuit is proposed. Using a high speed and low voltage multiplexer (MUX) and employing simple and efficient switches the proposed sample and hold circuit can operate with a sampling rates of about 5 GS/s making it suitable for high-speed analog-to-digital converters and wired-line communication systems. Designed in a 0.18μm CMOS process and under a supply voltage of 1 V, it consumes approximately 106 μW for a 125 MHz input sinusoidal signal with a 0.8V peak-to-peak swing at mentioned sampling rate using a 2.5 GHz clock signal.**

Keywords—**open-loop, sample and hold circuits, low-voltage multiplexer, high-speed switches**

I. INTRODUCTION

High performance ADCs have many applications in signal processing systems and broadband communication application. One of the important blocks in very high speed ADCs is sample and hold (S/H) circuit where high speed, high linearity and high precision characteristics are simultaneously required. The sample and hold circuits can be substantially divided in two categories; 1) closed-loop S/H and 2) open-loop S/H. The first type of S/Hs and their operations are explained in [1-3] while the second types are discussed in [3-5]. Both types have their own disadvantages and benefits comparing to each other. There is a main trade-off between linearity and speed in both types. Close-loop S/Hs are more linear than open-loop counterpart. Open-loop S/Hs, on the other hand, achieve higher speed, low power consumption and have lower circuit complexity [4]. However, in very high speed applications, linearity will be the main challenge if using open-loop S/H circuits. It can be shown that availability of switches with very low time constant can alleviate the trade-off between speed and linearity in the open-loop S/Hs.

This paper proposed a new and simple open-loop S/H circuit, which employs a low voltage 2:1 MUX to increase the speed and reduce the power. Fundamental limits of switches for S/H circuits are summarized in Section II. In Section III, usual methods for reducing these limitations are described. In Section IV two architectures of 2:1 MUX for high speed applications are compared and their benefits and shortcomings

are addressed. Section V describes the implementation of a novel open-loop S/H using a 0.18μm CMOS process followed by results and discussion provided in Section VI.

II. SWITCH LIMITATIONS FOR S/H CIRCUITS

A. Background

Fig. 1 shows a simple sampling switch consisting of a nMOS transistor as a switch and a holding capacitor. In this switch, when the CLK goes high (switch is on), V_{out} follows V_{in} and when goes low (switch is off) V_{out} remains constant freezing the instantaneous value of V_{in} across C_H.

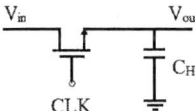

Fig.1. A simple sampling switch

When $V_{in} < V_{DD} - V_{th}$ (assuming CLK is a rail-to-rail signal), this transistor is in deep triode region and exhibits an on-resistance of

$$R_{on} = \frac{1}{\mu_n C_{ox} \frac{W}{L} (V_{DD} - V_{in} - V_{th})} \tag{1}$$

Now, we can represent V_{out} as a function of time for $t > 0$ s as follows:

$$V_{out} = \frac{V_{in}\left\{ \exp\left[(V_{DD} - V_{th} - V_{in})\mu_n \frac{C_{ox}}{C_H} . \frac{W}{L} t \right] + 1 \right\} - 2(V_{DD} - V_{th})}{\exp\left[(V_{DD} - V_{th} - V_{in})\mu_n \frac{C_{ox}}{C_H} . \frac{W}{L} t \right] - 1} \tag{2}$$

This equation reveals that V_{out} eventually reaches to V_{in}.

978-1-4673-2395-6/12 $31.00 © 2012 IEEE

B. Speed Limitations

Switch speed can be defined as the time required for the output voltage to go from zero to the maximum input level after the switch turns on [1]. In this situation, we can represent the speed with a simple time constant τ as follows:

$$\tau = R_{on} . C_H \qquad (3)$$

where R_{on} is the transistor on-resistance and C_H is holding capacitor, as shown in Fig. 1. Note that we can reduce holding capacitor or transistor on-resistance to decrease this time-constant thereby improving the speed.

C. Charge Injection Limitations

When a transistor is on, a finite amount of mobile charge there exists in its channel. When the transistor turns off, this charge will exit trough the source and drain terminals of the device [1]. The total amount of this charge in the inversion layer is

$$Q_{tot} = C_{ox} WL \left(V_{DD} - V_{in} - V_{th} \right) \qquad (4)$$

The charge that injected to the source terminal is stored on the holding capacitor and creates an error in the output voltage (Fig. 2). Assuming that the channel charges are divided equally between the drain and source side, the estimated voltage error can be obtained as

$$V_{error} = \frac{WLC_{ox} \left(V_{DD} - V_{in} - V_{th} \right)}{2C_H} \qquad (5)$$

Unlike to the speed, this equation shows that V_{error} is inversely proportional to holding capacitor and directly proportional to W and L.

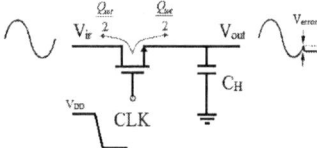

Fig.2. Effect of charge injection assuming the charges are divided equally between both terminals

D. Phase and Amplitude Distortion

Shown in Fig. 3, for defining phase and amplitude distortions, the simple nMOS switch must be assumed as a single-input single-output system. In this situation, the transfer function of the system is as follows:

$$H(S) = \frac{V_{out}(s)}{V_{in}(s)} = \frac{1}{R_{on} C_H S + 1} \qquad (6)$$

Representing this transfer function in amplitude and phase form, we have

$$H(jw) = \frac{1}{\sqrt{\left(R_{on} C_H \omega \right)^2 + 1}} e^{-j \arctan(R_{on} C_H \omega)} \qquad (7)$$

where ω is the angular frequency. As a result, at high frequencies, the switch attenuates the signal amplitude and shifts the input signal phase at the output.

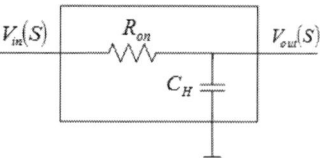

Fig.3. An on-switch as a single-input single-output system

III. USUAL METHODS FOR REDUCING SWITCH LIMITATIONS

Phase and amplitude difference between input and output of simple switch given in (6) and (7) are directly proportional to $R_{on} . C_H$ while switch speed from (3) is inversely proportional to $R_{on} . C_H$. Moreover, charge injection as a representative for sampling precision, from (5), is inversely proportional with C_H. Thus, with decreasing the amount of transistor on-resistance and holding capacitor, we can reduce the speed limitations, and phase and amplitude distortion. However, this will lower the sampling precision. To overcome this problem, we should choose an optimum amount for holding capacitor, and reduce the transistor on-resistance as much as possible. Consequently, improving the charge injection and speed limitations might be accomplished using following three methods:

A. Reducing Channel Length

According to (5), (1) and (3), this issue reduces the charge injection (and therefore enhances the sampling precision), and increases the speed.

B. Employing Differential Architecture

An alternative approach to reduce the effects of charge injection is using differential switch. In this approach, the charge injection would appear as a common mode distortion. As shown in Fig. 4 and from (4), we know that $q_1 = q_2$ only if $V_{in+} = V_{in-}$ [1]. Furthermore, with this method, the effect of clock feed-through due to the gate-source or gate-drain overlap capacitance of the MOS switch phenomenon can be reduced.

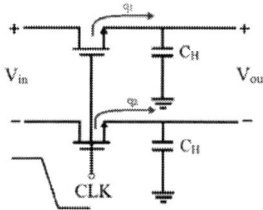

Fig.4. Differential switch

C. Using Dummy Transistor

Simply, a dummy transistor is a transistor that its drain and source terminals are shorted. Assuming the charge that is pushed away from the switch to each side is half of the total charge in the channel, a dummy transistor with a half size of the switch, if turned on when the switch turns off can absorb the emitted charges from switch and prevents them from generating an error-voltage on the sampling capacitor.

IV. 2:1 MUX ARCHITECTURE

High speed operation imposes a MUX that has minimum capacitance at its output nodes. As shown in Fig. 5(a), a tail current (I_{DD}) is switched between two circuit paths through the M5, M6. The CLK command determines whether the tail current steers through M1-M2 or M3-M4. After each transition of CLK, V_{out} will be a new value. It means that when CLK goes high, V_{out} follows V_{in1} and for low levels of CLK, V_{out} follows V_{in2} [6].

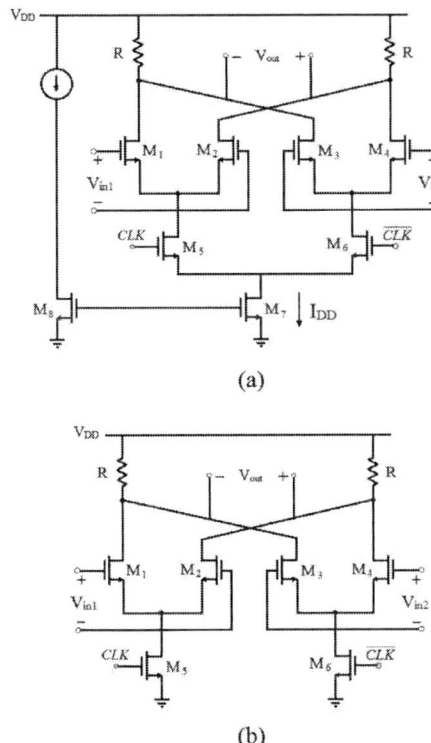

(a)

(b)

Fig.5. Two architecture for a 2:1 MUX (a) a MUX biased with a specific tail current (b) a MUX without a specific tail current

In fact, in Fig. 5(a), a biasing current mirror is used to discharge the output capacitances. Thus, this circuit is not very sensitive to PVT variations. Since in very high speed applications, maximum available speed of technology is needed, an amount of I_{DD} must be chosen such that unity gain frequency of M1-M4 is maximized. However, one of the main problems of this topology is the number of accumulated transistors in a path that limits functionality of this circuit, when the supply voltage is low. Moreover, these stacked transistors limit the output swing.

Fig. 5(b) shows an alternative topology, which is a modified version of the previous topology. In this circuit, the biasing current mirror transistors (M7-M8) are eliminated to reduce the number of stacked transistors. The operation of this circuit is similar to the previous circuit. In this circuit, M5 and M6 operate as switches and work in triode region. The recent topology has some advantages in comparison to the previous topology. By reducing the number of stacked transistors; the maximum voltage swing at the output node will be increased. On the other hand, owing to the fewer number of stacked transistors, the V_{GS} and V_{DS} of M1-M4 will be increased. Thus, according to the relation for transconductance of a MOS transistor

$$g_m = \mu_n C_{ox} \frac{W}{L}(V_{GS} - V_{th})(1 + \lambda V_{DS}) \qquad (8)$$

we can decrease transistor sizes to have the same value of g_m comparing to the previous topology. This means that we have smaller output capacitance and therefore, the latter topology is faster.

V. PROPOSED S/H CIRCUIT

Fig. 6 shows the proposed dual edge sampling S/H consisting of two track and hold circuits and one 2:1 multiplexer. This circuit is designed fully differential to increase robustness against noise and other distortions. In this figure, the transistors where the input signal is connected to their drain terminal work as switch, and the transistors that their drain and source terminals are shorted together, are dummy transistor.

In this circuit when, for example, CLK goes high, M_8 and M_{12} turn on and track the input signal while M_1, M_9 remain off. In other words, when CLK signal has its high level, output of M_1 and M_9 is a held level of input signal and for situations where CLK is low the input will be sampled by M_8 and M_{12}. As a result, by a 2:1 MUX we can choose one of the sampled levels of input signal on each clock edges. In addition, as shown in Fig. 6, when CLK command is high, M_{13} is on and M_{14} is off. Therefore, M3 and M4 pass to the output a sample of input at the output of M1 and M9.

As shown in Fig. 6, at the output of switches there is not any specific holding capacitor. Actually there is not any need to use explicit holding capacitors with large values. In fact, parasitic input capacitances of MUX stage were utilized as holding capacitors.

VI. SIMULATION RESULTS

An up to 5 GS/s dual edge triggered sampling S/H was designed in a 0.18µm CMOS technology under a 1V supply. Simulations for the MUX circuits were done under equal power consumption and equal transconductances for M1-M4. A differential 2.5GHz clock signal (CLK) was utilized and a sinusoidal 1.25GHz input signal was applied to both inputs of each MUX circuits. Fig. 7 reveals that the circuit of Fig. 5(b) achieves higher bandwidth in comparison to the circuit of Fig. 5(a).

Simulations of proposed S/H circuit under a 1 V supply voltage shows approximately 106 µW power consumption and

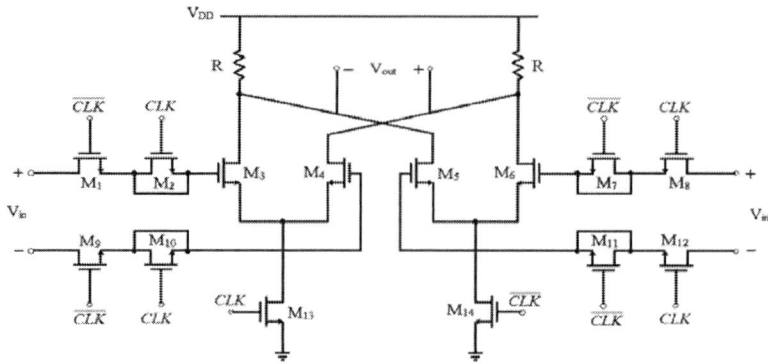

Fig. 6. Circuit schematic of the proposed S/H

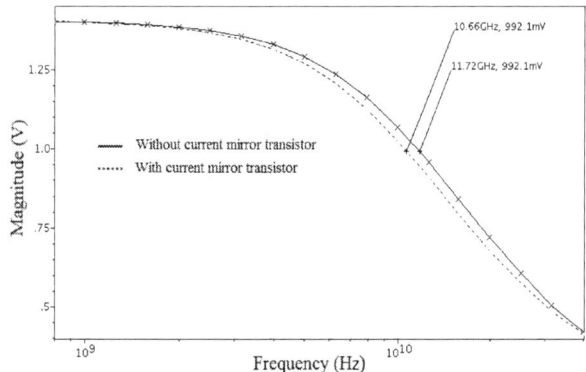

Fig.7. Comparison of 3dB bandwidths of both MUX architectures.

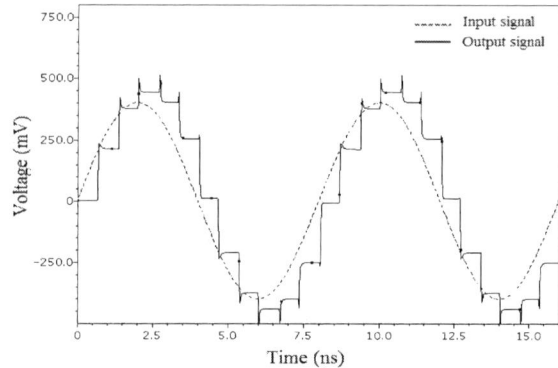

Fig.8. Input and output waveforms of proposed S/H

a differential signal swing of 0.8 V_{p-p}. To examine the operation of this circuit, a differential mode 125 MHz, 0.8 V peak-to-peak sinusoidal wave was applied to the inputs and then sampled at 1.5 GS/s sampling rate. Fig. 8 shows the input and output waveforms and Fig. 9 shows the spectrum of the output revealing a 34 dB SFDR. To justify the circuit operability at higher frequencies, input and output of the circuit for a 1.25GHz sinusoidal input and sampling rate of 5GS/s are shown in Fig. 10. Consequently, this circuit can be used in very high speed applications such as wired-line transceivers and other high speed communication systems.

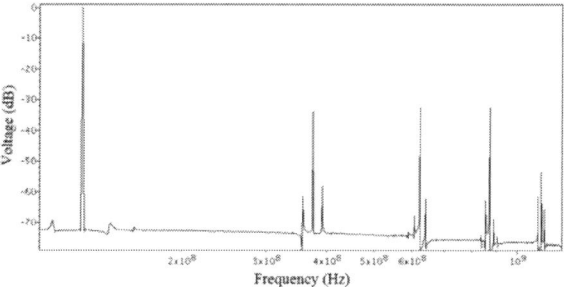

Fig.9. Proposed S/H output spectrum for 125 MHz input sampled at 1.5 GS/s.

REFERENCES

[1] B. Razavi, Design of analog CMOS integrated circuits vol. 212: McGraw-Hill Singapore, 2001.

[2] T. S. Lee, et al., "A very-high-speed low-power low-voltage fully-differential CMOS sample-and-hold circuit with low hold pedestal," in IEEE Int. Symposiom on Circuits and Systems Conf. (ISCAS), 2005, pp. 3111-3114 Vol. 4.

[3] M. Mousazadeh, et al., "A novel open-loop high-speed CMOS sample-and-hold," AEU-International Journal of Electronics and Communications, vol. 62, pp. 588-596, 2008.

[4] K. D. Sadeghipour, "A new passive sample and hold structure for high-speed, high-resolution ADCs," AEU-International Journal of Electronics and Communications, 2011.

[5] A. Shirazi, et al., "Linearity improvement of open-loop NMOS source-follower sample and hold circuits," Circuits, Devices & Systems, IET, vol. 5, pp. 1-7, 2011.

[6] B. Razavi, Design of integrated circuits for optical communications vol. 1: McGraw-Hill New York, 2003.

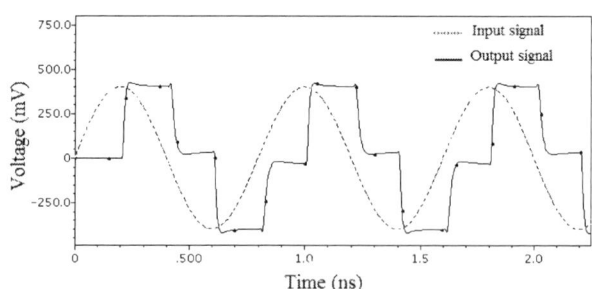

Fig.10. proposed circuit waveforms for 1.25 GHz input frequency and 5 GS/s sampling rate

978-1-4673-2395-6/12 $31.00 © 2012 IEEE   648

Development of Capacity Indices for Semiconductor Fabrication

Mohd Azizi Chik, Kader Ibrahim, Mohd Hazmuni Saidin, Faizah Md Yusof, G. Devandran & *U. Hashim
Silterra Malaysia Sdn. Bhd
Lot 8, Phase 2, Kulim Hi-Tech Park,
0900 Kulim, Kedah, Malaysia.
mohd_azizi@silterra.com
*Institue Nano Electronic Engineering (INEE)
University Malaysia Perlis,
Lot 106,108,110 Tingkat 1, Blok A, Taman Pertiwi Indah,
Jalan Kangar-Alor Setar, Seriab, 01000, Kangar, Perlis
uda@unimap.edu.my

Abstract – A typical semiconductor fabrication process contains 300 to 1000 steps and its variation depends on the product complexity. Most of the processes are re-entrance to same equipments especially at photolithography, etching, implanter, film deposition, chemical mechanical polishing (CMP) and cleaning. For example, photolithography steps for island, poly, and contact module will be processed at same equipment. Another complication is that the equipment types are different from one to another resulting in different approach for cycle time calculation, difference in availability and efficiency. The objective of this paper is to establish capacity indices to guide for monthly output in semiconductor fabrication facilities. The approach in this paper is using the waterfall chart for individual process and equipment types. Data extraction is being done through reporting systems of Advance Productivity Family (APF), an industrial standard software for data collection that is integrated with individual equipment and product processing historical data. The data was then analyzed using JMP to check for sanity. Results were used to develop capacity indices, which are wafer per hour (wph), manufacturing efficiency, and equipment availability. All these information will be later used to develop the final capacity figure. The final capacity number will then be used to guide the planning team to schedule product combination that will achieve monthly and quarterly wafer shipment goal to customers. This approach reached accuracy of 99% compared to actual throughput. In conclusion, this approach helps the company to provide planning guidelines in meeting the financial goals.

Index Terms—Advance Productivity Family (APF), Chemical mechanical polishing (CMP), Work In progress (WIP), wafer per hour (wph)

I. INTRODUCTION

Capacity is the main problem in semiconductor fabrication foundries and it always has challenges to meet the on time delivery to the customers [1]. The complexity happens when a foundry runs various product mix, due to the demand from customers with short cycle time [2]. Normally, in a semiconductor fabrication foundry the Overall Equipment Effectiveness (OEE) for the most constraint equipment is considered the built in capacity of that foundry [3]. OEE is used as a benchmark among the foundries in the world. OEE components can be classified as equipment availability, equipment performance and equipment quality. OEE components such as Scheduled Downtime, Unscheduled Downtime and Engineering Time will contribute to Equipment Availability. Situation when the equipment is utilized or unutilized, or utilized but not making any output will be considered in the Equipment Performance part of OEE. Scrap or rework from the equipment will fall into the Equipment Quality segment. Among the OEE segmentation, the equipment performance is the most important and difficult to determine. One of the factors that contribute to the Equipment Performance loss is the speed lost. Speed lost is hidden by many factors such as deterioration of robot movement, small lot sizes, inefficient cascading, integrated system failures, equipment alarms and others. OEE is also used to compare equipment to equipment performance, with the same manufacturer, model and configuration. The best equipment will then be used as the "golden equipment". Other equipments will be benchmarked on every single set-up with the golden equipment, resulting all equipments to perform at the best level.

Silterra started the OEE improvement initiative in 2008. The initiative was part of Silterra's quest for continuous

Improvement. This improvement contributed to the capacity gain which resulted in higher output. Our first aim was the most expensive and constraint equipment in the Fab, photolithography patterning equipment marked as DUV [4]. In 2008 the OEE for DUV group consisting of 7 DUV equipments was only 65.1 %. The inefficiency was due to speed loss at 19.4%. The team deep dived to understand the speed loss and observed variations in performance between similar equipments such as improper cascade level of the production lots, variation of equipment recipe and variation of robot movement. OEE result for DUV is summarized in figure 1.

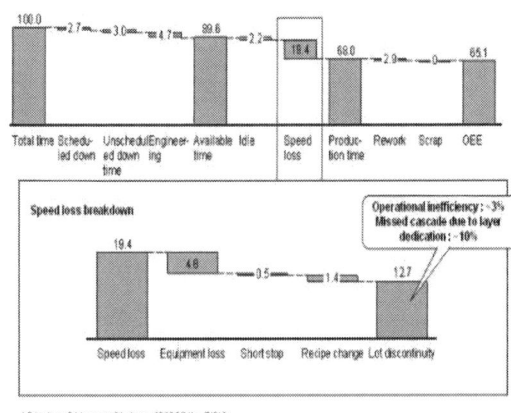

Fig. 1. Example of OEE Study for DUV Photolithography Equipments In Silterra.

II. METHODOLOGY

This research is based on actual 200mm semiconductor fabrication facility simulated using AutoSched AP simulation software. Fundamental information required in establishing the indices are product information, equipment information and the product mix [5]. The product information is the process plan, related steps and step cycle time. The equipments information consists of type of equipment, count, availability, efficiency and WPH [wafer per hour, throughput]. The product mix is the breakdown of the type of products and volume. Based on these informations, the capacity can be derived as below:-

- WPH * Equipment Count * Availability % * Efficiency % = Daily Capacity Moves
- (WSPW * Weighted Processing Steps) / 7 days = Daily Moves Required (WSPW is Wafer Starts Per Week)
- Daily Moves Required / Daily Capacity Moves = BNI (Bottle Neck Indices).
- BNI interpretations, if BNI 1.0; the demand is same as the capacity, if the BNI is < 1.0; the

demand is lower than the capacity and if the BNI is > 1.0; the demand is higher than capacity

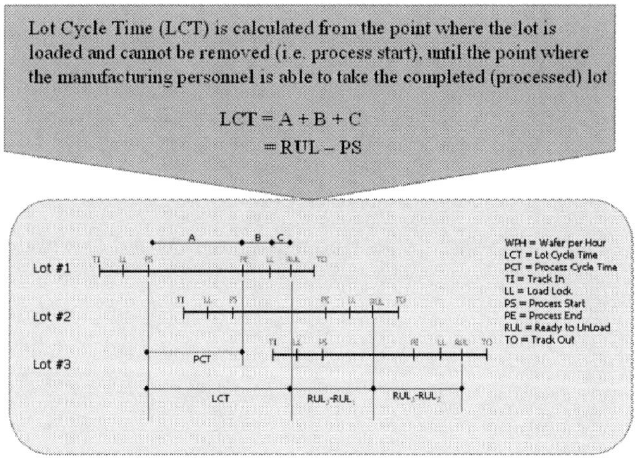

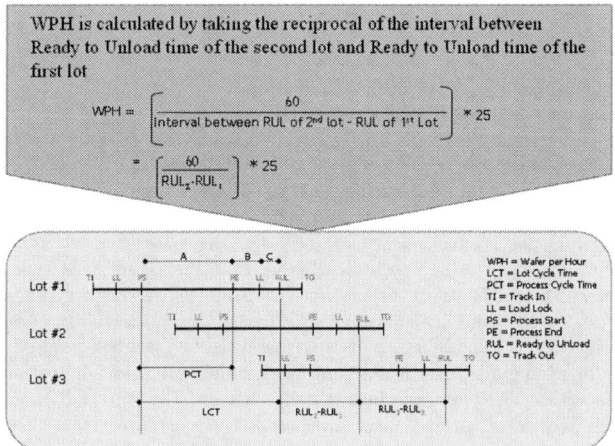

Fig. 2 Lot Cycle Time (LCT) Diagram

Fig. 3. WPH Calculations Diagram

The key input parameters for WPH calculation, the activity events and event time of equipments are extracted from equipment log files via customized software (ELGS, Electronic Logging Sheets). The information's from ELGS is available to end users for WPH and LCT [lot cycle time] analysis via Unix/windows servers. APF and JMP analysis software were used to validate and establish WPH and LCT indices.

The equipment availability, which is the key input for capacity indices. The equipment availability definition is based on SEMI E10 Standard.

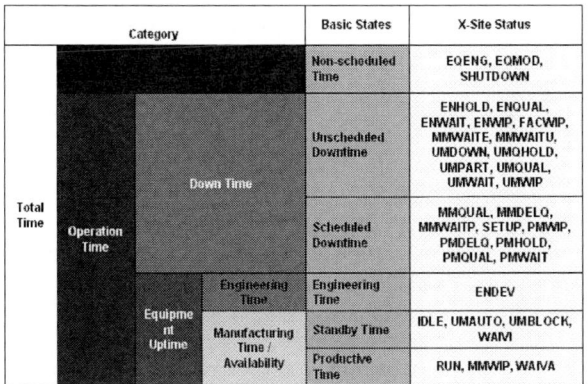

Fig. 4. Source: SEMI E10-0699 [3]

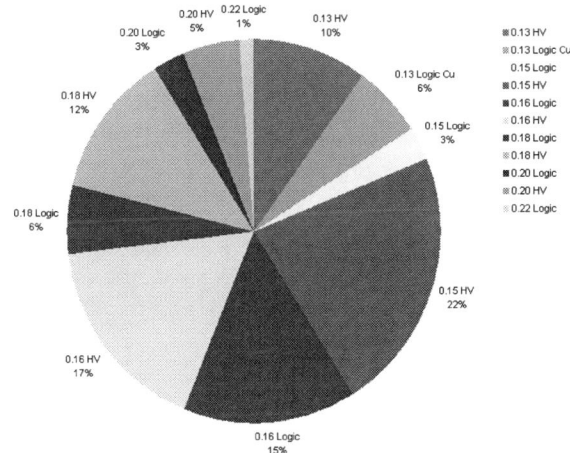

Fig. 5. Product Mix Chart Sample

One of the key challenges in capacity indices methodology is the product mix. The common challenge in product mix is product to product process step variability. This variability causes frequent changes of setup or running condition of the affected equipment; this directly affects the manufacturing efficiency of the affected equipment. The above chart is sample product mix summary by two main groups HV (high voltage) and Logic for each product.

III. DATA ANALYSIS

APF extracted data from EI [equipment interface] server in raw data format. These data are given necessary treatment to transform them into valuable information. Those data were analyzed using the combination of Microsoft Excel, JMP statistical software and Minitab statistical software.
Equipment cycle time and wafer per hour (WPH) distributions are plotted using statistical software. Any outliers in the distribution chart will be removed before further analysis is done to determine the best overall cycle

time and WPH. Figure 6 shows example of WPH distribution from a thin film deposition equipment

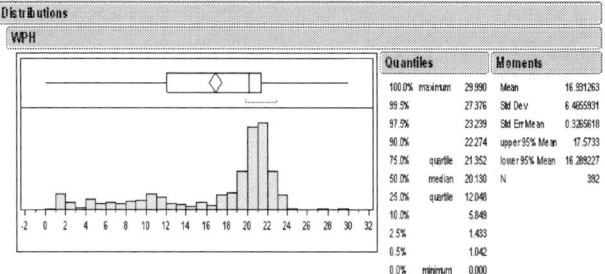

Fig. 6. WPH distribution from a thin film deposition equipment

Equipment count to process a step increases with the increase in wafer demand. In this case, it is critical for the same equipment type to have the same WPH. However variation may occur as a result of different equipment parameters and hardware setting. This variation will be analyzed by process and equipment engineering group to ensure continuous improvement program is alive.

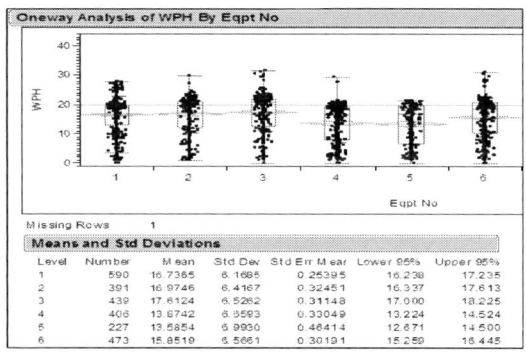

Fig. 7. Box plot showing equipment to equipment WPH variation

Besides equipment to equipment variation, ELGS data availability down to chamber level, has enabled the analysis of WPH increment with additional 1 chamber and 2 chambers added to an equipment. Example of this analysis is shown in Figure 8. From the analysis, Equipment X WPH does not increase linearly as additional chamber added to the main platform. This analysis can help to drive further analysis to determine the actual bottleneck unit in the equipment.

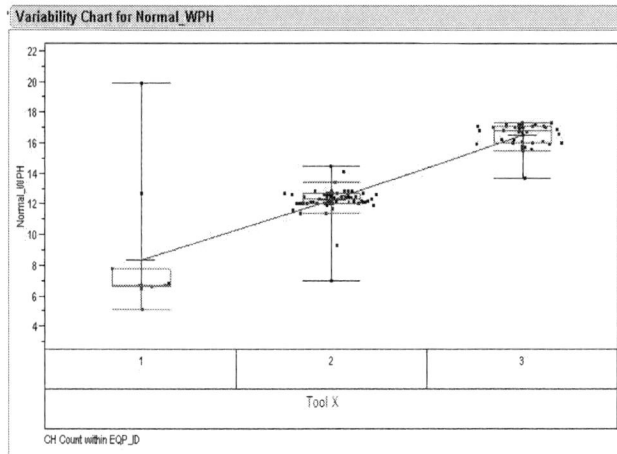

Fig. 8. WPH increment by adding up chambers at Equipment X

Availability analyzed from 3 months historical data is shown in figure 9.

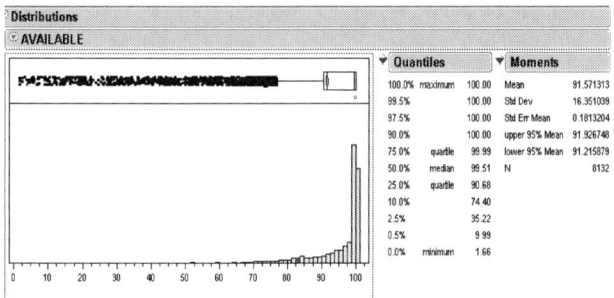

Fig. 9. Historical 3 months distribution of a lithogprahy equipment availability

Overall Equipment Efficiency (OEE) is carried out on an equipment group to determine the efficiency. The OEE result is then summarized using a waterfall chart that clearly shows the contribution of availability, performance and quality as shown in figure 10.

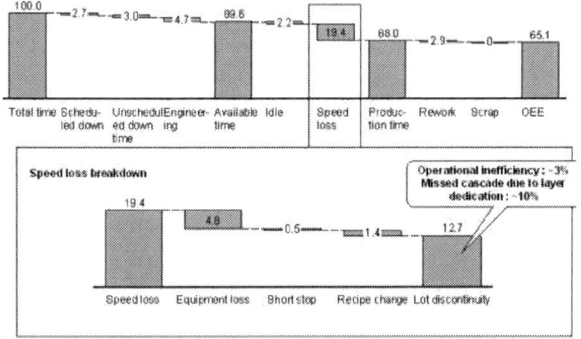

Fig. 10. OEE waterfall chart of a lithography equipment

To determine if the existing equipment sets are capable of meeting the customer demand - static and dynamic capacity analysis are made by inputting the WPH, cycle time, availability and efficiency figures in the model.

IV. CONCLUSION

Large variability of product mix in semiconductor fabrication requires capacity planning to support manufacturing [5]. This will impact the semiconductor manufacturing performance and support. Semiconductor manufacturing companies will allocate million of dollars for equipments purchase during ramping up or technology change. Typically the rate of change in products and technology create difficulty for companies to have good estimate of future equipment requirements. In order to achieve monthly and quarterly target, it is critical for a company to have precise capacity number as a guide. This help companies to have proper plan on the product mix with optimum cost and revenue.

This is unique and helps save time for data collection ahead by more than 6 months. Furthermore, this helps Industrial Engineering team to provide planning guidelines for the company.

In conclusion the accuracy for planning is now at 99% compare to actual. Figure 11 shows the actual monthly fab utilization which shows that the planned Fab-capacity is on par with the actual Fab-capacity.

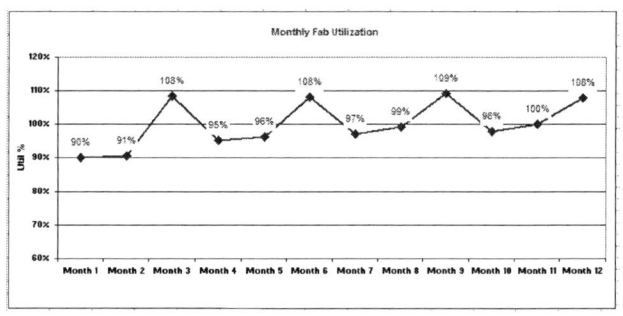

Fig. 11. Monthly Fab Utilization

ACKNOWLEDGEMENT

Authors would like to thank Silterra Managements and Industrial Engineering, for their continued support in this research.

REFERENCE

[1] Mohd Azizi Chik, Mohd Hazmuni Saidin and Uda Hashim, "Industrial Engineering Roles in Semiconductor Fabrication,", *11ᵗʰ Asia Pacific Industrial Engineering & Management Systems Conference,* 2010.

[2] Mohd Azizi Chik, Ve Chun Yung, Puvaneswaran Balakrishna, Uda Hashim, Ibrahim Ahmad & Bashir Mohamad "A study for Optimum Productivity in 0.16μm Products in Wafer Fabrication Facility," *IEEE*

International Conference on Semiconductor Electronics (ICSE), 2010, pp377-380.

[3] SEMI E10-0699 Standard for Definition and Measurement of Equipment Reliability, Availability, and Maintainability (RAM), May 3rd 1999

[4] Kader Ibrahim, Mohd Azizi Chik and Uda Hashim "Managing Demand Variabiliy to Achieve Optimum Cost and revenue in Wafer Foundry" *11th Asia Pacific Industrial Engineering & Management Systems Conference,* 2010.

[5] Wen-Chih Chen, Shu-Hsing Chung and Chih-Wei Lai "Determine product family mix and priority mix for semiconductor fabrication considering demand variation," *International Conference on 40th Computers and Industrial Engineering (CIE)*, 2010, pp 1-6.

Oxidation on Copper Lead Frame Surface Which Leads to Package Delamination

Lai Chin Yung, Lee Chai Ying, Cheong Choke Fei, Aw Tiam Ann and Soellner Norbert
Infineon Technologies (Malaysia) Sdn. Bhd.
Email : ChinYung.Lai@infineon.com

Abstract - **Based on several studies, package delamination caused by copper oxide is one of the most common IC packaging defects detected in backend assembly processes. The possible root causes for the copper oxide formation could be due to leadframe surface oxidation, incompatible process temperature and ineffectiveness of surface cleaning prior to molding. In order to effectively investigate the contribution of copper oxide formation to package delamination, this study have been initiated with the application of relevant reliability stress tests and surface interface analytical techniques. Reliability stress test of HTS was carried out to accelerate the copper oxide formation after molding and the delamination occurrence being monitored by SAM scanning after defined stress test intervals. To further on the study, a detailed elemental low KeV EDX analysis was performed to assess the delaminated interfaces after mechanical decapsulation of the failure units. In addition, AES analysis including depth profiling was also applied to determine the copper oxide layer. A good correlation of copper oxide formation to package delamination was demonstrated and inappropriate temperature used in wirebonding clamping was identified to be the most probable root cause of the package delamination.**

INTRODUCTION

Copper material is widely used for leadframe material in intergrated circuits packaging nowadays due to its superior electrical performance, high thermal conductivity and lower cost [1]. However, copper is susceptible to thermal oxidation when exposed to elevated temperature in semiconductor manufacturing processes [2]. With the copper oxide formation on the leadframe surface, poor adhesion at the interface of copper leadframe and EMC had been reported in several studies [1, 2]. Yoshioka et al [3] reported poor adhesion of copper oxide to mold compound that caused corrosion due to the moisture ingress through package from external. Kim [4] studied the adhesion properties between copper oxide and mold compound at the copper oxide surface. Therefore, the understanding of the copper oxide formation mechanism is important to ensure good adhesion quality of the interfaces that assure the reliability of a semiconductor.

According to Ying Zheng, there are two stages of copper oxidation mechanism as shown in the Figure 1. In the first stage, a thin layer of Cu_2O will readily be oxidized on the surface of copper bulk at ambient temperature. Later, the CuO will slowly form from the first stage by interdiffusion of oxygen into the oxide [5]. In this study, a series of analysis methods were applied in order to identify such copper oxidation in PSSO package after delamination detection at package robustness screening of the new developed product.

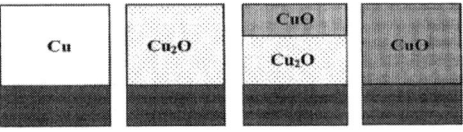

FIGURE 1: Copper oxidation mechanism

ANALYSIS METHODOLOGY AND

EXPERIMENT DESIGN

2.1 Scanning Acoustic Microscopy (SAM)

Sonoscan D9000 SAM machine was used to check the package robustness performance before subjecting to reliability tests for product qualification. 3 interfaces of the package were examined by using 30MHz transducer. The interfaces are EMC/die top, EMC/ die pad & EMC/lead as shown in Figure 2.

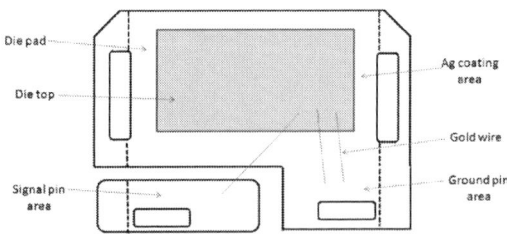

FIGURE 2: Schematic that shown the 3 interfaces: 1: EMC/die top; 2: EMC/die pad; 3: EMC/lead.

2.2 New Developed Decapsulation Method: Laser + Mechanical Decapsulation

In order to maximize the probability of preserving the suspected contaminants that might cause package delamination, chemical decapsulation was excluded in this study. This is because acids used in chemical decapsulation to etch away the EMC will clean the contamination evident at the same time. Therefore, mechanical decapsulation was selected and this method was combined with laser ablation as the first step to ease the later mechanical decapsulation.

The model of laser ablation machine used was Rofin. Laser ablation current, voltage and the repeating time are some parameters that need to be considered for not to overly thin the sample. Once the opening reached the area near to interest, the sample was then placed on a customized mechanical decapsulation tool. With the correct positioning to the intended separation interface, clamping force would be applied manually and slowly to the sample until the separation was achieved. Figure 3 gives the separated pieces of die pad and EMC after the mechanical decapsulation.

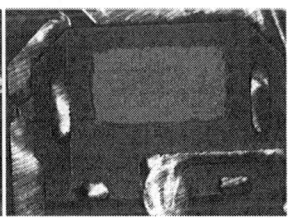

FIGURE 3: PSSO package separation between copper leadframe and mold compound after laser + mechanical decapsulation.

2.3 Low KeV EDX Analysis

Energy dispersive X-ray or EDX was used to find out some hints of the delamination elements. With the conscious that the contamination was only on the surface, EDX analysis with low KeV set-up was recommended to get shallow electron beam penetration into the sample surface. The details impact with different KeV is demonstrated in Figure 4 at results and discussion section.

2.4 Auger Electron Spectroscopy (AES)

Auger electron spectroscopy was applied to further assess the elements on the top most surface in the range of 3-5nm region near the surface with 0.1-1% higher detection sensitivity than EDX. Additional advantage of AES is its depth profiling function that provides the layers information of the interested elements detected on the surface. Ion Argon source was used for surface sputtering in depth profiling, with the rate of 8nm/s using

SiO2 standard. The model used was PHI and energizer used was in the range of 10kV.

2.5 High Temperature Storage Test (HTS)

In order to accelerate the failure of delamination in the package, HTS at 200°C for 168 hours was used. Espec oven was applied instead of the common 175°C to shorten the reliability stress time. As reference to mold compound supplier that HTS 200°C is an accelerate stress condition to surface out delamination defect in short cycle time.

RESULTS AND DISCUSSION

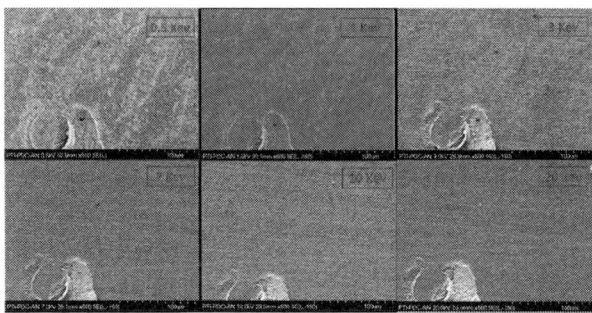

FIGURE 4: Impact of applying different kV on same analysis surface. High accelerated voltage used can cause deep penetration of electron beam down to the sample and covered the weak signals from the surface, which is not wanted in this analysis.

In order to determine the proper kV to be used in the analysis, the famous Kanaya-Okayama equation was used to estimate the electron beam penetration depth of the material surface as given in the Equation 1.

$$H = \frac{0.0276 \, A \, E^{1.67}}{(Z^{0.89} D)} \, um$$

(1)

A = atomic weight, g/mole, D = density, g/cm^3, z= atomic number, E = accelerating voltage, H = penetration depth (μm).

The estimation of the penetration depth can be done easily nowadays by using the Monte-Carlo software. The inputs needed are Line of Interest and the kV (Accelerating Voltage) for the penetration depth estimation as shown in Figure 5.

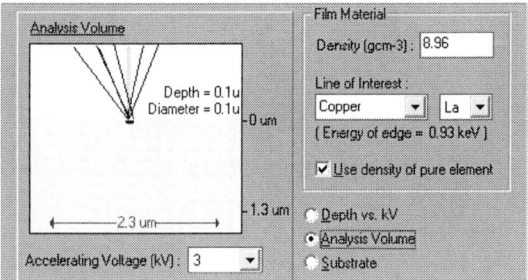

FIGURE 5: Interaction volume activated by 3kV accelerating voltage on copper bulk material.

A few kV were assessed and based on the Figure 6, 3kV was selected for this analysis to assess the surface contaminant elements on the delamination area. The penetration depth is about 0.1μm.

Penetration depth Vs Accelerating voltage

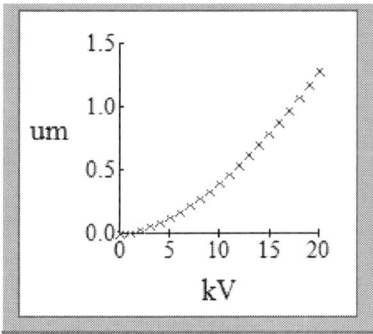

FIGURE 6: Graph of penetration depth vs kV applied for Cu element.

3.1 C-SAM Results after HTS defined intervals

The package was first assessed by SAM analysis in C-mode to check for whether there was any delamination or not in the package. Figure 7 shown the delaminated interface between signal pin and mold compound is shown below. The delamination signal that shown in red surrounded by yellow was then be confirmed by the waveform capture in negative signal in point 4. Another reference signal of non-delaminated area was captured the waveform as well in point 3 that shown positive signal as expected. The negative waveform is due to the impedance difference of the EMC to the air/gap and the same principle applied to the positive signal in vice versa.

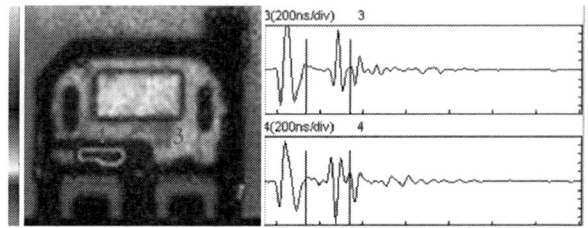

FIGURE 7: SAM analysis observed negative signal shown in red color indicating the delamination in the interface of EMC to signal pin.

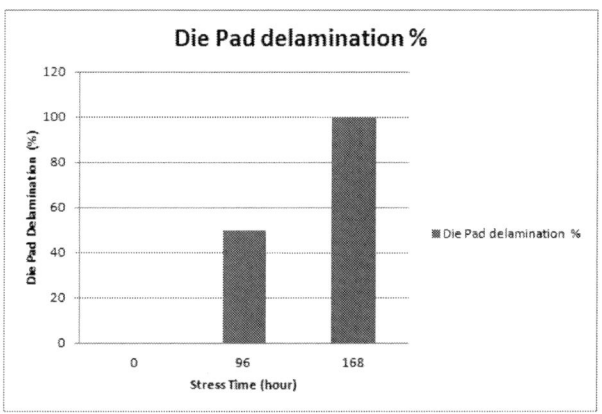

TABLE 1: Shown an example of delamination % Vs HTS 200°C at different stress time interval.

3.2 Laser + Mechanical Decapsulation Observation & SEM/EDX Results

After the laser + mechanical de-capsulation, the package was separated into 2 pieces. One of the pieces was the diepad with some remnants of glue left and the stitch of ground bond. The other piece was the separated EMC with embedded wirebonded die. Further zoom in optically at the delaminated area of the diepad, a stretch of brownish contaminant observed at the edge of the ground pin (Figure 8).

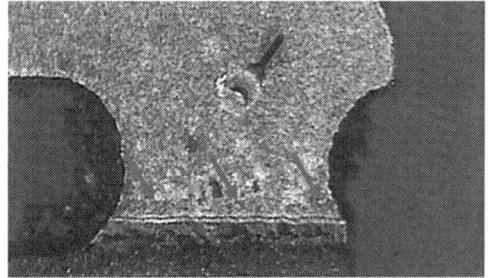

FIGURE 8: Mechanical de-capsulation interface at signal pin area shown a brownies color due to WB clamping effect.

Later the decapsulated unit was brought to SEM inspection that revealed the rough surface morphology of the contaminated area suggesting metal oxidation that caused the package delamination Figure 9. Effort of mapping the delamination position to the manufacturing processes gives the wirebond clamping to be the most probable process to cause such package delamination. The investigation was later continued with EDX point ID analysis at the brownish area (Point A) shown significant peaks of copper and oxygen Figure 9. EDX line scan analysis across the delamination had shown overlapping elements of copper and oxygen that confirmed the copper oxidation formation at the delamination area as shown in Figure 9 (c).

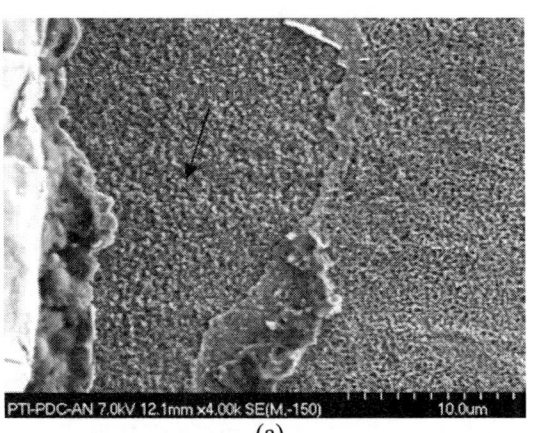

(a)

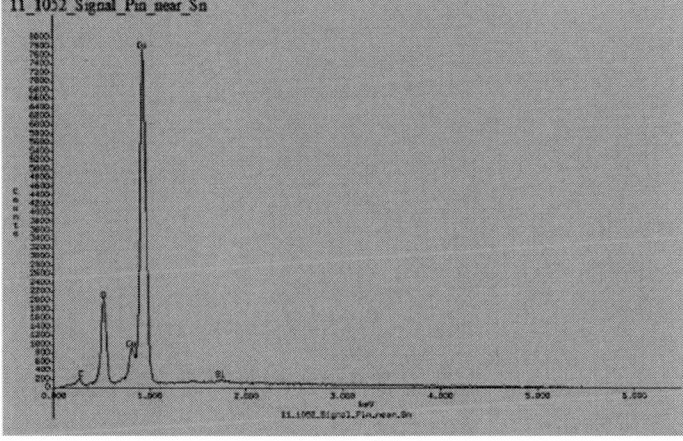

(b)

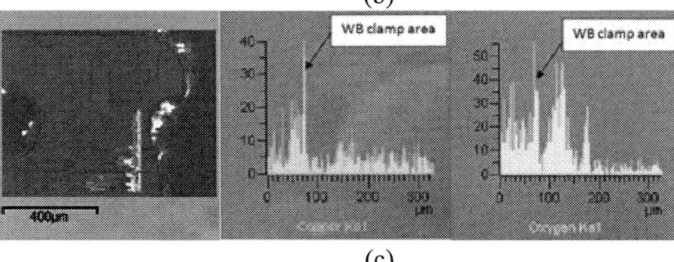

(c)

FIGURE 9: (a) SEM image of the delamination shown to have rough surface morphology; (b) EDX Point ID analysis at Point A

(c) EDX line scan at signal pin area that observed significant peaks of copper and oxygen at the brownish area.

To further confirm the EDX linescan results of delaminated area, a linescan analysis also applied at a reference of non-delaminated area, the ground pin. The EDX results shown that low oxygen peak intensity observed as compared to the signal pin Figure 10. With all these results, it is explainable that WB clamping area with high temperature exposure at long heating time would increase the probability of copper oxidation that caused delamination at this interface. In some of the marginal cases, WB clamping that later caused copper oxidation might not be easily shown up after completing molding process. Therefore, in order to accelerate the copper oxide formation, HTS test with 200 degree C was carried out.

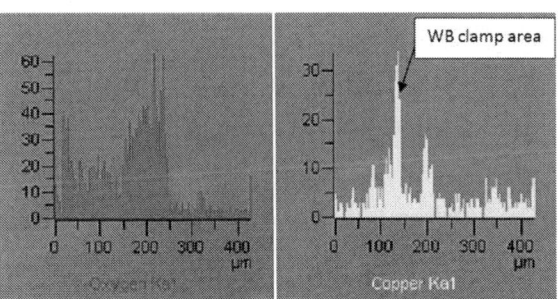

FIGURE 10: EDX line scan at Ground pin area and low oxygen signal peak observed at WB clamping area.

3.3 Auger Electron Spectroscopy Results

The investigation of the probable root cause finding continued in the vertical direction after all the analysis done in lateral. In AES analysis was carried out to find out the thickness of the copper oxide. Figure 11 shown the AES spectrum obtained from the copper oxide area at as-received condition. High copper and oxygen peaks were observed once again that confirmed the possibility of copper oxidation. The high carbon of 59% atomic percentage was believed contributed from the EMC residues from the mechanical decapsulation. In the AES depth profiling (shown in Figure 12) up to 10 min sputtering with in Argon ion, the copper oxide thickness was estimated to be 200-230nm. This thickness is correlated well to the low kV results that shown in the earlier section.

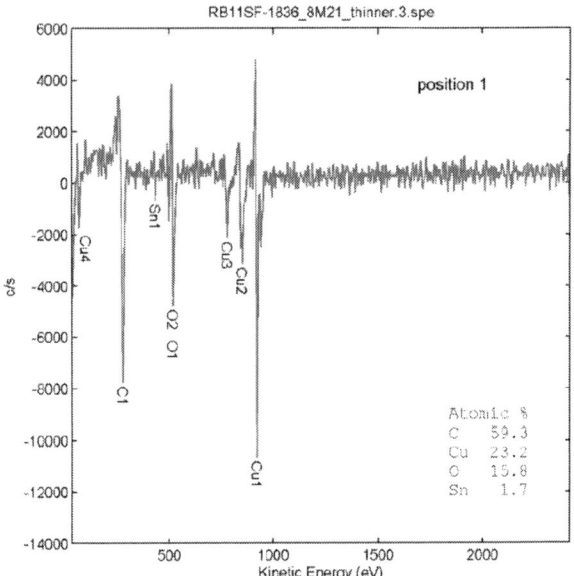

FIGURE 11: AES spectrum on WB clamping area at 0 second sputtering.

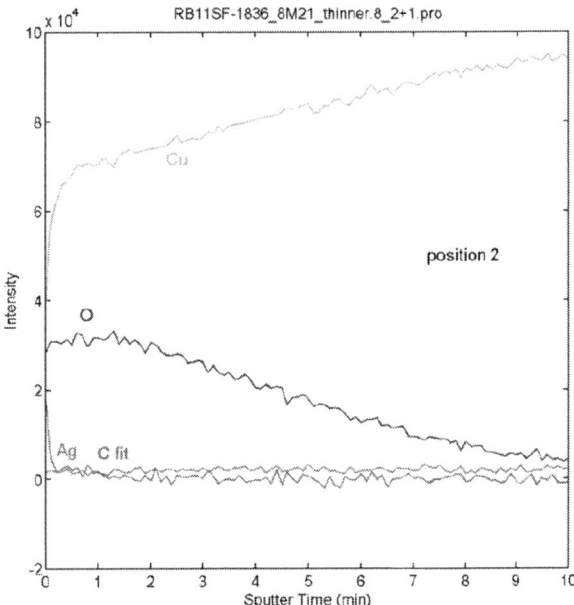

FIGURE 12: AES depth profile spectrum on WB clamping area observed a thick Cu oxide.

CONCLUSION

Careful and systematic analytical approach is important in ensuring the preservation of evident in surface analysis. As shown in this investigation, several analytical methods were complementally used to find out the most probable root cause of

copper oxidation that leads to package delamination. The cause of the copper oxide shown a good correlation to the improper temperature used in the clamping area during the wirebonding process.

ACKNOWLEDGEMENT

The authors would like to take this opportunity to thank the support from PSSO development team for the sample build and the FA support from Regensburg team, especially to Mr. Renner Frank for his detailed explanation on AES analysis.

REFERENCES

[1] Jiaqing Xi et.al, Effects of Trace Element Ag on Oxidation Failure of Copper Lead Frame. 7th International Conference on Electronics Packaging Technology, 2006.

[2] Soon-Jin Cho et.al, The Effects of the Oxidation of Cu-Base Leadframe on the Interface Adhesion Between Cu Metal and Epoxy Molding Compound. IEEE Transactions on Components, Packaging, and Manufacturing Technology, Vol 20, 1997.

[3] Osame Yoshioka, Norio Okabe Ryozo Yamagishi, Sadao Nagayama, Gen Murakami, " Improvement of Moisture Resistance in Plastic Encapsulants MOS-IC By surface Finishing Copper Leadframe", E1CTC' 89 pp.464-471.

[4] Samuel Kim, "The Role of Plastic Package Adhesion in IC Performance", ECTC'91 pp.750-758.

[5] "ASM handbook corrosion" 9th ed. Vol. 13, (ASM international, 1978), pp 610-640.

Measurement and Characterization of Hot Carrier Safe Operating Area (HCI-SOA) in 24V n-type Lateral DMOS Transistors

N. Soin, S.S. Shahabuddin, *Member, IEEE* and K.K. Goh

Faculty of Engineering, University of Malaya, 50603 Kuala Lumpur, Malaysia

Silterra (M) Sdn. Bhd., Kulim Hi-Tech Park, 09000 Kulim, Malaysia

Email: sharifah_shafini@silterra.com

Abstract - **Due to strong demand on smart-power technologies, LDMOS device has been widely used because of its compatibility with standard CMOS process. For most fabrication companies, this is an attractive reason to extend the existing technology applications. However, the device is vulnerable to hot carrier (HCI) damage. SOA in HCI is one of the major criteria to address the concern in LDMOS device design. There are a lot of reliability studies focusing on the HCI failure mechanism but not many literatures available on SOA characterization. This paper will focus on the HCI-SOA test methodology and characterization for n-type LDMOS device by adopting the conventional CMOS HCI test method. This will be a useful guideline for the industry on the device characterization process.**

I. INTRODUCTION

Lateral DMOS transistors are widely used in various smart-power applications, as it can be easily integrated within existing CMOS technologies without significant process changes [1,2]. Due to the high drain voltages being applied to these LDMOS, these devices are more sensitive to hot carrier degradation than low voltage transistors. As a result, new degradation mechanisms appear due to the architecture of these devices [3-6]. This has become a serious concern with regard to reliability.

One of the criteria to address reliability concern in LDMOS devices is Safe Operating Area (SOA), which can be described as the operating conditions of a device in which the device can be expected to operate without self-damage [7]. SOA is usually presented in device datasheets as a graph or a table. The area under the curve is referred to as the safe area. The SOA of LDMOS transistors cover a wide span in which different phenomena may lead to failure [8]. It is an important criterion for device and circuit designers. In general, the SOA boundary is specified by maximum voltage above which a mechanism such as avalanche breakdown which will lead to loss of electrical control. In the long term, SOA is ultimately limited by hot-carrier injection that is slowly degrading the current characteristics.

In this paper, the focus will be on the test methodology to determine SOA in HCI for n-type LDMOS transistor. The test characterization is established using conventional n-type CMOS HCI test procedure as a reference. But due to the differences in device architecture between both devices, the degradation mechanism will be varied. Thus, it is necessary to validate the degradation and lifetime models for the n-type LDMOS transistor using n-type CMOS transistor.

Power LDMOS for smart applications typically operate under two static bias conditions (ON/OFF) state. These requirements must be taken into account and therefore, the linear drain current ($I_{D,lin}$) will be selected as monitoring parameter to obtain SOA lifetime plot.

II. DEVICE DESCRIPTION

The schematic cross section of the n-type LDMOS transistor used in this paper is shown in Fig. 1 illustrating Lch and Ldr region. The device features an STI in n-drift region near the drain with buried body implant. This device is integrated into a 0.18μm CMOS compatible process with self-aligned gate on the source side and the drain side. The gate-length and gate-width of the device are 0.4μm and 20μm, respectively. The gate oxide is 115-angstrom. The device operational voltages are 24V for V_{DS} and 5V for V_{GS}. The STar Scorpio hiVIP HCE equipment is used to perform hot carrier stress test. Constant voltage stressing for a large set of V_{DS} - V_{GS} condition and is performed at room temperature with the source and substrate connected to ground. The stress is applied until the targeted failure criterion or maximum 100 hours. At selected time intervals, the device electrical parameters are measured to monitor the degradation of the device. In this work, the focus will be on the drain current performance; i.e. linear current ($I_{D,lin}$ @ V_{DS} = 0.1V and V_{GS} = 5V) and saturation current ($I_{D,sat}$ @ V_{DS} = 24V and V_{GS} = 5V). Hot carrier failure criterion is fixed at maximum 10% of electrical parameter shift.

III. EXPERIMENTS

A. Test Characterization

By adopting the conventional n-type CMOS HCI test method as a reference, a sufficiently large set of V_{DS} - V_{GS} condition is determined to obtain n-type LDMOS SOA. In conventional n-type CMOS device, the gate voltage stress is generally determined at peak substrate current which usually produce maximum degradation [9]. Equation (1) shows the Substrate-Current model

978-1-4673-2395-6/12 $31.00 © 2012 IEEE

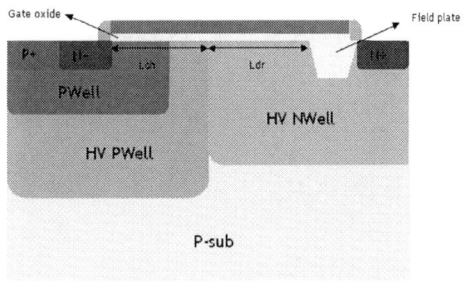

Fig. 1. Schematic cross section of the n-type LDMOS transistor used in this paper. The length of channel region (Lch) and the length of drift region between channel and STI (Ldr) are illustrated in the figure.

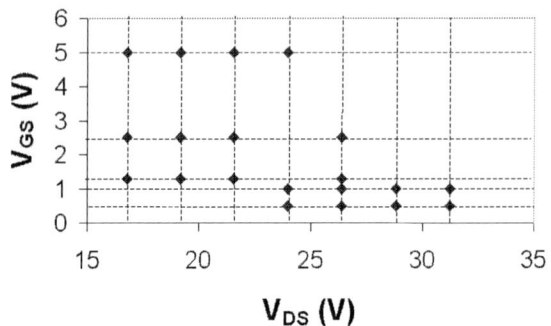

Fig. 2. Test conditions for the DC HCI-SOA measurement and characterization. For every stress condition (indicated by a dot), several samples were stressed.

for HCI lifetime.

$$t_{TAR} = C(I_{sub}/W)^{-m} \qquad (1)$$

where t_{TAR} is the device lifetime and I_{sub} is the device substrate current. The parameter C and m in (1) are the lifetime acceleration factor and exponent, respectively. In conventional n-type CMOS, the lifetime exponent m is a function of the amount of energy required to create an interface trap divided by the impact ionization energy and typically has a value of 3 [6].

However, the substrate current characteristic of an n-type LDMOS as a function of gate bias does not exhibits a bell-shape curve [10,11]. Recent studies shows that the device reveals a double hump characteristic when biased at low drain voltages [12] and vanishes at high drain voltage. In this experiment however, the double hump characteristic was not observed. Thus, the test conditions to obtain SOA in HCI are determined as in Table 1.

The stress conditions at the gate and drain terminal is selected based on the device degradation linearity adopted from the conventional CMOS procedure [13]. Several test conditions are normally tested using a few package units and the selection is based on the device degradation data. The reason for choosing $V_{GS}/2$ as one of the V_{GS} voltage is that the I_{sub} peak point normally happened at this condition [13]. However, for LDMOS device, it is quite difficult to see obvious I_{sub} peak point, and therefore $V_{GS}/2$ is selected for this experiment.

Fig. 2 schematically shows the different gate/drain bias configurations that are measured in this work. For each stress condition, five devices are measured in order to check for the repeatability of the result. In this experiment, the drain current degradation in linear region was evaluated.

TABLE 1
HCI-SOA TEST CONDITIONS

VGS (V)	VDS (V)
VGS	VDD, 1.1 VDD, 1.2 VDD
VGS @Isubmax or VGS /2	VDD, 1.1 VDD, 1.2 VDD
Vt+0.3	VDD, 1.1 VDD, 1.2 VDD

III. RESULTS AND DISCUSSION

A. Drain-current degradation analysis

Fig. 3 and 4 show a typical degradation characteristics of the drain current for the n-type LDMOS for linear current ($I_{D,lin}$) and saturation current ($I_{D,sat}$) respectively after hot carrier stressing. The device is stressed under static bias condition of $V_{DS} = 24V$ for $I_{D,lin}$ monitoring and $V_{DS} = 36V$ for $I_{D,sat}$ monitoring with $V_{GS} = 5V$. It is observed that the I_{DS}-V_{DS} sweep curve is reduced and the magnitude of reduction will increase with increment in stress time.

The linear mode of operation defined the threshold voltage of the n-type LDMOS. As shown in Fig. 3, a positive shift in the threshold voltage is observed. The reduction in the drain current will impact the speed of turning on the device, as it takes higher gate voltage to reach the same drain current. The increase in threshold voltage is attributed to the increase in interface state density and the fixed charge at the Si-SiO2 interface. This is ascribed to the hot-carrier induced electron trapping at lightly doped source region, adjacent to the source region [12].

Fig. 4 shows the same decrease trend in the saturation drain current. However, the saturation region current is affected to a lesser extent compared to the linear drain current. The reduction in drain current in saturation mode can be attributed to increase in interface state density and the fixed charges at the Si-SiO2 interface as well. This will impact the mobility of electrons in the inversion channel as per (1).

$$I_{DS} = I_{D,sat} = \mu n C_{ox} W/2L(V_{GS} - V_{TH})^2(1+\lambda V_{DS}) \qquad (1)$$

where μn is the charge-carrier effective mobility, W is the gate width, L is the gate length and C_{ox} is the gate oxide capacitance per unit area and λ, the channel-length modulation parameter.

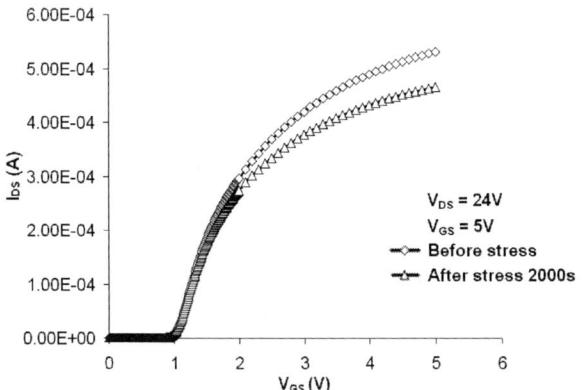

Fig. 3. Typical drain current vs. gate voltage characteristics of an n-type LDMOS transistor before and after hot-carrier induced oxide damage.

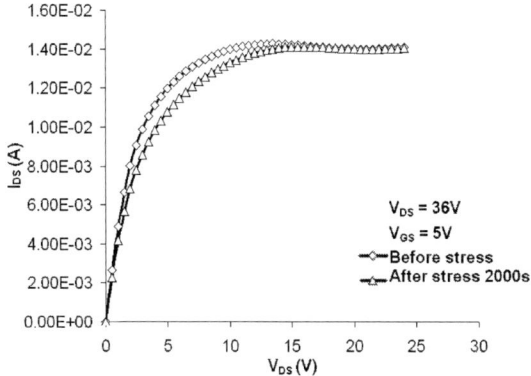

Fig. 4. Typical drain current vs. drain voltage characteristics of an n-type LDMOS transistor before and after hot-carrier induced oxide damage.

In general, the failure criterion for hot carrier reliability experiment is determined as 10% changes in forward drain current. The percent change values are calculated as (2).

$$\Delta I_D = [I_D(t) - I_D(0) / I_D(0)]*100 \qquad (2)$$

where $I_D(0)$ is the initial parameter value and $I_D(t)$ is the parameter value at time t.

The most commonly used time dependency function describing device degradation under hot carrier stress is the power law [9]. This is where the absolute value of the percentage change Y(t) in a parameter as a function of time t.

$$[Y(t)] = Ct^n \qquad (3)$$

A linear regression of Log Y(t) versus log(t) provides the fit coefficients n and C. The slope, n, in a log-log plot is strongly dependent on V_{GS} but less dependence on V_{DS}. This suggests that n changes according to hot-carrier injection mechanisms.

In conventional CMOS where the gate stress voltage corresponding to maximum substrate current ($I_{sub,max}$), a degradation exponent n of 0.5-0.7 indicates that interface trap generation (N_{it}) plays a dominant role in the drain current degradation [6]. Previous studies also suggest that other types of damage exist under different gate voltage conditions. For example, at $V_{GS}=V_{DS}$, oxide traps (N_{ot}) will be created with n value of 0.2-0.3. While at low gate voltages, it is shown that interface traps, hole traps and electron traps are created at high lateral fields with n values, which vary between 0.15 and 0.35. While at low gate voltages, it is shown that interface traps, hole traps and electron traps are created at high lateral fields with n values, which vary between 0.15 and 0.35 [14,15,16].

For LDMOS device, linear relationship between log of the drain degradation versus log of stress time indicates power law behavior. Fig. 5 shows the hot carrier degradation for V_{GS} = 5V at different V_{DS} for n-channel LDMOS. The degradation increases with increasing V_{DS}. The degradation model is valid for LDMOS device. In this experiment, the value of n varies between 0.07 and 0.13. This is assumed to be due to the different types of damage that occurred during the stress [13,15].

B. DC HCI Lifetime Analysis

The time to reach 10% degradation in $I_{D,lin}$ is the chosen fail criteria. Fig. 6 displays the cumulative probability plot for the experimental data for three splits. This is similar to standard CMOS analysis. There is no significant variation in the sigma value for the three splits indicating the failure mechanism at all three conditions is the same. The extrapolated time to 0.1% failure is obtained for each stress level and used in lifetime predictions to follow [17].

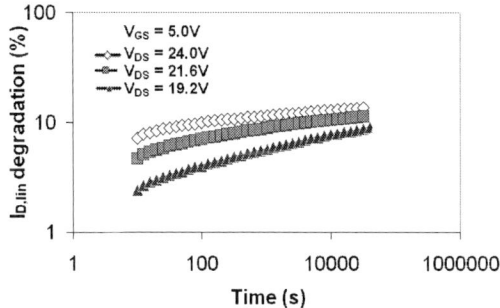

Fig. 5. $I_{D,lin}$ shift as a function of stress time for the device stressed under various V_{DS} with V_{GS} = 5V.

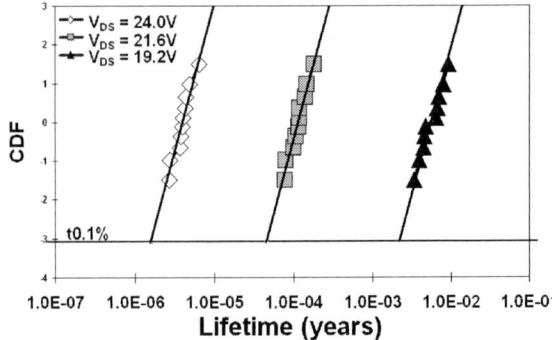

Fig. 6. Cumulative probability plot for different V_{DS}.

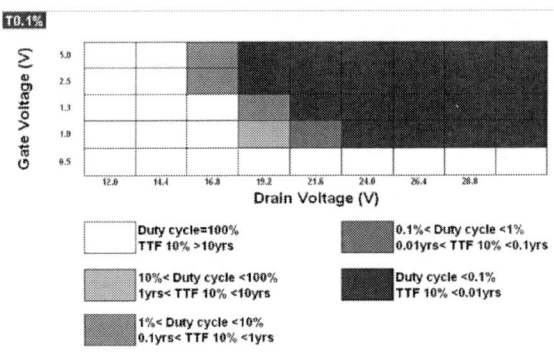

Duty cycle=100% TTF 10% >10yrs	0.1%< Duty cycle <1% 0.01yrs< TTF 10% <0.1yrs
10%< Duty cycle <100% 1yrs< TTF 10% <10yrs	Duty cycle <0.1% TTF 10% <0.01yrs
1%< Duty cycle <10% 0.1yrs< TTF 10% <1yrs	

Fig. 7. HCI-SOA plot for 24V n-channel LDMOS transistor for various duty cycles.

ACKNOWLEDGMENT

The authors would like to thanks Goh Kee Kuang for assisting on the HCI experiments.

In conventional CMOS, there are three analysis methods commonly used: Substrate/drain current ratio method, Drain-source voltage acceleration method and Substrate current method. In this work, we use Drain-source voltage acceleration method [18]. The simplicity of this model allows quick and easy extrapolation of device lifetime under stressing conditions to the real life basis [9]. Furthermore, the device does not observed I_{sub} peak. Based on the stress conditions selected in Fig. 2, the time to reach the failure criteria is obtained based on (4).

$$T_{TAR} = t0\ exp(B/V_{DS}) \qquad (4)$$

where $t0$ and B are fit parameters.

To present SOA of the device, the HCI lifetime at each condition will be plotted in the V_{DS} - V_{GS} plane. Fig. 7 shows the HCI-SOA lifetime boundary chart based on the lifetime characterization on various V_{DS} - V_{GS} conditions [7]. This is based on time to 10% $I_{D,lin}$ degradation in 10 years at $t0.1\%$. The data will be useful for the circuit designer. Different color coding or shading is used to identify the regions suitable for analog and digital operation. It is then further sub divided depending upon allowed duty cycle for digital operating region.

IV. CONCLUSION

In this paper, the test characterization method for HCI-SOA in n-type LDMOS device has been presented. The overall characterization process was performed based on conventional n-type CMOS HCI test method. The degradation and lifetime models adopted from conventional n-type CMOS transistor are valid for n-type LDMOS device. The test condition characterization was modified from the original method in conventional CMOS, as I_{sub} peak was not observed. Furthermore, to obtain SOA in HCI for n-type LDMOS device, a large set of V_{GS}-V_{DS} test conditions is needed. The drain current degradation characteristics have been discussed and it shows similar trend as conventional CMOS. Finally, HCI-SOA lifetime plot was generated in a chart, which will be useful for device & circuit designer.

REFERENCES

[1] J. F. Chen et al., "Mechanism and Improvement of On-Resistance Degradation Induced by Avalanche Breakdown in Lateral DMOS Transistors", *IEEE Transactions on Electron Devices*, Vol. 55, No. 8, 2008.

[2] W. Kanert, "Reliability Challenges for Power Devices under Active Cycling", in *Proceedings of the IRPS*, pp. 409-415, 2009.

[3] G.Cao, M.M. de Souza, *IEEE 04CH3753, 42nd Annual International Reliability Physics Symposium, Phoenix.* pp. 283, 2004.

[4] R. Versari, A. Pieracci, "Experimental Study of Hot Carrier Effects in LDMOS Transistors" *IEEE, 46*, pp.1228-1233, June 1999.

[5] D. Brisbin, A. Strachan, P. Chaparala, *IEEE 04CH3753, 42nd Annual International Reliability Physics Symposium, Phoenix*, pp.265, 2004

[6] D.Brisbin, A. Strachan, P. Chaparala, "Hot Carrier Reliability of N-LDMOS Transistor Arrays for Power BICMOS Applications", *IEEE 2002*, pp 105-109.

[7] P. L Hower and S. Pendharkar, "Short and Long-term Safe Operating Area Considerations in LDMOS Transistors," in *Proc. IRPS*, pp. 545–550, 2005.

[8] P. Hower, "Safe operating area - a new frontier in LDMOS design", *Proc. ISPSD*, Santa Fe, pp. 1-8, 2002.

[9] E. Takeda, N. Suzuki, "An Empirical Model for Device Degradation due to Hot-Carrier Injection", *IEEE, 4*, pp. 111-113, 1983.

[10] E. Takeda, C. Y. Yang, and A. Hamada, "Hot-Carrier Effects in MOS Devices", San Diego, Academic Press, 1995.

[11] Y. Tsividis, "Operation and Modeling of the MOS Transistor", *McGraw-Hill International Editions Electrical Engineering Series*, Second Edition, 1999, pp. 181-188.

[12] R. Versari, A. Pieracci, S. Manzini, C. Contiero, B. Ricco, "Hot Carrier Reliability in Submicrometer Lateral DMOS Transistors", *IEDM*, pp 371-374, 1997.

[13] JESD28-A, "Procedure for Measuring N-Channel MOSFET Hot-Carrier-Induced Degradation Under DC Stress", *JEDEC Publication*, December 2001.

[14] B. S. Doyle, M. Bourcerie, J.-C. Marchetaux, and A. Boudou, "Interface State Creation and Charge Trapping in the Medium-to-High Gate Voltage Range (Vd=2>Vg>Vd) during Hot Carrier Stressing of NMOS Transistors," *IEEE Trans. Electron Devices*, vol. 37, pp. 744–755, 1990.

[15] P. Bellens, P. Heremans, G. Groeseneken, and H. E. Maes, "Hot Carrier Effects in N-Channel MOS Transistors under Alternate Stress Conditions," *IEEE Electron Device Lett.*, vol. 9, pp. 232–234, 1988.

[16] B. S. Doyle, M. Bourcerie, C. Bergonzoni, R. Benecci, A. Bravais, K. R. Mistry, and A. Boudou, "The Generation and Characterization of Electron and Hole Traps Created by Hole Injection during Low Gate Voltage Hot

Carrier Stressing of N-MOS Transistors," *IEEE Trans. Electron Devices*, vol. 37, pp. 1969–1877, 1990.

[17] O'Donovan, V., Whiston, S., Deignan, A. & Chleirigh, C.N., "Hot Carrier Reliability of Lateral DMOS Transistors", *IEEE International Reliability Physics Symposium Proceedings*, pp. 174-179, 2000.

[18] JESD28-A, "N-Channel MOSFET Hot Carrier Data Analysis", *JEDEC Publication*, September 2001.

Characteristic Analysis of 1024-Point Quantized Radix-2 FFT/IFFT Processor

Rozita Teymourzadeh, *Member IEEE/IET*, Memtode Jim Abigo, Mok Vee Hong, *Member IET*
Faculty of Engineering, Technology & Built Environment,
UCSI University, 56000, Malaysia
rozita@ucsi.edu.my

Abstract-The precise analysis and accurate measurement of harmonic provides a reliable scientific industrial application. However, the high performance DSP processor is the important method of electrical harmonic analysis. Hence, in this research work, the effort was taken to design a novel high-resolution single 1024-point fast Fourier transform (FFT) and inverse fast Fourier transform (IFFT) processors for improvement of the harmonic measurement techniques. Meanwhile the project is started with design and simulation to demonstrate the benefit that is achieved by the proposed 1024-point FFT/IFFT processor. Pipelined structure is incorporated in order to enhance the system efficiency. As such, a pipelined architecture was proposed to statically scale the resolution of the processor to suite adequate trade-off constraints. The proposed FFT makes use of programmable fixed-point/floating-point to realize higher precision FFT.

Keywords -- DFT, IDFT, Fast Fourier Transform (FFT), IFFT, quantized, floating point, Radix

I. Introduction

Discrete Fourier Transform (DFT) is amongst the most fundamental operations in digital signal processing. However, the widespread uses of DFTs make its computational requirements an important issue. The direct computation of the DFT requires approximately N^2 operations where N is the transform size. The breakthrough of Cooley-Tukey (CT) FFT comes from the fact that it reduces the complexity to an order of $N\log_2 N$ operations. The FFT is therefore an efficient algorithm to compute the DFT and its inverse (IDFT). It has several applications in the field of signal processing including the real-time processing of wireless time-domain and frequency-domain signals especially for use in Orthogonal Frequency Division Multiplexing (OFDM) systems such as Digital Video Broadcasting (DVB), Digital Subscriber Line (xDSL) and WiMAX (IEEE 802.16) [1-4]. These applications require large-point FFT processing, such as 1024/2048/8192-point, FFTs for multiple carrier modulation.

Many FFT algorithms based on the CT decomposition such as radix-2^2, radix-2^3, radix-4, radix-(4+2), prime-factor as well as split-radix algorithms, have been proposed using the complex mathematical relationship to reduce the hardware complexity [2-3]. For example, in [4] one butterfly unit is used for all computations and $N+N.\log_2 N$ clock cycles are required for the computation of the FFT. A second implementation approach is for speed demanding

applications, where one butterfly unit is used for each decimation stage of a radix-2 FFT [5]. A pipeline architecture based on the constant geometry radix-2 FFT algorithm, which uses $\log_2 N$ complex-number multipliers (more precisely butterfly units) and is capable of computing a full N-point FFT in N/2 clock cycles has been proposed in 2009 [8]. All these developments have introduced their own disadvantages, in addition to the age-long finite word-length effects of digital circuitry [7-9]. This paper thus, uses the pipeline architecture [6] to propose a model for the analysis of important design constraints like the finite word-length effects and amount of resolution needed to achieve the appropriate SNR [8-10] for the desired design needs using the statistical tools for the analysis of a range of feasible resolution.

II. Architecture Development

A. Algorithm development of the decimation-in-time (DIT) radix-p FFT

The DFT of an N-point sequence x[n] is given by:

$$X[k] = \sum_{n=0}^{N-1} x[n] W_N^{kn} \quad \text{For k = 0, 1, 2,...,N-1} \quad (1)$$

Where $W_N = e^{-j\left(\frac{2\pi}{N}\right)}$.

Consider the general formula of the DIT Radix-p FFT as follows:

$$X\left[k + r\left(\frac{N}{p}\right)\right] = \sum_{n=0}^{\frac{N}{p}-1}\left(\sum_{j=0}^{p-1} x[pn+j] . W_p^{jr} W_N^{jk}\right) W_{\frac{N}{p}}^{nk} \quad (2)$$

for k = 0,1,2,...,N/p-1 and R = 0,1,2,...,p-1. Using the above decomposition, the DFT can be reduced successively to *N/p* p-point DFTs. In general, this process can be repeated m times and therefore there are totally *m* stages in the implementation of the DFT.

B. Parallel Architecture

The computational structure of a butterfly unit is shown in Fig. 1. It is the fundamental computational of the parallel architecture

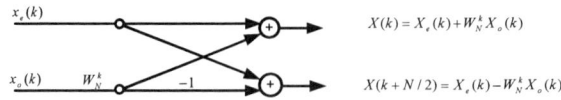

Fig. 1. Radix-2 butterfly unit

The butterfly unit requires a complex multiply and two complex additions. Therefore, it takes a total of $(N/2) \text{Log}_2 N$ complex multiplies and $N\log_2 N$ complex additions to compute all N-point DFT samples. An 8-point Radix-2 DIT FFT requires $N/2$ butterfly units per stage for all m stages [11-15]. For larger butterflies ($N > 2^6$), the processor becomes extremely complex and slow. Hence, a simpler and faster architecture is then required. Therefore, the proposed system was designed and simulated by MATLAB software. Fig. 2, shows the overall pipelined system structure and its designed control signals.

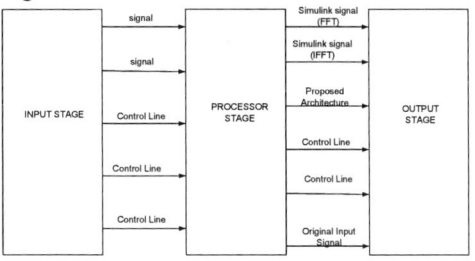

Fig. 2. Proposed pipelined system algorithm

C. Pipeline Architecture

The same butterfly unit can perform the $N/2$ butterfly operations computed in every stage sequentially. Since the two inputs of a next butterfly unit of a stage are provided from the output of the butterfly unit of the previous stage at different time points, a shuffling unit is inserted between two successive butterfly units in order to route these outputs to the corresponding inputs of the next stage. To increase the system efficiency in the Radix butterfly algorithm, the pipeline registers are located after each addition, subtraction blocks that is the end of each stage. Hence, the pipeline butterfly algorithm keeps the final result in the register to be transferred to next step by the next calculation cycle. However the measurement of system efficiency after applying pipeline structure only can be evaluated after the hardware implementation. Fig. 3 shows the inner layer of proposed FFT processor design where pipelining is applied in signal input logic block.

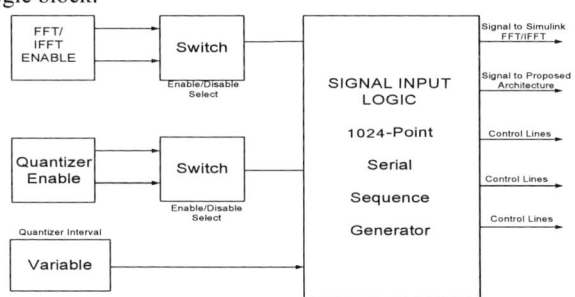

Fig. 3 Proposed inner layer of FFT/IFFT Processor

The signal input is inserted at the control signal to program the processor functionality. The control signals are to select FFT or IFFT calculation, while the other enables and disables the quantization of the twiddle factors. Fig. 4 illustrates the

10 stages butterfly for 1024-point pipeline FFT/IFFT processor.

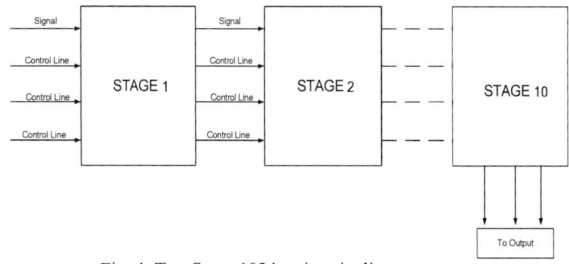

Fig. 4. Ten-Stage 1024-point pipeline processor

In the proposed design, excluding the input stage, the rest of stages consist of the twiddle factors, the shuffling unit and a floating-point quantize model. The interval of the quantize unit for each stage is preset statically and this is used to vary the bit-resolution of the processor. Fig. 5 shows the flowchart of overall system operation while quantizing and pipelining are applied.

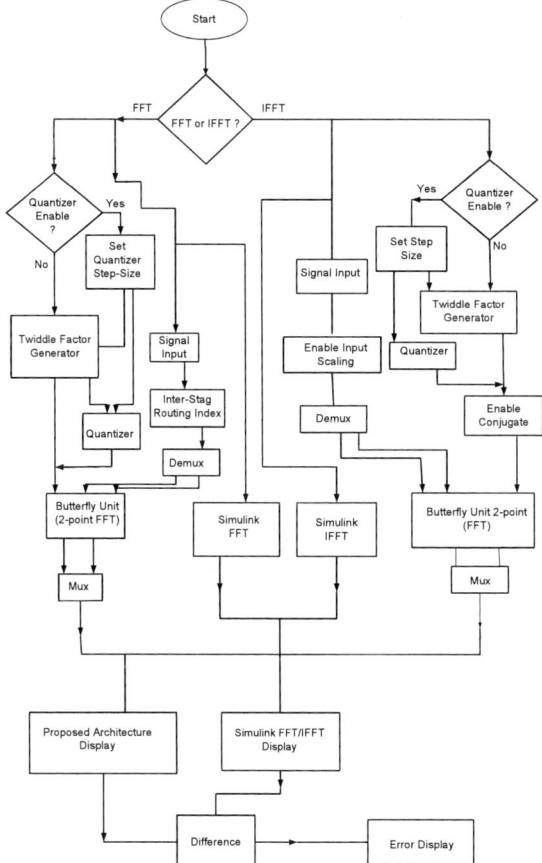

Fig. 5. Proposed overall system operation

The IFFT computation uses the same fundamental Radix-2 DIT Butterfly unit. However, the input is scaled by the factor of N (1024). These discrete input values are then sent through the processor stage, which performs the same operation except that the conjugate of the twiddle factors are used instead. The output stage simply compares the results of the proposed FFT/IFFT Processor with the idle FFT/IFFT processor and their difference is observed as system error that will be analysed in the next chapter.

III. STATISTICAL THEORY OF QUANTIZATION

A. Uniform Quantization

One would expect that quantization has a similar effect on functions of the amplitude as sampling has on functions of time. Quantization is an operation on signals that is represented as a "staircase" function. Each input value is rounded toward the nearest allowable discrete level. The probability of each discrete output level equals the probability of the input signal occurring within the associated quantum band [16]. For example, the probability that the output signal has the value zero equals the probability that the input signal falls between $\pm q/2$, where q is the quantization box size [8]. Fig. 6 shows the model of quantizing.

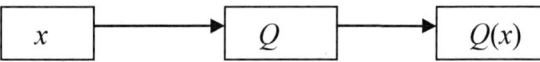

Fig. 6. Uniform Quantize Model

The quantize error (h) is given as

$$h = x - Q(x) \tag{3}$$

If x and h are real, with probability density function (PDF) as $P_x(.)$, then the quantization error variance is

$$
\begin{aligned}
\sigma_h^2 = E\{h^2\} &= \int_{-\infty}^{\infty} h^2 P_h(h)dh \\
&= \sum_{k=1}^{L} \int_{x_k}^{x_{k+1}} \left(x - Q(x)\right)^2 p_x(x)dx
\end{aligned} \tag{4}
$$

$$\sigma_h^2 = \int_{-\frac{q}{2}}^{\frac{q}{2}} h^2 \frac{1}{q} dq = \frac{q^2}{12} = \frac{1}{3} x_{max}^2 \, 2^{-2b} \tag{5}$$

$$SNR(dB) = 10 log_{10}\left(\frac{\sigma_x^2}{\sigma_h^2}\right) \tag{6}$$

where q is the quantization interval, b is the number of bits. Quantization noise is defined as the difference between the output and input of the quantized signal. Since the quantized unit is designed in the proposed processor to enhance the calculations, Fig. 7 illustrates a plot of the error versus the number of bits for the uniform quantized, while Fig. 8 shows a comparison of the mean, standard deviation as well as variance of the uniform quantization model achieved by the proposed FFT Processor.

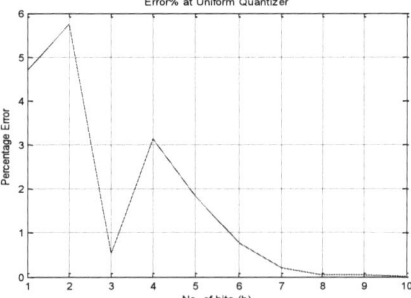

Fig. 7. Error of the Uniform Quantization

The probability of getting a given error value is the sum of probabilities of all the quantization boxes. The uniform quantize model performs uniform quantization on the signal input and thus, a linear signal is required at the input. Fixed-point numbers are considered linear since the Radix point remains fixed. However, this research was focused on floating-point numbers hence; the response of the uniform quantize to floating point input was observed. As result the quantization noise was increased. Equation (6) gives an expression for the SNR of the quantizer using the ratio of the variances of the input to noise.

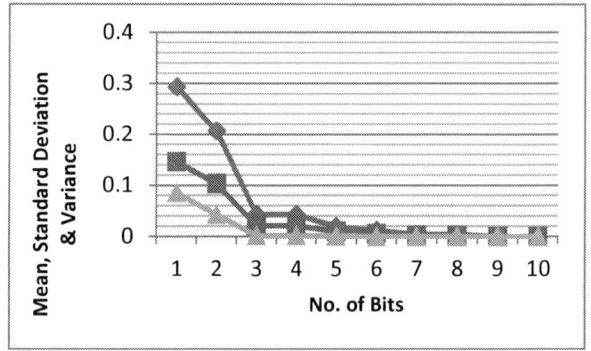

Fig. 8 Measured dispersion of uniform quantize

B. Non-Uniform Quantization

The uniform quantization model is usually not used for floating-point quantization due to the overall non-uniform characteristic of the latter. Quantization of floating-point numbers is carried out only on the mantissa hence; it is more relevant to consider the relative error ε caused by the quantization process. The relative error defined in terms of the numerical values of the quantized floating-point number $Q(x) = 2^e \, Q(M)$ and the un-quantized number $x = 2^e \, M$ is given as

$$\varepsilon = \frac{(Q(x)-x)}{x} = \frac{(Q(M)-M)}{M} = \frac{\alpha}{M} \tag{7}$$

It is possible however, to represent the floating-point quantizer using a combination of a compressor, a uniform quantize and an expander. Fig. 9 shows the non-uniform quantized model.

Fig. 9. Non-uniform quantization model

$$
\begin{aligned}
\sigma_\varepsilon^2 = E\{\varepsilon^2\} &= \frac{2}{q} \int_{1/2}^{1} \int_{-q/2}^{q/2} \frac{\alpha^2}{M^2 \, d\alpha} dM \\
&= q^2/6 = (0.167) \, 2^{-2b}
\end{aligned} \tag{8}
$$

An expression for the variance is shown in (8). The variance from the floating-point quantization equals half that obtained from the uniform quantization which is a generally preferred characteristic. Fig. 10 determines that the stability of the

978-1-4673-2395-6/12 $31.00 © 2012 IEEE

processor performance such that no variation occurs when the number of bits is 7 bit, unlike that of the uniform quantization which attains this stability at bit position eight 8, as shown in Fig. 7. That is the advantage of the system while modeling the floating-point structure.

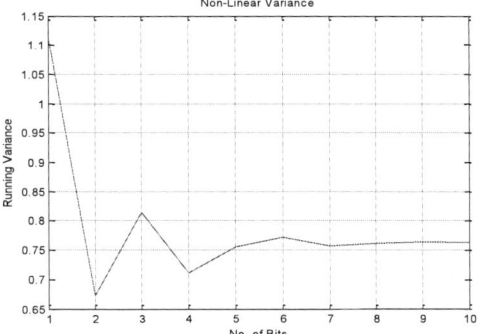

Fig.10 Error variance of the non-uniform quantizer

In addition, bit position 2 of Fig. 10 gives minimum swing before stability, contrary to that of Fig. 7 which occurs at bit position 3. This minimum swing gives a false minimum error position and can be used for less sensitive applications in which minimum error is not important.

Fig. 11 illustrates the comparison between mean, standard deviation when number of bit increased.

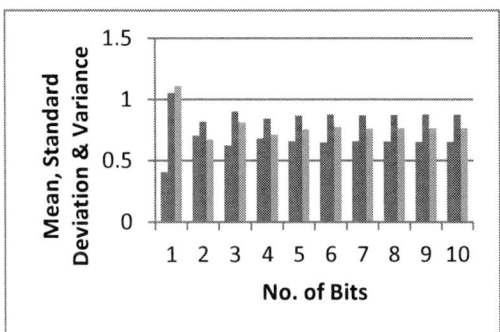

Fig. 11   Comparison of the measured dispersion of non-uniform quantization

IV.   DISCUSSION

Statistical parameters like the mean, standard deviation and variance were used to analytically develop expressions for the variation in error as a function of the quantization interval. These parameters were also known to have relationships with the SQNR. The percentage error of the non-uniform quantization generally decreased with an increase in the quantizer interval.

As such, the SQNR is increased with respect to the quantization step size. The same general trend was observed in the uniform quantization, as well as the FFT and IFFT results. Fig. 12 shows the error variation when the input data increased.

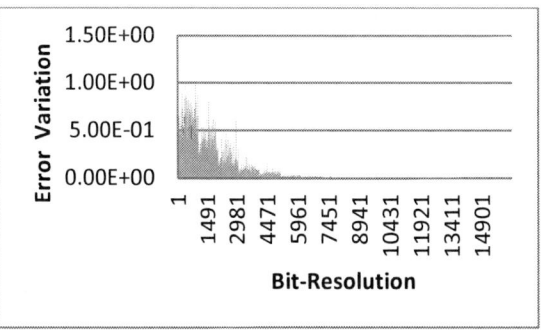

Fig. 12. Error variation of the FFT processor

The general trend observed from the results indicates that the measured dispersion can only be valuable when they are used alongside the mean since the mean actually provides the benchmark for understanding the decreasing trend. Fig. 13 shows the mean standard deviation and variance when the number of bit increased.

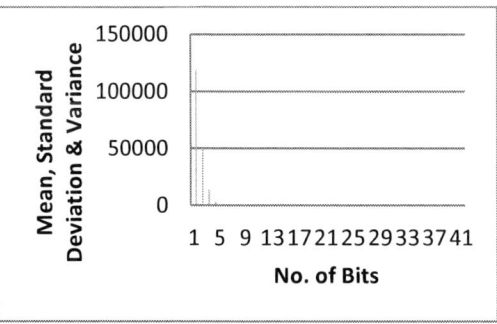

Fig. 13. Measured dispersion of FFT/IFFT processor

Therefore, the variance is decreased as quantization interval increased. Hence, the variance is inversely proportional to the percentage error, and as such, inversely proportional to the SQNR. This provides experimental proof to the theoretical models given earlier and provides a benchmark for the trade-off between the SQNR and the resolution. As shown in Fig. 14 the error variation also decreased as bit resolution increased.

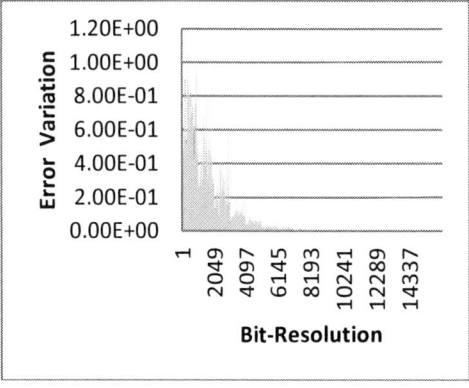

Fig. 14. Error variation of IFFT processor

978-1-4673-2395-6/12 $31.00 © 2012 IEEE   667

V. CONCLUSION

The 1024-point radix-2 FFT/IFFT processor is designed and simulated using MATLAB Simulink toolbox. The proposed processor satisfies the FFT size requirement of the 1024-points for a quantized pipeline structure. The percentage error and all the measured dispersion were found to decrease as the bit-resolution increased. This shows how the SQNR improves with bit-resolution. Although the power requirement for such SQNR systems are high, the proposed architecture provides an ease in the trade-off decision between the SQNR, power requirement and bit-resolution of the Radix-2 FFT/IFFT processor.

REFERENCES

[1] Meletis, C. Bougas, P. Economakos, G. Kalivas, Kaimal Pekmestzi, P. 2005. High Speed Pipeline Implementation of Radix-2 DIF Algorithm, *Proceedings of World Academy of Science, Engineering and Technology*. Vol. 2. ISSN 1307-6884, pp. 1-4

[2] Curtis, T. Curtis, M. 2004. High Performance Digital Signal Processing, *IOA Conference on Sonar Signal Processing*, Lough borough, pp. 1-10.

[3] M. Mohamed Ismail, M.J.S Rangachar, Ch. D. V. Paradesi Rao.2010. VLSI Implementation and Performance Analysis of Efficient Mixed-Radix 8-2 FFT Algorithm with Bit Reversal for the Output Sequences. *Journal of Theoretical and Applied Information Technology, pp. 164-168.*

[4] Luis Diaz de Cerio, Miguel Valero-Garcia and Antonio Gonzalez. 1985. Efficient FFT on Torus Multicomputers: A Performance Study. Universitat Politecnica de Catalunya, Barcelona (Spain) pp. 1 -10.

[5] Chih-Peng and Guo-An Su. 2006. A Grouped Fast Fourier Transform Algorithm Design for Selective Transformed Outputs. Depart of Electrical, National Chung Hsing University, Taiwan, pp 1 – 8.

[6] Rozita Teymourzadeh, Yazan Samir Algenabi, Nooshin Mahdavi, Masuri Bin Othman. 2010. On-Chip Implementation of High Resolution High Speed Floating Point Adder/Subtractor with Reducing Mean Latency for OFDM. American Journal of Engineering and Applied Sciences. 3(1): 25-30. ISSN: 1941-7020. DOI: 10.3844/ajeassp.2010.25.30.

[7] Jonathan Greene, Robert Cooper. 2005. A Parallel 64K Complex FFT Algorithm for the IBM/Son/Toshiba Cell Broadband Engine Processor. Mercury Computer Systems, pp 1 – 7.

[8] Ahmed Saeed, M. Elbably, G. Abdelfadeel, and M-I. Eladawy. 2009. Efficient FPGA Implementation of FFT/IFFT Processor. International Journal of Circuits, Systems and Signal Processing, Chelmsford MA, 3(3): 103-110.

[9] E. H. Satorius, M. J. Grimm, G. A. Zimmerman and H. C. Wilck.1986. Finite Word length Implementation of a Megachannel Digital Spectrum Analyzer. TDA Progress Report, pp 42 –86.

[10] Pel-Yun Tsai, Chia-Wei Chen and Meng-Yuan Huang. 2011. Automatic IP Generation of FFT/IFFT Processors with Word-Length Optimization for MIMO-OFDM Systems, Vol 2011, National Central University, Taiwan, pp 1 – 15.

[11] Wei-Hsin Chang and Truong Nguyen. 2007. Integer FFT Optimized Coefficient Sets, Department of Electrical and Computer Engineering, UCSD, La Jolla, California 92093 – 0407, pp 1 – 4.

[12] Cheng-Yeh Wang, Chih-Bin Kuo, and Jing-Yang Jou. 2007, Hybrid Word-Length Optimization Methods of Pipelined FFT Processors IEEE Transactions on Computers, vol 56(8):1 – 14.

[13] Shannon, C. E. 1949. Communication in the presence of noise, *Proc. IRE*,vol. 47, pp. 10-21.

[14] Sripad, B. and Snyder, D. L. A. 1917. Necessary and sufficient condition for quantization errors to be uniform and white, *IEEE Trans. Acoust, Speech, Signal Processing,* vol. ASSP- 25(5):442-448.

[15] Kontro, J. Kalliojiirvi, K. and Neuvo, Y. 1992, *Floating-point arithmetic in signal processing. In Proc. IEEE Int. Symp. Circuits and Systems*, San Diego, CA, May 10-13, 1992, 92CH3139-3, vol. 4, pp. 1784-1791.

[16] IEEE Standard.1987. The Institute of Electrical and Electronics Engineers. "IEEE Standard for binary floating-point nithmetic," ANSVIEEE Standard 754-1985, New York, Aug. "IEEE Standard for radix-independent floating-point arithmetic," ANSUIEEE Standard 854-1987, New York.

Effect of Platinum Catalyst Loading on Membrane Electrode Assembly (MEA) in Proton Exchange Membrane Fuel Cell (PEMFC)

Norfarhanim Mohd Zahari[1] and Azlan Abdul Aziz[1,2]
[1]School of Physics
Universiti Sains Malaysia (USM)
11800 USM, Pulau Pinang, Malaysia
[2]NanoBiotechnology Research and Innovation, INFORMM,
Universiti Sains Malaysia,
11800 Pulau Pinang, Malaysia
Email: farhanim_zahari@yahoo.com

Abstract- **Catalyst layer distribution of platinum particles by brush coating method on a gas diffusion layer (GDL) for proton exchange membrane fuel cell (PEMFC) was studied. The performance of proton exchange membrane fuel cells (PEMFC) was characterized at different catalyst loadings (0.3mg/cm^2, 0.35mg/cm^2, 0.40mg/cm^2, 045mg/cm^2 and 0.50 mg/cm^2 on a carbon support). The best fuel cell performance was found for cell which has 0.50 mg/cm^2 platinum loading in both anode and cathode with 0.513V at an operating current of 10A at 36.8°C with ambient pressure. It is due to more platinum loading will have more reaction with hydrogen gas. The characterization of the particle size and identification of individual surface atoms was performed by means of Xray diffraction (XRD), scanning electron microscopy (SEM) and atomic force microscopy (AFM).**

I. INTRODUCTION

Fuel cell-an electrochemical power plant which directly converts chemical energy to electrical energy, is getting more attentions due to growing concernment to the energy and environmental issues since it Furthermore, it is resulting in high energy conversion efficiency and high environmental affinity.

One of the challenges facing proton exchange membrane fuel cells (PEMFC) is to improve the utilization of platinum within the catalyst layer that should reduce the platinum loading in the electrodes. Ideally, all the platinum should be active for hydrogen and oxygen gases reaction for both anode and cathode side in the fuel cell.

Conventionally, brush coating method is used to obtain platinum loading on electrodes. In order to get better performance for fuel cell, pulsed electrodeposition (PED) method is used in PEM fuel cells. A low platinum loading on the electrode was obtained by PED method without any loss in fuel cell performance compared with electrodes prepared by conventional brush coating method.

The fuel cell performance of the electrodes prepared by brush coating method has been presented in these studies due to variation of platinum catalyst loading on each of the electrodes. The typical Pt loading used for PEMFC has been between 20 and 40wt.% Pt/Carbon. However, 60wt.% Pt loading will be favorable in this studies which means relative amount of supporting carbon will be reduces due to high Pt loading.

II. THEORY

Proton Exchange Membrane (PEM) fuel cells also known as Polymer exchange membrane fuel cells. Typically, PEM fuel cells operate on pure (99.999%) hydrogen gas. This fuel cell combines the pure hydrogen and oxygen gases from atmosphere at anode and cathode respectively to produce water, heat and generate electricity.

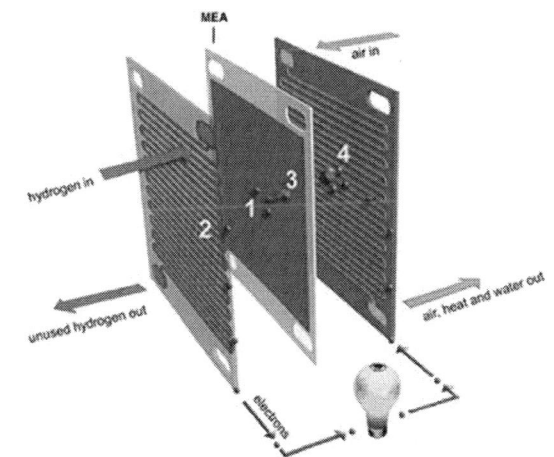

Fig. 1 PEM Fuel Cell

PEM fuel cell consist of membrane electrode assembly (MEA) and bipolar plate which has hydrogen flow field and oxygen flow field on anode and cathode side respectively. Hydrogen gas is purge into flow field plates to the anode side while oxygen gas from the air is channeled to the cathode side.

978-1-4673-2395-6/12 $31.00 © 2012 IEEE

At the anode, platinum catalyst cause the hydrogen gas to split into positive ions (protons) and negative electrons as showed in Fig 1.

The Proton Exchange Membrane (PEM) in the middle of the fuel cell allows only positively charged ions pass through it to the cathode while electron must travel along external circuit to the cathode. At cathode side, positively charged ions and electron will combine with oxygen gas to form water and creating an electrical current.

For each cells, it is compulsory to connect it in series, in order to have fuel cell stack. In this fuel cell stack, the voltage from each cells will be accumulate, thus will shows the performance of fuel cell stack. The accumulation of voltage calculated by using Equation (1).

$$V_T = V_1 + V_2 + \ldots + V_n \qquad (1)$$

The power produced from fuel cell stack can be calculated according equation (2)

$$P = I \, V \qquad (2)$$

Where P is power output in Watts, I is current in Ampere which is amount of load, and V is voltage in Volt.

III. METHODOLOGY

In this work, four MEA with each different platinum loading was fabricate using brush coating method. The discussion is divided into three sections.

First section is, preparation of Platinum solution. The platinum solution consists of Platinum, Nafion 30% and de-ionized water. The mixture was sonicated to ensure it is well-mixed.

Second section is to prepare the Gas Diffusion Electrode (GDE). First, the Gas Diffusion Layer (GDL), which is the carbon paper with Polytetrafluoroethylene (PTFE) is weighted. Next, platinum solution is brushed on the GDL for uniform distribution and consistent layering. The distributed platinum on the GDL, is now called GDE. GDE is then left to dry so that all water evaporates. GDE is again weighted to obtain the amount of solution on the surface and calculation the exact platinum loading on the GDE is performed.

Last section is to prepare the Membrane Electrode Assembly (MEA). The Proton Exchange Membrane (PEM) is sandwiched with the GDE on aluminum foil. After that, the Proton Exchange Membrane (PEM) and GDE is hot pressed at 200°C to become MEA.

IV. RESULTS AND DISCUSSION

The graph of current density versus voltage in Fig. 2 shows the performance of 5 cell stack with the different platinum loading 0.30mg/cm^2, 0.35mg/cm^2, 0.40mg/cm^2, 0.45mg/cm^2 and 0.50mg/cm^2. To examine the effect of both anode and cathode catalyst loading on cell performance, V-I curves or polarization curve were measured for the single cells. The

performance of each platinum loading was tested with pressure 0.3 bar with load current from 0A until 10A.

In order to investigate the performance of fuel cell with different supported platinum loading on the both anode and cathode, the bipolar plate design was fixed. The voltage of the each cell is drop abruptly at voltage region above 0.7V.

Abrupt voltage drop is due to activation of electrochemical reaction between surface areas of platinum catalyst with hydrogen gas. To achieve the stabilization of measured voltage, the current is load until 2A. Voltage will decreases as the load current increases. After load current with 2A it called ohmic losses region because the voltage will decrease gradually with increasing of load current. Consequently, it occurs caused by ohmic loss and mass transport loss. Ohmic losses occur by reason of electron resistance that flow inside the bipolar plate show from previous study[6].

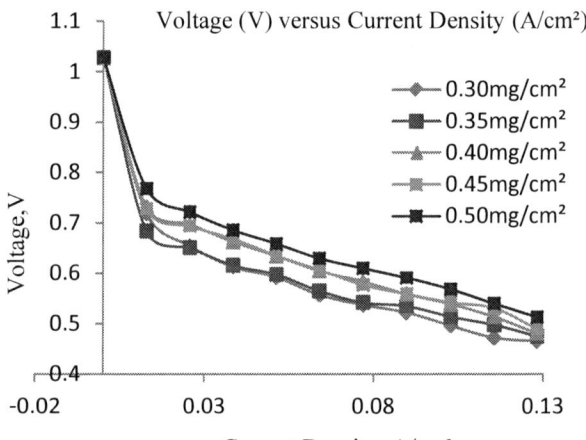

Fig. 2 Voltage dependence on current density

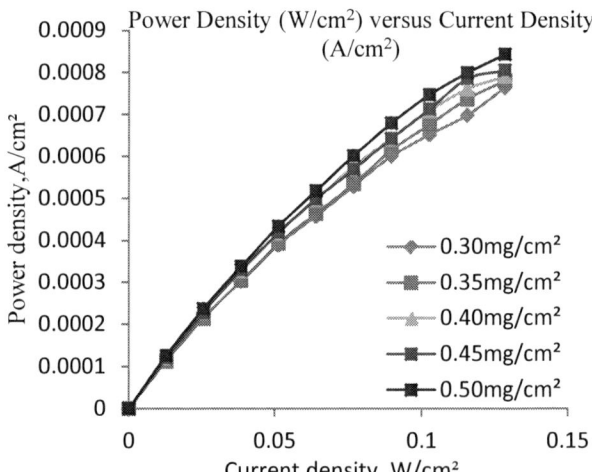

Fig. 3 Graph of Power Density versus Current Density

Furthermore, mass transport loss from the depletion of reactants which is hydrogen gas that being trapped at the platinum catalyst sites under high loads of current. Hence, the causing rapid loss of voltage thus will decrease the performance of fuel cell. Cell with 0.5mg/cm^2 platinum loading exhibit better performance due to more platinum reaction with hydrogen reactant. For the single cell that has very high platinum loading which is more than 0.5mg/cm^2, the voltage will drop abruptly at high current densities. This observation might due to mass transport resistance caused by the thick electrode layer according to the previous studies [1]. As a suggestion, the utilization of platinum loading should be optimized to 0.5mg/cm^2.

Fig. 3 is graph of current density versus power density that shows how the performance is related to the platinum loading for the electrode that have different platinum loading for both anode and cathode side. As the platinum loading further increased, the power density will increase when load current increased. Large increases of power density were observed when the platinum loading increased to 0.5mg/cm^2. The highest power density achieved was 65.7 mW/cm^2 at 0.5mg/cm^2, while the lowest power density achieved was 59.6 mW/cm^2 at 0.3mg/cm^2. It is obvious that the higher platinum loading on a supported carbon is, the higher power density will be. It is also clearly shown that the power density has a linear relationship with the current density. If load current increased further, it is predicted that the power density will drop off as observed by other worker [2] .This effect is due to insufficient supply of hydrogen on the active area surface. To achieve higher current load, more hydrogen is needed to be supplied to the cell. To find out the influence of each cell with respect to different platinum loading, the bipolar plate design parameter need to be constant. The cell performance curves increases by about 10% under operating voltage of 0.51V.

Fig. 4 shows sample's surface morphology and composition of the brush coated deposits corresponding to a MEA with 30wt% Nafion and various 60wt% Pt loading of 0.30mg/cm^2, 0.35mg/cm^2, 0.40mg/cm^2, 0.45mg/cm^2 and 0.50mg/cm^2. In this studies, the used of 30wt% Nafion contents which optimized to that value obtained for the best performance in the previous studies [3]. The image of (0.30mg/cm^2 - 0.45mg/cm^2) shows that there are many pores with large diameter. The diameter is thousand nanometers and few pores in the micrometer range. However, the electrode containing 0.50mg/cm2 platinum loading exhibit very small porous like sponge strucutre. The platinum loading has porosity in hundred nanometers range. This is due to well distributed of platinum loading all over the area.

Images in Fig. 4 also shows the granularity of each platinum loading. For 0.30mg/cm^2 to 0.45mg/cm^2 catalysts coarse-grained properties which consist of few pores in a very large diameter was observed. On the other hand, 0.50mg/cm^2 catalyst exhibited finer grain in which small well distributed pores all over the active area. The finer the granularity, the greater the power density of fuel cell and hence speed up the reaction of platinum catalyst with hydrogen gas.

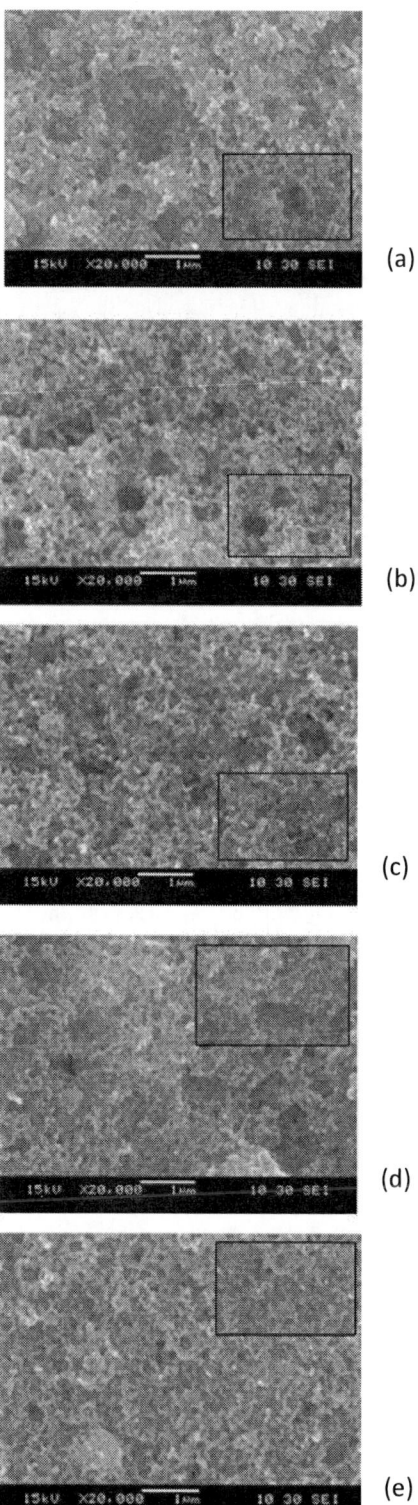

Fig. 4 SEM micrographs of brush coated catalyst layers corresponding to a fixed Nafion content of 30% and platinum loading of (a)0.30mg/cm^2 (b)0.35mg/cm^2 (c)0.40mg/cm^2 (d)0.45mg/cm^2 (e)0.50mg/cm^2. The micron bar in figures in figure (a) - (e) corresponds to 1μm length for the main and the enlarged inset is 0.5μm in length.

In order to attain the best performance of catalyst is necessary to optimize the size of pores. Decreasing the size of pores will increase the total surface area of catalyst. Thus, available areas that participate in the reactions per unit volume of platinum used also increased. If the granularity is too fine, the performance will increase the overhead. On the other hand, if the granularity is too coarse, the performance of fuel cell will suffer from the increasing of load current. The results indicate that the porosity can have a significant effect on the performance of MEA in the fuel cell. Even though the $0.50mg/cm^2$ platinum loading gives the best performance for the fuel cell, the cathode electrode appears to cause increasing the resistance of mass transfer. This confirmed by electrochemical impedance analysis by Litster et al [4] where smallest pores that contain in $0.50mg/cm^2$ platinum loading might cause water flooding in cathode electrode.

Fig. 5 shows analysis of EDX for $0.5mg/cm^2$ platinum loading. At 20 000x magnification the catalyst layers clearly showed to be platinum rich in the distribution of platinum clustered areas. An EDX spectrum of platinum surrounding area that not observed is also included for comparison shown in Fig 5(b). Fig 5(b) shows the presence of F and Si has also been identified by EDX.

The use of silicone piping and the sealing gaskets in stacking the fuel cells are believed to have introduced the contamination on MEA. The F indicated by the EDX is introduced by the Polytetrafluoroethylene (PTFE), which is coated on the carbon paper, before the platinum is brushed on the GDL. PTFE is used to increase the diffused of gases in the sense that it will provides hydrophobicity to an open pore to prevent it from being clogged with water. The fluoride can be transported along the fuel cell stack with the water or gas and can be accumulated in the last cell of the stacks. F loss measured is probably from degradation of the PTFE ionomer acts as a binder in the platinum layer. The rate release of fluoride can be measured by using the in-situ method using a fluoride ion selective electrode as showed by Hong Chu et al [5].

VI. CONCLUSIONS

Proton Exchange Membrane Fuel Cell (PEMFC) at different catalyst loading ($0.3mg/cm^2$, $0.35mg/cm^2$, $0.40mg/cm^2$, $0.45mg/cm^2$ and 0.50 mg/cm^2 on a carbon support) were MEA fabricated by a simple brush coating method and their distribution morphology was examined by SEM. The highest cell performance in this study was found to be 65.7 mW/cm^2 at $0.5mg/cm^2$.

The 60wt% Pt/C applied to both electrode side of membrane assembly (MEA), the cell performance decreased gradually in high current density.

Comparing the polarization curves in Fig. 2 for single cell with different platinum loading, the MEA with $0.3mg/cm^2$ showed a slightly higher ohmic resistance. However, these due to small platinum loading is less expensive than $0.50mg/cm^2$. As a conclusion, the ratio of catalyst loading and Nafion

solution relative to the carbon support and the different loading of platinum on MEA will affect the fuel cell performance.

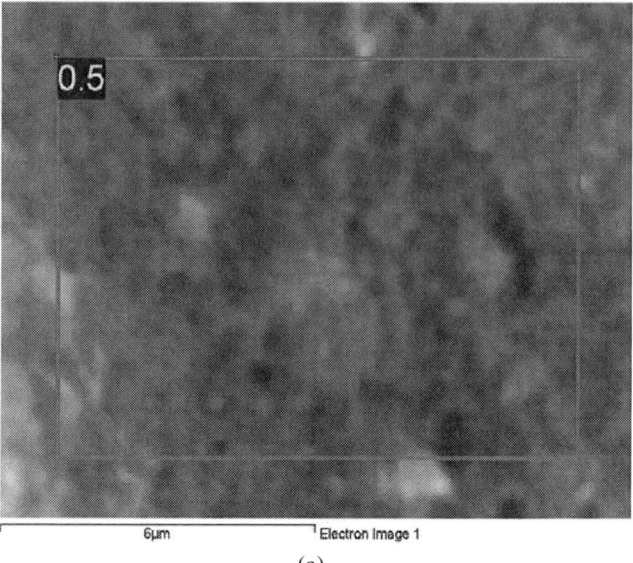

(a)

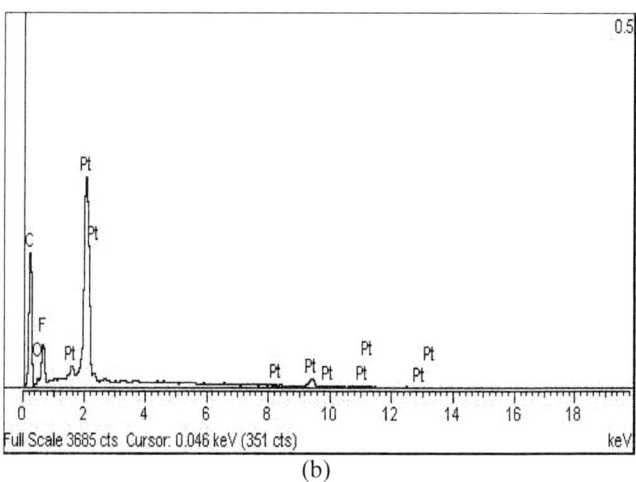

(b)

Fig. 5 (a) SEM and associated (b) EDX of platinum loading on membrane electrode assembly (MEA)

ACKNOWLEDGEMENT

This study was financially supported by Universiti Sains Malaysia under grant 1001/PFIZIK/814113. Special thanks to Yeap Tea Sin, Kelvin Tan and Rashila of G-Energy Sdn Bhd for MEA samples and tools for loading characterization. The assistance of Nano Optoelectronic Research and Technology Lab (NOR Lab) staff for the SEM and EDX measurement are also gratefully acknowledged.

REFERENCES

[1] M. Prasanna, E.A. Cho, H.-J. Kim,I.-H. Oh, T.-H Lim, S.-A Hong, "Performance of proton exchange membrane fuel cells the catalyst gradient electrode technique," J.Power Source 166(2007) 53-58.

[2] David M. Smiadak, "The Characteristic Curve of a Fuel Cell and Parameters Influencing the Characteristic Curve," Experiment report EGR 380 – Renewable and Sustainable Energy , School of Engineering Grand Valley State University.

[3] R.A. Silva, T. Hashimoto, G.E. Thompson, C.M. Rangel, "Characterization of MEA degradation for an open air cathode PEM fuel cell," Int. Journal of Hydrogen Energy xxx (2012) 1e10.

[4] S. Litster, G. McLean, Review PEM fuel cell electrodes, J. Power Source 130 (2004) 61-76.

[5] Yong-Hun Cho, Hyun-Seo Park, Yoon-Hwan Cho, Dae-Sik Jung, Hee- Young Park, Yung-Eun Sung,"Effect of platinum amount in carbon supported platinum catalyst on performance of polymer electrolyte membrane fuel cell", J. Power Source 172 (2007) 89-93

[6] S.Martin, P.L. Garcia-Ybarra, J.L Castillo, "High Platinum Utilization in ultra low Pt loaded PEM fuel cell cathodes prepared by electrospraying," Int. journal of hydrogen energy 35(2010)10446e10451

[7] Wan R.W. Daud, Abu Bakar Mohamad, Abdul Amir H.Kadhum, Rachid Chebbi,Sunny E.Iyuke, " Performance optimization of PEM fuel cell during MEA fabrication," Energy Conversion and Management 45 (2004) 3239–3249

[8] Zhigang Qi, Arthur Kaufman, "Low Pt loading high performance cathodes for PEM fuel cells," Journal of Power Sources 113 (2003) 37–43

Optimized Flow Field Bipolar Plate Design In Proton Exchange Membrane Fuel Cell

Mohd Ikhwan Mohd Isa[2] and Azlan Abdul Aziz[1,2]
[1]School of Physics
Universiti Sains Malaysia (USM)
11800 USM, Pulau Pinang, Malaysia
[2]NanoBiotechnology Research and Innovation, INFORMM,
Universiti Sains Malaysia,
11800 Pulau Pinang, Malaysia
Email: muhammadikhwan_i@yahoo.com

Abstract- **Proton Exchange Membrane Fuel Cell (PEMFC) performance is related to the bipolar plate design and their channels pattern. To study the effect of these parameters on output performance, four design were prepared: (i) Twin Vertical serpentine, (ii) Multiple Horizontal serpentine, (iii) Large channel parallel, and (iv) Pin flow field. Results shows that Multiple Horizontal serpentine produced highest output current and better thermal stability. Reason for these improvement are mainly due its ability to induces large pressure drop between adjacent flow channels over the entire electrode surface. Thus, it increase mass transport rates of reactants and reduce the amount liquid water that is entrapped. It was found that, after load 15A from the cell, this flow field produced 0.594V, higher output was obtained as compared to other design measured at 38.6°C.**

I. INTRODUCTION

A hydrogen fuel cell generates electron from electrochemical reaction between hydrogen and platinum catalyst. This electron then will flow to create electricity. The Proton Exchange Membrane Fuel Cell (PEMFC) is a zero-emission power source, which has high efficiency, low temperature operation, and high power density. Bipolar Plate is one of the vital part of PEMFC which supplies hydrogen (reactant) and oxygen (oxidant) to reactive area, remove water, collect produced current and provides mechanical support to Membrane Electrode Assembly (MEA) in fuel cell stack.

One of the challenges in PEMFC is to obtain effective flow field bipolar plates. An optimal flow field is critical to obtain high power density in fuel cell. Early development of fuel cell has conventional design such as parallel or straight channels. In modern days, there generally three major flow field design such as parallel, serpentine and interdigitated flow fields. Even these three design is a major design, serpentine flow fields remain most commonly used because of its effective reactant distribution. Many researchers also investigate the design based on a bio inspired pattern such as fluid distribution pattern on banana leaf [1].

II. THEORY

Fuel Cells

Fuel Cell has many types. This project discuss on Proton Exchange Membrane Fuel Cell (PEMFC), particularly flow field bipolar plate. PEMFC consist of Bipolar Plate, Current collector, and Membrane Electrode Assembly (MEA). MEA is a solid polymer membrane (nafion) disposed between two electrodes formed of porous, electrically conductive sheet material, typically carbon paper. The Platinum catalyst is coated on each carbon paper to produced desired electrochemical reaction. Nafion used to only allow H^+ to flow through.

Fuel Cells Operation

Fuel Cell operation is very simple. At anode, hydrogen gas will be purge into the bipolar plate from inlet and flow through the flow field channels on the bipolar plate. Before hydrogen flow out to the outlet, the hydrogen will be permeates into the porous electrode material and reacts with platinum catalyst to form electron and hydrogen proton. The solid Polymer membrane only allows hydrogen proton to permeate across membrane, while the electron is prohibited to flow across, it will flow through the external circuit to the cathode simultaneously generating electricity. At the cathode, oxygen is supplied, combines with hydrogen proton migrated from anode and electron from the external circuit to produce water and heat. Reaction at electrodes as follows:

Anode : $2H_2 = 4H^+ + 4e^-$
Cathode : $O_2 + 4H^+ + 4e^- = 2H_2O + Heat$
Overall reaction : $2H_2 + O_2 = 2H_2O + Electricity + Heat$

This cells is then stacked together which mean each cells is connected in the series, the voltage will be accumulated, such in equation (1).

$$V_T = V_1 + V_2 + \ldots + V_n \qquad (1)$$

The power produced from fuel cell stack can be calculated according equation (2)

$$P = I\,V \qquad (2)$$

978-1-4673-2395-6/12 $31.00 © 2012 IEEE

Where P is power output in Watts, I is current in Ampere which is amount of load, and V is voltage in Volt.

Bipolar Plate

This work focused on the flow field bipolar plate, where the hydrogen flow through the channel in active area reacted with platinum on the MEA. The main function of Bipolar plate is to distribute the hydrogen gas on the active area uniformly so that all area will have same hydrogen mass transport and high electron conductivity for current collection. An optimized design would have wider channels and narrow lands [2]. Uniform current density for serpentine flow fields can be obtain with shorter path lengths or larger number of channels [3].

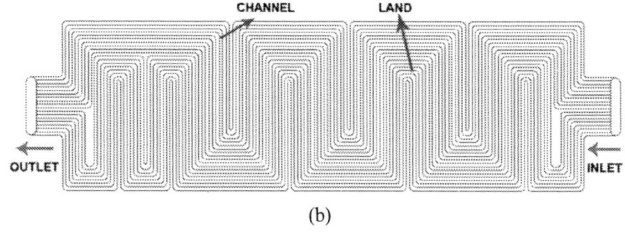

(b)

III. EXPERIMENT METHODOLOGY

In this work, 4 different design flow fields on the graphite bipolar plate were constructed. The software used to design flow field was AutoCAD. For first flow field, Twin vertical serpentine, it is design so that it is flow field is vertical direction which is up and down. For second flow field, it is the same with the first (serpentine) but in the horizontal direction. In the third design, the channel is like normal parallel flow, only the width of channel is much larger. Lastly in fourth design, these pin type flow fields only have four areas in contact with MEA. Fig. 1 shows four flow field design.

Next, check whether all the line connected together or not. This procedure help prevent the machine cutting wrong line thus break the bipolar plate because the bipolar plate is very fragile. After that, construct the cutter path line. This cutter path line helps drive the machine cutter engrave the channel precisely as in the design.

Then, the graphite with 78 cm^2 active area is placed perfectly on the machine plate. Machine cutter must be 1 mm cutter because all four designs have 1mm width channel. The cutter path line constructed is then plotted on the bipolar plate. Set up the cutter so that it has 0.4 mm depth.

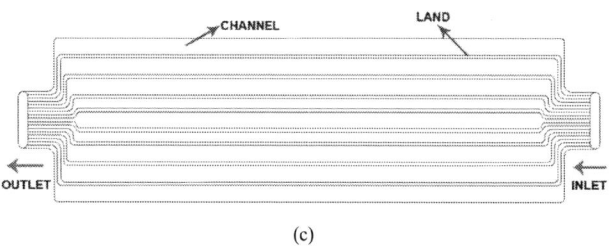

(c)

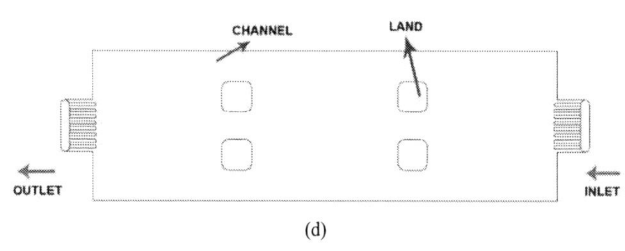

(d)

IV. RESULTS AND DISCUSSION

Experiment was performed on each different bipolar plate flow field design. The 7 cell stack was tested with pressure 0.3 bar at initial temperature 28°C is load current 1A until 15A and the voltage output recorded. The MEA has fixed Pt loading which is 0.3 mg/cm^2. The active area is 78 cm^2.

Fig. 1 Flow Field design (a) Twin Vertical serpentine, (b) Multiple Horizontal serpentine, (c) Large channel parallel, and (d) Pin flow field

Fig. 2 shows the performance and the voltage output of the four flow field design under high humidity condition. The polarization curve indicates cell voltage is decrease with the increase of current density due to increase overall potential loss.

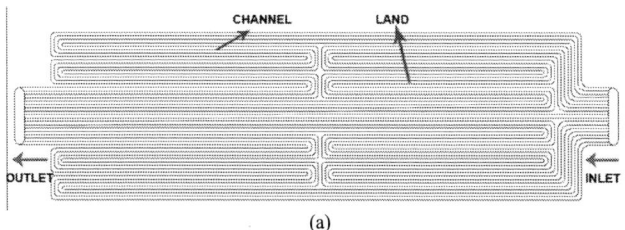

(a)

978-1-4673-2395-6/12 $31.00 © 2012 IEEE

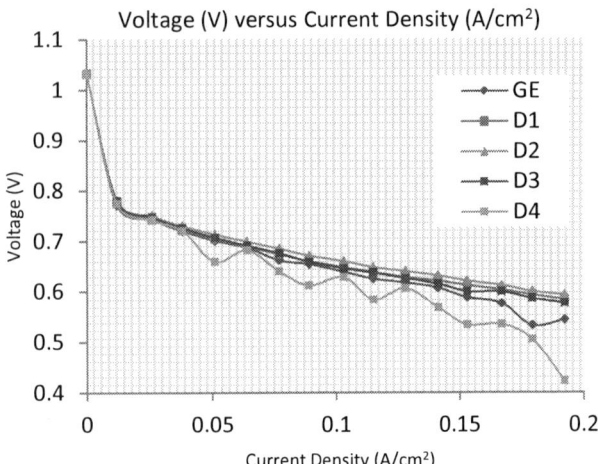

Fig. 2 Bipolar Plates Geometry influence on voltage output

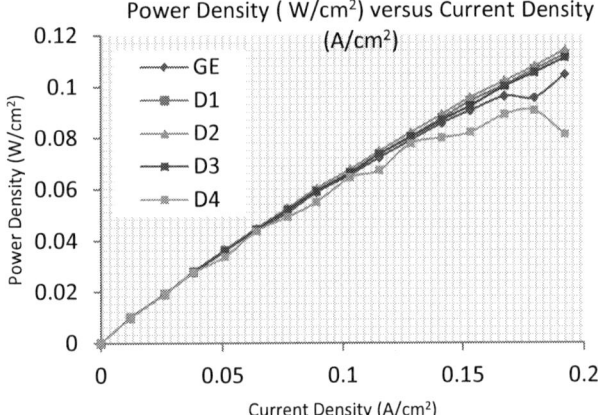

Fig 3 Bipolar Plate Geometry influence on power density

The region at the 0A until 1A is called activation region. Activation region is directly proportional to the rates of electrochemical reactions. Above the activation region, it is called ohmic losses, where occur because of resistance to flow of electron in the electrodes and also resistance in to flow of ions in the membrane.

Design 2, the Multiple Horizontal Serpentine exhibit better performance as compared to the others. It shows the no fluctuations on the graph therefore this design show high stability of voltage. Previous studies conducted by Jeon et al [4] showed that single serpentine flow fields forces the reactant to flow the entire active area thus rate of electrochemical reaction increase. Hence, produce higher output. The total channel area over total active area is 46.8% and the total land area is 53.2%. The land area provides contact with the Membrane Electrode Assembly (MEA), which also act as the conductivity area for electron. This clearly shows that the total conductivity area is larger than the total channel area.

The second highest performance shown by design 1, which is the Twin Vertical Serpentine flow fields, shows no fluctuation and high voltage stability. This design allows the reactant to distribute uniformly all over the active area. Each four portions will have same reactant mass transport. The total channel area calculated in this design is 52.4% form the total active area and the total conductivity area is 47.6%. The land or area in contact with the MEA is well uniformly distributed all over the active area. From the calculations, the channel area is larger than the conductivity area.

The third high output voltage is exhibit by design 3, the Large Channel Parallel flow fields. This design have large parallel channel where more reactant can enter the channel thus resulted in more reaction with platinum catalyst. The total channel area is 80.3% over the active area and land area is 19.7%. The surface conductivity is small even though it has large channel area, which shows that even it has more hydrogen and platinum reaction but the conductivity area is inadequate for electron flow.

Design 4 shows lowest output voltage and it tends to fluctuates above current load 3A. The design has only 5.2% conductivity area and 94.8% the channel area. It has lowest voltage because of the conductivity area is not well distributed on the bipolar plate. It only in contact with MEA at the specific area thus has high resistance. The channel area is very large, the hydrogen is free to flow without directional motion, and thus the reaction will occur at certain area leaving some area without reaction. This is indicates by the fluctuation of the curve.

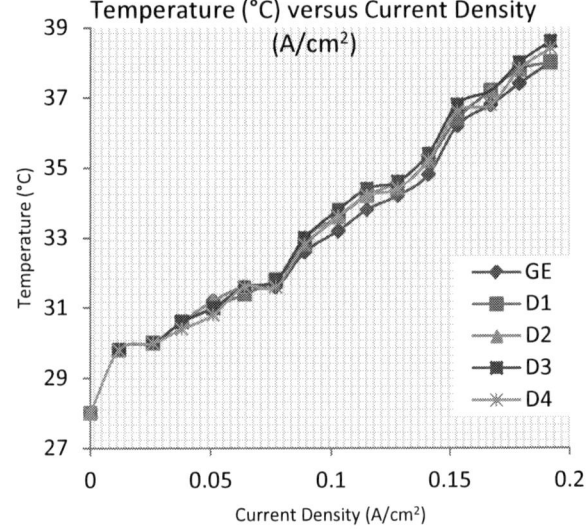

Fig. 4 Graph of temperature versus current density

Fig. 3 shows power density curve for different bipolar plate flow field design. Power is in term of current and voltage, manipulating the current to get the voltage. According to the equation (2), to obtain high power output, constant the current thus reduce the voltage. Power density for design 1, design 2, and design 3 is stable until the current load 15A. This is due to the uniform distribution of hydrogen on the active area. But design 4 exhibit unusual anomaly which the power density start to fluctuates at 4A until 14A, the power density drops slightly. The slight drop is because of at higher current, the it needs more hydrogen reaction with platinum to maintain the voltage output, but design 4 has very large channel, which the hydrogen tend to flow path where there is less resistance and the hydrogen has less directional flow. It tends to flow everywhere but some areas there no reaction occur. If we further load the current, performance would be drop. That is why the fuel cells operating below or at the maximum power.

Fig. 4 shows the temperature curve for different flow field design. Design 2 has the higher temperature increase with increasing the current density. Ii is confirming by its high output voltage where more hydrogen and platinum reaction will increase the temperature. The overall reaction is known as exothermic reaction, which reaction that release energy in form of heat from the system.

V. CONCLUSIONS

The design of the flow field channels on the bipolar plates can significantly affected the output power in PEMFC. The Horizontal Serpentine design exhibited better performance than the other 3 design.

The analysis shows that cell performance is dependent on the size and distribution of conductivity area of electron. This also prove that, larger channel does not necessarily provide more hydrogen reaction with the platinum on active area. It is because larger channel area caused the hydrogen motion less directional, thus it is not distributed uniformly all over the active area surface.

ACKNOWLEDGEMENT

The study was supported by School of Physics, Universiti Sains Malaysia under grant 1001/PIZIK/814113. Special thanks to Yeap Tea Sin and Kelvin Tan of G-Energy Sdn. Bhd. for kind assistance in design and model development.

REFERENCES

[1] R. Roshandel, F. Arbabi, G. Karimi M., "Simulation flow field design based on a bio inspired pattern for PEM fuel cells," Renewable Energy 41(2012) 86-95

[2] Li X, Sabir I. "Review of bipolar plates in PEM fuel cells: flow field designs," Int J Hydrogen Energy 2005;30:359-71

[3] Shimpalee S, Greenway S, Van Zee J. "The impact of channel path length on PEMFC flow-field Design," J Power Source 2006;160:398-406w

[4] Jeon DH, Greenway S, Shimpalee S, Van Zee JW. "The Effect of serpentine flow-field design on PEM fuel cell performance," Int. Journal of Hydrogen Energy 2008;33:1052-66

978-1-4673-2395-6/12 $31.00 © 2012 IEEE

Fabrication of Cu_2ZnSnS_4 Thin film Solar Cells by the Spin Coating technique

M.A. Olopade[a,*], O.E. Awe[b], A.M. Awobode[b], A. Oberafo[c], M.G. Zebaze Kana[c].

a Department of Physics, University of Lagos, Nigeria.
b Department of Physics, University of Ibadan, Nigeria.
c Physics Advanced Laboratory, Sheda Science and Technology Complex, Abuja, Nigeria.
Email: molopade@unilag.edu.ng

Abstract— **This study presents a novel investigation on thin film Cu_2ZnSnS_4 (CZTS) solar cells in the superstrate structure in which all the semiconductor layers were prepared under non-vacuum conditions. The solar cell structure (SLG)/FTO (Fluorine doped Tin-Oxide/Ag/CdS/CZTS/Al had the SnO_2:F (window), CdS (buffer) and CZTS (absorber) layers deposited by APCVD, Sol-gel and Sol-gel sulphurizing methods respectively. Each of the layers of the solar cells was first optimized before fabricating the solar cells. As a result of our investigations, the most efficient solar cell showed an open circuit voltage of 230mV, a short circuit current density of 4.40mA/cm^2, a fill factor of 0.277 and a conversion efficiency of 0.28%.**

Index Terms— **Sol-gel preparation, Spin coating, Superstrate, Thin films**

I. Introduction

Cu_2ZnSnS_4 (CZTS) is one of the promising materials for absorber layers of thin film solar cells [1]. CZTS thin films have suitable optical band-gap energy of 1.4-1.5eV and a large absorption co-efficient of ~10^4cm^{-1}.There is considerable interest in this compound due to the abundance of its constituents in the earth crust and its non-toxicity as compared to $CuIn_{1-x}GaSe_2$ (CIGS). Currently, CZTS thin films are prepared under vacuum conditions which make the process rather expensive and complicated. Kunihiko et al. investigated CZTS thin film solar cells on SLG substrate under a non-vacuum condition by the sol-gel sulfurization method and obtained an efficiency of 1.01% [2]. Later, Moritake et al. [3] built on this through the optimization of the CdS buffer layer to produce a conversion efficiency of 1.61%. Furthermore, Zhihua et al. [4] were able to fabricate CZTS on flexible polymide substrates by screen printing leading to an efficiency of 0.49%. Qin Miao Chen et al. [5] fabricated a superstrate structure of CZTS thin film solar cell by the screen printing process with an efficiency of 0.53%. Also, Chen Qin-Miao et al. [6] used the doctor-blade method to investigate the preparation of superstrate CZTS film for low cost solar cell and had an efficiency of 0.55%.

In this study, we adopted the sol-gel spin coating method to fabricate CZTS solar cells in the superstrate structure.

This is to enable these solar cells to be used as top cells in tandem structures of thin film solar cells to further improve their efficiencies.

II. Experiment

In this study solar cells with (SLG)/FTO/Ag/CdS/CZTS/Al structure were prepared. The SnO_2:F(FTO) film on soda lime glass were prepared using atmospheric pressure chemical vapour deposition(APCVD) machine developed at the Sheda Science and Technology Complex, Abuja, Nigeria with Tin(IV) chloride($SnCl_4$) and Hydrogen Fluoride(HF) as the precursor and dopant. High purity N_2 was used as the carrier gas and H_2O was used as the activator. $SnCl_4$ (99% pure) and HF (99% pure) were gasified in volumetric flasks by bubbling with N_2 and heating at 52^0C respectively. The flow rates were $0.54 \pm 0.1 L/m$ for $SnCl_4$ and $0.090 \pm 0.1 L/m$ for HF. The activator gas (H_2O vapour) was carried with air and gasified at 30^0C using a pump attached to a regulator. Its flow rate was $0.767 \pm 0.1 L/m$. The carrier gas (N_2) flow rate was $0.5 \pm 0.1 L/m$. The susceptor of the APCVD machine can be moved, allowing multiple, or graded deposition across the same substrate. The temperature of the glass surface (substrate) during deposition was varied from 320^0C-360^0C while the deposition time was varied from 60s to 140s. The most efficient film gave a transmittance of 80% and sheet resistance of 15Ω/sq [7].

CdS thin films were deposited onto glass substrates by sol-gel spin-coating method, from a precursor solution of cadmium acetate in 2-methoxy ethanol and Poly ethylene glycol (PEG). We dissolved 0.4ml of poly ethylene glycol in 20ml of 2-methoxy ethanol and stirred for 1 hour. Thereafter, 0.0186M of CdAc was added and stirred continuously for another 30mins, after which 0.01582M of thiourea was dissolved in 5ml of 2-Methoxy ethanol and added drop wise to the initially stirred solution. Thereafter, it was filtered and aged for 48hours. After depositions, the films were heated at a temperature of 200^0C in order to remove unwanted materials .The films have high transparency (more than 75%) in the spectral range from

978-1-4673-2395-6/12 $31.00 © 2012 IEEE

450nm to 1300nm .The analysis of the absorbance spectra showed that the optical band gap energy ranged between 2.2eV and 2.5 eV.

The depositions of CZTS thin films was carried out by the sol-gel spin coating method on the Soda Lime Glass Substrate (SLG) before sulphurizing in the furnace. The sol-gel solutions were prepared from Copper Chloride (BDH), Tin (II) chloride (BDH), Zinc Chloride (BDH), 2-Methoxy ethanol, Thiourea (BDH) and Acetic acid as the stabilizer. 2-Methoxy ethanol and Acetic acid were used as the solvent and the stabilizer respectively. 0.8ml Acetic acid was mixed in 25ml of 2-methoxy ethanol and stirred for 1 hour. Thereafter, 0.0480M of $CuCl_2$, 0.0293M of $ZnCl_2$ and 0.0277M of $SnCl_2$ were added and stirring continuously for another 30mins. After which 0.1846M of thiourea was dissolved in 5ml of 2-Methoxy ethanol and added drop-wisely to the initially stirred solution. The entire solution precipitated and 2 drops of Acetyl acetone was added to have a clear solution .The stirring continued for another 30minutes. Thereafter, it was filtered and aged for 48hours. The solution was spin coated onto SLG substrates at speed of 2400 rpm for 30seconds respectively. These films were dried in air at 250℃ for 3mins using a hot plate

CZTS solar cells with active area of $2.25cm^2$ were prepared and labeled cell A and cell B respectively. Each of the cells having a buffer layer thickness of 40nm and 60nm respectively. The crystallinity of CZTS thin films was analyzed with an X-ray Diffractometer (MD-10 Precision Mini X-ray Diffractometer). The surface morphology of the films was determined by a scanning electron microscope (SEM, MA10, Zeiss) and the layer thickness by Dektak 8 Profilometer. Elemental composition was characterized by an energy dispersive spectrometer (EDS) attached to the ZEISS SEM. Optical properties of the prepared films were recorded on a spectrophotometer (AvaSpec-2048 Standard Fiber Optic spectrometer). A solar simulator that has a 500W Xe lamp and a Keithley 2400 digital sourcemeter was used to determine the conversion efficiency and current density-Voltage (J-V) characteristics of the solar cells under illumination of AM 1.5 and power density of $1000W/m^2$.

III. RESULTS AND DISCUSSION

X-ray diffraction pattern of the CZTS thin film deposited on SLG is shown in Fig.1. The X-ray diffraction pattern shows peaks appearing at 28.55^0 and 48.61^0. These peaks are due to diffraction from the (1 1 2) and (2 2 0) crystal planes. The peak of (1 1 2) is more prominent than the peak of (2 2 0) and this indicates that the film was oriented to (1 1 2) which agrees with reported data [8, 9].

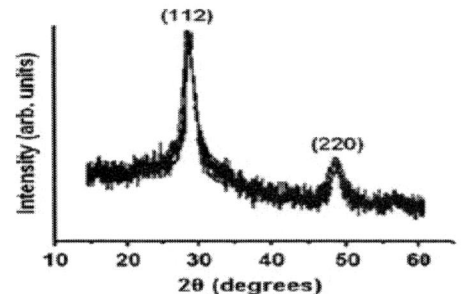

Fig.1: X-ray diffraction pattern of CZTS films grown on SLG

Fig. 2 is the surface image of the CZTS film which indicates the formation of a continuous and smooth film. Our film had smaller grain size as compared to those found in literature [10] and there are no large voids on the surface of the film.

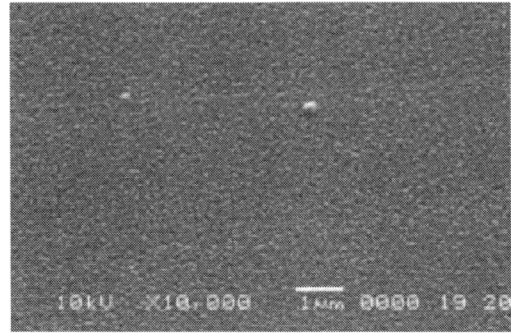

Fig.2: Surface SEM image of the CZTS thin film

The optical band gap of the samples were deduced from the plot of $(\alpha h v)^2$ vs hv as shown in Fig.3., by extrapolating the straight line portion of the graph in the high absorption regime, where α and hv are absorption co-efficient and photon energy. It was observed that our CZTS film had a band gap of ~1.51eV. This band gap is comparable with the result from reported values [1]. The absorption co-efficient in the visible region was larger than 10^4cm^{-1}. The optical properties of the CZTS thin film suggest that they are suitable for absorber layers in thin film solar cells.

TABLE I
CHEMICAL COMPOSITIONS OF CZTS THIN FILMS

	Chemical composition (at %)				Ratio of Composition		
	Cu	Zn	Sn	S	Cu/(Zn+Sn)	Zn/Sn	S/metal
CZTS thin film	26	14	13	47	0.96	1.08	1.13

The chemical composition of the CZTS thin film was almost stoichiometric but slightly Sulphur poor.

The Open circuit voltage (V_{oc}), Short circuit current (J_{sc}), Fill factor (FF) and Efficiency (η) of our most efficient solar cell deduced from the J-V characteristics as shown in Fig. 4 under light irradiation were 230mV, $4.40mA/cm^2$, 0.277 and 0.28% respectively.

This efficiency of 0.28% is less than the reported efficiencies of CZTS superstrate thin films prepared under non-vacuum conditions in which the least is 0.53% [5] and the highest is 0.55% [6].

The reduction in efficiency mentioned above is possibly due to the following:

(i) Insufficient incorporation of Na into the absorber layer from the SLG substrate likely contributed to the reduction in the efficiencies. This is because in ref. [11, 13], it was reported that inclusion of Na enhances conductivity since Na is a highly ionic species.

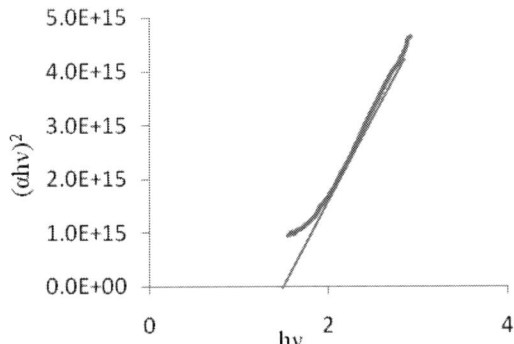

Fig.3: Absorption spectrum of CZTS thin film

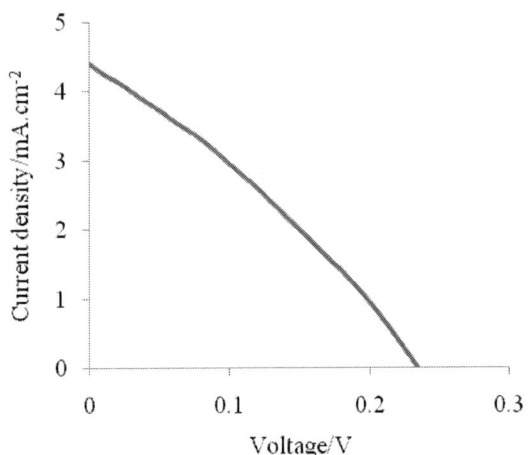

Fig.4: J-V characteristics of the most efficient CZTS photovoltaic cell with an active area of $2.25 cm^2$

(ii) The quality of the heterojunction interface between FTO and CZTS thin films possibly contributed to the observed reduction in efficiencies. This is in agreement with the submission of Haug et al. in ref. [12] on the effect of heterojunction between FTO and CIGS thin films.

(iii) The smallness of the grain size of the CZTS film possibly contributed to the reduction in efficiencies of the solar cells as suggested by Kunihiko et al. in ref. [10].

(iv)The thickness of CdS layer likely contributed to the low efficiencies of our solar cells.

(v) We also suspect that the standard of our clean room must have contributed to the poor efficiencies of our solar cells.

It is to be noted that the transmittance of our most efficient FTO films was 80% in the visible and near infra red regions. Thus in our opinion, the transmittance of our window layer was not responsible for the cause of low efficiencies of the grown solar cells.

IV. CONCLUSION

The thin film solar cells with (SLG)/FTO/Ag/CdS/CZTS/Al superstrate structure were prepared. The semiconductor layers of the CZTS thin film solar cells, FTO, CdS and CZTS, were deposited under non-vacuum conditions. The superstrate structure thin film of CZTS solar cells of active area $2.25 cm^2$ was examined under irradiation of AM 1.5 and $1000 W/m^2$. Our solar cells with buffer layers of 40nm and 60nm had efficiencies of 0.28% and 0.13% respectively.

ACKNOWLEDGEMENT

This work was supported primarily by the Sheda Science and Technology Complex, Abuja, Nigeria under the STEP-B project (a World Bank assisted grant).

REFERENCES

[1] Tooru Tanaka, Akihiro Yoshida, Daisuke Saiki, Katsuhiko Saito, Qixin Guo, Mitsuhiro Nishio, Toshiyuki Yamaguchi, Influence of composition ratio on properties of Cu2ZnSnS4 thin films fabricated by co-evaporation , Thin Solid Films (2010), doi:10.1016/j.tsf.2010.03.026

[2] T.Kunihiko, O. Masatoshi, M. Noriko, H. Uchiki, Cu_2ZnSnS_4 thin film solar cells prepared by non-vacuum processing, Solar Energy Material and Solar Cells 93(2009)583–587.

[3] N. Moritake, Y. Fukui, M. Oonuki, K. Tanaka, H. Uchiki, Preparation of Cu_2ZnSnS_4 thin film solar cells under non-vacuum condition, Physica Status Solidi C 6(2009) 1233-1236.

[4] Zhihua Zhou, Yanyan Wang, Dong Xu, Yafei Zhang, Fabrication of Cu_2ZnSnS_4 screen printed layers for solar cells, Solar Energy Material and Solar Cells 94(2010) 2042-2045.

[5] Qin Miao Chen, Xiao Ming Dou, Zhen Qing Li, Shu Yi Cheng, Song Lin Zhuang, Preparation of CZTS film by printing process for low-cost solar cell, Advanced Materials Research, (335-336) 1406-1411.

[6] Chen Qin-Miao, Li Zhen-Qing, Ni Yi, Cheng Shu-Yi, and Dou Xiao-Ming, Doctor-bladed Cu_2ZnSnS_4 light absorption layer for low-cost solar cell application, Chinese Physics B, Vol. 21, No. 3 (2012) 038401.

[7] M.A. Olopade, O.E. Awe, A.M. Awobode, N.Alu, Characterization of Sno_2:F films deposited by Atmospheric Pressure Chemical Vapour Deposition for Optimum Performance Solar Cells, The African Review of Physics (2012) 7:0018.

[8] H.katagiri, N. Ihigaki, T. Ishida, K. Saito, Characterization of Cu_2ZnSnS_4 thin films prepared by vapour phase sulfurization, Japanese Journal of Applied Physics 40 (2001) 500-504.

[9] T.Tanaka, D.Kawasaki, M.Nishio, Q.Guo, H.Ogawa, Fabrication of Cu_2ZnSnS_4 thin films by co-evaporation, Physica Status Solidi C3 (2006) 2844-2847.

[10] Kunihiko Tanaka, Noriko Moritake, Hisao Uchiki, Preparation of Cu_2ZnSnS_4 thin films by sulphurizing sol-gel deposited precursors, Solar Energy Materials & Solar cells 91 (2007) 1199-1201.

[11] T. Prabhakar, N. Jampana, Effect of Na diffusion on the structural and electrical properties of CZTS thin films, Solar Energy Material and Solar Cells (2010) doi: 10.1016/j.solmat.2010.12.012.

[12] F.J. Haug, D. Rudmann, A. Pomev, H. Zogy, and A.N. Tiwan , Electrical properties of the heterojunction in $Cu(In,Ga)Se_2$ superstrate solar cells. Proceedings of 3rd world conference in photovoltaic solar energy conversion, Osaka, (2003).

[13] Su-Huai Zhang Wei, Alex S.B. Zunger, Effects of Na on the electrical and structural properties of $CuInSe_2$, Journal of Applied Physics 85(1999) 7214-7218.

[14] Kunihiko Tanaka, Yuki Fukui, Noriko Moritake, and Hisao Uchiki , Chemical composition dependence of morphological and optical properties of Cu_2ZnSnS_4 thin films deposited by sol–gel sulfurization and Cu_2ZnSnS_4 thin film solar cell efficiency, Solar Energy Materials & Solar cells 95 (2011) 838-842.

Structural Study and Sensitivity Measurements of ZnO based Ammonia (NH₃) Sensor

S. Ahmad[1], N. D. Md Sin[2], M. H. Mamat[2], M.Salina[2], M. N. Berhan[1], M. Rusop[2]

[1] Faculty of Mechanical Engineering
[2] NANO-Electronic Centre, Faculty of Electrical Engineering
Universiti Teknologi MARA (UiTM)
40450 Shah Alam, Malaysia
Email: samsiah81@gmail.com

Abstract. Zinc Oxide (ZnO) thin films were prepared on thermally oxidized SiO₂ by radio frequency (RF) magnetron sputtering at various substrate temperature ranging from room temperature (25°C) to 500°C. The surface morphology and crystallinity were analyzed by field emission scanning electron microscopy (FESEM) and X-Ray Diffractometer (XRD) respectively. The grain size measured by FESEM is increasing with the increased of substrate temperature used. All films grown were c-axis oriented and the film deposited at 300°C exhibit the highest crystallinity. The film deposited at room temperature exhibit the highest sensitivity due the smallest grain size and the highest surface to volume ratio.

Keywords- ZnO; Magnetron sputtering; Substrate temperature; Structural properties; Sensitivity

I. INTRODUCTION

Ammonia (NH₃) is a colorless gas with a pungent odor. It is widely used in industry however it is hazardous, therefore the monitoring of its leakage is very important. Metal-oxides (SnO₂, ZnO and TiO₂) based ammonia gas sensors are being developed extensively as they are more sensitive to NH₃ and also easy to fabricate [1].

Among the listed metal oxides materials, ZnO is an attractive since it is chemically and thermally stable n-type semiconductor with a large exciton binding energy of 60 meV and large bandgap energy of 3.37 eV at room temperature [2, 3]. These properties allow the usage of ZnO in detectors sensitive to toxic and combustible gases [4].

ZnO thin films were prepared by various techniques [5-8]. In this study we choose RF magnetron sputtering method since it requires a simple apparatus and it gives higher film orientation with low growth temperature [9]. RF sputtering techniques also offers other advantages such as it gives uniform films on various substrates [10], high purity films, better controlled composition, homogeneity and it permits better control of the film thickness [11].

Although many researcher have grown ZnO on various substrates but the study on the effect of substrate temperature on SiO₂/Si substrate is still lack. Therefore in this paper, we study the effect of substrate temperature on the structural and sensitivity of ZnO thin film deposited on SiO₂/Si substrate.

The sensitivity (S) i.e. the percent reduction of sensor resistance is given by (1) [12];

$$S = \frac{Ra - Rg}{Ra} \times 100\% \quad (1)$$

where R_a is the sample resistance measured at ambient environment while R_g is that under the test gas.

II. METHODOLOGY

The ZnO thin film were deposited on 60 nm thermally grown SiO₂ on p-Si using high purity (99.999%) ZnO target. During the growth, argon (Ar) gas with 99.99% purity was introduced into the reaction system and the Ar flow rate was kept at 45 sccm. The pressure inside the chamber, the total sputtering time and RF power used were 8 mTorr, 60 min and 50 W respectively. The substrate temperature were varies from room temperature, 100, 200, 300 , 400 qnd 500 °C. The thickness of ZnO thin films was measured using surface profiler (Dektak 150+). A JEOL JSM-7000F field emission scanning electron microscope (FESEM) operating at 5 keV was used to characterize the morphology while the crystalinity have been characterized using XRD (Rigaku Ultima IV). 60 nm thick metal contacts were deposited using thermal evaporator (VCP-1100) using Aluminium (Al) metal. The device configuration is shown in Fig. 1.

The sensitivity of the ammonia (NH₃) sensor were measured using the simple gas sensor setup as shown in Fig. 2. The ammonia solution was heated at 50°C to produce NH₃ gaseous and the flow rate of carrier gas (Ar) is 300ml/min at 1 bar while the voltage was varied from -10 to 10 V. The sample box and the solution container were sealed to prevent any leakage. The measurement were done at room temperature.

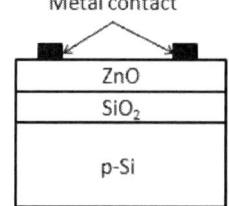

Fig. 1. Schematic of the fabricated NH₃ sensor structure

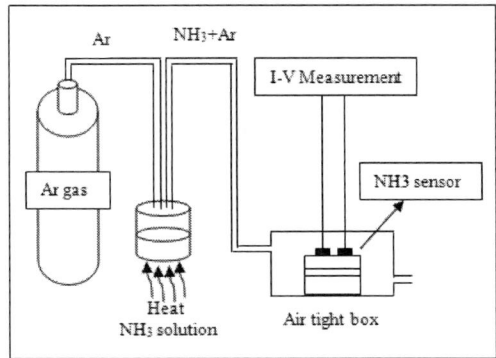

Fig. 2. The schematic NH₃ sensing properties measurement

III. RESULT AND DISCUSSION

A. Structural Properties

The average thickness of ZnO thin film deposited at room temperature, 100, 200, 300, 400 and 500°C were less than 100 nm while the deposition rate were less than 2 nm/min for all the samples. Singh et al. also reported that the thickness and deposition rate of ZnO deposited on quartz substrate also did not varies much with the variation of substrate temperature [13].

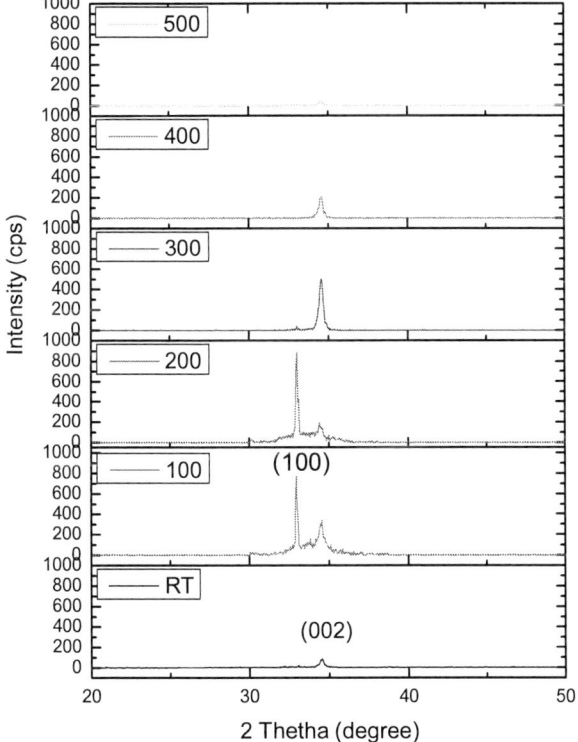

Fig. 3. XRD spectra of ZnO thin films at different substrate temperature

The XRD spectra for the ZnO thin film deposited at different substrate temperature is shown in Fig. 3. All deposited films show a peak at 2θ = 34.4° that correspond to the (002) hexagonal wurtzite structure of ZnO with JCPDS Card no. 36-1451. Overall we can conclude that as the substrate temperature increases, the (002) peak increases however if the substrate temperature is further increases to 400 and 500°C the (002) peak decreases. The highest (002) was recorded for the sample deposited at 300°C. The sample deposited at 100 and 200 °C show a strong (100) peak at 2θ = 32.9° [14].

Fig. 4(a-f) shows the plane view-FESEM images of the ZnO thin films at different substrate temperature. Fig. 4(a) shows a homogenous microstruture with sphere shape grains with average grain size of 27.84 nm. As the substrate temperature increased to 100°C as in Fig. 4(b), the grain become coarser with an average grain size of 32.44 nm. Fig. 4(c) which correspond to the sample deposited at 200°C shows that the microstructure consists of fine sphere grain and coarse agglomerated grain. The fine particle has an average diameter of 34 nm while the coarse grain have an average diameter of 69 nm. Fig. 4(d) also shows a combination of a fine and coarse grain with average diameter of 46 and 77 nm respectively. Fig. 4(e) shows that the sample consists of an elongated particles with average grain size of 61 nm. Finally for the sample deposited at 500°C as in Fig. 4(f), the sample consists of coarse grain with an average diameter of 85 nm.

From these FESEM observations, we noticed that the microstructure of the ZnO thin film change with increasing substrate temperature. The grain size becomes coarser and some of the grain agglomerate when deposited at high substrate temperature. Lee et al. [15] also reported that grain size increases with increasing substrate temperature. At low substrate temperature, a smaller particles is deposited due to the limited mobility the of the arrived atoms. As the substrate temperature increases, the mobility of the atoms also increases [16]. It was also noticed that the sample deposited at room temperature is more dense compared to the sample deposited at the higher temperature. ZnO deposited onto Corning glass also develops a dense structure with decreasing substrate temperature [17].

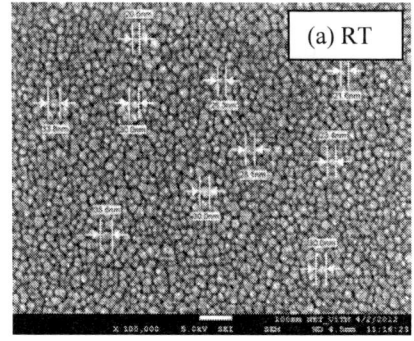

(a) RT

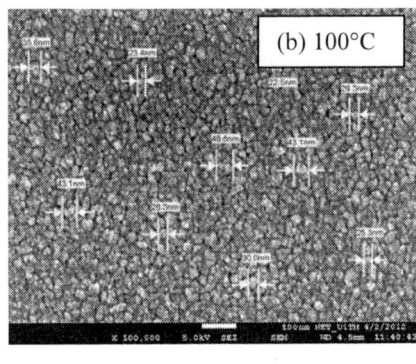

(b) 100°C

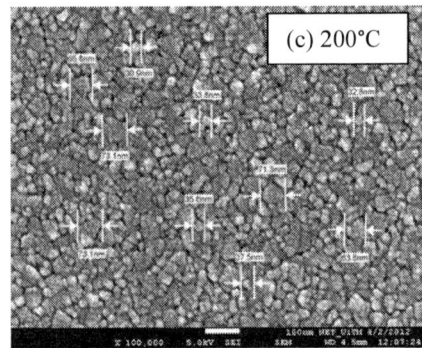

(c) 200°C

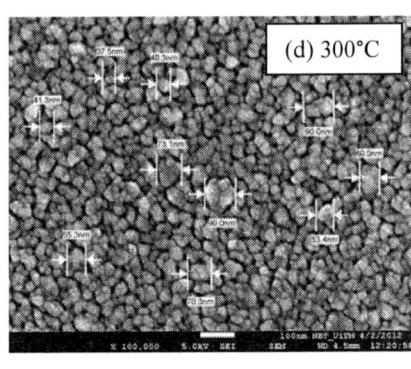

(d) 300°C

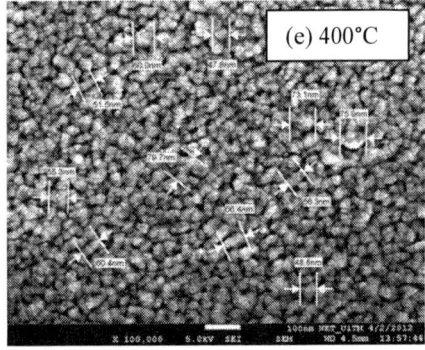

(e) 400°C

(f) 500°C

Fig. 4(a-f). FESEM images of the ZnO thin films measured at 100,000 times magnification with 5 kV. The films were deposited at various substrate temperature

B. Sensitivity

Fig. 5 shows the variation of sensitivity as a function of substrate temperature. It was found that the highest sensitivity was recorded for the sample deposited at room temperature with the sensitivity value of 37.80%. The sensitivity for the sample deposited at room temperature is the highest due to the smallest grain as compared to the other sample. Several papers [18, 19] also reported that sensitivity value increases with the decrease in the grain size. This is due to the fact that smaller grain size materials have a high surface to volume ratio. The high surface to volume ratio materials provide more active site on the surface of materials for physical or chemical interactions therefore it is expected to have high sensitivity [20]. The sample deposited at 500°C also displays a quite high sensitivity which is 34.54%. It is expected that the roughness of the materials also give a significant effect on the sensitivity of the film. According to Farmakis et. al. [21], a high surface roughness of ZnO would improve the chemical active area of the film and hence increase the sensitivity of the film.

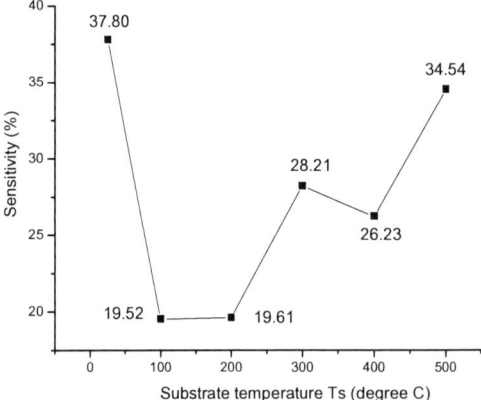

Fig 5. Variation of sensitivity as a function of substrate temperature

CONCLUSION

ZnO thin films were deposited by RF magnetron sputtering method at various substrate temperature. All ZnO films show (002) preferred orientations with the highest peak at 300°C. In summary, the sample deposited at room temperature shows the smallest grain size that gives the highest surface to volume ratio which in turns gives the highest sensitivity (36%) when exposed to NH_3 gas at room temperature.

ACKNOWLEDGMENT

The authors wish to thank the Research Management Institute (RMI), UiTM and Ministry of Higher Education (MOHE) for their financial support. Special thanks also to NANO-SciTech Centre, Institute of Science (IOS), UiTM for laboratory facilities.

REFERENCES

[1] B.Renganathan, D.Sastikumar, G.Gobi, N.R.Yogamalar, and A.C.Bose, "Nanocrystalline ZnO coated fiber optic sensor for ammonia gas detection," *Optics & Laser Technology,* vol. 43, pp. 1398-1404, 2011.

[2] Z.Bai, C.Xie, S.Zhang, L.Zhang, Q.Zhang, W.Xu, and J.Xu, "Microstructure and gas sensing properties of the ZnO thick film treated by hydrothermal method," *Sensors and Actuators B: Chemical,* vol. 151, pp. 107-113, 2010.

[3] A. M.Soleimanpour and A.H.Jayatissa, "Effect of UV irradiation on gas sensing behavior of nanocrystalline ZnO thin films," in *Nanotechnology Materials and Devices Conference (NMDC), 2010 IEEE,* 2010, pp. 225-229.

[4] S.J.Chang, W.Y.Weng, C.L.Hsu, and T.J.Hsueh, "High sensitivity of a ZnO nanowire-based ammonia gas sensor with Pt nano-particles," *Nano Communication Networks,* vol. 1, pp. 283-288, 2010.

[5] J.N.Zeng, J.K.Low, Z.M.Ren,T.Liew, and Y.F.Lu, "Effect of deposition conditions on optical and electrical properties of ZnO films prepared by pulsed laser deposition," *Applied Surface Science,* vol. 197–198, pp. 362-367, 2002.

[6] A. J. C.Fiddes, K.Durose, A. W.Brinkman, J.Woods, P. D.Coates, and A.J.Banister, "Preparation of ZnO films by spray pyrolysis," *Journal of Crystal Growth,* vol. 159, pp. 210-213, 1996.

[7] M. N.Kamalasanan and S.Chandra, "Sol-gel synthesis of ZnO thin films," *Thin Solid Films,* vol. 288, pp. 112-115, 1996.

[8] K.B. Sundaram and A. Khan, "Characterization and optimization of zinc oxide films by r.f. magnetron sputtering," *Thin Solid Films,* vol. 295, pp. 87-91, 1997.

[9] C.W.Hsu, T.C.Cheng,C.H.Yang, Y.L.Shen, J.S.Wu, and S.Y. Wu, "Effects of oxygen addition on physical properties of ZnO thin film grown by radio frequency reactive magnetron sputtering," *Journal of Alloys and Compounds,* vol. 509, pp. 1774-1776, 2011.

[10] M.Selmi, F.Chaabouni, M.Abaab, and B.Rezig, "Studies on the properties of sputter-deposited Al-doped ZnO films," *Superlattices and Microstructures,* vol. 44, pp. 268-275, 2008.

[11] M. D. J.Ooi, A. A.Aziz, and M. J.Abdullah, "Recent development in the growth of ZnO nanoparticles thin film by magnetron sputtering," in *Semiconductor Electronics, 2008. ICSE 2008. IEEE International Conference on,* 2008, pp. 514-518.

[12] J.F.Chang, H.H.Kuo, I.C.Leu, and M.H.Hon, "The effects of thickness and operation temperature on ZnO:Al thin film CO gas sensor," *Sensors and Actuators B: Chemical,* vol. 84, pp. 258-264, 2002.

[13] S.Singh, R.S.Srinivasa, and S.S.Major, "Effect of substrate temperature on the structure and optical properties of ZnO thin films deposited by reactive rf magnetron sputtering," *Thin Solid Films,* vol. 515, pp. 8718-8722, 2007.

[14] M.K.Hossain, S.C.Ghosh, Y.Boontongkong, C.Thanachayanont, and J.Dutta, "Growth of zinc oxide nanowires and nanobelts for gas sensing applications," *Journal of Metastable and Nanocrystalline Materials,* vol. 23, pp. 27-30, 2005.

[15] C.M.Lee, C.M.Kim, S.J.Kim, and Y.K.Park, "Enhancement of the Quality of the ZnO Thin Films by Optimizing the Process Parameters of High-Temperature RF Magnetron Sputtering," *Key Engineering Materials,* vol. 336, pp. 581-584, 2007.

[16] J.F.Chang and M.H.Hon, "The effect of deposition temperature on the properties of Al-doped zinc oxide thin films," *Thin Solid Films,* vol. 386, pp. 79-86, 2001.

[17] S.S.Lin, J.L.Huang, and D.F.Lii, "The effects of r.f. power and substrate temperature on the properties of ZnO films," *Surface and Coatings Technology,* vol. 176, pp. 173-181, 2004.

[18] X.L.Cao, "The Controllable Flexible Features of ZnO Thin Film Gas Sensor," *Advanced Materials Research,* vol. 335, pp. 478-482, 2011.

[19] L.F.Dong, Z.L.Cui, and Z.K.Zhang, "Gas sensing properties of nano-ZnO prepared by arc plasma method," *Nanostructured materials,* vol. 8, pp. 815-823, 1997.

[20] C.Li, Z.Du, H.Yu, and T.Wang, "Low-temperature sensing and high sensitivity of ZnO nanoneedles due to small size effect," *Thin Solid Films,* vol. 517, pp. 5931-5934, 2009.

[21] F.V.Farmakis, T.Speliotis, K.P.Alexandrou, C.Tsamis, M.Kompitsas, I.Fasaki,P.Jedrasik, G.Petersson, and B.Nilsson, "Field-effect transistors with thin ZnO as active layer for gas sensor applications," *Microelectronic Engineering,* vol. 85, pp. 1035-1038, 2008.

Design and Characterization of Bandgap Voltage Reference

Yuzman Yusoff, Hanif Che Lah, Nabihah Razali, Siti Noor Harun and Tan Kong Yew
Integrated Circuit Development, MIMOS Berhad
Technology Park Malaysia,
57000 Kuala Lumpur, Malaysia.
Email: yuzman@mimos.my

Abstract- **This paper presents the design and characterization of 1.8V bandgap voltage reference fabricated using Siltera's 0.18um CMOS process technology. The proposed bandgap voltage reference employed two-stage amplifier, start-up and power down circuit. The paper focuses on circuit analysis using SPECTRE and Monte Carlo, layout design technique for reducing mismatch and silicon characterization. The result shows the designed bandgap voltage reference generates a stable voltage reference at 1.204V with average temperature coefficient of 6.5ppm/°C. The power dissipation for this bandgap voltage reference is 150uW under 1.8V supply voltage and it occupies silicon area of 370umx300um.**

I. INTRODUCTION

A bandgap voltage reference is used extensively in most analog mixed signal circuit like data converters, smart sensors and power management circuits [1]. In these applications, an accurate and a stable biasing voltage are very critical to the accuracy and performance of the overall systems. Thus, it is important for bandgap voltage reference to provide a nearly constant reference voltage output irrespective of process, voltage and temperature (*PVT*) variations.

There have been a number of approaches reported to realize a temperature independent bandgap voltage reference in analog circuits. The most popular approach is using base-emitter voltage, (V_{BE}), of bipolar junction transistor (*BJT*) [2]. The characteristic of the bipolar transistors is unique as it can produce positive and negative temperature coefficients. It is well understood that V_{BE} is nearly complementary to absolute temperature (*CTAT*), that it decreases linearly with temperature [3]. Meanwhile, if two *BJTs* operate with unequal current densities, then the difference in base emitter voltages, ΔV_{BE}, is found to be proportional to absolute temperature (*PTAT*). The standard equation for the collector current for a bipolar transistor is shown in (1), where I_S is the saturation current and V_T is the thermal voltage equal to kT/q. From (1), the temperature dependence of the V_T is very apparent.

$$I_C = I_S \exp\left(\frac{V_{BE}}{V_T}\right) \quad (1)$$

Writing the base-emitter voltage as a function of collector current and temperature can be expressed as

$$V_{BE} = V_{G0}\left(1 - \frac{T}{T_0}\right) + V_{BE}\frac{T}{T_0} + \frac{mkT}{q}\ln\left(\frac{T_0}{T}\right) + \frac{kT}{q}\ln\left(\frac{J_C}{J_{C0}}\right) \quad (2)$$

where,

V_{G0} : Bandgap voltage of silicon (~1.205V)
T : Absolute temperature
T_0 : Reference temperature
V_{BE0} : V_{BE} at T_0
M : Temperature constant approximately 2.3
k : Boltzmann's constant
q : Electronic charge
J_C : Collector current density
J_{C0} : J_C at T_0

For a constant collector current, V_{BE} will have approximately -2.2mV/°C temperature dependence at room temperature. It can be shown by (2) that voltage difference of two base-emitter junctions biased at fixed but different current densities is proportional to absolute temperature [1]. The difference in their junction voltage is given by

$$\Delta V_{BE} = V_{BE1} - V_{BE2} = \frac{kT}{q}\ln\left(\frac{J_1}{J_2}\right) \quad (3)$$

This proportionality is quite accurate even when the collector currents are temperature dependent, as long as their ratio remains fixed [2].

The reference voltage that now can be obtained by multiplying the PTAT voltage by a factor K and adding it to the CTAT voltage. Therefore we can write

$$V_{REF} = V_{BE2} + K\Delta V_{BE} \quad (4)$$

or,

$$V_{REF} = V_{G0} + \frac{T}{T_0}\left(V_{BE2} - V_{G0}\right) + (m-1)\frac{kT}{q}\ln\left(\frac{T_0}{T}\right) + K\frac{kT}{q}\ln\left(\frac{J_2}{J_1}\right) \quad (5)$$

These equations can be used to estimate the temperature dependence at temperatures different from the reference temperature. A simplified block diagram of bandgap voltage reference is shown symbolically in Fig.1 [2].

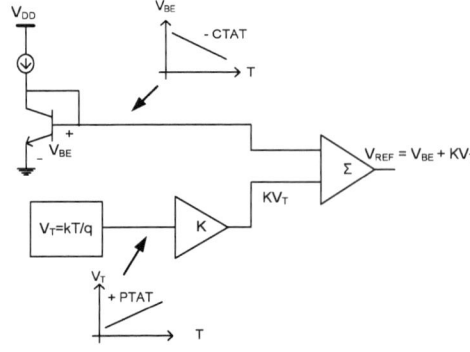

Fig. 1. Basic diagram of bandgap voltage reference

II. DESIGN IMPLEMENTION

A. Circuit Design

The proposed circuitry for bandgap voltage reference is shown in Fig. 2. It is consists of three parts; bandgap voltage reference core, start-up and power down circuit.

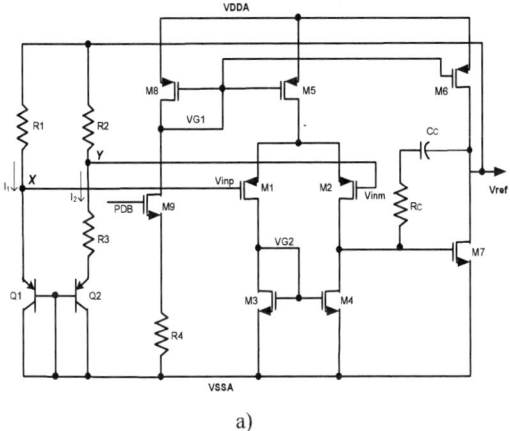

a)

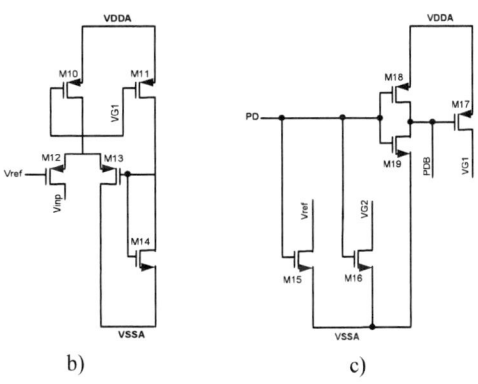

b)   c)

Fig. 2. Circuit schematic of a) Bandgap voltage reference core b) Start-up circuit c) Power down circuit

Fig. 2(a) shows the main circuitry for bandgap voltage reference. It uses operational amplifier to force nodes X and Y to be identical through the relationship (6).

$$I_1 R_1 \approx I_2 R_3 \qquad (6)$$

This architecture introduces some error on output voltage because of operational amplifier's input offset. In a real case, it is very difficult to design operational amplifier with zero offset. Thus, to lower input offset large transistor input devices are used.

The generated reference voltage is describes in (7). Eq. (7) includes the effect of input offset (V_{OS}) from operational amplifier.

$$V_{REF} = V_{BE2} + \left(1 + \frac{R_2}{R_3}\right)\left(V_T \ln n - V_{OS}\right) \qquad (7)$$

i. Two-Stage Operational Amplifier

The operational amplifier used for proposed bandgap voltage reference is two-stage amplifier with high DC gain.

This characteristic helps to enhance Power Supply Rejection Ratio (PSRR) of bandgap voltage reference and reduces some errors on the voltage reference, V_{REF}. The first stage (formed by transistors M1 - M5) and second stage (formed by transistor M6-M7) are standard differential amplifier with Miller compensation. The compensation resistor R_C is added to improve the stability performance. Fig. 3 shows the simulated bode plots of this two-stage operational amplifier in 44 conditions across process, voltage and temperature (PVT) variation. Table I summarizes its performances.

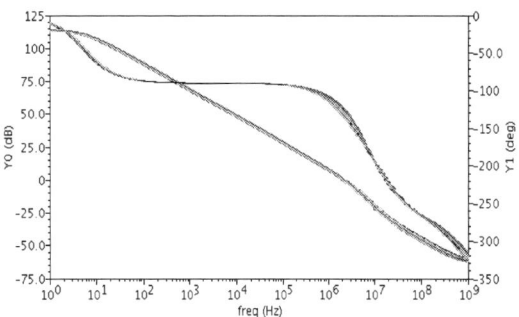

Fig. 3. Simulated gain and phase for operational amplifier

TABLE I
SUMMARY OF TWO-STAGE OPERATIONAL AMPLIFIER PERFORMANCES

Parameter	Simulated Result
DC Gain	113dB
Unity Gain Bandwidth	2.5MHz
Phase Margin	62 degree
PSRR	- 99dB
Input Offset	90uV
Current Consumption	65uA

ii. Start-up Circuit

In order to make sure the circuit works properly, a mechanism to provide small current flow through amplifier is needed. This mechanism is also known as start-up circuit. This start-up circuit is presented in Fig. 2(b). It consists of transistors M10 to M14. In this circuit, transistor M10 and M12 are used to inject current to node $Vinp$ if node $Vinp$ goes below threshold voltage. When the voltage reference V_{REF} established, transistor M13 turns on and transistor M12 turns off.

iii. Power Down Circuit

The power down (PD) signal is an active high signal that can switch ON or OFF the whole circuit whenever necessary. This power down circuit is shown in Fig. 2(c). When PD=1, M16 and M17 are ON. Due to M16, the gate of M3 and M4 goes LOW and they becomes OFF. Meanwhile, when M17 is ON, the gate of transistor M5, M6 and M8 is goes HIGH and they become OFF and, thereby, ceases the operation of the core circuit. Transistor M9 is used to cut off the branch that provides bias current to the operational amplifier. The voltage reference V_{REF} is also pulled to ground during the power down mode.

978-1-4673-2395-6/12 $31.00 © 2012 IEEE   687

B. Physical Layout Design

The layout arrangement and technique is the key point for the precision of the output of bandgap reference voltage. The main important point that should consider in the bandgap layout is the parasitic resistance rather than parasitic capacitance. Things that had been taken care in this bandgap layout are matching the transistor, resistor and also BJTs.

For operational amplifier, the matching of net for both the input is the important thing. Mismatching means amount of signal reaching at particular point at given time will be different which impact the circuit performance. The common centroid and current mirror techniques are applied for both inputs net to achieve best matching. It is beneficial to precision of the reference since the equality of the MOS pair will determines the equality of the both input current.

Then the inter-digitized matching was applied for the resistor part with the dummies on the ends of rows. To improve resistor matching, all should be made up with the same material and have the same widths. Since some effects of process such like the lateral diffusion will introduce systematic mismatches to resistors having different widths. It is much better to constitute the resistors by groups of parallel or serial resisters with identical geometries, this will reduce the corner effects.

The other parts are *BJTs*, this use the common centroid technique. The single *BJT* is put in the middle and surrounded by the others number of *BJTs*, fully covered by a guard-ring and also make wide metal connection. Same goes for capacitor.

The complete physical layout of this bandgap voltage reference is shown in Fig. 3. It occupies $(370 \times 300) \mathrm{um}^2$ silicon area without I/O pads. Most of the area is taken up by the two-stage operational amplifier that uses large transistor devices.

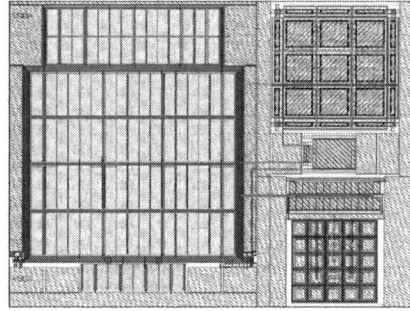

Fig. 4. Layout of bandgap voltage reference

II. SIMULATION AND MEASUREMENT RESULTS

Simulations have been carried out using 180nm Siltera's CMOS technology under the operating voltage of 1.8V and simulation results are obtained by *SPECTRE*. Fig. 5 shows the simulated output reference voltage V_{REF} (1.204V) of the proposed bandgap reference as a function of temperature over the range -20°C to 90°C. The variation of the V_{REF} for 5 different process corners (i.e. tt, ss, ff, fs and sf) are shown in the same figure over the aforesaid temperature range. The curves exhibit a variation of 659uV, 661uV, 666uV, 678uV and 654uV for tt, ss, ff, fs and sf corners respectively. The corresponding temperature coefficients are

5.99ppm/°C, 6.01ppm/°C, 6.05ppm/°C, 6.16ppm/°C and 5.95ppm/°C respectively.

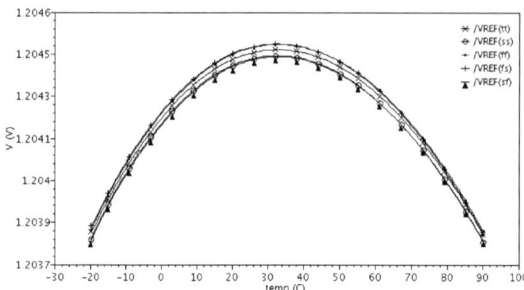

Fig. 5. Variation of V_{REF} over 5 different process corners

The time required for the bandgap voltage reference to produce a stable output is ~9us (27°C, 1.8V). It is found from the transient simulation of the bandgap voltage reference output voltage when the supply voltage V_{DD} is given as a ramp signal. This simulation of the proposed bandgap voltage reference has been carried out for ramp input that goes from 0V to V_{DD} within 1uS and the corresponding output voltage waveforms are shown in Fig. 6.

Fig. 7 shows the output V_{REF} variation with three different power supply; 1.7V, 1.8V and 1.9V over temperature range of -20°C to 90°C. The simulations are performed at typical process corner. As seen from the plot, the variation of V_{REF} is within 800uV.

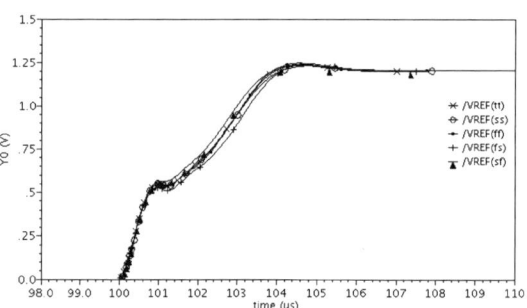

Fig. 6. Start-up behavior of the bandgap voltage reference output with a V_{DD} ramp

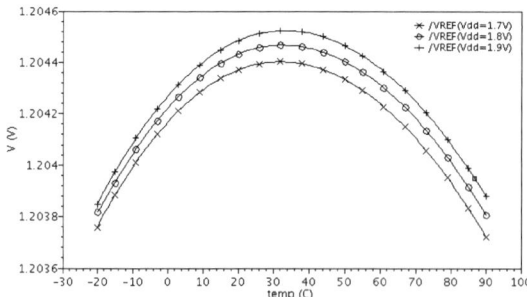

Fig. 7. Variation of V_{REF} with different power supply V_{DD} and temperature at typical process corner.

The PSRR Vs. frequency plot is shown in Fig. 8, from which it can be found that PSRR of the proposed bandgap voltage reference is 84.6dB at tt process corner at 27°C at low frequency. The values of PSRR over different process corners at 30°C and 90°C are given in the Table II.

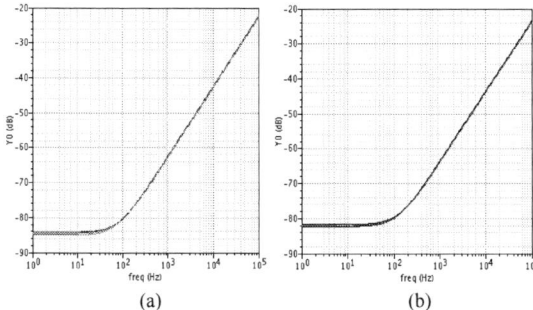

(a)   (b)

Fig. 8. Variation of V_{REF} with different power supply V_{DD} and temperature (a) 30ₒC (b) 90°C at 5 different process corners

TABLE II
SUMMARY OF PSRR OVER PROCESS CORNERS

Process Corners	PSRR (30°C)	PSRR (90°C)
tt	84.6dB	82.3dB
ss	84.1dB	81.7dB
ff	85.1dB	82.4dB
fs	84.4dB	81.6dB
sf	84.5dB	82.1dB

Monte Carlo (MC) simulations were used to determine the process and devices (transistor, resistor and BJT) mismatch affects on the bandgap voltage reference output. This analysis is performed under 1.8V supply at 27°C in typical (tt) process. As shown in Fig. 9, the output reference varied by less than 4mV across 1000 runs. The mean value for is V_{REF} 1.2044V.

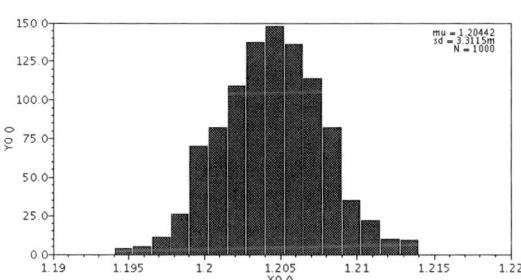

Fig. 9. Simulation result for V_{REF} on 1000 Monte Carlo runs

The performance metrics of different state-of-the-art bandgap voltage reference circuits along with the proposed one are shown in Table III. From the table, it can be seen that the proposed bandgap voltage reference has got a very low temperature coefficient and a high PSRR compared to other architectures.

TABLE III
BANDGAP VOLTAGE REFERENCE PERFORMANCES COMPARISON

PARAMETERS	[4]	[5]	[6]	This work
Technology	180nm	180nm	180nm	180nm
Power Supply, V_{DD}	1.5V	3.3V	1.8V	1.8V
V_{REF}	1.116V	1V	712mV	1.204V
Power Dissipation	0.22mW	0.9mW	-NA-	150uW

Temp. Range	0°C to 80°C	15°C to 100°C	-40°C to 80°C	-20°C to 90°C
TC	37 ppm/°C	9.5 ppm/°C	8 ppm/°C	6.1 ppm/°C
PSRR	45dB	-NA-	81dB	84dB

Fig. 10 shows the measurement results for V_{REF} that were captured from 6 chips. Due to equipment limitation, the fabricated bandgap voltage reference can be measured at 5 different temperatures; 10°C, 30°C, 50°C and 70°C. The maximum variation for V_{REF} is 1.3mVgiven by chip 5.

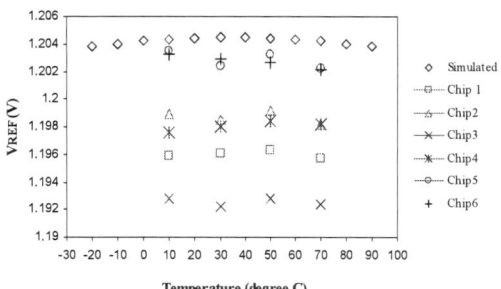

Fig. 10. Measurement result on V_{REF} for 6 different chips.

IV. CONCLUSION

A bandgap voltage reference with high PSRR and low temperature coefficient is presented. Using 1.8V supply, a reference voltage V_{REF} of 1.204V has been generated, which shows a temperature coefficient less than 6.5ppm/°C over a wide temperature range from -20°C to 90°C. The proposed circuit registers a PSRR of more than 80dB at low frequency at typical process corner. It consumes 150uW and occupies (370x300)um² silicon area.

V. REFERENCES

[1] P. E Allen and D. R Holberg, "Bandgap Reference," in *CMOS Analog Circuit Design*, 2ⁿᵈ Ed. Oxford University Press, 2002.

[2] B. Razavi, "Bandgap References," in *Design of Analog CMOS Integrated Circuits*, McGraw-Hill, 2001.

[3] T. H. Lee, "Voltage References and Biasing," in The Design of CMOS Radio-Frequency Integrated Circuits, 2ⁿᵈ Ed. Cambridge University Press, 2004.

[4] M. Jingwen, C. Tingqian, C. Cheng, R. Junyan and Y. Li, "CMOS 1.5V Bandgap Voltage Reference," ASIC, 2005. *ASICON 2005. 6ᵗʰ International Conference*, vol. 1, pp. 469-472, October 2005.

[5] D. Li, J. Ye and Z. Hong, "A Current Mode CMOS Bandgap Reference for Differential Signal Processing," ASIC, 2005. *ASICON 2005. 6ᵗʰ International Conference*, vol. 1, pp. 454-458, October 2005.

[6] W. Xichuan, S. Cuiying and X. Xing, "Curvature Compensated CMOS Bandgap Reference with 1.8-V Operation," *High Density Microsystem Design and Packaging and Component Failure Analysis 2006*, pp.20-23, 2006.

An Improved P+/N Diode Leakage Current in BiCMOS Technologies with Fluorine Co-implant

Siti Zubaidah Md Saad[a,c,*], Tan Chan Lik[a], Marhanis Abu Othman[a], Poehle Holger[b], and Sukreen Hana Herman[c]

[a] *Technology Department, Infineon Technologies (Kulim) Sdn Bhd, Lot 10 & 11, Jalan Hi-Tech 7, Industrial Zone Phase II, Kulim Hi-Tech Park, 09000 Kulim, Kedah, Malaysia.*
[b] *Technology Department, Infineon Technologies AG, Am Campeon 1-12, 85579 Neubiberg, Bavaria, Germany.*
[c] *Faculty of Electrical Engineering, Universiti Teknologi MARA, 40450 Shah Alam, Selangor, Malaysia.*
*sitizubaidah_mdsaad@yahoo.com

Abstract- **Fluorine (F) is a co-implant species known to have numbers of beneficial effects to the semiconductor device. In this study, we demonstrate the effect of fluorine to the p+/n-junction leakage current improvement in BiCMOS technologies. In which, fluorine and boron fluoride (BF_2) were used instead of fluorine-boron (F-B) or BF_2 only. By changing the implant sequence at P+ region from F followed by BF_2 to BF_2 followed by fluorine (BF_2-F), the leakage current improved by one decade with a higher breakdown voltage also observed in the reverse sequence, BF_2-F.**

I. INTRODUCTION

The influence of fluorine as a co-implant species has been studied by many researcher recently. Many of these papers, study the effect of fluorine in the formation of shallower pn-junction due to the Transient Enhanced Diffusion (TED) effects by using BF_2 or F-B [1-2]. A. Nakada *et al.* [3] investigated the effect of fluorine in the BF_2 to the leakage current at different annealing temperature. But not many studies have been done on the effect of fluorine to the leakage current upon implant sequence at the P+ region.

In most of the previous studies, fluorine's effect has been studied in the form of boron fluoride (BF_2) [1,3] and in some works, fluorine and boron (B) were implanted separately [4,5]. However in this study, F-BF_2 is used instead of F-B or BF_2 only. The main purpose of the fluorine implant prior to the BF_2 is to suppress the Negative Bias Temperature Instability (NBTI) effect of this device [6] and not to serve as co-implant in creating shallower pn-junction. Therefore, this paper will discuss more on the effect of fluorine to the leakage current.

End-of-Range defect (EOR) is one of the common defects introduced by ion implantation that is usually situated in the depletion region of the junction, causing a leakage current [7]. However, Haizhou Yin *et al* [8] have demonstrated that junction leakage current can be minimized by shallow-implant-induced EOR defects due to the fact that these defects are outside of depletion region junction in Direct Silicone Bonded (DSB) technology. Hence, in this experiment the effect of EOR on the improvement of the leakage current is investigated.

II. EXPERIMENT

The experiments were performed in a BiCMOS product but only focus at the p+/n-well junction of PMOS device. In order to investigate the effect of the implant sequences, we fabricated two types of sample. For the first one, $3x10^{15}$ ion/cm^2 fluorine was implanted with the implantation energy of 30 keV followed by the source/drain (S/D) implant with $5x10^{15}$ ion/cm^2 BF_2 at 40keV (F-BF_2). The second sample was implanted with a reverse sequence, where BF_2 is implanted prior to the fluorine implant (BF_2-F) with the same energy and dosage as the first one. Table 1 summarizes the implant conditions.

Agilent 4156C Semiconductor Device Analyzer and Cascade Microtech Probe station were used for the electrical characterization of the devices at temperatures of 25°C, 100°C and 175°C on the completed product. The voltage is sweep from -10V to 10V to study on the current-voltage (IV) curves. Leakage current is defined at -6V and breakdown voltage at 1μA. While for the physical characterization, Secondary Ion Mass Spectroscopy (SIMS) profiling is used to compare the implant profile for both samples and Transmission Electron Microscopy (TEM) for the EOR defects verification.

Table 1: Implant conditions

Process Parameters\Sample	F-BF_2	BF_2-F
First implant (Dopant/Dose/Energy)	F/3.00E+15/30keV	BF2/5.00E+15/40keV
Second implant (Dopant/Dose/Energy)	BF_2/5.00E+15/40keV	F/3.00E+15/30keV

III. RESULTS AND DISCUSSIONS

The IV- curves for both samples in reversed and forward bias are shown in Figures 1 (a) and (b) below. The electrical results of the reverse bias depict a reduction of leakage current by one decade on the BF_2-F sequence compared to the F-BF_2 at -6V. Fig. 1 (b) however shows a comparable leakage in the forward bias. Forward bias is mainly contributed by lightly doped area, which is N-well. Since there is no change has been done in N-well, we should expect there is no effect at the forward bias for both sequences.

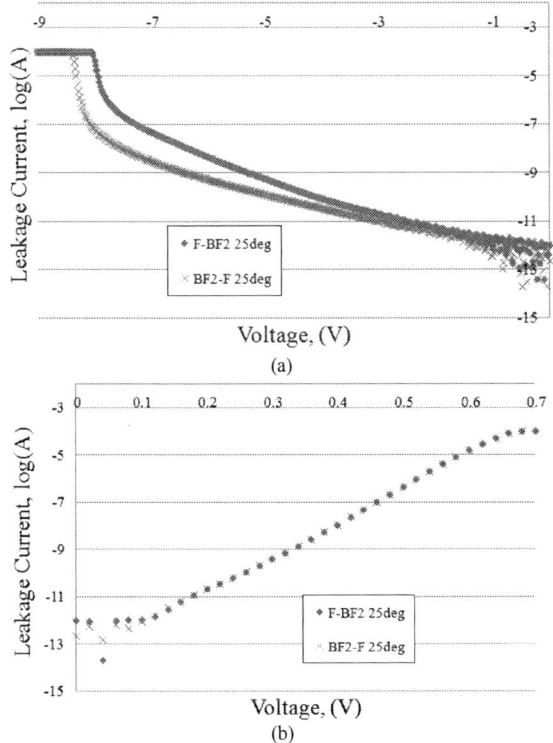

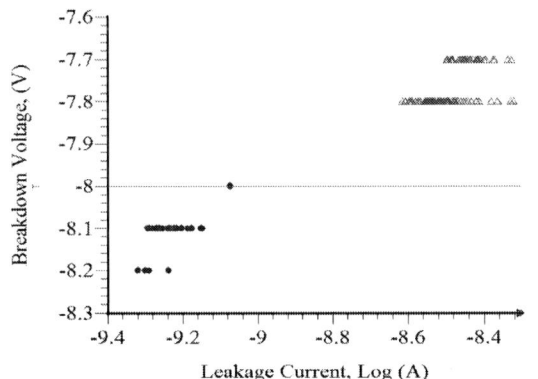

Fig. 1: IV-curve of BF_2-F and F-BF_2 (a) reversed bias and (b) forward bias [9].

Fig. 2: Breakdown voltage vs. leakage current for a number of samples in BF_2-F (●) and F-BF_2 (△) [9].

The breakdown voltage also improved for the BF_2-F sequence, as seen in Fig. 2.

The SIMS plots of boron and fluorine profiles for both samples are shown in Fig. 3 together with an overlay of phosphorus-doped of N-well profile to portray the p+/n-junction depth. The p+/n-junction, which is defined at the intercept of B and N-well profile, shows BF_2-F is slightly shallower by 14nm compared to the F-BF_2. This is probably due to more extensive amorphization of the P+ area by BF_2 compared to that of F, thus when BF_2 was implanted prior to the fluorine (BF_2-F), BF_2 has fully amorphized the P+ area and therefore, slightly shallower junction is observed.

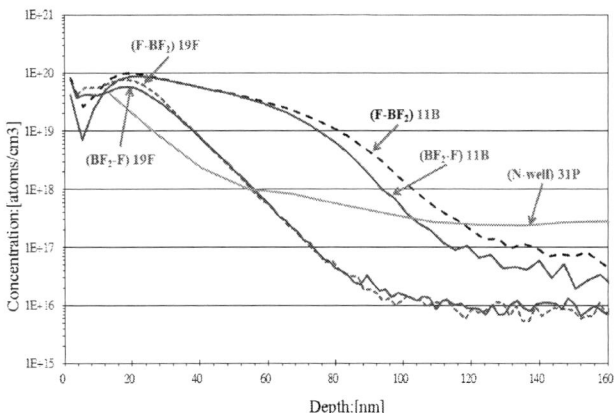

Fig. 3: Boron B (denoted as 11B), fluorine F (19F) and phosphorous P (31P) SIMS profiles. The p+/n-junction of the sample is defined as the intercept of B and P.

From the SIMS plots as well, lower dopant concentration by ~12% of boron and ~27% of fluorine in BF_2-F compared to F-BF_2 is seen. This explains both the higher breakdown voltage and the lower leakage current in the BF_2-F as can be explicated by the device physics itself. In the thermal equilibrium condition, the total negative charge per unit area in the *p*-side must be equal to the total positive charge per unit area in *n*-side [10]. Since there is no change in the N-well concentration for both sequences, the lower boron level in BF_2-F sequence which can be translated to lower negative charge concentration on the p-side results in a wider depletion region, thus the breakdown voltage also higher compared to that of F-BF_2 sequence.

The lower leakage current in BF_2-F, however, can be explained by the lower fluorine level. In an ideal pn-junction, most of the minority carriers pass across the depletion region but in reality there will be some generation and recombination in the trapping centers. Equation (1) below, current in pn-junction is the sum of an ideal or diffusion current (I_{DIFF}) and generation-recombination current (I_{R-G}) [11]. This I_{R-G} is the leakage current in pn-junction.

$$I = I_{DIFF} + I_{R-G} \qquad (1)$$

We assume when the fluorine level is lower in BF_2-F, fewer traps are created by fluorine at the depletion region, therefore will be less generation-recombination which contributing to lower leakage current.

Further, as can be seen from Fig. 4, for the temperature dependency of the reverse current, leakage current is consistently lower for the BF_2-F at temperature 25°C, 100°C and 175°C. At room temperature, the BF_2-F leakage current improved by one decade compared to F-BF_2. When the temperature is increases, the difference between BF_2-F and F-BF_2 is smaller. Most probably at high temperature, the leakage current in the depletion region is dominated by the generation-recombination effect. So, the effects of the process change, such as fluorine and boron concentration, are less severe at high temperature.

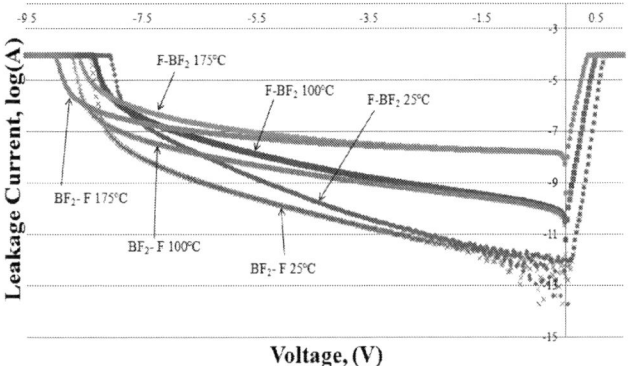

Fig. 4: IV-curve for temperature dependency at 25°C, 100°C and 175°C.

From the temperature test results above (Fig. 4), an Arrhenius plot as Fig. 5 is plotted. From the equation (2), a plot of $\ln(I_s/T^2)$ versus $1/T$ gives a straight line and the activation energy (E_a) is calculated from the slope. Also from the intercept on the y-axis, the effective Richardson constant (A^*) can be determined [12].

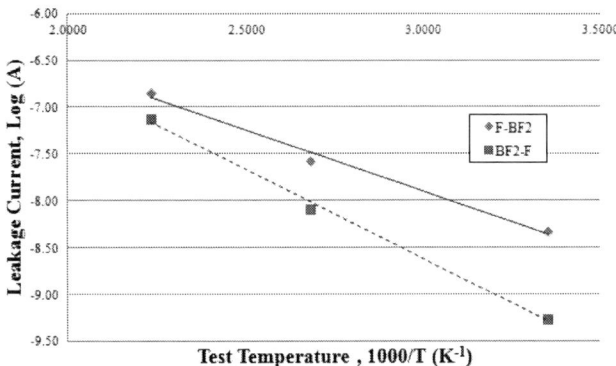

Fig. 5: Temperature dependency of leakage current at 25°C, 100°C and 175°C at applied voltage of -6V [9].

$$\ln\left(\frac{I_s}{T^2}\right) = -\frac{1}{T}\frac{q}{k}\Phi_B + \ln(AA^*) \qquad (2)$$

The E_a of the BF_2-F is 0.16eV, while F-BF_2 is 0.11eV. From the Arrhenius plot, for ideal diode, higher E_a means the pn-junction is more sensitive to the temperature effect, thus will give higher A^*. However in this experiment, the E_a is higher but the A^* is lower in the BF_2-F compared to the F-BF_2. These results explain that the lower leakage current for the BF_2-F is due the lower A^*. From (3), A^* is a function of the effective mass (m^*) [10]. Based on the SIMS results (Fig. 3), since the BF_2-F has fully amorphized the P+ region, the effective mass will be less which contributes to the lower leakage current.

$$A^* = \left(\frac{4\pi e m^* k^2}{h^3}\right) \qquad (3)$$

The TEM results are shown in Fig. 6. There is no significant different between F-BF_2 and BF_2-F samples. Therefore the reduction in the leakage current of BF_2-F is not due to the EOR effect.

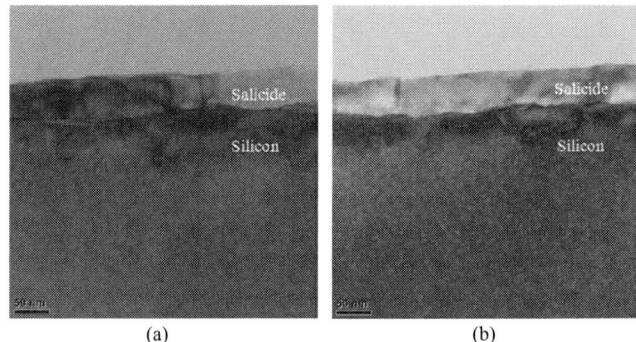

Fig. 6: TEM images of (a) F-BF_2 and (b) BF_2-F.

IV. CONCLUSION

The effect of the implant sequence at the P+ region of a BiCMOS device was investigated. We demonstrated a reduction of the leakage current by 1 decade by implanting BF_2 followed F (BF_2-F) compared by the reverse sequence implant process (F-BF_2). This is due to heavier Atomic Mass Unit (AMU) of BF_2 than fluorine which has fully amorphized the P+ region and creates slightly shallower pn-junction in BF_2-F compared to that of F-BF_2. This at the same time creates a lower effective mass, thus reduced the leakage current in BF_2-F.

Lower boron level in the BF_2-F SIMS profile explains on the improvement of the breakdown voltage by creating a wider depletion region, whereas the lower fluorine level in the BF_2-F SIMS profile possibly explains the improvement of the leakage current by forming less traps for generation-recombination.

ACKNOWLEDGEMENTS

Work supported in part by Infineon Technologies (Kulim) Sdn Bhd and FRGS/1/2012/TK02/UITM/03/8.

REFERENCES

[1] J. Liu, D. F. Downey, K. S. Jones and E. Ishida, "Fluorine effect on boron diffusion: chemical or damage?", *Proc. IEEE Int. Conf. on Ion Implantation Tech.* 2, pp. 951 – 954 (1999).

[2] G. Impellizzeri, S. Mirabella, F. Priolo, E. Napolitana and A. Carnera, *J. Appl. Phys.*, 99, 103510 (2006).

[3] A. Nakada, K. Kanemoto, M. Massazumi Oka, Y. Tamai and T. Ohmi, "Influence of fluorine in BF2+ implantation on the formation of ultrashallow and low-leakage silicon p+n junctions by 450–500 °C annealing", *J. Appl. Phys.*, 82, 2560-2565 (1997).

[4] L. Y. Krasnobaev, N. M. Omelyanovskaya, and V. V. Makarov , "The effect of fluorine on the redistribution of boron in ion-implanted silicon", *J. Appl. Phys.*, 74, 6020-6022 (1993).

[5] H. Kinoshita, T. H. Huang and D. L. Kwong, "Modeling of suppressed dopant activation in boron- and BF2-implanted silicon", *J. Appl. Phys.*, 75, 8213-8215 (1994).

[6] T. B. Hook, E. Adler, F. Guarin, J. Lukaitis, N. Rovedo, and K. Schruefer, "The effects of fluorine on parametrics and reliability in a 0.18-μm 3.5_6.8 nm dual gate oxide CMOS technology", *IEEE Electron Devices,* 48, pp. 1346-1353 (2001).

[7] S. Acco, J.S. Custer and F.W. Saris, "Avoiding end-of-range dislocations in ion-implanted silicon", *Materials Science and Engineering B34*, pp. 168-174 (1995).

[8] H. Yin *et al.*, "Effect of End-of-Range Defects on Device Leakage in Direct Silicon Bonded (DSB) Technology", *IEEE VLSI-TSA*, pp. 34-35 (2008).

[9] S.Z.M Saad, C. L. Tan, M. A. Othman, P. Holger and S. H. Herman, "A study of fluorine implant in the formation of low leakage p+/n junction in BiCMOS technologies", *Proc. IEEE Int. Conf. on ESciNano*, pp. 1-2 (2012).

[10] S. M. Sze and Kwok K. Ng, "Physics of Semiconductor Devices", *Wiley*, pp. 81, 156 (2007).

[11] Dieter K. Schroder, "Semiconductor Material and Device Characterization", *Wiley*, pp. 186 (2006).

[12] Muhammad Yusuf Ali, "Passivation of Si (100) Surface and Fabrication of Doping-Free MOSFET", PhD Thesis University of Texas at Arlington, pp.20(2008).

Oxygen Uptake During the MBE Growth of $Al_xGa_{1-x}As$ Epitaxial Layers

[1]A. A. RahmanOthman, *Student Member, IEEE*, [1]B. F. Usher, [1]A. Loykaew and [2]D. Nelson

[1]Department of Electronics Engineering,
La Trobe University, Bundoora,
Victoria 3086, Australia.
[2]School of Natural Sciences,
University of Western Sydney,
Penrith 2751, New South Wales.
Email: b.usher@latrobe.edu.au, D.Nelson@uws.edu.au

Abstract- **This paper presents the results of a study of oxygen incorporation in $Al_xGa_{1-x}As$ epitaxial layers during MBE growth. Controlled, low levels of high purity oxygen have been introduced during growth and the structural properties of the layers measured by high resolution x-ray diffraction (HRXRD), while Secondary Ion Mass Spectrometry (SIMS) has been used to measure the oxygen content in the layers. The observations are interpreted by comparing these lattice parameter measurements with $Al_xGa_{1-x}As$ layers grown without admitting oxygen to the growth chamber. In the presence of oxygen, the $Al_xGa_{1-x}As$ lattice parameter exhibits contraction for Al fractions x up to 0.65 while layers with x fractions greater than 0.65 show an expansion in the $Al_xGa_{1-x}As$ lattice parameter. The x-fraction oxygen incorporation dependence, as well as the lattice parameter behavior, are accounted for by a model which discriminates between the propensities for O_2 to incorporate at As sites by (i) bonding to pairs of Ga atoms in the layer below, (ii) to pairs of Al atoms in the layer below or (iii) a combination of one Ga and one Al atom in the layer below.**

I. Introduction

Epitaxial layers of group III-V compound semiconductors such as AlGaAs have found wide application in high speed electronic and optoelectronic devices. It is of considerable importance to consider factors that may compromise device reliability as this is a key factor in materials choice, particularly if replacement due to device failure is expensive, as it is for example in long haul undersea fibre-optic transmission systems. Due to the high reactivity of oxygen with Al [1,3-8], the choice to employ AlGaAs layers in a device structure should be made in the light of an understanding of the factors that govern the rate of oxygen incorporation. These considerations are important, irrespective of the growth technique [4].

It has been observed that the presence of a vacuum leak, even at the lowest detectable level, in a molecular beam epitaxy (MBE) growth chamber can cause the lattice constant of a grown AlAs layer to change significantly from what it would have been if it was grown with no leak [5]. While this is readily noticeable in the binary compound AlAs, it is likely to be an issue in the ternary $Al_xGa_{1-x}As$ also. This can result in considerable errors when using x-ray diffraction to characterise the structural properties of a layer and therefore infer the Al

content. Such a process depends on having reliable knowledge of the endpoint binary lattice parameters as well as the dependence of the lattice constants of the ternary AlGaAs across the range of Al compositions. Equally importantly, Al containing semiconductor alloys are known to be sensitive to oxygen contamination particularly when grown at higher Al fractions and while the presence of a small amount of oxygen in a growth chamber is unavoidable and perhaps tolerable, at higher levels the Al oxidizes readily and forms nonradiative recombination centers. Oxygen contamination in this and other Al containing materials is therefore one of the most important determinants of material quality in the context of device performance and reliability [1-8].

High resolution x-ray diffraction (HRXRD) is a nondestructive technique requiring no sample preparation that is routinely used to characterise semiconductor materials [9-20,22], to determine lattice parameters, layer thicknesses and compositions. The application of the HRXRD technique to the analysis of AlGaAs materials requires the AlAs lattice parameter to be known with high accuracy, however a wide range of values have been reported, between 5.6605 Å and 5.6622 Å [2,10,12-20]. AlAs does not exist in bulk form due to its reactivity with oxygen. Since the lattice parameter of homo-epitaxial AlAs grown on an AlAs substrate would still depend on the cleanliness of the growth environment, standardization of the lattice parameter of AlAs is difficult. This problem can give rise to significant errors in determining $Al_xGa_{1-x}As$ compositions and so this issue remains a matter of technological importance and scientific interest.

In this study we report on the uptake of oxygen in AlGaAs epitaxial layers when grown by the MBE technique. We used SIMS to measure oxygen levels in the layers and HRXRD to measure the change in the perpendicular lattice parameter as a result of the incorporation of oxygen. To our knowledge, no study has been reported of the effect on the AlGaAs lattice parameters in layers grown by the MBE technique, of oxygen incorporation due to a deliberate oxygen leak. We have measured the effects of such an oxygen leak as a function of Al fraction x, including changes in the AlGaAs lattice constant due to oxygen uptake in epitaxial layers as measured by HRXRD. Our experimental data show that oxygen incorporation into a growing AlGaAs layer results in significant lattice parameter changes and the effects of oxygen incorporation are nonlinear across the full range of Al fractions.

II. EXPERIMENT

Before the effects of oxygen incorporation could be measured, a series of AlGaAs standard samples were grown along the lines of the structure shown in Fig.1, at compositions between $x \cong 0$ and 1.0, without any deliberate exposure of the growth surface to oxygen. Growths were performed in a highly modified solid source Varian MBE-360 system with a residual background pressure in the growth chamber better than 10^{-10} Torr. Prior to growth, the GaAs substrates were degreased, etched using semiconductor grade chemicals, rinsed in de-ionized water and blown dry with N_2. They were then soldered onto Mo blocks using high purity Indium and introduced into the MBE system through a vacuum load lock. The surface oxides were thermally removed in vacuum following which the substrate temperature was reduced by about ~10°C and held for 15min in order to ensure complete removal. A GaAs buffer layer was grown and the surface temperature was calibrated by making RHEED observations of the surface reconstruction change from a 2x4 to a 2x2 pattern. A second GaAs buffer layer was then grown for 5min at a substrate temperature of 600°C to provide a smooth and clean surface for subsequent growth. Reflection high energy electron diffraction (RHEED) oscillation observations, primarily the persistence of the oscillations, allowed a check to be made regarding the optimum As_4 flux for the desired growth rate [21]. The setting of furnace and substrate temperature and their control and the shutter switching sequences were computer-controlled, allowing reproducible growth of precise structures.

Returning to Figure 1, the 10 period AlAs/GaAs MQW was designed to allow precise determination of Al and Ga fluxes in each sample, thus allowing determination of the Al composition with precision as described in [22]. The AlAs reference perpendicular lattice constant was measured by including an AlAs single layer in each sample grown to provide an internal calibration for one of the end-point binaries associated with the AlGaAs system, the other being GaAs, the growth substrate. The AlGaAs layer was designated as the layer of interest and

| 200Å GaAs cap |
| 4000Å Al$_x$Ga$_{1-x}$As |
| 200Å GaAs |
| 4000Å AlAs |
| [AlAs/GaAs] MQW (x10) |
| GaAs (001) substrate + buffer layer |

Fig. 1. The generic sample structure for flux rate calibration and AlGaAs lattice parameter measurements.

the designed structures XRD spectra were simulated at different compositions $x \cong 0.15, 0.30, 0.50, 0.75$ and 1.0 using dynamical simulation software to ensure the fringes arising from the MQW and the two single layer peaks did not overlap. To inhibit oxidation of the Al containing layers a GaAs layer capped the structure. For a particular target composition, each AlGaAs sample was grown at different Al and Ga furnace temperatures which were determined from Arrhenius plots of historical data and those temperatures were kept constant to within 0.1°C to ensure constant group III fluxes. The total group III growth rate was maintained at 1 ML/s during the growth of the AlGaAs layers in each sample, regardless of the Al fraction in the layer.

Following growth of the set of standard samples, a set of five AlGaAs samples as shown in Fig. 2, each containing two AlGaAs layers, were grown with one of the AlGaAs layers of interest grown while oxygen was admitted to the growth chamber through a precision leak valve. The leak rate was set and controlled by monitoring the output signal of a mass spectrometer tuned to the 32 amu oxygen peak, at an indicated ion current of $\cong 0.4 \times 10^{-12}$ Amps.

III. RESULTS AND DISCUSSIONS

All samples grown were measured by high resolution x-ray diffraction, using a conventional x-ray source with a Cu target. The source was operated at 45kV and 40mA and an x-ray mirror was used in conjunction with an asymmetric reflection Ge (220) Bartels style monochromator to give incident x-ray fluxes of the order of 7×10^5 cps. For all samples studied, (004) spectra were collected at azimuthal Phi angles of 0°, 90°, 180° and 270° to allow tilt effects to be detected and accounted for in the analysis. XRD rocking curves were collected for both the standard samples and the samples grown with an oxygen leak into the growth chamber. Each spectra in the set of standard samples included a set of MQW satellite fringes and two individual peaks arising from the Al$_x$Ga$_{1-x}$As layer of interest and the AlAs reference layer. The error associated with determination of the peak positions from XRD rocking curves

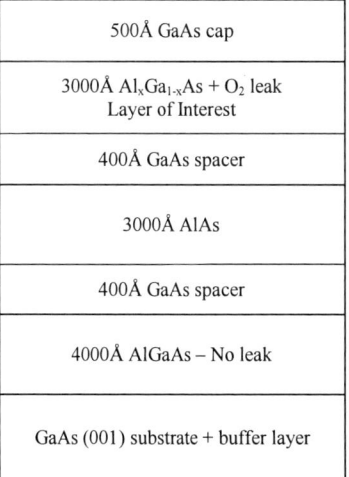

| 500Å GaAs cap |
| 3000Å Al$_x$Ga$_{1-x}$As + O$_2$ leak Layer of Interest |
| 400Å GaAs spacer |
| 3000Å AlAs |
| 400Å GaAs spacer |
| 4000Å AlGaAs – No leak |
| GaAs (001) substrate + buffer layer |

Fig. 2. The generic sample structure grown with a deliberate oxygen leak for AlGaAs lattice parameter measurements.

were ±3 arc seconds, which resulted in an Al composition error following analysis of the MQW structures of ±0.005. To establish a standard curve for comparison with samples grown with an oxygen leak, we developed relationships between the $Al_xGa_{1-x}As$ perpendicular lattice constant and the Al mole fraction x by considering four cases. The first two cases assumed Vegard's law, that the natural lattice constant can be linearly interpolated between the end-point binaries according to:

$$a_x = xa_{AlAs} + (1-x)a_o \qquad (1)$$

while the second two cases included a bowing parameter in the lattice constant-composition relationship so that:

$$a_x = xa_{AlAs} + (1-x)a_o + cx(1-x) \qquad (2)$$

where a_{AlAs} and a_o are the AlAs and GaAs lattice constants, a_x is the natural lattice constant of $Al_xGa_{1-x}As$, x is the Al mole fraction and c is a bowing parameter. Of the first two possible cases which assume Vegard's law, one assumes Poisson's ratio is linear while the other allows for a bowing factor. The relevant equation for Poisson's ratio mirror equations (1) and (2) in their form. For the second two cases they differ again as to whether the compositional dependence of Poisson's ratio is assumed to be linear or non-linear. (004) x-ray measurements of as-grown samples do not yield the natural lattice constant, rather they measure the perpendicular lattice constant which can be related to the natural and in-plane lattice constants via linear elasticity theory in the context of an isotropic solid, so that:

$$a_x = a_x^\perp \left(\frac{1-v_x}{1+v_x} \right) + a_o \left(\frac{2v_x}{1+v_x} \right) \qquad (3)$$

where $a_x^\perp$ and v_x are the perpendicular lattice constant and Poisson's ratio of AlGaAs, respectively.

For each possible case, we develop a relationship for the perpendicular lattice constant versus Al mole fraction of AlGaAs, as shown in Table I, where $\Delta a = a_{AlAs} - a_o$, v_o and v_{AlAs} are Poisson's ratio of GaAs and AlAs respectively, $\Delta v = v_{AlAs} - v_o$ and b is a bowing parameter to allow for non-linearity of Poisson's

ratio of AlGaAs with composition. The separations between GaAs and $Al_xGa_{1-x}As$ peaks in HRXRD spectra can be determined by substituting $a_x^\perp$ in Table I into Braggs equation (4) to yield the peak separation between GaAs and AlGaAs layers which can be expressed, in the case of (004) reflections, as:

$$\Delta\omega = \theta_o - \theta_x = \sin^{-1}\left(\frac{2\lambda}{a_o}\right) - \sin^{-1}\left(\frac{2\lambda}{a_x^\perp}\right) \qquad (4)$$

where θ_o and θ_x are the Bragg angles associated with GaAs and $Al_xGa_{1-x}As$, respectively. Then, the peak separations between GaAs and $Al_xGa_{1-x}As$ as a function of Al fraction x were plotted and compared with the four model curves using a chi-square (χ^2) test on each curve. It was found that the model III curve was the best fit to our standard sample data, the model that assumed Vegard's law is not true and that Poisson's ratio is linear, which is in agreement with previously reported studies [12,17,19-20]. The bowing parameter c was found to be 2.2×10^{-3}, in good agreement with most reported studies of the validity of Vegard's law [12,17,19-20]. The data and the model fit are shown in Fig. 3. The AlAs perpendicular lattice constant measured was generally different for each sample, so we take an average which suggested the AlAs lattice constant was 5.66071Å. The GaAs lattice constant, a_o was assumed to be 5.65325Å, while the Poisson ratio of AlAs and GaAs were assumed to be 0.326 and 0.312 [19-20] respectively.

The samples containing AlGaAs layers grown with and without exposure to oxygen allowed two important observations to be made. Firstly, establishment of the standard curve allowed the Al mole fraction to be determined from the peak separation of the layer grown without oxygen. Secondly, the peak separations of the layers grown with a high purity oxygen leak were generally different to those of the layer grown without the leak, but in the same sample, and these differences in peak separations are shown in Fig. 4.

Negative peak separation differences represent a smaller perpendicular lattice constant and positive differences represent an expansion in the perpendicular lattice constant of $Al_xGa_{1-x}As$ due to the uptake of oxygen. The HRXRD AlGaAs layer peaks showed no broadening as a result of oxygen incorporation when compared with the standard layer peaks.

TABLE I
PERPENDICULAR LATTICE CONSTANT EXPRESSIONS FOR EACH OF FOUR CASES.

Model Assumptions	Expressions for the Al fraction dependence of the Perpendicular Lattice Constant of $Al_xGa_{1-x}As$.
Model I Vegard's law is true and Poisson's ratio is linear	$a_x^\perp = a_o + \left[\dfrac{\Delta a \Delta v x^2 + \Delta a(1 + v_o)x}{1 - v_o - \Delta v x} \right].$
Model II Vegard's law is true and Poisson's ratio is non-linear	$a_x^\perp = a_o + \left[\dfrac{b\Delta a x^3 - \Delta a(b + \Delta v)x^2 - \Delta a(1 + v_o)x}{-bx^2 + (b + \Delta v)x - (1 - v_o)} \right].$
Model III Vegard's law is not true and Poisson's ratio is linear	$a_x^\perp = a_o + \left[\dfrac{x}{1 - v_o - \Delta v x} \right] . [-c\Delta v x^2 - [c(1 + v_o) - \Delta v(c + \Delta a)]x + (c + \Delta a)(1 + v_o)].$
Model IV Vegard's law is not true and Poisson's ratio is non-linear	$a_x^\perp = a_o + \left[\dfrac{bc}{(1 - v_o) - (b + \Delta v)x + bx^2} \right] . \left[x^4 - \left[\dfrac{(b + \Delta v)}{b} + \dfrac{(c + \Delta a)}{c} \right]x^3 - \left[\dfrac{(1 + v_o)}{b} - \dfrac{(c + \Delta a)(b + \Delta v)}{bc} \right]x^2 + \dfrac{(c + \Delta a)(1 + v_o)}{bc} \right]x.$

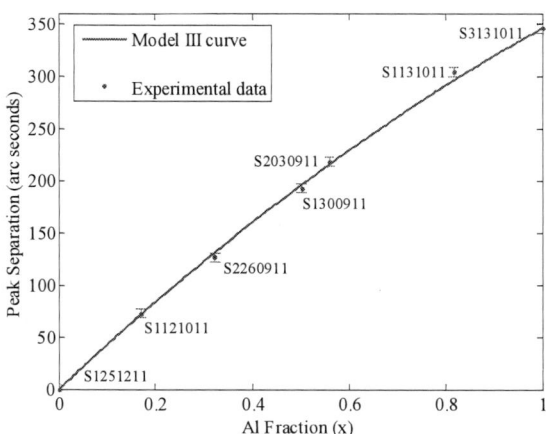

Fig. 3. Standard sample GaAs-Al$_x$Ga$_{1-x}$As peak separations plotted against Al fraction, x.

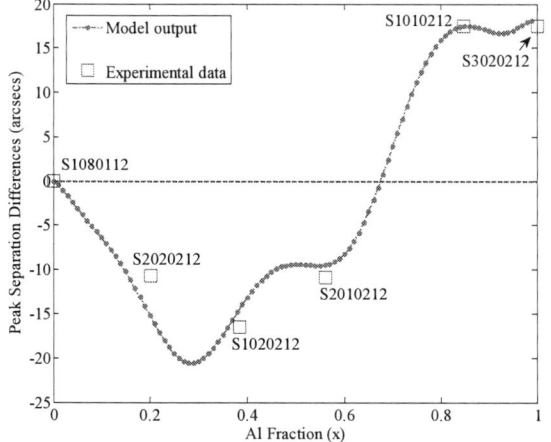

Fig. 4. Experimental data and calculated differences in peak separations between standard Al$_x$Ga$_{1-x}$As samples and samples grown with a high purity oxygen leak.

The SIMS measurements made on all AlGaAs samples grown during exposure to oxygen are shown in Fig. 5, where the oxygen levels have been normalised to the arsenic yield. The high purity oxygen leak data shows there are two peaks across the Al fraction range, the largest at an Al fraction around 0.3 and the second at 0.8. This is in agreement with our XRD results in the sense that at those same compositions the negative and positive changes in lattice parameter are at their extremes. To explain our results, we have developed a model that assumes O$_2$ molecules occupy As sites substitutionally at the growing surface. We also assume that O$_2$ molecules will not replace As atoms at the growing surface if the bonds to the lattice below are to two Ga atoms. In this we recognize the experimental observation that the uptake of O$_2$ in GaAs is significantly less than that in AlGaAs. However we allow an O$_2$ molecule to incorporate into the lattice if one or two of the bonds to the lattice below are to Al atoms. This is because the bond between Al and O is much stronger than that between Ga and O because of the high reactivity of Al with O. Additionally, GaO$_2$ is

volatile compared to AlO$_2$, which is very stable and nonvolatile [23] at typical growth temperatures, suggesting that GaO$_2$ would evaporate should it form. By assuming that the concentrations of O$_2$ molecules sitting on As sites when bonded to (i) one Al and (ii) two Al atoms are proportional to the population statistics of As sites bonded to one or two Al atoms in the layer below, we obtain expressions for the O$_2$ populations in all cases as shown in Table II, where C$_T$ is the total volume concentration of As sites. The $\propto$ in the term $\propto I_{O_2}$ scales our experimental oxygen concentration data as measured by SIMS. The population statistics for 2-Ga's, 1-Al and 1-Ga and 2-Al's in Table II are associated with the lattice constants of GaAs, Al$_{0.5}$Ga$_{0.5}$As and AlAs unit cells respectively.

Therefore the natural lattice constant of a layer containing O$_2$ in various types of As sites is given by:

$$a_{O_2}(x) = a_o(1-x)^2 + a_{Al_{0.5}Ga_{0.5}As}\left[\frac{2(1-x)}{C_T}\left[xC_T\right.\right.$$
$$\left.\left. -\propto I_{O_2}(x)\cdot\frac{1}{2-x}\right]\right] + \left(a_{Al_{0.5}Ga_{0.5}As} + \Delta_1\right)\left[\frac{2\propto I_{O_2}(x)}{C_T}\cdot\frac{1-x}{2-x}\right]$$
$$+ a_{AlAs}\left[\frac{x}{C_T}\left[xC_T - \propto I_{O_2}(x)\cdot\frac{1}{2-x}\right]\right] + \left(a_{AlAs} + \Delta_2\right)\left[\frac{\propto I_{O_2}(x)}{C_T}\cdot\frac{x}{2-x}\right]$$

$$(5)$$

where Δ_1 and Δ_2 allow for lattice parameter changes due to the presence of an O$_2$ molecule on the two difference types of As sites. In equation (5) the x dependence of the SIMS O$_2$ concentration data has been made explicit and Vegard's Law has been assumed with negligible error for simplicity. The perpendicular lattice parameter of AlGaAs containing O$_2$ is then given by:

$$a_{O_2}^{\perp}(x) = a_{O_2}(x)\left(\frac{1+v_x}{1-v_x}\right) - a_{GaAs}\left(\frac{2v_x}{1-v_x}\right)$$

$$(6)$$

and the perpendicular lattice constant of the same x fraction AlGaAs grown in the absence of an oxygen leak is given by the Model III perpendicular lattice constant expression given in Table I.

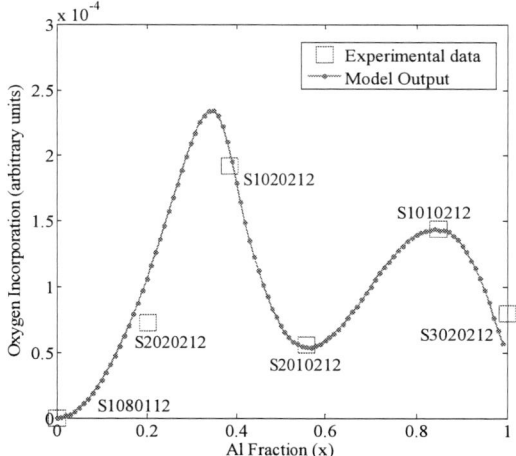

Fig. 5. Experimental data and theoretical curve for oxygen levels in AlGaAs epitaxial layers exposed to a high purity oxygen leak during growth.

978-1-4673-2395-6/12 $31.00 © 2012 IEEE

TABLE II
POPULATION STATISTICS ASSOCIATED WITH EACH TYPE OF ARSENIC SITE.

Site Type	Statistics for Each Site Type
2-Ga's	$C_T(1-x)^2$.
1-Al and 1-Ga	$2(1-x)\left[xC_T - \propto I_{O_2} \cdot \frac{1}{2-x}\right] + 2\propto I_{O_2} \cdot \frac{1-x}{2-x}$.
2-Al's	$x\left[xC_T - \propto I_{O_2} \cdot \frac{1}{2-x}\right] + \propto I_{O_2} \cdot \frac{x}{2-x}$.

Fig. 4 shows the difference between the standard curve and the curve relevant to the situation when O_2 incorporation is active. Δ_1 and Δ_2 were found to be -24×10^{-3} and $+30 \times 10^{-3}$, to achieve the best fit to the data. The specific details associated with the model described briefly above will be published elsewhere. Our observations are in agreement with the results obtained by Wagner et al. who reported a contraction in the $Al_{0.39}Ga_{0.61}As$ lattice constant due to oxygen incorporation in a layer grown by liquid phase epitaxy [24]. They explained that oxygen incorporation caused a contraction in the $Al_xGa_{1-x}As$ lattice constant at $x \sim 0.39$ due to electronic and ion size effects. However, their study did not quantitatively identify the levels of oxygen incorporation, nor account for the processes that caused oxygen to be incorporated.

IV. CONCLUSIONS

Our experimental data show that oxygen uptake in $Al_xGa_{1-x}As$ epilayers caused lattice parameter changes, due to O_2 molecules substituting on As sites. Lattice parameter changes were influenced by whether O_2 molecules bonded to a single or a double Al site, the former resulting in lattice contraction relative to an $Al_{0.5}Ga_{0.5}As$ unit cell and the latter resulting in expansion relative to an AlAs unit cell. The strengths of the two effects were accounted for by determining the populations of these different bonding sites as a function of Al fraction. Further work on oxygen incorporation in $Al_xGa_{1-x}As$ epitaxial layers grown when the growth chamber was exposed to various air leaks and the modeling of oxygen incorporation as a result is in progress.

ACKNOWLEDGEMENTS

We would like to thank the Public Service Department of Malaysia and AINSE for financial support as well as Ashland for the supply of semiconductor grade chemicals. Their support is gratefully acknowledged.

REFERENCES

[1] N. Chand, "Growth of High Quality AlGaAs/GaAs heterostructures by molecular beam epitaxy for photonic and electronic device applications," *Thin Solid Films* vol. 231, p. 15, 1993.

[2] S. Adachi, "GaAs, AlAs, and $Al_xGa_{1-x}As$: Material parameters for use in research and device applications," *Journal of Applied Physics* vol. 58, pp. R1-R29, 1985.

[3] T. Achtnich, G. Burri, and M. Ilegems, "Study of oxygen incorporation in AlGaAs layers grown by molecular-beam epitaxy," *Journal of Vacuum Science & Technology A: Vacuum, Surfaces, and Films* vol. 7, pp. 2537-2541, 1988.

[4] K. A. Prior, G. J. David, and R. Heckingbottom, "The Thermodynamics of Oxygen Incorporation into III-V Semiconductor Compounds and Alloys in MBE," *Journal of Crystal Growth*, vol. 66, pp. 55-62, 1984.

[5] B. F. Usher, Private Communication, June, 2010.

[6] O. Kobayashi. S. Naritsuka, K. Mitsuda, and T. Nishinaga, "Oxygen incorporation mechanism in AlGaAs layers grown by molecular beam epitaxy," *Journal of Crystal Growth*, vol. 254, pp. 310-315, 2003.

[7] S. Naritsuka, O. Kobayashi, and T. Maruyama, "Numerical Model for Oxygen Incorporation into AlGaAs Layer Grown by Molecular Beam Epitaxy," *Japanese Journal of Applied Physics*, vol. 42, p. 3, 2003.

[8] T. Achtnich, G. Burri, M. A. Py, and M. Ilegems, "Secondary ion mass spectrometry study of oxygen accumulation at GaAs/AlGaAs interfaces grown by molecular beam epitaxy," *Applied Physics Letters*, vol. 50, pp. 1730-1732, 1987.

[9] M. T. Asom, M. Geva, R. E. Leibenguth, and S. N. G. Chu, "Interface disorder in AlAs/(Al)GaAs Bragg reflectors," *Applied Physics Letter*, vol. 59, pp. 976-978, 1991.

[10] M. Leszczynskmi, M. Micovic, C. A. C. Mendoncaa, A. Ciepielewska, and P. Ciepielewski, "Lattice Constant of AlAs," *Crystal Research Technology*, vol. 27, pp. 97-100, 1992.

[11] M. A. G. Halliwell, M. H. Lyons, and M. J. Hill, "The Interpretation of X-ray Rocking Curves from III-V Semiconductor Device Structures," *Journal of Crystal Growth*, vol. 68, p. 9, 1984.

[12] G. S. Solomon, D. Kirillov, H. C. Chui, and J. S. Harris, "Determination of AlAs mole fraction in $Al_xGa_{1-x}As$ using Raman spectroscopy and x-ray diffraction," *Journal of Vacuum Science and Technology B*, vol. 12, pp. 1078-1081, 1994.

[13] R. P. Leavitt and F. J. Towner, "Determination of the lattice parameter and Poisson ratio for AlAs via high-resolution x-ray diffraction studies of epitaxial films," *Physical Review B*, vol. 48, pp. 9154-9157, 1993.

[14] M. Krieger, H. Sigg, N. Herres, K. Bachem, and K. Ko¨hler, "Elastic constants and Poisson ratio in the system AlAs–GaAs," *Applied Physic Letters*, vol. 66, pp. 682-684, 1995.

[15] M. S. Goorsky, T. F. Kuech, M. A. Tischler, and R. M. Potemski, "Determination of epitaxial AlGaAs composition from x-ray diffraction measurements," *Applied Physics Letter*, vol. 59, pp. 2269 -2271, 1991.

[16] B. K. Tanner, A. G. Turnbull, C. R. Stanley, A. H. Kean, and M. McElhinney, "Measurement of aluminum concentration in epitaxial layers of AlGaAs on GaAs by double axis x-ray diffractometry," *Applied Physics Letter* vol. 59, p. 3, 1991.

[17] Z. R. Wasilewski et al., "Composition of AlGaAs," *Journal of Applied Physics*, vol. 81, pp. 1683-1694, 1997.

[18] S. Gehrsitz, H. Sigg, N. Herres, K. Bachem, K. Ko¨hler, and F. K. Reinhart, "Compositional dependence of the elastic constants and the lattice parameter of $Al_xGa_{1-x}As$," *Physical Review B*, vol. 60, pp. 11 601 - 11 610, 1999.

[19] D. Zhou and B. F. Usher, "Deviation of the AlGaAs lattice constant from Vegard's law," *Journal of Physics D: Applied Physics*, vol. 34, pp. 1461-1465, 2001.

[20] D. Zhou, B. F. Usher, T. Warminski, R. Absin, and M. Madebo, "Poisson's Ratio of AlAs," *Proceeding of SPIE*, vol. 4086, pp. 168-173, 2000.

[21] T. E. Harvey, K. A. Bertness, R. K. Hickernell, C. M. Wang, and J. D. Splett, "Accuracy of AlGaAs growth rates and composition determination using RHEED oscillations," *Journal of Crystal Growth* vol. 251, pp. 73-79, 2003.

[22] B. F. Usher and D. Zhou, "Thickness and composition determination of MBE grown strained multiple quantum well structures by X-ray diffraction," *Proceedings of SPIE* vol. 4086, pp. 76-81, 2000.

[23] A. Taguchi and H. Kageshima, "Diffusion and stability of oxygen in GaAs and AlAs," *Physical Review B*, vol. 60, pp. 5383-5391, 1999.

[24] W. R. Wagner, N. E. Schumaker, J. L. Zilka, P. J. Anthany, and V. Swaminathan, "Lattice Parameter Changes in $Al_{0.39}Ga_{0.61}As$ Due to O, Ge, Si and S Doping," *Journal of Electrochemical Society: Solid-State Science and Technology*, vol. 130, p. 670, 1983.

Characterization of NBTI by Evaluation of Hydrogen Amount in the Si/SiO$_2$ Interface

Surya Kris Amethystna[1], Karuna Nidhi[1], Shao-Ming Yang[1*], Gene Sheu[1,2], and Jung-Ruey Tsai[1,2], Md Imran Siddiqui[1]

[1]Department of Computer Science and Information Engineering, Asia University, Taichung, Taiwan, ROC
[2]Department of Photonic and Communication Engineering, Asia University, Taichung, Taiwan, ROC
500, Lioufeng Rd., Wufeng, Taichung 41354, Taiwan, R. O. C
Phone: +886-4-2332-3456 ext. 1784, Fax: +886-4-2332-0718, *E-mail: rickyyang121@asia.edu.tw

Abstract- **In this work, the behavior of Si-H bond generating interface trap was studied by experiment and TCAD simulation. Its behavior is responsible for the increasing of PMOSFET absolute threshold voltage due to negative bias temperature instability (NBTI) stress for device reliability issue. It was found that the temperature stress has more significant influence on initial interface trap generation as compare to the electric field stress. In addition, fast triangular pulse measurement with elevated temperature was applied to eliminate the NBTI recovery effect and gives a better evaluation of the number of generated interface trap, as compared to the conventional DCIV measurement under NBTI stress.**

Keyword-NBTI, Interface Trap Distribution, TCAD Simulation, Recovery Effect, Fast Triangular Pulse.

I. INTRODUCTION

Nowadays, negative bias temperature instability (NBTI) has reintroduced as a major reliability concern especially for p-type metal-oxide-semiconductor (PMOS) devices [1-3]. Aggressive scaling of oxide thickness compare to slower scaling of operating voltage has increase the effective electric field across the oxide which leads into the device instability operation due to NBTI.

Hydrogen model [4] explain the interface trap generation during NBTI process as a reaction of hydrogen atom from Si-H bond in the Si/SiO$_2$ interface to become electrically active species while remaining empty traps at the Si/SiO$_2$ interface. These defects are treated as charge trapping centres with an energy distribution throughout the silicon band gap. Threshold voltage shift due to generated interface trap can be represented as

$$\Delta V_T = -\Delta N_{it} \cdot \phi_s / C_{ox} \qquad (1)$$

Where N_{it} is a number of interface traps ϕ_s is the surface potential, and C_{ox} is the oxide capacitance of the device. In the PMOS device, negative bias stress generates donor states in the lower half of the band gap. As the results, interface traps are positively charged and leading to negative threshold voltage shift [5].

Till now, Reaction-Diffusion (R-D) model is the most prevalent model [6, 7] to interpret the power law dependence of interface trap generation during NBTI stress. There are two main interface trap generation rate. First, initial creation of interface traps (reaction phase) which is depends on the amount of trapped hydrogen atoms which able to react with silicon

atoms during trapping/de-trapping process and later on interface trap generation rate depends on hydrogen diffusion into the oxide. However, the total amount of hydrogen trapped initially at the Si/SiO$_2$ interface is still not well understood.

Transient behaviour of hydrogen cannot be well explained by using the conventional direct current I-V (DCIV) measurement. Trapping hydrogen species back so called as recovery effect will underestimate the NBTI influence on the device.

This study also employed a new measurement technique for NBTI characterization on PMOS by using triangle shape pulse to minimize recovery effect [8-11] with elevated temperature to evaluate the amount of hydrogen in the Si/SiO$_2$ interface more accurate. In addition, electric field and temperature stress effects will first be studied to distinguish which stress parameter has more influence on initial interface trap generation.

II. EXPERIMENTAL SETUP

This experiment was done using standard 0.25μm 5V BCD PMOS process fabrication and the threshold voltage shift extracted and plotted using conventional direct current current-voltage (DCIV) measurement. In addition, TCAD Sentaurus simulator was employed to simulate the process flow and extract the electrical parameter as well. The Id-Vg result from TCAD simulation shows a good agreement compare to silicon measurement data as shown in Fig. 1.

Conventional DCIV stress without any elevated temperature is used to observe long time electric field stress effect on generated interface traps. Conventional DCIV schematic diagram is shown in Fig. 2. Device was stressed at a room temperature with constant 5MV/cm negative electric field across the gate oxide.

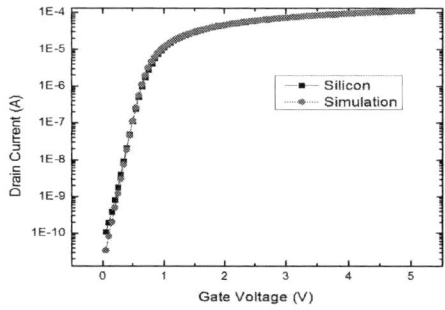

Fig. 1. Device I_D-V_G curve comparison between silicon measurement and TCAD Sentaurus Simulation. It shows a good agreement in initial threshold voltage of the fresh sample.

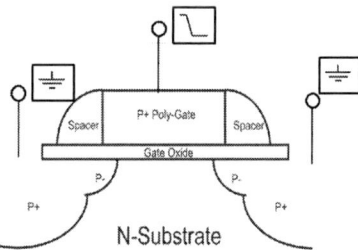

Fig. 2. Schematic diagram of direct current current-voltage (DCIV) measurement. Gate voltage stressed at constant bias while source, drain, and substrate were grounded.

Generated interface trap behaviour due to electric field stress will be analyzed by varying electric field ranging from 2.5 to 7.5 MV/cm. In the other hand, its behaviour due to temperature stress will also be analyzed by varying temperature at 300 and 350 K. Second experiment proposed a triangle pulse as a measurement technique with an elevated temperature. The schematic diagram for proposed measurement is shown in Fig. 3. Device treated at temperature ranging from 300 to 400K. Fast triangle shape pulse with 100ns rising and falling time was applied on the gate contact while, drain contact was biased at 0.1V. The schematic waveform on gate and drain is shown in Fig. 4.

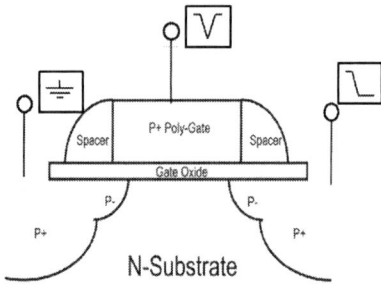

Fig. 3. Schematic diagram of fast triangle pulse measurement. Drain voltage was biased at low voltage while source and substrate were grounded. Gate terminal then biased with fast triangular pulse to measure NBTI degradation

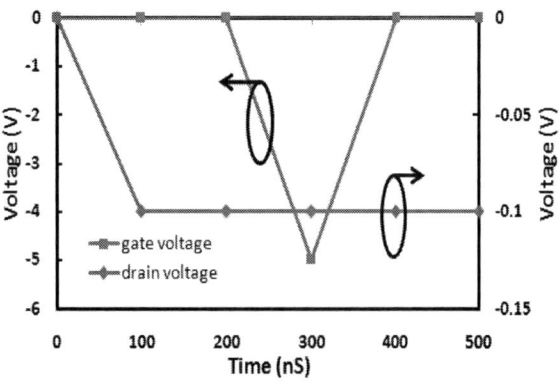

Fig. 4. Schematic gate and drain voltage waveform of the fast triangular pulse measurement. Rising part of the pulse was employed as initial I_D-V_G measurement while falling part was employed as degraded I_D-V_G

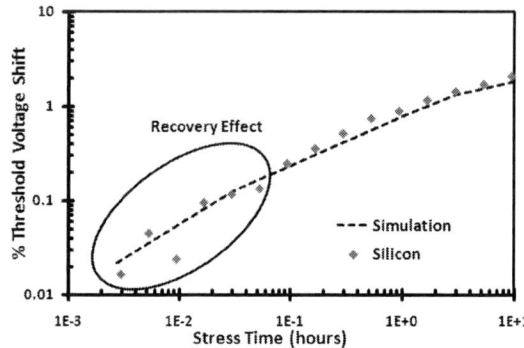

Fig. 5. Threshold voltage degradation due to NBTI-DCIV for 10 hours stress. It shows a good agreement between silicon measurement and TCAD simulation results.

III. RESULT AND DISCUSSION

A. Direct current current-voltage (DCIV)

Figure 5 shows the degradation trend for conventional DCIV after 10 hours stress time for both silicon and simulation result. Simulation result using TCAD Sentaurus was obtained by employing the trap formation kinetics in the Si/SiO_2 interface under the influence of negative gate bias. Degradation model invoked and specified to generate donor state interface trap. It shows a good agreement with the silicon data.

When short stress time applied, it observed that silicon measurement exhibit non-consistent threshold voltage shift trends. This phenomenon is explained as a recovery effect. Hydrogen that diffused into the oxide, passivate back to the silicon dangling bonds during the transition from stress mode to measure mode of the measurement machine.

This simulation results gives not only a degradation trend as a function as stress time but also the number of generated interface trap as well as its distribution over the Si/SiO_2 interface.

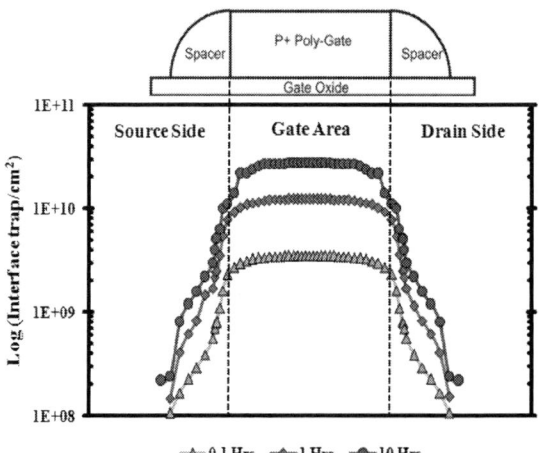

Fig. 6. Interface trap evolution and its distribution over the Si/SiO_2 interface for 0.1, 1, and 10 hours NBTI-DCIV stress. The interface traps were generated mainly under the gate area.

Fig 6, shows the distribution of total interface trap due to NBTI stress at Si/SiO$_2$ interface during DCIV stress at room temperature. The device assumed has 1e8 cm^{-2} initial interface trap. After 10 hours stress total interface traps increase to 2.8e10 cm^{-2} and located mainly under the gate area.

From the experimental result in Fig. 7, the evolution of total interface trap can be divided into three regions. First region is called as reaction limited. It started from the beginning until few seconds after stress applied. Interface trap generation rate depends on the Si-H bond dissociation (reaction phase). After that, generation rate controlled by the diffusion phase until it began to reach the saturation phase. It was observed that varying the electric field stress from 2.5 MV/cm, 5MV/cm, and 7.5MV/cm give less significant influence on initial interface trap generation. The effect will obviously observed after the device entering the diffusion phase.

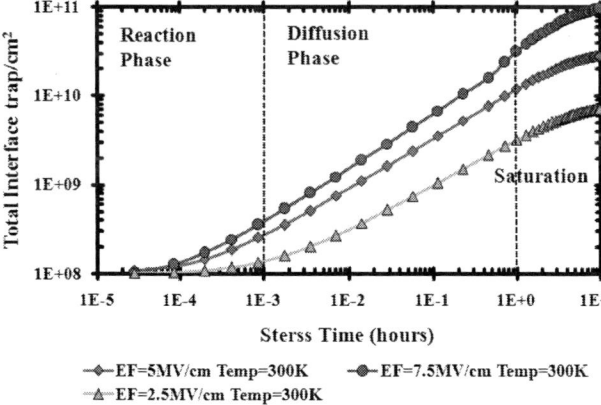

Fig. 7. Behavior of generated interface trap by the influence of electric field stress. Increasing the electric field stress does not have a significant influence of initial interface trap generation (reaction phase) but it will affect more in the diffusion phase at the stress time larger than few seconds.

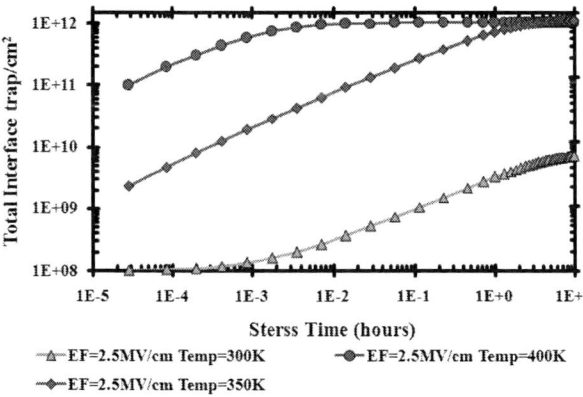

Fig. 8. Behavior of generated interface trap by the influence of temperature stress. Increasing the ambient temperature stress will give a significant influence of the initial interface trap generation (reaction phase) due to weaker Si-H bond as the effect of temperature stress.

In the other hand, as it observed form Fig 8, temperature stress will produce a significant increment on the initial interface trap generation even at low electric field stress applied. In this case, 2.5MV/ cm were selected in order to minimize the electric field effect on interface trap generation. Three different ambient temperatures 300K, 350K, and 400K were applied to the device. After 0.1s stress, total interface trap increase to 1.08×10^8, 1.97×10^9, and 9.74×10^{10} cm^{-2} respectively. As the stress time increase, total interface trap become saturates more easily which indicates this measurement technique is overestimating the NBTI degradation.

B. Fast Pulse Measurement with Elevated Temperature

As it mention in the previous section, conventional DCIV measurement technique at room temperature suffers from NBTI recovery effect and become a major limitation to understand NBTI mechanism. As a result, it will underestimate the total number of interface traps. Thus, we proposed a triangle pulse with 100ns rising and falling time as an I-V extraction to overcome such recovery effect problem. Those pulse rising and falling time was fast enough to be considered as a free NBTI recovery measurement. Elevating the temperature is used to increase the initial interface trap generation.

We have done an experiment with three different temperatures. First experiment was done in room temperature, second and third done at 350K And 400K respectively. Elevated temperature absolutely needed in order to weaken and break as many Si-H bonds as possible creating the initial interface trap on the reaction phase. In the simulation result on Fig. 9, it observed 2.98mV (0.38%) and 3.46mV (0.48%) threshold voltage shift at 350K and 400K measurement temperature respectively. According to (1), this threshold voltage shift was caused by generation of 5×10^{11} cm^{-2} and 6×10^{11} cm^{-2} interface traps concentration for 350 and 400 K temperature stress, respectively. In addition, the increasing of initial generation of interface traps will be saturated as the temperature stress goes beyond 400K.

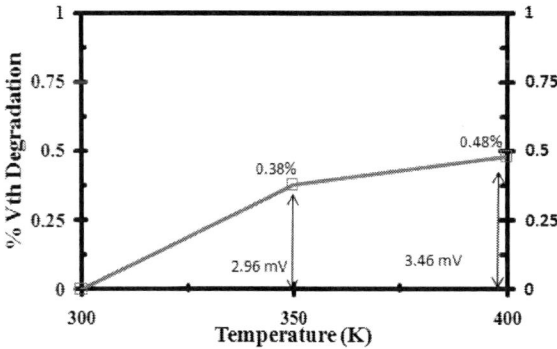

Fig. 9. Threshold voltage degradation measured by fast triangular pulse at elevated temperature. The degradation seems to be saturated when the ambient temperature elevated beyond 400K.

IV. CONCLUSION

This paper successfully characterizes NBTI on PMOS device by evaluating the amount of hydrogen in Si/SiO$_2$ interface. Electric field stress will not give a significant influence on the initial interface trap generation (reaction phase) and suffer from recovery effect. In the other hand, DCIV-temperature stress will overestimate the NBTI degradation. Therefore fast triangular pulse with 100ns rise and fall time with elevated temperature was considered to elucidate the transient de-trapping behaviour of hydrogen atom from Si-H bond generating interface trap in the Si/SiO$_2$ interface.

ACKNOWLEDGMENT

The authors are grateful to Vanguard International Semiconductor Corporation for the valuable discussion and for the financial support in a part on this topic, and to the National Center, Taiwan for supporting the TCAD licenses in this study.

REFERENCES

[1] P. Chaparala, J. Shibley, P. Lim, "Threshold voltage drift in PMOSFETS due to NBTI and HCI," *In: Proc Int Reliability Workshop*, pp.95–97, 2000.

[2] M. Makabe, T. Kubota, T Kitano, "Bias-temperature degradation of pMOSFETs: mechanism and suppression," *IEEE AIRPS*, pp.205–209, 2000.

[3] S. Mahapatra and M.A Alam, "A predictive reliability model for PMOS bias temperature degradation," *IEEE IEDM*, pp.505–508, 2002.

[4] D. Schroder and J. Babcock, "Negative bias temperature instability: Road to cross in deep submicron silicon semiconductor manufacturing," *J. Appl. Phys*, vol. 94, pp.1-18, 2003.

[5] D. Schroder, "Negative bias temperature instability: What do we understand?," *Microelectronic Reliability*, vol. 47, pp.841–852, 2007.

[6] M.A Alam and S. Mahapatra, "A comprehensive model of PMOS NBTI degradation," *Microelectronic Reliability*, vol. 45, pp.71-78, 2005.

[7] K.O Jeppson and C.M Svensson, "Negative bias stress of MOS devices at high electric fields and degradation of MOS devices," *J. Appl. Phys*, vol. 48, pp.2004-2014, 1977.

[8] S. Mahapatra, M. Alam, P. Kumar, T. Dalei, and D. Saha, "Mechanism of negative bias temperature instability in CMOS devices: degradation, recovery and impact of nitrogen," *IEEE IEDM*, pp.105-108, 2004.

[9] M.F Li, *et al*, "Understanding NBTI mechanism by developing novel measurement techniques," *IEEE Trans. On Dev. And Mat. Rel.*, vol. 8, pp.62-71, 2008.

[10] G.A Du, D.S Ang, Z.Q Teo, and Y.Z Hu, "Ultrafast measurement on NBTI," *IEEE Electron Device Letter*, vol. 30, pp.275-277, 2009.

[11] M. Denais, *et al*, "New methodologies of NBTI characterization eliminating recovery effects," *IEEE ESSDERC*, pp.265-268, 2004.

[12] V. Huard, F. Ivfonsieur, G. Ribes, S. Bruyere, "Evidence for hydrogen-related defects during NBTI stress in p-MOSFETs," *IEEE AIRPS*, pp 178–182, 2003.

[13] O. Penzin, A. Haggag, W. McMahon, E. Lyumkis, K. Hess, "MOSFET degradation kinetics and its simulation," *IEEE Trans. On Electron Devices*, vol. 50, pp. 1445–1450, 2003.

Effect of Annealing Duration on the Memristive Behavior of Pt/TiO$_2$/ITO Memristive Device

[*]N.S Kamarozaman, Z. Aznilinda, [**]S.H Herman, R. A Bakar and M. Rusop

NANO-ElecTronic Centre (NET), Faculty of Electrical Engineering, Universiti Teknologi Mara (UiTM), 40450 Shah Alam, Selangor, Malaysia

Email: [*]sheeraee@yahoo.com, [**]hana1617@salam.uitm.edu.my

Abstract— A titanium dioxide (TiO$_2$) based memristive device was fabricated and investigated for its memristive behavior. In this paper, the effect of annealing duration on the memristive behavior of device was studied. TiO$_2$ thin films were deposited on ITO substrate using RF magnetron sputtering method and then annealed in nitrogen at 450°C for 10, 30 and 60min. Characterization of current-voltage (I-V) measurement and surface morphology using scanning electron microscopy (FESEM) between annealed and nonannealed samples were investigated. From the result, it shows that sample annealed at 450°C for 60min resulted in switching loop with high conductivity when negative bias is applied probably due to the presence of high oxygen vacancies.

Keywords- Titanium dioxide;memristor;RF-magnetron sputtering method;annealing

I.   INTRODUCTION

Memristor is known as the fourth fundamental passive circuit element that has attracted much attention from many researches due to its ability to be used as a nonvolatile memory [1-6] and biologically inspired computing [2]. It was first proposed by L. Chua in the year 1971 [2], but was only fabricated by HP labs in the year 2008 [2] owing to the advancement of nanotechnology. Memristive devices have the advantages of nonvolatility, fast programming, small cell size and low power consumption [3]. A wide range of materials such as organic monolayers [7], silicon [8,9] and titanium dioxide [3,4,6] have been used in fabricating memristor. However, titanium dioxide (TiO$_2$) is chosen as the most popular material because it exhibit good memristive behavior [1,2]. The memristive behavior depends on the oxide layer thickness and distribution of oxygen vacancies. The oxide layer thickness is favorably to be in 3-30nm for a memristive device [2].

Most have reported Pt/TiO$_2$/Pt structure exhibit good memristive behavior. In this work we investigated switching behavior using Pt/TiO$_2$/ITO structure. Indium tin oxide or ITO (90wt% In$_2$O$_3$ and 10wt% SnO$_2$) as bottom electrode provide an excellent transparency in visible light, improved functional reproducibility and also shows bipolar switching behavior [10,11] which promises wider applications for future electronic devices.

Generally, memristor consists of two layers, one is mainly an insulating layer and another one is conductive layer due to oxygen vacancies. The structure in this work only contains one TiO$_2$ layer, thus in this work after deposition of TiO$_2$ thin film, annealing process is performed to create the necessary oxygen vacancies within the TiO$_2$ layer [3] to act as an active layer for memristor. Effect of annealing duration to the memristive behavior of the device is studied.

II.   METHODOLOGY

A.   Memristor Fabrication

ITO as a substrate and bottom electrode for the device was cleaned using acetone, methanol and distilled water, followed by drying with nitrogen gases. Then, 50 nm of TiO$_2$ thin film was deposited on the ITO substrate by RF magnetron sputtering method using TiO$_2$ (99.999% purity) as a target. The deposition process was performed at 300W RF power, with applied heat, heating time and bias power at 200°C, 120s and 20W, respectively. During deposition process of 5 minutes, 50sccm of argon gas and 1sccm of oxygen gas were introduced into the chamber. The working pressure of 5mTorr was fixed during deposition process. After that, the sample was annealed in nitrogen at 450°C for 10, 30 and 60 min. Then, Pt as a top electrode was sputtered onto the film for 60nm thickness. Finally, the electrical and physical properties of the device structure were investigated.

B.   Memristor Characterization

I-V measurement for each sample was performed by the two point probe method using Keithley 4200 semiconductor characterization system connected to a probe station at room temperature. To test the memristive behavior, the electric potential is applied to the top electrode. The sample was characterized by sweeping the bias voltage from 0 to 5V, 5V to -5V and back to 0, while simultaneously measuring the current. Fig. 1 shows the schematic diagram of the memristor device structure and the I-V measurement arrangement used in this work.

978-1-4673-2395-6/12 $31.00 © 2012 IEEE

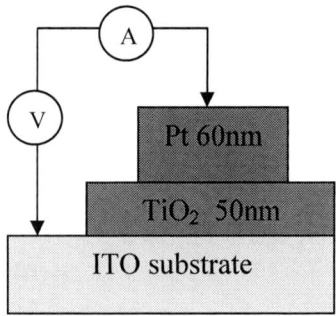

Fig. 1. Schematic diagram of I-V measurement for a memristor device structure with ITO and Pt used as bottom and top electrode.

III. RESULT AND DISCUSSION

A. Electrical Properties

Fig. 2 (a) to (d) show the current-voltage (I-V) measurement for four samples; which are nonannealed, and annealed for 10, 30 and 60 mins, respectively.

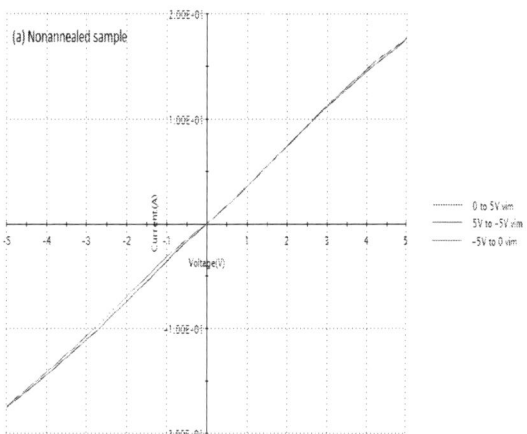

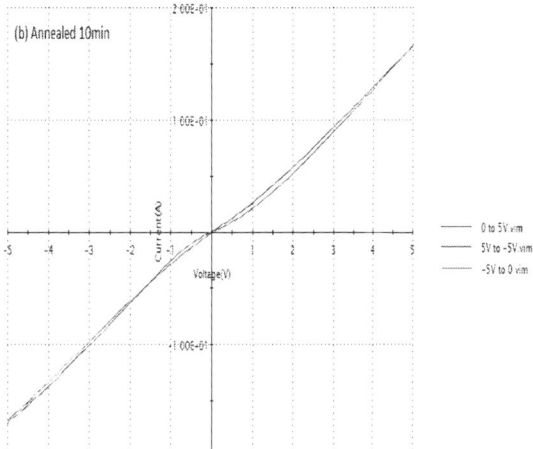

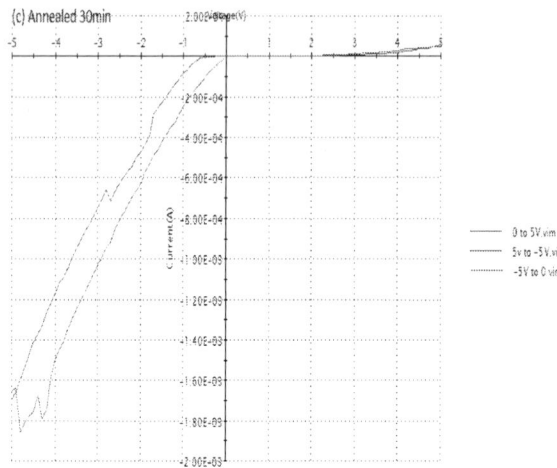

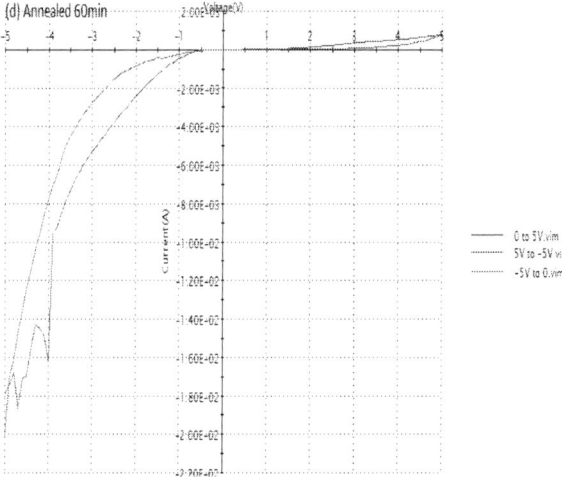

Fig. 2. Current-voltage measurement for Pt/TiO$_2$/ITO memristive device for (a) nonannealed sample (b) annealed sample at 450°C for 10min (c) 30min and (d) 60min.

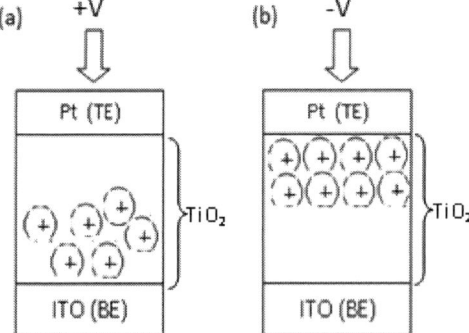

Fig. 3. Schematic diagram for a device behavior with (a) positive bias and (b) negative bias. The thickness of the TiO$_2$ active layer is drawn excessively thicker for a clearer view and thus not to scale.

978-1-4673-2395-6/12 $31.00 © 2012 IEEE

From the result, it was observed that both samples for nonannealed and annealed at 10 mins exhibit almost undistinguishable switching loops of a memristive behavior. The sample annealed for 10 min, shows a rectifying curve compare to the nonannealed sample. In contrast with these, the samples in Fig. 2(c) and (d) show the switching loops at negative bias indicating that the device was only conductive when negative voltage was applied to the sample. The sample annealed for 60 min was more conductive compared to sample annealed for 30 min probably due to the presence of more oxygen vacancies within TiO_2 layer. It is reported that the presence of oxygen vacancies increased the conductivity of the TiO_2 layer [2]. This in return suggests that the oxygen in the TiO_2 layer was released in the annealing process, due to which, the longer annealing duration, the more oxygen vacancies were created. As reported by Chris Yakopcic et. al [12], memristor is a device where the hysteresis loop present in the I-V curve where the intersection of current-voltage occur at zero point. From the result, all sample show a clear hysteresis loop intersects at zero point that proved the memristive behavior.

As shown in Fig. 2(c) and (d), when positive voltage was applied to the sample, the graph shows an insulative I-V curve. We hypothesize that this might be due to the presence of insulating layer of TiO_2 layer on top of the surface. Oxygen vacancies is positively charged, thus when positive bias is applied to the device, the positively charged oxygen vacancies will be pushed downward and cause the insulating region become wider. So, the more positive bias is applied, the more insulative the sample becomes. When negative bias is applied to the device, the positively charged oxygen vacancies will be attracted to the surface, thus the device will be more conductive. This behavior is illustrated as shown in figure 3.

B. Physical Properties

The cross-section and the surface morphology of the TiO_2/ITO are shown in Figs. 4 and 5, respectively. From Fig. 4, it can be seen that the grains are denser after the annealing process (Fig. 4(b)) compared to before the annealing (Fig. 4(a)). This is also can be seen from the surface morphology shown in Fig. 5 (a) and (b).

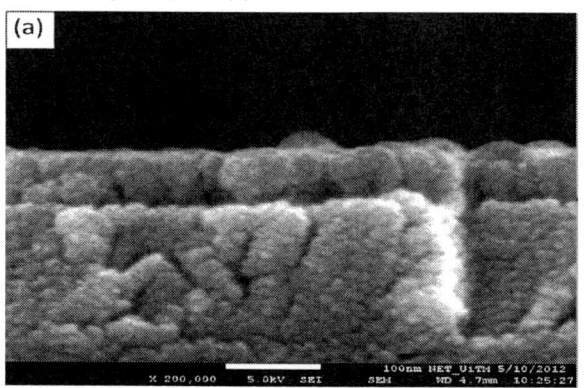

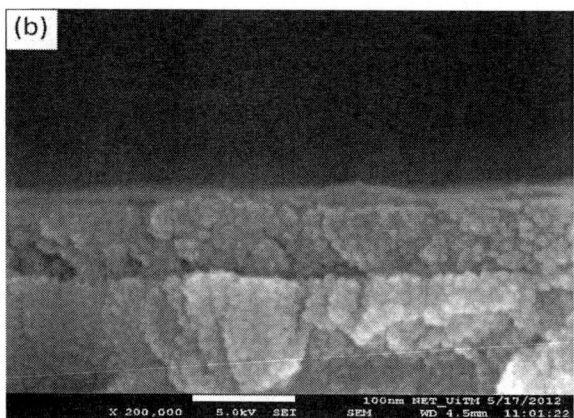

Fig. 4. Cross-section of the sample (a) before annealed (b) annealed at 450°C for 60min.

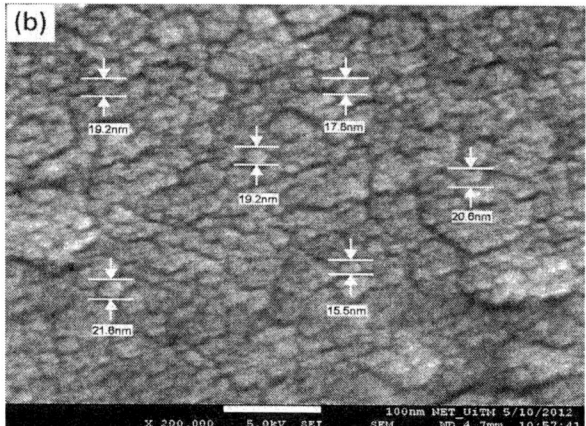

Fig. 5. Surface morphology of the sample (a) before annealed (b) annealed at 450°C for 60min.

The analyses of oxygen content in the samples were measured using EDS as shown in Fig. 6 respectively.

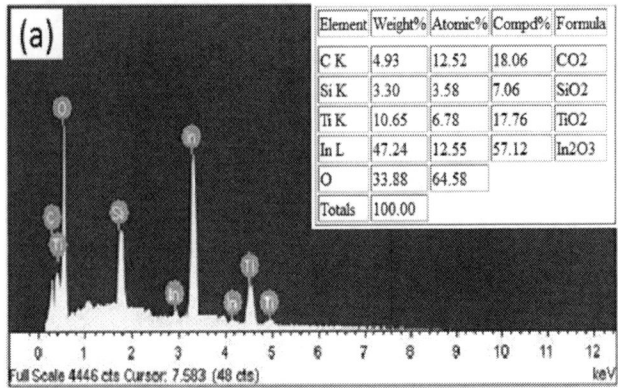

Element	Weight%	Atomic%	Compd%	Formula
C K	4.93	12.52	18.06	CO2
Si K	3.30	3.58	7.06	SiO2
Ti K	10.65	6.78	17.76	TiO2
In L	47.24	12.55	57.12	In2O3
O	33.88	64.58		
Totals	100.00			

Full Scale 4446 cts Cursor: 7.583 (48 cts)

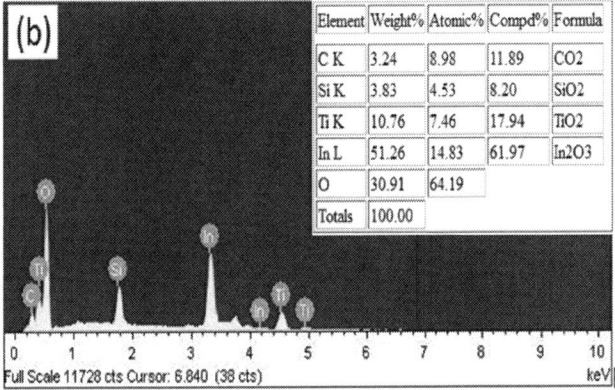

Element	Weight%	Atomic%	Compd%	Formula
C K	3.24	8.98	11.89	CO2
Si K	3.83	4.53	8.20	SiO2
Ti K	10.76	7.46	17.94	TiO2
In L	51.26	14.83	61.97	In2O3
O	30.91	64.19		
Totals	100.00			

Full Scale 11728 cts Cursor: 6.840 (38 cts)

Fig. 6. EDS of the sample (a) before annealed (b) annealed at 450°C for 60min.

The EDS result indicated that the nonannealed sample shows higher oxygen content compared to annealed sample at 450°C for 60 min. Thus, it proved that annealing induce extra oxygen vacancies to the sample as it shows that sample annealed has higher oxygen vacancies compared to nonannealed sample.

IV. CONCLUSION

In conclusion, the device that has been fabricated by sandwiching the TiO_2 films between n-type conducting ITO substrate and Pt electrodes exhibit memristive behavior. It was observed that sample annealed in longer times result in the better memristive behavior which may be due to the existence of more oxygen vacancies. The sample annealed in longer times show higher conductivity when negatively biased.

ACKNOWLEDGEMENT

The authors would like to thank all members of NANO-ElecTronic Centre (NET) Universiti Teknologi MARA Malaysia for their cooperation and support. Also thanks to Ministry of Higher Education (MOHE) Malaysia for supporting the research under SLAB and Excellence Fund Research Management Institute (RMI) Universiti Teknologi Mara, Shah Alam (Project Code: 600-RMI/ST/DANA 5/3/Dst (415/2011)) for their financial support.

REFERENCES

[1] L. O. Chua, "Memristor- The Missing Circuit Element," IEEE Trans. Circuit Theory, vol. CT-18, no. 5, pp.507-519, Sep. 1971.
[2] R. S. Williams, " How We Found The Missing Memristor," IEEE Spectr., vol. 45, no. 12, pp. 28-35, Dec. 2008.
[3] K. Miller, K.S. Nalwa, A. Bergerud, N.M. Neihart and S. Chaudhary, " Memristive Behavior in Thin Anodic Titania," IEEE Electron Device Lett., vol. 31, no. 7, pp.737-739, July 2010.
[4] D. Panda, A. Dhar and S.K. Ray, "Nonvolatile Memristive Switching Characteristics of TiO2 Films Embedded with Nickel Nanocrystals," IEEE Trans. on Nanotechnology, vol. 11, no. 1, pp. 51-55, 2012.
[5] O. Jambois, P. Carreras, A. Antony, J. Bertomeu, C. Martinez-Boubeta, "Resistance Switching in Transparent Magnetic MgO Films," Solid State Communications, vol. 151, pp.1856-1859, 2011.
[6] J. J. Yang, J. P. Strachan, F. Miao, M-X. Zhang, M.D. Pickett, W. Yi, D.A.A. Ohlberg, G. Medeiros-Ribeiro, R.S. Williams, "Metal/TiO2 Interfaces for Memristive Switches," Appl Phsy A 102, pp. 785-789, 2011.
[7] D.R. Stewart, D.A.A. Ohlberg, P.A. Beck, Y. Chen, R.S. Williams, J.O. Jeppesen, K.A. Nielsen, J.F. Stoddart, Nano Lett, vol.4, 1, pp. 133-136 2004.
[8] S.H. Jo, W. Lu, Nano Lett, vol.8, 2, p.392-397, 2008.
[9] Mehonic Adnan, Cueff Sebastian, Wojdak Maciej, Hudziak Stephen, Jambois Olivier, Labbe Christophe, Garrido Blas, Rizk Richard, Kenyon Anthony J., "Resistive Switching in Silicon Suboxide Film", J.Appl. Phys., vol. 111, no. 7, 2012.
[10] L. Z-Yu, Z. P-Jian, M. Yang, L. Dong, M. Q-Yu, L. J-Qi and Z. H-Wu, "The Influence of Interfacial Barrier Engineering on the Resistance Switching of In2O3:SnO2/TiO2/In2O3:SnO2 Device," Chin. Phys. B, vol. 21, no. 4, 2012.
[11] M. Yang, Z. P-Jian, L. Z-Yu, L. Z-Liang, P. X-Yu, L. X-Jin, Z. H-Wu and C. D-Min, "Enhanced Resistance Switching Stability of Transparent ITO/TiO2/ITO Sandwiches, Chin. Phys. B, vol. 19, no. 3, 2010.
[12] Yakopcic, C., Shin E., Taha, T.M., Subramanyam G., Murray, P.T., Rogers S., "Fabrication and Testing of Memristive Devices," Neural Networks (IJCNN), The 2010 International Joint Conference.

Effective Heat Dissipation of High Power LEDs Mounted on MCPCBs with Different Thickness of Aluminium Substrates

Soon Bee Law, Anithambigai Permal, Mutharasu Devarajan

Nano Optoelectronics Research Laboratory, Universiti Sains Malaysia, 11800 Minden, Penang, Malaysia
E-mail: lsb11_fkp005@student.usm.my

Abstract – This paper mainly discusses the thermal characterization of high power Light Emitting Diodes (LEDs). The Metal Core Printed Circuit Board (MCPCB) was designed with the aid of the recommended solder pad given in the product datasheet. Aluminium Nitride (AlN) was employed as the dielectric layer. Industrial grade aluminium (grade 5052) was utilized as the substrate material of the MCPCBs. The samples under test were varied in terms of substrate thickness on which the LEDs were mounted. The temperature rise at the junction and junction to ambient thermal resistance of the LEDs was investigated. The device performances were investigated with various substrate thicknesses under different ambient temperature at a fixed current input. The experimental results show that the increase in the substrate thickness greatly improved the thermal and optical performance of the LED packages. There were increments in the percentage of decrease in junction temperature rise and the percentage of decrease in the total thermal resistance which is from 19.8 % to 21.2 % and 15.2 % to 17.2 % respectively. The decreases in wall plug efficiency were reduced from 0.9% to 0.2%. Transient dual interface method was employed to determine the junction to board thermal resistance of the sample.

Keywords – Metal core printed circuit board (MCPCB), dielectric material, aluminium substrate thickness, junction temperature rise, thermal interface material, junction to board thermal resistance, junction to ambient thermal resistance.

I. Introduction

The introduction of the high brightness Light Emitting Diodes (LEDs) with white lights and monochromatic colours leads towards a tremendous progress of the industry. Today, one may observe a significant change in the worldwide lighting industry as these LEDs are competing the traditional lighting sources with the increasing energy conversion efficiency [1]. Successful thermal management ensures the reliability and efficiency of LEDs hence the increased electrical currents use to drive LEDs have focused more attention on the thermal management [2].

Modifications on the printed circuit boards (PCB) significantly improves the thermal performance of an LED package, conveniently allows an elimination of an external heat sink [3]. Thus, the design of the board is ultimately crucial to optimize the heat dissipation performance. Previous studies have shown that the metal core printed circuit boards (MCPCBs) have a realistic value of thermal conductivity

compare to PCBs on basis of FR4 or ceramic based solutions [4].

A typical MCPCB consist of number of layers including a dielectric sandwiched between copper foil layer act as a circuit layer for electrical connection, the other layer such as aluminium alloy or copper alloy serve as heat spreader. Studies have shown that Aluminium Nitride (AlN) had the highest dielectric constant of 10 Dk and highest thermal conductivity of 117 W/mK among other dielectric like alumina, boron nitride and beryllium oxide [5].

In this work, MCPCB has been custom designed to have Aluminium Nitride (AlN) as the dielectric layer and aluminium alloy as the substrate material with variable thickness. The LED chip was mounted on the MCPCB and a combined thermal and radiometric measurement was carried out to analyze the thermal characteristic of the LED packages with different thickness of MCPCB substrate. Transient dual interface method using thermal paste and mylar tape was carried out to identify the junction to board thermal resistance [6].

II. Theoretical Background

A. Thermal Resistance

According to the JEDEC standard 51-1 [7], thermal resistance of a single semiconductor device is defined as shown in (1)

$$R_{JX} = \frac{T_J - T_X}{P_H} \qquad (1)$$

where R_{JX} is the thermal resistance from device junction to the specific environment, T_J is the junction temperature of the device at steady state condition, T_X is the reference temperature for the specific environment and P_H is the power dissipation in the device. The dependence of thermal resistance with optical power and temperature of an LED is given in (2),

$$R_{th} = \frac{T_J - T_X}{P_{el} - P_{opt}} \qquad (2)$$

where P_{el} is the electrical power and P_{opt} is the optical power. Considering optical power in the thermal resistance calculation according to (2) yields the real thermal resistance [8].

B. Newton's Law of Cooling

Heat transfers by convection is an exchange among nearly

stationary molecules adjacent to the solid surface, as occurs in the heat conduction, and transport the heat away from the solid surfaces by the bulk motion of the fluid [9]. The relationship of heat transfer by convection is referred to as Newton's Law of Cooling as given in (3):

$$q = hA(T_s - T_f) \qquad (3)$$

where q is the convective heat transfer, h is the heat transfer coefficient (W/m^{2}K), A is the surface area of convection, T_s is the surface temperature and T_f is the bulk temperature of the nearby fluid.

III. EXPERIMENT METHOD

Three pieces of cool white Cree XLamps MX6 LED were mounted on three different thickness of MCPCB. The sample size of all the three samples had MCPCB dimensions of 1.5 cm width x 1.5 cm length. The bases of the MCPCB board were made of aluminium (Al) and the thicknesses were 0.9, 1.6 and 2.1 mm respectively.

A. Optical and Thermal Measurement

The samples were mounted on the Peltier cooled-fixture which was then attached to an integrating sphere. The Peltier cooled fixture served as cooled plate for thermal measurement as well as stabilize the LED temperature during optical measurement [10]. The ambient temperature was varied between 25, 35, 45, 55, 65 and 75 °C with the fixed current of 350 mA during the measurement. The optical measurement is performed and the optical power is determined using TERALED. After the optical measurement, the LED was switched off and was left to cool down for 900 seconds, thus cooling transient was captured by using the T3Ster equipment.

B. Transient Dual Interface Method (TDIM)

Transient Dual Interface Method (TDIM) was adapted to determine the junction to board thermal resistance, R_{thJ-B}. The dual interface method which allows the R_{thJ-B} with higher accuracy and better reproducibility had recently been accepted by JEDEC standard as JESD51-14 [11]. Two types of materials were used which were thermal paste and mylar tape to attach the LED on the Peltier cooled plate.

IV. RESULTS AND DISCUSSION

A. Optical and Thermal Measurement

The temperature rise at the junction, T_{RISE} for all three

samples were plotted against increasing ambient temperature, T_A in order to study the influence of the various thicknesses of the aluminium substrate to the LED samples.

It was observed from Fig.1 that T_{RISE} decreased as the T_A increased. T_{RISE} is the difference between junction temperature and ambient temperature. The decrease in the T_{RISE} is mainly due to the insufficient rise of the junction temperature when the T_A was increased. The increase in the T_A leads to the lower heat dissipation of the LED at a constant current. This will contribute to the decrease in the T_{RISE} of the LED. Fig.1 describes the temperature rise at junction, T_{RISE} versus ambient temperature, T_A for LED with the aluminium substrate thickness of 0.9, 1.6 and 2.1 mm at a fixed current of 350 mA.

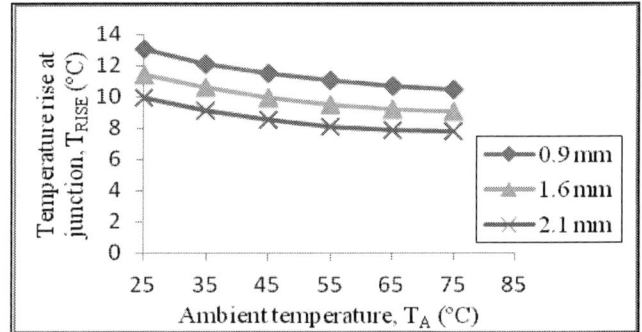

Fig.1. Temperature rise at junction, T_{RISE} versus ambient temperature, T_A for LED with the aluminium substrate thickness of 0.9, 1.6 and 2.1 mm at fixed current of 350 mA

It can also be observed that the T_{RISE} decreased with the increase in the thickness of the aluminium substrate. In this study, all the samples were not attached to any external heat sink, so the aluminium substrate of the MCPCB is eventually acted as a heat sink to dissipate heat to the ambient.

Theoretically, this follows the Newton's Law of Cooling, by referring to (3), it is obviously shown that the increased thickness of the aluminium substrate led to the widening of the surface area, further led to the increase in the convective heat transfer.

As observed form the Fig.2, the mechanism can be explained as the layer of the surrounding air adjacent to the hot surface becomes warmer, its density decrease and become buoyant. A cooler air near the surface replaces the warmer air and the pattern of circulation form. The increased convective heat transfer to the surrounding area will reduce the heat which generated at the junction further reduce the T_{RISE}.

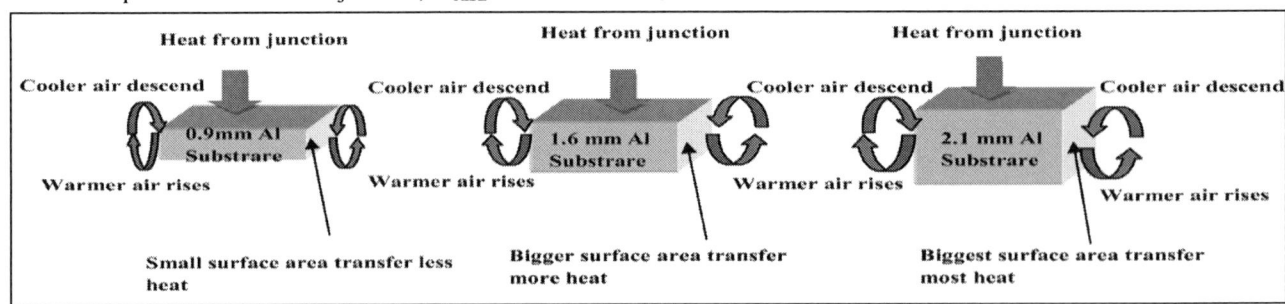

Fig.2. The increased thickness of the aluminium substrate led to the widening of the surface area and increase in the convective heat transfer

Fig.3 describes the thermal resistance from the junction to ambient, R_{thJ-A} versus ambient temperature, T_A at the fixed current of 350 mA for the aluminium substrate thickness of 0.9, 1.6 and 2.1 mm.

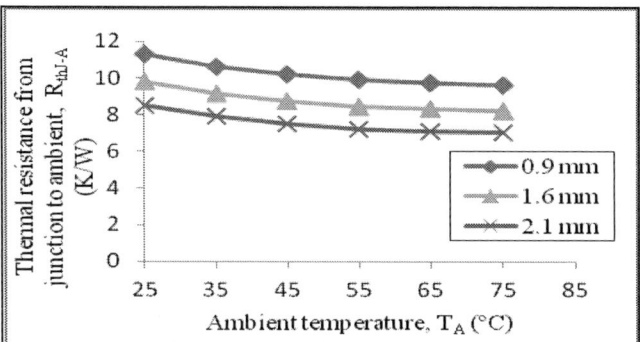

Fig.3. Thermal resistance from junction to ambient, R_{thJ-A} (K/W) versus ambient temperature, T_A for LED with the aluminium substrate thickness of 0.9, 1.6 and 2.1 mm at fixed current of 350 mA

As the T_A is increased, the R_{thJ-A} is reduced. Besides, it can be seen that increase in the thickness of the substrate caused a decrease in the R_{thJ-A}. This is mainly due to the low T_{RISE}. Referring to the (2), the T_{RISE} is directly proportional to the R_{thJ-A}.

Besides, the increase in the aluminium substrate thickness will increase the volume of the aluminium substrate, thus, induce higher thermal conductivity whereby absorb and transfer the heat faster and therefore attribute to reduced in thermal resistance. The thermal resistance, R_{th} can be expressed in terms of

$$R_{th} = \frac{L}{kA} \qquad (4)$$

where L is the thickness, k is the thermal conductivity and A is the cross-sectional area normal to the heat flow [12].

The numerical values of the temperature rise at junction and thermal resistances from junction to ambient of 0.9, 1.6 and 2.1 mm thickness of the samples with ambient temperature in the order of 10 °C from 25 °C to 75 °C were summarized in the Table I and Table II respectively.

TABLE I

THE NUMERICAL VALUES OF THE TEMPERATURE RISE AT JUNCTION OF 0.9, 1.6 AND 2.1 MM THICKNESS OF THE ALUMINIUM SUBSTRATE OF LED WITH AMBIENT TEMPERATURE IN THE ORDER OF 10 °C FROM 25 °C TO 75 °C

Ambient Temperature, T_A (°C)	Thickness of the aluminium substrate (mm)		
	0.9	1.6	2.1
	Temperature rise at junction, T_{RISE} (°C)		
25	13.1	11.4	9.9
35	12.0	10.6	9.1
45	11.5	9.9	8.5
55	11.0	9.5	8.1
65	10.7	9.2	7.9
75	10.5	9.1	7.8

TABLE II

THE NUMERICAL VALUES OF THE THERMAL RESISTANCES FROM JUNCTION TO AMBIENT OF 0.9, 1.6 AND 2.1 MM THICKNESS OF THE ALUMINIUM SUBSTRATE OF LED WITH AMBIENT TEMPERATURE IN THE ORDER OF 10 °C FROM 25 °C TO 75 °C

Ambient Temperature, T_A (°C)	Thickness of the aluminium substrate (mm)		
	0.9	1.6	2.1
	Junction to ambient thermal resistance, R_{thJ-A} (K/W)		
25	11.34	9.77	8.49
35	10.63	9.17	7.90
45	10.23	8.72	7.48
55	9.91	8.46	7.22
65	9.74	8.31	7.09
75	9.62	8.22	7.03

Refer to Table I and Table II, from 25 °C to 75 °C, the percentages of decrease in T_{RISE} for 0.9, 1.6 and 2.1 mm of substrate thickness were 19.8, 20.2 and 21.2 % respectively.

Whereas for the percentages of decrease in total thermal resistance or thermal resistances from junction to ambient were 15.2, 15.9 and 17.2% respectively for 0.9, 1.6 and 2.1 mm of substrate thickness, from 25 °C to 75 °C. The results show that an increase in the thickness of the substrate will enhance the overall thermal performance.

On the other hand, the wall plug efficiency, WPE has been calculated. Wall plug efficiency was defined as ratio of radiant flux, also known as radiant efficiency, was typically the energy conversion efficiency where the electrical power is converting to the optical power. It was found that the decrease in WPE from 25 °C to 75 °C were 0.9, 0.3 and 0.2% for 0.9, 1.6 and 2.1 mm of the substrate thickness respectively. The results show the improvement of the optical performance with the increase of the thickness of the substrate.

B. Transient Dual Interface Method (TDIM)

Apart from obtaining the junction to ambient thermal resistance, R_{thJ-A}, the partial thermal resistance which is thermal resistance from junction to board of the package, R_{thJ-B} was determined employing TDIM method.

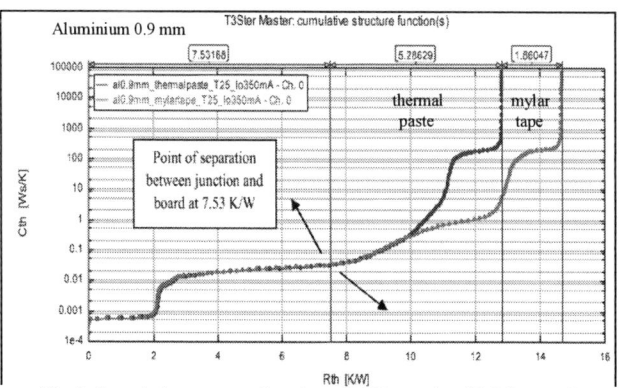

Fig.4. Cumulative structure function of LED sample with 0.9 mm of aluminium substrate thickness using TDIM method

978-1-4673-2395-6/12 $31.00 © 2012 IEEE

Mylar tape and thermal paste was employed as the Thermal Interface Material (TIM). Mylar tape and thermal paste was placed as the interface layer between the MCPCB and the Peltier cooled plate. Fig.4. shows the cumulative structure function of LED sample with 0.9 mm of aluminium substrate thickness using TDIM method.

As seen from Fig.4, it was observed that there was a significant deviation point on the graph of cumulative structure function using TDIM method. By comparing both the conditions of thermal paste and mylar tape as TIM, the deviation point on the graph indicates the point of separation between junction and board which were also known as thermal resistance from junction to the board, R_{thJ-B}. This point assumed to be the R_{thJ-B} as the heat spreading at the board of the package is more significant.

The corresponding numerical values of thermal resistance of junction to board and thermal resistance of junction to ambient using TDIM method of all the three LED samples with 0.9, 1.6 and 2.1mm were summarized in the Table III.

TABLE III
THE NUMERICAL VALUES OF THERMAL RESISTANCES FROM JUNCTION TO BOARD AND THERMAL RESISTANCES FROM JUNCTION TO AMBIENT USING TDIM METHOD.

Thermal resistance, R_{th} (K/W)	Aluminium substrate thickness (mm)		
	0.9	1.6	2.1
R_{thJ-B}	7.53	6.46	5.00
R_{thJ-A}	12.82	11.15	9.87

Based on Table III, the obtained results show that the both of the values of R_{thJ-A} and R_{thJ-B} decreases as the substrate thickness increases. This observation indicates there is a significant influence of the thickness of the aluminium substrate to the R_{thJ-A} and R_{thJ-B}. The thermal performance has been improved by increasing the thickness of the aluminium substrate.

As observed from Table III, the experimental values of the R_{thJ-B} fall within the range of 5 to 8 K/W. This compensates the junction to case thermal resistance datasheet value provided, which is 5 K/W [13]. Moreover, from Fig.4, the inconsistence and the higher value of R_{thJ-A} using mylar tape as the TIM can be observed compared to the value of R_{thJ-A} using thermal paste. Studies had shown that TIM material act as filler and fill the micro gaps, thus enabling the efficient heat flow from the device [14].

For the mylar tape, although the surface is smooth and flat, however it will form micro air gaps. This issues arises due to the fact that no matter how smooth and flat surface which observed by the naked eyes will have asperities. Hence, when two surfaces with the asperities come into contact, micro gaps are formed at interface [15]. Whereas for the thermal paste, the semi solid form of the material will eventually filled all the micro air gaps and create an perfect contact between the aluminium substrate and the Peltier cooled plate. This explains the higher value of R_{thJ-A} obtained using mylar tape as TIM.

V. CONCLUSION

In this paper, thermal characteristic of cool white Cree Xlamp MX6 LEDs were presented using different thickness of MCPCB substrate. MCPCB has been designed and AlN was employed as dielectric layer to enhance the thermal conductivity. The LED samples were differed in terms of thickness of the aluminium substrate of the MCPCB. In terms of measurement, optical and thermal measurement was carried out in the condition of fixed current and various ambient temperatures. As observed, increases in the substrate thickness enhance the thermal and optical performance of the LED packages significantly. The decrease in the temperature rise at the junction of the LED has been increased from 19.8 % to 21.2 % and 15.2 % to 17.2 % as for the total thermal resistance. The decreases in wall plug efficiency were reduced from 0.9 % to 0.2 %. In addition, the TDIM measurement results show the well agreement of junction to board thermal resistance between the measured values and the one in the datasheet.

REFERENCES

[1] G. Marosy, Z. Kovács, G. Molnár, A. Poppe, "Diagnostics of LED-based Street Lighting Luminaires By Means of Thermal Tranient Method", *Proceedings of IEEE, THERMINIC*, 2010, pp. 1-6.

[2] W.R. Hyun, S.S. Kwang, W.OK. Chi, and B.H. Yoon, "Heat Transfer Behaviour of High Power Light Emitting Diode Packages", *Korean J. Chem. Eng.*, vol. 24, No. 2, 2007, pp. 197-203.

[3] www.cree.com/products/pdf/Xlamp_PCB_Thermal.pdf

[4] K. Oliver, "High Power LED Arrays Special Requirements on Packaging", *Proc. of SPIE*, vol. 6134, 2006, pp. 613404-1-8.

[5] W.K.C. Yung, "Using Metal Core Printed Circuit Board as a Solution for Thermal Management", Journal of the HKPA, Q2, No. 24, 2007, pp. 12-16.

[6] P. Anithambigai, N. Teeba, K. Dinash, D. Mutharasu and K.L. Choon, "Influence of Thermal Interface Material on the Thermal Resistance of High Power LED", *IEEE 2nd International Conference on Photonics (ICP)*, Kota Kinabalu, 2011, pp. 1-5.

[7] www.jedec.org/sites/default/files/docs/jesd51-1.pdf

[8] P.Anithambigai, K. Dinash, D. Mutharasu, S. Shanmugana and K.L. Choon, "Thermal Analysis of Power LED Employing Dual Interface Method and Water Flow as a Cooling System", *Thermochimica Acta*, vol. 523, Iss. 1-2, 2011, pp. 237-244.

[9] Rao R. Tummala, "Fundamental of Microsystem Packaging", Mc Graw Hill, 2001

[10] A. Poppe, G. Farkas and Gy. Horvát, "Electrical, Thermal and Optical Characterization of Power LED Assemblies", *Therminic*, France, 2006.

[11] www.jedec.org/sites/default/files/docs/JESD5114_1.pdf

[12] R.H. Horng, C.C. Chiang, Y.L. Tsai, C.P. Lim, K. Kan, H.I. Lin and D.S. Wu, "Thermal Management Design from Chip to Package for High Power InGaN/Sapphire LED Applications", *Electrochemical and Solid-State Letters*, 12(6), 2009, pp. H222-H225.

[13] http://www.cree.com/products/pdf/XLampsMX-6.pdf

[14] K. Dinash, D. Mutharasu and Y.T. Lee, "Paper Study on Thermal Conductivity of Al2O3 Thin Film of Different Thicknesses on Copper Substrate under Different Contact Pressures", *IEEE, ISIEA*, 2011, pp. 620-623.

[15] K. Dinash, D. Mutharasu and Y.T. Lee, "Thermal Measurements of Two Layered Thin Films of Different Orders on Copper", *IEEE (ISIEA)*, 2011, pp. 606-609.

978-1-4673-2395-6/12 $31.00 © 2012 IEEE

Rhombohedral In₂O₃ Thin Films Preparation from In Metal Film using Oxygen Plasma

Subramani Shanmugan* Devarajan Mutharasu and Ibrahim Kamarulazizi

Nano Optoelectronics Research Laboratory, School of Physics, Universiti Sains Malaysia (USM), Minden, Pulau Penang, 11800, Malaysia

* Corresponding author E-mail: subashanmugan@gmail.com (S.Shanmugan)

Tel: +60-04-6533672; Fax: +60-04-6579150.

Abstract- **Rhombohedral (*rh*-In₂O₃) In₂O₃ thin films were synthesized by Oxygen plasma process of RF sputtered In metal film. The formation of (110) oriented *rh*-In₂O₃ was confirmed by XRD analysis and well matched with JCPDS File no. 73-1809. The effect of process parameters on the growth of *rh*-In₂O₃ thin film and their optical properties was studied. The observed band gap was in between 2.45 and 2.92 eV. High process power and high gas flow rate affect the growth of *rh*-In₂O₃ and its band gap considerably. FTIR spectra analysis was performed to confirm the synthesis of *rh*-In₂O₃ processed by O₂ plasma. The elemental composition of processed films was studied by EDAX spectra.**

I. INTRODUCTION

Indium oxide (In₂O₃) is a potential material for use in solar cells, for ultraviolet lasers and sensor applications [1]. In thin film form, their wide ranges of applications include transparent windows in liquid crystal displays, anti-reflection coatings [2], optoelectronic and electro-chromic devices [3] have shown remarkable prospects in the field of upcoming nanoelectronic building blocks and nanosensors. The observance of both high optical transmittance and high electrical conductivity simultaneously in Indium Oxide makes a suitable transparent conducting oxide material for many device developments [4].

In₂O₃ can appears in two stable modifications as body-centered (bcc) cubic (a=10.118 Å) and Rhombohedral (*rh*) (a=5.478 Å and c=14.51 Å [5]). The band gap of both is currently a subject of discussion and it seems to be that this material exhibits indirect transitions [6]. The values for the indirect and direct gaps of the cubic phase are between 2.1 and 3.1 eV and between 3.1 and 3.7 eV, respectively [7]. In the case of the rhombohedral In₂O₃ phase, the indirect gap was estimated to be in the range from 3.0 to 3.3 eV, whereas the direct gap can be expected 3.3 and 3.4 eV [7]. The rhombohedral structure of In₂O₃ is a high pressure phase [8] and consequently has been rarely produced. This rhombic phase exhibits better physical properties than the cubic one, such as a more stable conductivity. The difference seems to be the result of a better packing of the anion layers in the rhombohedral In₂O₃.

Therefore, lot of works have been published on investigation of *c*-In₂O₃ growth conditions and optimizing their properties in dependence on the synthesis methods ranging from such as DC and RF sputtering [9], reactive evaporation[10], evaporation of metallic indium and subsequent oxidation [11], chemical vapor deposition[12], spray pyrolysis[13], sol–gel [14] and laser ablation[15].

However, the stoichiometric form of indium oxide, In₂O₃, is of a dielectric material [16]. In the case of *rh*-In₂O₃, there are only a few reports concerning the growth and characterization of *rh*-In₂O₃ in the literature like hydrothermal [17] and Metal Organic Chemical Vapor Deposition [18], because *rh*-In₂O₃ has been reported as can only be synthesized through high pressure and high temperature processes [19]. However, the growth of *rh*-In₂O₃ thin films at room temperature has not been reported in the literature so far. Thus, further research on the properties and applications of this material is hindered due to lack of appropriate deposition methods.

It is also widely reported in literature that oxygen plasma is effective at causing oxidation of treated materials, with Reactive Ion Etching (RIE), a common technique used for the treatment of synthesized films such as $YBa_2Cu_3O_{7-x}$ thin films [20] and CuO thin films [21]. Here, the plasma process technique is employed to prepare the *rh*-In₂O₃ thin film which is new and limited research work has been carried on the synthesis of rhombohedral In₂O₃. This paper presents a systematic study on the effect of plasma power, oxygen flow-rate and plasma duration in the formation of the *rh*-In₂O₃ thin films and more importantly, the attained structural and optical properties have been compared with the published literatures.

II. EXPERIMENTAL TECHNIQUES

Indium (In) thin film was deposited on soda lime glass substrates in Ar atmosphere at ambient temperature by RF Magnetron Sputtering (Edwards make, Model-Auto 500). Pure In (99.999%) and high pure argon gas were used as the sputtering target and the work gas, respectively. The base pressure of the chamber was ~2 x 10⁻⁷ torr. During the deposition, a gas flow rate of 14 sccm and a gas pressure P_{Ar} of 1.4 x 10⁻² torr were employed. RF power of 40 W was used for all In coatings. To get the uniform thickness, rotary drive system was used and 25 rpm was fixed as rotary speed for entire coatings. Sputtering duration was adjusted to yield In thicknesses of about 100 nm. All Indium films were coated at the deposition rate of 0.09 nm/sec.

The sputtered films were then subjected to oxygen plasma in the Inductively Coupled Plasma and Reactive Ion Etching System (ICP-RIE, Oxford Plasmalab 80 Plus) for various plasma power (100W and 200 W), gas flow rate (15 and 20 sccm) and various process time (5 and 10 min). The samples names are identified for different process conditions throughout this text as follows:

TABLE.I

COMPARISON OF OBSERVED (O) 2θ AND d SPACE VALUE WITH STANDARD (S) JCPDS CARD NO. 731809 OF RH-In_2O_3

Sample	Orientation	2θ (O)	2θ (S)	d space (O)	d space (S)
S2	1 1 0	32.70	32.59	2.740	2.745
S3	1 1 0	32.72	32.59	2.735	2.745
S4	1 1 0	32.75	32.59	2.733	2.745
S5	1 1 0	32.71	32.59	2.735	2.745
S6	1 1 0	32.77	32.59	2.730	2.745
S7	1 1 0	32.76	32.59	2.731	2.745
S8	1 1 0	32.76	32.59	2.731	2.745
S9	1 1 0	32.73	32.59	2.734	2.745

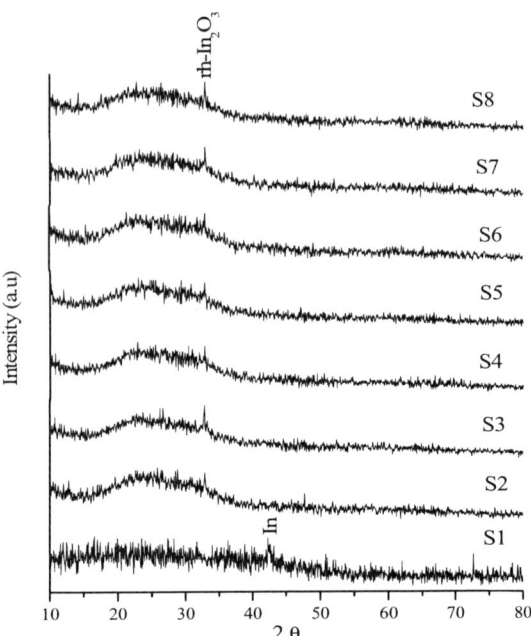

Fig. 1. XRD spectra of as grown (S1) and O_2 plasma processed Indium thin films

S1- as grown, S2-100W;15 sccm;5 min, S3- 100W;15 sccm;10 min, S4- 200W;15 sccm;5 min, S5- 200W;15 sccm;10 min, S6- 100W;20 sccm;5 min, S7- 100W;20 sccm;10 min, S8- 200W;20 sccm;5 min, S9- 200W;20 sccm;10 min. All thin films were processed at a pressure of 80mT.

The synthesized samples were characterized by X-ray diffraction (XRD) with a General Area Detector Diffraction System (PANalytical's X'Pert PRO). The observed results were compared with the JCPDS files. Transmission curves were also obtained for wavelengths of 300–1100 nm, using a Shimadzu UV-1800 UV/Vis/NIR Scanning Spectrophotometer. The FTIR spectra were recorded in the range 400–4000 cm^{-1}, with a spectral resolution of 4 cm^{-1} using Perkin Elmer spectrum GX. The film composition was analyzed using EDAX spectrum measured by Scanning Electron Microscope (Model - JSM-6460 LV).

III. RESULTS AND DISCUSSIONS

3.1 XRD analysis

Films deposited on substrates held at room temperature are expected to be highly amorphous [22]. Fig. 1 shows the X-ray pattern of O_2 plasma processed In thin films at various process conditions. The broad shoulder was observed for as grown film (S1). It reveals that the samples are in the amorphous phase.

From the Fig. 1, it also reveals that O_2 plasma processed samples (S2-S9) show peaks observed nearly at $2\theta=32.70°$ which correspond to the rhombohedral structure of In_2O_3 films with (1 1 0) orientation. The observed results are well matched with the JCPDS Data card No.73-1809 of rhombohedral In_2O_3.

The observed 2θ and d space values are compared with standard JCPDS file and given in Table-1.Fig.1 also shows that all thin films exhibit only rh- In_2O_3 with low intensity except processed at S2 conditions. It is also found that the (110) In_2O_3 peak was observed at very low intensity when Indium film processed at high plasma power (200W), gas flow rate (20 sccm) and process duration (10 min).

Figure 1 also shows that a mixture of In and rh- In_2O_3 is obtained for the film processed at 100W power and 15 sccm for 5 min. It also reveals that the intensity of rh- In_2O_3 peaks varies with respect to the plasma power and also O_2 flow rate. The observed intensity of rh- In_2O_3 peak for the sample S3 shows high value when compared to other samples. It seems to be the nature of conversion from amorphous to crystalline during O_2 plasma process. It is also the evidence of crystalline nature of the film when processed at condition of S3 sample. Using the Scherrer equation [21], the calculated average crystallite size of the processed samples gives a mean value of 35 nm, with a distribution in the range of 18–57 nm. It is also revealed the synthesis of nano crystallites using the O_2 plasma process. Low value of about 18 nm in crystallite size is observed with the samples S2 and S6.

From the Fig.1, it is also found that the films processed at high power and high oxygen flow rate do not show any residual Indium. This can be qualitatively explained by noting that dissociated oxygen is the primary chemically reactive species in the plasma [22]. The equilibrium amount of dissociated oxygen within the plasma is thus dependent on the power supplied. Reducing the power would reduce the amount of dissociated oxygen and affect the oxidation accordingly and hence the In metal peaks are existing at low power with low oxygen flow rate.

Additionally, increasing the O_2 flow-rate from 15 sccm to 20 sccm slowed the oxidation of Indium. This is because an increased flow-rate means that molecular oxygen is replaced at a faster rate. As such, the equilibrium amount of dissociated oxygen would be lower for a higher flow-rate and the oxidation rate would be reduced. But it would be compensated by increasing the power from 100 W to 200W as well as the time from 5 min to 10 min. Hence the rate of oxidation is same. It is also observed that the area under (110) peak increases as gas flow rate increases from 15 sccm to 20 sccm with respective to plasma power.

3.2 Optical Properties
3.2.1 Transmission

978-1-4673-2395-6/12 $31.00 © 2012 IEEE

Fig. 2 shows the transmittance spectra of O_2 plasma processed Indium films. The spectra show a lower transmittance in between 8% and 18% for all films and also the films processed at low plasma power with high O_2 flow rate show very low transmittance <10% at the band gap region.

The low optical transmittance for S3 and S4 may be due to insufficient oxygen available for reaction of indium with oxygen to form indium oxide. The un-oxidized indium acts as scattering centers for light hence of low optical transmittance at lower oxygen partial pressures. The higher transmittance (~18%) observed in the films is attributed to less scattering effects, structural homogeneity and better crystallinity, whereas, the lower transmittance (~8%) might be due to the less crystallinity leading to more light scattering [23]. A moderately lower transparency may be due to the existence of donor levels as a result of oxygen vacancies or in other words an excess of free metal, which makes it stable and sensitive gas sensor [24]. From the Fig. 2, it is found that the transmittance decreases with the order as S9,S5,S2,S8,S3,S4,S6,S7 and hence the sample S9 shows high transmittance than the other samples.

3.2.2 Absorption coefficient

Fig. 3 shows the spectral dependence of absorption coefficient in rh- In_2O_3 films. The absorption coefficient was determined in the range between 7.5×10^4 and 1.25×10^5 cm^{-1} at the absorption edge. From the Fig. 3, it is also observed that the absorption coefficient decreases with the increase of wavelength as well as oxygen partial pressure. Fig. 3 (inset) shows the variation of absorption coefficient of rh- In_2O_3 thin films with respect to plasma conditions measured at 550 nm. From the Fig. 3, it is observed that the films exhibit a higher absorption coefficient of 1.06×10^5 cm^{-1} at oxygen flow rate of 15 sccm. Then it decreases to 8.6×10^4 cm^{-1} at a higher oxygen flow rate of 20 sccm. The Fig. 3 also shows that the increase in absorption coefficient for S3, S4, S6, S7 and S8 is due to the variation of process power and process duration.

3.2.3 Band gap

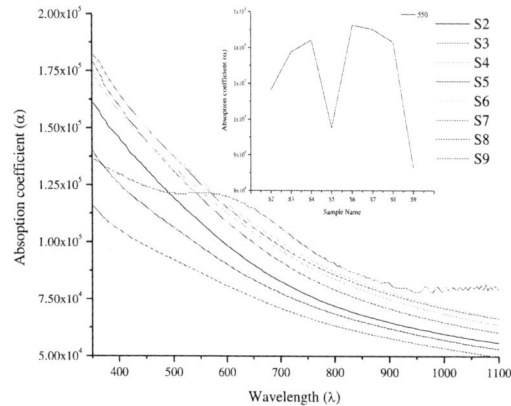

Fig. 3. Spectral dependence of absorption coefficient of O_2 plasma processed Indium thin films

The relation between the absorption coefficient (α) and the incident photon energy ($h\nu$) is given by $\alpha h\nu = A (h\nu - E_g)^{1/2}$ for direct allowed transitions, where A is a constant and E_g is the optical band gap energy. To obtain E_g from this relation, $(\alpha h\nu)^2$ as a function of photon energy is calculated and plotted. The optical band gap is calculated from $(\alpha h\nu)^2$ versus $h\nu$ plot (Fig. 4). The optical band gap of the films is evaluated by extrapolating the linear portion of the above plots to $\alpha = 0$. The optical band gap of the films measured between 2.45 and 2.92 eV. The observed values are close to the published results in the literatures [25]. From the Fig. 4, it is found that the band gap of rh- In_2O_3 thin films decreases with increase of plasma power and O_2 gas flow rate. From the plot, it is also found that rh- In_2O_3 thin films with the high band gap energy is observed with low plasma power (100 W) and high oxygen flow rate (20 sccm). It is attributed to lower density of oxygen vacancies, which act as donor states near the conduction band. To support this argument, EDAX spectra are recorded and the elemental composition of O_2 plasma processed Indium thin films are presented in Table – 2. From this table, it is also revealed that the very low band gap of about 2.45 eV is observed with S2 which is having 77.59% of O_2 and 22.41 % of In.

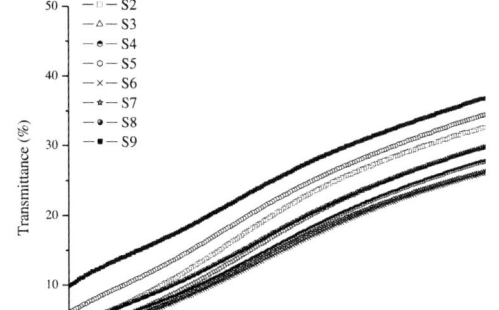

Fig. 2. Transmittance spectra of O_2 plasma processed Indium thin films

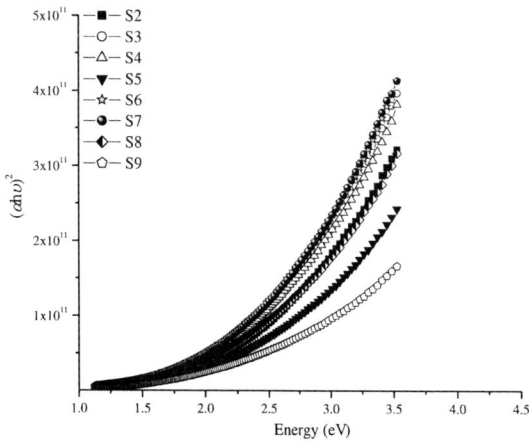

Fig. 4. Variation of band gap of O_2 plasma processed Indium thin films

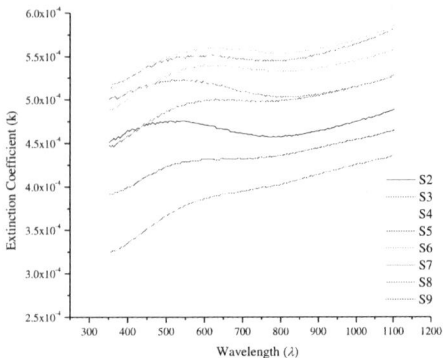

Fig. 5. Spectral dependence of extinction coefficient of O_2 plasma processed Indium thin films

It shows that the Indium film processed at 200W, 20 sccm and 10 min enhance the rate of oxidation and detailed explanation has been given in XRD results.

3.2.4 Extinction coefficient

The spectral variation of extinction coefficient (k) as a function of wavelength (λ) is shown in Fig. 5. The observed value lies in between 3.25×10^{-4} and 5.75×10^{-4}. It indicates a fall in the value of k at lower wavelengths and reaches a minimum at a wavelength close to the band gap energy for all the films. The value of extinction coefficient is found to be very low for all the films that may be due to the less crystallographic defects such as grain boundaries, void free layers and seems to be varying with the process conditions. Such low value of k in the order of 10^{-4} in the visible and near infra red region is a qualitative indication of excellent surface smoothness of the rh- In_2O_3 thin films. From the Fig.5, it is also found that the high k value is observed with low plasma power and high O_2 gas flow rate at band gap region.

It is also found that the low k value seems to be the improvement in the crystallinity with the film processed at high (200W) plasma power and low (15 sccm) O_2 gas flow rate leading to minimum imperfections.

3.2.5 FTIR analysis

IR spectra of processed Indium thin films are presented in Fig.6. The region of interest is below 800 cm^{-1}, where In–O–In vibrations are generally observed [26]. The appearance of two bands peaking at 402 and 538 cm^{-1} can be assigned to the phonon vibrations of In–O bonds and well agreed with the results published elsewhere [27].

TABLE II

ELEMENTAL COMPOSITION AND BANDGAP OF O_2 PLASMA PROCESSED INDIUM THIN FILM MEASURED FROM EDAX ANALYSIS

Process parameters	Atomic % (O_2)	Atomic % (In)	Band gap (eV)
S4	82.07	17.93	2.67
S2	77.59	22.41	2.45
S5	78.82	21.18	2.62
S9	80.68	19.32	2.82

From the Fig. 6, a strong band at near 1195 cm^{-1} and a weak band at near 801 cm^{-1} are the characteristics peaks of Si-O-Si stretching and Si-O bending vibration respectively [28] which is observed as a result of SiO_2 content in glass substrates. Figure 7 (a) and (b) show the changes in peak position and intensity of absorption of rh- In_2O_3 at their characteristics peak of nearby 402 cm^{-1} and 538 cm^{-1} respectively. It is observed that the film processed at 100 W power, 15 sccm gas flow rate and 5 min duration shows high intensity. It is also found that the peak position at 538 cm^{-1} shows high absorption than the peak observed at 402 cm^{-1}. It also reveals that the peak position of 538 cm^{-1} band shifts toward higher frequency which may be due to the change of stress applied during the process with high power as well as O_2 flow rate. But no shift in peak corresponding to 402 cm^{-1} band is observed for all process conditions other than S6 and S8.

IV. CONCLUSIONS

Indium thin films were prepared using RF sputtering and processed under O_2 plasma at various power, O_2 gas flow rate and time. XRD analysis confirmed the presence of rhombohedral In_2O_3 with (110) orientation in all processed samples. The optical properties measured from the transmittance spectra showed the effects of O_2 flow rate, plasma power and time duration on the transmission as well as band gap of the rh- In_2O_3. The observed band gap was low (2.45 eV - 2.92 eV) when compared to pure rh- In_2O_3. FTIR analysis was also supported to confirm the formation of rh-In_2O_3 using O_2 plasma process at room temperatures. EDAX studies revealed the composition of In_2O_3 thin film as non stoichiometry. It is suggested that oxygen plasma could be used to synthesize rh- In_2O_3 films from pure Indium via a relatively rapid and simple oxidation process. Finally, variation on indium oxidation rate was observed with O_2 flow-rate and plasma power with irrespective to process duration.

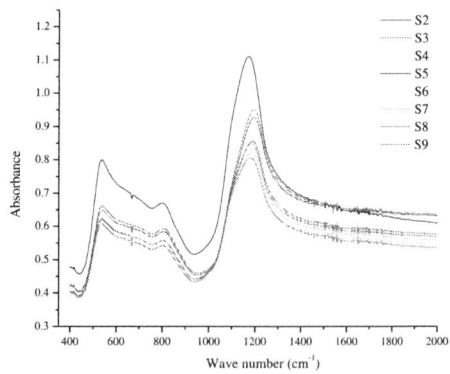

Fig. 6. FTIR spectra of O_2 plasma processed Indium thin films

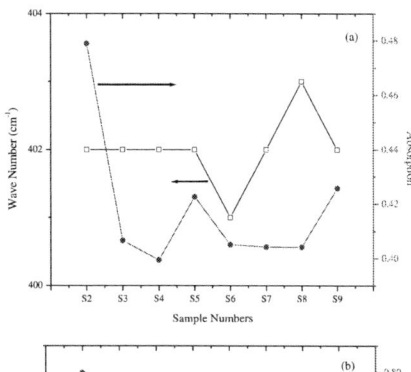

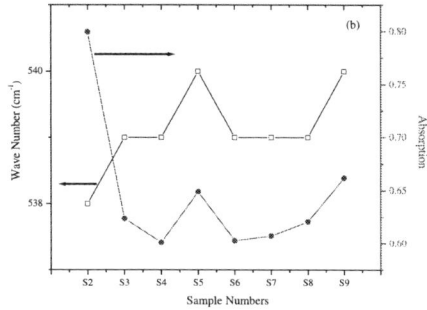

Fig. 7. Change in peak position and intensity of absorption of *rh*-In$_2$O$_3$ at their characteristics peak of nearby (a) 402 cm^{-1} and (b) 538 cm^{-1}

ACKNOWLEDGEMENT

The authors would like to thank the Solid State Laboratory and Nano Optoelectronics Research Laboratory in School of physics who has provided the RF sputtering system to coat In metal films and the ICP-RIE, SEM, AFM and four probe apparatus to perform the plasma process and characterization for the prepared film respectively.

REFERENCES

[1] J. Xu, Y. Chen and J. Shen, "Ethanol sensor based on hexagonal indium oxide nanorods prepared by solvothermal methods," *Mater. Lett.* vol. 62, pp. 1363 – 1365, 2008.

[2] K.L. Chopra, S. Major and D.K. Pandya, "Transparent conductors-A status review," *Thin Solid Films* vol. 102, pp. 1 – 46, 1983.

[3] C.G. Granquist, "Electrochromic tungsten oxide films: Review of progress 1993–1998," *Sol. Energy Mater. Sol. Cells,* vol. 60, pp. 201 – 262, 2000.

[4] A.N.H. Ajili and S.C. Bayliss, "A study of the optical, electrical and structural properties of reactively sputtered InO$_x$ and ITO$_x$ thin films," *Thin solid Films* vol. 305, pp. 116 – 123, 1997.

[5] ICDD PDF-2 Data base, JCPDS-Int. Center for Diffraction Data, Pensylvania, 1994.

[6] R.L. Weiher and R.P. Ley, "Optical Properties of Indium Oxide," *J. Appl. Phys.* vol. 37, pp. 299 – 302, 1966.

[7] P.D.C. King, T.D. Veal, F. Fuchs, Ch.Y. Wang, D.J. Payne, et.al., "Band gap, electronic structure, and surface electron accumulation of cubic and rhombohedral In$_2$O$_3$," *Phys. Rev. B,* vol. 79, 205211, 2009.

[8] C. T. Prewitt, R. D. Shannon, D. B. Rogers and A. W. Sleight, "The C rare earth oxide-corundum transition and crystal chemistry of oxides

having the corundum structure," *Inorg. Chem.* vol. 8, pp. 1985 – 1992, 1969.

[9] G. Kiriakidis, M. Bender, N. Katsarakis, E. Gagoudaskis, E. Hourdakis, et. al., "Ozone Sensing Properties of Polycrystalline Indium Oxide Films at Room Temperature," *Phys. Stat. Sol. (a)* vol. 185, pp. 27–32, 2001.

[10] K.G. Copchandran, B. Joseph, J.T. Abraham, P. Koshy and V.K. Vaidyan, "The preparation of transparent electrically conducting indium oxide films by reactive vacuum evaporation," *Vacuum,* vol. 48, pp. 547 – 550, 1997.

[11] V. Damodara Das, S. Kirupavathy, L. Damodare and N. Lakshminarayan, "Optical and electrical investigations of indium oxide thin films prepared by thermal oxidation of indium thin films," *J. Appl. Phys.* vol. 79, pp. 8521–8530, 1996.

[12] A.P. Mammana, E.S. Braga, I. Torriani and R.P. Anderson, "Structural characterization of transparent semiconducting thin films of SnO$_2$ and In$_2$O$_3$," *Thin Solid Films,* vol. 85, pp. 355 – 359, 1981.

[13] W. Siefert, "Properties of thin In$_2$O$_3$ and SnO$_2$ films prepared by corona spray pyrolysis, and a discussion of the spray pyrolysis process," *Thin Solid Films,* vol. 120, pp. 275-282, 1984.

[14] R.B.H. Tahar, T. Ban, Y. Ohya and Y. Takahashi, "Optical, structural, and electrical properties of indium oxide thin films prepared by the sol-gel method," *J. Appl. Phys.* vol. 82, pp. 865 – 870, 1997.

[15] S. Mailis, C. Grivas, D. Gill, L. Boutsikaris, N.A. Vainos, et al., "Dynamic holography in indium oxide and indium-tin-oxide thin films" *Opt. Memory Neural Networks,* vol. 5, pp. 191–196, 1996.

[16] S. Kasiviswanathan and G. Rangarajan, "Direct current magnetron sputtered In$_2$O$_3$ films as tunnel barriers" *J. Appl. Phys.* vol. 75, pp. 2572-2577, 1994.

[17] M. Sorescu, L. Diamandescu, D. Tarabasanu-Mihaila and V. S. Teodorescu , "Nanocrystalline rhombohedral In$_2$O$_3$ synthesized by hydrothermal and postannealing pathways," *J. Mater. Sci.,* vol. 39, pp. 675 – 677, 2004.

[18] Ch. Y. Wang, Y. Dai, J. Pezoldt, B. Lu, Th. Kups, et. al., "Phase Stabilization and Phonon Properties of Single Crystalline Rhombohedral Indium Oxide," *Crystal Growth & Design,* vol. 8, pp. 1257–1260, 2008.

[19] R.D. Shannon, "New high pressure phases having the corundum structure' *Solid State Commun.* vol. 4, pp. 629–630, 1966.

[20] B.G. Bagley, L.H. Greene, J.M. Tarascon and G.W.Hull, "Plasma oxidation of the high T$_c$ superconducting perovskites," *Appl. Phys. Lett.* vol. 51, pp. 622-624, 1987.

[21] C. Ooi and G.K.L. Goh, "Formation of Cuprous Oxide Films via Oxygen Plasma" *Thin Solid Films,* vol. 518, pp. 98-100, 2010.

[22] F. O. Adurodija, H. Izumi, T. Ishıhara, H. Yoshioka, H. Matsui et. al., "High-quality indium oxide films at low substrate temperature" *Appl. Phys. Lett.* vol. 74, pp. 3059-3061, 1999.

[23] N. Bellakhal, K. Draou and J.L. Brisset, "Electrochemical investigation of copper oxide films formed by oxygen plasma treatment," *J. Appl. Electrochem.* vol. 27, pp. 414–421, 1997.

[24] P. Luzeau, X.Z.Xu, M. Lagues, N.Hess, J.P. Contour, et. al., "Copper oxide thin-film growth using an oxygen plasma source" *J. Vac. Sci. Technol.* vol. A8, pp. 3938–3940, 1990.

[25] Ch. Y. Wang, V. Cimalla, H. Romanus, Th. Kups, G. Ecke, et. al., "Phase selective growth and properties of rhombohedral and cubic indium Oxide", *Appl. Phy. Lett.* vol. 89, 011904, 2006.

[26] N.G. Patel, K. K. Makhija, C.J. Panchal, D.B. Dave, and V.S. Vaishnav, "Fabrication of carbon tetrachloride gas sensors using indium tin oxide thin films" *Sens. & Act.* B vol. 23, pp. 49–53, 1995.

[27] K. Ulutas, D. Deger and Y. Skarlatos, "Thickness dependence of optical properties of amorphous indium oxide thin films deposited by reactive evaporation" *phys. stat. sol. (a),* vol. 203, pp. 2432–2437, 2006.

[28] M. Anwar, I.M. Ghauri and S. A. Siddiqi, "The study of optical properties of In$_2$O$_3$ and of mixed oxides In$_2$O$_3$–MoO$_3$ system deposited by coevaporation" *J. Mater. Sci.* vol. 41, pp. 2859-2867, 2006

Capacitive micro-sensor for the detection of dextrose

Q.Humayun, U.Hashim, M.Kashif

Nano Structure Lab-On-Chip Research Group,Institute Nano Electronic Engineering (INEE),
Universiti Malaysia Perlis (UniMap), 01000 Kangar,Perlis Malaysia
uda@unimap@gmail.com, qhumayun2@gmail.com

Abstract— **This work demonstrates the fabrication and electrical characterization of aluminum micro-gap on silicon substrate. For low cost and time conventional photolithography technique has been applied to fabricate the aluminum micro-gap. The fabricated device was tested for different molarities of dextrose. For the micro-gap fabrication chrome mask was design using AutoCAD software. The size of the fabricated micro-gap was around 81.7μm. The electrical measurements were carried out using Alpha high frequency dielectric analyzer. It was observed that when the dextrose concentration varied from 250μM to 1μM the capacitance value fluctuates from 330nF to 10nF. The limit of detection for the fabricated device was 1μM.**

Keywords-; micro-gap; chrome mask; molarities; dextrose; capacitance;

I. INTRODUCTION

During past two decades, micro and nano fabrications were progressively being applied to create ultra miniature sensors for characterizing and detecting of bio molecules [1]. The reduction in sensor size can result in lower materials cost, reduced weight, and lower power consumption, which are the key factors driving new opportunities for sensors in the market place [2]. Micro and nano-sensors of highly reduced power consumption suitable for integration into wireless communication devices [3]. Low power micro and nano-sensors would also be beneficial for use as battery-operated handheld or wearable sensors [4]. Logical and promising sensing application areas for micro and nano-sensors including medical (e.g., blood analysis, patient monitoring and diagnostic testing), bio warfare detection, genetic analysis, drug discovery, food inspection/testing, environmental monitoring, and industrial chemical process monitoring [5]. Micro and nano-gap structures fabricated by semiconductor materials have been used quite often for such sensors [6]. In this research conventional photolithography technique has been used to fabricate aluminum micro-gap onto silicon substrate and employed as an electrochemical micro sensor to measure the dextrose at different concentration. As the development in medicine steadily progresses toward diagnostics based on molecular markers, and highly specific therapies aimed at molecular targets, the necessity for high-throughput methods for the detection of biomolecules, and their abundance, concomitantly increases [7]. Technology platforms that provide the reliable, rapid, quantitative, low cost and multi-channel identification of biomarkers such as genes and proteins are the fact to the rate-limiting steps for the clinical deployment of personalized medicine in domains such as the early detection and the treatment of malignant diseases [8]. Early detection is particularly important in the case of cancer because in the early stages of disease it is quite possible to treat it with success [11].

II. RESEARCH METHODOLOGY

A. Mask Design

In this research, p-type; 100 mm in diameter (4 inch wafer) silicon substrate wafer has been used to fabricate the aluminum microgap structure. For the fabrication of the aluminum microgap, first mask was designed and gently applied on silicon (Si) substrate. Chemical etching is used to create the aluminum micro-gap structure. As for the lithography process, photo mask was employed to fabricate the aluminum micro-gap using conventional photolithography. Commercial chrome mask was used in this research for the better photo-masking process. The photo-mask was designed using AutoCAD and then prepared by printing onto the chrome glass surface. Fig.1 a, b shows AutoCAD design mask for micro-gap formation with length 4000 μm width 2750μm and the distance between two micro-gaps is 1000μm and the actual arrangement of micro-gap design on chrome mask. Chrome glass mask consists of 22 dies with 60, 80 and 100μm gap designs as shown in Fig. 1(b).

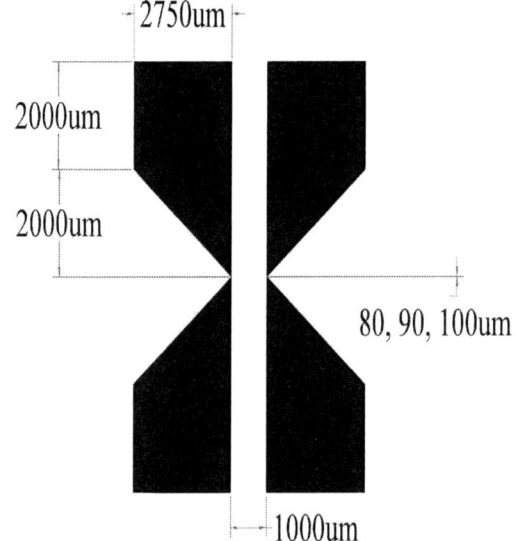

Figure 1.   (a) Design specification of the first Mask (b) Schematic design of the actual mask on chrome glass

B.   Micro -gap Fabrication

A clean Si wafer is deposited with 150nm aluminum using an auto 360 vacuum thermal evaporator. Next, in the photolithography process, 1.2μm layer of positive photoresist is first applied onto the aluminum surface, and then exposed to ultraviolet light through mask as shown in Fig. 3 (e). After development only the unexposed resist will remain and then chemical etching process of Al layer is done before removing the resist. Finally, the wafer was rinsed in developer to remove the photoresist from the surface of aluminum micro-gap. The fabricated aluminum micro-gap structure can be seen in the process flow of Fig. 2.

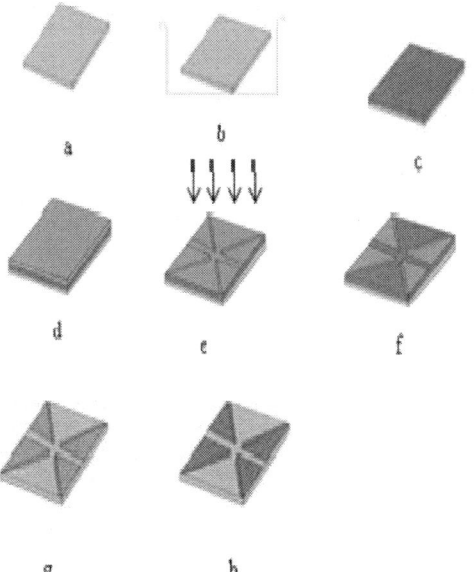

Figure 2.   Fabrication process flow of microgap structure (a) Silicon wafer sample (b) Wafer sample cleaning (c) Aluminum deposition (d) Coating PR-1 2000 resist (e) Exposure of ultra violet light (f) Resist removing (g) Aluminum etching (i) Resist stripping.

C.   Microgap Mesearument and Characterization

The surface morphology of fabricated micro-gap structure was characterized using scanning electron microscope (SEM, JEOL 6460LA. Fig. 3 shows the final device structures used for the electrical characterization using alpha high dielectric analyzer from Novacontrol. The micro-gap will act as a capacitor between two aluminum pads that were 81.7μm apart from each other.

Figure 3.   Capacitance measurement set up with a dielectric analyzer.

D.   Dextrose Measurment and Characterization

Before starting the dextrose measurements on the prepared micro-gap, the fabricated device was cleaned with deionized water followed by dry spinning and then baked at less than 100ºC for few minutes. After that different dextrose solutions were tested drop wise across the micro-gap structure and a change in the capacitance was measured. The obtained data is shown in result and discussion part.

III. RESULT AND DISCUSION

Fig. 4 shows the SEM image of the aluminum micro-gap, it was observed that the gap size of the fabricated device was approximately 81.7μm.

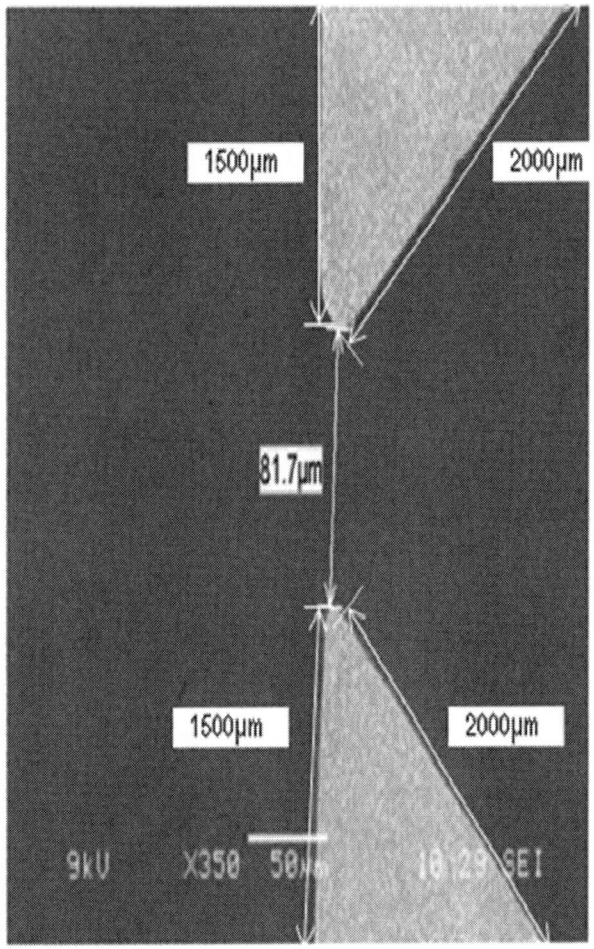

Figure 4. SEM image of the aluminum microgap structure

Fig.9 shows the capacitive curve as a function of frequency at different molarities of dextrose concentrations; 250μM, 200μM, 150μM, 100μM, 25μM, 20μM, 10μM, 5μM, 1μM. All the concentrations were tested at a gap size of 81.7μm. The measurements were carried out in the frequency range of 10Hz to 10MHz.

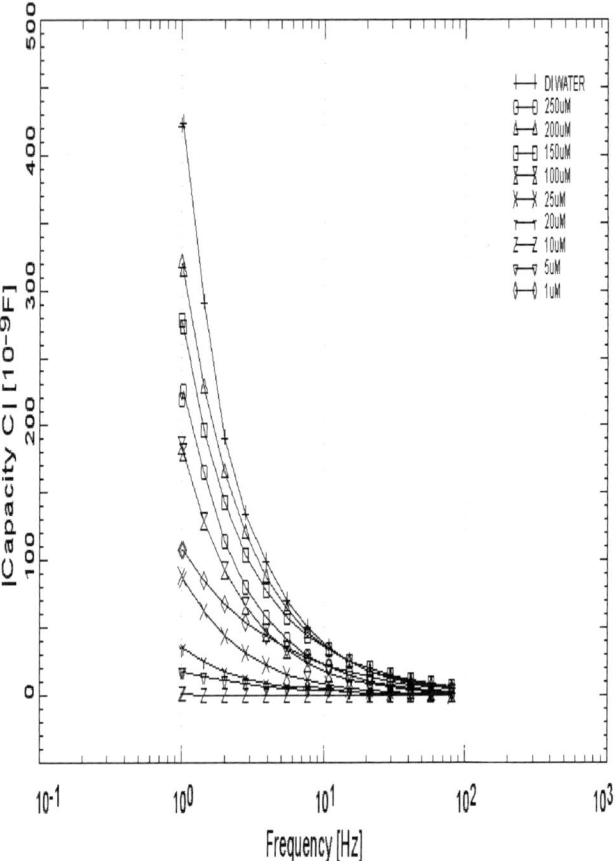

Figure 5. Capacitance curves at different concentration of dextrose.

The capacitive curves revealed that the measured capacitance with DI water is higher than 1μM molar dextrose concentration. As the dextrose concentration increased from 1μM to 250μM, there was a measureable change in the capacitance of the fabricated device. From the capacitance values listed in table1, it was concluded that the capacitance value recorded at 1μM was 1nF whereas at concentration value of 250μM the fabricated device showed 330nF capacitance. It means by varying the concentration from 1μM till 250μM the device showed approximately 30% change in capacitance value. From the capacitance curves we can conclude that our fabricated device was suitable for the detection of dextrose. The limit of detection (LOD) for the fabricated device was 1μM. Further work is in progress to reduce the gap size in order to increase the sensitivity as well as to drop the limit of detection to lower levels.

TABLE I

CAPACITANCE VLAUES AT DIFFERENT DEXTORSE CONCENTRATION

Micro-gap size	Liquid	Capacitance (nF)
	DI Water 430nF	DI Water 430nF
	Dextrose Molarities	
	250μM 330nF	250μM 330nF
	200μM 280nF	200μM 280nF
	150μM 230nF	150μM 230nF
81.7μm	100μM 190nF	100μM 190nF
	25μM 110nF	25μM 110nF
	20μM 95nF	20μM 95nF
	35nF	35nF
	15nF	15nF
	10μM	10μM

III. CONCLUSION

We have successfully demonstrated the fabrication and characterization of aluminum micro-gap as a dextrose sensor. We have applied simple conventional lithographic method using chrome mask to fabricate the micro-gap dextrose sensor. Capacitances were measured when the fabricated micro-gap sensor was exposed to different dextrose concentrations. The noticeable changes in the capacitance curves of the fabricated devices showed promising application of fabricated micro-gap sensor. The results present a potential development towards future practical applications. The limit of detection of the device is still higher and further work is in progress to reduce the limit of detection by reducing the gap size.

ACKNOWLEDGEMENT

The authors acknowledge the financial support from ministry of higher education (MOHE). The authors also would like to thank all of the team members and technical staff in the Institute of Nano Electronic Engineering.

REFRENCES

[1]. Uda Hashim, Siti Fatimah Abd. Rahman, M. Nuzaihan, Md. Nor, Shahrir Salleh," Design and Process Development of Silicon Nanowire Base DNA Biosensor using Electron Beam Lithography" *2008 International Conference on Electronic Design*, December1-3, 2008, Penang, Malaysia.

[2]. Th. S. Dhahi, U. Hashim, M. E. Ali, N. M. Ahmed and T. Nazwa Fabrication of Lateral Polysilicon Gap of Less than 50nm Using Conventional Lithography, *Journal of Nano materials*. Volume 2011, Article ID 250350.

[3]. Dhahi. T. S, Hashim. U, Ahmed, N. M. 2011.Improvement in Processing of Nano Structure Fabrication Using O2

[4]. Dhahi. T. S, Hashim. U, Ahmed. N.M, Taib. A.M. 2010. A review on the Electrochemical Sensors and Biosensors Composed of Nanogaps as Sensing Material. *J. Optoelectr. Adv. Materials*, 12 (29):1857-1862.

[5]. Dhahi. T. S, Hashim .U, Ahmed, N.M. 2011. Fabrication and Characterization of 50nm Silicon Nanogap Structure. *J. Sci. Adv. Materials* 3(2): 233–238.

[6]. Dhahi. T. S, Hashim. U, Ali. M.E, Nazwa. T, Ahmed, N. M. 2011. Fabrication and characterization of lateral polysilicon gap less than 50nm using conventional lithography process. *J. Nano Materials*, Article in Press.

[7]. H. Namatsu, Y. Watanabe, K. Yamazaki. "Fabrication of Si single electron transistors with precise dimensions by electron-beam nanolithography, "*Journal of Vacuum Science and Technology B*, vol.21, no1,pp.1 –5, 2003.

[8]. R. Sasajima, K. Fujimaru, and H. Matsumura, "A metal/insulator tunnel transistor with 16 nm channel length" *Applied Physics Letters*, vol.74, no 21, pp.3215–3217, 1999.

[9]. A. Marchi, E. Gnani, S. Reggiani, M. Rudan, and G. Baccarani G. "Investigation the performance limits of silicon-nanowire and carbon nanotube FETs" *Solid-State Electronics*, vol.50, no.1, pp.78–85, 2006.

[10]. B. L. Allen, P. D. Kichambare, and A. Star, "Carbon nanotube field effect-transistor-based biosensors," *Advanced Materials*, vol. 19, no.11, pp. 1439–1451, 2007.

[11]. Y. Ding, Y. Dong, A. Bapat. "Single nanoparticle semiconductor devices," *IEEE Transactions on Electron Devices*, vol. 53, no. 10, pp. 2525–2531, 2006.

978-1-4673-2395-6/12 $31.00 © 2012 IEEE

Thermoelectric properties and devices of p-type $Bi_{0.4}Sb_{1.6}Se_{2.4}Te_{0.6}$ and n-type $Bi_2Se_{0.6}Te_{2.4}$ prepared by solid state microwave synthesis

Arej Kadhim Abbas, Arshad Hmood Abd Al Kadhim, Haslan Abu Hassan

Nano-Optoelectronic Research and Technology Laboratory (N.O.R.)

School of physics, Universiti Sains Malaysia

11800, Pulau Penang, Malaysia

Email: areejkadhim@yahoo.com

Abstract- This study reports on the fabrication of a chalcogen-based thermoelectric power generation (TEG) device using p-type $Bi_{0.4}Sb_{1.6}Se_{2.4}Te_{0.6}$ and n-type $Bi_2Se_{0.6}Te_{2.4}$ legs. Electrical power generation characteristics were monitored by changing both the temperature conditions and the number of p-n couples required to generate maximum power. The significance of the resistances, including the internal resistance (R_{in}) and contact resistance (R_c) between legs and electrodes, are discussed. The maximum open circuit voltage (V_{oc}) and output power (P_{max}) obtained with the 18 p-n couples device were 1480.5 mV and 273.2 mW, respectively, under the thermal condition of T_H=523 K hot-side temperature and $\Delta T = 184$ K temperature difference.

I. INTRODUCTION

A thermoelectric power generation (TEG) device produces voltage between the hot and the cold sides when a temperature difference (ΔT) between the two sides is present as a result of thermoelectric effect (TE). TE includes Seebeck (S), Peltier, and Thomson effects. TE is also associated with other effects such as Joulean and Fourier effects [1].

The following are the advantages of these generators: no moving parts, small and lightweight, maintenance-free, acoustically silent and electrically "quiet," same-module heating and cooling, wide-ranging operating temperature, and environmentally friendly [2].

A practical TEG device generally consists of two or more elements of n-type- and p-type-doped semiconductor materials that are electrically connected in a series and thermally connected in parallel. The extra electrons in the n-material and the "holes" resulting from the deficiency of electrons in the p-material are the carriers that move heat energy through the TE material. Most TEG devices are fabricated using an equal number of n-type and p-type elements, where a pair of n and p elements forms a thermoelectric "couple" [3].

Yet, enhancing in the TE properties is strongly required for empirical applications. One approach to enhance the TE prop-

erties is optimization of the doping effective route. For example, recent devices use Bi_2Te_3, a semiconductor, which when alloyed with antimony (Sb) or selenium (Se) becomes an efficient TE material for refrigeration or for portable power generation [4]. As an alternate approach, researchers have attempted to improve the efficiency of materials based on Bi_2Te_3 by creating structures with one or more reduced dimensions [5]. In one case, an n-type Bi_2Te_3 has been shown to have improved S. However, S and electrical conductivity (σ) have a trade-off; a higher S results in decreased carrier concentrations and decreased σ [6].

In our study, we fabricated 9 and 18 couples of TEG devices using solid-state microwave synthesis. We focused on the properties of two TEG devices that use $Bi_{0.4}Sb_{1.6}Se_{2.4}Te_{0.6}$ as p-type and $Bi_2Se_{0.6}Te_{2.4}$ as n-type. Fundamental physical parameters of the p-type and the n-type samples, such as S, σ, and power factor (P_{factor}), were subsequently examined.

II. EXPREIMENTAL

Bi, Sb, Se, and Te that were used in this study were highly pure powders (99.999%). The typical element ratio for the preparation of p-type $Bi_{0.4}Sb_{1.6}Se_{2.4}Te_{0.6}$ is as follows: 0.3071g Bi, 0.7156g Sb, 0.6961g Se, and 0.2812g Te. The n-type $Bi_2Se_{0.6}Te_{2.4}$ was prepared by mixing 1.0834g Bi, 0.1228g Se, and 0.7938g Te. The p-type and n-type ingots were fabricated through a solid-state microwave synthesis that was described in a previous literature [7]. The samples were irradiated in an 800 W (100% power) MS2147C microwave oven (LG) at 2.45 GHz for 10 min. Bright, whitish-blue plasma was observed emerging from both the ampules from the first minute of microwave exposure. The temperature of the samples was measured using an OS524E infrared thermometer (OMEGA SCOPE) with values 898K for p-type sample, whereas 873K for n-type sample. Selected portions of the ingots were imaged using field emission scanning electron microscopy (FESEM) (Leo-Supra 50VP,Carl Zeiss, Germany).

978-1-4673-2395-6/12 $31.00 © 2012 IEEE

After grinding, the samples were then characterized to determine their crystallization via X-ray diffraction (XRD, PANalytical X'Pert PRO MRD PW3040, Almelo, The Netherlands). The resultant powders were pressed into disk shapes (5 mm diameter and 3.5 mm thickness) through cold pressing at 10 tons. The Seebeck coefficient (S) was determined by the slope of the linear relationship between the thermoelectromotive force and the temperature difference between the two ends of each sample. The electrical conductivity (σ) was measured using the standard four-probe dc method under a vacuum of 10^{-3} mbar within temperatures ranging from room temperature to about 523 K.

The assembly of 9 and 18 couples from these pellets was placed between two alumina plates with the corresponding dimensions of 50 mm $\times$ 25 mm and 50 mm $\times$ 50 mm, respectively, which served as hot and cold ends for the relevant TE pellets. By using Ag paste and Cu plates, the Ag paste–Cu plates–Ag paste electrodes were made on the inner surface of the alumina substrate. The devices were then dried at room temperature for one day to metalize the electrodes on the devices. To evaluate device performance, the bottom alumina plate was heated up to 523 K by one brass block as heater for the device, and the top plate was cooled by another brass block with circulated cooling water. ΔT between the hot and the cold sides was measured by two digital K-type $E^{©}$ Sun (ECS820C) thermocouples near the inner surface of the alumina substrates. ΔT along the length of the devices was approximately equal to the difference in interface temperatures between the hot alumina plate T_H and the cold alumina plate T_C. The current-voltage (I-V) lines and the current-power (I-P) curves of power generation were performed in the air by sweeping the load resistance (R_L) using the variable resistance box. The open circuit voltage and many other voltages at the condition of power generation were measured by a voltage meter (Keithley 197).

III. RESULTS AND DISCUSSION

A typical FESEM image of the surface morphology of the ingots $Bi_{0.4}Sb_{1.6}Se_{2.4}Te_{0.6}$ prepared through solid-state microwave synthesis is shown in Fig. 1(a). Hexagonal rods with polished surfaces and different widths and lengths were observed. The addition of Se to Bi_2Te_3 compound, the FESEM observations revealed the appearance of a typical layered and well-packed structure, which indicating that Se alloying is an effective approach for crystalline refinement, as shown in Fig. 1(b). XRD experiments were carried out to determine the structure of the powder samples, and the results are shown in Figs. 2. In Fig. 2(a) the XRD pattern of $Bi_2Se_{0.6}Te_{2.4}$ powder was showed. All the peaks in this pattern can be indexed according to JCPDS15-0863 and 33-0214 for Bi_2Te_3 and Bi_2Se_3, respectively, rhombohedral structure (R3m). In Fig. 2(b), all the XRD peaks of $Bi_{0.4}Sb_{1.6}Se_{2.4}Te_{0.6}$, which can be

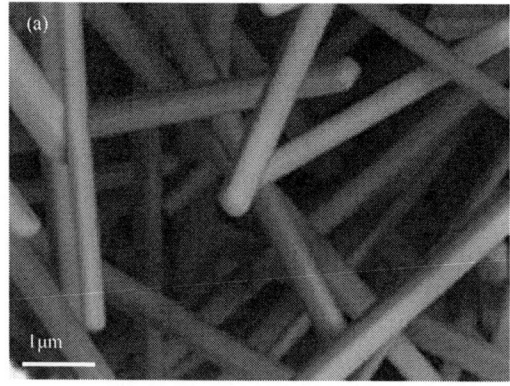

Fig. 1. Felid emission scanning electron microscopy image (FESEM) of (a) $Bi_{0.4}Sb_{1.6}Se_{2.4}Te_{0.6}$ and (b) of $Bi_2Se_{0.6}Te_{2.4}$ ingots.

indexed as the rhombohedral phase of Bi_2Te_3, Sb_2Te_3, and Bi_2Se_3 (JCPDS15-0863, 15-0874, and 33-0214, respectively) with a small amount of Sb_2Se_3 phase, appeared as (400), (331), (060), and (412). The diffraction peaks (015) displayed an apparent shift to a higher angle with increasing Se content, which is mainly due to the smaller atomic radius of Se (1.15 Å) compared with that of Te (1.4 Å).

The transport properties of p- and n-type samples in terms of σ, S, and P_{factor} were investigated from 300 to 523 K (Figs. 3 to 5). Both sample types had nearly the same behavior as σ, which gradually decreased as the experimental temperature increased, that is, a degenerate semiconductor (Fig. 3). σ for the p-type $Bi_{0.4}Sb_{1.6}Se_{2.4}Te_{0.6}$ sample (σ_p) varied from 4.96×10^5 S/m at 300 K to 1.79×10^5 S/m at 523 K, whereas σ for the n-type $Bi_2Se_{0.6}Te_3$ sample (σ_n) was from 1.99×10^4 S/m at 300 K to 1.74×10^4 S/m at 523 K. The behavior of the σ of the two types contrary to its S. S for both types of the samples gradually incr-

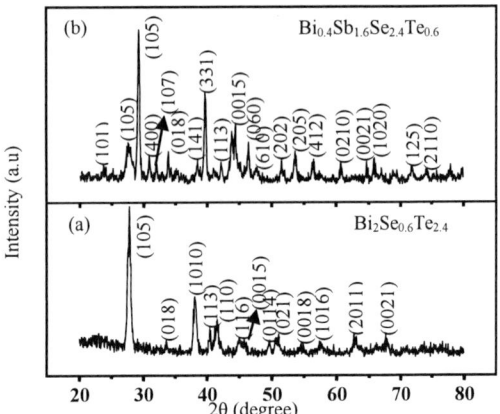

Fig. 2. X-ray diffraction of (a) $Bi_{0.4}Sb_{1.6}Se_{2.4}Te_{0.6}$ and (b) of $Bi_2Se_{0.6}Te_{2.4}$ powders.

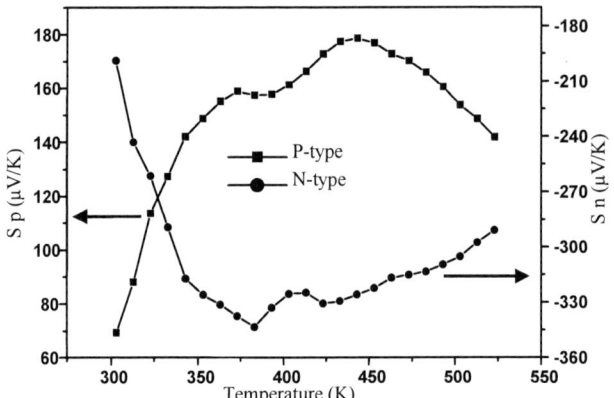

Fig. 4. Temperature dependence values of the Seebeck coefficient of p- and n-type samples.

ease as the temperature increases. As shown in Fig. 4, S of the p-type sample (S_p) was 178.5 at 443 K, and for n-type (S_n) was -330.6 μV/K at 423 K. σ and S were influenced by the incorporation of Se atoms into the crystal lattice, thus changing the formation energy of the lattice defects in the mixed crystals [4]. As evident in Fig. 5, the temperature behaviors of P_{factor} for the p-type and n-type samples were similar: increasing with increasing temperature, achieving a maximum, and then decreasing with further increases in temperature. It indicated that p- and n-type samples can be appropriately used at low temperatures. $P_{factor-P}$ values obtained for p-type samples were larger than those for n-type sample within the entire temperature range investigated. The maximum $P_{factor-P}$ measured was 7.47 mW/mK^2 at 373 K, as previously reported [7], whereas the

maximum $P_{factor-n}$ was 2.22 mW/mK^2 at 383 K. The output voltage and the output power of the fabricated 9(D_1) and 18 (D_2) couples versus the current were measured by sweeping R_L at several temperature conditions, as shown in Figs. 6 and 7. The open-circuit voltage (V_{oc}) that is equal to the intercept of the I-V line reached 586 mV and 1480.5 mV for D1 and D2, respectively, at ΔT of 184 K and T_H of 523 K, which are in agreement with the expression $V=V_{oc}-R_L I$. It is lower than that calculated S–T curves of both p-and n-type legs ($V_{calculated} = (S_p-S_n) \times \Delta T \times n$, where n is the number of couples). This voltage loss could have originated from many factors including low thermal conductivity of alumina substrate and unfavorable junctions between the TE legs and the electrodes [8]. I-P curves illustrated in Figs. 6 and 7 exhibit the parabolic curves of the output power (P_{out}); an analysis to plots of I-V lines

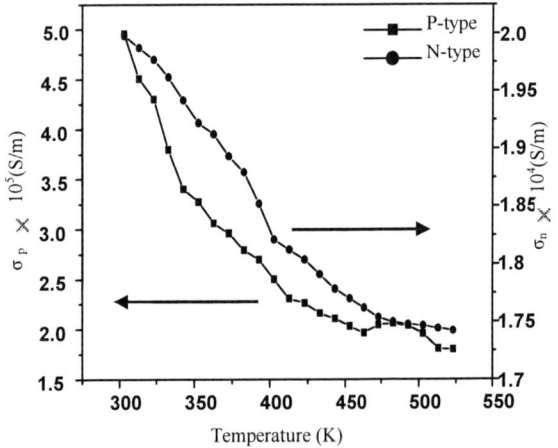

Fig. 3. Temperature dependence values of the electrical conductivity of p- and n-type samples.

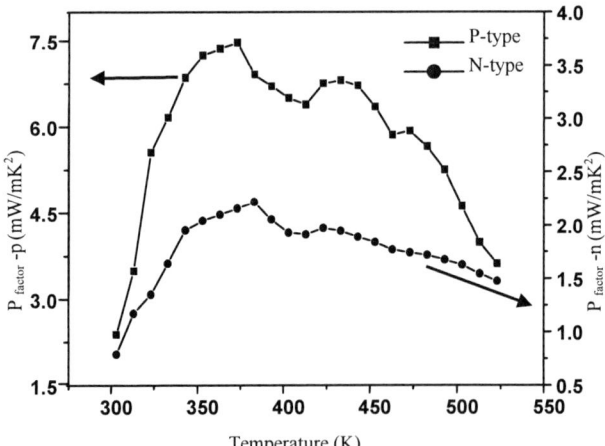

Fig. 5. Temperature dependence values of the power factor of p- and n-type samples.

978-1-4673-2395-6/12 $31.00 © 2012 IEEE

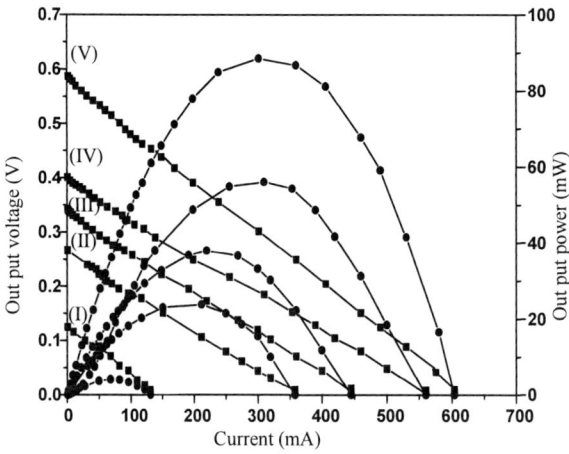

Fig. 6. The power generation characteristics of the TEG device that comprises 9, where (I)ΔT=27,(II)ΔT=66(III)ΔT=104(IV)ΔT= 145,and (V) ΔT=184K.

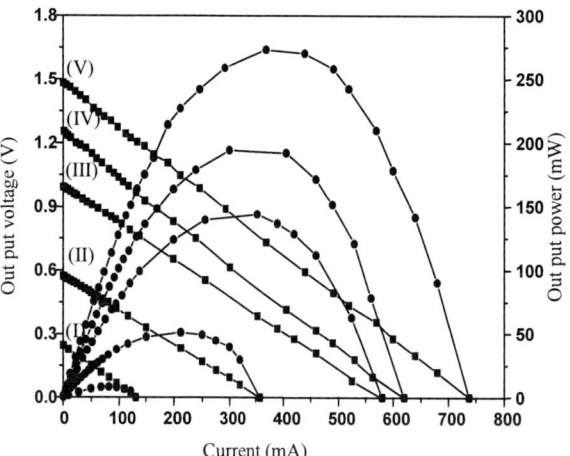

Fig. 7. The power generation characteristics of the TEG device that comprises 18, where (I)ΔT=27,(II)ΔT=66(III)ΔT=104(IV)ΔT= 145,and (V) ΔT=184K.

allows the observation of an increasing in the P_{out} with the ΔT. The explanation for this observation results from the rise of the ΔT, whose consequence is an increase in the output voltage(V_{out}).As high is this V_{out}, the high will be the output current (I_{out}) for a given R_L, and therefore will be the dissipated power in the external load, e.g. $P_{out} = R_L I_{out}^2$.Considering several values for the R_L, it was also possible to obtain for the P_{out}, versus the I_{out} (or versus the R_L). The maximum output power (P_{max}) values were 88.5 and 273.2 mW for D1 and D2, respectively, at the thermal condition of 523 K T_H and $\Delta T = 184$ K, which means these results could be comparable with the results of Wang et al. [4]. It was investigated that the powers of the devices improved by increasing the number of couples between the hot and the cold sides. Based on these results, V_{oc} and P_{max} systematically increased with the number of p-n couples, indicating that TE power could be simply controlled by a change in the module design.

The internal resistance (R_{in}) of each device corresponding to the slope of the I-V lines was directly obtained by the measured system. The ideal internal resistance (R_{id}) was calculated by the sum of the resistance values of p-type and n-type samples. With R_{in} and R_{id}, contact resistance (R_c) can be obtained by $R_c = R_{in} - R_{id}$ [2]. The resistance values of D_2 ($R_{in}=2\Omega$ and $R_c=1.57\ \Omega$) were two times larger than those of D_1 ($R_{in} =0.97\ \Omega$ and $R_c=0.754\ \Omega$), which could be attributed to the differences in size and electrode contact areas among elements [9]. These results demonstrate that, with the relationship between R_c and P_{max}, R_c should be minimized for each device because it plays a key role in TEG device performance. Two methods were adopted to optimize the device performance. First, the surface of the alumina plates was treated with NaOH solution to increase roughness and to enhance both mechanical strength and electrical contact between the alumina plates and the Cu electrodes. Second, the ends of p-type and n-type samples were

grooved to increase the surface area, also improving the mechanical and the electrical properties of the contacts [2]. Electrical and thermal contacts are known to play an important key role in improving the device-manufacturing factor (MF). MF represents the cumulative influence of various parameters in the fabrication process on the quality of the devices, and it is defined as $MF = R_{id}/R_{in}$ [8]. In the present research, MF at ΔT of 184 K reached 22.3% and 21.5% for D1 and D2, respectively, which could be attributed to the value of the Rc as previously mentioned. Based on the data obtained, the good TEG properties originally came from the relatively high electrical properties of p-type and n-type samples that were prepared by solid-state microwave synthesis.

IV. CONCLUSIONS

TE materials p-type $Bi_{0.4}Sb_{1.6}Se_{2.4}Te_{0.6}$ and n-type $Bi_2Se_{0.6}Te_{2.4}$ were prepared via solid-state microwave synthesis. TEG devices were fabricated and characterized in terms of high V_{oc} and P_{max}. A maximum V_{oc} and P_{max} of 1480.5 mV and 273.2 mW, respectively, were achieved in the device with 18 p-n couples with T_H of 523 K and ΔT of 184 K. V_{oc} and P_{max} increased with ΔT and also systematically increased with the number of p-n couples. Based on these results, the device using p-type $Bi_{0.4}Sb_{1.6}Se_{2.4}Te_{0.6}$ and n-type $Bi_2Se_{0.6}Te_{2.4}$ worked successfully and it was stable with satisfactory TE performances.

ACKNOWLEDGMENT

This work was supported by the (PRGS) 1001/PFIZIK/844091 of the Universiti Sains Malaysia (USM).

REFERENCES

[1] Y. Jiang, J. Zhou and X. Zhang, "Analytical model of parallel thermoelectric generator", Appl. Energy, vol. 88,pp. 5193–5199, 2011.

[2] L. Han, Y. Jiang, S. Li, H. Su, X. Lan, K. Qin et al., "High temperature thermoelectric properties and energy transfer devices of $Ca_3Co_{4-x}Ag_xO_9$ and $Ca_{1-y}SmyMn_{O3}$", J Alloys Compd. vol. 509, pp. 8970– 8977, 2011.

[3] A.G. Agwu Nnanna, W. Rutherford, W. Elomar, B. Sankowski," Assessment of thermoelectric module with nanofluid heat exchanger", Appl. Therm. Eng. vol. 29, pp.491–500, 2009.

[4] S. Wang, W. Xie, H. Li, X. Tang,"Enhanced performances of melt spun Bi2(Te,Se)3 for n-type thermoelectric legs", Intermetallics, vol. 19, pp.1024–1031.2011.

[5] J. Kang, J.-S. Noh and W. Lee," Simple two-step fabrication method of Bi_2Te_3 nanowires", Nanoscale Res. Lett., vol. 6, pp. 277–280, 2011.

[6] B.Y. Yoo, C.-K. Huang, J.R. Lim, J. Herman, M.A. Ryan, J.-P. Fleurial et al.," Electrochemically deposited thermoelectric n-type Bi_2Te_3 thin films", Electrochim. Act., vol. 50, pp. 4371–4377, 2005.

[7] A. Kadhim, A. Hmood, H. Abu Hassan," Novel hexagonal rods and characterization of $Bi_{0.4}Sb_{1.6}Se_{3x}Te_{3(1-x)}$ using solid-state microwave synthesis", Mater. Lett., vol. 81,pp.31–33, 2012.

[8] D. Zhao, C. Tian, S. Tang, Y. Liu, L. Jiang, L. Chen," FabricationofaCoSb3-based thermoelectricmodule", Mater. Sci. Semicond. Process., vol. 13, pp. 221–224, 2010.

[9] W. Shin, N. Murayama, K. Ikeda, S. Sago. Thermoelectric power generation using Li-doped NiO and (Br, Sr)PbO₃ module. J. Power Sour. 103 (2001) 80-85.

Fabrication and characterization of $Pb_{1-x}Yb_xTe$ based alloy thin film thermoelectric generators using thermal evaporation method

Arshad Hmood Abd Al Kadhim, Arej Kadhim Abbas, Haslan Abu Hassan

Nano-Optoelectronic Research and Technology Laboratory (N.O.R.)

School of physics, Universiti Sains Malaysia

11800, Pulau Penang, Malaysia

Email: Arshad.phy73@gmail.com

Abstract- **In this work fabricated p-$Pb_{0.925}Yb_{0.075}Te$:Te and n-$Pb_{0.925}Yb_{0.075}Te$ thin films thermoelectric devices are composed of 20-pair and 10-pair deposited on a glass substrate using simple thermal evaporation method. Overall size of thin films thermoelectric generators which consist of 20-pairs and 10-pair of legs connected by aluminum electrodes (Al-electrode) was 23 mm×20 mm and 12 mm×10 mm, respectively. The 20-pair p–n thermocouples in series device generated output maximum open-circuit voltage of 742.7 mV and a maximum output power up to 0.657 µW at temperature difference $\Delta T = 162$ K, and 467.9 mV and 0.346 µW at $\Delta T = 162$ K, for 10-pair, respectively.**

I. INTRODUCTION

Recently, the improvements in thermoelectric (TE) thin films have strongly increased in over a wide temperature range have spurred interest in micro-scale TE generators [1]. Although these TE thin films devices produced few micro-watts of power at relatively high-voltage there have been several applications calling for the operation of small electric devices and systems in the microelectronics industry [2, 3]. Micro thermoelectric generators can be used as a power source of small electronic devices, wireless sensors, and wearable electronics [4]. These include equipment used in military, aero space, medical, industrial, and consumer and scientific institutions [2, 5]. However, recent significant advances in the scientific understanding of quantum well and nanostructure effects on TE properties and modern thin layer and nano-scale manufacturing technologies have combined to create the opportunity of advanced TE materials with potential conversion efficiencies of over 15% [6]. The advent of these advanced TE materials offers new opportunities to recover waste heat more efficiently and economically with highly reliable and relatively passive systems that produce no noise and vibration[7, 8]. In the same time, the interest in producing micro-thermoelectric devices opens new opportunity in the field of micro power generation. Micro thermoelectric converters are a promising technology due to the high reliability, quiet operation and are usually environmentally friendly [9]. Recent developments in micro-thermoelectric devices using thin film deposition different growth methods

based on MBE [10], MOCVD [11], RF co-sputtering [5, 12], a simple vacuum thermal evaporation [6], flash evaporation [2], and co-evaporation [4] have been used to grow single layers and super-lattices on various substrates [2] have shown that energy scavenged from human environment. TE generation have primarily focused on increasing the material figure of merit (ZT), which is the standard measure of a material's TE performance, and defined as $ZT = S^2\sigma T/\kappa$, where S is the Seebeck coefficient, σ the electrical conductivity, κ the thermal conductivity, and T is the absolute temperature. The product $S^2\sigma$ is defined as the thermoelectric power factor [13]. The power factor should be maximized and the thermal conductivity should be minimized in order to achieve high efficiency thermoelectric materials. In present study, we focus on the characterized and enhance micro-fabrication methods to improve the performance and integration of micro-scale TE generators. We prepare $Pb_{0.925}Yb_{0.075}Te$:Te as p-type ingot and $Pb_{0.925}Yb_{0.075}Te$ as n-type ingot for the fabrication 20-pair and 10-pair thermoelectric micro-devices (23mm × 20mm and 12mm × 10 mm, respectively) of the thin films by a thermal evaporation method deposited onto glass substrates. First, investigate the intrinsic properties of each of the constituent thin films, and then we measure the output voltage and estimate the maximum output power of a complete generator near room temperature as functions of the temperature difference between hot and cold junctions.

II. EXPERIMENTAL

The n-type $Pb_{0.925}Yb_{0.075}Te$ was synthesized as a ternary compound by solid-state microwave standard. Weighted 2 g amounts, according to the stoichiometric ratio (1-x) : x : 1, were prepared from three high-purity element powders (Pb, Te ≥99.999% 100 meshes, and Yb ≥99.9% 157 µm) as described elsewhere [14]. The polycrystalline alloys of p-type of $Pb_{0.925}Yb_{0.075}Te$:Te prepared in above technique after added excess tellurium element. The thin film thermoelectric generators are deposited onto clean glass substrates by thermal evaporation of 10^{-6} mbar at 300 K using silicon monoxide source design (SM) SO-20 series by (R. D. Mathis Co. USA).

978-1-4673-2395-6/12 $31.00 © 2012 IEEE

The p-type ($Pb_{0.925}Yb_{0.075}Te:Te$) and n-type ($Pb_{0.925}Yb_{0.075}Te$) powder is positioned in the boat in load cavity, when heated it follows an indirect path through a series of baffles and then out the vertical chimney, which the height's and diameter were 12 mm and 6.3 mm, respectively. So the substrates cannot see the bulk $Pb_{1-x}Yb_xTe$ material at any time, this, essentially, eliminates any chance of spitting and streaming, which causes pinholes as shown in Fig. 1. The distance between the tantalum bout and the substrate is 180 mm. The thicknesses of the thin films were determined to be approximately 0.947 μm using an optical reflectometer (Filmatric F20, USA) measurement system. The patterned shadow masks fabricated 20-pair and 10-pair thermoelectric micro-devices (23 mm×20 mm and 12 mm×10 mm, respectively) for the p and n-legs thin films and their junctions. Tracks on square substrates had 400 μm in width and 20 mm in length, on rectangular ones – 400 μm and 10 mm, respectively. The dimension of p- and n-legs was 20 mm (l) × 400 μm (w) × 0.947 μm (t), and the spacing between both legs was arranged to be 150 μm are shown in Fig. 2. A diffusion barrier layer (Al-electrode) was also deposited between p- and n-type of the thin film thermoelectric generator at the junctions. The fundamental physical parameters of semi-magnetic semiconductor lead chalcogenide thin films such as lattice constant, electrical conductivity, and the Seebeck coefficient are referred to in our previous papers [14]. We first measure the output voltage of the thin film thermoelectric generators while imposing a temperature gradients $\Delta T = T_h - T_c$ between hot and cold junctions of the generators. The schematic diagram of the measurement for the output voltage is shown in Fig. 3. The output voltage V_{out}, and the respective current, I_{out} are measured at the silver paint pads connected to the thermoelectric legs.

Measurement pairs of the voltage and current, $\{V_{out}, I_{out}\}$, are acquired while the load resistance, R_{load}, is manually adjusted. We also measure the overall resistance of the thin film thermoelectric generators by a two-wire method. The internal resistance, R_{in}, of the thermoelectric generator is calculated as follows: $R_{in} = V_{oc}/I_{sc}$, where V_{oc} is the open-circuit voltage and I_{sc} is the short-circuit current. The maximum output power of the thin film thermoelectric generators is estimated from the output voltage and the overall resistance of the generators.

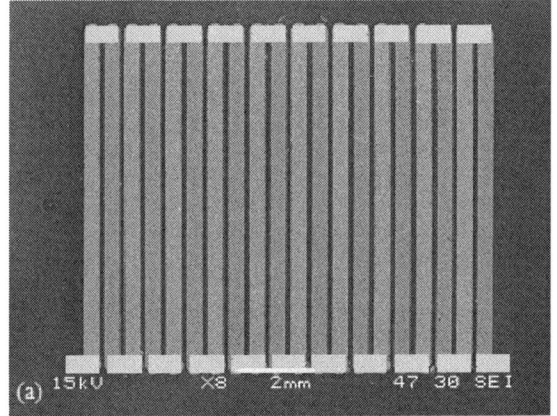

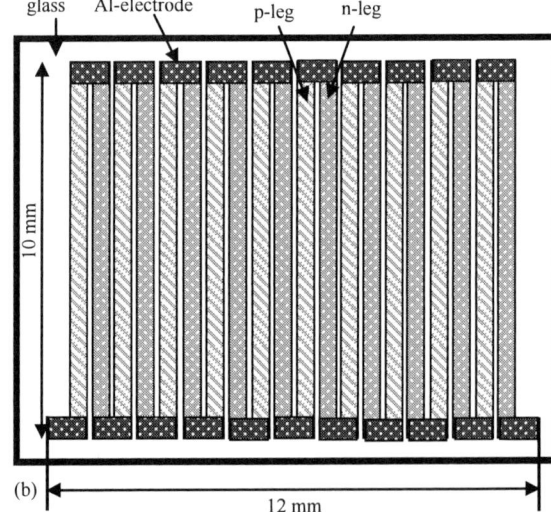

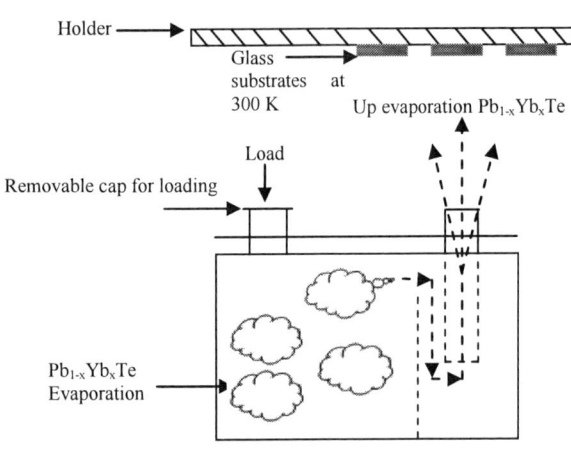

Fig.1. Scheme of SM source used to prepare polycrystalline superlattice $Pb_{1-x}Yb_xTe$ thin films.

Fig.2. (a) Photograph of fabricated thin film thermoelectric generator on glass substrate, and (b) Schematic of thin film thermoelectric generator.

III. RESULTS AND DISCUSSION

The performance of lead–ytterbium–telluride-based micro-generators is investigated at temperature range of 298-532 K. Thus, a temperature difference ΔT was induced between hot and cold junctions of the micro-generators. The first set of tests consisted in the measurement the output voltage versus the output current output characteristic of the thermoelectric converter, and estimates the maximum output power. Fig. 3 shows the load characteristics of the two micro-generators, namely, the output voltage V_{out} versus the output current I_{out} as functions of the temperature difference ΔT for (a) 20-pair and (b) 10-pair micro-generators. As expected the output voltage increases with the temperature gradient, ΔT increased, and this conclusion is valid for the plot of the Fig. 3 and for similar plots, since the gradient temperatures are such that the thermoelectric converter is not behind a certain thermal saturation point. It can be observed a high linearity in all V_{out} v.s I_{out} plots and almost the same slope. This means that the internal resistance $R_{in} = V_{oc}/I_{sc}$ of the thermoelectric converter still constant with the gradient temperature and load resistance. From this analysis, it possible to conclude that the internal resistance of the analyzed thermoelectric device is equal to R_{in} = 209.8 KΩ for 20-pair, and 157 KΩ for 10-pair, respectively. An analysis to both plots of the Fig. 3 (a) and (b) allows the observation of an increasing in the output power, P_{out}, with the temperatures gradient. Thus, the observation results from the rise of the temperature gradient, ΔT, whose consequence is an increase in the output voltage, V_{out}. As high is this output voltage, the high will be the output current I_{out} (considering several values for the load resistance) and therefore will be the dissipated power in the external load resistance, e.g. $P_{out} = R_{Load}I_{out}^2$. The alternative set of plots the output power, P_{out}, versus the output current, I_{out} (or versus the load resistance, R_{Load}). Fig. 3 illustrates six plots for the six difference temperatures, an alternative set of plots, but for the output power, *Pout*, versus the output voltage, V_{out}. The voltage, or thermoelectric electromotive force (emf), produced by Seebeck effect is defined as $V_{out} = S_{ab}ΔT$, where S_{ab} indicates the relative Seebeck coefficient for the material pair a–b [9]. In order to maximize the generated output voltage, several thermocouples are connected electrically in series each other and thermally in parallel to form a thermopile, which is able to generate *n* times the output voltage of one thermocouple (if n is the number of thermocouples in series) and a maximum output electric power (with optimal impedance matching) can be expressed as $P_{max} = (nS_{ab}ΔT)^2/4R_{in}$, where R_{in} internal electrical resistance of generator, the internal electrical resistance is calculated from their dimensions, the number of thermocouples in series, and the electrical resistance [15, 16]. Assuming that the maximum output power is achieved when $R_{Load} = R_{in}$, so the maximum output power of micro-generators estimated as functions of the temperature difference in this measured temperature region [15]. At the temperature difference of 162 K, the output voltage reaches 742.7 mV and 0.657 μW, respectively, for 20-pairs and 467.9 mV and 0.346 μW,

respectively for 10-paris. In fact, the thermoelectric generators are generally based on heavily doped semiconductors use the Seebeck effect to produce electrical energy [14-18]. In a thermoelectric generator, another performance factor is more appropriate, which is the power factor, $S^2σ$ (W/K^2 m^{-1}). The $S^2σ$ is defined as the electric power per unit of area through which the heat flows, per unit of temperature gradient between the hot and the cold sides [19]. This is attributed to the structure not having been optimized and to the high contact resistance caused by the non-optimized bonding process [11]. It is well known that the electrical contact and thermal contact will play important roles in improving the device power-generation performance and need to be minimized.

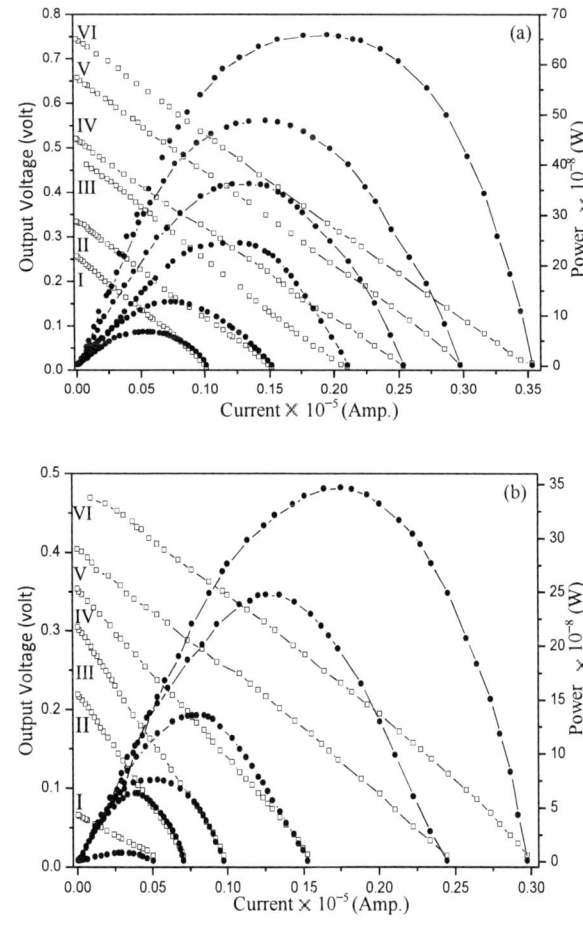

Fig. 3. Output power of the generator versus the output voltage and the output current of (a) 20–pair and (b) 10–pair Pb$_{1-x}$Yb$_x$Te micro-devices as a different temperature gradients ΔT at (I) ΔT=75 K, (II) ΔT=92 K, (III) ΔT=112 K, (IV) ΔT=122 K, (V) ΔT=148 K, and (VI) ΔT=162 K.

IV. CONCLUSIONS

Lead−ytterbium−telluride-based micro-generators are successfully fabricated by a thermal evaporation method, and their performance is improved by micro scale design. We measure the performance of the micro-generators at 298-523 K. The high output voltage of 742.7 mV, 467.9 mV and estimated output power of 0.657 μW, 0.346 μW for 20-pair and 10-pair, respectively, are obtained from a temperature difference ΔT = 162 K. Seebeck voltage and electrical power have shown a high voltage and low power generate for each device geometry. This means that these configurations are not enough well adapted to supply high power but are interesting for temperature sensors thanks to a good sensitiveness in voltage. We can conclude that thermoelectric micro-devices exhibit scalability in output power per unit volume. These power values are enough low, mainly due to high electrical contact resistances, and so due to device geometry.

ACKNOWLEDGMENT

The authors would like to the Nano-Optoelectronics Research and Technology Laboratory (N.O.R.) of the School of Physics for the help extended during the research. This work was supported by the Postgraduate Research Grant Scheme (PRGS) 1001/PFIZIK/844134 of the Universiti Sains Malaysia.

REFERENCES

[1] M. Takashiri, T. Shirakawa, K. Miyazaki, H. Tsukamoto, Sens. Actuators A 138 (2007) 329–334.

[2] G. Savelli, M. Plissonnier, J. Bablet, C. Salvi, J.M. Fournier, MEMS/MOEMS - DTIP (2006) ISBN: 2-916187-03-0.

[3] Il-Ho Kim, Mater. Lett. 43 (2000) 221–224.

[4] Nam-Ho Bae, Seungwoo Han, Kwang Eun Lee, Byeongil Kim, Seon-Tai Kim, Current Appl. Phys. 11 (2011) S40−S44.

[5] N. Kaiwa, M. Hoshino, T. Yaginuma, R. Izaki, S. Yamaguchi, A. Yamamoto, Thin Solid Films 515 (2007) 4501–4504.

[6] Ming Tan, Yao Wang, Yuan Deng, Zhiwei Zhang, Bingwei Luo, Junyou Yang, Yibin Xu, Sens. Actuators A 171 (2011) 252– 259.

[7] Xing Niu, Jianlin Yu, ShuzhongWang, J. Power Sou. 188 (2009) 621–626.

[8] Degang Zhao, ChangwenTian, ShouqiuTang, YuntengLiu, LikunJiang, LidongChen, Mater. Sci. Semicond. Process. 13 (2010) 221–224.

[9] L. Francioso, C. De Pascali, I. Farella, C. Martucci, P. Creti, P. Siciliano, A. Perrone, J. Power Sou. 196 (2011) 3239–3243.

[10] Gehong Zeng, Je-Hyeong Bahk, John E. Bowers, Hong Lu, Arthur C. Gossard, Suzanne L. Singer, Arun Majumdar, Zhixi Bian, Mona Zebarjadi, Ali Shakouri, Appl. Phys. Lett. 95 (2009) 083503.

[11] S.D. Kwon, B.K. Ju, S.J. Yoon, J.S. Kim, J. Electron. Mater. 38 (7) (2009) 920–924.

[12] Ryohei Izaki, Masayuki Hoshino, Tadashi Yaginuma, Nakaba Kaiwa, Shigeo Yamaguchi, Atsushi Yamamoto, Microelectronics Journal 38 (2007) 667–671.

[13] A. Hmood, A. Kadhim, H. Abu Hassan, Superl. Microst. 51 (2012) 825–833.

[14] A. Hmood, A. Kadhim, H. Abu Hassan, J. Alloys Compd. 520 (2012) 1–6.

[15] Y. Pei, A. D. La Londe, S. Iwanaga, G. J. Snyder, Energy Environ. Sci. 4 (2011) 2085-2089.

[16] Yu. I. Ravich, S. A. Nemov, Semiconductors 36 (2002) 1–20.

[17] H.S. Dow, M.W. Oh, B.S. Kim, S.D. Park, B.K. Min, H.W. Lee, D.M. Wee1, J. Appl. Phys. 108 (2010) 113709.

[18] H. Wang, Y. Pei, A. D. LaLonde, G. J. Snyder, Adv. Mater. 23 (2011) 1366–1370.

[19] J.P. Carmo, Joaquim Antunes, M.F. Silva, J.F. Ribeiro, L.M. Goncalves, J.H. Correia, Measurement 44 (2011) 2194–2199.

Growth and Fabrication of AlGaN/GaN HEMT on SiC Substrate

Yuen-Yee Wong, Yu-Sheng Chiu, Tien-Tung Luong, Tai-Ming Lin, Yen-Teng Ho, Yue-Chin Lin, Edward Yi Chang,
Senior Member, IEEE
Department of Materials Science and Engineering
National Chiao Tung University
1001 University Rd., Hsinchu, 30010 Taiwan
Email: yuenyee98.mse94g@nctu.edu.tw

Abstract- AlGaN/GaN high electron mobility transistor (HEMT) was grown on silicon carbide substrate by metalorganic chemical vapor deposition technique. The AlGaN/GaN structure was optimized by tuning the growth conditions such as AlN buffer thickness, and the Al composition and thickness of AlGaN barrier layer. As a result, the X-ray rocking curve widths were 277 arcsec and 324 arcsec for the GaN (002) and (102) planes, respectively, indicating a high crystalline quality. Hall measurement showed that the AlGaN/GaN structure has a high electron mobility of 1840 (cm^2/V-s) and a sheet electron concentration of 9.85 $x10^{12}$ cm^{-2}. Besides, HEMT device with sub-micron gate-length (0.7 μm) was also successfully fabricated. DC measurement showed that the HEMT device has a saturated current of 800 mA/mm, a transconductance of 257 mS/mm and an off-state breakdown voltage larger than 100 V. For RF performance, the device has achieved an output power density (P_{out}), gain, and power added efficiency (PAE) of 7 W/mm, 23.5 dB and 61.7%, respectively, measured at 2 GHz frequency. On the other hand, the RF performance measured at 8 GHz showed that the device could achieve P_{out}, gain and PAE of 5.01 W/mm, 14.9 dB and 26.23% respectively.

I. INTRODUCTION

Rapid development has been seen on the compound semiconductor technologies recently especially for the GaN related materials and devices. Due to the advantages of GaN material such as high energy bandgap, high breakdown field, high electron saturated velocity and good thermal conductivity (as shown in Table 1), the GaN devices are thus potential for high power and high frequency applications. Currently, the most commonly used substrates for GaN growth include Si, sapphire and silicon carbide (SiC). Although the Si has several advantages like lower cost and larger size, but high crystal quality GaN material is hard to grow on it due to the large lattice constant and thermal expansion coefficient (TEC) mismatches. The sapphire, which is the most commonly used substrate for GaN light emitting diodes growing, is not suitable for GaN electronic devices, especially for high power applications, due to its poor thermal conductivity. On the other hand, the SiC substrate has the smallest lattice constant and TEC mismatches with that of GaN (Table 2) and is therefore

suitable for high crystalline quality GaN growth. Furthermore, owing to the high thermal conductivity, the GaN-on-SiC devices have the best electrical properties and device reliability.

TABLE 1 MATERIAL PROPERTIES OF SOME COMMON SEMICONDUCTORS.

Material	Si	GaAs	SiC	GaN
Band Gap Energy (eV)	1.1	1.4	3.2	3.4
Breakdown Field ($MVcm^{-1}$)	0.3	0.4	3.0	3.0
Thermal Conductance (W/cm/K)	1.5	0.5	4.9	1.5
Mobility ($cm^2V^{-1}.S$)	1300	6000	600	1500
Saturated Velocity ($x10^7cms^{-1}$)	1.0	1.3	2.0	2.7

TABLE 2. MATERIAL PROPERTIES OF DIFFERENT SUBSTRATES

Substrate	Si	Sapphire	SiC
Thermal conductivity (W/cm/K)	1.5	0.3	4.9
Lattice mismatch with GaN (%)	17	16	4
Thermal expansion coefficient mismatch (%)	54	39	3.2

Despite the advantages of SiC, there are still some growth issues that needed to be attended. For example, the GaN cannot be grown directly on SiC due to poor wetting ability [1]. Three dimensional islands will form at the initial stages of growth and then deteriorates the GaN film quality. In an effort to avoid this problem, L. Lie et al. has proposed to use AlN as buffer layer [2]. This is because the AlN is a good wetting agent on the SiC. Besides, the AlN, which has an intermediate lattice constant between the GaN and the SiC, is also proven to improve significantly the GaN film quality grown on SiC. The AlN buffer is also played an important role to prevent the crack of GaN grown on SiC due to the TEC mismatch. The 3.2% TEC mismatch between GaN and SiC (Table 2) will cause the cracking of GaN film grown on SiC if the tensile stress is not handled properly [2]. Apart from the AlN, other buffer materials such as AlGaN can also be used [3, 4].

The current study is divided into two parts: (1) the growth of GaN HEMT structure on SiC substrate by metalorganic chemical vapor deposition (MOCVD) and (2) the fabrication

978-1-4673-2395-6/12 $31.00 © 2012 IEEE

of GaN HEMT devices. Both the material and electrical properties of the GaN HEMT will be presented.

II. EXPERIMENTAL

The AlGaN/GaN HEMT structures were grown by MOCVD (EMCORE; D-180) on 2-inch semi-insulating SiC substrate. The main goal was to achieve a high crystalline quality material by optimizing every epi-structure layers. The effects of AlN buffer thickness on the GaN film quality were first investigated. A thin AlN spacer layer was also inserted in between the AlGaN barrier and GaN layers to improve the electrical properties of the GaN HEMT structure. Finally, the influence of Al composition in the AlGaN barrier layer on the HEMT device characteristic was also discussed. The material properties, such as film thickness, surface morphology, and crystalline quality, were determined using scanning electron microscopy (SEM) and high resolution X-ray diffraction (HRXRD). On the other hand, the basic electrical properties such as electron mobility and sheet electron density were characterized from the Hall effect measurement.

After the material growth, GaN HEMT devices were fabricated. First of all, the device active region was defined using an optical lithography and device isolation was achieved using an inductive-coupling-plasma (ICP) etching. Ohmic contact to the AlGaN/GaN structure was achieved by depositing Ti/Al/Ni/Au metal stack using electron-beam evaporation and followed by high temperature annealing at 850°C for 30 sec. On the other hand, gate contact was formed using the as-deposited Ni/Au. A double-layer photo-resist (PMMA/P(MMA-MAA)) was used in order to achieve sub-micron gate length. By using the optical lithography, 0.7 micron gate was successfully fabricated. After that, Si_3N_4 passivation layer was deposited on the whole device by using the plasma enhanced chemical vapor deposition (PECVD). Finally, multi-finger gate device was connected using air-bridge technique. The HEMT device DC and high RF characteristics were determined using an Agilent E5270 and Load-pull measurement systems, respectively.

III. RESULTS AND DISCUSSION

A. The effect of AlN buffer thickness

In order to understand the effect of AlN buffer layer thickness on the GaN crystalline quality grown on SiC substrate, AlN with three difference thicknesses were prepared. Fig. 1 shows the GaN film X-ray rocking curve (XRC) widths grown on 70, 100, and 140 nm AlN buffer layers. It can be seen that the GaN XRC width on both (002) and (102) planes decreased with the increased of AlN thickness, implying that the GaN film quality improved with the buffer layer thickness. Furthermore, when grown on the 70-nm-AlN buffer, cracks on the GaN films occurred as shown in inset of Fig. 1. This pointed out that this buffer

layer was too thin to accommodate the stress induced from the TEC mismatch between the GaN film and SiC substrate. Table 3 summaries the Hall effect measurement results for GaN film grown on 100 and 140 nm AlN buffer layers. The results are quite similar for these two samples and suggesting that, for good quality GaN HEMT device grown on SiC, an AlN buffer layer with at least 100 nm is required.

B. The effect of AlN spacer layer

In the AlGaN/GaN HEMT structure, the electrical performances of the device are critically affected by the electron properties in the two dimensional electron gas (2DEG) channel [5]. The 2DEG formed in the quantum well at the AlGaN/GaN interface. Therefore, the abruptness of the AlGaN/GaN interface is essential. In this study, a thin AlN spacer layer (1 nm) was inserted prior to the growth of AlGaN barrier layer in order to enhance the heterostructure interface. As shown in Fig. 2, interference fringes can be observed for the X-ray reflectivity (XRR) scanned on the sample with AlN spacer. This indicates that, as a result of the AlN spacer, the AlGaN/GaN has a sharp interface. In fact, by fitting the XRR result, the AlGaN properties such as interface roughness, thickness, and Al composition can be estimated. On the other hand, when the AlN spacer layer was not inserted, no interference fringe was observed during the XRR scan (not shown here). The Hall measurement results for samples with and without the AlN spacer layer were presented in Table 4. The result clearly shows that the electron mobility was improved significantly after using the AlN spacer layer. Besides, the structure sheet resistance was also reduced. In contrast, for structure without the spacer layer, the electrical properties was degraded due to the poor AlGaN/GaN interface and also the alloy scattering effect induced by the ternary compound in AlGaN barrier layer.

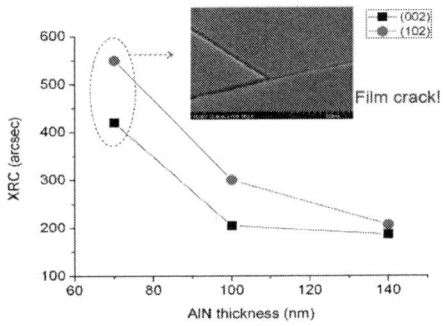

Fig. 1. Dependence of GaN XRC on the AlN buffer thickness.

TABLE 3. EFFECT OF ALN BUFFER THICKNESS ON THE ELECTRICAL CHARACTERISTIC OF GAN HEMT

AlN thickness (nm)	Sheet resistance (ohm/sq)	Mobility (cm²/V-s)	Sheet carrier conc. (x10¹² cm⁻²)
100	344	2060	8.79
140	343	2130	8.54

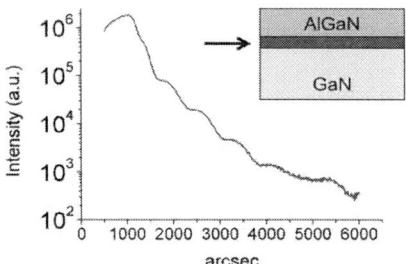

Fig. 2. XRR scan reveal the sharp interface of AlGaN/GaN heterostructure.

TABLE 4. Effect of AlN Spacer on the Electrical Characteristic of GaN HEMT

AlN spacer	Sheet resistance (ohm/sq)	Mobility (cm^2/V-s)	Sheet carrier conc. (x10^{12}cm^{-2})
No	609	1260	8.16
Yes	343	2130	8.54

C. The effect of Al composition in the AlGaN buffer layer

From the previous reports [6, 7], it is understood that the electron carrier in the 2DEG channel of GaN HEMT is induced by the material electrical polarization effect and surface states. For the case of AlGaN/GaN structure, the AlGaN is formed under tensile stress on the GaN film due to lattice mismatch. The lattice mismatch increase with the increase of Al composition in the AlGaN and thus the resulted larger piezoelectric polarization will also increase the carrier density in the 2DEG channel. However, if the Al composition is larger than 35%, serious degradation of the 2DEG properties may occur. Micro cracks may be formed in the AlGaN film due to large tensile stress. Thus, both the AlGaN thickness and Al composition have to the carefully controlled for a good quality GaN HEMT. In this study, the AlGaN barrier thickness was fixed at approximately 25 nm but two different Al compositions were used. As can be seen in Table 5, the electron density in the 2DEG was significantly increased (despite a slight decrease of the carrier mobility) by increased the Al composition from 19% to 25%. As a consequence, the overall sheet resistance of the structure was also decreased and the result is in accordance with that from the literature [8].

After optimizing all the layers as discussed above, AlGaN/GaN HEMT structure was grown on the semi-insulation SiC substrate. The HEMT structure consists of an AlN buffer (100 nm), a GaN film (2.5 μm), an AlN spacer (~1 nm) and an AlGaN barrier (25% Al, 25 nm) layers. The GaN XRC has a width of 277 arcsec and 324 arcsec for the (002) and (102) planes, respectively (Fig. 3). This structure also shows an electron mobility of 1840 (cm^2/V-s) and a sheet carrier density of 9.85 x10^{12} cm-^{2} from the Hall effect measurement.

TABLE 5. Effect of Al Composition in AlGaN Barrier Layer on the Electrical Characteristic of GaN HEMT

Al%	Sheet resistance (ohm/sq)	Mobility (cm^2/V-s)	Sheet carrier density. (x10^{12}cm^{-2})
19	479	1980	6.54
25	344	1840	9.85

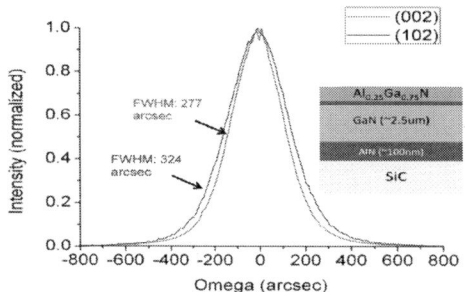

Fig. 3. XRC of GaN film grown on SiC. Inset shows the structure of AlGaN/GaN.

Finally, the grown structure was fabricated into GaN HEMT device. The device has source drain spacing of 7-μm, gate length of 0.7μm and gate width of 100 μm. Both the DC and RF performances of the HEMT device also characterized. As shown in Fig. 4 and Fig. 5, the saturated current (I$_{DSS}$) and transconductance (g$_m$) are 800 mA/mm and 257 mS/mm, respectively. The three-terminal breakdown voltage is larger than 100 V (Fig. 6). From the RF performances measured at 2 GHz frequency (S-band), the HEMT device shows an output power (P$_{out}$) of 28.49 dBm (equivalent to 7W/mm), a gain of 23.5 dB and power added efficiency (PAE) of 61.7 %. Besides, when measured at 8 GHz frequency (X-band), a P$_{out}$ of 27 dBm (equivalent to 5.01W/mm), a gain of 14.9 dB and a PAE of 26.23% has been achieved. The results suggest that good quality AlGaN/GaN HEMT structure has been grown on the SiC substrate.

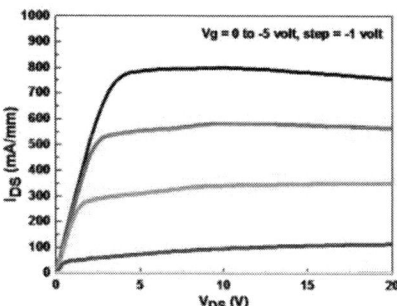

Fig. 4. I$_D$-V$_D$ curves of AlGaN/GaN HEMT

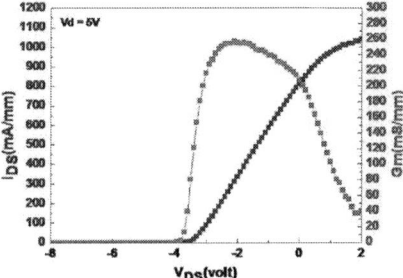

Fig. 5. I_D-g_m curves of AlGaN/GaN HEMT

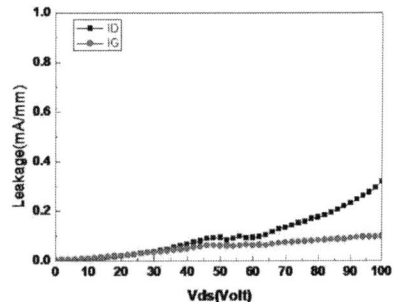

Fig. 6 Off-state characteristics of AlGaN/GaN HEMT

IV. CONCLUSIONS

AlGaN/GaN HEMT structure has been successfully grown on the SiC substrate. The effects of AlN buffer thickness, AlN spacer, and Al composition in AlGaN barrier layers were discussed. After optimizing the AlGaN/GaN structure, Hall electron mobility of 1840 cm^2/V-s and sheet carrier density of 9.85 x10^{12} cm^{-2} have been achieved. HEMT devices were also fabricated using the in-house facilities. Load-pull power performances measured at 2GHz frequency shows that the device has a P_{out} of 28.49 dBm (7W/mm), a gain of 23.5 dB and a PAE of 61.7 %. When measured at 8 GHz frequency, the device has a P_{out} of 27 dBm (5.01W/mm), a gain of 14.9 dB and a PAE of 26.23 %. The results suggested that high quality GaN HEMT device has been successfully fabricated for the RF power applications such as base station, satellite communication, phase array radar and power switching.

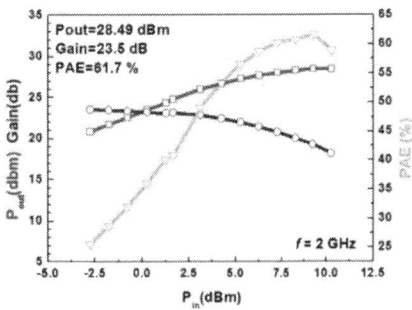

Fig. 7. RF power performance of AlGaN/GaN HEMT measured at 2-GHz

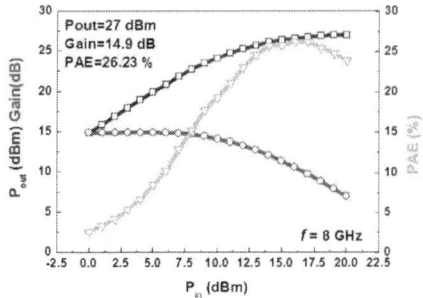

Fig. 8. RF power performance of AlGaN/GaN HEMT measured at 8-GHz

ACKNOWLEDGMENT

This work was financially supported by the Taiwanese National Council under Research Grants NFC100-2623-E-009-008-D. The authors would like to thanks CSIST for the Load-pull measurements.

REFERENCES

[1] H. Lahrèche, et al., "Buffer free direct growth of GaN on 6H-SiC by metalorganic vapor phase epitaxy," J. Appl. Phys., vol. 87, pp. 577-583, 2000.

[2] L. Liu and J. H. Edgar, "Substrates for gallium nitride epitaxy," Materials Science and Engineering: R: Reports, vol. 37, pp. 61-127, 2002.

[3] C. F. Lin, et al., "Growth and characterizations of GaN on SiC substrates with buffer layers," J. Appl. Phys., vol. 82, pp. 2378-2382, 1997.

[4] N. Onojima, et al., "High Off-state Breakdown Voltage 60-nm-Long-Gate AlGaN/GaN Heterostructure Field-Effect Transistors with AlGaN Back-Barrier," Jpn. J. Appl. Phys., vol. 48, p. 094502, 2009.

[5] R. Tulek, et al., "Comparison of the transport properties of high quality AlGaN/AlN/GaN and AlInN/AlN/GaN two-dimensional electron gas heterostructures," J. Appl. Phys., vol. 105, p. 013707, 2009.

[6] J. P. Ibbetson, et al., "Polarization effects, surface states, and the source of electrons in AlGaN/GaN heterostructure field effect transistors," Appl. Phys. Lett., vol. 77, pp. 250-252, 2000.

[7] M. Miyoshi, et al., "Metalorganic Chemical Vapor Deposition and Material Characterization of Lattice-Matched InAlN/GaN Two-Dimensional Electron Gas Heterostructures," Appl. Phys. Express, vol. 1, p. 081102, 2008.

[8] H. Tang and J. B. Webb, "Molecular Beam Epitaxy for III-N Materials," in III-Nitride Semiconductor Materials, Z. C. Feng, Ed., ed: Imperial College Press, 2006, pp. 117-160.

Structural, Morphological and Photoluminescence Studies of SnO$_2$ Microparticles

Karkeng Lim[1], Muhammad Azmi Abdul Hamid[1], Roslinda Shamsudin[1], Azman Jalar[2] and N. H. Al-Hardan[2]

[1]School of Applied Physics, Faculty of Science and Technology
[2]Institute of Microengineering and Nanoelectronics
Universiti Kebangsaan Malaysia
43600 UKM Bangi, Selangor, Malaysia
E-mel: azmi@ukm.my

Abstract- **Tin dioxide (SnO$_2$) microparticles have been grown on p-type Si (100) substrate by thermal evaporation method. The experiment was conducted at 1080°C, under 1.6% oxygen (O$_2$) gas in atmospheric ambient. The prepared film were characterized using X-ray diffraction (XRD), field emission scanning electron microscopy (FESEM) equipped with energy dispersive X-ray spectroscopy (EDX) and photoluminescence (PL) measurement. The growth particles were crystalline with size ranging from 100 nm to 500 nm. The PL spectrum of the SnO$_2$ microparticles exhibits a broad visible light emission with a peak centered at around 611 nm.**

I. Introduction

Revolution of metal oxide thin films deposition has occurred rapidly since past four decades. Metal oxides are classified as the basic of smart and functional materials that have tunable and multifaceted technological application. By varying either or both of their structural characteristics, namely cations with mixed valence state and/or anions with deficiencies/vacancies: the electrical, optical, magnetic and chemical properties of the products can be altered. These will provide the feasibility of fabricating smart devices that harnessing the semiconductivity, superconductivity, ferroelectricity and also magnetism offered by the oxides [1]. Variety metal oxides have been widely study in term of their structural, morphological, and PL properties, namely ZnO [2], MgO [3], SnO$_2$ [4], CdO [5] and even ZnMgO alloy [6]. SnO$_2$ is an important and stable n-type semiconductor material with a wide band gap of 3.62 eV at 300 K, with a distinction photoelectric properties and gas sensitivities [7], which make it a promising material for electronic and optoelectronic devices. Numeral synthesis methods have been employed to fabricate SnO$_2$ such as redox reaction, laser ablation, thermal decomposition, thermal evaporation and rapid oxidation [8]. Thermal evaporation has been extensively used due to its simplicity, inexpensive, and can produce high quality structures [9-11]. It was found that the produced thin films is strongly depends on its microstructure, composition, atomic structure, local chemistry of interfaces and crystal defects which form during the fabrication process [12].

In this study, we report the synthesis of SnO$_2$ microparticles via thermal evaporation method using a three zone furnace. The carrier and the reactant flow gases (Ar and 1.6 O$_2$ balanced with Ar) were kept constant at 0.3 L/min with the temperature at 1080°C. Since there is little information available on PL properties especially on SnO$_2$ structures growth using low O$_2$ concentration (1.6%). Under these situations, it is worthy to conduct this investigation.

II. Experimental Method

p-type Si wafer with orientation of (100) was used as a substrate for the growth of SnO$_2$. High purity tin (Sn) powder (<99.5%, Aldrich) with particle size of 150 μm was used as a starting material. The Si substrate with dimension 10 mm x 15 mm was first cleaned using Radio Corporation of America (RCA) method. Three zone furnace was used to prepared the SnO$_2$ structures. Fig. 1 depicts the configuration of the thermal evaporation method. The substrate is kept apart at top of the alumina plate that contained the source material at 10 mm distance. The substrate and the source material were inserted in the tube furnace at room temperature, and then the temperature was increased to 1080°C at a rate of 10°C/min. A mixed gas of argon 98.4% (Ar 99.999%) and oxygen 1.6% (O$_2$ 99.999%) was used to purge the furnace with a constant flow rate of 0.3 L/min. After an hour, the furnace was shut off and the temperature was allowed to cool down to room temperature under a 15 min flow of argon in order to purge out the residue O$_2$ gas inside the tube.

The morphology, chemical composition and crystal structure of as-grown SnO$_2$ microparticles were examined using field emission scanning electron microscopy (FESEM, Supra 55 VP Zeiss) coupled with energy dispersive X-ray spectroscopy (EDX, oxford instruments) and X-ray diffraction (XRD, D8 advance XRD diffractometer). The photoluminescence (PL) property of the as-grown SnO$_2$ microparticles were investigated at room temperature using Renishaw inVia Raman Microscope with a He-Cd laser as the excitation source with 325 nm line at a power of 0.5 mW.

978-1-4673-2395-6/12 $31.00 © 2012 IEEE

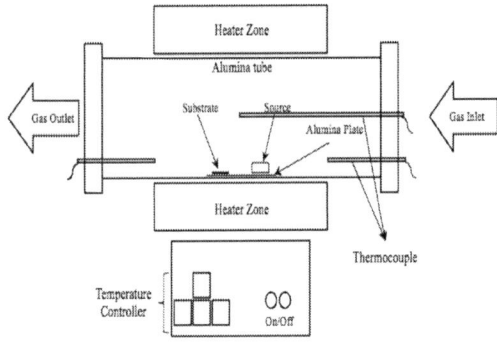

Fig. 1. Schematic diagram of thermal evaporation setup

III. RESULTS AND DISCUSSIONS

Fig. 2 illustrates the XRD pattern of the prepared SnO_2 film which can readily indexed to the tetragonal rutile structure by JCPDS card no. 01-077-0450. The sharp peaks observed indicate good crystalline structures were obtained. XRD also shows the present of pure SnO_2 as no other characteristics peak of impurities was detected in the pattern. The crystallite size, D of the grown SnO_2 particles is 47 nm computed using Scherrer's formula using the full width at half maximum (FWHM) of the (110) peak [13]:

$$D = 0.9\lambda/(\beta\cos\theta) \qquad (1)$$

where λ, β and θ are the X-ray wavelength (Cu Kα line of 0.154 nm), FWHM and Bragg diffraction angle, respectively.

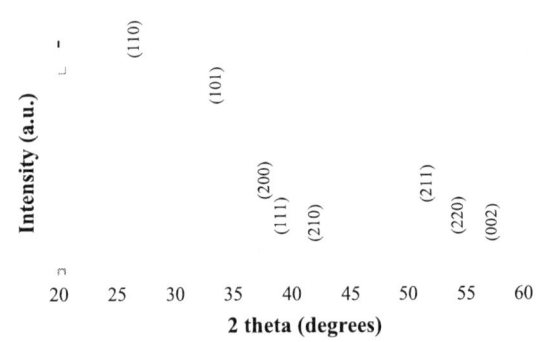

Fig. 2. XRD pattern of as-grown SnO_2 microparticles

The FESEM image (Fig. 3a) shows that the SnO_2 formed irregular shape microparticles with individual particles size ranging from 100 nm to 500 nm. The particles have a defined facet surface, which is common for a crystalline structure and this observation were consistent with XRD results. The particles are mostly agglomerate to form a bigger cluster most

probably through Ostwald ripening process while maintaining facetted surface structure (Fig. 3b). The morphology obtained are different from that reported by previous researcher [14] using similar O_2 concentration most probably due to different flow gas rate and growth temperature. The growth of SnO_2 microstructure follows a typical vapor transport mechanism since no catalyst is used. The Sn atoms vaporize when the melting point is reached, and react with O atoms, which supply by 1.6% O_2 gas to form metastable SnO and then decomposes into SnO_2 [15]. However, further work is on going to clarify the detailed growth mechanism.

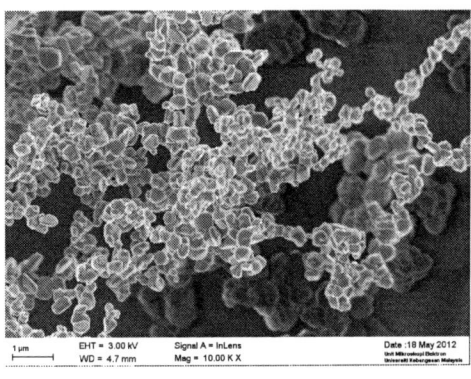

(a)

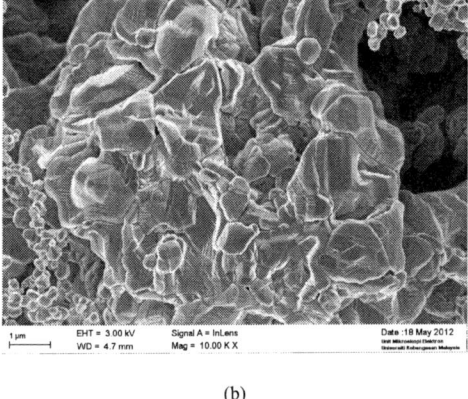

(b)

Fig. 3. FESEM micrographs (a) SnO_2 microparticles and (b) agglomerated of SnO_2 microparticles

Fig.4 depicts the EDX spectrum of the structures was composed of Sn and O with percentage ratio of Sn:O is 2:1, which is in accord with the stoichiometric ratio of SnO_2.

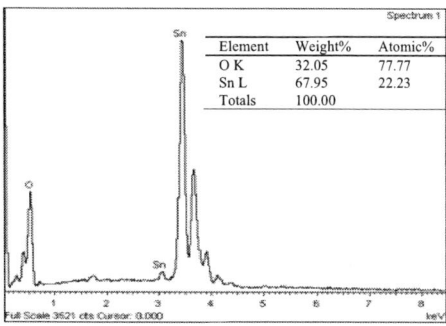

Element	Weight%	Atomic%
O K	32.05	77.77
Sn L	67.95	22.23
Totals	100.00	

Fig. 2. XRD pattern of as-grown SnO_2 microparticles

The room temperature PL spectrum of SnO_2 microparticles is depicts in Fig. 5. A broad visible emission with wavelength peak centered at around 611 nm is observed. This visible light emission is due to oxygen vacancies or Sn interstitials that formed during synthesis process, which know to be related to defects levels with the band gap of SnO_2 [8, 16,17].

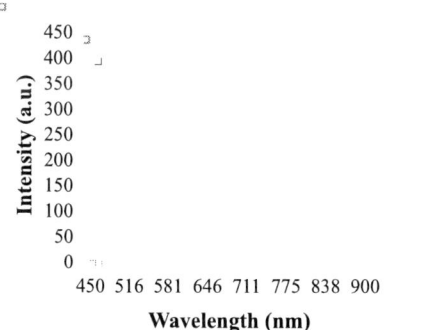

Fig. 5. PL spectrum of SnO_2 microparticles at 0.3 L/min and 1.6 % O_2 gas

Similar PL emission was reported earlier. The reported emission was around 595 nm for different SnO_2 nanostructurs [8,14,16]. The difference of the PL emission of the prepared SnO_2 with the published may be due to the defects in the crystal's structure [8].

III. CONCLUSIONS

SnO_2 microparticles were successfully synthesized by thermal evaporation method with 0.3 L/min gas flow rate of 1.6% O_2 and Ar. Structural and morphological characterization shows that SnO_2 microparticles were crystalline with rutile structure with diameter ranging from 100 nm to 500 nm. PL measurement depicts that the SnO_2 microparticles emits visible light that may have potential to apply in electronic or optoelectronic devices.

ACKNOWLEDGEMENTS

Authors appreciate the support of Universiti Kebangsaan Malaysia, UKM for facilities and financial assistance under research university grant (UKM-DIP-2012-014) and Research University Fellowship.

REFERENCES

[1] J. H. He, et al., "Beaklike SnO2 Nanorods with Strong Photoluminescent and Field-Emission Properties," Small, vol. 2, pp. 116-120, 2006.

[2] Y. Tian, H. B. Lu, J. C. Li, Y. Wu, and Q. Fu, "Synthesis, characterization and photoluminescence properties of ZnO hexagonal pyramids by the thermal evaporation method," Physica E, vol. 43, pp. 410-414, 2010.

[3] H. W. Kim, M. H. Kong, and J. -H. Yang, "Catalyst- Free Growth of Magnesium Oxide Whiskers and Their Characteristics," Acta Physica Polonica A, vol. 113, pp. 1021-1024, 2008.

[4] P. G. Li, X. Guo, X. F. Wang, and W. H. Tang, "Synthesis, photoluminescence and dielectric properties of O-deficient SnO2 nanowires," Journal of Alloys and Compounds, vol. 479, pp. 74-77, 2009.

[5] M. Zaien, K. Omar, and Z. Hassan, " Synthesis of dendrite-like petals of CdO nanostructure," Optoelectronics and Advanced Materials-Rapid Communications, vol. 5, pp. 982-984, 2011.

[6] R. Yousefi and M.R. Muhamad, "Effects of gold catalysts and thermal evaporation method modifications on the growth process of $Zn_{1-x}Mg_xO$ nanowires," Journal of Solid State Chemistry, vol. 183, pp. 1733-1739, 2010.

[7] M. Salavati-Niasari, N. Mir, and F. Davar, "Synthesis, characterization and optical properties of tin oxide nanoclusters prepared from a novel precursor via thermal decomposition route," Inorganica Chimica Acta, vol. 363, pp. 1719-1726, 2010.

[8] S. Park, C. Hong, J. Kang, N. Cho, and C. Lee, "Growth of SnO2 nanowires by thermal evaporation on Au-coated Si substrates," Current Applied Physics, vol. 9, pp. S230-S233, 2009.

[9] Z. Chen, et al., "Effect of N2 flow rate on morphology and structure of ZnO nanocrystals synthesized via vapor deposition," Scripta Materialia, vol. 52, pp. 63-67, 2005.

[10] A. Khan, S. N. Khan, W. M. Jadwisienczak, and M. E. Kordesch, "Growth and optical properties for non-catalytically grown ZnO micro-tubules by simple thermal evaporation," Materials Letters, vol. 63, pp. 2019-2021, 2009.

[11] H. W. Kim, S. H. Shim, and J. W. Lee, "Growth of MgO thin films with subsequent fabrication of ZnO rods: Structural and photoluminescence properties," Thin Solid Films, vol. 515, pp. 6433-6437, 2007.

[12] Z. W. Chen, et al., "Microstructural evolution of oxides and semiconductor thin films," Progress in Materials Science, vol. 56, pp. 901-1029, 2011.

[13] H. I. Abdulgafour, Z. Hassan, N. H. Al-Hardan, and F. K. Yam, "Growth of high-quality ZnO nanowires without a catalyst," Physica B: Condensed Matter, vol. 405, pp. 4216-4218, 2010.

[14] S. N. F. Hasim, M. A. A. Hamid, R. Shamsudin, S. Radiman, and A. Jalar, "Surface morphology of metal oxide SnO2 under different concentrations of oxygen by thermal evaporation method," Advanced Materials Research vol. 501, pp. 266-270, 2012.

[15] H. W. Kim, S. H. Shim, and C. Lee, "SnO2 microparticles by thermal evaporation and their properties," Ceramics International, vol. 32, pp. 943-946, 2006.

[16] H. W. Kim and S. H. Shim, "Synthesis and characteristics of SnO2 needle-shaped nanostructures," Journal of Alloys and Compounds, vol. 426, pp. 286-289, 2006.

[17] H. W. Kim, S. H. Shim, and J. H. Myung, "Synthesis and characteristics of SnO2 nanorods on Pd-coated substrates," Brazilian Journal of Physics, vol. 35, pp. 1006-1009, 2005.

Fabrication and Characterization of IDE ZnO Thin Films Using Sol-Gel Method for PBS Solution Measurement

K.L.Foo[1], U.Hashim[1], Haarindra Prasad s/o RajintraPrasat[2] and M.Kashif[1]

[1] Nano Biochip Research Group, Institute of Nano Electronic Engineering (INEE), UniversitiMalaysia Perlis (UniMAP), 01000 Kangar, Perlis, Malaysia

[2] Microelectronic Engineering, University Malaysia Perlis (UNIMAP), 01000 Kangar, Perlis , Malaysia

Email: elitefoo@yahoo.com

Abstract- **Microelectronics technologies have contributed a lot of facilities to fabricate small scale devices. With the enhancement of microelectronics technologies, IDE in micro-gap size is successfully fabricated on the SiO_2/Si substrate. In this paper, ZnO thin films, which were prepared using low-cost sol-gel method, were coated on the IDE device. As to study the influence of different solvent to the morphology, particles size and quality of the ZnO thin films, two types of solvent had been chosen, namely ethanol and methanol. For the study of morphology, grain size and uniformity distribution of the ZnO on the substrate, AFM had been chosen. XRD result shows that all annealed polycrystalline ZnO thin film exhibit highly crystalline hexagonal wurtzite structure. The optical characteristic of the ZnO thin films had been determined using UV-Vis spectroscopy. The capacitance and conductance of the device at certain frequency had been conducted using dielectrical analyzer. The influence of PBS solution on the IDE ZnO device had been studied using dielectrical analyzer.**

I. INTRODUCTION

Zinc oxide (ZnO), which has noble physical and electrical properties, can be used for various applications in the micro technology field. One of the main applications that can be used is as sensor. The main purpose of using metal oxide (ZnO) thin film is due to cheap, small size and the accuracy results, which can improve the output results if compared to others material. Therefore, ZnO thin film has widely contributed in various types of application, such as gas and chemical sensor[1-2], bio-molecular sensor[3-4], light emitting diode[5] and solar cell[6-7].

For the ZnO films preparation, various types of method has been used, such as sol-gel spin coatin[8], physical vapor deposition[9], chemical vapor deposition (CVD)[10], spray pyrolysis[11], sputtering[12] and ink-jet printing[2]. Due to the advantages of chemical composition control, and homogeneity of the sol solution, scalable and low cost given by sol-gel method, it has been chosen in this project.

Generally, the sol-gel process is the transformation of a solution system from a liquid "sol" (mostly colloidal) into a solid "gel" phase. Typically, the material stared to use in the preparation of "sol" are inorganic metal salts or metal organic compounds (metal alkoxides)[13]. Phosphate buffer solution (PBS), which is very sensitive with its concentration to the bimolecular in biomedical sensor application, had been studied in this work by using interdigit electrode (IDE) ZnO thin films.

The aim of this study is to produce uniform and nanoparticles of ZnO on the SiO_2/Si substrate with IDE etectrode. In this project, atomic force microscope (AFM) were used to study the surface morphologies and microstructures of the ZnO films, while X-ray diffraction (XRD) were used to examine the crystallinity and structural properties of the ZnO films. By the way, optical properties of the ZnO had been determined with ultraviolet-visible spectroscopy (UV-Vis). Then the capacitance and conductivity test for the IDE ZnO were conducted using dielectrical analyzer and source meter. Besides that, the capacitance of the IDE ZnO device was tested under different condition.

II. METHODS AND MATERIALS

The ZnO solution was prepared by using the sol- gel method. ZnO solution was prepared by ensuring the ratio of zinc acetate and solvent is 1:1. The concentration of the ZnO solution was maintained at 0.2mol/L. Zinc acetate dehydrate $[Zn(CH_3COO)_2 \cdot 2H_2O]$ was dissolved in two different solvents (ethanol and methanol) and stirred at 1000rpm with the temperature 60°C on the hot plate . The mixture solution was further stirred for 20 minutes without changing the temperature and stirring speed. After 20 minutes, monoethanolamine (MEA) was drop little by little to the mixed solvent for the next two hours as to get a homogeneous clear solution. The solution was left for ageing at room temperature for more than 24 hours.

SiO_2/Si sample with orientation of Si (100) was used for the deposition of ZnO thin film. As to eliminate the particles which were presence on the wafer surface, the wafer samples must be cleaned prior the deposition process. The silicon wafer samples were first cleaned with acetone, BOE and nitride acid. Then, the samples were further ultrasonic cleaned in IPA and Hydrochloric acid (HCl) solution for 5 min in each process. The cleaned silicon wafer was then undergoing wet oxidation process as to get the SiO_2 insulating layer with the thickness

~180nm. Prior to the ZnO thin films deposition, an aluminium (Al) IDE with the gap between the finger is 250µm was form on the SiO$_2$/Si by using conventional top-down lithography process.

The prepared ZnO solutions were then deposited on the cleaned substrates with spin coating technique with the spin speed of 3000rpm for 30s.The deposited ZnO thin film was then dried with hot plate for 20 minutes at 150°C. This procedure was repeated 3 times as to get 3 layers of ZnO on the substrate. The ZnO thin films were then annealed with furnace at 300°C for 2 hours as to get the crystallization of ZnO. For the electrode formation, a transparency mask had been used for the formation of the electrode area. The complete fabrication process is shown in Fig. 1.

The characteristic of the Zinc Oxide was studied by examining the surface topologies and grains size by using Atomic Force Microscopy (AFM, SPA400, SII Nanotechnology). X-ray diffraction (XRD, Bruk D8) was used for examining the crystal structure of sol-gel derived ZnO film and the Ultraviolet-Visible (UV-Vis) spectroscopy was used to study the optical properties and band-gap of the ZnO films. For the IDE ZnO thin films, a real-time α-high dielectric analyzer from Nova Control was used to study the capacitance and conductance of the ZnO thin film. With the same equipment, the capacitance of IDE ZnO device under different condition had been measured.

III. RESULTS AND DISCUSSION

Atomic Force Microscopy (AFM) is used to analyze the surface roughness and the topographies of the samples which are coated with two different solutions. The results obtained from the AFM test shows that the ZnO solution prepared by ethanol solvent gave higher value of root mean square (RMS) if compared to the methanol solvent, which are 7.48nm and 6.62nm respectively. The result also shows that the surface roughness of the ZnO for ethanol is the highest. It can be concluded that the value of RMS is directly proportional to the value of surface roughness. The grains size of the ZnO thin films for both solvent was distributed in nanometer-sized. By the way, the result shows that ZnO synthesis with methanol solvent provide smaller particles size if compared to the ethanol solvent.

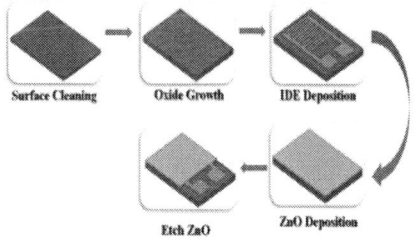

Fig. 1. ZnO films with IDE deposition process

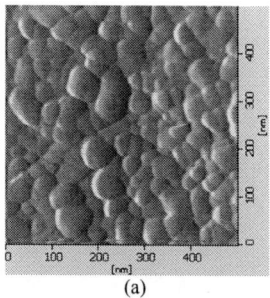

(a)

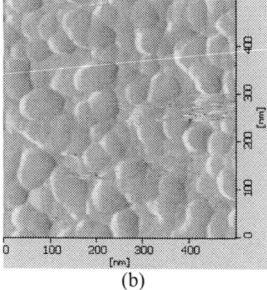

(b)

Fig. 2. AFM image of ZnO thin films synthesized with (a) methanol and (b)ethanol

In order to study the crystallinity, crystallographic and phase evaluation of the ZnO films, XRD analysis had been carried out. The XRD patern for the ZnO thin films prepared under methanol and ethanol is demonstrated in Fig. 3. All the diffraction peaks in the result is according to the standard card (JCPDS 36-1451). From the XRD result, it shows that all sol-gel derived ZnO thin film, which have the crystal growth orientation of the ZnO thin films has preferential growth along (002) planes. Besides that, both solvent also have the high intensities peak at (100) and (101) planes and low intensities peak at (110), (103) and (112) plane. Anywhere, methanol provides an extra peak at (102) plane, which is not indicated in ethanol solvent. Those sharp and narrow diffraction peaks in XRD patern indicates that the ZnO material exhibit high crystallinity. It also shows that these thin films are polycrystalline with a hexagonal wurtzite structure.

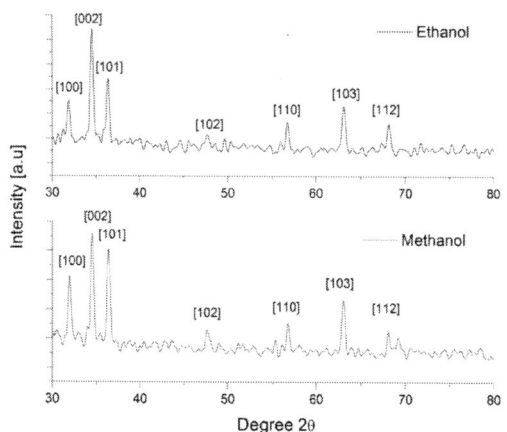

Fig. 3. X-ray diffraction patterns of ZnO thin film with (a) methanol (b) ethanol

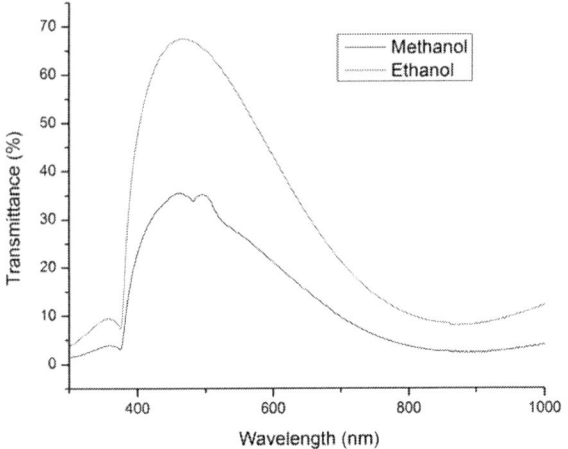

Fig. 4. Optical transmittance spectra of sol-gel derived ZnO thin films with methanol and ethanol

The optical properties of the sol-gel spin coated ZnO thin films with different solvent had been studied using ultraviolet-visible (UV-Vis) spectroscopy. The wavelength for the transmittance measurement of the ZnO thin films was set in the range of 300-1100nm at room temperature. Form the result shows in Fig. 4, it indicates that ZnO thin films has a dramatically drops of transmission in the ultraviolet region, which is in the range of around 370nm for both methanol and ethanol solvent. This is due to the sharp absorption edge of ZnO, which is very close to the intrinsic band-gap of ZnO (3.3eV)[14]. The results also indicate that ethanol solvents has higher optical transmittance (~67%) if compare to the methanol solvent (~36%).

The capacitance and conductivity behavior of the ZnO had been investigated using dielectrical analyzer. The measurement was investigated at frequency range of 100Hz to 1MHz, $0V_{DC}$ bias and $1V_{AC}$. Fig. 5 shows the result of capacitance vs frequency for methanol and ethanol solvent. The result shows that the capacitances for both solvent decreases from 1MHz to 0.2MHz; then increase slightly at lower frequency. A high capacitance value of 71pF was obtained at 1MHz for methanol solvent. In other way, the capacitance value for ethanol is only 33.6pF at 1MHz. This can be concluded that methanol solvent obtains higher capacitance if compared to ethanol solvent

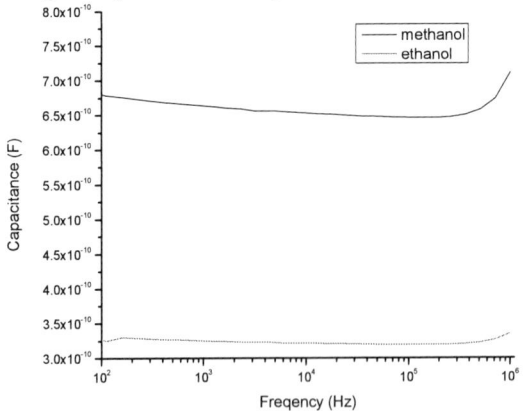

Fig. 5. Dependence of capacitance on frequency for different solvent used.

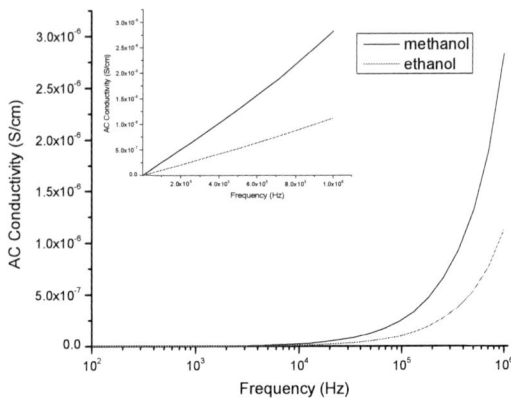

Fig. 6. Dependence of ac conductivity on the frequency for IDE ZnO thin films

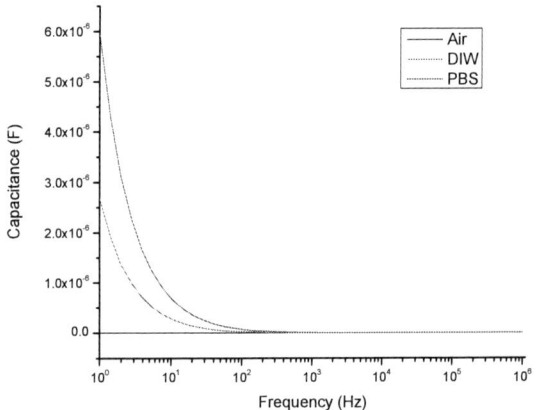

Fig. 7. Capacitance of IDE ZnO device under various conditions.

The dependence of ac electrical conductivity on frequency for different solvent is shown in Fig. 6. The result indicates that the conductivity for both solvents is nearly constant at lower frequency. When going to high frequency (>10KHz), it exhibits a huge dispersion on the conductivity. From the graph, it can be concluded that methanol has a higher conductivity ($2.83X10^{-6}$S/cm at 1MHz) if compared to the ethanol solvent, which has the conductivity of $1.12X10^{-6}$S/cm at 1MHz.

Fig. 7 shows the capacitance of IDE ZnO in air condition, with deionized water (DIW) and PBS solution. The result indicates that IDE ZnO device gives the lowest capacitance in air condition. This might due to no ion contain in the air, therefore electron is hardly transfer between the electrodes. When DIW was dropped on the device, the electric current is increased dramatically. This is believed due to the higher conductivity of water if compared to the air. However, the electric current shows highest value with PBS solution. This might due to the thickness of the double layer become smaller when PBS solution is dropped on the device.

IV. CONCLUSIONS

ZnO thin films with different solvetn (methanol and ethanol), which had been synthesized using low-cost sol-gel spin coating method giving a high crytallinity. AFM results shows that both solvent provide ZnO particles size in nano ranges. The hexagonal wurzite structure of ZnO films had been indicated by XRD and UV-Vis shows that the ZnO have the intrinsic band-gap of 3.3eV. For the electrical charaterization of IDE ZnO thin films, methanol shows a better capacitance and condutivity if campared to the ethanol solvent. Besides that, the PBS solution gives the highest capacitance value if compared to the DIW and air.

ACKNOWLEDGEMENTS

The authors are very grateful to ministry of high education (MOHE) for providing the FRGS grant to conduct this research. Besides that, the authors also would like to thank all of the team members in the Institute of Nano Electronic Engineering especially the member in Nano Biochip Research Group.

REFERENCES

[1] C. Baratto, *et al.*, "ZnO nanocrystals by chemical route for optical gas sensing," in *Sensors, 2008 IEEE*, 2008, pp. 1293-1296.

[2] W. Shen, *et al.*, "The preparation of ZnO based gas-sensing thin films by ink-jet printing method," *Thin Solid Films*, vol. 483, pp. 382-387, 2005.

[3] A. Fulati, *et al.*, "An intracellular glucose biosensor based on nanoflake ZnO," *Sensors and Actuators B: Chemical*, vol. 150, pp. 673-680, 2010.

[4] S. M. Usman Ali, *et al.*, "A fast and sensitive potentiometric glucose microsensor based on glucose oxidase coated ZnO nanowires grown on a thin silver wire," *Sensors and Actuators B: Chemical*, vol. 145, pp. 869-874, 2010.

[5] K. Kim, *et al.*, "Light-emitting diodes composed of n-ZnO and p-Si nanowires constructed on plastic substrates by dielectrophoresis," *Solid State Sciences*, vol. 13, pp. 1735-1739, 2011.

[6] E. Guillen, *et al.*, "ZnO solar cells with an indoline sensitizer: a comparison between nanoparticulate films and electrodeposited nanowire arrays," *Energy & Environmental Science*, vol. 4, pp. 3400-3407, 2011.

[7] K. Matsubara, *et al.*, "ZnO transparent conducting films deposited by pulsed laser deposition for solar cell applications," *Thin Solid Films*, vol. 431-432, pp. 369-372, 2003.

[8] M. Kashif, *et al.*, "Morphological, optical, and Raman characteristics of ZnO nanoflakes prepared via a sol–gel method," *physica status solidi (a)*, vol. 209, pp. 143-147, 2012.

[9] L. Wang, *et al.*, "Synthesis of well-aligned ZnO nanowires by simple physical vapor deposition on c-oriented ZnO thin films without catalysts or additives," *Applied Physics Letters*, vol. 86, pp. 024108-024108-3, 2005.

[10] B. S. Li, *et al.*, "Effects of RF power on properties of ZnO thin films grown on Si (001) substrate by plasma enhanced chemical vapor deposition," *Journal of Crystal Growth*, vol. 249, pp. 179-185, 2003.

[11] A. Ashour, *et al.*, "Physical properties of ZnO thin films deposited by spray pyrolysis technique," *Applied Surface Science*, vol. 252, pp. 7844-7848, 2006.

[12] B. Deng, *et al.*, "AFM characterization of nonwoven material functionalized by ZnO sputter coating," *Materials Characterization*, vol. 58, pp. 854-858, 2007.

[13] E. Comini, *et al.*, *Solid State Gas Sensing*: Springer, 2008.

[14] X. Zhao, *et al.*, "Dependence of the properties of hydrothermally grown ZnO on precursor concentration," *Physica E: Low-dimensional Systems and Nanostructures*, vol. 41, pp. 1423-1426, 2009.

Surface Defect on SiC Ohmic Contact During Thermal Annealing

Izhan Abdullah[1], Azman Jalar[1], *Member, IEEE*, Mohammad Azmi Abdul Hamid[2], Ishak Mansor[3], & Burhanuddin Yeop Majlis[1], *Senior Member, IEEE*

[1]Institute of Micro Engineering and Nanoelectronics (IMEN)
[2]School of Applied Physics, Faculty of Science and Technology,
Universiti Kebangsaan Malaysia,
43600 Bangi, Selangor, Malaysia.
[3]Nuclear Malaysia Agency, Bangi,
Selangor, Malaysia.
Email: eizhan@eng.ukm.my

Abstract- **Capability and reliability of Silicon Carbide (SiC) material for semiconductor power devices are influence by surface defects. 4H-SiC have been measured to investigate the surface defect on 4H-SiC epitaxial in order to obtain the defects information related to electrical performances. Ohmic contact was prepared using Aluminium and Platinum with thickness 90 nm and 150 nm respectively and annealed at 400 °C at three various time . Surface morphology was observed using both surface profile technique and Scanning Electron Microscope (SEM) prior to ohmic deposition surface. The mapping defect studies of 4H-SiC surface have revealed that the large area of micropipes and shallow pit will exposed defects exist.**

Keywords: **Silicon carbide, surface defect, ohmic contact, thermal annealing**

1. INTRODUCTION

Silicon Carbide (SiC) has been recognized as an ideal semiconductor material for applications that require high strength, good thermal conductivity and wider frequency [1, 2]. It was used widely in high-power industry due to operating at high temperatures and harsh environment. Compare to conventional Si semiconductor, SiC has superior electrical and thermal performance [3]. Due to wide bandgap rather than the well-known of semiconductor, it is suitable applied to the radiation detector because of capability various radiation detection [4, 5]. However, the defects from SiC growth such as micro-pipes and hollow screw were the presence on SiC surface will influence the potential of devices, particularly in performances and reliability [6]. In previous research, the research was found that SiC bipolar devices have major issues at performance failures such as large leakage current and premature switching cause the effect of the surface defects [7].

In this work, the relationship between defect and electric properties of 4H-SiC Schottky diode has been investigated. Many studies have found that the large surface defect such as micro pipes and screw hollow dislocation would reduce the blocking voltage, and the leakage current increased [8-9].

This paper has focused to investigation on 4H-SiC wafers in order to determine the correlation of surface defect and electric properties of 4H-SiC for radiation detector application. In a preliminary study, the surface defects on deposited metal contact have been performed by profiler meter, digital optical microscope and Scanning Electron Microscope (SEM) at MEMS Laboratory. Different types of defects will give distinct values on electrical properties of 4H-SiC. 4H-SiC wafers were selected to Schottky diode fabrication using aluminium and platinum as surface contact through the metallization process.

2. FABRICATION METHODS AND MEASUREMENT

In this experimental work, Three samples of 7 mm x 5 mm p-type 4H-SiC wafer dices with 1000 orientations, the thickness approximately 380 μm with the resistivity of 10-18 ohm-cm were used to fabricate back of Schottky diode. 4H-SiC of ohmic contact were fabricated on silicon face, p-type 4H-SiC as shown in Figure 1. Prior to metallization process, the sample of 4H-SiC wafers was prepared to clean the particles, fingerprints and grease using acetone and methanol for reducing the particles' defect of electrical performance and also prevent the contaminant. 10 percent of Hydrofluoric (HF) concentration with 100 ml of deionized water (DI) was applied to remove the stubborn of silicon oxide ($SiO2$) which growth naturally on the 4H-SiC surface.

Later cleaning process was performed, 90 nm of thin film on the C-face (Ohmic) was formed by aluminium deposition using thermal evaporator under a background pressure around 1 x 10-6 torr. 150 nm thick with platinum were deposited as surface contact using sputtering coater. Before the thermal annealing process was performed, the samples cool down in 5 minutes before inserted in the furnace. The furnace was set to 400 °C at three variant times, i.e. 15 minutes, 20 minutes and 25 minutes. The annealing was performed to allow the formation of silicide between 4H-SiC and Al/Pt layer for lower contact resistivity. The protective layer was fabricating to prevent the scratch from machines and human before developing the Schottky contact at the top surface (Si face).

Defect's densities on these samples were determined by optical microscopy using the Nikon microscope lens with magnification up to 50× (scale of 10 um). Defect distribution was reported in order to obtain a detailed of mapping defect on the 4H-SiC wafer surface. The highlighted defects were seen through Infinite Focus Microscope (Alicona) and Scanning Electron Microscope (SEM). Schottky diodes were realized, and electrical properties were correlated to presence of defects.

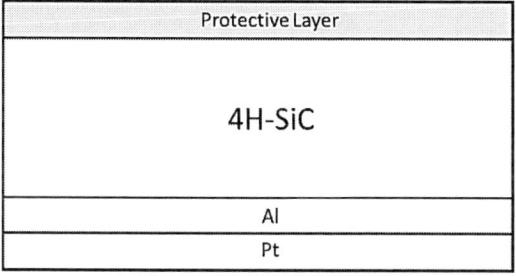

Fig. 1. 4H-SiC Ohmic contact fabrication structure and materials.

3. RESULTS AND DISCUSSION

By using a microscope and Scanning Electron Microscope (SEM), the surface contact was observed. Lots of defect and bubble structures were observed on the surface. As a result, it was a cause of an increase of contact. Three samples that were deposited have shown that remains micro-pipes have been reduced because those tiny holes have implanted during sputter within Al/Pt thin films. There is several of defects' effect that occurred during metalization process for ohmic contact formation. Figure 2 showed the digital optical microscope images of the surface defects on the 4H-SiC single layer during anneal at 20 minutes (sample 2). It was shown that the large micro-pipes area would not reduce because aluminium and platinum particles from the source were simply sputtered on the substrate with uniform thickness. Several of small holes were sealed by metal particles. That surface becomes rough and darkest due to thermal reaction on deposited metals. Sample 1 would not show the different effect during annealed at 15 minutes.

After 25 minutes of thermal annealing (sample 2), the surface contact showed bubble structures and desolated due to surface at morphology change on thermal reaction [10a]. Since this incident, Aluminium atoms will diffuse into Platinum's atoms. Maybe it will become alloyed metal. It can be affected to interface adhesive between 4H-SiC and Pt thin film. The surface roughness decreased to almost half value compared to that of the un-annealed contact. As a result of annealing process, SiC surface roughness decreased. Due to

annealing process, the surfaces resistant were increased than un-annealed sample [11b].

(a)

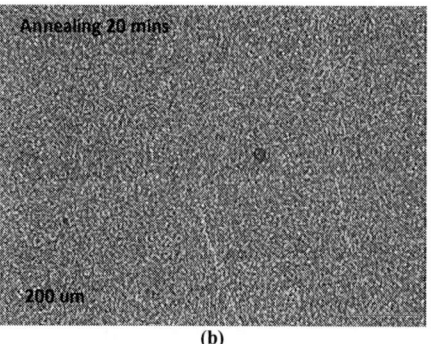

(b)

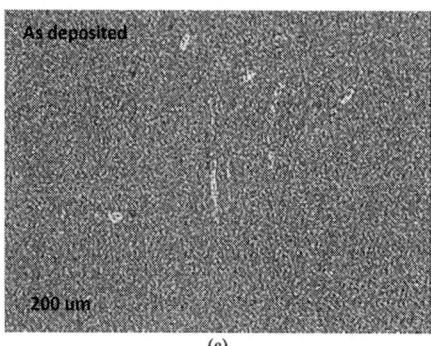

(c)

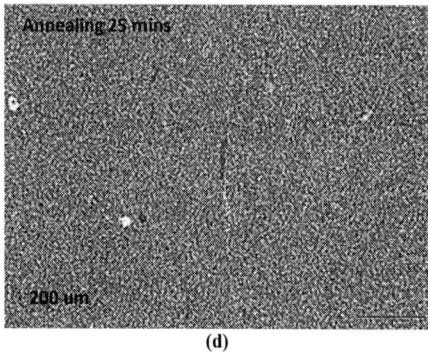

(d)

Fig. 2. Surface morphology on ohmic surface contact during thermal annealing stage after deposited by Al/Pt: (a) Sample 1 as deposited. (b)

978-1-4673-2395-6/12 $31.00 © 2012 IEEE   741

Sample 1 annealed in 20 minutes. (c) Sample 2 as deposited. (d) Sample 2 annealed in 25 minutes.

From three samples, the investigation was accomplished by digital optical microscope and SEM for determine the location of defects and mapping the defect areas. The surface results have shown that the large areas of defects were appeared in the 4H-SiC samples after deposited and annealed. Figure 2 also shows the mapping defect on the 4H-SiC diced of the wafer by using optical microscopy magnification 5x. The typical morphological defects observed on the grown surface: micro-pipes, small growth pit and core screw dislocation. Another back surface defect observed was scratched, which is line-shaped surface damage introduced by a polishing and machining process.

Defect A from figure 2(a) and 2(b) (as shown in figure 3) on SEM image present the large micro-pipes areas can be disrupted the operating switch and will affect the reverse bias voltage due to over than 50 percents of breakdown voltage reduction and the increased leakage currents [12]. Micro-pipes are documented to cause of premature reverse failure in high-frequency devices. Therefore, SiC power devices that uniformly distribute to break down current over the entire junction area exhibit much greater reliability than silicon devices that manifest localized breakdown behaviour [7, 12-13]. Defect B from figure 2(c) and 2(d) (as shown in figure 4) was present the small indentation shape which gives the impact to Schottky rectifying properties [14]. In observed of impact on electrical properties, core screw dislocation would occur in 10 ~ 30 percentages of breakdown voltage reduction and reduced the carrier lifetime [15].

(a)

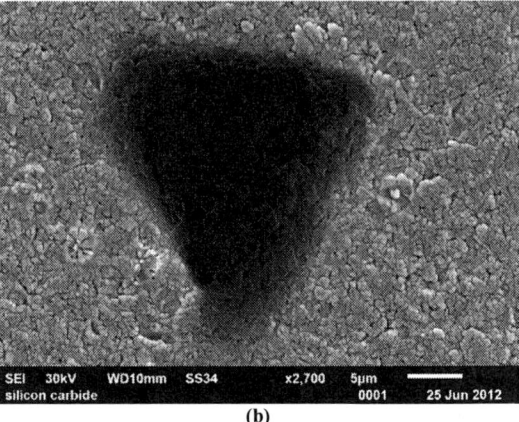

(b)

Fig. 4. SEM image of small growth pits at Sample 2 after annealed in 25 minutes.

4. CONCLUSION

The correlation between I-V characteristics and 4H-SiC surface defect were investigated and studied. Surface defects have been identified and shown the presence of different defects on the 4H-SiC wafer. Surface morphology technique was used to observe typical defects would be impacted to Schottky diode performances. The 4H-SiC Schottky diodes were realized using aluminium and platinum as metal contact. It was shown that the large of micro-pipes areas maybe lead the presences of defects for give the instability of reverse voltage on defect surface contact of Schottky diode during operating stage. Thermal annealing on surface contact can be increased the surface resistant and reduces micro-pipe's density than unannealed.

ACKNOWLEDGEMENT

This work has been sponsored by National University of Malaysia, Nuclear Malaysia Agency and under research university grant; UKM-GUP-2011-219 and OUP-2012-120.

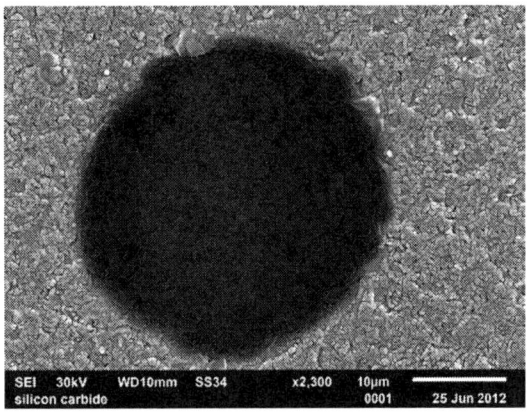

Fig. 3. SEM image of micro-pipe at Sample 1 after annealed in 20 minutes.

REFERENCES

[1] Nitin P. Padture, Christopher J. Evans, Hockin H. K. Xu, and Brian R. Lawn., "Enhanced Machinability of Silicon Carbide via Microstructural Design," J. Am. Ceramic . Soc., 78, Jan. 1995.

[2] G. Pensl, H. Morkoc, B. Monemar and E. Janze'n., eds., *Silicon Carbide, III-Nitrides and Related Materials, Part I and II,* Tech Publications, Switzerland, 1998.

[3] Katsutoshi Komeya, Yi-Bing Cheng, Junichi Tatami and Mamoru Mitomo., "Mechanical and Thermal Properties of Silicon Carbide Composites with Chopped Si-Al-C Fiber Addition," Key Engineering Materials, vol. 403, pp. 257-260, Dec. 2008.

[4] P. J. Sellin., "Recent advances in compound semiconductor radiation detectors", Nuclear Instruments and Methods in Physics Research A 513, 332-339, 2003

[5] F. Nava, G. Wagner, C. Lanzieri, P. Vanni and E. Vittone., " Investigation of Ni/4H-SiC diodes as radiation detectors with low doped n-type 4H-SiC epilayers ", Nuclear Instruments and Methods in Physics Research Section A510(3): 273-280, 2003.

[6] Neudeck, P.G. and Powell, J.A., "Performance-Limiting Micropipe Defects Identified in SiC Wafers", IEEE Trans. Electron Device Letters 15(2): 63 – 65, 1994.

[7] P.G. Muzykov, A.V. Bolotnikov, T.S. Sudarshan. " Study of leakage current and breakdown issues in 4H–SiC unterminated Schottky diodes", Solid-State Electronics 53: 14 – 17, Jan. 2009.

[8] P. G. Neudeck. *Encyclopedia of Materials: Science and Technology 9*, Elsevier Science, Oxford, 2001.

[9] S. Ferrero, S. Porro, F. Giorgis, C. F. Pirri, P. Mandracci, C. Ricciardi, L. Scaltrito, C. Sgorlon, G. Richieri and L. Merlin. , "Defect characterization of 4H-SiC wafers for power electronic device applications " J. Phys.: Condens. Matter 14(48), 2002.

[10] Hsu, C.-Y., Lan, W.-H., Wu, Y.C.S., "Effect of thermal annealing of Ni/Au ohmic contact on leakage current of GaN based light emitting diodes". Appl. Phys. Lett. 83 (12), 2447–2449, 2003.

[11] J.H. Ha, S.M. Kang, S.H. Park, H.S. Kim, Y.H. Cho, J.H. Lee, N.H. Lee, J.B. Kim and Y.K. Kim., "Annealing effect of the 6H-SiC semiconductor detector for alpha particles". Radiation Measurements 43: 1140 – 1143, 2008.

[12] T. Kimoto, N. Miyamoto and H. Matsunami., "Effects of surface defects on the performance of 4H– and 6H–SiC pn junction diodes", Material Science and Engineering B61-62: 349 – 352, 1999.

[13] F. Roccaforte, S. D. Franco, F. Giannazzo, F. L. Via, S. Libertino, V. Raineri, M. Saggio and E. Zanetti., "Silicon carbide: Defects and devices", Solid State Phenomena 108 – 109: 663-670, Sept. 2005.

[14] M. Ben Karoui, R. Gharbi, N. Alzaied, M. Fathallah, E. Tresso, L. Scaltrito and S. Ferrero., " Effect of defects on electrical properties of 4H-SiC Schottky diodes ", Materials Science and Engineering: C, 28: 799 – 804, 2008.

[15] S. I. Maximenko, J. A. Freitas, Jr., R. L. Myers-Ward, K.-K. Lew, B. L. VanMil, C. R. Eddy, Jr., D. K. Gaskill, P. G. Muzykov, and T. S. Sudarshan., " Effect of threading screw and edge dislocations on transport properties of 4H–SiC homoepitaxial layers",J. Appl. Phys. 108: 013708, July 2010.

ICSE2012 Proc. 2012, Melaka, Malaysia

Fabrication of AlGaN/GaN HEMTs with Slant Field Plates by Using Deep-UV Lithography

Ting-En Hsieh, Lu-Che Huang, Yueh-Chin Lin, Chia-Hua Chang, Huan-Chung Wang and Edward Yi Chang[*]
[1] National Chiao-Tung University, Hsinchu, Taiwan, R.O.C.
Department of Materials Science and Engineering
Tel : 886-3-5712121 ext.31536, Fax : 886-3-5745497, Email : edc@mail.nctu.edu.tw

Abstract- **In this work, AlGaN/GaN HEMTs with slant plate have been successfully fabricated using deep-UV lithography. By using an angle exposure technique, submicron T-shaped gates with slant sidewalls were achieved. The method is simple of cost effective. The $0.6 \times 100 \mu m^2$ slant-field-plated AlGaN/GaN HEMT on silicon substrate exhibited a peak value of transconductance higher than 200 mS/mm and a breakdown voltage higher than 100 V. Through high-frequency measurements, the device revealed a current gain cut-off frequency (fT) of 24 GHz, a maximum oscillation frequency (fmax) of 49 GHz.**

I. Introduction

AlGaN/GaN-based high electron mobility transistors (HEMTs) are expected to be widely used for high power and high frequency applications owing to their outstanding properties, such as high electron mobility and high breakdown electric field. The use of field plates (FP) in GaN HEMTs enhances breakdown voltage and boosts power performance due to the mitigation of crowded electric field at the gate edges [1-3]. However, the induction of additional gate capacitance due to field plate degrades high frequency performance [4]. A solution to the trade-off between breakdown voltage and frequency response is the slant field plate structure. In 2006, Y. Dora et al. demonstrated that the slant field plate possess superior performance in electric field suppression compared with the conventional field plate [5].

In order to improved GaN HEMTs RF performances, the T-shaped slant gate is introduced. The deep-UV lithography process with PMMA/P(MMA-MAA)/PMMA tri-layer resists is used to form the T-shaped resist cavity. While the sensitivity of P(MMA-MAA) is just a little higher than that of PMMA at deep-UV wavelength, it is difficult to achieve a well T-shaped resist cavity by deep-UV lithography. An angle exposure technique was introduced to improve this process realized on GaN-on-Si HEMTs in this work. With the assistance of angle exposure technique, the T-shaped gate is carefully

designed as an integrated slant field plate to further improve the power performance.

II. Experimental

The AlGaN/GaN HEMT structure used in this work is grown on silicon substrate by MOCVD. The epitaxial structure consisted of a 30 nm-thick AlN nucleation layer, a 1 μm-thick GaN buffer layer, and a 20 nm-thick AlGaN interlayer. Fabrication of the GaN HEMTs started with ohmic-contact formation. Ti/Al/Ni/Au metal stacks were evaporated as ohmic metals and subsequently annealed in N_2 ambient at 850℃ for 30 seconds. Mesa etching was performed by ICP-RIE system with Cl_2/Ar gas for device isolation.

A tri-layer resist system (PMMA/ copolymer P(MMA-MAA)/PMMA) was used to fabricate sub-micron slant T-shaped gates by using deep-UV lithography. Fig. 1 shows the proposed tri-layer process for enlarging the top to bottom ratio of the T-shaped gate. A Ti metal layer of 20nm was deposited on top of the tri-layer resists followed by patterned and DHF wet etch. the patterned Ti metal layer played the role of mask for the following tilted deep-UV lithography process. The wafer was tilted to perform an angle exposure. As shown in Fig. 1 (b) (c), two of angle exposures were applied from the opposite sides. For each exposure, only half of exposure-dose was applied. The Ti metal layer was follow-up removed by DHF before development. Then, Ni/Au metal stack was evaporated and lifted off as gate contact for those two kinds of devices. PECVD silicon nitride was deposited as a passivation layer. DC (Agilent E5270), pulse IV (Accent DIVA 225) and S-parameters (HP85112A vector network analyzer), and continuous wave load pull measurements were taken in this work.

III. Results and discussion

Fig. 2 shows the cross-section profile of the T-shaped gate fabricated by angle exposure. It has a small footprint of 0.6 μm, a large upper layer of 2.5 μm and a slant sidewall of 30°. This T-shaped gate is integrated slant field plate and suitable for high-voltage RF operation.

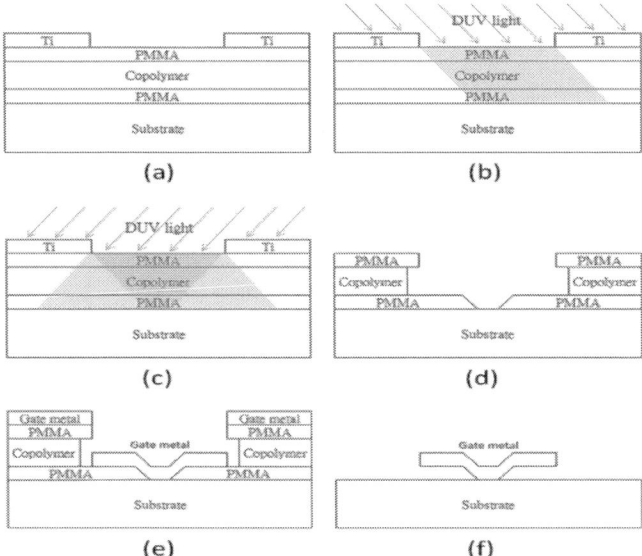

Fig. 1 Fabrication process of the AlGaN/GaN HEMT with slant field-plate structure using deep-UV lithography with angle exposure.

Fig. 3 shows the ID-VD characteristics of the unpassivated AlGaN/GaN HEMT. At Vgs = 0 V, it has a saturation current density higher than 550 mA/mm and a knee voltage of 4 V. The transconductance curve is plotted in Fig. 4, it can be seen that the device has a pinch-off voltage of 3.2 V, a maximum drain current density higher than 700mA/mm and a peak value of transconductance close to 200 mS/mm. Fig. 5 shows the result of the off-state breakdown measurement, a breakdown voltage higher than 100 V is obtained.

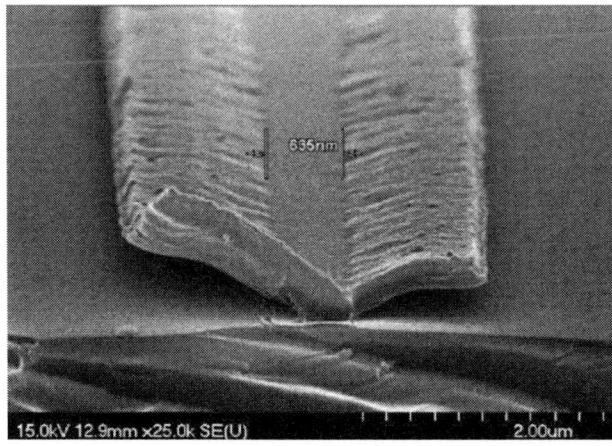

Fig. 2 Cross-section profile of T-shaped gate with slant sidewall

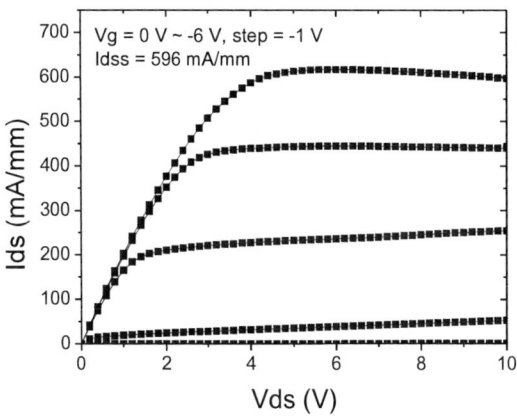

Fig. 3 ID-VD characteristics of unpassivated AlGaN/GaN HEMT

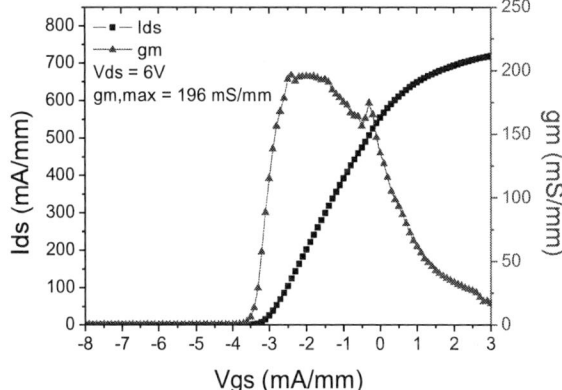

Fig. 4 Transconductance characteristics of unpassivated AlGaN/GaN HEMT

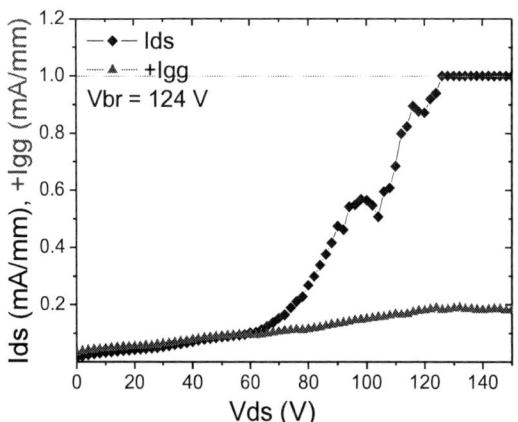

Fig. 5 Three-terminal off-state breakdown of unpassivated AlGaN/GaN HEMT

978-1-4673-2395-6/12 $31.00 © 2012 IEEE

A load-pull system was used to measure the power performance of the field plated GaN HEMTs at 8 GHz. In order to obtain the maximum output power in this device, various drain-source voltages were then applied. Attributed to the slant field plate, the device could be biased to 60 V for load-pull measurement. The output power density was saturated at V_{DS} = 50 V and the self-heating effect started to dominate the output power density at V_{DS} = 60 V. At V_{DS} = 50 V, an output power density higher than 5.0 W/mm was achieved. The small signal RF characteristics of unpassivated AlGaN/GaN HEMT are measured. Fig. 6 shows the de-embedded results which exclude the parasitic effects due to probing pads, the current gain cut-off frequency (fT) and maximum oscillation frequency (fmax) were extrapolated as 39.3 GHz and 63 GHz, respectively, using -20 dB/dec regression.

References

[1] Shreepad Karmalkar and Umesh K. Mishra, IEEE Trans. Electron Devices **48** (2001) 499.

[2] Huili Xing et al., IEEE Electron Device Lett. **25** (2004) 161.

[3] N. –Q. Zhang et al., IEEE Electron Device Lett. **21** (2000) 421.

[4] Alessio Pantellini et al., Proceedings of the 4th European Microwave Integrated Circuits Conference (2009) 140.

[5] Y. Dora et al., IEEE Electron Device Lett. **27** (2006) 713.

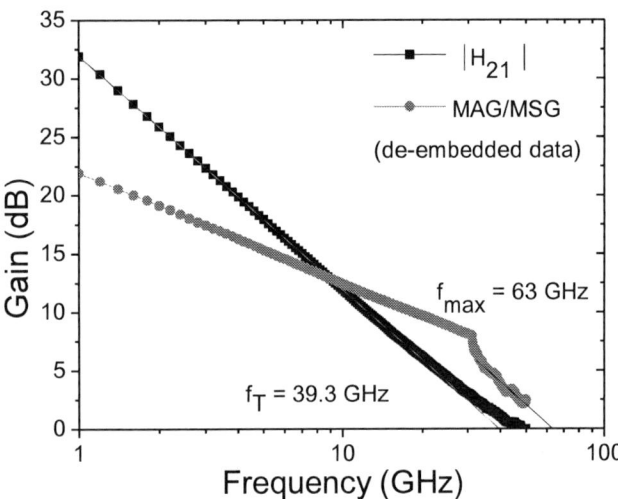

Fig. 6 RF performance at VDS = 10 V and VGS = -1.7 V for AlGaN/GaN HEMT. Extrapolation with -20 dB/dec yields fT = 24 GHz and fmax = 49 GHz.

IV. Conclusion

In this work, the improved deep-UV lithography processed for T-shaped gate fabrication is discussed. By using an angle exposure technique, a well-defined T-shaped gate with slant sidewalls has been successfully fabricated on AlGaN/GaN/Si HEMT. Through surface passivation, the slant field plate formed as the gaps beneath T-shaped gates filled with PECVD silicon nitride. The slant field-plated AlGaN/GaN HEMT exhibited a f_T of 24 GHz and a f_{max} of 49 GHz, and could be biased at 50 V for RF power measurement. An output power of 5.0 W/mm was achieved at 8 GHz. This result demonstrated the power capability of AlGaN/GaN HEMTs on silicon substrate in x-band.

978-1-4673-2395-6/12 $31.00 © 2012 IEEE

Influence of post deposition annealing temperatures on electrical properties of Al$_2$O$_3$/InSb MOSCAPs

Hai-Dang Trinh,[1] Yue-Chin Lin,[1] Edward Yi Chang,[1,2] *Senior member, IEEE*, Hong-Quan Nguyen,[1] Shin-Yuan Wang,[1] Yuen-Yee Wong,[1] Binh-Tinh Tran,[1] Quang-Ho Luc,[1] Chi-Lang Nguyen,[1] and Chang-Fu Dee[1]

[1]Department of Materials Science and Engineering, National Chiao Tung University, 1001, Univ. Rd., Hsinchu, Taiwan, ROC
[2] Department of Electronics Engineering, National Chiao Tung University, 1001, Univ. Rd., Hsinchu, Taiwan, ROC

Abstract- **The influence of post deposition annealing (PDA) temperatures on electrical characteristics of Al$_2$O$_3$/InSb metal-oxide-semiconductor capacitor (MOSCAP) structures is investigated. Low frequency C-V responses with strong inversion behavior in the whole range of measured frequency (100 Hz-1 MHz) are observed, indicating very short minority carrier response time in InSb. The PDA temperature of 300°C and above would result in the reduction of maximum capacitance. At the PDA temperature of above 300°C the C-V hysteresis, frequency dispersion and stretch out increases significantly, indicating the degradation of the MOSCAP structures. The degradation might relate to the interdiffusion between Al$_2$O$_3$ and InSb during thermal steps.**

I. INTRODUCTION

High carrier mobility, narrow-gap InSb has been raised the attention for future extremely high speed, low power transistors. Among III-V compounds, InSb has highest electron mobility of 7.7×10^4 cm^{2}V^{-1}s^{-1} and hole mobility of 840 cm^{2}V^{-1}s^{-1},[1] which promise of both n- and p-channel high performance transistors. Both p- and n-channels InSb quantum-well transistors have been demonstrated very-high-speed performance at low supply voltage of 0.5 V.[2, 3] For high k/InSb structure, the study is still relatively unexplored.[4, 5] Due to low thermal budget, the properties of InSb based devices are very sensitive to thermal processes. In this work, we investigate the influence of post deposition annealing temperature on the electrical properties of atomic layer deposition Al$_2$O$_3$/InSb MOSCAPs.

II. EXPERIMENT

Figure 1 shows the Al$_2$O$_3$/InSb MOSCAP structure and fabrication process flow. Wafers used in this study was undoped InSb(100) subtrate. From Hall measurement, the substrate showed n-type behavior with doping concentration of 2×10^{16} cm^{-3}. The wafers were rinsed in acetone and isopropanol for 2 min each, followed by dipping into HCl 3.5% for 2min and DI water rinsed. The samples were then loaded into atomic layer (ALD) deposition chamber (Cambridge NanoTech Fiji202 DSC). In ALD chamber, the samples were pre-cleaned by using 10 pulses of trimethyl aluminum (TMA)-

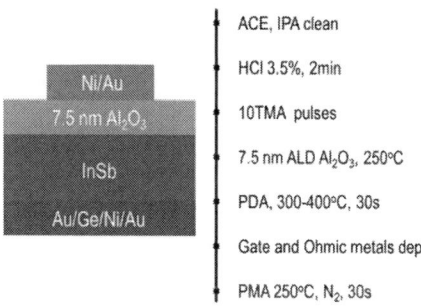

Figure 1. Al$_2$O$_3$/InSb MOSCAP structure and process flow

/Ar[6, 7] before the deposition of 7.5 nm Al$_2$O$_3$ at 250°C, using TMA and H$_2$O as precursors. After that, the samples were PDA at different temperatures in N$_2$ for 30 s. Ni/Au metal gate was formed via photolithography/e-beam evaporation /lift-off process. Finally, Au/Ge/Ni/Au was deposited for backside ohmic contact followed by post metal annealing at 250°C in N2 for 30s.

III. RESULT AND DISCUSSION

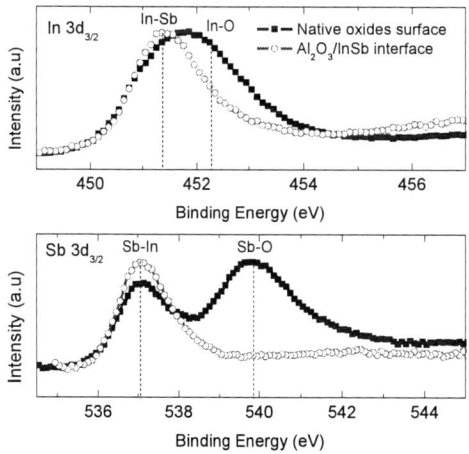

Figure 2. In 3d$_{3/2}$ and Sb3d$_{3/2}$ XPS spectra of bare InSb native oxide surface and as deposited 2 nm Al$_2$O$_3$/HCl-treated InSb interface show the significant reduction of InSb native oxides after HCl treatment and Al$_2$O$_3$ deposition.

Fig.2 shows the $In3d_{3/2}$ and $Sb3d_{3/2}$ X-ray photo electron spectroscopy (XPS) spectra of InSb native oxides surface and 2nm Al_2O_3/HCl plus TMA treated InSb interface. The use of HCl treatment, and TMA pretreatment before the deposition of Al_2O_3 resulted in significant reduction of both In-O and Sb-O bonds.

Figure 3 shows typical multifrequency C-V responses of the MOSCAPs. The C-V reponses were measured by using an HP4284A *LCR* meter. Strong inversion responses are observed in the whole range of measured frequencies (100 Hz - 1 MHz) due to very short minority carrier response time in InSb. The frequency dispersion in conduction band side is always lager that in valence band side for all samples [see also Fig. 4(b)]. This indicates lager amount of boder traps located in conduction band side.

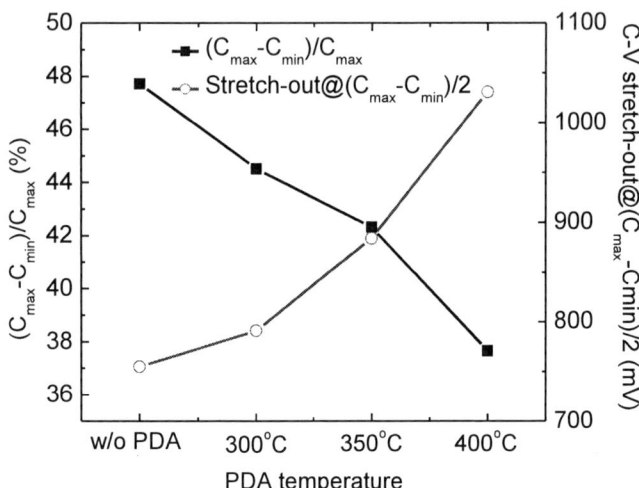

Figure 5. Maximum capacitance, hysteresis, and frequency dispersions vary PDA temperature of Al_2O_3/InSb structures.

Figure 4(a) shows the reduction of maximum capacitance at the PDA temperature of 300°C and above. The hysteresis in conductance band side increased from 70 mV to 100 mV while PDA temperature increased to 350°C and above. The frequency dispersion in both side of InSb bandgap also increase with the increase of PDA temperatures as shown in Fig 4(b).

Figure 5 shows the comparison of $(C_{max}-C_{min})/C_{max}$ and the C-V stretch-out values of the samples. The $(C_{max}-C_{min})/C_{max}$ value decreases from 47.7% for the sample without PDA to 37.7% for the sample PDA at 400°C. This indicates that samples without or with low PDA temperatures are easily to get inversion state as compared to samples with higher PDA temperatures. The C-V stretch-out value increases from 755 mV (w/o PDA sample) to 1030 mV (400°C PDA sample) indicate the increase of border traps inside the oxide when PDA temperature increased. The performing degradation of the MOSCAP samples with increasing PDA temperatures could attribute to the interdiffusion between Al_2O_3 and InSb during thermal process. In fact, the transmission electron microscopy (TEM) graphs and energy-dispersive X-ray spectroscopy (EDX) exhibit the extension of interdiffusion regions (data not shown). The interdiffusion would result in the degradation of both gate oxide as well as InSb layer near the Al_2O_3/InSb interface

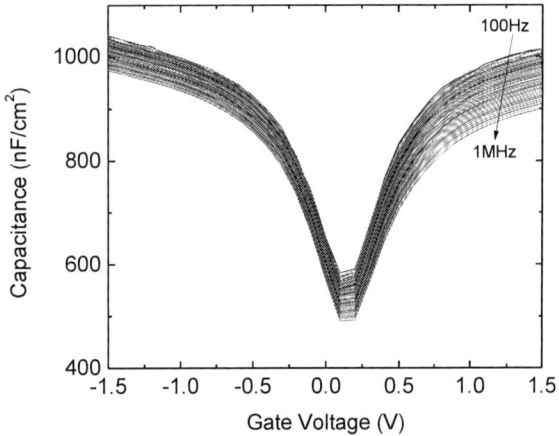

Figure 3. A typical multi-frequency C-V responses of Al_2O_3/InSb MOSCAPs.

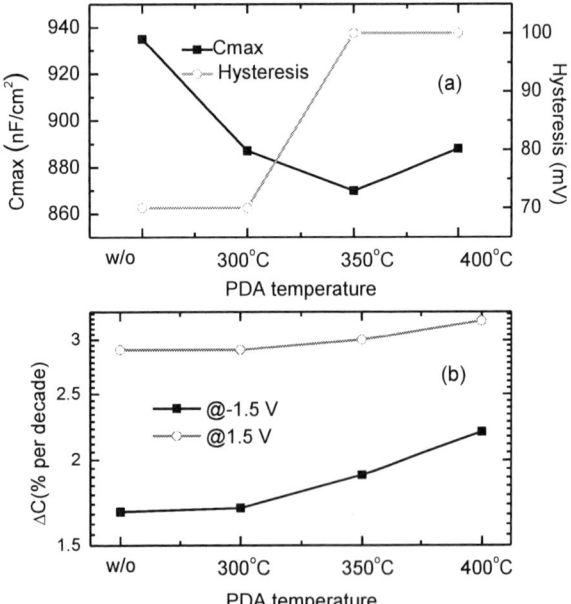

Figure 4. (a) Maximum capacitance, hysteresis, and (b) frequency dispersions vary PDA temperature of Al_2O_3/InSb structures.

IV. CONCLUSIONS

The electrical properties of Al_2O_3/InSb MOSCAPs with different PDA temperatures have been studied. XPS showed the significant reduction of InSb native oxides by using HCl plus TMA treatment before the deposition of Al_2O_3. Nice C-V responses with strong inversion behavior in whole range of measured frequency were observed. However, the electrical properties of the MOSCAPs were strongly affected by PDA temperatures. The PDA temperature of above 300°C would result in the significant degradation of electrical properties of the MOSCAPs. The degradation could attribute to the interdiffusion between Al_2O_3 and InSb during thermal treatment process.

AKNOWLEDGMENT

The authors would like to thank the Ministry of Education and the National Science Council of the Republic of China for supporting this research under contract Nos. 98-2923-E-009-002-MY3 and 100-2120-M-009-010.

REFERENCES

[1] *Hanbook series on semiconductor parameter*, vol. 1, M. Levinshtein, S. Rumyantsev and M. Shur, Eds. World Scientific Publishing Co. Pte. Ltd, Singapore, 1996.

[2] S. Datta, et al., "85nm gate length enhancement and depletion mode InSb quantum well transistors for ultra high speed and very low power digital logic applications," *IEDM Tech. Dig.*, 2005.

[3] M. Radosavljevic, et al., "High-performance 40nm gate length InSb p-channel compressively strained quantum well field effect transistors for low-power (VCC=0.5V) logic applications," *IEDM Tech. Dig.*, 2008.

[4] C. H. Hou, M. C. Chen, C. H. Chang, T. B. Wu and C. D. Chiang, "Interfacial cleaning effects in passivating InSb with Al_2O_3 by atomic layer deposition," *Electrochem. Solid-State Lett.*, vol. 11, pp.D60-D63, April 2008.

[5] C. H. Hou, M. C. Chen, C. H. Chang, T. B. Wu, C. D. Chiang and J. J. Luo, "Effects of surface treatments on interfacial self-cleaning in atomic layer deposition of Al_2O_3 on InSb" *J. Electrochem. Soc.*, vol. 155, pp.G180-G183, July 2008.

[6] H.-D. Trinh et al., "Effects of wet chemical and trimethyl aluminum treatments on the interface properties in atomic layer deposition of Al_2O_3 on InAs," *Jpn. J. Appl. Phys.*, vol. 49, pp. 111201, November 2010.

[7] H.-D. Trinh et al., "Electrical Characterization of Al_2O_3/n-InAs Metal–Oxide–Semiconductor Capacitors With Various Surface Treatments," *IEEE Electron Device Lett.*, vol. 32, pp. 572-574, June 2011.

Organic Field-Effect Transistors for Nonvolatile Memory Devices using Charge-Acceptor Layers

Khairul Anuar Mohamad[1*], Afishah Alias[1], Ismail Saad[1], Bablu Kumar Gosh[1]
Katsuhiro Uesugi[2], Hisashi Fukuda[2]

[1]Nano Engineering & Materials (NEMs) Research Group, School of Engineering and Information Technology,
Universiti Malaysia Sabah, 88400 Kota Kinabalu, Sabah, Malaysia
[2]Division of Engineering for Composite Functions, Muroran Institute of Technology,
27-1 Mizumoto, Muroran 050-8585 Hokkaido, Japan
[*]Email: khairul@ums.edu.my

Abstract- **We introduce a charge-accepting layer on a gate dielectric to investigate the reversible threshold voltage (V_{th}) shifts in both p-channel and n-channel organic field-effect transistors (OFETs) using organic semiconductors of pentacene and poly-naphthalene dicarboximide [P(NDI2OD-T2)], respectively. Bottom gate with top drain-source contact structure of both devices exhibited a unipolar property of field-effect transistor behavior. Furthermore, the existence of fullerene (PCBM) and poly(3-hexylthiophene) (P3HT) films as a charge-accepting-like storage layers in p-channel and n-channel devices, respectively, resulted in a reversible V_{th} shifts upon the application of external gate bias (V_{bias}). Hence, p-channel OFETs exhibited a memory window of 2.4 V and n-channel OFETs exhibited a memory window of 10.7 V for program and erase electrically upon application of gate bias.**

I. INTRODUCTION

Research on organic field-effect transistors (OFETs) for their broad range of applications in the electronic industry has attracted scientific and technological interest. OFETs have been the subject of interest for the past few years [1]. During this decade, the use of OFETs in electronic applications has seen a spectacular evolution which provides a unique opportunity to enable low-cost, large areas, flexibility, and easier fabrication procedure. However, organic semiconductors became more than a curiosity when it was recognized to have excellent electronic properties compared with inorganic semiconductors. Recently, there has been impressive progress in the development of electronic devices, particularly high performance organic transistors which been utilized in inverters and logic elements, sensors, light-emitting displays, photovoltaic cells, and integrated circuits [2-6]. While the Silicon-based nonvolatile memory has emerged as the most mature nonvolatile semiconductor memory [7], it has only recently been demonstrated to have potential in organic transistors [8]. In conventional silicon technology, research and development for silicon nonvolatile memory is fueled by the serious drawback in the physical limitation of the device structure toward nano-scale transistors. Such spectacular evolution in organic-based devices offers a significant advantages over silicon technology in numerous and innovative applications where the performance level of silicon is not

essential. Hence, this could be an alternative or complement technology to the conventional semiconductor technology in the nano-scale devices.

Initial attempts at nonvolatile organic memory were reported using p-channel OFET memory with charge storage in polymer electret, dielectric with embedded metallic or semiconductor nanoparticles, and ferroelectric gate insulators with permanent or switchable electrical dipoles [9-11]. Such devices can have memory characteristics if either reversible charge trapping or detrapping mechanism can be made to occur in the gate dielectric layer. Recently, nonvolatile n-channel organic memory using a block copolymer-nanoparticle hybrid system has been reported [12]. The n-channel memory device demonstrated programmable-erasable properties with a large memory window (~9–11 V) using n-channel (perfluorinated copper phthalocyanine) OFETs memories where in-situ synthesized gold (Au) nanoparticle in self-assembled polystyrene-block-poly-4-vinylpyridine (PS-b-P4VP) block copolymer nano-domains as charge storage elements. Thus, the evolution of organic semiconductors has become important for the continuous development of OFETs with memory element.

In this study, we demonstrate fabrication and characterization of p-channel and n-channel OFETs with heterojunction structure thin films with charge-acceptor layers, respectively. Moreover, we also present the characteristics of memory element in p-channel and n-channel OFETs using charge-acceptor layers for nonvolatile memory application.

II. DEVICE CONCEPT AND FABRICATION

A. Fabrication of p-channel organic field-effect transistors

A (100)-oriented Sb-doped n^+-type Si wafer (< 0.1 Ω cm) was used as the substrate and gate contact. The substrates were rinsed by ultrasonic in deionized water, ethyl alcohol, acetone, and then methyl alcohol, and cleaned by standard RCA cleaning procedure. A thermally oxidized 200-nm-thick SiO_2 layer formed by dry oxidation was used as a gate dielectric. A thin fullerene layer; the acceptor layer, was deposited onto the SiO_2 layer by spin coating from dichloromethane solution at 1500 rpm and followed by an annealing treatment inside a glove box under the N_2 atmosphere. Pentacene (p-channel organic semiconductor) was then deposited by vacuum evaporation to form a 50-nm-thick pentacene thin film onto the fullerene layer. Finally, a gold (Au) film was deposited through a designated shadow mask in a vacuum chamber for the top source and drain contact. The channel dimension were $W/L = 5000~\mu m~/100~\mu m$. The schematic diagram of the bottom gate with top drain-source contact structure for p-channel OFETs is shown in Fig. 1(a). The transistor characteristics of the devices were measured using a computer-controlled automatic electrical analyzer at room temperature in the dark under atmospheric pressure.

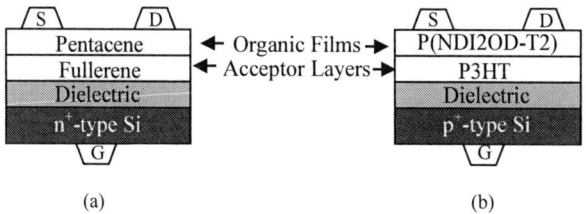

(a)   (b)

Fig. 1. Schematic diagram of bottom gate with top drain-source contact structure for (a) p-channel and (b) n-channel OFETs

B. Fabrication of n-channel organic field-effect transistors

A (100)-oriented B-doped p^+-type Si wafer (< 0.1 Ω cm) was used as the substrate and gate contact. The substrates were rinsed ultrasonically in deionized water, ethyl alcohol, acetone, and then methyl alcohol, and cleaned by the standard RCA cleaning procedure. Poly(methyl methacrylate) (PMMA) was diluted in ethyl acetate and then spin-coated directly onto the silicon substrate to form an organic dielectric layer. Poly(3-hexylthiophene) (P3HT); the acceptor layer, was deposited by spin-coating from dichloromethane solution at 5000 rpm onto the PMMA dielectric layer, followed by spin coating of poly-naphthalene dicarboximide [P(NDI2OD-T2)] from chloroform solution at 5000 rpm to form the n-channel active semiconducting layer. Finally, a gold (Au) film was deposited through a designated shadow mask in a vacuum chamber for the top source and drain contacts. The channel dimension of these devices was $W/L = 2500~\mu m/50~\mu m$. The schematic diagram of the bottom gate with top drain-source contact structure n-channel OFETs is shown in Fig. 1(b). The transistor characteristics were measured in a probe station with a computer-controlled automatic electrical analyzer at room temperature in the dark under the ambient atmosphere.

III. Device Characterization Results

A. Transistor characteristics of p-channel OFETs

Fig. 2 shows representative output and transfer characteristics of pentacene/fullerene OFETs on the highly doped p-type silicon substrate. The output curve of the device shows typically p-channel field-effect transistor (FET) as shown in Fig. 2(a). This indicates that pentacene/fullerene OFETs did not change the nature of charge transport, as the heterojunction device was found to behave as p-channel transistors. Similar behavior has been observed in top source-

drain contact structure of copper phthaocyanine and hexadeca-fluoro phthalocyanine (CuPc/F_{16}CuPc) based OFETs [13]. However, most heterojunction structure devices exhibit ambipolar transport and OFETs based on fullerene/pentacene heterojunction with bottom drain-source contact structure have been previously reported [14,15]. Thus, this indicates that charge transport mode is strongly determined by the interface characteristics on layering order in OFETs. The performances of p-channel OFETs are generally defined by several important parameters, such as the field-effect mobility (μ), on/off current ratio (I_{on}/I_{off}), and threshold voltage (V_{th}) were extracted from Fig. 2(b). The field-effect mobility was a 3.2×10^{-2} cm^{2}V^{-1}s^{-1}, a threshold voltage of −1.1 V, and an on/off current ratio of 10^4 were obtained. In fact, the mobility was obtained using the standard transistor equation for FETs [16]. The parameters, especially the field-effect mobility is low comparable with the best mobilities reported for pentacene OFETs [17] due to the variation with the molecular structures of semiconductor films [18].

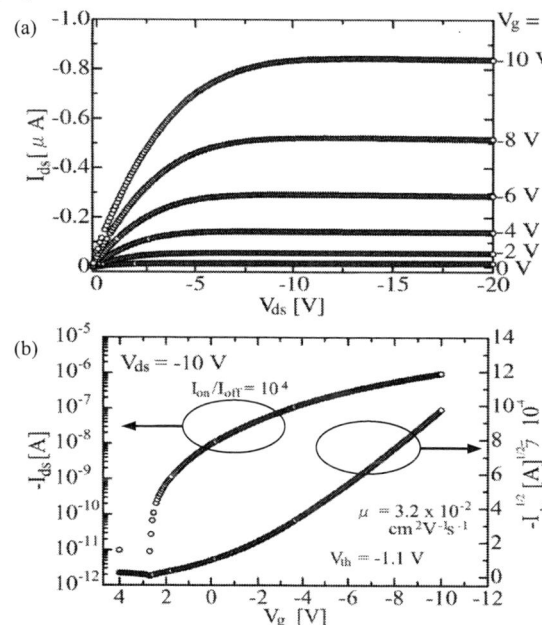

Fig. 2. (a) Output characteristics and (b) semi-logarithmic plots of both drain current and square root of drain current versus gate voltage showing the transfer characteristics of p-channel OFET

B. Transistor characteristics of n-channel OFETs

Fig. 3 shows representative output and transfer characteristics of P(NDI2OD-T2)/P3HT OFETs on the highly doped p-type silicon substrate. The device shows a typical output curve of an n-channel FET, as shown in Fig. 3(a). A bottom source-drain with a bottom drain-source structure ambipolar OFETs based on a bulk heterojunction layer of P(NDI2OD-T2) and P3HT have been reported [19], in which the p-channel and n-channel semiconductor layers have to be deposited using orthogonal solvents in order to prevent mixing of the layers. In fact, the top drain-source contact structure device based on heterojunction thin films did not interfere with the nature of charge transport, as the devices were found to behave as n-channel operation mode similarly to the unipolar P(NDI2OD-T2) based OFETs [20]. In addition, transistor parameters such as field-effect mobility (μ), threshold voltage (V_{th}), and on/off current ratio (I_{on}/I_{off}) were 2.8×10^{-4} cm^{2}V^{-1}s^{-1}, 30.6 V, and 10^2 extracted from Fig. 3(b), respectively. The mobility was extracted using the standard transistor equation for FETs [16]. In fact, the mobility was in the range of the reported values for other unipolar OFETs based on NDI-polymer [20], and typically, spin-coated n-channel polymer devices displayed mobility in the range of 10^{-3}-10^{-5} cm^{2}V^{-1}s^{-1}. The on-off current ratio was comparable with other OFETs [21].

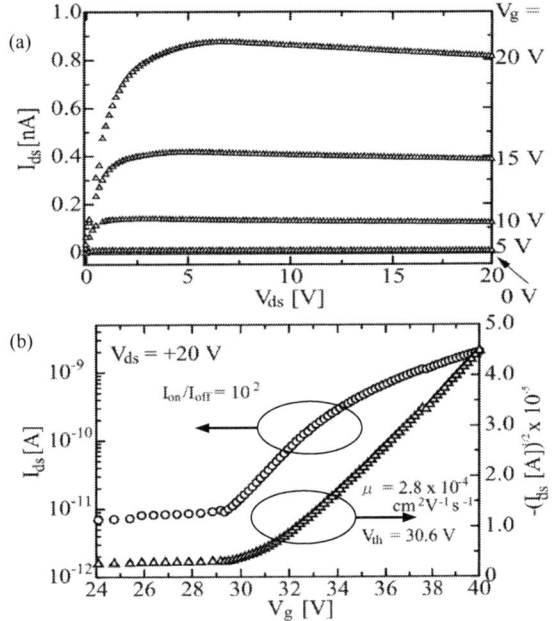

Fig. 3. (a) Output characteristics and (b) semi-logarithmic plots of both drain current and square root of drain current versus gate voltage showing the transfer characteristics of n-channel OFET

C. Memory characteristics in p-channel and n-channel OFETs

A gate bias (V_{bias}) was applied to the gate contact to cause reversible shifts in threshold voltage in order to determine whether the device can sense for memory state, i.e. program and erase states, which is similar to the characteristics described by S. Tiwari *et al.* for nonvolatile memory [22]. Fig. 4 shows the memory characteristics of both p-channel and n-channel OFETs with acceptor layers following the application of V_{bias}. From Fig. 4(a), p-channel OFETs exhibit that the initial transfer characteristic was clearly shifted in a negative direction after the application of external V_{bias} = +10 V (programming voltage) for T_{bias} = 10 min. The memory window (ΔV_{th}) was observed to be 2.4 V. Thereafter, the application of an external V_{bias} = −30 V (erasing voltage) for T_{bias} = 2 min caused the positively shifted transfer characteristic to return completely to its initial position. On the other hand, as

shown in Fig. 4(b), n-channel OFETs show that the initial transfer characteristic was clearly shifted in a negative direction after the application of external V_{bias} = −60 V (programming voltage) for T_{bias} = 5 min. The memory window of 10.7 V was obtained. Thereafter, the application of an external V_{bias} = +60 V (erasing voltage) for T_{bias} = 5 min caused the positively shifted transfer characteristic to return nearly to its initial position. The memory characteristics in both devices were obtained without any degradation of charge transport properties.

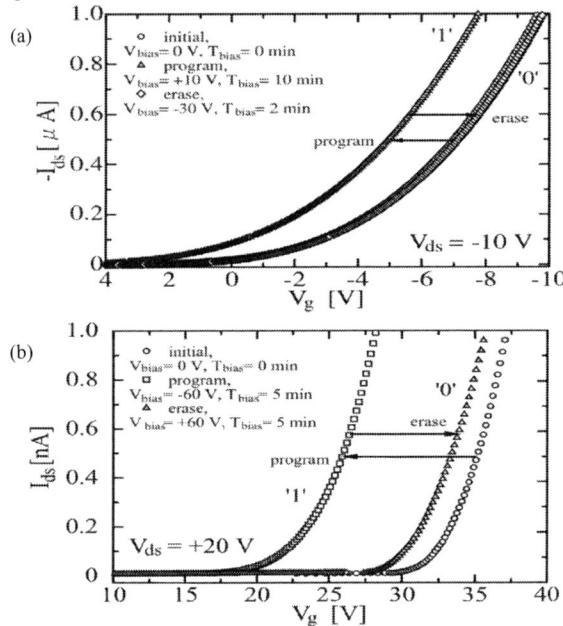

Fig. 4. Memory characteristics obtained upon application of different external gate bias conditions at constant drain-source voltage (V_{ds}) for both p-channel and n-channel OFETs. (a) p-channel OFETs: V_{bias} = 0 V (initial state), V_{bias} = +10 V at T_{bias} = 10 min (program or '1' state), and V_{bias} = −30 V at T_{bias} = 2 min (erase or '0' state). (b) n-channel OFETs: V_{bias} = 0 V (initial state), V_{bias} = −60 V at T_{bias} = 5 min (program or '1' state), and V_{bias} = +60 V at T_{bias} = 5 min (erase or '0' state)

The programming voltage caused direction shifts in V_{th}, which indicates that charges are stored or trapped locally from the channel into the thin film of acceptor layers; fullerene and P3HT. The dependence of programming voltage in both p-channel and n-channel devices could be explained by the difference of energy barriers for a charge transfer and trapping between the organic and the acceptor layers. The energy barrier for a charge transfer and trapping is defined as an energy gap between the organic semiconductor and the acceptor layer [23]. On the other hand, the reversible direction in V_{th} shift in an opposite direction was caused by an erasing voltage. This indicates that the detrapping of charges was occurred locally at the organic/acceptor layer interface to completely restore the transfer curve to its initial state. Moreover, much higher gate bias was required to detrap due to the large injection barrier height for charge between the energetic level of organic semiconductor and acceptor layers [23]. Therefore, the degree of position shift of the transfer curve or threshold voltage could be controlled by the proper selection of an external gate bias for a certain period of time. The investigation of the lowest unoccupied molecular orbital (LUMO) and highest occupied molecular orbital (HOMO) for both organic semiconductor and acceptor materials will be studied later, which might give relationship between the relative energy barrier and applied gate bias. Here, one should distinguish the memory behavior upon bias-stress effects in OFETs by defect states or impurities in semiconductor and/or gate dielectrics, which also lead to threshold voltage shifts [24]. Although this external effect would enhance to be opened the memory window, it cannot be controllable to use as a practical memory as well as not permanent [25].

IV. CONCLUSIONS

We fabricated and characterized both p-channel and n-channel OFETs with charge-acceptor layers. Both OFETs exhibited a unipolar property of field-effect transistor behavior. Furthermore, we also demonstrated memory characteristics in both p-channel and n-channel OFETs in which fullerene and P3HT functioned as the acceptor or charge storage layers, respectively. The p-channel OFETs with fullerene-acceptor layer manifested a memory window of 2.4 V (ΔV_{th} = 2.4 V) whereas the n-channel OFETs with P3HT-acceptor layer manifested a memory window of 10.7 (ΔV_{th} = 10.7 V) upon the application of gate bias (programming voltage). The incorporation of acceptor layers as charge storage sites resulted in a reversible shift in the threshold voltage. The recovery of the threshold voltage was achieved by the application of higher reverse gate bias (erasing voltage). Thus, OFETs with charge trap (charge-acceptor layer) sites could be used for nonvolatile memory application. However, there are still some critical issues to be addresses such as stability and degradation issues, before organic transistor memory can be used in practical devices. Nevertheless, promising progress has been made in designing next-generation nonvolatile memories, and the organic transistor memory has a bright prospect for future electronics.

ACKNOWLEDGMENT

This study was supported by a Grant-in-Aid for Scientific Research (No. 17510107) from the Ministry of Education, Culture, Sports, Science and Technology of Japan and FRGS (FRG0306-TK-1/2012) fund from Ministry of Higher Education (MOHE) of Malaysia.

REFERENCES

[1] H. Inokuchi, "The discovery of organic semiconductors: Its light and shadow," *Org. Electron.*, vol. 7, pp. 62-76, April 2006.

[2] E. J. Meijer, D. M. de Leeuw, S. Setayesh, E. Van Veenendaal, B. H. huisman, P. W. M. Blom, J. C. Hummelen, U. Scherf, and T. M. Klapwijk, "Solution-processed ambipolar organic field-effect transistors and inverters," *Nat. Mater.*, vol. 2, pp. 678-684, September 2003.

[3] Th. B. Singh, N. Marjanovic, G. J. Matt, N. S. Sariciftci, R. Schwodiauer, and S. Bauer, "Nonvolatile organic field-effect transistor memory

element with a polymeric gate electret," *Appl. Phys. Lett.*, vol. 85, pp. 5409-5411, October 2004.

[4] G. Darlinski, U. Bottger, R. Wasser, H. Klauk, M. Halik, U. Zschieschang, G. Schmid, and C. Dehm, "Mechanical force sensors using organic thin-film transistors," *J. Appl. Phys.*, vol. 97, pp. 093708-093711, April 2005.

[5] M. A. McCarthy, B. Liu, E. P. Donoghue, I. Kravchenko, D. Y. Kim, F. So, and A. G. Rinzler, "Low-voltage, low-power, organic light-emitting transistors for active matrix displays," *Science*, vol. 332, pp. 570-573, March 2011.

[6] A. C. Hubler, G. C. Schmidt, H. Kempa, K. Reuter, M. Hambsch, and M. Bellmann, "Three-dimensional integrated circuit using printed electronics," *Org. Electron.*, vol. 12, pp. 419-423, March 2011.

[7] S. M. Sze, *Semiconductor Devices: Physics and Technology*, New York: Wiley, 1985, pp. 6-7, 216-218, 507-510.

[8] H. E. Katz, X. M. Hong, A. Dodabalapur, and R. Sarpeshkar, "Organic field-effect transistors with polarizable gate insulators," *J. Appl. Phys.*, vol. 91, pp. 1572-1576, October 2001.

[9] H. E. Katz, X. M. Hong, A. Dodabalapur, and R. Sarpeshkar, "High-performance n-channel organic thin-film transistor for CMOS circuits using electron-donating self-assembled layer," *J. Appl. Phys.*, vol. 91(3), pp. 1572-1576, February 2002.

[10] K. J. Baeg, Y. Y. Noh, H. Sirringhaus, and D. Y. Kim, "Controllable shifts in threshold voltage of top-gate polymer field-effect transistors for applications in organic nano floating gate memory," *Adv. Funct. Mater.*, vol. 20(2), pp. 224-230, December 2010.

[11] Q. D. Ling, D. J. Liaw, C. Zhu, D. S. H. Chan, E. T. Kang, and K. G. Neoh, "Polymer electronic memories: Materials, devices and mechanism," *Prog. Polym. Sci.*, vol. 33, pp. 917-978, October 2008.

[12] W. L. Leong, N. Mathews, S. Mhaisalkar, Y. M. Lam, T. Chen, and P. S. Lee, "Micellar poly(styrene-b-4-vinylpyridine)-nanoparticle hybrid system for non-volatile organic transistor memory," *J. Mater. Chem.*, vol. 19, pp. 7354-7361, September 2009.

[13] J. Wang, H. Wang, X. Yan, H. Huang, and D. Yan, "Organic heterojunction and its application for double channel field-effect transistors," *Appl. Phys. Lett.*, vol. 87, pp. 093507-093509, August 2005.

[14] S. D. Wang, K. Kanai, Y. Ouchi, and K. Seki, "Bottom contact ambipolar organic thin film transistor and organic inverter based on C60/pentacene heterostructure," *Org. Electron.*, vol. 7, pp. 457-464, December 2006.

[15] S. J. Kang, Y. Yi, C. Y. Kim, S. W. Cho, M. Noh, K. Jeong, and C. N. Whang, "Energy level diagrams of C60/pentacene/Au and pentacene/C60/Au," *Synth. Metal.*, vol. 156, pp. 32-37, January 2006.

[16] S. M. Sze, *Physics of Semiconductor Devices*, 2nd ed., New York: Wiley, 1981, pp. 440.

[17] L. A. Majewski, R. Schroeder, and M. Grell, "One volt organic transistor," *Adv. Mater.*, vol. 17, pp. 192-196, January 2005.

[18] M. Kano, T. Minari, K. Tsukagoshi, and H. Maeda, "Control of device parameters by active layer thickness in organic field-effect transistors," *Appl. Phys. Lett.*, vol. 98, pp. 073307-073309, February 2011.

[19] K. Szendrei, D. Jarzab, Z. Chen, A. Facchetti, and M. A. Loi, "Ambipolar all-polymer bulk heterojunction field-effect transistors," *J. Mater. Chem.*, vol. 20, pp. 1317-1321, December 2009.

[20] J. H. Oh, S. Liu, Z. Bao, R. Schmidt, and F. Wurthner, "Air-stable n-channel organic thin-film transistors with high field-effect mobility based on N,N'-bis(heptafluorobutyl)-3,4:9,10-perylene diimide," *Appl. Phys. Lett.*, vol. 91, pp. 212107-212109, November 2007.

[21] S. Scheinert, G. Paasch, M. Scrodner, H.-K. Roth, S. Sensfus, and Th. Doll, "Subthreshold characteristics of field effect transistors based on poly(3-dodecylthiophene) and an organic insulator," *J. Appl. Phys.*, vol. 92, pp. 330-337, April 2002.

[22] S. Tiwari, F. Rana, H. Hanafi, A. Hartstein, E. F. Crabbe, and K. Chan, "A silicon nanocrystals based memory," *Appl. Phys. Lett.*, vol. 68, pp. 1377-1379, December 1995.

[23] K. J. Baeg, Y. Y. Noh, and D. Y. Kim, "Charge transfer and trapping properties in polymer gate dielectrics for non-volatile organic field-effect transistor memory applications," *Solid-State Electron.*, vol. 53, pp. 1165-1168, August 2009.

[24] A. Salleo and R. A. Street, "Kinetics of bias-stress and bipolaron formation in polythiophene," *Phys. Rev. B*, vol. 70, pp. 235324-235331, December 2004.

[25] P. Heremans, G. H. Gelinck, R. Muller, K. J. Baeg, D. Y. Kim, and Y. -Y. Noh, "Polymer and organic nonvolatile memory devices," *Chem. Mater.*, vol. 23, pp. 341-358, October 2010.

Nanoindentation Creep Analysis of Gold Ball Bond

Muhammad Nubli Zulkifli[1], Azman Jalar[1], Shahrum Abdullah[2], Norinsan Kamil Othman[3], and Muhammad Azmi Abdul Hamid[3]

[1]Institute of Microengineering and Nanoelectronic (IMEN),
[2]Department of Mechanical & Materials Engineering,
[3]School of Applied Physics, Faculty of Science and Technology,
Universiti Kebangsaan Malaysia,
43600 UKM, Selangor, Malaysia.
Email: azmn@ukm.my

Abstract- **The analysis of indentation creep of gold, Au ball bond was carried out by using nanoindentation approach. 3 X 4 arrays of indentation were indented at three location of Au ball bond namely gold, Au Zone, intermetallic compounds, IMC Zone and Silicon, Si Zone. It was observed that Au and IMC have higher creep behavior compared to that of Si. The responsible indentation creep mechanism for Au and IMC of ball bond that have been subjected 1000 hours of HTS was the dislocation glide. It was noted that the lower plastic deformation or creep effect of IMC was due to the higher hardness value which demonstrated the strain hardening effect compared to that of Au.**

I. INTRODUCTION

The quality and reliability assessment of Au wire bond is a crucial part in order to evaluate the performance of electronic packaging. The conventional tests that has been widely used namely tensile, wire pull, and ball bond shear tests provide limited data regarding the quality of wire bond [1-8]. In addition, today's trend that requires the smaller size of wire bond introduces a lot of technology challenges in terms of process and quality assessment [9].

Creep is one of the issues that have been used to characterize the performance of materials over time [10-15]. Creep is the time-dependent deformation under a constant load or constant stress [10]. Li et al. [10] reported that indentation creep can happen in materials during nanoindentation even when the test is conducted in the temperatures ranging from room temperature or 25 °C to the half of melting temperature $(0.5T_m)$.

The creep analysis of Au ball bond is still lacking due to the difficulty in performing creep analysis by using conventional tests [9,16]. In the present analysis, nanoindentation approach is proposed in order to analyse the creep behaviour of Au ball bond. Nanoindentation test has the capability to characterize the micromechanical properties of material in detailed and localized manner [17,18]. In addition, nanoindentation test also provides continuous measurement of indentation depth over applied load and this will allow the determination of creep behaviour.

In this paper, creep behaviour of Au ball bond encapsulated in the Quad Flat No-Lead (QFN) package that has been exposed to the high temperature storage (HTS) was analysed by using the nanoindentation test. The constant load method introduced by Goodall and Clyne et al. [11] was used in order to evaluate the creep behaviour of Au ball bond. The stress exponent, *n* and strain rate sensitivity, *m* of Au ball bond were obtained by using constant load method to determine the responsible mechanism of creep and the creep behaviour of Au ball bond after subjected to HTS.

II. EXPERIMENTAL WORKS

The as-received Quad Flat No-Lead (QFN) package with dimension of 4 mm X 4 mm, built up with double stack dies and also bonded with 25 μm diameter of Au wire bond was used as the sample. High temperature storage, HTS was carried out on the QFN package sample with temperature of 175 °C and 1000 hours of time duration based on the condition C of JESD22-A103C JEDEC standard [19].

The sample for nanoindentation test was prepared by resin mounting the QFN package. After mounting and curing, wet grinding was carried out to reveal the Au ball bond from the QFN package using 600, 800, and 1200 grit of abrasive papers followed by polishing. Polishing was conducted using 6 μm, 3 μm, and 0.25 μm of diamond suspension on silk cloths.

Nanoindentation test was performed using Micro Materials Nanotest™ indenter, equipped with a Berkovich diamond tip. Nanoindentation test was carried out at twelve different locations with 3 X 4 arrays of indentations on Au ball bond. The distance between each indentation was maintained around 6 μm. Nanoindentation was performed at room temperature with loading and unloading rate of 0.5 mn/s, 10 second of hold time at the peak load and 60 second of hold time at 90 % unload for thermal drift correction. The data obtained throughout the nanoindentation test were based on Oliver and Phar method [17]. This method determines the hardness and reduced Young's modulus from depth sensing indentation (DSI) of load-displacement data.

III. RESULTS AND DISCUSSION

Fig. 1 shows the indentation location with its respective number on Au ball bond subjected to 1000 hours of HTS. Whereas, Fig. 2 shows the *P-h* profiles for Au ball bond subjected to 1000 hours of HTS.

978-1-4673-2395-6/12 $31.00 © 2012 IEEE

Fig. 4 show the depth versus dwell time graph for the indentations in the IMC Zone. Figs. 5 and 6 show the log-log graph of strain rate versus stress for the indentations in the Au and IMC Zones, respectively.

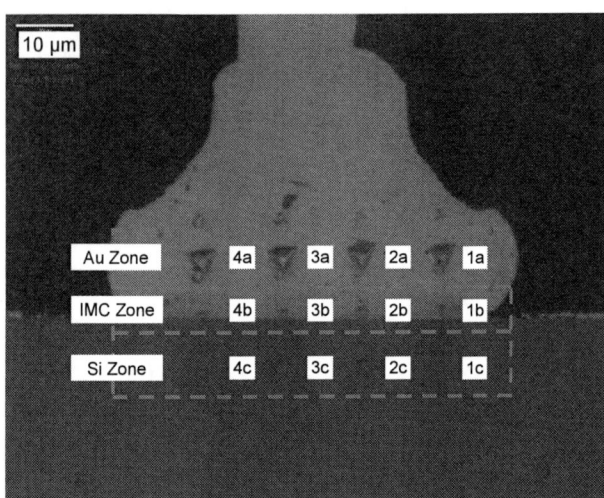

Fig. 1 Indentation location on Au ball bond subjected to 1000 hours of HTS

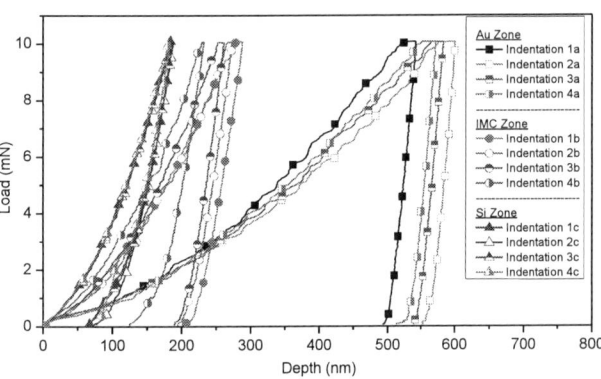

Fig. 2 *P-h* profile for Au ball bond subjected to 1000 hours of HTS

In Fig. 1 the 3 X 4 arrays of indentation are assigned at three Zones namely gold, Au Zone, intermetallic compounds, IMC Zone and Silicon, Si Zone. The *P-h* profiles for the indentations on Au and IMC Zones display pronounce creep effect compared to that of indentations on Si Zone as shown in Fig. 2. The occurrence of creep can be observed through the increase of depth at the maximum applied load [18]. In the present analysis, the creep behavior is measured for the indentions on Au and IMC Zones using the constant load method introduced by Goodall and Clyne [11]. This technique has been chosen because it requires a simple method of measurement and several researchers have successfully utilized this method in analyzing the creep behavior [12-15]. The creep behavior for the Si is not measured due to the least creep effect observed as shown in Fig. 2. Fig. 3 shows the depth versus dwell time graph for the indentations in the Au Zone. While,

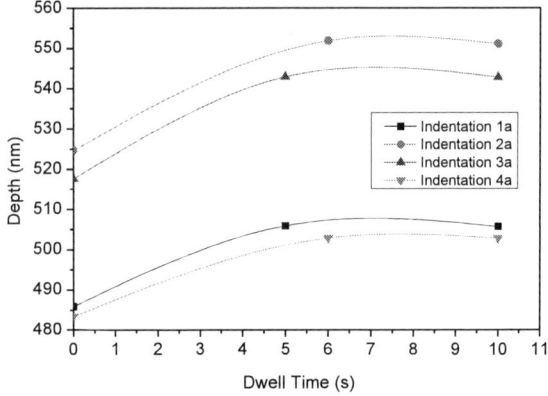

Fig. 3 Graph of depth versus dwell time for the indentations in the Au Zone

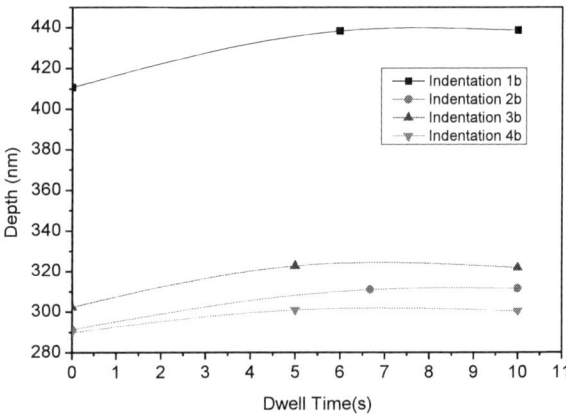

Fig. 4 Graph of depth versus dwell time for the indentations in the IMC Zone

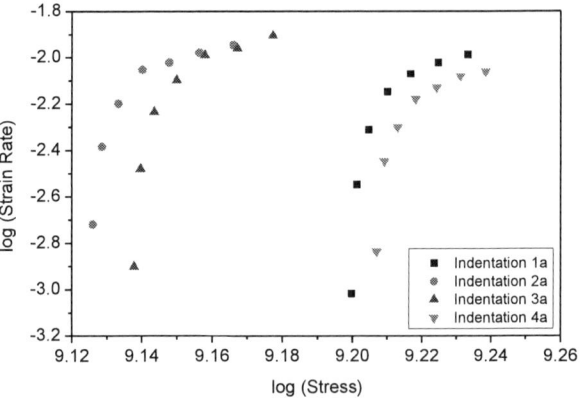

Fig. 5 Graph of log strain rate versus log stress for indentations in the Au Zone

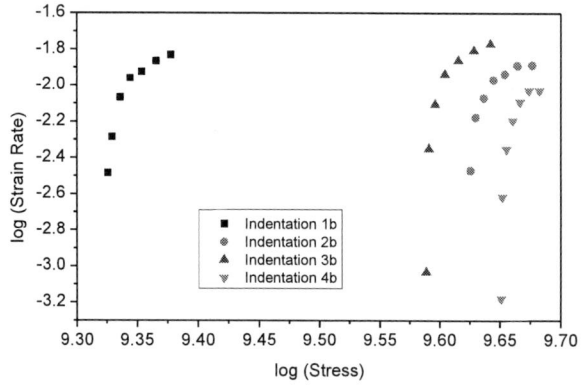

Fig. 6 Graph of log strain rate versus log stress for indentations in the IMC Zone

Figs. 3, 4, 5, and 6 are used to measure the creep parameters namely stress exponent, n and strain rate sensitivity, m based on the constant load method. The plot of log-log graph of strain rate versus stress is created by measuring the strain rate and stress value based on the depth versus dwell time graph [11]. Following are the equations for the stress and strain rate, respectively [11]:

$$\sigma = \frac{F}{A_p} \qquad (1)$$

$$\dot{\varepsilon} = \frac{1}{h}\frac{dh}{dt} \qquad (2)$$

where F is applied load, A_p is indentation projected area, h is depth, dh is depth gradient and dt is time gradient. A_p for Berkovich indenter is equals to $24.5h^2$. The stress exponent, n value is measured by getting the slope of the log-log graph of strain rate versus stress as shown in Figs. 5 and 6. The strain rate sensitivity, m value is obtained by using the following equation [20]

$$m = \frac{1}{n} \qquad (3)$$

Tables I and II show the stress exponent, n value and strain rate sensitivity, m measured for indentations on Au and IMC Zones subjected to 1000 hours of HTS. Table III shows the responsible indentation creep mechanism with its respective stress exponent, n value as reported by Mahmudi et al. [21].

TABLE I STRESS EXPONENT, n VALUES AND STRAIN SENSITIVITY, m VALUE FOR INDENTATIONS ON AU ZONE SUBJECTED TO 1000 HOURS OF HTS

Indentation	Stress Exponent, n	Strain Rate Sensitivity, m
1	23.714	0.042
4	15.560	0.064
7	19.688	0.051
10	19.443	0.051
Average	19.601	0.052

TABLE II STRESS EXPONENT, n VALUES AND STRAIN SENSITIVITY, m VALUE FOR INDENTATIONS ON IMC ZONE SUBJECTED TO 1000 HOURS OF HTS

Indentation	Stress Exponent, n	Strain Rate Sensitivity, m
2	10.98	0.09
5	9.38	0.11
8	16.75	0.06
11	27.87	0.04
Average	16.24	0.07

TABLE III RESPONSIBLE INDENTATION CREEP MECHANISMS WITH ITS RESPECTIVE STRESS EXPONENT, n VALUE [21]

Stress Exponent, n Value	Indentation Creep Mechanism
1	Diffusional
2	Grain Boundary Sliding
4 to 6	Dislocation Climb
More than 6	Dislocation Movement / Dislocation Climb

From Tables I, II, and III it is shown that the stress exponent, n values for indentations on Au and IMC Zones range from 9.383 to 27.868. This shows that the dislocation glide is the responsible indentation creep mechanism in Au and IMC. In addition, it is noted that the stress exponent, n for Au with average value of 19.601 is higher than IMC with value of 16.245. While, the strain rate sensitivity, m for Au with average value of 0.052 is lower than IMC with value of 0.073. The Au and IMC have different crystal structure and consequently different response towards creep [22-24]. Au is a polycrystalline metal that has multiple grain boundaries. Thus, the occurrence of creep in Au is due to the dislocation of grains in gliding manner [22,23]. Meanwhile, the occurrence of creep of IMC is due to the increase of strain hardening effect [24]. Different crystal structure of IMC phases will be created when the HTS time interval is increased [25]. Consequently, different value of applied load is required for each IMC phase to initiate the occurrence of sliding and dislocation [24,25]. This will make each of the IMC phases that possess different crystal structures to have different values of micromechanical properties and creep behavior.

Fig. 7 shows the variation of hardness towards indentation location.

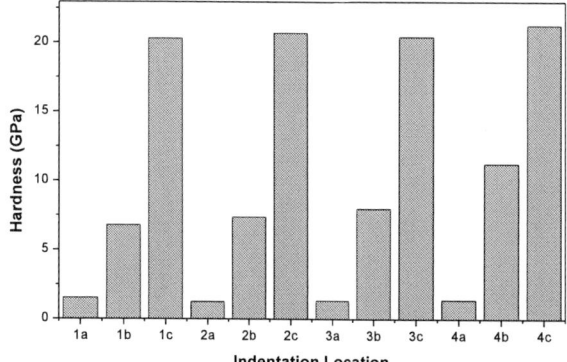

Fig. 7 Variation of hardness towards indentation location

From Fig. 9 it is indicated that the hardness value of indentations in the IMC Zone (indentations 1b, 2b, 3c, and 4b) is higher than that of indentations in Au Zone (indentations 1a, 2a, 3a, and 4a). As mentioned earlier, the increase of HTS time duration will form different phases of IMC with higher value of Vickers hardness [25]. However, the hardness value for Au-Al IMC obtained from the present analysis does not represent the hardness value for individual phase of IMC formed after Au ball bond was subjected to HTS. This is because the indentation projected area in the IMC Zone is covered mostly by the IMC thickness as shown in Fig. 1. Based on preliminary study, the applied load with value of 10 mN that creates the indentation depth of 200 nm until 400 nm has been used to avoid the indentation size effect and the work hardening effect occurred on the surface of cross-section of Au ball bond [26]. For the purpose of comparison, it is believed that higher hardness value for the indentations in the IMC Zone is due to the phases with higher hardness has been formed during 1000 hours of HTS compared to that of indentations in Au zone. The difficulty of plastic deformation also increases with the increment of the hardness or strain hardening effect [24]. Thus, the higher value of hardness possess by IMC will reduce the occurrence of plastic deformation or creep effect compared to that of Au.

IV. CONCLUSION

The nanoindentation test has been carried out to analyse the creep effect of Au ball bond. Au and IMC showed creep behaviour compared to that of Si. The occurrence of creep for Au was due to the dislocation of grain in gliding manner while IMC was due to the increase of strain hardening effect. The responsible indentation creep mechanism for Au and IMC was dislocation glide mechanism based on the value of stress exponent obtained that ranging from 9.38 to 27.868.

ACKNOWLEDMENT

This work was sponsored by National University of Malaysia under research university grants (ERGS/1/2011/STG/UKM/02/10, OUP-2012-120, and UKM-RRR1-07-FRGS0257-2010).

REFERENCES

[1] W.D.V. Driel, R.B.R.V Silfhout, and G.Q. Zhang, "Reliability of wirebonds in micro-electronic packages," *Microelectron. Int.*, vol. 25/2, pp.15-22, 2008.

[2] Z.W. Zhong, "Fine and ultra-fine pitch wire bonding: challenges and solutions," *Microelectron. Int.*, vol. 26/2, pp. 10-18, 2009.

[3] V. Sundaraman, D.R. Edwards, W.E. Subido, H.R. Test, "Wire Pull on Fine Pitch Pads: An Obsolete Test for First Bond Integrity," *IEEE 50th Electronic Components and Technology Conf.*, pp. 416-420, 2000.

[4] G.G. Harman, C.A. Cannon, "The Microelectronic Wire Bond Pull Test-How to Use It, How to Abuse It," *IEEE T. Compon. Hybr.*, vol. 3, pp. 203-210, 1978.

[5] M. Petch, D. Barker, P. Lall, "Development of an Alternative Wire Bond Test Technique," *IEEE T. Compon. Pack. A*, vol. 17, pp. 610-615, 1994.

[6] Z.N Liang, F.G. Kuper M.S. Chen, "A concept to relate wire bonding parameters to bondability and ball bond reliability," *Microelectron. Reliab.*, vol. 38, pp. 1287-1291, 1998.

[7] R. Pantaleon, J. Sanchez-Mendoza, M. Mena, "Rationaliztion of Gold Ball Bond Shear Strengths," *IEEE 44th Electronic Components and Technology Conf.*, pp. 733-740, 1994.

[8] S.Murali, N, Srikanth, C.J. Vath III, "Effect of wire diameter on the thermosonic bond reliability," *Microelectron. Reliab.*, vol. 46, pp. 467-475, 2006.

[9] C.D Breach, F.W Wulff, "A brief review of selected aspects of the materials science of ball bonding," *Microelectron. Reliab.*, vol. 50, pp. 1-20, 2010.

[10] W.B. Li, J.L. Henshall, R.M. Hooper, K.E. Easterling, "The Mechanisms of Indentation Creep," *Acta Metall. Mater.*, vol. 39, 3099-3110, 1991.

[11] R. Goodall, T.W. Clyne, "A critical appraisal of the extraction of creep parameters from nanoindentation data obtained at room temperature," *Acta Mater.*, vol. 54, pp. 5489-5499, 2006.

[12] J. Alkorta, J.M. Martinez-Esnaola, J.G. Sevillano, "Critical examination of strain-rate sensitivity measurement by nanoindentation methods: Application to severely deformed niobium," *Acta Mater.*, vol. 56, pp. 884-893, 2008.

[13] Z.S. Ma, S.G. Long, Y.C. Zhou, Y. Pan, "Indentation scale dependence of tip-in creep behavior in Ni thin films," *Scripta Mater.*, vol. 59, pp. 195-198, 2008.

[14] Z.H. Cao, P.Y. Li, & X.K. Meng, "Nanoindentation creep behaviors of amorphous, tetragonal, and bcc Ta films," *Mat. Sci. Eng. A-Struct.*, vol. 516, pp. 253-258, 2009.

[15] Y. Liu, C. Huang, H. Bei, X. He, W. Hu, "Room temperature nanoindentation creep of nanocrystalline Cu and Cu alloys," *Mater. Lett.*, vol. 70, pp. 26-29, 2012.

[16] A. Jalar, M.N. Zulkifli, N.K. Othman, and S. Abdullah, "The Re-Evaluation of Mechanical Properties of Wire Bonding," *IEEE International Symposium on Advanced Packaging Materials (APM)*, PP. 226-233, 2012.

[17] W.C. Oliver, G.M. Pharr, "An improved technique for determining hardness and elastic modulus using load and displacement sensing indentation experiments," *J. Mater. Res.*, Vol. 7, pp. 1564-1583, 1992.

[18] A.C. Fischer-Cripps, Introduction to Contact Mechanics, Springer, New York, 2000.

[19] JEDEC Standard, JESD22-AIO3C, High Temperature Storage Life

[20] B.N. Lucas, W.C. Oliver, "Indentation Power-Law Creep of High-Purity Indium," *Metall. Mater. Trans. A*, vol. 30A, pp. 601-610, 1999.

[21] R. Mahmudi, R. Roumina, B. Raeisinia, "Investigation of stress exponent in the power-law creep of Pb-Sb alloys," *Mat. Sci. Eng. A-Struct.*, vol. 382, pp. 15-22, 2004.

[22] J. Chen, K. Lu, "Hardness and strain rate sensitivity of nanocrystalline Cu," *Scripta Mater.*, vol. 54, pp. 1913-1918, 2006.

[23] X.Y. Zhu, X.J. Liu, F. Zeng, F. Pan, "Room temperature nanoindentation creep of nanoscale Ag/Fe multilayers," *Mater. Lett.*, vol. 64, pp. 53-56, 2010.

[24] J. Song, Y. Shen, C. Su, Y. Lai, Y. Chiu, "Strain Rate Dependence on Nanoindetation Responses of Interfacial Intermetallic Compounds in Electronic Solder Joints with Cu and Ag Substrates," *Mater. T.*, vol. 50, pp. 1231-1234, 2009.

[25] G. Harman, Wire Bonding in Microelectronics, third ed., McGraw-Hill, New York, 2010.

[26] A. Jalar, M.N. Zulkifli, S. Abdullah, "Nanoindentation Test for the Strength Distribution Analysis of Bonded Au Ball Bonds," *Adv. Mat. Res.*, vol. 148-149, pp. 1163-1166, 2011.

The Effects of Mixed Electroluminescent (EL) Polymer Layer Thickness on the Single Layer Organic Light Emitting Diode (OLED) Performance

Mohd Shahrul Akram Mohd Mokhtar[1], Chi Chin Yap[2], Muhamad Mat Salleh[1], *Member, IEEE*, Akrajas Ali Umar[1], *Member, IEEE*, Muhammad Yahaya[2], *Member, IEEE*

[1]Institute of Microengineering and Nanoelectronics (IMEN)
Universiti Kebangsaan Malaysia (UKM)
43600 UKM Bangi, Selangor, Malaysia
[2]School of Applied Physics, Faculty of Science and Technology
Universiti Kebangsaan Malaysia (UKM)
43600 UKM Bangi, Selangor, Malaysia
E-mail: mms@ukm.my

Abstract- Single layer OLEDs by using polymer poly(9,9-di-n-hexylfluorenyl-2,7-diyl) (PHF) mixed with organic salt tetrabutylammonium hexafluorophosphate (TBAPF$_6$) as the electroluminescent (EL) layer were fabricated. This paper reports the effects of variation of that EL layer thickness on the OLED performance. OLEDs performance are determined in terms of their turn-on voltage and luminance. The PHF:TBAPF$_6$ EL layers with weight ratios of 100:20 were used and prepared by spin coating technique using chloroform as solvent. The thickness being incremented by adding layer-by-layer during the spin coating process. The thickness effects to the EL layer surface roughness were also studied. The best device performance was given by the thinnest layer which also has the lowest average surface roughness (0.557 nm), it gave the lowest turn-on voltage (7.0 V) and highest maximum luminance (9.12 cd/m^2 at 10 V). It emitted greenish-yellow color at it maximum luminance with CIE coordinate (0.30, 0.55).

I. INTRODUCTION

Organic light emitting diodes (OLEDs) are electronic thin film devices that sandwiched organic-based EL films layer between two charged electrodes, one as a cathode and one as a transparent anode. It will emit light when electrical voltage is applied through the layers. The attractive features of this technology are thin device structure, low weight, easy manufacturing process, flexibility, low cost, low operating voltages, large viewing angles, tunability of the color emission, fast response time and ease of forming large area, promising innovative electronic products, hence, gained a lot of attention and interest [1,2, 3]. OLEDs are widely used in large and small area full color flat-panel screen displays in many commercial products (computer, television, cellular phones, PDA and so on) and recently as lighting [4, 5, 6].

OLEDs can be fabricated in multilayer or single layer structure. Single layer OLED is a device containing only one organic EL layer sandwiched between a transparent conducting anode and a metal cathode. Even though the multilayer structure OLED gives better performance in terms of turn-on voltage and brightness, it requires higher cost in fabrication process as compared to the single layer structure OLED. The number of layers is little, so it decreases the cost of manufacturing and integration [7]. Therefore, it is valuable to improve the performance of low cost single layer structure OLED. Besides, the understanding of fundamental concept of OLED could be further strengthened by investigating the single layer structure OLED. The previous works have shown that organic salt doping [8, 9] and annealing process [10, 11] are among the approaches to improve the OLEDs performance in terms of turn on voltage, luminance, power efficiency and lifetimes. The usage of polymer give advantages over small molecule materials, such as it can be processed in solution form which results into simpler and cheaper fabrication cost [12].

Our previous work also proved that the mixing of OLED based on a blue light polymer emitting material, poly(9,9-di-n-hexylfluorenyl-2,7-diyl) (PHF) with organic salt, tetrabutylammonium hexafluorophosphate (TBAPF$_6$) has reduced the turn on voltage and increased the luminance [13]. This paper reports the effects of EL layer thickness variation on the ITO/PHF:TBAPF$_6$/Al OLEDs performances. With the increasing of EL layer thickness, the threshold voltage (V_{th}), turn on voltage (V_{on}), and the voltage needed to reach the maximum luminance (L_{max}) increased. The highest L_{max} was gave by the thinnest EL layer. Besides, the EL spectra's shape and the CIE coordinates were not much affected by the difference of EL layer thickness.

II. METHODOLOGY

The polymer emitting material, PHF and the organic salt, TBAPF$_6$ were purchased from Sigma-Aldrich Corporation and used as received without further purification.

OLEDs with structure of ITO/PHF:TBAPF$_6$/Al were fabricated (Figure 1). The ratio used for PHF:TBAPF$_6$ was 100:20. EL layer thickness was incremented by adding the number of layers deposited (during spin coating process). Five different EL layer thickness of PHF:TBAPF$_6$ were used, noted by BL1, BL2, BL3, BL4 and BL5, which correspond to the number of layers deposited. Glass substrates coated with indium tin oxide (ITO) were patterned by exposing it to the acid vapor, and then they were cleaned in acetone and 2-propanol for 15 minutes each in ultrasonic bath. The PHF:TBAPF$_6$ thin films were deposited on ITO anode using spin coating technique at 3000 rpm for 30 seconds. All thin films depositions were prepared in a glove box system with humidity level of below 18%. Finally the 150 nm thick aluminium as the cathode layer was deposited onto the emitting layer by electron-gun evaporation technique at a chamber pressure of 2.5 x 10^{-5} mbar. They were patterned using masks to create the emitting areas of 0.07 cm^2.

The thin films surface morphology's images and roughnesses were analyzed using Atomic Force Microscopic (AFM). Meanwhile, the EL layer thicknesses were measured using Surface Profiler DEKTAK 150. The current density-voltage (J-V) measurements were taken using Keithley 238 source measurement unit and the electroluminescence (EL) spectra of the OLEDs were obtained by Ocean Optic HR2000 spectrometer. The photoluminescence (PL) spectra of PHF and PHF:TBAPF$_6$ were collected using Perkin Elmer LS 55 Luminescence Spectrometer. The devices characterizations processes were carried out in a dark room.

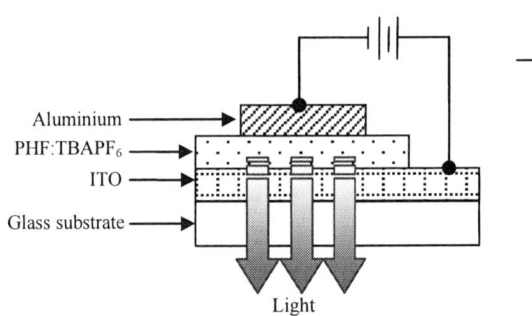

Fig. 1. The schematic structure of a single-layer OLED device

III. RESULTS AND DISCUSSION

Figure 2 shows the photoluminescence spectra of PHF and PHF:TBAPF$_6$ thin film. It was obtained that the emissions for all solutions are in blue-range with the emission peak of PHF and PHF:TBAPF$_6$ located at 443 nm and 441 nm respectively. This result indicated that the presence of TBAPF$_6$ gave no significant effect to the emission peak.

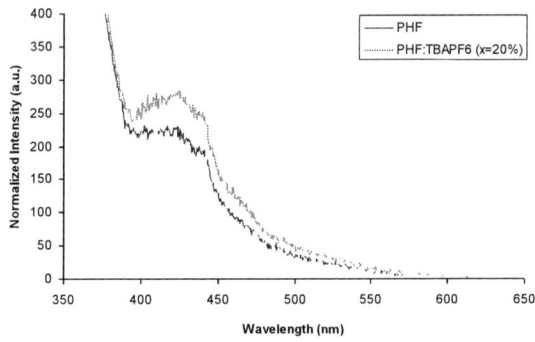

Fig. 2. Photoluminescence (PL) spectra of PHF and PHF:TBAPF$_6$ thin film

Thickness of BL1, BL2, BL3, BL4 and BL5 EL layers were varied from 38nm to 67nm. The J-V curves of all fabricated devices are shown in Fig. 3, which exhibit rectifier diode behavior. The threshold voltage (V$_{th}$) of the BL1 to BL5 devices were 7.0V, 13.0V, 15.5V, 16.0V and 12.0V, respectively. This means that the best performance in term of V$_{th}$ was given by the thinnest EL layer which is BL1. Generally, the V$_{th}$ decreased accordingly with the EL layer thickness.

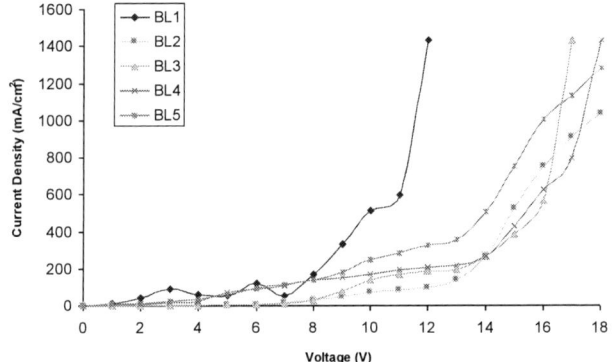

Fig. 3. Current density-voltage (J-V) curves of ITO/PHF:TBAPF$_6$/Al devices with variation of EL layer thickness

Figure 4 shows the luminance-voltage (L-V) characteristics for each device. From this L-V curve, we can determine the V$_{on}$ and L$_{max}$ of each device. V$_{on}$ was defined as the voltage applied when the luminance reach 1 cd/m^2. A line was put at 1 cd/m^2 on the graph. The V$_{on}$ of BL1 to BL5 devices were 7V, 11V, 13V, 14V and 17V, respectively. The device with the thinnest EL layer (BL1) gave the best V$_{on}$ (lowest) and also highest L$_{max}$ (9.12 cd/m^2 at 10 V), followed by BL3 (2.68 cd/m^2 at 16 V), BL5 (1.75 cd/m^2 at 18 V), BL4 (1.57 cd/m^2 at 17 V), and BL2 (1.33 cd/m^2 at 12 V). After reaching the maximum luminance, all the curves dropped tremendously, indicating the degradation of the devices. The thinnest EL film gave lowest voltage needed to reach the L$_{max}$.

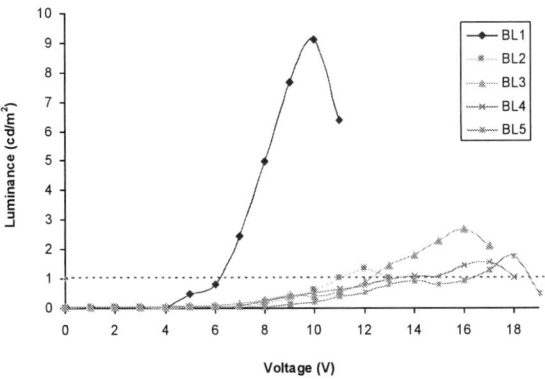

Fig. 4. Luminance-voltage (L-V) curves of ITO/PHF:TBAPF$_6$/Al devices with variation of EL layer thickness

Figure 5 shows the EL spectral of all devices measured at L$_{max}$. The thinnest EL layer, which is BL1, gave the EL peak at 491 nm. The result indicated that the thickness greater than BL1 (38 nm) has shifted the EL peaks to the right (in range 550 nm – 600 nm). Table 1 summarizes the performance of OLEDs with five variations of EL thickness. OLED achieve the optimum performance when the thinnest EL layer was deposited, which is BL1. It gave the best performance in terms of V$_{th}$, V$_{on}$ and L$_{max}$.

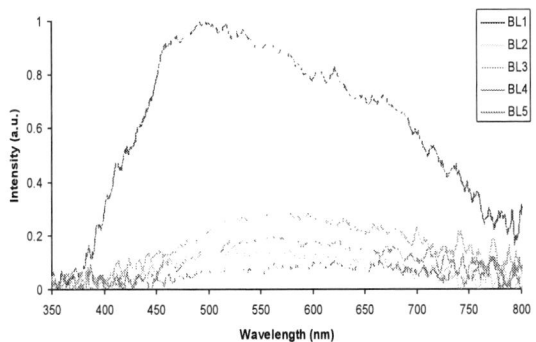

Fig. 5. EL spectra of ITO/PHF:TBAPF$_6$/Al devices at maximum brightness with different of EL layer thickness

TABLE 1
COMPARISON RESULTS OF THE ITO/PHF:TBAPF$_6$/AL DEVICES WITH DIFFERENT WEIGHT RATIOS.

Sample	V$_{th}$ (V)	V$_{on}$ (V)	L$_{max}$ (cd/m^2)	CIE coordinates (at L$_{max}$)	Color (at L$_{max}$)
BL1	7.00	7.00	9.12	0.30, 0.55	Greenish-yellow
BL2	13.00	11.00	1.33	0.46, 0.47	Orange
BL3	15.50	13.00	2.68	0.47, 0.46	Orange
BL4	16.00	14.00	1.57	0.48, 0.46	Orange
BL5	12.00	17.00	1.75	0.48, 0.47	Orange

Device performance of ITO/PHF:TBAPF$_6$/Al has deteriorated with increasing thickness of the thin film, because the thickness increment led to deterioration morphology of the film and increase the electrical resistivity. Deterioration morphology of the film is said to occur when the average roughness (R$_{ave}$) of the film surface increased. R$_{ave}$ of the first

film layer influences the formation of the second layer, and so on to the next layer - which affected the R$_{ave}$ to increase with the increment of the EL layer thickness, generally. Figure 6 shows the film morphology of five different EL layer thicknesses, taken on 3.0 × 3.0 μm^2 surface area, sample: a) BL1, b) BL2, c) BL3, d) BL4 and e) BL5.

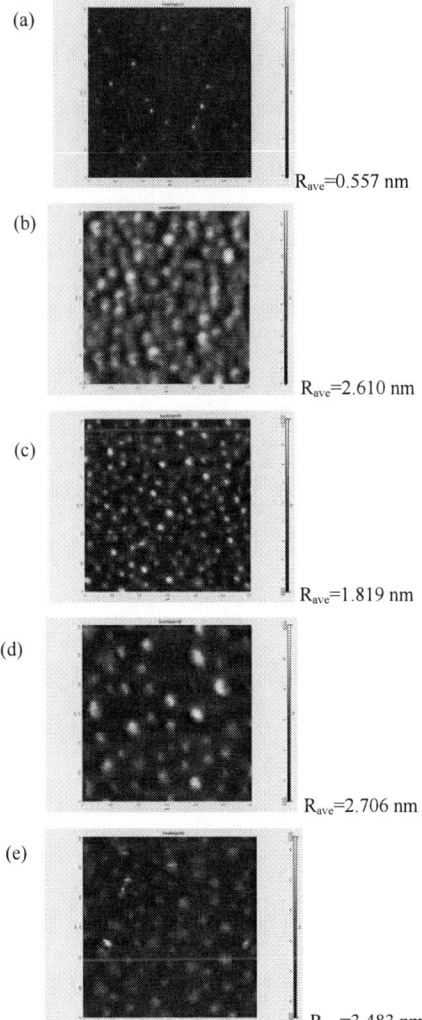

Fig. 6. AFM images of EL thin film (PHF:TBAPF$_6$) surface of five different thicknesses

The mechanism of the effect of organic salt mixing in the OLED device may be explained through Ionic space charge and Fowler-Nordheim tunneling injection [6]. At the ITO anode contact, the accumulation of separated negative PF$_6^-$ ions can assist hole injection from ITO to the emitting layer. At the same time, the accumulation of the positive TBA$^+$ ions near the cathode aids the injection of electrons from the Al cathode to the emitting layer by reducing the tunneling barrier of the cathode interface. Figure 6 represents the schematic diagrams of charge injection mechanism at cathode (Al) and organic layer interface for (a) unmixed device and (b) mixed devices with variation of TBAPF$_6$ mixing ratios, respectively in forward-bias state. Figure 6(b) energy level of emitting layer corresponding to TBAPF$_6$ mixing ratios of 100:20. The presence of the TBAPF$_6$ gives smaller shaded area and thus smaller electron injection barrier area. Therefore, the electron tunneling probability increases and more electrons are able to overcome the injection barrier and tunnel into the emitting layer. The same mechanism happens to the holes at the anode. It resulted into more electrons and holes tunnel into the emitting layer of the doped device for the same value of voltage applied. When more recombination of electron and holes occur, the luminance is significantly higher. The reduction of barrier at interface may also reduce the turn-on voltage.

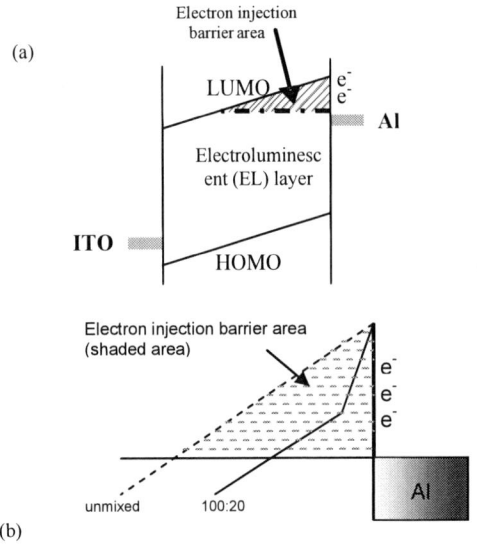

(a)

(b)

Fig. 6. Schematic diagram of charge injection mechanism for (a) unmixed and (b) mixing ratios of 100:20 devices.

IV. CONCLUSION

Single layer OLED devices with structure of ITO/PHF:TBAPF$_6$ /Al were fabricated with five different EL layer thickness. It was found that the thinnest EL layer gave the best OLED performances. Lower threshold voltage, turn on voltage and higher brightness were achieved. Meanwhile, there was no significant effect on the color emitted. The optimum ITO/PHF:TBAPF$_6$(100:20)/Al device gave the lowest turn on voltage (7 V) and the highest brightness (9.12cd/m^2 at 10V). The increase in film thickness has also caused a decrease

quality in film morphology and increase the electrical resistivity. Deterioration morphology of the film is said to occur when there is an increase in average surface roughness of the film.

ACKNOWLEDGMENTS

The authors wish to thank to the Malaysian Ministry of Higher Education and Universiti Kebangsaan Malaysia for financial assistance through grants of UKM-GUP-BTT-07-26-178 and UKM-RRRI-07-FRGS0037-2009.

REFERENCES

[1] S. R. Forest, *Organic Electronics*, 4, 45-48 (2003).
[2] B. Young, *Information Display* 25(9), 14-17 (2009). [1] S. R. Forest, *Organic Electronics*, 4, 45-48 (2003).
[3] Klaus Meerholz, Christoph-David Müller and Oskar Nuyken, "Crosslinkable organic semiconductors for use in organic light-emitting diodes (OLEDs)", in *Organic light-emitting devices: synthesis properties and application*, Klaus Müllen and Ullrich Scherf, Eds. Germany: Wiley-VCH, 2006, pp.293-318.
[4] M. Eritt, C. May, K. Leo, M. Toerker and C. Radehaus, *Thin Solid Films* 518, 3042-3045 (2010).
[5] Xiong Gong, Daniel Moses and Alan J. Heeger, "Polymer-based light-emitting diodes (PLEDs) and displays fabricated from arrays of PLEDs", in *Organic light-emitting devices: synthesis properties and application*, Klaus Müllen and Ullrich Scherf, Eds. Germany: Wiley-VCH, 2006, pp.151-180.
[6] Hong Meng and Norman Herron, "Organic small molecule materials for organic light-emitting diodes", in *Organic light-emitting materials and devices*, Zhigang Li and Hong Meng, Eds. Boca Raton: CRC Press Taylor & Francis Group, 2007, pp.295-412.
[7] Sue A. Carter, "Print-based manufacturing technologies for organic light-emitting displays", in *Organic light-emitting materials and devices*, Zhigang Li and Hong Meng, Eds. Boca Raton: CRC Press Taylor & Francis Group, 2007, pp.567-581.
[8] C.C. Yap, M. Yahaya and M.M. Salleh, *Current Applied Physics* **8**, 637–644 (2008).
[9] C.C. Yap, M. Yahaya and M.M. Salleh, *Current Applied Physics* **9**, 722–726 (2009).
[10] Y.-L. Hua, J. Li, X.-L. Feng and X.-M. Wu, *Guangdianzi Jiguang/Journal of Optoelectronics Laser* 15, 516-519 (2004).
[11] S. Sepeai, M.M. Salleh, M. Yahaya and A.A.Umar, *Thin Solid Films*, 517, 4679-4683 (2009).
[12] Gang Yu and Jian Wang, "Organic light-emitting devices and their applications for flat-panel displays", in *Organic light-emitting materials and devices*, Zhigang Li and Hong Meng, Eds. Boca Raton: CRC Press Taylor & Francis Group, 2007, pp.567-581.
[13] M.S.A.M. Mokhtar, M.M. Salleh, A.A. Umar, C.C. Yap and M. Yahaya, "Effect of Organic Salt Doping on The Performance of Poly(9,9-di-n-hexylfluorenyl-2,7-diyl) Organic Light Emitting Diode, OLED" in *The 3rd Nanoscience and Nanotechnology Symposium 2010, edited by* M. Abdullah et al., AIP Conference Proceedings 1284, American Institute of Physics, Melville, NY, 2010, pp. 80-82.

Fabrication of CuGaO₂ Films by Sol-gel Method for UV Detector Application

Afishah Alias[1*], Khairul Anuar Mohamad[1], Bablu Kumar Gosh[1]
Masato Sakamoto[2], Katsuhiro Uesugi[2]

[1]Nano Engineering & Materials (NEMs) Research Group, School of Engineering and Information Technology,
Universiti Malaysia Sabah, 88400 Kota Kinabalu, Sabah, Malaysia
[2]Faculty of Engineering, Muroran Institute of Technology,
27-1 Mizumoto, Muroran 050-8585 Hokkaido, Japan
*Email: afishah79@yahoo.co.jp

Abstract- **Cu-based conductive oxide such as CuGaO₂ is promising for transparent p-type oxide material. In this work, the CuGaO₂ film has been fabricated using liquid-phase sol-gel method. The sol-gel derived CuGaO₂ films showed p-type conductivity, and high transparency with transmittance of about 80% in the visible light region. The energy gap for direct allowed transition was about 3.6 eV. Furthermore, we also investigated the potential application of CuGaO₂ films by measure the drain current in the dark and under a UV light illumination. Upon illumination with UV light, a significant increase in the drain current was observed which indicates that the charge carrier was excited when energy with more than energy gap was applied.**

I. INTRODUCTION

Transparent oxide semiconductors such as ZnO and SnO₂ have been widely used in optoelectronics areas owing to their high optical transparency, low resistivity, and wide energy band gap [1]. The notable application is for transparent electrodes and thin-film transistors (TFTs) of flat panel displays and light-emitting diodes [1,2]. However, most of the oxide semiconductors have n-type conductivity [3]. The development of p-type materials is much desired as this would open up the applications of new transparent devices. Recently, Cu-based oxide semiconductors such as delafossite CuMO₂ (M=Al, Ga, In) have been widely studied for p-type materials using various deposition methods [4]. Ueda et al. reported that CuGaO₂ films were epitaxially grown on α-Al₂O₃(001) using pulsed laser deposition (PLD), and the band gap energy and hole concentration of CuGaO₂ films were about ~3.6 eV and 1.7×10^{18} cm^{-3}, respectively [2]. Mine et al. reported the control of hole concentration in the films from 10^{14} to 10^{17} cm^{-3} by varying the O₂ pressure during PLD [5]. However, the fabrication of CuGaO₂ films using vapor- and liquid-phase growth methods such as chemical vapor deposition, metal-organic molecular beam epitaxy, and sol-gel methods has not been reported yet because stable Cu sources are not available.

In this study, we demonstrated the sol-gel processing of CuGaO₂ films using 2-propanol solutions of copper(II) acetate monohydrate [Cu(CH₃COO)₂·H₂O] and tris(acetylacetonato) gallium(III) [Ga(C₅H₇O₂)₃]. The temperature dependences of the stabilization of sol solutions and the crystallization of sol-gel-derived films were investigated to fabricate p-type CuGaO₂ films. Further, we investigated the potential application of CuGaO₂ films as ultraviolet (UV) detector by measure the drain current in the dark and under a UV light illumination.

II. DEVICE CONCEPT AND FABRICATION

A. Materials

The precursor for the growth of CuGaO₂ films was prepared by mixing two separates 2-propanol solutions of copper(II) acetate monohydrate [Cu(CH₃COO)₂·H₂O] and tris(acetylacetonato) gallium(III) [Ga(C₅H₇O₂)₃]. Both [Cu(CH₃COO)₂·H₂O] and [Ga(C₅H₇O₂)₃] powders were purchased from Kanto Chemical Co. and used as received. Monoethanolamine (MEA) was also purchased from Kanto Chemical Co. and used without further purification.

B. Fabrication of CuGaO₂ films by sol-gel method

The substrates used were n⁺-Si(100), with a resistivity of 0.1 Ω cm, 200-nm-thick SiO₂ layers thermally grown on the n⁺-Si(100) and silica glass substrates. CuGaO₂ films were fabricated on the substrates by the sol-gel method. Initially, Cu-O and Ga-O sol solutions were prepared separately. 0.4 M Cu(CH₃COO)₂·H₂O and Ga(C₅H₇O₂)₃ were dissolved in 10 ml 2-propanol aqueous solution. As a sol stabilizer, 2 M MEA was also added into the solutions. After each solution was stirred at 50 °C for 1 h, they were kept at room temperature (23 °C) for 24 h. Stable medium blue Cu-O and light yellow Ga-O sol solutions were obtained. Sol solutions for CuGaO₂ growth were fabricated by the mixing of these Cu-O and Ga-O sol solutions in Cu/Ga=1 atomic ratio. After the Cu-Ga-O sol solution was spin-coated on the substrates at room temperature, they were prebaked in two steps: 100 °C for 5 min, followed by 300 °C for 10 min under N₂ ambient for the gelation process. After cooling, these processes were repeated two times. The gel films were crystallized by postbaking at a temperature between 650 and 950 °C under N₂ ambient. The typical film thickness prepared in this way was about 80-90 nm.

C. Characterization

X-ray diffraction (XRD) spectrometer (RIGAKU) was used to characterize the crystallographic orientation of the films. The

978-1-4673-2395-6/12 $31.00 © 2012 IEEE

X-ray beam was generated by a Cu Kα radiation, using a tube voltage of 40 kV at an electron beam current of 40 mA. The scanning angle was increased in steps of 0.02° over the range of 0-60°. Meanwhile, the surface morphology of the films was observed by atomic force microscopy (AFM) (Seiko SPA-300). The optical property such as the transmittance measurement of the films was performed using an 1800 ultraviolet-visible (UV-VIS) absorption spectrometer (Seiko). The films were scanned between 200 and 1100 nm.

The transistor structure was fabricated on (100)-oriented Sb-doped n^+-type Si wafer (< 0.1 Ω cm), which was used as the substrate and gate contact. The substrates were rinsed by ultrasonic in deionized water, ethyl alcohol, acetone, and then methyl alcohol, and cleaned by standard RCA cleaning procedure. A thermally oxidized 200-nm-thick SiO_2 layer formed by dry oxidation was used as a gate dielectric. Source- and drain-Au electrodes were fabricated on sol-gel-derived $CuGaO_2$ films by the vacuum deposition method (channel width, W=5mm and channel length, L=100 μm). The schematic diagram of the bottom gate with top drain-source contact structure for $CuGaO_2$ TFT is shown in Fig. 1. The transistor characteristics of the $CuGaO_2$ TFTs were measured at room temperature in a shield box using a computer-controlled automatic electrical analyzer (Measure Jig MI-494) in the dark and under UV light illumination.

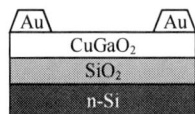

Fig. 1. Schematic diagram of bottom gate with top drain-source contact structure for $CuGaO_2$ TFT

III. CHARACTERIZATION RESULTS

A. Characterization of $CuGaO_2$ films

Fig. 2 shows the XRD patterns of the films crystallized at 800 °C for (a) 1 and (b) 5 h. The XRD peak of $CuGaO_2$ (012) improved with longer postbake duration, but a broad peak indicated by the arrow was observed. This peak is unidentified, but may be due to the formation or degradation of undesirable phases by the longer postbake duration. The $CuGa_2O_4(222)$ peak was not dependent on the annealing duration.

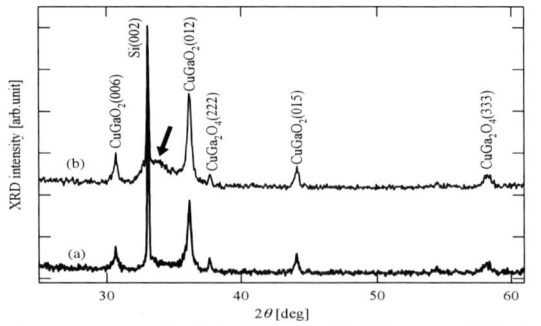

Fig. 2. XRD patterns of $CuGaO_2$ films crystallized for (a) 1 and (b) 5 h at 800 °C

Fig. 3 shows typical AFM images of these films. The surfaces were covered with grains and the root-mean-square (RMS) values for the films postbaked (a) 1 and (b) 5 h were 1.7 and 2.2 nm, respectively. The 1-h-postbaked sample showed uniform grains, while the 5-h-postbaked sample showed the formation of large grains, as shown in Fig. 3(b), which may be caused by the degradation on the surface with longer postbake, which is assumed as a $CuGaO_2$ phase and also an undesirable phase that is unidentified.

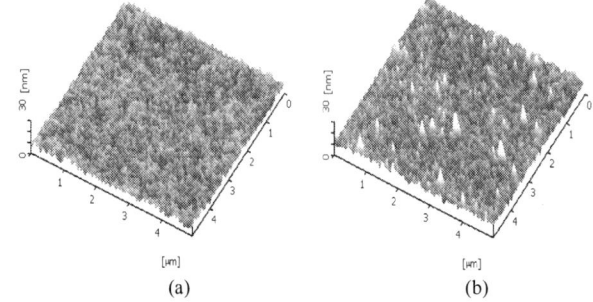

Fig. 3. AFM images of $CuGaO_2$ films postbaked for (a) 1 and (b) 5 h

Fig. 4 shows a typical transmission spectrum of the sol-gel derived $CuGaO_2$ film on a silica substrate postbaked at 800 °C for 1 h. The films show a high transparency with a transmittance of about 80% in the visible range. The inset to Fig. 4 shows a $(\alpha E)^2$ versus energy plot for the estimation of absorption edge energy. The energy gap of the films was estimated to be about 3.6 eV, which is consistent with that of the films fabricated by the PLD method [2]. As shown in Fig. 4, the absorption edge of the films is less sharp, which is due to the large activation energy of 0.18-0.22 eV and the existence of the $CuGa_2O_4$ phase [6]. To improve the optical and electrical properties of sol-gel derived $CuGaO_2$ films, the formation of a $CuGa_2O_4$ phase in the films with an energy gap of 1.38 eV has to be suppressed [7].

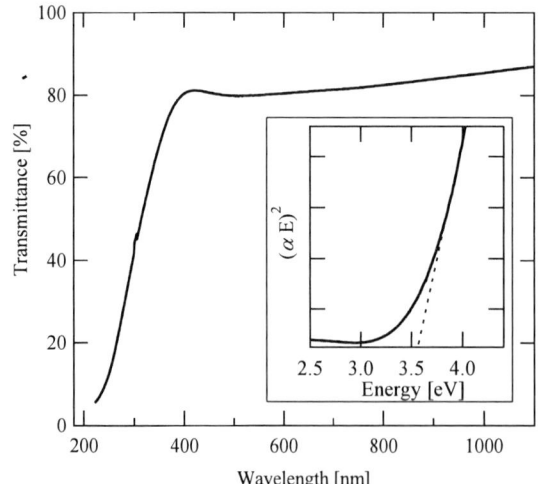

Fig. 4. Transmission spectrum of the sol-gel derived $CuGaO_2$ film on the silica substrate

B. Transistor characteristics of CuGaO₂ films

The CuGaO₂ films on the SiO₂/n⁺-Si substrates were crystallized at 800 °C for (a) 1 and (b) 5 h. The resistivities obtained for the films were (a) 1.1×10^6 and (b) 5.5×10^5 Ω cm, respectively. The Hall measurement was not carried out owing to the high resistivity. Figure 4 shows the electrical-transfer characteristics of the CuGaO₂-based TFT structure at a constant source-drain voltage V_{ds} of −10 V. The property in the negative gate bias V_g shows that the hole dominated transport properties. This strongly proved that p-type CuGaO₂ was fabricated using this sol-gel processing. The field effect mobility, μ was estimated using the equation [8]

$$\left.\frac{\delta I_{ds}}{\delta V_{ds}}\right|_{V_{ds}=\text{constant}} = \frac{WC_i\mu V_{ds}}{L} \qquad (1)$$

where I_{ds} is the drain-source current, V_g is the gate voltage, W is the channel length, C_i is the capacitance per unit area of the SiO₂ layer, V_{ds} is the drain-source voltage, and L is the channel length. The field-effect mobility estimated from the linear region in Fig. 5 was (a) 6.3 and (b) 8.9 cm² V⁻¹s⁻¹. The 5-h-postbaked sample showed slightly higher mobility and lower resistivity, which may be due to the improvement of the CuGaO₂(012) crystal quality, as shown in Fig. 2.

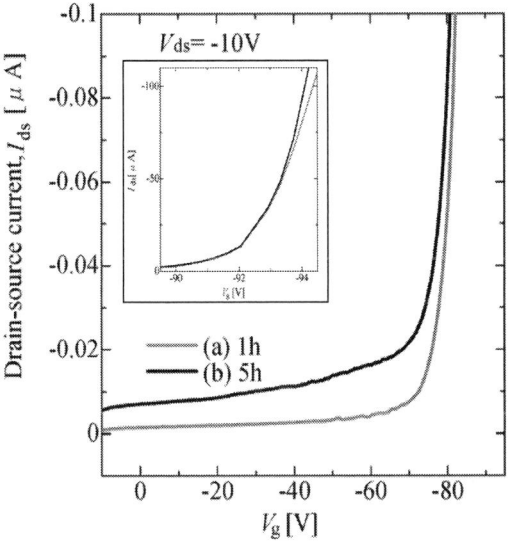

Fig. 5. Transfer characteristics of the CuGaO₂-based TFT structure at a constant source-drain voltage V_{ds} of -10 V

Fig. 6 shows the output characteristic of CuGaO₂ films measured at room temperature in dark and under UV light illumination. UV light with 254 nm wavelength, which about 4.88 eV was supplied. Significant different was seen in drain current when measured under UV light illumination. Indicating the carrier is excited when energy with more than energy gap is supplied. This shows that the CuGaO₂ films have a potential as UV detector as it can detect the UV light with more that its energy gap.

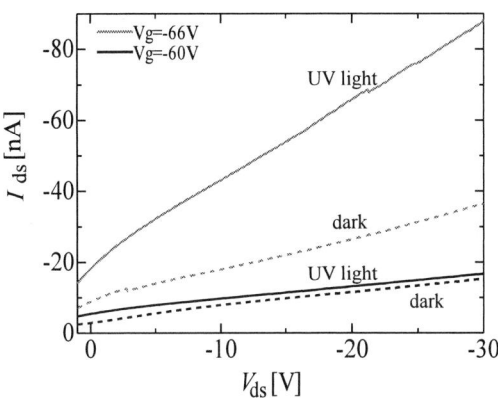

Fig. 6. The output characteristic of CuGaO₂ films at $V_g = -60$ and -66 V in the dark and under UV light illumination

IV. CONCLUSIONS

We fabricated CuGaO₂ films using sol-gel method. CuGaO₂ films with energy gap 3.6 eV were obtained by annealing at 800 °C for 1 h under N₂ ambient at 1 atm. Transmittance of CuGaO₂ films was about >80% at visible region. The output characteristic has proved its potential as UV detector in the future as it can detect the UV lights with more that its energy gap.

ACKNOWLEDGMENT

This study was supported by ERGS (ERGS0023-TK-1/2012) fund from Ministry of Higher Education (MOHE) Malaysia. The author is thankful to the Muroran Institute of Technology at Hokkaido, Japan for providing research facility in which to complete this work.

REFERENCES

[1] A. N. Banerjee and K. K. Chattopadhyay, "Recent developments in the emerging field of crystalline p-type transparent conducting oxide thin films," *Prog. Cryst. Growth Charact. Mater.*, vol. 50, pp. 52-105, November 2005.

[2] K. Ueda, T. Hase, H. Yanagi, H. Kawazoe, H. Hosono, H Ohta, M.Orita, and M. Hirano, "Epitaxial growth of transparent p-type conducting CuGaO₂ thin films on sapphire (001) substrates by pulsed laser deposition," *J. Appl. Phys.*, vol. 89, pp. 1790-1793, November 2000.

[3] Z. J. Fang, C. Fang, L. J. Shi, Y. H. Liu, and M. C. He, "First-principles study of defects in CuGaO₂," *Chin. Phys. Lett.*, vol. 25, pp. 2997-3000, April 2008.

[4] H. Kawazoe, M. Yasukawa, H. Hyodo, M. Kurita, H. Yanagi, and H. Hosono, "P-type electrical conduction in transparent thin films of CuAlO₂," *Nature*, vol. 389, pp. 939-942, September 1997.

[5] T. Mine, H. Yanagi, K. Nomura, T. Kamiya, M. Hirano, and H. Hosono, "Control of carrier concentration and surface flattening of CuGaO2 epitaxial films for a p-channel transparent transistor," *Thin Solid Films*, vol. 516, pp. 5790-5794, July 2008.

[6] N. A. Ashmore and D. P. Cann, "Electrical and structural characteristics of non-stoichiometric Cu-based delafossites," *J. Mater. Sci.*, vol. 40, pp. 3891-3896, August 2005.

[7] K. Gurunathan, J. O. Baeg, S. M. Lee, E. Subramaniam, S. J. Moon, and K. J. Kong, "Visible light active pristine and Fe³⁺ doped CuGa₂O₄ spinel photocatalysts for solar hydrogen production," *Int. J. Hydrogen Energy*, vol. 33, pp. 2646-2652, June 2008.

[8] C. R. Kagan and P. Andry, *Thin-Film Transistors*, Marcel Dekker, Inc. New York, 2003, pp. 343.

The Parasitic Reaction During the MOCVD Growth of AlInN Material

Wei-Ching Huang, Yuen-Yee Wong, Kusan-Shin Liu, Chi-Feng Hsieh, Edward Yi Chang, *Senior Member, IEEE*

Department of Materials Science and Engineering
National Chiao Tung University
1001 University Rd., Hsinchu, 30010 Taiwan
Email: weiching0928@gmail.com

Abstract- For observation of parasitic reaction effect during InAlN material growth by Matel-Organic Chemical Vapor Deposition (MOCVD), the growth parameters include temperature and pressure had been varied to investigate it. The pressure would be kept at 50torr, 100torr and 150torr and temperature was varied from 700 °C to 780 °C by 20 °C step in each growth pressure. The experimental results appeared that higher pressure gave rise to more serious parasitic reaction during material growth. It made the less Al atoms incorporate into the AlInN. In addition the 100-torr growth pressure shows the best efficiency of Al atom incorporation compare with other two growth pressures. By the AFM analysis, the morphology of Al0.82In0.18N grown with three different pressures was also examined.

I. INTRODUCTION

In recent years, the AlInN material had been attracted more and more attentions due to the capability of lattice match with GaN and larger spontaneous polarization. When the In composition at 18%, its lattice constant can perfectly match to GaN. This feature would reduce undesirable degradations of performance in the AlInN/GaN HEMT. On the other hand, larger spontaneous polarizations also can provide the higher carrier concentration in the two-dimensional election gas (2DEG) channel even without the piezoelectric polarization. Therefore, it had became the most potential candidate for replacing the AlGaN barrier in the AlGaN/GaN high electron mobility transistor (HEMT) to be the next generation high power and frequency device. However, the extremely differences of the growth condition between AlN and InN make a difficulty in material growth. In general, during the growth of Al-contained material, one issue was appeared. That was the parasitic effect. The parasitic effect was a vapor reaction between the ammonia
(NH3) and trimethylaluminum (TMAl)[1]. When this effect occurred, pre-reaction of AlN will be formed before reaching the reaction surface and didn't contribute to the growth[2]. In this article, the parasitic effect during the growth of AlInN by mocvd was investigated. By adjusting the growth pressure and growth temperature, the influence of parasitic effect on the alloy composition in the AlInN is observed. Through AFM analysis, the morphology of a fixed-composition InAlN was also examined.

II. EXPERIMENTAL

The AlInN material was regrown on the GaN templates by metal-organic chemical vapor deposition (MOCVD). The GaN templates were grown on the c-plane sapphire substrate by MOCVD. It consisted of 25-nm thick low temperature GaN, which was grown at 530 °C and 2-μm undoped GaN, which was grown at 1020 °C. H_2 was the carrier gas for GaN template growth and N_2 was the carrier gas for AlInN material growth. During the growth of AlInN, except the temperature and pressure, all other parameters were kept at a constant, respectively. The variation of growth pressure was from 150torr to 50torr with 50torr in each step. In addition, temperature were also changed from 780°C to 700°C with 20°C under each growth pressure. The aluminum composition and crystal quality of AlInN was estimated in ω-2Θ and ω (rocking curve) scan mode by the high-resolution X-ray diffraction (HRXRD). By estimating the aluminum composition in the AlInN under different growth pressure and temperature, the influence of the parasitic effect on the growth of AlInN could be understood.

III. RESULT AND DISCUSSION

In order to discuss the parasitic reaction during AlInN material growth, the influence of growth parameter include pressure and temperature would be investigated. Fig.1 showed the dependence of Al composition with the growth pressure. The XRD results indicated that less Al atoms could be incorporated into the AlInN because of enhancement of parasitic reaction as growth pressure was increased. At higher growth pressure, the boundary layer above the heated-substrate became thicker[3] and more serious parasitic reaction would occur inside the boundary layer. Therefore many Al atoms reacted with nitrogen in this boundary before reaching the growth surface and did not incorporate into the AlInN.

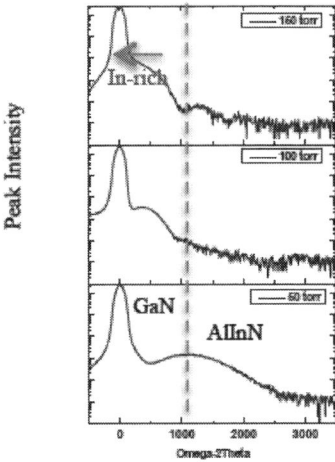

Fig. 1 The dependence of Al composition with growth pressure. Parasitic effect was dominated at higher growth pressure.

Beside the growth pressure, the effect of different temperature under the growth pressures of 50,100 and 150 torr was shown in the Fig.2. At growth pressure of 50torr, the Al composition under all temperature range was the higher than the other two samples grown at 100 torr and 150 torr, respectively. It pointed out that the influence of parasitic effect is less at lower growth pressure. Because the surface kinetically mechanism can overcome the gas phase reaction and becomes dominated at lower growth pressure[4]. Thus more Al atoms could avoid the pre-reaction to approach surface and contributed to the growth of AlInN. On the other hand, by observation of slope of the line in the Fig.2 the experimental results appeared that the largest changes of Al% with temperature at the growth pressure of 100torr. This achievement indicated that the incorporation of Al atoms would be easier under 100torr pressure.

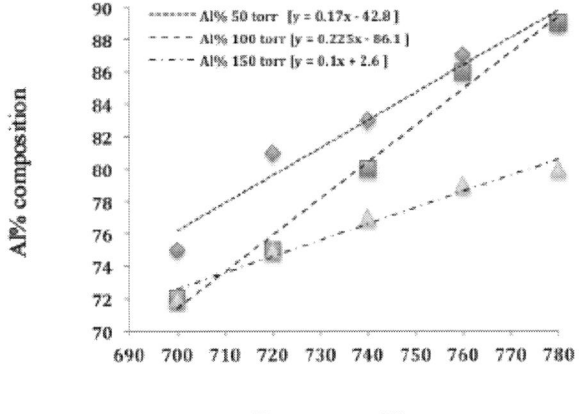

Fig. 2 The Al composition of AlInN with different growth temperature, which were grown under the pressure of 50,100 and 150 torr, respectively.

Fig.3 indicates the crystal quality of AlInN in (0002) plane. Under these three different growth pressure, the results show that the better crystal quality always appears at 17%~20%

indium composition. For AlInN material, as the indium composition at 18% the lattice of AlInN can match with GaN. Thus the defects and strain were least. Therefore, the better crystal quality can be obtained in this composition range.

(a)

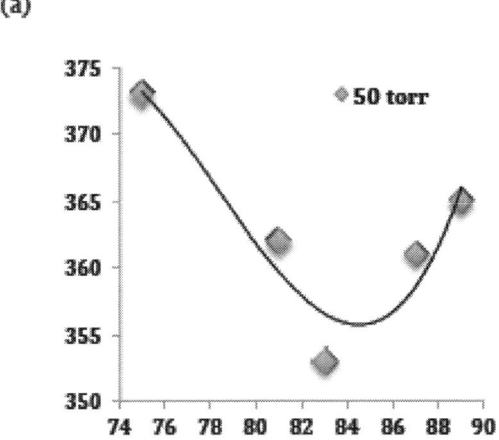

(b)

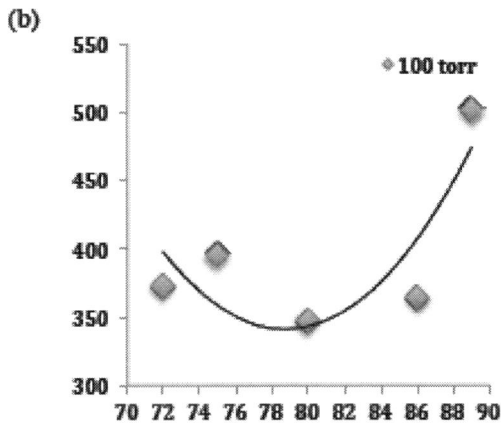

(c)

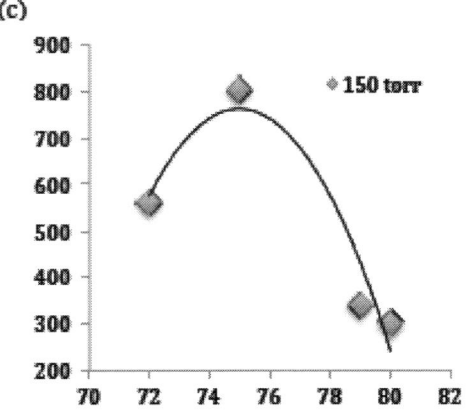

Fig. 3 (0002) rocking curve of AlInN with different Al composition under 50torr, 100torr and 150 torr growth pressure. (a) 50 torr (b)100 torr (c) 150 torr

The morphology of Al0.82In0.18N surface grown with different growth pressures were also examined by the AFM 5μm×5μm scans. As Fig.4 showed, the circle indicated that the quantum-dot like grain appeared on the surface. At a fixed Al composition, the density and size of these grains increased with the growth pressure. These sizes of grains were examined by the extract line scans in the AFM. Table I show the results of grain size analysis. The size of grain could be increased by enhancing pressure, especially for the size of lateral direction. While the pressure was increased to 150torr the size of lateral direction would be increased from 156.25nm to 507.81nm, however in the size of vertical was only slightly increased from 1.062 nm to 1.913nm. It could be concluded that the higher pressure promoted the 2D(lateral direction) growth mode.

TABLE I
THE SIZE OF GRAIN IN LATERAL AND VERTICAL DIRECTION WITH 50-TORR, 100-TORR AND 150-TORR PRESSURE. WITH 150-TORR GROWTH PRESSURE SHOWS THE LARGE ENHANCEMENT OF LATERAL SIZE AND SLIGHT INCREASE IN VERTICAL SIZE.

Growth pressure	Lateral(nm)	Vertical(nm)
50 torr	156.25	1.062
100 torr	224.61	1.065
150 torr	507.81	1.913

IV. CONCLUSION

In conclusion, the influence of parasitic effect on theAl-InN material growth by MOCVD has been investigated by adjusting growth pressure and temperature. Lower pressure could suppress the parasitic reaction effectively and make the more Al atoms incorporated into AlInN. In addition, it was found that the pressure of 100torr had the highest Al incorporation efficiency and the crystal quality was always better as indium composition at 17%~20%, because the ideal match in lattice between AlInN and GaN. At a fixed composition, the morphology of AlInN tended to have 2D growth mode as pressure increased. For these quantum-dot like grain, further study to investigate the principle of formation of these quantum-dot like grain would be proceed

REFERENCE

[1] Chen, C.H., et al., *A study of parasitic reactions between NH3 and TMGa or TMAl.* Journal of Electronic Materials, 1996. **25**(6): p. 1004-1008.

[2] Kondratyev, A.V., et al., *Aluminum incorporation control in AlGaN MOVPE: experimental and modeling study.* Journal of Crystal Growth, 2004. **272**(1-4): p. 420-425.

[3] F. C. Everstyn, P.J.W.S., C. H. J. van den Brekel, and H. L. Peek, J. Electrochem. Soc., 1970. **119**: p. 925.

[4] Zhao, D.G., et al., *Parasitic reaction and its effect on the growth rate of AlN by metalorganic chemical vapor deposition.* Journal of Crystal Growth, 2006. **289**(1): p. 72-75.

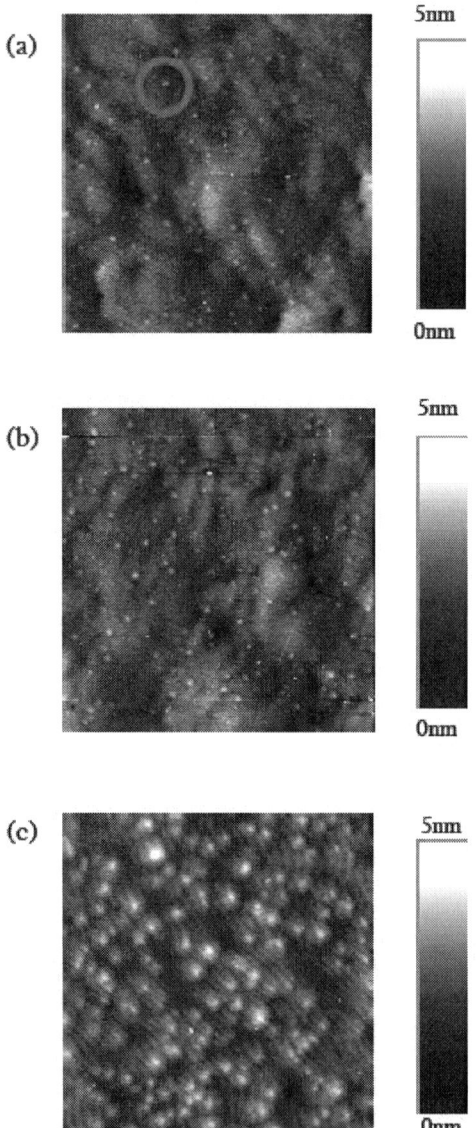

Fig.4 The AFM analysis of Al0.82In0.18N surface by 5×5 scans with (a)50torr, (b)100torr and (c)150torr. The quantum-dot like grain was indicate by circle in (a).

Analysis of Energy Harvesters for Powering a Wireless Sensor Node Device

Asral Bahari Jambek[1], Choo Pey See[2], Uda Hashim[3]

[1,2]School of Microelectronic Engineering, Universiti Malaysia Perlis, Malaysia

[3]Institute of Nanoelectronic Engineering, Universiti Malaysia Perlis, Malaysia

[1]asral@unimap.edu.my, [2]xiaoxun0222@yahoo.com [3]uda@unimap.edu.my

Abstract

This study analyses energy harvesters used to power a wireless sensor node. The sensor node measures the ambient temperature and transmits the data to a coordinator. Three main energy sources are investigated: solar, radio frequency and thermal. The energy produced by these devices is investigated under various environmental conditions to ensure that it can reliably supply the required amount of power to the wireless sensor node. The results show that these energy sources can provide power to the wireless sensor node at different transmission rates with an average power of 0.16 W during each data transmission.

Keywords

energy harvester, solar, radio frequency, thermal energy, ultra-low power, wireless sensor node

1. Introduction

Wireless sensor nodes will become increasingly important in the future. These devices allow users to monitor and collect data regarding their surrounding environment and use it to improve services, life quality and safety. The devices monitor the environment remotely and transmit the data to a base station. Existing wireless sensor nodes are operated using batteries, which limits the operation of devices to only a few years. For mission-critical applications, these devices must be able to operate continuously with minimal human intervention. Thus, harvesting energy from an ambient source is important to ensure the continuous, long-term operation of these devices.

The power requirement of the sensor nodes must be known beforehand in order for energy harvesting to power the wireless sensor nodes. Once the power requirement is known, the design of the proper energy harvester can be made. This ensures that the peak power requirement of the device can be achieved in any situation.

The work described in this paper forms part of our effort to design a multi-energy harvester to power a wireless sensor node. This study analyses several energy harvesters that have a high potential to power wireless sensor nodes. The harvested energy is investigated in various conditions so that

the limit of each energy provider can be known. The generated energy is then used to power a wireless sensor node device.

This paper is organised as follows. Section 2 reviews existing methods used to harvest energy from various sources. Section 3 discusses our approach to analysing energy harvesters and wireless sensor nodes. The results of the experiment are discussed in Section 4, and Section 5 presents the conclusion.

2. Literature Review

This section discusses existing methods used to harvest energy from various energy sources. Since they are abundant in our surroundings, solar, radio frequency and thermal energy sources are reviewed here.

In [1], solar energy is used to power a LPR2430ERA wireless sensor node operated at 3.3 V. The solar panel produces an output of 3 V, which is stored in the Li-Ion battery. Since charging the battery requires 4.5 V, a booster circuit is use to increase the voltage output from 3 V to 4.75 V. In addition, a 1F super capacitor is used in the circuit to ensure that the peak energy requirement during data transmission can be provided sufficiently. The maximum power point tracker (MPPT) discussed in [2] is able to achieve maximum power transfer from a solar panel to a battery. In this method, the maximum energy transfer from the solar cell to the battery can be obtained even in non-optimal weather conditions.

In [3], a thermal energy generator to power a wireless sensor node, ABBRF03, is discussed. Since a temperature difference is necessary for thermal power conversion, a heat sink is connected to a window. The temperature difference is obtained between the outside temperature and the inside temperature. The highest current output of 7 mA is generated by the thermal energy harvester when there is a temperature difference of 38 °C. Since the output of the energy harvester is too low to operate the sensor, a super capacitor is used to power the sensor directly. The sensor transmits the data every hour, and each transmission requires 27 mA. In [4] the use of thermal energy to power a wireless sensor network is investigated. For continuous operation, a temperature difference higher than 15 °C is required to power a 60 mW wireless sensor node without interruption.

Paper [5] investigates a multiple wireless sensor network arranged in a multi-hop network powered by a 3 W RF energy transmitter. The RF energy transmitter allows a maximum achievable distance between the RF energy sources and receiver of 12 metres. The sensor node must harvest at least 7 uW before it can operate properly. Once it has enough energy, the node 'wakes up' and communicates with the synchronizer. The farther away the sensor is from the energy transmitter, the longer the node needs to gather enough energy to wake up. In [6] radio frequency energy harvesting is used to power a wireless device without a battery. The device can operate with 3 W of power up to 90 feet from the radio frequency power source.

3. Methodology

This section discusses the work done to evaluate several energy harvesters used to power a wireless sensor node. Three types of energy harvesters are evaluated: solar energy, RF and thermal energy. The first part of this study analyses the output of the energy harvesters, and the second part analyses the behaviour of the wireless sensor node when powered by the energy harvester. The overall block diagram of our experiments is shown in Figure 1.

To evaluate the solar energy harvester, a solar panel 2 cm x 6 cm is used to convert the light energy to electrical energy. A small solar panel is used here to ensure that the panel can be easily integrated into a portable system. The generated energy is converted to a useable power by a power management to ensure a stable voltage at the output. The power management circuit has two 50 uAh solid state batteries that store the harvested energy [7].

Harvesting energy using RF is performed using a RF energy harvester [8]. The RF energy harvester consists of a transmitter and a receiver. The transmitter is powered through USB and emits a RF signal of 13.56 Hz frequency. The receiver converts the RF into useable electric and store the energy into a solid state battery of 50 uAh.

A thermo generator (TEG) is used to convert available waste heat to electric energy [9]. The device can output voltage whenever there is a temperature difference between two surfaces. The electric generated by the TEG is converted to a usable voltage by a power management circuit. The thermo generator used in this experiment can output voltage between 1.8 V to 4.5 V, with a temperature ranging from 0 °C to 105 °C [9].

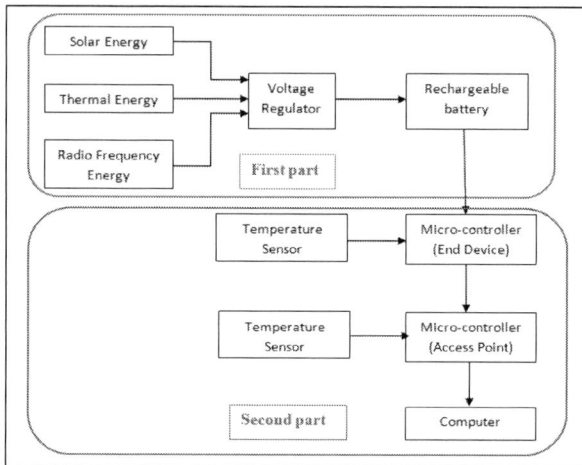

Figure 1: The block diagram of temperature sensor power by rechargeable battery using multi-energy sources.

Once the useable energy has been generated by the energy harvester, the energy is used to power a wireless sensor node. In this work, a wireless sensor node from Texas Instruments [10] was used. The device consists of two sets of devices that work as a coordinator and an end device. Each set consists of microprocessors, temperature sensors and wireless transmitters/receivers. The transmitter behaves as a coordinator that receives and displays the data on a PC. The receiver performs as an end device measuring the ambient temperature and transmitting the data with the current voltage level to the coordinator. The coordinator, which is connected directly to the PC through USB cable, then displays the data on the PC monitor.

4. Results and Discussion

In this section, we perform experiments to analyse the characteristic of the energy harvester from various energy sources. The results of the analysis will be used to power a wireless sensor node.

To analyse the solar energy harvester, the solar panel was first evaluated to determine the amount of electrical energy generated from it. Figure 2 shows various output voltages produced by the solar panel. The measurement was done for various environment conditions: direct sunlight; indoors with a light source; indoors without a light source; with a direct secondary light source. As expected, the solar panel generated maximum power when the solar panel was located under direct sunlight, with 1.5 V produced by the solar panel. 0.3 V was obtained in the condition of indoors without a light source.

Figure 3 shows the output voltage when it is connected to the power management circuit. The power management circuit takes the output voltage from the solar panel and stores it into a solid state battery so that a stable output voltage can be obtained. As shown in Figure 3, a voltage of 3.3 V is

978-1-4673-2395-6/12 $31.00 © 2012 IEEE   770

produced from the power management circuit regardless of the environment. The power management circuit ensures that stable output voltage can be produced at the output before it is connected to a device. Since the device is equipped with a battery, a stable 3.3 V can be obtained. However, the time it takes for the power management circuit to provide useful energy is affected by the amount of sunlight available.

Figure 4 shows the results of measuring the output voltage using the RF energy harvester. As mentioned in Section 3, the transmitter emits the RF signal, and the receiver converts the signal to electrical energy. The results show that the output from the energy harvester can produce 3.3 V. For the RF energy harvester, a constant voltage output at the receiver can be obtained as long as the RF source is powered by a constant voltage.

Figure 5 shows the results of harvesting energy from thermal energy harvester. In this experiment, several temperatures were applied to the device and the time taken to produce useable energy was recorded. The maximum output voltage for this device was 2.4 V. As shown in Figure 5, at 50 ^{0}C, the device required 15 seconds to generate the maximum output voltage. As the temperature increases, the time it takes to produce the output voltage decreases. At 90 ^{0}C, it requires less than 5 seconds to produce 2.4 V.

To investigate the wireless sensor node operation, the sensor node was first connected to batteries, and the current consumption by the device was then monitored. Figure 7 shows that the device transmitted the data every 1 second. The device required 7 ms to complete each transmission. Figure 8 shows a detailed current profile for each transmission. As shown in the figure, each transmission consisted of a sub-operation that began when the device woke up, communicated with the coordinator, and then returned the device to sleep mode. Table 1 presents a summary of the power required by the sensor node during each transmission. Each transmission required 0.16 W of energy with a peak voltage of 3.2 V and an average voltage of 1.41 V. From the above results showed that the solar, RF and thermal harvesters can reliably provide the voltage required by the wireless sensor node.

Table 2 shows the effectiveness of each energy harvester in powering the wireless sensor node as the transmission rate is increased. Using AA batteries, the wireless sensor node transmits the data every 1 second. The RF and solar energy harvesters can transmit the data up to every 5 seconds and 10 seconds, respectively. The results showed that the RF energy harvester can provide better energy compared to the solar harvester as long as the energy source of the RF harvester is constant. However, in the case where an RF energy source is not available, the solar energy harvester is a better choice although it has a lower transmission rate compared to the RF energy harvester.

In this experiment, the thermal energy harvester was not evaluated to power the wireless sensor node since the output voltage from the device was lower than the required operating voltage of the wireless sensor node. Work is in progress to modify the thermal device to increase the output voltage to 3.3 V.

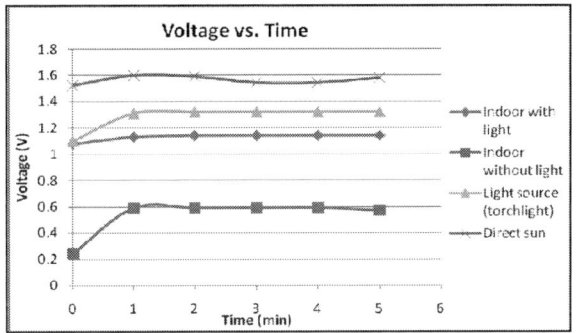
Figure 2: Output voltage for solar panel

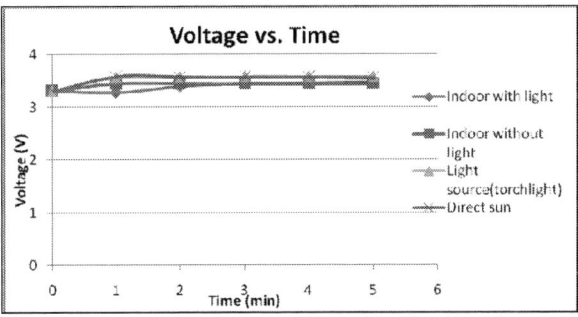
Figure 3: Output voltage of CBC-EVAL-09 with solar panel

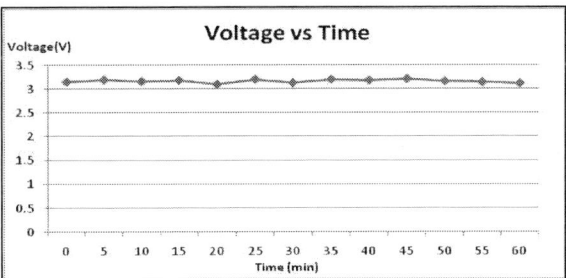
Figure 4: Output voltage of CBC-EVAL-11

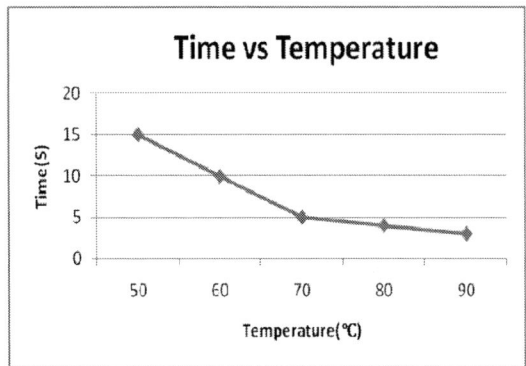

Figure 5: Output voltage of TE-CORE7

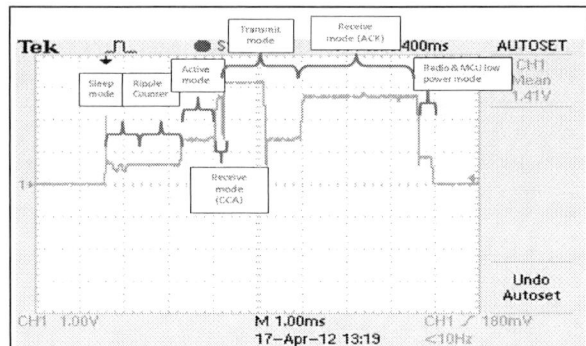

Figure 8: End device transmission current

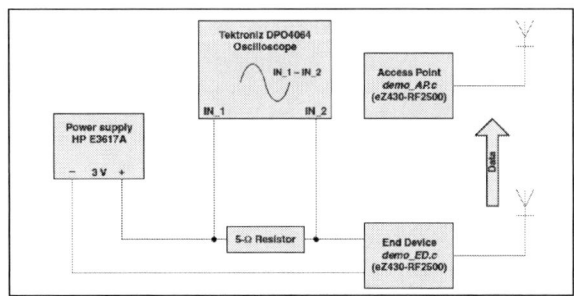

Figure 6: Experimental setup to measure current consumption by the wireless sensor node [11].

Figure 9: End device current over 50 seconds (10 seconds delay for solar)

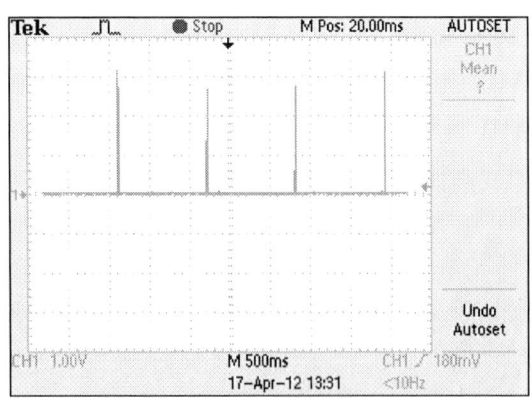

Figure 7: End device current over 5 seconds. (1 second delay)

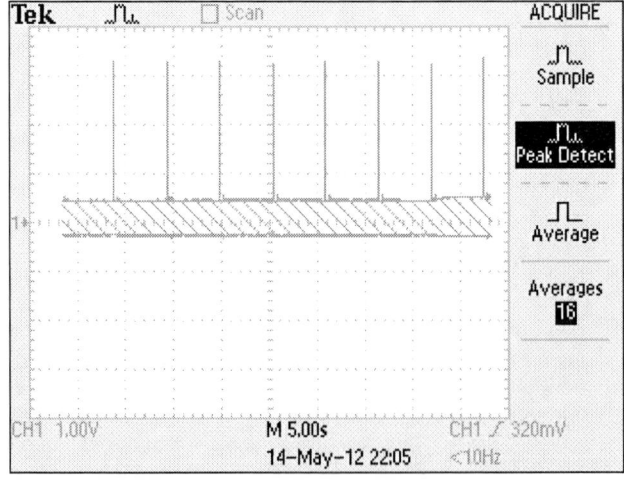

Figure 10: End device current over 50 seconds (5 seconds delay for RF)

978-1-4673-2395-6/12 $31.00 © 2012 IEEE

	Voltage (V)	Resistor (Ω)	$P = V^2/R(W)$
Peak	3.2	12.5	853 m
Average	1.41	12.5	160 m

Table 1: Power consumption

	Battery	RF	Solar	Thermal
Time (s)	1 s	5 s	10 s	NA

Table 2: Minimum interval time for the wireless sensor node to operate when powered by different energy harvesters

5. Conclusion

Several energy harvesters were used to power wireless sensor nodes. Energy generated by a solar panel, RF and thermal were evaluated to measure their effectiveness as energy sources for a wireless sensor node. Our experiment showed that these energy sources are potential sources for wireless devices. However, since harvesting the energy from the environment takes some time, the wireless sensor node operation has to be adjusted to match the available energy. Our results showed that for a wireless sensor node that requires average power of 0.16 W, the RF and solar harvesters are reliable sources of energy. In future studies, the integration of these energy sources for powering a single device will be investigated.

6. Reference

[1] RF Monolithics Inc, "A Solar Energy Harvester for Wireless Sensor Networks," Application note M1001, 2010.

[2] Cesare Alippi and Cristian Galperti, "An Adaptive System for Optimal Solar Energy Harvesting in Wireless Sensor Network Nodes," Regular Papers, VOL. 55, NO.6, July 2008.

[3] Reza Abbaspour, "A Practical Approach to Powering Wireless Sensor Nodes by Harvesting Energy from Heat Flow in Room Temperature," International Congress on Ultra Modern Telecommunications and Control System and Workshops, pp178-181, 2010.

[4] Piotr Dziurdzia and Jacek Stepien, "Autonomous Wireless Link Powered with Harvested Heat Energy," Technical Paper for AGH University of Science and Technology Department of Electronics, Krakow, 30-059 Poland, 2009.

[5] J.P.Olds and Winston K.G.Seah, "Design of an Active Radio Frequency Powered Multi-Hop Wireless Sensor Work," Technical Report, School of Engineering and Computer Science, 2011.

[6] Harry Ostaffe, Powercast Corporation, "RF Energy Harvesting Perpetually Power Wireless Sensor," article from ECN, 18 July 2011.

[7] Cymbet Corporation, EnerChip™ EP Universal Energy Harvester Eval Kit, CBC-EVAL-09 Data Sheet, 2011.

[8] Cymbet Corporation, EnerChip™ CC Inductive Charging Evaluation Kit, CBC-EVAL-11 Data Sheet, 2010.

[9] Micropelt Power generation, Thermo Harvesting Power Module, TE-CORE7 Data Sheet, 2011.

[10] Texas Instruments, eZ430-RF2500 Development Tool User's Guide, 2009.

[11] Texas Instruments, Wireless Sensor Monitor Using the eZ430-RF2500, Application Report, 2008.

Development of Microstructure on Polysilicon Substrate by Reactive Ion Etching (RIE) for future Reproductivity of Nanogap

Q.Humayun, U.Hashim

Nano Structure Lab-On-Chip Research Group,Institute Nano Electronic Engineering (INEE),
Universiti Malaysia Perlis (UniMap), 01000 Kangar,Perlis Malaysia
uda@unimap@gmail.com, qhumayun2@gmail.com

Abstract— **This article is a study about dry etching as applied to etch substrate directionality, which is the important feature of RIE. The biosensors based on polysilicon material should be sensitive and selective for the detection of bio molecule. The objective of this research is to design, and fabricate polysilicon microstructure. The proposed microstructure was designed initially by using AutoCAD software and then transferred to commercial chrome mask. Standard CMOS photolithography process coupled with RIE dry etching is used for fabrication of proposed microstructure. The fabrication process start by microstructure formation on resist and than by reactive ion etching (RIE) the proposed polysilicon microstructure was created onto samples wafer. Future work will focuse to reduce the microstructure width and finally break the microstructure to create nanogap by size reduction technique using thermal oxidation. For biomolecules sensing applications, the size of the gap must be small enough to allow the biomolecule inserted into the gap space to connect both leads to keep the molecules in a relaxed and undistorted state.**

Keywords-; biosensor; photolithography; size reduction technique; nanogap.

I. INTRODUCTION

Recently, the biological and medical fields have seen great advances in the development of biosensors and biochips capable of characterizing and quantifying biomolecules [1]. The RIE equipment used in this research was a Vacutech parallel-plate system [2]. The lower electrode is powered by a 13.56 MHz-RF generator coupled through an automatic tuning network [3]. Each electrode is 200 mm in diameter and the distance between them is 23 mm [4]. The RF electrodes are made of an odised aluminum [5].The chamber volume is 131 and the system is pumped by a 350 l/rain lurbo molecular pump backed by a mechanical rotary vane pump [6]. The base pressure before each run was less than 5 x10-2 Torr [7]. A two-level factorial design of experiments was used to find the main and interaction effects governing etch rate and sidewall slope [8]. Even through the RIE process performance is influenced by the interaction effects between different factors, such as, oxygen content, power density, pressure and loading, at a chosen set of etching parameters [9]. Chemical etch; laser sculpturing and plasma etching have been used to texture the structures that can be realized are limited [16]. In previous reports, different techniques for narrow nanogap biosensor fabrication has been demonstrated: electron beam lithography [17] [18] electromigration [19], mechanical breakjunction [20],

sacrificial layer-assisted silicon and gold nanogaps [21], and surface-catalyzed chemical deposition [22]. However, except for electron beam lithography and sacrificial layer assisted nanogaps, all other techniques have several problems in nanogap formation and commercialization because of the complex steps and difficulties in fabricating reproducible nanogap and their compatibility with other semiconductor circuits and processes. Therefore, new approaches and integration [23] methods for fabricating nanogap arrays need to be developed in order to overcome these problems [20]. On the other hand metallic nanoparticles have been used to establish self-assembly nanostructure of which physical and chemical properties have been investigated in recent years [19]. In particular, gold nano particles can be easily prepared and have the characteristic of biocompatibility [16]. Some new devices have been fabricated for the application of immune assay by making use of the latest properties of nano structure that is self-assembled by gold nanoparticles [23]. However, the properties of gold nanostructure can vary significantly with the size of gold nanoparticles and the pitch between gold [23].

II. RESEARCH METHODOLOGY

A. Mask Design

The starting material used in this research is a P-type, 100 mm in diameter (4 inch) polysilicon wafer The first process is to check the wafer type from its specification, measure wafer thickness measure the sheet resistance. After that, lightly scribe the backside of each wafer, protect the top surface, using the scribe tool provided. Mark gently but make it visible and place scribed wafer in container. Wafer is cleaned before each process. As for the lithography process, photomask were employed to fabricate the polysilicon microstructure using conventional photolithography and reactive ion etching (RIE) techniques. Commercial chrome mask was used in this research for better photo masking process and to pattern microstructure onto polysilicon substrate. The photomasks were designed using AutoCAD and then printed onto a chrome glass surface. Figure. 1 shows the first mask for microstructure formation of length and the width of 27000μm and 200μm respectively.

978-1-4673-2395-6/12 $31.00 © 2012 IEEE

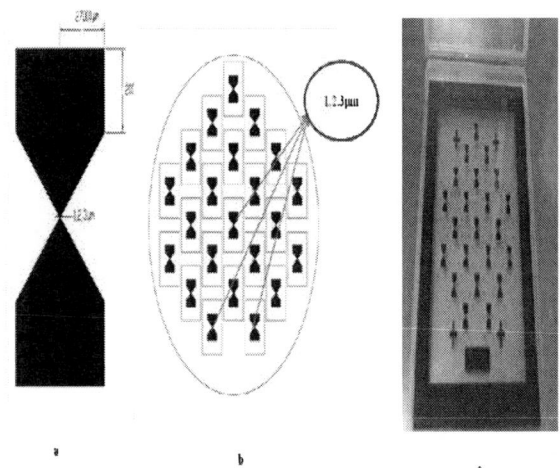

Fig.1. (a) Microstructure design specification (b) Microstructure AutoCAD design mask (c) Microstructure chrome mask.

B. Micro Structure Fabrication

Polysilicon wafers were cleaned in acetone, methanol, DI water, and dried on hot plate. PR-1 2000A was spin coated onto polysilicon at different RPM's to study the effect of rotation speed on the thickness of photoresist. After coating the PR-1 2000A photoresist on samples, the samples were soft soft-baked, and exposed to UV light through a chrome mask. All the samples were then baked on a hot plate to cross-link the exposed photoresist. To etch photoresist from exposed area, develop the wafer samples in resist developer for 30s, finally by (RIE) the polysilicon substrate was etched to create polysilicon microstructure as shown in Fig.2f. The Fig.2 demonstrates the fabrication process flow of microstructure formation by RIE.

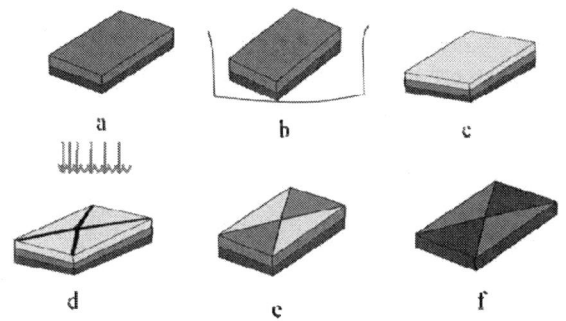

Fig.2. Fabrication process flow of micro structure formation by RIE (a) Polysilicon wafer sample (b) Wafer sample cleaning (c) Coating PR-1 2000 resist (d) Exposure of ultra violet light (e) Resist stripping (f) Polysilicon etching by RIE.

C. Micro Structure Width Verses Etching Time Mesurement

Before starting the polysilcon etching measurements process all the fabricated wafer samples surfaces should be cleaned by air blower. After that the wafer samples were placed inside the chamber of reactive ion etching equipment (RIE) consecutively at different etching time for each sample,

and record the microstructures width verses etching time data by using High power Microscope (HPM) and Scanning Electron Microscope (SEM).

III. RESULT AND DISCUSION

Fig. 4 shows the high power microscope (HPM) and scanning electron microscope (SEM) images of different wafer samples after using RIE. It was observed that the widths of microstructures after RIE was approximately 0.16, 1.4, 1.2, 0.10, 0.09, 0.07, 0.05, and 0.02μm with the etching time of 1, 2, 3, 4, 5, 6, 7, and 8 mint.

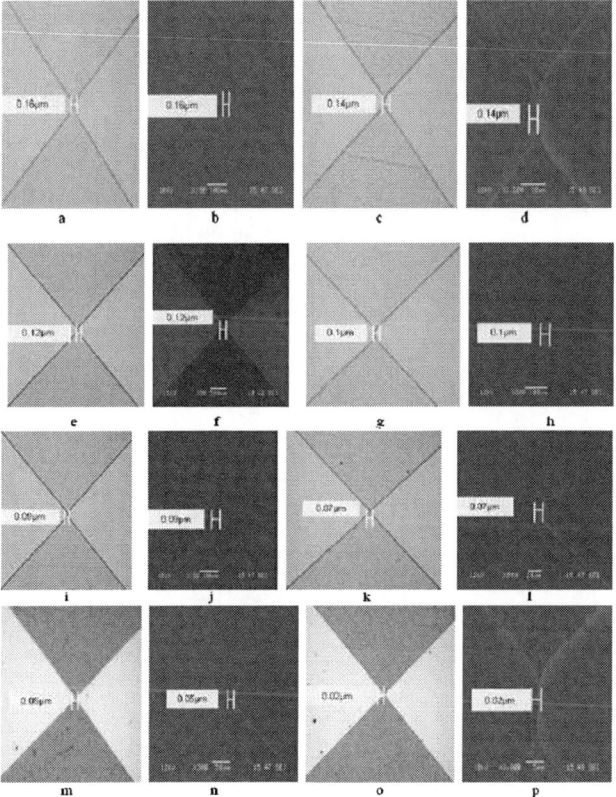

Fig. 3. Shows the high power microscope (HPM) and Scanning Electron Microscope (SEM) images, it was observed that after etching the polysilicon layer from the wafers surfaces the polysilicon microstructure was approximately 0.16, 0.14, 0.12, 0.10, 0.09, 0.07, 0.05, and 0.02μm at etching time of 1,2,3,4,5,6,7,and 8 mint respectively.

The etching time verses microstructures widths curve revealed that the measured microstructures widths with 1mint are wider than 8mint. As the etching time for all samples wafers increased from 1mint to 8mint, there was a measureable change in the widths of microstructures. From the values of microstructures width listed in table2, it was observed that the microstructures widths values recorded at 1mint etching time was 0.16μm whereas at 8mint was 0.02μm. From the tabulated values of table2 it was concluded that by varying the etching time from 1min up to 8mint the samples showed 90% change in microstructures widths values. It was concluded from etching time curve that samples were suitable for producing the smallest microstructures widths by increasing the etching time.

978-1-4673-2395-6/12 $31.00 © 2012 IEEE

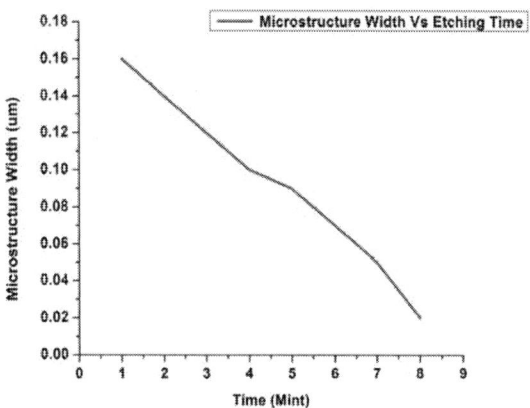

Fig. 4. Shows the etching time curve as a function of microstructure widths. All the wafer samples of microstructure widths of 0.16, 0.14, 0.12, 0.10, 0.09, 0.07, 0.05, and 0.02μm developed at etching time of 1,2,3,4,5,6,7, and 8mint.

TABLE 1

RECIPE FOR RIE PROCESS.

Gas	Pressure (Torr)
f4	0
CHF3	0
SF6	50
O₂	0
Ar	30
Voltage	Volts(V)
Bais	Bais
Power	Watt (W)
Power ICP power	650
APC/control(Pa)	4.00

TABLE 2

ETCHING TIME VERSES MICROSTRUCTURE WIDTH.

Wafer Sample	Etching Time(Mint)	Micro Structure (μm)
Sample1	1	0.16
Sample2	2	0.14
Sample3	3	0.12
Sample4	4	0.10
Sample5	5	0.09
Sample6	6	0.07
Sample7	7	0.05
Sample8	8	0.02

The results of the experimental designs clearly attest to the significance of microstructures widths and etching time as crucial determinants for microstructure initial structure after RIE [6]. Etching time verses microstructures widths are statistically significance for all wafers samples while using RIE [8]. Unfortunately, extrapolating the exact results for thickness of these photoresist coated samples due to photoresist properties and photoresist coating techniques to be critical for reactive ion etching process [9]. However, it is reasonable to expect the same microstructures widths value at mentioned etching time as calculated by past researchers [3]. The detrimental effects of extended spin times and spin speeds are probably ubiquitous for photoresist thickness [7].Resist coating on wafer is the dominance of radial coating variations on alignment error [10]. Since polysilicon wafers were used for all experiments, it is probable that the observed effects will be greater on six and eight inch wafers [5]. Although a site-by-site alignment technique was used in this study [9].

CONCLUSION

The proposed microstructure was designed, fabricated, and optimized in this paper by conventional photolithography process coupled with RIE dry etching technique respectively. The proposed method has been experimentally demonstrated by fabricating microstructures of widths 0.02, 0.05, 0.07, 0.09, 0.10, 0.12, 0.14, and 0.16μm. The next part of this research is to reduce the widths of microstructures by size reduction technique using thermal oxidation process until it finally break and create nanogap. Future experiments will focus on building up nanogap biosensors by coupling biomolecules within the nanogap with dimensions close to their persistence length.

ACKNOWLEDGEMENT

The authors acknowledge the financial support from ministry of higher education (MOHE). The authors also would like to thank all of the team members and technical staff in the Institute of Nano Electronic Engineering.

REFRENCES

[1]. S. Winderbaum, O. Reinhold, F. Yun."Reactive ion etching (RIE) as method for texturing polycrystalline silicon solar cells". Solar Energy Materials and Solar Cells 46(1997) 239-248.
[2]. E. Yablonovitch. "Inhibited Spontaneous Emission in Solid-State Physics and Electronics". Phys. Rev. Lett. 58 (1987) 2059.
[3]. V.V. Poborchii. "Photonic-band-gap properties of two-dimensional lattices of Si nano pillars". J. Appl. Phys. 91 (2002) 3299.
[4]. F. Pommereau, L. Legouezigou, G.H. Duan,B. Lombardet. "Fabrication of low loss two dimensional
[5]. In P photonic crystals by inductively coupled plasma etching". J. Appl. Phys. 95(2004) 2242.
[6]. H.C. Guo, D. Nau, A. Ranke, H. Giessen."Large-area metallic photonic crystal fabrication
[7]. with interference lithography and dry etching".Appl. Phys. B. 81 (2005) 271.
[8]. I. Soten, H. Miguez, S.M. Yang, S. Petrov,N. Coombs, N. Tetreault, N. Matsuura, H.E.Ruda, G.A. Ozin. "Barium Titanate Inverted Opals - Synthesis, Characterization, and Optical Properties". Adv. Funct. Mater. 12 (2002) 71.
[9]. T. Aoki, M. Kuwabara. "Micro patterne depitaxial (Pb,La) (Zr,Ti)O3 thin films on NbdopedSrTiO3 substrates by a chemical solution deposition process with resist molds". Appl .Phys. Lett. 27 (2004) 2580.
[10]. Process with Resist Molds". J. Appl. Phys. 45(2006) 350.
[11]. J. E. Jang, S.N. Cha, Y. Choi, G.A.J. Ameratunga, D.J. Kang. "Nanoscale capacitors based on metal-insulator-carbon nanotube-metal structures". Appl. Phys. Lett. 87 (2005) 263103.
[12]. A.G. Rinzler, J.H. Hafner, P. Nikolaev, L.Lou, R.E. Smalley. "Unraveling nanotubes: Field emission from an atomic wire". Science269 (1995) 1550.
[13]. V. Mizeikis, S. Juodkazis, J.Y. Ye, A.Rode, S. Matsuo, H. Misawa. "Silicon surface processing techniques for micro-systems fabrication". Thin Solid Films 438 (2003) 445.
[14]. V. V. Poborchii. "A visible–near in fraredrange photonic crystal made up of Si nano pillars". Appl. Phys. Lett. 75 (1999) 3276.
[15]. H. Toyota, K. Takahara, M. Okano, T.Yotsuya, H. Kikuta. "Fabrication of Microcone Array for Anti reflection Structured Surface Using Metal

Dotted Pattern". J. Appl. Phys. 40 (2001) 747D.H. Macdonald, A. Cuevas, M.J. Kerr, C. Samund sett, D. Ruby. "Texturing industrial multi crystalline silicon solar cells". SolarEnerg.76 (2004) 277.

[16]. Minoru Mizuhata, Takuya Miyake, Yuki Nomoto, Shigehito Deki. "Deep reactive ion etching (Deep-RIE) process for fabrication of ordered structural metal oxide thin films by the liquid phase infiltration method".

[17]. Y.X. Li, M.R. Wolffen buttel, P.J French,M. Laros, P.M. Sarro and R.F. Wolffen buttel."Reactive ion etching (RIE) techniques for micromachining applications". Sensors and Actuators A, 41-42 (1994) 317-323.

[18]. P. B. Fischer and S. Y. Chou. "10 nm Electron Beam Lithography and Sub-50 nm Overlay Using a Modified Scanning Electron Microscope". Appl. Phys. Lett., Vol. 62, No. 23,(1993)2989-2991.

[19]. S. Itousa, Han Young Yu, Chil Seong Ah,"Fabrication and AFM Characterization of a Coplanar Tunnel Junction with a Less Than 30nm Inter electrode Gap" Nanotechnology, vol. 5(1994)19-25.

[20]. In P photonic crystals by inductively coupled plasma etching". J. Appl. Phys. 95(2004) 2242.

[21]. Y.X. Li, M.R. Wolffen buttel, P.J French,M. Laros, P.M. Sarro and R.F. Wolffen buttel."Reactive ion etching (RIE) techniques for micromachining applications". Sensors and Actuators A, 41-42 (1994) 317-323.

[22]. DDD In P photonic crystals by inductively coupled plasma etching". J. Appl. Phys. 95(2004) 2242.

[23]. Irudayaraj, J., and Reh, C. 2008. Nondestructive Testing of Food Quality, 4: 978-0-8138-2885-5. James J.A. 2005. Micro Electro Mechanical System Design. Dekker Mechanical Engineering, 0824758242/0-8247-5824-2.

AUTHORS INDEX

A. A. Jasim	381
A. A. RahmanOthman	694
A. B. Suriani	141
A. Bag	285
A. H. You	366
A. Hadi	302
A. Ishak	78
A. Loykaew	694
A. N. Arshad	65, 73
A. N. Fadzilah	52, 102
A. Oberafo	678
A.Azlinda	214
A.M. Awobode	678
A.R. Zainun	128
A.Zaharim	219
Abdul Aziz. A	34
Abijith Prakash	462
Abrar Ismardi	306
Abu Bakar AR	177, 249
Abu Khari A'ain	503
Afifah Maheran A.H.	173
Afifah Maheran A.Hamid	219
Afiq Hamzah	298, 396
Afishah Alias	407, 750, 763
Aftanasar Md. Shahar	583
AHM Zahirul Alam	254
Ahmad Afif Safwan Mohd Radzi	399
Ahmad Al Ali	6
Ahmad Jalaluddin Yusof	583
Ahmad Jamal Salim	512
Ahmad Sudin	238
Akhlesh Lakhtakia	A1
Akrajas Ali Umar	344, 352, 759
Akshay Salimath	69, 486
Akzhigitova Meruyert	363
Alireza Bahadorimehr	280, 333
Andreas König	229, 517
Anees Abdul Aziz	293
Ang Boon Chong	431
Anis Nurashikin Nordin	254
Anithambigai Permal	707
Arej Kadhim Abbas	720, 725
Arshad Hmood Abd Al Kadhim	720, 725
Asral Bahari Jambek	324, 769

Asrulnizam Abd Manaf	234
Aun Shih Teh	14
Aw Tiam Ann	654
Aziz, M.H.A.	311, 316
Azlan Abdul Aziz	669, 674
Azlan Sulaiman	378
Azlan Zakaria	200
Azlin Bahador	298
Azman Jalar	242, 592, 604, 733, 740, 755
Azri Husni Hasani	583
Azrul Azlan Hamzah	186, 348
B. F. Usher	694
B. P. Singh	526
B.Vasuki	271
B.Y. Majlis	173, 302
B.Yeop Majlis	219
Bablu Ghosh	177, 249
Bablu K. Ghosh	407
Bablu Kumar Gosh	750, 763
Badariah Bais	168, 336, 340
Bahniman Ghosh	61, 69, 478, 486
Beena Pandey	538, 565
Benyamin Davaji	333
Bhupesh Bishnoi	61, 69, 486
Binghai Liu	436
Binh Tinh Tran	246
Binh-Tinh Tran	747
Burhanuddin Yeop Majlis	6, 123, 168, 186, 195, 205, 210, 280, 306, 333, 336, 340, 344, 348, 352, 357, 360, 422, 740
C. C. Sun	366
C. F. Dee	302
C. K. Maiti	285
C. Mahata	285
C.M.Firdaus	158
C.V.Krishna Reddy	490
Chang-Fu Dee	246, 747
Chee-Pun, Ooi	543, 556
Chen Chen Chung	246
Cheong Choke Fei	458, 654
Chew Pei Yi	426
Chi Chin Yap	30
Chi Lang Nguyen	246
Chia-Hua Chang	744
Chia-Ta Chang	411
Chien-Hung Kuo	574

Chi-Feng Hsieh	766
Chih-Peng Hsu	414, 418
Chi-Lang Nguyen	747
Ching Hsiang Hsu	246
Chin-Peng Ching	384, 388
Choo Pey See	769
Chu Tsui Ping	578
Chung-Yu Lu	411
ChunLei Wu	440, 444, 495
Chun-Yen Chang	414, 418
Chyi-Shiang Hoo	448
D. Nelson	694
Daniel C. S. Bien	14
Deb Kumar Pal	588
Dee Chang Fu	306
Deepa Yagain	551
Devarajan Mutharasu	132, 711
Dhafer Abdul-Ameer Shnawah	453
Dharmendra Hiranandani	69
Diao Yuan Chiou	246
Divya Pogaku	177, 249
Diwakar Agrawal	486
DiWei Fan	440, 444
E. K. Wong	366
E.S.Shajahan	276
Edward Yi Chang	A2, 246, 411, 729, 744, 747, 766
Eng Siew Kang	97
Eskendirov Sharipzhan	363
F. Razaghian	302
F. Salehuddin	173, 219
F.A.Hamid	219
F.S.S. Zahid	57, 153
Faisal Mohd-Yasin	186, 608
Faizah Md Yusof	649
Fatemeh Kohani Khoshkbijari	288, 600
Fatimah K. A. Hamid	86, 298, 396
Fazrena Azlee Hamid	636
Furat A. Aldaamee	470
G. Devandran	649
G.Uma	271
Gan Leong Kit	636
Gandi Sugandi	360
Gaojie Wen	436, 495
Gene Sheu	462, 699
Grace Song	436, 495
H Meidia	259

H. Ahmad	381
H.A. Elgomati	173, 219
Haarindra Prasad s/o RajintraPrasat	191, 736
Hafzaliza Erny Zainal Abidin	348
Hai-Dang Trinh	246, 747
Hamam Maher Abd	517
Hanif Che Lah	686
Hanim A.R	422
Harikrishnan Ramiah	448
Harith Ahmad	378
Hasanah, L.	422
Hashim, M.N.	316
Haslan Abu Hassan	720, 725
Haslinda Abdul Hamid	30
Hazura H.	422
Himadri Singh Raghav	526
Hing Wah Lee	18
Hisashi Fukuda	750
Hong-Quan Nguyen	246, 747
Huan-Chung Wang	744
Hung Wei Yu	246
I. Ahmad	1, 173, 219
I. Saurdi	82
Ian Grout	503
Ibrahim Ahmad	403, 627
Ibrahim Kamarulazizi	711
Irfan Anjum Badruddin	453
Irman Abdul Rahman	604
Ishak Hj. Abd. Azid	18
Ishak Mansor	592, 740
Iskhandar Md Nasir	263
Ismail Saad	177, 249, 407, 750
Izhan Abdullah	740
J. Karamdel	302
J.R. Rusli	482
Jackson Kong	583
Jafar Alvankarian	280, 333
Jalil. Md. Desa	378
Jamilah Karim	254
Jatmiko E Suseno	86
Jet-Rung Chang	414, 418
Joe Yu	436, 495
Johan Stiens	374
John F. McDonald	618, 622
Jubayer Jalil	466
Jui-Chien Huang	411

Jumril Yunas	195, 205, 210, 357
Jung-Ruey Tsai	699
K. Dayana	52, 102
K.K. Goh	659
K.L.Foo	191, 736
Kader Ibrahim	649
Kamarulzaman Mohamed Zin	A6
Kamaruzzaman Sopian	38
Kanesan Jeevan	448
Kang Cheng Wei	466
Karkeng Lim	733
Karuna Nidhi	699
Katsuhiro Uesugi	750, 763
Khairul A.M	177, 249
Khairul Anuar Abd Wahid	18
Khairul Anuar Mohamad	407, 750, 763
Khaw Mei Kum	608
Kobchai Dejhan	522
Kuan-Hsun Wang	574
Kusan-Shin Liu	766
L.N. Ismail	65, 73, 182
Labonnah F. Rahman	466
Lai Chin Yung	654
Lau Chyun Wenn	324
Lee Chai Ying	458, 654
Lesley Wong Ying Ying	588
Li Tian	436, 440, 444
Luay Yassin Taha	6
Lu-Che Huang	744
M S Bhat	613
M. Ain Zubaidah	94, 149
M. Amirul	78
M. B. I. Reaz	466
M. Devarajan	499
M. H. Fadzilah Suhaimi	149
M. H. Mamat	73, 682
M. H. Wahid	73
M. Khairizal	65
M. Maryam	141
M. N. Berhan	682
M. Rusop	34, 52, 57, 65, 73, 78, 82, 94, 102, 111, 128, 141, 149, 153, 158, 182, 214, 328, 682, 703
M. S. Bahrudin	1
M. T. Ahmadi	86
M. Z. Muhammad	381

M.A. Baqir	392
M.A. Olopade	678
M.Amir	90
M.G. Zebaze Kana	678
M.H. Abdullah	82
M.H. Mamat	34, 82
M.H. Wahid	65
M.Kashif	191, 716, 736
M.N. Amalina	128
M.Othman	123
M.S. Alias	267
M.S. Shamsudin	141
M.S.Bhuyan	123
M.S.P. Sarah	57, 153
M.Salina	682
M.T.Ahmadi	298
M.Y.Sulaiman	38
M.Z. Musa	34, 82
M.Z. Sahdan	267
Maan M. Alkaisi	119
Mahesh Kumar Talari	200
Mahmudin, D.	422
Mai Woon Lee	14
Man, B.	641
Mardhiah Mohd Nor	340
Mardiana B.	422
Marhanis Abu Othman	690
Maria Abu Bakar	242
Marianah Masrie	205
Marlia Morsin	352
Maryam Mousavi	168
Masato Sakamoto	763
Masuri Othman	470
Md Hanif Md Nasir	475
Md Imran Siddiqui	699
Md. Imran Siddiqui	462
Md. Shabiul Islam	123, 470, 547
Md. Syedul Amin	466
Mehdi Tajaldini	370
Memtode Jim Abigo	664
Menon, P.S	173, 422
Miao Wu	440, 444
Michaelina Ong	588
Mimiwaty Mohd Noor	210, 360
Mohamad Hasan-Sagha	645
Mohamad Rusop	293, 399

Mohammad Azmi Abdul Hamid	740
Mohammad Javad Kiani	298
Mohammad Redzuan Zahariman	596
Mohammadmahdi Vakilian	168
Mohammed Sadique Anwar	462
Mohd Azizi Chik	627, 649
Mohd Fadzlie Bin Ahmad	596
Mohd Faizul Mohd Sabri	453
Mohd Hazmuni Saidin	649
Mohd Hazrul Zakaria	336
Mohd Hezri Abu Bakar	263
Mohd Ikhwan Mohd Isa	674
Mohd Norzaidi Mat Nawi	234
Mohd Rizal Arshad	234
Mohd Rosydi Zakaria	224
Mohd Shahrul Akram Mohd Mokhtar	759
Mohd Zainizan Sahdan	596
Mohd Zaki Mohd Yusoff	30
Mohd Zubir Khalid	631
Mohd Zubir Mat Jafri	370
Mohd Zuhir H.	177, 249
Mohd Zuhir Hamzah	407
Mohsen Jalali	645
Mok Vee Hong	664
Mola Srinivasarao	526
Montree Kumngern	522
Muhamad Amri Ismail	263
Muhamad Khairol Ab Rani	631
Muhamad Mat Salleh	344, 352, 759
Muhammad Akmal Johar	229
Muhammad Azmi Abdul Hamid	242, 592, 604, 733, 755
Muhammad Nubli Zulkifli	755
Muhammad Sadiq Sahari	503
Muhammad Yahaya	344, 759
Mulyanti, B.	422
Munawar A Riyadi	86
Mutharasu Devarajan	115, 384, 388, 532, 707
N. Bolong	177, 249
N. D. Md Sin	682
N. H. Al-Hardan	733
N. H. Mahzan	111
N. Nafarizal	267
N. Soin	659
N.A. Asli	94
N.A. Rasheid	128

N.D. Md Sin	34
N.N. Hafizah	65, 182
N.S Kamarozaman	703
Nabihah Razali	475, 686
Nadia Md Razib	14
Nadzri, N.S.	641
Nadzril Sulaiman	357
Nafarizal Nayan	596
Narayan Sahoo	47
Nazaliza Othman	475
Nazwa Taib	238
Nennie Farina Mahat	570
Ng Hong Seng	588
Nico F. de Rooij	A3
Nishant Singh	478
Noraini Marsi	186
Norazlin Bahador	86, 396
Norazreen Abd Aziz	340
Norfarariyanti Parimon	407
Norfarhanim Mohd Zahari	669
Norhaimi, W.M.W.	316
Norhaslinawati Ramli	90
Norhayati Abu Bakar	344
Norihan Abdul Hamid	210
Norinsan Kamil Othman	604, 755
Norizam Saad	592
Norsuzlin Bt Mohad Sahar	560
Nur Humaira Md Salleh	238
Nur Raihana Samsudin	512
Nurhafizah Zainal Abidin	293
Nurmin Bolong	407
Nurul Izrini Ikhsan	399
Nurul Syahidah Sabri	200
Nurul Zayana Yahya	293
O.Anjaneyulu	490
O.E. Awe	678
Ong Hang See	10
Ooi Boon Siew	A5
Othman Sidek	90, 234
P A Chen	462
P. Dhavachelvan	507
P.K. Choudhury	392
P.N A. Diyana	26
P.R.Apte	219
P.Susthitha Menon	205, 219, 403
Palianysamy, M.	311

Pedro Torruella	229
Poehle Holger	690
Po-Min Tu	414, 418
Q. Humayun	22, 716, 774
Quang-Ho Luc	246, 747
Quanxi Cao	42
R. A Bakar	73, 328, 703
R. Abu Bakar	111
R.H.Salimin	158
R.M. Sidek	482
Rabab Khalid Sendi	163
Rahman, N.A.Z.	311, 641
Raja Mohd Fuad Tengku Aziz	475
Rajeev Komar	613
Ramzan Mat Ayub	224
Raymond Tan	588
Razak, H.A.	311
Razali Ismail	86, 97, 177, 249, 298, 396
Retnasamy, V.	311, 316, 320, 641
Reza Barkhordari	288, 600
Reza Fouladi	288, 600
Reza Kohani Khoshkbijari	288, 600
Riyaz Ahmad Mohamed Ali	596
Robert Freier	517
Rohaya Abdul Wahab	475
Roslinda Shamsudin	242, 592, 604, 733
Rosminazuin Ab. Rahim	336, 340
Rozaimah Baharim	475
Rozita Teymourzadeh	470, 560, 664
Ru Han	137
S Mahajan	259
S. Abdullah	94, 149
S. Ahmad	34, 682
S. B. Hashim	111
S. F. M. Yusop	149
S. H. Herman	65, 111, 328, 703
S. Mallik	285
S.F. Abdullah	1
S.Fazlili Abdullah	403
S.Isaak	86
S.Kalthom Tasirin	403
S.Noorjannah Ibrahim	119
S.R.M.S.Baki	158
S.S. Shahabuddin	659
S.W. Harun	381
S.Y. Lee	499

Saat Shukri Embong	18
SachchidaNand Shukla	538, 565
Sachin Maheshwari	526
Saifollah Abdullah	399
Salah Hasan Ibrahim	547
Saleem H.Zaidi	38
Sangeeta Palo	47
Sani Irwan Md Salim	512
Sauli. Z.	311, 316, 320, 641
Sawal Hamid Md Ali	123, 547
Seng Teik Ten	238
Sew-Kin, Wong	543, 556
Shafinaz Sobihana Shariffudin	293
Shahrom Mahmud	163
Shahrul Azam Abdullah	14
Shahrum Abdullah	755
Shamsul Faez Mohd Yusop	399
Shankaranarayana M Bhat	276
Shao-Ming Yang	699
Shapri, A.H.M.	320
Sharifah Saleh	475
Sharipah Nadzirah SAA	200
Sharipah Nadzirah Syed Ahmad Ayob	145
Sheetal Chennapnoor	551
Shide Nejati	288, 600
Shih-Cheng Huang	414, 418
Shin-Yuan Wang	747
Shiva Nejati	288, 600
Shun-Kuei Yang	414, 418
Siti Aisyah Zawawi	30
Siti Noor Harun	686
Siti Zubaidah Md Saad	690
Soellner Norbert	654
Soo Kien Chen	14
Soon Bee Law	707
Subramani Shanmugan	132, 711
Suhaila Isaak	396
Suhaila Sepeai	38
Suhana Binti Mohd Said	453
Sukreen Hana Herman	293, 690
Sulaiman Rabbaa	374
Sulaiman Wadi Harun	378
Surya Kris Amethystna	699
Susmrita Srivastava	538, 565
Syahril Ridzuan Ab Rahim	627

Syed Khaleel Ahmed	636
Sze-Yen Lee	388
T Laxminidhi	613
T.Pradeep	490
Tai Zhi Ling	10
Tai-Ming Lin	411, 729
Tan Chan Lik	690
Tan Kong Yew	686
Taniselass, S	316, 320
Tee Pei Ling	578
Teh Kim Ting	608
Thanh Hoa Phan Van	246
Thikra S.Dhahi	238
Tien-Tung Luong	729
Tijjani Adam	26, 107
Ting-En Hsieh	744
Tiong Teck Yaw	306
Tong Gee Hong	588
Trinath Sahu	47
U. Hashim	22, 107, 145, 177, 191, 200, 249, 254, 267, 649, 716, 736, 774
U. M. Noor	52, 57, 102, 153
U. Mohd Noor	111
Uda Hashim	26, 224, 238, 324, 596, 769
Uma.Ramadass	507
Umarova Zhanat	363
Ummikalsom Abidin	195
Vairavan, R.	320, 641
Velappa Ganapathy	448
Vengdasalam, K.	320
Vijaya Krishna A.	551
Vijayaram Thoguluva Raghavan	426
Vikas Nandal	61, 69, 486
W.H. Wan Zuha	482
Wai Yee Lee	18
Wai-Leong, Pang	556
Waleed Ahmed AL Garidi	560
Wan Yusmawati Wan Yusoff	604
Wei Ching LIEW	532
Wei Sun Leong	86
Wei-Ching Huang	766
Winter Wang	436, 440, 444, 495
Wiranto, G.	422
Wong Jian Sang	588
Xuelian Liu	618, 622
Y.Sujan	271

Yang Peng	578
Ya-wen Lin	414, 418
Yazan Samir Algnabi	470
Yen-Teng Ho	729
Yewguan Soo	512
Yinhua Yao	42
Yoon Soon Fatt	A4
You Ah Heng	426
Yu Chan Thien	97
Yue-Chin Lin	729, 747
Yueh-Chin Lin	744
Yuen-Yee Wong	246, 729, 747, 766
Yun-Fen, Yong	543
Yu-Sheng Chiu	411, 729
Yu-Ting Chou	411
Yuzman Yusoff	686
Z. Aznilinda	328, 703
Z. Habibah	65, 73
Z. Khusaimi	214
Z. Zulkifli	57
Zaharah Johari	86
Zarimawaty Zailan	224
Zhi-Yin Lee	115, 388
Zi-Yi, Lam	543, 556

CURRAN ASSOCIATES INC.
proceedings
.com

9781467323956